PITTCON Conference and Expo 2016

Abstracts

Atlanta, Georgia, USA
6 - 10 March 2016

Volume 1 of 2

ISBN: 978-1-5108-2394-5

Printed from e-media with permission by:

Curran Associates, Inc.
57 Morehouse Lane
Red Hook, NY 12571

CURRAN ASSOCIATES INC. proceedings.com

Some format issues inherent in the e-media version may also appear in this print version.

Printed by Curran Associates, Inc. (2016)

Pittsburgh Conference
300 Penn Center Boulevard
Suite 332
Pittsburgh, PA 15235-5503
USA

Phone: (412) 825-3220
(800) 825-3221
Fax: (412) 825-3224

info@pittcon.org

Additional copies of this publication are available from:

Curran Associates, Inc.
57 Morehouse Lane
Red Hook, NY 12571 USA
Phone: 845-758-0400
Fax: 845-758-2633
Email: curran@proceedings.com
Web: www.proceedings.com

SUNDAY, MARCH 6, 2016
AFTERNOON

The Wallace H. Coulter Lecture — Session 10

The Wallace H. Coulter Lecture

Sunday Afternoon, Sidney Marcus Auditorium, Bldg A, Level 4

5:00	(10-1)	**How Optical Single-Molecule Detection in Solids Led to Super-Resolution Nanoscopy in Cells and Beyond** W.E. (WILLIAM ESCO) MOERNER, Stanford University

AWARDS — Session 20

The Pittcon Heritage Award

Sunday Afternoon, Sidney Marcus Auditorium, Bldg A, Level 4

4:45		**Presentation of the 2016 Pittcon Heritage Award posthumously awarded to Kenji Kazato and Kazuo Ito, founders of JEOL - accepted by Gon-emon Kurihara, President of JEOL**

SYMPOSIUM — Session 30

ACS-ANYL - Tracing the Metabolome: Application of Stable-Isotope Tracers in Bioanalytical Chemistry

arranged by Cynthia K Larive, University of California-Riverside and Gregory A Barding, California State Polytechnic University

Sunday Afternoon, Room B308
Cynthia K Larive, University of California-Riverside, Presiding

1:30		**Introductory Remarks - Cynthia K Larive and Gregory A Barding**
1:35	(30-1)	**Stable Isotope Labeling and UHPLC/MS Strategies for Probing Dynamics of Plant Specialized Metabolism** A DANIEL JONES, Michigan State University, Zhenzhen Wang, Xiaoxiao Liu, Banibrata Ghosh
2:10	(30-2)	**Developments in ^{13}C-based Metabolomics** ARTHUR S EDISON, University of Georgia
2:45	(30-3)	**Chemical Isotope Labeling LC-MS for Quantitative Metabolomics with High Metabolomic Coverage** LIANG LI, University of Alberta
3:20		**Recess**
3:35	(30-4)	**Lipid Metabolic Pro°°filing Using Stimulated Raman Scattering Microscopy** MENG WANG, Baylor College of Medicine
4:10	(30-5)	**Monitoring the Incorporation of Stable-Isotope Labeled Alcohols by Bacteria with GC-MS** GREGORY A BARDING, California State Polytechnic University, Rakesh Mogul, Nicole G Perkins

SYMPOSIUM — Session 40

Emerging Leaders in Biological Mass Spectrometry

arranged by Amanda B Hummon, University of Notre Dame and Heather R Desaire, University of Kansas

Sunday Afternoon, Room B302
Amanda B Hummon, University of Notre Dame, Presiding

1:30		**Introductory Remarks - Amanda B Hummon and Heather R Desaire**
1:35	(40-1)	**New Mass Spectrometry Methods Show HIV Vaccine Candidates' Protein Structures are Misfolded** HEATHER R DESAIRE, University of Kansas
2:10 Madison	(40-2)	**Novel Strategies in Top-down Proteomics** YING GE, University of Wisconsin
2:45	(40-3)	**Multi-Tier Approach to Understand the Biology of Alzheimer's Disease** RENÃ ROBINSON, University of Pittsburgh
3:20		**Recess**
3:35	(40-4)	**Old Photochemistry Brings New Capabilities in Unsaturated Lipid Analysis** YU XIA, Purdue University, Zheng Ouyang, Xiaoxiao Ma, Craig Stinson
4:10	(40-5)	**Evaluating Small Molecule Histone Inhibitors with High Resolution Mass Spectrometry and 3D Cell Cultures** AMANDA B HUMMON, University of Notre Dame, Simone Sidoli, Peter E Feist, Monica M Schroll, Benjamin A Garcia

SYMPOSIUM — Session 50

Enabling Sample Preconcentration Methods for Bioanalysis

arranged by Adam T Woolley, Brigham Young University

Sunday Afternoon, Room B304
Adam T Woolley, Brigham Young University, Presiding

1:30		**Introductory Remarks - Adam T Woolley**
1:35	(50-1)	**Bottom-up Proteomics of E.coli Using Dynamic pH Junction Preconcentration and CZE-ESI-MS/MS** NORMAN J DOVICHI, University of Notre Dame, Guijie Zhu, Liangliang Sun
2:10	(50-2)	**Electrokinetic Sample Preconcentration and Hydrodynamic Sample Injection for Capillary Electrophoresis Using a Pneumatic Microvalve** RYAN T KELLY, Pacific Northwest National Laboratory, Yongzheng Cong, Katipamula Shanta, Tang Keqi
2:45	(50-3)	**Solid Phase Microextraction in Bioanalysis** JANUSZ PAWLISZYN, University of Waterloo
3:20		**Recess**
3:35	(50-4)	**Microfluidic Integration of Solid-Phase Extraction with Fluorescence Labeling for Microfluidic Analysis** ADAM T WOOLLEY, Brigham Young University, Suresh Kumar, Vishal Sahore, Radim Knob, Mukul Sonker
4:10	(50-5)	**Liquid Biopsies: Microfluidic Enabling the Clinical Utility of These Markers** STEVEN A SOPER, University of North Carolina

SYMPOSIUM — Session 60

Frontiers of In Situ and In Vivo Spectroscopic Imaging

arranged by Ji-Xin Cheng and Mikhail Slipchenko, Purdue University

Sunday Afternoon, Room B305
Ji-Xin Cheng, Purdue University, Presiding

1:30		**Introductory Remarks - Ji-Xin Cheng and Mikhail Slipchenko**
1:35	(60-1)	**Correlated Raman and Mass Spectrometric Chemical Imaging of Multiscale Spatiotemporal Signaling in Microbial Communities** PAUL W BOHN, University of Notre Dame, Nameera Baig, Nydia Morales-Soto, Sage B Dunham, Jonathan V Sweedler, Joshua D Shrout, Sneha Polisetti
2:10	(60-2)	**Nanotechnology for In-Vivo and Intraoperative Cancer Detection and Image-Guided Surgery** SHUMING NIE, Emory University
2:45	(60-3)	**Nonlinear Microscopy to Detect and Grade Cancer** WARREN S WARREN, Duke University
3:20		**Recess**
3:35	(60-4)	**Multi-Parametric and Multi-Spectral Photoacoustic Microscopy** SONG HU, University of Virginia
4:10	(60-5)	**Vibrational Spectroscopic Imaging of Molecular Dynamics in Living Systems** JI-XIN CHENG, Purdue University

SYMPOSIUM **Session 70**

Wearable and Point-of-Care Sensor Technologies for Biomonitoring
arranged by Ian Papautsky and William R Heineman, University of Cincinnati

Sunday Afternoon, Room B401
Ian Papautsky, University of Cincinnati, Presiding

1:30		**Introductory Remarks - Ian Papautsky and William R Heineman**
1:35	(70-1)	**Personal Exposure Monitoring Using Portable Samplers and Paper Analytical Devices** CHARLES HENRY, Colorado State University
2:10	(70-2)	**Pencil and Paper Diagnostic Devices** ANDRES W MARTINEZ, California Polytechnic State University
2:45	(70-3)	**Electrochemical Metal Determination for Point-of-Care Assessment of Environmental Exposure** IAN PAPAUTSKY, University of Cincinnati
3:20		**Recess**
3:35	(70-4)	**Ionic Liquid Electrochemistry and Biointerface Design for Reliable and Smart Sensors** XIANGQUN ZENG, Oakland University
4:10	(70-5)	**Monitoring Corrosion of Biodegradable Magnesium Implants with Hydrogen Gas Sensors** WILLIAM R HEINEMAN, University of Cincinnati, Tingting Wang, Daoli Zhao, Zhongyun Dong, William Hoagland, David Benson

WORKSHOPS **Session 80**

Light Sources in Analytical Chemistry: Solid State Light Sources and Beyond
arranged by Mirek Macka, University of Tasmania

Sunday Afternoon, Room B402
Mirek Macka, University of Tasmania, Presiding

1:30		**Introductory Remarks - Mirek Macka**
1:35	(80-1)	**Gatherers and Foragers? Analytical Scientists in the Quest for Better Light Sources** MIREK MACKA, University of Tasmania
2:05	(80-2)	**Applications of Cavity Ring-Down and Cavity-Enhanced Absorption Spectroscopy in Atmospheric Chemistry** HANS DIETER OSTHOFF, University of Calgary, Charles A Odame-Ankrah, Youssef M Taha, Jason Pak, Connie Z Ye, Nick R Yordanov
2:35	(80-3)	**A Fourier-Domain Fluorescence Excitation Emission Matrix Spectrometer and Its Applications in Measuring Oxidative Stability of Industrial Liquids** HANS-PETER LOOCK, Queen's University, Nicholas L Andrews, James Z Fan, Oliver Reich, Alexander Dudelzak, Hengameh Omrani
3:05		**Recess**
3:20	(80-4)	**Deep UV-LEDs in Detectors for HPLC and Capillary Electrophoresis** PETER CH HAUSER, University of Basel, Duy Anh Bui
3:50	(80-5)	**Solid State Light Sources in Capillary Electrophoresis** DAN XIAO, Sichuan University N/A
4:20	(80-6)	**Quantum Cascade Lasers: How to Revolutionize Mid-Infrared Gas and Liquid Phase Diagnostics** BORIS MIZAIKOFF, Ulm University

ORGANIZED CONTRIBUTED SESSIONS **Session 90**

Advances in Mass Spectrometry of RNA - Half Session
arranged by Norman Chiu, University of North Carolina at Greensboro

Sunday Afternoon, Room B403
Norman Chiu, University of North Carolina at Greensboro, Presiding

1:30	(90-1)	**Toward Detection of Isomeric/Isobaric MicroRNA Biomarkers** NORMAN CHIU, University of North Carolina at Greensboro
1:50	(90-2)	**New LC-MS/MS Strategies for Modified Oligonucleotides and RNAs** PATRICK A LIMBACH, University of Cincinnati, Robert L Ross
2:10	(90-3)	**Mass Spectrometry, RNA and Infectious Disease: tRNA Reprogramming and a Second Genetic Code Control Mycobacterial Dormancy** PETER DEDON, Massachusetts Institute of Technology
2:30	(90-4)	**Identification and Quantitation of RNA Post-Transcriptional Modifications by Nano-flow Liquid Chromatography - Tandem Mass Spectrometry and Database Searching.** HIROSHI NAKAYAMA, RIKEN CSRS, Masato Taoka, Nobuhiro Takahashi, Toshiaki Isobe

ORGANIZED CONTRIBUTED SESSIONS **Session 100**

Ionophore-Based Chemical Sensors I
arranged by Philippe Buhlmann, University of Minnesota and Eric Bakker, University of Geneva

Sunday Afternoon, Room B404
Philippe Buhlmann, University of Minnesota, Presiding

1:30	(100-1)	**pH Independent Fluorescent Ion Sensors** ERIC BAKKER, University of Geneva
1:50	(100-2)	**Measurement of Carbon Dioxide in Urine to Guide the Treatment of Patients in Severe Sepsis or Septic Shock** ERNO LINDNER, The University of Memphis, James G Atherton, Marcin Guzinski, Jasinski Artur, Bradford D Pendley
2:10	(100-3)	**Potentiometric Nanosensors Towards Direct Detection of Nucleic Acids** RÓBERT E GYURCSÁNYI, Budapest University of Technology and Economics, Istvan Makra, Gyula Jágerszki, Alexandra Brajnovits, Peter Fürjes, Márton Bojtár
2:30	(100-4)	**Potentiometric Detection of Biomacromolecules Based on Surface Molecular Imprinting** RONGNING LIANG, Yantai Institute of Coastal Zone Research, CAS, Wei Qin N/A
2:50		**Recess**
3:05	(100-5)	**Voltammetric Ionophore-Based Electrode for Protamine in Human Blood** SHIGERU AMEMIYA, University of Pittsburgh
3:25	(100-6)	**Conducting Polymer-Coated Electrode and Its Application to the Thin Layer Electrolysis Cell for Coulometric Determination** YUMI YOSHIDA, Kyoto Institute of Technology, Mao Fukuyama, Kohji Maeda
3:45	(100-7)	**Biofouling of Ionophore-Doped Ion-Selective Electrode Membranes Revisited** PHILIPPE BUHLMANN, University of Minnesota, Adam Dittmer
4:05	(100-8)	**Metastable Photoacids Towards Activatable and Controllable Ion Sensing Bulk Membranes for Cations Detection** KARIN Y CHUMBIMUNI-TORRES, University of Central Florida, Parth K Patel

ORGANIZED CONTRIBUTED SESSIONS **Session 110**

R&D to QC: Bridging the Gap - Half Session
arranged by Justin Shearer, Dow AgroSciences

Sunday Afternoon, Room B312
Justin Shearer, Dow AgroSciences, Presiding

1:30	(110-1)	**Back to the Future: Bringing Life to an Old Crop Protection Product** JENNIFER HOLLIE JONES, Dow AgroSciences, Tammy Hamilton, Grant Von Wald, Maciej Turowski, Binghe Gu, Chengli Zu, Scott Kelso, Edward Bellinger, Justin Shearer N/A
1:50	(110-2)	**Effective Method Transfer: Ensuring End Results** MARY ELLEN P MCNALLY, DuPont Crop Protection, Stephen Platz
2:10	(110-3)	**Transferring Analytical Methods to a Quality Control Laboratory** AMANDA DARLAND, Dow AgroSciences N/A
2:30	(110-4)	**R&D to QC: An R&D Perspective in Chromatographic Method Development** JUSTIN SHEARER, Dow AgroSciences, Suresh Annangudi Palani, Daniel Knueppel, Rose Nelson

ORAL SESSIONS **Session 120**

Bioanalytical: LC Techniques - Half Session

Sunday Afternoon, Room B315
Sean M Burrows, Oregon State University, Presiding

1:30	(120-1)	**Characterization of New 2 µm Particle Size, 25 nm Pore Size Analytical Size Exclusion Chromatography Column with Larger Exclusion Limit Useful for the Separation of Biomolecules Using UHPLC and HPLC** CRYSTAL BENNER, Tosoh Bioscience LLC, Atis Chakrabarti
1:50	(120-2)	**Studies of Drug Interactions with Alpha$_1$-Acid Glycoprotein by Using On-Line Immunoextraction and High Performance Affinity Chromatography** CONG BI, University of Nebraska-Lincoln, Ryan Matsuda, Zitha Isingizwe, Chenhua Zhang, David Hage
2:10	(120-3)	**HPAE-PAD Applications for Stepwise Biosimilar Development Processes: Monosaccharide and Sialic Acid Determinations** HUA YANG, Thermo Fisher Scientific, Linda Lopez
2:30	(120-4)	**Analysis of Hormone-Protein Binding in Solution by Ultrafast Affinity Extraction** XIWEI ZHENG, University of Nebraska-Lincoln, Cong Bi, Marrisa Brooks, David Hage

Author and presider lists as of January 20, 2016, are available at www.pittcon.org

TECHNICAL PROGRAM

ORAL SESSIONS **Session 130**

Bioanalytical: LC/MS Techniques - Half Session

Sunday Afternoon, Room B315
Sean M Burrows, Oregon State University, Presiding

3:05 (130-1) **Discovery of Brown Recluse Spider Sex Pheromones Using Advanced Ultrafast Liquid Chromatography – Tandem Mass Spectrometry Methodologies** CASEY BURTON, Missouri University of Science and Technology, Ariel Donovan, Jennifer Parks, William Stoecker, Honglan Shi

3:25 (130-2) **Analysis of Herceptin Oxidation Variants Using a Supermacroporous Reverse Phase Column Coupled with an Orbitrap Mass Spectrometer** SHANHUA LIN, Thermo Fisher Scientific, Ilze Birznieks, Christopher Pohl, Xiaodong Liu, Jessica Wang, Terry Zhang, Jonathan Josephs

3:45 (130-3) **Quantitation of Endogenous Adenosine in Mouse Blood, Cell Lysate and Lysosomes by LC-MS/MS Using SurrogateMatrix Method** XIAO DING, Genentech

4:05 (130-4) **Isomeric Separation of Glycopeptides Using Porous Graphitic Carbon (PGC) LC at High Temperature** RUI ZHU, Texas Tech University, Jingfu Zhao, Yehia Mechref

ORAL SESSIONS **Session 140**

Chemical Methods - Half Session

Sunday Afternoon, Room B316
Richard Bormett, Renishaw, Presiding

1:30 (140-1) **Bromide-Assisted Anisotropic Growth of Surfactant-Free Gold Nanoparticles** MELISSA KERR, North Carolina Central University, Fei Yan

1:50 (140-2) **Simultaneous Reduction of Metal Ions by Multiple Reducing Agents Initiate the Asymmetric Growth of Metallic Nanocrystals** MAHMOUD MAHMOUD, Georgia Institute of Technology

2:10 (140-3) **Novel Optical Properties of Segmented Au-Ag Nanocylinders: Effects of Metallic Junctions on Surface Plasmon Resonance** VINEET KUMAR, North Carolina State University, Gufeng Wang

2:30 (140-4) **Measurement of Oxidative Potential of Particulate Matter by DTT assay** SHIORI OTA, Tokai University, Kazuhiro Misawa, Yoshika Sekine

ORAL SESSIONS **Session 150**

Gas Chromatography Innovations

Sunday Afternoon, Room B310
Olujide T Akinbo, Butler University, Presiding

1:30 (150-1) **Method Translation in Gas Chromatography to Get the Same Chromatogram** JAAP DE ZEEUW, Restek, Chris English, Chris Nelson, Jack Cochran

1:50 (150-2) **Chemometric Treatment of GC-VUV Data - Samples in a New Light** JAMES J HARYNUK, University of Alberta, Keisean Stevenson, Seo Lin Nam, Lawrence A Adutwum

2:10 (150-3) **All 5 Phases are Not the Same! Considerations for Method Development and Selection** RAMKUMAR DHANDAPANI, Phenomenex, Tim Anderson, Kristen Parnell

2:30 (150-4) **Flow-through Microfluidic Photoionization Detectors for Rapid and Highly Sensitive Vapor Detection** HONGBO ZHU, University of Michigan, Robert Nidetz, Menglian Zhou, Jiwon Lee, Sanketh Buggaveeti, Katsuo Kurabayashi, Xudong Fan

2:50 **Recess**

3:05 (150-5) **Safeguarding a Mass Detector from Difficult Sample Components** AMANDA B DLUGASCH, Waters Corporation, Thomas E Wheat, Patricia R McConville

3:25 (150-6) **Wide Range Inert Dilution System for Gas Standard Generation** STANLEY D STEARNS, Valco Instruments Co. Inc., Alex Plistil, Sunny Sinlapadech, David McCharthy, Huamin Cai

3:45 (150-7) **Multimode Plasma Emission Detector** YVES GAMACHE, Analytical Flow Products

4:05 (150-8) **Polyionic Ionic Liquid GC Stationary Phase Evaluations** LEONARD M SIDISKY, Supelco/Sigma-Aldrich, Greg A Baney, James L Desorcie, Gustavo Serrano, Daniel Shollenberger

ORAL SESSIONS **Session 160**

GC Optimization - Half Session

Sunday Afternoon, Room B309
David Benanou, Veolia, Presiding

1:30 (160-1) **Evaluation and Application of Sample Preparation Techniques for the Determination of Residual Volatiles in Biological Matrices Using GC-FID** ADRIENE MALSBURY, Bristol-Myers Squibb, William Fish, Frank Tomasella

1:50 (160-2) **A New Web-Based Application for Modelling Gas Chromatographic Separations** REBECCA STEVENS, Restek Corporation, Jaap de zeeuw, Amanda Rigdon, Linx Waclaski, Dan Li

2:10 (160-3) **Overcoming New Problems With Old Solutions: Enhanced Separations With GC Retention Gap Columns** RAMKUMAR DHANDAPANI, Phenomenex, Tim Anderson, Kristen Parnell

2:30 (160-4) **Improved Inertness Performance for Polyethylene Glycol GC Columns** KENNETH G LYNAM, Agilent Technologies, Ngoc-A Dang, Allen Vickers, Yun Zou

ORAL SESSIONS **Session 170**

Instrument Innovations - Half Session

Sunday Afternoon, Room B316
Richard Bormett, Renishaw, Presiding

3:05 (170-1) **Rapid Analysis of SO_2 to Determine Catalyst Efficiency** DEBBIE ALCORN, INFICON

3:25 (170-2) **Portable Gas Analyzer for Continuous Monitoring of Sulfur Dioxide in Gas Streams** SAYED A MARZOUK, UAE University, Mohamed Alnaqbi, Muna Bufaroosha, Mohamed Al-Marzouqi

3:45 (170-3) **Triple Mode of Action of L-tyrosine Derived Probes: Solvent Mediated Flip-Flop Halide (iodide/fluoride) Sensors and Reversible Chromogenic pH Indicators** SANJAY KUMAR MANDAL, IISER Mohali

4:05 (170-4) **Saving Time and Improving Accuracy by Eliminating the Need for Standards and Calibration in GC/FID Analyses** JONES ANDREW, Activated Research Company, Charlie Spanjers

ORAL SESSIONS **Session 180**

Measurement Strategies - Sensors and Spectroscopy

Sunday Afternoon, Room B313
Stephanie Archer-Hartmann, Complex Carbohydrate Research Center, Presiding

1:30 (180-1) **Real-Time Voltammetric Characterization of Non-Electroactive Metal Complexation** THUSHANI SIRIWARDHANE, University of South Carolina, Shawn McElmurry, Parastoo Hashemi

1:50 (180-2) **Phthalocyanine Based Microfluidic Sensors for the Detection of Oxidative Stress** KEVIN J KLUNDER, Colorado State University

2:10 (180-3) **Multiple Light Scattering to Characterize Emulsions with Polymers** JONATHAN DENIS, Formulaction, Yoann Lefeuvre, Pascal Bru, Christelle Tisserand, Gérard Meunier

2:30 (180-4) **A Practical Approach for Maximizing ICP-MS Data: A Closer Look at Microwave Sample Prep, The Steps that Precede It, and How to Minimize External Factors that Affect Data Quality** JOHAN NORTJE, Milestone Inc.

2:50 **Recess**

3:05 (180-5) **Obtaining Maximum Information from Fast Chemical Reactions Using a Photodiode Array (PDA) UV-Visible Spectrophotometer** IAN ROBERTSON, PerkinElmer Limited, Steve Upstone, Christopher Lynch

3:25 (180-6) **Microfluidic Visual Rheometer** CHRISTELLE TISSERAND, Formulaction, Patrick Abgrall, Patrycja Adamska, Gérard Meunier

3:45 (180-7) **Real-time Investigation of Antibiotics-induced Oxidative Stress and Superoxide Release in Bacteria Using an Electrochemical Biosensor** XIAOBO LIU, Clarkson University, Mouna Marrakchi, Michael Jahne, Shane Rogers, Silvana Andreescu

4:05 (180-8) **Fabrication of Micro Ir/IrO_x pH Sensor for Dental Applications** CHINDANAI RATANAPORNCHAROEN, Tokyo Medical and Dental University, Miyuki Tabata, Yuji Miyahara, Tatsuro Goda, Akira Matsumoto, Junji Tagami, Yuichi Kitasako, Masaomi Ikeda

Author and presider lists as of January 20, 2016, are available at www.pittcon.org

ORAL SESSIONS **Session 190**

Process Analytical Techniques - Half Session

Sunday Afternoon, Room B309
David Benanou, Veolia, Presiding

3:05	(190-1)	**A Model Study of Pseudo-Absolute Quantitative Analysis Using Gas Chromatography – Vacuum Ultraviolet Spectroscopy** LING BAI, The University of Texas at Arlington, Jonathan Smuts, Phillip Walsh, Kevin A Schug
3:25	(190-2)	**Rapid Process Control Using GC-ION Mobility Spectrometry** CHANDRASEKHARA HARIHARAN, ION-GAS GmbH, Wolfgang Vautz
3:45	(190-3)	**Facilitating Complex Analysis Using Multiple Fast Temperature Programming Zones** DALE ASHWORTH, Valco Instruments Co. Inc., Huamin Cai, Martin Brisbin, Chris Bishop, William Coontz, Andrew Rochon, Steve Werner, Stanley D Stearns
4:05	(190-4)	**Analyzing Trace Level Nitric Oxide in a Flammable Gas Matrix** KENNETH WONG, Air Liquide

SUNDAY POSTER SESSION **Session 200**

Sunday posters will be on display from 3:30 PM to 7:30 PM with authors present from 5:30 PM to 7:30 PM. The location for the Sunday posters is Room A412.

Sunday Poster Session
Sunday Afternoon, Room A412

(200-1 P)	**Influence of Particle Size Distribution on Protein Digestibility and Functionality** ALBERTA ARYEE, Agriculture & Agri-Food Canada, Joyce I Boye N/A
(200-2 P)	**Increasing Stability of a Core-Shell Particle** MARK WOODRUFF, Fortis Technologies, Ken Butchart
(200-3 P)	**Rapid Screening of Virgin and Recycled Polymer Resins Using FTIR and Raman Libraries of Pre-Computed Mixture Spectra** WILLIAM COSTA, Fiveash Data Management, Inc., Bill McCarthy, Todd Strother
(200-4 P)	**The Complimentary Nature of NMR and SPME GCMS for the Analysis of Fermented Beverages** NEIL FITZGERALD, Marist College, Samantha E Soprano, Sarah R Johnson, John C Edwards
(200-5 P)	**Evaluation of an Integrated Optic-Fiber-Microfluidic Analyzer for Polyphenols Measurement** MARIA CAÑIZARES-MACÍAS, Universidad Nacional Autónoma de México, Oscar Sandoval-Ventura, Luis F Olguín-Contreras N/A
(200-6 P)	**Selective Separation and Detection of Catecholamines with Capillary Electrophoresis** MAOJUN GONG, Wichita State University, Qiyang Zhang
(200-7 P)	**Cold EI – Approaching the Ideal GC-MS Interface and Ion Source** AVIV AMIRAV, Tel Aviv University, Bogdan Belgorodsky, Uri Keshet, Alexander Fialkov, Tal Alon
(200-8 P)	**New Sampling Device for the Analysis of Lung Cancer Biomarkers in Exhaled Breath** MOHAMED ABDEL-REHIM, Stockholm University N/A
(200-9 P)	**Ceramic Ion Source for Catalytic Pyrolysis and Catalytic Combustion Ionization Detection of Selected Constituents in Complex Petroleum and Biofuel Samples** PAUL L PATTERSON, Detector Engineering & Technology, Inc., Jennifer Seroy
(200-10 P)	**Using Microscopy to Measure Rates of Heterogeneous Reactions** WALTER BOWYER, Hobart and William Smith Colleges, Gabriella Mylod
(200-11 P)	**Crowding in Cell-Like Environments Alters Diffusion and Enzyme Kinetics** KRISTIN M SLADE, Hobart and William Smith Colleges, Michael Conroy, Sophia Melvin
(200-12 P)	**Effects of Macromolecular Crowding on YADH Enzyme Mechanism** KRISTIN M SLADE, Hobart and William Smith Colleges, Allison Wilcox
(200-13 P)	**Mixed Valence Mn,La,Sr-oxide based Magnetic Nanoparticles Coated with Silica Nanoparticles for Immunosensor Fabrication** AMOS MUGWERU, Rowan University
(200-14 P)	**Hydrophilic Interaction Chromatography (HILIC) and Enzymatic/Spectrometric Methods for the Determination of Uric Acid and Creatinine in Human Biofluids** YUEGANG ZUO, University of Massachusetts Dartmouth, Si Zhou, Xiaofei Lu, Ningning Zhang, Faten Albalawi
(200-15 P)	**Profiling/Monitoring Activated Carbons for Drinking Water** HENRY G NOWICKI, PACS Inc
(200-16 P)	**Profiling/Monitoring Gas Phase Activated Carbon Systems** HENRY G NOWICKI, PACS Inc
(200-17 P)	**Lignin Waste Transformed to High Quality Activated Carbons** HENRY G NOWICKI, PACS Inc
(200-18 P)	**A Rapid Hydrophilic Interaction Liquid Chromatography (HILIC) Method for Determination of Trace Nitrate and Nitrite in Snow and Rain Samples** XIAOFEI LU, University of Massachusetts Dartmouth, Yuegang Zuo
(200-19 P)	**Occurrence and Identification of Bisphenol A and Other Alkylphenols in Sea Shore Crabs** JOSEPH MICHAEL, University of Massachusetts Dartmouth, Yuegang Zuo
(200-20 P)	**Evaluation of Porous Layer Thickness of Core Shell Particle for Separation of Proteins** NORIKAZU NAGAE, ChromaNik Technologies Inc., Tomoyasu Tsukamoto

ACS POSTER **Session 210**

ACS posters will be on display from 3:30 PM to 7:30 PM with authors present from 5:30 PM to 7:30 PM. The location for the ACS posters is Room A412.

ACS Division of Analytical Chemistry Poster Session
Sunday Afternoon, Room A412

(210-1 P)	**Phospholipid/Aromatic Thiol Hybrid Bilayers** CHAO LI, Auburn University, Mingming Wang, Matthew Ferguson, Wei Zhan
(210-2 P)	**Dynamic Records of Spatiotemporal Pb: Lichens versus Trees and Sediments** NATHAN W BOWER, Colorado College, Ben Greydanus, Sam Kramer, Eric Wolatz, Stephen Getty, Craig Lundstrom
(210-3 P)	**Dating Paintings and Prints Using ATR-FTIR: A Philatelic Case Study from Lodz Ghetto** NATHAN W BOWER, Colorado College, Conor J Blanchet, Michael S Epstein
(210-4 P)	**SPME/GC-MS Analysis of Dynamic Compositional Changes in the Volatile Components in Red Wine After "Breathing"** AMBER I HILLS, East Stroudsburg University, Jon S Gold, Richard S Kelly
(210-5 P)	**Development of Portable Elisa System for Infectious Disease Diagnosis** KAZUHIRO MORIOKA, Tokyo Metropolitan University, Harpal Singh, Hizuru Nakajima, Akihide Hemmi, Masayuki Shimojima, Le Van An, Sazaly AbuBakar, Hulie Zeng, Shungo Kato, Masami Sugamata, Ming Yang, Katsumi Uchiyama
(210-6 P)	**Applications of a Quartz Crystal Microbalance for Monitoring Bacterial Biofilm Growth and Removal** HUNTER J SISMAET, Northeastern University, Pegah N Abadian, Edgar D Goluch
(210-7 P)	**A Capillary Electrophoresis Approach to the Characterization and Application of Graphene Quantum Dots as Sensing Agents** LEONA SIRKISOON, Wake Forest University, Honest Makamba, Shingo Saito, Christa Colyer
(210-8 P)	**A pH Sensor for Non-Invasive In Vivo Detection and Imaging of Implant Associated Infection** UNAIZA UZAIR, Clemson University
(210-9 P)	**Scanning Electrospray Microscopy with Nanopipettes** ELIZABETH M YUILL, Indiana University, Wenqing Shi, Lane A Baker
(210-10 P)	**Phosphoproteomics Studies of Human Immunodeficiency Virus-1** KEVIN MARK, LaGuardia Community College City University of NY, Pratikkumar Rathod, Emmanuel Chang, Hsin-Pin Ho, Xu Yu, Mathias Lichterfeld
(210-11 P)	**Segmented Flow Sampling with Theta Pipettes** ANUMITA SAHA-SHAH, Indiana University, Curtis M Green, Lane A Baker
(210-12 P)	**Mapping Local Permeability Change During Degradation with Scanning Ion Conductance Microscopy-Scanning Electrochemical Microscopy (SICM-SECM)** WENQING SHI, Indiana University, Lane A Baker
(210-13 P)	**Profiling N, N'-Dibutylbenzimidazolium Salt and Its Derivatives Using CYCLIC Voltammetry** HUGGINS Z MSIMANGA, Kennesaw State University, Andrew Montalvo, Daniela Tapu
(210-14 P)	**Comparative Study of Elemental Nutrients in Organic and Conventional Vegetables by Laser Induced Breakdown Spectroscopy (LIBS)** CHET R BHATT, Mississippi State University
(210-15 P)	**Photoacoustic Spectroscopy with SF6, An Optically Thick Greenhouse Gas** HAN PARK, University of Tennessee at Chattanooga
(210-16 P)	**Integrating Authentic Research Experiences into Undergraduate Analytical Chemistry** FEI YAN, North Carolina Central University, Melissa Kerr
(210-17 P)	**Proteomic Analysis of Tetrahymena Thermophila Using MALDI-TOF/TOF** DOUGLAS BEUSSMAN, St. Olaf College, Harrison VanDolah
(210-18 P)	**Possible Age and Location Differences in Human Scent Profiles** DOUGLAS BEUSSMAN, St. Olaf College, Laura Muehlbauer
(210-19 P)	**LC-QTOF Detection of Methamphetamine in Packaging Residue** DOUGLAS BEUSSMAN, St. Olaf College, Matthew Bock
(210-20 P)	**Environmental Toxins from Decorative Candles** DOUGLAS BEUSSMAN, St. Olaf College, Allison Sager
(210-21 P)	**Analysis of Fibers Via Isotope Ratio Mass Spectrometry** DOUGLAS BEUSSMAN, St. Olaf College, Elaine Macon, Hannah Brown
(210-22 P)	**Withdrawn**
(210-23 P)	**HPLC Method Development for Analysis of Dissolution Samples of a Highly Soluble Drug Substance in a Hydrophobic Matrix** MATTHEW LOUCKS, Banner Life Sciences, Angela Moore, Wayne Craig, Sara Draper
(210-24 P)	**Confocal Raman Microscopy Investigation of the Kinetic Barrier to PAH Partitioning into Individual C_{18}–Silica Particles** DAVID A BRYCE, University of Utah, Jay P Kitt, Joel M Harris
(210-25 P)	**Development of a Flow-Based Electrogenerated Chemiluminescent Detection of Biogenic Amines on a Microfluidic Chip** ERIN GROSS, Creighton University, Leah Schaffer, Emily Lowry, Charles Henry, Rachel M Feeny, John Wydallis

TECHNICAL PROGRAM

MONDAY, MARCH 7, 2016
MORNING

AWARDS — **Session 220**

Chromatography Forum of Delaware Valley Dal Nogare Award
arranged by Mary Ellen McNally, E.I. DuPont de Nemours and Company

Monday Morning, Room B312
Mary Ellen McNally, E.I. DuPont de Nemours and Company, Presiding

8:30		**Introductory Remarks - Mary Ellen McNally**
8:35		**Presentation of the 2016 Dal Nogare Award to Stephen Weber, University of Pittsburgh, by Mary Ellen McNally, E.I. DuPont de Nemours and Company**
8:40	(220-1)	**Opportunities and Challenges with Capillary Liquid Chromatography** STEPHEN WEBER, University of Pittsburgh
9:15	(220-2)	**How Much Performance is Lost in LC by Optimizing Only Velocity and Column Length Using Limited Number of Particle Size?** PETER W CARR, University of Minnesota, Adam Matula, Dwight R Stoll
9:50	(220-3)	**UHPLC-MS for Multiplexed Neurochemical Analysis** ROBERT T KENNEDY, University of Michigan
10:25		**Recess**
10:40	(220-4)	**Integrated Microfluidic Separations Devices Interfaced to Mass Spectrometry** J MICHAEL RAMSEY, University of North Carolina at Chapel Hill, Erin Redman, William Black
11:15	(220-5)	**The Second Dimension is a Strange Place - Fundamental Aspects of the Optimization of the Second Dimension in Two-Dimensional Liquid Chromatography** DWIGHT R STOLL, Gustavus Adolphus College, John Halvorson, Carston Dammann, Eli Larson, Alex Wilson, David C Harmes, Monika Dittmann, Sarah C Rutan, Abraham Lenhoff

AWARDS — **Session 230**

The Pittsburgh Conference Achievement Award
arranged by Amit Ghosh, Bayer Materials Science LLC

Monday Morning, Room B314
Amit Ghosh, Bayer Materials Science LLC, Presiding

8:30		**Introductory Remarks - Amit Ghosh**
8:35		**Presentation of the 2016 Pittsburgh Conference Achievement Award to Jared L Anderson, Iowa State University, by Elias S Absey, Chair, Society for Analytical Chemists of Pittsburgh**
8:40	(230-1)	**Exploiting Ionic Liquids, Magnetic Ionic Liquids, and Polymeric Ionic Liquids in Sample Preparation and Multidimensional Gas Chromatography** JARED L ANDERSON, Iowa State University
9:15	(230-2)	**The Impact and Evolution of High Efficiency Chiral and Achiral Separations** DANIEL W ARMSTRONG, University of Texas at Arlington
9:50	(230-3)	**Use of Metal-Organic Frameworks in Sample Preparation Analytical Schemes: An Overview of Their Performance in Solid-Phase Extraction Modes** VERÓNICA PINO, University of La Laguna, Priscilla Rocio-Bautista, Juan H Ayala, Jorge Pasán, Catalina Ruiz-Pérez, Ana M Afonso
10:25		**Recess**
10:40	(230-4)	**Achiral LSER Investigation of HPLC Retention on Cinchona Alkaloid-Based Chiral Stationary Phases** APRYLL M STALCUP, Dublin City University
11:15	(230-5)	**Applications of Conductive Ionic Liquids in Chemical Analysis** JON R KIRCHHOFF, University of Toledo, Joshua A Young, Amila M Devasurendra, LM Viranga Tillekeratne, Jared L Anderson, Cheng Zhang

SYMPOSIUM — **Session 240**

ACS-ANYL - BRAIN Initiative Advancements in Neurochemical and Physiological Measurements
arranged by Michael L Heien, University of Arizona

Monday Morning, Room B308
Michael L Heien, University of Arizona, Presiding

8:30		**Introductory Remarks - Michael L Heien**
8:35	(240-1)	**New Platforms for Multiplexing Electrochemical Measurements In Vivo** MICHAEL L HEIEN, University of Arizona
9:10	(240-2)	**Simultaneous Detection of Dopamine Release and Multiple Single-Unit Activity in Awake and Behaving Rats** STEPHEN COWEN, University of Arizona, Michael L Heien, Kate L Parent, Daniel F Hill, Jean-Paul Wiegand, Michael A Miller, Christopher W Atcherley
9:45	(240-3)	**New Nano Tools for Molecular Sensing and Imaging of Single Neuron-Neuron Communication** X NANCY XU, Old Dominion University, Pavan K Cherukuri, Preeyaporn Songkiatisak
10:20		**Recess**
10:35	(240-4)	**In Vivo Multiphoton Microscopy of Mouse Brain** CHRIS XU, Cornell University
11:10		**Open Discussion**

SYMPOSIUM — **Session 250**

Advances in Raman Spectroscopy
arranged by Sanford A Asher, University of Pittsburgh

Monday Morning, Room B304
Sanford A Asher, University of Pittsburgh, Presiding

8:30		**Introductory Remarks - Sanford A Asher**
8:35	(250-1)	**Micro-Scale Spatially Offset Raman Spectroscopy (Micro-SORS)** PAVEL MATOUSEK, Rutherford Appleton Laboratory, Claudia Conti, Marco Realini, Chiara Colombo, Giuseppe Zerbi
9:10	(250-2)	**Recent Advances in Raman Optical Activity (ROA) Methodology and Applications** LAURENCE A NAFIE, Syracuse University, Rina K Dukor
9:45	(250-3)	**Fast Raman Imaging Using Optimized Binary Compressive Detection** DOR BEN-AMOTZ, Purdue University, Owen G Rehrauer, Bharat R Mankani, Gregery T Buzzard, Bradley J Lucier, Stanley Chan
10:20		**Recess**
10:35	(250-4)	**Stimulated Raman Scattering Microscopy of Vibrational Tags** WEI MIN, Columbia University
11:10	(250-5)	**Nanoparticle Based Analysis of Cells, Molecules and Tissue by SERS** DUNCAN GRAHAM, University of Strathclyde, Samuel Mabbott, Lee Barrett, Steven Asiala, Kirsten Gracie, Karen Faulds

SYMPOSIUM — **Session 260**

Innovative Applications of Mass Spectrometry in Biopharmaceutical and Diagnostic Development
arranged by William Christopher Lamanna, Sandoz

Monday Morning, Room B305
William Christopher Lamanna, Sandoz, Presiding

8:30		**Introductory Remarks - William Christopher Lamanna**
8:35	(260-1)	**Characterization of Filgrastim Using Intact and Top-Down MS** WILLIAM CHRISTOPHER LAMANNA, Sandoz, Johann Holzmann, Hansjoerg Toll
9:10	(260-2)	**Tandem Mass Spectrometry as an Excellent Quantification Tool to Guide Process Development for Optimal Host Cell Protein Removal in Biopharmaceutical Products** DONALD WALKER, Genentech
9:45	(260-3)	**Antibody Characterization with Ultrahigh-Resolution Mass Spectrometry: Selecting Well Behaved Molecules for Optimal Product Quality Attributes** LISA MARZILLI, Pfizer, Inc., Mellissa Ly, Elaine Stephens, Heather S DeGruttola, Andrew E Saati, Jason C Rouse
10:20		**Recess**

10:35	(260-4)	**Advanced Process Control Using Fast Analytics to Monitor Multiple Critical Quality Attributes of a Monoclonal Antibody** LI ZANG, Biogen
11:10	(260-5)	**Analysis of Glycosaminoglycan Non-Reducing End Structures by Liquid Chromatography Tandem Mass Spectrometry and Its Uses in Discovering New Disease-Specific Biomarkers** ROGER LAWRENCE, BioMarin, Heather Prill, Evelyn Wang, Raymond Y Wang, Jeffery D Esko, Brett Crawford, Marzia Pasquali, Toni Pollock, William Christopher Lamanna, Nancy Pryer

SYMPOSIUM **Session 270**

Neurotransmission at Single and Nano-Resolved Bio-Structures
arranged by Mei Shen, University of Illinois at Urbana-Champaign and Christian A Amatore, CNRS & ENS

Monday Morning, Room B303
Mei Shen, University of Illinois at Urbana-Champaign, Presiding

8:30		**Introductory Remarks - Mei Shen and Christian Amatore**
8:35	(270-1)	**Electrochemical Cytometry of the Contents of Nanometer Vesicles Out and Inside Cells** ANDREW EWING, University of Gothenburg and Chalmers University
9:10	(270-2)	**Detection of Ionic Neurotransmitters at Biological Nanostructures via Versatile Nanosensor Probes** MEI SHEN, University of Illinois at Urbana-Champaign
9:45	(270-3)	**Electrochemical Approaches to Monitoring Neural Activity** R MARK WIGHTMAN, University of North Carolina at Chapel Hill
10:20		**Recess**
10:35	(270-4)	**Full Fusion During Vesicular Exocytosis: A Real Release Stage or a Hoax?** CHRISTIAN A AMATORE, CNRS & ENS
11:10	(270-5)	**Coulter Counter Voltage Trapping of Nanoparticles with Sub-Nanometer Size Resolution** HENRY WHITE, University of Utah, Sean German, Martin Edwards, Yulun Zhang

SYMPOSIUM **Session 280**

Omics for Environmental and Public Health Protection
arranged by Jeanette Van Emon, U.S. EPA and Hercules Moura, CDC

Monday Morning, Room B309
Jeanette Van Emon, U.S. EPA, Presiding

8:30		**Introductory Remarks - Jeanette Van Emon and Hercules Moura**
8:35	(280-1)	**Applications of Accelerator Mass Spectrometry to Human Health Research** TED OGNIBENE, Lawrence Livermore National Laboratory
9:10	(280-2)	**Proteomic Approaches for Molecular Epidemiological Studies** ROLF U HALDEN, Arizona State University, Nicole Hansmeier, Tzu-Chiao Chao, Julie B Herbstman, Lynn R Goldman, Frank R Witter
9:45	(280-3)	**Proteomic Applications in Microbiology** HERCULES MOURA, CDC NCEH-DLS, John R Barr
10:20		**Recess**
10:35	(280-4)	**Development of Metagenomic Tools for Identification and Characterization of Infectious Agents from Clinical Samples and for Microbiome Association Studies** JAMES POSEY, CDC
11:10	(280-5)	**Proteomics for Adverse Outcome Pathway Discovery for Environmental Contaminants** JEANETTE VAN EMON, U.S. EPA

SYMPOSIUM **Session 290**

SEAC - Nano-Electroanalysis for a Sustainable World
arranged by Joaquin Rodriguez Lopez, University of Illinois

Monday Morning, Room B310
Joaquin Rodriguez Lopez, University of Illinois, Presiding

8:30		**Introductory Remarks - Joaquin Rodriguez Lopez**
8:35	(290-1)	**Chemically Modified 3D Energy-Storage Architectures** DEBRA R ROLISON, U.S. Naval Research Laboratory, Jeffrey W Long, Christopher N Chervin, Joseph F Parker, Megan B Sassin
9:10	(290-2)	**Sunlight-Driven Hydrogen Formation by Membrane-Supported Photoelectrochemical Water Splitting** NATHAN S LEWIS, Caltech
9:45	(290-3)	**Elucidating Charge Transfer on Polymer Nano-Structures for a New Concept in Energy Storage** JOAQUIN RODRIGUEZ LOPEZ, University of Illinois
10:20		**Recess**
10:35	(290-4)	**Ion Transport in 1D and 3D Networked Porous Nanostructure Electrodes** SANG BOK LEE, University of Maryland, Eleanor Gillette
11:10	(290-5)	**Operando Methods for the Characterization Energy Materials** HECTOR D ABRUNA, Cornell University

ORGANIZED CONTRIBUTED SESSIONS **Session 300**

Infrared Spectroscopy Beyond the Diffraction Limit
arranged by Ellen Miseo, Hamamatsu Corporation and Craig Prater, Anasys Instruments

Monday Morning, Room B311
Ellen Miseo, Hamamatsu Corporation, Presiding

8:30	(300-1)	**Advancing the Field of Infrared Nanospectroscopy for Materials and Life Sciences** CRAIG PRATER, Anasys Instruments
8:50	(300-2)	**Resonance Enhanced AFM-IR Induced by Quantum Cascade Laser** ALEXANDRE DAZZI, University Paris-Sud, Jeremie Mathurin, Ariane Deniset-Besseau, Johanna Saunier, Najet Yagoubi, Kevin Kjoller
9:10	(300-3)	**Unprecedented Nanoscale Chemical Characterization of Materials Using AFM-Based Infrared Spectroscopy** CURTIS MARCOTT, Light Light Solutions, Dillon Eoghan, Kevin Kjoller
9:30	(300-4)	**Withdrawn**
9:50		**Recess**
10:05	(300-5)	**AFM-IR Spectroscopy and Imaging of Biodegradable Polymers** JOHN F RABOLT, University of Delaware, Liang Gong, Isao Noda, Bruce Chase
10:25	(300-6)	**Nanoscale Molecular Imaging of Polymer Systems Using AFM-IR** MARK RICKARD, Dow Chemical, Gregory Meyers, Carl Reinhardt, Jamie Stanley
10:45	(300-7)	**Absorption Spectroscopy and Imaging from the Visible through Mid-IR with 20 nm Resolution Using AFM Probes** ANDREA CENTRONE, NIST
11:05	(300-8)	**Open Discussion**

ORGANIZED CONTRIBUTED SESSIONS **Session 310**

Ionophore-Based Chemical Sensors II
arranged by Philippe Buhlmann, University of Minnesota and Eric Bakker, University of Geneva

Monday Morning, Room B404
Philippe Buhlmann, University of Minnesota, Presiding

8:30	(310-1)	**Magnetic Nanoparticles as Dispersible Electrodes** JUSTIN J GOODING, University of New South Wales, Roya Tavallaie, Elizabeth Morago, Saimon M Silva, Kyloon Chuah
8:50	(310-2)	**Development of an Ultra-Selective Optode Nanosensor for Potassium Imaging** ALI SAHARI, Northeastern University, Tim Ruckh, Richard Hutchings, Heather Clark
9:10	(310-3)	**Low Detection Limit of Ion-Selective Electrodes; Is the Story Really Over?** ALEKSANDAR RADU, Keele University, Lukasz K Mendecki, Sergio Granados-Focil, Christina McGraw, Peter Dillingham
9:30	(310-4)	**Multifunctional Detection and Delivery** ELIZABETH (LISA) HALL, University of Cambridge, Nadia Tsao
9:50		**Recess**

10:05	(310-5)	**Printed Paper-Based Ion-Selective Optode Devices** DANIEL CITTERIO, Keio University, Hiroyuki Shibata, Terence G Henares, Nobutoshi Komuro, Koji Suzuki
10:25	(310-6)	**Conducting Polymers Nanospheres for Sensors – Receptor or Transducers?** AGATA MICHALSKA, University of Warsaw, Katarzyna Klucinska, Anna Kisiel, Krzysztof Maksymiuk N/A
10:45	(310-7)	**Signal Transduction Based on Constant-Potential Coulometry for Solid-Contact ISEs** JOHAN BOBACKA, Åbo Akademi University, Ulriika Vanamo, Elisa Hupa, Ville Yrjänä
11:05	(310-8)	**Plasticizer-Free Paper-Based Ion-Selective Optodes** XUEWEI WANG, University of Michigan, Mark Meyerhoff, Yu Qin

ORGANIZED CONTRIBUTED SESSIONS — Session 320

New Perspectives on the History of Chromatography

arranged by Richard Ulrych, Chemical Heritage Foundation

Monday Morning, Room B407
Sarah Reisert, Chemical Heritage Foundation, Presiding

8:30	(320-1)	**Building the Market for Ion Chromatography: The Early Days** ARTHUR WILLIAM FITCHETT, Thermo Fisher Scientific
8:50	(320-2)	**Early Days in GC 1957-1962** HAROLD MONROE MCNAIR, Virginia Tech
9:10	(320-3)	**Evolution of Capillary Chromatography and Capillary Electrophoresis: A Personal Perspective** MILOS V NOVOTNY, Indiana University
9:30	(320-4)	**The Development of HPLC Method Development** LLOYD R SNYDER, LC Resources Inc.
9:50		Recess
10:05	(320-5)	**Pioneering Days of Chromatography in Lab Automation** JACK GILL, Chemical Heritage Foundation
10:25	(320-6)	**The Role of Temperature Programming in F&M's Success** AARON J MARTIN, Marlabs, Retired
10:45		**Liquid Chromatography Before It Became HPLC** JAMES WATERS, Chemical Heritage Foundation N/A
11:05	(320-7)	**Chromatography In, or Next to Chemistry? Discipline Formation and Identity of Chromatography Practitioners in the 1960s and 1970s** APOSTOLOS GERONTAS, Coburg University of Applied Sciences

ORAL SESSIONS — Session 330

Biomedical: Advances in Point-of-Care Technologies

Monday Morning, Room B313
Mustafa Culha, Yeditepe University, Presiding

8:30	(330-1)	**Low-Cost Molecular Diagnostics on a Smartphone** HYUNGSOON IM, Massachusetts General Hospital, Cesar M Castro, Huilin Shao, Monty Liong, Jun Song, Divay Pathania, Lioubov Fexon, Changwook Min, Rosemary H Tambouret, Misha Pivovarov, Ralph Weissleder, Hakho Lee
8:50	(330-2)	**Smartphone Based Detection of Stress Biomarkers in Saliva** AADHAR JAIN, Cornell University, Elizabeth Rey, David Erickson, Dakota O'Dell, Seoho Lee
9:10	(330-3)	**A Smartphone Platform for Quantitative Point-of-Care Detection of Micronutrient Deficiencies** DAKOTA O'DELL, Cornell University, Seoho Lee, Jessica Hohenstein, David Erickson
9:30	(330-4)	**Bed-Side Detection of Volatile Human Metabolites for Medical Applications** CHANDRASEKHARA HARIHARAN, ION-GAS GmbH, Wolfgang Vautz, Sascha Liedtke
9:50		Recess
10:05	(330-5)	**Improved Optical Cavity Based Biosensor with Differential Detection Method Through Simultaneous Detection** TONY BUJANA, Letourneau University, DongGee Rho, Cody Joy, Peter Cowles, Seunghyun Kim
10:25	(330-6)	**A PDMS/Paper Hybrid Microfluidic Biochip for Multiplexed Instrument-Free Bacterial Meningitis Diagnosis** MAOWEI DOU, University of Texas El Paso, Sanjay Sharma Timilsina, Juan Sanchez, Delfina Dominguez, Xiujun James Li
10:45	(330-7)	**Measuring Biochemical Effects of Pulmonary Rehabilitation and OMT on COPD Patients through LC-MS-MS Analysis of Plasma Metabolites** CHEN ZHANG, Michigan State University, Sherman Gorbis, John Wang, A Daniel Jones
11:05	(330-8)	**Surface-Enhanced Raman Scattering for In Situ Analysis of Living Systems** MUSTAFA CULHA, Yeditepe University

ORAL SESSIONS — Session 340

Capillary Electrophoresis

Monday Morning, Room B302
Christopher R Harrison, San Diego State University, Presiding

8:30	(340-1)	**Quantitation of Kinase Activity in a Social Amoeba Using Capillary Electrophoresis and a Peptide Substrate Reporter** MICHELLE L KOVARIK, Trinity College, Kunwei Yang, Allison J Tierney
8:50	(340-2)	**Development of a Degradation Resistant Peptide Reporter for Monitoring E3 Ligase and Proteasome Activity** GREGERY WOSS, University of North Carolina at Chapel Hill, Nancy Allbritton, Marcey Waters, Adam Melvin, Kaiulani Houston
9:10	(340-3)	**A New Analytical Technique to Profile Multiple Steroids in Individual Female Zebrafish to Study Endocrine Disruption** VINCENT T NYAKUBAYA, West Virginia University, Paige A Reed, William J Feeney, Regina Rockwell, Amber Kantes, Jennifer Ripley-Stueckle, Lisa A Holland
9:30	(340-4)	**A Miniaturized Immunoaffinity Capillary Electrophoresis Based-Biomarker Analyzer for Bioanalytical Applications** NORBERTO A GUZMAN, Princeton Biochemicals, Inc., Daniel E Guzman
9:50		Recess
10:05	(340-5)	**Electrophoretic Technologies for Catecholamine Detection** MAOJUN GONG, Wichita State University, Qiyang Zhang, Maddukuri Naveen
10:25	(340-6)	**Integrating Microscale Enzyme Reactions Into Capillary Separation** SRIKANTH GATTU, West Virginia University, Cassandra Crihfield, Lisa A Holland, Anthony Moncrief
10:45	(340-7)	**Investigation of Oxidative Metabolism by Electrochemistry/Capillary Electrophoresis/Mass Spectrometry Using a Novel Sheathless Interface Design** NHAN TO, University of Kansas, Ryan T Johnson, John Stobaugh, Craig E Lunte
11:05	(340-8)	**Manipulation of the EOF in Phospholipid Coated Capillaries Through the Incorporation of Different Metal Cations** CHRISTOPHER R HARRISON, San Diego State University, Shane Wells, Eduardo De La Toba, Srilatha Vydha

ORAL SESSIONS — Session 350

Data Analysis and Manipulation

Monday Morning, Room B301
A Peter Snyder, Retired Government, Presiding

8:30	(350-1)	**Quantitative Evaluation of Spectral Interferences in Atomic Emission Spectroscopy** MATTHIEU BAUDELET, University of Central Florida, Jessica Chappell, Brandon Seesahai, Martin Richardson, Michael Sigman
8:50	(350-2)	**Comparison of Various Methods for the Determination of Uncertainty due to Long Term Stability** DANIEL BIGGERSTAFF, o2si Smart Solutions, Huichen Stavros, Mark Filla
9:10	(350-3)	**Lifecycle of Analytical Methods: Development of Equivalent Dissolution Methods for Immediate-Release Oral Dosage Forms Post-Approval** IVELISSE COLON, Vertex Pharmaceuticals, Taryn Ryan, Joseph Medendorp
9:30	(350-4)	**Investigating Robustness and Ruggedness of Analytical Methods Employing aQBD Principles** PETER ANDREAS LUND JACOBSEN, Fertin Pharma
9:50		Recess
10:05	(350-5)	**Chemometrics for Big Data** ROBERT A LODDER, University of Kentucky, Anne Brooks
10:25	(350-6)	**Air Quality Networks- Results from Validation and Lessons on Calibration** JOHN R SAFFELL, Alphasense Ltd., Roderic L Jones
10:45	(350-7)	**Using Prior Probabilities to Increase the Confidence of Chemical Identification** TYLER A ZIMMERMAN, NIST, Tytus D Mak, W Gary Mallard, Nirina R Andriamaharavo, Dmitrii V Tchekhovskoi, Stephen E Stein
11:05	(350-8)	**Applied Analytics: Using Performance-Based Analytical Test Methodology for Monitoring Laboratory Methods Required by EPA Tier III Standards** JOHN MAURER, Valero

ORAL SESSIONS — Session 360

Environmental LC

Monday Morning, Room B408
Ariel Donovan, Missouri University Science and Technology, Presiding

8:30	(360-1)	**Intrigues of Analyte Peak Distortion in Ion Chromatography by Overloaded Matrix Ions –Trace Analysis of Bromate in High Ionic Strength Samples** MICHAEL K PAPPOE, University of Alberta, Mohammad H Naeeni, Charles A Lucy
8:50	(360-2)	**Multi-Dimensional Ion Chromatography for Trace Ion Analysis** RONG LIN, Thermo Fisher Scientific, Kannan Srinivasan, Herb Wagner
9:10	(360-3)	**Assessment of Amino Acid Stability in the Early Oceans by Liquid Chromatography-Mass Spectrometry** ERIC T PARKER, Georgia Institute of Technology, Karen L Britnon, Aaron S Burton, Daniel P Glavin, Jason P Dworkin, Jeffrey L Bada
9:30	(360-4)	**Increase Sample Productivity in the Environmental Lab with High Pressure Ion Chromatography** FRANK HOEFLER, Thermo Fisher Scientific, David G Moore, Yan Liu
9:50		**Recess**
10:05	(360-5)	**A Novel Cation Exchange Stationary Phase for Analysis of Common Cations and Amines Using Ion Chromatography** MANI JAYARAMAN, Thermo Fisher Scientific, Christopher Pohl, Charanjit Saini, Yan Liu
10:25	(360-6)	**Effects of Titanium Dioxide Nanoparticles on Endocrine Disruption in Zebrafish** MARRIAH ELLINGTON, West Virginia University, Cassandra Crihfield, Sara Melow, Lisa A Holland
10:45	(360-7)	**Trace Analysis of Guanidine Compounds in Surface Water with Resorcinarene-Based Ion Chromatography Columns** ROGER G HARRISON, Brigham Young University, Tayyabeh Panahi, John Lamb
11:05	(360-8)	**Analysis of Molecular Markers of Animal Waste by LC-MS/MS** SREE HARSHITHA VELAGA, Tennessee Technological University, Sreedharan Lakshmi Narayanan, John Harwood

ORAL SESSIONS — Session 370

Fluorescence and Luminescence Advances

Monday Morning, Room B315
Chenzhong Li, Florida International University, Presiding

8:30	(370-1)	**Measuring Displacement on the Micron Scale: Novel Luminescent Spectral Rulers for Orthopedic Applications** MELISSA M ROGALSKI, Clemson University, Donald Benza, Hunter Pelham, Nakul Ravikumar, Dakotah Anderson, Johnathan Heath, John D DesJardins, Jeffrey N Anker
8:50	(370-2)	**Simplified Two-Photon Synchronous Scanning Fluorescence for Resolution of Co-Localized Emitters** CHRISTOPHER K ALMLIE, Oregon State University, Sean M Burrows, Kuan-Jen Chen, Karan A Patel
9:10	(370-3)	**Development of Chitosan-Modified Near Infrared Fluorescent Graphene Oxide Nanocomposite** MINH H DUONG, University of North Dakota, Steve Xu Wu, Julia Xiaojun Zhao
9:30	(370-4)	**Studying Chemical Reaction, Mass Transport and Their Coupling in 3D Multi-layer Catalysts at Single-Molecule Level** BIN DONG, Georgia State University, Chen Y Pei, Xiao X Chao, Wen Y Huang, Ning Fang
9:50		**Recess**
10:05	(370-5)	**Slowing Down of Nanoparticle Diffusion on Sub-Micrometer Oil Droplet-Aqueous Buffer Interface Studied with Three Dimensional Tracking** YANING ZHONG, North Carolina State University, Luyang Zhao, Gufeng Wang
10:25	(370-6)	**Two-Photon Metal Enhanced Fluorescence Properties of Gold Nanostars** LIXIA ZHOU, Oregon State University, Sean M Burrows
10:45	(370-7)	**DNA-Aligner Controlled Nicking and Extension for Isothermal Amplification of Nucleic Acids** TAO ZHANG, Zhejiang University N/A
11:05	(370-8)	**Construction of Convenient Biosensors Based on Optical Techniques** QIAOLI YUE, Liaocheng University, Chenzhong Li

ORAL SESSIONS — Session 380

FTIR and Terahertz Applications

Monday Morning, Room B316
Chin I Shyr, The Pittsburgh Conference, Presiding

8:30	(380-1)	**EPA Methods 320 and 18 by Portable GC/FTIR for Total Source Emission Determination** MARTIN L SPARTZ, Prism Analytical Technologies, Inc., Anthony S Bonanno, Charles Mark Phillips, Peter P Behnke, Kelly R McPartland
8:50	(380-2)	**A Novel Infrared Interferometer Suitable for 3D Infrared Hyperspectral Imaging** RYUJI TAO, Kagawa University, Akira Nishiyama, Kenji Wada, Ichiro Ishimaru
9:10	(380-3)	**New Methodology for Finding Optimal Spectral Matches in Reference Databases** GREGORY BANIK, Bio-Rad Laboratories, Inc., Ty Abshear, Karl Nedwed
9:30	(380-4)	**Experimental Optimization of IR pMAIRS Using A New Analytical Concept** TAKESHI HASEGAWA, Kyoto University, Nobutaka Shioya, Takafumi Shimoaka, Miyako Hada
9:50		**Recess**
10:05	(380-5)	**ATR-FTIR Spectroscopic Imaging of Pharmaceutical Formulations Under Continuous Flow** SERGEI KAZARIAN, Imperial College London, Andrew Ewing
10:25	(380-6)	**Advances in Pattern Recognition for the Remote Detection of Sulfur Dioxide by Passive Infrared Spectrometry** BRIAN W DESS, University of Iowa, Gary W Small
10:45	(380-7)	**Demonstration of Detection of Hidden Persons and Illegal Substances with an Array of Quantum Cascade Lasers and Cantilever Enhanced Photoacoustic Spectroscopy** SAULI SINISALO, Gasera Ltd., Ismo Kauppinen
11:05	(380-8)	**Measurement of Trace HF in Clean Rooms and Ambient Air** ISMO KAUPPINEN, Gasera Ltd., Tuomas Hieta, Timo Rajamäki, Sauli Sinisalo

ORAL SESSIONS — Session 390

GCMS of Environmental Analysis

Monday Morning, Room B401
Archana Kumar, Genentech, Presiding

8:30	(390-1)	**Continuous Monitoring of Polycyclic Aromatic Hydrocarbons Using Automatic Thermal Desorption-Gas Chromatography** FRANCK AMIET, Chromatotec Inc., Michel Robert, Damien Bazin
8:50	(390-2)	**Method Development for Evaluation of Pesticides Residue in Lake Lanier-Georgia Gwinnett County's Drinking Water Resource Using Disposable Pipette Extraction (DPX) and Gas Chromatography—Mass Spectrometry (GC-MS)** HONGXIA GUAN, Georgia Gwinnett College, Huang Wenlin, Xiaoping Li, Simon Mwongela, Rashad Simmons
9:10	(390-3)	**Overcoming Cost and Supply: Let's Use Nitrogen** LEE MAROTTA, PerkinElmer, Jacob Rebholz, Roger Bardsley, Tom Hartlein
9:30	(390-4)	**Analytical Method Development for 2,4,6,8-Tetrachlorodibenzothiohiophene (TCDT) in River Sediments Utilizing GCxGC-TOFMS and APGC-TQS** MAURA K MCGONIGAL, Penn State, Frank Dorman, Kari Organtini, Robert Parette, Wendy Pearson, Doug Stevens, Adam Ladak
9:50		**Recess**
10:05	(390-5)	**A Kendrick Mass Defect Approach Towards the Characterization of Hydraulic Fracturing Fluids** PAULINA PIOTROWSKI, The Pennsylvania State University, Frank Dorman, Jonathan D Byer, Joseph E Binkley
10:25	(390-6)	**Collaboration of Government (EPA), Industry, Academia to Update EPA Method 625 to Solid Phase Extraction for Drinking and Wastewater Analysis Combining Stir-Bar Sorptive Extraction and EPA Method 6800** ANIL SRINIVAS CHAITANYA VISHNUVAJJHALA, Duquesne University, Weier Hao, Matt Pamuku, Andrew Boggess, Mizanur Rahman, Skip Kingston, David Singer, Ed Pfannkoch
10:45	(390-7)	**Analysis of Vapor Mercury in Ambient Air and Flue Gas Emissions Using Thermal Desorption Trap Hyphenated with Gas Chromatography/Mass Spectrometry** PAOLO BENEDETTI, IIA - CNR, Carlo Crescenzi, Ettore Guerriero
11:05	(390-8)	**Soil Gas: Are Targets Being Missed?** LEE MAROTTA, PerkinElmer, Roberta Provost

TECHNICAL PROGRAM

ORAL SESSIONS — Session 400

Innovative Approaches to Science Education - Half Session

Monday Morning, Room B409
Hubert MacDonald, The Pittsburgh Conference, Presiding

8:30	(400-1)	**Teaching Through Research (Mini) Projects of Industrial Relevance** JURICA BAUER, Inholland, Rosalba Bellini, Gertjan Heijne, Mark Jansen, Iris Kuiper, Maarten Kuiper, Lieke van Hemert, Mark Verheij, John Vessies, Niek Persoon
8:50	(400-2)	**Promote Science Education Through Collaborative Learning** YI HE, John Jay College/CUNY, Sandra Swenson
9:10	(400-3)	**Open Source Instruments for Environmental Monitoring and Science Education** JACK SUMMERS, Western Carolina University, Benjamin Hickman
9:30	(400-4)	**Application of Environmental Data for Public Health Response Actions** JAMES S HOLLER, ATSDR

ORAL SESSIONS — Session 410

Microfluidics/Lab-on-a-Chip - Bioanalytical and Others

Monday Morning, Room B406
Vishal Sahore, Brigham Young University, Presiding

8:30	(410-1)	**Droplet Microfluidic Device: Application of Nucleosome Preparation to Nucleosome Analysis** YI XU, University of Illinois at Urbana-Champaign, Richard M Graybill, Mallika V Modak, Jeong-Heon Lee, Tamas Ordog, Ryan C Bailey
8:50	(410-2)	**3D-Printed Analytical Devices Facilitate Investigation of Stored Erythrocytes Used in Transfusion Medicine** CHENGPENG CHEN, Saint Louis University, Dana Spence, R Scott Martin
9:10	(410-3)	**Investigation of Bacterial Behavior in Water Filtration Processes by Using Nanofluidic Devices** NIL TANDOGAN, Northeastern University, Kai-Tak Wan, Edgar D Goluch
9:30	(410-4)	**A Novel Chemiluminescence Signal Amplification Strategy Based on Microchip Electrophoresis Platform for Highly Sensitive MicroRNA Detection** SHULIN ZHAO, Guangxi Normal University, Jian Li, Jingjin Zhao, Yong Huang, Yi-Ming Liu
9:50		Recess
10:05	(410-5)	**Singe Molecule Nanoelectrophoresis within Thermoplastics** COLLEEN ONEIL, University of North Carolina at Chapel Hill, Swathi Pullagurla, Steven A Soper
10:25	(410-6)	**Parallel Functionality Determination of shRNA Knockdown Constructs Using Microfluidic Technologies** KRISTEN ENTWISTLE, Michigan State University, Dana Spence
10:45	(410-7)	**A Universal Droplet Microfluidic Strategy for On-Chip Operations Including Reagent Injection, Sample Washing, and Droplet Tagging** STEVEN R DOONAN, University of Illinois at Urbana-Champaign, Richard M Graybill, Dongkwan Lee, Yi Xu, Ryan C Bailey
11:05	(410-8)	**Reversible Wettability Switching in Fabric-Based Microfluidic Devices** TANYA NARAHARI, Northeastern University, Dhananjaya Dendukuri, Tripurari Choudhary, Shashi K Murthy

ORAL SESSIONS — Session 420

Neurochemistry

Monday Morning, Room B402
Rose Ann Clark, Saint Francis University, Presiding

8:30	(420-1)	**Fast-Scan Cyclic Voltammetry Measurements of Serotonin Release and Reser Pools in R6/2 Huntington's Disease Model Mice** RACHEL C GEHRINGER, The University of Kansas, Sam V Kaplan, Sarah Fantin, Michael A Johnson
8:50	(420-2)	**Combining Voltammetry, Mathematical Modeling and 2-Photon Microscopy to Correlate In Vivo Serotonin Chemistry to Physiology** AYA ABDALLA, University of South Carolina, Christopher W Atcherley, Yunju Jin, Michael L Heien, Janet Best, Michael C Reed, David Linden, Parastoo Hashemi
9:10	(420-3)	**Coregulation of Serotonin and Histamine in the Mouse Premammilary Nucleus** SRIMAL A SAMARANAYAKE, University of South Carolina, Aya Abdalla, Rhiannon Robke, H Frederick Nijhout, Michael C Reed, Janet Best, Parastoo Hashemi
9:30	(420-4)	**Application of In Vivo Fast-scan Cyclic Voltammetry to Examine the Impact of (+)-Methamphetamine of the Regulation of Norepinephrine and Dopamine** KEN WAKABAYASHI, State University of New York at Buffalo, Jinwoo Park
9:50		Recess
10:05	(420-5)	**Characterization of Spontaneous Transient Adenosine in Rat Brain Slices** SCOTT T LEE, University of Virginia, B Jill Venton
10:25	(420-6)	**Comparing Contralateral Catecholamine Release with Simultaneous Voltammetric Measurements** MEGAN E FOX, University of North Carolina at Chapel Hill, R Mark Wightman
10:45	(420-7)	**Insights into the Effects of Non-stimulant ADHD Drugs on Catecholamine Transmission in the Rat Brain Using In Vivo Fast-scan Cyclic Voltammetry** ROHAN BHIMANI, The State University of New York at Buffalo, Jinwoo Park
11:05	(420-8)	**Investigation of Enkephalin-Evoked Catecholamine Secretion in Adrenal Tissue by Fast-Scan Cyclic Voltammetry** LARS DUNAWAY, North Carolina State University, Leslie A Sombers

ORAL SESSIONS — Session 430

Pharmaceutical-GC and LC

Monday Morning, Room B403
Sam Subramaniam, Miles College, Presiding

8:30	(430-1)	**Analyte Stability and Side Reactions During Pharmaceutical Analysis** LEAH XIONG, Merck Co., Nathan D Contrella, Leih-Shan Yeung, Cory Bottone, Brian Regler, Justin Pennington, Eugenia Muschajew
8:50	(430-2)	**A New Tool for Residual Solvent Analysis of Technical Grade Hexanes** KENNETH G LYNAM, Agilent Technologies, John Oostdijk, Johan Kuipers, Ramaprasad Ganni
9:10	(430-3)	**Development and Validation of a Fast Ion-Paired Reversed Phase Stability-Indicating Method for the Assay of Thiabendazole and Estimation of its Related Compounds** PENG ZHANG, Merial, A Sanofi Company, Jingzhi Tian, Abu Rustum
9:30	(430-4)	**Determining Lithium in Pharmaceutical Products** SACHIN P PATIL, Thermo Fisher Scientific, Jeffrey Rohrer
9:50		Recess
10:05	(430-5)	**Two Dimensional Liquid Chromatography (2D-LC), A "Must-Have" Tool in Pharmaceutical Development, When 1D-LC is Inadequate** IMAD A HAIDAR AHMAD, Novartis, Adrian Clarke, James Tam, Xue Li, Thomas Tarara, Andrei Blasko
10:25	(430-6)	**Enantioresolution of Several Amino Alcohol Drugs Containing Multiple Stereogenic Centers Using Immobilized Polysaccharide-Based HPLC Chiral Stationary Phases** MOHAMED HEFNAWY, King Saud University N/A
10:45	(430-7)	**Introducing Modern LC Column Technology into a Research-Led Pharmaceutical Teaching Environment** WILLIAM J LOUGH, University of Sunderland
11:05	(430-8)	**Improved Gas Chromatographic (GC) Quantification of Acidic and Neutral Cannabinoids** AMANDA RIGDON, Restek Corporation, Jack Cochran, Joan Serdar, Corby Hilliard, Linx Waclaski, Rebecca Stevens

ORAL SESSIONS — Session 440

Synthesis and Characterization of Nano Particles

Monday Morning, Room B405
Ashish Tripathi, Leidos, Inc., Presiding

8:30	(440-1)	**Antibacterial Activities and Cytotoxicity of Green Synthesized Stable Gold Nanoparticles from Flavonoid Derivatives** FRANCIS JUMA OSONGA, Binghamton University, Idris Yazgan, David C Luther, Apryl P Jimenez, Phuong N Lee, Omowunmi Sadik
8:50	(440-2)	**Wavelength Selective Photocatalysis Using Gold-Platinum Nano-Rattles** MAHMOUD MAHMOUD, Georgia Institute of Technology, Batyr Garlyyev, Mostafa A El-Sayed
9:10	(440-3)	**Reversible Electron Delocalization of Molecule-Like CdSe Nanoclusters Using Z-Type Ligand Functionalization** KATIE N LAWRENCE, Indiana University-Purdue University Indianapolis
9:30	(440-4)	**Characterization Techniques for Nanomaterials – An Overview** CHADY STEPHAN, PerkinElmer
9:50		Recess

10:05 (440-5) **A Novel Nanoparticle Tracking Analysis System for Improved Determinations of Nanoparticle Concentration and Size Distribution** DARIUSZ STRAMSKI, Manta Instruments Inc, Kuba Tatarkiewicz, Rick A Reynolds, Monette Karr, Rick Cooper

10:25 (440-6) **Coupled Calorimetric-Manometric Technique for the Study of Sorption and Thermodynamic Properties of Macroscopic and Nanosized Materials** KRISTINA LILOVA, Setaram Inc., Link Brown

10:45 (440-7) **Exploring the Effects of Surface Ligand Structural Parameters on Exciton Delocalization of CdSe Nanocrystals** MEGHAN TEUNIS, Indiana University-Purdue University Indianapolis

11:05 (440-8) **Metallic and Hybrid Nanomaterials: Fabrication and Applications** SIMONA HUNYADI MURPH, Savannah River National Lab

POSTER SESSION — Session 450

All posters are to be mounted by 10:00 AM and remain on display until 4:00 PM. Authors must be at their posters from 10:00 AM to 12:00 PM. Location of the morning posters is on the Exposition Floor, 400 Aisle. PLEASE NOTE: You cannot get onto the Exposition Floor until after 9:00 AM.

Agriculture

Monday Morning, Exposition Floor, 400 Aisle

(450-1 P) **Growth Performance of Chicory to Plant Growth Regulators** RAMESHBHAI P DABHI, J & J College of Science, Alpesh M Patel, Maheshkumar B Chauhan N/A

(450-2 P) **Analysis of Sugarcane Growing Soils at Kakamega North District Kenya for Micro and Macro Nutrients** ONDITI O ANAM, Jomo Kenyatta University of Agriculture and Technology N/A

(450-3 P) **Fast and Precise Nitrogen and Carbon Determination Using an Elemental Analyzer** GUIDO GIAZZI, Thermo Fisher Scientific, Liliana Krotz, Francesco Leone, Dominique Chevalier

(450-4 P) **Carbon/Nitrogen Ratio in Soils and Plants Using an Elemental Analyzer** GUIDO GIAZZI, Thermo Fisher Scientific, Liliana Krotz

(450-5 P) **Development of Electrochemical Detection System Combining with Nitrocellulose Membrane for Quantitative Immunochromatography** WATARU IWASAKI, Natl. Inst. Adv. Ind. Sci. Technol. (AIST), Mizuki Ryu, Ramachandra Rao Sathuluri, Ryoji Kurita, Osamu Niwa, Masaya Miyazaki

POSTER SESSION — Session 460

All posters are to be mounted by 10:00 AM and remain on display until 4:00 PM. Authors must be at their posters from 10:00 AM to 12:00 PM. Location of the morning posters is on the Exposition Floor, 400 Aisle. PLEASE NOTE: You cannot get onto the Exposition Floor until after 9:00 AM.

Clinical/Toxicology

Monday Morning, Exposition Floor, 400 Aisle

(460-1 P) **Benzo[a]pyrene Levels in Mainstream Smoke from Spectrum Research Cigarettes** JARED HUGHES, Centers for Disease Control and Prevention, Wayne Chen, Bryan Hearn, Shirley Ding, Clifford Watson

(460-2 P) **Selected Carbonyl Levels in Mainstream Smoke from Spectrum Research Cigarettes** MICHELE CHAN, Centers for Disease Control and Prevention, Shirley Ding, Clifford Watson, Xizheng Yan

(460-3 P) **Absolute Quantification of Apolipoproteins in Serum and the Efficacy of Trypsin While Utilizing Ultra Performance Liquid Chromatography - Isotope Dilution Mass Spectrometry (UPLC-IDMS)** MICHAEL L ANDREWS, Centers for Disease Control and Prevention, Bryan A Parks, Christopher Toth, Jeffrey Jones, Kuklenyik Zsuzsanna, Michael S Gardner, Jon Rees, David Schieltz, John R Barr

(460-4 P) **Mainstream Smoke Deliveries of Tobacco Specific Nitrosamines in Spectrum Clinical Research Cigarettes** PATRICK CHEN, Centers for Disease Control and Prevention, Clifford Watson, Shirley Ding, Liqin Zhang, Josh Wong

(460-5 P) **Separation of Vitamin D2 and D3 for Clinical Application** MARK WOODRUFF, Fortis Technologies, Ken Butchart

(460-6 P) **Measurement of Ammonia Emanating from Human Skin as a Possible Biomarker for Physical/Mental Stress Responses** SHOTA FURUKAWA, Tokai University, Minami Takahashi, Keita Kimura, Shiro Ikeda, Yoshika Sekine, Asai Satomi, Umezawa Kazuo, Hayato Miyachi

(460-7 P) **Measurement of Acetic Acid Emanating from Human Skin as a Potential Biomarker for Quality of Sleep** MINAMI TAKAHASHI, Tokai University, Furukawa Shota, Keita Kimura, Yoshika Sekine, Satomi Asai, Kazuo Umezawa, Miyachi Hayato

(460-8 P) **LC-MS/MS Determination of Interactions Between Sunitinib and Green Tea Polyphenol by Equilibrium Dialysis** MATTHEW VERGNE, Lipscomb University, Lincoln Shade

(460-9 P) **Identification of Hypocretin-1 in Cerebrospinal Fluid: A Potential Diagnostic Biomarker for Narcolepsy** HEMASUDHA CHATRAGADDA, Duquesne University, Skip Kingston, Birgitte R Kornum, Matt Pamuku

(460-10 P) **ICP-MS – A Perfect Tool for the Bio-monitoring of Trace Elements in Body Fluids** EWA M PRUSZKOWSKI, PerkinElmer, Inc.

(460-11 P) **Incorporation of Amphiphilic Dendrimers in Supported Lipid Bilayers to Enhance Stability and Functionalization** CHARLES J RUIZ, University of California, Riverside, Samuel S Hinman, Quan Cheng, Ling Peng

(460-12 P) **Chiral Capillary Electrophoresis-Mass Spectrometry: Turning an Analytical Technique into High Throughput Screening of Chiral Compounds Using Novel Polymerized beta-D-Glucopyranoside Surfactants** LIU YIJIN, Georgia State University, Shahab S Shamsi

(460-13 P) **Measurement of Diacetyl and 2-Nonenal Emanating from Human Skin by Passive Flux Sampler** KEITA KIMURA, Tokai University, Shota Furukawa, Minami Takahashi, Yoshika Sekine, Kazuo Umezawa, Satomi Asai, Hayato Miyachi

(460-14 P) **Aptamer-Modified Gold Nanoparticles Coupled with Nitrocellulose Membranes for Detection of Thrombin by LDI-MS** CHIA-YIN CHANG, National Taiwan Ocean University

(460-15 P) **On-line Membrane Assisted Distillation Coupled with Ion Chromatography: A Novel Approach to Determine Trace Fluoride in Serum** LOU CHAOYAN, Zhejiang University

(460-16 P) **SERS Probe for Rapid Detection of Erythropoietin in Urine** UGUR TAMER, Gazi University, Yesim Selbes, Gokhan M Caglayan, Nejdet Saglam N/A

(460-17 P) **Electrochemical Measurement of Thyroid Hormone for Rapid Diagnostics Technology** BARBARA CATA, Northern Kentucky University, Celeste A Morris

(460-18 P) **Forensics Meets Green Chemistry: Removing Known Carcinogens from Blood Alcohol Content Protocols for Safer Applied Spectroscopy Laboratories** SARAH E GRAY, Armstrong State University, Mathew Holmes

(460-19 P) **Preparation and Evaluation of Nifedipine-Cyclodextrin Complex Microspheres with In-Vitro Studies** SUNILKUMAR H CHAUDHARI, Mantri Health Care, Shailesh Jain N/A

(460-20 P) **A Rapid and Sensitive SERS Based Measles Detection** RAMESH KATTUMENU, Argent Diagnostics Inc

(460-21 P) **The Efficiency of Multi-Sample Analysis Using Dual Gradient LCMS System** WATANABE SATORU, Shimadzu Corporation, Nakayama Daisuke, Yamaguchi Tadayuki, Inohana Yusuke

POSTER SESSION — Session 470

All posters are to be mounted by 10:00 AM and remain on display until 4:00 PM. Authors must be at their posters from 10:00 AM to 12:00 PM. Location of the morning posters is on the Exposition Floor, 400 Aisle. PLEASE NOTE: You cannot get onto the Exposition Floor until after 9:00 AM.

Drug Discovery

Monday Morning, Exposition Floor, 400 Aisle

(470-1 P) **Automated Mass-Directed Purification of Bioactive Peptides** LAINE STEWART, Gilson, Inc., Luke Roenneburg, Karen Kleman N/A

(470-2 P) **Synthesis of Nanosized Poly B-aminoester Holding Chlorambucil Drug as Slow Release Drug System for Antitumer** FAHIMA MOSAAD HELALY, National Research Centre (NRC)

(470-3 P) **Synthesis, Characterization and Antibacterial Studies of Cobalt Complexes of Isomeric Aminophenol Schiff Bases** TOLULOPE M FASINA, University of Lagos, Felicia N Ejiah, Oluwole B Familoni, Neerish Revaprasadu

(470-4 P) **Mass Spectral and Chromatographic Studies on Substituted Cathinones: Bath Salt-type Aminoketone Designer Drugs** YOUNIS ABIEDALLA, Auburn University, Randall Clark, Jack DeRuiter, Karim Abdel-Hay

(470-5 P) **An In Vitro Microfluidic Platform to Unravel Mechanisms of Action of Drug Therapies used in Multiple Sclerosis** TIFFANY BELL, Michigan State University, Dana Spence

(470-6 P) **Streamlining Compound Isolation Automatically with UPLC to Prep Chromatography Using Mass-Directed Auto Purification** JO-ANN M JABLONSKI, Waters Corporation, Andrew J Aubin, Wendy Harrop, Thomas E Wheat

(470-7 P) **Phytochemical Screening and Antimicrobial Activity of *Boerhavia Verticillata Poir.* and *Fagonia Schweinfurthii Hadidi*** YAMINI A JOSHI, M G Science Institute N/A

(470-8 P) **Aluminum-Based MOF Composite as Polymer Monoliths for Microextraction of Sulfonamides** YUNG-HAN SHIH, Chung Yuan Christian University, Kuen-Yun Wang, Hsi-Ya Huang N/A

(470-9 P) **Preparation and Study of Tramadol Imprinted Micro-and Nanoparticles by Precipitation Polymerization: Microwave Irradiation and Conventional Heating Method** MARYAM HASSANPOUR MOGHADAM, Pharmaceutical Research Center, School of Pharmacy N/A

(470-10 P) **Spectral Characterization of Cytochrome P450cam Active Site Using NMR Methods Including ^{13}C-Doubled Filtered ^{1}H-^{1}H Noesy Experiments for Mapping Distances** REMIGIO USAI, Marquette University, James R Kincaid, Daniel S Sem, Daniel Kaluka

(470-11 P) **Synthesis, Anti-TB and Crystallography of Benzothiazole Analogues** KATHARIGATTA NARAYANASWAMY VENUGOPALA, King Faisal University, Al-Dhubiab Bander N/A

(470-12 P) **Inhibition of Human Cytochrome P450 Enzymes by Rottlerin, A Naturally Occurring Constituent of Mallotus Philippensis Using an In-Vitro Cocktail Approach and Liquid Chromatography-Tandem Mass Spectrometry** ATUL S RATHORE, Bharati Vidyapeeth Deemed University, L Sathiyanarayanan, Kakasaheb R Mahadik N/A

(470-13 P) **Photostability of Pharmaceutical Drug Substance as Free Acid and Salt** JENNY WANG, Genentech Inc., Geoffrey Yeh, Lulu Dai, Christine Gu, Kelly Zhang

POSTER SESSION — Session 480

All posters are to be mounted by 10:00 AM and remain on display until 4:00 PM. Authors must be at their posters from 10:00 AM to 12:00 PM. Location of the morning posters is on the Exposition Floor, 400 Aisle. PLEASE NOTE: You cannot get onto the Exposition Floor until after 9:00 AM.

Food Safety

Monday Morning, Exposition Floor, 400 Aisle

(480-1 P) **Analysis of Local Brews in Jua, Kiambu County, Kenya, for Methanol and Ethanol by Gas Chromatography** ONDITI O ANAM, Jomo Kenyatta University of Agriculture and Technology, Momanyi M Moraa N/A

(480-2 P) **Cannabinoids and Residual Solvents by Headspace GC** TIM ANDERSON, Phenomenex, Kristen Parnell, Ramkumar Dhandapani

(480-3 P) **Melamine in Pet Food** TIM ANDERSON, Phenomenex, Kristen Parnell, Ramkumar Dhandapani

(480-4 P) **Impact of HPLC Stationary Phase Selection on Matrix Effects During LC-MS/MS Analysis of Multiple Mycotoxins in Corn** EMILY R BARREY, Supelco/Sigma-Aldrich, Lynne Perez-Blanco, Olga I Shimelis, Michael Ye, Jennifer Claus

(480-5 P) **UHPLC/MS/MS Analysis of Lipophilic Marine Toxins from Homogenized Shellfish** EMILY R BARREY, Supelco/Sigma-Aldrich, Olga I Shimelis, Michael Ye, Jennifer Claus

(480-6 P) **Utilizing Mass Spectrometry for Gluten Detection for Use in Gluten-Free Foods** SOPHIE BROMILOW, University of Manchester, Lee Gethings, Peter Shewry, Michael Buckley, Michael Bromley, Phil Johnson, Clare Mills

(480-7 P) **Easy Development of an MRM Method for Analysis of Environmental Contaminants in Milk Using SPME and GC-MS/MS** NICOLE M LOCK, Shimadzu Scientific Instruments, Di Wang, Laura Chambers, Brahm Prakash, Robert Clifford, Shilpi Chopra N/A

(480-8 P) **Analysis of Target Pesticides in Essential Oils Using a Novel GC/MS/MS System** THOMAS DILLON, PerkinElmer, Samuel Tolley, Adam J Patkin, Sharanya Reddy

(480-9 P) **A Raman Spectroscopic Method for Determination of Erucic Acid in Canola Oils** ELIF ERCIOGLU, Hacettepe University, Ismail H Boyaci, Tumay H Temiz, Serap Durakli Velioglu, Murat H Velioglu

(480-10 P) **Multiple Foodborne Pathogen Analysis Using a 96-Well Assay in Less than 5 Hours** STUART FARQUHARSON, Real-Time Analyzers, Inc, Chetan Shende, Kathryn Dana

(480-11 P) **Stand-Off Raman Detection of Adulteration in Honey** KENNETH GARCIA, Alabama A&M University, Carlton Farley, Aschalew Kassu, Anup Sharma

(480-12 P) **Evaluating the Extraction Efficiency of Food Borne Pathogens on Automated Homogenizer Platforms** SHARI GARRETT, Omni International, James Atwood, Brandon Easparro

(480-13 P) **Nitrogen/Protein Determination in Infant Food by Dumas Combustion Method in Alternative to Kjeldahl Method** GUIDO GIAZZI, Thermo Fisher Scientific, Liliana Krotz

(480-14 P) **Non-Targeted Screening of Nutritional Supplements with GC, GC×GC, TOFMS, and HR-TOFMS** ELIZABETH M HUMSTON-FULMER, Leco Corporation, David E Alonso, Jonathan D Byer, Joseph E Binkley, Lorne E Fell

(480-15 P) **Synthesis, Physico-Chemical Characterization and Potential Biological Activity of Newly Synthetised Cu(II) and Ni(II) Complexes of N-(Benzyloxycarbonyl)-1H-pyrazOle-1-Carboxamidine (HL) ligand** ZELJKO JACIMOVIC, Faculty of Metallurgy and Technology, Milica Kosovic, Goran Bogdanovic, Sladjana Novakovic, Nedeljko Latinovic, Gerald Giester N/A

(480-16 P) **Combining SERS with Liquid-Liquid Extraction Method for Simple, Rapid Detection of Rhodamine B in Raw Food** HUAIZHI KANG, Xiamen University

(480-17 P) **Concurrent High Sensitivity Conductometric Detection of Sulfide and Cyanide in a Suppressed Anion Chromatography System** HONGZHU LIAO, University of Texas at Arlington, Purnendu Dasgupta, Akinde Kadjo

(480-18 P) **A Computational Study Assisted Nano-Aptasensor for Detection of Tetracycline in Honey with a DNA Aptamer** SAI WANG, Beijing Univerisity of Chemical Technology

(480-19 P) **Simple, Rapid Extraction of Chlorinated Pesticides in Poultry Fat by Solid Phase Extraction and GC/ECD** ALLEN MISA, Phenomenex, Ramkumar Dhandapani, Tim Anderson

(480-20 P) **Approaches to Measuring both Trace and Nutritional Elements in Food in a Single Analysis by ICP-MS** KENNETH NEUBAUER, PerkinElmer, Stan Smith

(480-21 P) **Comparative Evaluation of the Antibacterial Activities of the Essential Oils of Rosmarinus Officinalis L. Obtained by Hydrodistillation and Solvent Free Microwave Extraction Methods** OMOBOLA OLURANTI OKOH, University of Fort Hare N/A

(480-22 P) **Impact of Thermal Processing on the Solubility and Detection of Peanut Allergens Using LC-MS/MS based Targeted Proteomics with Multiple-Reaction Monitoring (MRM)** REBEKAH L SAYERS, University of Manchester, Phil E Johnson, Justin T Marsh, Clare Mills, Helen Brown

(480-23 P) **Measurement of Arsenic in Wine by Hydride Generation – Flame Atomic Absorption** NICK SPIVEY, PerkinElmer Inc., Kenneth Neubauer, Stan Smith

(480-24 P) **Determination of Organic Tin Pesticides in Fruits and Vegetables by Gas Chromatography Coupled to Tandem Mass Spectrometry** XIAOBO LIU, Shimadzu

(480-25 P) **Determination of 198 Pesticide Residues in Eggplant Using Gas Chromatography Tandem Mass Spectrometry/Mass Spectrometry (GCMS/MS)** WANG YAN, Shimadzu

(480-26 P) **A Direct Ultra-Performance Liquid Chromatography-Mass Spectrometry Method for the Simultaneous Quantitation of Fatty Acids in Olive Oils of Different Origins** ZEID ABDULLAH ALOTHMAN, King Saud University N/A

(480-27 P) **Development of On-Line SFE-SFC System and Its Application for Food Contaminant Analysis** SHIN-ICHI KAWANO, Shimadzu (China) Co., Ltd., Xiaohua Liu, Lingling Shen, Yan Wang, Taohong Huang, Naoki Hamada, Yuki Hashi

(480-28 P) **High Resolution Accurate Mass (HRAM) Collision Energy Profile of Residues of Concern for Food Safety** DANIEL BIGGERSTAFF, o2si Smart Solutions, Huichen Stavros, Min Cai

(480-29 P) **Quantification of Pesticide Residues in Fruits and Vegetables by Gas Chromatography - Mass Spectrometry** KAELYB SUCHEVITS, California University of PA

(480-30 P) **Study of Sugar and Humectant Profiles in Smokeless Tobacco Products Using an LC-ESI-MS/MS Method** LIQUN WANG, Battelle, Roberto Bravo, Stephen Stanfill, Liza Valentin, Clifford Waston

(480-31 P) **High-Throughput Screening of Domoic Acid in Shellfish by Laser Ablation Electrospray Ionization (LAESI)-MS** PEARSE MCCARRON, National Research Council, Kelley Reeves, Callee M Walsh, Pamela Cantrell, Wade A Rourke, Sinead O'Brien, Daniel Beach

(480-32 P) **Development and Implementation of a Fast, Reliable and Sensitive Analytical Test for Determining Methylmercury in Fish** ANA MARIA MUÑOZ, Lasallian University , Claudio Jiménez, Daniel E León

(480-33 P) **Characterization of Adulterated Argan Oil Using a Portable Gas Chromatography Mass Spectrometer** PRESHIOUS REARDEN, 1st Detect Corporation, Parminder Kaur N/A

(480-34 P) **A New Insight Into Fish Meat Freshness: ZnO /PPy Modified Biosensor** BUKET SAHYAR (YALCIN), Indesit Company, Merve Kaplan, Ahmet Yavas, Mehmet Ozsoz, Erdal Celik, Semih Otles, Ömer Mindivanli, Metin Ozgul

POSTER SESSION — Session 490

All posters are to be mounted by 10:00 AM and remain on display until 4:00 PM. Authors must be at their posters from 10:00 AM to 12:00 PM. Location of the morning posters is on the Exposition Floor, 400 Aisle. PLEASE NOTE: You cannot get onto the Exposition Floor until after 9:00 AM.

Microfluidics/Lab-on-a-Chip

Monday Morning, Exposition Floor, 400 Aisle

(490-1 P) **A Simple and Sensitive Paper-Based Device Coupling Electrochemical Sample Pretreatment and Colorimetric Detection** WILLIAM REIS DE ARAUJO, USP, Thalita G Silva, Thiago Paixao

(490-2 P) **Mixing Reaction on Paper-Based Analytical Devices** CHING MAN CHOY, California State Polytechnic University, Pomona, Yan Liu

(490-3 P) **Combining Vibrational Spectroscopy with Microfluidics** KATHLEEN E BERG, Colorado State University, Scott D Noblitt, Monpichar Srisa-Art, Amber T Krummel, Charles Henry

(490-4 P) **Prostate Cancer Biomarker Detection Using a 16-Sensor Electrochemical Microfluidic Immunoarray** ABBY JONES, University of Connecticut, Brunah A Otieno, Colleen Krause, Mohammed Sherafeldin, Amit Joshi, James F Rusling

(490-5 P) **Real-Time Profiling of Pancreatic Hormone Secretion Dynamics Using an in Flow Fluorescence Polarization Immunoassay** NIKITA MUKHITOV, Florida State University, Adrian M Schrell, Michael G Roper

(490-6 P) **Alternating Current Driven Electroosmotic Pumping through Conical Pore Membrane** XIAOJIAN WU, University of Florida, Charles R Martin

(490-7 P) **Two-Fold Control of Pressure and Flow-Rate for Flow Control and Quality Management in Fluidic Processes** ANNE LE NEL, Fluigent, Thibaut Thupnot, Benjamin Rouffet, Nicolas Petit

(490-8 P) **A 3D-printed Device for High-Throughput Membrane-Based Cellular Analysis** RUIPENG MU, Michigan State University, Dana Spence

(490-9 P) **3D-Printed Tools to Enhance Targeted Drug Therapy** CODY WAYNE PINGER, Michigan State University, Dana Spence

(490-10 P) **Fluorescence-Based Quantification of Oxygen in Paper-Based Cultures of Mammalian Cells** MATTHEW W BOYCE, University of North Carolina at Chapel Hill, Andrew S Truong, Rachael M Kenney, Matthew R Lockett

(490-11 P) **Methods for Quantifying Hypoxia and the Hypoxic Responses of Cells in Paper-Based Invasion Assays** ANDREW S TRUONG, University of North Carolina at Chapel Hill, Rachael M Kenney, Matthew W Boyce, Christian A Lochbaum, C Chad Lloyd, Matthew R Lockett

(490-12 P) **A Microfluidic Copper Detection System Incorporating a Ratiometric Fluorescent Quantum Dot Pair** SUMATE PENGPUMKIAT, Oregon State University, Yuanyuan Wu, Anukul Boonloed, Chandima Bandara, Vincent T Remcho

(490-13 P) **Ultra-Thin Layer Chromatography with Integrated Silver Colloid-based SERS Detection** RYAN A WALLACE, University of Tennessee, Knoxville, Nickolay V Lavrik, Michael J Sepaniak

(490-14 P) **Affinity Cell Separation Based on Surface Antigen Expression Difference in a Sequential Concentration Microfluidic Chip** YE ZHANG, Texas Tech University, Dimitri Pappas

(490-15 P) **Microscale Size-Based Sorting with Capillary Electrophoresis and Phospholipid Additives** CASSANDRA CRIHFIELD, West Virginia University, Lisa A Holland

(490-16 P) **Kinetic Characterization of Neurotransmitters Release from Neurons Cells Using MicroChip Electrophoresis–ESI/Mass Spectrometry** XIANGTANG LI, Jackson State University, Shulin Zhao, Yiming Liu N/A

(490-17 P) **High-Throughput Microfluidic Isolation and Analysis of Exosomes** KRISTINA M HERRERA, University of North Carolina at Chapel Hill, Steven A Soper

(490-18 P) **Quick Production of Microfluidic Devices by Laser Engraving of Wax-Coated Glass Slides** MAURO SERGIO FERREIRA SANTOS, Clemson University, Eric T da Costa, Hong Jiao, Claudimir L do Lago, Ivano G R Gutz, Carlos D Garcia

(490-19 P) **Microfluidic Platform for Mass Spectrometry-Based Monitoring of Protein-ligand Binding Dynamics** YONGZHENG CONG, Pacific Northwest National Laboratory, Cameron Trader, Ryan T Kelly, Daniel Orton, Erin S Baker, Tao Geng

(490-20 P) **Low-Cost Microfluidic Devices for the Determination of Renal Health** CHRISTOPHER A HEIST, Oregon State University, Vincent T Remcho, Joel C Pommerenck

POSTER SESSION — Session 500

All posters are to be mounted by 10:00 AM and remain on display until 4:00 PM. Authors must be at their posters from 10:00 AM to 12:00 PM. Location of the morning posters is on the Exposition Floor, 400 Aisle. PLEASE NOTE: You cannot get onto the Exposition Floor until after 9:00 AM.

Process Analytical Techniques

Monday Morning, Exposition Floor, 400 Aisle

(500-1 P) **Characterization and Monitoring of Cavitation Through Its Acoustic Emission** NOEMIE CAILLOL, IDEEL, Franck Baco-Antoniali, Sebastien Leinardi, Sylvain Charquet, Serge Henrot, Pascal Pitiot, Davide Zonca, Christine Villard

(500-2 P) **Monitoring of Tablet Coating Using Raman Spectroscopy** HOEIL CHUNG, Hanyang University, Jaejin Kim, Young Ah Woo

(500-3 P) **Simple and Rapid Determination of Polyanion-Polycation Binding Ratio Using Pulsed Chronopotentiometry with Polyion-Selective Electrodes** EMMA GORDON, Northern Kentucky University, Kebede L Gemene

(500-4 P) **How Can Your Process Benefit From External Flow Cells and Optimized Pump Heads?** KATHRYN E MONKS, Knauer, Ingo Piotrowski

(500-5 P) **Applications of a New Wear Resistant, Chemically Inert Coating that Improves Reliability, Lifetime and Accuracy of Process, Analytical and Sampling Systems** LUKE PATTERSON, SilcoTek Corporation, Min Yuan, David Smith

(500-6 P) **Withdrawn**

(500-7 P) **Online Analysis Using LIBS for Industrial Process** RONALD BERGER-LEFÉBURE, IVEA

(500-8 P) **Non-Porous, No Glass, Leak-Free Reference Electrode** ZIAD H TAHA, Innovative Instruments, Inc.

(500-9 P) **Suggested QC Practices for On-line Analyzers** WILLIAM LIPPS, Shimadzu Scientific Instruments

(500-10 P) **A Novel Approach to Cleaning Validation for Pharmaceutical Manufacturing by Online SFE-SFC** KENICHIRO TANAKA, Shimadzu Scientific Instruments, Inc., William Hedgepeth, Daisuke Nakayama, Hidetoshi Terada, Minori Nakashima, Tadayuki Yamaguchi

(500-11 P) **Advances in Raman Analyzers for In Situ Studies of Small Volume Liquid-phase Reactors** LISA GANSTER, Kaiser Optical, Ian Lewis

(500-12 P) **Establishing Raman Spectroscopy for the Process Environment** LISA GANSTER, Kaiser Optical, Karen Esmonde-White, Harry Owen, Ian Lewis

(500-13 P) **In situ Raman Measurements of Pharmaceutical Solids During Process Unit Operations** LISA GANSTER, Kaiser Optical, Karen Esmonde-White, Carsyen Uerpmann, Sean Gilliam, Ian Lewis

(500-14 P) **Recovery of Challenging Compounds in Cleaning Validation Using Total Organic Carbon (TOC) Analysis** JENNY G WATSON, GE Analytical Instruments, David Wayne

POSTER SESSION — Session 510

All posters are to be mounted by 10:00 AM and remain on display until 4:00 PM. Authors must be at their posters from 10:00 AM to 12:00 PM. Location of the morning posters is on the Exposition Floor, 400 Aisle. PLEASE NOTE: You cannot get onto the Exposition Floor until after 9:00 AM.

Safety

Monday Morning, Exposition Floor, 400 Aisle

(510-1 P) **A Cheap and Simple Approach to Monitor Sun Exposure based on Photocatalytic Properties of Titanium Dioxide** PARISA SOWTI, University of New South Wales

(510-2 P) **A Wide Selection of New Psychoactive Substances Investigated with Proton Transfer Reaction – Mass Spectrometry within the Marie Curie Training Network Proton Ionization Molecular Mass Spectrometry (PIMMS)** MATTEO LANZA, IONICON Analytik, W Joe Acton, Philipp Sulzer, Kostiantyn Breiev, Simone Jürschik, Alfons Jordan, Eugen Hartungen, Gernot Hanel, Jens Herbig, Lukas Märk, Christian Lindinger, Chris A Mayhew, Tilmann D Maerk

POSTER SESSION — Session 520

All posters are to be mounted by 10:00 AM and remain on display until 4:00 PM. Authors must be at their posters from 10:00 AM to 12:00 PM. Location of the morning posters is on the Exposition Floor, 400 Aisle. PLEASE NOTE: You cannot get onto the Exposition Floor until after 9:00 AM.

Sampling and Sample Preparation - Pharmaceutical, Clinical/Toxicology, Food Safety, and Others

Monday Morning, Exposition Floor, 400 Aisle

(520-1 P) **Preserving the Free-radical Scavenger Activity of Key Bioactives During Extraction and Purification Processes** VALERIE DESYROY, SiliCycle, Vincent Bedard, Genevieve Gingras, Denis Boudriau, Aurelie Meng

(520-2 P) **Determination of Quinolones in Bovine Kidney Using Hollow-Fiber Supported Liquid Membrane (HF-SLM) and LC-ESI+MS/MS** SIMISO DUBE, University of South Africa N/A

(520-3 P) **Working with Challenging Samples, Sampling Systems** YVES GAMACHE, Analytical Flow Products

(520-4 P) **Analytical Method Development for Cleaning Verification of Manufacturing Equipment - Exploring the Effect of the Cleanliness of Stainless Steel Coupons on Sample Recovery** IMAD A HAIDAR AHMAD, Novartis, James Tam, Xue Li, Thomas Tarara, Andrei Blasko

(520-5 P) **Evaluation of Antibacterial and Wound Healing Properties of Hydro-ethanolic Extracts of Gossypium Barbadense Leaves** NWAMAKA H IGBOKWE, University of Lagos, Eugene E Ikobi, Cecilia I Igwilo

(520-6 P) **Evaluating the Potential for Cross Contamination when Performing 96-Well Sample Preparation Prior to LC-MS/MS Analysis** HELEN LODDER, Biotage GB Limited, Lee Williams, Victor Vandell N/A

(520-7 P) **Acrylamide from Coffee Using a Simplified Liquid** ALLEN MISA, Phenomenex, Matthew Brusius, Zeshan Aqeel

(520-8 P) **Analysis of Food Grains Using Automated Block Digestion** MICHAEL A RUTZKE, Cornell University, Suhas Narkhede, Nick McLeod

(520-9 P) **Does Weather Effect Pipetting?** GEORGE W RODRIGUES, Artel, Emily Avis, Alf Price, Verena Gerbe

TECHNICAL PROGRAM

MONDAY, MARCH 7, 2016
AFTERNOON

AWARDS — **Session 530**

LCGC Lifetime Achievement and Emerging Leader in Chromatography Award
arranged by Laura Bush, LCGC & Spectroscopy

Monday Afternoon, Room B314
Laura Bush, LCGC & Spectroscopy, Presiding

1:30		**Introductory Remarks - Laura Bush**
1:35		**Presentation of the LCGC 2016 Lifetime Achievement in Chromatography Award to Milton L Lee, Brigham Young University, by Laura Bush, LCGC & Spectroscopy**
1:40	(530-1)	**Columns in Small-Scale Chromatography** MILTON L LEE, Brigham Young University
2:15	(530-2)	**Recent Chromatographic and Mass Spectrometric Developments Applied to the Characterization of Recombinant Proteins, Monoclonal Antibodies and Antibody-Drug Conjugates** PAT SANDRA, RIC, Koen Sandra
2:50	(530-3)	**Analytical Glycoscience: Quo Vadis?** MILOS V NOVOTNY, Indiana University
3:25		**Recess**
3:40		**Presentation of the LCGC Emerging Leader in Chromatography Award to Debby Mangelings, Vrije Universiteit Brussel, by Laura Bush, LCGC**
3:45	(530-4)	**Generic Chiral Separation Strategies for Pharmaceutical Compounds Using Chromatographic and Electrophoretic Techniques** DEBBY MANGELINGS, Vrije Universiteit Brussel, Yvan Vander Heyden
4:20	(530-5)	**Recent Trends in High-Performance Liquid Chromatographic Separation of Enantiomers** BEZHAN CHANKVETADZE, Tbilisi State University N/A

AWARDS — **Session 540**

SEAC - Charles N Reilley and Royce W Murray Award
arranged by Richard M Crooks, University of Texas at Austin

Monday Afternoon, Room B312
Shelley Minteer, University of Utah, Presiding

1:30		**Introductory Remarks - Shelley Minteer**
1:35		**Presentation of the 2016 SEAC - Charles N Reilley Award to Reginald M Penner, University of California, Irvine by Shelley Minteer, SEAC President**
1:40	(540-1)	**Electrodeposited Nanowires for Faster and More Sensitive Hydrogen Gas Detection** REGINALD M PENNER, University of California, Irvine
2:15	(540-2)	**A Membrane-Based AC Electroosmotic Pump** CHARLES R MARTIN, University of Florida, Pradeep Ramiah Rajasekaran, Xiaojian Wu
2:50	(540-3)	**Electrocatalytic Amplification of Single Nanoparticle Collisions Using DNA-Modified Surfaces** RICHARD M CROOKS, University of Texas at Austin, Timothy M Alligrant, Radhika Dasari, Keith J Stevenson
3:25		**Recess**
3:40		**Presentation of the 2016 SEAC - Royce W Murray Award to Ryan J White, University of Maryland Baltimore County, by Shelley Minteer, SEAC President**
3:45	(540-4)	**Imaging Molecular Flux Using Protein Channel Based Scanned Probe Microscopy** RYAN J WHITE, University of Maryland Baltimore County, Florika Caling Macazo
4:20	(540-5)	**Imaging with Nanopipettes** LANE A BAKER, Indiana University

SYMPOSIUM — **Session 550**

Emerging Platforms for Lab-on-a-Chip Analyses
arranged by Ryan T Kelly, Pacific Northwest National Laboratory

Monday Afternoon, Room B301
Ryan T Kelly, Pacific Northwest National Laboratory, Presiding

1:30		**Introductory Remarks - Ryan T Kelly**
1:35	(550-1)	**Rapid Screening for Infectious Diseases Using Paper-Analytic Devices** CHARLES HENRY, Colorado State University
2:10	(550-2)	**High-Throughput Microfluidic Experimentation One Drop at a Time** ANDREW J DEMELLO, ETH Zürich
2:45	(550-3)	**Simple, Microfluidic Flow Distance-Based Determination of Biomolecule Concentrations** ADAM T WOOLLEY, Brigham Young University, Chatterjee Debolina, Sahore Vishal
3:20		**Recess**
3:35	(550-4)	**Automated Droplet Manipulation, Analysis and Screening Based on Sequential Operation Droplet Array Technique** QUN FANG, Zhejiang University, Zhu Ying
4:10	(550-5)	**Microfluidic Platforms with Integrated Microvalves for Novel Biochemical Analyses** RYAN T KELLY, Pacific Northwest National Laboratory, Yongzheng Cong, Tao Geng, Sachin Jambovane, Katipamula Shanta, Spencer Prost, Michael Russcher N/A

SYMPOSIUM — **Session 560**

Innovative Applications of Surface-Enhanced Raman Spectroscopy
arranged by Bhavya Sharma, University of Tennessee, Knoxville

Monday Afternoon, Room B303
Bhavya Sharma, University of Tennessee, Knoxville, Presiding

1:30		**Introductory Remarks - Bhavya Sharma**
1:35	(560-1)	**Ultrafast Surface-Enhanced Raman Spectroscopy** RENEE R FRONTIERA, University of Minnesota, Nathaniel C Brandt, Emily L Keller, Alyssa A Cassabaum, James L Brooks
2:10	(560-2)	**Detecting Small Molecules Using Probe-Mediated SERS Schemes** JON CAMDEN, University of Notre Dame
2:45	(560-3)	**Utilization of SERS Nanoparticles as Contrast Agents for Molecular Imaging in Cancer** CRISTINA ZAVALETA, Stanford University
3:20		**Recess**
3:35	(560-4)	**Gels, Inkjet, and Laser Ablation: A New Toolkit for Surface-Enhanced Raman Spectroscopy** MARCO LEONA, The Metropolitan Museum of Art N/A
4:10	(560-5)	**Development of Surface Enhanced Spatially Offset Raman Spectroscopy (SESORS) for Neuroscience** BHAVYA SHARMA, University of Tennessee

SYMPOSIUM — **Session 570**

Miniature Mass Spectrometers
arranged by R Graham Cooks and Zheng Ouyang, Purdue University

Monday Afternoon, Room B304
R Graham Cooks, Purdue University, Presiding

1:30		**Introductory Remarks - R Graham Cooks and Zheng Ouyang**
1:35	(570-1)	**Introduction of a Multi-Optic Coaxial Ring Ion Trap (MoCRIT) for External Ionization in Portable Mass Spectrometry** GUIDO F VERBECK, University of North Texas
2:10	(570-2)	**From 2G to 3G - Quantitative Analysis and Biomarker Profiling Using Miniature Mass Spectrometry System** ZHENG OUYANG, Purdue University, Ren Yue, Ran Zou, Yuan Su, Graham Cooks, Yu Xia
2:45	(570-3)	**Describing and Optimizing Toroidal Trapping Fields for the Development of Miniature Mass Spectrometers** STEPHEN A LAMMERT, PerkinElmer, Edgar D Lee, Daniel E Austin, Karl R Warnick
3:20		**Recess**
3:35	(570-4)	**Practical Applications of Outside-the-Lab Mass Spectrometry** MITCH WELLS, FLIR Detection, Inc.
4:10	(570-5)	**The Development and Performance Enhancement of A Mini-Mass Spectrometer with Continuous Atmospheric Pressure Interface** WEI XU, Beijing Institute of Technology, YanBing Zhai, Muyi He

Author and presider lists as of January 20, 2016, are available at www.pittcon.org

SYMPOSIUM **Session 580**

Nanomedicine, From Diagnostics to Large Animal Therapy
arranged by Raoul Kopelman, University of Michigan

Monday Afternoon, Room B305
Raoul Kopelman, University of Michigan, Presiding

1:30		Introductory Remarks - Raoul Kopelman
1:35	(580-1)	The Plasmonic Photo-Thermal Death of Cancer in Cells(1) and in Animals(2,3) Using Gold Nano-rod MOSTAFA A EL-SAYED, Georgia Institute of Technology
2:10	(580-2)	Platinum(II) and Gold(III) Compounds for Cancer Diagnosis and Therapy CHI-MING CHE, The University of Hong Kong N/A
2:45	(580-3)	Nanophotonics-Based Diagnostics and Therapy: From Deep In-Vivo Photo-Acoustic Chemical Imaging of Tumors to Photodynamic Treatment of Heart Disease RAOUL KOPELMAN, University of Michigan
3:20		Recess
3:35	(580-4)	Interfacial Assembly of Functional Mesoporous Nanospheres with Multi-Level Architectures for Bioapplications DONGYUAN ZHAO, Fudan University
4:10	(580-5)	Stickyflares: Tracking the Amount and Location of RNA in Single Cells CHAD A MIRKIN, Northwestern University

SYMPOSIUM **Session 590**

Non-Traditional Human Biometrics for Threat Assessment: Using Chemical Forensics for National Security and Intelligence Applications
arranged by Joachim Dieter Pleil, US EPA and Kevin O'Connell, InQTel

Monday Afternoon, Room B308
Joachim Dieter Pleil, US EPA, Presiding

1:30		Introductory Remarks - Joachim Dieter Pleil and Kevin O' Connell
1:35	(590-1)	Human Biomonitoring and In Vitro Toxicity Testing Applications for Covert Threat Analysis and Security Applications JOACHIM DIETER PLEIL, US EPA, William E Funk
2:10	(590-2)	Noninvasive Infectious Disease Monitoring for Health Applications MICHAEL SCHIVO, University of California, Davis, Cristina E Davis
2:45	(590-3)	Comprehensive Analysis of the Chemicals Within: The Human Exposome and Applications for Threat Assessment GARY W MILLER, Emory University
3:20		Recess
3:35	(590-4)	The US Army Edgewood Chemical Biological Center's Chemical Threat Analysis Within the Framework of the Organization for the Prohibition of Chemical Weapons STANLEY A OSTAZESKI, US Army ECBC, Ethan A Jestel N/A
4:10	(590-5)	Feasibility of Using Breath to Predict Exposure to Ionizing Radiation TERENCE H RISBY, Johns Hopkins University

SYMPOSIUM **Session 600**

Novel Mass Spectrometric Approaches and Applications to Polymer Analysis
arranged by Charles L Wilkins, University of Arkansas

Monday Afternoon, Room B309
Charles L Wilkins, University of Arkansas, Presiding

1:30		Introductory Remarks - Charles L Wilkins
1:35	(600-1)	Characterization of Synthetic Polymers Using as Little as a Small Molecule Matrix and the Vacuum of the Mass Spectrometer for Ionization SARAH TRIMPIN, Wayne State University, Casey D Foley, Joshua Fischer, Zachary Devereaux, Sashiprabha M Vithanarachchi, Matthew J Allen, Barbara S Larsen
2:10	(600-2)	Model Polymer Systems: Studies by Mass Spectrometry, Ion Mobility, and Computational Strategies DAVID M HERCULES, Vanderbilt University, Sarah M Stow, Tiffany M Onifer, John A McLean
2:45	(600-3)	Shape Selective Studies of Macromolecular Systems JAMES HOWARD SCRIVENS, Teesside University
3:20		Recess
3:35	(600-4)	Shape Sensitive Multidimensional Mass Spectrometry of Synthetic Polymers and Hybrid Materials CHRYS WESDEMIOTIS, The University of Akron
4:10	(600-5)	MALDI-TOF and MALDI-FTICR of Challenging Polymer Analysis Problems CHARLES L WILKINS, University of Arkansas, Evegenia Tisdale

SYMPOSIUM **Session 610**

The Twenty-Seventh James L Waters Symposium on Super-Resolution Microscopy
arranged by W Richard Howe, University of Pittsburgh

Monday Afternoon, Room B405
W Richard Howe, University of Pittsburgh, Presiding

1:30		Introductory Remarks - W Richard Howe
1:35	(610-1)	Imaging Life at High Spatiotemporal Resolution ERIC BETZIG, Janelia Research Campus
2:10	(610-2)	Accessing the Emerging Imaging Technologies at HHMI Janelia TENG-LEONG CHEW, Janelia Research Campus
2:45	(610-3)	Closing the Gap Between First Implementation and Product – Update on Lattice Light Sheet Commercialization Effort ALEX SOELL, Carl Zeiss Microscopy, LLC
3:20		Recess
3:35	(610-4)	Life Inside the Cell: STORM, CRISPR and Imagenomics BO HUANG, University of California, San Francisco
4:10	(610-5)	Structured Scanned Plane Bessel Microscopy for Super-Resolution Neuroanatomy TIMOTHY HARRIS, HHMI Janelia Research Campus

SYMPOSIUM **Session 620**

Trials and Tribulations of Dietary Supplement Analysis: Authentication, Adulteration and Contaminant Testing
arranged by Lowri DeJager, U.S. Food and Drug Administration

Monday Afternoon, Room B311
Lowri DeJager, U.S. Food and Drug Administration, Presiding

1:30		Introductory Remarks - Lowri DeJager
1:35	(620-1)	Protecting Consumers One Analysis at a Time: Identifying Harmful Adulterants in Dietary Supplements TRAVIS M FALCONER, U.S. Food & Drug Administration
2:10	(620-2)	Authentication of Foods and Botanical Supplements Using Chemometric Methods JAMES HARNLY, USDA
2:45	(620-3)	DNA Authentication and Adulterant Detection: Dispelling the Myths and Facing the Facts DANICA HARBAUGH REYNAUD, Authen Technologies N/A
3:20		Recess
3:35	(620-4)	Authenticity of Herbal Dietary Supplements: Comparison of Chemical and DNA Barcoding Methods RAHUL PAWAR, CFSAN/FDA, Sara M Handy, Erich Grundel, Raymond Cheng, Nicole Shyong
4:10	(620-5)	Spectroscopic Detection of Adulteration in Botanical Dietary Supplements: What is Licorice? CHARLOTTE SIMMLER, University of Chicago, Guido Pauli, Shao-Nong Chen

TECHNICAL PROGRAM

ORGANIZED CONTRIBUTED SESSIONS — Session 630

Cell Phone Spectroscopy - Handheld Spectroscopy for the Citizen
arranged by Mark Druy, Druy Consulting and Richard Crocombe, PerkinElmer

Monday Afternoon, Room B402
Mark Druy, Druy Consulting, Presiding

1:30	(630-1)	**Ready When You Are: Cell Phone Spectrometry When Cellcams Yield RAW Data** ALEXANDER SCHEELINE, SpectroClick
1:50	(630-2)	**Biomedical Applications of Cellphone Spectroscopy** ANSHUMAN DAS, MIT Media Lab
2:10	(630-3)	**Optical Smartphone Biosensing Techniques** BRIAN T CUNNINGHAM, University of Illinois at Urbana-Champaign, Kenneth Long
2:30	(630-4)	**Withdrawn**
2:50		**Recess**
3:05	(630-5)	**Quantum Dots and Smartphone Fluorescence Imaging: A Perfect Marriage for Bioassays** RUSS ALGAR, University of British Columbia, Eleonora Petryayeva
3:25	(630-6)	**Point-of-Care Colorimetric Detection with a Cell Phone** IAN PAPAUTSKY, University of Cincinnati
3:45	(630-7)	**Mobile Technologies for Personalized Diagnostics and Global Health** DAVID ERICKSON, Cornell University
4:05	(630-8)	**A Handheld Optoelectronic Nose** KENNETH S SUSLICK, University of Illinois at Urbana-Champaign, Zheng Li, Maria K LaGasse, Jon R Askim

ORGANIZED CONTRIBUTED SESSIONS — Session 640

Isolation and Characterization of Impurities/Degradation Product: Understanding Your Impurity Profile Throughout the Development Process
arranged by Andy Miles and Paul Wrezel, Regis Technologies

Monday Afternoon, Room B313
Andy Miles, Regis Technologies, Presiding

1:30	(640-1)	**Risk Analysis of Impurities in Drug Substances** DAN WEISSMUELLER, Regis Technologies
1:50	(640-2)	**Low level Impurity Isolations for Impurity Profiling and Structure Elucidation** TONY (QI) YAN, Pfizer
2:10	(640-3)	**Case Studies Involving Method Development for Trace-Level Impurities** PAUL WREZEL, Regis Technologies
2:30	(640-4)	**Impurity Identification for API Process Development** ZHAO YANQUN, AbbVie, Inc., Wayne Pritts
2:50		**Recess**
3:05	(640-5)	**Emerging Techniques for the Identification of Impurities and Degradation Products in Agricultural Research** JEFFRIE A GODBEY, Dow AgroSciences, Jesse L Balcer, Jeffrey R Gilbert, Yelena A Adelfinskaya, Mary D Evenson
3:25	(640-6)	**The Development of SFC Stationary Phases for the Optimized Separation of Chemical Mixtures Containing a Wide Range of Polarities** MATTHEW PRZYBYCIEL, ES Industries
3:45	(640-7)	**Isolation and Characterization of Impurities in the Synthesis of Drug Substance to Support Drug Development** QIFENG XUE, Theravance Biopharma US, Inc., Kanaka Hettiarachchi, Ken Ngim, Zhengtian (Titan) Gu
4:05	(640-8)	**Open Discussion**

ORAL SESSIONS — Session 650

Biomedical: Advances in Detection and Therapeutics of Cancer

Monday Afternoon, Room B302
Rabih E Jabbour, US Army Edgewood Chemical Biological Center, Presiding

1:30	(650-1)	**KS-Detect: A Portable, Solar-Thermal, Polymerase Chain Reaction System for the Point-of-Care Diagnosis of Kaposi's Sarcoma** RYAN SNODGRASS, Cornell University, Andrea Gardner, Li Jiang, Ethel Cesarman, David Erickson
1:50	(650-2)	**Nano-Plasmonic Exosome Platform (nPLEX) for Label-Free Detection and Molecular Profiling of Exosomes** HYUNGSOON IM, Massachusetts General Hospital, Huilin Shao, Park Yongil, Vanessa Peterson, Cesar M Castro, Ralph Weissleder, Hakho Lee
2:10	(650-3)	**Generate DNA Aptamers Against Glypican 3 with Expanded Genetic Systems** LIQIN ZHANG, University of Florida
2:30	(650-4)	**Systematic and Quantitative Analysis of Surface N-Sialoglycoproteins in Cancer Cells with Distinct Invasiveness** RONGHU WU, Georgia Institute of Technology
2:50		**Recess**
3:05	(650-5)	**DNA "AND" Logic Platform Integrated on Nanoparticle for Programmed Recognition and Therapy of Cancer** ZHENBAO LIU, University of Florida, Weihong Tan N/A
3:25	(650-6)	**A Core/Shell Structure of Reduced Graphene Oxide/ Mesoporous Silica with Oligonucleotide Gates for Cancer Treatment** XIAO LIU, University of North Dakota, Xu Wu, Xuefei Zhang, Yuqian Xing, Julia Xiaojun Zhao
3:45	(650-7)	**Simultaneous Photothermo-/Chemotherapy Using Reduced Graphene Oxide Based Nanocomposites** YUQIAN XING, University of North Dakota, Xu Wu, Xiao Liu, Julia Xiaojun Zhao
4:05	(650-8)	**Imaging of Cancer Protein-Protein Interactions and Small Molecule Inhibitions by a Surface Plasmon Resonance Microarray** CHARUKSHA WALGAMA, Oklahoma State University, Zainab Hussain Al Mubarak, Bing Zhang, Mayowa Akinwale, Anuruddha Pathiranage, Junpeng Deng, Darrell K Berlin, Doris M Benbrook, Sadagopan Krishnan

ORAL SESSIONS — Session 660

Environmental Applications of Elemental Analysis

Monday Afternoon, Room B315
Eugene F Barry, University of Massachusetts Lowell, Presiding

1:30	(660-1)	**Analyzing Mercury from Contaminated Mining Sites Using a Direct Mercury Analyzer** SUMEDH PHATAK, Milestone Inc.
1:50	(660-2)	**ICP-MS for the Analysis of High Salinity Samples** ERICA CAHOON, PerkinElmer, Daniel H Jones
2:10	(660-3)	**Nanoparticle Removal During Alum and Ferric Coagulation Characterized by Single Particle ICP-MS** ARIEL DONOVAN, Missouri University of Science and Technology, Honglan Shi, Yinfa Ma, Craig Adams, Chady Stephan, Todd Eichholz
2:30	(660-4)	**Applications of Single Particle ICP-MS to Environmental Matrices** CHADY STEPHAN, PerkinElmer, Kenneth Neubauer
2:50		**Recess**
3:05	(660-5)	**Issues in Deep Ocean LIBS: Internal Calibration and the Effect of Suspended Particles** JOSEPH BONVALLET, University of South Carolina, S Michael Angel
3:25	(660-6)	**Selective Measurement of Metal Ions at Covalently Functionalized Carbon-Fiber Microelectrodes** YUANYUAN YANG, Wayne State University, Ahmad A Ibrahim, Jennifer L Stockdill, Parastoo Hashemi
3:45	(660-7)	**MWCNT/Bi Composite Film Modified PGE for Voltammetric Determination of Lead and Cadmium** YESIM TUGCE YAMAN, Hacettepe University, Serdar Abaci N/A
4:05	(660-8)	**Application of Mössbauer spectrometry in Environmental Studies of Fly Ashes and Road Dusts** TADEUSZ SZUMIATA, RADWAG Balances and Scales, Tadeusz Szumiata, Malgorzata Gzik-Szumiata, Katarzyna Brzozka, Bogumil Gorka, Michal Gawronski, Ryszard Swietlik, Marzena Trojanowska

ORAL SESSIONS **Session 670**

Environmental GC

Monday Afternoon, Room B406
Walter B Wilson, National Institute of Standards and Technology, Presiding

1:30	(670-1)	**Multi-Dimensional Micro Gas Chromatography Device for the Rapid and Sensitive Analysis of Volatile Organic Compounds** JIWON LEE, University of Michigan, Menglian Zhou, Hongbo Zhu, Robert Nidetz, Katsuo Kurabayashi, Xudong Fan
1:50	(670-2)	**SPME On-Fiber Derivatization Using a Stable and Reusable Pentafluorophenyl Hydrazine Standard Gas Generating Vial** JUSTEN J POOLE, University of Waterloo, Jonathan J Grandy, German A Gomez-Rios, Emanuela Gionfriddo, Janusz Pawliszyn
2:10	(670-3)	**Gas Chromatography-Vacuum Ultraviolet Detection (GC-VUV) for Analysis of Polychlorinated Biphenyls (PCBs)** CHANGLING QIU, University of Texas at Arlington, Jonathan Smuts, Kevin A Schug
2:30	(670-4)	**Gas Chromatographic Retention Behavior on Select Groups of Isomeric Polycyclic Aromatic Compounds and Their Alkyl-Substituted Derivatives on Stationary Phases of Different Selectivity** WALTER B WILSON, National Institute of Standard and Technology, Federica Nalin, Lane C Sander, Leonard M Sidisky, Stephen A Wise
2:50		**Recess**
3:05	(670-5)	**Optimization of Phthalate Analysis by Gas Chromatography-Mass Spectrometry Using Computer Modeling Software** DAN LI, Restek Corporation, Rebecca Stevens, Jack Cochran, Amanda Rigdon, Chris English
3:25	(670-6)	**Analysis of Greenhouse Gases Using the Dielectric Barrier Discharge Detector** MATTHEW MONAGLE, AIC LLC N/A
3:45	(670-7)	**Fast, Very Fast and Ultra-Fast Forensic and Homeland Security GC-MS** AVIV AMIRAV, Tel Aviv University, Uri Keshet, Alexander Fialkov, Tal Alon
4:05	(670-8)	**Passive Monitoring – A Guide to Sorbent Tube Sampling for EPA Method 325** CAROLINE WIDDOWSON, Markes International

ORAL SESSIONS **Session 680**

Food Product Quality and Component Characterization

Monday Afternoon, Room B316
Ed Guthrie, Agilent Technologies, Presiding

1:30	(680-1)	**Analysis of Fat Crystallization Thanks to Microrheology** MAXIME BAZIN, Formulaction, Giovanni Brambilla, Roland Ramsch, Mathias Fleury, Gérard Meunier
1:50	(680-2)	**Atomic Absorption and Potentiometric Analysis of Electrolytes in Food Substances Used for Rapid Muscle Cramp Relief by Athletes** STEPHANIE HOOPER MAROSEK, Methodist University, Taylor Tipton
2:10	(680-3)	**Microrheological Analyses for Dairy Formulations** ROLAND RAMSCH, Formulaction, Giovanni Brambilla, Gérard Meunier
2:30	(680-4)	**Analysis of Aroma Compounds in Beer by TD–GC–TOF MS with Soft Electron Ionization** LAURA MCGREGOR, Markes International Ltd, Caroline Widdowson, Massimo Santoro, Chris Hall, Ken Umbarger
2:50		**Recess**
3:05	(680-5)	**Food Analysis Using Laser Desorption GS-Ion Mobility Spectrometry – Olive Oil as an Example** WOLFGANG VAUTZ, ISAS, Sascha Liedtke, Joachim Franzke
3:25	(680-6)	**Monitoring of Protein Changes in Pasteurized Liquid Egg Using Capillary Electrophoresis** REYHAN S UYSAL, Hacettepe University, Ismail H Boyaci, Esra Acar, Nusret Ertas
3:45	(680-7)	**Exploiting Polymeric Ionic Liquids-Based SPME Sorbents Coupled to Gas-Chromatography/Mass Spectrometry for Food Quality and Safety Assessment** ERICA A SOUZA-SILVA, Universidade Federal do Rio Grande do Sul, Emanuela Gionfriddo, Nathaly Reyes-Garces, German A Gomez-Rios, Jared L Anderson, Janusz Pawliszyn
4:05	(680-8)	**Detection, Identification, and Pattern Recognition of Microbial Volatile Organic Compounds from Virulent and Hypo-Virulent Cryphonectria Parasitica Species by Headspace-SPME-GC-MS and Chemometrics** JINYAN SHE, Mississippi State University

ORAL SESSIONS **Session 690**

LC Method Development - Half Session

Monday Afternoon, Room B403
William E Barber, Agilent Technologies, Presiding

1:30	(690-1)	**Method Transfer and Column Scalability with Superficially Porous Particles** WILLIAM LONG, Agilent Technologies, Anne Mack, Stephen Luke, Jason Link
1:50	(690-2)	**Isolation of Impurities in Biopharmaceutical Formulations** THOMAS E WHEAT, Waters Corporation, Amanda B Dlugasch, Patricia R McConville
2:10	(690-3)	**New Technological Solutions to Maximizing Uptime on an Ion Chromatography System** DAVID G MOORE, Thermo Fisher Scientific, Pranathi Perati, Sally Eastman
2:30	(690-4)	**Accurate Measurement of Analyte Dispersion through Connecting Tubes used in Fast Very High-Pressure Liquid Chromatography** FABRICE GRITTI, Waters Corporation, Martin Gilar, Thomas McDonald

ORAL SESSIONS **Session 700**

LC Optimization - Half Session

Monday Afternoon, Room B404
Rachael Wilson, University of Pittsburgh, Presiding

1:30	(700-1)	**Dynamic Temperature Control in Capillary Liquid Chromatography: Increasing Analysis Sensitivity, Speed, and Peak Capacity** STEPHEN R GROSKREUTZ, University of Pittsburgh, Rachael Wilson, Stephen Weber
1:50	(700-2)	**Towards the Development of a Refractive Index-Based Optical Microcavity Mass Concentration Detector Compatible with Gradient Elution Liquid Chromatography** ZACHARY S WIERSMA, University of Illinois at Urbana-Champaign, Todd O Pangburn, Alexandria L Stanton, James H Wade, David Mark Meunier, Ryan C Bailey
2:10	(700-3)	**Increasing the Peak Capacity of Quantitative Liquid Chromatography Using Chemometric Curve Resolution** DANIEL W COOK, Virginia Commonwealth University, Sarah C Rutan, Dwight R Stoll
2:30	(700-4)	**Improving Sensitivity, Resolution, and Peak Capacity of Gradient Elution Capillary Liquid Chromatography with Temperature-Assisted On-Column Solute Focusing** RACHAEL WILSON, University of Pittsburgh, Stephen R Groskreutz, Stephen Weber

ORAL SESSIONS **Session 710**

Neurochemistry

Monday Afternoon, Room B401
Chi Leng Leong, Imperial College London, Presiding

1:30	(710-1)	**Measurement of Phasic Dopamine Signals in the Rat Nucleus Accumbens Core and Shell in Response to Noxious Stimuli** CHRISTOPHER W ATCHERLEY, Mayo Clinic, Edita Navratilova, Yanhua Xie, Levi Lazarus, Michael L Heien, Frank Porreca N/A
1:50	(710-2)	**Oxygen Changes and Dopamine Release during Spreading Depression** CADDY N HOBBS, University of North Carolina at Chapel Hill, Justin A Johnson, R Mark Wightman
2:10	(710-3)	**Spontaneous, Transient Adenosine Release from Brain During Ischemia-Reperfusion Injury** MALLIKARJUNARAO GANESANA, University of Virginia, B Jill Venton
2:30	(710-4)	**Treatment on Carbon Nanotube Yarn Microelectrode for Sensitive and Rapid Dopamine Detection In Vivo** CHENG YANG, University of Virginia, B Jill Venton, Ilia N Ivanov, Michael D Nguyen, Christopher B Jacobs
2:50		**Recess**
3:05	(710-5)	**Neurochemical Investigation of Epilepsy Using Microdialysis Sampling to Study Multiple Seizure Rat Models** AMANDA M FURNESS, University of Kansas
3:25	(710-6)	**PKC-b Inhibitors Attenuate Amphetamine and Cocaine Stimulated Dopamine Release** ALEXANDROS G ZESTOS, University of Michigan, Robert T Kennedy, Margaret E Gnegy

3:45 (710-7) **Clinical Measurements at the Bedside: Dynamic Neurochemical Changes in the Injured Human Brain Monitored Using an Online Microdialysis System** MICHELLE L ROGERS, Imperial College London, Chi Leng Leong, Sally A Gowers, Sharon Jewell, Shumaila Khan, Anthony J Strong, Martyn G Boutelle

4:05 (710-8) **Online Clinical Microdialysis: Detecting the Neurochemical Consequences of Spreading Depolarization** CHI LENG LEONG, Imperial College London, Michelle L Rogers, Sally A Gowers, Sharon Jewell, Anthony J Strong, Shumaila Khan, Martyn G Boutelle

ORAL SESSIONS — Session 720

Unique Developments in Spectroscopy - Half Session

Monday Afternoon, Room B403
William E Barber, Agilent Technologies, Presiding

3:05 (720-1) **Simultaneous Measurement of N_2O and CH4 Emissions from Agriculture Using Photoacoustic Detection and QCL Laser** ARTO BRANDERS, Gasera Ltd., Ismo Kauppinen, Jaakko Lehtinen

3:25 (720-2) **Infrared Spectroscopic Remote Sensing of Pulsed Signals from Nearby Stars** ROBERT A LODDER, University of Kentucky, Anne Brooks

3:45 (720-3) **High Resolution Coherent Multidimensional Spectroscopy** PETER CHEN, Spelman College

4:05 (720-4) **Biologics Starting Materials Identified through Opaque Containers by Spatially Offset Raman Spectroscopy (SORS)** MATTHEW BLOOMFIELD, Cobalt Light Systems, Darren Andrews, Pavel Matousek

POSTER SESSION — Session 730

All posters are to be mounted by 10:00 AM and remain on display until 4:00 PM. Authors must be at their posters from 1:00 PM to 3:00 PM. Location of the afternoon posters is on the Exposition Floor, 400 Aisle. PLEASE NOTE: You cannot get onto the Exposition Floor until after 9:00 AM.

Art and Archaeology

Monday Afternoon, Exposition Floor, 400 Aisle

(730-1 P) **Maximizing the Information Obtained from Small Archaeological Samples by Sequencing DART-MS and Plasma-Chemical Oxidation for AMS Radiocarbon Dating** RUTH ANN ARMITAGE, Eastern Michigan University, Kathryn Jakes, Suzanne Baker

(730-2 P) **The Dose Makes the Poison: Quantitation of Pollutant VOCs From Materials Used in A Museum Environment** MICHAEL J SAMIDE, Butler University, Gregory D Smith, Jericha Mill

(730-3 P) **Using Reflectance Spectroscopy to Determine the Rate of Formation of Prussian Blue Pigment** JACOB APPLEGARTH, Butler University, Michael J Samide

(730-4 P) **Elemental Profiling of Archaeometallurgical Artefacts by ICP Spectrometry** SANDA RONCEVIC, University of Zagreb, Ivan Nemet N/A

POSTER SESSION — Session 740

All posters are to be mounted by 10:00 AM and remain on display until 4:00 PM. Authors must be at their posters from 1:00 PM to 3:00 PM. Location of the afternoon posters is on the Exposition Floor, 400 Aisle. PLEASE NOTE: You cannot get onto the Exposition Floor until after 9:00 AM.

Chemical Methods

Monday Afternoon, Exposition Floor, 400 Aisle

(740-1 P) **Single Nanoparticle to 3D Supercage: Framing for an Artificial Enzyme System** REN CAI, University of Florida, Weihong Tan

(740-2 P) **Synthesis and Characterization of Superparamagnetic Manganese Ferrite Nanoparticles** SIMONAS RAMANAVICIUS, Vilnius University, Marija Kurtinaitiene, Kestutis Mazeika, Vidas Pakstas, Arunas Jagminas

(740-3 P) **Optimization of Ibuprofen Micronization by the Rapid Expansion of a Supercritical Solution (RESS)** ROLF SCHLAKE, Applied Separations, Al Kaziunas, Brian Day

(740-4 P) **DNA-Aligner-Controlled Nicking-Based Isothermal Exponential Amplification Reaction for High Sensitive Detection of Nucleic Acids** WU WANGHUA, Zhejiang University, Zhou Jianguang, Zhang Tao, Yu Dongdong

(740-5 P) **Electrochemical, Spectroscopic and Chromatographic Techniques for Monitoring the Active Pharmaceutical Ingredients Degradation Kinetics: Which Methodology Fulfills More Principles of the Green Analytical Chemistry?** MOHAMED K ABD EL-RAHMAN, Cairo University, Amr M Mahmoud

(740-6 P) **Investigation of Heterogeneous Reaction Mechanism Between Formaldehyde and MnO_2/CeO_2 at Room Temperature by Gas Analysis Approach** HAYASHI HIROKI, Tokai University, Sekine Yoshika

(740-7 P) **Automated, Low-Level Distillation of Phenolics for Use in Environmental and Manufacturing Applications** BRANT HOEKSTRA, OI Analytical, Jonathan Howerton N/A

(740-8 P) **Identification of 6-chlorotestosterone and Other Designer Anabolic Steroids in Dietary Supplements with Semi-Quantitative Content Determination Using Surrogate Compounds** SARAH ELIZABETH VOELKER, U.S. FDA, Forensic Chemistry Center, Lisa M Lorenz, Mary B Jones, Travis M Falconer, Rick A Flurer

POSTER SESSION — Session 750

All posters are to be mounted by 10:00 AM and remain on display until 4:00 PM. Authors must be at their posters from 1:00 PM to 3:00 PM. Location of the afternoon posters is on the Exposition Floor, 400 Aisle. PLEASE NOTE: You cannot get onto the Exposition Floor until after 9:00 AM.

Commercial Products Characterization

Monday Afternoon, Exposition Floor, 400 Aisle

(750-1 P) **Considerations in Sample Preparation and Method Development Using ICP-OES** KENNETH NEUBAUER, PerkinElmer, Nick Spivey, Stan Smith, Laura Thompson

(750-2 P) **Chemical Fingerprinting of Tobacco and Related Products by TD–GC–TOF MS** LAURA MCGREGOR, Markes International Ltd, Chris Hall, Caroline Widdowson, Nicola Watson, Ken Umbarger

(750-3 P) **Levels of Mercury and Methylmercury in Fish (Prochilodus Magdalenae) Ayapel Marsh (Colombia)** EDINELDO LANS CEBALLOS, University of Cordoba, Mauricio Lora Agamez, Amira Padilla Jimenez

(750-4 P) **Mercury Residual and Methylmercury in Canned Tuna Distributed in Monteria –Colombia** EDINELDO LANS CEBALLOS, University of Cordoba, Mauro Lombana Agamez, Basilio Diaz Pogutá

(750-5 P) **Automated SPME for the Analysis of Environmental Contaminants in Milk** NICOLE M LOCK, Shimadzu Scientific Instruments, Robert Clifford, Di Wang, Laura Chambers, Brahm Prakash, Shilpi Chopra

(750-6 P) **Estimation of Dietary Intake of Essential and Non-Essential Metals through the Consumption of Dietary Supplements Available on Nigerian Retail Market** IMAOBONG UDOUSORO, University of Uyo, Ikem Abua, Olujide T Akinbo

(750-7 P) **Analysis of Aflatoxins in Pet Food by UHPLC Using PDA and Fluorescence Detection** CATHARINE LAYTON, PerkinElmer, Jason Weisenseel, Wilhad M Reuter

(750-8 P) **Antioxidant Activity of Flavored Tea and Its Content of Phenol, Flavonoid and Tannins** ABD EL-MONEIM M AFIFY, Cairo University

(750-9 P) **High Sensitive LC/MS Analysis of Stevia Sweeteners Using Polymer-Based Amino Column** RONALD BENSON, Shodex/Showa Denko America, Inc., Junji Sasuga, Satoko Sakai, Tomokazu Umezawa

(750-10 P) **Assigning Unknown Fingernail Polish to Known Manufacturers by Raman Spectroscopy and Multivariate Statistics** GARY H NAISBITT, Utah Valley University, Kelsey Hartt N/A

(750-11 P) **Characterizing the Rheological Properties of Wax Emulsions used as Carriers for Biopesticides in Agricultural Pest Management** KRISTEN JORDAN, Western Carolina University, Cynthia Atterholt

(750-12 P) **Multivariate Quantification of API Release from Combination Tablets in the Presence of Matrix Effects Using Fiber Optic Dissolution** JOSEPH MEDENDORP, Vertex Pharmaceuticals, Taryn Ryan, Mahidhar Shapally, Ivelisse Colon

(750-13 P) **Simultaneous Analysis of Methylisothiazolinone, Salicylic Acid and Parabens in Antidandruff Hair Shampoos by Monolithic Silica High-performance Liquid Chromatographic Column** ABDULRAHMAN ALMAJED, King Saud Univesriy, College of Pharmacy

(750-14 P) **Simultaneous Analysis Red Wine Absorbance, CIE Lab Color Indices and Fluorescence Excitation-Emission Matrices** ADAM MATTHEW GILMORE, Horiba, Sakiko Akaji

(750-15 P) **Extractables and Leachables Analysis of IV Bag Systems Using Thermal Desorption and Stir Bar Sorptive Extraction with GC Single Quad and GC Time-of-Flight Mass Spectrometric Detection** ANDREAS HOFFMANN, Gerstel GmbH & Co.KG, Thomas Albinus, Kurt Thaxton, Elizabeth Almasi

POSTER SESSION **Session 760**

All posters are to be mounted by 10:00 AM and remain on display until 4:00 PM. Authors must be at their posters from 1:00 PM to 3:00 PM. Location of the afternoon posters is on the Exposition Floor, 400 Aisle. PLEASE NOTE: You cannot get onto the Exposition Floor until after 9:00 AM.

Data Analysis and Manipulation

Monday Afternoon, Exposition Floor, 400 Aisle

(760-1 P) **Automatic, High-Throughput Motion Classification of Surface-Attached E. Coli Cells Using Bright Field Microscopy Data** SIMON SHEN, Arizona State University, Karan Syal, Nongjian Tao, Shaopeng Wang N/A

(760-2 P) **Ranking of Variables in the Dataset Using the Degree of Separation: Pure Statistical Method** WALEED MASWADEH, US Army, Jason A Guicheteau, A Peter Snyder

(760-3 P) **Procurement and Distribution Channels of Commonly Used Drugs in Nigeria: A Case Study of the Pharmaceutical Industry in Abia State** LILIAN I OGUGUO, National Malaria Elimination Programme FMoH, Chidiebere A Odike-Aduaka, Ifeoma Agwo

(760-4 P) **Effective Data Management in the Analytical Laboratory** TOSHINOBU YANAGISAWA, Shimadzu Corporation, Kazuhito Wakabayashi, Masayuki Shibata, Keisuke Yoshizawa, Ryuji Nishimoto

(760-5 P) **Combining PLS Classification and Regression Analyses for Robust Monitoring of Nuclear Materials Reprocessing** ROBERT LASCOLA, Savannah River National Laboratory, Patrick O'Rourke

POSTER SESSION **Session 770**

All posters are to be mounted by 10:00 AM and remain on display until 4:00 PM. Authors must be at their posters from 1:00 PM to 3:00 PM. Location of the afternoon posters is on the Exposition Floor, 400 Aisle. PLEASE NOTE: You cannot get onto the Exposition Floor until after 9:00 AM.

Education/Teaching

Monday Afternoon, Exposition Floor, 400 Aisle

(770-1 P) **Development of Interactive Learning Modules used in Teaching Instrumental Analysis** YI HE, John Jay College/CUNY, Sandra Swenson, Wong Tiffany, Colleen McNamara

(770-2 P) **Conservation Science Tutorial Website: Analysis of VOCs Emitted from Museum Construction Materials** JERICHA MILL, Butler University, Michael J Samide, Gregory D Smith

(770-3 P) **Analytical Chemistry Course Embedded STEM Projects with Trolox Equivalent Antioxidant Capacity Assay** XIAOPING LI, Georgia Gwinnett College, Rashad Simmons

(770-4 P) **Inexpensive Teaching Instruments for Atomic Emission and Molecular Spectroscopy** ABD AL-KARIM ALI, Miami University, Taryn L Winner, Neil D Danielson

(770-5 P) **The Antibacterial Effects of Nanoparticles** NEYDA BRIDGITTE CHACON, SUNO N/A

(770-6 P) **Strategies for Designing Calibration Curves that Extend over Three Decades of Concentration for Ultraviolet-Visible Absorption Spectroscopy** LAUREN GRABOWSKI, University of South Carolina, Scott Goode

(770-7 P) **Analytical Chemistry Experiments with a Forensic Flavor** ROBERT Q THOMPSON, Oberlin College

(770-8 P) **HPLC Refurbishment** CONNOR PUTNAM, St. John Fisher College

(770-9 P) **Analytical Chemistry 2.1: An Open-Access Digital Resource for Undergraduate Education in Analytical Chemistry** DAVID T HARVEY, DePauw University

(770-10 P) **Introducing Undergraduate Chemists to Chemometrics: Using Microsoft Excel to Illustrate the NIPALS Algorithm used in Principal Component Analysis and Partial Least Squares Regression** MARK T STAUFFER, University of Pittsburgh - Greensburg

(770-11 P) **Applying Analytical Chemistry to Solve Problems in the Developing World** RHONDA L GROSSE, Chemists Without Borders, Bego Gerber, Marya Lieberman, Steve Chambreau, Satinder Ahuja

(770-12 P) **Interdisciplinary Undergraduate Research in Chemometrics: The Faculty Perspective** HELEN M BOYLAN, Westminster College, Carolyn Cuff, Stephanie Homitz, Christopher Caroff, Keilah Ireland, Domenic DiSanti

(770-13 P) **Development of an Undergraduate Laboratory Experiment to Determine Arsenic in Sinus Wash and Tap Water by Inductively Coupled Plasma-Mass Spectrometry** ANNA M DONNELL, University of Cincinnati, Keaton Nahan, Dawone Holloway, Anne P Vonderheide

POSTER SESSION **Session 780**

All posters are to be mounted by 10:00 AM and remain on display until 4:00 PM. Authors must be at their posters from 1:00 PM to 3:00 PM. Location of the afternoon posters is on the Exposition Floor, 400 Aisle. PLEASE NOTE: You cannot get onto the Exposition Floor until after 9:00 AM.

Electrochemistry

Monday Afternoon, Exposition Floor, 400 Aisle

(780-1 P) **Laser-based Heating for Nanopore DNA Duplex Analysis** CHRISTOPHER E ANGEVINE, Virginia Commonwealth University, Sarah Seashols-Williams, Joseph E Reiner

(780-2 P) **A Novel Sandwiched Electrochemiluminescence Immunosensor for Detection of Carcinoembryonic Antigen Based on Carbon Quantum Dots** WANG HUAI-SHENG, Liaocheng University N/A

(780-3 P) **Label-Free DNA Sensors for Detection of BRCA1 Mutation** BRUNO C JANEGITZ, Federal University of Sal Carlos, Lais Ribovski, Valtencir Zucolotto N/A

(780-4 P) **Conductivity as Sensor for Real-time Monitoring of the Solution Soluble Corrosion Products under Cell Culture Conditions During Corrosion of High Purity Magnesium** KOLADE O OJO, University of Cincinnati, Tracy Hopkins, Sarah Pixley, William R Heineman, Vesselin Shanov, Madhura Joshi, Pravahan Salunke, Keaton Nahan, Guangqi Zhang

(780-5 P) **Gold Nanoparticles Enzymatic Sensors for Determination of Glucose** ARUNAS RAMANAVICIUS, Vilnius University, Povilas Genys, German Natalija, Almira Ramanaviciene

(780-6 P) **RC Constant Based Label Free Biomarkers Detection** PRADEEP RAMIAH RAJASEKARAN, University of Florida, Charles R Martin, Jennifer Ottoy

(780-7 P) **Electrochemical Detection of Evoked Dopamine Release in Zebrafish** MIMI SHIN, University of Kansas, Thomas Field, Michael A Johnson, Mia Furgurson

(780-8 P) **Quantification of Ion Selectivity of Single Asymmetric Nanopore-Channels for Better Energy Harvesting from Salinity Gradients** WARREN D BROWN, Georgia State University, Yan Li, Dengchao Wang, Maksim Kvetny, Gangli Wang

(780-9 P) **Charge Transfer Mechanism of Organic Molecules Associated with Dendrimers at Polarized Liquid/Liquid Interfaces** HIROKI SAKAE, Kanazawa University, Hirohisa Nagatani, Hisanori Imura

(780-10 P) **Study of Electrochemical Hydrogen Evolution and Oxygen Evolution Reactions at Ir and Ru Oxide Alloys with Scanning Electrochemical Microscopy** YUN-BIN CHO, Ewha Womans University, Chongmok Lee, Youngmi Lee

(780-11 P) **Automated, Accurate pH and Conductivity Measurements Using a Discrete Photometric Analyzer with an ECM Module** MARI KIVILUOMA, Thermo Fisher Scientific, Annu Suoniemi-Kahara

(780-12 P) **Atmospheric Corrosion Study of Metals in an Industrial Environment of Ahmedabad** PAREKH P SUNILKUMAR PUNAMBHAI, CU Shah Science College N/A

(780-13 P) **Synthesis and Characterization of Silver Coated Electrospun Cobalt Nanotubes and Their Electrocatalytic Activity** AREUM YU, Ewha Womans University, Chongmok Lee, Myung Hwa Kim, Youngmi Lee

(780-14 P) **Single Microwire Electrodeposition by Electrochemical Liquid-Liquid-Solid Growth** TIM ZHANG, University of Michigan

(780-15 P) **Simultaneous Detection Heavy Metals by Anodic Stripping Voltammetry Using Carbon Nanotube Thread** DAOLI ZHAO, University of Cincinnati, David R Siebold, Tingting Wang, Noe R Alvarez, Vesselin Shanov, William R Heineman

(780-16 P) **Low Concentration Detection by Electrogenerated Chemiluminescence Using Bipolar Electrochemistry in a Thin Layer Manner** SONGYAN YU, Auburn University, Mark D Holtan, Sanjun Fan, Curtis Shannon

(780-17 P) **Atypical Induction of Membrane Potential in Ion Selective Membranes** DEMETRA M PANTELIS, University of Florida, Pradeep R Rajasekaran

(780-18 P) **Characterization and Spectroelectrochemical Sensing with a Boron Doped Diamond Optically Transparent Electrode Coated with a Charge Selective Polymer Film** CORY A RUSINEK, University of Cincinnati, William R Heineman, Michael Becker, Robert Rechenberg, Daoli Zhao, Necati Kaval

(780-19 P) **Simultaneous Sensing of α-Tocopherol and Retinol in Micellar Media by Using Poly(2,21-(1,4 phenylenedivinylene) bis-8-hydroxyquinaldine)/MWCNTs Modified Electrode** HAYATI FILIK, Istanbul University

(780-20 P) **Electrochemical Investigation of a Series of Uranyl Salen Complexes: Effect of Ligand Conjugation on the U(V)/U(VI) Redox Couple** EMILY E HARDY, Auburn University, Madeleine A Eddy, Anne E Gorden

(780-21 P) **Predictable Standard Potential of Solid-Contact Ion-Selective Electrodes by Using Prussian Blue Analogues as Solid Contacts** YU ISHIGE, Hitachi Ltd., Stefan Klink, Wolfgang Schuhmann

(780-22 P) **Electrochemical Behaviour of Highly- and Poorly-doped p-Si-AuNP Electrodes** MEHRAN KASHI, University of New South Wales, Yanfang Wu, Vinicius Goncales, Moinul Choudhury, Simone Ciampi, Justin J Gooding

Monday Afternoon

(780-23 P) **Withdrawn**

(780-24 P) **The Effect of Excited Fluorophore on Vesicle Fusion at the Surface of the Electrode** NEDA NAJAFINOBAR, Chalmers University of Technology, Jelena Lovric, Johan Dunevall, Hoda M Fathali, Andrew Ewing, Ann-Sofie Cans N/A

(780-25 P) **Novel Electrode PtCr/PAA (Polyamic Acid) for Efficient Ethanol Oxidation Reaction** JING ZHANG, Binghamton University

(780-26 P) **Voltammetric Serotonin Measurements in Mouse Models of Depression** RACHEL A SAYLOR, University of South Carolina, Aya Abdalla, Parastoo Hashemi

(780-27 P) **Kinetic Size-Spectra of Gas Molecules at Ionic Liquid (IL)-Metal Interface and Its Application for Highly Selective Gas Sensing** ZHE WANG, Xavier University

(780-28 P) **New Arginine-Acetaminophen Incorporation for Selective Determination in Serum with Graphene-based Biosensor** ZHE WANG, Xavier University

(780-29 P) **Core-shell Nanoparticles Supported on Carbon Nanotubes as Promising Catalyst for Methanol Electro-oxidation** MANZAR SOHAIL, Center Of Excellence In Nanotechnology, KFUPM N/A

POSTER SESSION — Session 790

All posters are to be mounted by 10:00 AM and remain on display until 4:00 PM. Authors must be at their posters from 1:00 PM to 3:00 PM. Location of the afternoon posters is on the Exposition Floor, 400 Aisle. PLEASE NOTE: You cannot get onto the Exposition Floor until after 9:00 AM.

Materials Science

Monday Afternoon, Exposition Floor, 400 Aisle

(790-1 P) **Non-Toxic Corrosion Inhibitor from Tagetes Patula L. for Aluminum in Acid Fluids used in Industrial Operations** OLUSEGUN KEHINDE ABIOLA, Federal University of Petroleum Resources, A O Aliyu, E E Elemike

(790-2 P) **Synthesis of Fluorine Substituted $LiFeBO_3$ Composite Material as a Cathode Material for a Lithium Secondary Battery and Characterization Using the Solid State NMR** YOUNGIL LEE, University of Ulsan, JaeMin Bak, Hansol Lee

(790-3 P) **Innovative Instrument Design for Highly Precise Determination of Sulfur and Carbon Concentration** MARKUS JUNG, Elementar Analysensysteme GmbH, David Nennstiel, Christian Schmidt N/A

(790-4 P) **Analytical Methods to Estimate Biodurability of Engineered Nanomaterials** MARY-LUYZA AVRAMESCU, Health Canada, Pat E Rasmussen, Marc Chénier, H David Gardner

(790-5 P) **Intensity Ratio of Resonant Raman Modes for Chirality Enriched Carbon Nanotubes** YANMEI PIAO, National Institute of Standards and Technology, Jeffrey Simpson, Jason Streit, Stephanie Lam, Geyou Ao, Jeffrey Fagan, Angela Hight Walker

(790-6 P) **Liquid-Phase Laser Ablation as a Controlled Method to Produce Graphene Quantum Dots** ROSEMARY LYNN EASTERDAY, University of Kentucky, Wenjin Cao, Yiyang Liu, Doo Young Kim, Dong-Sheng Yang

(790-7 P) **Physicochemical Characterization and Nanotoxicity of Polishing Slurries During CMP Process** EDUARD DUMITRESCU, Clarkson University, Dinusha Karunaratne, Kenneth Wallace, Silvana Andreescu

(790-8 P) **Polycapillary X-Ray Optics to Play a Key Role in NASA Mars 2020 Mission** NING GAO, XOS, Jared Sachs, George Allen, Genevieve DeMarco, George Allen, Larry Wade, Peng Lu, Igor Ponomarev, Jay Burdett, Robert Sharrow, Jaime Luna, Douglas Dawson

(790-9 P) **Preparation and Characterization of Bleached Ground Peanut Hulls** HOLLY TRULUCK, Western Carolina University, Melisa J Glatte, Carmen L Huffman

(790-10 P) **Real Time Measurement of Layer Thickness, Erosion Rates and Crater Depth in Glow Discharge Optical Emission Spectrometry** SOFIA GAIASCHI, Horiba Jobin Yvon, Simon Richard, Patrick Chapon, Kayvon Savadkouei, Philippe Hunault

(790-11 P) **HH-XRF and HH-LIBS for Alloy Analysis** JIYAN GU, Bruker

(790-12 P) **X-Ray Photoelectron Spectroscopy, Low Energy Ion Scattering, and Time-of-Flight Secondary Ion Mass Spectrometry, including Chemometrics Analysis, of Display Glass Surfaces** MATTHEW R LINFORD, Brigham Young University, Cody V Cushman, Barry M Lunt, Philipp Brüner, Julia Zakel, Thomas Grehl, Nicholas J Smith

(790-13 P) **The Particle Size Paradox** JACK G SAAD, Micromeritics, Paul A Webb

(790-14 P) **Gas Chromatography-Vacuum Ultraviolet Absorbance Spectroscopy for Quantitation of Trace and Bulk Water in Organic Solvents: An Emerging Alternative to Karl Fischer Titration** LINDSEY NICHOLE SHEAR, VUV Analytics, Leonard M Sidisky, Dale Harrison

(790-15 P) **Complete Size Characterization of Diatomaceous Earth** JACK G SAAD, Micromeritics

POSTER SESSION — Session 800

All posters are to be mounted by 10:00 AM and remain on display until 4:00 PM. Authors must be at their posters from 1:00 PM to 3:00 PM. Location of the afternoon posters is on the Exposition Floor, 400 Aisle. PLEASE NOTE: You cannot get onto the Exposition Floor until after 9:00 AM.

Physical Measurements

Monday Afternoon, Exposition Floor, 400 Aisle

(800-1 P) **Structural Studies of Co-Spinel Ferrite Synthesized by an Auto Combustion Method** DIPAL P CHAUDHARY, Mehsana Urban Institute of Sciences

(800-2 P) **Fabrication of A Novel Fiber-Optic based Single-Cell Temperature Sensor** QINGBO YANG, Missouri University of Science and Technology, Ke Li, Hai Xiao, Honglan Shi, Yinfa Ma

(800-3 P) **Withdrawn**

POSTER SESSION — Session 810

All posters are to be mounted by 10:00 AM and remain on display until 4:00 PM. Authors must be at their posters from 1:00 PM to 3:00 PM. Location of the afternoon posters is on the Exposition Floor, 400 Aisle. PLEASE NOTE: You cannot get onto the Exposition Floor until after 9:00 AM.

Polymer Characterization and Analysis

Monday Afternoon, Exposition Floor, 400 Aisle

(810-1 P) **Analyzing the Vibration-Rotation Spectrum of Hydrochloric Acid Generated with a Thermogravimetric Analyzer Coupled to a Fourier Transform Infrared Spectrophotometer** ANTHONY J LANG, PerkinElmer, Jack Botting

(810-2 P) **Analysis of PLGA Molecular Weight and Structure by the Latest Advanced Multi-Detector GPC Systems** MARK RICHARD POTHECARY, Malvern Instruments, Carrie Schindler, Ulf Nobbmann

(810-3 P) **Molecular Level Understanding of Etching Chemistry at Polymer Interfaces** JOHN MYERS, University of Michigan

(810-4 P) **Dissolution Dynamic Nuclear Polarization Study of Living Anionic Polymerization Reaction** YOUNGBOK LEE, Hanyang University

(810-5 P) **Extractables Analysis of SPE Frits** KIRSTEN KINNEBERG, Porex Corporation, Gary Li

(810-6 P) **Automated Microwave Sample Preparation of Difficult Petroleum and Polymer Matrices** ROBERT LOCKERMAN, CEM Corporation, Dan Iversen, Tina Restivo, Ian Goldstein, Austin Thornton

(810-7 P) **Spectroelectrochemical and Electrochemical Impedance Spectroscopy Analysis of Coronene and Perylene Dimide Copolymers** SAMIR C PAUL, Auburn University, Vince Cammarata N/A

(810-8 P) **Microwave, r0 Structural Parameters, Conformational Stability and Vibrational Assignment of (Chloromethyl)fluorosilane** DATTATRAY K SAWANT, University of Missouri Kansas City, James R Durig

(810-9 P) **Micro/Nano-Structured Flexible Foils for Anti-Counterfeiting Purposes** NASTASIA OKULOVA, Danapak Flexibles A/S and Technical University of Denmark, Rafael Taboryski, Lars Christensen

(810-10 P) **Self-Cleaning Properties of Nanostructured Polypropylene Foils Fabricated by Roll-to-Roll Extrusion Coating** AGNIESZKA TELECKA, Danish Technical University, Rafael Taboryski, Ling Sun

(810-11 P) **Synthesis and Characterization of Poly(p-methylstyrene) Spiropyran Conjugates** MATTHEW J PRICE, California University of PA, Zachary Sullenberger

(810-12 P) **Advantages of Ion Mobility Mass Spectrometry for Extractables Testing** BAIBA CABOVSKA, Waters Corporation, Eleanor Riches

(810-13 P) **A Novel Device for DART-MS System** MICHAEL J CHURCHILL, BioChromato, Chikako Takei

(810-14 P) **Characterization of Inorganic Fillers in Complex Polymer Matrices Using Mid-IR and Far-IR Spectroscopy** WILLIAM WIHLBORG, Thermo Fisher Scientific, Ronald Rubinovitz

POSTER SESSION — Session 820

All posters are to be mounted by 10:00 AM and remain on display until 4:00 PM. Authors must be at their posters from 1:00 PM to 3:00 PM. Location of the afternoon posters is on the Exposition Floor, 400 Aisle. PLEASE NOTE: You cannot get onto the Exposition Floor until after 9:00 AM.

Portable Instruments

Monday Afternoon, Exposition Floor, 400 Aisle

(820-1 P) **Portable Optical Detection System for Determining Physical and Chemical Changes on Bioimplant Surfaces** DONALD BENZA, Clemson University, Fenglin Wang, Jeffrey N Anker

(820-2 P) **Field-Portable LED Array Based Multi-Wavelength Photothermal Lens Spectrometer** MICAH W ELLER, Tennessee Technological University, Andrew Callender

(820-3 P) **Translating Molecular Recognition into Pressure Signal for Rapid, Sensitive, and Portable Biomedical Analysis** ZHI ZHU, Xiamen University, Yang Chaoyong N/A

(820-4 P) **Development of Optical "Clamp-meter" Using Silicone Optical Technology for In-Situ Absorption Spectroscopy** HIROAKI NOMADA, Kyushu University, Hirokazu Higuchi, Hiroaki Yoshioka, Morita Kinichi, Yuji Oki

(820-5 P) **Development of a Portable Optical Cavity to Enhance 1-GHz Mode-Locked Laser Pulses for Broadband Absorption Spectroscopy** YUTARO ITO, Kyushu University, Zaitsu Shin-ichi, Imasaka Totaro

(820-6 P) **Characterization of a Miniaturized Benchtop Fast Gas Chromatograph Mass Spectrometer** COREY STEDWELL, 1st Detect Corporation, Daniel DeBord N/A

(820-7 P) **Enzymes and Photometers** ELLEN RUTH CAMPBELL, NECi Superior Enzymes, Bill Campbell, Justin Walbeck, David A Squires

(820-8 P) **Personal Monitoring of Ozone Exposure: A Fully Portable Device for Under $150 USD Cost** TINGTING CAO, Texas Tech University, Jonathan E Thompson

(820-9 P) **Producing a Miniaturized High-Performance Liquid Chromatography System for Exploration-Based Research** KYLE B LYNCH, University of Oklahoma

POSTER SESSION — Session 830

All posters are to be mounted by 10:00 AM and remain on display until 4:00 PM. Authors must be at their posters from 1:00 PM to 3:00 PM. Location of the afternoon posters is on the Exposition Floor, 400 Aisle. PLEASE NOTE: You cannot get onto the Exposition Floor until after 9:00 AM.

Quality/QA/QC

Monday Afternoon, Exposition Floor, 400 Aisle

(830-1 P) **A Novel Approach to Specialty Gas Certifications by Means of GC/FTIR Analysis** PETER P BEHNKE, Prism Analytical Technologies, Inc., Martin L Spartz, Charles Mark Phillips, Anthony S Bonanno, Kelly R McPartland

(830-2 P) **Evaluation of the Flavor of Strawberry Preparation Using Gas Chromatography Electronic Nose** ANDREW COWELL, Alpha MOS, Jean-Christophe Mifsud, Herve Lechat, Valérie Vabre, Fatma Ayouni, Marion Bonnefille

(830-3 P) **Developments in the Automatism for CHNS and Oxygen Determination Using an Elemental Analyzer for Chemical Characterization** GUIDO GIAZZI, Thermo Fisher Scientific, Liliana Krotz, Francesco Leone, Walter Galotta

(830-4 P) **Running Legacy HPLC and Optimized UPLC Methods on a Single UPLC Platform, Comparative Studies and Steps Towards Implementation in a QC Environment** CHRISTOPHER HENRY, Waters Corporation, Mark Wrona, Richard Ladd, Andy Boughey

(830-5 P) **Quantitative Analysis of Calcite in Desulfurization Gypsum Using Raman Spectroscopy** YOUNG TAEK MA, Hanyang University

(830-6 P) **The Use of a Triple Detection System, (UV, ELSD, MS) for Pharmaceutical Degradation Studies** AARON D PHOEBE, Waters Corporation, Paula Hong, Patricia R McConville

(830-7 P) **Fast Analysis of Short Chain Fatty Acids in Feeds By SFC/MS** JINCHUAN YANG, Waters Corporation, Carrie Snyder, BJ Bench, Jessica Lance, Jayant Shringarpure

(830-8 P) **Application of FTIR Spectroscopy to Study the Effect of Processing on the Secondary Structure of Lentil Proteins** ALBERTA ARYEE, Agriculture & Agri-Food Canada, Joyce I Boye N/A

(830-9 P) **Accelerated Oxidation Tests on Olive Oil Stored in Plastic Packaging and Submitted to Autoclaving** STEFANO CASIRAGHI, VELP Scientifica, Stefania Corti, Claudia Mancinelli

(830-10 P) **Fast Analytical Method for Essential Oil Flavor Characterization Using a Polar Stationary Phase for Optimized Selectivity** RAMKUMAR DHANDAPANI, Phenomenex, Tim Anderson, Kristen Parnell

(830-11 P) **Steep Time and Temperature Effects on Flavor and Flavonoid Extraction of Black Tea** ANNE JUREK, EST Analytical, Lindsey Pyron, Kelly Cravenor, Justin Murphy

(830-12 P) **A Novel Approach to the Analysis of Multivitamin in Foodstuff by Online Supercritical Fluid Sample Extraction/Supercritical Fluid Chromatography** QIANG LI, Shimadzu (China) Co., LTD

(830-13 P) **Simultaneous Measurement of L-Lactate and Ethanol in Tomato-Based Products** WILLIAM MILLER, YSI, Inc, June Klingensmith

(830-14 P) **Iron Translocation in Adzuki Beans Sprouts: Enrichment Effects** ALINE P OLIVEIRA, Federal University of São Paulo, Juliana Naozuka, Alexandrina C Carvalho, Cassiana S Nomura

(830-15 P) **The Qualitative and Quantitative Analysis of α-Acids in Hops and Beers by UHPLC with UV Detection** WILHAD M REUTER, PerkinElmer, Jason Weisenseel

(830-16 P) **SFC Analytical Method Development for Vitamin D3 and Related Compounds** TAKASHI SATO, YMC Co., Ltd., Roland Spaegele, Junko Kawabata, Saoko Nozawa, Toshikazu Adachi, Noritaka Kuroda

(830-17 P) **Fast and Cost-effective Sugar Analysis Using HPAE-PAD** HUA YANG, Thermo Fisher Scientific, Linda Lopez, David G Moore

(830-18 P) **Improving Analytical Performance with Enhanced Matrix Removal - Lipid** DERICK LUCAS, Agilent Technologies, Limian Zhao, Bruce Richter, Megan Juck, Joan Stevens

(830-19 P) **Food Inspection Using NIR Reflectance and Transmittance Imaging** SATORU TSUCHIKAWA, Nagoya University, Hikaru Kobori, Te Ma, Norihisa Katayama N/A

(830-20 P) **A Fast Method for Predicting the Phenolic Content of Whisky Malts** ANDREW COWELL, Alpha MOS, Jean-Christophe Mifsud, Herve Lechat, Valérie Vabre, Fatma Ayouni, Marion Bonnefille

(830-21 P) **Combining FTIR and Mass Spectrometric Detection in the Gas Chromatographic Analysis of Fragrances** TRACY PHILLPOTT, DANI Instruments, Roberta Lariccia, Matthew S Klee, Daniele Recenti

(830-22 P) **The Effect of Food Additives on Hippocampus NMDA Receptor Subunits and Oxidative Stress** ZAFER YONDEN, Mustafa Kemal University ZAFER YONDEN, Mustafa Kemal University, Oguzhan Ozcan, Yeşim Göçmen, Namık Delibaş N/A

(830-23 P) **Holographic Characterization of Multicomponent Colloidal Suspensions** JAROSLAW M BLUSEWICZ, Spheryx, Inc., David B Ruffner, Laura Philips

(830-24 P) **Mitigating Electrostatic Effects Improves Measurement Accuracy** GREG GUMKOWSKI, NRD LLC, Arnold Steinman

(830-25 P) **Safe Approach to Gas Purification** BRIAN WARRICK, ARM, Inc, Daniel Spohn

(830-26 P) **Advanced UHPLC Instrument to Instrument Method Transfer** GREGORY HUNLEN, Agilent Technologies, Michael Woodman

POSTER SESSION — Session 840

All posters are to be mounted by 10:00 AM and remain on display until 4:00 PM. Authors must be at their posters from 1:00 PM to 3:00 PM. Location of the afternoon posters is on the Exposition Floor, 400 Aisle. PLEASE NOTE: You cannot get onto the Exposition Floor until after 9:00 AM.

Sensors

Monday Afternoon, Exposition Floor, 400 Aisle

(840-1 P) **An Ultrasensitive Impedance Biosensor based on Immunomagnetic Separation and Urease Catalysis for Rapid Detection of Foodborne Pathogens Using an Immobilization-Free Interdigitated Microelectrode** LIN JIANHAN, China Agricultural University N/A

(840-2 P) **Enabling Tools to Combat Antibiotic Resistant Bacteria** ANDREW HELLER, Michigan State University, Dana Spence

(840-3 P) **Optimizing the Scintillation Cascade in Nano-Scintillation Proximity Assay (nanoSPA) for Multiplexed Detection of Small Biomolecules** ZEINAB MOKHTARI, University of Arizona, Isen Andrew C Calderon, Colleen Janczak, Craig A Aspinwall

(840-4 P) **Dendritic Gold Structures for Glucose Biosensor Design** ALMIRA RAMANAVICIENE, Vilnius University, Laura Sakalauskiene, Anton Popov, Natalija German, Asta Kausaite-Minkstimiene, Arunas Ramanavicius

(840-5 P) **High-Throughput Thiamine Quantification in Environmental Matrices Using Periplasmic-Binding Protein Biorecognition** KATIE A EDWARDS, Cornell University, Seth Feder, Antje J Baeumner, Cliff Kraft

(840-6 P) **Multi-Purpose Individual Air Monitor – Conception to Proof of Concept Laboratory Prototype** NICHOLAS FITZGERALD, Defence Science and Technology Group, Karl Pavey

(840-7 P) **Characterization of Self-Oscillating Reaction Catalyzed by Metal Porphyrin Complexes** TAKASHI ARIMURA, AIST Interdisciplinary Research Center, Masaru Mukai, Jung Hee Do, Kenichi Tominaga

(840-8 P) **Simultaneous Detection of Dopamine Release and Neural Activity** KATE L PARENT, University of Arizona, Daniel F Hill, Jean-Paul Wiegand, Michael A Miller, Christopher W Atcherley, Stephen Cowen, Michael L Heien

(840-9 P) **Development of a Novel Position-Sensitive MCP Detector** BLAKE WIGGINS, Indiana University, Davinder Siwal, Romualdo T deSouza

(840-10 P) **Real Time Monitoring Magnesium Alloys Corrosion by Electrochemical H2 Sensor In Vivo** DAOLI ZHAO, University of Cincinnati, Tingting Wang, Zhongyun Dong, William R Heineman, Vesselin Shanov

(840-11 P) **A Cellulose Acetate Membrane-Based Colorimetric Device to Discriminate Bacteria** LIGIA BUENO, Instituto de Química - Universidade de Sao Paulo, Subrayal M Reddy, Alison Cottel, Thiago Paixao

(840-12 P) **A Smart Wearable and Autonomous Negative Pressure Device for Wound Monitoring** FABIO DI FRANCESCO, Università di Pisa, Bernardo Melai, Pietro Salvo, Andrea Pucci, Vincenzo Mollica, Anna Maria Raspolli, Valentina Dini, Marco Romanelli, Beatrice Lazzerini, Aldo Paolicchi, Valter Castelvetro, Roger Fuoco N/A

(840-13 P) **A Smart Polymer Hydrogel as a Chemical Sensor on Biomedical Implant** MOHAMMED ARIFUZZAMAN, Clemson University, Caleb Behrend, Jeffrey N Anker

(840-14 P) **Oxygen Sensitive Probes** PETER GENNARO, Clemson University

(840-15 P) **Graphene Oxide as an Efficient Antibacterial Agent in Macrophages and in Mice** XU WU, University of North Dakota, Xiao Liu, Yuqian Xing, Julia Xiaojun Zhao, Min Wu

(840-16 P) **pH Indicator to Enhance Surface Plasmon Resonance Imaging Detection of Small Organic Molecules** ZAINAB HUSSAIN AL MUBARAK, Oklahoma State University, Gayan Premaratne, Cassandra Rodenbaugh, Lucy Lehoczky, Sadagopan Krishnan

(840-17 P) **Membrane Based Electrodes for Electrochemical Applications in Biology** TANYA BAKMAND, DTU Nanotech, Dorota Kwasny, Fatima Al-Zahraa Al Atraktchi, Helle Waagepetersen, Winnie E Svendsen, Maria Dimaki

(840-18 P) **Paper-Based Colorimetric Glucose Determination Using Smartphone** HAKAN CIFTCI, Kirikkale University, Nazlı Ayyıldız, Ugur Tamer

POSTER SESSION — Session 850

All posters are to be mounted by 10:00 AM and remain on display until 4:00 PM. Authors must be at their posters from 1:00 PM to 3:00 PM. Location of the afternoon posters is on the Exposition Floor, 400 Aisle. PLEASE NOTE: You cannot get onto the Exposition Floor until after 9:00 AM.

Thermal Analysis

Monday Afternoon, Exposition Floor, 400 Aisle

(850-1 P) **New TGA-FT-IR Library for Polymers** EKKEHARD POST, NETZSCH Geraetebau GmbH, Carolin Fischer, Bob Fidler

(850-2 P) **Beyond Classical Dynamic Mechanical Analysis - Using A New High-Force, High-Temperature DMA to Characterize Advanced Materials** BOB FIDLER, NETZSCH Instruments NA LLC, Horst Deckmann, Juergen Blumm, Ekkehard Post, Tobias Pflock

(850-3 P) **Exploring the Properties of Textiles with Thermal Analysis, Infrared and UV/Vis Spectroscopy** ANTHONY J LANG, PerkinElmer, Aaron H Adams, Jack Botting

TUESDAY, MARCH 8, 2016
MORNING

AWARDS — Session 860

Pittsburgh Analytical Chemistry Award

arranged by Joanne Smith, The Pittsburgh Conference

Tuesday Morning, Room B312

Joanne Smith, The Pittsburgh Conference, Presiding

8:30		**Introductory Remarks - Joanne Smith**
8:35		**Presentation of the 2016 Pittsburgh Analytical Chemistry Award to Sanford A Asher, University of Pittsburgh, by Elias S Absey, Chair, Society for Analytical Chemists of Pittsburgh**
8:40	(860-1)	**Photonic Crystal Hydrogel and Organogel Sensors for Chemical and Biological Analytes** SANFORD A ASHER, University of Pittsburgh, Natasha L Smith, Zhongyu Cai, Andrew E Coukouma
9:15	(860-2)	**Structural Characterization of Methylenedianiline Regioisomers by Ion Mobility-Mass Spectrometry, Tandem Mass Spectrometry, and Computational Strategies - MALDI Spectra** DAVID M HERCULES, Vanderbilt University, Sarah M Stow, Tiffany M Onifer, John A McLean
9:50	(860-3)	**Liposomal Spherical Nucleic Acids: A New Approach to Immunomodulatory Therapy** CHAD A MIRKIN, Northwestern University
10:25		**Recess**
10:40	(860-4)	**Measuring DNA Hybridization Kinetics by Single-Molecule Fluorescence Imaging** JOEL M HARRIS, University of Utah, Eric M Peterson, Michael Manhart, Frances Morris
11:15	(860-5)	**SERS-Based Metabolic Profiling for Diagnostics and Forensics** LAWRENCE ZIEGLER, Boston University

AWARDS — Session 870

The Coblentz Society/ABB - Bomem-Michelson Award

arranged by Brooks Pate, University of Virginia

Tuesday Morning, Room B314

Brooks Pate, University of Virginia, Presiding

8:30		**Introductory Remarks - Brooks Pate**
8:35		**Presentation of the 2016 Coblentz Society/ABB- Bomem-Michelson Award to Shaul Mukamel, University of California Irvine, by Henry L. Buijs, ABB**
8:40	(870-1)	**Coherent Ultrafast Multidimensional Spectroscopy of Molecules with Optical, X-Ray, and Quantum Light** SHAUL MUKAMEL, University of California, Irvine
9:15	(870-2)	**Photosynthetic Light Harvesting from Individual Complexes to the Grana Membrane** GRAHAM RICHARD FLEMING, University of California at Berkeley N/A
9:50	(870-3)	**A Few Lessons from Non-Adiabatic Excited State Dynamics Simulations of Large Molecules** SERGEI TRETIAK, Los Alamos National Laboratory
10:25		**Recess**
10:40	(870-4)	**Photosynthetic Light Harvesting and Ultrafast Energy Transfer** GREG D SCHOLES, Princeton University
11:15	(870-5)	**Elucidation of Chemical Reactions by Two-Dimensional Resonance Raman Spectroscopy** ANDREW MORAN, University of North Carolina

SYMPOSIUM **Session 880**

ACS-ANLY - Chemometrics: A New Dimension in Chromatography
arranged by Amber M Hupp, College of the Holy Cross

Tuesday Morning, Room B308
Amber M Hupp, College of the Holy Cross, Presiding

8:30		**Introductory Remarks - Amber M Hupp**
8:35	(880-1)	**Applying the Hotelling Trace Criterion to Optimize Chromatogram Alignment of Biodiesel Diesel Blended Fuels** AMBER M HUPP, College of the Holy Cross, Gopal Yalla, John O'Connor, Kevin Walsh, Edward Soares
9:10	(880-2)	**Forensic Signatures for the Source Attribution of Chemical Threat Agents Using Chemical Profiling, Stable Isotope Ratios and Chemometrics** CARLOS FRAGA, Pacific Northwest National Laboratory
9:45	(880-3)	**Chemometrics: An Old Dimension in Chromatography – Application to New Dimensions** SARAH C RUTAN, Virginia Commonwealth University, Daniel W Cook, Melanie M Sinanian
10:20		**Recess**
10:35	(880-4)	**Data Reduction and Processing Tools for GCxGC-TOFMS** JAMES J HARYNUK, University of Alberta, Lawrence A Adutwum, A Paulina de la Mata
11:10	(880-5)	**Chemometric Approaches to Maximize Interpretation of GCxGC - TOFMS Data for Discovery-Based Analyses** ROB SYNOVEC, University of Washington

SYMPOSIUM **Session 890**

Advances in Analytical Methodologies for the Detection of Food Allergens and Gluten
arranged by Lowri DeJager and Christine Parker, U.S. Food and Drug Administration

Tuesday Morning, Room B302
Christine Parker, U.S. Food and Drug Administration, Presiding

8:30		**Introductory Remarks - Lowri DeJager and Christine Parker**
8:35	(890-1)	**Approaches to Multianalyte Allergen Analysis in Food** CLARE MILLS, The University of Manchester, Karine Adel-Patient, Sabine Baumgartner
9:10	(890-2)	**Effects of Cross-Reactivity in Food Allergy Detection and Diagnosis** SOHEILA J MALEKI, USDA-Agricultural Research Service
9:45	(890-3)	**Novel Approaches to Identifying Amadori Products in Peanut Extract** GEOFFREY MUELLER, National Institute of Environmental Health Sciences, Jason Williams, Katina Johnson
10:20		**Recess**
10:35	(890-4)	**Overview of Analytical Methods for Food Allergen and Gluten Analysis: Challenges and Trends from a Regulatory Perspective** TERRY KOERNER, Health Canada
11:10	(890-5)	**Effects of a Proline Endopeptidase on the Detection and Quantification of Gluten During the Fermentation of Beer** RAKHI PANDA, Food and Drug Administration, Katherine L Fiedler, Chung Y Cho, Eric A Garber

SYMPOSIUM **Session 900**

Advances in Two-Dimensional Liquid Chromatography Separations of Biopharmaceuticals
arranged by Dwight R Stoll, Gustavus Adolphus College

Tuesday Morning, Room B303
Dwight R Stoll, Gustavus Adolphus College, Presiding

8:30		**Introductory Remarks - Dwight R Stoll**
8:35	(900-1)	**Column Technology for the Chromatographic Characterization of Biopharmaceuticals** DAVY GUILLARME, University of Geneva, Jean-Luc Veuthey, Szabolcs Fekete
9:10	(900-2)	**Optimization of Two-Dimensional Liquid Chromatography Separations of Therapeutic Monoclonal Antibodies Involving Ion-Exchange and Reversed-Phase Separation Modes** DWIGHT R STOLL, Gustavus Adolphus College, David C Harmes, Jacob Bush, Matthew Sorensen, Gregory Staples, Szabolcs Fekete, Davy Guillarme, Alain Beck
9:45	(900-3)	**Application of a Multi-Dimensional Approach for Quantification of Free Drug in Antibody Drug Conjugates** BROOKE M KOSHEL, Waters Corporation, Robert Birdsall, Alain Beck, Sean M McCarthy
10:20		**Recess**
10:35	(900-4)	**2-Dimensional Liquid Chromatography for Rapid Parallel Separation of Biopharmaceuticals** YAN HE, Pfizer, Ian Hartzel, Michael T Jones
11:10	(900-5)	**Antibody-Drug-Conjugates (ADC) Drug Product Profiling by Multi-dimensional HPLC** KELLY ZHANG, Genentech

SYMPOSIUM **Session 910**

Graphene Nanomaterials for Bio/Sensing Applications
arranged by Ashok Mulchandani, University of California, Riverside and Chenzhong Li, Florida International University

Tuesday Morning, Room B304
Ashok Mulchandani, University of California, Riverside, Presiding

8:30		**Introductory Remarks - Ashok Mulchandani and Chenzhong Li**
8:35	(910-1)	**New Concepts in Biosensing Using Single Walled Carbon Nanotubes and Graphene** MICHAEL S STRANO, Massachusetts Institute of Technology
9:10	(910-2)	**Graphene Electrodes for Bio/chemical Sensors** ASHOK MULCHANDANI, University of California, Riverside
9:45	(910-3)	**A Novel Wireless Biosensing Platform Enabled by Graphene Varactors** STEVEN J KOESTER, University of Minnesota
10:20		**Recess**
10:35	(910-4)	**Graphene-Enabled Nano/Bio Hybrids for Chemical Detection and Medical Diagnostics** A T CHARLIE JOHNSON, University of Pennsylvania
11:10	(910-5)	**Impedance Sensing of Nanotoxicity of Graphene at the Cellular and Tissue Level** CHENZHONG LI, Florida International University

SYMPOSIUM **Session 920**

JAIMA - The State-of-the-Art Technologies from Japan: Analytical Instruments with/for Nano-Chemistry Technology and Advanced Diagnosis (I)
arranged by Koichiro Matsuda, JAIMA

Tuesday Morning, Room B305
Koichiro Matsuda, JAIMA, Presiding
Takeshi Kawamoto, JAIMA, Presiding

8:30		**Introductory Remarks - Gon-emon Kurihara, President of JAIMA**
8:35	(920-1)	**Creation of Bio/Chemical Sensing Probes** KOJI SUZUKI, Keio University, Daniel Citterio
9:10	(920-2)	**Ab Initio Powder Structure Determination Opening Up New Research Fields** MASAKI KAWANO, Pohang University of Science and Technology
9:45	(920-3)	**Nanodroplet Formation and Chemical Analysis in Microfluidic Devices** AKIHIDE HIBARA, Tokyo Institute of Technology
10:20		**Recess**
10:35	(920-4)	**Development of Mass Microscope and Applications in Drug and Oncometabolite Visualization** SHUICHI SHIMMA, Osaka University
11:10	(920-5)	**Plasma Phospholipids and Prevalence of Mild Cognitive Impairment/ Dementia** DANNI LI, University of Minnesota

TECHNICAL PROGRAM

SYMPOSIUM — Session 930

Ultrahigh-Resolution Mass Spectrometry: A New Frontier
arranged by Alan G Marshall, Florida State University

Tuesday Morning, Room B301
Alan G Marshall, Florida State University, Presiding

8:30		**Introductory Remarks - Alan G Marshall**
8:35	(930-1)	**High-Resolution TOFMS via a Folded Flight Path: You Can Have Your Cake and Eat It Too** VIATCHESLAV ARTAEV, Leco Corporation, Michael Mason, Peter Willis, Anatoly Verenchikov
9:10	(930-2)	**21 Tesla FT-ICR Mass Spectrometer for Top-Down Protein Identification and Characterization** CHRISTOPHER L HENDRICKSON, National High Magnetic Field Laboratory, Lissa C Anderson, John P Quinn, Nathan K Kaiser, Donald F Smith, Greg T Blakney, Alan G Marshall
9:45	(930-3)	**Exploring the Microbiome via Advanced Mass Spectrometry** LJILJANA PASA-TOLIC, EMSL, PNNL, Jared B Shaw, Malak Tfaily, Nikola Tolic, Errol Robinson, David W Koppenaal, Nancy J Hess
10:20		**Recess**
10:35	(930-4)	**Complex Mixture Analysis by Ultra-High Resolution Mass Spectrometry** RYAN P RODGERS, Florida State University, Amy C Clingenpeel, Steven M Rowland, Vladislav V Lobodin, David C Podgorski, Yuri E Corilo, Priscila M Lalli, Jonathan C Putman, Winston K Robbins
11:10	(930-5)	**Orbitrap Mass Analyzers: The Teenage Years** MICHAEL W SENKO, Thermo Fisher Scientific, Jesse D Canterbury, Eduard Denisov, Alexander Makarov

ORGANIZED CONTRIBUTED SESSIONS — Session 940

High Performance SFC for the Analysis of Pharmaceuticals, Nutraceuticals, Natural Products and Metabolomics
arranged by David Kohler, ES Industries and Patrick G Fromal, Shimadzu Scientific Instruments Inc

Tuesday Morning, Room B309
David Kohler, ES Industries, Presiding

8:30	(940-1)	**Analysis of Mycotoxins in Various Food Matrices via SFE/SFC/MS** TODD ANDERSON, Shimadzu Scientific
8:50	(940-2)	**Evaluation of a Liquid Carbon Dioxide Based Flash Chromatography System** RAY MCCLAIN, Merck
9:10	(940-3)	**Fluoro-Methylphenylcarbamates of Cellulose as Chiral Stationary Phases for Supercritical Fluid Chromatography** WILLIAM FARRELL, Pfizer, Inc., Matthew Przybyciel
9:30	(940-4)	**Potential of Supercritical Fluid Extraction and Separation Technologies in Metabolomics** BAMBA TAKESHI, Kyushu University
9:50		**Recess**
10:05	(940-5)	**SFC – An Essential "Green" Separation Technique for the Chromatographic Analysis of a Nutraceuticals** MATTHEW PRZYBYCIEL, ES Industries
10:25	(940-6)	**SFC, A Walk on the Wild Side** BRENT MURPHY, Genentech, Mengling Wong, Joseph H Pease, Michael Hayes, Amber Guillen, Christopher Hamman
10:45	(940-7)	**Understanding the Value of Preparative Chiral Separations in Advancing Pharmaceutical R&D Projects** MICKEY REGO, Averica Discovery Services, Jeffrey Kiplinger, Paul Lefebvre, Keith Galyan, Emily Showell-Rouse
11:05	(940-8)	**Open Discussion**

ORGANIZED CONTRIBUTED SESSIONS — Session 950

Molecular Modelling and Quantum Mechanical Calculations: From Small Molecules to Large Multimeric Protein Complexes
arranged by Iain Campuzano and Michael D Bartberger, Amgen

Tuesday Morning, Room B310
Iain Campuzano, Amgen, Presiding

8:30	(950-1)	**Vibrational Optical Activity of Helical Peptides** JAMES R CHEESEMAN, Gaussian, Inc.
8:50	(950-2)	**Interplay of Theory and Experiment: Small Molecule Ion Mobility and Spectroscopic Characterization of Metabolites** MICHAEL D BARTBERGER, Amgen
9:10	(950-3)	**Efficient Photostitching of Peptide Ion Complexes in the Gas Phase with the Photo-Leucine Zipper** FRANTISEK TURECEK, University of Washington, Christopher J Shaffer, Akis P Andrikopoulos, Lubomir Rulisek
9:30	(950-4)	**Chiroptical Spectroscopy for Molecular Structure Determination** PRASAD L POLAVARAPU, Vanderbilt University
9:50		**Recess**
10:05	(950-5)	**The Use of Molecular Modelling for Ion Mobility Drift Time and Fragment Ion Prediction in Ion Mobility and Mass Spectrometry** CRIS LAPTHORN, University of Greenwich, Frank S Pullen, Babur Chowdhry, Patricia Wright, George Perkins, Yanira Heredia, Trevor Dines
10:25	(950-6)	**Dynamical Networks of Protein Residue-Residue Contacts Provide Insights into Enzyme Function** DONALD HAMELBERG, Georgia State University
10:45	(950-7)	**A Molecular Dynamics and Ion Mobility Study of Protein Structure Collapse in the Gas-phase** IAIN CAMPUZANO, Amgen, Morgan Lawrenz, Carlos Larriba-Andaluz
11:05	(950-8)	**Hell Bent on Opening: Structural Basis for Estrogen Modulation of Potassium Recycling During Epithelial Secretion** BRETT KRONCKE, Vanderbilt University

ORGANIZED CONTRIBUTED SESSIONS — Session 960

SEAC Young Investigator Session
arranged by Hitomi Mukaibo, University of Rochester

Tuesday Morning, Room B311
Hitomi Mukaibo, University of Rochester, Presiding

8:30	(960-1)	**Fluorescence-Enabled Electrochemistry** BO ZHANG, University of Washington
8:50	(960-2)	**Designing Functional Nanostructured Gels for Electrochemical Energy Storage and Biosensors** GUIHUA YU, University of Texas at Austin
9:10	(960-3)	**Inhibiting Electrochemical Processes in Li-ion Batteries at High Temperatures Using Responsive Polymers** MARK E ROBERTS, Clemson University, Jesse C Kelly, Nick L DeGrood
9:30	(960-4)	**Revealing the Mechanism of Electron Uptake in Methanogenic Biofilm Community Using a Solid-Phase Electron Donor** SOFIA BABANOVA, JCVI, Kayla Carpenter, Sujal Phadke, Tony Phan, Shunichi Ishii, Shino Suzuki, Michael Salvacion, Michael Flynn, John Hogan, Orianna Bretschger
9:50		**Recess**
10:05	(960-5)	**Reagentless and Reusable Electrochemical Metal Ion Sensors** REBECCA Y LAI, University of Nebraska-Lincoln
10:25	(960-6)	**Electrochemical Imaging of Ionic Reactivity at Operating Ion Batteries** JOAQUIN RODRIGUEZ LOPEZ, University of Illinois
10:45	(960-7)	**In Vivo Serotonin Chemistry and Local Cytoarchitecture: A Combined Voltammetric, Mathematical and Microscopy Study** PARASTOO HASHEMI, University of South Carolina, Aya Abdalla
11:05	(960-8)	**Ec-LLS at the Micro- and Macroscale: Electrodeposition of Semiconductor Films and Nanocrystals** STEPHEN MALDONADO, University of Michigan

ORGANIZED CONTRIBUTED SESSIONS **Session 970**

Supercritical CO_2 – SFE/SFC: Advances in Extraction and Purification for Pharmaceutical and Natural Products
arranged by Andy Miles, Regis Technologies and Gregory K. Webster, AbbVie

Tuesday Morning, Room B313
Andy Miles, Regis Technologies, Presiding

8:30	(970-1)	**Supercritical Fluid Chromatography in Support of Pharmaceutical – A Study of Scale-Up from Analytical to Preparative Scale with Isocratic Conditions** MIRLINDA BIBA, Merck, Jinchu Liu, Lindsey Jacobs, Judy Morris, Ingrid Mergelsberg
8:50	(970-2)	**Strategies for Increasing Throughput of Chiral Separations by Supercritical Fluid Chromatography** ERIN JORDAN, AbbVie, Philip A Searle
9:10	(970-3)	**Withdrawn**
9:30	(970-4)	**SFC-MS as the Technique of Choice for Small Molecules Purification** GERARD ROSSE, Dart Neuroscience
9:50		**Recess**
10:05	(970-5)	**Practical Approaches to SFE and SFC in Drug Discovery** JOSEPH H PEASE, Genentech, Brent Murphy, Mengling Wong, Michael Hayes, Amber Guillen
10:25	(970-6)	**One Phase, Three Techniques: Utilizing One SFC Stationary Phase Across Three Chromatographic Techniques** CHRISTINE AURIGEMMA, Pfizer, Inc., Perrine Hoerter, William Farrell
10:45	(970-7)	**Online SFE-SFC Purification Method Development Using SFC Solubility Determination** WES BARNHART, Amgen, Kyung Gahm
11:05	(970-8)	**Utilizing SFE and SFC for Extraction and Isolation of Cannabinoids** CHRISTOPHER HUDALLA, ProVerde Laboratories

ORAL SESSIONS **Session 980**

Biomedical: Advances in Glucose Monitoring and Therapeutics of Diabetes - Half Session

Tuesday Morning, Room B408
Barbara Manner, The Pittsburgh Conference, Presiding

8:30	(980-1)	**Skin Interface and Spectrometer Development for Noninvasive Glucose Measurements in People over Combination Near-Infrared Wavelengths** ARIEL BOHMAN, University of Iowa, Mark A Arnold, Gary W Small, Michael J Miller
8:50	(980-2)	**Advances in Development of Glucose Biosensors** ARUNAS RAMANAVICIUS, Vilnius University, Natalija German, Asta Kausaite-Minkstimiene, Inga Vilkonciene, Povilas Genys, Jaroslav Voronovic, Jurate Petroniene, Almira Ramanaviciene
9:10	(980-3)	**A Microfluidic Cell Culture Device for Automated Sample Preparation and Improved Biomimetic Modeling in Diabetes Metabolomics** LAURA FILLA, Saint Louis University, James L Edwards
9:30	(980-4)	**Analysis of Drug-Protein Interactions During Diabetes by High-Performance Affinity Chromatography** ZHAO LI, University of Nebraska-Lincoln, Ryan Matsuda, David Hage

ORAL SESSIONS **Session 990**

Capillary Electrophoresis

Tuesday Morning, Room B315
Elizabeth Harris, Mannkind Corporation, Presiding

8:30	(990-1)	**Monitoring Neurotransmitter Secretion from Islets of Langerhans** KIMBERLY EVANS, Florida State University, Xue Wang, Michael G Roper
8:50	(990-2)	**Observing Peptide Folding Intermediates Using Capillary Electrophoresis** ALISON E HOLLIDAY, Moravian College, John D Barr, Liuqing Shi, David E Clemmer
9:10	(990-3)	**PDMS Micro-cross Junction for Online Nanoliter Heart-cut for Two Dimensional CE Separation** VITALY AVILOV, University of Illinois at Chicago, Scott A Shippy
9:30	(990-4)	**Real Time Monitoring of Amino Acid Secretions from Islets of Langerhans Using a Microfluidic Device** XUE WANG, Florida State University, Michael G Roper, Lian Yi
9:50		**Recess**
10:05	(990-5)	**Single-Cell Chemical Cytometry of Akt Activity within Primary Rheumatoid Arthritis Synovial Fibroblasts Illustrates Heterogeneity in Patient Responses to TNFα** EMILIE R MAINZ, University of North Carolina at Chapel Hill, Stephen Serafin, David Lawrence, Teresa Tarrant, Christopher Sims, Nancy Allbritton
10:25	(990-6)	**The Determination of Oxidative Stress Biomarkers of Lipid Peroxidation Using a Novel Capillary Electrophoresis-Mass spectrometry Sheathless Interface Design** RYAN T JOHNSON, University of Kansas, Craig E Lunte, Nhan To, John Stobaugh
10:45	(990-7)	**Analysis of Metabolite Biomarkers in Prostrate Cancer Tissues by Capillary Electrochromatography Mass Spectrometry** YANG LU, Georgia State University, Shahab S Shamsi, Dean Troyer
11:05	(990-8)	**Rapid Determination of Cyanide in Human Urine by Capillary Electrophoresis Coupled with Laser-Induced Fluorescence Detection** QIYANG ZHANG, Wichita State University, Naveen Maddukuri, Maojun Gong

ORAL SESSIONS **Session 1000**

Electrochemistry - New Approaches and Techniques

Tuesday Morning, Room B316
Lawrence A Bottomley, Georgia Institute of Technology, Presiding

8:30	(1000-1)	**Stochastic Electrochemistry of TiO_2 Nanoparticles and Their Agglomerates** MARIO ALPUCHE-AVILES, University of Nevada, Reno, Pushpa Chhetri, Krishna K Barakoti, Ganesh Rana, Andrew Recinos
8:50	(1000-2)	**The Development of an Automated NanoElectrode Array Sensor to Detect Evaporation and Changes in Cellular Bioenergetics in a Submicroliter Chamber from an Organ-on-a-Chip System** ANNA N DAVIS, Vanderbilt University, John Wikswo, David Cliffel
9:10	(1000-3)	**Scanning Electrochemical Microscopy (SECM): A New Tool to Study Microbial Metabolism** DIPANKAR KOLEY, Oregon State University, Vrushali Joshi
9:30	(1000-4)	**Drug Metabolism Assays and Metabolite Synthesis Using Microsomes Based Bioreactor** RAJASEKHARA NERIMETLA, Oklahoma State University, Sadagopan Krishnan
9:50		**Recess**
10:05	(1000-5)	**High-Density Microelectrode Arrays as Electrochemical Imaging Platforms** RACHEL M FEENY, Colorado State University, John Wydallis, Lang Yang, Stacy Willett, Tom Chen, Stuart A Tobet, Charles Henry
10:25	(1000-6)	**Imaging of Immobilized Enzymes and Yeast Cells by Scanning Electrochemical Microscopy** ARUNAS RAMANAVICIUS, Vilnius University, Inga Vilkonciene, Rita Sareikaite, Aura Kisieliute, Povilas Genys, Jurate Petroniene, Almira Ramanaviciene
10:45	(1000-7)	**Rapid Temperature Control of Electrochemical and Biochemical Systems Using Microfabricated Heaters** NICHOLAS M CONTENTO, Biomolecular Measurement Division, Sarah M Robinson, Kurt D Benkstein, Herman O Sintim, Steve Semancik
11:05	(1000-8)	**A Simple, High Yield Method to Microfabricate Planar Microelectrode Arrays** MOHAMED M MAREI, University of Louisville, Mark Crain, Richard P Baldwin, Robert S Keynton

ORAL SESSIONS **Session 1010**

Enhancements in Pharmaceutical and Environmental Separations

Tuesday Morning, Room B402
Richard Henry, Consultant, Presiding

8:30	(1010-1)	**Sub-ppm Level Formaldehyde Measurement in Complex Sample Matrices Using a Variety of Analytical Methods – Method Comparison and Practical Considerations** PEILIN YANG, The Dow Chemical Company, Francois Huby, James N Alexander IV
8:50	(1010-2)	**Optimization of a Single-Stage, Consumable Free Thermal Modulator for GC x GC** HALEIGH BOSWELL, University of Waterloo, Tadeusz Gorecki
9:10	(1010-3)	**Thermo Tuning of Redox Potential on Nanostructured Adsorbent for a Reagent-less Recovery of Pollutant Oxoanions** MANUEL VALIENTE, Universitat Autonoma de Barcelona, He Liu, Liu Tong, Gustavo Perez
9:30	(1010-4)	**Super/subcritical Fluid Chromatographic Chiral Separations with High Efficiency Chiral Stationary Phases** ZACHARY BREITBACH, The University of Texas at Arlington, Chandan L Barhate, M Farooq Wahab, Daniel W Armstrong

9:50		Recess
10:05	(1010-5)	**Novel Solid-Phase Microextraction and Capillary Electrochromatographic Column Techniques for Pharmaceutical Analysis** ZILIN CHEN, Wuhan University
10:25	(1010-6)	**Retention in Porous Layer Pillar Array Planar Separation Platforms** DANIELLE R LINCOLN, University of Tennessee, Knoxville, Nickolay V Lavrik, Michael J Sepaniak
10:45	(1010-7)	**Carbon Nanofibers Decorated with Magnetic Nanoparticles as a New Sorbent for the Magnetic Solid Phase Extraction of Selected Polycyclic Aromatic Hydrocarbons from Water Samples** ALI SARAFRAZ YAZDI, Ferdowsi University of Mashhad N/A
11:05	(1010-8)	**Analytical Method Development: Are We Solving the Right Problem? A Systematic Approach to Select an Appropriate RPLC Column and to Optimize Separation** IMAD A HAIDAR AHMAD, Novartis, James Tam, Xue Li, Thomas Tarara, Andrei Blasko

ORAL SESSIONS — Session 1020

Environmental Air Quality and Analysis

Tuesday Morning, Room B403
John Kokosa, Research Consultant, Presiding

8:30	(1020-1)	**On-site Determination of Formaldehyde Using SPME and a Portable GC-TMS** JUSTEN J POOLE, University of Waterloo, Jonathan J Grandy, German A Gomez-Rios, Janusz Pawliszyn
8:50	(1020-2)	**Fast, Accurate, and Precise: How to Comply with EPA Method 325b (Fence Line Monitoring for Benzene)** LEE MAROTTA, PerkinElmer, Roberta Provost, Amy Jacobson, Mariah Peronto
9:10	(1020-3)	**Investigation and Modelling of the Sampling Process in a PDMS-Based Permeation Passive Sampler** FATEN SALIM, University of Waterloo, Marios Ioannidis, Tadeusz Gorecki
9:30	(1020-4)	**QCL-Based Perfluorocarbon Emission Monitoring** LUIS H ESPINOZA-NAVA, Alcoa Inc
9:50		Recess
10:05	(1020-5)	**Laser Derivitization for Soot Source Identification** RANDY VANDER WAL, Penn State University, Chethan K Gaddam
10:25	(1020-6)	**Analysis of Damaged Floor Coverings Emissions in Indoor Air Quality with Cantilever-Enhanced Photoacoustic Spectroscopy** JUSSI RAITTILA, Gasera Ltd., Ismo Kauppinen, Sauli Sinisalo, Jaakko Lehtinen
10:45	(1020-7)	**Ship Emissions Monitoring with Laser-Based Cantilever-Enhanced Photoacoustic Detection** JAAKKO LEHTINEN, Gasera Ltd., Ismo Kauppinen, Jussi Raittila
11:05	(1020-8)	**Hand Held Detector Based on an Ion Mobility Spectrometer and an Additional Detector (Electrochemical Cell or Alternatively a Photo Ionization Detector) for the Detection of Toxic Gases in Chemical Industries** ANDREAS WALTE, Airsense Analytics, Bert Ungethuem, Wolf Muenchmeyer

ORAL SESSIONS — Session 1030

Magnetic Resonance - Half Session

Tuesday Morning, Room B408
Barbara Manner, The Pittsburgh Conference, Presiding

10:05	(1030-1)	**Quantitative Determination of Short-Chain Branching and Co-Monomer Content in High Density Polyethylene by NMR Spectroscopy** RAVINDRA KUMAR, Indian Oil Corporation Ltd, Veena Bansal, Sujit Mondal, GS Kapur, V Kagdiyal, Deepak Saxena N/A
10:25	(1030-2)	**GPC-NMR Analysis of Polymeric Mixtures** NIKKI H LAFEMINA, Arkema N/A
10:45	(1030-3)	**Quantitative Analysis of Phosphorous Containing Drug Molecules Encapsulated in pH-Sensitive Nanoparticle Formulations to Establish Drug Loading and Drug Releasing Profile by Utilizing 31P Solid State and Solution State NMR Spectroscopy** SUDHAUNSHU S PUROHIT, University of Missouri Kansas City, Jianing Meng, Vivek Agrahari, Bi Botti Celestin Youan, Nathan Oyler
11:05	(1030-4)	**Nanoparticle-Assisted Removal of Proteins in Human Serum for Metabolomics Studies** BO ZHANG, The Ohio State University, Mouzhe Xie, Lei Bruschweiler-Li, Rafael Brüschweiler

ORAL SESSIONS — Session 1040

Mass Spectrometry - Bioanalytical

Tuesday Morning, Room B404
Daniel E Austin, Brigham Young University, Presiding

8:30	(1040-1)	**Electrospray Ionization-High Pressure Mass Spectrometry for Peptide and Protein Analysis** RUSSELL E BORNSCHEIN, University of North Carolina at Chapel Hill, William M Gilliland, J Michael Ramsey
8:50	(1040-2)	**Current Analytical Techniques for Glycoprotein Characterization by Mass Spectrometry** PARASTOO AZADI, Complex Carbohydrate Research Center
9:10	(1040-3)	**Nanopatterning Ligands to Enable Cell-Based Assays Using SAMDI-Mass Spectrometry** MARIA D CABEZAS, Northwestern University, Milan Mrksich, Chad A Mirkin
9:30	(1040-4)	**Shotgun Lipidomic Analysis of Human Meibum by MS/MS^{all} with Successive Switching between Positive and Negative Detection Modes** JIANZHONG CHEN, University of Alabama at Birmingham, Kelly Nichols
9:50		Recess
10:05	(1040-5)	**Development, Characterization, and Application of Coated Blade Spray Ionization** GERMAN A GOMEZ-RIOS, University of Waterloo, Nathaly Reyes-Garces, Ezel Boyaci, Janusz Pawliszyn
10:25	(1040-6)	**Discrimination of Carbohydrate Isomers as Transition Metal Adducts Using Ion Mobility Spectrometry and Tandem Mass Spectrometry** YUTING HUANG, University of Nebraska-Lincoln, Eric D Dodds, Lauren M Petrosh
10:45	(1040-7)	**Scan-by-Scan Analysis of Orbitrap Fine Isotope Structures for Unique Elemental Composition Determination** YONGDONG WANG, Cerno Bioscience, Ming Gu
11:05	(1040-8)	**Mass Spectra of Analytical Derivatives of Amino Acids and Small Peptides** NINO G TODUA, NIST, Stephen E Stein, Anzor I Mikaia

ORAL SESSIONS — Session 1050

Materials Characterization and Engineering

Tuesday Morning, Room B405
Cecil Dybowski, University of Delaware, Presiding

8:30	(1050-1)	**Chemical Analysis Applications and Optical Properties of 3D Printed <100 µm Dimension Microfluidic Channels** MICHAEL BEAUCHAMP, Brigham Young University, Hua Gong, Steven Perry, Greg Nordin, Adam T Woolley
8:50	(1050-2)	**Crack-Free Three-Dimensionally Ordered Macroporous (3DOM) Structure in Microfluidic Reactor** XIAORAN ZHANG, Michigan State University, Gary Blanchard
9:10	(1050-3)	**Interfacial Structure-Function Correlation of Perovskite Solar Cell** MINYU XIAO, University of Michigan
9:30	(1050-4)	**Optical Spectroscopy Analyses of Perovskite Nanomaterials** DANIEL J FREPPON, Iowa State University, Long Men, Ujjal Bhattacharjee, Feng Zhu, Jacob W Petrich, Emily A Smith, Javier Vela
9:50		Recess
10:05	(1050-5)	**Utilizing Thermal Analysis Methods to Develop and Optimized Porous Structural Ceramics** JOHN P SANDERS, Clemson University, Nathaniel Huygen N/A
10:25	(1050-6)	**Carbazole-Dye Conjugate - Derived Group of Uniform Materials Based on Organic Salts (GUMBOS) for Optoelectronic Applications** DEEPTHIKA DE SILVA, Louisiana State University, Noureen Siraj, Isiah M Warner
10:45	(1050-7)	**Molecular-Scale IR Thermometer Reveals Sub-Molecular Photo-Plasticization in Azomaterials** CHRISTIAN PELLERIN, University of Montreal, Audrey Laventure, Jaana Vapaavuori, Geraldine Bazuin, Olivier Lebel
11:05	(1050-8)	**Complete Characterization of the UV-Visible Properties of Optical Materials Using a Total Absolute Measurement System (TAMS)** IAN ROBERTSON, PerkinElmer Limited, Steve Upstone, Christopher Lynch

ORAL SESSIONS **Session 1060**

Pharmaceutical-MS, UV-VIS and Others

Tuesday Morning, Room B406
Jason N Payne, Western Kentucky University, Presiding

8:30	(1060-1)	**Speciation of Elemental Impurities for Compliance with USP <232>** JON L SIMS, Perkin Elmer, Helmut Ernstberger, Kenneth Neubauer
8:50	(1060-2)	**Oxidative Degradation in Pharmaceuticals: Mechanism and Stabilization of Spray Dried Amorphous Drug - A Case Study** ARCHANA KUMAR, Genentech, Raghendhar Kotha
9:10	(1060-3)	**Evaluation of Antimicrobial and Neutraceutical Properties of Plukenetia Conophora (Walnut) Leaves** CHUKWUEMEKA PAUL AZUBUIKE, University of Lagos, Cecilia I Igwilo, Karamot O Suara
9:30	(1060-4)	**Novel Self-Patented Gold Nanoparticles for Antineoplastic Activity** JASON N PAYNE, Western Kentucky University, Rajalingam Dakshinamurthy
9:50		**Recess**
10:05	(1060-5)	**Expanding the Analytical Toolbox for Material Verification: Spectroscopic Screening of Raw Ingredients Using Portable Spectrometers** JASON D RODRIGUEZ, FDA Division of Pharmaceutical Analysis, Hirsch Srivastava, Fabiola Semidei Ortiz
10:25	(1060-6)	**Image Directed Identification of Sub-visible Particles in Protein Based Therapeutics, Classification According USP<787> of Intrinsic, Inherent and Extrinsic Particulate Matter on the Sub-visible Level** OLGA LASKINA, rap.ID, Kathryn A Lee, Markus Lankers, Oliver Valet
10:45	(1060-7)	**Coupling Chemical Analysis to High Resolution Dark Field Microscopy for Enhanced Physicochemical Characterization of Complex Drug Formulations** KATHERINE TYNER, Food and Drug Administration, Sheetal D'Mello, Sau (Larry) Lee
11:05	(1060-8)	**Therapeutic Potential of Polyamine-Tethered Low Generation PAMAM Dendrimer Derivatives for Nucleic Acid Delivery** RUBY BANSAL, CSIR-Institute of Genomics and Integrative Biology N/A

ORAL SESSIONS **Session 1070**

Sensors - Biomedical

Tuesday Morning, Room B407
Emil Ciurczak, Doramaxx Consulting, Presiding

8:30	(1070-1)	**Disposable Sensors for the Remote Monitoring of Chronic Wounds** FABIO DI FRANCESCO, Università di Pisa, Nicola Calisi, Bernardo Melai, Pietro Salvo, Clara Paoletti, Letizia Moni, Consuelo Politino, Alessio Ceccarini
8:50	(1070-2)	**Real-Time Monitoring Urinary Encrustation Using Quartz Crystal Microbalance Sensor** PEGAH N ABADIAN, Northeastern University, Jun Li, Edgar D Goluch, John Victor, Jonathan Zhang
9:10	(1070-3)	**A Fiber Optic Biosensor for Noninvasive Transdermal Glucose Sensing Based on the Glucose Binding Protein** CRISTINA E TIANGCO, University of Maryland Baltimore County, Dieudonne Fon, Yordan Kostov, Govind Rao, Leah Tolosa, Fortunato Sevilla, Dayanand Bagdure
9:30	(1070-4)	**Direct Measurement of Total Concentration of Major Physiological Anions, Chloride and Bicarbonate, Using Pulsed Chronopotentiometry with Ion-Selective Electrodes (Pulstrode)** KEBEDE L GEMENE, Northern Kentucky University, Adaeze Stella Iloegbunam, Sara Keshtvarz, Simon Segal
9:50		**Recess**
10:05	(1070-5)	**Non-Invasively Interrogating Chemical and Mechanical Sensors on Implanted Medical Devices** JEFFREY N ANKER, Clemson University, Jeremy Tzeng, Fenglin Wang, Donald Benza, Peter Gennaro, Yash Raval, Mohammed Arifuzzaman
10:25	(1070-6)	**Detection of MicroRNA Presence or Absence with Dual Functioning Signal-On/Off Fluorescent Biosensors** NICHOLAS E LARKEY, Oregon State University, Lulu Zhang, Sean M Burrows
10:45	(1070-7)	**Single Nanoparticle Plasmonic Spectroscopy and Biosensors for Imaging of Efflux Functions of Single Live Cells** X NANCY XU, Old Dominion University, Kerry J Lee, Tao Huang, Prakash d Nallathamby, Feng Ding
11:05	(1070-8)	**Detection of Chemotherapeutic-Induced Damage in Genomic DNA Using Integrated Thermoplastic Nanofluidic Sensor Devices** KUMUDITHA MADUSHANKA WEERAKOON-RATNAYAKE, University of North Carolina at Chapel Hill, Franklin I Uba, Robert Schotzinger, Steven A Soper

POSTER SESSION **Session 1080**

All posters are to be mounted by 10:00 AM and remain on display until 4:00 PM. Authors must be at their posters from 10:00 AM to 12:00 PM. Location of the morning posters is on the Exposition Floor, 400 Aisle. PLEASE NOTE: You cannot get onto the Exposition Floor until after 9:00 AM.

Environmental Air Quality and Analysis

Tuesday Morning, Exposition Floor, 400 Aisle

(1080-1 P)	**Tea Polyphenols Reduces Toxicity of PM2.5 in Human Alveolar Epithelial A549 Cells** YING ZHANG, Shijiazhuang CDC, Jie Jiang, Yan He, Yanzhong Chang
(1080-2 P)	**Testing of Gas Purifiers for VOC Removal Down to the PMOL/MOL Level** ANNARITA BALDAN, VSL, Janneke van Wijk, Jianrong Li, Rina Wortman, Jennifer Englert, Christian Plass-Duelmer, Stefan Persijn, Guido Sassi, Alessia Demichelis, Mariapaola Sassi
(1080-3 P)	**An Analytical Method for the Measurement of Reduced Sulfur Gases** JÜRGEN MICHAEL LOBERT, Entegris, Inc., Charles M Miller, Emily C Zaloga N/A
(1080-4 P)	**Developing and Field Tests of an Automatic Impinger System for Continuous Sampling of Volatile Amines in the Environment** CHIA-JUNG LU, National Taiwan Normal University, Wang Chih-Chia, Rih-Sheng Jian, Sung Lung-Yu
(1080-5 P)	**SIFT-MS: A One-Stop Analytical Tool for Detection of Fumigation Chemicals** DANIEL MILLIGAN, Syft Technologies Ltd, Vaughan Langford, Barry Prince, Thomas McKellar, David Hera, Murray McEwan
(1080-6 P)	**SIFT-MS: A Complete Solution for Analysis of Ambient Air** DANIEL MILLIGAN, Syft Technologies Ltd, Vaughan Langford, Thomas McKellar, David Hera, Barry Prince, Murray McEwan
(1080-7 P)	**Real Time Detection and Identification of Chemical Releases via GCMS** PARMINDER KAUR, 1st Detect Corporation, Corey Stedwell, Daniel DeBord
(1080-8 P)	**Continuous Fenceline Monitoring Using a Miniature Mass Spectrometer** PRESHIOUS REARDEN, 1st Detect Corporation, Madonna Marie Mamerow, Parminder Kaur
(1080-9 P)	**Changes in Tobacco-Specific Nitrosamine Cigarette Smoke Deliveries in Unburned, Recycled Portions of Roll-Your-Own Cigarettes** BENJAMIN WADE ALVERSON, Centers for Disease Control and Prevention, Jose J Perez, Morgan E Larango, Roberto Bravo, Liqin Zhang, Patrick Chen, Shirley Ding, Mary Halstead, R Steven Pappas, Clifford Watson, Liza Valentin-Blasini
(1080-10 P)	**A Method of Performing In-Trap Photoionization in a Miniature Ion Trap Mass Spectrometer** COREY STEDWELL, 1st Detect Corporation, Daniel DeBord, Michael Spencer, David Rafferty

POSTER SESSION **Session 1090**

All posters are to be mounted by 10:00 AM and remain on display until 4:00 PM. Authors must be at their posters from 10:00 AM to 12:00 PM. Location of the morning posters is on the Exposition Floor, 400 Aisle. PLEASE NOTE: You cannot get onto the Exposition Floor until after 9:00 AM.

Environmental and Geochemical Analysis: Soils, Minerals, and Agriculture

Tuesday Morning, Exposition Floor, 400 Aisle

(1090-1 P)	**Automated Analysis of Explosives in Soil Samples** WILLIAM HEDGEPETH, Shimadzu Scientific Instruments, Kenichiro Tanaka
(1090-2 P)	**Mechanisms for Controlling Soil Organic Matter Decompositions: An Application of Pyrolysis –Cryogenic –GC/MS to Molecular Characterizations of Organic Matter in Grass and Forestry Soils** XIANZHI (AMANDA) SONG, Young Harris College
(1090-3 P)	**A New Analytical Technique for the Determination of Carbonate in Soil Samples** TACETTIN OZTURK, Lita Analytical, Onur Iscan N/A
(1090-4 P)	**Ion Selective Electrodes – A Cheaper, Simpler and more Robust Analytical Method for Monitoring of Nitrate and Ammonium in Water and Soil** TOLULOPE A FAYOSE, Keele University
(1090-5 P)	**Multiple Surface-Science Techniques to Elucidate the Reactive Nature of a Metal Phosphide Mineral** DANNA QASIM, Kennesaw State University, Aaron Pital, Thomas Beckman, Heather Abbott-Lyon
(1090-6 P)	**A Systems Chemistry Investigation into Nucleoside Phosphorylation Mechanisms by Prebiotic Meteoritic Materials** MIKE A MOJICA, Georgia Institute of Technology, Maheen Gull, Matthew Pasek, Charles Liotta, Thomas Orlando, Aaron McKee, Facundo M Fernandez N/A
(1090-7 P)	**Current Advances in Instrumentation of Soil Elemental Analyzers** DOMINIK MARGRAF, Elementar Analysensysteme GmbH, Christian Schmidt, Tony Szuppa, Sabine Kraus, Lutz Lange, Hans-Peter Sieper N/A
(1090-8 P)	**Analysis of PAHs in Soil by Online SFE-SFC** KENICHIRO TANAKA, Shimadzu Scientific Instruments, Inc., William Hedgepeth N/A
(1090-9 P)	**Determination of PAHs in Durban City Road Dusts** ABDULMUMEEN ABDULKADIR, University of Kwazulu-Natal N/A

(1090-10 P) **Speciation of Metal-Binding Proteins in Marine Environment** PILAR BERMEJO-BARRERA, University of Santiago de Compostela, Spain, Maria del Carmen Barciela-Alonso, Natalia García-Otero, Olga Cristina Vázquez-Padín, Antonio Moreda-Piñeiro, Raquel Dominguez-Gonzalez, Elena Pena-Vazquez N/A

(1090-11 P) **A New Advancement in the Automated Preparation of Pressed Pellets for XRF Analysis** DAVID COLER, FLSmidth, Lukas Bruzenak, Ian Campbell

(1090-12 P) **Multifunctional Ligand Platform for Detection, Capturing and Removal of Cerium Oxide Nanoparticles** ALI OTHMAN, Clarkson University, Silvana Andreescu

POSTER SESSION — Session 1100

All posters are to be mounted by 10:00 AM and remain on display until 4:00 PM. Authors must be at their posters from 10:00 AM to 12:00 PM. Location of the morning posters is on the Exposition Floor, 400 Aisle. PLEASE NOTE: You cannot get onto the Exposition Floor until after 9:00 AM.

Environmental Applications of Elemental Analysis and Speciation

Tuesday Morning, Exposition Floor, 400 Aisle

(1100-1 P) **Rare Earth Elements – How to Accurately Determine Contamination Levels and Remove Spectral Interferences Created by Them** EWA M PRUSZKOWSKI, PerkinElmer, Inc., Cynthia Bosnak

(1100-2 P) **Determination of Heavy Metals In and Around Lake Ontario** SIMRAN SANDHU, St. John Fisher College, Kimberly Chichester

(1100-3 P) **The Analysis of Flue Gas Desulfurization Fluids by ICP-MS Using Universal Cell and FastFIAS Technology** MICHELLE M COKER, SCE&G, Daniel H Jones, George W Eargle

(1100-4 P) **Determination of Cadmium in Environmental Water Samples Collected in Superfund Sites in New York City** YI HE, John Jay College/CUNY, Kathleen Lopez, Sandra Swenson, Kate Good

(1100-5 P) **Withdrawn**

(1100-6 P) **Evaluation of Chromium Stability on Filters** TAMUTSIWA M MUTUTUVARI, High Purity Standards, Svetlana Uzunova, Kim-Phuong Tran

(1100-7 P) **Withdrawn**

(1100-8 P) **Online Analysis and Speciation of Antimony in Various Wastewater Streams Using Hydride Generation-AFS** BIN CHEN, P S Analytical, Warren T Corns, Peter B Stockwell

(1100-9 P) **An Efficient Recovery of Rare Metal Ions with Calix[4]arene Derivatives from Acidic Media Using Droplet-Based Microreactor System** MASAYA MIYAZAKI, National Institute of Advanced Industrial Science and Technology, Keisuke Ohto, Masatoshi Maeki, Ramachandra Rao Sathuluri

(1100-10 P) **Breakthrough Development for Quantitative Analysis of Total Metals in Soil by Portable High Definition X-ray Fluorescence** ZEWU CHEN, XOS, Shenghua Song, Danhong Li, Kyle Kuwitzky

(1100-11 P) **Chromogenic and Fluorogenic Chemosensors for the Selective Detection of Nd(III) Ion and Their Electrochemical Properties** ASHOK K SINGH, IIT Roorkee, Neha Gupta N/A

(1100-12 P) **Collaborative Certification of a New Low-Level Hexavalent Chromium Standard Reference Material in a Soil Matrix** JAMES E HENDERSON, Duquesne University, Patrick Benecewicz, Anil Srinivas Chaitanya Vishnuvajjhala, Weier Hao, Logan T Miller, Matt Pamuku, Jennifer Crawford, Teresa Switzer, Vasile Furdui, Pam Wee, Francine Walker, Bob O'Brien, Skip Kingston

(1100-13 P) **Impact of Coexistent Elements and Its Concentration in the Quantification of Strontium-90 Using ICP-MS with Cascade Separation System** YOSHITAKA TAKAGAI, Fukushima University, Takahiro Suzuki, Aya Yokoyama, Makoto Furukawa, Yutaka Kameo, Katz Suzuki

POSTER SESSION — Session 1110

All posters are to be mounted by 10:00 AM and remain on display until 4:00 PM. Authors must be at their posters from 10:00 AM to 12:00 PM. Location of the morning posters is on the Exposition Floor, 400 Aisle. PLEASE NOTE: You cannot get onto the Exposition Floor until after 9:00 AM.

Environmental Organic Analysis: VOCs, Pesticides, and Others

Tuesday Morning, Exposition Floor, 400 Aisle

(1110-1 P) **Determination of VOCs by US EPA Method 8260 with Extended Dynamic Range Using Fast, Sensitive Capillary GC/MS** BRAHM PRAKASH, Shimadzu Scientific Instruments, Inc, William Lipps, Di Wang, Nicole M Lock, Shilpi Chopra, Laura Chambers

(1110-2 P) **Determination of Organochlorine Pesticides and Polychlorinated Biphenyls Using GC-MS/MS Operated in the MRM Mode** BRAHM PRAKASH, Shimadzu Scientific Instruments, Inc., William Lipps, Di Wang, Nicole M Lock, Shilpi Chopra, Laura Chambers

(1110-3 P) **Analysis of Terpenes Using Gas Chromatography with Vacuum Ultraviolet Detection** CHANGLING QIU, University of Texas at Arlington, Jonathan Smuts, Phillip Walsh, Kevin A Schug

(1110-4 P) **Use of a Boron Doped Diamond Electrode Sensor for Carbamate Pesticide Classification Using a Chemometric Approach** THIAGO SELVA, University of Sao Paulo, Thiago Paixao

(1110-5 P) **PCBs and DDTs in Bluefin Tuna From the Adriatic Sea** DARIJA KLIN I, Institute for Med Research and Occupational Health, Snjažana Herceg Romanić, Zorana Kljakovi -Gašpi , Vjekoslav Tičina

(1110-6 P) **From Freon to PAHs - A New Generation of Multipurpose Thermal Desorption Tubes** PAOLO BENEDETTI, IIA - CNR, Carlo Crescenzi, Ettore Guerriero

(1110-7 P) **Search for Organic Substance Sources in the 1B Mine Pool of Sydney Coalfield Abandoned Mines** ALLEN BRITTEN, Cape Breton University, Ceilidh MacDonald, Judy MacInnis, Martin Mkandawire N/A

(1110-8 P) **Analysis of Volatile Organic Compounds in Wastewater by Purge and Trap GC/TOF-MS According to EPA Method 624** MOIRA ZANABONI, DANI Instruments, Roberta Lariccia, Matthew S Klee

(1110-9 P) **Degradation of Environmental Contaminants Using Chlorine Dioxide** SUSHMA APPALA, Middle Tennessee State University, Samanwi Munagala, Megan Z Chong, Ooi G Beng, Ngee Sing Chong

(1110-10 P) **Automating Solid Phase Extraction and Florisil Clean-Up for Organichlorine Pesticides and PCB Aroclors** PHILIP BASSIGNANI, Fluid Management Systems, Rudolf Addink

(1110-11 P) **Evaluation of a Novel Hand-Held and Easy-to-Use GC-PID Prototype for Fast and Selective On-site Analysis of Benzene and VOC** MATTHIAS SCHMITTMANN, Bentekk GmbH

(1110-12 P) **Direct Determination of Glyphosate, Glufosinate, and AMPA in Egg by Liquid Chromatography/Tandem Mass Spectrometry** NARONG CHAMKASEM, FDA, Cynthia Morris, Krystle L Hargrove

(1110-13 P) **Direct Coupling of Active and Passive Samplings of Organics with Microwave Assisted Thermal Desorption as an Innovative Solvent-Free Method** WILLIAMS ESTEVE, INRS

(1110-14 P) **Deuterated Monitoring Compounds for Better Accuracy and Precision Measurement of GC/MS Environmental Data** CHARLES G APPLEBY, U.S. Environmental Protection Agency

(1110-15 P) **Withdrawn**

(1110-16 P) **Wastewater Monitoring by Fluorescence Excitation and Emission Matrix with Parallel Factor Analysis** SAM LI, NUS, Baisheng Chen

(1110-17 P) **New Plastic In-Syringe Based Ultrasound Assisted Salt-induced Liquid-liquid Microextraction Technique for the Rapid Analysis of Triclosan in Aqueous Samples** VINOTH KUMAR PONNUSAMY, National Chung Hsing University, Jen Fon Jen

POSTER SESSION **Session 1120**

All posters are to be mounted by 10:00 AM and remain on display until 4:00 PM. Authors must be at their posters from 10:00 AM to 12:00 PM. Location of the morning posters is on the Exposition Floor, 400 Aisle. PLEASE NOTE: You cannot get onto the Exposition Floor until after 9:00 AM.

Environmental Water Quality and Analysis

Tuesday Morning, Exposition Floor, 400 Aisle

(1120-1 P) **Electrochemical Determination of the Levels of Some Potentially Toxic Metals in Groundwater and Leachate Samples from Solous II and III Landfill, Lagos Nigeria** WESLEY OHIFEME OKIEI, University of Lagos, Modupe M Ogunlesi, Jeffery Undiandeye N/A

(1120-2 P) **Determination of N-Nitrosamines by USEPA Method 521 Using Triple Quadrupole Mass Spectrometry** BRAHM PRAKASH, Shimadzu Scientific Instruments, Inc, William Lipps, Di Wang, Nicole M Lock, Shilpi Chopra, Laura Chambers

(1120-3 P) **Optimizing Treatment of Reclaimed Water at a Drinking Water Plant by Online Monitoring of Organic Carbon Levels** MARK MULLET, GE Analytical Instruments, Dondra Biller

(1120-4 P) **Development of an Arduino Shield for Water Quality Analysis Probes** MICHAEL CHIA, Northern Kentucky University, Kelley Weigman, Grant Foreman, Richard Durtsche, Celeste A Morris

(1120-5 P) **Analysis of Peanut Hulls as an Alternative to Ion Exchange Resins** CARMEN L HUFFMAN, Western Carolina University, Kanika O Davis, Melisa J Glatte, Holly Truluck, Tyler S Cook

(1120-6 P) **Analytical Strategies for Monitoring Ionic Liquids Breakdown by Electro-Fenton Process** ELISA GONZÁLEZ-ROMERO, University of Vigo, Elvira Bocos, Jessica Meijide, Aida Díez, Marta Pazos, María Ángeles Sanromán

(1120-7 P) **A Study of Quality of Water and Effect of Various Coagulants on Water of Lodra Lake of Gandhinagar District** JAIMIN K DESAI, Gujarat University, Parekh P Paresh N/A

(1120-8 P) **Determination of the Toxins Found in Lake Ontario** MARTA LABECKI, St. John Fisher College, Kimberly Chichester

(1120-9 P) **EPA Method 557 Quantitation of Haloacetic Acids, Bromate and Dalapon in Drinking Water Using Ion Chromatography and Tandem Mass Spectrometry** JONATHAN BECK, Thermo Fisher Scientific, Terri T Christison, Charles Yang, Hans Schweingruber

(1120-10 P) **Assessment of Water Quality Parameters from the Lowber Abandoned Mine Drainage Treatment Facility, Part 2: Further Studies and Results** MARK T STAUFFER, University of Pittsburgh - Greensburg, Tell M Lovelace, Aaron K Hirshka, Luke J Metzler

(1120-11 P) **Breaking the Biofouling Code: Towards Reliable In-Pipe Water Quality Sensors** ROBERT E WILSON, Imperial College London, Ivan Stoianov, Danny O'Hare

(1120-12 P) **Analysis of the Residual Oil in Water with Excitation Emission Matrix and Parallel Factor Analysis** KAWAGUCHI YOSHIHIKO, HORIBA Advanced Techno, Co., Ltd., Kojima Reiji

(1120-13 P) **Sensitive Determination of Arsenate and Phosphate by Molybdenum Blue Method with Membrane Filter Extraction Using a Portable 8-Channel LED-Based Reflective Photometer** YASUTADA SUZUKI, University of Yamanashi, Susumu Kawakubo

(1120-14 P) **Assessment of Physico-Chemical Analysis of Drinking Water of Kheda District Gujarat (INDIA) and its Impact on Human Health** MAHESHKUMAR B CHAUHAN, J & J College of Science, Dipakkumar K Bhoi, Rameshbhai P Dabhi N/A

(1120-15 P) **Nano-Sensing Approach for the Determination of Phosphate Ions** MADELEINE JOHNSON, University of Central Florida N/A

(1120-16 P) **Study of Enhancement Effects of Functionalized Gold Nanorods in Quantitative Analysis of 1-H Benzotriazole by Surface Enhanced Raman Spectroscopy (SERS)** UTTAM SHARMA PHUYAL, Tennessee Technological University, Andrew Callender

(1120-17 P) **Seasonal Variations in Water Quality Parameters at Thol Bird Sanctuary, Gujarat India** AMRUTLAL B PARMAR, J & J College of Science, Rameshkumar V Parmar, Arunkumar H Dholakia N/A

(1120-18 P) **A Study of Water Quality Index (W.Q.I) of Pariyej Lake District: Kheda - Gujarat** BUNTY R PATEL, J & J College of Science, Alpesh M Patel, Fulaji J Thakor N/A

(1120-19 P) **Colorimetric and Electrochemical Sensor for Monohydrogen Phosphate Ion Based on 1,3-bis(2-formylphenoxy) Propane Bis(2,4-dinitrophenyl hydrazine)** DIVYA SINGHAL, IIT Roorkee, Ashok K Singh N/A

(1120-20 P) **The Effect of Slaughter House Waste on the Water Quality of Okpu River Aba** CHIDIEBERE A ODIKE-ADUAKA, Abia State Teaching Hospital Aba, Lilian I Oguguo

(1120-21 P) **Biosensor for Toxic Compounds in Wastewater Based on Microbial Electrochemistry** SAM LI, NUS

(1120-22 P) **Rapid Determination of Endocrine Disrupting Bisphenol A (BPA) in Drinking Water by Solid Phase Nano-Extraction and Room-Temperature Fluorescence Spectroscopy** MAHA AL-TAMEEMI, University of Central Florida, Bassam Alfarhani, Andres Campiglia, Jung Jong Seok

(1120-23 P) **GC/MS Screening of Water Samples for Organic Pollutants by Stir Bar Sorptive Extraction (SBSE)** OLIVER LERCH, Gerstel GmbH & Co. KG, Jasmin Zboron, Andreas Hoffmann, Chris Sandy

POSTER SESSION **Session 1130**

All posters are to be mounted by 10:00 AM and remain on display until 4:00 PM. Authors must be at their posters from 10:00 AM to 12:00 PM. Location of the morning posters is on the Exposition Floor, 400 Aisle. PLEASE NOTE: You cannot get onto the Exposition Floor until after 9:00 AM.

Environmental, Food and Elemental Analyses - Atomic Spectroscopy

Tuesday Morning, Exposition Floor, 400 Aisle

(1130-1 P) **Analysis of Radium-226 in Shale-Gas Wastewater Using Inductively Coupled Plasma Mass Spectrometry** YUQIANG BI, University of Michigan, Wenjia Fan, Thomas P Yavaraski, Brian R Ellis, Kim F Hayes

(1130-2 P) **Speciation of Organic Mercury in Water Samples by Alkylation, Organic Solvent Extraction and GC-AFS Detection – A Comparison of Ethylation, Propylation and Phenylation** CORNELIUS BROMBACH, P S Analytical, Bin Chen, Warren T Corns, Jun Yoshinaga

(1130-3 P) **Models for Predicting Atmospheric Mercury Concentrations Using Meteorological Data and Mercury Concentrations in *Salix* (Willow) Leaves** DAVID LEHMPUHL, Colorado State University Pueblo, Lauren Bartolo

(1130-4 P) **Assessment of Titanium Dioxide Nanoparticles in Aquatic Tanks** SARA MELOW, Elmira College, Marriah Ellington, Lisa A Holland

(1130-5 P) **Feasibility Use of Ceramics as Solid Support for Cr(III) Measurement in Water by LIBS** CASSIANA S NOMURA, Institute of Chemistry - University of Sao Paulo, Alexandrina C Carvalho, Daniel M Silvestre, Flávio O Leme, Danielle P Intima, Juliana Naozuka

(1130-6 P) **Lead Determination in Soil from a Recreational Shooting Range Built on a Reclaimed Strip Mine: Effects of Oxidant Flow Rate on PB Measurements Obtained by FAAS, and Other Considerations** MARK T STAUFFER, University of Pittsburgh - Greensburg, Luke J Metzler

(1130-7 P) **Determination of Metals in Three Types of Loose-Leaf Tea: Can Metal Content Indicate the Type of Tea?** MARK T STAUFFER, University of Pittsburgh - Greensburg, Aaron K Hirshka

(1130-8 P) **Determination of Selected Metals and Nonmetals in Pre-Workout Supplements: Results from Initial and Ongoing Investigations** MARK T STAUFFER, University of Pittsburgh - Greensburg, Kelly M Boyles, Nicholas E Glotfelty N/A

(1130-9 P) **Speciation of Mercury in Rice with a New Online Pre-Concentration HPLC-CV-AFS Method** CORNELIUS BROMBACH, P S Analytical, Piumi K Dona, Bin Chen, Warren T Corns, Eva Krupp, Joerg Feldman

(1130-10 P) **Optimizing a Total Protein Combustion Instrument for Maximum Sample Throughput and Lowest Cost-Per-Analysis** JEFFERY GAST, LECO Corporation, Mason Marsh

(1130-11 P) **Laser Induced Breakdown Spectroscopy (LIBS) of Food Samples: Case Study of Tortillas** CHARLES GHANY, Mississippi State University, Herve Sanghapi, Chet R Bhatt, Bader Alfarraj, Fang Y Yueh, Jagdish P Singh

(1130-12 P) **Development of a Method For the Determination of Titanium Dioxide Nanoparticles in Food Products Using SP-ICPMS** ANTONIO MOREDA-PINEIRO, University of Santiago de Compostela, Maria del Carmen Barciela-Alonso, Olga Cristina Vázquez-Padín, Francisco Javier Vilariño-Páxaro, Manuel Aboal-Somoza, Pilar Bermejo-Barrera, Elena Pena-Vazquez

(1130-13 P) **Evaluation of Optical Depths of Ca Emission Lines in Laser Induced Breakdown Spectroscopy (LIBS)** BADER ALFARRAJ, Mississippi State University, Herve Sanghapi, Chet R Bhatt, Charles Ghany, Fang Y Yueh, Jagdish P Singh

(1130-14 P) **Standard Dilution Analysis for the Determination of Calcium by Flame Atomic Emission Spectrophotometry** CLIFTON P CALLOWAY, Winthrop University, Katja A Hall, Emily A Watson

(1130-15 P) **Cyanide Detection in Blood Using Indirect Atomic Absorption Spectroscopy** JEFFREY ROSENTRETER, Idaho State University, Matt Kirkham, Jeff Kuhlmeier

(1130-16 P) **Advanced Application of Speciation Analysis Using ICP-MS detection** DANIEL KUTSCHER, Thermo Fisher Scientific, Monika Verma, Shona McSheehy Ducos

(1130-17 P) **Multi-trophic Analysis of Lead Using a Flame Atomic Absorption Spectrometer** MICHAEL DECAROLIS, St. John Fisher College, Christopher Collins, Kimberly Chichester

(1130-18 P) **Atomic and Molecular Laser Induced Breakdown Spectroscopy for Detection of Chlorine in Concrete** WILL B JONES, University of Florida, Ebo Ewusi-Annan, Tobias Guenther, Ben Smith, Nico Omenetto

(1130-19 P) **A Modern Multi-Excitation Concept for Total Reflection X-Ray Spectrometry** ARMIN GROSS, Bruker Nano GmbH, Ulrich Waldschlaeger N/A

(1130-20 P) **Determination of Lead in Keratin Tissues: Method Development and Candidate Reference Material Production Using Caprine Horns** MINA TEHRANI, Wadsworth Center/ SUNY Albany, Patrick J Parsons N/A

(1130-21 P) **Lead Concentrations in Lagos Groundwater** AMALACHUKWU OLIVE ANADI, Hydrochrom Resources Ltd, Chimezie A Anyakora N/A

(1130-22 P) **Fast Monitoring Processed Manure Using WD-XRF Spectroscopy for Nutrients and Metals** ALEXANDER SEYFARTH, Rock River Axs LLC, Aicado Roa-Espinosa

(1130-23 P) **Onsite Monitoring of Manure Using XRF for Nutrients and Metals** ALEXANDER SEYFARTH, Bruker Nano Analytics, Aicado Roa-Espinosa

POSTER SESSION — Session 1140

All posters are to be mounted by 10:00 AM and remain on display until 4:00 PM. Authors must be at their posters from 10:00 AM to 12:00 PM. Location of the morning posters is on the Exposition Floor, 400 Aisle. PLEASE NOTE: You cannot get onto the Exposition Floor until after 9:00 AM.

Sampling and Sample Preparation - Environmental

Tuesday Morning, Exposition Floor, 400 Aisle

(1140-1 P)	**The Measurement of Formaldehyde in Drinking Water Using Automated SPE and HPLC** ALICIA CANNON, Horizon Technology, Chris Shevlin, Michael Ebitson
(1140-2 P)	**Analysis of Extract Drying Criteria for Oil and Grease Method 1664A/B** MICHAEL EBITSON, Horizon Technology, David Gallagher, William Jones
(1140-3 P)	**Fundamentals and Comparisons for Organic Sample Extract Evaporation** ZOE GROSSER, Horizon Technology, Robert Johnson
(1140-4 P)	**Determination of 2-Methylisoborneol and Geosmin in Water Using Solid Phase Micro Extraction** ANNE JUREK, EST Analytical
(1140-5 P)	**The Advantages of Automated Sample Preparation** ANNE JUREK, EST Analytical, Lindsey Pyron, Kelly Cravenor, Justin Murphy
(1140-6 P)	**Highly Robust Polymeric Ionic Liquid Coatings for Solid Phase Microextraction: Multiclass Determinations with Application of Direct-Immersion-Headspace Mode Using Gas Chromatography-Mass Spectrometry** JOSIAS MERIB, Iowa State University, Honglian Yu, Carasek Eduardo, Jared L Anderson
(1140-7 P)	**Development of Dispersive Liquid-Liquid Microextraction for the Determination of Six Steroidal Hormones in Wastewater Using High Pressure Liquid Chromatography-Charged Aerosol Detector** MATHEW MUZI NINDI, Unisa, Cecilia O Osunmakinde, Simiso Dube
(1140-8 P)	**Extraction of Analytes of Forensic Toxicological Interest from Plasma with Enhanced Matrix Removal-Lipid Material** JOAN STEVENS, Agilent Technologies, Derick Lucas, Limian Zhao, Megan Juck, William Long
(1140-9 P)	**Modified Sample Clean-up for Combined POPs Using Automated Multi-Column Fractionation and Analytical Optimization** PHILIP BASSIGNANI, Fluid Management Systems, Rudolf Addink
(1140-10 P)	**Ice Concentration Linked with Extractive Stir Bar** NUJUD O MASLAMANI, South Dakota State University
(1140-11 P)	**The Best Sample Preparation for High Throughput Quantitative X-Ray Diffraction of Mineral Mixtures** ROGER MEIER, FLSmidth A/S, Ian Campbell, Lukas Bruzenak, Detlev Götz
(1140-12 P)	**Nanoporous Solid Phase Microextraction (SPME) Fibers by Sputtering Silicon** MASSOUD KAYKHAII, Brigham Young University, Anubhav Diwan, Bhupinder Singh, Matthew R Linford

TUESDAY, MARCH 8, 2016
AFTERNOON

AWARDS — Session 1150

Pittsburgh Spectroscopy Award

arranged by Joseph Grabowski, University of Pittsburgh

Tuesday Afternoon, Room B312
Fu-Tyan Lin, The Pittsburgh Conference, Presiding

1:30		**Introductory Remarks - Fu-Tyan Lin**
1:35		**Presentation of the 2016 Pittsburgh Spectroscopy Award to Jürgen Popp, Friedrich-Schiller University Jena, by Joseph Grabowski, Chair, Spectroscopy Society of Pittsburgh**
1:40	(1150-1)	**Overcoming Unmet Medical Needs: Advances in Raman Spectroscopy** JÜERGEN POPP, Friedrich-Schiller University Jena
2:15	(1150-2)	**Developing Deep UV Raman Standoff Spectrometers for Trace Explosives** SANFORD A ASHER, University of Pittsburgh, Sergei V Bykov, Katie L Gares, Kyle T Hufziger
2:50	(1150-3)	**Enhanced Vibrational Circular Dichroism: Moving Beyond Established Applications of Vibrational Circular Dichorism (VCD)** LAURENCE A NAFIE, Syracuse University
3:25		**Recess**
3:40	(1150-4)	**Raman Big Data Analysis for Automatic and Objective Living Cell Discrimination/Diagnosis** HIROO HAMAGUCHI, National Chiao Tung University, Masahiro Ando
4:15	(1150-5)	**Raman Scattering from Single, Laser-Trapped Microparticles: A Review** WOLFGANG KIEFER, University of Würzburg

AWARDS — Session 1160

RSC - JAAS Emerging Investigator Lectureship Award

arranged by Philippa Ross, Royal Society of Chemistry

Tuesday Afternoon, Room B314
Philippa Ross, Royal Society of Chemistry, Presiding

1:30		**Introductory Remarks - Philippa Ross**
1:35		**Presentation of the 2016 RSC - JAAS Emerging Investigator Lectureship Award to Gerardo Gamez, Texas Tech University, by Philippa Ross, Royal Society of Chemistry**
1:40	(1160-1)	**Revealing Surface Elemental Landscapes with Ultra-High Throughput via GDOES** GERARDO GAMEZ, Texas Tech University
2:15	(1160-2)	**Laser Ablation-Based Chemical Analysis Techniques: A Short Review** JHANIS JOSE GONZALEZ, Lawrence Berkeley National Laboratory
2:50	(1160-3)	**Direct Determination of Trace Antimony and Arsenic in Natural Waters by Photochemical Vapor Generation ICPMS** LU YANG, National Research Council Canada
3:25		**Recess**
3:40		**Presentation of the 2016 RSC - JAAS Emerging Investigator Lectureship Award to Lara Lobo Revilla, University of Oviedo, by Philippa Ross, Royal Society of Chemistry**
3:45	(1160-4)	**Pulsed Glow Discharge Time-of-Flight Mass Spectrometry (Positive and Negative Ionization Modes) for Elemental Depth Profiling of Innovative Materials and Polymer Fingerprinting** LARA LOBO REVILLA, University of Oviedo, R Muniz, B Fernandez, R Pereiro, A Sanz-Medel
4:20	(1160-5)	**What is XRF Doing in a Mass Spectrometry Award Symposium?** GEORGE HAVRILLA, Los Alamos National Laboratory

SYMPOSIUM **Session 1170**

ACS-ANYL - New Approaches to Nuclear Safeguards and Forensics Analysis
arranged by Douglas Duckworth, Pacific Northwest National Laboratory

Tuesday Afternoon, Room B308
Douglas Duckworth, Pacific Northwest National Laboratory, Presiding

1:30		**Introductory Remarks - Douglas Duckworth**
1:35	(1170-1)	**Advances in Online Spectroscopic Monitoring for Process Control and Safeguarding of Radiochemical Streams** SAM A BRYAN, Pacific Northwest National Laboratory, Amanda J Casella, Amanda M Lines, Gilbert L Nelson, Job M Bello
2:10	(1170-2)	**New XRF Applications to Nuclear Safeguards and Nuclear Forensics** GEORGE HAVRILLA, Los Alamos National Laboratory
2:45	(1170-3)	**Fieldable Mass Spectrometry: Sample Preparation and Rapid Field Analysis for Nuclear Safeguards** PETER STARK, Los Alamos National Laboratory, Elizabeth Judge, Keri Campbell, Lisa Meyers, Chris Leibman, Ning Xu, Thomas Yoshida, Matthew R Dirmyer
3:20		**Recess**
3:35	(1170-4)	**New Analytical Methods for Trace Elemental and Isotopic Analysis of Nuclear Fuel Cycle Materials** GREGORY C EIDEN, Pacific Northwest National Laboratory, Andrew M Duffin, Michael P Dion, April J Carman, Orville T Farmer, Jesse D Ward, Carmen S Menoni, Martin Liezers
4:10	(1170-5)	**Contribution and Impact of Laser Ablation Inductively Coupled Plasma Mass Spectrometry (LA-ICP-MS) and Laser Induced Breakdown Spectroscopy (LIBS) to Forensic Analysis** JHANIS JOSE GONZALEZ, Lawrence Berkeley National Laboratory

SYMPOSIUM **Session 1180**

Emerging Mass Spectrometry-Based Techniques for Biomolecular Analysis
arranged by Ronghu Wu, Georgia Institute of Technology

Tuesday Afternoon, Room B302
Ronghu Wu, Georgia Institute of Technology, Presiding

1:30		**Introductory Remarks - Ronghu Wu**
1:35	(1180-1)	**Searching for Biomarkers Using Ambient Ionization Mass Spectrometry** GRAHAM COOKS, Purdue University
2:10	(1180-2)	**An Inside-Outside Strategy to Study Cell Communication** CATHERINE FENSELAU, University of Maryland, Sitara Chauhan, Lucia Geis-Asteggiante, Avantika Dhabaria, Nathan Edwards, Suzanne Ostrand-Rosenberg
2:45	(1180-3)	**Developing Cross-linking Mass Spectrometry to Define Protein-Protein Interactions** LAN HUANG, University of California, Irvine
3:20		**Recess**
3:35	(1180-4)	**Innovative Instrumentation and Methods for the Identification of Intact Proteins in Mixtures and for Sequence Analysis of Antibodies and Posttranslationally-Modified, Intact Proteins on a Chromatographic Time-Scale** DONALD F HUNT, University of Virginia
4:10	(1180-5)	**Charge Detection Mass Spectrometry for Single Ions** EVAN WILLIAMS, University of California, Berkeley, Andrew Elliot, Zijie Xia

SYMPOSIUM **Session 1190**

Emerging Pollutants in the Environment – from Sources to Effects
arranged by Rudolf J Schneider, BAM Federal Institute for Materials Research and Testing

Tuesday Afternoon, Room B303
Rudolf J Schneider, BAM Federal Institute for Materials Research and Testing, Presiding

1:30		**Introductory Remarks - Rudolf J Schneider**
1:35	(1190-1)	**Using High Resolution Mass Spectrometry to Uncover New, Emerging Iodinated and Nitrogen-Containing Disinfection Byproducts** SUSAN D RICHARDSON, University of South Carolina, Cristina Postigo, Christina M Joseph, Hannah K Liberatore, Jessie Kadlec, Amy Cuthbertson, Friedrich Wendel, Christian Luetke-Eversloh, Thomas A Ternes, Edward Machek, Stephen Duirk, Elizabeth Wagner, Michael J Plewa
2:10	(1190-2)	**Measuring Plant Uptake and Effects of Pharmaceuticals Using Liquid Chromatography/Mass Spectrometry** DIANA S AGA, University at Buffalo, Rachel Mullen
2:45	(1190-3)	**Analysis of Hydraulic Fracturing Additives by LC/Q-TOF-MS** IMMA FERRER, University of Colorado, Michael Thurman
3:20		**Recess**
3:35	(1190-4)	**Uptake and Disposition of Pharmaceuticals by Bluegill Exposed at Constant Concentrations in a Flow-Through Aquatic Exposure System** EDWARD FURLONG, U.S. Geological Survey, Jian-Liang Zhao, Heiko L Schoenfuss, Dana W Kolpin, Kyle L Bird, David J Feifarek, Eric A Schwab, Guang-Guo Ying
4:10	(1190-5)	**Pharmaceuticals in Surface Waters - Analysis and Effects** RUDOLF J SCHNEIDER, BAM

SYMPOSIUM **Session 1200**

Emerging Technologies for Disease Biomarker Detection
arranged by Xiujun James Li, University of Texas at El Paso

Tuesday Afternoon, Room B304
Xiujun James Li, University of Texas at El Paso, Presiding

1:30		**Introductory Remarks - Xiujun James Li**
1:35	(1200-1)	**Phase Separated Droplets Enable Multiplexing of Difficult ELISA Panels** SHUICHI TAKAYAMA, University of Michigan
2:10	(1200-2)	**Looking for Rare Cells via High-throughput Single Cell Mass Spectrometry Profiling** JONATHAN V SWEEDLER, University of Illinois
2:45	(1200-3)	**Biomarker Detection Using Paper/PDMS Hybrid Microfluidic Platforms for Low-cost Disease Diagnosis** XIUJUN JAMES LI, University of Texas at El Paso, Maowei Dou, Sharma T Sanjay
3:20		**Recess**
3:35	(1200-4)	**'Cytology-on-a-Chip' Based Sensors for Monitoring of Potentially Malignant Oral Lesions** JOHN T MCDEVITT, New York University
4:10	(1200-5)	**Biomarker Discovery Using DNA Aptamers** WEIHONG TAN, University of Florida

SYMPOSIUM **Session 1210**

JAIMA - The State-of-the-Art Technologies from Japan: Analytical Instruments with/for Nano-Chemistry Technology and Advanced Diagnosis (II)
arranged by Koichiro Matsuda, JAIMA

Tuesday Afternoon, Room B305
Koichiro Matsuda, JAIMA, Presiding
Koji Suzuki, Keio University, Presiding

1:30		**Introductory Remarks - Koji Suzuki, President of JSAC**
1:35	(1210-1)	**Electrochemical Application of Boron-Doped Diamond Electrodes** YASUAKI EINAGA, Keio University
2:10	(1210-2)	**Plasmonic Nanomaterials** TETSU TATSUMA, University of Tokyo
2:45	(1210-3)	**The Unique Combination of Nanotechnology with Raman and SPRi Platforms Offers Innovative and Ultrasensitive Solutions for Diagnostics** MARINELLA SANDROS, HORIBA Scientific
3:20		**Recess**
3:35	(1210-4)	**Medicinal Cannabinomics and Mass Spectrometry Applications to Cannabis Testing Laboratories** SCOTT KUZDZAL, Shimadzu Scientific Instruments, Di Wang, Jonathan Edwardsen, William Lipps, Jeff H Dahl
4:10	(1210-5)	**Application of Laser/desorption Ionization Mass Spectrometry as a Novel Surface Analytical Tool** TAKAYA SATOH, JEOL Ltd.

TECHNICAL PROGRAM

SYMPOSIUM — **Session 1220**

New Advances in Analytical Mass Spectrometry
arranged by Gary Martin Hieftje, Indiana University

Tuesday Afternoon, Room B309
Gary Martin Hieftje, Indiana University, Presiding

1:30		**Introductory Remarks - Gary Martin Hieftje**
1:35	(1220-1)	**21 Tesla Fourier Transform Ion Cyclotron Resonance Mass Spectrometry: New Capabilities for Ultrahigh-Resolution Mass Analysis** ALAN G MARSHALL, Florida State University, John P Quinn, Nathan K Kaiser, Donald F Smith, Greg T Blakney, Tong Chen, Steven C Beu, Christopher L Hendrickson
2:10	(1220-2)	**The Unique Analytical Capabilities of Distance-of-Flight Mass Spectrometry** STEVEN J RAY, SUNY-Buffalo, Elise Dennis, David W Koppenaal, Christie G Enke, Gary Martin Hieftje
2:45	(1220-3)	**Plasma Mass Spectrometry: A Tool for and a Source of Chemical Reactions** JACOB T SHELLEY, Kent State University, Sunil P Badal, Yi You, Kelsey Williams, Garett MacLean
3:20		**Recess**
3:35	(1220-4)	**New Paths for Mass Spectrometry based upon Structures for Lossless Ion Manipulations (SLIM)** RICHARD D SMITH, Pacific Northwest National Laboratory
4:10	(1220-5)	**Improvements in Velocity-Based Mass Analysis by Use of Constant-Momentum Acceleration** GARY MARTIN HIEFTJE, Indiana University, Elise Dennis, Steven J Ray, Christie G Enke, Alexander W Gundlach-Graham

SYMPOSIUM — **Session 1230**

SEAC - New Trends in Electrochemical Neurochemistry
arranged by Adrian C Michael, University of Pittsburgh

Tuesday Afternoon, Room B310
Ann-Sofie Cans, Chalmers University, Presiding

1:30		**Introductory Remarks - Ann-Sofie Cans**
1:35	(1230-1)	**Fast-scan Cyclic Voltammetry Reveals Dopamine Spikes to Food Reward that are Tuned by Physiological State and Its Proxies** MITCHELL ROITMAN, University of Illinois at Chicago
2:10	(1230-2)	**Electroenzymatic Detection of Basolateral Amygdala Glutamate Release During Reward Seeking** KATE M WASSUM, University of California Los Angeles, Melissa Malvaez, Venuz Y Greenfield, Allison M Yorita, Lili Feng, Harold G Monbouquette
2:45	(1230-3)	**Improving Temporal Resolution of Enzyme Based Electrochemical Sensors for Detection of Non-Electroactive Analytes Important in Brain Chemistry** ANN-SOFIE CANS, Chalmers University of Technology, Yuanmo Wang, Jenny Bergman, Joakim Wigström
3:20		**Recess**
3:35	(1230-4)	**Building a 'Well-tempered' Biosensor for Real-Time Neurochemical Monitoring in the Intensive Care Unit** MARTYN G BOUTELLE, Imperial College London, Isabelle C Samper, Chu Wang, Thomas Watts, Michelle L Rogers, Chi Leng Leong, Sally A Gowers
4:10	(1230-5)	**Optogenetic-Control of Glutamate Release in the Rat Hippocampus and Frontal Cortex Measured Using Enzyme-Coated Ceramic Based Microelectrode Arrays** GREG A GERHARDT, University of Kentucky Medical Center

SYMPOSIUM — **Session 1240**

The Challenge of Detection for Drugged Driving
arranged by Herbert H Hill, Washington State University

Tuesday Afternoon, Room B311
Herbert H Hill, Washington State University, Presiding

1:30		**Introductory Remarks - Herbert H Hill**
1:35	(1240-1)	**Police Officer Difficulties with Drug-Impaired Driver Arrests** NICHOLAS PETER LOVRICH, Washington State University
2:10	(1240-2)	**Human Cannabinoid Metabolism and Disposition in Biological Matrices after Controlled Cannabis Administration** MARILYN ANN HUESTIS, NIDA
2:45	(1240-3)	**Analytical Methods for the Detection of Marijuana in Biological Fluids in a Forensic Toxicology Laboratory** BRIANNA PETERSON, Washington State Patrol
3:20		**Recess**
3:35	(1240-4)	**Detection of Drug Consumption in Europe** WOLFGANG VAUTZ, ISAS
4:10	(1240-5)	**Detection of Marijuana from Human Breath by Breathalyzer-IMS** HERBERT H HILL, Washington State University, Jessica A Tufariello

WORKSHOPS — **Session 1250**

Core-Shell versus Fully Porous HPLC Particles – The Current State of the Art in HPLC Columns
arranged by Jason Anspach and Lawrence Loo, Phenomenex

Tuesday Afternoon, Room B313
Jason Anspach, Phenomenex, Presiding

1:30		**Introductory Remarks - Jason Anspach and Lawrence Loo**
1:35	(1250-1)	**Perspectives on the Development and Future of Monodisperse Fully-Porous Silica Supports** DAVID S BELL, Supelco/Sigma-Aldrich
2:05	(1250-2)	**Fully-Porous vs. Core-shell Particles — The Past, the Present and the Future** LAWRENCE LOO, Phenomenex, Jason Anspach, Tivadar Farkas, Mike Chitty, Thuylinh Tran
2:35	(1250-3)	**Advantages and Limitations of Superficially Porous Particles** KEN BROECKHOVEN, Vrije Universiteit Brussel, Gert Desmet
3:05		**Recess**
3:20	(1250-4)	**Next Generation Superficially Porous Particle Technology: Highly Ordered Pore Structures Formed by Pseudomorphic Transformation** WILLIAM E BARBER, Agilent Technologies, Ta-Chen Wei, Wu Chen, Anne Mack, Jia Liu, Monika Dittmann, Xiaoli Wang
3:50	(1250-5)	**Clarifying the Difference Between Columns Packed with Solid Core and Fully Porous Particles** JACOB FAIRCHILD, Waters Corporation, Jonathan Turner, Bonnie Alden, Kevin Wyndham, Babajide Okandeji

ORGANIZED CONTRIBUTED SESSIONS — **Session 1260**

LIMS Live @ Pittcon: Best Practices and Lessons Learned From The Laboratory
arranged by Ken Ochi and Devender Gandhi, Accelerated Technology Laboratories

Tuesday Afternoon, Room B315
Ken Ochi, Accelerated Technology Laboratories, Presiding

1:30	(1260-1)	**Where Do We Start? A Roadmap to LIMS Success** CHRISTINE PASZKO, Accelerated Technology Laboratories
1:50	(1260-2)	**The LIMS Needs Assessment: Your Secret Weapon to a Successful Deployment** ALAN SERRERO, Gwinnett County Department of Public Utilities
2:10	(1260-3)	**Finding the Perfect LIMS: Keys to a Successful RFP** ASTER TEKLE, Alexandria Renew Enterprises
2:30	(1260-4)	**Preparing for a LIMS – The Importance of Proper Planning** KEITH KEESEE, Oklahoma Department of Agriculture, Cassandra Kontas
2:50		**Recess**
3:05	(1260-5)	**Implementing a LIMS is a Project – Treat It Like One** ROY D JONES, Duke Energy
3:25	(1260-6)	**Life with a LIMS: What It's Meant for the City of Clearwater** MARIA DE LA CANTERA, City of Clearwater (FL)
3:45	(1260-7)	**Leveraging LIMS for Streamlining Next Generation Sequencing Data** JENNIFER WELLER, University of North Carolina - Charlotte
4:05	(1260-8)	**What Have We Learned? Final Thoughts On the Road to LIMS Success (Getting the Most From Your LIMS)** DEVENDER GANDHI, Accelerated Technology Laboratories

ORGANIZED CONTRIBUTED SESSIONS **Session 1270**

Quantifying the Tumor Microenvironment
arranged by Matthew R Lockett, University of North Carolina at Chapel Hill

Tuesday Afternoon, Room B316
Matthew R Lockett, University of North Carolina at Chapel Hill, Presiding

1:30	(1270-1)	**Phosphoproteomics in Prostate Extracellular Vesicles** W ANDY TAO, Purdue University
1:50	(1270-2)	**Exploring the Permissive Stromal Microenvironment** AMANDA B HUMMON, University of Notre Dame, Pinar Zorlutuna, Eric M Weaver
2:10	(1270-3)	**Modulating Drug Resistance in Hypoxia Tumors** DIMITRI PAPPAS, Texas Tech University
2:30	(1270-4)	**Paper-Based Assays for the Study of Cancer Cell Biology, Invasion, and Metastasis** MATTHEW R LOCKETT, University of North Carolina at Chapel Hill
2:50		**Recess**
3:05	(1270-5)	**Examining Small Molecule Cellular Signaling in Complex Environments with Microscale Systems** ASHLEIGH B THEBERGE, University of Washington
3:25	(1270-6)	**Microengineered Physiological Biomimicry: Human Organ-on-Chips** D DAN HUH, University of Pennsylvania
3:45	(1270-7)	**Direct Optical Microscopy of Biological Interfaces** CHARLES R MACE, Tufts University, Jenna A Walz, Irene Lui, Daniel J Wilson
4:05	(1270-8)	**Deconvolving Glycans in Metastasis** LARA K MAHAL, New York University

ORGANIZED CONTRIBUTED SESSIONS **Session 1280**

SEAC - The Student Session in Electroanalysis
arranged by Stephen Maldonado, University of Michigan

Tuesday Afternoon, Room B401
Stephen Maldonado, University of Michigan , Presiding

1:30	(1280-1)	**Developing an Innovative Bio-Inspired Scanning Probe Microscopy (Bio-SPM) Approach to Map Specific Molecular Flux** FLORIKA CALING MACAZO, University of Maryland Baltimore County, Ryan J White
1:50	(1280-2)	**Electrodeposition of Semiconductor Thin Films Using Electrochemical Liquid-Liquid-Solid (ec-LLS) Deposition** JOSHUA DEMUTH, University of Michigan
2:10	(1280-3)	**Alternating Current Potentiometric Scanning Ion Conductance Microscopy (AC-PSICM)** LUSHAN ZHOU, Indiana University Bloomington, Yi Zhou, Wenqing Shi, Lane A Baker
2:30	(1280-4)	**Toward the Electrochemical Detection of Single Atoms and Ions** JEFFREY E DICK, The University of Texas at Austin
2:50		**Recess**
3:05	(1280-5)	**Miniaturized Potentiometric Ion-Sensing Systems: From Bulk Electrodes to Paper-Based Ion-Sensing Devices** JINBO HU, University of Minnesota, Andreas Stein, Philippe Buhlmann
3:25	(1280-6)	**Spontaneous Photoelectrochemical Growth of Nanopatterned Semiconductor Films Driven by Anisotropic Interfacial Light Collection** AZHAR I CARIM, California Institute of Technology, Nicolas A Batara, Anjali Premkumar, Harry A Atwater, Nathan S Lewis
3:45	(1280-7)	**New Enzymes for the Hybrid Enzymatic and Organic Electrocatalytic Cascade for the Complete Oxidation of Glycerol** SOFIENE ABDELLAOUI, University of Utah, Shelley D Minteer
4:05	(1280-8)	**The Unique Electrochemical Reactivity of Small Metal Nanoparticles** RAFAEL MASITAS, University of Louisville, Francis Zamborini

ORGANIZED CONTRIBUTED SESSIONS **Session 1290**

Specialty Gas Analysis
arranged by Tracey Jacksier, Air Liquide

Tuesday Afternoon, Room B402
Tracey Jacksier, Air Liquide, Presiding

1:30	(1290-1)	**Real-Time, Selective Analysis of Air and Specialty Gases** DANIEL MILLIGAN, Syft Technologies Ltd, Vaughan Langford, Barry Prince, Murray McEwan
1:50	(1290-2)	**Cavity Ring-Down Spectroscopy Analyzer for Trace Moisture Detection in Ultra-Pure Ammonia** HELEN WAECHTER, Tiger Optics, Florian Adler, Marten Beels, Brian Siller, Bill West, Yu Chen
2:10	(1290-3)	**Novel FTIR/GC Detector for Analyzing Impurities in Gas Standards** MARTIN L SPARTZ, Prism Analytical Technologies, Inc., Charles Mark Phillips, Anthony S Bonanno, Peter P Behnke, Kelly R McPartland
2:30	(1290-4)	**Optimization of a Cavity Ring-Down Spectrometer for the Measurement of Trace Ammonia Contamination in Semicondutor Cleanroom Environments** GRAHAM A LEGGETT, Picarro, Mark Camenzind
2:50		**Recess**
3:05	(1290-5)	**Analytical Challenges of Measuring Impurities in Biogas** JANNEKE VAN WIJK, VSL, Annarita Baldan, Adriaan van der Veen, Stefan Persijn, Jeanrong Li
3:25	(1290-6)	**Analysis and Stability of Low Concentration HCl Standards** NATHALIE LUU , Air Liquide, Steve Hagen, Anthony Schleisman
3:45	(1290-7)	**Analytical QA/QC Measures to Validate Your Analysis Method Using FTIR** SYLVIE BOSCH-CHARPENAY, MKS Instruments
4:05	(1290-8)	**Open Discussion**

ORAL SESSIONS **Session 1300**

Biomedical: Nanotechnology - Half Session

Tuesday Afternoon, Room B301
Jagdish Singh, Mississippi State University, Presiding

1:30	(1300-1)	**Green Synthesis, Characterization of Saccharide Coated Gold Nanoparticles for Catalytic Applications** HARSH MOOLANI, Western Kentucky University, Jason N Payne, Rajalingam Dakshinamurthy
1:50	(1300-2)	**NanoCluster Beacons for Detection of a Single N6-Methyladenine Epigenetic Modification** TIM YEH, University of Texas at Austin
2:10	(1300-3)	**Self-assembly Approach to Integrated Nanozymes: Rational Design and Biomedical Applications** HUI WEI, Nanjing University
2:30	(1300-4)	**High Resolution Separation of Oligonucleotides and DNA Fragments Using a New Polymer-Based Reversed Phase Column** JULIA BAEK, Thermo Fisher Scientific, Shanhua Lin, Xiaodong Liu, Jessica Wang

ORAL SESSIONS **Session 1310**

Consumer Products Characterization - Half Session

Tuesday Afternoon, Room B408
William R LaCourse, University of Maryland Baltimore County, Presiding

1:30	(1310-1)	**Characterization of Metallic Nanoparticles in Tattoo Ink Using Asymmetrical Flow Field-Flow Fractionation Coupled with ICP-MS** SOHEYL TADJIKI, Postnova Analytics Inc., Evelin Moldenhauer, Tony Pfaffe, Trevor Havard, Thorsten Klein
1:50	(1310-2)	**Developing a Color Matching Database in Supercritical CO_2 for Waterless Textile Dyeing** ROLF SCHLAKE, Applied Separations, Susan Crowe, Madhu Anand, Rob Dorrycott
2:10	(1310-3)	**Detection and Quantification of Allergens in Personal Care Products by GC and GC×GC Paired with TOFMS** ELIZABETH M HUMSTON-FULMER, Leco Corporation, David E Alonso, Jonathan D Byer, Joseph E Binkley, Lorne E Fell
2:30	(1310-4)	**Carbohydrate-Rich Microalgae Characterization Using High-Performance Anion-Exchange Chromatography** PARUL ANGRISH, Thermo Fisher Scientific N/A

ORAL SESSIONS **Session 1320**

Detection of Illicit Drugs - Half Session

Tuesday Afternoon, Room B407
Nathaniel R Gomer, ChemImage Sensor Systems, Presiding

1:30	(1320-1)	**The Detection of Illicit Drugs and Cutting Agents Using Shortwave Infrared Hyperspectral Imaging** NATHANIEL R GOMER, ChemImage Sensor Systems, Jeffrey Beckstead, Oksana Olkhovyk, Matthew P Nelson
1:50	(1320-2)	**Characterization of Synthetic Phenethylamines Using High-Resolution GC-TOFMS and Mass Defect Filters** RUTH SMITH, Michigan State University, Alexandria Anstett, Fanny Chu, David E Alonso
2:10	(1320-3)	**The Development of a Novel Color Test for Improved Detection of Synthetic Cathinones** TSUNGHSUEH WU, University of Wisconsin-Platteville, Charles Cornett, Brook Tashner, Nicole Kloepfer
2:30	(1320-4)	**Potentiometric Sensor for Forensic Analysis: The Detection of the 'Undetectable Poison' Succinylcholine and Study of Its Enzymatic Degradation Kinetics** MOHAMED K ABD EL-RAHMAN, Cairo University, Amr M Mahmoud

ORAL SESSIONS **Session 1330**

Environmental and Instrumentation Application of LC/MS - Half Session

Tuesday Afternoon, Room B408
William R LaCourse, University of Maryland Baltimore County, Presiding

3:05	(1330-1)	**A New Type of Electron Ionization LC-MS and Its Applications** AVIV AMIRAV, Tel Aviv University, Boaz Seemann, Svetlana Tsizin, Alexander Fialkov, Tal Alon
3:25	(1330-2)	**Assessing the Occurrence and Fate of UV Filters in Seawater Swimming Pools by UPLC-Q-ToF-MS** TAREK MANASFI, Aix-Marseille University, Ravier Sylvain, Bruno Coulomb, Jean-Luc Boudenne
3:45	(1330-3)	**Expanded Analysis of Human Hormones in Drinking Water Using Solid Phase Extraction and Liquid Chromatography Tandem Mass Spectrometry** CARL FISHER, Thermo Fisher Scientific, Claudia Martins, Pranathi Perati
4:05	(1330-4)	**Multiclass Determination of New Psychoactive Substances in Municipal Wastewater** IVAN SENTA, Rudjer Boskovic Institute, Ivona Krizman, Marijan Ahel, Senka Terzic

ORAL SESSIONS **Session 1340**

Environmental Applications of Electrochemistry and Sensors - Half Session

Tuesday Afternoon, Room B409
Susan S Marine, Miami University Middletown, Presiding

1:30	(1340-1)	**Pretreatment and Spectroelectrochemical Sensing of Re(I)-Carbonyl Complexes** SHIRMIR D BRANCH, University of Cincinnati, Amanda D French, Amanda M Lines, Brian M Rapko, Sam A Bryan, William R Heineman
1:50	(1340-2)	**Wearable Gas Sensors: Shrinking Electrochemical Cells** JOHN R SAFFELL, Alphasense Ltd., Joseph R Stetter
2:10	(1340-3)	**Cloud Point Extraction for Electroanalysis: Anodic Stripping Voltammetry of Lead** CORY A RUSINEK, University of Cincinnati, William R Heineman, Adam Bange, Ian Papautsky, Mercedes Warren
2:30	(1340-4)	**Development of an Electrochemical Sensor for Detection of Dissolved Polycyclic Aromatic Hydrocarbons in Water** ABRA PENEZIC, Rudjer Boskovic Institute, Blazenka Gasparovic, Andrew Nelson, Drazenka Stipanicev

ORAL SESSIONS **Session 1350**

Food Product Quality and Component Characterization II

Tuesday Afternoon, Room B403
Alexander J Krynitsky, US FDA, Presiding

1:30	(1350-1)	**A Green Sample Preparation Device for Complex Biological, Environmental, Food, Pharmaceutical and Toxicological Samples** ABUZAR KABIR, Florida International University, Rodolfo Mesa, Rayma Blanko, Kenneth G Furton
1:50	(1350-2)	**Cannabinoids and Terpenes in Food** TIM ANDERSON, Phenomenex, Kristen Parnell, Ramkumar Dhandapani
2:10	(1350-3)	**Determination of Whey Adulteration in Milk Powder by Using Laser Induced Breakdown Spectroscopy** GONCA BILGE, Hacettepe University, Banu Sezer, Kemal E Eseller, Halil Berbero lu, Ali Topcu, Ismail H Boyaci
2:30	(1350-4)	**Bioactive Compounds and Antioxidant Activity of Guava Processed Byproducts and Wastes** NEELA EMANUEL, NIFTEM, Sao Khushbu, Aman Kaushik
2:50		**Recess**
3:05	(1350-5)	**Analysis of Non Volatile Congeners in Spirits by Creation of an Ion Fragmentation Database for Use with Time of Flight LC MS** RITA STEED, Agilent Technologies, Joni Stevens, Tarun Anumol, Gregory Hunlen, Luke Adam, Sue Dantonio
3:25	(1350-6)	**A Simple Field Test Kit for the Detection of Iodine in Food Grade (Table) Salt** RUFUS SHA'ATO, University of Agriculture, Nasiru L Usman
3.45	(1350-7)	**Chromatography Advancements in Nutraceuticals and Dietary Supplements Testing** ALLEN MISA, Phenomenex, Zeshan Aqeel
4:05	(1350-8)	**Raman Microspectroscopy and Chemometrics: A Combined Approach to Perform A Rapid Untargeted Screening of Bacteria Present in Food Samples** ALI ASSAF, University of Nantes, Emilie Grangé, Christophe Cordella, Douglas Rutledge, Michele Lees, Gerald Thouand N/A

ORAL SESSIONS **Session 1360**

Forensic Trace Analysis - Half Session

Tuesday Afternoon, Room B407
Nathaniel R Gomer, ChemImage Sensor Systems, Presiding

3:05	(1360-1)	**Analysis and Comparison of Fatty Acid Compositions in Latent and Smudged Fingermarks via Gas Chromatography-Mass Spectrometry (GC-MS) and Comprehensive Two-Dimensional Gas Chromatography-Time-of-Flight Mass Spectrometry (GCxGC-MS)** CAITLIN COBERN, Penn State, Frank Dorman, Seth Michalski
3:25	(1360-2)	**Mathematical Model to Predict Evaporation of Ignitable Liquids for Forensic Applications** VICTORIA L MCGUFFIN, Michigan State University, Rebecca Brehe, John W McIlroy, Ruth Smith
3:45	(1360-3)	**Analysis of Smoke Residues from Illicit Drugs as a Potential Source of Forensic Evidence** JULIE BITTER, National Institute of Standards and Technology, Matthew Staymates
4:05	(1360-4)	**LIBS Instrumentation for Fast Quantitative Analysis of Soil and Forensic Investigation of Nuclear Materials** ALAIN BLOUIN, National Research Council Canada, Paul Bouchard, Josette El Haddad, Aissa Harhira, Mohamad Sabsabi

ORAL SESSIONS **Session 1370**

Laser Induced Breakdown Spectroscopy (LIBS) and Glow Discharge in Atomic Spectroscopy - Half Session

Tuesday Afternoon, Room B301
Jagdish Singh, Mississippi State University, Presiding

3:05	(1370-1)	**Standoff LIBS Using a Spatial Heterodyne Spectrometer with Sub-Microsteradian Collection Optics** PATRICK D BARNETT, University of South Carolina, Nirmal Lamsal, S Michael Angel
3:25	(1370-2)	**Study of Matrix Effects for Reproducible LIBS Analysis of Powders** MATTHIEU BAUDELET, University of Central Florida, Sudeep Jung Pandey, Richard Locke, Brandon Seesahai, Romain Gaume, Martin Richardson
3:45	(1370-3)	**Particulate Identification Using Image Directed Laser Induced Breakdown Spectroscopy (LIBS) with Enhanced Spectral and Spatial Resolution for Pharmaceutical and Industrial Applications** MARK SULLIVAN, rap.ID Inc, Oliver Valet
4:05	(1370-4)	**Spectroscopy and Imaging Studies of a Solution-Cathode Glow Discharge** MICHAEL R WEBB, University of North Carolina Wilmington, Christian G Decker, Denise E Moon

ORAL SESSIONS **Session 1380**

LC/MS Biological Applications

Tuesday Afternoon, Room B404
Mary Ellen McNally, E.I. DuPont de Nemours and Company, Presiding

1:30	(1380-1)	**Distinguishing Isomeric Acylsugar Metabolites: Strategies for Labeling Using ^{13}C-amino Acid Precursors and LC-MS/MS** XIAOXIAO LIU, Michigan State University, Banibrata Ghosh, A Daniel Jones
1:50	(1380-2)	**Oxidative Techniques for Pteridine Bioanalysis and Implications for ESI-MS Applications** CASEY BURTON, Missouri University of Science and Technology, Honglan Shi, Yinfa Ma
2:10	(1380-3)	**Intact Histones Separation by Using Submicron Particles with RPLC-MS** XIMO ZHANG, Purdue University, Mary J Wirth
2:30	(1380-4)	**High Throughput Analysis of TCA Metabolites Using Column Switching and IC-HRAM Mass Spectroscopy** TERRI T CHRISTISON, Thermo Fisher Scientific, Junhua Wang, John E Madden, Monika Verma
2:50		**Recess**
3:05	(1380-5)	**Unified Drug Testing by Online SPE-LC/MS/MS with Focus on Productivity Achieved Through Ease of Use by Lab Technicians: One Totally Automated Method Measures ALL the Drugs in Urine and/or Oral Fluids** MARK HAYWARD, ITSP Solutions, Rick Youngblood, Kim Gamble, Martin Johnson, Matthew Hardison
3:25	(1380-6)	**High-Throughput Mass Spectrometric Analysis of Covalent Protein-Inhibitor Adducts for the Discovery of Irreversible Inhibitors: A Complete Workflow** IAIN CAMPUZANO, Amgen, Tisha San Miguel, Todd Rowe, Daniel Onea, Victor Cee, Tara Arvedson, John McCarter
3:45	(1380-7)	**Efficient Use of pH Control in Developing LC/UV/MS Methods** THOMAS E WHEAT, Waters Corporation, Amanda B Dlugasch, Patricia R McConville
4:05	(1380-8)	**Selectivity and Column Choices in HPLC Method Development** WILLIAM LONG, Agilent Technologies, Anne Mack, Stephen Luke, Jason Link

ORAL SESSIONS **Session 1390**

Polymer Characterization and Applications

Tuesday Afternoon, Room B405
Christopher Henry, Waters Corporation, Presiding

1:30	(1390-1)	**Transferring a GPC Method to A More Efficient SEC Method for Zoladex Co-Polymer Using an Advanced Polymer Chromatography Based U(H)PLC System Coupled with RI Detection** CHRISTOPHER HENRY, Waters Corporation, Jeanette Bowden, Mark Wrona, Richard Ladd, Andy Boughey
1:50	(1390-2)	**Characterization of Polyacrylamide at Different Ionic Strength and pH Conditions Using Asymmetrical Flow FFF and Multi-Angle Light Scattering Detector** SOHEYL TADJIKI, Postnova Analytics Inc., Trevor Havard, Japan Trivedi, Thorsten Klein
2:10	(1390-3)	**Patterning of Polycaprolactone-Impregnated Glass Microfiber Membranes: A Novel Approach to Fabrication of Microfluidic Devices** GAYAN C BANDARA, Oregon State University, Vincent T Remcho
2:30	(1390-4)	**Advanced Polymer Chromatography - Method Development Tools for SEC Analysis of PEG** MICHAEL OLEARY, Waters Corporation, Damian Morrison
2:50		**Recess**
3:05	(1390-5)	**Determination of Minor Component Differences and Additives in Polyethylene Using Thermal Desorption, Heart-Cutting EGA, Reactive Pyrolysis and GC/MS Techniques** TERRY RAMUS, Diablo Analytical, Dave Randle, Itsuko Iwai, Robert R Freeman
3:25	(1390-6)	**Applying Automatic Polymer Identification Capability to DSC Thermograms** BOB FIDLER, NETZSCH Instruments NA LLC, Ekkehard Post, Tobias Pflock, Stefan Schmoelzer, Gabriele Kaiser, Alexander Schindler
3:45	(1390-7)	**Characterization and Determination of Irganox 1076 and 1010 in Polyethylene Using Thermal Desorption and Reactive Pyrolysis – GC/MS** DAVE RANDLE, Frontier Lab USA, Itsuko Iwai, Terry Ramus, Robert R Freeman, Aki Hosaka, Ichi Watanabe
4:05	(1390-8)	**Simple Analysis Method of Degradation for The Molded Articles Polymer by FTIR and SEM** YOKO KATO, Advantest N/A

ORAL SESSIONS **Session 1400**

Sampling and Sample Preparation-Environmental and Food (ID, Safety and Contaminants)

Tuesday Afternoon, Room B406
Garry Lynch, Bechtel Marine Propulsion Corporation, Presiding

1:30	(1400-1)	**Sampling Condensed Hookah Smoke with C-18 SPME Fibers** AMBERLIE A CLUTTERBUCK, University of Cincinnati, Joseph Caruso, William Wetzel, Julio Landero N/A
1:50	(1400-2)	**Bacterial DNA Analysis Using Solid-Phase Microextraction by Different Polymeric Ionic Liquid-Based Sorbent Coatings** OMPRAKASH NACHAM, Iowa State University, Kevin D Clark, Matthew P Bommarito, Andrea E Tsatalis, Jared L Anderson
2:10	(1400-3)	**ICE Concentration Linked with Extractive Stir Bar (ICECLES): A Novel Sample Preparation Technique for Ultratrace Analysis** BRIAN LOGUE, South Dakota State University
2:30	(1400-4)	**Analysis of Pesticides in Foods by Direct Immersion SPME Using an Overcoated Fiber** LEONARD M SIDISKY, Supelco/Sigma-Aldrich, Katherine Stenerson, Robert Shirey, Yong Chen, Tyler Young
2:50		**Recess**
3:05	(1400-5)	**Recent Advances in Sample Preparation for Extraction and SPE** SM RAHMAT ULLAH, Thermo Fisher Scientific, Kannan Srinivasan, Mike McAdams, Glenn Kuse, Aaron Kettle
3:25	(1400-6)	**Development of a Carbon Mesh Supported Thin Film Microextraction Membrane as a Means to Lower the Detection Limits of Bench-top and Portable GC-MS Instrumentation** JONATHAN J GRANDY, University of Waterloo, Janusz Pawliszyn
3:45	(1400-7)	**Investigation of Polymeric Ionic Liquid Sorbent Coatings in Solid-Phase Microextraction Coupled to High-Performance Liquid Chromatography for the Analysis of Polar Compounds** HONGLIAN YU, Iowa State University, Josias Merib, Jared L Anderson
4:05	(1400-8)	**Justification of Kinetic Calibration in Pre-Equilibrium Solid Phase Microextraction with a Mathematical Model and Computational Simulation** MD NAZMUL ALAM, University of Waterloo, Fardin Ahmadi, Luis Ricardez-Sandoval, Janusz Pawliszyn

POSTER SESSION — Session 1410

All posters are to be mounted by 10:00 AM and remain on display until 4:00 PM. Authors must be at their posters from 1:00 PM to 3:00 PM. Location of the afternoon posters is on the Exposition Floor, 400 Aisle. PLEASE NOTE: You cannot get onto the Exposition Floor until after 9:00 AM.

Advances in Biomedical Applications

Tuesday Afternoon, Exposition Floor, 400 Aisle

(1410-1 P) **Strategies for Glycomics, Glycoproteomics, and Glycosaminoglycan Research at the Complex Carbohydrate Research Center** STEPHANIE A ARCHER-HARTMANN, Complex Carbohydrate Research Center, Roberto Sonon, Parastoo Azadi, Christian Heiss, Mayumi Ishihara

(1410-2 P) **Down Regulation of Smad-2 and VEGF Transcription and TGF-B1 Signaling in Nano Sized Titanium Dioxide-Induced Liver Injury in Mice by Potent Antioxidants** SAMY A ABDEL AZIM, Cairo University, Abd El-Moneim M Afify

(1410-3 P) **One Hydrothermal Processing of 1D Hydroxyapatite for Biomedical Application** ZORAN S STOJANOVIC, Institute of Technical Sciences of SASA, Nenad L Ignjatovic, Vuk D Uskokovic, Miroslav M Miljkovic, Vojka Zunic

(1410-4 P) **DNA Micelle Flares: Thermodynamic Stability and Cellular Internalization** YANYUE WANG, University of Florida, Weihong Tan

(1410-5 P) **Study of Various Cationic CPEs' Interaction with Mammalian Cells** SHANSHAN WANG, University of Florida, Zhiliang Li, Yun Huang, Kirk S Schanze

(1410-6 P) **Generation of T-cell Specific Aptamers Using a Novel Cell-Selex Method: Antibody Guided Cell-SELEX Technology** HASAN E ZUMRUT, City University of New York, The Graduate Center, Mallikaratchy Prabodhika, Shomi Chakrabarti, Mst Naznin Ara, George Maio

(1410-7 P) **Adeno-Associated Virus** YUAN WU, University of Florida

(1410-8 P) **Continuous Electroporation Through a Mesoporous Gold Membrane** JULIETTE EXPERTON, University of Florida, Aaron Wilson, Charles R Martin

(1410-9 P) **Analyzing the Effect of Beverages and Fluoride on Tooth Enamel with a FAAS and Dissolution** ANDREA GERCHMAN, St. John Fisher College, Kimberly Chichester

(1410-10 P) **Profiling Volatile Organic Compounds in Exhaled Breath by TD–GC–TOF MS** LAURA MCGREGOR, Markes International Ltd, Caroline Widdowson, Nicola Watson, Chris Hall, Ken Umbarger

(1410-11 P) **Purification of Pharmaceutical Proteins Including Antibody and Peptides Using Ion-Exchange Bulk Media Designed for High-Throughput Purification** TAKASHI SATO, YMC CO., Ltd., Munehiro Shoda, Chiaki Iwata, Noriko Shoji, Takatomo Takai

(1410-12 P) **An Analysis of the Protective Effects of Selenium on Porcine Jejunal Epithelial Cells Following Cadmium-induced Oxidative DNA Damage** SARAH JOANNE LYNCH, Dublin City University, Blánaid White, Dermot Walls, Karina Horgan

(1410-13 P) **Imaging Mass Spectrometry Reveals the Chemistry in Chemically Fixed Adrenal Cells Prepared for Transmission Electron Microscopy Analysis** JELENA LOVRIC, Chalmers University of Technology, Per Malmberg, Bengt R Johansson, John S Fletcher, Andrew Ewing

(1410-14 P) **Measurement of the Correlation Between Type of Mutation and Conditions that Alter Aging in Yeast** ANDREEA P MUSTEATA, Rensselaer Polytechnic Institute, Patrick Maxwell

(1410-15 P) **Self-Assembly Aptamer–Graphene Oxide Nanosheets as an Anticoagulant** PEI-XIN LAI, National Taiwan Ocean University

(1410-16 P) **Synthesis of Protein-Capped Gold Nanoparticles with Specific Protein Orientation as a DNA Transfection Vehicle** JU-YI MAO, National Taiwan Ocean University

(1410-17 P) **Recovering the Electrocatalytic Activity of Pt Nanoparticle-DNA Collisions via Nuclease Digestion for Sensing Applications** ALMA CASTANEDA, University of Texas at Austin, Donald Robinson, Richard M Crooks, Keith Stevenson

(1410-18 P) **Study of Stressed Monoclonal Antibody (mAb) Pharmaceuticals by Using Deep-UV Resonance Raman (DUVRR) Spectroscopy** CHEN QIU, US Food and Drug Administration, Sergey Arzhanstev, John Kauffman

(1410-19 P) **Paper Membrane Based SERS Platform for Rapid Bacteria Enumeration** UGUR TAMER, Gazi University, Aysen Gumustas, Bozkurt Akif, Merve Eryilmaz, Esra Acar, Demet Cetin, Zekiye Suludere, Ismail H Boyaci

(1410-20 P) **Microfluidics for the Detection of Minimal Residual Disease in Acute Myeloid Leukemia Patients Using Circulating Leukemic Cells Selected from Blood** JAMES TAYLOR, University of North Carolina at Chapel Hill, Joshua Jackson, Malgorzata Witek, Paul Armistead, Steven A Soper

POSTER SESSION — Session 1420

All posters are to be mounted by 10:00 AM and remain on display until 4:00 PM. Authors must be at their posters from 1:00 PM to 3:00 PM. Location of the afternoon posters is on the Exposition Floor, 400 Aisle. PLEASE NOTE: You cannot get onto the Exposition Floor until after 9:00 AM.

Advances in Metabolomics, Proteomics, Lipidomics

Tuesday Afternoon, Exposition Floor, 400 Aisle

(1420-1 P) **Universal Derivatization of Metabolites for Improved Sensitivity in Electrospray Ionization Mass Spectrometry** TIANJIAO HUANG, Saint Louis University, Maria Toro, James L Edwards

(1420-2 P) **Identification of a Biological Fear Cue in Blue Crab Urine Via 1H NMR-Based Metabolomics** REMINGTON X POULIN, Georgia Institute of Technology, Marc Weissburg, Julia Kubanek

(1420-3 P) **Analysis of Urine SRMs Using Solid Phase Micro Extraction, Dynamic Headspace and Liquid Injection with Comprehensive Two-Dimensional Gas Chromatography (GC×GC)-High Resolution Time-of-Flight Mass Spectrometry** DAVID E ALONSO, Leco Corporation, Joseph E Binkley, Elizabeth M Humston-Fulmer, Lorne E Fell, Jonathan D Byer

(1420-4 P) **Feasibility of Early Detection of Acute Pulmonary Exacerbations by Exhaled Breath Condensate Metabolomics** XIAOLING ZANG, Georgia Institute of Technology, Maria E Monge, Nael A McCarty, Arlene Stecenko, Facundo M Fernandez

(1420-5 P) **Prebiotic Peptidomics: An Ultra-Performance Liquid Chromatography-Ion Mobility-Tandem Mass Spectrometry (UPLC-IM-MS/MS) Workflow Applied to Origins-of-Life Chemistry** JAY G FORSYTHE, Georgia Institute of Technology, Sheng-Sheng Yu, Ramanarayanan Krishnamurthy, Martha A Grover, Nicholas V Hud, Facundo M Fernandez

(1420-6 P) **Development of a High Throughput Organelle Extraction Procedure from Rat Tissues** BRANDON EASPARRO, Omni International, Shari Garrett, James Atwood

(1420-7 P) **Serum Lipidomics Identifies Biomarkers of Acute Traumatic Brain Injury** SCOTT HOGAN, Georgia Institute of Technology, David A Gaul, Melissa A Velez, Michelle C LaPlaca, Facundo M Fernandez

(1420-8 P) **An Accelerated Protein Sample Preparation Method for LC-MS-Based Proteomics** SUJATHA CHILAKALA, Cleveland State University, Yan Xu

(1420-9 P) **High-Throughput Proteomics Analysis by LC-MS with AJS-CESI Technology** SUJATHA CHILAKALA, Cleveland State University, Yan Xu

(1420-10 P) **Response of NEIL1 to Oxidatively Damaged G-Quadruplexes via Affinity Purification-Mass Spectrometry** JONATHAN ASHBY, University of California, Davis, Brittany Anderson, Sheila David N/A

(1420-11 P) **Expression and Spectroscopic Characterization of Allene Oxide Synthase: A Cytochrome P450 for the Rearrangement of Small Molecule Hydroperoxides** JULIE C MCINTOSH, University of North Carolina at Chapel Hill, Nathan A Whitman, Matthew R Lockett

(1420-12 P) **Practical Considerations for Quantifying Protein Therapeutics in Biological Matrices by LC-MS/MS** ERIN E CHAMBERS, Waters Corporation, Mary Lame, Paula Orens, Hua Yang N/A

POSTER SESSION — Session 1430

All posters are to be mounted by 10:00 AM and remain on display until 4:00 PM. Authors must be at their posters from 1:00 PM to 3:00 PM. Location of the afternoon posters is on the Exposition Floor, 400 Aisle. PLEASE NOTE: You cannot get onto the Exposition Floor until after 9:00 AM.

Bioanalytical and Neurochemistry

Tuesday Afternoon, Exposition Floor, 400 Aisle

(1430-1 P) **Using Effective Conductivity to Study Brain Tissue Morphology** JENNA DEVIVO, University of Pittsburgh, Yangguang Ou, Erika Varner, Adrian C Michael, Stephen Weber

(1430-2 P) **Serotonin and Histamine Coregulation in the Mouse Premammilary Nucleus** RHIANNON ROBKE, University of South Carolina, Srimal A Samaranayake, Aya Abdalla, H Frederick Nijhout, Michael C Reed, Janet Best, Parastoo Hashemi

(1430-3 P) **Development of a Nanoscale Calcium-Selective Electrode** THERESA M RUWE, Northern Kentucky University, Edward A Dobrzykowski, Teri Rae Armstrong, Celeste A Morris

(1430-4 P) **Optimized Determination of L-Dopa at a Glassy Carbon Electrode Modified with Electrodeposited Films of Caffeic Acid** AHMAD ROHANI FAR, The University of Toledo, Amila M Devasurendra, Joshua A Young, Jon R Kirchhoff

(1430-5 P) **Mass Spectrometry Imaging for Targeted Metabolomics of Medulloblastoma** MARTIN R PAINE, Georgia Institute of Technology, Jingbo Liu, Danning Huang, Tobey MacDonald, Facundo M Fernandez

(1430-6 P) **Ultra-High Performance Liquid Chromatography Mass Spectrometry Metabolic Fingerprinting of a Medulloblastoma Mouse Model** DANNING HUANG, Georgia Institute of Technology, Martin R Paine, Jingbo Liu, Tobey MacDonald, Facundo M Fernandez

(1430-7 P) **Biophysical Evaluation of Surfactant Effects on Nanoparticle Toxicity a Lipid Model of the Blood-Brain Barrier** ADAM L HOFFMANN, Northern Kentucky University, Darcy Poor, Rolf Fowee, Marcus Jones, Andrew Hall, Kristi L Haik, Celeste A Morris

(1430-8 P) **Novel Graphene-Modified Graphite Pencil Electrode for the Trace Quantification of L-Tyrosine in Human Urine** ABDEL-NASSER KAWDE, King Fahd University of Petroleum and Minerals

(1430-9 P) **Online Liquid Chromatography - Surface Enhanced Raman Detection for Metabolic Profiling** ANH H NGUYEN, University of Notre Dame, Zachary D Schultz

(1430-10 P) **Combining Microchip Electrophoresis, Mass Spectrometry, and Standard Addition to Identify N-glycan Structures in Serum** XIAOMEI ZHOU, Indiana University, Christa Snyder, Margit I Campos, Milos V Novotny, Stephen C Jacobson

(1430-11 P) **Multichannel Impedance-Based Biosensing Using Virus-Polyethylenimine Films for Bladder Cancer Detection** ALANA OGATA, University of California Irvine, Reginald M Penner, Crystin Eggers N/A

(1430-12 P) **Development of a Wearable Device for Neurochemical Monitoring of Energy Availability in the Injured Brain Using On-Line Microdialysis** ISABELLE C SAMPER, Imperial College London, Sally A Gowers, Chu Wang, Martyn G Boutelle

(1430-13 P) **Label-Free Profiling of O-Linked Glycans by HPLC with Charged Aerosol Detection** IAN N ACWORTH, Thermo Fisher Scientific, David Thomas, Rainer Bauder, William Kopaciewicz

(1430-14 P) **Amperometric Quantitative Measurements of Individual Vesicular Transmitters in Single Adrenal Chromaffin Cells with Nano-Tip Conical Carbon Fiber Microelectrodes** SOODABEH MAJDI, Chalmers University of Technology N/A

(1430-15 P) **In Vitro Electrochemical Investigation of ATP: Catecholamine Interactions** ZAHRA TALEAT, Chalmers University, Johan Dunevall, Ricardo Borges, Judith Estevez, Andrew Ewing

(1430-16 P) **Quantifying the Progression of Amyotrophic Lateral Sclerosis** AIDAN P WICKHAM, Imperial College London, Christopher E Shaw, Kerry R Mills, James Bashford, Emmanuel M Drakakis, Martyn G Boutelle

POSTER SESSION **Session 1440**

All posters are to be mounted by 10:00 AM and remain on display until 4:00 PM. Authors must be at their posters from 1:00 PM to 3:00 PM. Location of the afternoon posters is on the Exposition Floor, 400 Aisle. PLEASE NOTE: You cannot get onto the Exposition Floor until after 9:00 AM.

Bioanalytical: Miscellaneous Analytical Techniques

Tuesday Afternoon, Exposition Floor, 400 Aisle

(1440-1 P) **An Analytic Approach to the Laser-Assisted Microscopy of Erythrocyte Deformability** FARAMARZ RAHNAMA, University of Calcutta, Azamat Vaseghi N/A

(1440-2 P) **Fabrication of Passive Microfluidic Diodes with Tunable Breakthrough Junctions** MARK D HOLTAN, Auburn University, Christopher J Easley

(1440-3 P) **Microfluidic Thermofluorimetric Analysis (µTFA) for Protein Quantification in Nanoliter and Picoliter Volumes** JUAN HU, Auburn University, Joonyul Kim, Mark D Holtan, Christopher J Easley

(1440-4 P) **Analysis of Fluorinated Lidocaine Derivatives Under Varying Conditions and Its Application in the Body** CYANN CICCONI, Seton Hill University

(1440-5 P) **Non-Enzymatic Modification of Human Serum Albumin: A Study Focusing On Advanced Glycation End Products by D-Galactose and D-Glucose** MENASHI A COHENFORD, Marshall University

(1440-6 P) **Study Interactions Between 21 Proteins and Nanoparticles** YAOKAI DUAN, University of California Riverside, Wenwan Zhong

(1440-7 P) **Development of Conductive Polymeric Ionic Liquid-Based Electrodes** AMILA M DEVASURENDRA, The University of Toledo, Joshua A Young, Ahmad Rohani Far, LM Viranga Tillekeratne, Jon R Kirchhoff

(1440-8 P) **Plasmonic Applications in the Mid-IR: Spectroscopic Surface Plasmon Resonance (SPR) Detection of N_2O Gas and Hexadecanethiol Self-Assembled Monolayer on a Low Loss, Plasmonic Tunable Novel Material Dy:CdO** HNIANG KHAMH, North Carolina State University

(1440-9 P) **Orientation of Membrane-Bound Cytochrome P450's in Nanodiscs** IVAN LENOV, University of Illinois, Stephen Sligar

(1440-10 P) **A Comparison of Binding Constants for FITC-Labeled Single Stranded DNA with Anti-FITC Antibody** QIAN LIU, Wake Forest University, Keith Bonin, Jason Gagliano, Kathryn Riley

(1440-11 P) **A Micro Pathogenic Microorganism Detector Applied for Mobile Phone Devices** LIANG SIJIA, Zhejiang University, Yu Dongdong, Zhou Jianguang

(1440-12 P) **The Preservation of DNA Using Magnetic Ionic Liquids** MATTHEW SORENSEN, Gustavus Adolphus College, Kevin D Clark, Omprakash Nacham, Jared L Anderson

(1440-13 P) **Automatic Data Analysis for Nanopore Detection** ZHEN GU, East China University of Science and Technology, Chan Cao, Yi-Lun Ying, Yi-Tao Long N/A

(1440-14 P) **Experiments and Modeling of Sequential Enzyme Activity in Biphasic Reaction Media** BRADLEY W DAVIS, Waynesburg University, William M Aumiller, Negar Hashemian, Antonios Armaou, Christine D Keating

(1440-15 P) **Fast Multipoint Immobilized MOF Bioreactor** HSI-YA HUANG, Chung Yuan Christian University, Wan-Ling Liu, Stephen Lirio, Chia-Her Lin

(1440-16 P) **HPLC Determination of the Rosmarinic Acid Content of Common Household Herbs** BRANDT A WOOD, California University of Pennsylvania N/A

(1440-17 P) **Molecular Modeling of Binding of Coumarins to DNA Gyrase** SARA MARIYAM ABDULLA, College of Chemical Sciences N/A

(1440-18 P) **Monitoring Cluster Ions Derived from Aptamer-Modified Gold Nanofilms under Laser Desorption/Ionization for the Detection of Circulating Tumor Cells** WEI-JANE CHIU, National Taiwan Ocean University, Chih-Ching Huang N/A

(1440-19 P) **Same Nucleotide Composition and High Sequence Similarity of MicroRNAs - An Analytical Challenge to RNA Studies?** JOSEPH N MWANGI, University of North Carolina at Greensboro, Norman Chiu

(1440-20 P) **Travelling Wave Ion Mobility Mass Spectrometry of Isomeric MicroRNA Biomarkers** NORMAN CHIU, University of North Carolina at Greensboro, Joseph Mwangi N/A

(1440-21 P) **Ultrafast Spectroscopic Studies of Molecular Interactions and Vibrational Energy Relaxation Dynamics in Binary Solvents** CHEN QIU, Michigan State University, Gary Blanchard N/A

(1440-22 P) **Surface-enhanced Raman Spectroscopy Based Total Protein Assay** MERVE ERYILMAZ, Gazi University, Adem Zengin, Ugur Tamer

(1440-23 P) **Ionic Liquid Crosslinkers for Chiral Imprinted NanoGUMBOS** SUZANA HAMDAN, Louisiana State University, Leonard Moore, Jason LeJeune, Farhana Hasan, Trevor K Carlisle, Jason E Bara, Douglas Gin, Andrew L LaFrate, Richard D Noble, David Spivak, Isiah M Warner

(1440-24 P) **Fabricated Multi-functional DNA Nanogel to Delivery Drug to Target Cells** XIGAO CHEN, University of Florida N/A

(1440-25 P) **Withdrawn**

(1440-26 P) **A Novel Bioassay Platform Using Silica Core-Stabilized Liposome Shell Microparticles for Ligand Discovery** KENDALL E SANDY, University of Arizona, Jinyan Wang, Mark T Agasid, Craig A Aspinwall

(1440-27 P) **Detection of Disease Associated MicroRNA Combinations with a Smart AND Sensor** LULU ZHANG, Oregon State University, Sean M Burrows

(1440-28 P) **Withdrawn**

(1440-29 P) **E-spun Collagen-CNT/Silk-CNT Composite Fibers to Transmit Electrical Signals for Cell Stimulation** NAIWEI CHI, Illinois Institute of Technology

(1440-30 P) **Sensitive Quantification of Protease Activity in Biological Samples by Using MCE-LIF** LI PAN, Jackson State University, Yiming Liu, Shulin Zhao N/A

POSTER SESSION — Session 1450

All posters are to be mounted by 10:00 AM and remain on display until 4:00 PM. Authors must be at their posters from 1:00 PM to 3:00 PM. Location of the afternoon posters is on the Exposition Floor, 400 Aisle. PLEASE NOTE: You cannot get onto the Exposition Floor until after 9:00 AM.

Bioanalytical: Separation Techniques

Tuesday Afternoon, Exposition Floor, 400 Aisle

(1450-1 P) **Surface Modified Nylon Capillary-Channeled Polymer (C-CP) Fibers for Protein Ion-Exchange Separations** LIUWEI JIANG, Clemson University, R Kenneth Marcus

(1450-2 P) **SEC Analysis of a Monoclonal Antibodies Using a Hybrid Silica Based Stationary Phase** JEFFREY KAKALEY, YMC America Inc., Ernest Sobkow N/A

(1450-3 P) **HPLC Method Transfer for Biopharmaceutical Analysis** BROOKE M KOSHEL, Waters Corporation, Sean M McCarthy

(1450-4 P) **Ratio of Different Fatty Acids Determined by GC-MS in Exosomes Purified Through Size Exclusion Chromatography** RUI XU, Jackson State University, Yiming Liu, Joseph Fernandes, Radhika Pochampally

(1450-5 P) **Analysis of Chromium Species in Dietary Supplements Using ICP-MS and Speciated Isotope Dilution Mass Spectrometry (SIDMS)** KAITLIN MILLER, Duquesne University, Logan T Miller, Jennifer Crawford, Stuart Procter, Matt Pamuku, Skip Kingston

(1450-6 P) **Metabolomic Signatures from Early Stage Ovarian Cancer Patients** DAVID A GAUL, Georgia Institute of Technology, Christina M Jones, Long Q Tran, John F McDonald, Facundo M Fernandez

(1450-7 P) **Reducing Adhesion of Proteins on Stainless Steel Components by the Application of a Carboxysilane Coating** LUKE PATTERSON, SilcoTek Corporation, Alfredo Narvaez, David Daghfal, Vaidya Shyam, Min Yuan, David Smith

(1450-8 P) **Optimizations of Proteomic Sample Preparation Method for *Xenopus Laevis* Embryonic Proteomics** ELIZABETH H PEUCHEN, University of Notre Dame, Liangliang Sun, Norman J Dovichi

(1450-9 P) **Determination of the Constituent Compounds in the Essential Oil from the Stem Bark of Ficus Capensis, A Multipurpose Phytomedicine, by GCMS and their Relevance to the Bioactivity of the Plant** MODUPE M OGUNLESI, University of Lagos, Christianah T Aleshinloye

(1450-10 P) **GC-MS Analysis of the Essential Oil from the Stem Bark of Tetrapleura Tetrapetra, a Multipurpose Medicinal Plant, and Bioactivities of some Constituent Compounds** MODUPE M OGUNLESI, University of Lagos, Christianah T Aleshinloye

(1450-11 P) Quantification of Trehalose and Other Sugars in Submergence Resistant Rice ELIZABETH N MARTINEZ, California State Polytechnic University, Pomona, Rejbana Alam, Julia Bailey-Serres, Endang M Septiningsih, Gregory A Barding

(1450-12 P) **Exploring SFC for the Separation of Peptides and Small Proteins** CECILIA MAZZA, AkzoNobel PPC AB, Joakim Högblom, Peter Gidlund

(1450-13 P) **Automated Solid Phase Extraction Method for the Assessment of Human Exposure to Polycyclic Aromatic Hydrocarbons Using the Biomarker Metabolite 1-Hydroxypyrene in Urine** MICHAEL JOE TANNER, J2 Scientific, Jeff Wiseman

(1450-14 P) **High Speed SDS–PAGE of Proteins** PARUL MODI, Thermo Fisher Scientific, Stephen Roemer

(1450-15 P) **Withdrawn**

(1450-16 P) **Optimized Wide Pore Superficially Porous Particles by One-Step Coating Process for Fast and Efficient Separation of Large Biomolecules** WU CHEN, Agilent Technologies, Anne Mack, Xiaoli Wang

(1450-17 P) **Fast Quantification of Immunoglobulin G Using A New Protein A Analytical HPLC Column** KOSUKE ARAKI, Tosoh, Satoshi Fujii, Shigeru Nakatani, Atis Chakrabarti

(1450-18 P) **Charging YOYO-1 on Capillary Wall for Online DNA Intercalation and Integrating This Approach with Multiplex PCR and Bare Narrow Capillary–Hydrodynamic Chromatography for Online DNA Analysis** HUANG CHEN, University of Oklahoma, Zaifang Zhu, Joann Lu, Shaorong Liu

(1450-19 P) **From Peptide Fractions to Pure, Dry Powders: Development of a Novel Automated Chromatographic Purification Process Supported by Solid-Phase Trapping** YAMAZAKI TOMOYUKI, Shimadzu Corporation, Okoba Tsutomu, Matsuo Eiichi, Masuda Junichi, Iwata Yosuke, Robert E Buco, Nishimura Masayuki

(1450-20 P) **Coupling Ion Exchange Chromatography with Reverse Phase Liquid Chromatography for High-Throughput Analysis of Intact Proteins** ZAIFANG ZHU, University of Oklahoma, Joann Lu, Huang Chen, Shaorong Liu

POSTER SESSION — Session 1460

All posters are to be mounted by 10:00 AM and remain on display until 4:00 PM. Authors must be at their posters from 1:00 PM to 3:00 PM. Location of the afternoon posters is on the Exposition Floor, 400 Aisle. PLEASE NOTE: You cannot get onto the Exposition Floor until after 9:00 AM.

High-Throughput Chemical Analysis

Tuesday Afternoon, Exposition Floor, 400 Aisle

(1460-1 P) **Withdrawn**

(1460-2 P) **Characterization and Use of a Microspectrophotometer for Quantitative Bio-Applications** THOMAS M SPUDICH, Maryville University, Bradley Postier, Ronald Mills

(1460-3 P) **California Chlor-Alkali Production Facility Monitors Organic Carbon for Increased Reliability and Equipment Protection** MARK MULLET, GE Analytical Instruments, Gary Erickson

(1460-4 P) **New Analysis Technology of Ultra-Trace Yellow Components in a Transparent Film** HOKO SUTO, Hitachi Chemical Co., Ltd., Kosuke Iwamoto, Akihiro Unno

(1460-5 P) **Leaning out Stage 1 Conductivity** JENNY G WATSON, GE Analytical Instruments

(1460-6 P) **Theoretical Simulation of a Helium DC Glow Discharge Used as an Ambient Desorption/Ionization Source for Mass Spectrometry** WADE C ELLIS, Brigham Young University, Paul B Farnsworth, Ross L Spencer

(1460-7 P) **Mathematical Modeling and Computational Simulation of Matrix Effect on Uptake Kinetics in Solid Phase Microextraction** MD NAZMUL AALM, University of Waterloo, Luis Ricardez-Sandoval, Janusz Pawliszyn

(1460-8 P) **Simple Imager for Multi-Well Plates** THAYUMANASAMY SOMASUNDARAM, Florida State University, Michael Zawrotny

POSTER SESSION — Session 1470

All posters are to be mounted by 10:00 AM and remain on display until 4:00 PM. Authors must be at their posters from 1:00 PM to 3:00 PM. Location of the afternoon posters is on the Exposition Floor, 400 Aisle. PLEASE NOTE: You cannot get onto the Exposition Floor until after 9:00 AM.

Surface and Microscopic Characterization of Nanostructures and Biological Materials

Tuesday Afternoon, Exposition Floor, 400 Aisle

(1470-1 P) **Characterizing Nanoparticle Size and Particle-Surface Interactions Using Nanophotonic Force Microscopy** DAKOTA O'DELL, Cornell University, Perry Schein, Summer Saraf, David Erickson

(1470-2 P) **Tuning Localized Surface Plasmon Resonance Wavelengths of Nanoparticles by Mechanical Deformation** FATHIMA S AMEER, Clemson University, Shilpa Varahagiri, Fenglin Wang, Hannah Mack, Marian Kennedy, Jeffrey N Anker

(1470-3 P) **Comparison of Color Pigment Removal between Graphitized Carbon Black and Zirconia-Based Adsorbents for QuEChERS Process** PATRICK MYERS, Supelco/Sigma-Aldrich, Katherine Stenerson, Tyler Young, Jennifer Claus, Michael Ye

(1470-4 P) **Potential-Dependent Adsorption of Water-Soluble Porphyrins at Liquid/Liquid Interfaces Studied by Polarization-Modulation Total Internal Reflection Fluorescence Spectroscopy** SHO YAMAMOTO, Kanazawa University, Hirohisa Nagatani, Hisanori Imura, Kotaro Morita

(1470-5 P) **The Concept of Lipobeads in the Context of Encapsulated Drug Delivery: Technological Challenges vs. Potential Advantages** SERGEY V KAZAKOV, Pace University

(1470-6 P) **Combination of Surface Plasmon Resonance - Surface Enhanced Raman Scattering Spectroscopy in the Kretschmann Configuration** JU-YOUNG KIM, University of Notre Dame, Zachary D Schultz

(1470-7 P) **Preparation and Characterization of Photo-Patterned Amorphous Carbon Films with Thiol-Click Reactions** CATHERINE G MCKENAS, University of North Carolina at Chapel Hill, Matthew R Lockett

(1470-8 P) **Electrophoretic Separation of Carbon Dots** KARINA M TIRADO-GONZÁLEZ, University at Buffalo, SUNY, Zuqin Xue, Luis A Colón

(1470-9 P) **Preparation and Separation of Highly Fluorescent Carbon Dots** ZUQIN XUE, University at Buffalo, SUNY, Luis A Colón, Karina M Tirado-González

(1470-10 P) **Microscopy in Analysis of Erythrocyte Shape Changes** VALIEV HAMMAT, Institute Applied Mechanics RAS, Karnet Yulia, Yumashev Oleg, Snegireva Nataliya N/A

(1470-11 P) **High Temperature In-Situ Reaction Monitoring of CdS Quantum Dots Using Spectrophotometers with Peltier Cell Holders** KYUNBAE LEE, Scinco R&D Center, In-Sung Kang, Kyung-Won Ro

(1470-12 P) **Determination of Airborne Concentration of Single-Wall Carbon Nanotubes and Metals by Wet Electrostatic Precipitation and Inductively Coupled Plasma Mass Spectroscopy** PETER ANDERSEN, Elemental Scientific, John Aumen, Matt Anderson, Grant Josh

(1470-13 P) **Impacts of Mesoporous Silica Shells on Reactivity of Metal-Semiconductor Hybrid Nanocatalysts** FEI ZHAO, Georgia State University, Bin Dong

(1470-14 P) **Investigating Chemical Reactivity of Nanoparticles Using Nano-Impact Electrochemistry** ANAHITA KARIMI, Clarkson University

(1470-15 P) **Advanced Analysis of LIB and Related Materials** KEIJI SUMIYA, Hitachi Chemical Co., Ltd., Hiroki Hirano

(1470-16 P) **Evaluation of Apples Browning Using a Camera-Imaging Visual Analyzer** ANDREW COWELL, Alpha MOS, Jean-Christophe Mifsud, Herve Lechat, Marion Bonnefille

(1470-17 P) **Ligand-Coated Zinc Oxide Nanoparticle Adsorption to Cellulose** STEPHEN R PRINTZ, Western Carolina University, Carmen L Huffman N/A

(1470-18 P) **Gemcitabine Conjugated Cysteine Modified Gold Nanoparticles: Improved Potency and Targeted Nanocarriers for Cancer Treatment** NIKUNJKUMAR NARAYANBHAI VALAND, Gujarat University, Manish B Patel, Kalpesh B Solanki, Shobhana K Menon N/A

(1470-19 P) **Investigation of Nitrogen-Doped Graphene Quantum Dots: Temperature Dependent Nitrogen Incorporations and their Effect on Optical Properties** TIMOTHY PILLAR-LITTLE, University of Kentucky, Doo Young Kim

(1470-20 P) **Industrial Characterization of Nano-scale Roughness on Polished Surfaces** NIKOLAJ A FEIDENHANS'L, Danish Fundamental Metrology, Poul-Erik Hansen, Lukas Pilny, Morten H Madsen, Giuliano Bissacco, Jan C Petersen, Rafael Taboryski

(1470-21 P) **High Throughput Fabrication of Nano and Micro Structured Polymer Foils by Roll-to-Roll-Extrusion Coating** SWATHI MURTHY, Danish Tchnical University/ Inmold, Maria Matschuk, Henrik Pranov, Henrik Pedersen, Guggi Kofod, Rafael Taboryski N/A

(1470-22 P) **Holographic Characterization of Particles in Complex Suspensions** DAVID B RUFFNER, Spheryx, Inc., Jaroslaw M Blusewicz, Priya Kasimbeg, David G Grier, Laura Philips

(1470-23 P) **Withdrawn**

(1470-24 P) **Impacts of Mesoporous Silica Shells on Reactivity of Metal-Semiconductor Hybrid Nanocatalysts** FEI ZHAO, Georgia State University, Bin Dong N/A

(1470-25 P) **Characterization of Carbon Nanomaterial Dispersions for Printed Electronics** QIHUA WU, Brewer Science, Carissa Jones, Kay Mangelson, Christopher Landorf, Stephen Gibbons

(1470-26 P) **Quantum Dots Embedded Silica Nanoparticles Functionalized for Interface Partitioning** SAFYAN AKRAM KHAN, King Fahd University of Petroleum and Minerals, Zain H Yamani, Mohammed H Aljabri N/A

WEDNESDAY, MARCH 9, 2016
MORNING

AWARDS **Session 1480**

Satinder Ahuja Award for Young Investigators in Separation Sciences

arranged by Neil D Danielson, Miami University of Ohio

Wednesday Morning, Room B312

Neil D Danielson, Miami University of Ohio, Presiding

8:30		**Introductory Remarks - Neil D Danielson**
8:35		**Presentation of the 2016 ACS Division of Analytical Chemistry Award for Young Investigators in Separation Sciences to Matthew D Miller, The Dow Chemical Company, by Neil D Danielson, Miami University of Ohio**
8:40	(1480-1)	**Fully Unlocking Polyolefin Chemical Composition Distributions: Break-through Separations Using Graphitic Carbon** MATTHEW D MILLER, The Dow Chemical Company, Rongjuan Cong, Willem deGroot, Chanda Klinker, Dean Lee, John W Lyons, David Mark Meunier, Abhishek Roy, Freddy Van Damme, Bill Winniford, Zhe Zhou
9:15	(1480-2)	**Demystifying Flow Modulated Comprehensive Two Dimensional Gas Chromatography (GC x GC) as a Practical Problem Solving Tool** BILL WINNIFORD, Dow Chemical, James Griffith, Anna Sandlin, Jim Luong, Chris Siegler, Kefu Sun
9:50	(1480-3)	**Separation of Building Blocks from Block Copolymers by High Performance Liquid Chromatography with Preloaded Adsorption Barriers** DAVID MARK MEUNIER, Dow Chemical Company, Yongfu Li, Tirtha Chatterjee, Todd O Pangburn, Eric Pearce, Mark Rickard, John W Lyons
10:25		**Recess**
10:40	(1480-4)	**Macromolecule and Nanoparticle Analyses: Beyond Molecular Weight and Size Measurements** S KIM R WILLIAMS, Colorado School of Mines
11:15	(1480-5)	**Uncertainties in Analyte Measurements in Dried Blood Spots** PURNENDU DASGUPTA, University of Texas at Arlington, Brian Stamos, Jordan Berg

SYMPOSIUM **Session 1490**

ACS-ANYL - Supported Bilayers in Bio/Chemical Analysis

arranged by Craig A Aspinwall, University of Arizona

Wednesday Morning, Room B308

Craig A Aspinwall, University of Arizona, Presiding

8:30		**Introductory Remarks - Craig A Aspinwall**
8:35	(1490-1)	**Spectroscopic Studies of the Formation, Structure, and Applications of Hybrid Supported Phospholipid Bilayers** JOEL M HARRIS, University of Utah, Jay P Kitt
9:10	(1490-2)	**Surface-Sensitive Imaging of Supported Membranes and Single Lipid Vesicles for Medical Applications** FREDRIK HÖÖK, Chalmers University of Technology
9:45	(1490-3)	**Mobile, Oriented Proteinaceous Supported Bilayers Made Directly from Cell Plasma Membranes for Bioanalytical Assays** SUSAN DANIEL, Cornell University
10:20		**Recess**
10:35	(1490-4)	**Stabilized Lipid Bilayers as a Platform for Fabrication of Ion Channel Functionalized Sensors** CRAIG A ASPINWALL, University of Arizona, S Scott Saavedra, Leonard K Bright, Mark T Agasid
11:10	(1490-5)	**Supported Lipid Bilayers and Scanning Ion Conductance Microscopy** LANE A BAKER, Indiana University

TECHNICAL PROGRAM

SYMPOSIUM — **Session 1500**

Analytical Applications of Terahertz Time Domain Spectroscopy (THz-TDS)
arranged by Mark A Arnold, University of Iowa

Wednesday Morning, Room B302
Mark A Arnold, University of Iowa, Presiding

8:30		**Introductory Remarks - Mark A Arnold**
8:35	(1500-1)	**Exploration of Interlayer Chemistry in Clay Minerals by Terahertz Spectroscopy** INGRID WILKE, Rensselaer Polytechnic Institute
9:10	(1500-2)	**Analytical Applications of Terahertz Spectroscopy in Nanotechnology and Biotechnology** ANIS RAHMAN, Applied Research & Photonics
9:45	(1500-3)	**PAT Measurements to Enhance Small-Molecule Drug Development and Manufacturing** HUIQUAN WU, Food and Drug Administration
10:20		**Recess**
10:35	(1500-4)	**Understanding Bonding in Cocrystals and Identifying Polymorphs in Small-Molecule Drugs** TIMOTHY MICHAEL KORTER, Syracuse University
11:10	(1500-5)	**Analytical Measurements and Dielectric Properties of Organic Cocrystals by Time-Domain Terahertz Spectroscopy** MARK A ARNOLD, University of Iowa

SYMPOSIUM — **Session 1510**

Frontiers of Plasmonics
arranged by Zachary D Schultz, University of Notre Dame and Jean-Francois Masson, Université de Montréal

Wednesday Morning, Room B303
Zachary D Schultz, University of Notre Dame, Presiding

8:30		**Introductory Remarks - Zachary D Schultz and Jean-Francois Masson**
8:35	(1510-1)	**Real-Time Tunable Emission from Plasmonic Nanolasers** TERI W ODOM, Northwestern University
9:10	(1510-2)	**Nanoplasmonic Sensors for Rapid Concentration and Sensitive Detection of Biomolecules** SANG-HYUN OH, University of Minnesota
9:45	(1510-3)	**Plasmonic Rectification** ZACHARY D SCHULTZ, University of Notre Dame
10:20		**Recess**
10:35	(1510-4)	**Trends and Challenges of Nanoplasmonic Biosensors for Clinical Use in Diagnostics** LAURA M LECHUGA, ICN2. CSIC & CIBER-BBN
11:10	(1510-5)	**Plasmonic Nanobiosensors: From Therapeutic Drug Monitoring to the Detection of Molecules Secreted by Cells** JEAN-FRANCOIS MASSON, Universite de Montreal

SYMPOSIUM — **Session 1520**

IAEAC: International Association of Environmental Analytical Chemistry - Upconverting Nanocrystals: Near Infrared Excitable Probes for Background-Free Luminescent Sensing
arranged by Antje J Baeumner, University of Regensburg

Wednesday Morning, Room B304
Antje J Baeumner, University of Regensburg, Presiding

8:30		**Introductory Remarks - Antje J Baeumner**
8:35	(1520-1)	**Optical Nanotransformers for In-Situ Upconversion: From Design to Functional Imaging and Sensing** PARAS N PRASAD, SUNY Buffalo, Guanying Chen, Tymish Y Ohulchanskyy
9:10	(1520-2)	**Photodynamic Therapy, Drug Delivery, Persistent and Photo-Stimulated Emission Using Low Excitation Photons** JOHN A CAPOBIANCO, Concordia University
9:45	(1520-3)	**Printing Enhanced Upconverting Nanocrystals on Solid Supports** STANLEY MAY, University of South Dakota, Aravind Baride, Jeevan Meruga, Mary Berry, Jon Kellar, William Cross, Grant Crawford
10:20		**Recess**
10:35	(1520-4)	**Surface Modifications of Upconverting Nanoparticles and Their (Bio)analytical Applications** THOMAS HIRSCH, University of Regensburg
11:10	(1520-5)	**Upconversion Nanoparticles as Active Elements in Optical Sensing for Development of Protein and Oligonucleotide Bioassays** ULRICH J KRULL, University of Toronto Mississauga, Samer Doughan, Anna Shahmuradyan, Feng Zhou, Yi Han

SYMPOSIUM — **Session 1530**

Ion Mobility/Mass Spectrometry for Metabolomics and Clinical Analysis
arranged by Richard Alan Yost, University of Florida

Wednesday Morning, Room B305
Richard Alan Yost, University of Florida, Presiding

8:30		**Introductory Remarks - Richard Alan Yost**
8:35	(1530-1)	**Structural Mass Spectrometry for Systems, Synthetic, and Chemical Biology** JOHN A MCLEAN, Vanderbilt University
9:10	(1530-2)	**Enhancing Ion Mobility-Mass Spectrometry Metabolomic Analyses with High Throughput Front End Separations** ERIN S BAKER, Pacific Northwest National Laboratory, Xing Zhang, Kristin E Burnum-Johnson, Jennifer E Kyle, Cameron P Casy, Young-Mo Kim, Erika M Zink, Dennis Mehinagic, Matthew E Monroe, Yehia M Ibrahim, Daniel Orton, Justin G Teeguarden, Thomas O Metz, Richard D Smith
9:45	(1530-3)	**Ion Mobility and Metabolomics, Two New Tools for Current Drug Discovery and Drug Development** ROB J VREEKEN, Janssen Pharmaceutica
10:20		**Recess**
10:35	(1530-4)	**Cystic Fibrosis Breathomics by Transmission-Mode Direct Analysis in Real Time-Traveling Wave Ion Mobility-Mass Spectrometry** FACUNDO M FERNANDEZ Georgia Institute of Technology, Facundo M Fernandez, Christina M Jones, Maria E Monge, Nael A McCarty, Arlene Stecenko, Jose J Perez
11:10	(1530-5)	**Ion Mobility/Mass Spectrometry for Metabolomics and Clinical Analysis: Progress and Prospects** RICHARD ALAN YOST, University of Florida, Christopher D Chouinard, Christopher R Beekman, Michael T Costanzo, Jared J Boock, Robin Kemperman, Wei S Michael, Timothy J Garrett, Christopher A Beecher

SYMPOSIUM — **Session 1540**

Precision Bioanalytical Measurements
arranged by Steven A Soper, St. Louis University and Susan M Lunte, University of Kansas

Wednesday Morning, Room B309
Steven A Soper, St. Louis University, Presiding

8:30		**Introductory Remarks - Steven A Soper and Susan M Lunte**
8:35	(1540-1)	**Microdialysis Sampling and Separations: A Tribute** SUSAN M LUNTE, University of Kansas
9:10	(1540-2)	**Expanding the Application Space of In Vivo Microdialysis Sampling in the Areas of Drug Metabolism, Free Radical Chemistry, Neurochemistry, and Tissue Engineering** JULIE STENKEN, University of Arkansas
9:45	(1540-3)	**The Development of New Tools Based on Whispering Gallery Mode Sensing** ROBERT C DUNN, University of Kansas, Sarah Wildgen, Daniel Kim, Judith Flores
10:20		**Recess**
10:35	(1540-4)	**Utilizing Oxidative DNA Damage to Explore the Mode of Action of Oxidative Events and Antioxidative Responses** BLÁNAID WHITE, Dublin City University, Dermot Walls, Sarah Joanne Lynch, Karina Horgan, Sinead Loughran, Roya Hakimjavadi
11:10	(1540-5)	**Precision Medicine: Enabled by Single Cell Analysis** STEVEN A SOPER, University of North Carolina

Author and presider lists as of January 20, 2016, are available at www.pittcon.org

SYMPOSIUM — Session 1550

Standoff Detection Methods for Security Applications
arranged by Sanford A Asher, University of Pittsburgh
and Michael Shepard, Dept. Homeland Security, S&T Directorate

Wednesday Morning, Room B310
Michael Shepard, Dept. Homeland Security, S&T Directorate, Presiding

8:30		**Introductory Remarks - Michael Shepard**
8:35	(1550-1)	**Novel Approaches to Standoff Hyperspectral Imaging Based Detection of Explosives and Other Threats** CHARLES W GARDNER, ChemImage Sensor Systems, Matthew P Nelson, Nathaniel R Gomer, Oksana P Klueva
9:10	(1550-2)	**Standoff Detection of Explosives Residues Using Integrated Quantum Cascade Laser Arrays** MARK F WITINSKI, Pendar Technologies, Romain Blanchard, Daryoosh Vakhshoori
9:45	(1550-3)	**Standards for Optical Based Standoff Detection Fabricated Using Inkjet Printing** GREG GILLEN, NIST
10:20		**Recess**
10:35	(1550-4)	**Recent Advances in Standoff Chemical Threat Detection Using Deep-Ultraviolet Raman Spectroscopy** ADAM J HOPKINS, Alakai Defense Systems, Kenneth R Pohl, Rob Waterbury, Edwin Dottery
11:10	(1550-5)	**Deep UV Standoff Raman Detection of Explosives: Fundamentals and Methodologies** SERGEI V BYKOV, University of Pittsburgh, Katie L Gares, Kyle T Hufziger, Sanford A Asher

SYMPOSIUM — Session 1560

Sum Frequency Generation (SFG) Vibrational Spectroscopic Studies on Proteins and Peptides at Interfaces
arranged by Zhan Chen, University of Michigan

Wednesday Morning, Room B311
Zhan Chen, University of Michigan, Presiding

8:30		**Introductory Remarks - Zhan Chen**
8:35	(1560-1)	**Determining the Structure of Surface Bound Proteins** DAVID G CASTNER, University of Washington
9:10	(1560-2)	**The Interactions of Ions with Peptides and Lipid Bilayers** PAUL CREMER, Penn State University
9:45	(1560-3)	**Proteins at Interfaces: Structure and Platform for Novel Biomaterials** MISCHA BONN, Max Planck Institute for Polymer Research, Tobias Weidner
10:20		**Recess**
10:35	(1560-4)	**Proteins at Interfaces Probed by Chiral Sum Frequency Generation** ELSA CY YAN, Yale University
11:10	(1560-5)	**Structure-Function Relationships of Surface Immobilized Peptides and Enzymes** ZHAN CHEN, University of Michigan

WORKSHOPS — Session 1570

Analytical Information Markup Language (AnIML) Data Standards
arranged by Burkhard Schaefer, BSSN Software GmbH

Wednesday Morning, Room B313
Burkhard Schaefer, BSSN Software GmbH, Presiding

8:30		**Introductory Remarks - Burkhard Schaefer**
8:35	(1570-1)	**Generating AnIML Technique Definitions** MARK FAULKNER BEAN, GSK
9:05	(1570-2)	**Filling the Automation and Enterprise Gap with Data and Device Standardization** CARMEN CONDRAU, SiLA Consortium, Devon Johnston
9:35	(1570-3)	**Instant JChem and AnIML: Curation, Searching, and Visualization of MS Library and Empirical Data** MICHAEL BRADEN, ChemAxon N/A
10:05		**Recess**
10:20	(1570-4)	**Use of AnIML and Other Methods for Software Visualization and Automation** DAVID M COX, SCIEX, John Gibbons
10:50	(1570-5)	**Cloud-Based Analytical Data Management Using the AnIML Standard** BURKHARD SCHAEFER, BSSN Software GmbH

ORGANIZED CONTRIBUTED SESSIONS — Session 1580

Food Contaminant Methods
arranged by Wendy Young, Food and Drug Administration

Wednesday Morning, Room B315
Wendy Young, Food and Drug Administration, Presiding

8:30	(1580-1)	**Quantitative HPLC-MS/MS Analysis of Metabolites of Hypoglycin A and Methylenecyclopropylglycine (MCPG) in Human Urine** SAMANTHA L ISENBERG, Centers for Disease Control and Prevention, Melissa D Carter, Leigh A Graham, Thomas P Mathews, Darryl Johnson, Jerry D Thomas, James Pirkle, Rudolph C Johnson
8:50	(1580-2)	**LC-MS/MS Detection of Tetrodotoxin in Fresh/Frozen and Salt-Dried Fish Products** SARA C MCGRATH, FDA/CFSAN, Jonathan R Deeds
9:10	(1580-3)	**Fast Liquid Chromatography - Tandem Mass Spectrometry Analysis of >150 Drug Residues including Aminoglycosides in Food Animal Tissues** STEVEN J LEHOTAY, USDA Agricultural Research Service, Alan R Lightfield
9:30	(1580-4)	**Determination of Diglycolic Acid in Foods Containing Carboxymethyl Cellulose** WENDY YOUNG, Food and Drug Administration, Lowri DeJager
9:50		**Recess**
10:05	(1580-5)	**Sulfite Determination in Food by Liquid Chromatography-Mass Spectrometry** KATHERINE S ROBBINS, US FDA, Shaun MacMahon, Lowri de Jager
10:25	(1580-6)	**Pesticide Screen Method Development Using Micro Flow LC and High Resolving Power Mass Spectrometry** MARK CROSSWHITE, Florida Department of Agriculture, Walter Hammack, Gerard Ghislain, Fadi Aldeek, Matthew Standland, Daniel Canzani N/A
10:45	(1580-7)	**Application of Raman Imaging for the Analysis of Food Packaging Stability** RIC GONZALEZ, ConAgra Foods, Inc., Indarpal Singh
11:05	(1580-8)	**Open Discussion**

ORGANIZED CONTRIBUTED SESSIONS — Session 1590

On-Site Detection of THC and Related Drugs
arranged by Maggie Tam, Canada Border Services Agency and Charles S Harden, US Army ECBC

Wednesday Morning, Room B316
Maggie Tam, Canada Border Services Agency, Presiding

8:30	(1590-1)	**Challenges of Drug Detection in the Field from the Enforcement Perspective** MAGGIE TAM, Canada Border Services Agency
8:50	(1590-2)	**Forensic Screening Using High-Performance Ion Mobility Spectrometry** CHING WU, Excellims Corporation, Mark Osgood, Anthony Midey
9:10	(1590-3)	**Handheld Differential Mobility Spectrometry for Drug Interdiction** PAUL J RAUCH, Chemring Detection Systems, Eric Lynch
9:30	(1590-4)	**Detection of Tetrahydrocannabinol and Related Compounds in Human Breath Using High-Field Asymmetric Waveform Ion Mobility Spectrometry** JARED J BOOCK, University of Florida, Michael T Costanzo, Richard Alan Yost
9:50		**Recess**
10:05	(1590-5)	**Detection of Illicit Drugs of Abuse Using Existing Military Chemical Detection Equipment** CHARLES S HARDEN, US Army ECBC
10:25	(1590-6)	**Innovative and Rapid Detection of Marihuana Consumption from Direct Breath Analysis** CHANDRASEKHARA HARIHARAN, ION-GAS GmbH, Oliver Kayser
10:45	(1590-7)	**Understanding Canine Detection of Explosives and Narcotics Using a 3D Printed Artificial Dog Nose** MATTHEW STAYMATES, NIST, William MacCrehan, Jessica Staymates, Greg Gillen, Brent Craven
11:05	(1590-8)	**Open Discussion**

TECHNICAL PROGRAM

ORGANIZED CONTRIBUTED SESSIONS — Session 1600

PAI-NET - Characterization of Micro/Nano Liquid Phases
arranged by Akihide Hibara, Tokyo Institute of Technology and Kenji Kojima, PAI-NET

Wednesday Morning, Room B301
Akihide Hibara, Tokyo Institute of Technology, Presiding

8:30	(1600-1)	**Development of Microfluidic Lattices for High-Performance Cell/Particle Separations** MASUMI YAMADA, Chiba University, Minoru Seki
8:50	(1600-2)	**Microfluidic-Based Approach for Producing Diffraction Quality Protein Crystals** MAEKI MASATOSHI, Hokkaido University, Pawate Ashtamurthy, Sugishima Masakazu, Watanabe Keiichi, Tokeshi Manabu, Kenis J Paul, Miyazaki Masaya
9:10	(1600-3)	**Sample Pretreatments for Droplet-Based Microanalytical System by Using Nanodroplet Formation** MAO FUKUYAMA, Kyoto Institute of Technology, Akihide Hibara, Yumi Yoshida, Kohji Maeda
9:30	(1600-4)	**Development of DNA and/or RNA Extraction Method from Single Cell** YUKIHIRO OKAMOTO, Osaka University
9:50		**Recess**
10:05	(1600-5)	**In Vivo Detection and Quantification of Circulating Cells and Nanoparticles in Whole Blood Using Photoacoustic-Fluorescence Flow Cytometry** DMITRY A NEDOSEKIN, University of Arkansas for Medical Sciences, Vladimir P Zharov
10:25	(1600-6)	**Microfluidic Construction of Artificial Cell-Like Reactors** MASAHIRO TAKINOUE, Tokyo Technology
10:45	(1600-7)	**Parallel Lipid Bilayer Formation by Microfabrication and Its Applications** RYUJI KAWANO, Tokyo University of Agriculture and Technology
11:05	(1600-8)	**Unique Liquid Properties by Surface Silanol Groups in Extended-Nano Spaces** YUTAKA KAZOE, The University of Tokyo, Keisuke Ikeda, Kazuma Mawatari, Takehiko Kitamori

ORAL SESSIONS — Session 1610

Biomedical: New Technologies for Breath Analysis (Half Session)

Wednesday Morning, Room B409

8:30	(1610-1)	**Heart-Cutting Multidimensional Micro Gas Chromatography (μGC) System for Breath Analysis** MENGLIAN ZHOU, University of Michigan, Jiwon Lee, Hongbo Zhu, Robert Nidetz, Kevin Ward, Xudong Fan, Katsuo Kurabayashi
8:50	(1610-2)	**Real-time PTR-TOF-MS Measurements Reveal Effects of Respiratory Maneuvers onto Exhaled Breath Biomarker Profiles** PRITAM SUKUL, University Medicine of Rostock, Peter Oertel, Phillip Trefz, Khushman Taunk, Jochen K Schubert, Wolfram Miekisch, Svend Kamysek
9:10	(1610-3)	**High Altitude Respiratory Research Using Quadrupole Mass Spectrometry** CHARLES W DE CARLO, Extrel CMS, James R Brenner, Zbigniew Krieger, Frank DeThomas
9:30	(1610-4)	**VOC Detection in Animal Models for Medical Research** WOLFGANG VAUTZ, ISAS, Liedtke Sascha, Nils Kunze, Thorsten Perl, Ursula Telgheder

ORAL SESSIONS — Session 1620

Clinical/Toxicology

Wednesday Morning, Room B402
Tyler Davis, West Virginia University, Presiding

8:30	(1620-1)	**Determination of Manganese Using Cathodic Stripping Voltammetry on a Platinum Electrode** WENJING KANG, University of Cincinnati, Cory A Rusinek, Adam Bange, Erin Haynes, William R Heineman, Ian Papautsky
8:50	(1620-2)	**Impact of Amphiphilic Poly(amido)amine Dendrimers on the Biophysical and Biorecognition Properties of Bilayer Membranes** SAMUEL S HINMAN, University of California, Riverside, Charles J Ruiz, Ling Peng, Quan Cheng
9:10	(1620-3)	**Analysis of Drugs in Saliva During Treatment of Post-Traumatic Stress Disorder Patients** KATHRYN DANA, Real-Time Analyzers, Inc, Chetan Shende, Stuart Farquharson, Albert J Arias
9:30	(1620-4)	**G-Quadruplex: A Biocompatible Additive for Enhancing the Antibacterial Activity of H_2O_2** YUQIAN XING, University of North Dakota, Xiao Liu, Minh H Duong, Julia Xiaojun Zhao
9:50		**Recess**
10:05	(1620-5)	**Capillary Electrophoresis Investigation of Multiwalled Carbon Nanotube-Biomolecule Binding** TYLER DAVIS, West Virginia University, Lisa A Holland, Julia Ann Mouch
10:25	(1620-6)	**Determining Variability in Potency of Marijuana for Potential Medical Application** REBECCA PLESSEL, Pennsylvania State University, Frank Dorman, Amanda Rigdon
10:45	(1620-7)	**Design and Development of a Portable Aptasensor for Toxicity Monitoring of Field Samples** GONCA BULBUL, Clarkson University, Akhtar Hayat, Silvana Andreescu
11:05	(1620-8)	**Quantitative Assessment of Nanoparticles Exposure Potential Effects in Embryonic Zebrafish** XIAOBO LIU, Clarkson University, Eduard Dumitrescu, Kenneth Wallace, Silvana Andreescu

ORAL SESSIONS — Session 1630

Electrochemistry - Biological Applications

Wednesday Morning, Room B403
Yinfa Ma, Missouri University of Science and Technology, Presiding

8:30	(1630-1)	**Development of an Optrode for In Vivo Neurochemical Studies** THOMAS FIELD, University of Kansas, Meng Sun, Peter M Ruggles, Michael A Johnson
8:50	(1630-2)	**Adenosine Transiently Modulates Vasodilation in Caudate-Putamen** YING WANG, University of Virginia, B Jill Venton
9:10	(1630-3)	**Amperometric and Voltammetric Measurements in the Cell Cytosol Using Conical Carbon Nanoelectrodes** EDWIN MITCHELL, North Carolina State University, James Roberts, Leslie A Sombers
9:30	(1630-4)	**Tunable Electroosmotic Push-Pull Perfusion Shows Higher Aminopeptidase N Activity in CA1 than CA3 of the Rat Hippocampus** YANGGUANG OU, University of Pittsburgh, Bocheng Yin, Jenna DeVivo, German Barrionuevo, Stephen Weber
9:50		**Recess**
10:05	(1630-5)	**Optogenetic Control of Octopamine Release in Drosophila Melanogaster Larval Ventral Nerve Cord and Detection with Fast Scan Cyclic Voltammetry (FSCV)** POOJAN PYAKUREL, University of Virginia, B Jill Venton
10:25	(1630-6)	**Clarifying the Complex Chemical Mechanisms that Underlie the Voltammetric Detection of Hydrogen Peroxide** LESLIE A SOMBERS, North Carolina State University, Samantha K Smith, Leyda Z Lugo-Morales, James Roberts, Maxim A Voinov, Tatyana I Smirnova
10:45	(1630-7)	**FSCV Measurements of Neurotransmitters in Daphnia** MATT N JACKSON, Wayne State University, Srimal A Samaranayake, David Pitts, Shawn McElmurry, Annette R Tremonti, Parastoo Hashemi
11:05	(1630-8)	**Glutamate Receptor Influence on Localized Oxygen Metabolism** LINDSAY WALTON, University of North Carolina at Chapel Hill, Nick Boustead, Susan Carroll, R Mark Wightman

ORAL SESSIONS **Session 1640**

Environmental, Pharm and Nano Methods Development in Atomic Spectroscopy

Wednesday Morning, Room B404
David Clarke, Teledyne CETAC Technologies, Presiding

8:30	(1640-1)	**Novel Automation Streamlining Microwave Digestion to Detection for Elemental Analysis** DAVID CLARKE, Teledyne CETAC Technologies, Matthew Nigro, James Block
8:50	(1640-2)	**Use of Atomic Fluorescence Spectrometry (AFS) for Element Specific Hg Detection Combined with Combustion** WARREN T CORNS, P S Analytical
9:10	(1640-3)	**Development of Online Method for Simultaneous Preconcentration of Cd, Cu, Ni and Zn in Environmental Samples Using Modified Alumina** ZAHEER A KHAN, SBBU, SBA
9:30	(1640-4)	**Dual Analyte Analysis of Bimetallic and Monometallic Nanoparticle Mixtures Using Field Flow Fractionation Separation Coupled to spICP-MS** CHADY STEPHAN, PerkinElmer, Ruth Merrifield
9:50		**Recess**
10:05	(1640-5)	**Analysis of Trace Metals in Pharmaceutical Formulations - Issues with USP 232 and Q3D Regulations** PHILIP RIBY, Liverpool John Moores University, Samar Thiab, Philip Salmon, Matt Roberts, Philip Riby N/A
10:25	(1640-6)	**Development and Validation of a New Method to Measure Activity of the Na+, K+ ATPase Using ICP-MS QQQ** CORY A STINER, University of Cincinnati, Julio Landero, Judith Heiny
10:45	(1640-7)	**Arsenic Speciation in Water Samples – Development of a New ISO/CEN Method** CORNELIUS BROMBACH, P S Analytical, Bin Chen, Warren T Corns, K C Thompson
11:05	(1640-8)	**Exploiting the Limits of Single Particle ICP-MS – From Particle Size to Particle Number** CHADY STEPHAN, PerkinElmer

ORAL SESSIONS **Session 1650**

Food Contaminants

Wednesday Morning, Room B314
Joan M Stevens, Agilent Technologies, Inc., Presiding

8:30	(1650-1)	**The Analysis of Chlorinated Dioxins and Difurans in Pet Food** PHILIP BASSIGNANI, Fluid Management Systems, Rudolf Addink
8:50	(1650-2)	**Selective Lipid Removal from Complex Samples for Multi-Residue Analysis** DERICK LUCAS, Agilent Technologies, Limian Zhao, Bruce Richter, Joan Stevens, Megan Juck
9:10	(1650-3)	**Arsenic Contamination and the Emergence of Speciation in the Food Chain** HELMUT ERNSTBERGER, Perkin Elmer, Kenneth Neubauer
9:30	(1650-4)	**Elemental Profile of Tobacco used in Counterfeit Cigarettes** YI HE, John Jay College/CUNY, Carrie Green, Rufus Chaney, Fidelis Tan, Ye Hua, Victoria Mei, Marin Kurti, Klaus von Lampe
9:50		**Recess**
10:05	(1650-5)	**Determination of Antimony in Food Samples by Slurry Sampling Hydride Generation Atomic Absorption Spectrometry** JERZY MIERZWA, Tennessee State University
10:25	(1650-6)	**The Analysis of Chlorinated Dioxins, Difurans and Polychlorinated Biphenyls in Edible Oils** PHILIP BASSIGNANI, Fluid Management Systems, Rudolf Addink
10:45	(1650-7)	**Application of Mid-Infrared Portable Spectrometry in Determination of Trans-fatty Acid Content in Bakery Products** MEI SHOTTS, The Ohio State University, Luis Rodriguez-Saona
11:05	(1650-8)	**Rapid Detection of Processed Uranium in Food** ABDUR-RAFAY SHAREEF, FDA, Lin Zhichao, Emanuele Kathryn, Stephanie Healey, Patrick Regan, William Cunningham, Brian Baker

ORAL SESSIONS **Session 1660**

LIMS – No One Size Fits All

Wednesday Morning, Room B405
Hubert MacDonald, The Pittsburgh Conference, Presiding

8:30	(1660-1)	**Quantum Time Savings with LIMS Deployment** DEVENDER GANDHI, Accelerated Technology Laboratories
8:50	(1660-2)	**Informatics for Externalization** GRAHAM A MCGIBBON, ACD/Labs, Pranas Japertas, Ryan Sasaki
9:10	(1660-3)	**Going Mobile with LIMS for Field Data Collection** KEN OCHI, Accelerated Technology Laboratories
9:30	(1660-4)	**Pay Now or Later: Creating Solid System Application User Requirements** KATHERINE TEMPLE, CSOLS, Inc., Daniel Freel
9:50		**Recess**
10:05	(1660-5)	**Key Factors to Consider in Transitioning to a New LIMS** SONJA STUTSMAN, Accelerated Technology Laboratories
10:25	(1660-6)	**LIMS Implementation: "Big Bang" or "Phased" Approach** HOWARD J ROSENBERG, CSols, Inc.
10:45	(1660-7)	**Migrating from a Legacy LIMS to a Modern Platform** LAURA LEE WILLIFORD, Accelerated Technology Laboratories, Devender Gandhi
11:05	(1660-8)	**Integrated Informatics: Single Vendor vs. "Best of Breed"** HOWARD J ROSENBERG, CSols, Inc.

ORAL SESSIONS **Session 1670**

Mass Spectrometry-Environmental, ICP-MS and Others

Wednesday Morning, Room B406
Yongbo Dan, Missouri University of Science and Technology, Presiding

8:30	(1670-1)	**Ultra-Trace Analysis of Mercury Species in Drinking Water Using Ion Chromatography and Speciated Isotope Dilution Mass Spectrometry (IC-SIDMS)** PATRICK BENECEWICZ, Duquesne University, Skip Kingston, Matt Pamuku, Stuart Procter, Christopher Loran
8:50	(1670-2)	**Ion Mobility-Mass Spectrometry Screening Reveals Small Molecules Capable of Chemical and Structural Modulation of Amyloidogenic Protein** RICHARD A KERR, University of Michigan, Jeffrey Derrick, Michael Beck, Hyuck Jin Lee, Mi Hee Lim, Brandon T Ruotolo
9:10	(1670-3)	**Transformation Kinetics of Metallic Nanoparticles in Environmental and Cell Culture Exposure Media as Measured by spICP-MS** CHADY STEPHAN, PerkinElmer, Ruth Merrifield
9:30	(1670-4)	**Reading Information Digitally-Encoded in Synthetic Polymers: A Sequencing Approach by Tandem Mass Spectrometry** LAURENCE CHARLES, Aix-Marseille University, Jean-François Lutz
9:50		**Recess**
10:05	(1670-5)	**Enzymatic Digestion-Single Particle ICP-MS Method to Characterize Nanoparticle Uptake by Plants** DAN YONGBO, Missouri University of Science and Technology, Weilan Zhang, Runmiao Xue, Ma Xingmao, Chady Stephan, Honglan Shi
10:25	(1670-6)	**Study of the Transmission of Megadalton-Sized Ions from Atmospheric Pressure to Vacuum in a Q-TOF Charge Detection Mass Spectrometer** STACI N ANTHONY, Indiana University, Benjamin E Draper, Martin F Jarrold
10:45	(1670-7)	**Characterization of TiO_2-Nanoparticles Using Asymmetrical Flow- and Centrifugal Field-Flow-Fractionation Coupled with MALS, DLS and ICP-MS Detection** SOHEYL TADJIKI, Postnova Analytics Inc., Florian Meier, Evelin Moldenhauer, Trevor Havard, Thorsten Klein
11:05	(1670-8)	**Enhanced Real-Time Gas Analysis with SIFT-MS Using Negative Reagent Ions** DANIEL MILLIGAN, Syft Technologies Ltd, Vaughan Langford, Thomas McKellar, Barry Prince, Murray McEwan

ORAL SESSIONS — Session 1680

Pharmaceutical Applications of Liquid Chromatography

Wednesday Morning, Room B401
Paulina Piotrowski, The Pennsylvania State University, Presiding

8:30	(1680-1)	**Method Development Strategies and Applications of 2D LC for Pharmaceutical Analysis** PANKAJ AGGARWAL, Pfizer Inc., David T Fortin, Angel R Diaz N/A
8:50	(1680-2)	**Coupling Mass Detection with UV to Improve Method Sensitivity for Esters of Benzenesulfonic and p-Toluenesulfonic Acids in Analysis of Genotoxic Impurities** MARGARET MAZIARZ, Waters Corporation, Christopher Henry, Mark Wrona
9:10	(1680-3)	**Method Development for Non-Chromophoric Pharmaceutical Analytes Using Alternative Chromatographic and Detection Techniques** ZONGYUN HUANG, Bristol-Myers Squibb, William Fish
9:30	(1680-4)	**The Impact of LC Instrument Characteristics on HPLC and UPLC Method Migration and Method Transfer** PAULA HONG, Waters Corporation, Patricia R McConville
9:50		**Recess**
10:05	(1680-5)	**Development and Validation of A Novel Stability-Indicating Reversed-Phase High-Performance Liquid Chromatography Method for Assay of Thiabendazole and Estimation of its Related Compounds** JIANGTAO HE, Merial, A Sanofi Company, Huang Junmin, Abu Rustum
10:25	(1680-6)	**Fast Centrifugal Partitioning Chromatography** ROB DRISCOLL, Robatel Inc.
10:45	(1680-7)	**Development and Validation of a Reversed Phase Chiral HPLC Method for Verification of Afoxolaner as a Racemic Mixture** NILUSHA PADIVITAGE, Merial, A Sanofi Company, Satish Kumar, Abu Rustum
11:05	(1680-8)	**Development for a Sensitive Method for the Determination of Diphenylphosphoryl Azide and Hydrogen Azide in Active Pharmaceutical Compounds** XUEJUN XU, Bristol-Myers Squibb, Martin Nunez, Yun K Ye, Thomas V Raglione

ORAL SESSIONS — Session 1690

Portable Instruments - Half Session

Wednesday Morning, Room B409
Robert W Baudoux, Sr., RWB Convention Mgt, Presiding

10:05	(1690-1)	**Development of Portable Instrumentation Using the Arduino Microcontroller Platform for Field-Ready Electrochemical Experimentation** DREW C FARRELL, University of Arizona, Michael L Heien
10:25	(1690-2)	**A Novel Instrument for Microscale IR Thermography in High Temperature Applied to Solar Salts** JUNKO MORIKAWA, Tokyo Institute of Technology, Massimiliano Zamengo, Yukitaka Kato
10:45	(1690-3)	**A New GC/FTIR Detection Method as Applied to Inline Monitoring of Siloxanes in Biogas** CHARLES MARK PHILLIPS, Prism Analytical Technologies, Inc., Martin L Spartz, Anthony S Bonanno, Peter P Behnke
11:05	(1690-4)	**Next-Generation Handheld XRF Analyzer – Smarter, Smaller and Faster** ESA NUMMI, Thermo Fisher Scientific

ORAL SESSIONS — Session 1700

Sensors - Bioanalytical and Homeland Security/Forensics

Wednesday Morning, Room B407
Katie Edwards, Cornell University, Presiding

8:30	(1700-1)	**Photonic Crystal Protein Hydrogel Sensor for Candida Albicans** ZHONGYU CAI, University of Pittsburgh, Daniel H Kwak, David Punihaole, Zhenmin Hong, Sachin S Velankar, Xinyu Liu, Sanford A Asher
8:50	(1700-2)	**Development of a Wireless Microfluidic Biosensor System for Real-Time Monitoring of Human Transplant Organs in Transit** SALLY A GOWERS, Imperial College London, Isabelle C Samper, Chu Wang, Thomas Watts, Bynvant Sandhu, Vassilios Papalois, Martyn G Boutelle
9:10	(1700-3)	**Ion-Selective Electrodes with PEDOT(PSS) as Solid Contact: Influence of the PEDOT(PSS) Thickness on the Equilibration Time** MARCIN GUZINSKI, The University of Memphis, Jennifer M Jarvis, Erno Lindner, Bradford D Pendley
9:30	(1700-4)	**Evaluating Real Time Binding Interactions in Insulin Immunoassay for Diagnosis of Type of Diabetes by Surface Plasmon Resonance** VINI SINGH, Oklahoma State University, Sadagopan Krishnan
9:50		**Recess**
10:05	(1700-5)	**Extended Nanopore Residence Times via Metallic Clusters** JOSEPH E REINER, Virginia Commonwealth University, Amy E Chavis, Kyle T Brady, Nuwan Kothalawala
10:25	(1700-6)	**Kinetics Quantification of MicroRNAs as Disease Biomarkers on Microelectrode Point-of-Care Sensors at Attomoles within Minutes** TANYU WANG, Georgia State University, Gangli Wang
10:45	(1700-7)	**Hyphenated Inverted Mesa E-QCM Sensors for Explosives Detection** ABDUL REHMAN, King Fahd University of Petroleum and Minerals N/A
11:05	(1700-8)	**A New Miniaturized Sensor for Real-Time Suit Penetration Assessment in the Man-In-Simulant-Test (MIST) Protocol** NICHOLAS FITZGERALD, Defence Science and Technology Group, Karl Pavey

ORAL SESSIONS — Session 1710

Thermal Analysis

Wednesday Morning, Room B408
Fu-mei Lin, The Pittsburgh Conference, Presiding

8:30	(1710-1)	**Using Isothermal Titration Calorimetry to Measure Thermodynamic Parameters of Adsorption on Chromatographic Media** ANTHONY R HORNER, University of Pittsburgh, Stephen R Groskreutz, Stephen Weber
8:50	(1710-2)	**Thermal Conductivity Measurement of Solar Salt in High Temperature Using the Temperature Wave Method** JUNKO MORIKAWA, Tokyo Institute of Technology, Yukitaka Kato, Massimiliano Zamengo
9:10	(1710-3)	**TGA-GC-MS Analysis of Different Tobacco Types** EKKEHARD POST, NETZSCH Geraetebau GmbH, Bob Fidler, Jan Hanss
9:30	(1710-4)	**Pyrolysis Gases of Polycarbonate Identified by TGA-FT-IR and TGA-GC-MS** EKKEHARD POST, NETZSCH Geraetebau GmbH, Bob Fidler
9:50		**Recess**
10:05	(1710-5)	**Understanding Auto-Catalysis Using Scanning, Isothermal and Adiabatic Calorimetry** PETER RALBOVSKY, Netzsch Instruments, Bob Fidler
10:25	(1710-6)	**Developing Databases and Optimized Spectral Searching from TGA-IR Hyphenation Experiments** IAN ROBERTSON, PerkinElmer Limited, Justin Lang, Jack Botting
10:45	(1710-7)	**Thermal Excitation, Optical Response: A Novel Approach to Thermal Analysis by TMOR** SARAH SCHWARZ G HENRIQUES, Anton Paar OptoTec GmbH, Tobias Husemann, Jens Kruse, Nils Bertram
11:05	(1710-8)	**Characterization of Nanomaterials with Thermal Analysis and Hyphenated Techniques** CHADY STEPHAN, PerkinElmer, Jun Wang

POSTER SESSION — Session 1720

All posters are to be mounted by 10:00 AM and remain on display until 4:00 PM. Authors must be at their posters from 10:00 AM to 12:00 PM. Location of the morning posters is on the Exposition Floor, 400 Aisle. PLEASE NOTE: You cannot get onto the Exposition Floor until after 9:00 AM.

Applications of LC/MS

Wednesday Morning, Exposition Floor, 400 Aisle

(1720-1 P) **Determination of Fentanyl in Canine Plasma Using HPLC-MS Detection** JOAN B BAILEY, University of Tennessee, Sherry K Cox, Molly White, Kristen Gordon, Reza Seddighi

(1720-2 P) **Targeted Metabonomic Study of Plasma from Rats with Acute Colitis Using LCMS-IT-TOF Based Metabonomics** LINGLING SHEN, Shimadzu Global COE, Shimadzu Co., Ltd, Jingting Yao, Taohong Huang, Xiaojun Zhang, Qisheng Zhong

(1720-3 P) **Optimization of On-column Trypsin Digestion Coupled with LC-MS/MS for Analysis of Apolipoproteins in Serum** CHRISTOPHER TOTH, Centers for Disease Control and Prevention, Zsuzsanna Kuklenyik, Jeffrey Jones, Bryan A Parks, Michael S Gardner, Jon Rees, Yulanda Williamson, David Schieltz, Lisa McWilliams, James Pirkle, John R Barr

(1720-4 P) **A Simple and Sensitive LC-MS/MS Method for the Determination of Free 8-Hydroxy-2'-deoxyguanosine in Human Urine** ZUWEI WANG, JES Tech, Scott M Smith

(1720-5 P) **Evaluation of Streamlined SPE Processing Using Novel Column based Components prior to LC-MS/MS** LEE WILLIAMS, Biotage GB Limited, Helen Lodder, Victor Vandell

(1720-6 P) **Multi-Class Screening of Drug Abuse in Hair by Matrix Solid Phase Dispersion – Ultrasound Extraction and HPLC-MS/MS Detection** ANTONIO MOREDA-PIÑEIRO, University of Santiago de Compostela, Mercedes Saavedra-Suarez, Juan Sanchez-Gonzalez, Pilar Bermejo-Barrera, Maria del Carmen Barciela-Alonso, Elena Pena-Vazquez, Raquel Dominguez-Gonzalez

(1720-7 P) **Optimization of QuEChERS Sample Preparation Method for the Determination of Bisphenol A in Carrots** OLUJIDE T AKINBO, Butler University, Hugh Kestufskie

(1720-8 P) **Quantification of Iodine-Containing Hormones Present in Dietary Supplements by Liquid Chromatography Tandem Mass Spectrometry (LC-MS/MS)** ENRIQUE G YANES, U.S. Food and Drug Administration (FDA), Robert A Wilson, James A Turner, Ryan Saadawi

(1720-9 P) **Ultra High Performance Liquid Chromatography Tandem Mass Spectrometry Method for Simultaneous Determination of Multiple Bioactive Constituents in Fruit Extracts of Myristica Fragrans and its Marketed Poly Herbal Formulations Using a Polarity Switch Technique** RENU PANDEY, Sophisticated Analytical Instrument Facility, CSIR, Brijesh Kumar N/A

(1720-10 P) **A Targeted Multidimensional Approach with MS Detection (SPE-RPLC/MS) for the Assessment of Trace Free Drug Species in Unadulterated Antibody-Drug Conjugate (ADC) Samples with Improved Specificity and Sensitivity** ROBERT BIRDSALL, Waters Corporation, Sean M McCarthy, Scott Berger, Weibin Chen, Alain Beck

(1720-11 P) **Method Development and Optimization for the Combined Analysis of Synthetic Cannabinoids and Designer Cathinones in Urine** HOLLY CASTELLANO, Duquesne University, Stephanie Wetzel N/A

(1720-12 P) **Separation and Identification of Lentil (Lens Culinaris) Proteins by Mass Spectrometry** ALBERTA ARYEE, Agriculture & Agri-Food Canada, Joyce I Boye N/A

(1720-13 P) **Withdrawn**

(1720-14 P) **Quantitation of Proto-Peptide Building Blocks in Complex Model Prebiotic Mixtures via Liquid Chromatography-Tandem Mass Spectrometry** ERIC T PARKER, Georgia Institute of Technology, Jeffrey L Bada, Facundo M Fernandez

POSTER SESSION — Session 1730

All posters are to be mounted by 10:00 AM and remain on display until 4:00 PM. Authors must be at their posters from 10:00 AM to 12:00 PM. Location of the morning posters is on the Exposition Floor, 400 Aisle. PLEASE NOTE: You cannot get onto the Exposition Floor until after 9:00 AM.

Capillary Electrophoresis

Wednesday Morning, Exposition Floor, 400 Aisle

(1730-1 P) **Capillary Electrophoresis Method to Detect Circulating Steroids in Individual Zebrafish Plasma** PAIGE A REED, West Virginia University, Vincent T Nyakubaya, William J Feeney, Lisa A Holland, Jennifer Ripley-Stueckle

(1730-2 P) **Fluorogenic Derivatization of Amino Acids for Laser-Induced Fluorescence Detection in Capillary Electrophoresis** NAVEEN MADDUKURI, Wichita State University, Qiyang Zhang, Maojun Gong

(1730-3 P) **Investigating Electrospray Behavior in Capillary Electrophoresis Coupled Mass Spectrometry** JARED A LAMP, University of Notre Dame

(1730-4 P) **Validated Capillary Zone Electrophoretic Method for Simultaneous Determination of Some Antihypertensive Drugs in Their Single-pill Combination Therapy** FAWZY A EL-YAZBI, Alexandria University, Hytham M Ahmed, Tarek S Belal, Rasha A Shaalan, Sohaila M Elonsy

(1730-5 P) **Single Enzyme Molecule Studies with CE-LIF and CE-MS** EMILY A AMENSON, University of Notre Dame, Norman J Dovichi

(1730-6 P) **Xenopus Laevis Metabolomic Profiling via CZE-ESI-MS/MS and MALDI-TOFMS** JENNIFER ARCEO, University of Notre Dame, Nicole Schiavone, Danielle Boley, Elizabeth H Peuchen, Norman J Dovichi

(1730-7 P) **Insights into Chiral Recognition Mechanisms for Acryloyl Terminated Polymeric Surfactants: Application of Linear Solvation Energy Relationship in Micellar Electrokinetic Chromatography and Capillary Electrochromatography** YANG LU, Georgia State University, Shahab S Shamsi

(1730-8 P) **Study of Electrooxidation Products of Primary Alcohols by EC-CE-C4D: Assessment of the Conversion Efficiency of Alcohols into Their Carboxylates on Gold and Platinum Electrodes in Different Media** MAURO SERGIO FERREIRA SANTOS, USP, Fernando S Lopes, Ivano G R Gutz

(1730-9 P) **Microscale Quantification of Nanoparticle-Biomolecule Interactions with Capillary Electrophoresis** JULIA ANN MOUCH, Bethany College, Tyler Davis, Lisa A Holland

(1730-10 P) **Comparative Validation of Sofosbuvir Determination in Pharmaceuticals by Several Chromatographic, Electrophoretic and Spectrophotometric Methods** AMIRA F EL-YAZBI, Alexandria University

(1730-11 P) **Quantification of Amino Acids in Cordyceps by MEKC** XIN WEI, Wuhan University Zhongnan Hospital, Hankun Hu, Yiming Liu, Yue Xu

(1730-12) Isolation of Intact Cell Populations by Preparative Capilary Zone Electrophoresis Sarh N. Lum, Bonnie Huge, Matthew Champion, Norman J. Dovichi

POSTER SESSION — Session 1740

All posters are to be mounted by 10:00 AM and remain on display until 4:00 PM. Authors must be at their posters from 10:00 AM to 12:00 PM. Location of the morning posters is on the Exposition Floor, 400 Aisle. PLEASE NOTE: You cannot get onto the Exposition Floor until after 9:00 AM.

Forensics and Homeland Security

Wednesday Morning, Exposition Floor, 400 Aisle

(1740-1 P) **Fingerprinting of Falsified Artemisinin Combination Therapies (ACTs) via DART Ionization Coupled to a Compact Single Quadrupole** Mass Spectrometer MATTHEW C BERNIER, Georgia Institute of Technology, Joseph LaPointe, Brian Musselman, Facundo M Fernandez

(1740-2 P) **Lead-Free Gunshot Residues as Forensic Evidence** CHRISTOPHER R DOCKERY, Kennesaw State University, Lashaundra Fambro, Ethan Miller, Deidre VanDenbos, Wassim Abdul Khalek

(1740-3 P) **Forensic Analysis of Textile Fibers Exposed to Cigarette Smoke Using Non-Destructive Room Temperature Excitation-Emission Fluorescence Microscopy** HUGH HAYES, University of Central Florida, Andres Campiglia, Arsenio Munoz de la Pena, Hector C Goicoechea N/A

(1740-4 P) **Forensic Analysis Of Textile Fibers Exposed To Laundry Detergents Using Fluorescence Excitation-Emission Spectroscopy** NIRVANI MUJUMDAR, University of Central Florida

(1740-5 P) **Eye-safe, Wide-Area Hyperspectral Raman Imaging Using a Spatial Heterodyne Raman Spectrometer** NATHANIEL R GOMER, ChemImage Sensor Systems, Matthew P Nelson

(1740-6 P) **Development and Optimization of Solid Phase Extraction (SPE) Method for Determination of Benzodiazepines in Wastewater and Surface Water by High-Performance Liquid Chromatography (HPLC)** HONGXIA GUAN, Georgia Gwinnett College, Qingsong Cai

(1740-7 P) **Determination and Measurement of Wildland Fire Markers in Residential Structures Using TD-GCMS** MARY C MARTIN, Prism Analytical and Central Michigan University, Alice Delia, Dale LeCaptain

(1740-8 P) **Differentiate Delta-9-tetrahydrocannabinol (Δ9-THC) and Delta-8-tetrahydrocannabinol (Δ8-THC)** KEN TSENG, Nacalai USA, Toshi Ono, Tsunehisa Hirose, Kazuhiro Kimata

(1740-9 P) **Identification and Quantification of Explosives and Their Residues in Water Using a Novel Surfactant in MEKC** CHRISTINE COPPER, U.S. Naval Academy, Marlene Perez, Alexis Clark, Karen Brensinger, Christopher Rollman, Jacqueline Rine, Ashton Genzman, Ira Lurie, Mehdi Moini

(1740-10 P) **Novel Concept of Biomarker Analysis in Forensic Analysis** JAN HALAMEK, SUNY Albany N/A

(1740-11 P) **Evaluation of GC-IRD Methods for Sensitivity and Selectivity in Forensic Drug Identification** RANDALL CLARK, Auburn University, Younis Abiedalla, Jack DeRuiter N/A

(1740-12 P) **Screening of Drugs of Abuse Using DART-MS and Real Time Reverse Library Search** FREDERICK LI, IonSense, Inc., Brian Musselman, Joe Tice, Stephen Shrader

(1740-13 P) **Sol-gel Sorbent Beds for All-in-One Sampling, Preconcentration, and Separation of Trace Explosive Vapors** MICHELLE CERRETA, U.S. Naval Research Laboratory, Braden Giordano, Kevin Johnson

(1740-14 P) **Techniques for Analyzing Volatile Organic Compounds Emitted During Aerobic Decomposition of Pig Carcasses and Swine Tissues** MASOUMEH DALILIAN, Middle Tennessee State University, Ngee Sing Chong, Samantha Keene, Lydia Rickman

(1740-15 P) **High Frequency, High Pressure Tandem Mass Spectrometry** ANDREW HAMPTON, University of North Carolina at Chapel Hill, J Michael Ramsey

(1740-16 P) **Gun Residue Analysis Using Paper Microfluidics** CHASTITY PAREDES-RODRIGUEZ, Penn State Berks, James Karlinsey

POSTER SESSION Session 1750

All posters are to be mounted by 10:00 AM and remain on display until 4:00 PM. Authors must be at their posters from 10:00 AM to 12:00 PM. Location of the morning posters is on the Exposition Floor, 400 Aisle. PLEASE NOTE: You cannot get onto the Exposition Floor until after 9:00 AM.

Liquid Chromatography

Wednesday Morning, Exposition Floor, 400 Aisle

(1750-1 P) **Determination of Pyrethrins in Pyrethrum Oil Extracts** IAN N ACWORTH, Thermo Fisher Scientific, David Thomas, Alan Wong, Jan Glinski

(1750-2 P) **Pesticide Residue Analysis of Cereal Grains Treated with Traditional Fungicides or a Tannic Acid Biopesticide** BROOKE M BIEN, Western Carolina University, Cynthia Atterholt

(1750-3 P) **Microwave-Assisted Extraction of Triketone and Pyrazole Corn Herbicides from Agricultural Soil** SANJA STIPIČEVI , Institute for Medical Research and Occupational Health, Milena Milaković, Marija Dvoršćak, Sanja Fingler, Gordana Mendaš

(1750-4 P) **Development and Validation of a Method for the Simultaneous Extraction and Separated Measurement of Diflubenzuron and Azamethiphos from the Soft Tissue of Mussel M. Chilensis** LUIS NORAMBUENA, Instituto de Fomento Pesquero, Sergio Contreras-Lynch

(1750-5 P) **Determination of Anions and Cations in Water Matrices Using Non-suppressed Ion Chromatography** JOSEPH PAUL ROMANO, Waters Corporation, Mark E Benvenuti, Kenneth Rosnack

(1750-6 P) **Application of Factorial Design for the Determination of Ammonium in Saline Waters by Ion Chromatography Using an Ultrasound-assisted Purge-and-Trap System** MANUEL C CARNEIRO, CETEM, Fernanda N Ferreira, Fernanda Veronesi M Pontes, Julio C Afonso, Arnaldo A Neto, Maria Ines Couto Monteiro N/A

(1750-7 P) **Determination of Nitrogen Compounds in Saline Waters by Matrix Interference-Free Ion Chromatography** MARIA INES COUTO MONTEIRO, CETEM, Fernanda N Ferreira, Fernanda Veronesi M Pontes, Julio C Afonso, Manuel C Carneiro, Arnaldo A Neto N/A

(1750-8 P) **Micellar and Sub-Micellar UHPLC of Aromatic Acid Geometric Isomers** JENNIFER M FASCIANO, Miami University, Neil D Danielson

(1750-9 P) **Hydrophilic Interaction Liquid Chromatography of Aromatic Acid Isomers on a Plain Silica Stationary Phase Using a Ternary Mobile Phase** ASHLEY E RICHARDSON, Miami University, Neil D Danielson

(1750-10 P) **Understanding the Importance of Instrument Design To Take Full Advantage of 1.0mm Internal Diameter (ID) Columns when Running UPLC** JENNIFER SIMEONE, Waters Corporation, Paula Hong, Patricia R McConville

(1750-11 P) **Investigation of Reverse-Phase - HILIC Continuous Analysis Using a One Column** RONALD BENSON, Shodex/Showa Denko America, Inc., Junji Sasuga, Tomokazu Umezawa, Hideyuki Kondo

(1750-12 P) **One Diode Array Detector for Analytical, Semi-Preparative, Preparative and Biocompatible HPLC** KATHRYN E MONKS, Knauer

(1750-13 P) **A Novel Diphenyl Stationary Phase for Metabolite Profiling** MARK WOODRUFF, Fortis Technologies, Ken Butchart

(1750-14 P) **Silica Hybrid Monoliths with a Carbonaceous Surface for Liquid Chromatography** AMARIS C BORGES-MUNOZ, University at Buffalo SUNY, Luis A Colón

(1750-15 P) **Evaluation of the Performance of Ultra-high Pressure Chromatography (UPLC) Systems** IMAD A HAIDAR AHMAD, Novartis, Frank Hrovat, Adrian Clarke, James Tam, Xue Li, Thomas Tarara, Andrei Blasko

(1750-16 P) **Optimizing Reagent Free Ion Chromatography; Electrolytic Water Purification** JOHN M RIVIELLO, Trovion Company, Archava Siriraks, Daniel Khor

(1750-17 P) **Functionalized Carbon Nanotubes as Pseudo-Stationary Phases in Capillary Electrokinetic Chromatography - Evaluation of Retention Energetics and Analysis of a Wide Range of Neutral and Charged Species** SARAH ALHARTHI, Oklahoma State University, Ziad El Rassi

(1750-18 P) **Determination of Hydroperoxides Using High Performance Liquid Chromatography with Reductive Electrochemical Detection** JUN CHENG, Thermo Fisher Scientific, Yan Liu, Khalil Divan

(1750-19 P) **Label-Free Analysis by HPLC with Charged Aerosol Detection of N-linked Glycans Separated by Charge** IAN N ACWORTH, Thermo Fisher Scientific, David Thomas, William Kopaciewicz

(1750-20 P) **Multi-Modal Analyte Detection of Cyclodextrin and Ketoprofen Inclusion Complex Using UV and CAD on an Integrated UHPLC System** BRUCE BAILEY, Thermo Fisher Scientific, Marc Plante, Ian N Acworth, Evert-Jan Sneekes, Frank Steiner

(1750-21 P) **RP-HPLC Method Development and Validation for the Analysis of Pharmaceutical Drugs-Linezolid** RAJESHKUMAR H CHAUDHARI, B/H Saikrishna Hospital, Vadilal G Patel N/A

(1750-22 P) **Purification of Chlorogenic Acid from Green Coffee Using Core-shell Technology in Axia Preparative Format** MARC JEAN JACOB, Phenomenex

(1750-23 P) **HILIC HPLC Separation of Oxymorphone for Assay and Purity** BRADLEY KUMAGAI, Theravance Biopharma

(1750-24 P) **Transfer of the Method for Related Substances Analysis of Metoclopramide HCl Between Different LC Systems** MARGARET MAZIARZ, Waters Corporation, Christopher Henry, Mark Wrona

(1750-25 P) **RP-HPLC Method Development and Validation for the Analysis of Pharmaceutical Drugs-60% Sodium Lactate** VADILAL G PATEL, B/H Rajpath Hotel, Rajeshkumar H Chaudhari N/A

(1750-26 P) **Polar Stationary Phases for Capillary Liquid Chromatography based on Metallic Oxides** CARLA GRAZIELI AZEVEDO DA SILVA, Unicamp, Carol H Collins, Carla B Bottoli

(1750-27 P) **Selection of Chromatographic Columns by Supercritical Fluid Chromatography** CARLA GRAZIELI AZEVEDO DA SILVA, IQ/UNICAMP, Carol H Collins, Isabel Cristina S Jardim

(1750-28 P) **Stability Evaluation of Core Shell C18 with Encapsulated Type End-Capping** NORIKAZU NAGAE, ChromaNik Technologies Inc., Tomoyasu Tsukamoto

(1750-29 P) **Development of an Interactive Counter Current Extraction Simulation** SEAN A REED, Westminster College, Erin Wilson, James Anthony

POSTER SESSION Session 1760

All posters are to be mounted by 10:00 AM and remain on display until 4:00 PM. Authors must be at their posters from 10:00 AM to 12:00 PM. Location of the morning posters is on the Exposition Floor, 400 Aisle. PLEASE NOTE: You cannot get onto the Exposition Floor until after 9:00 AM.

Pharmaceutical-IC, LC, and SFC

Wednesday Morning, Exposition Floor, 400 Aisle

(1760-1 P) **Determination of Aluminum in OTC Products** MANALI AGGRAWAL, Thermo Fisher Scientific, Jeffrey Rohrer

(1760-2 P) **Analysis of Antibiotics Sold in Pharmaceutical Market in Idumota, Lagos Using HPLC** SIXTUS I AMADI, Hydrochrom Resources Ltd

(1760-3 P) **A Sensitive Method for Direct Analysis of Impurities in Apramycin and Other Aminoglycoside-Antibiotics Using Hydrophilic Interaction Liquid Chromatography and Charged Aerosol Detection** ZHEN LONG, Thermo Fisher Scientific, Qi Zhang, Lina Liang, Yan Jin, Ian N Acworth, Evert-Jan Sneekes, Frank Steiner

(1760-4 P) **Chemically Stable Reversed Phase Chromatography Material Scalable from UHPLC to Semi Prep and Production** CECILIA MAZZA, AkzoNobel PPC AB, Fredrik Lime

(1760-5 P) **Characterization and Lot-to-Lot Variability of Complex Surfactants by High Performance Liquid Chromatography and Charged Aerosol Detection** MARC PLANTE, Thermo Fisher Scientific, Ian N Acworth, Bruce Bailey, Evert-Jan Sneekes, Frank Steiner

(1760-6 P) **Direct-Determination of Underivatized Carbohydrates in Biopharmaceutical Formulations Using Heart-Cut, 2D HPLC-HILIC and Charged Aerosol Detection** MARC PLANTE, Thermo Fisher Scientific, Bruce Bailey, Ian N Acworth, Evert-Jan Sneekes, Frank Steiner

(1760-7 P) **Fast Method Screening for Chromatographic Separation of Enantiomers Utilizing Polysaccharides Type Chiral Stationary Phases** TAKASHI SATO, YMC CO., Ltd., Noritaka Kuroda, Saoko Nozawa, Keiko Kihara, Noriko Shoji, Takatomo Takai

(1760-8 P) **Label-Free Measurement of Sialic Acids Released From Glycoproteins, by High Performance Liquid Chromatography and Charged Aerosol Detection** QI ZHANG, Thermo Fisher Scientific, Ian N Acworth, Bruce Bailey, Evert-Jan Sneekes, Frank Steiner

(1760-9 P) **Challenges in Developing Analytical Methods for Cleaning Validations in a GMP Regulated Environment** XIAOHUI YANG, Baxter Healthcare, Robert Garber, Walter Wasylenko, Lakshmy Nair, Jane Werling, George Monen, Beena Uchil

(1760-10 P) **SFC Analysis of Nutraceuticals and Pharmaceuticals Using SFC Optimized Stationary Phases** MATTHEW PRZYBYCIEL, ES Industries

(1760-11 P) **Analysis of Metoprolol and Select Impurities Using a Hydrophilic Interaction Chromatography Method with Combined UV and Charged Aerosol Detection** BRUCE BAILEY, Thermo Fisher Scientific, Ian N Acworth, Evert-Jan Sneekes, Frank Steiner

(1760-12 P) **Benefits of Using Mass Detection for Assessing Quality and Purity of Cetrimonium Bromide Pharmaceutical Raw Material** MARGARET MAZIARZ, Waters Corporation, Christopher Henry, Mark Wrona, Dominic Moore, Chengxia O'Shea

(1760-13 P) **Characterization of a Biologic Therapeutic: Reversed Phase/HILIC Analysis of Protein and Excipients** BRUCE BAILEY, Thermo Fisher Scientific, Marc Plante, Ian N Acworth, Evert-Jan Sneekes, Frank Steiner

(1760-14 P) **Efficient, Effective and Proven Approach to Chiral Method Development for Purification Scale Up** SEAN ORLOWICZ, Phenomenex, Michael McCoy

(1760-15 P) **Guidelines for Method Transfer and Optimization of the Newest Charged Aerosol Detector** MARC PLANTE, Thermo Fisher Scientific, Bruce Bailey, Ian N Acworth, Paul Gamache, Evert-Jan Sneekes, Frank Steiner

(1760-16 P) **Chiral Separation Using SFC and HPLC** HIDETOSHI TERADA, Shimadzu, Takato Uchikata, Funada Yasuhiro, Tanaka Kenichiro, Arao Yohei

(1760-17 P) **Determination of Ascorbic Acid in Citrus Fruits and Pharmaceutical Formulations by Hydrophilic Interaction Chromatography (HILIC)** YUEGANG ZUO, University of Massachusetts Dartmouth, Ruiting Zuo, Si Zhou, Yiwei Deng

(1760-18 P) **HPLC Method Development and Validation for the Assay and Organic Impurities of Naproxen in Naproxen API, Naproxen Delayed-Release Tablets and Naproxen Oral Suspension** JENNIFER FEDOROWSKI, United States Pharmacopeial Convention, Joshua Bhattacharya, Arindam Ganguly, Natalia Kouznetsova, Jennifer Belsky

(1760-19 P) **Sample Analysis of Compounds with Multiple Chiral Centers by Two-Dimensional HPLC** CHI TSANG, Genentech, Kelly Zhang

(1760-20 P) **Mobile Phase Effects in Reversed-Phase Chromatography of Monoclonal Antibodies at High Temperature** HILLEL K BRANDES, Sigma-Aldrich/Supelco

(1760-21 P) **Separation of Aminoglycoside Antibiotics by Using Hydrophilic Interaction Liquid Chromatography** YU LONG, Dalian Institute of Chemical Physics, Wei Jie, Shen Aijin, Guo Zhimou, Liang Xinmiao N/A

(1760-22 P) **Utilization of HPTLC and Diffuse Reflectance Spectroscopy to Quickly Evaluate Product Quality of Cotrimoxazole Tablets from Tanzania** DAVID WAYNE JENKINS, FHI 360, Eliangiringa Kaale, Samuel Hope, Thomas Layloff

(1760-23 P) **Amino Acid Analysis for Qualitative and Quantitative Composition of Pharmaceutical Products** NATALIA BELIKOVA, SGS Life Science Services

(1760-24 P) **Separation and Detection of Small Molecules by Low-Flow Liquid Chromatography Mass Spectrometry for Pharmacokinetic Studies** JAMES N MARR, Merck & Co., Rena Zhang, Gary Adamson

POSTER SESSION — Session 1770

All posters are to be mounted by 10:00 AM and remain on display until 4:00 PM. Authors must be at their posters from 10:00 AM to 12:00 PM. Location of the morning posters is on the Exposition Floor, 400 Aisle. PLEASE NOTE: You cannot get onto the Exposition Floor until after 9:00 AM.

Pharmaceutical-MS, LC/MS and Others

Wednesday Morning, Exposition Floor, 400 Aisle

(1770-1 P) **A New Approach to the Automated Identification of Metabolites in Multi-Vendor Datasets** RICHARD LEE, ACD/Labs, Vitaly Lashin, Andrey Paramonov, Alexandr Sakharov, Rytis Kubilius

(1770-2 P) **Analysis of Atropine Sulfate by HPLC Using Mass Spec HPLC Mobile Phase** JEFFREY KAKALEY, YMC America Inc., Ernest Sobkow

(1770-3 P) **Alternative Oxidation Technique for Pharmaceutical Forced Degradations** KEN NGIM, Theravance Biopharma US, Inc., Claudine Ooi, Bradley Kumagai

(1770-4 P) **Investigating Transdermal Diffusion of Vitamin D and 25-Hydroxyvitamin D** MARCEL MUSTEATA, Albany College of Pharmacy, Isaac Mall

(1770-5 P) **Calibration Free, Semi-Quantitative Analysis of Defined Drug Formulations Using FTIR Pre-computed Mixture Spectra** WILLIAM COSTA, Fiveash Data Management, Inc., Bill McCarthy, Todd Strother

(1770-6 P) **Novel Self-Patented Gold Nanoparticle Synthesis, Characterization and Antibacterial Susceptibility Testing** WILLIAM HAMILTON, Western Kentucky University, Jason N Payne, Fenil Chavda, Rajalingam Dakshinamurthy

(1770-7 P) **Unattended, Representative Sampling of a Wide Range of Chemical Reactions** VASO VLACHOOS, Mettler Toledo, Jane Riley N/A

(1770-8 P) **Square Wave Adsorptive Stripping Voltammetric Determination of Ketoconazole Drug in the Saudi Market** ABDEL-NASSER KAWDE, King Fahd University of Petroleum and Minerals, Mohamed A Morsy

(1770-9 P) **Capillary Electrophoresis: MEKC Assay Method for Simultaneous Determination of Olmesartan Medoxomil, Amlodipine Besylate and Hydrochlorothiazide in Tablets** MAHESH V ATTIMARAD, College of Clinical Pharmacy, Sree N Harsha, Bander E Al Dubaib, Anroop B Nair N/A

(1770-10 P) **A Comparison of Complete Dissolution Versus Leach of the Target Analytes by Using and Omitting HF** DAN IVERSEN, CEM, Tyler Edwards

(1770-11 P) **Standardization of Experimental Conditions of USP Melting Point Reference Standards in DSC Applications** OSOMWONKEN IGBINOSUN, United States Pharmacopeial Convention, Guillermo A Casay, Antonio Hernandez-Cardoso, Steven T Rau, Kanda K Balasubramanian

(1770-12 P) **Identification of Persistent Pd-containing Impurities Using LC-MS/MS and LC-ICP-MS** RENEE K DERMENJIAN, Merck N/A

(1770-13 P) **A Co-Crystal of Febuxostat and Isonicotinamide: Synthesis and Characterization** YANLEI KANG, Zhejiang University, Xiurong Hu, Jianming Gu, Dongdong Yu, Jianguang Zhou

(1770-14 P) **Real Time Stability of NSAIDs in Aqueous Solutions by Infrared Spectroscopy** ANUMEHA P MUTHAL, Seton Hall University, Vrushali Bhawtankar, Nicholas Snow

(1770-15 P) **Headspace Sampling of Residual Solvents Per USP 467 Using a Gas Tight Syringe** ANNE JUREK, EST Analytical, Lindsey Pyron, Kelly Cravenor, Justin Murphy

(1770-16 P) **Comparative Evaluation of Physicochemical Properties of Some Commercially Available Brands of Metformin Hydrochloride Tablets in Lagos, Nigeria** ADERONKE A ADEPOJU-BELLO, University of Lagos, Olawale S Bisiriyu N/A

(1770-17 P) **A Simple and Rapid LC-MS/MS Method for the Determination of BMCL26, a Novel Anti-Parasitic Agent, in Rat Plasma** RAMAKRISHNA REDDY VOGGU, Cleveland State University, Xiang Zhou, Bin Su, Baochuan Guo

(1770-18 P) **Derivative and Chemometric Spectrophotometric Methods for the Determination of Aripiprazole in Presence of its Related Impurities** SALLY TAREK, Cairo University N/A

(1770-19 P) **Drug-Herb Interaction: A Crossover Study of the Effect of a Polyherbal Formulation on Metroinidazole Pharmacokinetic Profile** GRACE EIGBIBHALU UKPO, University of Lagos, Teddy S Ehianeta, Steve O Ogbonnia, Idris O Balogun

(1770-20 P) **Supersaturation of Spray-Dried-Dispersion (SDDs) - Development and Evaluation of a Characterization Method** BENJAMIN H WU, Bristol-Myers Squibb

(1770-21 P) **Headspace Grade Solvents for Trace Level Analyte Detection** SUBHRA BHATTACHARYA, Thermo Fisher Scientific, Eric Oliver, Stephen Roemer

(1770-22 P) **Analytical Strategies in the Development of Generic Drug Products: Excipient Quantitation** ALEXANDER WILLIAM GARNER, Mayne Pharma

(1770-23 P) **Quantitative Laser Diffraction Method for the Assessment of Subvisible Protein Particles** ROBERT E BUCO, Shimadzu Corporation, Matthew Ferrarelli, Ariadna Martos, Andrea Hawe, Michael Wiggenhorn, Shinichiro Totoki, Haruo Shimaoka, Markus Ortlieb

(1770-24 P) **3D-Printed LTP Ionization Source for the Direct Analysis of Biomolecules** SANDRA MARTINEZ JARQUIN, Centro de Investigation Y de Estudios Avanzados Del IPN, Robert Winkler

(1770-25 P) **Demonstrating the Uptake Mechanism of Cisplatin in Cells by Single Cell ICP-MS** CHADY STEPHAN, PerkinElmer

(1770-26 P) **The Practice and Challenges of Ultrafast Chiral Separations in UHPLC and Super/Subcritical Fluid Chromatography (SFC)** CHANDAN L BARHATE, University of Texas at Arlington, M Farooq Wahab, Daniel W Armstrong

(1770-27 P) **Isolation of Pharmaceutical Impurities Utilizing MS-Directed Purification Platforms** ERIC STRECKFUSS, Merck N/A

(1770-28 P) **Determining Equivalency of Generic and Name Brand Oral Suspensions Using Zeta Potential** JACK G SAAD, Micromeritics, Myke Scoggins, Danielle Sowle

(1770-29 P) **Analysis of the Photodegradation Products of Nifedipine in Photoprotective Topical Formulations by HPLC and Mass Spectrometry** ELLEN K WASAN, University of Saskatchewan, Joshua Poteet, Ed Krol, Jacqueline Cawthray, Anas El-Aneed, Kevin Soulsbury, Munawar Mohammed N/A

POSTER SESSION — Session 1780

All posters are to be mounted by 10:00 AM and remain on display until 4:00 PM. Authors must be at their posters from 10:00 AM to 12:00 PM. Location of the morning posters is on the Exposition Floor, 400 Aisle. PLEASE NOTE: You cannot get onto the Exposition Floor until after 9:00 AM.

Practical Chromatography in Today's Laboratory

Wednesday Morning, Exposition Floor, 400 Aisle

(1780-1 P) **A Rapid Isocratic LC Separation of Soft Drink Additives Using an Environmentally Friendly Mobile Phase with UV Detection** MARK E BENVENUTI, Waters Corporation, Joseph Paul Romano

(1780-2 P) **High Efficiency Chiral Separations Using Sub-2 µm Monodisperse Chiral Stationary Phases** ZACHARY BREITBACH, The University of Texas at Arlington, Chandan L Barhate, M Farooq Wahab, David S Bell, Daniel W Armstrong

(1780-3 P) **Advances in Chiral HPLC Column Technology: Superficially Porous Particle Based Chiral Stationary Phases** ZACHARY BREITBACH, The University of Texas at Arlington, M Farooq Wahab, Daniel W Armstrong

(1780-4 P) **A New Standard in Analytical Workflow Design** WILLIAM HEDGEPETH, Shimadzu Scientific Instruments, Kenichiro Tanaka

(1780-5 P) **Conductivity and pH Monitoring as an Effective Tool for Multi-Step Biological Separations** LUKE ROENNEBURG, Gilson, Inc., Laine Stewart, Karen Kleman N/A

(1780-6 P) **Centrifugal Partition Chromatography: A Preparative Tool for Isolation and Purification of Xylindein from Chlorociboria Aeruginosa** ANUKUL BOONLOED, Oregon State University, Genevieve L Weber, Sumate Pengpumkiat, Vincent T Remcho

(1780-7 P) **Analysis of Petroleum Products Using Comprehensive Two-Dimensional Gas Chromatography (GC×GC) with Both Time of Flight MS and Flame Ionization Detectors** JOSEPH E BINKLEY, Leco Corporation, Christina N Kelly, Jonathan D Byer, David E Alonso, Lorne E Fell

(1780-8 P) **Evaluation of Polycyclic Aromatic Hydrocarbon Standard Reference Materials 869b and 1647f on Different Stationary Phases for Liquid Chromatography** WALTER B WILSON, National Institute of Standards and Technology, Jorge O Oña-Ruales, Lane C Sander, Stephen A Wise

(1780-9 P) **Liquid Chromatographic Retention Behavior of Polycyclic Aromatic Sulfur Heterocycles and Their Alkyl-Substituted Derivatives** WALTER B WILSON, National Institute of Standard and Technology, Lane C Sander, Stephen A Wise

(1780-10 P) **Evaluation of 25-Hydroxy Vitamin D Extraction Using Phospholipid Depletion Plate Technology and Method Comparison Using Automated Sample Preparation** KERRY CHALLENGER, Biotage GB Limited, Lee Williams, Helen Lodder, Victor Vandell

(1780-11 P) **Effects of Ultracentrifugation on HDL and LDL Size Distribution** JEFFREY JONES, Centers for Disease Control and Prevention, Zsuzsanna Kuklenyik, Christopher Toth, Bryan A Parks, Michael S Gardner, Jon Rees, Yulanda Williamson, David Schieltz, Lisa McWilliams, John R Barr, James Pirkle

(1780-12 P) **Biomolecular Separations through Tunable Nanoporous Gold Membranes** DANIEL A MCCURRY, University of Illinois at Urbana-Champaign, Ryan C Bailey

(1780-13 P) **Non-Contact Pd Separation based on Laser-Induced Particle Formation for Determination of ^{107}Pd with ICP-MS** TAKUMI YOMOGIDA, Japan Atomic Energy Agency, Shiho Asai, Morihisa Saeki, Yukiko Hanzawa, Fumitaka Esaka, Hironori Ohba, Yoshihiro Kitatsuji

(1780-14 P) **Surfactant-Pluronic Gel Phases for Electrophoresis** ASHLEY E RICHARDSON, Miami University, Elise M Leonard, Neil D Danielson

(1780-15 P) **Insights into the Effect of the PDMS-layer on the Kinetics and Thermodynamics of Analytes Sorption onto the PDMS-overcoated Coating** ERICA A SOUZA-SILVA, Universidade Federal do Rio Grande do Sul, Emanuela Gionfriddo, Janusz Pawliszyn

(1780-16 P) **A New Anion Exchange Column for Fast Ion Chromatographic Separation of Monosaccharides and Disaccharides in Biofuel, Food, and Beverage Samples** YAN LIU, Thermo Fisher Scientific, Andy Woodruff, Charanjit Saini, Yury Agroskin, Christopher Pohl

(1780-17 P) **Column Robustness Challenges during HPLC Method Development** JUN WANG, Takeda Boston, Laila Kott, Elizabeth Hewitt, Scott Zugel

(1780-18 P) **There are Problems Associated with Gradient and Method Transfer in HPLC and UHPLC – Are There Explanations and Usable Workarounds? Part 1 of 2** MICHAEL WOODMAN, Agilent Technologies, Gregory Hunlen

(1780-19 P) **Understanding the Causes and Minimizing the Impact of LC Carryover in Most LC Systems** MICHAEL WOODMAN, Agilent Technologies, Gregory Hunlen

SEAC POSTER SESSION — Session 1790

All posters are to be mounted by 10:00 AM and remain on display until 4:00 PM. Authors must be at their posters from 10:00 AM to 12:00 PM. Location of the morning posters is on the Exposition Floor, 400 Aisle. PLEASE NOTE: You cannot get onto the Exposition Floor until after 9:00 AM.

SEAC Poster Session

Wednesday Morning, Exposition Floor, 400 Aisle

(1790-1 P) **New Portable Electrochemical Instrument for *In Situ* Analysis** PABLO FANJUL-BOLADO, DropSens S.L, David Hernández-Santos, Laura Fernández-Llano, Marta Neves, Pablo Bobes-Limenes, Alejandro Pérez-Junquera, Begoña González-García, Carla Navarro-Hernández

(1790-2 P) **Aqueous and Non-Aqueous Electrochemical Activities of Mercaptosuccinic Acid Stabilized Au11-13 Clusters** JONATHAN PADELFORD, Georgia State University, Gangli Wang

(1790-3 P) **Copolymerized Triazole Based Ionic Liquid as New Sensing Material in Ion-Selective Sensors** LUKASZ K MENDECKI, Keele University

(1790-4 P) **Investigation of Cloud-Point Extraction and UV-Vis for Determining Copper and Cadmium in Vegetables** ASADUZZAMAN NUR, Tennessee Technological University, Andrew Callender

(1790-5 P) **Potentiometric Ion-Selective Electrodes based on Metastable Photoacid for Cation Detection** PARTH K PATEL, University of Central Florida, Karin Y Chumbimuni-Torres

(1790-6 P) **Signal Amplification of a Highly Selective Universal MicroRNA Electrochemical Sensor for Single Nucleotide Polymorphism Detection** DAWN MILLS, University of Central Florida, Jeffer Pinzon, Percy Calvo-Marzal, Dmitry M Kolpashchikov, Karin Y Chumbimuni-Torres

(1790-7 P) **Self-Reference Single Strip Paper Based Sensors for Ion Detection** ANDREW J MANHAN, University of Central Florida, Stephanie Armas, Karin Y Chumbimuni-Torres

(1790-8 P) **Development and Characterization of an Ion Selective Microsensor for the Detection and Monitoring of Zinc Levels in Citrus Plants** COURTNEY HULCE, University of Central Florida, Jared Church, Swadeshmukul Santra, Woo Hyoung Lee, Karin Y Chumbimuni-Torres

(1790-9 P) **Poly(3-octylthiophene)-Based Solid Contact Ion-Selective Electrodes with Improved Potential Stability** JENNIFER M JARVIS, University of Memphis, Marcin Guzinski, Bradford D Pendley, Erno Lindner

(1790-10 P) **Urine Carbon Dioxide in Septic Shock** JAMES G ATHERTON, University of Memphis, Artur Jasinski, Erno Lindner, Bradford D Pendley, Marcin Guzinski, William King

(1790-11 P) **Disposable Paper-Based Electrochemical Ion-Sensing Platform** JINBO HU, University of Minnesota, Andreas Stein, Philippe Buhlmann

(1790-12 P) **Development of Calibration-Free Electrochemical Sensors Using Novel Redox Polymers** XUE ZHEN, University of Minnesota, Philippe Buhlmann

(1790-13 P) **Ultrasensitive Detection of Dopamine with Carbon Nanopipettes** KEKE HU, Queens College CUNY, Michael Mirkin, Yun Yu, Min Zhou

(1790-14 P) **Avoiding Errors in Electrochemical Measurements: Effect of Frit Material on the Performance of Reference Electrodes with Porous Frits** MARAL PS MOUSAVI, University of Minnesota, Stacey Saba, Evan Anderson, Marc A Hillmyer, Philippe Buhlmann

WEDNESDAY, MARCH 9, 2016
AFTERNOON

AWARDS — **Session 1800**

Ralph N Adams Award
arranged by Norman J Dovichi, University of Notre Dame

Wednesday Afternoon, Room B312
Norman J Dovichi, University of Notre Dame, Presiding

1:30		**Introductory Remarks - R Mark Wightman**
1:35		**Presentation of the 2016 Ralph Adams Award to David R Walt, Tufts University, by Norman J Dovichi, University of Notre Dame**
1:40	(1800-1)	**Clinical Applications of Single Molecule Arrays (Simoa)** DAVID R WALT, Tufts University
2:15	(1800-2)	**Microengineered Devices for Biomedical Research** NANCY ALLBRITTON, University of North Carolina at Chapel Hill
2:50	(1800-3)	**Micro-and Nanofabricated Molecular Measurement Devices** J MICHAEL RAMSEY, University of North Carolina at Chapel Hill N/A
3:25		**Recess**
3:40	(1800-4)	**Paper-Based Microfluidic Devices for Point-of-Need Bioanalysis** CHARLES R MACE, Tufts University, Syrena C Fernandes, Samuel Berry
4:15	(1800-5)	**Developmental Proteomics** NORMAN J DOVICHI, University of Notre Dame, Liangliang Sun, Matthew M Champion, Guijie Zhu, Paul Huber

AWARDS — **Session 1810**

The Coblentz Society - Williams-Wright Award
arranged by Nancy Jestel, Sabic Innovative Plastics

Wednesday Afternoon, Room B314
Nancy Jestel, Sabic Innovative Plastics, Presiding

1:30		**Introductory Remarks - Nancy Jestel**
1:35		**Presentation of the 2016 Coblentz Society's Williams-Wright Award to D Warren Vidrine, Vidrine Consulting, by Nancy Jestel, Sabic Innovative Plastics**
1:40	(1810-1)	**Industrial Analysis Utilizing Vibrational Spectrometry** D WARREN VIDRINE, Vidrine Consulting
2:15	(1810-2)	**Biological Infrared Spectroscopy: An Overnight Success Story 64 Years in the Making** BOB MESSERSCHMIDT, Nueon Inc.
2:50	(1810-3)	**Reflections of a Chemometric Spectroscopist** DAVID M HAALAND, Spectral Resolutions, David A Melgaard, Howland D Jones
3:25		**Recess**
3:40	(1810-4)	**The Laminated Card** STUART YANIGER, ITW Fluids North America N/A
4:15	(1810-5)	**Open Discussion**

SYMPOSIUM — **Session 1820**

ACS-ANYL - Advances in Instrumentation for Ion Mobility Mass Spectrometry
arranged by Matthew F Bush, University of Washington

Wednesday Afternoon, Room B308
Matthew F Bush, University of Washington, Presiding

1:30		**Introductory Remarks - Matthew F Bush**
1:35	(1820-1)	**Ultra-High Resolution Ion Mobility Separations based upon Long Path Length Structures for Lossless Ion Manipulations (SLIM)** RICHARD D SMITH, Pacific Northwest National Laboratory, Ahmed Hamid, Sandilya Garimella, Yehia M Ibrahim, Liulin Deng, Ian K Webb, Aleksey V Tolmachev, Erin S Baker
2:10	(1820-2)	**Ion Mobility Mass Spectrometers for Structural Biology and Biophysics** MATTHEW F BUSH, University of Washington
2:45	(1820-3)	**Toward Protein Ion Surface Characterization with IMS/HDX-MS/MS Techniques** STEPHEN J VALENTINE, West Virginia University, Mahdiar Khakinejad, Gregory Donohoe, Samaneh Ghassabi Kondalaji
3:20		**Recess**
3:35	(1820-4)	**Maximizing Fragmentation Duty Cycle for Ion Mobility-Ion Trap Instrumentation Using the Fourier Transform and Photodissociation** BRIAN CLOWERS, Washington State University, Kelsey Morrison
4:10	(1820-5)	**A Multi-Pass Cyclic Ion Mobility Separator: Design and Performance** KEVIN GILES, Waters Corporation, Jason Wildgoose, Steve Pringle

SYMPOSIUM — **Session 1830**

Advancing Strategies for Chronic In Vivo Sensing
arranged by R Mark Wightman, University of North Carolina at Chapel Hill
and Adrian C Michael, University of Pittsburgh

Wednesday Afternoon, Room B302
R Mark Wightman, University of North Carolina at Chapel Hill, Presiding

1:30		**Introductory Remarks - R Mark Wightman and Adrian C Michael**
1:35	(1830-1)	**Modulating Blood-Brain Barrier Healing Around Intracortical Electrode Implants** RAVI BELLAMKONDA, Georgia Institute of Technology, Varun Yarabarla, Alexus Clark, Brianna Gresham, Robert Kretschmar, Shoba Paul, Jessica Falcone
2:10	(1830-2)	**Is Microdialysis Specifically Monitoring the Tonic Modality of Dopamine Transmission?** GAETANO DI CHIARA, University of Cagliari
2:45	(1830-3)	**Long-Term Monitoring of Dopamine** PAUL EM PHILLIPS, University of Washington
3:20		**Recess**
3:35	(1830-4)	**Calibration of In Vivo Voltammetry** R MARK WIGHTMAN, University of North Carolina at Chapel Hill
4:10	(1830-5)	**Advancing the Possibilities for Chronic Brain Microdialysis** ADRIAN C MICHAEL, University of Pittsburgh, Erika Varner, Khanh Ngo, Andrea Jaquins-Gerstl, Stephen Weber

SYMPOSIUM — **Session 1840**

Analytical Challenges Relating to the Discovery, Development, Manufacturing and Use of Cancer Immunotherapy Medicines
arranged by Maribel Beaumont and Christopher Welch, Merck Research Laboratories

Wednesday Afternoon, Room B303
Maribel Beaumont, Merck Research Laboratories, Presiding

1:30		**Introductory Remarks - Maribel Beaumont**
1:35	(1840-1)	**Concepts of Cancer Immunotherapy** ROBERT KASTELEIN, Merck & Co.
2:10	(1840-2)	**Identifying and Profiling Tumor Specific T Cells Using Mass Cytometry and Highly Multiplexed Peptide-MHC Tetramer Staining** EVAN W NEWELL, Singapore Immunology Network
2:45	(1840-3)	**Analytical Challenges in the Discovery, Development and Commercialization of Keytruda** MARIBEL BEAUMONT, Merck Research Laboratories
3:20		**Recess**
3:35	(1840-4)	**Development of Microtools for Immune Therapy Applications** CHRISTOPHER LOVE, Koch Institute for Integrative Cancer Research at MIT
4:10	(1840-5)	**Analytical Challenges During Development of Monoclonal Antibody Therapeutics** JOHN T STULTS, Genentech, Inc, David Michels

TECHNICAL PROGRAM

SYMPOSIUM — Session 1850

Analytical Chemistry of Oil and Gas Prospecting in Brazil
arranged by Maira Menezes, NurnbergMesse Brasil

Wednesday Afternoon, Room B304
Jailson de Andrade, Universidade Federal da Bahia-UFBA, Presiding

1:30		**Introductory Remarks -** Jailson de Andrade
1:35	(1850-1)	**Analytical Chemistry Challenges of Oil and Gas Prospection and Exploration** JAILSON B DE ANDRADE, UFBA N/A
2:10	(1850-2)	**Compound Specific $\delta^{13}C$ Determination of Light Hydrocarbons (n-alkanes and Olefins from C1 to C5) at Low Concentration for Oil and Gas Prospection** ARTHUR DE LEMOS SCOFIELD, PUC-RIO, Angela R Wagener, Laura R Morales, Lilian F Almeida
2:45	(1850-3)	**Mass Spectrometry by FT-ICR and Orbitrap: Analysis of Crude Oil and Its Derivatives** EDUARDO MORGADO SCHMIDT, State University Campinas

SYMPOSIUM — Session 1860

Big Data in Analytical Sciences - Challenges and Solutions
arranged by Hang Lu, Georgia Tech and Andriana San Miguel, North Carolina State University

Wednesday Afternoon, Room B305
Hang Lu, Georgia Tech, Presiding

1:30		**Introductory Remarks -** Hang Lu
1:35	(1860-1)	**Data Analysis Challenges in Plant Biology** PHILIP BENFEY, Duke University
2:10	(1860-2)	**First Steps into Big Data Chemistry- Ultra-High-Throughput Screening** SPENCER D DREHER, Merck & Co., Inc.
2:45	(1860-3)	**Big, But Small, Data: Network Analysis in Small Sample Size Systems Biology** MARK P STYCZYNSKI, Georgia Institute of Technology
3:20		**Recess**
3:35	(1860-4)	**Clustering and Differential Alignment Algorithm: Identification of Early Stage Regulators in the Arabidopsis Thaliana Iron Deficiency Response** CRANOS WILLIAMS, North Carolina State University, Terri A Long, Alexandr Koryachko, Anna Matthiadis, James Tuck, Durreshahwar Muhammad, Joel Ducoste, Siobhan Brady, Jessica Foret
4:10	(1860-5)	**Machine Learning for Big Nonlinear Problems in Science and Engineering** LE SONG, Georgia Institute of Technology

SYMPOSIUM — Session 1870

Electrical and Electrochemical Sensing and Detection based on Nucleic Acid Recognition
arranged by Rebecca Lai, University of Nebraska-Lincoln
and Ryan J White, University of Maryland Baltimore County (UMBC)

Wednesday Afternoon, Room B309
Rebecca Lai, University of Nebraska-Lincoln, Presiding

1:30		**Introductory Remarks -** Rebecca Lai and Ryan J White
1:35	(1870-1)	**A New Approach to POC Serology** KEVIN W PLAXCO, University of California Santa Barbara
2:10	(1870-2)	**Detection of Hepatitis B Virus DNA with a Paper Electrochemical Sensor** RICHARD M CROOKS, University of Texas at Austin, Xiang Li, Karen Scida
2:45	(1870-3)	**Using Widely Available Electrochemical Device for Sensing or Diagnostics** YI LU, University of Illinois at Urbana-Champaign, JingJing Zhang, Yu Xiang
3:20		**Recess**
3:35	(1870-4)	**Electrochemical Analysis of Clinically-Relevant Biomolecular Analytes Using Nanostructured Microelectrodes** SHANA KELLEY, University of Toronto
4:10	(1870-5)	**Signal Amplification for Biosensing Based on Nucleic Acid Recognition** JU HUANGXIAN, Nanjing University

SYMPOSIUM — Session 1880

Vibrational Spectroscopy of Biodegradable Plastics: Evolution, Revolution or Back to the Future
arranged by John F Rabolt, University of Delaware

Wednesday Afternoon, Room B310
John F Rabolt, University of Delaware, Presiding

1:30		**Introductory Remarks -** John F Rabolt
1:35	(1880-1)	**Spectroscopic Study of Structural Evolution Dynamics of Bio-Based and Biodegradable Poly(hydroxyalkanoate) Copolymers** ISAO NODA, University of Delaware, Brian Sobieski, Liang Gong, Bruce Chase, John F Rabolt
2:10	(1880-2)	**Degradation Mechanisms of Poly(lactic acid)-Nanoparticle Composite and Phthalate Plasticized Poly(vinyl chloride)** ZHAN CHEN, University of Michigan
2:45	(1880-3)	**Vibrational Spectroscopic Studies on Biodegradable Polymer Microstructures** SHAW L HSU, University of Massachusetts
3:20		**Recess**
3:35	(1880-4)	**2D IR Correlation Study of Thin Films of Biodegradable PHA/PEG Blend** YOUNG MEE JUNG, Kangwon National University, Yeonju Park, Yujing Chen, Isao Noda
4:10	(1880-5)	**Characterization of Single Electrospun Biopolymer Nanofibers Using AFM-IR and Selected Area Electron Diffraction** JOHN F RABOLT, University of Delaware, Liang Gong, Isao Noda, Bruce Chase, Curtis Marcott, C J McBrin, David Martin, Jinglin Liu, Chao Ni

WORKSHOPS — Session 1890

CACA - How to be Successful in Your Career
arranged by Tao Jiang, Mallinckrodt Pharmaceuticals and Michael Ye, Supelco/Sigma-Aldrich

Wednesday Afternoon, Room B311
Tao Jiang, Mallinckrodt Pharmaceuticals, Presiding

1:30		**Introductory Remarks -** Tao Jiang and Michael Ye
1:35	(1890-1)	**Insights on Job Searching** MICHAEL W DONG, MWD Consulting
2:05	(1890-2)	**Recognizing and Managing Career Passover** ROBERT L STEVENSON, American Laboratory
2:35	(1890-3)	**Advancing Academic Career as a Faculty** YI HE, John Jay College/CUNY
3:05		**Recess**
3:20	(1890-4)	**Career Development Workshop** NAIDONG WENG, JNJ
3:50	(1890-5)	**From a Bench Chemist to an Entrepreneur** BRUCE LIU, ACB Scitech, Inc. N/A
4:20	(1890-6)	**Go Wider and Higher** CHUPING LUO, Waters Corporation

WORKSHOPS — Session 1900

Natural Health Products: Scientific Approaches to Securing Product Quality and Safety
arranged by Rob O'Brien, ISURA and Bob Chapman, National Research Council of Canada

Wednesday Afternoon, Room B313
Rob O'Brien, ISURA, Presiding

1:30		**Introductory Remarks -** Rob O'Brien and Bob Chapman
1:35	(1900-1)	**Development of a Chemical Barcoding Methodology to Identify and Support the Quality and Safety of Functional Ingredients** BOB CHAPMAN, National Research Council of Canada, Fabrice Berrue, Junzeng Zhang, Ian Burton, Joseph Hui, Sabrena MacKenzie, Camilo Martinez-Farina, Aissa Harhira, El Haddad Josette, Mohamad Sabsabi, Alain Blouin, Yuan-Chun Ma, Rob O'Brien
2:05	(1900-2)	**Utility and Limitations of DNA Barcoding and Next Generation Sequencing for Herbal Product Authentication** JONATHAN VAN HAMME, Thompson Rivers University
2:35	(1900-3)	**Developing Metabolomics Approaches for Natural Product Authentication** LIANG LI, University of Alberta
3:05		**Recess**
3:20	(1900-4)	**Natural Health Product Quality Control: Addressing the Current Issue of Adulterants** YUAN-CHUN MA, Canadian Phytopharmaceuticals Corporation
3:50	(1900-5)	**Practical Gas Chromatographic based Approaches to Confirm Identity of Powdered Herbal Products** ROB O'BRIEN, ISURA, Anderson Smith
4:20:50	(1900-6)	**TLC and HPLC Fingerprinting for Authentication of Natural Health Products** RUDOLF BAUER, University of Graz

ORGANIZED CONTRIBUTED SESSIONS **Session 1910**

High Throughput Analysis for Food Safety and Cosmetics: Challenges and Validation
arranged by Perry G Wang, US FDA and Xiaogang Chu, China Academy of Inspection and Quarantine

Wednesday Afternoon, Room B315
Perry G Wang, US FDA, Presiding

1:30	(1910-1)	**Rapid Determination of Non-Allowed Active Pharmaceutical Ingredients for the Treatments of Hair Loss in Cosmetics Using UHPLC-HRMS** WANLONG ZHOU, US FDA, Perry G Wang, James B Wittenberg, Maria A Dionisio De Sousa, Alexander J Krynitsky
1:50	(1910-2)	**Advances in High-Throughput Analysis for Determination of Marine Biotoxins in Seafood** PEARSE MCCARRON, National Research Council, Daniel Beach
2:10	(1910-3)	**Determination of Prostaglandin Analogs in Eye Area Cosmetic Products by High Performance Liquid Chromatography with Tandem Mass Spectrometry** JAMES B WITTENBERG, Food and Drug Administration, Wanlong Zhou, Perry G Wang, Alexander J Krynitsky
2:30	(1910-4)	**Withdrawn**
2:50		**Recess**
3:05	(1910-5)	**Analysis of Color Additives (Permitted and Non-Permitted) in Different Food Matrices by a HPLC Method** SNEH D BHANDARI, Merieux NutrSciences, Tiffany Gallegos-Peretz
3:25	(1910-6)	**A Comprehensive Approach on Food Safety Analysis, Screening and Quantitation by Using Data Independent Acquisition (DIA) and DDMS2 on HR/AM Q Exactive System** JAMES S CHANG, Thermo
3:45	(1910-7)	**Separation of Aminoglycoside Antibiotics by Using Hydrophilic Interaction Liquid Chromatography** YU LONG, Dalian Institute of Chemical Physics, Wei Jie, Shen Aijin, Guo Zhimou, Liang Xinmiao
4:05	(1910-8)	**Open Discussion**

ORGANIZED CONTRIBUTED SESSIONS **Session 1920**

Precision Bioanalytical Measurements
arranged by Robert Dunn, University of Kansas

Wednesday Afternoon, Room B316
Robert Dunn, University of Kansas, Presiding

1:30	(1920-1)	**Analytical Precision in the Age of Metabolomics – Focus on the Fundamentals** HOWARD HENDRICKSON, University of Arkansas for Medical Sciences, Lin Song
1:50	(1920-2)	**Biocompatible Self-Tuning Nanomaterials Improve the Precision of Biomolecule Processing and Separation** LISA A HOLLAND, West Virginia University, Srikanth Gattu, Brandon C Durney, Cassandra Crihfield
2:10	(1920-3)	**Analytical Approaches for Environmental Metabolomics and Ecotoxicity Modeling** CYNTHIA K LARIVE, University of California - Riverside, Corey M Griffith, Melissa M Morgan
2:30	(1920-4)	**Extending the Free Drug Hypothesis: Physical Properties Driving Asymmetric Tissue Distribution** DENNIS O SCOTT, Pfizer
2:50		**Recess**
3:05	(1920-5)	**Barbeques and Tornadoes: Analytical Strategies for Metal Ion Determinations Using Separations and Sensors** FIONA REGAN, Dublin City University
3:25	(1920-6)	**A Mass Spectrometry Based High Throughput Screening Approach with Exquisite Selectivity** HEATHER R DESAIRE, University of Kansas, Imaduwage Kasun
3:45	(1920-7)	**Microfluidic Devices with Integrated Electrodes for Monitoring Cellular Systems** R SCOTT MARTIN, Saint Louis University
4:05	(1920-8)	**Novel Applications of Microdialysis Sampling: Where No Probe has Gone Before!** SARA R THOMAS, Kansas University, Craig E Lunte, Susan M Lunte

ORGANIZED CONTRIBUTED SESSIONS **Session 1930**

Technology Strategies for Explosives Sensing
arranged by Michael Shepard, Dept. Homeland Security, S&T Directorate
and Samar K Guharay, The MITRE Corporation

Wednesday Afternoon, Room B401
Samar K Guharay, The MITRE Corporation, Presiding

1:30	(1930-1)	**Overview of Operational Challenges to Domestic Security Screening** MICHAEL SHEPARD, US Dept. of Homeland Security
1:50	(1930-2)	**Generalized Systems Analysis Framework** SAMAR K GUHARAY, The Mitre Corporation
2:10	(1930-3)	**Statistical Methods for Constructing Chemometric Signatures in Noisy Environments** CHRISTOPHER P SAUNDERS, South Dakota State University
2:30	(1930-4)	**Operational Outlook for Remote Trace Explosives Detection** RODERICK KUNZ, MIT Lincoln Laboratory
2:50		**Recess**
3:05	(1930-5)	**Pushing the Limits of Trace Detection Technology** STEFAN LUKOW, Morpho Detection
3:25	(1930-6)	**Optimized Sampling and Analysis Strategies for Trace Explosives Detection** GREG GILLEN, NIST
3:45	(1930-7)	**Chemical Attribution Signatures of Homemade Explosives** JOE CHIPUK, Signature Science N/A
4:05	(1930-8)	**Open Discussion**

ORAL SESSIONS **Session 1940**

Advances in Fuel and Petrochemical Analyses

Wednesday Afternoon, Room B402
John Baltrus, US Department of Energy-NETL, Presiding

1:30	(1940-1)	**Chemical Fingerprinting of Crude Oils: Gaining an Extra Dimension from GCxGC–TOF MS** LAURA MCGREGOR, Markes International Ltd, Nicola Watson, Kevin Collins, Chris Hall, Ken Umbarger
1:50	(1940-2)	**Fast and Accurate Analysis of Extended Natural Gas Composition and Physical Properties Using a Temperature Programmable Gas Analyzer** DEBBIE ALCORN, INFICON
2:10	(1940-3)	**Application of Polymeric Ionic Liquids as Highly Robust and Selective Stationary Phases for Comprehensive Two-Dimensional Gas Chromatography** CHENG ZHANG, Iowa State University, Rodney A Park, Jared L Anderson
2:30	(1940-4)	**Solving Industrial Problems by Determining Compound Classes in Refinery Streams and Products** CHRIS GOSS, Alberta Innovates Technology Futures, Dan Wispinski, Lee Marotta
2:50		**Recess**
3:05	(1940-5)	**Optical Sensors for the Detection of Nitrogen Compounds in Aviation Fuels** ROBERTO A FEDERICO-PEREZ, University of Tennessee, Ziling (Ben) Xue
3:25	(1940-6)	**Fast Profiling of Petrochemical Samples by Thermal Analysis- Soft Ionization Mass Spectrometry: From Source Rock and Kerogen via Crude Oil to Petrochemical Products** RALF ZIMMERMANN, University Rostock /HMGU, Thorsten Streibel, Mohammad Saraji, Michael Fischer , Sebastian Wohlfahrt, Andreas Walte, Thomas Denner
3:45	(1940-7)	**Micro-Hollow Discharge Analysis of Carbonaceous Materials** RANDY VANDER WAL, Penn State University, Chethan K Gaddam N/A
4:05	(1940-8)	**On-Site Fuel Analysis Using a Portable Near-Infrared Spectrometer** STUART FARQUHARSON, Real-Time Analyzers, Inc, Carl Brouillette, Wayne Smith, Chetan Shende

TECHNICAL PROGRAM

ORAL SESSIONS **Session 1950**

Bioanalytical: Using Microfluidics/Lab-on-a-Chip Techniques

Wednesday Afternoon, Room B404
Casey Burton, Missouri University of Science and Technology, Presiding

1:30	(1950-1)	**Pressure-Actuated Integrated Microfluidic Devices for Biomarker Analysis** VISHAL SAHORE, Brigham Young University, Suresh Kumar, Adam T Woolley
1:50	(1950-2)	**Real-Time Imaging of Cancer Cell Chemotaxis in Paper-Based Scaffolds** RACHAEL M KENNEY, University of North Carolina at Chapel Hill, Matthew W Boyce, Andrew S Truong, Matthew R Lockett
2:10	(1950-3)	**Observation of Rapid Changes in Drug Susceptibility of Tumor Cells in a Hypoxia Microfluidic Culture Device** TODD GERMAIN, Texas Tech University, Dimitri Pappas
2:30	(1950-4)	**Magnetic Microbead Based Capture and Labeling of DNA from Carbapenemase Resistance Genes** RILEY K MILLS, Brigham Young University, Radim Knob, Adam T Woolley N/A
2:50		Recess
3:05	(1950-5)	**Screening Small Molecule Modulators of Cellular Chemotaxis in Paper-Based Invasion Assays** C CHAD LLOYD, University of North Carolina at Chapel Hill, Matthew R Lockett, Rachael M Kenney, Andrew S Truong, Matthew W Boyce, Christian A Lochbaum
3:25	(1950-6)	**Microchip Affinity Monoliths for Solid Phase Extraction of DNA for Bacteria Infection Detection** RADIM KNOB, Brigham Young University, Riley K Mills, Adam T Woolley
3:45	(1950-7)	**MALDI-IMS Evaluation of 3D Cell Cultures Treated with Combination Chemotherapeutics by a 3D Printed In-Vitro PK/PD Microfluidic Platform** GABRIEL J LABONIA, University of Notre Dame, Amanda B Hummon
4:05	(1950-8)	**Fully Automated Microfluidic Input/Output Multiplexer for Endocrine Tissue Culturing and Hormone Secretion Sampling** XIANGPENG LI, Auburn University, Jessica C Brooks, Katarena Ford, Christopher J Easley

ORAL SESSIONS **Session 1960**

Environmental Water Quality and Analysis

Wednesday Afternoon, Room B403
John P Auses, University of Pittsburgh, Presiding

1:30	(1960-1)	**Targeted Discovery of Disinfection By-Products in Swimming Pools and Hot Tubs** SUSAN D RICHARDSON, University of South Carolina, Jonathan D Byer, Eric J Daiber, Sridevi A Ravuri, Joseph E Binkley, Christina M Joseph
1:50	(1960-2)	**Evaluation of Iodo-, Bromo-, and Chloro-Acetic Acids Formation by Peracetic Acid Disinfection Using a Newly Developed Rapid HPIC-MS/MS Method** RUNMIAO XUE, Missouri University of Science and Technology, Honglan Shi, Yinfa Ma, John Yang, Bin Hua, Enos Inniss, Craig Adams, Todd Erichholz
2:10	(1960-3)	**Removal of Pharmaceutical Products from Wastewater Using Magnetic Bio-Char** AKILA G KARUNANAYAKE, Mississippi State University, Olivia A Todd, Todd E Mlsna
2:30	(1960-4)	**Determination of Priority Water Contaminants by Solid-Phase Extraction and UFLC-MS/MS Method** HAITING ZHANG, Missouri University of Science and Technology, Danielle West, Honglan Shi, Yinfa Ma, Craig Adams, Todd Erichholz
2:50		Recess
3:05	(1960-5)	**Characterization of Oil-based Pollutants Using Webcam-based Spectrometer** YAGIZ SUTCU, InfoScope Research, Aysegul Ergin
3:25	(1960-6)	**Coupling of Thin Film Microextraction Techniques to Portable GC-TMS Instrumentation for the On-Site, Sub-ppb, Detection of Pesticides From Environmental Waters** JONATHAN J GRANDY, University of Waterloo, Janusz Pawliszyn
3:45	(1960-7)	**SPE in US EPA Method 625, A Step Closer to Reality after Good Performance in Two Round Robins** ZOE GROSSER, Horizon Technology, William Jones, David Gallagher, Michael Ebitson, Alicia Cannon
4:05	(1960-8)	**Adsorption of Pb^{2+} from Aqueous Solution Using Low-Cost Chitosan-Modified Biochar, A Green Adsorbent** NARADA W BOMBUWALA DEWAGE, Mississippi State University, Todd E Mlsna

ORAL SESSIONS **Session 1970**

GC Fuels, Energy and Petrochemical

Wednesday Afternoon, Room B405
Patricia Ranaivo, Indiana University Southeast, Presiding

1:30	(1970-1)	**Zeolite-Loaded Metal-Catalyzed Hydrotreatment of Lignin to Aromatic Monomers Using Subcritical Water** ERIC A BOAKYE, South Dakota State University, Douglas Raynie
1:50	(1970-2)	**Olefins in Refinery Streams by GC-VUV** DAN WISPINSKI, Alberta Innovates Technology Futures, Chris Goss, Phillip Walsh
2:10	(1970-3)	**GC-Ion Mobility Spectrometry for Determination of Ageing of Mineral Oil Impregnated Presspaper Isolation Systems** WOLFGANG VAUTZ, ISAS, Liedtke Sascha, Torben Muth, Frank Jenau
2:30	(1970-4)	**High Level Fixed Gas Analysis, Including Hydrogen, with one Detector and One Carrier** MATTHEW MONAGLE, AIC LLC N/A
2:50		Recess
3:05	(1970-5)	**Analysis of Dissolved Hydrocarbon Gases in Water – Pitfalls and Improvements** MARK L BRUCE, TestAmerica
3:25	(1970-6)	**Performance Evaluation of Modern Stainless Steel Capillary GC Columns** REBECCA STEVENS, Restek Corporation, Corby Hilliard, Amanda Rigdon, Linx Waclaski, Jaap de zeeuw
3:45	(1970-7)	**Determination of Polycyclic Aromatic Sulfur Heterocycles and Their Alkyl-Substituted Derivatives in Standard Reference Material 1597a** WALTER B WILSON, National Institute of Standards and Technology, Stephen A Wise
4:05	(1970-8)	**Vehicle Interior Air Quality - (S)VOC Emission from Materials: Regulation, Standard Methods and Analytical Implementation** CAROLINE WIDDOWSON, Markes International

ORAL SESSIONS **Session 1980**

Mass Spectrometry - Bioanalytical and Omics

Wednesday Afternoon, Room B406
Logan T Miller, Duquesne University, Presiding

1:30	(1980-1)	**Mass Spectrometry in Discovery of Lipid Markers for Alzheimer's Disease** SATYA GIRISH CHANDRA AVULA, Cleveland State University, Christine Reece, Jagan A Pillai, Yan Xu
1:50	(1980-2)	**Novel Cationization Strategies for Improved Separation of Metabolite Isomers with Ion Mobility – Mass Spectrometry** CHRISTOPHER D CHOUINARD, University of Florida, Robin Kemperman, Harrison King, Christopher R Beekman, Richard Alan Yost
2:10	(1980-3)	**Development of Dried Matrix Card Quantification by Speciated Isotope Dilution Mass Spectrometry Using Elution and Laser Desorption Techniques** LOGAN T MILLER, Duquesne University, Kaitlin Miller, Sarah Sheffield, Skip Kingston, Matt Pamuku, Scott Faber, Mark Little, Silverio Iacono
2:30	(1980-4)	**Protease-Containing Membranes for Rapid Antibody Digestion Prior to Mass Spectrometry Analysis** YONGLE PANG, Michigan State University, Wei-Han Wang, Gavin Reid, Donald F Hunt, Merlin Bruening
2:50		Recess
3:05	(1980-5)	**Quantitative Environmental Human Health Assessment of Inorganic Elements in Dried Blood Spots Using Direct Isotope Dilution Laser Ablation and Elution Mass Spectrometry** SARAH SHEFFIELD, Duquesne University, Logan T Miller, Scott Faber, Matt Pamuku, Skip Kingston
3:25	(1980-6)	**Identification of the Cell Surface N-Glycoproteome by MS-based Proteomics** JOHANNA SMEEKENS, Georgia Institute of Technology, Weixuan Chen, Ronghu Wu
3:45	(1980-7)	**Probing the Cell-Surface N-Glycoproteome with Metabolic Chemical Reporters (MCRs)** HAOPENG XIAO, Georgia Tech
4:05	(1980-8)	**Quantitative Characterization of Protein Content from HDL and LDL Size Fractions** BRYAN A PARKS, Centers for Disease Control and Prevention, Zsuzsanna Kuklenyik, David Schieltz, Michael S Gardner, Jon Rees, Lisa McWilliams, Yulanda Williamson, John R Barr

ORAL SESSIONS **Session 1990**

Metabolomics, Proteomics, and Genomics

Wednesday Afternoon, Room B407
Qingbo (Roger) Yang, Missouri University of Science and Technology, Presiding

1:30	(1990-1)	**A Combination of Multidimensional Chromatography and High Resolution Mass Spectrometry for Chemical Exposure Analysis** DAVID E ALONSO, Leco Corporation, Joseph E Binkley, Elizabeth M Humston-Fulmer, Jonathan D Byer, Lorne E Fell
1:50	(1990-2)	**Metabolomics of Nonalcoholic Fatty Liver Disease via LC-MS** RAINEY E PATTERSON, University of Florida, Richard Alan Yost, Timothy J Garrett, Nishanth E Sunny, Srilaxmi Kalavalapalli, Kenneth Cusi
2:10	(1990-3)	**Application of PDMS-Overcoated Matrix-Compatible Solid Phase Microextraction Fiber Coupled to Comprehensive Two-Dimensional Gas Chromatography–Time-of-Flight Mass Spectrometry for Improved Exploitation of Chromatographic Space in Global Chemical Profiling of Brazilian Cachaça** ERICA A SOUZA-SILVA, Universidade Federal do Rio Grande do Sul, Fernando C Fontanive, Claudia A Zini
2:30	(1990-4)	**Application of the Isotopic Ratio Outlier Analysis Phenotypic Protocol for Metabolomic Biomarker Discovery in Type 1 Diabetes Using T Cells** CANDICE Z ULMER, University of Florida, Richard Alan Yost, Timothy J Garrett, Christopher A Beecher, Jing Chen, Clayton Matthews
2:50		**Recess**
3:05	(1990-5)	**Secretome of Murine Islets of Langerhans** ANDREW W SCHMUDLACH, University of Notre Dame, Jeremy Felton, Robert T Kennedy, Norman J Dovichi
3:25	(1990-6)	**Optimizing Sampling Protocols for the Identification and Quantitation of Neuropeptides from Brain Tissues** NING YANG, University of Illinois at Urbana-Champaign, Stanislav S Rubakhin, Jonathan V Sweedler, Krishna D Anapindi
3:45	(1990-7)	**A Study of Protein Degradation and the Effect of Proteasome Inhibition by Combining Biorthogonal Noncanonical Amino Acid Labeling with Click Chemistry** ZHENYU ZHOU, Georgia Institute of Technology N/A
4:05	(1990-8)	**Benchmarking of DNA-seq and RNA-seq** Variant Detection Software GANG FENG, Northwestern University, Zhujun Huang, Lei Huang, Riyue Bao, Hongmei Jiang, Yuchen Bai, Xiaoyong Sun N/A

ORAL SESSIONS **Session 2000**

Raman, SERS, UVRR Applications

Wednesday Afternoon, Room B408
Matt Cerutti, Thermo Scientific, Presiding

1:30	(2000-1)	**UV Resonance Raman (UVRR) Structural Studies of Polyglutamine (polyQ) Side Chains and Fibrils** DAVID PUNIHAOLE, University of Pittsburgh, Zhenmin Hong, Ryan Jakubek, Elizabeth Dahlburg, Riley Workman, Jeffry Madura, Sanford A Asher
1:50	(2000-2)	**Sheath-Flow Microfluidic Approach for Combined Surface Enhanced Raman Scattering and Electrochemical Detection** MATTHEW R BAILEY, University of Notre Dame, Amber Pentecost, Asmira Selimovic, R Scott Martin, Zachary D Schultz
2:10	(2000-3)	**Raman Spectroscopy Monitors Glutamine and Asparagine Side Chain OCCC Dihedral Angles** RYAN JAKUBEK, University of Pittsburgh, David Punihaole, Zhenmin Hong, Elizabeth Dahlburg, Steven Geib, Sanford A Asher
2:30	(2000-4)	**Assessment of the Protein–Protein Interactions in a Highly Concentrated Antibody Solution by Using Raman Spectroscopy** CHIKASHI OTA, Horiba, Ltd., Shintaro Noguchi, Satoru Nagatoishi, Kouhei Tsumoto
2:50		**Recess**
3:05	(2000-5)	**Raman Imaging of Samples with Complex Surface Topographies** TIM BATTEN, Renishaw plc, Tim Smith
3:25	(2000-6)	**Improving Raman and Surface Enhanced Raman Spectroscopy Through Effective Sampling** DONGMAO ZHANG, Mississippi State University N/A
3:45	(2000-7)	**Spatially Offset vs Conventional Raman for Through-Barrier Material Identification** ROBERT STOKES, Cobalt Light Systems, Paul Loeffen, Darren Andrews, Ken Mann, Oliver Presly, Pavel Matousek
4:05	(2000-8)	**Degradation Analysis of International Space Station Medications by Raman Spectroscopy** STUART FARQUHARSON, Real-Time Analyzers, Inc, Chetan Shende, Alexander May, Carl Brouillette, Joseph Cosgrove

ORAL SESSIONS **Session 2010**

Surface and Microscopic Characterization of Nanostructures and Biological Materials

Wednesday Afternoon, Room B409
Brian R Strohmeier, United States Steel Corporation, Presiding

1:30	(2010-1)	**Fundamental Understanding of the Synergy Between Electroactive Poly (amic) Acid Membranes and Their Interaction with Nanoparticles** VICTOR KARIUKI, Binghamton University
1:50	(2010-2)	**Optical Property and Catalytic Activity of Gold Nanorods End-capped with a Second Metal** GUFENG WANG, North Carolina State University, Nathalia Ortiz, Vineet Kumar
2:10	(2010-3)	**Autocorrelation Function Analysis of Rotational Dynamics of Gold Nanorod** KUANGCAI CHEN, Iowa State University/Georgia State University, Ning Fang
2:30	(2010-4)	**Antibody-Like Biorecognition Sites for Proteins from Surface Imprinting on Nanoparticles** SNEHASIS BHAKTA, University of Connecticut, Saiful Seraji, James F Rusling, Steven L Suib
2:50		**Recess**
3:05	(2010-5)	**Selective Raman Imaging of Integrin Receptors Through Coupled Plasmonic Nanostructures** LIFU XIAO, University of Notre Dame, Zachary D Schultz, Hao Wang
3:25	(2010-6)	**Deep and High-Resolution Three-Dimensional Tracking of Single Particles Using Nonlinear and Multiplexed Illumination** TIM YEH, University of Texas at Austin
3:45	(2010-7)	**Development of a Dual Microscope System for Integration of Intracellular Calcium Imaging with Monitoring Insulin Secretion from Islets of Langerhans** LIAN YI, Florida State University, Xue Wang, Michael G Roper
4:05	(2010-8)	**Establishing Statistical Criteria for Detecting Molecular Motion in Single-molecule Diffusional Trajectories** MOUSSA BARHOUM, University of Utah, Karl-Heinz Gericke, Joel M Harris N/A

ORAL SESSIONS **Session 2020**

Trace Explosives Detection - Half Session

Wednesday Afternoon, Room B301
Maria K Ferguson, PA Department of Environmental Protection, Presiding

1:30	(2020-1)	**Analysis of the Decomposition of Hexamethylene Triperoxide Diamine (HMTD) as Determined by SPME-GC/MS and LC/MS** LAURYN DEGREEFF, Naval Research Laboratory, Christopher Katilie, Frank L Steinkamp
1:50	(2020-2)	**Separation of Inorganic Ions and Neutral Organic Nitroaromatic Compounds by Electrokinetic Chromatography** JULIE R MCGETTRICK, University of Montana, Christopher P Palmer
2:10	(2020-3)	**Comparison of Inter-Instrument Relative Response Factors with Thermal Desorption Internal Standards** H MITCHELL RUBENSTEIN, USAF, Maomian Fan, Claude C Grigsby, Kathy Fullerton, Brian Geier, Darrin Ott
2:30	(2020-4)	**Through-Barrier Explosives and Hazardous Material Detection Using a Handheld Spatially Offset Raman Spectrometer** PAUL LOEFFEN, Cobalt Light Systems, Robert Stokes, Ken Mann, Darren Andrews, Oliver Presly, Pavel Matousek

UNDERGRADUATE POSTER SESSION — Session 2030

All posters are to be mounted by 10:00 AM and remain on display until 4:00 PM. Authors must be at their posters from 1:00 PM to 3:00 PM. Location of the afternoon posters is on the Exposition Floor, 400 Aisle. PLEASE NOTE: You cannot get onto the Exposition Floor until after 9:00 AM.

Undergraduate Poster Session

Wednesday Afternoon, Exposition Floor, 400 Aisle

(2030-1 P) **Methods for Improving Cytokine in Vivo Calibration During Microdialysis Sampling** PATRICK M PYSZ, University of Arkansas, Tina M Poseno, Julie Stenken N/A

(2030-2 P) **Analysis of Nicotine Levels in Electronic Cigarettes** MARTIN ERIC MILLER, Austin Peay State University, Jacob Williams, Dillon Burrow

(2030-3 P) **Thermal Behavior of Barium and Strontium Carbonates** CHARLES MANSFIELD EARNEST, Berry College, Ethan Miller

(2030-4 P) **Improved Quantification of Gibbsite in Bauxite Ores** CHARLES MANSFIELD EARNEST, Berry College, Karla Gann, Britney Stong

(2030-5 P) **Signal Enhancement Compensation in ICP-MS Analysis for Arsenic in the Presence of Organics** ADAM KAGEL, University of California, Davis, Carla Kagel, Richard Kagel, Nels Worden

(2030-6 P) **Isotope Labeling in Astrobiology: Ethanol as a Carbon Source** NICOLE G PERKINS, California State Polytechnic University, Pomona, Gregory A Barding, Rakesh Mogul

(2030-7 P) **Sulfidation of Silver Nanoparticles** NATHANIEL D FLETCHER, College of Charleston, Katherine M Mullaugh

(2030-8 P) **Photodegradation and Ecotoxicity Studies of Sertraline, Fluoxetine, and their Photodegradants** SYLVIA C DAVILA, College of Charleston, Jessica R Hinson, Allison Welch, Wendy C Cory

(2030-9 P) **Photodegradation of Bupropion and Gabapentin** NEHA V MUPPALA, College of Charleston, Kristina K Tran, Wendy C Cory

(2030-10 P) **Accelerated Degradation of Diphenhydramine** JAMES L SOLOMON, College of Charleston, Michael T Blanton, Wendy C Cory N/A

(2030-11 P) **Preliminary Investigation of the Geographical Distribution of Synthetic UV Filters in Bahamian Surface Waters** AEJIN KIM, Elmira College, Christopher B Hall, Emily J Feldpausch, Betsy A Smith, Jared S Baker

(2030-12 P) **Surface Plasma Polaritons from Voltage Charged Gold Nanotubes** SAM KONCHAN, University of Florida, Pradeep Ramiah Rajasekaran, Charles R Martin, Alec Talin

(2030-13 P) **Electroporation Facilitation via Gold-Plated Membranes** AARON WILSON, University of Florida, Juliette Experton, Charles R Martin

(2030-14 P) **The Analysis of Electronic Cigarette Solutions by ICP-MS for USP Regulated Contaminates** CHLOE E FERNANDES, Georgia Gwinnett College, Daniel H Jones

(2030-15 P) **Comparison of Cellular Response to Atorvastatin Among Selected Cell Lines by MS-Based Proteomics** GEORGE TANG, Georgia Institute of Technology N/A

(2030-16 P) **Discovery Metabolomics of Early-Stage Ovarian Cancer in a Dicer-Pten Double Knockout Murine Model** LAURA C WINALSKI, Georgia Institute of Technology, Christina Jones, Maria E Monge, Jaeyeon Kim, Martin M Matzuk, Facundo M Fernandez

(2030-17 P) **Colorimetric Detection of Pyrocatechol as a Model for Urushiol Analysis in Poison Ivy** COLLIN J STEEN, Kalamazoo College, Kari Anderson

(2030-18 P) **Quantification of Salicylates in Stomach Relief Aids Using GC-MS and UV-Vis** MARINA CHRISTINA KOETHER, Kennesaw State University, Kimberly C Powers

(2030-19 P) **Correlation Between Different Extraction Methods and the Ratio of Neral to Geranial in the Essential Oils of Lemongrass by GC-MS** MARINA CHRISTINA KOETHER, Kennesaw State University, Skyler Mize

(2030-20 P) **Determining Dissolution Testing Time of Potassium in Potassium Gluconate Tablets by Conductivity with Confirmation by Flame Atomic Absorption Spectrometry** MARINA CHRISTINA KOETHER, Kennesaw State University, Minwoo Lee

(2030-21 P) **Study of the Implementation of the Systematic Method in General Chemistry II** VICTORIA Y REINDERS, Maryville University Saint Louis, Thomas M Spudich, Natalie Ulrich

(2030-22 P) **Characterization and Identification of Biosurfactants for Oil Remediation** KAYDREN B ORCUTT, Mercer University, Justis E Ward, Joseph W Kloepper, Garland L Crawford, Kathryn D Kloepper

(2030-23 P) **Green Diesel and Biodiesel Fuel Additives and Their Effects on VOC Emissions** CHRISTOPHER MOORE, Middle Tennessee State University, Joseph Close N/A

(2030-24 P) **Investigating Dopamine Fluctuations Associated with Impulsive Decision Making Using Fast Scan Cyclic Voltammetry** BRENNEN GUZIK, North Carolina State University, Xiaohu Xie, Leslie A Sombers

(2030-25 P) **Tracking Cellular Invasion and Death in Three Dimensional Paper-Based Cultures with Quantitative PCR** CHRISTIAN A LOCHBAUM, University of North Carolina at Chapel Hill, Andrew S Truong, Matthew R Lockett

(2030-26 P) **Real-Time Measurements of Oxidative Stress During Chronic L-DOPA Treatment for Parkinson's Disease** CATHERINE F MASON, North Carolina State University

(2030-27 P) **Pulsed Chronopotentiometry with Asymmetric Cellulose Triacetate Membrane-Based Ion-Selective Electrodes for Kinetic Discrimination of Lipophilic Ions During Measurement of Chloride** SIMON SEGAL, Northern Kentucky University, Jeremy Meyers, Kebede L Gemene

(2030-28 P) **Water Quality Analysis of the Chattahoochee River** KIZGEL DAVIS-DESOUZA, Oglethorpe University, Kelly Jacobson, Md H Kabir

(2030-29 P) **Physicochemical Characterization of Biodiesel from Different Vegetable Oils and Seeds** JOSEPH GOODWIN, Oglethorpe University, Michelle Huang, Grace Djokoto, Markus Germann, Md H Kabir N/A

(2030-30 P) **Colloidal CdSe Quantum Dots: Synthesis, Characterization, and Applications** KELLY JACOBSON, Oglethorpe University, Md H Kabir, Michael Rulison

(2030-31 P) **Nanoparticle Toxicity on the Development of Brine Shrimp and Zebrafish** CAITLIN MAY, Oglethorpe University, Gregory Gabriel, Michael Rulison, Md H Kabir

(2030-32 P) **Evaluation of Phthalate Wiping Protocols for Estimation of Dermal Exposure from Consumer Products** ALEXANDRIA VAN GROUW, University of Portland, Richard Kagel, Carla Kagel, Stephen McWeeney

(2030-33 P) **Gaseous Molecular Analysis by High Resolution Coherent Multidimensional Spectroscopy (HRCMDS)** ANGELAR KANINI MUTHIKE, Spelman College, Peter Chen, Jessica Robinson

(2030-34 P) **Spectral Imaging of Plasma Optical Emission Via Compressed Sensing** JOHN D USALA, Texas Tech University, Gerardo Gamez

(2030-35 P) **Analysis of Melamine in Solid Pet Food Samples Using Gold Nanoparticles and Reversed Phase Chromatography** AARON HUMMERT, Washburn University, Seid Adem

(2030-36 P) **Using a Solvent-Free Synthesis-Grafting Method to Attach Phenanthroline and Dimethoxybenzene Groups to Glassy Carbon Electrodes** SOPHIA L MELNYK, Wittenberg University, Kristin K Cline N/A

(2030-37 P) **Does Grinding Glassy Carbon Electrodes in Diazonium Salts Lead to Covalently Bonded Groups?** CHELSEA L HORVATH, Wittenberg University, Kristin K Cline

(2030-38 P) **Comparison of Grafted and Untreated Activated Carbon as Solid Phase Extraction Media for Preconcentrating Copper and Lead Ions** MARGARET COLE, Wittenberg University, Kristin K Cline

(2030-39 P) **Barium Leaching in Ceramic Glazes** JASON HALMO, Hampden-Sydney College, Joshua Chamberlin, Paul Mueller

(2030-40 P) **Spectroelectrochemical Urinalysis: A Kinetic Assay for Uric Acid** PAUL FLOWERS, University of North Carolina at Pembroke, Sean Downes, Sonvia Brown

(2030-41 P) **Determination of Formaldehyde Concentration in Electronic Cigarettes** ROLAND LANDERS, Cumberland University, Sarah Pierce

(2030-42 P) **Matrix Targeting Peptide Impact on Tertiary Folding of Cargo Proteins** TYLER J SMITH, Truman State University, Brian P Adams, Bethany P Manning

(2030-43 P) **Electrospray Mass Spectrometry and Density Functional Theory Studies of the Estrone Fragmentation Mechanisms** YASSIN JEILANI, Spelman College, Daphney Sihwa, Gabrielle Webb, Nasrin Aweis

(2030-44 P) **Metals in Mushrooms of Western Pennsylvania** KAELYN MARGARET GRESKO, California University of PA, Kimberly A Woznack N/A

(2030-45 P) **Method Development for the Analysis of Pesticide Degradates by GC-ECD** JESSICA REILLY, Saint Francis University, Samantha Radford

(2030-46 P) **Investigation of Iron Dissolution from Pyrite Electrodes Using Electrochemical and Atomic Absorption Methods** KATELYN NUSBAUM, Saint Francis University, Brandyn Pryce, Rose A Clark

(2030-47 P) **The Efficacy of Duckweed in Reducing the Concentration of Manganese in Abandoned Mine Drainage (AMD) through Phytoremediation** REBECCA ANNE BRADNAM, Westminster College, Helen M Boylan

(2030-48 P) **Analysis of Water Quality in Western Pennsylvania near Hydraulic Fracturing Sites** KELSEY ANN KILBANE, Westminster College, Helen M Boylan, Jamie Linderman, Christina Mauri

(2030-49 P) **Electrophoretic Character of Borate Buffers in Capillary and Microfluidic Channels** LAUNICK SAINT-FORT, Pennsylvania State University-Berks Campus, James Karlinsey

(2030-50 P) **Interdisciplinary Undergraduate Research in Chemometrics: The Students' Perspective** STEPHANIE HOMITZ, Westminster College, Helen M Boylan, Christopher Caroff, Keilah Ireland, Carolyn Cuff

(2030-51 P) **Analysis of the Toluene Efflux Pumps in Microorganisms Through Bioinformatics** MATHILDA WILLOUGHBY, Westminster College, Samantha Tower, Sarah Kennedy

(2030-52 P) **Analysis of Urine Organic Acids via GC/MS-based Metabolomics to Determine the Effect of Diet on Urine Composition** JESSICA MINNICK, Georgia College and State University, Catrena Lisse N/A

(2030-53 P) **A Cascade SERS Signal Amplification Approach for Telomerase Activity at Single-cell Level** MULING SHI, Hunan University

Author and presider lists as of January 20, 2016, are available at www.pittcon.org

THURSDAY, MARCH 10, 2016
MORNING

SYMPOSIUM **Session 2040**

ACS-ANLY - Advances in Electrokinetic Methods for Bioanalysis
arranged by Alexandra Ros, Arizona State University

Thursday Morning, Room B308
Alexandra Ros, Arizona State University, Presiding

8:30		**Introductory Remarks - Alexandra Ros**
8:35	(2040-1)	**Nanofluidic Devices for Single-Particle Analysis of Virus Assembly** STEPHEN C JACOBSON, Indiana University, Zachary D Harms, Lisa Selzer, Adam Zlotnick
9:10	(2040-2)	**Amyloid Oligomers Analysis Using Microchannel Electrophoresis** CHRISTA HESTEKIN, University of Arkansas, Sadia Paracha, Melissa Moss
9:45	(2040-3)	**Surface Isoelectric Focusing (SIEF) for Therapeutic Protein Separations** ADRIENNE R MINERICK, Michigan Technological University, Zhichao Wang
10:20		**Recess**
10:35	(2040-4)	**Microfluidic-Based Electrokinetic Methods for Protein Separation and Sensing and Manipulation of Particles and Droplets** CAROLYN L REN, University of Waterloo
11:10	(2040-5)	**Towards Organelle Separation Exploiting Deterministic Absolute Negative Mobility** ALEXANDRA ROS, Arizona State University, Jinghui Luo, Edgar Arriaga, Katherine Muratore N/A

SYMPOSIUM **Session 2050**

Advances in Vibrational Spectroscopy for Medical Diagnostics
arranged by Igor K Lednev, University at Albany, SUNY

Thursday Morning, Room B302
Igor K Lednev, University at Albany, SUNY, Presiding

8:30		**Introductory Remarks - Igor K Lednev**
8:35	(2050-1)	**Stimulated Raman Scattering Microscopy as a Tool for Brain and Skin Cancer Tissue Diagnoses** SUNNEY XIE, Harvard University
9:10	(2050-2)	**Raman Spectroscopy for Clinical Cell Analysis** JÜERGEN POPP, Friedrich-Schiller University Jena
9:45	(2050-3)	**Classification of Lung Cancers by Infrared Spectral Histopathology (SHP)** MAX DIEM, Northeastern University
10:20		**Recess**
10:35	(2050-4)	**Label-Free Spectroscopic Imaging for Molecular Diagnosis** JI-XIN CHENG, Purdue University
11:10	(2050-5)	**Raman Spectroscopy of Blood for Alzheimer's Disease Diagnostics** IGOR K LEDNEV, University at Albany, SUNY, Elena Ryzhikova, Oleksandr Kazakov, Lenka Halamkova, Eric Molho, Earl Zimmerman

SYMPOSIUM **Session 2060**

Computational Chemistry Coupled to Analytical Measurements: A Synergistic Relationship
arranged by Bruce Chase, University of Delaware

Thursday Morning, Room B303
Bruce Chase, University of Delaware, Presiding

8:30		**Introductory Remarks - Bruce Chase**
8:35	(2060-1)	**Advancing the Understanding of Rigid Rod Polymers with Statistical Mechanics and Analytical Chemistry** STEVE LUSTIG, DuPont, Steven Allen, Juan David Londono, Christopher Seay
9:10		**Binding Free Energy of DNA on Graphite and Carbon Nanotubes by Single-Molecule Peeling** ANAND JAGOTA, Lehigh University N/A
9:45	(2060-3)	**Theory and Simulations of Macromolecular Soft Materials: Linking Molecular Design to Macroscale Morphology and Function** ARTHI JAYARAMAN, University of Delaware, Tyler B Martin
10:20		**Recess**
10:35	(2060-2)	**Recent Computational Studies Related to the Use of Plasmonic Materials for Analytical Applications** GEORGE C SCHATZ, Northwestern University
11:10	(2060-4)	**Differentiation of Alpha and Beta Crystalline Polymorphs in Biodegradable Poly Hydroxybutyrate** BRUCE CHASE, University of Delaware, Brian Sobieski, Isao Noda

SYMPOSIUM **Session 2070**

Identification and Analysis for Food Safety
arranged by Perry G Wang, US FDA

Thursday Morning, Room B304
Perry G Wang, US FDA, Presiding

8:30		**Introductory Remarks - Perry G Wang**
8:35	(2070-1)	**Validation and Challenge for the Determination of Chemical Components in Cosmetic Products Using LC-MS and GC-MS** PERRY G WANG, US FDA, Wanlong Zhou, Alexander J Krynitsky
9:10	(2070-2)	**Development of Multi-Functional Ambient Mass Spectrometry for Food Safety Screening and Characterizing Polymers in Packing Materials** JENTAIE SHIEA, National Sun Yat-sen University
9:45	(2070-3)	**Emerging Disinfection Byproducts Halobenzoquinones in Treated Drinking Water** XING-FANG LI, University of Alberta
10:20		**Recess**
10:35	(2070-4)	**Effects of Different Dietary Doses of Copper and High Fructose Feeding on Rat Fecal and Liver Metabolome** XIANG ZHANG, University of Louisville, Ming Song, Xiaoli Wei, Xinmin Yin, Aminul Prodhan, Craig McClain
11:10	(2070-5)	**Identification and Confirmation of Chemical Residues in Foods for Regulatory Purposes** STEVEN J LEHOTAY, USDA Agricultural Research Service

SYMPOSIUM **Session 2080**

Integrated Microfluidics
arranged by R Scott Martin, Saint Louis University

Thursday Morning, Room B305
R Scott Martin, Saint Louis University, Presiding

8:30		**Introductory Remarks - R Scott Martin**
8:35	(2080-1)	**Integrated Microfluidics for Forensic Analysis: Creating Simple, Portable and Cost-effective Systems** JAMES LANDERS, University of Virginia
9:10	(2080-2)	**Microfluidic Technology for Protein Crystallization and Pharmaceutical Solid form Screening** PAUL KENIS, University of Illinois at Urbana Champaign
9:45	(2080-3)	**Integrated Microfluidic Systems for Measuring Secretion from Cellular Networks** MICHAEL G ROPER, Florida State University, Lian Yi, Adrian M Schrell, Xue Wang, Nikita Mukhitov, Basel Bandak, Kimberly Evans
10:20		**Recess**
10:35	(2080-4)	**New Strategies for Enhancing Human-on-Chip Systems** DANA SPENCE, Michigan State University
11:10	(2080-5)	**Integrated Microfluidic Platform for Mass Spectrometry based Metabolomics** JAMES L EDWARDS, Saint Louis University

SYMPOSIUM — Session 2090

New Bioanalytical Separations for Molecular Mechanisms of Disease
arranged by B Jill Venton, University of Virginia and Lisa A Holland, West Virginia University

Thursday Morning, Room B309
B Jill Venton, University of Virginia, Presiding

8:30		**Introductory Remarks - B Jill Venton and Lisa A Holland**
8:35	(2090-1)	**Capillary Separations that Unravel Molecular Mechanisms of Endocrine Dysfunction** LISA A HOLLAND, West Virginia University, Vincent T Nyakubaya, Jennifer R Stueckle
9:10	(2090-2)	**Separation-Based Methods for Measuring Reactive Oxygen and Nitrogen Species in Biological Samples** SUSAN M LUNTE, University of Kansas
9:45	(2090-3)	**How Separations—Both High and Low Resolution—Enable Selection of Clinically Useful Aptamers** REBECCA WHELAN, Oberlin College
10:20		**Recess**
10:35	(2090-4)	**Merging Microfluidics, Electrophoresis, and Mass Spectrometry for Protein Assays** ROBERT T KENNEDY, University of Michigan
11:10	(2090-5)	**CE-FSCV for Determining Neurotransmitter Tissue Content in Drosophila Disease Models** B JILL VENTON, University of Virginia, Madelaine E Denno, Ryan Borman

SYMPOSIUM — Session 2100

Overcoming the Obstacles to Making Measurements in the Brain
arranged by Stephen Weber, University of Pittsburgh

Thursday Morning, Room B310
Stephen Weber, University of Pittsburgh, Presiding

8:30		**Introductory Remarks - Stephen Weber**
8:35	(2100-1)	**Probing the Spatial and Temporal Dynamics of Signaling Peptides in the Nervous Systems by a Multi-Faceted MS Approach** LINGJUN LI, University of Wisconsin
9:10	(2100-2)	**Overcoming Obstacles to Understanding Voltammetric Measurements of Dopamine in the Brain** ADRIAN C MICHAEL, University of Pittsburgh, Seth H Walters, Elaine M Robbins, Brendan P Sestokas, Andrea Jaquins-Gerstl
9:45	(2100-3)	**Simultaneous Optical Imaging of Neuronal, Glia, and Hemodynamic Waves During Seizures** HONGTAO MA, Weill Cornell Medical College, Andy Daniel, Eliza Baird-Daniel, Philippe Laffont, Mingrui Zhao, Theodore H Schwartz
10:20		**Recess**
10:35	(2100-4)	**Placing New Pieces in the Puzzle of Human Traumatic Brain Injury Using Multimodal Monitoring** MARTYN G BOUTELLE, Imperial College London, Michelle L Rogers, Sally A Gowers, Chi Leng Leong, Vassilios Kontojannis, Anthony J Strong, Sharon Jewell
11:10	(2100-5)	**Electroosmotic Perfusion in Brain Tissue for Determining Ectopeptidase Activity** STEPHEN WEBER, University of Pittsburgh, Stephen R Groskreutz, Khanh Ngo, Yangguang Ou, Rachael Wilson, Jenna DeVivo, Bocheng Yin

SYMPOSIUM — Session 2110

SAS - Handheld Spectrometers
arranged by Richard A Crocombe, PerkinElmer and Mark A Druy, Technology Assessment & Partnering

Thursday Morning, Room B311
Richard A Crocombe, PerkinElmer, Presiding

8:30		**Introductory Remarks - Richard A Crocombe and Mark A Druy**
8:35	(2110-1)	**Handheld Laser-Induced Breakdown Spectroscopic Instruments** AMY JO RAY BAUER, TSI, Incorporated
9:10	(2110-2)	**Chemometrics in Action: Moving the Lab to the Field** SUZANNE SCHREYER, Thermo Fisher Scientific, Michael Hargreaves
9:45	(2110-3)	**Mass Spectrometry in Miniature** CHRISTOPHER BROWN, 908 Devices
10:20		**Recess**
10:35	(2110-4)	**Next Generation Portable Spectrometers: Spectroscopy Solutions Wherever You Want Them** KATHERINE A BAKEEV, B&W Tek, Inc, Ken Li, Sean Wang, Jing Li, Jack Zhou
11:10	(2110-5)	**High Sensitivity Measurements in Liquids Using Mid-IR Lasers** DON KUEHL, RedShift Systems, Rick Sharp, Eugene Ma, Jinghong Kim, Chip Marshall

SYMPOSIUM — Session 2120

Single Cell Molecular Analysis
arranged by Milan Mrksich, Northwestern University

Thursday Morning, Room B312
Milan Mrksich, Northwestern University, Presiding

8:30		**Introductory Remarks - Milan Mrksich**
8:35	(2120-1)	**Using Single Molecule Arrays (Simoa) to Measure Proteins and Nucleic Acids in Single Cells** DAVID R WALT, Tufts University, Stephanie Walter, Stephanie Schubert, Payal Ghatak
9:10	(2120-2)	**Semiconducting Polymer Dots (Pdots) for Single-Cell Sensing and Molecular Analysis** DANIEL T CHIU, University of Washington
9:45	(2120-3)	**Assaying Single Cells as a Diagnostic Tool** NANCY ALLBRITTON, University of North Carolina at Chapel Hill
10:20		**Recess**
10:35	(2120-4)	**Using SAMDI Mass Spectrometry to Measure Enzyme Activity in Single Cell Lysates** MILAN MRKSICH, Northwestern University
11:10	(2120-5)	**A Novel Tool for Detection and Enumeration of Circulating Tumor Cells** DAVID A GILJOHANN, AuraSense

ORGANIZED CONTRIBUTED SESSIONS — Session 2130

Recent Advances in Ion Analysis
arranged by Kannan Srinivasan, Thermo Fisher Scientific

Thursday Morning, Room B313
Kannan Srinivasan, Thermo Fisher Scientific, Presiding

8:30	(2130-1)	**Anion Analysis Using Capillary Ion Chromatography of Steam Cycle Water and Anion Analysis of Reactor Water by Matrix Elimination in a Nuclear Power Plant** RICHARD WALLWORK, Pacific Gas and Electric
8:50	(2130-2)	**New Developments in Multidimensional Analysis for Drinking Water Applications** HERB WAGNER, Independent Contractor
9:10	(2130-3)	**Recent Advances in Suppressor Technology in Ion Chromatography for Achieving Low Noise Performance** KANNAN SRINIVASAN, Thermo Fisher Scientific, Brittany Omphroy, Sheetal Bhardwaj, Rong Lin, Christopher Pohl
9:30	(2130-4)	**Indirect Pulsed Electrochemical Detection following Anion-Exchange Chromatography** BILL LACOURSE, University of Maryland Baltimore County N/A
9:50		**Recess**
10:05	(2130-5)	**Revisiting the Many Facets of Ion Exclusion Chromatography** C PHILLIP SHELOR, University of Texas Arlington, Purnendu Dasgupta
10:25	(2130-6)	**Factors Underlying Recent Advances in High Speed and High Resolution Ion Chromatography** CHARLES A LUCY, University of Alberta
10:45	(2130-7)	**Peak Shapes in High Efficiency and Fast Chromatography: Contributions from the Slurry Packing Process and Detector Settings** M FAROOQ WAHAB, University of Texas at Arlington, Purnendu Dasgupta, Daniel W Armstrong
11:05	(2130-8)	**Recent Developments in Stationary Phases for Ion Chromatography** CHRISTOPHER POHL, Thermo Fisher Scientific, Charanjit Saini, Mani Jayaraman, Maria Rey, Andy Woodruff

ORAL SESSIONS **Session 2140**

Bioanalytical: Fluorescence/Luminescence Techniques

Thursday Morning, Room B405
Nathan Whitman, University of North Carolina at Chapel Hill, Presiding

8:30	(2140-1)	**Ultrasensitive Detection of Ricin Toxin in Multiple Sample Matrixes Using Single-Domain Antibodies** TRINH DINH, Tufts University, Shonda Gaylord, David R Walt, Kevin Ngan
8:50	(2140-2)	**Unusual Red Emission of Graphene Quantum Dots at Extremely High pH** YIYANG LIU, University of Kentucky, Doo Young Kim
9:10	(2140-3)	**Bright Large Stokes' Shift NIR Fluorescent Silica Nanoparticle Labels and Probes** GABOR PATONAY, Georgia State University, Gala Chapman, Maged Henary, Kyle Emer
9:30	(2140-4)	**Development of Fluorescent Magnetic Particles for "On-Off" Switching based Detection of Various Lectin-Saccharide Interactions** SUZUKI YOSHIO, AIST
9:50		**Recess**
10:05	(2140-5)	**Cell-free Expression of Cytochrome P450-containing Liposomes for Drug Metabolism Screens** NATHAN A WHITMAN, University of North Carolina at Chapel Hill, Julie C McIntosh, Jeffrey B Penley, Matthew R Lockett
10:25	(2140-6)	**Tethered Cationic Lipoplex Nanoparticle Biochip for Fast Disease Detection** JIAMING HU, The Ohio State University, L J Lee N/A
10:45	(2140-7)	**Cell Surface Engineering with Lipid-Molecular Beacon Aptamer for Real Time Probing of Proteins in Cellular Microenvironment** WEIJIA HOU, University of Florida
11:05	(2140-8)	**Ratiometric Fluorescence Assay based Quantum Dots for Recognition of RRE RNA Ligands Using A New Fluorescence Indicator** ZHIQI ZHANG, Shaanxi Normal University N/A

ORAL SESSIONS **Session 2150**

Electrochemistry - New Methods and Applications

Thursday Morning, Room B314
Leslie A Sombers, North Carolina State University, Presiding

8:30	(2150-1)	**Elucidating the Structure/Function Relationship of Conductive Polymer Microelectrodes for Use in Fast-Scan Cyclic Voltammetry of Neurotransmitters** ADAM R MEIER, University of Arizona, William Bahureksa, Michael L Heien
8:50	(2150-2)	**Electrochemically Reduced Graphene Oxide as an Electrocatalyst Support for H_2S Detection** JASON A BENNETT, Penn State Erie - The Behrend College
9:10	(2150-3)	**Amperometric Determination of Aurocyanide for Hydrometallurgical Gold Processing** WAYNE DICKINSON, Kemira
9:30	(2150-4)	**Method for Removal of Non-Faradaic Contributions to Fast-Scan Cyclic Voltammetry Recordings** JUSTIN A JOHNSON, University of North Carolina at Chapel Hill, R Mark Wightman
9:50		**Recess**
10:05	(2150-5)	**Quantitative Analysis of Microiontophoresis Drug Delivery** DOUGLAS KIRKPATRICK, University of North Carolina at Chapel Hill, R Mark Wightman
10:25	(2150-6)	**Fast Scan Cyclic Voltammetry of Metals at Carbon-Fiber Microelectrodes: Correlation Between FSCV Response and Solution Dynamics** PAVITHRA PATHIRATHNA, Wayne State University, Thushani Siriwardhane, Shawn McElmurry, Parastoo Hashemi
10:45	(2150-7)	**Construction of Training Sets for Valid Calibration Using Principal Component Analysis** NATHAN T RODEBERG, University of North Carolina at Chapel Hill, Justin A Johnson, R Mark Wightman
11:05	(2150-8)	**Electrochemical Characterization and Catalytic Activity of Ultrasmall Gold Nanoparticles** FRANCIS ZAMBORINI, University of Louisville, Rafael Masitas, Stacy Allen, Dhruba Pattadar

ORAL SESSIONS **Session 2160**

Food Safety Evaluations - Half Session

Thursday Morning, Room B407
Alice Chen, The Pittsburgh Conference, Presiding

8:30	(2160-1)	**Characterization of the Adulteration, Counterfeiting and Contamination of Spices, Spice Products and Supplements by the Detection of Toxic and Banned Organic Chemicals in Commercial Botanical Products** PATRICIA ATKINS, SPEX CertiPrep
8:50	(2160-2)	**Using SPE to Adjust Sensitivity to Analytical Requirements for Food Safety** CHRIS SHEVLIN, Horizon Technology, William Jones
9:10	(2160-3)	**Ensuring a Safe and Stable Food Supply Using ICP-OES for Elemental Monitoring** NICK SPIVEY, PerkinElmer Inc., Kenneth Neubauer, Stan Smith, Laura Thompson
9:30	(2160-4)	**Automated Solid Phase Extraction and Quantitative UHPLC Analysis of Cannabis Compounds in Food Matrices** CHRIS SHEVLIN, Horizon Technology, Robert E Buco

ORAL SESSIONS **Session 2170**

Liquid Chromatography Column Chemistry

Thursday Morning, Room B406
A Carl Sanchez, Phenomenex, Presiding

8:30	(2170-1)	**Functionalized Octatetrayne as Novel Carbon Media for Capillary Liquid Chromatography** JIAYI LIU, The Ohio State University, Susan Olesik
8:50	(2170-2)	**Stationary Phases Based on the Thiol-ene Reaction on Mercaptopropylsilane-Modified Nonporous Silica** ERIN SHIELDS, University of Pittsburgh, Kayla Thomas, Stephen Weber
9:10	(2170-3)	**Using 5 Micron Superficially Porous Particles in Capillary and Microfluidic LC Columns** JAMES P GRINIAS, University of Michigan, Robert T Kennedy
9:30	(2170-4)	**Selectivity in Reversed-Phase Liquid Chromatography: Impact of Stationary Phase Chemistry** DAVID S BELL, Supelco/Sigma-Aldrich
9:50		**Recess**
10:05	(2170-5)	**A Specialty Column for High Resolution Separation of Aminoglycoside Antibiotics by HPLC** XUEFEI SUN, Thermo Fisher Scientific, Xiao Cui, Yoginder Singh, Xiaodong Liu
10:25	(2170-6)	**Liquid Chromatography, Hydrophilic Interaction Chromatography, HILIC** DAVID S BELL, Supelco/Sigma-Aldrich
10:45	(2170-7)	**The Benefits of 1mm ID UHPLC Columns Made Real** STEPHEN LUKE, Agilent Technologies, Norwin Von Doehren, William Long, Jason Link
11:05	(2170-8)	**Comparison of the Practical Kinetic Performance Limits of Core-Shell and Fully Porous (U)HPLC Sorbents Using Commercially Available Column Formats** A CARL SANCHEZ, Phenomenex, Gareth Friedlander, Jason Anspach, Tivadar Farkas

ORAL SESSIONS **Session 2180**

Microfluidics/Lab-on-a-Chip - Bioanalytical I

Thursday Morning, Room B403
Hubert MacDonald, The Pittsburgh Conference, Presiding

8:30	(2180-1)	**A Method for Measurement of Temporally Resolved Insulin Secretion from Islets of Langerhans in Response to Fatty Acid Hydroxy Fatty Acids** BASEL BANDAK, Florida State University, Lian Yi, Nikita Mukhitov, Michael G Roper
8:50	(2180-2)	**Macro-to-Microfluidic Interfacing for Primary Endocrine Cell Culture and Sampling Using 3D Printed Device Templates and Fluidic Manifolds** JESSICA C BROOKS, Auburn University, Mark D Holtan, Katarena Ford, Dylan Holder, Christopher J Easley
9:10	(2180-3)	**A Simple Droplet Microfluidic Capillary Viscometer Based on Droplet Frequency for Rheological Measurements of Proteins** MICHAEL F DELAMARRE, University of Illinois at Chicago, Scott A Shippy
9:30	(2180-4)	**Improving Detection Sensitivity in Microchip Electrophoresis-Laser Induced Fluorescence Assays by Target-Induced Exonuclease Assisted Strand Circle Signal Amplification** SHULIN ZHAO, Guangxi Normal University, Yingfeng Qin, Liangliang Zhang, Yong Huang, Yi-Ming Liu

9:50		Recess
10:05	(2180-5)	**A Complementary Method to CD4 Counting: Measurement of CD4+/CD8+ T Lymphocytes Ratio in a Serial Microfluidic System** WENJIE LI, Texas Tech University, Dimitri Pappas
10:25	(2180-6)	**PDMS-Based Injection Valves for SPE-MS Analysis of Biomolecules** JAMES P GRINIAS, University of Michigan, Colleen E Dugan, Robert T Kennedy
10:45	(2180-7)	**Development of an Online Microchip Electrophoresis with LED-Induced Fluorescence System for In Vivo Detection of Excitatory Amino Acid Neurotransmitters Following a Traumatic Brain Injury** MICHAEL L HOGARD, University of Kansas, Nathan Oborny, Elton E Melo Costa, Susan M Lunte, Craig E Lunte
11:05	(2180-8)	**Determination of Amplification of Cellular Effects by Hormones Derived from Different Tissues** KESHAVARZ HAMIDEH, Michigan State University, Dana Spence

ORAL SESSIONS — **Session 2190**

Novel Applications with Gas Chromatography Mass Spectrometry

Thursday Morning, Room B315
Rudolf Addink, Toxic Report, Presiding

8:30	(2190-1)	**How GC-MS with Cold EI Improves NIST Library Identification** AVIV AMIRAV, Tel Aviv University, Uri Keshet, Tal Alon
8:50	(2190-2)	**Use of Automated Column Chromatography Clean Up with Reduced Solvent Volume in POPs Analysis** RUDOLF ADDINK, Toxic Report, Philip Bassignani
9:10	(2190-3)	**Effective QuEChERS Cleanup and Quantitation of Planar Pesticides from Spinach and Other Food Matrices Using a Novel Graphitized Carbon Black and a Zirconia-Based Adsorbent** PATRICK MYERS, Supelco/Sigma-Aldrich, William Betz, Bill Ozanich, Jennifer Claus, Michael Ye
9:30	(2190-4)	**Use of Micro Scale Solid Phase Extraction and Automated Clean Up in POPs Analysis of Human Milk and Serum** RUDOLF ADDINK, Toxic Report
9:50		Recess
10:05	(2190-5)	**Multidimensional Comprehensive Gas Chromatography Multireflection High Resolution Time-of-Flight Mass Spectrometry: Combining Accurate Mass Information with Ultra-High Chromatographic Resolution** RALF ZIMMERMANN, University Rostock/HMGU, Thomas Groeger, Benedikt Weggler, Juergen Wendt
10:25	(2190-6)	**Rapid, New Methods for the Analysis of 3-MCPD and 1,3 DCP in Soy Sauce** SUSAN GENUALDI, US FDA, Lowri DeJager, Patsy Nyman
10:45	(2190-7)	**A Method Development Software Tool for Comprehensive Two-Dimensional Gas Chromatography Evaluated for GCxGC-TOFMS** MARK F MERRICK, LECO Corporation, Viatcheslav Artaev, Leonid M Blumberg
11:05	(2190-8)	**Innovative TG-GC-MS Methods for Thermal Degradation Studies of Polymers** KRISTINA LILOVA, Setaram Inc., Link Brown

ORAL SESSIONS — **Session 2200**

Novel Synthesis and Applications of Nanomaterials

Thursday Morning, Room B316
David Pensenstadler, The Pittsburgh Conference, Presiding

8:30	(2200-1)	**Nanofibers from Hydrothermal Treatment of Cellulose Nanocrystals** YIMEI WEN, Clemson University
8:50	(2200-2)	**Graphene Nanoribbons: Engineering, Characterization of Edge Defects and Sensor Applications** PANKAJ RAMNANI, University of California, Riverside, Ashok Mulchandani
9:10	(2200-3)	**Modeling of Seed-Mediated Nanoparticle Growth** HANS MUSGRAVE, University of North Dakota, Julia Xiaojun Zhao N/A
9:30	(2200-4)	**Single-Molecule Tracking Studies of the Effects of Solvent Swelling on the Properties of Cylindrical Block Copolymer Microdomains** TAKASHI ITO, Kansas State University, Dol R Sapkota, Khanh-Hoa Tran-Ba, Daniel A Higgins
9:50		Recess
10:05	(2200-5)	**Characterization of Single and Multi-Walled Carbon Theranostic Nanovectors** MARKUS MARTINCIC, Institut de Ciencia de Materials de Barcelona, Elzbieta Pach, Belen Ballesteros, Gerard Tobias
10:25	(2200-6)	**Investigating 4-Wheeled Nanocar Diffusion Kinetics on Differently Modified Solid Surface with Single Molecule Fluorescence Microscopy (SMFM)** FANG CHEN, North Carolina State University, Víctor García-López, James Tour, Gufeng Wang
10:45	(2200-7)	**Sealing and Opening of Metallic Nanotubes with a Laser Beam: A Potential Drug Delivery Vehicle** NATHALIA ORTIZ, North Carolina State University, Gufeng Wang
11:05	(2200-8)	**Optically Transparent Carbon Nanotube Film Electrode for Thin Layer Spectroelectrochemistry** TINGTING WANG, University of Cincinnati

ORAL SESSIONS — **Session 2210**

Sampling and Sample Preparation - Bioanalytical, Neurochemistry, and Material Science

Thursday Morning, Room B404
Denise Wilkins, Bechtel Bettis Inc., Presiding

8:30	(2210-1)	**Rapid Protein Purification and Digestion with Membrane-Containing Pipette Tips** WENJING NING, Michigan State University, Merlin Bruening
8:50	(2210-2)	**DNA Extraction and Analysis Using Magnetic Ionic Liquid Solvents** KEVIN D CLARK, Iowa State University, Melissa Yamsek, Omprakash Nacham, Jared L Anderson
9:10	(2210-3)	**Localized Laser Ablation Sample Transfer for Tissue Proteomics** FABRIZIO DONNARUMMA, Louisiana State University, Michael E Pettit, Touradj Solouki, Kermit K Murray
9:30	(2210-4)	**An Ultra Sensitive Sample Preparation Approach that Eliminates the Need to Dry Down and Reconstitute** SHAHANA HUQ, Phenomenex, Matthew Brusius, Jessica Detsch, Zeshan Aqeel, Ramkumar Dhandapani
9:50		Recess
10:05	(2210-5)	**Pulled Low Flow Push-Pull Perfusion Probe Tips for Sampling from Tissue Slices** MARISSA R BECKER, University of Illinois at Chicago, Scott A Shippy, David E Featherstone
10:25	(2210-6)	**Design of Protein-Binding Membranes through Adsorption of Star-Shaped Polyelectrolytes in Membrane Pores** WEIJING LIU, Michigan State University, Salinda Wijeratne, Merlin Bruening
10:45	(2210-7)	**Direct Coupling of Solid Phase Microextraction to Mass Spectrometry Via Nano-Electrospray Ionization: Development and Applications in Bioanalysis** GERMAN A GOMEZ-RIOS, University of Waterloo, Nathaly Reyes-Garces, Ezel Boyaci, Barbara Bojko, Janusz Pawliszyn
11:05	(2210-8)	**Molecularly Imprinted Polymer-Sol-Gel for Dispersive Micro-Solid Phase Extraction** MOHAMED ABDEL-REHIM, Stockholm University, Aziza El-Beqqali N/A

ORAL SESSIONS — **Session 2220**

Sensors - Others

Thursday Morning, Room B401
Gufeng Wang, North Carolina State University, Presiding

8:30	(2220-1)	**PANi Electrospun Fibers and Drop Cast Film Sensor Array for the Detection of Small Chained Alcohols** KELVIN TRAN, University of California Riverside, Andrew J Burris, Quan Cheng
8:50	(2220-2)	**Application of Thermo-Reversible Interpenetrating Poly(Vinyl Alcohol) Networks to Stabilize Mechanically Fragile Hydrogel Sensors** ANDREW E COUKOUMA, University of Pittsburgh, Sanford A Asher N/A
9:10	(2220-3)	**Improving the Sustainability of Drinking Water Systems Using Nanostructured Biosensors for Escherichia Coli** HEATHER A CRAPO, Binghamton University, Melissa McDonald, Idris Yazgan, Omowunmi Sadik
9:30	(2220-4)	**An Innovative Biosensor to Assess Quickly the Biodegradable Organic Fraction** SULIVAN JOUANNEAU, University of Nantes, Marie-José Durand, Gerald Thouand, Ali Boukabache, Yves Primault N/A
9:50		Recess
10:05	(2220-5)	**Surface Plasmon Resonance Immunosensor Using Au Nanoparticle Modified Antibody** DULAL C KABIRAZ, Hokkaido University, Kinichi Morita, Toshikazu Kawaguchi
10:25	(2220-6)	**QCM Virtual Multisensor Array for Detection of Gasoline Adulterants** NICHOLAS SPELLER, Louisiana State University, Noureen Siraj, Isiah M Warner, Stephanie Vaughan

10:45 (2220-7) **Development of an Aptamer Functionalized Electrode Array for Real-Time In Vivo Cocaine Detection Using Square Wave Voltammetry** IAN MITCHELL TAYLOR, University of Pittsburgh, Emma Bigelow, Tracy Cui

11:05 (2220-8) **Optomechanical Switching by Plasmonic Nanoparticle Monolayers on Elastic Substrate Induced by Stretching** MAHMOUD MAHMOUD, Georgia Institute of Technology

ORAL SESSIONS — Session 2230

Vibrational Spectroscopy Instrumentation and Applications

Thursday Morning, Room B402
Jane Chan, Bechtel Bettis Inc., Presiding

8:30 (2230-1) **Experimental Studies and Electromagnetic Modeling of Localized Plasmon Surface on a New SERS Platform** MATHIEU EDELY, IMMM Institut des Molécules et Matériaux du Mans, Guy Louarn, Jean-François Bardeau, Nicolas Delorme, Ludovic Douillard N/A

8:50 (2230-2) **The Effect of Molecular Polarity and Solubility on Adsorption Rates and Equilibrium Constants for Molecules on Noble Metal Surfaces Using Surface-Enhanced Raman Spectroscopy** ERIK DAVID EMMONS, US Army, Ashish Tripathi, Neal D Kline, Jerry Cabalo, Jason A Guicheteau, Augustus W Fountain, Steven D Christesen

9:10 (2230-3) **Room Temperature Freezing and Orientation Control of Surface Immobilized Biomoelcules in Air** YAOXIN LI, University of Michigan, Zhang Xiaoxian, Somayesadat Badieyan, Zhan Chen

9:30 (2230-4) **Fiber Spectroscopy for Process Control and Medical Diagnostics** VIACHESLAV ARTYUSHENKO, Art Photonics GmbH

9:50 **Recess**

10:05 (2230-5) **Characterization of Polymer/Epoxy Buried Interfaces with Silane Adhesion Promoters Before and after Hygrothermal Aging for the Elucidation of Molecular Level Details Relevant to Adhesion** NATHAN W ULRICH, The University of Michigan, John Myers, Zhan Chen

10:25 (2230-6) **Surface Interaction of Nitrogen Containing Aromatic Molecules with Gold Investigated with Surface Enhanced Raman Spectroscopy (SERS)** ASHISH TRIPATHI, Leidos, Inc., Erik David Emmons, Augustus W Fountain, Jason A Guicheteau, Steven D Christesen, Martin Moskovits

10:45 (2230-7) **Transmission Raman Spectroscopy as a Regulatory-Approved Method for Content Uniformity Analysis – Replacing HPLC** DARREN ANDREWS, Cobalt Light Systems, Julia Griffen, Matthew Bloomfield, Andrew Owen, Mark Mabry, Pavel Matousek

11:05 (2230-8) **A New Microscope for FT-IR Microspectrometry** DAVID W SCHIERING, Czitek, Gregg Ressler

POSTER SESSION — Session 2240

All posters are to be mounted by 10:00 AM and remain on display until 2:00 PM. Authors must be at their posters from 10:00 AM to 12:00 PM. Location of the posters is on the Exposition Floor, 400 Aisle. PLEASE NOTE: You cannot get onto the Exposition Floor until after 9:00 AM.

Application of Mass Spectrometry

Thursday Morning, Exposition Floor, 400 Aisle

(2240-1 P) **Variations on a Theme: The Detection of NBOMe Designer Drugs on Blotter Paper by High Resolution Time-of-flight Mass Spectrometry (TOFMS) with and without Chromatography** DAVID BARAJAS, Boston University School of Medicine, Frank A Kero, Noelle Elliot, Bogdan Bogdanov, Craig Young, Jason Weisenseel, Sabra Botch

(2240-2 P) **Proton Transfer Reaction – Mass Spectrometry: Automated Measurement and Evaluation** JENS HERBIG, IONICON Analytik, Klaus Winkler, Johann Seehauser, Lukas Mäerk, Christian Lindinger, Alfons Jordan

(2240-3 P) **Functionalized Gold Surface for SPR-MS Determination of Enzymatic Activities and Specificity of Lectins** HYOJIK YANG, University of California, Riverside, Quan Cheng

(2240-4 P) **High-Resolution Atmospheric Pressure Drift Tube Ion Mobility Spectrometry Coupled with High-Resolution Accurate Mass Orbitrap Mass Spectrometry** JOEL D KEELOR, Georgia Institute of Technology, Facundo M Fernandez, Brian Clowers

(2240-5 P) **Real-Time Metabolome Analysis by Probe Electrospray Ionization-Tandem Mass Spectrometry (PESI-MS/MS): Preliminary Challenge to Real-Time Metabolomics** ZAITSU KEI, Nagoya University, Hayashi Yumi, Murata Tasuku, Nakajima Hiroki, Nakajima Tamie, Kusano Maiko, Tsuchihashi Hitoshi, Ishikawa Tetsuya, Ishii Akira

(2240-6 P) **Use of High Speed/High Resolution Size Based Chromatographic Separation of Surfactants and Oligomeric Materials with Single Quadrupole Mass Spectrometry Detection** MICHAEL OLEARY, Waters Corporation

(2240-7 P) **Coupling Surface Acoustic Wave Nebulization (SAWN) with Vacuum-Assisted Plasma Ionization (VaPI) Mass Spectrometry for Enhanced Ionization and Transmission Efficiency** STEPHEN C ZAMBRZYCKI, Georgia Institute of Technology, Matthew C Bernier, Joel D Keelor, Fernandez M Facundo, Sung H Yoon, David Goodlett

(2240-8 P) **Withdrawn**

(2240-9 P) **Withdrawn**

(2240-10 P) **Visualizing the Distribution of Volatile and Semi-Volatile Compounds by Low Temperature Plasma Mass Imaging (LTP-MSI)** ROBERT WINKLER, CINVESTAV Unidad Irapuato, Sandra Martinez Jarquin, Abigail Moreno Pedraza

POSTER SESSION — Session 2250

All posters are to be mounted by 10:00 AM and remain on display until 2:00 PM. Authors must be at their posters from 10:00 AM to 12:00 PM. Location of the posters is on the Exposition Floor, 400 Aisle. PLEASE NOTE: You cannot get onto the Exposition Floor until after 9:00 AM.

Fluorescence and Luminescence

Thursday Morning, Exposition Floor, 400 Aisle

(2250-1 P) **Determination of Thiamine by Flow Analysis System Based on the Chemiluminescence Inhibition Using Multicommutation** DEBORAH AZZI, Federal University of São Carlos, Marina Batistão, Bruno C Janegitz, Marina Baccarin, Geiser Oliveira, Orlando Fatibello-Filho

(2250-2 P) **Selection of Aptamers Targeting B-Cell Receptor (BCR) Using Antibody Guided Cell-Selex Method - A Novel Approach** SHOMI CHAKRABARTI, City University of New York, The Graduate Center, Hasan E Zumrut, George Maio, Prabodhika Mallikaratchy, Mst Naznin Ara

(2250-3 P) **Spectral Encoders: Detecting Position-Dependent Luminescent Spectra Through Tissue** MELISSA M ROGALSKI, Clemson University, Bobby Smith, Hunter Pelham, Nakul Ravikumar, John D DesJardins, Jeffrey N Anker

(2250-4 P) **Photoluminescence of Novel Osmium and Ruthenium Complexes in the Presence of Polyanions** MEHRUN UDDIN, St. John's University, Cody Piotrowski, Besiana Kurti, Armando Seitllari, Elise Megehee, Enju Wang

(2250-5 P) **Quantifying Free Fatty Acid Uptake Dynamics in Primary Adipocytes Using Customized Micro-Wells Made with 3D-Printed Templates** TESFAGEBRIEL M HAGOS, Auburn University, Jessica C Brooks, Christopher J Easley

(2250-6 P) **Toward Non-Invasively Detecting Radiolabeled Analytes Near Implanted Medical Devices Coated in Radioluminescent Phosphors** GRETCHEN B SCHOBER, Clemson University, Jeffrey N Anker

(2250-7 P) **Development of Silicone Filled Optical Module for Laser Fluorescence Trace Molecular Detection** HIROKAZU HIGUCHI, Kyushu University, Hiroaki Nomada, Hiroaki Yoshioka, Kinichi Morita, Yuji Oki

(2250-8 P) **Recombinant Tobacco Peroxidase: A 100-Fold More Effective Luminescent Label Than Horseradish Peroxidase** IRINA GAZARYAN, Pace University, Galina Zakharova, Andrey Poloznikov, Dmitry Hushpulian, Vladimir Tishkov

(2250-9 P) **Using Diffusional Motion to Gauge Fluidity and Interfacial Adhesion Strength of Supported Octadecylphosphonic Acid (ODPA) Monolayers** STEPHEN BAUMLER, Michigan State University

(2250-10 P) **Improving Selectivity of Fluorimetric Water Sensing in Aprotic Solvents** KATARZYNA KŁUCIŃSKA, University of Warsaw, Patryk Rzepiński, Michał Cyrański, Krzysztof Maksymiuk, Agata Michalska N/A

(2250-11 P) **Irradiation of Gold Nanodots by Ultraviolet Light: Modulation of Ligand Density and Photoluminescence** YU-TING TSENG, National Taiwan University

(2250-12 P) **A Preliminary Study of Factors Affecting Quenching or Enhancement of Fluorescence of 1,10-Phenantrololine Toward Quantitative Determination of Selected Metals in mine Drainage and Related Samples** MARK T STAUFFER, University of Pittsburgh - Greensburg, Tell M Lovelace N/A

(2250-13 P) **Traceable Mercury Gas Phase Calibrations Based Upon Gravimetry** ANNARITA BALDAN, VSL, Janneke van Wijk, Hugo Ent

(2250-14 P) **Detection of Caffeine Using a Field-Portable Fluorescence Device** HALEY CURTIS, Tennessee Technological University, Andrew Callender

(2250-15 P) **The Quenching of Riboflavin Fluorence by Nicotine in Bicontinuous Microemulsion** MAURICE O IWUNZE, Morgan State University

(2250-16 P) **Optimizing Fluorescence Tagging Strategy to Study Single Molecule Diffusion on Air-Solid Surface** TAO JIN, North Carolina State University, Fang Chen, James Tour, Victor Lopez, Gufeng Wang

TECHNICAL PROGRAM

POSTER SESSION — Session 2260

All posters are to be mounted by 10:00 AM and remain on display until 2:00 PM. Authors must be at their posters from 10:00 AM to 12:00 PM. Location of the posters is on the Exposition Floor, 400 Aisle. PLEASE NOTE: You cannot get onto the Exposition Floor until after 9:00 AM.

Fuels, Energy and Petrochemical

Thursday Morning, Exposition Floor, 400 Aisle

(2260-1 P) **Increasing Resolution of Propylene Glycol Impurities with High-Efficiency GC Columns** RAMKUMAR DHANDAPANI, Phenomenex, Tim Anderson, Kristen Parnell

(2260-2 P) **Analysis of Impurities in Propane/Propylene Streams Using a Pulsed Flame Photometric Detector (PFPD)** CYNTHIA ELMORE, OI Analytical, Michael Duffy, Brian Mistovich, J Garrett Slaton

(2260-3 P) **High C6+ Analysis** MATTHEW MONAGLE, AIC LLC N/A

(2260-4 P) **High Performance Chromatographic Diatomaceous Earth (DE)** KATARINA ODEN, Restek, Jaap de zeeuw, Barry Burger, Linx Waclaski, Rebecca Stevens

(2260-5 P) **Methanizer – A Simple Solution for CO/CO2 Gas Analysis by Gas Chromatography (GC)** KATARINA ODEN, Restek, Jaap de zeeuw, Barry Burger, Linx Waclaski, Rebecca Stevens

(2260-6 P) **Analysis of A Sulfur Mixture in Hydrocarbon Standard Cylinder in ppb Level** YANG QIN, Air Liquide Specialty Gases

(2260-7 P) **Reactive Pyrolysis-GC/MS of Polymers in a Steam Environment Used to Study Potential Bio-oil** KAREN SAM, CDS Analytical, Gary Deger, Steve Wesson

(2260-8 P) **Extended Lower Detection Range for Hydrogen Sulfide and Carbonyl Sulfide with Metal Surface Deactivated Sample Inlet for Micro Gas Chromatography** REMKO VAN LOON, Agilent Technologies, Thomas Szakas, Coen Duvekot

(2260-9 P) **The Improvement of ASTM D3612 TOGA Analysis** MAX WANG, Shimadzu Scientific Instrument Inc., Clifford M Taylor, Marty Smith

(2260-10 P) **Accurate and Reproducible Determination of Halogens in Coal Using Combustion Ion Chromatography** CARL FISHER, Thermo Fisher Scientific, Daniel Khor, Mark Manahan, Adelon Augustin, Kirk Chassaniol

(2260-11 P) **Nitrogen Determination in Lubricant by Elemental Analyzer** GUIDO GIAZZI, Thermo Fisher Scientific, Liliana Krotz, Francesco Leone

(2260-12 P) **New Methodology for the Analysis of Silicon in Petrochemical Samples by ICP-MS** ANTHONY PALERMO, PerkinElmer, Kenneth Neubauer, Stan Smith

(2260-13 P) **Artificial Photosynthesis** MINGMING WANG, Auburn University, Wei Zhan, Chao Li

(2260-14 P) **Rapid Measurement of Xylose and Glucose for Monitoring Corn Stover Fermentation in Bioethanol Production** WILLIAM MILLER, YSI, Inc, June Klingensmith

(2260-15 P) **Safety and Performance Testing of Li-ion Cells Using Thermal Analysis** PETER RALBOVSKY, Netzsch Instruments, Bob Fidler

(2260-16 P) **The Determination of Mercury in Liquefied Petroleum Gas – A Comparison of Sampling Techniques** MATTHEW A DEXTER, P S Analytical, C A Rogers, Warren T Corns

(2260-17 P) **Salen Quinoxalinol Ligands for Selective Coordination and Sensors** ANNE E GORDEN, Auburn University

(2260-18 P) **A New Hydroxide Selective Anion Exchange Phase for Ion Chromatography** CHARANJIT SAINI, Thermo Fisher Scientific, Christopher Pohl, Yan Liu

(2260-19 P) **Physicochemical Characterization of Microalgal Biodiesel** KIZGEL DAVIS-DESOUZA, Oglethorpe University, Grace Djokoto, Md H Kabir

(2260-20 P) **Graphene Quantum Dots Immobilized Nanoporous N-TiO2 Thin Films for Efficient Photocatalytic Water Splitting** NAMAL WANNINAYAKE, University of Kentucky, Syed Z Islam, Allen Reed, Stephen E Rankin, Doo Young Kim

(2260-21 P) **Techniques for Polysaccharide Research at the Complex Carbohydrate Research Center** ROBERTO SONON, Complex Carbohydrate Research Center, Stephanie A Archer-Hartmann, Zhirui Wang, Ian Black, Radnaa Naran, Mayumi Ishihara, Christian Heiss, Dandan Zhou, Artur Muszynski, Scott Forsberg, Asif Shajahn, Justyna Dobruchowska, Parastoo Azadi

(2260-22 P) **Application of Laser Induced Breakdown Spectroscopy (LIBS) in Analysis of Out Crop Samples from the Marcellus Shale** JINESH JAIN, AECOM, Alexander Bol'shakov, Herve Sanghapi, Christina Lopano, Dustin McIntyre, Richard Russo

(2260-23 P) **Direct Detection of Hydrocarbons from Microalgae Using Low Temperature Plasma – Mass Spectrometry (LTP-MS)** ABIGAIL MORENO PEDRAZA, Cinvestav, Robert Winkler

(2260-24 P) **Comprehensive Online Real-Time Analysis of Natural Gas Using VUV Absorption Spectroscopy** JAMES ANTHONY DIEKMANN III, VUV Analytics, Jonathan Smuts, Michael Roecker

(2260-25 P) **IR-Spectroscopic Analyses of Chemical Adaptation Dynamics of Live Microalgal Biomass to Shifting Nutrient Conditions** FRANK VOGT, University of Tennessee N/A

(2260-26 P) **Inorganic Salt Doped with Nanoparticulate Additives for Thermal Energy Storage with Improved Radiative Heat Transfer** PHILIP D MYERS, University of South Florida Clean Energy Research Center, D Yogi Goswami N/A

(2260-27 P) **Transformer Oil Gas Analysis Using Gas Chromatography – Vacuum Ultraviolet Absorbance Spectroscopy** JONATHAN SMUTS, VUV Analytics, Lindsey Nichole Shear, James Anthony Diekmann III, Jeff Tenney, Andy Shkolnik

(2260-28 P) **Ionic Liquids as Electrolytes for Electrochemical Double-Layer Capacitors: Structures that Optimize Specific Energy** MARAL PS MOUSAVI, University of Minnesota, Benjamin E Wilson, Sadra Kashefolgheta, Evan Anderson, Andreas Stein, Philippe Buhlmann

(2260-29 P) **Using Time Resolved FT-IR-ATR Spectroscopy to Study Biodiesel Fuel Diffusion in Flexible Elastomer Materials** JAMES M SLOAN, U.S. Army Research Laboratory N/A

(2260-30 P) **Engine State Monitoring Technology based on Engine Exhaust Test** WEIKUI WANG, Beijing Research Institute of Telemetry, Xiantao Yang, Yi Zheng N/A

POSTER SESSION — Session 2270

All posters are to be mounted by 10:00 AM and remain on display until 2:00 PM. Authors must be at their posters from 10:00 AM to 12:00 PM. Location of the posters is on the Exposition Floor, 400 Aisle. PLEASE NOTE: You cannot get onto the Exposition Floor until after 9:00 AM.

New Developments in GC

Thursday Morning, Exposition Floor, 400 Aisle

(2270-1 P) **Introduction of a New, High-Quality, Cost Efficient Headspace GC Autosampler** MAX WANG, Shimadzu Scientific Instrument Inc., Clifford M Taylor, Marty Smith

(2270-2 P) **Unique Selectivity: The Power of Ionic Liquid GC Columns** LEONARD M SIDISKY, Supelco/Sigma-Aldrich, Greg A Baney, James L Desorcie, Gustavo Serrano, Daniel Shollenberger, Michael D Buchanan

(2270-3 P) **Use of Traditional/Apex Track Integration in Commercially Available Software for GC Data Generated for EPA Methods 8082 and 8015** TOM KWOKA, PerkinElmer, Bill Hahn, Sharanya Reddy

(2270-4 P) **Evaluation of GC Conditions Using Inert Nano Stationary Phase GC Columns** ALLEN BRITTEN, Cape Breton University, Krishnat Naikwadi, Amy Clemens, Robyn Novorolsky, Allison Clarke N/A

(2270-5 P) **Evaluation of Gas Chromatographic Liner Deactivation when Exposed to Various Solvents and Extracts** LINX WACLASKI, Restek, Jack Cochran, Scott Adams, Rebecca Stevens, Dan Li

(2270-6 P) **Development of an Innovative New Thermal Modulator for Comprehensive Multidimensional Gas Chromatography** MATTHEW S KLEE, DANI Instruments, Roberta Lariccia, Vincenzo Casilli N/A

(2270-7 P) **Modification of a GC for High Level Tritium Exposure** WILLIAM SPENCER, SRNL, Jose Cortes Concepcion, Jacob Schaufler, Robert Lascola

(2270-8 P) **Micro-Scale Vapor Extractor for Micro-GC Analysis of VOCs in Biofluids** JUNQI WANG, University of Michigan, Joseph A Potkay, Edward T Zellers

POSTER SESSION — Session 2280

All posters are to be mounted by 10:00 AM and remain on display until 2:00 PM. Authors must be at their posters from 10:00 AM to 12:00 PM. Location of the posters is on the Exposition Floor, 400 Aisle. PLEASE NOTE: You cannot get onto the Exposition Floor until after 9:00 AM.

UV/VIS

Thursday Morning, Exposition Floor, 400 Aisle

(2280-1 P) **Qualitative Colorimetric and Quantitative Flow Injection Determination of Alkyl Nitrite** ABD AL-KARIM ALI, Miami University, Neil D Danielson

(2280-2 P) **Multi-order Visible Absorption and Reflectance Spectrometry: Parallels to Atomic Emission Line Interferences** ALEXANDER SCHEELINE, SpectroClick, Mark D Ginsberg

(2280-3 P) **Nano-Embedded Optochemical Sensors for In-Vivo Photo-acoustic Chemical Imaging of Potassium Ions** WULIANG ZHANG, University of Michigan, Chang Lee, Jeffery Folz, Xueding Wang, Raoul Kopelman

(2280-4 P) **Colorimetric Determination of Sulfate Using Barium Ion and the Chromate/Dichromate Equilibrium: Preliminary Results and Comparison with the Classical Gravimetric Method** MARK T STAUFFER, University of Pittsburgh - Greensburg, Jarrod W Qualk, Jeremiah C Jamrom N/A

(2280-5 P) **Derivatives of 4,5-Diazafluorene for Chelation and Colorimetric Determination of FE(II) and CU(I): Better FE(II) and CU(I) Chelators Than 1,10-Phenanthroline?** MARK T STAUFFER, University of Pittsburgh - Greensburg, Luke J Metzler, Matthew R Luderer N/A

(2280-6 P) **Simultaneous Determination of Iron and Aluminum by Spectrophotometry and Partial Least Squares Regression: Comparison of Two Potential Ligands and Application to Mine Drainage and Related Water Samples** MARK T STAUFFER, University of Pittsburgh - Greensburg

(2280-7 P) **A Highly Selective Naked Eye Anion Detector Based on Chromone Derivatives and their Electroanalytical Studies** NEHA GUPTA, IIT Roorkee, Ashok K Singh N/A

(2280-8 P) **TBAF and It's Spectral Interference** IAN ADAMS, The University of Alabama

POSTER SESSION — Session 2290

All posters are to be mounted by 10:00 AM and remain on display until 2:00 PM. Authors must be at their posters from 10:00 AM to 12:00 PM. Location of the posters is on the Exposition Floor, 400 Aisle. PLEASE NOTE: You cannot get onto the Exposition Floor until after 9:00 AM.

Various Applications of GCMS

Thursday Morning, Exposition Floor, 400 Aisle

(2290-1 P) **Evaluation of Methylisothiazolinone (MI) Extraction from Sunscreen Using Supported Liquid Extraction prior to GC/MS Analysis** RHYS JONES, Biotage GB Limited, Lee Williams, Victor Vandell

(2290-2 P) **Extending the Range of GC-MS Applications with Cold EI** AVIV AMIRAV, Tel Aviv University, Alexander Fialkov, Uri Keshet, Tal Alon

(2290-3 P) **New Sampling Device for Early Cancer Screening by Non-Invasive Detection of VOCs Biomarkers in Exhaled Breath** PAOLO BENEDETTI, IIA - CNR, Ettore Guerriero, Federico Marini, Mark Ragusa, Maria Cristina Zappa, Carlo Crescenzi

(2290-4 P) **Quantitative and Qualitative Multi-Residue Analysis of Chemical Contaminants in Food and Feed by GC-HRTOFMS** JONATHAN D BYER, Leco Corporation, Joseph E Binkley, David E Alonso N/A

(2290-5 P) **Chemical Warfare Agents (CWAVX) by Large Volume Injection/Programmable Temperature Vaporization (LVI/PTV) - Gas Chromatography/Quadrupole Mass Spectrometry** TOM FOWLER, CSS-Dynamac, Kelly Head, Julia Capri

(2290-6 P) **Quantification of Persistent Organic Pollutants in Dietary Supplements Using Stir-bar Sorptive Extraction- Thermal Desorption- GC/MS and Isotope Dilution Mass Spectrometry** WEIER HAO, Duquesne University, Andrew Boggess, Skip Kingston

(2290-7 P) **Automated Sampling of Methanol Extractions** ANNE JUREK, EST Analytical, Lindsey Pyron, Kelly Cravenor, Justin Murphy

(2290-8 P) **Complex Sample Characterization by GC×GC-TOF** ROBERTA LARICCIA, DANI Instruments, Vincenzo Casilli, Matthew S Klee N/A

(2290-9 P) **Pesticide Analysis in Drinking Water and Beverages Using Multiple Techniques** XIAOPING LI, Georgia Gwinnett College, Zoe Goldstein, Hongxia Guan, Michelle Huang, Rashad Simmons, Simon Mwongela

(2290-10 P) **Identification of Contaminants in Powdered Foods by Direct Extraction Thermal Desorption GC/MS** RONALD SHOMO, Scientific Instrument Services, Christopher Baker, John Manura

(2290-11 P) **Methods Development for Sampling and Analysis of Biogenic Volatile Organic Compounds Released from Plants** PRITHVIRAJ SRIPATHI, Middle Tennessee State University, Christopher Moore, Kathleen Kuklewicz, James G Milstead, Beng Guat Ooi, Ngee Sing Chong

(2290-12 P) **Pyrolysis-GC/MS as a Screening Tool for Phthalate Esters and Brominated Flame Retardants in the RoHS Directive** NICOLE M LOCK, Shimadzu Scientific Instruments, Shilpi Chopra, Di Wang, Laura Chambers, Mark Janeczko

(2290-13 P) **Withdrawn**

(2290-14 P) **Improved Flavor Profile of Italian Wine and Scotch Whiskey Using an Aqueous-Stable Polyethylene Glycol Stationary Phase** RAMKUMAR DHANDAPANI, Phenomenex, Tim Anderson, Kristen Parnell

(2290-15 P) **Separation Solutions for Triglycerides in Food Fat and Oil by High Temperature GC** RAMKUMAR DHANDAPANI, Phenomenex, Kristen Parnell, Tim Anderson

(2290-16 P) **Volatile Organic Compounds in Energy Drinks as Determined by GC/MS with Purge and Trap Sample Concentration** CYNTHIA ELMORE, OI Analytical, J Garrett Slaton

(2290-17 P) **What's in Your Morning Drink? Comprehensive Characterization of Coffee and Tea Extracts by GCxGC–TOF MS** LAURA MCGREGOR, Markes International Ltd, Nicola Watson, Massimo Santoro, Chris Hall, Ken Umbarger

(2290-18 P) **Research of Polychlorinated Biphenyls (PCBs) in Vegetables by GC-MS/MS** XIZHI WANG, Shimadzu, Shiheng Luo, Feifei Tian, Jun Fan, Guixiang Yang, Taohong Huang, Shin-ichi Kawano, Yuki Hashi

(2290-19 P) **Analysis of FAMEs Using Cold EI GC/MS for Improved Molecular Ion Information** ADAM J PATKIN, PerkinElmer, Sharanya Reddy

(2290-20 P) **Basmati or Not Basmati? That is the Question** KENNETH ROSNACK, Waters Corporation, Gareth Cleland, Adam Ladak, Jennifer Burgess, Steven Lai

POSTER SESSION — Session 2300

All posters are to be mounted by 10:00 AM and remain on display until 2:00 PM. Authors must be at their posters from 10:00 AM to 12:00 PM. Location of the posters is on the Exposition Floor, 400 Aisle. PLEASE NOTE: You cannot get onto the Exposition Floor until after 9:00 AM.

Vibrational Spectroscopy Advances

Thursday Morning, Exposition Floor, 400 Aisle

(2300-1 P) **Development of a Non-Invasive Probing Method for Pharmaceutical Analysis Using Spatially Offset Raman Spectroscopy** HYUNG MIN KIM, Kookmin University

(2300-2 P) **Spatial Heterodyne Raman Spectrometer with LED Sources** WILLIAM HUNTINGTON, University of South Carolina, S Michael Angel

(2300-3 P) **Silver-Coated Self-Assembled Polystyrene Nanospheres for Surface-enhanced Raman Spectroscopy** LARA MIKAC, Rudjer Boskovic Institute, Mile Ivanda, Marijan Gotic, Vesna Janicki, Hrvoje Zorc, Tibor Janci, Sanja Vidacek N/A

(2300-4 P) **Second Harmonic Generation of Gold Nanorods Coupled with Atomic Force Microscopy** DARBY NELSON, University of Notre Dame, Zachary D Schultz

(2300-5 P) **Chemical Detection and Tracking of c-RGD Peptide-Conjugated Gold Nanoparticles Interacting with αVβ3 Integrins in Living Cells** HAO WANG, University of Notre Dame, Lifu Xiao, Zachary D Schultz

(2300-6 P) **Molecular Interactions between Nanoparticles and Model Cell Membranes Determined via Combined Vibrational Spectroscopic Studies** PEIPEI HU, The University of Michigan at Ann Arbor

(2300-7 P) **A Stable, Disposable Nanostructured Substrate for Surface Enhanced Raman Scattering (SERS) Detection of Drugs with Environmental Applications** HONEY MADUPALLI, Central Michigan University, Brandon Russell, Mary M Tecklenburg N/A

(2300-8 P) **A New Diamond ATR Video Microscopy Accessory** DAVID W SCHIERING, Czitek

(2300-9 P) **From Mars to Mission Critical Process Control** DAN WOOD, Keit Ltd

(2300-10 P) **Measurement of Component Distribution in "Soft Chew" Formulations by ATR FTIR Imaging** RONALD RUBINOVITZ, Thermo Fisher Scientific, William Wihlborg

THURSDAY, MARCH 10, 2016
AFTERNOON

SYMPOSIUM **Session 2310**

ACS-ANLY - Ultrasensitive Bioanalysis on the Pico-to Femtoliter Scales
arranged by Yong Zeng, University of Kansas

Thursday Afternoon, Room B308
Yong Zeng, University of Kansas, Presiding

1:30		**Introductory Remarks - Yong Zeng**
1:35	(2310-1)	**Bioanalysis in Ultrasmall Volumes** DANIEL T CHIU, University of Washington N/A
2:10	(2310-2)	**Engineering Hydrogels for Sensitive miRNA Assays** PATRICK DOYLE, Massachusetts Institute of Technology
2:45	(2310-3)	**Microfluidic Devices with Integrated Nanochannel Arrays to Study Development and Aging of Individual Bacteria** STEPHEN C JACOBSON, Indiana University, Joshua D Baker, David T Kysela, Yves V Brun
3:20		**Recess**
3:35	(2310-4)	**Single Cell Genomic and Proteomic Analysis** ANUP K SINGH, Sandia National Laboratories
4:10	(2310-5)	**Ultrasensitive and Broad-Range Microfluidic Immunoassays** YONG ZENG, University of Kansas

SYMPOSIUM **Session 2320**

Analytical Techniques in Neuroscience
arranged by Hang Lu, Georgia Tech and Andriana San Miguel, North Carolina State University

Thursday Afternoon, Room B309
Hang Lu, Georgia Tech, Presiding

1:30		**Introductory Remarks - Hang Lu**
1:35	(2320-1)	**The Chemical Characterization of the Brain: From New MS-Based Measurement Tools to New Insights** JONATHAN V SWEEDLER, University of Illinois
2:10	(2320-2)	**Scalable Proteomic Imaging of Intact Biological Systems** KWANGHUN CHUNG, MIT
2:45	(2320-3)	**In Situ RNA Profiling by FISH SCALYS (Sequential Coding anALYSis)** LONG CAI, Caltech N/A
3:20		**Recess**
3:35	(2320-4)	**Expansion Sequencing (ExSEQ): Comprehensive In Situ Transcriptome Characterization Throughout Intact Brain Circuits** SHAHAR ALON, Massachusetts Institute of Technology
4:10	(2320-5)	**Genetic Approaches to Brain Circuit Mapping and Cell Type Characterization** HONGKUI ZENG, Allen Institute for Brain Science

SYMPOSIUM **Session 2330**

Bioanalytical Chemistry Using the Next Generation of Nanomaterials
arranged by Isiah M Warner, Louisiana State University and Chieu D Tran, Marquette University

Thursday Afternoon, Room B310
Isiah M Warner, Louisiana State University, Presiding

1:30		**Introductory Remarks - Isiah M Warner and Chieu D Tran**
1:35	(2330-1)	**GUMBOS at the Nanoscale: Size Control, Characterization, and Applications** ISIAH M WARNER, Louisiana State University, Suzana Hamden
2:10	(2330-2)	**Discovering "Genetic Codes" for Nanomaterials Morphologies and Employing the DNA-Encoded Nanomaterials for Sensing and Imaging** YI LU, University of Illinois at Urbana-Champaign, Li Huey Tan, Peiwen Wu, Nitya Sai Reddy Satyavolu
2:45	(2330-3)	**Free-Standing Gold and Silver Nanoparticles Films as Flexible Sensing Platforms** JENNIFER S SHUMAKER-PARRY, University of Utah
3:20		**Recess**
3:35	(2330-4)	**Ultrafast and Nonlinear Spectroscopy of Plasmonic Nanoparticles for Drug Delivery and Photothermal Applications** LOUIS H HABER, Louisiana State University, Raju R Kumal, Tony E Karam, Holden T Smith, Corey R Landry, Mohammad Abu-Laban, Daniel J Hayes
4:10	(2330-5)	**Biocompatible Nanoparticle Composite Materials: Green Synthesis and Applications** CHIEU D TRAN, Marquette University

SYMPOSIUM **Session 2340**

Bioinformatics: Metabolite Identification and Quantification
arranged by Xiang Zhang, University of Louisville

Thursday Afternoon, Room B311
Xiang Zhang, University of Louisville, Presiding

1:30		**Introductory Remarks - Xiang Zhang**
1:35	(2340-1)	**New Methods for Improved NMR Quantitation and Reliable MS Coverage in Metabolomics** DANIEL RAFTERY, University of Washington
2:10	(2340-2)	**Point Matching Global Peak Alignment Algorithms for Comprehensive Two-Dimensional Gas Chromatography Coupled with Mass Spectrometry (GC×GC-MS)** KIM SEONGHO, Wayne State University/Karmanos Cancer Institute
2:45	(2340-3)	**Computational and Database Assisted Structure Identification Tools for Untargeted Metabolomics** DAVID GRANT, University of Connecticut
3:20		**Recess**
3:35	(2340-4)	**Mass Informatics of Quantitative Metabolomics by Integrating LCxLC-MS and GCxGC-MS Data** XIANG ZHANG, University of Louisville, Xiaoli Wei, Imhoi Koo, Pawel Lorkiewicz
4:10	(2340-5)	**Translating Big Data from HR Imaging MS Data into Molecular Knowledge** THEODORE ALEXANDROV, EMBL

SYMPOSIUM **Session 2350**

Micro and Nano-Scale Optofluidic Lasers for Biological Applications
arranged by Xudong Fan, University of Michigan and Seok-Hyun Yun, Harvard Medical School

Thursday Afternoon, Room B312
Xudong Fan, University of Michigan, Presiding

1:30		**Introductory Remarks - Xudong Fan and Seok-Hyun Yun**
1:35	(2350-1)	**Optofluidic Laser as a New Bioanalytical Tool** XUDONG FAN, University of Michigan
2:10	(2350-2)	**Latest Progress in Spasers** MARK I STOCKMAN, Georgia State University
2:45	(2350-3)	**Recent Advances in Biolasers: Stimulated Emission from Solid-State Fluorescent Protein and Lasing Inside Live Cells** MALTE C GATHER, University of St Andrews
3:20		**Recess**
3:35	(2350-4)	**Speckle-Free Lasers** HUI CAO, Yale University
4:10	(2350-5)	**Bio-Lasers** S H ANDY YUN, Harvard University N/A

Thursday Afternoon

SYMPOSIUM **Session 2360**

Nanofiber Materials Overcome Enduring (Bio) Analytical Challenges
arranged by Margaret W Frey, Cornell University and Antje J Baeumner, University of Regensburg

Thursday Afternoon, Room B313
Margaret W Frey, Cornell University, Presiding

1:30		**Introductory Remarks - Margaret W Frey and Antje J Baeumner**
1:35	(2360-1)	**Nanofiber Chemistry and Synthesis and the Impact on Analytical Systems** MARGARET W FREY, Cornell University, Larissa M Shepherd, Edurne Gonzalez, Nidia Trejo
2:10	(2360-2)	**Electrospun Fibers for Electrochemical Analysis** GREGORY RUTLEDGE, Massachusetts Institute of Technology, Xianwen Mao, Yuxi Zhang, T Alan Hatton, Harry Tuller
2:45	(2360-3)	**Carbon Nanotube–Nanocrystalline Diamond Hybrid Electrodes: A Route for Development of a Highly Sensitive Neurochemical Microsensor** PRABHU ARUMUGAM, Louisiana Tech University
3:20		**Recess**
3:35	(2360-4)	**Piezoelectric Nanofiber Platform for Cell Monitoring** CAROLINE L SCHAUER, Drexel University
4:10	(2360-5)	**New Concepts for Lab-on-a-Chip Systems Using Electrospun Nanofibers** ANTJE J BAEUMNER, University of Regensburg

SYMPOSIUM **Session 2370**

SEAC - Nanoengineered Biosensors
arranged by Adrian C Michael, University of Pittsburgh

Thursday Afternoon, Room B314
Reginald M Penner, University of California Irvine, Presiding

1:30		**Introductory Remarks - Reginald M Penner**
1:35	(2370-1)	**Biosensors for Early Cancer Detection Based Upon an Electrical Interface to Virus Particles** REGINALD M PENNER, University of California Irvine
2:10	(2370-2)	**Organic Electronics Biosensors for Label-Free Femtomolar Protein Detection** LUISA TORSI, Università degli Studi di Bari "A. Moro", Maria Magliulo, Gerdo Palazzo
2:45	(2370-3)	**Single Nanoparticle SPR Imaging and Plasmonic Nanocone Arrays: Smart Materials and Smart Chemistries for Advanced Optical Biosensors and Biomimetic Devices** ROBERT M CORN, University of California Irvine, Adam M Maley, H W Millie Fung
3:20		**Recess**
3:35	(2370-4)	**Ultrasensitive Biomolecular Detection Using Nanostructured Microelectrodes** SHANA KELLEY, University of Toronto
4:10	(2370-5)	**DNA Nanostructures and Networks** WEIHONG TAN, University of Florida

ORGANIZED CONTRIBUTED SESSIONS **Session 2380**

Biosensing Devices for Neuron Mapping
arranged by Chenzhong Li, Florida International University

Thursday Afternoon, Room B315
Chenzhong Li, Florida International University, Presiding

1:30	(2380-1)	**Integration of CNS and PNS Components with Silicon Devices via Surface Microengineering for Neuronal Mapping Applications** JAMES J HICKMAN, University of Central Florida
1:50	(2380-2)	**Biomimetic Strategies for Seamless Integration of Neural Interface Technology** TRACY CUI, University of Pittsburgh
2:10	(2380-3)	**Automated Micro- and Nanoscale Systems for Single Neuronal Activity** JIT MUTHUSWAMY, Arizona State University, Swathy Sampath Kumar, Sindhu Anand
2:30	(2380-4)	**Electronic Biosensing Devices for Recording Neuron Transmitter Expression at Single Cell and 3D Tissue Level** CHENZHONG LI, Florida International University
2:50		**Recess**
3:05	(2380-5)	**Visualization of Nanoscale Neuron Surface Topography and Detection of Neurotransmitter Release Using Nano Electrochemical Microscopy** YASUFUMI TAKAHASHI, Tohoku University, Hitoshi Shiku, Tomokazu Matsue
3:25	(2380-6)	**Trans-Synaptic In Vitro Mapping Using Microfluidic Approaches for Neuroscience Discovery** ANNE MARION TAYLOR, University of North Carolina at Chapel Hill and North Carolina State
3:45	(2380-7)	**CMOS Technology Enabled Brain Machine Interface (BMI) For Chronic Neuronal Mapping** MUHAMMAD MUSTAFA HUSSAIN, King Abdullah University of Science and Technology (KAUST), Aftab M Hussain, Amir N Hanna
4:05	(2380-8)	**Wireless Stimulation and Recording for In-Vivo Electrophysiology** JAMES MORIZIO, Triangle BioSystems Inc.

ORAL SESSIONS **Session 2390**

Bioanalytical: Electrochemical Techniques

Thursday Afternoon, Room B305
Leslie Wilson, North Carolina State University, Presiding

1:30	(2390-1)	**Compatibility of Nitric Oxide Release Coatings with Implantable Enzymatic Glucose Sensors Based on Osmium(III/II) Mediated Electrochemistry** KYOUNG HA CHA, University of Michigan
1:50	(2390-2)	**Application of Nanopipette Electrodes for Real-Time Measurement of Thyroid Hormones to Evaluate Thyrotoxic Storm** CELESTE A MORRIS, Northern Kentucky University, Barbara Cata, Theresa M Ruwe, Edward A Dobrzykowski, Teri Rae Armstrong
2:10	(2390-3)	**Neurochemical and Behavioral Analysis of Post-Chemotherapy Cognitive Impairment** MICHAEL A JOHNSON, University of Kansas, Sam V Kaplan, David P Jarmolowicz, Rachel C Gehringer, Michael J Sofis, Ryan Limbocker, Mimi Shin, Meng Sun
2:30	(2390-4)	**In-Vitro Amperometric Sensing of Dynamic Changes of Endogenous NO and CO Gases for Co-cultured Endothelial and Neuronal Cells** HA YEJIN, Ewha Womans University, Heo Chaejeong, Woo Juhyun, Suh Minah, Lee Youngmi
2:50		**Recess**
3:05	(2390-5)	**Real-Time Measurements of Oxidative Stress During Chronic L-DOPA Treatment For Parkinson's Disease** LESLIE WILSON, North Carolina State University, Christie Lee, Catherine F Mason, Leslie A Sombers
3:25	(2390-6)	**Dual Function Ion Selective Microelectrodes for Scanning Electrochemical Microscopy** GANESH UMMADI, Oregon State University, Corey Downs, Dipankar Koley
3:45	(2390-7)	**Label-Free Electrochemical microRNA Detection based on Different Modifier: Conducting Polymer and Graphene on the Surface of Pencil Graphite Electrode** MEHMET OZSOZ, Gediz University, Merve Kaplan
4:05	(2390-8)	**Quantitative Measurement of Transmitters in Individual Vesicles with Microelectrodes** XIANCHAN LI, University of Gothenburg, Soodabeh Majdi, Andrew Ewing

ORAL SESSIONS **Session 2400**

Bioanalytical: Sampling and Sample Preparation - Half Session

Thursday Afternoon, Room B304
Jinesh Jain, NETL, Presiding

1:30	(2400-1)	**Aptamer Functionalized Solid Phase Microextraction for Selective Enrichment of Thrombin** MD NAZMUL ALAM, University of Waterloo, Fuyou Du, Janusz Pawliszyn
1:50	(2400-2)	**New Generation of Solid SPME Coatings for Complementary Gas- and Liquid- Phase Separation: A Step Toward Integration of Metabolomics Platforms** EMANUELA GIONFRIDDO, University of Waterloo, Ezel Boyaci, Janusz Pawliszyn
2:10	(2400-3)	**Solid Phase Microextraction as Sample Preparation Tool in Brain Tumors Analysis** NATHALY REYES-GARCES, University of Waterloo, Barbara Bojko, Janusz Pawliszyn
2:30	(2400-4)	**MEMS Based Pre-Concentrator GC-Ion Mobility Spectrometry for Trace Gas Analysis** WOLFGANG VAUTZ, ISAS, Sascha Liedtke, Stefano Zampolli, Chandrasekhara Hariharan

TECHNICAL PROGRAM

ORAL SESSIONS — **Session 2410**

Bioanalytical: Techniques Using Sensors

Thursday Afternoon, Room B302
Ke Li, Missouri University of Science and Technology, Presiding

1:30	(2410-1)	**Core-Shell Nanoparticle Scintillator Probes for Low-Energy Radionuclide Quantification in Aqueous Media** COLLEEN JANCZAK, University of Arizona, Isen Andrew C Calderon, Zeinab Mokhtari, Craig A Aspinwall
1:50	(2410-2)	**Use of Silicon Photonic Microring Resonators for the High-Throughput Analysis of Multi-Protein Complex Formation in the Blood Coagulation Cascade** ELLEN MUEHL, University of Illinois at Urbana-Champaign, Josh M Gajsiewicz, Ivan Lenov, Jim H Morrissey, Ryan C Bailey
2:10	(2410-3)	**Label-Free RNA Probes for Live Cell Dual-Color Imaging of EGFR** XIAOHONG TAN, Carnegie Mellon University
2:30	(2410-4)	**Integration of Whispering Gallery Mode Detectors into Fluidic Platforms for Clinical Diagnostics** DANIEL KIM, University of Kansas, Robert C Dunn
2:50		Recess
3:05	(2410-5)	**Ultra-High Spatial Resolution Detection of Localized pH within a Single Live Cell** QINGBO YANG, Missouri University of Science and Technology, Hai Xiao, Honglan Shi, Xiaobei Zhang, Yinfa Ma
3:25	(2410-6)	**Real Time Analysis of Hepatitis B Virus Assembly with Multi-Pore Nano-fluidic Devices for Enhanced Resolution of Particle Size and Electrophoretic Mobility** PANAGIOTIS KONDYLIS, Indiana University, Jinsheng Zhou, Zachary D Harms, Lisa Selzer, Adam Zlotnick, Stephen C Jacobson
3:45	(2410-7)	**Graphene-Based Chemiresistive Nanobiosensors for Detection of Citrus Greening Disease** THIEN-TOAN H TRAN, University of California, Riverside, Clark Kelley, Jinxia Shi, Wenbo Ma, Ashok Mulchandani
4:05	(2410-8)	**A Quantum Dot Based Fiber-Optic Micro-Sensor for Niche-Environment Temperature Monitoring** KE LI, Missouri University of Science and Technology, Qingbo Yang, Honglan Shi, Yinfa Ma, Hai Xiao, Qi Zhang

ORAL SESSIONS — **Session 2420**

Computers in Chemistry - Half Session

Thursday Afternoon, Room B301
Dean Tzeng, The Pittsburgh Conference, Presiding

1:30	(2420-1)	**An Inexpensive, Programmable System for Prototyping Instruments, Making Short Run Specialty Measurement Systems, and Computerizing Outdated Hardware** SCOT D ABBOTT, Phoenix Frist Response
1:50	(2420-2)	**Is SAP the Only System You Need to Operate Your QC Lab?** GEOFF R TURNBULL, CSols, Inc. N/A
2:10	(2420-3)	**Maximizing the Information from the Infrared Spectra of Mixtures Using Advanced Software Algorithms** IAN ROBERTSON, PerkinElmer Limited, Jerry Sellors, Justin Lang
2:30	(2420-4)	**Driving Governance and Organizational Change in Large and Complex Informatics Projects** ADAM S BORENSTEIN, LabAnswer, Brian Brunner, Terryl Kibodeaux

ORAL SESSIONS — **Session 2430**

Glycan Analysis - Half Session

Thursday Afternoon, Room B303
Christa M Snyder, Indiana University, Presiding

1:30	(2430-1)	**Identification of Serum N-Glycans as Cancer-Specific Biomarkers by Microchip Electrophoresis and Mass Spectrometry** CHRISTA SNYDER, Indiana University, Xiaomei Zhou, Margit I Campos, Milos V Novotny, Stephen C Jacobson
1:50	(2430-2)	**Improved Separation of Saccharide Standards by EFL-HILIC** RAFFEAL BENNETT, The Ohio State University, Susan Olesik
2:10	(2430-3)	**Receptor for Advanced Glycation End Products Diffusion and Ligand-Binding Events Studied by Fluorescence Recovery after Photobleaching and Surface Plasmon Resonance** QIAOCHU ZHU, Iowa State University, Aleem Syed, Chamari S Wijesooriya, Emily A Smith
2:30	(2430-4)	**Comprehensive Quantitative and Structural Analysis of Permethylated N-glycan Using PGC-LC-MS/MS** SHIYUE ZHOU, Texas Tech University, Yehia Mechref

ORAL SESSIONS — **Session 2440**

LC and Sample Matrix Solutions - Half Session

Thursday Afternoon, Room B304
Jinesh Jain, NETL, Presiding

3:05	(2440-1)	**Analysis of Biofluids Using Solid Phase Microextraction Devices Made on Plastic Support** NATHALY REYES-GARCES, University of Waterloo, Barbara Bojko, Janusz Pawliszyn
3:25	(2440-2)	**Studies of Matrix Effects on the Determination of Hydrogen Peroxide by Using High Performance Anion Exchange Chromatography with Pulsed Amperometric Detection** JUN CHENG, Thermo Fisher Scientific, Yan Liu, Kannan Srinivasan, Christopher Pohl
3:45	(2440-3)	**HPLC Purity Method Development Using Low pH Mobile Phase and Ion-Pairing Reagent: Application to GMP Tert-Leucine Analysis** JOHN VINCI, AbbVie, Inc., Clifford Mitchell
4:05	(2440-4)	**Combining Orthogonal Separation Modes for Analysis of Multiple Sample Components** THOMAS E WHEAT, Waters Corporation, Amanda B Dlugasch, Patricia R McConville

ORAL SESSIONS — **Session 2450**

Microfluidics/Lab-on-a-Chip - Bioanalytical II

Thursday Afternoon, Room B316
Michael Woodman, Agilent Technologies, Presiding

1:30	(2450-1)	**Separation of Preterm Birth Biomarkers Using Capillary and Microchip Electrophoresis** ANNA V NIELSEN, Brigham Young University, Radim Knob, Adam T Woolley
1:50	(2450-2)	**Analysis of Nitrosative Stress in Macrophage Cells Using Microchip Electrophoresis with Electrochemical Detection** JOSEPH M SIEGEL, University of Kansas, Manjula B Wijeshinghe, Kelci M Schilly, Susan M Lunte
2:10	(2450-3)	**Ultrasensitive Electrochemical Microfluidic Immunoarray for Assessment of Aggressive vs Indolent forms of Prostate Cancer Biomarkers** BRUNAH A OTIENO, University of Connecticut, Colleen E Krause, Abby Jones, Amit Joshi, Mohammed Sherafeldin, James F Rusling
2:30	(2450-4)	**Microfluidic and 3-D Printed Devices for Near-Real-Time and Simultaneous Detection of Neurotransmitters** ALEXANDRA D TOWNSEND, Saint Louis University, R Scott Martin
2:50		Recess
3:05	(2450-5)	**Electrokinetically Operated Integrated Microfluidic Platform for Preterm Birth Biomarker Analysis** MUKUL SONKER, Brigham Young University, Radim Knob, Vishal Sahore, Adam T Woolley
3:25	(2450-6)	**Discrete Stimulation of Lymph Node Slices on Chip** ASHLEY E ROSS, University of Virginia, Jacob F Woodroof, Rebecca R Pompano
3:45	(2450-7)	**Ultrasensitive ELISA for Detection of Infectious Diseases on Surface Modified PMMA Microfluidic Microplates** SANJAY SHARMA TIMILSINA, University of Texas at El Paso, Maowei Dou, Xiujun James Li
4:05	(2450-8)	**Optical Formulation of 3D Printer Resin for Minimum Microfluidic Flow Channel Size** GREG NORDIN, Brigham Young University, Hua Gong, Michael Beauchamp, Steven Perry, Adam T Woolley

PITTCON® CONFERENCE & EXPO 2016

TECHNICAL PROGRAM

AGENDA OF SESSIONS

ABSTRACT PDF

March 6-10, 2016

Atlanta, Georgia

www.pittcon.org

TO SEARCH THIS DOCUMENT, PLEASE USE CTRL+F

Session # 10 Abstract # 10-1 **The Wallace H. Coulter Lecture**

Session Title The Wallace H. Coulter Lecture

Abstract Title **How Optical Single-Molecule Detection in Solids Led to Super-Resolution Nanoscopy in Cells and Beyond**

Date: Sunday, March 06, 2016 - Afternoon
Time: 05:00 PM
Room: Sidney Marcus Auditorium, Bldg A, Le

Primary Author W.E. (William Esco) Moerner
Stanford University

Co-Author(s)

Abstract Text
More than 25 years ago, low temperature experiments aimed at establishing the ultimate limits to optical storage in solids led to the first optical detection and spectroscopy of a single molecule in the condensed phase. At this unexplored ultimate limit, many surprises occurred where single molecules showed both spontaneous changes (blinking) and light-driven control of emission, properties that were also observed in 1997 at room temperature with single green fluorescent protein variants. In 2006, PALM and subsequent approaches showed that the optical diffraction limit of ~200 nm can be circumvented to achieve super-resolution fluorescence microscopy, or nanoscopy, with relatively nonperturbative visible light. Essential to this is the combination of single-molecule fluorescence imaging with active control of the emitting concentration and sequential localization of single fluorophores decorating a structure. Super-resolution microscopy has opened up a new frontier in which biological structures and behavior can be observed in live cells with resolutions down to 20-40 nm and below. Examples range from protein superstructures in bacteria to bands in actin filaments to details of the shapes of amyloid fibrils and much more. Current methods development research addresses ways to extract more information from each single molecule such as 3D position and orientation, in thick cells. Still, it is worth noting that in spite of all the focus on super-resolution, even in the "conventional" single-molecule tracking regime where the motions of individual biomolecules are recorded in solution or in cells rather than the shapes of extended structures, much can still be learned about biological processes.

Keywords: Molecular Spectroscopy
Application Code: Bioanalytical
Methodology Code: Molecular Spectroscopy

Session # 30 **Abstract # 30-1** **Symposia**

Session Title: ACS-ANYL - Tracing the Metabolome: Application of Stable-Isotope Tracers in Bioanalytical Chemistry

Abstract Title: **Stable Isotope Labeling and UHPLC/MS Strategies for Probing Dynamics of Plant Specialized Metabolism**

Date: Sunday, March 06, 2016 - Afternoon
Time: 01:35 PM
Room: B308

Primary Author: A Daniel Jones
Michigan State University

Co-Author(s): Banibrata Ghosh, Xiaoxiao Liu, Zhenzhen Wang

Abstract Text

The plant kingdom synthesizes an enormous suite of specialized metabolites that contribute economic value and inspire developments of modern medicines. However, limited knowledge about identities of biosynthetic pathway intermediates, genes responsible for metabolic transformations, and mechanisms that regulate metabolite accumulation often stymies engineering of plant metabolic networks. These knowledge gaps can be addressed through the use of stable isotope tracers, either through photosynthetic fixation of [sup]13[/sup]CO[sub]2[/sub] or isotopically-labeled central metabolites including amino acids. However, limitations of mass spectrometry strategies for measurements of isotope incorporation into complex metabolites have hindered adoption of these approaches.

We recently developed data-independent mass-unselective collision-induced dissociation (CID) mass spectrometry strategies for measuring stable isotope incorporation into both intact metabolites and fragment ion masses that reflect labeling in metabolite substructures. Quasi-simultaneous parallel spectrum acquisition using multiple collision potentials on a time-of-flight mass spectrometer enabled resolution of otherwise overlapping isotopolog envelopes. This analytical approach was applied to characterize specialized metabolite dynamics in tomato (Solanum lycopersicum) seedlings grown for five days under an atmosphere containing [sup]13[/sup]CO[sub]2[/sub], followed by return of the plants to ambient atmosphere. Dynamics of metabolite accumulation and turnover were compared in tomato and an introgression line (IL5-3) to probe the functions of putative hydrolase genes in accumulation of acylsucrose metabolites. In addition, mass-unselective CID and MS/MS spectra were used with [sup]13[/sup]C,[sup]15[/sup]N-labeled amino acid precursors to distinguish isomeric metabolites that could not be resolved by MS/MS and to reveal metabolic transformations that contribute to acylsucrose chemical diversity.

Keywords: Liquid Chromatography/Mass Spectroscopy, Natural Products, Time of Flight MS
Application Code: Bioanalytical
Methodology Code: Liquid Chromatography/Mass Spectrometry

Session # 30 **Abstract # 30-2** **Symposia**

Session Title ACS-ANYL - Tracing the Metabolome: Application of Stable-Isotope Tracers in Bioanalytical Chemistry

Abstract Title **Developments in [sup]13[/sup]C-based Metabolomics**

Date: Sunday, March 06, 2016 - Afternoon
Time: 02:10 PM
Room: B308

Primary Author Arthur S. Edison
University of Georgia

Co-Author(s)

Abstract Text

Reliable compound identification and quantification are some of the biggest challenges in metabolomics. Studies using NMR typically use 1D [sup]1[/sup]H methods, which can have extensive overlap, leading to uncertainty in resonance assignments. In LC-MS studies, an accurate mass and retention time is oftentimes not enough information for a reliable compound identification. Utilization of [sup]13[/sup]C can improve the ability to reliably identify compounds using both NMR and LC-MS.[1] Using NMR, we have developed approaches using both natural abundance [sup]13[/sup]C and isotopically enriched samples to obtain more reliable carbon backbone maps of metabolites in a complex mixture. With natural abundance samples, we collect [sup]13[/sup]C and [sup]1[/sup]H 1D spectra of all the samples in a study to obtain [sup]13[/sup]C-[sup]13[/sup]C and [sup]13[/sup]C-[sup]1[/sup]H statistical correlation networks, which can be used to query databases. With isotopically enriched samples, we collect 2D INADEQUATE spectra to directly obtain [sup]13[/sup]C-[sup]13[/sup]C networks, which can be automatically identified and matched to databases or analyzed de novo. With LC-MS we have used isotopic ratio outlier analysis (IROA) to analyze samples that have been properly labeled with [sup]13[/sup]C, and this technology allows for the determination of the number of carbons in a molecule and can provide relative quantification. I will discuss recent advances on enabling technology for these methods, including specialized NMR probe development for [sup]13[/sup]C detection.[2] I will also show strategies for combining [sup]13[/sup]C NMR methods with IROA.

1.□lendinen, C.S., et al., An Overview of Methods using 13C for Improved Compound Identification in Metabolomics and Natural Products. Frontiers in Plant Science, 2015. 6.
2.□amaswamy, V., et al., Development of a (13)C-optimized 1.5-mm high temperature superconducting NMR probe. Journal of Magnetic Resonance, 2013. 235C: p. 58-65.

Keywords: Liquid Chromatography/Mass Spectroscopy, Method Development, Metabolomics, Metabonomics
Application Code: Genomics, Proteomics and Other 'Omics
Methodology Code: Magnetic Resonance

Session # 30 **Abstract # 30-3** **Symposia**

Session Title: ACS-ANYL - Tracing the Metabolome: Application of Stable-Isotope Tracers in Bioanalytical Chemistry

Abstract Title: **Chemical Isotope Labeling LC-MS for Quantitative Metabolomics with High Metabolomic Coverage**

Date: Sunday, March 06, 2016 - Afternoon
Time: 02:45 PM
Room: B308

Primary Author: Liang Li
University of Alberta

Co-Author(s)

Abstract Text

Chemical isotope labeling (CIL) liquid chromatography mass spectrometry (LC-MS) uses differential isotope mass tags to label a metabolite in two comparative samples (e.g., 12C-labeling of an individual sample and 13C-labeling of a pooled sample or control), followed by mixing the light-labeled sample and the heavy-labeled control and LC-MS analysis of the resultant mixture. Individual metabolites are detected as peak pairs in the mass spectra. The MS or chromatographic intensity ratio of a peak pair can be used to measure the relative concentration of the same metabolite in the sample vs. the control. For a metabolomics study working with many different samples, the same heavy-labeled control is spiked to all the light-labeled individual samples. Thus, the intensity ratios of a given peak pair from LC-MS analyses of all the light-/heavy-labeled mixtures reflect the relative concentration changes of a metabolite in these samples. CIL LC-MS can overcome the technical problems such as matrix effects, ion suppression and instrument drifts to generate more precise and accurate quantitative results, compared to conventional LC-MS. CIL LC-MS can also significantly increase the detectability of metabolites by rationally designing the labeling reagents to target a group of metabolites (e.g., all amine-containing metabolites or amine submetabolome) to improve both LC separation and MS sensitivity. In this presentation, recent advances in CIL LC-MS for quantitative metabolomics will be described. In particular, new labeling chemistries targeting difference groups of submetabolomes will be presented. A robust ample amount normalization method based on UV measurement of labeled metabolites will be shown for improving metabolome quantification. Finally, some recent applications of CIL LC-MS metabolomics for disease biomarker discovery research will be discussed.

Keywords: Derivatization, Isotope Ratio MS, Liquid Chromatography/Mass Spectroscopy, Metabolomics, Metab
Application Code: Genomics, Proteomics and Other 'Omics
Methodology Code: Liquid Chromatography/Mass Spectrometry

Session # 30 **Abstract # 30-4** **Symposia**

Session Title: ACS-ANYL - Tracing the Metabolome: Application of Stable-Isotope Tracers in Bioanalytical Chemistry

Abstract Title: **Lipid Metabolic Profiling Using Stimulated Raman Scattering Microscopy**

Primary Author: Meng Wang
Baylor College of Medicine

Date: Sunday, March 06, 2016 - Afternoon
Time: 03:35 PM
Room: B308

Co-Author(s)

Abstract Text

Lipid molecules are crucial for various cellular responses, and their misregulation results in such human diseases as metabolic disorders, neurodegenerative diseases, and cancers. As in proteins, the physiological and pathological activities of lipid molecules are tightly associated with their spatial distribution and temporal dynamics. The ability to track specific lipid molecules in vivo is essential for understanding their physiological impacts and regulatory mechanisms. Using stimulated Raman scattering (SRS) microscopy, we quantitatively imaged lipid distribution at the whole organism level in live Caenorhabditis elegans, and conducted genome-scale analysis to identify new genes and pathways in regulating lipid metabolism. Using hyperspectral SRS microscopy, we profiled lipid composition at the single lipid droplet level in living cells and revealed the changes of lipid composition in association with obesity and hepatic steatosis. Through incorporation of isotope labeling/tracing into SRS microscopy, we directly visualized the temporal dynamics of specific lipid molecules during their transportation between cells and tissues, and revealed metabolic heterogeneity among different lipid molecules. These studies demonstrate the biological application of SRS microscopy, and reveal its power in discovery of new regulatory mechanisms that govern lipid dynamics under both physiological and pathological conditions.

Keywords: Biomedical, Infrared and Raman, Lipids, Metabolomics, Metabonomics
Application Code: Biomedical
Methodology Code: Microscopy

Session # 30 Abstract # 30-5 Symposia

Session Title ACS-ANYL - Tracing the Metabolome: Application of Stable-Isotope Tracers in Bioanalytical Chemistry

Abstract Title **Monitoring the Incorporation of Stable-Isotope Labeled Alcohols by Bacteria with GC-MS**

Primary Author Gregory A. Barding
California State Polytechnic University

Date: Sunday, March 06, 2016 - Afternoon
Time: 04:10 PM
Room: B308

Co-Author(s) Nicole G. Perkins, Rakesh Mogul

Abstract Text

Spacecraft cleanliness is strictly controlled in efforts to minimize the microbial contamination of extraterrestrial environments and to bolster the integrity of life detection missions. Despite these efforts, the assembly facilities for Mars spacecraft harbor a diverse yet low abundant microbial inventory, which possess tolerances towards extreme conditions. Isolated from the Mars Odyssey prior to launch, Acinetobacter radioresistens 50v1 was found to exhibit extreme tolerances towards hydrogen peroxide and ultraviolet radiation. Furthermore, A. radioresistens 50v1 was found to grow on ethanol as a sole carbon source, thus suggesting that the alcohol cleaning solvents used during spacecraft assembly also serve as carbon and energy sources. However, it is unknown if and how A. radioresistens metabolically incorporates ethanol as a sole carbon source. To obtain evidence of the metabolic intake of ethanol, whole cells and cellular extracts of A. radioresistens 50v1 were analyzed by GC-MS using 1,2-13C ethanol, 1-13C ethanol, and 2-13C ethanol as a sole carbon source. GC-MS is a powerful and sensitive detection platform for studying changes in metabolism as well as the incorporation of isotope labels to elucidate specific metabolic pathways associated with a stress. Interestingly, evidence of gluconeogenesis and extensive metabolite labeling of many classes of metabolites demonstrates the flexibility of A. radioresistens to thrive in extreme conditions. These results demonstrate that the cleaning protocols for spacecraft may favor certain members within the microbial inventory of the NASA spacecraft assembly facilities, and hence impact the probability of contamination of Mars.

Keywords: Bioanalytical, Biological Samples, Gas Chromatography/Mass Spectrometry, GC-MS
Application Code: Bioanalytical
Methodology Code: Gas Chromatography/Mass Spectrometry

Session # 40 **Abstract # 40-1** **Symposia**

Session Title: Emerging Leaders in Biological Mass Spectrometry

Abstract Title: **New Mass Spectrometry Methods Show HIV Vaccine Candidates' Protein Structures are Misfolded**

Primary Author: Heather R. Desaire
University of Kansas

Date: Sunday, March 06, 2016 - Afternoon
Time: 01:35 PM
Room: B302

Co-Author(s)

Abstract Text

Numerous HIV vaccine design strategies that have advanced towards clinical trials include boosting immune response with a recombinant, truncated form of Env, the trimeric surface antigen on the HIV virus. Production of this protein in quantities large enough to support clinical trials is a challenge, and vaccine development efforts now favor generation of the soluble, monomeric gp120 component of the protein. Thus far, gp120 has not been effective at eliciting neutralizing antibodies in humans, and we hypothesized that a contributing cause for this lackluster performance was that the recombinant, truncated preparations of gp120 are ineffective mimics of native Env.

To test this hypothesis, we have spent years developing analytical technology to measure whether or not a native protein structure is present in these candidates. Specifically, we have developed an LC-MS workflow for disulfide bonding analysis that is robust, 100% reproducible, and effective for small quantities of heavily glycosylated proteins. We also developed relative quantitative methods for the disulfide bonded peptides. We used the methods to determine whether current HIV vaccine candidates destined for clinical trials are well designed to mimic viral Env.

We have conducted disulfide bonding analyses of many gp120 vaccine candidates and compared their molecular profiles to trimeric, native-like Env. The results show that a large fraction, in many cases the majority of proteoforms present in several HIV vaccine candidates, are not disulfide bonded in the same fashion as native Env. We have furthermore used these methods to identify vaccine design and purification methods that produce almost exclusively properly connected, folded, and glycosylated Env. These results emphasize that many vaccine design and production efforts need reconsideration, if developers intend to deliver recombinant Env in a native conformation as part of their vaccine strategy.

Keywords: Biopharmaceutical, Data Analysis, Mass Spectrometry, Protein
Application Code: Bioanalytical
Methodology Code: Mass Spectrometry

Session # 40 **Abstract # 40-2** **Symposia**

Session Title: Emerging Leaders in Biological Mass Spectrometry

Abstract Title: **Novel Strategies in Top-down Proteomics**

Date: Sunday, March 06, 2016 - Afternoon
Time: 02:10 PM
Room: B302

Primary Author: Ying Ge
University of Wisconsin Madison

Co-Author(s)

Abstract Text

A comprehensive analysis of all proteoforms that arise from genetic variations, alternative splicing, and post-translational modifications (PTMs), is essential for gaining a transformative understanding of disease mechanisms and identifying new therapeutic targets. Top-down mass spectrometry (MS)-based proteomics is arguably the most powerful method to comprehensively characterize proteoforms. Despite the significant recent advances, top-down proteomics still faces significant challenges. Herein, we aim to develop novel strategies to address the challenges in top-down proteomics in a comprehensive manner by developing new MS-compatible surfactants for protein solubilization, new methods for chromatography separation, and novel nanomaterials for enrichment of low-abundance proteins.

To address the protein solubility challenge, we are developing new degradable surfactants that can effectively solubilize proteins and is compatible with top-down MS. We have discovered an MS-compatible degradable surfactant that solubilizes all categories of proteins with performance comparable to SDS and significantly improved the detection of membrane protein. To address the proteome complexity challenge, we are developing novel strategies for multi-dimensional liquid chromatography (MDLC) to separate intact proteins. We have developed a novel 3DLC strategy by coupling HIC with ion exchange chromatography (IEC) and reverse phase chromatography (RPC) for intact protein separation for top-down proteomics. To address the proteome dynamic range, we have been developing novel nanomaterials that can bind low abundance proteins with PTMs with high specificity. The current focus is to develop multivalent nanoparticle reagents for capturing phosphoproteins globally out of the human proteome for top-down proteomics. We have designed and synthesized the first generation of functionalized NPs and showed the high specificity in enriching phosphoproteins from complex cell and tissue lysates.

Keywords: Biological Samples, Mass Spectrometry, Nanotechnology, Proteomics
Application Code: Genomics, Proteomics and Other 'Omics
Methodology Code: Mass Spectrometry

Session # 40 **Abstract # 40-3** **Symposia**

Session Title: Emerging Leaders in Biological Mass Spectrometry

Abstract Title: **Multi-Tier Approach to Understand the Biology of Alzheimer's Disease**

Date: Sunday, March 06, 2016 - Afternoon
Time: 02:45 PM
Room: B302

Primary Author: Renã Robinson
University of Pittsburgh

Co-Author(s)

Abstract Text

'Omics-based approaches are increasingly being recognized for offering powerful ways to investigate biological processes, especially in the context of Alzheimer's disease. Advances in mass spectrometry instrumentation and bioinformatics have increased the throughput with which we can begin to ask and answer difficult biological questions and study complex systems. There are many levels (i.e., epidemiological, clinical, genetic, proteomic, metabolomic, etc.) with which one can answer biological questions about Alzheimer's disease, and each of these levels is important to guide research directions, facilitate hypothesis generation, and inform diagnostic or therapeutic strategies. Synthesizing information from these multiple levels however can be challenging due to the amount and complexity of the data that is generated, lack of tools for data integration, and missing pieces of data. This presentation will give examples of our multi-tier proteomics and information synthesis approach to understanding aspects of biology in Alzheimer's disease.

Keywords: Bioanalytical, Mass Spectrometry, Proteomics
Application Code: Genomics, Proteomics and Other 'Omics
Methodology Code: Mass Spectrometry

Session # 40 **Abstract # 40-4** **Symposia**

Session Title: Emerging Leaders in Biological Mass Spectrometry

Abstract Title: **Old Photochemistry Brings New Capabilities in Unsaturated Lipid Analysis**

Primary Author: Yu Xia, Purdue University

Date: Sunday, March 06, 2016 - Afternoon
Time: 03:35 PM
Room: B302

Co-Author(s): Craig Stinson, Xiaoxiao Ma, Zheng Ouyang

Abstract Text

Mass spectrometry (MS)-based analysis has become a primary tool in lipidomics for providing global lipid identification and quantitation capabilities. Among many notable analytical figures of merits, high molecular specificity is a distinct feature of mass spectrometry. Despite the fact that multi-levels of structural information for lipids are readily achieved from MS and tandem mass spectrometry (MS/MS) such as molecular weight information, lipid class, and fatty acyl composition, the location information of a carbon-carbon double bond (C=C), however, is rarely obtainable using commercially available MS platforms. This is fundamentally routed in that cleavages of a C-C or C=C require significantly higher energies than other possible fragmentation channels (neutral of or charged head-group and acyl chain losses), preventing forming fragment ions useful for C=C location determination. To meet the above analytical challenge, our group has investigated several well established photochemical reactions for their utility in lipid C=C location determination and isomer quantitation. We found that Paternò–Büchi (PB) reaction and UV-induced ozonolysis are good candidates since they can be implemented on-line with MS or MS/MS analysis and provide confident C=C location information for a broad spectrum of lipid classes. The methods to couple these reactions with electrospray ionization (ESI) or nanoESI will be introduced. The potential and scope of these methods are demonstrated with lipid extracts from biological samples (tissues, cell lines, and plasma) for simultaneous unsaturated lipid characterization and quantitation.

Keywords: Bioanalytical, Lipids, Mass Spectrometry
Application Code: Bioanalytical
Methodology Code: Mass Spectrometry

Session # 40 **Abstract # 40-5** **Symposia**

Session Title: Emerging Leaders in Biological Mass Spectrometry

Abstract Title: **Evaluating Small Molecule Histone Inhibitors with High Resolution Mass Spectrometry and 3D Cell Cultures**

Date: Sunday, March 06, 2016 - Afternoon
Time: 04:10 PM
Room: B302

Primary Author: Amanda B. Hummon
University of Notre Dame

Co-Author(s): Benjamin A. Garcia, Monica M. Schroll, Peter E. Feist, Simone Sidoli

Abstract Text

Three dimensional cell cultures are simplified models of human organs or tumors. Similar to in vivo systems, they often contain complex physiological and biochemical gradients, such as regions of enhanced cellular viability or proliferation and increasing or decreasing pH. For the first time, we are characterizing the histone post-translational modifications (PTMs) in different layers of 3D cell cultures. Using high-resolution tandem mass spectrometry, we have characterized pronounced gradients in the histone PTMs that correlate with spatial regions of pathophysiology in the 3D cell mass. For example, histone H3 lysine 27 (H3K27) is highly methylated in the necrotic cells in the center of the 3D cell cultures. This methylation is a transcriptionally repressive modification, and is correlated with many cancers with corresponding interest therapeutically. In addition to characterizing the endogenous post-translational status of the histones, we are also taking advantage of the 3D cell culture system to evaluate the effectiveness of small molecule inhibitors for altering epigenetic histone modification profiles. Enhancer of zeste homolog 2 (EZH2) is the catalytic subunit of the polycomb repressive complex 2 that catalyzes methylation of H3K27. Using imaging mass spectrometry approaches, we have characterized the ability of the EZH2 inhibitor, UNC 1999, to penetrate 3D cell culture masses and alter the methylation status for the H3K27 mark. While we demonstrate the proof-of-principle with UNC1999, this approach is ideal to evaluate a larger range of molecular inhibitors of epigenetic modifiers in a tumor-like environment.

Keywords: Bioanalytical, Biomedical, Mass Spectrometry, Protein
Application Code: Biomedical
Methodology Code: Mass Spectrometry

Session # 50 Abstract # 50-1 **Symposia**

Session Title: Enabling Sample Preconcentration Methods for Bioanalysis

Abstract Title: **Bottom-up Proteomics of E.coli Using Dynamic pH Junction Preconcentration and CZE-ESI-MS/MS**

Date: Sunday, March 06, 2016 - Afternoon
Time: 01:35 PM
Room: B304

Primary Author: Norman J. Dovichi
University of Notre Dame

Co-Author(s): Guijie Zhu, Liangliang Sun

Abstract Text

We report the use of dynamic pH junction based CZE-ESI-MS/MS for bottom-up proteomics an electrokinetically-pumped sheath-flow nanospray CE-MS interface and both an LTQ-ion trap and a Orbitrap-Velos mass spectrometers. Conventional injection of 20 nL of a 1 mg/mL BSA digest identified 37 peptides and produced 66% sequence coverage. In contrast, pH junction injection of 130 nL (or larger) of a 0.05 mg/mL BSA digest identified 40 peptides and produced 70% coverage using a pH 6.5 sample buffer and the LTQ. A 20 nL conventional injection of a 1 mg/mL E. coli digest identified 508 peptides and 199 proteins with the LTQ. A 400 nL pH junction injection of a 0.1 mg/mL E. coli digest identified 527 peptides and 179 proteins with the LTQ. Triplicate technical replicates of a 0.01 mg/mL sample with 400-nL injection volume using a pH junction identified 288 ± 9 peptides and 121 ± 5 proteins with the LTQ. The Orbitrap-Velos analysis of a 10-nL injection of a 0.1 mg/mL E. coli digest identified of 185 proteins and 828 peptides, which appears to be the state-of-the art performance for single shot analysis of a 1-ng prokaryote digest. There was outstanding concordance in migration time between the pH junction and normal injection. The pH junction produced narrower peaks and significant concentration for all but the most acidic components in the sample. We also applied the pH junction to three intact standard proteins and observed a >10x increase in peak intensity compared to conventional injection.

Keywords: Electrophoresis, Peptides, Proteomics
Application Code: Bioanalytical
Methodology Code: Capillary Electrophoresis

Session # 50 **Abstract # 50-2** **Symposia**

Session Title: Enabling Sample Preconcentration Methods for Bioanalysis

Abstract Title: **Electrokinetic Sample Preconcentration and Hydrodynamic Sample Injection for Capillary Electrophoresis Using a Pneumatic Microvalve**

Date: Sunday, March 06, 2016 - Afternoon
Time: 02:10 PM
Room: B304

Primary Author: Ryan T. Kelly
Pacific Northwest National Laboratory

Co-Author(s): Katipamula Shanta, Tang Keqi, Yongzheng Cong

Abstract Text

Online analyte preconcentration is a preferred method to increase the limited detection sensitivity of miniaturized capillary electrophoresis (CE) resulting from the required small sample injection volumes. We have developed a microfluidic platform that employs a pneumatic microvalve to perform online preconcentration and controlled sample injection for CE. The pneumatic microvalve, created using multilayer soft lithography, is precisely aligned at a T-junction on a PDMS chip and can serve as a permselective membrane under an applied potential. For sample preconcentration, potentials are set such that analytes are driven to and focused at the closed valve by concentration polarization. Once analytes are concentrated, the valve is briefly opened and the stacked sample is hydrodynamically injected onto a separation column. Preconcentration and injection of cationic and anionic species have been separately demonstrated using appropriate microchannel surface treatments. Enrichment factors exceeding 1000 have been achieved, and the preconcentration/injection functionality has been coupled with both microchip and capillary-based separations. The latter enabled straightforward coupling with electrospray ionization mass spectrometry via an electrokinetically driven sheath flow interface. To preserve the high resolution made possible by the rapid and precise actuation of the microvalve for capillary-based separations, a method was developed to interface microchannels with fused silica capillaries with no detectable dead volume. Additionally, analyte preconcentration can take place while an electrophoretic separation is ongoing, which makes this a promising method for coupling two-dimensional separations, such as liquid chromatography coupled to CE with no loss of analyte between dimensions.

Keywords: Bioanalytical, Capillary Electrophoresis, Electrospray, Lab-on-a-Chip/Microfluidics
Application Code: Bioanalytical
Methodology Code: Microfluidics/Lab-on-a-Chip

Session # 50 **Abstract # 50-3** **Symposia**

Session Title: Enabling Sample Preconcentration Methods for Bioanalysis

Abstract Title: **Solid Phase Microextraction in Bioanalysis**

Date: Sunday, March 06, 2016 - Afternoon
Time: 02:45 PM
Room: B304

Primary Author: Janusz Pawliszyn
University of Waterloo

Co-Author(s)

Abstract Text

Recent trends in sample preparation include shift towards automation, high-throughput, miniaturization, and extraction methodology with low or no solvent consumption with preferably integrated with on-site sampling ("green chemistry"). In the presentation the ways in which solid-phase microextraction (SPME) and related techniques can complement currently used techniques will be discussed. The most promising application of these approaches, which distinguishes this method from other extraction and sample preparation techniques, is their applicability to on-site and in vivo sampling with no need of sample withdrawal. In this presentation, an automated, high-throughput method based on thin-film solid phase microextraction and liquid chromatography mass spectrometry will be introduced for simultaneous quantitative analysis of 110 drugs chosen from ten classes and varying in physical and chemical properties. SPME does not require any sample collection because extraction takes place in situ by inserting a biocompatible microfiber directly into tissue, blood or other biological matrix for a short period of time. Alternatively, the same device can be used for ex vivo analysis using a small amount of the studied sample. Despite the multiple advantages of SPME, few efforts have been done regarding its direct coupling to MS towards the analysis of drugs and biomarkers in vivo and in complex matrices. This work presents multiple strategies recently developed for the direct coupling of SPME to MS. In order to have a broader range of applications, different SPME geometries such coated fibers and meshes, as well as ionization approaches such DART and ESI, were studied.

Keywords: Clinical Chemistry, Extraction, Sample Handling/Automation, Sample Preparation

Application Code: Bioanalytical

Methodology Code: Sampling and Sample Preparation

Session # 50 **Abstract # 50-4** **Symposia**

Session Title: Enabling Sample Preconcentration Methods for Bioanalysis

Abstract Title: **Microfluidic Integration of Solid-Phase Extraction with Fluorescence Labeling for Microfluidic Analysis**

Date: Sunday, March 06, 2016 - Afternoon
Time: 03:35 PM
Room: B304

Primary Author: Adam T. Woolley
Brigham Young University

Co-Author(s): Mukul Sonker, Radim Knob, Suresh Kumar, Vishal Sahore

Abstract Text

Advances in biomolecular analysis are continuing to push detection limits lower, and sample preconcentration offers a key tool for further improving these limits of detection. We are developing microfluidic systems that integrate solid-phase extraction with fluorescence labeling in a microfluidic format. Our microdevices address two important human health issues: preterm birth and sepsis. We are integrating affinity capture, solid-phase extraction and microchip electrophoresis in microfluidic devices for rapid determination of preterm birth biomarker concentrations. We have observed an increase in signal of nearly 100-fold in microchip electrophoresis after an on-chip solid-phase extraction preconcentration step. We are also forming monolithic columns in microfluidic devices for the selective capture of nucleic acids from bacteria associated with blood infections, as well as genes related to antibiotic resistance. We have demonstrated the ability to capture and elute model DNA sequences related to the KPC gene. Present efforts are focusing on the extraction of targeted nucleic acids from bacterial lysate purified from blood. Our microfluidic systems for preterm birth biomarker analysis and sepsis diagnosis have promising capabilities for improving health outcomes.

We are grateful to the National Institutes of Health for funding (R01 EB006124 and R01 AI116989).

Keywords: Bioanalytical, Capillary Electrophoresis, Lab-on-a-Chip/Microfluidics, Solid Phase Extraction
Application Code: Bioanalytical
Methodology Code: Microfluidics/Lab-on-a-Chip

Session # 50 **Abstract # 50-5** **Symposia**

Session Title: Enabling Sample Preconcentration Methods for Bioanalysis

Abstract Title: **Liquid Biopsies: Microfluidic Enabling the Clinical Utility of These Markers**

Primary Author: Maggie Witek
University of North Carolina

Date: Sunday, March 06, 2016 - Afternoon
Time: 04:10 PM
Room: B304

Co-Author(s): Steven A. Soper

Abstract Text

Circulating markers can be used for a variety of disease management scenarios including prenatal diagnostics, cardiovascular diseases, bacterial infections and stroke. In addition, it is now being realized that many cancer-related diseases may also be managed using circulating markers (i.e., liquid biopsies) instead of relying exclusively on imaging-based technologies. While the utility of circulating markers has been recognized, the hardware and assays for their selection from clinical samples has been less than appealing. For example, many plasma-based markers are selected from whole blood using a variety of centrifugation and/or magnetic bead techniques. In addition, most assays are geared toward selecting only a single marker type instead of the plethora of markers that exist in blood (biological cells, cell free DNA, exosomes, proteins, etc.). In this presentation, exciting new microfluidic technologies will be discussed that can process whole blood directly and search for a number of different types of clinically relevant biomarkers to allow processing of large inputs and search for rare targets. The technology platform consists of task-specific fluidic modules made from plastics using micro-replication that can be generated in a single step irrespective of the module architecture and structure size, which is conducive to producing large number of modules at low cost and with tight tolerances. The modules that will be discussed in this presentation include modules to: (1) Selection of biological cells from whole blood; (2) isolation of plasma; (3) solid-phase extraction of cell free DNA; and (4) affinity selection of exosomes. The selection of these markers can be accomplished from a single sample requiring short processing times (<30 min even for sample inputs ranging to 10 mL) excellent recoveries (>90%) and with high purity (>80%).

Keywords: Biological Samples, Biotechnology, Medical, Sample Preparation
Application Code: Biomedical
Methodology Code: Microfluidics/Lab-on-a-Chip

Session # 60 **Abstract # 60-1** **Symposia**

Session Title: Frontiers of In Situ and In Vivo Spectroscopic Imaging

Abstract Title: **Correlated Raman and Mass Spectrometric Chemical Imaging of Multiscale Spatiotemporal Signaling in Microbial Communities**

Date: Sunday, March 06, 2016 - Afternoon
Time: 01:35 PM
Room: B305

Primary Author: Paul W. Bohn
University of Notre Dame

Co-Author(s): Jonathan V. Sweedler, Joshua D. Shrout, Nameera Baig, Nydia Morales-Soto, Sage B. Dunham, Sneha Polisetti

Abstract Text

Correlated chemical imaging is an emerging strategy for acquisition of images by combining information from multiplexed measurement platforms to track, visualize, and interpret in situ changes in the structure, organization, and activities of interesting biological systems, frequently spanning multiple decades in space and time. Acquiring and correlating information from complementary imaging experiments has the potential to expose complex chemical behavior in ways that are simply not available from single methods applied in isolation. However, correlating image information across platforms presents a number of challenges. Signals are obtained from disparate experiments with fundamentally different figures of merit, including pixel size, spatial resolution, dynamic range and acquisition rates. In addition, images are often acquired on different instruments in different locations, so the sample must be registered spatially so that the same area of the sample landscape is addressed. The signals acquired must be correlated in both spatial and temporal domains, and the resulting information has to be presented in a way that is readily understood. We are exploring the potential of heterocorrelated mass spectrometric (MS) and confocal Raman microscopy (CRM) chemical imaging, as targeted to problems in microbial community development. This talk will center on the use of correlated CRM-MSI in order to understand the community behavior in the development of biofilms in Pseudomonas seruginosa and will illustrate how integrating the tools of modern molecular/cellular biology with advanced chemical imaging concepts can yield heretofore unknown (and unknowable) features of the collective behavior of bacteria.

Keywords: Imaging, Infrared and Raman
Application Code: Bioanalytical
Methodology Code: Vibrational Spectroscopy

Session # 60 Abstract # 60-2 **Symposia**

Session Title: Frontiers of In Situ and In Vivo Spectroscopic Imaging

Abstract Title: **Nanotechnology for In-Vivo and Intraoperative Cancer Detection and Image-Guided Surgery**

Date: Sunday, March 06, 2016 - Afternoon
Time: 02:10 PM
Room: B305

Primary Author: Shuming Nie
Emory University

Co-Author(s)

Abstract Text

Nanotechnology is an area of considerable current interest in biomedical engineering because of its broad applications in biomedical imaging, in-vitro diagnostics, and targeted therapy. The basic rationale is that nanometer-sized particles such as quantum dots, colloidal gold, and polymeric nanomicelles have functional and structural properties that are not available from either discrete molecules or bulk materials. When conjugated with targeting ligands such as monoclonal antibodies, peptides or small molecules, these nanoparticles can be used to target malignant tumor cells and the tumor microenvironment (such as tumor stroma and tumor vasculatures) with high specificity and affinity. In the "mesoscopic" size range of 10-100 nm, nanoparticles also have large surface areas for conjugating to multiple diagnostic and therapeutic agents, opening new possibilities in imaging, therapy, and surgery. At the present, however, there are several fundamental problems and technical barriers that must be understood and overcome. In this talk, I will discuss the major challenges and opportunities in the development of nanomedicine for intraoperative cancer detection and image-guided surgery.

This work was supported by grants from the US National Institutes of Health (U54 CA119338, RC2 CA148265, and R01CA163256).

Keywords: Biomedical, Biospectroscopy, Fluorescence, Spectroscopy

Application Code: Nanotechnology
Methodology Code: Biospectroscopy

Session # **60** **Abstract #** **60-3** **Symposia**

Session Title: Frontiers of In Situ and In Vivo Spectroscopic Imaging

Abstract Title: **Nonlinear Microscopy to Detect and Grade Cancer**

Date: Sunday, March 06, 2016 - Afternoon
Time: 02:45 PM
Room: B305

Primary Author: Warren S. Warren
Duke University

Co-Author(s)

Abstract Text

Enhanced control over optical sources (femtosecond laser pulse shaping and pulse train modulation) turn out to be powerful techniques to drastically improve contrast in optical imaging. For example, nonlinear optical signatures can be explored with very modest laser powers (less average power than a laser pointer), and these signatures for biologically important endogenous chromophores such as melanin are exquisitely sensitive to aggregation, trace metal concentration, and composition changes. This means that nonlinear imaging produces clinically relevant contrast, reflecting metastatic potential. Applications to cutaneous, ocular, and vulvar melanoma have been extensively explored

Keywords: Biomedical, Laser, Near Infrared, Ultra Fast Spectroscopy
Application Code: Clinical/Toxicology
Methodology Code: Near Infrared

Session # **60** **Abstract #** **60-4** **Symposia**

Session Title: Frontiers of In Situ and In Vivo Spectroscopic Imaging

Abstract Title: **Multi-Parametric and Multi-Spectral Photoacoustic Microscopy**

Date: Sunday, March 06, 2016 - Afternoon
Time: 03:35 PM
Room: B305

Primary Author: Song Hu
University of Virginia

Co-Author(s)

Abstract Text

Photoacoustic microscopy (PAM), providing the unique contrast of optical absorption and the superb scalability of spatial resolution and tissue penetration, has emerged as an enabling technology in basic and translational biomedicine. Recent technical advances in our laboratory have opened new application niches for PAM. Specifically, our multi-parametric PAM enables—for the first time—simultaneous quantification of hemoglobin concentration, oxygen saturation, and blood flow down to single capillaries [i]in vivo[/i]. This technical innovation, in combination with fluorescence microscopy, permits dynamic interrogation of the molecular basis of ischemia-induced anatomical and functional vascular remodeling. Moreover, PAM of the three hemodynamic parameters at the same spatiotemporal scale lays the foundation for high-resolution mapping of the metabolic rate of oxygen, which may shed new light on the metabolic dysfunction in cancer and brain ischemia. Our multi-spectral PAM, based on a novel optical-acoustic objective, provides a near-constant lateral resolution (i.e., 2.8 m) over a broad spectral range of 270–1300 nm. The consistent performance over the ultraviolet, visible, and near-infrared range enables label-free PAM of cell nucleus (DNA/RNA contrast at 270 nm), blood vessel (hemoglobin contrast at 532 and 559 nm), and sebaceous gland (lipid contrast at 1260 nm) at the microscopic level [i]in vivo[/i]. Concurrent imaging of irregular nuclear morphology, disrupted oxygen metabolism, and abbarent lipogenesis in experimental cancer models holds great potential to reveal new biomarkers for early cancer detection and targeted therapy.

Keywords: Biomedical, Imaging, Microscopy, Photoacoustic

Application Code: Biomedical

Methodology Code: Microscopy

Session # 60 **Abstract # 60-5** **Symposia**

Session Title: Frontiers of In Situ and In Vivo Spectroscopic Imaging

Abstract Title: **Vibrational Spectroscopic Imaging of Molecular Dynamics in Living Systems**

Primary Author: Ji-Xin Cheng
Purdue University

Date: Sunday, March 06, 2016 - Afternoon
Time: 04:10 PM
Room: B305

Co-Author(s)

Abstract Text

Vibrational fingerprint spectroscopy has been extensively applied to study molecules in gas phase, condensed phase, and at interfaces. The transition from spectroscopy to spectroscopic imaging of living systems is opening new opportunities to unveil cellular machinery and to enable molecule-based diagnosis. Such transition is, however, not a simple combination of spectrometer and microscope. My talk will review recent efforts that pushed the boundary of the vibrational spectroscopic imaging field in terms of spectral acquisition speed, detection sensitivity, spatial resolution, and imaging depth. I will further highlight significant applications in functional analysis of single cells and in label-free detection of diseases. An outlook of the field will be presented.

Keywords: Infrared and Raman, Microscopy, Microspectroscopy, Vibrational Spectroscopy
Application Code: Bioanalytical
Methodology Code: Molecular Spectroscopy

Session # 70 **Abstract # 70-1** **Symposia**

Session Title: Wearable and Point-of-Care Sensor Technologies for Biomonitoring

Abstract Title: **Personal Exposure Monitoring Using Portable Samplers and Paper Analytical Devices**

Primary Author: Charles Henry
Colorado State University

Date: Sunday, March 06, 2016 - Afternoon
Time: 01:35 PM
Room: B401

Co-Author(s)

Abstract Text

Exposure to ambient particulate matter (PM) in both indoor and outdoor settings has a significant impact on human health. The most recent Global Burden of Disease study published in conjunction with the World Health Organization suggests PM exposure as the single largest environmental contributor to disability-adjusted life year (DALY) loss. While on-going studies utilizing traditional analytical instrumentation coupled with fixed site monitoring have yielded important insights into PM exposure, the approach cannot account for the heterogeneity of an individual's exposure because of the highly variable nature of PM composition and concentration. Attempts at personal exposure monitoring have been limited by the use of bulky, expensive, loud sampling systems and traditional analytical techniques that, while sensitive and selective, require large sample volumes resulting in poor sensitivity. We have been developing alternative battery powered, wearable PM sampling systems that generate filter-based as an alternative to traditional samplers. To improve chemical measurements, we are using paper-based analytic devices to determine composition and reactivity because the technology works well with small volumes and complex samples, can ultimately be used at the point of need, and provides rapid results. This talk will focus on recent results that have coupled the innovative sampling and analytic methods enabling accurate measurements of personal exposure to indoor and outdoor PM pollutants. Specific examples of detecting metals and oxidative load in different samples will be shown.

Keywords: Aerosols/Particulates, Electrochemistry, Lab-on-a-Chip/Microfluidics, Sampling
Application Code: Environmental
Methodology Code: Microfluidics/Lab-on-a-Chip

Session # 70 **Abstract # 70-2** **Symposia**

Session Title: Wearable and Point-of-Care Sensor Technologies for Biomonitoring

Abstract Title: **Pencil and Paper Diagnostic Devices**

Primary Author: Andres W. Martinez
California Polytechnic State University

Date: Sunday, March 06, 2016 - Afternoon
Time: 02:10 PM
Room: B401

Co-Author(s)

Abstract Text

Paper-based fluidic devices, also known as microPADS, offer a promising platform for the development of point-of-care diagnostic assays for use in remote, resource-limited settings. MicroPADs are inexpensive to fabricate, portable, simple to operate, and can complete an assay without relying on electrical power or supporting equipment. The reagents required to perform the assays on paper-based devices are typically added to the devices as solutions during fabrication. The solutions are then dried on the devices, and the reagents are stored in dry form until they are rehydrated when the device is used to perform a test. This presentation will describe our work on the development of an alternative method for depositing and storing reagents on paper-based devices using custom-made pencils containing reagents dispersed in a solid matrix. These reagent pencils enable rapid and solvent-free deposition of reagents onto membrane-based fluidic devices in a method that is as simple as drawing with the reagent pencils on the devices. When aqueous samples are added to the devices, the reagents dissolve from the pencil matrix and become available to react with analytes in the sample. We evaluated the capabilities of the pencils for depositing reagents onto paper-based devices and also for storing reagents in a dry and convenient platform. We believe reagent pencils will offer a new option for preparing and customizing diagnostic tests at the point of care without the need for specialized equipment, which will provide an alternative option to the conventional approach of depositing reagents on the devices during fabrication.

Keywords: Bioanalytical, Clinical Chemistry, Enzyme Assays, Lab-on-a-Chip/Microfluidics
Application Code: Bioanalytical
Methodology Code: Microfluidics/Lab-on-a-Chip

Session # 70 Abstract # 70-3 **Symposia**

Session Title: Wearable and Point-of-Care Sensor Technologies for Biomonitoring

Abstract Title: **Electrochemical Metal Determination for Point-of-Care Assessment of Environmental Exposure**

Primary Author: Ian Papautsky
University of Cincinnati

Date: Sunday, March 06, 2016 - Afternoon
Time: 02:45 PM
Room: B401

Co-Author(s)

Abstract Text

Metals are ubiquitous in the environment and have long been recognized to pose significant threat to human health. Blood lead (Pb) has been consistently associated with deficits in IQ and academic achievement in numerous controlled studies. Manganese (Mn) is an essential element, yet neurotoxic in excess, capable of crossing the blood-brain barrier and accumulating in the brain. Another ubiquitous metal, cadmium (Cd), has the potential to cause kidney, liver, bone, brain, and vascular damage. Current approaches to measuring exposure to these metals often suffer from high cost and lengthy turnaround. This talk will discuss recent developments in electrochemical metal determination for point-of-care assessment of these metals. While anodic stripping of Pb has been well-reported in literature, stripping analysis of Mn and Cd on microscale remains a critical challenge due to the strong negative potential of stripping peak or complexation with other metals. Necessity for low limits of detection, high reproducibility, and low (disposable) sensor costs present additional challenges. Our ultimate goal is to demonstrate rapid, point-of-care, multi-analyte assessment of Pb, Mn and Cd in a finger prick of blood.

Keywords: Bioanalytical, Electrochemistry, Environmental/Biological Samples, Metals
Application Code: Bioanalytical
Methodology Code: Electrochemistry

Session # 70 **Abstract # 70-4** **Symposia**

Session Title: Wearable and Point-of-Care Sensor Technologies for Biomonitoring

Abstract Title: **Ionic Liquid Electrochemistry and Biointerface Design for Reliable and Smart Sensors**

Date: Sunday, March 06, 2016 - Afternoon
Time: 03:35 PM
Room: B401

Primary Author: Xiangqun Zeng
Oakland University

Co-Author(s)

Abstract Text

Following the 1980s PC revolution and the 1990s internet revolutions, recent decades, have experienced a revolution in sensor research which promises to have a significant impact on a broad range of applications including national security, health care, environment, energy, industry and food safety, and manufacturing. Sensor research is cross-disciplinary involving chemistry, physics, material science, engineering, biology and medicine. This presentation will discuss the notable contributions to the chemical sensor and biosensor fields from our laboratory. Our innovative approach stems from a systematic design process across all design layers, with judicious choices in materials and transducers to uniquely overcome all performance challenges, especially those related to sensing capability and miniaturized sensor system implementation. By molecular design and control of unique interface materials including ionic liquids, conductive polymers, carbohydrates, proteins and cells, we have fabricated new sensor interfaces and synergistically integrated them with low cost and low power piezoelectric/QCM and electrochemical transducers to yield forward-thinking solutions to many sensors challenges, especially miniaturization for wearable sensors and robustness for field uses. The presentation will include three parts: (1) Challenges for current chemical sensor and biosensors; (2) Ionic liquid electrochemistry and gas sensor applications; (3) label free biosensors for real time and non-invasive characterization of biomolecular interactions and for studying cell biology at the whole cell level. The simplicity and the high performance of the demonstrated detection principles allows for easy integration with engineering advancements for lab on a chip and wearable sensors with the performance, cost, power, and operational lifetime characteristics to suit a broad range of applications.

Keywords: Biosensors, Electrochemistry, Environmental/Air, Sensors
Application Code: Other
Methodology Code: Sensors

Session # 70 **Abstract # 70-5** **Symposia**

Session Title: Wearable and Point-of-Care Sensor Technologies for Biomonitoring

Abstract Title: **Monitoring Corrosion of Biodegradable Magnesium Implants with Hydrogen Gas Sensors**

Primary Author: William R. Heineman
University of Cincinnati

Date: Sunday, March 06, 2016 - Afternoon
Time: 04:10 PM
Room: B401

Co-Author(s): Daoli Zhao, David Benson, Tingting Wang, William Hoagland, Zhongyun Dong

Abstract Text

Magnesium and its alloys are being developed as biodegradable metallic implant materials for bone repair, stents and other medical applications because they exhibit properties such as strength, light weight, and corrosion in aqueous environments. The corrosion of magnesium and its alloys can be monitored by measuring the concentrations of the solution soluble corrosion products magnesium ion, hydroxyl ion and hydrogen gas. Hydrogen has proven to be especially useful in this regard because electrochemical and optical sensors are available that detect hydrogen at low levels with little interference. Of special importance is the ability to detect hydrogen produced by a magnesium implant through the skin. Monitoring hydrogen transdermally with a sensor provides a means for monitoring biodegration in point of care situations.

Keywords: Bioanalytical, Electrochemistry, Sensors
Application Code: Biomedical
Methodology Code: Sensors

Session # 80 **Abstract # 80-1** **Workshops**

Session Title: Light Sources in Analytical Chemistry: Solid State Light Sources and Beyond

Abstract Title: **Gatherers and Foragers? Analytical Scientists in the Quest for Better Light Sources**

Primary Author: Mirek Macka
University of Tasmania

Date: Sunday, March 06, 2016 - Afternoon
Time: 01:35 PM
Room: B402

Co-Author(s)

Abstract Text

This talk will introduce the topic of the workshop "Light Sources in Analytical Chemistry: Solid State Light Sources and Beyond". Advances in many areas of analytical chemistry depend to a significant degree on advances in light source technology. Solid state light sources (SSLS), namely light emitting diodes (LEDs), laser diodes, quantum cascade lasers etc.) provided an immense impetus for further developments and most importantly to portable analytical instrumentation. This talk will highlight some of the most important developments especially in the area of light emitting diodes, and illustrate them by examples primarily from own research in areas including optical detection in separation techniques and imaging. Areas and directions of future progress will be discussed.

Keywords: Analysis, Characterization, Instrumentation, Portable Instruments
Application Code: Other
Methodology Code: Chemical Methods

Session # **80** Abstract # **80-2** **Workshops**

Session Title: Light Sources in Analytical Chemistry: Solid State Light Sources and Beyond

Abstract Title: **Applications of Cavity Ring-Down and Cavity-Enhanced Absorption Spectroscopy in Atmospheric Chemistry**

Primary Author: Hans D. Osthoff
University of Calgary

Date: Sunday, March 06, 2016 - Afternoon
Time: 02:05 PM
Room: B402

Co-Author(s): Charles A. Odame-Ankrah, Connie Z. Ye, Jason Pak, Nick R. Yordanov, Youssef M. Taha

Abstract Text

High-finesse optical absorption techniques such as cavity ring-down spectroscopy (CRDS) and cavity-enhanced absorption spectroscopy (CEAS) are sensitive and hence popular tools for accurate quantification of gases. Here, applications of CRDS and CEAS to the analysis of trace gases of interest to the atmospheric chemistry community will be presented. A 6-channel CRDS for quantification of nitrogen oxides and a CEAS powered by a cyan (505 nm) LED for quantification of nitrogen dioxide (NO[sub]2[/sub]) and I[sub]2[/sub] mixing ratios were constructed. These molecules play many important roles in the troposphere, including the catalysis of photochemical ozone production; hence, knowledge of their abundances and transformation pathways is of interest. Mixing ratios of NO[sub]2[/sub] and of the nitrate radical (NO[sub]3[/sub]) are directly quantified by blue (405 nm) and red (662 nm) diode laser CRDS with 1-minute detection limits of 20 parts-per-trillion (pptv, 10[sup]-12[/sup]) and 2 pptv, respectively. Other nitrogen oxides, including nitric oxide (NO), dinitrogen pentoxide (N[sub]2[/sub]O[sub]5[/sub]), total peroxyacyl nitrates (PN = RC(O)O[sub]2[/sub]NO[sub]2[/sub]), and total alkyl nitrates, (AN = RONO[sub]2[/sub]) are quantified by selective conversion to either NO[sub]2[/sub] or NO[sub]3[/sub]. To improve the detection limits for PN and AN to below 10 pptv and to minimize interferences, the instrument was equipped with an automated purge-and-trap preconcentration setup. Results from the Ozone-depleting reactions in a coastal atmosphere (ORCA) campaign, which took place at the Amphitrite Point Observatory in Ucluelet, BC, Canada, July 8 - 31, 2015, are presented. Current work focuses on the quantification of total odd nitrogen (NO[sub]y[/sub]) and aromatic nitrates by catalytic conversion to NO[sub]2[/sub].

Keywords: Environmental/~~Air~~, ~~Laser~~, Specialty Gas Analysis, Spectrometer
Application Code: Environmental
Methodology Code: UV/VIS

Session # 80 **Abstract # 80-3** **Workshops**

Session Title: Light Sources in Analytical Chemistry: Solid State Light Sources and Beyond

Abstract Title: **A Fourier-Domain Fluorescence Excitation Emission Matrix Spectrometer and Its Applications in Measuring Oxidative Stability of Industrial Liquids**

Date: Sunday, March 06, 2016 - Afternoon
Time: 02:35 PM
Room: B402

Primary Author: Hans-Peter Loock
Queen's University

Co-Author(s): Alexander Dudelzak, Hengameh Omrani, James Z. Fan, Nicholas L. Andrews, Oliver Reich

Abstract Text

Internal combustion engines and turbines are ubiquitous in today's society, and it may appear surprising that there exists not a single commercial sensor that could reliably monitor in real-time, the condition of the lubricants that are used in these engines. We previously established fluorescence excitation emission matrix (EEM) spectroscopy as a method capable of determining the antioxidant concentration - a good indicator species for lubricant quality. By scanning the excitation wavelength and recording a fluorescence spectrum at every wavelength a three dimensional excitation emission matrix (EEM) spectrum can be obtained. EEM spectra are typically generated by step-scanning a grating to 50-100 positions for each wavelength of the emission spectrum while a second grating is scanned to positions corresponding to each excitation wavelength. A new fiber probe permits these measurements in situ, i.e. in a running engine. The fluorescent components of lubricants, i.e. mostly antioxidants and other additives can be quantified using multivariate analysis techniques such as Parallel Factor (PARAFAC) analysis.

Unfortunately, the sequential scanning of EEM spectrometers corresponds to a point by point generation of the EEM spectrum and is quite slow. The acquisition of a single EEM spectrum takes typically between 10-40 minutes and cannot be applied to fast phenomena.
We demonstrate a technique whereby multiplexing such a process has the potential of speeding up data acquisition by a factor of up to 5,000. Using a digital micromirror array each excitation event consists not of a single colour but a "barcode spectrum" of colours. The sample subsequently fluoresces and the fluorescence is dispersed in a conventional array spectrometer. After decoding the spectral array the whole EEM spectrum is obtained. Utilizing a Hadamard modulation approach, the technique makes use of the throughput and multiplex advantages that are inherent with Fourier-transform-based methods.

Keywords: Fluorescence, Instrumentation, Spectrometer
Application Code: Process Analytical Chemistry
Methodology Code: Fluorescence/Luminescence

Session # 80 **Abstract # 80-4** **Workshops**

Session Title: Light Sources in Analytical Chemistry: Solid State Light Sources and Beyond

Abstract Title: **Deep UV-LEDs in Detectors for HPLC and Capillary Electrophoresis**

Date: Sunday, March 06, 2016 - Afternoon
Time: 03:20 PM
Room: B402

Primary Author: Peter C. Hauser
University of Basel

Co-Author(s): Duy Anh Bui

Abstract Text

Optical absorbance detection in HPLC and capillary electrophoresis usually requires wavelengths below 300 nm as many molecules show absorbance bands in this region but not at longer wavelengths. The arrival of LEDs for the deep UV-range therefore opened up the possibility of constructing inexpensive and compact detectors for these common analytical separation methods. These devices were found to perform as well as state-of-the art conventional detectors.

Design considerations for the construction of detectors for standard and narrow bore HPLC will be discussed as well as for the more demanding platform of capillary electrophoresis. Important aspects are the avoidance of stray light, achieving high stability in intensity, and mechanical robustness. The implementation of Beer's law with analog electronic circuitry will be detailed and the effect of non-chromaticity of the LEDs on the signal will be examined.

Keywords: Capillary Electrophoresis, Capillary LC, Detector, HPLC Detection
Application Code: General Interest
Methodology Code: Molecular Spectroscopy

Session # 80 **Abstract # 80-6** **Workshops**

Session Title: Light Sources in Analytical Chemistry: Solid State Light Sources and Beyond

Abstract Title: **Quantum Cascade Lasers: How to Revolutionize Mid-Infrared Gas and Liquid Phase Diagnostics**

Date: Sunday, March 06, 2016 - Afternoon
Time: 04:20 PM
Room: B402

Primary Author: Boris Mizaikoff
Ulm University

Co-Author(s)

Abstract Text

In recent years, chem/bio sensing platforms increasingly benefit from miniaturized and integrated optical technologies providing direct access to molecular information. Since in-situ analytical strategies are becoming more prevalent e.g., in harsh environments or for point-of-care diagnostics, detection schemes that do not require reagents or labels are of particular interest providing localized on-site information in – or close to - real-time.

Mid-infrared (MIR; 3-20 μm) sensor technology is progressively more adopted in environmental analysis, process monitoring, and biodiagnostics owing to the inherent molecular specificity. Thereby, discrimination of molecular constituents at ppm-ppb concentration levels in condensed and vapor phase media is enabled. While recently emerging technologies include innovative waveguide structures such as mid-infrared transparent fiberoptics, substrate-integrated hollow waveguides, and planar semiconductor waveguide, the true revolution in MIR sensing is based on next-generation laser light sources - broadly tunable quantum cascade lasers (QCLs). Given these developments, compact yet robust MIR chem/bio sensors and diagnostics are on the horizon for applications in extreme environments such as the deep sea, but also for advanced breath monitoring in clinical analysis. Selected examples and recent developments will highlight the potential of QCL-based MIR sensing technologies.

Keywords: Biomedical, Biosensors, Environmental Analysis, Laser
Application Code: General Interest
Methodology Code: Sensors

Session # 90 **Abstract # 90-1** **Organized Contributed Sessions**

Session Title: Advances in Mass Spectrometry of RNA - Half Session

Abstract Title: **Toward Detection of Isomeric/Isobaric MicroRNA Biomarkers**

Primary Author: Norman Chiu
University of North Carolina at Greensboro

Date: Sunday, March 06, 2016 - Afternoon
Time: 01:30 PM
Room: B403

Co-Author(s)

Abstract Text

In the human body, there are more than 2,000 different microRNA (miRNA). Approximately, half of the human miRNA have the same size and consist of 22 nuclecotides. With the limit on four different nucleobases (A, U, G, C), it is not surprising to have multiple isomeric miRNA, i.e. nucleotide composition is identical, that may originate from the same type of tissue. Some of isomeric human miRNA can also have high percentage of sequence similarity. Together with the increasing interest in using specific miRNA as diagnostic biomarkers or potential drug targets, it has raised the demand for more accurate miRNA detection. One way to improve the accuracy is by using mass spectrometry (MS) to measure miRNA directly. Matrix-assisted laser desorption/ionization (MALDI) MS stands apart from other MS techniques due to the fact that a MALDI matrix is required for sample preparation. In this study, by exploiting the acidity of MALDI matrix and its mixing with miRNA prior to MS measurements, a simple method to generate RNA sequencing ladders is developed. The method utilizes the MALDI matrix to hydrolyze RNA at high temperature. The resulting sequencing ladders are ready to be measured without any desalting. By using MALDI SpiralTOF MS, the monoisotopic mass of each RNA fragment was measured. The RNA sequence was determined by sequentially comparing nucleotide compositions that were calculated from measured monoisotopic masses. The use of nucleotide compositions to assist the spectral interpretation has the advantages on distinguishing the complementary sequencing ladders, and allows the nucleotide identity at each position to be crosschecked multiple times. Together with the analysis of both complementary sequencing ladders, 100% sequence coverage and sequence accuracy were achieved in a blinded study.

Keywords: Bioanalytical, Mass Spectrometry, Method Development, Nucleic Acids
Application Code: Bioanalytical
Methodology Code: Mass Spectrometry

Session # 90 **Abstract # 90-2** **Organized Contributed Sessions**

Session Title: Advances in Mass Spectrometry of RNA - Half Session

Abstract Title: **New LC-MS/MS Strategies for Modified Oligonucleotides and RNAs**

Date: Sunday, March 06, 2016 - Afternoon
Time: 01:50 PM
Room: B403

Primary Author: Patrick A. Limbach
University of Cincinnati

Co-Author(s): Robert L. Ross

Abstract Text

Liquid chromatography tandem mass spectrometry (LC-MS/MS) methods have proven to be quite effective at determining oligonucleotide purity, sequence and modification status. The standard approaches have been to use C18 reverse phase chromatography for sample separation, followed by electrospray ionization (ESI) for sample introduction. Despite the success of these standard approaches, issues related to chromatographic background and carryover, along with ion suppression, currently limit high sensitivity measurements of modified oligonucleotides and RNAs. In this presentation, we will discuss new strategies for LC-MS/MS of modified oligonucleotides including the use of droplet-based sample introduction. Advantages of these alternative methods will be illustrated, and future areas of improvement will be discussed.

Keywords: Bioanalytical, Liquid Chromatography/Mass Spectroscopy, Nucleic Acids, Software
Application Code: Bioanalytical
Methodology Code: Mass Spectrometry

Session # 90 Abstract # 90-3 **Organized Contributed Sessions**

Session Title: Advances in Mass Spectrometry of RNA - Half Session

Abstract Title: **Mass Spectrometry, RNA and Infectious Disease: tRNA Reprogramming and a Second Genetic Code Control Mycobacterial Dormancy**

Primary Author: Peter Dedon
Massachusetts Institute of Technology

Date: Sunday, March 06, 2016 - Afternoon
Time: 02:10 PM
Room: B403

Co-Author(s)

Abstract Text

The past decade has seen a resurgence of interest in the biological function of the >120 known chemical modifications of all types of RNA in all forms of life. Every cell possesses 25-40 different modified ribonucleosides most frequently occurring in tRNA. While the modifications are known to affect tRNA stability and translational fidelity, their higher-level biological function has eluded definition. Here we look at the role of tRNA modifications in controlling cellular adaptation to stress. Cells respond to environmental changes by altering gene expression at several levels, with translational control of cell response being poorly understood. Using a novel bioanalytical platform, we recently discovered a new systems-level mechanism in yeast, in which the dozens of tRNA modifications reprogram in response to stress to control selective translation of families of codon-biased mRNAs from stress response genes. We now report that this translational mechanism is a general feature of both prokaryotes and eukaryotes, with a demonstration that it governs the dormancy response of mycobacteria during the hypoxic stress characteristic of tuberculous granulomas. Focusing on [i]Mycobacterium bovis[/i] BCG, an [i]M. tuberculosis[/i] surrogate, we observed signature reprogramming of 40 tRNA modifications at each stage of hypoxic dormancy and aerobic resuscitation. This reprogramming was linked to selective translation of transcripts from families of codon-biased genes. For example, early hypoxia caused increases in wobble cmo[sup]5[/sup]U in tRNA[sup]Thr(UGU)[/sup] that paralleled translation of transcripts biased in the use of its cognate codon, ACG, such as the DosR master regulator of hypoxic dormancy. Codon reengineering of [i]dosR[/i] produced corresponding changes in fitness during hypoxia and recovery. Ongoing studies illustrate the generality of this mechanism in human cancer, the life cycle of malaria parasites, and dengue virus infection.

Keywords: Biomedical, Liquid Chromatography/Mass Spectroscopy, Mass Spectrometry, Proteomics
Application Code: Biomedical
Methodology Code: Liquid Chromatography/Mass Spectrometry

Session # 90 Abstract # 90-4 **Organized Contributed Sessions**

Session Title: Advances in Mass Spectrometry of RNA - Half Session

Abstract Title: **Identification and Quantitation of RNA Post-Transcriptional Modifications by Nano-flow Liquid Chromatography - Tandem Mass Spectrometry and Database Searching.**

Date: Sunday, March 06, 2016 - Afternoon
Time: 02:30 PM
Room: B403

Primary Author: Hiroshi Nakayama
RIKEN CSRS

Co-Author(s): Masato Taoka, Nobuhiro Takahashi, Toshiaki Isobe

Abstract Text

Post-transcriptional modifications (PTMs) play important roles in the structure and function of RNA. Mass spectrometry has been proven to be an invaluable tool for the direct analysis of PTMs because almost all PTMs accompany the mass shifts from the equivalent, unmodified molecules and because highly sensitive genomics-based technologies can characterize only several types of known PTMs. We are developing a platform toward the comprehensive chemical analysis of RNA, which consists of direct nano-flow liquid chromatography and high resolution tandem mass spectrometry (nLC-MS/MS) coupled with a genome-oriented search program, Ariadne (Nakayama H. et al. 2009 Nucleic Acids Res, Taoka M. et al. 2009 Nucleic Acids Res). To date, this analytical system has been applied successfully to discover a novel metabolic pathway that ensures the quality of U snRNAs important for pre-mRNA splicing and isoform expression (Ishikawa H. et al. 2014 Nucleic Acids Res) and to the direct identification of human cellular miRNAs of around 22 nucleotides (Nakayama H. et al. 2015 Anal Chem). More recently, we have developed a method, termed SILNAS, for the comprehensive, quantitative identification of RNA PTMs, which utilizes an in vitro transcribed 13C-coded RNA as a reference standard to the biological RNA to be analyzed, and determined the complete chemical structure of eukaryotic ribosomal RNAs for the first time (Taoka M. et al. 2015 Nucleic Acids Res). Thus, our current system allows the identification of all RNA PTMs listed in public databases, i.e. ~130 PTMs, and can even discriminate a “mass-silent” PTM from uridine to pseudouridine at a single nucleotide resolution. In this presentation, we will present the current status of our system and discuss its potential application to the analysis of biological RNAs, as well as of synthetic nucleic acids produced for the pharmaceutical purpose.

Keywords: Bioinformatics, Biological Samples, Biopharmaceutical, Nucleic Acids
Application Code: Genomics, Proteomics and Other 'Omics
Methodology Code: Liquid Chromatography/Mass Spectrometry

Session # 100 **Abstract # 100-1** **Organized Contributed Sessions**

Session Title: Ionophore-Based Chemical Sensors I

Abstract Title: **pH Independent Fluorescent Ion Sensors**

Primary Author: Eric Bakker
University of Geneva

Date: Sunday, March 06, 2016 - Afternoon
Time: 01:30 PM
Room: B404

Co-Author(s)

Abstract Text

Optical sensors based on ionophore assisted competitive or cooperative extraction are today well established, but they typically exhibit a pH cross-sensitivity that is inherent in the sensing mechanism. Finding fundamental solutions to overcome this limitation is of utmost practical importance.

This talk explores to what extent the use of voltage sensitive dyes can help to design ion opposed that work independently of sample pH. These systems function by the partitioning of a solvatochromic charged dye as a function of the interfacial potential, although it can also be understood on the basis of simple ion-exchange or coextraction with the ion of interest. This type of detection must be performed with short diffusion distances, since the dye most equilibrium partition, and this is achieved with emulsion based sensor systems. While the sample pH is confirmed not to influence the sensor readout, the volume ratio of sample to sensor phase does impact on the signal.

This talk will explore whether a partial molecular transfer of the chromophore unit of the dye, while keeping the bulky lipophilic tail near the organic phase, will reduce this sample volume dependence to make this new sensing principle truly promising.

Keywords: Bioanalytical, Ion Selective Electrodes, Sensors
Application Code: Bioanalytical
Methodology Code: Sensors

Session # 100 **Abstract # 100-2** **Organized Contributed Sessions**

Session Title: Ionophore-Based Chemical Sensors I

Abstract Title: **Measurement of Carbon Dioxide in Urine to Guide the Treatment of Patients in Severe Sepsis or Septic Shock**

Date: Sunday, March 06, 2016 - Afternoon
Time: 01:50 PM
Room: B404

Primary Author: Erno Lindner
The University of Memphis

Co-Author(s): Bradford D. Pendley, James G. Atherton, Jasinski Artur, Marcin Guzinski

Abstract Text

It has been hypothesized that patients in severe sepsis or septic shock possess urine carbon dioxide levels significantly different from healthy patients. If the hypothesis could be proved true and the CO[sub]2[/sub] can be monitored during the treatment of these patients it is expected to lead to better outcomes. Today the mortality rate of patients in septic shock using the "early goal-directed therapy protocol"of Rivers[sup]1[/sup] is 30.5%.

To investigate the utility of urine carbon dioxide as a prognostic indicator for severe sepsis and septic shock we have collected urine samples from human subjects in an IRB approved protocol at the Methodist University Hospital, Memphis, and determined their CO[sub]2[/sub] levels using a home made wall-jet type flow through manifold in combination with a Severinghaus type carbon dioxide sensor. In the frame of our studies we followed the CO[sub]2[/sub] levels in urine samples collected from a Foley catheter before treatment and 12 hours and 24 hours after treatment.

The data collected from 12 intensive care unit patients does not allow drawing unambiguous conclusion. In certain patients, in agreement with the expectations, the urinary CO[sub]2[/sub] levels gradually decreased during the patient's recovery. However, in other patients it hardly changed or even increased. We are convinced that the uncertainties are related to uncertainties in the sampling. To establish that urine CO[sub]2[/sub] may indeed correlate with patient's hemodynamic status the sampling should be synchronized with the treatment. Towards this goal we are working on the implementation of a miniaturized CO[sub]2[/sub] sensor into the Foley catheter.

Related to the precision and accuracy of the determination of CO[sub]2[/sub] levels in urine we discuss the challenges of sampling, sample storage, and the influence of urine pH, temperature and ionic strength on the results.

1. Rivers, E. et. al. New Engl. J. Med. 2001, 345, 1368.

Keywords: Clinical Chemistry, Ion Selective Electrodes, Potentiometry, Sampling
Application Code: Bioanalytical
Methodology Code: Electrochemistry

Session # 100 Abstract # 100-3 **Organized Contributed Sessions**

Session Title: Ionophore-Based Chemical Sensors I

Abstract Title: **Potentiometric Nanosensors Towards Direct Detection of Nucleic Acids**

Primary Author: Róbert E. Gyurcsányi
Budapest University of Technology and Economics

Date: Sunday, March 06, 2016 - Afternoon
Time: 02:10 PM
Room: B404

Co-Author(s): Alexandra Brajnovits, Gyula Jágerszki, Istvan Makra, Márton Bojtár, Peter Fürjes

Abstract Text

We are going to report on our most recent results regarding the application of nanomaterials for potentiometric sensing. This includes single nanopore-based potentiometric sensors for direct determination of ions and nucleic acids as well as ion-selective electrodes based on nanoparticle-ionophore conjugates. In terms of nanopore-based potentiometric sensors for nucleic acid determination we will report the detailed mechanism of the potentiometric response formation. The theoretical treatment includes the effect of the location and surface concentration of the receptor molecule in the nanopore environment as well as the binding kinetics of the ligand to the surface confined receptor molecules.

Keywords: Biosensors, Nanotechnology, Nucleic Acids, Potentiometry
Application Code: Bioanalytical
Methodology Code: Sensors

Session Title Ionophore-Based Chemical Sensors I

Abstract Title **Voltammetric Ionophore-Based Electrode for Protamine in Human Blood**

Primary Author Shigeru Amemiya
University of Pittsburgh

Date: Sunday, March 06, 2016 - Afternoon
Time: 03:05 PM
Room: B404

Co-Author(s)

Abstract Text

In this presentation, I will demonstrate the first biomedical application of voltammetric ion-selective electrodes based on double-polymer membranes. Specifically, plasticized poly(vinyl chloride) membrane is doped with dinonylnaphthalenesulfonate to selectively and reversibly detect protamine as an antidote of heparin in human blood. The application of the new protamine sensor for titration of heparin is also discussed.

Keywords: Electrochemistry, Ion Selective Electrodes
Application Code: Biomedical
Methodology Code: Electrochemistry

Session # 100 Abstract # 100-6 **Organized Contributed Sessions**

Session Title Ionophore-Based Chemical Sensors I

Abstract Title **Conducting Polymer-Coated Electrode and Its Application to the Thin Layer Electrolysis Cell for Coulometric Determination**

Primary Author Yumi Yoshida
Kyoto Institute of Technology

Date: Sunday, March 06, 2016 - Afternoon
Time: 03:25 PM
Room: B404

Co-Author(s) Kohji Maeda, Mao Fukuyama

Abstract Text

A conducting polymer-coated electrode has been investigated as an inner electrode of the all-solid ion selective electrode. However, its stability and reproducibility are rather low, and practical use of the all-solid ion selective electrode with the conducting polymer-coated electrode has not been realized yet because potential stability significantly affects determination based on the potentiometric response. To examine the potential response of the all-solid ion selective electrode, the potential response of the conducting polymer-coated electrode in the organic solution, which is involved in the PVC membrane of the ion selective electrode, should be confirmed. But, there is few report regarding to potential response in the organic solution immiscible in water. In the present work, electrode response of the conducting polymer-coated electrode in the organic phase is reported. Moreover, in order to apply the conducting polymer-coated electrode to voltammetry measurement, the effect of current flow on depolarization of the electrode potential was also examined. The conducting polymer electrode adjusted by authors was applied to the thin layer electrolysis cell for coulometric determination, which was a two-electrode system. Results of the flow injection method and the stripping method will be indicated as coulometric measurements [1, 2].

[1] Y. Yoshida, S. Nakamura, J. Uchida, A. Hemmi, K. Maeda, J. Electroanal. Chem.,707 (2013), 95–101, [2] Y. Yoshida, J. Uchida, S. Nakamura, S. Yamaguchi, K. Maeda, Anal. Sci. 30 (2014) 351-357

Keywords: Electrochemistry, Electrodes, Extraction, Stripping Analysis
Application Code: General Interest
Methodology Code: Electrochemistry

Session # 100 Abstract # 100-7 **Organized Contributed Sessions**

Session Title Ionophore-Based Chemical Sensors I

Abstract Title **Biofouling of Ionophore-Doped Ion-Selective Electrode Membranes Revisited**

Date: Sunday, March 06, 2016 - Afternoon
Time: 03:45 PM
Room: B404

Primary Author Philippe Buhlmann
University of Minnesota

Co-Author(s) Adam Dittmer

Abstract Text

The primary focus of this research is the investigation of the mechanism of biofouling in current fluorous phase ion-selective electrode (ISE) systems. With prolonged exposure to biological samples, potentiometric measurements using conventional polymeric-membrane ISEs exhibit a breakdown of selectivity and response. Therefore, extensive washing procedures and frequent recalibrations are needed for many clinical and biological applications. Initial work with fluorous phase ISEs has shown significant improvements in selectivity and limits of detection over conventional polymeric-membrane ISEs.1 Moreover, experiments with fluorous pH electrodes have shown that long term serum exposure does not affect the electrode selectivity but stirred serum solutions appeared to cause a transient EMF drift. To explore this effect more systematically, a potentiometric stir tests was developed. Both conventional polymeric membrane and fluorous membrane electrodes were exposed to solutions stirred intermittently. Both types of membranes exhibited an EMF response to stirring when exposed to of 10% v/v solutions equine blood serum but not when exposed to simple electrolyte solutions. The transient potentiometric response depends on the hydrophobicity of the ionic sites incorporated into the ISE membranes; specifically, a lower hydrophobicity results in a larger effect of stirring on the EMF. For the fluorous electrodes, synthesis of a more hydrophobic ionic site and its use along with fluorophilic $H+$ ionophores successfully mitigated the effect of sample stirring on the emf. The use of a fairly simple phase boundary model confirms that the effect of stirring is caused by loss of ionic sites into the serum containing sample.

Keywords: Clinical Chemistry, Electrochemistry, Electrode Surfaces, Electrodes

Application Code: Clinical/Toxicology

Methodology Code: Electrochemistry

Session # 100 Abstract # 100-8 **Organized Contributed Sessions**

Session Title: Ionophore-Based Chemical Sensors I

Abstract Title: **Metastable Photoacids Towards Activatable and Controllable Ion Sensing Bulk Membranes for Cations Detection**

Date: Sunday, March 06, 2016 - Afternoon
Time: 04:05 PM
Room: B404

Primary Author: Karin Y. Chumbimuni-Torres
University of Central Florida

Co-Author(s): Parth K. Patel

Abstract Text

An entirely new approach in activatable and controllable sensing is presented here using metastable photoacids (mPAHs) within the sensor that may be activated by visible light or near-infrared (NIR). This could broaden and significantly enhance the current biological, biomedical and environmental sensing applications. Most of the activatable and controllable platforms for sensing require ultraviolet (UV) irradiation, which promotes cellular damage and limits sensing in biological and biomedical applications. Additional limitations are irreversibility and photo-fatigue of the sensor, which is inherent to the systems that use UV-irradiation for activation. These limitations could be overcome by using mPAHs that do not require UV irradiation.

A carbanion mPAH was reported to generate a large photo-induced proton concentration by visible light and was compatible towards organic media. This carbanion mPAH displays two absorption peaks that corresponds to carbanion mPAH protonated and deprotonated forms, which allows for ratiometric measurements to indirectly correlate the ion of interest. Furthermore, when this mPAH is introduced into the calcium sensing membrane, no photo-fatigue was observed after repeated activation (turn ON/OFF) cycles. This sensing membrane was also exposed to increasing concentrations of calcium ions without activation, resulting in negligible change in absorbance. When activated the membrane with visible light, the sensing membrane showed a change in the absorbance for different concentrations of calcium ions, allowing detection at the micromolar level, indicating that the photodissociated proton exchanged with the calcium ions.

Keywords: Sensors
Application Code: Bioanalytical
Methodology Code: Sensors

Session # 110 Abstract # 110-2 **Organized Contributed Sessions**

Session Title R&D to QC: Bridging the Gap - Half Session

Abstract Title **Effective Method Transfer: Ensuring End Results**

Date: Sunday, March 06, 2016 - Afternoon
Time: 01:50 PM
Room: B312

Primary Author Mary Ellen P. McNally
DuPont Crop Protection

Co-Author(s) Stephen Platz

Abstract Text

In a regulated industrial environment, analytical methods are typically developed in R&D and transferred widely to manufacturing and formulation sites, alliance partners, and CRO's. These transfers readily determine the ruggedness of the methods. The methods themselves determine the purity of the active ingredients or technical material, the level of the active ingredient in the formulated products and the level of impurities. Since these analytical methods are also submitted to regulatory bodies during the process of product registration, the regulators have the right to use them to analyze products pulled from the market place. It is essential therefore that industry knows how rugged the methods are, and has the ability to evaluate that via an evaluation of the effectiveness of a method transfer. We have a process that describes an effective protocol for method transfer and an evaluation of the data after that process is conducted. The process is used world-wide. This presentation will highlight this process and show results of effective transfers.

Keywords: Agricultural, HPLC, Liquid Chromatography, Quality
Application Code: Quality/QA/QC
Methodology Code: Liquid Chromatography

Session # 110 **Abstract # 110-4** **Organized Contributed Sessions**

Session Title: R&D to QC: Bridging the Gap - Half Session

Abstract Title: **R&D to QC: An R&D Perspective in Chromatographic Method Development**

Date: Sunday, March 06, 2016 - Afternoon
Time: 02:30 PM
Room: B312

Primary Author: Justin Shearer
Dow AgroSciences

Co-Author(s): Daniel Knueppel, Rose Nelson, Suresh Annangudi Palani

Abstract Text

One of the primary roles many analytical scientists take on in the industrial chemistry setting is the development of methods that will permit reliable quantitative analyses. Many of these methods will employ some form of chromatography to enable quantitation of the analytes of interest. There have been many advancements in the field of separation science recently, that include UHPLC, nano-LC, and advances in column technology, and these technologies would be advantageous for of developing quantitative methods for many projects. This talk will focus on the considerations that will go into developing primary methods to enable the accurate, precise, and robust analysis of analytes in complex matrices. The goal of this work is to demonstrate the balance required to develop methods using the highest level science and transferring that to quality control facilities.

Keywords: HPLC, Liquid Chromatography, Method Development, Quality Control
Application Code: Other
Methodology Code: Liquid Chromatography

Session # 120 Abstract # 120-1 **Oral Sessions**

Session Title Bioanalytical: LC Techniques - Half Session

Abstract Title **Characterization of New 2 [micro]m Particle Size, 25 nm Pore Size Analytical Size Exclusion Chromatography Column with Larger Exclusion Limit Useful for the Separation of Biomolecules Using UHPLC and HPLC**

Date: Sunday, March 06, 2016 - Afternoon
Time: 01:30 PM
Room: B315

Primary Author Crystal Benner
Tosoh Bioscience LLC

Co-Author(s) Atis Chakrabarti

Abstract Text

To improve the separation performance of Size Exclusion Chromatography columns, improvement of the packing materials is essential. It is important to compromise between the particle size and pore volume, in order to get the desired separation. For large biomolecules, such as therapeutic proteins and monoclonal antibodies (mAbs), a larger exclusion limit will yield better separation. With the increased use of UHPLC instruments in laboratories today, it is important to have a column which is compatible with both UHPLC and conventional HPLC instruments.

Here we introduce two new silica based, 2 µm particle Size, 25 nm pore size analytical size exclusion chromatography columns with larger exclusion limits; 4.6 mm I.D. x 15 cm column for ultra-high speed separation and 4.6 mm I.D. x 30 cm column for ultra-high resolution. Both columns are modified with diol groups on the surface. The slope around molecular weight region of □-globulin is shallow. Pore characteristics were optimized to have the high resolution of mAb monomer from dimer and higher order aggregates besides monomer/fragment with this shallower calibration slope. Surface characteristics are almost equivalent to existing SW-type columns. The columns are compatible to both UHPLC and HPLC. For best performance it is necessary to use semi-micro HPLC system with low dead volume.

In this presentation, we will discuss the characterization of these two columns in the analysis of therapeutic mAbs and proteins. We have evaluated different peak parameters such as retention time, peak asymmetry, column efficiency, peak resolution, run time, loading capacity etc. for the characterization.

Keywords: Biopharmaceutical, HPLC Columns, Liquid Chromatography, Protein
Application Code: Bioanalytical
Methodology Code: Liquid Chromatography

Session # 120 **Abstract # 120-2** **Oral Sessions**

Session Title: Bioanalytical: LC Techniques - Half Session

Abstract Title: **Studies of Drug Interactions with Alpha[sub]1[/sub]-Acid Glycoprotein by Using On-Line Immunoextraction and High Performance Affinity Chromatography**

Date: Sunday, March 06, 2016 - Afternoon
Time: 01:50 PM
Room: B315

Primary Author: Cong Bi
University of Nebraska-Lincoln

Co-Author(s): Chenhua Zhang, David Hage, Ryan Matsuda, Zitha Isingizwe

Abstract Text

The binding of drugs with serum proteins, such as alpha[sub]1[/sub]-acid glycoprotein (AGP) and human serum albumin (HSA), is important in determining the transport, excretion and metabolism of such drugs in the body. AGP can bind and transport numerous basic and neutral drugs in the blood stream. It is also an acute phase protein and has levels that may be altered during many acute and chronic disease states, such as systemic lupus erythematosus (SLE), a chronic autoimmune disease. Higher plasma levels and a differential glycosylation pattern of AGP have been noted in patients with SLE. This suggests that the binding of some drugs with AGP may be changed during SLE due to the variation in its concentration or glycosylation pattern. However, little information exists on the changes in the drug-binding of AGP under these conditions. Recently, a method combining on-line immunoextraction and high-performance affinity chromatography has been developed and shown to be a rapid means for examining drug interactions with HSA. In this work, AGP isolated directly from serum was used to investigate changes in drug-binding with this protein in SLE. Both frontal analysis and zonal elution were conducted on the isolated AGP with various drugs, with some of these showing significant changes in binding versus that with normal AGP. The glycosylation patterns of the AGP samples were also examined by capillary electrophoresis to compare the structural variations in AGP with the changes in its drug interactions. The same chromatographic method should also be applicable to drug-protein binding studies in other disease states, as could be used for clinical studies or personalized medicine.

Keywords: Bioanalytical, Drugs, HPLC, Protein
Application Code: Bioanalytical
Methodology Code: Liquid Chromatography

Session # **120** **Abstract #** **120-3** **Oral Sessions**

Session Title: Bioanalytical: LC Techniques - Half Session

Abstract Title: **HPAE-PAD Applications for Stepwise Biosimilar Development Processes: Monosaccharide and Sialic Acid Determinations**

Date: Sunday, March 06, 2016 - Afternoon
Time: 02:10 PM
Room: B315

Primary Author: Hua Yang
Thermo Fisher Scientific

Co-Author(s): Linda Lopez

Abstract Text

Biosimilar development is increasing because of expiring patents. The U.S. Food and Drug Administration (US FDA) recently recommended that biosimilar sponsors use a stepwise approach to develop the evidence needed to demonstrate biosimilarity. The stepwise approach starts with extensive structural and functional characterization of both the proposed product and the reference product. This regulatory guidance for biosimilar development is driving demand for monosaccharide and sialic acid analyses, which is used for monitoring biosimilar glycosylation.

High Performance Anion-Exchange with Pulsed Amperometric Detection (HPAE-PAD) is a direct analysis method for monosaccharides, sialic acids, and other carbohydrates. It is sensitive and selective for a large variety of carbohydrates. Here HPAE-PAD is used to directly determine the carbohydrates present in glycoproteins, without additional labeling steps that are often needed for other analysis methods, saving time and reagent costs.

Keywords: Biopharmaceutical, Carbohydrates, Electrochemistry, Ion Chromatography
Application Code: Bioanalytical
Methodology Code: Liquid Chromatography

Session # 120 Abstract # 120-4 **Oral Sessions**

Session Title: Bioanalytical: LC Techniques - Half Session

Abstract Title: **Analysis of Hormone-Protein Binding in Solution by Ultrafast Affinity Extraction**

Date: Sunday, March 06, 2016 - Afternoon
Time: 02:30 PM
Room: B315

Primary Author: Xiwei Zheng
University of Nebraska-Lincoln

Co-Author(s): Cong Bi, David Hage, Marrisa Brooks

Abstract Text

Testosterone circulating in the bloodstream can be either in a free form or a protein-bound form, both of which have been reported to play important roles in tissue uptake and biological activity. To better understand the interactions of testosterone with plasma protein, and the resulting free fraction of this hormone in blood or serum, it is important to have the information regarding the overall binding and kinetics that govern this interaction. Two serum proteins that bind to testosterone are human serum albumin, the most abundant serum protein, and sex hormone-binding globulin, the main steroid hormone binding protein. In this study, a novel approach was developed for this type of study based on ultrafast affinity extraction and high performance affinity chromatography (HPAC). This approach was used to simultaneously and rapidly determine both the global association equilibrium constants and dissociation rate constants for these testosterone-protein interactions while also providing an automated system that required only small amounts of protein and hormone. This method was also used in a multi-dimensional affinity system to measure the free fractions of testosterone in samples that contained physiological concentrations of this hormone and its binding proteins. The results indicated that ultrafast affinity extraction and HPAC can be powerful tools for studying hormone-protein interactions and can provide information regarding both the rate constants and binding strengths for a hormone with serum proteins, as well as the free fractions for such a target in samples with clinically-relevant concentrations.

Keywords: Analysis, Bioanalytical, HPLC, Pharmaceutical
Application Code: Bioanalytical
Methodology Code: Liquid Chromatography

Session # 130 **Abstract # 130-1** **Oral Sessions**

Session Title: Bioanalytical: LC/MS Techniques - Half Session

Abstract Title: **Discovery of Brown Recluse Spider Sex Pheromones Using Advanced Ultrafast Liquid Chromatography – Tandem Mass Spectrometry Methodologies**

Date: Sunday, March 06, 2016 - Afternoon
Time: 03:05 PM
Room: B315

Primary Author: Casey Burton
Missouri University of Science and Technology

Co-Author(s): Ariel Donovan, Honglan Shi, Jennifer Parks, William Stoecker

Abstract Text

The brown recluse spider (BRS), [i]Loxosceles reclusa[/i], inhabits much of the Midwestern and Southern US, where it presents a serious public health challenge owing to the risks associated with its venomous bite that include hemolysis, renal failure, multiple organ dysfunction, and uncommonly, death. These risks are exacerbated by widespread home infestations and a lack of effective population control practices. This inspired us to seek novel BRS sex pheromones from spider silk using advanced ultrafast liquid chromatography – tandem mass spectrometry (UFLC-MS[sup]n[/sup]) methodologies for use in BRS trap applications. Spider silk from both sexually mature males and females were biologically assayed using male courtship behavioral monitoring to determine presence of sex pheromone followed by subsequent pheromone extraction. Silk extracts were separated with an isocratic UFLC methodology utilizing a C18 column. State-of-the-art silk profiling was performed using high-resolution enhanced mass spectrum scans coupled with information dependent acquisition of enhanced product ion scans (EMS-IDA-EPI) for compounds with mass-to-charge ratios (m/z) ranging from 50 to 800 m/z. Silk components were identified according to their fragmentation patterns with integrated MassBank support. A total of 24 unique compounds were identified on pheromone-active silk that were not found on pheromone-inactive silk from both female and male spiders. Moreover, our results demonstrate marked differences in silk composition between males and females that may be further used for improved BRS trap designs. The detailed UFLC-MS[sup]n[/sup] EMS-IDA-EPI method, extraction protocol, and results will be presented at the conference.

This study was supported by a National Science Foundation Graduate Research Fellowship.

Keywords: Identification, Liquid Chromatography/Mass Spectroscopy, Semi-Volatiles, Volatile Organic Compoun
Application Code: Bioanalytical
Methodology Code: Liquid Chromatography/Mass Spectrometry

Session # 130 Abstract # 130-2 **Oral Sessions**

Session Title: Bioanalytical: LC/MS Techniques - Half Session

Abstract Title: **Analysis of Herceptin Oxidation Variants Using a Supermacroporous Reverse Phase Column Coupled with an Orbitrap Mass Spectrometer**

Primary Author: Shanhua Lin
Thermo Fisher Scientific

Date: Sunday, March 06, 2016 - Afternoon
Time: 03:25 PM
Room: B315

Co-Author(s): Christopher Pohl, Ilze Birznieks, Jessica Wang, Jonathan Josephs, Terry Zhang, Xiaodong Liu

Abstract Text

The monoclonal antibody (mAb) therapeutics market is growing at a rapid rate owing to increased demand for targeted treatments. Therapeutic mAbs, such as Rituxan, Herceptin, Remicade, and Avastin, are mostly produced from mammalian cell. These biological products are heterogeneous due to post-translational modifications. Additional modifications such as oxidation can be introduced during the manufacturing process. A comprehensive characterization of mAb purity, aggregates, and variants is required for the final biopharmaceutical product approval and subsequent manufacturing processes.

Herceptin was partially oxidized with 0.01% H2O2 overnight. The mAb fragments were then generated by subsequent DTT reduction, or papain digestion, or IdeS protease digestion. These fragments and their oxidation variants were separated on a supermacroporous reverse phase (SMP-RP) column coupled to a Q Exactive Plus Orbitrap mass spectrometer.

There is a growing trend to obtain intact mass information as well as the glycan profile in the QC of mAbs using high resolution mass spectrometers. The most commonly employed LC/MS method is to desalt mAbs via reversed phase chromatography and perform an MS analysis. Further MS analysis of mAb fragments such as the heavy chain (HC), light chain (LC), Fab, and Fc can quickly reveal the location as well as nature of the modification. In the current study, we are presenting a fast separation method for mAb fragments and their oxidation variants from Herceptin using a novel supermacroporous reverse phase column. Baseline separation of LC and HC, Fc and Fab fragments, scFc and F(ab')2, was achieved in all cases using a 10-min gradient with water/acetonitrile/TFA eluents. Using an Orbitrap mass spectrometer accurate masses of mAb fragments were measured and the presence of oxidation variants was detected. The fast LC/MS approach described here is generally applicable to mAb variant characterization.

Keywords: Biopharmaceutical, Characterization, Liquid Chromatography/Mass Spectroscopy, Protein
Application Code: Bioanalytical
Methodology Code: Liquid Chromatography/Mass Spectrometry

Session # 130 **Abstract # 130-3** **Oral Sessions**

Session Title: Bioanalytical: LC/MS Techniques - Half Session

Abstract Title: **Quantitation of Endogenous Adenosine in Mouse Blood, Cell Lysate and Lysosomes by LC-MS/MS Using Surrogate Matrix Method**

Date: Sunday, March 06, 2016 - Afternoon
Time: 03:45 PM
Room: B315

Primary Author: Xiao Ding
Genentech

Co-Author(s)

Abstract Text

Lysosomal Storage Diseases (LSDs) is a group of heterogeneous disorders that result from lysosomal dysfunction. Symptoms of LSDs include splenic and liver enlargement, joint stiffness, etc. It was proposed that equilibrative nucleoside transporter 3 (ENT3) deficiency perturbs lysosomal function by preventing the efflux of nucleosides such as adenosine. Experiment showed that ENT3 KO mice developed spontaneous splenic enlargement, lymphadenopathy, etc. Do these KO mice have increased adenosine accumulation in blood, lysosomes, or macrophages? To help understand it, we developed a LC-MS/MS method to determine the concentration of adenosine in mouse blood, cell lysate and lysosomes. Adenosine was ionized using a turbo-ionspray source operated in the positive ionization mode on a TSQ Vantage triple-quadrupole mass spectrometer (Thermo). MRM transitions were m/z 268.1/136.1 for adenosine and 283.1/146.1 for adenosine-13C1015N5 (internal standard). Hydrophilic interaction liquid chromatographic separation was performed on an Accela HPLC system (Thermo) using a HALO Penta-HILIC (2.1 x 50 mm) column with a gradient elution at flow rate 0.5 mL/min. Mobile phases were 5 mM ammonium acetate in water (A) and in ACN/MeOH, 95:5, v/v (B). Adenosine eluted at 0.83 min. Standards were prepared using water, a surrogate matrix, due to the endogenous nature of adenosine. Extraction recovery was 91.6% and no significant matrix effect was observed. Enzymatic instability of adenosine was prevented during sample collection. It was found that concentration of adenosine in splenic macrophage whole cell lysate and lysosomes was higher in ENT3 KO than in WT, proving adenosine accumulation in the cell lysate and lysosomes.

Keywords: Bioanalytical, Drug Discovery, Liquid Chromatography/Mass Spectroscopy, Method Development
Application Code: Bioanalytical
Methodology Code: Liquid Chromatography/Mass Spectrometry

Session # 130 **Abstract # 130-4** **Oral Sessions**

Session Title: Bioanalytical: LC/MS Techniques - Half Session

Abstract Title: **Isomeric Separation of Glycopeptides Using Porous Graphitic Carbon (PGC) LC at High Temperature**

Primary Author: Rui Zhu
Texas Tech University

Date: Sunday, March 06, 2016 - Afternoon
Time: 04:05 PM
Room: B315

Co-Author(s): Jingfu Zhao, Yehia Mechref

Abstract Text

The interpretation of isomeric heterogeneity is one of the major technical challenges of glycoprotein and glycopeptide analysis. Herein, we firstly report temperature plays a critical role in the isomeric separation of glycopeptide isomers on porous graphitic carbon (PGC) LC systems. The temperature condition of glycopeptide isomers separation on PGC LC was described. Preliminarily, model glycoproteins (ribonuclease B and fetal calf serum fetuin) were digested with site specific proteases (trypsin and endoproteinaseglu-C) yielding glycopeptide containing consistent peptide moiety. The resulting glycopeptide mixture was separated on a PGC column and detected by mass spectrometry. The isomeric separation of glycopeptide on PGC column was significantly increased as the LC temperature increased. Base peak isomeric separation of glycopeptide mixture was attained at 75 oC. For the glycopeptides derived from ribonuclease B, three Man 7 and three Man 8 isomers were separated and detected, which agreed with previous NMR study. For the glycopeptides derived from fetal calf serum fetuin, more than 20 glycopeptide isomers corresponding to 6 compositional N-glycopeptides were separated and detected. Fetuin results were also in agreement with previous NMR data. These results suggested that PGC LC/MS could be a potential platform for the analysis of the isomeric microheterogeneity of glycoproteins. By using site specific proteases, both quantitative and qualitative analysis of glycopeptide isomers can be achieved through a single PGC LC/MS analysis.

Keywords: Bioanalytical, Carbohydrates, Liquid Chromatography/Mass Spectroscopy, Protein
Application Code: Bioanalytical
Methodology Code: Liquid Chromatography/Mass Spectrometry

Session # 140 **Abstract # 140-1** **Oral Sessions**

Session Title: Chemical Methods - Half Session

Abstract Title: **Bromide-Assisted Anisotropic Growth of Surfactant-Free Gold Nanoparticles**

Date: Sunday, March 06, 2016 - Afternoon
Time: 01:30 PM
Room: B316

Primary Author: Melissa Kerr
North Carolina Central University

Co-Author(s): Fei Yan

Abstract Text

An uncomplicated approach that uses the ideals of a Green Chemistry for shape-controlled synthesis of gold nanoparticles (AuNPs) using 5-hydroxyindoleacetic acid (5-HIAA) as the reducing agent has been reported. Using the reported procedure as a base, AuNPs of various shapes (triangles, hexagons, and semi-spheres, etc.) were prepared. Attempts to isolate a singular AuNP shape was made by supplementing the 5-HIAA with another reducing agent, namely Potassium Bromide. This synthesis was then carried out using various concentrations and temperatures, and a detailed characterization of the resulting AuNPs was performed using ultraviolet–visible (UV-Vis) spectroscopy, scanning electron microscopy (SEM), X-ray diffraction (XRD), and Raman spectroscopy. The as-prepared AuNPs exhibited excellent surface-enhanced Raman scattering (SERS) properties, which make them very attractive for the development of SERS-based chemical and biological sensors. Being able to isolate a particular AuNP shape or structure will tailor the resulting properties. This may lead to simple and effective changes to the design of many analytical devices.

Keywords: Material Science, Nanotechnology, Raman, Surface Enhanced Raman
Application Code: Material Science
Methodology Code: Chemical Methods

Session # 140 **Abstract # 140-2** **Oral Sessions**

Session Title: Chemical Methods - Half Session

Abstract Title: **Simultaneous Reduction of Metal Ions by Multiple Reducing Agents Initiate the Asymmetric Growth of Metallic Nanocrystals**

Primary Author: Mahmoud Mahmoud
Georgia Institute of Technology

Date: Sunday, March 06, 2016 - Afternoon
Time: 01:50 PM
Room: B316

Co-Author(s)

Abstract Text

Thermodynamically unfavorable metallic nanocrystals can be prepared only by the growth of the nanocrystals under kinetically controlled experimental conditions. The common technique to drive the growth of metallic nanocrystals under kinetic control is to adjust the rate of the generation of metal atoms to be slower than the rate of deposition of such atoms onto the surface of nanocrystal nuclei, which form in the first step of the nanoparticles synthesis. The kinetically controlled growth leads to the formation of seeds with crystal defects, which are needed for the growth of anisotropic nanocrystals such as silver nanodisks (AgNDs). The simultaneous multiple asymmetric reduction technique (SMART) is introduced here to successfully prepare AgNDs of controllable sizes and in large scale within a few seconds. SMART is simply based on the simultaneous reduction of silver ions with a strong reducing agent such as borohydride (redox potential of 1.24 V) and a weak reducing agent such as L-ascorbic acid (redox potential of 0.35 V) in the presence of a polyvinyl pyrrolidone capping agent. The random formation and deposition of silver atoms by the two different reducing agents generated stacking faults in the growing nanocrystal. The hexagonal close-packed {111} layers of silver atoms were then deposited on the surface of the growing nanocrystal containing stacked faults along the [111] plane. This initiated asymmetric growth necessary for the formation of plate-like seeds with planar twin defects, which is required for the formation of anisotropic AgNDs.

Keywords: Materials Characterization, Material Science, Metals, Nanotechnology
Application Code: Nanotechnology
Methodology Code: Chemical Methods

Session # **140** **Abstract #** **140-3** **Oral Sessions**

Session Title: Chemical Methods - Half Session

Abstract Title: **Novel Optical Properties of Segmented Au-Ag Nanocylinders: Effects of Metallic Junctions on Surface Plasmon Resonance**

Date: Sunday, March 06, 2016 - Afternoon
Time: 02:10 PM
Room: B316

Primary Author: Vineet Kumar
North Carolina State University

Co-Author(s): Gufeng Wang

Abstract Text

Metal nanoparticles (MNPs) are fascinating materials for their plasmonic applications. Introduction of another type of metal segments may expand their optical response to the whole UV-Vis-NIR range. However, fabrication of bimetallic systems give rise to the formation of a metal-metal interface between different metals. This interface leads to a change in multiple factors that could affect the optical response of such systems. Mainly, the electron density and the dielectric function around the interface. To date, there is no set theory in the literature which predicts the optical response of such bimetallic systems. Thus, this study will explore the interface effect on the surface plasmon resonance (SPR) and the electric field around the MNPs. We simulated Au-Ag bimetallic nanocylinders using Finite-difference time-domain (FDTD), a computational electrodynamics modeling technique. The FDTD simulation demonstrated that the bimetallic nanoparticles have certain advantages over single particle systems. Further, the electric field study was shows how a constituent segment could affect the |E|2 around the other metal segment in segmented nanowires. We then synthesized Au-Ag segmented nanowires (NWs) by electrochemical plating method to deposit metals into the templates. Results show the metal-metal interfaces enhance the luminescence and second-harmonic generation (SHG) signal in segmented nanowires. In conclusion, using single particle measurements, we show that enhanced optical properties can be obtained in bimetallic segmented nanoparticles

Keywords: Electrochemistry, Spectrophotometry, UV-VIS Absorbance/Luminescence
Application Code: Process Analytical Chemistry
Methodology Code: Chemical Methods

Session # 140 Abstract # 140-4 **Oral Sessions**

Session Title Chemical Methods - Half Session

Abstract Title **Measurement of Oxidative Potential of Particulate Matter by DTT assay**

Primary Author Shiori Ota
Tokai University

Date: Sunday, March 06, 2016 - Afternoon
Time: 02:30 PM
Room: B316

Co-Author(s) Kazuhiro Misawa, Yoshika Sekine

Abstract Text

The rate of consumption of dithiothreitol (DTT) is widely used in epidemiological studies to measure the oxidative potential of particulate matter (PM), which has been linked to adverse health effects of PM. However, little is known for present status of the oxidative potential of PM in Asia. This study then aimed to evaluate the toxicity level by particle size of PM collected at Kanagawa, Japan, where annual mean concentration of PM2.5 (50% cut-off at 2.5 µm) was 14.6 µg/m3 in 2014. PM2.5 and SPM (100% cut-off at 10 µm) were simultaneously collected on a quartz fiber filter by high-volume air sampler at a flow rate of 500 L/min for 24 hours both indoor and outdoor of the school building. Collected PM was extracted in pure water and the extract was added to DTT solution at 310 K. At known times, aliquots of the mixed solution were removed and the reaction was terminated by adding trichloroacetic acid. When all time points were quenched, the residual DTT was determined by spectrophotometry. The results showed the dependence of particle size on the oxidative potential was different between indoor and outdoor samples. As for outdoor samples, toxicity of PM2.5 was quite greater than that of SPM. Meanwhile, SPM showed higher toxicity level than PM2.5 in indoor samples. This may be because of difference in chemical compositions of PM. Therefore, coarse particles should be carefully considered as a possible oxidative stressor in indoor environment.

Keywords: Aerosols/Particulates, Air, Environmental Analysis, Environmental/Air
Application Code: Environmental
Methodology Code: Chemical Methods

Session # 150 **Abstract # 150-1** **Oral Sessions**

Session Title: Gas Chromatography Innovations

Abstract Title: **Method Translation in Gas Chromatography to Get the Same Chromatogram**

Date: Sunday, March 06, 2016 - Afternoon
Time: 01:30 PM
Room: B310

Primary Author: Jaap de zeeuw, Restek

Co-Author(s): Chris English, Chris Nelson, Jack Cochran

Abstract Text

In Gas chromatography there is often a need to optimize separations using different column dimensions, different linear gas velocity, using a different detector or using a different carrier gas. If you want to get the same peak elution order (same chromatogram), you must make sure that the elution temperatures of components is kept the same. This is only possible using a different oven temperature program. To calculate this program, there are free calculation programs available on the web. In this course we will discuss the details of conversion of methods so you get the same chromatograms with the new method.

The basics of converting existing GC methods into a new (mostly faster) method, and aiming for the same separation / peak elution order. If column dimensions, linear velocity or pressure drop over a capillary column is changed, and without changing temperature program, the separation of many components will change. Some separations will be better, some will go worse. In order to keep the separation similar, one needs to adjust the oven temperature, to get the same elution temperatures. For this, one can use free calculation programs, available on the web. The relevance and options of these calculation programs will be explained.

Keywords: Capillary GC, Gas Chromatography, GC Columns, Teaching/Education
Application Code: General Interest
Methodology Code: Gas Chromatography/Mass Spectrometry

Session # 150 Abstract # 150-2 **Oral Sessions**

Session Title Gas Chromatography Innovations

Abstract Title **Chemometric Treatment of GC-VUV Data - Samples in a New Light**

Primary Author James J. Harynuk
University of Alberta

Date: Sunday, March 06, 2016 - Afternoon
Time: 01:50 PM
Room: B310

Co-Author(s) Keisean Stevenson, Lawrence A. Adutwum, Seo Lin Nam

Abstract Text

The new vacuum-ultraviolet (VUV) detection strategy for gas chromatography presents unique opportuinites for studying samples amenable to gas chromatography. Already, several new methods for petroleum analysis have been published, and many more are sure to follow across a variety of fields. The VUV detector also presents a unique data structure that is highly amenable to chemometric processing. This is due to all wavelengths in the spectrum being collected simultaneously, and the data surface being fundamentally smooth in both the time and spectral dimensions. In this work, we will explore the application of chemometric tools to a variety of problems in GC-VUV, including classification and regression problems in petroleum and other fields.

Keywords: Chemometrics, Gas Chromatography, GC Detectors, Spectroscopy
Application Code: General Interest
Methodology Code: Gas Chromatography

Session # 150 **Abstract # 150-3** **Oral Sessions**

Session Title: Gas Chromatography Innovations

Abstract Title: **All 5 Phases are Not the Same! Considerations for Method Development and Selection**

Primary Author: Ramkumar Dhandapani, Phenomenex

Co-Author(s): Kristen Parnell, Tim Anderson

Date: Sunday, March 06, 2016 - Afternoon
Time: 02:10 PM
Room: B310

Abstract Text

5 % phenyl type stationary phases are very popular in gas chromatographic separation. Among the 5 % phenyl phases, there are countless options to choose from as a starting point for method development. Choosing the right gas chromatographic stationary phase is a challenge, considering the number of selections available. Though many GC methods employ the use of a 5 % phenyl type phase, not all 5 phases are the same! Presented in this work is are guidelines for choosing the best-suited 5 % phenyl phase based on analytical and application goals. A systematic approach has been laid to identify the optimal phase based on several parameters, including selectivity, temperature stability, and inertness to active analytes. Also explored are the impacts of 5 % phenyl phase efficiency through the use of specific dimensions, to provide an additional vehicle for enhanced resolution in addition to selectivity.

Keywords: Gas Chromatography, Gas Chromatography/Mass Spectrometry, GC, GC Columns

Application Code: General Interest

Methodology Code: Gas Chromatography

Session # 150 Abstract # 150-4 **Oral Sessions**

Session Title Gas Chromatography Innovations

Abstract Title **Flow-through Microfluidic Photoionization Detectors for Rapid and Highly Sensitive Vapor Detection**

Primary Author Hongbo Zhu
University of Michigan

Date: Sunday, March 06, 2016 - Afternoon
Time: 02:30 PM
Room: B310

Co-Author(s) Jiwon Lee, Katsuo Kurabayashi, Menglian Zhou, Robert Nidetz, Sanketh Buggaveeti, Xudong Fan

Abstract Text
Photoionization detector is a widely used sensor for volatile organic compounds detection due to its high sensitivity and large dynamic range. In the micro-gas chromatography (micro-GC) devices, because of its lightweight and small footprint, PID has distinct advantages over other vapor detectors. However, due to its tardy response, which results from the relatively large ionization chamber and dead volume, the peaks in chromatogram are broadened, which compromises the micro-GC To solve this problem, an extremely high flow rate (30 mL/min) or make-up gas (20 mL/min) rates is generally used, neither of which is desired due to complicated fluidic design and/or significant reduction in sensitivity. Here we developed a microfluidic PID that is micro-fabricated directly on a conductive silicon wafer and has a significantly reduced ionization chamber volume of only 1.3 µL with virtually zero dead volume (~2 nL) owing to its flow-through design. Consequently, the microfluidic PID can considerably shorten the response time (~100 ms, 10-20X faster than the conventional PID) while maintaining excellent sensitivity (pico-gram) and large dynamic range (~6 orders of magnitude). In this talk, we will describe the details of fabrication and characterization of the microfluidic PID, and show how it is used in a GC or micro-GC system.

Keywords: Detector, Gas Chromatography, GC Detectors, Lab-on-a-Chip/Microfluidics
Application Code: General Interest
Methodology Code: Gas Chromatography

Session # 150 Abstract # 150-5 **Oral Sessions**

Session Title: Gas Chromatography Innovations

Abstract Title: **Safeguarding a Mass Detector from Difficult Sample Components**

Date: Sunday, March 06, 2016 - Afternoon
Time: 03:05 PM
Room: B310

Primary Author: Amanda B. Dlugasch
Waters Corporation

Co-Author(s): Patricia R. McConville, Thomas E. Wheat

Abstract Text

The mass spectrometer can provide valuable identification information in the chromatographic analysis of all types of samples. The composition of the sample, however, can compromise the quality of the analysis in a variety of ways. Two major classes of problems can be encountered. First, sample components can damage the MS source. Such elements might include high concentrations of salt or buffers that would block the orifice or corrode the source. Second, sample constituents at extreme molar ratios can compromise the quantitative and resolution of the results obtained, for example as might be encountered with assays of genotoxic impurities. Both challenges are often addressed by some form of physical and chemical sample preparation. As an alternative, the flow from the column can be diverted to waste undesirable materials. In this study we evaluate a divert valve. The suitability of the valve will be tested for its function in solving both types of problems and will be assessed by its ability to eliminate the problem material without reducing recovery, resolution or chromatographic peak shape. Two classes of potentially damaging material will be evaluated. One sample is a high concentration of salt and buffer that would elute near the solvent front potentially occluding the source and suppressing ionization. The second sample is a hydrophobic material that would elute at the end of the gradient and leave solid residue in the source. The class of problems associated with extreme molar ratios will be analyzed with a trace component eluting either before or after the main peak. The larger peak will be diverted to waste. These experiments show the general utility of incorporating a divert valve into an LC-MS system.

Keywords: Detector, Liquid Chromatography/Mass Spectroscopy, Mass Spectrometry, Quadrupole MS
Application Code: General Interest
Methodology Code: Liquid Chromatography/Mass Spectrometry

Session # 150 Abstract # 150-6 **Oral Sessions**

Session Title Gas Chromatography Innovations

Abstract Title **Wide Range Inert Dilution System for Gas Standard Generation**

Date: Sunday, March 06, 2016 - Afternoon
Time: 03:25 PM
Room: B310

Primary Author Stanley D. Stearns
Valco Instruments Co. Inc.

Co-Author(s) Alex Plistil, David McCharthy, Huamin Cai, Sunny Sinlapadech

Abstract Text

The use of single or dual linear pump with large volumetric range can easily generate serial gas standards. The pump was originally designed to delivering a bidirectional flow of liquid to over six orders of magnitude. It has no separate fill cycle during delivery and has a high volumetric accuracy (0.5%). These features make the pump an excellent choice for delivering gas for the purpose of generating a series of gas standards from a single concentration. In this presentation, we will discuss different configurations and their test results.

Keywords: Calibration, Gas, GC
Application Code: General Interest
Methodology Code: Gas Chromatography

Session # 150 **Abstract # 150-7** **Oral Sessions**

Session Title: Gas Chromatography Innovations

Abstract Title: **Multimode Plasma Emission Detector**

Primary Author: Yves Gamache
Analytical Flow Products

Date: Sunday, March 06, 2016 - Afternoon
Time: 03:45 PM
Room: B310

Co-Author(s)

Abstract Text

Several types of gas detectors for detecting, measuring and/or analyzing constituents of a gas sample are known in the art. For example, in the context of chromatographic systems, it is known to select a detector based on the application at hand, the type of carrier gas and impurities to be detected, the desired information, the required precision of the results, price considerations, etc. Gas detectors suitable for some chromatography applications include Flame Ionization Detectors (FID), Electron Capture Detectors (ECD), Thermal Conductivity Detectors (ECD), Photoionization Detectors (PID) and Mass spectrometers (MS), to name only a few Plasma-based detectors using optical spectroscopic techniques for analyzing the constituents of gas samples will be explained. In accordance with some embodiments, the plasma-based detector may be provided with one or more features allowing its use for different applications and in different operating conditions.Such features include mechanisms for plasma generation and stability improvement. In one aspect, a mechanism for generating a localizing field in the plasma chamber is provided. Such a mechanism may for example include localizing electrodes or magnets configured to generate an electrostatic, magnetic or electromagnetic field transversally to the plasma-generating field. In another aspect, electron-injecting electrodes may be provided in the plasma chamber. Pressure control mechanisms may also be provided and used to improve plasma control or enable plasma generation in hard to ionize gases.

Keywords: Chromatography, Detector, Gas Chromatography, GC Detectors
Application Code: Other
Methodology Code: Gas Chromatography

Session # 150   Abstract # 150-8   **Oral Sessions**

Session Title   Gas Chromatography Innovations

Abstract Title   **Polyionic Ionic Liquid GC Stationary Phase Evaluations**

Primary Author   Leonard M. Sidisky
Supelco/Sigma-Aldrich

Date: Sunday, March 06, 2016 - Afternoon
Time: 04:05 PM
Room: B310

Co-Author(s)   Daniel Shollenberger, Greg A. Baney, Gustavo Serrano, James L. Desorcie

Abstract Text

Ionic liquids are a class of nonmolecular ionic solvents with low melting points. These liquids are unique combination of cations and anions and can provide a variety of different selectivities when used as stationary phases in capillary gas chromatography. The majority of the polyionic ionic liquid phases that we have been evaluating all provide polar and highly polar selectivities similar to polyethylene glycol based our biscyanopropylpolysiloxane phases. These phases will provide unique selectivity for the evaluation of a number of variety of different samples including petrochemical, environmental and food and beverage. The purpose of our studies is to determine the effects changing the cation and spacer groups on the selectivity and stability of the phases. Selectivity was determined and compared using various isothermal and temperature programmed test mixes. Particular cation and anion combinations appear to provide very unique selectivity by the shifting of normal alkane relative to aromatic and other chemical species. New combinations of cations and anions have been evaluated which provide unique selectivity and stability to the phases. We will also demonstrate the effects of the various functional group combinations on the overall stability of the ionic liquid stationary phases.

Keywords: Capillary GC, Gas Chromatography
Application Code: Other
Methodology Code: Gas Chromatography

Session # **160** **Abstract #** **160-1** **Oral Sessions**

Session Title GC Optimization - Half Session

Abstract Title **Evaluation and Application of Sample Preparation Techniques for the Determination of Residual Volatiles in Biological Matrices Using GC-FID**

Date: Sunday, March 06, 2016 - Afternoon
Time: 01:30 PM
Room: B309

Primary Author Adriene Malsbury
Bristol-Myers Squibb

Co-Author(s) Frank Tomasella, William Fish

Abstract Text

The advancement of novel antibody drug conjugates (ADCs) into the drug development pipeline has introduced new challenges in analytical analysis. The determination and quantification of residual solvents is common practice in the registration of a new small molecule drug product. However, the conjugation of a novel synthetic drug to an antibody increases the complexity of the residual solvent analysis of the final ADC drug product which is typically formulated in an aqueous buffer solution. The aqueous buffered sample matrix of the final product poses a significant challenge for residual solvent analysis when using the common direct inject technique as the expansion of water in the GC liner only allows for small amounts of sample to be injected, impacting sensitivity. This presentation will discuss the limitations of both traditional direct inject and headspace analysis as well as alternative sample preparation techniques used to overcome the challenges associated with the quantification of residual solvents in ADCs.

Keywords: Bioanalytical, Capillary GC, Quantitative, Sample Preparation
Application Code: Bioanalytical
Methodology Code: Gas Chromatography

Session # 160 **Abstract # 160-2** **Oral Sessions**

Session Title: GC Optimization - Half Session

Abstract Title: **A New Web-Based Application for Modelling Gas Chromatographic Separations**

Date: Sunday, March 06, 2016 - Afternoon
Time: 01:50 PM
Room: B309

Primary Author: Rebecca Stevens
Restek Corporation

Co-Author(s): Amanda Rigdon, Dan Li, Jaap de zeeuw, Linx Waclaski

Abstract Text

In previous work, our lab has demonstrated the utility of computer modeling for gas chromatography (GC) method development and the high accuracy of our time-summation modeling approach. While our modeling program was able to provide solutions for given separation problems it lacked flexibility.

Based on comments from end users of the program we have directed our efforts to making the computer modeling technique more powerful. We have added features that allow users to directly control every aspect of the chromatography being modeled including carrier gas, column stationary phase, column dimension, and temperature program. The modeling program now constitutes a complete "dry-lab" GC melding computer optimization with end user's intuition.

These new features will be discussed with a focus on experimental agreement for both atmospheric and vacuum outlet GC, investigation of the analysis time to resolution tradeoff, and migration of methods between carrier gasses without loss of resolution.

Another particularly interesting use of the program is for reverse peak identification. This means that for a chromatogram of known compounds under any set of instrument conditions the precise elution order can be obtained instantly without the need to run several confirming standards.

Keywords: Capillary GC, GC Columns, Method Development, Optimization
Application Code: Other
Methodology Code: Gas Chromatography

Session # 160 **Abstract #** 160-3 **Oral Sessions**

Session Title: GC Optimization - Half Session

Abstract Title: **Overcoming New Problems With Old Solutions: Enhanced Separations With GC Retention Gap Columns**

Date: Sunday, March 06, 2016 - Afternoon
Time: 02:10 PM
Room: B309

Primary Author: Ramkumar Dhandapani
Phenomenex

Co-Author(s): Kristen Parnell, Tim Anderson

Abstract Text

Gas chromatographic guard columns have dual functions. When used as a guard column, it protects the analytical column by retaining non-volatile compounds and preventing them from entering the analytical column. Apart from this function, the guard column also offers peak focusing of the analytes when used as a retention gap. Multiple experiments were conducted to demonstrate the benefits of guard columns used as a retention gap. Different injection techniques, including splitless mode and large sample volumes, were explored to illustrate the enhancement in resolution of early eluting analytes. The difference in the chromatographic performance by using a guard column and an analytical column having integrated guard column was also explored. The results not only proved that the GC guard column offered enhanced peak focusing and improved resolution by acting as a retention gap but also helped as a guard against nonvolatile impurities entering the analytical column.

Keywords: Gas Chromatography, Gas Chromatography/Mass Spectrometry, GC, GC Columns
Application Code: General Interest
Methodology Code: Gas Chromatography

Session # 160 **Abstract # 160-4** **Oral Sessions**

Session Title: GC Optimization - Half Session

Abstract Title: **Improved Inertness Performance for Polyethylene Glycol GC Columns**

Date: Sunday, March 06, 2016 - Afternoon
Time: 02:30 PM
Room: B309

Primary Author: Kenneth G. Lynam
Agilent Technologies

Co-Author(s): Allen Vickers, Ngoc-A Dang, Yun Zou

Abstract Text

Polyethylene glycol (PEG) or wax based GC columns are staples in laboratories analyzing alcohols, acids, essential oils, fragrances, flavors, spirits, solvents, and fatty acids. The high polarity of this phase type is used to resolve similar components based on their polarities. Sharp peak shapes, consistent resolution and retention of closely eluting compounds on these phases can be elusive particularly when using columns where inertness has not been probed critically and/or effectively. This presentation demonstrates how improved inertness performance on an inert wax column leads directly to improved chromatographic performance, better column consistency, and ultimately better quantitation. Problematic analytes on PEG phases include diols, glycols, and ethyl maltol to name just a few. The chromatographic behavior of these compounds and other challenging analytes illustrate just how peak shape and run to run repeatability improve when an inert wax column is used vs. more traditional wax columns from several GC column suppliers. The difference between inert and traditional wax based columns is striking for many of these challenging analytes, showing convincingly that all wax based columns are not equal.

Continuous improvements in column manufacturing processes, rigorous testing procedures and a relentless commitment to improve the inertness performance of our products have all been key facets in the successful development of our new inert wax columns.

Keywords: Beverage, Flavor/Essential Oil, Food Science, GC Columns
Application Code: Food Science
Methodology Code: Gas Chromatography

Session # 170 **Abstract # 170-1** **Oral Sessions**

Session Title: Instrument Innovations - Half Session

Abstract Title: **Rapid Analysis of SO2 to Determine Catalyst Efficiency**

Date: Sunday, March 06, 2016 - Afternoon
Time: 03:05 PM
Room: B316

Primary Author: Debbie Alcorn
INFICON

Co-Author(s)

Abstract Text

Sulfuric acid is one of the most produced chemicals in the world. Almost 180 million tons are consumed per year on a worldwide basis. In the United States alone, billions of pounds are produced and sold for a variety of uses including the formulation of fertilizers, insecticides and detergents. To generate sulfuric acid, sulfur dioxide (SO2) is oxidized to generate sulfur trioxide (SO3), which when reacted with water, forms sulfuric acid (H2SO4). Catalysts facilitate the oxidation of SO2. The analysis of SO2 at the inlet and outlet of the catalytic bed determines the conversion efficiency and performance of the catalyst. Since sample integrity may be compromised due to the delay encountered transporting the sample to an analysis lab, it is preferable to obtain accurate results quickly and reliably on-site. Micro GC Fusion is a small, transportable GC capable of analyzing SO2 across a broad linear range. The microelectromechanical systems (MEMS) based injector and detector in Micro GC Fusion miniaturize the GC, making it suitable for on-site SO2 analysis.

Keywords: Chemical, Gas Chromatography, GC, Portable Instruments
Application Code: General Interest
Methodology Code: Gas Chromatography

Session # 170   Abstract # 170-2   **Oral Sessions**

Session Title: Instrument Innovations - Half Session

Abstract Title: **Portable Gas Analyzer for Continuous Monitoring of Sulfur Dioxide in Gas Streams**

Primary Author: Sayed A. Marzouk
UAE University

Date: Sunday, March 06, 2016 - Afternoon
Time: 03:25 PM
Room: B316

Co-Author(s): Mohamed Al-Marzouqi, Mohamed Alnaqbi, Muna Bufaroosha

Abstract Text

The construction, optimization and use of simple and inexpensive portable gas analyzer for real time measurement of sulfur dioxide in gas stream are described. The analyzer is based on stabilized-gravity-driven carrier/stripping solution flowing through a bundle of hollow fibers which acts as gas contactor/diffusion scrubber to absorb SO[sub]2[/sub] molecules from the gas stream. The produced sulfite ions -in the carrier/stripping solution- are then detected by anodic amperometry at a carbon nanotube electrode – downstream– polarized at 0.35 V vs Ag/AgCl reference electrode. A novel compact cell design that integrates the diffusion scrubber (gas sampler) with the 3-electrode amperometric flow detector along with the other careful construction aspects provided a light weight portable analyzer (~900 g) with small footprint. The analyzer provided linear amperometric response to SO[sub]2[/sub] concentration up to 2000 ppm. Excellent SO[sub]2[/sub] recoveries were obtained in the presence of several folds of CO[sub]2[/sub] and H[sub]2[/sub]S. Under optimized conditions, the analyzer offered several favorable performance characteristics such as (i) reasonably fast response and recovery times, (ii) excellent signal stability and reproducibility (RSD = 0.5%), (iii) high selectivity in presence of common non-ionogenic gases, e.g., CH[sub]4[/sub], N[sub]2[/sub], O[sub]2[/sub], CO and electroinactive species such as CO[sub]2[/sub] and (iv) standalone operation for 6 hours prior to the need for carrier solution refill. The suggested analyzer was applied successfully in monitoring the removal of SO[sub]2[/sub] from SO[sub]2[/sub]-CO[sub]2[/sub]-N[sub]2[/sub] gas mixtures as well as in the determination of sulfites in lemon juice and dried apricot.

Keywords: Automation, Electrodes, Gas, Instrumentation
Application Code: General Interest
Methodology Code: Sensors

Session # 170 Abstract # 170-3 **Oral Sessions**

Session Title: Instrument Innovations - Half Session

Abstract Title: **Triple Mode of Action of L-tyrosine Derived Probes: Solvent Mediated Flip-Flop Halide (iodide/fluoride) Sensors and Reversible Chromogenic pH Indicators**

Date: Sunday, March 06, 2016 - Afternoon
Time: 03:45 PM
Room: B316

Primary Author: Sanjay K. Mandal
IISER Mohali

Co-Author(s)

Abstract Text

In recent years, the progress in supramolecular chemistry has opened new avenues in the field of sensing multiple analytes and physical observable (pH, temperature, etc.) using a single-molecular probe. However, only few multiple sensing fluorescent probes have been reported for the ratiometric detection of di□erent cations or di□erent anions or cation with anions or metal-ions along with proteins or amino acids or sugar along with physical parameters like temperature. While few solvent mediated multiple analyte sensing have been reported in the literature, several probes acting as pH indicators are also documented. To the best of our knowledge, a single-molecular probe acting as a solvent mediated differential halide sensor as well as an optical pH indicator is unprecedented. The design of such new multiple-action molecular probes is important for budding new tools for sensing multiple analytes and observables along with the basic understanding of the mechanism of their action. We initiated our work towards this direction by formulating L-tyrosine derived fluorescent probes and exploring their role as halide sensors and pH indicators. Two fluorophores (phenol and 3- or 4-nitrobenzyl groups) judiciously put together in a single molecule are attributed to its triple mode of action. Herein, we report simple L-tyrosine derived fluorescent probes (S)-3-(4-hydroxyphenyl)-2-((4-nitrobenzyl)amino)propanoic acid (H2Tyr-4-nitro, [b]1[/b]) and (S)-3-(4-hydroxyphenyl)-2-((3-nitrobenzyl)amino)propanoic acid (H2Tyr-3-nitro, [b]2[/b]) that are synthesized via the combination of L-tyrosine and 4-nitrobenzaldehyde or 3-nitrobenzaldehyde moieties, respectively, for their abilities to act as a solvent mediated differential sensor for iodide in methanol and fluoride in DMSO as well as a reversible chromogenic pH indicator in DMSO (see Figure 1).

Keywords: Amino Acids, Analysis, Fluorescence, Sensors
Application Code: General Interest
Methodology Code: Fluorescence/Luminescence

Session # 170 Abstract # 170-4 **Oral Sessions**

Session Title: Instrument Innovations - Half Session

Abstract Title: **Saving Time and Improving Accuracy by Eliminating the Need for Standards and Calibration in GC/FID Analyses**

Date: Sunday, March 06, 2016 - Afternoon
Time: 04:05 PM
Room: B316

Primary Author: Jones Andrew
Activated Research Company

Co-Author(s): Charlie Spanjers

Abstract Text

The accurate quantification of carbon-containing compounds using GC/FID requires painstaking calibration using carefully prepared commercial standards. In many industries, accurate standards are either prohibitively expensive, difficult to synthesize, or simply not available, and the required calibrations can sometimes take enormous amounts of time. The conversion of carbon molecules in GC effluents to methane before FID analysis leads to identical response factors, thereby eliminating the need for calibrations and the standards they rely upon. Here we discuss this approach to the quantitative analysis of carbon-containing compounds using the newly developed, catalytic Polyarc™ reactor. Relative response factors collapse to unity for all molecules, allowing for the quantification of molecules without standards or their related calibrations. Compared to FID-only analyses, the Polyarc[sup]TM[/sup] reactor demonstrates improved response and repeatability and is useful for all industries that currently use gas chromatography.

Keywords: Chemical, Gas Chromatography, GC Detectors, Quantitative

Application Code: General Interest

Methodology Code: Gas Chromatography

Session # 180 **Abstract # 180-1** **Oral Sessions**

Session Title: Measurement Strategies - Sensors and Spectroscopy

Abstract Title: **Real-Time Voltammetric Characterization of Non-Electroactive Metal Complexation**

Date: Sunday, March 06, 2016 - Afternoon
Time: 01:30 PM
Room: B313

Primary Author: Thushani Siriwardhane
University of South Carolina

Co-Author(s): Parastoo Hashemi, Shawn McElmurry

Abstract Text

Accumulation of trace metals in the environment has increased rapidly during the past few decades. Toxicity and bioavailability of these harmful metals depends on their speciation which can change rapidly during environmental processes. Most traditional analytical and electrochemical methods fail to analyze metals during such rapid environmental events because they do not have a high temporal resolution. Recently we developed fast scan cyclic voltammetry (FSCV) on carbon fiber microelectrodes (CFMs) for real-time monitoring of copper and lead with a temporal resolution of 100 ms. Further we developed a paradigm to monitor metal complexation and to calculate complexation constants (Ks) in real time using a thermodynamic and a hydrodynamic model. However, this paradigm does not facilitate the detection of metals which have their redox potentials out of the potential window of CFMs. Here, we describe FSCV to detect non-electroactive metals based on changes in the double layer capacitance on CFM surfaces. We correlate capacitative changes to metal concentration and metal binding with different known ligands. Finally we use an empirical model to calculate the unknown K of Aluminum binding with L-cysteine. Our ultimate goal is to develop this technique to rapidly calculate fundamental thermodynamic parameters of unknown non-electroactive metal complexes.

Keywords: Electrochemistry, Microelectrode, Voltammetry
Application Code: General Interest
Methodology Code: Electrochemistry

Session # 180 Abstract # 180-2 **Oral Sessions**

Session Title: Measurement Strategies - Sensors and Spectroscopy

Abstract Title: **Phthalocyanine Based Microfluidic Sensors for the Detection of Oxidative Stress**

Date: Sunday, March 06, 2016 - Afternoon
Time: 01:50 PM
Room: B313

Primary Author: Kevin J. Klunder
Colorado State University

Co-Author(s)

Abstract Text

Environmental oxidative stress, according to the World Health Organization, has been directly linked to increased morbidity and mortality in a concentration dependent manner. Measuring the ability of particulate matter (PM) to cause oxidative stress, a parameter referred to as the oxidative load, is challenging because of the low relative concentrations of oxidants coupled with the heterogeneity of the samples. The use of dithiothreitol (DTT) as a model oxidant is a common method to gauge oxidative load. Polymeric Cobalt tetraaminophthalocyanine have long been used in the oxidation of thiols. In this work we continue this investigation of cobalt tetraaminophthalocyanine (CoTAPc) for thiol oxidation, with a specific focus on DTT in microfluidics. The lesser studied nickel, copper and iron derivatives are also investigated. Additionally, for the first time tetranitrophthalocyaine complexes are examined for this application. Impedance spectroscopy and rotating ring disk electrochemistry are used to understand the underlying electron transfer process of the oxidation of DTT with phthalocyanines.

Cyclic voltammetry experiments show that polymeric CoTAPc has a 200 mV lower onset oxidation of DTT, when compared to a traditional cobalt phthalocyanine modified electrode. This lower oxidation potential is important to minimize the impact of sample interferences and/or inclusion of an internal standard. Initial tests show that polymeric tetraaminophthalocyanine are stable on the time scale of days, with hours of continuous use. Work is ongoing to improve the temporal resolution and sensitivity of these phthalocyanine based microfluidics to measure real-time changes in oxidative load associated with ambient PM.

Keywords: Chemical, Chemically Modified Electrodes, Electrochemistry, Electrodes
Application Code: General Interest
Methodology Code: Sensors

Session # **180** **Abstract #** **180-3** **Oral Sessions**

Session Title: Measurement Strategies - Sensors and Spectroscopy

Abstract Title: **Multiple Light Scattering to Characterize Emulsions with Polymers**

Date: Sunday, March 06, 2016 - Afternoon
Time: 02:10 PM
Room: B313

Primary Author: Jonathan Denis
Formulaction

Co-Author(s): Christelle Tisserand, Gérard Meunier, Pascal Bru, Yoann Lefeuvre

Abstract Text

Polymers are widely used in the industry as a tool to increase the stability. Depending on their concentration, they can act as depletion agents or gel agent. The stability of these systems is driven by the polymers and the structure of the network of droplets and can lead to collapse of the emulsions.

In this work, Multiple Light Scattering device is used to monitor the behaviour of w/o emulsions stabilized with polymers. The heart of the optical scanning analyser is a detection head, which moves up and down along a flat-bottomed cylindrical glass cell (see figure). The detection head is composed of a pulsed near infrared light source (wavelenght = 880 nm) and two synchronous detectors. The transmission detector (at 180°) receives the light, which goes through the sample, while the backscattering detector (at 45°) receives the light scattered backward by the sample. The detection head scans the entire height of the sample, acquiring transmission and backscattering data every 40 µm.

We propose a description of the behaviour of o/w emulsions stabilized with different polysaccharides, we will show the advantages of using Multiple Light Scattering (MLS) to monitor their stability and propose a method to predict stability of these emulsions thanks to their size evolution in the first days after preparation.

Keywords: Characterization, Chemical, Food Science, Particle Size and Distribution
Application Code: General Interest
Methodology Code: Physical Measurements

Session # 180 Abstract # 180-4 **Oral Sessions**

Session Title: Measurement Strategies - Sensors and Spectroscopy

Abstract Title: **A Practical Approach for Maximizing ICP-MS Data: A Closer Look at Microwave Sample Prep, The Steps that Precede It, and How to Minimize External Factors that Affect Data Quality**

Date: Sunday, March 06, 2016 - Afternoon
Time: 02:30 PM
Room: B313

Primary Author: Johan Nortje
Milestone Inc.

Co-Author(s)

Abstract Text

A continuing trend in trace metals analysis is the transition to more sensitive analytical techniques like ICP-MS. The use of such techniques requires an increased emphasis on the handling and preparation of samples (typically performed via microwave digestion) prior to analysis. Utilizing a robust sample prep technique is critical but nuanced factors, such as acid cleanliness and vessel contamination, precede this step and are often overlooked. Controlling these variables can be cost- and time prohibitive but a variety of solutions now exist to minimize them in a practical, cost-effective manner. This presentation will discuss microwave sample prep, the steps preceding it, and how they affect the quality of analysis. We will present useful solutions that enable labs to minimize the effects of these variables and ultimately enhance their ICP-MS data.

Keywords: ICP-MS, Metals, Sample Preparation, Trace Analysis
Application Code: General Interest
Methodology Code: Sampling and Sample Preparation

Session # 180 **Abstract # 180-5** **Oral Sessions**

Session Title: Measurement Strategies - Sensors and Spectroscopy

Abstract Title: **Obtaining Maximum Information from Fast Chemical Reactions Using a Photodiode Array (PDA) UV-Visible Spectrophotometer**

Date: Sunday, March 06, 2016 - Afternoon
Time: 03:05 PM
Room: B313

Primary Author: Ian Robertson
PerkinElmer Limited

Co-Author(s): Christopher Lynch, Steve Upstone

Abstract Text

UV-visible spectroscopy has been widely used for the monitoring of chemical reactions. The principle is that the absorbance values at known wavelengths are monitored as the reaction progresses. A reaction curve can be plotted for each of the wavelengths measured from which it is possible to determine the reaction order and rate. Such measurements require up-front knowledge of the materials and the wavelengths at which they have absorptions.

Photodiode Array (PDA) spectrophotometers are capable of collecting complete UV-Visible spectra over the entire UV-Visible wavelength range from 190-1100nm at significantly faster rates than conventional scanning dispersive UV-Visible instruments. This allows the PDA spectrophotometer to be used to measure complete spectral information from fast chemical reactions that are completed within seconds. Time-resolved information for multiple wavelengths can also be extracted from the spectra.

Example applications that demonstrate the significant information that can be obtained from fast chemical reactions will be shown, including stopped-flow reactions. Advantages of PDA instruments for these types of measurements will also be discussed.

Keywords: Chemical, Flow Injection Analysis, Ultra Fast Spectroscopy, UV-VIS Absorbance/Luminescence
Application Code: General Interest
Methodology Code: UV/VIS

Session # 180 Abstract # 180-6 **Oral Sessions**

Session Title: Measurement Strategies - Sensors and Spectroscopy

Abstract Title: **Microfluidic Visual Rheometer**

Date: Sunday, March 06, 2016 - Afternoon
Time: 03:25 PM
Room: B313

Primary Author: Christelle Tisserand
Formulaction

Co-Author(s): Gérard Meunier, Patrick Abgrall, Patrycja Adamska

Abstract Text

Rheological analysis is made fast and easy with a novel instrument based on a simple microfluidic flow comparator. Using only tiny amount of samples, the technology allows flow viscosity measurements of liquid products from water-like inks to thick cosmetic formulations, under a wide range of shear rates (including high values up to 105 s-1). Additionally, the user-friendly visual principle makes possible routine optical control of both samples and disposable flow cells to assess measurement quality. A sample and a viscosity standard are pushed together through a microfluidic comparator (Y-junction) at controlled flow rates. Images of the resulting laminar co-flow are acquired via an integrated optical system and the position of the interface position is measured.

The interface position is simply related to the viscosity and the flowrates ratio between the sample and the reference. Using dedicated algorithms, sample viscosity is automatically extracted over a controlled range of shear rates and temperatures.

Benefits

-User-friendly, fast and simple measurement

-Visual quality control

-Accuracy over a wide range of viscosity (0.1 to 10000 mPa/s) and shear rates (up to 105 s-1)

-Automated analysis as a function of both shear rates and temperature (4-80 °C)

-Small sample volume

-No user intervention (e.g. resampling) required between measurement repetitions

Applications

-Inkjet (printers, flexible electronics, biochips)

-Spray (cosmetics, food, paints)

-Drug injection (pharma)

-Blood, synovia, etc. (biomedical diagnostics)

-Oil industry

-Lubricants (mechanics, cutting, wire drawing)

Keywords: Characterization, Chemical, Food Science, Rheology

Application Code: General Interest

Methodology Code: Physical Measurements

Session # 180 Abstract # 180-7 **Oral Sessions**

Session Title: Measurement Strategies - Sensors and Spectroscopy

Abstract Title: **Real-Time Investigation of Antibiotics-induced Oxidative Stress and Superoxide Release in Bacteria Using an Electrochemical Biosensor**

Primary Author: Xiaobo Liu
Clarkson University

Date: Sunday, March 06, 2016 - Afternoon
Time: 03:45 PM
Room: B313

Co-Author(s): Michael Jahne, Mouna Marrakchi, Shane Rogers, Silvana Andreescu

Abstract Text

The involvement of oxidative stress in the mechanism of antibiotics-meditated cell death is unclear and subject to debate. This presentation will describe development and application of a cytochrome c electrochemical biosensor for measuring the release of superoxide radicals (O[sub]2[/sub][sup].-[/sup]), a major contributor to ROS, in antibiotics-treated bacterial cultures. The specificity of the electrochemical measurements was established by the addition of superoxide dismutase (SOD). Measurements using a general ROS-specific fluorescence dye and colony forming units (CFU) assays were performed side-by-side to determine the total ROS and establish the relationship between ROS and the degree of lethality. Exposure of Escherichia coli and Listeria monocytogenes cultures to antibiotics increased the release of O[sub]2[/sub][sup].-[/sup] in a dose-dependent manner, suggesting that the transmembrane generation of ROS may occur as part of the antibiotic action. The study provides a quantitative methodology and fundamental knowledge to further explore the role of oxidative stress in antibiotics-meditated bacterial death and to assess physiological changes associated with the complex metabolic events related to oxidative stress and bacterial resistance.

Keywords: Bioanalytical, Biosensors, Biotechnology, Electrochemistry
Application Code: Bioanalytical
Methodology Code: Electrochemistry

Session # 180 Abstract # 180-8 **Oral Sessions**

Session Title: Measurement Strategies - Sensors and Spectroscopy

Abstract Title: **Fabrication of Micro Ir/IrO[sub]x[/sub] pH Sensor for Dental Applications**

Primary Author: Chindanai Ratanaporncharoen
Tokyo Medical and Dental University

Date: Sunday, March 06, 2016 - Afternoon
Time: 04:05 PM
Room: B313

Co-Author(s): Akira Matsumoto, Junji Tagami, Masaomi Ikeda, Miyuki Tabata, Tatsuro Goda, Yuichi Kitasako, Yuji Miyahara

Abstract Text

A new way to diagnose the carious lesions which combine bio-sensing technology and dental knowledge to improve the new invasive carious lesion detection. This study aims to diagnose carious lesion by pH with the measurement system using metal oxide films electrode, Ir/IrO[sub]x[/sub] as a measuring electrode. We have developed the simplest pH measurement instrument which has fast response, high precision and accuracy system. The micro electrode detects the concentration of proton which directly related to pH, which also related to the state of mineralization cycle.

The potentiometric response to pH showed -57.4 mV/pH in evaluated pH range from 4 to 8, which demonstrated that our electrode possesses the excellent proton sensitivity (Nernst slope; -59.2 mV/pH). Subsequently, the pH of the surface of an extracted carious tooth was measured 3 times, and the average pH of the carious lesion and the sound enamel were 6.2 and 7.0, respectively. From these results, we confirmed the possibility of quantitative diagnosis of dental caries using Ir/IrO[sub]x[/sub] electrode based on the pH index.

Our examination on the measurement system with a large number of extracted tooth samples will ensure a detail understanding between pH and dental caries, we are expecting the significantly different between sound tooth and carious lesion. The quantitative method used in this research aims to classify the sound teeth and carious lesion which can further diagnose for active or arrested condition. The development of this system is expected to be great innovation for dental caries sensor.

Keywords: Biosensors, Microelectrode, Portable Instruments, Quantitative
Application Code: Biomedical
Methodology Code: Sensors

Session # 190 **Abstract # 190-1** **Oral Sessions**

Session Title: Process Analytical Techniques - Half Session

Abstract Title: **A Model Study of Pseudo-Absolute Quantitative Analysis Using Gas Chromatography - Vacuum Ultraviolet Spectroscopy**

Date: Sunday, March 06, 2016 - Afternoon
Time: 03:05 PM
Room: B309

Primary Author: Ling Bai
The University of Texas at Arlington

Co-Author(s): Jonathan Smuts, Kevin A. Schug, Phillip Walsh

Abstract Text

While different types of GC detectors, either selective or not, can be used to generate signals, relationships between detector responses and analyte quantities is very important to be known and be precisely controlled. The concentrations of eluted solutes are proportional to the areas or the height under the recorded peaks integrated in GC system. In addition to conventional methods of quantification (internal and external standard), gas chromatography-vacuum ultraviolet spectroscopy also provides a new means for pseudo-absolute quantitation, which is based on recorded molecular cross sections, without the need for traditional calibration. This means that once all sample loss and recovery sources associated with sample introduction into the instrument are addressed, the number of molecules registering an absorption event can be directly determined and related back to sample concentration. Standard samples of benzene and natural gas have been used to assess error or sample loss for the analysis of liquid and gaseous samples. Results indicate that error introduction and sample loss would occur in many steps. For example, column installation, split ratio accuracy, sampling times for splitless analysis, detector scan rate, and make-up gas flow can all possibly cause sample discrimination. Thus, the capability of VUV detector has been evaluated and results indicate that it provides an excellent means for carrying out system performance checks and solving challenging quantitative analytical problems.

Keywords: GC, Method Development, Quantitative
Application Code: Process Analytical Chemistry
Methodology Code: Gas Chromatography

Session # 190 Abstract # 190-2 **Oral Sessions**

Session Title Process Analytical Techniques - Half Session

Abstract Title **Rapid Process Control Using GC-ION Mobility Spectrometry**

Primary Author Chandrasekhara Hariharan
ION-GAS GmbH

Co-Author(s) Wolfgang Vautz

Date: Sunday, March 06, 2016 - Afternoon
Time: 03:25 PM
Room: B309

Abstract Text

Ion mobility spectrometry (IMS) is a method for the characterisation of trace substances using gas-phase mobility in a weak electrical field. If a complex and humid mixture is subject of an IMS analyses, rapid gas-chromatographic pre-separation is coupled to the IMS. Besides the higher selectivity of a GC-IMS it furthermore avoids negative effects such as clustering. With this method, complex mixtures can be analysed in few minutes including identification of the available analytes by comparison with a data bases and quantification by comparison with a calibration carried out earlier.

On behalf of significant examples, the suitability of IMS for quality and process control of a broad range of processes and bio-processes will be demonstrated. IMS was already applied for characterisation of the development of polymerisation processes. The monomer concentration could be determined on-line via the headspace a of continuous sample flow in sufficient accuracy. Furthermore for fermentation, the most time consuming part of beer brewing, a suitable on-line process control could be developed based on Diacetyl and Pentanedione. Various other compounds detected moreover could be used to characterise the beer itself. Recent developments demonstrated, that not only headspace analysis of volatiles as in the examples above but also semi-volatiles could be detected when laser desorption is applied for sampling, e.g. for the characterisation of olive oils.

Finally, the potential of the method with regard to optimisation of the analysis time down to few seconds, to miniaturisation of the equipment and to other relevant applications will be discussed.

Keywords: Process Analytical Chemistry, Process Control, Process Monitoring, Quality Control
Application Code: Process Analytical Chemistry
Methodology Code: Gas Chromatography

Session # 190 Abstract # 190-3 **Oral Sessions**

Session Title: Process Analytical Techniques - Half Session

Abstract Title: **Facilitating Complex Analysis Using Multiple Fast Temperature Programming Zones**

Primary Author: Dale Ashworth
Valco Instruments Co. Inc.

Date: Sunday, March 06, 2016 - Afternoon
Time: 03:45 PM
Room: B309

Co-Author(s): Andrew Rochon, Chris Bishop, Huamin Cai, Martin Brisbin, Stanley D. Stearns, Steve Werner, William Coontz

Abstract Text

Difficult analysis can often require traps, programmable rate vaporizers, multiple columns, and zones for focusing, and refocusing to enhance separations. By using high accuracy, high precision, high rate, temperature controlled zones the separations can be improved dramatically. The technology discussed in this presentation can be applied to existing instrumentation to improve separation techniques, and enhance resolution of current methods. Examples of the variety of ways to apply this technology in both conventional and experiment hardware will be presented. This technology is applied with a single wire approach to heating and sensing of temperature, and is significantly more efficient in it use of power than conventional heating techniques. It can be applied to columns and devices in ways that both enhance separation and reduce time of analysis.

Keywords: Gas Chromatography, Hydrocarbons, Method Development, Trace Analysis
Application Code: Process Analytical Chemistry
Methodology Code: Gas Chromatography

Session # 190 **Abstract # 190-4** **Oral Sessions**

Session Title: Process Analytical Techniques - Half Session

Abstract Title: **Analyzing Trace Level Nitric Oxide in a Flammable Gas Matrix**

Primary Author: Kenneth Wong
Air Liquide

Date: Sunday, March 06, 2016 - Afternoon
Time: 04:05 PM
Room: B309

Co-Author(s)

Abstract Text

Single digit ppb-mole nitric oxide concentration in nitrogen can easily be measured by a chemiluminescence analyzer. However, a chemiluminescence analyzer cannot be used for measuring trace nitric oxide in a flammable gas matrix containing carbon monoxide and hydrogen because such an analyzer is not made for use in a Class 1 Div 1 environment. Furthermore, a chemiluminescence technique is severely interfered by the presence of high concentration of carbon dioxide and moisture that might be present in a real-life sample. FTIR has a detection limit that is close to that of chemiluminescence technique. FTIR, however, is a more universal detector than chemiluminescence method. Infrared active species such as carbon monoxide, methane, carbon dioxide and ethylene in high concentrations can give rise to severe spectral interferences. Chemometric techniques can be applied to correct these spectral interferences. But due to the large concentration differences between the matrix gases and nitric oxide, chemometric techniques were found to give rise to high margin of uncertainty. This presentation discusses a wet colorimetric method that had been applied successfully to measure trace nitric oxide in several flammable gas matrices. Except for a few known interferences, the method had been found to be more reliable than some current instrumental methods. Unlike FTIR, the wet method does not require extensive data interpretation.

Keywords: FTIR, Gas, Process Analytical Chemistry, Wet Chemical Methods
Application Code: General Interest
Methodology Code: Process Analytical Techniques

Session Title: Sunday Poster Session

Abstract Title: **Increasing Stability of a Core-Shell Particle**

Date: Sunday, March 06, 2016 - Afternoon
Time:
Room: A412

Primary Author: Mark Woodruff
Fortis Technologies

Co-Author(s): Ken Butchart

Abstract Text

Much has been made of the ability of core-shell particles to improve speed of analysis and increase sensitivity in HPLC. In this poster we look at the use of a core-shell column that can withstand high pH conditions. Up until now core-shell columns have had a very limited pH range, which means that the use of pH to alter selectivity, and in particular retain polar basic analytes with good peak shape has been limited.

Now with a new surface grafted technology the core-shell particle pH range has been extended. We can look at the entire pH range and show how this technology works, allowing method development options not previously available. Utilising pH from 1-12 should allow us to screen acidic, basic and neutral compounds in order to obtain the optimum pH in which to run our LC method. We discuss how this affects our robustness and reproducibility in method development.

Keywords: High Throughput Chemical Analysis, HPLC Columns, Modified Silica, Chromatography
Application Code: General Interest
Methodology Code: Separation Sciences

Session # 200 Abstract # 200-3 **Sunday Poster Session**

Session Title: Sunday Poster Session

Abstract Title: **Rapid Screening of Virgin and Recycled Polymer Resins Using FTIR and Raman Libraries of Pre-computed Mixture Spectra**

Date: Sunday, March 06, 2016 - Afternoon
Time:
Room: A412

Primary Author: William Costa
Fiveash Data Management, Inc.

Co-Author(s): Bill McCarthy, Todd Strother

Abstract Text

Raman and Fourier transform infrared (FTIR) spectroscopy have become analysis methods of choice for the polymer and plastics industry because of their shared significant advantages such as non-destructive nature, fast analysis times and availability of spectral libraries. The rapid screening, simple sample preparation, and linear absorbance of attenuated total reflectance (ATR) accessories for FTIR spectrometers is particularly useful for identifying polymers, plastics and resins. Plastic pellets and scrap are often a mix of different polymers. In the past, spectroscopists utilized spectral libraries of pure compounds as an indicator of the chemical make-up of a sample, but still relied heavily on subjective interpretation. Another method uses proprietary algorithms that are not always clearly understood in an attempt to programmatically reduce the effects of mixed spectra. A third method of creating samples of all likely permutations is not feasible because of the high cost, substantial investment in time, and specialized labor involved.

We will present the development of a spectral library of pre-computed mixture spectra of binary and ternary polymers from a collection of neat polymer spectra. This spectral library can be searched using well established, and commonly available search methods that are widely accepted by the scientific community. The efficiency and accuracy of search results will be evaluated. Results of Raman and FTIR analysis for virgin polymers (materials that were never made into a finished product) and recycled polymers will be compared. Advantages and disadvantages of Raman and FTIR mixture searches for the polymer industry will be discussed.

Keywords: FTIR, Infrared and Raman, Polymers & Plastics, Raman
Application Code: Polymers and Plastics
Methodology Code: Molecular Spectroscopy

Session # 200 **Abstract #** 200-4 **Sunday Poster Session**

Session Title: Sunday Poster Session

Abstract Title: **The Complimentary Nature of NMR and SPME GCMS for the Analysis of Fermented Beverages**

Date: Sunday, March 06, 2016 - Afternoon
Time:
Room: A412

Primary Author: Neil Fitzgerald
Marist College

Co-Author(s): John C. Edwards, Samantha E. Soprano, Sarah R. Johnson

Abstract Text

The craft alcoholic beverage market has experienced dramatic growth over the past few years, and the Brewers Association estimates an 18% increase in sales volume of craft beer in 2014. Craft distilleries and cideries have also increased. Small fermentation beverage companies and home brew enthusiasts provide an increased demand for chemical analysis to demonstrate consistencies between batches, highlight deficiencies in their products, and gain a better understanding of the fermentation process. Nuclear Magnetic Resonance (NMR) and Solid Phase Micro Extraction (SPME) combined with Gas Chromatography Mass Spectrometry (GCMS) are examples of techniques that can provide useful chemical information to the fermentation beverage industry. The techniques have distinct advantages and limitations. In this work, a large number of beverage samples of varying types were analyzed by both techniques. The complimentary nature of the techniques is demonstrated. NMR is shown capable of detecting a number of compounds at the minor or semi-trace level in the liquid phase such as ethanol, fusel alcohols, sugars, dextrins, and organic acids. Quantitative analysis is readily achievable using a single internal standard. SPME GCMS is capable of identifying chemicals at the trace level. SPME GCMS of the headspace is particularly useful for the identification of aroma compounds such as esters, ethers, alcohols, aldehydes, ketones, and volatile sulfur compounds. Quantitation is less straightforward than NMR. When used together, the two methods give a good indication of the chemical composition of the fermented beverage that can be interpreted to provide useful information.

Keywords: Beverage, GC-MS, NMR, SPME
Application Code: Food Science
Methodology Code: Gas Chromatography/Mass Spectrometry

Session # 200 **Abstract # 200-6** **Sunday Poster Session**

Session Title: Sunday Poster Session

Abstract Title: **Selective Separation and Detection of Catecholamines with Capillary Electrophoresis**

Date: Sunday, March 06, 2016 - Afternoon
Time:
Room: A412

Primary Author: Maojun Gong
Wichita State University

Co-Author(s): Qiyang Zhang

Abstract Text

Catecholamines are a group of correlated compounds including dopamine, norepinephrine and epinephrine, which are important neurotransmitters and biomarkers of pheochromocytoma, a rare but deadly cancer. The detection method mainly depends on electrochemistry considering their redox activity. However, interferences such as vitamin C and other redox reactive species set challenging to the detection selectivity. In addition, the levels of these compounds are low in biological fluids such as cerebrospinal fluid (CSF), blood plasma, and urine. Therefore, it is challenging to determine their concentrations in real time. Here we report a sensitive fluorescent strategy to rapidly separate and detect fluorescent catecholamine derivatives via capillary electrophoresis. First, catecholamines and other amines were fluorogenically derivatized through online mixing and derivatization. Then, the reaction mixture was selectively injected to the separation capillary by taking advantage of the difference of their electrophoretic mobilities in a high electric field. Finally, there were separated and detected with laser-induced fluorescence. As typical examples, NDA (Naphthalene-2,3-Dicarboxaldehyde)and NBD-F (4-fluoro-7-nitro-2,1,3-benzoxadiazole) were used as the model derivatization reagents which produced neutral derivatives, while other charged derivatives were partially or completely rejected from entering the capillary during the electrokinetic injection process. The baseline cleanup of the electropherograms effectively facilitated the detection of catecholamines at low concentrations. This strategy has been applied to the detection of neurotransmitters, especially dopamine and norepinephrine, in CSF, and they will be used for in vivo neurotransmitter monitoring and cancer diagnosis.

Keywords: Amino Acids, Bioanalytical, Capillary Electrophoresis, Separation Sciences

Application Code: Bioanalytical

Methodology Code: Capillary Electrophoresis

Session # 200 Abstract # 200-7 **Sunday Poster Session**

Session Title Sunday Poster Session

Abstract Title **Cold EI – Approaching the Ideal GC-MS Interface and Ion Source**

Date: Sunday, March 06, 2016 - Afternoon
Time:
Room: A412

Primary Author Aviv Amirav
Tel Aviv University

Co-Author(s) Alexander Fialkov, Bogdan Belgorodsky, Tal Alon, Uri Keshet

Abstract Text

Cold EI is electron ionization of very cold molecules in supersonic molecular beams (SMB). The Aviv Analytical 5975-SMB GC-MS with Cold EI is based on interfacing an Agilent 7890 GC and 5975 or 5977 MS with SMB while replacing the EI ion source with Cold EI fly-through ion source. Cold EI approaches the ideal ion source as in combination with the SMB interface it significantly improves all the major performance aspects of GC-MS:

A) Improved sample identification: Enhanced molecular ions are provided while retaining the fragment ions. Thus NIST library identification probabilities are improved due to the enhancement of the molecular ions that serve to better reject incorrect candidates.

B) Molecule identifier software (TAMI) was developed that improves the quadrupole mass accuracy and inverts the enhanced molecular ion isotope pattern into elemental formula.

C) Significantly extended range of low volatility, polar and thermally labile compounds are amenable for analysis.

D) Superior sensitivity is provided particularly for difficult to analyze compounds via improved signal, enhanced molecular ion and the elimination of vacuum background noise.

E) Much faster analysis is provided, from few minutes down to few seconds with the help of low thermal mass fast GC. Real time analysis with separation is provided when Open Probe Fast GC inlet is used.

F) Uniform compound independent response is exhibited for improved quantitation and provision of chemical reaction yields.

G) GC-MS with Cold EI is seamlessly coupled with pulsed flow modulation GCxGC for the combination of ultimate MS information and improved GC separation.

Keywords: Gas Chromatography/Mass Spectrometry, GC-MS
Application Code: Other
Methodology Code: Gas Chromatography/Mass Spectrometry

Session # 200 Abstract # 200-9 **Sunday Poster Session**

Session Title: Sunday Poster Session

Abstract Title: **Ceramic Ion Source for Catalytic Pyrolysis and Catalytic Combustion Ionization Detection of Selected Constituents in Complex Petroleum and Biofuel Samples**

Primary Author: Paul L. Patterson
Detector Engineering & Technology, Inc.

Co-Author(s): Jennifer Seroy

Date: Sunday, March 06, 2016 - Afternoon
Time:
Room: A412

Abstract Text

An electrically heated, cylindrically shaped ion source element made of catalytically active ceramic provides unique GC detection of selected constituents in complex Petroleum and Biofuel samples. When used in a detector gas environment of Nitrogen, the ion source provides extraordinarily large selective responses to electronegative heteroatom compounds such as Oxygenates, Halogenates, Nitro compounds, and compounds containing the Pyrrole versus Pyridine functional group, among others. A Nitrogen detector gas environment is also characterized by a Catalytic Pyrolysis Ionization Detection (CPID) process which provides selective responses to high concentrations of Alkane Hydrocarbons having multiple branched Methyl (CH3) groups. By changing the detector gases to Air or Oxygen, the same equipment is converted to Catalytic Combustion Ionization Detection (CCID) which is selective to compounds containing chains of Methylene (CH2) functional groups. Neither CPID nor CCID are responsive to Aromatic or Cyclo-Hydrocarbons so these modes of selective detection provide unique simplified GC chromatograms of complex samples such as Gasoline, Auto Diesel, B20 Diesel, Kerosene, Fuel Oil, Shale Oil, etc. CPID and CCID can be combined in series to provide two different signals simultaneously from the same incoming sample.

Keywords: GC Detectors, Instrumentation, Petroleum
Application Code: Fuels, Energy and Petrochemical
Methodology Code: Gas Chromatography

Session # **200** Abstract # **200-10** **Sunday Poster Session**

Session Title: Sunday Poster Session

Abstract Title: **Using Microscopy to Measure Rates of Heterogeneous Reactions**

Primary Author: Walter Bowyer
Hobart and William Smith Colleges

Date: Sunday, March 06, 2016 - Afternoon
Time:
Room: A412

Co-Author(s): Gabriella Mylod

Abstract Text

We describe the use of microscopy for determining rates of both diffusion and reaction during indium mediated allylations. As the reaction proceeds, the metal surface retreats. Photomicroscopy is used to measure the rate of retreat as a function of convection(in this case, rotation) rate. Because the reactions exhibit mixed diffusion/kinetic control over the range of available rotation rates (100 to 3000 rpm), the plots of reaction rate vs. rotation rate allow determination of both diffusion coefficients and heterogeneous rate constants. Precision and error of the results are emphasized. The strategy is applied to determining rates in a variety of solvents. Potential applications of the strategy to other systems are explored.

Keywords: Electrochemistry, Microscopy, Surface Analysis
Application Code: Other
Methodology Code: Microscopy

Session # 200 **Abstract # 200-11** **Sunday Poster Session**

Session Title: Sunday Poster Session

Abstract Title: **Crowding in Cell-Like Environments Alters Diffusion and Enzyme Kinetics**

Primary Author: Kristin M. Slade
Hobart and William Smith Colleges

Date: Sunday, March 06, 2016 - Afternoon
Time:
Room: A412

Co-Author(s): Michael Conroy, Sophia Melvin

Abstract Text

Cells consist of a heterogeneous mixture of macromolecules that are tens to hundreds of times more concentrated than the dilute conditions used in most biophysical experiments. Evidence suggests that this macromolecular crowding alters enzyme behavior, resulting in changes to stability, diffusion, aggregation, association, and ultimately function. To better understand how crowding affects enzymes, high concentrations of the glucose polymer, dextran were used to simulate the crowded cellular environment during in vitro spectrophotometric kinetic assays of three enzymes: yeast alcohol dehydrogenase, lactate dehydrogenase, and citrate synthase. For both dehydrogenases, crowding enhances the substrate binding affinity, while decreasing the maximum reaction rate. We have used electrochemical methods to link this decreased rate with slowed diffusion. For citrate synthase, low concentrations of dextran enhanced the rate and weakened binding affinity, while higher concentrations decreased the rate and enhance substrate binding. Thus, for all three enzymes, the effects from crowding depend on the dextran concentration but not the size of the polymers.

Keywords: Enzyme Assays, UV-VIS Absorbance/Luminescence
Application Code: Bioanalytical
Methodology Code: UV/VIS

Session Title Sunday Poster Session

Abstract Title **Effects of Macromolecular Crowding on YADH Enzyme Mechanism**

Primary Author Kristin M. Slade
Hobart and William Smith Colleges

Co-Author(s) Allison Wilcox, Micaela LoConte

Date: Sunday, March 06, 2016 - Afternoon
Time:
Room: A412

Abstract Text

The intracellular environment is crowded with high concentrations of carbohydrates, proteins, nucleic acids, and other macromolecules. Traditionally, biophysical studies are conducted in dilute conditions, yet experimental and theoretical evidence indicate that enzyme behavior is affected by the presence of these macromolecules, resulting in slowed diffusion, enhanced enzyme-substrate binding, and altered enzyme conformation. To begin to characterize macromolecular crowding's effects on enzyme kinetics, Michaelis-Menten parameters were obtained from spectrophotometric assays conducted on the enzyme yeast alcohol dehydrogenase (YADH) in the absence and presence of crowding agents. These crowding agents range from the inert glucose polymer, dextran, to charged proteins such as BSA and lysozyme. Our results show that when the primary substrate, ethanol, is used and the rate-limiting step is the release of the NADH product, crowding decreases the maximum rate of the reaction by approximately 40%. Conversely, when the alternative substrate, isopropanol, is used and the rate-limiting step becomes the chemical hydride transfer, a 20-40 % increase in maximum rate of the reaction is observed, likely due to the compression of the active site. Although further investigation is required, these studies provide insight into how crowding impacts individual steps of the enzyme mechanism.

Keywords: Enzyme Assays
Application Code: Bioanalytical
Methodology Code: UV/VIS

Session # 200 **Abstract # 200-13** **Sunday Poster Session**

Session Title: Sunday Poster Session

Abstract Title: **Mixed Valence Mn,La,Sr-oxide based Magnetic Nanoparticles Coated with Silica Nanoparticles for Immunosensor Fabrication**

Date: Sunday, March 06, 2016 - Afternoon

Time:

Room: A412

Primary Author: Amos Mugweru
Rowan University

Co-Author(s)

Abstract Text

A mixed valence manganite nanoparticles (NPs) of the type La0.67Sr0.33MnO3 for use in electrochemical immunoassays is described. The NPs were synthesized using the reverse micelle method and their surface was then functionalized with silica nanoparticles. The resulting NPs (La0.67Sr0.33MnO3@SiO2) were characterized by X-ray diffraction, scanning electron microscopy and physical property measurements. The La0.67Sr0.33MnO3@SiO2 NPs were used to develop an electrochemical immunoassay for human IgG. Antibody (anti-human IgG) was immobilized on the silica NPs on the magnetic core, and this process was monitored by cyclic voltammetry and square-wave voltammetry. The assay has a linear working range up to an IgG concentration of 5 ng mL , and the detection limit is 0.6 ng mL .

Keywords: Bioanalytical, Biomedical, Biosensors, Immunoassay

Application Code: Bioanalytical

Methodology Code: Electrochemistry

Session # 200 Abstract # 200-14 Sunday Poster Session

Session Title Sunday Poster Session

Abstract Title **Hydrophilic Interaction Chromatography (HILIC) and Enzymatic/Spectrometric Methods for the Determination of Uric Acid and Creatinine in Human Biofluids**

Date: Sunday, March 06, 2016 - Afternoon
Time:
Room: A412

Primary Author Yuegang Zuo
University of Massachusetts Dartmouth

Co-Author(s) Faten Albalawi, Ningning Zhang, Si Zhou, Xiaofei Lu

Abstract Text

Abnormal concentrations of uric acid (UA) and creatinine (Cr) in plasma and urine are associated with various diseases and are routinely examined in clinical and biomedical laboratories using enzymatic/photometric techniques. However, many endogenous and exogenous compounds interfere with these photometrical measurements. In this study, a hydrophilic interaction chromatography (HILIC) [Zuo, Y, Ed., High-Performance Liquid chromatography (HPLC): Principles, Procedures and Practices. Nova Science Publishers, Inc., New York (2014)] was developed for the simultaneous determination of uric acid and creatinine in human fluids. The retention of two analytes (UA and Cr) and the internal standard on the column was found to increase with increasing of both pH (range from 3.75 to 5.75) and the percentage of acetonitrile (ACN) in mobile phase. The developed HILIC method has proved fast, selective, and accurate when compared with commonly used enzymatic/photometrical methods. The enzymatic/spectrometric methods compared were based on different reaction principles for these two analytes. The product of the each reaction was quantitatively analyzed by UV-Vis spectrometry.

Keywords: Bioanalytical, Clinical Chemistry, Forensics, HPLC
Application Code: Bioanalytical
Methodology Code: Liquid Chromatography

Session # 200	Abstract # 200-15	Sunday Poster Session

Session Title Sunday Poster Session

Abstract Title **Profiling/Monitoring Activated Carbons for Drinking Water**

Primary Author Henry G. Nowicki
PACS Inc

Date: Sunday, March 06, 2016 - Afternoon
Time:
Room: A412

Co-Author(s)

Abstract Text

Activated carbons (AC) are the most used sorbents at drinking water plants to remove trace organic contaminants from our drinking water supplies. AC is mandated by U.S. EPA as the best available technology for this application.

Drinking water plant managers have a difficult decision when to replace used AC or spent AC with un-used AC or fresh. They also have difficulty in distinguishing reactivated AC from virgin- or un-used AC. Picking the best AC for each drinking water plant could be improved. There are about 1,000 drinking water plants in the U.S. These defined problems for drinking water plants is universal and on a global scale.

We provide solutions for these drinking water plant problems. We provide an example of characterizing the AC filters at a drinking water plant.

ASTM and advanced activated carbon test methods were employed. A water plant with eight large activated carbon filters, 20-25,000 pounds AC per filter. Test methods revealed the best to worst of eight filters. Laboratory results provided recommendations for which filters needed replaced and estimated time others may need replaced. This level of client information is above industry standards and useful to plant managers.

We will present some of the ASTM data and concentrate of the advanced testing data from Gravimetic Adsorption Energy Distribution or GAED aqueous full characterizations. GAED provides characterization for both aqueous- and vapor-phase AC applications. GAED enables determination of isotherms for any compound of interest at any desired operating temperature. GAED is a home made instrument with software not yet commercialized.

We will present the characteristic adsorption curves for all eight AC analytics and interpretation of data. It has capability of solving several present refractory problems for drinking water plants and other AC users.
More information PACS web site: www.pacslabs.com.

Keywords: Contamination, Materials Characterization, Process Analytical Chemistry, Quality Control

Application Code: Process Analytical Chemistry
Methodology Code: Process Analytical Techniques

Session # 200 **Abstract # 200-16** **Sunday Poster Session**

Session Title: Sunday Poster Session

Abstract Title: **Profiling/Monitoring Gas Phase Activated Carbon Systems**

Primary Author: Henry G. Nowicki
PACS Inc

Co-Author(s)

Date: Sunday, March 06, 2016 - Afternoon
Time:
Room: A412

Abstract Text

Gas phase activated carbon adsorbers are used by municipal wastewater plants to control obnoxious odors. Gas phase adsorbers are used in a large variety of other applications: control of specific off-gases, solvent recovery, residential and business office buildings.

PACS has developed profiling and monitoring of gas emissions from wastewater plants. In-coming fresh carbons are validated by standard and advanced test methods. Hydrogen sulfide capacity is determined on fresh carbon and used carbons. Monitoring remaining service life to estimate time for carbon media replacement is provided by: monitoring elemental sulfur build-up in the carbon pores, the heat-of-immersion at different stages of carbon use, and changes in the apparent density of media with use. Examples of monitoring will be provided and explanations on the basis of the new test methods employed to solve problems.

Solvent recovery is monitored with standard methods on the carbons used. GC/MS is used to identify the adsorbates on fouled carbons. Activated carbon adsorbates are recovered by using competitive desorbtion. A stronger adsorbate in a solvent is used to take off the accumulated organics in AC pores for GC-MS analysis. Heat-of-immersion (HOI) is used to estimate remaining service time left in the carbon.

Adsorbates are described and identified and quantified with GC/MS analysis. An example will be provided.

Heat of immersion developed at PACS Laboratories and is now an ASTM standard method. HOI is based on comparison of adsorption heat from the starting un-used carbon and the used carbon. This information is useful for changing out the used- with fresh AC.

These techniques have capability of solving refractory problems for gas phase application of activated carbons.

Keywords: Contamination, Environmental/Air, Fuels\Energy\Petrochemical, Gas Chromatography/Mass Spectro
Application Code: Process Analytical Chemistry
Methodology Code: Process Analytical Techniques

Session # 200 Abstract # 200-17 **Sunday Poster Session**

Session Title Sunday Poster Session

Abstract Title **Lignin Waste Transformed to High Quality Activated Carbons**

Date: Sunday, March 06, 2016 - Afternoon

Time:

Room: A412

Primary Author Henry G. Nowicki
PACS Inc

Co-Author(s)

Abstract Text

Lignin is a natural product of polymerized propyl-aromatic moieties, present in plants to provide its rigid structure and cell walls. During the process of recovering carbohydrates from biomass lignin is a major volume recovered residue. Lignin is treated as waste and used for heating by forest products firms.

We have discovered a better use for lignin by transforming it to high quality activated carbon. We built a small laboratory sized rotary kiln to simulate large commercial production sized kilns. Lab sized kiln provides a quick low-cost way to evaluate feedstocks for making activated carbons and maximizing process parameters.

We will describe our small furnace construction and their products performance using ASTM standard test methods and advanced test methods including GAED full characterizations.

The best product made at our lab from lignin has an adsorption pores or adsorption space equal to or better than best commercially available coconut shell based activated carbons. Coconut shell activated carbons need to be imported, thus they are negative to U.S. GNP. Being able to substitute lignin for coconut carbons would improve the U.S. GNP and decrease reliance on foreign manufacturers.

Coconut and lignin are considered green chemistry compared to coal family activated carbon production which is the major feedstock for production of activated carbons. It is considered green, because it releases carbon dioxide which was assimilated in present time compared to coal family members who release carbon dioxide that were incorporated eons ago. Typically five pounds of feedstock are used to make one pound of activated carbon, thus you generate a lot of carbon dioxide.

Presently we are writing grant proposals to develop the lignin towards commercial scale production and determining its real-world applications.

Keywords: Adsorption, Agricultural, Carbohydrates, Energy

Application Code: Agriculture

Methodology Code: Thermal Analysis

Session Title Sunday Poster Session

Abstract Title **A Rapid Hydrophilic Interaction Liquid Chromatography (HILIC) Method for Determination of Trace Nitrate and Nitrite in Snow and Rain Samples**

Date: Sunday, March 06, 2016 - Afternoon
Time:
Room: A412

Primary Author Xiaofei Lu
University of Massachusetts Dartmouth

Co-Author(s) Yuegang Zuo

Abstract Text

Nitrite and nitrate are two common ions in natural waters. It is of great significance to determine the concentration of these ions because they can put potential risk to human and environment health. A rapid, specific, and sensitive method for determining nitrite and nitrate in snow and rain water has been developed based on high performance liquid chromatography. A hydrophilic amino column was used to separate the nitrite and nitrate which were then detected by an UV-Vis detector at wavelength of 215 nm [Zuo, Y. Ed., High-Performance Liquid Chromatography (HPLC): Principles, Procedures and Practices. Nova Science Publishers, Inc., New York (2014)]. The mobile phase consisted of acetonitrile and acetate buffer solution. It was found that the retention times of nitrite and nitrate increased with decrease of pH of mobile phase or decrease of proportion of buffer solution in mobile phase. The effect of acetate concentration on retention time is more complicated. The retention times of nitrite and nitrate increased with the increase of concentration in low acetate concentrations range (below 2.5 mM), whereas in the high concentration range (2.5 mM to 10 mM), they decreased with increase of concentration. The results may indicate an ion exchange mechanism to the separation of nitrite and nitrate on amino column. The LOQ (limit of quantitation) of the two compounds were both 0.01 mg/L with linear range of 0.01 mg/L to 60 mg/L. The average recoveries of the two compounds were 97.86% and 98.39%, respectively. Nitrate concentration varied from 0.072 to 1.39 mg/L in the in snow or rain sample. The nitrite in most of the samples was below the LOQ and was quantified only in one snow sample with concentration of 0.012 mg/L.

Keywords: Environmental Analysis, Environmental/Water, HPLC, Separation Sciences
Application Code: Environmental
Methodology Code: Liquid Chromatography

Session # 200 Abstract # 200-19 **Sunday Poster Session**

Session Title: Sunday Poster Session

Abstract Title: **Occurrence and Identification of Bisphenol A and Other Alkylphenols in Sea Shore Crabs**

Primary Author: Joseph Michael
University of Massachusetts Dartmouth

Date: Sunday, March 06, 2016 - Afternoon
Time:
Room: A412

Co-Author(s): Yuegang Zuo

Abstract Text

The global industrialization in recent decades has led to an increase in environmental pollutants from the development of commercial products, especially those from plastic components. Plastics are organic materials that are used in everyday households for a variety of functions. With the increase in volume of plastics in common everyday items like baby bottles, food cans and drinking bottles, there is a concern over the health effects that come with the exposure of plasticizers to humans and wild lives because plasticizers, such as phthalates, alkylphenols and bisphenol A, are endocrine-disrupting chemicals (EDCs). Determining the levels of these EDCs in the environment and animals is important so that certain thresholds can be set to limit the health effects to the wildlife and humans. Both GC and HPLC techniques have been developed for determining BPA in the environment. The purpose of this experiment is to identify and quantify alkylphenols and bisphenol A in a local Asian Shore Crab species using GC-FID and GC-MS techniques. BPA and two of its analogues, 2,4-bis-(dimethylbenzyl)phenol and 4-cumylphenol, have been found in Asian crab samples. Further details will be discussed at the presentation.

Keywords: Biological Samples, Environmental, Food Science, Gas Chromatography
Application Code: Bioanalytical
Methodology Code: Gas Chromatography/Mass Spectrometry

Session # 200 **Abstract # 200-20** **Sunday Poster Session**

Session Title: Sunday Poster Session

Abstract Title: **Evaluation of Porous Layer Thickness of Core Shell Particle for Separation of Proteins**

Date: Sunday, March 06, 2016 - Afternoon

Time:

Room: A412

Primary Author: Norikazu Nagae
ChromaNik Technologies Inc.

Co-Author(s): Tomoyasu Tsukamoto

Abstract Text

The feature of superficially porous (core shell) particle used as a highly efficient material is existence of a core, a thin porous layer and narrow particle size distribution, which lead to higher efficiency than totally porous particle. Recently a core shell particle with wide pore for biomacromolecular separations has developed by a few manufacturers. It has been said that thin porous layer of core shell particle have an advantage for separation of large molecules such proteins because a diffusion coefficient becomes small to proportional to a molecular weight and a mass transfer speed also decreases. In this paper, thickness of porous layer of core shell particle was evaluated to separate proteins.

2 kinds of thickness of porous layer such as 0.2 □m and 0.5 □m thickness were applied for separation of standard protein samples. One was 3.4 □m particle size, 0.2 □m porous layer and 15 m2/g specific surface area and another was 2.6 □m, 0.5 □m and 15 m2/g. Separation was achieved using a gradient elution with 0.1% trifluoroacetic acid and acetonitrile including 0.08% trifluoroacetic acid.

On fast separation, 0.2 □m of porous layer showed sharper peaks than 0.5 □m of porous layer. However at 80 degree Celsius and using 60 min gradient time program, 0.5 □m of porous layer showed much sharper peaks than 0.2 □m of porous layer. It was considered that 0.5 □m of porous layer had the a wider specific surface area than 0.2 □m of porous layer and this wider specific surface area leaded separation efficiency concerning the partition interaction on the stationary phase to be large.

Better separation of proteins contributes not only the thin porous layer but also the large surface area.

Keywords: HPLC Columns
Application Code: General Interest
Methodology Code: Liquid Chromatography

Session # 210 Abstract # 210-1 **ACS Poster**

Session Title: ACS Division of Analytical Chemistry Poster Session

Abstract Title: **Phospholipid/Aromatic Thiol Hybrid Bilayers**

Primary Author: Chao Li
Auburn University

Co-Author(s): Matthew Ferguson, Mingming Wang, Wei Zhan

Date: Sunday, March 06, 2016 - Afternoon
Time:
Room: A412

Abstract Text

Gold-supported hybrid bilayers comprising phospholipids and alkanethiols have been found to be highly useful in biomembrane mimicking as well as biosensing ever since their introduction by Plant in 1993 (Plant, A. L. Langmuir 1993, 9, 2764?2767). Generalizing the mechanism (i.e., hydrophobic/hydrophobic interaction) that primarily drives bilayer formation, we have reported such a bilayer structure could also be successfully obtained when aromatic thiols were employed in place of alkanethiols. Four aromatic thiols were studied (thiophenol, 2-naphthalene thiol, biphenyl-4-thiol, and diphenylenevinylene methanethiol), all affording reliable bilayer formation when 1-palmitoyl-2-oleoyl-sn-glycero-3-phosphocholine liposomes were incubated with self-assembled monolayers of these thiols. Significant differences in electrochemical blocking and mechanical characteristics of these new bilayers were identified in comparison to their alkanethiol counterparts. Taking advantage of these new features, a convenient and sensitive voltammetric scheme for the straightforward bio-recognition of a lipolytic enzyme (phospholipase A2) is introduced, which revealed important differences between aromatic- and alkane-based hybrid bilayers. Then, chemical force microscopy (CFM) with hydrophobic tips to mimic phospholipid aliphatic chains is used to measure hydrophobic forces on the aromatic thiol SAMs and alkane thiol SAMs, which supports the voltammetric bio-recognition analysis.

Keywords: Electrochemistry, Lipids, Microscopy, Statistical Data Analysis
Application Code: Bioanalytical
Methodology Code: Microscopy

Session # 210 Abstract # 210-2 ACS Poster

Session Title ACS Division of Analytical Chemistry Poster Session

Abstract Title **Dynamic Records of Spatiotemporal Pb: Lichens versus Trees and Sediments**

Date: Sunday, March 06, 2016 - Afternoon
Time:
Room: A412

Primary Author Nathan W. Bower
Colorado College

Co-Author(s) Ben Greydanus, Craig Lundstrom, Eric Wolatz, Sam Kramer, Stephen Getty

Abstract Text

Lead (Pb) isotopic and concentration analyses of epiphytic lichens are often employed in the development of spatial records of anthropogenic Pb releases, and 210Pb and 14C dated sediments from lakes, rivers, estuaries and peat bogs have been used to construct temporal records. These profiles across space and time help identify specific Pb sources. They are also valuable for developing spatiotemporal isoscapes useful in forensic science and for migration studies. But in arid regions such as the Rocky Mountain West, lake sediments and peat bogs are relatively uncommon. Alternatives are needed for developing isoscapes with more spatial resolution. While some tree species have been successfully tested as biomarkers of temporal Pb release, many studies have given ambiguous results. Similarly, although epilithic lichens collected at different times have been used, the temporal variations within individual lichens have not been fully explored. In this study we employ ICP-OES and multi-collector inductively-coupled mass-spectrometry (MC-ICP-MS) to measure the concentration and isotopic profiles of Pb in Ponderosa pine tree rings and in individual Xanthoparmelia epilithic lichens, exploring the conditions under which they are most useful as biomarkers for measuring atmospheric Pb deposition over time.

Keywords: Atomic Spectroscopy, Forensics, Geochemistry, Mass Spectrometry
Application Code: Homeland Security/Forensics
Methodology Code: Mass Spectrometry

Session # 210 Abstract # 210-3 **ACS Poster**

Session Title: ACS Division of Analytical Chemistry Poster Session

Abstract Title: **Dating Paintings and Prints Using ATR-FTIR: A Philatelic Case Study from Lodz Ghetto**

Primary Author: Nathan W. Bower
Colorado College

Date: Sunday, March 06, 2016 - Afternoon
Time:
Room: A412

Co-Author(s): Conor J. Blanchet, Michael S. Epstein

Abstract Text

The determination of dates of production for questioned paintings and printed works would be useful in authentication and art historical studies. Rates of decay of radioactive elements such as Pb-210 and C-14 can sometimes be employed, and the profile of pigments used at different times has also provided insights. But relatively little use has been made of the changes that happen over time as the binding media ages. Using samples of printed and painted works with known dates, we calibrate the drying and/or loss of oil binders in paintings and printing inks using changes in their attenuated total reflectance infra-red (ATR-IR) spectra. We demonstrate the efficacy of this non-destructive method by dating authentic and forged postage stamps from a World War II Jewish Ghetto in Occupied Poland and an oil painting of an alchemist from early in the 20th century.

Keywords: Art/Archaeology, Forensics, FTIR, Portable Instruments
Application Code: Art/Archaeology
Methodology Code: Vibrational Spectroscopy

Session # **210** **Abstract #** **210-4** **ACS Poster**

Session Title: ACS Division of Analytical Chemistry Poster Session

Abstract Title: **SPME/GC-MS Analysis of Dynamic Compositional Changes in the Volatile Components in Red Wine After "Breathing"**

Primary Author: Amber I. Hills
East Stroudsburg University

Co-Author(s): Jon S. Gold, Richard S. Kelly

Date: Sunday, March 06, 2016 - Afternoon
Time:
Room: A412

Abstract Text

Extended exposure to oxygen in air can be detrimental to the flavors in wine, and most wine makers limit the amount of exposure during the production process. It is common practice, however, to allow some wines to "breathe" upon opening to fully develop the volatiles on the "nose" and express the character within the juice. While much has been reported on the identities of flavor-related volatiles in the vapor phase of some wines, there is less literature dedicated to the dynamic detection of volatile components in wine after it is opened and exposed to air. In this work, solid phase microextraction (SPME) coupled with gas chromatography-mass spectrometry (GC-MS) was used to identify the major components in the vapor phase above red wine from the Côtes du Rhône, and to measure the concentrations of these components immediately after opening, after being exposed to air without stirring for a period of 112 minutes, and after being stirred for 10 seconds every 15 minutes for a total of 225 minutes. Among the results were increases in alcohol concentrations (isoamyl alcohol, for example) over time with concomitant decreases in the concentrations of certain esters (isoamyl acetate, for example). These results suggest an acid catalyzed hydrolysis of esters to alcohols in the vapor phase during the "breathing" process.

Keywords: Beverage, Food Science, GC-MS, SPME
Application Code: Food Science
Methodology Code: Gas Chromatography/Mass Spectrometry

Session # 210 Abstract # 210-5 **ACS Poster**

Session Title ACS Division of Analytical Chemistry Poster Session

Abstract Title **Development of Portable Elisa System for Infectious Disease Diagnosis**

Date: Sunday, March 06, 2016 - Afternoon
Time:
Room: A412

Primary Author: Kazuhiro Morioka
Tokyo Metropolitan University

Co-Author(s): Akihide Hemmi, Harpal Singh, Hizuru Nakajima, Hulie Zeng, Katsumi Uchiyama, Le Van An, Masami Sugamata, Masayuki Shimojima, Ming Yang, Sazaly AbuBakar, Shungo Kato

Abstract Text

A pandemic outbreak of infectious diseases is threatening our lives. The prevention of infection disease by appropriate vaccination based on viral antibody test is very important to prevent the pandemic outbreak of infectious disease. Enzyme-linked immunesorbent assay (ELISA) is commonly used for the antibody test. Conventional ELISA requires complicated operations and long analysis times. In addition, the microplate reader used for detection is large-size and expensive. Therefore, it is difficult to perform the antibody test rapidly in the field.

The objective of this study is to develop a portable ELISA system that can perform the antibody test rapidly in the field. For the purpose, we developed a small and inexpensive ELISA system consisted of a PDMS microtiter plate and a fluorescence microplate reader using LEDs and photodiodes. The developed system is palm-size and lightweight. We can measure the fluorescence intensity wirelessly using a notebook PC or tablet PC since a Bluetooth module is built into the system.

The basic performance of the developed microplate reader was evaluated using resorufin as model sample. The calibration curve for resorufin solutions at concentrations under 6.25 M showed good linearity with a correlation coefficient of 0.999. The limit of detection (LOD) defined as 3was estimated to be 50.7 fmol. On the other hand, LOD obtained on a 96-well microtiter plate and a commercially available microplate reader was estimated to be 42.7 fmol. The sensitivity of the developed microplate reader was comparable to that of the commercially available microplate reader. The developed ELISA system was applied to the measurement of measles IgG in human serum. The test results obtained on the present ELISA system were identical to that on the conventional ELISA system. Compared to conventional ELISA system, the present ELISA system shortened the analysis time to half and reduced the amounts of the reagent and sample to one third, respectively.

Keywords: Biosensors, Enzyme Assays

Application Code: Bioanalytical

Methodology Code: Sensors

Session # 210 **Abstract # 210-6** **ACS Poster**

Session Title: ACS Division of Analytical Chemistry Poster Session

Abstract Title: **Applications of a Quartz Crystal Microbalance for Monitoring Bacterial Biofilm Growth and Removal**

Date: Sunday, March 06, 2016 - Afternoon
Time:
Room: A412

Primary Author: Hunter J. Sismaet
Northeastern University

Co-Author(s): Edgar D. Goluch, Pegah N. Abadian

Abstract Text

Biofilms are matrices of extracellular polymeric substance that are readily formed by bacteria, allowing them to adhere to multiple surfaces and grow in the most extreme environments. As a result, highly-resistant bacterial species have emerged in the hospital setting, where they are known to persist on hospital surfaces and contribute to infection transmission. To address this concern, this study focuses on using a quartz crystal microbalance (QCM) to investigate bacterial biofilm growth and removal. Developing platforms and techniques for studying clinically-relevant bacterial species as they adhere and grow on surfaces as well as strategies for their removal from surfaces can provide significant insight in preventing biofilm-associated infections.

Bacterial species [i]Pseudomonas aeruginosa[/i] and [i]Staphylococcus aureus[/i] were grown overnight at 37 °C in lysogeny broth growth media. The bacteria were diluted and loaded into the QCM, where they were allowed to adhere and grow on the gold-coated quartz surface over the course of several hours under low-flow conditions. Real-time changes in both frequency and damping were recorded to monitor bacterial accumulation on the surface over time. Once the bacterial cells were stably attached to the surface, an SDS detergent was loaded into the system to study the effect of biofilm removal. Biofilm removal was determined by the sensor's response as it returned to the original baseline. This study validates the use of a QCM for studying bacterial adhesion and removal from surfaces as well as investigating the effectiveness of biofilm removal agents.

Keywords: Biomedical, Biosensors, Detection, Monitoring
Application Code: Biomedical
Methodology Code: Sensors

Session # 210 Abstract # 210-7 **ACS Poster**

Session Title ACS Division of Analytical Chemistry Poster Session

Abstract Title **A Capillary Electrophoresis Approach to the Characterization and Application of Graphene Quantum Dots as Sensing Agents**

Primary Author Leona Sirkisoon
Wake Forest University

Date: Sunday, March 06, 2016 - Afternoon
Time:
Room: A412

Co-Author(s) Christa Colyer, Honest Makamba, Shingo Saito

Abstract Text

Graphene quantum dots (GQDs) are safe, convenient, and effective alternatives to organic dyes, inorganic quantum dots or metal nanoparticles for molecular sensing. The goal of the present work is to use capillary electrophoresis (CE) to characterize new polymer-modified GQDs (PM-GQDs) and to compare the utility of PM-GQDs to unmodified dots for the separation of analytes by CE. The modified and unmodified dots were synthesized by a 'bottom up' approach involving the carbonization of citric acid as the organic precursor with or without added sodium polyacrylate (PAAS), followed by extraction in water or an NaOH solution. Preliminary characterization experiments revealed differences in excitation spectra upon polymer modification (with □[sub]ex,max[/sub] = 345 nm for unmodified GQDs and 365 nm for PM-GQDs), although □[sub]em,max[/sub] was relatively unaffected. Whereas the emission intensities of PM-GQDs and GQD samples were comparable in aqueous solution, the PM-GQD emission intensity was quenched in acetonitrile. To demonstrate the applicability of GQDs as separation adjuvants, we analyzed a mixture of holo (metallated) and apo (demetallated) forms of transferrin (Tf, an iron transport protein). In the absence of GQDs, the proteins were not resolved by a simple CE method; however, upon addition of GQDs to the separation buffer, multiple forms of Tf were resolved. Additionally, enhanced fluorescence of the GQDs was observed upon the addition of increasing quantities of Tf, allowing for improved sensitivity and selectivity. It is anticipated that the PM-GQDs will interact differentially with metallated and demetallated Tf or other metalloprotein targets, facilitating further analytical method development.

Keywords: Characterization, Electrophoresis, Fluorescence
Application Code: Process Analytical Chemistry
Methodology Code: Capillary Electrophoresis

Session # 210 **Abstract # 210-8** **ACS Poster**

Session Title: ACS Division of Analytical Chemistry Poster Session

Abstract Title: **A pH Sensor for Non-Invasive In Vivo Detection and Imaging of Implant Associated Infection**

Date: Sunday, March 06, 2016 - Afternoon

Time:

Room: A412

Primary Author: Unaiza Uzair
Clemson University

Co-Author(s)

Abstract Text

We develop a pH sensor based on use of X-ray Excited Chemical Luminescence Imaging (XECLI) to non-invasively detect and image changes in pH of a surface with high spatial and pH resolution while minimizing tissue scattering effects. Diagnosis and treatment of infections associated with implanted medical devices is a challenge, as clinical symptoms of implant associated infection are often delayed and can sometimes be completely absent till infection reaches a later stage. Early diagnosis of implant associated infection and non-invasive continuous monitoring of infection to evaluate eradication and success of treatment has not been established yet. Treatment of implant infection without implant removal is possible if infection can be diagnosed at its onset. Our pH sensor can be attached to the implant surface to non-invasively diagnose and monitor implant associated infection in situ. Bacteria and inflammatory responses cause a pH drop in the area and pH shifts to acidic from in situ pH (~ 7.3). Our pH sensor consists of a layered structure of a pH sensitive polymer film over radioluminescent particles. The pH sensor is characterized for reversibility, sensitivity and resolution. XECLI provides high spatial resolution images mainly limited by X-ray beam width with minimum increase from X-ray scattering in the tissue. It allows point by point mapping of the surface with minimum background. We studied pH changes during the formation of biofilm on the pH sensitive sensor film. In summary, our sensor provides a novel approach to non-invasively image surface pH to diagnose implant infection and assess treatment.

Acknowledgements:This research was supported by NSF CAREER award CHE1255535, a Fulbright Scholarship award to Unaiza Uzair, and animal studies funded through SC Bioengineering Center of Regeneration and Formation of Tissues (SCBioCRAFT) under NIH grant R15EB014560-01A1

Keywords: Bioanalytical, Chemiluminescence, Imaging, Sensors
Application Code: Bioanalytical
Methodology Code: Sensors

Session # 210 **Abstract # 210-9** **ACS Poster**

Session Title ACS Division of Analytical Chemistry Poster Session

Abstract Title **Scanning Electrospray Microscopy with Nanopipettes**

Primary Author Elizabeth M. Yuill
Indiana University

Date: Sunday, March 06, 2016 - Afternoon
Time:
Room: A412

Co-Author(s) Lane A. Baker, Wenqing Shi

Abstract Text

Electrospray from nanopipettes is used to realize a mode of scanning probe microscopy introduced here as scanning electrospray microscopy (SESM). Nanopipettes, or pulled glass capillaries with inner diameters ranging from tens to hundreds of nanometers, have been recently developed and characterized as emitters for electrospray ionization mass spectrometry. Further characterization of electrospray current from nanopipettes confirmed a well-known distance-dependent relationship, which has been utilized here as a feedback mechanism for SESM. SESM provides an ambient, non-contact method to investigate surface topography. Images can be generated over both conductive and insulative features and feature size agrees well with atomic force microscopy of the same substrate. As a consequence of electrospray, scanning a surface with SESM results in deposition of salt solution on the substrate. This controlled deposition presents opportunities for application in the realm of electrodeposition, nanofabrication, and electrospinning, and will be further investigated.

Keywords: Electrospray, Imaging, Method Development, Microscopy
Application Code: Bioanalytical
Methodology Code: Surface Analysis/Imaging

Session # 210 **Abstract # 210-10** **ACS Poster**

Session Title: ACS Division of Analytical Chemistry Poster Session

Abstract Title: **Phosphoproteomics Studies of Human Immunodeficiency Virus-1**

Primary Author: Kevin Mark
LaGuardia Community College City University of NY

Date: Sunday, March 06, 2016 - Afternoon
Time:
Room: A412

Co-Author(s): Emmanuel Chang, Hsin-Pin Ho, Mathias Lichterfeld, Pratikkumar Rathod, Xu Yu

Abstract Text

Protein phosphorylation is a ubiquitous post-translational modification involved in proliferation,signal transduction, energy metabolism and other key biochemical pathways. The hijacking of host cell kinases by viruses represents a powerful means to reduce the complexity of viral genomes, eliminating the need for virus- encoded kinases and allowing individual viral polypeptides to exist in diverse functional states. The hypothesis, supported by preliminary data, suggests that host cell kinase phosphorylation of HIV proteins drives processes critical to viral propagation, and that an abundance of such phosphorylation sites in HIV remain uncharacterized. In response to this need for comprehensive HIV determination, we propose a top-down strategy to unearth the entire complement of HIV phosphorylation sites, to discover the kinases associated with each phosphorylation site, and to generate and test hypotheses for the function of novel sites.

We have determined more than 25 different potential phosphorylation sites for different HIV-proteins including Gag, Pol, Env, Nef, Rev, Tat, Vif, Vpr and Vpu using phosphorylation prediction algorithms such as NetPhosK, GPS and KinasePhos. Based on high scores from multiple prediction algorithms, and sequence conservation we have evaluated 14 different kinases including Cyclin-Dependant Kinase (CDK2/Cyclin A, CDK2/Cyclin E, CDK1/Cyclin B, CDK4/Cyclin D, CDK9/Cyclin T1 and CDK7/Cyclin H), Protein Kinase C (alpha, beta and delta), Protein Kinase A, Casein Kinase I and II, MAPK and GSK3 in this study. Synthetic HIV-peptides corresponding to bioinformatics prediction were chosen to perform kinase assay and to evaluate in vitro phosphorylation of the predicted sites using MALDI mass spectrometry.Our preliminary data shows that there are 11 potential phosphorylation sites predicted using phosphorylation prediction algorithms of HIV-protein, Vif (Figure 1).

Keywords: Bioanalytical, Informatics, Mass Spectrometry, Peptides
Application Code: Bioanalytical
Methodology Code: Mass Spectrometry

Session # 210 **Abstract # 210-11** **ACS Poster**

Session Title: ACS Division of Analytical Chemistry Poster Session

Abstract Title: **Segmented Flow Sampling with Theta Pipettes**

Primary Author: Anumita Saha-Shah
Indiana University

Date: Sunday, March 06, 2016 - Afternoon
Time:
Room: A412

Co-Author(s): Curtis M. Green, Lane A. Baker

Abstract Text

Nano- and micro pipettes have been widely used as probes for local electrochemical analysis and manipulation of samples at small scales. In this report, we have used theta pipettes as a mobile microfluidic device that can capture chemical information in sub-nanoliter plugs and delivers them to a mass spectrometer for dilution free analysis of molecular signals. In this method, multiple analyte plugs were sandwiched between immiscible fluorous (perfluorodecalin (PFD)) solvent phases to minimize dispersion of analyte. The pipette also served as an electrospray emitter enabling fast and convenient analysis of segments as small as few hundreds of picoliters. To our knowledge these segment are the smallest segments subjected to ESI-MS analysis. Small pipette tip size and flow rates in the order of 5-20 nL/min enables ESI-MS analysis of such small segments. These approaches may find application in mapping local reactivity of a surface, creation of chemical maps, and performing single cell analysis. Further, a novel strategy to regularize and automate the segmented flow sampling will be discussed and application of the new method in monitoring local chemical changes will be demonstrated.

Keywords: Biological Samples, Lab-on-a-Chip/Microfluidics, Mass Spectrometry, Sampling

Application Code: Bioanalytical

Methodology Code: Sampling and Sample Preparation

Session # 210 **Abstract # 210-12** **ACS Poster**

Session Title: ACS Division of Analytical Chemistry Poster Session

Abstract Title: **Mapping Local Permeability Change During Degradation with Scanning Ion Conductance Microscopy-Scanning Electrochemical Microscopy (SICM-SECM)**

Primary Author: Wenqing Shi
Indiana University

Co-Author(s): Lane A. Baker

Date: Sunday, March 06, 2016 - Afternoon
Time:
Room: A412

Abstract Text

In this work, accelerated aging experiments of Nafion® 212 (N212) membrane were carried out using Fenton's reagent under a series of degradation conditions. Morphological change after degradation was examined by scanning electron microscopy (SEM) analyses. X-ray photoelectron spectroscopy (XPS) mapping was performed to further investigate the heterogeneity of chemical composition in the degraded membrane samples. Furthermore, a hybrid scanning ion conductance-scanning electrochemical microscopy (SICM-SECM) technique was developed and further employed to study the heterogeneous local permeability of the degraded membranes. Results obtained from the aforementioned SEM analyses, XPS mapping and SICM-SECM measurements in combination can be utilized to gain insights of the membrane degradation mechanism, which can further provide important guidelines to improve membrane stability and further increase the lifetime of the PEMFCs.

Keywords: Imaging, Materials Characterization, Membrane, Surface Analysis
Application Code: Material Science
Methodology Code: Chemical Methods

Session # 210 Abstract # 210-13 **ACS Poster**

Session Title ACS Division of Analytical Chemistry Poster Session

Abstract Title **Profiling N, N'-Dibutylbenzimidazolium Salt and Its Derivatives Using CYCLIC Voltammetry**

Date: Sunday, March 06, 2016 - Afternoon
Time:
Room: A412

Primary Author Huggins Z. Msimanga
Kennesaw State University

Co-Author(s) Andrew Montalvo, Daniela Tapu

Abstract Text

Carbenes are molecules that contain a neutral carbon atom with a valence of two and two unshared electrons. Stable nucleophilic carbenes are actively being studied due to their role as ligands in homogeneous catalysis. They have high affinity for a wide range of main group elements and transition metals. Novel o-phenylenediamine-fused imidazolium salts have been successfully synthesized and characterized in Dr. Daniela Tapu's laboratory at Kennesaw State University. These salts are used as precursors in the synthesis of the corresponding N-heterocyclic carbenes. o-Phenylenediamine-fused imidazol-2-ylidenes are a particularly attractive class of ligands. The o-Phenylenediamine moiety is electroactive. Fusion of the phenylenediamine with the imidazol-2-ylidenes may alter the electronic properties of the resulting carbine center. Those properties may in turn be used to fine-tune the carbene center and offer the possibility to control the stoichiometric and catalytic reactivity of the corresponding transition metal complexes. To this end, the electronic properties of 1,3-dibutylbenz-imidazolium iodide and its derivatives where amine and nitro-groups were substituted at the o-position of the benzene moiety, were studied using cyclic voltammetry. Voltammograms were acquired using a three-electrode electrochemical cell consisting of a silver/silver chloride reference, glassy carbon working, and a platinum wire auxiliary electrodes. Dry dichloromethane was used in 0.1 M [Bu4N][PF6] as the support electrolyte. Without the amine and nitro substituents, N, N'-dibutylbenzimidazolium salt showed a quasi-reversible behavior around 587 mV, and irreversibility around -350 mV. Of the substituents, 5,6-damine-1,3-dibutylbenzimidazolium salt shifted the first reduction potential from 587 mV to 955 mV. The nitro-substituted salt did not cause much change from 587 mV. O-Phenylenediamine-fused imidazolium salts thus increase the nucleophilic properties of these salts.

Keywords: Electrochemistry, Electrode Surfaces, Metals, Voltammetry

Application Code: Other

Methodology Code: Electrochemistry

Session # 210 **Abstract # 210-14** **ACS Poster**

Session Title: ACS Division of Analytical Chemistry Poster Session

Abstract Title: **Comparative Study of Elemental Nutrients in Organic and Conventional Vegetables by Laser Induced Breakdown Spectroscopy (LIBS)**

Date: Sunday, March 06, 2016 - Afternoon

Time:

Room: A412

Primary Author: Chet R. Bhatt
Mississippi State University

Co-Author(s)

Abstract Text

In this study, LIBS technique is used to compare the presence of major nutrient elements (Ca, Na, K, and Mg) in organic and conventional vegetables. Different parts of Cauliflower and Broccoli were used as working samples. Optimum values of laser energy, gate delay, and gate width were used to acquire the LIBS spectra from those samples. The intensity ratios of different elemental lines present on the similar location of the organic Cauliflower and Broccoli samples were compared with those of conventional ones. The intensity ratio of elemental lines between different parts of the Cauliflower and Broccoli are also compared. The detailed analysis of the elemental nutrients in Cauliflower and Broccoli will be presented.

Keywords: Atomic Emission Spectroscopy, Spectroscopy
Application Code: Material Science
Methodology Code: Atomic Spectroscopy/Elemental Analysis

Session # 210 **Abstract # 210-15** **ACS Poster**

Session Title: ACS Division of Analytical Chemistry Poster Session

Abstract Title: **Photoacoustic Spectroscopy with SF6, An Optically Thick Greenhouse Gas**

Date: Sunday, March 06, 2016 - Afternoon
Time:
Room: A412

Primary Author: Han Park
University of Tennessee at Chattanooga

Co-Author(s)

Abstract Text

Photoacoustic spectroscopy was used to test the photoacoustic properties of sulfur hexafluoride, an optically thick and potent greenhouse gas. Detection of trace amounts of the gas was also implemented. The conditions in which the gas was tested, gas cell length, temperature, concentration, and power of the laser, were varied in order to determine their effect on the photoacoustic signal, and an ideal condition to detect trace gas amounts. A detection limit of 2.86 ppb was determined for SF6.

Keywords: Detection, Environmental/Air, Gas, Spectroscopy
Application Code: Environmental
Methodology Code: Physical Measurements

Session # 210 **Abstract # 210-16** **ACS Poster**

Session Title: ACS Division of Analytical Chemistry Poster Session

Abstract Title: **Integrating Authentic Research Experiences into Undergraduate Analytical Chemistry**

Date: Sunday, March 06, 2016 - Afternoon
Time:
Room: A412

Primary Author: Fei Yan
North Carolina Central University

Co-Author(s): Melissa Kerr

Abstract Text

Course-based undergraduate research experiences (CUREs) are garnering increasing attention as one of the desirable pedagogical strategies to facilitate science learning in college. The laboratory components of two undergraduate chemistry courses, namely Quantitative Chemical Analysis and Instrumental Analysis, have been revamped to incorporate CUREs into the curricula, thereby offering students an opportunity to explore problems and challenges that have real-world applications. Traditional cookbook-type laboratory exercises are integrated with student-initiated research projects that have the elements of scientific inquiry. Each CUREs project involves identifying a problem, designing a study, locating an appropriate method via literature search, collecting samples, measuring variables, analyzing data, and presenting the results in a formal report and an oral presentation. Our evaluation of the preliminary impacts of CUREs integration suggests that the majority of students become more excited about careers in chemistry, and science in general, as an outcome of this exercise.

Acknowledgements: This work was partially supported by the National Science Foundation (NSF) Grant No. #1137462 (FY), the Department of Education Grant No. P120A150022(PD: C. R. Jackson), and a DREAM STEM Mini-Grant to FY under the NSF Grant No. #1238547(PI: C. R. Jackson).

Keywords: Detection, Education, Environmental Analysis, Forensics
Application Code: General Interest
Methodology Code: Education/Teaching

Session # 210 Abstract # 210-17 **ACS Poster**

Session Title ACS Division of Analytical Chemistry Poster Session

Abstract Title **Proteomic Analysis of Tetrahymena Thermophila Using MALDI-TOF/TOF**

Date: Sunday, March 06, 2016 - Afternoon
Time:
Room: A412

Primary Author Douglas Beussman
St. Olaf College

Co-Author(s) Harrison VanDolah

Abstract Text

Matrix-assisted laser desorption/ionization mass spectrometry (MALDI-MS) has become an invaluable tool in modern proteomics research, with broad applications in medicine and at the crossroads of biology and chemistry. We used MALDI coupled with a tandem Time-of-flight (TOF/TOF) mass spectrometer to identify several proteins isolated from Tetrahymena thermophila, a model single-cell organism. Proteins were separated using SDS-PAGE with individual protein bands of interest excised. Bands were reduced and alkylated and steps were taken to digest the protein into smaller peptides using Trypsin. After cleanup, the peptides were then mixed with □-Cyano-4-hydroxycinnamic acid, spotted on the metal target, and allowed to dry. After MALDI-MS analysis, the resulting peptide masses were compared with theoretical spectra using BioTools and the online MASCOT database. Matched proteins were analyzed further by tandem mass spectrometry of specific peaks in order to unambiguously determine their identities, again using BioTools and MASCOT. We were able to identify proteins in 28 different samples, finding evidence of 11 unique proteins.

Keywords: Mass Spectrometry, Proteomics, Tandem Mass Spec, Time of Flight MS

Application Code: Genomics, Proteomics and Other 'Omics

Methodology Code: Mass Spectrometry

Session # 210 Abstract # 210-18 **ACS Poster**

Session Title ACS Division of Analytical Chemistry Poster Session

Abstract Title **Possible Age and Location Differences in Human Scent Profiles**

Date: Sunday, March 06, 2016 - Afternoon

Time:

Room: A412

Primary Author Douglas Beussman
St. Olaf College

Co-Author(s) Laura Muehlbauer

Abstract Text

Using analytical instrumentation to study differentiable human scent profiles is of forensic interest because many applications exist, especially the validation of canine evidence in criminal investigations. A scent profile is composed of the volatile organic compounds (VOCs) that are released by dead epithelial cells, bacterial cells, and eccrine and sebaceous gland excretions. These VOCs can be present in varying combinations and abundances between people, allowing dogs to track a specific scent among numerous profiles. We used GC-MS and cryfocusing headspace preconcentration to identify VOCs from the epithelial cells of the unwashed hands of fifty-eight subjects, ages 0-82. The samples were sealed in glass jars for 12-24 hours to allow the VOCs to accumulate in the headspace. The relative amount of ethanol was found to be a potential marker for age and/or gender, as the average amount of ethanol decreased as the age of male donors increased, with young boys under the age of 16 having a higher average amount than any other age or gender group. A general decreasing trend was observed with female donors. Another potential age marker could be 6-methyl-5-hepten-2-one, with very small amounts in young donors and a sharp increase in amounts from donors in the middle range of ages. Additionally, the samples were divided into six groups from different geographic locations, and certain trends emerged with the presence or absence of particular compounds. These data suggest that age, gender, and location might have an effect on scent profiles and aid in the identification of possible suspects.

Keywords: Forensics, Gas Chromatography/Mass Spectrometry, Trace Analysis, Volatile Organic Compounds
Application Code: Homeland Security/Forensics
Methodology Code: Gas Chromatography/Mass Spectrometry

Session # 210 Abstract # 210-19 **ACS Poster**

Session Title: ACS Division of Analytical Chemistry Poster Session

Abstract Title: **LC-QTOF Detection of Methamphetamine in Packaging Residue**

Date: Sunday, March 06, 2016 - Afternoon
Time:
Room: A412

Primary Author: Douglas Beussman
St. Olaf College

Co-Author(s): Matthew Bock

Abstract Text

When a bag is seized for suspected drug paraphernalia, there might only be a powder residue remaining on the inside surface. While analysis via GC/MS is appropriate in many circumstances, the residue might not have enough powder to examine. Furthermore, the residue might not be volatile, requiring derivatization prior to analysis. Liquid Chromatography/Mass Spectrometry (LC/MS) is one possible solution to these problems because the residue can be readily dissolved in a liquid and analyzed directly. This research attempted to determine the ability of LC/MS to detect drug residues of adulterated methamphetamine and quantify the results. Methamphetamine was mixed with acetaminophen, caffeine, dextrose, lactose, lidocaine, mannitol, phenobarbital, procaine, or sucrose at various concentrations. Residues were obtained by putting the adulterated methamphetamine into 1x1 bags and then dumping them. The residues were dissolved in water and analyzed via LC/MS to determine limits of detection and quantitation. Furthermore, tandem MS/MS was used to further support the identification of methamphetamine residues. Through this method, single-adulterated methamphetamine residues could be detected at concentrations of 0.15% drug or lower, well below levels found in actual street samples.

Keywords: Drugs, Forensics, Liquid Chromatography/Mass Spectroscopy, Tandem Mass Spec
Application Code: Homeland Security/Forensics
Methodology Code: Liquid Chromatography/Mass Spectrometry

Session # 210 **Abstract # 210-20** **ACS Poster**

Session Title: ACS Division of Analytical Chemistry Poster Session

Abstract Title: **Environmental Toxins from Decorative Candles**

Date: Sunday, March 06, 2016 - Afternoon
Time:
Room: A412

Primary Author: Douglas Beussman
St. Olaf College

Co-Author(s): Allison Sager

Abstract Text

Candles have been around for millennia, and still are a staple of daily life today. Companies who produce candles today claim that candles made of certain types of material will burn differently or burn 'cleaner' than other types. They claim certain candles produce a greater amount of toxic chemical compounds when they burn. There has been very little research done to test these claims. Understanding the differences in chemical output of different types of unscented candles is important for both human and environmental health. Our procedure used a cryogenic headspace preconcentrator and GC/MS to identify volatile organic compounds (VOCs) emitted from burning candles. Samples were collected from fourteen candles of five different types: beeswax, gel, palm oil, paraffin, and soywax. Candles were burned for approximately 1 hour to collect the VOCs in 500-600 mL headspace containers. VOCs were identified by comparison of the mass spectra with a database. The VOCs were divided into eleven chemical categories and studied. Candle types were evaluated based on the relative abundance of the eleven classes and on the number of distinct VOCs found in each class for each candle. Results show that gel candles produce the largest number of distinct alkenes. In addition, soywax candles produced, overall, a smaller amount of VOCs than other candle types. Paraffin and palm oil candles were found to produce the largest amount of VOCs with known LD50 values. Further research is needed on the toxicity implications to humans at these concentrations.

Keywords: Environmental/Air, Gas Chromatography/Mass Spectrometry, Trace Analysis, Volatile Organic Comp
Application Code: Environmental
Methodology Code: Gas Chromatography/Mass Spectrometry

Session # 210 Abstract # 210-21 **ACS Poster**

Session Title: ACS Division of Analytical Chemistry Poster Session

Abstract Title: **Analysis of Fibers Via Isotope Ratio Mass Spectrometry**

Date: Sunday, March 06, 2016 - Afternoon
Time:
Room: A412

Primary Author: Douglas Beussman, St. Olaf College

Co-Author(s): Elaine Macon, Hannah Brown

Abstract Text

Cloth fibers found at the scene of a crime can be used as circumstantial evidence in a court of law. Current methodology can only determine the color and chemical composition of the sample, giving limited information about its origins. Isotope ratio mass spectrometry (IRMS), however, can be used to analyze the ratio of carbon, hydrogen, nitrogen, and oxygen isotopes, further differentiating fibers of the same color and chemical composition. For the purpose of this study, carpet, synthetic fabrics, and blood stained fibers were analyzed using IRMS. Additionally, the lower limit of detection for this analysis method was studied using plain white cotton and denim fibers.

The synthetic fibers studied all had unique isotope ratios that differentiate them from fibers of similar color and composition. These unique isotope ratios likely resulted from differing petroleum sources and purchasing years, although all fibers were obtained from the same manufacturer. Similarly, all of the carpet fibers studied had unique and distinguishable isotope ratios. Blood stains were found to have a statistically significant impact on the isotope ratios of threads. With respect to the investigation of limit of detection, fibers weighing more than ~50 micrograms possessed statistically similar isotope ratios, while at masses smaller than ~50 micrograms the isotope ratios became differentiable. This research enhances IRMS analysis by expanding differentiation into synthetic fibers, including carpet, exploring the effects of blood stains on the isotope ratios of fibers, and preliminarily identifying the limit of detection of IRMS to ~50 micrograms.

Keywords: Forensics, Isotope Ratio MS
Application Code: Homeland Security/Forensics
Methodology Code: Mass Spectrometry

Session # 210 **Abstract # 210-23** **ACS Poster**

Session Title: ACS Division of Analytical Chemistry Poster Session

Abstract Title: **HPLC Method Development for Analysis of Dissolution Samples of a Highly Soluble Drug Substance in a Hydrophobic Matrix**

Date: Sunday, March 06, 2016 - Afternoon
Time:
Room: A412

Primary Author: Matthew Loucks
Banner Life Sciences

Co-Author(s): Angela Moore, Sara Draper, Wayne Craig

Abstract Text

High Pressure Liquid Chromatography (HPLC) for Dissolution testing is a critical component to development, as well as, stability of a manufactured drug product.

Dissolutions incorporating hydrophilic matrices dissolve more readily in water, aiding in drug substance solubility. However, drug products formulated in hydrophobic matrices lack an affinity for water, inhibiting solubility. Dissolution medium additives (surfactants) lower the surface tension between the matrix and medium, aiding in solubility. The solubilized solution is measured to quantify the amount of drug substance present.

Surfactants can be problematic for sample analysis. An HPLC method showed an unknown extraneous peak, resolved from the peak of interest, but losing resolution with each proceeding injection (regardless of blank, standard or sample). Throughout the run, the extraneous peak would elute later and later, eventually after the peak of interest, which appeared to be losing retention itself.

Initial method conditions were isocratic using a high flow rate (1.5 mL/min), high column temperature (60°C) and a C8 column. Experiments were conducted to determine the extraneous peak source, as well as, resolution and retention loss. Dissolution medium injections showed the extraneous peak, and the shifting retention times. Differing isocratic compositions, surfactant in buffer, and a gradient profile using differing flow rates, various column temperatures, and differing columns were evaluated. The extraneous peak and retention shifting was determined to be from a surfactant necessary for sample dissolution.

A new method using different mobile phases and gradient elution, flow rate, column, and column temperature, eliminated the shift in retention for both peaks and provided consistent method specificity and analytical chromatography.

Keywords: Dissolution, Liquid Chromatography, Surfactants
Application Code: Pharmaceutical
Methodology Code: Liquid Chromatography

Session # 210 Abstract # 210-24 **ACS Poster**

Session Title ACS Division of Analytical Chemistry Poster Session

Abstract Title **Confocal Raman Microscopy Investigation of the Kinetic Barrier to PAH Partitioning into Individual C[sub]18[/sub]–Silica Particles**

Date: Sunday, March 06, 2016 - Afternoon
Time:
Room: A412

Primary Author David A. Bryce
University of Utah

Co-Author(s) Jay P. Kitt, Joel M. Harris

Abstract Text

Porous silica supports are employed in many applications in chemistry, including stationary phase supports in chromatography and solid phase extraction (SPE). Recently, confocal Raman microscopy was employed as an in-situ detection method within individual C[sub]18[/sub]-functionalized porous silica particles, bringing SPE pre-concentration methods to the femtoliter volume scale. Here we report on efforts to interrogate the accumulation kinetics which control partitioning of polycyclic aromatic hydrocarbons into individual chromatographic particles under high retention conditions. Specifically, accumulation of 2µM pyrene was measured into particles of radii from 2.2 - 7.3µm from 20% methanol: 80% water solution. Strong partitioning of pyrene leads to 30 mM concentrations within the C[sub]18[/sub]-silica particle representing a 15,000-fold concentration gain relative to the source phase. The observed accumulation kinetics are consistent with a model where the rate-limiting step is partitioning of the analyte in a boundary layer at the particle exterior. The accumulation rate is proportional to 1/r, governed by flux across the particle boundary (4□r[sup]2[/sup]) divided by the volume of the collector (4/3□r[sup]3[/sup]). This model allows determination of the heterogeneous rate constant for partitioning into the particle, 4.8±0.1x10[sup]-3[/sup]cm/sec. For comparison, the partitioning rate of pyrene onto planar C18 surface was also determined, 2.6±0.4x10[sup]-4[/sup] cm/sec, which is 20-times slower than the capture rate of a porous particle due to the higher surface area and greater collision frequency of molecules at the porous particle-solution interface.

Keywords: Chromatography, Microspectroscopy, Raman, Solid Phase Extraction
Application Code: Material Science
Methodology Code: Molecular Spectroscopy

Session # 210 **Abstract # 210-25** **ACS Poster**

Session Title: ACS Division of Analytical Chemistry Poster Session

Abstract Title: **Development of a Flow-Based Electrogenerated Chemiluminescent Detection of Biogenic Amines on a Microfluidic Chip**

Primary Author: Erin Gross
Creighton University

Date: Sunday, March 06, 2016 - Afternoon
Time:
Room: A412

Co-Author(s): Charles Henry, Emily Lowry, John Wydallis, Leah Schaffer, Rachel M. Feeny

Abstract Text

Carbon paste microelectrodes electrodes are easily fabricated, inexpensive and can be used with electroanalytical-based microfluidic devices. In this work, the electrochemiluminescent (ECL) reaction between tris(2,2'-bipyridyl)ruthenium(II) and biogenic amines was used to detect these amines in a microfluidic flow system. The ability to measure these biogenic amines is important for food safety applications. The linear range and optimal parameters for the reaction were determined. The ECL reaction was able to detect spermine, spermidine, and putrescine with detection limits of 1.8, 4.3 and 28 micromolar, respectively. A milk sample was spiked with spermine and the method was able to detect spermine in the sample. Work was initiated to develop a microchip capillary electrophoresis separation of the amines with ECL detection.

Keywords: Bioanalytical, Chemiluminescence, Electrochemistry, Food Safety
Application Code: Bioanalytical
Methodology Code: Electrochemistry

Session # 220 Abstract # 220-1 **Awards**

Session Title: Chromatography Forum of Delaware Valley Dal Nogare Award

Abstract Title: **Opportunities and Challenges with Capillary Liquid Chromatography**

Primary Author: Stephen Weber
University of Pittsburgh

Date: Monday, March 07, 2016 - Morning
Time: 08:40 AM
Room: B312

Co-Author(s)

Abstract Text

Capillary liquid chromatography (cLC) is attractive, in principle, for its small dimensions. The small diameter permits the analysis of small samples and provides an opportunity for on-column concentration of solutes in larger volumes of injected samples. The small diameter favors thermal equilibration across the column which permits high mobile phase velocities without suffering from the bandspreading that radial thermal gradients create. The lower flow rates used decrease significantly the cost and environmental burden of liquid chromatography. However, cLC does have limitations. Extracolumn volume plays a more significant role in bandspreading in cLC in comparison to standard LC.

Because the column diameter is a critical variable in defining both the benefits and the problems of cLC, we have created an optimization approach for maximizing sensitivity with a minimum number of theoretical plates and a defined separation time. This has been applied to the determination of neurotransmitters by online analysis of microdialysate once per minute. We have also attacked the extracolumn dispersion and volume overload problems by using temperature-assisted solute focusing.

Keywords: Bioanalytical, Capillary LC, Neurochemistry, Temperature
Application Code: Bioanalytical
Methodology Code: Liquid Chromatography

Session # 220 Abstract # 220-2 **Awards**

Session Title: Chromatography Forum of Delaware Valley Dal Nogare Award

Abstract Title: **How Much Performance is Lost in LC by Optimizing Only Velocity and Column Length Using Limited Number of Particle Size?**

Primary Author: Peter W. Carr
University of Minnesota

Date: Monday, March 07, 2016 - Morning
Time: 09:15 AM
Room: B312

Co-Author(s): Adam Matula, Dwight R. Stoll

Abstract Text

It has been known for over 40 years that the maximum possible plate count per unit time in LC is obtained by simultaneously adjusting the particle size, column length and eluent velocity. We refer to this as a three-parameter optimization or Knox-Saleem optimization. However, particle size is not a continuous variable and thus performance optimization is done by picking an appropriate particle size for the timescale of the separation and then adjusting the column length and eluent velocity. This optimization is referred to here as two-parameter optimization or Poppe optimization. Of course, the column length is not a continuous variable but one can achieve column lengths within 1 cm of the optimum by combining commercially available column lengths. This issue is only a problem for extremely fast analyses. Thus the question naturally arises as to how much performance is lost due to the availability of only a limited number of different particle sizes. We have solved this problem by observing that the maximum loss in performance occurs at the crossing point of the two curves which define the optimized plate count as a function of column dead time for the two different size particles. One can solve the mathematical problem of finding the appropriate column dead time, column length and eluent velocity at the crossover point. Our work shows that provided one changes from the smaller to the larger particle size at the column dead time the maximum loss in plate count with currently available particle sizes amounts to less than 11%. Interestingly, this loss in performance is independent of the dynamic characteristics of the column, the operating pressure, the diffusion coefficients of the analyte and the operating temperature. It depends only on the ratio of the two particle sizes. Since resolving power varies with the square root of the plate count this loss in performance is essentially insignificant.

Keywords: Liquid Chromatography
Application Code: General Interest
Methodology Code: Liquid Chromatography

Session # 220 Abstract # 220-3 **Awards**

Session Title: Chromatography Forum of Delaware Valley Dal Nogare Award

Abstract Title: **UHPLC-MS for Multiplexed Neurochemical Analysis**

Date: Monday, March 07, 2016 - Morning
Time: 09:50 AM
Room: B312

Primary Author: Robert T. Kennedy
University of Michigan

Co-Author(s)

Abstract Text

Metabolomics based on HPLC-MS has become a powerful tool for investigating biological systems. We have applied metabolomics to the brain by coupling sampling probes to assays for neurotransmitters and peptides. Such measurements require ability to retain and resolve many, often polar compounds. We have developed methods based on derivatization of neurotransmitters and metabolites to enhance retention and single in LC-MS assays. For peptides we have developed improved sampling approaches that improve sensitivity by over 100-fold based on reducing adsorption. Finally, we are investigating advanced ultra high pressure LC systems to improve metabolomic coverage for neurochemistry.

Keywords: Liquid Chromatography, Neurochemistry, Sampling
Application Code: Bioanalytical
Methodology Code: Liquid Chromatography/Mass Spectrometry

Session # 220 **Abstract # 220-4** **Awards**

Session Title: Chromatography Forum of Delaware Valley Dal Nogare Award

Abstract Title: **Integrated Microfluidic Separations Devices Interfaced to Mass Spectrometry**

Date: Monday, March 07, 2016 - Morning
Time: 10:40 AM
Room: B312

Primary Author: J Michael Ramsey
University of North Carolina at Chapel Hill

Co-Author(s): Erin Redman, William Black

Abstract Text

The first demonstration of micromachined devices that emulate the functions of laboratory chemical instrumentation, i.e., the silicon gas chromatograph (GC), is now over three decades old. Due largely to the modest performance of these early devices, further developments in MEMS-based chemical instrumentation were slow to materialize. Microfabricated fluidic devices that accomplished chemical measurement strategies were first reported in the early 1990s and have several demonstrated significant advantages over conventional approaches. Since that time there has been rapidly growing interest in microfabricated fluidic devices (microchips). The diversity of biochemical measurement techniques that have been implemented on microchips includes various electrophoretic and chromatographic separations, chemical and enzymatic reactions, noncovalent recognition interactions, sample concentration enhancement, and cellular manipulations. In addition, the types of samples addressed by microchips has been broad in scope, e.g., small ions and molecules, single and double stranded DNA, amino acids, peptides, and proteins. These devices have low cost and small footprints while consuming miniscule quantities of reagents and can rapidly produce precise results. All of these features suggest the possibility to perform chemical and biochemical experimentation on a massive scale at low cost on a bench top, a goal being pursued by many laboratories around the world. We will discuss microfluidic devices that provide state-of-the-art separations performance while efficiently coupled to mass spectrometry through integrated nano-electrospray ionization. In addition, the ability to integrate sample processing functionalities on these devices will also be discussed.

Keywords: Bioanalytical, Biopharmaceutical, Capillary Electrophoresis, Mass Spectrometry
Application Code: Bioanalytical
Methodology Code: Microfluidics/Lab-on-a-Chip

Session # 220 Abstract # 220-5 **Awards**

Session Title: Chromatography Forum of Delaware Valley Dal Nogare Award

Abstract Title: **The Second Dimension is a Strange Place - Fundamental Aspects of the Optimization of the Second Dimension in Two-dimensional Liquid Chromatography**

Primary Author: Dwight R. Stoll
Gustavus Adolphus College

Date: Monday, March 07, 2016 - Morning
Time: 11:15 AM
Room: B312

Co-Author(s): Abraham Lenhoff, Alex Wilson, Carston Dammann, David C. Harmes, Eli Larson, John Halvorson, Monika Dittmann, Sarah C. Rutan

Abstract Text

Literature on the optimization of online two-dimensional liquid chromatography (2D-LC) points to the need for conditions in the second dimension of the system that enable much faster (10 to 100-fold faster) separations compared to the separation speed in the first dimension. This theoretical perspective prompts us to think about separation speeds (e.g., < 20 seconds) that are rare in other contexts where liquid chromatography is used. Modern instrumentation and column technologies allow us to realize such high speed separations in practice, with sometimes impressively high performance, however the characteristics of the resulting chromatograms place extraordinary demands on the instrument if we are to avoid serious performance losses due to peak dispersion outside of the column. Specifically, as separation times become short, the optimal particle size and column length also become very small, and these factors lead to peaks that are very narrow in time, and have small volumes. To make matters even more difficult, in the context of 2D-LC we typically want to inject volumes of first dimension column effluent that are on the order of the second dimension column volume to enable sensitive detection in the second dimension.

In this presentation we will discuss our efforts to address these issues, using both experiments and theory and simulations. First, we will describe work aimed at modelling pressure drops in narrow capillaries that we would like to use in the second dimension with high eluent velocities. Second, we will describe the development of a numerical approach to modelling large injections of first dimension column effluent into short, narrow second dimension columns. We will summarize the findings of set of survey simulations aimed at finding optimal combinations of particle size, and column length and diameter for use with very fast second dimension separations, and compare these findings with the results of selected experiments under similar conditions.

Keywords: Food Science, HPLC, Optimization, Pharmaceutical
Application Code: Pharmaceutical
Methodology Code: Liquid Chromatography

Session # 230 **Abstract # 230-1** **Awards**

Session Title: The Pittsburgh Conference Achievement Award

Abstract Title: **Exploiting Ionic Liquids, Magnetic Ionic Liquids, and Polymeric Ionic Liquids in Sample Preparation and Multidimensional Gas Chromatography**

Date: Monday, March 07, 2016 - Morning
Time: 08:40 AM
Room: B314

Primary Author: Jared L. Anderson
Iowa State University

Co-Author(s)

Abstract Text

Ionic liquids (ILs) can be designed to exhibit unique properties for their use in a number of applications in analytical chemistry. This talk will focus on the design and synthesis of ILs, magnetic ionic liquids (MILs) and polymeric ionic liquids (PILs) and the use of these materials in a number of applications within sample preparation and multidimensional gas chromatography. A series of monocationic and dicationic ionic liquid (IL)-based stationary phases were evaluated as secondary columns in comprehensive two-dimensional gas chromatography (GC×GC) for the separation of aliphatic hydrocarbons from kerosene. In order to further understand the role that structural features of ILs play on the selectivity of nonpolar analytes, a series of dicationic IL-based stationary phases were evaluated using GC×GC. PIL-based sorbent coatings in solid-phase microextraction (SPME) will be discussed in their use for the extraction of a number of analytes ranging from pesticides, insecticides, and food products within a variety of matrices. The SPME applications involve both GC-SPME and HPLC-SPME. Finally, the ability to perform rapid DNA extraction from biological samples using custom-designed MILs will be presented. MIL solvents exhibit a number of advantages that allow them to be used in downstream DNA analysis including polymerase chain reaction (PCR). The ability to customize components within a PCR buffer permitting the direct analysis of DNA-enriched MIL will be discussed.

Keywords: Nucleic Acids, Sample Preparation, Separation Sciences
Application Code: Bioanalytical
Methodology Code: Separation Sciences

Session # 230 Abstract # 230-2 **Awards**

Session Title The Pittsburgh Conference Achievement Award

Abstract Title **The Impact and Evolution of High Efficiency Chiral and Achiral Separations**

Primary Author Daniel W. Armstrong
University of Texas at Arlington

Date: Monday, March 07, 2016 - Morning
Time: 09:15 AM
Room: B314

Co-Author(s)

Abstract Text

The development of rapid, efficient enantiomeric separations over the last 25 years has had a tremendous impact on several areas of science and technology. Discoveries in just a few academic laboratories changed the way drugs were regulated world-wide and forever altered the pharmaceutical industry. Biological and analytical aspects of closely related compounds and enantiomers will be outlined and discussed in terms of high throughput analysis and limitation of current instrumentation.

Keywords: Chiral Separations, HPLC, HPLC Columns, Supercritical Fluid Chromatography
Application Code: Pharmaceutical
Methodology Code: Liquid Chromatography

Session # 230 Abstract # 230-3 **Awards**

Session Title: The Pittsburgh Conference Achievement Award

Abstract Title: **Use of Metal-Organic Frameworks in Sample Preparation Analytical Schemes: An Overview of Their Performance in Solid-Phase Extraction Modes**

Date: Monday, March 07, 2016 - Morning
Time: 09:50 AM
Room: B314

Primary Author: Verónica Pino
University of La Laguna

Co-Author(s): Ana M. Afonso, Catalina Ruiz-Pérez, Jorge Pasán, Juan H. Ayala, Priscilla Rocío-Bautista

Abstract Text

Metal-organic frameworks (MOFs) are new class of hybrid inorganic-organic micro-porous crystalline materials self-assembled straightforwardly from metal ions with organic linkers via coordination bonds. The availability of various building blocks of metal ions and organic linkers makes it possible to prepare an infinite number of new MOFs with diverse structure, topologies and porosity. The porosity of MOFs spans from ultra-microporous to mesoporous materials, well bridging the gap between the small pore size of zeolites and large pore size of mesoporous silicate materials. MOFs can be prepared with a minimum amount of metal ions and organic ligands while achieving maximum surface areas with predictable, controllable, tailorable and post-modifiable pores and cavities. Owing to their fascinating structures and unusual properties, such as permanent nanoscale porosity, high surface area, good thermostability, and uniform structure cavities, MOFs have attracted tremendous interest, and an enormous amount of MOF structures have been designed and synthesized for various applications such as hydrogen storage, gas separation, catalysis, sensing and bioimaging. Clearly, the use of MOFs as materials in analytical chemistry has started to receive important attention, as novel stationary phases in chromatography or as novel extractant materials for a number of analytes in a variety of samples. The main aim of this work is to give an overview of main uses of MOFs in analytical sample preparation techniques, particularly in dispersive solid-phase extraction approaches.

Keywords: Environmental/Water, GC, HPLC, Sample Preparation
Application Code: Environmental
Methodology Code: Sampling and Sample Preparation

Session # 230 Abstract # 230-4 **Awards**

Session Title: The Pittsburgh Conference Achievement Award

Abstract Title: **Achiral LSER Investigation of HPLC Retention on Cinchona Alkaloid-Based Chiral Stationary Phases**

Date: Monday, March 07, 2016 - Morning
Time: 10:40 AM
Room: B314

Primary Author: Apryll M. Stalcup
Dublin City University

Co-Author(s)

Abstract Text

Cinchona alkaloid-based chiral stationary phases incorporate a variety of potential interaction modalities that have proven particularly effective for the resolution of acidic chiral compounds under reversed phase and polar-ionic phase conditions. In this work, we examine the retention of achiral probe solutes on immobilized quinine stationary phase with under a range of mobile phase conditions in a Linear Solvation Energy Relationship (LSER) analysis to delineate the impact of mobile phase composition on retention. In this study, a modified LSER model that has proven very effective in describing retention on multimodal surface-confined ionic liquid phases will be used to simultaneously model retention of neutral, weakly acidic and weakly basic solutes.

Keywords: Characterization, Chiral Separations, HPLC
Application Code: Other
Methodology Code: Liquid Chromatography

Session # 230 Abstract # 230-5 **Awards**

Session Title: The Pittsburgh Conference Achievement Award

Abstract Title: **Applications of Conductive Ionic Liquids in Chemical Analysis**

Date: Monday, March 07, 2016 - Morning
Time: 11:15 AM
Room: B314

Primary Author: Jon R. Kirchhoff
University of Toledo

Co-Author(s): Amila M. Devasurendra, Cheng Zhang, Jared L. Anderson, Joshua A. Young, LM Viranga Tillekeratne

Abstract Text

Ionic liquids (ILs) have been shown to possess many unique features such as tunable viscosity and solubility, low vapor pressure, and excellent thermal stability, which have facilitated their increasing use in chemical analysis. However, as liquids there is a tendency for ILs to flow when used as thin-film coatings to modify surface structures thus minimizing their utility in some applications. To this end polymeric ILs have been developed to provide additional structural stability on substrate surfaces.

In this work, electropolymerizable thiophene-based ionic liquids (ILs) were synthesized and characterized as potential candidates for developing selective extraction media for chemical analysis. Electropolymerization of the bis[(trifluoromethyl)sulfonyl]imide ([NTf2]-) analogues successfully produced uniform polymeric thin-films on macro- and microelectrode substrates from both vinyl and methylimidazolium IL monomer derivatives. The resultant conducting polymer IL (CPIL) films were characterized by electrochemical methods and found to exhibit attractive behavior towards anionic species while simultaneously providing an exclusion barrier toward cationic species. The methylimidazolium thiophene-based IL monomer also exhibited high thermal stability and subsequently was examined as a sorbent coating for solid-phase microextraction (SPME). The SPME coating was easily prepared and provided high selectivity for gas chromatographic analysis of polar analytes. The normalized response of the methylimidazolium CPIL sorbent coating exhibited higher extraction efficiency compared to an 85 µm polyacrylate fiber and excellent fiber-to-fiber reproducibility. Therefore, the electropolymerizable thiophene-based ILs were found to be viable new materials for the preparation of sorbent coatings for SPME.

Keywords: Bioanalytical, Chemically Modified Electrodes, Separation Sciences, SPME
Application Code: General Interest
Methodology Code: Electrochemistry

Session # 240 **Abstract # 240-1** **Symposia**

Session Title: ACS-ANYL - BRAIN Initiative Advancements in Neurochemical and Physiological Measurements

Abstract Title: **New Platforms for Multiplexing Electrochemical Measurements In Vivo**

Date: Monday, March 07, 2016 - Morning
Time: 08:35 AM
Room: B308

Primary Author: Michael L. Heien
University of Arizona

Co-Author(s)

Abstract Text

Fast scan cyclic voltammetry has been used extensively to monitor dynamic changes in neurotransmitters (e.g. dopamine, serotonin, norepinephrine) in real time. However, these measurements are limited in terms of the number of implantable sensors. This does not allow the coordinated chemical communication and interactions between groups of neurons to be mapped in the same animal. Here, we present arrays of electrodes with biocompatible conducting polymers capable of measuring biogenic amines. Metal wires are coated with the electrochemical polymerization of the monomer 3,4-ethylenedioxythiophene (EDOT). After baking and rinsing, a film with a positively charged polymer backbone balanced by negatively charged tosylate ions is formed. These electrodes can then be easily integrated in mass-producible arrays, brining voltammetry on par with multiple probe recordings common in electrophysiology. These voltammetric recordings also require custom instrumentation that integrates low-noise operational amplifiers with high-performance data acquisition cards and software. The system presented is capable of recording from eight channels (with appropriate instrumentation). By electrically isolating each channel, they can be independently addressed and a different neurotransmitter can be measured at each sensor. This greatly increases the ability to map neurotransmitter release in the brain of animals in real time. Data measuring both dopamine and serotonin in different brain regions is presented.

Keywords: Bioanalytical, Biosensors, Electrochemistry, Neurochemistry
Application Code: Neurochemistry
Methodology Code: Electrochemistry

Session # 240 **Abstract # 240-2** **Symposia**

Session Title ACS-ANYL - BRAIN Initiative Advancements in Neurochemical and Physiological Measurements

Abstract Title **Simultaneous Detection of Dopamine Release and Multiple Single-Unit Activity in Awake and Behaving Rats**

Primary Author: Stephen Cowen
University of Arizona

Date: Monday, March 07, 2016 - Morning
Time: 09:10 AM
Room: B308

Co-Author(s): Christopher W. Atcherley, Daniel F. Hill, Jean-Paul Wiegand, Kate L. Parent, Michael A. Miller, Michael L. Heien

Abstract Text

The neurotransmitter dopamine plays a central role in learning, decision making, motivation, and the control of movement. It is thought that dopamine release regulates such behaviors through the coordination of the firing of groups of neurons and by altering local-field oscillations. This assumption has not been tested directly as no instrument exists for the real-time measurement of both dopamine release and the activities of groups of individual neurons in freely-moving/deciding animals. This is largely due to difficulties in integrating electrophysiological and electrochemical hardware. We have engineered a measurement platform capable of tandem measurements of dopamine and neural ensemble activity in awake and behaving rats. This platform consists of high-density electrode arrays for measurement of multiple single-neuron activity and local-field potentials in conjunction with a separate carbon electrode for measurement of real-time dopamine release using fast-scan cyclic voltammetry. Fast and low-noise solid-state relays combined with improved electrochemical instrumentation allows a common reference to be utilized for both electrophysiological and electrochemical measurements, resulting in a higher signal-to-noise ratio for all measurements. Tests in anesthetized and awake rats revealed that single-unit action potentials (300 – 6000 Hz) and local-field activity in the hippocampus at frequencies > 50 Hz could be recovered while measuring dopamine concentrations in the nucleus accumbens.

Keywords: Biomedical, Biosensors, Electrodes, Neurochemistry
Application Code: Neurochemistry
Methodology Code: Sensors

Session # 240 Abstract # 240-3 **Symposia**

Session Title ACS-ANYL - BRAIN Initiative Advancements in Neurochemical and Physiological Measurements

Abstract Title **New Nano Tools for Molecular Sensing and Imaging of Single Neuron-Neuron Communication**

Primary Author X Nancy Xu
Old Dominion University

Date: Monday, March 07, 2016 - Morning
Time: 09:45 AM
Room: B308

Co-Author(s) Pavan K. Cherukuri, Preeyaporn Songkiatisak

Abstract Text

A human brain comprises a large number of neurons (~85 billion neurons) and can direct specific communication between their cells via synaptic connections through specific molecular interactions with high precision of spatial and temporal resolutions. Current technologies are unable to real-time detect, image and study these wide ranges of molecular interactions simultaneously with sufficient spatial and temporal resolutions and over an extended period of time. We are developing new nano photonic sensing and imaging tools for capturing functions of vital individual neurotransmitters at individual live neuron with single-molecule (SM) sensitivity for a wide variety of potential applications. The detailed experimental design and updated results will be presented.

The work is supported in part by NSF (CBET 0507036 and CBET 1450936) and NIH (R01 GM0764401; 3R01 GM076440-04S1).

Keywords: Bioanalytical, Biomedical, Biosensors, Nanotechnology

Application Code: Nanotechnology

Methodology Code: Biospectroscopy

Session # 240 **Abstract # 240-4** **Symposia**

Session Title ACS-ANYL - BRAIN Initiative Advancements in Neurochemical and Physiological Measurements

Abstract Title **In Vivo Multiphoton Microscopy of Mouse Brain**

Date: Monday, March 07, 2016 - Morning
Time: 10:35 AM
Room: B308

Primary Author Chris Xu
Cornell University

Co-Author(s)

Abstract Text

Over the last two decades, multiphoton microscopy has created a renaissance in the brain imaging community. It has changed how we visualize neurons by providing high-resolution, non-invasive imaging capability deep within intact brain tissue. Multiphoton imaging will likely play an essential role in understanding how the brain works at the level of neural circuits, which will provide a bridge between microscopic interactions at the neuronal level and the macroscopic structures that perform complex computations.

In this talk, the fundamental challenges of deep tissue, high-resolution optical imaging are discussed. New technologies for in vivo imaging of mouse brain using three-photon excitation at the long wavelength spectral windows of 1300 nm and 1700 nm will be presented. We will show results of high spatial resolution, deep imaging of structure and function of intact mouse brain in vivo. We will further discuss the requirements for imaging the dynamic neuronal activity at the cellular level over a large area and depth in awake and behaving animals.

Keywords: Fluorescence, Imaging, Laser, Microscopy
Application Code: Neurochemistry
Methodology Code: Microscopy

Session # 250 Abstract # 250-1 **Symposia**

Session Title: Advances in Raman Spectroscopy

Abstract Title: **Micro-Scale Spatially Offset Raman Spectroscopy (Micro-SORS)**

Date: Monday, March 07, 2016 - Morning
Time: 08:35 AM
Room: B304

Primary Author: Pavel Matousek
Rutherford Appleton Laboratory

Co-Author(s): Chiara Colombo, Claudia Conti, Giuseppe Zerbi, Marco Realini

Abstract Text

The analysis of diffusely scattering media at depths is a topical and fast evolving area of Raman spectroscopy spurred by the recent advent of Spatially Offset Raman Spectroscopy (SORS). The accessible depths can be up to two orders of magnitude larger than those with conventional confocal Raman microscopy enabling, for example, non-invasive interrogation several mm's, and in some cases several cm's, deep inside biological tissues.

This presentation will focus on the development of micro-scale variant of SORS, micro-SORS, permitting to resolve thin, tens of micrometres thick, highly turbid multilayers, which are often beyond the reach of conventional confocal Raman microscopy and spatially unresolvable by traditional macro-scale SORS. The concept is beneficial in a number of areas including the non-destructive interrogation of painted layers on objects of art, an application of particular relevance to the conservation and valorisation of cultural heritage avoiding highly undesirable cross sectional analysis. Other application areas will also be discussed.

Keywords: Raman
Application Code: Other
Methodology Code: Vibrational Spectroscopy

Session # 250 Abstract # 250-2 **Symposia**

Session Title: Advances in Raman Spectroscopy

Abstract Title: **Recent Advances in Raman Optical Activity (ROA) Methodology and Applications**

Date: Monday, March 07, 2016 - Morning
Time: 09:10 AM
Room: B304

Primary Author: Laurence A. Nafie
Syracuse University

Co-Author(s): Rina K. Dukor

Abstract Text

Raman optical activity (ROA) is now an established spectroscopic methods with many instruments world-wide carrying out a variety of applications[1]. In this talk we explore recent advances in ROA methodology and briefly describe a variety of new applications. Among the advances in ROA methodology to be discussed is the simultaneous measruement of all four forms of circular polarization (CP) ROA[2]. Among the many novel applications of ROA, we will features observations of resonance ROA (RROA) in which small but significant departures fomr the single-electronic-state (SES) theory[3] of RROA is extended in a variety ways. An additional application of broad significane is the use of ROA in the biopharmaceutical industry to test formulations of protein-based drugs for long-term stability and the early onset of signs of protein aggregation and degradation[4].

[1] Laurence A. Nafie, Vibrational Optical Activity: Principles and Applications, John Wiley & Sons, Ltd., Chichester, 2011.
[2] Honggang Li and Laurence A. Nafie , J. Raman Spectrosc. 43, 89-94 (2012).
[3] Laurence A. Nafie, Chem. Phys. 205, 309-322 (1996).
[4] Geetha Thiagarajan, Effendi Wedjaja, Jun Hyuk Heo, Jason K. Cheung, Busolo Wabuyele, Xiaodun Mou and Mohammed Shameem, J. Raman Spectrosc. 46, 531-536 (2015).

Keywords: Bioanalytical, Infrared and Raman, Molecular Spectroscopy, Pharmaceutical
Application Code: Bioanalytical
Methodology Code: Biospectroscopy

Session # 250 **Abstract # 250-3** **Symposia**

Session Title: Advances in Raman Spectroscopy

Abstract Title: **Fast Raman Imaging Using Optimized Binary Compressive Detection**

Primary Author: Dor Ben-Amotz
Purdue University

Date: Monday, March 07, 2016 - Morning
Time: 09:45 AM
Room: B304

Co-Author(s): Bharat R. Mankani, Bradley J. Lucier, Gregery T. Buzzard, Owen G. Rehrauer, Stanley Chan

Abstract Text

Optimized binary compressive detection (OB-CD) makes it possible to collect hyperspectral Raman images in real time. Here we demonstrate the capabilities of this strategy as well as head-to-head comparisons with results obtained using conventional full-spectral measurements performed using an electron multiplying charge coupled device (EM-CCD) detector. We further demonstrate the use of OB-CD for fluorescence suppression, and the combined application of OB-CD and image denoising to obtain high quality chemical images at speeds exceeding one frame per second (or 10 microseconds per pixel).

Keywords: Chemical, Chemometrics, Forensics, Raman
Application Code: Pharmaceutical
Methodology Code: Vibrational Spectroscopy

Session # 250 **Abstract # 250-4** **Symposia**

Session Title: Advances in Raman Spectroscopy

Abstract Title: **Stimulated Raman Scattering Microscopy of Vibrational Tags**

Date: Monday, March 07, 2016 - Morning
Time: 10:35 AM
Room: B304

Primary Author: Wei Min
Columbia University

Co-Author(s)

Abstract Text

Innovations in optical imaging technology have significantly impacted modern biology and medicine. Here we will discuss an emerging chemical imaging platform, stimulated Raman scattering (SRS) microscopy, which is capable of generating concentration maps of targeted chemical bonds in living systems with high sensitivity, specificity and resolution. When coupled with stable isotopes (e.g., deuterium and 13C) or bioorthogonal chemical moieties (e.g., alkynes), SRS microscopy is well suited for probing in vivo metabolic dynamics of small bio-molecules which cannot be labeled by bulky fluorophores. Physical principle of the underlying optical spectroscopy and exciting biomedical applications such as imaging lipid metabolism, protein synthesis, DNA replication, protein degradation, RNA synthesis, glucose uptake, drug trafficking and tumor metabolism will be presented.

Keywords: Imaging, Microscopy, Raman, Vibrational Spectroscopy
Application Code: Biomedical
Methodology Code: Microscopy

Session # 250 **Abstract # 250-5** **Symposia**

Session Title: Advances in Raman Spectroscopy

Abstract Title: **Nanoparticle Based Analysis of Cells, Molecules and Tissue by SERS**

Primary Author: Duncan Graham
University of Strathclyde

Date: Monday, March 07, 2016 - Morning
Time: 11:10 AM
Room: B304

Co-Author(s): Karen Faulds, Kirsten Gracie, Lee Barrett, Samuel Mabbott, Steven Asiala

Abstract Text

Metallic nanoparticles offer many opportunities in terms of detection including light scattering, surface plasmon resonance and surface enhanced Raman scattering (SERS). We are interested in the optical properties of metal nanoparticles and their potential application in a range of different biological studies. We can make use of the optical properties of nanoparticles in two ways.

1.The nanoparticle can act as an extrinsic label for a specific biomolecular target in the same way as a fluorescent label is used. The advantage of using the nanoparticle is its optical brightness (typically several orders of magnitude more than fluorophores) and the lack of background vibrational signals. Functionalisation of the nanoparticle with a specific targeting species such as an antibody or peptide aptamer allows this approach to be used in a wide range of studies including cell, tissue and in vivo analysis.

2.Nanoparticles can be designed to contain a specific recognition probe designed to cause a change in the aggregation status of the nanoparticles resulting in a discernible optical change when it interacts with its biomolecular target. This allows separation free analysis of specific biomolecular interactions and can be applied to a range of different probe/target interactions such as DNA-DNA, peptide-protein and sugar-protein.

We have been making use of nanoparticles in both of these approaches in conjunction with SERS which is an advanced vibrational spectroscopy.

Keywords: Bioanalytical, Biomedical, Raman, Surface Enhanced Raman
Application Code: Biomedical
Methodology Code: Vibrational Spectroscopy

Session # 260 Abstract # 260-1 **Symposia**

Session Title Innovative Applications of Mass Spectrometry in Biopharmaceutical and Diagnostic Development

Abstract Title **Characterization of Filgrastim Using Intact and Top-Down MS**

Date: Monday, March 07, 2016 - Morning
Time: 08:35 AM
Room: B305

Primary Author William C. Lamanna
Sandoz

Co-Author(s) Hansjoerg Toll, Johann Holzmann

Abstract Text

In March of 2015 the Food and Drug Administration (FDA) approved Sandoz's Zarxio as the first biosimilar in the United States (U.S.). Detailed characterization using a number of analytical techniques – most prominently mass spectrometry (MS) - provided the foundation for approval by demonstrating the high similarity between Zarxio and the reference product Neupogen. With the continuous improvement in MS instrumentation and data analysis software, intact and top-down MS are becoming highly attractive methods for the characterization of protein drugs. Using a benchtop Exactive MS we identified several modifications in filgrastim with high sensitivity on the intact level and subsequently performed site assignment using the instruments all-ion fragmentation mode. In our presentation we will show some examples of successful top-down MS experiments such as the site assignment of methionine oxidized variants as low as 0.1 percent.

Keywords: Biopharmaceutical, Mass Spectrometry
Application Code: Bioanalytical
Methodology Code: Mass Spectrometry

Session # 260 **Abstract # 260-2** **Symposia**

Session Title: Innovative Applications of Mass Spectrometry in Biopharmaceutical and Diagnostic Development

Abstract Title: **Tandem Mass Spectrometry as an Excellent Quantification Tool to Guide Process Development for Optimal Host Cell Protein Removal in Biopharmaceutical Products**

Primary Author: Donald Walker
Genentech

Date: Monday, March 07, 2016 - Morning
Time: 09:10 AM
Room: B305

Co-Author(s)

Abstract Text

Residual host cell protein (HCP) impurities in biotherapeutics are a potential risk factor for patient safety, and health authorities expect appropriate detection and adequate removal. As a result, the detection of a host cell protein in a drug product could result in significant delays in process development, clinical testing, or drug approval. When a host cell protein is detected, process development engineers often need quick access to analytical methods that report the critical quantitative information necessary to guide process modifications required for reduction of HCP levels. Overall HCP content is typically determined by a platform immunoassay, but quantification of an individual HCP often requires development of a specific ELISA. Expression and purification of a recombinant HCP standard for immunization, as well as generation and characterization of the antibodies necessary for ELISA development requires several months, leaving a large gap of time in which immunoassays to measure HCP levels are unavailable. Increasingly used in the identification of residual HCP, Liquid Chromatography-Tandem Mass Spectrometry (LC-MS/MS) can be also be used as a quantification tool to bridge the gap between HCP discovery and the development of a qualified immunoassay. This process starts with initial discovery of the HCP contaminant, development of LC-MS/MS methods for both relative and absolute quantification of HCP in in-process pools to guide process design. The design and implementation of the LC-MS/MS methods used for the HCP assays will be discussed, along with problems and solutions encountered during their development.

Keywords: Biopharmaceutical, Liquid Chromatography, Mass Spectrometry, Peptides
Application Code: Pharmaceutical
Methodology Code: Liquid Chromatography/Mass Spectrometry

Session # 260 Abstract # 260-3 **Symposia**

Session Title: Innovative Applications of Mass Spectrometry in Biopharmaceutical and Diagnostic Development

Abstract Title: **Antibody Characterization with Ultrahigh-Resolution Mass Spectrometry: Selecting Well Behaved Molecules for Optimal Product Quality Attributes**

Date: Monday, March 07, 2016 - Morning
Time: 09:45 AM
Room: B305

Primary Author: Lisa Marzilli
Pfizer, Inc.

Co-Author(s): Andrew E. Saati, Elaine Stephens, Heather S. DeGruttola, Jason C. Rouse, Mellissa Ly

Abstract Text

Ultrahigh-resolution mass spectrometers offer advanced capabilities for rapid, thorough, and definitive structural characterization of protein therapeutics, especially for monoclonal antibodies (mAbs). A case study on the mass spectrometric detection, site-localization and quantification of Asn deamidation in a complementarity determining region (CDR) of light chain will be presented, as well as the implications for this mAb project and others with CDR hotspots. Consequently, a new cross-functional team was formed to systematically catalog the CDR hotspots in >40 mAbs and then develop a sequence screening algorithm to ensure selection of well-behaved molecules without CDR liabilities. In an effort to generate accurate information, a novel peptide mapping method was developed to minimize method-related deamidation and more reliably separate Asp isomerization products, which will be described.

Keywords: Biopharmaceutical, Liquid Chromatography/Mass Spectroscopy, Mass Spectrometry, Protein
Application Code: Pharmaceutical
Methodology Code: Liquid Chromatography/Mass Spectrometry

Session # 260 Abstract # 260-4 **Symposia**

Session Title: Innovative Applications of Mass Spectrometry in Biopharmaceutical and Diagnostic Development

Abstract Title: **Advanced Process Control Using Fast Analytics to Monitor Multiple Critical Quality Attributes of a Monoclonal Antibody**

Date: Monday, March 07, 2016 - Morning
Time: 10:35 AM
Room: B305

Primary Author: Li Zang
Biogen

Co-Author(s)

Abstract Text

A monoclonal antibody as a therapeutic agent has to be monitored for multiple quality attributes in order to ensure its consistent quality profile to provide the intended therapeutic effect and safety. The quality attributes typically include aggregation, degradation, N-glycosylation, charge heterogeneity among others. Traditional assays such as chromatography or electrophoresis methods used in release testing or characterization are focused on individual attribute and can be lengthy and labor-intensive, which make these assays less amenable to apply to process control. This talk will present alternative analytical approaches developed using LC-MS to potentially report multiple quality attributes for real-time monitoring monoclonal antibody production as process analytical tools. These new approaches have high potentials to replace traditional assays for either in-process testing or release testing to minimize the number of assays performed for monoclonal antibody biopharmaceuticals.

Keywords: Bioanalytical
Application Code: Bioanalytical
Methodology Code: Mass Spectrometry

Session # 260 **Abstract # 260-5** **Symposia**

Session Title: Innovative Applications of Mass Spectrometry in Biopharmaceutical and Diagnostic Development

Abstract Title: **Analysis of Glycosaminoglycan Non-Reducing End Structures by Liquid Chromatography Tandem Mass Spectrometry and Its Uses in Discovering New Disease-Specific Biomarkers**

Date: Monday, March 07, 2016 - Morning
Time: 11:10 AM
Room: B305

Primary Author: Roger Lawrence
BioMarin

Co-Author(s): Brett Crawford, Evelyn Wang, Heather Prill, Jeffery D. Esko, Marzia Pasquali, Nancy Pryer, Raymond Y. Wang, Toni Pollock, William C. Lamanna

Abstract Text

The analysis of disease-specific metabolites is important for the diagnosis and determination of disease severity in many different conditions. Mucopolysaccharidoses (MPS) result from the deficiency of lysosomal catabolic enzymes leading to the accumulation of glycosaminoglycan metabolites containing characteristic non-reducing end carbohydrate biomarkers which can be monitored and quantified by liquid chromatography/mass spectrometry. Sensi-Pro is a technology developed to take advantage of these distinctive non-reducing end structures to diagnose and monitor therapy for MPS disorders and it is currently being used in a clinical setting to follow patients with MPS I and MPS II. Morquio syndrome type A (MPS IVA) is a disorder caused by deficiency of a lysosomal aryl-sulfatase (N-acetylgalactosamine-6 sulfatase) that is required for the degradation of both Keratan Sulfate (KS) and Chondroitin Sulfate (CS). Urinary Keratan Sulfate (uKS) levels are routinely used to grade disease severity and to monitor therapeutic efficacy in patients with MPS IVA. However, uKS levels are obscured in older patients by age-dependent structural changes to KS such as increased fucosylation. We used Sensi-Pro to identify a more specific and sensitive non-reducing end biomarker derived from CS for MPS IVA. This new biomarker accurately detects disease in all patients, independently from age, and its levels decrease with therapy. Overall, the Sensi-Pro approach is a very effective means of diagnosing and evaluating disease severity and response to therapy for an increasing number of MPS disorders and is expected to have growing clinical applications.

Keywords: Bioanalytical, Biomedical, Carbohydrates, Liquid Chromatography/Mass Spectroscopy
Application Code: Bioanalytical
Methodology Code: Liquid Chromatography/Mass Spectrometry

Session # 270 **Abstract # 270-1** **Symposia**

Session Title: Neurotransmission at Single and Nano-Resolved Bio-Structures

Abstract Title: **Electrochemical Cytometry of the Contents of Nanometer Vesicles Out and Inside Cells**

Primary Author: Andrew Ewing
University of Gothenburg and Chalmers University

Date: Monday, March 07, 2016 - Morning
Time: 08:35 AM
Room: B303

Co-Author(s)

Abstract Text

Electrochemical cytometry is a new method that can be used to separate nanometer vesicles, lyse them on an electrode surface, and amperometrically detect the active contents of each vesicle in a high throughput manner. First, a hybrid capillary-microfluidic device surrounding the electrode was developed to rapidly determine levels of aminergic transmitters in vesicles. More recently, we have developed a new method of electrochemical cytometry we call vesicle impact cytometry to measure the total content of single neurotransmitter vesicles.1 The electrochemical response to single adrenal chromaffin vesicles filled with hormone transmitters as they impact a 33-um diameter disk-shaped carbon electrode will be shown. The vesicles appear to adsorb onto the electrode surface and sequentially spread out over the electrode surface trapping their contents against the electrode. These contents are then oxidized and a peak results for each vesicle that bursts. A large number of current transients can be observed if the concentration of vesicles is high relative to the area of the electrode. We have also been able to accomplish this type of cytometry in the cytoplasm of living PC12 and adrenal cells.2 Comparison of the contents of these biological vesicles to the release of catecholamine from single cells supports the concept that only a fraction of transmitter is released during exocytosis.

REFERENCES

1. J. Dunevall, H. Fathali, N. Najafinobar, J. Lovric, J. Wigström, A.-S. Cans, A. G. Ewing, J. Am. Chem. Soc. 137: 4344-4346 (2015).
2. X. Li, S. Majdi, J. Dunevall, H. Fathali, A. G. Ewing, Angew. Chem. Int. Ed. (2015): DOI: 10.1002/anie.201504839.

Keywords: Bioanalytical, Electrochemistry, Electrode Surfaces
Application Code: Neurochemistry
Methodology Code: Electrochemistry

Session # **270** **Abstract #** **270-2** **Symposia**

Session Title: Neurotransmission at Single and Nano-Resolved Bio-Structures

Abstract Title: **Detection of Ionic Neurotransmitters at Biological Nanostructures via Versatile Nanosensor Probes**

Date: Monday, March 07, 2016 - Morning
Time: 09:10 AM
Room: B303

Primary Author: Mei Shen
University of Illinois at Urbana-Champaign

Co-Author(s)

Abstract Text

The Nano-Interface between Two Immiscible Electrolyte Solutions (ITIES) provides a unique analytical platform for the study of ionic species of biological interest such as neurotransmitters on living bio-structures. A typical nano-ITIES consists of a laser pulled pipette with an orifice in the few nm range that can be filled with an immiscible organic solution for its immersion into biologically-relevant fluids, thus forming a liquid/liquid interface. Upon electrochemical polarization, charged neurotransmitters can be transferred from one phase to another, which is the basis for quantitative ionic species sensing. This is particularly useful for the detection of non-redox active neurotransmitters, especially those whose detection on carbon microelectrodes is challenging. Nano-ITIES probes thus possess significant advantages over the state of the art carbon microelectrodes, which need to be enzymatically modified, for the detection of non-redox active neurotransmitters.

Scanning electrochemical microscope coupled with nanoelectrodes have been a very useful technique providing chemical information of nanostructures with unprecedented nanometer spatial resolution, e.g. ion transfer properties across a single nanopore. We employed SECM for accurate imaging of neurotransmitter flux from the synapses of model neurons, Aplysia californica. We studied on the nanopipette based sensor probe for the detection of neurotransmitters that pose a significant analytical challenge for state-of-the-art carbon microelectrodes, i.e. acetylcholine. Our results show that our nanoprobes, with a typical diameter of 30 nm, can detect acetylcholine both qualitatively and quantitatively, with excellent signal to noise ratios and in biologically-relevant fluids. The strategy presented here will provide an unprecedented view on the mechanisms of neurotransmission.

Acknowledgement: This research was supported by the National Institutes of Health under Award Number R21NS085665.

Keywords: Electrochemistry, Electrodes, Nanotechnology, Neurochemistry
Application Code: Neurochemistry
Methodology Code: Electrochemistry

Session # **270** **Abstract #** **270-3** **Symposia**

Session Title: Neurotransmission at Single and Nano-Resolved Bio-Structures

Abstract Title: **Electrochemical Approaches to Monitoring Neural Activity**

Date: Monday, March 07, 2016 - Morning
Time: 09:45 AM
Room: B303

Primary Author: R Mark Wightman
University of North Carolina at Chapel Hill

Co-Author(s)

Abstract Text

Chemical signaling through the release of neurotransmitters into the extracellular space is the primary means of communication between neurons. Because many neurotransmitters are easily oxidized, electrochemical methods have been shown to be useful for the study of dopamine, norepinephrine, and serotonin and their metabolites. In addition, molecules of metabolic interest such as oxygen and hydrogen peroxide can also be monitored with microelectrodes. Thus, the implantation of microelectrodes into the brain of an intact animal or in a brain slice s provides considerable information concerning the ongoing chemical activity. Electrochemical techniques enable spatially resolved recordings of rapid neurotransmitters dynamics in a variety of biological preparations spanning from single cells to the intact brain of behaving animals. Furthermore, the combination of electrochemistry with iontophoresis and single unit recording enhances the scope of these measurments.

Keywords: Electrodes, Neurochemistry, Voltammetry
Application Code: Neurochemistry
Methodology Code: Electrochemistry

Session # 270 Abstract # 270-4 **Symposia**

Session Title: Neurotransmission at Single and Nano-Resolved Bio-Structures

Abstract Title: **Full Fusion During Vesicular Exocytosis: A Real Release Stage or a Hoax?**

Date: Monday, March 07, 2016 - Morning
Time: 10:35 AM
Room: B303

Primary Author: Christian A. Amatore
CNRS & ENS

Co-Author(s)

Abstract Text

Vesicular exocytosis involves a connection between nanometric vesicles and the cell membrane through a nanopore across the two membranes. Single exocytotic events are recorded by the 'artificial synapse' amperometric method [1] at carbon fiber microelectrodes. Analysis of the current time dependence and intensity afford topological, energetic and dynamic information about these nanopores [2-3]. However, the main physicochemical parameters characterizing a spike are not known a priori. Yet, complete reconstruction requires that at least one independent entry. To this end, we used initial fusion nanopore radius values (1.2±0.35 nm) from patch-clamp measurements [4] to determine the mean neurotransmitter diffusion rate (□= D/Rves2) within the vesicle, so the fusion nanopore dynamics could be reconstructed for any given spike.

This afforded statistically significant analysis of size distributions of initial and final fusion pores [4]. This evidences that the "full fusion" stage does not end into full fusion but stops after less than ca. 1% of the vesicle membrane surface area is exposed. This shows that fusion mechanism is more complex than thought and most involves also the actin cytoskeleton or proteins. In functional. Probing inside neuronal synapses with nano-conical carbon fiber electrodes supported these views [5,6].

References

1.□. Amatore, S. Arbault, M. Guille, F. Lemaître, Chem. Rev. 108, 2585-2621 (2008).
2.□. Amatore, A. I. Oleinick, I. Svir, ChemPhysChem 11, 149-158 (2010).
3.□. Amatore, A. I. Oleinick, I. Svir, ChemPhysChem 11, 159-174 (2010).
4.□. Oleinick, F. Lemaître, M. Guille Collignon, I. Svir, C. Amatore, Faraday Discuss. 164, 33–55 (2013).
5.□-T. Li, S.H. Zhang, L. Wang, R.-R. Xiao, W. Liu, X.-W. Zhang, Z. Zhou, C. Amatore, W.-H. Huang, Angew. Chem. Int. Ed., 53, 12456-12460 (2013).
6.□T. Li, S.H. Zhang, X.Y. Wang, X.W. Zhang, A.I. Oleinick, I. Svir, C. Amatore, W.H. Huang. Angew. Chem., 54, 9313–9318 (2015).

Keywords: Microelectrode, Neurochemistry
Application Code: Biomedical
Methodology Code: Electrochemistry

Session # 270 Abstract # 270-5 Symposia

Session Title: Neurotransmission at Single and Nano-Resolved Bio-Structures

Abstract Title: **Coulter Counter Voltage Trapping of Nanoparticles with Sub-Nanometer Size Resolution**

Primary Author: Henry White
University of Utah

Date: Monday, March 07, 2016 - Morning
Time: 11:10 AM
Room: B303

Co-Author(s): Martin Edwards, Sean German, Yulun Zhang

Abstract Text

Resistive-pulse sensing has generated considerable interest as a technique for characterizing nanoparticle suspensions. The size, charge, and shape of individual particles can be estimated from features of the resistive pulse, but the technique suffers from an inherent variability due to the stochastic nature of particles translocating through a small orifice or channel. Here we report a method, and associated automated instrumentation, that allows repeated voltage-driven translocation of individual particles back and forth across the orifice of a conical nanopore, greatly reducing uncertainty in particle size that results from streamline path distributions, particle diffusion, particle asphericity, and electronic noise.

Keywords: Particle Size and Distribution
Application Code: Bioanalytical
Methodology Code: Electrochemistry

Session # 280 **Abstract # 280-1** **Symposia**

Session Title: Omics for Environmental and Public Health Protection

Abstract Title: **Applications of Accelerator Mass Spectrometry to Human Health Research**

Date: Monday, March 07, 2016 - Morning
Time: 08:35 AM
Room: B309

Primary Author: Ted Ognibene
Lawrence Livermore National Laboratory

Co-Author(s)

Abstract Text

Accelerator mass spectrometry (AMS) quantifies attomole (10[sup]-18[/sup]) amounts of [sup]14[/sup]C-labeled compounds in sub-milligram sized samples. This sensitivity is used to trace nutrients, toxins and therapeutics in humans and animals at less than mg/kg doses containing 1-100 nCi of [sup]14[/sup]C. AMS can also be coupled with standard separation methodologies in the identification and quantification of metabolites and target molecules. Recent advances have allowed for smaller systems with subsequent reductions in complexity and costs. At the Center for Accelerator Mass Spectrometry at Lawrence Livermore National Laboratory, we have developed an on-line combustion interface, coupled to our CO[sub]2[/sub] gas-accepting AMS ion source, to enable the direct analysis of nonvolatile carbonaceous samples. Our interface allows for both the analysis of discrete small samples, as well as for continuous flow applications to directly measure the output of a coupled HPLC in real time, thereby eliminating the complexity and costs associated with traditional AMS sample preparation techniques. Additionally, the sensitivity for [sup]14[/sup]C quantitation has been increased by a least a factor of 20. Liquid sample AMS provides a new technology to expand our biomedical AMS program by enabling the capability to measure low-level biochemicals in extremely small samples that would otherwise be inaccessible.

Work performed at the Research Resource for Biomedical AMS, which is operated at LLNL under the auspices of the U.S. DOE under contract DE-AC52-07NA27344, is supported by the National Institutes of Health, National Institute of General Medical Sciences, Biomedical Technology Research Resources under grant number 8P41GM103483.

Keywords: Instrumentation, Isotope Ratio MS, Quantitative
Application Code: Biomedical
Methodology Code: Mass Spectrometry

Session # 280 Abstract # 280-2 **Symposia**

Session Title Omics for Environmental and Public Health Protection

Abstract Title **Proteomic Approaches for Molecular Epidemiological Studies**

Primary Author Nicole Hansmeier
University of Osnabruck

Date: Monday, March 07, 2016 - Morning
Time: 09:10 AM
Room: B309

Co-Author(s) Frank R. Witter, Julie B. Herbstman, Lynn R. Goldman, Rolf U. Halden, Tzu-Chiao Chao

Abstract Text

Environmental exposures have been associated with diverse adverse health effects such as cancer, diabetes, obesity, cardiovascular, respiratory and kidney diseases. Recent studies show that 24% of global diseases are caused by environmental exposures. Preventing or minimizing these environmental risks could improve health and reduce therapeutic expenses.

A major challenge is a lack of knowledge on mechanisms linking environment to health. Furthermore, individuals may respond differently to exposures, complicating the identification of exposure-associated biological effects.

Integrating molecular omics-data into population-wide studies can provide the missing mechanistic link between exposures and health outcomes by highlighting disruptions of molecular networks due to harmful exposures. Despite the large benefits of using omics-analyses, there are significant challenges in the integration and interpretation of high-dimensional molecular data. However, omics-based epidemiological studies hold the promise to improve the risk assessment of potentially harmful exposure by improving our understanding of toxic mechanisms and disease progression. Moreover, new developments in quantitative proteomics also enable us to monitor molecular changes on the individual level, which can identify groups with an increased risk of adverse health outcomes. Ideally, these studies can lead to the development of biomarker-based tools for accurately quantifying the toxicant or disease burden within a population, thereby providing decision-makers with a tool for data-driven action and policies.

In this presentation, we will discuss the benefits and challenges of integrating proteomic investigations into epidemiological studies and provide insights from pilot studies on infant and adult health, leveraging a body of new knowledge in the rapidly growing field of environmental proteomics.

Keywords: Environmental Analysis, Environmental/Biological Samples, Mass Spectrometry, Protein
Application Code: Environmental
Methodology Code: Mass Spectrometry

Session # 280 **Abstract # 280-3** **Symposia**

Session Title: Omics for Environmental and Public Health Protection

Abstract Title: **Proteomic Applications in Microbiology**

Primary Author: Hercules Moura
CDC NCEH-DLS

Co-Author(s): John R. Barr

Date: Monday, March 07, 2016 - Morning
Time: 09:45 AM
Room: B309

Abstract Text

Proteomics, the study of protein expression within cells and tissues, is currently an area of great interest. In the area of functional genomics, it is an array of complex technologies that incorporates protein separation methods, mass spectrometry and bioinformatics on a massive scale. Different proteomic platforms have been used for biomarker discovery, from gel-based to liquid chromatography (LC)-based separations of proteins and peptides, alongside the corresponding detection by mass spectrometry (MS). We describe here a combination of top-down and bottom-up proteomics approaches applied by our group to the study of microorganisms and toxins. We applied gel-based proteomics and a top-down whole organism for rapid identification of biomarkers using MALDI-TOF MS. We then applied bottom-up gel-independent shotgun proteomics approaches based on LC-MS/MS studies as the next step for biomarker identification. The complexity of the proteomes is due to sequence polymorphisms, posttranslational modifications and other protein-processing mechanisms. In addition, a proteome's proteins span a concentration range that exceeds the dynamic range of any single analytical method or instrument. Of the many fractionation technologies available, we applied three: affinity purification, LC-based and GeLC-based separation. The different fractions were digested and the resulting peptides were analyzed by MS and tandem-MS; label-free quantitative data was obtained. MS/MS data were searched against the entire NCBI database using the Mascot search algorithm to generate a list of valid proteins. Either common or unique biomarkers detected were validated. Proteomics will profoundly improve the diagnosis, prognosis, treatment and prevention of diseases.

Keywords: Liquid Chromatography/Mass Spectroscopy, Protein, Proteomics
Application Code: Genomics, Proteomics and Other 'Omics
Methodology Code: Mass Spectrometry

Session # 280 Abstract # 280-4 **Symposia**

Session Title: Omics for Environmental and Public Health Protection

Abstract Title: **Development of Metagenomic Tools for Identification and Characterization of Infectious Agents from Clinical Samples and for Microbiome Association Studies**

Primary Author: James Posey
CDC

Date: Monday, March 07, 2016 - Morning
Time: 10:35 AM
Room: B309

Co-Author(s)

Abstract Text
Infectious disease detection, characterization, and surveillance for public health traditionally rely on methodologies that identify etiologic agents based on previously characterized associations to known clinical syndromes. In the case of novel emerging pathogens, however, attempting to detect previously uncharacterized infectious agents can prove extremely challenging. The field of metagenomics has the potential to revolutionize pathogen detection in public health laboratories through the use of next generation sequencing (NGS) on primary clinical samples. Metagenomics allows for the simultaneous detection of all microorganisms in a sample without a priori knowledge of their identities. However, a significant challenge with this approach is the vast majority of NGS information generated from a single complex clinical specimen is derived from the host genome and/or commensal organisms The increasing clinical adoption of culture independent diagnostics threatens the availability of isolated pathogens, and clutter mitigation will be a necessary first step when processing nucleic acid from specimens. To compare methodologies, we are evaluating two phases: pre- and post-extraction. We are using a previously characterized specimen sets (blood, sputum, and stool) spiked with an array of whole intact pathogens for pre-extraction, and we constructed synthetic sample sets consisting of mixtures of pathogen DNA/RNA in a clinically relevant nucleic acid background for the post-extraction. We are assessing a myriad of CDC-developed and commercially available host/commensal depletion and targeted enrichment strategies that chemically, physically, or enzymatically increase the signal-to-noise ratio, and the resulting clutter mitigation and enrichment protocols from this systematic study are being applied to real-world outbreak samples and will assist public health programs to produce and analyze clinical data for improved pathogen detection and characterization.

Keywords: Biological Samples, Characterization, Genomics, Nucleic Acids
Application Code: Genomics, Proteomics and Other 'Omics
Methodology Code: Sampling and Sample Preparation

Session # 280 **Abstract # 280-5** **Symposia**

Session Title: Omics for Environmental and Public Health Protection

Abstract Title: **Proteomics for Adverse Outcome Pathway Discovery for Environmental Contaminants**

Primary Author: Jeanette Van Emon
U.S. EPA

Date: Monday, March 07, 2016 - Morning
Time: 11:10 AM
Room: B309

Co-Author(s)

Abstract Text

An Adverse Outcome Pathway (AOP) is a conceptual framework to apply molecular pathway-based data for use in risk assessments and to support regulatory decisions. The development of AOPs requires data on the effects of chemicals on biological processes (i.e., molecular initiating events, key intermediate toxicity events, and changes in protein and metabolite expression) to link exposure to health effects. Proteomics analysis on human cell cultures was conducted to support AOP development for the compounds chlorpyrifos and 3,5,6-trichloro-2-pyridinol (TCP). Human embryonic kidney cells (HEK293) were cultured in DMEM supplemented with FBS and exposed to chlorpyrifos at 10, 30, 60, and 90 µg/mL, or 3,5,6-trichloro-2-pyridinol (TCP) at 50, 100, 150, and 200 µg/mL. After 24 h of exposure, the cells were removed from the culture plates, stained with trypan blue, counted and tested for cytotoxicity. These range finding experiments indicated the dose at which each of the compounds exerted a toxic effect on the cells based on morphology and apoptosis. Concentrations from these experiments were chosen to study potential proteomic effects exhibited by the exposed cells. A stable isotope labeling cell culture approach was used followed by proteomic determination. Over 1000 peptides were identified and assigned to 372 proteins in the cell pellet preparations using high-resolution mass spectrometry. Groups of proteins were identified that consistently appeared to be up-regulated or down-regulated in response to the exposure. One group of 25 proteins was detected after the cells were exposed to either chlorpyrifos or TCP, including cDNA FLJ54290, ADP-ribosylation factor 3 and Serpinei mRNA binding proteins. Another group of 19 proteins were detected after exposure to TCP alone, while three proteins were detected after exposure to just chlorpyrifos. The identification and relevance of the proteins are being determined for application to the construction of an AOP.

Keywords: Bioanalytical, Environmental/Biological Samples, Immunoassay, Proteomics
Application Code: Genomics, Proteomics and Other 'Omics
Methodology Code: Chemical Methods

Session # 290 **Abstract # 290-1** **Symposia**

Session Title: SEAC - Nano-Electroanalysis for a Sustainable World

Abstract Title: **Chemically Modified 3D Energy-Storage Architectures**

Date: Monday, March 07, 2016 - Morning
Time: 08:35 AM
Room: B310

Primary Author: Debra R. Rolison
U.S. Naval Research Laboratory

Co-Author(s): Christopher N. Chervin, Jeffrey W. Long, Joseph F. Parker, Megan B. Sassin

Abstract Text

Any future success in the global effort to shift energy usage away from fossil fuels will need high-performance energy storage in batteries and electrochemical capacitors (ECs) [1]. A marked improvement in the energy density of these power sources is critical, yet both are mature technologies. By remembering the 20[sup]th[/sup]-century lessons of chemically modified electrodes [2], 21[sup]st[/sup]-century energy researchers can integrate the requisite multifunctionmass and charge transport, electronic and ionic conductivity, and electron-transfer kineticsnto 3D architectures with marked improvement in performance [3,4]. The design and fabrication of size- and energy-scalable 3D multifunctional architectures from the appropriate micro-to-nanoscale building blocks will be highlighted, including the use of "nothing" (void space) and deliberate disorder as design components. We build energy-storage devices relying on aperiodic, monolithic foam or sponge form-factors including carbon aerogel-like nanofoam papers and zinc sponges. The former when painted with 10-nm–thick MnO[i]x[/i] introduces pulse-power capability into air cathodes [5] and the latter solves the historic inability to deeply discharge and cycle zinc anodes in alkaline electrolyte without forming dendrites [6]. Our work creates low-cost and scalable 3D battery components with the properties that enable next-generation energy-storage devices.

[1] D.R. Rolison, L.F. Nazar, [i]MRS Bull[/i]. 36[7] (2011) 486.
[2] R.W. Murray, [i]Molecular Design of Electrode Surfaces[/i], 1992, John Wiley & Sons (New York).
[3] D.R. Rolison, J.W. Long, [i]Acc. Chem. Res[/i]. 40 (2007) 854.
[4] D.R. Rolison, J.W. Long, J.C. Lytle, A.E. Fischer, C.P. Rhodes, T.M. McEvoy, M.E. Bourg, A.M. Lubers, [i]Chem. Soc. Rev[/i]. 38 (2009) 226.
[5] J.W. Long, C.N. Chervin, N.W. Kucko, E.S. Nelson, D.R. Rolison, [i]Adv. Energy Mater[/i]. 3 (2013) 584.
[6] D.R. Rolison, J.F. Parker, J.W. Long, U.S. Patent Appl. No. 20140147757.[sup]

Keywords: Electrochemistry, Electrodes, Energy, Nanotechnology
Application Code: Fuels, Energy and Petrochemical
Methodology Code: Electrochemistry

Session # **290** **Abstract #** **290-2** **Symposia**

Session Title: SEAC - Nano-Electroanalysis for a Sustainable World

Abstract Title: **Sunlight-Driven Hydrogen Formation by Membrane-Supported Photoelectrochemical Water Splitting**

Date: Monday, March 07, 2016 - Morning
Time: 09:10 AM
Room: B310

Primary Author: Nathan S. Lewis
Caltech

Co-Author(s)

Abstract Text

We are developing an artificial photosynthetic system that utilizes sunlight and water as input and produces hydrogen and oxygen as the outputs. We are taking a modular, parallel development approach in which three distinct primary components-the photoanode, the photocathode, and the product-separating but ion-conducting membrane-are fabricated and optimized separately before assembly into a complete water-splitting system. Design principles incorporate two separate, photosensitive semiconductor/liquid junctions that will collectively generate the 1.7-1.9 V at open circuit necessary to support both the oxidation of H[sub]2[/sub]O (or OH[sup]-[/sup]) and the reduction of H[sup]+[/sup] (or H[sub]2[/sub]O). The photoanode and photocathode will consist of rod-like semiconductor components, with attached heterogeneous multi-electron transfer catalysts, which are needed to drive oxidation or reduction reactions at low overpotentials. The high aspect-ratio semiconductor rod electrode architecture allows for use of low cost, earth abundant materials without sacrificing energy conversion efficiency due to orthogonalization of light absorption and charge-carrier collection. Additionally, the high surface-area design of the rod-based semiconductor array electrode inherently lowers the flux of charge carriers over the rod array surface relative to the projected geometric surface of the photoelectrode, thus lowering the photocurrent density at the solid/liquid junction and thereby relaxing demands on the activity (and cost) of any electrocatalysts. A flexible composite polymer film will allow for electron and ion conduction between the photoanode and photocathode while simultaneously preventing mixing of the gaseous products. Separate polymeric materials will be used to make electrical contact between the anode and cathode, and also to provide structural support. Interspersed patches of an ion conducting polymer will maintain charge balance between the two half-cells.

Keywords: Fuels\Energy\Petrochemical
Application Code: Fuels, Energy and Petrochemical
Methodology Code: Electrochemistry

Session # 290 Abstract # 290-3 **Symposia**

Session Title SEAC - Nano-Electroanalysis for a Sustainable World

Abstract Title **Elucidating Charge Transfer on Polymer Nano-Structures for a New Concept in Energy Storage**

Date: Monday, March 07, 2016 - Morning
Time: 09:45 AM
Room: B310

Primary Author Joaquin Rodriguez Lopez
University of Illinois

Co-Author(s)

Abstract Text

We introduce voltammetric studies and single-particle methods for the elucidation of redox dynamics in novel polymer nanoparticle energy storage materials. Redox Active Polymers (RAPs) and Redox Active Colloids (RACs) are promising charge storage materials for a new type of size-selective Non-aqueous Redox Flow Battery (NRFB) intended for large-scale grid energy storage (see Figure). RAPs and RACs consist of a polymeric unconjugated backbone that is decorated with high energy density and highly stable redox active pendants that display facile electron transfer.

We will present our advances in the fundamental voltammetric and chronocoulometric considerations that impact RAP/RAC charge transfer performance when in solution and when forming films. Viologen-, nitrobenzene-, and ferrocene- based polymers will be discussed. RAPs and RACs are highly versatile materials that display sizes between few to hundreds of nm, and show high solubility, reversible electron transfer, and near-quantitative charge/discharge efficiency [1].

The application of new electrochemical and ionic imaging methods based on nano-pipettes and nano-electrodes developed in our group has allowed us to better understand the careful balance between RAP size and ion and electron mobility. We will discuss how these guidelines have significantly aided our molecular design and how they impact the performance of this new type of materials for flow batteries.

This work was supported as part of the Joint Center for Energy Storage Research, an Energy Innovation Hub funded by the U.S. Department of Energy, Office of Science, Basic Energy Sciences.

[1] N. Gavvalapalli, J. Hui, K. Cheng, T. Lichtenstein, M. Shen, J.S. Moore and J. Rodríguez-López. J. Impact of Redox Active Polymer Molecular Weight on the Electrochemical Properties and Transport Across Porous Separators in Non-Aqueous Solvents. J. Am. Chem. Soc. 136 (2014) 16309.

Keywords: Chemically Modified Electrodes, Energy, Imaging, Polymers & Plastics
Application Code: Fuels, Energy and Petrochemical
Methodology Code: Electrochemistry

Session # 290 Abstract # 290-4 **Symposia**

Session Title SEAC - Nano-Electroanalysis for a Sustainable World

Abstract Title **Ion Transport in 1D and 3D Networked Porous Nanostructure Electrodes**

Date: Monday, March 07, 2016 - Morning
Time: 10:35 AM
Room: B310

Primary Author Sang Bok Lee
University of Maryland

Co-Author(s) Eleanor Gillette

Abstract Text

While interest in nanostructured materials for batteries has been growing enormously in recent years, limited systematic studies have been carried out on controlled architectures to explore the impact of structure on ion transport and resulting charge-discharge rate performance in these electrodes. Here we utilize a combination of well-defined porous 1D and 3D networked nanostructures and atomic layer deposition to fabricate a variety of systematically variable electrode architectures. The structural control and electrode design are described in detail. Then, analysis of the rate performance, with a focus on distinguishing between diffusion and charge transfer limited reaction mechanisms, is carried out for two distinct electrode systems, focusing on different issues which face advanced electrode architectures. First, we study a quantitative understanding of the balance between the loss of capacity due to resistance increases and improvements due to surface area increases. The kinetics of the charge storage reaction strongly influences the magnitude of this trade off. Second, we analyze the impact of transitioning from 1D to 3D structured electrodes. This opens up opportunities for rationally designed advanced electrode architectures to optimize the performance of electrochemical energy storage devices in the future.

Keywords: Electrochemistry, Energy
Application Code: Nanotechnology
Methodology Code: Electrochemistry

Session # 290 **Abstract # 290-5** **Symposia**

Session Title SEAC - Nano-Electroanalysis for a Sustainable World

Abstract Title **Operando Methods for the Characterization Energy Materials**

Primary Author Hector D. Abruna
Cornell University

Date: Monday, March 07, 2016 - Morning
Time: 11:10 AM
Room: B310

Co-Author(s)

Abstract Text
This presentation will deal with the development of operando methods for the study and characterization of fuel cell and battery materials. The presentation will begin with a brief overview of the methods employed. Particular emphasis will be placed on the use of X-ray diffraction (XRD) and X-ray absorption spectroscopy (XAS), transmission electron microscopy (TEM) under active potential control, confocal Raman and differential electrochemical mass spectrometry (DEMS). The utility of these methods will be illustrated by selected examples including the use of Ge nanowires as lithium ion battery anodes, spectroscopic studies of Li/S batteries and the use of DEMS to characterize electrolyte systems for LIBs. The use of operando TEM will be illustrated by studies of fuel cell catalyst degradation and coalescence and lithiation/de-lithiation dynamics of $LiFePO_4$ via energy-filtered TEM. The presentation will conclude with an assessment of future directions.

Keywords: Electrochemistry, Electrode Surfaces, Energy
Application Code: Other
Methodology Code: Electrochemistry

Session # 300 **Abstract # 300-1** **Organized Contributed Sessions**

Session Title: Infrared Spectroscopy Beyond the Diffraction Limit

Abstract Title: **Advancing the Field of Infrared Nanospectroscopy for Materials and Life Sciences**

Date: Monday, March 07, 2016 - Morning
Time: 08:30 AM
Room: B311

Primary Author: Craig Prater
Anasys Instruments

Co-Author(s)

Abstract Text

The combination of atomic force microscopy and infrared spectroscopy (AFM-IR) has grown rapidly in recent years due to its ability to provide chemical analysis and compositional imaging with very high spatial resolution. These capabilities have led to successful applications in many fields in the materials and life sciences. For example AFM-IR has been used to map component distribution in polymer blends and nanocomposites, visualize the distribution of biomolecules in tissue, cells and membranes, reveal secondary structure in proteins, and even measure molecular orientation and conformation in single nanofibers.

The presentation will overview key technologies that have enabled the field of infrared nanospectroscopy with specific focus on recent advances in the field that have dramatically (1) improved the measurement speed and sensitivity; and (2) extended the range of samples and materials that can be successfully measured. We will also survey recent applications and discoveries that have been made using AFM-IR and related techniques. The presentation will also focus specifically on research problems where the combination of nanometer scale spatial resolution, composition mapping and high spectral resolution spectroscopy is providing new research insights. We will also discuss scanning near field optical microscopy (SNOM) techniques that are enabling new insights into optical and electronic properties of 2D materials (e.g. graphene, hexagonal boron nitride) and photonic structures.

Keywords: Infrared and Raman, Microscopy, Polymers & Plastics, Protein
Application Code: Material Science
Methodology Code: Vibrational Spectroscopy

Session # 300 Abstract # 300-2 **Organized Contributed Sessions**

Session Title: Infrared Spectroscopy Beyond the Diffraction Limit

Abstract Title: **Resonance Enhanced AFM-IR Induced by Quantum Cascade Laser**

Primary Author: Rolando Rebois
University Paris-Sud

Date: Monday, March 07, 2016 - Morning
Time: 08:50 AM
Room: B311

Co-Author(s): Alexandre Dazzi, Ariane Deniset-Besseau, Jeremie Mathurin, Johanna Saunier, Kevin Kjoller, Najet Yagoubi

Abstract Text

We present results obtained with Resonance Enhanced AFM-IR, a new generation of AFM-IR technique[1] allowing analysis of much thinner samples by making use of a Quantum Cascade Laser (QCL) as the IR source. With this new technique, it is possible to increase significantly the sensitivity and the spatial resolution of the AFM-IR technique. Imaging of self-assembled monolayers was realized [2] proving the high sensitivity and resolution of this new technique which is achieved by measuring molecular expansion using force detection.
The first demonstration will be illustrated by a direct application in pharmaceutics domain [3]. The technique is used to detect some traces of organics materials (lubricant) on polymer (polyurethane) used in hospital to make implantable catheters (fig.1). Surface state is one of the most important parameter on the biocompatibility of an implantable medical device. By using the resonance enhanced mode, it is possible to identify with high resolution topographic and chemical maps the chemical nature of very thin film observed on the surface.

The second illustration will deal with proteins spectra obtained on biomaterials. Streptomyces is a filament bacteria that growths naturally into the earth. Usually the mycelium evolves from basal to aerial state with an autophagy process. The result of such a process is to find ghosts inside the cell culture which are "empty filaments" composed only of cell walls and membranes. The resonance enhanced technique will allow us to detect the proteins by their infrared spectra directly on the ghost filament.

[1] A.Dazzi, R.Prazeres, F.Glotin, J.M.Ortega, Opt. Lett., Vol. 30, Issue 18, 2388-2390 (2005).
[2] F. Lu, M. Jin, M.A. Belkin, Nat. Photon. 8, 307–312 (2014).
[3] A. Dazzi, J. Saunier, K. Kjoller, N. Yagoubi, International Journal of Pharmaceutics Volume 484, Issues 1–2, pp 109–114, (2015).

Keywords: Imaging, Nanotechnology, Spectroscopy
Application Code: Material Science
Methodology Code: Surface Analysis/Imaging

Session # 300 **Abstract # 300-3** **Organized Contributed Sessions**

Session Title: Infrared Spectroscopy Beyond the Diffraction Limit

Abstract Title: **Unprecedented Nanoscale Chemical Characterization of Materials Using AFM-Based Infrared Spectroscopy**

Date: Monday, March 07, 2016 - Morning
Time: 09:10 AM
Room: B311

Primary Author: Curtis Marcott
Light Light Solutions

Co-Author(s): Dillon Eoghan, Kevin Kjoller

Abstract Text

Atomic force microscope-based infrared spectroscopy (AFM-IR) has been developed in recent years providing extremely high spatial resolution chemical characterization and imaging. The technique is based on the combination of a tunable infrared laser with an atomic force microscope that can locally map and measure thermal expansion of nanoscale regions of a sample resulting from the absorption of infrared radiation. Because the AFM probe tip can map the thermal expansion on very fine length scales, the AFM-IR technique provides a robust way to obtain interpretable IR absorption spectra at spatial resolution scales well below the diffraction limit. Several applications of AFM-IR spectroscopy and imaging to problems in the materials sciences will be presented, including the chemical identification of components in multilayer films and polymer defect analysis.

Keywords: Infrared and Raman, Material Science, Microspectroscopy, Polymers & Plastics
Application Code: Material Science
Methodology Code: Vibrational Spectroscopy

Session # **300** **Abstract #** **300-5** **Organized Contributed Sessions**

Session Title: Infrared Spectroscopy Beyond the Diffraction Limit

Abstract Title: **AFM-IR Spectroscopy and Imaging of Biodegradable Polymers**

Date: Monday, March 07, 2016 - Morning
Time: 10:05 AM
Room: B311

Primary Author: John F. Rabolt
University of Delaware

Co-Author(s): Bruce Chase, Isao Noda, Liang Gong

Abstract Text

The combination of atomic force microscopy (AFM) and infrared (IR) spectroscopy is an extremely powerful tool that can provide topographic information that can be correlated with chemical, conformational and molecular orientation information at a spatial resolution of 50-100 nm. Using an AFM-IR instrument, we have explored the correlation between structure, processing and chain orientation/crystallinity in biodegradable polymer nanofibers and thin films and tested the hypothesis that different processing protocols can alter the concentration of different crystalline polymorphs. The ability to obtain IR spectral images at high spatial resolutions has allowed us to probe, for example, crystalline populations as a function of nanofiber diameter and as a function of location within the fiber.

Keywords: Characterization, Infrared and Raman, Nanotechnology, Polymers & Plastics
Application Code: Polymers and Plastics
Methodology Code: Vibrational Spectroscopy

Session # **300** Abstract # **300-6** **Organized Contributed Sessions**

Session Title: Infrared Spectroscopy Beyond the Diffraction Limit

Abstract Title: **Nanoscale Molecular Imaging of Polymer Systems Using AFM-IR**

Date: Monday, March 07, 2016 - Morning
Time: 10:25 AM
Room: B311

Primary Author: Mark Rickard
Dow Chemical

Co-Author(s): Carl Reinhardt, Gregory F. Meyers, Jamie Stanley

Abstract Text

Material design Improvement often requires tailoring the chemistry and structure of materials at sub-micron scales. Thin film interfaces, phase-separated morphologies and confined or constrained geometries are examples of these situations. Characterizing the spatial distributions of chemical functionalities by conventional methods is difficult or even impossible in these cases. Therefore, new measurement capabilities are required to fully investigate and understand the relationships between chemical functionality, structure and performance of nanoscale materials.

Within the last five years, spectroscopic technologies have become commercially available that exceed the diffraction limit of light and enable nanoscale spectroscopy and spectral imaging/mapping. These technologies include photothermal induced resonance (AFM-IR), scattering near-field infrared microscopy (IR s-SNOM) and tip-enhanced Raman spectroscopy (TERS). In all cases, these technologies are mediated by an atomic force microscopy (AFM) tip. The interaction area of the tip with the surface dictates the spatial resolution of the spectral response, not the wavelength of light.

In this talk we will describe our validation of spatial resolution of the AFM-IR method using selected polymer multilayer materials. We will demonstrate the utility and potential of AFM-IR to provide spatially resolved chemical information in phase separated blends, membranes, functionalized resins, and composites. The method now allows us to 'see' where the chemistry goes in the morphology.

Keywords: Imaging, Molecular Spectroscopy, Nanotechnology, Polymers & Plastics
Application Code: Polymers and Plastics
Methodology Code: Microscopy

Session # 300 Abstract # 300-7 **Organized Contributed Sessions**

Session Title: Infrared Spectroscopy Beyond the Diffraction Limit

Abstract Title: **Absorption Spectroscopy and Imaging from the Visible through Mid-IR with 20 nm Resolution Using AFM Probes**

Date: Monday, March 07, 2016 - Morning
Time: 10:45 AM
Room: B311

Primary Author: Andrea Centrone
NIST

Co-Author(s)

Abstract Text

Correlated nanoscale composition and optical property maps are important to engineer nanomaterials in applications ranging from photovoltaics to sensing and therapeutics. Light in the visible and near-IR probes electronic transitions in materials, providing information on the band gap and defects. Light in mid-IR probes vibrational transitions and provides information on chemical composition. However, diffraction limits the lateral resolution of conventional micro-spectroscopic techniques to approximately half the wavelength of light, which is insufficient for characterizing materials at the nanoscale. Additionally, the wavelength-dependent resolution impedes direct comparison of spectral maps from different spectral ranges.

Photo Thermal Induced Resonance (PTIR) is a novel technique that circumvents light diffraction by employing an AFM tip as a local detector to measure light absorption with a wavelength-independent resolution. Our PTIR setup combines an AFM microscope with three lasers providing wavelength-tunability from 500 nm to 16000 nm continuously.

In the first part of the talk I will discuss the PTIR working principles, its wavelength-independent lateral resolution (□20 nm) and its sensitivity. Recently examples of PTIR characterization from my lab include: i) plasmonic nanomaterials, ii) metal-organic frameworks crystals and iii) tri-halide perovskites. The latter will be the focus of this presentation. Tri-halide perovskites attract interest in photovoltaic application because they combine the high efficiency typical of inorganic semiconductors with low material cost and ease of fabrication. However, the knowledge of how the local material properties, such as the chemical composition, the bandgap and the defect density are related to tri-halide perovskite devices operation is still limited. It will be shown that PTIR can provide unique information to characterize and engineer these materials.

Keywords: Imaging, Material Science, Microspectroscopy, Spectroscopy
Application Code: Material Science
Methodology Code: Microscopy

Session # 310 **Abstract # 310-1** **Organized Contributed Sessions**

Session Title: Ionophore-Based Chemical Sensors II

Abstract Title: **Magnetic Nanoparticles as Dispersible Electrodes**

Date: Monday, March 07, 2016 - Morning
Time: 08:30 AM
Room: B404

Primary Author: Justin J. Gooding
University of New South Wales

Co-Author(s): Elizabeth Morago, Kyloon Chuah, Roya Tavallaie, Saimon M. Silva

Abstract Text

Nanoparticles have attracted enormous interest in biosensing related to a range of unique properties they possess. Here we will present our findings on using gold coated magnetic nanoparticles as dispersible electrodes that can be sent out into the sample, collect the analyte and bring it back to an macroelectrode for detection. This strategy is shown to give highly sensitive sensors that have much faster response times than classical macroscopic sensors. Here how the particles are made is discussed followed by how they perform as electrodes before demonstrating their surface modification strategies to convert these nanoparticles into selective nanoelectrodes for sensing. Application of these dispersible nanoelectrodes are demonstrated for the sensing of metal ions, proteins such as prostrate specific antigen and small organic molecules such as the antibiotic enrofloxacin. These three examples all demonstrates different transducing formats in which these dispersible electrodes can be employed, from straight amperometric detection, to labelling approaches and finally employing these particles in biochemresistors. In our most recent work we have demonstrated how this concept can be used for the detection of nucleic acids, for multiplex sensing and to make calibration free devices.

Keywords: Biosensors, Electrochemistry, Electrode Surfaces, Ion Selective Electrodes
Application Code: Bioanalytical
Methodology Code: Electrochemistry

Session # 310 Abstract # 310-2 **Organized Contributed Sessions**

Session Title: Ionophore-Based Chemical Sensors II

Abstract Title: **Development of an Ultra-Selective Optode Nanosensor for Potassium Imaging**

Primary Author: Heather Clark
Northeastern University

Co-Author(s): Ali Sahari, Richard Hutchings, Tim Ruckh

Date: Monday, March 07, 2016 - Morning
Time: 08:50 AM
Room: B404

Abstract Text

The existing fluorescent indicators synthesized for measuring extracellular changes in potassium concentration suffer from the interference of background cations especially sodium. Nanoparticle sensors have shown potential to overcome this issue experienced by molecular indicators for measuring potassium ions in a biological environment. Top-down nanoemulsion methods provide a fast and easy synthesis of nanoparticle sensors compared to the earlier fabrication strategies that first polymerize the nanospheres and then incorporate the sensing phase post-fabrication. However, the previous attempts for achieving adequate selectivity for potassium measurement with chromoionophore-based sensors, made using nanoemulsion procedures, fell short compared to the results from alternative fabrication methods.

Here, we refined the formulation of the existing nanoscale optode-based sensor by substituting a new pH-sensitive quencher molecule, blueberry-C6-ester-652 (synthesized by Berry and Associates), for the commonly used chromoionophore III to enhance the sensor selectivity. In addition, a static fluorophore was paired with the quencher molecule to produce a fluorescent signal from the nanosensor construct. The particle diameter and charge was measured to be 128 nm and -47.25 ± 2.24 mV. The nanosensor was characterized against calibration solutions of potassium (as well as sodium for the selectivity study) and the fluorescence/absorbance characteristics of the nanosensors were measured by fluorometer. The nanosensor measured potassium with nearly one order of magnitude higher selectivity compared to our chromoionophore-based optode nanosensors. In addition to the improved selectivity, the nanosensor responded to the physiological concentration of potassium at a relevant ionic strength, and detected serum potassium levels. The potassium nanosensor presented in this study is envisioned to have applications in cellular imaging and drug screening after being modified to enable ratiometric measurement with a built-in correction factor for sensor concentration, photobleaching and environmental changes.

Keywords: Bioanalytical, Biosensors, Fluorescence, Nanotechnology
Application Code: Bioanalytical
Methodology Code: Sensors

Session # 310 **Abstract # 310-3** **Organized Contributed Sessions**

Session Title: Ionophore-Based Chemical Sensors II

Abstract Title: **Low Detection Limit of Ion-Selective Electrodes; Is the Story Really Over?**

Primary Author: Aleksandar Radu
Keele University

Date: Monday, March 07, 2016 - Morning
Time: 09:10 AM
Room: B404

Co-Author(s): Christina McGraw, Lukasz K. Mendecki, Peter Dillingham, Sergio Granados-Focil

Abstract Text

Ion-selective electrodes (ISEs) are a rare example of an experimental technique whose limit of detection (LOD) is not defined as signal-to-noise (S/N) ratio. Rather, it is defined as the intersection of two straight lines; the first line represents the high-concentration region of Nernstian response and the second line represents the low-concentration, baseline region where the electrode is unresponsive to primary ions. At the intersection of these lines, 50% of primary ions in the membrane are replaced with interfering ions. However, in practice, this definition results in signals less than 17.8 mV/z¬I above the baseline to be neglected. Moreover, due to the curvilinear response at the LOD, the estimated concentration in unknown samples LOD is significantly biased.

In this work, we present a Bayesian methodology for the calibration of ISEs. This new methodology improves the precision of the estimated analyte concentrations and reduces bias in the curvilinear segment near the LOD. The Bayesian calibration can also be used to define LOD according to the standard S/N definition, leading to substantially reduced LODs.

In the presentation, we will give some fundamental information on the Bayesian methodology and demonstrate its utility for environmental and biomedical analysis.

Keywords: Bioanalytical, Chemical, Environmental Analysis, Ion Selective Electrodes
Application Code: Environmental
Methodology Code: Sensors

Session # 310 Abstract # 310-4 **Organized Contributed Sessions**

Session Title Ionophore-Based Chemical Sensors II

Abstract Title **Multifunctional Detection and Delivery**

Date: Monday, March 07, 2016 - Morning
Time: 09:30 AM
Room: B404

Primary Author Elizabeth (Lisa) Hall
University of Cambridge

Co-Author(s) Nadia Tsao

Abstract Text

A multifunctional silica-containing particle sensor will be reported which has the capacity to (a) carry a therapeutic payload (b) perform an ion-selective measurement (c) can be visualised by ultrasound and stimulated to deliver a payload.

The vehicles are fabricated through a sacrificial template, self-assembly mechanism, and comprise of a hollow core with a layered enzyme-hydrolysable polymer and a mesoporous organosilica shell. The microcapsules are sensitive to ultrasound (US). Upon insonation at low acoustic pressures, they resonate non-linearly. This results in a large backscatter echo, which allows for their visualisation using conventional US imaging equipment.

They can also be modified with ionophores, so that an ion-selective response can be interrogated optically in the cell. The geometry overcomes one of the limitations in cell imaging of using nanoparticle-labelled probes, whose structures are below the resolution limits for conventional confocal microscopy. Nano-particles are useful core agents, since their small size suggests a fast time response, but without good spatial resolution of individual particles, this advantage is lost.

A simple method for encapsulating high concentrations (approx~1 x 10[sup]8[/sup] mol/capsule) of materials in the cavity has also been developed and a theoretical model for both loading and release is presented. The system is tuned and demonstrated for docetaxel delivery to C4-B2 human prostate cancer cells at a normal ultrasound contrast agent concentration of 10[sup]6[/sup] capsule/mL. The effect on cell viability is demonstrated.

Keywords: Biosensors, Fluorescence, Modified Silica
Application Code: Bioanalytical
Methodology Code: Sensors

Session # 310 **Abstract # 310-5** **Organized Contributed Sessions**

Session Title Ionophore-Based Chemical Sensors II

Abstract Title **Printed Paper-Based Ion-Selective Optode Devices**

Date: Monday, March 07, 2016 - Morning
Time: 10:05 AM
Room: B404

Primary Author Daniel Citterio
Keio University

Co-Author(s) Hiroyuki Shibata, Koji Suzuki, Nobutoshi Komuro, Terence G. Henares

Abstract Text

Ion-selective optodes (ISOs) relying on the ion-exchange between a lipophilic bulk membrane phase and the aqueous sample solution are well-studied and widely reported optical chemical sensing systems. In combination with a lipophilic colorimetric pH-indicator (chromoionophore), they make use of the vast variety of highly selective ligands (ionophores) available for almost any species of cation. One disadvantage is their signal dependency on pH, making buffering of sample solutions or simultaneous pH determination necessary.

Since 2008, our research group has been working on the development of a variety of inkjet-printed microfluidic paper-based analytical devices (µPADs). In the present work, the concept of cation-exchange optodes has been implemented into printed µPAD devices with integrated pH control, making the buffering of sample solutions unnecessary. Ion concentrations are converted into a colorimetric signal, which after digital color analysis, results in semiquantitative data. Plasticized PVC cation-exchange ISO membranes are inkjet deposited on 6 mm diameter disks of filter paper and placed onto a microfluidically patterned paper structure. Control of pH is achieved by placing an additional paper disk impregnated with a pH buffer component. A completed µPAD is obtained by laminating the entire structure, preventing the evaporation of the sample liquid that would result in a disturbance of the ion-exchange equilibrium. The developed "double-shunt" (buffer-shunt and optode-shunt) system allows for the flexible variation of target cation and pH in the system. The robustness of the devices against changes of the sample pH has been experimentally confirmed. The buffer-shunt system provides an amount of pH buffering compounds sufficient to maintain a constant pH value during the measurement. Calibration curves for sodium and potassium ions based on digital color analysis resulted in the typical sigmoidal response of ISOs, covering a range from 10⁻⁴ to 10⁻¹ M. The ionophore-dependent selectivity (DD16C5 for sodium; valinomycin for potassium) known from conventional membrane-based ion-exchange optodes has been maintained. A simple exchange of pH-buffer shunt allowed for the shift of the dynamic response range on the basis of ISOs.

Keywords: Clinical Chemistry, Lab-on-a-Chip/Microfluidics, Membrane, Sensors
Application Code: Clinical/Toxicology
Methodology Code: Sensors

Session # 310 **Abstract # 310-7** **Organized Contributed Sessions**

Session Title: Ionophore-Based Chemical Sensors II

Abstract Title: **Signal Transduction Based on Constant-Potential Coulometry for Solid-Contact ISEs**

Date: Monday, March 07, 2016 - Morning
Time: 10:45 AM
Room: B404

Primary Author: Johan Bobacka
Åbo Akademi University

Co-Author(s): Elisa Hupa, Ulriika Vanamo, Ville Yrjänä

Abstract Text

Signal transduction based on constant-potential coulometry was recently introduced and applied to solid-contact ISEs (SC-ISEs) [1]. In this method, the potential between the SC-ISE and the reference electrode is kept constant while the current and charge are recorded between the SC-ISE and the counter electrode, by using a potentiostat. This novel transduction principle has been evaluated for SC-ISEs utilizing poly(3,4-ethylenedioxythiophene) (PEDOT) as the ion-to-electron transducer (solid contact). The charge recorded for a given change in ion activity in solution is proportional to the amount (redox capacitance) of the PEDOT-layer of the solid contact. This means that it is possible to amplify the analytical signal (charge) by increasing the capacitance of the solid contact layer, as schematically illustrated in Figure 1. Advantages and limitations of the new signal transduction method will be presented and discussed based on our recent experimental results.

References:

[1] E. Hupa, U. Vanamo, J. Bobacka, Electroanalysis, 27 (2015) 591-594.

Keywords: Electrochemistry, Ion Selective Electrodes, Membrane, Potentiometry
Application Code: General Interest
Methodology Code: Electrochemistry

Session # 310 **Abstract # 310-8** **Organized Contributed Sessions**

Session Title: Ionophore-Based Chemical Sensors II

Abstract Title: **Plasticizer-Free Paper-Based Ion-Selective Optodes**

Date: Monday, March 07, 2016 - Morning
Time: 11:05 AM
Room: B404

Primary Author: Xuewei Wang
University of Michigan

Co-Author(s): Mark Meyerhoff, Yu Qin

Abstract Text

In this work, cellulose paper is shown to be an excellent substrate for strong adsorption of hydrophobic sensing components (chromoionophores, ionophores, and ion-exchangers) required to prepare ionophore-based ion-selective optodes. The adsorbed components form a water-immiscible phase and enable optode sensing based on the classical ion-selective optode mechanism (heterogeneous ion-exchange or co-extraction) in the absence of any plasticizer or organic polymer phase. A camera phone can be used to take pictures of the optode paper and Hue-Saturation-Value color coordinates are employed for ion quantification. This methodology works well for all main electrolytes in body fluids (sodium, potassium, calcium, and chloride ions) using commercially available sensing components. In contrast to the traditional polymeric membrane optodes which are highy hydrophobic, the new paper-based optode retains the capability of the cellulose paper to wick fluids effectively and is thereby compatible with pump-free paper microfluidics. By using a piece of patterned filer paper with multiple optode spots for different ions, simultaneous multi-ion analysis that only requires a single drop of aquoues sample can be developed. Combined with paper-based blood separation techniques, the direct analysis of multiple electrolytes in blood or other samples can be fulfilled. Such paper-based ion sensing platform holds great promise for low cost, point-of-care tesing in resource-limited settings such as in a patient's home, at accident scenes, and in underdeveloped countries/areas.

Keywords: Biological Samples, Clinical Chemistry, Ion Selective Electrodes, Sensors
Application Code: Clinical/Toxicology
Methodology Code: Sensors

Session # 320 **Abstract # 320-1** **Organized Contributed Sessions**

Session Title: New Perspectives on the History of Chromatography

Abstract Title: **Building the Market for Ion Chromatography: The Early Days**

Primary Author: Arthur W. Fitchett
Thermo Fisher Scientific

Date: Monday, March 07, 2016 - Morning
Time: 08:30 AM
Room: B407

Co-Author(s)

Abstract Text

Forty years ago the technique of Ion Chromatography was introduced to the marketplace and the first commercial IC was sold in October 1975. From its humble beginnings at Dow Chemical by researchers Hamish Small, Tim Stevens and Bill Bauman to the commercialization by the newly formed Dionex Corporation, Ion Chromatography has grown into a very significant player in the field of liquid chromatography.

This presentation will take a nostalgic look back at those early days of Ion Chromatography and the history behind what it took to make this technique what it has become today.

Keywords: History of Chemistry, Ion Chromatography, Trace Analysis
Application Code: Environmental
Methodology Code: Liquid Chromatography

Session # 320 **Abstract # 320-2** **Organized Contributed Sessions**

Session Title: New Perspectives on the History of Chromatography

Abstract Title: **Early Days in GC 1957-1962**

Primary Author: Harold M. McNair
Virginia Tech

Date: Monday, March 07, 2016 - Morning
Time: 08:50 AM
Room: B407

Co-Author(s)

Abstract Text

I was introduced to GC in the summer of 1957 at the Amoco Refinery. I was building their first GC from simple pieces, 4 foot packed columns, a variety of liquid phases and crushing firebricks into small particles. I was assisting Dr. A.J.P. Martin install his newly invented Gas Density Balance. I was using a Gow-Mac TCD. I wanted a new hot PhD topic. He convinced me to do GC. I returned to Purdue and was able to do that. I believe my PhD thesis on selective liquid phases was the first PhD thesis in the USA. I introduced TRIS, which was the most selective liquid phase for about 40years. I spent the next two summers working in GC at DuPont (Dr. Steve dal Nogare 1957) and Esso R&D(1958) in New Jersey and in 1959/60 I enjoyed a Fulbright Post-doctoral position with Pro.Dr.Ing. A.I.M.Keulemans in Eindhoven, the Netherlands. It was one of the best years of my life. I worked, dined and visited with Dr.Marcel Golay. Dr. A.J.P.Martin, Dennis Desty, Dr.R.P.W. Scott and Dr. Jim Lovelock. This presentation is stories from those days.

Keywords: GC
Application Code: General Interest
Methodology Code: Gas Chromatography

Session # 320 Abstract # 320-3 **Organized Contributed Sessions**

Session Title: New Perspectives on the History of Chromatography

Abstract Title: **Evolution of Capillary Chromatography and Capillary Electrophoresis: A Personal Perspective**

Primary Author: Milos V. Novotny
Indiana University

Date: Monday, March 07, 2016 - Morning
Time: 09:10 AM
Room: B407

Co-Author(s)

Abstract Text

Today's capillary separations represent powerful approaches to analyze complex mixtures of natural or synthetic origin. From the concept of open tubular (capillary) column introduced by Marcel Golay in 1956, capillary gas chromatography (GC) evolved through the 1960s and 1970s to demonstrate high resolving power and its numerous applications: first in petrochemistry, but increasingly in environmental and biochemical analysis. New column technologies (surface treatments, coating procedures, bonded phases) were important, while the initially used metal and glass capillaries yielded to fused silica columns. New detection technologies, including coupling of capillary GC with mass spectrometry (MS), demanded advances in instrumental design, to which instrument companies gradually responded. The fields of capillary liquid chromatography (LC) and capillary electrophoresis (CE) were initiated in the late 1970s and early 1980s in response to the needs to analyze complex nonvolatile biological mixtures. Diffusional characteristics of the solutes in the liquid phase necessitated significant decreases in column diameters and particle size, and the term "capillary" became no longer synonymous with "open tubular". While progress in these fields was initially slow, capillary LC, CE, and ultrahigh-pressure LC have developed to revolutionize some of the most important fields of human endeavor (Human Genome Project and systems biology). Miniaturization of instrumental design has greatly facilitated certain trends in both gas-phase and liquid-phase separations such as microfluidics and multidimensional separations. Their combination with MS (as "the ultimate detector") is vital to reach successful applications of these tools. Capillary separation techniques will further thrive with advances in separation matrixes, column design and instrumental improvements.

Keywords: Capillary Electrophoresis, Capillary GC, Capillary LC, Chromatography
Application Code: Biomedical
Methodology Code: Separation Sciences

Session # 320 **Abstract # 320-4** **Organized Contributed Sessions**

Session Title: New Perspectives on the History of Chromatography

Abstract Title: **The Development of HPLC Method Development**

Primary Author: Lloyd R. Snyder
LC Resources Inc.

Date: Monday, March 07, 2016 - Morning
Time: 09:30 AM
Room: B407

Co-Author(s)

Abstract Text

The resolution of an isocratic separation has long been known to depend on sample retention (k), column efficiency (N) and selectivity (). The general dependence of k and N on separation conditions has been understood since the 1960s, allowing their contributions to resolution to be conveniently optimized. This was not the case for selectivity at the beginning of HPLC. Several separation conditions have a major effect on selectivity, including the choice of column type, mobile phase solvents, and mobile-phase pH. Temperature and solvent strength have received less attention because of their apparently minor effect on selectivity. However with increases in column efficiency these latter separation conditions can also play an important role in optimizing selectivity and resolution, especially when computer simulation is used. Advances in our understanding and control of selectivity since 1970 will be reviewed.

The use of gradient elution seems to provide a further level of complexity in the control of separation selectivity. However the use of the linear-solvent-strength model of gradient elution eliminates this complexity, and allows selectivity in either isocratic or gradient elution to be controlled in essentially the same way.

The preponderance of this presentation will relate to reversed-phase systems.

Keywords: Liquid Chromatography, Method Development
Application Code: General Interest
Methodology Code: Liquid Chromatography

Session # 320 Abstract # 320-5 **Organized Contributed Sessions**

Session Title New Perspectives on the History of Chromatography

Abstract Title **Pioneering Days of Chromatography and Lab Automation**

Primary Author Jack M. Gill
Vanguard Ventures (Retired)

Date: Monday, March 07, 2016 - Morning
Time: 10:05 AM
Room: B407

Co-Author(s)

Abstract Text

Early commercial GC Instrumentation in the 1950's and 1960's had serious design flaws and limited performance characteristics as typical of an emerging technology. The author, performing PhD thesis research in synthetic organic chemistry at Indiana university, followed by research at Monsanto's Hydrocarbon Division in St. Louis, created major design and performance improvements in GC inlets, capillary splitter designs, capillary columns, flame ionization detectors, instrument design and computerized methods for quantification of chromatography data. Soon thereafter, Gill became Director of Research and Engineering at age 28 at Wilkens Instruments (later Varian Aerograph) in Walnut Creek, California. Working with the legendary Keene P. Dimick and colleagues, Dr. Harold McNair, T.Z. Chu, and Roger Sant, Dr. Jack Gill was charged with the responsibility to develop the next generation of GC's, plus pioneer the development of digital integrators, computing integrators and computerized lab automation systems (LIMS) at Varian Aerograph and later Autolab (Spectra Physics). These products were the industry's first to bring Intel microprocessor-based instrumentation to the Chromatography market. Such products revolutionized the standards of quantitation and accuracy in Chromatography Automation, both in GC and HPLC. Anecdotal stories and technical contributions of many pioneering colleagues, customers and manufacturers will be shared in this presentation.

Keywords: Gas Chromatography, Laboratory
Application Code: Other
Methodology Code: Liquid Chromatography

Session # 320 **Abstract # 320-6** **Organized Contributed Sessions**

Session Title: New Perspectives on the History of Chromatography

Abstract Title: **The Role of Temperature Programming in F&M's Success**

Date: Monday, March 07, 2016 - Morning
Time: 10:25 AM
Room: B407

Primary Author: Aaron J. Martin
Marlabs, Retired

Co-Author(s)

Abstract Text

Two DuPont analytical chemists, Steven DalNogare and Eugene Bennett early learned of the usefulness of newly-discovered gas chromatography before commercial instruments were available. They designed two chromatographs and had them built by their resident glassblower, Frank Martinez. The instruments worked so well that other DuPont researchers wanted copies. Soon requests were received from outside the Company, and Martinez was given permission to make and sell them from a company he operated in his home basement, F&M Scientific Glassware Co. By word-of-mouth and advertizing, sales increased to overwhelm his free time. He wanted to leave DuPont, but could not leave his technical support friends. Aaron Martin, Martinez's supervisor, and Bennett recognized his situation, and the three decided to leave DuPont and enlarge on Frank's instrument business: On January 1, 1959, they started under the name, F&M Scientific Corporation with Frank responsible for manufacturing, Bennett for Sales and Marketing, and Martin for Research and Development.

They wanted to introduce a new instrument at the 1959 Pittcon Conference and chose programmed temperature chromatography which DalNogare had shown to bring meaningful advantages. They were too late to rent space on the Exhibit floor, took a salesman's-demonstration room on the forth floor. They set up their new instrument there. Potential customers had been solicited to bring in their samples for analysis, and were amazed at what information could be gotten from the first analysis using programmed temperature.

Sales of the instrument grew rapidly worldwide. An assembly factory was set up in Amsterdam in 1963. The need for more capital limited growth and opportunity, so in 1965, F&M merged into Hewlett Packard Co. who has brought new features and continues to be a dominant supplier. The feature of Programmed Temperature, and F&M's short-term exclusive provision of it, was very important to F&M's success.

Keywords: Gas Chromatography, GC, History of Chemistry, Instrumentation
Application Code: Other
Methodology Code: Gas Chromatography

Session # 320 Abstract # 320-7 **Organized Contributed Sessions**

Session Title: New Perspectives on the History of Chromatography

Abstract Title: **Chromatography In, or Next to Chemistry? Discipline Formation and Identity of Chromatography Practitioners in the 1960s and 1970s**

Date: Monday, March 07, 2016 - Morning
Time: 10:45 AM
Room: B407

Primary Author: Apostolos Gerontas
Coburg University of Applied Sciences

Co-Author(s)

Abstract Text

Chromatography is a cluster of analytical techniques widely recognized as among the most powerful group of analytical tools for modern chemistry. All through the 1960s and 1970s, chromatography knew an explosive growth, due mainly to the appearance of ultra-fast, efficient and flexible mechanized versions of it (specifically gas chromatography and high performance liquid chromatography). This ran parallel to an attempt of the chromatography practitioners to delineate a distinct "separation science" from chemistry and –in some cases— to proclaim chromatography itself an independent scientific field. In this talk, we shall see the strategies used by the chromatography specialists towards the building of the identity of their field, and discuss the possible reasons for the eventual failure of the attempt to create a separation science altogether independent from chemistry. Furthermore the claim of several chromatographists to the existence of an independent "chromatographic science" will be presented, and the possible reasons for such a claim will be discussed.

Keywords: GC, GC-MS, History of Chemistry, HPLC
Application Code: Other
Methodology Code: Chemical Methods

Session # 330 **Abstract # 330-1** **Oral Sessions**

Session Title: Biomedical: Advances in Point-of-Care Technologies

Abstract Title: **Low-Cost Molecular Diagnostics on a Smartphone**

Primary Author: Hyungsoon Im
Massachusetts General Hospital

Date: Monday, March 07, 2016 - Morning
Time: 08:30 AM
Room: B313

Co-Author(s): Cesar M. Castro, Changwook Min, Divay Pathania, Hakho Lee, Huilin Shao, Jun Song, Lioubov Fexon, Misha Pivovarov, Monty Liong, Ralph Weissleder, Rosemary H. Tambouret

Abstract Text

Smartphones with their integrated image sensors and communication capabilities have emerged as a promising platform for point-of-care diagnostics in resource-limited settings. For on-site cancer diagnosis, obtaining molecular phenotypes on cells is critical to enhance the detection accuracy. Here, we developed a smartphone-based portable diffraction imaging system to detect and molecularly profile cancer cells. By attaching a small snap-on module, we converted a smartphone into a screening tool. Cells were immunolabeled with microbeads, which transducer the molecular information into diffraction signatures detectable by the smartphone. For effective point-of-care operation, we adopted a cloud computing system in which acquired images were transferred to a remote server through a secure cloud storage for computationally intensive image processing and analysis. Analytical readouts were sent back to the smartphone. The system accurately detected target cells with immunobeads and enabled quantitative cellular analysis for not only cancer cell counts but also the expression levels of molecular markers. We also showed that the system can be used to screen for pre-cancerous or cancerous cells in clinical cervical specimens. The system generated readouts within 45 min and showed excellent agreement with gold-standard pathology. With its capacity for imaging and wireless communication, the smartphone-based imaging system would be a promising point-of-care tool for cancer screening in geographically and/or socioeconomically limited settings with pathology bottlenecks.

Keywords: Biomedical, Imaging, Immunoassay, Microscopy
Application Code: Biomedical
Methodology Code: Separation Sciences

Session # 330 Abstract # 330-2 **Oral Sessions**

Session Title Biomedical: Advances in Point-of-Care Technologies

Abstract Title **Smartphone Based Detection of Stress Biomarkers in Saliva**

Date: Monday, March 07, 2016 - Morning
Time: 08:50 AM
Room: B313

Primary Author Aadhar Jain
Cornell University

Co-Author(s) Dakota O'Dell, David Erickson, Elizabeth Rey, Seoho Lee

Abstract Text

We are developing a point of care diagnostic device that can be integrated with smartphones or tablets for the measurement of cortisol. Cortisol is a stress-related biomarker found in saliva which is known to be strongly involved in body's natural response to a physical or emotional stressor, and has therefore been used as a measurement of stress of an individual. A device that utilizes personal diagnostics in order to help identify stressed individuals would therefore be greatly beneficial for workforce productivity as anxiety disorders are estimated to affect around 40 million people, in addition to costing the US industry up to $300 billion as a result of lower productivity and absenteeism. The technology involves a competitive lateral flow assay which creates a distinct colorimetric signal which can be read and quantified through an accessory compatible with a smartphone. In this presentation, we present the results of a study conducted by us which investigates the diurnal patterns of cortisol levels and their relation to the alertness, circadian rhythm and sleep habits of an individual, where the cortisol levels are measured using a smartphone and a related accessory. We plan to use the insight provided by combining data concerning stress-related chemical biomarkers and day-to-day alertness and sleep patterns of an individual to lead to a better-informed and optimized activity schedule for maximized work output.

Keywords: Biomedical, Biosensors, Imaging, Lab-on-a-Chip/Microfluidics
Application Code: Biomedical
Methodology Code: Portable Instruments

Session # 330 **Abstract # 330-3** **Oral Sessions**

Session Title: Biomedical: Advances in Point-of-Care Technologies

Abstract Title: **A Smartphone Platform for Quantitative Point-of-Care Detection of Micronutrient Deficiencies**

Date: Monday, March 07, 2016 - Morning
Time: 09:10 AM
Room: B313

Primary Author: Dakota O'Dell
Cornell University

Co-Author(s): David Erickson, Jessica Hohenstein, Seoho Lee

Abstract Text

Micronutrient deficiencies are a “hidden hunger” which affect billions of people worldwide. According to the World Health Organization, vitamin A, zinc, and iron deficiencies are responsible for more than a million deaths annually. Nevertheless, micronutrient deficiencies are often asymptomatic and can be difficult to diagnose, particularly in the developing world where centralized medical infrastructure is limited. To address this need, we have developed the NutriPhone, a low-cost hardware device for point-of-care monitoring of micronutrients in whole blood which attaches directly to a smartphone. By using the smartphone camera as an optical sensor and a custom app for image processing, the NutriPhone device can quantitatively analyze rapid diagnostic test strips with no need for expensive laboratory devices or highly trained personnel. Hardware accessories are rapidly prototyped using 3D printing technology for multiple smartphone and tablet geometries. By modifying only the processing software, it is also possible to measure many different micronutrients and biomarkers with the same hardware accessory. To demonstrate this multiplexed capability, we evaluated the use of both smartphone and iPad devices on a number of test strips—both commercially available rapid diagnostic tests and custom strips developed in-house. We report on quantitative analysis of vitamins A, B12, and D, as well as other biomarkers including cholesterol, C reactive protein, and ferritin. Human field trials implementing the vitamin B12 test were also conducted at a rural clinic in Madanapalle, India.

Keywords: Biomedical, Immunoassay, Lab-on-a-Chip/Microfluidics, Portable Instruments
Application Code: Biomedical
Methodology Code: Portable Instruments

Session # 330 **Abstract # 330-4** **Oral Sessions**

Session Title: Biomedical: Advances in Point-of-Care Technologies

Abstract Title: **Bed-Side Detection of Volatile Human Metabolites for Medical Applications**

Date: Monday, March 07, 2016 - Morning
Time: 09:30 AM
Room: B313

Primary Author: Chandrasekhara Hariharan
ION-GAS GmbH

Co-Author(s): Sascha Liedtke, Wolfgang Vautz

Abstract Text

There is general agreement, that human breath or skin volatiles are carrier of significant information on the status of the underlying metabolism, including health, diet, environmental influences and the consumption of medical or illicit drugs. The comprehensive analysis of the human metabolic profile could make this information accessible to the medical staff e.g. for diagnosis and therapy control. However, the design of the related analytical method must consider all requirements of the end user including mobility, time of operation, analysis time and presentation of the results but also sensitivity and selectivity.

Ion mobility spectrometers, when appropriately coupled to rapid gas-chromatographic pre-separation (GC/IMS), offer an enormous potential for the exploration of the human metabolome and thereby cover a broad range of possible applications. Beside the typical experimental conditions and alternative setups leading to a full analysis after few minutes, we will be present significant examples on on-site non-invasive breath and sweat analysis applications in the fields of animal models, pharmacokinetics as well as medical diagnosis and therapy control:

Quantification of remedies by non-invasive breath analysis – on-line quantification of anaesthetics as an example.

Medical diagnosis by non-invasive breath analysis in Nephrology and Diabetes.

Potential of the rapid on-site analysis of human sweat.

Applications in animal models.

Keywords: Gas Chromatography, Medical, Metabolomics, Metabonomics
Application Code: Biomedical
Methodology Code: Chemical Methods

Session # 330	Abstract # 330-5	Oral Sessions

Session Title: Biomedical: Advances in Point-of-Care Technologies

Abstract Title: **Improved Optical Cavity Based Biosensor with Differential Detection Method Through Simultaneous Detection**

Date: Monday, March 07, 2016 - Morning
Time: 10:05 AM
Room: B313

Primary Author: Tony Bujana
Letourneau University

Co-Author(s): Cody Joy, DongGee Rho, Peter Cowles, Seunghyun Kim

Abstract Text

We report an optical cavity based sensor using a differential detection method towards point-of-care diagnostics. The optical cavity is composed of two partially reflective surfaces made of thin silver layers with a small gap. A commercial simulation tool is employed for simulating its characteristics, and the samples are fabricated by SU-8 to PMMA bonding. Two collimated lasers at different wavelengths propagate through the same path, and pass through the functionalized optical cavity. Then the lasers are separated and their intensities are measured by two CMOS cameras simultaneously. The cavity is designed for the two wavelengths' intensities to change in opposite directions as specific biomolecules are immobilized. The PMMA surface is exposed to ultraviolet light, treated with a mixture of 1-ethyl-3-(3-dimethylaminopropyl) carbodiimide hydrochloride (EDC) and N-hydroxysuccinimide (NHS) which allows Avidin binding, and biotinylated BSA is immobilized on Avidin. The measured intensities, changing upon avidin-biotin interaction, are used for the differential detection method to determine the concentration of biotinylated BSA in a sample fluid. The differential detection method, which divides the difference of the individual intensities by the sum of them, provides greater responsivity than the intensity change of the individual lasers. Since both lasers propagate through the same path, random variations are expected to be cancelled out with the differential detection method, thus enhancing the sensitivity through higher responsivity and noise cancellation. The proposed system contains no moving components and can be used for real time display.

This work was supported by NSF grant CBET-1350648.

Keywords: Biosensors
Application Code: Biomedical
Methodology Code: Sensors

Session # 330 Abstract # 330-6 **Oral Sessions**

Session Title: Biomedical: Advances in Point-of-Care Technologies

Abstract Title: **A PDMS/Paper Hybrid Microfluidic Biochip for Multiplexed Instrument-Free Bacterial Meningitis Diagnosis**

Primary Author: Maowei Dou
University of Texas El Paso

Date: Monday, March 07, 2016 - Morning
Time: 10:25 AM
Room: B313

Co-Author(s): Delfina Dominguez, Juan Sanchez, Sanjay Sharma Timilsina, Xiujun James Li

Abstract Text

Bacterial meningitis remains the most serious form of meningitis disease. [i]Neisseria meningitidis[/i] ([i]N. meningitidis[/i]), [i]Streptococcus pneumoniae[/i] ([i]S. pneumoniae[/i]), and [i]Haemophilus influenzae[/i] type B (Hib) are three most common pathogens accounting for most of bacterial meningitis. Due to the high fatality rate and the damaging effect caused by the untreated disease, immediate and early diagnosis of bacterial meningitis is in urgent need. Herein, we have developed a PDMS/paper hybrid microfluidic biochip (Figure 1) integrated with loop mediated isothermal amplification (LAMP) for multiplexed bacterial meningitis diagnosis. The limit of detection of [i]N. meningitidis[/i], [i]S. pneumoniae[/i] and Hib was a few copies per LAMP zone within 1 hour. In addition, by using artificial cerebrospinal fluid (ACSF) samples, our instrument-free direct detection of pathogenic microorganisms was proved effective without laborious sample preparation process or use of centrifuges. This hybrid microfluidic biochip with the introduction of paper for LAMP reaction enabled stable testing results over a much longer period than that of the paper-free microfluidic biochip. This work provides a low-cost, fast, highly sensitive and instrument-free microfluidic approach for multiplexed bacterial meningitis diagnosis, which has great potential for point of care (POC) detection of meningitis and other diseases in resource-limited settings.

Financial support from NIH, UT STARS Award, MRAP, BBRC, IDR2 and URI award from UTEP is gratefully acknowledged.

Keywords: Detection, Fluorescence, Lab-on-a-Chip/Microfluidics

Application Code: Bioanalytical

Methodology Code: Microfluidics/Lab-on-a-Chip

Session # 330 Abstract # 330-7 **Oral Sessions**

Session Title: Biomedical: Advances in Point-of-Care Technologies

Abstract Title: **Measuring Biochemical Effects of Pulmonary Rehabilitation and OMT on COPD Patients through LC-MS-MS Analysis of Plasma Metabolites**

Date: Monday, March 07, 2016 - Morning
Time: 10:45 AM
Room: B313

Primary Author: Chen Zhang
Michigan State University

Co-Author(s): A Daniel Jones, John Wang, Sherman Gorbis

Abstract Text

Chronic obstructive pulmonary disease (COPD) is a major cause of morbidity and mortality throughout the world. Pulmonary rehabilitation (PR) is often prescribed to improve the quality of life of COPD patients. Osteopathic manipulative treatment (OMT) has long been used for improving pulmonary function. However, research on the biochemical mechanisms underlying the effects of PR and OMT on COPD has been limited. Bioactive lipid mediators including oxylipins and endocannabinoids are endogenous metabolites playing essential roles in many physiologic and pathophysiologic events. Here we developed a targeted LC-MS-MS method to simultaneously quantify bioactive lipid mediators from multiple classes in plasma. Metabolites were extracted from human plasma using solid phase cartridges and analyzed using a Waters Xevo TQ-S instrument and multiple reaction monitoring (MRM). Measurements of oxylipins and fatty acids employed electrospray ionization in negative mode, and the same aliquot was analyzed using positive mode electrospray ionization for endocannabinoids by switching polarity within the analysis method. From these results, we identified biochemical alternations of COPD patients in a 12-week pulmonary rehabilitation program before, 1 h post, and 48 h post treatment and exercises. Comparison between patients revealed that their basal oxylipin concentrations differ greatly. Hydroxyl, epoxide, and diol products from the lipoxygenase and cytochrome P450 pathways were detected, and remarkable differences were observed in some metabolites before and after one pulmonary rehabilitation treatment. Both exercises and OMT had significant impact on epoxide oxylipins. Our findings will help improve our understanding of the effects of pulmonary rehabilitation on wellness in COPD patients.

Keywords: Bioanalytical, Clinical Chemistry, Liquid Chromatography/Mass Spectroscopy
Application Code: Bioanalytical
Methodology Code: Liquid Chromatography/Mass Spectrometry

Session # 330 **Abstract # 330-8** **Oral Sessions**

Session Title: Biomedical: Advances in Point-of-Care Technologies

Abstract Title: **Surface-Enhanced Raman Scattering for In Situ Analysis of Living Systems**

Date: Monday, March 07, 2016 - Morning
Time: 11:05 AM
Room: B313

Primary Author: Mustafa Culha
Yeditepe University

Co-Author(s)

Abstract Text

Surface-enhanced Raman scattering (SERS) continues to evolve into a unique technique to collect molecular level information from dynamic biological systems. In our effort to push the technique to its limits, the spectral information obtained from dynamic biological systems is attempted to correlate with conventional molecular biology techniques. In this context, a number of systems from biofilm formation to determination of cytotoxicology of nanomaterials on eukaryotic cells are evaluated using the technique. For each case, a method employing gold (AuNPs) or silver (AgNPs) nanoparticles is developed. Either the plasmonic NPs are introduced into the living cells or brought to the contact with them. It is found that the technique can provide very rich molecular label-free information, which can replace the conventional techniques or provide complementary information. For example, cytotocicity of nanomaterials can be determined faster, cheaper and more accurate compared to conventional colorimetric viability tests. In the second example, biofilm formation in 2D or 3D scaffold can be monitored to shed light onto the response of microorganisms to the external stimuli such as antibiotics and temperature changes.

The authors acknowledge the financial support of The scientific and Technological Council of Turkey (TUBITAK) through projects 113Z554 and 214Z129 and Yeditepe University.

Keywords: Biological Samples, Biomedical, Surface Enhanced Raman, Vibrational Spectroscopy

Application Code: Biomedical

Methodology Code: Vibrational Spectroscopy

Session # 340 Abstract # 340-1 **Oral Sessions**

Session Title: Capillary Electrophoresis

Abstract Title: **Quantitation of Kinase Activity in a Social Amoeba Using Capillary Electrophoresis and a Peptide Substrate Reporter**

Date: Monday, March 07, 2016 - Morning
Time: 08:30 AM
Room: B302

Primary Author: Michelle L. Kovarik
Trinity College

Co-Author(s): Allison J. Tierney, Kunwei Yang

Abstract Text

Peptide substrate reporters are molecular tools used to measure the activity of an enzyme of interest. Briefly, a peptide sequence is identified that acts as a substrate and is also resistant to degradation in cells and cell lysates. This peptide is fluorescently labeled and incubated in intact cells or lysates. Electrophoretic separation is then used to resolve the unmodified reporter from any enzymatic products. These experiments provide quantitative measurements of enzyme activity and complement studies of enzymes based on mRNA transcripts or antibody labeling. However, development of peptide reporters is challenging because the peptide sequence must be tailored to obtain stable substrates that have good phosphorylation kinetics and are specific to enzymes-of-interest. We report on our work to apply a peptide reporter substrate, developed in human cells for protein kinase B, to an important model organism, the social amoeba [i]Dictyostelium discoideum[/i]. PKB is highly conserved threonine kinase involved in cell proliferation and survival in human cells and in chemotaxis during [i]Dictyostelium[/i] social development. We find that the peptide reporter is at least as resistant to degradation by peptidases in [i]Dictyostelium[/i] lysates as in human cell lysates and is phosphorylated at rates comparable to those in intact human cells (~0.03-0.05 zmol pg[sup]-1[/sup] s[sup]-1[/sup]). Phosphorylation increases with stimulation of the PI3K-PKB pathway with cyclic AMP, suggesting at least partial specificity for this signaling cascade. In our on-going work, we are measuring PKB activity throughout the course of social development and further evaluating specificity of the reporter using mutant cell line controls that are readily available in this genetically tractable organism. This validation of a tool developed in human cell lines to a new organism will open possibilities for studying signaling pathways via peptide reporters in diverse cell types.

Keywords: Bioanalytical, Capillary Electrophoresis
Application Code: Bioanalytical
Methodology Code: Capillary Electrophoresis

Session # 340 Abstract # 340-2 **Oral Sessions**

Session Title Capillary Electrophoresis

Abstract Title **Development of a Degradation Resistant Peptide Reporter for Monitoring E3 Ligase and Proteasome Activity**

Primary Author Gregery Woss
University of North Carolina at Chapel Hill

Date: Monday, March 07, 2016 - Morning
Time: 08:50 AM
Room: B302

Co-Author(s) Adam Melvin, Kaiulani Houston, Marcey Waters, Nancy Allbritton

Abstract Text

Dysregulation of the ubiquitin proteasome system (UPS) is suspected to play an important role in several diseases, including cancers such as acute myeloid leukemia, and neurodegenerative diseases such as Parkinson's disease. Inhibitors targeting a key component of the UPS, E3 Ligases, represent a burgeoning new branch of targeted therapeutics for treating these diseases; however, there is currently no assay capable of measuring E3 Ligase activity in patient samples. Patient samples present unique obstacles to conventional analysis, especially given their small sample size.

Here we present a novel bioanalytical tool capable of measuring E3 ligase activity in small biological samples. We have developed a degradation resistant fluorescent peptide reporter which is ubiquitinated in cell lysates. Capillary electrophoresis with laser induced fluorescence detection (CE-LIF) enables separation and quantification of modified and unmodified reporters in small sample sizes. We show results of degradation studies demonstrating the peptide's robust nature and cell lysate studies demonstrating ubiquitination. We also report on current efforts focused on adapting reporters to analysis by CE-LIF, as well as efforts in determining the minimum number of cells required for analysis.

Keywords: Bioanalytical, Capillary Electrophoresis, Fluorescence, Peptides
Application Code: Bioanalytical
Methodology Code: Capillary Electrophoresis

Session # 340 **Abstract # 340-3** **Oral Sessions**

Session Title: Capillary Electrophoresis

Abstract Title: **A New Analytical Technique to Profile Multiple Steroids in Individual Female Zebrafish to Study Endocrine Disruption**

Date: Monday, March 07, 2016 - Morning
Time: 09:10 AM
Room: B302

Primary Author: Vincent T. Nyakubaya
West Virginia University

Co-Author(s): Amber Kantes, Jennifer Ripley-Stueckle, Lisa A. Holland, Paige A. Reed, Regina Rockwell, William J. Feeney

Abstract Text

Exposure to endocrine disrupting chemicals is associated with reproductive impairment as well as health issues including cancer, diabetes, and obesity. A new capillary electrophoresis method has been developed to analyze circulating steroids in plasma volumes less than 5 microliters. This research is significant because it is enables rapid analysis of multiple circulating steroids, which generates more information about the mechanism of endocrine disruption. This is innovative because for the first time multiple steroids are analyzed in limited volume plasma samples from single fish. The capillary electrophoresis method can analyze 5 natural steroids in 5 minutes from 5 microliters of plasma. With a technique called pH-mediated stacking, limits of detection ranging from 0.2 to 2 ng/mL (0.8 to 6 nM) for the steroids are achieved with ultraviolet-visible absorbance detection. This steroid assay was utilized to assess circulating steroids in zebrafish exposed to 17□-estradiol, a positive control for estrogenic activity, using protocol outlined by the Organisation for Economic Co-operation and Development Test No. 229: Fish Short Term Reproduction Assay. Plasma analysis revealed that exposure to 17□-estradiol leads to an increase in the level of circulating estrone in female zebrafish. Estrone is produced from 17□-estradiol in the steroid synthetic pathway. Thus, by monitoring the circulating steroids using this new technique, more information is obtained about the mechanism of disruption. An increase in the level of circulating estrone is also observed in fish that has been exposed to acetone, which is used as a delivery solvent for toxicity studies. Therefore, care must be taken when choosing the solvent vehicle to be used to expose the fish to the endocrine disrupting chemical. The use of capillary electrophoresis in endocrine disruption toxicity study is a fast method that gives insight into the mechanism of disruption in individual fish.

Keywords: Capillary Electrophoresis, Cyclodextrin, Environmental, Toxicology
Application Code: Bioanalytical
Methodology Code: Capillary Electrophoresis

Session # 340 Abstract # 340-4 **Oral Sessions**

Session Title Capillary Electrophoresis

Abstract Title **A Miniaturized Immunoaffinity Capillary Electrophoresis Based-Biomarker Analyzer for Bioanalytical Applications**

Primary Author Norberto A. Guzman
Princeton Biochemicals, Inc.

Co-Author(s) Daniel E. Guzman

Date: Monday, March 07, 2016 - Morning
Time: 09:30 AM
Room: B302

Abstract Text

Over the years analytical chemistry and immunology have contributed significantly to the field of clinical diagnosis by introducing quantitative techniques that can detect crucial and distinct chemical, biochemical and cellular biomarkers present in biosamples. Currently, two-dimensional hybrid immuno-analytical separation technologies are emerging as powerful tools for the simultaneous capture, isolation, enrichment, quantification, identification and characterization of several selected proteins simultaneously, including those with subtle structural changes such as variants, isoforms, peptide fragments, and post-translational modifications. One such technique to perform this challenging task is immunoaffinity capillary electrophoresis (IACE), which combines the use of antibodies and/or other affinity ligands as highly selective capture agents with the high resolving power of capillary electrophoresis. Since affinity ligands can be polyreactives, binding and capturing more than one molecule, they may generate false positive results when tested under mono-dimensional techniques, such as ELISA. IACE, on the other hand, is a two-dimensional technique that captures, isolates, concentrates, separates, quantifies, identifies and characterizes each component of a sample, when coupled to one or more detectors simultaneously, without the presence of false positive or false negative data. This disruptive technique, capable of preconcentrate on-line analytes present in simple or complex matrices, may change the traditional system of testing biomarkers to obtain more accurate diagnose of diseases, ideally before symptoms of a disease are manifested.

In this presentation, we will discuss examples of the determination of biomarkers by IACE and the design of a miniaturized multi-dimensional IACE apparatus capable of improved sensitivity, specificity and throughput, with potential of being used as a point-of-care instrument.

Keywords: Biological Samples, Capillary Electrophoresis, Immunoassay, Portable Instruments
Application Code: Bioanalytical
Methodology Code: Capillary Electrophoresis

Session # 340 **Abstract # 340-5** **Oral Sessions**

Session Title: Capillary Electrophoresis

Abstract Title: **Electrophoretic Technologies for Catecholamine Detection**

Date: Monday, March 07, 2016 - Morning
Time: 10:05 AM
Room: B302

Primary Author: Maojun Gong
Wichita State University

Co-Author(s): Maddukuri Naveen, Qiyang Zhang

Abstract Text

Catecholamines including dopamine, norepinephrine, epinephrine, and their metabolites are important neurotransmitters and biomarkers of pheochromocytoma, a rare but deadly cancer. The detection of these compounds is challenging due to their low levels in biological fluids such as cerebrospinal fluid (CSF), blood plasma, and urine. The objective of this project is to develop sensitive fluorescence strategies to be used for rapid separation and determination of catecholamines via capillary electrophoresis. First, o-phthalaldehyde (OPA)-based derivatization of primary amines at the presence of a fluorescent thiol was developed. The fast reaction kinetics of OPA, amine and thiol was used to facilitate online derivatization while the sensitive fluorescence based on Fluorescein or Rhodamine provided high fluorescence quantum yield and sensitivity. The limit of detection was lowered to pico-molar range. Second, an online sample pre-concentration strategy was developed to facilitate the detection of catecholamines. Specifically, catecholamines were electrokinetically injected into the separation capillary and were concentrated at the assistance of borate. This novel concentration method is different from other techniques based on sweeping, pH-junction, or electric field gradient. Both techniques have been applied to the detection of neurotransmitters, especially dopamine and norepinephrine, in CSF, and they will be used for in vivo neurotransmitter monitoring and cancer diagnosis.

Keywords: Amino Acids, Capillary Electrophoresis, Derivatization, Fluorescence
Application Code: Bioanalytical
Methodology Code: Capillary Electrophoresis

Session # 340 **Abstract # 340-6** **Oral Sessions**

Session Title: Capillary Electrophoresis

Abstract Title: **Integrating Microscale Enzyme Reactions Into Capillary Separation**

Date: Monday, March 07, 2016 - Morning
Time: 10:25 AM
Room: B302

Primary Author: Srikanth Gattu
West Virginia University

Co-Author(s): Anthony Moncrief, Cassandra Crihfield, Lisa A. Holland

Abstract Text

Enzymes are used to determine the sequence of glycans. Studying glycan sequences is important because change in glycosylation pattern is a hallmark of cancer and analyzing the glycan composition is essential to therapeutics [1-2]. Neuraminidase is an important enzyme used to evaluate and control sialic acid content. Neuraminidase catalytically cleaves the sialic acid from the non-reducing end of the oligosaccharide. Traditional methods of enzyme characterization use milliliters of enzyme which require several hours for incubation. By using capillary electrophoresis, the amount of enzyme required for analysis is reduced to nanoliter levels and enzymatic processing is completed in minutes. Phospholipid additives were used to suppress the electro osmotic flow and to create a stationary enzyme plug. These preparations have fluid like properties at lower temperatures and become a viscous gel at higher temperatures [4]. Generally oligosaccharides do not absorb in the UV region. So, the UV label 2-amino benzoic acid is utilized to label the sugar and detect the substrate and product after the enzyme reaction. The labeling reaction is driven to completion and the labeled substrate is purified to avoid preferential injection. UV absorbance detection is employed to work at substrate concentrations needed to achieve the Km value of neuraminidase for 3'-Siallylactose and 6'-Siallylactose as reported in the literature [3]. Currently the neuraminidase being studied preferentially cleaves 2,3 and 2,6 bonds. The goal of the present study is to utilize enzyme specificity and to evaluate the catalytic efficiency of neuraminidases specific to different linkages with 3' and 6'- Siallylactose.

Keywords: Capillary Electrophoresis, Lipids, Protein, UV-VIS Absorbance/Luminescence
Application Code: Bioanalytical
Methodology Code: Capillary Electrophoresis

Session # 340 Abstract # 340-7 **Oral Sessions**

Session Title: Capillary Electrophoresis

Abstract Title: **Investigation of Oxidative Metabolism by Electrochemistry/Capillary Electrophoresis/Mass Spectrometry Using a Novel Sheathless Interface Design**

Date: Monday, March 07, 2016 - Morning
Time: 10:45 AM
Room: B302

Primary Author: Nhan To
University of Kansas

Co-Author(s): Craig E. Lunte, John Stobaugh, Ryan T. Johnson

Abstract Text

Coupling electrochemistry (EC) and mass spectrometry (MS) provides a simulation of metabolic pathways for biomolecules and drugs. EC oxidizes analytes of interest to produce short-lived intermediates that may form adducts with antioxidants possessing thiol moieties. These adducts can then be characterized using MS to better understand the detoxification process of analytes in the body. Additionally, the integration of capillary electrophoresis (CE) with MS provides a separation method that has low sample consumption, high separation efficiencies, and selectivity complementary to liquid chromatography. CE-MS has the potential resolve isomeric adducts and minimize ion suppression with electrospray ionization (ESI). However, one of the largest difficulties when integrating CE with ESI-MS is that the ESI current must be supplied at the tip of the capillary, while the separation current is supplied throughout the capillary. In this work, a cellulose acetate cast decoupler was used to interface CE with ESI-MS. The application of this method has been used to study acetaminophen, with the potential to investigate peptides and larger biomolecules.

Keywords: Capillary Electrophoresis, Electrochemistry, Instrumentation, Mass Spectrometry
Application Code: Bioanalytical
Methodology Code: Capillary Electrophoresis

Session # 340 **Abstract # 340-8** **Oral Sessions**

Session Title: Capillary Electrophoresis

Abstract Title: **Manipulation of the EOF in Phospholipid Coated Capillaries Through the Incorporation of Different Metal Cations**

Primary Author: Christopher R. Harrison
San Diego State University

Date: Monday, March 07, 2016 - Morning
Time: 11:05 AM
Room: B302

Co-Author(s): Eduardo De La Toba, Shane Wells, Srilatha Vydha

Abstract Text

Phospholipids have been exploited by many doing bioanalytical separations by capillary electrophoresis due to their inherent compatibility with biomolecules such as proteins. Commonly, the phospholipids have been used as protective coatings on the inner surface of the capillary, forming an effective barrier against unwanted protein adsorption. A unique aspect of the bilayers formed by the phospholipids is the ability to alter the electroosmotic flow (EOF) not only through their adsorption to the capillary surface, but also through their interaction with separation buffer components. Most notable of these is the impact of calcium on the EOF in capillaries with phosphocholine head groups, such as 1,2-dilauroyl-sn-glycero-3-phosphocholine (DLPC). In the presence of separation buffers with modest calcium concentrations DLPC coated capillaries exhibit a significant reversed EOF (~ □2×10-4 cm2/V•s), yet in the absence of calcium the EOF is weakly normal (~ 3 ×10-5 cm2/V•s).

We have been exploring how the choice of metal cation in the separation buffer can influence the magnitude of the EOF obtained with DLPC coated capillaries. We have been particularly interested to see if different divalent cations, with different sizes, and affinities for phosphocholine, will alter the EOF in DLPC coated capillaries. To this end we have identified divalent cations that can slow the reversed EOF by at least an order of magnitude. As the EOF can be altered by simply filling the capillary with a buffer containing a different metal cation, we have been further exploiting this feature to develop means of performing sample stacking. By using different metal cations in the sample and separation buffers regions of different EOF magnitudes can be achieved and manipulated for the stacking of analytes such as proteins.

Keywords: Bioanalytical, Capillary Electrophoresis, Metals, Separation Sciences
Application Code: Bioanalytical
Methodology Code: Capillary Electrophoresis

Session # 350 **Abstract # 350-1** **Oral Sessions**

Session Title: Data Analysis and Manipulation

Abstract Title: **Quantitative Evaluation of Spectral Interferences in Atomic Emission Spectroscopy**

Primary Author: Matthieu Baudelet
University of Central Florida

Date: Monday, March 07, 2016 - Morning
Time: 08:30 AM
Room: B301

Co-Author(s): Brandon Seesahai, Jessica Chappell, Martin Richardson, Michael Sigman

Abstract Text

Quantitative analysis in any Optical Emission Spectroscopy starts with a high level of confidence in the spectral line assignment from databases. Nevertheless, the resolution of the spectra is usually degraded by the instrument and/or the line broadening. This low spectral resolution makes it possible that spectral interferences occur for the majority of the lines. Such interference makes the elemental profile uncertain.

This study shows the development of a factor to quantify the level of confidence attributed to line assignment in LIBS as an example and its extension to a complete quantitative atomic profiling. This factor is combining a physical understanding of the plasma emission with a statistical analysis of the spectrum. The possible applications and outcomes of using such a quantitative factor to evaluate line assignment in plasma spectroscopy will be discussed in the context of forensic science.

The work presented is funded by the US National Institute of Justice (2012-DN-BX-K027: "Level of Confidence in Elemental Analysis by LIBS") and the State of Florida.

Keywords: Atomic Emission Spectroscopy, Automation, Forensics, Spectroscopy
Application Code: Homeland Security/Forensics
Methodology Code: Data Analysis and Manipulation

Session # 350 **Abstract # 350-2** **Oral Sessions**

Session Title: Data Analysis and Manipulation

Abstract Title: **Comparison of Various Methods for the Determination of Uncertainty due to Long Term Stability**

Primary Author: Daniel Biggerstaff
o2si Smart Solutions

Date: Monday, March 07, 2016 - Morning
Time: 08:50 AM
Room: B301

Co-Author(s): Huichen Stavros, Mark Filla

Abstract Text

One requirement for a reference material (RM) to be classified as certified reference material (CRM) is the determination of the total combined uncertainty of the certified property value. One of the components of the total combined uncertainty is the long term stability and how it relates to the assigned shelf life of the product. o2si has produced certified reference materials for calibration of inorganic and organic chemical measurements for 18 years, and has been accredited to the requirements of ISO Guide 34 for 4 years. In this period o2si has accumulated stability data for many CRMs for organic and inorganic compounds using accelerated stability models, with testing under repeatability conditions, and real time stability data gathered using the same method under conditions of long term within-laboratory precision. Several methods have been developed over the years based on the Arrheneous Equation and the application of various assumptions about the activation energy to estimate an expiration period by stressing the product with applicable environmental variables. Various approaches for accelerated stability testing (accelerated isochronus testing) will be compared to real time data (classic stability testing) for solutions of organic compounds which show where some approaches are more accurate and where caution should be exercised in applying the methods.

Keywords: Quality, Quality Control, Reference Material, Validation
Application Code: Validation
Methodology Code: Data Analysis and Manipulation

Session # 350 Abstract # 350-3 **Oral Sessions**

Session Title: Data Analysis and Manipulation

Abstract Title: **Lifecycle of Analytical Methods: Development of Equivalent Dissolution Methods for Immediate-Release Oral Dosage Forms Post-Approval**

Date: Monday, March 07, 2016 - Morning
Time: 09:10 AM
Room: B301

Primary Author: Ivelisse Colon
Vertex Pharmaceuticals

Co-Author(s): Joseph Medendorp, Taryn Ryan

Abstract Text

Monitoring the performance and capability of quality control methods is a key component of the analytical methods lifecycle in the pharmaceutical industry. It is an integral part of the Analytical Quality by Design (AQbD) paradigm and promotes continuous improvement of analytical methods. However, the evaluation and implementation of newer technology can be challenging during the commercial phases of the product lifecycle due to regulatory & GMP requirements, laboratory capabilities and the demonstration of equivalency. Appropriate demonstration of equivalency is not only important from a regulatory standpoint, but also prevents false trends, false product failures and increases the confidence in the analytical capability of the method and its degree of variability.

This study presents different statistical approaches for demonstration of equivalency between reference and test profiles across instruments and dissolution methods. These principles were applied to a case study for the development of alternate dissolution methods for an immediate release tablet in order to decrease analytical variability (e.g. use of fiber-optic UV analysis and automatic sampling stations). Using known variability from actual tablet dissolution profiles, theoretical reference and test batches were simulated and evaluated for similarity. Global and local similarity rejection rates will be presented for the tests of interest in the context of regulatory relevance for dissolution testing cases referenced in the SUPAC-IR FDA Guidance for Industry. While focused primarily on similarity between dissolution methods, lifecycle management of analytical methods suggests the need for principles such as these to apply equally to the evaluation of tablets manufactured at different sites, tablets with minor differences in formulation or manufacturing processes, and as a gauge of analytical variability across labs.

Keywords: Dissolution, Fiber Optics, Pharmaceutical, Statistical Data Analysis

Application Code: Pharmaceutical
Methodology Code: Data Analysis and Manipulation

Session # 350 **Abstract # 350-4** **Oral Sessions**

Session Title: Data Analysis and Manipulation

Abstract Title: **Investigating Robustness and Ruggedness of Analytical Methods Employing aQBD Principles**

Date: Monday, March 07, 2016 - Morning
Time: 09:30 AM
Room: B301

Primary Author: Peter A. Jacobsen
Fertin Pharma

Co-Author(s)

Abstract Text

Investigation of an analytical methods robustness and ruggedness are in recent years becoming increasingly important within the pharmaceutical area. Risk assessment and investigation of robustness has become an integrated part on the newest FDA guideline for method validation. The principles are also integrated in SixSigma and analytical Quality by Design (aQBD), which is a proposed systematic approach for developing, evaluating, optimizing and maintaining analytical methods. Even though aQBD is focused on pharma, the principles for evaluating and improving method robustness and ruggedness are relevant throughout the analytical area. A multivariate approach to investigating robustness and ruggedness of an analytical method based on risk assessment, where interactions between factors are investigated, will give insight into method performance at variating operating conditions. This knowledge can then be used to pinpoint factors that need to be better controlled, in order to have a sufficiently robust method that will work reliable during routine analysis. The future method performance can also be estimated from the multivariate robustness and ruggedness study. This presentation will give an introduction to performing robustness and ruggedness investigations based on a risk assessment with examples of how to use the data based on real world data from the pharmaceutical industry.

Keywords: Chemometrics, Data Analysis, Method Development, Statistical Data Analysis
Application Code: Other
Methodology Code: Data Analysis and Manipulation

Session # 350 **Abstract # 350-5** **Oral Sessions**

Session Title: Data Analysis and Manipulation

Abstract Title: **Chemometrics for Big Data**

Primary Author: Robert A. Lodder
University of Kentucky

Co-Author(s): Anne Brooks

Date: Monday, March 07, 2016 - Morning
Time: 10:05 AM
Room: B301

Abstract Text

Science needs new tools and algorithms that reflect the continuing convergence of High Performance Computing and Big Data. Big Data are common in the Search for Extra Terrestrial Intelligence (SETI) because of the broad bandwidth of the receivers and fast data rates. Fig. 1 shows the data collection screen of our NIR SETI telescope with a test beacon and data packet.

As a result of the Breakthrough Listen project, the small SETI community will before long be inundated with a gush of microwave spectral data, too - potentially as much in a day as earlier SETI projects collected in a year. The data will be publicly available to allow other data scientists to join the search. Breakthrough Listen will also work with the SETI@home project that connects home computers and uses them to analyze data when the screensaver is activated. In a statement, the Greenbank observatory said: "This will likely constitute the largest amount of scientific data ever made publicly available." Big data of this scale require new: (1) Extraction, Transform, and Load (ETL) tools, and (2) Query tools, both described in this paper.

1. ETL applications integrate data developed and supported by different vendors or hosted on separate computer hardware managed and operated by different employees, as is the case with current SETI projects worldwide. Contemporary ETL tools still require too much human interaction when a new application is built using the databases resulting from the use of those ETL tools.

2. There are two general concepts used in data federation, both have strengths and weaknesses.
The first concept is a database-centric approach, used by relational database (RDBMS) vendors like Teradata (QueryGrid) and IBM (FluidQuery) or by specialty technologies like the former Composite Software
The second concept is a query tool-centric concept, used by Tableau, Qlik, etc. These implementations of the approach allow end users to mashup many sources, but they may not scale to big data volumes because data are often mashed up on the user's desktop computer or web browser rather than in a scalable big data backend like Apache Spark.

Keywords: Biosensors, Biospectroscopy, Chemometrics, Statistical Data Analysis
Application Code: General Interest
Methodology Code: Data Analysis and Manipulation

Session # 350 Abstract # 350-6 **Oral Sessions**

Session Title: Data Analysis and Manipulation

Abstract Title: **Air Quality Networks- Results from Validation and Lessons on Calibration**

Primary Author: John R. Saffell
Alphasense Ltd.

Date: Monday, March 07, 2016 - Morning
Time: 10:25 AM
Room: B301

Co-Author(s): Roderic L. Jones

Abstract Text

Air quality is an increasingly important global concern. Air quality monitoring stations (AQMs) are very good point sampling analysers but the purchase and maintenance costs limit their deployment. Recent developments have led to low cost urban air quality networks with from 50 to thousands of sites in large cities. The underlying technologies have larger errors than AQMs so our recent efforts have focused on validating and using new techniques for maintaining calibration across these networks.

Validation involves two steps:
1 Lab validation of the calibration facilities where concentrations as low as 10 ppb must be generated with known error
2 Field validation of the long term zero and sensitivity for each measurement for periods of months and years

Network calibration uses three tools:
1 Laboratory calibration of the sensor performance which is then included in the correction algorithms
2 Use of statistics to correct mostly for the zero current with temperature
3 Understanding of local and non-local pollution sources to correct the baseline across the network.

Overall network performance will be reviewed.

Keywords: Calibration, Environmental Analysis, Environmental/Air, Monitoring
Application Code: Environmental
Methodology Code: Data Analysis and Manipulation

Session # 350 **Abstract # 350-7** **Oral Sessions**

Session Title: Data Analysis and Manipulation

Abstract Title: **Using Prior Probabilities to Increase the Confidence of Chemical Identification**

Date: Monday, March 07, 2016 - Morning
Time: 10:45 AM
Room: B301

Primary Author: Tyler A. Zimmerman
NIST

Co-Author(s): Dmitrii V. Tchekhovskoi, Nirina R. Andriamaharavo, Stephen E. Stein, Tytus D. Mak, W Gary Mallard

Abstract Text

Chemical identification through matching spectral properties to a spectral database is a common practice. One factor that is overlooked by many algorithms is the "prior probability" of finding the matched compound in the experiment. Mass spectral matching is not always adequate, therefore the combination of spectral matching with prior probability improves chemical identification. In this study, prior probability information is calculated from several sources (including Bing.com, ChEBI, ChemExper, ChemSpider, HMDB, Jochem, KEGG, LIPID MAPS, MetaCyc, MolPort, NIST WebBook, PubChem, and UniChem) and is used to improve the compound identification for GC-MS spectra (gas chromatography-mass spectrometry) searched against the NIST EI (electron-ionization) spectral library. The prior probability algorithms improve compound identification for the two GC-MS datasets of essential oils and urine metabolites. The developed prior probability algorithms are generalizable to any type of spectral identification (including mass spectrometry, infrared, Raman, and other spectroscopies).

Keywords: Data Analysis, Database, GC-MS, Standards
Application Code: General Interest
Methodology Code: Data Analysis and Manipulation

Session # 350 Abstract # 350-8 **Oral Sessions**

Session Title: Data Analysis and Manipulation

Abstract Title: **Applied Analytics: Using Performance-Based Analytical Test Methodology for Monitoring Laboratory Methods Required by EPA Tier III Standards**

Primary Author: John Maurer
Northwest Analytics, Inc.

Date: Monday, March 07, 2016 - Morning
Time: 11:05 AM
Room: B301

Co-Author(s)

Abstract Text

The new EPA Tier III requirements are putting more demands on testing laboratories to ensure that their analytical procedures meet published standards. 40CFR80.47 defines a "Performance-based Analytical Test Method Approach" detailing specific QA/QC requirements for each method. The rules prescribe the use of Statistical Quality Control and other statistical techniques referencing multiple ASTM standards (ASTM 6299 in particular) covering both VCSB and non-VCSB methods. This presentation reviews the effects of the Tier III standard on analytical laboratory operations, QA/QC methodology, and the application of SQC and other statistical techniques required to establish the performance-based methodologies necessary to meet the standard.

Keywords: Data Analysis, Lab Management, Petroleum, Quality Control
Application Code: Laboratory Management
Methodology Code: Data Analysis and Manipulation

Session # 360 Abstract # 360-1 **Oral Sessions**

Session Title: Environmental LC

Abstract Title: **Intrigues of Analyte Peak Distortion in Ion Chromatography by Overloaded Matrix Ions –Trace Analysis of Bromate in High Ionic Strength Samples**

Date: Monday, March 07, 2016 - Morning
Time: 08:30 AM
Room: B408

Primary Author: Michael K. Pappoe
University of Alberta

Co-Author(s): Charles A. Lucy, Mohammad H. Naeeni

Abstract Text

Ion chromatography (IC) is perhaps the most well-established and robust analytical technique for the analysis of inorganic and organic anions. IC is routinely utilized in water and environmental monitoring, the power industry, and pharmaceutical analysis. Due to the wide range of selectivities needed for these diverse analysis, over 70 IC columns with different packing materials and functionalities have been manufactured. Complex samples may contain ions of widely differing concentrations. Overload effects can be observed when high concentrations of analyte or matrix are injected into the column, resulting in distorted (non-Gaussian) peaks. This situation is even worse when analyzing samples with a high concentration of matrix ions and trace concentration of the analyte of interest. The matrix ion peaks can severely distort the analyte peak. Previous studies established that overload effects and their occurrence can best explained on the basis of competitive Langmuir isotherm.(1)

We present our investigation into the intriguing effects of major matrix ions, chloride and sulfate, on the environmentally important trace analyte, bromate, using hydroxide- and carbonate-selective columns in isocratic IC with eluent suppression and the post-column UV analysis of bromate. Strategies to tune the overload chloride peak to reduce its effect on the bromate peak for quantitative analysis will also be discussed.

(1). Wahab M.F.; Anderson, J.K.; Abdelrady, M.; Lucy, C.A. Anal. Chem. 2014, 86, 559-566

Keywords: Environmental/Water, Ion Chromatography, Ion Exchange, Liquid Chromatography
Application Code: Environmental
Methodology Code: Liquid Chromatography

Session # 360 Abstract # 360-2 **Oral Sessions**

Session Title Environmental LC

Abstract Title **Multi-Dimensional Ion Chromatography for Trace Ion Analysis**

Date: Monday, March 07, 2016 - Morning
Time: 08:50 AM
Room: B408

Primary Author Rong Lin
Thermo Fisher Scientific

Co-Author(s) Herb Wagner, Kannan Srinivasan

Abstract Text

Ion chromatography with suppressed conductivity detection has been used routinely for monitoring the contaminants in various water matrices. Water samples around the world are diverse and contain varying amounts of matrix components therefore analytical methods have to be flexible enough to handle the demands of this varying sample.

Also as the demand for sensitivity increases, the interest in multi-dimensional ion chromatography has increased significantly due to the inherent resolution capability of this technology. In this presentation, we will go over different strategies for implementing the multi-dimensional ion chromatography. We will then show the application of this technology in drinking water analysis, especially how this technology can greatly lower the quantitation limits for trace contaminants in the presence of large matrices.

Keywords: Environmental/Water, Ion Chromatography, Liquid Chromatography

Application Code: Environmental

Methodology Code: Liquid Chromatography

Session # 360	Abstract # 360-3	Oral Sessions

Session Title: Environmental LC

Abstract Title: **Assessment of Amino Acid Stability in the Early Oceans by Liquid Chromatography-Mass Spectrometry**

Date: Monday, March 07, 2016 - Morning
Time: 09:10 AM
Room: B408

Primary Author: Eric T. Parker
Georgia Institute of Technology

Co-Author(s): Aaron S. Burton, Daniel P. Glavin, Jason P. Dworkin, Jeffrey L. Bada, Karen L. Britnon

Abstract Text

Protein- and non-protein-amino acids (AAs) likely occupied the oceans at the time of the origin and evolution of life. Primordial soup-, hydrothermal vent-, and meteoritic-processes likely contributed to this chemical inventory. Prebiotic synthesis and carbonaceous meteorite studies suggest non-protein AAs were likely more abundant than protein AAs. Amino acid preservation before abiotic and biotic destruction remains an important uncertainty.

To constrain primitive AA lifetimes, a 1992 archived seawater/beach sand mixture was spiked with D,L-alanine, D,L-valine (Val), □-aminoisobutyric acid (□-AIB), D,L-isovaline (Iva), and glycine (Gly). Analysis by high performance liquid chromatography with fluorescence detection (HPLC-FD) showed only D-Val and non-protein AAs were abundant after 2250 days.

The mixture was re-analyzed in 2012 using HPLC-FD and a triple quadrupole mass spectrometer (QqQ-MS). The analytical results after 20 years were strikingly similar to those after 2250 days.

To confirm that viable microorganisms were still present, the mixture was re-spiked with Gly in 2012. Aliquots were collected upon spiking and at 5- and 9-month intervals thereafter. Final HPLC-FD/QqQ-MS analyses were performed in 2014.

The 2014 analyses revealed only □-AIB, D,L-Iva, and D-Val remained abundant. The disappearance of Gly indicated that microorganisms still lived in the mixture and were capable of consuming protein AAs. These findings demonstrate that non-protein AAs are minimally impacted by biological degradation and thus have very long lifetimes under these conditions.

Primitive non-protein AAs from terrestrial synthesis, or meteorite in-fall, likely experienced greater preservation than protein AAs. Such robust molecules may have reached a steady state concentration dependent on ocean circulation through hydrothermal systems and synthetic input processes. We are presently trying to estimate this concentration.

Keywords: Amino Acids, Liquid Chromatography, Mass Spectrometry, Trace Analysis
Application Code: Environmental
Methodology Code: Liquid Chromatography/Mass Spectrometry

Session # 360 Abstract # 360-4 **Oral Sessions**

Session Title: Environmental LC

Abstract Title: **Increase Sample Productivity in the Environmental Lab with High Pressure Ion Chromatography**

Date: Monday, March 07, 2016 - Morning
Time: 09:30 AM
Room: B408

Primary Author: Frank Hoefler
Thermo Fisher Scientific

Co-Author(s): David G. Moore, Yan Liu

Abstract Text

Ion chromatography evolved over the last 40 years from low pressure separations at less than 2000 psi to high pressure operation up to 5000 psi; all with metal-free, polymeric flowpath ion chromatography systems. High pressure ion chromatography system capabilities allow the use of longer columns, smaller particle columns with diameters down to 4.0 m, and operation at higher flow rates. In particular, the use of smaller particle columns allows the maintenance of efficient separations with either shorter columns or increased flow rates. For complex sample matrices, fast gradient separations can be applied to further improve the analysis time. In this presentation we discuss strategies how to speed up and optimize methods using high pressure ion chromatography (HPIC) for routine environmental applications such as anion separations in bottled water, cation determinations in waste water, and disinfection byproducts in tap water. Examples are presented how throughput in a routine lab can be increased and how methods can be transferred to HPIC.

Keywords: Environmental Analysis, Environmental/Water, Ion Chromatography
Application Code: Environmental
Methodology Code: Liquid Chromatography

Session # **360** **Abstract #** **360-5** **Oral Sessions**

Session Title: Environmental LC

Abstract Title: **A Novel Cation Exchange Stationary Phase for Analysis of Common Cations and Amines Using Ion Chromatography**

Date: Monday, March 07, 2016 - Morning
Time: 10:05 AM
Room: B408

Primary Author: Mani Jayaraman
Thermo Fisher Scientific

Co-Author(s): Charanjit Saini, Christopher Pohl, Yan Liu

Abstract Text

A new cation exchange phase consisting of a novel tri-polymer system containing sulfonic, phosphonic and carboxylic acid groups has been developed for use in suppressed ion chromatography. This new hybrid phase utilizes multiple types of cation-exchange sites homogeneously distributed throughout the stationary phase.

The present work combines the best aspects of both carboxylic and sulfonic acid based stationary phases. This new material incorporates multiple types of ion exchange sites in the stationary phase in such a way as to achieve improved selectivity for monovalent species while at the same time enabling the facile elution of divalent species. Furthermore, this new material has a unique selectivity which makes it optimal for eluting most amines after the alkali metals but before the alkaline earth metals using gradient ion chromatography. The utility of the new material will be demonstrated with a number of real and simulated samples.

Keywords: Environmental Analysis, Fuels\Energy\Petrochemical, Ion Chromatography, Liquid Chromatography
Application Code: Environmental
Methodology Code: Liquid Chromatography

Session # 360 **Abstract # 360-6** **Oral Sessions**

Session Title: Environmental LC

Abstract Title: **Effects of Titanium Dioxide Nanoparticles on Endocrine Disruption in Zebrafish**

Primary Author: Marriah Ellington
West Virginia University

Date: Monday, March 07, 2016 - Morning
Time: 10:25 AM
Room: B408

Co-Author(s): Cassandra Crihfield, Lisa A. Holland, Sara Melow

Abstract Text

Nanoparticles have unique functionality, with health effects that are vastly unexplored. These effects can be enhanced depending on the surrounding matrix, exposure duration, and localization at specific target organs. Titanium dioxide nanoparticles in particular are known to concentrate in tissue and interact with pharmaceuticals. Combining these known characteristics, the purpose of the research has been to understand the complexation of titanium dioxide nanoparticles with various biologics and/or endocrine disrupting chemicals for the chief aim of studying reproductive failure following exposures. The steps leading to this goal include: binding affinity studies using pH-mediated capillary electrophoresis, homogeneity studies and flow- through system validation (both utilizing ICP-OES), and the 21-day exposure studies performed on zebrafish in accordance with OECD guidelines, whereupon blood analysis is conducted using capillary electrophoresis to quantify circulating steroid hormones. With these combined techniques, the impact of titanium dioxide nanoparticles interactions is fully realized, starting from pristine water samples to model protein mixtures, and finally in naturally occurring organic matter and blood plasma.

This research was supported by the NSF Integrative Graduate Education and Research Traineeship Program.

Keywords: Bioanalytical, Capillary Electrophoresis, Clinical/Toxicology, ICP
Application Code: Bioanalytical
Methodology Code: Capillary Electrophoresis

Session # 360 **Abstract # 360-7** **Oral Sessions**

Session Title: Environmental LC

Abstract Title: **Trace Analysis of Guanidine Compounds in Surface Water with Resorcinarene-Based Ion Chromatography Columns**

Date: Monday, March 07, 2016 - Morning
Time: 10:45 AM
Room: B408

Primary Author: Roger G. Harrison
Brigham Young University

Co-Author(s): John Lamb, Tayyabeh Panahi

Abstract Text

Trace levels of pharmaceuticals have been detected in surface water and may pose a health risk to humans and other organisms. Chromatographic materials will help identify and quantify these contaminants. We report a ion chromatographic (IC) material designed to separate cationic pharmaceuticals and report its ability to separate a group of guanidine compounds. Guanidine moieties are strongly basic and protonated under acidic conditions, and therefore can be separated on the newly designed stationary phase and detected by ion exchange chromatography. The new column packing material is based on glutamic acids bonded to resorcinarene moieties that in turn are bound to divinylbenzene macroporous resin. Detection limits in the range of 5 - 30 ppb were achieved using integrated pulsed amperometric detection (IPAD) for guanidine (G), methylguanidine (MG), 1,1-dimethylbiguanide (DMG), agmatine (AGM), guanidinobenzoic acid (GBA) and cimetidine (CIM). Suppressed conductivity (CD) and UV-vis detection resulted in limits of detection similar to IPAD. Three water sources, river, lake, and marsh, were analyzed and despite matrix effects, sensitivity for guanidine compounds was in the 100 ppb range and apparent recoveries were 80-96 %.

Keywords: Ion Chromatography
Application Code: Environmental
Methodology Code: Liquid Chromatography

Session # **360** **Abstract #** **360-8** **Oral Sessions**

Session Title: Environmental LC

Abstract Title: **Analysis of Molecular Markers of Animal Waste by LC-MS/MS**

Date: Monday, March 07, 2016 - Morning
Time: 11:05 AM
Room: B408

Primary Author: Sree Harshitha Velaga
Tennessee Technological University

Co-Author(s): John Harwood, Sreedharan Lakshmi Narayanan

Abstract Text

Contamination of water by animal wastes degrades water quality and may cause disease in humans and animals. Determining the individual sources of the contamination is important both to allow effective remediation of polluted waters and to identify waters where human disease vectors may be present. Bile acids and fecal sterols are molecular markers which can identify which animal species may have contributed to contamination of a particular water. The profiles of relative concentrations of bile acids and fecal sterols are characteristic of the waste of individual animal species. LC-MS/MS affords a potentially much superior means of analyzing these compounds in comparison with the traditional GC-MS methods. We have developed and validated an LC-MS/MS method for analysis of animal bile acids and fecal sterols. The method employs a standard C-18 column and a gradient of methanol:acetonitrile and 0.2 mM ammonium acetate in water. We have also developed preconcentration discs of the markers using C-18 extractions discs. The method has been field tested by analysis of local sewage treatment plant effluent and a farm stream polluted by cattle waste.

Keywords: Environmental/Waste/Sludge, Liquid Chromatography, Mass Spectrometry, Optimization
Application Code: Environmental
Methodology Code: Liquid Chromatography/Mass Spectrometry

Session # 370 **Abstract # 370-1** **Oral Sessions**

Session Title: Fluorescence and Luminescence Advances

Abstract Title: **Measuring Displacement on the Micron Scale: Novel Luminescent Spectral Rulers for Orthopedic Applications**

Date: Monday, March 07, 2016 - Morning
Time: 08:30 AM
Room: B315

Primary Author: Melissa M. Rogalski
Clemson University

Co-Author(s): Dakotah Anderson, Donald Benza, Hunter Pelham, Jeffrey N. Anker, John D. DesJardins, Johnathan Heath, Nakul Ravikumar

Abstract Text

Measuring strain in vivo is critical for assessing load-sharing between implanted fixation devices and healing bone fracture. Fracture callus stiffness, a mechanical indicator of healing is difficult to quantify in vivo. Radiographs are commonly used to image the fracture callus, however do not provide biomechanical stiffness and have insufficient sensitivity for determining whether weight bearing is safe. We have developed novel luminescent spectral rulers capable of measuring micron scale displacements non-invasively through 6 mm of tissue, a clinically relevant depth for assessing strain on tibial fixation implants. Our sensor, similar in design to an optical encoder, contains two overlaid substrates: a surface patterned with alternating stripes of luminescent materials (the encoder) and a mask containing opaque regions and transparent windows. Displacement of the encoder with respect to the mask is mechanically confined to a single axis and can be measured based on color visible through the transparent regions. By acquiring spectra rather than imaging the surface requiring position of the luminescence to be resolved, we are able to overcome many of the challenges associated with optical imaging through tissue and have measured reproducible 14.5 ± 0.7 µm displacements. We have developed two variations of the sensor, an x-ray excited optical luminescent spectral ruler where signal from a gadolinium oxysulfide film is modulated by an encoder patterned with bromocresol purple dye, and a fluorescent ruler patterned with spectrally distinct CdSeS/ZnS core shell quantum dots.

Keywords: Biomedical, Fluorescence, Luminescence, Spectroscopy
Application Code: Biomedical
Methodology Code: Fluorescence/Luminescence

Session # **370** **Abstract #** **370-2** **Oral Sessions**

Session Title: Fluorescence and Luminescence Advances

Abstract Title: **Simplified Two-Photon Synchronous Scanning Fluorescence for Resolution of Co-Localized Emitters**

Primary Author: Christopher K. Almlie
Oregon State University

Date: Monday, March 07, 2016 - Morning
Time: 08:50 AM
Room: B315

Co-Author(s): Karan A. Patel, Kuan-Jen Chen, Sean M. Burrows

Abstract Text

Fluorescence spectroscopy is a ubiquitous analytical method due to its simplicity, robustness, and sensitivity. However, a major limitation is the number of co-localized fluorescent emitters that can be resolved in a given spectral window, typically no more than three. Current techniques to reduce spectral crosstalk have their own drawbacks, such as: 1) multiple dye-specific detection channels with dedicated interference filters and detectors, or 2) the need for detailed spectral data from each emitter for labor-intensive deconvolution of overlapping spectra. Use of Liquid Crystal and Acousto-Optical Tunable Filters are complicated, have limited light throughput, and require precise alignment of optical equipment to achieve optimum performance. To advance multicolor detection, we sought to resolve emitters that have strong overlap in the excitation and emission spectrum. A simplified two-photon synchronous scanning fluorescence system and method for resolution of three co-localized fluorescent emitters will be described. Simple short-wavepass and long-wavepass linear variable filters were combined in series to create a bandpass filter with a customizable center wavelength and bandwidth. Several synchronous scanning methods with different excitation and emission wavelength offsets, scanning trajectories, and different emission bandwidths will be discussed. Each methods ability to reduce spectral cross-talk will be revealed. Complex chemical and biological systems will be better understood with improved resolution of several dyes from the same spatial location. The simple two-photon synchronous scanning method has broad analytical applicability across many fields such as biochemistry, genetics, pharmaceuticals, and food science.

Keywords: Automation, Bioanalytical, Fluorescence, Spectroscopy

Application Code: Bioanalytical

Methodology Code: Fluorescence/Luminescence

Session # 370 Abstract # 370-3 **Oral Sessions**

Session Title: Fluorescence and Luminescence Advances

Abstract Title: **Development of Chitosan-Modified Near Infrared Fluorescent Graphene Oxide Nanocomposite**

Date: Monday, March 07, 2016 - Morning
Time: 09:10 AM
Room: B315

Primary Author: Minh H. Duong
University of North Dakota

Co-Author(s): Julia Xiaojun Zhao, Steve Xu Wu

Abstract Text

Graphene-based near-infrared fluorescent (NIRF) nanomaterials have great potential for bioimaging and bioanalysis. However, the two major limitations exist: poor water solubility and low fluorescence intensity. The objective of this work was to synthesize chitosan-attached graphene oxide (CS-GO) nanocomposites to obtain a new graphene-based NIRF nanomaterial with high water solubility and high fluorescence intensity. The method includes two major steps: maximizing the amount of CS molecules attached to GO sheets and optimizing the conditions for breakdown of CS-GO composites. Since CS attached to GO via the interactions between the amine groups of CS and the carboxyl groups of GO, most of hydroxyl and epoxy groups in the GO sheets were converted to carboxyl groups by reacting with chloroacetic acid under strong basic condition. Then the breakdown of the GO sheets (size 500 – 1000 nm) was optimized by ultrasonicating the CS-GO solution at 25[degree]C for different periods of time. The interactions between CS and GO were initiated in two ways. The first way was to form amide bonds between CS and GO in 0.1 M of 2-(N-morpholino)ethanesulfonic acid (MES) hydrate buffer solution in the presence of N-hydroxysuccinimide (NHS) and 1-ethyl-3-(3-dimethylaminopropyl)carbodiimide•HCl (EDC). The second way was to form electrostatic interactions between CS and GO in 0.1 M acetic acid. By applying either of these routes, CS-GO nanocomposites with size less than 20 nm, high water solubility, and relatively high fluorescence intensity were successfully synthesized.

The Neuroscience COBRE Pilot Grant from NIH

Keywords: Fluorescence, Nanotechnology, Near Infrared, UV-VIS Absorbance/Luminescence

Application Code: Nanotechnology
Methodology Code: Fluorescence/Luminescence

Session # 370 Abstract # 370-4 **Oral Sessions**

Session Title: Fluorescence and Luminescence Advances

Abstract Title: **Studying Chemical Reaction, Mass Transport and Their Coupling in 3D Multilayer Catalysts at Single-Molecule Level**

Primary Author: Bin Dong
Georgia State University

Date: Monday, March 07, 2016 - Morning
Time: 09:30 AM
Room: B315

Co-Author(s): Chen Y. Pei, Ning Fang, Wen Y. Huang, Xiao X. Chao

Abstract Text

The emergence and advance of super-resolution and super-localization fluorescence microscopy in the past decades had enabled researchers to reveal more detailed molecular dynamics and structural information from single molecule imaging. The application of single molecule localization based super-resolution imaging techniques in imaging chemical reactions had brought new insights of the reaction mechanism at single particle level void of the averaging effect in classical ensemble experiments. Heterogeneous properties of porous materials had been revealed in single molecule fluorescence microscopy. Here we synthesized a multilayer nanocatalysts composed of platinum nanoparticles sandwiched between an optically transparent solid core and a mesoporous shell with aligned pores. A model fluorogenic oxidation reaction of a non-fluorescent reactant molecule amplex red into a highly fluorescent product molecule resorufin was used to study the catalytic properties. The 3D multilayer catalyst provides a uniquely well-defined structure to study diffusion and adsorption at the single molecule level because a fluorescent product molecule is generated on metal NPs located at the surface of the core silica sphere, which provides an unambiguous starting point (both in time and space) of the molecular transport out of a single reactive site. Factors, including pore size, pore length, pore structure, and mass transport, etc. affecting the catalytic properties and efficiency were systematically investigated at single particle and single molecule level.

Keywords: Imaging, Material Science, Microscopy, Nanotechnology
Application Code: Nanotechnology
Methodology Code: Fluorescence/Luminescence

Session Title: Fluorescence and Luminescence Advances

Abstract Title: **Slowing Down of Nanoparticle Diffusion on Sub-Micrometer Oil Droplet-Aqueous Buffer Interface Studied with Three Dimensional Tracking**

Date: Monday, March 07, 2016 - Morning
Time: 10:05 AM
Room: B315

Primary Author: Yaning Zhong
North Carolina State University

Co-Author(s): Gufeng Wang, Luyang Zhao

Abstract Text

The attachment process of nanoparticles to the fluid interface is considered to play a crucial role in understanding dynamic behavior of complicated multiphase system and modeling behavior of interface absorbed molecules conveniently. The diffusion coefficient of the spherical polystyrene nanoparticles on oil droplet-water interface is measured by our three dimensional single particle tracking technique. This technique is based on astigmatic imaging by adding a cylindrical lens to the light path. The tracking precision is 20 nm, 20 nm, and 30 nm in x, y, and z directions, respectively. The 3D single particle tracking technique provides us a powerful tool to study the movement of nanoparticles on curved interface. Our study shows that polystyrene particles tend to attach onto and diffuse on oil-water interface. Surprisingly, the diffusion coefficient slows down significantly by a factor 3 when the oil droplet size decreases to below ~1 micrometer. Our calculation suggests that this size effect on diffusion coefficient is caused by the changes of their three-phase contact angle, which decreases as the curvature of the oil-water interface increase. This study reveals an unexpected decrease of the diffusion coefficient and will shed light on the study of dynamic properties of absorbed particles in complex media, emulsion stabilization and mass transport on biological cells.

Keywords: Adsorption, Analysis, Fluorescence, Surfactants
Application Code: Nanotechnology
Methodology Code: Fluorescence/Luminescence

Session # 370 Abstract # 370-6 **Oral Sessions**

Session Title: Fluorescence and Luminescence Advances

Abstract Title: **Two-Photon Metal Enhanced Fluorescence Properties of Gold Nanostars**

Date: Monday, March 07, 2016 - Morning
Time: 10:25 AM
Room: B315

Primary Author: Lixia Zhou
Oregon State University

Co-Author(s): Sean M. Burrows

Abstract Text

Gold nanostars are gaining popularity due to their sharp spikes localizing the enhanced electric field on a single nanoparticle. As a result, the need for a reproducible colloidal solution is relaxed for signal enhancement applications. While nanostars have gained popularity for Surface Enhanced Raman Spectroscopy, little work has been done to learn about the Metal Enhanced Fluorescence (MEF) properties of nanostars. MEF is a distance-dependent phenomenon requiring determination of the optimal fluorophore–to-nanostructure distance to achieve the brightest signal enhancement. Here, different length DNA strands were used to vary the distances between the nanostars and Cy3. The nanostars were functionalized with avidin, then a biotin-TEG-polyT-DNA linker containing biotin, triethyleneglycol (TEG), different length poly-Thymine (polyT) regions, and a region that was complementary to another DNA strand labeled with Cy3 (Cy3-DNA). Nanosight and Dynamic Light Scattering were used to confirm the attachment of the nanostars to avidin and biotin-TEG-polyT-DNA sequences. We investigated how the MEF was influenced by the concentration of nanostars, the mole ratio of avidin to nanostars, and the mole ratio of biotin-TEG-polyT-DNA to avidin. Furthermore, we were curious about how the orientation of Cy3 influenced the enhancement by using a 3-prime and a 5-prime biotin-TEG-polyT-DNA linker to orient the dye far and close to the nanostar surface, respectively.

Keywords: Bioanalytical, Fluorescence, Nanotechnology, Spectroscopy
Application Code: Bioanalytical
Methodology Code: Fluorescence/Luminescence

Session # 370 Abstract # 370-8 **Oral Sessions**

Session Title: Fluorescence and Luminescence Advances

Abstract Title: **Construction of Convenient Biosensors Based on Optical Techniques**

Primary Author: Qiaoli Yue, Liaocheng University

Co-Author(s): Chenzhong Li

Date: Monday, March 07, 2016 - Morning
Time: 11:05 AM
Room: B315

Abstract Text

Biosensors are greatly significant because they are capable to resolve a potentially large number of analytical problems and challenges in very diverse areas such as defense, homeland security, agriculture and food safety, environmental monitoring, medicine, pharmacology, industry, etc. Among the current biosensors, optical ones, as a powerful alternative to conventional analytical and biochemical techniques, enable the highly sensitive, real-time, and high-frequency monitoring of analyte without extensive sample preparation. In concept, optical biosensors are those based on the detection of changes on absorption of UV/visible/Infrared light when chemical reactions occur or on the quantity of light emitted by some luminescent process. Our research is focused on developing precise, sensitive, specific, rapid, and easy-to-use biosensor using functional biorecognition materials like aptamers, proteins, mitochondrion and whole cells based on the enhancement of nanostructured substances. These materials provide specificity through use of specific receptors and enhance sensitivity through optical amplification, and they employ materials that can integrate naturally with tissue, such as nanoclusters and nanoparticle suspensions. While aimed primarily toward the long-term goal of biosensors for the maintenance of health, these systems may provide opportunities for advanced basic research as well as potential clinical applications. Towards this goal, with an emphasis on monitoring of cancer cells and the mitochondrial diseases, we have demonstrated hydrogel-based biochemical sensors that change optical properties as measured by luminescence intensity, lifetime, and fluorescence correlated spectrum. This talk will describe several examples of these materials and the underlying motivation for their design, particularly highlighting the major challenges to long-term monitoring.

Keywords: Biosensors, Fluorescence, Nanotechnology, Protein
Application Code: Bioanalytical
Methodology Code: Fluorescence/Luminescence

Session # 380 Abstract # 380-1 **Oral Sessions**

Session Title FTIR and Terahertz Applications

Abstract Title **EPA Methods 320 and 18 by Portable GC/FTIR for Total Source Emission Determination**

Date: Monday, March 07, 2016 - Morning
Time: 08:30 AM
Room: B316

Primary Author Martin L. Spartz
Prism Analytical Technologies, Inc.

Co-Author(s) Anthony S. Bonanno, Charles M. Phillips, Kelly R. McPartland, Peter P. Behnke

Abstract Text

On-site analysis of source emissions is routinely performed using EPA Method 320. Under this method an FTIR gas analyzer utilizing a multiple pass "White" cell can analyze many chemicals from percent to low ppm levels. For ppb MDL requirements, many times a direct FTIR measurement is not sensitive enough.

In the case of VOCs, EPA Method 18 is regularly performed to qualify and quantify the source emissions. This Method is traditionally performed using on-site sample collection (thermal desorption tubes (TDT) or Tedlar bag) followed by an off-site laboratory GC or GC/MS analysis. While this testing is widely used it prevents the customer, source tester and/or regulator from knowing the near-real-time source emissions and acting on that information.

Coupling a gas chromatography (GC) system to a multiple pass FTIR gas analyzer, while maintaining the direct FTIR measurement functionality, allows for EPA Method 18 measurements to be performed in the field in an unprecedented manner. An EPA Method 320 test can be performed while analytes are concentrated onto TDTs for an EPA Method 18 analysis. Once the Method 320 testing is complete, novel GC/FTIR techniques perform the Method 18 analysis. This new GC/FTIR technology traps the GC effluents in the gas cell and measures them continuously allowing for integration and averaging of each species. This allows the FTIR to provide low ppb detection limits for each compound. Due to the nature of spectroscopic detection, a vast, pre-calibrated, compound library can be used to fully speciate individual components of the emission stream.

Keywords: Environmental Analysis, FTIR, Gas Chromatography, Instrumentation
Application Code: Environmental
Methodology Code: Vibrational Spectroscopy

Session # 380 **Abstract # 380-2** **Oral Sessions**

Session Title: FTIR and Terahertz Applications

Abstract Title: **A Novel Infrared Interferometer Suitable for 3D Infrared Hyperspectral Imaging**

Date: Monday, March 07, 2016 - Morning
Time: 08:50 AM
Room: B316

Primary Author: Ryuji Tao
Kagawa University

Co-Author(s): Akira Nishiyama, Ichiro Ishimaru, Kenji Wada

Abstract Text

Handheld FTIR spectrometers have been becoming more and more popular as infrared hyperspectral imagers included in the most promising applications of which are material characterization and surface analysis. Although only two dimensional (2D) images are obtained in principle, three dimensional (3D) images can be reconstituted by coupling the 2D infrared images with the corresponding 2D visible images. To do this reconstitution, one must deduce what the 3D structures look like from the 2D visible images by making use of accumulated knowledge. However, there remain some cases when 3D infrared images cannot readily be reconstituted like gas distribution in chambers or in the open air.

We have been developing a novel infrared interferometer for handheld near infrared or mid infrared hyperspectral imagers where, unlike the conventional Michelson interferometer, interference takes place only when split light beams are combined again. Therefore, light beams originating solely from a focal plane in the object interfere with each other that helps obtain 3D hyperspectral images when an additional capability is implemented which drives the focal plane in the direction along the optical axis.

One of the most appropriate applications of this 3D infrared hyperspectral imager is the gas monitoring in clean rooms in the semiconductor manufacturing facilities. Although it depends on the optical characteristics of the actual implementation, a 1 m spatial resolution can be obtained along the optical axis for the position 10 m from the imager. Detecting gas leakage with the information about where the leakage is taking place should help take immediate measures and minimize the damages caused by the leakage. Another application is the flame detection in bright rooms referring to the specific wavelengths of the flames.

Keywords: Environmental Analysis, Imaging, Monitoring, Vibrational Spectroscopy
Application Code: Safety
Methodology Code: Vibrational Spectroscopy

Session # 380 Abstract # 380-3 **Oral Sessions**

Session Title FTIR and Terahertz Applications

Abstract Title **New Methodology for Finding Optimal Spectral Matches in Reference Databases**

Date: Monday, March 07, 2016 - Morning
Time: 09:10 AM
Room: B316

Primary Author Gregory Banik
Bio-Rad Laboratories, Inc.

Co-Author(s) Karl Nedwed, Ty Abshear

Abstract Text

Optimized Curve Matching and Display is a novel and valuable curve matching and visualization methodology. It allows users to identify optimal spectral matches within reference databases and visualize the comparative results in a way that is more discernible to the human eye. We will discuss multiple corrections that can be applied automatically to compensate for differences between spectral instruments, environmental conditions, sample concentration, ATR correction, and others to optimize the match between spectral curves.

Keywords: Data Analysis, FTIR, Infrared and Raman, Raman
Application Code: General Interest
Methodology Code: Vibrational Spectroscopy

Session # 380 Abstract # 380-4 **Oral Sessions**

Session Title: FTIR and Terahertz Applications

Abstract Title: **Experimental Optimization of IR pMAIRS Using A New Analytical Concept**

Primary Author: Takeshi Hasegawa, Kyoto University

Co-Author(s): Miyako Hada, Nobutaka Shioya, Takafumi Shimoaka

Date: Monday, March 07, 2016 - Morning
Time: 09:30 AM
Room: B316

Abstract Text

Infrared p-polarized multiple-angle incidence resolution spectrometry (IR p-MAIRS) is a recently developed powerful tool to reveal the molecular orientation in a thin film deposited on an IR transparent substrate made of such as silicon, germanium and calcium fluoride. IR pMAIRS yields a set of in-plane (IP) and out-of-plane (OP) spectra of the thin film, which correspond to the conventional transmission (Tr) and reflection-absorption (RA) spectra, respectively, at a time. Although this technique is powerful to study a thin film at hand, experimental optimization is necessary to discuss the molecular orientation quantitatively. Since the molecular orientation is reflected as an intensity ratio of the IP and OP spectra, a standard sample to check pMAIRS has thus been considered to be a film of "oriented molecules" as found in an LB film and SAM. The preparation of such an oriented film is difficult to prepare for the manufacture of the pMAIRS equipment and the users.

In the present study, the "band position" has first been found to be quite useful to check the quantitative reliability of IR pMAIRS without using the band intensity. To do that, the TO-LO splitting of a strongly-absorbing IR band of an isotropic sample should be used. A perfluoroalkyl-containing polymer liquid is found to be suitable for this purpose. As a result, the experimental optimization has quite conveniently been established for the substrates: Ge, Si, ZnSe and CaF2. This would be an important basis for the use of IR pMAIRS.

Keywords: FTIR, Surface Analysis
Application Code: Material Science
Methodology Code: Vibrational Spectroscopy

Session # 380 **Abstract # 380-5** **Oral Sessions**

Session Title: FTIR and Terahertz Applications

Abstract Title: **ATR-FTIR Spectroscopic Imaging of Pharmaceutical Formulations Under Continuous Flow**

Primary Author: Sergei Kazarian
Imperial College London

Co-Author(s): Andrew Ewing

Date: Monday, March 07, 2016 - Morning
Time: 10:05 AM
Room: B316

Abstract Text

We have demonstrated the potential of ATR-FTIR spectroscopic imaging for in situ screening of pharmaceutical systems. Spectroscopic imaging data collected as a function of time reveal spatially resolved and chemically specific information about the samples. Previous work has demonstrated that it is possible to obtain spectroscopic images of dissolving tablets and to determine polymorphic and structural changes in real time, study their behaviour in pH modified environments and validate mathematical models. These studies have provided an understanding of how drug release can be influenced by the selection of appropriate excipients, highlighted the specific conditions that induce structural changes of the drug and measured the effectiveness of controlled release formulations. Here, we present exciting results which combine the use of ATR-FTIR spectroscopic imaging with microfluidic devices for the screening of pharmaceutical systems. Firstly, in these novel imaging studies, high-throughput analysis of release of ibuprofen from several micro-formulations under flowing conditions has been introduced. Moreover, it was possible to investigate how aqueous flows with different pH values affect the release of ibuprofen from micro-formulations in different channels simultaneously. Further investigations report the behaviour of a dissolved drug that crystallised upon contact with an acidic solution. This resulted in precipitation and subsequently agglomeration of the solid particles in the channels. Innovative technologies for high-throughput analysis of formulations in situ are of great interest for pharmaceutical development. The success of combining microfluidics with spectroscopic imaging to study pharmaceutical formulations under flow will impact applications in the fields across pharmaceutical science, biological chemistry and medicine.

Keywords: Dissolution, FTIR, High Throughput Chemical Analysis, Infrared and Raman
Application Code: Drug Discovery
Methodology Code: Vibrational Spectroscopy

Session # 380 Abstract # 380-6 **Oral Sessions**

Session Title: FTIR and Terahertz Applications

Abstract Title: **Advances in Pattern Recognition for the Remote Detection of Sulfur Dioxide by Passive Infrared Spectrometry**

Date: Monday, March 07, 2016 - Morning
Time: 10:25 AM
Room: B316

Primary Author: Brian W. Dess
University of Iowa

Co-Author(s): Gary W. Small

Abstract Text

The U.S Environmental Protection Agency (EPA) Airborne Spectral Photometric Environmental Collection Technology (ASPECT) program is designed in part to provide first responders with accurate chemical awareness at sites of interest. This program utilizes an aircraft fitted with a Fourier transform infrared spectrometer in a passive detection configuration. The challenges present in detecting an infrared signal remotely are strongly tied to the signal-to-noise ratio of the collected interferogram and the specific signal processing and pattern recognition methods used in the data analysis. Depending on the atmospheric effects on the signal of the target analyte, design of a targeted classifier must be tailored to separate the signal from a complex background. Using both laboratory data and data supplied by the ASPECT program, advances have been made in the detection of sulfur dioxide at field locations. New signal processing methods and optimized classification techniques have been applied to achieve these results. In this presentation, the developed data analysis protocols will be demonstrated for use in the automated detection of sulfur dioxide. This classification methodology makes use of a background correction procedure in conjunction with digital filtering and piecewise linear discriminant analysis. The optimization of the individual techniques will be described, as well as the data requirements for their implementation. The performance of the developed classifiers will be assessed by use of interferograms collected from airborne mapping of field locations in which sulfur dioxide is present, most notably a coal-fired powerplant.

Keywords: Chemometrics, Environmental Analysis, FTIR, Pattern Recognition
Application Code: Environmental
Methodology Code: Chemometrics

Session # 380 Abstract # 380-7 **Oral Sessions**

Session Title: FTIR and Terahertz Applications

Abstract Title: **Demonstration of Detection of Hidden Persons and Illegal Substances with an Array of Quantum Cascade Lasers and Cantilever Enhanced Photoacoustic Spectroscopy**

Date: Monday, March 07, 2016 - Morning
Time: 10:45 AM
Room: B316

Primary Author: Sauli Sinisalo
Gasera Ltd.

Co-Author(s): Ismo Kauppinen

Abstract Text

Border security is one of the key security challenges to be taken up in the following years. In particular, the deployment of practical efficient means to detect hidden persons and illegal substances at border crossing points is instrumental in avoiding terrorism, human trafficking or smuggling. The required gas analyzer, which is handheld size, simple to use, sensitive, and capable of providing selective screening over a large number of compounds, is demonstrated to be a reachable target in the near future.

The proposed solution combines cantilever enhanced photoacoustic spectroscopy with an array of distributed feedback quantum cascade lasers (DFB-QCL), which is capable of measuring infrared gas phase spectra of the analyte substances. The high sensitivity in a wide dynamic range is achieved with a silicon MEMS cantilever sensor coupled with an optical interferometric readout system [1,2]. Simultaneous detection of a high number of substances is achieved by measuring the infrared spectra of the sample gas utilizing widely tunable DFB-QCL array technology. The use of these two technologies allows the development of a handheld size device.

The research work, funded by the European Commission within the FP7 project called DOGGIES, has started in 2012 and lasted till November 2015. The project results are given in the presentation along with the description of the DFB-QCL array photoacoustic sensor technology. Also the results with measurements of hidden person detection, illegal drugs and explosives from the proof-of-principle demonstrators will be presented.

[1] T. Kuusela, J. Kauppinen. Appl. Spectrosc. Rev., 42, (2007)
[2] V. Koskinen, J. Fonsen, J. Kauppinen, I. Kauppinen, Vibr. Spectrosc. 42, (2006)
[3] J. Fonsen, V. Koskinen, K. Roth, J. Kauppinen, Vibr. Spectrosc. 50, (2009)

Keywords: Laser, Photoacoustic, Trace Analysis, Vibrational Spectroscopy
Application Code: Homeland Security/Forensics
Methodology Code: Vibrational Spectroscopy

Session # 380 **Abstract # 380-8** **Oral Sessions**

Session Title: FTIR and Terahertz Applications

Abstract Title: **Measurement of Trace HF in Clean Rooms and Ambient Air**

Date: Monday, March 07, 2016 - Morning
Time: 11:05 AM
Room: B316

Primary Author: Ismo Kauppinen
Gasera Ltd.

Co-Author(s): Sauli Sinisalo, Timo Rajamäki, Tuomas Hieta

Abstract Text

During the micro-fabrication process in clean rooms, yield loss is caused for example by Airborne Molecular Contamination (AMC). One of the most critical contaminants is Hydrogen fluoride (HF). Monitoring of HF spills and background levels is extremely demanding due to the high sensitivity (sub-ppb) and fast response time (real-time) requirement from the monitoring equipment.

HF is a common industrial compound used in various fields of industry, such as aluminum smelting, glassware production, manufacturing of pharmaceuticals etc.

HF is non-flammable, highly corrosive and very toxic with strong irritating odor. HF is a very strong acid causing respiratory problems and irritation at low concentrations or at higher concentrations more serious lung and heart damage. This could be fatal. Exposure to HF in the atmosphere can have an adverse effect on the infrastructure. Dissolved HF will cause damage to building materials, such as concrete, limestone and metals. Therefore, it is essential to continuously monitor possible HF sources such as aluminum smelters for detecting and preventing HF leakage to the ambient air.

Gasera addresses the trace HF measurement need with tunable diode
laser-based photoacoustic detection [1,2]. In addition to utilizing the novel cantilever sensor for achieving ultimate detection sensitivity, semiconductor optical amplifier (SOA) and multi-pass optical arrangement is employed. Special arrangement is made for the gas sampling in order to overcome adsorption issues related to the sticky nature of HF gas. Ultimately sub-ppb level detection limits are demonstrated which proves high suitability for the purpose of HF spills and background contamination monitoring in clean room micro-fabrication processes and monitoring of plant emissions in ambient air.

[1] T. Kuusela, J. Kauppinen. Appl. Spectrosc. Rev., 42, (2007)
[2] C. B. Hirschmann, J. Lehtinen, J. Uotila, S. Ojala, R. L. Keiski, Appl. Phys. B (2013)

Keywords: Environmental/Air, Photoacoustic, Semiconductor, Trace Analysis
Application Code: Environmental
Methodology Code: Vibrational Spectroscopy

Session # 390 Abstract # 390-1 **Oral Sessions**

Session Title GCMS of Environmental Analysis

Abstract Title **Continuous Monitoring of Polycyclic Aromatic Hydrocarbons Using Automatic Thermal Desorption-Gas Chromatography**

Date: Monday, March 07, 2016 - Morning
Time: 08:30 AM
Room: B401

Primary Author Franck Amiet
Chromatotec Inc.

Co-Author(s) Damien Bazin, Michel Robert

Abstract Text

Polycyclic Aromatic Hydrocarbons (PAHs) are a group of over 100 different chemicals that are known to be formed typically during incomplete combustion of organic matter at high temperature. Their major sources in the atmosphere include industrial processes, vehicle exhausts, waste incinerations, and domestic heating emissions. Due to their carcinogenic/mutagenic effects, 16 PAHs are currently listed as priority air pollutants. Actual analytical methods dedicated to monitor PAHs require multistep sampling preparations and are not suited for continuous monitoring. Automatic Thermal Desorption-Gas chromatography equipped with flame ionization detector (AUTO TDGC-FID/MS) is the standard method for the monitoring of volatile and semi-volatile hydrocarbons. This technique allows for identifying and quantifying continuously hydrocarbons from ethane to naphthalene. The main goal of this work was to implement a new and simple method for sampling and determination of PAHs in gas and solid phase in air by using thermal desorption technique followed by gas chromatography equipped with two detectors: a flame ionization detector and a Mass spectrometer. A detailed study was carried out to optimize the experimental method in each of its phases, including sampling, thermal desorption, analytical separation, and detection. First, the limits of use of the analytical system were determined during the laboratory phase using liquid standards of the 16 PAHs. Then the applicability of the novel methodology was tested in real environment, namely, at 200 meters from a highway.

Keywords: Gas Chromatography/Mass Spectrometry, PAH, Sample Handling/Automation, Thermal Desorption
Application Code: Environmental
Methodology Code: Gas Chromatography/Mass Spectrometry

Session # 390 **Abstract # 390-2** **Oral Sessions**

Session Title: GCMS of Environmental Analysis

Abstract Title: **Method Development for Evaluation of Pesticides Residue in Lake Lanier-Georgia Gwinnett County's Drinking Water Resource Using Disposable Pipette Extraction (DPX) and Gas Chromatography Mass Spectrometry (GC-MS)**

Date: Monday, March 07, 2016 - Morning
Time: 08:50 AM
Room: B401

Primary Author: Hongxia Guan
Georgia Gwinnett College

Co-Author(s): Huang Wenlin, Rashad Simmons, Simon Mwongela, Xiaoping Li

Abstract Text

Pesticides have become ubiquitous environmental and human health hazards. There is clear evidence that long term exposure to pesticides can cause serious diseases such as neurological disorders, endocrine disruption, birth defects, and cancer. Regulatory and public concern over pesticide residues in water supplies has been increasing. Monitoring of pesticide exposure requires rapid and selective sample preparation so that preventive and treatment protocols can be initiated promptly. Conventional solid-phase extraction (SPE) methods generally require multiple steps due to the need for conditioning and wash steps. In this study, disposable pipette extraction (DPX) method used reversed phase (RP) mechanism has been found to be a rapid and reliable SPE method for pesticides extraction from water. Sample solutions are mixed with the DPX sorbent to provide efficient extractions without concerns of channeling or solution flow rates. Recoveries and relative standard deviation (%RSDs) for the target pesticides are greater than 80% and less than 10%, respectively.

Keywords: GC-MS, Pesticides, Validation, Water
Application Code: Environmental
Methodology Code: Gas Chromatography/Mass Spectrometry

Session # 390 **Abstract # 390-3** **Oral Sessions**

Session Title GCMS of Environmental Analysis

Abstract Title **Overcoming Cost and Supply: Let's Use Nitrogen**

Date: Monday, March 07, 2016 - Morning
Time: 09:10 AM
Room: B401

Primary Author Lee Marotta
PerkinElmer

Co-Author(s) Jacob Rebholz, Roger Bardsley, Tom Hartlein

Abstract Text

The United States Environmental Protection Agency (EPA) has developed several volatile methods designed to protect public health and the environment. The sample introduction technique used in volatile methods is purge and trap (P & T). This technique has many uses including the ability to concentrate a sample to enable low detection limits. The effluent from the P & T enters a gas chromatograph (GC) and the compounds are separated on a GC column. Detection is performed by mass spectrometry (MS), and thus the overall method is referred to as P&T/GC/MS.

Traditionally, these methods use helium as the source for the P & T purge gas and for the GC column carrier gas. However, due to increasing cost and potential limited supply of helium, alternative carrier gases, such as hydrogen and nitrogen are being considered.

Therefore, the goal of this research was to rigorously investigate the use of alternative carrier gases while robustly meeting EPA method criteria.

This presentation will review all method parameters used with the three carrier gases, and the results using nitrogen and hydrogen as alternate carrier gases compared to results attained with the referee carrier gas helium. The EPA methods discussed are 524.3, 624 and 8260. Success was attained with both nitrogen and hydrogen; however, data and reasoning are presented to show that nitrogen could be a more suitable and desirable carrier gas.

Keywords: Automation, Environmental/Water, Purge and Trap, Water
Application Code: Environmental
Methodology Code: Gas Chromatography/Mass Spectrometry

Session # 390 Abstract # 390-4 **Oral Sessions**

Session Title: GCMS of Environmental Analysis

Abstract Title: **Analytical Method Development for 2,4,6,8-Tetrachlorodibenzothiohiophene (TCDT) in River Sediments Utilizing GCxGC-TOFMS and APGC-TQS**

Date: Monday, March 07, 2016 - Morning
Time: 09:30 AM
Room: B401

Primary Author: Maura K. McGonigal
Penn State

Co-Author(s): Adam Ladak, Doug Stevens, Frank Dorman, Kari Organtini, Robert Parette, Wendy Pearson

Abstract Text

A number of environmental studies have been performed to identify contaminants in the Passaic River in New Jersey. The contaminants of interest that emerged from these studies were 2,4,6,8-tetrachlorodibenzothiphene (2,4,6,8-TCDT), 2,3,7,8-tetrachlorodibenzo-p-dioxin (2,3,7,8-TCDD) and polychlorinated dibenzo-p-dioxin and dibenzofuran (PCDD/F) species.

One tool for source identification is calculating the ratio of 2,4,6,8-TCDT to 2,3,7,8-TCDD in the river sediment samples. Early studies on Passaic River sediment struggled with misidentification of 2,4,6,8-TCDT as 1,2,8,9-TCDD due to the two compounds being isobaric (both have a nominal mass of 322). Therefore, chromatographic resolution and access to reference standards for mass spectral comparison is crucial.

Using GCxGC-TOFMS, the mass spectra of 2,3,7,8-TCDD and 2,4,6,8-TCDT were collected. The fragmentation patterns were observed to be different between the two compounds, allowing for the two species to be distinguished from one another. Considering the chromatography of the pair of compounds, 2,4,6,8-TCDT was found to elute close to the tetrachlorodibezno-p-dioxin isomer elution window. A RTX-Dioxin2 column, was utilized to resolve 2,4,6,8-TCDT from the closest tetrachloro dioxin isomer (1,2,8,9-TCDD). GCxGC-TOFMS analysis was performed on the sediment samples and, using isotope dilution, an initial quantification of the samples was performed.

The levels of the dioxins, furans, and dibenzothiophenes were not high enough in all of the river sediments to be observed using GCxGC-TOFMS. Therefore, the river sediment extracts were also analyzed with an APGC-TQS. This analysis was performed for more accurate quantitation as well as selective MRM transitions were used to again resolve the two compound groups from one another.

Keywords: Environmental Analysis, Gas Chromatography, Tandem Mass Spec, Time of Flight MS
Application Code: Environmental
Methodology Code: Gas Chromatography/Mass Spectrometry

Session # 390 **Abstract # 390-5** **Oral Sessions**

Session Title: GCMS of Environmental Analysis

Abstract Title: **A Kendrick Mass Defect Approach Towards the Characterization of Hydraulic Fracturing Fluids**

Primary Author: Paulina Piotrowski
The Pennsylvania State University

Date: Monday, March 07, 2016 - Morning
Time: 10:05 AM
Room: B401

Co-Author(s): Frank Dorman, Jonathan D. Byer, Joseph E. Binkley

Abstract Text

Hydraulic fracturing is an increasingly common technique for the extraction of natural gas entrapped in shale formations. This technique has been highly criticized by some due to the possibility of environmental contamination, among other concerns. The chemical modifiers used in the process in addition to the materials which are solubilized in the well bore and subsequently returned to the surface are most likely contaminants. In this study, we present an environmental forensics approach towards the characterization of these solubilized hydrocarbons in hydraulic fracturing flowback fluids and drinking well water by chromatographic separation using GCxGC coupled to a high-resolution time-of-flight mass spectrometer. This unique instrumentation combination allows for hydrocarbon fingerprinting using a Kendrick Mass Defect approach. Analysis of flowback waters from geographically distinctive locations in Pennsylvania has previously revealed unique hydrocarbon patterns, and Kendrick diagrams were also used to further distinguish well waters impacted by fracking. We hypothesize these findings are a result of the distinctive geologic formations associated with each site. Development of such characterization technologies can aid in determining the sources of contamination implicated by unconventional gas development.

Keywords: Environmental/Water, Gas Chromatography/Mass Spectrometry, Geochemistry, Hydrocarbons
Application Code: Environmental
Methodology Code: Gas Chromatography/Mass Spectrometry

Session # 390 **Abstract # 390-6** **Oral Sessions**

Session Title: GCMS of Environmental Analysis

Abstract Title: **Collaboration of Government (EPA), Industry, Academia to Update EPA Method 625 to Solid Phase Extraction for Drinking and Wastewater Analysis Combining Stir-Bar Sorptive Extraction and EPA Method 6800**

Date: Monday, March 07, 2016 - Morning
Time: 10:25 AM
Room: B401

Primary Author: Anil Srinivas Chaitanya Vishnuvajjhala
Duquesne University

Co-Author(s): Andrew Boggess, David Singer, Ed Pfannkoch, Matt Pamuku, Mizanur Rahman, Skip Kingston, Weier Hao

Abstract Text

The original EPA Method 625 organic pollutants in drinking water is approximately 30 years old and is a liquid-liquid extraction method that requires liters of solvent in sample preparation and replicates. Updating this method to a greener solid phase extraction method was undertaken as a collaboration among solid phase extraction manufacturers and industrial instrument manufactures in collaboration with EPA and Academia. Newer and greener solid phase extraction methods, namely, disk, column, and polydimethylsiloxane (PDMS) coated stir bars followed by automated thermal desorption which uses little solvents, were tested, validated, and compared. Newer methods of direct digital calibration were incorporated eliminating calibration curves and combining EPA Method 625 with EPA Method 6800 for quantification and calibration. Automated digital transfer from laboratory to laboratory was evaluated and found to be within 3-5% between laboratories and method improvements of an order of magnitude in precision and accuracy were observed. Comparative results between and within laboratories for the PDMS stir bars and Method 6800 direct mathematical calibration are compared. An academic and two industrial labs teamed up using the PDMS stir bars without and with the 6800 isotope dilution mass spectrometry in manual and fully automated green technology using thermal desorption reducing solvents in the solid phase extraction to a minimum. These methods are much more environmentally friendly, faster, cost effective, less labor intensive, and automated. These data from the collaborative study and method validation are demonstrated and discussed.

Keywords: Environmental Analysis, Gas Chromatography/Mass Spectrometry, Isotope Ratio MS, Solid Phase Ext
Application Code: Environmental
Methodology Code: Gas Chromatography/Mass Spectrometry

Session # 390 Abstract # 390-7 **Oral Sessions**

Session Title GCMS of Environmental Analysis

Abstract Title **Analysis of Vapor Mercury in Ambient Air and Flue Gas Emissions Using Thermal Desorption Trap Hyphenated with Gas Chromatography/Mass Spectrometry**

Date: Monday, March 07, 2016 - Morning
Time: 10:45 AM
Room: B401

Primary Author Paolo Benedetti
IIA - CNR

Co-Author(s) Carlo Crescenzi, Ettore Guerriero

Abstract Text

Mercury is a potent neurotoxin regarded as globally widespread persistent pollutant. Very recently many countries finalized the negotiations and signed the Minamata convention on Mercury, which commits participating countries to reduce emissions and use of mercury. Successful implementation of the treaty will require adequate verification through global monitoring. The sustainability of a global scale atmospheric monitoring network for mercury in ambient air is limited by the lack cost-effective and reliable analytical methods. For this reason a simple TD-GC-MS method for monitoring and characterizing mercury compounds in the atmosphere was developed using a sampling device based on an innovative sorbent material. Well known advantages of TD-GC technique are reduced risks of contaminations and reduced sample manipulation. The whole analytical procedure can be automatized and, once collected, samples can be safely stored for long periods before the analysis. Air samples are collected on thermal desorption tubes packed with an innovative sorbent. The mercury is then quantitatively thermally stripped and analyzed by GC-MS. The sampling device is suitable for both active and passive collection. The robustness of the method was investigated for major factors, such as humidity and air temperature, potentially affect trapping system and analytical performances. The risk of false positives is reduced by the extreme selectivity of GC-MS detection. Results demonstrate that LOD as low as 30 pg for cubic meter can be easily achieved with high precision (5 %RSD at 3 LOQ).

Keywords: Environmental/Air, GC-MS, Mercury, Volatile Organic Compounds
Application Code: Environmental
Methodology Code: Gas Chromatography/Mass Spectrometry

Session # 390 **Abstract # 390-8** **Oral Sessions**

Session Title: GCMS of Environmental Analysis

Abstract Title: **Soil Gas: Are Targets Being Missed?**

Primary Author: Lee Marotta
PerkinElmer

Co-Author(s): Roberta Provost

Date: Monday, March 07, 2016 - Morning
Time: 11:05 AM
Room: B401

Abstract Text

Compounds found in the interstitial space of soil are referred to as soil gas (or vapor). These compounds may be toxic if located near or at a contaminated water and/or soil source where the compounds volatilized and entered this space. Soil vapor intrusion (SVI) is when these toxic compounds find a pathway into a dwelling located near or at the site causing a negative impact on human health.

The goal of this research was to determine if compounds above the boiling point of naphthalene were present in soil gas and being missed using Environmental Protection Agency (EPA) method TO-15, which is a canister based sampling technique that has poor to no recovery of compounds above the boiling point of naphthalene.

A new EPA TO-17 sorbent tube (XRO-440) has been developed to analyze both volatile and semi-volatile compounds. This tube has an analyte range from C4 to C40, and is capable of accurately measuring compounds from 1,3-butadiene to benzo(g,h,i)perylene, including all 16 regulated polynuclear aromatic hydrocarbons (PAHs).

Research included side by side experiments were conducted at three suspected SVI sites to determine if semi-volatile targets were being missed, thereby comparing TO-17 using XRO-440 tubes to TO-15 using canisters. Yes, targets were being missed by EPA method TO-15. In addition, the results from different sorbent tubes were compared.

In addition, discussed are all method parameters, sampling procedures, analytical parameters, reporting limits and comparison of the results from the site studies will be discussed. An introduction to thermal desorption technique will be provided.

Keywords: Air, Environmental/Air, Gas Chromatography/Mass Spectrometry, Thermal Desorption
Application Code: Environmental
Methodology Code: Gas Chromatography/Mass Spectrometry

Session # 400 Abstract # 400-1 **Oral Sessions**

Session Title Innovative Approaches to Science Education - Half Session

Abstract Title **Teaching Through Research (Mini) Projects of Industrial Relevance**

Date: Monday, March 07, 2016 - Morning
Time: 08:30 AM
Room: B409

Primary Author Jurica Bauer
Inholland

Co-Author(s) Gertjan Heijne, Iris Kuiper, John Vessies, Lieke van Hemert, Maarten Kuiper, Mark Jansen, Mark Verheij, Niek Persoon, Rosalba Bellini

Abstract Text

The Life Sciences and Chemistry (LSC) cluster within Inholland University of Applied Sciences, located in Amsterdam, The Netherlands, offers a four-year BSc study with three majors: biotechnology, biomedical research and chemistry. The study shapes the students into hands-on analists skilled to conduct research in one of the three fields, making them ready to face the challenges of the job market or to pursue further education.

In close collaboration with the industrial partners in the region, the LSC cluster also conducts research in many diverse fields like plant breeding, seed technology and bee diseases. The research is supported by a center of excellence for chemical analysis and carried out by both the staff and the students. In this environment the students learn through conducting research and get to experience the academic curiosity as well as the industrial demand. Inholland strives to maintain a strong connection and balance between education and applied research of interest to the industrial sector in the region. This is accomplished by means of research miniprojects carried out in small groups as well as through more elaborate and demanding individual student internship and graduation projects.

This contribution aims to present the chemistry curriculum at Inholland along with several representative examples of student research projects. The successes and limitations of teaching chemistry through research (mini)projects of industrial relevance will be analyzed and discussed.

The work presented is funded by Inholland and the indutrial partners.

Keywords: Chemometrics, Education, Laboratory, Teaching/Education
Application Code: General Interest
Methodology Code: Education/Teaching

Session # 400 **Abstract # 400-2** **Oral Sessions**

Session Title: Innovative Approaches to Science Education - Half Session

Abstract Title: **Promote Science Education Through Collaborative Learning**

Primary Author: Yi He
John Jay College/CUNY

Co-Author(s): Sandra Swenson

Date: Monday, March 07, 2016 - Morning
Time: 08:50 AM
Room: B409

Abstract Text

A collaborative learning model was created and implemented at John Jay College of the City University of New York to promote science education. This model involves student cohorts at different academic levels, for example, research students, science major students and non-science major students. All students are connected through a research topic that is of common interest. Three levels of collaboration, i.e. research students with non-research students, science major with non-science major students, and collaboration within a cohort, can be developed. Students are engaged and motivated to learn science by working with peers, doing research and applying their knowledge to solve real-world problems that are relevant to their daily life. At John Jay College/CUNY, this model was demonstrated by three groups of undergraduate students, i.e. science research students, Instrumental Analysis students and non-science major Environmental Science students, to study organic and inorganic pollutants in EPA Superfund Sites in New York City. By choosing proper research topics and coupling proper student cohorts, this model can be transferred to other disciplines to promote undergraduate STEM education.

Keywords: Education
Application Code: Other
Methodology Code: Education/Teaching

Session # 400 Abstract # 400-3 **Oral Sessions**

Session Title Innovative Approaches to Science Education - Half Session

Abstract Title **Open Source Instruments for Environmental Monitoring and Science Education**

Date: Monday, March 07, 2016 - Morning
Time: 09:10 AM
Room: B409

Primary Author Jack Summers
Western Carolina University

Co-Author(s) Benjamin Hickman

Abstract Text

Barriers to working with microcontrollers have dropped substantially since the introduction of the Arduino in 2005. Academic labs can now contribute to development of low cost, open source instruments. Once developed, such instruments can be used by citizen groups to monitor environmental pollution and by educators to provide more affordable hands-on, experiences for their students. This talk will cover the development and open publication of two instruments, the WheeStat, a three electrode potentiostat, and the Titraumatic, an automated pH titrator. Enabling technologies like 3d printing technology, free software for coding microcontrollers and graphic user interfaces and electronics design will be discussed. The benefits and pitfalls of starting a business based on open-source technology developed in an academic environment will also be discussed.

Keywords: Education, Electrochemistry, Titration
Application Code: Environmental
Methodology Code: Education/Teaching

Session # **400** **Abstract #** **400-4** **Oral Sessions**

Session Title: Innovative Approaches to Science Education - Half Session

Abstract Title: **Application of Environmental Data for Public Health Response Actions**

Date: Monday, March 07, 2016 - Morning
Time: 09:30 AM
Room: B409

Primary Author: James S. Holler
ATSDR

Co-Author(s)

Abstract Text

The Agency for Toxic Substance and Disease Registry (ATSDR) is the federal public health agency identified to respond to public health issues in the event of acute chemical releases. The Emergency Response Program of ATSDR is composed of specialists in toxicology, chemistry, risk assessment and other sciences. In the event of a chemical release, the ATSDR staff can collaborate with the organization leading the response. The organization assessing and managing the event may be state or local public health officials, other state officials or a federal On-Scene Coordinator from the US Environmental Protection Agency (EPA) or US Coast Guard. The coordinated response often involves the review and assessment of environmental data and a comparison with related reference values. The environmental data is often the observed concentration of toxicants in environmental media-air, water or soil.. Reference values most often used are the ATSDR Minimal Risk Levels or the EPA Reference Dose or Reference Concentration. Such a review allows an estimate of possible public health risks in the community and a determination of public health actions to mitigate such risks. Recent collaborative response actions illustrate how such an effective partnership works to enhance an effective environmental public health response.

Keywords: Clinical/Toxicology, Environmental Analysis, Medical, Toxicology
Application Code: Environmental
Methodology Code: Data Analysis and Manipulation

Session # 410 **Abstract #** 410-1 **Oral Sessions**

Session Title: Microfluidics/Lab-on-a-Chip - Bioanalytical and Others

Abstract Title: **Droplet Microfluidic Device: Application of Nucleosome Preparation to Nucleosome Analysis**

Primary Author: Yi Xu
University of Illinois at Urbana-Champaign

Date: Monday, March 07, 2016 - Morning
Time: 08:30 AM
Room: B406

Co-Author(s): Jeong-Heon Lee, Mallika V. Modak, Richard M. Graybill, Ryan C. Bailey, Tamas Ordog

Abstract Text

Nucleosome preparation is important for many epigenetic studies including chromatin immunoprecipitation (ChIP). ChIP is the gold standard for probing epigenetic protein-DNA interactions that plays critical roles in cell fate and function, aging and carcinogenesis. While developed and applied for many years, traditional ChIP protocols require a large cellular input (10[sub]6[/sub]-10[sub]7[/sub] cells) to prepare target nucleosomal DNA, which limits their utility to study biopsies, rare cells such as cancer and stem cells, and to assess tumor heterogeneity. The nucleosome preparation in ChIP is also laborious and time-consuming, and user-dependent. To address these disadvantages, we have developed an automated, droplet-based microfluidic device that prepares nucleosomal DNA directly from cells within half an hour, more rapidly than traditional protocols. With this device, starting number of cells and reaction time are flexible. Percentages of mono-, di- and tri-nucleosomes are adjustable to satisfy the requirement of different downstream ChIP processing with the capability to accurately control on-chip incubation time. This enabling device is also suitable for other epigenetic study assays, including nucleosome positioning assay and DNase I hypersensitive sites assay, as the type and concentration of enzyme are changeable. This device will provide unprecedented opportunity to prepare samples for studies of multiple epigenetic profiles down to the level of single cells allowing both the assessment of cell heterogeneity within complex clinical samples and the application of cost-effective epigenetic testing to very small samples in individualized medicine settings directly at the point of care.

Keywords: Bioanalytical, Biomedical, Sample Handling/Automation, Sample Preparation
Application Code: Bioanalytical
Methodology Code: Microfluidics/Lab-on-a-Chip

Session # 410 Abstract # 410-2 **Oral Sessions**

Session Title Microfluidics/Lab-on-a-Chip - Bioanalytical and Others

Abstract Title **3D-Printed Analytical Devices Facilitate Investigation of Stored Erythrocytes Used in Transfusion Medicine**

Primary Author Chengpeng Chen
Saint Louis University

Date: Monday, March 07, 2016 - Morning
Time: 08:50 AM
Room: B406

Co-Author(s) Dana Spence, R Scott Martin

Abstract Text

3D-printing has been recently used to fabricate novel microfluidic devices that are rugged, reusable, reproducible and integrated. In this work, a set of 3D-printed analytical devices were developed to investigate stored ERYs used in transfusion medicine, with the goal of understanding ERY storage lesions. A reusable 3D-printed circulation-mimic device enabled the evaluation of stored ERYs on the same device during the whole storage period of 36 days, the results of which suggest that the currently approved hyperglycemic ERY storage solutions irreversibly diminish ERY derived ATP release after 15 days storage. The mechanism by which hyperglycemia reduces ERY derived ATP was also investigated using a 3D-printed cell filtration device. This device minimizes the overall dead volume and simplifies the experimental operation. Results show that hyperglycemia permanently alters ERYs deformability after 1 to 2 week storage, which may explain for the diminished ATP release. An injection device and a -cell-on-chip platform were developed with 3D-printing technology to investigate the response of stored ERYs to the endocrine function. It was discovered that unlike fresh cells or normoglycemia-stored ERYs, hyperglycemia stored ERYs fail to respond to the endocrine mimic correctly. All these results suggest that the currently approved ERY storage solutions can lead to multiple lesions of stored ERYs, which can increase the risk of transfusion related complications. Moreover, 3D-printing enables the fabrication of demand-based analytical devices that facilitate in vitro studies of physiological processes.

Keywords: Biological Samples, Biomedical, Lab-on-a-Chip/Microfluidics, Sampling

Application Code: Biomedical

Methodology Code: Microfluidics/Lab-on-a-Chip

Session # 410 Abstract # 410-3 **Oral Sessions**

Session Title: Microfluidics/Lab-on-a-Chip - Bioanalytical and Others

Abstract Title: **Investigation of Bacterial Behavior in Water Filtration Processes by Using Nanofluidic Devices**

Date: Monday, March 07, 2016 - Morning
Time: 09:10 AM
Room: B406

Primary Author: Nil Tandogan
Northeastern University

Co-Author(s): Edgar D. Goluch, Kai-Tak Wan

Abstract Text

Finding resources for clean water has become a big concern. It is essential to have effective water filtration processes. Slow sand filtration is an economical water filtration technique using sand grains to form pores and filters pathogens as water flows through the sand. The behavior of bacteria in these pores is still not well-known. With the advances in nanotechnology, it was realized that bacteria could pass through structures that are as small as half of their own diameter. Bacteria could change their morphology to move into these structures, which is facilitated under applied pressure. In this study, we investigate the effects of applied pressure on bacterial movement in nanochannels, which mimic the pores. We compared two microorganisms: [i]Escherichia coli[/i] and [i]Pseudomonas aeruginosa[/i]. Nanochannels were fabricated with different widths. The finalized device contains two microchannels, an inlet where the pressure is applied from and an outlet channel open to atmosphere. Inlet and outlet microchannels are connected via these nanochannels, where the bacterial movement is hindered. Electron-beam lithography was used to pattern the nanochannels on a silicon wafer, and inlet and outlet channels were aligned to them with photolithography. The completed features were transferred onto a PDMS polymer through soft lithography and the devices were permanently bonded to glass coverslips using oxygen plasma. Bacterial strains were cultured in Lysogeny Broth overnight at 37 °C, and diluted prior to experiments to avoid clogging of the channels. Our results illustrated that bacterial species show different behavior in narrow structures. Although [i]E. coli[/i] and [i]P. aeruginosa[/i] have similar cell morphology, [i]P. aeruginosa[/i] entered narrow structures at lower pressure values than [i]E. coli[/i] did. These observations might be due to their motility and changes in their size with growth phases.

Keywords: Biotechnology, Environmental/Biological Samples, Lab-on-a-Chip/Microfluidics
Application Code: Environmental
Methodology Code: Microfluidics/Lab-on-a-Chip

Session # 410 Abstract # 410-4 **Oral Sessions**

Session Title: Microfluidics/Lab-on-a-Chip - Bioanalytical and Others

Abstract Title: **A Novel Chemiluminescence Signal Amplification Strategy Based on Microchip Electrophoresis Platform for Highly Sensitive MicroRNA Detection**

Date: Monday, March 07, 2016 - Morning
Time: 09:30 AM
Room: B406

Primary Author: Shulin Zhao
Guangxi Normal University

Co-Author(s): Jian Li, Jingjin Zhao, Yi-Ming Liu, Yong Huang

Abstract Text

A micro analysis platform with chemiluminescence (CL) signal amplification based on nicking endonuclease, G4 DNAzyme and microchip electrophoresis (MCE) was developed for microRNA detection. G-quadruplex DNAzyme as one of DNAzymes can promote hydrogen peroxide to oxidize luminol for CL detection, and easy to conbine with nicking endonuclease for signal amplification (NESA). In this work, we established a NESA-G4-MCE-CL platform for highly sensitive microRNA detection. Using miR-30b as a model target, the levels of miR-30b in cell lysamples were determined within 90 s, with a limit of detection of 8.9 pM.

Keywords: Bioanalytical, Capillary Electrophoresis, Chemiluminescence, Lab-on-a-Chip/Microfluidics
Application Code: Bioanalytical
Methodology Code: Microfluidics/Lab-on-a-Chip

Session # 410 Abstract # 410-5 **Oral Sessions**

Session Title: Microfluidics/Lab-on-a-Chip - Bioanalytical and Others

Abstract Title: **Singe Molecule Nanoelectrophoresis within Thermoplastics**

Primary Author: Colleen ONeil
University of North Carolina at Chapel Hill

Co-Author(s): Steven A. Soper, Swathi Pullagurla

Date: Monday, March 07, 2016 - Morning
Time: 10:05 AM
Room: B406

Abstract Text

Electrophoresis is a rapid and efficient technique for the separation of molecules based on their charge to size ratio. Microchip electrophoresis has proven to be useful for the automated separation of various molecules; however, new efforts have been devoted to reducing electrophoresis columns to nanometer dimensions, because unique phenomena specific to nano-confined domains can be exploited for new separations. These phenomena include, electric double layer (EDL) overlap, increased surface area-to-volume ratios and transverse electromigration (TEM). Because these flow profiles can be generated and with the action of TEM, separations can be undertaken that are not possible in microscale electrophoresis. Conventionally, electrophoresis is carried out using glass-based devices, which have a well characterized surface chemistry that is highly ordered and homogenous; however, fabrication of such devices is costly and time-consuming. In contrast, multiple thermoplastic nanocolumns can be replicated from a single master mold by nanoimprint lithography (NIL), thereby reducing cost and simplifying fabrication. Electrophoresis within thermoplastics nanochannels is an exciting prospect due to these fabrication advantages; however, little is known regarding the effects of thermoplastic surfaces on nanoelectrophoresis separations. For example, activation of thermoplastic surfaces through O[sub]2[/sub] plasma exposure generates surface confined carboxylic acid functional groups. This allows for a tunable surface charge density; thus unique control of surface properties. This research presents the separation of fluorescently labeled single deoxynucleotide monophosphate molecules within thermoplastic nanochannels and how factors such as field strength, channel dimension, surface charge, buffer composition and ionic strength affect electrophoretic mobility and resolution within thermoplastic nanochannels.

Keywords: Capillary Electrophoresis, Microscopy, Nanotechnology, Polymers & Plastics
Application Code: Nanotechnology
Methodology Code: Capillary Electrophoresis

Session # 410 **Abstract # 410-6** **Oral Sessions**

Session Title: Microfluidics/Lab-on-a-Chip - Bioanalytical and Others

Abstract Title: **Parallel Functionality Determination of shRNA Knockdown Constructs Using Microfluidic Technologies**

Date: Monday, March 07, 2016 - Morning
Time: 10:25 AM
Room: B406

Primary Author: Kristen Entwistle
Michigan State University

Co-Author(s): Dana Spence

Abstract Text

Recently, the zinc transporter protein, ZnT-8, has been implicated in the etiology of diabetes. ZnT-8 is present on secretory granules in the pancreatic beta cells, and is responsible for the transport of zinc from the cytoplasm of the beta cells into the insulin granule. Loss of function mutations of the gene encoding for ZnT-8 that result in truncated protein have been linked to increased susceptibility to type 2 diabetes in humans, and ZnT-8 has also been found to be downregulated in hyperglycemic mice. Our group has successfully created variants of the INS-1 cell line (rat insulinoma cells) with differing levels of ZnT-8 expression using standard shRNA technology. While these cells showed varying level of expression (from ~40-65% knockdown depending on the construct transfected into the cells) using classical tools such as gel electrophoresis and western blotting, nothing is known about the level of functionality of the INS-1 cells following transfection and knockdown of the ZnT-8 protein. Here, we report the use of a fluidic device with multiple parallel channels for simultaneous determination of such secretogues as insulin, C-peptide, and zinc from each of the knockdown cells. While used for an insulin-secreting cell line in this study, the platform could be integrated with any cell line system for which a functional assay is known.

Keywords: Biological Samples, Biotechnology, Lab-on-a-Chip/Microfluidics, Protein
Application Code: Biomedical
Methodology Code: Microfluidics/Lab-on-a-Chip

Session # 410 **Abstract # 410-7** **Oral Sessions**

Session Title: Microfluidics/Lab-on-a-Chip - Bioanalytical and Others

Abstract Title: **A Universal Droplet Microfluidic Strategy for On-Chip Operations Including Reagent Injection, Sample Washing, and Droplet Tagging**

Date: Monday, March 07, 2016 - Morning
Time: 10:45 AM
Room: B406

Primary Author: Steven R. Doonan
University of Illinois at Urbana-Champaign

Co-Author(s): Dongkwan Lee, Richard M. Graybill, Ryan C. Bailey, Yi Xu

Abstract Text

Droplet microfluidics for small sample or even single cell analysis has brought the potential of personalized medicine closer to realization. This technique compartmentalizes samples into picoliter-sized droplets carried by immiscible oil. Due to encapsulation samples do not interact with device walls, which limits loss to nonspecific adsorption. Small volumes also facilitate rapid mixing. In order to process samples, microfluidic devices must form droplets, inject reagents, and decant waste. These operations each proceed through combinations of specific geometries. On the other hand, this means a given device is typically limited to one set of processes and conditions. To overcome this limitation we have recently developed the "K Junction," a versatile geometry for many droplet processes.

The K Junction provides tunable reagent injection or volume removal on a single device. It operates via a parallel fluid stream in an electric field that influences passing droplets at the junction. Integration with antibody-conjugated magnetic beads even allows for selective on-chip washing steps and sample concentration via a device-embedded magnet. Other K Junction processes include droplet splitting, droplet respacing, and programmable droplet tagging. By altering applied pressures and electric field, we can modify performance or switch a single K Junction device from one operation to another. Through characterizing this architecture, we have successfully engineered a generalized geometry that can be rationally applied and quickly optimized for a range of applications.

Keywords: Automation, Lab-on-a-Chip/Microfluidics, Sample Handling/Automation, Sample Preparation
Application Code: High-Throughput Chemical Analysis
Methodology Code: Microfluidics/Lab-on-a-Chip

Session # 410 **Abstract # 410-8** **Oral Sessions**

Session Title: Microfluidics/Lab-on-a-Chip - Bioanalytical and Others

Abstract Title: **Reversible Wettability Switching in Fabric-Based Microfluidic Devices**

Primary Author: Tanya Narahari
Northeastern University

Date: Monday, March 07, 2016 - Morning
Time: 11:05 AM
Room: B406

Co-Author(s): Dhananjaya Dendukuri, Shashi K. Murthy, Tripurari Choudhary

Abstract Text

Controlling the wettability of a solid surface is important in microfluidics. For instance, the sensitivity of an over-the-counter pregnancy test (a membrane-based lateral flow test) varies inversely with sample flow rate. Such devices lack a precise, active means of flow control and are limited to analytes located in the high concentration detection window. In prior work, we developed a textile weaving approach to manufacture lateral flow devices. In our approach, yarns are pre-selected based on their surface properties, functionalized with the appropriate reagents and weaved into a testing device. The inherent benefits of a weaving based approach are its low cost, scalability and the fact that devices are assembled in a quasi-single step. Liquid flow was tuned passively by tuning the twist frequency, thickness and composition of yarns. We now present an active tuning approach. Polypyrrole(s) are a class of conductive polymers that switch reversibly between non-wetting (oxidized) and wetting (reduced) states. For the present device, yarns were coated by polymerizing Pyrrole monomer in the presence of Iron (III) Chloride and incorporated into the fabric sensor along with metal wire electrodes. When subjected to a reducing potential, the Water contact angle in the polymer-coated region switches from superhydrophobic (137 °) to superhydrophilic (0 °). With a subsequent oxidizing potential the region reverts to hydrophobic (> 100 °). The polymer behaves as a microfluidic valve, initially acting as a barrier to sample (plasma or urine) flow and allowing ample time for analyte-reagent interaction. Using electron dispersive x-ray spectroscopy, we show that the hydrophobic-to-hydrophilic transition is likely due to an efflux of Chloride ions from the polymer, and vice versa. The major improvement over the state of the art was the avoidance of an additional oxidizing dopant. The switch was achieved within short time scales (<2 min) by applying a small potential (<20V/cm).

Keywords: Lab-on-a-Chip/Microfluidics, Material Science, Microelectrode, Polymers & Plastics
Application Code: Material Science
Methodology Code: Microfluidics/Lab-on-a-Chip

Session # 420 Abstract # 420-1 **Oral Sessions**

Session Title Neurochemistry

Abstract Title **Fast-Scan Cyclic Voltammetry Measurements of Serotonin Release and Reser Pools in R6/2 Huntington's Disease Model Mice**

Date: Monday, March 07, 2016 - Morning
Time: 08:30 AM
Room: B402

Primary Author Rachel C. Gehringer
The University of Kansas

Co-Author(s) Michael A. Johnson, Sam V. Kaplan, Sarah Fantin

Abstract Text

Huntington's disease (HD) is an autosomal dominant, neurodegenerative disorder that causes uncontrolled movements, cognitive impairment, mood disorders, and ultimately, death. Up to 60 percent of patients with HD suffer from depression. Much attention has been given to the dopaminergic system in HD due to the massive degeneration of the striatum, a region heavily innervated by dopamine neurons. However, due to the well-established connection between depression and serotonin, we chose to investigate serotonin alterations due to HD. Fast-scan cyclic voltammetry (FSCV) is an electrochemical technique that allows for µm spatial resolution, sub-second temporal resolution, and nM detection limits when used with carbon fiber microelectrodes. Furthermore, FSCV can be performed [i]ex vivo [/i] in brain slices in order to measure changes in serotonin concentration in the extracellular space in real time. Here, using FSCV at carbon fiber microelectrodes, we measured serotonin release due to an electrical stimulus in two different regions of R6/2 HD model mice: the dorsal raphe (DR) and the substantia nigra, pars reticulata (SNr). In addition to release, we also measured serotonin reserve pool content in both the DR and SNr. Our results show that in R6/2 mice aged 12-14 weeks, serotonin release was diminished to 18±3% of wild-type controls in the SNr region, and 28±8% of wild-type controls. However, our data show that reserve pool serotonin content was not altered in the DR of 12-14 week-old R6/2 model mice.

Keywords: Bioanalytical, Electrochemistry, Neurochemistry, Voltammetry
Application Code: Bioanalytical
Methodology Code: Electrochemistry

Session # 420 **Abstract #** 420-2 **Oral Sessions**

Session Title: Neurochemistry

Abstract Title: **Combining Voltammetry, Mathematical Modeling and 2-Photon Microscopy to Correlate In Vivo Serotonin Chemistry to Physiology**

Date: Monday, March 07, 2016 - Morning
Time: 08:50 AM
Room: B402

Primary Author: Aya Abdalla
University of South Carolina

Co-Author(s): Christopher W. Atcherley, David Linden, Janet Best, Michael C. Reed, Michael L. Heien, Parastoo Hashemi, Yunju Jin

Abstract Text

Serotonin is one of the most important neuromodulators in the brain and its role in mood and emotional processes as well as its association with diseases like depression make it an important pharmacology target. However many serotonin targeting agents (e.g antidepressants) have notoriously low efficacy rates primary because the principle mechanisms that regulate extracellular serotonin in vivo are not well-understood. In defining the fundamentals of serotonin related disorders, it is important to understand how the physiology of different brain areas controls serotonin's chemistry. The dynamics of serotonin release and reuptake can be studied in depth using fast scan cyclic voltammetry (FSCV), but because of technological limitations (background subtraction), FSCV cannot be used to study ambient serotonin levels. The ability to measure ambient, in vivo, serotonin levels is critical to understand serotonin's chemistry.

In this paper, we first describe a novel method, fast scan controlled adsorption voltammetry (FSCAV), which depends on controlled adsorption, to measure serotonin's steady-state, extracellular chemistry. We investigate three serotonin circuitries via the medial forebrain bundle terminating in the prefrontal cortex, in the hippocampus and in the substantia nigra, pars reticulata. Using FSCV, FSCAV, mathematical modeling and 2 photon imaging, we show how the local cytoarchitechture of different brain areas controls serotonin chemistry hence how chemical measurements can give information about local physiology.

Our study is the first to combine a powerful interdisciplinary tool set that provides fundamental information on serotonin chemistry in different brain regions.

Keywords: Bioanalytical, Electrochemistry, Neurochemistry, Voltammetry
Application Code: Neurochemistry
Methodology Code: Electrochemistry

Session # 420 Abstract # 420-3 **Oral Sessions**

Session Title Neurochemistry

Abstract Title **Coregulation of Serotonin and Histamine in the Mouse Premammilary Nucleus**

Primary Author Srimal A. Samaranayake
University of South Carolina

Date: Monday, March 07, 2016 - Morning
Time: 09:10 AM
Room: B402

Co-Author(s) Aya Abdalla, H Frederick Nijhout, Janet Best, Michael C. Reed, Parastoo Hashemi, Rhiannon Robke

Abstract Text

Neurodegeneration cause incurable disorders such as Parkinson's and Alzheimer's diseases. These disorders are incurable and difficult to treat. To develop more effective therapies for neurodegenerative diseases, it is critical to better understand the dynamic [i]in vivo[/i] chemistry of neurotransmitters. Serotonin and histamine are co-localized as in many brain regions and are thought to be involved in many neurological disorders, in particular, Parkinson's disease. In this paper we develop a novel voltammetric method that can simultaneously and selectively monitor both serotonin and histamine in real time. We evoke histamine release in mouse brain at premammillary nucleus via electrical stimulation of medial forebrain bundle. Histamine release causes rapid inhibition of serotonin in a concentration dependent manner. Mathematical models illustrate an active uptake mechanism for histamine, and indicate that serotonin inhibition is an autoreceptor mediated process. We therefore pharmacologically explore active histamine reuptake via monoamine transporters and serotonin inhibition via histamine autoreceptors. This fundamental mechanistic information can be used to examine full extent of interconnectivity between histamine and serotonin in brain.

Keywords: Bioanalytical, Electrochemistry, Neurochemistry, Voltammetry
Application Code: Neurochemistry
Methodology Code: Electrochemistry

Session # 420 **Abstract # 420-4** **Oral Sessions**

Session Title: Neurochemistry

Abstract Title: **Application of In Vivo Fast-scan Cyclic Voltammetry to Examine the Impact of (+)-Methamphetamine of the Regulation of Norepinephrine and Dopamine**

Date: Monday, March 07, 2016 - Morning
Time: 09:30 AM
Room: B402

Primary Author: Ken Wakabayashi
State University of New York at Buffalo

Co-Author(s): Jinwoo Park

Abstract Text

(+)-Methamphetamine (METH) is a widely abused psychostimulant drug with profound effects on dopaminergic transmission in the mesolimbic pathway in the brain, mainly by inducing dopamine efflux by reversing the presynaptic transporter, influencing neurons that originate in the ventral tegmental area (VTA) and project to the nucleus accumbens (NAc). The effects of METH on norepinephrine transmission, another major catecholamine system in the brain, is less understood. The impact of METH on norepinephrine neurons may be critical, as METH has been reported to be more potent at releasing norepinephrine via a similar presynaptic mechanism than at releasing dopamine. To explore this issue in vivo, we employ fast-scan cyclic voltammetry coupled with carbon-fiber microelectrodes, a monitoring technique that provides both sub-second temporal resolution and high spatial resolution. This approach allows real-time measurement of METH-induced changes in extracellular NE concentration in the bed nucleus of the stria terminalis (BNST), a region that receives dense NA innervation. This result elucidates how METH effects norepinephrine regulation in comparison to its impact on DA in an intact brain, which may underlie its mechanism of action and possibly contribute to its addictive properties.

Keywords: Drugs, Electrochemistry, Neurochemistry, Voltammetry
Application Code: Neurochemistry
Methodology Code: Electrochemistry

Session # 420 Abstract # 420-5 **Oral Sessions**

Session Title: Neurochemistry

Abstract Title: **Characterization of Spontaneous Transient Adenosine in Rat Brain Slices**

Primary Author: Scott T. Lee
University of Virginia

Date: Monday, March 07, 2016 - Morning
Time: 10:05 AM
Room: B402

Co-Author(s): B Jill Venton

Abstract Text

Adenosine is an important neurochemical responsible for protecting tissue during hypoxic events, such as stroke, and for modulating the activity of other neurotransmitters. Previous work has primarily looked at variations in adenosine concentration over the course of several minutes to hours. Fast-scan cyclic voltammetry (FSCV) has been used to measure neural adenosine activity with sub-second temporal resolution. Recently, spontaneous, transient adenosine events have been characterized by the Venton lab in anesthetized rats. In vivo, the concentration of these adenosine events are 0.19 (+/- 0.01) μM on average and last about 2.9 seconds in the pre-frontal cortex (PFC). The release of transient adenosine is regulated in part by the inhibitory adenosine A1 receptor. However, there are several other possible receptors and channels that could affect the release and clearance of transient adenosine events. Unfortunately, many pharmacological agents that could be used to probe these pathways are either not readily available or don't cross the blood-brain barrier. Experimenting in brain slices allows us to supplement in vivo work and combat the latter issue. In this work, adenosine transients were characterized in brain slices of the PFC and compared to results found in vivo. Transient adenosine in slices had concentrations of 0.20 (+/- 0.02) μM and lasted for 5.0 (+/- 0.3) seconds. The concentration agrees well with in vivo measurements, though transients seem to last longer in situ. Transients in brain slices also seem to occur more frequently than those found in vivo, with an event occurring every 70 seconds in slices on average and every 108 seconds in vivo. Knowing the characteristics of transient adenosine events in brain slices allow for a more complete analysis of the biological pathways that control them.

Keywords: Bioanalytical, Electrochemistry, Microelectrode, Neurochemistry
Application Code: Neurochemistry
Methodology Code: Electrochemistry

Session # 420 **Abstract # 420-6** **Oral Sessions**

Session Title: Neurochemistry

Abstract Title: **Comparing Contralateral Catecholamine Release with Simultaneous Voltammetric Measurements**

Date: Monday, March 07, 2016 - Morning
Time: 10:25 AM
Room: B402

Primary Author: Megan E. Fox
University of North Carolina at Chapel Hill

Co-Author(s): R Mark Wightman

Abstract Text

Although there has been significant work characterizing monoamine projections throughout the brain, our knowledge of its connections rely heavily on early tracing and lesioning studies. There is ongoing debate regarding the existence and contribution of contralateral projections to catecholamine release at nerve terminals. Newer tracing studies provide evidence for a small number of dopamine neurons that cross the midline. Here we examine the functional connectivity of these neurons by measuring contralateral catecholamine release with fast-scan cyclic voltammetry. By employing a 2-electrode, simultaneous recording strategy, we can compare electrically evoked release in both hemispheres to the same stimulation. We demonstrate sites of contralateral dopamine release in both dorsal and ventral striatum, and compare it with contralateral norepinephrine release in the bed nucleus of the stria terminalis. These measurements provide insight into functional connectivity and provide a new context for unilateral manipulations in the brain.

Keywords: Electrochemistry, Neurochemistry
Application Code: Neurochemistry
Methodology Code: Electrochemistry

Session # 420 Abstract # 420-7 **Oral Sessions**

Session Title: Neurochemistry

Abstract Title: **Insights into the Effects of Non-stimulant ADHD Drugs on Catecholamine Transmission in the Rat Brain Using In Vivo Fast-scan Cyclic Voltammetry**

Primary Author: Rohan Bhimani
The State University of New York at Buffalo

Co-Author(s): Jinwoo Park

Date: Monday, March 07, 2016 - Morning
Time: 10:45 AM
Room: B402

Abstract Text

Perturbations of catecholamine systems in the brain are prominent in many psychological disorders. Attention deficit/hyperactivity disorder (ADHD) is of particular importance as abnormal levels of dopamine and norepinephrine lead to a negative phenotype. Noradranergic and dopaminergic receptors and transporters are ubiquitously distributed throughout the central nervous system, making them prime drug targets to modulate the levels of catecholamines. Unfortunately, dopamine and norepinephrine differ by only a single hydroxyl group and selective drugs for targeting central catecholamines have an effect on the contra-catecholamine albeit at a lower affinity. Until recently, first response treatments to ADHD have often included psychostimulants (e.g. methylphenidate and dextroamphetamine) that have a relatively high abuse potential. Here, we present the effects of non-stimulant ADHD drugs targeting noradrenergic receptors and transporters; guanfacine, a norepinephrine alpha-2 autoreceptor agonist and atomoxetine, a norepinephrine uptake transporter inhibitor on both norepinephrine and dopamine transmission in the anesthetized rat brain. For this study, multi-channel fast-scan cyclic voltammetry at two separate carbon-fiber microelectrodes was used to examine how these drugs specifically modulate the dynamics of catecholamine release and clearance in discrete brain regions simultaneously (i.e. nucleus accumbens shell and ventral bed nucleus of the stria terminalis) in a single animal. Fast-scan cyclic voltammetry provides the temporal and spatial resolutions necessary to investigate both the changes in catecholamine release from ADHD medications and understand how they alter off-target catecholamine concentrations.

Keywords: Electrochemistry, Electrodes, Neurochemistry, Voltammetry
Application Code: Neurochemistry
Methodology Code: Electrochemistry

Session # 420 **Abstract # 420-8** **Oral Sessions**

Session Title: Neurochemistry

Abstract Title: **Investigation of Enkephalin-Evoked Catecholamine Secretion in Adrenal Tissue by Fast-Scan Cyclic Voltammetry**

Date: Monday, March 07, 2016 - Morning
Time: 11:05 AM
Room: B402

Primary Author: Lars Dunaway
North Carolina State University

Co-Author(s): Leslie A. Sombers

Abstract Text

The adrenal glands regulate physiological responses to stressors, in part by secretion of the catecholamines, epinephrine and norepinephrine. A variety of peptides, including opioid peptides, are thought to be co-stored with catecholamines in dense core vesicles in adrenal cells. The opioid system is strikingly complex, and the precise interaction between opioid peptides and catecholamines is not well understood. We have used fast-scan cyclic voltammetry coupled to carbon-fiber microelectrodes to study the kinetic properties of the opioid/catecholamine interaction in a rat adrenal slice preparation. These experiments are complicated by the fact that many peptides, both endogenous and synthetic, foul the electrode surface and interfere with catecholamine detection. To address this issue, a sawhorse waveform was developed to clean the electrode surface, enabling reproducible measurements. This allows simultaneous monitoring of the application and clearance of exogenous opioid peptide (met-enkephalin or its synthetic analog DAMGO, a mu opioid receptor agonist), and the effects of this manipulation on local catecholamine dynamics. We found that an acute application of M-ENK or DAMGO evokes catecholamine release that is sensitive to blockade by naltrexone, a non-selective opioid receptor antagonist. DPDPE, a delta opioid receptor agonist, does not elicit a physiological response. These results demonstrate that M-ENK evokes catecholamine secretion in the adrenal medulla by binding to mu opioid receptors, providing a chemical mechanism by which opioid peptides can regulate an organism's response to physiological and environmental demands.

Keywords: Electrochemistry, Neurochemistry, Voltammetry
Application Code: Neurochemistry
Methodology Code: Electrochemistry

Session # 430 **Abstract #** 430-1 **Oral Sessions**

Session Title Pharmaceutical-GC and LC

Abstract Title **Analyte Stability and Side Reactions During Pharmaceutical Analysis**

Primary Author Leah Xiong
Merck Co.

Date: Monday, March 07, 2016 - Morning
Time: 08:30 AM
Room: B403

Co-Author(s) Brian Regler, Cory Bottone, Eugenia Muschajew, Justin Pennington, Leih-Shan Yeung, Nathan D. Contrella

Abstract Text

Pharmaceutical analyses of drug substance and drug product are critical to ensure the safety and quality of medications. Common pharmaceutical analyses include purity by HPLC/UPLC, dissolution by USP apparatus, residual solvents by GC-FID, and moisture content by Karl-Fischer (KF) titration. During the method development, the Design of Experiment should consider analyte stability during the analysis to minimize any potential method bias. Three case studies of analyte degradation and the corresponding mitigation strategy will be discussed in details:

• Compound A is prone to Peroxy radical mediated oxidative degradation. Dissolution media containing Tween 80 and iron contamination could generate peoxy radicals and therefore limited sample stability. Investigation of antioxidant and chelating agent identified the solution and effectively improved the sample stability.

• Product B is a Fixed Combination Product that contains three Active Pharmaceutical Ingredients. During the transfer of oven-based KF method, different carrier gas – air or nitrogen was found to result in significantly different moisture content. Upon further investigation, two of the APIs formed a condensation degradate and released water in the presence of oxygen. Inert nitrogen was chosen as the carrier gas in the final oven KF method.

• Residual solvent of compound C was measured using headspace GC-FID. Significant amount of methanol was detected in compound C even though methanol was not used as a process-related solvent. In depth investigation revealed that the carbamate group of compound C degraded in the head space oven and released methanol as a side product. Headspace GC method was optimized to eliminate this side reaction successfully.

Keywords: Dissolution, Drugs, Gas Chromatography, Mass Spectrometry
Application Code: Pharmaceutical
Methodology Code: Liquid Chromatography/Mass Spectrometry

Session # 430 **Abstract # 430-2** **Oral Sessions**

Session Title: Pharmaceutical-GC and LC

Abstract Title: **A New Tool for Residual Solvent Analysis of Technical Grade Hexanes**

Date: Monday, March 07, 2016 - Morning
Time: 08:50 AM
Room: B403

Primary Author: Kenneth G. Lynam
Agilent Technologies

Co-Author(s): Johan Kuipers, John Oostdijk, Ramaprasad Ganni

Abstract Text

Residual solvent testing according to United States Pharmacopeia chapter <467> is a requirement for drugs substances either produced or imported into the US. This test is done by static headspace GC FID and specifies types and allowable limits for the solvents used in the manufacture of drug substances. Permissible limits for individual solvents are based on their toxicity. Chlorinated solvents are to be avoided where ever possible based on their toxicity. Resolving hexane isomers in technical grade hexane from dichloromethane isomers using FID detection is challenging using traditional cyano-propyl phenyl (G43 or 624) based GC stationary phases. This presentation highlights the challenge of routinely achieving sufficient chromatographic resolution with traditional 624 phases. A new tool is introduced for addressing this challenge in the form of a 624 phase optimized for resolution of hexane and dichloromethane isomers. Example chromatograms illustrating the issue along with improved results with the optimized phase are shown. Tips and tricks for successful separation of technical grade hexane isomers from dichloromethane isomers are discussed. Best practices for achieving consistent static headspace results will also be part of the discussion.

Keywords: Gas Chromatography, GC Columns, Pharmaceutical, Sample Handling/Automation
Application Code: Pharmaceutical
Methodology Code: Gas Chromatography

Session # 430 **Abstract # 430-3** **Oral Sessions**

Session Title: Pharmaceutical-GC and LC

Abstract Title: **Development and Validation of a Fast Ion-Paired Reversed Phase Stability-Indicating Method for the Assay of Thiabendazole and Estimation of its Related Compounds**

Date: Monday, March 07, 2016 - Morning
Time: 09:10 AM
Room: B403

Primary Author: Peng Zhang
Merial, A Sanofi Company

Co-Author(s): Abu Rustum, Jingzhi Tian

Abstract Text

Thiabendazole is a benzimidazole based antihelmintic active pharmaceutical ingredient (API) which is used to treat of intestinal pinworm and strongyloides infections. Its mechanism of action is via selective binding to beta-tubulin of parasitic worms, causing their immobilization and death. Thiabendazole is also used as fungicide to control fungal diseases in fruits and vegetables.

A stability-indicating reversed-phase ion-paired HPLC method was developed for the assay of thiabendazole and estimation of its related compounds. Chromatographic separation of thiabendazole and its related compounds was achieved by using an isocratic elution at a flow rate of 1.5 mL/minute using ACE 5 (C18, 4.6x50 mm, 5um particle size) as the primary column and Phenomenex Luna (C18, 4.6x50 mm, 5um particle size) as the equivalent column at 35°C. The mobile phase consists of 25% acetonitrile and 75% 10 mM sodium 1-octanesulfonate (as an ion-pairing reagent) aqueous solution containing 0.1% of methanesulfonic acid. A UV detector at 300 nm was used to detect the analytes. The total run time for this method is only 4 minutes. The new method was successfully validated according to International Conference on Harmonization (ICH) guidelines and was found to be specific, linear, accurate, precise, robust and sensitive. The stability indicating capability of the method was demonstrated through adequate separation of all potential thiabendazole related compounds (from thiabendazole and from each other) that are present in aged and stressed degraded samples under heat, light, base, acid and oxidation. This fast ion-paired reversed phase HPLC method is ideal for QC labs to conduct routine test for the assay of thiabendazole and estimation of its related compounds.

Keywords: HPLC, Ion Chromatography, Method Development, Validation
Application Code: Pharmaceutical
Methodology Code: Separation Sciences

Session # 430 Abstract # 430-4 **Oral Sessions**

Session Title Pharmaceutical-GC and LC

Abstract Title **Determining Lithium in Pharmaceutical Products**

Primary Author Sachin P. Patil
Thermo Fisher Scientific

Date: Monday, March 07, 2016 - Morning
Time: 09:30 AM
Room: B403

Co-Author(s) Jeffrey Rohrer

Abstract Text

In response to the USP's effort to modernize existing monographs across all compendia, this work describes alternative methods for lithium analysis that are automated, fast, and use an aqueous mobile phase (eluent). Ion chromatography (IC) offers a significant improvement to the existing assays because it can simultaneously determine lithium and common cation impurities in a single injection. The two methods described here for lithium determination in lithium citrate and lithium hydroxide were validated in accordance with guidelines prescribed in the USP General Chapter <1225> for following parameters:

1)Separation: A representative chromatogram demonstrates separation of six cation mix using the method proposed for lithium citrate. Lithium is well resolved from nearest cation i.e. sodium under proposed gradient conditions with other cations eluting quickly (Figure 1).

2)Sample Analysis: Recoveries of commercially available lithium salts were 100.4% (lithium citrate) and 99.2% (lithium hydroxide) of the label claim respectively, indicating that the methods are capable of determining lithium concentration within the USP specification.

3)Linearity: Analysis of lithium based on proposed methods yielded a linear relationship of peak area to concentration. The linearity range for lithium was 0.1 to 15 mg/L for lithium citrate assay and 0.3 to 20 mg/L for lithium hydroxide assay.

4)Sample Accuracy and Precision: Recovery of all three spike levels of tested, yielded was satisfactory (Tables 1A and 1B). Replicate injections at three different concentration levels of lithium resulted in the retention time RSDs of □0.05% and the peak area RSDs of □0.37% (Table 1) indicating excellent assay precision.

5)Robustness: No Major effect of small (±10%) variations in procedural parameters on chromatographic performance of the methods was observed. (data not shown)

The assays described here are candidates to replace the existing lithium assays in the USP monograph.

Keywords: Drugs, Ion Chromatography, Pharmaceutical, Quality Control

Application Code: Pharmaceutical

Methodology Code: Liquid Chromatography

Session # 430 **Abstract # 430-5** **Oral Sessions**

Session Title: Pharmaceutical-GC and LC

Abstract Title: **Two Dimensional Liquid Chromatography (2D-LC), A "Must-have" Tool in Pharmaceutical Development, When 1D-LC is Inadequate**

Primary Author: Imad A. Haidar Ahmad
Novartis

Date: Monday, March 07, 2016 - Morning
Time: 10:05 AM
Room: B403

Co-Author(s): Adrian Clarke, Andrei Blasko, James Tam, Thomas Tarara, Xue Li

Abstract Text

One-dimensional liquid chromatography (1D-LC) is not always capable of efficiently separating complex samples efficiently. This drawback is not solely because of the lack of column efficiency but is mainly due to insufficient selectivity and the need to separate with orthogonal retention mechanisms. In this regard, two-dimensional liquid chromatography (2D-LC) is currently attracting much interest due to its markedly higher resolving power compared to one dimensional separations. In this work, accurate mass spectrometry (MS) data were needed to identify multiple degradation product peaks observed with a method that used a non-volatile mobile phase. The MS data were collected with a MS friendly volatile buffer mobile phase using the same LC column. 2D-LC was used to facilitate the correlation of MS data to impurity peaks collected with the non-volatile buffer method. Peaks of interest were collected from the first dimension, via heart-cutting mode and re-injected into the second dimension, where a volatile LC-MS compatible buffer was used. Using 2D-LC, we were able to correlate the retention times of the peaks of both separation methods and assign accurate MS results to the peaks of interest in the original method. Without 2D-LC, much more time consuming fraction collection experiments would be been needed to achieve the goal.

*Due to confidentiality reasons, no MS results can be shown, only chromatograms.

Keywords: Chromatography, Liquid Chromatography, Liquid Chromatography/Mass Spectroscopy
Application Code: Pharmaceutical
Methodology Code: Separation Sciences

Session # 430 **Abstract # 430-7** **Oral Sessions**

Session Title: Pharmaceutical-GC and LC

Abstract Title: **Introducing Modern LC Column Technology into a Research-Led Pharmaceutical Teaching Environment**

Date: Monday, March 07, 2016 - Morning
Time: 10:45 AM
Room: B403

Primary Author: William J. Lough
University of Sunderland

Co-Author(s)

Abstract Text

Of the challenges facing the pharmaceutical industry, a shortage of skilled analysts could be perceived as not being the most pressing or as high profile as others. Nonetheless it is one that should command attention and not be underestimated. Accordingly in teaching separation science on pharmaceutical science programmes it is imperative not just to provide students with appropriate knowledge and skills for employment in the pharmaceutical industry but also to inspire them to go on to become tomorrow's expert separation scientists. While working in a research-led teaching environment is conducive to this, it is necessary to go beyond basing teaching around research interests, in this case in LC, to move to having students conduct research through their 'taught' classes and laboratory exercises, focussing on exploiting the latest developments in modern LC column technology. Such activities have involved (a) evaluating and attempting to exploit the selectivity of recent commercial LC stationary phase introductions (b) developing and exploiting ballistic uHPLC gradients by adapting them for method development and (c) optimising and minimising chiral LC screening. Similar strategies have previously been successful, but only time will tell if this latest incarnation fulfils its aims. However, an important outcome has been the evolving symbiotic relationship between teaching and research. While not quite going as far as teaching-led research, it has become easier to sustain research in a teaching-intensive background and the leads developed through teaching have fed back into research (applications of chiral uHPLC) and have been adapted for use in routine QC laboratories.

Keywords: Chiral Separations, HPLC Columns, Liquid Chromatography, Pharmaceutical

Application Code: Pharmaceutical

Methodology Code: Liquid Chromatography

Session # 430 **Abstract # 430-8** **Oral Sessions**

Session Title: Pharmaceutical-GC and LC

Abstract Title: **Improved Gas Chromatographic (GC) Quantification of Acidic and Neutral Cannabinoids**

Primary Author: Amanda Rigdon
Restek Corporation

Date: Monday, March 07, 2016 - Morning
Time: 11:05 AM
Room: B403

Co-Author(s): Corby Hilliard, Jack Cochran, Joan Serdar, Linx Waclaski, Rebecca Stevens

Abstract Text

Quantification of cannabinoids is required for labeling purposes in most states where medical and/or recreational cannabis is legally dispensed. Accurate quantification of cannabinoid content is important for dosage purposes and assurance of product uniformity, and is required in most states where cannabis is available. The most straightforward way to accurately quantify both acidic and neutral cannabinoids is liquid chromatography (LC), however some laboratories do not have access to LC instrumentation and are limited to GC techniques.

Neutral cannabinoids may be quantified using GC, but acidic cannabinoids cannot be separately quantified via GC because they are decarboxylated in the hot GC injection port, resulting in the acidic and neutral cannabinoids being reported as a sum. Another drawback of GC cannabinoid quantification is the phenomenon of irreproducible decarboxylation, which results in inaccurate reporting.
A fast and simple derivatization method using BSTFA + 1% TMCS was developed to address both of the drawbacks associated with GC cannabinoid testing. This method was tested for linearity and stability using different concentrations of acidic and neutral cannabinoids, as well as different plant matrix concentrations. Because some of the derivatization sites are sterically hindered, mass spectrometry was employed to ensure derivatization went to completion during the course of the experiment. Results show that the derivatization process was completed in all samples.

This presentation will explore the phenomenon of incomplete decarboxylation, introduce a fast derivatization method for cannabinoids in plant matrix, and explore the feasibility of using this method in alternative matrices such as food.

Keywords: Derivatization, Drugs, Food Safety, Pharmaceutical
Application Code: General Interest
Methodology Code: Gas Chromatography/Mass Spectrometry

Session # 440   Abstract # 440-1   **Oral Sessions**

Session Title   Synthesis and Characterization of Nano Particles

Abstract Title   **Antibacterial Activities and Cytotoxicity of Green Synthesized Stable Gold Nanoparticles from Flavonoid Derivatives**

Date: Monday, March 07, 2016 - Morning
Time: 08:30 AM
Room: B405

Primary Author   Francis J. Osonga
Binghamton University

Co-Author(s)   Apryl P. Jimenez, David C. Luther, Idris Yazgan, Omowunmi Sadik, Phuong N. Lee

Abstract Text

The fundamental goal of sustainable nanotechnology is tailored towards nanoscale control of synthesis and processing of matter without footprints that give rise to environmental degradation. Therefore, currently there is a search for synthetic methods that utilize fewer amounts of materials, water, and energy; while reducing or replacing the need for organic solvents. We hereby present a novel approach for synthesis of gold nanoparticles using naturally-derived water soluble flavonoids including Quercetin pentaphosphate (QPP), Quercetin sulfonic acid (QSA) and Apigenin Triphosphate (ATRP) which were utilized both as reducing agent and stabilizer. The synthesis was achieved at room temperature using water as a solvent and it requires no capping agents. In this case the synthesis of gold nanoparticles using naturally derived flavonoids contributes immensely in promoting ideals of green synthesis and nanotechnology by eliminating the use of hazardous and toxic organic solvents and adopting the use of water as a solvent. The synthesized nanoparticles were characterized using Uv-visible spectroscopy (Uv-Vis),X-ray diffraction (XRD), Transmission electron microscopy (TEM),High resolution transmission electron microscopy (HR-TEM) and selected area electron diffraction (SAED). The gold nanoparticles were spherical in shape with average particle size of 10.45 nm, 12.66 nm and 13.54 nm for the nanoparticles derived from QPP, ATRP and QSA respectively. The surface plasmon resonance peak of the synthesized gold nanoparticles derived from QSA, ATRP and QPP was observed at 541 nm, 544 nm and 547 nm respectively which correspond to gold nanoparticles peak. The synthesized AuNPs exhibit good antibacterial activities and hence can be used in water treatment plants for purification of water since the nanoparticles are able to inhibit the bacterial growth.

Keywords: Characterization, Surface Analysis, UV-VIS Absorbance/Luminescence, X-ray Diffraction
Application Code: Nanotechnology
Methodology Code: UV/VIS

Session # 440 Abstract # 440-2 **Oral Sessions**

Session Title: Synthesis and Characterization of Nano Particles

Abstract Title: **Wavelength Selective Photocatalysis Using Gold-Platinum Nano-Rattles**

Date: Monday, March 07, 2016 - Morning
Time: 08:50 AM
Room: B405

Primary Author: Mahmoud Mahmoud
Georgia Institute of Technology

Co-Author(s): Batyr Garlyyev, Mostafa A. El-Sayed

Abstract Text

The selectivity of a thermal catalyzed reaction with multiple products, catalyzed on the surface of a catalyst exhibiting different surface facets and surface energy, is controlled by the temperature. Manipulating the temperature makes one product more dominant than the other. The selectivity of the photochemical reaction on the surface of plasmonic nanocatalyst of multiple plasmon modes is controlled by changing the wavelength of the exciting light. Gold nanospheres (AuNSs) located inside of gold-platinum double shell nanoparticles in a rattle structure were prepared with different sizes and showed two plasmon spectral modes. The high energy plasmon mode corresponds to the photo-excitation of the small nanosphere, while the low energy plasmonic mode is related to both the gold-platinum double shell plasmon and the inside nanosphere, as assigned by calculation using the discrete dipole approximation (DDA) simulation technique. Photodimerization of 4-nitrothiophenol (4NTP) adsorbed on the surface of gold platinum nanorattles (AuPtNRTs) was studied using the surface-enhanced Raman spectroscopy technique. When the AuPtNRTs are photo-excited at the high energy band at 532 nm, which selectively excites the AuNS, 4NTP is photodimerized into an azo compound only on the surface of AuNS. The 4NTP adsorbed on the surface of the outer gold-platinum double shell did not react. Despite the fact that 785 nm photons excite both the AuNS and the outer shell of the AuPtNRTs, no photodimerization is observed with such low energy photons.

Keywords: Raman, Spectrophotometry, Surface Enhanced Raman
Application Code: Nanotechnology
Methodology Code: Physical Measurements

Session # 440   Abstract # 440-3   **Oral Sessions**

Session Title   Synthesis and Characterization of Nano Particles

Abstract Title   **Reversible Electron Delocalization of Molecule-Like CdSe Nanoclusters Using Z-Type Ligand Functionalization**

Primary Author   Katie N. Lawrence
Indiana University-Purdue University Indianapolis

Date: Monday, March 07, 2016 - Morning
Time: 09:10 AM
Room: B405

Co-Author(s)

Abstract Text

Precise understanding of mechanisms underlying the exciton relaxation pathways of semiconductor nanocrystals is extremely important for the rational design of solid state devices using these materials. Unlike quantum dots (QDs, >3.0 nm diameter), which contain a low percentage of surface atoms with their photophysical properties dominated by core atoms, a relatively larger number of atoms occupies the surface of semiconductor nanoclusters (<2.5 nm diameter, SNCs). Therefore, the surface structure of SNCs substantially dominates the mechanism of photo-excited carrier relaxation. Neutral ligands, such as amines, only bind to surface metal centers leaving the chaocogenide sites unpassivated. This creates nonradiative trap states, which directly affect the inter-band exciton relaxation pathways. Here we show that passivation of Se sites of metal chalocogenide SNCs by Z-type ligands (e.g., M(O[sub]2[/sub]CR)[sub]2[/sub]) facilitates the delocalization of electron wavefunctions into ligand monolayers and increases the excited state lifetimes. We have also observed that the electron delocalization process and exciton relaxation dynamics are controlled by the energetic alignment between SNC and ligand orbitals and formation of hybrid orbitals. M(O[sub]2[/sub]CR)[sub]2[/sub] is a sigma-accepting Z-type ligand, thus it will bind to the Se sites in the SNCs causing a decrease in defect sites and an increase in direct recombination. During this time, the QY increases (From 1% to 73% PL-QY). We have proven that our shift is directly related to the band gap alignment by reversing the electronic delocalization using TMEDA and through control experiments in which no shift occurred. The qualitative molecular orbital modeling of the electron delocalization through band alignment is provided. This first example of fully reversible electron delocalization will have a great impact on material science and more specifically materials for optoelectronic and photovoltaic devices.

Keywords: Adsorption, Fluorescence, Nanotechnology, Semiconductor
Application Code: Nanotechnology
Methodology Code: UV/VIS

Session # 440 Abstract # 440-4 **Oral Sessions**

Session Title: Synthesis and Characterization of Nano Particles

Abstract Title: **Characterization Techniques for Nanomaterials – An overview**

Primary Author: Chady Stephan
PerkinElmer

Date: Monday, March 07, 2016 - Morning
Time: 09:30 AM
Room: B405

Co-Author(s)

Abstract Text

Nanomaterials characterization is complex and requires many analytical platforms in order to achieve the task at hand. This talk connects nanomaterials parameters that most commonly need to be measured with corresponding measurement techniques. It is a brief overview for nanotechnology scientists to rapidly identify the appropriate technique needed for measuring nanomaterials. Among the techniques that will be discussed;

Single Particle ICP-MS is a new advancement in ICP-MS devoted to the analysis of individual metallic nanoparticles ranging from single digit nm up to a few µm. It is element specific, allows the differentiation between ionic (M+) and particulate signals (particles) in a wide variety of matrices without any prior separation. In one sample analysis, SP-ICP-MS provides ionic and particle concentration, particle composition, size and size distribution.

Direct Sampling Analysis (DSA) Time of Flight (TOF) mass spectrometer is an ambient ionization technique coupled to an accurate mass spectrometer allowing the analysis of capping agent. The organic capped ligands are released from the nanoparticles due to thermo-lability of the covalent bonds. Using accurate mass and isotope profile information provided by the TOF, we were able to confirm the presence of the different types of ligands attached to nanoparticles. Besides just identifying and confirming one type of organic monolayer covalently bound to the nanoparticle, we were also able to identify bilayers wherein, one organic monolayer is covalently modified with a second type of organic ligand.

Thermal Gravimetric Analysis (TGA) – Infra Red (IR) – Gas Chromatography Mass Spectrometry (GC/MS) is a very powerful combination of analytical techniques all operated together in synergy towards the identification of various organics residing on the surface of Nanostructure.

Keywords: ICP-MS, Mass Spectrometry, Nanotechnology, Thermal Analysis

Application Code: Nanotechnology

Methodology Code: Atomic Spectroscopy/Elemental Analysis

Session # 440 Abstract # 440-5 **Oral Sessions**

Session Title: Synthesis and Characterization of Nano Particles

Abstract Title: **A Novel Nanoparticle Tracking Analysis System for Improved Determinations of Nanoparticle Concentration and Size Distribution**

Date: Monday, March 07, 2016 - Morning
Time: 10:05 AM
Room: B405

Primary Author: Dariusz Stramski
Manta Instruments Inc

Co-Author(s): Kuba Tatarkiewicz, Monette Karr, Rick Cooper, Rick A. Reynolds

Abstract Text

Although nanoparticle uses are widespread and growing rapidly, there is an unmet need for a routine, automated, and user-friendly method that reproducibly provides accurate characterization of concentration and size distribution for a wide diversity of nanoparticles. Accurate determinations of nanoparticle number concentration and size distribution are particularly challenging for samples of mixed nanoparticle sizes. Improved knowledge of these properties, from innovations in particle characterization instruments, will contribute to future advancements of nanoparticle-enabled products in many industries. We describe a novel Nanoparticle Tracking Analysis (NTA) instrument for routine determination of nanoparticle concentrations and sizes based on the analysis of individual nanoparticles undergoing Brownian motion in liquid medium. Such individual particle measurement systems offer significant advantages over methods that measure an ensemble of particles. The invention is based on innovative, multi-spectral illumination and detection techniques that enable video recording of scattered light from wide-ranging sizes of individual nanoparticles simultaneously. A key advancement of the system, and its novel particle tracking analysis software, is the reduction of artifacts and uncertainties common in other techniques, especially those resulting from the very large dynamic range of scattered light intensity produced by differently-sized nanoparticles co-existing in a polydisperse sample. The instrument has successfully characterized different nanoparticle materials in various liquids including samples from several independent third parties. We will present results of validation testing of our system with various nanoparticle samples.

The development of the NTA system was funded by NSF, Scripps Institution of Oceanography UCSD, and Triton Technology Fund.

Keywords: Instrumentation, Nanotechnology, Particle Size and Distribution
Application Code: Nanotechnology
Methodology Code: Physical Measurements

Session # 440 **Abstract # 440-6** **Oral Sessions**

Session Title: Synthesis and Characterization of Nano Particles

Abstract Title: **Coupled Calorimetric-Manometric Technique for the Study of Sorption and Thermodynamic Properties of Macroscopic and Nanosized Materials**

Date: Monday, March 07, 2016 - Morning
Time: 10:25 AM
Room: B405

Primary Author: Kristina Lilova
Setaram Inc.

Co-Author(s): Link Brown

Abstract Text

Understanding the thermodynamics of the adsorption is essential for the practical applications of different porous catalysts and among all the enthalpy of adsorption (or desorption) is a key parameter. The surface properties of the nanomaterials cannot be properly studied without taking into account the energetics of the physi and chemisorbed gases. The combination of Sievert's technique (to quantify the amount of gas absorbed/released) and calorimetry was successfully applied for direct measurement of enthalpy of formation per mole of gas.
The gas sorption Sievert's technique has proven to have many advantages for the evaluation of the ad- or ab- sorbed amount of gas by porous materials in a wide range of temperature and pressure. Coupling calorimetry and manometry allows a full characterization of different gas adsorption on porous bulk and nanomaterials and provides the following thermodynamic values: 1) Adsorption isotherms; 2) Integral heats of adsorption; 3) Isosteric heat of adsorption; 4) Differential heat of adsorption as a function of the adsorbed amounts. A gas sorption calorimetry method, using Calvet type microcalorimeter (Sensys evo DSC) coupled with PCTPro-2000 Sievert's manometric system at high pressure other types of manometric dosing system at atmospheric pressure will be presented. Several examples including Cu-BTC, MOF-5, and CD-MOF2 were selected to demonstrate the technique. Additionally, the water adsorption energetics of nanosized oxides will be discussed and illustrated with several examples.

Keywords: Adsorption, Nanotechnology, Other Hyphenated Techniques, Thermal Analysis
Application Code: Nanotechnology
Methodology Code: Thermal Analysis

Session # 440 Abstract # 440-7 **Oral Sessions**

Session Title: Synthesis and Characterization of Nano Particles

Abstract Title: **Exploring the Effects of Surface Ligand Structural Parameters on Exciton Delocalization of CdSe Nanocrystals**

Date: Monday, March 07, 2016 - Morning
Time: 10:45 AM
Room: B405

Primary Author: Meghan Teunis
Indiana University-Purdue University Indianapolis

Co-Author(s)

Abstract Text
A complete understanding of the mechanism of exciton delocalization of semiconductor nanocrystals is critical for the design and optimization of solid-state devices. Ultrasmall semiconductor nanocrystals, with atomically precise core composition, are an ideal system to study these mechanisms due to their large number of surface atoms that control the photophysical properties of the nanocrystals. We investigate how exchanging the native oleylamine ligands with various chalcogenol ligands, with different levels of electronic interaction (mode of binding, binding head group, and pi-electron conjugation), affect the optoelectronic properties of CdSe nanocrystals. Specifically, we used optical spectroscopic techniques to investigate the ground-state exciton delocalization process. We observe an unprecedented red shift of 650 meV of the first excitonic peak of (CdSe)[sub]34 [/sub] nanocrystals when exchanged with pyrenedithiocarbamate (Py-DTC). We propose the "molecular orbitals" model to rationalize the delocalization of strongly confined excitons..

Keywords: Material Science, Nanotechnology, Semiconductor, UV-VIS Absorbance/Luminescence
Application Code: Material Science
Methodology Code: UV/VIS

Session # 440 Abstract # 440-8 **Oral Sessions**

Session Title: Synthesis and Characterization of Nano Particles

Abstract Title: **Metallic and Hybrid Nanomaterials: Fabrication and Applications**

Date: Monday, March 07, 2016 - Morning
Time: 11:05 AM
Room: B405

Primary Author: Simona Hunyadi Murph
Savannah River National Lab

Co-Author(s)

Abstract Text

The development of functional nanomaterials with multi-detection modalities opens up new avenues for creating multi-purpose technologies for many applications. Multifunctional nanoparticles take advantage of the physicochemical properties of two or more materials to create a new multifunctional composite nanostructure. The "ideal" nanoparticle would combine multiple detection modalities (magnetic, plasmonic, elastic or inelastic light scattering, fluorescence, storage, tracking, delivery), be stable under biological/industrial conditions, could be easily functionalized, environmentally friendly, and could be prepared in large quantities. Noble metal nanoparticles, particularly gold and silver, have exciting physical and chemical properties that are entirely different from the bulk and meet many of these characteristics. By combining these metallic nanomaterials with either magnetic (Fe_2O_3) or photocatalytic (TiO_2) materials one could create tunable multifunctional nanostructures with multi detection-capabilities, e.g. plasmonic, magnetic, catalytic functionalities. The objective of this talk is to discuss our recent synthesis and characterization techniques used to create different size, shape, composition and morphology multifunctional nanoparticles and their corresponding applications.

Keywords: Material Science, Metals, Nanotechnology, Sensors
Application Code: Material Science
Methodology Code: Chemical Methods

Session # 450 Abstract # 450-3 **Poster Sessions**

Session Title Agriculture

Abstract Title **Fast and Precise Nitrogen and Carbon Determination Using an Elemental Analyzer**

Primary Author Guido Giazzi
Thermo Fisher Scientific

Date: Monday, March 07, 2016 - Morning
Time:
Room: Exposition Floor, 400 Aisle

Co-Author(s) Dominique Chevalier, Francesco Leone, Liliana Krotz

Abstract Text

Nitrogen and Carbon analysis by combustion analysis is very commonly used for soils, sediments, deposit on filters, plants, animal tissues analysis and marine samples. The classical configuration of the Elemental Analyzer showed in this paper is based on a double reactors system: one reactor for combustion and the second one for reduction. The high performances of the Dynamic Flash Combustion method used by the FLASH 2000 allows the NC determination using a single reactor in which both steps, combustion and reduction are present.

In this way the chemicals consume is reduced, the analysis is faster and it is possible to analyze the same number of samples. This paper presents data on NC determination in different soils and plants samples to demonstrate the excellent repeatability of this new configuration without matrix effect.

Keywords: Agricultural, Elemental Analysis, Soil
Application Code: Agriculture
Methodology Code: Atomic Spectroscopy/Elemental Analysis

Session # 450 **Abstract # 450-4** **Poster Sessions**

Session Title: Agriculture

Abstract Title: **Carbon/Nitrogen Ratio in Soils and Plants Using an Elemental Analyzer**

Date: Monday, March 07, 2016 - Morning
Time:
Room: Exposition Floor, 400 Aisle

Primary Author: Guido Giazzi
Thermo Fisher Scientific

Co-Author(s): Liliana Krotz

Abstract Text

The Carbon Nitrogen Ratio of the organic material added to the soil influences the rate of decomposition of organic matter and this results in the release (mineralization) or immobilization of soil nitrogen. If the added organic material contains more Nitrogen in proportion to the Carbon, then Nitrogen is released into the soil from the decomposing organic material, while if the organic material has a less amount of Nitrogen in relation to the Carbon then the microorganisms will utilize the soil Nitrogen for further decomposition and the soil nitrogen will be immobilized and will not be available. In few words, cellular Carbon and Nitrogen metabolism must be tightly coordinated to sustain optimal growth and development for plants. Furthermore, C/N balance is also critical for the ecosystem response to elevated atmospheric CO_2. For these reasons the use of an accurate instrumental analytical techniques is required. As the demand for improved sample throughput, reduction of operational costs and minimization of human errors is becoming every day more notable, it is very important apply a simple and automatic technique which allows the fast analysis with excellent reproducibility. A new Elemental Analyzer based on the dynamic flash combustion of the sample, copes effortlessly with the wide array of laboratory requirements such as accuracy, day to day reproducibility and high sample throughput. This paper presents data on Nitrogen and Carbon determination in soils and plants and the relative ratio calculated automatically by the software Eager Xperience, to show the repeatability of the results obtained.

Keywords: Agricultural, Characterization, Elemental Analysis, Soil
Application Code: Agriculture
Methodology Code: Atomic Spectroscopy/Elemental Analysis

Session # 450 Abstract # 450-5 **Poster Sessions**

Session Title: Agriculture

Abstract Title: **Development of Electrochemical Detection System Combining with Nitrocellulose Membrane for Quantitative Immunochromatography**

Primary Author: Wataru Iwasaki
Natl. Inst. Adv. Ind. Sci. Technol. (AIST)

Date: Monday, March 07, 2016 - Morning
Time:
Room: Exposition Floor, 400 Aisle

Co-Author(s): Masaya Miyazaki, Mizuki Ryu, Osamu Niwa, Ramachandra Rao Sathuluri, Ryoji Kurita

Abstract Text

Immunochromatography is one of rapid and simple bioanalytical device. However it is not suitable for quantitative detection of biomarkers. Therefore, we have been developing quantitative immunochromatography by applying electrochemical method to immunochromatography. When we measure redox current of electrochemical species flowing through a nitrocellulose membrane, the contact between the membrane and the electrode is very important to improve the sensitivity. In this regards, we have fabricate electrochemical detector with micropyramid array electrode to improve the contact between the membrane and the electrode. In this study, we developed combining system of nitrocellulose membrane and the electrochemical detector at constant and low contact force. Furthermore, we evaluated the combining system and electrochemical detector by measuring potassium ferricyaninde.

The combining system consists of plastic jig, polymer block, USB connector, load cell and pinch cock. The electrochemical detector was inserted into USB connector and placed onto plastic jig fabricated by 3D printer. Load cell is built in the plastic jig to measure contact force between the membrane and electrode. The nitrocellulose membrane is placed onto detection area of electrochemical detector and pressed with PDMS polymer block, another plastic jig and pinchcock. Screwing pinchcock enables to apply and control the contact force. Elastic polymer block contributes to precise control of contact force. We evaluated the electrochemical immunochromatography with potassium ferricyanide.

Keywords: Biosensors, Electrochemistry, Lab-on-a-Chip/Microfluidics
Application Code: Agriculture
Methodology Code: Electrochemistry

Session # 460 Abstract # 460-1 **Poster Sessions**

Session Title: Clinical/Toxicology

Abstract Title: **Benzo[a]pyrene Levels in Mainstream Smoke from Spectrum Research Cigarettes**

Date: Monday, March 07, 2016 - Morning

Time:

Room: Exposition Floor, 400 Aisle

Primary Author: Jared Hughes
Centers for Disease Control and Prevention

Co-Author(s): Bryan Hearn, Clifford Watson, Shirley Ding, Wayne Chen

Abstract Text

The objective of this research was to quantitate benzo[a]pyrene (BAP) levels in different formulations of Spectrum research cigarettes. BAP is a well-known carcinogen in the family of compounds polycyclic aromatic hydrocarbons. The significance of the study is that it shows levels of BAP in mainstream smoke as cigarette design is varied significantly to give a wide range of tar and nicotine levels. The total particulate matter from each cigarette was collected onto a filter pad as the cigarette was smoked with a commercial smoking machine. Solvent extraction of the filter pad was followed by solid-phase extraction for sample cleanup. Analysis was done by isotope dilution GC/MS in the selective ion mode. The mainstream smoke yields of BAP for the different cigarette formulations were obtained for two common smoking regimes. The Spectrum cigarettes contained amounts of BAP that are similar to those found in 3R4F Kentucky Reference Cigarette. The smoke deliveries of BAP for each regime were found to be less than the deliveries of a popular leading brand cigarette in the United States. The concentration of BAP had a positive linear relationship with the stated amount of tar in the cigarette. There was no statistically significant relationship between BAP and the reported nicotine or menthol.

This study was supported by internal funds at the Centers for Disease Control and Prevention.

Keywords: Clinical/Toxicology, GC-MS, PAH

Application Code: Clinical/Toxicology

Methodology Code: Gas Chromatography/Mass Spectrometry

Session # 460 **Abstract # 460-2** **Poster Sessions**

Session Title: Clinical/Toxicology

Abstract Title: **Selected Carbonyl Levels in Mainstream Smoke from Spectrum Research Cigarettes**

Date: Monday, March 07, 2016 - Morning
Time:
Room: Exposition Floor, 400 Aisle

Primary Author: Michele Chan
Centers for Disease Control and Prevention

Co-Author(s): Clifford Watson, Shirley Ding, Xizheng Yan

Abstract Text

Carbonyls are a group of highly volatile and reactive organic compounds found in mainstream cigarette smoke. They are compounds of interests due to their short and long term health effects when inhaled and therefore are included on the FDA's Harmful and Potentially Harmful Constituents list. This study looks at the carbonyl levels in Spectrum reduced nicotine research cigarettes, which are specially designed with various tar and nicotine levels. These cigarettes will be used in clinical studies, but to date have not been characterized for other common tobacco constituents. A linear smoking machine is used to smoke cigarettes using two smoking regimens: ISO (35ml puff volume at 60s intervals with filter tip ventilation open) and CI (55 ml puff volume at 30s intervals and closed ventilation). Seven carbonyls (formaldehyde, acetaldehyde, acetone, acrolein, propionaldehyde, crotonaldehyde and methyl ethyl ketone) were derivatized and trapped "in-situ" on filter pads. The derivatized carbonyls are extracted from the pads and subsequently quantified by isotope dilution ultra-performance liquid chromatography tandem mass spectrometry. From currently collected data, we observed high ventilation spectrum cigarettes delivered lower levels of carbonyls compared to the ones with low ventilation under ISO regimen. Under CI regimen, levels of carbonyls from no or low nicotine variants are similar regardless of tar level. We observed that smoke formaldehyde levels were much lower in spectrum cigarettes compared to 3R4F research cigarettes. Smoke acrolein and crotonaldehyde also demonstrated a similar trend. Accurate assessment of cigarette yields are important to help assess potential exposures to harmful smoke constituents.

Keywords: Clinical/Toxicology, Liquid Chromatography/Mass Spectroscopy
Application Code: Clinical/Toxicology
Methodology Code: Liquid Chromatography/Mass Spectrometry

Session # 460 **Abstract # 460-3** **Poster Sessions**

Session Title: Clinical/Toxicology

Abstract Title: **Absolute Quantification of Apolipoproteins in Serum and the Efficacy of Trypsin While Utilizing Ultra Performance Liquid Chromatography - Isotope Dilution Mass Spectrometry (UPLC-IDMS)**

Date: Monday, March 07, 2016 - Morning

Time:

Room: Exposition Floor, 400 Aisle

Primary Author: Michael L. Andrews
Centers for Disease Control and Prevention

Co-Author(s): Bryan A. Parks, Christopher Toth, David Schieltz, Jeffrey Jones, John R. Barr, Jon Rees, Kuklenyik Zsuzsanna, Michael S. Gardner

Abstract Text

Apolipoprotein A-1 (ApoA-1) and B-100 (ApoB-100) serum levels are two of the strongest measurable predictors for cardiovascular disease (CVD) risk. ApoA-1 and ApoB-100 are the primary protein components of high density lipoprotein (HDL) and low density lipoprotein (LDL), respectively. Our laboratory seeks to develop a reference method to quantify ApoA-1 and ApoB-100 using liquid chromatography (UPLC) coupled isotope dilution mass spectrometry (IDMS) approach in a manner that is traceable to universal peptide and protein reference standards. The method uses 30 µL (100 fold diluted) serum for trypsin digestion in the presence of acid labile detergent followed by UPLC-IDMS analysis using peptide calibrators (i.e., absolute quantification). The efficacy of the trypsin digest was evaluated with timed incubation and monitoring the maximum concentration of three cleavage peptides relative to isotope labeled peptide analogs. Two cleavage peptides (limit peptides) with highest digestion efficacy in 4 hours were selected for both ApoA-I and ApoB-100 quantification. The agreement between the measured ApoA-I and ApoB-100 concentration in primary and secondary standard serum materials and their certified value, suggest close to 95% digestion efficacy. The ruggedness of the UPLC-IDMS analysis was tested by modifying eluent parameters (pH and column temperature) and MS scanning cycle time optimal for 15 min sample run time. The intra-day and inter-day method precision with CVs of □10% was obtained in 20 independent runs by triplicate analysis of serum samples collected from individual donors with low, medium and high HDL/LDL levels. The method throughput is high enough for application in epidemiologic studies.

Keywords: Biological Samples, Clinical Chemistry, Lipids, Liquid Chromatography/Mass Spectroscopy
Application Code: Clinical/Toxicology
Methodology Code: Liquid Chromatography/Mass Spectrometry

Session # 460 **Abstract # 460-4** **Poster Sessions**

Session Title: Clinical/Toxicology

Abstract Title: **Mainstream Smoke Deliveries of Tobacco Specific Nitrosamines in Spectrum Clinical Research Cigarettes**

Date: Monday, March 07, 2016 - Morning
Time:
Room: Exposition Floor, 400 Aisle

Primary Author: Patrick Chen
Centers for Disease Control and Prevention

Co-Author(s): Clifford Watson, Josh Wong, Liqin Zhang, Shirley Ding

Abstract Text

Tobacco Specific Nitrosamines (TSNAs) are formed from the nitrosation of alkaloids (e.g. nicotine) which are found commonly in tobacco products. Analysis of TSNA levels is of interest because they are potential human carcinogens. The Centers for Disease Control and Prevention Tobacco Product Laboratory conducted a chemical characterization of Spectrum Research cigarettes. These cigarettes are made in various brand variants with different tar and nicotine levels. These cigarettes will be used in clinical studies, but to date have not been characterized for other common tobacco constituents. Therefore it is important to quantitatively characterize TSNA deliveries in Spectrum cigarettes' mainstream smoke.

Cigarettes were smoked using a linear smoking machine following ISO and Canadian Intense machine smoking regimens. Mainstream smoke total particulate matter was collected on a filter pad. TSNAs were extracted from the pad and analyzed by isotopic dilution using LC/MS/MS.

The levels of NNN ranged from 26.636-185.83 ng/cig and the levels of NNK ranged from 7.8882-96.223 ng/cig under ISO conditions. The levels of NNK for all the cigarettes remained lower than a leading US brand cigarette. For NNN, one of the Spectrum variants was higher than this leading US brand cigarette.

We analyzed the TSNAs deliveries with respect to nicotine and tar deliveries. A correlation was observed between TSNA concentrations and nicotine levels. We found no statistically significant correlation between TSNA concentrations and tar levels. The lack of correlation between TSNA and tar is important as tar has historically been used as a surrogate for carcinogenicity.

Keywords: Clinical/Toxicology, Liquid Chromatography/Mass Spectroscopy, Quantitative
Application Code: Clinical/Toxicology
Methodology Code: Liquid Chromatography/Mass Spectrometry

Session # 460 **Abstract # 460-5** **Poster Sessions**

Session Title: Clinical/Toxicology

Abstract Title: **Separation of Vitamin D2 and D3 for Clinical Application**

Date: Monday, March 07, 2016 - Morning

Time:

Room: Exposition Floor, 400 Aisle

Primary Author: Mark Woodruff
Fortis Technologies

Co-Author(s): Ken Butchart

Abstract Text

In this poster we discuss the separation of Vitamin D2 and D3, two crucial vitamins either ingested as supplements or synthesised naturally by the skin. They allow the adsorption of several minerals calcium, phosphate, iron, magnesium and zinc in the human body, which help prevent the disease osteomalacia and rickets, a weakening of the bones due to defective bone mineralization.

The separation chromatographically is important before detection of these compounds as they have the same molecular weight, meaning that MS detection cannot be relied on to separate them alone. We highlight a rapid highly sensitive method in which a simple polar-endcapped column and mobile phase combination separates the two forms of Vitamin D, allowing high qualitative and quantitative results to be obtained.

Keywords: Biopharmaceutical, Chromatography, Clinical/Toxicology, Pharmaceutical

Application Code: Clinical/Toxicology

Methodology Code: Separation Sciences

Session # 460 Abstract # 460-6 **Poster Sessions**

Session Title: Clinical/Toxicology

Abstract Title: **Measurement of Ammonia Emanating from Human Skin as a Possible Biomarker for Physical/Mental Stress Responses**

Date: Monday, March 07, 2016 - Morning

Time:

Room: Exposition Floor, 400 Aisle

Primary Author: Shota Furukawa
Tokai University

Co-Author(s): Asai Satomi, Hayato Miyachi, Keita Kimura, Minami Takahashi, Shiro Ikeda, Umezawa Kazuo, Yoshika Sekine

Abstract Text

Bio-gases emanating from human skin are potentially non-invasive biomarkers of individual physical or physiological status. Among the human skin gases, we are focusing on ammonia in relation to our daily life. In previous studies, we have developed a passive flux sampler (PFS) – Ion Chromatography methodology and determined the diurnal variation of emission flux of ammonia from human skin surface. The results suggested the human skin ammonia is indicative of fatigue and stress response of tested volunteers. Recently, stress-related diseases are becoming a great social problem. Then, this study aimed to investigate a relationship of human skin ammonia and physical/mental stress response. Firstly, a short-term stressor was loaded to healthy volunteers. Emission flux of ammonia was determined by the PFS method before and after Uchida- Kraepelin test (20 min), together with salivary amylase tests. In such a short term test, salivary amylase showed good response to the stressor, but skin ammonia did not. Secondary, a long-term stressor was loaded to smoking volunteers: smoking is prohibited. In this case, emission flux of ammonia increased with an increase in non-smoking duration, whilst salivary amylase did not correspond to the status of the volunteers. Thirdly, variations of emission flux of ammonia were measured for volunteers during sleep. Skin ammonia decreased with time when the volunteers had a good quality of sleep, whilst it irregularly increased with time when the quality of sleep was poor. These results suggested human skin ammonia may be indicative of accumulated physical/mental stress responses.

Keywords: Biological Samples, Clinical/Toxicology, Ion Chromatography

Application Code: Clinical/Toxicology

Methodology Code: Chemical Methods

Session # 460 **Abstract #** 460-7 **Poster Sessions**

Session Title Clinical/Toxicology

Abstract Title **Measurement of Acetic Acid Emanating from Human Skin as a Potential Biomarker for Quality of Sleep**

Primary Author Minami Takahashi
Tokai University

Date: Monday, March 07, 2016 - Morning
Time:
Room: Exposition Floor, 400 Aisle

Co-Author(s) Furukawa Shota, Kazuo Umezawa, Keita Kimura, Miyachi Hayato, Satomi Asai, Yoshika Sekine

Abstract Text
Volatile compounds emanating from human skin are potentially non-invasive biomarkers of individual physical or physiological status. Acetic acid has been known as one of the human skin gases whose major emission route is a sweat grand. Sweating during sleep is related to the quality of sleep and can be a sign of illness. This study then aimed to investigate variations of emission flux (emission rate per unit area) of acetic acid from skin surface, comparing with the sleep quality of volunteers while sleeping. Emission flux of acetic acid was measured by Passive Flux Sampler (PFS) – Ion Chromatography methodology, and level of sleep was monitored by polysomnography (brain and pulse waves) and Halter electrocardiography (heart rate) during the tests. A questionnaire survey was separately conducted for the volunteers using a Pittsburgh Sleep Quality Index as a subjective assessment. The continuous measurements were carried out from 23:00pm to 6:00am. The results showed volunteers with a good sleep repeated rapid eye motion (REM) sleep and non-REM sleep during the tests. Emission flux of acetic acid increased during the first non-REM sleep period and gradually decreased toward awakening because of higher frequency of REM sleep in the morning. This time course was very similar to typical variation of amount of sweat during sleep. On the other hand, no cyclic and/or regular patterns in emission flux of acetic acid were found for volunteers with poor sleep. These results suggest monitoring of acetic acid from human skin may work for diagnosing quality of sleep.

Keywords: Ion Chromatography, Medical, Sampling
Application Code: Clinical/Toxicology
Methodology Code: Chemical Methods

Session # **460** **Abstract #** **460-8** **Poster Sessions**

Session Title: Clinical/Toxicology

Abstract Title: **LC-MS/MS Determination of Interactions Between Sunitinib and Green Tea Polyphenol by Equilibrium Dialysis**

Primary Author: Matthew Vergne
Lipscomb University

Date: Monday, March 07, 2016 - Morning
Time:
Room: Exposition Floor, 400 Aisle

Co-Author(s): Lincoln Shade

Abstract Text

Tea is the second most consumed beverage in the world behind water. Catechins are polyphenols with antioxidant properties found in high amounts in tea, especially green tea. Green tea beverages and dietary supplements containing green tea extracts are often marketed for health benefits. The most abundant catechin in tea is epigallocatechin-3-gallate (EGCG). While EGCG has a high amount of antioxidant activity compared to other catechins in tea, it is also believed to have the least bioavailability due to protein binding. Recent studies indicate that the drug sunitinib, a tyrosine kinase inhibitor, may interact with components in green tea. It is believed that sunitinib and certain catechins bind non-covalently via hydrogen bonding. We sought to determine if the bioavailability of sunitinib is reduced in the presence of EGCG in serum by measuring serum protein binding using an in vitro rapid equilibrium dialysis method. Post dialysis, sunitinib concentrations will be measured with a reverse phase gradient elution liquid chromatography-tandem mass spectrometry (LC/MS/MS) method.

Keywords: Liquid Chromatography/Mass Spectroscopy, Pharmaceutical, Quantitative, Tandem Mass Spec
Application Code: Clinical/Toxicology
Methodology Code: Liquid Chromatography/Mass Spectrometry

Session # 460 **Abstract # 460-9** **Poster Sessions**

Session Title: Clinical/Toxicology

Abstract Title: **Identification of Hypocretin-1 in Cerebrospinal Fluid: A Potential Diagnostic Biomarker for Narcolepsy**

Primary Author: Hemasudha Chatragadda
Duquesne University

Date: Monday, March 07, 2016 - Morning

Time:

Room: Exposition Floor, 400 Aisle

Co-Author(s): Birgitte R Kornum, Matt Pamuku, Skip Kingston

Abstract Text

Although the etiology of Narcolepsy (NA) remains unclear, there seems to be mounting evidence hypothesizing it as an autoimmune disease, resulting from the loss of hypothalamic neurons producing Hypocretins (Hcrt) by an unknown autoimmune mechanism. NA affects approximately 1 in 2000 individuals in US and the onset of disease is often around adolescence. The current diagnostic tests used are based on clinical history and examination of objective measures such as Polysomnograph, Multiple sleep latency test, Human Leukocyte Antigen typing rather than the pathological conditions leading to false negatives and false positives in the clinical practice. With key pathophysiological findings of NA being associated with deficiency of (Hcrt-1) neurotransmission in the brain, measurement of post translationally modified biologically active Hcrt-1 in Cerebrospinal Fluid (CSF) can be proven as a diagnostic biomarker.

There have been several reports of measurement of Hcrt-1 in CSF using a radioimmunoassay procedure. Most of these studies with disparate results have not been verified, validated and translated to clinical diagnosis accurately due to lack of sensitivity to measure the entire Hcrt-1 fragment essential for biological activity. This is due to low abundance, lack of specific antibody and rapid degradation. These critical analytical challenges can be overcome by using nanoparticles containing bait, which not only prevent degradation but amplify the concentration, increasing the sensitivity of Mass spectrometry detection. The proposed study focuses on identification of Hcrt-1 in CSF using nanoparticles and Isotope dilution Mass Spectrometry allowing unambiguous measurement with greater accuracy and precision.

Keywords: Liquid Chromatography, Mass Spectrometry, Proteomics, Sample Preparation

Application Code: Clinical/Toxicology

Methodology Code: Liquid Chromatography/Mass Spectrometry

Session # **460** Abstract # **460-10** **Poster Sessions**

Session Title Clinical/Toxicology

Abstract Title **ICP-MS – A Perfect Tool for the Bio-monitoring of Trace Elements in Body Fluids**

Primary Author Ewa M. Pruszkowski
PerkinElmer, Inc.

Date: Monday, March 07, 2016 - Morning
Time:
Room: Exposition Floor, 400 Aisle

Co-Author(s)

Abstract Text

For many years the ICP-MS was a tool of choice for the trace analysis of elements like Pb, As, Hg, and Cu in body fluids such as urine, blood, serum and saliva. Single toxic or nutritional elements or panels of elements were run in the proper matrices. Recently, due to the popularity of implants, elements like Ti, Co, and Mn that provide information on implant degradation were added to the common list of tested analytes.

The goal of this poster is to demonstrate the capability of the current ICP-MS technology for trace element analysis in body fluids for research applications. A winning combination of reaction/collision spectral interference removal allows for the accurate determination of the low levels of analytes of interest.

It will be shown that the ICP-MS, in combination with an optimized sample introduction system, is a perfect technique for the analysis of diverse types of matrices including urine, serum and blood. One simple sample preparation technique, the appropriate diluent, and panels or individual analytes can be measured quickly and precisely. The results in reference materials will be shown and discussed.

* For research use only. Not intended for diagnostic procedures.

Keywords: Clinical/Toxicology, ICP-MS
Application Code: Clinical/Toxicology
Methodology Code: Atomic Spectroscopy/Elemental Analysis

Session # 460 Abstract # 460-11 **Poster Sessions**

Session Title Clinical/Toxicology

Abstract Title **Incorporation of Amphiphilic Dendrimers in Supported Lipid Bilayers to Enhance Stability and Functionalization**

Date: Monday, March 07, 2016 - Morning

Time:

Room: Exposition Floor, 400 Aisle

Primary Author Charles J. Ruiz
University of California, Riverside

Co-Author(s) Ling Peng, Quan Cheng, Samuel S. Hinman

Abstract Text

The development of robust and functionalizable solid-supported lipid bilayer (SLB) systems for use as biosensors presents a practical platform for modifying the biorecognition properties of SLB's while retaining the inherent membrane fluidity. Polyamidoamine (PAMAM) amphiphilic dendrimers (ADs) have been extensively investigated as nonviral vectors for effective delivery of siRNA therapeutics. While PAMAM amphiphilic dendrimers have proven their efficacy as drug delivery vectors, their incorporation into phospholipid-containing SLBs for increasing membrane stability on hydrophilic surfaces and introduction of coupling handles has not been studied. By premixing varying ratios of PAMAM ADs and POPC into small unilamellar vesicles, we demonstrated that SLBs spontaneously form onto hydrophilic nanoglassified gold SPR substrates via vesicle fusion. Lateral mobilities of PAMAM ADs were examined via confocal fluorescence microscopy using fluorescence recovery after photobleaching (FRAP) and yielded comparable membrane fluidity to traditional SLBs comprised solely of POPC. The stability of PAMAM AD/POPC SLBs was investigated via SPR in the presence of surfactants and revealed that the stability of the membrane was proportional to the concentration of PAMAM ADs in the mixture. The biorecognition properties of PAMAM AD/POPC SLBs were evaluated via a biotin/streptavidin binding scheme after in situ covalent modification of biotin onto the primary amines found on the hydrophilic PAMAM AD head group. The incorporation of PAMAM ADs into SLBs allows for the expanded utility of SPR biosensors by enhancing stability and functionalization. Given the PAMAM ADs applications in drug delivery, these studies serve as a valuable tool for gaining insight on lipid-based drug carrier designs.

Keywords: Bioanalytical, Biospectroscopy, Lipids, Membrane

Application Code: Clinical/Toxicology

Methodology Code: Biospectroscopy

Session # 460 **Abstract # 460-12** **Poster Sessions**

Session Title: Clinical/Toxicology

Abstract Title: **Chiral Capillary Electrophoresis-Mass Spectrometry: Turning an Analytical Technique into High Throughput Screening of Chiral Compounds Using Novel Polymerized beta-D-Glucopyranoside Surfactants**

Date: Monday, March 07, 2016 - Morning

Time:

Room: Exposition Floor, 400 Aisle

Primary Author: Liu Yijin
Georgia State University

Co-Author(s): Shahab S. Shamsi

Abstract Text

Increasing concerns on the improving sensitivity and throughput of chiral compounds screening boosted the development of different types of high molecular mass polymeric chiral surfactants (aka. molecular nanomicelles, (MNM)) for capillary electrophoresis (CE). The advantages of using MNM as chiral selector over conventional low-molecular-mass chiral selectors such as cyclodextrins, macrocyclic antibiotics is that they are covalently stabilized micellar aggregates, which cannot be fragmented in electrospray ionization (ESI)-MS [1,2,3].
Sugar based surfactants forming unpolymerized micelles have been successfully used as pseudophase in chiral MEKC with UV detection [4] but there are no reports of using either unpolymerized or the polymerized version in MEKC-MS.

In this work, beta-D-glucopyranoside MNM with different head groups and chain lengths have been successfully synthesized, characterized and applied to evaluate multiple chiral cationic compounds (ephedrine alkaloids, □-blockers, homatropine, terbutaline etc.) and neutral or negatively charged binaphthyl derivatives in MEKC-MS. Results will be presented for high throughput separation of these chiral compounds at different pH and background electrolyte using sugar based MNM in in MEKC-MS.

[1] Wang, X., Hou, J, Jann, M., Hon, Y.Y., Shamsi, S.A., J. Chromatogr. A 2013, 127, 207-216
[2] Hou, J., Zheng, J., Shamsi, S. A., J. Chromatogr. A 2007, 1159, 208-216
[3] Rizvi, S. A. A., Zheng, J., Apkarian, R. P., Dublin, S. N., Shamsi, S. A., Anal. Chem. 2007, 79, 879-898
[4] D.C. Tickle, N.G. Okaf, P. Camilleri, R.F.D. Jones, A.J. Kirby, Anal.Chem. 1994, 66, 4121-4126.

Keywords: Capillary Electrophoresis, Mass Spectrometry
Application Code: Clinical/Toxicology
Methodology Code: Capillary Electrophoresis

Session # 460 **Abstract # 460-13** **Poster Sessions**

Session Title: Clinical/Toxicology

Abstract Title: **Measurement of Diacetyl and 2-Nonenal Emanating from Human Skin by Passive Flux Sampler**

Primary Author: Keita Kimura
Tokai University

Date: Monday, March 07, 2016 - Morning
Time:
Room: Exposition Floor, 400 Aisle

Co-Author(s): Hayato Miyachi, Kazuo Umezawa, Minami Takahashi, Satomi Asai, Shota Furukawa, Yoshika Sekine

Abstract Text

Volatile compounds emanating from human skin are potentially non-invasive biomarkers of individual physical or physiological status. Among the compounds, diacetyl and 2-nonanal have been known as a source of body odor specific to aged people. While they were previously detected in outgases from used underwear and/or sebum, there has been no information on emission flux (emission rate per unit area). Then, this study aimed to develop a passive flux sampler (PFS) method for the measurement of emission fluxes of diacetyl and 2-nonenal from human skin. The PFS was constructed with sampler body, trapping media and stopper. A disk type monolithic silica containing activated charcoal was employed as an absorbent. The PFS was tested for ten healthy volunteers at six sampling positions. Considering the analytical sensitivity, sampling time was set at 7 hours. After sampling, the analytes were extracted by dichloromethane with ultrasonic waves. The extraction solution was subsequently served for GC-MS with SIM mode. Greater emission fluxes of both compounds were found at occipital of all volunteers, where there is a cutaneous gland that secretes sebum for lubricating hair and skin. Based on the odor threshold of each gas in air, we had estimated thresholds for the emission flux by a simple gas diffusion model. While emission fluxes were very low for volunteers under 30 years old, they exceeded the thresholds for volunteers from 30 to 60's. This tendency was more apparent for male volunteers. These results show the PFS is useful for determining the age-specific odor gases.

Keywords: GC-MS, Sampling, Volatile Organic Compounds

Application Code: Clinical/Toxicology

Methodology Code: Chemical Methods

Session # 460 **Abstract # 460-14** **Poster Sessions**

Session Title: Clinical/Toxicology

Abstract Title: **Aptamer-Modified Gold Nanoparticles Coupled with Nitrocellulose Membranes for Detection of Thrombin by LDI-MS**

Date: Monday, March 07, 2016 - Morning

Time:

Room: Exposition Floor, 400 Aisle

Primary Author: Chia-Yin Chang
National Taiwan Ocean University

Co-Author(s)

Abstract Text

Thrombin (activated blood-coagulation factor II) plays an important role in the blood coagulation cascades such as hydrolysis of soluble fibrinogen into insoluble strands of fibrin, feedback amplification of factors V, VIII, XI, and XII, initiation of anticoagulation by activating protein C and platelets. Monitoring thrombin in physiological conditions is useful for diagnosing various health issues related to hypocoagulability and hyper-coagulability. Here, we report a simple and rapid assay to detect thrombin using aptamer-modified gold nanoparticles (Apt□AuNPs) coupled with nitrocellulose membrane (NCM) through laser desorption/ionization mass spectrometry (LDI-MS). In order to increase the binding affinity of Apt□AuNPs to thrombin, we designed a DNA aptamer having base pair 8-mer complementary 15-mer thiolated-thrombin-binding-aptamer (TBA15) and 29-mer thiolated-thrombin-binding-aptamer 29 (TBA29) (TBA15/29-AuNPs), to bind with exosite I and exosite II of thrombin, respectively. The Apt□AuNPs were sunk into NCM deeper after thrombin were specific bound to the particles. Hence, the signal intensity of Au cationic cluster ions ([Aun]+; n = 1□3) were decreased unpon increasing the thrombin under the analysis by LDI-MS. We employed Apt□AuNPs/NCM to monitor thrombin with good selectivity (>100-fold toward thrombin with respect to other proteins) and sensitivity (limit of detection for thrombin of 100 pM in human serum samples). Thus, the remarkable performance of Apt□AuNPs/NCM to detect thrombin in real samples suggests its possible practical applications for monitoring thrombin related diseases.

Keywords: Mass Spectrometry, Nanotechnology, Protein

Application Code: Nanotechnology

Methodology Code: Mass Spectrometry

Session # 460 **Abstract # 460-15** **Poster Sessions**

Session Title: Clinical/Toxicology

Abstract Title: **On-Line Membrane Assisted Distillation Coupled with Ion Chromatography: A Novel Approach to Determine Trace Fluoride in Serum**

Primary Author: Lou Chaoyan
Zhejiang University

Date: Monday, March 07, 2016 - Morning
Time:
Room: Exposition Floor, 400 Aisle

Co-Author(s)

Abstract Text

Fluorine is one of the essential trace elements in human body. It plays an important part in the formation of bone issues, enamel and metabolism of calcium and phosphorus.The determination of fluoride in human serum, plasma and urine has been an essential task in chemistry and biology field.Ion Chromatography(IC) is a very promising alternative for the determination of trace fluoride due to its excellent reproducibility, accuracy and sensitivity. However, due to the complexity of sample matrix, it is nearly impossible for the direct determination of fluoride in biological samples.

The common process to extract fluoride from complicated matrix sample is distillation. Fluoride is easily acidified into hydrogen fluoride, and for the purification of this halogen acids(HF), the conventional distillation method has been utilized for various types of matrices. But these obtained distillates are not applicable for ion chromatography due to its residuals and by-products. Due to the principle of distillation, we proposed an on-line membrane assisted distillation coupled with ion chromatography system to automatically analyze trace fluoride in serum and urine.

Keywords: Biological Samples, Chromatography, Environmental/Biological Samples, Isolation/Purification
Application Code: Bioanalytical
Methodology Code: Sampling and Sample Preparation

Session # 460 Abstract # 460-17 **Poster Sessions**

Session Title Clinical/Toxicology

Abstract Title **Electrochemical Measurement of Thyroid Hormone for Rapid Diagnostics Technology**

Date: Monday, March 07, 2016 - Morning

Time:

Room: Exposition Floor, 400 Aisle

Primary Author Barbara Cata
Northern Kentucky University

Co-Author(s) Celeste A. Morris

Abstract Text

Development of rapid-analysis technology for thyroid hormones is of particular interest due to its clinical relevance as a marker for thyroid disease, autoimmune diseases and evaluation of thyrotoxic storm. In this study, we employed principles of electrochemistry to observe oxidation and reduction of thyroxine at the surface of glassy carbon and gold electrodes. Micro and nanoscale electrodes utilized for electrochemical measurement of thyroxine are especially suited for point-of-care testing due to rapid analysis times, simple calibration, and ability to provide hormone concentration levels in blood for patients experiencing thyrotoxicosis. The selective detection and measurement of thyroxine in blood serum has been demonstrated via cyclic voltammetry with M sensitivity. Improvements in sensitivity and dynamic range of thyroxine measurement was achieved through formation of a self-assembled monolayer with (11-mercaptoundecyl)-N,N,N-trimethylammonium bromide on gold microelectrodes. In conclusion, we determined thyroxine reduction could be achieved with gold microelectrodes and nanopipettes while developing a technology for real-time thyrotoxic storm evaluation.

Keywords: Bioanalytical, Chemically Modified Electrodes, Clinical Chemistry, Nanotechnology

Application Code: Bioanalytical

Methodology Code: Electrochemistry

Session # 460 Abstract # 460-18 **Poster Sessions**

Session Title: Clinical/Toxicology

Abstract Title: **Forensics Meets Green Chemistry: Removing Known Carcinogens from Blood Alcohol Content Protocols for Safer Applied Spectroscopy Laboratories**

Primary Author: Sarah E. Gray
Armstrong State University

Co-Author(s): Mathew Holmes

Date: Monday, March 07, 2016 - Morning
Time:
Room: Exposition Floor, 400 Aisle

Abstract Text

The importance of Beer's Law in spectroscopy is covered throughout the chemistry curriculum, particularly in general chemistry, analytical chemistry and instrumental analysis. The use of real world applications can help increase the engagement of students in this topic, as well as increase their retention of new knowledge. One application that has shown high interest from students is forensics, especially with the prevalence of forensic analyses on television. The analysis of blood alcohol content (BAC) by UV/VIS absorption is one such applied spectroscopy method. The standard wet chemical method, however, uses potassium dichromate, a known carcinogen. Here, we present an optimized version of the standard method that replaces potassium dichromate with potassium permanganate. Several method conditions are presented so that the laboratory can be implemented as a troubleshooting exercise for students, rather than a cookbook method. After experimenting and deciding on an optimized method for using potassium permanganate to determine BAC, students go on to use their self-designed procedure to determine the BAC of a suspect during a multi-week forensics scenario in instrumental analysis.

Keywords: Forensic Chemistry, Molecular Spectroscopy, Teaching/Education, UV-VIS Absorbance/Luminescence
Application Code: Other
Methodology Code: Education/Teaching

Session # 460 Abstract # 460-20 **Poster Sessions**

Session Title Clinical/Toxicology

Abstract Title **A Rapid and Sensitive SERS Based Measles Detection**

Date: Monday, March 07, 2016 - Morning
Time:
Room: Exposition Floor, 400 Aisle

Primary Author Ramesh Kattumenu
Argent Diagnostics Inc

Co-Author(s)

Abstract Text

Measles outbreak and the testing that follows are very critical in the identification of the virus. The turnaround time for current serological test results is adequate in accordance to WHO Measles/Rubells Laboratory Network, but specimens collected in remote areas can be delayed due to poor infrastructure for collection, storage and transportation. Hence, a rapid response to measles outbreak is very critical, and diagnostic assays that can be performed in field conditions may be necessary for rapid outbreak confirmation. The availability of a quick diagnostic test would improve the timing of the public health response to an outbreak. The test being developed is a portable spectroscopic diagnostic assay based on surface enhanced Raman scattering (SERS) to detect measles virus that is sensitive and highly specific to measles virus. The SERS based assay relies on the fabrication of silver nanorod array substrates (by oblique angle deposition) that are extremely SERS-active allowing for magnification of spectral signal more than million times, that is a quick identification technique specific to measles. The spectroscopic based assay along with a supervised chemometric method, partial least squares-discriminant analysis (PLS-DA) was able to differentiate measles viral RNA from the controls with 96% sensitivity and 90% specificity for a viral concentration as low as 0.9 ug/ul based solely on the inherent SERS spectra. These results demonstrate that SERS, in combination with multivariate statistical analysis methods, is a highly sensitive and rapid viral identification method and can be used for pathogen detection central to human health care.

Keywords: Chemometrics, Detection, Raman
Application Code: Biomedical
Methodology Code: Vibrational Spectroscopy

Session # 460 **Abstract # 460-21** **Poster Sessions**

Session Title: Clinical/Toxicology

Abstract Title: **The Efficiency of Multi-Sample Analysis Using Dual Gradient LCMS System**

Primary Author: Watanabe Satoru
Shimadzu Corporation

Date: Monday, March 07, 2016 - Morning
Time:
Room: Exposition Floor, 400 Aisle

Co-Author(s): Inohana Yusuke, Nakayama Daisuke, Yamaguchi Tadayuki

Abstract Text

Higher throughput is of the utmost importance to laboratory efficiency and profitability. Conventional analysis requires performing various processes such as column washing, equilibration at initial mobile phase concentrations, and the next sample injection by the autosampler during the analysis. We developed and evaluated the dual gradient LCMS system which incorporates a special flow line structure and instrument control system, and performs overlap control of sample injection by using two analysis systems (streams) alternately.
The dual gradient LCMS system has two analytical streams which contain pump units and analytical columns. By using these, after one stream completes data acquisition, the other stream starts data acquisition immediately without interruption, making it possible to use nearly the whole time of LCMS operation for data acquisition. We evaluated the analytical cycle time and the repeatability in the analysis of four biomarker compounds for the four major molecular species in the Cytochrome P450 family, using an ultra-high speed reversed phase column and LCMS-8060 triple quadrupole mass spectrometer.

The dual gradient LCMS system is evaluated for four compounds. This system showed that the cycle time is twice or more faster compared with conventional LCMS system. The repeatability of the peak area results good for all compounds.

The evaluation results showed that the dual gradient LCMS system can increase the number of samples and maximize throughput. it is quite effective to laboratory efficiency and profitability.

Keywords: Clinical/Toxicology, Drug Discovery, HPLC, Liquid Chromatography/Mass Spectroscopy
Application Code: Clinical/Toxicology
Methodology Code: Liquid Chromatography/Mass Spectrometry

Session # 470 **Abstract # 470-2** **Poster Sessions**

Session Title: Drug Discovery

Abstract Title: **Synthesis of Nanosized Poly B-aminoester Holding Chlorambucil Drug as Slow Release Drug System for Antitumer**

Primary Author: Fahima M. Helaly
National Research Centre (NRC)

Date: Monday, March 07, 2016 - Morning
Time:
Room: Exposition Floor, 400 Aisle

Co-Author(s)

Abstract Text

The current research deal to synthesize nanosized poly B-aminoester containing chlorambucil drug and evaluate its efficiency as slow sustained release drug system for long term antitumer.Spectrophotometric analysis was used to measure the release of chlorambucil in aquatic media.Poly B-aminoester was prepared from 1,4 butanediol diacrylate and piprazine by Mecheal addition polymerization reaction.The nanosized polymer was produced by controlled solvent displacement (ethanol/water) system .Chlorambucil was holded on the investigated polymer during stirring at the end of the reaction.the resulting product was formulated into circular disks.The nano size was controlled by different factors as:speed of material addition,stirring speed,and the concentration of nonionic surfactant.The size of nanoparticles was about 160 nm as measured by transmetion electron microscope(TEM).The release analysis of chlorambucil illustrated that,the comulative release percent was about,70,65,and50 for the composites containing 10%,20%and 30% respectivellyafter 150hrs.Also,the release rate in distilled water was ranged from 0.15 to 0.35 mg/unite area(cm)and was extended to 50 days.The release rate in acidic water(pH2.5)was greater than in alkaline media(pH8.5) and distilled water(pH6.9).The efficiency of the released chlorambucil onthe human liver cancer cell line showed promissing results,due to the growth inhibition of the cell line observed.

Keywords: Identification, Material Science, Nanotechnology, Spectrophotometry

Application Code: Drug Discovery

Methodology Code: UV/VIS

Session # 470 Abstract # 470-3 **Poster Sessions**

Session Title: Drug Discovery

Abstract Title: **Synthesis, Characterization and Antibacterial Studies of Cobalt Complexes of Isomeric Aminophenol Schiff Bases**

Date: Monday, March 07, 2016 - Morning

Time:

Room: Exposition Floor, 400 Aisle

Primary Author: Tolulope M. Fasina
University of Lagos

Co-Author(s): Felicia N. Ejiah, Neerish Revaprasadu, Oluwole B. Familoni

Abstract Text

The present study was carried out to investigate the antibacterial activities of isomeric aminophenol Schiff bases and their metal complexes. The burden of health-care associated infections experienced by both developed and developing countries is of great concern particularly with the growing reports of bacterial resistance to currently used antibacterial agents. There is therefore an urgent need to synthesize and investigate new antibacterial agents. Schiff bases are a class of versatile ligands with variable coordination modes. The coordination of these ligands with transition metal ions has resulted in formation of complexes with interesting biological and photochemical properties.
In this study, new cobalt (II) complexes of isomeric Schiff bases derived from aminophenols and substituted benzaldehydes were synthesized using condensation method.

The compounds were fully characterized using elemental analysis, infrared spectroscopy (IR), electronic absorption spectroscopy, magnetic susceptibility measurements, differential scanning calorimetry and thermal gravimetry analysis. The Schiff bases and their metal complexes were screened for in-vitro antibacterial activities against 6 human pathogenic bacteria ; Escheriachia coli, Staphylococcus aureus, Pseudomonas aeruginosa, Bacillus cereus, Enterococcus faecalis and Kribsella pneumonia. Ampicillin was used as a reference compound.
Results indicate that all metal complexes had a 1:1 metal ligand ratio with magnetic moments characteristic of octahedral and tetrahedral geometry around the metal ion. The complexes exhibited higher antibacterial activity than the free ligands.

Our results show that these complexes can be employed as active ingredients in development of broad spectrum antibacterial agents.

The authors acknowledge support from National Research Foundation, University of Zululand, South Africa

Keywords: Atomic Absorption, Elemental Analysis, FTIR, UV-VIS Absorbance/Luminescence

Application Code: Drug Discovery

Methodology Code: Chemical Methods

Session # 470 **Abstract # 470-4** **Poster Sessions**

Session Title: Drug Discovery

Abstract Title: **Mass Spectral and Chromatographic Studies on Substituted Cathinones: Bath Salt-type Aminoketone Designer Drugs**

Date: Monday, March 07, 2016 - Morning

Time:

Room: Exposition Floor, 400 Aisle

Primary Author: Younis Abiedalla, Auburn University

Co-Author(s): Jack DeRuiter, Karim Abdel-Hay, Randall Clark

Abstract Text

This presentation will describe our research efforts to evaluate the structure-retention, structure-fragmentation and other structure-property analytical relationships for a large series of substituted aminoketones related to the cathinone-type drugs. The synthesis, GC-MS, GC-MS/MS, GC-IR and related spectroscopic properties will be presented for several series of substituted cathinone derivatives.

Our research has focused on the development of regioisomer specific methods for the identification of ring substituted aminoketone compounds (cathinone derivatives). The work includes the chemical synthesis of all regioisomeric forms of selected aromatic ring substituted aminoketones; generation of analytical profiles for each compound; chromatographic studies to separate/resolve all regioisomeric aminoketones having overlapping analytical profiles, and design and validation of confirmation level methods to identify individual compounds.

Based on the structure of the unsubstituted cathinone molecule, designer modifications are possible in three distinct regions of the molecule: the aromaric ring, the alkyl side chain and the amino group. Example compounds from all three of these areas of designer modification have been reported as components of clandestine drug samples. Commercially available precursor aldehydes, alkyl halides and amines can be converted to a wide variety of cathinone-type compounds. Legal control of a specific molecule often provides the driving force for clandestine development of additional substituted cathinone designer molecules.

Keywords: Analysis, Gas Chromatography/Mass Spectrometry, Infrared and Raman, Ion Trap

Application Code: Drug Discovery

Methodology Code: Gas Chromatography/Mass Spectrometry

Session # 470   Abstract # 470-5   **Poster Sessions**

Session Title: Drug Discovery

Abstract Title: **An In Vitro Microfluidic Platform to Unravel Mechanisms of Action of Drug Therapies used in Multiple Sclerosis**

Primary Author: Tiffany Bell
Michigan State University

Co-Author(s): Dana Spence

Date: Monday, March 07, 2016 - Morning
Time:
Room: Exposition Floor, 400 Aisle

Abstract Text

Multiple Sclerosis (MS) is a disease of the central nervous system affecting over 2.3 million people worldwide. MS is characterized by blood brain barrier (BBB) damage and demyelination of axons, which causes brain lesions that are detected through magnetic resonance images. Medications, such as steroids and interferon beta, are available to help treat symptoms of MS although the mechanisms of action of these therapies are not completely understood. Here, data will be presented showing the effects of various MS therapies (e.g., estriol, prednisolone, and interferon beta) on red blood cells (RBCs). A 3D-printed fluidic device designed by our group has been used for all studies reported. The device mimics various aspects of the bloodstream circulation by using Tygon tubing and a peristaltic pump to create a closed loop system. Our results to date have shown that RBCs obtained from people with MS release significantly more ATP than control RBCs (MS: 197 +/- 20 nM Control: 276 +/- 35 nM). Importantly, this increased ATP release is reduced to that of healthy, non-MS controls when incubated at physiological levels with the steroids estriol or prednisolone. This presentation will also explore the cell pathways by which each therapy is exerting its cellular effects.

Keywords: Bioanalytical, Biomedical, Biopharmaceutical, Lab-on-a-Chip/Microfluidics

Application Code: Drug Discovery

Methodology Code: Microfluidics/Lab-on-a-Chip

Session Title Drug Discovery

Abstract Title **Streamlining Compound Isolation Automatically with UPLC to Prep Chromatography Using Mass-Directed Auto Purification**

Date: Monday, March 07, 2016 - Morning
Time:
Room: Exposition Floor, 400 Aisle

Primary Author Jo-Ann M. Jablonski
Waters Corporation

Co-Author(s) Andrew J. Aubin, Thomas E. Wheat, Wendy Harrop

Abstract Text

Drug discovery laboratories that process large numbers of samples must develop efficient purification protocols, thereby reducing the solvent, time, and effort necessary to make critical strategic decisions about therapeutic candidates and their progression to the next phase of testing. Combining UPLC-driven compound screening and fraction analysis with mass-directed purification effectively streamlines the compound isolation process. Sub-2[micro]m particle size columns, combined with the fluidically-optimized flow path of the UPLC, increase sensitivity and resolution and reduce chromatographic run times, making it ideal for fast sample analysis. AutoPurify[circumflex O], a feature included in the FractionLynx[circumflex O] software, assesses the purity of the target compound in the crude mixture, suggests a purification strategy, and generates the sample list for the isolation on the preparative LC system. The sample list is imported and run on the preparative LC system. Once the isolation is completed, AutoPurify reports the purity of the fractions and creates a sample list for fraction analysis on the UPLC system. In this study, we illustrate how dedicating the UPLC for compound screening and fraction analysis, reserving the preparative LC system for compound isolation, and using AutoPurify to coordinate and execute the workflow effectively manages the purification process, increasing efficiency and throughput.

Keywords: Drug Discovery, Isolation/Purification, Liquid Chromatography/Mass Spectroscopy, Prep Chromatogr
Application Code: Drug Discovery
Methodology Code: Liquid Chromatography/Mass Spectrometry

Session # 470 Abstract # 470-10 **Poster Sessions**

Session Title: Drug Discovery

Abstract Title: **Spectral Characterization of Cytochrome P450cam Active Site Using NMR Methods Including [sup]13[/sup]C-Doubled Filtered [sup]1[/sup]H-[sup]1[/sup]H Noesy Experiments for Mapping Distances**

Primary Author: Remigio Usai
Marquette University

Date: Monday, March 07, 2016 - Morning
Time:
Room: Exposition Floor, 400 Aisle

Co-Author(s): Daniel Kaluka, Daniel S. Sem, James R. Kincaid

Abstract Text

Cytochrome P450 superfamily is one of the most diverse enzyme systems in nature and has been subject of intense research for over six decades. In this presentation an efficient approach for the labeling of the prosthetic group in heme proteins has been exploited to produce an analogue of the soluble bacterial cytochrome P450cam (P450cam) that contains a [sup]13[/sup]C labeled prosthetic group. The HU227 strain of [i]E. coli[/i], which lacks the [delta]-aminolevulinic acid ([delta]-ALA) synthase gene, was employed in the heterologous expression of P450cam harboring a prosthetic group labeled with [sup]13[/sup]C at the Cm and C[alpha] positions by growing cells in the presence of [5-[sup]13[/sup]C] [delta]-ALA, which was synthesized in four steps from [2-[sup]13[/sup]C] glycine. NMR spectrometry was used to confirm labeling of the hemes at the Cm and C[alpha] positions. This system has been utilized as proof of principle for the strategy of defining active site structure in cytochrome P450cam, including iron-to-proton distances on bound substrates, using NMR methods. Such data are potentially of great use in furnishing necessary experimental restrictions in docking routines, which are commonly employed in determining the relative affinities of drug candidates. NOESY, a 2D NMR technique which utilizes through-space interaction of protons has been employed and resonances obtained from the [sup]13[/sup]C labeled reference positions and substrate. To confirm these resonances camphor, a normal substrate for P450cam was deuterated at the 5n and 5x positions. GC/MS was employed in characterization of intermediates during the synthesis of D[sub]2[/sub]-camphor.

Keywords: GC-MS, Infrared and Raman, NMR, Protein
Application Code: Drug Discovery
Methodology Code: Magnetic Resonance

Session # 470 Abstract # 470-13 **Poster Sessions**

Session Title: Drug Discovery

Abstract Title: **Photostability of Pharmaceutical Drug Substance as Free Acid and Salt**

Primary Author: Jenny Wang
Genentech Inc.

Date: Monday, March 07, 2016 - Morning
Time:
Room: Exposition Floor, 400 Aisle

Co-Author(s): Christine Gu, Geoffrey Yeh, Kelly Zhang, Lulu Dai

Abstract Text

Small molecule drugs can be developed in free acid (or free base) and salt forms. Different forms may have different stability. Two main degradation products formed under photolytic conditions were observed in GNE-A new drug substance forced degradation testing. The photolytic degradants were determined to be the Z-isomer of drug substance, and a dehydrogenated des-chloro compound. In order to support new drug discovery and development, to assess the need for manufacturing, handling, and storage controls due to the risk of drug substance instability, comprehensive photolysis experiments were executed, and the two degradants were used as metrics of photolability. Photostability testing of pharmaceutical salt and free acid was investigated by exposure to various light sources including ICH light condition, white fluorescent light and natural sunlight. Additionally, milled drug substance was compared to unmilled drug substance to identify precautionary measures needed in manufacturing or in formulation of free acid drug product.

Both GNE-A salt and free acid exhibited measurable decomposition under all conditions, but the extent of degradation of free acid was substantially greater than that of pharmaceutical salt. There was no significant difference in the extent of degradation of milled free acid form drug powder versus unmilled free acid. GNE-A drug substance exposed to hood light or natural light formed relatively greater amounts of RRT 0.88 than drug substance exposed to ICH light, showing that the illumination spectrum dictates decomposition branching pathways.

Keywords: Drugs, Pharmaceutical, Process Analytical Chemistry, Quality Control
Application Code: Pharmaceutical
Methodology Code: Process Analytical Techniques

Session # 480 **Abstract # 480-2** **Poster Sessions**

Session Title: Food Safety

Abstract Title: **Cannabinoids and Residual Solvents by Headspace GC**

Date: Monday, March 07, 2016 - Morning

Time:

Room: Exposition Floor, 400 Aisle

Primary Author: Tim Anderson
Phenomenex

Co-Author(s): Kristen Parnell, Ramkumar Dhandapani

Abstract Text

Organic solvents are commonly used by cannabis product manufactures to extract cannabinoids and terpenes from cannabis flowers prior to infusion in edibles. Unlike pharmacopeia products, there are no standardized monographs by which to monitor residual solvents. Nor are there standardized methods to determine the concentration of cannabinoids in headspace. This paper explores a method by which cannabis is substituted with hops, spiked with cannabinoids and organic solvents. The terpenes from the hops, cannabinoids, and solvents influence each other in terms of concentration in the headspace, resulting in complex headspace sampling and chromatographic analysis. This paper explores how headspace and GC can be optimized to provide reproducibly chromatography for the emerging cannabis industry.

Keywords: Drugs, Gas Chromatography, Headspace, Optimization
Application Code: Food Safety
Methodology Code: Gas Chromatography

Session # 480 Abstract # 480-3 **Poster Sessions**

Session Title: Food Safety

Abstract Title: **Melamine in Pet Food**

Date: Monday, March 07, 2016 - Morning
Time:
Room: Exposition Floor, 400 Aisle

Primary Author: Tim Anderson
Phenomenex

Co-Author(s): Kristen Parnell, Ramkumar Dhandapani

Abstract Text

Pet food quality and safety is on on-going challenge for manufacturers, and melamine is a particularly common analyte of interest. Solid Phase Extraction (SPE) and GC/MS analysis is an easy and reproducible method by which to determine the concentration of melamine and related compounds in pet food. Early methods were primarily liquid-liquid extraction and filtration prior to analysis. The use of SPE can isolate and concentrate the analytes of interest in a complex matrix, leading to improved recovery and chromatography. This poster discusses a technique by which pet food spiked with melamine can be prepared, cleaned with SPE, and analyzed with an improved GC column.

Keywords: Food Safety, Gas Chromatography/Mass Spectrometry, Sample Preparation
Application Code: Food Safety
Methodology Code: Gas Chromatography/Mass Spectrometry

Session # **480** Abstract # **480-4** **Poster Sessions**

Session Title Food Safety

Abstract Title **Impact of HPLC Stationary Phase Selection on Matrix Effects During LC-MS/MS Analysis of Multiple Mycotoxins in Corn**

Date: Monday, March 07, 2016 - Morning

Time:

Room: Exposition Floor, 400 Aisle

Primary Author Emily R. Barrey
Supelco/Sigma-Aldrich

Co-Author(s) Jennifer Claus, Lynne Perez-Blanco, Michael Ye, Olga I. Shimelis

Abstract Text

LC-MS methods allow for rapid analysis of multiple analytes, increased selectivity and sensitivity. Recently a number of advancements have been made toward LC-MS analysis of mycotoxins in foods and feeds. This study investigated the use of C18 and Phenyl-hexyl HPLC column stationary phases for the analysis of twelve mycotoxin compounds. While C18 is a commonly used column, stronger retention of most analytes was observed on phenyl-hexyl stationary phase using the same gradient method. The separation of analytes from matrix is important to reduce the impact of matrix effects, as often limited or no sample cleanup is applied during mycotoxin analysis by LC-MS. We will present the results of our investigation of matrix effects in corn meal using both C18 and phenyl-hexyl column chemistries, and discuss the recommended LC conditions for best overall method performance.

Keywords: Food Contaminants, Food Safety, Liquid Chromatography/Mass Spectroscopy

Application Code: Food Safety

Methodology Code: Liquid Chromatography/Mass Spectrometry

Session # 480 Abstract # 480-5 **Poster Sessions**

Session Title Food Safety

Abstract Title **UHPLC/MS/MS Analysis of Lipophilic Marine Toxins from Homogenized Shellfish**

Date: Monday, March 07, 2016 - Morning

Time:

Room: Exposition Floor, 400 Aisle

Primary Author Emily R. Barrey
Supelco/Sigma-Aldrich

Co-Author(s) Jennifer Claus, Michael Ye, Olga I. Shimelis

Abstract Text

Biotoxins in shellfish have recently come under increased scrutiny. The European Union (EU) has issued regulation changes requiring all shellfish commodities to be tested for the toxins. These toxins pose a health risk to humans with complications such as headaches and gastrointestinal side effects. This class of compounds presents challenges to LC/MS/MS analysis due to the need for both positive and negative ionization techniques, as well as the need for chromatographic resolution of several isobaric compounds. A robust and highly sensitive method is needed in order to reliably detect and quantitate these compounds from marine food products. This study will demonstrate a rapid and reproducible method for the analysis of lipophilic marine toxins using certified reference materials. Separation was achieved in less than 7 minutes, and using matrix-matched calibration standards for quantitation, recoveries from shellfish ranged from 80 to120%.

Keywords: Food Contaminants, Food Safety, Liquid Chromatography/Mass Spectroscopy
Application Code: Food Safety
Methodology Code: Liquid Chromatography/Mass Spectrometry

Session # 480 Abstract # 480-6 **Poster Sessions**

Session Title Food Safety

Abstract Title **Utilizing Mass Spectrometry for Gluten Detection for Use in Gluten-Free Foods**

Primary Author Sophie Bromilow
University of Manchester

Date: Monday, March 07, 2016 - Morning
Time:
Room: Exposition Floor, 400 Aisle

Co-Author(s) Clare Mills, Lee Gethings, Michael Bromley, Michael Buckley, Peter Shewry, Phil E. Johnson

Abstract Text

Coeliac disease (CD) is an autoimmune condition that is classified as a non-IgE mediated food intolerance to gluten with an estimated prevalence of 1% in the United States and Europe. In the absence of a cure, patients with CD must adhere to a strict gluten-free diet, the Codex standard defines "gluten-free" as foods containing <20mg/Kg gluten. However, methods for detecting and quantifying gluten in foods often return ambiguous and inaccurate results. Mass spectrometry (MS) may provide an alternative novel method with the specificity and sensitivity required for unambiguously detecting gluten in free-from foods.

This research is novel by using a knowledge based approach, requiring additional research to support MS method development. This study includes the development of a curated database of unique gluten protein sequences with celiac toxicity data that the MS data can be mined against with an optimized gluten protein extraction buffer. Extracts from a variety of wheat grains and flours were prepared using various extraction methods and digested with chymotrypsin. LC/MS data were acquired using data independent analysis, incorporating ion mobility into the workflow (DIA-IM-MS). Data were processed and searched against a complete protein database using Progenesis QI for Proteomics providing label-free quantitation, resulting in 620 gluten proteins being identified. Normalisation of the data was based on all proteins with 493 proteins having been quantified. Over 80% of the proteins observed are common proteins shared between seven extraction methods, with additional unique proteins identified in each extraction. MRM targets were picked which target proteins that are highly abundant and highly coeliac toxic.

Keywords: Food Identification, Food Safety, Mass Spectrometry, Proteomics

Application Code: Food Safety

Methodology Code: Mass Spectrometry

Session # 480 **Abstract # 480-8** **Poster Sessions**

Session Title: Food Safety

Abstract Title: **Analysis of Target Pesticides in Essential Oils Using a Novel GC/MS/MS System**

Primary Author: Charles Schmidt
PerkinElmer

Date: Monday, March 07, 2016 - Morning
Time:
Room: Exposition Floor, 400 Aisle

Co-Author(s): Adam J. Patkin, Samuel Tolley, Sharanya Reddy, Thomas Dillon

Abstract Text

Essential or botanical oils are used as flavors and fragrances in aromatherapy and alternative medicine. The presence of pesticide residues in these essential oils is of growing concern because of the widespread commercial usage of these oils. Analysis of pesticides using a single quadrupole mass spectrometer in the complex matrix of botanical oil normally requires extensive sample cleanup prior to analysis. We present a study of the analysis of target pesticides spiked in lemon oil using a novel hybrid GC/MS/MS system, with no additional sample cleanup other than a simple dilution and an online Swafer backflush system to remove high-boiling matrix components. Unlike a traditional triple quadrupole system, this novel GC/MS/MS system has the ability to measure "all product ions all the time" with no loss in sensitivity. Besides method simplification, monitoring for multiple product ions improves confidence in results. The other big advantage of full product spectrum acquisition is realized when the unanticipated interference of a matrix ion with a target product ion can be eliminated in post-run processing by simply selecting and quantifying an interference-free product ion - without having to reacquire the calibration and sample chromatograms. This saves both time and expense. The pesticide detection limits were determined to be well within the USP regulatory limits.

Keywords: Agricultural, Gas Chromatography/Mass Spectrometry, GC-MS
Application Code: Food Safety
Methodology Code: Gas Chromatography/Mass Spectrometry

Session # 480 Abstract # 480-9 **Poster Sessions**

Session Title Food Safety

Abstract Title **A Raman Spectroscopic Method for Determination of Erucic Acid in Canola Oils**

Date: Monday, March 07, 2016 - Morning
Time:
Room: Exposition Floor, 400 Aisle

Primary Author Elif Ercioglu
Hacettepe University

Co-Author(s) Ismail H. Boyaci, Murat H. Velioglu, Serap Durakli Velioglu, Tumay H. Temiz

Abstract Text

Erucic acid is the principal fatty acid in the rapeseed which is commonly used for edible oil production. However, in vivo experiments reported that consuming erucic acid containing oils could lead to serious health problems. Due to these health concerns, a new strain of rapeseed known as canola which contains low erucic acid was developed using selective breeding. Today, canola is often used with other vegetable oils for the production of edible oils, margarine or shortenings. However, studies related to the toxicity of erucic acid are still being conducted. Regarding the safety of vegetable oils, European Council requires that canola oil must contain erucic acid less than 5%, while US legislations specify a maximum level as 2% in total fatty acids. Hence, establishing a rapid, simple and reliable method for determination of erucic acid is of great importance in terms of consumer safety and food legislations. Chromatographic methods namely, GC, GC-MS, and HPLC are commonly used for erucic acid determination. However, these methods mostly require time and chemical consuming pre-treatment procedures. To eliminate the disadvantages of traditional analysis, vibrational methods have been widely used in food analysis. The aim of this study was to evaluate the potential of Raman spectroscopy for determination of erucic acid percentage in total fatty acids of the canola oil. The erucic acid content of canola oils, between 0% and 33.56% (w/w) were determined by means of partial least squares (PLS) analysis. High coefficient of determination values were obtained for both calibration and validation graphs, 0.990 and 0.982 respectively. This study presents a rapid (45 s), non-destructive and accurate method for determination of erucic acid content in canola oil.

Keywords: Analysis, Chemometrics, Food Safety, Spectroscopy
Application Code: Food Safety
Methodology Code: Vibrational Spectroscopy

Session # **480** **Abstract #** **480-10** **Poster Sessions**

Session Title: Food Safety

Abstract Title: **Multiple Foodborne Pathogen Analysis Using a 96-Well Assay in Less than 5 Hours**

Date: Monday, March 07, 2016 - Morning
Time:
Room: Exposition Floor, 400 Aisle

Primary Author: Stuart Farquharson
Real-Time Analyzers, Inc

Co-Author(s): Chetan Shende, Kathryn Dana

Abstract Text

Each year foodborne pathogen infections result in ~50 million illnesses, ~130,000 hospitalizations, and 3000 deaths in the USA alone. Preventing the distribution and consumption of contaminated foods is challenging, because just 100 bacterial cells can rapidly multiply to millions, reaching infectious doses within a few days. While polymerase chain reaction analyzers (PCR) are replacing culture growth to reduce the analysis times, they are only effective if there are 10,000 to100,000 pathogenic cells/g present. For this reason numerous samples are "pooled" together and added to enrichment broths. Not only does analysis still take 20-30 hours, but if a sample tests positive, all of the individual samples (typically 20) must now be analyzed. Consequently, there is a critical need for a multiplexable analyzer that can rapidly detect foodborne pathogens at 1 colony forming units per gram (cfu/g) of food in a few hours (not days). In an effort to meet this need, we have been developing an assay in a 96 micro-well format that extracts such pathogens from food, selectively binds these pathogens, and produces surface-enhanced Raman spectra (SERS) when read by a Raman analyzer. Measurements of Campylobacter jejuni in chicken, Escherichia coli in ground beef, Listeria monocytogenes in cheese and Salmonella typhimurium in chocolate at <10 cfu/g using this assay will be presented.

Keywords: Food Safety, Immunoassay, Process Monitoring, Surface Enhanced Raman
Application Code: Food Safety
Methodology Code: Vibrational Spectroscopy

Session # 480   Abstract # 480-11   **Poster Sessions**

Session Title: Food Safety

Abstract Title: **Stand-Off Raman Detection of Adulteration in Honey**

Date: Monday, March 07, 2016 - Morning

Time:

Room: Exposition Floor, 400 Aisle

Primary Author: Kenneth Garcia
Alabama A&M University

Co-Author(s): Anup Sharma, Aschalew Kassu, Carlton Farley

Abstract Text

Pure, natural honey is a labor intensive product as beekeepers must care for the bees year-round and manually harvest the honey. This contributes to a higher cost per volume than comparable syrups and sweeteners such as corn syrup and rice malt syrup. Due to the limited supply and high cost, honey is a prime target for dilution with less expensive adulterants. Viability of standoff Raman technique is demonstrated for detecting contamination/adulteration of honey with sweeteners like Maltose and Fructose syrups from a distance of up to one meter. We have characterized standoff Raman technique with reference to parameters like, nature and concentration of contaminants, standoff distance, ability to make quantitative measurements, and sensitivity. The equipment involves a portable Raman system with a 785 nm laser. The Raman system is coupled to a 2-inch refracting telescope and can be operated in the field either with batteries or a gas-powered generator. This set-up has been used to detect adulteration in Honey through the glass walls of the containers. Results include detection of adulterating syrups by their distinct spectral fingerprints. Adulterating syrups in honey can be detected from a distance of up to one meter. Stand-off Raman technique can detect food contaminants/adulterants which could be toxic or laced with biological pathogens from a safe, non-contact distance of several meters. The results show a potential to solve issues related to monitoring food-supply chain and characterize the nature of vulnerability.

This work is supported by the U.S. Department of Homeland Security under award no. 2010-ST-061-FD0001.

Keywords: Contamination, Food Safety, Molecular Spectroscopy, Raman

Application Code: Food Safety

Methodology Code: Molecular Spectroscopy

Session # **480** **Abstract #** **480-12** **Poster Sessions**

Session Title: Food Safety

Abstract Title: **Evaluating the Extraction Efficiency of Food Borne Pathogens on Automated Homogenizer Platforms**

Date: Monday, March 07, 2016 - Morning

Time:

Room: Exposition Floor, 400 Aisle

Primary Author: Shari Garrett
Omni International

Co-Author(s): Brandon Easparro, James Atwood

Abstract Text

There is an ever increasing demand for food and beverage manufacturers to adhere to rigorous food safety regulations. As part of the global need for safe food, food and beverage manufacturers test their products for the presence of food borne pathogens. While safety regulations describe what pathogens should be controlled through testing there is no predefined method of sample preparation or detection. Typically, the first step in a food borne pathogen detection study is the disaggregation of the food product or the washing of the sample to release bacteria from the food surface using a Stomacher. While the later method is great for large masses of food stuffs, some researchers may require small amounts of product for analysis. Bead mill homogenizers, as well as rotor-stator homogenizers, can be used to aid in the sample preparation process. Processing can be completed in less than one minute while maintaining bacterial cell and analyte integrity. Depending on the pathogen of interest and the food matrix, the downstream results, including sensitivity can vary greatly based on the sample preparation method chosen. With this is mind, we analyzed two sample preparation techniques on a common contaminated food. Spinach samples were inoculated with known levels of recombinant E.coli expressing GFP. The inoculated samples were processed on a bead mill homogenizer and an automated rotor-stator homogenizer. The homogenates were grown on nutrient agar and the proportion of viable cell recovery was quantified. LODs were determined by lysis of recovered homogenates, DNA purification and PCR amplification.

Keywords: Biotechnology, Detection, Food Safety, Food Science

Application Code: Food Safety

Methodology Code: Sampling and Sample Preparation

Session # 480 **Abstract # 480-13** **Poster Sessions**

Session Title: Food Safety

Abstract Title: **Nitrogen/Protein Determination in Infant Food by Dumas Combustion Method in Alternative to Kjeldahl Method**

Date: Monday, March 07, 2016 - Morning

Time:

Room: Exposition Floor, 400 Aisle

Primary Author: Guido Giazzi
Thermo Fisher Scientific

Co-Author(s): Liliana Krotz

Abstract Text

In the market the available selection of infant products is very wide and the decision of the product choice of consumers depends of a set of information comprise the quality valuation. New regulations regarding all processed food and most raw foods include a series of tests determine what their food contains and also how they relate to a healthy diet for infant and children. One of the most important parameter of the quality is the protein content, calculated through the Nitrogen determination, used also for R&D purpose. So, for this the use of an accurate and automatic analytical techniques which allows the fast analysis with an excellent reproducibility is required. An new Elemental Analyzer, based on the dynamic flash combustion of the sample, copes effortlessly with the wide array of laboratory requirements such as accuracy, day to day reproducibility and high sample throughput. This alternative to the classical Kjeldahl method, based on Dumas (combustion) method, has been developed and approved by different associations. This paper presents data on Nitrogen/Protein determination in different infant food samples, obtained with the analyzer to demonstrate the validity of the method without matrix effect.

Keywords: Elemental Analysis, Food Safety, Protein, Quality Control
Application Code: Food Safety
Methodology Code: Atomic Spectroscopy/Elemental Analysis

Session # 480 **Abstract # 480-14** **Poster Sessions**

Session Title: Food Safety

Abstract Title: **Non-Targeted Screening of Nutritional Supplements with GC, GC×GC, TOFMS, and HR-TOFMS**

Primary Author: Elizabeth M. Humston-Fulmer
Leco Corporation

Date: Monday, March 07, 2016 - Morning
Time:
Room: Exposition Floor, 400 Aisle

Co-Author(s): David E. Alonso, Jonathan D. Byer, Joseph E. Binkley, Lorne E. Fell

Abstract Text

Understanding the composition of nutritional supplements that are commonly taken by many consumers and athletes is important. Nutritional supplements are more loosely regulated by the FDA and not subjected to the same processes and requirements as pharmaceutical drugs, so questions of efficacy and safety can occur. Typically, manufacturers and distributors are responsible for ensuring that all claims and information on their packaging material are truthful and not misleading and the consumer relies on their compliance. Athletes are often concerned about unintentional consumption of unreported substances in supplements that are performance enhancing or otherwise banned within their sport. Thus, it is not uncommon for athletes to pay to have their supplements screened for banned substances. The work reported here aims to explore what else is present in supplements through complementary analytical technologies. We performed a general non-targeted extraction of a variety of nutritional powders and pills followed by non-targeted analyses with gas chromatography (GC), two-dimensional GC (GC×GC), mass spectrometry (MS), and high resolution MS (HR-MS). GC×GC allowed for the chromatographic separation of analytes that coelute in a one-dimensional separation and high resolution mass spectrometry added confidence to identifications through formulae determinations from accurate mass information. Further insight to these complex supplements was gained and is reported.

Keywords: Food Safety, GC-MS, Time of Flight MS
Application Code: Food Safety
Methodology Code: Gas Chromatography/Mass Spectrometry

Session # 480 **Abstract # 480-16** **Poster Sessions**

Session Title: Food Safety

Abstract Title: **Combining SERS with Liquid-Liquid Extraction Method for Simple, Rapid Detection of Rhodamine B in Raw Food**

Primary Author: Huaizhi Kang
Xiamen University

Date: Monday, March 07, 2016 - Morning

Time:

Room: Exposition Floor, 400 Aisle

Co-Author(s)

Abstract Text

Although Rhodamine B has been declared toxic and a cancer-causing substance, it is still illegally used commonly in food industry as an additive based on its bright color, high dyeing efficiency and low cost. In response, many analytical and physical chemical methods have been developed to analyze Rhodamine B in food samples, including GC-MS, HPLC, and UV/vis. However, most of them are lab-based methods and both time- and labor- consuming. A (Surface-enhanced Raman spectroscopy-) SERS-based method has been developed in recent years and is attractive due to its high sensitivity, intrinsic selectivity and nondestructive data acquisition. However, lack of an accompanying high-efficient extraction method displays unsatisfied overall performance and has restricted its applications. Here we introduce a SERS-based detection method using gold nanoparticles as the Raman signal amplification substrate and a novel liquid-liquid extraction route to detect Rhodamine B in raw paprika. The entire detection process can be accomplished in few simple steps within 10 minutes and can detect Rhodamine B down to 0.2ppm. More significantly, the results show no noticeable false positives or false negatives, and the method can be easily used on-site for the examination of suspicious food products.

Keywords: Food Safety, Instrumentation

Application Code: Food Safety

Methodology Code: Portable Instruments

Session # 480 Abstract # 480-17 **Poster Sessions**

Session Title Food Safety

Abstract Title **Concurrent High Sensitivity Conductometric Detection of Sulfide and Cyanide in a Suppressed Anion Chromatography System**

Primary Author Hongzhu Liao
University of Texas at Arlington

Date: Monday, March 07, 2016 - Morning
Time:
Room: Exposition Floor, 400 Aisle

Co-Author(s) Akinde Kadjo, Purnendu Dasgupta

Abstract Text

Sulfide and cyanide derive from the corresponding highly toxic acids that play important roles in different fields, including corrosion in petrochemical plants, mine wastes, wastewater treatment. Suppressed conductometric ion chromatography (SCIC) has become the benchmark in anion analysis in general. However, very weak acids like sulfide and cyanide cannot be determined by SCIC because these very weak acids do not ionize sufficiently to be conductive. By far the most popular technique to measure sulfide and cyanide in conjunction with IC has been pulsed amperometric detection. However, in order to achieve optimum performance, many special steps are required. Efforts to detect these weakly dissociated acids better by conductometric detection have also been made for some time. The most promising involved introduction of a base at a constant concentration after the suppressed conductivity detector and monitoring the conductivity again. All anions, including very weak acids show up as negative peaks, because hydroxide anions are replaced by analyte anions. In present work, a laboratory made volatile acid transfer device transfers and preconcentrates the H2S and HCN in the IC effluent. The baseline noise remains the same as that of the suppressed IC system even though the background is far higher. The limits of detection for sulfide and cyanide were sub-[micro]M the linear dynamic range extended from 0.5 to 100 [micro]M.

Keywords: Detection, Food Safety, Ion Chromatography, Method Development
Application Code: Food Safety
Methodology Code: Liquid Chromatography

Session # 480 Abstract # 480-18 Poster Sessions

Session Title: Food Safety

Abstract Title: **A Computational Study Assisted Nano-Aptasensor for Detection of Tetracycline in Honey with a DNA Aptamer**

Primary Author: Sai Wang
Beijing Univerisity of Chemical Technology

Date: Monday, March 07, 2016 - Morning

Time:

Room: Exposition Floor, 400 Aisle

Co-Author(s)

Abstract Text

A 40 mer T-rich aptamer was used in the present study to develop a label-free AuNPs-based nano-aptasensor for tetracycline (TC) residue detection in honey sample. Color of unmodified AuNPs will change from red to purple when the AuNPs aggregate from dispersion state. Adsorption of aptamer onto the surface of provides the AuNPs from aggregation at high ionic condition, while aggregation of AuNPs will specifically induced by desorption of aptamer from the surface of AuNPs as a result of the aptamer-TC binding. The length of the anti-TC aptamer is desirable for the easy adsorption of aptamer to the surface of AuNPs since the 40 mer aptamer has simple and flexible conformational structure. The rich T bases are helpful for the sensitivity of the biosensor, because T base has less affinity to the surface of AuNPs. Moreover, computational analysis was applied to help us study the molecular recognition between the 40 mer aptamer and TC. A spectrophotometer was used for precise quantitation both in buffer and honey sample. The present study provides a computation-experiment-combined concept for biosensor development. The label-free nano-aptasensor provides a limit of detection (LOD) of 12.4 ng/mL for TC detection in honey. This assay can be a simple and efficient method for basic on-site screening for TC residue in honey samples, both visibly and quantatively.

Keywords: Bioanalytical, Biosensors, Detection, Food Safety

Application Code: Food Safety

Methodology Code: Sensors

Session # 480 **Abstract # 480-19** **Poster Sessions**

Session Title: Food Safety

Abstract Title: **Simple, Rapid Extraction of Chlorinated Pesticides in Poultry Fat by Solid Phase Extraction and GC/ECD**

Date: Monday, March 07, 2016 - Morning
Time:
Room: Exposition Floor, 400 Aisle

Primary Author: Allen Misa
Phenomenex

Co-Author(s): Ramkumar Dhandapani, Tim Anderson

Abstract Text

Pesticide contamination is not limited to fresh produce; poultry fat used for food consumption is also exposed to contaminants at levels that can pose harm to the human population. Pesticide extraction from poultry fat using conventional liquid-liquid extraction (LLE) techniques can be time-consuming and create excess solvent waste. Additionally, the fat matrix is complex due to the presence of proteins and lipids. Non-selective extraction methods such as LLE do not eliminate all interferences, and eventually result in decreased column lifetime and increased system maintenance.

Presented is a simple, rapid solid phase extraction (SPE) method developed to selectively extract chlorinated pesticides from poultry fat using minimal solvents, utilizing Strata® Alumina-N solid phase extraction cartridges. Following the extraction, GC analysis is performed using a Zebron™ ZB-MultiResidue™-1 column and electron capture detector (ECD). The column employs an extensively cross-linked stationary phase that offers selectivity necessary to separate structurally similar chlorinated pesticides. The optimized GC method results in a 19 min total run time for all the chlorinated pesticides, eluting all analytes within 15 min. Analytically, the method presents cleaner chromatograms that are free from matrix impurities and suitable for quantitative analysis. Comparatively, the SPE GC/ECD method for pesticide analysis from poultry fat outperformed the traditional procedure with decreased laboratory space requirements, reduction of hazardous waste, and significant reduction of labor consumption, leading to greater overall laboratory productivity.

Keywords: Chromatography, Food Safety, GC
Application Code: Food Safety
Methodology Code: Gas Chromatography

Session # 480 **Abstract # 480-20** **Poster Sessions**

Session Title: Food Safety

Abstract Title: **Approaches to Measuring both Trace and Nutritional Elements in Food in a Single Analysis by ICP-MS**

Primary Author: Kenneth Neubauer
PerkinElmer

Co-Author(s): Stan Smith

Date: Monday, March 07, 2016 - Morning
Time:
Room: Exposition Floor, 400 Aisle

Abstract Text

With the increasing global population, both the availability and quality of food are of growing concern. In order to monitor the nutritional content and contamination in foods, both trace-level toxic elements and high-level nutritional elements must be measured.

Inductively coupled plasma mass spectrometry (ICP-MS) is a trace elemental technique capable of measuring both ultra-trace and high level elements. Typically ICP-MS is used for low-level analyses; measuring very high analyte levels (such as nutritional elements) can shorten the detector's lifetime. Additionally, because of the complex nature of food products, a wide variety of spectral interferences exist.

One common method to deal with these interferences is the use of collision mode, where physical collisions and an energy barrier both remove interferences and decrease analyte sensitivity. This has the benefit of simplicity, but also significantly reduces analyte signal. Although beneficial for nutritional analytes, it may also affect the ability to measure trace level contaminants.

An alternative approach is reaction mode, where controlled chemistry is used to remove spectral interferences without affecting analyte sensitivity, thereby allowing the lowest levels to be accurately measured. With a properly designed reaction cell, analyte sensitivity can be tuned per-mass, which effectively extends the dynamic range without sacrificing detector lifetime. The significance is that high-level nutritional elements can be measured during the same analysis as trace level elements without sacrificing sensitivity of the low-level analytes.

This work will focus on the analysis of both toxic and trace elements in a variety of food matrices, comparing both collision and reaction modes.

Keywords: Elemental Analysis, Food Safety, ICP-MS
Application Code: Food Safety
Methodology Code: Atomic Spectroscopy/Elemental Analysis

Session # 480 **Abstract # 480-22** **Poster Sessions**

Session Title: Food Safety

Abstract Title: **Impact of Thermal Processing on the Solubility and Detection of Peanut Allergens Using LC-MS/MS based Targeted Proteomics with Multiple-Reaction Monitoring (MRM)**

Date: Monday, March 07, 2016 - Morning
Time:
Room: Exposition Floor, 400 Aisle

Primary Author: Rebekah L. Sayers
University of Manchester

Co-Author(s): Clare Mills, Helen Brown, Justin T. Marsh, Phil E. Johnson

Abstract Text

Peanut allergy is considered one of the most prominent food allergies, reactions to which are often severe and can be fatal. Peanuts typically undergo thermal processing, which can lead to a reduction in protein solubility and therefore extractability, making accurate detection and quantification problematic. In this study the robustness of proteomic analysis to determine the allergenic content of thermally processed peanuts was assessed.

Thermally processed peanuts were defatted and the extraction efficiencies of five buffers determined. Specific analyte targets representing the major allergenic proteins in peanuts were selected and isotopically labelled peptides synthesised. MRM experiments were designed to detect target peptides using 3 transitions. Calibration curves were generated from stable isotope dilutions in buffer and peanut matrix. Analytical response of light relative to heavy peptide was used to interpolate the resulting linear regression analysis allowing absolute quantification of allergenic proteins.

Thermal processing decreases protein solubility and optimal extraction was achieved using harsh denaturing conditions. The thermal stability of target peptides was closely associated with the proteins biochemical properties. In extensively processed samples peptides derived from cupin proteins proved less robust than prolamins which provided a more reliable target. In addition peptides flanked by arginine residues proved less susceptible to thermal processing than lysine, possibly linked to Maillard-reactions.

This work highlights the need for improved MS compatible extracts and more efficient proteolysis in order to accurately quantify protein. While this work focuses on understanding the thermal dependence of allergen profiles of peanut the application of this study will aid detection of peanut in foods.

Keywords: Proteomics, Quadrupole MS, Quantitative, Sample Preparation
Application Code: Food Safety
Methodology Code: Mass Spectrometry

Session # 480 Abstract # 480-23 **Poster Sessions**

Session Title Food Safety

Abstract Title **Measurement of Arsenic in Wine by Hydride Generation – Flame Atomic Absorption**

Date: Monday, March 07, 2016 - Morning
Time:
Room: Exposition Floor, 400 Aisle

Primary Author Nick Spivey
PerkinElmer Inc.

Co-Author(s) Kenneth Neubauer, Stan Smith

Abstract Text

Recent concern about arsenic in the food supply has mainly focused on fruit juice and rice, but other substances have been undergoing testing. There have been published reports of arsenic being found in wine at levels higher than the EPA limit for drinking water (10 ppb). As a result, wine producers are rapidly adopting the ability to monitor arsenic in their products.

The easiest and most accurate way to measure arsenic at levels of 10 ppb or less is with ICP-MS. However, the instrumentation is costly, the result of the ability of the ICP-MS to rapidly and accurately measure trace elements across the entire atomic mass the range. For small wineries that are primarily concerned with arsenic levels, ICP-MS can be prohibitively expensive.

Other atomic spectrometry techniques which may be considered are ICP-OES, graphite furnance AA (GFAA), and flame AA. Unfortunately, reliable arsenic measurements at 10 ppb are either not possible or are pushing the limits of ICP-OES and flame AA. While GFAA can measure sub 10 ppb levels, it is a complex technique.

Coupling hydride generation to a flame AA (HG-FAA) or ICP-OES (HG-ICP-OES) provides a simple solution: generation of the arsenic hydride eliminates matrix effects and allows low levels of arsenic to be easily measured. This work will discuss the measurement of arsenic in wine by HG-FAA and HG-ICP-OES.

Keywords: Atomic Absorption, Beverage, Food Safety
Application Code: Food Safety
Methodology Code: Atomic Spectroscopy/Elemental Analysis

Session # 480 **Abstract # 480-24** **Poster Sessions**

Session Title: Food Safety

Abstract Title: **Determination of Organic Tin Pesticides in Fruits and Vegetables by Gas Chromatography Coupled to Tandem Mass Spectrometry**

Date: Monday, March 07, 2016 - Morning

Time:

Room: Exposition Floor, 400 Aisle

Primary Author: Xiaobo Liu
Shimadzu

Co-Author(s)

Abstract Text

Organic tin pesticides are often used to sterilize and kill of mites in fruit and vegetables.This kind of pesticides is dangerous substance with high toxicity ,and some of them can cause cancers.The latest implementation of the food safety standards of china sets the maximum residue limit of azacyclotin and fenbutatin oxide in apple,organges,tomato etc, which were range from 0.1 mg/kg to 2.0 mg/kg. The EU, Japan, the United States and other countries also have a limited level of the food. Therefore, it is very important to establish a method for the determination of organic tin pesticides in food. At present, the methods of detecting organic tin pesticides in China and abroad are mainly by gas chromatography and GC/MS. Because of the complexity of vegetable and fruit matrix, the method of gas and GC/MS is very complicated. A method for the simultaneous determination of azacyclotin, triphenylhydroxytin and fenbutatin oxide residues in fruits and vegetables was developed by pentylmagnesium bromide derivatization and gas chromatography coupled to tandem mass spectrometry(GC-MS/MS).The samples were firstly extracted by hexane and acetone,followed derivatization with pentylmagnesium bromide.Then after purification using Envi-Carb and florisil SPE columns,the sample extracts were finally analyzed by GC-MS/MS. This method showed good linearity within the range of 0.005~0.5mg/kg. The recoveries of the method ranged from 74.1% to 124.5%.This method shows strong anti-interference capability, accuracy and high sensitivity to detect Organic tin pesticides in fruits and vegetables.

Keywords: Analysis, Food Safety, GC-MS
Application Code: Food Safety
Methodology Code: Gas Chromatography/Mass Spectrometry

Session # **480** **Abstract #** **480-25** **Poster Sessions**

Session Title Food Safety

Abstract Title **Determination of 198 Pesticide Residues in Eggplant Using Gas Chromatography Tandem Mass Spectrometry/Mass Spectrometry (GCMS/MS)**

Primary Author Wang Yan
Shimadzu

Date: Monday, March 07, 2016 - Morning
Time:
Room: Exposition Floor, 400 Aisle

Co-Author(s)

Abstract Text

A fast method for simultaneous analysis of 198 pesticide residues in the substrate of eggplant by Gas Chromatography tandem Mass Spectrometry/Mass Spectrometry (GCMS/MS) in multiple reaction monitoring(MRM) acquisition mode was established. The eggplant samples were weighted, extracted by acetonitrile, and separated with liquid-liquid partitioning from water in the sample by salting out with sodium acetate and magnesium sulfate (MgSO4). The supernatant liquid was purified by primary-secondary amine (PSA) to remove most of the pigments in samples, vaporized to near dryness by nitrogen, then dissolved in ethylacetate. The treated samples were then subjected to MRM analysis for 198 pesticides using GC-MS/MS, every compound has 2 MRM transitions (primary for quantification, secondary for qualification), totally 396 transitions in 38 minutes. These results showed that the average recoveries were 70%~120% of target compounds and the relative standard deviations (n = 6) were below 12% in eggplant spiked levels at 0.005mg/kg. The Limits of determination were range from 0.01 to 0.5 [micro]g/kg. The results show that this method is rapid, simple, easy to operate, and it is feasible to determine simultaneous pesticide residues in eggplant.

Keywords: Gas Chromatography/Mass Spectrometry
Application Code: Food Safety
Methodology Code: Gas Chromatography/Mass Spectrometry

Session # 480 Abstract # 480-27 **Poster Sessions**

Session Title: Food Safety

Abstract Title: **Development of On-Line SFE-SFC System and Its Application for Food Contaminant Analysis**

Primary Author: Shin-ichi Kawano
Shimadzu (China) Co., Ltd.

Date: Monday, March 07, 2016 - Morning
Time:
Room: Exposition Floor, 400 Aisle

Co-Author(s): Lingling Shen, Naoki Hamada, Taohong Huang, Xiaohua Liu, Yan Wang, Yuki Hashi

Abstract Text

Sample pretreatment is always one of the key issues in a series of analytical procedures. Items such as efficiency of sample extraction, sample throughput, reproducibility, etc. are evaluated through the analytical validation. On-line sample pretreatment with an automated chromatographic system is often used for effective and reliable extraction and analysis. Supercritical fluid extraction (SFE) has been a commonly used substantial extraction technique and applied for extraction of bioactive compounds, drugs, lipids, and contaminants from natural products. Compared with other extraction methods, SFE with supercritical carbon dioxide (SC-CO2) has advantages such as low consumption of organic solvent and low toxicity. SC-CO2 is also readily removed after use. Supercritical fluid chromatography (SFC) has the mutual advantages with SFE. Compared with HPLC, reduction of the viscosity of mobile phase and the improvement of the diffusion coefficient make better separation performance in SFC. For the enhancement of productivity and quality of analysis, we have developed an on-line SFE-SFC system. Devices like CO2 pump, SFE unit, BPR (back pressure regulator) were newly designed for the system. Mass spectrometers can be combined so that the system achieves the maximum performance. The system was evaluated through the applications such as residual pesticide and antibiotic analysis. The system performance was compared with that of the on-line pretreatment HPLC system.

Keywords: Food Contaminants, Mass Spectrometry, SFC, SFE
Application Code: Food Contaminants
Methodology Code: Supercritical Fluid Chromatography

Session # 480 **Abstract # 480-28** **Poster Sessions**

Session Title: Food Safety

Abstract Title: **High Resolution Accurate Mass (HRAM) Collision Energy Profile of Residues of Concern for Food Safety**

Date: Monday, March 07, 2016 - Morning

Time:

Room: Exposition Floor, 400 Aisle

Primary Author: Daniel Biggerstaff
o2si Smart Solutions

Co-Author(s): Huichen Stavros, Min Cai

Abstract Text

High resolution accurate mass (HRAM) fragmentation patterns for hundreds of pesticides, toxins, and drugs have been determined using at least ten different collision energies from 10 to 105 eV using a Thermo Scientific Exactive Orbitrap. The fragmentation patterns of representative compounds are displayed. The fragmentation patterns provide useful information for multi-residue method development. These can be especially useful for co-eluting compounds where one compound may contribute an interfering ion for the quantitation of another. A method for the quantification of isomers that coelute and have the same mass fragments is demonstrated. In the development of multicomponent residue mixes for the use in rapid triple quad LC/MS/MS source optimization solutions, having all parent and fragment ions be unique by unit mass resolution at all energies is essential.

Keywords: Analysis, Food Safety, Liquid Chromatography/Mass Spectroscopy

Application Code: Food Safety

Methodology Code: Liquid Chromatography/Mass Spectrometry

Session # 480 Abstract # 480-29 **Poster Sessions**

Session Title Food Safety

Abstract Title **Quantification of Pesticide Residues in Fruits and Vegetables by Gas Chromatography - Mass Spectrometry**

Date: Monday, March 07, 2016 - Morning

Time:

Room: Exposition Floor, 400 Aisle

Primary Author Kaelyb Suchevits
California University of PA

Co-Author(s)

Abstract Text

In this project, fresh fruits and vegetables including apples, peaches, green beans and celery were analyzed for pesticide residues using a Gas Chromatography Mass Spectrometry. The fruit and vegetable samples were extracted using the QuEChERS method for sample concentration and purification. Prior to the extraction, an internal standard of triphenyl phosphate was spiked to all samples. About 60 pesticide compounds were identified in the samples by authentic multi-residue standard mixtures. A recovery study of all 60 pesticide compounds was conducted on randomly selected samples (2 per fruit and vegetable). Quantification of each of the 60 pesticide compounds was performed with their relative response to triphenyl phosphate and three quantification m/z ions.

Keywords: Biological Samples, Food Science, Gas Chromatography/Mass Spectrometry, GC-MS

Application Code: Food Science

Methodology Code: Gas Chromatography/Mass Spectrometry

Session # 480 Abstract # 480-30 **Poster Sessions**

Session Title Food Safety

Abstract Title **Study of Sugar and Humectant Profiles in Smokeless Tobacco Products Using an LC-ESI-MS/MS Method**

Date: Monday, March 07, 2016 - Morning

Time:

Room: Exposition Floor, 400 Aisle

Primary Author Liqun Wang
Battelle

Co-Author(s) Clifford Waston, Liza Valentin, Roberto Bravo, Stephen Stanfill

Abstract Text

Smokeless tobacco (ST) products are used broadly and their consumption is increasing in the world. To better understand the chemical constituents of ST, mono- and disaccharide sugars (fructose, glucose, mannose; sucrose, maltose), alditols (xylitol, sorbitol, and myo-inositol), and humectants (glycerol, propylene glycol, and triethylene glycol) in 13 major categories/subcategories of ST were characterized, including chewing (loose leaf, plug, and twist), US moist snuff loose, Sweden moist snus loose/pouch, creamy snuff, dry snuff without pouch, US moist snus, dissolvable tobacco (pellet, stick, and strip), and tobacco stick. The highest mean sugar level was detected in chewing tobacco (9.3□27.5%), followed by dissolvable tobacco (2.1%). All others had a mean sugar level lower than 1%. Creamy snuff had the highest mean alditol level detected (22.6%), followed by dissolvable tobacco (15.4%). All others had a mean alditol level lower than 1%. The detected mean humectant levels ranged from non-detectable to 5.9%. This study demonstrates the broad diversity among ST. This research may aid researchers and public health advocates investigating the risk of using ST.

Keywords: Analysis, Characterization, Liquid Chromatography/Mass Spectroscopy

Application Code: Food Safety

Methodology Code: Liquid Chromatography/Mass Spectrometry

Session # 480 Abstract # 480-31 **Poster Sessions**

Session Title Food Safety

Abstract Title **High-Throughput Screening of Domoic Acid in Shellfish by Laser Ablation Electrospray Ionization (LAESI)-MS**

Date: Monday, March 07, 2016 - Morning
Time:
Room: Exposition Floor, 400 Aisle

Primary Author Pearse McCarron
National Research Council

Co-Author(s) Callee M. Walsh, Daniel Beach, Kelley Reeves, Pamela Cantrell, Sinead O'Brien, Wade A. Rourke

Abstract Text

Domoic Acid (DA) is a potent neurotoxin produced by marine diatoms that is accumulates in shellfish. Regulatory shellfish safety testing programs play an important role in minimizing the public health and economic impacts of DA outbreaks in shellfish harvesting areas. We recently showed that Laser Ablation Electrospray Ionization (LAESI)-MS/MS could detect and quantify DA directly from mussel tissue homogenates without sample extraction, cleanup or chromatographic separation (Beach, Walsh and McCarron, Toxicon, 2014). The decrease in run time from 20 min for LC methods to approximately 10 sec/sample for LAESI-MS is of interest to regulatory labs carrying out shellfish safety testing. For example, the Canadian Food Inspection Agency currently runs about 10,000 shellfish samples annually testing for DA, the vast majority of which are negative. Here, we assess the suitability of LAESI-MS as a high-throughput screening or quantitation tool for DA in a variety of shellfish matrices. The method was first optimized for use with high resolution MS detection. Samples included 190 real shellfish samples previously analyzed by regulatory labs as well as mussel matrix reference materials certified for DA. LAESI-MS showed good agreement with LC-UV and was capable of differentiating samples above and below 5 mg/kg, one quarter of the regulatory limit. This makes LAESI-MS promising as a high-throughput screening tool and its implementation in routine monitoring labs could lead to significant cost/time savings and expanded sample throughput capability.

Keywords: Food Safety, High Throughput Chemical Analysis, Mass Spectrometry, Natural Products
Application Code: High-Throughput Chemical Analysis
Methodology Code: Mass Spectrometry

Session # 480 **Abstract # 480-32** **Poster Sessions**

Session Title: Food Safety

Abstract Title: **Development and Implementation of a Fast, Reliable and Sensitive Analytical Test for Determining Methylmercury in Fish**

Primary Author: Ana M. Muñoz
Lasallian University

Date: Monday, March 07, 2016 - Morning
Time:
Room: Exposition Floor, 400 Aisle

Co-Author(s): Claudio Jiménez, Daniel E. León

Abstract Text

Determining organic mercury species requires cumbersome procedures often leading to analyte losses and changes, low sensitivity and alterations of the accuracy of the method. The aim of this study was to develop and apply a method for the determination of methylmercury in fish. The developed method is applied in different brands of canned tuna, and frozen fish, all obtained in supermarkets in Medellin – Colombia. Development of the method includes a comparison of processing techniques shown by solid phase microextraction (SPME) and liquid-liquid microextraction (LLME). The response surface methodology was used to obtain a polynomial expression to describe the optimum combination of variables that influence the extraction. The analysis is performed on a gas chromatograph with mass detector. The optimized method was validated according to the directive 2002/657 / EC1 of the European Union. Reference material used NIST 2976 y NMiJ CRM 7402-a. Decision limit and detection capability of CC□=13.88 µg/L and CC□=14.55µg/L was obtained respectively, recoveries between 82.14% and 101.71%, linearity between 20-900µg/L. From 13 samples, only three from different brands of canned tuna showed MeHg ranging between 109.63 and 497.10µg/kg. In addition other fish samples showed MeHg concentrations ranging between 42.32 and 59.46□g/Kg. These values do not exceed the reference level (RL) recommended by the Food and Agriculture Organization (FAO).

Keywords: Gas Chromatography/Mass Spectrometry, SPME, Trace Analysis, Validation
Application Code: Validation
Methodology Code: Gas Chromatography/Mass Spectrometry

Session # 480 **Abstract # 480-34** **Poster Sessions**

Session Title Food Safety

Abstract Title **A New Insight Into Fish Meat Freshness: ZnO /PPy Modified Biosensor**

Date: Monday, March 07, 2016 - Morning
Time:
Room: Exposition Floor, 400 Aisle

Primary Author Buket Sahyar (Yalcin)
Indesit Company

Co-Author(s) Ahmet Yavas, Erdal Celik, Mehmet Ozsoz, Merve Kaplan, Metin Ozgul, Ömer Mindivanli, Semih Otles

Abstract Text

Biosensors offer superior advantages over conventional food toxicity analysis techniques due to their low assy time, ease-of-use, portability and low cost. Electrochemical biosensors is the major subgroup employed for food analysis covering a wide range of applications from detection of microbial contaminants of food to detection of veterinary drugs in foodstuffs of animal origin. Electrochemical sensors have been designed for decades to assess the freshness of meat, more specially for fish based on detection of biogenic amines via enzyme electrode transducers.

In the proposed work, an electrochemical enzyme based biosensor was designed for the assessment of fish freshness. The three electrode system consisting of, zinc oxide (ZnO) nanoparticles were entrapped to the electropolymerized polypyrolle (PPy) covered pencil graphite electrode as working electrode, Ag/AgCl as reference electrode and Pt as counter electrode. The detection was based on enzymatic conversion of xanthine via xanthine oxidase that was physically adsorbed on working electrode and measured by chronoamperometry and electrochemical impedance spectroscopy. The proposed freshness biosensor provides high sensitive detection with its label-free and rapid detection scheme.

Keywords: Biosensors, Electrochemistry, Food Safety, Material Science
Application Code: Food Science
Methodology Code: Electrochemistry

Session # 490 Abstract # 490-1 **Poster Sessions**

Session Title Microfluidics/Lab-on-a-Chip

Abstract Title **A Simple and Sensitive Paper-Based Device Coupling Electrochemical Sample Pretreatment and Colorimetric Detection**

Date: Monday, March 07, 2016 - Morning

Time:

Room: Exposition Floor, 400 Aisle

Primary Author William R. de Araujo
USP

Co-Author(s) Thalita G. Silva, Thiago Paixao

Abstract Text

Microfluidic paper-based analytical devices (□PADs) attract tremendous attention as an economical tool for in-field diagnosis, food safety, environmental monitoring and homeland security [1]. The paper-based devices can provide a totally integration for chemical analysis process (sample pretreatment, transport and mixture of solutions due the paper microfluidic properties) eliminating complex laboratory structure and well-trained persons [2]. In this study, we report the development of a simple, portable and low cost colorimetric paper-based analytical device able to detect a cutting agent, procaine, in seized cocaine samples. The interference of most common cutting agents found in cocaine samples was verified and a novel electrochemical approach was used as a sample pretreatment in order to increase the selectivity. Under the optimum experimental conditions, a linear analytical curve was obtained for procaine concentrations ranging from 5 to 60 □mol L□1, with a detection limit of 0.9 μmol L□1. The accuracy of the proposed method was evaluated using seized cocaine samples and an addition and recovery protocol. The recovery values obtained ranged from 86 to 104□%, which highlight the accuracy and robustness of the proposed method. Financial support: FAPESP, CNPq and CAPES.

References:

[1] Lab Chip, 2015, 15, 1642; [2] Anal. Chem. 2010, 82, 3–10

Keywords: Drugs, Forensic Chemistry, Lab-on-a-Chip/Microfluidics, Sensors

Application Code: Homeland Security/Forensics

Methodology Code: Microfluidics/Lab-on-a-Chip

Session # 490 **Abstract # 490-2** **Poster Sessions**

Session Title: Microfluidics/Lab-on-a-Chip

Abstract Title: **Mixing Reaction on Paper-Based Analytical Devices**

Primary Author: Ching Man Choy
California State Polytechnic University, Pomona

Date: Monday, March 07, 2016 - Morning
Time:
Room: Exposition Floor, 400 Aisle

Co-Author(s): Yan Liu

Abstract Text

Microfluidic devices have been applied to the analysis of a variety of samples due to its minimal reagent consumption and waste production while still maintaining high performance. Different materials such as quartz, polydimethylsiloxane, and paper have been adopted for microfluidic device fabrication. Among them, paper-based device is cheaper in fabrication, lighter in weight, easier in storage and transportation, and friendlier to environment. However, the flow of aqueous solutions on paper-based devices has not been thoroughly investigated; the focus of this project was to investigate mixing reactions on paper-based analytical devices using different flow patterns.
Two different patterns of microfluidic network, single channel and three-channel mixing, fabricated by wax printing were investigated, respectively. A 15 microlitter of water was introduced into the reservoir connected to the single channel, and the time required for water to travel through the channel was recorded. The flow rate of water was calculated to be 0.03068 mm[sup]2[/sup]/s. The mixing reactions of a salt solution and water were investigated by introducing a salt solution into the middle channel and water into two side channels. The flow rate of 1 mM copper sulfate was determined to be 0.08412 mm[sup]2[/sup]/s, while that of 1 mM potassium permanganate was determined to be 0.07719 mm[sup]2[/sup]/s. On-chip Indophenol blue reaction by mixing all three reagents loading from three different channels was tested and will be applied toward semi-quantitative analysis of ambient ammonia.

This project is supported by the Department of Chemistry and Biochemistry at California State Polytechnic University Pomona.

Keywords: Analysis, Environmental/Air, Lab-on-a-Chip/Microfluidics
Application Code: Environmental
Methodology Code: Microfluidics/Lab-on-a-Chip

Session # 490 Abstract # 490-3 **Poster Sessions**

Session Title: Microfluidics/Lab-on-a-Chip

Abstract Title: **Combining Vibrational Spectroscopy with Microfluidics**

Primary Author: Kathleen E. Berg
Colorado State University

Date: Monday, March 07, 2016 - Morning
Time:
Room: Exposition Floor, 400 Aisle

Co-Author(s): Amber T. Krummel, Charles Henry, Monpichar Srisa-Art, Scott D. Noblitt

Abstract Text

Microfluidic devices have employed many different detection methods depending upon the application. One of these methods is infrared (IR) spectroscopy, a useful qualitative, quantitative, and label-free analytical technique for identifying and characterizing chemical species. Coupling IR microscopy containing a focal plane array (FPA) detector with microfluidics provides simultaneous spatial and temporal data; however, analysis is challenging because many traditional substrates (e.g. poly(dimethylsiloxane), glass, plastics) used to fabricate devices are not IR transparent. One solution is to use thin (< 100-micron) layers of even incompatible materials because there is not a substantial IR radiation absorption, enabling the use of IR to analyze chemical compounds. Building on this approach, our lab has fabricated microfluidic devices compatible with in situ IR spectroscopy. Both traditional and novel device materials have been explored. Studying chemical kinetics with IR-compatible devices can provide useful information for determining mechanistic models and optimizing experimental parameters, such as reagent structure, concentration, and solvent. As an example application of these devices, we have examined various aqueous and organic reactions and reaction kinetics. Examples of diffusional mixing, proton exchange, and metal-ligand interactions will be discussed.

Keywords: FTIR, Microscopy, Polymers & Plastics, Vibrational Spectroscopy
Application Code: General Interest
Methodology Code: Microfluidics/Lab-on-a-Chip

Session # 490 **Abstract # 490-4** **Poster Sessions**

Session Title: Microfluidics/Lab-on-a-Chip

Abstract Title: **Prostate Cancer Biomarker Detection Using a 16-Sensor Electrochemical Microfluidic Immunoarray**

Date: Monday, March 07, 2016 - Morning
Time:
Room: Exposition Floor, 400 Aisle

Primary Author: Abby Jones
University of Connecticut

Co-Author(s): Amit Joshi, Brunah A. Otieno, Colleen Krause, James F. Rusling, Mohammed Sherafeldin

Abstract Text

Prostate cancer is the most common malignancy in men throughout the world. In the United States, the lifetime risk of men developing prostate cancer is 1 in 7. Current strategies in detection and staging of prostate cancer include prostate specific antigen (PSA) test, digital rectal exam (DRE) and biopsy. These practices often fall short in terms of sensitivity, specificity and inability to distinguish between aggressive and indolent forms of prostate cancer. These limitations lead to unnecessary treatments that adversely affect the patients' quality of life with minimal gain. Measurement of small panels of molecular biomarkers in serum holds tremendous potential for cancer diagnostics and personalized therapy. Herein, we describe a semi-automated multiple biomarker-based diagnostic microfluidic device for on-line capture and detection of prostate cancer biomarkers. The protein panel includes PSA, CD-14, ERG, GOLM-1, PEDF-1, IGF-1, VEGF-D and IGFBP-3, many of which are thought to be specific for aggressive prostate cancer. The protein analytes are captured from serum samples in a two channel on-line microfluidic chamber by magnetic beads labeled with antibodies and signal transducing elements. Captured analytes are then magnetically separated, washed, and introduced into a two channel detection chamber housing an array of 8 or 16 nanostructured, antibody-labeled electrodes. Ultralow detection limits in the sub fg ml-1 range was achieved for the protein panel. Using this strategy, 8-16 proteins can be detected simultaneously in 30 minutes. Measurements of this panel of selected biomarkers will be tested with prostate cancer patient samples in future to assess its diagnostic capability.

Keywords: Bioanalytical, Electrochemistry, Immunoassay, Sensors
Application Code: Bioanalytical
Methodology Code: Microfluidics/Lab-on-a-Chip

Session # 490 Abstract # 490-5 **Poster Sessions**

Session Title Microfluidics/Lab-on-a-Chip

Abstract Title **Real-Time Profiling of Pancreatic Hormone Secretion Dynamics Using an in Flow Fluorescence Polarization Immunoassay**

Primary Author Nikita Mukhitov
Florida State University

Co-Author(s) Adrian M. Schrell, Michael G. Roper

Date: Monday, March 07, 2016 - Morning
Time:
Room: Exposition Floor, 400 Aisle

Abstract Text

Fluorescence polarization immunoassays (FPIAs) are an appealing readout scheme due to the lack of required washing steps and the ability to conduct the assay in flow without separation. While single analyte FPIAs have been demonstrated, the translation of the method to multi-analyte assays has been restricted due to the scaling of the optical complexity. In this study, we present a means to conduct simultaneous FPIAs through the use of frequency encoding of the individual signals from each IA.

As a proof of concept, the assays were designed for measuring glucagon and insulin secretion from islets of Langerhans. The immunoassays were performed in a competitive scheme with glucagon and insulin probes labeled with fluorescein isothiocyanate and cyanine-5 maleimide, respectively. A 488 and 635 nm laser were made coaxial with a dichroic mirror, filtered with a linear polarizer and free space aligned to make incident upon a point in a microfluidic channel. The emission was collected by epi-fluorescence and directed into a photometer, where the parallel and perpendicular polarizations were split with a polarizing beam splitter into two separate PMTs. During operation, the 488 nm laser and 635 nm laser were pulsed at 73 and 153 Hz, respectively. The individual parallel and perpendicular polarization signals were then extracted for both channels using Fourier analysis and the depolarization was subsequently determined. Besides the ability to encode more channels, frequency modulation also demonstrated an improvement in the LODs compared to conventional operation, with LODs of 1 and 5 nM for insulin and glucagon respectively.

Keywords: Bioanalytical, Fluorescence, Immunoassay, Lab-on-a-Chip/Microfluidics
Application Code: Bioanalytical
Methodology Code: Microfluidics/Lab-on-a-Chip

Session # **490** Abstract # **490-6** **Poster Sessions**

Session Title Microfluidics/Lab-on-a-Chip

Abstract Title **Alternating Current Driven Electroosmotic Pumping through Conical Pore Membrane**

Date: Monday, March 07, 2016 - Morning

Time:

Room: Exposition Floor, 400 Aisle

Primary Author Xiaojian Wu
University of Florida

Co-Author(s) Charles R. Martin

Abstract Text

Electroosmotic flow (EOF) is an electrokinetic phenomenon that occurs when an ionic current is passed through a channel or porous material that contains excess surface charge. While EOF is typically driven by passing a direct current (DC) through the membrane or device, there is considerable practical interest in developing alternating current (AC) driven electroosmotic pumps, since applying AC can eliminate unwanted Faradaic reactions. We have found that net flow through a conical-pore membrane in the direction base-to-tip can be achieved by AC electroosmotic pumping. This was accomplished by applying a sinusoidal voltage across polyethylene terephthalate (PET) membranes that contain conically shaped pores. This phenomenon was based on EOF rectification which we have recently shown in such membranes. The effects of magnitude and frequency of the AC voltage on flow rate were investigated. A maximum flow of 3.5 ⎕L min[sup]-1[/sup] was achieved with a root-mean-square voltage of 3.5 V[sub]rms[/sub] at a frequency of 20 Hz. The effect of electrolyte concentration was also studied. We report the results of these investigations here.

Keywords: Electrochemistry, Membrane, Nanotechnology

Application Code: Nanotechnology

Methodology Code: Microfluidics/Lab-on-a-Chip

Session # 490 **Abstract # 490-7** **Poster Sessions**

Session Title: Microfluidics/Lab-on-a-Chip

Abstract Title: **Two-Fold Control of Pressure and Flow-rate for Flow Control and Quality Management in Fluidic Processes**

Date: Monday, March 07, 2016 - Morning
Time:
Room: Exposition Floor, 400 Aisle

Primary Author: Anne Le Nel
Fluigent

Co-Author(s): Benjamin Rouffet, Nicolas Petit, Thibaut Thupnot

Abstract Text

Conventional flow control systems, such as syringe and peristaltic pumps, are not well adapted to the control of flow in microchannels. They often result in long equilibration times, hysteresis and low stability. Herein, we present a new method to control the flows in microchannels based on pressure actuation, by pressurization of reservoirs filled with fluids to be injected in the microsystem. The regulated pressure within the reservoirs generates pulse-free and very stable flows through the microchannels with short settling times. To control the flow-rates with pressure actuation, highly precise flow sensors are implemented in the fluidic system and an algorithm has been developed to automatically adjust the applied pressure to reach the targeted flow-rates. Unlike conventional PID regulation which is very sensitive to any transient behavior, our algorithm deals with any coupling effects between the differ-ent channels, and is designed to deliver the fastest and the most stable flow response. It calculates a matrix image of the microsystem with the relationships between each actuated pressure channel, and the measured flow-rates. Furthermore, the system can cope with any external disturbances of the system (presence of air bubble, partial clogging, variation of viscosity or temperature, etc...), and continuously re-adjust the applied pressures. The technology is perfectly suited for droplet manipulation experiments (among other applications) where we can generate 2pL water-in-oil droplets with very high monodispersity (1.63% CV) at up to 12 kHz frequency. Only few seconds are needed to stop the droplets flow, reducing costs by a huge factor.

Keywords: Flow Injection Analysis, Instrumentation, Process Control, Sample Handling/Automation
Application Code: Other
Methodology Code: Microfluidics/Lab-on-a-Chip

Session # 490 **Abstract # 490-8** **Poster Sessions**

Session Title: Microfluidics/Lab-on-a-Chip

Abstract Title: **A 3D-Printed Device for High-Throughput Membrane-Based Cellular Analysis**

Date: Monday, March 07, 2016 - Morning
Time:
Room: Exposition Floor, 400 Aisle

Primary Author: Ruipeng Mu
Michigan State University

Co-Author(s): Dana Spence

Abstract Text

Red blood cell (RBC) transfusion has become a highly organized healthcare activity. The most recent national reports on blood banking suggest that hospitals and other institutions were handling blood donations in a most efficient manner. In fact, a major concern was post-transfusion complications, as opposed to an overall "blood shortage". Previously, our group and others have shown that RBCs stored in standard storage solutions become rigid, hindering their ability to pass through the microcirculation. It is well known that deformability is related to the ability of the RBC to release ATP, a now-recognized determinant of blood flow, in vivo. Investigating RBC deformability from all current FDA approved collection and storage solutions, and combinations thereof, would be time-consuming. Therefore, we have developed a multi-port, 3D-printed, membrane based device for RBC deformability measurements. RBC deformability is quantified by counting the number of cells with a hemocytometer that pass through a 3 □m membrane. The device is designed to contain multiple ports, controllable by a single switch or individually, allowing high-throughput studies and individual port control. Preliminary results have shown that the deformability of RBCs stored in our group's modified collection and storage solutions remains unchanged throughout 35 days of storage. However, RBCs stored in FDA-approved standard solutions lose about 20% of their deformability. Further applications of this device, such as protein cleanup from biological samples and drugs affecting deformability, will also be explored.

Keywords: Bioanalytical, Biomedical, Biotechnology, Lab-on-a-Chip/Microfluidics
Application Code: Biomedical
Methodology Code: Microfluidics/Lab-on-a-Chip

Session # 490 Abstract # 490-9 **Poster Sessions**

Session Title: Microfluidics/Lab-on-a-Chip

Abstract Title: **3D-Printed Tools to Enhance Targeted Drug Therapy**

Primary Author: Cody W. Pinger
Michigan State University

Co-Author(s): Dana Spence

Date: Monday, March 07, 2016 - Morning

Time:

Room: Exposition Floor, 400 Aisle

Abstract Text

Our group has recently shown that C-peptide and zinc delivery to erythrocytes (ERYs) requires albumin, an important hormone and drug transporter in the bloodstream. We have also shown that ERYs exhibit a "resistance" to C-peptide when suspended in hyperglycemic conditions often seen in diabetic patients. However, it is not known if this "resistance" or inability of C-peptide binding in hyperglycemic conditions is caused by glycated ERYs or glycated albumin. When albumin is incubated in a hyperglycemic environment, as seen in diabetes, it undergoes structural and functional changes, often resulting in altered binding capability of a wide range of species. Here we have studied the delivery of C-peptide to ERYs in the presence of in vitro glycated albumin in both a static environment (using ELISA), and on a 3D-printed blood circulation mimicking device. The device has six parallel channels in which ERYs flow in a buffer containing either normal albumin or glycated-albumin. In order to mimic exogenous C-peptide administration (as would be administered to a person with type I diabetes), the device is also integrated with a 3D-printed rubberized, septum-like injection port that allows a bolus syringe injection directly to the flowing ERYs for sample introduction, or aliquot removal for off-chip analysis. This strategy enables us to effectively control the glycated species (molecules or cells) and determine the role of each in targeted drug therapy.

Keywords: Bioanalytical, Biotechnology, Drug Discovery, Lab-on-a-Chip/Microfluidics

Application Code: Bioanalytical

Methodology Code: Microfluidics/Lab-on-a-Chip

Session # 490 Abstract # 490-10 **Poster Sessions**

Session Title: Microfluidics/Lab-on-a-Chip

Abstract Title: **Fluorescence-Based Quantification of Oxygen in Paper-Based Cultures of Mammalian Cells**

Primary Author: Matthew W. Boyce
University of North Carolina at Chapel Hill

Date: Monday, March 07, 2016 - Morning

Time:

Room: Exposition Floor, 400 Aisle

Co-Author(s): Andrew S. Truong, Matthew R. Lockett, Rachael M. Kenney

Abstract Text

Paper-based scaffolds are an attractive material for cell culture because, with the addition of hydrogels, they form a three-dimensional cellular microenvironment of defined thickness. Paper is also readily available to all laboratories, is easily patterned and cut to meet specific experimental setups, requires minimal investment, and supports prolonged culture of mammalian cells. Paper-based invasion assays have been previously described, and found that cells exposed to an oxygen gradient selectively invade regions of higher oxygen tension. Despite this finding, no method has been developed to quantify oxygenation in paper-based cultures; moreover few analytical sensors have been incorporated into paper-based cultures, but are needed to better understand the evolving cellular microenvironment. In this work, we characterize an oxygen-sensing polystyrene thin film to spatially and temporally quantify oxygen tensions in paper-based cultures. For thin film preparation, a polystyrene support matrix was mixed with Pd(II) [i]meso[/i]-tetrakis(pentafluorophenyl)porphrine, a porphyrin dye quenched by oxygen, and spin coated onto glass coverslips. The oxygen sensor's response was characterized by flowing different ratios of an oxygen and nitrogen gas, in the presence of 5% carbon dioxide to mimic the conditions regularly used in cell culture, across the thin film. A Stern-Volmer relationship was determined via fluorescence microscopy and was found to be linear to 160 mmHg O[sub]2[/sub] with an I[sub]0[/sub]/I[sub]160[/sub] = 35. These sensors were found to exhibit no significant cytoxic effects, but were capable of measuring real-time changes in oxygen tension. We plan to further implement these films for simultaneous measurement of cellular invasion and extracellular oxygen tension.

Keywords: Bioanalytical, Biotechnology, Microscopy, Sensors

Application Code: Bioanalytical

Methodology Code: Microfluidics/Lab-on-a-Chip

Session # 490 Abstract # 490-11 **Poster Sessions**

Session Title: Microfluidics/Lab-on-a-Chip

Abstract Title: **Methods for Quantifying Hypoxia and the Hypoxic Responses of Cells in Paper-Based Invasion Assays**

Primary Author: Andrew S. Truong
University of North Carolina at Chapel Hill

Date: Monday, March 07, 2016 - Morning
Time:
Room: Exposition Floor, 400 Aisle

Co-Author(s): C Chad Lloyd, Christian A. Lochbaum, Matthew R. Lockett, Matthew W. Boyce, Rachael M. Kenney

Abstract Text

Many solid tumors contain areas of decreased oxygen tension due to limitations in tumor-associated vasculature. Hypoxia often results in cellular phenotypes displaying increased invasiveness and an increased resistance to chemo- and radiation-based therapies. While two-dimensional cultures are widely used to study cancer biology and potential pharmacotherapies, they cannot mimic the complex [i]in vivo[/i] environment needed to accurately predict cellular responses. Consequently there is a need for [i]in vitro[/i] models that can mimic tumor-like environments. Paper-based scaffolds not only support the prolonged culture of mammalian cells in three-dimensional solid tumor-like environments but are also easy to prepare. We use a paper-based invasion assay to determine the effect of decreasing oxygen tensions on the invasiveness of MDA-MB-231 breast adenocarcinoma cells. In this assay, cells suspended in a hydrogel are seeded into a single sheet of paper and subsequently sandwiched between other sheets containing only hydrogel. The stack is placed in a custom-built holder in which we can control the direction of the oxygen gradients formed in the culture. Using this invasion assay, we show that oxygen acts as a chemoattractant for MDA-MB-231 cells. We developed immunofluorescence, confocal microscopy, protein- and DNA-based profiling methods that are compatible with the paper-based scaffolds to tease out the biomolecular responses that distinguish subpopulations of noninvasive cells from those that undergo chemotactic invasion. Our results confirm the presence of oxygen gradients in these invasion assays, and the role of oxygen in directly cellular invasion and promoting an invasive phenotype.

Keywords: Bioanalytical, Biomedical, Detection, Lab-on-a-Chip/Microfluidics
Application Code: Bioanalytical
Methodology Code: Microfluidics/Lab-on-a-Chip

Session # **490** Abstract # **490-12** **Poster Sessions**

Session Title: Microfluidics/Lab-on-a-Chip

Abstract Title: **A Microfluidic Copper Detection System Incorporating a Ratiometric Fluorescent Quantum Dot Pair**

Date: Monday, March 07, 2016 - Morning
Time:
Room: Exposition Floor, 400 Aisle

Primary Author: Sumate Pengpumkiat
Oregon State University

Co-Author(s): Anukul Boonloed, Chandima Bandara, Vincent T. Remcho, Yuanyuan Wu

Abstract Text

Described is a new approach to building a microfluidic quantum dot ratiometric sensor for detection of copper in biological samples, indicating copper toxicity or canine copper toxicosis. CdTe quantum dots of two different sizes, emitting green and red light, were synthesized, characterized and utilized as fluorophores. The green dot is used as a constant emitter and encapsulated in a silica shell. The red dot, which is immobilized on the silica surface, is static quenched in the presence of copper ion. The dual emission of the quantum dot ratiometric sensor results in a fluorescence color change from red to green by visual identification corresponding to the absence and the presence of copper ion in the sample, respectively. The ratiometric sensor was diluted and mixed with microcrystalline cellulose and dropcast on the detection zone of a novel microfluidic chip, which was made from poly(methyl methacrylate) (PMMA) assembled by polycaprolactone (PCL). Capillary action is used to transfer the liquid sample from the introduction zone to the detection zone, where the ratiometric sensor is located. Copper ion was quantitatively determined by constructing a Stern-Volmer plot. Red and green intensity values from the RGB color system were used as analytical signals for the calibration curve. The fluorescence signal ratio of red to green quantum dot was optimized in order to detect copper ion concentration in a clinical important range. The microfluidic format of the quantum dot ratiometric sensor makes rapid, low-cost testing feasible and convenient for copper ion detection specific to the diagnostic goal.

Keywords: Environmental/Biological Samples, Lab-on-a-Chip/Microfluidics, Metals
Application Code: Clinical/Toxicology
Methodology Code: Microfluidics/Lab-on-a-Chip

Session # 490 Abstract # 490-13 **Poster Sessions**

Session Title Microfluidics/Lab-on-a-Chip

Abstract Title **Ultra-Thin Layer Chromatography with Integrated Silver Colloid-Based SERS Detection**

Date: Monday, March 07, 2016 - Morning

Time:

Room: Exposition Floor, 400 Aisle

Primary Author Ryan A. Wallace
University of Tennessee, Knoxville

Co-Author(s) Michael J. Sepaniak, Nickolay V. Lavrik

Abstract Text

Simplified lab-on-a-chip techniques are desirable for quick and efficient detection of analytes of interest in the field. The following work involves the use of photolithographically fabricated deterministic pillar arrays on the micro-scale as a planar separation platforms. Separations are performed by simple low volume (2-5 μL) spotting of samples and using simple capillary-action driven flow to develop. While these platforms offer advantages for separations, detection can be complicated by their small dimensions. In this work, Ag colloid is deposited within the arrays as a source of increased signal via surface enhanced Raman spectroscopy (SERS). The colloid is shown to be stable to flow during the development process. Ag colloid is easily produced; however a common problem traditionally seen with SERS surfaces containing Ag colloid is oxidation. Our platforms are superhydrophobic inhibiting water vapor contact. We will show that oxidation is reduced and shelf-life is increased. This work Includes the successful separation and SERS detection of fluorescent anti-tumor drugs (Adriamycin and Daunomycin), as well as the DNA bases (adenine, cytosine, guanine, and thymine). Included in the study is an evaluation of the reproducibility and limit of detection of the studied compounds via fluorescence and SERS detection.

Keywords: Chromatography, Raman, Spectroscopy
Application Code: Bioanalytical
Methodology Code: Microfluidics/Lab-on-a-Chip

Session # 490 **Abstract # 490-14** **Poster Sessions**

Session Title: Microfluidics/Lab-on-a-Chip

Abstract Title: **Affinity Cell Separation Based on Surface Antigen Expression Difference in a Sequential Concentration Microfluidic Chip**

Date: Monday, March 07, 2016 - Morning
Time:
Room: Exposition Floor, 400 Aisle

Primary Author: Ye Zhang
Texas Tech University

Co-Author(s): Dimitri Pappas

Abstract Text

Antigen expression plays a significant role in clinical studies, pathology, biology and chemistry. Different types of expressed antigens and different amounts of certain expressed antigens can provide information for diagnosis, classification, and monitoring of diseases. Analyses based on antigen expression are easily operated with high specificity and sensitivity. Therefore, investigations of antigen expression at different levels are meaningful in disease detection and diagnosis. In this work, an affinity separation method was developed based on antigen expression difference in a single microfluidic chip. This 2-region microfluidic chip was coated with two different concentrations of a capture antibody to capture two cell types based on differences in antigen expression. We observed that the capture ratio of Ramos and HuT 78 cells matched the expression ratio of CD71 for the two cell lines when lower concentrations of antibody were used (R2 = 0.98). Using herringbone-modified capture channels, a separation purity of 95% and a capture efficiency of 15% were achieved under continuous-flow conditions. This chip allows differentiation of cell types using a single antigen and a single device. To further validate our analytical method, Ramos B lymphocytes were spiked into blood samples to demonstrate performance with a complex sample. Expression ratios were measured over a 1-40-fold difference, and the sample enrichment was 9.5X. This method has proven to be a robust system to measure cell types of differing antigen expression levels, and can be used to isolate cells without having a unique surface antigen if the expression level is sufficiently high in one cell type.

Keywords: Analysis, Bioanalytical, Lab-on-a-Chip/Microfluidics
Application Code: Biomedical
Methodology Code: Microfluidics/Lab-on-a-Chip

Session # 490 Abstract # 490-15 **Poster Sessions**

Session Title: Microfluidics/Lab-on-a-Chip

Abstract Title: **Microscale Size-Based Sorting with Capillary Electrophoresis and Phospholipid Additives**

Date: Monday, March 07, 2016 - Morning
Time:
Room: Exposition Floor, 400 Aisle

Primary Author: Cassandra Crihfield
West Virginia University

Co-Author(s): Lisa A. Holland

Abstract Text

Phospholipid additives have been utilized in the past to create a thermally responsive pseudo-gel for size-based capillary electrophoresis separations of DNA[1,2]. Performing phospholipid separations in a microfluidic device has the potential to greatly increase the throughput of the separations since multiple channels can be monitored simultaneously. Size separations in a microfluidic is a step toward a device capable of rapidly screening heterogeneous mixtures for catalytic activity. This research develops a method to sort molecules in the nanometer regime by size in a microfluidic platform utilizing the phospholipid pseudo-gel. Models with known sizes, such as proteins, DNA, and nanoparticles, are used to validate the method and determine resolution.

1. Durney BC, Lounsbury JA, Poe BL, Landers JP, Holland L. A Thermo-responsive Phospholipid Pseudo-gel: Tunable DNA Sieving with Capillary Electrophoresis Analytical Chemistry. 2013;85(14):6617-25.

2. Durney BC, Bachert BA, Sloane HS, Lukomski S, Landers JP, Holland LA. Reversible phospholipid nanogels for deoxyribonucleic acid fragment size determinations up to 1500 base pairs and integrated sample stacking. Analytica Chimica Acta. 2015;880(0):136-44.

Keywords: Analysis, Bioanalytical, Electrophoresis, Lab-on-a-Chip/Microfluidics
Application Code: Bioanalytical
Methodology Code: Microfluidics/Lab-on-a-Chip

Session # 490 Abstract # 490-17 **Poster Sessions**

Session Title Microfluidics/Lab-on-a-Chip

Abstract Title **High-Throughput Microfluidic Isolation and Analysis of Exosomes**

Primary Author Kristina M. Herrera
University of North Carolina at Chapel Hill

Date: Monday, March 07, 2016 - Morning
Time:
Room: Exposition Floor, 400 Aisle

Co-Author(s) Steven A. Soper

Abstract Text

Exosomes are small vesicles (30-100 nm diameter) that are shed from cells with cargo containing nucleic acids that can indicate the cells from which they originated. It is of clinical interest to isolate disease-specific exosomes to provide material from which biomarkers can be used for guiding therapeutic decisions. Exosomes, like circulating tumor cells and cell-free DNA, can be isolated from the blood of patients, forming the basis of a liquid biopsy for disease management. Recently, microfluidics have been used to isolate exosomes, but are limited by their low-throughput capabilities, which limits the quantity of disease-specific exosomes that can be isolated. We have designed a microfluidic device for the positive affinity selection of exosomes that can rapidly (<30 min) process 1 mL of plasma to search for rare disease-specific exosomes. The device is manufactured via hot embossing of a cyclic olefin copolymer (COC), and features an array of pillars in each extraction bed that are 5 μm both diameter and spacing, and 20 μm deep. Several extraction beds can be placed in parallel to increase processing throughput without sacrificing recovery of target exosomes. The antibodies used for exosomes selection can either be anti-CD63 antibodies or a disease specific affinity agent, such as anti-EpCAM. The antibodies are covalently linked to the plastic microfluidic surface via EDC/NHS coupling chemistry of a UV/O3 activated COC. The design of the chip, including pillar spacing and number and length of capture beds was computationally optimized using fluidic resistance and diffusion-based capture models. The embossing master was fabricated using UV-LiGA, which allowed for the fabrication of high aspect ratio features designed to maximize the recovery of exosomes. From the exosome isolate, we were able to extract mRNA, which was reverse transcribed to allow for expression profiling and/or mutation scanning, which provides information indicative of a patient's disease.

Keywords: Biological Samples, Biomedical, Lab-on-a-Chip/Microfluidics, Solid Phase Extraction

Application Code: Biomedical

Methodology Code: Microfluidics/Lab-on-a-Chip

Session # 490 Abstract # 490-18 **Poster Sessions**

Session Title Microfluidics/Lab-on-a-Chip

Abstract Title **Quick Production of Microfluidic Devices by Laser Engraving of Wax-Coated Glass Slides**

Date: Monday, March 07, 2016 - Morning

Primary Author Mauro S. Santos
Clemson University

Time:

Room: Exposition Floor, 400 Aisle

Co-Author(s) Carlos D. Garcia, Claudimir L. do Lago, Eric T. da Costa, Hong Jiao, Ivano G R. Gutz

Abstract Text

Due to its characteristics, glass is one of the most convenient materials for the development of microfluidic devices. However, most of the fabrication protocols, e.g. photolithographic methods, require long processing times, expensive facilities, and/or harsh chemical conditions. Polymeric materials like PMMA and PDMS, despite its limited resistance to organic solvents, higher permeability and lower thermal and mechanical stability have been extensively used due to the availability of fast, low-cost production methods such as casting and laser ablation [1]. While convenient, laser ablation is difficult to extend to the fabrication of glass devices, because the local heating by the laser beam causes surface cracking. We herein describe a simple procedure to produce microfluidic devices for capillary electrophoresis from microscope glass slides.

The channels and reservoirs were fabricated on standard soda-lime glass microscope slides (1.0mm thick) coated with paraffin wax before ablation. Firstly, some paraffin was melted at 60oC and a thin and uniform layer was deposited on the top and bottom faces of the slides (0.40g and 0.20g respectively). All ablations were carried out with a commercial CO2 laser engraver. The channels were ablated using 80% of the laser's power and 10% of maximum travel speed of laser's carriage while the reservoirs were cut by repeatedly scanning (5x) the area on the slide at maximum power and low speed to (2%).

In this context, the wax works as an effective "heat sink" once solid-liquid-vapor phase transitions absorb energy, reducing the propagation of the local temperature rise caused by laser beam, thus minimizing thermal stress and, consequently, crack formation on the substrate. With this method, it was possible to obtain channels of semi-circle profile of around 400μm width and 70μm depth (Figure) and holes (used as reservoirs) of 3mm diameter. The surface roughness of ablated channels was the same for glass and polymeric materials, although microscope images showed some small cracks around the glass channels that became irrelevant after sealing together two glass slides (or a PDMS layer on the glass) to form the CE microchip. Also, no high voltage leakage was observed during electrophoretic separations.

Acknowledgments

The financial support provided by NASA STTR and CNPq (Brazil) and FACES (Clemson University) for the microscope images.

References

[1] Gabriel, E.F.M., et.al., Electrophoresis 2014, 35, 2325.

Keywords: Instrumentation, Lab-on-a-Chip/Microfluidics, Laser, Method Development

Application Code: General Interest

Methodology Code: Microfluidics/Lab-on-a-Chip

Session # 490 Abstract # 490-19 **Poster Sessions**

Session Title: Microfluidics/Lab-on-a-Chip

Abstract Title: **Microfluidic Platform for Mass Spectrometry-Based Monitoring of Protein-ligand Binding Dynamics**

Primary Author: Yongzheng Cong
Pacific Northwest National Laboratory

Date: Monday, March 07, 2016 - Morning

Time:

Room: Exposition Floor, 400 Aisle

Co-Author(s): Cameron Trader, Daniel Orton, Erin S. Baker, Ryan T. Kelly, Tao Geng

Abstract Text

Protein-ligand binding interactions are fundamental to many biological processes including signal transduction, enzymatic catalysis and immune response. Determination of protein-ligand binding affinities and dynamics is crucial to understanding protein function and the development of new therapeutic agents. We have designed a novel microfluidic platform that combines rapid mixing of protein and ligand, variable incubation times controlled by integrated microvalves, and an on-chip electrospray ionization emitter to perform label-free and solution-based monitoring of protein-ligand binding kinetics. The platform incorporates a multi-lamellar flow mixer for rapid mixing in which protein and ligand are distributed on-chip to 28 coflowing channels of alternating composition that subsequently merge into a single narrow channel. The small volume and the sub-micrometer diffusion distances established by the mixer produce mixing times of around 15 ms at 3 µL/min total flow rate. The mixture is then incubated on-chip, with the time determined by the volume of the eight different flow paths at a constant flow rate. By stepping through each path, a range of incubation times spanning a factor of ~30 is achieved such that reactions can be monitored from <100 ms to several seconds. After incubation, the mixture is electrosprayed for MS determination. Device performance was characterized by measuring the carbonic anhydrase-furosemide binding kinetics in a millisecond time scale. The kinetic profile monitors the conversion of unbound to drug-bound protein over time until equilibrium is reached. This platform can be used for monitoring a variety of rapid, label-free reactions in addition to protein-ligand binding.

Keywords: Mass Spectrometry, Protein

Application Code: Bioanalytical

Methodology Code: Microfluidics/Lab-on-a-Chip

Session # 490 Abstract # 490-20 **Poster Sessions**

Session Title: Microfluidics/Lab-on-a-Chip

Abstract Title: **Low-Cost Microfluidic Devices for the Determination of Renal Health**

Primary Author: Christopher A. Heist, Oregon State University

Co-Author(s): Joel C. Pommerenck, Vincent T. Remcho

Date: Monday, March 07, 2016 - Morning
Time:
Room: Exposition Floor, 400 Aisle

Abstract Text

Recently, microfluidic paper-based analytical devices (PADS) have garnered increasing interest as platforms for analysis of a variety of complex biological and environmental systems. PADS have many benefits over traditional systems due to the unique properties of paper including capillary action, reagent compatibility, widespread availability, and low cost. Common fabrication methods include photolithography, wax printing/dipping, and inkjet printing; all suffer drawbacks including high cost, difficult post processing, unstable hydrophobic materials, and non-biodegradability.

This work makes use of polycaprolactone (PCL) a low-cost, biodegradable polyester with favorable melting and glass transition temperatures of 60°C and -60°C respectively. PCL was used as both a hydrophobic and adhesive material for the fabrication of multilayer hybrid microfluidic devices. Fabrication methodology consisted of simple cut and stack followed by thermal bonding to ensure a fluidic seal.

Kidney health was targeted due to the number of diseases that have a negative effect on renal health and the importance of monitoring kidney function during various disease states. Creatinine and blood-urea-nitrogen (BUN) are often used as markers of kidney health, but enzymatic-based detection methods are either complex or require the detection of ammonia leading to difficulties with traditional PADs. Using the unique properties of PCL saturated paper, ammonia is able to diffuse through the paper and be quantitatively determined in a colorimetric fashion through the use of a simple pH indicator. Colorimetric assays for glucose and albumin were also incorporated due to their importance in the determination of kidney function.

Keywords: Clinical Chemistry, Lab-on-a-Chip/Microfluidics, Polymers & Plastics

Application Code: Bioanalytical

Methodology Code: Microfluidics/Lab-on-a-Chip

Session # **500** **Abstract #** **500-1** **Poster Sessions**

Session Title: Process Analytical Techniques

Abstract Title: **Characterization and Monitoring of Cavitation Through Its Acoustic Emission**

Primary Author: Noemie Caillol, IDEEL

Date: Monday, March 07, 2016 - Morning

Time:

Room: Exposition Floor, 400 Aisle

Co-Author(s): Christine Villard, Davide Zonca, Franck Baco-Antoniali, Pascal Pitiot, Sebastien Leinardi, Serge Henrot, Sylvain Charquet

Abstract Text

As the need for online diagnostics and characterization of chemical processes is increasing, we explore in this presentation the potential of acoustic emission for online analysis. This technic is especially attractive as it can be completely non-intrusive. All it requires is the contact of a small piezoelectric microphone with the structure to monitore just as a doctor's stethoscope. It can be implemented on almost all scales of devices from micro lab flows to tectonic plates. After an introduction to acoustic emission and presenting its potential for online analysis in the chemical and petroleum industries, we will show results of a study where different Venturi systems were monitored in pressure and flow measurments and acoustic emission recorded. The cavitation phenomenon corresponds to the spontaneous creation of bubbles of vapors where pressure drops. Their implosion as pressure increases generate micro jets and pressure waves which are responsible for major material damages in valves, pumps, etc. The appearance of the phenomena was found to have specific emissions that could be easily monitored on a flowing online venturi of a pilot plant. Signals were analyzed both in their temporal and frequency domain. Most effective parameters for the specific characterization of the different phenomena occurring were quantified to establish the bases for a monitoring methodology of the systems studied.

Keywords: On-line, Process Monitoring, Sensors, Water

Application Code: Process Analytical Chemistry

Methodology Code: Process Analytical Techniques

Session # 500 Abstract # 500-2 **Poster Sessions**

Session Title: Process Analytical Techniques

Abstract Title: **Monitoring of Tablet Coating Using Raman Spectroscopy**

Date: Monday, March 07, 2016 - Morning

Time:

Room: Exposition Floor, 400 Aisle

Primary Author: Hoeil Chung, Hanyang University

Co-Author(s): Jaejin Kim, Young Ah Woo

Abstract Text

A method based on Raman spectroscopy to monitor tablet coating has been studied. In Raman measurement, the relative intensity of coating to a tablet could vary depends on the laser illumination angles. This is a critical issue to address to make the method reliable for monitoring of tablet coating. Therefore, Raman spectra of a coated table were obtained by changing the laser illumination angles as well as illumination spots, and the resulting spectral features were examined. As expected, the peak intensity of coating vary depends on the illumination angles, so a spectrum acquired at a given illumination angle on a coated tablet is unable to represent entire tablet coating. Therefore, to achieve correct sample representation, Raman spectra of tablets were collected with shaking to make their orientations random for the spectral acquisition. Also, to collect more reproducible spectra, a wide area illumination (WAI) scheme that applies a laser beam to a sample in a circular fashion (6 mm diameter) with a long focal length was employed. The prediction performance a PLS model developed using the spectra of tablets with repeated shakings was acceptable to monitor tablet coating as an on-line manner.

Keywords: Molecular Spectroscopy, Process Analytical Chemistry, Raman

Application Code: Pharmaceutical

Methodology Code: Process Analytical Techniques

Session # 500 Abstract # 500-3 **Poster Sessions**

Session Title: Process Analytical Techniques

Abstract Title: **Simple and Rapid Determination of Polyanion-Polycation Binding Ratio Using Pulsed Chronopotentiometry with Polyion-Selective Electrodes**

Date: Monday, March 07, 2016 - Morning
Time:
Room: Exposition Floor, 400 Aisle

Primary Author: Emma Gordon
Northern Kentucky University

Co-Author(s): Kebede L. Gemene

Abstract Text

The use of biological polyions is common in biomedical applications. The polysaccharides heparin, LMWHs, DexS and PPS are used as anticoagulant and therapeutic agents. On the other hand, protamine and other polypeptides are used as antidotes of the widely used heparin and can effectively bind the other polysaccharides as well. Thus, developing simple, rapid and inexpensive detection methods of the binding affinities and ratios of the anticoagulant polysaccharides and the antidote polypeptides has significant biomedical applications including for screening efficient and safe antidotes for new anticoagulants such as LMWHs and for continuous monitoring of the anticoagulants and the antidotes in blood before, during and after surgical procedures. We have demonstrated here a simple and rapid detection of polycation-polyanion binding ratio using pulsed chronopotentiometry with reversible polyion-selective membrane electrodes. As model systems, we have determined the binding ratio of protamine with heparin, dextran sulfate (DexS) and pentosan polysulfate (PPS). The binding ratios were found to be 1.2:1.0 (protamine:heparin), 1.5:1.0 (protamine:DexS) and 1.55:1.0 (protamine:PPS). The binding ratio of protamine with heparin and PPS determined by our method are in good agreement with the published data. We report here the protamine-DexS binding ratio for the first time and the value is reasonable based on the charge density of the polysaccharide.

Keywords: Electrochemistry, Ion Selective Electrodes, Membrane, Potentiometry
Application Code: Biomedical
Methodology Code: Process Analytical Techniques

Session # 500 Abstract # 500-4 **Poster Sessions**

Session Title: Process Analytical Techniques

Abstract Title: **How Can Your Process Benefit From External Flow Cells and Optimized Pump Heads?**

Date: Monday, March 07, 2016 - Morning

Time:

Room: Exposition Floor, 400 Aisle

Primary Author: Kathryn E. Monks
Knauer

Co-Author(s): Ingo Piotrowski

Abstract Text

KNAUER has been manufacturing world-class HPLC instruments for the past five decades. The increased demand for precise measurements in hazardous, explosive or toxic work processes has initiated a number of technical optimizations particularly on our high pressure pump heads and UV detector flow cells. The key focus in these optimizations has been enhancing security and adaption to applicational demands whilst maintaining performance.

Keywords: Chromatography, Fiber Optics, HPLC Detection, Optimization
Application Code: Process Analytical Chemistry
Methodology Code: Process Analytical Techniques

Session # 500 Abstract # 500-5 **Poster Sessions**

Session Title Process Analytical Techniques

Abstract Title **Applications of a New Wear Resistant, Chemically Inert Coating that Improves Reliability, Lifetime and Accuracy of Process, Analytical and Sampling Systems**

Date: Monday, March 07, 2016 - Morning
Time:
Room: Exposition Floor, 400 Aisle

Primary Author Luke Patterson
SilcoTek Corporation

Co-Author(s) David Smith, Min Yuan

Abstract Text

The internal and external component surfaces of Analyzers and process sampling systems can be exposed to challenging environments.. Many sample streams are corrosive or contain active compounds that reduce component lifetimes or require extended preventative maintenance. For systems that are required to give accurate, reliable and repeatable data, the cost of upkeep is much larger than systems in benign environments. Current inert coatings are often challenged and limited in some of these applications due to acidic or basic conditions, erosive affects or physical wear demands.

This presentation will report data on the newest generation of inert and wear resistant surface treatments that greatly reduce maintenance cycles and improve analytical reliability. The applications include sampling and analytical systems found in off-shore analyzers, refineries, petrochemical plants, continuous emission monitors and more. Data from the analysis of sulfurs containing streams and other active indicators will be discussed to illustrate the inertness of the new coating in comparison to existing inert coatings available on the market.

Keywords: Adsorption, Gas Chromatography, Material Science, Process Monitoring
Application Code: Process Analytical Chemistry
Methodology Code: Process Analytical Techniques

Session # 500 Abstract # 500-7 **Poster Sessions**

Session Title Process Analytical Techniques

Abstract Title **Online Analysis Using LIBS for Industrial Process**

Primary Author Ronald Berger-Lefébure
IVEA

Co-Author(s)

Date: Monday, March 07, 2016 - Morning
Time:
Room: Exposition Floor, 400 Aisle

Abstract Text

Whatever their fields of activity, industrial companies require expertise to fulfil their responsibilities. Some of those obligations can be legal by respecting the environmental and/or human impacts regulations. Improving their manufacturing process also permit industrial companies to increase their profitability. To that end, laser-induced breakdown spectroscopy (LIBS) appears to be an efficient technique. Indeed, it is quantitative, fast (< 1 min), requires no sample preparation and can be performed at remote distance and in a harsh environment.
Since 2010, a French consortium works on the development and optimization of LIBS technique in order to develop specific analytical systems dedicated to industrial measurements. Amongst them the centres of expertise IDEEL and ISA from Lyon University, industrial companies like SOLVAY, ARKEMA, IFP-EN and IVEA a LIBS instrumentation manufacturer. One of these developments is devoted to elementary control on heavy grade oils and preliminary results will be presented and were achieved with a modular multi-purpose LIBS sub-system. Due to the viscosity of the sample, a dedicated sampling cell has also been developed in order to generate steady stream necessary for LIBS analysis. Calibration curves were obtained by considering standard samples resulting from the process. Those curves shows good linear trends and detection limits in agreement with industrial specifications. Finally, the LIBS system was implemented during one month on an industrial pilot site during a monitoring campaign for hydro treating heavy oil products. Fully automatic quantitative LIBS measurements were realized once or twice per day. The results obtained were compared to standard measurements with a WDXRF laboratory system (ASTM D2622). Good correlation between online LIBS system and WDXRF standard measurements are obtained confirming the capability of LIBS technique to make fast and online analysis for specific industrial needs.

Keywords: Elemental Analysis, Instrumentation, On-line, Process Analytical Chemistry
Application Code: Process Analytical Chemistry
Methodology Code: Process Analytical Techniques

Session # 500 **Abstract # 500-8** **Poster Sessions**

Session Title: Process Analytical Techniques

Abstract Title: **Non-Porous, No Glass, Leak-Free Reference Electrode**

Primary Author: Ziad H. Taha
Innovative Instruments, Inc.

Date: Monday, March 07, 2016 - Morning
Time:
Room: Exposition Floor, 400 Aisle

Co-Author(s)

Abstract Text

The properties and performance of an all-plastic, Ag/AgCl/sat KCl, leak-free reference electrode are presented. This class of reference electrodes utilizes a highly conductive, non-porous junction, eliminating the need for the filling electrolyte to leak to provide ionic conductivity. This electrode does not need electrolyte refill and can store dry, offering long term use and stability. It does not contaminate the sample. This electrode offers many advantages over conventional reference electrodes. It can be used in hydrofuoric acid, concentrated hydroxide , concentrated acid, perchlorate, cyanide and many other solutions where conventional electrode use is limited or problematic. Moreover, it can be used in common organic solvents. This electrode is very useful in biological research, energy research, ion selective electrodes, semiconductors, corrosion in addition to common electrochemical experiments. This electrode can be prepared in any size. Most common available sizes are 0.5, 1, 1.6, 2, 3.2, 5 and 6.4 mm diameter at any desired length. Moreover, it can be prepared flexible and in many shapes. The electrode has been coupled with conventional glass pH electrode to eliminate the problems of associated with porous junctions. The use of this electrode with maintenance-free, multi-parameter devices is under investigation.

Keywords: Analysis, Electrodes, Process Analytical Chemistry, Sensors
Application Code: General Interest
Methodology Code: Electrochemistry

Session # 500 Abstract # 500-9 **Poster Sessions**

Session Title: Process Analytical Techniques

Abstract Title: **Suggested QC Practices for On-line Analyzers**

Date: Monday, March 07, 2016 - Morning
Time:
Room: Exposition Floor, 400 Aisle

Primary Author: William Lipps
Shimadzu Scientific Instruments

Co-Author(s)

Abstract Text

The USEPA regulates by establishing a contaminant, an MCL, and a method, or methods, that are required to use for compliance purposes. These methods are almost all laboratory methods and include sampling, sample preservation, holding times, and batch QC criteria. Even if an on-line analyzer uses the same measurement technique as a laboratory method, you cannot assume equivalence of an on-line test to a laboratory method because the on-line lacks some of the required QC. This presentation suggests QC practices that could be applied to new on-line analyzer methods.

Keywords: Environmental Analysis, Environmental/Water, Monitoring, Process Monitoring
Application Code: Environmental
Methodology Code: Process Analytical Techniques

Session # 500 **Abstract # 500-10** **Poster Sessions**

Session Title: Process Analytical Techniques

Abstract Title: **A Novel Approach to Cleaning Validation for Pharmaceutical Manufacturing by Online SFE-SFC**

Primary Author: Kenichiro Tanaka
Shimadzu Scientific Instruments, Inc.

Date: Monday, March 07, 2016 - Morning
Time:
Room: Exposition Floor, 400 Aisle

Co-Author(s): Daisuke Nakayama, Hidetoshi Terada, Minori Nakashima, Tadayuki Yamaguchi, William Hedgepeth

Abstract Text

Cleaning validation is necessary to establish the quality and safety of pharmaceutical drug products. In cleaning validation protocols, direct sampling is performed with swabs, which are sticks with textiles at one side. The sample in the swab after swabbing the surface of equipment is extracted with water or organic solvent before analyses with a TOC analyzer and HPLC. Although swabbing is the preferable method to validate cleaning, it has some problems. The TOC analyzer is not applicable to hydrophobic compounds because ethanol is needed for swabbing. Sample condensation is also sometimes required before HPLC. Thus, we developed a novel sample pretreatment and analysis method for cleaning validation using supercritical fluid for both extraction and analysis.

A commercially available detergent containing alkylbenzene sulfonates was used as a standard sample. For the test of sample extraction, the sample was dropped onto a swab (ITW Texwipe, USA). The Nexera UC system (Shimadzu Corporation, Japan) was used for both the screening of the method using supercritical fluid chromatography (SFC) and the supercritical fluid sample extraction (SFE) followed by SFC directly (SFE/SFC). Shim-pack UC series columns were screened and used for subsequent analysis.

Keywords: Pharmaceutical, SFC, SFE, Supercritical Fluid Chromatography
Application Code: Pharmaceutical
Methodology Code: Supercritical Fluid Chromatography

Session # 500 **Abstract # 500-11** **Poster Sessions**

Session Title: Process Analytical Techniques

Abstract Title: **Advances in Raman Analyzers for In Situ Studies of Small Volume Liquid-Phase Reactors**

Primary Author: Lisa Ganster
Kaiser Optical

Co-Author(s): Ian Lewis

Date: Monday, March 07, 2016 - Morning
Time:
Room: Exposition Floor, 400 Aisle

Abstract Text

In the last 15 years Raman spectroscopy has emerged as an important in situ analytical and process control tool. The early years of this period were dominated by improvements in Raman spectrometer components including the development of high power, small footprint NIR lasers, high performance holographic laser rejection filters, and low-noise CCD array detectors. The later years have seen a significant increase in the type and quantity of published in situ application successes. At the core of this emergence has been developments in sampling and sampling interfaces. The ability to flexibly configure the optical sampling interface using fiber-optic delivery allows Raman analyzers to be integrated to reactors ranging from the micro-scale to large volume manufacturing reactors. In this presentation, examples of improvements in sampling for in situ liquid-phase Raman applications will be shown, as well as applications of Raman spectroscopy for the study and control of small reactor systems including sealed microwave systems, continuous flow reactors, NeSSI platform devices, and small volume thermal reactors.

Keywords: Laser, Process Control, Sampling, Spectroscopy
Application Code: Process Analytical Chemistry
Methodology Code: Process Analytical Techniques

Session # 500 **Abstract # 500-12** **Poster Sessions**

Session Title: Process Analytical Techniques

Abstract Title: **Establishing Raman Spectroscopy for the Process Environment**

Primary Author: Lisa Ganster, Kaiser Optical

Date: Monday, March 07, 2016 - Morning
Time:
Room: Exposition Floor, 400 Aisle

Co-Author(s): Harry Owen, Ian Lewis, Karen Esmonde-White

Abstract Text

Raman spectroscopy is an established tool in research analytical laboratories because of its sampling versatility, minimal sample preparation, and compatibility with aqueous systems. For the past 20 years, we have applied the measurement principles of Raman spectroscopy in a manufacturing environment for understanding, monitoring, and controlling continuous processes or unit operations. Raman spectroscopy in a process or manufacturing environment required an integrated approach, with careful attention to the reliability of the analyzer and sampling probe optics and transferability of the data analysis method. Some manufacturing environments have additional environmental or regulatory requirements, which also impacted our technology development and product manufacturing. We broadly discuss these requirements in the context of customer applications and provide examples that illustrate the challenges, successes and benefits of Raman spectroscopy in the process and manufacturing environments. Our approach includes an integrated hardware, optical probe, control software and model development platform, that addresses the challenges of Raman spectroscopy in a process environment. We demonstrate successful application of Raman-based process analytical technology (PAT) and Quality by Design (QbD) in pharmaceutical, specialty chemical, petrochemical and polymer manufacturing environments. We show Raman spectroscopy as a valuable technique for process monitoring and control.

Keywords: Atomic Spectroscopy, Quality, Raman, Spectroscopy
Application Code: Process Analytical Chemistry
Methodology Code: Atomic Spectroscopy/Elemental Analysis

Session # 500 Abstract # 500-13 **Poster Sessions**

Session Title: Process Analytical Techniques

Abstract Title: **In situ Raman Measurements of Pharmaceutical Solids During Process Unit Operations**

Date: Monday, March 07, 2016 - Morning
Time:
Room: Exposition Floor, 400 Aisle

Primary Author: Lisa Ganster
Kaiser Optical

Co-Author(s): Carsyen Uerpmann, Ian Lewis, Karen Esmonde-White, Sean Gilliam

Abstract Text

Understanding, monitoring and controlling solid phase unit operations in pharmaceutical manufacturing is a critical component in ensuring a drug product's safety, bioavailability and efficacy. Consistent quality is a key requirement in the FDA's process analytical technology (PAT) and Quality by Design (QbD) initiatives. In situ Raman spectroscopy can improve control over the solid-phase unit operation, reduce batch rejections and improve process quality. Continuous, in situ, Raman measurements align with these FDA initiatives and provide direct process feedback without the need for sample removal or preparation. We describe laboratory and manufacturing examples where Raman spectroscopy provides monitoring and control of a critical solid-phase pharmaceutical process. In our examples, a Kaiser Optical Systems RamanRXN system analyzer, operating at 785 nm, was coupled directly into the processing equipment using the Kaiser PhAT probe. The PhAT probe delivered approximately 100-200 mW onto the sample and signal was collected for 1-3 seconds. In one example, the PhAT probe was used to measure signal from the coating and the API during the tablet coating process. Ratiometric analysis of the coating-to-API signal was used to generate a Raman-defined endpoint, which was used on subsequent batches to control coating thickness. Through these examples, we have demonstrated Raman spectroscopy as a valuable technique in measuring, monitoring and controlling pharmaceutical solids during process unit operations.

Keywords: Process Analytical Chemistry, Process Control, Raman, Spectroscopy
Application Code: Process Analytical Chemistry
Methodology Code: Process Analytical Techniques

Session # 500 **Abstract # 500-14** **Poster Sessions**

Session Title: Process Analytical Techniques

Abstract Title: **Recovery of Challenging Compounds in Cleaning Validation Using Total Organic Carbon (TOC) Analysis**

Primary Author: Dondra Biller
GE Analytical Instruments

Date: Monday, March 07, 2016 - Morning
Time:
Room: Exposition Floor, 400 Aisle

Co-Author(s): Andy Young, David Wayne, Jenny Watson

Abstract Text

Biologics, pharmaceutical, cosmetics, and nutraceutical manufacturing companies spend significant time cleaning and verifying the cleanliness of production equipment. In many cases, non-specific analytical methods such as total organic carbon (TOC) have the potential to help release clean equipment back to production more quickly. TOC has become a very common analytical method for cleaning validation in recent years, primarily due to ease of validation, speed of analysis, and ability to be automated as either an at-line or on-line measurement. In order to successfully replace compound-specific analytical methods with TOC, recovery studies must be performed to show that the compounds of concern can be measured. Traditional wisdom is that compounds with complex molecular structures or low solubility in water may be difficult to sample and measure using currently available TOC instrumentation. This study involved the selection of multiple difficult to oxidize or low solubility organic compounds, recovery testing on TOC instrumentation, as well as physical recovery of each compound from stainless steel coupons using swab sampling. Results from this study demonstrate the usefulness of TOC analysis for the detection of a broad range of organic residues that are of interest to industry.

Keywords: Biopharmaceutical, Pharmaceutical, Total Organic Carbon, Validation
Application Code: Validation
Methodology Code: Sampling and Sample Preparation

Session # 510 Abstract # 510-1 **Poster Sessions**

Session Title Safety

Abstract Title **A Cheap and Simple Approach to Monitor Sun Exposure based on Photocatalytic Properties of Titanium Dioxide**

Primary Author Parisa Sowti Khiabani
University of New South Wales

Date: Monday, March 07, 2016 - Morning
Time:
Room: Exposition Floor, 400 Aisle

Co-Author(s) Alexander H. Soeriyadi, J Justin Gooding, Peter J. Reece

Abstract Text

Overexposure to UV-radiation of sun is the main cause of skin cancer. The duration that is required to exceed the exposure limit varies with the intensity of solar UVR and skin type of the person who is being exposed to solar UV. This makes it challenging to judge the appropriate amount of time that is safe under UV sun exposure. Therefore, there is a need for some disposable sun exposure sensor which is simple, easy to use and inexpensive. Another important aspect is that it should not be only cheap but also composed of entirely benign materials. This is what we seek to achieve with the technology described herein using entirely materials that are already approved for human use.

For this purpose, the decolouration of common food dyes by TiO_2 when exposed to UV radiation is applied to fabricate an easy to fabricate, simple to use, disposable sun exposure sensor that can be tuned to different skin types. This single layer sticker was fabricated by inject printing of a suspension containing a FDA approved food dye (e.g. brilliant blue FCF), TiO_2 and polyvinylpyrrolidone (PVP) as a binder on paper. As a result of decomposition of this food dye by TiO_2 in presence of UV, the film will lose the inherent colour of the dyes. The decolouration can be easily observed by the naked eye. Quantitative analysis of decolouration change was performed using UV-Vis reflectance spectra of films fabricated using ink jet printing. Finally, decolouration of the films was calibrated to match UV exposure time of different skin types, by using different UV neutral density filter with the ability of transmit between 1.5 to 70% of UV radiation from the sources to the photoactive film. The decolouration time could be adjusted most effectively by coating the device using filters although fabrication parameters such as the actual dye, the ratio of dye to TiO_2 or the film thickness also gave some control over the decolouration time.

Keywords: Detection, Nanotechnology
Application Code: Safety
Methodology Code: Sensors

Session # 510 **Abstract # 510-2** **Poster Sessions**

Session Title: Safety

Abstract Title: **A Wide Selection of New Psychoactive Substances Investigated with Proton Transfer Reaction – Mass Spectrometry within the Marie Curie Training Network Proton Ionization Molecular Mass Spectrometry (PIMMS)**

Date: Monday, March 07, 2016 - Morning

Time:

Room: Exposition Floor, 400 Aisle

Primary Author: Matteo Lanza, IONICON Analytik

Co-Author(s): Alfons Jordan, Chris A. Mayhew, Christian Lindinger, Eugen Hartungen, Gernot Hanel, Jens Herbig, Kostiantyn Breiev, Lukas Märk, Philipp Sulzer, Simone Jürschik, Tilmann D. Maerk, W Joe Acton

Abstract Text

12 full and 4 associate partners from all over Europe (renowned universities, research institutes and industry) have trained a total of 15 Early Stage Researchers within the Marie Curie Training Network PIMMS during the past four years. PIMMS has focused on the application of PTR-MS to four key areas: environmental, food and health sciences and homeland security. Here we present work on one of the successful projects within PIMMS: the analysis of New Psychoactive Substances (NPS), i.e. compounds that do not belong to the common group of illicit drugs (e.g. cocaine, ecstasy, LSD, etc.) but mimic their intoxicating effects. Utilizing a high-resolution and high-sensitivity PTR-TOF 8000 (PTR-MS instrument equipped with a time-of-flight mass analyzer and switchable reagent ions capability) we studied the reduced electric field dependence of the reactions of a wide selection of NPS with the reagent ions H[sub]3[/sub]O[sup]+[/sup], NO[sup]+[/sup], O[sub]2[/sub][sup]+[/sup] and Kr[sup]+[/sup].

We present an overview of these results and highlight key findings, and in particular the interesting results we obtained for the popular NPS ethylphenidate (EPH, C[sub]15[/sub]H[sub]21[/sub]NO[sub]2[/sub]), a drug closely related to the prescription drug methylphenidate, better known as Ritalin. Over a period of three years we purchased a total of six EPH samples via the Internet. All samples contained the advertised active ingredient EPH and some impurities, which could be assigned to residuals from the synthesis process, as illustrated in the figure below, which shows the product ions originating from various samples (those from EPH marked with an E). Although purchased from the same vendor, we found that the impurities differed between batches, indicating that different routes of synthesis had been used in the production processes.

PIMMS is supported by the EC 7th Framework Programme
(GA 287382). WJA received a BBSRC-Industrial CASE studentship.

Keywords: Drugs, Mass Spectrometry, Method Development, Time of Flight MS
Application Code: Safety
Methodology Code: Mass Spectrometry

Session # 520 Abstract # 520-1 **Poster Sessions**

Session Title	Sampling and Sample Preparation - Pharmaceutical, Clinical/Toxicology, Food Safety, and Others
Abstract Title	**Preserving the Free-Radical Scavenger Activity of Key Bioactives During Extraction and Purification Processes**

Primary Author: Valerie Desyroy, SiliCycle

Date: Monday, March 07, 2016 - Morning

Time:

Room: Exposition Floor, 400 Aisle

Co-Author(s): Aurelie Meng, Denis Boudriau, Genevieve Gingras, Vincent Bedard

Abstract Text

Scutellaria baicalensis, commonly known as Chinese scullcap or Baikal scullcap, is a member of the mint family, grown in China, Mongolia, Korea and Russia. It is one of the 50 fundamental herbs used in traditional Chinese medicine and it has been used for centuries. Recently, it was found to block MAP Kinase by growth factors and estrogen in human breast cancer cells. We herein demonstrate how to Preserving the free-radical scavenger activity of key bioactives during the extraction and purification process. We herein demonstrate how to preserve the free-radical scavenging activity of such key bioactives during the extraction and purification process.

Keywords: Extraction, Isolation/Purification, Sample Preparation, Solid Phase Extraction

Application Code: Drug Discovery

Methodology Code: Sampling and Sample Preparation

Session # 520 **Abstract # 520-3** **Poster Sessions**

Session Title: Sampling and Sample Preparation - Pharmaceutical, Clinical/Toxicology, Food Safety, and Others

Abstract Title: **Working with Challenging Samples, Sampling Systems**

Date: Monday, March 07, 2016 - Morning
Time:
Room: Exposition Floor, 400 Aisle

Primary Author: Yves Gamache
Analytical Flow Products

Co-Author(s)

Abstract Text

When dealing with challenging samples like corrosive, toxic, flammable and the like samples (Silane, HF, ClF3, phosphine, ammonia, chlorine, boron trichloride, nitrogen trifluoride, fluorine, bromine, hydrogen etc.), there is a high risk for user safety and equipment damage. We will show how that AFP design philosophy is used to build sampling systems and define methods in order to get a safe and intelligent system while having long term repeatability and accurate measurement. A new sealed GC valve concept will also be introduced.

Keywords: Analysis, Chromatography, Gas, Trace Analysis
Application Code: Other
Methodology Code: Sampling and Sample Preparation

Session # 520 Abstract # 520-4 **Poster Sessions**

Session Title: Sampling and Sample Preparation - Pharmaceutical, Clinical/Toxicology, Food Safety, and Others

Abstract Title: **Analytical Method Development for Cleaning Verification of Manufacturing Equipment - Exploring the Effect of the Cleanliness of Stainless Steel Coupons on Sample Recovery**

Primary Author: Imad A. Haidar Ahmad
Novartis

Date: Monday, March 07, 2016 - Morning
Time:
Room: Exposition Floor, 400 Aisle

Co-Author(s): Andrei Blasko, James Tam, Thomas Tarara, Xue Li

Abstract Text

The aim of this work was to identify the parameters that affect the recovery of pharmaceutical residues from the surface of stainless steel coupons. A series of factors were assessed, including drug product spike levels, spiking procedure, drug/excipient ratios, analyst-to-analyst variability, intraday variability, and cleaning procedure of the coupons. The low recoveries were narrowed down to the cleaning procedure of the coupons as the major contributor. Assessment of acid, base, and peroxide washes, as well as the order of treatment, showed that a base-water-acid-water wash procedure resulted in consistently high spiked recovery (>90%) and reproducible results. By applying this cleaning procedure to the previously used coupons that failed the cleaning acceptance criteria, Multiple analysts were able to obtain consistent recoveries from day-to-day for different drugs, and drug/excipient ratios at various spike levels. We successfully applied our approach for cleaning verification of small molecules (MW < 1000 Da) as well as large biomolecules (MW up to 50 KDa). The analytical techniques used were either Total Organic Carbon or liquid chromatography with UV detection.

Keywords: Biopharmaceutical, Total Organic Carbon, Trace Analysis, Validation
Application Code: Pharmaceutical
Methodology Code: Sampling and Sample Preparation

Session # 520 **Abstract # 520-5** **Poster Sessions**

Session Title: Sampling and Sample Preparation - Pharmaceutical, Clinical/Toxicology, Food Safety, and Others

Abstract Title: **Evaluation of Antibacterial and Wound Healing Properties of Hydro-Ethanolic Extracts of Gossypium Barbadense Leaves**

Date: Monday, March 07, 2016 - Morning
Time:
Room: Exposition Floor, 400 Aisle

Primary Author: Nwamaka H. Igbokwe
University of Lagos

Co-Author(s): Cecilia I. Igwilo, Eugene E. Ikobi

Abstract Text

This study was designed to evaluate the antibacterial and wound healing properties of hydro-ethanolic extract of Gossypium barbadense leaves, a breakthrough in an era of multi-drug resistance. The in-vitro antimicrobial activities of different concentrations of the extract (10mg/ml, 20mg/ml and 30mg/ml) were evaluated on multi-drug resistant wound isolates (Pseudomonas aeruginosa, Staphylococcus aureus, Escherichia coli, Proteus mirabilis and Shigella sonnei) from patients at the Lagos University Teaching Hospital, Lagos Nigeria, using the well diffusion method. Propylene glycol was used as negative control while Bacitracin/Neomycin was used as positive control. Wound healing properties of the extract were also investigated on 25 Wister albino rats using the excision model. Five groups (A, B, C D and E) of five Wister albino rats each, induced with sized wounds were treated with 10mg/ml, 20mg/ml and 30mg/ml of the extract for groups A, B and C. Group D was treated with Bacitracin/Neomycin positive control while group E rats were treated with sterile distilled water as negative control

After ten days of treatment, 20mg/ml of the extract gave 91% healing while Bacitracin/Neomycin gave about 80%. Distilled water produced 36% healing Hydro-ethanolic extract of Gossypium barbadense leaves compared favorably with commercial Bacitracin/Neomycin in antimicrobial activity and also showed faster and better wound healing than the commercial brand.

Keywords: Analysis, Extraction, Natural Products
Application Code: Pharmaceutical
Methodology Code: Sampling and Sample Preparation

Session # 520   Abstract # 520-7   Poster Sessions

Session Title: Sampling and Sample Preparation - Pharmaceutical, Clinical/Toxicology, Food Safety, and Others

Abstract Title: **Acrylamide from Coffee Using a Simplified Liquid**

Primary Author: Allen Misa
Phenomenex

Co-Author(s): Matthew Brusius, Zeshan Aqeel

Date: Monday, March 07, 2016 - Morning

Time:

Room: Exposition Floor, 400 Aisle

Abstract Text

According to the American Cancer society, cooking at high temperatures causes a chemical reaction between certain sugars and
asparagine which causes acrylamide to form. Acrylamide is commonly found in foods that are made from plants such as potato
products, grain products, and coffee whose preparation often requires longer cooking times and higher temperatures. In this poster we explore how to use Novum SLE tubes to clean up a coffee matrix in order to quantitate known acrylamide levels, demonstrating that the SLE technique can be applied to a variety of compounds and sample matrices outside of the clinical research industry.

Keywords: Food Safety, Food Science, Liquid Chromatography

Application Code: Food Safety

Methodology Code: Sampling and Sample Preparation

Session # 520 Abstract # 520-8 **Poster Sessions**

Session Title: Sampling and Sample Preparation - Pharmaceutical, Clinical/Toxicology, Food Safety, and Others

Abstract Title: **Analysis of Food Grains Using Automated Block Digestion**

Primary Author: Michael A. Rutzke
Cornell University

Co-Author(s): Nick McLeod, Suhas Narkhede

Date: Monday, March 07, 2016 - Morning
Time:
Room: Exposition Floor, 400 Aisle

Abstract Text

Cereal grains represent an important food class consumed by all living organisms on a daily basis. Ensuring trace level amounts of micronutrients obtained from rice, wheat or maize is imperative as deficiencies in Fe, Zn and I are on the rise.[superscript 1] Alternatively, the presence of toxic heavy metals present concerns as short or long term exposure can lead to adverse effects including nervous system and/or organ damage.[superscript 2] High-throughput trace metal analysis in food grains is therefore key in regulating their respective accepted levels. Sample preparation, in particular acid digestion, consumes the largest amount of time during these analyses.[superscript 3] Improved techniques such as microwave digestion have significantly reduced digestion times but are limited to small sample sizes and are labor intensive. Automated open vessel digestions are capable of handling large sample sizes and dispensing accurate amounts of reagent while increasing analyst efficiency. Through the use of automated block digesters and automated dilution work-stations, approximately 320 samples could be processed in a short period of time using minimal labour. We describe the nitric/perchloric digestion procedure used in the analysis of various food grains, as well as reporting the recoveries of elements; measured by ICP-AES.

1. Tekli□ T.; Lon□ari□ Z.; Kova□evi□ V.; Singh, B. R. [i]Food and Energy Security[/i] [b]2013[/b], [i]2[/i](2), 81-95.
2. Singh, R.; Gautam, N.; Mishra, A.; Gupta, R. [i]Indian J. Pharmacol.[/i] [b]2011[/b], [i]43[/i](3), 246–253.
3. Korn, M. G. A. et al. [i]Appl. Spectrosc. Rev.[/i] [b]2008[/b], [i]43[/i], 67-92.

Keywords: Food Contaminants, ICP, Sample Handling/Automation, Sample Preparation
Application Code: Food Safety
Methodology Code: Sampling and Sample Preparation

Session # 520 Abstract # 520-9 **Poster Sessions**

Session Title: Sampling and Sample Preparation - Pharmaceutical, Clinical/Toxicology, Food Safety, and Others

Abstract Title: **Does Weather Effect Pipetting?**

Primary Author: George W. Rodrigues
Artel

Co-Author(s): Alf Price, Emily Avis, Verena Gerber

Date: Monday, March 07, 2016 - Morning
Time:
Room: Exposition Floor, 400 Aisle

Abstract Text

Automated liquid handling performance can be optimized by adjusting system operating parameters, but the effect of environmental conditions on liquid handling quality is often under-appreciated.

This study uses a two factor design (temperature and humidity) to investigate volumetric performance (precision and trueness) at environmental conditions from 15 °C to 30 °C and 30% to 80% relative humidity. Nine environmental conditions were tested, the mid point, four corner cases and four edge cases. Replicate testing was performed at the mid point and corner cases.

The Tecan QC Kit, which is based on Artel's ratiometric photometry, was used to make measurements at each set of environmental conditions. The precision (%CV), and trueness (%SE) is reported for each environmental condition using default liquid classes.

The precision of the Tecan Freedom EVO was found to be insensitive to environmental conditions and performed within manufacturer's limits over the entire range tested in this study (15 °C to 30 °C and 30% to 80% relative humidity).

Over the range of this test, the trueness (systematic error) does depend on environmental conditions. Increased temperature and decreased relative humidity resulted in lower delivered volume. For the greatest accuracy in laboratory work, and particularly when operated outside of recommended temperatures, the automated liquid handling system should be tested and optimized for the local environment.

The environmental effect on trueness exhibited a linear correlation with the "evaporation potential" (the driving force for evaporation of the sample). The evaporation potential is dependent on both temperature and relative humidity. Use of this concept permits the experiment to be reduced from two factors (temperature and relative humidity) to one factor (evaporation potential).

Keywords: Calibration, Laboratory Automation, Quality Control, Validation
Application Code: General Interest
Methodology Code: Sampling and Sample Preparation

Session # 530 Abstract # 530-1 **Awards**

Session Title LCGC Lifetime Achievement and Emerging Leader in Chromatography Award

Abstract Title **Columns in Small-Scale Chromatography**

Primary Author Milton L. Lee
Brigham Young University

Date: Monday, March 07, 2016 - Afternoo
Time: 01:40 PM
Room: B314

Co-Author(s)

Abstract Text

Miniaturization of columns in chromatography began with gas chromatography (GC) and continued steadily over the past half century. The main driving forces have been prospects for increasingly higher chromatographic resolution and simpler coupling of chromatography to mass spectrometry (MS). Major breakthroughs in GC include flexible fused silica columns, robust column deactivation methods, high viscosity stationary phases, understanding of Raleigh instability, rapid static coating, free-radical crosslinking, and thermally stable stationary phases. In liquid chromatography (LC), developments in small particle morphology and uniformity have been most critical, with optimized packing methods following close behind. Development of supercritical fluid chromatography (SFC) introduced its own unique requirements that contributed to improvements in column technology for the other chromatographic techniques as well. Although continually exciting to consider, efforts to incorporate chromatographic separation channels in microfluidic platforms have been relatively slow. This presentation will describe the author's first-hand experience in developing capillary column technology for chromatography over the past 40 years, followed by speculation of emerging opportunities that may provide even better chromatographic performance in the future.

Keywords: Capillary GC, Capillary LC, GC Columns, HPLC Columns
Application Code: General Interest
Methodology Code: Separation Sciences

Session # 530 Abstract # 530-2 **Awards**

Session Title LCGC Lifetime Achievement and Emerging Leader in Chromatography Award

Abstract Title **Recent Chromatographic and Mass Spectrometric Developments Applied to the Characterization of Recombinant Proteins, Monoclonal Antibodies and Antibody-drug Conjugates**

Primary Author Pat Sandra
RIC

Co-Author(s) Koen Sandra

Date: Monday, March 07, 2016 - Afternoo
Time: 02:15 PM
Room: B314

Abstract Text

In recent years within the pharmaceutical industry and also in our research activities related to pharmaceutical analysis, a remarkable shift from small to large molecules was noticed. Being on the market since early 1980s, protein biopharmaceuticals have seen an enormous growth in the last decade. It is even expected that within the current decade, more than 50% of new drug approvals will be of biological nature. A dominant role is thereby played by protein biopharmaceuticals (recombinant proteins, monoclonal antibodies-mAbs and antibody-drug conjugates-ADCs). The sales top 10 of pharmaceuticals is currently heavily populated with mAbs and a substantial number of mAbs has reached blockbuster status.

Protein biopharmaceuticals are large and heterogeneous and their in-depth analysis during development, during lifetime and in QA/QC requires the best of both chromatography and mass spectrometry. Moreover, with the patents of the first generation protein biopharmaceuticals expired and blockbuster mAbs becoming open to the market, activities in biosimilars exploded in recent years. Biosimilar developers are also confronted with huge analytical challenges as opposed to generic versions of small molecules since exact copies of recombinant proteins cannot be produced due to differences in cell clone and processes used.

In this presentation, analytical platforms for full characterization and analysis of protein biopharmaceuticals will be presented and illustrated. From a chromatographic point of view, one-dimensional (1 D) and comprehensive two-dimensional liquid chromatography (LC×LC) will be compared with emphasis on their application in a QA/QC environment. For in-depth characterization of protein biopharmaceuticals, mass spectrometry is of utmost importance and the features of high-end MS (QTOF, QQQ, ion mobility MS) will be illustrated with crucial compound characteristics.

Keywords: Biopharmaceutical, Liquid Chromatography, Mass Spectrometry
Application Code: Pharmaceutical
Methodology Code: Liquid Chromatography/Mass Spectrometry

Session # **530** **Abstract #** **530-3** **Awards**

Session Title LCGC Lifetime Achievement and Emerging Leader in Chromatography Award

Abstract Title **Analytical Glycoscience: Quo Vadis?**

Date: Monday, March 07, 2016 - Afternoo
Time: 02:50 PM
Room: B314

Primary Author Milos V. Novotny
Indiana University

Co-Author(s)

Abstract Text

The enormous complexity of most glycoconjugate natural pools challenges even the best available analytical tools and methodologies developed for the other areas of systems biology. Both eukaryotic and prokaryotic systems present unique analytical problems of their own. Among the distinct and important tasks of contemporary glycobiology is to identify and quantify components of human glycomes and distinguish "normal" glycan levels from disease-related glycosylation aberrations. In analytical terms, this often means resolution of many different glycans, including various isomeric species, and reaching even the trace components of complex glycan mixtures. Mass-spectrometric (MS) glycan profiling is conveniently performed through the use of derivatization (e.g., permethylation) and MALDI-MS or LC- ESI-MS. To deal with mixtures containing isomeric glycans, it is beneficial to employ capillary electrophoresis (CE) or chromatographic separations, as has been shown by the recent applications in biomedical field. In addition to describing the glycomes, it is essential to characterize the proteins conjugating their respective glycans. The current glycoproteomics, which has evolved from the proteome-centered approaches, must add unique considerations of glycobiology in the further developments of effective analytical platforms. Importantly, the field also includes new analytical methodologies in the rapidly developing area of recombinant glycoprotein biopharmaceuticals. Advanced understanding of glycan-protein binding interactions necessitates the use of complementary tools such as glycan arrays and lectin arrays.

Keywords: Bioanalytical, Biological Samples, Carbohydrates, Separation Sciences
Application Code: Bioanalytical
Methodology Code: Separation Sciences

Session # 530 **Abstract # 530-4** **Awards**

Session Title: LCGC Lifetime Achievement and Emerging Leader in Chromatography Award

Abstract Title: **Generic Chiral Separation Strategies for Pharmaceutical Compounds Using Chromatographic and Electrophoretic Techniques**

Date: Monday, March 07, 2016 - Afternoo
Time: 03:45 PM
Room: B314

Primary Author: Debby Mangelings
Vrije Universiteit Brussel

Co-Author(s): Yvan Vander Heyden

Abstract Text

Enantiomers of chiral pharmaceutical drug compounds can exhibit different pharmacological and pharmacokinetic activities in the human body. Therefore, regulatory authorities demand that, if possible, a single enantiomer drug is developed which only contains the therapeutically active enantiomer. In addition, methods for the separation and quantification of the enantiomers must be presented in the registration procedure of a chiral drug molecule. However, in early drug development, industry mostly prefers to synthesize a racemate and to separate it afterwards. A fast screening for separation conditions is already performed in this stage to reduce the method development time at later stages. Therefore, generic separation strategies can be very useful. A fast screening experiment gives an idea about the enantioselectivity, and the optimization steps can be used afterwards to enhance the obtained separation. Generic strategies are developed in such a way that they are applicable on large sets of structurally diverse molecules. Previously, polysaccharide-based chiral stationary phases (CSP) proved to be well suited to define such strategies, as they show a broad enantioselectivity range. Different strategies were already developed in normal-phase liquid chromatography, polar organic solvent chromatography, reversed-phase liquid chromatography and capillary electrochromatography. For supercritical fluid chromatography, only a screening step was defined. A new challenge concerned the evaluation of recently introduced CSP with other types of polysaccharide selectors for their applicability in generic analysis. In a first step, the applicability of the screening step was evaluated on the new CSP for all considered techniques. When the enantioselectivity of these new CSP was higher, the screening step was altered by replacement of some CSP to achieve a higher success rate. In a second part of the project, the optimization steps were adapted or defined.

Keywords: Chiral Separations, Chromatography, Electrophoresis, Pharmaceutical
Application Code: Pharmaceutical
Methodology Code: Separation Sciences

Session # 540 Abstract # 540-1 **Awards**

Session Title SEAC - Charles N Reilley and Royce W Murray Award

Abstract Title **Electrodeposited Nanowires for Faster and More Sensitive Hydrogen Gas Detection**

Date: Monday, March 07, 2016 - Afternoo
Time: 01:40 PM
Room: B312

Primary Author Reginald M. Penner
University of California, Irvine

Co-Author(s)

Abstract Text

Hydrogen (H2) gas sensors that are sensitive, rapid-responding, stable, selective, compact, and inexpensive are needed to support the operation of fuel cell-electric automobiles. Palladium absorbs hydrogen to form a hydride (PdHx) with x saturating at 0.67 and since 1869 it has been known that the electrical resistivity of this hydride increases linearly with x by a factor of 1.8 – 1.9 over the range from x = 0 to 0.67. This property of PdHx was first exploited for hydrogen sensing by Hughes and Schubert in 1992. As hydrogen gas sensors, Pd thin film resistors are elegant in their simplicity and cheap, but they are much too slow. The solution is to use a Pd nanowire as a resistive H2 sensor instead of a film. In this talk, we discuss recent innovations in nanowire-based H2 detection, focusing attention on platinum-modified palladium nanowires (or Pd@Pt nanowires) that are prepared with control of the Pt coverage. We assess the influence of the Pt surface layer on various H2 detection metrics. Pd nanowires with dimensions of 40 nm(h) x 100 nm(w) x 50 µm(l) fabricated using lithographically-patterned nanowire electrodeposition (LPNE). Then a thin Pt surface layer is electrodeposited conformally onto the Pd nanowire at coverages of 0.10 monolayer (ML), 1.0 ML, and 10 ML. The resistance of a single Pd@Pt nanowire is measured during the exposure of these nanowires to pulses of hydrogen gas in air at concentrations ranging from 0.05 to 5.0 vol%. Both Pd nanowires and Pd@Pt nanowires show a prompt, reversible increase in resistance upon exposure to H2 in air, caused by the conversion of Pd to more resistive PdHx. But significant accelerations in response and recovery are observed with the addition of just 1.0 ML of Pt to the Pd nanowire surface.

Keywords: Electrochemistry, Metals, Microelectrode, Nanotechnology
Application Code: Nanotechnology
Methodology Code: Electrochemistry

Session # 540 Abstract # 540-2 **Awards**

Session Title SEAC - Charles N Reilley and Royce W Murray Award

Abstract Title **A Membrane-Based AC Electroosmotic Pump**

Date: Monday, March 07, 2016 - Afternoo
Time: 02:15 PM
Room: B312

Primary Author Charles R. Martin
University of Florida

Co-Author(s) Pradeep Ramiah Rajasekaran, Xiaojian Wu

Abstract Text

Microfluidic devices are of considerable research interest in areas such as analytical chemistry, biochemistry, and cell biology. Here we present our recent work on the design and development of a new type of pump that can be used in microfluidic devices. We have developed a membrane-based AC electroosmotic pump with no moving parts that can generate a flow rate on the order of 4 mL min-1 using a sinusoidal voltage waveform of 3.5 Vrms at 20 Hz. The polymeric membrane employed contains conically shaped pores, and the asymmetric nature of these pores enables electroosmotic flow rectification – the phenomenon responsible for generating net flow from a symmetrical sinusoidal voltage wave. This AC operation offers a number of potential advantages including lower power consumption. The pumping performance of the device was experimentally investigated by analyzing the flow rate as a function of frequency and magnitude of the voltage wave.

Keywords: Automation, Biomedical, Electrochemistry, Environmental/Biological Samples
Application Code: Biomedical
Methodology Code: Electrochemistry

Session # 540 **Abstract # 540-3** **Awards**

Session Title: SEAC - Charles N Reilley and Royce W Murray Award

Abstract Title: **Electrocatalytic Amplification of Single Nanoparticle Collisions Using DNA-Modified Surfaces**

Primary Author: Richard M. Crooks
University of Texas at Austin

Date: Monday, March 07, 2016 - Afternoo
Time: 02:50 PM
Room: B312

Co-Author(s): Keith J. Stevenson, Radhika Dasari, Timothy M. Alligrant

Abstract Text

Here we report on the effect of DNA modification on individual collisions between Pt nanoparticles (PtNPs) and ultramicroelectrode (UME) surfaces. These results extend recent reports of electrocatalytic amplification (ECA) arising from collisions between naked surfaces, and they are motivated by our interest in using ECA for low-level biosensing applications. In the present case, we studied collisions between naked PtNPs and DNA-modified Au and Hg UMEs and also collisions between DNA-modified PtNPs and naked Au and Hg UMEs. In all cases, the sensing reaction is the catalytic oxidation of N2H4. The presence of ssDNA (5-mer or 25-mer) immobilized on the UME surface has little effect on the magnitude or frequency of ECA signals, regardless of whether the electrode is Au or Hg. In contrast, when DNA is immobilized on the PtNPs and the electrodes are naked, clear trends emerge. Specifically, as the surface concentration of ssDNA on the PtNP surface increases, the magnitude and frequency of the current transients decrease. This trend is most apparent for the longer 25-mer. We interpret these results as follows. When ssDNA is immobilized at high concentration on the PtNPs, the surface sites on the NP required for electrocatalytic N2H4 oxidation are blocked. This leads to lower and fewer ECA signals. In contrast, naked PtNPs are able to transfer electrons to UMEs having sparse coatings of ssDNA.

Keywords: Electrochemistry
Application Code: Bioanalytical
Methodology Code: Electrochemistry

Session # 540 Abstract # 540-4 **Awards**

Session Title SEAC - Charles N Reilley and Royce W Murray Award

Abstract Title **Imaging Molecular Flux Using Protein Channel Based Scanned Probe Microscopy**

Date: Monday, March 07, 2016 - Afternoo
Time: 03:45 PM
Room: B312

Primary Author Ryan J. White
University of Maryland Baltimore County

Co-Author(s) Florika C. Macazo

Abstract Text

The use of scanning electrochemical (SECM) and scanning ion conductance microscopy (SICM) has enabled chemical imaging of a variety of surface and transport processes. Coupling these scanned probed techniques with the sensitivity afforded by ion channel stochastic single molecule detection has the potential to enable chemical imaging of a wide variety of new molecules with exquisite sensitivity. Here I discuss the development of a DC-based scanning ion conductance measurement with imaging probes incorporating the protein channel -hemolysin. Using these probes we are able to image the molecular flux of the small molecule, -cyclodextrin, through porous membranes. The measurement uses both channel conductance and channel activity (i.e., blocking) to provide both conductance and cyclodextrin images of the membrane surface simultaneously. Naturally occurring protein channels represents a new method of producing reproducible nanometer-scaled pores for scanned probe microscopy.

Keywords: Bioanalytical, Biosensors, Electrochemistry, Sensors
Application Code: Bioanalytical
Methodology Code: Electrochemistry

Session # 540 Abstract # 540-5 **Awards**

Session Title SEAC - Charles N Reilley and Royce W Murray Award

Abstract Title **Imaging with Nanopipettes**

Primary Author Lane A. Baker
Indiana University

Date: Monday, March 07, 2016 - Afternoo
Time: 04:20 PM
Room: B312

Co-Author(s)

Abstract Text

Nanopipettes provide an interesting tool for collection of local information from sample surfaces. Here, we describe electrospray from nanopipettes to realize a mode of electrochemical scanning probe microscopy introduced here as scanning electrospray microscopy (SESM). This technique provides an ambient, non-contact method to investigate surface topography with distance-dependence of electrospray current as a means of feedback for imaging. Approach curves, line scans, and images generated with SESM are reported. Deposition of salt on the sample surface from electrospray imaging is characterized.

Keywords: Electrochemistry, Electrospray, Imaging
Application Code: Bioanalytical
Methodology Code: Electrochemistry

Session # 550 Abstract # 550-1 **Symposia**

Session Title Emerging Platforms for Lab-on-a-Chip Analyses

Abstract Title **Rapid Screening for Infectious Diseases Using Paper-Analytic Devices**

Primary Author Charles Henry
Colorado State University

Date: Monday, March 07, 2016 - Afternoo
Time: 01:35 PM
Room: B301

Co-Author(s)

Abstract Text

Infectious diseases are a major threat to human and animal health worldwide, and are responsible for killing millions of people each year and costing billions of dollars in economic losses. From 1995-2008, the estimated losses from zoonotic infectious diseases was $120B. As a result, there has been a long standing interest in rapid screening tools that enable detection and even identification of bacterial and viral pathogens. Traditional microbiological methods rely on culturing, which, despite being slow and time intensive, are effective for those species which can be readily cultured. Molecular techniques such as RT-PCR and ELISA have improved accuracy and speed but are rarely performed at the point of need, largely due to the need for multiple processing steps and benchtop instrumentation. We have been developing new methods for molecular detection of infectious diseases making use of recent developments in paper-based analytic devices. Multiple methods for detecting infectious diseases using paper-based devices will be presented. The first relies on detection of enzymes produced by bacteria using electrochemical paper-based analytic devices (ePADs). ePADs provide better sensitivity than colorimetric methods. Discussion will focus on detection of Salmonella and E. Coli using this approach. The second method detects viral and bacterial DNA colorimetrically using peptide nucleic acids coupled with silver nanoparticles in a simplified aggregation assay. Examples of detecting DNA from human papilloma virus, MERS, and tuberculosis will be shown.

Keywords: Bioanalytical, Biosensors, Lab-on-a-Chip/Microfluidics
Application Code: Biomedical
Methodology Code: Microfluidics/Lab-on-a-Chip

Session # 550 Abstract # 550-2 **Symposia**

Session Title: Emerging Platforms for Lab-on-a-Chip Analyses

Abstract Title: **High-Throughput Microfluidic Experimentation One Drop at a Time**

Date: Monday, March 07, 2016 - Afternoo

Time: 02:10 PM

Room: B301

Primary Author: Andrew J. deMello
ETH Zürich

Co-Author(s)

Abstract Text

The relevance of microfluidic technology is modern experimental science is significant and driven by including the ability to process ultra-small volumes of fluid, enhanced analytical performance, reduced instrumental footprints, facile integration of functional components and the capacity to exploit atypical fluid behaviour to control chemical and biological entities in both time and space. Droplet-based microfluidic systems allow the generation and manipulation of discrete pL-volume droplets contained within an immiscible continuous phase. They leverage immiscibility to create discrete volumes that reside and move within a continuous flow. Such segmented-flows allow for the production of monodisperse droplets at rates in excess of tens of KHz and independent control of each droplet in terms of size, position and chemical makeup. Moreover, the use of droplets in complex chemical and biological processing relies on the ability to perform a range of integrated, unit operations such as droplet generation, merging/fusion, sorting, splitting, dilution, storage and sampling. I will provide examples of how droplet-based microfluidic systems can be used to perform a range of chemical and biological experiments (including nanomaterial synthesis and cell-based assays) in a rapid and efficient manner and also introduce the use of stroboscopic epifluorescence imaging in the extensive characterization of enzyme–inhibitor reaction kinetics and high-throughput imaging flow cytometry.

Keywords: Fluorescence, Lab-on-a-Chip/Microfluidics, Microscopy, Multichannel Spectrometry (CCD CID array)
Application Code: Bioanalytical
Methodology Code: Microfluidics/Lab-on-a-Chip

Session # 550 Abstract # 550-3 Symposia

Session Title: Emerging Platforms for Lab-on-a-Chip Analyses

Abstract Title: **Simple, Microfluidic Flow Distance-Based Determination of Biomolecule Concentrations**

Date: Monday, March 07, 2016 - Afternoo
Time: 02:45 PM
Room: B301

Primary Author: Adam T. Woolley
Brigham Young University

Co-Author(s): Chatterjee Debolina, Sahore Vishal

Abstract Text

We have designed and tested a simple, distance-based microfluidic analysis system for determining concentrations of biomolecules. Our microchips do not need detection hardware; instead, they utilize visualization of fluid flow distances in microfluidic channels, wherein a given flow distance corresponds to a specific analyte concentration.(1,2) We fabricated the microchannels in an elastically deformable polymer (PDMS) and linked receptor molecules on the microchannel surfaces. When fluid containing the analyte(s) of interest flows in a microchannel, receptor pairs are crosslinked, which causes constriction of the channel and stoppage of flow. The solution flow distance (before stoppage) can then be used to infer the target analyte concentration. We have evaluated our flow distance assays on streptavidin in solution, with a limit of detection of 1 ng/mL.(1) We determined a linear relationship between the logarithm of streptavidin concentration and the solution flow distance in the microchannel.(1) We have also utilized these flow assays to measure the concentrations of nucleic acids in solutions, including synthetic urine. We determined the concentrations of two disease-related microRNA analogues in synthetic urine with detection limits of 10 pg/mL,(2) which are very close to the levels of microRNAs in urine that occur in a diseased state (5 pg/mL). We are working to further improve the limits of detection and expand the range of microRNAs that can be determined in our flow devices. We expect that these microfluidic systems should have broad applicability for various point-of-usage bioassays.

References

1.Chatterjee, D.; Mansfield, D.S.; Anderson, N.G.; Subedi, S.; Woolley, A.T. Analytical Chem. 84, 7057-7063 (2012).
2.Chatterjee, D.; Mansfield, D.S.; Woolley, A.T. Analytical Methods 6, 8173-8179 (2014).

Keywords: Bioanalytical, Lab-on-a-Chip/Microfluidics
Application Code: Bioanalytical
Methodology Code: Microfluidics/Lab-on-a-Chip

Session # 550 Abstract # 550-4 **Symposia**

Session Title: Emerging Platforms for Lab-on-a-Chip Analyses

Abstract Title: **Automated Droplet Manipulation, Analysis and Screening Based on Sequential Operation Droplet Array Technique**

Date: Monday, March 07, 2016 - Afternoo
Time: 03:35 PM
Room: B301

Primary Author: Qun Fang
Zhejiang University

Co-Author(s): Zhu Ying

Abstract Text

In 2013, we developed a sequential operation droplet array (SODA) system [1] for performing automated picoliter to nanoliter-scale droplet manipulation, analysis and screening. The SODA system consists of a tapered capillary connected with a syringe pump and a two-dimensional oil-covered droplet array fixed on a x-y-z translational stage [1]. It can achieve multiple liquid handling manipulations including liquid droplet assembling, generation, indexing, transferring, splitting and fusion, under control of a computer program. All of these operations are completed using different combinations of three elemental operations as capillary-based liquid aspirating and depositing, and the moving of the oil-covered droplet array, which endows it with substantial high versatility and flexibility.

We have applied the SODA system in enzyme inhibitor screening [1], cell-based drug combination screening [2], and protein crystallization condition screening [3]. Recently, we applied the SODA system in quantification of microRNA in single cells to perform in-droplet single cell encapsulation, cell lysis, reverse transcription, and real-time PCR [4]. We also coupled the nanoliter-scale droplet array with high-resolution separation systems such as capillary electrophoresis, liquid chromatography and electrospray ionization mass spectrometry [5].
The above results demonstrated that the SODA strategy has the potential to provide a versatile and flexible micro-liquid manipulation and treatment approach, especially for analysis of small amount of samples.

References

1. Y. Zhu, et al., Anal. Chem., 2013, 85, 6723.
2. G-S. Du, et. al, Anal. Chem. 2013, 85, 6740.
3. Y. Zhu, et al, Sci. Rep., 2014, 4, 5046.
4. Y. Zhu, et al., Sci. Rep., 2015, 5, 9551.
5. D-Q. Jin, et al., Anal. Chem., 2014, 86, 10796.

Keywords: Lab-on-a-Chip/Microfluidics
Application Code: Bioanalytical
Methodology Code: Microfluidics/Lab-on-a-Chip

Session # **560** **Abstract #** **560-1** **Symposia**

Session Title: Innovative Applications of Surface-Enhanced Raman Spectroscopy

Abstract Title: **Ultrafast Surface-Enhanced Raman Spectroscopy**

Date: Monday, March 07, 2016 - Afternoo
Time: 01:35 PM
Room: B303

Primary Author: Renee R. Frontiera
University of Minnesota

Co-Author(s): Alyssa A. Cassabaum, Emily L. Keller, James L. Brooks, Nathaniel C. Brandt

Abstract Text

Plasmonic materials offer a promising approach for developing new light harvesting devices and for directing the outcome of chemical reactions. Their ability to concentrate light to subwavelength volumes leads to the formation of hot spots and hot electrons, which can drive energy flow, modify potential energy surfaces, and enable new photochemical and photophysical processes. Currently, a fundamental mechanistic understanding of how photoreactions occur at plasmonic surfaces is lacking. We utilize a novel time-resolved surface-enhanced Raman technique to follow structural dynamics in molecules adsorbed to or proximal to plasmonic materials, providing picosecond timescale information on bond-making and bond-breaking processes.

This talk will focus on the use of ultrafast surface-enhanced Raman spectroscopies to follow chemical reaction dynamics of molecular plasmonic systems. We use a femtosecond photoexcitation pulse to excite plasmon-molecule systems, and a picosecond spontaneous Raman probe to follow the transient molecular response. We discuss two specific systems of interest. In the first, we track the dimerization of 4-nitrobenzenethiol on silver and gold plasmonic substrates. We see a transient response with a lifetime of 7 picoseconds, which we assign to strong coupling and energy flow between the plasmonic particle and the chemisorbed molecules. In the second system, we examine p-nitrophenol on gold colloids, in which the transient response corresponds to a tenfold increase in the SERS signal. Our results should lead to rational design in the use of plasmonic systems to drive photochemical and photophysical processes.

This work was supported by the Air Force Office of Scientific Research grant FA9550-15-1-0022.

Keywords: Nanotechnology, Raman, Surface Enhanced Raman, Ultra Fast Spectroscopy
Application Code: Nanotechnology
Methodology Code: Vibrational Spectroscopy

Session # **560** **Abstract #** **560-2** **Symposia**

Session Title Innovative Applications of Surface-Enhanced Raman Spectroscopy

Abstract Title **Detecting Small Molecules Using Probe-Mediated SERS Schemes**

Date: Monday, March 07, 2016 - Afternoo
Time: 02:10 PM
Room: B303

Primary Author Jon Camden
University of Notre Dame

Co-Author(s)

Abstract Text

Surface-enhanced Raman spectroscopy (SERS) has extraordinary promise for thedetection of trace anaylte because of its extreme sensitivity, down to the single-molecule level, and itsfreedom from water interferences. This powerful combination makes it a candidate for a range of analytical targets ranging from bioanalysis to environmental remediation and process monitoring. In this talk we present several SERS based, probe-mediated schemes for the detection of small molecules with inherently weak Raman signals. In particular we focus on environmental contaminants and nuclear forensic targets.

Keywords: Environmental Analysis, Fuels\Energy\Petrochemical, Nanotechnology, Vibrational Spectroscopy
Application Code: Nanotechnology
Methodology Code: Molecular Spectroscopy

Session # 560 Abstract # 560-3 **Symposia**

Session Title Innovative Applications of Surface-Enhanced Raman Spectroscopy

Abstract Title **Utilization of SERS Nanoparticles as Contrast Agents for Molecular Imaging in Cancer**

Primary Author Cristina Zavaleta
Stanford University

Date: Monday, March 07, 2016 - Afternoo
Time: 02:45 PM
Room: B303

Co-Author(s)

Abstract Text

Several preclinical molecular imaging modalities have been used to study cellular and molecular processes that have the potential to provide important functional information about various cancer types. Some of these imaging modalities include positron emission tomography, single photon emission computed tomography, optical bioluminescence and fluorescence, photoacoustics, as well as several other emerging modalities. Recently, there has been an overwhelming interest from the biomedical community to sensitively track nanoparticles for their diagnostic and therapeutic potential. As a result, new strategies optimized for nanoparticle imaging have been developed, further expanding the field of molecular imaging. One of the newer molecular imaging approaches we have spent considerable effort in developing involves the use of Raman spectroscopy in conjunction with tumor targeting contrast agents known as surface enhanced Raman scattering (SERS) nanoparticles. Raman spectroscopy is based on an inelastic scattering effect that was discovered back in 1928, but has just recently generated interest in the molecular imaging community with its ability to detect femtomolar concentrations along with its unique ability to multiplex using SERS nanoparticles. We have developed both a preclinical Raman imaging platform capable of non-invasive deep tissue imaging in small animals and a Raman endoscope capable of clinical imaging in patients both to be used in conjunction with Raman nanoparticles. This talk will cover the preclinical evaluation of our small animal imaging platform to image deep tissues and multiplex using various SERS nanoparticles, as well as localize the biodistribution properties of our Raman nanoparticles in preclinical models after different administration routes. I will also discuss our plans to move this novel strategy into the clinics with the recent development of our Raman endoscope and its potential to improve cancer detection during routine screening.

Keywords: Biomedical, Nanotechnology, Raman, Surface Enhanced Raman
Application Code: Biomedical
Methodology Code: Vibrational Spectroscopy

Session # 560 Abstract # 560-5 **Symposia**

Session Title Innovative Applications of Surface-Enhanced Raman Spectroscopy

Abstract Title **Development of Surface Enhanced Spatially Offset Raman Spectroscopy (SESORS) for Neuroscience**

Date: Monday, March 07, 2016 - Afternoo
Time: 04:10 PM
Room: B303

Primary Author Bhavya Sharma
University of Tennessee

Co-Author(s)

Abstract Text

Surface-enhanced Raman spectroscopy (SERS) is a powerful vibrational spectroscopy that allows for label-free, highly sensitive and selective detection of low concentration analytes. We combine the power of SERS with the advantages of spatially offset Raman spectroscopy (SORS) for biomedical applications. SORS allows Raman measurements to be made through distinctly different layers within a diffusely scattering medium. This combined surface-enhanced SORS is designated as SESORS.

We present progress on developing SESORS for the detection of neurotransmitters in the brain through the skull. Most measurements of chemicals in the brain involve drilling holes through the skull to access the soft tissue and little is known about the local concentrations of these chemicals. We are working to develop a Raman spectroscopy-based, non-invasive methodology for the detection and measurement of key neurochemicals in vivo.

Keywords: Bioanalytical, Neurochemistry, Surface Enhanced Raman
Application Code: Bioanalytical
Methodology Code: Vibrational Spectroscopy

Session # 570 **Abstract # 570-1** **Symposia**

Session Title: Miniature Mass Spectrometers

Abstract Title: **Introduction of a Multi-Optic Coaxial Ring Ion Trap (MoCRIT) for External Ionization in Portable Mass Spectrometry**

Primary Author: Guido F. Verbeck
University of North Texas

Date: Monday, March 07, 2016 - Afternoo
Time: 01:35 PM
Room: B304

Co-Author(s)

Abstract Text

Here we present a multi-optic coaxial ring ion trap (MoCRIT) developed for external ionization-coupled to miniature portable mass spectrometry. Performance characterization will be shown by introduction of laser desorption ions (high energy) through a multiport pulse valve, thermalized, and trapped within the MoCRIT. The portable platform will also be introduced. Applications to analysis of illicit drugs and organophosphates will be shown. Preliminary addition of a resistive glass kinetic filter in front of the multi-optic coaxial ring ion trap will be introduced to show future directions. Lastly, preliminary data for nanospray introduction will be shown, specifically to the analysis of illicit drugs.

Keywords: Drugs, Forensic Chemistry, Mass Spectrometry, Portable Instruments
Application Code: Homeland Security/Forensics
Methodology Code: Mass Spectrometry

Session # 570 Abstract # 570-2 **Symposia**

Session Title: Miniature Mass Spectrometers

Abstract Title: **From 2G to 3G - Quantitative Analysis and Biomarker Profiling Using Miniature Mass Spectrometry System**

Date: Monday, March 07, 2016 - Afternoo
Time: 02:10 PM
Room: B304

Primary Author: Zheng Ouyang
Purdue University

Co-Author(s): Graham Cooks, Ran Zou, Ren Yue, Yu Xia, Yuan Su

Abstract Text

At Purdue we have developed two generations of miniature mass spectrometry (MS) systems and are designing the 3rd generation. The 1st generation (1G), regardless of the size of the individual systems (ranging from Mini 5 of 65 kg to Mini 11 of 4 kg), was characterized with analysis of volatile or semi-volatile organic compounds and internal ionization. With the incorporation of the discontinuous atmospheric pressure interface on Mini 10 and Mini 11, we developed the 2nd generation (2G) miniature MS systems. The coupling of ambient ionization for direct sampling ionization was a main focus and enabled the analysis of non-volatile organic/biological compounds in complex samples.
For the 2G miniature MS systems, a major effort was put in the simplification of the complete analysis procedures, which is a key to enable the use of the miniature MS analytical systems by non-expert end users. We purposely focused on the applications where single compound could be used as the marker, which is quite applicable for many regulatory and clinical applications. We have demonstrated that through the combination of optimized ambition ionization, real-time reactions and MS/MS, adequate sensitivity could be achieved using miniature MS systems for target analysis. Incorporation of internal standards into the use with disposable sample cartridges was the strategy for achieving the quantitation accuracy. The developed technologies are being used in the development of a commercial miniature MS product (PURSPEC).

The 3rd generation miniature MS systems is targeting clinical analysis, with biological compounds such as fatty acids and lipids as the potential biomarkers to be analyzed. Analysis of a chemical/biological profile, instead of a single compound, need to be executed. Widened m/z ranges, wider dynamic range, faster MS/MS scan and advanced data processing are required for the 3G miniature MS systems.

Keywords: Biological Samples, Lipids, Mass Spectrometry
Application Code: Clinical/Toxicology
Methodology Code: Mass Spectrometry

Session # 570 Abstract # 570-3 **Symposia**

Session Title: Miniature Mass Spectrometers

Abstract Title: **Describing and Optimizing Toroidal Trapping Fields for the Development of Miniature Mass Spectrometers**

Date: Monday, March 07, 2016 - Afternoo
Time: 02:45 PM
Room: B304

Primary Author: Stephen A. Lammert
PerkinElmer

Co-Author(s): Daniel E. Austin, Edgar D. Lee, Karl R. Warnick

Abstract Text

Quadrupole mass analyzers (2D and 3D) are the overwhelming choice as mass analyzers for researchers developing portable mass spectrometry systems. And as the analyzer dimensions decrease in size, an increased reliance on the fundamental mathematical underpinnings of these devices is required to correct for field imperfections introduced by compromises in the machining or degradation in the accuracy of the fields formed by intentionally approximating or compromising 'ideal' electrodes geometries. Fortunately, researchers developing these miniature devices can rely on the theoretical trapping stability and field equations that were co-developed with the the devices in the 1950's. Toroidal ion traps are another novel approach to reduced size mass analyzers. And while the conception of the toroidal mass analyzer proceeded from quadrupole devices, they cannot be described by the quadrupole device fundamental mathematical equations. This barrier originates from the fact that 2D/3D quadrupole device fields can be described in Cartesian (or cylindrical) coordinates while the toroidal trapping field cannot. For example, the sublinear even-order (mostly octapole) contribution to the trapping field added by opening slits in the original Finnigan ion trap was compensated by a adding a positive octapole component (stretching the endcap electrode distance). There currently is no analog process known in the toroidal coordinate system.

We have recently begun an effort to develop the field description and analysis tools needed to describe, analyze and optimize trapping fields in the toroidal coordinate system. For example, we have solved the Laplace equations for the first 10 'poles' in a toroidal coordinate system and have used these solutions in an attempt to describe the current PerkinElmer TRIDION mass analyzer trapping field. This example is part of a multifaceted approach using fundamental/theoretical, computational and empirical methods which will be presented.

Keywords: Instrumentation, Ion Trap, Mass Spectrometry, Spectrometer
Application Code: General Interest
Methodology Code: Mass Spectrometry

Session # 570 **Abstract # 570-4** **Symposia**

Session Title: Miniature Mass Spectrometers

Abstract Title: **Practical Applications of Outside-the-Lab Mass Spectrometry**

Primary Author: Mitch Wells
FLIR Detection, Inc.

Date: Monday, March 07, 2016 - Afternoo
Time: 03:35 PM
Room: B304

Co-Author(s)

Abstract Text

Recent developments in sample collection have expanded the utility of mass spectrometry to data gathered at the site of action, reducing the time and logistics of sample analysis versus a traditional laboratory. Efficiency of data collection in the field requires sampling techniques that reduce the sample preparation. The work demonstrated here highlights the integration of a prepless sampling introduction technique capable of directly sampling materials of multiple phases of matter, including solids and aqueous liquids, into a mobile GC/MS platform.

In addition to sample preparation, ionization plays a key role in the information that can be extracted from a sample. Fielded mass spectrometry (MS) is traditionally limited to electron ionization (EI) and positive ionization, or chemical ionization (CI) with ammonia or methane reagents which are considered toxic and/or flammable. FLIR Systems has developed a low pressure dual polarity CI source that uses non-hazardous carbon dioxide as the reagent gas, capable of running in positive, negative or even combined modes, whereby rapid polarity switching permits the near simultaneous analysis of both signals.

Through the cooperation of multiple law enforcement agencies, analysis of suspect materials from actual forensic scenarios has been carried out, including the analysis of both drug and explosive chemical classes. Among those scenarios includes clandestine laboratories involved in the suspected production of designer drugs. Emphasis will be placed on techniques associated with sample collection and presentation for a number of native surfaces, and the high fidelity data produced from the PSI Probe technique linked to a mobile GC/MS platform through the implementation of stringent searching techniques and use of a simplified user interface to expedite data collection and decision making.

Keywords: Forensics, GC-MS, Mass Spectrometry, Portable Instruments
Application Code: Homeland Security/Forensics
Methodology Code: Mass Spectrometry

Session # 570 Abstract # 570-5 **Symposia**

Session Title Miniature Mass Spectrometers

Abstract Title **The Development and Performance Enhancement of A Mini-Mass Spectrometer with Continuous Atmospheric Pressure Interface**

Date: Monday, March 07, 2016 - Afternoo
Time: 04:10 PM
Room: B304

Primary Author Wei Xu
Beijing Institute of Technology

Co-Author(s) Muyi He, YanBing Zhai

Abstract Text

Owing to its portability, miniature mass spectrometer (MS) could be used for in-situ analysis of trace amount of chemicals and biological samples. Performance, size, weight and power consumption of a mini MS are greatly determined by vacuum system and the atmospheric pressure interface. Continuous gas inlet coupled with internal ionization source have been used in most mini MS. With the cost of scanning speed, discontinuous atmospheric pressure interface (DAPI) has been developed, which enabled efficient ion transfer from atmospheric pressure ionization source to the mass analyzer in vacuum. Herein, we describe the development of a mini MS, which is small in size but has a continuous atmospheric pressure interface (API). The mini-MS was characterized and optimized in terms of stability, sensitivity, mass range, mass resolution and scan speed. Rapid analysis (5 Hz) of mixtures was demonstrated by coupling the mini-MS with paper spray.

Keywords: Instrumentation, Ion Trap, Mass Spectrometry, Portable Instruments
Application Code: Biomedical
Methodology Code: Mass Spectrometry

Session # **580** **Abstract #** **580-1** **Symposia**

Session Title: Nanomedicine, From Diagnostics to Large Animal Therapy

Abstract Title: **The Plasmonic Photo-Thermal Death of Cancer in Cells(1) and in Animals(2,3) Using Gold Nano-Rod**

Date: Monday, March 07, 2016 - Afternoo
Time: 01:35 PM
Room: B305

Primary Author: Mostafa A. El-Sayed
Georgia Institute of Technology

Co-Author(s)

Abstract Text

Cancer cells were photo thermally destroyed with gold nano-spheres in vitro (1) and with gold nano-rods in-vivo in mice (2), in cats and in dogs(3). Interesting results will be shown and discussed, including the larger difference of the outcome of the normal surgery carried out on breast cancer known for humans and the observed photothermal treatment carried out on a female cat.

1) Aioub, M.; El-Sayed, M. A. A Real-Time Surface Enhanced Raman Spectroscopy Study of Plasmonic Photothermal Cell Death in vivo Using Targeted Gold Nanoparticles. JACS, In Review,.
2) Dickerson, E. B.; Dreaden, E. C.; Huang, X.; El-Sayed, I. H.; Chu, H.; Pushpanketh, S.; McDonald, J. F.; El-Sayed, M. A. Gold nanorod assisted near-infrared plasmonic photothermal therapy (PPTT) of squamous cell carcinoma in mice. Cancer letters 2008, 269, 57-66.
3) Abdoon et al., Normal Pregnancy and Lactation in a Cat after Treatment of Mammary Gland Tumor When Using Photothermal Therapy with Gold Nanorods: A Case Report. Nanomed and Nanotechnology, 2015, 6, 3241

Keywords: Medical
Application Code: Nanotechnology
Methodology Code: Mass Spectrometry

Session # 580 **Abstract # 580-3** **Symposia**

Session Title: Nanomedicine, From Diagnostics to Large Animal Therapy

Abstract Title: **Nanophotonics-Based Diagnostics and Therapy: From Deep In-Vivo Photo-Acoustic Chemical Imaging of Tumors to Photodynamic Treatment of Heart Disease**

Primary Author: Raoul Kopelman
University of Michigan

Date: Monday, March 07, 2016 - Afternoo
Time: 02:45 PM
Room: B305

Co-Author(s)

Abstract Text

In recent years, the use of nanoparticles has advanced from its original utilization to study cells in vitro to application in animal models for cancer diagnostics and therapy. No such advances have been reported concerning the number one killer, heart disease. Here we report on in vivo photon based diagnostics as well as therapy for both diseases. It is well known that the chemistry of the extracellular tumor environment differs from that of healthy organs in two major respects: oxygen depletion and acidosis, i.e. lower pH. The latter may affect the efficacy of therapy, e.g. radiation therapy, chemotherapy and photodynamic therapy. It would thus be advantageous for precision medicine to have this information. We have shown that nanoparticle based photoacoustic imaging can quantify both tissue oxygen as well as pH, and note that this photon and ultrasound based imaging technique is both non-invasive and relatively inexpensive. Regarding photon based therapy, photodynamic therapy (PDT) has been used clinically for skin cancers. Previous animal model tests have also indicated its potential use for internal organs and even brain cancer. We now show that PDT can also serve to fix arrhythmia, possibly the most critical aspect in heart disease. Starting with in vitro tests, continuing with rodent models and culminating with large animal models (sheep), we have shown that cell selective photoablation has major advances compared to the traditional modes of treatment. The latter approach is based on the use of targeted photoactive nanoparticles. Notably, the latter have to be of a much smaller size (below 10 nm) compared to the photodynamic nanoparticles used for the treatment of cancer, which can be of the order of 100 nm.

Keywords: Nanotechnology
Application Code: Biomedical
Methodology Code: Sensors

Session # 580	Abstract # 580-4	Symposia

Session Title Nanomedicine, From Diagnostics to Large Animal Therapy

Abstract Title **Interfacial Assembly of Functional Mesoporous Nanospheres with Multi-Level Architectures for Bioapplications**

Date: Monday, March 07, 2016 - Afternoo
Time: 03:35 PM
Room: B305

Primary Author Dongyuan Zhao
Fudan University

Co-Author(s)

Abstract Text

Functional mesoporous materials possess uniform large pore size, unique ordered mesostructure, high surface area and large pore volume, showing great potential applications on catalysis, adsorption, separation petroleum oil industry, especially in drug delivery and biomedicines. Here we present the development and progress of the synthesis of the spherical mesoporous materials from surfactant assembly approach for the drug carriers, fluorescence detection and diagnosis. We focus on the development of new synthesis approaches, including the liquid-liquid biphase synthesis, evaporation induced aggregation assembly (EIAA), and interface driven orientation arrangement to create novel mesoporous nanospheres with one-level and multi-level architectures, such as the core-shell, yolk-shell structures for silica, TiO2, carbon spheres and hemispheres, Janus nanoparticles and single hole nanocages. These spherical materials with uniform large pore channels (> 3.0 nm), high surface area (~ 1150 m2/g), large pore volume (1.5 ~ 3.5 cm3/g) and open frameworks are non-toxic, easy degradation, and can be used to remove body toxin, as carriers for controlling release of drugs and agents for photothermal therapy (elevated temperature as high as 85 [sup]o[/sup]C at 808 nm).

References

1. Y. Wan, D. Y. Zhao, Chem. Rev., 2007, 107, 2821.
2. Y. Deng, et al., J. Am. Chem. Soc., 2008, 130, 28; Y. Fang, et al., Angew. Chem. Int. Ed., 2010, 49, 7987; W. Li, et al., J. Am. Chem. Soc., 2012, 134, 1186; W. Li, et al., Adv. Mater., 2013, 25, 5129; Y. Fang, et al., Angew. Chem. Int. Ed., 2014, 53, 5366. X. Li, et al., J. Am. Chem. Soc., 2014, 136, 15086; D. Shen, et al., Nano Lett., 2014, 14, 923; Y. Fang, et al., J. Am. Chem. Soc., 2015, 137, 2803; W. Li, et al., Nano Lett., 2015, 15, 2186. L. Zhou, et al., Nat. Commun., 2015, 6, 6938; Y. Liu, et al., Science Adv., 2015, 1, e1500166; X. Li, et al., J. Am. Chem. Soc., 2015, 137, 5903; B. Kong, et al., J. Am. Chem. Soc., 2015, 137, 4260-4266.

Keywords: Biomedical, Nanotechnology, Wet Chemical Methods
Application Code: Material Science
Methodology Code: Chemical Methods

Session # 580 Abstract # 580-5 **Symposia**

Session Title Nanomedicine, From Diagnostics to Large Animal Therapy

Abstract Title **Stickyflares: Tracking the Amount and Location of RNA in Single Cells**

Primary Author Chad A. Mirkin
Northwestern University

Date: Monday, March 07, 2016 - Afternoo
Time: 04:10 PM
Room: B305

Co-Author(s)

Abstract Text

Proper function of RNA is critical to the health and maintenance of a cell, and its misregulation plays a critical role in the development of many disorders. Despite this, the ability to study RNA has been severely limited. Many analytical techniques are only capable of quantifying expression levels of transcripts, and do not offer insight into the dynamics of RNA transport or localization. Recently, the Mirkin group has developed a novel nanoconjugate, termed the "Stickyflare", capable of reporting on both of these critical components in live cells, with the intent to enable a more complete picture of RNA function than any other analytical techniques to date. Based on the spherical nucleic acid (SNA) platform, the Stickyflares enter live cells without the need for harmful transfection agents, quantify target RNA expression with single cell resolution, and allow for real-time analysis of the transport and localization of endogenous RNA. We believe this nanoconjugate will be valuable for studying the function of endogenous RNA in healthy and diseased cells, as well as offer insight into how a change of environment (drug treatments, hypoxia, starvation, etc.) affect the dynamics of RNA expression.

Keywords: Biotechnology, Fluorescence, Nanotechnology, Nucleic Acids
Application Code: Bioanalytical
Methodology Code: Fluorescence/Luminescence

Session # **590** **Abstract #** **590-1** **Symposia**

Session Title: Non-Traditional Human Biometrics for Threat Assessment: Using Chemical Forensics for National Secu

Abstract Title: **Human Biomonitoring and In Vitro Toxicity Testing Applications for Covert Threat Analysis and Security Applications**

Date: Monday, March 07, 2016 - Afternoo
Time: 01:35 PM
Room: B308

Primary Author: Joachim D. Pleil
US EPA

Co-Author(s): William E. Funk

Abstract Text

Non-traditional biometrics takes security-oriented biomonitoring beyond traditional identification (e.g. fingerprinting, retinal scan, DNA profile) into the realm of threat assessment. Rather than answering "Who are you?", we want to know "Are you a threat?". This presentation suggests methods for medical and environmental chemical assessments that could be adapted to identify persons posing a security threat. Specifically, exhaled breath can be tested for exogenous chemicals including jet fuels, volatile ingredients of illicit drug manufacture, and explosives handling, as well as for endogenous chemicals indicative of infectious status. Whole blood and dried blood spots can be tested for protein adducts of various chemical exposures. High-throughput (in vitro) immunochemical instrumentation can be deployed at risk points for biological screening. Although these techniques are currently based on deliberate sampling and analysis, they could also be adapted for covert and rapid detection. The advantage of such chemical interrogation is that the body-borne signature is unambiguous and difficult to obscure.

Keywords: Biological Samples
Application Code: Homeland Security/Forensics
Methodology Code: Gas Chromatography/Mass Spectrometry

Session # 590 **Abstract # 590-2** **Symposia**

Session Title: Non-Traditional Human Biometrics for Threat Assessment: Using Chemical Forensics for National Secu

Abstract Title: **Noninvasive Infectious Disease Monitoring for Health Applications**

Date: Monday, March 07, 2016 - Afternoo
Time: 02:10 PM
Room: B308

Primary Author: Michael Schivo
University of California, Davis

Co-Author(s): Cristina E. Davis

Abstract Text

Human breath is a complex mixture of water, gases, and small molecules that represent metabolic products. The detection and identification of these lightweight compounds may be important to advance non-invasive, point-of-care monitoring systems for humans and animals. While there are many potential applications of noninvasive monitoring, infectious diseases are important targets for early diagnostic systems.

We have taken a bottom-up approach to define metabolic signatures in breath as an avenue toward developing monitoring systems for infectious diseases. We used gas chromatography-mass spectrometry (GC/MS) to show that influenza-infected immune cells in vitro demonstrate reproducibly unique metabolic signatures based on virus strain. We then extrapolated this data to the human airway by showing that rhinovirus-infected primary human respiratory cells emit unique compounds compared to uninfected cells. Last, our breath collection studies serve as excellent platforms to apply our in vitro experience for biomarker discovery.

Through careful collaboration between engineers and clinicians to ultimately conduct clinical studies, noninvasive detection systems are becoming a reality that may help change the landscape of health monitoring.

Keywords: Bioanalytical, Biomedical, GC-MS, Headspace
Application Code: Biomedical
Methodology Code: Gas Chromatography/Mass Spectrometry

Session # **590** **Abstract #** **590-3** **Symposia**

Session Title: Non-Traditional Human Biometrics for Threat Assessment: Using Chemical Forensics for National Secu

Abstract Title: **Comprehensive Analysis of the Chemicals Within: The Human Exposome and Applications for Threat Assessment**

Date: Monday, March 07, 2016 - Afternoo
Time: 02:45 PM
Room: B308

Primary Author: Gary W. Miller
Emory University

Co-Author(s)

Abstract Text

Many of the threats we face are unknown. Focusing on suspected chemicals has to potential to overlook unknown adverse exposures. The exposome is an emerging concept that takes a holistic look at exposure. By examining as many chemicals as possible while also assessing alterations to the underlying biology, the exposome can capture perturbations in function without knowing the precise cause of the disruption. Such approaches could be useful in assessing unknown threats. While suspected exposure to a cholinesterase inhibitor has a simple biological marker, other chemicals target multiple biological components. High-resolution metabolomics allows the identification of tens of thousands of chemical and biological species in a single drop of human blood. The processes that occur downstream from the initial interactions with exogenous compounds influences the biological impact of exposures. The pathway level analysis that can be conducted with exposome studies allows one to peer into nearly all biological pathways. In addition, as many exposures are transient, the chemicals underlying a specific exposure may be long gone from the body. However, these past chemical exposures often leave molecular fingerprints that may be able to provide information on these past exposures and these can be detected with high-resolution metabolomics. The metabolomics-based approaches used to support exposome research could advance our ability to detect chemical and biological threats as well as identifying adverse health outcomes. Exposome-based approaches can also help develop strategies aimed at mitigating adverse health outcomes that occur due to such exposures.

Keywords: Detection, Toxicology
Application Code: High-Throughput Chemical Analysis
Methodology Code: Mass Spectrometry

Session # **590** **Abstract #** **590-5** **Symposia**

Session Title Non-Traditional Human Biometrics for Threat Assessment: Using Chemical Forensics for National Secu

Abstract Title **Feasibility of Using Breath to Predict Exposure to Ionizing Radiation**

Date: Monday, March 07, 2016 - Afternoo

Time: 04:10 PM

Room: B308

Primary Author Terence H. Risby
Johns Hopkins University

Co-Author(s)

Abstract Text

Free radical formation is the key mechanism responsible for injury when tissues are exposed to ionizing radiation. It has been postulated that free radicals are responsible for two-thirds of the indirect damage to the cell. Many biomolecules, including DNA, are targets of attack by these reactive species. The unsaturated fatty acids present in cell membranes are readily peroxidized when electrons generated by ionizing radiation are carried by oxygen-containing species. Oxidative status of the cell and the levels of readily available intracellular free radical scavengers determine the degree of damage caused by oxygen free radicals to biomolecules.

Lipid peroxidation has been most widely studied and is a chain reaction with initiation, propagation, and chain termination steps. Specifically, ROS, typically the hydroxyl radical, abstracts an allylic hydrogen atom from an unsaturated lipid to produce a carbon centered radical and water. This radical is conjugated, peroxidized by molecular oxygen and undergoes a variety of reactions. Lipid peroxidation is measured by quantifying the stable products of this chain reaction. Examples of these stable products are: hydrocarbons (ethylene, ethane, 1-pentane, and branched chain hydrocarbons), conjugated dienes, aldehydes (including malondialdehyde), lipid peroxides, and isoprostanes. The identities and concentrations of these products are dependent upon the identities of the source fatty acid substrates and the efficiencies of their production (typically<1%) and finally the biological specimen (tissue, blood, urine, breath or breath condensate) will determine which stable product can be determined. This discussion will be limited to volatile markers of exposure to ionizing radiation in exhaled breath.

Keywords: Biomedical, Characterization

Application Code: Bioanalytical

Methodology Code: Gas Chromatography/Mass Spectrometry

Session # 600 **Abstract # 600-1** **Symposia**

Session Title: Novel Mass Spectrometric Approaches and Applications to Polymer Analysis

Abstract Title: **Characterization of Synthetic Polymers Using as Little as a Small Molecule Matrix and the Vacuum of the Mass Spectrometer for Ionization**

Date: Monday, March 07, 2016 - Afternoo

Time: 01:35 PM

Room: B309

Primary Author: Sarah Trimpin
Wayne State University

Co-Author(s): Barbara S. Larsen, Casey D. Foley, Joshua Fischer, Matthew J. Allen, Sashiprabha M. Vithanarachchi, Zachary Devereaux

Abstract Text

Abundant, multiply charged gas-phase ions of synthetic polymers are produced by exposing the sample with a suitable ionizing matrix and salt to the vacuum of a mass spectrometer in the absence of high energy sources such as lasers and high voltages (Trimpin et al., IJMS 2015, DOI:10.1016/j.ijms.2014.07.033). A fundamental understanding of the role of these unique matrix compounds and their self-ionizing characteristics as well as the importance of salt additives for enhancing the efficiency and applicability of this novel ionization process for polymer characterization are presented here. Although lasers are not necessary, they can be employed to obtain the typical spatial resolution and imaging of surfaces from these multiply charged polymer ions. These novel ionization processes for use in mass spectrometry depend on the proper temperature and pressure and are ideally suited for surface analyses from atmospheric and intermediate pressure conditions of low and high performance mass spectrometers. These polymer surface characterization methods are simple and fast. We demonstrate wide applicability to polymer characterization including spatial information for polymer complexes synthetized for use as magnetic resonance imaging contrast agents to commercially important branched polymers and how the measurement is enhanced by including ion mobility spectrometry dimension, especially from polymer surfaces that are limited to direct separation technologies.

Keywords: Mass Spectrometry, Polymers & Plastics, Portable Instruments, Surface Analysis
Application Code: Polymers and Plastics
Methodology Code: Mass Spectrometry

Session # 600 Abstract # 600-2 **Symposia**

Session Title: Novel Mass Spectrometric Approaches and Applications to Polymer Analysis

Abstract Title: **Model Polymer Systems: Studies by Mass Spectrometry, Ion Mobility, and Computational Strategies**

Date: Monday, March 07, 2016 - Afternoo
Time: 02:10 PM
Room: B309

Primary Author: David M. Hercules
Vanderbilt University

Co-Author(s): John A. McLean, Sarah M. Stow, Tiffany M. Onifer

Abstract Text

There has been considerable interest recently in using mass spectrometry and related methods to address structures of complex systems, including block copolymers. A significant problem with polyurethanes (PURs), polymers that contain multiple hard blocks and soft blocks, is that different group sequences along the chain are possible. The ultimate goal for studying such materials would be the ability to provide a sequence analysis for oligomers, both linear and cyclic. We have established a protocol on smaller molecules that will be used to address this issue using a combination of mass spectrometry, ion mobility spectrometry, oligomer synthesis, and computational strategies. A key component of the project is the synthesis of PUR oligomers having discreet molecular weights (~1900) from methylenediphenyldiisocyanate (MDI) and bis-diol terminated poly(butylene adipate) (PBA), each containing 3 MDIs, 4 butanediols, and 4 polyester repeat units, but having the components arranged in different orders. These syntheses are accomplished by using protecting group chemistry. The main collision-induced fragmentation mechanisms involving the MDI groups are 1,5 H-shift reactions between the PUR carbonyl group and a butanediol H atom, and 1,3 H-shift reactions involving the PUR N-H group and a butanediol oxygen atom. A key component for distinguishing different structures are the 1,5 H-shift reactions that occur within a polyester segment, yielding products different from those produced by MDI groups. Computational strategies are important for calculating collisional cross sections, identifying different species in the ion mobility spectra and identifying the specific atoms involved in fragmentation reactions. They also serve to identify for detailed study those oligomers that show the greatest differences in IMS based on their structures. Computations are also reported for several species from our earlier PUR paper to expand the interpretation of those results.

Keywords: Characterization, Electrospray, Mass Spectrometry, Time of Flight MS
Application Code: Polymers and Plastics
Methodology Code: Mass Spectrometry

Session # 600 Abstract # 600-3 **Symposia**

Session Title: Novel Mass Spectrometric Approaches and Applications to Polymer Analysis

Abstract Title: **Shape Selective Studies of Macromolecular Systems**

Primary Author: James H. Scrivens
Teesside University

Date: Monday, March 07, 2016 - Afternoo
Time: 02:45 PM
Room: B309

Co-Author(s)

Abstract Text

Modern macromolecular product formulations may be characterized by the complex multi-component nature of their composition. In order to achieve the desired, increasingly stringent, structure-property requirements components of widely different molecular weight ranges, volatility, polarity and reactivity are present in the product. There is an increasing need to develop measurement approaches that can cope with this challenge whilst operating on a timescale relevant to synthesis requirements. Ion mobility mass spectrometry (IMMS) which incorporates rapid, shape selective separation with the ability to characterize components utilising tandem (MS/MS) and to be interfaced with ambient ionisation techniques has been shown to provide information rich experimental data enabling complex formulations to be characterised.

Mechanically interlocked polymers are an example of novel polymer architectures that can possess significant additional physical properties in addition to those associated with their constituent parts. Their unique properties make them attractive for a range of potential applications, such as biomaterials and molecular machines. Their efficient and reproducible synthesis is therefore of much interest. The properties of mechanically interlocked polymeric systems depend not only on the properties of their individual components but also on the topology of the subsequent product.

IMMS approaches, coupled with varied MS/MS experiments can be used both to simplify the, often complex data sets that macromolecular systems produce and also to reveal the shape that the individual molecules adopt. Examples from a number of polymeric systems will be presented which show the useful additional data that may be gained from making shape selective measurements and the experimental, structural and synthetic inferences that can be made.

Keywords: Data Analysis, Mass Spectrometry, Method Development, Polymers & Plastics
Application Code: Polymers and Plastics
Methodology Code: Mass Spectrometry

Session # 600 **Abstract # 600-4** **Symposia**

Session Title: Novel Mass Spectrometric Approaches and Applications to Polymer Analysis

Abstract Title: **Shape Sensitive Multidimensional Mass Spectrometry of Synthetic Polymers and Hybrid Materials**

Primary Author: Chrys Wesdemiotis
The University of Akron

Date: Monday, March 07, 2016 - Afternoo
Time: 03:35 PM
Room: B309

Co-Author(s)

Abstract Text

Advanced polymeric materials designed for industrial, pharmaceutical, and biomedical applications are usually complex blends that cannot be characterized by single-stage mass spectrometry (MS). For these systems, MS must be interfaced with a separation method, such as liquid chromatography (LC) or ion mobility (IM) spectrometry. Particularly powerful are ultrahigh performance LC which reduces elution times, and the IM dimension which enables separation by charge, size, and shape and also minimizes solvent use. IM-MS is ideally suitable for the analysis of labile or reactive macromolecular mixtures which cannot be subjected to LC, because they would be destroyed or permanently retained by the stationary phase. The analytical power of IM-MS will be demonstrated with the complete characterization of amphiphilic copolymers, hybrid materials containing peptides conjugated with synthetic polymers, and supramolecular materials bound via noncovalent interactions. The IM step disperses the desired products into a unique mobility space, from which they can be extracted and identified by MS and tandem MS without interference from byproducts and remaining reactants. Moreover, from the IM-MS data, collision cross-sections (CCSs) can be derived for the analyzed materials, which unveil insight about the corresponding macromolecular architectures and 3-D geometries. CCSs have been used in this study to differentiate isomeric peptide-polymer hybrids and to confirm the occurrence of fission and fusion phenomena in noncovalently bonded smart materials.

Keywords: Identification, Ion Chromatography, Mass Spectrometry, Materials Characterization
Application Code: Material Science
Methodology Code: Mass Spectrometry

Session # 600 **Abstract # 600-5** **Symposia**

Session Title: Novel Mass Spectrometric Approaches and Applications to Polymer Analysis

Abstract Title: **MALDI-TOF and MALDI-FTICR of Challenging Polymer Analysis Problems**

Date: Monday, March 07, 2016 - Afternoo
Time: 04:10 PM
Room: B309

Primary Author: Charles L. Wilkins
University of Arkansas

Co-Author(s): Evegenia Tisdale

Abstract Text

In a recent study, MALDI analysis of pristine low molecular weight polyvinyl acetate (a common chewing gum base) and a commercial chewing gum sample were analyzed(1). Sample preparation was optimized using MALDI-TOF mass spectrometry. The optimized sample preparation conditions were used to perform high resolution FTMS analysis for chemical composition assignments. It was found that best results were obtained for the lowest matrix/analyte ratio used with ethyl acetate as solvent. Under FTMS conditions, polyvinyl acetate looses acetic acid molecules from the backbone. In a second study, the challenging MALDI-TOF analysis of a styrene-pentafluorostyrene copolymer was undertaken (2). Here, also, sample preparation optimization was carried out using MALDI-TOF to investigate the influence of matrix, matrix/analyte and analyte/salt ratios. It was determined the 2,5-DHB was the matrix of choice for these analyses and that when a matrix:analyte:salt ratio of 40:5:1 was employed, the highest analyte signals were obtained. Detailed information about the influence of polymerization conditions on copolymer composition was revealed in this study. This presentation will discuss the relative advantages of MALDI-TOF and MALDI-FTMS for chewing gum analysis and describe details of the MALDI-TOF styrene-PFS copolymer analyses.

References

(1) E. Tisdale, C.Wilkins, Anal. Chim.Acta, 820, 92-103 (2014).
(2) E. Tisdale, X. Xu,D. Kennedy,C. Wilkins,Anal. Chim.Acta, 808, 151-162(2014).

Keywords: Characterization, Ion Cyclotron Resonance, Mass Spectrometry, Polymers & Plastics
Application Code: Polymers and Plastics
Methodology Code: Mass Spectrometry

Session # 610 **Abstract # 610-1** **Symposia**

Session Title: The Twenty-Seventh James L Waters Symposium on Super-resolution Microscopy

Abstract Title: **Imaging Life at High Spatiotemporal Resolution**

Date: Monday, March 07, 2016 - Afternoo
Time: 01:35 PM
Room: B405

Primary Author: Eric Betzig
Janelia Research Campus

Co-Author(s)

Abstract Text

As our understanding of biological systems as increased, so has the complexity of our questions and the need for more advanced optical tools to answer them. For example, there is a hundred-fold gap between the resolution of conventional optical microscopy and the scale at which molecules self-assemble to form sub-cellular structures. Furthermore, as we attempt to peer more closely at the dynamic complexity of living systems, the actinic glare of our microscopes can adversely influence the specimens we hope to study. Finally, the heterogeneity of living tissue can seriously impede our ability to image at high resolution, due to the resulting warping and scattering of light rays. I will describe three areas focused on addressing these challenges: super-resolution microscopy for imaging specific proteins within cells down to near-molecular resolution; plane illumination microscopy using non-diffracting beams for noninvasive imaging of three-dimensional dynamics within live cells and embryos; and adaptive optics to recover optimal images from within optically heterogeneous specimens.

Keywords: Biological Samples, Biomedical, Microscopy
Application Code: Biomedical
Methodology Code: Microscopy

Session # 610   Abstract # 610-2   Symposia

Session Title: The Twenty-Seventh James L Waters Symposium on Super-resolution Microscopy

Abstract Title: **Accessing the Emerging Imaging Technologies at HHMI Janelia Research Campus**

Date: Monday, March 07, 2016 - Afternoo
Time: 02:10 PM
Room: B405

Primary Author: Teng-Leong Chew
Janelia Research Campus

Co-Author(s)

Abstract Text

Visualizing and understanding complex biological processes demands the integrated efforts of biologists and physicists. The mission of the Advanced Imaging Center (AIC) is to make cutting-edge imaging technologies developed at Janelia widely accessible, and at no cost, to scientists before the instruments are commercially available. Operating strategically at the interface of engineering and biological applications, the AIC is positioned to drastically reduce the time between instrument development and widespread use in the increasingly technology-intensive field of biology. The AIC will expand the number and diversity of biologists who have access to the unique, state-of-the-art optical imaging microscopes developed at Janelia years before they become commercially available. This unique imaging center is thus uniquely positioned to empower investigators with tools currently not widely available elsewhere. In alignment with Janelia's philosophy of encouraging bold and risky science, the AIC welcomes proposals with high-risk-high-gain projects that may challenge the current paradigm. In fact, it serves as an ideal platform for researchers to test out their novel ideas with the emerging microscopy technologies, fully supported by Janelia's in-house imaging experts and research infrastructure.

Keywords: Bioanalytical, Microscopy
Application Code: Bioanalytical
Methodology Code: Microscopy

Session # **610** **Abstract #** **610-3** **Symposia**

Session Title: The Twenty-Seventh James L Waters Symposium on Super-resolution Microscopy

Abstract Title: **Closing the Gap Between First Implementation and Product - Update on Lattice Light Sheet Commercialization Effort**

Date: Monday, March 07, 2016 - Afternoo

Time: 02:45 PM

Room: B405

Primary Author: Alex Soell
Carl Zeiss Microscopy, LLC

Co-Author(s)

Abstract Text

Commercializing a new technology is difficult and risky. Many questions need to be considered: Does this technology offer a tangible and important advantage to a large enough group of customers and does it align with future research trends? Do the advantages outweigh the disadvantages? What are the barriers to adoption and commercial success?

These are just some of the questions which need to be answered. At Carl Zeiss Microscopy, a dedicated group of physicists, engineers and application scientists ("Advanced Development Team") works together with the inventor(s) of the technique in a clearly defined collaborative process to prepare a decision regarding potential commercialization.

The talk will review how Carl Zeiss Microscopy is approaching the commercializing challenge for the Lattice Light Sheet technology licensed from Eric Betzig.

Keywords: Bioanalytical, Education, Microscopy
Application Code: Biomedical
Methodology Code: Education/Teaching

Session # 610 Abstract # 610-4 **Symposia**

Session Title The Twenty-Seventh James L Waters Symposium on Super-resolution Microscopy

Abstract Title **Life Inside the Cell: STORM, CRISPR and Imagenomics**

Date: Monday, March 07, 2016 - Afternoo
Time: 03:35 PM
Room: B405

Primary Author Bo Huang
University of California, San Francisco

Co-Author(s)

Abstract Text

Cellular processes are carried out by coordinated participation of many biomolecules in a tiny volume. Many people have been dreaming to see clear pictures of these processes in order to understand how these molecules work together. Taking on this challenge, we are developing new imaging techniques and imaging probes, including the use of super-resolution microscopy to dissect the architecture of centrosome, as well as fluorescent probes based on the CRISPR-Cas9 system to visualize the spatial organization of the genome.

Keywords: Biotechnology, Fluorescence, Genetic Engineering, Microscopy
Application Code: Biomedical
Methodology Code: Microscopy

Session # 610 **Abstract # 610-5** **Symposia**

Session Title: The Twenty-Seventh James L Waters Symposium on Super-resolution Microscopy

Abstract Title: **Structured Scanned Plane Bessel Microscopy for Super-Resolution Neuroanatomy**

Date: Monday, March 07, 2016 - Afternoo
Time: 04:10 PM
Room: B405

Primary Author: Timothy Harris
HHMI Janelia Research Campus

Co-Author(s)

Abstract Text

Understanding the function of a brain requires determination of both the activity of neurons and the connections between them. The size range important in this tissue creates a challenging problem for structure determination. Synapses ~ 100 nm wide and dendrites as small as 50 nm diameter can be part of a neuron many millimeters long. Resolution is needed in three dimensions for large tissue volumes, 0.1-1 mm3, at the least. Electron microscopy (EM) has been the method of choice for neuroanatomy, providing clear synaptic resolution. However, EM imaging such large tissue volumes needs years of scanning and manual tracing. Super-resolution optical microscopy approaches the required resolution and is many orders of magnitude faster. I will discuss the design and construction of a microscope designed to offer maximum resolution for volume imaging neural tissue. The core question is the synaptic connection between labeled neurons. The size and crowding of synapses and other processes in Drosophila is even more challenging than typical mammalian brain. Achieving instrument limited super-resolution requires that sample aberrations do not blur the images. This requires either adaptive optics, difficult for volume imaging, or preparation of tissue without aberrations. We have prepared whole Drosophila brains without detectable aberrations and built a custom objective, structured illumination fluorescence microscope (SIM) using scanned Bessel beams to achieve 0.18 x 0.18 x 0.23 um voxels. This three objective, three color, immersion microscope uses all available methods for linear super-resolution. Examples of images from this system, labeling strategies to use the unique sample preparation, and prospects for performance of this system will be used to discuss the performance and the general utility of a complex instrument.

Keywords: Biological Samples, Fluorescence, Imaging, Instrumentation
Application Code: Neurochemistry
Methodology Code: Microscopy

Session # 620 **Abstract # 620-1** **Symposia**

Session Title Trials and Tribulations of Dietary Supplement Analysis: Authentication, Adulteration and Contaminant

Abstract Title **Protecting Consumers One Analysis at a Time: Identifying Harmful Adulterants in Dietary Supplements**

Date: Monday, March 07, 2016 - Afternoo
Time: 01:35 PM
Room: B311

Primary Author Travis M. Falconer
U.S. Food & Drug Administration

Co-Author(s)

Abstract Text

It has been reported that more than half of American adults used dietary supplements in the period 2003 to 2006, creating a massive market that generated roughly $32 billion in revenue in 2012. Unscrupulous manufacturers sell products that are adulterated through the presence of ingredients that are either unlabeled or not "generally recognized as safe." These ingredients may include new plant extracts; unlabeled prescription drugs, such as the inclusion of tadalafil in sexual performance enhancing supplements; unapproved drugs, such as the inclusion of sibutramine in weight-loss supplements; and analogs or derivatives thereof, such as anabolic steroids in body-building supplements. To keep pace with this rapidly changing landscape, our lab employs a combination of analytical techniques in a "semi-targeted" fashion for adulterant detection and identification. Examples of such analyses will be given and additional challenges discussed, including diverse sample matrices, difficulty in obtaining reference materials, etc.

Keywords: Food Contaminants, Food Safety, Gas Chromatography/Mass Spectrometry, Liquid Chromatography/
Application Code: Food Contaminants
Methodology Code: Chemical Methods

Session # 620 Abstract # 620-2 **Symposia**

Session Title Trials and Tribulations of Dietary Supplement Analysis: Authentication, Adulteration and Contaminant

Abstract Title **Authentication of Foods and Botanical Supplements Using Chemometric Methods**

Date: Monday, March 07, 2016 - Afternoo

Time: 02:10 PM

Room: B311

Primary Author James Harnly

USDA

Co-Author(s)

Abstract Text

The variability of the chemical composition of plant materials is largely unexplored. Conventional statistics (e.g., t-test, F-test, ANOVA, and many others) are commonly used to determine concentrations and similarities of individual compounds, but when we consider a whole plant material with thousands of chemical components, multivariate analysis, frequently referred to as chemometrics, becomes the tool of the trade. My lab has found that multivariate analysis, in combination with direct analysis (no chromatography) using MS, NMR, NIR, IR, and UV, is a powerful tool for characterizing the total variance and sources of variance associated with food plants and botanical supplements. In essence, we use chemometrics for the analysis of chemical fingerprints. This information and these methods are critical for establishing identity, authenticity, adulteration, contamination, and differences arising from genetic, environmental, and processing conditions. For foods, we have used chemometric methods to characterize differences arising from plant varieties, growing location, growing year, growing conditions (conventional vs organic), and time of harvest. We have shown that skim milk powders coming from different world-wide locations are sufficiently different to prevent the use of a global standard for detection of adulteration. For botanical supplements, we have identified different species and growing locations and demonstrated that raw materials and processed, commercial supplements cannot be directly compared. Thus, chemometric methods allow us to pull meaningful information from the mountains of data we get from our instrumental methods.

Keywords: Characterization, Chemometrics, Identification, Mass Spectrometry

Application Code: Other

Methodology Code: Chemometrics

Session # 620 **Abstract # 620-4** **Symposia**

Session Title: Trials and Tribulations of Dietary Supplement Analysis: Authentication, Adulteration and Contaminant

Abstract Title: **Authenticity of Herbal Dietary Supplements: Comparison of Chemical and DNA Barcoding Methods**

Primary Author: Rahul Pawar
CFSAN/FDA

Date: Monday, March 07, 2016 - Afternoo
Time: 03:35 PM
Room: B311

Co-Author(s): Erich Grundel, Nicole Shyong, Raymond Cheng, Sara M. Handy

Abstract Text

Dietary supplements are a growing, multi-billion dollar, global business. The consumer perceptions that these products are natural and safe, may prevent disease, may replace prescription medicines, or may make up for a poor diet, has led to their growing popularity. At the same time, some of these businesses have been criticized for problems related to, among others, poor quality control, safety, misbranding, and adulteration. Investigations have shown that some of these products were found to be unauthentic as they lack chemical markers associated with the particular plant. Other investigations have also found that botanical supplements were devoid of the labeled ingredients based on not finding their genetic signature according to DNA barcoding. These investigations have brought into the discussion the suitability of traditional DNA barcoding methods for analyzing finished dietary supplements. In our study, we investigated five popular botanical ingredients (Valerian, soy isoflavones, yohimbe, St. John's wort and Ginkgo) for their authenticity by chemical and DNA barcoding methods. The results of the HPLC quantification of the chemical markers with respect to the label claims are discussed and compared to the results of the DNA barcoding study. Based on the findings, the advantages and drawbacks of the two methods are discussed.

Keywords: Liquid Chromatography, Natural Products, Quantitative
Application Code: Bioanalytical
Methodology Code: Liquid Chromatography

Session # 620 Abstract # 620-5 **Symposia**

Session Title: Trials and Tribulations of Dietary Supplement Analysis: Authentication, Adulteration and Contaminant

Abstract Title: **Spectroscopic Detection of Adulteration in Botanical Dietary Supplements: What is Licorice?**

Date: Monday, March 07, 2016 - Afternoo
Time: 04:10 PM
Room: B311

Primary Author: Guido F. Pauli
UIC College of Pharmacy

Co-Author(s): Charlotte Simmler, Shao-Nong Chen

Abstract Text

Botanical Dietary Supplements (BDSs) are products derived from dried raw plant materials. Due to the complexity of plants as chemical factories and the known impact of their growth conditions and (microbiological) environment, BDSs can exhibit considerable chemical variation, even when they are produced from authentic specimens of the desired source plant. Distinction of this "natural" variation of authentic from truly adulterated material is an additional level of challenge in botanical analysis.

As spectroscopic methods are probing the chemistry of samples, any spectroscopic adulteration assay relies on a combination of three main factors: (a) the availability of authentic botanical standards, which provide the reference metabolomes; (b) reproducible sampling of the plant metabolome; and (c) a spectroscopic analysis with adequate sensitivity and specificity, capable of detecting a characteristic sub-portion of the metabolome ("markers) that enable the distinction of authentic vs adulterated material.

Licorice is one of the most widely used BDSs worldwide. However, the definition of its authenticity is complicated by the concurrent use of at least three Glycyrrhiza species. Recent research in our UIC/NIH Botanical Center has shown that these species are non-equivalent, neither chemically nor biologically. Using 50+ DNA authenticated accessions, chemometrics (PCA and SIMCA) were performed on 1H NMR spectra of standardized extracts, and compared with CDA results performed with AUC of UHPLC-UV profiles. The presentation explains the (a/b/c) workflow and parameters of the NMR spectroscopic authentication concept. Demonstrating the practical feasibility of spectroscopic botanical authentication that targets bioactive markers without the need for reference standards, provides new insights towards the definition of what licorice represents.

Keywords: Bioanalytical, Chemometrics, Natural Products, NMR
Application Code: Other
Methodology Code: Magnetic Resonance

Session # 630 Abstract # 630-1 **Organized Contributed Sessions**

Session Title: Cell Phone Spectroscopy - Handheld Spectroscopy for the Citizen

Abstract Title: **Ready When You Are: Cell Phone Spectrometry When Cellcams Yield RAW Data**

Date: Monday, March 07, 2016 - Afternoo
Time: 01:30 PM
Room: B402

Primary Author: Alexander Scheeline
SpectroClick

Co-Author(s)

Abstract Text

As soon as cameras were built into cellular telephones and the phones became programmable, it was clear to many that portable spectrometry could be carried out on the phones. So why is it, many years after the introduction of smart phones, that cell phone spectrometry (CPS) is not yet common? There are a number of issues. First is that the cell cams are designed to compress raw data to save memory. While the distortion in making JPEG files is acceptable for selfies and landscape photos, it is deadly for doing photometry. Second is the white balancing and autoexposure controls that allow selfies in sunlight and selfies at night under street lights to both look good to people. Spectrometry requires that exposure be under the control of the instrument, and that intensity be reported without bias. While iPhones have had exposure controls for some time and Android has exposure control hooks, getting raw data is far from easy. There are additional challenges with wavelength calibration and, potentially, compensation for lens distortion. Finally, most cell phone owners are not trained spectroscopists, so having a working instrument is only part of the battle. The chemistry and sample handling must also be non-scientist friendly. We report progress on overcoming the various limitations of cell phone cameras and non-technical users, leading towards, but not yet delivering, a "totally awesome" and acceptably accurate CPS analytical spectrometry experience.

Keywords: Instrumentation, Portable Instruments, Spectrometer, UV-VIS Absorbance/Luminescence
Application Code: Consumer Products
Methodology Code: UV/VIS

Session # 630	Abstract # 630-2	Organized Contributed Sessions

Session Title Cell Phone Spectroscopy - Handheld Spectroscopy for the Citizen

Abstract Title **Biomedical Applications of Cellphone Spectroscopy**

Primary Author Anshuman Das
MIT Media Lab

Date: Monday, March 07, 2016 - Afternoo
Time: 01:50 PM
Room: B402

Co-Author(s)

Abstract Text

Spectrometers have evolved from being large table top devices to miniature palm-top devices. This has been possible due to the advent of advanced integration techniques like nano-imprint lithography and MEMS fabrication. Consequently, the size of the constituent components of the spectrometer like slits, gratings, lenses and electronics has diminished to give rise to ultra-compact spectrometers. Since the technologies involved are inherently large scale, the cost of the devices has rapidly come down as well. As a result, many applications that were not possible to implement due to the size and cost of spectrometers, can now be realized. Furthermore, when combined with cellphone and wireless technologies, compact spectrometers become truly portable, standalone, hand-held devices and can deliver powerful data from a wide range of applications. E.g. these devices can sense food quality; they can be used as probes for biomedical point-of-care screening or even as educational tools. This surge in the development of compact spectrometers is driving its transformation from a research tool to a consumer device. This paper presents some of the recent advances in cellphone spectroscopy and its applications in biomedical imaging.

Keywords: Biomedical, Fluorescence, Spectrometer, Spectroscopy
Application Code: Biomedical
Methodology Code: Portable Instruments

Session # 630 **Abstract # 630-3** **Organized Contributed Sessions**

Session Title: Cell Phone Spectroscopy - Handheld Spectroscopy for the Citizen

Abstract Title: **Optical Smartphone Biosensing Techniques**

Date: Monday, March 07, 2016 - Afternoo
Time: 02:10 PM
Room: B402

Primary Author: Brian T. Cunningham
University of Illinois at Urbana-Champaign

Co-Author(s): Kenneth Long

Abstract Text

The field of medical diagnostics is rapidly expanding, and is expected to reach a value of $54.5 billion this year. When coupled with the progression from hospital-based emergency care to small clinic-based preventative care, the importance of providing accessible point-of-care testing (POCT) technologies is rapidly increasing. In addition to being of significant utility in situations when portability and immediate results are key, such as food safety, water quality, animal diagnostics, and global health situations, advances in smartphone technology present a key component of the field's rapid development. Since the introduction of the iPhone, there has been a rapid development of fundamental computational, communication, and sensing capabilities that have transformed the smartphone into a complex, miniature computer, with integrated high-resolution cameras, GPS and GIS capabilities, and cloud integration. We have demonstrated the spectrometric capabilities of a smartphone-based system for absorption, fluorescence, and photonic-crystal-based label-free detection with sensitivity and LODs either comparable or better than traditional laboratory equipment can provide. A 2nd-generation device was designed with custom components to portablize and maximize sensitivity of the instrument, while allowing for future multi-modal absorptive, fluorescent, and label-free biosensing on a unified platform. A 10x increase in utilization of the non-spectral dimension of the CMOS chip, and an 8% increase in pixel resolution were demonstrated, providing the foundation for a portable optical biosensing instrument, with future potential for the integration of similarly portablized fluid-handling hardware. By miniaturizing the detection instrument for a broadly used immunologic test, this work seeks to meet a crucial, unmet need for the adoption of POCT as a ubiquitous tool in consumer testing, food safety, and both rural and global health arenas.

Keywords: Biomedical, Portable Instruments, Spectroscopy

Application Code: Biomedical

Methodology Code: Portable Instruments

Session # 630 **Abstract # 630-5** **Organized Contributed Sessions**

Session Title: Cell Phone Spectroscopy - Handheld Spectroscopy for the Citizen

Abstract Title: **Quantum Dots and Smartphone Fluorescence Imaging: A Perfect Marriage for Bioassays**

Date: Monday, March 07, 2016 - Afternoo
Time: 03:05 PM
Room: B402

Primary Author: Russ Algar
University of British Columbia

Co-Author(s): Eleonora Petryayeva

Abstract Text

Smartphones offer many exciting possibilities as platforms for bioassays and point-of-care diagnostics. These mass-produced consumer electronic devices are already ubiquitous throughout much of the globe, and have advantageous features such as portability, connectivity, data handling capabilities, and built-in cameras that are suitable for spectroscopy and imaging. In turn, the unique optical properties of semiconductor quantum dots (QDs) have attracted considerable interest for assay and biosensor development. Properties such as excellent brightness, spectrally broad absorption spectra, and spectrally narrow and tunable emission spectra make QDs an excellent partner for smartphones in the context of fluorescence measurements and imaging. This presentation will describe our research toward smartphone fluorescence assays with QDs, including ratiometric F[small o with diaresis]rster resonance energy transfer (FRET)-based assays of proteolytic enzyme activity on paper test strips, the adaptation of these test strips to assays with serum and whole blood, and triply multiplexed homogeneous assays of proteolytic activity in bulk solution. In addition, recent results demonstrating fully integrated smartphone fluorescence assays with the built-in LED flash and camera for excitation and emission measurement will be discussed, along with ongoing efforts to transition to chip-based assays. Overall, the properties of QDs and the features of smartphones are a near-perfect match for one another, and suggest a bright future for fluorescence-based point-of-care diagnostics.

Keywords: Bioanalytical, Enzyme Assays, Fluorescence, Imaging

Application Code: Bioanalytical

Methodology Code: Fluorescence/Luminescence

Session # **630** **Abstract #** **630-6** **Organized Contributed Sessions**

Session Title: Cell Phone Spectroscopy - Handheld Spectroscopy for the Citizen

Abstract Title: **Point-of-Care Colorimetric Detection with a Cell Phone**

Date: Monday, March 07, 2016 - Afternoo

Time: 03:25 PM

Room: B402

Primary Author: Ian Papautsky
University of Cincinnati

Co-Author(s)

Abstract Text

Paper-based immunoassays are becoming powerful and low-cost diagnostic tools, especially in resource-limited settings. Inexpensive methods for quantifying these assays have been shown using desktop scanners, which lack portability, and cameras, which suffer from the ever changing ambient light conditions. In this talk we will discuss a novel approach of quantifying colors in colorimetric diagnostic assays with a cell phone, which allows high accuracy measurements in a wide range of ambient conditions, making it a truly portable system. Instead of directly using the red, green, and blue intensities of the color images, we use chromaticity values to construct calibration curves of analyte concentrations. With this approach, we demonstrate a number of paper-based measurements, from simple pH strips to paper-based devices for detection of human performance biomarkers.

Keywords: Bioanalytical, Biomedical, Environmental/Biological Samples

Application Code: Bioanalytical

Methodology Code: Microfluidics/Lab-on-a-Chip

Session # **630** **Abstract #** **630-7** **Organized Contributed Sessions**

Session Title Cell Phone Spectroscopy - Handheld Spectroscopy for the Citizen

Abstract Title **Mobile Technologies for Personalized Diagnostics and Global Health**

Date: Monday, March 07, 2016 - Afternoo
Time: 03:45 PM
Room: B402

Primary Author Dakota O'Dell
Cornell University

Co-Author(s) David Erickson

Abstract Text
Smartphones and other mobile technologies will be transformative to the deployment of biomolecular diagnostics both domestically and worldwide. In this talk, I will review the existing commercial and technical roadblocks to the deployment molecular diagnostics to the consumer market and how they can be fundamentally altered by taking advantage of the now ubiquitous installed base of smartphones. I will discuss two technologies in this talk. The first is our NutriPhone technology which is designed to detect micronutrient and vitamin deficiencies both in both individuals and populations. The second is our KS-Detect system which is a solar-powered PCR system currently targeted towards the diagnosis of Kaposi's sarcoma in sub-Saharan Africa. In addition to covering the basic engineering science advancements that led to the development of these technologies, I will also discuss our strategies for deployment and commercialization.

Keywords: Bioanalytical, Biomedical, Biosensors, Lab-on-a-Chip/Microfluidics
Application Code: Biomedical
Methodology Code: Microfluidics/Lab-on-a-Chip

Session # 630 **Abstract #** 630-8 **Organized Contributed Sessions**

Session Title: Cell Phone Spectroscopy - Handheld Spectroscopy for the Citizen

Abstract Title: **A Handheld Optoelectronic Nose**

Date: Monday, March 07, 2016 - Afternoo
Time: 04:05 PM
Room: B402

Primary Author: Kenneth S. Suslick
University of Illinois at Urbana-Champaign

Co-Author(s): Jon R. Askim, Maria K. LaGasse, Zheng Li

Abstract Text

There is an increasingly urgent need for rapid, portable, and inexpensive detection of a wide range of volatiles, ranging from toxic industrial chemicals (TICs) to homemade explosives (HMEs) to environmental pollutants. We have developed the colorimetric sensor array as a general "optoelectronic nose" for the detection and identification of a wide range of VOCs, for both single analytes and complex mixtures. We will report on a robust and low-cost handheld device for analysis of colorimetric sensor arrays. The device makes use of a contact image sensor (CIS), technology typically used in business card scanners, to collect colorimetric data. The lack of moving parts and one dimensional imaging of a linear colorimetric sensor array allows for lower noise and improved scan rates compared to other digital imaging techniques (e.g., cell phones, CMOS cameras, flatbed scanners); signal-to-noise ratios are a factor of 3-10 higher than currently used methods and scan rates are up to 50 Hz. As an example of its use, we have examined the forensic identification of the source of production of HMEs, which poses a difficult analytical challenge, especially for in□field evaluations. The ease of synthesis and difficulty of detection of peroxides (most notably triacetone triperoxide, TATP, and hexamethylene triperoxide diamine, HMTD) make them explosives of ideal choice for terrorists. In this work, we report the use of a handheld reader and a simple colorimetric sensor array, with a field□ready sampling protocol, for the forensic identification of peroxide HMEs based on their source or manufacturing details. HCA, PCA and SVM analysis show excellent discrimination among twelve peroxide explosives produced by a range of synthetic methods. This method may prove to be a useful supplement to other available detecting technologies used in security checks and forensic evaluation of improvised explosives.

Keywords: Chemometrics, Forensics, Integrated Sensor Systems, UV-VIS Absorbance/Luminescence
Application Code: Homeland Security/Forensics
Methodology Code: Sensors

Session # 640 Abstract # 640-1 **Organized Contributed Sessions**

Session Title: Isolation and Characterization of Impurities/Degradation Product: Understanding Your Impurity Profile

Abstract Title: **Risk Analysis of Impurities in Drug Substances**

Date: Monday, March 07, 2016 - Afternoo
Time: 01:30 PM
Room: B313

Primary Author: Dan Weissmueller
Regis Technologies

Co-Author(s)

Abstract Text

Quality and purity of drug substances are essential in pharmaceuticals. Impurities in pharmaceutical products do not offer any therapeutic benefit for the patient and sometimes they are potentially toxic. By the identification and quantification of the impurities and related substances, the risk of side effect can be avoided or minimized. Control of impurities in drug substance and drug product is described in ICH Q3A, Q3B and Q3C. This presentation will address different case studies involving impurities in drug substances from a Quality Assurance prospective.

Keywords: Characterization, HPLC, Quality
Application Code: Quality/QA/QC
Methodology Code: Liquid Chromatography/Mass Spectrometry

Session # **640** Abstract # **640-2** **Organized Contributed Sessions**

Session Title Isolation and Characterization of Impurities/Degradation Product: Understanding Your Impurity Profile

Abstract Title **Low level Impurity Isolations for Impurity Profiling and Structure Elucidation**

Date: Monday, March 07, 2016 - Afternoo
Time: 01:50 PM
Room: B313

Primary Author Tony (Qi) Yan
Pfizer

Co-Author(s)

Abstract Text

Low level impurity isolation from the drug substance or an excipient in the drug product has been proven as a challenge task. A frequent goal is to isolate a quantity of each impurity for impurity profiling and structure elucidation. In this presentation, we will provide a general overview for impurity isolation strategy including various purification technologies, column chromatography, and analytical method development for optimizing the resolution as well as the discussions on impurity structure work flow. Three case studies were discussed in this presentation. The first study discusses a Pfizer research compound contains two impurities at the level of ~0.05-0.1%. The impurities are generated by first enriched from a chemical reaction, and followed by the chromatographic resolution. The second study used a late stage Pfizer compound. Several of impurities associated with this compound are resolved using SFC, normal phase and reverse phase HPLC. The third study involved with a drug product (capsule). Three desired impurities are isolated from ~30 impurities presented in this drug product. The structures are elucidated and filed for regulatory submissions.

Keywords: Chromatography, Isolation/Purification
Application Code: Pharmaceutical
Methodology Code: Liquid Chromatography

Session # 640 Abstract # 640-3 **Organized Contributed Sessions**

Session Title Isolation and Characterization of Impurities/Degradation Product: Understanding Your Impurity Profile

Abstract Title **Case Studies Involving Method Development for Trace-Level Impurities**

Date: Monday, March 07, 2016 - Afternoo
Time: 02:10 PM
Room: B313

Primary Author Igor Likhotvorik
Regis Technologies

Co-Author(s) Paul Wrezel

Abstract Text
In the manufacturing of pharmaceutical drug substances, it may be necessary to demonstrate that impurities are below an appropriate safety threshold. The threshold is defined by the International Conference of Harmonization (ICH) and determined by the target dose and route of administration. Robust methods are needed to perform impurities testing on the final drug substance as well as starting material, intermediates, and critical control points in a synthesis. This presentation will review case studies involving method development of trace-level impurities in drug substances as well as starting materials and intermediate. Analytical techniques, such as chemical derivatization, HPLC, and MS, which are used in developing methods for the control and monitoring of impurities and degradation products will be discussed.

Keywords: Characterization, Chromatography, HPLC, Pharmaceutical
Application Code: Pharmaceutical
Methodology Code: Liquid Chromatography/Mass Spectrometry

Session # 640   Abstract # 640-4   Organized Contributed Sessions

Session Title: Isolation and Characterization of Impurities/Degradation Product: Understanding Your Impurity Profile

Abstract Title: **Impurity Identification for API Process Development**

Primary Author: Zhao Yanqun
AbbVie, Inc.

Co-Author(s): Wayne Pritts

Date: Monday, March 07, 2016 - Afternoo
Time: 02:30 PM
Room: B313

Abstract Text

LC/MS has become a widely used technique in pharmaceutical development. In this presentation, applications of LC/MS for impurity identification, impurity profiling and problem solving for the API process development and understanding will be discussed.

Keywords: Liquid Chromatography/Mass Spectroscopy, Pharmaceutical, Process Analytical Chemistry
Application Code: Process Analytical Chemistry
Methodology Code: Liquid Chromatography/Mass Spectrometry

Session # 640 **Abstract # 640-5** **Organized Contributed Sessions**

Session Title: Isolation and Characterization of Impurities/Degradation Product: Understanding Your Impurity Profile

Abstract Title: **Emerging Techniques for the Identification of Impurities and Degradation Products in Agricultural Research**

Date: Monday, March 07, 2016 - Afternoo
Time: 03:05 PM
Room: B313

Primary Author: Jeffrie A. Godbey
Dow AgroSciences

Co-Author(s): Jeffrey R. Gilbert, Jesse L. Balcer, Mary D. Evenson, Yelena A. Adelfinskaya

Abstract Text

The successful development of agrochemical products presents ever increasing challenges requiring continual evaluation and implementation of new analytical technologies. The applications which will be shared in this presentation include metabolite identification from complex environmental matrices, chromatographic challenges in the analysis of chiral compounds, and lipids using SFC-MS.

We will describe several new approaches using high resolution LC/MS on Sciex 5600 QqTOF and Thermo Fusion QqITOT instruments, as well as SFC/MS on the Agilent 6530 Q-TOF. We will also present the application of ion mobility separations into the metabolite ID process.

Finally, we will describe the use of advanced software tools to separate, detect, and identify the unique isotopic fine structure of isotopically-labeled metabolites.

METHODS:

LC/MS

Samples were analyzed on the: Sciex 5600 (QqTOF), Thermo Fusion (QqITOT), and Sciex 5600+ (QqTOF) with SelexION (DMS). Accurate MS and MS/MS data were acquired and data processed using software packages; Compound Discoverer (Thermo), PeakView (Sciex), and MsXelerator (MsMetrix).

SFC/MS

Several columns were evaluated including, Silica, C18, chiral, and SBAQ with a variety of modifiers to determine optimal SFC conditions for sample type. Sets were analyzed by both positive and negative ESI as appropriate.

PRELIMINARY DATA

When isotopically-labeled application solutions are blended to produce a unique isotopic pattern, modern metabolite identification programs can readily detect related metabolites. Sciex's PeakView provides a powerful isotope filtering tool & interactive MS, MS/MS, and structure/fragmentation windows for structure elucidation. Thermo's new Compound Discoverer offers a node-based approach for compound detection, and isotope filtering.

SFC can provide a valuable and unique chromatographic tool for a wide range of samples including separation of several classes of lipids, and providing novel separation of stereoisomers.

Keywords: Chiral Separations, Liquid Chromatography/Mass Spectroscopy, SFC, Software
Application Code: Agriculture
Methodology Code: Mass Spectrometry

Session # 640 Abstract # 640-6 **Organized Contributed Sessions**

Session Title: Isolation and Characterization of Impurities/Degradation Product: Understanding Your Impurity Profile

Abstract Title: **The Development of SFC Stationary Phases for the Optimized Separation of Chemical Mixtures Containing a Wide Range of Polarities**

Date: Monday, March 07, 2016 - Afternoo
Time: 03:25 PM
Room: B313

Primary Author: Matthew Przybyciel
ES Industries

Co-Author(s)

Abstract Text

Reversed-phase HPLC is widely used for separation and analytical analysis of many pharmaceutical compounds. Unfortunately, there are mixtures that are not well separated by HPLC leading to incomplete analytical analysis. An alternative separation technique maybe required such as SFC (supercritical fluid chromatography) to effect a complete separation of many mixtures. In addition, SFC can be utilized as an orthogonal separation technique to HPLC for many separations. SFC provides many unique features including producing high pressure/high speed separations "green" separations. These features suit SFC well to the separation of chemical mixtures containing of wide range of polarities. We will provide examples that show the benefits of stationary phases designed for SFC applications. We will demonstrate how these SFC columns can provide for the high resolution separations over a wide variety conditions.

Keywords: Pharmaceutical, SFC, Supercritical Fluid Chromatography
Application Code: Pharmaceutical
Methodology Code: Supercritical Fluid Chromatography

Session # **640** **Abstract #** **640-7** **Organized Contributed Sessions**

Session Title: Isolation and Characterization of Impurities/Degradation Product: Understanding Your Impurity Profile

Abstract Title: **Isolation and Characterization of Impurities in the Synthesis of Drug Substance to Support Drug Development**

Date: Monday, March 07, 2016 - Afternoo
Time: 03:45 PM
Room: B313

Primary Author: Qifeng Xue
Theravance Biopharma US, Inc.

Co-Author(s): Kanaka Hettiarachchi, Ken Ngim, Zhengtian (Titan) Gu

Abstract Text

Drug substance-related impurities are a major category of potential impurities in Active Pharmaceutical Ingredients (APIs) and regulatory authorities emphasize the importance of characterization and control of such impurities. Identification often provides a better understanding of how such impurities are generated and improves quality control of the manufacturing process. It is also important for evaluating whether genotoxic impurities are present in the API and for developing a strategy to eliminate such impurities as early as possible. This presentation focuses on Theravance Biopharma's strategy of using HPLC / SFC in impurity isolation for early development compounds and late stage drug candidates. A roadmap for identification, structure elucidation and quantification of key manufacturing impurities with LC-MS, MS/MS and NMR is also presented.

Keywords: HPLC, Liquid Chromatography, Mass Spectrometry, Pharmaceutical

Application Code: Pharmaceutical

Methodology Code: Liquid Chromatography/Mass Spectrometry

Session # 650 Abstract # 650-1 **Oral Sessions**

Session Title: Biomedical: Advances in Detection and Therapeutics of Cancer

Abstract Title: **KS-Detect: A Portable, Solar-Thermal, Polymerase Chain Reaction System for the Point-of-Care Diagnosis of Kaposi's Sarcoma**

Date: Monday, March 07, 2016 - Afternoo
Time: 01:30 PM
Room: B302

Primary Author: Ryan Snodgrass
Cornell University

Co-Author(s): Andrea Gardner, David Erickson, Ethel Cesarman, Li Jiang

Abstract Text

Kaposi's Sarcoma (KS) is one of the most common AIDS-related cancers. Incidence of KS is highest in sub Saharan Africa, where almost nine of ten cases result in death within five years of diagnosis. Early identification of the disease has been shown to greatly improve survival, but is challenging in resource limited settings because of reliance on benchtop laboratory equipment. We have developed a portable diagnostic device (termed KS-Detect) that combines microfluidics and solar-thermal heating to perform the polymerase chain reaction (PCR) for the identification of KSHV (Kaposi's Sarcoma herpesvirus) in skin biopsies. The portable device operates independently of traditional energy sources, using focused sunlight and a continuous-flow microchannel to achieve a thermal profile appropriate for PCR. To test the device, imitation biopsies were made by clotting together different ratios of cultured KSHV-positive and KSHV-negative human cells. DNA was extracted from our imitation biopsies using HotSHOT, a simple two-step lysis procedure. We report successful KSHV identification with samples containing 100% KSHV-positive cells down to 1% KSHV-positive cells: a physiologically relevant range, as confirmed by histological analysis on KS-positive human biopsies. The device is operated through use of a smartphone or tablet, making it accessible to a wide audience and convenient for point-of-care diagnosis. Our custom smartphone application allows for temperature monitoring, microfluidic control, and analysis of PCR results through fluorescent image measurements. With its portability, smartphone integration, and independence from traditional energy sources, KS-Detect allows for disease diagnosis at the point-of-care in limited resource settings.

This work was supported by a grant from the National Institutes of Health (NIH).

Keywords: Biomedical, Lab-on-a-Chip/Microfluidics, Nucleic Acids, Portable Instruments
Application Code: Biomedical
Methodology Code: Portable Instruments

Session # 650 Abstract # 650-2 **Oral Sessions**

Session Title Biomedical: Advances in Detection and Therapeutics of Cancer

Abstract Title **Nano-Plasmonic Exosome Platform (nPLEX) for Label-Free Detection and Molecular Profiling of Exosomes**

Date: Monday, March 07, 2016 - Afternoo
Time: 01:50 PM
Room: B302

Primary Author Hyungsoon Im
Massachusetts General Hospital

Co-Author(s) Cesar M. Castro, Hakho Lee, Huilin Shao, Park Yongil, Ralph Weissleder, Vanessa Peterson

Abstract Text

Exosomes present new opportunities for cancer diagnoses and treatment monitoring. These cell-derived membrane-bound vesicles (50–200 nm in diameter) are abundant in biological fluids (>10[sup]9[/sup] vesicles per mL of blood and ascites) and carry cell-specific cargos (lipids, proteins and genetic materials), which can be harnessed as a minimally invasive means to probe the molecular status of tumors. Despite such clinical potential, routine exosome analysis is still a challenging task especially due to their small sizes that require large sample volumes and extensive processing. Here, we describe a label-free, high-throughput approach for quantitative analyses of exosomes directly from clinical samples. We specifically developed a nano-plasmonic exosome (nPLEX) sensor, which comprised multiple arrays of periodic nanoholes patterned in a gold film. The sensor surface was functionalized with antibodies to capture target-specific exosomes. Binding of exosomes induced spectral shifts in the resonance wavelength; the magnitude of shifts is proportional to the abundance of cancer antigen, thereby enabling quantitative molecular profiling. Compared to conventional analytical methods (e.g., Western blot and ELISA), the nPLEX demonstrated >100-fold improved sensitivity and enabled fast (< 60 min), portable operations. We used the nPLEX assay to screen exosomes across various cancer cell lines and showed showed excellent correlation between the protein profiles of exosomes and their corresponding cell lines. We applied the nPLEX to detect ovarian cancer exosomes in patients' ascites. For 30 tested ascites (20 from ovarian cancer patients and 10 from noncancerous cirrhosis controls), the detection accuracy was 97%. We also showed that the nPLEX assay could differentiate treatment responses to cancer therapies. With its capability of high-throughput, label-free exosome detection, the nPLEX would facilitate comprehensive exosomal analyses for clinical trial testing.

Keywords: Biomedical, Biosensors, Nanotechnology, Spectroscopy
Application Code: Biomedical
Methodology Code: Sensors

Session # **650** **Abstract #** **650-3** **Oral Sessions**

Session Title Biomedical: Advances in Detection and Therapeutics of Cancer

Abstract Title **Generate DNA Aptamers Against Glypican 3 with Expanded Genetic Systems**

Date: Monday, March 07, 2016 - Afternoo
Time: 02:10 PM
Room: B302

Primary Author Liqin Zhang
University of Florida

Co-Author(s)

Abstract Text

It was proposed 25 years ago that analogous Darwinism might be applied in vitro with nucleic acid libraries to deliver aptamers, which are nucleic acid BMODs selected to bind to a specific target. Termed as "SELEX" (Systematic Evolution of Ligands by EXponential enrichment), this method is now commonly used to generate aptamers targeting ranging from small molecules, biomacromolecule to various cancer cells. However, as SELEX matured, it became clear that the power and binding diversity of nucleic acid aptamers built from standard nucleotides could not match those powers and diversities in antibodies. Artificial Expanded Genetic Information Systems (AEGIS) are DNA-like molecules whose nucleotides display different hydrogen bonding patterns. AEGIS pairs are Watson-Crick complementary, allowing six-nucleotide DNA to be supported by a full molecular biology. Here, we are aiming to select a panel of aptamers targeting a potential liver cancer biomarker--Glypican 3 (GPC3). GPC3 is a membrane protein that is found to be over-expressed on primary liver cancer cells, which is related to several important signaling pathways. GPC3 has become an attractive therapeutic and detection target in liver cancer treatment. In this experiment, the GPC3 was overexpressed on a GPC3 negative cell line, 1MEA cells. GPC3 positive and GPC3 negative 1MEA cells were employed as a pair in the laboratory in vitro evolution process to deliver AEGIS DNA aptamers. Several aptamers have been selected against GPC3 with high affinity, and the binding ability was confirmed on living liver cancer cells as well as recombinant human GPC3. We will also demonstrate that these aptamers can be used for tumor imaging and serum biomarker detection. All these experiments show that GPC3 could be used as a useful target and our AEGIS DNA aptamers could have great potential in liver cancer theranostics.

Keywords: Nucleic Acids, Protein
Application Code: Biomedical
Methodology Code: Fluorescence/Luminescence

Session # 650 Abstract # 650-4 **Oral Sessions**

Session Title: Biomedical: Advances in Detection and Therapeutics of Cancer

Abstract Title: **Systematic and Quantitative Analysis of Surface N-Sialoglycoproteins in Cancer Cells with Distinct Invasiveness**

Primary Author: Ronghu Wu
Georgia Institute of Technology

Date: Monday, March 07, 2016 - Afternoo
Time: 02:30 PM
Room: B302

Co-Author(s)

Abstract Text

Glycoproteins on the cell surface are ubiquitous and essential for cells to communicate with other cells and interact with environment. Sialic acid is very unique among monosaccharides because it is very hydrophilic and carries one negative charge under physiological conditions. Sialoglycoproteins on the cell surface can dramatically impact cell properties and represent different cellular statuses. However, the global analysis of sialoglycoproteins only on the cell surface is extraordinarily challenging. An effective method integrating bio-orthogonal sugar analog labeling and click chemistry was developed to selectively enrich surface sialoglycoproteins for mass spectrometric analysis. Systematic and quantitative analysis of the surface N-sialoglycoproteome in cancer cells with distinct invasiveness was performed, and the experimental results demonstrated that there were dramatic differences for surface N-sialoglycoproteins among breast cancer cells with minimally or highly invasiveness. Many more N-sialoglycosylation sites were identified in highly invasive cancer cells and a number of N-sialoglycoproteins were up-regulated in invasive cells, including quite a few CDs. Among up-regulated N-sialoglycoproteins, the majority of them contain cell adhesion-related domains, which suggests that the sialoglycosylation of these proteins plays a critically important role in regulating cell-cell interactions.

Keywords: Bioanalytical, Biomedical, Mass Spectrometry, Protein
Application Code: Biomedical
Methodology Code: Mass Spectrometry

Session # 650 **Abstract # 650-6** **Oral Sessions**

Session Title: Biomedical: Advances in Detection and Therapeutics of Cancer

Abstract Title: **A Core/Shell Structure of Reduced Graphene Oxide/ Mesoporous Silica with Oligonucleotide Gates for Cancer Treatment**

Date: Monday, March 07, 2016 - Afternoo
Time: 03:25 PM
Room: B302

Primary Author: Xiao Liu
University of North Dakota

Co-Author(s): Julia Xiaojun Zhao, Xu Wu, Xuefei Zhang, Yuqian Xing

Abstract Text

Due to the limitations and side effect of traditional cancer therapies, Chemo-photothermal therapy are conceived to set up a new nanoparticle designed for drug delivery system. The nanoparticle composes of a core of reduced graphene oxide, a shell of mesoporous silica, and an outmost layer of double strand DNAs used for sealing the pores. The drug was doped in the silica layer. When the nanoparticles are irradiated by a near infrared laser, the reduced graphene oxide will produce heat to cause the dsDNAs unwinding. Therefore, the drug was released through the pores. The core was synthesized by reduction of graphene oxide in aqueous solution. The silica layer was coated on the core with through polymerization of TEOS (Tetraethyl orthosilicate) and APTES((3-Aminopropyl)triethoxysilane). The final step is using a cross-linking agent to combine the double string DNA molecules with the core/shell structure. The average of the mesoporous silica pores is about 3 or 4 nm detected by Autosorb-iQ, the length of the dsDNA is about 5 nm. The drug loading efficiency is about 70% while DNA linking efficiency is up to 60%. Both are detected by a Jobin Yvon Horiba Flourologspectrofluorometer machine.

The Neuroscience COBRE Pilot Grant from NIH

Keywords: Biomedical, Modified Silica, Nanotechnology, Near Infrared
Application Code: Biomedical
Methodology Code: Near Infrared

Session # 650 Abstract # 650-7 **Oral Sessions**

Session Title Biomedical: Advances in Detection and Therapeutics of Cancer

Abstract Title **Simultaneous Photothermo-/Chemotherapy Using Reduced Graphene Oxide Based Nanocomposites**

Date: Monday, March 07, 2016 - Afternoo
Time: 03:45 PM
Room: B302

Primary Author Yuqian Xing
University of North Dakota

Co-Author(s) Julia Xiaojun Zhao, Xiao Liu, Xu Wu

Abstract Text
Currently, simultaneous multiple-therapies are becoming promising approaches for cancer treatments. Herein, based on hyphenating the photothermal property of reduced graphene oxide (RGO) and the thermo-responsive property of the polymer (PNIPAM-AAm), we proposed a bi-functional nanocomposite for effective cancer treatments. The RGO core was first coated with a layer of mesoporous silica, and then the polymer was in situ synthesized on the surface of the silica layer. The polymer here plays two significant roles, providing the space for drug loading and squeezing the drug out under near infrared radiation. Doxorubicin, a clinical antitumor drug, was loaded into the nanocomposite for chemotherapy with a loading content of 18% and an entrapment efficiency of 71%. The nanocomposite shows excellent photo-thermal and thermo-responsive properties and is possible for in vitro and in vivo cancer therapy.

Keywords: Biomedical, Nanotechnology, Near Infrared
Application Code: Biomedical
Methodology Code: Near Infrared

Session # **650** **Abstract #** **650-8** **Oral Sessions**

Session Title Biomedical: Advances in Detection and Therapeutics of Cancer

Abstract Title **Imaging of Cancer Protein-Protein Interactions and Small Molecule Inhibitions by a Surface Plasmon Resonance Microarray**

Primary Author Charuksha Walgama
Oklahoma State University

Date: Monday, March 07, 2016 - Afternoo
Time: 04:05 PM
Room: B302

Co-Author(s) Anuruddha Pathiranage, Bing Zhang, Darrell K. Berlin, Doris M. Benbrook, Junpeng Deng, Mayowa Akinwale, Sadagopan Krishnan, Zainab H. Al Mubarak

Abstract Text

We developed a SPRi (Surface Plasmon Resonance Imaging) based platform model for the identification of small molecule inhibition of cancer protein interactions. In vitro interaction between tumor suppressor p53 and its negative regulator MDM2 was successfully imaged on a 16 spot SPRi multi-array chip with high sensitivity and specificity over mutant controls. We could detect 50 nM of HDM2 oncoprotein via a rapid microfluidic system attached to the SPR imager. Instant and real time percentage reflectivity changes of the array spots due to p53-HDM2 interaction were further utilized to calculate apparent binding kinetics. We utilized simple and stable construction of the p53 transactivation domain peptide on a gold surface via in situ self-assembled thiol linkages to facilitate observation of Nutlin-3a inhibition of the p53-HDM2 interaction, which produced an apparent IC50 value of 90 nM. Additionally, QCM (quartz crystal microbalance) experiments validated our SPRi results.

Keywords: Array Detectors, Bioanalytical, Imaging, Plasma
Application Code: Biomedical
Methodology Code: Surface Analysis/Imaging

Session # **660** **Abstract #** **660-1** **Oral Sessions**

Session Title Environmental Applications of Elemental Analysis

Abstract Title **Analyzing Mercury from Contaminated Mining Sites Using a Direct Mercury Analyzer**

Date: Monday, March 07, 2016 - Afternoo
Time: 01:30 PM
Room: B315

Primary Author
Sumedh Phatak
Milestone Inc.

Co-Author(s)

Abstract Text

With over 220 million pounds of mercury produced between 1851 and 1981 in California, it had been widely used in gold mining operations to enhance gold recovery. The repercussions of this heavy mercury usage are seen even today in the form of contaminated sites which consequently have adverse health effects on human and wildlife alike. There are several remedial measures that can be taken to clean these mercury contaminated sites such as solidification/stabilization, soil washing, thermal treatment etc. but the process first starts with accurately analyzing the mercury concentrations from these sites.

The challenge with conventional analysis tools such as CVAA and ICP is that these techniques need the sample to be digested in acid prior to analysis. Consequently, this increases cost of analysis. Additionally, the concentration of mercury at these sites is usually in the high ppm range, which can cause memory effects in these instruments. Direct mercury analysis with a wide range configuration allows a user to eliminate the sample prep step while generating cost savings with respect to maintenance of digestion equipment, acid usage, waste disposal and labor costs. Another advantage of using a direct mercury analyzer with a wide range configuration is that it gives the users the ability analyze mercury from single digit ppb's to up to 1000 ppm accurately, reproducibly and within 6 minutes per sample.

Keywords: Atomic Spectroscopy, Environmental Analysis, Mercury, Soil
Application Code: Environmental
Methodology Code: Atomic Spectroscopy/Elemental Analysis

Session # **660** **Abstract #** **660-2** **Oral Sessions**

Session Title: Environmental Applications of Elemental Analysis

Abstract Title: **ICP-MS for the Analysis of High Salinity Samples**

Date: Monday, March 07, 2016 - Afternoo
Time: 01:50 PM
Room: B315

Primary Author: Erica Cahoon
PerkinElmer

Co-Author(s): Daniel H. Jones

Abstract Text

Sea water and other high salinity water samples pose many challenges to the ICP-MS user. Sample preparation is time consuming and increases the possibility for errors. Regardless of the challenges, the environmental laboratories must provide high-quality data with a fast turnaround time on a diverse array of samples. Environmental samples require the removal of unknown interferences or applications requiring the best performance with the lowest detection limits. ICP-MS allows for multi-element trace metal analysis, a wide linear dynamic range while still reaching low limits of detection.

Generally, there is a limitation of dissolved content allowed into the ICP-MS. This presentation will cover ICP-MS instrument optimization for the reduction of polyatomic interferences in different high salinity matrices. Strategies to increase the robustness and stability of ICP-MS to handle higher total dissolved solid content and dramatically improve long-term stability for high matrix solutions will be discussed.

Keywords: Atomic Spectroscopy, Environmental Analysis, Environmental/Water, ICP-MS
Application Code: Environmental
Methodology Code: Atomic Spectroscopy/Elemental Analysis

Session # 660 Abstract # 660-3 **Oral Sessions**

Session Title: Environmental Applications of Elemental Analysis

Abstract Title: **Nanoparticle Removal During Alum and Ferric Coagulation Characterized by Single Particle ICP-MS**

Date: Monday, March 07, 2016 - Afternoo
Time: 02:10 PM
Room: B315

Primary Author: Ariel Donovan
Missouri University of Science and Technology

Co-Author(s): Chady Stephan, Craig Adams, Honglan Shi, Todd Eichholz, Yinfa Ma

Abstract Text

It is well-known that the manufacturing and applications of engineered metal-containing nanoparticles (NPs) are increasing each year, resulting in higher concentrations of NPs reaching environmental systems. When introduced to water systems, some particles settle into the sedimentation, but many will remain in suspension potentially resulting in health risks to human, animal, plant, and aquatic life. In this study, alum and ferric coagulation were simulated in river water to evaluate the removal of spiked Au, Ag, TiO[sub]2[/sub], CeO[sub]2[/sub], and ZnO NPs. Charge neutralization (zone 2) and sweep floc (zone 4) coagulation processes were conducted using three different coagulants: Al[sub]2[/sub](SO[sub]4[/sub])[sub]3[/sub], Fe[sub]2[/sub](SO[sub]4[/sub])[sub]3[/sub], and FeCl[sub]3[/sub]. The fate of NPs was monitored by single particle (SP)-ICP-MS using a state-of-the-art NexION 300/350D ICP-MS with Syngistix Nano Application software from PerkinElmer. This instrument has the capability to simultaneously detect NP size, size distribution, particle concentration, and dissolved metal concentration. Detection limits for the monitored NPs range from 18-65 nm diameter and 0.10-0.55 µg/L for different dissolved ions when using surface water as the calibration matrix. Each method was validated in surface water with spike recoveries between 70-130% for dissolved ions and citrate-stabilized Au and Ag NPs. Over 85% of TiO[sub]2[/sub], CeO[sub]2[/sub], and ZnO NPs were removed during zone 2 and zone 4 coagulation with each type of coagulant. Citrate-stabilized Au and Ag NPs exhibited varied removal efficiencies during zone 2 and zone 4 coagulation processes with each type of coagulant, ranging from 48-95%.

This project was supported by PerkinElmer and the Missouri Department of Natural Resources.

Keywords: Elemental Mass Spec, Environmental/Water, ICP-MS, Nanotechnology
Application Code: Environmental
Methodology Code: Atomic Spectroscopy/Elemental Analysis

Session # 660 **Abstract # 660-4** **Oral Sessions**

Session Title Environmental Applications of Elemental Analysis

Abstract Title **Applications of Single Particle ICP-MS to Environmental Matrices**

Date: Monday, March 07, 2016 - Afternoo
Time: 02:30 PM
Room: B315

Primary Author Chady Stephan
PerkinElmer

Co-Author(s) Kenneth Neubauer

Abstract Text

The recent surge in the use of engineered nanomaterials (ENMs) in consumer products has resulted in concerns about their release into the environment. In order to properly assess their impact on ecological and human health, it is necessary not only to predict exposure through modelling, but also to perform quantitative physical and chemical measurements. Historically, metrics like particle size have been carried out using techniques like dynamic light scattering (DLS), nanoparticle tracking analysis (NTA) and transmission electron microscopy (TEM), while dissolved content has been typically measured by ultrafiltration. However, these traditional techniques have known limitations when it comes to quantifying very low levels of ENMs in the presence of natural colloidal species in complex waters, and as a result, are not ideally-suited for real-world environmental matrices where nanoparticle concentrations are typically extremely low. Single particle inductively coupled plasma mass spectrometry (SP-ICP-MS) is an exciting new technique for detecting and characterizing metal nanoparticles (NP) at very low concentrations. It is fast and can provide significantly more information than other traditional techniques, including particle number concentration, particle size, and size distribution, in addition to the concentration of dissolved metals in solution. The added benefit of using ICP-MS is that it can distinguish between particles of different elemental compositions. The study will investigate the use of SP-ICP-MS to track the release of ENMs into the environment and to better understand their fate and behavior specifically in drinking, surface and wastewater samples.

Keywords: Dissolution, Environmental/Water, ICP-MS, Nanotechnology
Application Code: Environmental
Methodology Code: Mass Spectrometry

Session # **660** **Abstract #** **660-5** **Oral Sessions**

Session Title Environmental Applications of Elemental Analysis

Abstract Title **Issues in Deep Ocean LIBS: Internal Calibration and the Effect of Suspended Particles**

Date: Monday, March 07, 2016 - Afternoo
Time: 03:05 PM
Room: B315

Primary Author Joseph Bonvallet
University of South Carolina

Co-Author(s) S Michael Angel

Abstract Text
Laser induced breakdown spectroscopy (LIBS) is being investigated for in situ underwater LIBS measurements of dissolved species at deep-ocean hydrothermal vent sites. In previous work, we demonstrated LIBS at oceanic pressures, using single pulse LIBS, of many elements that are relevant to vent chemistry. In recent studies, the use of in situ internal standards for underwater LIBS measurements has been investigated. It was found that O (I) and H (I) emission lines are useful at increasing calibration dynamic range and reducing measurement variability. In other recent studies, the effect of suspended particles causes the emission intensity, of dissolved elements, to decrease in proportion to the scattering coefficient of the solution, following a simple scattering model. This talk will present the latest studies of high-pressure underwater LIBS of dissolved species as well as future applications of this technique to measurements in supercritical CO2, which is relevant to future Venus lander instruments.

Keywords: Atomic Spectroscopy, Laser, Plasma, Spectroscopy
Application Code: Environmental
Methodology Code: Atomic Spectroscopy/Elemental Analysis

Session # 660 Abstract # 660-6 **Oral Sessions**

Session Title Environmental Applications of Elemental Analysis

Abstract Title **Selective Measurement of Metal Ions at Covalently Functionalized Carbon-Fiber Microelectrodes**

Primary Author Yuanyuan Yang
Wayne State University

Date: Monday, March 07, 2016 - Afternoo
Time: 03:25 PM
Room: B315

Co-Author(s) Ahmad A. Ibrahim, Jennifer L. Stockdill, Parastoo Hashemi

Abstract Text

Real-time detection of trace metal ions such as Cu(II) and Pb(II) is desirable in a number of environmental and biological studies. We previously described fast scan cyclic voltammetry (FSCV) at carbon fiber microelectrodes (CFMs) for real-time detection.1, 2 However, FSCV's application for trace metal detection in complex systems has been limited to low analytical selectivity in presence of other interfering electroactive substances. To more effectively improve the selectivity without compromising time resolution, we developed a novel approach to functionalize the CFMs.3 Diazonium reduction and click chemistry were employed to achieve a covalent grafting of metal-selective ionophores onto CFMs' surfaces. This affords an extremely high level of selectivity towards a certain metal analyte of interest without response delay. As proof of principle, Cu(II) ionophore grafted CFMs were fully characterized with FSCV for sensitivity, selectivity and stability. Our unique functionalization method makes fast metal voltammetry extremely powerful as a device that can report trace metal levels instantaneously and continuously with high sensitivity selectively.

1. P. Pathirathna, Y. Yang, K. Forzley, S. P. McElmurry and P. Hashemi, Anal Chem, 84, 6298-6302.
2. Y. Yang, P. Pathirathna, T. Siriwardhane, S. P. McElmurry and P. Hashemi, Analytical Chemistry, 2013, 85, 7535-7541.
3. Y. Yang, A. A. Ibrahim, J. L. Stockdill and P. Hashemi, Analytical Methods, 2015, DOI: 10.1039/C5AY00501A.

Keywords: Chemically Modified Electrodes, Electrochemistry, Electrodes, Environmental Analysis
Application Code: Environmental
Methodology Code: Electrochemistry

Session # 660 Abstract # 660-8 **Oral Sessions**

Session Title Environmental Applications of Elemental Analysis

Abstract Title **Application of Mössbauer Spectrometry in Environmental Studies of Fly Ashes and Road Dusts**

Date: Monday, March 07, 2016 - Afternoo
Time: 04:05 PM
Room: B315

Primary Author Tadeusz Szumiata
RADWAG Balances and Scales

Co-Author(s) Bogumil Gorka, Katarzyna Brzozka, Malgorzata Gzik-Szumiata, Marzena Trojanowska, Michal Gawronski, Ryszard Swietlik, Tadeusz Szumiata

Abstract Text

Mössbauer spectrometry utilizing a recoilless, resonant absorption of gamma rays by atomic nuclei in solids is a popular experimental technique in physics and materials science. Recently a significant increase of interest of this method has been observed as an effective research tool in chemistry and environmental studies. The most frequently used variant of Mössbauer spectrometry is based on [sup]57[/sup]Fe isotope, thus it is suitable for the qualitative and quantitative analysis of iron-containing phases (both paramagnetic and ferromagnetic ones). It is regarded as a complementary method to the X-ray diffraction, NMR and to the techniques for the elemental analysis. Present work gives an exemplary review of Mössbauer spectrometry application in phase analysis of fly ashes from different coal combustions systems and of the road dusts from express ways. The fly ash samples came from stoker-fired boiler of municipal heating plant and from pulverized coal boiler of power plant. The performed Mössbauer studies showed that proportion of the content of aluminosilicate glass (with Fe[sup]3+[/sup] ions) to hercynite spinel phase (with Fe[sup]2+[/sup] ions) strongly depends not only on the combustion temperature but also on the grain fraction. The higher concentration of glassy phase was observed in samples of ash with finer grains, which were collected in successive sections of multicyclon device or electrostatic precipitator. Moreover, the fly ashes contained hematite and magnetite iron oxides. The investigated road dusts were collected from the acoustic barriers of European E-77 dual carriageway. The elemental analysis pointed to the presence of (rather nontoxic) Fe, Al and Zn metals of the road traffic origin. The occurrence of these elements was accompanied by carcinogenic Pb, Cr and Ni heavy metals. The Mössbauer spectrometry revealed a dominance of paramagnetic fine iron particles and iron carbonates in the dusts.

Keywords: Characterization, Coal, Environmental Analysis, Environmental/Waste/Sludge
Application Code: Environmental
Methodology Code: Atomic Spectroscopy/Elemental Analysis

Session # 670 Abstract # 670-1 **Oral Sessions**

Session Title: Environmental GC

Abstract Title: **Multi-Dimensional Micro Gas Chromatography Device for the Rapid and Sensitive Analysis of Volatile Organic Compounds**

Date: Monday, March 07, 2016 - Afternoo
Time: 01:30 PM
Room: B406

Primary Author: Jiwon Lee
University of Michigan

Co-Author(s): Hongbo Zhu, Katsuo Kurabayashi, Menglian Zhou, Robert Nidetz, Xudong Fan

Abstract Text

We developed multi-channel multi-dimensional micro gas chromatography ([micro]GC) systems for the rapid and sensitive analysis of volatile organic compounds (VOCs) for various applications in healthcare, environmental protection, and homeland security. This [micro]GC device includes a pre-concentrator, thermal injectors, micro-separation columns, non-destructive on column gas detectors, and a flow routing system in each dimension and channel. The micrfabricated pre-concentrator and thermal injectors consists of a deep-reactive-ion-etched silicon chamber, an integrated platinum heater and temperature sensor and micro fluidic channels. The chamber is packed with graphitized carbons that can trap the analytes and thermally inject them within a second. The micro fluidic photoionization detector ([micro]PID) is installed after the microfabricated columns as a non-destructive, and highly sensitive on column gas detector that does not interfere with the flow and analytes. Between the two adjacent separation columns, a flow routing module is used to divert the analytes into different channels in the downstream dimension with an automated algorithm. The unique design of the [micro]GC architecture significantly enhances the GC peak capacity while remaining compact in size, thus enabling automated, rapid, and sensitive detection of VOCs within 20 min at the ppb or sub-ppb level. In this talk, we will describe the details of the multi-dimensional [micro]GC design and its components, and then demonstrate the separation and detection of 50 VOCs within 20 mins using the above GC system.

Keywords: Chromatography, GC
Application Code: Environmental
Methodology Code: Gas Chromatography

Session # 670 Abstract # 670-2 **Oral Sessions**

Session Title Environmental GC

Abstract Title **SPME On-Fiber Derivatization Using a Stable and Reusable Pentafluorophenyl Hydrazine Standard Gas Generating Vial**

Date: Monday, March 07, 2016 - Afternoo
Time: 01:50 PM
Room: B406

Primary Author Justen J. Poole
University of Waterloo

Co-Author(s) Emanuela Gionfriddo, German A. Gomez-Rios, Janusz Pawliszyn, Jonathan J. Grandy

Abstract Text

SPME on-fiber derivatization methods have facilitated the achievement of lower detection limits and targeted analysis of substances with poor chromatographic behavior, thermal stability, or high reactivity. However, previously developed methods to load derivatization reagents onto SPME fibers have been either single use and/or provide poor loading reproducibility over time. In this study, a novel standard gas generating vial containing pentafluorophenyl hydrazine (PFPH) that can be reused for on-fiber derivatization methods is presented. Intra-batch vial reproducibility, along with PFPH headspace stability over a period of 11 weeks was monitored. Four randomly selected, newly formulated vials demonstrated strong intra-batch agreement (n=3, %RSD=2), while no vial sampled throughout the 11 week stability study was observed to show significant deviation in SPME fiber loadings (RSD < 9%). In addition, headspace extractions utilizing PFPH pre-loaded SPME fibers above aqueous solutions of 10-200ppb/v of C4-C9 linear aldehydes demonstrated strong response correlation (R2 > 0.992) for derivatization reaction products. Reproducibility of the derivatization reaction was also monitored over 7 days using an aldehyde containing standard gas generating vial (C4-C9), demonstrating derivatization reaction product reproducibility over seven days (RSD <9%). The proposed standard gas generating vial represents the first documented method for the long term storage of a usable headspace standard for a reactive and otherwise unstable molecule.

Keywords: Derivatization, Gas Chromatography, Gas Chromatography/Mass Spectrometry, SPME
Application Code: Environmental
Methodology Code: Gas Chromatography

Session # 670 Abstract # 670-3 **Oral Sessions**

Session Title: Environmental GC

Abstract Title: **Gas Chromatography-Vacuum Ultraviolet Detection (GC-VUV) for Analysis of Polychlorinated Biphenyls (PCBs)**

Date: Monday, March 07, 2016 - Afternoo
Time: 02:10 PM
Room: B406

Primary Author: Changling Qiu
University of Texas at Arlington

Co-Author(s): Jonathan Smuts, Kevin A. Schug

Abstract Text

Polychlorinated biphenyls (PCBs) are a group of synthetic chlorinated compounds that are widely used as dielectric fluids in capacitors and transformers. Due to their toxicity, persistence, and bioaccumulation in the food chain, PCBs are an environmental concern and among the most analyzed compounds in environmental analysis. The most common analytical methods for analysis of PCBs are based on GC-ECD and GC-MS. However, number of possible congeners (209), similarities of physical and chemical properties, and complexity of sample matrices make it difficult to distinguish and accurately quantify PCB congeners using existing methods. This study presents a new method for analysis of PCBs using gas chromatography with vacuum ultraviolet detection, which offers full scan absorption detection in the range of 115-240 nm, where all chemical species have absorption. The VUV absorption spectra for a majority of the PCB congeners were collected and shown to be differentiable. VUV was verified for both qualitative and quantitative analyses of PCBs with great specificity, high sensitivity, and fast data acquisition. The capability of VUV data analysis software for deconvolution of co-eluting signals was also demonstrated.

Keywords: Detector, Environmental Analysis, Gas Chromatography
Application Code: Environmental
Methodology Code: Gas Chromatography

Session # 670 **Abstract # 670-4** **Oral Sessions**

Session Title: Environmental GC

Abstract Title: **Gas Chromatographic Retention Behavior on Select Groups of Isomeric Polycyclic Aromatic Compounds and Their Alkyl-Substituted Derivatives on Stationary Phases of Different Selectivity**

Date: Monday, March 07, 2016 - Afternoo
Time: 02:30 PM
Room: B406

Primary Author: Walter B. Wilson
National Institute of Standard and Technology

Co-Author(s): Federica Nalin, Lane C. Sander, Leonard M. Sidisky, Stephen A. Wise

Abstract Text

The retention behavior for many key isomeric groups of polycyclic aromatic compounds (PAC) and their alkyl-substituted derivatives were evaluated by using high-resolution capillary gas chromatography (GC) coupled with mass spectrometry using three different stationary phases: a ~50% (mole fraction) phenyl-substituted methylpolysiloxane column, smectic liquid crystalline column and a ionic liquid column. Retention data were obtained for the following groups of PAC: (1) polycyclic aromatic hydrocarbons (PAHs) with molecular mass (MM) 252, 276, 278, 300 and 302 Da; (2) alkylated-PAHs with MM 192, 216, 242 and 266 Da; (3) polycyclic aromatic sulfur heterocycles (PASHs) with MM 184, 234, 258 and 284 Da; (4) alkylated-PASHs with MM 198, 212, 226 and 248 Da. Molecular descriptors (length, breadth, thickness (T) and length-to-breadth (L/B) ratio were calculated for all the compounds studied. Correlations for retention on both stationary phases and PASH geometry (L/B and T) ratios were investigated.

Keywords: Environmental Analysis, Gas Chromatography, GC Columns, Method Development

Application Code: Environmental

Methodology Code: Gas Chromatography

Session # 670 **Abstract # 670-5** **Oral Sessions**

Session Title: Environmental GC

Abstract Title: **Optimization of Phthalate Analysis by Gas Chromatography-Mass Spectrometry Using Computer Modeling Software**

Primary Author: Dan Li
Restek Corporation

Date: Monday, March 07, 2016 - Afternoo
Time: 03:05 PM
Room: B406

Co-Author(s): Amanda Rigdon, Chris English, Jack Cochran, Rebecca Stevens

Abstract Text

Phthalates are esters of 1,2-benzenedicarboxylic acid, the phthalic acid esters. They are commonly used as plasticizers, and ubiquitous in the environment. The presence of phthalates has attracted considerable attention due to their potential impacts on ecosystem and public health. These health concerns have resulted in the regulation of phthalates in products and in the environment. There are sixteen phthalates listed in method 8061A by US Environmental Protection Agency (EPA) and the European regulations specify seven phthalates. The six phthalates that are regulated in both USA and EU are: di-n-butyl phthalate, benzyl butyl phthalate, bis(2-ethylhexyl) phthalate, di-n-octyl phthalate, diisononyl phthalate, and diisodecyl phthalate.

Libraries for 41 phthalates have been generated on 7 different phases. A gas chromatographic computer program was used to determine the best stationary phase and conditions. One column could separate 37 phthalates in less than 40 min, and 17 EPA and EU listed phthalates in 6 min. Non-regulated technical-grade phthalates contain EPA or EU contaminants at relatively high levels, including di-n-butyl phthalate, di-n-hexyl phthalate, bis(2-ethylhexyl) phthalate, dicyclohexyl phthalate, and di-n-octyl phthalate. The concentration of the impurities in non-regulated phthalates was measured and will be presented on the most appropriate stationary phase. The analysis of the EPA chlorinated pesticides using an Electron Capture Detector poses many challenges for the analyst. One issue with using this non-universal detector is interference from compounds that respond on the ECD but are not target analytes of interest. Phthalates are a good example of compounds that are extracted and interfere with quantification of chlorinated pesticides. Using computer program predictions of phthalates as potential interference by EPA Method 8081A will be compared to actual analytical results.

Keywords: Environmental, GC-MS, Method Development, Optimization
Application Code: Environmental
Methodology Code: Gas Chromatography/Mass Spectrometry

Session # **670** **Abstract #** **670-7** **Oral Sessions**

Session Title: Environmental GC

Abstract Title: **Fast, Very Fast and Ultra-Fast Forensic and Homeland Security GC-MS**

Date: Monday, March 07, 2016 - Afternoo
Time: 03:45 PM
Room: B406

Primary Author: Aviv Amirav
Tel Aviv University

Co-Author(s): Alexander Fialkov, Tal Alon, Uri Keshet

Abstract Text

Forensic and homeland security applications can clearly benefit from much faster GC-MS analysis. GC-MS with Cold EI was explored in several forensic applications with emphasis on reduced time of analysis. This unique GC-MS is based on interfacing the GC and MS with supersonic molecular beams (SMB) along with electron ionization of vibrationally cold sample compounds in SMB in a fly-through ion source (hence the name Cold EI). The use of short column with flow programming enables universal drugs of abuse screening with two minutes chromatography time and three minutes full analysis cycle time. We also developed an Open Probe Fast GC-MS for obtaining real time analysis with separation. Open Probe Fast GC-MS can be combined with standard GC-MS or with GC-MS with Cold EI that enables improved sample identification via the availability of enhanced molecular ions and TAMI software which provides elemental formulae. GC-MS with Cold EI enables faster analysis of all types of forensic GC-MS applications including:

A) Full range of organic explosives in several minutes analysis time.

B) Universal very fast GC-MS analysis method for illicit drugs (heroin, cocaine) in under 3 minutes total analysis cycle time, using flow programming.

C) High throughput forensic analyses with Open Probe fast GC-MS for real time analysis (30s) with separation and library identification.

D) Improved fast unknown sample identification with enhanced molecular ions, extended range of compounds amenable for GC-MS analysis and the TAMI software.

E) Fast arson analysis with isomer distribution analysis for fuels and oils source characterization.

Keywords: Forensics, Gas Chromatography/Mass Spectrometry, GC-MS

Application Code: Homeland Security/Forensics

Methodology Code: Gas Chromatography/Mass Spectrometry

Session # 670 Abstract # 670-8 **Oral Sessions**

Session Title: Environmental GC

Abstract Title: **Passive Monitoring – A Guide to Sorbent Tube Sampling for EPA Method 325**

Primary Author: Caroline Widdowson
Markes International

Date: Monday, March 07, 2016 - Afternoo
Time: 04:05 PM
Room: B406

Co-Author(s)

Abstract Text

EPA Method 325 is required to comply with the new US federal regulation (CFR 40), similar regulations are being developed in multiple countries. This method entails the monitoring of volatile organic compounds (VOC) around the boundary of refineries and requires 2-week passive sampling and TD-GC (or GC-MS) analysis. Whilst benzene is the primary target compound, the sampling and analysis methodology and apparatus can also be used to determine other VOCs including other Hazardous Air Pollutants (HAPs) without further method developments. All petroleum refineries seeking compliance with US EPA regulations should adhere to EPA Method 325, together with federal, regional and independent test laboratories.

Diffusive monitoring has been used in a range of air monitoring scenarios, i.e. occupational hygiene, as well as indoor air and ambient air monitoring. This presentation will discuss the application of passive sampling with industry-standard sorbent tubes and factors that need to be taken in to account when deploying them.

Keywords: Environmental Analysis, Environmental/Air, Fuels\Energy\Petrochemical, Volatile Organic Compound
Application Code: Environmental
Methodology Code: Gas Chromatography/Mass Spectrometry

Session # 680 **Abstract # 680-1** **Oral Sessions**

Session Title: Food Product Quality and Component Characterization

Abstract Title: **Analysis of Fat Crystallization Thanks to Microrheology**

Date: Monday, March 07, 2016 - Afternoo
Time: 01:30 PM
Room: B316

Primary Author: Maxime Bazin
Formulaction

Co-Author(s): Gérard Meunier, Giovanni Brambilla, Mathias Fleury, Roland Ramsch

Abstract Text

The crystalline form of fats in chocolate, butter and vegetable oils was studied thanks to microrheology. Passive microrheology studies the mobility and displacement of micron sized particles [1]: we used Multi Speckle Diffusing Wave Spectroscopy (MS-DWS) coupled with a temperature ramp in order to probe the particle displacement to analyze the viscoelastic properties of an opaque product. Under heating or cooling conditions, particle movements can be related to the crystalline form of the fat: the rearrangements occurring during melting or during crystallization provide crucial data about the fat's polymorphic transitions. Crystalline form and melting temperature of fats are important data for the elaboration of new products or for quality control of finished products. In the case of chocolate, the microrheology analysis during melting can identify the crystalline form of finished chocolate products, and so help to predict its stability against blooming. Moreover, microrheology can be used to study the impact of formulation and process on melting temperatures of low-fat butters. In addition to the analyses of crystalline forms of fat, the MS-DWS provides data on viscoelastic property changes.

[1] D. A. Weitz, D. J. Pine, in: Dynamic Light Scattering, W. Brown (Ed.) (Oxford Univ. Press, New York, 1993), Chap. 16

Keywords: Characterization, Food Science, Instrumentation
Application Code: Food Science
Methodology Code: Physical Measurements

Session # **680** **Abstract #** **680-2** **Oral Sessions**

Session Title: Food Product Quality and Component Characterization

Abstract Title: **Atomic Absorption and Potentiometric Analysis of Electrolytes in Food Substances Used for Rapid Muscle Cramp Relief by Athletes**

Date: Monday, March 07, 2016 - Afternoo
Time: 01:50 PM
Room: B316

Primary Author: Stephanie Hooper Marosek
Methodist University

Co-Author(s): Taylor Tipton

Abstract Text

The sudden onset of muscle cramps can be a debilitating and painful issue for many athletes in competition today. Studies suggest that the mechanism involved with relieving such cramps is two-fold: the first is primarily related to dehydration and loss of electrolytes, but the second involves a neuronal reaction triggered by sour acids that reduces alpha motor neuron pool activity in cramping muscles. Consumption of some common food items such as pickle juice and mustard has been utilized by athletes as a rapid remedy for cramps. Cramping ceases within minutes of ingestion of these substances, most likely due to the high potassium and acetic acid content. In this work, several food substances (dill and sweet pickle juice, yellow mustard, and white distilled and apple cider vinegar) were analyzed by potentiometry and atomic absorption spectroscopy for four common electrolytes (potassium, sodium, magnesium, and calcium) that are essential to physiological function. The corresponding acetic acid concentration was also determined through volumetric analysis. Each electrolyte was quantified and evaluated with the associated acetic acid concentration for each particular food substance. Sodium was by far the largest electrolyte present, but does not seem to play a significant role in relieving cramps. Potassium and calcium content varied moderately for each substance, while magnesium was noticeably present in the mustard. Recommended intake values for each analyte were also incorporated in determining that either dill pickle juice or apple cider vinegar would be the most suitable substance for athletes to consume for muscle cramp relief.

Keywords: Atomic Absorption, Bioanalytical, Food Science, Potentiometry

Application Code: Food Science
Methodology Code: Atomic Spectroscopy/Elemental Analysis

Session Title: Food Product Quality and Component Characterization

Abstract Title: **Microrheological Analyses for Dairy Formulations**

Date: Monday, March 07, 2016 - Afternoo
Time: 02:10 PM
Room: B316

Primary Author: Roland Ramsch
Formulaction

Co-Author(s): Gérard Meunier, Giovanni Brambilla

Abstract Text

This work presents studies on yoghurt preparation using microrheology. Passive microrheology studies the mobility and displacement of micron sized particles which results from Brownian motion [1]. The motion of particles induces local deformations of the sample, which are directly related to its viscoelastic properties.

Our technique is based on Multi Speckle Diffusing Wave Spectroscopy (MS-DWS), which consists of Dynamic Light Scattering (DLS) extended to an opaque media. With a patented algorithm, the backscattered interfering light can be analysed in terms of Mean Square Displacement (MSD), which is directly related to the viscoelastic properties of a sample. Moreover, the optical method allows to study especially weak gels without any applied shear, which avoids perturbation of the sample.

Nowadays, yogurt formation has caught much attention, as it is a growing market. The choice of proteins and other components are very important. This work shows how passive microrheology can be used to follow up yoghurt preparation and the influence of proteins type and content on the gelation time, and the viscoelastic properties of the yogurt. Gel time was determined by a new rescaling method, namely Time-Cure Superposition (TCS) [2,3]. This data processing determines the gel point according to the Winter-Chambon criterion [4].

[1] D. A. Weitz, D. J. Pine, in: Dynamic Light Scattering, W. Brown (Ed.) (Oxford Univ. Press, New York, 1993), Chap. 16
[2] T. H. Larsen, E. M. Furst, Phys. Rev. Letters, 2008, 100, 14600
[3] K. M. Schultz, E. M. Furst, Soft Matter, 2012, 8, 6198
[4] H. H. Winter, F. Chambon, J. Rheology 1986, 30, 364-382

Keywords: Beverage, Food Science, Rheology
Application Code: Food Science
Methodology Code: Physical Measurements

Session # 680 Abstract # 680-4 **Oral Sessions**

Session Title Food Product Quality and Component Characterization

Abstract Title **Analysis of Aroma Compounds in Beer by TD–GC–TOF MS with Soft Electron Ionization**

Primary Author Laura McGregor
Markes International Ltd

Date: Monday, March 07, 2016 - Afternoo
Time: 02:30 PM
Room: B316

Co-Author(s) Caroline Widdowson, Chris Hall, Ken Umbarger, Massimo Santoro

Abstract Text

Beer contains hundreds of organic ingredients, with concentrations spanning many orders of magnitude. Mono- and sesquiterpenes (C10 and C15 respectively) are aromatic hydrocarbons found in the essential oils of various plants, and (most notably for the brewing industry) in hops. Hops provide much of the characteristic flavouring of the finished beer, so the terpene content has a major impact on the final aroma and flavour. These compounds have very low odour thresholds, making them challenging to detect analytically. The ability to apply quality control to the raw ingredients and the finished product offers desirable cost and time saving to breweries. A rugged, quantitative technique has been developed for the determination of these flavour compounds in beer, and is based on sorptive extraction coupled with thermal desorption–gas chromatography–time-of-flight mass spectrometry (TD–GC–TOF MS). Sorptive extraction is an enrichment technique designed for the extraction of non-polar, semi-volatile constituents from aqueous samples. Like solid-phase micro-extraction (SPME), the approach is simple, quick and economical, but sorptive extraction media are physically much larger and can therefore allow higher levels of enrichment for amenable compounds. Coupling this with highly sensitive TOF MS detection and novel soft ionization technology ensures that a comprehensive flavour profile can be collected in a single sequence.

Keywords: Beverage, Semi-Volatiles, Thermal Desorption, Time of Flight MS
Application Code: Food Science
Methodology Code: Gas Chromatography/Mass Spectrometry

Session # 680 Abstract # 680-5 **Oral Sessions**

Session Title: Food Product Quality and Component Characterization

Abstract Title: **Food Analysis Using Laser Desorption GS-Ion Mobility Spectrometry – Olive Oil as an Example**

Primary Author: Wolfgang Vautz
ISAS

Co-Author(s): Joachim Franzke, Sascha Liedtke

Date: Monday, March 07, 2016 - Afternoo
Time: 03:05 PM
Room: B316

Abstract Text

Conventional methods like multi-dimensional GC- or LC-MS for food analysis such as e.g. the classification of olive oils are time consuming and expensive. Therefore, innovative approaches to characterise not only olive oils but liquid samples in general are still required. Headspace analysis of liquid samples using ion mobility spectrometry with fast GC pre-separation delivers results within a few minutes. However, besides the time consuming heating of the sample to ensure the equilibrium in the headspace which is required during sample preparation – this method provides only little information about low volatile compounds.

In order to increase the information about characteristic volatiles, including low volatile compounds as well, we investigated if the formation of a gas-phase vapour by laser desorption of a liquid sample. It was expected that this provides advantages compared to traditional headspace analyses, in particular for GC-IMS analyses of complex samples such as olive oils. For this purpose, a specially designed laser desorption chamber was implemented in the sampling line of a GC-IMS for the analysis of the generated vapours.

It could be demonstrated that sampling is fast and reproducible in comparison to headspace analysis. Furthermore, different olive oils showed distinct differences in the substance patterns detected, in particular with additional signals of low volatile substances which could not be detected by common headspace analysis. Therefore, laser desorption coupled to GC-ion mobility spectrometry (LD-GC-IMS) is a suitable method for the characterisation of complex liquid samples like olive oils.

Keywords: Food Identification, Food Safety, Gas Chromatography, Laser Desorption
Application Code: Food Science
Methodology Code: Chemical Methods

Session # 680 Abstract # 680-6 **Oral Sessions**

Session Title Food Product Quality and Component Characterization

Abstract Title **Monitoring of Protein Changes in Pasteurized Liquid Egg Using Capillary Electrophoresis**

Date: Monday, March 07, 2016 - Afternoo
Time: 03:25 PM
Room: B316

Primary Author Reyhan S. Uysal
Hacettepe University

Co-Author(s) Esra Acar, Ismail H. Boyaci, Nusret Ertas

Abstract Text

Egg is a nutritious food and consumed all over the world. Liquid egg products are valuable for food industry due to their ease of use, high protein content, and low cost. However, liquid eggs have a favorable environment for the development of pathogens, due to their rich nutritive value. Heat pasteurization is considered the best solution for controlling these pathogens in liquid egg products. Functional properties of egg, which make it a crucial ingredient in food industry, are related to protein quality. The protein quality of eggs is severely affected when heated, due to protein denaturation. The aim of this study is to reveal changes in protein structure due to heat treatments. In the first step, egg samples to be analyzed were prepared. Heat treatment was applied at different pasteurization parameters (time and temperature). Pasteurized samples were diluted and loaded onto the capillary column. The analysis was performed using uncoated fused-silica capillaries of 50□m i.d. Electropherogram images of untreated and heat treated eggs were obtained in 20 minutes. Differences in protein bands were observed between untreated and treated samples depending on different heat effects. These differences were obtained considering changes in both band area and height. It was found to be more changes in the pattern of protein bands by increasing heat treatment. Despite the complexity of the matrices, high reproducibility and interference-free electropherograms were obtained. The results indicate that changing of protein structure in liquid egg-induced different heat treatment parameters can be monitored and revealed by capillary zone electrophoresis.

Keywords: Electrophoresis, Food Science, Monitoring, Protein
Application Code: Food Science
Methodology Code: Capillary Electrophoresis

Session # 680 Abstract # 680-7 **Oral Sessions**

Session Title: Food Product Quality and Component Characterization

Abstract Title: **Exploiting Polymeric Ionic Liquids-Based SPME Sorbents Coupled to Gas-Chromatography/Mass Spectrometry for Food Quality and Safety Assessment**

Primary Author: Erica A. Souza-Silva
Universidade Federal do Rio Grande do Sul

Date: Monday, March 07, 2016 - Afternoo
Time: 03:45 PM
Room: B316

Co-Author(s): Emanuela Gionfriddo, German A. Gomez-Rios, Janusz Pawliszyn, Jared L. Anderson, Nathaly Reyes-Garces

Abstract Text

Solid phase microextraction process is governed by the shape, size and chemistry of the extracting phase that also plays an essential role in the selectivity of the extraction. Thus, efforts are currently focused in the development of new coatings amenable to different sampling objectives and with exclusive characteristics. In recent years, ionic liquids (ILs) have gained increasing interest in the analytical chemistry community because their unique physical and chemical properties. Due to their tunable physical and chemical properties, polymeric ionic liquids PILs have been recently used as extracting phases for SPME with promising potentialities compared to commercial SPME coatings.
The present work describes the evaluation of two different PILs coatings, poly([ViBHDIM][NTf2]) and Poly([DDMGlu][MTFSI]), and the comparison of their extraction performances with commercially available SPME coatings (PA and PDMS). The study can be divided in two parts: (a) the determination of organosphophorous pesticides in grapes, which accounts for the implementation of PILs coatings in safety evaluation of foodstuff, and (b) extraction of selected analytes representing different chemical classes of metabolites commonly found in fruit, which accounts for the implementation of PILs in metabolomics studies of foodstuff. The two PIL-based fibers exhibited satisfactory performances in terms of linearity and reproducibility compared to commercial SPME fibers. In spite of the important differences in film thicknesses among the fibers, the extraction performances could be considered satisfactory using the PIL coatings. The poly([VBHDIM][NTf2]) coating showed good selectivity towards aromatic compounds and its extraction efficiency is comparable to the performance of PA and PDMS coatings.

Keywords: Contamination, Gas Chromatography/Mass Spectrometry
Application Code: Food Contaminants
Methodology Code: Sampling and Sample Preparation

Session # 680 Abstract # 680-8 **Oral Sessions**

Session Title Food Product Quality and Component Characterization

Abstract Title **Detection, Identification, and Pattern Recognition of Microbial Volatile Organic Compounds from Virulent and Hypo-Virulent Cryphonectria Parasitica Species by Headspace-SPME-GC-MS and Chemometrics**

Primary Author Jinyan She
Mississippi State University

Date: Monday, March 07, 2016 - Afternoo
Time: 04:05 PM
Room: B316

Co-Author(s)

Abstract Text

American chestnut (Castanea dentata) played an important ecological role and was virtually eliminated by 1940s due to the chestnut blight fungus (Cryphonectria parasitica). Natural control of the blight involves the usage of debilitated virus-infected strains of hypo-virulent (HV) isolates to convert the lethal virulent (V) forms into HV's. Traditional identification methods of HV strains are labor intensive and time consuming. Studies showed that Microbial volatile organic compounds (MVOCs) are intermediate or end products of organism metabolic pathways, and can be used as bio-markers of a disease.Herein, we report the determination of MVOCs patterns by Headspace Solid Phase Microextraction (HS-SPME) with an 85 µm Carboxen/Polydimethylsiloxane (CAR/PDMS) fiber coupled with Gas Chromatography-Mass Spectrometry (GC-MS). From this study, the VOC patterns of seven virulent C.parasitica isolates and three hypo-virulent C.parasitica isolates were identified. Facts that can alter the MVOCs profile such as temperature and growth media are also studied. Moreover, the MVOCs profiles associated with fungi life circle were also detected. Finally, data pattern recognition was achieved by applying chemometrics.

Keywords: Chemometrics, Gas Chromatography/Mass Spectrometry, SPME, Statistical Data Analysis
Application Code: Food Science
Methodology Code: Gas Chromatography/Mass Spectrometry

Session # 690 Abstract # 690-1 **Oral Sessions**

Session Title: LC Method Development - Half Session

Abstract Title: **Method Transfer and Column Scalability with Superficially Porous Particles**

Date: Monday, March 07, 2016 - Afternoo
Time: 01:30 PM
Room: B403

Primary Author: William Long
Agilent Technologies

Co-Author(s): Anne Mack, Jason Link, Stephen Luke

Abstract Text

Superficially porous particles offer improved efficiency and performance over similarly sized traditional totally porous particles. This is primarily due to a shorter mass transfer distance and substantially narrower particle size distribution of the particles in the column. Higher efficiency leads to improved resolution and possible time savings with superficially porous particles, hence their growing popularity for HPLC analyses. Columns using superficially porous particles are currently available in a several particle sizes, pore sizes and stationary phase chemistries to meet most analysts' needs. This work will address method transfer and scalability from totally porous particles, as well as among different varieties of superficially porous particles and when to use which particle and column configurations. Overcoming common barriers of method transfer with high efficiency columns, including instrument configuration, method settings, and proper connections will also be addressed.

Example applications are transferred from traditional totally porous particle HPLC columns to more modern superficially porous particle columns. Methods are geometrically scaled, as column volumes are adjusted, with the intent to preserve chromatographic integrity. As methods are adjusted to higher efficiency columns with smaller column volumes, instrument hardware and software is also appropriately adjusted. Some example applications include compendial USP methods, which are transferred to superficially porous particle HPLC columns from their prescribed totally porous particle columns. Results are compared against the permissible adjustments found in the recently revised USP General Chapter 621.

Keywords: Clinical Chemistry, HPLC, Method Development
Application Code: Pharmaceutical
Methodology Code: Liquid Chromatography

Session # 690 Abstract # 690-2 **Oral Sessions**

Session Title: LC Method Development - Half Session

Abstract Title: **Isolation of Impurities in Biopharmaceutical Formulations**

Date: Monday, March 07, 2016 - Afternoo
Time: 01:50 PM
Room: B403

Primary Author: Thomas E. Wheat
Waters Corporation

Co-Author(s): Amanda B. Dlugasch, Patricia R. McConville

Abstract Text

The rapid development and acceptance of protein biopharmaceuticals has created a need for improved analytical techniques. In contrast to small molecules, proteins are best analyzed with a variety of tools including separation using several different chromatographic techniques. Each mode of chromatography is selected to analyze a particular class of chemical or physical modification of the protein. The advent of UPLC separation chemistries suitable for proteins has brought improvements in resolution and sensitivity to these assays. The ability to recognize heterogeneity in a protein sample still requires that the multiple molecular forms be characterized with respect to both structure and their impact of the biological properties of the formulation. Such assessments often require that the impurity observed chromatographically be available in a pure form for further analysis. The simplest approach to isolation is the collection of the peak as it elutes from the column. The use of UPLC techniques, however, adds some constraints to this approach. The desired peaks are difficult to collect because they are narrow in both time and volume. Collection must be very exact to ensure good yield of pure material. The collector must account for any transfer tubing and valve volumes that represent a significant fraction of peak volume. The peaks of interest may be well-resolved from an analytical perspective, but the times and volumes separating the several analytes are also small. The collection must provide for the elimination of carryover and cross-contamination. We have developed a small volume fraction collector for use with UPLC. We will show the function and optimal use of this device for the isolation of impurities from reversed-phase, ion exchange, and size exclusion separations of biopharmaceutical protein formulations.

Keywords: HPLC, Liquid Chromatography, Prep Chromatography, Protein
Application Code: Pharmaceutical
Methodology Code: Liquid Chromatography

Session # **690** **Abstract #** **690-3** **Oral Sessions**

Session Title LC Method Development - Half Session

Abstract Title **New Technological Solutions to Maximizing Uptime on an Ion Chromatography System**

Date: Monday, March 07, 2016 - Afternoo
Time: 02:10 PM
Room: B403

Primary Author David G. Moore
Thermo Fisher Scientific

Co-Author(s) Pranathi Perati, Sally Eastman

Abstract Text

The financial performance of laboratories utilizing ion chromatography systems is strongly tied to the percentage of time that the system is operational. Events such as consumables replacement, preventive maintenance, and instrument failure can remove the system from productive use for durations ranging from minutes to days. This presentation describes the use of modern technological and educational solutions to reduce the duration of such events, allow planning of predictable events to coincide with the optimal timing for the laboratory, and minimize the occurrence of unexpected events. Additionally, various tools to estimate the savings associated with maximizing uptime are discussed and used within a case study of the impact on a routine analysis, high-throughput water analysis laboratory.

Keywords: Ion Chromatography, Lab Management, Laboratory Automation, Software
Application Code: Laboratory Management
Methodology Code: Liquid Chromatography

Session # 690 **Abstract #** 690-4 **Oral Sessions**

Session Title LC Method Development - Half Session

Abstract Title **Accurate Measurement of Analyte Dispersion through Connecting Tubes used in Fast Very High-Pressure Liquid Chromatography**

Date: Monday, March 07, 2016 - Afternoo
Time: 02:30 PM
Room: B403

Primary Author Fabrice Gritti
Waters Corporation

Co-Author(s) Martin Gilar, Thomas McDonald

Abstract Text

Analyte dispersion through short connecting tubes used in fast very high-pressure liquid chromatography (vHPLC) limits the performance of modern 2.1 mm i.d. columns packed with sub-2 [micro]m particles. Its accurate measurement is required so that the loss of column efficiency and the intrinsic performance of the column can be assessed.

The complete theory of sample dispersion through tubes has been established by Taylor [1] and Aris [2] in the mid 1950s. It is only valid in the spatial domain and it cannot predict the observed temporal dispersion data departing from the classical linear Taylor-Aris behavior [3]. The chromatographer measure only temporal dispersion data which are often inaccurate and irreproducible due to imperfections (large detection and injection volumes, pressure surge) of classical vHPLC systems.

In this presentation, we address unresolved issues regarding temporal dispersion through tubes used in vHPLC: how can we measure it accurately at very high linear velocity? Beyond which velocity the observed temporal dispersion data start deviating from the spatial dispersion data predicted by Taylor-Aris theory? How does the observed dispersion data change with increasing further the eluent velocity?

The answers to these questions are based on the fabrication of a virtually dispersion-less instrument enabling the accurate measurement of time-based dispersion data through 75 and 100 [micro]m x 30.5 cm tubes. Dispersion through the injection device was reduced to less than 0.03 [micro]L[sup]2[/sup] at 0.5 mL/min and detection was performed in situ across the inner diameter of the capillary using UV-Vis light guided along a fiber optic. Errors made from the classical methods (theories and vHPLC set-up) are discussed.

[1] G. Taylor, Proc. R. Soc. Lond. A 225 (1954) 473-477.
[2] R. Aris, Proc. R. Soc. Lond.A 235 (1956) 67-77.
[3] A. Alizadeh, C. Nieto de Castro, W. Wakeham, Int. J. Therm. 1 (1980) 243-284.

Keywords: Capillary LC, Drug Discovery, HPLC, HPLC Detection
Application Code: General Interest
Methodology Code: Liquid Chromatography

Session # 700 Abstract # 700-1 **Oral Sessions**

Session Title LC Optimization - Half Session

Abstract Title **Dynamic Temperature Control in Capillary Liquid Chromatography: Increasing Analysis Sensitivity, Speed, and Peak Capacity**

Primary Author Stephen R. Groskreutz
University of Pittsburgh

Co-Author(s) Rachael Wilson, Stephen Weber

Date: Monday, March 07, 2016 - Afternoo
Time: 01:30 PM
Room: B404

Abstract Text

In this work we present on a series of technical improvements to the instrumentation used to dynamically control column temperature in capillary liquid chromatography. The low thermal mass and high thermal conductivity of fused silica capillary columns (75-150 □m ID) makes the application of rapid temperature gradients, >750 °C/min, a chromatographically useful tool for improving analysis performance. Column temperature can be regulated precisely in space and in time within a chosen temperature range of about 70 °C with a minimum of about -10 and a maximum of about 100 °C by using an array of independently controlled 1.0 cm x 1.0 cm Peltier devices. This design offers significant system flexibility. For example, the system can be operated to enhance sensitivity by cooling the first Peltier to ca. 0 °C while samples are loaded onto the column. Cooling the first segment during loading enhances sample preconcentration at the head of the column and significantly reduces valve-induced precolumn dispersion. In this application the remaining Peltiers are initially maintained at a higher temperature, 40 °C. Following loading, the first Peltier is rapidly heated to release the focused band into the segment heated by the downstream Peltiers. These Peltiers can be held at a constant temperature or subject the column to temperature gradients in space or in time. We report on improvements in peak height, peak shape, analysis time and peak capacity offered by this technology for a series of p-hydroxybenzoates (parabens), monoamine neurotransmitters and neuropeptides.

Keywords: Capillary LC, Chromatography, HPLC, Liquid Chromatography
Application Code: Bioanalytical
Methodology Code: Liquid Chromatography

Session # **700** Abstract # **700-2** **Oral Sessions**

Session Title LC Optimization - Half Session

Abstract Title **Towards the Development of a Refractive Index-Based Optical Microcavity Mass Concentration Detector Compatible with Gradient Elution Liquid Chromatography**

Primary Author Zachary S. Wiersma
University of Illinois at Urbana-Champaign

Date: Monday, March 07, 2016 - Afternoo
Time: 01:50 PM
Room: B404

Co-Author(s) Alexandria L. Stanton, David M. Meunier, James H. Wade, Ryan C. Bailey, Todd O. Pangburn

Abstract Text

There are many liquid chromatography applications that require quantitative mass concentration detection for molecules lacking chromophoric or fluorogenic functional groups. Numerous industrially important polymers (e.g., polyolefins) fall into this category and cannot be detected through conventional UV-VIS methods. Moreover, commercial differential refractive index (RI) detectors are limited by small RI dynamic range and long temperature equilibration times, making them essentially incompatible with solvent gradient liquid chromatography.

Microring resonators are a class of RI detectors that in recent years have been largely applied to the detection of chemical and biochemical analytes localized through specific binding interactions. However, they are also sensitive to bulk RI changes, and therefore have applicability to liquid chromatography detection. Specifically, the compact design, enormous RI dynamic range, and real-time temperature fluctuation compensation capability makes this a promising approach for HPLC detection. Considering the great industrial need for a detector capable of quantitatively characterizing polymers lacking chromophoric or fluorogenic groups, we have focused our attention on gel permeation chromatography (GPC) separations.

This research highlights recent efforts in the application of silicon photonic microring resonators for the characterization of synthetic polymers in both isocratic and gradient elution separations. Specifically, this emerging technology is directly and quantitatively compared with commercial RI detectors through an ASTM protocol. Proof of concept studies on industrial polymer separations, including separations of defined polymer standards in THF, are also presented in the context of quantitative analysis of GPC.

Keywords: Characterization, HPLC Detection, Liquid Chromatography, Polymers & Plastics

Application Code: Polymers and Plastics

Methodology Code: Liquid Chromatography

Session # 700 **Abstract # 700-3** **Oral Sessions**

Session Title: LC Optimization - Half Session

Abstract Title: **Increasing the Peak Capacity of Quantitative Liquid Chromatography Using Chemometric Curve Resolution**

Date: Monday, March 07, 2016 - Afternoo
Time: 02:10 PM
Room: B404

Primary Author: Daniel W. Cook
Virginia Commonwealth University

Co-Author(s): Dwight R. Stoll, Sarah C. Rutan

Abstract Text

Quantitation is a crucial step in chromatographic analyses. This requires that analyte peaks are sufficiently resolved to allow for integration, calibration, and concentration prediction for unknown samples. Often the optimized separation of moderately complex samples entails increased analysis times and complicated methods (i.e., multistep gradients, etc.). Curve resolution techniques are able to mathematically resolve peaks, facilitating quantification. One such technique is multivariate curve resolution alternating least squares (MCR-ALS) which separates signals based on differences between the analyte and background spectra. Previous works have suggested that MCR-ALS and other curve resolution techniques are powerful tools for quantifying compounds in crowded chromatograms; however, many of these works simulate data under ideal conditions, where background signals are absent, which can significantly affect the evaluation of these methods. The current work demonstrates the quantitative performance of MCR-ALS by performing Monte Carlo simulations varying backgrounds derived from experimental data. The performance of MCR-ALS is explored under varied conditions by changing conditions such as spectral similarity, chromatographic resolution, and peak intensity (signal-to-background and signal-to-signal). These results show that MCR-ALS is indeed a powerful tool for the chromatographer under a wide range of conditions. Experimental results from a quantitative analysis of furanocoumarins in apiaceous vegetables show that with MCR-ALS, quantitation with good precision (< 2 %) is possible even at low chromatographic resolution (< 0.5). Finally, these strategies are extended to comprehensive two-dimensional liquid chromatography.

Keywords: Chemometrics, Food Science, Liquid Chromatography, Quantitative
Application Code: General Interest
Methodology Code: Liquid Chromatography

Session # 700 **Abstract # 700-4** **Oral Sessions**

Session Title: LC Optimization - Half Session

Abstract Title: **Improving Sensitivity, Resolution, and Peak Capacity of Gradient Elution Capillary Liquid Chromatography with Temperature-Assisted On-Column Solute Focusing**

Primary Author: Rachael Wilson
University of Pittsburgh

Date: Monday, March 07, 2016 - Afternoo
Time: 02:30 PM
Room: B404

Co-Author(s): Stephen Weber, Stephen R. Groskreutz

Abstract Text

Capillary HPLC offers increased sensitivity for dilute samples typical of complex biological mixtures. However, small tubing and column volumes make capillary systems especially vulnerable to volume overload and extracolumn broadening processes. TASF uses sub-ambient temperatures to minimize broadening during injection, concentrating solutes into narrow bands at the head of the column. This allows the use of large volume injections by decreasing peak width and increasing peak height compared to isothermal gradients alone. We have demonstrated this computationally and experimentally with small molecules as well as peptides. Using a software-controlled Peltier thermoelectric device (TEC), solutes were focused at -7.5 °C during the injection time plus an additional five seconds then the temperature was rapidly elevated to the separation temperature of 65 °C. This resulted in up to four times greater peak heights and five times smaller peak widths for early eluting solutes in a mixture of acetanilide, parabens, and alkyl phenones with an injection volume of 146% of the column volume. Injection of bovine serum albumin (BSA) tryptic digest at over 290% of the column volume also resulted in improved peak shape and resolution compared to gradient elution alone. These results demonstrate the effectiveness of TASF with gradient elution and the ability to improve sensitivity, resolution, and peak capacity of volume overloaded samples containing a range of functional groups.

Keywords: Bioanalytical, Capillary LC, HPLC, Peptides
Application Code: Bioanalytical
Methodology Code: Liquid Chromatography

Session # 710 Abstract # 710-2 **Oral Sessions**

Session Title Neurochemistry

Abstract Title **Oxygen Changes and Dopamine Release during Spreading Depression**

Date: Monday, March 07, 2016 - Afternoo
Time: 01:50 PM
Room: B401

Primary Author Caddy N. Hobbs
University of North Carolina at Chapel Hill

Co-Author(s) Justin A. Johnson, R Mark Wightman

Abstract Text

Spreading depression (SD) is a pathophysiological event that is known to increase the area of damaged brain tissue in cases of traumatic brain injury such as ischemia, stroke, and epilepsy. As waves of SD propagate through grey matter a massive depolarization of brain cells occurs, followed by a decrease of neuronal activity. Ions passively distribute along their gradients across the cell membrane and excitatory neurotransmitters are released. Tissue responds to spreading depression by increasing blood flow which delivers oxygen and glucose so that the compromised region has the energy it requires to reestablish ion gradients and return to normal function. The deleterious effects of SD and need for treatment methods have inspired great interest in the hemodynamic and metabolic changes that occur during a wave of SD. Typical recordings of SD include large rises in extracellular potassium by ion selective microelectrodes; decreased glucose and increased neurotransmitter concentrations monitored with microdialysis; increased blood flow recorded with laser Doppler flowmetry; and widespread brain cell depolarization and decreased activity with electrocorticography. Rapid measurement of neurotransmitters has historically been a struggle since it requires off-line analysis. The first studies showing on-line analysis of neurotransmitter release during SD in the nucleus accumbens were carried out in the Ralph Adams group using chronoamperometry, which lacks optimal compound specificity. Also troublesome are direct measurements of oxygen, requiring Clark electrodes which perturb and damage tissue upon implantation. We present an alternative, minimally invasive method that uses a single carbon fiber microelectrode to simultaneously detect rapid and selective oxygen and neurotransmitter events during SD with fast-scan cyclic voltammetry (FSCV) coupled with single unit recording electrophysiology.

Keywords: Electrochemistry, Neurochemistry, Voltammetry

Application Code: Neurochemistry

Methodology Code: Electrochemistry

Session # 710 **Abstract # 710-3** **Oral Sessions**

Session Title: Neurochemistry

Abstract Title: **Spontaneous, Transient Adenosine Release from Brain During Ischemia-Reperfusion Injury**

Date: Monday, March 07, 2016 - Afternoo
Time: 02:10 PM
Room: B401

Primary Author: Mallikarjunarao Ganesana
University of Virginia

Co-Author(s): B Jill Venton

Abstract Text

Adenosine is an important neuroprotective modulator that regulates blood flow and neurotransmission. Previous studies have shown that tissue adenosine levels increases during ischemic events and attenuates the excitotoxic neuronal injury. Also adenosine is known to decrease the neurotransmitter release and neuronal firing during ischemia. Animal studies have shown protective effect of adenosine through various drugs during ischemia. Having these profound tissue protective effects in situations of ischemia, measuring adenosine concentrations in real-time would provide us a crucial information during the progression of the disease. Adenosine has traditionally been studied using slower resolution techniques, and proven to act on a longer time scale ranging from minutes to hours. Recently, our lab developed an electrochemical fast-scan cyclic voltammetry (FSCV) method using carbon-fiber microelectrodes to directly measure adenosine changes on a sub-second time scale. In this study, we discuss the spontaneous, transient adenosine release in vivo, in the caudate-putamen of anesthetized rats during the progression of ischemia-reperfusion (I-R) injury in a rat model of stroke for the first time. We will present the spontaneous, transient adenosine release during normoxia, ischemia and reperfusion. Our results demonstrates a significant difference in the frequency of transient release and inter event time during ischemia and reperfusion compared to normoxia. These findings provide us an initial understanding on the time course release of transient adenosine during I-R periods. Extending these studies by using pharmacological drugs to manipulate adenosine levels could help us to understand the specific role of adenosine and its release in the tissue protection and can be applied in the clinical arena of stroke.

Keywords: Electrochemistry, Neurochemistry, Voltammetry
Application Code: Neurochemistry
Methodology Code: Electrochemistry

Session # 710 **Abstract # 710-4** **Oral Sessions**

Session Title: Neurochemistry

Abstract Title: **Treatment on Carbon Nanotube Yarn Microelectrode for Sensitive and Rapid Dopamine Detection In Vivo**

Primary Author: Cheng Yang
University of Virginia

Date: Monday, March 07, 2016 - Afternoo
Time: 02:30 PM
Room: B401

Co-Author(s): B Jill Venton, Christopher B. Jacobs, Ilia N. Ivanov, Michael D. Nguyen

Abstract Text

Carbon nanotube yarn microelectrodes (CNTYMEs) exhibit rapid and selective detection of dopamine with fast-scan cyclic voltammetry (FSCV). However, the sensitivity limits their application in vivo. In this study, we introduce Laser treatment and oxygen plasma etching as a simple, reliable and efficient approach to improve the sensitivity of CNTYMEs by about three fold while maintaining high temporal resolution. The effect of Laser treatment on microelectrode surface was characterized by scanning electron microscopy, Raman spectroscopy, energy dispersion spectroscopy, and Laser confocal microscopy. Laser treatment increases surface area and oxygen containing functional groups on the surface which provides more adsorption sites to dopamine than unmodified CNTYMEs. Moreover, similar to unmodified CNTYMEs, dopamine signal at Laser treated CNTYMEs barely dropped with increasing scan frequency compared to a significant decrease at carbon fiber microelectrodes (CFMEs). This is presumably caused by the significantly larger surface roughness which would trap dopamine-o-quinone within each scan and amplify dopamine signal without influencing temporal resolution. In addition, CNTYMEs were applied as an in vivo sensor with FSCV for the first time and Laser treated CNTYMEs exhibited sensitive detection of stimulated dopamine in anesthetized, male, Sprague-Dawley rats with the scan frequency of 50 Hz, which is five-fold faster than the frequency conventionally applied at CFME. CNTYMEs with Laser treatment which are advantageous of easy fabrication, high reproducibility, fast electron transfer kinetics, and rapid in vivo measurement of dopamine with high sensitivity would be expected to be a potential alternative of CFMEs in the future.

Keywords: Bioanalytical, Microelectrode, Neurochemistry, Sensors
Application Code: Neurochemistry
Methodology Code: Electrochemistry

Session # 710 Abstract # 710-5 **Oral Sessions**

Session Title Neurochemistry

Abstract Title **Neurochemical Investigation of Epilepsy Using Microdialysis Sampling to Study Multiple Seizure Rat Models**

Date: Monday, March 07, 2016 - Afternoo
Time: 03:05 PM
Room: B401

Primary Author Amanda M. Furness
University of Kansas

Co-Author(s)

Abstract Text

Patients are diagnosed with epilepsy after experiencing two or more non-provoked seizures; therefore, animal models in which the animal experiences multiple seizures within one experiment more accurately represent human epileptic patients compared to ¬in vitro or in slicio studies. In these studies, a microdialysis probe was used to locally deliver an epileptic agent into the hippocampus of an anesthetized rat while simultaneously collecting dialysate for analysis. Samples were analyzed for the excitatory and inhibitory amino acids, glutamate and GABA as well as catecholamine neurotransmitters norepinephrine and dopamine. Increases in glutamate and decreases in GABA indicate that a seizure has occurred, and provide information on its strength and duration. Norepinephrine and dopamine are of interest because they are involved in reducing oxidative stress and are believed to have antiepileptic effects. Samples were analyzed using liquid chromatography with fluorescence detection (amino acids) or electrochemical detection (catecholamines). Several dosing time regimens were explored to study seizure strength and duration in response to the epileptic agent. It was determined that there was approximately a twofold decrease in glutamate response to a secondary epileptic episode compared to the first episode. This could be indicative of desensitization of the neurons to the epileptic agent as a protective response or, alternatively, due to changes in neuronal plasticity caused by oxidative stress. To determine the cause of the attenuation of glutamate concentration during the second epileptic event, awake animal studies were performed. By moving to awake animal experiments, it was possible to add third and fourth seizures to the epilepsy model to further evaluate the neuronal response to the epileptic agent. This also allowed the time between seizures to be extended, ruling out glutamate depletion as a possible cause of the attenuated response.

Keywords: Bioanalytical, Liquid Chromatography, Neurochemistry, Sampling
Application Code: Neurochemistry
Methodology Code: Liquid Chromatography

Session # 710 Abstract # 710-6 **Oral Sessions**

Session Title Neurochemistry

Abstract Title **PKC-b Inhibitors Attenuate Amphetamine and Cocaine Stimulated Dopamine Release**

Date: Monday, March 07, 2016 - Afternoo
Time: 03:25 PM
Room: B401

Primary Author Alexandros G. Zestos
University of Michigan

Co-Author(s) Margaret E. Gnegy, Robert T. Kennedy

Abstract Text

We have developed a novel technique to measure basal changes of dopamine and its analytes in response to pharmacological agents. Activation of protein kinase C-b enhances extracellular dopamine in the presence of amphetamine by enhancing the reverse transport of dopamine and internalizing the D2 autoreceptor.In this study, we utilized in vivo microdialysis with LCMS for detection in live, behaving rats to assess the effect of the PKCb inhibitors enzastaurin and ruboxistaurin on amphetamine-stimulated increases in monoamines (and over 25 neurochemicals) and their metabolites. Previous techniques can only measure one analyte at a time using HPLC or voltammetry. A 30 min perfusion of the nucleus accumbens core with 1 uM enzastaurin or 1 uM ruboxistaurin reduced amphetamine-stimulated efflux of dopamine and its metabolite 3-methoxytyramine by approximately 50%. The inhibitors also significantly reduced extracellular levels of norepinephrine and its metabolite normetanephrine after amphetamine. The stimulation of locomotor behavior by amphetamine, measured simultaneously with the analytes, was comparably reduced by the PKCb inhibitors. These inhibitors are specific for dopamine and their metabolites. Ruboxistaurin also attenuated cocaine stimulated extracellular dopamine, a process that would not be dependent upon DAT reversal. In order to see if this process was D2 autoreceptor mediated, we examined the effect of ruboxistaurin on cocaine activation when D2 receptors were blocked with raclopride. The inhibitory effect of ruboxistaurin was reduced in the presence of cocaine and raclopride, suggesting that ruboxistaurin action involved D2 autoreceptors. Using a stable isotope label retrodialysis procedure, we determined that ruboxistaurin had no effect on basal levels of dopamine, norepinephrine, glutamate, or GABA. Our results support the utility of using PKCb inhibitors to reduce the effects of amphetamine and cocaine.

Keywords: Bioanalytical, HPLC, Liquid Chromatography, Mass Spectrometry
Application Code: Neurochemistry
Methodology Code: Liquid Chromatography/Mass Spectrometry

Session # **710** **Abstract #** **710-7** **Oral Sessions**

Session Title Neurochemistry

Abstract Title **Clinical Measurements at the Bedside: Dynamic Neurochemical Changes in the Injured Human Brain Monitored Using an Online Microdialysis System**

Date: Monday, March 07, 2016 - Afternoo
Time: 03:45 PM
Room: B401

Primary Author Michelle L. Rogers
Imperial College London

Co-Author(s) Anthony J. Strong, Chi Leng Leong, Martyn G. Boutelle, Sally A. Gowers, Sharon Jewell, Shumaila Khan

Abstract Text

It has been reported that the tissue surrounding the ischaemic core of a traumatic brain injury, in some patients, suffers further insults resulting from events known as Spreading Depolarisations (SDs) [1]. This results in rapid metabolic changes leading to a crisis of energy delivery to the tissue. It has been reported that patients with a higher occurrence of SDs have a poorer health outcome [2]. SDs can occur in clusters, where an event circles throughout the tissue hitting the same area of tissue repeatedly. This can drive down the local levels of glucose and oxygen, to a point where the tissue is no longer viable, increasing the injured area.

We have designed a bedside monitoring system to analyse the dialysate from a microdialysis probe implanted in the "at-risk" brain tissue. This incorporates combined needle electrodes, designed to fit into an analysis chamber on a microfluidic chip [3]. The electrodes can then be fabricated into sensors by applying a protective mPD film followed by an enzyme layer [4].

The clinical system is run continuously for a minimum of 24 hours, by the patient bedside. Every 3 hours, an automated calibration system switches in to ensure accurate and reliable data. When an SD event is picked up on an adjacent ECoG probe, the microdialysis data is scrutinised. Rapid metabolic changes can be observed, where the local level of glucose falls and a dynamic increase of lactate is seen. Clinical data will be shown, investigating how the tissue copes with various physiological insults.

1. Consensus Statement from the 2014 International Microdialysis Forum. Intensive Care Medicine (2015)
2. Hartings JA et al. The Lancet Nuerology, 10, 12, 1058-1064 (2011)
3. Rogers, ML. et al. ACS Chem Neurosci 4, 5, 799-807 (2013)
4. Vasylieva, N. et al Biosens. Bioelectron. 26, 10, 3993-4000 (2011)

Keywords: Electrochemistry, Lab-on-a-Chip/Microfluidics, Neurochemistry, Sensors
Application Code: Neurochemistry
Methodology Code: Sensors

Session # 710 Abstract # 710-8 **Oral Sessions**

Session Title Neurochemistry

Abstract Title **Online Clinical Microdialysis: Detecting the Neurochemical Consequences of Spreading Depolarization**

Primary Author Chi Leng Leong
Imperial College London

Date: Monday, March 07, 2016 - Afternoo
Time: 04:05 PM
Room: B401

Co-Author(s) Anthony J. Strong, Martyn G. Boutelle, Michelle L. Rogers, Sally A. Gowers, Sharon Jewell, Shumaila Khan

Abstract Text

Traumatic brain injury (TBI) is the major cause of disability and morbidity of adults under the age of 35. After the development of the irreversible primary damage at impact, 40% of the TBI patients develop secondary brain damage in the following days. The international collaboration, COSBID, has found a link between poor patient outcomes with spontaneous depolarization waves, termed spreading depolarization (SD) [1]. These waves propagate from the injury core to nearby healthy tissue, sending mass of neurons into depolarization, hence suppressing further activity. The recovery from this silencing phase is extremely energy demanding, relying heavily on the supply and delivery of metabolites to the tissue. Repeated episodes of SD can create a net energy deficit, resulting in the death of the neighboring tissue and expansion of the lesion [2].

Our goal is to detect SD chemically in real-time using online microdialysis in the intensive care unit. A microdialysis probe is placed in proximity to the injury core in the brain, sampling the neurochemicals in the nearby "at-risk" tissue. The continuous dialysate sample is fed into our bedside monitoring trolley, housing both microfluidic flow cells and custom-built electronics, for real-time quantification of ions and metabolites [3]. Typically, patients are monitored post-operatively for 24 hours up to days. To ensure correct interpretation of the data, an automated calibration system is implemented to track the performance of the sensors at 3-hours intervals. Using this system, we have been able to capture the chemical signature of dynamic SD events with high resolution, and linked it with electrophysiology data using electrocorticography.

1. Hartings, J. et al. [i]Brain[/i] [b]134[/b], 1529-1540 (2011)
2. Feuerstein, D. et al. [i]J Cereb Blood Flow Metab[/i] [b]30[/b], 1343-1355 (2010)
3. Rogers, M.L et al., [i]ACS Chem. Neurosci. [/i] [b]4[/b], 799-807 (2013)

Keywords: Bioanalytical, Electrochemistry, Lab-on-a-Chip/Microfluidics, Neurochemistry
Application Code: Neurochemistry
Methodology Code: Sensors

Session # 720 Abstract # 720-1 **Oral Sessions**

Session Title Unique Developments in Spectroscopy - Half Session

Abstract Title **Simultaneous Measurement of N2O and CH4 Emissions from Agriculture Using Photoacoustic Detection and QCL Laser**

Date: Monday, March 07, 2016 - Afternoo
Time: 03:05 PM
Room: B403

Primary Author Arto Branders
Gasera Ltd.

Co-Author(s) Ismo Kauppinen, Jaakko Lehtinen

Abstract Text

Agriculture contributes significantly to the anthropogenic emissions of non-CO2 greenhouse gases (GHG) methane (CH4) and nitrous oxide (N2O). GHG emissions comes from livestock such as cows, agricultural soils, and rice production. Simultaneous monitoring of these GHG compounds is demanding due to the low ambient background level and interference from water vapor. Single digit ppb level sensitivity is required for accurately detecting changes in the ambient background level in real time.

Cantilever-enhanced photoacoustic spectroscopy (CEPAS) combined with tunable mid-IR laser sources can meet the demanding requirements for sensitivity and selectivity [1,2]. One tunable distributed feedback quantum cascade laser (DFB-QCL) source can cover the spectral lines of both CH4 and N2O, and therefore, both target molecules can be simultaneously measured with one single instrument. The proposed solution combines the tunable DFB-QCL source with photoacoustic detection based on a silicon MEMS cantilever sensor coupled with an optical interferometric readout system [1].

[1] T. Kuusela, J. Kauppinen. Appl. Spectrosc. Rev., 42, (2007)
[2] C. B. Hirschmann, S. Sinisalo, J. Uotila, S. Ojala, and R. L. Keiski, Vib. Spectrosc. 68, 170–176 (2013)

Keywords: Agricultural, Environmental/Soils, Photoacoustic, Spectroscopy
Application Code: Agriculture
Methodology Code: Molecular Spectroscopy

Session # **720** **Abstract #** **720-2** **Oral Sessions**

Session Title: Unique Developments in Spectroscopy - Half Session

Abstract Title: **Infrared Spectroscopic Remote Sensing of Pulsed Signals from Nearby Stars**

Primary Author: Robert A. Lodder
University of Kentucky

Co-Author(s): Anne Brooks

Date: Monday, March 07, 2016 - Afternoo
Time: 03:25 PM
Room: B403

Abstract Text

At the Royal Society in London on July 20, 2015, the Breakthrough Listen initiative, funded by Russian high tech billionaire Yuri Milner, was announced. Milner also released an open letter backing the idea of intensifying the search for extraterrestrial intelligence, and the letter was cosigned by many scientists including Stephen Hawking. Three days later, astronomers using NASA's Kepler Space Telescope announced the discovery of the most Earth-like planet yet orbiting a distant star similar to our own sun, bringing to 12 the number of small worlds spotted elsewhere in the galaxy that are potentially suitable for life. The newly discovered planet, called Kepler-452b, is the smallest planet found to date orbiting in the habitable zone of a distant star. The steady discovery of new Earthlike planets and the 10-year Breakthrough Listen project have injected new life into astrobiology and the Search for ExtraTerrestrial Intelligence (SETI). Breakthrough Listen will use radio telescopes at Greenbank in West Virginia, the Parkes Observatory in Australia, and the Lick Observatory's optical telescope in San Jose, California, to scan approximately 1 million stars in the Milky Way and 100 nearby galaxies. It is worth noting that the Breakthrough Listen initiative is no longer relying upon simple microwave spectrometric searching, but instead includes a significant optical telescope for spectroscopy. The absorption and scattering of visible light by dust in the plane of the galaxy argues for the use of near-infrared and infrared wavelengths for galactic communication, much as fiber optics on earth frequently use near-infrared light. This paper describes the construction and use of a new near-IR telescope with first surface gold optics, near-IR pulse detector, electronic assemblies, data preprocessing system, and a Dobsonian mount. The telescope constructed by Anne appears in Fig. 1.

Keywords: Biosensors, Charge Transfer Devices (CID CCD), Near Infrared, Spectrophotometry

Application Code: General Interest

Methodology Code: Near Infrared

Session # 720 Abstract # 720-3 **Oral Sessions**

Session Title: Unique Developments in Spectroscopy - Half Session

Abstract Title: **High Resolution Coherent Multidimensional Spectroscopy**

Primary Author: Peter Chen
Spelman College

Date: Monday, March 07, 2016 - Afternoo
Time: 03:45 PM
Room: B403

Co-Author(s)

Abstract Text

High resolution molecular spectroscopy is a venerable technique that suffers from congestion and entanglement when used on large and complex molecules as well as simple mixtures. This problem can be addressed by expanding the technique to the second and third dimensions. In this talk, high resolution coherent 2D and 3D spectroscopies will be discussed, and applications demonstrating the advantages of going to higher dimensions will be presented.

Keywords: Molecular Spectroscopy
Application Code: General Interest
Methodology Code: Molecular Spectroscopy

Session # 720 Abstract # 720-4 **Oral Sessions**

Session Title: Unique Developments in Spectroscopy - Half Session

Abstract Title: **Biologics Starting Materials Identified through Opaque Containers by Spatially Offset Raman Spectroscopy (SORS)**

Date: Monday, March 07, 2016 - Afternoo
Time: 04:05 PM
Room: B403

Primary Author: Matthew Bloomfield
Cobalt Light Systems

Co-Author(s): Darren Andrews, Pavel Matousek

Abstract Text

The increase in the number of biologic, “large molecule” active pharmaceutical ingredients (API), new chemical entities (NCE), biosimilars and “biobetters” registered has been well-reported and are expected to match the revenues from traditional drugs within a few years.

The manufacturing process for these complex biological materials, often proteins and monoclonal antibodies, bears little resemblance to the synthesis of “small molecule” APIs. Commonly, living unicellular systems like bacterial microorganisms and cells are used in the manufacturing process. The conditions required for successful fermentation are a delicate blend of nutrients, pH, temperature and other factors that all affect the quality and yield of the product. A key requirement is that each component, for example, the buffers, surfactants and media blends, are sterile when added to the bioreactor. Simple exposure of the reagents and excipients to moist ambient air could be sufficient to introduce adventitious virus or other unwanted bacteria.

This necessity for raw material sterility must be balanced with the regulatory requirement for identity verification. Spectroscopic techniques are increasingly popular for identifying materials due to their speed, convenience and accuracy. It has emerged that the most versatile technique is Raman spectroscopy due to its high chemical specificity and simple operation. However, in its conventional form, Raman measurements require a clear line of sight to the material to be measured, limiting their application to clear plastic bags. Sampling from the original raw material containers to this secondary container clearly represents an area of risk in biologics manufacturing. This report focuses on the application of Spatially Offset Raman Spectroscopy (SORS) to accurate, robust ID through the original, often opaque or absorbing, packaging.

Keywords: Biopharmaceutical, Portable Instruments, Quality Control, Raman
Application Code: Quality/QA/QC
Methodology Code: Vibrational Spectroscopy

Session # 730 Abstract # 730-1 **Poster Sessions**

Session Title: Art and Archaeology

Abstract Title: **Maximizing the Information Obtained from Small Archaeological Samples by Sequencing DART-MS and Plasma-Chemical Oxidation for AMS Radiocarbon Dating**

Primary Author: Ruth Ann Armitage
Eastern Michigan University

Co-Author(s): Kathryn Jakes, Suzanne Baker

Date: Monday, March 07, 2016 - Afternoo
Time:
Room: Exposition Floor, 400 Aisle

Abstract Text

In the analysis of organic archaeological materials like textiles and rock paintings, the very small samples available is a significant limitation. To maximize the amount of information that can be obtained from these fragmentary and microgram quantity samples, we have developed methods for applying direct mass spectrometric analysis to the washes collected in the preparation of the samples for accelerator mass spectrometric radiocarbon dating. With only a few milligrams of textile from the Seip Mound Group in Ohio, we used DART-MS to identify the red anthraquinone dye colorants in the wash solutions as consistent with those from bedstraw ([i]Galium[/i] species) roots. The solid material was then prepared for AMS radiocarbon analysis using the plasma-chemical oxidation method, yielding the first ever radiocarbon date on a textile artifact from this important site. In order to measure a radiocarbon age for rock paintings, there must be either charcoal or an organic binding medium present. The black images of the Guara Pictograph Region of Cuba consist of geometric, animal, and human figures, some of which were drawn in charcoal. DART-MS on the residue from radiocarbon sample preparation clearly showed that others contained plant resins, which was promising for possibility of dating these images. The radiocarbon ages for these samples indicate that subfossil resins or petroleum were present in these paintings, yielding dates that do not correlate with the time at which people created the images. These examples show the importance and utility of combining direct mass spectral analysis with sample preparation for AMS dating.

Keywords: Art/Archaeology, Mass Spectrometry, Materials Characterization, Sample Preparation

Application Code: Art/Archaeology
Methodology Code: Mass Spectrometry

Session # 730 **Abstract #** 730-2 **Poster Sessions**

Session Title Art and Archaeology

Abstract Title **The Dose Makes the Poison: Quantitation of Pollutant VOCs From Materials Used in A Museum Environment**

Primary Author Michael J. Samide
Butler University

Date: Monday, March 07, 2016 - Afternoo
Time:
Room: Exposition Floor, 400 Aisle

Co-Author(s) Gregory D. Smith, Jericha Mill

Abstract Text

Pollutant volatile organic compounds (VOCs) off-gassed from various materials can pose a threat to artworks and objects of cultural heritage. Currently, materials to be used in a museum setting are screened for potential problems using the Oddy test, a 28-day test based on subjective determination of the level of corrosion on lead, silver and copper coupons caused during accelerated emission of VOCs. Instrumental methods centered on qualitative analysis by gas chromatography-mass spectrometry (GC-MS) are being developed where the VOCs are sampled using headspace, SPME, thermal desorption, or evolved-gas analysis. Once separated, the identity of the VOCs can be determined and evaluated for their potential to damage artworks. However, knowing that "the dose makes the poison," it is important to know quantity as well as identity. In this work, quantitation of VOCs off-gassed during headspace sampling and evolved-gas analysis will be presented. Acetic acid emitted from poly(vinyl acetate) glues and cellulose acetate was quantified using an external calibration. Amount of acetic acid per mass of sample was related to glue type and cure time. In a similar fashion, residual n-butylmethacrylate monomer emitted from poly(n-butylmethacrylate) and n-alkylmorpholines in polyurethane foams were examined. The impact of sample surface area and EGA conditions will be discussed. A real-world case study for rigid poly(vinyl chloride) boards will be presented.

Keywords: Art/Archaeology, Gas Chromatography/Mass Spectrometry, Quantitative
Application Code: Art/Archaeology
Methodology Code: Gas Chromatography/Mass Spectrometry

Session # 730 Abstract # 730-3 **Poster Sessions**

Session Title: Art and Archaeology

Abstract Title: **Using Reflectance Spectroscopy to Determine the Rate of Formation of Prussian Blue Pigment**

Date: Monday, March 07, 2016 - Afternoo

Time:

Room: Exposition Floor, 400 Aisle

Primary Author: Jacob Applegarth
Butler University

Co-Author(s): Michael J. Samide

Abstract Text

The cyanotype process begins with the photo-induced reduction of an iron(III) carboxylate salt to iron(II), along with subsequent oxidation of the carboxylate anion. The iron(II) ion is not particularly stable and will undergo a redox reaction with the ferricyanide anion present in the sensitizer to produce insoluble iron(III) ferrocyanide Prussian Blue pigment. While much is known about the formation of Prussian Blue, the rate of the reaction has not been fully examined. In this work, reflectance spectroscopy has been employed to monitor the rate of formation of Prussian Blue using different sensitizer formulations on different papers. By measuring and plotting color values (CIE L*a*b*) with respect to time, an initial rate could be determined and these rates were used to derive a rate equation. The reaction appears to be 1st order with respect to ammonium iron (III) oxalate, but the overall printing process appears to be limited to a very narrow concentration range. The dependence of rate on (1) identity of the counter ions, (2) sensitizer concentration, and (3) paper type will be discussed. In addition, the rate at which the Prussian Blue fades from its deepest blue (as determined by CIE) back to Prussian White will also be presented.

Keywords: Art/Archaeology, Spectroscopy, UV-VIS Absorbance/Luminescence
Application Code: Art/Archaeology
Methodology Code: UV/VIS

Session # 740 Abstract # 740-1 **Poster Sessions**

Session Title: Chemical Methods

Abstract Title: **Single Nanoparticle to 3D Supercage: Framing for an Artificial Enzyme System**

Date: Monday, March 07, 2016 - Afternoo

Time:

Room: Exposition Floor, 400 Aisle

Primary Author: Ren Cai
University of Florida

Co-Author(s): Weihong Tan

Abstract Text

A facile strategy has been developed to fabricate Cu(OH)2supercages (SCs)as an artificial enzyme system with intrinsic peroxidase mimic activities (PMA). SCs with high catalytic activity and excellent recyclability were generated via direct conversion of amorphous Cu(OH)2 nanoparticles (NPs) at room temperature. More specifically, the process that takes a single nanoparticle to a three-dimensional (3D) supercage involves two basic steps. First, with addition of a copper-ammonia complex, the Cu2+ ions, which located on the surface of amorphous Cu(OH)2 NPs, would evolve into a fine lamellar structure by coordination and migration, and eventually converts to1D nanoribbons around the NPs. Second, accompanied by the migration of Cu2+, a hollow cavity is generated in the inner NPs, such that a single nanoparticle eventually becomes a nanoribbon-assembled 3D hollow cage. These Cu(OH)2 SCs were then engineered as an artificial enzymatic system with higher efficiency for intrinsic PMA than that of a natural enzyme - horseradish peroxidase□

Keywords: Adsorption, Analysis, Drugs, FTIR

Application Code: Material Science

Methodology Code: Chemical Methods

Session # 740 Abstract # 740-2 **Poster Sessions**

Session Title Chemical Methods

Abstract Title **Synthesis and Characterization of Superparamagnetic Manganese Ferrite Nanoparticles**

Primary Author Simonas Ramanavicius
Vilnius University

Date: Monday, March 07, 2016 - Afternoo
Time:
Room: Exposition Floor, 400 Aisle

Co-Author(s) Arunas Jagminas, Kestutis Mazeika, Marija Kurtinaitiene, Vidas Pakstas

Abstract Text

Superparamagnetic iron oxide nanoparticles, composed of magnetite, Fe_3O_4, or maghemite, core and biocompatible polymer shell, such as dextran or chitozan, recently are widely used for magnetic resonance imaging, contrast enhancement, hyperthermia and therapy [1,2].
This study is focused on the synthesis of manganese ferrite ($MnFe_2O_4$) nanoparticles by most commonly employed chemical way seeking to better control their size, purity and magnetic properties. In this way, hydrothermal and microwave-assisted co-precipitation syntheses were performed using aqueous alkaline solutions of Mn(II) and Fe(III) salts and NaOH within a wide pH range at various experimental regimes. Different additives, such as citric acid, cysteine, glicine, polyetylene glycol, triethanolamine, chitozan, etc., were also tested seeking to obtain pure phase and monodisperse nanoparticles in average size of 20 nm at a good yield. Transmission electron microscopy (TEM), X-ray diffraction, Mössbauer spectroscopy, magnetic measurements and inductively coupled plasma mass spectrometry were employed in this study.

References

[1] R.C. O'Handley. Modern magnetic materials: Principles and Applications, Wiley, New York, 2000.
[2]K. Mazeika, A. Jagminas, M. Kurtinaitiene. Mossbauer studies of superparamagnetic ferrite nanoparticles for functional applications. Hyperfine Interactions 218 (2013) 89-94.

Keywords: Materials Characterization, Particle Size and Distribution, Wet Chemical Methods, X-ray Diffraction
Application Code: Nanotechnology
Methodology Code: Chemical Methods

Session # **740** Abstract # **740-3** **Poster Sessions**

Session Title: Chemical Methods

Abstract Title: **Optimization of Ibuprofen Micronization by the Rapid Expansion of a Supercritical Solution (RESS)**

Primary Author: Rolf Schlake
Applied Separations

Co-Author(s): Al Kaziunas, Brian Day

Date: Monday, March 07, 2016 - Afternoo
Time:
Room: Exposition Floor, 400 Aisle

Abstract Text

The RESS process is used for the micronization of chemical compounds, typically for biomedical applications. It is done by placing a sample in a high pressure vessel and filling this vessel with CO2, which is then brought into a supercritical state by pressurizing and heating it beyond its critical point. In this state the compound is readily dissolved. By allowing the supercritical fluid to expand through a capillary nozzle into a heated collection vessel, one can achieve high levels of supersaturation, resulting in the rapid precipitation of the sample compound into nanoparticles.

This poster describes the optimization of conditions, equipment, and operations in the formation of fine particles of Ibuprofen by the rapid expansion of a supercritical solution. Rigorous temperature control of the dissolution vessel, associated tubing, valves, nozzles, and collector was necessary to prevent unwanted precipitation. In addition, nozzle design integrated into the collector/filter vessel was necessary for enhanced particle collection. Operational improvements included a dissolution vessel with a bypass line that allowed for trouble free operation. These enhancements resulted in reproducible particle formation with a reduction in the chronic plugging associated with the RESS process.

Keywords: Biomedical, Biopharmaceutical, Particle Size and Distribution, SFE
Application Code: Biomedical
Methodology Code: Chemical Methods

Session # 740 Abstract # 740-4 **Poster Sessions**

Session Title Chemical Methods

Abstract Title **DNA-Aligner-Controlled Nicking-Based Isothermal Exponential Amplification Reaction for High Sensitive Detection of Nucleic Acids**

Date: Monday, March 07, 2016 - Afternoo
Time:
Room: Exposition Floor, 400 Aisle

Primary Author Wu Wanghua
Zhejiang University

Co-Author(s) Yu Dongdong, Zhang Tao, Zhou Jianguang

Abstract Text

Improved detection of nucleic acids holds great promise in the early diagnosis of cancer and genetic diseases. Isothermal exponential amplification reaction (EXPAR) is a simple, low cost and highly sensitive method possessing high amplification efficiency and rapid kinetics. This approach, however, is unable to detect the sequence that is not adjacent to a specific recognition site of a certain endonuclease, and is hard to identify the single-nucleotide mutation. Herein, to address these limitations, we have developed a new method called DNA-aligner-controlled nicking-based isothermal amplification reaction (DACN-EXPAR). This method involves two isothermal stages by a combination of polymerase and nicking endonuclease (nickase). In the first stage, the nickase, directed by a hairpin-shaped DNA-aligner containing the nickase recgonition site in its stem, performs specific nicking on target strands. Then, in the second stage, the cleaved target strands serve as triggers to initiate the exponential amplification reaction catalyzed by both polymerase and nickase. Thanks to DNA-aligner, the DACN process does not require any special sequence in target strand, thus endowing the newly proposed DACN-EXPAR with excellent generality, which in theory, can be used to assay the target strand of any sequence. Moreover, the proposed method exhibits high sensitivity with a detection limit of 100 aM, a broad range of 8 orders of magnitude, and as expected, high specificity to discriminate the single-nucleotide mutation from the wild type. It also performs well in 5% human serum, demonstrating a good compatibility with complex environment. Therefore, the proposed DCAN-EXPAR assay shows a great potential in early clinical diagnosis.

Supported by National Science Foundation of China (21275129) and National Key Technology Support Program (2012BAB19B07).

Keywords: Bioanalytical, Detection, Enzyme Assays, Nucleic Acids
Application Code: Bioanalytical
Methodology Code: Chemical Methods

Session # 740 Abstract # 740-5 **Poster Sessions**

Session Title Chemical Methods

Abstract Title **Electrochemical, Spectroscopic and Chromatographic Techniques for Monitoring the Active Pharmaceutical Ingredients Degradation Kinetics: Which Methodology Fulfills More Principles of the Green Analytical Chemistry?**

Date: Monday, March 07, 2016 - Afternoo

Time:

Room: Exposition Floor, 400 Aisle

Primary Author Mohamed K. Abd El-Rahman
Cairo University

Co-Author(s) Amr M. Mahmoud

Abstract Text

This work addresses the promising junction of two areas of recent and increasing concern, the chemical kinetics of active pharmaceutical ingredients (APIs) and Green Analytical Chemistry (GAC) principles. Importantly, stability studies ensuring the maintenance of product quality, safety and efficacy throughout the shelf life are considered as pre-requisite for both, the approval of any pharmaceutical product and saving countless lives of humankind. Distinguishably, monitoring of the degradation kinetics of APIs can be achieved through different analytical strategies; off-line, at-line and in-line according to the degree of integration between the two units. This work compares different analytical techniques used for monitoring of degradation kinetics of APIs with respect to the 12 principles of GAC. For a meaningful comparison, distgmine bromide (DB) was chosen as a hydrolysable anti-cholinesterase drug and its degradation kinetics were monitored by different strategies. The first in-line strategy is achieved by the incorporation of an in-site DB selective electrode constructed using PVC membrane to track the hydrolysis kinetics of DB by continuous measurement of the decrease in the produced emf over time. The second at-line strategy by UV spectrophotometry via continuous tracking either the decrease of DB peak at 269 nm or the increase of 3-hydroxy- N-methylpyridinum (THMP) peak at 320 nm over time. The third off-line strategy utilizes separation-based chromatographic HPLC and TLC methods. The advantages and shortcomings of each strategy considering GAC principles are highlighted. A systematic approach for the development of an environmentally benign stability-indicating method is addressed.

Keywords: Electrochemistry, HPLC, Pharmaceutical, Spectrophotometry
Application Code: Pharmaceutical
Methodology Code: Chemical Methods

Session # 740 Abstract # 740-6 **Poster Sessions**

Session Title: Chemical Methods

Abstract Title: **Investigation of Heterogeneous Reaction Mechanism Between Formaldehyde and MnO_2/CeO_2 at Room Temperature by Gas Analysis Approach**

Date: Monday, March 07, 2016 - Afternoo
Time:
Room: Exposition Floor, 400 Aisle

Primary Author: Hayashi Hiroki
Tokai University

Co-Author(s): Sekine Yoshika

Abstract Text

Exposure to formaldehyde (HCHO) in air has been known to cause human health disorders. Manganese dioxide (MnO_2) is currently used for a major ingredient of air cleaning materials, because it reacts with HCHO to give carbon dioxide (CO_2) even at room temperature. However, the removal efficiency decreases by long-time uses. This is probably due to accumulation of intermediates on the surface caused by changes in oxidation state of manganese. Takao et al. found blending cerium oxide (CeO_2) with MnO_2 increased the life-time of removal of HCHO. However, the mechanism is unknown. Then, this study aimed to investigate the reaction process of the heterogeneous reaction by gas analysis. The test samples were simply prepared by mechanical mixing of the MnO_2 and CeO_2 powders. The mixed oxide was placed in a reaction vessel with a constant gas flow system of 500 ppm of HCHO. Concentrations of interest gases were determined by a conventional gas detector tube method. The results showed the breakthrough easily occurred when the sample was only MnO_2. Meanwhile, the effect of blending CeO_2 was remarkable with the appearance of steady state removal performance for HCHO under purified air. However, when employing nitrogen as the background gas, the steady-state performance did not appear. These results showed the mechanism can be explained by self-redox cycle between MnO_2 and CeO_2; the reduced manganese is re-oxidized into MnO_2 by CeO_2, and the reduced cerium oxide is thus regenerated to CeO_2 by oxygen in air.

Keywords: Chemical, Environmental, Environmental/Air, Oxygenates
Application Code: Environmental
Methodology Code: Chemical Methods

Session # 740 **Abstract # 740-8** **Poster Sessions**

Session Title: Chemical Methods

Abstract Title: **Identification of 6-Chlorotestosterone and Other Designer Anabolic Steroids in Dietary Supplements with Semi-quantitative Content Determination Using Surrogate Compounds**

Date: Monday, March 07, 2016 - Afternoo

Time:

Room: Exposition Floor, 400 Aisle

Primary Author: Sarah E. Voelker
U.S. FDA, Forensic Chemistry Center

Co-Author(s): Lisa M. Lorenz, Mary B. Jones, Rick A. Flurer, Travis M. Falconer

Abstract Text

Several dietary supplements recently submitted to the FDA's Forensic Chemistry Center were found to contain steroid-like compounds that could not be readily identified due to the lack of library reference spectra or commercially available standards. Analytical-scale high performance liquid chromatography with fraction collection was used to isolate and collect several new "designer steroids" observed in these supplements, including 6-chlorotestosterone. The isolated compounds were identified using a multi-technique analytical approach including analysis by nuclear magnetic resonance (NMR) spectroscopy and high resolution accurate mass-mass spectrometry (HRAM-MS); however, quantitative analysis could not be achieved using typical assay methods by high performance liquid chromatography with ultraviolet detection (HPLC-UV) due to the lack of commercially available reference standards.

Due to the increasing frequency of laboratory requests for content determination, it has become necessary to obtain quantitative or semi-quantitative data for dietary supplement samples containing steroid and steroid-like compounds. After evaluation of structural characteristics of compounds of interest, molar absorptivity data was collected for several common steroids using ultraviolet/visible spectroscopy, and compared to those of the new "designer steroids" isolated and identified by the Forensic Chemistry Center. This approach was employed to find suitable compounds for use as reference materials in the semi-quantitative determination of these compounds. Using laboratory fortified matrix (LFM) samples and evaluation of actual dietary supplement samples, a method for the estimation of these emerging substances by HPLC-UV is demonstrated.

Keywords: Forensic Chemistry, Gas Chromatography/Mass Spectrometry, Liquid Chromatography/Mass Spectro
Application Code: General Interest
Methodology Code: Chemical Methods

Session # 750 **Abstract # 750-1** **Poster Sessions**

Session Title: Commercial Products Characterization

Abstract Title: **Considerations in Sample Preparation and Method Development Using ICP-OES**

Date: Monday, March 07, 2016 - Afternoo
Time:
Room: Exposition Floor, 400 Aisle

Primary Author: Kenneth Neubauer
PerkinElmer

Co-Author(s): Laura Thompson, Nick Spivey, Stan Smith

Abstract Text

With the growing popularity of electronics and their shorter lifecycles as existing equipment is continuously being replaced by more advanced models, a growing concern is the leaching of toxic metals into the environment when electronics are disposed. Even though electronics recycling is becoming more common, many electronic products are still being disposed of.

To address this concern, the Restriction of Hazardous Substances (RoHS) directive was created. This directive limits the amount of cadmium, hexavalent chromium, arsenic, and lead which can be present in devices. As such, manufacturers of electronic devices must change the composition of their products to comply.

There are a variety of techniques capable of performing these analyses, but ICP-OES strikes the best balance between cost, speed, and simplicity. With its multi-elemental capability, it is faster than flame AA and significantly less expensive than ICP-MS. In addition, it can easily handle matrices with high acid concentrations and high levels of dissolved solids.

One of the challenges of the RoHS directive is sample preparation: with such a wide variety of products and components involved, the possibility of an enormous amount of sample preparation procedures exists. However, by using a microwave digestion system and carefully selecting parameters, the number of sample preparation methods can be minimized.

This work focuses on the preparation and analysis of a variety of materials covered under the RoHS directive using microwave digestion in conjunction with ICP-OES.

Keywords: Consumer Products, Elemental Analysis, ICP, Sample Preparation
Application Code: Consumer Products
Methodology Code: Atomic Spectroscopy/Elemental Analysis

Session # 750 **Abstract # 750-2** **Poster Sessions**

Session Title: Commercial Products Characterization

Abstract Title: **Chemical Fingerprinting of Tobacco and Related Products by TD–GC–TOF MS**

Primary Author: Laura McGregor
Markes International Ltd

Date: Monday, March 07, 2016 - Afternoo
Time:
Room: Exposition Floor, 400 Aisle

Co-Author(s): Caroline Widdowson, Chris Hall, Ken Umbarger, Nicola Watson

Abstract Text

The hazardous constituents of cigarette smoke have attracted considerable media attention, especially with increasing regulation around the world limiting or banning smoking in public places – and even in private cars if children are present. Furthermore, the recent surge in tobacco-replacement devices, such as e-cigarettes, is driving the development of fast and efficient quality control procedures. E-cigarette solutions may contain potentially harmful chemicals, including nitrosamines and polycyclic aromatic hydrocarbons (PAHs). The presence of such chemicals naturally gives rise to some concern, and confident chemical fingerprinting is required for both research and development and regulatory purposes. Although e-cigarettes emit less particulate matter than regular tobacco cigarettes (since no combustion takes place), they still produce a wide range of compounds at trace levels. Organic constituents of tobacco smoke have historically been analysed by gas chromatography coupled with quadrupole mass spectrometry (GC–MS). However, quadrupoles are mass filters, with a high percentage of ions being wasted, which limits sensitivity. Moreover, in selected ion monitoring (SIM) mode, only target compounds can be monitored, meaning that full characterisation of the sample is not possible in a single run and retrospective searching of data is limited. The use of time-of-flight mass spectrometry (TOF MS) overcomes this issue by providing highly sensitive detection whilst acquiring full-range mass spectra, to allow both target and unknown identification in a single, rapid analysis. This poster explores the use of a multi-functional thermal desorption (TD)–GC–TOF MS system to capture and identify whole e-cigarette emissions using a single, highly-automated platform.

Keywords: Gas Chromatography/Mass Spectrometry, Thermal Desorption, Time of Flight MS, Volatile Organic C
Application Code: Consumer Products
Methodology Code: Gas Chromatography/Mass Spectrometry

Session # 750 **Abstract # 750-3** **Poster Sessions**

Session Title: Commercial Products Characterization

Abstract Title: **Levels of Mercury and Methylmercury in Fish (Prochilodus Magdalenae) Ayapel Marsh (Colombia)**

Date: Monday, March 07, 2016 - Afternoo

Time:

Room: Exposition Floor, 400 Aisle

Primary Author: Edineldo Lans Ceballos
University of Cordoba

Co-Author(s): Amira Padilla Jimenez, Mauricio Lora Agamez

Abstract Text

Determinig the levels of total mercury(T-Hg) and methylmercury (Me-Hg)in fish(prochilodus magdalenae) to establish the risk associated with their daily in take. Two samplings were carried out during the months August and Semptember 2014. 40 specimens depending of size and weight were selected and purchased to fishermen of region. The T-Hg was determinated with a direct mercury analizer (DMA) by thermal descomposition,amalgamation and atomic absortion TDA-AAS and Me-Hg by gas chromatography with electron capture detector (GC-ECD). The T-Hg concentrations are between 50.7 to 302.3 mg Kg-1 and Me-Hg between 40.28 to 243.29 mg Kg-1. The Me-Hg/T-Hg reasons range from 0.46 to 0.95 increasing this ratio based on the size and weight of the fish. The hazard index (HI) for children and adults are between 1.73 to 12.97 and 0.99 to 7.40 for T-Hg and Me-Hg respectively. The T-Hg concentrations of the analyzed samples did not exceed 500 mg kg-1 established by Colombian legislation in fish for human consumption.

There are significant differences between the concentration of T-Hg and Me-Hg in relation to the size and weight of the fish (P[greater than]0.05). HI indicates that consumption of 0.148 kg of fish per day in the people of the region could increase the risk and susceptibility of poisoning these pollutans. The results indicate a high risk for human population specially children who are most vulnerable.

This project was funded by the University of Cordoba.

Keywords: Atomic Absorption, Gas Chromatography
Application Code: Consumer Products
Methodology Code: Gas Chromatography

Session # 750 Abstract # 750-4 **Poster Sessions**

Session Title: Commercial Products Characterization

Abstract Title: **Mercury Residual and Methylmercury in Canned Tuna Distributed in Monteria –Colombia**

Primary Author: Edineldo Lans Ceballos
University of Cordoba

Date: Monday, March 07, 2016 - Afternoo
Time:
Room: Exposition Floor, 400 Aisle

Co-Author(s): Basilio Diaz Pogutá, Mauro Lombana Agamez

Abstract Text

Determining total mercury (T-Hg) and Methylmercury (Me-Hg) in canned tuna distributed in the city of Monteria –Colombia. 40 samples were analyzed from August to September 2014, taking 10 samples from different batches and identified as L, M, N and O. Me-Hg was determined by gas chromatography with electron capture detector (Perkin Elmer XL Autsystem) Column Restek RTX-1701 USA corp. (30 m x 0.53 [micro] m ID and 3[micro] m film thickness). The T-Hg was determined with a mercury analyzer DMA-80. The L, N and O marks showed concentrations of T-Hg and Me-Hg, on Maximum Residual Limit established by the Ministry of Social Protection, Colombian. The highest concentration of T-Hg and Me-Hg was detected in L with values (1157.17 ± 229.06 mg / kg) and (1046.19 ± 209.19 mg / kg) respectively. In contrast, M presented the lowest values (253.65 ± 46.15 mg / Kg) and (223.57 ± 41.16 mg / kg) respectively. The evaluation of the health risk from eating canned tuna based on percentage of total allowable weekly intake suggests that tuna consumption of L, N and O marks, can be considered a risk to the health of the population and especially the children are most vulnerable. This study allows to know the risk to which the population is exposed by consuming canned tuna distributed in the city.

The authors thank the University of Cordoba in the financing of this project.

Keywords: Atomic Absorption, Contamination, Gas Chromatography

Application Code: Food Contaminants

Methodology Code: Gas Chromatography

Session # 750 **Abstract # 750-5** **Poster Sessions**

Session Title: Commercial Products Characterization

Abstract Title: **Automated SPME for the Analysis of Environmental Contaminants in Milk**

Primary Author: Nicole M. Lock
Shimadzu Scientific Instruments

Date: Monday, March 07, 2016 - Afternoo
Time:
Room: Exposition Floor, 400 Aisle

Co-Author(s): Brahm Prakash, Di Wang, Laura Chambers, Robert Clifford, Shilpi Chopra

Abstract Text

Contamination of milk is prevalent and is the consequence of decades of inadequately controlled pollution of the environment by toxic chemicals. The reports of residues in milk is increasing consistently which raises important issues for the practice of public health, and for the environmental health research community. The analysis of the environmental contaminants in complex matrices matrices traditionally requires several steps of extraction and pre-concentration for the analytes and clean-up procedures. Moreover, these extraction methods need expensive and hazardous solvents that are undesirable for health and disposal reasons. Solid phase microextraction (SPME) has been introduced as an alternative to traditional extraction techniques. This work will demonstrate the optimization of automated SPME for the extraction of environmental contaminants in milk. SPME efficiency is directly related to fiber material, temperature, extraction time interval, sample amount, desorption time. Half factorial design will be used for simultaneous investigation of the effect of several variables over the evaluated response (e.g., analytical sensitivity) requiring a reduced number of experiments hence providing best optimized parameters for SPME extraction.

Keywords: Food Safety, GC-MS, Sample Preparation, SPME
Application Code: Food Contaminants
Methodology Code: Sampling and Sample Preparation

Session # 750 Abstract # 750-6 **Poster Sessions**

Session Title: Commercial Products Characterization

Abstract Title: **Estimation of Dietary Intake of Essential and Non-Essential Metals through the Consumption of Dietary Supplements Available on Nigerian Retail Market**

Date: Monday, March 07, 2016 - Afternoo
Time:
Room: Exposition Floor, 400 Aisle

Primary Author: Imaobong Udousoro
University of Uyo

Co-Author(s): Ikem Abua, Olujide T. Akinbo

Abstract Text

The continuing rise in consumption of dietary supplements (Vitamin, mineral supplements, and herbal/botanical supplements) is a global phenomenon. This is in spite of concern over their quality safety and potential side effect. Dietary supplements are readily available and commonly sold without prescription in Nigeria. However, it has been reported that medical doctors in Lagos are becoming more favorably disposed to prescribing them. To this end, some authors have recommended further studies to minimize potential risk. There is still paucity of data on human exposure to trace and essential metals through consumption of dietary supplements sold in Nigerian markets. Also, quality of existing data is questionable due to limited or no information on quality control samples during analysis. The goal of this work is to evaluate dietary supplements from Nigeria and (1) compare the levels of essential metals in them to Recommended Daily Allowance (RDA) (2) compare the concentration of the elements to the levels that are declared on the labels (3) assess if the levels of non-essential metals in the supplements exceeds the Tolerable Upper Intake (UL) and (4) use principal component analysis to establish potential for classification of the supplement based on the elemental profiles. Eighteen dietary supplements (including vitamins, mineral and herbal supplements) from the Nigerian market were analyzed using ICPMS (Li, V, Cr, Mn, Fe, Co, Ni, Cu, Zn, As, Se, Sr, Mo, Cd, Ba, Pb, Th and U), ICPOES (Na, Mg, Ca, Al), and mercury analyzer (Hg). Results of the analysis will be presented.

Keywords: Atomic Spectroscopy, Environmental Analysis, ICP, ICP-MS
Application Code: Food Contaminants
Methodology Code: Atomic Spectroscopy/Elemental Analysis

Session # **750** **Abstract #** **750-7** **Poster Sessions**

Session Title Commercial Products Characterization

Abstract Title **Analysis of Aflatoxins in Pet Food by UHPLC Using PDA and Fluorescence Detection**

Date: Monday, March 07, 2016 - Afternoo
Time:
Room: Exposition Floor, 400 Aisle

Primary Author Amanda Prior
PerkinElmer

Co-Author(s) Jason Weisenseel, Wilhad M. Reuter

Abstract Text

Commercially prepared pet foods are an easy and economical way to fulfill the nutritional requirements for pets. Dry pet food is produced with grains and cereal by-products rejected for human consumption. The contamination of these by-products with toxigenic fungal metabolites, called mycotoxins, pose a serious health threat to pets.

Aflatoxins B1, B2, G1 and G2 are some of the most carcinogenic mycotoxins known. In particular, aflatoxin B1 is considered to be the most genotoxic of the mycotoxins, and, when ingested by farm animals, can contaminate dairy, eggs and meat products intended for human consumption.[sup]1[/sup]

With the above mind, this work describes a method for monitoring B1, B2, G1 and G2 aflatoxins at ppb to ppt levels, quantitatable well below regulated limits, without the need for post-column derivatization. This method takes advantage of a simple solid phase extraction (SPE) procedure, followed by UHPLC analysis, using a sub-3 µm particle column and fluorescence (FL) detection.

1. Hudler, George W., Magical Mushrooms, Mischievous Molds: The Remarkable Story of the Fungus Kingdom and Its Impact on Human Affairs, 1998.

Keywords: Food Contaminants, Liquid Chromatography
Application Code: Food Contaminants
Methodology Code: Liquid Chromatography

Session # 750 **Abstract #** 750-8 **Poster Sessions**

Session Title: Commercial Products Characterization

Abstract Title: **Antioxidant Activity of Flavored Tea and Its Content of Phenol, Flavonoid and Tannins**

Primary Author: Abd El-Moneim M. Afify
Cairo University

Date: Monday, March 07, 2016 - Afternoo
Time:
Room: Exposition Floor, 400 Aisle

Co-Author(s)

Abstract Text

This investigation concern the benefit of tea products, used in the Giza area of Egypt. The antioxidant activity of cold and hot water extracts of six flavored (i.e. raspberry, apple plus fruit pieces, blackcurrant, vanilla pods, cinnamon and peppermint plus lemon flavored tea) and one unflavored tea products were dtermined. The extracts were screened for total phenol, flavonoids and tannins content and antioxidant. Antioxidant activities were carried out using DPPH, ABTS methods. Hot water extract of tea flavored with blackcurrant had the highest total phenol and tannins content. Cold water extract of tea flavored with blackcurrant had the highest total flavonoid and tannins content. In contrast tea flavored with peppermint plus lemon had the lowest tannins content in both hot and cold water extract. Cinnamon flavored tea extracted with hot and cold water had the highest antioxidant activity against DPPH. In addition hot water extract of cinnamon flavored black tea and cold water extract of raspberry flavored and unflavored tea had the highest antioxidant activity against ABTS. Generally, hot extracts showed higher antioxidant activities than cold extracts with different tea samples. Generally, this flavored tea extracts had high level of antioxidant activity and phenolic contents. Therefore the investigation proved that flavored tee had more antioxidant activity which will be benifite for human consuming On thae same time changing tradition of social drinking tea drinking could have significant positive primary health implications, which would ultimately lighten the economic health burden on the state.

Keywords: Chemical, Consumer Products, Food Identification
Application Code: Food Identification
Methodology Code: Chemical Methods

Session # 750 Abstract # 750-9 **Poster Sessions**

Session Title: Commercial Products Characterization

Abstract Title: **High Sensitive LC/MS Analysis of Stevia Sweeteners Using Polymer-Based Amino Column**

Primary Author: Ronald Benson
Shodex/Showa Denko America, Inc.

Date: Monday, March 07, 2016 - Afternoo
Time:
Room: Exposition Floor, 400 Aisle

Co-Author(s): Junji Sasuga, Satoko Sakai, Tomokazu Umezawa

Abstract Text

In recent years, such as Rebaudioside A, a glycoside derived from stevia, is added to a variety of processed foods as a sweetener for reduced calories. It is important to quantify the stevia sweeteners in processed food for quality control purposes. Herein we describe the highly sensitive and highly selective analytical separation by using polymer-based amino column with detection by LC/MS. Chromatographic separation was conducted using polymer-based amino column with isocratic mixture of (A) 0.1% NH3 aq. / (B) CH3CN as the eluent. The flow rate was 0.2mL/min and the temperature was 30 °C. ESI was selected for ionization method and SIM (-) for detection mode. Eight components, Rebaudioside A/B/C/D/F, Isosteviol, Stevioside, Steviolbioside, were mixed in each 50ng/mL as a standard sample. Injection volume was 5L. Eight components were almost separated within 20 minutes and were detected with sufficient S/N in the condition of 78% CH3CN by HILIC mode. Rebaudioside B and Steviolbioside were strongly adsorbed to the column when the eluent (A) was only water since they have a carboxyl group. It became possible to adjust the appropriate retention time by alkaline with NH3. Result of measuring commercially available energy drink containing stevia sweeteners as a real sample, Rebaudioside A was detected, and additionally the other additives such as Sucralose, Acesulfame, Erythritol, Carnitine, Leucine, Isoleucine, Varine and Pyridoxine were also detected at the same time with good peak shape. The analysis condition show that the stevia sweetener in processed food and other high polarity food additives can be separated by HILIC mode and quantify with high sensitive.

Keywords: Food Identification, Food Science, Liquid Chromatography/Mass Spectroscopy
Application Code: Food Identification
Methodology Code: Liquid Chromatography/Mass Spectrometry

Session # 750	Abstract # 750-11	Poster Sessions

Session Title Commercial Products Characterization

Abstract Title **Characterizing the Rheological Properties of Wax Emulsions used as Carriers for Biopesticides in Agricultural Pest Management**

Date: Monday, March 07, 2016 - Afternoo
Time:
Room: Exposition Floor, 400 Aisle

Primary Author Kristen Jordan
Western Carolina University

Co-Author(s) Cynthia Atterholt

Abstract Text

The purpose of this research is to characterize the rheological properties of various wax emulsions developed as carriers for biopesticides used for agricultural pest management. Wax emulsions have been developed as carriers for the controlled release of non-toxic materials such as insect pheromones and essential oils used in crop protection as part of Integrated Pest Management programs (IPM) (Atterholt et al., 1998; Ballew, 2011; McCracken, 2006). However, the rheological properties of these wax emulsion carriers have not been evaluated and reported in the literature. Therefore, to obtain a more thorough evaluation of these wax emulsions, the rheological properties will be characterized. The data collected can be used to determine the best rheological properties for different field application methods. In this research, wax emulsions will be prepared using different formulations, such as different waxes, emulsifiers, and biopesticides, and different concentrations of each. Each of these emulsions will be tested using the Physica MCR 101 rheometer. The goal is to determine the rheological properties that will facilitate the application of these wax emulsions to various plant surfaces.

Geranyl propionate will be used as a substitute for insect pheromones because it is chemically and structurally similar to many insect sex pheromones and is less expensive (Atterholt et al., 1998). Wax emulsions will also be prepared with a small concentration of spearmint oil, which has been shown to be an effective feeding deterrent for white-tailed deer. Each of the emulsions prepared for this research will be tested for their rheological properties, such as viscosity and viscoelasticity. By gaining a better understanding of the rheological properties of these wax emulsions, formulations could be designed to improve the desired outcomes.

Keywords: Agricultural, Characterization, Pesticides, Rheology
Application Code: Agriculture
Methodology Code: Surface Analysis/Imaging

Session # 750 Abstract # 750-12 **Poster Sessions**

Session Title Commercial Products Characterization

Abstract Title **Multivariate Quantification of API Release from Combination Tablets in the Presence of Matrix Effects Using Fiber Optic Dissolution**

Date: Monday, March 07, 2016 - Afternoo
Time:
Room: Exposition Floor, 400 Aisle

Primary Author Joseph Medendorp
Vertex Pharmaceuticals

Co-Author(s) Ivelisse Colon, Mahidhar Shapally, Taryn Ryan

Abstract Text

The purpose of this research was to develop a fiber optic (FO) dissolution method for combination pharmaceutical tablets. FO dissolution allows direct API quantification in the vessel, obviating the need for error prone facets of standard dissolution methods. However, FO dissolution is potentially challenged by matrix effects: UV active excipients, API interactions with excipients and media, undissolved components attenuating the UV signal among others. These obstacles might render FO dissolution method development more complex than LC-end dissolution.

This study tackles an example with the added complexity of a combination product, where the two actives have similar release kinetics and UV spectra. Multiple methods were employed for the quantification of actives, including: single wavelength per active, a modified classical least squares (CLS) approach modeling matrix effects as a third component, and partial least squares for multivariate calibration and prediction using LC-end dissolution as reference data. Single wavelength quantification requires unique features for the actives of interest, which are not always readily available. The initial pass with the CLS approach requires that linear combinations of standards adequately describe the behavior of the actives in situ. And the multivariate approach requires manual reference data for calibration. Since a combination tablet typically demonstrates collinear API release, individual quantification is non-trivial. The advantages of each of these methods of quantification will be discussed in the context of the test systems under investigation. Additionally, some guidelines will be suggested for the development of FO methods for other test systems.

Keywords: Chemometrics, Dissolution, Pharmaceutical, UV-VIS Absorbance/Luminescence
Application Code: Pharmaceutical
Methodology Code: Chemometrics

Session # 750 Abstract # 750-13 Poster Sessions

Session Title Commercial Products Characterization

Abstract Title **Simultaneous Analysis of Methylisothiazolinone, Salicylic Acid and Parabens in Antidandruff Hair Shampoos by Monolithic Silica High-performance Liquid Chromatographic Column**

Date: Monday, March 07, 2016 - Afternoo

Time:

Room: Exposition Floor, 400 Aisle

Primary Author Abdulrahman AlMajed
King Saud Univesriy, College of Pharmacy

Co-Author(s)

Abstract Text

A highly selective, precise and rapid high-performance liquid chromatography (HPLC) method has been developed and validated for the determination of methylisothiazolinone (MIT), salicylic acid (SA), methyl paraben (MP), ethyl paraben (EP) and propyl paraben (PP) in antidandruff hair shampoos. Caffeine (CAF) was used as an internal standard (IS) to guarantee a high level of quantitative performance. The chromatographic separation was achieved on a monolithic column with a mobile phase consisted of acetonitrile: 10 mM phosphate buffer (pH 3.0) as gradient elution pumped at a flow rate of 1.0 mL min-1. The chromatographic behaviour of these compounds was studied to demonstrate their chromatographic efficiency, retention, and peak symmetry. The developed method was validated for its specificity, linearity, accuracy, precision and robustness. An experimental design was used during validation to evaluate method robustness. The calibration curves showed excellent linearity (r = 0.998) over concentrations ranging from 1.0 to 100.0 µg mL-1 for all analytes. The mean relative standard deviation (RSD) of the results of inter- and intra-day precision and accuracy of all compounds were 5%. The overall recoveries of all compounds from antidandruff hair shampoos were in the range 97.0–101.5% with %RSD ranging from 0.98 to 3.61%, which were in line with ICH guidelines. The assay was successfully applied in a large number of cosmetic products containing one/and all of these studied compounds obtained from Saudi Arabia market. The developed method can be considered as the quality control of antidandruff hair shampoos.

Keywords: Analysis, Detector, HPLC Columns, Validation

Application Code: Pharmaceutical

Methodology Code: Liquid Chromatography

Session # 750 Abstract # 750-14 **Poster Sessions**

Session Title: Commercial Products Characterization

Abstract Title: **Simultaneous Analysis Red Wine Absorbance, CIE Lab Color Indices and Fluorescence Excitation-Emission Matrices**

Date: Monday, March 07, 2016 - Afternoo

Time:

Room: Exposition Floor, 400 Aisle

Primary Author: Adam M. Gilmore
Horiba

Co-Author(s): Sakiko Akaji

Abstract Text

This study describes the unique capacity of simultaneous absorbance and fluorescence excitation-emission matrix (EEM) spectral instrument and chemometric analysis technology for rapid, precise determination of a wide range of parameters important to commercial wine processing and quality characterization. The instrument method acquires a complete UV-VIS spectrum including the industry standard absorbance wavelength values at 280, 420, 520 and 620 nm which are routinely used to evaluate a wine's phenolic content, hue and intensity. The method also reports the transmission spectrum which can be used to determine a complete array of CIELab Tri-Coordinate Color Descriptions, also valuable for wine color and flavor evaluation. Both the absorbance and CIE Lab analyses yielded significant spectral resolution of several red wine varieties as well as the effects of oxidation on the wine samples. Most importantly, however, the instrument method reports a National Institute of Standards and Technology (NIST) traceable EEM which can be evaluated using a variety of chemometric methods including Parallel Factor Analysis (PARAFAC) and Principal Component Analysis (PCA). For all wines evaluated, the EEM chemometric analyses resolved significant qualitative and quantitative composition parameters that were not discernable with the routine absorbance or CIE Lab data analyses. In conclusion, it is proposed that the absorbance, transmission and EEM chemometric data can be used synergistically to evaluate lot-to-lot, regional, and varietal characteristics as well as sensing the effects of oxidation and sulfite treatment thus making the instrument and analysis methods potentially valuable tools for industrial wine characterization.

Keywords: Chemometrics, Fluorescence, Food Science, UV-VIS Absorbance/Luminescence

Application Code: Food Science

Methodology Code: Fluorescence/Luminescence

Session # 750 Abstract # 750-15 **Poster Sessions**

Session Title: Commercial Products Characterization

Abstract Title: **Extractables and Leachables Analysis of IV Bag Systems Using Thermal Desorption and Stir Bar Sorptive Extraction with GC Single Quad and GC Time-of-Flight Mass Spectrometric Detection**

Primary Author: Andreas Hoffmann
Gerstel GmbH & Co.KG

Date: Monday, March 07, 2016 - Afternoo

Time:

Room: Exposition Floor, 400 Aisle

Co-Author(s): Elizabeth Almasi, Kurt Thaxton, Thomas Albinus

Abstract Text

To assure the safety of pharmaceutical products, the monitoring of potentially harmful contaminants originating from the product itself or its packaging is necessary. The assessment, regulatory guidance, and safety thresholds of the so called "leachables / extractables" from packaging is addressed by the U.S. Food and Drug Administration (FDA), the Product Quality Research Institute (PQRI), the United States Pharmacopeia (USP 1663 and USP 1664) and the International Organization for Standardization (ISO 10993).

In this study, IV bag components were analyzed for extractables using direct thermal desorption/thermal extraction combined with unit resolution GC/MS system. The results were compared to those obtained for leachables by stir bar sorptive extraction of an aqueous simulant stored in the exact same type of IV bag, again combined with GC/MS determination of the leached compounds. In addition, high resolution GC/QTOF mass spectrometer was used to confirm or refute some of the analyte identifications obtained using commercial library searches with the unit resolution MSD system.

Keywords: Gas Chromatography/Mass Spectrometry, Pharmaceutical, Thermal Desorption, Time of Flight MS

Application Code: Pharmaceutical

Methodology Code: Gas Chromatography/Mass Spectrometry

Session # 760 Abstract # 760-2 **Poster Sessions**

Session Title: Data Analysis and Manipulation

Abstract Title: **Ranking of Variables in the Dataset Using the Degree of Separation: Pure Statistical Method**

Primary Author: Waleed Maswadeh
US Army

Date: Monday, March 07, 2016 - Afternoo
Time:
Room: Exposition Floor, 400 Aisle

Co-Author(s): A Peter Snyder, Jason A. Guicheteau

Abstract Text

A dataset can consist of variable responses over multiple classes or groups. Variables are removed that contain very little information. There is no formal algorithm to arrive at a degree of separation (DS) between two distributions of data. The DS90 is defined as the average of the sum of the areas from two probability density functions (PDFs) that contain a >= 90% of A and/or B. To arrive at a DS value, two synthesized PDFs or very large experimental datasets are required. Experimentally it is common practice to generate relatively small datasets. Therefore, the challenge was to find a statistical parameter that can be used on small datasets to estimate and highly correlate with the DS90 parameter. Established statistical methods include the overlap area of the two data distribution profiles, Welch's t-test, Kolmogorov-Smirnov (K-S) test, Mann-Whitney-Wilcoxon test, and the area under the Receiver Operating Characteristics (ROC) curve (AUC). The area between the ROC curve and diagonal (ACD) and the length of the ROC curve (LROC) are introduced. The established, ACD, and LROC methods were correlated to the DS90 when applied on many pairs of synthesized PDFs. The LROC method provided the best linear correlation with, and estimation of, the DS90. The estimated DS90 from the LROC (DS90-LROC) is applied to a database of three Italian wines and three species of iris flowers. An important highlight of the DS90-LROC method is utilizing the LROC curve methodology to test all variables one-at-a-time with all pairs of classes in a dataset.

Keywords: Chemometrics, Data Analysis, Data Mining, Statistical Data Analysis

Application Code: General Interest

Methodology Code: Data Analysis and Manipulation

Session # 760 Abstract # 760-3 **Poster Sessions**

Session Title Data Analysis and Manipulation

Abstract Title **Procurement and Distribution Channels of Commonly Used Drugs in Nigeria: A Case Study of the Pharmaceutical Industry in Abia State**

Date: Monday, March 07, 2016 - Afternoo
Time:
Room: Exposition Floor, 400 Aisle

Primary Author Lilian I. Oguguo
National Malaria Elimination Programme FMoH

Co-Author(s) Chidiebere A. Odike-Aduaka, Ifeoma Agwo

Abstract Text

The objective of the project is to study the distribution and procurement channels of commonly used drugs using selected indicator essential drugs. Considering the unavailability of genuine and essential drugs in the Country which has led to a devastated healthcare with infant mortatlity rate 217/100,000, a crude death rate of 16/ 1000, and Malaria prevalence rate of 919/100,000, this study is crucial.

Significance of the Study: This study will assess the channel of distribution of medicines in the third largest market in Nigeria for sale of medicines and make recommendations applicable to serve similar chaotic situations in the entire country and outside. Experimental procedures / Tools: 162 questionnaire respondents participated, representing 89,5% response, comprising of 28 Pharmacists, 35 hospitals and 118 Patent Medicine vendors as sample population using stratified random sampling method. Core issues and factors determining procurement and distribution such as Source, Quality, Cost per price, availability and wide range were assessed.

Results: Tables and percentages were used to analyze the data. The Chaotic drug distribution system were found to permit unethical practices and so result in the presence of fake and adulterated drugs in the channel. Quality was found to be the most important issue followed by Source while a glaring 27. 7% would compromise Cost for a source they are 100% sure of.

Conclusions: Open markets and porous boarders are found to have resulted from non implementation of drug laws and poor NAFDAC performance towards regulation. A recommendation to sanitize the drug channels and enforce drug laws are made.

Keywords: Drugs, Pharmaceutical, Quality, Quality Control

Application Code: Safety

Methodology Code: Data Analysis and Manipulation

Session # 760 **Abstract # 760-4** **Poster Sessions**

Session Title: Data Analysis and Manipulation

Abstract Title: **Effective Data Management in the Analytical Laboratory**

Primary Author: Toshinobu Yanagisawa
Shimadzu Corporation

Date: Monday, March 07, 2016 - Afternoo
Time:
Room: Exposition Floor, 400 Aisle

Co-Author(s): Kazuhito Wakabayashi, Keisuke Yoshizawa, Masayuki Shibata, Ryuji Nishimoto

Abstract Text

It is very common to support security and data integrity in chromatography data system (CDS). This situation is slightly different for other analytical instruments such as spectroscopy equipment and electronic balances because they are not as many as HPLC and GC in the laboratory and they are often standalone system. A separate data management system is implemented in the software to store, backup, retrieve and archive data for each instrument to comply with regulations. As having separate data management system increases maintenance cost, a unified data management system is required. A solution for this requirement is to support a generic data management interface in the analytical data system to realize unified access control and data integrity for various instruments. This system also enables the user to make summary report from various instruments and to connect to the LIMS system through single interface. As a result, it decreases human errors and total operation cost. In this presentation, it is shown how the unified analytical data system achieves effective data management in the laboratory.

Keywords: Lab Management, Laboratory Informatics
Application Code: Laboratory Management
Methodology Code: Laboratory Informatics

Session # 760 **Abstract #** 760-5 **Poster Sessions**

Session Title: Data Analysis and Manipulation

Abstract Title: **Combining PLS Classification and Regression Analyses for Robust Monitoring of Nuclear Materials Reprocessing**

Date: Monday, March 07, 2016 - Afternoo
Time:
Room: Exposition Floor, 400 Aisle

Primary Author: Robert Lascola
Savannah River National Laboratory

Co-Author(s): Patrick O'Rourke

Abstract Text

Finite signal-to-noise in a data set limits the amount of meaningful variance that can be extracted with a chemometric model. This is problematic if the physical system being modeled has many sources of variation. For example, the process for recovery and purification of plutonium in a nuclear materials processing facility can generate at least 9 different forms of Pu nitrate, depending on the Pu oxidation state and nitric acid content, each of which has unique visible absorption spectra. Solution temperature and other physical parameters will also influence the spectra. Global partial least-squares (PLS) models for total Pu based on this data and requiring 9 or more principal components tend to correlate substantial predictive capacity with a comparatively small and noisy spectral signal. Here, we describe a scheme where spectral classification permits the use of simpler models that are localized to a region of the overall range of process conditions. Wavelength selection is enforced through observation of solution absorbance, and binning based on acidity reduces the underlying sources of variation that must be covered by a single model from 9-10 to 3-6. Overall accuracy for total Pu is improved by a factor of 3, long-term drift is greatly reduced, and the dynamic range of the measurement is improved by ~3x. Potential drawbacks, such as the risks associated with misclassification and the added computational overhead, are shown to be small.

Keywords: Chemometrics, Nuclear Analytical Applications, Process Monitoring, UV-VIS Absorbance/Luminescen
Application Code: Nuclear
Methodology Code: Chemometrics

Session # 770 Abstract # 770-1 **Poster Sessions**

Session Title: Education/Teaching

Abstract Title: **Development of Interactive Learning Modules used in Teaching Instrumental Analysis**

Date: Monday, March 07, 2016 - Afternoo
Time:
Room: Exposition Floor, 400 Aisle

Primary Author: Yi He
John Jay College/CUNY

Co-Author(s): Colleen McNamara, Sandra Swenson, Wong Tiffany

Abstract Text

Hands-on experience on instruments in lab is a critical part in teaching the course of Instrumental Analysis, however, this practice is often restricted by the availability of instruments to students. In order to solve this problem, this work developed a virtual learning module to simulate the use of a specific instrument through touch-screen of an iPad. Students get familiar with and learn how to use the instrument before they access the actual instrument. This practice significantly reduced the learning time that has to be spent on an actual instrument. In addition, this learning module could be accessed anytime anywhere. A demonstration module, gas chromatography-mass spectrometry (GC-MS), was developed and used in teaching undergraduate Instrumental Analysis lab at John Jay College, the City University of New York (CUNY).

Keywords: Education
Application Code: Other
Methodology Code: Education/Teaching

Session # **770** **Abstract #** **770-2** **Poster Sessions**

Session Title: Education/Teaching

Abstract Title: **Conservation Science Tutorial Website: Analysis of VOCs Emitted from Museum Construction Materials**

Primary Author: Jericha Mill
Butler University

Date: Monday, March 07, 2016 - Afternoo
Time:
Room: Exposition Floor, 400 Aisle

Co-Author(s): Gregory D. Smith, Michael J. Samide

Abstract Text

Conservation science applies the techniques of analytical chemistry to the study and preservation of art and objects of cultural heritage. One area of current research is the development of methodology to more quickly assess the suitability of materials for use in a museum space (shipping, display, construction, etc.). In this project, volatile organic compounds (VOCs) emitted from plastics commonly found in a museum environment were examined. Qualitative data was collected using headspace gas chromatography-mass spectrometry (HS-GC-MS) and solid-phase microextraction-GC-MS (SPME-GC-MS) on materials commonly found in packing crates, display cases, and gallery space. Major VOCs detected include unreacted monomers, plasticizers, heat stabilizers, and fire retardants. As data was collected, it was incorporated into an online educational module that is designed to inform a general population on how science can apply to art conservation. Chromatograms and associated mass spectra from 12 common materials are displayed and users are able to interact with the data to learn more about the processes used during this project and the hidden danger posed by pollutant VOCs. This tutorial will be hosted on the Indianapolis Museum of Art's website.

Keywords: Art/Archaeology, Education, Gas Chromatography/Mass Spectrometry, Headspace
Application Code: Art/Archaeology
Methodology Code: Education/Teaching

Session # 770 **Abstract # 770-3** **Poster Sessions**

Session Title: Education/Teaching

Abstract Title: **Analytical Chemistry Course Embedded STEM Projects with Trolox Equivalent Antioxidant Capacity Assay**

Date: Monday, March 07, 2016 - Afternoo

Time:

Room: Exposition Floor, 400 Aisle

Primary Author: Xiaoping Li
Georgia Gwinnett College

Co-Author(s): Rashad Simmons

Abstract Text

Georgia Gwinnett College (GGC) is one of eight institutions participating in the Board of Regents' USG-STEM Initiative. In order to enhance student engagement and student learning, Course embedded STEM research projects are popular in GGC. Several STEM projects have been embedded into analytical chemistry course over the years. In this presentation, we will focus on the STEM project targeting trolox equivalent antioxidant capacity (TEAC) assay. We will showcase a few interesting undergraduate research work, which include the TEAC assay for different juice, wine, tea, coffee, chocolate, beer, vegetable as well as supplement capsule. Different storage condition, extracting condition, solvent effect and brand difference will be discussed as well.

Keywords: Analysis, Food Science, Quantitative, UV-VIS Absorbance/Luminescence
Application Code: Food Science
Methodology Code: Education/Teaching

Session # 770 Abstract # 770-4 **Poster Sessions**

Session Title Education/Teaching

Abstract Title **Inexpensive Teaching Instruments for Atomic Emission and Molecular Spectroscopy**

Primary Author Abd al-karim Ali
Miami University

Co-Author(s) Neil D. Danielson, Taryn L. Winner

Date: Monday, March 07, 2016 - Afternoo
Time:
Room: Exposition Floor, 400 Aisle

Abstract Text

For atomic spectroscopy, we have designed and characterized (using off-the-shelf components) two flame emission instruments of differing complexity for teaching instrumental analysis to our chemistry and biochemistry majors. The first instrument uses a pharmaceutical nebulizer, a camping propane torch for emission, plastic cuvette with colored solution as a filter (e.g. copper sulfate for Na), and a simple photodiode detector with a breadboard current-to-voltage converter circuit to permit voltage measurements. This instrument is easily constructed and portable with a total investment of less than $200. A more sophisticated instrument uses an airbrush as a nebulizer, a short hard plastic tube as a spray chamber, the same propane torch as the excitation source, and a photodiode array spectrometer permitting sub ppm detection limits for the alkali metals.

All three forms of molecular spectroscopy (spectrophotometry, fluorescence, and chemiluminescence) can be effectively demonstrated in our instrumental analysis class using Lego blocks to align the cuvette with the light source (if needed) and the same photodiode detector as previously described for the first flame emission instrument. For spectrophotometry, the appropriate colored LED source mounted on a Lego block is used for measurement of a wide variety of colored acid-base indicators. Due to the flexible design, the effect of pathlength and stray light can be easily taught. The response linearity is comparable to a commercial LED colorimeter. The determination of peroxide in hair lightener products has been determined using luminol chemiluminescence using a similar Lego block arrangement but without the LED source. Current work is comparing UV LED sources for fluorescence of quinine with the photodiode now positioned at right angles to the source. All the Lego spectroscopy instruments cost less than $100; the only real initial expense is the breadboard and power supply for the circuit.

Keywords: Atomic Emission Spectroscopy, Molecular Spectroscopy, Teaching/Education
Application Code: Consumer Products
Methodology Code: Education/Teaching

Session # 770 Abstract # 770-6 **Poster Sessions**

Session Title Education/Teaching

Abstract Title **Strategies for Designing Calibration Curves that Extend over Three Decades of Concentration for Ultraviolet-Visible Absorption Spectroscopy**

Primary Author Lauren Grabowski
University of South Carolina

Date: Monday, March 07, 2016 - Afternoo
Time:
Room: Exposition Floor, 400 Aisle

Co-Author(s) Scott Goode

Abstract Text

Ultraviolet-visible absorption spectroscopy is a widely used technique for determining the concentration of an analyte by measuring the absorbance. Although there have been many studies examining the relative concentration error associated with ultraviolet-visible measurements, there is little that addresses the use of modern instruments capable of measuring absorbances that vary of three decades with high accuracy and precision. There is no information to date optimizing calibration curves to minimize spectrophotometric imprecision.

The major source of this error is detector noise, which obscures and degrades the ability to interpret the response and is dominate at low concentrations. Random instrumental noise in spectrophotometric measurements can be divided into three classes: (1) sources that are completely independent of response; (2) sources with variance that is directly proportional to the response; and (3) sources of noise with variances that are related to the square of the response. The magnitudes of these noise sources affect the signal-to-noise ratio of ultraviolet-visible absorbance measurements.

The goal of this study was to design calibration strategies that span over three decades to optimize precision over the calibration graph for ultraviolet-visible spectrochemical analyses. Different calibration strategies, composed of different concentrations and number of replicates, have been evaluated to try to determine the calibration design that will minimize imprecision as measured by the average relative concentration error integrated over the entire calibration graph.

Keywords: Calibration, Education, UV-VIS Absorbance/Luminescence
Application Code: Quality/QA/QC
Methodology Code: Education/Teaching

Session # 770 Abstract # 770-7 **Poster Sessions**

Session Title Education/Teaching

Abstract Title **Analytical Chemistry Experiments with a Forensic Flavor**

Date: Monday, March 07, 2016 - Afternoo
Time:
Room: Exposition Floor, 400 Aisle

Primary Author Robert Q. Thompson
Oberlin College

Co-Author(s)

Abstract Text

The laboratory portion of the undergraduate analytical chemistry course can be made more relevant and interesting by incorporating experiments that mimic those conducted in a crime laboratory. The experiments, developed by the author over the past decade, can introduce instrumental techniques common to the analytical course – atomic and molecular (UV/Vis/IR) spectrophotometry, gas and liquid chromatography, mass spectrometry – as well as a few less common techniques – optical microscopy, scanning electron microscopy, Raman spectrophotometry – with a decidedly forensic flavor. Experiments include the discovery of ignitable liquid residues by gas chromatography-mass spectrometry, identification of bulk drugs by attenuated total reflectance infrared spectrophotometry, matching of inks from pens and paper by liquid chromatography with multi-wavelength detection, and the determination of the metals content of gunshot residue by atomic absorption spectrophotometry. Inexpensive devices are described for sampling the headspace above simulated blood for ethanol, producing pyrolyzate from automotive clearcoats, and measuring the refractive index of glass shards by the immersion method. In addition, details are given of the preparation of simulated samples such as blood, urine, and bulk drugs, and the collection of real samples such as automotive paint, fibers, glass, and gunshot residue. Typical student-generated results and helpful instructor notes are included with each experiment.

The work was supported by Oberlin College and its research and development funds.

Keywords: Education, Forensic Chemistry, Laboratory, Teaching/Education
Application Code: Homeland Security/Forensics
Methodology Code: Education/Teaching

Session # **770** **Abstract #** **770-8** **Poster Sessions**

Session Title Education/Teaching

Abstract Title **HPLC Refurbishment**

Primary Author Connor Putnam

St. John Fisher College

Date: Monday, March 07, 2016 - Afternoo

Time:

Room: Exposition Floor, 400 Aisle

Co-Author(s)

Abstract Text

High performance liquid chromatography (HPLC) is a widely used analytical technique used in all aspects of chemical analysis. The availability of these instruments to chemistry students and researchers allows for efficient analysis of various liquid chemical samples. It is essential that students have significant experience with HPLC as it is a critical technique used in all fields of chemistry. Having more than one instrument allows for increased time using the instrument and strongly benefits the student by allowing them to become more familiar with the technique. This study included the refurbishing of an HP 1050 Series HPLC and interfacing the instrument with a computer. Parts were acquired from Phoenix Equipment and Agilent Technologies. The parts included a vial tray and inner cabinet, and it was assembled in its entirety. Capillaries were connected between various components of the instrument including the auto-sampler, UV/VIS detector, and degasser. Calibration of the instrument may be performed and will be ready for use to fellow undergraduate students. This projection benefited me by providing hands on experience with working on an analytical experience. Also, it allowed me to experience the logistics side of working on an instrument by contacting various companies for parts that were needed for the instrument.

Keywords: HPLC

Application Code: Other

Methodology Code: Liquid Chromatography

Session # 770 **Abstract # 770-9** **Poster Sessions**

Session Title: Education/Teaching

Abstract Title: **Analytical Chemistry 2.1: An Open-Access Digital Resource for Undergraduate Education in Analytical Chemistry**

Primary Author: David T. Harvey
DePauw University

Date: Monday, March 07, 2016 - Afternoo
Time:
Room: Exposition Floor, 400 Aisle

Co-Author(s)

Abstract Text
The rising cost of textbooks, both print and digital, and the slow pace of change in the publishing industry have created a growing interest in re-envisioning the way textbooks are produced and distributed. Open-access resources are an attractive alternative to the traditional marketplace for publishing, providing faculty members and students with freely available pedagogical materials. This poster will review the open-access electronic textbook [i]Analytical Chemistry 2.1[/i] as well as associated resources, such as case studies and applications written in R, that support the teaching of undergraduate courses in analytical chemistry.

Keywords: Education, Teaching/Education
Application Code: General Interest
Methodology Code: Education/Teaching

Session # **770** Abstract # **770-10** **Poster Sessions**

Session Title Education/Teaching

Abstract Title **Introducing Undergraduate Chemists to Chemometrics: Using Microsoft Excel to Illustrate the NIPALS Algorithm used in Principal Component Analysis and Partial Least Squares Regression**

Primary Author Mark T. Stauffer
University of Pittsburgh - Greensburg

Date: Monday, March 07, 2016 - Afternoo
Time:
Room: Exposition Floor, 400 Aisle

Co-Author(s)

Abstract Text

This paper will examine introduction of undergraduate chemists to chemometrics by illustrating, via Microsoft Excel, the operation of a well-known algorithm for multivariate calibration as well as pattern recognition, two topics covered in chemometrics texts. The algorithm, known as nonlinear iterative partial least squares (NIPALS), is widely used for obtaining principal components (PCs) from arrays of data. NIPALS has been applied to pattern recognition techniques such as principal component analysis (PCA) and multivariate calibration methods such as partial least squares regression (PLSR). The NIPALS algorithm iteratively extracts scores and loadings for the first PC (PC1) of a data matrix that has been preprocessed (at least mean-centered), and calculates a residuals matrix obtained from the preprocessed data matrix. The residuals matrix is then subjected to the NIPALS algorithm to obtain scores and loadings for the next PC, and so on. Use of Excel to perform PCA or PLSR via NIPALS allows students to work through the steps of this algorithm and thus see how NIPALS works. Some examples of the application of NIPALS to real and synthetic datasets, using Excel, will be provided, along with the NIPALS algorithm pseudocode. Results obtained by NIPALS from the example datasets via Excel will be compared with results obtained for the same datasets using singular value decomposition (SVD), another matrix decomposition technique, and MATLAB, a highly utilized software package for chemometrics and statistics. The significance of using Excel to illustrate the operation of NIPALS for PCA and PLSR, along with background information on PCA and PLSR, will also be presented and discussed.

Keywords: Calibration, Chemometrics, Statistical Data Analysis, Teaching/Education
Application Code: Other
Methodology Code: Chemometrics

Session # 770 **Abstract # 770-11** **Poster Sessions**

Session Title: Education/Teaching

Abstract Title: **Applying Analytical Chemistry to Solve Problems in the Developing World**

Date: Monday, March 07, 2016 - Afternoo

Time:

Room: Exposition Floor, 400 Aisle

Primary Author: Ronda L. Grosse
Chemists Without Borders

Co-Author(s): Bego Gerber, Marya Lieberman, Satinder Ahuja, Steve Chambreau

Abstract Text

This poster will describe volunteer work in progress to solve humanitarian problems with analytical chemistry by mobilizing the resources and expertise of the global chemistry community and its networks. Work to date has included clean water initiatives, science education in developing countries and inexpensive tests for analysis of medicines. This poster presentation will review recent projects to measure arsenic concentrations in water sources in Bangladesh, and interactions with local universities to implement testing. Projects in Africa include development of paper analytical devices to provide high quality chemical analysis of pharmaceutical samples and prevent falsified or substandard medications. Validation methods used for field test verification include High Performance Liquid Chromatography and Inductively Coupled Plasma – Mass Spectrometry. The status of these initiatives, technical progress, and ongoing opportunities and challenges will be discussed.

Keywords: Contamination, Environmental/Water, Teaching/Education, Wet Chemical Methods

Application Code: Safety

Methodology Code: Education/Teaching

Session # 770 **Abstract # 770-12** **Poster Sessions**

Session Title: Education/Teaching

Abstract Title: **Interdisciplinary Undergraduate Research in Chemometrics: The Faculty Perspective**

Date: Monday, March 07, 2016 - Afternoo

Time:

Room: Exposition Floor, 400 Aisle

Primary Author: Helen M. Boylan
Westminster College

Co-Author(s): Carolyn Cuff, Christopher Caroff, Domenic DiSanti, Keilah Ireland, Stephanie Homitz

Abstract Text

At Westminster College we have established an ongoing undergraduate research project in the field of chemometrics. This research is a collaboration involving an analytical chemist, a statistician, undergraduate chemistry students, and undergraduate mathematics students. The language barrier between the fields of chemistry and math presents a significant challenge for cross-communication between the two disciplines, but regular interdisciplinary meetings allow us to overcome this difficulty. The statistical analysis of data sets that both disciplines can make sense of (body fat, solar energy production) have been key to our ability to progress to complex chemical data sets (laser induced breakdown spectroscopy, multi-element analysis by ICP-OES). We utilize the open source package RStudio, as well as commercial chemometrics software, to process and analyze our data. In an era of big data, we feel it is critical to provide our students with training and research in the tools and techniques necessary to study complex data sets. This presentation will highlight our use of principal component analysis on a wide range of data sets as a means of collaborating with undergraduate research students in chemistry and math.

Keywords: Chemometrics, Education, Plasma Emission (ICP/MIP/DCP/etc.)
Application Code: Other
Methodology Code: Education/Teaching

Session # 770 **Abstract # 770-13** **Poster Sessions**

Session Title: Education/Teaching

Abstract Title: **Development of an Undergraduate Laboratory Experiment to Determine Arsenic in Sinus Wash and Tap Water by Inductively Coupled Plasma-Mass Spectrometry**

Date: Monday, March 07, 2016 - Afternoo

Time:

Room: Exposition Floor, 400 Aisle

Primary Author: Anna M. Donnell
University of Cincinnati

Co-Author(s): Anne P. Vonderheide, Dawone Holloway, Keaton Nahan

Abstract Text

Arsenic is a toxic element that humans are primarily exposed to through food and water; it occurs as a result of human activities and naturally from the earth's crust. An experiment was developed for a senior level analytical laboratory utilizing an Inductively Coupled Plasma-Mass Spectrometer (ICP-MS) for the analysis of arsenic in household sinus wash and tap water at The University of Cincinnati. This powerful technique can be used to elucidate the elemental composition of a complex sample while offering the benefits of high-precision quantitative analysis.

The chosen matrices further provided the opportunity to demonstrate two important analytical concepts. First, the sinus wash samples displayed signal suppression, illustrating the necessity for employment of the internal standard calibration technique. Secondly, the high chloride matrix of the sinus wash led to the formation of the $ArCl^+$ polyatomic ion in the argon plasma, which overlapped with arsenic's single isotope at $m/z = 75$. Analysis was performed in no gas mode and also with the collision cell in helium mode, allowing the students to observe firsthand the false positive results for arsenic by comparing 75 m/z results in the two modes. In addition, in a third objective, students were able to compare the sinus wash and tap water to current federal drinking water regulations. This laboratory exercise provided a novel and engaging application of ICP-MS and allowed the students to use modern instrumentation and analytical techniques to investigate the current and relevant issue of arsenic in our environment.

The ICP-MS experiment has become part of the Analytical Laboratory course which enrolls roughly 120 students in five sections each spring, many of which are Chemical Engineering majors. It is the hope that this laboratory experiment will lead to an increase in the use of ICP-MS in the undergraduate curriculum across institutions.

Keywords: Education, Elemental Analysis, Environmental/Water, ICP-MS

Application Code: Consumer Products

Methodology Code: Education/Teaching

Session # 780 Abstract # 780-1 **Poster Sessions**

Session Title Electrochemistry

Abstract Title **Laser-Based Heating for Nanopore DNA Duplex Analysis**

Primary Author Christopher E. Angevine
Virginia Commonwealth University

Date: Monday, March 07, 2016 - Afternoo
Time:
Room: Exposition Floor, 400 Aisle

Co-Author(s) Joseph E. Reiner, Sarah Seashols-Williams

Abstract Text

DNA fragment analysis is used in a number of applications from forensic science to diagnostic medicine. The most thorough analytical technique involves sequencing the DNA molecules, but this can prove difficult, time-consuming and costly. Throughput and effectiveness could be improved by first characterizing the relative proportion of different sized DNA sequences in solution. For example, forensic DNA analysis could be streamlined and both time and cost of analysis reduced if the number of contributors in a sample could be estimated prior to sequencing. Methods for "pre-screening" a DNA sample are numerous, but nanopore sensing offers a number of intriguing advantages; namely, the ability to rapidly analyze DNA mixtures in a label-free manner.

This presentation will describe efforts to utilize nanopore sensing as a pre-screening method for characterizing DNA mixtures of different sized DNA fragments. Our approach extends traditional nanopore sensing by incorporating an infrared laser to locally heat the solution in and around a pore. Heating partially unzips the double stranded DNA (dsDNA) confined in the pore volume, which facilitates transit of the DNA through the pore. The time the DNA spends in the pore, or residence time, is directly related to the size of the DNA, and our measurements show that laser-based heating can adjust the dsDNA residence times in an alpha hemolysin pore over several orders of magnitude. This enables better discrimination between different sized DNA. We will demonstrate proof-of-concept characterization of various mixture ratios of a mixture of homopolymer DNA molecules up to 30 basepairs in length.

Keywords: Bioanalytical, Forensic Chemistry, Laser, Nanotechnology
Application Code: Bioanalytical
Methodology Code: Electrochemistry

Session Title Electrochemistry

Abstract Title **Conductivity as Sensor for Real-Time Monitoring of the Solution Soluble Corrosion Products under Cell Culture Conditions During Corrosion of High Purity Magnesium**

Date: Monday, March 07, 2016 - Afternoo
Time:
Room: Exposition Floor, 400 Aisle

Primary Author Kolade O. Ojo
University of Cincinnati

Co-Author(s) Guangqi Zhang, Keaton Nahan, Madhura Joshi, Pravahan Salunke, Sarah Pixley, Tracy Hopkins, Vesselin Shanov, William R. Heineman

Abstract Text
Because controlling the corrosion rate of magnesium metal will be crucial to the success of biomedical implants containing pure magnesium or magnesium alloys, many ways have been sought to improve in vitro tests to analyze corrosion rates, and also to identify new methods of preparing or post-processing magnesium. In this work, for an in vitro assay, we explored the use of a commercially available conductivity sensor to study magnesium corrosion under cell culture conditions that duplicate many physiologically appropriate parameters. With this sensor, we studied the corrosion of two previously untested magnesium single crystal samples that differed in surface treatments that could alter corrosion rates. The results, show that the conductivity changes in (mS/cm) over the total time of immersion, were proportional to the corrosion rates in (mm/y), and also to the total magnesium released detected by inductively coupled plasma mass spectrometry (ICP-MS).

Keywords: Biomedical, Electrochemistry, Elemental Analysis, FTIR
Application Code: Biomedical
Methodology Code: Electrochemistry

Session # 780 **Abstract # 780-5** **Poster Sessions**

Session Title: Electrochemistry

Abstract Title: **Gold Nanoparticles Enzymatic Sensors for Determination of Glucose**

Primary Author: Arunas Ramanavicius
Vilnius University

Date: Monday, March 07, 2016 - Afternoo
Time:
Room: Exposition Floor, 400 Aisle

Co-Author(s): Almira Ramanaviciene, German Natalija, Povilas Genys

Abstract Text

Determination of glucose is very important direction in bioanalytical chemistry due to increasing number of diabetic patients. Enzymatic determination is still dominating in this technological area and the most of commercial glucose biosensors are based on electrochemical activity of glucose oxidizing enzymes. Therefore electron transfer (ET) between enzymes and electrodes is very important issue in the development of biosensors. To overcome the limitations of the most redox enzymes to transfer electrons to usual, a variety of attempts have been done to promote the electron transfer between redox enzymes and electrodes [1], however this issue is still the most critical in the development of electrochemical enzymatic biosensors. In this direction gold nanoparticles (AuNPs) are very promising as redox mediator [2]. In this research 13 nm diameter AuNPs and glucose oxidase (GOx) modified electrode is reported and electrochemically evaluated. Electrochemical glucose biosensor based on graphite electrodes modified by electrochemically deposited gold nanoparticles, immobilized glucose oxidase and Ppy layer were evaluated. It was demonstrated that electrochemical biosensors based on such modified electrodes were applicable for the determination of glucose in human serum samples in the presence of interfering species.

References
1. Ramanavicius A. Ramanaviciene A, Malinauskas A. (2006) Electrochemical Sensors Based on Conducting polymer – Polypyrrole (Review)" Electrochimica Acta 51, 6025-6037.
2. Ramanaviciene A., Nastajute G., Snitka V., Kausaite A., German N., Barauskas-Memenas D., Ramanavicius A. (2009) Spectrophotometric Evaluation of Gold Nanoparticles as Red-Ox Mediator for Glucose Oxidase Sensors and Actuators B-Chemical 137 483–489.

Keywords: Bioanalytical, Biological Samples, Nanotechnology, Sensors
Application Code: Bioanalytical
Methodology Code: Electrochemistry

Session # 780 Abstract # 780-6 **Poster Sessions**

Session Title: Electrochemistry

Abstract Title: **RC Constant Based Label Free Biomarkers Detection**

Date: Monday, March 07, 2016 - Afternoo
Time:
Room: Exposition Floor, 400 Aisle

Primary Author: Pradeep Ramiah Rajasekaran
University of Florida

Co-Author(s): Charles R. Martin, Jennifer Ottoy

Abstract Text

Biomarker sensing is a rapidly growing area as it plays a key role in detection of potentially harmful diseases. Current techniques for bio marker sensing involve labeling with fluorescent dyes, nanoparticles, immune assays etc. Many of the currently existing techniques are labor intensive, expensive or require extreme skill. Here we propose a label free electrochemical technique based on ion current measurement taking into account the poly ionic nature of the biomolecules (proteins and DNA). The conductive natures of the biomolecules are exploited to sense them after immobilizing them between micro electrodes. Upon surface functionalization with an appropriate molecule, biomarkers (proteins or DNA) of interest will be preferentially bound to the surface and provide a conductive pathway for the propagation of ionic current. Standard biochemical immobilization techniques can be employed and they can be easily adapted for any bio marker system. A simple charging curve and an associated RC constant will be able to identify the conductive state of the surface and thus sense the binding of the molecule of interest. Extremely simple fabrication and experimental setup in addition to its label free sensing mechanism are the main features of this technique. This technique has the potential to be extremely versatile and adaptable for many different biological systems.

Keywords: Biological Samples, Biosensors, Electrochemistry
Application Code: Clinical/Toxicology
Methodology Code: Electrochemistry

Session # 780 Abstract # 780-7 **Poster Sessions**

Session Title Electrochemistry

Abstract Title **Electrochemical Detection of Evoked Dopamine Release in Zebrafish**

Primary Author Mimi Shin
University of Kansas

Date: Monday, March 07, 2016 - Afternoo
Time:
Room: Exposition Floor, 400 Aisle

Co-Author(s) Mia Furgurson, Michael A. Johnson, Thomas Field

Abstract Text

Zebrafish have emerged as a useful vertebrate model in neuroscience, toxicology, drug screening, and developmental biology research. Immunohistochemistry, imaging, and electrophysiology have been used to understand dopaminergic neuron function in zebrafish; however, the direct electrochemical measurement of dopamine release has not been accomplished. In this study, we harvested zebrafish brains and maintained them under conditions that kept them viable for 6 hours. Electrically-evoked dopamine release in the telencephalon was measured using fast-scan cyclic voltammetry (FSCV) at carbon fiber microelectrodes. Electrical stimulation parameters, including number of pulses, recovery time, and stimulation current were optimized. Additionally, pharmacological agents were applied in order to confirm the presence of dopamine. To our best knowledge, this is the first time dopamine release has been electrochemically measured in zebrafish brain.

Keywords: Bioanalytical, Neurochemistry
Application Code: Neurochemistry
Methodology Code: Electrochemistry

Session # 780 **Abstract # 780-8** **Poster Sessions**

Session Title: Electrochemistry

Abstract Title: **Quantification of Ion Selectivity of Single Asymmetric Nanopore-Channels for Better Energy Harvesting from Salinity Gradients**

Date: Monday, March 07, 2016 - Afternoo

Time:

Room: Exposition Floor, 400 Aisle

Primary Author: Warren D. Brown
Georgia State University

Co-Author(s): Dengchao Wang, Gangli Wang, Maksim Kvetny, Yan Li

Abstract Text

Membranes allowing ion exchange are an integral part of energy devices such as fuel cells and supercapacitors, separation applications such as water desalination, and energy harvesting from salinity gradients that have attracted significant research interest recently. Ion selectivity is an important parameter characterizing the efficiency of energy conversion or separation in ion-exchange membranes. However the interactions between the mobile ions and the membrane surfaces at atomic and nanometer scale are not well understood. It is well known that ensemble description and analysis fail to explain many emerging phenomena of nanoscale transport. To address the uncertainty in studying transport in ion-exchange membranes which are constituted of ensemble pores with variability in pore size and geometry, we study ion transport through single conical quartz nanopipettes with defined geometry. An analytical method is established to determine the transference number of cations with different mobility, i.e. the ion selectivity of a single conical nanodevice/interface. Aqueous electrolytes with common anion but different monovalent cations such as lithium, sodium, potassium and tetra alkyl ammonium ions were studied. The ion concentration and salinity gradient across the nanopipettes were systematically varied in current-potential measurements and current clamping potentiometry. These electroanalytical measurements were combined with Comsol simulation to quantify the transference number of different cations through a single asymmetric nanopore with negative surface charges. The magnitude of the transference number for the same cation in different nanopipettes reveals the impacts of nanogeometry and nanointerfaces. Understanding the ion transport selectivity and mechanism will provide insights to further improve the performance of ensemble membrane devices for better energy and separation applications.

Keywords: Electrochemistry, Fuels\Energy\Petrochemical, Ion Exchange, Nanotechnology
Application Code: Nanotechnology
Methodology Code: Electrochemistry

Session # 780 Abstract # 780-9 **Poster Sessions**

Session Title Electrochemistry

Abstract Title **Charge Transfer Mechanism of Organic Molecules Associated with Dendrimers at Polarized Liquid/Liquid Interfaces**

Date: Monday, March 07, 2016 - Afternoo
Time:
Room: Exposition Floor, 400 Aisle

Primary Author Hiroki Sakae
Kanazawa University

Co-Author(s) Hirohisa Nagatani, Hisanori Imura

Abstract Text

Charge transfer reactions at interfaces between two immiscible electrolyte solutions (ITIES) have been investigated as a model reaction for mass-transport and enzymatic reactions on biomembrane systems. The ion partitioning property at ITIES is useful to evaluate the pharmacokinetic distribution of drugs Dendrimers are unique and nontraditional polymers which have a number of advantages as drug carrier or container. Therefore, the dendrimers have been attracted much attention to develop novel drug delivery systems (DDS). Recently, we reported the molecular encapsulation behavior of anionic molecules in the polyamidoamine (PAMAM) dendrimers at the polarized water|1,2-dichloroethane (DCE) interface [1], and the generation of the dendrimer, electrostatic and hydrophobic interactions, and the axial coordination to the metal center play important roles for the stability of ion associates at the interface.

In order to evaluate the functionality of the dendrimer as a molecular container in DDS, we investigated the interfacial behavior of bio-related materials at ITIES in the presence of the dendrimers by means of potential modulated fluorescence (PMF) spectroscopy [2]. For instance, the PMF responses for the adsorption process of dipyridamole (vasodilator, DIP) were significantly modified in the presence of the dendrimer under acidic conditions. This result indicates that positively charged dendrimers adsorbed at the interface prevent the coadsorption of protonated DIP through the electrostatic repulsion. PMF analysis demonstrated that the dendrimers have a potential ability as a modifier for the pharmacokinetic properties of organic molecules on a biomembrane.

[1] H. Sakae, H. Nagatani, K. Morita, H. Imura, [i]Langmuir[/i] [b]30[/b] (2014) 937.
[2] H. Nagatani, T. Sagara, [i]Anal. Sci.[/i] [b]23[/b] (2007) 1041.

Keywords: Drugs, Separation Sciences, Spectroelectrochemistry, Surface Analysis
Application Code: Drug Discovery
Methodology Code: Electrochemistry

Session # 780 Abstract # 780-10 **Poster Sessions**

Session Title Electrochemistry

Abstract Title **Study of Electrochemical Hydrogen Evolution and Oxygen Evolution Reactions at Ir and Ru Oxide Alloys with Scanning Electrochemical Microscopy**

Primary Author Yun-Bin Cho
Ewha Womans University

Co-Author(s) Chongmok Lee, Youngmi Lee

Date: Monday, March 07, 2016 - Afternoo
Time:
Room: Exposition Floor, 400 Aisle

Abstract Text

Hydrogen is an important energy source. The electrochemical hydrogen evolution reaction (HER) is attractive due to its usage in fuel cells without concerns of contaminants. The electrochemical oxygen evolution reaction (OER) has been of great interest in a wide range of industrial applications such as energy conversion/storage devices. Many works have proposed efficient catalyst systems for HER and OER. These reactions are catalyzed most effectively by Pt group metals. Among them, Ir and Ru oxides have exhibited their excellent activities for HER and OER, respectively, showing high ohmic conductivity and chemical/thermal stability. In this study, in order to make an optimized catalyst system facilitating both HER and OER, Ir and Ru oxide alloys are fabricated by electrochemical co-deposition. For the composition change of the relative ratios between Ir and Ru oxides, the contents of Ir and Ru precursor concentrations are varied in the deposition solutions. Quantitative analyses for their activities are performed with scanning electrochemical microscopy (SECM) which can provide much specific and direct information toward OER and HER. SECM has been proven to be useful in a quantitative study associated with electrode reactions because this technique can detect the electrode reaction intermediates and/or products sensitively. The collection efficiency of the electrode reaction intermediates/products in SECM is mainly affected by displacement between the tip and substrate electrodes, and these electrode diameters, and RG value (the ratio of the metal radius to the tip radius including a glass sheath). These factors are also adjusted to obtain nearly ~100% collection efficiency for quantitative study of HER and OER with a better precision.

This work was finally supported by the National Research Foundation of Korea (NRF) funded by the Ministry of Science, ICT and Future Planning (2014R1A2A2A05003769).

Keywords: Electrochemistry, Electrode Surfaces, Microelectrode

Application Code: Fuels, Energy and Petrochemical

Methodology Code: Electrochemistry

Session # 780 Abstract # 780-11 **Poster Sessions**

Session Title Electrochemistry

Abstract Title **Automated, Accurate pH and Conductivity Measurements Using a Discrete Photometric Analyzer with an ECM Module**

Primary Author Mari Kiviluoma
Thermo Fisher Scientific

Date: Monday, March 07, 2016 - Afternoo
Time:
Room: Exposition Floor, 400 Aisle

Co-Author(s) Annu Suoniemi-Kahara

Abstract Text

In this study pH and electrical conductivity (EC) were measured from water and beverage samples using an Electro Chemical Measurement (ECM) unit integrated into an automated discrete photometric analyzer. This integrated ECM unit is capable of simultaneously measuring both pH and conductivity alongside the photometric testing. Accuracy of the results was verified by parallel testing with manual conductivity and pH meters.

In the ECM unit, the conductivity measurement is performed via two electrodes. In between these electrodes a sine wave potential gives rise to a measureable current from which the sample conductivity can be calculated. pH is measured using a two-electrode galvanic cell consisting of an indicator pH electrode and a reference electrode. Both conductivity and pH are measured at 37 °C however, conductivity results can also be automatically reported at 25 °C because the discrete analyzer software has a robust system in which the sample result may be corrected to correspond to a reference analyzer result. The correction factor (Cfactor) may be used in combination with the measured EC to calculate the corrected ECcorr. Published Cfactor values are available for a variety of sample types. This test has been developed to measure conductivity within a range from 20 [illegible]S/cm to 112 mS/cm. The pH test measures a pH range from 2 to 12. Results show very good precision (0.5% or better for pH and 1.8% or better for EC) and good correlation to the manual methods with an r2 ranging from 0.910 to 1.000.

Keywords: Electrochemistry, Environmental Analysis, Instrumentation, Laboratory Automation

Application Code: Environmental

Methodology Code: Electrochemistry

Session # 780 Abstract # 780-13 **Poster Sessions**

Session Title Electrochemistry

Abstract Title **Synthesis and Characterization of Silver Coated Electrospun Cobalt Nanotubes and Their Electrocatalytic Activity**

Date: Monday, March 07, 2016 - Afternoo
Time:
Room: Exposition Floor, 400 Aisle

Primary Author Areum Yu
Ewha Womans University

Co-Author(s) Chongmok Lee, Myung Hwa Kim, Youngmi Lee

Abstract Text

Electrospinning has gained much attention due to not only its diversity of the electrospinning materials but also its simplicity. In this study, we prepare Co nanotubes by electrospinning, followed by calcination and reduction. The Co nanotubes undergo galvanic replacement reaction (GRR) with Ag precursor in aqueous media to produce AgCo nanotubes. In order to optimize the synthetic condition of AgCo nanotubes, the reduction time of Co[sub]3[/sub]O[sub]4[/sub] nanotubes and the GRR time between Co nanotubes and Ag precursor are changed. The morphologies and compositions of the prepared AgCo nanotubes are characterized by field-emission scanning electron microscopy, transmission electron microscopy, X-ray diffraction and X-ray photoelectron spectroscopy. For the electrochemical characterization, three-electrode cell is used with a glassy carbon electrode loaded with the synthesized AgCo nanotubes as the working electrode, a saturated calomel electrode as the reference electrode, and a coiled platinum wire as the counter electrode. Electrochemical activities of these nanomaterials for oxygen reduction reaction (ORR) are characterized with rotating disk electrode voltammetry. As prepared AgCo nanotubes show better electrocatalytic activity (onset potential and current density) for ORR than either Ag nanowires or Co nanotubes.

This research was supported by Basic Science Research Program through the National Research Foundation of Korea (NRF) funded by the Ministry of Science, ICT& Future Planning (2014R1A2A2A05003769).

Keywords: Electrochemistry, Materials Characterization, Material Science, Voltammetry

Application Code: Material Science

Methodology Code: Electrochemistry

Session # 780 Abstract # 780-14 **Poster Sessions**

Session Title: Electrochemistry

Abstract Title: **Single Microwire Electrodeposition by Electrochemical Liquid-Liquid-Solid Growth**

Date: Monday, March 07, 2016 - Afternoo

Time:

Room: Exposition Floor, 400 Aisle

Primary Author: Tim Zhang
University of Michigan

Co-Author(s)

Abstract Text

The electrochemical liquid-liquid-solid growth technique was investigated through the electrodeposition of individual crystalline germanium microwires. Experiments tested whether the current transient from potentiostatic electrodeposition can provide predictive information about the microwire morphology. A photoresist pattern of a single hole of defined diameter and depth was fabricated onto a silicon substrate, creating an ultramicroelectrode. The hole was filled with liquid metal, which acted as both the crystallization solvent and the working electrode. Temperature, electrode diameter and liquid metal identity were varied as a germanium microwire was electrodeposited under constant potential onto the liquid metal. The surface tension and curvature of the liquid metal before and after the germanium electrodeposition were observed to determine if a contact angle threshold for nucleation exists. The current transients were examined for correlations to the microwires' structural features to determine if real-time adjustment of microwires during the growth process is feasible.

Funding was provided by the National Science Foundation.

Keywords: Electrochemistry, Material Science, Microelectrode, Semiconductor

Application Code: Material Science

Methodology Code: Electrochemistry

Session # **780** **Abstract #** **780-15** **Poster Sessions**

Session Title: Electrochemistry

Abstract Title: **Simultaneous Detection Heavy Metals by Anodic Stripping Voltammetry Using Carbon Nanotube Thread**

Date: Monday, March 07, 2016 - Afternoo

Time:

Room: Exposition Floor, 400 Aisle

Primary Author: Daoli Zhao
University of Cincinnati

Co-Author(s): David R. Siebold, Noe R. Alvarez, Tingting Wang, Vesselin Shanov, William R. Heineman

Abstract Text

Carbon nanotube (CNT) threads are a type of CNT arrays that consist of super long CNTs. CNT threads inherit the advantages of CNTs, while avoiding the potential toxicity caused by individual CNTs. Electrodes based on CNT threads were fabricated and used for simultaneous detection of trace levels of Cu2+, Pb2+ Cd2+ and Hg2+ by anodic stripping voltammetry (ASV). The operational parameters such as deposition potential and deposition time were optimized in 0.1 M acetate buffer solution (pH = 4.5). The CNT thread electrode gives well-defined, reproducible and sharp stripping signals for individual and simultaneous detection of heavy metals. The detection limits are far below the requirement of WHO in water. The attractive behavior of the novel CNT thread sensing provides promise for onsite environmental and biomonitoring of heavy metals.

Keywords: Electrochemistry, Electrodes, Sensors

Application Code: Environmental

Methodology Code: Electrochemistry

Session # 780 **Abstract # 780-16** **Poster Sessions**

Session Title: Electrochemistry

Abstract Title: **Low Concentration Detection by Electrogenerated Chemiluminescence Using Bipolar Electrochemistry in a Thin Layer Manner**

Date: Monday, March 07, 2016 - Afternoo
Time:
Room: Exposition Floor, 400 Aisle

Primary Author: Songyan Yu
Auburn University

Co-Author(s): Curtis Shannon, Mark D. Holtan, Sanjun Fan

Abstract Text

Bipolar electrochemistry generates an asymmetric behavior on conductive objects wirelessly. Electrogenerated chemiluminescence(ECL), provides us a way of indirect detection instead of current readout. The thin layer(gap) electrochemistry can enhance the signal, even for single molecule detection. Taking advantage of these methods, our goal is to detect low concentration electrochemical active species by ECL using a thin layer method via bipolar electrodes, especially with simple readout instrument. We first began our project by employing 3D printing technique to build up devices which match the requisite configuration. We found that the ECL intensity was linear versus the redox couple concentration, such as K3Fe(CN)6/K4Fe(CN)6, at a certain range in either closed bipolar or open bipolar cell with distinct external potential requirement. Many parameters, such as timing, degassing, conductance of the electrolyte solution, ionic strength of the sample were studied during research with record of the potential profile, demonstrating that they need to be controlled carefully. However, these devices are restricted to their resolution, reproducibility, consistency, sample replacement, minimizing and precise control of gap size. Hence, we are going to combine the microfluidics methodology with the 3D printed template. By doing this, better fluid dynamics, less contamination on the surface, less energy loss, consistent gap size and expedient and continuous measurement will be realized. By way of electron-beam lithography-based microfabrication, it may be capable of detecting single molecule in the future. Overall, our thin layer bipolar electrochemical cell will separate the small with necessary ECL solution, making it in an intact situation, in a steady-state wireless behavior and offer us a significant signal enhancement with simple instrumentation. It would become a flexible, high throughput, cost-effective and energy saving and reusable method for analytical detection.

Keywords: Chemiluminescence, Electrochemistry, High Throughput Chemical Analysis, Sensors
Application Code: High-Throughput Chemical Analysis
Methodology Code: Electrochemistry

Session # 780 **Abstract # 780-17** **Poster Sessions**

Session Title: Electrochemistry

Abstract Title: **Atypical Induction of Membrane Potential in Ion Selective Membranes**

Date: Monday, March 07, 2016 - Afternoo

Time:

Room: Exposition Floor, 400 Aisle

Primary Author: Demetra M. Pantelis
University of Florida

Co-Author(s): Pradeep R. Rajasekaran

Abstract Text

Within every eukaryotic cell, a membrane potential associated with an ion selective membrane enables signal transduction between neighboring cells. Driven by a concentration gradient, ions flow across the membrane. This flow of ions in turn generates a non-zero resting membrane potential. Biological membranes can be simulated under laboratory conditions using an ion exchange membrane. We are investigating the effect of a current driven concentration gradient on the resting membrane potential of an ion exchange membrane. In our experimentation, a quaternary ammonium anion exchange membrane is used. The concentration gradient required for generation of a membrane potential is produced in an unconventional way by passing a current through a membrane, which separates an isotonic solution of sodium nitrate. The application of a current/potential across a membrane placed between two isotonic solutions forces nitrate ions to move in the direction of the applied current in order to support the applied current. This leaves the membrane with a non-zero resting potential upon termination of the applied potential. This method allows us to gradually generate and maintain a finite potential in perm-selective membranes with an initial membrane potential of zero. With the support and funding of the University of Florida Department of Chemistry, our current and future research will help us to better understand the transfer of ions, subsequent potential generation and signal transduction across biological membranes. This may have therapeutic applications for a wide variety of nervous disorders.

Keywords: Electrochemistry, Ion Exchange, Membrane
Application Code: Process Analytical Chemistry
Methodology Code: Electrochemistry

Session Title: Electrochemistry

Abstract Title: **Characterization and Spectroelectrochemical Sensing with a Boron Doped Diamond Optically Transparent Electrode Coated with a Charge Selective Polymer Film**

Primary Author: Cory A. Rusinek
University of Cincinnati

Date: Monday, March 07, 2016 - Afternoo
Time:
Room: Exposition Floor, 400 Aisle

Co-Author(s): Daoli Zhao, Michael Becker, Necati Kaval, Robert Rechenberg, William R. Heineman

Abstract Text

Spectroelectrochemistry is a well-known technique used to achieve multimode selectivity in one sensor. The sensor usually consists of an optically transparent electrode (OTE) coated with a charge selective polymer film. These polymer films are employed to pre-concentrate analyte at the OTE surface so that it can be sufficiently detected optically through electrochemical modulation. OTEs such as Indium Tin Oxide (ITO) have been used extensively in this method but little is known about the applicability of such sensors using other OTE materials, such as Boron Doped Diamond (BDD). One distinct advantage of BDD OTEs over ITO OTEs is their significant increase in sensitivity for organic compounds, such as para-aminophenol and hydroquinone. With this, we have developed an absorbance and fluorescence-based sensor with a BDD OTE coated with a sulfonated ionomer film, Nafion. This is demonstrated with tris(2,2'-bipyridyl)ruthenium(II) ion [Ru(bpy3)2+] using an attenuated total reflectance (ATR) flow cell setup for both absorbance and fluorescence. With a Nafion coated BDD optically transparent thin layer electrode (OTTLE), we developed a fluorescence based sensor for a common PAH, 1-hydroxypyrene (1-pyOH), achieving a detection limit of 398 nM (87 ppb). Using these sensing techniques we were able to manifest new applications while broadening the use of spectroelectrochemistry, OTEs, and BDD as an electrode material.

Keywords: Electrodes, PAH, Spectroelectrochemistry, Voltammetry
Application Code: General Interest
Methodology Code: Electrochemistry

Session # 780	Abstract # 780-19	Poster Sessions

Session Title Electrochemistry

Abstract Title **Simultaneous Sensing of -Tocopherol and Retinol in Micellar Media by Using Poly(2,2-(1,4 phenylenedivinylene) bis-8-hydroxyquinaldine)/MWCNTs Modified Electrode**

Primary Author Hayati Filik
Istanbul University

Date: Monday, March 07, 2016 - Afternoo
Time:
Room: Exposition Floor, 400 Aisle

Co-Author(s)

Abstract Text

Vitamins are categorized into two main classes, fat-soluble and water-soluble. Some fat-soluble vitamins include vitamin A (retinol), vitamin E (-tocopherol), vitamin D (radiostol), and vitamin K (antihemorrhagic vitamins). Fat-soluble vitamins are found mainly in animal products, fatty foods, and pharmaceuticals. The ensure of vitamins depends on the diet; however, even foods that bear the essential vitamins can have reduced vitamin content after storage, processing, or cooking. Therefore, many people take multivitamin tablets and/or lead in milk powder and vitamin-fortified beverages to supplement their diet. To supply that these foods and multivitamin tablets contain the labeled quantities of vitamins, there requires to be a quality control assay for them.

A novel method was developed for the simultaneous electrochemical determination of -tocopherol and retinol (vitamin A) using poly(2,2´-(1,4-phenylenedivinylene) bis-8-hydroxyquinaldine)/multi-walled carbon nanotubes composite modified glassy carbon electrode. Simple, quick and credible voltammetric approach for the simultaneous electrochemical determination of -tocopherol and retinol in real samples using micellar media (Triton X-100) have been evaluated. The use of nonionic surfactants in voltammetry of -tocopherol and retinol provides their solubilization and allows to perform detection in solvent-free media. The calibration graph was linear in the range 8–100 μM of -tocopherol with the detection limit of 0.1 μM in the presence of 80 μM retinol. But also, the peak intensity is linear with the concentration of retinol from 5 μM to 200 μM with a detection limit of 0.8 μM (S/N = 3) in the presence of 40 μM -tocopherol (Fig. 1.). The developed method is simple and rapid and it was successfully implemented in the determination of the two essential vitamins in pharmaceutical samples.

Keywords: Electrochemistry, Food Contaminants, Trace Analysis, Voltammetry
Application Code: Food Science
Methodology Code: Electrochemistry

Session # 780 Abstract # 780-20 **Poster Sessions**

Session Title: Electrochemistry

Abstract Title: **Electrochemical Investigation of a Series of Uranyl Salen Complexes: Effect of Ligand Conjugation on the U(V)/U(VI) Redox Couple**

Date: Monday, March 07, 2016 - Afternoo

Time:

Room: Exposition Floor, 400 Aisle

Primary Author: Emily E. Hardy
Auburn University

Co-Author(s): Anne E. Gorden, Madeleine A. Eddy

Abstract Text

Although nuclear power is an integral part of the United States energy policy, the calculated risk of environmental exposure and the presence of radioactive waste is still a common concern. Uranium, when introduced into the environment, is stable in aqueous solution and in wastewater streams as the uranyl (UO[sub]2[/sub][sup]2+[/sup]) ion, with uranium in the hexavalent oxidation state. Many examples of salen type ligands, with soft imine Schiff base donors, have been used to sense uranyl contamination. Here, a series of these scaffolds and uranyl complexes have been characterized by cyclic voltammetry, in particular the U(V)/U(VI) redox couple. Understanding the stabilization of this redox couple can aide in fundamental actinide chemistry and the covalency of the bound uranium.

Keywords: Voltammetry
Application Code: Nuclear
Methodology Code: Electrochemistry

Session # 780 Abstract # 780-21 **Poster Sessions**

Session Title Electrochemistry

Abstract Title **Predictable Standard Potential of Solid-Contact Ion-Selective Electrodes by Using Prussian Blue Analogues as Solid Contacts**

Primary Author Yu Ishige
Hitachi Ltd.

Date: Monday, March 07, 2016 - Afternoo
Time:
Room: Exposition Floor, 400 Aisle

Co-Author(s) Stefan Klink, Wolfgang Schuhmann

Abstract Text

Solid-contact ion-selective electrodes (SC-ISEs) have already been applied for point-of-care diagnostics according to their small sizes and low costs. Nonetheless, they need improvement to achieve as much reproducible, stable and predictable potential as classical ISEs with inner filling solutions. Choice of a solid contact which exists between ion-selective membrane and underlying solid contact is crucial for SC-ISEs' potential. We considered that intercalation compounds which are recently applied to the solid contact should satisfy the requirement to define the standard potential of SC-ISEs due to the narrow range of standard potential of intercalation compounds. We focused on Prussian Blue analogues (PBAs) because of their tunability of standard potential by altering their redox active transition metal, by changing their state of charge or by choosing intercalating cations. PBAs of potassium intercalated iron hexcacyanoferrate (K-FeHCF), nickel hexacyanoferrate (K-NiHCF) and copper hexacyanoferrate (K-CuHCF) were prepared and showed standard potential around 250, 500 and 650 mV vs Ag/AgCl. PBAs of sodium and calcium intercalated nickel hexacyanoferrate (Na-NiHCF, Ca-NiHCF) were also prepared and showed standard potential around 450 and 250 mV vs Ag/AgCl. SC-ISEs were fabricated by combining these PBAs with corresponding ion-selective membranes. These SC-ISEs showed standard potentials correlated with those of PBAs (r[sup]2[/sup]=0.95). These results clearly indicated that standard potentials of SC-ISEs using PBAs were dominated by those of PBAs. These SC-ISEs also showed high reproducibility (±0.6 mV, n=4) and high stability (±2.0 mV in 8 months) in potential.

Keywords: Bioanalytical, Electrochemistry, Electrodes, Ion Selective Electrodes

Application Code: Bioanalytical

Methodology Code: Electrochemistry

Session # **780** **Abstract #** **780-22** **Poster Sessions**

Session Title Electrochemistry

Abstract Title **Electrochemical Behaviour of Highly- and Poorly-Doped p-Si-AuNP Electrodes**

Primary Author Mehran Kashi
University of New South Wales

Date: Monday, March 07, 2016 - Afternoo
Time:
Room: Exposition Floor, 400 Aisle

Co-Author(s) Justin J. Gooding, Moinul Choudhury, Simone Ciampi, Vinicius Goncales, Yanfang Wu

Abstract Text

Light-activated electrochemistry of silicon enables us to switch electrochemistry on and off using light. This concept has been used to fabricate electrode arrays versatile in terms of electrode position and geometry. Moreover, high-density electrochemical arrays would be possible in a way that electrochemistry can be performed where we want, when we want and as dense as we want, but only using a single connecting wire. Herein, to expand the capability of this concept, electrochemical behaviour of highly- and poorly-doped p-type silicon electrodes modified with gold nanoparticles was studied. For this purpose, Si electrodes were modified with 1,8-nonadiyne and then, AuNPs were attached to the surface by incubating the electrode in AuNP dispersion after clicking an amine-terminated self-assembled monolayer to the acetylene end. Cyclic voltammetry results for a highly-doped silicon show that while the amine-terminated surface blocks the electrochemistry of hexamine ruthenium chloride, electron transfer is switched on after depositing the gold nanoparticles. Interestingly, electron transfer on a poorly-doped Si-AuNP electrode behaves differently from a highly-doped due to presence of a depletion layer at a certain potential range. Also, the electron transfer was shown to be negligible in dark while the peaks appeared distinctly under illumination. Finally, Si oxide evolution was studied by X-ray photoelectron spectroscopy before and after the voltammetry tests.

Keywords: Electrochemistry, Microelectrode, Nanotechnology, Semiconductor
Application Code: Material Science
Methodology Code: Electrochemistry

Session # 780 **Abstract # 780-25** **Poster Sessions**

Session Title: Electrochemistry

Abstract Title: **Novel Electrode PtCr/PAA (Polyamic Acid) for Efficient Ethanol Oxidation Reaction**

Primary Author: Jing Zhang
Binghamton University

Date: Monday, March 07, 2016 - Afternoo
Time:
Room: Exposition Floor, 400 Aisle

Co-Author(s)

Abstract Text

Among different types of fuel cells, direct alcohol fuel cells (DAFC), are recently attracting more attention and become an interesting and clean alternative energy source, due to their great application potential [1].Direct ethanol fuel cells can work at low temperature, possess high theoretical mass energy density, and are environmentally friendly. Ethanol Oxidation Reaction (EOR) can be operated both in acidic or alkaline solutions. Fundamental studies of ethanol electro-oxidation in acidic media have been mostly performed on platinum. Deposit platinum and chromium on the surface of carbon electrode as catalyst for EOR[2-4]. The electrodeposited PtCr nanoparticles will help to improve the efficiency of EOR but they are very easily lost into solution. An efficient electrode for a fuel cell must be conductive, hydrophobic, requires high surface areas and high porosity to enable mass transport of H+ as well as be corrosion-resistant. We hereby present the use of □-conjugated poly(amic) acid (PAA) [5-8] as novel materials for EOR. PAA is a yellow viscous solution. The catalyst was protected by PAA in order to improve stability. The electrode stability was recorded at varying potential cycling. Based on the study, the PAA is not a catalyst itself, but a PAA layer permits the diffusion of ethanol towards the PtCr nanoparticles in acidic solution.

Keywords: Electrochemistry, Material Science
Application Code: Material Science
Methodology Code: Electrochemistry

Session # 780 Abstract # 780-26 **Poster Sessions**

Session Title Electrochemistry

Abstract Title **Voltammetric Serotonin Measurements in Mouse Models of Depression**

Date: Monday, March 07, 2016 - Afternoo
Time:
Room: Exposition Floor, 400 Aisle

Primary Author Rachel A. Saylor
University of South Carolina

Co-Author(s) Aya Abdalla, Parastoo Hashemi

Abstract Text
Depression is a debilitating disorder, affecting a significant global population. Because the chemistry underlying depression is not well defined, pharmacological therapies are variable and have low efficacy rates. In addition, because there exist no reliable preclinical screening tools for antidepressant efficacy, major pharmaceutical companies have dramatically toned down their drug discovery efforts. Given that the WHO predicts that depression will be the leading cause of disability worldwide by 2030, there is a critical need to address the fundamental chemistry that underpins depression to provide new paths for drug discovers towards more effective therapies. In this work, we combine [i]in vivo [/i] fast scan cyclic voltammetry (FSCV), fast scan controlled adsorption voltammetry (FSCAV), mathematical modeling, and animal behavior to ask a fundamentally important question: "how do serotonin dynamics differ between normal and behaviorally depressed mice?" We employ a potent model, the social defeat model, to behaviorally depress mice over 10 days. After behavioral tests that confirm a depressive state, we apply [i]in vivo[/i] FSCV and FSCAV in the mouse hippocampus to make real-time, quantitative measures of electrically stimulated and ambient serotonin. We finally apply a 12 differential equation model to experimental data to decipher mechanistic differences in the chemical regulatory mechanisms controlling serotonin between healthy and depressed mice. Our work reveals previously unknown, important perturbations to the serotonin system in a depression model that has the potential to better focus therapeutic efforts.

Keywords: Electrochemistry, Microelectrode, Neurochemistry, Voltammetry
Application Code: Neurochemistry
Methodology Code: Electrochemistry

Session # 780 **Abstract # 780-27** **Poster Sessions**

Session Title Electrochemistry

Abstract Title **Kinetic Size-Spectra of Gas Molecules at Ionic Liquid (IL)-Metal Interface and Its Application for Highly Selective Gas Sensing**

Date: Monday, March 07, 2016 - Afternoo

Time:

Room: Exposition Floor, 400 Aisle

Primary Author Zhe Wang
Xavier University

Co-Author(s)

Abstract Text

Ionic liquids are widely used as nonvolatile solvents and electrolytes. In the presence of electric fields, their structure and properties at the electrode interface resemble crystalline solids that are significantly different from traditional electric double layers. Currently, the electrified IL-electrode interfaces with adsorbed gas molecules have not been experimentally studied. In this work, we systematically measured the differential capacitances of electrified IL-electrode interface in the presence of eight different gas molecules. The observed change of differential capacitance at the IL-electrode interface in the presence of gas molecules has a local maximum at a certain potential. This potential was found to be unique for each specific gas molecule and the amplitude of the differential capacitance change at the potential is related to the concentration of the gas molecules. We analyzed the effect of various factors (a dipole moment of the gas molecule, an applied electric field, availability of the free space at the IL-electrode interface) on the experimentally measured differential capacitance values. This allows us to establish a reversible and sensitive gas sensor method that is based on a voltage-dependent change of the distribution of gas molecules of a given kinetic diameter within a double-layer region adjacent to the electrode. We validate this new sensor method by characterizing SO2 detection. A reversible quantification of SO2 at ppb levels with less than 1.8% signal from other interfering species (i.e. CO2, O2, NO2, NO, SO2, H2O, H2 and cyclohexane when they are tested at the same concentration as SO2) was achieved. This study opens a new avenue of utilizing tunable electrified IL-electrode interfaces for selective sensing of molecules with a kinetic size resolution of 0.1 Å.

Keywords: Electrochemistry, Electrode Surfaces, Environmental Analysis, Material Science

Application Code: Environmental

Methodology Code: Electrochemistry

Session # 780 Abstract # 780-28 **Poster Sessions**

Session Title Electrochemistry

Abstract Title **New Arginine-Acetaminophen Incorporation for Selective Determination in Serum with Graphene-based Biosensor**

Date: Monday, March 07, 2016 - Afternoo
Time:
Room: Exposition Floor, 400 Aisle

Primary Author Zhe Wang
Xavier University

Co-Author(s)

Abstract Text

An electrochemical sensor based on arginine (Arg) functionalized graphene (Arg-G) nanocomposite was fabricated for sensitive detection of acetaminophen. The nanocomposite was characterized by transmission electron microscopy (TEM), scanning electron microscope (SEM),fourier transform infrared spectroscopy (FTIR), atomic fluorescence spectroscopy (AFS) and ultraviolet (UV) spectra etc. The electrochemical behaviors of acetaminophen on Arg-G composite film modified glassy carbon electrode (GCE) were investigated by cyclic voltammetry (CV) differential pulse voltammetry (DPV) and Electrochemical Quartz Crystal Microbalance (EQCM). The experimental results indicated that the incorporation of arginine and graphene greatly enhanced the electrochemical response of acetaminophen, which was represented by density function. This fabricated sensor displayed excellent analytical performance for acetaminophen detection over a range from 0.1 to 3000 M with a detection limit of 0.05 mol L-1 (S/N=3) in serum. Moreover, the proposed electrochemical sensor also exhibited good reproducibility and stability, and has been used to detect acetaminophen in tablets.

Keywords: Bioanalytical, Biosensors, Electrochemistry, Electrode Surfaces
Application Code: Bioanalytical
Methodology Code: Electrochemistry

Session # 790 Abstract # 790-1 **Poster Sessions**

Session Title: Materials Science

Abstract Title: **Non-Toxic Corrosion Inhibitor from Tagetes Patula L. for Aluminum in Acid Fluids used in Industrial Operations**

Primary Author: Olusegun K. Abiola
Federal University of Petroleum Resources

Date: Monday, March 07, 2016 - Afternoo
Time:
Room: Exposition Floor, 400 Aisle

Co-Author(s): A O. Aliyu, E E. Elemike

Abstract Text

Acid fluids play important roles in industrial processes such as oil well acidizing, acid pickling, acid cleaning and acid descaling, where metals are exposed to the corrosive action of the acid fluids. HCl solution is the acid fluid of choice for these industrial processes. There has been increased interest in employing plant extracts as corrosion inhibitors for metals in acid fluids for sustainable development. The ability of Tagetes Patula L. flower extract (TPFE) as green corrosion inhibitor for acid corrosion of aluminium in acid solution (1M HCl) was studied using chemical technique. TPFE extract inhibited the acid corrosion of Al significantly, with about 90 % efficiency at the highest concentration of the extract. The adsorption of the inhibitor molecules on aluminium surface was found to obey Langmuir adsorption isotherm with Goads value of 13.01 k J mol-1.This extract could find possible applications as green corrosion inhibitor for aluminium in acid fluids used in industrial operations.

This research was sponsored by Tertiary Education Trust Fund.

Keywords: Adsorption, Chemical, Material Science, Metals

Application Code: Material Science

Methodology Code: Chemical Methods

Session # 790 Abstract # 790-2 **Poster Sessions**

Session Title: Materials Science

Abstract Title: **Synthesis of Fluorine Substituted LiFeBO[sub]3[/sub] Composite Material as a Cathode Material for a Lithium Secondary Battery and Characterization Using the Solid State NMR**

Date: Monday, March 07, 2016 - Afternoo
Time:
Room: Exposition Floor, 400 Aisle

Primary Author: Youngil Lee
University of Ulsan

Co-Author(s): Hansol Lee, JaeMin Bak

Abstract Text

In lithium rechargeable battery, LiMBO[sub]3[/sub] (M=Fe, Mn and Co) is increasing to interest due to its strong boron-oxygen covalent bond, high theoretical specific capacity, environmental friendly benefit, and less expensive. The fluorine substituted LiFe(BO[sub]3[/sub])[sub]1-x[/sub]F[sub]3x[/sub] composites have been synthesized by solid-state reaction using planetary mill method without carbon coating. It was characterized by XRD, SEM, solid-state NMR and CV. The XRD results of fluorine composite have been successfully substituted to BO[sub]3[/sub][sup]3-[/sup] sites without structural modification. Electrochemical analysis has been also performed with various C-rates and cycleability.

Keywords: Electrochemistry, NMR
Application Code: Material Science
Methodology Code: Magnetic Resonance

Session # **790** **Abstract #** **790-4** **Poster Sessions**

Session Title: Materials Science

Abstract Title: **Analytical Methods to Estimate Biodurability of Engineered Nanomaterials**

Date: Monday, March 07, 2016 - Afternoo
Time:
Room: Exposition Floor, 400 Aisle

Primary Author: Mary-Luyza Avramescu
Health Canada

Co-Author(s): H David Gardner, Marc Chénier, Pat E. Rasmussen

Abstract Text

The main risks of exposure to engineered nanomaterials (ENMs) have been associated with inhalation, and effects also have been linked to gastrointestinal intake. Risk assessments of many engineered nanomaterials are difficult to undertake because of the current lack of reliable information on their physicochemical characteristics. Solubility testing is a critical component of physicochemical characterisation of ENMs. Methodologies are needed to determine their dissolution rates and biodurability in biological fluids.

High throughput assays were developed to measure the solubility of metallic ENMs (nano-ZnO and nano-TiO2) in simulated lung and gastric environments. NexION 300s ICP-MS and/or Optima 5300V ICP-OES were used to quantify the soluble and total metal fraction in the sample digests. XRD and SAXS analyses were carried out with a Rigaku Ultima IV diffractometer to determine the crystallographic structure, sample purity and size for materials used in the study.

The dissolution parameters determined for ENMs were compared with those of their micron-sized analogues for both inhalation and ingestion exposure pathways. Experimental results demonstrated that nanomaterials may display greater solubility then their micron-size analogues in both gastric and lung environments. Solubility of nanomaterials is also influenced by crystalline form (e.g., anatase vs rutile).

The approaches developed during this study provide nanomaterial dissolution and biodurability information which is essential for screening risks associated with human exposure to nanomaterials.

Keywords: Characterization, ICP-MS, Metals, Nanotechnology
Application Code: Nanotechnology
Methodology Code: Atomic Spectroscopy/Elemental Analysis

Session # 790 Abstract # 790-5 **Poster Sessions**

Session Title Materials Science

Abstract Title **Intensity Ratio of Resonant Raman Modes for Chirality Enriched Carbon Nanotubes**

Date: Monday, March 07, 2016 - Afternoo
Time:
Room: Exposition Floor, 400 Aisle

Primary Author Yanmei Piao
National Institute of Standards and Technology

Co-Author(s) Angela Hight Walker, Geyou Ao, Jason Streit, Jeffrey Fagan, Jeffrey Simpson, Stephanie Lam

Abstract Text

Relative intensities of resonant Raman spectral features, specifically the radial breathing mode (RBM) and G modes, of eleven chirality-seperated single-wall carbon nanotube (SWCNT) species were established under second-order optical transition excitation. The results demonstrate a significantly under-recognized complexity in the evaluation of Raman spectra for the assignment of (n,m) population distributions. The strong chiral angle and mod dependencies affect the intensity ratio of RBM to G band and can result in misleading interpretations. Additionally, we validate our results on chirality dependent G+ and G- positions, supporting accuracy in literature values for these peak positions and further extend the trend to the small diameter regime by including the first (5,4) second-order resonance Raman spectra. Together, the Raman spectra library is demonstrated to be sufficient for decoupling multiple species via a spectral fitting process, to enable fundamental characterization even in mixed chiral population samples.

Keywords: Characterization, Chiral Separations, Chromatography, Raman
Application Code: Material Science
Methodology Code: Separation Sciences

Session # 790 Abstract # 790-6 **Poster Sessions**

Session Title Materials Science

Abstract Title **Liquid-Phase Laser Ablation as a Controlled Method to Produce Graphene Quantum Dots**

Date: Monday, March 07, 2016 - Afternoo

Time:

Room: Exposition Floor, 400 Aisle

Primary Author Rosemary L. Easterday
University of Kentucky

Co-Author(s) Dong-Sheng Yang, Doo Young Kim, Wenjin Cao, Yiyang Liu

Abstract Text

Graphene quantum dots (GQDs) are an emerging class of carbon nanomaterials. Due to their tunable chemical structures and optical propeties, GQDs are considered as proming materials for wide-range applicatons in biosensing, photovoltaics, and catalysis. One common way to produce GQDs is chemical oxidation in which harsh chemicals break source materials into small graphene pieces. The chemically synthesized GQDs have a significant inhomogeneity in sizes and surface functional groups, which makes it difficult to track the origin of photoluminescence. Therefore, alternative ways to synthesize GQDs with a smaller use of chemicals, less impurity production, and a greater control of sizes and functional groups are highly desirable. Laser ablation in the liquid phase is an attractive and new way for generating GQDs. In this method, carbon plasma plumes formed by the absorption of high-power laser pulses are expanded, cooled, and condensed into carbon dots by the surrounding liquid. In our study, a high-power nanosecond pulsed laser beam was focused on the pellet of carbon sources immersed in water. FT-IR results showed that the resultant GQDs had fewer carboxylic groups than the GQDs produced by chemical oxidations. The size and morphology of GQDs were characterized by atomic force and transmissive electron microscopy. The effects of laser ablation parameters such as laser power, wavelength, and duration on sizes, chemical structures, and optical properties of GQDs were studied.

This research was supported by the Kentucky Science & Engineering Foundation.

Keywords: Materials Characterization, Method Development, Nanotechnology, UV-VIS Absorbance/Luminescen

Application Code: Material Science

Methodology Code: Fluorescence/Luminescence

Session # 790 Abstract # 790-7 **Poster Sessions**

Session Title: Materials Science

Abstract Title: **Physicochemical Characterization and Nanotoxicity of Polishing Slurries During CMP Process**

Primary Author: Eduard Dumitrescu, Clarkson University

Date: Monday, March 07, 2016 - Afternoo
Time:
Room: Exposition Floor, 400 Aisle

Co-Author(s): Dinusha Karunaratne, Kenneth Wallace, Silvana Andreescu

Abstract Text

Nanosized materials have extensive applications in large-scale industrial processes due to their remarkable properties. Chemical mechanical planarization (CMP) is an essential step of the manufacturing process of integrated circuit chips in semiconductor industry. During CMP, excess deposited material is removed from metallic substrates by mechanical polishing. The major component affecting the polishing process is the CMP slurry, an aqueous dispersion of engineered nanoparticles (ENPs) and other chemical additives. The chemical additives selectively dissolve the materials present on the wafer surface, while nanoparticles, such as silica, ceria or alumina, mechanically remove the chemically modified surface through abrasion. None of these slurry components are incorporated into the semiconductor product, but they are eliminated through CMP effluents, along with dissolved and particulate materials removed by the polishing process. Therefore, it is essential to assess the environmental impact of the released nanoparticles and reactive additives. Here, our work is focused on describing the compositional change of a CMP slurry throughout a particular CMP process. For this purpose, the nanoparticles in the polishing slurries are characterized before and after the CMP process by means of particle sizing, zeta potential and SEM. The interaction between nanoparticles and chemical additives are studied using FTIR, TGA and XRD. In order to assess the toxicity of individual slurry components and their mixtures, zebrafish embryos were used as biological model. The mortality rate was evaluated using viability assays. We suggest that the chemical additives interact with nanoparticles, changing their physicochemical properties, which consequently affects their toxicity profile.

Keywords: Environmental Analysis, Materials Characterization, Nanotechnology
Application Code: Environmental
Methodology Code: Thermal Analysis

Session # 790 **Abstract # 790-8** **Poster Sessions**

Session Title: Materials Science

Abstract Title: **Polycapillary X-Ray Optics to Play a Key Role in NASA Mars 2020 Mission**

Primary Author: Ning Gao
XOS

Date: Monday, March 07, 2016 - Afternoo
Time:
Room: Exposition Floor, 400 Aisle

Co-Author(s): Douglas Dawson, Genevieve DeMarco, George Allen, Igor Ponomarev, Jaime Luna, Jared Sachs, Jay Burdett, Larry Wade, Peng Lu, Robert Sharrow

Abstract Text

For the very first time, a polycapillary x-ray optic will be sent to Mars as a key component in a micro XRF analysis instrument being developed by JPL under the NASA's Mars 2020 program. As the world leader of the polycapillary x-ray optic technology, XOS has been collaborating with JPL and others to develop the micro XRF instrument PIXL (Planetary Instrument for X-ray Lithochemistry) that has been selected for flight on the upcoming Mars rover in year 2020. The use of the polycapillary x-ray optic in PIXL instrument will allow elemental analysis of rocks and soils on Mars at the resolution and detection sensitivity significantly higher than being achieved from previous Mars missions. While the polycapillary x-ray optic technology has been successfully used in many research and commercial x-ray analytical systems, the special requirements for integrating an optic in the flight mission post unique challenges. The preliminary research and development work involving the polycapillary optic design, characterizations, and performance evaluation will be reviewed and discussed in this presentation.

Keywords: Nuclear Analytical Applications, Spectrometer, X-ray Diffraction, X-ray Fluorescence
Application Code: Material Science
Methodology Code: X-ray Techniques

Session # 790 Abstract # 790-9 **Poster Sessions**

Session Title Materials Science

Abstract Title **Preparation and Characterization of Bleached Ground Peanut Hulls**

Primary Author Holly Truluck
Western Carolina University

Date: Monday, March 07, 2016 - Afternoo
Time:
Room: Exposition Floor, 400 Aisle

Co-Author(s) Carmen L. Huffman, Melisa J. Glatte

Abstract Text

Heavy metals are a persistent source of water contamination, particularly in industrial settings such as mining and electroplating. Heavy metals are toxic, so they must be removed from wastewater before the water is discharged into natural systems. Removal of heavy metals typically involves the use of expensive ion exchange resins. An alternative material, ground peanut hulls, has been shown to be an economically viable biomaterial for the adsorption of heavy metal ions in water. Uptake is significantly increased upon simple modification of the hulls using a bleaching treatment involving alkaline peroxide. The bleaching method has been optimized to balance cost, safety and effectiveness. Effectiveness was characterized by the degree of oxidation, which is thought to be a key component in the metal binding process. Oxidation was quantified by the adsorption of methylene blue, a cationic dye that binds strongly to carbonyl groups. Porosity of the hulls was also examined, as the bleaching process is believed to dissolve lignin. This leads to an increased surface area of the hulls and consequently, a greater number of metal binding sites. Results from these experiments will be presented.

Keywords: Adsorption, Characterization, Infrared and Raman, UV-VIS Absorbance/Luminescence
Application Code: Material Science
Methodology Code: UV/VIS

Session # 790 **Abstract # 790-10** **Poster Sessions**

Session Title: Materials Science

Abstract Title: **Real Time Measurement of Layer Thickness, Erosion Rates and Crater Depth in Glow Discharge Optical Emission Spectrometry**

Date: Monday, March 07, 2016 - Afternoo

Time:

Room: Exposition Floor, 400 Aisle

Primary Author: Sofia Gaiaschi
Horiba Jobin Yvon

Co-Author(s): Kayvon Savadkouei, Patrick Chapon, Philippe Hunault, Simon Richard

Abstract Text

Pulsed Radio Frequency Glow Discharge Optical Emission Spectrometry provides Ultra Fast Elemental Depth Profile of thin and thick films. The technique relies on the sputtering of a representative area of the material of interest by a plasma which also excites the sputtered species. The quantitative information on the elements present in the film can be achieved through the real time analysis of the light emitted from the de-excitation of the excited elements.

A new function giving the capability to measure, in real time, the depth of the sputtered crater is being introduced. Such measurement, based on differential interferometry with nanometer sensitivity, is able to provide direct information about the thickness of the layers and their erosion rates. This is crucially important, notably when the investigated materials are non transparent, as in this case an alternative technique such as ellipsometry cannot be used.

Several examples of applications of this new development - on both non transparent and transparent layers - will be presented.

Keywords: Elemental Analysis, Material Science

Application Code: Material Science

Methodology Code: Atomic Spectroscopy/Elemental Analysis

Session # 790 **Abstract # 790-11** **Poster Sessions**

Session Title Materials Science

Abstract Title **HH-XRF and HH-LIBS for Alloy Analysis**

Date: Monday, March 07, 2016 - Afternoo

Time:

Room: Exposition Floor, 400 Aisle

Primary Author Jiyan Gu
Bruker

Co-Author(s)

Abstract Text

Handheld-XRF is a method of Energy Dispersive X-ray Fluorescence in which the radiation produced by a miniature X-ray tube strikes the sample surface and causes ionizations of the inner shell of the atoms constituting the sample. The technique is widely used in the industry for metal sorting and identification. Recently, handheld-LIBS as an emerging new technique have been introduced as a complementary tool for alloy analysis. Instead of using X-ray as the excitation source, HH-LIBS systems focused a laser beam onto the sample and generate a plasma plum in the temperature range of 5,000-20,000K. Atomic emission lines are collected by a miniature CCD spectrometer and then analyzed. A detailed comparison study is conducted to compare the HH-LIBS and HH-XRF performance for Aluminium analysis.

Keywords: Atomic Spectroscopy, Elemental Analysis, X-ray Fluorescence
Application Code: Material Science
Methodology Code: Atomic Spectroscopy/Elemental Analysis

Session # **790** **Abstract #** **790-12** **Poster Sessions**

Session Title Materials Science

Abstract Title **X-ray Photoelectron Spectroscopy, Low Energy Ion Scattering, and Time-of-Flight Secondary Ion Mass Spectrometry, including Chemometrics Analysis, of Display Glass Surfaces**

Primary Author Matthew R. Linford
Brigham Young University

Date: Monday, March 07, 2016 - Afternoo
Time:
Room: Exposition Floor, 400 Aisle

Co-Author(s) Barry M. Lunt, Cody V. Cushman, Julia Zakel, Nicholas J. Smith, Philipp Brüner, Thomas Grehl

Abstract Text

Display glasses are widely used to manufacture consumer electronics such as smart phones, tablets, and televisions. The surface compositions of these glasses significantly influences the manufacturing process for making displays and the performance of the finished devices. This composition can be altered during manufacturing by exposure to various wet-chemical or gas-phase treatments. We have applied angle-resolved X-ray photoelectron spectroscopy (AR-XPS), low-energy ion scattering (LEIS), and time-of-flight secondary ion mass spectrometry (ToF-SIMS) to evaluate the surface composition of a model display glass, Corning Eagle XG, under different information depths. Samples analyzed included the as formed glass, a fracture surface, and surfaces exposed to a series of aqueous chemistries, including HF, HCl, and tetramethylammonium hydroxide. Both ToF-SIMS static surface spectra and depth profiles were obtained. The data were analyzed manually and with a suite of chemometric techniques that included principle components analysis (PCA), multivariate curve resolution (MCR), and cluster analysis. These analyses, including radar plots of the ToF-SIMS intensities of the cationic glass components, showed that the elemental composition of the as-formed surface varied substantially from that of the fracture surface, and that the wet chemical treatments significantly altered the surface composition. Brief HF and HCl treatment leached network-modifying components from the glass, creating similar surface compositions despite significantly different extents of network dissolution. Depth profiles showed a gradient in the near-surface composition of the materials. In good agreement with the AR-XPS analysis, LEIS showed significant variations in the concentrations of network-modifying elements at the outermost surface as a function of surface preparation.

Keywords: Chemometrics, Material Science, Surface Analysis, Time of Flight MS
Application Code: Material Science
Methodology Code: Surface Analysis/Imaging

Session # 790 Abstract # 790-13 **Poster Sessions**

Session Title Materials Science

Abstract Title **The Particle Size Paradox**

Date: Monday, March 07, 2016 - Afternoo
Time:
Room: Exposition Floor, 400 Aisle

Primary Author Jack G. Saad
Micromeritics

Co-Author(s) Paul A. Webb

Abstract Text

Particle size is an important piece of information for research and development, quality control, and quality assurance as well as understanding the small physical details in a milling or powder compaction process. These details can contribute to potential desired and undesired products. With recent technological advances, particles are now measured using various analytical techniques and instrumentation. Different analytical techniques seldom provide the same value for particle size. The "paradox" of particle sizing is that all the different values are the correct value. The four analytical particle sizing instrument techniques discussed and compared include dynamic image analysis, sedimentation, light-scattering, and electric sensing zone. The result of the comparison is that particle size data is specific to the analytical technique used to collect that size data and particle shape is an important characteristic to consider.

Keywords: Characterization, Microscopy, Particle Size and Distribution, Process Analytical Chemistry
Application Code: Material Science
Methodology Code: Process Analytical Techniques

Session # 790 Abstract # 790-14 **Poster Sessions**

Session Title: Materials Science

Abstract Title: **Gas Chromatography-Vacuum Ultraviolet Absorbance Spectroscopy for Quantitation of Trace and Bulk Water in Organic Solvents: An Emerging Alternative to Karl Fischer Titration**

Primary Author: Lindsey N. Shear
VUV Analytics

Co-Author(s): Dale Harrison, Leonard M. Sidisky

Date: Monday, March 07, 2016 - Afternoo
Time:
Room: Exposition Floor, 400 Aisle

Abstract Text

Vacuum ultraviolet absorbance (VUV) spectroscopy is a powerful tool for characterizing the vibronic transitions of organic and inorganic compounds in the wavelength region of 125-240nm. Until recently, this technique has been limited to niche research applications, mostly due to difficulty obtaining high quality absorbance spectra in this wavelength region. Recent advances in vacuum ultraviolet spectroscopy have allowed for the first ever application of this technology for chromatographic systems. The GC-VUV detector produces highly characteristic absorbance spectra for nearly all chemical species in the wavelength region of 125-240 nm. This system allows not only for identification but also for robust quantitation of a variety of compounds separable by gas chromatography, including water. Accurate determination of the water content of organic solvents is critical for applications in synthetic organic chemistry, specialty chemical manufacturing and quality assurance. Traditionally the water content of materials including organic solvents has been determined using Karl Fischer Titration or GC-TCD, however both of these techniques are limited in sensitivity, reproducibility and ease of use. GC-VUV offers a powerful, yet simple alternative for the determination of water in organic solvents while also allowing for unique spectral identification of other chemical components in the sample via the measurement of characteristic VUV absorbance spectra. We present here a robust, rapid and validated method for water determination in a variety of organic solvents on a variety of GC column stationary phases. This methodology has the potential to transform the way routine analysis of water content is conducted.

Keywords: Detector, Gas Chromatography, Molecular Spectroscopy, Water
Application Code: Validation
Methodology Code: Gas Chromatography

Session # 790 **Abstract # 790-15** **Poster Sessions**

Session Title: Materials Science

Abstract Title: **Complete Size Characterization of Diatomaceous Earth**

Primary Author: Jack G. Saad
Micromeritics

Date: Monday, March 07, 2016 - Afternoo
Time:
Room: Exposition Floor, 400 Aisle

Co-Author(s)

Abstract Text

Diatomaceous earth is commonly used in many manufacturing and production processes as a filtering agent. The quality of the diatomaceous earth greatly affects the effectiveness of the filter. One of the primary characteristics of determining diatomaceous earth quality is particle size. Traditional methods, like using sieves, can be tedious and may not offer enough information to completely characterize the material. Automated sizing techniques, such as sedimentation analysis or static light-scattering, are limited in scope since diatomaceous earth is not a uniform material, but a composite of shapes, sizes, densities, and colors. To completely characterize the particle size of diatomaceous earth, dynamic image analysis and dynamic light scattering (DLS) analysis are used to compliment to each other to accomplish this goal. Dynamic image analysis uses shape factors to collect size data in the micron range while DLS is used to determine the size of nanoparticles that remain suspended in the medium. Testing is performed on two types of diatomaceous earth commonly used in the beer brewing industry.

Keywords: Characterization, Food Science, Imaging, Process Analytical Chemistry
Application Code: Process Analytical Chemistry
Methodology Code: Process Analytical Techniques

Session # 800 Abstract # 800-1 **Poster Sessions**

Session Title Physical Measurements

Abstract Title **Structural Studies of Co-Spinel Ferrite Synthesized by an Auto Combustion Method**

Date: Monday, March 07, 2016 - Afternoo
Time:
Room: Exposition Floor, 400 Aisle

Primary Author Dipal P. Chaudhary
Mehsana Urban Institute of Sciences

Co-Author(s)

Abstract Text

Nanocrystallite magnetic particles are attractive due to many importants applications like magnetic drug delivery, hyperthermia for cancer treatment, ferro fluids, magnetic storage data etc. Nanosized particles of cobalt ferrite have been synthesized by an auto combustion method. The citrate – nitrate solution was prepared under molar ratio of 1:1. The final product has been received by self ignition then as burnt powder was calcined at 950 ºC. The formation and phase identification of the synthesized particles were confirmed by FTIR and XRD. The particle size was investigated by a Scanning Electron Microscope.

Keywords: Characterization, Drugs, Particle Size and Distribution, Pharmaceutical

Application Code: Pharmaceutical

Methodology Code: Physical Measurements

Session # 800 **Abstract # 800-2** **Poster Sessions**

Session Title: Physical Measurements

Abstract Title: **Fabrication of A Novel Fiber-Optic based Single-Cell Temperature Sensor**

Primary Author: Qingbo Yang
Missouri University of Science and Technology

Co-Author(s): Hai Xiao, Honglan Shi, Ke Li, Yinfa Ma

Date: Monday, March 07, 2016 - Afternoo
Time:
Room: Exposition Floor, 400 Aisle

Abstract Text

Comprehensive understanding of heterogeneous cell behavior in response to niche environmental changes is becoming the foundation of next generation, single-cell based biomedical research. Subtle oscillations of extra- and Intra-cellular temperature, for instance, is the basis of many cell activities such as division, gene expression, enzyme reaction, metabolism processes as well as reflections of cell response to external stimuli. However, these small scale, single-cell level changes were highly localized and transiently occurred, rendering it rather difficult to measure using conventional methods. In this study, a fiber-optic based micro-temperature sensor for intracellular measurement was designed and fabricated using sub-picoliter quantum dots (QDs) as sensing material. A 442 nm argon laser source was used to excite QD fluorescence, and a 650 nm peak was acquired and quantified to linearly correlate with temperature that ranges from 30 to 45 °C. The newly developed single cell temperature sensor is easy to fabricate and operate, fast response and highly reproducible, with a high spatial resolution and biological-relevant sensitivity up to ~0.1 °C. This novel temperature sensor thus may allow researchers to obtain cellular temperature changes when cell responses to immediate stimuli. The detailed results will be presented at the conference.

This project was supported by the National Institute of Health (1R21GM104696-01).

Keywords: Bioanalytical, Biosensors, Fluorescence, Sensors

Application Code: Bioanalytical

Methodology Code: Sensors

Session # 810 **Abstract # 810-1** **Poster Sessions**

Session Title: Polymer Characterization and Analysis

Abstract Title: **Analyzing the Vibration-Rotation Spectrum of Hydrochloric Acid Generated with a Thermogravimetric Analyzer Coupled to a Fourier Transform Infrared Spectrophotometer**

Date: Monday, March 07, 2016 - Afternoo
Time:
Room: Exposition Floor, 400 Aisle

Primary Author: Anthony J. Lang
PerkinElmer

Co-Author(s): Jack Botting

Abstract Text

A well-known undergraduate laboratory experiment is the determination of the hydrogen-chlorine bond length in hydrochloric acid (HCl). This experiment is often achieved by trapping HCl in a gas cell and measuring its spectrum with a Fourier Transform Infrared Spectrophotometer (FTIR). This measurement is performed on a sealed gas cell. Concerns are often raised about the sensitivity of Thermogravimetric Analysis coupled to FTIR (TGA-FTIR). In an attempt to demonstrate the power of this technique, the vibration-rotation spectra of HCl will be collected using a gas flowing at 20mL min-1. Utilizing the decomposition gases of polyvinyl chloride, the HCl spectrum can easily be obtained with the interface of the two instruments. A unique feature of this methodology is the temperature control of the flowing gas, which adds additional experimental flexibility. The bond lengths and fundamental frequencies for the common isotopes of chlorine will be determined using the spectrum collected.

Keywords: FTIR, Polymers & Plastics, Teaching/Education, Thermal Analysis
Application Code: Polymers and Plastics
Methodology Code: Thermal Analysis

Session # 810 Abstract # 810-2 **Poster Sessions**

Session Title: Polymer Characterization and Analysis

Abstract Title: **Analysis of PLGA Molecular Weight and Structure by the Latest Advanced Multi-Detector GPC Systems**

Date: Monday, March 07, 2016 - Afternoo

Time:

Room: Exposition Floor, 400 Aisle

Primary Author: Carrie Schindler
Malvern Instruments

Co-Author(s): Mark R. Pothecary, Ulf Nobbmann

Abstract Text

Poly(D,L-lactide-co-glycolic acid), PLGA, is a biodegradable and biocompatible polymer, that has found use in a number of medical devices such as grafts and sutures as well as in drug delivery applications. Drug release process is dependent on the distribution of molecular weight, molecular structure and composition of the copolymer being used. Gel-permeation chromatography (GPC) is widely used for the measurement of molecular weight and molecular weight distribution polymers. Historically, the elution volume of an unknown sample was compared with that of known standards to estimate molecular weight. However, this 'conventional calibration' is limited by the structural differences between standards and samples, meaning that the measured molecular weight is only comparative. This is particularly true for PLGA where both structure and composition will affect the elution volume of different products of similar molecular weight. Static light scattering detectors allow the direct measurement of the sample molecular weight independent of its elution volume. A viscosity detector can also be used as part of a GPC system to measure intrinsic viscosity. In combination these data allow detailed structural information of a polymer to be generated in a single GPC measurement which can be compared with other samples in Mark-Houwink plots. In this paper, we analysed different samples of commercially available PLGA to compare their absolute molecular weight from light scattering to those quoted with the product using Malvern's latest GPC/SEC system, OMNISEC. Additionally, we compared the Mark-Houwink plots of different examples containing different ratios of the two co-monomers. Structural and molecular weight differences are clearly visible which will result in changes in drug release and delivery profile. More detailed analysis of these parameters can be used to better control the end-properties of the PLGA and its release rate of drugs in delivery applications.

Keywords: HPLC Detection, Instrumentation, Light Scattering, Pharmaceutical
Application Code: Polymers and Plastics
Methodology Code: Liquid Chromatography

Session # 810 **Abstract # 810-3** **Poster Sessions**

Session Title: Polymer Characterization and Analysis

Abstract Title: **Molecular Level Understanding of Etching Chemistry at Polymer Interfaces**

Date: Monday, March 07, 2016 - Afternoo

Time:

Room: Exposition Floor, 400 Aisle

Primary Author: John Myers
University of Michigan

Co-Author(s)

Abstract Text

Etching of polymers is a common processing step that occurs during the fabrication of microelectronic components used in microelectronic devices. However, controllable etching of polymers used in microelectronics remains challenging due to a lack of fundamental understanding of etching chemistry at the molecular level. In particular, as etching initially occurs at material surfaces, an understanding of etching chemistry at polymer interfaces is important. In this research, a nonlinear optical laser spectroscopic technique, sum frequency generation (SFG) vibrational spectroscopy, will be used together with complimentary surface analytical techniques to elucidate the molecular structure of polymer interfaces before, during, and after etching processes. In situ information about etching chemistry will be achieved by acquiring SFG spectra from polymer/etching solution interfaces in real time. In addition, the effects of dry plasma treatment on wet etching rates will be investigated by exposing the polymers to different plasma treatments prior to wet etching. Pristine and etched polymer interfaces will be characterized in terms of interfacial functional groups, functional group orientation, and number density of functional groups. By characterizing the molecular structure of polymer interfaces during etching processes, a molecular level understanding of etching chemistry will be developed. Understanding etching chemistry at the molecular level will aide in design and development of new etching techniques for the fabrication of advanced microelectronic components.

Keywords: Characterization, Polymers & Plastics, Surface Analysis, Vibrational Spectroscopy

Application Code: Polymers and Plastics

Methodology Code: Surface Analysis/Imaging

Session # 810 Abstract # 810-4 **Poster Sessions**

Session Title Polymer Characterization and Analysis

Abstract Title **Dissolution Dynamic Nuclear Polarization Study of Living Anionic Polymerization Reaction**

Primary Author Youngbok Lee
Hanyang University

Date: Monday, March 07, 2016 - Afternoo
Time:
Room: Exposition Floor, 400 Aisle

Co-Author(s)

Abstract Text

NMR has long played a pivotal role in studying reaction mechanisms in the field of polymer chemistry, since the technique reveals molecular structures and interactions with atomic resolution. In the overwhelming majority of uses, NMR is applied as a steady-state method for analyzing products after completion of a reaction. It is then often necessary to use sophisticated synthetic strategies such as the selective incorporation of stable isotopes, to infer information about the reaction mechanism. However, a more direct and potentially more powerful approach to characterizing a reaction is the monitoring of species that arise as the reaction occurs. Using dissolution dynamic nuclear polarization, a hyperpolarization technique that can enhance NMR signal by several orders of magnitude, it is possible to study chemical reactions in real time. Here, we use this unique feature of a hyperpolarized polymerization reaction to enable the detection of intermediate species as they arise during the synthesis of polystyrene by anionic polymerization of hyperpolarized styrene monomer.

Keywords: NMR, Polymers & Plastics
Application Code: Polymers and Plastics
Methodology Code: Magnetic Resonance

Session # 810 Abstract # 810-5 **Poster Sessions**

Session Title Polymer Characterization and Analysis

Abstract Title **Extractables Analysis of SPE Frits**

Primary Author Kirsten Kinneberg
Porex Corporation

Co-Author(s) Gary Li

Date: Monday, March 07, 2016 - Afternoo

Time:

Room: Exposition Floor, 400 Aisle

Abstract Text

Sample processing accounts for 61% of time spent on chromatographic analysis and is the source of 30% of chromatographic errors.[1] Accordingly, there is motive to optimize SPE processes and reduce interferences. [2-4] The objective of this study was to evaluate the extractables profile and the heavy metals profile of a polyethylene (PE) resin and a PE frit, respectively, utilized in SPE applications.

GC-MS: Approx. 1 g of PE resin was added to 5 mL of DCM. Samples were mixed for 30 minutes, then analyzed using a HP 6890 gas chromatograph in conjunction with a 5973 mass selective detector using liquid injection. Data acquisition was accomplished using chemstation software. The major components detected were long chain alkanes (C10 ~ C20) which are characteristic peaks for polyethylene oligomers (Fig. 1). The PE resin did not contain antioxidants or siloxanes.

ICP-MS: SPE frits were tested using a Perkin Elmer Elan DRC II equipped with a Cetac ASX-520 auto-sampler. Samples were digested by an inductively coupled plasma source; samples were burned and the resulting components were identified using a highly sensitive mass spectrometer. Elemental concentration of metals known to cause potential interference with HPLC [5] were found to be below their detection limit (Table 1).

The results presented here demonstrate that the application-specific PE material is SPE and HPLC compatible. This study highlights the importance of utilizing clean resins in manufacturing of porous plastics commonly used in chromatographic analysis, particularly in sample preparation.

References

1. R.E. Majors, LC/GC Magazine, 2002
2. Seb□k, Á., et al. J Chromatogr. A 1216 (2009) 2288-2301.
3. Stiles, R., et al. J Sep. Sci. 30 (2007) 1029-1036.
4. Rodushkin, I., et al. Sources of contamination and remedial strategies in the multi-elemental trace analysis laboratory.
5. Williams, S. J of Chromatorgr. A 1052 (2004) 1-11.

Keywords: Elemental Analysis, Extraction, GC-MS, Solid Phase Extraction

Application Code: Polymers and Plastics

Methodology Code: Mass Spectrometry

Session # 810 **Abstract # 810-6** **Poster Sessions**

Session Title: Polymer Characterization and Analysis

Abstract Title: **Automated Microwave Sample Preparation of Difficult Petroleum and Polymer Matrices**

Primary Author: Robert Lockerman
CEM Corporation

Date: Monday, March 07, 2016 - Afternoo
Time:
Room: Exposition Floor, 400 Aisle

Co-Author(s): Austin Thornton, Dan Iversen, Ian Goldstein, Tina Restivo

Abstract Text

Sample preparation for metals analysis of petroleum and plastic products provide many challenges. Materials are difficult to sample and completely digest. Incomplete digestion can lead to background interferences. In addition the typical sample size of 0.1 grams provides challenges for homogeneity as well as detection limits. Finally, the preparation is normally done in a batch environment that requires a lot of handling and no opportunity to automate the process. We will show a novel automated microwave digestion system and prepare samples such as bunker oil, Kevlar and other thermoplastics at the 0.25 gram range using only nitric acid. Methodologies will be presented and results will be discussed

Keywords: Atomic Emission Spectroscopy, High Temperature, Laboratory Automation, Microwave
Application Code: Polymers and Plastics
Methodology Code: Sampling and Sample Preparation

Session # **810** **Abstract #** **810-8** **Poster Sessions**

Session Title: Polymer Characterization and Analysis

Abstract Title: **Microwave, r0 Structural Parameters, Conformational Stability and Vibrational Assignment of (Chloromethyl)fluorosilane**

Date: Monday, March 07, 2016 - Afternoo

Time:

Room: Exposition Floor, 400 Aisle

Primary Author: Dattatray K. Sawant
University of Missouri Kansas City

Co-Author(s): James R. Durig

Abstract Text

The FT-microwave spectrum (6.5 – 26 GHz) of (chloromethyl)fluorosilane, ClCH[sub]2[/sub]-SiH[sub]2[/sub]F has been recorded and 250 transitions for the parent species along with [sup]13[/sup]C, [sup]37[/sup]Cl, [sup]29[/sup]Si and [sup]30[/sup]Si isotopologues have been assigned for [i]trans[/i] conformer. Infrared spectra (3,100 to 400 cm[sup]-1[/sup]) of gas, solid and the variable temperature (-60 to -100 °C) studies of the infrared spectra of the sample dissolved in xenon have been recorded. Additionally the variable temperature (-133 to -153 °C) studies of the Raman spectra of the sample dissolved in krypton have been recorded. The enthalpy difference between the [i]trans[/i] and [i]gauche[/i] conformers in xenon solutions has been determined to be 109 + 15 cm[sup]-1[/sup] (1.47 + 0.16 kJ mol[sup]-1[/sup]) and in krypton solution the enthalpy difference has been determined to be 97 + 16 cm[sup]-1[/sup] (1.16 + 0.19 kJ mol[sup]-1[/sup]) with the [i]trans[/i] conformer as the more stable form. Approximately 46 + 2 % of the [i]trans[/i] form is present at ambient temperature. By utilizing the microwave rotational constants of five isotopologues for trans and the structural parameters predicted from MP2(full)/6-311+G(d,p) calculations, adjusted r0 parameters have been obtained for trans conformer. The r0 structural parameter values for the trans form are for the heavy atom distances (Å): Si-F = 1.608 (3); C-Cl = 1.771 (3); Si-C = 1.884 (3); and angles (°):[capital eth]FSiC = 108.9 (5); [capital eth]ClCSi = 104.9 (5). The obtained results are compared with the corresponding properties of some related molecules.[sub][sub]

Keywords: FTIR, Microwave, Molecular Spectroscopy, Optimization

Application Code: General Interest

Methodology Code: Molecular Spectroscopy

Session # 810 **Abstract # 810-9** **Poster Sessions**

Session Title: Polymer Characterization and Analysis

Abstract Title: **Micro/Nano-Structured Flexible Foils for Anti-Counterfeiting Purposes**

Primary Author: Nastasia Okulova
Danapak Flexibles A/S and Technical University of Denmak

Co-Author(s): Lars Christensen, Rafael Taboryski

Date: Monday, March 07, 2016 - Afternoo
Time:
Room: Exposition Floor, 400 Aisle

Abstract Text

Up to date there have been found many ways of producing samples with functional nano-pattering, however for mass production of such samples the method of fabrication of the surface structure must be fast and cheap. A recently developed method suggests using extrusion coating of polymer materials in a roll-to roll process, where the functional micro-/nano- surface structures are imprinted directly onto the surface of a polymer foil. This new technology can both answer the requirement for being fast and cheap. The aim of this project is implementation of the technology for mass production and it is carried out in collaboration between technical Universuty of Denmark and Danapak Flexibles A/S. In a roll to roll extrusion coating a molten polymer film is extruded through a flat nozzle, then stretched in air, and finally laminated onto a carrier foil. The lamination process takes place as the melt curtain is squeezed between a structured cooling roller and a rubber counter roller. A force is exerted on the compliant counter roller to form a so-called nip region where the molten polymer solidifies and adheres to the carrier foil. The extrusion coating process is fast, mainly due to the fact that the polymer is molten to begin with, and cools rapidly by contact with the cooling roller. Previously a large area replication at high throughput of patterns both on micrometer- and nanometer scale in thermoplastic foils using standard industrial extrusion coating equipment and standard thermoplastic polymers has been demonstrated. The focus of this study lies on the reproduction of the previous results for nano- or micro-structures and implementation of this technology for mass production of such patterned foils for the use in packaging. An interesting application is production of holograms with build in anti-counterfeiting designs on micro- or nanoscale.

Keywords: Method Development, Nanotechnology, Polymers & Plastics, Process Control
Application Code: Polymers and Plastics
Methodology Code: Surface Analysis/Imaging

Session # **810** **Abstract #** **810-10** **Poster Sessions**

Session Title Polymer Characterization and Analysis

Abstract Title **Self-Cleaning Properties of Nanostructured Polypropylene Foils Fabricated by Roll-to-Roll Extrusion Coating**

Primary Author Agnieszka Telecka
Danish Technical University

Date: Monday, March 07, 2016 - Afternoo
Time:
Room: Exposition Floor, 400 Aisle

Co-Author(s) Ling Sun, Rafael Taboryski

Abstract Text

We present systematic wetting properties study of nanostructured polypropylene, functional foils, fabricated by roll – to – roll extrusion coating process (R2R EC) [1]. It is a fast and effective manufacture method, widely uses for smooth polymer films, which allows to large – scale replication of micro – and nanometric scale arrays onto polymer substrates. Metal templates used for patterns imprint were prepared through a one step, maskless, black silicon etching process [2] , what led to covering of the whole 4 inch silicon wafer area, and consequent NiV electroplating. By the manipulation of reactive SF6 and O2 gases, diverse nanograss topographies were generated. R2R extrusion was performed under varied imprinting temperatures and extrusion speeds to control structures replication quality. Wetting properties of fabricated foils were characterized by contact angle measurements of water sessile drop in static and dynamic method. We recorded values of static contact angles above 150□and contact angle hysteresis in a range of 10 - 20□what indicates on superhydrophobic surfaces with self – cleaning potential.

References

[1] S. Murthy and et. al., "Fabrication of Nanostructures by Roll-to-Roll Extrusion Coating," Adv. Eng. Mat., 2015.
[2] J. Shieh and et al., "Plasma-made silicon nanograss and related nanostructures," J. Phys. D: Appl. Phys., vol. 44, no. 17, p. 174010 (6pp), 2011.
[3] M. Schnell , R. Ludemann and S. Schaefer , "Plasma surface texturization for multicrystalline silicon solar cells," CONFERENCE RECORD OF THE TWENTY-EIGHTH IEEE PHOTOVOLTAIC SPECIALISTS CONFERENCE - 2000 Book Series: IEEE PHOTOVOLTAIC SPECIALISTS CONFERENCE, pp. 367-370, 2000.

Keywords: Characterization, Nanotechnology, Polymers & Plastics, Surface Analysis
Application Code: Polymers and Plastics
Methodology Code: Surface Analysis/Imaging

Session # 810 **Abstract # 810-11** **Poster Sessions**

Session Title: Polymer Characterization and Analysis

Abstract Title: **Synthesis and Characterization of Poly(p-methylstyrene) Spiropyran Conjugates**

Primary Author: Matthew J. Price
California University of PA

Date: Monday, March 07, 2016 - Afternoo
Time:
Room: Exposition Floor, 400 Aisle

Co-Author(s): Zachary Sullenberger

Abstract Text

Multiple methodologies regarding the synthesis of spiropyran polymer conjugates using the poly(p-methylstyrene) template were developed. The initial approach utilized NBS to undergo benzylic bromination of the homopolymer poly(p-methylstyrene) generating reactive sites for 2,3,3-trimethyl-3H-indolinine activation during spiropyran synthesis. The subsequent methodology produced a spiropyran linked p-methylstyrene monomer unit that constructed a statistical photochromic copolymer when mixed with p-methylstyrene under radical polymerization techniques using azobisisobutyronitrile as the initiator. Proton NMR and size exclusion chromatography were applied to determine an average molecular weight of the statistical photochromic copolymers. FT-IR spectroscopy was employed to observe functional group transformations between reactants and products ensuring the reaction proceeded to completion. UV-vis spectroscopy was utilized to determine the photochromic properties of the statistical copolymer conjugates.

Keywords: Derivatization, NMR, Polymers & Plastics, UV-VIS Absorbance/Luminescence
Application Code: Polymers and Plastics
Methodology Code: Education/Teaching

Session # 810 Abstract # 810-12 **Poster Sessions**

Session Title Polymer Characterization and Analysis

Abstract Title **Advantages of Ion Mobility Mass Spectrometry for Extractables Testing**

Date: Monday, March 07, 2016 - Afternoo

Time:

Room: Exposition Floor, 400 Aisle

Primary Author Baiba Cabovska
Waters Corporation

Co-Author(s) Eleanor Riches

Abstract Text

Extractable and leachable components, which are potentially harmful to human health, are of great concern to manufacturing industries, particularly manufacturers of food contact materials, pharmaceutical packaging and devices as well as cosmetics packaging. Globally, much legislation exists to try to mitigate exposure to these components, which results in a significant demand for rapid, accurate, and reliable analytical methodologies. One such method is targeted screening using LC-MS techniques. Typically, in identification of compounds, retention time, accurate mass, and fragmentation ion information is used. However, if different chromatographic methods are used, the retention times will vary. In this work, we demonstrate how the inclusion of collisional cross section (CCS) values, acquired using ion mobility-mass spectrometry, can provide increased confidence in compound identification. Collisional cross section is a key physicochemical property of compounds. The CCS depends on an ion's size, shape and charge. For example, in the case of two ions with the same m/z but different shapes, the less compact, straight-chain species will have a longer drift time than the smaller, more compact species. Solvent standards, solvent extracts of various plastic packaging materials, and spiked extracts of the same materials were analysed using LC-ion mobility-mass spectrometry (LC-IM-MS). Extracts were prepared by cutting the packaging materials into 5 mm2 pieces then sonicating 1 g of the pieces in 10 mL of 2-propanol for nine hours, at an average temperature of 45 oC. Solvent standards of representative packaging components such as dyes, were prepared in 100% methanol. A generic, 10-minute LC gradient was used with different column chemistries.

Keywords: Identification, Mass Spectrometry, Pharmaceutical, Polymers & Plastics

Application Code: Polymers and Plastics

Methodology Code: Mass Spectrometry

Session # 810 Abstract # 810-13 **Poster Sessions**

Session Title: Polymer Characterization and Analysis

Abstract Title: **A Novel Device for DART-MS System**

Primary Author: Michael J. Churchill
BioChromato

Date: Monday, March 07, 2016 - Afternoo
Time:
Room: Exposition Floor, 400 Aisle

Co-Author(s): Chikako Takei

Abstract Text

Direct analysis in real-time mass spectrometry (DART-MS) is a powerful method for rapid analysis of mixtures. However, this method is unsuitable for polymer analysis because polymers are difficult to volatilize. To overcome this disadvantage, we developed the ionRocket temperature gradient system for DART-MS (Fig. 1). On elevating the temperature at a controlled rate, vapor phase was generated from the analytes, which was then ionized and introduced into the mass spectrometer.

[b]Application 1: Identification of Nylons[/b]
Nylon, an aliphatic polyamide, is a well-known polymer material. There are several types of nylon such as nylon-6,6, nylon-6,10, and nylon-6,12. Although several analysis methods are in practice for identification of polymer structure, it is difficult to distinguish the nylon types easily because of their structural complexity. Using ionRocket, the pyrolysis products of nylons (monomers and multimers) were detected at over 300 °C. Monomers and dimers that were produced in the pyrolysis reaction were key compounds for identification of nylon types. The nylon types and individual nylon products were determined easily by analyzing the alteration of the spectral pattern with time.

[b]Application 2: Material Analysis of Vulcanized Rubbers[/b]
Vulcanized rubbers contain several additives and different kinds of polymers. For material analysis by Mass Spec various kinds of pretreatment were required; however, pretreatment for insoluble samples were difficult. ionRocket makes DART-Ms analysis of vulcanized rubbers possible without any sample pretreatment (Fig. 2). Comparing both new and aged rubber bands, it was observed that benzothiazole content decreased and the amount of degradation compounds of isoprene increased after the degradation test.

Keywords: Forensic Chemistry, Mass Spectrometry, Polymers & Plastics, Rubber
Application Code: Polymers and Plastics
Methodology Code: Mass Spectrometry

Session # 810 **Abstract # 810-14** **Poster Sessions**

Session Title: Polymer Characterization and Analysis

Abstract Title: **Characterization of Inorganic Fillers in Complex Polymer Matrices Using Mid-IR and Far-IR Spectroscopy**

Date: Monday, March 07, 2016 - Afternoo

Time:

Room: Exposition Floor, 400 Aisle

Primary Author: William Wihlborg
Thermo Fisher Scientific

Co-Author(s): Ronald Rubinovitz

Abstract Text

Inorganic powders are often added to polymers to alter the mechanical and optical properties of the composite material. Spectroscopically characterizing these composite materials requires the use of a multi-range FTIR system. The mid-infrared spectral range is used here to identify the polymers present, as well as the fillers and additives. The far-infrared range is needed in this study of filled-polymers because some inorganic fillers only absorb in the far-IR or have mid-infrared bands that overlap with other components of the composite material. In this paper we use Attenuated Total Reflectance (ATR) in both the mid-IR and far-IR ranges to characterize neat inorganic fillers as well as the components of filled polymer composites. These measurements were made using a purged FTIR system, a single infrared source and one infrared detector, which obviates the need for a costly vacuum FTIR system. The successful identification of both the polymer matrix and filler materials will be established. Distinguishing between the different crystal forms of calcium carbonate and titanium dioxide when dispersed in a polymer matrix will also be discussed.

Keywords: FTIR, Infrared and Raman, Materials Characterization, Polymers & Plastics

Application Code: Material Science

Methodology Code: Vibrational Spectroscopy

Session # 820 Abstract # 820-1 **Poster Sessions**

Session Title: Portable Instruments

Abstract Title: **Portable Optical Detection System for Determining Physical and Chemical Changes on Bioimplant Surfaces**

Primary Author: Donald Benza
Clemson University

Date: Monday, March 07, 2016 - Afternoo
Time:
Room: Exposition Floor, 400 Aisle

Co-Author(s): Fenglin Wang, Jeffrey N. Anker

Abstract Text

More than 2 million fracture fixation surgeries are performed each year in the US. When revisions are necessary, causes include hardware failures and infection which come at a large cost per year. It is essential to be able to detect these failures as quickly and noninvasively as possible. Disclosed here is a method to determine physical and chemical changes on implant surfaces. To do this, two photomultiplier tubes (PMTs) are used to detect single photons. Low levels of light are collected with PMTs in photon counting mode. An image is formed by scanning an x ray, which is used as a light source, point by point with an x-y moveable stage. Sensors consisting of nanoparticles are excited by the x ray source to luminesce which is modulated by chemical sensitive films or cross polarizers (physical changes). The light is then split by a 50/50 beam splitter after which the light is band pass filtered at 620nm and 700nm. Each PMT will detect one of these wavelengths and the ratio of the intensities of each wavelength will form the final image. Future work will include small animal studies.

This work was supported by an NSF CAREER grant under award CHE 1255535 and by NIGMS of the National Institutes of Health under award number 5P20GM103444-07.

Keywords: Biomedical, Biosensors, Imaging, Portable Instruments
Application Code: Bioanalytical
Methodology Code: Portable Instruments

Session # **820** Abstract # **820-2** **Poster Sessions**

Session Title: Portable Instruments

Abstract Title: **Field-Portable LED Array Based Multi-Wavelength Photohermal Lens Spectrometer**

Primary Author: Micah W. Eller
Tennessee Technological University

Co-Author(s): Andrew Callender

Date: Monday, March 07, 2016 - Afternoo
Time:
Room: Exposition Floor, 400 Aisle

Abstract Text

Photothermal lens spectroscopy, coupled with microscale liquid-liquid extraction techniques, has been previously demonstrated as an ultrasensitive method for the determination of trace metals in water. The development of inexpensive, field-portable PTLS instrumentation, along with field viable extraction techniques, could allow the ultrasensitive determination of trace metal concentrations in samples of natural and drinking water on site.

PTLS normally utilizes expensive high power lasers or vapor lamps as excitation sources, neither of which are appealing for use in a rugged, portable instrument. LED arrays are an inexpensive, rugged alternative, appealing for portable instrumentation. Coupled with inexpensive electronic hardware and rugged housing, white LED arrays could allow for the development of an inexpensive, portable, broad spectrum PTLS instrument.

Progress toward a field portable/ luggable, single-beam photothermal lens spectrometer, utilizing a white LED array as the pump-probe source with optical filter wavelength selection, is presented. The photothermal lens effect is demonstrated in dye solutions utilizing a white LED array. The determination of trace metals in water samples with the instrument after cloud point extraction of the 1-(2-pyridylazo)-2-napthol complexes is demonstrated.

Keywords: Environmental/Water, Metals, Molecular Spectroscopy, Portable Instruments
Application Code: Environmental
Methodology Code: Portable Instruments

Session # 820 Abstract # 820-4 **Poster Sessions**

Session Title: Portable Instruments

Abstract Title: **Development of Optical "Clamp-Meter" Using Silicone Optical Technology for In-Situ Absorption Spectroscopy**

Date: Monday, March 07, 2016 - Afternoo
Time:
Room: Exposition Floor, 400 Aisle

Primary Author: Hiroaki Nomada
Kyushu University

Co-Author(s): Hiroaki Yoshioka, Hirokazu Higuchi, Morita Kinichi, Yuji Oki

Abstract Text

Conventional spectrophotometers are constructed by air paths, optical components and optical traps to reduce an internal scattering noise. Also, a robust base for optical alignment accuracy requires the heavy and large equipment size. Traditionally, the sampling procedure must be needed for the absorption measurement due to the closed black chamber[1]

Recently, we proposed novel optical design concept "SOT (silicone optical technology)", in which PDMS (polydimethylsiloxane) is adopted as optical material to realize the compact system.[2] A light channel by SOT can suppress internal scattering light. It has a structure of the core of PDMS and the clad of PDMS compound with black pigment. This "light trapping" optical channel passes only the light entering with channel axis.

Light trapping performance were evaluated by bending the channel. When carbon nanotube was dispersed, optical density of 4.5 was obtained by the curving angle of 30 degree. This channel seemed useful such as a chamber free optical system. As an application, we also propose coupled probe for in-situ and open loop optical measurement such as fiber coupling[3]. It is also another demonstration of our proposed concept "labo-on-tablet" [4] It is coupled probes like clamp-meter and applicable to any objective in situ without sampling procedure. The source probe emits light from PDMS pickup on the tablet's LCD via a fiber. The measuring probe guides incident light to the tablet's camera via the optical trap channel. When the objective was clamped with the coupled probe, only light transmitted via objective from source probe was selectively guided to the tablet's camera and measured.

The measurement accuracy was evaluated with glass filter. Experimentally, the signal reproducibility about 0.17% can be estimated.

1.A. R. H. Cole et al.: Applied Optics 9(1970) 23
2.K. Morita et al.: Flow Analysis XIII (2015) P25
3.http://www.expedeon.com/VersaWave.aspx
4.H. Nomada et al.: Pittcon (2014) 1450-1

Keywords: Fiber Optics, Lab-on-a-Chip/Microfluidics, Portable Instruments, Spectrometer
Application Code: Biomedical
Methodology Code: Integrated Sensor Systems

Session # 820	Abstract # 820-5	Poster Sessions

Session Title Portable Instruments

Abstract Title **Development of a Portable Optical Cavity to Enhance 1-GHz Mode-Locked Laser Pulses for Broadband Absorption Spectroscopy**

Date: Monday, March 07, 2016 - Afternoo
Time:
Room: Exposition Floor, 400 Aisle

Primary Author Yutaro Ito
Kyushu University

Co-Author(s) Imasaka Totaro, Zaitsu Shin-ichi

Abstract Text

On-site multi-gas monitoring with high sensitivity to collect trace species data is essential for applications such as volcanic gas sensing for eruption prediction or infinitesimal radioactive gas monitoring in nuclear power plant. Mode-locked cavity-enhanced absorption spectroscopy (ML-CEAS) is expected to be a promising approach due to several remarkable features; high sensitivity (ppb~ppt), fast acquisition (>1s), broad spectral width for multispecies detection, and high resolution based on the thousands of laser lines. However, this technique has a trade-off between sensitivity and the size of optical cavity that is a main component of ML-CEAS. In previous research, the cavity size has tended to be large at the expense of portability (cavity length reported ever is 0.5~1.2 m), because sensitivity has been put on high priority for trace analysis in laboratory. Here, we focus on downsizing of the cavity and the improvement of portability for on-site gas analysis. The use of mode-locked laser with the repetition rate of 1-GHz as a light source allows us to design a palm-size cavity (length:0.072 m) portable enough to be lifted with one hand. We believe compact cavity is a hopeful method to balance between maintaining and downsizing the cavity. This achievement will open the door to utilize ML-CEAS with high performance in on-site gas monitoring.

Keywords: Gas, Laser, Portable Instruments, Spectroscopy
Application Code: Environmental
Methodology Code: Molecular Spectroscopy

Session # **820** Abstract # **820-7** **Poster Sessions**

Session Title: Portable Instruments

Abstract Title: **Enzymes and Photometers**

Primary Author: Ellen R. Campbell
NECi Superior Enzymes

Date: Monday, March 07, 2016 - Afternoo
Time:
Room: Exposition Floor, 400 Aisle

Co-Author(s): Bill Campbell, David A. Squires, Justin Walbeck

Abstract Text

Modern analytical techniques flow toward "Faster, Better, Cheaper". We develop wet chemistry methods that replace metals with enzyme-based reagents. These reagents are recombinant, produced in tightly controlled protein expression systems for lot to lot consistency: they can be regarded as any other reagent-grade chemical. Our Nitrate Reductase (NaR) reagents for nitrate detection have been on the market for over ten years. Validations include USGS, ASTM, and USEPA Clean Water Act, Safe Drinking Water Act and Standard Methods are in process.

Enzyme-based methods are "Fast": NaR reacts near the rate of diffusion. They're "Better": enzymes have unmatched selectivity, so assays work in complex mixtures with reduced requirement for sample prep. Plus, they're proteins – nonhazardous materials for reduced operator exposure and lower shipping and disposal costs. Sample and reaction volumes can be tiny – microplate range to one milliliter total assay volume for our test kits. Compare costs to handle cadmium, which NaR replaces. When these factors are taken into account, enzymes might even be "Cheaper".

We simplified the reagent system for onsite test kit applications, and find that even though the reaction results are visible over a broad color intensity range, users want a digital readout: they want a number. Handheld photometers on the market are based on vendors' test kit products. Reaction vessels are round and range from 5 to 20 mL volume. Here we show development of a portable photometer based on a standard cuvette holder. Data is sent to smartphones via Bluetooth. Open Source and commercial versions are available.

Keywords: Bioanalytical, Environmental/Biological Samples, Enzyme Assays, Wet Chemical Methods
Application Code: Environmental
Methodology Code: UV/VIS

Session # 820	Abstract # 820-8	Poster Sessions

Session Title: Portable Instruments

Abstract Title: **Personal Monitoring of Ozone Exposure: A Fully Portable Device for Under $150 USD Cost**

Date: Monday, March 07, 2016 - Afternoo
Time:
Room: Exposition Floor, 400 Aisle

Primary Author: Tingting Cao
Texas Tech University

Co-Author(s): Jonathan E. Thompson

Abstract Text

An inexpensive and portable device for monitoring personal ozone exposure is described and its performance characterized. The device is built from commercially available components, exhibits time resolution of approx. 60 sec, and highest analytical sensitivity under 100 ppbv ozone. The sensor has been employed to provide insights into ozone exposure for 8 volunteers living in Lubbock, Texas during the winter months of 2015. Consistent with previous literature, the results indicate the volunteers were exposed to highest levels of ozone when outdoors during daylight hours. Exposure to ozone indoors was typically only a fraction (0.3-0.7) of the dose observed during times spent outdoors. The sensing system described requires minimal technical skills to assemble and use at a cost of approximately $150 USD per unit. The device's batteries provide power for 8-10 hours on a single charge and the sensor can be re-used many times after recharging the battery pack. A major advantage of the sensor over chromogenic filters for exposure monitoring is the collection of time-series data that allows users to better understand when and where individuals are exposed to highest ozone concentrations. The device may prove useful for industries requiring a low-cost solution to monitor employee exposure to ozone for specific work environments.

Keywords: Environmental, Environmental/Air, Portable Instruments
Application Code: Environmental
Methodology Code: Portable Instruments

Session # 820 **Abstract # 820-9** **Poster Sessions**

Session Title: Portable Instruments

Abstract Title: **Producing a Miniaturized High-Performance Liquid Chromatography System for Exploration-Based Research**

Date: Monday, March 07, 2016 - Afternoo
Time:
Room: Exposition Floor, 400 Aisle

Primary Author: Kyle B. Lynch
University of Oklahoma

Co-Author(s)

Abstract Text

One main hurdle to overcome when trying to develop a HPLC system is the gradient generation that drives the separation within the column. This system needs to be both robust as reproducible in order to serve the analytical needs of chemists. As a whole, these systems (not including the detector) can be rather large and fairly expensive ($20,000-$30,000). In recent years, the drive for smaller and cheaper systems has been a topic of research with hopes of replacing the conventional commercial systems. The development and continued research into capillary electrochromatography (CEC) in recent years has allowed for both the decrease in size and cost of high-pressure pumps and gradient generation systems. It is with our hopes in finding a system that proves to be as reliable and reproducible as the conventional HPLC system at a fraction of the cost and size.

Presented are two gradient generation systems that are under current development within our laboratory. The first involves a combination of a selector valve, a series of different eluents, and a low-pressure pump to create the gradient that is used in separation. The second system greatly mimics a conventional HPLC and utilizes a proportional valve with a low-pressure pump along with a low and high concentration eluent to generate a gradient through the use of Labview programming. Both systems then use a miniaturized high-pressure electroosmotic pumping system developed in laboratory to drive separation of the sample of interest. It is our hopes with these miniaturized systems to create a low-cost, reliable, and portable system that can be coupled with a variety of detectors including UV-Vis and laser-induced spectroscopy detectors and mass spectrometers to name a few. Also, with the reduced size of the overall system, dwell volume is decreased by ~300% leading to a drastic increase in throughput compared to a conventional HPLC system.

Keywords: Capillary Electrophoresis, High Throughput Chemical Analysis, HPLC, Liquid Chromatography/Mass Sp
Application Code: High-Throughput Chemical Analysis
Methodology Code: Liquid Chromatography/Mass Spectrometry

Session # **830** **Abstract #** **830-1** **Poster Sessions**

Session Title Quality/QA/QC

Abstract Title **A Novel Approach to Specialty Gas Certifications by Means of GC/FTIR Analysis**

Primary Author Peter P. Behnke
Prism Analytical Technologies, Inc.

Date: Monday, March 07, 2016 - Afternoo
Time:
Room: Exposition Floor, 400 Aisle

Co-Author(s) Anthony S. Bonanno, Charles M. Phillips, Kelly R. McPartland, Martin L. Spartz

Abstract Text

The current state of the specialty gas market requires a diverse group of gas analysis methodologies to adequately quantify and certify the concentration of components for a given mixture. This requirement drives the need for a suite of analyzers along with their own particular maintenance and calibration requirements prior to each mixture analysis. A new instrument utilizing GC/FTIR technology has been developed that allows for faster, more sensitive and accurate quantification without the need for multiple instrumentation and continuous customer calibration. Instrument transferrable calibrations are supplied with the GC/FTIR and can be verified on setup, as QA/QC procedures require. Deconvolution of absorption spectra via classical least squares methodology ensures that compounds not fully chromatographically resolved still produce quantitatively accurate results. The control of environmental factors within the GC/FTIR (such as temperature and pressure) guarantees that calibrations are constant over both time and system conditions. This allows one instrument to test a variety of different mixtures without continuous calibration or interference, while reducing analysis time and maintaining current levels of precision and accuracy for any given mixture. Representative data and analytical results of an EPA Compendium Method TO-15 cylinder utilizing a GC/FTIR system equipped with a pre-concentrator will be discussed.

Keywords: FTIR, Gas Chromatography, Process Analytical Chemistry, Specialty Gas Analysis
Application Code: Quality/QA/QC
Methodology Code: Vibrational Spectroscopy

Session # 830 **Abstract # 830-2** **Poster Sessions**

Session Title: Quality/QA/QC

Abstract Title: **Evaluation of the Flavor of Strawberry Preparation Using Gas Chromatography Electronic Nose**

Primary Author: Andrew Cowell
Alpha MOS

Date: Monday, March 07, 2016 - Afternoo
Time:
Room: Exposition Floor, 400 Aisle

Co-Author(s): Fatma Ayouni, Herve Lechat, Jean-Christophe Mifsud, Marion Bonnefille, Valérie Vabre

Abstract Text

One of the major concerns of food producers is to avoid consumer claims by applying strict controls on products. Among the different aspects under scrutiny, sensory features have an important part.

This study presents the investigation of a sensory defect in strawberry preparations using a fast gas chromatography electronic nose. The objective is to explain the cause of off-odors in a claimed product and create tools to rapidly monitor the overall sensory quality of production batches. The comparison of chromatograms shows significant differences of volatile profiles between good and bad fruit preparation batches.

Additionally, an odor map based on Principal Components Analysis on all peaks clearly differentiates batches depending on quality, the distribution on the map being influenced by volatile compounds concentration. The nature of the most discriminant volatile compounds involved in the aroma of strawberry fruit preparation is investigated using their Kovats index and the AroChemBase database. It appears that the molecules responsible for odor differences are mostly esters (hexyl butyrate, hexyl propanoate...), aldehydes (benzaldehyde) and alcohol (cis-3-hexenol). To rapidly monitor the quality of fruit preparations at the production level, a quality control card based on Statistical Quality Control model has finally been set up by taking into account the good samples as the reference quality.

Keywords: Chemometrics, Food Science, Gas Chromatography
Application Code: Quality/QA/QC
Methodology Code: Chemometrics

Session # 830 **Abstract # 830-3** **Poster Sessions**

Session Title Quality/QA/QC

Abstract Title **Developments in the Automatism for CHNS and Oxygen Determination Using an Elemental Analyzer for Chemical Characterization**

Date: Monday, March 07, 2016 - Afternoo
Time:
Room: Exposition Floor, 400 Aisle

Primary Author Guido Giazzi
Thermo Fisher Scientific

Co-Author(s) Francesco Leone, Liliana Krotz, Walter Galotta

Abstract Text

Carbon, Nitrogen, Hydrogen, Sulfur by combustion analysis, and Oxygen determination by pyrolysis are very commonly used for the characterization of solid and liquid samples as raw and final products in pharmaceutical, cosmetics, universities and material industries for quality control and R&D purposes. The use of an accurate and automatic analytical techniques which allows the fast analysis with an excellent reproducibility is required. The Elemental Analyzer showed in this paper is equipped with two totally independent furnaces allowing the installation of two analytical circuits – CHNS and Oxygen - which are used alternatively and completely automatic through hardware and software improvements. Each analytical circuit can received is own autosampler. In this way the system copes effortlessly with the wide array of laboratory requirements such as accuracy, day to day reproducibility and high sample throughput. This paper introduces the data on CHNS/O determination to illustrate the high performance of the system.

Keywords: Automation, Characterization, Elemental Analysis, Pharmaceutical
Application Code: Quality/QA/QC
Methodology Code: Atomic Spectroscopy/Elemental Analysis

Session # 830 Abstract # 830-4 Poster Sessions

Session Title Quality/QA/QC

Abstract Title **Running Legacy HPLC and Optimized UPLC Methods on a Single UPLC Platform, Comparative Studies and Steps Towards Implementation in a QC Environment**

Date: Monday, March 07, 2016 - Afternoo

Time:

Room: Exposition Floor, 400 Aisle

Primary Author Christopher Henry
Waters Corporation

Co-Author(s) Andy Boughey, Mark Wrona, Richard Ladd

Abstract Text

With QC Departments looking for tools to update and modernize AstraZeneca Macclesfield updated their LC platforms by implementing Waters ACQUITY UPLC H-Class system in their pharmaceutical development department to enable the development of new methods for new products using UPLC technology. While needing to future proof the QC department to receive these UPLC methods it was critical to retain the ability to robustly run the legacy HPLC methods. The technology of choice was the Waters ACQUITY UPLC H-Class which has now been deployed in the AstraZeneca QC department based at Macclesfield, UK. AstraZeneca is successfully running all registered QC methods on the ACQUITY UPLC H-Class Systems with three high throughput products having been successfully transferred to UPLC and validated. The novel UPLC methods employ sub 2µm particle analytical columns combined with the low dispersion UPLC instrument to significantly reduce runtime while at the same time improving peak shape and resolution when compared to the corresponding legacy HPLC method. This reduction in runtime has shown potential instrument usage savings of between approximately 70-97%, which will be highlighted along with corresponding significant solvent usage reduction. Within this body of work we will give examples of two legacy HPLC methods, compounds B and C which have been successfully transferred to H-Class from the legacy LC platform. We will also demonstrate the successful development and validation of UPLC methods for three high volume products, compounds A, B and C, highlighting and contrasting the workflow efficiency savings possible with implementation of these methods.

Keywords: Analysis, Liquid Chromatography, Method Development, Quality Control

Application Code: Quality/QA/QC

Methodology Code: Liquid Chromatography

Session # **830** **Abstract #** **830-5** **Poster Sessions**

Session Title Quality/QA/QC

Abstract Title **Quantitative Analysis of Calcite in Desulfurization Gypsum Using Raman Spectroscopy**

Primary Author Young Taek Ma
Hanyang University

Date: Monday, March 07, 2016 - Afternoo
Time:
Room: Exposition Floor, 400 Aisle

Co-Author(s)

Abstract Text

The WAI(Wide Area Illumination) scheme provides unique advantages over the conventional Raman collection scheme to improve the performance of quantitative Raman analysis. It can improve the reliability of Raman measurements by significantly increasing surface coverage area. In this study, It purpose that thermal gravity analysis (TGA) method of desulfurization gypsum impurity replace new analysis method using WAI Raman spectroscopy. Usually quantitative analysis of Calcite contents used TGA method. But TGA take long time for analysis and affected organic contents. Raman spectroscopy is possible to quick analysis and nondestructive analysis, high sensitivity of $CaCO_3$ and nondestructive analysis. Generally gypsum board made of desulfurization gypsum. And it has some impurity like a Calcite($CaCO_3$). Calcite concentration influence mechanical and chemical property of gypsum board. Using variety sample (0, 0.1, 0.5, 1, 2, 3, 5 % seven STD samples and 12 deferent desulfurization gypsum made by chinese thermoelectric power plant) compare RAMAN and TGA result. This study establishes the great potential for the use of Raman spectroscopy in quantitative analysis of calcite concentration in variety constructive material.

Keywords: Analysis, Quality Control, Quantitative, Raman
Application Code: Quality/QA/QC
Methodology Code: Vibrational Spectroscopy

Session # **830** Abstract # **830-6** **Poster Sessions**

Session Title: Quality/QA/QC

Abstract Title: **The Use of a Triple Detection System, (UV, ELSD, MS) for Pharmaceutical Degradation Studies**

Primary Author: Aaron D. Phoebe
Waters Corporation

Date: Monday, March 07, 2016 - Afternoo
Time:
Room: Exposition Floor, 400 Aisle

Co-Author(s): Patricia R. McConville, Paula Hong

Abstract Text

In the pharmaceutical industry, there is a need to fully understand the stability of the active pharmaceutical ingredient (API) and characterize any impurities that may be formed, including those found in forced degradation studies. Reversed-phase liquid chromatography UV based techniques are often used for these types of analyses. However, separating and detecting the related impurities and other components can be challenging. Separation can be challenging because of the structural similarities of the components, and detection can prove to be difficult due to the fact that not every compound may be detected using a single detection technique. By combining a UV detector, an evaporative light scattering detector, and a mass spectrometer, it is possible to detect compounds with different chemical properties.
In the following study, we will conduct a forced degradation study of glimepiride. The degradation reaction was monitored using a UHPLC system, with a photodiode array detector, evaporative light scattering detector and a mass spectrometer. In this example we will illustrate the characterization and quatitation of the related impurities. We will also illustrate the advantage of using mass detection and evaporative light scattering for analysis of by-products that do not contain a chromophore and therefore cannot be detected by UV.

Keywords: Light Scattering, Liquid Chromatography/Mass Spectroscopy, Pharmaceutical
Application Code: Quality/QA/QC
Methodology Code: Liquid Chromatography/Mass Spectrometry

Session # 830 **Abstract #** 830-7 **Poster Sessions**

Session Title Quality/QA/QC

Abstract Title **Fast Analysis of Short Chain Fatty Acids in Feeds By SFC/MS**

Primary Author Jinchuan Yang
Waters Corporation

Date: Monday, March 07, 2016 - Afternoo
Time:
Room: Exposition Floor, 400 Aisle

Co-Author(s) BJ Bench, Carrie Snyder, Jayant Shringarpure, Jessica Lance

Abstract Text

SCFAs may have a positive role in Salmonella mitigation in chickens (Ref 1). The analysis of SCFAs as additives in feeds by GC-MS is about 60 minute long. In order to shorten the run time, UltraPerformance Convergence ChromatographyTM (UPC2) has been investigated. UPC2 is a next-generation supercritical fluid chromatography. UPC2 leverages the unique properties of compressed CO2 at or near its supercritical state, such as low viscosity and high diffusivity, and sub-two micron particle packed columns to improve separation efficiency, speed, and selectivity. A rapid and sensitive UPC2-MS method for the quantitative determination of seven short-chain fatty acids (SCFAs) in animal feeds is presented. The SCFAs include acetic acid, propionic acid, butyric acid, isobutyric acid, valeric acid, isovaleric acid, and hexanoic acid. The extraction of SCFAs in feeds was a liquid-liquid extraction using tetrahydrofuran. The extract was directly analyzed by UPC2-MS on two diol columns (ACQUITY UPLC BEH 125 4.6x150 mm, 1.7 um). Selected Ion Recording (SIR) mode in mass spectrometry was used to selectively collect each of the SCFAs molecular ions. For the two pairs of the isobaric ions (valeric and isovaleric acids, and butyric and isobutyric acids), chromatographic baseline separation was achieved (Rs larger than 1.6). The chromatography run time was shortened to 12 min as compared to the run time of 60 min by GC-MS. The performance of this method, such as calibration, limit of detection, precision, and accuracy will be presented. This method has been transferred and evaluated in a production lab, and the evaluation results will also be presented.

Keywords: Lipids, Supercritical Fluid Chromatography
Application Code: Quality/QA/QC
Methodology Code: Supercritical Fluid Chromatography

Session # 830 Abstract # 830-9 **Poster Sessions**

Session Title Quality/QA/QC

Abstract Title **Accelerated Oxidation Tests on Olive Oil Stored in Plastic Packaging and Submitted to Autoclaving**

Date: Monday, March 07, 2016 - Afternoo
Time:
Room: Exposition Floor, 400 Aisle

Primary Author Stefano Casiraghi
VELP Scientifica

Co-Author(s) Claudia Mancinelli, Stefania Corti

Abstract Text

Food packaging technology has undergone fast and significant developments in the past three decades, and the smartest are yet to be seen. Plastic materials can now be used for applications once restricted to metals and glass, allowing weight saving and design versatility benefits. In this context, an important step to take into account during the production process is the resistance to autoclaving, a sterilization process at high pressure and temperature. Indeed an alteration of the properties of the packaging material would affect the shelf-life of the food product to be stored. The purpose of the investigation was to evaluate the suitability of two polypropylene-based materials with oxygen barrier, as an alternative to metal or glass, for canned oily products. To this aim, the oxidative stability of an olive oil sample contained in the two different packaging materials was tested before and after autoclaving. Results were compared with those obtained on the same oil stored in glass. The oxidative stability of the oil samples was investigated by accelerating the oxidation process using the OXITEST (VELP Scientifica, Italy), a reactor based on the use of high temperatures and over-pressure of oxygen. Between the two polypropylene-based materials, the one showing higher stability after autoclaving was then subjected to an accelerated ageing process using a climate chamber in order to evaluate the effects of a shelf-life greater than 6 months at room temperature. The results obtained confirm that the Oxitest is particularly useful for determining the packaging material that maintains the product in the freshest condition.

Keywords: Food Science, Lipids, Quality Control
Application Code: Food Science
Methodology Code: Chemical Methods

Session # 830 Abstract # 830-10 **Poster Sessions**

Session Title Quality/QA/QC

Abstract Title **Fast Analytical Method for Essential Oil Flavor Characterization Using a Polar Stationary Phase for Optimized Selectivity**

Date: Monday, March 07, 2016 - Afternoo
Time:
Room: Exposition Floor, 400 Aisle

Primary Author Ramkumar Dhandapani
Phenomenex

Co-Author(s) Kristen Parnell, Tim Anderson

Abstract Text

Essential oils are fragrant plant essences primarily composed of terpenes, their derivatives, and other aromatic compounds. Variation in plant location and growing conditions produces natural differences in essential oil components, and due to their high price, premium oils are subject to adulteration with cheaper terpenes or other agents. Characterization of essential oils is therefore necessary, but testing is complex due to the number of compounds and their presence at trace levels. Long GC columns are therefore traditionally used to provide adequate resolution and sensitivity for these volatile flavor compounds.

The present study employs a high polarity GC column and optimized methods for the flavor characterization of cold-pressed orange oil. Initially, samples were analyzed using 60 m x 0.25mm x 0.25 μm dimensions, with a run time of 21 min. A modified, lean method was then developed using optimized parameters and dimensions. The analysis was run at Golay's optimum carrier gas flow rate to reduce the run time to 7 min; simultaneously, the small internal diameter and thin film used resulted in sharper peaks due to faster mass transfer. The analytical method presented thus resulted in a 65% reduction in analysis time, significantly improving overall productivity without sacrificing resolution or selectivity to key analytes.

Keywords: Flavor/Essential Oil, Food Science, Gas Chromatography, GC Columns
Application Code: Food Science
Methodology Code: Gas Chromatography

Session # 830 Abstract # 830-11 **Poster Sessions**

Session Title Quality/QA/QC

Abstract Title **Steep Time and Temperature Effects on Flavor and Flavonoid Extraction of Black Tea**

Primary Author Anne Jurek
EST Analytical

Date: Monday, March 07, 2016 - Afternoo
Time:
Room: Exposition Floor, 400 Aisle

Co-Author(s) Justin Murphy, Kelly Cravenor, Lindsey Pyron

Abstract Text

The steeping time of tea can have a profound effect on flavor. Most tea products have a recommended steeping time and temperature. The steeping temperature of the water can vary from “bring water to a rolling boil” to a “very light steam”, while the steeping time can fluctuate from two to five minutes. If a tea is steeped for too long or at too high of a temperature; it can become bitter. However, longer steeping times enhance the health benefits of tea by increasing the amount of flavonoids extracted from the tea. This application will look at the effects of time and temperature on the extraction of tea flavors and flavonoids.

Keywords: Food Science, Gas Chromatography/Mass Spectrometry, SPME
Application Code: Food Science
Methodology Code: Gas Chromatography/Mass Spectrometry

Session Title Quality/QA/QC

Abstract Title **A Novel Approach to the Analysis of Multivitamin in Foodstuff by Online Supercritical Fluid Sample Extraction/Supercritical Fluid Chromatography**

Primary Author Qiang Li
Shimadzu (China) Co., LTD

Date: Monday, March 07, 2016 - Afternoo
Time:
Room: Exposition Floor, 400 Aisle

Co-Author(s)

Abstract Text
Nexera UC system (Shimadzu corporation, Japan) was used for both the screening of the method using supercritical fluid chromatography (SFC) and the supercritical fluid sample extraction (SFE) followed by SFC directly (SFE/SFC). A new method was developed for the analysis of a series vitamins and its derivatives in foodstuff by online SFE/SFC . Foods as the sample were pretreated by SFE with the mixture of supercritical CO_2 and organic solvent like acetonitrile and methanol. Extracted components were separated directly by SFC which combined with SFE. The separation was carried out on C18 column and vitamins were detected via triple quadrupole mass spectrometer. Comparing with conventional pretreatment, online -SFE/SFC provide a new way to extract the analyte from complex matrix and avoid oxidation of unstable compounds. The Nexera UC enabled significant reduction of analysis time and cost without sacrificing quality of analyses. This method is applicable for the determination of fat-soluble vitamins and their derivatives in foodstuff.

Keywords: Chromatography, Food Science, SFC, SFE
Application Code: Food Science
Methodology Code: Supercritical Fluid Chromatography

Session # 830 Abstract # 830-13 **Poster Sessions**

Session Title Quality/QA/QC

Abstract Title **Simultaneous Measurement of L-Lactate and Ethanol in Tomato-Based Products**

Primary Author William Miller
YSI, Inc

Date: Monday, March 07, 2016 - Afternoo
Time:
Room: Exposition Floor, 400 Aisle

Co-Author(s) June Klingensmith

Abstract Text

In the manufacture and packaging of ketchup and related tomato products, both lactate and ethanol concentrations can be used to quantify microbial testing of ingredients, in-process and finished products. Lactate producing bacteria and ethanol-producing yeast may both contribute to microbial load. When mixed with broth and grown in cultures for 1-5 days, the lactate and ethanol concentrations may be correlated with the degree of microbial load. In this study, Lactate and ethanol were measured in supernatants of diluted ketchup samples that were obtained from a commercially available source. Precision of replicate samples was determined from selected samples, and percent recovery was determined for samples spiked with both lactate and ethanol. Spiked samples of diluted ketchup were measured using the YSI Biochemistry Analyzer within 30 minutes of spiking and mixing. The results of this study demonstrated that the YSI Biochemistry Analyzer can simultaneously measure lactate and ethanol in a tomato matrix with adequate precision and accuracy to make process and quality assurance decisions in tomato product manufacturing. This method of measuring lactate and/or ethanol can reduce QC test time by hours and provide results that better predict potential flavor issues and incipient spoilage compared to traditional microbial methods.

Keywords: Biosensors, Food Safety, Food Science, Process Monitoring
Application Code: Food Science
Methodology Code: Process Analytical Techniques

Session # 830 Abstract # 830-14 **Poster Sessions**

Session Title Quality/QA/QC

Abstract Title **Iron Translocation in Adzuki Beans Sprouts: Enrichment Effects**

Primary Author Aline P. Oliveira
Federal University of São Paulo

Date: Monday, March 07, 2016 - Afternoo
Time:
Room: Exposition Floor, 400 Aisle

Co-Author(s) Alexandrina C. Carvalho, Cassiana S. Nomura, Juliana Naozuka

Abstract Text

The sprouts are considered pesticide-free foods that should be highlighted in human nutrition and grows considerably in different countries. The sprouts are fully natural, using only the stored reserves in seeds to germinate and to achieve size to be consumed, in addition to being a good source of minerals, vitamins, proteins, and present low calorie. This way, when the present power supply potential nutritional deficiencies. Beside this, it is possible to enrich the sprouts with essential elements. So, in this work the Fe translocation was verified analysing roots, stems and cotyledons after germinated process of adzuki bean sprouts. Beside this, the effect of Fe enrichment in the elements essential (Ca, Cu, K, Mg, P, S and Zn) concentration was studied. Germination of seeds (13 days) was done at 25 °C in dark environment, wetting absorbent cotton with deionized water (control group) or different solutions containing 125 and 500 [micro]g of FeCl[sub]3[/sub] and irrigating the sprouts all days with deionized water. The elemental determination was done by ICP OES, after the acid digestion (HNO[sub]3[/sub] + H[sub]2[/sub]O[sub]2[/sub] + H[sub]2[/sub]O) in a closed-vessel microwave oven. Iron concentration increased in different beans parts, depending on the iron mass added to the growth medium: roots (47-202%); stem (0.6-11.2%); cotyledons (18-111%). The Ca, Cu, K, Mg, P, S and Zn concentrations in the different beans parts were very close to the control group. Iron fortification process was efficient to supply the Fe deficiency and the enrichment did not affect the elements essential concentration.

ACKNOWLEDGMENT: FAPESP(2015/01128-6)/CNPq/CAPES.

Keywords: Elemental Analysis, Food Science, Spectrophotometry
Application Code: Food Science
Methodology Code: Atomic Spectroscopy/Elemental Analysis

Session # 830 **Abstract # 830-15** **Poster Sessions**

Session Title: Quality/QA/QC

Abstract Title: **The Qualitative and Quantitative Analysis of α-Acids in Hops and Beers by UHPLC with UV Detection**

Date: Monday, March 07, 2016 - Afternoo
Time:
Room: Exposition Floor, 400 Aisle

Primary Author: Charles Schmidt
PerkinElmer

Co-Author(s): Wilhad M. Reuter

Abstract Text

Alpha acids (α-acids) are a class of chemical compounds primarily of importance to the production of beer, providing beers with their aroma and bitter taste.

The α-acids found in hop resins are isomerized to form the iso-α-acids during prolonged boiling in the wort. The degree of isomerization and the amount of bitter taste produced by the addition of hops is highly dependent on the type of hop and the length of time the hops are boiled.[sup]1[/sup]

Since the quality and quantity of α-acids is so important in consistently providing individual beers with their recognizable taste, it is essential to monitor their amount in hops and beers and to monitor the formation of the iso-α-acids during the beer brewing process. The focus of this application note is to provide a simplified robust analytical method for establishing the type and amount of α-acids in hops pellets, as well as determining the amount of α-acids and iso-α-acids in various beers. In particular, this method is novel in that it chromatographically separates humulone and adhumulone, which, in the past, have been reported to coelute.

1. International Calibration Standards, American Society of Brewing Chemist, 2009

Keywords: Beverage, Carbohydrates, Food Science, Liquid Chromatography
Application Code: Food Science
Methodology Code: Liquid Chromatography

Session # 830 **Abstract # 830-16** **Poster Sessions**

Session Title Quality/QA/QC

Abstract Title **SFC Analytical Method Development for Vitamin D3 and Related Compounds**

Date: Monday, March 07, 2016 - Afternoo

Time:

Room: Exposition Floor, 400 Aisle

Primary Author Takashi Sato
YMC Co., Ltd.

Co-Author(s) Junko Kawabata, Noritaka Kuroda, Roland Spaegele, Saoko Nozawa, Toshikazu Adachi

Abstract Text

By utilizing the advantages of Supercritical Fluid Chromatography (SFC) such as high fluid permeability and rapid adsorption/desorption, we could achieve higher resolution analysis in a shorter run time than HPLC. It is known that a combination of achiral columns and SFC is effective for reducing analysis cycle time of fat-soluble vitamins and terpenoids.

In this research, we intended to develop a rapid SFC analysis method for Vitamin D3 and three related compounds (Pre-cholecalciferol, 5,6-trans-Cholecalciferol, Tachysterol3), especially in nutritional products. At first, method screening by combination of five achiral phases and three types of modifier were conducted. This proposed that Triart Diol (3 [micro]m, 12 nm), column dimension of 250x4.6 mmI.D., and ethanol were the best combination. Then, fine-tuning of the method (e.g. modifier ratio in a mobile phase and column temperature) was conducted so that a concentration of Vitamin D3 in the nutritional products can be determined.

The established method can separate Vitamin D3, its related compounds, and impurities in the nutritional products. Analysis time was decreased to ten minutes, which is one fourths of a conventional HPLC method. Furthermore, analysis reproducibility and linearity offered reliability which is commonly required as a quantitative analysis method.

Keywords: Food Science, High Throughput Chemical Analysis, SFC, Supercritical Fluid Chromatography

Application Code: Food Science

Methodology Code: Supercritical Fluid Chromatography

Session # 830 Abstract # 830-17 **Poster Sessions**

Session Title Quality/QA/QC

Abstract Title **Fast and Cost-effective Sugar Analysis Using HPAE-PAD**

Primary Author Hua Yang
Thermo Fisher Scientific

Date: Monday, March 07, 2016 - Afternoo
Time:
Room: Exposition Floor, 400 Aisle

Co-Author(s) David G. Moore, Linda Lopez

Abstract Text

Mono- and disaccharide sugar determinations are often used in the food and beverage industry to comply with the mandatory nutrition declaration (EU regulation No.1169/2011), to ensure the quality of a formulated product, to maintain or select for desired sweetness, and to characterize and confirm the source of the carbohydrates. Carbohydrates have poor chromophores and are hard to detect by absorbance detectors without lengthy and costly derivitization. However, carbohydrates can be determined directly by High Performance Anionic-exchange Pulsed Amperometric Detection (HPAE-PAD). This is a well-established method that does direct detections.

Here demonstrates the use of HPAE-PAD for direct determination of the carbohydrates in food samples. With the new Thermo Scientific™ Dionex™ CarboPac™ SA10-4µm column, the sugar analysis is completed in less than 10 minutes. The new Thermo Scientific™ Dionex™ Integrion™ HPIC™ system introduces more easy-of-use features for this application such as the Thermo Scientific™ Dionex™ Reagent-Free™ IC Eluent Generation (RFIC™-EG) and the palladium reference electrode (pdH).

Keywords: Carbohydrates, Electrochemistry, Food Identification, Ion Chromatography
Application Code: Food Science
Methodology Code: Liquid Chromatography

Session # 830	Abstract # 830-18	Poster Sessions

Session Title Quality/QA/QC

Abstract Title **Improving Analytical Performance with Enhanced Matrix Removal - Lipid**

Primary Author Derick Lucas
Agilent Technologies

Co-Author(s) Bruce Richter, Joan Stevens, Limian Zhao, Megan Juck

Date: Monday, March 07, 2016 - Afternoo
Time:
Room: Exposition Floor, 400 Aisle

Abstract Text

The analysis of complex samples often requires unwanted sample preparation steps to extract analytes of interest at appropriate concentrations and eliminate unwanted co-extractives, which can result in chromatographic interferences, matrix effects in mass spectrometry, and accumulation in chromatographic flow paths. Developing concise and efficient sample preparation protocols for lipid rich matrices are of particular interest as current techniques can be time consuming, indiscriminately retain lipids and analytes of interest, and/or provide insufficient lipid removal. Agilent Bond Elut Enhanced Matrix Removal – Lipid (EMR-Lipid) is a novel matrix removal sorbent for highly selective lipid removal from complex, high fat samples without unwanted analyte retention. This work implements EMR-Lipid as a dispersive solid phase extraction (dSPE) cleanup for fatty samples in multi-class, multi-analyte analysis workflows such as QuEChERS. Data will demonstrate thc impact of superior cleanliness on analyte recovery, reproducibility, and instrumental performance. Furthermore, comparisons between EMR-Lipid and conventionally used sorbent materials demonstrate dramatic improvements in matrix removal and analyte recovery and reproducibility. The ease of use, time and cost savings, minimal method development, and dramatically cleaner samples make this an attractive cleanup option for laboratories currently performing or moving into contaminant analysis in complex sample types.

Keywords: Gas Chromatography/Mass Spectrometry, Lipids, Liquid Chromatography/Mass Spectroscopy, Sampl

Application Code: Food Contaminants

Methodology Code: Sampling and Sample Preparation

Session # 830 **Abstract # 830-20** **Poster Sessions**

Session Title: Quality/QA/QC

Abstract Title: **A Fast Method for Predicting the Phenolic Content of Whisky Malts**

Primary Author: Andrew Cowell
Alpha MOS

Date: Monday, March 07, 2016 - Afternoo
Time:
Room: Exposition Floor, 400 Aisle

Co-Author(s): Fatma Ayouni, Herve Lechat, Jean-Christophe Mifsud, Marion Bonnefille, Valérie Vabre

Abstract Text

Phenolic compounds, mainly coming from the smoking of barley, the thermal degradation and the maturation processes, are strongly related with the peaty and smoky flavor of a whisky. Most often, the concentration in phenolic compounds is determined by UV/visible spectroscopy, HPLC or GC-MS.

This study proposes to use a fast gas chromatography based electronic nose to predict the total phenolic content in barleys, by correlating the measurement to known data on phenolic scores and creating a predictive model for future samples. Six samples of smoked malts of known phenolic content were analyzed with HERACLES e-nose. After water addition, each malted barley is analyzed directly and the chromatography separation of samples headspace is achieved in less than two minutes. The evaluation of the overall flavor profiles on an odor map (Principal Components Analysis applied on the chromatography data) clearly shows that the various malts have different sensory features. On the graph, the samples are distributed along an axis linked with the total phenolic content. This confirms that these compounds strongly influence the aroma. A quantitative model based on Partial Least Square (PLS) was then built by correlating the known phenolic scores and the HERACLES measurements. As a high level of correlation was obtained (correlation coefficient > 89%), the model could be successfully used for predicting the total phenolic content in four blind samples of malt. Therefore, this analytical method can be an alternative solution for rapidly estimating the total phenolic content in malts.

Keywords: Beverage, Food Identification, Gas Chromatography, Headspace
Application Code: Food Identification
Methodology Code: Gas Chromatography

Session # **830** **Abstract #** **830-21** **Poster Sessions**

Session Title Quality/QA/QC

Abstract Title **Combining FTIR and Mass Spectrometric Detection in the Gas Chromatographic Analysis of Fragrances**

Date: Monday, March 07, 2016 - Afternoo

Time:

Room: Exposition Floor, 400 Aisle

Primary Author Tracy Phillpott
DANI Instruments

Co-Author(s) Daniele Recenti, Matthew S. Klee, Roberta Lariccia

Abstract Text

Characterization of fragrances is extremely important for the confirmation of the origin of the product and for quality control. While gas chromatography mass spectrometry is most commonly used for characterization and quantification of compounds in fragrances, presence of isomers sometimes makes the complete characterization more challenging from both GC separation and mass spectral differentiation standpoints. Characterization of isomers is important because they can have different sensorial characteristics. Linalool for instance, is a positional isomer of geraniol but with completely different sensorial characteristics. Isomers can often be easily differentiated by their infrared spectra, so by adding FTIR detection in a format that is optimized specifically for GC significantly improves the information that can be acquired in the same time frame as the typical GC-MS experiment.

Results of fragrance analysis from a system that combines FTIR detection and Time of Flight (TOF) mass spectrometry will be presented. High quality FTIR spectra will be shown that were generated using a GC FTIR detector technology optimized for use with capillary GC. This technology relies on deposition of eluting sample on a cryogenically cooled disk followed by post-run acquisition of solid-phase infrared spectra. This approach provides best-in-class resolution and sensitivity. As a choice of mass spectrometer, TOF-MS combines benefits of fast acquisition rate and full mass range spectra. Several example analyses of products used in the perfume industry will be used to demonstrate the power of combining GC, TOF-MS, and FTIR to provide meaningful and important information on the composition of complex samples.

Keywords: Flavor/Essential Oil, FTIR, Instrumentation, Mass Spectrometry

Application Code: Food Identification

Methodology Code: Molecular Spectroscopy

Session # **830** **Abstract #** **830-23** **Poster Sessions**

Session Title Quality/QA/QC

Abstract Title **Holographic Characterization of Multicomponent Colloidal Suspensions**

Primary Author Jaroslaw M. Blusewicz
Spheryx, Inc.

Date: Monday, March 07, 2016 - Afternoo
Time:
Room: Exposition Floor, 400 Aisle

Co-Author(s) David B. Ruffner, Laura Philips

Abstract Text

Holographic video microscopy characterizes colloidal microspheres with great accuracy and precision. Lorenz-Mie theory is used to determine the size, index of refraction, and 3D position of particles in the difficult to measure size range of 200nm to 20[micro]m. A major advantage of this technique is individual measurement of each particle in the suspension, which enables the detection and analysis of multicomponent suspensions without special sample preparation or the application of multiple technologies. The experimental suspensions were created using combinations of various colloidal particles (silica, polymethyl methacrylate (PMMA), polystyrene (PS)) varying in size from 400nm to 7[micro]m. Holographic video microscopy separately identifies each particle population even in samples with particles of the same size but different refractive index. We demonstrate this capability in samples with 7 distinct particle populations coexisting in the same sample. Every particle is characterized multiple times as it flows through the microscope. The results of these characterizations are tracked and statistically analyzed to provide high precision analysis of each particle population.

Keywords: Characterization, Light Scattering, Microscopy, Particle Size and Distribution
Application Code: Quality/QA/QC
Methodology Code: Microscopy

Session # **830** Abstract # **830-24** **Poster Sessions**

Session Title Quality/QA/QC

Abstract Title **Mitigating Electrostatic Effects Improves Measurement Accuracy**

Date: Monday, March 07, 2016 - Afternoo

Time:

Room: Exposition Floor, 400 Aisle

Primary Author Colleen Clancy
NRD LLC

Co-Author(s) Arnold Steinman

Abstract Text

Abstract: Electrostatic charge affects the accuracy of instruments making precision weighing measurements. Accurate measurements are vital, whether in a research facility exploring new opportunities, or in the production of an existing product. This paper will discuss electrostatic phenomena and their interaction with weighing operations.

Charge is generated primarily by contact and separation of dissimilar materials and is unavoidable throughout the weighing process. The materials involved are almost certain to be charged, and this includes samples, transport media, and weighing equipment parts or enclosures.

Once generated, charge affects both the instruments and materials being weighed. Electrostatic forces interact with the mechanisms of weighing machines, making precise measurements all but impossible. Electrostatic forces of attraction and repulsion affect light weight sample materials, causing unwanted movement, losses during transfers, and clinging of unwanted particles to measurement surfaces. Charge problems are not limited to weighing applications, occurring whenever small physical quantities or objects need to be measured, as with atomic force microscopes, force and mass measurements, and electrochemical measurements.

The electronics industry has developed mitigation methods to protect sensitive integrated circuits from the effects of static charge. Grounding of conductors, static dissipative materials, and air ionization are the primary static control methods. Air ionization is of particular importance in weighing operations as equipment parts, samples, and transport media are often insulators or isolated conductors. This paper discusses how static control may be applied in precision weighing operations, both in equipment and in sample transport, to neutralize static charge and improve accuracy and repeatability of measurements.

Keywords: Instrumentation, Laboratory, Pharmaceutical, Sample & Data Management

Application Code: Laboratory Management

Methodology Code: Physical Measurements

Session # 830 **Abstract # 830-25** **Poster Sessions**

Session Title Quality/QA/QC

Abstract Title **Safe Approach to Gas Purification**

Date: Monday, March 07, 2016 - Afternoo

Time:

Room: Exposition Floor, 400 Aisle

Primary Author Brian Warrick
ARM, Inc

Co-Author(s) Daniel Spohn

Abstract Text

Ultra-high purity gases are a necessity in many industries, including fabrication of electronics, photovoltaic, and lighting component as well as generation of high purity carrier gases for laboratory use. In many cases, the end users may be unaware of the risks associated with selecting, installing and operating a gas purifier.

Purification of industrial gases often requires use of highly reactive media. Select media may react exothermically with contaminates within the feed gas stream resulting in elevated temperature sufficient to breach the stainless steel containment vessel. Each year manufacturing facilities around the world are impacted by such failures. In many cases the damage is localized to an analyzer, tool, or laboratory, but when coupled with an on-site air separation plant, exothermic events may result shutdown of the entire manufacturing facility, with a typical recovery time of one month.

This study provides an historical review of gas purification technology ranging from methods of filling high purity gas cylinders, to point-of-use purification of laboratory gases, to purification of bulk and on-site atmospheric and specialty gases. The authors then introduce novel approaches to creating ultra high purity gases using safer methods and media. In many applications, use of advanced VPSA and TSA methods, coupled with safe catalysts, can be used to generate 7N, 8N, and 9N pure gases which historically required use of cryogenic separation or highly reactive media. The authors' goal is to introduce the concept of safer gas purification and offer alternative solutions for creation of ultra high purity gases.

Keywords: Adsorption, Chemical, Gas, Material Science

Application Code: Safety

Methodology Code: Chemical Methods

Session # 830 Abstract # 830-26 **Poster Sessions**

Session Title Quality/QA/QC

Abstract Title **Advanced UHPLC Instrument to Instrument Method Transfer**

Primary Author Gregory Hunlen
Agilent Technologies

Co-Author(s) Michael Woodman

Date: Monday, March 07, 2016 - Afternoo
Time:
Room: Exposition Floor, 400 Aisle

Abstract Text

The transfer of methods from conventional liquid chromatography to ultra high performance liquid chromatography systems in a mult-vendor world may result in significant differences in retention time, resolution, and even selectivity. Emulation technology, comprised of unique computer algorithms and interfaced with advance UHPLC pumps, characterizes critical parameters such as dwell volume and mixing behavior. The concept allows the system "adaptability" for nearly universal instrument to instrument method transfers. The results of this study show that validation parameters between a targeted and an emulated HPLC system are robust for retention time and resolution and correlate within method validation acceptance criteria.

Keywords: HPLC
Application Code: Validation
Methodology Code: Separation Sciences

Session # 840 **Abstract # 840-2** **Poster Sessions**

Session Title Sensors

Abstract Title **Enabling Tools to Combat Antibiotic Resistant Bacteria**

Primary Author Andrew Heller
Michigan State University

Date: Monday, March 07, 2016 - Afternoo
Time:
Room: Exposition Floor, 400 Aisle

Co-Author(s) Dana Spence

Abstract Text

Nearly a year ago, in response to a Presidential strategy, the CDC, NIH, and FDA chaired an 84-point action plan to combat antibiotic resistance. An antibiotic resistant bacterium can quickly reproduce and in turn negate the functionality of a given antibiotic. Thus, the need for fast and accurate measures of a bacterium's resistance to a particular antibiotic is a high priority action point. Here, we present multiple strategies that will facilitate a measure of antibiotic resistance with enhanced rapid and/or parallel diagnostics. First, the release of ATP during the bacterial growth cycle is exploited to decrease time associated with the Kirby-Bauer test which requires time for growth of bacteria. Aliquots of a bacterial pathogen are removed from a small batch of bacteria growing in media. The levels of ATP are measured at various time increments up to 20 minutes and the slope analyzed to determine the rate of growth (which is typically correlated with an increase of ATP secretions from the bacteria). Our current method allows for single digit nanomolar levels of ATP to be determined in less than 20 minutes after a bacterial sample has been removed from an initial growth colony. Secondly, we will demonstrate how a fluidic device is being utilized to expose bacteria to pharmacokinetic (PK) profiles of antibiotic followed by live/dead assays to determine drug resistance.

Keywords: Bioanalytical, Biological Samples, Biotechnology, Chemiluminescence
Application Code: Bioanalytical
Methodology Code: Sensors

Session # 840 Abstract # 840-3 **Poster Sessions**

Session Title Sensors

Abstract Title **Optimizing the Scintillation Cascade in Nano-Scintillation Proximity Assay (nanoSPA) for Multiplexed Detection of Small Biomolecules**

Primary Author Zeinab Mokhtari
University of Arizona

Date: Monday, March 07, 2016 - Afternoo
Time:
Room: Exposition Floor, 400 Aisle

Co-Author(s) Colleen Janczak, Craig A. Aspinwall, Isen Andrew C. Calderon

Abstract Text

Detection of small biomolecules such as glucose, which are involved in cellular signaling pathways but lack optical/electrochemical activity, is challenging in living cells. To detect such species, polystyrene/silica core/shell nanoparticle composites were prepared that are doped with scintillant fluorophores to facilitate the detection and measurement of 3H-labeled analytes. Beta particles emitted from 3H-labeled analytes are detected as visible light emission via energy transfer through the scintillants. The resulting nanoparticle composite, referred to as nanoSPA, yields specific response to 3H-labeled analytes. In this work, we sought to optimize the nanoSPA response to obtain a more efficient scintillation cascade and to shift the emission to red/near-IR wavelengths in order to avoid background fluorescence and allow for simultaneous detection of multiple analytes (multiplexed detection). To shift emission wavelength, we characterized the energy transfer efficiencies of fluorophore dyes with large Stokes' shifts and explored the use of quantum dots due to their broad absorption and size-dependent narrow emission spectrum. Core/shell QDs were also explored, as previous studies indicate that they provide more stable fluorescence due to the protection of the surface which prevents nonradiative relaxation of excitons. Ultimately optimization of scintillant composition will yield the most efficient energy transfer among the fluorophores enabling scintillation detection of 3H-labeled analytes with temporal and spatial resolution, properties that are currently lacking.

Keywords: Bioanalytical, Biosensors, Nanotechnology, Radiochemical Methods
Application Code: Bioanalytical
Methodology Code: Sensors

Session # **840** Abstract # **840-4** **Poster Sessions**

Session Title Sensors

Abstract Title **Dendritic Gold Structures for Glucose Biosensor Design**

Primary Author Almira Ramanaviciene
Vilnius University

Date: Monday, March 07, 2016 - Afternoo
Time:
Room: Exposition Floor, 400 Aisle

Co-Author(s) Anton Popov, Arunas Ramanavicius, Asta Kausaite-Minkstimiene, Laura Sakalauskiene, Natalija German

Abstract Text

Gold micro- and nano-structures have received considerable attention due to their attractive physical and chemical properties [1]. Some challenging bioanalytical problems, such as sensitivity, specificity, reproducibility, duration and cost of analysis can be resolved by micro- and nanostructure-based electrochemical biosensors [2-4].
The aim of this work was to synthetize dendritic gold structures on the graphite rod electrode surface and to develop sensitive and convenient amperometric glucose biosensor. Several enzyme immobilization on graphite electrode premodified with dendritic gold structures were tested. Electrode modified with dendritic gold structures, self-assembled monolayer and covalently immobilized glucose oxidase showed the highest analytical signal and the best limit of glucose detection.

Acknowledgment
This research was funded by a grant (No. SEN-15095) from the Research Council of Lithuania.

References

[1] N. Li, P. Zhao, D. Astruc. Anisotropic gold nanoparticles: synthesis, properties, applications, and toxicity. Angew. Chem. Int. Ed. 53 (2014) 1756-1789.
[2] Y. Li, Y. Li, M. Hong, Q. Bin, Z. Lin, Z. Lin, Z. Cai, G. Chen. Highly sensitive protein molecularly imprinted electro-chemical sensor based on gold microdendrites electrode and prussian blue mediated amplification. Biosens. Bioelectron. 42 (2013) 612-617.
[3] N. German, A. Ramanavicius, A. Ramanaviciene*. Electrochemical deposition of gold nanoparticles on graphite rod for glucose biosensing. Sensor Actuat. B: Chem. 203 (2014) 25–34.
[4] N. German, A. Kausaite-Minkstimiene, A. Ramanavicius, T. Semashko, R. Mikhailova, A. Ramanaviciene*. The use of different glucose oxidases for the development of an amperometric reagentless glucose biosensor based on gold nanoparticles covered by polypyrrole. Electrochim. Acta 169 (2015) 326-333.

Keywords: Biosensors, Electrochemistry, Immobilization, Nanotechnology
Application Code: Bioanalytical
Methodology Code: Sensors

Session # 840 **Abstract # 840-5** **Poster Sessions**

Session Title Sensors

Abstract Title **High-Throughput Thiamine Quantification in Environmental Matrices Using Periplasmic-Binding Protein Biorecognition**

Primary Author Katie A. Edwards
Cornell University

Date: Monday, March 07, 2016 - Afternoo
Time:
Room: Exposition Floor, 400 Aisle

Co-Author(s) Antje J. Baeumner, Cliff Kraft, Seth Feder

Abstract Text

Thiamine (vitamin B1) deficiency is a major cause for high mortality in fish of commercial and sport-fishing interest in the Great Lakes and the Baltic Sea. Its deficiency has been associated with poor immune system development and survival rate of fry, as well as reduced visual acuity, neurological symptoms, and muscle weakness in adult fish leading to their inability to migrate during spawning. Its deficiency is also known to cause neurological and cardiovascular problems in other species, including humans. This molecule has recently also been identified as an important nutrient in algal bloom proliferation. Current procedures for thiamine quantification rely on extensive sample preparation and formation of the fluorescent oxidation product, thiochrome, prior to chromatographic analysis. The cost and time required for these procedures limits data available for informed decisions critical to fish health by the fisheries industry and researchers alike. A high-throughput assay for thiamine quantification with specificity provided by biorecognition via periplasmic binding proteins is the objective of this work. The thiamine periplasmic binding protein (TBP) exhibits exceptional specificity to thiamine and its phosphate derivatives. Here, TBP has been applied in a competitive microtiter-plate assay for thiamine using fluorescent dye-encapsulating liposomes for signal enhancement. Results and challenges in the development of a total thiamine assay in environmental matrices will be presented.

Keywords: Bioanalytical, Environmental Analysis, Fluorescence, Immunoassay
Application Code: Environmental
Methodology Code: Sensors

Session # 840 Abstract # 840-6 **Poster Sessions**

Session Title Sensors

Abstract Title **Multi-Purpose Individual Air Monitor – Conception to Proof of Concept Laboratory Prototype**

Date: Monday, March 07, 2016 - Afternoo
Time:
Room: Exposition Floor, 400 Aisle

Primary Author Nicholas Fitzgerald
Defence Science and Technology Group

Co-Author(s) Karl Pavey

Abstract Text

Many toxic chemical vapours are invisible to the eye and by the time you smell them or feel the physical effects, it can be too late. What if someone had dosed the room you are in with something you can't see, you can't smell but are breathing it in right now? How would you know?

These are thoughts that must, at times, enter the minds of first responders, law enforcement, and military personnel as they attend to their often dangerous and stressful work. The authors will outline the development of a new prototype vapour detection device, dubbed Black Canary (BC), which brings colorimetry into the 21st century in a small, lightweight, flexible, autonomous package which delivers to a clear capability gap in the current generation of toxic chemical vapour detection devices.

BC introduces a new concept in hand held detection combining a suite of replaceable detection cartridges with a smart mobile phone sized electronics unit capable of identifying chemical threats at or below the Permissible Exposure Limit and having an automatic alarm capability.

From reactive chemical element identification, test and manufacture to integration with a purpose designed hardware solution; the authors will chronicle the development of the BC detector from proof of concept to working laboratory prototype and beyond.

Keywords: Air, Detection, Portable Instruments, Spectrophotometry
Application Code: Homeland Security/Forensics
Methodology Code: Sensors

Session # 840 Abstract # 840-7 **Poster Sessions**

Session Title: Sensors

Abstract Title: **Characterization of Self-Oscillating Reaction Catalyzed by Metal Porphyrin Complexes**

Date: Monday, March 07, 2016 - Afternoo

Time:

Room: Exposition Floor, 400 Aisle

Primary Author: Takashi Arimura
AIST Interdisciplinary Research Center

Co-Author(s): Jung Hee Do, Kenichi Tominaga, Masaru Mukai

Abstract Text

One dissipative structure that exhibits temporal and spatiotemporal oscillation is known to be chemically generated by the famous Belousov-Zhabotinsky (BZ) reaction. BZ reaction has been investigated as self-oscillating system toward a key feature of space-time self-assemble of nonequilibrium systems. We have developed a self-oscillating gel actuator in the wake of BZ reaction catalyzed by ferroin for the first time,1) which causes autonomous mechanical oscillation without an external control in a completely closed solution. So far all self-oscillating cross-linked gel systems undergoing expanding-contracting oscillation are catalyzed only by using the expensive Ru(bpy)3 complex and iron complex ferroin, which are expensive and toxic. The thing is to develop a less expensive catalyst of the BZ reaction for industrial applications such as nursing care equipment. Porphyrin derivatives play an important role in biochemistry and possess a small load on the experiment and a high safety to the human body. Herein, rhythmic oscillation of the metal porphyrin solution by BZ reaction was demonstrated without any on-off switching of external stimuli for the first time. The reaction was performed in the presence of bromate ion, malonic acid, and the metal catalyst in sulfuric acid solution. In the course of the reaction, the oscillating waveform of the metal porphyrin was affected by the concentration of the components, especially the metal catalyst.

References

1.� Arimura and M. Mukai, Chem. Commun., 2014, 50, 5861-5863.

Keywords: Biotechnology, Chemical, Material Science, Nanotechnology

Application Code: Nanotechnology

Methodology Code: Sensors

Session # 840 **Abstract # 840-8** **Poster Sessions**

Session Title: Sensors

Abstract Title: **Simultaneous Detection of Dopamine Release and Neural Activity**

Primary Author: Kate L. Parent
University of Arizona

Date: Monday, March 07, 2016 - Afternoo
Time:
Room: Exposition Floor, 400 Aisle

Co-Author(s): Christopher W. Atcherley, Daniel F. Hill, Jean-Paul Wiegand, Michael A. Miller, Michael L. Heien, Stephen Cowen

Abstract Text

Complex behaviors rely on coordinated firing of neurons in local networks which are dynamically regulated by neuromodulators such as dopamine. However, tools for simultaneous monitoring of neural assemblies and dopamine dynamics have not previously been developed due to difficulties in integrating electrophysiological and electrochemical hardware. We have engineered a measurement platform capable of tandem measurements of dopamine and neural activity. This platform consists of high-density electrode arrays for measurement of single-neuron activity and local-field potentials in conjunction with a separate carbon electrode for measurement of real-time dopamine release using fast-scan cyclic voltammetry. Electrochemical instrumentation was improved to allow a common reference to be utilized for both electrophysiological and electrochemical measurements resulting in a higher signal-to-noise ratio for all measurements. To prevent damage to sensitive instrumentation, current between the implanted electrodes and the neural recording amplifier circuitry is interrupted during the application of the voltammetric waveform via a solid-state relay. In vitro testing was carried out in a porcine gelatin to mimic brain conductivity and 800 µV sine waves with various frequencies (10 – 3000 Hz) were applied near the electrode array to simulate neural oscillations. It was found that frequencies above 50 Hz were rapidly recovered (< 20 ms) following resumption of neural recordings in relation to the time between waveforms (200 ms). Stimulated dopamine release and multiple single-unit activities were measured in an anesthetized animal.

Keywords: Bioanalytical, Biosensors, Neurochemistry, Voltammetry
Application Code: Neurochemistry
Methodology Code: Sensors

Session # 840	Abstract # 840-9	Poster Sessions

Session Title Sensors

Abstract Title **Development of a Novel Position-Sensitive MCP Detector**

Date: Monday, March 07, 2016 - Afternoo
Time:
Room: Exposition Floor, 400 Aisle

Primary Author Blake Wiggins
Indiana University

Co-Author(s) Davinder Siwal, Romualdo T. deSouza

Abstract Text

Position-sensitive microchannel plate (MCP) detectors are useful tools to measure the position of an incident electron, ion, or photon. Several applications in the fields of medical imaging, neutron imaging, and astronomy benefit from achieving good position resolution at the lowest level of signal-to-background possible. Recently, a novel detector, which utilizes an induced signal approach to provide position sensitivity at the single electron level, has been developed. This detector has the capability of resolving two simultaneous but spatially separated incident electrons. In the prototype detector, using only the zero-crossing point of the inherently bipolar signals, a position resolution of 466 □m (FWHM) has been achieved. Initial implementation of a differential readout has improved this resolution to 400 □m (FWHM). To further improve the resolution using the differential readout, a better understanding of the dependence of the induced signal shape on the position of the electron cloud is required. To characterize the dependence of the induced signal shape on position, a resistive anode (RA) has been incorporated into the detector. Use of the RA allows for the measurement of the centroid of the electron cloud. Factors impacting the position resolution obtained with the RA will be discussed and the achieved position resolution of 157 □m (FWHM) will be presented.

Keywords: Characterization, Detector, Imaging, Instrumentation
Application Code: Validation
Methodology Code: Sensors

Session # 840 **Abstract # 840-10** **Poster Sessions**

Session Title: Sensors

Abstract Title: **Real Time Monitoring Magnesium Alloys Corrosion by Electrochemical H2 Sensor In Vivo**

Date: Monday, March 07, 2016 - Afternoo
Time:
Room: Exposition Floor, 400 Aisle

Primary Author: Daoli Zhao
University of Cincinnati

Co-Author(s): Tingting Wang, Vesselin Shanov, William R. Heineman, Zhongyun Dong

Abstract Text

Biodegradable implant especially magnesium and its alloys have attracted a lot interest due to their good metallic strength, light weight, low toxicity and rapid corrosion rate in aqueous solutions. One of way to evaluate the development of the alloy is by monitoring the corrosion rate. To evaluate the corrosion rate of the various alloys in vivo, we develop the method for the real time monitoring the corrosion by electrochemical H2 sensors. We also established and developed a protocol for in vivo corrosion studies to obtain the surface composition of explanted Mg alloy by surface spectroscopies such as XPS; the distribution of corrosion products in the tissue surrounding an implant and in critical organs of test animals using bioanalytical techniques involving extraction and ICP-MS.

Keywords: Analysis, Detector, Electrochemistry, Sensors
Application Code: Material Science
Methodology Code: Sensors

Session # **840** **Abstract #** **840-11** **Poster Sessions**

Session Title Sensors

Abstract Title **A Cellulose Acetate Membrane-Based Colorimetric Device to Discriminate Bacteria**

Date: Monday, March 07, 2016 - Afternoo
Time:
Room: Exposition Floor, 400 Aisle

Primary Author Ligia Bueno
Instituto de Química - Universidade de Sao Paulo

Co-Author(s) Alison Cottel, Subrayal M. Reddy, Thiago Paixao

Abstract Text

Microorganisms are associated with disease and infection besides being culprits in food poisoning cases. The latter is a big concern since contamination may be caused by bacteria and viruses. Although viruses account for a large proportion of all foodborne illnesses, most hospitalizations and deaths related to foodborne infections are due to bacterial agents. Enterobacteria are associated with food adulteration either due to lack of hygiene or lack of specific care, and also within natural contamination. Bacteria produce volatile organic compounds (VOCs) in their growth process and such VOCs or sets of them are unique to each genus or species. The latter can be viewed as a chemical fingerprint of these organisms and can help to detect them. In order to demonstrate this, we report the use of a colorimetric plastic-based device to discriminate four pathogenic bacteria: Klebsiella pneumonia (KP), Proteus vulgaris (PV), Proteus mirabilis (PM) and Escherichia coli (EC). Cellulose acetate membranes were prepared with 5 entrapped pH indicators and Tween 20 as a plasticizer; they were placed in contact with the latter mentioned bacteria species in order to analyse volatile organic compounds released from the microorganisms during an incubation process at two temperatures (37°C and 25°C). The colour changes of the membranes were analysed with RGB values extracted from membrane images obtained with a smartphone and used as input data for non-supervised pattern recognition methods. With subsequent mathematical treatment, it was possible to see a clear discrimination amongst the studied bacteria, without any misclassification.

Financial support: FAPESP, CNPq and CAPES, Royal Society, UK.

Keywords: Chemometrics, Food Contaminants, Quality Control
Application Code: Food Contaminants
Methodology Code: Sensors

Session # 840 **Abstract # 840-13** **Poster Sessions**

Session Title: Sensors

Abstract Title: **A Smart Polymer Hydrogel as a Chemical Sensor on Biomedical Implant**

Date: Monday, March 07, 2016 - Afternoo

Time:

Room: Exposition Floor, 400 Aisle

Primary Author: Mohammed Arifuzzaman
Clemson University

Co-Author(s): Caleb Behrend, Jeffrey N. Anker

Abstract Text

Bacteria are often localized near implanted medical device (IMD) surfaces specifically at early stages, which bring challenges to current detection methods. In order to detect infections around biomedical implants a noninvasive chemical sensor is required. In vivo detection of low pH around IMD is an indication of bacterial infection. Therefore, the sensor, which can detect low and/or high pH in the target, needs to be coupled with the IMD. In addition, it should be biocompatible and stable material. To address this issue building a smart IMD sensor out of soft, wet hydrogels would be highly acceptable. A hydrogel, poly acrylic acid (PAAc) was synthesized to fabricate a pH sensor for IMD. After synthesis, a piece of PAAc hydrogel swollen in water (pH~5) was placed on an IMD attached to a cadaver femur. When phosphate buffer (pH~7.4) was applied for swelling, the mass/volume of the gel increased remarkably and significant displacement of a tungsten dial placed on the gel surface was clearly observed in a radiograph. The tungsten dial was attached to the IMD in such a way that the dial touched the top surface of the gel and was displaced when the gel mass/volume increased. Based on the results of this experiment we are claiming a successful "simple and smart IMD sensor" design, where a piece of hydrogel will sense bacterial infection level in the target by changing mass/volume of the gel due to swelling.

Keywords: Biomedical, Detection, Polymers & Plastics, Sensors
Application Code: Biomedical
Methodology Code: Sensors

Session # 840 Abstract # 840-14 **Poster Sessions**

Session Title Sensors

Abstract Title **Oxygen Sensitive Probes**

Primary Author Peter Gennaro
Clemson University

Co-Author(s)

Date: Monday, March 07, 2016 - Afternoo
Time:
Room: Exposition Floor, 400 Aisle

Abstract Text

A battery-powered device was developed to measure oxygen tension in tissue surrounding medical implants using an implanted LED as a fluorescent light source and an oxygen-sensitive fluorescent dye as an indicator. The ultimate goal is to detect hypoxia during implant infection. The fluorescent oxygen indicator dye, Oxyphor G4, was incorporated into a thin film of PDMS and uniformly coated onto the exterior of a 633 nm LED. Oxygen quenches the fluorescence reducing the intensity of 800 nm fluorescence passing through tissue, and is collected with a PMT. To account for variation in optical penetration through tissue and collection efficiency, a second nearby 800 nm LED is used as a reference. Although the Oxyphor G4 and reference LEDs emit light at the same wavelength, the two light sources are distinguished because they blink at different frequencies and times. The light source coated with the OSP showed great response to hypoxic and hyperoxic environments modeled by nitrogen purging. To conserve battery, a reed switch turns the circuit on only when an external magnetic field is applied. The fluorescent intensity of the in situ light source was also observed through tissue samples.

Keywords: Bioanalytical, Biomedical, Biosensors, Fluorescence
Application Code: Biomedical
Methodology Code: Sensors

Session # 840 **Abstract # 840-15** **Poster Sessions**

Session Title: Sensors

Abstract Title: **Graphene Oxide as an Efficient Antibacterial Agent in Macrophages and in Mice**

Date: Monday, March 07, 2016 - Afternoo
Time:
Room: Exposition Floor, 400 Aisle

Primary Author: Xu Wu
University of North Dakota

Co-Author(s): Julia Xiaojun Zhao, Min Wu, Xiao Liu, Yuqian Xing

Abstract Text

Graphene oxide (GO) has recently attracted significant attention due to broad applications in biomedical fields, such as biosensors, bioimaging, drug delivery, and antimicrobial. GO has been shown to kill bacteria in bacterial culture; however, how GO impacts bacterial invasive potential in vivo remains elusive. In this study, we characterized the antibacterial activity of GO in both cell culture and animal models. Klebsiella pneumoniae (Kp) is one of the most common multidrug resistant (MDR) pathogens in causing persistent nosocomial infections and is very difficult to eradicate once established in the host. First, we demonstrated that GO exerted direct killing of Kp in agar dishes and afforded the protection of alveolar macrophages (AM) from Kp infection in culture. We then evaluated the mortality, tissue damage, polymorphonuclear neutrophil (PMN) penetration, and bacterial dissemination in Kp-infected mice. Our results revealed that GO can counteract the invasive ability of Kp in vivo, resulting in lessened tissue injury, significant but subdued inflammatory response, and prolonged mouse survival. These findings indicate that GO may be an alternative agent for controlling MDR pathogens in clinics.

Keywords: Biomedical, Imaging, Nanotechnology, Toxicology
Application Code: Biomedical
Methodology Code: Sensors

Session # **840** Abstract # **840-16** **Poster Sessions**

Session Title: Sensors

Abstract Title: **pH Indicator to Enhance Surface Plasmon Resonance Imaging Detection of Small Organic Molecules**

Date: Monday, March 07, 2016 - Afternoo
Time:
Room: Exposition Floor, 400 Aisle

Primary Author: Zainab H. Al Mubarak
Oklahoma State University

Co-Author(s): Cassandra Rodenbaugh, Gayan Premaratne, Lucy Lehoczky, Sadagopan Krishnan

Abstract Text

pH is an essential parameter in biological, medical and industrial applications which reflects the formation of new chemicals in a deliberated medium. For instance, monitoring the pH level in biological fluids reflects the abnormal conditions in human body[1]. Accordingly, highly sensitive, rapid and continuous measurement of pH is extremely desirable. Surface plasmon resonance imaging (SPRi) is advantageous over other methods because it is label free, highly sensitive to refractive index changes (~ 3.75x10-5 RI change = 1 pixel), offers high throughput, and a small volume (~ 200 nL) of sample is sufficient per spot [2]. Changing the color of chemical substance such as pH indicator reveals the changes in the refractive index of the material, which is hypothesized to be sensitive for detection by SPRi. Thus, SPRi technique using a pH indicator to measure different low concentrations of H+ will offer a simple tool to monitor biological events as well as detection of small molecules. Herein, we demonstrate that SPRi could detect low acid concentrations (ppm-ppb range) via the refractive index changes of added pH indicators such as methyl red and bromothymol blue, which can be considered as a new method to study and investigate many biological and chemical proton coupled reactions.

Acknowledgments
Oklahoma State University and the National Institute of Health (NIH/NIDDK) supported this project.

[1] Sim, J.; Kwon, D.-S.; Kim, J., Acid-sensitive pH Sensor Using Electrolysis and a Microfluidic Channel for Read-out Amplification. RSC Advances 2014, 4 (75), 39634-39638.

[2] Fasoli, J. B.; Corn, R. M., Surface Enzyme Chemistries for Ultrasensitive Microarray Biosensing with SPR Imaging. Langmuir 2015, 10.1021/la504797z.

Keywords: Bioanalytical, Biosensors, Detector, Surface Analysis
Application Code: Bioanalytical
Methodology Code: Sensors

Session # 840 Abstract # 840-17 **Poster Sessions**

Session Title: Sensors

Abstract Title: **Membrane Based Electrodes for Electrochemical Applications in Biology**

Primary Author: Winnie E. Svendsen, DTU Nanotech

Date: Monday, March 07, 2016 - Afternoo
Time:
Room: Exposition Floor, 400 Aisle

Co-Author(s): Dorota Kwasny, Fatima Al-Zahraa Al Atraktchi, Helle Waagepetersen, Maria Dimaki, Tanya Bakmand

Abstract Text

Traditionally the state of cells and tissue is evaluated at the end of the culturing/maturation period by a visual inspection. By introducing continuous real time monitoring of cells and tissue in culture the culturing process can be optimized and the homogeneity of the samples can be ensured. Conventionally electrochemical measurements on cells and tissue are performed as end-point analysis in order to provide information on specific events. Here we present an electrochemical sensor system that enables monitoring of cell populations and tissue during the entire culturing/maturation period. Traditionally electrodes for electrochemical measurements on cells and tissue consist of either simple wires probing the area of interest or thin film electrodes covering the bottom surface of culture ware. The presented sensor system is based on commercially available tissue culturing membranes that are already widely used. The presented sensor system has potential not only in standard culturing methods but also within microfluidic culturing and Lab-on-Chip.

In this work we demonstrate how to fabricate the membrane based electrochemical sensors and how to incorporate them in polymer microfluidic devices and traditional cultureware. We present the latest results of the integration of the membrane electrode in both conventional steady state cell culturing and microfluidic culturing with flow. Furthermore, we discuss possibilities, advantages and drawbacks of introducing the membrane based electrodes in areas such as e.g. migration studies or neurobiology as a substitute to conventional electrodes.

Figure 1. Picture of a membrane sensor inserted in a costume made chip for determination of sensitivity and stability.

Keywords: Biosensors, Electrochemistry, Lab-on-a-Chip/Microfluidics, Sensors
Application Code: Bioanalytical
Methodology Code: Integrated Sensor Systems

Session # **840** **Abstract #** **840-18** **Poster Sessions**

Session Title: Sensors

Abstract Title: **Paper-Based Colorimetric Glucose Determination Using Smartphone**

Date: Monday, March 07, 2016 - Afternoo

Time:

Room: Exposition Floor, 400 Aisle

Primary Author: Hakan Ciftci
Kirikkale University

Co-Author(s): Nazl Ayydz, Ugur Tamer

Abstract Text

Colorimetric biosensing has attracted much attention due to its low cost, simplicity, and practicality. Colorimetric biosensing does not require expensive or sophisticated instrumentation and can be applied to field analysis and point-of-care diagnosis [1,2]. Paper-based sensors are becoming powerful and low-cost diagnostic tools, especially in resource-limited settings. Inexpensive methods for quantifying these assays have been shown using desktop scanners, which lack portability, and cameras, which suffer from the ever changing ambient light conditions. The advantages of inexpensive and convenient paper-based colorimetry and the ubiquitous smartphone are tied to achieve a ready-to-go POC diagnosis [3].

In this study, modified paper was prepared by poly(2-aminothiazole) entrapped bi-enzyme system and it was used for colorimetric glucose determination. GOx/HRP bienzyme system was utilized to amplify the color signal. Our proposed method gives direct outcomes, which can be observed by a smartphones in converting the colour changes to digitize values. The results show good use of a smartphone as an analytical instrument for glucose measurement.

*This project is supported by TUBITAK. Project No:114Z715

References

[1] Sun, J., Ge, J., Liu, W., Lan, M., Zhang, H., Wang, P., Wang, Y., Niu, Z. (2014) Nanoscale, 6, 255.
[2] Liu, Q., Jia, Q., Zhu, R., Shao, Q., Wang, D., Cui, P., Ge, J. (2014) Materials Science and Engineering C, 42, 177.
[3] Hong, J., Chang B.Y. (2014)Lab Chip, 14, 1725

Keywords: Biosensors, Portable Instruments, Sensors
Application Code: Biomedical
Methodology Code: Sensors

Session # 850 Abstract # 850-1 **Poster Sessions**

Session Title: Thermal Analysis

Abstract Title: **New TGA-FT-IR Library for Polymers**

Date: Monday, March 07, 2016 - Afternoo

Time:

Room: Exposition Floor, 400 Aisle

Primary Author: Ekkehard Post
NETZSCH Geraetebau GmbH

Co-Author(s): Bob Fidler, Carolin Fischer

Abstract Text

During the pyrolysis of polymers, the evolved gases can be a fingerprint for the identification of these materials. Several libraries for mass spectrometer or GC-MS are already available. But these were often produced by a fast-pyrolysis process of the samples. The gases from this fast-pyrolysis process differ very often significantly in the amount or even in the gaseous products itself from those received by a TGA analysis. For TGA-FT-IR some older booklets were still on the market, but are not suitable for an electronically library. In this poster contribution a TGA-FT-IR library will be introduced with the most important polymers included. It can be integrated into the Bruker OPUS library search and is also expandable by the user with their own measured data.

Keywords: Data Analysis, FTIR, Polymers & Plastics, Thermal Analysis
Application Code: Polymers and Plastics
Methodology Code: Thermal Analysis

Session # 850 Abstract # 850-2 **Poster Sessions**

Session Title Thermal Analysis

Abstract Title **Beyond Classical Dynamic Mechanical Analysis - Using A New High-Force, High-Temperature DMA to Characterize Advanced Materials**

Primary Author Bob Fidler
NETZSCH Instruments NA LLC

Date: Monday, March 07, 2016 - Afternoo
Time:
Room: Exposition Floor, 400 Aisle

Co-Author(s) Ekkehard Post, Horst Deckmann, Juergen Blumm, Tobias Pflock

Abstract Text

Dynamic Mechanical Analysis (DMA) is a powerful technique to analyze mechanical properties of many materials in research, development and quality control. Traditionally, DMA has been limited in terms of temperature and force which, in turn, limits the scope of materials that can be successfully characterized. Besides the traditional applications involving rubber, plastics, and composites, this poster will detail newly expanded applications of DMA made possible by applying both high force and high temperature characteristics to examine materials. Variations include high force capability up to 8000N, whereas in other configurations, high temperatures up to 1500 °C care are now possible for determination of absolute values of E modulus, for example. Fields of application include dynamic component testing, durability tests, analysis of visco-elastic material properties, dynamic fatigue testing, relaxation and retardation, creep testing, and more. In addition to using very high forces to characterize polymers and elastomers, this poster will also demonstrate unique examples of high temperature DMA now applied to glasses, ceramics, advanced composites, and

Keywords: Materials Characterization, Polymers & Plastics, Rubber, Thermal Analysis
Application Code: Material Science
Methodology Code: Thermal Analysis

Session # **850** **Abstract #** **850-3** **Poster Sessions**

Session Title Thermal Analysis

Abstract Title **Exploring the Properties of Textiles with Thermal Analysis, Infrared and UV/Vis Spectroscopy**

Primary Author Anthony J. Lang
PerkinElmer

Date: Monday, March 07, 2016 - Afternoo
Time:
Room: Exposition Floor, 400 Aisle

Co-Author(s) Aaron H. Adams, Jack Botting

Abstract Text

Textiles come in many forms, ranging from clothing to umbrellas. It is not an easy task to assign the identity of the fibers or predict the performance of a material. These determinations are often achieved through the application of analytical instrumentation. Throughout this project, materials will be subjected to a variety of experimentation to determine parameters such as identity and ultraviolet protection factor (UPF). Differential Scanning Calorimetry (DSC) will be implemented to characterize the synthetic fiber melting points, glass transitions, and detect possible mixtures. Dynamic Mechanical Analysis (DMA) will assist in the determination of a fiber's response to the relative humidity of the air around it. Determination of a fiber identity can often be achieved quickly with a Fourier Transform Infrared spectrometer (FTIR); however, blends of fibers tend to be more challenging. Advanced algorithms will be implemented to identify mixtures of fibers. Color, reflectivity, and transmittance are key parameters for the application of the textile. These values will be determined through the use of a spectrophotometer equipped with an integrating sphere.

Keywords: FTIR, Spectrophotometry, Spectroscopy, Thermal Analysis
Application Code: Material Science
Methodology Code: Molecular Spectroscopy

Session # **860** **Abstract #** **860-1** **Awards**

Session Title: Pittsburgh Analytical Chemistry Award

Abstract Title: **Photonic Crystal Hydrogel and Organogel Sensors for Chemical and Biological Analytes**

Date: Tuesday, March 08, 2016 - Morning
Time: 08:40 AM
Room: B312

Primary Author: Sanford A. Asher
University of Pittsburgh

Co-Author(s): Andrew E. Coukouma, Natasha L. Smith, Zhongyu Cai

Abstract Text

In the work here, we describe sensing motifs that utilize 2-D and 3-D arrays and monolayers of particles embedded onto a molecular recognition polymer hydrogel networks. The 3-D and 2-D arrays alter their visually evident diffraction colors because the polymer networks swell or shrink in response to analyte concentration changes. We recently developed easily fabricated 2-D photonic crystals for molecular recognition and chemical sensing applications. We prepared close packed 2-D polystyrene particle arrays by solvent evaporation of an assembling monolayer on a mercury surface or by lifting off the array from a water surface. We then transfer the 2-D arrays onto a hydrogel thin film. These responsive hydrogel and organogels show volume phase transitions (VPT) in response to specific analytes. These VPT alter the array spacings, changing the array diffraction wavelengths. These 2-D and 3-D photonic crystals exhibit ultrahigh diffraction efficiencies that enable them to be used for visual detection of analyte concentrations. The 3-D arrays undergo 3-D volume changes to shift the diffraction wavelength. We developed novel protein hydrogels that are highly selective for charged species binding. These protein hydrogels act as Coulometers that sense binding of individual charged species. These protein hydrogel photonic crystal sensors also respond to conformational changes induced by protein ligand binding.

Keywords: Biosensors, Polymers & Plastics, Protein, Sensors
Application Code: Bioanalytical
Methodology Code: Sensors

Session # 860 Abstract # 860-2 **Awards**

Session Title Pittsburgh Analytical Chemistry Award

Abstract Title **Structural Characterization of Methylenedianiline Regioisomers by Ion Mobility-Mass Spectrometry, Tandem Mass Spectrometry, and Computational Strategies - MALDI Spectra**

Date: Tuesday, March 08, 2016 - Morning
Time: 09:15 AM
Room: B312

Primary Author David M. Hercules
Vanderbilt University

Co-Author(s) John A. McLean, Sarah M. Stow, Tiffany M. Onifer

Abstract Text

Methylenedianiline (MDA) regioisomers, containing 2, 3 and 4 aromatic rings, were characterized by MALDI mass spectrometry for comparison with our earlier studies of MDA by electrospray (ESI) mass spectrometry. Because both MALDI and ESI are "soft" ionization methods, frequently their MS and MS/MS spectra are very similar. However, the MDA regioisomer spectra are quite different. In the present presentation we will deal mostly with the 2-ring isomers having 2,2'-, 2,4'-, and 4,4'- amino group substituents; typically the 3- and 4-ring compounds behave similarly to the 4,4' 2-ring isomer . In ESI, the 4,4' isomer (M+H)+ (199 Da) is the base peak in the parent-ion region of the spectrum while a fragment ion peak at 106 Da is base for the other two 2-ring isomers. In comparable MALDI spectra the (M-H)+ peak at 197 Da is dominant for all three isomers and is barely visible in the ESI spectra. It is proposed that the 197 Da peak in MALDI results from formation of a diamine substituted 9-H fluorene ring structure. MS/MS and computational data are presented to support this assignment. We also studied the collision-induced dissociation curves as a function of collision energy. For the ESI produced (M+H)+ ions complete fragmentation occurred at low collision energy (1.5 eV) for the 2,2'- and 2,4'- isomers but not the 4,4'- isomer (8 eV). For the MALDI species, all three isomers showed very similar collision-induced dissociation behavior. Collision cross section (CCS) values were measured for both ESI and MALDI species and compared with results from computational methods. Typically good agreement was found between computed and measured values. Some values for different isomers differed significantly. For example, in the 2,4'- and 2,2'- species, two protonation sites exist in the molecules, which give rise to differences in IMS spectra and fragmentation mechanisms. Because the MALDI species have greater similarity, this effect is not seen.

Keywords: Analysis, Mass Spectrometry, Organic Mass Spectrometry, Polymers & Plastics

Application Code: Polymers and Plastics

Methodology Code: Mass Spectrometry

Session # **860** **Abstract #** **860-3** **Awards**

Session Title: Pittsburgh Analytical Chemistry Award

Abstract Title: **Liposomal Spherical Nucleic Acids: A New Approach to Immunomodulatory Therapy**

Date: Tuesday, March 08, 2016 - Morning

Time: 09:50 AM

Room: B312

Primary Author: Chad A. Mirkin
Northwestern University

Co-Author(s)

Abstract Text

Immunomodulatory oligonucleotides are short strands of nucleotides containing motifs that activate disease-fighting, innate immune signaling networks. Spherical nucleic acid (SNA) is attractive as a novel immunotherapy platform, providing a safe and effective delivery of immune-activating oligonucleotides in vitro and in vivo. SNAs consist of nanoparticles (5-30 nm) functionalized with a dense, spherical shell of radially oriented oligonucleotides. SNAs efficiently enter cells without the need for transfection agents or chemical modifications, demonstrate enhanced resistance towards degradation by nucleases relative to linear oligonucleotides, and exhibit minimal toxicity and off-target immunogenicity.

In comparison to linear, free oligonucleotides, SNAs with a 30 nm liposomal core functionalized with immune-activating oligonucleotides show up to 80-fold increase in efficacy activating endosomal TLRs in vitro. Furthermore, SNAs conjugated with a model antigen demonstrate up to 700-fold higher antibody titers, 400-fold higher cellular response, and significantly improve median survival in a murine model lymphoma through antigen-specific tumor inhibition. The tailorability of the SNA platform, therefore, is advantageous to develop therapeutics to treat various cancer types as well as immune and inflammation diseases. Combined with their at-scale synthesis from readily available FDA-approved materials, minimal toxicity profile and potent delivery via well-defined chemical nature, liposomal SNAs hold great promise for the development of immunotherapies.

Keywords: Biomedical, Biopharmaceutical, Drug Discovery, Nanotechnology
Application Code: Nanotechnology
Methodology Code: Chemical Methods

Session # 860 Abstract # 860-4 **Awards**

Session Title Pittsburgh Analytical Chemistry Award

Abstract Title **Measuring DNA Hybridization Kinetics by Single-Molecule Fluorescence Imaging**

Primary Author Joel M. Harris
University of Utah

Date: Tuesday, March 08, 2016 - Morning
Time: 10:40 AM
Room: B312

Co-Author(s) Eric M. Peterson, Frances Morris, Michael Manhart

Abstract Text

Quantitative imaging of individual fluorescently-labeled molecules is a powerful and extremely sensitive method to characterize equilibria and dynamics of bio-recognition events at liquid-solid interfaces. Total-internal-reflection excitation combined with high-efficiency imaging of fluorescence allows surface-bound molecular populations to be quantified in situ by counting individual molecules. We have applied this methodology to characterize oligonucleotide hybridization, the chemistry of which is integral to the design and understanding of modern chip-based DNA screening. DNA hybridization is challenging to measure by single-molecule techniques because of nonspecific adsorption, which is typically mitigated by FRET methods that compromise detection efficiency. In this work, we describe a scheme for immobilizing probe single-stranded DNA at a glass interface passivated against nonspecific adsorption with anionic blocking groups, yielding surfaces where complementary binding dominates the population of fluorescently-labeled target DNA on the surface. Surface populations of complementary target DNA are nearly 3-orders-of-magnitude greater than those of a scrambled sequence. The surface density of immobilized probe DNA can be controlled by the probe concentration used in the immobilization reaction. At low surface densities, association and dissociation kinetics are measured under equilibrium conditions where individual probe molecules can be resolved so that hybridization and dissociation rates can be measured directly from single-molecule trajectories. At high probe capture site densities, the substrates are efficient at capturing target strands from solution, making them excellent ssDNA sensors with very low detection limits. These selective substrates can also be used to investigate fundamental factors influencing the rates DNA hybridization at surfaces. They can also be used to unravel the kinetics of DNA split aptamer assembly and analyte recognition.

Keywords: Fluorescence, Imaging, Microscopy
Application Code: Bioanalytical
Methodology Code: Surface Analysis/Imaging

Session # 860 **Abstract # 860-5** **Awards**

Session Title: Pittsburgh Analytical Chemistry Award

Abstract Title: **SERS-Based Metabolic Profiling for Diagnostics and Forensics**

Date: Tuesday, March 08, 2016 - Morning
Time: 11:15 AM
Room: B312

Primary Author: Lawrence Ziegler
Boston University

Co-Author(s)

Abstract Text

Surface enhanced Raman spectroscopy (SERS) spectra excited at 785 nm on Au and Ag nanostructured substrates are found to provide sensitive and specific vibrational signatures for bacterial diagnostics, cancer cell identification, measures of blood aging and trace body fluid identification. A SERS based methodology for rapid, growth-free diagnostics of bacterial infections has been demonstrated. SERS spectra of bacteria enriched from infected body fluids, combined with multivariate data analysis techniques and a previously determined SERS library, results in rapid, antibiotic specific bacterial diagnostics. These bacterial SERS spectra are due to the metabolic degradation of purine containing biologically active molecules. Isotope effects, SERS spectra of non-cell containing extra-cellular regions, spectral fitting and mass spectrometry establish the molecular origins of these bacterial SERS signatures. The effects of gene knockouts and co-enzymes are consistent with this mechanism for the origin of the bacterial SERS signals. SERS signal kinetics also provide measures of the relevant metabolic pathway enzyme rates. Tumor cells which are well-known to exhibit high metabolic rates compared to normal, non-pathogenic cells, also exhibit SERS spectra with different purine degradation metabolites. Characteristic SERS vibrational signatures due to purines, AMP and NADH appear over the course of several hours from single cancer cells and the time dependence of these signatures are different for pathogenic and nonpathogenic cells. Thus these SERS may be used for in vitro single cell cancer detection as well as fundamental studies of the effects of genetic or proteomic manipulation for cancer therapy efficacy evaluation. Several of these purine metabolites are also identified in the SERS spectra of human body fluids which contain cellular components and contribute to the SERS identification of these fluids for forensic science applications.

Keywords: Biomedical, Surface Enhanced Raman, Vibrational Spectroscopy, Metabolomics
Application Code: Biomedical
Methodology Code: Vibrational Spectroscopy

Session # 870 Abstract # 870-1 **Awards**

Session Title The Coblentz Society/ABB - Bomem-Michelson Award

Abstract Title **Coherent Ultrafast Multidimensional Spectroscopy of Molecules with Optical, X-Ray, and Quantum Light**

Primary Author Shaul Mukamel
University of California, Irvine

Date: Tuesday, March 08, 2016 - Morning
Time: 08:40 AM
Room: B314

Co-Author(s)

Abstract Text

Multidimensional spectroscopy uses sequences of optical pulses to study dynamical processes in complex molecules through correlation plots involving several time delay periods. Recent extensions of these techniques to the x-ray regime as well as by utilizing the quantum nature of light will be discussed.

Ultrafast nonlinear x-ray spectroscopy is made possible by newly developed free electron laser and high harmonic generation sources. The attosecond duration of X-ray pulses and the atomic selectivity of core X-ray excitations offer a uniquely high spatial and temporal resolution. We demonstrate how stimulated Raman detection of an X-ray probe may be used to monitor the phase and dynamics of the nonequilibrium valence electronic state wavepacket created by e.g. photoexcitation, photoionization and Auger processes. Applications will be presented to long-range charge transfer in proteins and to excitation energy transfer in porphyrin arrays.

Many important photophysical and photochemical molecular processes take place via conical intersections (COIS) where nuclear and electronic degrees of freedom become strongly coupled and the adiabatic Born Oppenheimer approximation beaks down. A new technique, TRUE-CARS, Transient Redistribution of Ultrafast Electronic Coherences in Attosecond Raman Signals, is proposed that can detect the passage through a COIS. Entangled photons provide novel nonlinear spectroscopic probes of excitation-energy-transfer and charge-separation processes in chromophore aggregates. Signals that utilize the quantum nature of the optical field by varying parameters of the photon wavefunction rather than classical field delays and frequencies will be presented. The unusual spectral and temporal characteristics of entangled photon pairs combined with interferometric detection make it possible to manipulate and control two photon absorption and Raman signals and extract information not available with classical light.

Keywords: Ultra Fast Spectroscopy
Application Code: Other
Methodology Code: Computers, Modeling and Simulation

Session # 870 **Abstract # 870-3** **Awards**

Session Title: The Coblentz Society/ABB - Bomem-Michelson Award

Abstract Title: **A Few Lessons from Non-Adiabatic Excited State Dynamics Simulations of Large Molecules**

Primary Author: Sergei Tretiak
Los Alamos National Laboratory

Date: Tuesday, March 08, 2016 - Morning
Time: 09:50 AM
Room: B314

Co-Author(s)

Abstract Text
Modelling of non-adiabatic dynamics in extended molecular systems and solids is a next frontier of atomistic electronic structure theory. The underlying numerical algorithms should operate only with a few quantities (that can be efficiently obtained from quantum chemistry), provide a controlled approximation (which can be systematically improved) and capture important phenomena such as branching (multiple products), detailed balance and evolution of electronic coherences. This talk will overview recently developed theoretical methodologies applicable for simulating large molecules underlying an efficient Non-Adiabatic Excited State Molecular Dynamics (NA-ESMD) framework incorporating non-adiabatic quantum transitions. Our calculations of several molecular systems show intricate details of photoinduced vibronic relaxation and identify the conformational degrees of freedom leading to ultrafast dynamics and energy transfer. This theoretical modeling allows us to understand and to potentially manipulate energy transfer pathways in molecular materials suitable for solar energy conversion.

Keywords: Electron Spectroscopy, Energy, Material Science, Molecular Spectroscopy
Application Code: Material Science
Methodology Code: Computers, Modeling and Simulation

Session # 870 **Abstract # 870-4** **Awards**

Session Title: The Coblentz Society/ABB - Bomem-Michelson Award

Abstract Title: **Photosynthetic Light Harvesting and Ultrafast Energy Transfer**

Date: Tuesday, March 08, 2016 - Morning
Time: 10:40 AM
Room: B314

Primary Author: Greg D. Scholes
Princeton University

Co-Author(s)

Abstract Text

Photosynthetic solar energy conversion occurs on an immense scale across the earth, influencing our biosphere from climate to oceanic food webs. Photosynthetic light-harvesting complexes are sophisticated multichromophoric assemblies used to regulate and concentrate photo-excitations for delivery to reaction centers under wide-ranging incident irradiances [1]. They provide wonderful model systems to study energy transfer mechanisms in well-defined structures. I will describe recent examples of ultrafast energy transfer in photosynthetic light harvesting and discuss critically the role of coherence. I will show new studies revealing how certain algae switch excitons on and off in light harvesting.

[1] "Lessons from nature about solar light harvesting" Nature Chem. 3, 763–774 (2011).

Keywords: Laser, Mass Spectrometry
Application Code: General Interest
Methodology Code: Biospectroscopy

Session # 870 **Abstract # 870-5** **Awards**

Session Title The Coblentz Society/ABB - Bomem-Michelson Award

Abstract Title **Elucidation of Chemical Reactions by Two-Dimensional Resonance Raman Spectroscopy**

Date: Tuesday, March 08, 2016 - Morning
Time: 11:15 AM
Room: B314

Primary Author Andrew Moran
University of North Carolina

Co-Author(s)

Abstract Text

Two-dimensional (2D) Raman spectroscopies were proposed by Mukamel and Loring in1985 as a method for resolving line broadening mechanisms of vibrational motions in liquids. Significant technical issues challenged the development of both five- and seven-pulse 2D Raman spectroscopies. For this reason, 2D Raman experiments were largely abandoned in 2002 following the first demonstrations of 2D infrared spectroscopies (i.e., an alternate approach for obtaining similar information). We have recently shown that 2D Raman experiments conducted under electronically resonant conditions are much less susceptible to the problems encountered in the earlier 2D Raman work, which was carried out off-resonance. In effect, Franck-Condon activity obviates the problematic selection rules encountered under electronically off-resonant conditions. In this presentation, I will discuss applications of 2D resonance Raman spectroscopies to photodissocation reactions of triiodide and myoglobin. It will be shown that vibrational resonances of the reactants and products can be displayed in separate dimensions of a 2D resonance Raman spectrum when the photo-dissociation reaction is fast compared to the vibrational period. Such 2D spectra expose correlations between the nonequilibrium geometry of the reactant and the distribution of vibrational quanta in the product, thereby yielding insight in the photo-dissociation mechanism. Our results suggest that the ability of 2D resonance Raman spectroscopy to detect correlations between reactants and products will generalize to other ultrafast processes such as electron transfer and energy transfer.

Keywords: Infrared and Raman, Spectroscopy, Vibrational Spectroscopy
Application Code: General Interest
Methodology Code: Molecular Spectroscopy

Session # 880 Abstract # 880-1 **Symposia**

Session Title ACS-ANLY - Chemometrics: A New Dimension in Chromatography

Abstract Title **Applying the Hotelling Trace Criterion to Optimize Chromatogram Alignment of Biodiesel Diesel Blended Fuels**

Date: Tuesday, March 08, 2016 - Morning
Time: 08:35 AM
Room: B308

Primary Author Amber M. Hupp
College of the Holy Cross

Co-Author(s) Edward Soares, Gopal Yalla, John O'Connor, Kevin Walsh

Abstract Text

Chemometric methods, such as principal component analysis (PCA), can be applied to complex data sets with multiple variables and multiple sample types in an effort to reveal trends that otherwise may not be observed. Biodiesel-diesel blended fuels represent such a complex data set. Biodiesel is created through the transesterification of triglycerides in oils and fats (both plant and animal) with methanol and a sodium hydroxide catalyst to form the fatty acid methyl esters otherwise known as biodiesel. Diesel is created through the process of refining petroleum oil and contains a mixture of aliphatic and aromatic components, typically C14 to C20. Blending these two fuels creates a very complex mixture. Additional differences arise from biodiesel feedstock sources (soybean, canola, palm, animal) and oil distillation process. As such, the methods used to process the data prior to chemometric analysis must make like samples as similar as possible and maximize the differences between samples that are not alike.

In this talk, I will present our efforts to develop a chromatographic method that is ideal for isolating components in the biodiesel-diesel blends using a polar gas chromatography column. This step is key to producing a rich, diverse data set. I will also present a methodology for optimization of chromatogram alignment using a class separability measure called the Hotelling trace criterion (HTC). This metric is a multi-class distance measure that accounts for within-class and between-class variation. A correlation optimized warping algorithm was used for alignment with the HTC used to judge the effectiveness of the alignment based on the algorithm parameters. The entire data set was baseline corrected, aligned, normalized, and mean-centered prior to PCA. The results demonstrated that the parameters derived from maximizing the HTC more effectively aligned the data, as evidenced by better clustering of the biodiesel-diesel blends in the scores plots.

Keywords: Biofuels, Chemometrics, Fuels\Energy\Petrochemical, Gas Chromatography/Mass Spectrometry
Application Code: Fuels, Energy and Petrochemical
Methodology Code: Chemometrics

Session # 880 **Abstract # 880-2** **Symposia**

Session Title ACS-ANLY - Chemometrics: A New Dimension in Chromatography

Abstract Title **Forensic Signatures for the Source Attribution of Chemical Threat Agents Using Chemical Profiling, Stable Isotope Ratios and Chemometrics**

Primary Author Carlos Fraga
Pacific Northwest National Laboratory

Date: Tuesday, March 08, 2016 - Morning
Time: 09:10 AM
Room: B308

Co-Author(s)

Abstract Text

Chemical Threat Agents (CTAs) like sarin (GB), nitrogen mustard (HN3), and cyanide salts (NaCN, KCN) pose a serious threat to national security. In the event that a CTA is used in a crime or act of terrorism, there will likely be a need to provide evidence for attribution. Pacific Northwest National Laboratory and other labs are developing forensic analysis methods in order to provide chemical evidence for attribution following a chemical attack. Herein, we report our recent research efforts using chromatographic, mass spectrometric, and chemometric methods for matching the above mentioned CTAs to their specific reagent stocks or countries of origins using stable isotope ratios, organic and ionic impurities, and trace elemental constituents.

Keywords: Chemometrics, GC-MS, Ion Chromatography, Isotope Ratio MS
Application Code: Homeland Security/Forensics
Methodology Code: Chemometrics

Session # 880 **Abstract # 880-3** **Symposia**

Session Title: ACS-ANLY - Chemometrics: A New Dimension in Chromatography

Abstract Title: **Chemometrics: An Old Dimension in Chromatography – Application to New Dimensions**

Date: Tuesday, March 08, 2016 - Morning
Time: 09:45 AM
Room: B308

Primary Author: Sarah C. Rutan
Virginia Commonwealth University

Co-Author(s): Daniel W. Cook, Melanie M. Sinanian

Abstract Text

The application of chemometrics to chromatography has been mentioned in the literature as early as 1977 and a Web of Science search of these two topic key words comes up with 1,819 citations. Still, it is very true that routine application of chemometric tools in everyday chromatography practice is still somewhat limited. In this presentation I will discuss the possible reasons for this. It is possible that this barrier will be broken by the need to analyze complex data resulting from newer multidimensional chromatography methods and the high computational demands for high resolution mass spectrometric data. The focus of this talk will be on the use of a chemometric technique, specifically multivariate curve resolution alternating least squares, to improve the quantitative performance of chromatographic analyses, specifically with respect to comprehensive two-dimensional liquid chromatography and liquid chromatography coupled to high resolution mass spectrometry.

Acknowledgement: This research is supported via a grant from NSF (CHE-1507332).

Keywords: Chemometrics, HPLC, Mass Spectrometry
Application Code: General Interest
Methodology Code: Chemometrics

Session # 880 **Abstract # 880-4** **Symposia**

Session Title ACS-ANLY - Chemometrics: A New Dimension in Chromatography

Abstract Title **Data Reduction and Processing Tools for GCxGC-TOFMS**

Date: Tuesday, March 08, 2016 - Morning
Time: 10:35 AM
Room: B308

Primary Author James J. Harynuk
University of Alberta

Co-Author(s) A Paulina de la Mata, Lawrence A. Adutwum

Abstract Text

Comprehensive two-dimensional gas chromatography (GCxGC) is simultaneously a blessing and a curse for the analytical chemist. This technique permits the analysis of increasingly complex samples for metabolomics, fuels, the environment, forensics, foods, flavours, etc etc... Analysts can ask questions about complex systems, collect and analyze samples carefully, and get very detailed data about the samples. We routinely track 2000-4000 peaks across a set of samples. The curse of GCxGC is in handling these data to quickly obtain relevant information. "What are the peaks?" and "Which peaks are important?" (i.e. to distinguish healthy from diseased, or "normal" from "abnormal") are two central questions. These have not been easy questions to answer in the past. Some of the automated, objective tools we have developed will be presented, as we work towards data processing tools that can match the power of GCxGC-TOFMS to generate data.

Keywords: Chemometrics, Gas Chromatography/Mass Spectrometry, Separation Sciences, Statistical Data Analy
Application Code: General Interest
Methodology Code: Gas Chromatography/Mass Spectrometry

Session # 880 Abstract # 880-5 **Symposia**

Session Title ACS-ANLY - Chemometrics: A New Dimension in Chromatography

Abstract Title **Chemometric Approaches to Maximize Interpretation of GCxGC - TOFMS Data for Discovery-Based Analyses**

Primary Author Rob Synovec
University of Washington

Date: Tuesday, March 08, 2016 - Morning
Time: 11:10 AM
Room: B308

Co-Author(s)

Abstract Text

For the analysis of complex samples, comprehensive two-dimensional gas chromatography coupled with time-of-flight mass spectrometry (GC x GC - TOFMS) is a powerful instrumental platform. Chemometric approaches play a pivotal role in the analytical workflow for the translation of the raw data into useful information. We are developing advanced approaches for non-targeted discovery-based analysis of cross-sample comparisons of GC x GC - TOFMS data, coupled with robust deconvolution, identification and quantification of meaningful analytes. Recent advances in Fisher-ratio (F-ratio) analysis will be presented, a statistically-based data mining technique to discover analytes that distinguish sample classes based upon the experimental design. Recently developed tile-based F-ratio software will be presented, which substantially improves chemical selectivity in the discovery process for the determination of an analyte "hit list." Important features discovered by F-ratio analysis are further analyzed using complementary deconvolution methods to readily provided analyte deconvolution, identification, and quantification. This presentation will also focus on rapid and high-quality extraction of the most useful information from metabolomics and forensic studies.

Keywords: Chemometrics, Data Mining, Gas Chromatography, Time of Flight MS
Application Code: General Interest
Methodology Code: Chemometrics

Session # 890 **Abstract # 890-1** **Symposia**

Session Title: Advances in Analytical Methodologies for the Detection of Food Allergens and Gluten

Abstract Title: **Approaches to Multianalyte Allergen Analysis in Food**

Date: Tuesday, March 08, 2016 - Morning
Time: 08:35 AM
Room: B302

Primary Author: Clare Mills
The University of Manchester

Co-Author(s): Karine Adel-Patient, Sabine Baumgartner

Abstract Text

The underpinning philosophy of food allergen management is quality assurance, which means systems that span the food chain from primary production through to point-of-sale, are designed and maintained with the aim of ensuring that untoward events do not occur. Central to implementing such an approach is the availability of tools able to determine the levels of allergenic food protein(s) to monitor either factory cleaning, or ingredients and finished products. The currently available tests can give highly variable, matrix dependent results. In addition the methods themselves have only undergone limited validation. Consequently the validation and monitoring of food allergen management plans is an uncertain process and means that food manufacturers and retailers may have to resort to the use of precautionary 'may contain' labelling if there is uncertainty with regards the presence of unintended allergens in a product. Analytical methods need to target biologically relevant molecules in foods ("the hazard") and detect them in a manner relevant to their allergenic potency, thus aiding interpretation of test results. This is a challenge given the variation in performance of current test methods, which are particularly affected by processing-induced changes to allergens, such as conformational changes, aggregation and thermally-induced chemical modifications. As well as altering the intrinsic allergenicity, processing also modifies the effectiveness of extraction methods for determination of allergens in foods. The allergen analysis module of the iFAAM project aims to address these problems and to link the development of effective multianalyte immuno-based tools suitable for in-factory testing with confirmatory in-laboratory mass spectrometry-based analysis. This is being undertaken for 5 of the 7 most prevalent allergenic foods, peanut, tree nuts (hazelnut and walnut) together with egg and milk in three food matrices (mousse-style dessert, cookie and chocolate).

Keywords: Food Science, Immunoassay, Mass Spectrometry, Protein
Application Code: Food Safety
Methodology Code: Mass Spectrometry

Session # 890 Abstract # 890-2 **Symposia**

Session Title: Advances in Analytical Methodologies for the Detection of Food Allergens and Gluten

Abstract Title: **Effects of Cross-Reactivity in Food Allergy Detection and Diagnosis**

Date: Tuesday, March 08, 2016 - Morning
Time: 09:10 AM
Room: B302

Primary Author: Soheila J. Maleki
USDA-Agricultural Research Service

Co-Author(s)

Abstract Text

Cross-reactivity among peanuts and tree nuts is a major problem in accurate detection of nut allergens and in diagnosis. While much is known about the sequences and structural details of the peanut proteins that contribute to IgE based response, substantially less work has been done to determine why nuts from different botanical sources have such a high degree of cross reactivity. We have developed a computational and experimental approach to identify common epitopes of peanut and tree nut allergens that could contribute to cross-reactivity. The protein extracts of peanuts and select tree nuts were all normalized according to protein content and subjected to SDS-PAGE and immunoassays. Immunoblots of proteins and synthetic peptides were performed using specific antibodies and sera from patients with elevated specific IgE and a positive food challenge or convincing known history of peanut or tree nut allergy. Specific antibody or IgE-reactive bands were excised, digested and identified using mass spectroscopy. Potentially cross-reactive peptides were identified with peptide-microarray analysis and computationally with tools from the Structural Database of Allergenic Proteins (SDAP). Immunoblot and microarray analysis showed antibody cross-reactivity, both at protein and peptide levels between the nuts. Several protein bands that showed cross-reactivity with various antibodies were previously known allergens. Both microarray analysis and computational methods were able to identify, strongly cross-reactive peptides with low sequence identity, often found in proteins that are not in the same protein families (mostly 7S, 2S and 11S Pfams). As expected, extensive IgE cross-reactivity was seen between peanuts and tree nuts within the known major allergen protein families. However, previously unidentified cross-reactive proteins seem to also exist. Novel epitopes and repeated peptide sequences were identified that could account for cross reactivity among nuts.

Keywords: Bioinformatics, Biomedical, Identification, Protein

Application Code: Biomedical

Methodology Code: Computers, Modeling and Simulation

Session # 890 Abstract # 890-3 **Symposia**

Session Title: Advances in Analytical Methodologies for the Detection of Food Allergens and Gluten

Abstract Title: **Novel Approaches to Identifying Amadori Products in Peanut Extract**

Date: Tuesday, March 08, 2016 - Morning
Time: 09:45 AM
Room: B302

Primary Author: Geoffrey Mueller
National Institute of Environmental Health Sciences

Co-Author(s): Jason Williams, Katina Johnson

Abstract Text

Peanut allergy affects 1-2% of the population of the U.S. It is hypothesized that roasting of peanuts influences the allergenicity of the peanut proteins due to formation of advanced glycation end products (AGE). In efforts to evaluate the AGE composition of peanut proteins we performed nanoLC-ESI-MS and MS/MS analyses of peanut extracts as well as recombinant peanut proteins. Initial experiments showed that in both raw and roasted peanut extracts one of the common modifications was the Amadori product (mass gain of 162 Da). The ions matched in mass to expected peptides +162 Da and yielded fragments in the MS/MS that were included neutral loss of 2, 3, and 4 waters and a loss of 3 waters and HCHO. Due to the absence of b- and y-ions from the CID MS/MS of these glycated peptides, standard search algorithms do not perform well in the identification of these peptides. Hence, we developed an algorithm that applies filters based upon information content (Shannon entropy) and the loss of water and HCHO. Results with a test data set show that the algorithm finds the correct spectra with high precision. The flagged spectra contained all of the Amadori product spectra with a high (66%) false positive rate. Isotopic labeling with 13C xylose and 13C glucose confirmed that these spectra contained Amadori products. If dry roasting enhances allergenicity, identification of the chemical differences between raw and roasted peanuts will lead to a better understanding of the chemical pathways that skew immune responses toward allergy.

Keywords: Food Science, Identification, Mass Spectrometry, Pattern Recognition

Application Code: Food Science

Methodology Code: Mass Spectrometry

Session # 890 Abstract # 890-4 **Symposia**

Session Title: Advances in Analytical Methodologies for the Detection of Food Allergens and Gluten

Abstract Title: **Overview of Analytical Methods for Food Allergen and Gluten Analysis: Challenges and Trends from a Regulatory Perspective**

Date: Tuesday, March 08, 2016 - Morning
Time: 10:35 AM
Room: B302

Primary Author: Terry Koerner
Health Canada

Co-Author(s)

Abstract Text

Food allergies and intolerance to gluten are important health concerns worldwide with estimates of approximately 4% of adults and 6% of children having food allergies and another 1% having gluten intolerance. The only prescription is the complete avoidance of the specific food proteins in order to eliminate the risk of an adverse reaction. Unfortunately, inadvertent cross-contamination is a real possibility in a modern food system and good industrial controls and labelling practices are essential to protect the food sensitive consumer. Analytical methods are a key component in allergen control programs and in the enforcement of regulatory requirements. Immunological based methods such as enzyme-linked immunosorbent assays (ELISA) and lateral flow devices (LFD) are important tools for the detection and quantification of food allergens, but these methods are not perfect and misleading results can be obtained in some instances. Due to the limitations of the ELISA there is a need for confirmatory, and in some cases, alternative methods of analysis for food allergens and gluten. Both PCR and mass spectrometry methods have been developed as confirmatory tools for food allergen analysis. This talk will provide an overview of the analytical methods used at Health Canada for food allergen analysis and provide examples of questionable ELISA results and how confirmatory methods were used to assess the safety concerns for food sensitive individuals. A discussion of the challenges of these methods with regard to determining the level of allergen present in the sample will also be discussed.

Keywords: Food Science, Mass Spectrometry
Application Code: Food Safety
Methodology Code: Mass Spectrometry

Session # 890 **Abstract # 890-5** **Symposia**

Session Title: Advances in Analytical Methodologies for the Detection of Food Allergens and Gluten

Abstract Title: **Effects of a Proline Endopeptidase on the Detection and Quantification of Gluten During the Fermentation of Beer**

Primary Author: Rakhi Panda
Food and Drug Administration

Co-Author(s): Chung Y. Cho, Eric A. Garber, Katherine L. Fiedler

Date: Tuesday, March 08, 2016 - Morning
Time: 11:10 AM
Room: B302

Abstract Text

Approximately 1 in 141 people in the US are affected by celiac disease and adherence to a strict gluten free diet is the only option to prevent inflammatory symptoms in sensitive individuals. In 2013, the FDA issued a regulation requiring food bearing the claim "gluten-free" must contain less than 20 ppm gluten. It was also recognized at the time that scientifically valid analytical methods did not exist for the quantification of fermented and hydrolyzed gluten. Although several antibody-based and mass spectrometric methods have been used to detect and characterize gluten proteins and peptides in fermented foods such as beer, very limited information is available on whether the hydrolyzed peptides remain immunopathogenic to celiac patients. The use of proteases to enzymatically hydrolyze and detoxify gluten has recently gained popularity. However, it is unclear how the enzymes affect the gluten proteins and peptides containing immunopathogenic epitopes during the beer brewing process. This presentation will focus on a study evaluating antibody-based and mass spectrometric methods to detect and quantify hydrolyzed gluten, using brewing of beer as a model for a fermentation process that involves gluten hydrolysis. The effects of proline endopeptidase (PEP), an enzyme marketed to degrade immunopathogenic sequences suspected of causing celiac disease, on the detection and quantification of gluten were also examined and the findings will be included in the presentation.

Keywords: Immunoassay, Proteomics
Application Code: Food Safety
Methodology Code: Mass Spectrometry

Session # 900 Abstract # 900-1 **Symposia**

Session Title: Advances in Two-Dimensional Liquid Chromatography Separations of Biopharmaceuticals

Abstract Title: **Column Technology for the Chromatographic Characterization of Biopharmaceuticals**

Primary Author: Davy Guillarme, University of Geneva

Co-Author(s): Jean-Luc Veuthey, Szabolcs Fekete

Date: Tuesday, March 08, 2016 - Morning
Time: 08:35 AM
Room: B303

Abstract Text

Therapeutic proteins are large and heterogeneous molecules subjected to a variety of enzymatic and chemical modifications during expression, purification and long-term storage. Regulatory bodies require a detailed characterization (e.g., verifying primary structure and appropriate post-translational modifications, secondary and tertiary structure), lot-to-lot and batch-to-batch comparisons, stability studies, impurity profiling, glycoprofiling, determination of related proteins and excipients as well as determination of protein aggregates.
Today, one of the most widely used analytical technique for therapeutic protein characterization is liquid chromatography, probably due to the remarkable developments of the past few years, enabling a new level of chromatographic performance. Recent developments in LC, such as ultra-high-pressure LC (UHPLC), columns packed with wide-pore superficially porous particles and organic monolith columns now allow a dramatic increase in separation efficiency, even with large intact biomolecules.

The aim of this presentation will be to review the possibilities and trends of current state-of-the-art LC column technology applied for different modes of chromatography for the characterization of therapeutic proteins. Therefore, the recent trends in column technology for reversed phase liquid chromatography (RPLC), size exclusion chromatography (SEC), ion exchange chromatography (IEX) and hydrophobic interaction chromatography (HIC) for analysis, at the protein level, of biopharmaceuticals including therapeutic proteins, monoclonal antibodies and antibody-drug conjugates will be critically discussed.

Keywords: Biopharmaceutical, HPLC, Pharmaceutical
Application Code: Pharmaceutical
Methodology Code: Liquid Chromatography

Session # 900 Abstract # 900-2 **Symposia**

Session Title Advances in Two-Dimensional Liquid Chromatography Separations of Biopharmaceuticals

Abstract Title **Optimization of Two-Dimensional Liquid Chromatography Separations of Therapeutic Monoclonal Antibodies Involving Ion-Exchange and Reversed-Phase Separation Modes**

Primary Author Dwight R. Stoll
Gustavus Adolphus College

Date: Tuesday, March 08, 2016 - Morning
Time: 09:10 AM
Room: B303

Co-Author(s) Alain Beck, David C. Harmes, Davy Guillarme, Gregory Staples, Jacob Bush, Matthew Sorensen, Szabolcs Fekete

Abstract Text

Two-dimensional liquid chromatography (2D-LC) is a powerful and versatile tool for the characterization of therapeutic monoclonal antibodies (mAbs) that, when coupled with mass spectrometry, can provide information about protein charge variants, hydrophobicity, and molecular weight in a single analysis. Obtaining such information efficiently is becoming increasingly important as mAbs are being made in increasingly complex ways (e.g., especially with conjugation chemistry), and as the biopharmaceutical community seeks more effective ways to characterize biosimilar and biobetter mAbs. Whereas there is a large and deep literature on the optimization of separation speed and selectivity for separation of small molecules by LC, relatively little work has been done on the same topics for large biomolecules. Furthermore, the commercialization of new stationary phase chemistries and particle architectures (e.g., superficially porous) for large biomolecules over the past few years presents us with more and diverse options to consider when developing a 2D-LC method for the characterization of mAbs. A thorough understanding of separation kinetics and selectivity is very important in this context, as the selectivities of the phases used in the two dimensions should be highly complementary, and the first and second dimension separations should be optimized for relatively low and high speeds, respectively.

In this presentation we will discuss results of experiments aimed at better understanding the kinetics of both cation-exchange and reversed-phase separations of mAbs, within the context of 2D separations. This is an important distinction, because in this context serious compromises are often made to allow the coupling of the first and second dimensions together, forcing operation of each of them under sub-optimal conditions. We will demonstrate the impact of optimizing the first and second dimensions on overall separation quality using closely related mAbs as a case study.

Keywords: Biopharmaceutical, Liquid Chromatography, Liquid Chromatography/Mass Spectroscopy, Pharmaceu
Application Code: Pharmaceutical
Methodology Code: Liquid Chromatography/Mass Spectrometry

Session # 900 **Abstract # 900-3** **Symposia**

Session Title: Advances in Two-Dimensional Liquid Chromatography Separations of Biopharmaceuticals

Abstract Title: **Application of a Multi Dimensional Approach for Quantification of Free Drug in Antibody Drug Conjugates**

Date: Tuesday, March 08, 2016 - Morning
Time: 09:45 AM
Room: B303

Primary Author: Brooke M. Koshel
Waters Corporation

Co-Author(s): Alain Beck, Robert Birdsall, Sean M. McCarthy

Abstract Text

Biopharmaceuticals are rapidly expanding as a within the pharmaceutical pipeline and subsequent release for commercial sale. It is well understood that the complexity of biopharmaceuticals, and their inherent heterogeneity, far exceeds small molecule compounds. In addition to the inherent complexity associated with protein based therapeutics themselves, modified proteins are becoming more common. In particular, antibody drug conjugates (ADCs) are a very promising new class of biotherapeutic due to their high specificity towards their target and potent cytotoxic agents. The use of potent cytotoxic drugs leads to a need for analytical techniques characterize residual free drug present during a variety of stages in drug development. In addition, these techniques are intended to be used as routine assays highlighting the need for methods and associated instrumentation must be robust, transferrable, and easy to deploy.

The current study addresses these challenges through the development of an SPE-RPLC/MS approach that is specific, sensitive, and enables method control in both dimensions. The proposed method was evaluated using a clinically relevant valine-citrulline surrogate molecule based on brentuximab vedotin. Assay sensitivity was found to be two orders of magnitude more sensitive using MS detection in comparison to UV based detection with an LOQ of 0.30 ng/mL. Free-drug species were present in an unadulterated ADC surrogate sample at concentrations below 7.0 ng/mL, levels not detectable by UV alone. The proposed 2DLC method provides a high degree of specificity and sensitivity in the assessment of trace free drug species with improved control over each dimension enabling straightforward integration into existing or novel workflows.

Keywords: Biological Samples, Biopharmaceutical, Liquid Chromatography/Mass Spectroscopy, Pharmaceutical
Application Code: Pharmaceutical
Methodology Code: Liquid Chromatography/Mass Spectrometry

Session # 900 **Abstract # 900-4** **Symposia**

Session Title: Advances in Two-Dimensional Liquid Chromatography Separations of Biopharmaceuticals

Abstract Title: **2-Dimensional Liquid Chromatography for Rapid Parallel Separation of Biopharmaceuticals**

Date: Tuesday, March 08, 2016 - Morning
Time: 10:35 AM
Room: B303

Primary Author: Yan He
Pfizer

Co-Author(s): Ian Hartzel, Michael T. Jones

Abstract Text

Multiple assays are usually required to monitor different quality attributes of complex biopharmaceuticals. The work on the evaluation of 2-dimensional liquid chromatography (2-D LC) for parallel separation of biopharmaceuticals is presented. The first application is on the use of 2-D LC for simultaneous analysis of size variants and charge variants of monoclonal antibody (mAb) by size exclusion chromatography (SEC) and cation exchange chromatography (CEX). The second application is on the use of 2-D LC for simultaneous analysis of free polysaccharide and free protein of polysaccharide-protein conjugate by SEC and reversed phase liquid chromatography (RPLC). CEX or RPLC separation is not interfered by SEC separation. However, SEC separation is interfered by both CEX and RPLC separation. Approaches to eliminate interference in SEC separation were evaluated. Under optimal condition, 2-D LC provided two fold enhancement in analytical throughput without compromising separation capability.

Keywords: Biopharmaceutical, High Throughput Chemical Analysis, HPLC
Application Code: Pharmaceutical
Methodology Code: Liquid Chromatography

Session # **900** **Abstract #** **900-5** **Symposia**

Session Title Advances in Two-Dimensional Liquid Chromatography Separations of Biopharmaceuticals

Abstract Title **Antibody-Drug-Conjugates (ADC) Drug Product Profiling by Multi-Dimensional HPLC**

Primary Author Kelly Zhang
Genentech

Date: Tuesday, March 08, 2016 - Morning
Time: 11:10 AM
Room: B303

Co-Author(s)

Abstract Text

Antibody-drug-conjugate (ADC) is the new horizon of drug delivery. ADC combines the potency of small molecule drugs and the target specificity of antibodies. Fully understand and characterize the ADC drug product is critical for drug research and development. Profiling the components in antibody-drug-conjugates is highly challenging due to the complexity nature of ADC drug products. An ADC drug product could contain the conjugates, the unconjugated free small molecules and free antibodies, formulation excipients (such as polysorbate, histidine, sucrose, sodium succinate etc.), process related impurities and degradation products. Multiple methods involved with indirect analysis and complicated sample preparations are required to characterize and quantify the ADC drug purity profile. 2DLC is a powerful tool that offers the selectivity and peak capacity to enable the complete profiling of ADC drug products. Hyphenated detection is used to ensure all ADC components can be captured no matter they have UV chromophores or not. Different separation mechanisms are used in each dimension to significantly enhance the selectivity.

Keywords: HPLC, HPLC Detection, Method Development, Pharmaceutical
Application Code: Pharmaceutical
Methodology Code: Liquid Chromatography

Session # **910** **Abstract #** **910-1** **Symposia**

Session Title: Graphene Nanomaterials for Bio/Sensing Applications

Abstract Title: **New Concepts in Biosensing Using Single Walled Carbon Nanotubes and Graphene**

Date: Tuesday, March 08, 2016 - Morning
Time: 08:35 AM
Room: B304

Primary Author: Michael S. Strano
Massachusetts Institute of Technology

Co-Author(s)

Abstract Text

Our lab at MIT has been interested in how the 1D and 2D electronic structures of carbon nanotubes and graphene respectively can be utilized to advance new concepts in molecular detection. We introduce CoPhMoRe or corona phase molecular recognition1 as a method of discovering synthetic antibodies, or nanotube-templated recognition sites from a heteropolymer library. We show that certain synthetic heteropolymers, once constrained onto a single-walled carbon nanotube by chemical adsorption, also form a new corona phase that exhibits highly selective recognition for speci c molecules. To prove the generality of this phenomenon, we report three examples of heteropolymers–nanotube recognition complexes for ribo avin, L-thyroxine and estradiol. The platform opens new opportunities to create synthetic recognition sites for molecular detection. We have also extended this molecular recognition technique to neurotransmitters, producing the first fluorescent sensor for dopamine. Another area of advancement in biosensor development is the use of near infrared fluorescent carbon nanotube sensors for in-vivo detection2. Here, we show that PEG-ligated d(AAAT)7 DNA wrapped SWNT are selective for nitric oxide, a vasodilator of blood vessels, and can be tail vein injected into mice and localized within the viable mouse liver. We use an SJL mouse model to study liver inflammation in vivo using the spatially and spectrally resolved nIR signature of the localized SWNT sensors. Lastly, we discuss graphene as an interfacial optical biosensor, showing that it possesses two pKa values in alkaline and basic ranges.

1. Zhang, JQ et. al. Molecular recognition using corona phase complexes made of synthetic polymers adsorbed on carbon nanotubes. Nature Nanotechnology, 8, 12, 2013, 959-968
2. Iverson, NM, et. al. In vivo biosensing via tissue-localizable near-infrared-fluorescent single-walled carbon nanotubes. Nature Nanotechnology, 8, 11, 2013, 873

Keywords: Nanotechnology
Application Code: Nanotechnology
Methodology Code: Molecular Spectroscopy

Session # 910 Abstract # 910-2 Symposia

Session Title: Graphene Nanomaterials for Bio/Sensing Applications

Abstract Title: **Graphene Electrodes for Bio/chemical Sensors**

Primary Author: Ashok Mulchandani
University of California, Riverside

Date: Tuesday, March 08, 2016 - Morning
Time: 09:10 AM
Room: B304

Co-Author(s)

Abstract Text

Graphene is a single sheet of carbon atoms with outstanding electrical and physical properties being exploited for applications in electronics, sensors, fuel cells, photovoltaics and energy storage. We have synthesized different graphene electrode systems having metallic and semiconducting properties and evaluated their applications as bioanalytical sensors.

The zero bandgap of semimetal graphene still limits its application as an effective field-effect transistor device or a chemiresistor biosensor. It has been shown theoretically and experimentally that graphene nanoribbons (GNRs) or nanomeshes (GNMs) can attain a bandgap that is large enough for a transistor device, and hence would show high sensitivity. Towards this goal, we have synthesized large-area mono- and bilayer graphene films by chemical vapor deposition (CVD) technique using ethanol as carbon source. Results revealed the graphene film to be polycrystalline, contained about 3 atomic% carboxylic groups and exhibited a p-type semiconductor behavior with an on-off ratio of ~2. Additionally, simple reactive ion etching (RIE) combined with e-beam and nanosphere lithography were performed on the synthesized CVD-grown monolayer graphene platform to fabricate large area GNRs and GNMs with specific dimension and periodicity to further enhance the on-off ratio of field-effect transistor and an improved detection sensitivity was demonstrated.

Furthermore, a three-dimensional (3D) carbon electrode in the form of vertically aligned metallic carbon nanotubes (CNTs) on a graphene floor was synthesized and modified with biocatalysts such as heme peptide, horseradish peroxidase and glucose oxidase. Experimental results demonstrated an enhanced efficiency of the 3D graphene-carbon nanotubes hybrid film of high surface area as support for biocatalysts for highly sensitive and selective detection of hydrogen peroxide and glucose.

Keywords: Bioanalytical, Biosensors, Electrodes, Nanotechnology
Application Code: Bioanalytical
Methodology Code: Sensors

Session # 910 Abstract # 910-3 **Symposia**

Session Title: Graphene Nanomaterials for Bio/Sensing Applications

Abstract Title: **A Novel Wireless Biosensing Platform Enabled by Graphene Varactors**

Primary Author: Steven J. Koester
University of Minnesota

Date: Tuesday, March 08, 2016 - Morning
Time: 09:45 AM
Room: B304

Co-Author(s)

Abstract Text

Graphene has the potential to form the basis of a powerful sensing platform due to its unique combination of properties, including high surface sensitivity, chemical stability, mechanical strength, and biocompatibility. However, nearly all sensor concepts based upon graphene require direct electrical connections to the graphene. This can limit the range of applications suitable for these devices, including many biological sensing functions, where wire leads can be cumbersome or impractical. In this talk, I will describe a novel sensor concept that utilizes a unique property of graphene, the quantum capacitance effect, to form a passive sensor that can be interrogated wirelessly. This effect allows the creation of variable capacitors, or varactors, that, when integrated with an inductor, form a variable LC resonator circuit whose oscillation frequency can vary with the concentration of an adsorbed analyte. We have developed a fabrication process for graphene varactors and demonstrated devices with capacitance tuning ratios as high as 1.6-to-1, with excellent uniformity and yield. We have also demonstrated this wireless sensing concept using water vapor as a test analyte and have developed further understanding of the effect of water on the sensor performance. Finally, I will show progress toward utilization of graphene varactors for a wide variety of biological sensing applications, including in vivo glucose sensors for closed-loop diabetes treatment and wireless vapor sensors for breath analysis.

Keywords: Chemical, Nanotechnology, Sensors
Application Code: Biomedical
Methodology Code: Sensors

Session # 910 **Abstract # 910-4** **Symposia**

Session Title: Graphene Nanomaterials for Bio/Sensing Applications

Abstract Title: **Graphene-Enabled Nano/Bio Hybrids for Chemical Detection and Medical Diagnostics**

Date: Tuesday, March 08, 2016 - Morning
Time: 10:35 AM
Room: B304

Primary Author: A T Charlie Johnson
University of Pennsylvania

Co-Author(s)

Abstract Text

We have explored all-electronic chemical detectors based on bio-nano hybrids, where the biomolecule (DNA or protein) provides chemical recognition and a carbon nanotube (NT), graphene, or monolayer molybdenum disulfide (MoS2) transistor enables electronic readout. This sensor class represents a promising approach towards sensitive and selective detection of liquid- and vapor-phase anayltes. Rapid advances in nanomaterials research and biomolecular engineering offer the prospect of optimizing the performance of each component of the hybrid to yield sensors with enhanced sensitivity and specificity for arbitrary chemical targets. Coupling chemistries have been developed that allow for the creation of a nanoelectronic and/or nanophotonic interface to a variety of different proteins including: antibodies and antibody fragments, optically active proteins, G-protein coupled receptors (GPCRs) housed in nanoscale membrane analogues, and GPCR variants that are engineered for enhanced solubility and stability in aqueous solution. We demonstrated the use of these bio-nano hybrids for detection of protein cancer biomarkers, antigen from various pathogens, and small molecule receptor targets at concentrations ~ 1 pg/mL. Single stranded DNA can be coupled to nanotubes and graphene through non-covalent functionalization. The DNA is used for its chemical recognition for small molecule analytes rather than recognition of complementary DNA. Vapor sensors based on this approach were able to discriminate between highly similar compounds such as enantiomers, as well as very similar complex mixtures of vapors characteristic of humans. More recently we have shown the promise of this system for diagnosis of disease based on volatile biomarkers.

Keywords: Array Detectors, Chemical, Material Science, Sensors
Application Code: Nanotechnology
Methodology Code: Integrated Sensor Systems

Session # 910 **Abstract # 910-5** **Symposia**

Session Title Graphene Nanomaterials for Bio/Sensing Applications

Abstract Title **Impedance Sensing of Nanotoxicity of Graphene at the Cellular and Tissue Level**

Primary Author Chenzhong Li
Florida International University

Date: Tuesday, March 08, 2016 - Morning
Time: 11:10 AM
Room: B304

Co-Author(s)

Abstract Text

The interest in graphene for biomedical applications has grown substantially in the past few years creating a need for biocompatibility testing. Biomedical engineering applications using graphene such as biosensing devices, microbial detection, disease diagnosis and drug delivery systems are progressing rapidly, perhaps overlooking any possible hazards as graphene nanomaterials may interact with biological materials differently than other graphitic materials such as carbon nanotubes and fullerenes. As a potential application for graphene is drug delivery, the toxicity of graphene was tested against an in vitro model of the blood brain barrier (BBB) by measuring the trans-endothelial electrical resistance (TEER). Conventional biological methods for measurement of cytotoxicity including mitochondrial reduction of tetrazolium salts into an insoluble dye (the MTT test) and enzyme lactate dehydrogenase (LDH) release tests measure toxicity at a final time point. As a result, the kinetic model of nanoparticle uptake in the cells usually cannot be observed leading to a lack of knowledge of the total nanoparticle interaction with the cell. Recent advancements in cytotoxicity studies have lead to the use of biosensors for real-time monitoring of cellular activity when subjected to various compounds such as nanoparticles.

In this presentation, a new approach in terms of electrical impedance sensing will be introduced. The novel biosensing technology enable us kinetically analyzing the cytotoxicity of graphene nanomaterials towards the BBB model's individual components, rat astrocytes (CRL- 2006™) and mouse endothelial cells (CRL-2583™), in real time by measuring the impedemetric response. Graphene showed little or no toxicity towards both cell types showing the biocompatibility of graphene and the broad potential of using these new nanomaterials for biomedical applications.

Keywords: Biosensors, Nanotechnology, Toxicology
Application Code: Nanotechnology
Methodology Code: Integrated Sensor Systems

Session # 920 **Abstract #** 920-1 **Symposia**

Session Title JAIMA - The State-of-the-Art Technologies from Japan: Analytical Instruments with/for Nano-Chemistr

Abstract Title **Creation of Bio/Chemical Sensing Probes**

Primary Author Koji Suzuki
Keio University

Date: Tuesday, March 08, 2016 - Morning
Time: 08:35 AM
Room: B305

Co-Author(s) Daniel Citterio

Abstract Text

The creation of original chemical sensing materials and bio/chemical sensors is one of the main concerns in our research group. To obtain bright fluorescent chemical probes, we have developed a set of fluorescent dyes (BODIPY-based KFL series dyes), which have excellent optical properties like sharp fluorescence spectra with high quantum yields, and moreover, the wavelength is finely tunable over a wide spectral range including the NIR region by introducing proper electron-donating groups into the furan moieties of the chromophore. By linking an ion recognizing ionophore with a fluorescent dye as a transducer, a chemical probe (fluorescent probe) is obtained, transforming a simple molecular recognition ligand into a sensor ligand. For instance, a KFL fluorescent dye combined with a BAPTA chelating group resulted in a bright fluorescent probe for calcium ions. We also have developed a set of chemiluminescent (CL) dyes (BODIPY-based KCL and KBI series dyes) with excellent CL properties. The luciferin-based CL probe KBI is a useful probe for highly sensitive detection of ROS such as O2-

Several bioluminescent (BL) systems have been investigating based on synthetic coelenterazine (CTZ) derivatives as substrates in combination with Renilla luciferase (Rluc) variants or artificial luciferases (Aluc) as the enzyme. It was found that extending the conjugated system at the 6-carbon position of CTZ is more effective compared to extensions at the 2-, 5-, or 8-carbon positions. The 6-position carbon variants of CTZ combined with the known Rluc mutant Rluc8.6 resulted in the most intense bioluminescence in the blue spectral region. In addition, with the system consisting of a CTZ derivative and the artificial luciferase Aluc, we have succeeded in the development of the most high-intensity artificial bioluminescence system.

We thank all members of our research roup at Keio University.

Keywords: Sensors

Application Code: Bioanalytical
Methodology Code: Sensors

Session # 920 Abstract # 920-2 **Symposia**

Session Title JAIMA - The State-of-the-Art Technologies from Japan: Analytical Instruments with/for Nano-Chemistr

Abstract Title **Ab Initio Powder Structure Determination Opening Up New Research Fields**

Date: Tuesday, March 08, 2016 - Morning
Time: 09:10 AM
Room: B305

Primary Author
Masaki Kawano
Pohang University of Science and Technology

Co-Author(s)

Abstract Text

Powder diffraction technique has been widely used for qualitative analysis over half a century. In sharp contrast, the history of ab initio powder structure analysis is quite short especially in soft materials fields. I will introduce the applications of ab initio powder structure analysis for exploring a new class of materials and science. One of representative methods for structure analysis is single crystal structure analysis. However, in reality there are a number of crystalline powder materials rather than single crystals in industrial fields. Recently we developed the preparation method of interactive porous materials using kinetic/thermodynamic control. Construction of interactive pores is a key to production of functional porous materials. In this talk, I will introduce interactive porous materials, the applications, and new science unveiled by crystallography.

Keywords: Vibrational Spectroscopy, X-ray Diffraction
Application Code: Material Science
Methodology Code: X-ray Techniques

Session # 920 **Abstract # 920-3** **Symposia**

Session Title: JAIMA - The State-of-the-Art Technologies from Japan: Analytical Instruments with/for Nano-Chemistr

Abstract Title: **Nanodroplet Formation and Chemical Analysis in Microfluidic Devices**

Primary Author: Akihide Hibara
Tokyo Institute of Technology

Date: Tuesday, March 08, 2016 - Morning
Time: 09:45 AM
Room: B305

Co-Author(s)

Abstract Text

This paper reports the condensation and separation method of microdroplet contents by aqueous nanodroplet formation in organic continuous phase. This method utilizes spontaneous emulsification at the microdroplets' interface. During the spontaneous emulsification, water in the aqueous microdroplet is transferred to nanodroplet, and the microdroplet shrinks.

We have conceived to integrate sample pretreatment methods utilizing the shrinkage. When solute molecules stay in the microdroplet during the emulsification, it is condensed in the microdroplet. When the solute does not partition to the nanodroplet side at all, hundreds to a thousand times condensation is expected with a 10-times diameter-decrease. On the other hand, when solutes are partitioned without any bias (the concentration in the nanodroplet is equal to that in the microdroplet), the solute is not condensed.

We found that the partition characteristics of the solute molecules depend on their size and hydrophobicity. For example, large hydrophilic proteins stay in the microdroplet while small hydrophobic dyes partition to the nanodroplet side. In this presentation, basic conception of the analytical pretreatment methods utilizing the spontaneous emulsification (nanodroplet formation) is introduced. Then, some applications of the methods such as biotin-avidin bioassay, and protein crystallization will be demonstrated.

Keywords: Lab-on-a-Chip/Microfluidics
Application Code: Bioanalytical
Methodology Code: Microfluidics/Lab-on-a-Chip

Session # 920 Abstract # 920-4 **Symposia**

Session Title JAIMA - The State-of-the-Art Technologies from Japan: Analytical Instruments with/for Nano-Chemistr

Abstract Title **Development of Mass Microscope and Applications in Drug and Oncometabolite Visualization**

Date: Tuesday, March 08, 2016 - Morning
Time: 10:35 AM
Room: B305

Primary Author Shuichi Shimma
Osaka University

Co-Author(s)

Abstract Text
Conventional mass spectrometric analysis provides quantitative information on pharmaceuticals in blood or normal/diseased tissues such as cancer. However, tissue extraction precludes determining information on their spatial distribution. To bring precise spatial distribution of molecules in specimens, imaging mass spectrometry (IMS) based on Mass Microscope was developed. Especially, cancer tissues have complicated morphology due to heterogeneity, therefore to perform IMS under the microscopic view with high sensitivity is beneficial to evaluate the pharmacokinetics. This feature attracts interest in early phase drug developments, especially in oncology. On the other hand, there are a lot of challenges to apply IMS for drug developments, for example, avoidance of artifacts, sensitivity, reproducibility, throughput and quantification. Therefore, I consider methodological optimizations from tissue sampling to data analysis is absolutely essential for practical application even though IMS is cutting-edge technology. In this presentation, I will introduce the IMS specific instrument "Mass Microscope" and provide the methodologies for the practical use. The main drugs in this presentation will be anticancer agents. Drug administered tumor block was resected from human lung cancer and mice xenograft model of non-small cell lung cancer (NSCLC). In addition, I will provide oncometabolites visualization results using an on-tissue derivatization technique.

Keywords: Drugs, Imaging, Instrumentation, Mass Spectrometry
Application Code: Clinical/Toxicology
Methodology Code: Mass Spectrometry

Session # 920 **Abstract # 920-5** **Symposia**

Session Title JAIMA - The State-of-the-Art Technologies from Japan: Analytical Instruments with/for Nano-Chemistr

Abstract Title **Plasma Phospholipids and Prevalence of Mild Cognitive Impairment/Dementia**

Primary Author Danni Li
University of Minnesota

Date: Tuesday, March 08, 2016 - Morning
Time: 11:10 AM
Room: B305

Co-Author(s)

Abstract Text

Plasma phospholipids may be useful in the prediction and clinical diagnosis of mild cognitive impairment (MCI) and dementia. Using targeted metabolomics, we measured 188 plasma metabolites in 441 participants who underwent a detailed neuropsychological assessment and diagnosed as normal (n=153), MCI (n=145) or dementia (n=143) by expert adjudication. Our results showed that some phospholipids and metabolites were altered in MCI and dementia. However, their ability for aid in diagnosis of MCI and dementia remains to be determined.

Keywords: Liquid Chromatography/Mass Spectroscopy
Application Code: Genomics, Proteomics and Other 'Omics
Methodology Code: Liquid Chromatography/Mass Spectrometry

Session # 930 Abstract # 930-1 **Symposia**

Session Title: Ultrahigh-Resolution Mass Spectrometry: A New Frontier

Abstract Title: **High-Resolution TOFMS via a Folded Flight Path: You Can Have Your Cake and Eat It Too**

Date: Tuesday, March 08, 2016 - Morning
Time: 08:35 AM
Room: B301

Primary Author: Viatcheslav Artaev
Leco Corporation

Co-Author(s): Anatoly Verenchikov, Michael Mason, Peter Willis

Abstract Text

Use of high resolution mass spectrometry has grown significantly in the recent past with availability of new instrumentation being a significant contributor to that growth. Recognition of the value of high mass accuracy and accurate isotope abundance in compound detection and identification has become ever more prominent. Simultaneously, challenges faced by the modern analytical chemist in the need for fast and accurate analysis of various compounds in complex matrices have created pressure on analysts to apply comprehensive tools in everyday routine.

The novel ion optics concept - Folded Flight Path (FFP) - allows significant improvements in resolving power of TOF analyzers (up to 200,000 and more) while keeping overall dimensions of the instrument within practical limits. Additionally, FFP optics provides very high ion transmission despite elongated ion path. A custom high speed data acquisition system (up to 200 full spectra/s) was developed along with comprehensive automatic data processing algorithms to address challenges of dealing with large amounts of high precision and high dynamic range data generated by FFP TOFMS and multi-dimensional separation methods. The challenges of low duty cycle, arisen from a long TOF employed in FFP analyzers, were addressed via multiplexing method of Encoded Frequency Pulsing, which results in increasing duty cycle in 10-100 times and brings FFP TOF sensitivity on par with most efficient single reflecting TOFMS systems. Thus the FFP technology combined with high data acquisition speed and high duty cycle operation mode creates unprecedented combination of characteristics significantly expanding the tool box of the modern analyst.

Keywords: GC-MS, Instrumentation, Mass Spectrometry, Time of Flight MS
Application Code: Other
Methodology Code: Mass Spectrometry

Session # **930** **Abstract #** **930-2** **Symposia**

Session Title: Ultrahigh-Resolution Mass Spectrometry: A New Frontier

Abstract Title: **21 Tesla FT-ICR Mass Spectrometer for Top-Down Protein Identification and Characterization**

Date: Tuesday, March 08, 2016 - Morning
Time: 09:10 AM
Room: B301

Primary Author: Christopher L. Hendrickson
National High Magnetic Field Laboratory

Co-Author(s): Alan G. Marshall, Donald F. Smith, Greg T. Blakney, John P. Quinn, Lissa C. Anderson, Nathan K. Kaiser

Abstract Text

Fourier transform ion cyclotron resonance (FT-ICR) mass spectrometry offers the highest achievable broadband mass resolving power and mass accuracy of any mass analyzer, which is especially important for characterization of large molecules such as biopolymers. We report here top-down protein analysis with a 21 tesla FT-ICR mass spectrometer, which is the highest field ICR system to date. The instrument is a part of the National High Field FT-ICR User Facility and is available to all qualified users. Unit mass resolving power is achieved for proteins as large as bovine serum albumin (BSA, 66 kDa) by use of a 0.38 second detection interval, which facilitates top-down proteomic analysis of large proteins by online LC/MS. Measured mass errors are routinely less than 1 ppm for proteins as large as 30 kDa, which increases protein identification confidence. External fragment ion accumulation and high magnetic field combine to facilitate rapid acquisition of high resolution tandem mass spectra. Electron-transfer and collisional dissociation combine to produce hundreds of fragment ions, facilitating site-localization of post-translational modifications.

We will report progress toward extending the mass range for efficient top-down LC/MS beyond 30 kDa, by optimization of chromatographic and mass spectrometric parameters. We demonstrate that, coupled with front-end protein fractionation, hundreds of spectral matches are obtained from single top-down LC-MS experiments of samples derived from human cancer cell lines. High-throughput characterization of proteins approaching 50 kDa will soon be a reality.

Work supported by the National Science Foundation through DMR-1157490 and CHE-1019193, and the State of Florida.

Keywords: Ion Cyclotron Resonance, Liquid Chromatography/Mass Spectroscopy, Proteomics, Tandem Mass Sp
Application Code: Genomics, Proteomics and Other 'Omics
Methodology Code: Liquid Chromatography/Mass Spectrometry

Session # 930 **Abstract # 930-3** **Symposia**

Session Title: Ultrahigh-Resolution Mass Spectrometry: A New Frontier

Abstract Title: **Exploring the Microbiome via Advanced Mass Spectrometry**

Primary Author: Ljiljana Pasa-tolic
EMSL, PNNL

Date: Tuesday, March 08, 2016 - Morning
Time: 09:45 AM
Room: B301

Co-Author(s): David W. Koppenaal, Errol Robinson, Jared B. Shaw, Malak Tfaily, Nancy J. Hess, Nikola Tolic

Abstract Text

Fourier transform ion cyclotron resonance (FTICR) offers the highest mass resolving power and accuracy of any mass analyzer; nevertheless, even higher resolution and accuracy is required to capture the full range of information for increasingly complex natural mixtures (e.g., aerosols, soil organic matter) and biological complexity (e.g., proteomics, metabolomics). Herein, we will describe advanced mass spectrometry capabilities available at EMSL, a DOE national scientific user facility located at PNNL, to address the key knowledge gaps in functional understanding of how complex microbiomes influence and are influenced by their environment and present selected applications to oceanic, freshwater and soil microbiome field studies. Special emphasis will be on the 21T FTICR mass spectrometer recently brought online at EMSL. This capability will arguably provide that next level of performance needed to understand chemistry and dynamics of the belowground carbon cycle. Identification and quantitation of intact molecular structures derived from higher plants and soil microbiome within their native inorganic soil matrix is essential for advancing molecular-scale mechanistic understanding of the role of physical, geochemical, and biological processes belowground. 21T FTICR platform will dramatically increase mass resolving power, accuracy, dynamic range and attainable sensitivity, and thus enable characterization of intact molecular structures (metabolites, peptides, proteins) in complex environmental matrices while addressing spatial relationships and heterogeneity, particularly in combination with other imaging modalities, e.g. FISH and/or nanoSIMS.

Keywords: Lipids, Mass Spectrometry, Metabolomics, Metabonomics, Method Development
Application Code: Genomics, Proteomics and Other 'Omics
Methodology Code: Mass Spectrometry

Session # 930 Abstract # 930-4 **Symposia**

Session Title: Ultrahigh-Resolution Mass Spectrometry: A New Frontier

Abstract Title: **Complex Mixture Analysis by Ultra-High Resolution Mass Spectrometry**

Primary Author: Ryan P. Rodgers
Florida State University

Date: Tuesday, March 08, 2016 - Morning
Time: 10:35 AM
Room: B301

Co-Author(s): Amy C. Clingenpeel, David C. Podgorski, Jonathan C. Putman, Priscila M. Lalli, Steven M. Rowland, Vladislav V. Lobodin, Winston K. Robbins, Yuri E. Corilo

Abstract Text

High field FT-ICR mass spectrometry has changed the utility and expectations of complex mixture analysis by mass spectrometry over the past decade. The inherent high resolving power and high mass accuracy enable direct determination of elemental compositions to tens of thousands of individual components by mass measurement alone. Modern ionization methods facilitate the selective ionization of components based coarsely on chemical functionality, which combined with FT-ICR MS, reveals acidic, basic, and aromatic contributions to complex mixtures at a molecular level. Here, we expand on previous work in the field to expose chromatographic, non-covalent adduct, and novel ionization methods that further refine the ability to selectively monitor species in complex mixtures at the molecular level. The primary targets of these efforts are high heteroatom-containing species (N, O, and S) that are either naturally occurring, or are produced by abiotic / biotic modification of the source material. We will highlight the latest analytical tools and their contribution to basic and applied research advances in petroleum science. Applications include the detailed characterization of environmental contaminants, interfacial material, and naphthenic acids. Data reduction, molecular formula assignment, data visualization, and chemometric comparisons of the data are handled by a custom software package (Petro-Org©).

Keywords: Energy, Environmental Analysis, Hydrocarbons, Mass Spectrometry
Application Code: Fuels, Energy and Petrochemical
Methodology Code: Mass Spectrometry

Session # **930** **Abstract #** **930-5** **Symposia**

Session Title Ultrahigh-Resolution Mass Spectrometry: A New Frontier

Abstract Title **Orbitrap Mass Analyzers: The Teenage Years**

Date: Tuesday, March 08, 2016 - Morning
Time: 11:10 AM
Room: B301

Primary Author Michael W. Senko
Thermo Fisher Scientific

Co-Author(s) Alexander Makarov, Eduard Denisov, Jesse D. Canterbury

Abstract Text

Fifteen years ago, the Orbitrap mass analyzer was newly born. Five years of growth and learning were necessary before this new analyzer was ready to move out into the world. The ten years since have shown tremendous growth in sensitivity, mass accuracy and resolution. As with any teenager, the Orbitrap is running up against new challenges. This presentation will describe some of those challenges and the developments that are necessary for the Orbitrap to become a unimpeachable leader in the mass spectrometry community.

Keywords: Ion Cyclotron Resonance, Ion Trap, Liquid Chromatography/Mass Spectroscopy, Mass Spectrometry
Application Code: Genomics, Proteomics and Other 'Omics
Methodology Code: Mass Spectrometry

Session # 940 **Abstract # 940-1** **Organized Contributed Sessions**

Session Title: High Performance SFC for the Analysis of Pharmaceuticals, Nutraceuticals, Natural Products and Meta

Abstract Title: **Analysis of Mycotoxins in Various Food Matrices via SFE/SFC/MS**

Date: Tuesday, March 08, 2016 - Morning
Time: 08:30 AM
Room: B309

Primary Author: Todd Anderson
Shimadzu Scientific

Co-Author(s)

Abstract Text

Mycotoxins are a metabolite of certain fungi that are produced throughout the harvest, transport and storage process. These toxins can therefore be found in everyday food products and can exhibit significant toxicity to humans and animals. Our objective is to develop a methodology by which we can detect these toxins from food products in a single instrument. Typically analysis consists of sample homogenization, some form of extraction technique followed by derivatization and finally analytical analysis. This talk will discuss the means by which we utilize a single instrument to perform all these functions, except for the sample homogenization. We will utilize samples of commercial commodities as a source for the mycotoxins and analyze for several mycotoxins using SFC CO2 for extraction, and SFC/MS for analysis.

Keywords: Chromatography, Natural Products, SFE, Supercritical Fluid Chromatography
Application Code: Food Contaminants
Methodology Code: Supercritical Fluid Chromatography

Session # 940 Abstract # 940-2 **Organized Contributed Sessions**

Session Title: High Performance SFC for the Analysis of Pharmaceuticals, Nutraceuticals, Natural Products and Meta

Abstract Title: **Evaluation of a Liquid Carbon Dioxide Based Flash Chromatography System**

Date: Tuesday, March 08, 2016 - Morning
Time: 08:50 AM
Room: B309

Primary Author: Ray McClain
Merck

Co-Author(s)

Abstract Text

Advances in SFC instrumentation, commercialization of widely applicable stationary phases, and the emergence of new areas of application using the powerful technique have made SFC mainstream in today's chromatography community. The green aspect of the using a carbon dioxide based mobile phase has been appealing to many audiences, particularly preparative chromatographers supporting drug discovery and development. A commercially available flash chromatography system utilizing such a mobile phase has been widely sought after for many years. This presentation will showcase the evaluation of a flash chromatography system utilizing liquid carbon dioxide as the weak mobile phase component for use in pharmaceutical research labs. In addition to the typical system evaluation data using standard components, real world data from several developmental compounds will be shared.

Keywords: Chromatography, Drug Discovery, Prep Chromatography, Supercritical Fluid Chromatography
Application Code: Drug Discovery
Methodology Code: Supercritical Fluid Chromatography

Session # 940 Abstract # 940-3 **Organized Contributed Sessions**

Session Title: High Performance SFC for the Analysis of Pharmaceuticals, Nutraceuticals, Natural Products and Meta

Abstract Title: **Fluoro-Methylphenylcarbamates of Cellulose as Chiral Stationary Phases for Supercritical Fluid Chromatography**

Date: Tuesday, March 08, 2016 - Morning
Time: 09:10 AM
Room: B309

Primary Author: William Farrell
Pfizer, Inc.

Co-Author(s): Matthew Przybyciel

Abstract Text

The use of aromatic halogenated substituents within carbohydrate chiral stationary phases is not a new concept, with early work dating back the mid to late 1990's [1, 2]. Chloro-, bromo, and fluoro-methylphenyl carbamates all showed chiral recognition and there are several chloro-phenyl substituted versions on available commercially. However, with the resurgence of fluorinated chemistry being used for drug development, the need for phases suitable for chiral recognition of fluorinated compounds is facilitated by re-introduction of these types of halogenated phases. Introduction of a fluorophilic retention mechanism can be particularly useful in support of medicinal chemistry drug discovery efforts, where more than a third of newly approved small molecule drugs contain fluorine (3, 4). Two new fluorinated phases, 4-fluoro-3-methylphenylcarbamate and 2-fluoro-5-methylphenylcarbamate were prepared on cellulose and evaluated for use with drug-like compounds containing fluorine constituents. Separations of a wide variety of these compounds will be presented as well as comparative separations using other halogenated stationary phases.

References

1.Enantioseparation on fluoro-methylphenylcarbamates of cellulose and amylose as chiral stationary phases for high performance liquid chromatography, Yashima et.al, Polymer Journal, vol 27, No. 8, pp 856-861 (1995)
2.3-Fluoro-, 3-chloro- and 3-bromo-5-methylphenyl carbamates of cellulose and amylose as chiral stationary phases for high performance liquid chromatographic enantioseparation, B. Chankvetadze, et al, Journ. Chrom. A, 787, (1997) 67-77.
3.Fluorous Reverse Phase Silia Gel. A New Tool for Preparative Separations in Synthetic Organic and Organofluorine Chemistry, Curran, D. P., Synlett 2001, (9), 1488-1496. http://dx.doi.org/10.1055/s-2001-16800
4.A Bountiful Year, Jarvis, L. M., Chemical & Engineering News 2013, 91, (5), 15-17 http://dx.doi.org/10.1021/cen-09105-bus1

Keywords: Chiral, Chromatography, Pharmaceutical, Supercritical Fluid Chromatography
Application Code: Pharmaceutical
Methodology Code: Supercritical Fluid Chromatography

Session # 940 Abstract # 940-4 **Organized Contributed Sessions**

Session Title: High Performance SFC for the Analysis of Pharmaceuticals, Nutraceuticals, Natural Products and Meta

Abstract Title: **Potential of Supercritical Fluid Extraction and Separation Technologies in Metabolomics**

Date: Tuesday, March 08, 2016 - Morning
Time: 09:30 AM
Room: B309

Primary Author: Bamba Takeshi
Kyushu University

Co-Author(s)

Abstract Text

Supercritical fluids have desirable properties such as high density, low viscosity, and high diffusivity that make it suitable for use as a solvent in supercritical fluid extraction (SFE), an effective and environment-friendly analytical method, and as a mobile phase for supercritical fluid chromatography (SFC), which facilitates high-throughput, high-resolution analysis. Because supercritical carbon dioxide (SCCO[sub]2[/sub]), which is generally used as the mobile phase in SFC, is automatically emitted at room temperature, SFC is most commonly used as a preparative method. However, SFC can also be used to perform high-precision biomolecular analysis, especially for hydrophobic metabolites, because of the low polarity of SCCO[sub]2[/sub]. Combining mass spectrometry (MS) with SFC can widen the scope of application of SFC to bioanalysis. Therefore, we have tried to apply SFC/MS to metabolic profiling.

We developed a useful workflow using SFC, a quadrupole orbitrap mass spectrometer, which can perform high-resolution full scanning and product ion scanning in both the positive-ion and negative-ion mode within a practical cycle time, and an automated lipid identification system. Additionally, we developed a polar lipid profiling method with trimethylsilylation and methylation and an oxidized-lipid-analysis system by applying SFC/MS effectively. Furthermore, we tried to apply SFC to hydrophilic compounds, even though SFC is recognized as a suitable chromatographic technology for hydrophobic compounds. The polarity of the mobile phase can be drastically changed in SFC. Thus, SFC can be applied to a wide range of compounds. Diverse bile acids, including their conjugates, could be simultaneously separated in a short time. Furthermore, we also successfully analyzed water- and fat-soluble vitamins in a single run.

Additionally, we developed technologies for combining SFC and supercritical fluid extraction (SFE). Exact information on labile metabolites in living bodies can be obtained, because online SFE/SFC can be used for the direct analysis of biological samples. Additionally, laborious extraction and purification processes are unnecessary with this technique. The reduced form of coenzyme Q[sub]10[/sub] in a photosynthetic bacterium was successfully analyzed (without oxidation taking place) by using online SFE/SFC/MS. SFE/SFC/MS allowed us to analyze many types of easily oxidized, fat-soluble vitamins and their metabolites in human serum.

Keywords: Mass Spectrometry, SFC, Metabolomics, Metabonomics
Application Code: Genomics, Proteomics and Other 'Omics
Methodology Code: Supercritical Fluid Chromatography

Session # 940	Abstract # 940-5	Organized Contributed Sessions

Session Title: High Performance SFC for the Analysis of Pharmaceuticals, Nutraceuticals, Natural Products and Meta

Abstract Title: **SFC – An Essential "Green" Separation Technique for the Chromatographic Analysis of a Nutraceuticals**

Primary Author: Matthew Przybyciel
ES Industries

Date: Tuesday, March 08, 2016 - Morning
Time: 10:05 AM
Room: B309

Co-Author(s)

Abstract Text

Nutraceuticals is an expansive term that describes any product resulting from food sources with extra health benefits in addition to basic nutritional value found in food. These nutraceuticals materials can be derived from various natural sources including plant materials, marine organisms and yeasts. Nutraceuticals contain bioactive components which have been discovered by epidemiological studies from diets enriched in bioactive components. These epidemiological studies have shown that certain bioactive components found can prevent diseases. The identification and quantification of bioactive components contained in natural sources can be challenging given their chemical complexity. The chemical complexity of natural product mixtures requires the use high performance chromatographic separation tools in order to reveal and measure bioactive components. In addition, bioactive active compounds possess a wide range of polarities from non-polar to extremely polar as well as potential thermal stability issues. All of these characteristics present a challenge to the well established chromatographic separation techniques such as HPLC and GC. To over the deficiencies of conventional chromatographic techniques we are employing SFC using carbon dioxide as a mobile phase for the high performance separation of bioactive components found in natural sources. SFC provides many unique features including the production of high pressure/high speed "green" separations. These features are well suited to SFC which can provide for the separation of chemical mixtures containing of wide range of polarities without creating thermal stability issues. We will demonstrate that recent improvements in SFC instrumentation and SFC optimized column technology have made SFC a primary and essential chromatographic technique for the separation and analysis of complex nutraceuticals mixtures.

Keywords: Natural Products, SFC, Supercritical Fluid Chromatography
Application Code: Other
Methodology Code: Supercritical Fluid Chromatography

Session # 940 **Abstract # 940-6** **Organized Contributed Sessions**

Session Title: High Performance SFC for the Analysis of Pharmaceuticals, Nutraceuticals, Natural Products and Meta

Abstract Title: **SFC, A Walk on the Wild Side**

Date: Tuesday, March 08, 2016 - Morning
Time: 10:25 AM
Room: B309

Primary Author: Brent Murphy
Genentech

Co-Author(s): Amber Guillen, Christopher Hamman, Joseph H. Pease, Mengling Wong, Michael Hayes

Abstract Text

Our group has experimented with altering SFC instrumentation and methods with the end goal of obtaining the best separation possible. Common problems encountered in preparative SFC include sample introduction, pressure gradients, resolution, etc. Here we present a wide array of things our lab has tried and the results we obtained, including grand successes, miserable failures, and everything in between.

Keywords: Chiral Separations, Chromatography, Method Development, SFC
Application Code: Pharmaceutical
Methodology Code: Supercritical Fluid Chromatography

Session # 940 Abstract # 940-7 **Organized Contributed Sessions**

Session Title: High Performance SFC for the Analysis of Pharmaceuticals, Nutraceuticals, Natural Products and Meta

Abstract Title: **Understanding the Value of Preparative Chiral Separations in Advancing Pharmaceutical R&D Projects**

Primary Author: Mickey Rego
Averica Discovery Services

Date: Tuesday, March 08, 2016 - Morning
Time: 10:45 AM
Room: B309

Co-Author(s): Emily Showell-Rouse, Jeffrey Kiplinger, Keith Galyan, Paul Lefebvre

Abstract Text

Chiral chromatography has been used to prepare samples of single stereoisomers for small to mid-scale efficacy, pharmacology, and toxicity testing. Conventional wisdom holds that chromatographic isolation of isomers of drugs becomes less cost-effective as the synthetic scale increases, so efforts to develop a stereospecific synthesis or a "batch resolution" approach are often initiated well before clinical trials begin.

The question of when – at what point in the R&D timeline, or at what scale-up level – chromatographic preparation becomes a cost burden is not often quantitatively evaluated. The choice often depends on balancing the costs of developing a non-chromatographic process (which will be wasted if the compound does not progress to clinical stage) against the cost-effectiveness of chromatography. Traditionally chromatographic efficiency is measured through loading studies, which provide an absolute productivity figure of merit in units such as KKD (kg feedstock/kg stationary phase/day). We suggest that an alternate approach, based on evaluating the relative efficiency of feedstocks produced by different chemistry options, offers a better assessment of cost-effectiveness. For example, partial enrichment of a desired isomer, changes in the impurity profile of a mixture, or protection/deprotection strategies can dramatically influence separation efficiency. Planning to compare production efficiency of different feeds is a rapid process, and offers more information than a final loading study on a synthetic product. Developing a chemistry strategy based on this feedback optimizes R&D program support during drug candidate profiling and de-risking, and facilitates cost-effective decision making prior to and during GMP production.

Keywords: Chiral, Chiral Separations, Drug Discovery, Prep Chromatography
Application Code: Pharmaceutical
Methodology Code: Supercritical Fluid Chromatography

Session # 950 Abstract # 950-1 **Organized Contributed Sessions**

Session Title Molecular Modelling and Quantum Mechanical Calculations: From Small Molecules to Large Multimeri

Abstract Title **Vibrational Optical Activity of Helical Peptides**

Date: Tuesday, March 08, 2016 - Morning
Time: 08:30 AM
Room: B310

Primary Author James R. Cheeseman
Gaussian, Inc.

Co-Author(s)

Abstract Text

Vibrational Circular Dichroism (VCD) and Raman Optical Activity (ROA) are powerful chiroptical spectroscopic methods for the structural characterization of chiral molecules. The theoretical prediction of chiroptical spectra, using [i]ab initio[/i] quantum chemical calculations, is becoming an important aspect to these methods. In addition to providing absolute configuration assignments, the comparison of calculated and experimental spectra can enhance the interpretation of the experimental spectra and provide insight into the structural sensitivity of these spectroscopic methods. Here I will present and discuss VCD and ROA spectra, obtained using DFT, for a pair of 22-residue antiparallel double-stranded [beta]-helical species and a series of alanine-rich peptides which form [alpha]-helices in water.

Keywords: Chiral, Peptides, Raman, Vibrational Spectroscopy
Application Code: General Interest
Methodology Code: Vibrational Spectroscopy

Session # **950** **Abstract #** **950-2** **Organized Contributed Sessions**

Session Title: Molecular Modelling and Quantum Mechanical Calculations: From Small Molecules to Large Multimeri

Abstract Title: **Interplay of Theory and Experiment: Small Molecule Ion Mobility and Spectroscopic Characterization of Metabolites**

Date: Tuesday, March 08, 2016 - Morning
Time: 08:50 AM
Room: B310

Primary Author: Michael D. Bartberger
Amgen

Co-Author(s)

Abstract Text

This presentation will outline some of the many potential roles of quantum chemistry when used in conjunction with experimental spectroscopic data such as ion mobility and NMR.

Quantum mechanical prediction of phenomena such as electronic and vibrational circular dichroism (ECD/VCD) and Raman optical activity (ROA), when used alongside corresponding experimental measurements, can furnish absolute configuration of enantiomeric drug like molecules and will be discussed briefly. In addition, we have recently undertaken studies involving the resolution and characterization of both constitutional isomers and diastereomers, via ionization or metal complexation, followed by comparison of experimental and theoretical cross sections based on QM-derived charge distribution.

Finally, the utility of NMR shielding tensor computation coupled with experiment in the characterization of human drug metabolites will be discussed. Such combinations of experiment and theory can be of great value in aiding the characterization of molecules from medicinal chemistry, process chemistry, or even PKDM campaigns.

Keywords: Characterization, Computers, Drug Discovery, Pharmaceutical
Application Code: Drug Discovery
Methodology Code: Computers, Modeling and Simulation

Session # 950 Abstract # 950-3 **Organized Contributed Sessions**

Session Title Molecular Modelling and Quantum Mechanical Calculations: From Small Molecules to Large Multimeri

Abstract Title **Efficient Photostitching of Peptide Ion Complexes in the Gas Phase with the Photo-Leucine Zipper**

Date: Tuesday, March 08, 2016 - Morning
Time: 09:10 AM
Room: B310

Primary Author Frantisek Turecek
University of Washington

Co-Author(s) Akis P. Andrikopoulos, Christopher J. Shaffer, Lubomir Rulisek

Abstract Text

Non-covalent protein-protein and protein-ligand interactions are the basis of molecular recognition of critical importance to many areas of biology. One of the methods to study non-covalent interactions relies on crosslinking by covalent bond formation between the protein and ligand at their contact sites upon chemical reaction or photolysis. Photochemical cross linking, in particular, utilizes highly reactive intermediates produced by photolysis of a chemically stable chromophore that is incorporated into the protein or ligand. We report a study of gas-phase complexes consisting of a hydrophobic neutral target peptide and a peptide ion furnished with the photoreactive L-2-amino-4,4-azi-pentanoic acid residue (L*) mimicking the natural leucine. Photodissociation of the diazirine ring in L* generates a highly reactive carbene that attacks the hydrophobic peptide by undergoing insertion into X-H bonds and forming new covalent bonds between the peptide moieties. The site of this "photostitching" is determined by MSn gas phase sequencing of the photo products. Complexes of GL*L*L*K and G(L2, L*)K with GLLLK, GLLLG and other hydrophobic pentapeptides were found to give high yields of covalently formed heterodimers. Extensive molecular dynamics modeling using a new semiempirical method with dispersion interactions revealed that the non-covalent gas-phase complexes are held together by polar interactions between the charged lysine side chain and polar groups of the neutral peptide counterpart. The complexes are fluxional, forming multiple contacts between the photolytically produced carbene and the groups on the other peptide in the course of trajectory simulations. Preference contacts were found for the neutral peptide carboxyl group, resulting in insertion into the O-H bond and formation of a covalent ester bond. The result of ion trajectory simulations were consistent with the sequence analysis of the photostitched products.

Keywords: Bioanalytical, Mass Spectrometry, Peptides, Tandem Mass Spec
Application Code: Bioanalytical
Methodology Code: Computers, Modeling and Simulation

Session # 950 **Abstract # 950-4** **Organized Contributed Sessions**

Session Title: Molecular Modelling and Quantum Mechanical Calculations: From Small Molecules to Large Multimeri

Abstract Title: **Chiroptical Spectroscopy for Molecular Structure Determination**

Date: Tuesday, March 08, 2016 - Morning
Time: 09:30 AM
Room: B310

Primary Author: Prasad L. Polavarapu
Vanderbilt University

Co-Author(s)

Abstract Text

The determination of absolute configuration and predominant conformations of chiral molecules and of the secondary structures of biomolecules are areas of active interest in chemical sciences, especially in pharmaceutical industry. Chiroptical spectroscopy is well suited for this purpose [1]. Four different methods, namely Electronic circular dichroism (ECD), optical rotatory dispersion (ORD), vibrational circular dichroism (VCD) and Raman optical activity (ROA), provide the needed information for elucidating chiral molecular structure. Facilitated by recent developments in chiroptical spectroscopic instrumentation and quantum chemical predictions of ECD, ORD, VCD and ROA properties, it is now becoming more or less routine to use combined experimental and computational studies for structural characterization of chiral molecules. In this talk I will provide a review of these methods, citing successful examples and limitations.

[1].Comprehensive Chiroptical Spectroscopy, 2 Volume Set; Editors: Nina Berova, Prasad L. Polavarapu, Koji Nakanishi, Robert W. Woody, John Wiley and Sons, 2012 (ISBN: 978-0-470-64135-4).

Keywords: Molecular Spectroscopy, Spectroscopy, Vibrational Spectroscopy
Application Code: Other
Methodology Code: Vibrational Spectroscopy

Session # 950 **Abstract # 950-5** **Organized Contributed Sessions**

Session Title: Molecular Modelling and Quantum Mechanical Calculations: From Small Molecules to Large Multimeri

Abstract Title: **The Use of Molecular Modelling for Ion Mobility Drift Time and Fragment Ion Prediction in Ion Mobility and Mass Spectrometry**

Date: Tuesday, March 08, 2016 - Morning
Time: 10:05 AM
Room: B310

Primary Author: Cris Lapthorn
University of Greenwich

Co-Author(s): Babur Chowdhry, Frank S. Pullen, George Perkins, Patricia Wright, Trevor Dines, Yanira Heredia

Abstract Text

Molecular modelling offers significant opportunities in ion mobility and mass spectrometry to 1) rationalise the drift-time observed for ions and 2) determine the identity of likely product (fragment) ions. These methods seek to utilise computational techniques to help rationalise experimental results in a time-frame useful for ion mobility mass spectrometry and mass spectrometry. Experimental ion mobility mass spectrometry data for some 'small molecule' compounds unexpectedly demonstrates the separation of at least two components, with significantly different collision cross section and distinct product ion spectra. Density functional theory calculations were utilised to generate geometry optimised structures which were used as inputs for calculating ion mobility collision cross sections. The theoretical structures of all protonation sites, using the projection approximation method that ignores charge distribution, revealed small differences in collision cross section consistent with the hypothesis that charge distribution plays a significant role in the separation of these components rather than the geometry. Further studies compare the theoretical collision cross sections with experimental collision cross sections which sought to benchmark the accuracy and feasibility for larger-scale databases to aid identification of small-molecule structures and sub-structures. The theoretical collision cross section was found to be within 3% of the experimental collision cross section for 90% of compounds when a simple correction was applied showing the potential for accurate prediction experimental collision cross sections. Both density functional theory and semi-empirical methods were used to rapidly predict the identity of product ions by observing changes in bond length following protonation. Polarised bond cleavage were predicted 100% correctly and over-estimation of bond cleavage was 34%, significantly lower than many alternative approaches to fragmentation prediction.

Keywords: Characterization, Mass Spectrometry
Application Code: High-Throughput Chemical Analysis
Methodology Code: Computers, Modeling and Simulation

Session # 950 Abstract # 950-6 **Organized Contributed Sessions**

Session Title: Molecular Modelling and Quantum Mechanical Calculations: From Small Molecules to Large Multimeri

Abstract Title: **Dynamical Networks of Protein Residue-Residue Contacts Provide Insights into Enzyme Function**

Date: Tuesday, March 08, 2016 - Morning
Time: 10:25 AM
Room: B310

Primary Author: Donald Hamelberg
Georgia State University

Co-Author(s)

Abstract Text

Although the relationship between structure and function in biomolecules is well established, it is not always adequate to provide a complete understanding of biomolecular function. It is now generally believed that dynamical fluctuations of biomolecules can also play an essential role in function. However, the precise nature of this dynamical contribution remains unclear. In this talk, I will discuss the development and use of theoretical and computational methods to understand how dynamical motions of enzymes are coupled to their function. I will present computational studies on members of a ubiquitous family of enzymes that catalyze peptidyl-prolyl bonds and regulate many sub-cellular processes. We map key dynamical features of the prototypical enzyme by defining dynamics in terms of residue-residue contacts. Analyzing large amount of time-dependent multi-dimensional data with a coarse-grained approach, we could capture the variation in contact dynamics during catalysis. I will present rationale of how enzyme dynamics is coupled to the reaction step and affects catalysis. Our results provide insights into the general interplay between enzyme conformational dynamics and catalysis from an atomistic perspective that have implications for structure based drug design and protein engineering.

Keywords: Biomedical, Computers, Method Development, Protein

Application Code: Biomedical

Methodology Code: Computers, Modeling and Simulation

Session # **950** **Abstract #** **950-7** **Organized Contributed Sessions**

Session Title Molecular Modelling and Quantum Mechanical Calculations: From Small Molecules to Large Multimeri

Abstract Title **A Molecular Dynamics and Ion Mobility Study of Protein Structure Collapse in the Gas-Phase**

Date: Tuesday, March 08, 2016 - Morning
Time: 10:45 AM
Room: B310

Primary Author Iain Campuzano
Amgen

Co-Author(s) Carlos Larriba-Andaluz, Morgan Lawrenz

Abstract Text

Being able to measure a protein or protein complex by MS and IM under native conditions has resulted in a new area of interest that is gas-phase structural biology. Generating an experimental value for a protein or protein complex of interest allows the researcher to then compare to a theoretically value, typically from an X-ray crystallography 3-D Cartesian coordinate set. However, to date, a detailed study comparing the gas-phase size and X-ray and NMR derived 3-D coordinate set of proteins, varying in size and quaternary structure has not been performed.

An RF-confining drift-cell was used to characterise proteins such as ubiquitin, cytochrome-C, lysozyme, albumin, alcohol dehydrogenase, serum amyloid protein and the humanised IgG1k NIST mAb, under native-IM/MS and buffer conditions. Experimental and theoretically derived values were generated in both He and N2 drift-gases. Gas-phase MD simulations were performed using Chem3D Pro and the MM2 force-field; Amber14 and the FF14SB force-field. An Adaptive Biasing Force was also implemented, biasing along the radius of gyration. Neutral and representative charge states were modelled using coordinates from available crystal structures. After equilibration in water at pH=7, protein charge states, the charge assignments were modified by adding hydrogen atoms to residues according to side-chain pKa, residue solvent accessibility and proximity to other like-charges for minimal Coulombic repulsion. Protein structures were minimized then simulated for a minimum of 10ns at 300K under vacuum conditions.

We will demonstrate how the majority of protein structures described herein, display some level of gas-phase collapse, when compared to their X-ray crystallographic derived coordinate set. This would imply that using an completely unprocessed protein coordinate set can lead to a significantly over estimated theoretical value.

Keywords: Biopharmaceutical, Mass Spectrometry
Application Code: Bioanalytical
Methodology Code: Computers, Modeling and Simulation

Session # 950	Abstract # 950-8	Organized Contributed Sessions

Session Title: Molecular Modelling and Quantum Mechanical Calculations: From Small Molecules to Large Multimeri

Abstract Title: **Hell Bent on Opening: Structural Basis for Estrogen Modulation of Potassium Recycling During Epithelial Secretion**

Date: Tuesday, March 08, 2016 - Morning
Time: 11:05 AM
Room: B310

Primary Author: Brett Kroncke
Vanderbilt University

Co-Author(s)

Abstract Text

Epithelial secretion in some tissues including lung and intestine is critically dependent upon recycling of potassium ions across basolateral membranes. In intestinal and tracheal epithelia, a potassium channel complex consisting of pore-forming (KCNQ1) and auxiliary (KCNE3) subunits are crucial for this ion recycling required for trans-epithelial chloride ion secretion. Human KCNE3 (MinK-related peptide 2, MiRP2) is a widely-expressed single span integral membrane protein that modulates the function and trafficking of several voltage-gated potassium channels, including KCNQ1, KCNQ4, hERG, Kv2.1, Kv3.1, and Kv3.2. KCNE3 converts KCNQ1 into a voltage-independent and constitutively active "leak" channel with enhanced conductance. KCNE3 has been implicated in gender-specific outcomes associated with salt homeostasis, such as occurs in cystic fibrosis (CF), where early epidemiological studies indicated females generally suffer significantly higher mortality rates than males, resulting in a shorter lifespan. Estrogen has been postulated to contribute to this gender gap in CF survival and is known to regulate the KCNE3–KCNQ1 channel complex, with KCNE3 Ser82 being critically involved. This suggests a mechanistic basis for the CF gender gap. In this work, we combined protein solution NMR for restraint generation and experimental flexibility characterization with Amber to refine and explain some interesting features, most notably the pronounced curvature and water accessibility of the KCNE3 transmembrane helix. With computational modeling, experimental structural biology, and electrophysiology we developed an integrative structural model that explains how KCNE3 modulates KCNQ1 function and how this modulation is attenuated by estrogen.

Keywords: Biomedical, Computers, NMR
Application Code: Biomedical
Methodology Code: Computers, Modeling and Simulation

Session # 960 **Abstract # 960-1** **Organized Contributed Sessions**

Session Title: SEAC Young Investigator Session

Abstract Title: **Fluorescence-Enabled Electrochemistry**

Date: Tuesday, March 08, 2016 - Morning
Time: 08:30 AM
Room: B311

Primary Author: Bo Zhang
University of Washington

Co-Author(s)

Abstract Text

In this talk, I will present fluorescence-enabled electrochemical microscopy (FEEM) as a new analytical method for imaging and studying dynamic redox events. FEEM is based on electrochemical coupling on a closed bipolar electrode which enables one to use optical readout to examine electrochemical reactions on numerous parallel microelectrodes. Here I will present new advances in FEEM focusing on the use of millions of micro- and nanoelectrodes to image dynamic redox events with exceedingly high spatial and temporal resolutions.

Keywords: Electrochemistry, Fluorescence, Imaging, Neurochemistry
Application Code: Neurochemistry
Methodology Code: Electrochemistry

Session # 960 Abstract # 960-2 **Organized Contributed Sessions**

Session Title: SEAC Young Investigator Session

Abstract Title: **Designing Functional Nanostructured Gels for Electrochemical Energy Storage and Biosensors**

Primary Author: Guihua Yu
University of Texas at Austin

Date: Tuesday, March 08, 2016 - Morning
Time: 08:50 AM
Room: B311

Co-Author(s)

Abstract Text

Nanostructured materials become critically important in many areas of technology, ranging from renewable energy, electronics, and photonics to biology and medicine, because of their unusual physical/chemical properties due to confined dimensions of such materials. This talk will present a new novel class of polymeric materials we developed recently: nanostructured electronic gels that offer an array of advantageous features such as intrinsic 3D nanostructured conducting framework, excellent electronic and electrochemical properties, synthetically tunable structures and chemical interfaces. These functional organic building blocks have been demonstrated powerful for a number of significant applications in energy, environmental and health-related technologies. Several examples on developing next-generation energy storage and biosensor devices will be discussed to illustrate 'structure-derived functions' of this special class of materials.

Keywords: Biosensors, Electrochemistry, Energy, Material Science
Application Code: Material Science
Methodology Code: Electrochemistry

Session # 960 **Abstract # 960-3** **Organized Contributed Sessions**

Session Title: SEAC Young Investigator Session

Abstract Title: **Inhibiting Electrochemical Processes in Li-ion Batteries at High Temperatures Using Responsive Polymers**

Primary Author: Mark E. Roberts
Clemson University

Date: Tuesday, March 08, 2016 - Morning
Time: 09:10 AM
Room: B311

Co-Author(s): Jesse C. Kelly, Nick L. DeGrood

Abstract Text

In recent years, remarkable progress in the development of Li-ion batteries has led to devices that deliver records in power and energy density. Their thermal instability, however, creates a major roadblock to large-format implementation for applications in transportation and energy storage from intermittent sources. Efforts to mitigate safety issues pertaining to flammability, reactivity, and thermal runaway typically involve the implementation of low conductivity solid-state materials or destructive safety devices, which prove problematic for high-power, large-format systems.

In this presentation, we will discuss a new approach to thermal stability in Li-ion batteries that uses an electrolyte system containing a polymer designed to phase separate from solution above a specific temperature. The phase separation causes a polymer coating on the electrode and separator, which increases the internal resistance and prevents current flow. This approach is advantageous because polymer phase separation is a local process, that is, if a hot spot forms in the battery, the polymer will phase separate and coat the electrode before catastrophic failure occurs.

Two responsive electrolyte systems will be described: poly(ethylene oxide) (PEO) in ionic liquids (ILs) and poly(benzyl methacrylate) PBMA in ILs. Both systems exhibit Li-ion dependent phase behavior, in which ion concentration affects the phase transition temperature and reversibility. Additionally, PEO-IL exhibits changes in solution conductivity and charge transfer resistance, while PBMA-IL only exhibits an increase in charge transfer resistance, but to a greater extent. Electrochemical impedance spectroscopy (EIS) is performed on planar and porous electrodes to correlate ionic and charge transfer resistances with temperature. The influence of ion and polymer concentration on the phase transition temperature will be discussed, along with the extent to which these properties can changed with temperature.

Keywords: Electrochemistry, Energy, Material Science, Polymers & Plastics
Application Code: Fuels, Energy and Petrochemical
Methodology Code: Electrochemistry

Session # 960 Abstract # 960-4 **Organized Contributed Sessions**

Session Title SEAC Young Investigator Session

Abstract Title **Revealing the Mechanism of Electron Uptake in Methanogenic Biofilm Community Using a Solid-Phase Electron Donor**

Date: Tuesday, March 08, 2016 - Morning
Time: 09:30 AM
Room: B311

Primary Author Sofia Babanova
JCVI

Co-Author(s) John Hogan, Kayla Carpenter, Michael Flynn, Michael Salvacion, Orianna Bretschger, Shino Suzuki, Shunichi Ishii, Sujal Phadke, Tony Phan

Abstract Text

The ability of microbes to produce electrical signals with chemical mediators was first reported in 1911. However, the concept of direct extracellular electron transfer (EET) for bacterial respiration was not discovered until 1988 through work conducted by Lovley (Geobacter) and Nealson (Shewanella). Immediately following these reports, the modern field of microbial fuel cells began to emerge. Fundamental and applied studies have revealed a mechanistic understanding for how microbes are able to utilize solid-phase substrates as electron acceptors during respiration. However, little is yet known about the mechanisms enabling microbes to utilize solid-phase minerals and

The presented study is focused on revealing the mechanism of electron uptake in methanogenic biofilm communities using a solid-phase electron donor. Methanogenesis is an essential part of the global carbon cycle and a key bioprocess for sustainable energy. Its underlying processes can be further explored as a tool for carbon dioxide fixation to methane as a renewable biofuel. The latter is described in the literature as microbial electrosynthesis - the process in which microorganisms use electrons derived from an electrode to reduce carbon dioxide to multicarbon, extracellular products. Data relative to the electron uptake as well as taxonomic and functional relationships that were observed for methanogenic biofilm in bioelectrochemical systems will be demonstrated. Duplicate bioelectrochemical systems were inoculated with rice paddy soil and subpassaged every 10-14 days to accelerate the enrichment of the cathode-associated methanogenic biofilm. Poised-potential electrodes (-500 mV vs SHE) were used as solid-phase electron donors and bioelectrochemical reactors were operated under anoxic conditions . Various electrochemical techniques were employed for revealing the EET mechanisms and the role of hydrogen exchange in electromethanogenic consortia.

Keywords: Biofuels, Biotechnology, Electrochemistry, Electrode Surfaces
Application Code: Fuels, Energy and Petrochemical
Methodology Code: Electrochemistry

Session # **960** **Abstract #** **960-5** **Organized Contributed Sessions**

Session Title SEAC Young Investigator Session

Abstract Title **Reagentless and Reusable Electrochemical Metal Ion Sensors**

Date: Tuesday, March 08, 2016 - Morning
Time: 10:05 AM
Room: B311

Primary Author Rebecca Y. Lai
University of Nebraska-Lincoln

Co-Author(s)

Abstract Text

A number of folding- and dynamics-based electrochemical biosensors have been developed in recent years. Targets reported include nucleic acids, proteins and small molecules. These sensors comprise an oligonucleotide probe attached to a gold electrode via an alkanethiol and modified with a redox reporter such as methylene blue (MB). Binding of an analyte to the redox-modified probe changes its conformation and/or flexibility, which, in turn, influences the efficiency of electron transfer to the interrogating electrode. We have recently expanded the application of this class of electrochemical biosensors for detection of metal ions. The targets of the electrochemical ion (E-ION) sensors include Hg(II), Ag(I), Au(III), and Cd(II). These sensors are sensitive, specific, and more importantly, selective enough to be employed directly in realistically complex samples such as soil extracts and aquifer water samples. Given that all of the sensing components are chemisorbed onto the electrode surface, they are readily regenerable and reusable. They are also compatible with paper-based gold-plated carbon electrodes and are thus well-suited for real-time, on-site analysis.

Keywords: Biosensors, Electrochemistry, Environmental/Water, Nucleic Acids

Application Code: Bioanalytical

Methodology Code: Electrochemistry

Session # 960 Abstract # 960-6 **Organized Contributed Sessions**

Session Title SEAC Young Investigator Session

Abstract Title **Electrochemical Imaging of Ionic Reactivity at Operating Ion Batteries**

Primary Author Joaquin Rodriguez Lopez
University of Illinois

Date: Tuesday, March 08, 2016 - Morning
Time: 10:25 AM
Room: B311

Co-Author(s)

Abstract Text

We introduce probe-based stripping voltammetric methods based on scanning electrochemical microscopy (SECM) for interrogating ionic and electronic reactivity at single reactive sites on operating ion-battery anodes. In spite of their significance in determining battery performance, a comprehensive understanding of how local surface (de)activation and site-specific differential reactivity impacts the dynamic ion-transfer capabilities of ion-battery anode interfaces has yet to be fully elucidated. New in-situ analytical tools are required for analyzing diverse ionic systems (e.g. Li[sup]+[/sup], Na[sup]+[/sup], K[sup]+[/sup], etc.) at interfacial and bulk nanostructures, especially in non-aqueous systems.

We present an experimental and simulation approach that integrates Hg cap electrodes, on which alkaline ions can be detected by means of anodic stripping voltammetry (see Figure), onto an SECM platform [1]. The probe potential provided chemical specificity while the limiting current for ion ingress into the Hg cap permitted to measure ionic fluxes with excellent stability and linearity for concentrations in the mM and sub-mM range. We also demonstrate the possibility of carrying out analysis on electrodes of ~100 nm. Additionally, we developed the simulation background for quantifying local charge/discharge curves by means of SECM ion collection. Here, we probed with sub-micron spatial resolution the alkaline ionic reactivity of patterned ultra-thin carbon samples. These techniques represent an unprecedented step in the analysis of anode ion insertion mechanisms, accessing aspects of surface reactivity that are lost during averaging in conventional electrochemical methods.

[1] Barton, Z.J.; Rodriguez-Lopez. Lithium Ion Quantification using Mercury Amalgams as In Situ Electrochemical Probes in Nonaqueous Media. Anal. Chem. 86 (2014) 10660.

Keywords: Energy, Imaging, Ion Selective Electrodes, Stripping Analysis
Application Code: Fuels, Energy and Petrochemical
Methodology Code: Electrochemistry

Session # 960 **Abstract # 960-7** **Organized Contributed Sessions**

Session Title: SEAC Young Investigator Session

Abstract Title: **In Vivo Serotonin Chemistry and Local Cytoarchitecture: A Combined Voltammetric, Mathematical and Microscopy Study**

Date: Tuesday, March 08, 2016 - Morning
Time: 10:45 AM
Room: B311

Primary Author: Parastoo Hashemi
University of South Carolina

Co-Author(s): Aya Abdalla

Abstract Text

Fast scan cyclic voltammetry (FSCV) can provide sub second in vivo measurements of evoked serotonin release and reuptake. We previously used FSCV to describe serotonin dynamics in the mouse substantia nigra, pars reticulate (SNr) and developed a mathematical model to fit experimental data corresponding to serotonin's two discrete reuptake mechanisms, Uptake 1 and Uptake 2. Here, we expand this work to two important regions, the CA2 region of the hippocampus and the prefrontal cortex (PFC). We describe two novel voltammetric circuitries for evoking and measuring serotonin in the CA2 region and the PFC. We compare evoked serotonin release and reuptake with FSCV and ambient serotonin levels (with a novel technique, fast scan adsorption controlled voltammetry (FSCAV)) between the SNr, CA2 region and PFC. Our models allow us to determine the average contribution of Uptake 1 and 2 in each region and correlate these contributions to local cytoarchitecture, as determined with 2-photon microscopy.

Keywords: Electrochemistry, Microelectrode, Neurochemistry, Voltammetry
Application Code: Neurochemistry
Methodology Code: Electrochemistry

Session # 960 Abstract # 960-8 **Organized Contributed Sessions**

Session Title: SEAC Young Investigator Session

Abstract Title: **Ec-LLS at the Micro- and Macroscale: Electrodeposition of Semiconductor Films and Nanocrystals**

Primary Author: Stephen Maldonado
University of Michigan

Date: Tuesday, March 08, 2016 - Morning
Time: 11:05 AM
Room: B311

Co-Author(s)

Abstract Text

This presentation will discuss advancements on two fronts with regards to the use of electrochemical liquid-liquid-solid crystal (ec-LLS) growth to prepare semiconductor materials at two extreme length scales. First, we will describe a methodology for the preparation of thin liquid metal electrodes that are useful for the electrodeposition of large-area, thin semiconductor films. The state of the art in process conditions as well as material quality will be discussed. Second, we present in-situ electron microscopy studies of the electrodeposition of individual Ge nanowires by ec-LLS. We will present data illustrating both factors that affect nanowire nucleation and growth as well as methodology advancements that advance the general use of in-situ transmission electron microscopy for electroanalytical studies.

Keywords: Electrochemistry, Electrodes, Material Science, Semiconductor
Application Code: Material Science
Methodology Code: Electrochemistry

Session # 970 **Abstract # 970-1** **Organized Contributed Sessions**

Session Title: Supercritical CO2- SFE/SFC: Advances in Extraction and Purification for Pharmaceutical and Natural Pr

Abstract Title: **Supercritical Fluid Chromatography in Support of Pharmaceutical – A Study of Scale-Up from Analytical to Preparative Scale with Isocratic Conditions**

Date: Tuesday, March 08, 2016 - Morning
Time: 08:30 AM
Room: B313

Primary Author: Mirlinda Biba
Merck

Co-Author(s): Ingrid Mergelsberg, Jinchu Liu, Judy Morris, Lindsey Jacobs

Abstract Text

Supercritical fluid chromatography (SFC) has been the preferred technology at Merck for routine enantiopurity analysis and preparative scale purifications in support of drug discovery and development for many years. The use of pressurized carbon dioxide, as opposed to organic solvents, leads to faster, more efficient, and greener separations compared to high performance liquid chromatography (HPLC). However, the advantages afforded by the SFC are coupled with various unknowns when scaling to preparative scale. Due to the compressible nature of the carbon dioxide mobile phase, scaling from analytical to preparative scale in SFC is considerably complicated and not widely understood. In these research studies, we investigated scaling from analytical SFC with 3 μm particle size chiral columns to preparative SFC with 5 μm chiral columns, using isocratic conditions. The results obtained from this systematic SFC scaling study illustrate the practical application and scaling rule from analytical to preparative SFC.

Keywords: Pharmaceutical, Supercritical Fluid Chromatography

Application Code: Pharmaceutical

Methodology Code: Separation Sciences

Session # 970 **Abstract # 970-2** **Organized Contributed Sessions**

Session Title: Supercritical CO2- SFE/SFC: Advances in Extraction and Purification for Pharmaceutical and Natural Pr

Abstract Title: **Strategies for Increasing Throughput of Chiral Separations by Supercritical Fluid Chromatography**

Date: Tuesday, March 08, 2016 - Morning
Time: 08:50 AM
Room: B313

Primary Author: Erin Jordan
AbbVie

Co-Author(s): Philip A. Searle

Abstract Text

The chirality of a drug can potentially have a large impact on its biological activity, metabolism and toxicity. Obtaining optically pure compounds has become increasingly important in the discovery of therapeutic compounds since one enantiomer can have positive therapeutic properties while the other can display non-therapeutic or negative biological activity. Due to the often difficult and limiting nature of achiral synthesis, chiral chromatography is typically applied to access stereochemically pure compounds. The application of Supercritical Fluid Chromatography (SFC) to chiral separations has proven very effective in recent years due to its many advantages over chiral HPLC, such as shorter retention times, higher efficiencies per unit time, and the reduction of organic solvent waste. At Abbvie, the Analytical and Purification Sciences (APS) group has provided a chiral preparative SFC service since 2009, which has steadily grown to impact over 30 projects and over 200 samples per year. This talk will discuss strategies to meet the growing need and variability of chiral separations within Drug Discovery, such as developing a streamlined scale-up approach, applying structure similarity software to minimize method development, and focusing on unique features offered in preparative SFC to address a broader range of chiral separations in a high-throughput laboratory.

Keywords: Automation, Chiral, Prep Chromatography, Supercritical Fluid Chromatography
Application Code: Drug Discovery
Methodology Code: Supercritical Fluid Chromatography

Session # 970 Abstract # 970-4 **Organized Contributed Sessions**

Session Title: Supercritical CO2- SFE/SFC: Advances in Extraction and Purification for Pharmaceutical and Natural Pr

Abstract Title: **SFC-MS as the Technique of Choice for Small Molecules Purification**

Primary Author: Gerard Rosse
Dart Neuroscience

Date: Tuesday, March 08, 2016 - Morning
Time: 09:30 AM
Room: B313

Co-Author(s)

Abstract Text

Productivity of modern medicinal chemistry requires automated synthesis and high throughput purification instrumentation that can process a large number of samples within a meaningful timeframe. The presentation will discuss our decision-making process for selecting SFC-MS as the prevailing method for compound purification. Instrumentation for analytical and preparative SFC-MS, infrastructure, logistics, workflows, and robotics to support the purification of >10,000 compounds per month will be presented. Challenges related to SFC-MS technology implementation and comparative studies for the application of SFC-MS and HPLC-MS techniques in small molecules analysis/purification will also be presented.

Keywords: Automation, Drug Discovery, Pharmaceutical, SFC
Application Code: Pharmaceutical
Methodology Code: Supercritical Fluid Chromatography

Session # 970 Abstract # 970-5 **Organized Contributed Sessions**

Session Title Supercritical CO2- SFE/SFC: Advances in Extraction and Purification for Pharmaceutical and Natural Pr

Abstract Title **Practical Approaches to SFE and SFC in Drug Discovery**

Primary Author Joseph H. Pease
Genentech

Date: Tuesday, March 08, 2016 - Morning
Time: 10:05 AM
Room: B313

Co-Author(s) Amber Guillen, Brent Murphy, Mengling Wong, Michael Hayes

Abstract Text

There have been many recent advances in SFC instrumentation and the latest systems not only improve performance but also make the technology more accessible to the non-expert. Here we present methods development and application of a new SFE-SFC system to resolve and identify both low level impurities in active pharmaceutical ingredient batches and metabolites from complex biological matrices. In particular, we will present separation strategies for structurally similar molecules, such as API and metabolites from plasma.

Keywords: Chiral Separations, Chromatography, Solid Phase Extraction, Supercritical Fluid Chromatography
Application Code: Pharmaceutical
Methodology Code: Supercritical Fluid Chromatography

Session # **970** **Abstract #** **970-6** **Organized Contributed Sessions**

Session Title: Supercritical CO2- SFE/SFC: Advances in Extraction and Purification for Pharmaceutical and Natural Pr

Abstract Title: **One Phase, Three Techniques: Utilizing One SFC Stationary Phase Across Three Chromatographic Techniques**

Date: Tuesday, March 08, 2016 - Morning
Time: 10:25 AM
Room: B313

Primary Author: Christine Aurigemma
Pfizer, Inc.

Co-Author(s): Perrine Hoerter, William Farrell

Abstract Text

The introduction of walk-up analytical SFC/MS in the medicinal chemistry labs was initially met with great enthusiasm, but usage quickly declined. Despite the excellent separations and fast cycle times by SFC, the chemists lacked the know-how to translate the analytical SFC methods to flash liquid chromatography, which is a common purification tool that chemists have at their disposal. This led the chemists to revert back to using traditional LC/MS as a guiding tool rather than changing their work habits to embrace the new (superior) technology. In an effort to facilitate the transition of SFC methodologies into their work habits, we explored the use of SFC-specific stationary phases as replacements for bare silica in TLC and flash liquid chromatographies. In our study, we evaluated both HA-pyridinyl and HA-morpholine phases as silica replacements. This presentation will explore the benefits and pitfalls of such an approach and how this can facilitate thesuccessful adoption of SFC methodologies to boost chemist efficiency and maximize productivity in the medicinal chemistry laboratory.

Keywords: Analysis, SFC, Supercritical Fluid Chromatography, Thin Layer Chromatography
Application Code: Pharmaceutical
Methodology Code: Supercritical Fluid Chromatography

Session # 970 **Abstract # 970-7** **Organized Contributed Sessions**

Session Title: Supercritical CO2- SFE/SFC: Advances in Extraction and Purification for Pharmaceutical and Natural Pr

Abstract Title: **Online SFE-SFC Purification Method Development Using SFC Solubility Determination**

Primary Author: Wes Barnhart
Amgen

Co-Author(s): Kyung Gahm

Date: Tuesday, March 08, 2016 - Morning
Time: 10:45 AM
Room: B313

Abstract Text

Supercritical fluid chromatography (SFC) has been in use for many years, providing certain advantages, such as high flow rates and reduced fraction volume, due to the presence of liquid carbon dioxide in the mobile phase [1]. Over the past 30 years, however, the process of introducing analyte has undergone little change. Recently, it was demonstrated that samples can be injected by coupling supercritical fluid extraction (SFE) to SFC on an analytical scale [2,3]. Solubility measurement data generated from this technique was then utilized for method development for a preparative chiral separation.

In this presentation, on-line SFE-SFC method development of varous samples will be discussed. Data obtained from the analytical scale will then be applied to preparative SFE-SFC purifications of these mixtures. It will be demonstrated that the method development process and purification can be mostly automated, allowing for a simple way to take advantage of this powerful technique.

[1] W. Barnhart, K. Gahm, S. Thomas, S. Notari, D. Semin, J Cheetham. J. Sep. Sci., 2005, 28, 619.
[2] K. Gahm, H. Tan, J. Liu, W. Barnhart, J. Eschelbach, S. Notari, S. Thomas, D. Semin, J. Cheetham, J. Pharm. and Biomed. Anal., 2008, 46, 831.
[3] K. Gahm, K. Huang, W. Barnhart, W. Goetzinger. Chirality, 2011, 23, 1E, E65.

Keywords: Drug Discovery, Extraction, SFE, Supercritical Fluid Chromatography
Application Code: Drug Discovery
Methodology Code: Supercritical Fluid Chromatography

Session # 970 Abstract # 970-8 **Organized Contributed Sessions**

Session Title Supercritical CO2- SFE/SFC: Advances in Extraction and Purification for Pharmaceutical and Natural Pr

Abstract Title **Utilizing SFE and SFC for Extraction and Isolation of Cannabinoids**

Primary Author Christopher Hudalla
ProVerde Laboratories

Date: Tuesday, March 08, 2016 - Morning
Time: 11:05 AM
Room: B313

Co-Author(s)

Abstract Text

The cannabis industry has been thriving for many years, with the use of cannabis tracing back thousands of years. However, the illicit status of the herb throughout much of the world has stifled commercialization and research, forcing most activities underground, often times with high risk and minimal accountability. As a result of this, technological advances in science and analytical instrumentation have found little to no application to this diverse field. Recent advances throughout the world in legislation, regulation and public acceptance have opened the door for legitimacy of this industry. This provides the new opportunity for the use of the latest advances in scientific instrumentation and methodologies to be applied to different aspects of this industry, including ensuring consumer safety, basic research, optimization of cultivation practices, and the design and development of Marijuana Infused Products (MIPs). Here we present the application of supercritical fluid technologies to the extraction and purification of cannabinoids for the preparation of cannabis based therapeutics. Supercritical Fluid Extraction (SFE), using liquid carbon dioxide, is used to extract the cannabinoids out of the raw plant material. Preparative Supercritical Fluid Chromatography (SFC) is used to further purify the extract, isolating the individual cannabinoids. The purified cannabinoids can be used to prepare custom formulations, providing the ability to generate precise dosing for cannabinoid therapies.

Keywords: Drug Discovery, Natural Products, SFC, SFE

Application Code: Other

Methodology Code: Supercritical Fluid Chromatography

Session # 980 Abstract # 980-1 **Oral Sessions**

Session Title: Biomedical: Advances in Glucose Monitoring and Therapeutics of Diabetes - Half Session

Abstract Title: **Skin Interface and Spectrometer Development for Noninvasive Glucose Measurements in People over Combination Near-Infrared Wavelengths**

Date: Tuesday, March 08, 2016 - Morning
Time: 08:30 AM
Room: B408

Primary Author: Ariel Bohman
University of Iowa

Co-Author(s): Gary W. Small, Mark A. Arnold, Michael J. Miller

Abstract Text

Treatment of diabetes demands glucose sensing technology that can provide the clinical data required to establish tight glycemic control. In practice, daily management of diabetes involves self-monitoring using test-strip technology. More recently, continuous glucose monitoring provides patterns over time as a means to refine insulin delivery strategies. Both technologies are invasive in the sense that either capillary blood must be collected or a sensing device must puncture the skin in order that the sensing device contacts interstitial fluid within subcutaneous tissue.

A noninvasive approach to glucose sensing involves passing electromagnetic radiation through a region of the body and extracting the concentration of glucose from an analysis of the resulting spectrum. This approach is painless and amenable to continuous monitoring. To date, however, noninvasive approaches have proven difficult owing to demands for chemical selectivity and spectral quality.

Our approach is to transmit near-infrared radiation (4000-5000 cm-1) through the dermis layer of skin. Feasibility of this approach has been demonstrated in an animal model where the near-infrared radiation was transmitted across a fold of rat skin held between two sapphire rods.

This presentation will focus on efforts to modify our procedures for noninvasive glucose measurements in people. Our efforts have focused on skin-interface and spectrometer performance, both of which are necessary to collect noninvasive spectra of sufficiently high quality to enable glucose measurement. Key elements of our noninvasive human measurements will be presented, including specific instrumental signal-to-noise ratios and effective aqueous optical path lengths necessary to discern glucose from the tissue background.

Keywords: Chemometrics, FTIR, Near Infrared, Spectroscopy
Application Code: Biomedical
Methodology Code: Near Infrared

Session # 980 Abstract # 980-2 **Oral Sessions**

Session Title: Biomedical: Advances in Glucose Monitoring and Therapeutics of Diabetes - Half Session

Abstract Title: **Advances in Development of Glucose Biosensors**

Date: Tuesday, March 08, 2016 - Morning
Time: 08:50 AM
Room: B408

Primary Author: Arunas Ramanavicius
Vilnius University

Co-Author(s): Almira Ramanaviciene, Asta Kausaite-Minkstimiene, Inga Vilkonciene, Jaroslav Voronovic, Jurate Petroniene, Natalija German, Povilas Genys

Abstract Text

Demographic changes, multiple and chronic diseases as well as possibilities provided by the modern information technology to compare and benchmark own health parameters at patient networks and databases are the drivers of coming health care innovations. To meet required safety and accuracy of medical services, at point-of-care units and, especially, at homes, emphasis must be put on the development of non-invasive, attachable, flexible, wireless, multiparameter, micro and nanosensors as well as biosensors. The benefit of health care at home has been highly appreciated by diabetic patients since 1987, when ExaTech introduced a pen-sized glucose biosensor for home use. Non-invasive glucose biosensing technologies have been proposed, however, still with a limited success. In this presentation experience of authors in the application of enzymes, including several different FAD dependent glucose oxidases [1], NAD and PQQ dependent dehydrogenases, is overviewed. Most successful immobilization methods for immobilization of these enzymes will be outlined. Electron transfer issues, which are very important in development of electrochemical enzymatic glucose biosensors will be presented and characteristic of some new redox mediators [2] will be presented. Applicability of some nanomaterials such as gold nanoparticles, graphene, carbon nanotubes will be discussed.

1. A. Ramanavicius, A. Kausaite-Minkstimiene, I. Morkvenaite-Vilkonciene, P. Genys, R. Mikhailova, T. Semashko, J. Voronovic, A. Ramanaviciene, Biofuel Cell Based on Glucose Oxidase from Penicillium Funiculosum 46.1 and Horseradish Peroxidase. Chemical Engineering Journal 2015, 264, 165–173.
2. R. Mikhailova, T. Semashko, O. Demeshko, A. Ramanaviciene, A. Ramanavicius, Effect of Some Redox Mediators on FAD Fluorescence of Glucose Oxidase from Penicillium Adametzii LF F-2044.1. Enzyme and Microbial Technology 2015, 72, 10-15.

Keywords: Bioanalytical, Biosensors, Electrochemistry, Electrode Surfaces
Application Code: Biomedical
Methodology Code: Sensors

Session # 980 **Abstract # 980-3** **Oral Sessions**

Session Title: Biomedical: Advances in Glucose Monitoring and Therapeutics of Diabetes - Half Session

Abstract Title: **A Microfluidic Cell Culture Device for Automated Sample Preparation and Improved Biomimetic Modeling in Diabetes Metabolomics**

Date: Tuesday, March 08, 2016 - Morning
Time: 09:10 AM
Room: B408

Primary Author: Laura Filla
Saint Louis University

Co-Author(s): James L. Edwards

Abstract Text

Diabetes is a metabolic disease characterized by a hyperglycemic state, which, if not effectively managed, can generate reactive oxygen species (ROS). When ROS attack endothelial cells (ECs), which line the walls of blood vessels, cardiovascular damage develops over time. Despite the strong vascular component of the disease, a suitable [i]in vitro[/i] system that models the metabolic consequences of hyperglycemia [i]in vivo[/i] has yet to be developed. Conventional cell culture-based methods for studying diabetes have proven inadequate both in their inability to mimic physiological conditions present [i]in vivo[/i] and in their lengthy sample preparation, which could lead to the loss of biological information prior to analysis. We describe here a microfluidic cell culture platform with embedded electrodes for rapid cell lysis and metabolite extraction. Microchannel dimensions are consistent with those of a blood vessel, allowing for reduced volumetric flow rates and reagent consumption. We have developed an automated cell lysis system which is coupled to the device; complete cell lysis is achieved within 15 seconds using only organic lysing solvents and within 4 seconds when combined with in-channel electroporation. Metabolic analysis of these samples has been undertaken using mass spectrometry. MALDI analysis of the extracted metabolites resulted in the detection of several species relevant to diabetic complications. This work will investigate the metabolic pathways that become damaged by hyperglycemia which will aid in the elucidation of potential therapeutic targets.

Funding for this project was provided by grants from the American Heart Association, National Institutes of Health, and Saint Louis University

Keywords: Bioanalytical, Lab-on-a-Chip/Microfluidics, Metabolomics, Metabonomics
Application Code: Bioanalytical
Methodology Code: Microfluidics/Lab-on-a-Chip

Session # 980 **Abstract # 980-4** **Oral Sessions**

Session Title: Biomedical: Advances in Glucose Monitoring and Therapeutics of Diabetes - Half Session

Abstract Title: **Analysis of Drug-Protein Interactions During Diabetes by High-Performance Affinity Chromatography**

Date: Tuesday, March 08, 2016 - Morning
Time: 09:30 AM
Room: B408

Primary Author: Zhao Li
University of Nebraska-Lincoln

Co-Author(s): David Hage, Ryan Matsuda

Abstract Text

High-performance affinity chromatography (HPAC) and small affinity columns were used to examine the changes in binding that occurred for chlorpropamide and tolazamide (i.e., two sulfonylurea drugs used to treated type II diabetes) with human serum albumin (HSA) at various stages of non-enzymatic glycation for HSA, as is produced during diabetes. Frontal analysis and competition studies, using warfarin and L-tryptophan as site-selective probes for Sudlow sites I and II of HSA, were carried out with these two drugs on columns that contained normal HSA or HSA with various levels of glycation. These two drugs were found to bind to both Sudlow sites I and II for normal HSA and glycated HSA. The approximate global affinity constants for these two drugs were 3.0 (± 0.7) × 10000 and 2.8 (± 0.5) × 100000 (1/M), respectively. An increase in affinity of 1.6- to 1.7- fold versus normal HSA was seen at Sudlow site I for these drugs when using HSA that had moderate to high levels of glycation. A larger increase of 1.3- to 2.3-fold in affinity was found at Sudlow site II when using the same preparations of glycated HSA. These results indicated that HPAC can be used as a useful tool for examining the interactions of sulfonylurea drugs like chlorpropamide and tolazamide with modified proteins, as can be used to provide a more comprehensive understanding of how glycation can change the protein binding of drugs in blood during diabetes.

Keywords: Bioanalytical, Drugs, HPLC, Immobilization
Application Code: Bioanalytical
Methodology Code: Liquid Chromatography

Session # 990 Abstract # 990-1 **Oral Sessions**

Session Title: Capillary Electrophoresis

Abstract Title: **Monitoring Neurotransmitter Secretion from Islets of Langerhans**

Primary Author: Kimberly Evans
Florida State University

Co-Author(s): Michael G. Roper, Xue Wang

Date: Tuesday, March 08, 2016 - Morning
Time: 08:30 AM
Room: B315

Abstract Text

Islets of Langerhans have been shown to release a variety of small molecules, such as amino acids and neurotransmitters, that help control glucose homeostasis. While initial work has focused on developing methods for measuring amino acid secretions from islets, there has been less research on incorporation of neurotransmitters. Therefore, our goal was to develop a method for resolution of secreted neurotransmitters and amino acids to help understand the role these molecules play in regulating physiology of islets of Langerhans.

The neurotransmitters, L-3,4-dihydroxyphenylalanine (L-DOPA), serotonin, dopamine, L-5-hydroxytryptophan and norepinephrine, and 18 amino acids were derivatized with 2,3-naphthalenedicarboxaldehyde and separated using micellar electrokinetic chromatography with laser induced fluorescence system. To fully resolve the neurotransmitters, the length of the capillary was optimized to 70 cm with 25 [micro]m i.d. and the separation voltage was optimized to 27 kV. During optimization, it was found that earlier peaks were better resolved at lower separation temperatures and later peaks at higher temperatures. To resolve the largest number of components in the sample, a temperature ramp was then incorporated during the separation. The final temperature ramp was 22[degree]C from 0-15 min, 25[degree]C from 15-20 min, and 27[degree]C 20-40 min.

Under these conditions, serotonin, dopamine, L-5-hydroxtrypophan, and L-DOPA were baseline resolved as well as 17 out of 18 amino acids. The increased resolution allows for the measurement of these neurotransmitters in the presence of amino acids from islet secretions.

Keywords: Amino Acids, Capillary Electrophoresis, Optimization, Separation Sciences
Application Code: Bioanalytical
Methodology Code: Capillary Electrophoresis

Session # 990 Abstract # 990-2 **Oral Sessions**

Session Title Capillary Electrophoresis

Abstract Title **Observing Peptide Folding Intermediates Using Capillary Electrophoresis**

Date: Tuesday, March 08, 2016 - Morning
Time: 08:50 AM
Room: B315

Primary Author Alison E. Holliday
Moravian College

Co-Author(s) David E. Clemmer, John D. Barr, Liuqing Shi

Abstract Text

Polyproline folds from an all-<i>cis</i> helical structure (PPI) to an all-<i>trans</i> helical structure (PPII) when immersed in aqueous solution. Due to high energetic barriers, this folding takes place over a much longer timescale than most peptide folding. Recently, intermediates along the folding transition pathway were observed using ion mobility spectrometry-mass spectrometry (IMS-MS) [1]. IMS is a gas-phase separation technique, and molecular dynamics simulations were required to provide a link between solution-phase structures and gas-phase structures produced with electrospray ionization. In contrast, capillary electrophoresis (CE) allows for direct separation and observation of charged species in solution.

Using CE, we have observed intermediates in the transition between PPI and PPII for polyproline-13. Polyproline samples were prepared in 1-propanol to obtain the PPI starting material. Samples were then transferred to an aqueous solution to initiate the transition to PPII. At various time points during the transition, aliquots of the sample were hydrodynamically injected for CE separation. Detection was by UV-Vis.
The transition process observed using CE occurs on the same time-scale as that reported for IMS using the same experimental protocol. Preliminary data analysis indicates that the occurrence of observed intermediates is consistent with IMS data. By bridging the gap between solution and the gas-phase, this research provides key evidence that ion mobility is able to provide information on solution-phase structures.

[1] Shi et al., <i>JACS</i>, <b>136</b>, 12702–12711 (2014).

Keywords: Capillary Electrophoresis, Peptides, Separation Sciences
Application Code: Bioanalytical
Methodology Code: Capillary Electrophoresis

Session # 990 **Abstract # 990-3** **Oral Sessions**

Session Title: Capillary Electrophoresis

Abstract Title: **PDMS Micro-cross Junction for Online Nanoliter Heart-cut for Two Dimensional CE Separation**

Primary Author: Vitaly Avilov
University of Illinois at Chicago

Date: Tuesday, March 08, 2016 - Morning
Time: 09:10 AM
Room: B315

Co-Author(s): Scott A. Shippy

Abstract Text

Two dimensional separations capillary electrophoresis provide great resolving power for complex separations. While there are a number of examples of microchip-based two dimensional electrophoresis separations, there has not been a simple means to utilize cylindrical fused-silica capillary for these separations due to low nanoliter zone volumes. The objective of this project is to develop a method for online CE heart-cut for a two-dimensional electrophoretic separation with a polydimethylsiloxane (PDMS) micro junction heart-cut transfer interface between cylindrical fused-silica capillaries. In this research a 50 µm inner diameter channel PDMS cross junction is molded from glued steel wires. Capillaries of 50 µm inner diameter are installed into the ends of the cross junction and the out-of-plane intersection of the two channels, providing a connection for performing a heart-cut by timing the migration of analytes and switching voltage strengths on the tips of four capillary ends. Different separation buffers can be loaded into the different channels for two dimensional separations. A standard solution of fluorescently labeled amino acids was analyzed through the junction devices to determine effects on peak migration time, peak efficiency, and signal-to-noise. Compared to a contiguous section of fused-silica capillary the PDMS junction provides efficient transfer through the junction compared to capillary
The internal volume of the micro junction was only 20 nL, which helped to avoid longer time migration and radial diffusion of analyte bands. Selected amino acid regions passed perpendicular angle through the micro junction with introduction into a second dimension in volume as small as 200 nL. Heart-cut efficiency in comparison to separation through a solid capillary showed only 5% time lag, 16 % loss in signal strength, and 35% decrease in number of theoretical plates. Two orthogonal buffers were introduced into separate columns of the device, to further separate the heart-cut region in a different buffer in a second dimension.
The proposed method of heart-cut using the developed micro cross junction gives the ability to heart-cut a two hundred nanoliters region of interest, anywhere on separation timeline, with minimal loss of signal, migration time, peak efficiency for continuing separation in a second dimension.

Keywords: Amino Acids, Capillary Electrophoresis, Sample Handling/Automation, Small Samples
Application Code: Bioanalytical
Methodology Code: Capillary Electrophoresis

Session # 990 Abstract # 990-4 **Oral Sessions**

Session Title Capillary Electrophoresis

Abstract Title **Real Time Monitoring of Amino Acid Secretions from Islets of Langerhans Using a Microfluidic Device**

Date: Tuesday, March 08, 2016 - Morning
Time: 09:30 AM
Room: B315

Primary Author Xue Wang
Florida State University

Co-Author(s) Lian Yi, Michael G. Roper

Abstract Text

Gamma-aminobutyric acid (GABA) and glutamate are hypothesized to play a paracrine role in controlling hormone release from islets of Langerhans, the endocrine portion of the pancreas. However, the secretion profile of GABA and Glu, as well as other amino acids (AA's) from islets remains unclear. To measure AA secretions from islets in a time resolved manner, we have developed a microfluidic device which is capable of culturing islets on chip, sampling the secretions, derivatizing the AA's, and performing fast separations to obtain high temporal resolution.

The secretion of AA's were sampled and mixed with naphthalene-2,3-dicarboxaldehyde (NDA) and cyanide by electroosmotic flow. The derivatized AA's were then separated and detected using laser-induced fluorescence. To prevent precipitation of the high salt buffer that islets are held in and the derivatization reagents' buffer, on-chip derivatization conditions were optimized, including the ratio of the reactants and the concentration of the labeling reagents. Ideal conditions were found at 1:1:1 (v:v:v) ratio of AA:NDA:cyanide. The separation conditions, including channel dimensions and separation voltage, were also optimized for resolution of the largest number of AA's. Optimal conditions were obtained with a 5 um deep and 10 cm long channel with a separation voltage of 15 kV. The channel design was also optimized to obtain an electric field of 770 V/cm in the separation channel and to prevent arcing under applied high voltage.

Under these conditions, 15 out of 18 standard AA's were detected and resolved within 2 min, whereas similar results were obtained with 20 min separation time on a commercial CE instrument. The faster separation gave rise to high temporal resolution to measure time-resolved secretion profiles of AA release from islets of Langerhans.

Keywords: Amino Acids, Electrophoresis, Optimization, Separation Sciences

Application Code: Bioanalytical

Methodology Code: Capillary Electrophoresis

Session # 990 Abstract # 990-5 **Oral Sessions**

Session Title Capillary Electrophoresis

Abstract Title **Single-Cell Chemical Cytometry of Akt Activity within Primary Rheumatoid Arthritis Synovial Fibroblasts Illustrates Heterogeneity in Patient Responses to TNF[alpha]**

Date: Tuesday, March 08, 2016 - Morning
Time: 10:05 AM
Room: B315

Primary Author Emilie R. Mainz
University of North Carolina at Chapel Hill

Co-Author(s) Christopher Sims, David Lawrence, Nancy Allbritton, Stephen Serafin, Teresa Tarrant

Abstract Text

The ability to directly screen individual cells for clinically relevant enzyme activity has long been a goal from a diagnostic and drug development standpoint. This is true of Rheumatoid Arthritis (RA), where 30% of patients are unresponsive to treatments targeting Tumor Necrosis Factor Alpha (TNF[alpha]). Akt kinase is a potential biomarker whose aberrant activity has been implicated with TNF[alpha] signaling in RA, and may act as a barometer of patient responses to biologic therapies. We describe an assay that employs peptide sensors and single-cell capillary electrophoresis (CE) to directly measure Akt activity in individual fibroblast like synoviocytes (FLS) from both RA and normal subjects. Peptide sensor was microinjected into single FLS and incubated to allow reporter phosphorylation. Individual cells were then selectively lysed and analyzed via chemical cytometry. Akt activity was quantified in each cell by the ratio of phosphorylated to total reporter. Analysis of Akt reporter in single RA FLS (n=11) showed significant (p = 0.043) elevation in reporter phosphorylation under Akt-stimulating conditions when compared to primary fibroblasts from healthy individuals (n=9). The effect of TNF[alpha] treatment on Akt activity was disparate both between and within multiple RA patients. In 2 subjects, a bimodal distribution of Akt activity ranging from 0-100%(□=44 and 50%, respectively) was observed, while one subject was nonresponsive (□=1%) to TNF[alpha] treatment, highlighting the heterogeneity in Akt activity under conditions mirroring the RA affected joint. The single-cell activity profile in a patient may represent a biomarker indicating which cells would respond favorably to anti-TNF[alpha] therapies. We expect that studies using peptide reporters paired with capillary electrophoresis will provide valuable data regarding aberrant kinase activity from small samples of clinical interest, including additional subjects with RA.

Keywords: Bioanalytical, Biosensors, Capillary Electrophoresis, Enzyme Assays
Application Code: Bioanalytical
Methodology Code: Capillary Electrophoresis

Session # 990 **Abstract # 990-6** **Oral Sessions**

Session Title: Capillary Electrophoresis

Abstract Title: **The Determination of Oxidative Stress Biomarkers of Lipid Peroxidation Using a Novel Capillary Electrophoresis-Mass spectrometry Sheathless Interface Design**

Date: Tuesday, March 08, 2016 - Morning
Time: 10:25 AM
Room: B315

Primary Author: Ryan T. Johnson
University of Kansas

Co-Author(s): Craig E. Lunte, John Stobaugh, Nhan To

Abstract Text

Oxidative stress occurs when there is an imbalance in the body between reactive oxygen species and the species that act to detoxify them. As a result, excess reactive oxygen species degrade important biological molecules such as DNA, lipids, and proteins. Our goal in this work was to develop a method which could investigate the pathways which degrade arachidonic acid, a polyunsaturated fat found in the membrane of the body's cells. Arachandoic acid can undergo degradation via cyclooxygenase (COX) and lipoxygenase (LOX) pathways, which are upregulated during oxidative stress. Biomarkers of the COX pathway are prostaglandins and thomboxanes, while those of the LOX pathway are HETEs and leukotrienes. In order to collect these biomarkers in vivo, microdialysis sampling using a rat model was used. Because dialysate sample volumes are small, a novel sheathless CE-MS/MS Interface design was employed in order to achieve high temporal resolution and specificity. In this design, the electrical connection for the ESI voltage is applied at holes ablated into the CE capillary wall that were then coated with cellulose acetate. The interface proved structurally robust while preventing any sample dilution present with most sheath liquid designs.

Keywords: Bioanalytical, Capillary Electrophoresis, Instrumentation, Method Development
Application Code: Bioanalytical
Methodology Code: Capillary Electrophoresis

Session # 990 Abstract # 990-7 **Oral Sessions**

Session Title: Capillary Electrophoresis

Abstract Title: **Analysis of Metabolite Biomarkers in Prostrate Cancer Tissues by Capillary Electrochromatography Mass Spectrometry**

Date: Tuesday, March 08, 2016 - Morning
Time: 10:45 AM
Room: B315

Primary Author: Yang Lu
Georgia State University

Co-Author(s): Dean Troyer, Shahab S. Shamsi

Abstract Text

Metabolomics or analysis of disease specific metabolite signature has been shown to be a promising discipline for comprehensive measurement of metabolites (e.g., <1.5 kDa) in biofluids. Majority of metabolites are not organism specific. Current Human Metabolome Data Base indicates nearly 40,000 metabolites are registered [1]. Some of the significant challenges in metabolomics includes: (a) large variations in physical and chemical properties of metabolites; (b) need to quantitate metabolite concentrations over a wide dynamic range (millimolar to picomolar); (c) allowing simultaneous quantitation of all metabolites of a certain biochemical group using a single method. Emerging technique of capillary electrophoresis (CE)-MS has recently attracted attention in various analyses due to its low cost and volume limited precious biological samples.

A capillary electrochromatography-tandem mass spectrometry (CEC–MS/MS) method was developed for the simultaneous determination and separation of eight proof of concept (POC) metabolites (betaine, malate, proline, N-acetylaspartate, N-acetylglucosamine, uracil, xanthine, and alanine) as potential prostate cancer diagnostic markers. A polymeric monolith column with hydrophilic crosslinker and strong anion-exchange mixed-mode has been fabricated by an in situ copolymerization of vinylbenzyl trimethylammonium chloride,and bisphenol A glycerolate dimethacrylate (BisGMA) in the presence of methanol and dodecyl alcohol as porogens and AIBN as initiator. After CEC separation, samples were analyzed by a triple–quadrupole mass spectrometer operated in positive ion mode. After optimization, the data showed that the CEC-MS/MS method using monolithic column achieved a much better chromatographic selectivity compared to coated columns and increased sensitivity than bare fused silica column. Future studies are planned to quantitate these metabolites in prostrate cancer tissues.

Keywords: Capillary Electrophoresis, Chromatography, Quadrupole MS
Application Code: Clinical/Toxicology
Methodology Code: Capillary Electrophoresis

Session # 990 **Abstract # 990-8** **Oral Sessions**

Session Title: Capillary Electrophoresis

Abstract Title: **Rapid Determination of Cyanide in Human Urine by Capillary Electrophoresis Coupled with Laser-Induced Fluorescence Detection**

Date: Tuesday, March 08, 2016 - Morning
Time: 11:05 AM
Room: B315

Primary Author: Qiyang Zhang
Wichita State University

Co-Author(s): Maojun Gong, Naveen Maddukuri

Abstract Text

Humans could be exposed to cyanide from smoking, industrial, environmental, and many other sources. For example, tobacco smoking is one of the common sources of cyanide, and the exposure may also occur from smoke due to fires. It is important to develop a method for rapid and accurate determination of cyanide exposure. Here we presented a rapid and sensitive method for the determination of cyanide ions in human urine samples. This method employed online mixing of samples and naphthalene-2,3-dicarboxaldehyde (NDA) and a primary amine (glycine). Mixtures were allowed to react for 2-4 minutes and then separated by an integrated CE system coupled with laser-induced fluorescence (LIF) detection. Conditions of derivatization and separation were optimized, and the separation was observed in 25 seconds. The method of standard addition was used to determine the cyanide levels in urine samples from smokers and nonsmoker. This method was rapid and accurate, and it is suitable for early diagnosis of cyanide exposure.

Keywords: Biological Samples, Capillary Electrophoresis, Fluorescence, Quantitative
Application Code: General Interest
Methodology Code: Capillary Electrophoresis

Session # 1000 Abstract # 1000-1 **Oral Sessions**

Session Title: Electrochemistry - New Approaches and Techniques

Abstract Title: **Stochastic Electrochemistry of TiO[sub]2[/sub] Nanoparticles and Their Agglomerates**

Date: Tuesday, March 08, 2016 - Morning
Time: 08:30 AM
Room: B316

Primary Author: Mario Alpuche-Aviles
University of Nevada, Reno

Co-Author(s): Andrew Recinos, Ganesh Rana, Krishna K. Barakoti, Pushpa Chhetri

Abstract Text

We present the effect of agglomerates on the stochastic photoelectrochemistry of anatase nanoparticles (NPs). Controlling the ionic strength and the concentration of the NPs we promote the formation of larger agglomerates of the nanoparticles in a colloidal suspension. The formation of NPs results in significantly larger photocurrents detected with a microelectrode. We will discuss the magnitude of the currents and its relationship with the 'particle size' obtained during dynamic light scattering. For example, the figure shows the larger spikes that are assigned to the collisions of large agglomerates to the surface of the microelectrode. After the agglomerates are discharged we can observe step-wise changes in the current that are assigned to the agglomerates breaking down into smaller units. We will discuss the possibility of these agglomerates being in equilibrium with NPs in solution and the effect that the illumination time and other parameters have on this equilibrium. The shape of the stochastic photocurrent is a strong function of the microelectrode material and we will present the methods that we use to prepare the microelectrodes for these measurements.

Keywords: Nanotechnology
Application Code: Material Science
Methodology Code: Electrochemistry

Session # 1000 Abstract # 1000-2 **Oral Sessions**

Session Title Electrochemistry - New Approaches and Techniques

Abstract Title **The Development of an Automated NanoElectrode Array Sensor to Detect Evaporation and Changes in Cellular Bioenergetics in a Submicroliter Chamber from an Organ-on-a-Chip System**

Date: Tuesday, March 08, 2016 - Morning
Time: 08:50 AM
Room: B316

Primary Author Anna N. Davis
Vanderbilt University

Co-Author(s) David Cliffel, John Wikswo

Abstract Text

Organ-on-a-chip (OoC) systems are designed to more realistically mimic [i]in vivo[/i] cellular responses compared to traditional two dimensional tissue culturing platforms (such as well plates and culture flasks). The Vanderbilt Institute for Integrative Biosystems Research and Education (VIIBRE), in collaboration with other institutions across the country, has been working to develop different OoCs to monitor responses to pharmaceuticals and environmental toxins. For meaningful information to be gained from long term studies on OoCs, the health of the cells needs to be monitored to observe real-time toxicological events and ensure the viability of additional results produced. Previous work has been shown using modified screen printed electrodes in a 26[micro]L chamber to detect changes in cellular glucose metabolism; however, the volume of effluent produced from smaller OoCs requires lower sample volumes for real-time analysis of cellular energetics. As devices become smaller and are run for longer periods of time, the risk of evaporation grows which could result in cells being supplemented with hypertonic or hyperosmotic media. Cells exposed to such conditions could experience apoptosis, irreversible DNA damage or inhibition of integral enzymatic pathways. When water evaporates from the media solutions, the salt concentrations increase and result in an increase in conductivity as well. A nanoelectrode array has been fabricated using electron beam deposition and soft lithography and modified with electrodeposition and ink-jet printing techniques. The modified sensor was utilized to perform electrochemical detection of glucose, lactate, oxygen, conductivity and acidification in a submicroliter multichannel PDMS sample chamber. This array can now be used to make automated offline measurements from small volume OoCs.

This work was supported by EPA R835736, NIH Grant U01 AI061223, NIH NCATS Grant UH2 TR000491, and the Vanderbilt Institute of Chemical Biology.

Keywords: Biosensors, Electrochemistry, Electrodes, Lab-on-a-Chip/Microfluidics
Application Code: Biomedical
Methodology Code: Electrochemistry

Session # 1000 Abstract # 1000-3 **Oral Sessions**

Session Title: Electrochemistry - New Approaches and Techniques

Abstract Title: **Scanning Electrochemical Microscopy (SECM): A New Tool to Study Microbial Metabolism**

Date: Tuesday, March 08, 2016 - Morning
Time: 09:10 AM
Room: B316

Primary Author: Dipankar Koley
Oregon State University

Co-Author(s): Vrushali Joshi

Abstract Text

Scanning electrochemical microscopy (SECM) has been used to study a wide variety of biological systems including bacterial biofilms. Herein, a new SECM based method has been developed to study the real-time bacterial metabolism at high spatial resolution. New peroxide sensor of submicromolar detection limit has been developed to detect and monitor the microbial metabolites such as hydrogen peroxide produced by Streptococcus Gordonii, an important species of bacteria for oral health. It has been observed that the peroxide is being produced by S. gordonii in a cyclic pattern in oxygen-limited condition, a condition similar to oral microenvironment. To correlate the peroxide production with the change in pH inside biofilm, a solid-state pH sensing SECM probe with near Nernstian slope has also been developed to map the pH inside and outside the gel-encapsulated biofilm. The local pH, both above and inside the biofilm, has been observed to shift towards more acidic as the peroxide production ceases. New findings of how the microbial metabolism is affected by the microenvironment and vice versa would be presented at the meeting.

Acknowledgement: Research reported in this publication was supported by the National Institute Of Dental & Craniofacial Research of the National Institutes of Health under Award Number R21DE025370. The content is solely the responsibility of the authors and does not necessarily represent the official views of the National Institutes of Health.

Keywords: Biosensors, Electrochemistry, Ion Selective Electrodes, Microscopy

Application Code: Biomedical
Methodology Code: Electrochemistry

Session # 1000	Abstract # 1000-4	Oral Sessions

Session Title: Electrochemistry - New Approaches and Techniques

Abstract Title: **Drug Metabolism Assays and Metabolite Synthesis Using Microsomes Based Bioreactor**

Primary Author: Rajasekhara Nerimetla
Oklahoma State University

Co-Author(s): Sadagopan Krishnan

Date: Tuesday, March 08, 2016 - Morning
Time: 09:30 AM
Room: B316

Abstract Text

Liver microsomes are subcellular fractions containing major drug metabolizing enzymes such as cytochrome P450 (CYP) and its redox partner, CYP-NADPH reductase. Liver microsomes are FDA approved in vitro systems to study drug-drug interactions, drug metabolism and inhibition and identify specific isoforms of CYP enzymes. Hence designing microsomes based bioreactors can help in identifying the toxicological and biochemical properties of new drugs under development via drug metabolites. Our prior studies identified surface defects and hydrophilic groups as optimal properties of electrodes to offer stable bioactive microsomal films; recently we also showed high turn-over rates in converting testosterone to 6□hydroxytestosterone stereoselectively by liver microsomes adsorbed on multiwalled carbon nanostructured modified graphite electrodes. Interestingly, the developed green bioreactor showed good stability and reusability. The objective of the present study is to scale-up the bioreactor design to enhance product yields and reusability with proportionate high electrocatalytic currents. Such biocatalytic reactors are useful for examining pharmacokinetic and pharmacological properties of new drugs in development.

Keywords: Electrochemistry
Application Code: Bioanalytical
Methodology Code: Electrochemistry

Session # 1000 Abstract # 1000-5 **Oral Sessions**

Session Title: Electrochemistry - New Approaches and Techniques

Abstract Title: **High-Density Microelectrode Arrays as Electrochemical Imaging Platforms**

Date: Tuesday, March 08, 2016 - Morning
Time: 10:05 AM
Room: B316

Primary Author: Rachel M. Feeny
Colorado State University

Co-Author(s): Charles Henry, John Wydallis, Lang Yang, Stacy Willett, Stuart A. Tobet, Tom Chen

Abstract Text

The ability to generate chemical images with high spatiotemporal resolution is critical to the understanding of biological processes such as chemotaxis and cancer metastasis. Many analytes that are difficult or impossible to image with optical techniques can be monitored electrochemically with high sensitivity, and enzyme-modified electrodes can further expand this library of biomolecules that can be imaged. Using amperometry provides temporal resolution to capture release of biomolecules from live tissue slices. High-density microelectrode arrays are capable of achieving good spatial resolution by using closely packed electrodes as "pixels" in the resulting electrochemical image. The array presented here consists of 8,192 individually addressable working electrodes in 2 mm × 2 mm. The microchip containing this array also houses an on-chip potentiostat, providing simultaneous control of 128 electrodes and the ability to cycle through the entire array at rates up to 30 Hz. The platform interfaces with an upright microscope enabling simultaneous optical and electrochemical imaging to correlate biological processes with chemical gradients. As a model neurotransmitter, norepinephrine distributions were generated through diffusion or microfluidic control and were electrochemically imaged using the high-density array. Oxidation at individual electrodes provides a current with high spatiotemporal resolution, which is expressed as color intensity to generate electrochemical images and videos. Real-time neurotransmitter release from mouse adrenal tissue has been imaged using this system and shows gradients in neurotransmitter release based on chemical stimulation. Electrochemical images and videos of biomolecule distributions and real-time release will be presented. This work is funded by the National Science Foundation.

Keywords: Array Detectors, Biosensors, Imaging, Microelectrode
Application Code: Bioanalytical
Methodology Code: Electrochemistry

Session # **1000** **Abstract #** **1000-6** **Oral Sessions**

Session Title: Electrochemistry - New Approaches and Techniques

Abstract Title: **Imaging of Immobilized Enzymes and Yeast Cells by Scanning Electrochemical Microscopy**

Date: Tuesday, March 08, 2016 - Morning
Time: 10:25 AM
Room: B316

Primary Author: Arunas Ramanavicius
Vilnius University

Co-Author(s): Almira Ramanaviciene, Aura Kisieliute, Inga Vilkonciene, Jurate Petroniene, Povilas Genys, Rita Sareikaite

Abstract Text

Localized evaluation of bio-electrocatalytic activity of redox enzymes immobilized on the surface could be very attractive for biosensor design or development of biofuel cells3. Scanning electrochemical microscopy (SECM) is an innovative method, which could be applied for the surface-activity analysis of enzymatic biosensors. Initially the SECM was designed as a method suitable for the investigation of electrochemically active surfaces. Applicability of SECM in the evaluation of biosensor and biofuel-cell surfaces will be justified. In this presentation experience of authors in the application of SECM in the evaluation of FAD dependent glucose oxidase [1] and yeast cells will be presented. Most efficient ways and SECM modes, which are suitable for the evaluation of enzymatic activity by SECM will be presented. Possibility to apply electrochemical impedance spectroscopy in SECM based evaluation of immobilized enzymes wil be introduced. Investigations of Yeasts Saccharomyces cerevisiae at generation-collection mode of SECM will be outlined. Details in application of two redox mediators based system, which was designed and adapted in order to evaluate redox activity of yeasts, will be sumarized. The most efficient concentrations of redox mediators suitable for the visualization of redox process of yeast cells will be determined.

1. I. Morkvenaite-Vilkonciene, A. Ramanaviciene, A. Ramanavicius, Redox Competition and Generation-Collection Modes Based Scanning Electrochemical Microscopy for the Evaluation of Immobilized Glucose Oxidase Catalysed Reaction. RSC Advances 2014, 4, 50064–50069.
2. I. Morkvenaite-Vilkonciene, P. Genys, A. Ramanaviciene, A. Ramanavicius Scanning Electrochemical Impedance Microscopy for Investigation of Glucose Oxidase Catalyzed Reaction Colloids and Surfaces B-Biointerfaces 2015, 126, 598–602.

Keywords: Electrochemistry, Electrode Surfaces, Electrodes, Enzyme Assays
Application Code: Bioanalytical
Methodology Code: Electrochemistry

Session # 1000 Abstract # 1000-7 **Oral Sessions**

Session Title: Electrochemistry - New Approaches and Techniques

Abstract Title: **Rapid Temperature Control of Electrochemical and Biochemical Systems Using Microfabricated Heaters**

Primary Author: Nicholas M. Contento
Biomolecular Measurement Division

Date: Tuesday, March 08, 2016 - Morning
Time: 10:45 AM
Room: B316

Co-Author(s): Herman O. Sintim, Kurt D. Benkstein, Sarah M. Robinson, Steve Semancik

Abstract Text

Temperature-dependent measurements can provide kinetic and thermodynamic information about chemical and biochemical systems. By determining, for example, DNA melting or protein denaturation temperatures, a biomolecule's thermal stability can be discerned, potentially revealing underlying relationships between molecular stability and structure. Common laboratory techniques, however, require relatively large sample volumes and have slow temperature control (~[degree]C/min).

We have developed micrometer-scale devices using standard microfabrication techniques that allow rapid temperature control (t ~ 1 s) of small sample volumes ([micro]L or smaller). These devices use micro-scale resistive heaters (microheaters) to controllably heat surfaces and proximal solution volumes. This approach is compatible with developing temperature-controlled sensor array platforms for high-throughput detection or (bio)chemical characterization methods.

To enable temperature-controlled electrochemical measurements, Pt disk electrodes (d = 50 [micro]m) were fabricated over Pt microheaters (area ~ 100 x 100 [micro]m[superscript 2]) with an insulating layer between the two active metallic layers. In a proof-of-concept demonstration using both bare and Au-modified electrodes, the electrocatalytic oxidation of hydrogen peroxide, an important chemical intermediate in many enzyme-mediated biosensing strategies, was studied. Temperature-dependent kinetic parameters were extracted from voltammetric data to quantify the performance of different electrode materials and determine their suitability for future biosensing devices.

These devices were also used to perform thermal-shift assays of a model protein, human IgG, both in solution and immobilized on a Au surface, using fluorescence and electrochemical monitoring. The use of a microscale device allows for temperature manipulation (ramping and cycling) at an unprecedented rate, enabling access to shorter timescale phenomena.

Keywords: Biosensors, Fluorescence, Temperature, Voltammetry
Application Code: Bioanalytical
Methodology Code: Electrochemistry

Session # 1000 **Abstract # 1000-8** **Oral Sessions**

Session Title: Electrochemistry - New Approaches and Techniques

Abstract Title: **A Simple, High Yield Method to Microfabricate Planar Microelectrode Arrays**

Date: Tuesday, March 08, 2016 - Morning
Time: 11:05 AM
Room: B316

Primary Author: Mohamed M. Marei
University of Louisville

Co-Author(s): Mark Crain, Richard P. Baldwin, Robert S. Keynton

Abstract Text

Electrochemical analysis is attractive for low cost decentralized applications involving µL samples; thus, large signal-to-noise ratios (S/N) are highly desirable. Although microelectrode arrays (MEAs) offer enhanced S/N, the microfabrication process for planar arrays typically involves photo-lithographic patterning and etching of an insulating layer (such as silicon nitride or silicon dioxide) to reveal selected geometries of the underlying larger area electrode. The final and limiting step of the MEA fabrication process is almost universally dry reactive ion etch (RIE) to anisotropically etch the inert insulating layer. The RIE process requires a complex system to fine-tune stoichiometry and pressure of etchant gases, power supplies to generate ionizing plasmas, vacuum pumps, and other equipment. Depending on the intended application, the size and number of microelectrodes is typically in the 2 to 50 µm range, whereas the number of microelectrodes can range from as few as a 3x3 grid of 9 electrodes to as many as several thousands. The number of active microelectrodes is rarely reported in the literature, and the very few cases reporting this number indicate the RIE process can result in 30% of the intended microelectrodes being electrochemically inactive. In this work, we will present an alternative method to RIE which utilizes HF vapor from a liquid reservoir (no gas cylinder needed) and can be performed in the laboratory fume hood. Despite its extreme simplicity, the process is capable of anisotropic etching of PECVD silicon nitride to yield arrays consisting of several thousand microelectrodes that are essentially 100% active with excellent control over geometry and dimensions. The presentation will overview the etching process and demonstrate the electrochemical activity of the microelectrode arrays. In addition, we look forward to reporting the results of investigations on the surface morphology, analytical performance, and durability of these arrays.

Keywords: Electrochemistry, Electrodes, Lab-on-a-Chip/Microfluidics, Trace Analysis
Application Code: General Interest
Methodology Code: Electrochemistry

Session # 1010 Abstract # 1010-1 **Oral Sessions**

Session Title: Enhancements in Pharmaceutical and Environmental Separations

Abstract Title: **Sub-ppm Level Formaldehyde Measurement in Complex Sample Matrices Using a Variety of Analytical Methods – Method Comparison and Practical Considerations**

Date: Tuesday, March 08, 2016 - Morning
Time: 08:30 AM
Room: B402

Primary Author: Peilin Yang
The Dow Chemical Company

Co-Author(s): Francois Huby, James N. Alexander IV

Abstract Text

Reducing formaldehyde emissions and minimizing formaldehyde levels in commercial products is a global challenge. Many analytical methods have been developed over the years to measure formaldehyde levels including several broadly adopted methods in the US such as the EPA 8315A method based on DNPH derivatization and ASTM D5910-05(2012) method based on post-column derivatization with the Nash reagent. In recent years, more and more strict regulations on formaldehyde were put in place in many countries and regions. We have seen formaldehyde specification on certain products set at low or sub-ppm levels. Due to the widespread occurrence of formaldehyde in the environment and the hydroscopic nature of the molecule, accurate measurement of formaldehyde at the sub-ppm level is very challenging and contamination in the formaldehyde analysis is a common problem. In this presentation, we will show a comparison of different formaldehyde methods and discuss the strengths and limitations of the different methods for low level formaldehyde measurement. A sample preparation guideline will be given based on the recent improvements we made to minimize formaldehyde contamination. The effect of sample matrix on the formaldehyde analysis will also be discussed.

Keywords: Derivatization, GC, HPLC, Sample Preparation
Application Code: Environmental
Methodology Code: Separation Sciences

Session # 1010 Abstract # 1010-2 **Oral Sessions**

Session Title: Enhancements in Pharmaceutical and Environmental Separations

Abstract Title: **Optimization of a Single-Stage, Consumable Free Thermal Modulator for GC x GC**

Date: Tuesday, March 08, 2016 - Morning
Time: 08:50 AM
Room: B402

Primary Author: Haleigh Boswell
University of Waterloo

Co-Author(s): Tadeusz Gorecki

Abstract Text

Comprehensive two-dimensional gas chromatography (GCxGC) has become a leader in the field of analytical separatory techniques for complex volatile and semi-volatile mixtures. A single-stage, consumable free thermal modulator has been developed in order to properly focus and reinject analytes into the second dimension column. Trapping in the modulator occurs with the help of a specially treated coated stainless steel capillary compressed between two passively or actively cooled ceramic pads. Desorption is accomplished via resistive heating using a capacitive discharge power supply. This design allows for rapid cooling after the capacitive heating stage to occur through continuous direct contact with the ceramic cooling pads. Various types of treatment of the coated stainless steel capillary trap have been evaluated to determine the optimal set of conditions for a wide variety of analytes. Investigated conditions include temperature, time, presence of oxygen and pulsed or constant heat. General classes of compounds including alkanes, alcohols, saturated hydrocarbons and aromatic hydrocarbons were examined to assess the overall performance of each set of treatment conditions. The results permit the analysis of a wide range of applications and samples utilizing a single instrumental setup that requires no consumables. The design of the modulator and the results obtained will be presented.

Keywords: Gas Chromatography, Instrumentation, Optimization
Application Code: Quality/QA/QC
Methodology Code: Separation Sciences

Session # 1010 Abstract # 1010-3 **Oral Sessions**

Session Title: Enhancements in Pharmaceutical and Environmental Separations

Abstract Title: **Thermo Tuning of Redox Potential on Nanostructured Adsorbent for a Reagent-less Recovery of Pollutant Oxoanions**

Date: Tuesday, March 08, 2016 - Morning
Time: 09:10 AM
Room: B402

Primary Author: Manuel Valiente
Universitat Autonoma de Barcelona

Co-Author(s): Gustavo Perez, He Liu, Liu Tong

Abstract Text

The present study reports a new reagentless method to clean separation (removal and recovery) of pollutant oxoanions (e.g, arsenic). The process is able to regenerate the adsorbent by tuning the temperature and redox conditions. Nanostructured adsorbent based on Super Paramagnetic Iron Oxide Nanoparticles, SPION, has been developed on a sponge support with convective properties. In the case of Arsenic oxoanions, the temperature-dependence of adsorption by SPION loaded sponge-has revealed adsorption of As(V) to take place at 20 C while at 70 C desorption process is produced at the same time of the regeneration of the nanostructured adsorbent. The effect of temperature on the adsorption of As(V) was studied in the continuous column mode by evaluating the adsorption at the temperature 10oC, 20oC and 70oC. The adsorption phenomenon increases with the decrease of temperature. On the other hand, As(III) shows very low adsorption and practically no temperature dependence in the range studied. Thus, tuning redox conditions can be used on the adsorption/desorption process. Results on the synergistic interaction of temperature and redox potential will be discussed. The results of adsorption are explained by the Langmuir, Freundilich models. Thermodynamics parameters of related ion exchange equilibrium have been evaluated including the equilibrium constant K as well as Gibbs free energy. The thermodynamic affinity defined by log K=4.20 and log K=1.02 under 20oC and 70oC were determined, respectively. The obtained results indicate the possible use of this system to both removal/recovery of oxoanions and the regeneration of the nanostructured adsorbent.

Keywords: Environmental/Water, Ion Exchange, Nanotechnology, Thermal Desorption
Application Code: Environmental
Methodology Code: Separation Sciences

Session # **1010** **Abstract #** **1010-4** **Oral Sessions**

Session Title: Enhancements in Pharmaceutical and Environmental Separations

Abstract Title: **Super/Subcritical Fluid Chromatographic Chiral Separations with High Efficiency Chiral Stationary Phases**

Primary Author: Zachary Breitbach
The University of Texas at Arlington

Date: Tuesday, March 08, 2016 - Morning
Time: 09:30 AM
Room: B402

Co-Author(s): Chandan L. Barhate, Daniel W. Armstrong, M Farooq Wahab

Abstract Text

Recently, we have developed a series of high efficiency HPLC chiral stationary phases based on macrocyclic glycopeptide and cyclofructan chiral selectors. The improved efficiencies observed in HPLC were afforded by the use of i) superficially porous particles (SPPs) or ii) sub-2um fully porous particles (FPPs) with narrow particle size distributions (NPSD). Both the SPP and NPSD chiral columns exhibited over 200,000 plates per meter when tested in HPLC. In this study, the chiral recognition capabilities and efficiencies of these high efficiency chiral columns were evaluated with supercritical and subcritical fluid mobile phases. The usage of carbon dioxide as a mobile phase component to replace either heptane (for transferring normal phase separations) or acetonitrile (for transferring polar organic separations) is evaluated. The efficiency of these columns will be compared to the state of the art commercial columns composed of the same chiral selectors. Additionally, the efficiency gains between the SPP based chiral phases and the sub-2um NPSD based phases will be discussed. Finally, the ability to perform ultra-fast (i.e. <1 min) chiral separations using these columns in SFC will be presented along with a discussion of instrumental concerns that must be addressed when performing such highly efficient and fast chiral separations.

Keywords: Chiral, Chiral Separations, Separation Sciences, Supercritical Fluid Chromatography
Application Code: Pharmaceutical
Methodology Code: Separation Sciences

Session # 1010 Abstract # 1010-5 **Oral Sessions**

Session Title: Enhancements in Pharmaceutical and Environmental Separations

Abstract Title: **Novel Solid-Phase Microextraction and Capillary Electrochromatographic Column Techniques for Pharmaceutical Analysis**

Primary Author: Zilin Chen
Wuhan University

Date: Tuesday, March 08, 2016 - Morning
Time: 10:05 AM
Room: B402

Co-Author(s)

Abstract Text

Solid-phase microextraction and capillary electrochromatographic column technology are very import research areas for pharmaceutical analysis. The major problems in pharmaceutical analysis are the low contents of target components and interference from complex matrix in real sample. Solid phase microextraction and capillary electrochromatographic separation are useful ways to solve these problems. Solid-phase microextraction can selectively extract the target analyte and cleanup the interference in matrix. Capillary electrochromatography can separate the complex multiple components with high performance and selectivity. Mussels attach to solid surfaces such as rocks in the sea. The tight adhesion results from the adhesive protein containing polydopamine secreted by mussels. Inspired by the composition of adhesive proteins in mussels and combined with the chemical self-assembly of multilayer grapheme and mental organic frame materials on chemically resistant plastic micro-tubes such as PEEK and stainless steel tubes, we have successfully developed several novel solid phase micro-extraction [1-5] and capillary electrochromatographic columns [6-8] and applied in environmental and pharmaceutical analysis. This talk will introduce the preparation, characterization, validation and application of novel solid-phase microextraction and open tubular capillary electrochromatographic columns based on mussel-inspired polydopamine functionalization and growth of metal organic frame materials in the inner wall of fused silica capillary using liquid-phase epitaxy.

Keywords: Capillary Electrophoresis, Pharmaceutical, Separation Sciences, Solid Phase Extraction
Application Code: Pharmaceutical
Methodology Code: Separation Sciences

Session # 1010 Abstract # 1010-6 **Oral Sessions**

Session Title: Enhancements in Pharmaceutical and Environmental Separations

Abstract Title: **Retention in Porous Layer Pillar Array Planar Separation Platforms**

Date: Tuesday, March 08, 2016 - Morning
Time: 10:25 AM
Room: B402

Primary Author: Danielle R. Lincoln
University of Tennessee, Knoxville

Co-Author(s): Michael J. Sepaniak, Nickolay V. Lavrik

Abstract Text

This work presents the retention capabilities and surface area enhancement of highly ordered, high-aspect ratio, open-platform pillar arrays when coated with a thin layer of porous silicon oxide (PSO). Separations were performed using capillary action-driven flow on these planar platforms. The photolithographically-fabricated pillar arrays were coated with 50-150 nm of PSO via plasma enhanced chemical vapor deposition and then functionalized in the liquid phase using octadecyltrichlorosilane or in the gas phase using n-butyldimethylchlorosilane. Theoretical calculations indicate that a 100 nm layer of PSO increases the surface area of a pillar nearly 100-fold, while experimental results indicate that increasing the PSO thickness from 50 to 100 nm increases the retentive surface by a factor of about 9. Retention capabilities were tested by observing analyte development under various conditions, as well as by running one-dimensional separations on varying thicknesses of PSO. Increasing the thickness of PSO on an array clearly resulted in greater retention of fluorescent dyes without deleterious effects on efficiency. A two-dimensional separation of fluorescent dyes and mycotoxins was also pursued.

Keywords: Chromatography, Modified Silica, Nanotechnology, Thin Layer Chromatography
Application Code: Nanotechnology
Methodology Code: Separation Sciences

Session # **1010** **Abstract #** **1010-8** **Oral Sessions**

Session Title Enhancements in Pharmaceutical and Environmental Separations

Abstract Title **Analytical Method Development: Are We Solving the Right Problem? A Systematic Approach to Select an Appropriate RPLC Column and to Optimize Separation**

Date: Tuesday, March 08, 2016 - Morning
Time: 11:05 AM
Room: B402

Primary Author Imad A. Haidar Ahmad
Novartis

Co-Author(s) Andrei Blasko, James Tam, Thomas Tarara, Xue Li

Abstract Text

A systematic rational approach for analytical method development was assessed as a better alternative to non-comprehensive approaches when selecting a chromatography column and optimizing the separation. The non-comprehensive approaches based trial and error, one-factor-at-a-time experimentation, are usually suitable for simple separations. They often do not consider column selectivity and the interplay between mobile phase gradient and column temperature, which requires the aid of specialized software to build a modeled matrix to optimize these conditions. This study demonstrates how to use the hydrophobic-subtraction model to carefully build a library of chromatography columns comprising multiple stationary phases, such as: C18, functionalized C18, polar embedded group, phenyl, phenyl hexyl, and biphenyl. Columns with different selectivities were used to achieve the best separation of the critical resolution pairs using ACD/Labs AutoChrom MS software with automated MS and UV peak tracking (Advanced Chemistry Development, Inc.). By screening the columns in our library along with the use of AutoChrom MS software, we were able to successfully optimize an extremely challenging 25-peak separation for a stability-indicating RPLC method in a 30 minute gradient.

Keywords: Chromatography, Liquid Chromatography
Application Code: Pharmaceutical
Methodology Code: Separation Sciences

Session # 1020 Abstract # 1020-1 **Oral Sessions**

Session Title: Environmental Air Quality and Analysis

Abstract Title: **On-Site Determination of Formaldehyde Using SPME and a Portable GC-TMS**

Date: Tuesday, March 08, 2016 - Morning
Time: 08:30 AM
Room: B403

Primary Author: Justen J. Poole
University of Waterloo

Co-Author(s): German A. Gomez-Rios, Janusz Pawliszyn, Jonathan J. Grandy

Abstract Text

Formaldehyde is a carcinogenic gaseous molecule which can originate from a wide variety of anthropogenic and natural matrices. Due to thermal instability it has proven difficult to quantify using classical analytical techniques, such as gas chromatography where dimerization can be observed. However, with increased concerns regarding human exposure to this difficult to analyze molecule, it is imperative that on-site analytical methods be developed for the screening and quantitation of formaldehyde in real time. To facilitate this goal, a pentafluorophenyl hydrazine (PFPH) gas generating vial was developed and used to pre-load PFPH onto solid phase microextraction (SPME) fibers in order to perform on-fiber derivatization of aldehydes. Initial experiments using C4-C9 linear aldehydes as markers demonstrated that the derivatization reaction was reproducible. The on-fiber derivatization was also shown to demonstrate a strong correlation when headspace extractions of aqueous solutions spiked with aldehydes C4-C9 at 10-200ppb/v. The developed method was then coupled to a portable GC-TMS and applied to the on-site, semi-quantitative determination of formaldehyde from car exhaust. A novel field portable vial heater was also utilized to control the temperature of the PFPH gas generating vial, facilitating rapid and reproducible SPME fiber loadings of PFPH. Using the proposed method, on-site determination of formaldehyde could be accomplished in less than 15 minutes.

Keywords: Derivatization, Gas Chromatography/Mass Spectrometry, Portable Instruments, SPME
Application Code: Environmental
Methodology Code: Gas Chromatography/Mass Spectrometry

Session # 1020	Abstract # 1020-2	Oral Sessions

Session Title: Environmental Air Quality and Analysis

Abstract Title: **Fast, Accurate, and Precise: How to Comply with EPA Method 325b (Fence Line Monitoring for Benzene)**

Date: Tuesday, March 08, 2016 - Morning
Time: 08:50 AM
Room: B403

Primary Author: Lee Marotta
PerkinElmer

Co-Author(s): Amy Jacobson, Mariah Peronto, Roberta Provost

Abstract Text

The Environmental Protection Agency (EPA) has developed a new method for the passive monitoring of several volatile components in air. Benzene is the compound that will be regulated at the fence line of most refineries, largely because of its impact on human health. There is a mandated sampling time of two weeks. The objectives of this new method and regulation will be discussed.

This presentation describes how to comply with the new EPA fence line regulations, and the steps taken to optimize this method for high sample throughput. Presented by the support team utilized by the EPA in their development of this method, validation steps will be discussed.

Refineries and testing laboratories will benefit by gaining valuable information on the optimum techniques to perform this method and what it takes to comply. An introduction to the theory and operation of thermal desorption will be included.

The data from several site studies will be reviewed. These samples were collected on the same tubes used by the EPA in their development, and analyzed using the same concentrator trap and automated thermal desorber.

Method criteria including minimum detection limits (MDLs), action limits, accuracy and precision will be demonstrated. The presentation will also include optimization of the method for high sample throughput, and elimination of false positives.

Keywords: Air, Environmental/Air, Fuels\Energy\Petrochemical, Thermal Desorption
Application Code: Environmental
Methodology Code: Gas Chromatography/Mass Spectrometry

Session # 1020 Abstract # 1020-3 **Oral Sessions**

Session Title: Environmental Air Quality and Analysis

Abstract Title: **Investigation and Modelling of the Sampling Process in a PDMS-Based Permeation Passive Sampler**

Date: Tuesday, March 08, 2016 - Morning
Time: 09:10 AM
Room: B403

Primary Author: Faten Salim
University of Waterloo

Co-Author(s): Marios Ioannidis, Tadeusz Gorecki

Abstract Text

Waterloo Membrane Sampler (WMS) is a permeation passive sampler utilizing a polydimethylsiloxane (PDMS) membrane as the uptake rate-determining barrier. The membrane covers the mouth of a small glass vial containing a sorbent material. The sampler has been used for measuring the time weighted average (TWA) concentrations of volatile organic compounds (VOCs) in air and soil gas. The TWA concentrations reflect the average concentrations over the time of exposure. The measurement of these concentrations is based on the assumption that the sampling rates are constant at a given temperature. In this ideal behaviour, the sorbent is assumed to behave as a perfect sink and the mass uptake is controlled by the resistance to mass transport in the membrane. Deviations from this linear behaviour have been thought to only occur as the sorbent approaches saturation. However, experiments revealed that during sampling, the concentrations of the collected analytes increase significantly near the interface between the membrane and the sorbent, while the rest of the sorbent bed remains relatively clean. This non-uniform distribution of the analytes within the sorbent bed would lead after a period of time to a change in the sampling rates as an additional resistance to mass transport is created within the sorbent bed itself. Therefore, a mathematical model of the mass transport inside the sampler has been developed to allow better understanding of the sampling process, optimization of the sampler design, and most importantly, prediction of any possible change in the sampling rates and its significance in different sampling conditions.

Keywords: Environmental Analysis, Sampling, Volatile Organic Compounds
Application Code: Environmental
Methodology Code: Sampling and Sample Preparation

Session # 1020 Abstract # 1020-4 **Oral Sessions**

Session Title: Environmental Air Quality and Analysis

Abstract Title: **QCL-Based Perfluorocarbon Emission Monitoring**

Date: Tuesday, March 08, 2016 - Morning
Time: 09:30 AM
Room: B403

Primary Author: Nicola Menegazzo
Alcoa Inc

Co-Author(s): Luis Espinoza-Nava

Abstract Text

Extractive sampling Fourier transform infrared s(FTIR) pectrometers are often employed for measuring perfluorocarbon (PFC) emissions at aluminum smelters during limited duration (as few days or weeks) test periods. The data from these limited duration measurement are then employed predict subsequent emissions using plant logs of pot voltage. Purpose-designed measurement systems could facilitate long-term in-situ stack measurement of CF4 of plants with one or few scrubber stacks. This work discusses paired in-plant measurements by FTIR and quantum cascade laser (QCL) based optical systems employed to develop future QCL measurement platforms to optimize the latter for long term stack monitoring.

Keywords: Environmental Analysis, Environmental/Air, FTIR, Specialty Gas Analysis
Application Code: Environmental
Methodology Code: Molecular Spectroscopy

Session # 1020 Abstract # 1020-5 **Oral Sessions**

Session Title Environmental Air Quality and Analysis

Abstract Title **Laser Derivitization for Soot Source Identification**

Date: Tuesday, March 08, 2016 - Morning
Time: 10:05 AM
Room: B403

Primary Author Randy Vander Wal
Penn State University

Co-Author(s) Chethan K. Gaddam

Abstract Text

Combustion produced soot is highly variable with details as dependent upon combustion conditions. Our prior studies have shown soot nanostructure to be dependent upon the source via quantification of HRTEM images for nanostructural parameters. In principle this permits identification of the soot source and its contribution to any particular receptor site. Yet many structural aspects are subtle and the chemistry of and between the lamella is unaddressed. How best to bring out small differences in nanostructure and other seemingly subtle differences in chemistry? We proposed the process of pulsed laser annealing to highlight compositional and structural differences thereby distinctively and uniquely identifying the source of the soot. Our overall objective is then to develop the laser-based heating as an analytical tool and identify the process conditions and operational parameters for optimal derivitization. Specific tasks directed towards achieving this goal include 1) Identifying optimal laser operational parameters for derivitization, 2) Defining the dependence upon nanostructure and composition using model soots while also identifying variability and range of products for uniform and homogeneous starting material, 3) Demonstrating differentiation upon combustion derived soots from real engines such as diesel, gasoline, boilers, etc., and 4) Quantifying the nanostructural changes and statistical differences between the derivatives of these actual samples to formulate statistical confidence intervals for nanostructure parameters of the derivatized soots. Results will be presented accordingly.

Keywords: Aerosols/Particulates, Analysis, Environmental Analysis, Environmental/Air

Application Code: Environmental

Methodology Code: Microscopy

Session # 1020 Abstract # 1020-6 **Oral Sessions**

Session Title: Environmental Air Quality and Analysis

Abstract Title: **Analysis of Damaged Floor Coverings Emissions in Indoor Air Quality with Cantilever-Enhanced Photoacoustic Spectroscopy**

Date: Tuesday, March 08, 2016 - Morning
Time: 10:25 AM
Room: B403

Primary Author: Jussi Raittila
Gasera Ltd.

Co-Author(s): Ismo Kauppinen, Jaakko Lehtinen, Sauli Sinisalo

Abstract Text

The damage in the floor coverings due to exceptional moisture is a common indoor air problem in both new and older buildings. The emissions of the damaged coverings lead often to several, mostly irritational symptoms to the users of the building. The exceptional moisture under the covering results in the chemical degradation of the flooring and its adhesive. Among other emissions, which are mostly Volatile Organic Compounds (VOC), there are some indicator compounds revealing the damage: 2-ethyl-1-hexanol in older coverings and C9 alcohols in newer coverings.

In case the flooring damage is not obvious by sensory observations or moisture measurements, the condition of the flooring is investigated by VOC sampling of the indoor air or the Field and Laboratory Emission Cell (FLEC) emission measurements on the surface of the covering. The sampling is very time-consuming and expensive and therefore the amount of samples are usually limited even though the surface area, where the damages are suspected is usually large. The laboratory analysis of the samples usually takes several weeks, which generates a need for a portable analyzer for fast analysis in the field.

The detection limits of the indoor air VOC samples depend on the investigated compound and are usually in the range of 0.02-0.2 ppb. Sub-ppb detection limits set a requirement for ultra-sensitive detection technique. Cantilever-enhanced photoacoustic spectroscopy (CEPAS) combined with widely tunable infrared sources can meet the requirements for sensitivity and the technology can be packed to portable, and eventually into hand-held size. The sensitivity of the analysis comes from the combination of optical cantilever microphone and powerful infrared laser sources and the selectivity comes from the high resolution of the lasers. Sub-ppb level detection limits for 2-ethyl-1-hexanol were demonstrated for a reliable floor coverings measurement.

Keywords: Environmental/Air, Trace Analysis, Volatile Organic Compounds
Application Code: Environmental
Methodology Code: Molecular Spectroscopy

Session # 1020 **Abstract #** 1020-7 **Oral Sessions**

Session Title: Environmental Air Quality and Analysis

Abstract Title: **Ship Emissions Monitoring with Laser-Based Cantilever-Enhanced Photoacoustic Detection**

Date: Tuesday, March 08, 2016 - Morning
Time: 10:45 AM
Room: B403

Primary Author: Jaakko Lehtinen
Gasera Ltd.

Co-Author(s): Ismo Kauppinen, Jussi Raittila

Abstract Text

The new environmental regulations for ship emissions monitoring in sulfur emissions control areas (SECAs) will generate a global need for new measurement tools in the near future. SECA includes the Baltic Sea, the North Sea, the North American ECA, including most of US and Canadian coast and the US Caribbean ECA. The new regulations force cargo ships to use fuel that contains notably lower levels of sulfur than previously. New regulations cost around $45 billion per year to the shipping industry and the shipping companies can save around $10000 per day per ship by using illegal higher sulfur level fuel. Emissions monitoring is required to detect the use of illegal fuel.

The sulfur concentration of the fuel used by a ship can be calculated by measuring the ratio of carbon dioxide (CO2) and sulfur dioxide (SO2) from the emissions. Current technologies that are used in this application are UV fluorescence spectroscopy for the detection of SO2 and cavity ringdown spectroscopy for the detection of CO2. Currently there is no standard method for the detection of the CO2/SO2 ratio and using two different methods increases the inaccuracy and unreliability in the measurement.

Gasera addresses the emissions monitoring need with laser-based photoacoustic detection (1,2). A combination of quantum cascade laser (QCL) and diode laser is used to accurately measure small changes in background sulfur dioxide and carbon dioxide concentrations. A narrow linewidth QCL is used to measure SO2 absorption lines in mid-infrared region. High resolution laser spectroscopy is needed as these lines are normally buried under stronger absorption lines of water. CO2 is measured in the near-infrared region. Sub-ppb level detection limits can be achieved with a combination of high power QCL and ultra-sensitive cantilever sensor.

[1] T. Kuusela, J. Kauppinen. Appl. Spectrosc. Rev., 42, (2007)
[2] C. B. Hirschmann, J. Lehtinen, J. Uotila, S. Ojala, R. L. Keiski, Appl. Phys. B (2013)

Keywords: Environmental/Air, Laser, Monitoring, Vibrational Spectroscopy
Application Code: Environmental
Methodology Code: Vibrational Spectroscopy

Session # 1020 Abstract # 1020-8 **Oral Sessions**

Session Title: Environmental Air Quality and Analysis

Abstract Title: **Hand Held Detector Based on an Ion Mobility Spectrometer and an Additional Detector (Electrochemical Cell or Alternatively a Photo Ionization Detector) for the Detection of Toxic Gases in Chemical Industries**

Date: Tuesday, March 08, 2016 - Morning
Time: 11:05 AM
Room: B403

Primary Author: Andreas Walte
Airsense Analytics

Co-Author(s): Bert Ungethuem, Wolf Muenchmeyer

Abstract Text

Toxic chemicals are very often needed in chemical industries. Therefore the detection of small leakages is needed in order to protect life and the environment. Several accidents in the past show the great potential of damage which is possible. A small handheld system which can monitor the chemicals processed is needed. Combinations of different gas detectors are often needed if more than one chemical has to be detected. By a specific combination of gas detectors the handheld system can be customized in order to detect only the toxic gases that are present in the specific industrial plant. Ion mobility spectrometers (IMS) are systems which can detect many different toxic industrial chemicals in very low concentrations. Unfortunately they cannot detect all toxic gases and also at the required exposure limits. Therefore additional detectors are often needed. Depending of the chemicals in use different detectors have to be added to the IMS based system.
We will discuss the advantages of using a hand held detection system (GDA-P) based on an ion mobility spectrometer (IMS) in combination with another orthogonal sensor such as a dedicated electrochemical cell (EC) or alternatively a photo ionization detector (PID). The PID can be used as an additional general VOC detector, capable to detect different hydrocarbons, such as aromatic compounds, which cannot be tested via IMS. Depending on the application also electrochemical cells can be integrated. Different EC are available, for example for the detection of CO, H_2S, NH_3, PH_3, AsH_3 or formaldehyde. It is desirable to detect leakages at an early stage by the hand held system based on the combination of the IMS with the second detector, which has to be chosen depending on the requirements. Measurements with the systems for different industries will be shown.

Keywords: Instrumentation, On-line, Portable Instruments, Volatile Organic Compounds
Application Code: Safety
Methodology Code: Portable Instruments

Session # 1030 Abstract # 1030-3 **Oral Sessions**

Session Title: Magnetic Resonance - Half Session

Abstract Title: **Quantitative Analysis of Phosphorous Containing Drug Molecules Encapsulated in pH-Sensitive Nanoparticle Formulations to Establish Drug Loading and Drug Releasing Profile by Utilizing 31P Solid State and Solution State NMR Spectroscopy**

Date: Tuesday, March 08, 2016 - Morning
Time: 10:45 AM
Room: B408

Primary Author: Sudhaunshu S. Purohit, University of Missouri Kansas City

Co-Author(s): Bi Botti Celestin Youan, Jianing Meng, Nathan Oyler, Vivek Agrahari

Abstract Text

The ability to accurately quantify the amount of drug being delivered in human body is a crucial requirement of a drug development process. Understanding the pharmacokinetics is vital when launching new drug-body profile as it offers a quantitative outline for the prospective optimization of therapeutic dosage treatments. We are focusing on phosphorous containing drugs that are being effectively used against diseases such as AIDS, Hepatitis, Cancer, Alzheimer etc. as a part of preventive treatment. We developed and implemented a general 31P qNMR method to achieve direct, real time quantification of in vitro drug release. We utilized both solution state and solid state 31P qNMR spectroscopic techniques to establish the kinetics of drug release and to determine the encapsulation efficiency of nano-formulation for a drug under study. Taking into account the principles of 31P qNMR, the proposed method can be extensively applied to all phosphorous containing drug molecules, signifying a subsequent huge scope of this method. The in vitro drug release profile will be studied in various human body fluids such as simulated vaginal and seminal fluids, blood, plasma etc.

Our study started with AIDS and Tenofovir (TFV), a well-tested antiretroviral drug with proven mettle against AIDS. We included a specific type of spray dried, mucoadhesive, pH sensitive nano-formulation, developed by Zhang T. et al. that will serve as an encasing for TFV. The respective nano-formulation is efficient in administering TFV in human body through the epithelial cells of human genitals. We introduced 31P qNMR spectroscopic method for quantification purposes. The preliminary results of method validation parameters for TFV in simulated vaginal, seminal fluids and plasma obtained by using 31P solution state qNMR spectroscopy and, the encapsulation efficiency of the respective nano-formulation encasing TFV in solid form was estimated by integrated 31P (144.597 MHz) Bloch decays are presented.

Keywords: Drugs, Identification, NMR, Pharmaceutical
Application Code: Pharmaceutical
Methodology Code: Magnetic Resonance

Session # 1030 Abstract # 1030-4 **Oral Sessions**

Session Title: Magnetic Resonance - Half Session

Abstract Title: **Nanoparticle-Assisted Removal of Proteins in Human Serum for Metabolomics Studies**

Date: Tuesday, March 08, 2016 - Morning
Time: 11:05 AM
Room: B408

Primary Author: Bo Zhang
The Ohio State University

Co-Author(s): Lei Bruschweiler-Li, Mouzhe Xie, Rafael Brüschweiler

Abstract Text

Metabolite identification and measurement of their concentrations and perturbations under different conditions is used routinely for biomarker identification, toxicity evaluation and disease diagnosis. Serum, as a body fluid, poses significant challenges for comprehensive and accurate identification of metabolites because its high protein content can strongly interfere with metabolite signal detection and quantitation. Considering the importance of human serum for diagnostic tests and for metabolomics analysis, the lack of standard operating procedure for the elimination of protein from human serum samples is rather urgent.

Here, we introduce novel procedures for the removal of protein from serum by the addition of nanoparticles. Nanoparticles have been found to have great potential in biomedical implications due to their size- and shape-tunable physicochemical properties. Recently, electrically charged silica nanoparticles have been applied in metabolomics as a tool to simplify NMR spectra of complex mixtures and to directly sense the charge of metabolites. Alternatively, the interactions of proteins and silica nanoparticles with a negatively charged surface cause significant attractive interactions with proteins, which leads to an efficient, cost-effective, and environmentally-friendly procedure for protein removal from serum in an aqueous environment at physiological pH. It is further shown that serum can be processed with nanoparticles prior to ultrafiltration or organic solvent induced protein precipitation complementarily for optimal protein removal. These approaches will be demonstrated and their efficiency will be discussed.

Keywords: Method Development, Nanotechnology, Metabolomics, Metabonomics
Application Code: Bioanalytical
Methodology Code: Magnetic Resonance

Session # 1040 Abstract # 1040-1 **Oral Sessions**

Session Title: Mass Spectrometry - Bioanalytical

Abstract Title: **Electrospray Ionization-High Pressure Mass Spectrometry for Peptide and Protein Analysis**

Date: Tuesday, March 08, 2016 - Morning
Time: 08:30 AM
Room: B404

Primary Author: Russell E. Bornschein
University of North Carolina at Chapel Hill

Co-Author(s): J Michael Ramsey, William M. Gilliland

Abstract Text

Peptides and proteins are the workhorses of biological systems performing major mechanical functions within organisms such as regulating transcellular transport and carrying oxygen through the blood stream to vital organs. These processes rely heavily on highly regulated equilibria. Many diseases shift these equilibria resulting in measurable changes in peptide and protein concentrations. Mass spectrometry is a central analytical method to proteomics, capable of monitoring relative concentrations and reliably identifying unknown peptides and proteins. Rapid detection of volatile organic compounds via handheld mass spectrometry is now possible due to the advent of high pressure mass spectrometry (HPMS), while peptide and protein identification for disease diagnostics relies upon transport of biological samples to laboratories for analysis. The goal of this work is to combine HPMS with electrospray ionization (ESI-HPMS) to provide field-deployable, in-clinic peptide and protein detection and identification.

HPMS utilizes ion trap technology operating a sub-millimeter cylindrical or stretched-length ion trap (CIT or SLIT) driven at relatively high frequency (>1 MHz) at □1 Torr. These operating conditions eliminate the need for turbomolecular pumps and reduce the size, weight, and power of the system. A microfluidic chip with an integrated ESI emitter is used. These devices are ideal for separation of complex mixtures prior to HPMS as they are compact, require minimal sample, and operate at low flow rates. Initial HPMS development focused on analysis of smaller compounds (<300 m/z); biological analytes are larger and require exploration of the parameter space for effective analysis. The focus of this work has been expanding the mass range to perform bottom-up proteomics analysis. Currently, analytes up to 1900 m/z are detected. Differences between analysis of smaller and larger compounds as well as unique considerations for the ESI-HPMS platform will be discussed.

Keywords: Electrophoresis, Electrospray, Ion Trap, Protein
Application Code: Bioanalytical
Methodology Code: Mass Spectrometry

Session # 1040 Abstract # 1040-2 **Oral Sessions**

Session Title: Mass Spectrometry - Bioanalytical

Abstract Title: **Current Analytical Techniques for Glycoprotein Characterization by Mass Spectrometry**

Primary Author: Parastoo Azadi
Complex Carbohydrate Research Center

Date: Tuesday, March 08, 2016 - Morning
Time: 08:50 AM
Room: B404

Co-Author(s)

Abstract Text

The Analytical Services Laboratory of the Complex Carbohydrates Research Center (CCRC) at The University of Georgia is a non-profit entity that offers services for structural characterization of glycoconjugates derived from animal, plant, or microbial origin. The Analytical Services Group routinely analyzes samples from a wide variety of institutions including universities, federal agencies, and industry groups from the US and other countries.

The analysis of the N/O-glycan portions of glycoproteins has become a crucial step in comparability studies as well as the quality control of therapeutic recombinant glycoproteins. We will discuss procedures necessary for the complete structural elucidation of any N/O-glycan mixture found in glycoprotein products as well as sites of glycosylation. Featured in this talk are new analytical developments in the structural elucidation of glycopeptides by LC-MS. This talk also serves to demonstrate both the benefits of glycoprotein analysis as well as number of challenges that are frequently experienced as part of glycopeptide mapping and glycan analysis.

Keywords: Biopharmaceutical, Carbohydrates, Liquid Chromatography/Mass Spectroscopy, Mass Spectrometry
Application Code: Bioanalytical
Methodology Code: Mass Spectrometry

Session # **1040** **Abstract #** **1040-3** **Oral Sessions**

Session Title: Mass Spectrometry - Bioanalytical

Abstract Title: **Nanopatterning Ligands to Enable Cell-Based Assays Using SAMDI-Mass Spectrometry**

Date: Tuesday, March 08, 2016 - Morning
Time: 09:10 AM
Room: B404

Primary Author: Maria D. Cabezas, Northwestern University

Co-Author(s): Chad A. Mirkin, Milan Mrksich

Abstract Text

Designing effective cell-based assays that allow cell culture preparation and protein analysis on the same chip would facilitate quantification of molecular activities and would significantly reduce the number of cells required for an assay. Achieving this goal remains a challenge for current methods used to measure enzyme activities, which often rely on time-consuming sample preparation steps, lead to loss of enzymatic activity and therefore limit the validity of the readout. We have been successful in developing a cell-based assay whereby a monolayer presenting both an RGD ligand and a peptide substrate for an enzyme can be used for the simultaneous culture of cells and analysis of enzyme activity. In this work, we expand the capabilities of our previous design to introduce extracellular matrix (ECM) attachment proteins, such as fibronectin, laminin and collagen, which would allow culture of a wide range of cell types. Polymer pen lithography was first used to pattern a template alkane thiolate on a gold-coated surface that would direct ECM protein attachment to discreet locations in an arrangement that is commensurate with the dimensions of a cell. The remaining surface was then functionalized to display short peptide ligands that would act as substrates for a desired enzyme reaction. In this manner, we combined surface chemistry amenable for cell culture, along with self-assembled monolayer laser desorption/ionization mass spectrometry (SAMDI-MS), thus enabling cell culture and analysis on the same monolayer-coated surface. Our results reveal that cell attachment is directed toward the patterned regions while the phosphatase reporter ligand displayed in the unpatterned area showed conversion to the dephosphorylated product following lysis and analysis with SAMDI-MS.

Keywords: Bioanalytical, Enzyme Assays, Mass Spectrometry, Nanotechnology
Application Code: Bioanalytical
Methodology Code: Mass Spectrometry

Session # 1040 Abstract # 1040-4 **Oral Sessions**

Session Title: Mass Spectrometry - Bioanalytical

Abstract Title: **Shotgun Lipidomic Analysis of Human Meibum by MS/MS[sup]all[/sup] with Successive Switching between Positive and Negative Detection Modes**

Date: Tuesday, March 08, 2016 - Morning
Time: 09:30 AM
Room: B404

Primary Author: Jianzhong Chen
University of Alabama at Birmingham

Co-Author(s): Kelly Nichols

Abstract Text

Meibum, the secretions from meibomian glands of the eye, is composed of lipids that form the outmost layer of tears and protect ocular surface of the eye. A good understanding of the lipid composition of meibum helps accurately diagnosis and/or treatment of eye diseases. However, detailed knowledge of the lipid composition of meibum remains lacking. Recently, we quantified many previously identified molecular species of major lipid classes including wax esters (WEs), cholesteryl esters (CEs) and diesters (DEs) based on optimized single stage mass spectrometry (Chen J, et al., Invest. Ophthalmol. Vis. Sci., 2013; 54:5730–5753). However, each peak of a certain m/z can correspond to a mixture of isomers with different combinations of moieties. In this study we determined the combinations of these isomers by utilizing shotgun lipidomics and the SWATH technology on a 5600 Triple TOF mass spectrometer (AB Sciex) with electrospray ionization. Product ion MS/MS spectra for all precursor ions in the range m/z 200 to 1200 at every one Dalton step were acquired. Each sample was analyzed sequentially under positive and negative detection modes, which not only saved time but also minimized the changes after the samples were prepared. Detailed information of meibum lipid composition including the isomeric components was obtained for the major molecular species of lipid classes including WEs, DEs, triacylglycerols, and O-acyl-□-hydroxy fatty acids.

This work was supported by unrestricted fund from University of Alabama at Birmingham School of Optometry (Chen J) and NIH grant (NEI R01EY015519, PI: Nichols KK).

Keywords: Bioanalytical, Electrospray, Lipids, Mass Spectrometry
Application Code: Bioanalytical
Methodology Code: Mass Spectrometry

Session # 1040 Abstract # 1040-5 **Oral Sessions**

Session Title Mass Spectrometry - Bioanalytical

Abstract Title **Development, Characterization, and Application of Coated Blade Spray Ionization**

Date: Tuesday, March 08, 2016 - Morning
Time: 10:05 AM
Room: B404

Primary Author German A. Gomez-Rios
University of Waterloo

Co-Author(s) Ezel Boyaci, Janusz Pawliszyn, Nathaly Reyes-Garces

Abstract Text

Coated Blades Spray (CBS) is a novel technology based on solid-phase microextraction (SPME) that efficiently integrates collection of analytes from complex matrices and direct ionization under ambient mass spectrometry conditions. Essentially, the device consists of a stainless steel sheet cut as a "sword" and coated with a biocompatible polymer (e.g. HLB-PAN). As a sample preparation method, the ultra-thin SPME coating simultaneously isolates and enriches small molecules present in the matrix, and allows for clean-up of undesirable artefacts that might provide ion suppression or enhancement. Whereas as an ambient ionization technique, CBS performs as a solid-substrate electrospray ionization source, where ions of the extracted analytes are generated by applying a high electric field to a device pre-wetted with a desorption solution. Analyte-enrichment and sample-clean-up is performed in exceedingly short times (i.e. 1 min or less), such the total analysis time does not exceed 3 minutes. Limits of detection at the low pg mL-1 levels, great accuracy, and outstanding reproducibility can be achieved for a broad range on analytes in complex matrices of clinical, forensic, and environmental relevance. Given the structural configuration of the apparatus, these can be used to perform extractions independently of the sample complexity (e.g. plasma) or its dimensions (i.e. few L up to liters when doing on-site analysis). In addition to the quantitation of target analytes in key matrices such blood and urine, this study presents a full characterization of CBS devices in terms of blade geometry and coating characteristics.

Keywords: Bioanalytical, Clinical Chemistry, Mass Spectrometry, SPME
Application Code: Bioanalytical
Methodology Code: Mass Spectrometry

Session # 1040 Abstract # 1040-6 **Oral Sessions**

Session Title: Mass Spectrometry - Bioanalytical

Abstract Title: **Discrimination of Carbohydrate Isomers as Transition Metal Adducts Using Ion Mobility Spectrometry and Tandem Mass Spectrometry**

Primary Author: Yuting Huang
University of Nebraska-Lincoln

Date: Tuesday, March 08, 2016 - Morning
Time: 10:25 AM
Room: B404

Co-Author(s): Eric D. Dodds, Lauren M. Petrosh

Abstract Text

The demand for rapid and comprehensive methods for carbohydrate analysis has arisen as the field of glycoscience has gained increasing attention in recent years. However, complete characterization of carbohydrate molecules remains challenging partly due to isomerism. Carbohydrate/metal ion interactions along with gas-phase ion chemistry have been found to be useful for rapid discrimination of isomeric carbohydrates by ion mobility spectroscopy (IMS) and tandem mass spectrometry (MS/MS).

Here, we explored the effects of five Period 4 transition metal ions (Fe[sup]2+[/sup], Co[sup]2+[/sup], Ni[sup]2+[/sup], Cu[sup]2+[/sup] and Zn[sup]2+[/sup]) and their corresponding gas-phase electron transfer (ET) products (Fe[sup]+•[/sup], Co[sup]+•[/sup], Ni[sup]+•[/sup], Cu[sup]+•[/sup] and Zn[sup]+•[/sup]) as charge carriers for carbohydrate isomers discrimination of by IMS and MS/MS. Several groups of carbohydrate isomers which have proven difficult to distinguish by both IMS and MS/MS were investigated, including pentasaccharide isomers (lacto-N-fucopentaose I, lacto-N-fucopentaose II, lacto-N-fucopentaose III, and lacto-N-fucopentaose V) and hexasaccharide isomers (lacto-N-difucohexaose I and lacto-N-difucohexaose II). Experiments were performed using a Waters Synapt G2-S quadrupole time-of-flight hybrid mass spectrometer coupled with traveling wave ion mobility spectrometer and equipped to carry out gas-phase ion-ion ET reactions. Ion-neutral collisional cross sections (CCSs) and MS/MS spectra under various vibrational activation energies of these carbohydrate/metal adducts have been obtained. With appropriate selection of charge carriers and gas-phase ion chemistry, differentiation of multiple isomers and conformations in IMS was achieved; and isomer-distinguishing features in the spectra of MS/MS fragmentation were observed.

Keywords: Bioanalytical, Carbohydrates, Mass Spectrometry, Tandem Mass Spec
Application Code: Bioanalytical
Methodology Code: Mass Spectrometry

Session # 1040   Abstract # 1040-7   **Oral Sessions**

Session Title   Mass Spectrometry - Bioanalytical

Abstract Title   **Scan-by-Scan Analysis of Orbitrap Fine Isotope Structures for Unique Elemental Composition Determination**

Date: Tuesday, March 08, 2016 - Morning
Time: 10:45 AM
Room: B404

Primary Author   Yongdong Wang
Cerno Bioscience

Co-Author(s)   Ming Gu

Abstract Text

Mass accuracy at the level of 1-2ppm or even sub-ppm can now be readily achieved on higher resolution MS systems such as Orbitrap to facilitate elemental composition determination of unknown compounds. Studies have shown that accurate isotope measurement and its proper utilization can make a significant difference in eliminating incorrect candidates whose exact monoisotope masses fall within a given accurate mass window, especially for unknown ions at m/z > 300 Da. This paper will quantitatively examine the measurement of isotope peaks, particularly the fine structures associated with higher (A+2 and A+3) isotopes, using an FT Orbital Trap MS (FT-OT-MS) system and explore key factors contributing to the isotope measurement accuracy.

Resolution and ion population (i.e., space charge) are two known factors contributing to MS measurement accuracy. While FT resolution can be easily adjusted from one experiment to another or even within a single experiment, the adjustment and control of ion population deserves more careful consideration. By observing and quantitatively comparing scan-by-scan FT mass spectrum across a chromatographic peak with and then without Automatic Gain Control (AGC), it is possible to observe the changes in FT spectrum and differentiate between resolution and space charge. The quantitative comparison of FT spectra across a chromatographic peak is achieved by a peak shape self-calibration performed on each scan individually, allowing for the examination of changes in resolution width and effects related to space charge.

Through this scab-by-scan analysis ultra-high resolution (resolving power > 240,000) Orbitrap MS data, one scan from each side of a typical chromatographic peak is found to be of the highest spectral accuracy to achieve unique elemental composition determination of true unknown compounds with reasonably open search conditions.

Keywords: Liquid Chromatography/Mass Spectroscopy, Mass Spectrometry, Organic Mass Spectrometry, Pharm
Application Code: Pharmaceutical
Methodology Code: Liquid Chromatography/Mass Spectrometry

Session # **1040** **Abstract #** **1040-8** **Oral Sessions**

Session Title: Mass Spectrometry - Bioanalytical

Abstract Title: **Mass Spectra of Analytical Derivatives of Amino Acids and Small Peptides**

Date: Tuesday, March 08, 2016 - Morning
Time: 11:05 AM
Room: B404

Primary Author: Nino G. Todua
NIST

Co-Author(s): Anzor I. Mikaia, Stephen E. Stein

Abstract Text

The NIST mass spectrometry data center is continuing to add reliable reference mass spectral and gas chromatography data for various derivatives of amino acids and peptides to the NIST/NIH/EPA mass spectral library (the Library). This will enable the Library to enhance its application to metabolomics. A systematic study of EI spectra for alkyl, acyl and N-alkoxycarbonyl derivatives of amino acids, di- and tripeptides has been done. Emphasis is made on interpretation of spectra for N-alkyloxyformates and comparative analysis of all data enabling location of branching centers and positions of functional groups. Advantages and disadvantages of a particular derivative for a specific compound are assessed. Special attention is paid to procedures for preparation of chemical derivatization products with more than two different functional groups; recommendations for their synthesis are given. The fragmentation pathways of compounds of interest and their derivatives, as well as characteristic ions in their mass spectra are discussed. Furthermore, evaluation methods of newly acquired mass spectral data of analytical derivatives for the addition to the Library are described.

Keywords: Amino Acids, Mass Spectrometry, Peptides
Application Code: Bioanalytical
Methodology Code: Mass Spectrometry

Session # 1050 Abstract # 1050-1 **Oral Sessions**

Session Title Materials Characterization and Engineering

Abstract Title **Chemical Analysis Applications and Optical Properties of 3D Printed <100 μm Dimension Microfluidic Channels**

Primary Author Michael Beauchamp
Brigham Young University

Date: Tuesday, March 08, 2016 - Morning
Time: 08:30 AM
Room: B405

Co-Author(s) Adam T. Woolley, Greg Nordin, Hua Gong, Steven Perry

Abstract Text

Most commercially available 3D printing resins used for stereolithography do not work for printing microfluidic features (<100 [micro]m). By studying the depth of polymerization at various exposure times we have determined an absorbance parameter and minimum exposure time that allow us to make predictions about the minimum size of channels that can be reliably printed, which agree well with experimental results. We used this approach with several resins: Clear (Full Spectrum Laser), PlasCLEAR (Asiga), PR48 (open sourced by Spark), and several polyethylene glycol diacrylate (PEGDA) formulations we have developed. We have optimized these parameters to print channels as small as 400 [micro]m x 190 [micro]m[sup]2[/sup] with PR48 and 60 [micro]m x 100 [micro]m[sup]2[/sup] with a PEGDA formula. Additionally, we are studying the visible absorbance and fluorescence of 3D printed devices for their potential use as microfluidic analyzers. We are working to create microfluidic channels for porous polymer monolith fabrication and for electrophoretic separation. Being able to 3D print true microfluidic devices (<100 [micro]m) may enable faster prototyping, encourage experimentation, and use cheaper materials than current methods to make microfluidic devices.

Keywords: Automation, Lab-on-a-Chip/Microfluidics, Materials Characterization, Polymers & Plastics
Application Code: Material Science
Methodology Code: Microfluidics/Lab-on-a-Chip

Session # 1050 Abstract # 1050-2 **Oral Sessions**

Session Title: Materials Characterization and Engineering

Abstract Title: **Crack-Free Three-Dimensionally Ordered Macroporous (3DOM) Structure in Microfluidic Reactor**

Date: Tuesday, March 08, 2016 - Morning
Time: 08:50 AM
Room: B405

Primary Author: Xiaoran Zhang
Michigan State University

Co-Author(s): Gary Blanchard

Abstract Text

We are interested in three dimensionally ordered macroporous (3DOM) structures, because of the characteristically high surface area of these materials and their ability to form flow-through structures with controllable void volumes and pore diameters. For the application of 3DOMs as microfluidic reactors, the material surface reactivity can allow for the chemical attachment of selected catalytic species.

One limitation of 3DOMs made using silica sol-gel chemistry is macroscopic cracking and defect formation. To address these issues, we report the formation of composite inverse opal 3DOM structures where the matrix used to form the inverse opal contains both silica, formed using sol□gel chemistry, and poly(ethylene glycol), PEG. We find that the morphology of the inverse opal structure depends on both the amount of PEG incorporated into the matrix and its molecular weight. The extent of organization in the inverse opal structure, characterized by scanning electron microscopy and optical reflectance data, is mediated by the chemical bonding interactions between the silica and PEG constituents in the hybrid matrix. The addition of PEG mitigates cracks and allows the formation of a matrix with fewer defects. Five types of defects can be quantitated. The resulting structures provide effective materials for catalyst support applications.

Keywords: Material Science, Modified Silica, Nanotechnology, Polymers & Plastics
Application Code: Material Science
Methodology Code: Microfluidics/Lab-on-a-Chip

Session # 1050 Abstract # 1050-3 **Oral Sessions**

Session Title: Materials Characterization and Engineering

Abstract Title: **Interfacial Structure-Function Correlation of Perovskite Solar Cell**

Date: Tuesday, March 08, 2016 - Morning
Time: 09:10 AM
Room: B405

Primary Author: Minyu Xiao
University of Michigan

Co-Author(s)

Abstract Text

Perovskite solar cell, an organic-inorganic hybrid material, has skyrocketed both in its efficiency and its popularity. However, for materials with such high quantum yield (reported 19.3% late last year), the lack of control in molecular orientation between active layer and its hole/electron extraction layer dampens charge carrier transfer between its layers, and ultimately decreases its overall power conversion efficiency (PCE). The aim is to combine a variety of state-of-the-art nonlinear optic interface-sensitive techniques in order to understand the interfacial molecular structure and efficiency correlation in a solar cell, for the purpose of improving cell efficiency via proper interfacial engineering, and increasing power conversion efficiency (PCE). A variety of organic hole-transport materials are investigated. P3HT, with its superior conductivity, solution processable property and its band gap matching, is an ideal candidate. In this study, combining sum frequency generation (SFG), a surface sensitive vibration spectroscopy, with device testing, we systematically investigated the correlation between P3HT/perovskite interfacial molecular orientation and its hole extraction efficiency. We concluded that molecular orientation of P3HT has a significant effect on PCE. Such studies on this interface will provide important insight regarding a molecular level understanding of layer coupling in real photovoltaic devices.

Keywords: Energy, Materials Characterization, Material Science, Vibrational Spectroscopy
Application Code: Material Science
Methodology Code: Vibrational Spectroscopy

Session # 1050 Abstract # 1050-4 **Oral Sessions**

Session Title: Materials Characterization and Engineering

Abstract Title: **Optical Spectroscopy Analyses of Perovskite Nanomaterials**

Primary Author: Daniel J. Freppon, Iowa State University

Date: Tuesday, March 08, 2016 - Morning
Time: 09:30 AM
Room: B405

Co-Author(s): Emily A. Smith, Feng Zhu, Jacob W. Petrich, Javier Vela, Long Men, Ujjal Bhattacharjee

Abstract Text

Inorganic-organic CH[sub]3[/sub]NH[sub]3[/sub]PbX[sub]3[/sub] (X = I, Br) hybrid perovskite materials suitable for photovoltaic devices must be stable in ambient conditions and under continuous illumination. They should also be free of surface defects. Photoluminescence (PL) spectroscopy is a useful tool for characterization of these materials. Initial studies of CH[sub]3[/sub]NH[sub]3[/sub]PbX[sub]3[/sub] nano-crystals showed more than one emission peak when exciting at 532 nm in the solid state. The presence of multiple emission peaks could be due to the presence of surface defects or photoinduced phase segregation. Using a modified precursor ratio, bromide and iodide perovskite nanoparticles are shown to have a single, stable PL peak under continuous illumination. The iodide-based perovskites emit light centered around 800 nm with a FWHM of 28 nm, and the bromide-based perovskites emit light centered around 535 nm with a FWHM of 15 nm. Mixed halide hybrid perovskites, specifically bromide-iodide mixtures, have previously been shown to be more stable in humid conditions compared to the pure bromide or iodide complementary materials. However, mixed halide perovskites have exhibited shifting emission peaks under continuous illumination using excitation power densities of 1.62 × 10[sup]5[/sup] W/cm[sup]2[/sup]. Preliminary XRD experiments show that the instability is likely not due to phase segregation of the two halide components of the perovskites.

This research is supported by the U.S. Department of Energy, Office of Basic Energy Sciences, Division of Chemical Sciences, Geosciences, and Biosciences through the Ames Laboratory. The Ames Laboratory is operated for the U.S. Department of Energy by Iowa State University under Contract No. DE-AC02-07CH11358.

Keywords: Energy, Luminescence, Materials Characterization, Nanotechnology
Application Code: Material Science
Methodology Code: Fluorescence/Luminescence

Session # **1050** Abstract # **1050-6** **Oral Sessions**

Session Title: Materials Characterization and Engineering

Abstract Title: **Carbazole-Dye Conjugate - Derived Group of Uniform Materials Based on Organic Salts (GUMBOS) for Optoelectronic Applications**

Date: Tuesday, March 08, 2016 - Morning
Time: 10:25 AM
Room: B405

Primary Author: Deepthika De Silva
Louisiana State University

Co-Author(s): Isiah M. Warner, Noureen Siraj

Abstract Text

Significantly lower energy consumption of organic optoelectronics have distinct advantages over conventional inorganic optoelectronics. We have recently introduced a new class of materials termed as GUMBOS (GROUP of UNIFORM MATERIALS BASED on ORGANIC SALTS). These compounds represent organic salts with melting points in the range of 25-250 ºC. We have designed carbazole based GUMBOS for optoelectronic applications. Carbazole possesses charge carrier properties and high triplet energies of approximately 3.0 eV. These carbazole based GUMBOS are potential candidates for use as emissive materials in optoelectronics, particularly in organic light emitting diodes (OLEDs) and organic field-effect transistors (OFETs). The synthetic procedures for production of these compounds are fairly simple using inexpensive starting materials that provide high product yields (70-90%). Carbazole based GUMBOS designed for OLEDs application show high quantum yields (above 90%) in the blue region of the electromagnetic spectrum. Evaluation of band gap values (3.1 eV), highest occupied molecular orbital (HOMO) and lowest unoccupied molecular orbital (LUMO) energy levels (-5.0 and -1.9 eV, respectively), photo and thermal stability data indicate that these carbazole derived GUMBOS can be successfully applied as emissive layers for blue OLEDs. Since generation of stable blue OLEDs is difficult, these materials exhibit great potential for serving as blue emissive layers in OLEDs for applications including TV and phone displays, computer screens, and home lighting.

Keywords: Material Science, Semiconductor
Application Code: Material Science
Methodology Code: Chemical Methods

Session # 1050 Abstract # 1050-7 **Oral Sessions**

Session Title: Materials Characterization and Engineering

Abstract Title: **Molecular-Scale IR Thermometer Reveals Sub-Molecular Photo-Plasticization in Azomaterials**

Date: Tuesday, March 08, 2016 - Morning
Time: 10:45 AM
Room: B405

Primary Author: Christian Pellerin
University of Montreal

Co-Author(s): Audrey Laventure, Geraldine Bazuin, Jaana Vapaavuori, Olivier Lebel

Abstract Text

We demonstrated experimentally for the first time that the photoinduced creation of free volume in azomaterials is heterogeneous at the sub-molecular level. This free volume gradient leads to localized photoplasticization within the molecules and helps to understand the puzzling phenomenon of photoinduced macroscopic material flow upon illumination far below the glass transition temperature (Tg). The findings stem from the correlation of infrared (IR) spectral band shifts measured upon illumination with those measured at controlled temperatures for two amorphous DR1-functionalized azo derivatives, a polymer, pDR1A, and a molecular glass, gDR1. This new approach reveals that IR spectroscopy can be used as an efficient label-free molecular-scale thermometer that allows the assignment of an effective temperature (T[sub]eff[/sub]) to each moiety in these compounds when irradiated. The direct measurement of T[sub]eff[/sub] offers a powerful probe of the local environment at the sub-molecular scale, paving the way towards better rationalization of the athermal malleability of azo-containing materials upon illumination below their Tg.

Keywords: FTIR, Materials Characterization, Material Science, Vibrational Spectroscopy
Application Code: Material Science
Methodology Code: Vibrational Spectroscopy

Session # 1050 Abstract # 1050-8 **Oral Sessions**

Session Title Materials Characterization and Engineering

Abstract Title **Complete Characterization of the UV-Visible Properties of Optical Materials Using a Total Absolute Measurement System (TAMS)**

Primary Author Ian Robertson
PerkinElmer Limited

Co-Author(s) Christopher Lynch, Steve Upstone

Date: Tuesday, March 08, 2016 - Morning
Time: 11:05 AM
Room: B405

Abstract Text

Optical materials are becoming ever more complex to meet a demanding range of new applications. Coating materials can be used to change the transmitting and reflecting properties of the substrate material, often glass. The optical properties of these materials need to be characterized across a wide spectral range, typically covering the UV/Vis/NIR region of the electromagnetic spectrum. Often these materials need to function at a wide range of non-normal angles of incidence.

Characterization of the reflectance and transmittance properties of these materials can be achieved in a fully automated method using a Total Absolute Measurement System (TAMS) on a UV/Vis/NIR spectrophotometer, measuring over a wide range of wavelengths, at the greatest range of sample angles, and using different polarizations.

Applications of this system for optical materials will be described showing the advantages of this type of measurement and the important information that can be obtained.

Keywords: Instrumentation, Material Science, Spectroscopy, UV-VIS Absorbance/Luminescence
Application Code: Material Science
Methodology Code: UV/VIS

Session # 1060 Abstract # 1060-1 **Oral Sessions**

Session Title: Pharmaceutical-MS, UV-VIS and Others

Abstract Title: **Speciation of Elemental Impurities for Compliance with USP <232>**

Date: Tuesday, March 08, 2016 - Morning
Time: 08:30 AM
Room: B406

Primary Author: Jon L. Sims
Perkin Elmer

Co-Author(s): Helmut Ernstberger, Kenneth Neubauer

Abstract Text

Testing requirements for elemental impurities in pharmaceuticals are currently undergoing a major shift away from long established protocols. The new standard defined in USP <232> specifies not only total elemental impurities but also refers to element species, taking into account the significant toxicity differences that exist between species for some elements. Species may differ in oxidation state, association with other elements, or extent of complexation with organic groups. Complying with the new USP testing requirements creates a need to determine element speciation for certain elements in order to risk assess cases where totals exceed the permitted daily exposure.

Highlighted in the USP <232> text is the case of arsenic and a distinction is made between inorganic and organic arsenic forms. The regulated levels apply to drug products, but also excipients and drug substances used in the manufacture of the product need to be monitored for metal impurities as stipulated by the regulations. Here we report arsenic speciation results for some common drug substances and excipients used in over the counter medicines using LC-ICP-MS. We employed ion interaction chromatography for adequate species separation in short run times. The results were validated using spike recoveries for inorganic arsenic species, demonstrating the suitability of the presented method for pharmaceutical analysis.

Keywords: HPLC, ICP-MS, Pharmaceutical
Application Code: Pharmaceutical
Methodology Code: Atomic Spectroscopy/Elemental Analysis

Session # 1060 Abstract # 1060-2 **Oral Sessions**

Session Title Pharmaceutical-MS, UV-VIS and Others

Abstract Title **Oxidative Degradation in Pharmaceuticals: Mechanism and Stabilization of Spray Dried Amorphous Drug - A Case Study**

Date: Tuesday, March 08, 2016 - Morning
Time: 08:50 AM
Room: B406

Primary Author Archana Kumar
Genentech

Co-Author(s) Raghendhar Kotha

Abstract Text

Overcoming poor aqueous solubility and oral bioavailability are common challenges in the pharmaceutical industry. Several formulations have been developed to address these issues. Spray-dried dispersion (SDD) is one such technique, which involves amorphous molecular dispersion of a drug in polymer matrix. Despite of improved solubility and bioavailability of SDD observed, resulting amorphous drug with higher energy and molecular mobility is inherently unstable and undergoes undesired degradation. Oxidative degradation is one of the most commonly observed degradation in formulated drug systems. Oxidation of active pharmaceutical ingredient (API) can be catalyzed by many sources such as radical chain initiators, molecular oxygen, peroxides, or trace metal contaminants, most of which are often carried over by either polymeric excipients or API synthesis.

In an attempt to improve solubility and bioavailability, one of the drugs in development was spray dried with 80% hydroxypropyl methylcellulose acetate succinate (HPMCAS). Three degradation products were observed on long-term storage conditions. Forced degradation studies and high-resolution mass spectrometry (HRMS) data suggested that the products are corresponding to oxidative degradation. It is very important to understand the source (s) and mechanism of degradation in order to stabilize the drug by using appropriate antioxidant(s) and packaging techniques.

We report here the results of various studies performed to understand the degradation mechanism, identification of appropriate antioxidant excipients (which may be added during formulation development), their synergistic effects and finally a practical way to minimize or completely mitigate the oxidative degradation during storage of amorphous drugs.

Keywords: Chromatography, Drug Discovery, Liquid Chromatography/Mass Spectroscopy, Pharmaceutical
Application Code: Pharmaceutical
Methodology Code: Liquid Chromatography/Mass Spectrometry

Session # 1060 **Abstract # 1060-3** **Oral Sessions**

Session Title: Pharmaceutical-MS, UV-VIS and Others

Abstract Title: **Evaluation of Antimicrobial and Neutraceutical Properties of Plukenetia Conophora (Walnut) Leaves**

Primary Author: Chukwuemeka P. Azubuike, University of Lagos

Co-Author(s): Cecilia I. Igwilo, Karamot O. Suara

Date: Tuesday, March 08, 2016 - Morning
Time: 09:10 AM
Room: B406

Abstract Text

The neutraceutical and antimicrobial properties of methanolic and ethylacetate extracts of the leaves of Plukenetia conophora were studied as well as physical properties of creams formulated with the extracts. Proximate analysis, mineral content and phytochemical analysis of the extracts were determined using standard methods. The polyphenolic contents were determined using ultraviolet spectroscopy while antimicrobial assay of the extracts was investigated. Standard methods were employed to determine the quality and stability of the creams formulated with the extracts. Mineral content analysis showed that both extracts contain the same minerals, however, higher values were observed for methanolic extract. Proximate analysis revealed 6.86% moisture, 11.78% protein, 8.57% total ash, 20.12% crude fibre, 1.56% total fat and 51.85% total carbohydrate. The phytochemical groups identified include alkaloids, cardenolides, flavonoids, sugars, and tannins. Polyphenolic content analysis also revealed the following in the methanolic extract versus the ethylacetate fraction: total flavonoids (78.27mg/g, 71.54mg/g), total proanthocyanidins (73.50mg/g, 77.75mg/g), and total phenolic acids (110.71mg/g, 64.71mg/g). The antimicrobial assay of the extracts displayed activity against Proteus mirabilis, Bacillus subtilis and Staphylococcus aureus with the ethylacetate fraction being more potent. The results indicate the potential of the extracts as a nutraceutical and the phytochemicals indicate the wide range of physiological and medicinal activities of the extract and thus support the folk use of Plukenetia conophora as a nutraceutical and antimicrobial. Furthermore, the antimicrobial activities of the methanolic extract and ethylacetate fraction suggest that they can be applied in various skin infections and wound implicated by the susceptible microorganisms.

Keywords: Drugs, Pharmaceutical, Spectroscopy

Application Code: Pharmaceutical
Methodology Code: Separation Sciences

Session # 1060 **Abstract # 1060-4** **Oral Sessions**

Session Title: Pharmaceutical-MS, UV-VIS and Others

Abstract Title: **Novel Self-Patented Gold Nanoparticles for Antineoplastic Activity**

Date: Tuesday, March 08, 2016 - Morning
Time: 09:30 AM
Room: B406

Primary Author: Jason N. Payne
Western Kentucky University

Co-Author(s): Rajalingam Dakshinamurthy

Abstract Text

Phloridzin, a natural hydoxychalcone constituent obtained from fruit trees is an antidiabetic and antineoplastic agent. Phloridzin was first isolated, for clinical pharmaceutical usage, from the pear tree bark of Pyrus communis in 1838 as the first sodium-glucose linked transport 2 (SGLT2) inhibitor. Additionally, phloridzin is also reported to have antineoplastic activity. Phloridzin had to take an exit from the pharmaceutical market due to its side effects and poor bioavailability when compared to other antidiabetic drug competitors. This limit of phloridzin's bioavailability is primary attributed to the degradation of the glyosidic bond of the drug to result in the formation of phloretin, the aglycone of pholoridzin. Phloretin displays a reduced capacity of SGLT2 inhibition, however this nutraceutical displays enhanced antineoplastic activity in comparison to phloridzin. Gold nanoparticles (AuNPs) have been studied for drug delivery applications for poorly bioavailable drugs. Hence, in order to tackle the bioavailability of these hydroxychalcones and study the unknown anticancer mechanism, we synthesized phloridzin and phloretin conjugated gold nanoparticles (Phl-AuNP and Pht-AuNP) in single step, rapid, biofriendly processes. The synthesized AuNPs morphology was characterized via transmission electron microscopy and UV-Vis spectroscopy. The presence of phloridzin or phloretin was confirmed using SEM-EDS. The percentage of organic component (phloridzin/phloretin) onto GNPs surface was characterized using TGA. Assessment of the antineoplastic potency of the hydroxylchalcone conjugated AuNPs against cancerous cell lines was accomplished through monitoring via flow cytometry. We hypothesize that functionalization of these chalcones onto the gold nanoparticles' surface may improve the pharmacokinetic profile of phloridzin and phloretin.

Keywords: Drugs, Nanotechnology, Natural Products, Pharmaceutical
Application Code: Pharmaceutical
Methodology Code: Fluorescence/Luminescence

Session # 1060 Abstract # 1060-5 **Oral Sessions**

Session Title: Pharmaceutical-MS, UV-VIS and Others

Abstract Title: **Expanding the Analytical Toolbox for Material Verification: Spectroscopic Screening of Raw Ingredients Using Portable Spectrometers**

Date: Tuesday, March 08, 2016 - Morning
Time: 10:05 AM
Room: B406

Primary Author: Jason D. Rodriguez
FDA Division of Pharmaceutical Analysis

Co-Author(s): Fabiola Semidei Ortiz, Hirsch Srivastava

Abstract Text

We report our results for a study designed to evaluate the performance of field-deployable Raman and near infrared methods to perform rapid raw pharmaceutical material verification. Identity testing of incoming raw materials has received much attention recently due to the increasing complexity arising from globalization of the supply chain. This complexity may increase the potential for economically-motivated adulteration of pharmaceutical materials. Portable spectroscopic instruments are capable of screening samples at a much higher throughput than can be achieved by chromatographic laboratory based methods. These tests can be carried out in the field and can typically be conducted in less than a minute without destroying the sample. We report a comparison of the sensitivity of the spectral-based screening methods with current compendial identification tests which rely largely on legacy methods which may be non-specific for certain adulterants or not amenable for rapid screening. Our results indicate that spectral based methods can be used to screen gross adulteration or mislabeling of pharmaceutical materials when using spectral library based correlation methods and the sensitivity of the spectral methods is further improved when utilizing multivariate algorithms such as principal component analysis.

Keywords: Near Infrared, Raman, Spectroscopy, Vibrational Spectroscopy
Application Code: Pharmaceutical
Methodology Code: Vibrational Spectroscopy

Session # 1060 Abstract # 1060-6 **Oral Sessions**

Session Title Pharmaceutical-MS, UV-VIS and Others

Abstract Title **Image Directed Identification of Sub-visible Particles in Protein Based Therapeutics, Classification According USP<787> of Intrinsic, Inherent and Extrinsic Particulate Matter on the Sub-visible Level**

Primary Author Olga Laskina
rap.ID

Date: Tuesday, March 08, 2016 - Morning
Time: 10:25 AM
Room: B406

Co-Author(s) Kathryn A. Lee, Markus Lankers, Oliver Valet

Abstract Text

Protein aggregation is a key quality attribute of bio-therapeutics. Aggregates hold the potential for adversely impacting production and patients.

The newly released USP<787> "Subvisible Particulate Matter in Therapeutic Protein Injections" defines particle types: Truly foreign particles are "extrinsic"; particles from the production environment or primary packaging are "intrinsic" and formulation particles are "inherent". In the visible inspection process inherent particles must be distinguished from the other two.

USP<787> furthermore states that membrane microscopy is not the preferred method and is suited only for other than inherent particles. USP<788> and USP<787> confirm this, stating MM gridded filter paper cannot isolate fragile or translucent particles, nor can they sufficiently be visualized in conventional microscopes.

We will show a method that allows the isolation of particles on a gold membrane with subsequent enumeration of particles by membrane microscopy utilizing a new illumination technique. Besides dark-field illumination, a UV light induces the auto-fluorescence of tryptophan and allows specific counting of inherent particles. In a second step, Raman spectroscopy can obtain fingerprint spectra that allow material identification and differentiation between extrinsic, intrinsic, and inherent particles. This method gives chemical composition of hundreds of particles per analysis and allows root cause investigation of contaminants to avoid further contamination.

With a novel sampling method we have overcome the limitation of the lower count number from the membrane method compared to micro flow imaging MFI. Image directed spectroscopy is used to obtain Raman spectra of the spherical silicone droplets specifically from a biopharmaceutical formulation between two quartz slides.

Keywords: Characterization, Materials Characterization, Pharmaceutical, Raman
Application Code: Pharmaceutical
Methodology Code: Microscopy

Session # **1060** **Abstract #** **1060-7** **Oral Sessions**

Session Title: Pharmaceutical-MS, UV-VIS and Others

Abstract Title: **Coupling Chemical Analysis to High Resolution Dark Field Microscopy for Enhanced Physicochemical Characterization of Complex Drug Formulations**

Date: Tuesday, March 08, 2016 - Morning
Time: 10:45 AM
Room: B406

Primary Author: Katherine Tyner
Food and Drug Administration

Co-Author(s): Sau (Larry) Lee, Sheetal D'Mello

Abstract Text

Controlling the quality of complex dosage forms relies on the ability to adequately characterize the drug product. Oftentimes information concerning the chemical microstructure of a drug product is unknown. This information becomes more critical from a quality perspective as the complexity of a formulation increases. The focus of this study, therefore, is to evaluate the distribution of different domains within complex drug formulations and obtain overall chemical microstructure by employing darkfield microscopy coupled with hyperspectral analysis.

Commercially available emulsions and creams were mounted onto a glass slide with a coverslip. An optical microscope equipped with a CytoViva® unit and a hyperspectral imaging spectrophotometer unit (Headwall Photonics) was used to acquire darkfield microscopy images and hyperspectral plots. ImageJ was used to analyze the different domains within the formulations. Hyperspectral images (400 nm -1000 nm) were acquired using a 0.5 s collection time and a halogen light source. Hyperspectral analysis of the acquired images was performed using ENVI 4.8. Spectral libraries were built with pure chemical components of the formulations and then applied to images of unmodified formulated products. The CytoViva® system allowed for the observation of domains into the nanoscale range. By constructing spectral libraries of reference materials, API and select excipient distribution within the formulation was able to be distinguished and discriminated from the liquid microenvironments.

Coupling chemical information with morphological information arising from high resolution imaging techniques can redefine the role of imaging complex drug products. This information will allow for specific questions about the morphology of complex formulations of interest to be addressed and would represent an improvement in our ability to understand complex microstructures and the critical quality attributes that govern these drug products.

Keywords: Characterization, Imaging, Microscopy, Pharmaceutical
Application Code: Pharmaceutical
Methodology Code: Microscopy

Session # **1070** **Abstract #** **1070-1** **Oral Sessions**

Session Title Sensors - Biomedical

Abstract Title **Disposable Sensors for the Remote Monitoring of Chronic Wounds**

Date: Tuesday, March 08, 2016 - Morning
Time: 08:30 AM
Room: B407

Primary Author Fabio Di Francesco
Università di Pisa

Co-Author(s) Alessio Ceccarini, Bernardo Melai, Clara Paoletti, Consuelo Politino, Letizia Moni, Nicola Calisi, Pietro Salvo

Abstract Text

Health care systems in Western countries are requiring increasing resources, and concerns about sustainability of universal health care are growing. The longer life expectancy resulting from successful public health interventions brings forth the new challenge of a growing burden of chronic illnesses. At present, about 1.5 - 2 % of people experiences a chronic wound in the course of life. Most of these wounds heal with normal treatments (e.g. compression bandaging in the case of venous leg ulcers or offloading in the case of diabetic foot ulcers), but about 25% of them remains stuck to one of the healing phases (haemostasis, inflammation, proliferation, and tissue remodelling or resolution) without an obvious reason.
Today, wearable sensors are creating great expectations for improving knowledge on the biochemical processes in action in these wounds and combining quality of treatment and low cost. SWAN-iCare is a project funded by the European Commission developing temperature, pH and metalloproteases activity sensors for monitoring and managing chronic wounds, mainly diabetic foot ulcers and venous leg ulcers.

We report here the fabrication, testing and validation of disposable sensors, namely a resistive sensor based on reduced graphene oxide for the measurement of temperature and a potentiometric sensor based on graphene oxide for the measurement of pH in the wound bed. In-vitro validation with model solutions and real samples established accuracies of ±0.5 °C (range 20-40 °C) and ±0.2 pH units (range 5.5-9 pH units). Issues concerning biocompatibility for the use in contact with the wound bed are also addressed.

Acknowledgements: The work presented in this paper was supported by the EU-funded FP7 ICT- 317894 SWAN-iCare project.

Keywords: Biomedical, Nanotechnology, Sensors

Application Code: Biomedical

Methodology Code: Sensors

Session # 1070 Abstract # 1070-2 **Oral Sessions**

Session Title Sensors - Biomedical

Abstract Title **Real-Time Monitoring Urinary Encrustation Using Quartz Crystal Microbalance Sensor**

Date: Tuesday, March 08, 2016 - Morning
Time: 08:50 AM
Room: B407

Primary Author Pegah N. Abadian
Northeastern University

Co-Author(s) Edgar D. Goluch, John Victor, Jonathan Zhang, Jun Li

Abstract Text

One of the major problems related to urological catheterization is intraluminal blockage and encrustation which happens in almost 50% of the patients undergoing long term bladder catheterization. If blockage is not detected early enough it can cause episodes of pyelonephritis, septicaemia, and endotoxic shock. Encrustation in urinary catheters usually starts with a urease producing bacteria such as Proteus mirabilis. The bacteria will colonize on the surface of the medical device and form a biofilm. This bacteria will then produce urease which will generate ammonia from urea, resulting in a pH increase in the environment. In this alkaline environment, crystals of magnesium and calcium salts (struvite and apatite) form in the urine and in the biofilm on the surface. The continuous process of crystal formation on the catheter will eventually result in catheter blockage.

For urological devices a substrate surface can effectively prevent early stage crystal formation to further prevent blockage. However, there are very limited models available to achieve a fast, real-time monitoring. In this work for the first time we developed a method to study in real-time the encrustation on the surface by using a Quartz Crystal Microbalance (QCM) biosensor. The QCM sensor is capable of label-free detection with very high sensitivity, which leads to rapid detection of a small amount of encrustation on the surface.

In order to provide a surface similar to the urinary catheters, the sensor surface was first spin coated with polyurethane. Then, various solutions were flown over the surface and the deposited mass of crystals was monitored. The results showed the pre-adsorption of urease has a significant direct effect on encrustation which preventing it can be an avenue to reduce encrustation. With this technique, we are able to monitor encrustation formation under different environmental factors, and compare different surface chemistries for their anti-encrustation performance.

Keywords: Analysis, Biomedical, Biosensors, Material Science
Application Code: Biomedical
Methodology Code: Sensors

Session # 1070	Abstract # 1070-3	Oral Sessions

Session Title: Sensors - Biomedical

Abstract Title: **A Fiber Optic Biosensor for Noninvasive Transdermal Glucose Sensing Based on the Glucose Binding Protein**

Primary Author: Cristina E. Tiangco
University of Maryland Baltimore County

Date: Tuesday, March 08, 2016 - Morning
Time: 09:10 AM
Room: B407

Co-Author(s): Dayanand Bagdure, Dieudonne Fon, Fortunato Sevilla, Govind Rao, Leah Tolosa, Yordan Kostov

Abstract Text

The ability to measure glucose levels is an important requirement for good clinical care in the intensive care unit. Current technologies for glucose measurement are enzyme-based sensors that work well in the mM glucose range of blood and interstitial fluid. These sensors require breaking the skin, which can be painful to the patient. Noninvasive monitoring of glucose could facilitate a more effective management of hyperglycaemic and hypoglycaemic episodes. Our group developed a painless, noninvasive method of collecting glucose passively diffusing through the skin. The transdermal glucose collected in this way has a concentration in the molar range so there is a need to develop a reliable glucose sensor that could measure glucose at these levels. Thus, we developed a fiber optic biosensor for glucose that is based on the glucose binding protein (GBP) H152C. GBP is highly specific and sensitive to glucose at M concentrations. GBP-labeled with BADAN was immobilized in Ni-NTA agarose beads via metal-histidine interaction. The portable, low-cost biosensor system consists of an optical fiber with the immobilized beads trapped on one end, and appropriate optics and electronics on the other end. The control software and the visual interface for the optical sensor is designed and implemented in LabVIEW and runs on tablet computer. The biosensor exhibited a stable response to the blank and with 10 M glucose for ~16 hours. Glucose responses are also reversible when washed with phosphate-buffered saline solution. Measured voltages resulting from fluorescence were recorded and the response time of the biosensor for 6 M glucose was approximately 70 seconds. A linear relationship (r2=0.9517) was observed between the sensor response and glucose standard solutions from 4 to 20 M. Different parameters like amount of beads and distance of the optical fiber to Ni-NTA-GBP beads were optimized. In vitro studies on pig skin showed faster diffusion of glucose with increasing glucose concentration in the bottom reservoir of a static Franz-type cell. This fiber optic sensor was also used in vivo to measure transdermal glucose from healthy adults. The collected results reveal the potential of GBP as a noninvasive glucose monitoring system and shows great potential for point of care use.

Keywords: Biosensors, Fiber Optics, Portable Instruments, Protein
Application Code: Biomedical
Methodology Code: Sensors

Session Title Sensors - Biomedical

Abstract Title **Direct Measurement of Total Concentration of Major Physiological Anions, Chloride and Bicarbonate, Using Pulsed Chronopotentiometry with Ion-Selective Electrodes (Pulstrode)**

Date: Tuesday, March 08, 2016 - Morning
Time: 09:30 AM
Room: B407

Primary Author Kebede L. Gemene
Northern Kentucky University

Co-Author(s) Adaeze Stella Iloegbunam, Sara Keshtvarz, Simon Segal

Abstract Text

Polymer membrane-based ion-selective electrodes (ISEs) have unparalleled applications in clinical analysis of blood electrolytes such as K+, Na+, Ca2+ and Cl-. However, bicarbonate ion, which is one of the major biological anions, does not have any known ionophore and its detection with ISEs has been challenging. We demonstrate here a simple pulsed chronopotentiometric ion-selective electrode (pulstrode) detection method for the total major physiological anions, chloride and bicarbonate. Chloride and bicarbonate have comparable lipophilicity, with chloride slightly more preferred when ionophore-free ion exchanger-based membranes are used. The membranes used in pulstrode are formulated with plasticized poly(vinyl chloride) containing lipophilic inert salts without added ion exchange site. Thus, the concentration of the extracted ions is controlled by the magnitude of the applied current pulse. By using an optimum current pulse (magnitude and duration) as well as optimum membrane composition, we have shown that the measured potential can be a function of the total concentration of chloride and bicarbonate in a solution. Near-Nernstian sensor responses were obtained when chloride and bicarbonate were measured separately as well in a 1:1 mixture. This simultaneous measurement of chloride and bicarbonate can be used for calculation of anion gap, after measuring the total measureable cations using the same electrode. Note that the same electrode can be used to measure cations by a mere change of the sign of the applied current. In addition, measuring the concentration of chloride alone in the same physiological sample using ionophore-based membrane electrode, can give the much-needed concentration of bicarbonate ions in the sample by simple subtraction of chloride concentration from the total concentrations.

Keywords: Biomedical, Electrochemistry, Ion Selective Electrodes, Sensors
Application Code: Biomedical
Methodology Code: Sensors

Session # 1070 Abstract # 1070-5 **Oral Sessions**

Session Title: Sensors - Biomedical

Abstract Title: **Non-Invasively Interrogating Chemical and Mechanical Sensors on Implanted Medical Devices**

Date: Tuesday, March 08, 2016 - Morning
Time: 10:05 AM
Room: B407

Primary Author: Jeffrey N. Anker
Clemson University

Co-Author(s): Donald Benza, Fenglin Wang, Jeremy Tzeng, Mohammed Arifuzzaman, Peter Gennaro, Yash Raval

Abstract Text

Detecting and monitoring implanted medical device infections is challenging at early stages and during antibiotic therapy, when the infection is localized on and near the device surface. Bacterial biofilms which grown on the device surface can resist the host's immune system and are antibiotics. Chemical changes near the device surface (e.g. low pH and oxygen tension) contribute to the antibiotic resistance and could also serve as a local infection indicator. However, detecting these chemical concentrations at the implant surface is difficult because the chemical sensor must provide a low background, surface-specific signal that can pass through tissue. We are developing a combination of X-ray and optical techniques to interrogate sensor films coated on implanted devices for high spatial resolution chemical sensing. One approach uses battery-powered LEDs as internal light sources. Another approach uses an X-ray scintillator film coated over the implant surface as an X-ray addressable light source. In either case, the internally generated light then passes through a sensor layer (e.g. pH indicator-loaded film that alters the spectrum according to pH), then passes through the tissue, and finally is detected by a spectrometer. Reference light sources account for tissue-induced spectral distortion. For the X-ray excited sensors, pH images are acquired by moving the X-ray relative to the sample and collecting a spectrum at each position, with a spatial resolution limited by the X-ray beam width. Using this technique, we detected a pH drop during bacterial growth on the sensor surface, and a pH restoration during antibiotic treatment, measured with millimeter resolution through 6 mm ex vivo porcine tissue. Overall, these methods provide noninvasive chemical measurements at the implant surface to detect and study implant infection.

Keywords: Bioanalytical, Biomedical, Biotechnology, Luminescence
Application Code: Biomedical
Methodology Code: Sensors

Session # **1070** Abstract # **1070-6** **Oral Sessions**

Session Title: Sensors - Biomedical

Abstract Title: **Detection of MicroRNA Presence or Absence with Dual Functioning Signal-On/Off Fluorescent Biosensors**

Date: Tuesday, March 08, 2016 - Morning
Time: 10:25 AM
Room: B407

Primary Author: Nicholas E. Larkey
Oregon State University

Co-Author(s): Lulu Zhang, Sean M. Burrows

Abstract Text

MicroRNAs are emerging as an early marker of disease because changes in their expression precede protein and morphological changes. Disease-associated microRNAs typically become over- or under-expressed during all stages of development, and this altered expression is cell type specific. These changes in microRNA concentrations related to altered expression can be in the femto- to nanomolar range. For these reasons, sensitive in situ biosensors that can detect both the presence and absence of small amounts of microRNA are in demand. Current sensing technologies typically look for just the presence of microRNA and are susceptible to false signals from nuclease degradation in cells. We have developed nucleic acid biosensors that can mitigate false-signals, as well as operate in either a signal-on or signal-off manner dependent on excitation wavelength. Different Förster Resonance Energy Transfer donor/acceptor dye and dye-quencher pairs were compared to determine the figures of merit (FOM) for both signal-on and signal-off mechanisms. Several chemical modifications were used to reduce the biosensors susceptibility to nuclease degradation, optimize sensitivity, and aid in keeping the biosensors in the cytoplasm of cells. The influence of polyethylene glycol spacers, locked nucleic acids, and conjugated peptide chemical modifications on the FOM for these biosensors will be addressed. These advances in nucleic acid biosensor design will aid in overcoming the challenges facing in situ microRNA

Keywords: Biosensors, Biospectroscopy, Fluorescence, Nucleic Acids
Application Code: Bioanalytical
Methodology Code: Sensors

Session # 1070 Abstract # 1070-7 **Oral Sessions**

Session Title: Sensors - Biomedical

Abstract Title: **Single Nanoparticle Plasmonic Spectroscopy and Biosensors for Imaging of Efflux Functions of Single Live Cells**

Date: Tuesday, March 08, 2016 - Morning
Time: 10:45 AM
Room: B407

Primary Author: X Nancy Xu
Old Dominion University

Co-Author(s): Feng Ding, Kerry J. Lee, Prakash d. Nallathamby, Tao Huang

Abstract Text

Multidrug membrane transporters (efflux pumps) can selectively extrude a variety of structurally and functionally diverse substrates (e.g., chemotoxics, antibiotics), leading to multidrug resistance and ineffective treatment of a wide variety of diseases. Current technologies are unable to effectively characterize molecular mechanisms of efflux function of membrane transporters. We have developed far-field photostable-optical-nanoscopy (PHOTON), which includes photostable single molecule nanoparticle optical biosensors and, single nanoparticle plasmonic microscopy and spectroscopy, for probing of interactions of membrane proteins of transporter with their substrates and quantitatively measuring of efflux dynamics of single membrane transporters. We have demonstrated that PHOTON can be used to characterize the efflux function of single membrane transporters in single live cells in real-time at single-molecule and nanometer resolutions for better understanding of multidrug resistance.

Keywords: Bioanalytical, Biomedical, Imaging, Nanotechnology
Application Code: Nanotechnology
Methodology Code: Sensors

Session # 1070 Abstract # 1070-8 **Oral Sessions**

Session Title Sensors - Biomedical

Abstract Title **Detection of Chemotherapeutic-Induced Damage in Genomic DNA Using Integrated Thermoplastic Nanofluidic Sensor Devices**

Primary Author Kumuditha M. Weerakoon-Ratnayake
University of North Carolina at Chapel Hill

Date: Tuesday, March 08, 2016 - Morning
Time: 11:05 AM
Room: B407

Co-Author(s) Franklin I. Uba, Robert Schotzinger, Steven A. Soper

Abstract Text

Damage can be induced in genomic DNA due to a number of processes, including therapeutic response to drugs used to treat a particular disease, such as chemotherapies used to arrest the activity of cancer cells. The damage that can occur in DNA can consist of double strand breaks, DNA crosslinking or the formation of abasic sites, which can be sensed using the appropriate chemistry and detection technology. Optical labeling of the abasic sites followed by far-field visualization is limited by the diffraction-limited optical resolution, which can be on the order of 250 nm or about 75 base pairs (bp). We propose a new sensing platform for the direct reading of abasic sites, which consists of optical quantification following DNA stretching to near its full contour length using a thermoplastic nanochannel that is imprinted into a polymer substrate as well as electrical sensing of labeled abasic sites while its passing through nanopore structures. Dual detection will ensure the accuracy of our proposed device. Our nanostructures are fabricated in poly(methyl) methacrylate (PMMA) using nanoimprint lithography (NIL). Imprinted channels <100 nm and single DNA molecules are electrokinetically driven into the nanochannels by an external applied electric field. Abasic sites are labeled with a biotinylated aldehyde-reactive-probe (ARP), which is then associated to streptavidin molecules with and without a fluorescent tag to assist dual detection. We are also detecting the presence of abasic sites in cancer cells secured from breast cancer patients that are undergoing chemotherapeutic treatment.

Keywords: Bioanalytical, Biosensors, Lab-on-a-Chip/Microfluidics, Nanotechnology

Application Code: Nanotechnology

Methodology Code: Fluorescence/Luminescence

Session # 1080 Abstract # 1080-1 **Poster Sessions**

Session Title Environmental Air Quality and Analysis

Abstract Title **Tea Polyphenols Reduces Toxicity of PM2.5 in Human Alveolar Epithelial A549 Cells**

Primary Author Ying Zhang
Shijiazhuang CDC

Date: Tuesday, March 08, 2016 - Morning
Time:
Room: Exposition Floor, 400 Aisle

Co-Author(s) Jie Jiang, Yan He, Yanzhong Chang

Abstract Text

Haze of fine particulate matter (Pariculate Matter 2.5, PM2.5) poses a huge threat to human health as an important environmental pollutant. Oxidative stress is suggested to be involved in PM2.5-induced cell injury and to induce cell apoptosis. The present research is designed to study both toxic effects of PM with an aerodynamic diameter of less than 2.5 um (PM2.5) and protective effects of tea polyphenols against PM2.5 on human alveolar epithelial A549 cells in vitro. Cytotoxic effects of the PM2.5 on A549 were measured by means of A549 cell viability and the generation of intracellular reactive oxygen species (ROS). The present results showed that PM2.5 (0.5–500 ug/ml) decreased A549 viability, and SOD levels, while results also showed an increased intracellular generation of ROS and malondialdehyde (MDA) in a concentration dependent manner. Tea polyphenols (0.1 and 80 ug/ml) diminished PM2.5 induced A549 cells viability and SOD levels, along with decreasing ROS and MDA generation. Meanwhile, PM2.5 induced A549 cells apoptosis that caspase-3 and bax expression increased, and deceased bcl-2 expression. Tea polyphenols could relieve the cells' apoptosis. These results suggested that tea polyphenols antagonize PM2.5-induced excess oxidative stress and apoptosis in A549 cells.

This work was supported by Hebei province postdoctoral research projects merit funding (B2015003023) and Hebei province key subjects funding of medical research program (20150170).

Kaywords: Tea polyphenols; PM2.5; Oxidative stress; apoptosis;
[micro]

Keywords: Bioanalytical, Environmental/Air, Toxicology
Application Code: Environmental
Methodology Code: Chemical Methods

Session # **1080** **Abstract #** **1080-2** **Poster Sessions**

Session Title: Environmental Air Quality and Analysis

Abstract Title: **Testing of Gas Purifiers for VOC Removal Down to the PMOL/MOL Level**

Primary Author: Annarita Baldan, VSL

Date: Tuesday, March 08, 2016 - Morning
Time:
Room: Exposition Floor, 400 Aisle

Co-Author(s): Alessia Demichelis, Christian Plass-Duelmer, Guido Sassi, Janneke van Wijk, Jennifer Englert, Jianrong Li, Mariapaola Sassi, Rina Wortman, Stefan Persijn

Abstract Text

Trace concentration of volatile organic compounds (VOCs) in air are typically measured by gas chromatography with pre-concentration techniques. In order to calibrate the measurement system at these trace levels high concentrated gas standards are dynamically diluted with a zero gas (nitrogen or air). Within the framework of the European KEY-VOCs project, different kinds of gas purifiers (including carbon and inorganic) were tested to remove residual VOC impurities which could bias the prepared calibration standard. The testing includes VOCs such as O3 precursors (C2-C9) and polar VOCs (including OVOCs). Results from the experiments are presented. Special attention is paid to the handling of instrumental artefacts such as memory effect which become relevant for measurements at such low levels.[sub][/sup]

Keywords: Calibration, Environmental/Air, GC, Volatile Organic Compounds
Application Code: Environmental
Methodology Code: Gas Chromatography

Session # 1080 Abstract # 1080-4 **Poster Sessions**

Session Title: Environmental Air Quality and Analysis

Abstract Title: **Developing and Field Tests of an Automatic Impinger System for Continuous Sampling of Volatile Amines in the Environment**

Primary Author: Chia-Jung Lu
National Taiwan Normal University

Co-Author(s): Rih-Sheng Jian, Sung Lung-Yu, Wang Chih-Chia

Date: Tuesday, March 08, 2016 - Morning
Time:
Room: Exposition Floor, 400 Aisle

Abstract Text

This research develops an automated and continuous sampling system which possesses high collection efficiency for lower-level concentration ambient amines in the environment. The key functional units of the system include two impingers, one peristaltic pump, a multi-way solenoid valve, a mass flow controller and a mini vacuum pump. All element function controls were connected to a 10.1-in tablet computer via an I/O module and software was written in LabVIEW (National Instrument). In the sampling mode, one of the impingers was filled with 25 mL DI-water by the peristaltic pump as the amines extraction media. The pump was then turned on and air samples were directly drawn though the impinger via the mass flow controller at an adjusted sampling rate of 1.0 L/min. After sampling completed, the multi-way valve was computer-switched and the sample was drained to glass vials. All collected samples were returned to the laboratory to be analyzed by ion chromatograph (IC). Seven amines including methylamine (MA), ethylamine (EA), dimethylamine (DMA), isopropylamine (IPA), propoylamine (PA), diethylamine (DEA) and trimethylamine (TMA) are selected as the testing compounds for this system. The amine collection efficiencies were ranged from 80.9 to 97.2 %. All calibration curves of the R-square values were > 0.99 with method detection limits (MDLs) of 0.1 to 0.5 ppb in the air. A field study using this sampling system was conducted inside a fish market. TMA was persistently found in both hand-operate impingers and automatic system. Five consecutive field samples showed that the concentration varied from 1.8 to 3.6 ppb inside a fish market. The analysis results of our auto-sampler agreed reasonably with hand-operate sampler in this field study. The next step is combining this sampling system with IC to be a fully automated amine analysis station that can provide the near-real time data for odor control in an environment.

Keywords: Automation, Environmental/Air, Ion Chromatography, Sampling

Application Code: Environmental

Methodology Code: Sampling and Sample Preparation

Session # 1080 **Abstract # 1080-5** **Poster Sessions**

Session Title: Environmental Air Quality and Analysis

Abstract Title: **SIFT-MS: A One-Stop Analytical Tool for Detection of Fumigation Chemicals**

Primary Author: Daniel Milligan
Syft Technologies Ltd

Date: Tuesday, March 08, 2016 - Morning
Time:
Room: Exposition Floor, 400 Aisle

Co-Author(s): Barry Prince, David Hera, Murray McEwan, Thomas McKellar, Vaughan Langford

Abstract Text

Fumigation chemicals (for example, hydrogen cyanide, methyl bromide, phosphine, cyanogen, and sulfuryl fluoride (VikaneTM)) are highly toxic species widely used to exterminate pests and prevent their transfer across international borders. They are commonly used in shipping containers and their presence poses a major risk to people working those containers. They are, however, very diverse in their chemical properties, which creates significant challenges for comprehensive detection and quantitation using traditional technologies. This is especially true when the chemically complex and variable air in shipping containers is considered.

Selected Ion Flow Tube Mass Spectrometry (SIFT-MS) is a real-time analytical technique that offers rapid analysis of volatile organic compounds (VOCs) and many inorganic gases to ultra-trace levels in air. Since 2006, SIFT-MS has been used as a front-line tool for fumigant analysis because it can detect a broad range of fumigants in a single analysis with high selectivity. The fumigant compounds have all been detected in the traditional SIFT-MS positive-ion mode, with the exception of sulfuryl fluoride, which has been detected using electron attachment in a flowing afterglow MS approach in the same instrument.

Recently, the option of using negatively charged ions (OH[sup]-[/sup], O[sup]-[/sup], O[sub]2[/sub][sup]-[/sup], and NO[sub]2[/sub][sup]-[/sup]) has become available on commercial SIFT-MS instruments. Application of negative-ion mode has enabled more selective and sensitive detection of sulfuryl fluoride than the flowing afterglow approach (low part-per-billion (by volume) rather than low part-per-million). Furthermore, most other common fumigants also benefit from enhanced selectivity through the addition of negative-ion-based detection. This paper will describe quantitative performance of the new method.

Keywords: Chemical Ionization MS, Gas, High Throughput Chemical Analysis, Industrial Hygiene
Application Code: Industrial Hygiene
Methodology Code: Mass Spectrometry

Session # 1080 Abstract # 1080-6 **Poster Sessions**

Session Title: Environmental Air Quality and Analysis

Abstract Title: **SIFT-MS: A Complete Solution for Analysis of Ambient Air**

Primary Author: Daniel Milligan, Syft Technologies Ltd

Date: Tuesday, March 08, 2016 - Morning

Time:

Room: Exposition Floor, 400 Aisle

Co-Author(s): Barry Prince, David Hera, Murray McEwan, Thomas McKellar, Vaughan Langford

Abstract Text

This paper presents key results obtained from evaluation of Selected Ion Flow Tube Mass Spectrometry (SIFT-MS) as a complete solution that detects VOCs and critical permanent gases, such as carbon dioxide, methane, nitrous oxide, ozone, sulfur dioxide, and water. Limits of detection, linearity range, and an appraisal of selectivity in the routine monitoring scenario will be discussed.

SIFT-MS is a real-time analytical technique that rapidly analyzes air to ultra-trace levels for volatile organic compounds (VOCs) and inorganic compounds (Prince et al., 2010). Traditional SIFT-MS utilizes three positively charged reagent ions, H[sub]3[/sub]O[sup]+[/sup], NO[sup]+[/sup], and O[sub]2[/sub][sup]+[/sup], which are created from a microwave discharge through moist air and subsequently mass-selected using a quadrupole mass filter.

Recently, the option of using negatively charged reagent ions (OH[sup]-[/sup], O[sup]-[/sup], O[sub]2[/sub][sup]-[/sup], and NO[sub]2[/sub][sup]-[/sup]) has become available on commercial SIFT-MS instruments. This enhancement enables detection of greenhouse gases and other environmental pollutants that were previously inaccessible to SIFT-MS, or were detectable at insufficient sensitivity.

Acknowledgement: This work was funded by Syft Technologies Ltd, New Zealand.
Prince, B.J., Milligan, D.B., & McEwan, M.J. (2010). Rapid Commun. Mass Spectrom., 24, 1763-1769.

Keywords: Chemical Ionization MS, Environmental/Air, High Throughput Chemical Analysis, Ultratrace Analysis

Application Code: Environmental

Methodology Code: Mass Spectrometry

Session # 1080 Abstract # 1080-7 **Poster Sessions**

Session Title: Environmental Air Quality and Analysis

Abstract Title: **Real Time Detection and Identification of Chemical Releases via GCMS**

Date: Tuesday, March 08, 2016 - Morning
Time:
Room: Exposition Floor, 400 Aisle

Primary Author: Parminder Kaur
1st Detect Corporation

Co-Author(s): Corey Stedwell, Daniel DeBord

Abstract Text

Real time monitoring of ambient air using gas chromatography mass spectrometry presents a host of technical changes from achieving stable, autonomous data collection to developing robust detection/identification algorithms for processing the substantial data throughput. A novel mass spectrometry system which is capable of autonomous pump down, acquisition startup, tuning, calibration, data export, and excursion detection has been developed for this application. Long term stability data (> 2 month deployment) will be presented along with the algorithms which enable automatic excursion detection. Excursions for each m/z are independently monitored based upon the probability density function (PDF) of its ion intensities. When an ion exhibits excursion conditions, the corresponding mass spectrum (MS) is compared against the NIST MS library using Pearson's correlation coefficient. The matches showing a p-value<0.05 and presence of GC peaks for the constituent ions of the compound are output as true excursion events. Ions from the matched spectra exhibit multiple consecutive excursions in time arising from the strong GC peaks. This methodology can be applied to real-time applications for generating an alarm when background levels exceed predetermined limits.

Keywords: Chemometrics, Data Mining, Gas Chromatography/Mass Spectrometry, High Throughput Chemical A
Application Code: High-Throughput Chemical Analysis
Methodology Code: Gas Chromatography/Mass Spectrometry

Session # 1080 Abstract # 1080-8 **Poster Sessions**

Session Title Environmental Air Quality and Analysis

Abstract Title **Continuous Fenceline Monitoring Using a Miniature Mass Spectrometer**

Primary Author Preshious Rearden
1st Detect Corporation

Co-Author(s) Madonna M. Mamerow, Parminder Kaur

Date: Tuesday, March 08, 2016 - Morning
Time:
Room: Exposition Floor, 400 Aisle

Abstract Text

Section 112 of the US EPA's Clean Air Act require industrial sites such as petroleum factories to monitor Hazardous Air Pollutants (HAP) levels at their boundaries and report any transient emissions above a certain threshold. Traditionally, passive samplers have been employed for monitoring HAPs due to their ease of deployment and low cost. The major disadvantages of passive methods is their inability to identify the emission source or provide data in real-time. Recent modifications to the regulations governing HAP emissions allow for the introduction and development of new technologies that would enable source attribution as well as real-time emissions tracking. In this study we demonstrate the capabilities of a miniature mass spectrometer as an alternate approach to monitoring transient emissions. The feasibility of real time monitoring of benzene and ethylbenzene fugitive emissions is reported. Results show that quantitative levels of benzene and ethylbenzene can be detected at distances of up to 200 feet with simultaneous detection of both compounds. Also, by factoring in meteorological data into our analysis, we can determine the direction and severity of transient HAP emissions. This approach offers enhanced data feedback to regulators and refineries which is invaluable in achieving prompt response to large-scale emission events and determining attribution in the case of shared fencelines.

Keywords: Environmental/Air, Mass Spectrometry, Petrochemical, Volatile Organic Compounds
Application Code: Regulatory
Methodology Code: Mass Spectrometry

Session # 1080 Abstract # 1080-9 **Poster Sessions**

Session Title: Environmental Air Quality and Analysis

Abstract Title: **Changes in Tobacco-Specific Nitrosamine Cigarette Smoke Deliveries in Unburned, Recycled Portions of Roll-Your-Own Cigarettes**

Primary Author: Benjamin W. Alverson
Centers for Disease Control and Prevention

Date: Tuesday, March 08, 2016 - Morning
Time:
Room: Exposition Floor, 400 Aisle

Co-Author(s): Clifford Watson, Jose J. Perez, Liqin Zhang, Liza Valentin-Blasini, Mary Halstead, Morgan E. Larango, Patrick Chen, R Steven Pappas, Roberto Bravo, Shirley Ding

Abstract Text

Roll-your-own cigarette (RYOC) use became increasingly popular among smokers between 2000–2009 because of significantly lower cost compared with factory-manufactured cigarettes (FMC). The equalization of taxes on FMC and RYO tobacco in 2009 shifted smokers towards using lower-taxed pipe tobacco instead of RYO tobacco to make RYOC. Some smokers perceive that smoking RYOC is a safer alternative to smoking FMC. However, RYOC smokers may have higher levels of harmful exposures resulting from reuse of the unburned tobacco portions of their cigarettes as a price-minimizing strategy. We analyzed carcinogenic tobacco-specific nitrosamines (TSNAs) using liquid chromatography-tandem mass spectrometry (LC-MS/MS) in smoke from RYOC made from fresh (1st generation) and recycled (2nd generation) pipe tobacco. Tobacco for 2nd-generation RYOC was obtained by smoking approximately two-thirds of the tobacco contained in multiple 1st-generation RYOC and collecting/pooling the remaining unburned tobacco. First-generation RYOC smoke data provided a baseline for relative comparison of 2nd-generation RYOC TSNA smoke deliveries. Of the twelve pipe tobacco brands tested, 2nd-generation RYOC smoke—collected using ISO and Canadian Intense smoking regimes—yielded statistically higher TSNA levels ($p < 0.05$) for most brands, suggesting that smokers who reuse the unburned fractions of RYO tobacco are at risk of increased exposure to TSNAs.

Keywords: Liquid Chromatography/Mass Spectroscopy
Application Code: Other
Methodology Code: Liquid Chromatography/Mass Spectrometry

Session # **1080** **Abstract #** **1080-10** **Poster Sessions**

Session Title Environmental Air Quality and Analysis

Abstract Title **A Method of Performing In-Trap Photoionization in a Miniature Ion Trap Mass Spectrometer**

Date: Tuesday, March 08, 2016 - Morning
Time:
Room: Exposition Floor, 400 Aisle

Primary Author Corey Stedwell
1st Detect Corporation

Co-Author(s) Daniel DeBord, David Rafferty, Michael Spencer

Abstract Text
Gas-phase ionization of molecules by electron ionization (EI) is one of the most widely used ionization methods in mass spectrometry. While EI is amenable to numerous types of chemicals, the energetic nature of the ionization can result in significant fragmentation, dilution of ion signal over many m/z values, and hindered interpretation of mass spectra. Many alternative ionization techniques (e.g., electrospray ionization, chemical ionization, photoionization) provide "softer" ionization conditions, resulting in reduced fragmentation. However, these ionization techniques are generally performed at atmospheric pressure, increasing the chance for ion losses crossing the vacuum barrier and are limited to laboratory equipment, and thus are not fieldable. Here, we present a method of performing photoionization on a series of volatile organic compounds (VOCs) within the confines of an ion trap mass analyzer.

Keywords: Mass Spectrometry
Application Code: Fuels, Energy and Petrochemical
Methodology Code: Mass Spectrometry

Session # 1090 Abstract # 1090-1 **Poster Sessions**

Session Title: Environmental and Geochemical Analysis: Soils, Minerals, and Agriculture

Abstract Title: **Automated Analysis of Explosives in Soil Samples**

Primary Author: William Hedgepeth
Shimadzu Scientific Instruments

Co-Author(s): Kenichiro Tanaka

Date: Tuesday, March 08, 2016 - Morning
Time:
Room: Exposition Floor, 400 Aisle

Abstract Text

There are a large number of explosives-contaminated sites in the US, Europe, and Asia. High levels of explosives in soil can threaten the health of humans, livestock, and wildlife. A number of remediation efforts are underway, which require the analysis of explosives in soil samples. Recently, a new technique was introduced the allows the automated supercritical extraction and SFC analysis of samples with minimal sample preparation and handling requirements to save analyst time and sample preparation expenses. This technique was applied to the analysis of explosives in soil samples and showed good recoveries of the explosives tested in a number of different soil samples. Automated analysis of up to 48 samples is possible without the need for manual sample preparation to allow quick screening of explosives in numerous soil samples.

Keywords: Automation, Environmental/Soils, Forensic Chemistry, SFE
Application Code: Environmental
Methodology Code: Separation Sciences

Session # 1090 Abstract # 1090-2 **Poster Sessions**

Session Title: Environmental and Geochemical Analysis: Soils, Minerals, and Agriculture

Abstract Title: **Mechanisms for Controlling Soil Organic Matter Decompositions: An Application of Pyrolysis –Cryogenic –GC/MS to Molecular Characterizations of Organic Matter in Grass and Forestry Soils**

Date: Tuesday, March 08, 2016 - Morning
Time:
Room: Exposition Floor, 400 Aisle

Primary Author: Xianzhi (Amanda) Song
Young Harris College

Co-Author(s)

Abstract Text

Soil organic matter consists of a mixture of plant and animal products in various stages of decomposition and humification. Although soil organic matter and associated humic substances plays an important role in many environmental processes such as soil carbon sequestration, contaminant fate and transportation, and carbon and nitrogen biogeochemical cycles, its molecular structural analyses remain challenging to analytical chemistry science. The primary objective of this study was to apply the pyrolysis –cryogenic GC/MS directly to raw soil samples for studying molecular structures of organic matter contained. The soil samples were collected from grass and forestry lands at the depths of 0-12 cm, 12-25 cm, and 25-38 cm. The secondary objective of the research was to investigate the mechanisms that control soil organic matter decomposition processes through comparing molecular structural changes over the soil profiles. The soil samples were first purged with helium and pyrolysized at the temperature programmed from 300 to 500°C with a temperature increasing rate of 10°C/min. The pyrolysates were condensed and collected in a cryogenic sample loop that was merged in a liquid nitrogen container. After a desired period of time, the pyrolysates were flashed out into GC/MS. And the pyrolysis GC/MS spectra were analyzed for derivatives from aliphatic hydrocarbon, lignin, polysaacharide, and heterocyclic N and non- heterocyclic N. The spectra were also compared with those obtained from humic and fulvic acids. Our preliminary results show distinguishable differences between grass and forestry soil samples and at different depths. With increasing soil depth, pyrolysate products of the soil samples resemble those from humic and fulvic acids. It is our hope that further analyses may offer new information of molecular structures of soil organic matter through a simple and direct analysis of soils.

Keywords: Environmental/Soils, Gas Chromatography/Mass Spectrometry, Quantitative, Speciation
Application Code: Environmental
Methodology Code: Gas Chromatography/Mass Spectrometry

Session # 1090 Abstract # 1090-4 **Poster Sessions**

Session Title: Environmental and Geochemical Analysis: Soils, Minerals, and Agriculture

Abstract Title: **Ion Selective Electrodes – A Cheaper, Simpler and more Robust Analytical Method for Monitoring of Nitrate and Ammonium in Water and Soil**

Date: Tuesday, March 08, 2016 - Morning
Time:
Room: Exposition Floor, 400 Aisle

Primary Author: Tolulope A. Fayose
Keele University

Co-Author(s)

Abstract Text

Mineral reactive nitrogen (nitrate-nitrogen and ammonium-nitrogen) are critical for soil fertility and crop productivity. Consequently the use of synthetic fertilizers and organic manures application has been increased tremendously to enhance and maintain high agricultural productivity. However, the increased use of fertilizers application to croplands often in excess of plant nutrient requirements has led to losses of reactive nitrogen (Nr) into surface, coastal and groundwater resources with implications on water quality, public health and health of aquatic ecosystems. Among others, one way to effectively match plan Nr requirement and to reduce losses of Nr to water resources, is to develop robust techniques that can routinely be used for quick, and low-cost monitoring of Nr in soil and water to aid end-user of fertilizers (farmers and regulatory agencies) achieve agricultural and environmental sustainability. Ion-selective electrodes (ISEs) are a promising analytical technique which offers robust sensitivity and selectivity for target ions in solution. ISEs are simple and low-cost, and are not affected by sample turbidity and/or matrix. Therefore, ISEs holds feasible promise for monitoring Nr in soil and water in support of the overall Nr management at various levels. In this abstract, we present data on the suitability, sensitivity, selectivity and robustness of ISEs for measuring Nr in varied environmental media including organic and mineral soils, slurry and water. The performance of ISEs in measuring Nr is validated and compared using laboratory-based ion chromatographic and spectrophotometric techniques.

Keywords: Electrochemistry, Electrodes, Environmental Analysis, Potentiometry
Application Code: Environmental
Methodology Code: Sensors

Session # **1090** **Abstract #** **1090-5** **Poster Sessions**

Session Title: Environmental and Geochemical Analysis: Soils, Minerals, and Agriculture

Abstract Title: **Multiple Surface-Science Techniques to Elucidate the Reactive Nature of a Metal Phosphide Mineral**

Date: Tuesday, March 08, 2016 - Morning
Time:
Room: Exposition Floor, 400 Aisle

Primary Author: Danna Qasim
Kennesaw State University

Co-Author(s): Aaron Pital, Heather Abbott-Lyon, Thomas Beckman

Abstract Text

A number of recent studies have been performed on the meteoritic mineral, schreibersite (Fe,Ni)[sub]3[/sub]P, which is hypothesized to be the ultimate source of phosphorous in life. Schreibersite is known to phosphorylate biomolecules such as glycerol, choline, adenosine and uracil in aqueous solution. However, there is little information on how these products are formed. Because these reactions occur in the presence of schreibersite, it is likely that the mineral surface has a direct role in product formation. This presentation discusses changes at the mineral interface related to corrosion and phosphorylation. Water, methanol, and a mixture of both were dosed onto the schreibersite surface at various coverages. A QMS and an FTIR were used to perform temperature programmed desorption (TPD) and reflection-absorption infrared spectroscopy (RAIRS), respectively. These techniques provide information about chemical structure changes on the surface as conditions change. SEM/EDS and XPS were also used to characterize the geometric structure and the composition of the surface. The combination of these surface-sensitive experiments helps illustrate the magnitude of the role of the schreibersite surface in the aqueous-phase reactions by providing details on the adsorbate-surface interaction.

This work was jointly supported by NSF and the NASA Astrobiology Program, under the NSF Center for Chemical Evolution, CHE-1004570.

Keywords: Mass Spectrometry, Microscopy, Surface Analysis, Vibrational Spectroscopy

Application Code: Environmental

Methodology Code: Surface Analysis/Imaging

Session # 1090 Abstract # 1090-11 **Poster Sessions**

Session Title: Environmental and Geochemical Analysis: Soils, Minerals, and Agriculture

Abstract Title: **A New Advancement in the Automated Preparation of Pressed Pellets for XRF Analysis**

Date: Tuesday, March 08, 2016 - Morning

Time:

Room: Exposition Floor, 400 Aisle

Primary Author: David Coler
FLSmidth

Co-Author(s): Ian Campbell, Lukas Bruzenak

Abstract Text

One of the single largest sources of error in XRF analysis is sample preparation. Minimizing error in the preparation of samples is therefore critical to obtaining high quality analytical results. This is particularly relevant for the preparation of pressed pellets which is one of the most common methods used for XRF analysis. Some of the major sources of sample preparation error include

1) Large or highly variable particle size distribution in the sample which can lead to heterogeneities in the sample relative to the sampling depth of the XRF and shadow effects on the surface of the sample.
2) Thin samples or those that vary in thickness can lead problems with infinite thickness and create errors relative to the energy of the elements being analyzed.
3) Incomplete compaction of the sample leading to small void spaces in the sample creating heterogeneities in the sample.
4) Contamination of the sample with residual material from previous samples in the milling and pressing process.
5) Variations in the dilution factor of samples caused by weighing errors of the sample and binder.

All of these errors can be magnified in by the variation induced by multiple lab personnel preparing samples for analysis.

We present data from the Centaurus, a new automated mill and press that demonstrates how this instrument addresses these main sources of error for the analysis of different geological materials.

Keywords: Automation, High Throughput Chemical Analysis, Process Control, X-ray Fluorescence
Application Code: Process Analytical Chemistry
Methodology Code: X-ray Techniques

Session # 1090 Abstract # 1090-12 **Poster Sessions**

Session Title: Environmental and Geochemical Analysis: Soils, Minerals, and Agriculture

Abstract Title: **Multifunctional Ligand Platform for Detection, Capturing and Removal of Cerium Oxide Nanoparticles**

Primary Author: Ali Othman
Clarkson University

Date: Tuesday, March 08, 2016 - Morning
Time:
Room: Exposition Floor, 400 Aisle

Co-Author(s): Silvana Andreescu

Abstract Text

Nanoceria is currently used in various catalytic processes due to their unique chemical and electronic configuration. These particles are characterized by high reactivity, and catalytic properties which make them useful for implementation in many practical applications. In addition to the recently discovered therapeutic applications, nanoceria has been used as fuel additive, as a fuel-borne catalyst and as abrasives in printed circuit manufacture to decrease the emission of particulate matter from diesel engines and to lower the generation of diesel exhaust particles (DEPs), but are emitted as cerium oxide nanoparticles (CeO2) along with DEP in the diesel exhaust. Studies show that these nanoparticles may induce lung injury and co-localized in the lung tissues after combined exposure. Moreover, CeO2 induced sustained inflammation and surfactant accumulation, and altered the balance of mediators involved in tissue repair process leading to excess collagen deposit and pulmonary fibrosis. Thus, the release of these nanoparticles into the environment may cause health concerns. Methods to determine the concentration of these nanoparticles under conditions relevant to environmental and biological systems are needed to determine the level of exposure and provide concentration limits for toxicological testing. In this presentation, we demonstrate design and development of a new method for the detection of nanoceria particles (CeO2). The method is based on the use of different organic ligands (chelating agents) such as ascorbic acid and catechol that are used to recognize and catalytically amplify signals, aiding in the detection of the nanoceria particles in the environment. The analytical capability of our approach and a potential implementation of this method for real world applications will be discussed.

Keywords: Adsorption, Detection, Environmental/Waste/Sludge, Sensors

Application Code: Environmental

Methodology Code: Chemical Methods

Session # 1100 Abstract # 1100-1 **Poster Sessions**

Session Title: Environmental Applications of Elemental Analysis and Speciation

Abstract Title: **Rare Earth Elements – How to Accurately Determine Contamination Levels and Remove Spectral Interferences Created by Them**

Date: Tuesday, March 08, 2016 - Morning
Time:
Room: Exposition Floor, 400 Aisle

Primary Author: Ewa M. Pruszkowski
PerkinElmer, Inc.

Co-Author(s): Cynthia Bosnak

Abstract Text

Rare earths have unique magnetic, luminescent, and electrochemical properties that are used in many technologies such as aerospace, health care, clean energy, electronics and transportation. However, often the functionality of high tech devices that incorporate REE depends on the cleanliness of the materials used. On the other hand, more and more REE (rare earth elements) are used in every day products and there is a growing concern that some of them can pollute the water supply and cause incorrect results for some elements that are routinely analyzed in drinking water due to spectral interferences. The objective of this poster is to investigate the capability of the current ICP-MS instrumentation for accurately determining contamination in high purity REE and in drinking water contaminated by REE. It will be shown that the ICP-MS with the universal cell, utilizing reaction/collision technology, can efficiently remove spectral interferences created by REE and also by Ar, O2, H2 and the acids added during sample preparation.

Keywords: Environmental Analysis, Environmental/Water, ICP-MS, Materials Characterization
Application Code: Environmental
Methodology Code: Atomic Spectroscopy/Elemental Analysis

Session # 1100 **Abstract # 1100-2** **Poster Sessions**

Session Title: Environmental Applications of Elemental Analysis and Speciation

Abstract Title: **Determination of Heavy Metals In and Around Lake Ontario**

Date: Tuesday, March 08, 2016 - Morning

Time:

Room: Exposition Floor, 400 Aisle

Primary Author: Simran Sandhu
St. John Fisher College

Co-Author(s): Kimberly Chichester

Abstract Text

The concentration of heavy metals in and around Lake Ontario was determined using Flame Atomic Absorption Spectroscopy (FAAS). Heavy metals can have negative developmental, reproductive, and neurological health effects on humans and the organisms of the aquatic ecosystem. The aim of the research was to quantify the concentrations of lead, nickel, cadmium, and mercury and to further determine how to remove these specific heavy metals from the water. It is imperative to preserve the aquatic ecosystems in Lake Ontario and the water quality since it is the largest source of fresh water in the Eastern Hemisphere. Water samples were collected from two beaches on Lake Ontario. Soil samples were also collected from the same two beaches along the shoreline and the water's edge. An extraction procedure was carried out multiple times on each soil sample with hydrochloric acid and nitric acid and the water samples were simply filtered before analysis by FAAS.

Keywords: Atomic Absorption, Atomic Spectroscopy, Environmental/Soils, Environmental/Water
Application Code: Environmental
Methodology Code: Atomic Spectroscopy/Elemental Analysis

Session # 1100 **Abstract # 1100-3** **Poster Sessions**

Session Title: Environmental Applications of Elemental Analysis and Speciation

Abstract Title: **The Analysis of Flue Gas Desulfurization Fluids by ICP-MS Using Universal Cell and FastFIAS Technology**

Primary Author: Michelle M. Coker
SCE&G

Date: Tuesday, March 08, 2016 - Morning
Time:
Room: Exposition Floor, 400 Aisle

Co-Author(s): Daniel H. Jones, George W. Eargle

Abstract Text

Gasses emitted from coal fired power plants have been the subject of much environmental scrutiny in recent years. These emissions from the smoke stack or Flue can contain greenhouse gasses such as sulfur dioxide and other pollutants. In order to reduce the level of these contaminates that are released into the atmosphere, power plants often employ a limestone oxidation scrubbing system to convert the gaseous sulfur dioxide into calcium sulfate. The solutions, often referred to as FGD (Flue Gas Desulfurization) solutions, used in this type of scrubbing system are very acidic and trap many other contaminates such as heavy metals, alkali earth metals, chlorides and dissolved organic compounds. In turn, FGD solutions must also be characterized for waste disposal, however the complex matrix of the FGD samples make analysis by ICP-MS difficult. First, the concentration of metals in these samples can cause polyatomic interferences that need to be eliminated for accurate ICP-MS analysis. This is accomplished using Universal Cell Technology employing a combination of Kinetic Energy Discrimination with Helium and the use of reactive gasses to completely remove the interferent. Secondly, the elevated amount of dissolved solids can cause long term stability issues by depositing matrix on the cones. This high matrix can also suppress the signal of the analyte below what an internal standard can accurately compensate. This issue can be resolved with fastFIAS sample introduction technology. FastFIAS injects the sample into the plasma in discrete microliter pulses. This eliminates the continuous flow of sample into the plasma during analysis while maintaining good sensitivity and vastly improved internal standard recoveries.

Keywords: Coal, Energy, Environmental/Air, ICP-MS
Application Code: Environmental
Methodology Code: Atomic Spectroscopy/Elemental Analysis

Session # 1100 Abstract # 1100-4 **Poster Sessions**

Session Title: Environmental Applications of Elemental Analysis and Speciation

Abstract Title: **Determination of Cadmium in Environmental Water Samples Collected in Superfund Sites in New York City**

Date: Tuesday, March 08, 2016 - Morning

Time:

Room: Exposition Floor, 400 Aisle

Primary Author: Yi He
John Jay College/CUNY

Co-Author(s): Kate Good, Kathleen Lopez, Sandra Swenson

Abstract Text

Cadmium, even at low concentration, impacts ecosystem and human health adversely. Monitoring Cd in environmental water, therefore, is important to ensure the levels of cadmium do not rise to harmful levels. In this project, preservation of water samples through acidification and filtration and measurement using graphite furnace atomic absorption spectroscopy (GFAAS) and inductively-coupled mass spectrometry (ICPMS) were investigated. Superfund sites around New York City, including the lower Hudson River, Gowanus Canal, and Newtown Creek were used as sampling sites for cadmium determination. These bodies of water have historically known to be contaminated with various pollutants, including cadmium. Analysis results revealed that cadmium concentrations for the Hudson River sampling site (59th St) was 1.98 ppb, Gowanus Canal was 2.95 ppb, and Newtown Creek was 17.56 ppb.

Keywords: Elemental Analysis, Environmental/Water

Application Code: Environmental

Methodology Code: Atomic Spectroscopy/Elemental Analysis

Session # 1100 Abstract # 1100-6 **Poster Sessions**

Session Title: Environmental Applications of Elemental Analysis and Speciation

Abstract Title: **Evaluation of Chromium Stability on Filters**

Primary Author: Tamutsiwa M. Mututuvari
High Purity Standards

Date: Tuesday, March 08, 2016 - Morning
Time:
Room: Exposition Floor, 400 Aisle

Co-Author(s): Kim-Phuong Tran, Svetlana Uzunova

Abstract Text

Hexavalent chromium (Cr (VI)) is commonly added to spray-paint to provide corrosion protection and to create specific colors. However, exposure to Cr (VI) creates health complications such as lung and nasal cancer. Therefore, air quality should be monitored to help protect workers from exposure to harmful levels of Cr (VI). Monitoring air quality is accomplished by using filters that adsorb Cr (VI) particulates from the air. However, the interactions between Cr (VI) and filter surfaces may induce inter-conversion between the more toxic Cr (VI) and the less toxic Cr (III) species before analysis of the filter sample. In addition, this interconversion may occur during extraction of the Cr (VI) from the filters prior to analysis. In the previous study, we investigated the stability of Cr (VI) on PVC filter medium. In the present study, we extended that study to include quartz filter medium. This was accomplished by storing filters under different environmental (temperature and humidity) conditions. Specifically, filters were stored under three different temperature conditions; ambient, 4 °C, and 40 °C. Preliminary results suggest that low temperature (4 °C) conditions stabilize Cr (VI) the best. This study will provide an insight into the chemistry of inter-conversion between Cr species. This will enable environmentalists to obtain more accurate values for the concentration of Cr (VI) in air.

Keywords: Chromatography, Environmental Analysis, Environmental/Waste/Sludge, ICP-MS
Application Code: Environmental
Methodology Code: Liquid Chromatography

Session # 1100 Abstract # 1100-8 **Poster Sessions**

Session Title: Environmental Applications of Elemental Analysis and Speciation

Abstract Title: **Online Analysis and Speciation of Antimony in Various Wastewater Streams Using Hydride Generation-AFS**

Primary Author: Bin Chen
P S Analytical

Date: Tuesday, March 08, 2016 - Morning
Time:
Room: Exposition Floor, 400 Aisle

Co-Author(s): Peter B. Stockwell, Warren T. Corns

Abstract Text

Antimony levels in drinking water are generally quite low and not considered a major health concern. However it is used in various industrial processes and is discharged in wastewater. Antimony compounds are used as catalysts in the production of polyesters and flame retardants and these ultimately cause releases of antimony to the environment during production and disposal. In June 2013, the EPA proposed a new rule "40 CFR Part 423" which relates to effluent limitation guidelines to limit the amount of toxic metals being discharged to surface waters. Antimony is one of the main elements of concern because of its high toxicity and potential impact to the environment and wildlife. Online analysis of Sb in wastewater streams enables the possibilities of precise monitoring of the complete water treatment process and ensures the compliance with discharge regulations. However the analytical challenges remain due to the complexity of the wastewater matrix. Hydride generation is also oxidation state dependent and therefore online chemistries have to be developed to ensure accurate measurements are achieved. In this presentation, various industrial wastewater streams are successfully analysed by online analysis based on hydride generation atomic fluorescence spectrometry (HG-AFS). Detection limits of 200 ppt with linearity to 100 ppm are achievable with the online instrumentation. The results were compared and cross examined by HPLC-HG-AFS and HG-AFS methodologies after the appropriate sample treatments.

Keywords: Atomic Spectroscopy, Environmental/Waste/Sludge, On-line, Trace Analysis
Application Code: Environmental
Methodology Code: Atomic Spectroscopy/Elemental Analysis

Session # 1100 Abstract # 1100-9 **Poster Sessions**

Session Title: Environmental Applications of Elemental Analysis and Speciation

Abstract Title: **An Efficient Recovery of Rare Metal Ions with Calix[4]arene Derivatives from Acidic Media Using Droplet-Based Microreactor System**

Date: Tuesday, March 08, 2016 - Morning

Time:

Room: Exposition Floor, 400 Aisle

Primary Author: Masaya Miyazaki
National Institute of Advanced Industrial Science and Techn

Co-Author(s): Keisuke Ohto, Masatoshi Maeki, Ramachandra Rao Sathuluri

Abstract Text

Precious metals are important rare metals for advanced materials. The supply, however, has been inconsistent due to poor natural abundance. Recycling from spent home appliances i.e. urban mine is complicated as it contains several other metals. Therefore, technique that enables recovery metal ions selectively is required. Calix[4]arenes are phenolic oligomers, which has the ability to discriminate metal ions making them suitable as specific receptors. However, the slow extraction rate is to be resolved in using calix[4]arene solvent extraction. We propose a droplet-based microfluidc reactor with larger surface/interface area per unit volume than conventional macroscale system. In this study, we investigated the potential of droplet-based reactors for recovery of silver and palladium from aqueous solutions as well as extraction from commercial metal waste. We fabricated microreactor (0.2 × 0.2 mm2) by micromachining process for droplet-based recovery. Silver nitrate, Palladium nitrates dissolved to a desired conc. in 0.1M HNO3 as aqueous solution, while organic solution is prepared by dissolving methylketonic, ethylamide calix[4]arene derivatives in chloroform. We evaluated the extraction percentage of metal ions by ICP-AES from aqueous phase by varying the extraction time (2 to 10 s). As a result, the time required to reach equilibrium for silver ion extraction was 4 s in the droplet-based microreactors, which is over 90% against 72 h in batch-wise method. Methylketonic calix[4]arene is selective for silver, while ethylamide type for palladium. These results show that an increasing liquid-liquid interface per unit volume is effective in solvent extraction of metal ion with calix[4]arene derivatives.

Keywords: Accelerated Solvent Extraction, Environmental/Waste/Sludge, Extraction, Lab-on-a-Chip/Microfluidic
Application Code: Environmental
Methodology Code: Separation Sciences

Session # **1100** **Abstract #** **1100-10** **Poster Sessions**

Session Title: Environmental Applications of Elemental Analysis and Speciation

Abstract Title: **Breakthrough Development for Quantitative Analysis of Total Metals in Soil by Portable High Definition X-ray Fluorescence**

Primary Author: Zewu Chen
XOS

Date: Tuesday, March 08, 2016 - Morning
Time:
Room: Exposition Floor, 400 Aisle

Co-Author(s): Danhong Li, Kyle Kuwitzky, Shenghua Song

Abstract Text

Heavy metal contamination from industrialization in developing countries is an important environmental issue today. Hand Held X-ray Fluorescence (HHXRF) analysis has recently developed into a powerful tool to screen heavy metal contamination at brownfield sites. However, HHXRF on-site analysis does not meet the needs in many contamination site survey and risk assessments due to three major reasons: 1) insufficient accuracy to provide good assessment of contaminated sites. 2) Poor detection limits during contamination surveys, especially for Hg, Cd, Ni, Cr, and As. 3) Widely varying measurement results due to the heterogeneous nature of soil samples. Currently, comprehensive site assessment is heavily dependent on field sample collection and lab analysis because of these limitations. The large scale sample collection and lab analysis is a severe bottle neck for environmental cleanup process in terms of cost, on-site decision making, and time management while waiting for off-site lab analysis.

Recently, new portable soil heavy metal analysis techniques using innovative, focused, and multiple monochromatic beam X-ray fluorescence technology have been developed to address industry concerns. By utilizing miniature doubly curved crystal optical systems coupled with a miniature x-ray tube and a sample rotation mechanism to increase analysis area and improve measurement repeatability, superior detection limits for heavy metal analysis and better comparability with laboratory methods have been achieved.

In this poster, in-depth analysis of a measurement data set of various NIST soil standards and soil samples from actual survey sites will be presented. Data from the new technique and laboratory analysis using US EPA ICP methods will be summarized. Results from the analysis demonstrate that on-site, quantitative analysis for low level heavy metals (1-10ppm range) in soil has been achieved for the first time.

Keywords: Environmental, Environmental/Soils, X-ray Fluorescence
Application Code: Environmental
Methodology Code: X-ray Techniques

Session # 1100 Abstract # 1100-12 **Poster Sessions**

Session Title: Environmental Applications of Elemental Analysis and Speciation

Abstract Title: **Collaborative Certification of a New Low-Level Hexavalent Chromium Standard Reference Material in a Soil Matrix**

Date: Tuesday, March 08, 2016 - Morning

Time:

Room: Exposition Floor, 400 Aisle

Primary Author: James E. Henderson
Duquesne University

Co-Author(s): Anil Srinivas Chaitanya Vishnuvajjhala, Bob O'Brien, Francine Walker, Jennifer Crawford, Logan T. Miller, Matt Pamuku, Pam Wee, Patrick Benecewicz, Skip Kingston, Teresa Switzer, Vasile Furdui,

Abstract Text

Several hexavalent chromium standards in soil, certified in the past decade, either have high concentrations of chromium or active matrices that render them unsuitable for validation studies that use natural soil untouched by industrial activity. NIST 2701, for example, is made of chromium processing ore residue (COPR) industrial waste material from New Jersey. It contains approximately four percent total chromium, with an unnaturally high Cr(VI) fraction (552.1 mg/kg). SRM-2701 also includes an active matrix that dominates the chemistry of the standard and causes significant Cr(III)/Cr(VI) species shift and biases during extraction. Presently, no low-background level standard containing an inert matrix is available for the appropriate validation of low level Cr(VI) in native or uncontaminated soil. An international collaboration is under way to produce a more appropriate series of low level soil certified standard reference materials for Cr(VI) analysis. Multiple laboratories in different locations are involved in certifying two new reference materials at ng/g levels. These were prepared by Sigma-Aldrich in larger batches and were distributed among the participating laboratories for collaborative certification. The methods used for certification are two EPA RCRA methods: alkaline extraction by EPA method 3060A and speciated isotope dilution mass spectrometry by EPA method 6800 (Update V, 2015). The methods used in the preparation of the materials and preliminary certification, including means and confidence limits, will be presented along with some discussion of the Eh and Ph phase diagram stability of the material.

Keywords: Elemental Analysis, ICP-MS, Reference Material, Validation
Application Code: Validation
Methodology Code: Mass Spectrometry

Session # 1100 Abstract # 1100-13 **Poster Sessions**

Session Title: Environmental Applications of Elemental Analysis and Speciation

Abstract Title: **Impact of Coexistent Elements and Its Concentration in the Quantification of Strontium-90 Using ICP-MS with Cascade Separation System**

Date: Tuesday, March 08, 2016 - Morning

Time:

Room: Exposition Floor, 400 Aisle

Primary Author: Yoshitaka Takagai
Fukushima University

Co-Author(s): Aya Yokoyama, Katz Suzuki, Makoto Furukawa, Takahiro Suzuki, Yutaka Kameo

Abstract Text

Radioactive strontium-90 ([sup]90[/sup]Sr) scattered by nuclear power plant accident was specifically quantified by conventional inductively coupled plasma quadrupole mass-spectrometry (ICP-QMS) preceded by on-line chelate column separation (based on lab-on-valve) and oxygen reaction (designated the cascade step). The proposed system overcomes the isobaric interference of [sup]90[/sup]Zr, whose soil concentration exceeds that of [sup]90[/sup]Sr by more than six orders of magnitude. In addition, the system requires no ultimate mass spectrometry or radioactive [sup]90[/sup]Sr standards. The modified ICP-QMS system yielded a precise, reproducible sharp [sup]90[/sup]Sr peak in the ICP-MS profile [sup][1][/sup]. However, the allowance of the upper concentration of coexistent element has been not known. In this presentation, we present the impact of coexistent elements and Its upper concentration in the quantification of [sup]90[/sup]Sr using ICP-MS with cascade separation system.

[1]Y Takagai, M. Furukawa, Y. Kameo, K. Suzuki, Analytical Methods, 6(2), 355-362 (2014).

Keywords: ICP-MS, Nuclear Analytical Applications
Application Code: Nuclear
Methodology Code: Atomic Spectroscopy/Elemental Analysis

Session # 1110 Abstract # 1110-1 **Poster Sessions**

Session Title: Environmental Organic Analysis: VOCs, Pesticides, and Others

Abstract Title: **Determination of VOCs by US EPA Method 8260 with Extended Dynamic Range Using Fast, Sensitive Capillary GC/MS**

Date: Tuesday, March 08, 2016 - Morning

Time:

Room: Exposition Floor, 400 Aisle

Primary Author: Brahm Prakash
Shimadzu Scientific Instruments, Inc

Co-Author(s): Di Wang, Laura Chambers, Nicole M. Lock, Shilpi Chopra, William Lipps

Abstract Text

The US EPA has developed and published multiple methods for the analysis of organic environmental pollutants, including Volatile Organic Compounds (VOCs), while standardizing on single-quadrupole gas chromatography-mass spectrometry (GC/MS). Over the years, GC/MS instrumentation has evolved with changes in MS detector design, resulting in improvements in sensitivity and reliability which help increase productivity of environmental laboratories, but there haven't been many significant advancements in the overall methodology since the mid-1980s.

US EPA Method 8260 is by far the most comprehensive method in terms of the number of VOCs included in the compound list, with as many as 100 or more RCRA compounds slated for testing. The method is used to determine VOCs in a variety of solid waste matrices, is applicable to nearly all types of samples, and is one of the most common VOC methods used by commercial testing laboratories today. The chemist is confronted with meeting regulatory compliance requirements for all compounds on a routine basis, which can be a challenging task.

This poster describes the effects of recent instrument improvements and method modifications on sensitivity for US EPA Method 8260. An extended calibration range minimizes the number of dilutions and re-analyses that are required for high-concentration compounds, while still reaching the required low detection limits. Analytical operating conditions including BFB tune parameters, calibration details, and a complete MDL and Precision and Accuracy study for almost 100 target compounds over an extended calibration range are described.

Keywords: Environmental, Gas Chromatography/Mass Spectrometry, Purge and Trap, Volatile Organic Compou
Application Code: Environmental
Methodology Code: Gas Chromatography/Mass Spectrometry

Session Title: Environmental Organic Analysis: VOCs, Pesticides, and Others

Abstract Title: **Determination of Organochlorine Pesticides and Polychlorinated Biphenyls Using GC-MS/MS Operated in the MRM Mode**

Primary Author: Brahm Prakash
Shimadzu Scientific Instruments, Inc

Date: Tuesday, March 08, 2016 - Morning
Time:
Room: Exposition Floor, 400 Aisle

Co-Author(s): Di Wang, Laura Chambers, Nicole M. Lock, Shilpi Chopra, William Lipps

Abstract Text

The determination of chlorinated pesticides (OCPs) and polychlorinated biphenyls (PCBs) in environmental matrices are common analyses in most environmental laboratories. These compounds are typically analyzed by employing solid phase or liquid liquid extraction with Methylene chloride, concentration, solvent exchange into hexane, and interference removal using acid, copper, or column chromatography. Analysis is done using gas chromatography (GC) with electron capture detection (ECD) and requires confirmation of every detected component on another, dissimilar GC column. GC-ECD techniques are prone to positive and negative bias in complex matrices resulting in unnecessary cleanup costs and/or violations of NPDES permits. Clearly, a new method for pesticides and PCBs, based on modern GCMS technology is needed.

This poster describes use of a tandem GC-MS/MS method using Multiple Reaction Monitoring (MRM) mode for sensitive and selective detection and quantitation of organochlorine pesticides and PCBs. A database with optimized MRM transitions for all of the OCPs and PCBs including relative retention times for all components makes method setup possible within minutes. The use of GCMSMS MRM mode provides enhanced selectivity, specificity and sensitivity in complex matrices with potential co-eluting interferences.

This poster presents all instrument operating conditions, and instrument method performance statistics including method linearity, accuracy, precision, and instrument detection limits for all compounds.

Keywords: PCB's, Pesticides, Semi-Volatiles, Tandem Mass Spec
Application Code: Environmental
Methodology Code: Gas Chromatography/Mass Spectrometry

Session # 1110 Abstract # 1110-3 **Poster Sessions**

Session Title: Environmental Organic Analysis: VOCs, Pesticides, and Others

Abstract Title: **Analysis of Terpenes Using Gas Chromatography with Vacuum Ultraviolet Detection**

Primary Author: Changling Qiu
University of Texas at Arlington

Date: Tuesday, March 08, 2016 - Morning
Time:
Room: Exposition Floor, 400 Aisle

Co-Author(s): Jonathan Smuts, Kevin A. Schug, Phillip Walsh

Abstract Text

The separation and identification of natural mixtures of terpenes is challenging and laborious. A new gas chromatography detector based on vacuum ultraviolet (VUV) spectroscopy, which collects full scan adsorption in the range of 115-240 nm, was developed and applied to analyze terpenes. VUV features deconvolution of co-eluting signals from different analytes, making it a great tool for analysis from complex matrices. In this study, the VUV absorption spectra for different terpenes (51) were investigated and compared. The spectra of terpenes were found to be unique and highly featured. VUV was applied to the analysis of turpentine and demonstrated excellent specificity for qualitative identification analysis of terpenes. Deconvolution of co-eluting signals of terpenes was achieved utilizing the VUV data analysis software.

Keywords: Capillary GC, Detector, Volatile Organic Compounds
Application Code: Environmental
Methodology Code: Gas Chromatography

Session # 1110 Abstract # 1110-4 **Poster Sessions**

Session Title: Environmental Organic Analysis: VOCs, Pesticides, and Others

Abstract Title: **Use of a Boron Doped Diamond Electrode Sensor for Carbamate Pesticide Classification Using a Chemometric Approach**

Date: Tuesday, March 08, 2016 - Morning
Time:
Room: Exposition Floor, 400 Aisle

Primary Author: Thiago Selva
University of Sao Paulo

Co-Author(s): Thiago Paixao

Abstract Text

Electrochemical boron doped diamond (BDD) sensors can be an interesting analytical devices for extracting pesticide information in water samples. However, this type of sensing requires low detection levels relevant for human health and ecosystem protection monitoring; moreover, pesticide classification also requires fast feedback based on the correct diagnostic of which pesticide is present [1]. To this end, this paper describes the development of an analytical method to classify five carbamates pesticides (aldicarb, carbaryl, carbofuran, methomyl and propoxur) using Principal Component Analysis (PCA) and a [i]k[/i]-nearest neighbours approach ([i]k[/i]-NN). Current values of the oxidation process of five carbamates in different concentrations (50 to 3000 [micro]mol L[sup]-1[/sup]) were recorded using a BDD sensor in a working potential range from 1.0 to 1.8 V and were pre-processed using a baseline correction and a normalization step to bring the current values between 0 and 1. These values were used as input data for PCA resulting in the discrimination of five clusters, each corresponding to the pesticides studied. Additionally, a [i]k[/i]-NN model was performed in order to test the capability of the proposed method to identify real samples of pesticide in water. The choice of [i]k[/i] was optimized by calculating the prediction ability of the calibration set using hold-out cross validation. The best cross-validation accuracy was achieved when [i]k[/i] = 1. To validate the proposed method three real pesticide samples, each one containing aldicarb, carbofuran and porpoxur, were used. No misclassification was noted for the validation set which demonstrates the reliability of the method to classify pesticides in water samples.

Financial support: FAPESP, CNPq and CAPES.

[1] Y. Ni, P. Qiu, S. Kokot, Anal. Chim. Acta 537 (2005) 321.

Keywords: Chemometrics, Electrochemistry, Environmental/Water, Pesticides
Application Code: Environmental
Methodology Code: Sensors

*** Formatting Issue - Best Available Paper/Graphic ***

Session # 1110 Abstract # 1110-5 **Poster Sessions**

Session Title: Environmental Organic Analysis: VOCs, Pesticides, and Others

Abstract Title: **PCBs and DDTs in Bluefin Tuna From the Adriatic Sea**

Primary Author: Darija Klin□□
Institute for Med Research and Occupational Health

Date: Tuesday, March 08, 2016 - Morning
Time:
Room: Exposition Floor, 400 Aisle

Co-Author(s): Snjažana Herceg Romani□ Vjekoslav Ti□na, Zorana Kljakovi□-Gašpi□

Abstract Text

Top marine predators with long life spans, such as tuna, can contain high concentrations of organochlorine contaminants (OCs). Tuna is a good bioindicator that serves for obtaining data on marine ecosystem contamination. Being an important foodstuff, it also represents a potential source of human exposure. Light meat samples of 9 wild Bluefin tuna (Thunnus thynnus) caught in the Adriatic Sea were freeze-dried and extracted with n-hexane:acetone=50:50 (20 mL) in a Microwave Accelerated Reaction System (CEM, USA). Extracts were further cleaned with 96% sulphuric acid, by adsorption chromatography on a multilayer silica column and finally on commercial tubes pre-packed with carbon. Analysis was done on a high-resolution gas chromatograph (Clarus 500, Perkin Elmer) with ECD detectors on two capillary columns simultaneously. The highest concentrations were found for p,p´-DDE, which is the main, highly stable metabolite of p,p´-DDT. Concentrations of p,p´-DDT were 2 to 20 times lower than those of p,p´-DDE. The sum of six indicator PCB congeners accounted for between 54% and 75% of total PCB concentrations. Higher chlorinated indicator PCBs (PCB-153, PCB-138, and PCB-180) were found in the highest concentrations. Positive correlations of PCB-138, -153, -180 and p,p´-DDE with tuna age and/or body weight confirmed that the most stable and most degradation-resistant organochlorine compounds accumulate in the organism of a tuna during its life. The OC concentrations found in this study should not raise concern regarding human health risk, but indicate the need for continuous monitoring in marine organisms, especially species such as farmed tuna, which is an important food and export product.

Keywords: Environmental/Biological Samples, GC, Monitoring, PCB's
Application Code: Environmental
Methodology Code: Gas Chromatography

Session # 1110 Abstract # 1110-6 **Poster Sessions**

Session Title: Environmental Organic Analysis: VOCs, Pesticides, and Others

Abstract Title: **From Freon to PAHs - A New Generation of Multipurpose Thermal Desorption Tubes**

Date: Tuesday, March 08, 2016 - Morning

Time:

Room: Exposition Floor, 400 Aisle

Primary Author: Paolo Benedetti
IIA - CNR

Co-Author(s): Carlo Crescenzi, Ettore Guerriero

Abstract Text

An innovative thermal desorption (TD) tubes was optimized using recently developed materials. The trapping device was applied to the TD-GC/MS analysis of several classes of compounds including: EPA TO-14, PAMS, oxygenated compounds several terpenoids and semi-volatile compounds (SVOCs). Results are compared with those of a multi-carbon thermal desorption tubes commonly used in the analysis of volatile and semi-volatile organic compounds (VOCs and SVOCs). Using just 30 mg of sorbent material the new trapping device demonstrates the ability of trapping more efficiently analytes even in harsh conditions. Excellent results from recovery experiment in controlled atmosphere are presented. Major demonstrated advantages are in terms of reduced impedance and inertness to water. Moreover the new trapping device has several additional features: higher thermal and mechanical stability (especially towards very fragile granular graphitized carbons) and "self-confining" properties (it does not require quartz wool frits). Methods validation was performed in conditions where conventional thermal desorption sorbent tubes fail such as wet vapor conditions. Finally the method was applied to the analysis of exhaust from a biomass combustion plant.

Keywords: Environmental/Air, GC-MS, Semi-Volatiles, Volatile Organic Compounds
Application Code: Environmental
Methodology Code: Gas Chromatography/Mass Spectrometry

Session # 1110 Abstract # 1110-8 **Poster Sessions**

Session Title: Environmental Organic Analysis: VOCs, Pesticides, and Others

Abstract Title: **Analysis of Volatile Organic Compounds in Wastewater by Purge and Trap GC/TOF-MS According to EPA Method 624**

Primary Author: Moira Zanaboni
DANI Instruments

Date: Tuesday, March 08, 2016 - Morning
Time:
Room: Exposition Floor, 400 Aisle

Co-Author(s): Matthew S. Klee, Roberta Lariccia

Abstract Text

Wastewater is any water that has been negatively affected in quality by anthropogenic influence. Wastewater can originate from a combination of industrial, commercial, domestic or agricultural activities, surface runoff or rainwater, and from sewer inflow or infiltration.

Wastewater may contain toxic compounds or compounds that potentially may be mutagenic or carcinogenic. To protect public health and the environment the analysis of wastewater is internationally regulated. For example the U.S. authority Environmental Protection Agency (EPA) has developed method 624, that covers the determination of a number of purgeable organics in municipal and industrial wastewater.

According to EPA 624 the Purge and Trap GC/TOF-MS technique allows a wide range of purgeable organics, commonly called Volatiles Organic Compounds (VOCs), to be determined.

In this work an innovative technology of in-vial purging is presented. This feature allows any risk of cross-contamination and carryover to be eliminated. In this way no additional workload of cleaning glassware or time-consuming line purging are requested. Moreover, the exclusive dew stop device provides an efficient removal of water regardless of the analytes, maintaining volatile compound recovery unaffected. The use of a TOF-MS guarantees high quality mass spectra and effective deconvolution algorithm to ensure a reliable identification, even with a complex matrix as wastewater. Results will be presented to show the efficiency, accuracy and reproducibility of this approach.

Keywords: Environmental/Water, Purge and Trap, Time of Flight MS, Volatile Organic Compounds
Application Code: Environmental
Methodology Code: Gas Chromatography/Mass Spectrometry

Session # 1110 Abstract # 1110-9 **Poster Sessions**

Session Title Environmental Organic Analysis: VOCs, Pesticides, and Others

Abstract Title **Degradation of Environmental Contaminants Using Chlorine Dioxide**

Primary Author Sushma Appala
Middle Tennessee State University

Date: Tuesday, March 08, 2016 - Morning
Time:
Room: Exposition Floor, 400 Aisle

Co-Author(s) Megan Z. Chong, Ngee Sing Chong, Ooi G. Beng, Samanwi Munagala

Abstract Text

Accumulation of environmental contaminants, especially persistent organic compounds, can pose adverse ecological and human health effects. Therefore, our research goal is to evaluate the efficiency of chlorine dioxide in removing or reducing harmful pollutants in different environmental media. An ingredient of birth control pill, 17-ethinylestradiol, is chosen for the study of its degradation kinetics when treated with ClO2. The oxidative degradation by-products are characterized by Fourier Transform infrared spectrometry (FTIR) and gas chromatography or liquid chromatography coupled to mass spectrometry (GC-MS and LC-MS). The LC-MS analysis of the estradiol solution treated with ClO2 shows a reduction in concentration of estradiol over a 20-hour period. The reason for studying the degradation of this estradiol compound is related to its adverse effects on aquatic life since it has been reported to show endocrine disrupting effects. It is capable of reducing male and female fertility as well as feminization of male fish. The degradation of pesticides in the presence of ClO2 is also investigated because it is often used to disinfect fruits and vegetables and it is expected that not only will ClO2 deactivate pathogenic bacteria in these food products, it may also yield the beneficial effects of breaking down toxic pesticides used for pest control. Degradation of pesticides including DDT, endrin, and chlordane have been studied and their degradation products characterized by GC-MS. Lastly, the degradation of odorous compounds including cadaverine and dimethyldisulfide has also been studied in gas phase using the GC-MS and FTIR techniques.

Keywords: Environmental, Gas Chromatography/Mass Spectrometry, Liquid Chromatography/Mass Spectrosco
Application Code: Environmental
Methodology Code: Liquid Chromatography/Mass Spectrometry

Session # 1110 Abstract # 1110-10 **Poster Sessions**

Session Title: Environmental Organic Analysis: VOCs, Pesticides, and Others

Abstract Title: **Automating Solid Phase Extraction and Florisil Clean-Up for Organichlorine Pesticides and PCB Aroclors**

Date: Tuesday, March 08, 2016 - Morning

Time:

Room: Exposition Floor, 400 Aisle

Primary Author: Philip Bassignani
Fluid Management Systems

Co-Author(s): Rudolf Addink

Abstract Text

Organochlorine Pesticides (OCPs) are a class of persistent contaminants that span the spectrum of EPA methodology. Typically analyzed by GC/ECD, florisil clean-up is often required prior to analysis to eliminate non-target interferences. The use of an automated SPE apparatus can enable the pairing of both extraction of aqueous samples with extract clean-up in a continuous process with no manual manipulation. A detailed examination of the procedure will be covered along with the fluidic pathways to accommodate the pairing of the extraction and clean-up steps. IPR, MDL and a variety of matrix specific validations were conducted testing the process. Final validation involved with participation in a commercial PT study by an accredited laboratory.

Keywords: Environmental/Biological Samples, Sample Handling/Automation, Solid Phase Extraction, Ultratrace
Application Code: Environmental
Methodology Code: Sampling and Sample Preparation

Session # 1110 Abstract # 1110-11 **Poster Sessions**

Session Title Environmental Organic Analysis: VOCs, Pesticides, and Others

Abstract Title **Evaluation of a Novel Hand-Held and Easy-to-Use GC-PID Prototype for Fast and Selective On-Site Analysis of Benzene and VOC**

Date: Tuesday, March 08, 2016 - Morning
Time:
Room: Exposition Floor, 400 Aisle

Primary Author Matthias Schmittmann
Bentekk GmbH

Co-Author(s)

Abstract Text

Environmental policies and threshold limit values (TLV) are the reasons for governmental and private organizations to use on-site analyses for detecting hazardous substances, protecting the environment and securing worker's safety. Modern applications increasingly demand methods and devices that are hand-held, fast and easy to use. In this paper, a device based on gas chromatographic (GC) separation and photo ionization detection (PID) of volatile organic compounds (VOC), in particular chlorinated hydrocarbons (CVOC) and benzene including its derivatives toluene, and xylene (BTX), is introduced for such applications. In particular, the separation of benzene and n-hexane relevant for refineries is discussed. The GC-PID prototype is characterized by short analysis cycles of less than a minute while matching required detection limits by industry of sub-ppm concentrations. Benzene can be analyzed in less than 15 seconds and down to 50 ppb. The device is using a short multi capillary column for increased volume flow and low retention times. Filtered ambient air serves as carrier gas. A second continuous mode measures the concentration of total aromatic compounds (TAC) and can be used to trigger the discontinuous GC analysis when TAC levels are above TLV. The device's software includes an algorithm for deconvolution of overlapping peaks and enables easy to use analyses including users not trained on gas chromatography.

Keywords: Environmental/Air, Petrochemical, Portable Instruments, Volatile Organic Compounds
Application Code: Environmental
Methodology Code: Gas Chromatography

Session # 1110 Abstract # 1110-12 **Poster Sessions**

Session Title: Environmental Organic Analysis: VOCs, Pesticides, and Others

Abstract Title: **Direct Determination of Glyphosate, Glufosinate, and AMPA in Egg by Liquid Chromatography/Tandem Mass Spectrometry**

Primary Author: Narong Chamkasem
FDA

Date: Tuesday, March 08, 2016 - Morning
Time:
Room: Exposition Floor, 400 Aisle

Co-Author(s): Cynthia Morris, Krystle L. Hargrove

Abstract Text

A simple high-throughput liquid chromatography/tandem mass spectrometry (LC-MS/MS) method was developed for the determination of glyphosate, aminomethylphosphonic acid (AMPA) and glufosinate in egg using a reversed-phase liquid chromatography column with weak anion/cation exchange stationary phase. Two grams of egg were shaken with 8.5 mL extracting solvent containing Na2EDTA and acetic acid along with 6 mL of methylene chloride for 10 min to precipitate the protein content and separate the fat portion. After centrifugation, the supernatant was filtered and passed through an Oasis HLB SPE to retain suspended particulates and phospholipids. The sample was directly injected and analyzed in 12 min, with no sample concentration or derivatization steps. Two multiple reaction monitoring (MRM) channels were monitored in the method for each target compound to achieve true positive identification. Three internal standards corresponding to each analyte were used to counter matrix suppression effects. The linearity of the detector response was demonstrated in the range of 5 to 500 ng/mL for each analyte, with a minimum coefficient of determination (R2) value of more than 0.9995. Through the use of this internal standard calibration method, the average recovery for all analytes at 0.05, 0.1, 0.5, and 1 µg/g (n = 7) are between 85-107% with a relative standard deviation of less than 6%.

Keywords: Food Contaminants
Application Code: Food Safety
Methodology Code: Liquid Chromatography/Mass Spectrometry

Session # 1110 Abstract # 1110-13 **Poster Sessions**

Session Title: Environmental Organic Analysis: VOCs, Pesticides, and Others

Abstract Title: **Direct Coupling of Active and Passive Samplings of Organics with Microwave Assisted Thermal Desorption as an Innovative Solvent-Free Method**

Date: Tuesday, March 08, 2016 - Morning
Time:
Room: Exposition Floor, 400 Aisle

Primary Author: Williams Esteve
INRS

Co-Author(s)

Abstract Text

Thermal desorption has been developed the past years because this solvent-free method perfectly fits in the current green chemistry trend. However, with classical thermal desorption systems, the use of activated charcoal as an adsorbent is not possible. Activated charcoal is a convenient, inexpensive, high trapping capacity material and compatible with a wide variety of organic, but the amount of provided energy during the heating desorption process is not sufficient to totally extract the most polar and low sterically hindered adsorbed compounds.

A new alternative microwave assisted thermal desorption technique allows the use of activated charcoal as sorbent. This technique provides almost instantly a high quantity of energy which enables a complete flash desorption of organics.

The goal of this study was to develop a sampling device usable in the microwave thermal desorption system directly after either active or passive sampling of organic compounds, without preparation steps. The main challenge was to find an inert diffusion media in the diffusive ceramic tubes for passive samplings resistant to high temperatures without suffering irreversible structural damages. A carbonaceous material has been chosen as a diffusion material. The stability of its structure and diffusive properties as a function of the number of desorption cycles have been studied by microscopic analyses and also by samplings of aromatic compounds.

Microscopic observations made on the diffusion media after various desorption cycles have shown no structural modification or damage between up to a 50-cycle diffusion material and a new one.

Then, three diffusion sampling experiments have been performed with three similar devices in a controlled atmosphere of toluene and m-xylene each at the threshold limit value TLV®. After the passive sampling step, the activated charcoal has been extracted with carbon disulfide and analyzed by GC-FID to determinate the mass of organics. The calculated uptake rates have shown less than 2% of inter-sampler variability, and only 7% of variability between the three experiments for both compounds (Figure 1).

The results have shown the developed sampler appears to be suitable for passive samplings directly followed by microwave thermal desorption. Besides the reliability of the diffusion media, another aspect of the study still in progress is the optimization of the quantification by external standard directly by the microwave thermal desorption system.

Keywords: Environmental/Air, Microwave, Sampling, Thermal Desorption
Application Code: Industrial Hygiene
Methodology Code: Thermal Analysis

Session # 1110 Abstract # 1110-14 **Poster Sessions**

Session Title: Environmental Organic Analysis: VOCs, Pesticides, and Others

Abstract Title: **Deuterated Monitoring Compounds for Better Accuracy and Precision Measurement of GC/MS Environmental Data**

Primary Author: Charles G. Appleby
U.S. Environmental Protection Agency

Date: Tuesday, March 08, 2016 - Morning
Time:
Room: Exposition Floor, 400 Aisle

Co-Author(s)

Abstract Text

The use of surrogate compounds to measure method performance in Gas Chromatography/Mass Spectroscopy (GC/MS) methods for environmental monitoring is not a new practice. All EPA-approved methods require the use of three to six compounds; however only a few are deuterated analogs of target analytes. Deuterated analogs are more representative of target analytes, thereby providing more information regarding matrix effects while measuring the accuracy and precision. Since 2001, the EPA's Office of Superfund Remediation and Technology Innovation's Contract Laboratory Program (CLP) has required laboratories to add over a dozen deuterated monitoring compounds (DMCs) to each sample, all analogs of target analytes. Developed to improve data quality used in decision-making processes, this approach ultimately reduced the cost to the Superfund Program. This presentation will show, with thousands of data points, how incorporating more DMCs into EPA-approved GC/MS methods has improved data quality, and provided cost savings to the Agency, and how it may benefit the entire analytical chemistry community.

Keywords: Gas Chromatography/Mass Spectrometry, Mass Spectrometry, Method Development, Quality
Application Code: Environmental
Methodology Code: Gas Chromatography/Mass Spectrometry

Session # 1110 Abstract # 1110-16 **Poster Sessions**

Session Title: Environmental Organic Analysis: VOCs, Pesticides, and Others

Abstract Title: **Wastewater Monitoring by Fluorescence Excitation and Emission Matrix with Parallel Factor Analysis**

Date: Tuesday, March 08, 2016 - Morning
Time:
Room: Exposition Floor, 400 Aisle

Primary Author: Sam Li
NUS

Co-Author(s): Baisheng Chen

Abstract Text

The potential of fluorescence analysis for on-line wastewater monitoring was investigated in this study. Parallel factor analysis (PARAFAC) was successfully applied in the analysis of fluorescence excitation and emission matrix (FEEM) data in wastewater treatment. Good correlations (R2=0.67-0.78) were obtained between the removal efficiencies of chemical oxygen demand (COD) and dissolved organic carbon (DOC) and FEEM by PARAFAC. The monitoring of treatment efficiency in a series of treatment processes in a wastewater treatment plant (WWTP) was also studied by FEEM. Moderately good correlation (R2=0.71-0.82) was found with COD removal efficiency. The correlations of removal efficiency of COD and DOC by FEEM were found to be comparable with those by absorbance spectroscopy. More insight in the composition changes could be obtained through the analysis of FEEM.

Keywords: Environmental Analysis, Fluorescence
Application Code: Environmental
Methodology Code: Fluorescence/Luminescence

Session # 1110 Abstract # 1110-17 **Poster Sessions**

Session Title: Environmental Organic Analysis: VOCs, Pesticides, and Others

Abstract Title: **New Plastic In-Syringe Based Ultrasound Assisted Salt-induced Liquid-Liquid Microextraction Technique for the Rapid Analysis of Triclosan in Aqueous Samples**

Primary Author: Vinoth Kumar Ponnusamy
National Chung Hsing University

Date: Tuesday, March 08, 2016 - Morning
Time:
Room: Exposition Floor, 400 Aisle

Co-Author(s): Jen Fon Jen

Abstract Text

A new plastic in-syringe based ultrasound assisted salt-induced liquid-liquid microextraction (IS-USA-SI-LLME) technique was developed for the determination of triclosan (TCS) in environmental water samples using high performance liquid chromatography-ultraviolet (HPLC-UV) detection. This method is based on the rapid phase separation of water-miscible organic solvent from the aqueous phase in the presence of high concentration of salt (salting-out phenomena) under ultrasonication. A simple home-made plastic syringe based extraction device (5-mL commercial plastic syringe coupled with a 100 microliter pipette tip) was adopted as the phase separation device for IS-USA-SI-LLME. After extraction, the upper extraction solvent layer was narrowed into the self-scaled capillary tip by pushing the plunger plug; thus making the collection and measurement of the upper organic solvent layer easy and convenient. Factors influencing the extraction efficiency such as selection of extraction solvent and salt, extraction solvent volume, salt amount, ultrasonication time, and pH were thoroughly investigated and optimized. Under the optimum conditions, the method showed good linearity in the concentration range from 0.4–100 ppb with correlation coefficient of 0.9991 for the target analyte. The limit of detection was 80 ng/L. The method was validated with real water samples and the relative recoveries of environmental water samples ranged between 95.6 -114.3% and relative standard deviations were ranged between 1.8-5.7%, making the proposed method highly reliable. The proposed method provides a rapid, simple, efficient, low cost, easy to handle and convenient environmental friendly sample preparation procedure to determine triclosan in aqueous samples.

Keywords: Environmental/Water, Liquid Chromatography, Sample Preparation, Ultratrace Analysis
Application Code: Environmental
Methodology Code: Liquid Chromatography

Session # 1120 Abstract # 1120-2 **Poster Sessions**

Session Title: Environmental Water Quality and Analysis

Abstract Title: **Determination of N-Nitrosamines by USEPA Method 521 Using Triple Quadrupole Mass Spectrometry**

Primary Author: Brahm Prakash
Shimadzu Scientific Instruments, Inc

Date: Tuesday, March 08, 2016 - Morning
Time:
Room: Exposition Floor, 400 Aisle

Co-Author(s): Di Wang, Laura Chambers, Nicole M. Lock, Shilpi Chopra, William Lipps

Abstract Text

Research indicates that N-nitrosamines are formed during chlorination of finished drinking water. These compounds are classified as possible human carcinogens by the International Agency for Research on Cancer (IARC), and were detected in about 25% of the drinking water systems monitored in the Unregulated Contaminant Monitoring Rule 2 (UCMR2).

Detection and quantitation of N-nitrosamines for contaminant monitoring requires using USEPA Method 521 (2004) employing an ion trap MS with chemical ionization (CI). Method 521 was originally developed using ion trap MS, and all of the validation data presented in the method are based on this technique, but unfortunately, ion trap mass spectrometers are no longer available. A new method is needed.

This poster describes development of a Multiple Reaction Monitoring (MRM) instrument method using a commercially available triple quadrupole GC-MS/MS for detection and quantitation of N-nitrosamines after extraction as described by USEPA Method 521. GC-MS/MS in MRM mode produces significant improvements in selectivity and specificity, as well as dramatically lower detection limits than single quadrupole GC/MS, especially in complex matrices producing background interferences. This poster presents final instrument configuration and operating conditions, as well as instrument validation results including LCMRL, and precision & accuracy as evaluated using spiked samples at various concentration levels. Use of the CI technique compared to electron impact ionization is also discussed.

Keywords: Environmental, GC-MS, Tandem Mass Spec, Water
Application Code: Environmental
Methodology Code: Gas Chromatography/Mass Spectrometry

Session # 1120 Abstract # 1120-3 **Poster Sessions**

Session Title: Environmental Water Quality and Analysis

Abstract Title: **Optimizing Treatment of Reclaimed Water at a Drinking Water Plant by Online Monitoring of Organic Carbon Levels**

Primary Author: Dondra Biller
GE Analytical Instruments

Co-Author(s): Mark A. Mullet

Date: Tuesday, March 08, 2016 - Morning
Time:
Room: Exposition Floor, 400 Aisle

Abstract Text

The Twin Oaks Valley drinking water treatment plant in San Diego County is a zero discharge facility and is the world's largest submerged membrane ultrafiltration water treatment plant producing 100 MGD. Ultrafiltration membranes can be fouled by high organic loading, therefore careful monitoring is essential.

Source water for the Twin Oaks Valley plant is 95% surface or Colorado River water that is mixed with reclaim water on site from an equalization (EQ) basin. The water from the EQ basin is primarily backwash from the ultrafiltration membrane banks, and as such, the EQ basin has high concentrations of contaminants. Contaminants in the EQ basin effluent not removed via centrifuging are then removed via coagulation and settling by addition of ferric chloride and a polymer. The treated EQ water goes to plate settlers where the coagulants can settle out. To optimize chemical treatment of water from the EQ basin and to determine the efficiency of the plate settlers, online monitoring of organic carbon was implemented.

The EQ basin effluent before chemical treatment and the effluent off the plate settlers were analyzed continually. Data from these two streams provided organic carbon levels before and after treatment of the EQ basin water. This information was beneficial for both optimizing treatment of the reclaimed water as well as understanding the organic loading of the membranes. Monitoring organic carbon using online TOC analysis was highly cost-effective and beneficial for the Twin Oaks Valley water treatment plant in preventing fouling of the costly ultrafiltration membranes.

Keywords: Instrumentation, On-line, Process Control, Total Organic Carbon
Application Code: Environmental
Methodology Code: Process Analytical Techniques

Session # 1120 Abstract # 1120-4 **Poster Sessions**

Session Title: Environmental Water Quality and Analysis

Abstract Title: **Development of an Arduino Shield for Water Quality Analysis Probes**

Primary Author: Michael Chia
Northern Kentucky University

Date: Tuesday, March 08, 2016 - Morning
Time:
Room: Exposition Floor, 400 Aisle

Co-Author(s): Celeste A. Morris, Grant Foreman, Kelley Weigman, Richard Durtsche

Abstract Text

We present application of Arduino technology to integrate solid-state pH probes with temperature, turbidity, dissolved oxygen and conductivity solid-state sensors for portable probes capable of assessing water quality. Individual sensors currently exist on the market; however, an integrated, efficient, and inexpensive probe is not currently available. According to the EPA, the most recent assessment of water quality in the Commonwealth of Kentucky was conducted nearly three years ago. Even then, only a small fraction of the total waterways in Kentucky was assessed with nearly two thirds of that water deemed impaired. We present development of an Arduino pH shield for integration of pH and temperature probes and longevity analysis of solid-state pH probes.

Keywords: Electrochemistry, Electrode Surfaces, Environmental Analysis, Environmental/Water
Application Code: Environmental
Methodology Code: Portable Instruments

Session # 1120 Abstract # 1120-5 **Poster Sessions**

Session Title: Environmental Water Quality and Analysis

Abstract Title: **Analysis of Peanut Hulls as an Alternative to Ion Exchange Resins**

Primary Author: Carmen L. Huffman
Western Carolina University

Date: Tuesday, March 08, 2016 - Morning
Time:
Room: Exposition Floor, 400 Aisle

Co-Author(s): Holly Truluck, Kanika O. Davis, Melisa J. Glatte, Tyler S. Cook

Abstract Text

Heavy metals are typically removed from contaminated wastewater through the use of ion exchange resins, which are expensive. Biosorption has long been pursued as an alternative method due to its economic and environmental advantages. One potential biosorbent is ground peanut hulls. They are a by-product of the food and agriculture industry, which makes them inexpensive. They are also a renewable and biodegradable product, which makes them environmentally friendly. Lastly, due to their high lignocellulosic content they have a moderate affinity for metal ions. Our studies have shown that the affinity can be enhanced by an alkaline peroxide bleaching process, which is believed to oxidize the hulls, creating additional binding sites. Simultaneously, the dissolution of lignin leads to an increased porosity and therefore greater surface area and increased number of binding sites. The adsorption of copper and multi-metal systems to both unmodified and modified hulls has been tested in both batch and continuous flow systems. In batch systems, hulls were mixed with solutions containing various concentrations of metal ions and allowed to come to equilibrium. After filtration, the supernatant liquid was analyzed for metal concentration using ICP-OES. The amount of bound metal was calculated by difference. In continuous flow experiments, metal containing solution was flushed through a column containing wet hulls. Aliquots of effluent were collected and analyzed for metal concentration using ICP-OES. Adsorption curves from batch studies and breakthrough curves from continuous flow studies will be presented. A comparison to ion exchange resin cost and performance will also be addressed.

Keywords: Adsorption, Environmental/Water, Materials Characterization, Plasma Emission (ICP/MIP/DCP/etc.)
Application Code: Environmental
Methodology Code: Atomic Spectroscopy/Elemental Analysis

Session # 1120 Abstract # 1120-6 **Poster Sessions**

Session Title: Environmental Water Quality and Analysis

Abstract Title: **Analytical Strategies for Monitoring Ionic Liquids Breakdown by Electro-Fenton Process**

Date: Tuesday, March 08, 2016 - Morning

Time:

Room: Exposition Floor, 400 Aisle

Primary Author: Elisa González-Romero
University of Vigo

Co-Author(s): Aida Díez, Elvira Bocos, Jessica Meijide, María Ángeles Sanromán, Marta Pazos

Abstract Text

Improving the framework of green chemistry principles, it should be necessary to decrease the major sources of waste as volatile organic solvents (VOSs) losses that end up in the atmosphere or in ground water, mainly caused by their use in industrial applications. Owing to their low volatility, the Ionic Liquids (ILs) have been regarded as greener replacements of the traditional VOSs, since their contribution to the air pollution can be effectively reduced. Unfortunately, their high solubility makes their entrance in water very easy. Furthermore, ILs has low biodegradability; this is the major reason why they have become a new threat for the aquatic ecosystems. This urges new efforts and the development a new strategies to eliminate the ILs from water at an acceptable cost, in a short time and with the least possible environmental and social impact. In this work, we illustrate the application of electro-Fenton for the degradation of several ILs. This technique is based on Fenton reaction by the action of a powerful oxidizing agent, hydroxyl radical, generated on the electrochemical process. In order to determine the pathways of the reactions that take places in the degradation process, it is necessary to determine the evolution of the composition of the samples along the treatment. For that, chromatographic analysis (Ionic/Ion-Exclusion HPLC and GC/MS) were performed. A number of others analytical techniques, including AAS, voltammetry, conductivity and potentiometry, are also used in our methodology to get insight into the degradation pathways of the ILs under study. The energy consumption and the measurement of the Total Organic Carbon (TOC) of the studied aqueous solution containing the ILs are also calculated.

Acknowledgements: Financial support from Spanish/European Institutions MINECO/FEDER (Project CTM2014-52471-R) and XUNTA (ReGaLIs R2014/015 and Bioauga R2014/030 Networks). The authors are grateful to Pepe Lamas for his helpful hands in the laboratory.

Keywords: Chromatography, Electrochemistry, Environmental/Water, Total Organic Carbon

Application Code: Environmental

Methodology Code: Electrochemistry

Session # 1120 Abstract # 1120-8 **Poster Sessions**

Session Title Environmental Water Quality and Analysis

Abstract Title **Determination of the Toxins Found in Lake Ontario**

Primary Author Marta Labecki
St. John Fisher College

Date: Tuesday, March 08, 2016 - Morning
Time:
Room: Exposition Floor, 400 Aisle

Co-Author(s) Kimberly Chichester

Abstract Text

Water samples from the New York side of the Lake Ontario watershed were collected to determine the presence and levels of cyanobacterial toxins. The specific sampling sites included: Webster Park, Oswego River, Sodus Bay, Salmon River in Mexico Point State Park Oak Orchard Creek, Eighteen Mile Creek, and Niagara River, which all feed into Lake Ontario. These locations were chosen because they provide a good range along the coast of Lake Ontario and they are places people frequent for fishing or other recreational activities exposing them to the water and toxins. The water samples were used to cultivate cyanobacteria in order to determine what types of toxins the cyanobacteria released and the quantity released. The toxins included in the study are saxitoxins, anatoxins, and microcystins. The cyanotoxins can be found in both the water and in the cyanobacteria, therefore cell lysis was performed. The toxins were studied and quantified using High-Performance Liquid Chromatography (HPLC). Calibration was performed using purchased pure toxin standards prior to testing with cell lysates and sample water. Ultimately, the aim of the study is to determine the amount of toxins produced and released by the cyanobacteria.

Keywords: Environmental Analysis, Liquid Chromatography, Method Development, UV-VIS Absorbance/Lumines
Application Code: Environmental
Methodology Code: Liquid Chromatography

Session # 1120 Abstract # 1120-9 **Poster Sessions**

Session Title: Environmental Water Quality and Analysis

Abstract Title: **EPA Method 557 Quantitation of Haloacetic Acids, Bromate and Dalapon in Drinking Water Using Ion Chromatography and Tandem Mass Spectrometry**

Primary Author: Jonathan Beck
Thermo Fisher Scientific

Co-Author(s): Charles Yang, Hans Schweingruber, Terri T. Christison

Date: Tuesday, March 08, 2016 - Morning
Time:
Room: Exposition Floor, 400 Aisle

Abstract Text

Introduction: Haloacetic Acids (HAAs) can be formed during drinking water purification in municipal water supplies during the chlorination, ozonation or chloramination of water. Reactions between chlorine and organic matter present in the water can create HAAs. There are health concerns regarding human consumption of HAAs. Because of these concerns, the US EPA has published Method 557 for the quantitation of HAAs using Ion Chromatography coupled to Tandem Mass Spectrometry (IC-MS/MS). The IC-MS/MS method bypasses derivitization of the HAAs and allows for direct analysis of the drinking water samples.

Method: Drinking water samples were collected and treated with preservatives. Calibration standards were prepared by spiking drinking water with 9 Haloacetic Acids, bromate, and dalapon stock solutions.

Preliminary Data: The following HAAs were analyzed, along with bromate and the pesticide dalapon: Bromochloroacetic acid (BCAA), Bromodichloroacetic acid (BDCAA), Chlorodibromoacetic acid (CDBAA), Dibromoacetic acid (DBAA), Dichloroacetic acid (DCAA), Monobromoacetic acid (MBAA), Monochloroacetic acid (MCAA), Tribromoacetic acid (TBAA) and Trichloroacetic acid (TCAA). While US regulations currently only require the monitoring of 5 of the HAAs, (MCAA, DCAA, TCAA, MBAA and DBAA), interest is growing in the additional 4 HAAs and were included in this analysis. Drinking water samples were tested using San Jose, CA municipal drinking water as well as bottled water for a variety of manufacturers to test for the presence of HAAs. The response of the HAAs, bromate, and dalapon over the concentration range was linear. An instrument detection limit (IDL) was calculated for each analyte based on replicate injections and the student's t-test.

Keywords: Environmental Analysis, Environmental/Water, Ion Chromatography, Mass Spectrometry
Application Code: Environmental
Methodology Code: Liquid Chromatography/Mass Spectrometry

Session # 1120 Abstract # 1120-10 **Poster Sessions**

Session Title: Environmental Water Quality and Analysis

Abstract Title: **Assessment of Water Quality Parameters from the Lowber Abandoned Mine Drainage Treatment Facility, Part 2: Further Studies and Results**

Primary Author: Mark T. Stauffer
University of Pittsburgh - Greensburg

Date: Tuesday, March 08, 2016 - Morning
Time:
Room: Exposition Floor, 400 Aisle

Co-Author(s): Aaron K. Hirshka, Luke J. Metzler, Tell M. Lovelace

Abstract Text

This presentation will provide an update on continued assessment of abandoned mine drainage (AMD) from the old Marchand coal mine, which is remediated at the Lowber AMD treatment facility in southwest Pennsylvania. This study is sponsored by the Rho Theta Chapter of the Gamma Sigma Epsilon Chemistry Honor Society, in collaboration with the Sewickley Creek Watershed Association (SCWA). The focus continues to be on determination of alkalinity, acidity, pH, conductivity, dissolved oxygen, sulfate, iron, aluminum, manganese, and calcium. The goals of this research continue to be: 1) assessment of selected analytes in AMD from the Lowber facility in the field and laboratory, and 2) lending assistance to the SCWA toward evaluation of AMD and development of remediation strategies for use at potential sites throughout the Sewickley Creek Watershed. Sample collection and analytical methodologies will be presented and discussed, as will results of determinations performed so far and their significance in light of the effectiveness of the passive treatment process used at Lowber, and future plans for this study.

Keywords: Environmental Analysis, Environmental/Water, Instrumentation, Wet Chemical Methods
Application Code: Environmental
Methodology Code: Education/Teaching

Session # 1120 Abstract # 1120-11 **Poster Sessions**

Session Title: Environmental Water Quality and Analysis

Abstract Title: **Breaking the Biofouling Code: Towards Reliable In-Pipe Water Quality Sensors**

Date: Tuesday, March 08, 2016 - Morning

Time:

Room: Exposition Floor, 400 Aisle

Primary Author: Robert E. Wilson, Imperial College London

Co-Author(s): Danny O'Hare, Ivan Stoianov

Abstract Text

Chlorination of drinking water has saved countless lives and remains vital to ensure that water is clean and safe. Adequate chlorination is essential to maintain an effective residual disinfectant concentration, which prevents water quality deteriorating as it travels to the consumer. The concentration of chlorine in drinking water must remain within a specific range, as both under and over chlorination pose serious risks to public health. Under-chlorination causes tragedies such as Walkerton and Milwaukee, where mass infections with water-borne diseases resulted in the deaths of consumers. Over-chlorination is responsible for many consumer complaints of poor water aesthetic qualities, and increases levels of potentially carcinogenic disinfectant by-products in the supply.

As not all bacteria are removed during treatment, biofilms are present within water distribution systems and deteriorate electrochemical disinfectant sensors. Biofilm deposition on the surfaces of these sensors begins within days of installation and within a week, they require removal for cleaning. Bacterial biofouling of in-pipe water quality sensors prevents reliable, reagentless, continuous, [i] in situ [/i], real-time monitoring of disinfectant in water distribution systems. The key challenges in this research are:

I) Understanding how biofilm formation effects electrochemical sensors
II) Removing biofilms [i] in situ [/i]
III) Checking reliability of sensors [i] in situ [/i]
IV) Achieving the reliable long-term operation of in-pipe reagentless water quality sensors

This research will involve a systematic investigation into electrochemical chlorine sensors with approaches that exploit recent developments in materials, signal processing, biocompatiblity, and water supply infrastructure to adapt biomedical and bio-inspired technologies to the field of environmental monitoring.

Keywords: Detection, Electrochemistry, Environmental/Water, Sensors
Application Code: Environmental
Methodology Code: Electrochemistry

Session # 1120	Abstract # 1120-12	Poster Sessions

Session Title Environmental Water Quality and Analysis

Abstract Title **Analysis of the Residual Oil in Water with Excitation Emission Matrix and Parallel Factor Analysis**

Primary Author Kawaguchi Yoshihiko
HORIBA Advanced Techno, Co.,Ltd.

Date: Tuesday, March 08, 2016 - Morning
Time:
Room: Exposition Floor, 400 Aisle

Co-Author(s) Kojima Reiji

Abstract Text
Generally, we use solvent extraction method when we measure residual oil concentration in water. But extraction efficiency is known to different by the difference of solvent and oil species. In addition, because of complicated experimental operation, it is difficult to rapid and on site analysis. Therefore we evaluated fluorescence analysis, for the purpose of developing a method of oil concentration. But, because internal absorption occurs by light absorption components in samples (Inner filter effect: IFE), we cannot be accurately measured fluorescence. In addition, fluorescence quenching occurs in halogen rich sample. Additionally, because environmental water contains many fluorescent materials, fluorescence spectra overlap. High precision measurement of the fluorescence from oil is not possible. Therefore, to correct reabsorption of absorption and the fluorescence of the excitation light, we used IFE correction function. And we measured the EEM of concentrate known oil and separated a fluorescence component using parallel factor analysis (PARAFAC). And we extracted relationship the intensity of fluorescence and concentration of oil. In addition, we obtained a quenching constant using a halogen ion and artificial seawater. It showed that it is possible to quantify the oil that is included in the environmental water (Figure). PARAFAC revealed that it is possible to eliminate the influence of other fluorescent components. In this study, because of oil from the environmental water was not detected, we added B heavy oil. It was possible the real environment underwater oil also detects as with pure water oil. Environmental water used two types of stream water and seawater. By performing quenching correction on fluorescence from oil in seawater, it was revealed that measurement in seawater oil is also possible. Finally the analysis technique is used and hopes that the better quality of the water is secured.

Keywords: Environmental/Water, Fluorescence, Water
Application Code: Environmental
Methodology Code: Fluorescence/Luminescence

Session # 1120 **Abstract # 1120-13** **Poster Sessions**

Session Title: Environmental Water Quality and Analysis

Abstract Title: **Sensitive Determination of Arsenate and Phosphate by Molybdenum Blue Method with Membrane Filter Extraction Using a Portable 8-Channel LED-Based Reflective Photometer**

Date: Tuesday, March 08, 2016 - Morning
Time:
Room: Exposition Floor, 400 Aisle

Primary Author: Yasutada Suzuki
University of Yamanashi

Co-Author(s): Susumu Kawakubo

Abstract Text

Arsenic in inorganic form is highly toxic and phosphorus is one of nutrient salts, therefore their precise and simple determination method is necessary. Although molybdenum blue methods are widely used for arsenate and phosphate determination, solvent extraction must be used to get high sensitivity. Membrane filter extraction is alternative to solvent extraction, and harmful organic solvent can be omitted if analyte collected by a filter could be directly analyzed without dissolution. We have developed a modified molybdenum blue method to produce ion-pairs of cation surfactant and molybdoarsenate and/or molybdophosphate, and then they are filtered and enriched on a membrane filter. Reflective absorbance and Kubelka-Munk function of the analyte in molybdenum blue form are measured at 740 nm by a laboratory-made LED-based 8-channel reflective colorimeter. Obtained calibration graphs for both phosphate and arsenate were linear between 0 and 100 µg/L and the detection limit at sub-ppb and ppb level could be attained for phosphate and arsenate, respectively. Fractional determination of phosphate, arsenate, and arsenite will be discussed.

This work was supported by JSPS KAKENHI Grand Number 24550098.

Keywords: Environmental/Water, Instrumentation, Trace Analysis, UV-VIS Absorbance/Luminescence
Application Code: Environmental
Methodology Code: Portable Instruments

Session # 1120 Abstract # 1120-16 **Poster Sessions**

Session Title: Environmental Water Quality and Analysis

Abstract Title: **Study of Enhancement Effects of Functionalized Gold Nanorods in Quantitative Analysis of 1-H Benzotriazole by Surface Enhanced Raman Spectroscopy (SERS)**

Date: Tuesday, March 08, 2016 - Morning
Time:
Room: Exposition Floor, 400 Aisle

Primary Author: Uttam Sharma Phuyal
Tennessee Technological University

Co-Author(s): Andrew Callender

Abstract Text

Surface enhanced Raman spectroscopy (SERS), by definition, is a technique of amplifying Raman signals from the Raman-active molecules that have been adsorbed onto specially prepared metal surfaces. The greater surface selectivity and sensitivity has extended the application of SERS to a wide variety of interfacial systems previously inaccessible to Raman spectroscopy. The SERS analysis of hydrophilic pollutants such as benzotriazoles in treated water has not been well-studied yet. Benzotriazoles are used as corrosion inhibitors in antifreeze formulations, cooling systems, hydraulic fluids and dishwasher detergents, UV stabilizers in plastics and drug precursors in pharmaceutical industries. The molecules are relatively stable and hydrophilic. The compounds are used widespread and likely to contaminate our drinking water systems with concentrations ranging from few ng/L to hundreds of □g/L. However, the current methods of analysis such as LC-MS and others are costly and tedious. The SERS selectivity and sensitivity of gold nanorods capped with three different capping agents: CTAB, mercapto-acetic acid and octadecanoic acid in detection of 1-H Benzotriazole will be compared and contrasted to develop the SERS method of analysis of 1-H Benzotriazole in treated water.

Keywords: Environmental/Water, Surface Enhanced Raman
Application Code: Environmental
Methodology Code: Vibrational Spectroscopy

Session # 1120 Abstract # 1120-20 **Poster Sessions**

Session Title Environmental Water Quality and Analysis

Abstract Title **The Effect of Slaughter House Waste on the Water Quality of Okpu River Aba**

Date: Tuesday, March 08, 2016 - Morning

Time:

Room: Exposition Floor, 400 Aisle

Primary Author Chidiebele A. Odike-Aduaka
Abia State Teaching Hospital Aba

Co-Author(s) Lilian I. Oguguo

Abstract Text

Study Objective : The aim is to investigate effect of the slaughter house waste water on the Okpu River in Aba. Significance of the research: The importance of clean water in healthy living cannot be over emphasized. This study will make recommendations on management of slaughter wastes that will form the bases for policy formulation to prevent the pollution of the Okpu River and others.

Experimental procedures and equipment used: Samples were collected from 4 stations using sterile 1.5 transparent plastic containers fitted with screw cap at a sampling depth of 15-30cm below water surface. Samples were analysed at the labs within 2hours. The following tests were done: Microbiologic analysis, heterotrophic bacteria /Fungi count, identification of bacteria and biochemical tests.

RESULTS: Results from the investigations showed that the effluent influences the microbiological and physicochemical properties of the river. The total microbial population count ranged from 0.7x 10 6 cfu/ml and 0.8 x10 3 to 6.5 x 10 3 cfu/ml for the bacteria and fungi isolates respectively. Microorganisms isolated include Staphylococus, Escheriche coli, Salmonella sp, Streptococcus sp, Proteus sp, Mucor sp, Yeast sp and Rhizopus sp. The values of the physiochemical parameters of the river samples collected from upstream, point of discharge, downstream and 200metres downstream were: PH ranging from 5.10 to 6.30, temperature 27C to 29C, Conductivity 64 to 80us/cm, dissolved oxygen 1.8 to 3.0mg/ml, biochemical oxygen demand 28.9 to 49.5mg/ml respectively. Moreso, the values of some physiochemical parameters of the river were higher than the WHO tolerant levels for surface water.

Discussion /Conclusions: The results show some level of contamination/ pollution of the surface water of Okpu River due to wastes discharged from the slaughter house in Aba. This could be hazardous to residents of Aba and Abia State at large.

Keywords: Quality, Sampling, Standards, Water

Application Code: Industrial Hygiene

Methodology Code: Microscopy

Session # **1120** **Abstract #** **1120-21** **Poster Sessions**

Session Title: Environmental Water Quality and Analysis

Abstract Title: **Biosensor for Toxic Compounds in Wastewater Based on Microbial Electrochemistry**

Date: Tuesday, March 08, 2016 - Morning
Time:
Room: Exposition Floor, 400 Aisle

Primary Author: Sam Li
NUS

Co-Author(s)

Abstract Text

Microbial fuel cell (MFC)-based biosensors are powered by water pollutants as renewable fuels. In this study, an MFC was constructed as an electrochemical biosensor for toxicity assessment of copper in contaminated domestic sewage. In the anode chamber, organic compounds present in real domestic sewage were bio-degraded to generate electricity. In the cathode chamber, cupric ions which are commonly discharged from heavy metal processing plants were reduced to elemental Cu to recover metallic resources. Based on linear sweep voltammetry (LSV) analysis, the external resistance of metallurgical MFC was optimized. The stabilized electrochemical system was utilized to sense the copper toxicity in wastewater. During the less than 1-h sensing period, copper concentrating ranging from 1 mg L-1 to 5 mg L-1 could be detected. A power output of around 100 Wh (kg Cu) -1 was achieved simultaneously. This study demonstrated that an MFC could be a promising candidate for monitoring copper toxicity in real domestic wastewater accompanied by wastewater treatment, energy generation and metallic copper recovery.

Keywords: Biosensors, Electrochemistry, Environmental/Water, Sensors
Application Code: Environmental
Methodology Code: Sensors

Session # 1120 Abstract # 1120-22 **Poster Sessions**

Session Title: Environmental Water Quality and Analysis

Abstract Title: **Rapid Determination of Endocrine Disrupting Bisphenol A (BPA) in Drinking Water by Solid Phase Nano-Extraction and Room-Temperature Fluorescence Spectroscopy**

Date: Tuesday, March 08, 2016 - Morning

Time:

Room: Exposition Floor, 400 Aisle

Primary Author: Maha Al-Tameemi
University of Central Florida

Co-Author(s): Andres Campiglia, Bassam Alfarhani, Jung Jong Seok

Abstract Text

A novel method for the determination of Bisphenol A (BPA) in drinking water samples is reported. BPA is pre-concentrated from the water sample with the aid of gold nanoparticles (Au NPs). After releasing BPA with a polar organic solvent (Mixture of 50 □ of acetonitrile and 12 □ of 1-hexanthiol), the analyte is determined in the supernatant via room-temperature fluorescence spectroscopy. The extraction efficiency of the new method is statistically equivalent to 100% with relative standard deviations lower than 2%. Complete analysis at the parts-per-trillion concentration level is made possible with milliliters of water sample and micro-litters of organic solvent. BPA average recoveries were 95.39 ± 2.19%. The simplicity of the experimental procedure, the low analysis cost, and the excellent analytical figures of merit demonstrate the potential of this approach for routine analysis of BPA in drinking water samples.

Keywords: Environmental Analysis, Fluorescence
Application Code: Environmental
Methodology Code: Fluorescence/Luminescence

Session # 1120 Abstract # 1120-23 **Poster Sessions**

Session Title: Environmental Water Quality and Analysis

Abstract Title: **GC/MS Screening of Water Samples for Organic Pollutants by Stir Bar Sorptive Extraction (SBSE)**

Date: Tuesday, March 08, 2016 - Morning
Time:
Room: Exposition Floor, 400 Aisle

Primary Author: Oliver Lerch
Gerstel GmbH & Co. KG

Co-Author(s): Andreas Hoffmann, Chris Sandy, Jasmin Zboron

Abstract Text

Rapid and low cost screening of surface and drinking water for contaminants is getting more and more important in times of water shortage and new emerging substances. We developed an easy screening method employing stir bar sorptive extraction (SBSE) for analyte enrichment, followed by thermal desorption-GC/MS detection. The method relies on a database of more than 1000 compounds including pesticides, fungicides, molluscicides, hydrocarbons, PAHs, emerging pollutants, industrial chemicals, metabolites, volatile solvents as well as pharmaceuticals and personal care products. A solvent free and "green" extraction by SBSE (stir bar coated with polydimethylsiloxane, PDMS) was developed to substitute dichloromethane extraction in the original method. A 50 mL water sample is extracted at pH 7 by a first stir bar for 1 h followed by a second stir bar extraction (1 h) at pH 1-2. Both stir bars are combined in a tube for automated thermal desorption and GC/full-scan-MS detection. Deconvoluted component spectra are searched against the target MS library. Retention time (retention time locking to fluorene), ion ratios and library match score are qualifiers for compound identification. Semi-quantitative results can be received by considering the response factor of a compound which is included in the library as well. A report is generated after optional reviewing of the identified compounds. Data processing of a single sample takes around 1 min. New compounds can be easily added to the target library allowing to adapt the method to current demands. Typical limits of detection are 0.1 µg/L. A number of different water samples (surface water, drinking water, sewage plant effluent) were screened with the developed method and results are presented.

Keywords: Environmental/Water, Gas Chromatography/Mass Spectrometry, Sample Preparation, Thermal Deso
Application Code: Environmental
Methodology Code: Gas Chromatography/Mass Spectrometry

Session # 1130 Abstract # 1130-1 **Poster Sessions**

Session Title: Environmental, Food and Elemental Analyses - Atomic Spectroscopy

Abstract Title: **Analysis of Radium-226 in Shale-Gas Wastewater Using Inductively Coupled Plasma Mass Spectrometry**

Date: Tuesday, March 08, 2016 - Morning
Time:
Room: Exposition Floor, 400 Aisle

Primary Author: Yuqiang Bi
University of Michigan

Co-Author(s): Brian R. Ellis, Kim F. Hayes, Thomas P. Yavaraski, Wenjia Fan

Abstract Text

The rapid development of horizontal drilling and hydraulic fracturing for natural gas production presents a significant challenge on the management of shale-gas wastewaters that return to the surface. Besides high total dissolved solids (TDS), shale-
gas water typically contains elevated concentrations of radioactive radium (Ra2+), which must be treated properly to minimize the potential environmental and health risks. The development of a rapid and precise method for analysis of 226Ra, the major naturally occurring radioactive isotope, is critical for addressing the regulatory and public concerns, as well as designing new brine treatment technologies.

In order to rapidly determine the concentration of 226Ra in shale-gas wastewater, an analytical method was established using inductively coupled plasma mass spectrometry (ICP-MS). Without complicated pre-treatment and separation, the concentration of 226Ra was accurately determined in the high salinity matrix up to 200,000 mg/L, with a detection limit of 100 pCi/L. The interference of coexisting metals (i.e., Ba2+, Ca2+, and Na+) in the high salinity solution was also investigated. The measurement by ICP-MS agreed well with the results from gamma spectrometry.

Keywords: Environmental, Fuels\Energy\Petrochemical, ICP-MS, Water
Application Code: Environmental
Methodology Code: Atomic Spectroscopy/Elemental Analysis

Session # 1130 Abstract # 1130-2 **Poster Sessions**

Session Title: Environmental, Food and Elemental Analyses - Atomic Spectroscopy

Abstract Title: **Speciation of Organic Mercury in Water Samples by Alkylation, Organic Solvent Extraction and GC-AFS Detection – A Comparison of Ethylation, Propylation and Phenylation**

Date: Tuesday, March 08, 2016 - Morning

Time:

Room: Exposition Floor, 400 Aisle

Primary Author: Cornelius Brombach
P S Analytical

Co-Author(s): Bin Chen, Jun Yoshinaga, Warren T. Corns

Abstract Text

Gas chromatography is an established method for mercury speciation coupled to ICP-MS or AFS. AFS offers a low cost alternative to ICP-MS with similar analytical performance characteristics. The EPA method 1630 uses distillation, ethylation, and purge and trap for speciation of MeHg. Distillation in combination with purge and trap is known to form artefacts of MeHg in sediments and is a time-consuming step. The use of Ethylation also masks the determination of ethylmercury which may be important in certain biogeochemical ecosystems such as the Florida Everglades and for wastewater discharges from industrial sources. In this work, we use the extraction of the alkylated species into an organic phase and then the direct injection of the extract into GC-AFS. This eliminates the purge and trap requirement thus simplifying instrument and reduces measuring times. This approach was tested with sodium tetraethylborate, sodium tetrapropylborate and sodium tetraphenylborate and the advantages and disadvantages of each alkylation reagent are presented. Various water samples including drinking water, river water, seawater and wastewater were tested using the methods developed and these findings will be reported.

Keywords: Atomic Spectroscopy, Environmental/Water, Gas Chromatography, Mercury
Application Code: Environmental
Methodology Code: Atomic Spectroscopy/Elemental Analysis

Session # 1130 Abstract # 1130-3 **Poster Sessions**

Session Title: Environmental, Food and Elemental Analyses - Atomic Spectroscopy

Abstract Title: **Models for Predicting Atmospheric Mercury Concentrations Using Meteorological Data and Mercury Concentrations in [i]Salix[/i] (Willow) Leaves**

Primary Author: David Lehmpuhl
Colorado State University Pueblo

Date: Tuesday, March 08, 2016 - Morning
Time:
Room: Exposition Floor, 400 Aisle

Co-Author(s): Lauren Bartolo

Abstract Text

Atmospheric mercury can be a significant source of mercury to water and land areas relatively uncontaminated by the heavy metal, yet it can be difficult and expensive to determine. The concentration of mercury ([Hg]) was determined in [i]Salix[/i] (Willow) leaves during the 2014 growing season as well as in the air at various points throughout the year in Pueblo, CO. Corresponding meteorological data was also collected. In the air, [Hg] correlated linearly with air temperature with an R[sup]2[/sup] value of 0.84. Some dependence on wind direction and speed was also found. The measured [Hg] in Willow leaves indicated two distinct regions; an accumulation phase during the active growing season (April 18-June 18), and a steady state phase once the leaf had matured (July 1-October 4). During the active growing season Willow leaf concentrations ranged from 9.4 ± 0.1 □g kg[sup]-1[/sup] to 18.7 ± 0.3 □g kg[sup]-1[/sup]. The range of the [Hg] in the mature leaves was 23.0 ± 0.3 □g kg[sup]-1[/sup] to 29.5 ± 0.6 □g kg[sup]-1[/sup]. Models that include temperature and progression into the growing season only, and temperature with corrections applied for corresponding meteorological data, are now being applied using a new set of Willow leaf samples and meteorological data collected during the 2015 growing season at the same locations. Corresponding measurements of actual atmospheric [Hg] are being used to assess the validity and robustness of the models developed, and facilitate the discussion of the applicability and limits of the models to actual [Hg] in air.

Keywords: Analysis, Environmental/Air, Mercury, Method Development
Application Code: Environmental
Methodology Code: Atomic Spectroscopy/Elemental Analysis

Session # **1130** Abstract # **1130-4** **Poster Sessions**

Session Title Environmental, Food and Elemental Analyses - Atomic Spectroscopy

Abstract Title **Assessment of Titanium Dioxide Nanoparticles in Aquatic Tanks**

Date: Tuesday, March 08, 2016 - Morning
Time:
Room: Exposition Floor, 400 Aisle

Primary Author Sara Melow
Elmira College

Co-Author(s) Lisa A. Holland, Marriah Ellington

Abstract Text

Titanium dioxide nanoparticles are widely used in makeup, sunscreens, and paint with an estimated 38,000 tons produced annually. Many of these nanoparticles enter waterways such as rivers and lakes. There is evidence that titanium dioxide nanoparticles impact the health of humans and wildlife; however, aquatic exposure cannot be accurately monitored without a clear understanding of the behavior of these nanoparticles in water samples. Understanding the distribution and stability of these nanoparticles in water samples is critical to quantitatively assay these chemicals to estimate environmental toxicity. Nanoparticles were dosed in a flow-through exposure tank over a 24-hour period and determined using inductively coupled plasma-atomic emission spectroscopy. An acid digestion of the titanium dioxide nanoparticles was validated to ensure the nanoparticles were solubilized and then analyzed by the inductively coupled plasma. With this study accomplished, the behavior of titanium dioxide nanoparticles in aquatic systems is better understood and exposure experiments can be performed more accurately. Animal exposure experiments are used to demonstrate the exacerbation of chemical toxicity through nanoparticles.

Keywords: Plasma Emission (ICP/MIP/DCP/etc.), Water
Application Code: Environmental
Methodology Code: Atomic Spectroscopy/Elemental Analysis

Session # 1130 Abstract # 1130-5 **Poster Sessions**

Session Title: Environmental, Food and Elemental Analyses - Atomic Spectroscopy

Abstract Title: **Feasibility Use of Ceramics as Solid Support for Cr(III) Measurement in Water by LIBS**

Date: Tuesday, March 08, 2016 - Morning

Primary Author: Cassiana S. Nomura
Institute of Chemistry - University of Sao Paulo

Time:

Room: Exposition Floor, 400 Aisle

Co-Author(s): Alexandrina C. Carvalho, Daniel M. Silvestre, Danielle P. Intima, Flávio O. Leme, Juliana Naozuka

Abstract Text

Laser Induced Breakdown Spectroscopy is a multi elemental technique which allows direct and in situ analysis. However, quantitative measurement is not easy due to strong matrix interference. In liquid sample analysis, loss of sensitivity is observed due to the laser and water interaction. To enable LIBS application in liquid samples, some strategies, such as liquid-to-solid transfer, have to be used. This work proposes water analysis by LIBS using ceramic as solid support for Cr(III) adsorption. The best adsorption condition was found on pH 7.0, contact time = 40min and no interference was observed even in the presence of high concentration of Na, K, Cu, Al, Fe, Zn, Ba, among others. A LIBS system (J200 Tandem, Applied Spectra) with a Nd:YAG laser operating at 266 nm, pulsed at 10 Hz and a high-resolution 6 channel-CCD spectrometer covers the spectrum ranging from 186 to 1044 nm was used to carried out the experiments. Instrumental parameters consisted of 403 accumulated laser pulses, 20 mJ/ pulse, 65 µm of spot size, 0.75 µs delay time and 1.05 ms gate width. Calibration was performed with ceramics spiked with increasing concentrations of Cr(III). The calibration equation obtained was y=790.51x + 16393 and R2=0.9589 when Cr I 425.435 nm emission signal was monitored. The use of Si as internal standard was investigated and the following equation was obtained: y = 0.0004x + 0.0011 and R2=0.9975. Fresh water sample spiked with 0.9 mg L-1 of Cr(III) was analyzed by the proposed method and recovery of 105±2% was observed.

Keywords: Atomic Emission Spectroscopy

Application Code: Environmental

Methodology Code: Atomic Spectroscopy/Elemental Analysis

Session # 1130 Abstract # 1130-6 **Poster Sessions**

Session Title: Environmental, Food and Elemental Analyses - Atomic Spectroscopy

Abstract Title: **Lead Determination in Soil from a Recreational Shooting Range Built on a Reclaimed Strip Mine: Effects of Oxidant Flow Rate on PB Measurements Obtained by FAAS, and Other Considerations**

Primary Author: Mark T. Stauffer
University of Pittsburgh - Greensburg

Date: Tuesday, March 08, 2016 - Morning
Time:
Room: Exposition Floor, 400 Aisle

Co-Author(s): Luke J. Metzler

Abstract Text

This investigation is a continuation of ongoing research into leaching of lead through soil at a recreational shooting range built on a reclaimed strip mine. The focus of the current investigation is to determine the optimum analysis conditions, particularly the oxidant flow rate used for determination of lead in soil from the shooting range by flame atomic absorption spectrometry (FAAS). Lower-than-expected Pb recoveries on spiked soil samples have generated high interest in determination of an optimum oxidant flow rate for the FAAS absorbance measurements for Pb. A preliminary experiment on a set of Pb-spiked soil samples yielded lower recoveries for Pb. It was hypothesized that not all of the ionic Pb present in the analysis solutions undergoes reduction to neutral Pb in the air-acetylene flame. Adjustment of oxidant flow rate to a value yielding a richer flame produced Pb results for spiked samples that indicated essentially complete Pb recoveries. The current objective is to study the effect of oxidant flow rate on the slopes of calibration curves used for Pb determinations, and subsequently on the results obtained for Pb in soil. In addition to the oxidant flow rate study, water from the man-made lake downhill from the shooting range will be analyzed for lead to determine if any migration actually occurs.

Results of the aforementioned investigation, along with details of sample collection, preparation, analytical methods, discussion of factors that might hinder Pb migration through the soil, and future directions for this research, will be presented and discussed.

Keywords: Atomic Absorption, Environmental/Soils, Lead, Water
Application Code: Environmental
Methodology Code: Atomic Spectroscopy/Elemental Analysis

Session # **1130** **Abstract #** **1130-7** **Poster Sessions**

Session Title: Environmental, Food and Elemental Analyses - Atomic Spectroscopy

Abstract Title: **Determination of Metals in Three Types of Loose-Leaf Tea: Can Metal Content Indicate the Type of Tea?**

Date: Tuesday, March 08, 2016 - Morning

Time:

Room: Exposition Floor, 400 Aisle

Primary Author: Mark T. Stauffer
University of Pittsburgh - Greensburg

Co-Author(s): Aaron K. Hirshka

Abstract Text

This presentation will examine the results of research performed on commercially available black, green, and white loose-leaf tea to determine if metal ion concentrations can be used as an indicator of the type of tea. The metal content of tea, just like any other plant, is susceptible to the metal content of the soil in which it is grown (1). Additionally, certain metals, e.g., manganese, are prevalent in tea. The hypothesis proposed in this study is that metal ion concentrations of selected metals (zinc, iron, copper, manganese, calcium, magnesium, chromium, aluminum, and lead) will differ among the three types of loose-leaf tea due to differences in the tea plants, growing conditions, and tea leaf processing prior to packaging. The tea samples will be decomposed by digestion with acid, e.g. nitric acid, with the aforementioned metals determined by flame atomic absorption spectrometry (FAAS). The results of the tea analyses for the selected metals will be analyzed statistically by, e.g., ANOVA and principal component analysis (PCA), to determine significant differences in metal ion concentrations among the three types of loose-leaf tea as well as identify trends in metals among the tea types. Experimental details, data and results obtained and their interpretation, and future plans for this study will be presented and discussed.

(1) P. L. Fernandez-Caceres, M. J. Martin, F. Pablos, and A.G. Gonzalez. J Agric Food Chem. 2001, 29, 4775

Keywords: Atomic Absorption, Chemometrics, Food Science, Metals
Application Code: Food Science
Methodology Code: Atomic Spectroscopy/Elemental Analysis

Session # 1130 Abstract # 1130-9 **Poster Sessions**

Session Title Environmental, Food and Elemental Analyses - Atomic Spectroscopy

Abstract Title **Speciation of Mercury in Rice with a New Online Pre-Concentration HPLC-CV-AFS Method**

Date: Tuesday, March 08, 2016 - Morning

Time:

Room: Exposition Floor, 400 Aisle

Primary Author Cornelius Brombach
P S Analytical

Co-Author(s) Bin Chen, Eva Krupp, Joerg Feldman, Piumi K. Dona, Warren T. Corns

Abstract Text

Rice is a global staple food and can provide up to 75 % of a person's dietary energy supply. It is now accepted that rice is a mercury accumulator. Methylmercury (MeHg) is powerful neurotoxin so the monitoring of this species in rice is crucial. We developed a speciation method for rice, which is based on an alkaline digestion and leaching extraction, the subsequent online pre-concentration and HPLC-CV-AFS detection. A market study was done, where 87 commercial rice products (from the United Kingdom, Germany, Switzerland, Taiwan and China), including baby-food rice, were analyzed for total Hg and methylmercury content. MeHg was analyzed with another independent method called species-specific isotope dilution GC-ICP-MS. The results for both methods were found to be in good agreement and a correlation coefficient of 0.972 and a slope of 1.013 was found. The MeHg concentrations in all rice products investigated range from 0.11 to 6.45 µg kg-1 with an average value of 1.91 ± 1.07 µg kg-1. Total Hg ranges from 0.53 to 11.1 µg kg-1 with an average of 3.04 ± 2.07 µg kg-1. MeHg is the main Hg species with 71 ± 26 % of total Hg but can be as low as 6 % and as high as 100 %. Whilst these results seem relatively low, the mercury dietary contribution to infants and populations consuming large rice quantities is significant especially if the person also consumes fish products or has additional exposures to Hg. This work clearly demonstrates the need for detailed food surveys to accurately determine MeHg in rice and rice products.

Keywords: Atomic Spectroscopy, Environmental Analysis, HPLC, Solid Phase Extraction
Application Code: Food Contaminants
Methodology Code: Atomic Spectroscopy/Elemental Analysis

Session # **1130** Abstract # **1130-10** **Poster Sessions**

Session Title Environmental, Food and Elemental Analyses - Atomic Spectroscopy

Abstract Title **Optimizing a Total Protein Combustion Instrument for Maximum Sample Throughput and Lowest Cost-Per-Analysis**

Date: Tuesday, March 08, 2016 - Morning
Time:
Room: Exposition Floor, 400 Aisle

Primary Author Fred Schultz
LECO Corporation

Co-Author(s) Jeffery Gast, Mason Marsh

Abstract Text

Total protein in foods and feeds is calculated using the total measured nitrogen content in the sample and a multiplier specific to the sample matrix. Nitrogen determination is commonly performed by one of two major methods, a classical wet chemistry (Kjeldahl) technique or combustion instrument-based technique, with the combustion technique gaining popularity due to several advantages including shorter analysis times, ease of operation, and improved safety characteristics.

Total nitrogen combustion instruments use a high temperature furnace with a pure oxygen environment to combust the sample. The nitrogen gases within the sample combustion gas are subsequently reduced to N2 gas, and detected using a thermo-conductivity (TC) detector with the excess oxygen and other sample combustion products being removed using multiple reagents within the instrument flow path. Total nitrogen combustion instruments manufactured by LECO Corporation utilize a system collecting and equilibrating the combustion gas, then sampling a small aliquot of the equilibrated combustion gas for nitrogen measurement, hereby reducing the reagent demand and cost associated with treating this gas for nitrogen measurement.

This poster presentation will cover the optimization of a total nitrogen combustion instrument for maximum sample throughput by optimizing the instrument method parameters, and concurrently reducing the cost-per-analysis by optimizing the aliquot gas volume while maintaining the instrument's performance requirements. Data will be examined that includes common foods, feeds, and reference materials analyzed with a LECO FP628 and TruMac® N instrument using helium and argon gas as a TC carrier gas.

Keywords: Elemental Analysis, Food Science, Method Development, Protein
Application Code: Food Science
Methodology Code: Atomic Spectroscopy/Elemental Analysis

Session # **1130** Abstract # **1130-11** **Poster Sessions**

Session Title: Environmental, Food and Elemental Analyses - Atomic Spectroscopy

Abstract Title: **Laser Induced Breakdown Spectroscopy (LIBS) of Food Samples: Case Study of Tortillas**

Primary Author: Charles Ghany
Mississippi State University

Date: Tuesday, March 08, 2016 - Morning
Time:
Room: Exposition Floor, 400 Aisle

Co-Author(s): Bader Alfarraj, Chet R. Bhatt, Fang Y. Yueh, Herve Sanghapi, Jagdish P. Singh

Abstract Text

Tortillas of various colors are cut into sizes comparable to slides (25x75x1.0mm) and studied with the help of LIBS. LIBS is presented as an effective tool for rapid in situ sample analysis of food samples with emphasis on tortillas due to its simplicity, fast response, little or no sample preparation. Spectroscopic analysis of the plasma generated by Nd:YAG laser irradiation of tortilla was carried out. A careful selection of spectral lines of Ca, Na and K which do not suffer from spectral interference was made. Spectral properties of the above mentioned elements such as peak intensities, intensity ratios, and area under spectral lines were analyzed. Optimization of laser pulse energy, detection gate width and gate delay for well resolved spectra with high signal-to-noise ratio was done and other parameters studied. Tortillas manufactured by Gruma Corporation of brown, orange, green and white colors as well as homemade tortillas were studied in this work and the results from both dried and undried samples will be presented.

Keywords: Analysis, Atomic Emission Spectroscopy, Spectrometer, Spectroscopy
Application Code: Food Safety
Methodology Code: Atomic Spectroscopy/Elemental Analysis

Session # 1130 Abstract # 1130-12 Poster Sessions

Session Title Environmental, Food and Elemental Analyses - Atomic Spectroscopy

Abstract Title **Development of a Method For the Determination of Titanium Dioxide Nanoparticles in Food Products Using SP-ICPMS**

Primary Author Antonio Moreda-Pineiro
University of Santiago de Compostela

Date: Tuesday, March 08, 2016 - Morning
Time:
Room: Exposition Floor, 400 Aisle

Co-Author(s) Antonio Moreda-Piñeiro, Francisco Javier Vilariño-Páxaro, Manuel Aboal-Somoza, Maria del Carmen Barciela-Alonso, Olga Cristina Vázquez-Padín, Pilar Bermejo-Barrera

Abstract Text

Nanotitanium (E-171) is one of the food additives that are being subjected to a re-evaluation program by the European Food Safety Authority in accordance with Commission Regulation (EU) No 257/2010 of the European Union, to collect more toxicological data and data relevant for the estimation of the human exposure. Nowadays, additives such us E-171 are not mandatorily qualified by the word "nano" in the list of food ingredients.

In this study, an analytical method for the detection and characterization of nanoparticles of titanium dioxide has been developed. The technique used was inductively coupled plasma-mass spectrometry with single particle detection (SP-ICP-MS). The instrumental operating conditions were selected (optimization of nebulization gas flow, torch position, deflector voltage, selection of isotopes, dwell time, readings, etc...). In optimal conditions, standards of ionic titanium and a certified reference material of gold nanoparticles were used for calibration and to calculate the concentration and size of the nanoparticles. Experiments for the preparation of suspensions were also performed using a ultrasonic probe or ultrasonic bath, different sonication times and nanoparticles provided by different manufacturers. Calcium, a possible interferent in the analysis, was also determined in several compounds that were tested as stabilizers (hexane, BSA, Triton-X100, isopropanol, lactose, citric acid, tergitol). The analytical characteristics of the method (limits of detection in size and concentration, reproducibility, recovery) were evaluated, and finally, different samples were analyzed (drinks, energetic drinks, pharmaceutical preparations for oral intake).

Keywords: Food Safety, ICP-MS, Nanotechnology

Application Code: Food Safety
Methodology Code: Atomic Spectroscopy/Elemental Analysis

Session # 1130 Abstract # 1130-13 **Poster Sessions**

Session Title: Environmental, Food and Elemental Analyses - Atomic Spectroscopy

Abstract Title: **Evaluation of Optical Depths of Ca Emission Lines in Laser Induced Breakdown Spectroscopy (LIBS)**

Primary Author: Bader Alfarraj
Mississippi State University

Date: Tuesday, March 08, 2016 - Morning
Time:
Room: Exposition Floor, 400 Aisle

Co-Author(s): Charles Ghany, Chet R. Bhatt, Fang Y. Yueh, Herve Sanghapi, Jagdish P. Singh

Abstract Text

Laser induced breakdown spectroscopy (LIBS) is widely used laser spectroscopic techniques in various fields, such as material science, forensic science, biological science, and the chemical and pharmaceutical industries. In most of LIBS work, the analysis is performed using radiative transitions from the atomic emissions. In this work, the focus is on the determination of plasma characterization, such as plasma temperature to obtain the optical depths of Ca lines. A binary mixture of calcium carbonate (CaCO[sub]3[/sub]) and strontium fluoride (NaCl[sub]2[/sub]) of different concentrations in powder form was used as sample. LIBS spectra were collected by varying various parameters, such as laser energy, gate delay, and gate width to optimize the LIBS signals. The atomic emission from Ca lines observed in LIBS spectra of different sample composition were used to characterize the laser induced plasma and evaluate of optical depths of LIBS. The details of this study will be presented in this paper.

Keywords: Atomic Emission Spectroscopy, Material Science, Plasma
Application Code: Material Science
Methodology Code: Atomic Spectroscopy/Elemental Analysis

Session # **1130** Abstract # **1130-14** **Poster Sessions**

Session Title: Environmental, Food and Elemental Analyses - Atomic Spectroscopy

Abstract Title: **Standard Dilution Analysis for the Determination of Calcium by Flame Atomic Emission Spectrophotometry**

Date: Tuesday, March 08, 2016 - Morning
Time:
Room: Exposition Floor, 400 Aisle

Primary Author: Clifton P. Calloway
Winthrop University

Co-Author(s): Emily A. Watson, Katja A. Hall

Abstract Text

Standard Dilution Analysis (SDA) is a novel spectroscopic calibration method that can be applied to most instrumental techniques that are capable of monitoring two wavelengths and will accept liquid samples. It combines traditional methods of calibration curve, standard addition and internal standard, therefore correcting for matrix effects and for fluctuations due to changes in sample size, orientation or instrumental parameters. SDA analysis time requires only about 200 seconds per sample with flame atomic emission spectrophotometry (AES). The preparation of a series of standard solutions and the construction of a universal calibration graph are not required. The analysis is performed by combining two solutions in a single container, the first solution containing 50% sample and 50% of a standard mixture (the analyte + the internal standard); the second solution containing 50% sample and 50% solvent. Data are collected in real time as the first solution is diluted by the second solution. The analyte concentration in the sample is determined from the ratio of the slope and the intercept of the resulting plot.

Phosphate is a well-known interferent in the determination of calcium in complex samples, such as vitamin tablets and calcium supplements. It is often recommended to add a complexing agent or matrix modifier for the determination of calcium by atomic spectrophotometry. Calcium has also been shown to produce non-linear standard addition plots in the extrapolated region when phosphate is present. SDA method of analysis has been applied to the determination of calcium in over-the-counter vitamin and calcium supplements, containing phosphate, using an inexpensive flame atomic emission spectrophotometer. Accuracy and precision are better than those observed for external calibration, standard addition or internal standard methods of analysis, even in the presence of significant concentrations of phosphate.

Keywords: Analysis, Atomic Spectroscopy, Elemental Analysis, Quantitative
Application Code: General Interest
Methodology Code: Atomic Spectroscopy/Elemental Analysis

Session # 1130 Abstract # 1130-15 Poster Sessions

Session Title: Environmental, Food and Elemental Analyses - Atomic Spectroscopy

Abstract Title: **Cyanide Detection in Blood Using Indirect Atomic Absorption Spectroscopy**

Date: Tuesday, March 08, 2016 - Morning

Time:

Room: Exposition Floor, 400 Aisle

Primary Author: Jeffrey Rosentreter
Idaho State University

Co-Author(s): Jeff Kuhlmeier, Matt Kirkham

Abstract Text

Cyanide is an important and useful compound used in large quantities in various mining and industrial processes. Cyanide, while useful, is also acutely toxic and has the potential to poison humans and animals through accidental or purposeful release to the environment. Cyanide proves difficult to detect after entering the bloodstream due to its tight binding with various biological iron complexes. In our research, various displacement techniques are investigated which allow the cyanide to be removed and rebound to a more preferable element for detection. By complexing the cyanide with silver one can facilitate simple analyte detection using atomic absorption spectroscopy. Complexation reactions have been performed under a series of variable conditions, including reaction with silver wire, silver mesh, and micro-porous silver filters. Varying reaction times and ppm concentrations of cyanide were reacted with the silver sources. Analysis of the formation of silver-cyanide complexes was done via flame atomic absorption spectroscopy. Reactions using the silver mesh supported on a glass rod aptly facilitate sample media of various viscosities and suspended solid concentrates and established a basis for the quantitative detection of cyanide. Using these results, ongoing research is investigating iron cyanide dissociation methods including visible and UV photodissociation, as well as, selective reaction kinetics. This will provide the methodology for detecting cyanide in blood that can differentiate free and complexed cyanide concentrations. Such a unique detector, providing cyanide speciation in blood would aid in toxicological determination of the environmental conditions that lead to the poising whether through accidental or purposeful exposure.

Keywords: Atomic Spectroscopy, Method Development, Toxicology
Application Code: Homeland Security/Forensics
Methodology Code: Atomic Spectroscopy/Elemental Analysis

Session # **1130** **Abstract #** **1130-16** **Poster Sessions**

Session Title: Environmental, Food and Elemental Analyses - Atomic Spectroscopy

Abstract Title: **Advanced Application of Speciation Analysis Using ICP-MS detection**

Date: Tuesday, March 08, 2016 - Morning
Time:
Room: Exposition Floor, 400 Aisle

Primary Author: Daniel Kutscher
Thermo Fisher Scientific

Co-Author(s): Monika Verma, Shona McSheehy Ducos

Abstract Text

It is well known that speciation analysis is a mandatory tool in order to correctly address the potential risks associated with for example As in fish or food or Cr in drinking water. Due to the ionic nature of the most commonly investigated species, ion chromatography hyphenated to inductively coupled plasma mass spectrometry (IC-ICP-MS) is normally the method of choice. The striking advantage of this combination is that modern IC systems are completely metal free, so that background contamination caused through the chromatographic system can be systematically avoided. However, also liquid chromatography can help to unravel the distribution of a given element over different chemical compounds, normally using reversed phase separations for non-ionic compounds with different polarities. A typical example for the combination of HPLC to ICP-MS is the analysis of Sulfur and Phosphorous containing peptides or proteins. Furthermore, recent hyphenation of e.g. Field Flow Fractionation (FFF or AF4) extends the capabilities of speciation analysis not only to investigate different chemical forms of an element, but can also answer the question whether the element is present in a nanomaterial. This presentation should highlight the advantages and applications of those chromatographic techniques when used in combination with ICP-MS as a powerful detection system for trace elements.

Keywords: HPLC, ICP-MS, Ion Chromatography, Speciation
Application Code: General Interest
Methodology Code: Atomic Spectroscopy/Elemental Analysis

Session # **1130** Abstract # **1130-17** **Poster Sessions**

Session Title Environmental, Food and Elemental Analyses - Atomic Spectroscopy

Abstract Title **Multi-Trophic Analysis of Lead Using a Flame Atomic Absorption Spectrometer**

Primary Author Michael DeCarolis
St. John Fisher College

Date: Tuesday, March 08, 2016 - Morning
Time:
Room: Exposition Floor, 400 Aisle

Co-Author(s) Christopher Collins, Kimberly Chichester

Abstract Text

Soil samples as well as hair samples from trapped mice and shrews from two sites in Rochester, N.Y. were analyzed for the presence and concentrations of lead using a Flame Atomic Absorption Spectrometer (FAAS). The research goal was to introduce a multi-sample mammal and soil baseline testing series of protocols that was pioneered in Europe to the United States. Multi-trophic analysis and monitoring techniques are effective tools in determining the concentrations of lead in an environment and determining if actions need to be taken to mitigate these toxins. Hair samples were used because they can serve as a non-lethal substitute for liver tissue as shown by Tête et al. in 2014. The target sampling site was a section of forested private property built atop a former landfill. A golf course was used as the control site due to its history of non-contamination. Soil samples were dried in an oven, ground into a uniform powder; then digested using a 3:1 ratio of hydrochloric-nitric acid. The solution was refluxed to extract lead, than filtered to remove contaminants. Hair samples were collected with a electric shaver from the live trapped mammals, than washed with alternating cycles of distilled water and acetone. Hair samples were digested in the same manner as the soil. All samples were subjected FAAS and the results showed statistical significance between lead in soil samples between sites, and also between the mammals and the soil at each site.

Keywords: Atomic Absorption, Environmental/Biological Samples, Environmental/Waste/Sludge, Lead
Application Code: Environmental
Methodology Code: Atomic Spectroscopy/Elemental Analysis

Session # **1130** **Abstract #** **1130-18** **Poster Sessions**

Session Title: Environmental, Food and Elemental Analyses - Atomic Spectroscopy

Abstract Title: **Atomic and Molecular Laser Induced Breakdown Spectroscopy for Detection of Chlorine in Concrete**

Date: Tuesday, March 08, 2016 - Morning
Time:
Room: Exposition Floor, 400 Aisle

Primary Author: Will B. Jones
University of Florida

Co-Author(s): Ben Smith, Ebo Ewusi-Annan, Nico Omenetto, Tobias Guenther

Abstract Text

Laser induced breakdown spectroscopy (LIBS) has been studied as a simple, fast and "in-situ" method of detecting chlorine in concrete samples. Both single pulse (SP) and double pulse (DP) experiments have been described in the literature [see, for example, 1]. In most studies, the neutral atomic line at 837.594 nm has been used, with variable success in terms of detection limits achieved. DP-LIBS has been demonstrated to be more sensitive, and an LOD of 50 ppm has been reported [2]. We will discuss here the results obtained in our laboratory using an ionic chlorine line at 479.454 nm as well as a molecular emission band due to CaCl around 593 nm. Detection limits for both SP and DP methods, obtained on cement mortar samples in the concentration range 0.06-0.6% Cl will be presented, together with a statistical evaluation of the data.

[1] R. Noll, "Laser Induced Breakdown Spectroscopy", Springer-Verlag, Berlin (2012).
[2] T. A. Labutin, A. M. Popov, S. N. Raikov, S. M. Zaytsoev, N. A. Labutina,a and N. B. Zorova, Determination of chlorine in concrete by laser-induced breakdown spectroscopy in air, Journal of Applied Spectroscopy, 80(3), 315-318 (2013).

Keywords: Atomic Emission Spectroscopy, Elemental Analysis, Laser
Application Code: Environmental
Methodology Code: Atomic Spectroscopy/Elemental Analysis

Session # 1130 Abstract # 1130-22 **Poster Sessions**

Session Title: Environmental, Food and Elemental Analyses - Atomic Spectroscopy

Abstract Title: **Fast Monitoring Processed Manure Using WD-XRF Spectroscopy for Nutrients and Metals**

Primary Author: Alexander Seyfarth, Rock River Axs LLC

Co-Author(s): Aicado Roa-Espinosa

Date: Tuesday, March 08, 2016 - Morning
Time:
Room: Exposition Floor, 400 Aisle

Abstract Text

The U.S. dairy industry produces more approximately 400,000 million pounds of manure each year. This poses a potential environmental problem as the excess amounts of phosphate and nitrogen in dairy manure can have negative impacts to the land and water systems around the dairy farms. It is therefore necessary to process dairy manure to remove excess nutrients and minimize the environmental impact. A typical manure processing plant often includes several units such as a rotating screen unit to remove excess fibers, and a digester unit in which manure is anaerobically fermented to produce biogases. Consequently it is crucial to monitor and evaluate the nutrient contents of the processed manure at each processing unit to determine the efficiency of these units in terms of solids and nutrients removal. The traditional methods for analysis of major and trace elements in manure are based on dissolution (digestion) of individual samples in acids and subsequent analysis of the homogenized solutions by atomic absorption spectrophotometry (AAS), or inductively coupled plasma optical emission spectrometry (ICP - OES). While these methods have been well established and characterized, the sample preparation procedure is time-consuming and hazardous. X-ray fluorescence (XRF) spectroscopy has gained popularity in the field of elemental analysis due to not only its comparable accuracy and precision to other spectroscopic methods, but also its simple sample preparation. Light Element analyzer crystals enable the quantification of Nitrogen, Carbon and Boron in manure in addition to the traditional major and trace elements. With ICP traceable matrix compliant reference samples as well as an optimized rapid sample preparation a method has been developed and is in routine operation using WD-XRF. The poster will present the performance especially for the light element analysis of N, C and Boron and compare the “whole” manure assay to commercially obtained atomic spectroscopy results.

Keywords: Agricultural, Environmental Analysis, Environmental/Soils, X-ray Fluorescence
Application Code: Agriculture
Methodology Code: X-ray Techniques

Session # **1130** **Abstract #** **1130-23** **Poster Sessions**

Session Title: Environmental, Food and Elemental Analyses - Atomic Spectroscopy

Abstract Title: **Onsite Monitoring of Manure Using XRF for Nutrients and Metals**

Date: Tuesday, March 08, 2016 - Morning
Time:
Room: Exposition Floor, 400 Aisle

Primary Author: Alexander Seyfarth
Bruker Nano Analytics

Co-Author(s): Aicado Roa-Espinosa

Abstract Text

The U.S. dairy industry produces more than 400,000 million pounds of manure each year. This poses a potential environmental problem as the excess amounts of phosphate and nitrogen in dairy manure can have negative impacts to the land and water systems around dairy farms. It is therefore necessary to process dairy manure to remove excess nutrients and to minimize the environmental impact. Large scale farm have typical manure processing plant which include various process steps. It is crucial to monitor and evaluate the nutrient contents of the processed manure at each processing step to determine the efficiency and tune the process. X-ray fluorescence (XRF) spectroscopy has gained popularity in the field of elemental analysis due to not only its comparable accuracy and precision to other spectroscopic methods, but also its simple sample preparation.The paper/poster illustrates a proprietary HHXRF solution which enables rapid and cost effective characterization of the various manure processing steps at line, as part of a patented manure solidification process. With a HH-XRF unit with a traceable calibration based on matrix matched and ICP and WD XRF characterized dairy manure reference samples it is possible to determine major and trace elements with a direct measurement of the solid manure process material in less than 1 minute measurement time.

Keywords: Agricultural, Environmental/Soils, Environmental/Waste/Sludge, X-ray Fluorescence
Application Code: Agriculture
Methodology Code: X-ray Techniques

Session # 1140 Abstract # 1140-1 **Poster Sessions**

Session Title: Sampling and Sample Preparation - Environmental

Abstract Title: **The Measurement of Formaldehyde in Drinking Water Using Automated SPE and HPLC**

Date: Tuesday, March 08, 2016 - Morning
Time:
Room: Exposition Floor, 400 Aisle

Primary Author: Alicia Cannon
Horizon Technology

Co-Author(s): Chris Shevlin, Michael Ebitson

Abstract Text

Formaldehyde is found in many products including particle board and permanent-press clothing. It has also been reported used in fracking fluid as a biocide or scale inhibitor. It is not unexpected that it might also be found in water supplies for drinking water. Although not currently regulated in drinking water it is included in Candidate Contaminant List which will be monitored for occurrence and further evaluated for regulation in the next few years. The EPA reason given for inclusion is "It is an ozoneation disinfection byproduct, can occur naturally and has been used as a fungicide."

Two EPA methods are available for preparation and analysis of the compound, US EPA Method 554 in the drinking water program and US EPA Method 8315A in the Office of Resource Conservation, using the same general methodology for collection on a C18 cartridge or disk and elution with ethanol. The sample is derivatized with 2, 4-dinitrophenyl hydrazine and detection with HPLC-UV is used for the analytical measurement. When cartridges are used for collection of the analytes, inconsistent flow rates or flow rates faster than specified can affect the recovery and precision. Performance using an automated system to provide consistent performance will be evaluated and compared with manual performance. A variety of concentration levels and types of drinking water will be shown.

Keywords: Environmental/Water, HPLC Detection, Solid Phase Extraction
Application Code: Environmental
Methodology Code: Sampling and Sample Preparation

Session # 1140 Abstract # 1140-2 **Poster Sessions**

Session Title: Sampling and Sample Preparation - Environmental

Abstract Title: **Analysis of Extract Drying Criteria for Oil and Grease Method 1664A/B**

Date: Tuesday, March 08, 2016 - Morning

Time:

Room: Exposition Floor, 400 Aisle

Primary Author: Michael Ebitson
Horizon Technology

Co-Author(s): David Gallagher, William Jones

Abstract Text

Oil and Grease is a simple measurement, used around the world to evaluate pollution, regulate release into sewer systems and ensure good wastewater treatment plant operation. There are a variety of options listed in the US EPA method 1664A/B to accomplish drying the hexane extract including sodium sulfate drying, phase separation paper and others. Drying is an important step because water left in the extract will make it hard for the extract to evaporate uniformly prior to gravimetric analysis. False negatives because light-end oil and grease components may evaporate off before the water is evaporated is one problem. False positives because water remains when the hexane is evaporated is another problem. Each of the drying options allowed by the method will be evaluated for performance, ease of use and cost.

Keywords: Environmental/Water, Sample Preparation, Solid Phase Extraction
Application Code: Environmental
Methodology Code: Sampling and Sample Preparation

Session # 1140 Abstract # 1140-3 **Poster Sessions**

Session Title Sampling and Sample Preparation - Environmental

Abstract Title **Fundamentals and Comparisons for Organic Sample Extract Evaporation**

Primary Author Zoe Grosser
Horizon Technology

Date: Tuesday, March 08, 2016 - Morning
Time:
Room: Exposition Floor, 400 Aisle

Co-Author(s) Robert Johnson

Abstract Text

Sample preparation is a key part of the analytical process, contributing to reproducibility and accuracy of the final results. Generally, sample preparation for organic analysis requires the analytes of interest to be first extracted from the matrix. Then cleanup of the extract may be required to remove interferences arising from the matrix. Water is removed during the drying step if it was introduced from the samples. Finally the extract is reduced in volume to accommodate the detection limits needed for the analysis and the ability of the instrument to accommodate a large-volume sample.

The evaporation/concentration step can be achieved with various technologies, including heat, vacuum, and blow-down. We will examine the parameters that go into each of these choices and describe criteria to consider in matching the sample to the technique. Further, solvent recovery has become increasingly important as the number of samples analyzed and the size of individual laboratory locations has increased. The implications for solvent recovery based on the type of evaporation will be discussed.

Keywords: Chromatography, Environmental, Sample Preparation
Application Code: Environmental
Methodology Code: Sampling and Sample Preparation

Session # 1140 Abstract # 1140-4 **Poster Sessions**

Session Title Sampling and Sample Preparation - Environmental

Abstract Title **Determination of 2-Methylisoborneol and Geosmin in Water Using Solid Phase Micro Extraction**

Date: Tuesday, March 08, 2016 - Morning
Time:
Room: Exposition Floor, 400 Aisle

Primary Author Anne Jurek
EST Analytical

Co-Author(s)

Abstract Text
The compounds 2-Methylisoborneol (2-MIB) and Geosmin are the primary source of the foul odor found in drinking water. Algal contamination is the primary cause of the formation of these compounds. Since Geosmin and 2-MIB have such a low odor threshold, even the slightest amount of them can produce an unpleasant odor and taste in drinking water. In order to detect 2-MIB and Geosmin at these low levels, the sampling and analysis of the water has to be optimized. Standard Method 6040D describes a procedure for the detection of 2-MIB and Geosmin using Solid Phase Micro Extraction (SPME) coupled with a Gas Chromatograph (GC) and Mass Spectrometer (MS). Selective Ion Monitoring (SIM) is used for compound detection down to the part per trillion (ppt) level. This examination will optimize the sampling and detection of 2-MIB and Geosmin.

Keywords: Environmental/Water, GC-MS, SPME
Application Code: Environmental
Methodology Code: Sampling and Sample Preparation

Session # 1140 Abstract # 1140-5 **Poster Sessions**

Session Title: Sampling and Sample Preparation - Environmental

Abstract Title: **The Advantages of Automated Sample Preparation**

Primary Author: Anne Jurek
EST Analytical

Date: Tuesday, March 08, 2016 - Morning
Time:
Room: Exposition Floor, 400 Aisle

Co-Author(s): Justin Murphy, Kelly Cravenor, Lindsey Pyron

Abstract Text

Sample preparation is one of the most important steps in analytical chemistry. Attention to detail and accuracy are essential. For this reason, many laboratories are interested in automating sample preparation procedures so as to limit the possibility of human error. Furthermore, quantitative analysis requires not only sample prep but also standard prep. When preparing a calibration curve, the analyst has to follow a recipe in order to ensure standard accuracy. Any mistake in the standard preparation and the calibration curve will need to be re-prepped and re-run. Thus, valuable time and resources are wasted. This application will explore automated standard preparation of Poly Aromatic Hydrocarbon (PAH) compounds.

Keywords: PAH, Sample Handling/Automation, Sample Preparation, Semi-Volatiles
Application Code: Environmental
Methodology Code: Sampling and Sample Preparation

Session # **1140** **Abstract #** **1140-6** **Poster Sessions**

Session Title: Sampling and Sample Preparation - Environmental

Abstract Title: **Highly Robust Polymeric Ionic Liquid Coatings for Solid Phase Microextraction: Multiclass Determinations with Application of Direct-Immersion-Headspace Mode Using Gas Chromatography-Mass Spectrometry**

Date: Tuesday, March 08, 2016 - Morning
Time:
Room: Exposition Floor, 400 Aisle

Primary Author: Josias Merib
Iowa State University

Co-Author(s): Carasek Eduardo, Honglian Yu, Jared L. Anderson

Abstract Text

Sample preparation is the most critical step in the analytical procedure. One of the techniques developed to solve some problems associated with sample preparation is the solid-phase microextraction (SPME). The SPME procedure is simple, relatively quick and does not require the use of organic extraction solvents. Ionic liquids (ILs) have emerged as a new class of sorbent coatings in SPME. ILs are organic salts which consist largely of organic cations paired with organic or inorganic anions. These compounds often possess melting points less than or equal to 100 °C. Important properties of ILs include their low vapor pressure, high thermal stability, variable viscosity, and the ability to interact with dissolved molecules through a multitude of solvation interactions. PILs are polymers synthesized from IL monomers. They possess a number of advantages over ILs when used as coatings in SPME. PILs often possess higher viscosity and greater mechanical strength compared to ILs while exhibiting similar extraction selectivity. Recently, the use of nitinol (NiTi) wires as highly robust support for polymeric ionic liquid coatings has been proposed. In this work, different PIL coatings chemically bonded to NiTi wires were applied for multiclass determination of potential water contaminants with different volatilities using GC-MS. In order to find a compromise in the extraction conditions among the various volatilities of the studied analytes, direct immersion and headspace extraction modes were used sequentially in the same extraction procedure. Central composite design was applied to determine the best extraction conditions. In addition, the extraction efficiency achieved by the PIL coatings will be compared with commercial SPME fibers and the determination of analytical figures of merit related to those PIL coatings that provide the best performance will be performed.

Keywords: Environmental Analysis, SPME, Volatile Organic Compounds
Application Code: Environmental
Methodology Code: Sampling and Sample Preparation

Session # 1140 Abstract # 1140-7 **Poster Sessions**

Session Title Sampling and Sample Preparation - Environmental

Abstract Title **Development of Dispersive Liquid-Liquid Microextraction for the Determination of Six Steroidal Hormones in Wastewater Using High Pressure Liquid Chromatography-Charged Aerosol Detector**

Primary Author Mathew M. Nindi
Unisa

Date: Tuesday, March 08, 2016 - Morning
Time:
Room: Exposition Floor, 400 Aisle

Co-Author(s) Cecilia O. Osunmakinde, Simiso Dube

Abstract Text

A dispersive liquid-liquid microextraction (DLLME) was developed as simple, rapid, cheap sample clean-up and/or pre-concentration method for steroid hormone. The DLLME method was used in combination with high performance liquid chromatography with charged aerosol detector (HPLC-CAD) for the determination of estriol (E3), bisphenol A (BPA), Beta estradiol (BetaE2), Alpha estradiol (Alpha E2), testosterone (T) and progesterone (P) in wastewater. A mixture of extraction solvent (35 µL tetrachloroethylene) and dispersive solvent (500 µL methanol) was used to extract the steroid hormones from 5 mL of wastewater samples under optimum conditions. Parameter such as the type and volume of the extracting and dispersive solvent was examined. Linearity was in the range (0.0002 – 0.02µg/L) with of R2 ranging from 0.9952 – 0.9996. Intra-day repeatability %RSD was between 2.8 to 7.6). Limit of detection (0.02–0.06µg/L), Limit of quantification (0.1- 0.4µg/L). All six steroid hormones were detected in the wastewater samples. Satisfactory recoveries reveal that numerous estrogens can be simultaneously detected in environments where steroids are in abundance.

Keywords: Environmental/Waste/Sludge, Environmental/Water, HPLC Detection, Sample Preparation

Application Code: Environmental

Methodology Code: Sampling and Sample Preparation

Session # 1140 **Abstract # 1140-8** **Poster Sessions**

Session Title: Sampling and Sample Preparation - Environmental

Abstract Title: **Extraction of Analytes of Forensic Toxicological Interest from Plasma with Enhanced Matrix Removal-Lipid Material**

Date: Tuesday, March 08, 2016 - Morning

Time:

Room: Exposition Floor, 400 Aisle

Primary Author: Joan Stevens
Agilent Technologies

Co-Author(s): Derick Lucas, Limian Zhao, Megan Juck, William Long

Abstract Text

Matrix interferences and ion suppression are two common analytical issues often encountered when using chromatographic techniques in drug analysis. Diluting the sample prior to analysis especially for LC/MS analysis is a common technique however it does not reduce matrix resulting in ion-suppression. Lack of sample preparation will reduce the method robustness, create chromatographic anomalies and increase instrument maintenance. Lipids are the major source of matrix interferences in plasma and will build up on the analytical column and cause ion-suppression. Removing lipids from a biological sample not only increases the robustness of the analytical method but improves recoveries and limits of quantitation/detection. Protein precipitation is a simple sample preparation technique that removes the proteins but the lipids remain in the sample. Solid phase extraction can produce a very clean sample removing proteins and lipids, however it requires technical expertise and additional laboratory equipment. Implementing protein precipitation with a novel enhanced matrix removal-lipid material offers the cleanliness of a solid phase extraction method with the simplicity of protein precipitation. Using a mixture of controlled substances and drugs of abuse at low ng/mL levels we will demonstrate the simple yet effective sample preparation technique incorporating EMR-Lipid.

Keywords: Biological Samples, Liquid Chromatography/Mass Spectroscopy, Sample Preparation, Toxicology
Application Code: Clinical/Toxicology
Methodology Code: Sampling and Sample Preparation

Session # 1140 Abstract # 1140-9 **Poster Sessions**

Session Title: Sampling and Sample Preparation - Environmental

Abstract Title: **Modified Sample Clean-up for Combined POPs Using Automated Multi-Column Fractionation and Analytical Optimization**

Date: Tuesday, March 08, 2016 - Morning

Time:

Room: Exposition Floor, 400 Aisle

Primary Author: Philip Bassignani
Fluid Management Systems

Co-Author(s): Rudolf Addink

Abstract Text

Analysis of Priority Organic Pollutants (POPs) often require complex, multi faceted clean-up procedures for low level analysis, including HRMS and tandem quadrapole MS. Traditionally separated as separate samples for independent analysis, the combination of samples often leads to complications with analyte separations and/or concentration variations. The utilization of automated, multi-column clean-ups in conjunction with optimization of analytical practices can overcome these problems, and result in streamlined sample prep for a variety of analyses. Analytes examined include PCDDs, PCDFs, PCBs, PBDEs, PCNs and OCPs to be combined in a single extract with optimized clean-up. Experimental trials on sorbent optimization, sample amounts and extract final and injection volumes are all variables of focus.

Keywords: Environmental/Biological Samples, Gas Chromatography/Mass Spectrometry, Sample Handling/Auto
Application Code: Environmental
Methodology Code: Sampling and Sample Preparation

Session # 1140 Abstract # 1140-10 **Poster Sessions**

Session Title Sampling and Sample Preparation - Environmental

Abstract Title **Ice Concentration Linked with Extractive Stir Bar**

Date: Tuesday, March 08, 2016 - Morning
Time:
Room: Exposition Floor, 400 Aisle

Primary Author Nujud O. Maslamani
South Dakota State University

Co-Author(s)

Abstract Text

Trace and ultra-trace analysis can be difficult to achieve, especially for volatile and/or thermally unstable analytes. A novel technique, coined ICE Concentration Linked with Extractive Stir bar (ICECLES), may address this problem. ICECLES combines stir bar sorpitive extraction (SBSE) with freeze concentration (FC), to extract analytes from aqueous solution by slowly freezing water to concentrate analytes into a polydimethlsiloxane (PDMS) coated stir bar. Two probe molecules, dimethyl trisulfide and benzaledhyde, were prepared from aqueous solutions using ICECLES. Thermal desorption gas–chromatography mass–spectrometry was then used to quantify these analytes. Parameters affecting the performance of ICECLES were evaluated, such as the initial concentration, stir speed and freeze rate. Extraction at low speeds resulted in higher extraction efficiency. However, the freeze rate and initial concentrations had a minor effect on ICECLES extraction efficiency. ICECLES produced concentration factors of over 1000x SBSE with excellent reproducibility 8% RSD. ICECLES also, provided higher extraction efficiencies and lower LODs. Overall, the ICECLES technique was excellent at preparing aqueous samples for trace analysis and shows promise as a novel analytical sample preparation technology.

Keywords: Extraction, GC-MS, Sample Preparation, Thermal Desorption

Application Code: Environmental

Methodology Code: Gas Chromatography/Mass Spectrometry

Session # 1140 **Abstract # 1140-11** **Poster Sessions**

Session Title: Sampling and Sample Preparation - Environmental

Abstract Title: **The Best Sample Preparation for High Throughput Quantitative X-Ray Diffraction of Mineral Mixtures**

Date: Tuesday, March 08, 2016 - Morning

Time:

Room: Exposition Floor, 400 Aisle

Primary Author: Roger Meier
FLSmidth A/S

Co-Author(s): Detlev Götz, Ian Campbell, Lukas Bruzenak

Abstract Text

The development in various industries towards more advanced processes and materials in combination with the recent X-ray diffraction technology asks for more and more quantitative full pattern XRD analysis at different stages of the process. The indispensible preconditions for a reproducible, repeatable and accurate result are the 'perfect' sample preparation, where the highest focus is required to minimize the human factor and to eliminate as far as possible the preferred orientation of the crystallites of different mineralogical phases during sample preparation. The last one of the most critical steps in the process is the pressing of the powder into a sample holder. A recent development of the automatic soft press machine (ASP100) opens an opportunity to combine optimal reproducibility, user-friendly operation and lowest preferred orientation, resulting in the best analytical results.

The unique soft press applies a four core step approach consisting of holder and bottom presentation, pouring of a defined sample amount, levelling including random crystal orientation and soft pressing of the sample. This approach results in a very statistical oriented presentation of the sample with a perfect defined surface and sample height. The subsequent XRD measurement shows the perfect data for a full pattern quantitative data analysis with the highest obtainable repeatability, reproducibility and accuracy, the ideal precondition for a reliable analytical result.

This article will demonstrate the revolutionary method and its advantages based on examples of different complex samples by comparing the various, in XRD mostly used, sample preparation methods. The in-depth data interpretation of the achieved results clearly proves the capabilities of the soft press in science, quality and process control.

Keywords: Quality, Quantitative, Sample Preparation, X-ray Diffraction
Application Code: Process Analytical Chemistry
Methodology Code: Sampling and Sample Preparation

Session # 1140 Abstract # 1140-12 **Poster Sessions**

Session Title: Sampling and Sample Preparation - Environmental

Abstract Title: **Nanoporous Solid Phase Microextraction (SPME) Fibers by Sputtering Silicon**

Primary Author: Massoud Kaykhaii, Brigham Young University

Co-Author(s): Anubhav Diwan, Bhupinder Singh, Matthew R. Linford

Date: Tuesday, March 08, 2016 - Morning

Time:

Room: Exposition Floor, 400 Aisle

Abstract Text

Solid phase microextraction (SPME) is a solventless, fast, easy, and relatively inexpensive sample preparation technique that integrates sampling, extraction and preconcentration into one step. It has widespread applicability in various fields that include environmental, food, drugs, in-vivo analyses etc. Commercial coatings may be expensive, show relatively short lifetimes, extract limited numbers of compounds, and have relatively low thermal and mechanical stability, and/or solvent incompatibility. We have endeavored to develop an SPME phase that is thinner, robust, longer lasting, has higher mechanical strength, greater thermal and solvent stability, and is devoid of the other drawbacks of commercial coatings. Silicon/silica-based nanoporous coatings were prepared via sputtering of high purity silicon onto a fiber substrate, which resulted in porous structures. The thicknesses of these fiber coatings were around 2 µm. To increase the density of –OH groups on the surfaces, the coatings were treated in piranha solution and then reacted with n-octadecyldimethylmonomethoxysilane to render the surfaces hydrophobic. Sputtering provides good control over coating thickness and high reproducibility. We compare our fibers against commercial PDMS 7 µm fiber for the extraction of saturated alkanes, primary alcohols, aldehydes, amines, carboxylic acids, and various real life samples. In general, noticeably higher signals are obtained for our 2 µm coatings. They also show very fast adsorption kinetics vis-à-vis a BTEX test mixture.

Keywords: GC, Sample Preparation, Sampling, SPME

Application Code: General Interest

Methodology Code: Sampling and Sample Preparation

Session # 1150 Abstract # 1150-1 Awards

Session Title: Pittsburgh Spectroscopy Award

Abstract Title: **Overcoming Unmet Medical Needs: Advances in Raman Spectroscopy**

Primary Author: Juergen Popp
Friedrich-Schiller University Jena

Date: Tuesday, March 08, 2016 - Afternoo
Time: 01:40 PM
Room: B312

Co-Author(s)

Abstract Text

Within the last years, Raman spectroscopy developed from a niche technology in analytical chemistry to a versatile biomedical analysis tool. The ability to obtain molecular fingerprint information labelfree makes Raman spectroscopy attractive for many applications in clinical diagnostics of bodily fluids, pathogens, cells, and tissue. This presentation reports about Raman spectroscopic approaches for the diagnosis and therapy of infectious diseases and for spectral histopathology.

Faster and more detailed diagnosis of acute life-threating human infections (like e.g. sepsis) represents an important unmet medical need. It will be shown that Raman spectroscopy holds great promise as point-of-care approach for a fast identification of pathogens and the determination of their antibiotic resistances, which is crucial for patient's survival. In this context, we will present innovative chip-based bacterial isolation strategies out of complex sample matrices (e.g. blood or urine).

The second part of this presentation reports about the Raman spectroscopic detection of tissue pathologies. Here, the medical focus predominantly lies on the determination of the tumor type and grade and a better delineation of tumor margins. In this context, it will be shown that the combination of Raman approaches with other spectroscopic technologies is very beneficial for addressing the aforementioned unmet medical needs. We will introduce a combined Raman /FLIM (fluorescence lifetime imaging microscopy) fiber optical probe for in-vivo tissue screening. Furthermore we demonstrate how the combination of CARS (coherent anti-Stokes Raman scattering), SHG (second harmonic generation) and two-photon excited autofluorescence (TPEF) enables the characterization of the morphochemistry of frozen section biopsy specimens.

Keywords: Biomedical, Biospectroscopy, Chemometrics, Infrared and Raman
Application Code: Biomedical
Methodology Code: Vibrational Spectroscopy

Session # **1150** **Abstract #** **1150-2** **Awards**

Session Title Pittsburgh Spectroscopy Award

Abstract Title **Developing Deep UV Raman Standoff Spectrometers for Trace Explosives**

Date: Tuesday, March 08, 2016 - Afternoo
Time: 02:15 PM
Room: B312

Primary Author Sanford A. Asher
University of Pittsburgh

Co-Author(s) Katie L. Gares, Kyle T. Hufziger, Sergei V. Bykov

Abstract Text

We are developing deep UV resonance Raman standoff spectrometers to detect trace explosives. For this effort we are developing novel compact lasers, ultrahigh efficient spectrometers and imaging spectrometers. We will discuss both rastering as well as an imaging instrument that utilizes a novel photonic crystal narrow wavelength band optic. We are developing methodologies to detect explosives at standoff distances greater than 2 m. We are exploring the spectroscopic signatures of explosives as well as their photochemistries, which gives rise to specific dynamic spectral signatures. Lastly, we will discuss the deep UV Raman spectral signatures of NaNO[sub]3[/sub], NH[sub]4[/sub]NO[sub]3[/sub], TNT, RDX, PETN and HMX.

Keywords: Detection, Raman, Spectrometer
Application Code: Homeland Security/Forensics
Methodology Code: Vibrational Spectroscopy

Session # 1150 Abstract # 1150-3 **Awards**

Session Title Pittsburgh Spectroscopy Award

Abstract Title **Enhanced Vibrational Circular Dichroism: Moving Beyond Established Applications of Vibrational Circular Dichorism (VCD)**

Date: Tuesday, March 08, 2016 - Afternoo
Time: 02:50 PM
Room: B312

Primary Author Laurence A. Nafie
Syracuse University

Co-Author(s)

Abstract Text

Vibrational circular dichroism (VCD) has emerged in recent years as a powerful technique for molecular structure analysis in chiral molecules[1]. Most applications have been carried out for natural products and organic molecules of pharmaceutical significance[2] and for molecules of biological significance. The principal advantage of VCD as a spectroscopic method is its combination of structural richness combined with chiroptical stereosensitivity. However, a disadvantage of VCD is the need for relatively high concentrations and long acquisition times due to the intensities, being roughly 4 to 5 orders of magnitude smaller than parent IR absorption intensities. Of interest are instances when VCD intensities are enhanced, not only for their relative ease of measurement but for understanding the origin of enhanced intensities which do not yet have the computational precision of non-enhanced VCD. Here we present several instances of enhanced VCD, where observed intensities are in the range of 2 to 3 orders of magnitude reduced from IR intensities. We will discuss VCD from protein amyloid fibrisl show unusually strong VCD intensities owing to the long-range supramolecular chiral structure of fibrils[3]. Additionally we discuss amplified VCD in molecules that have low-lying electronic states[4] which show promise to become localized probes of biomolecular structure.

[1] L. A. Nafie, Vibrational Optical Activity: Principles and Applications, John Wiley & Sons, Ltd., Chichester, 2011.
[2] Y. He, B. Wang, R. K. Dukor,L. A. Nafie, Appl. Spectrosc. 65, 699-723 (2011).
[3] D. Kurouski, X. Lu, L. Popova, W. Wan, M. Shanmugasundaram, G. Stubbs, R. K. Dukor, I. K. Lednev, L. A. Nafie, J. Am. Chem. Soc. 136, 2302-2312 (2014).
[4] S. R. Domingos, A. Huerta-Viga, L. Baij, S. Amirjalayer, D. A. E. Dunnebier, Annemarie J. C. Walters, M. Finger, L. A. Nafie, B. de Bruin, W. J. Buma, S. Woutersen, J. Am. Chem. Soc. 136, 3530-3535 (2014).

Keywords: Bioanalytical, Biospectroscopy, FTIR, Molecular Spectroscopy
Application Code: Bioanalytical
Methodology Code: Molecular Spectroscopy

Session # 1150 **Abstract # 1150-4** **Awards**

Session Title: Pittsburgh Spectroscopy Award

Abstract Title: **Raman Big Data Analysis for Automatic and Objective Living Cell Discrimination/Diagnosis**

Date: Tuesday, March 08, 2016 - Afternoo
Time: 03:40 PM
Room: B312

Primary Author: Hiroo Hamaguchi
National Chiao Tung University

Co-Author(s): Masahiro Ando

Abstract Text

We discuss the possibility of chemometrical multivariate curve resolution (MCR) for global analysis of Raman big data of living leukocyte cells. We analyze a total of 18400 Raman spectra (400 Raman spectra from 46 leukocyte cells; 21 neutrophils, 14 eosinophils, 7 lymphocytes, and 4 monocytes) to discriminate the cell type and diagnose the cell status by a small number of characteristic Raman spectra. We use MCR-alternating least squares (MCR-ALS) with L1-norm regularization to extract Raman spectra associated with physically plausible spatial distributions within a cell and across cells. Because of this physical constraint imposed by the L1-norm regularization, the extracted Raman spectra are ready to assign to molecules or combinations of molecules existing in a cell. Four types of leukocytes are clearly differentiated into four groups by using the four characteristic Raman spectra extracted by the MCR-ALS analysis; Spectra of myeloperoxidase, eosinophil peroxidase, proteins and proteins + myeloperoxidase. By the present Raman MCR-ALS analysis, it has become possible for the first time to automatically and objectively discriminate the leukocyte cell types at the molecular-level. It is noted that we are sure how the molecular-level differentiation is accomplished; we are using different Raman spectra of different types of heme proteins. In this regard, the present MCR-ALS approach contrasts with principal component analysis (PCA), which has been extensively used for cell type discrimination by Raman spectroscopy.PCA examines the spectral similarity among a large number of Raman spectra purely mathematically and therefore PCA retrieved spectra are often hard to assign to any definitive molecular origins.

Keywords: Bioinformatics, Biospectroscopy, Chemometrics, Raman
Application Code: Biomedical
Methodology Code: Chemometrics

Session # **1150** **Abstract #** **1150-5** **Awards**

Session Title Pittsburgh Spectroscopy Award

Abstract Title **Raman Scattering from Single, Laser-Trapped Microparticles: A Review**

Primary Author Wolfgang Kiefer
University of Würzburg

Date: Tuesday, March 08, 2016 - Afternoo
Time: 04:15 PM
Room: B312

Co-Author(s)

Abstract Text

Soon after the pioneering work by Arthur Ashkin in the nineteenseventies optical trapping techniques were successfully adapted to Raman spectroscopy. Such techniques allowed to obtain Raman spectra of single particles, whose sizes are of the order of or larger than the wavelength of the exciting light. However, in scattering systems with well defined geometries, e.g. dielectric spheres, the use of Raman spectroscopy as a diagnostic probe becomes complicated due to morphology-dependent resonances of the cavity (MDRs, Mie theory). Due to resonance enhancement of the Raman scattering, additional sharp peaks appear which are not present in bulk Raman spectra. On the other hand, such "Raman-Mie spectra" are very useful to precisely study the physical chemistry of single trapped microparticles, such as for example evaporation, phase transition, acid/base reaction, polymerization, etc. Additionally, stimulated Raman-Mie scattering applying external seeding allowed to study minority species. This lecture will review the huge amount of work in these fields the 2016 Pittsburgh Spectroscopy Awardee has contributed while he was in the author's laboratory. Also optical trapping of non-spherical microparticles has meanwhile become possible and was successfully applied for biomedical studies. A short review will be given on recent results of studies in this research area.

Keywords: Light Scattering, Microspectroscopy, Molecular Spectroscopy, Raman
Application Code: General Interest
Methodology Code: Vibrational Spectroscopy

Session # 1160	Abstract # 1160-1	Awards

Session Title: RSC - JAAS Emerging Investigator Lectureship Award

Abstract Title: **Revealing Surface Elemental Landscapes with Ultra-High Throughput via GDOES**

Date: Tuesday, March 08, 2016 - Afternoo
Time: 01:40 PM
Room: B314

Primary Author: Gerardo Gamez
Texas Tech University

Co-Author(s)

Abstract Text

Obtaining spatially-resolved chemical information is of utmost importance to understand the underlying mechanisms of natural and manufactured systems. Nonetheless, typical elemental mapping techniques are limited by low-throughput, requiring several hours to tens of hours for obtaining a full map of large-diameter surfaces with sufficient pixel density. Thus, the need for high-throughput elemental mapping techniques is evident. As such, routine elemental mapping for diagnostics or quality control, and even 3D surface elemental mapping, would become practical.

In the last few years, GDOES (operated with pulsed power and at higher pressures) has shown the capability to yield elemental composition landscapes with several orders of magnitude faster acquisition-time compared to traditional techniques. Recent developments in this area will be presented, including novel GD geometries that allow larger-diameter surfaces to be mapped, implementation of novel cost-effective imaging systems that allow even faster throughput and easier potential interfacing with commercial systems, as well as the latest application development.

Keywords: Atomic Emission Spectroscopy, High Throughput Chemical Analysis, Imaging, Plasma Emission (ICP/M
Application Code: Material Science
Methodology Code: Surface Analysis/Imaging

Session # 1160 Abstract # 1160-2 Awards

Session Title RSC - JAAS Emerging Investigator Lectureship Award

Abstract Title **Laser Ablation-Based Chemical Analysis Techniques: A Short Review**

Primary Author Jhanis J. Gonzalez
Lawrence Berkeley National Laboratory

Date: Tuesday, March 08, 2016 - Afternoo
Time: 02:15 PM
Room: B314

Co-Author(s)

Abstract Text
Laser ablation has become a dominant technology for direct solid sampling in analytical chemistry. Laser ablation refers to the process in which an intense burst of energy delivered by a short laser pulse is used to sample (remove a portion of) a material. The advantages of laser ablation chemical analysis include direct characterization of solids, no chemical procedures for dissolution, reduced risk of contamination or sample loss, analysis of very small samples not separable for solution analysis, and determination of spatial distributions of elemental composition. This presentation will describes the most common approaches Laser Induced Breakdown Spectroscopy LIBS) and Laser Ablation Inductively Coupled Plasma Spectrometry (LA-ICP-MS) and an introduction to Laser Ablation Molecular Isotopic Spectrometry (LAMIS)

Keywords: Laser
Application Code: General Interest
Methodology Code: Atomic Spectroscopy/Elemental Analysis

Session # 1160 Abstract # 1160-3 **Awards**

Session Title RSC - JAAS Emerging Investigator Lectureship Award

Abstract Title **Direct Determination of Trace Antimony and Arsenic in Natural Waters by Photochemical Vapor Generation ICPMS**

Date: Tuesday, March 08, 2016 - Afternoo
Time: 02:50 PM
Room: B314

Primary Author Lu Yang
National Research Council Canada

Co-Author(s)

Abstract Text

ABSTRACT: Antimony and arsenic are well-known environmental pollutants. The general population is usually exposed to antimony and arsenic from food and water1-2 which can cause adverse effects on human healthy. It has been reported that exposure to high levels of arsenic via drinking water can cause skin, lung and bladder cancers.2 More recent findings reveal that consumption of water with levels as low as 0.017 ng g-1 for As over long periods of time may lead to arsenicosis3. Consequently, monitoring of Sb and As concentrations in natural waters has become one of most frequently performed analyses.

Novel and sensitive methods for the accurate determination of Sb in natural waters and As in seawaters are presented using photochemical vapor generation (PVG) for sample introduction with ICPMS detection. Utilizing a unique flow-through photochemical reactor capable of subjecting the samples to deep UV (185 nm) radiation, generation efficiency was found to be independent of Sb or As species presenting in the waters, eliminating the shortcoming of Sb and As species depended sensitivity encountered during direct solution nebulization by ICPMS4. The accuracy of the proposed methods was demonstrated by analysis of several Certified Reference Materials (CRMs) of river water and seawaters, as well as spike recovery test with satisfying results. Results obtained in this study will be presented and discussed in details.

Keywords: Atomic Spectroscopy, Environmental/Water, ICP-MS
Application Code: Environmental
Methodology Code: Atomic Spectroscopy/Elemental Analysis

Session # 1160 Abstract # 1160-4 **Awards**

Session Title RSC - JAAS Emerging Investigator Lectureship Award

Abstract Title **Pulsed Glow Discharge Time-of-Flight Mass Spectrometry (Positive and Negative Ionization Modes) for Elemental Depth Profiling of Innovative Materials and Polymer Fingerprinting**

Date: Tuesday, March 08, 2016 - Afternoo
Time: 03:45 PM
Room: B314

Primary Author Lara Lobo Revilla
University of Oviedo

Co-Author(s) A Sanz-Medel, B Fernandez, R Muniz, R Pereiro

Abstract Text

Radiofrequency pulsed glow discharge time-of-flight mass spectrometry (rf-PGD-TOFMS) has received much attention during the last years for depth profiling analysis [1]. In fact, most of the investigations reported on this topic were performed using prototype instruments, being the most popular, the prototype formed by a GD bay developed by Horiba Scientific (France) and a TOF mass spectrometer by Tofwerk (Switzerland). At present, the commercial version of the prototype has been launched by Horiba Scientific including improvements to offer some additional features of interest for analytical applications, e.g. blanking (to reduce ion intensity of selected species such as Ar or matrix ions), negative ionization detection mode, among others.

In this work, performance and analytical potential of the commercial rf-PGD-TOFMS will be thoroughly evaluated, in both positive and negative detection modes,to achieve elemental and molecular specific information for different challenging materials, including fluorine detection during titania nanotubes synthesis, photovoltaic materials based on thin film solar cells, golden layers on ceramics and polymeric materials deposited on silicon wafers.

[1] J. Pisonero, N. Bordel, C. Gonzalez de Vega, B. Fernández, R. Pereiro, A. Sanz-Medel, Anal. Bioanal. Chem., 2013, 405, 5655-5662.

Keywords: Mass Spectrometry, Materials Characterization
Application Code: Material Science
Methodology Code: Atomic Spectroscopy/Elemental Analysis

Session # 1160 Abstract # 1160-5 **Awards**

Session Title RSC - JAAS Emerging Investigator Lectureship Award

Abstract Title **What is XRF Doing in a Mass Spectrometry Award Symposium?**

Primary Author George Havrilla
Los Alamos National Laboratory

Co-Author(s)

Date: Tuesday, March 08, 2016 - Afternoo
Time: 04:20 PM
Room: B314

Abstract Text

XRF (X-ray fluorescence) is the forgotten elemental analysis technique. Developed well before mass spectrometry, XRF offers unique capabilities not found in conventional methods of elemental analysis. Based on the excitation of core electrons, XRF provides elemental analysis covering a wide range of scientific problems, from probing the origins of the cosmos to the workings of biological processes in cells to uncovering hidden art works. Both synchrotron sources and lab-based X-ray tubes utilize X-ray optics to enable capabilities which offer high spatial and spectral resolution to probe these and many other problems. In addition, XRF is nondestructive in measuring the elemental composition of samples, thus allowing follow-on analyses such as isotopic measurements to be done. X-ray optics open high resolution elemental imaging into the 10's of nanometers as well as enabling the detection of the analyte oxidation state. By providing key information on the element present as well as its oxidation state new insights can be gleaned into the fundamental characteristics of many chemical and biological processes. Laboratory-based instrumentation can rival capabilities found in synchrotrons, particularly when coupled with X-ray optics and state-of-art detectors. Spatial resolution of lab-based systems approaches sub-10 micrometers along with the ability for 3D elemental imaging of a variety of samples which includes actinide contaminated soils, spent nuclear fuels and fusion targets. This presentation will provide a state-of-the-art survey of XRF covering these topics and many other XRF applications.

Keywords: Atomic Spectroscopy, Elemental Analysis, Microspectroscopy, X-ray Fluorescence
Application Code: General Interest
Methodology Code: X-ray Techniques

Session # 1170 Abstract # 1170-1 **Symposia**

Session Title ACS-ANYL - New Approaches to Nuclear Safeguards and Forensics Analysis

Abstract Title **Advances in Online Spectroscopic Monitoring for Process Control and Safeguarding of Radiochemical Streams**

Date: Tuesday, March 08, 2016 - Afternoo
Time: 01:35 PM
Room: B308

Primary Author Sam A. Bryan
Pacific Northwest National Laboratory

Co-Author(s) Amanda J. Casella, Amanda M. Lines, Gilbert L. Nelson, Job M. Bello

Abstract Text

There is renewed interest to promote the use of nuclear power and close the nuclear fuel cycle. The long term successful use of nuclear power is critically dependent upon adequate and safe processing and disposition of the spent nuclear fuel. Liquid-liquid extraction is a separation technique commonly employed for the processing of the dissolved spent nuclear fuel. The instrumentation used to monitor these processes must be robust, require little or no maintenance, and be able to withstand harsh environments such as high radiation fields and aggressive chemical matrices.

This paper describes the application of the electronic and vibrational spectroscopic techniques for radiochemical process monitoring. In this context, our team experimentally assessed the potential of Raman and UV-vis spectroscopic techniques for online real-time monitoring of the U(VI)/nitrate ion/nitric acid and Pu(VI/IV)/Np(VI/V/VI)/Nd(III) in solutions relevant to used nuclear fuel reprocessing. These techniques demonstrated robust performance in the repetitive batch measurements of each analyte in a wide concentration range using simulant and commercial dissolved used fuel solutions.

Currently this has been demonstrated on the macroscopic scale, using sample probes requiring large solution volumes. In an effort to minimize waste and reduce dose to personnel, we have modified this technique to allow measurement at the microfluidic scale using a Raman microprobe. Under the current sampling environment, Raman samples typically require upwards of 10 mL and larger. Using the new sampling system, we can sample volumes at 10 L or less, which is a scale reduction of over 1,000 fold in sample size. This will significantly reduce the requirements for handling hazardous materials, and greatly reduce the costs associated with worker protection from exposure to radioactive and hazardous materials, and costs due to waste disposal. This paper will summarize our current work in this area.

Keywords: Chemometrics, Process Analytical Chemistry, Raman, UV-VIS Absorbance/Luminescence
Application Code: Nuclear
Methodology Code: Process Analytical Techniques

Session # 1170 Abstract # 1170-2 **Symposia**

Session Title ACS-ANYL - New Approaches to Nuclear Safeguards and Forensics Analysis

Abstract Title **New XRF Applications to Nuclear Safeguards and Nuclear Forensics**

Date: Tuesday, March 08, 2016 - Afternoo
Time: 02:10 PM
Room: B308

Primary Author George Havrilla
Los Alamos National Laboratory

Co-Author(s)

Abstract Text

XRF offers unique capabilities for the characterization of nuclear materials as applied to both nuclear safeguards and nuclear forensics requirements. In both cases, the identification of actinide elements along with their quantification are key components in evaluating nuclear materials for both safeguards and forensics applications. Development of XRF as a new tool in the nuclear material characterization toolbox provides both opportunities and challenges for analytical methodology in answering the age old questions of “What is it” and “How much of it is there?” XRF when based upon X-ray optics offers new capabilities for high sensitivity, high selectivity and 3D elemental imaging not feasible with conventional elemental methods of analysis. hiRX or high resolution X-ray can detect sub-nanogram amounts of plutonium or uranium with high selectivity using doubly curved crystal optics. This new instrument has proven capable of analyzing small amounts of Pu in the presence of over 100 times uranium present in such samples as dissolved spent fuel found in nuclear fuel reprocessing plants. Micro X-ray fluorescence (MXRF) employing monolithic polycapillary optics can provide elemental maps of actinide distributions in soil samples with 10’s of micrometer spatial resolution. Adding a polycapillary optic on the detector enables confocal 3D elemental imaging which can identify the elemental spatial distribution within individual soil particles. In each instance, XRF offers new spatially resolved sensitive elemental analyses with the focus of identifying actinide elements of interest within the areas of nuclear safeguards and nuclear forensics.

Keywords: Forensics, Nuclear Analytical Applications, X-ray Fluorescence
Application Code: Nuclear
Methodology Code: X-ray Techniques

Session # 1170 Abstract # 1170-3 **Symposia**

Session Title ACS-ANYL - New Approaches to Nuclear Safeguards and Forensics Analysis

Abstract Title **Fieldable Mass Spectrometry: Sample Preparation and Rapid Field Analysis for Nuclear Safeguards**

Date: Tuesday, March 08, 2016 - Afternoo
Time: 02:45 PM
Room: B308

Primary Author Matthew R. Dirmyer
Los Alamos National Laboratory

Co-Author(s) Chris Leibman, Elizabeth Judge, Keri Campbell, Lisa Meyers, Ning Xu, Peter Stark, Thomas Yoshida

Abstract Text

Significant gaps exist in in-field sample preparation technology and suitable ion sources matched to sample type, target analyte(s), IAEA requirements, and suitability for non-expert operation. , Experimental work in the past two years has lead us to conclude that sample preparation technology must be developed in close integration with ion source development. This is especially true given the stringent demands of field portability, ease of operation, and operator/operational safety. Our strategy has been to develop new and improved sample preparation methods including 1) ammonium bifluoride to convert solid uranium oxides to uranyl fluorides and 2) UF6 precipitation to a solid form. We also will present our progress in developing a field portable ion source that is compatible with the new sample preparation methods and COTS MS technology for in-field uranium isotopic analysis. Focus will also be placed on methods that meet published IAEA International Target Values (ITVs).

Keywords: ICP-MS, Laser, Mass Spectrometry
Application Code: Nuclear
Methodology Code: Mass Spectrometry

Session # 1170 Abstract # 1170-4 **Symposia**

Session Title ACS-ANYL - New Approaches to Nuclear Safeguards and Forensics Analysis

Abstract Title **New Analytical Methods for Trace Elemental and Isotopic Analysis of Nuclear Fuel Cycle Materials**

Date: Tuesday, March 08, 2016 - Afternoo
Time: 03:35 PM
Room: B308

Primary Author Andrew Duffin
Pacific Northwest National Laboratory

Co-Author(s) April J. Carman, Carmen S. Menoni, Gregory Eiden, Jesse D. Ward, Martin Liezers, Michael P. Dion, Orville T. Farmer

Abstract Text

Recent progress in the development of new analytical methods for trace elemental and isotopic analysis of nuclear fuel cycle materials will be presented. These new methods include atomic mass-based separations as part of the sample preparation for radiometric counting ("MSRAD"), the use of femtosecond laser ablation and extreme UV photon sources for very high spatial resolution MS analysis of surfaces, the use of x-ray light sources to achieve very high spatial resolution in the analysis of nuclear fuel cycle materials, and finally the development of methods to create materials that mimic the chemical, isotopic, and morphologies of nuclear explosion debris. "MSRAD" represents a new paradigm in radiochemical analysis: separations along diagonal dimensions of the Chart of the Nuclides (within single or a few mass chains) versus along horizontal dimensions (single elements or groups of elements). Fs-LA and EUV light sources have demonstrated extremely good sensitivity, precision, and spatial resolution in the isotopic and elemental analysis of surfaces and particles. X-ray light sources enable us to probe the chemical nature of sample features down to the low nanometer scale where chemical transformations can be observed "in progress". Finally, the development of methods to produce synthetic nuclear explosion debris (SNED) is leading to new materials for use in both laboratory research and field exercises.

Keywords: Aerosols/Particulates, Elemental Analysis, ICP-MS, Ultratrace Analysis
Application Code: Nuclear
Methodology Code: Mass Spectrometry

Session # 1170 **Abstract # 1170-5** **Symposia**

Session Title: ACS-ANYL - New Approaches to Nuclear Safeguards and Forensics Analysis

Abstract Title: **Contribution and Impact of Laser Ablation Inductively Coupled Plasma Mass Spectrometry (LA-ICP-MS) and Laser Induced Breakdown Spectroscopy (LIBS) to Forensic Analysis**

Date: Tuesday, March 08, 2016 - Afternoo
Time: 04:10 PM
Room: B308

Primary Author: Jhanis J. Gonzalez
Lawrence Berkeley National Laboratory

Co-Author(s)

Abstract Text

Improvements in Laser Ablation for material sampling over the past few decades have led to the emergence of several applications of this technique (in its two modalities LIBS and LA-ICP-MS) to some important forensic measurements. These Laser Ablation modalities are commonly used for both elemental and isotopic analyses, and has multiple advantages compared to dissolution techniques, notably higher spatial resolution, easier and faster sample preparation, and for many applications it is a non-destructive method. However, LA-ICP-MS suffers of a significant limitation related to its elemental coverage. For example, non-metals such as H, N, O and halogens such as F, are difficult or impossible to analyze by conventional ICP-MS systems. On the other hand, one of the major benefits for using Laser Induced Breakdown Spectroscopy (LIBS) is its elemental coverage, and of course its ability to detect elements that are difficult or impossible to analyze by most analytical techniques and in particular those mentioned above. The combination of the two Laser Ablation modalities provides complementary measurements for elements that are separately unattainable due to low sensitivity and/or strong interferences. In this presentation we will discuss some of the recent advances and applications using Laser Ablation.

Keywords: Laser, Mass Spectrometry, Nuclear Analytical Applications, Spectroscopy
Application Code: General Interest
Methodology Code: Atomic Spectroscopy/Elemental Analysis

Session # 1180 Abstract # 1180-1 **Symposia**

Session Title Emerging Mass Spectrometry-Based Techniques for Biomolecular Analysis

Abstract Title **Searching for Biomarkers Using Ambient Ionization Mass Spectrometry**

Date: Tuesday, March 08, 2016 - Afternoo

Time: 01:35 PM

Room: B302

Primary Author Graham Cooks
Purdue University

Co-Author(s)

Abstract Text

Ambient mass spectrometry (MS) is used to ionize and characterize compounds present in samples examined in their natural state (no dilution or extraction or other chemical work-up). Desorption electrospray ionization (DESI) and paper spray (PS) ionization are typical methods. Sets of molecules detected by ambient MS can represent a snapshot of the state of the sample (organism, biofluid, tissue section, etc.) therefore such a set of molecules can serve a biomarker of disease and its progression. In contrast to the standard LC MS/MS method of characterizing one or a few molecular biomarkers, the ambient MS approach uses a pattern of markers. Two distinct approaches are described: (i) low resolution MS profiles, e.g. as used to characterize the disease state of tissue including brain and kidney by their lipid profiles and (ii) sets of multiple reaction monitoring (MRM) transitions as exemplified by searches for biomarkers for Parkinson's disease in cerebrospinal fluid.

Keywords: Biomedical, Instrumentation, Mass Spectrometry, Organic Mass Spectrometry

Application Code: Bioanalytical

Methodology Code: Mass Spectrometry

Session # 1180 Abstract # 1180-2 **Symposia**

Session Title Emerging Mass Spectrometry-Based Techniques for Biomolecular Analysis

Abstract Title **An Inside-Outside Strategy to Study Cell Communication**

Date: Tuesday, March 08, 2016 - Afternoo
Time: 02:10 PM
Room: B302

Primary Author Catherine Fenselau
University of Maryland

Co-Author(s) Avantika Dhabaria, Lucia Geis-Asteggiante, Nathan Edwards, Sitara Chauhan, Suzanne Ostrand-Rosenberg

Abstract Text

Exosomes are nanoscale membrane bound vesicles shed from cells via multivesicular bodies, which carry proteins, microRNA and lipids between cells. Exosomes may be viewed as a subset of the secretome, but a subset confined in a lipid bilayer that carries the addresses of intended receiver cells. We are studying both the internal cargo of the exosomes released by myeloid-derived suppressor cells (MDSC), and the cell surface glycoproteins that may direct their cellular interactions. Both bottom up and top down proteomic strategies have been applied to analysis of the protein cargo in MDSC exosomes. 105 proteins have been characterized as ubiquitin conjugates, of which 84 have not previously been reported to be ubiquitinated. Top-down analysis of cargo proteins has identified more than 80 isoforms of three members of the S100 calcium binding family. Two of these, S100 A8 and S100 A9 have been shown to exert chemotactic activity when carried by exosomes. The effect of inflammation on the abundances of the active proteins is being measured using top-down spectral counting and relative intensities. Cell surface chemistry, including oxidation of glycosidic diols and alkylation with tethered biotin, has been used to capture glycoproteins from the surfaces of both parental MDSC cells and their exosomes. Both tryptic peptides and PNGase products were recovered for analysis by LC-MS/MS from digestions carried out on biotinylated glycoproteins immobilized on streptavidin beads. 122 glycoproteins were identified from the MDSC surface, each supported by a PNGase peptide carrying a deamidated N-glycosylation motif and one other peptide. The exosomes appear to carry fewer cell surface proteins; 23 glycoproteins have been identified from the exosome surface using similar criteria. Functional roles for the inside and outside proteins will be briefly discussed.

Keywords: Bioanalytical, Biological Samples, Mass Spectrometry, Method Development
Application Code: Other
Methodology Code: Mass Spectrometry

Session # 1180 Abstract # 1180-3 **Symposia**

Session Title Emerging Mass Spectrometry-Based Techniques for Biomolecular Analysis

Abstract Title **Developing Cross-linking Mass Spectrometry to Define Protein-Protein Interactions**

Date: Tuesday, March 08, 2016 - Afternoo
Time: 02:45 PM
Room: B302

Primary Author Lan Huang
University of California, Irvine

Co-Author(s)

Abstract Text

Protein-protein interactions (PPIs) have been recognized as one of the major regulatory mechanisms for controlling protein functions in maintaining cell viability and homeostasis. Disruption of normal PPIs has been linked to various human disease and recent studies have described protein interaction interfaces as a new class of attractive therapeutic targets. Cross-linking mass spectrometry (XL-MS) represents a powerful technology for studying PPIs. Despite its great potential, XL-MS analysis remains challenging due to the difficulty in unambiguous identification of low abundant cross-linked peptides. In order to facilitate MS detection and identification of cross-linked peptides, we have developed a novel MS-cleavable cross-linker, disuccinimidyl sulfoxide (DSSO) (1). DSSO based XL-MS strategy enables fast and accurate identification of cross-linked peptides by multistage mass spectrometry (MSn), and has been successfully applied to define structural topologies of protein complexes (2) and protein-protein interaction interfaces. To further advance XL-MS studies, we have developed a series of DSSO derivatives to allow quantitative analysis of structural dynamics of protein complexes and in vivo profiling of protein interaction network topologies in living cells. In combination with affinity purification strategies, in vivo structural topologies of protein complexes can be determined (3). The analytical methods described here represent technological advancements in studying PPIs in vitro and in vivo.

This work is supported by NIH RO1GM074830 to L.H, and R01GM106003 to L.H. and S. R.. Refs:1.Kao, A., et. al. (2011) Mol Cell Proteomics 10, M110.002212; 2.Kao, A., et. al. (2012) Mol Cell Proteomics 11, 1566-1577; 3. Kaake, R. M., et. al. (2014) Mol Cell Proteomics 13, 3533-3543.

Keywords: Bioanalytical, Mass Spectrometry, Proteomics
Application Code: Genomics, Proteomics and Other 'Omics
Methodology Code: Mass Spectrometry

Session # 1180 Abstract # 1180-4 **Symposia**

Session Title: Emerging Mass Spectrometry-Based Techniques for Biomolecular Analysis

Abstract Title: **Innovative Instrumentation and Methods for the Identification of Intact Proteins in Mixtures and for Sequence Analysis of Antibodies and Posttranslationally-Modified, Intact Proteins on a Chromatographic Time-Scale**

Date: Tuesday, March 08, 2016 - Afternoo
Time: 03:35 PM
Room: B302

Primary Author: Donald F. Hunt
University of Virginia

Co-Author(s)

Abstract Text
This lecture will focus on data generated with a new ion source that facilitates simultaneous generation of positively charged sample ions by electrospray ionization and negatively charged reagent ions for both electron transfer dissociation (ETD) and ion-ion proton transfer (IIPT) reactions on Orbitrap mass spectrometers. Implementation of multiple C-trap fills for enhanced sensitivity will be discussed and both parallel peak parking, and ion ejection strategies to facilitate protein separation and enhanced sequence coverage of intact proteins will be described. Use of IIPT/ETD facilitates near complete sequence coverage on many intact proteins and is ideally suited for locating multiple posttranslational modifications on the same protein molecule. Sequence analysis of antibodies with an enzyme reactor that generates 3-10 KDa fragments in seconds will also be discussed. If time permits, the lecture will also provide an update on the use Class I MHC phosphopeptides for the immunotherapy of cancer.

Keywords: Instrumentation, Mass Spectrometry
Application Code: Genomics, Proteomics and Other 'Omics
Methodology Code: Mass Spectrometry

Session # 1180 Abstract # 1180-5 **Symposia**

Session Title Emerging Mass Spectrometry-Based Techniques for Biomolecular Analysis

Abstract Title **Charge Detection Mass Spectrometry for Single Ions**

Date: Tuesday, March 08, 2016 - Afternoo
Time: 04:10 PM
Room: B302

Primary Author Evan Williams
University of California, Berkeley

Co-Author(s) Andrew Elliot, Zijie Xia

Abstract Text

Obtaining information about structures of macromolecular complexes can be challenging when the complexes are large and/or heterogeneous. Heterogeneity can be a result of many different factors, including multiple different stoichiometries of the complex, heterogeneity of the molecular constituents within a complex, or the propensity of the complex to adduct salts and/or other molecules in solution. Although the complexes can often be readily ionized, the heterogeneity can lead to incomplete or no separation between the different charge states making assignment of charge states, and hence masses of the ions difficult or impossible. Strategies involving tandem MS have been used to obtain some information when charge state distributions are unresolved. Interferences between ions with overlapping m/z values can be eliminated by weighing individual ions. Individual ions of large synthetic polymers and biopolymers have been measured using both charge detection mass spectrometers and also FT-ICR mass spectrometry [1,2]. Here, results from a new charge detection mass spectrometer that uses multiple pick-up electrodes between cone electrodes to trap and measure individual ions multiple times will be presented. Progress on the development of this instrument and its use to measure the molecular masses of individual ions up to 100s of MDa will be presented.

[1] Fuerstenau, S. D.; Benner, W. H. Rapid Commun. Mass Spectrom. 1995, 9, 1528–1538.
[2] Smith, R. D.; Cheng, X.; Bruce, J. E.; Hofstadler, S. A.; Anderson, G. A. Nature 1994, 369, 137-139.

Keywords: Electrospray, Instrumentation, Mass Spectrometry, Protein
Application Code: Bioanalytical
Methodology Code: Mass Spectrometry

Session # 1190 Abstract # 1190-1 **Symposia**

Session Title: Emerging Pollutants in the Environment – from Sources to Effects

Abstract Title: **Using High Resolution Mass Spectrometry to Uncover New, Emerging Iodinated and Nitrogen-Containing Disinfection Byproducts**

Date: Tuesday, March 08, 2016 - Afternoo
Time: 01:35 PM
Room: B303

Primary Author: Susan D. Richardson
University of South Carolina

Co-Author(s): Amy Cuthbertson, Christian Luetke-Eversloh, Christina M. Joseph, Cristina Postigo, Edward Machek, Elizabeth Wagner, Friedrich Wendel, Hannah K. Liberatore, Jessie Kadlec, Michael J. Plewa, Stephen

Abstract Text

Disinfection by-products (DBPs) are an unintended consequence of using disinfectants to kill harmful pathogens in water. They are formed by the reaction of disinfectants with natural organic matter, bromide, and iodide, as well as with other anthropogenic contaminants, such as pharmaceuticals. DBPs have been associated with human health risk, including cancer, miscarriage, birth defects, and asthma. Only 11 DBPs are regulated in the U.S., but nearly 700 have been identified. Many of the unregulated DBPs are more toxic than the ones currently regulated, and these 'emerging DBPs' have been the focus of new chemistry, toxicology, and epidemiology studies. Several are suspected as causal agents in human health effects observed. Of these emerging DBPs, those containing nitrogen (so-called 'N-DBPs') and those containing iodine have risen to the top as being the most toxicologically important. N-DBPs and iodo-DBPs have been identified in drinking water and swimming pool waters, and they are expected to increase with climate change and population stresses on our fresh water supplies.

Mass spectrometry (MS) has been important for the identification of emerging DBPs, as well as the study of their mechanisms of formation, with high resolution-MS particularly important for providing accurate mass information to enable molecular formula determinations. This presentation will discuss the use of high resolution mass spectrometry, along with other analytical techniques, for helping to uncover and study new, emerging iodinated and N-DBPs. Iodinated DBPs discussed will include iodo-acids, iodo-trihalomethanes (iodo-THMs), and iodo-acetaldehyde that have been previously reported, as well as new iodo-DBPs not yet published. Their mechanisms of formation will also be addressed, including reactions with iodide from seawater intrusion, reactions with iodinated compounds used for medical imaging, and reactions with a new source of iodine just discovered.

Keywords: Environmental Analysis, Gas Chromatography/Mass Spectrometry, Liquid Chromatography/Mass Spe
Application Code: Environmental
Methodology Code: Mass Spectrometry

Session # **1190** **Abstract #** **1190-2** **Symposia**

Session Title: Emerging Pollutants in the Environment – from Sources to Effects

Abstract Title: **Measuring Plant Uptake and Effects of Pharmaceuticals Using Liquid Chromatography/Mass Spectrometry**

Primary Author: Diana S. Aga
University at Buffalo

Date: Tuesday, March 08, 2016 - Afternoo
Time: 02:10 PM
Room: B303

Co-Author(s): Rachel Mullen

Abstract Text

Human urine is a cost effective, renewable resource that can be used as a valuable source of fertilizer because it is rich in nitrogen, phosphorus and potassium. As fertilizers derived from urine become more widely used, it is important to understand the transport of the pharmaceuticals from urine to the environment. It is known that many pharmaceuticals are excreted from the human body in their native form; hence when urine is used as fertilizer pharmaceuticals can be released as contaminants in the environment. The goal of this study was to develop a sensitive analytical method to measure trace pharmaceuticals in urine, struvite, lysimeter water, soil, and food crops. Analysis of these matrices gave information about the pharmaceuticals present in urine that are carried over into the fertilizer, soil, ground water and eventually crops consumed by humans. Carrots and lettuce, grown in soil plots previously fertilized with urine and struvite, were analyzed when market ready using optimized extraction method and liquid chromatography with tandem mass spectrometry (LC-MS/MS). Lysimeter water samples were collected after rain events and the soil samples were tested pre- and post- fertilization. In addition to the urine and struvite fertilized plots, two control plots were monitored, one using synthetic fertilizer, and one with no fertilizer. Results showed detectable amounts of pharmaceuticals in lettuce and carrots grown in soil fertilized by urine. Pharmaceuticals were detected in lysimeter water, soil, and food crops at low part per billion levels. Hydroponic experiments using maize were conducted to examine possible detoxification of chlorinated pharmaceuticals via the glutathione pathway similar to the mechanism used by plants for herbicide detoxification. Results indicate preferential uptake and potential detoxification of some pharmaceutical compounds in maize.

Keywords: Agricultural, Environmental Analysis, Liquid Chromatography/Mass Spectroscopy
Application Code: Environmental
Methodology Code: Liquid Chromatography/Mass Spectrometry

Session # **1190** **Abstract #** **1190-3** **Symposia**

Session Title: Emerging Pollutants in the Environment – from Sources to Effects

Abstract Title: **Analysis of Hydraulic Fracturing Additives by LC/Q-TOF-MS**

Date: Tuesday, March 08, 2016 - Afternoo
Time: 02:45 PM
Room: B303

Primary Author: Imma Ferrer
University of Colorado

Co-Author(s): Michael Thurman

Abstract Text

Hydraulic fracturing, or simply "fracking' is the process of injecting water, sand, and various surfactant and biocide mixtures into deep wells, 5,000 to 10,000 feet deep in order to open the shales and release the oil and gas trapped inside. These waters present possible contamination to shallow and deep aquifers and to nearby surface water, if they should be released. About 30% of the injected water is returned to the surface. For these reasons, the topic of fracking is of prime environmental significance at the current time. The chemical additives used in fracturing fluids can be used as tracers of water contamination caused by hydraulic fracturing operations. For this purpose a complete chemical characterization is necessary using advanced analytical techniques. Liquid chromatography coupled with high resolution mass spectrometry (LC/Q-TOF-MS) was used to identify several chemical additives present in flowback and produced waters. Accurate mass measurements of main ions and fragments were used to discover the major components. Sodium adducts turned out to be the main molecular adduct ions detected for some additives due to oxygen rich structures. Among the several classes of chemical components identified by mass spectrometry include gels (guar gum), biocides (glutaraldehyde and alkyl dimethyl benzyl ammonium chloride) and surfactants (cocamidopropyl derivatives). The capabilities of accurate mass and MS-MS fragmentation are explored for the unequivocal identification of these compounds. A special emphasis is given to the mass spectrometry elucidation approaches used to identify a major class of hydraulic fracturing compounds, surfactants.

Keywords: Environmental, Liquid Chromatography/Mass Spectroscopy, Surfactants, Time of Flight MS
Application Code: Environmental
Methodology Code: Liquid Chromatography/Mass Spectrometry

Session # 1190 **Abstract # 1190-4** **Symposia**

Session Title: Emerging Pollutants in the Environment – from Sources to Effects

Abstract Title: **Uptake and Disposition of Pharmaceuticals by Bluegill Exposed at Constant Concentrations in a Flow-Through Aquatic Exposure System**

Date: Tuesday, March 08, 2016 - Afternoo
Time: 03:35 PM
Room: B303

Primary Author: Edward Furlong
U.S. Geological Survey

Co-Author(s): Dana W. Kolpin, David J. Feifarek, Eric A. Schwab, Guang-Guo Ying, Heiko L. Schoenfuss, Jian-Liang Zhao, Kyle L. Bird

Abstract Text

The widespread use of pharmaceuticals is reflected in their release from wastewater treatment plants into surface waters and and potentially has corresponding adverse effects upon aquatic organisms. Uptake, accumulation, and transformation of pharmaceuticals in fish have been less well studied than transport and distribution in aquatic ecosystems. In the present study, we investigated the uptake, disposition, and toxicokinetics of five pharmaceuticals—temazepam (neuroleptic), methocarbamol (muscle relaxant), sulfamethoxazole (antibiotic), diclofenac (anti-inflammatory), and rosuvastatin (antihyperlipidemic)—in bluegill (Lepomis macrochirus) tissues and fluids exposed to environmentally relevant concentrations (targeted at between1,000–4,000 ng/L) in a flow-through aquatic exposure system.

Temazepam and methocarbamol were consistently detected in bluegill tissue and fluid samples, with concentrations of up to 4,936 and 180 ng g□1 in bile, respectively. Sulfamethoxazole, diclofenac, and rosuvastatin were much less frequently detected in tissues and fluids. Over a 30-day exposure period, mean concentrations of temazepam and methocarbamol in tissues and fluids generally followed the order: bile □gut > liver and brain > muscle ~ plasma ~ gill. Short-term bioconcentration factors (BCFs) in different tissues ranged between 0.71–3,959 and 0.13–48.6 for temazepam and methocarbamol, respectively. Log BCFs were positively correlated to pH-adjusted log Kow (i.e. log Dow), especially for liver and brain (r2 = 0.92 and 0.99, respectively). These data suggest that bioaccumulation patterns of ionizable pharmaceuticals depend on the physicochemical properties and ionizability of each pharmaceutical within each tissue. A consistent pattern of rapid uptake and elimination of pharmaceuticals in bluegill tissues was observed, with uptake rate constants (Ku) and elimination rate constants (Ke) at 0.0066–0.0330 h□1 and 0.0075–0.0384 h□1, and elimination half-lives of 18.1–92.4 h.

Keywords: Environmental Analysis, Environmental/Biological Samples, HPLC, Tandem Mass Spec
Application Code: Environmental
Methodology Code: Mass Spectrometry

Session # 1190 Abstract # 1190-5 **Symposia**

Session Title Emerging Pollutants in the Environment – from Sources to Effects

Abstract Title **Pharmaceuticals in Surface Waters - Analysis and Effects**

Primary Author Rudolf J. Schneider
BAM

Co-Author(s)

Date: Tuesday, March 08, 2016 - Afternoo
Time: 04:10 PM
Room: B303

Abstract Text

Pharmaceuticals and other pharmacologically active compounds such as endocrine disruptors and stimulants are considered emerging pollutants being found in the environment since many years. Concentrations are usually low, in the nanogram-per-liter level in groundwater and drinking water to a maximum microgram-per-liter level in surface waters and wastewater. More recently it has been shown that after entering the water cycle they potentially may exert harmful effects on water-dwelling organisms. We developed, validated and implemented sensitive and fast analytical techniques, especially antibody-based immunochemical methods. These allow for high-throughput screening of environmental waters as well as an inexpensive monitoring of experiments with water-living species because they require a very small amount of sample. High sensitivity allows for studies at ambient concentrations as well as a quick distinction between pristine and contaminated areas since even low concentrations (ng/L to g/L), may provoke chronic effects in biota. Our main target analytes were pharmaceuticals such as carbamazepine and the hormones estradiol (E2), ethinylestradiol (EE2) and estrone (E1). Also ELISAs for anthropogenic markers such as caffeine – which at the same time is a psychoactive drug – have been developed. Toxicological studies with the compounds alone as well as in combination towards different species of aquatic life such as marine invertebrates, e.g. bivalves (clams) and benthic invertebrates, namely polychaetes, will be shown, the effects induced having been assessed through the application of a battery of biomarkers.

Keywords: Bioanalytical, Environmental Analysis, Liquid Chromatography, Water

Application Code: Environmental

Methodology Code: Liquid Chromatography/Mass Spectrometry

Session # 1200 Abstract # 1200-1 **Symposia**

Session Title: Emerging Technologies for Disease Biomarker Detection

Abstract Title: **Phase Separated Droplets Enable Multiplexing of Difficult ELISA Panels**

Primary Author: Shuichi Takayama
University of Michigan

Date: Tuesday, March 08, 2016 - Afternoo
Time: 01:35 PM
Room: B304

Co-Author(s)

Abstract Text

Despite the discovery and report of many potential protein-based disease biomarkers every year, only a handful become clinically useful. Typically, no one protein species is sensitive or specific enough to reliably diagnose a disease. What is needed, therefore, is a method to validate diagnostic protein panels. While mass spectrometry has significantly accelerated the discovery of potential panels of protein-based disease biomarkers, there is still a lack of efficient and effective tools for multiplex protein biomarker validation. While bead-based and microarray based multiplex immunoassays exist, they suffer from antibody cross-reactions and differences in the required dilution of patient samples and associated with insufficient dynamic range of assays. These limitations make development of novel assay panels difficult, and biomarker validation experiments difficult and time-consuming. This talk will present novel technologies based on aqueous two phase system microdroplets that overcome common limitations in multiplexing such as antibody cross-reaction. Specifically, the talk will present multiplexing of homogeneous immunoassays (AlphaLisa) that are typically not possible to multiplex beyond a color-based two-plex, as well as a microarray based heterogeneous immunoassay. Applications will focus on development of diagnostics for graft-versus-host disease (GVHD) and other diseases.

Keywords: Chemiluminescence, Immunoassay
Application Code: Biomedical
Methodology Code: Microfluidics/Lab-on-a-Chip

Session # **1200** Abstract # **1200-2** **Symposia**

Session Title Emerging Technologies for Disease Biomarker Detection

Abstract Title **Looking for Rare Cells via High-Throughput Single Cell Mass Spectrometry Profiling**

Date: Tuesday, March 08, 2016 - Afternoo
Time: 02:10 PM
Room: B304

Primary Author Jonathan V. Sweedler
University of Illinois

Co-Author(s)

Abstract Text

Cell-to-cell chemical heterogeneity is a fundamental property of many biological tissues. Chemically and functionally distinct subpopulations are found within so-called homogenous cell populations, and oftentimes, rare cells impart important aspects to disease progression and the pathological state or survival of an organism. To characterize rare cells, presorting approaches such as flow cytometry or other high-throughput methods that can perform large number of direct single cell measurements have advantages. For high-throughput mass spectrometry (MS) measurements, we perform optical microscopy-guided single cell profiling. The cells are deposited onto a microscope slide, their locations on the slide determined using microscopy, and these locations guide subsequent MS analysis to each cell. To validate these approaches, we study several well-characterized tissue samples including endocrine cells from the rat pituitary and pancreas. In the pituitary, the most common cell types contain POMC-related peptides. We detected rare cells with distinct MS peptide profiles. In the rat pancreas, we identified subpopulations of specific peptide containing cells at the expected yields, suggesting that we can work with heterogeneous populations of small cells. Intriguingly, MALDI MS does not use most of a single cell contents in a measurement, so the rare cells can be located and used for follow-up profiling such as immunohistochemical staining, metabolomics via capillary electrophoresis MS, and potentially even transcriptomic measurements. Hence MS becomes a high-throughput screening approach to locate rare cells that can be used for follow-up studies.

Keywords: Mass Spectrometry, Neurochemistry, Sample Preparation, Time of Flight MS
Application Code: Neurochemistry
Methodology Code: Mass Spectrometry

Session # 1200 Abstract # 1200-3 Symposia

Session Title Emerging Technologies for Disease Biomarker Detection

Abstract Title **Biomarker Detection Using Paper/PDMS Hybrid Microfluidic Platforms for Low-Cost Disease Diagnosis**

Date: Tuesday, March 08, 2016 - Afternoo
Time: 02:45 PM
Room: B304

Primary Author Xiujun James Li
University of Texas at El Paso

Co-Author(s) Maowei Dou, Sharma T Sanjay

Abstract Text

There is a great need for simple, affordable disease biomarker detection methods in low-resource settings that lack complicated and expensive diagnostic tools. Although numerous polydimethylsiloxane (PDMS) and paper-based microfluidic devices have been developed to address this issue, PDMS/paper hybrid systems that take advantage of both substrates are rarely reported. Each device substrate has its own advantages and disadvantages. Herein, we have developed different low-cost PDMS/paper hybrid microfluidic systems that take advantage of both PDMS and paper substrates for rapid and sensitive disease diagnosis, especially in low-resource settings. For instance, paper was used in a PDMS/ paper hybrid microfluidic system integrated with loop-mediated DNA isothermal amplification (LAMP) for rapid and sensitive multiplexed meningitis diagnosis, a global disease with high morbidity and mortality. The introduction of paper into the microfluidic device enables stable test results over a much longer period of time than a paper-free microfluidic system. Results can be observed by the naked eye. Although this hybrid system does not require expensive instruments, its sensitivity is even higher than conventional real-time PCR. Additionally, we have also developed a paper/PMMA hybrid microfluidic immunosensing microplate for hepatitis B diagnosis. The unique funnel-shaped microwell design enabled rapid antigen immobilization and efficient washing. Without any specialized equipment, the limit of detection of 1.60 ng/mL hepatitis B surface antigen was achieved within one hour, which is comparable to commercial kits using spectrophotometers. Financial support from NIH, UT STARS Award, MRAP, IDR2 and URI award from UTEP is gratefully acknowledged.

Keywords: Bioanalytical, Biosensors, Biotechnology, Lab-on-a-Chip/Microfluidics

Application Code: Bioanalytical
Methodology Code: Microfluidics/Lab-on-a-Chip

Session # 1200	Abstract # 1200-4	Symposia

Session Title: Emerging Technologies for Disease Biomarker Detection

Abstract Title: **'Cytology-on-a-Chip' Based Sensors for Monitoring of Potentially Malignant Oral Lesions**

Primary Author: John T. McDevitt
New York University

Date: Tuesday, March 08, 2016 - Afternoo
Time: 03:35 PM
Room: B304

Co-Author(s)

Abstract Text

Over past few decades, use of biomarkers has become increasingly intrinsic to practice of medicine and clinical decision-making. Diagnosis and management of oral cancer is a promising area whereby biomarker driven testing has potential to provide significant impact on patient care. Oral cancer is sixth most common cancer worldwide and has been marked by high morbidity and poor survival rates with little over the past few decades. Beyond prevention, early detection is the most crucial determinant for successful treatment and survival of oral cancer. This talk will feature details related to a new 'cytology-on-a-chip' platform capable of high-content single-cell measurements. This methodology permits concurrent analysis of molecular biomarker expression and cellular/nuclear morphology using over 200 fluorescence intensity and shape parameters for each region of interest extracted from multi-spectral fluorescence images. Molecular biomarkers: EGFR, □v□6, CD147, □-catenin, MCM2, and Ki67 were selected based on their capacity, through prior immunohistochemistry studies, to distinguish stages of disease progression towards oral cancer. Measurement time to complete this chip-based image analysis is approximately 20 minutes vs. about 1-3 days for gold standard pathology exam. This new clinical decision tool has been developed and validated in context of major clinical study involving 714 prospectively recruited patients. These efforts have led to collection of data across 6 diagnostic categories and assembly of one of largest well-qualified cytology database (confirmed by tissue biopsy) ever collected for prospectively recruited potentially malignant oral lesions. The application of statistical machine learning algorithms exploiting this large database has led to development of robust classification models with validated and stable parameters. High sensitivity and high specificity adjunctive diagnostic aids have been developed through these efforts.

Keywords: Bioinformatics, Biosensors, Lab-on-a-Chip/Microfluidics, Medical
Application Code: Biomedical
Methodology Code: Microfluidics/Lab-on-a-Chip

Session # 1200 Abstract # 1200-5 **Symposia**

Session Title Emerging Technologies for Disease Biomarker Detection

Abstract Title **Biomarker Discovery Using DNA Aptamers**

Date: Tuesday, March 08, 2016 - Afternoo
Time: 04:10 PM
Room: B304

Primary Author Weihong Tan
University of Florida

Co-Author(s)

Abstract Text

A full understanding of the molecular basis of diseases depends on the development of molecular probes able to recognize disease targets of interest. Until very recently, such tools have been absent from the clinical practice of medicine. The newest molecular probe, and one that holds most promise, is a new class of designer nucleic acids, termed aptamers, whichare single-stranded DNA/RNA able to recognize specific targets, such as single proteins and even small molecules. Recently, we applied a simple, fast and reproducible cell-based aptamer selection strategy called Cell-SELEX which uses whole, intact cells as the target for aptamer selection.This selection processthen generatesmultiple aptamers for the specific recognition of biological cells, but without the need for prior knowledge about the signature of target cell-surface molecules. The selected aptamers have dissociation constants in the nanomolar to picomolar range. Thus far, we have selected aptamer probes for many different diseases, andused them to carry out studiesat the vanguard of biomedicalscience, including ultrasensitive detection of tumors, molecular imaging, targeted drug delivery, and, mostcritically, cancer biomarker discovery. Taken together, these molecular level tools form a solid scientific platform from which to pursue advanced studies in molecular medicine. We will report our most recent progress in this exciting researcharea,especially the molecular elucidation of cancer biomarkers and targeted drug development.

Keywords: Bioanalytical, Biomedical, Drugs
Application Code: Biomedical
Methodology Code: Chemical Methods

Session # 1210 Abstract # 1210-1 **Symposia**

Session Title JAIMA - The State-of-the-Art Technologies from Japan: Analytical Instruments with/for Nano-Chemistr

Abstract Title **Electrochemical Application of Boron-Doped Diamond Electrodes**

Primary Author Yasuaki Einaga
Keio University

Date: Tuesday, March 08, 2016 - Afternoo
Time: 01:35 PM
Room: B305

Co-Author(s)

Abstract Text

Boron-doped diamond (BDD) electrodes are very attractive material, because of their wide potential window, low background current, chemical inertness, and mechanical durability[1]. In these years, we have reported several examples for electrochemical sensor applications[2]. Here, we report some recent examples of electrochemical sensor application of BDD such as ozone [3], pH [4], in vivo detection of neurotransmitter in monkey brain [5], and in vivo detection of glutathione for assessment of cancerous tumors [6] using BDD microelectrodes. Furthermore, other applications such as organic synthesis [7], ozone generation, and CO2 reduction [8] are also shown.

In vivo assessment of cancerous tumors

The in vivo electrochemical detection of the reduced form of glutathione (GSH) using BDD microelectrode for potential application in the assessment of cancerous tumors is presented. In vivo GSH detection measurements have been performed in human cancer cells inoculated in immunodeficient mice. These measurements have shown that the difference of GSH level between cancerous and normal tissues can be detected.

In vivo pH monitoring

The in vivo electrochemical monitoring of pH using BDD microelectrode and silver needles for potential application in medical diagnosis was studied. A quantitative analysis of the increase in stomach pH is also presented. It is proposed that the catheter-free pH monitoring system presented in this study could be potentially employed in any biological environment.

References

[1] Diamond Electrochemistry, Elsevier, 2005. [2] (a) Anal. Chem. 78, 3467 (2006), (b) Anal. Chem. 78, 6291 (2006)., (c) Anal. Chem. 80, 5783 (2008)., (d) Anal. Chem. 78, 7857 (2006)., (e) Anal. Chem. 79, 8608 (2007). [3] Anal. Chem. 85, 4284 (2013). [4] (a) Phys. Chem. Chem. Phys. 13, 16795 (2011)., (b) Sci. Rep. 3, 3257 (2013). [5] Neurosci. Res. 71, 49 (2011). [6] Sci. Rep. 2, 901 (2012). [7] Angew. Chem. Int. Ed. 51, 5443 (2012). [8] Angew. Chem. Int. Ed. 53, 871 (2014).

Keywords: Bioanalytical, Electrodes, Environmental Analysis, Sensors
Application Code: Environmental
Methodology Code: Electrochemistry

Session # **1210** **Abstract #** **1210-2** **Symposia**

Session Title JAIMA - The State-of-the-Art Technologies from Japan: Analytical Instruments with/for Nano-Chemistr

Abstract Title **Plasmonic Nanomaterials**

Date: Tuesday, March 08, 2016 - Afternoo

Time: 02:10 PM

Room: B305

Primary Author Tetsu Tatsuma
University of Tokyo

Co-Author(s)

Abstract Text

Plasmonic metal nanoparticles, which are characterized by strong light absorption, scattering, and their controllability, attract attention as new photofunctional materials. The lecture will focus on application of plasmonic nanoparticles to sensors, photovoltaic cells, photocatalysts, and advanced optical materials, as well as some basic aspects of localized surface plasmon resonance (LSPR). We found plasmon-induced charge separation (PICS) at the metal nanoparticle-semiconductor interface and applied it to photovoltaics and photocatalysis [1]. Now many other research groups are working on photosensors and photocatalysts based on PICS [2,3]. We also applied PICS to plasmonic sensors [4] for chemical and biological applications, image display [2], and data storage [5]. We also developed other plasmonic sensors [6] and color routers [7].

1. Y. Tian and T. Tatsuma, J. Am. Chem. Soc., 2005, 127, 7632.
2. T. Tatsuma, Bull. Chem. Soc. Jpn., 2013, 86, 1.
3. C. Clavero, Nat. Photonics, 2014, 8, 95.
4. T. Tatsuma, Y. Katagi, S. Watanabe, K. Akiyoshi, T. Kawawaki, H. Nishi, and E. Kazuma, Chem. Commun., 2015, 51, 6100.
5. I. Tanabe and T. Tatsuma, Nano Lett., 2012, 12, 5418.
6. H. Nishi, S. Hiroya, and T. Tatsuma, ACS Nano, 2015, 9, 6214.
7. K. Saito and T. Tatsuma, Adv. Opt. Mater., 2015, 3, 883.

Keywords: Electrochemistry, Metals, Nanotechnology, Sensors

Application Code: Nanotechnology

Methodology Code: Electrochemistry

Session # 1210 Abstract # 1210-3 **Symposia**

Session Title JAIMA - The State-of-the-Art Technologies from Japan: Analytical Instruments with/for Nano-Chemistr

Abstract Title **The Unique Combination of Nanotechnology with Raman and SPRi Platforms Offers Innovative and Ultrasensitive Solutions for Diagnostics**

Date: Tuesday, March 08, 2016 - Afternoo
Time: 02:45 PM
Room: B305

Primary Author Marinella Sandros
HORIBA Scientific

Co-Author(s)

Abstract Text

Raman Microscopy allows for simultaneous chemical identification and imaging through the distinctive vibrational signatures of chemical bonds and SPRi provides quantitative interaction data by measuring changes in refractive index on the sensor interface. Both optical non-destructive techniques provide unique diagnostic applications. During this lecture, we will highlight how Raman Microscopy and SPRi have enjoyed tremendous gains after integration with nanotechnology. Diagnostic biomarkers for many diseases at the early stage are often in low abundance presenting many challenges for their detection. Extending the application of SPRi and Raman Microscopy systems through the unique combination with nanomaterials provides ultra signal enhancement. Research developments in detecting biomarkers at ultralow levels (10-15 to 10-18 M) will be presented. Furthermore, cutting-edge techniques like Surface and Tip Enhanced Raman Spectroscopy offering nanoscale resolution with ultrasensitivity will also be reviewed.

Keywords: Array Detectors, Biosensors, Nanotechnology
Application Code: Biomedical
Methodology Code: Sensors

Session # **1210** **Abstract #** **1210-4** **Symposia**

Session Title JAIMA - The State-of-the-Art Technologies from Japan: Analytical Instruments with/for Nano-Chemistr

Abstract Title **Medicinal Cannabinomics and Mass Spectrometry Applications to Cannabis Testing Laboratories**

Primary Author Scott Kuzdzal
Shimadzu Scientific Instruments

Date: Tuesday, March 08, 2016 - Afternoo
Time: 03:35 PM
Room: B305

Co-Author(s) Di Wang, Jeff H. Dahl, Jonathan Edwardsen, William Lipps

Abstract Text

Thirty-seven states have pro- medical cannabis laws, including twenty-three states (and DC) that have medical laws, and four states that have full legalization. There are now over 2 million legal medical marijuana patients in the U.S.A., and nearly half of Americans are now living in states with some form of medical marijuana or CBD legislation. Cannabis testing laboratories have emerged to accurately determine cannabinoid potencies in cannabis products as well as ensure that these products are free from contaminants.

High performance liquid chromatography and gas chromatography are important cannabis testing laboratory separation techniques. Mass spectrometry is playing an increasingly important role in cannabis testing laboratories for the analysis of cannabinoids and terpenoids in cannabis products, including extracted oils. Mass spectrometry is also being employed for the ultra-low level quantitation of contaminants including pesticides and heavy metals. Application of ultra-fast electrospray ionization LC-MS/MS with continuous polarity switching for the simultaneous analysis of pesticides in positive and negative mode and approximately 200 pesticides were measured with over 500 MRM transitions per run. Additionally, data presented illustrates how a triple quadrupole GC-MS/MS operated in the MRM mode, can be used to analyze for trace-level pesticide residues in complex plant matrices such as medical cannabis flowers and extracted oils.

As mass spectrometry adoption continues to grow and its utility expands, it is critical that mass spectrometry knowledge gained from more established scientific disciplines, including food safety, clinical, pharmaceutical and environmental markets, be applied to cannabis testing laboratories. This presentation will also discuss future MS/MS-based opportunities for techniques like MALDI-TOF MS, SFE-SFC-MS, direct ionization methods and integrated “cannabis analyzers”.

Keywords: GC-MS, Liquid Chromatography/Mass Spectroscopy, Mass Spectrometry, Quality
Application Code: Quality/QA/QC
Methodology Code: Liquid Chromatography/Mass Spectrometry

Session # 1210 Abstract # 1210-5 **Symposia**

Session Title JAIMA - The State-of-the-Art Technologies from Japan: Analytical Instruments with/for Nano-Chemistr

Abstract Title **Application of Laser/desorption Ionization Mass Spectrometry as a Novel Surface Analytical Tool**

Primary Author Takaya Satoh
JEOL Ltd.

Date: Tuesday, March 08, 2016 - Afternoo
Time: 04:10 PM
Room: B305

Co-Author(s)

Abstract Text

Imaging mass spectrometry (IMS) for matrix-assisted laser desorption/ionization (MALDI) has been expanded during the last decade in biological applications, to assess the distribution of proteins, peptides, lipids, drugs, and metabolites in a tissue specimen. The MALDI-IMS also has an application potentiality to material science field; however, the studies in this field were limited. The laser desorption/ionization (LDI) IMS, including MALDI-IMS, has an advantage in organic compounds analysis due to an ability of separating by molecular weight, where other spectroscopy methods can obtain only elemental information. In this presentation, the potentiality of LDI-IMS as a new surface analytical tool is discussed. At first, we have developed high mass resolution time-of-fight mass spectrometer (TOFMS) for non-target IMS. It is necessary to extend flight path to improve mass resolution of TOFMS. We have developed a multi-turn type TOFMS with a spiral ion trajectory, SpiralTOF, to solve the issue. The total flight path of SpiralTOF was 17 m and achieved ultra-high mass resolution that could separate isobaric compounds. Secondly, we have studied the nano-particle surface assisted LDI (SALDI) for high lateral resolution IMS. The key technique for expanding the IMS will be establishment of methodology to apply the compounds to assist ionization homogeneously on a sample surface. The nano-particle SALDI-IMS was confirmed to be effective method for high lateral resolution IMS compared to a common matrix application method using spray. At last, we demonstrate the collaboration analysis with established surface analytical tools such as Raman spectroscopy, TOF-SIMS and XPS. The LDI-IMS has a strong point in analyzing organic molecular structure, however, it is a destructive-analytical tool and the available information regarding to depth direction was not well understood. Collaboration research reveals basic character and true advantages of LDI-IMS as surface analytical tools.

Keywords: Imaging, Mass Spectrometry, Surface Analysis, Time of Flight MS
Application Code: Material Science
Methodology Code: Mass Spectrometry

Session # 1220 Abstract # 1220-1 **Symposia**

Session Title New Advances in Analytical Mass Spectrometry

Abstract Title **21 Tesla Fourier Transform Ion Cyclotron Resonance Mass Spectrometry: New Capabilities for Ultrahigh-Resolution Mass Analysis**

Date: Tuesday, March 08, 2016 - Afternoo
Time: 01:35 PM
Room: B309

Primary Author Alan G. Marshall
Florida State University

Co-Author(s) Christopher L. Hendrickson, Donald F. Smith, Greg T. Blakney, John P. Quinn, Nathan K. Kaiser, Steven C. Beu, Tong Chen

Abstract Text

FT-ICR mass spectrometry offers the highest achievable broadband mass resolving power and mass accuracy of any mass analyzer. Resolving power and scan rate improve linearly, and mass accuracy and dynamic range improve quadratically with magnetic field strength [1], such that resolving power greater than 1 million and mass accuracy better than 1 ppm become routine at sufficiently high magnetic field strength. We describe the design and initial performance of the first 21 tesla Fourier transform ion cyclotron resonance mass spectrometer [2]. The 21 tesla magnet is the highest field superconducting magnet used for FT-ICR MS and features high spatial homogeneity, high temporal stability, and negligible liquid helium consumption. The instrument includes a commercial dual linear quadrupole trap front end that features high sensitivity, precise control of trapped ion number, and collisional and electron transfer dissociation. A third linear quadrupole trap offers high ion capacity and ejection efficiency, and rf quadrupole ion injection optics deliver ions to a novel dynamically harmonized ICR cell [3]. The instrument is part of the NSF High-Field FT-ICR User Facility and is available free of charge to qualified users Work supported by the National Science Foundation through DMR-1157490, CHE-1016942, CHE-1019193, and the State of Florida

(1) Marshall, A. G.; Hendrickson, C. L.; Jackson, G. S. Mass Spectrom. Rev. 1998, 17, 1-35.
(2) Hendrickson, C. L.; Quinn, J. P.; Kaiser, N. K.; Smith, D. F.; Blakney, G. T.; Chen, T.; Marshall, A. G.; Weisbrod, C. R.; Beu, S. C. JASMS. 2015, 26, 1626-1632.
(3) Chen, T.; Beu. S. C.; Kaiser, N. K.; Quinn, J. P.; McIntish, D. G.; Williams, V.; Marshall, A. G.; Hendrickson, C. L. 63rd ASMS Conference on Mass Spectrometry and Allied Topics, St. Louis, MO, 31 May – 5 June, 2015. Poster MP 081

Keywords: Fuels\Energy\Petrochemical, Ion Cyclotron Resonance, Mass Spectrometry, Protein
Application Code: Bioanalytical
Methodology Code: Mass Spectrometry

Session # 1220 Abstract # 1220-2 Symposia

Session Title New Advances in Analytical Mass Spectrometry

Abstract Title **The Unique Analytical Capabilities of Distance-of-Flight Mass Spectrometry**

Primary Author Steven J. Ray
SUNY-Buffalo

Date: Tuesday, March 08, 2016 - Afternoo
Time: 02:10 PM
Room: B309

Co-Author(s) Christie G. Enke, David W. Koppenaal, Elise Dennis, Gary M. Hieftje

Abstract Text

This presentation will describe a new type of mass spectrometer known as distance-of-flight mass spectrometry (DOFMS). The concept behind DOFMS is best explained by comparison with traditional time-of-flight mass spectrometry (TOFMS). Time-of-flight mass analyzers measure the m/z of an ion by imparting the same energy to all ions and then measuring the time required for each m/z to traverse a known distance and arrive at a single detector. In contrast, DOFMS measures the m/z of an ion by measuring the distance each ion travels during a set time period. Ions accelerated to a constant momentum separate in space, with ions of lower m/z traveling longer distances than ions of greater m/z. At a specific instant after acceleration, all m/z will achieve a sharp spatial focus and can then be directed onto the surface of a position-sensitive ion detector where their m/z is determined based upon location.

?

The DOFMS offers some intriguing opportunities for MS analyses. Because DOFMS physically separates ions according to m/z, is a naturally suited to ion collection via 'soft-landing' MS approaches. The DOFMS can also employ new solid-state ion detector arrays that measure ion charge directly, and thus show little or no mass bias. The DOFMS method also exploits an ion focusing strategy that is designed to reproduce the initial spatial distribution of ions upon the detector surface. Thus, surface ionization technique, such as MALDI, are well suited to DOFMS because the ions created by the source are initially confined at the sample surface. Taken together, these attributes suggest a new MS approach for the analysis and collection of ions over a very wide mass range (and without upper mass limit). Moreover, the shared architecture of TOFMS and DOFMS means that both techniques can be packaged in a single instrument. This new MS strategy will be discussed, and initial analytical performance of this new type of spectrometer described.

Keywords: Bioanalytical, Instrumentation, Mass Spectrometry, Time of Flight MS
Application Code: Bioanalytical
Methodology Code: Mass Spectrometry

Session # 1220 Abstract # 1220-3 **Symposia**

Session Title New Advances in Analytical Mass Spectrometry

Abstract Title **Plasma Mass Spectrometry: A Tool for and a Source of Chemical Reactions**

Date: Tuesday, March 08, 2016 - Afternoo
Time: 02:45 PM
Room: B309

Primary Author Jacob T. Shelley
Kent State University

Co-Author(s) Garett MacLean, Kelsey Williams, Sunil P. Badal, Yi You

Abstract Text

Plasma ionization sources and mass spectrometry have gone hand-in-hand since J.J. Thomson first separated ions produced by a gaseous discharge over 100 years ago. Reduced-pressure plasmas were often used as sampling and ionization tools for physical studies, such as isotope measurements, and analytical atomic MS. Later in the 20th century, the differentially pumped interface allowed the ion source to be external to the vacuum system, which simplified sample introduction for analytical purposes. Operating a plasma source at atmospheric-pressure also provides the possibility of softer ionization due to collisional cooling allowing detection of intact molecular species. Thus, there should be a large range of possible chemistries with these plasmas. This presentation will focus on the tunable chemistry of the plasma-based Flowing Atmospheric-Pressure Afterglow (FAPA) source. The FAPA relies on a DC atmospheric-pressure glow discharge. This source provides excellent sensitivity for molecular species (attomole detection limits), yet is easy to construct and use. Recently, our group has found that the ionization processes are highly tunable by altering the operating parameters or the plasma gas composition. For instance, the major ionization pathway can be switched between proton-transfer and charge transfer ionization. This rapidly adjustable mode of operation provides more comprehensive analyses due to the expanded range of detectable analytes. In one example, the progress of a liquid-crystal synthesis is monitored in real-time, where the reactants and products are highly non-polar and could not be detected with other MS approaches. Lastly, we have found that unique gas-phase chemistry can take place with a mixed-gas FAPA. The observed reactions usually require a catalyst in solution-phase, but occur on the sub-millisecond timescale here. Attempts to scale up these reactions as well as potential mechanisms for these effects will also be presented.

Keywords: Atomic Spectroscopy, Bioanalytical, Mass Spectrometry, Organic Mass Spectrometry
Application Code: General Interest
Methodology Code: Mass Spectrometry

Session # 1220 Abstract # 1220-4 **Symposia**

Session Title New Advances in Analytical Mass Spectrometry

Abstract Title **New Paths for Mass Spectrometry based upon Structures for Lossless Ion Manipulations (SLIM)**

Date: Tuesday, March 08, 2016 - Afternoo
Time: 03:35 PM
Room: B309

Primary Author Richard D. Smith
Pacific Northwest National Laboratory

Co-Author(s)

Abstract Text

Mass spectrometry (MS)-based proteomics measurements are having profound impacts on broad areas of biological research, and including applications involving proteomics, metabolomics, lipidomics and glycomics. Increasingly, advances in the quality, resolution, and the speed of e.g. polypeptide and protein separations are arguably as important as mass spectrometric developments in improving the sensitivity and coverage of proteomics measurements. Both liquid phase, e.g., using liquid chromatography, and increasingly gas phase ion mobility separations, respectively, provide a basis for increasing the quality of proteomics measurements, such as the completeness of protein coverage. While these capabilities are challenged by very small sample sizes, the recent development of more efficient nanoelectrospray ion sources and MS interfaces has helped enable ultra-sensitive measurements. Increasingly advances involve gas phase ion manipulations that are conducted between the ion source and m/z analyzer. These manipulations include: ion transport through regions of elevated pressure, trapping, reactions (both ion-molecule and ion-ion), and mobility-based separations. This presentation will discuss the utility of ion mobility separations for biological applications, and describe new developments based upon long path length Structures for Lossless Ion Manipulations (SLIM) that enable very fast high resolution separations, as well as the use of other gas phase ion manipulation approaches having broad utility for facilitating MS analysis capabilities. The SLIM developments will be discussed with regard to their sensitivity, measurement throughput, and their utility for both broad and targeted quantitative measurements. The presentation will conclude with consideration of pending developments enabled by SLIM and their potential impacts for mass spectrometry-based applications.

Keywords: Bioanalytical, Instrumentation, Mass Spectrometry
Application Code: Bioanalytical
Methodology Code: Mass Spectrometry

Session # 1220 Abstract # 1220-5 **Symposia**

Session Title New Advances in Analytical Mass Spectrometry

Abstract Title **Improvements in Velocity-Based Mass Analysis by Use of Constant-Momentum Acceleration**

Date: Tuesday, March 08, 2016 - Afternoo
Time: 04:10 PM
Room: B309

Primary Author Gary M. Hieftje
Indiana University

Co-Author(s) Alexander W. Gundlach-Graham, Christie G. Enke, Elise Dennis, Steven J. Ray

Abstract Text

Time-of-flight mass spectrometry (TOFMS) is the most common method for velocity-based mass analysis. In TOFMS, ions to be mass analyzed are all given the same kinetic energy by causing them to fall across the same voltage drop. The resulting kinetic energy (KE) is then equal to eV, where V is the voltage drop and e is the charge on the ion. Lighter ions therefore travel faster than heavier ones (KE = 0.5 mv2, where m is mass and v is velocity) and reach a distant detector sooner. Indeed, the time-resolved output of the detector is the mass spectrum itself.

There is an alternative approach for velocity-based mass analysis: give all ions the same momentum (= mv). Constant momentum is achieved by giving the ions all the same "kick"; that is, accelerating them for the same length of time. Again, the lightest ions fly the fastest, but here there is a linear proportionality between mass and velocity rather than the quadratic relationship in constant-energy acceleration (CEA). This linearity inherently yields better mass separation and consequently higher resolving power. Moreover, there is an important fundamental difference between the constant-energy and constant-momentum approaches. In CEA, ions of all m/z can be brought to focus at a specific location (where the detector is positioned) whereas constant-momentum acceleration (CMA) yields a different focal location for each m/z, but at m/z-dependent distances. CEA therefore lends itself to use in TOFMS whereas CMA seems ideal for distance-of-flight MS (DOFMS). In DOFMS, ions are therefore all measured at virtually the same moment, but at different locations with a position-sensitive detector or array.

Further, CMA does not suffer as seriously as CEA from differences in initial energy that all ions possess. It therefore provides even better resolving power than CEA, whether used in TOFMS or DOFMS. The practical and performance benefits of CMA will be highlighted with examples of where these benefits are most useful.

Keywords: Array Detectors, Instrumentation, Mass Spectrometry, Time of Flight MS
Application Code: Bioanalytical
Methodology Code: Mass Spectrometry

Session # 1230 Abstract # 1230-1 **Symposia**

Session Title SEAC - New Trends in Electrochemical Neurochemistry

Abstract Title **Fast-Scan Cyclic Voltammetry Reveals Dopamine Spikes to Food Reward that are Tuned by Physiological State and Its Proxies**

Primary Author Mitchell Roitman
University of Illinois at Chicago

Date: Tuesday, March 08, 2016 - Afternoo
Time: 01:35 PM
Room: B310

Co-Author(s)

Abstract Text

The neurotransmitter dopamine was identified as a key signal in reward close to 40 years ago. Yet its precise role in reward-directed behaviors remains unresolved, in part, because reward-directed behavior is complex and occurs on a sub-second timescale. More recently, fast-scan cyclic voltammetry at carbon fiber microelectrodes (CFMs) has been applied to sample fluctuations in dopamine concentration on a timescale commensurate with complex behavior. We acutely drive glass-insulated CFMs into brain regions of awake and behaving rats. Taking advantage of the relatively small sensor size, we focus on two subterritories of the nucleus accumbens, which are rich in dopamine terminals. By rapidly (400 V/s) altering the voltage of the electrode in a triangular fashion (-0.4 to +1.3 to -0.4 V vs Ag/AgCl reference) every 100 milliseconds, we measure dopamine based on its oxidative current and can separate it from other electroactive species. Using this approach, we have found that that phasic (1-2 s) spikes in dopamine concentration are evoked by food and, in trained rats, by cues that predict food (which elicit food-approach behaviors). We have also found that the physiological state of the animal and hormones that signal hunger and satiety modulate these dopamine spikes. If dopamine spikes play a causal role in food-directed behavior, a subject of current investigation in our lab using targeted opto- and chemogenetic approaches, then hormone receptors on dopamine neurons may represent novel therapeutic targets for disordered food-directed behavior such as the over-eating that leads to obesity.

Keywords: Neurochemistry, Peptides, Sensors, Voltammetry
Application Code: Neurochemistry
Methodology Code: Electrochemistry

Session # 1230 Abstract # 1230-2 **Symposia**

Session Title: SEAC - New Trends in Electrochemical Neurochemistry

Abstract Title: **Electroenzymatic Detection of Basolateral Amygdala Glutamate Release During Reward Seeking**

Date: Tuesday, March 08, 2016 - Afternoo
Time: 02:10 PM
Room: B310

Primary Author: Kate M. Wassum
University of California Los Angeles

Co-Author(s): Allison M. Yorita, Harold G. Monbouquette, Lili Feng, Melissa Malvaez, Venuz Y. Greenfield

Abstract Text

Adaptive reward seeking requires the ability to extract from the environment and mentally represent information about specific available rewards. The basolateral amygdala (BLA) participates in this cognitive process, but precisely how is unknown. There is a particular lack of information regarding how BLA input signals relate to and influence reward-seeking behavior. The BLA receives dense glutamatergic innervation from both cortical and thalamic structures, so we focused on measuring this excitatory input signal during several tests of reward seeking. To achieve this, we used an electroenzymatic glutamate biosensor technology that affords near-real time (<1s temporal resolution), sensitive, and selective measurement of glutamate concentration changes in the brains of freely-behaving rodents. Transient elevations in glutamate concentration were detected in the BLA during reward-related behavior. Targeted behavioral tests revealed that these glutamate transients occurred when rats encoded the value of specific rewards and also when the mental representation of those rewards was used to guide reward seeking. Interference methods were used to provide a complementary causal analysis to the correlational recording approach.

Keywords: Biosensors, Microelectrode
Application Code: Other
Methodology Code: Electrochemistry

Session # 1230 Abstract # 1230-3 **Symposia**

Session Title SEAC - New Trends in Electrochemical Neurochemistry

Abstract Title **Improving Temporal Resolution of Enzyme Based Electrochemical Sensors for Detection of Non-Electroactive Analytes Important in Brain Chemistry**

Date: Tuesday, March 08, 2016 - Afternoo
Time: 02:45 PM
Room: B310

Primary Author Ann-Sofie Cans
Chalmers University of Technology

Co-Author(s) Jenny Bergman, Joakim Wigström, Yuanmo Wang

Abstract Text

Acetylcholine and glutamate are highly important non-electroactive neurotransmitter in the mammalian central nervous system. A fast, sensitive method to detect the release of these molecules at the surface of a single cell is needed to gather data about the kinetics of exocytosis events in pathways involving these signaling molecules. To this end, carbon fiber electrodes have been modified with electrodeposited nanoparticles to increase the effective electrode surface area and provide a high curvature surface for enzyme attachment. For detection of acetylcholine, acetylcholine esterase and choline oxidase and for glutamate, glutamate oxidase were deposited onto the nanoparticle coated electrode surfaces to catalyze acetylcholine to hydrogen peroxide for electrochemical detection. The functionalized electrodes have been characterized to determine the KM and Vmax of the enzymes as well as the total enzyme coverage and gold nanoparticle surface area. This information was further used to evaluate the conditions for optimal retained enzyme activity of the sensor surface. Similarly, glutamate oxidize was placed onto the surface of electrodes plated with nanoparticles. The sensors were tested for analyte release from a synthetic cell model for exocytosis, and providing time resolved detection of single vesicle release events on the order of millisecond time scale.

Keywords: Bioanalytical, Biosensors, Microelectrode
Application Code: Bioanalytical
Methodology Code: Electrochemistry

Session # 1230 **Abstract # 1230-4** **Symposia**

Session Title: SEAC - New Trends in Electrochemical Neurochemistry

Abstract Title: **Building a 'Well-Tempered' Biosensor for Real-Time Neurochemical Monitoring in the Intensive Care Unit**

Date: Tuesday, March 08, 2016 - Afternoo
Time: 03:35 PM
Room: B310

Primary Author: Martyn G. Boutelle
Imperial College London

Co-Author(s): Chi Leng Leong, Chu Wang, Isabelle C. Samper, Michelle L. Rogers, Sally A. Gowers, Thomas Watts

Abstract Text

Beyond the few electroactive catecholamine neurotransmitters such as dopamine, detection of the majority of neurochemicals require the use of amperometric biosensors. Electrochemical biosensors are necessarily complex, typically comprising a carefully prepared electrode surface coated with an electropolymerised permselective layer (20 nm), an enzyme containing hydrophilic layer (20 microns) and potentially an outer layer to limit mass transport to the enzyme (extending the linear range of the biosensor). Response curves relating electrooxidation current and neurochemical concentration are well fitted by either the Michaelis-Menton or better the Hill equation. Such biosensors typically lose sensitivity by Vmax falling by many percent per hour. If this were not complex enough biosensors are craft items made by hand in small batches with substantial sensor to sensor variation.

All of these factors make biosensors difficult to use by non-experts (who are interested only in concentration), and importantly very challenging to integrate within expert detection systems as part of electroanalytical instruments (which detect concentration changes greater than some threshold). 30 years of biosensor research seeking the perfect, stable biosensor has not fundamentally changed the situation.

Inspired by the compromise of the well-tempered Clavier, playable in all 24 keys while not being perfect in any one [1], we have pursued a new approach in which a good (but imperfect) biosensor is coupled to an auto-calibration system, an ADC and a microcontroller to control the whole system. The system tracks falling sensitivity of the biosensor allowing prediction of sensitivity between calibrations, automatic adjustment of calibration intervals and prediction of biosensor failure. The output of the whole system appears to the user as a perfect, well-tempered biosensor giving a linear output of for example 1V per millimolar.

[1] Das wohltemperierte Klavier I, Bach JS, 1722, BWV 846-869

Keywords: Biosensors, Electrochemistry, Lab-on-a-Chip/Microfluidics, Neurochemistry
Application Code: Neurochemistry
Methodology Code: Electrochemistry

Session # 1230 Abstract # 1230-5 **Symposia**

Session Title SEAC - New Trends in Electrochemical Neurochemistry

Abstract Title **Optogenetic-Control of Glutamate Release in the Rat Hippocampus and Frontal Cortex Measured Using Enzyme-Coated Ceramic Based Microelectrode Arrays**

Date: Tuesday, March 08, 2016 - Afternoo
Time: 04:10 PM
Room: B310

Primary Author Greg A. Gerhardt
University of Kentucky Medical Center

Co-Author(s)

Abstract Text

Recent advancements in optogenetics, which introduces light sensitive proteins (opsins) into neurons that regulate transmembrane ion conductance, have allowed for more specific control of neurons in the CNS. Electrophysiological studies have shown that optical excitation or inhibition of neuronal activity is correlated with behavior, but to date, very few studies have examined neurotransmitter release combined with optical stimulation. We combined our expertise of electrochemical measurements of neurotransmitter release in vivo with optogenetics in order to examine glutamate release dynamics in the CNS. We infused (1 µl/each) AAV5-Syn-ChR2-EYFP (CHR2) into the right dentate gyrus (DG) of the hippocampus and the left infralimbic (IL) region of the frontal cortex. We attached an optical fiber (200 µm o.d. ~200 µm from the recording sites) to our ceramic-based microelectrode array (MEA) configured to directly record tonic and phasic glutamate release and lowered the assembly into regions of the DG or IL. We used constant light activation (DC: 488 nm, 1 to 10 mW) or pulses (train (TR): 10 ms; 40 Hz) to directly produce light-dependent glutamate release. Glutamate dynamics were in the same range as we have previously reported using other forms of stimulation or behavior. We observed highly reproducible glutamate release in the range of 1 to 70 µM and uptake rates of 0.1 to 10 µM/sec. Enhanced yellow fluorescent protein expression showed the distribution of CHR2 6 weeks after a single injection (1 µl) into the DG (Fig A, B and C). ChR2 distribution was not limited to the injection site DG (Fig A) and can be observed in the CA1-CA3 (Fig B) and the entorhinal cortex (Fig C). Figure D shows glutamate release by stimulating the transfected CHR2 locally using a fiber optic cannula with a variety of light pulse parameters. These results show the feasibility of directly measuring glutamate release in vivo while controlling glutamatergic neuronal systems using optogenetics.

Keywords: Electrochemistry, Electrodes, Method Development, Neurochemistry
Application Code: Biomedical
Methodology Code: Electrochemistry

Session # 1240 **Abstract # 1240-1** **Symposia**

Session Title: The Challenge of Detection for Drugged Driving

Abstract Title: **Police Officer Difficulties with Drug-Impaired Driver Arrests**

Primary Author: Nicholas P. Lovrich
Washington State University

Co-Author(s)

Date: Tuesday, March 08, 2016 - Afternoo
Time: 01:35 PM
Room: B311

Abstract Text

The rapid rate of substitution of alcohol-impaired driving with drug-impaired driving (both illicit drugs and prescription medications) poses a major challenge for law enforcement in the US and in many other counties in Europe and Asia. The need for point-of-contact documentation of impairment through physical evidence to complement a behavioral assessment of impairment is sorely needed. The presentation will entail noting some of the many steps to be taken to equip police officers with the tools necessary to collect, store, and retrieve such physical evidence in DUI/DWI traffic stop situations.

Keywords: Detector, Drug Discovery
Application Code: Drug Discovery
Methodology Code: Mass Spectrometry

Session # 1240 Abstract # 1240-2 **Symposia**

Session Title The Challenge of Detection for Drugged Driving

Abstract Title **Human Cannabinoid Metabolism and Disposition in Biological Matrices after Controlled Cannabis Administration**

Primary Author Marilyn A. Huestis
NIDA

Date: Tuesday, March 08, 2016 - Afternoo
Time: 02:10 PM
Room: B311

Co-Author(s)

Abstract Text

[delta]9-Tetrahydrocannabinol (THC) is the primary psychoactive compound in cannabis or marijuana. THC is oxidized in the liver primarily to the equipotent metabolite, 11-hydroxy-THC (11-OH-THC), and further to the inactive metabolite, 11-nor-9-carboxy-THC (THCCOOH). All three compounds are glucuronidated to increase polarity and improve excretion of the compounds. Other important analytes present in the cannabis plant are cannabinol (CBN), about 10% as active as THC, cannabidiol (CBD), a cannabis component that appears to offer multiple therapeutic benefits without psychoactive effects, cannabigerol (CBG), [delta]9-tetrahydrocannabivarin (THCV), and [delta]9-tetrahydrocannabinolic acid (THCAA), the precursor of THC. These analytes may be quantified in blood, plasma, oral fluid, urine & breath and their measurement may improve interpretation of cannabinoid test results. Controlled cannabinoid administration in occasional and chronic frequent cannabis smokers provides scientific evidence based data for interpreting individual cannabinoid concentrations. The metabolism and disposition of cannabinoids in different biological specimens following smoked, inhaled and oral cannabis will be presented. These data are critical for developing rational drug policy and legislation, especially important today with the increasing medicalization and legalization of cannabis in many US states.

Keywords: Biological Samples, Clinical/Toxicology, Drugs, Mass Spectrometry
Application Code: Clinical/Toxicology
Methodology Code: Liquid Chromatography/Mass Spectrometry

Session # 1240 Abstract # 1240-3 **Symposia**

Session Title The Challenge of Detection for Drugged Driving

Abstract Title **Analytical Methods for the Detection of Marijuana in Biological Fluids in a Forensic Toxicology Laboratory**

Date: Tuesday, March 08, 2016 - Afternoo
Time: 02:45 PM
Room: B311

Primary Author Brianna Peterson
Washington State Patrol

Co-Author(s)

Abstract Text

Forensic toxicology laboratories use a variety of analytical techniques for the detection of marijuana use in biological fluids. The compounds of interest are [delta] 9-tetrahydrocannabinol (THC), 11-nor-9-carboxy-[delta]9-THC (THCCOOH), and 11-OH-THC. These methods can range from immunoassay to gas chromatography-mass spectrometry (GC/MS) or liquid chromatography-mass spectrometry (LC/MS). Immunoassay is typically used as a screening technique to rapidly identify the presence of the metabolite THCCOOH. Confirmation is followed utilizing a more sensitive and specific methodology. Derivatization is required for effective GS/MS techniques.

The toxicology laboratory in Washington State utilizes an immunoassay with a cutoff for cannabinoids of 10 ng/mL for blood samples. Confirmation analysis is performed using a liquid-liquid extraction followed by analysis on LC/MS/MS. The limit of detection for THC is 0.5 ng/mL and 2.5 ng/mL for THCCOOH. The dynamic range is 1 to100 ng/mL and 5 to 500 ng/mL for THC and THCCOOH, respectively.

In 2012, the Washington legislature legalized possession and use of recreational marijuana. In addition, the state's driving under the influence (DUI) statutes were amended to include a per se level of 5 ng/mL THC in whole blood for drivers aged 21 years and older. The prevalence of both THC and THCCOOH detected in such cases pre-legalization was compared to the prevalence post-legalization. In 2009-2012, the average percentage of cases positive for THC and THCCOOH was 19.1 % (range 18.2-20.2 %) and 27.9 % (range 26.3-28.6 %), respectively. By 2014, the percentages had increased to 28 % and 36.3%, respectively. The need for accurate and sensitive testing needs is imperative for enforcement of changing legislation of marijuana.

Keywords: Drugs, Liquid Chromatography, Tandem Mass Spec, Toxicology
Application Code: Clinical/Toxicology
Methodology Code: Liquid Chromatography/Mass Spectrometry

Session # 1240 **Abstract # 1240-4** **Symposia**

Session Title The Challenge of Detection for Drugged Driving

Abstract Title **Detection of Drug Consumption in Europe**

Date: Tuesday, March 08, 2016 - Afternoo
Time: 03:35 PM
Room: B311

Primary Author Wolfgang Vautz
ISAS

Co-Author(s)

Abstract Text

Still, Marihuana is one of the most frequently consumed drugs of abuse in Europe. In the days of a world-wide increasing legalisation of Marihuana consumption, therefore the need of detection methods e.g. for traffic control by the responsible authorities is also increasing. However, many other drugs like heroin, cocain or amphetamines are also in the focus since many years.

Presently, methods for the on-site detection of THC – the psych-active component in Marihuana – are not available. Several provides offer test via wiping tests with cotton swabs for thermal desorption analysis or urine tests. However, all the on-site tests non-specific and are subject to numerous false positives and negatives. The same is true for many other common illicit drugs. Therefore, an – in the ideal case analytical – method for the on-site persuasive evidence of Marihuana and other drug consumption – analogous to the alcohol test devices – is still lacking.

The presentation will give an overview on the methods for drug detection available and in use by the authorities in Europe and on the general procedure with suspicious individuals. Furthermore, the challenges of such an approach will be discussed and an overview on recent developments in this field will be given.

Keywords: Drugs, Forensics, Portable Instruments, Validation
Application Code: Safety
Methodology Code: Chemical Methods

Session # **1240** **Abstract #** **1240-5** **Symposia**

Session Title: The Challenge of Detection for Drugged Driving

Abstract Title: **Detection of Marijuana from Human Breath by Breathalyzer-IMS**

Date: Tuesday, March 08, 2016 - Afternoo
Time: 04:10 PM
Room: B311

Primary Author: Herbert H. Hill
Washington State University

Co-Author(s): Jessica A. Tufariello

Abstract Text

Since 2012 four states, Washington, Colorado, Oregon, and Alaska, along with D.C. have legalized recreational marijuana use. Though the use of marijuana is becoming more prevalent there is still no field detection method for law enforcement to use when identifying intoxicated individuals. Ion mobility spectrometry has been used in the field for decades by military and law enforcement for the detection of chemical warfare agents, explosives and narcotics. Breath samples have been an excepted medium for law enforcement officers to collect for identification of intoxication. In the past year the Hill lab has investigated a hand held breathalyzer-IMS system for the detection of □9- tetrahydrocannabinol (THC), the main psychoactive compound in marijuana, in human breath. Numerous system checks and calibrations with THC standards were performed. The instrumental response was tested with an array of clean human breath samples and human breath with possible contaminates. Finally two separate field tests were performed, one spanning from March to May and another from September to November of 2015. They included over 30 different human subjects who recreationally used marijuana. The breathalyzer-IMS was able to identify THC from human breath from numerous human subjects in the field studies, contaminates studied did not to interfere with THC identification, and the breathalyzer-IMS prototype was optimized for improved identification with less false positives.

Keywords: Detection, Drugs, Portable Instruments, Sampling

Application Code: Clinical/Toxicology

Methodology Code: Portable Instruments

Session # 1250 Abstract # 1250-1 Workshops

Session Title Core-Shell versus Fully Porous HPLC Particles – The Current State of the Art in HPLC Columns

Abstract Title **Perspectives on the Development and Future of Monodisperse Fully-Porous Silica Supports**

Date: Tuesday, March 08, 2016 - Afternoo
Time: 01:35 PM
Room: B313

Primary Author David S. Bell
Supelco/Sigma-Aldrich

Co-Author(s)

Abstract Text

Over the past 15 years there have been a number of exciting advances in liquid chromatography particle technology. Initially, the development of commercial sub-2 µm particles in 2004 (along with the instrumentation required to effectively use them) provided improved efficiencies over common 3 µm and 5 µm materials. Novel 2.7 µm superficially porous particles (SPP) were then introduced in 2007 that provided greater efficiencies than similarly sized fully porous particles (FPP) and were comparable in performance to sub-2 µm particles. Columns packed with SPP particles could thus be utilized in place of smaller particles to obtain similar efficiencies but without the burden of high backpressures. Although the SPP architecture was initially designed to improve mass transfer kinetics, further research has revealed that the increased efficiencies observed for small molecules has been largely due to improvements in both eddy diffusion and longitudinal diffusion. This improvement is, at least, partly due to much narrower particle size distribution resulting from the SPP construction process. These discoveries have led to a revival of research and development concerning the importance of monodisperse particle technologies for liquid chromatography. This presentation highlights recent efforts to understand the importance of low particle size distribution and offers perspective regarding the future of monodisperse (< 10% RSD) particle technology in separation sciences. Contrasts and comparisons are provided against SPP and sub-2 µm platforms.

Keywords: HPLC, Liquid Chromatography, Method Development
Application Code: General Interest
Methodology Code: Liquid Chromatography

Session # 1250 **Abstract # 1250-2** **Workshops**

Session Title: Core-Shell versus Fully Porous HPLC Particles – The Current State of the Art in HPLC Columns

Abstract Title: **Fully-Porous vs. Core-Shell Particles -- The Past, the Present and the Future**

Date: Tuesday, March 08, 2016 - Afternoo
Time: 02:05 PM
Room: B313

Primary Author: Lawrence Loo
Phenomenex

Co-Author(s): Jason Anspach, Mike Chitty, Thuylinh Tran, Tivadar Farkas

Abstract Text

As the demand of higher productivity and the complexity of HPLC analyses increase, the quest to higher resolving power and reduced analysis time led to the development of HPLC sorbent with incrementally smaller particle size in the past decades. The use of HPLC columns packed with sub-2 µm fully porous silica sorbent has been shown to provide the desired benefits, but it requires the use of dedicated UHPLC systems because of the back pressure it generates. In the past 5-7 years, a new development approach to the silica particle morphology has led to the commercialization of core-shell materials which offer higher efficiencies than fully porous silica particles of the same size. Many publications have been dedicated to the comparison between the fully porous particles and the core-shell particles in their performances in different applications, as well as potential issues such as mechanical strength, sample loading capacity, particle scalability, ease of method transfer and required system optimization. It is the purpose of this presentation to summarize such comparisons, as well as to provide a glimpse into the future potential of these two approaches to particle morphology in various modes of separation.

Keywords: Chromatography, HPLC, HPLC Columns
Application Code: High-Throughput Chemical Analysis
Methodology Code: Liquid Chromatography

Session # **1250** **Abstract #** **1250-3** **Workshops**

Session Title Core-Shell versus Fully Porous HPLC Particles – The Current State of the Art in HPLC Columns

Abstract Title **Advantages and Limitations of Superficially Porous Particles**

Date: Tuesday, March 08, 2016 - Afternoo
Time: 02:35 PM
Room: B313

Primary Author Ken Broeckhoven
Vrije Universiteit Brussel

Co-Author(s) Gert Desmet

Abstract Text

The introduction of a new generation of superficially porous particles with a thick porous layer, has resulted in a large jump in separation performance comparable to the introduction of ultra-high pressure (UHPLC) instrumentation. The reduction of the minimum reduced plate height h from roughly 2, for fully porous particles, down to 1.4-1.6 for superficially porous particles, theoretically allows for two-fold decrease in analysis time for a fixed efficiency. Initially the enhanced performance was marketed as the result of faster mass transfer due to the smaller diffusion distance in the relative thin porous shell (C-term). However, further investigation revealed that the advantage mainly arose from the smaller longitudinal diffusion (B-term) and strongly reduced eddy-dispersion (A-term) due to the much more homogeneous column packing. In addition, the flow resistance [Phi] (based on the velocity of an unretained compound) is also smaller for superficially porous particles (~650) than for their fully porous counterparts (~800). The advantages of the decrease in h and [Phi] can best be represented in a so-called kinetic performance limit plot, as it provides a plot of time versus efficiency, under the conditions of both optimized flow rate and column length. The progress made in performance due to the introduction of both superficially particles and UHPLC systems can thus easily be visualized. Unfortunately, current state-of-the-art UHPLC instrumentation doesn't allow harnessing the full separation power of the most performant due to excessive performance loss as a result of extra-column band broadening. Reducing the inner diameter of the connection tubing can alleviate this partially, but results in excessive extra-column pressure drop. Further improvements in separation efficiency, by e.g. the introduction of even higher operating pressure and/or smaller particle, therefore first requires the availability of instruments with smaller extra-column volumes (<0.5μL^2).

Keywords: HPLC, HPLC Columns, Liquid Chromatography, Particle Size and Distribution
Application Code: General Interest
Methodology Code: Liquid Chromatography

Session # 1250 Abstract # 1250-4 **Workshops**

Session Title Core-Shell versus Fully Porous HPLC Particles – The Current State of the Art in HPLC Columns

Abstract Title **Next Generation Superficially Porous Particle Technology: Highly Ordered Pore Structures Formed by Pseudomorphic Transformation**

Date: Tuesday, March 08, 2016 - Afternoo
Time: 03:20 PM
Room: B313

Primary Author William E. Barber
Agilent Technologies

Co-Author(s) Anne Mack, Jia Liu, Monika Dittmann, Ta-Chen Wei, Wu Chen, Xiaoli Wang

Abstract Text

In recent years, superficially porous particles have drawn great interest because of their special characteristics and improvement in separation efficiency. Superficially porous particles are currently manufactured by adding silica nanoparticles onto solid cores either using a multi-layer or one-step coacervation process. The pore size is mainly controlled by the size of the silica nanoparticles and the tortuous pore channel geometry is determined by how those nanoparticles randomly aggregate. In this presentation, we report on the development of a next generation of superficially porous particles with a unique pore structure including thinner shell thickness and ordered pore channels oriented normal to the particle surface. The columns packed with these new superficially porous particles have reduced plate heights as low as 1.0.

The method of making the new superficially porous particles is a process/technology called pseudomorphic transformation (PMT), which is a form of micelle-templating. The porosity is no longer controlled by randomly aggregated nanoparticles but rather by micelles, which have an ordered liquid crystal structure. The new particles possess many advantageous features such as a narrower particle size distribution, thinner porous layers and, most importantly, highly-ordered, non-tortuous pore channels oriented normal to the particle surface.

The impact of the novel morphology of these particles, i.e., the pore structure and orientation, on the intra-particle diffusion of analytes and van Deemter coefficients will be discussed. Reduced plate heights as low as 1.0 have been achieved, which will be compared to reduced plate heights achieved with conventional totally porous and superficially porous particles. Moreover, the same technology can be applied to hybrid cores to form superficially porous hybrid particles. This new column technology has great potential for fast separations over a wide pH range with ultra-high efficiency.

Keywords: HPLC Columns
Application Code: Material Science
Methodology Code: Liquid Chromatography

Session # 1250 Abstract # 1250-5 **Workshops**

Session Title Core-Shell versus Fully Porous HPLC Particles – The Current State of the Art in HPLC Columns

Abstract Title **Clarifying the Difference Between Columns Packed with Solid Core and Fully Porous Particles**

Date: Tuesday, March 08, 2016 - Afternoo
Time: 03:50 PM
Room: B313

Primary Author Jacob Fairchild
Waters Corporation

Co-Author(s) Babajide Okandeji, Bonnie Alden, Jonathan Turner, Kevin Wyndham

Abstract Text

In recent years, there has been escalated interest in solid-core particles due to their lower reduced plate heights and lower back pressures, as compared to conventional fully porous particles of equivalent size. However, modern fully porous particles still play an important role in liquid chromatography in both analytical and preparative separations. Choosing the right particle and bonding technology can be difficult due to the high number of products available to a consumer. The user must consider their flexibility in altering or improving a particular method and the instrumentation available. We will show how the enhanced efficiency of solid-core particles impacts chromatography and high performance instrumentation is necessary. The wide range of selectivities currently available for columns containing both solid-core and fully porous particles will be shown as well. These results show that offering both solid-core and modern fully porous particles allow for the scientist to meet all their chromatographic goals.

Keywords: HPLC Columns, Liquid Chromatography, Liquid Chromatography/Mass Spectroscopy
Application Code: Pharmaceutical
Methodology Code: Liquid Chromatography

Session # 1260	Abstract # 1260-1	Organized Contributed Sessions

Session Title: LIMS Live @ Pittcon: Best Practices and Lessons Learned From The Laboratory

Abstract Title: **Where Do We Start? A Roadmap to LIMS Success**

Date: Tuesday, March 08, 2016 - Afternoo
Time: 01:30 PM
Room: B315

Primary Author: Christine Paszko
Accelerated Technology Laboratories

Co-Author(s)

Abstract Text

Implementing a Laboratory Information Management System (LIMS) can be a challenging and complex endeavor fraught with opportunities to derail, delay or add additional costs to the project. Because a LIMS has become an essential component of a laboratory's everyday operation, ensuring a successful deployment takes on an elevated level of importance. And once the LIMS is up and running, it is essential that a laboratory take steps to fully utilize its capabilities to realize the typical business benefits that a LIMS brings to the laboratory including higher data quality, faster turnaround on analysis and lower operational costs.

Thousands of organizations have implemented LIMS solutions since the early 1980s. While many of these implementations have been successful, a significant number of these deployments turned into costly and time-consuming projects that did not meet expectations or deemed failures and forced companies to start their LIMS selection process from scratch. During this presentation we will provide a "roadmap" that can guide an organization through a successful LIMS implementation project. We will begin with the scope definition phase and wind our way through gathering functional requirements and product selection. We'll then move on to the implementation phase which includes configuration of the system and training for the LIMS administrator and users. Throughout the presentation there will be an emphasis on sharing real-world deployment knowledge that can help those organizations whose goal is to achieve a successful LIMS implementation.

Keywords: Data Analysis, Laboratory Automation, Laboratory Informatics, LIMS
Application Code: Laboratory Management
Methodology Code: Laboratory Informatics

Session # 1260 Abstract # 1260-2 **Organized Contributed Sessions**

Session Title: LIMS Live @ Pittcon: Best Practices and Lessons Learned From The Laboratory

Abstract Title: **The LIMS Needs Assessment: Your Secret Weapon to a Successful Deployment**

Primary Author: Alan Serrero
Gwinnett County Department of Public Utilities

Date: Tuesday, March 08, 2016 - Afternoo
Time: 01:50 PM
Room: B315

Co-Author(s)

Abstract Text

For organizations seeking a successful LIMS implementation some will choose a LIMS vendor that provides a consulting service known as a LIMS needs assessment. The needs assessment is ideal for organizations that are planning to invest in a LIMS and see value in bringing in an external expert to assist with defining the business and technical requirements. The needs assessment can provide the business justification and a valuable roadmap to a successful LIMS implementation. It can also help avoid potential pitfalls along the way.

In this presentation we will highlight an environmental laboratory in Georgia who went from using a manual method of managing laboratory information to a LIMS. We'll focus on the journey this organization made to the LIMS and emphasize the following presentation topics:

•The benefits that the LIMS has brought to the laboratory over the previous solution.
•The role that the needs assessment played in the implementation strategy.
•The benefits that the needs assessment provided in ensuring a successful implementation.
•The reasons why this laboratory would recommend a needs assessment as part of the Request for Proposal (RFP) process and why the needs assessment should be performed by an independent LIMS consultant vs. an in-house resource.

Keywords: Data Analysis, Laboratory Automation, Laboratory Informatics, LIMS
Application Code: Laboratory Management
Methodology Code: Laboratory Informatics

Session # 1260 Abstract # 1260-3 **Organized Contributed Sessions**

Session Title LIMS Live @ Pittcon: Best Practices and Lessons Learned From The Laboratory

Abstract Title **Finding the Perfect LIMS: Keys to a Successful RFP**

Primary Author Aster Tekle
Alexandria Renew Enterprises

Date: Tuesday, March 08, 2016 - Afternoo
Time: 02:10 PM
Room: B315

Co-Author(s)

Abstract Text

A Laboratory Information Management System (LIMS) has become a critical tool for many organizations so the process of selecting the right LIMS should not be considered a simple task – the wrong choice could result in a failed project that costs time and money. For many organizations the process for selecting a LIMS is accomplished through a Request for Proposal (RFP). An RFP is a clear and detailed way to supply information on how the LIMS solution will meet the requirements of the laboratory. A well-designed RFP can be used as a meaningful tool that allows the laboratory to compare the capabilities of competing LIMS solutions based on the laboratory's functional requirements.

In this presentation we will focus on the proven fundamentals of writing a successful RFP. We'll focus on the experience of one organization based in the eastern US who had an existing LIMS that no longer was able to meet the lab's requirements. We will walk through the process of creating the RFP beginning with development of the business case and the functional requirements. We will talk about the structure of the RFP and the importance of allowing for easy and objective comparison between vendor responses. Once the RFP is written and responses are received we'll then focus on the process of evaluating responses and coming up with a scoring system that results in selecting a vendor. We'll share best practices and lessons learned from the RFP experience in the spirit of passing along valuable knowledge that will help Pittcon attendees who are currently evaluating LIMS solutions.

Keywords: Data Analysis, Laboratory Automation, Laboratory Informatics, LIMS
Application Code: Laboratory Management
Methodology Code: Laboratory Informatics

Session # 1260	Abstract # 1260-4	Organized Contributed Sessions

Session Title LIMS Live @ Pittcon: Best Practices and Lessons Learned From The Laboratory

Abstract Title **Preparing for a LIMS – The Importance of Proper Planning**

Primary Author Keith Keesee
Oklahoma Department of Agriculture

Date: Tuesday, March 08, 2016 - Afternoo
Time: 02:30 PM
Room: B315

Co-Author(s) Cassandra Kontas

Abstract Text

A pre-planning stage is inherent to the success of implementing a LIMS. Well-defined business practices, workflows, quality assurance criteria, and analytical testing procedures should be in place and documented prior to evaluating, acquiring, or implementing a LIMS to best streamline the process. A common problem when business practices and workflows are not evaluated for potential increases in efficiency is an attempt to recreate a legacy system. Current personnel may not even be aware what practices have been instituted to "work-around" a deficient legacy system. When laboratories do not have consistent and established procedures in place for analytical testing and quality assurance/control, articulating functional requirements for a LIMS becomes cumbersome to impossible. Attempts to implement a LIMS in tandem with establishing testing and quality requirements will inherently increase the project timeline due to multiple re-configurations or re-customizations.

In this presentation, the focus will be on the importance of proper planning while also pointing out some of the common pitfalls that can impact a typical implementation including failure to meet project deadlines, inability to achieve functional goals to increase productivity or experiencing significant cost overruns.

Keywords: Data Analysis, Laboratory Automation, Laboratory Informatics, LIMS
Application Code: Laboratory Management
Methodology Code: Laboratory Informatics

Session # 1260 Abstract # 1260-5 **Organized Contributed Sessions**

Session Title LIMS Live @ Pittcon: Best Practices and Lessons Learned From The Laboratory

Abstract Title **Implementing a LIMS is a Project – Treat It Like One**

Primary Author Roy D. Jones
Duke Energy

Co-Author(s)

Date: Tuesday, March 08, 2016 - Afternoo
Time: 03:05 PM
Room: B315

Abstract Text

Implementing a LIMS is a major endeavor for most organizations and usually requires a major capital expenditure along with a significant amount of time and resources to getting it deployed. For that reason, it is important to give the LIMS project the respect it deserves in terms of planning and execution. This presentation will provide real-world experiences and lessons learned throughout the LIMS implementation lifecycle. The advice given during the presentation will focus on:

- The importance of implementing a NEW LIMS vs. trying to re-implement to previous solution.
- Managing personalities on the project team – allocate people based on their strengths.
- People will be motivated to use the LIMS if you show them features that will make their job easier.
- Need to set management expectations – they want a one-button solution but need to be brought back to reality and also be told that a fully deployed LIMS may take 6 months to a year and possibly longer.
- Importance of using proven project management fundamentals throughout the implementation.
- Choosing a LIMS vendor who has expertise in your industry will pay dividends during the deployment.

Attendees will walk away from this presentation with proven tips and suggestions that will help ensure that their LIMS project is a success.

Keywords: Data Analysis, Laboratory Automation, Laboratory Informatics, LIMS
Application Code: Laboratory Management
Methodology Code: Laboratory Informatics

Session # 1260 Abstract # 1260-6 **Organized Contributed Sessions**

Session Title LIMS Live @ Pittcon: Best Practices and Lessons Learned From The Laboratory

Abstract Title **Life with a LIMS: What It's Meant for the City of Clearwater**

Date: Tuesday, March 08, 2016 - Afternoo
Time: 03:25 PM
Room: B315

Primary Author Maria de la Cantera
City of Clearwater (FL)

Co-Author(s)

Abstract Text

The Water Pollution Control Laboratory at the City of Clearwater, FL provides the community of 108,000 with safe drinking water and supports a proactive wastewater collection and treatment system. The WPC Laboratory tests drinking water, wastewater and solid materials for three wastewater plants and the reverse osmosis plant, performing approximately 52,000 tests annually. In addition, the lab performs quality control and regulatory compliance analysis and monthly bacteriological analyses are run for the community drinking water system.

The lab previously conducted their testing using bench sheets and a very labor-intensive manual entry process that was prone to data transcription errors. Eventually the lab was able to justify the purchase of a LIMS and the impact and the benefits to the organization and their residents has been enormous.

This presentation will focus on the many benefits that have been realized due to the implementation of the LIMS including the following:

• A significant increase in productivity throughout the laboratory. In many cases, data is entered only once, eliminating transcription errors due to entering data in multiple documents.
• Significantly reduced turnaround times for data analysis and reporting – end result are happy clients who get their reports quickly.
• Clients can actually access their reports 24/7 via access to a secure web-based portal. This solution not only enhances customer service but also allows lab staff to focus on analysis and not having to respond to simple customer requests for information.
• Ability to schedule tasks increases the efficiency of the lab operation by eliminating the need to perform these repetitive tasks.

Keywords: Data Analysis, Laboratory Automation, Laboratory Informatics, LIMS
Application Code: Laboratory Management
Methodology Code: Laboratory Informatics

Session # 1260 Abstract # 1260-7 **Organized Contributed Sessions**

Session Title: LIMS Live @ Pittcon: Best Practices and Lessons Learned From The Laboratory

Abstract Title: **Leveraging LIMS for Streamlining Next Generation Sequencing Data**

Date: Tuesday, March 08, 2016 - Afternoo
Time: 03:45 PM
Room: B315

Primary Author: Jennifer Weller
University of North Carolina - Charlotte

Co-Author(s)

Abstract Text

Managing various protocols and documenting all associated information in non-relational databases can be very challenging. Today's Next Generation Sequencing requires researchers manage numerous workflows to match their business rules, SOPs (Standard Operating Procedures), reagents, kits and all consumables, instrument data, analyst information, and much more. The ability to have all of the data in a powerful LIMS that leverages user configurable workflows with a user-friendly workflow designer, graphical reporting tool that can be used by non-IT staff and the ability to easily link data files and documents.

The presentation will review basic workflows of the illumina Next Gen Sequencing Platform. Key automation features of the presentation wil include the integration with barcoding, instruments, user defined workflows, graphical scheduling, automatic result flagging, full QA/QC support, integration with the statistical software SAS JMP, chain of custody, batch management, project management, data security, and regulatory compliance.

Keywords: Data Analysis, Laboratory Automation, Laboratory Informatics, LIMS
Application Code: Laboratory Management
Methodology Code: Laboratory Informatics

Session # 1260 Abstract # 1260-8 **Organized Contributed Sessions**

Session Title LIMS Live @ Pittcon: Best Practices and Lessons Learned From The Laboratory

Abstract Title **What Have We Learned? Final Thoughts On the Road to LIMS Success (Getting the Most From Your LIMS)**

Primary Author Devender Gandhi
Accelerated Technology Laboratories

Date: Tuesday, March 08, 2016 - Afternoo
Time: 04:05 PM
Room: B315

Co-Author(s)

Abstract Text

In this presentation, we will summarize the most important thoughts, ideas, best practices and lessons learned during this Organized Contributed Session. We'll also introduce additional resources that attendees can use as they return to their laboratories and look to enhancing their operation with the valuable knowledge they've learned here. Hopefully we will accomplish our goal of sharing real world experience and expertise when it comes to selecting, implementing and maintaining a LIMS. This wrapup will highlight the key learnings and send the audience out with new found wisdom that will help ensure LIMS success at their organizations. We will bring the speakers up to offer final thoughts and possibly take audience questions if time permits.

Keywords: Data Analysis, Laboratory Automation, Laboratory Informatics, LIMS
Application Code: Laboratory Management
Methodology Code: Laboratory Informatics

Session # 1270 Abstract # 1270-1 **Organized Contributed Sessions**

Session Title Quantifying the Tumor Microenvironment

Abstract Title **Phosphoproteomics in Prostate Extracellular Vesicles**

Date: Tuesday, March 08, 2016 - Afternoo

Time: 01:30 PM

Room: B316

Primary Author W Andy Tao
Purdue University

Co-Author(s)

Abstract Text

Clinically useful protein biomarkers can provide early diagnosis for effective treatments, improved prognosis for patient monitoring, and preventative screening that can identify patients at highest risk and subsequently offer intervention. While phosphoproteins are connected to onset and progression of numerous diseases, few phosphoproteins have been developed for biomarkers from biofluids.

Recent developments in biofluid analysis highlight the importance of cell-secreted extracellular vesicles (EVs). These typically include smaller size exosomes derived from multivesicular endosome-based secretions, and microvesicles derived from plasma membrane. The EVs are an effective and ubiquitous method for intercellular communication and removal of excess materials. As these are shed into virtually every type of biological fluid, and embody a good representation of their parent cells, analysis of the EV cargo has a great potential for biomarker discovery and disease diagnosis. We report an integrated platform technology to effectively isolate, enrich and detect phosphoproteins from biological matrices for proteomic analyses. EVs from control mice and patient-derived LuCaP35CR tumors xenograft mice were isolated for quantitative phosphorylation analyses.

Keywords: Mass Spectrometry, Proteomics

Application Code: Genomics, Proteomics and Other 'Omics

Methodology Code: Mass Spectrometry

Session # 1270 Abstract # 1270-2 **Organized Contributed Sessions**

Session Title Quantifying the Tumor Microenvironment

Abstract Title **Exploring the Permissive Stromal Microenvironment**

Primary Author Amanda B. Hummon
University of Notre Dame

Date: Tuesday, March 08, 2016 - Afternoo
Time: 01:50 PM
Room: B316

Co-Author(s) Eric M. Weaver, Pinar Zorlutuna

Abstract Text

Metastasis is the most formidable problem in cancer biology. In metastasis, a cell breaks away from the primary tumor and successfully invades the host stroma. To understand how the metastatic niche contributes to the success of a metastasizing cell, realistic in vitro models are needed. Animals can provide valuable models of the metastatic process, but it is difficult to capture the precise beginning of metastasis and to modulate the chemical and physical environment within the metastatic niche. In contrast, 3D cell cultures are easy to manipulate and they capture some of the spatial and molecular complexity of tumors. We are introducing 3D cell cultures into an in vitro microfabricated tissue engineered model to explore what constitutes a permissive stromal environment. Our stromal microenvironment includes co-cultures of multiple cell types in a tunable hydrogel to mimic the colon tumor microenvironment. 3D colon cancer cell cultures are introduced and chemically triggered to cause cells to detach and migrate. We are applying imaging mass spectrometry protocols developed in our laboratory to image the chemical and spatial gradients that are present in the stromal microenvironment following invasion. Additional samples will be processed for molecular profiling by nano-liquid chromatography tandem mass spectrometry to identify and quantify specific molecules present in the stromal niche. This platform represents an ideal testbed to evaluate the molecular and cellular components of a permissive stromal microenvironment.

Keywords: Biological Samples, Imaging, Mass Spectrometry, Proteomics
Application Code: Bioanalytical
Methodology Code: Mass Spectrometry

Session # 1270 Abstract # 1270-3 **Organized Contributed Sessions**

Session Title: Quantifying the Tumor Microenvironment

Abstract Title: **Modulating Drug Resistance in Hypoxia Tumors**

Primary Author: Dimitri Pappas
Texas Tech University

Date: Tuesday, March 08, 2016 - Afternoo
Time: 02:10 PM
Room: B316

Co-Author(s)

Abstract Text

The role of hypoxia in tumors is poorly understood, and serves as a therapeutic target from multiple approaches. In this talk we detail a microfluidic system for cancer cell and tumor aggregate hypoxia, and investigate changes in drug resistance due to rapid adaptation to hypoxia. We have shown that tumor cells respond to decreased oxygen within 30 minutes and exhibit remarkable drug resistance after this time. In this talk, we will discuss our latest efforts to target this mechanism as a method to restore drug susceptibility.

Keywords: Bioanalytical, Biomedical, Biotechnology, Lab-on-a-Chip/Microfluidics
Application Code: Biomedical
Methodology Code: Microfluidics/Lab-on-a-Chip

Session # 1270 Abstract # 1270-4 **Organized Contributed Sessions**

Session Title: Quantifying the Tumor Microenvironment

Abstract Title: **Paper-Based Assays for the Study of Cancer Cell Biology, Invasion, and Metastasis**

Primary Author: Matthew R. Lockett
University of North Carolina at Chapel Hill

Date: Tuesday, March 08, 2016 - Afternoo
Time: 02:30 PM
Room: B316

Co-Author(s)

Abstract Text

The tumor microenvironment contains a heterogeneous population of cancer cells and other cell types; these cells respond to environmental and physical constraints through cell-cell and cell-matrix interactions as well as through diffusion-based signaling. Microfluidic devices offer precise experimental control for assembling and maintaining a 3D culture, but have not been widely adopted by the biology community because they require specialized equipment, engineering expertise, and patience for experimental success. We are actively developing a three-dimensional culture system that utilizes paper-based scaffolds to engineer tissue-like constructs whose microenvironment can be easily controlled and readily measured. Sheets of paper are an ideal scaffold because they are commercially available, inexpensive, and easy to process and sterilize. Tissue-like cultures can be prepared by stacking the individual scaffolds, which are seeded with a particular cell type or extracellular matrix. We have developed a paper-based invasion assay to monitor the invasiveness of different cell types under experimentally controlled environments and chemical gradients. These cultures afford a system in which we can easily address questions about the role of the tissue microenvironment on cellular phenotype; we are particularly interested in answering questions about the role of oxygen in altering the proliferation, invasiveness, and chemotherapeutic sensitivity of endocrine-related cancers. Tumorigenic cells, particularly those found in the breast and prostate, express higher levels of hypoxia inducible factors than normal tissue, and there is a clear relationship between hypoxia, chemotherapy resistance, and poor prognosis in patients. By engineering cultures containing breast or prostate tumorigenic cells—of increasing complexity containing normal epithelial, endothelial, and other stromal cells—we aim to develop an in vitro culture system that can predict cellular responses in vivo.

Keywords: Bioanalytical, Biomedical, Biotechnology, Lab-on-a-Chip/Microfluidics
Application Code: Bioanalytical
Methodology Code: Microfluidics/Lab-on-a-Chip

Session # 1270 Abstract # 1270-5 **Organized Contributed Sessions**

Session Title Quantifying the Tumor Microenvironment

Abstract Title **Examining Small Molecule Cellular Signaling in Complex Environments with Microscale Systems**

Primary Author Ashleigh B. Theberge
University of Washington

Date: Tuesday, March 08, 2016 - Afternoo
Time: 03:05 PM
Room: B316

Co-Author(s)

Abstract Text

Small molecule signals provide a rich vocabulary for cellular communication. To better understand these signaling processes in both normal and disease states, there is a growing need for cell culture platforms that interface with analytical chemistry tools and recreate key features of the in vivo cell microenvironment. We have developed new microfluidic methods for cell culture that (1) integrate organic solvents for small molecule extraction enabling downstream metabolomic analysis and (2) accommodate the culture of multiple cell types in microfabricated compartments connected via channels to facilitate soluble factor signaling. Furthermore, our microscale culture systems allow a 50- to 1000-fold reduction in volume compared to conventional assays, enabling experiments with limited cells from patient samples. We used these platforms to study steroid synthesis in adrenal cells and oxylipin production in immune cells, demonstrating the ability to vary culture conditions in microchannels, perform on-chip liquid-liquid extraction, and quantify secreted compounds using liquid chromatography-mass spectrometry. To establish links between metabolomic profiles and biological function, we engineered functional readouts, such as assays for blood vessel formation, into the culture systems. These microfluidic models show great potential for disentangling complex relationships among cell types and understanding the chemistry responsible for these interactions.

Keywords: Bioanalytical, Liquid Chromatography/Mass Spectroscopy
Application Code: Biomedical
Methodology Code: Microfluidics/Lab-on-a-Chip

Session # 1270 Abstract # 1270-6 **Organized Contributed Sessions**

Session Title: Quantifying the Tumor Microenvironment

Abstract Title: **Microengineered Physiological Biomimicry: Human Organ-on-Chips**

Primary Author: D Dan Huh
University of Pennsylvania

Date: Tuesday, March 08, 2016 - Afternoo
Time: 03:25 PM
Room: B316

Co-Author(s)

Abstract Text

Human organs are complex living systems in which specialized cells and tissues are assembled in various proportions and patterns to carry out integrated functions essential to the survival of the entire organism. Lack of reliable model systems that recapitulate the complexity of living human organs poses major technical challenges in virtually all areas of life science and technology. This talk will present interdisciplinary research efforts focused on leveraging unique capabilities of microfluidics and microfabrication to develop microengineered biomimetic models that reconstitute complex structures, dynamic microenvironments, and physiological functionality of human organs. Specifically, I will talk about i) a bioinspired microsystem that mimics the structural and functional complexity of the alveolar-capillary interface in the living human lung, ii) a specialized in vitro human disease model that simulates pulmonary edema, and iii) a microengineered model of the ocular surface in the human eye.

Keywords: Biomedical, Biopharmaceutical, Environmental Analysis, Lab-on-a-Chip/Microfluidics
Application Code: Biomedical
Methodology Code: Microfluidics/Lab-on-a-Chip

Session # **1270** **Abstract #** **1270-7** **Organized Contributed Sessions**

Session Title: Quantifying the Tumor Microenvironment

Abstract Title: **Direct Optical Microscopy of Biological Interfaces**

Date: Tuesday, March 08, 2016 - Afternoo
Time: 03:45 PM
Room: B316

Primary Author: Charles R. Mace
Tufts University

Co-Author(s): Daniel J. Wilson, Irene Lui, Jenna A. Walz

Abstract Text

We have developed a microscope that allows for the direct observation of cell-substrate interactions. Using our imaging system, cell adhesion processes can be imaged in real-time and on any substrate—transparent, opaque, coated, or topographically patterned—without requiring provisions for labeling or specialized optical components. We have demonstrated the use of our microscope by studying the dynamic changes in cell morphology that occur upon the initial contact of a cell with a surface and using measurements of cell morphology (e.g., the contact angle) to characterize adhesion interactions. By measuring the rate of change in the contact angles of cells, we have established a means to quantitatively characterize cell morphology during adhesion, as well as a means to describe materials according to their ability to promote or resist adhesion. Using this approach, we are able to (i) rapidly differentiate invasive and non-invasive cancer cells, (ii) characterize the inherent heterogeneity within a population of cells at the resolution of a single cell, and (iii) identify new mechanisms that cancer cells use to recognize and adhere to surfaces.

Keywords: Bioanalytical, Biological Samples, Biomedical, Microscopy
Application Code: Bioanalytical
Methodology Code: Microscopy

Session # 1270 Abstract # 1270-8 **Organized Contributed Sessions**

Session Title Quantifying the Tumor Microenvironment

Abstract Title **Deconvolving Glycans in Metastasis**

Primary Author Lara K. Mahal
New York University

Date: Tuesday, March 08, 2016 - Afternoo
Time: 04:05 PM
Room: B316

Co-Author(s)

Abstract Text

Carbohydrates, which are altered in tumors, are involved in immune evasion, homing of cells to tissues, survival, and anchorage. Glycans are dendrimeric and epimeric structures and require an intricate biosynthetic pathway whose regulation is poorly understood. This talk will focus on using systems-based approaches, including lectin microarray technology and our microRNA proxy approach to map glycosylation in tumor metastasis.

Keywords: Bioanalytical, Bioinformatics, Carbohydrates, Genomics
Application Code: Genomics, Proteomics and Other 'Omics
Methodology Code: Microfluidics/Lab-on-a-Chip

Session # 1280 Abstract # 1280-1 **Organized Contributed Sessions**

Session Title SEAC - The Student Session in Electroanalysis

Abstract Title **Developing an Innovative Bio-Inspired Scanning Probe Microscopy (Bio-SPM) Approach to Map Specific Molecular Flux**

Primary Author Florika C. Macazo
University of Maryland Baltimore County

Date: Tuesday, March 08, 2016 - Afternoo
Time: 01:30 PM
Room: B401

Co-Author(s) Ryan J. White

Abstract Text

Successful reports on the utility of nanopore sensing and scanning ion conductance microscopy (SICM) in studying physiological systems are great motivations to pursue studies involving coupling of these two techniques to encompass broader applications. In this study, we combined the imaging ability of SICM with the sensitivity of protein nanopore sensing to develop a new, bio-inspired scanning ion conductance microscopy (bio-SICM) technique capable of quantitatively mapping specific molecular flux across membranes. We established the framework of this analytical platform using [alpha]-hemolysin ([alpha]HL) as a representative protein nanopore to map [beta]-cyclodextrin ([beta]CD) flux across a synthetic membrane. We fabricated [alpha]HL-based probes and used an in-house SPM to generate approach curves employing the distance-dependent current response as feedback. Continuous monitoring of the current fluctuations suggests successful detection of [beta]CD binding events as evidenced by typical current blockades (60 – 80%) for a single [alpha]HL channel inserted into a lipid bilayer. To demonstrate molecular flux imaging, we raster-scanned the [alpha]HL-based probe over a 25-[micro]m pore glass substrate, while recording the lateral positions and current fluctuations caused by the [beta]CD flux. Specifically, we utilized the frequency of [beta]CD single binding events occurring at a single [alpha]HL pore to generate a concentration profile of the [beta]CD flux across the membrane. When further optimized, we believe that this will provide a simple analytical methodology that is generalizable, which will lay the groundwork for pursuing other molecular flux-related studies especially in the areas of neuroscience and biology (e.g. mapping ATP flux from astrocyte cells).

Keywords: Biosensors, Electrochemistry, Imaging, Nanotechnology
Application Code: Bioanalytical
Methodology Code: Sensors

Session # 1280 Abstract # 1280-2

Organized Contributed Sessions

Session Title SEAC - The Student Session in Electroanalysis

Abstract Title **Electrodeposition of Semiconductor Thin Films Using Electrochemical Liquid-Liquid-Solid (ec-LLS) Deposition**

Date: Tuesday, March 08, 2016 - Afternoo
Time: 01:50 PM
Room: B401

Primary Author Joshua DeMuth
University of Michigan

Co-Author(s)

Abstract Text

Depositing thin films of group IV and III-V semiconductors from aqueous solutions has remained a difficult task in the field of electrodeposition. However, a new hybrid technique that combines aspects of electrodeposition and solution-based crystal growth, termed electrochemical liquid-liquid-solid (ec-LLS) deposition, has been developed for growing crystalline semiconductors from aqueous solutions. [1] In ec-LLS a potential is applied to a liquid metal electrode in order to reduce aqueous precursor ions. Dissolution into the liquid metal causes a supersaturation of the reduced precursor to develop which then leads to the growth of crystalline material. This presentation entails the use of ec-LLS to electrodeposit thin films of crystalline semiconductors. Previously, heterogeneous crystal growth has been demonstrated through ec-LLS on the nano- and microscale (forming nano/micro-wires, respectfully) using droplets of liquid metal as the electrode. [2, 3] However, a method to adapt ec-LLS for the heterogeneous growth of films has remained elusive. In this study attempts to amend ec-LLS growth to produce large area thin films of crystalline germanium from aqueous electrolyte are described using a custom electrochemical cell that enables the use of a thin film liquid metal electrode. The films are characterized through X-ray diffraction techniques and scanning electron microscopy.

References

1. Carim, A. I.; Collins, S. M.; Foley, J. M.; Maldonado, S., Journal of the American Chemical Society 2011, 133 (34), 13292-13295.
2. Fahrenkrug, E.; Biehl, J.; Maldonado, S., Chemistry of Materials 2015, 27 (9), 3389-3396.
3. Fahrenkrug, E.; Gu, J.; Jeon, S.; Veneman, P. A.; Goldman, R. S.; Maldonado, S., . Nano Letters 2014, 14 (2), 847-852.

Keywords: Electrochemistry, Material Science, Semiconductor
Application Code: Material Science
Methodology Code: Electrochemistry

Session # 1280 Abstract # 1280-3 **Organized Contributed Sessions**

Session Title SEAC - The Student Session in Electroanalysis

Abstract Title **Alternating Current Potentiometric Scanning Ion Conductance Microscopy (AC-PSICM)**

Primary Author Lushan Zhou
Indiana University Bloomington

Date: Tuesday, March 08, 2016 - Afternoo
Time: 02:10 PM
Room: B401

Co-Author(s) Lane A. Baker, Wenqing Shi, Yi Zhou

Abstract Text

Alternating current potentiometric scanning ion conductance microscopy (AC-PSICM) is a new tool that can be utilized to study ion transport at small length scales and at wide ranges of time scales. AC-PSCIM makes use of a nanopipet to measure the local potential over different features within a sample as an AC perturbation is applied across the sample. Phase and amplitude of the measured local potential deflections were analyzed for a range of frequencies (5 Hz – 50 kHz) with nanoporous membranes to detail the basics of this method. Measurements obtained with a single nanopore membrane demonstrated that phase is sensitive to local conductive pathways (nanopore) and can be fitted to an equivalent electrical circuit to yield the single nanopore resistance. The use of phase to differentiate heterogeneous conductive pathways within a multiple-pore membrane was also demonstrated and was found to be dependent on frequency. Therefore a "frequency selection rule" was developed by studying phase approach curves over a single nanopore and lateral phase distribution at a constant height around a single nanopore at various frequencies. Four distinct frequency ranges for resolving heterogeneous nanopores were defined and were confirmed with phase line profile measurements in membranes with different sized nanopores. AC-PSICM shows promise for a phase mapping technique, which can be utilized to visualize heterogeneity of conductive pathways in biological systems such as cell layers and tissues.

Keywords: Bioanalytical, Microscopy, Nanotechnology, Quantitative
Application Code: Bioanalytical
Methodology Code: Surface Analysis/Imaging

PITTCON Conference and Expo 2016

Abstracts

Atlanta, Georgia, USA
6 - 10 March 2016

Volume 2 of 2

ISBN: 978-1-5108-2394-5

Printed from e-media with permission by:

Curran Associates, Inc.
57 Morehouse Lane
Red Hook, NY 12571

Printed by Curran Associates, Inc. (2016)

For permission requests, please contact Pittsburgh Conference
at the address below.

Pittsburgh Conference
300 Penn Center Boulevard
Suite 332
Pittsburgh, PA 15235-5503
USA

Phone: (412) 825-3220
(800) 825-3221
Fax: (412) 825-3224

info@pittcon.org

Additional copies of this publication are available from:

Curran Associates, Inc.
57 Morehouse Lane
Red Hook, NY 12571 USA
Phone: 845-758-0400
Fax: 845-758-2633
Email: curran@proceedings.com
Web: www.proceedings.com

SUNDAY, MARCH 6, 2016
AFTERNOON

The Wallace H. Coulter Lecture — Session 10

The Wallace H. Coulter Lecture

Sunday Afternoon, Sidney Marcus Auditorium, Bldg A, Level 4

5:00	(10-1)	**How Optical Single-Molecule Detection in Solids Led to Super-Resolution Nanoscopy in Cells and Beyond** W.E. (WILLIAM ESCO) MOERNER, Stanford University

AWARDS — Session 20

The Pittcon Heritage Award

Sunday Afternoon, Sidney Marcus Auditorium, Bldg A, Level 4

4:45		**Presentation of the 2016 Pittcon Heritage Award posthumously awarded to Kenji Kazato and Kazuo Ito, founders of JEOL - accepted by Gon-emon Kurihara, President of JEOL**

SYMPOSIUM — Session 30

ACS-ANYL - Tracing the Metabolome: Application of Stable-Isotope Tracers in Bioanalytical Chemistry

arranged by Cynthia K Larive, University of California-Riverside and Gregory A Barding, California State Polytechnic University

Sunday Afternoon, Room B308

Cynthia K Larive, University of California-Riverside, Presiding

1:30		**Introductory Remarks - Cynthia K Larive and Gregory A Barding**
1:35	(30-1)	**Stable Isotope Labeling and UHPLC/MS Strategies for Probing Dynamics of Plant Specialized Metabolism** A DANIEL JONES, Michigan State University, Zhenzhen Wang, Xiaoxiao Liu, Banibrata Ghosh
2:10	(30-2)	**Developments in ^{13}C-based Metabolomics** ARTHUR S EDISON, University of Georgia
2:45	(30-3)	**Chemical Isotope Labeling LC-MS for Quantitative Metabolomics with High Metabolomic Coverage** LIANG LI, University of Alberta
3:20		**Recess**
3:35	(30-4)	**Lipid Metabolic Pro°°filing Using Stimulated Raman Scattering Microscopy** MENG WANG, Baylor College of Medicine
4:10	(30-5)	**Monitoring the Incorporation of Stable-Isotope Labeled Alcohols by Bacteria with GC-MS** GREGORY A BARDING, California State Polytechnic University, Rakesh Mogul, Nicole G Perkins

SYMPOSIUM — Session 40

Emerging Leaders in Biological Mass Spectrometry

arranged by Amanda B Hummon, University of Notre Dame and Heather R Desaire, University of Kansas

Sunday Afternoon, Room B302

Amanda B Hummon, University of Notre Dame, Presiding

1:30		**Introductory Remarks - Amanda B Hummon and Heather R Desaire**
1:35	(40-1)	**New Mass Spectrometry Methods Show HIV Vaccine Candidates' Protein Structures are Misfolded** HEATHER R DESAIRE, University of Kansas
2:10	(40-2)	**Novel Strategies in Top-down Proteomics** YING GE, University of Wisconsin Madison
2:45	(40-3)	**Multi-Tier Approach to Understand the Biology of Alzheimer's Disease** RENÃ ROBINSON, University of Pittsburgh
3:20		**Recess**
3:35	(40-4)	**Old Photochemistry Brings New Capabilities in Unsaturated Lipid Analysis** YU XIA, Purdue University, Zheng Ouyang, Xiaoxiao Ma, Craig Stinson
4:10	(40-5)	**Evaluating Small Molecule Histone Inhibitors with High Resolution Mass Spectrometry and 3D Cell Cultures** AMANDA B HUMMON, University of Notre Dame, Simone Sidoli, Peter E Feist, Monica M Schroll, Benjamin A Garcia

SYMPOSIUM — Session 50

Enabling Sample Preconcentration Methods for Bioanalysis

arranged by Adam T Woolley, Brigham Young University

Sunday Afternoon, Room B304

Adam T Woolley, Brigham Young University, Presiding

1:30		**Introductory Remarks - Adam T Woolley**
1:35	(50-1)	**Bottom-up Proteomics of E.coli Using Dynamic pH Junction Preconcentration and CZE-ESI-MS/MS** NORMAN J DOVICHI, University of Notre Dame, Guijie Zhu, Liangliang Sun
2:10	(50-2)	**Electrokinetic Sample Preconcentration and Hydrodynamic Sample Injection for Capillary Electrophoresis Using a Pneumatic Microvalve** RYAN T KELLY, Pacific Northwest National Laboratory, Yongzheng Cong, Katipamula Shanta, Tang Keqi
2:45	(50-3)	**Solid Phase Microextraction in Bioanalysis** JANUSZ PAWLISZYN, University of Waterloo
3:20		**Recess**
3:35	(50-4)	**Microfluidic Integration of Solid-Phase Extraction with Fluorescence Labeling for Microfluidic Analysis** ADAM T WOOLLEY, Brigham Young University, Suresh Kumar, Vishal Sahore, Radim Knob, Mukul Sonker
4:10	(50-5)	**Liquid Biopsies: Microfluidic Enabling the Clinical Utility of These Markers** STEVEN A SOPER, University of North Carolina

SYMPOSIUM — Session 60

Frontiers of In Situ and In Vivo Spectroscopic Imaging

arranged by Ji-Xin Cheng and Mikhail Slipchenko, Purdue University

Sunday Afternoon, Room B305

Ji-Xin Cheng, Purdue University, Presiding

1:30		**Introductory Remarks - Ji-Xin Cheng and Mikhail Slipchenko**
1:35	(60-1)	**Correlated Raman and Mass Spectrometric Chemical Imaging of Multiscale Spatiotemporal Signaling in Microbial Communities** PAUL W BOHN, University of Notre Dame, Nameera Baig, Nydia Morales-Soto, Sage B Dunham, Jonathan V Sweedler, Joshua D Shrout, Sneha Polisetti
2:10	(60-2)	**Nanotechnology for In-Vivo and Intraoperative Cancer Detection and Image-Guided Surgery** SHUMING NIE, Emory University
2:45	(60-3)	**Nonlinear Microscopy to Detect and Grade Cancer** WARREN S WARREN, Duke University
3:20		**Recess**
3:35	(60-4)	**Multi-Parametric and Multi-Spectral Photoacoustic Microscopy** SONG HU, University of Virginia
4:10	(60-5)	**Vibrational Spectroscopic Imaging of Molecular Dynamics in Living Systems** JI-XIN CHENG, Purdue University

SYMPOSIUM **Session 70**

Wearable and Point-of-Care Sensor Technologies for Biomonitoring
arranged by Ian Papautsky and William R Heineman, University of Cincinnati

Sunday Afternoon, Room B401
Ian Papautsky, University of Cincinnati, Presiding

1:30		**Introductory Remarks - Ian Papautsky and William R Heineman**
1:35	(70-1)	**Personal Exposure Monitoring Using Portable Samplers and Paper Analytical Devices** CHARLES HENRY, Colorado State University
2:10	(70-2)	**Pencil and Paper Diagnostic Devices** ANDRES W MARTINEZ, California Polytechnic State University
2:45	(70-3)	**Electrochemical Metal Determination for Point-of-Care Assessment of Environmental Exposure** IAN PAPAUTSKY, University of Cincinnati
3:20		**Recess**
3:35	(70-4)	**Ionic Liquid Electrochemistry and Biointerface Design for Reliable and Smart Sensors** XIANGQUN ZENG, Oakland University
4:10	(70-5)	**Monitoring Corrosion of Biodegradable Magnesium Implants with Hydrogen Gas Sensors** WILLIAM R HEINEMAN, University of Cincinnati, Tingting Wang, Daoli Zhao, Zhongyun Dong, William Hoagland, David Benson

WORKSHOPS **Session 80**

Light Sources in Analytical Chemistry: Solid State Light Sources and Beyond
arranged by Mirek Macka, University of Tasmania

Sunday Afternoon, Room B402
Mirek Macka, University of Tasmania, Presiding

1:30		**Introductory Remarks - Mirek Macka**
1:35	(80-1)	**Gatherers and Foragers? Analytical Scientists in the Quest for Better Light Sources** MIREK MACKA, University of Tasmania
2:05	(80-2)	**Applications of Cavity Ring-Down and Cavity-Enhanced Absorption Spectroscopy in Atmospheric Chemistry** HANS DIETER OSTHOFF, University of Calgary, Charles A Odame-Ankrah, Youssef M Taha, Jason Pak, Connie Z Ye, Nick R Yordanov
2:35	(80-3)	**A Fourier-Domain Fluorescence Excitation Emission Matrix Spectrometer and Its Applications in Measuring Oxidative Stability of Industrial Liquids** HANS-PETER LOOCK, Queen's University, Nicholas L Andrews, James Z Fan, Oliver Reich, Alexander Dudelzak, Hengameh Omrani
3:05		**Recess**
3:20	(80-4)	**Deep UV-LEDs in Detectors for HPLC and Capillary Electrophoresis** PETER CH HAUSER, University of Basel, Duy Anh Bui
3:50	(80-5)	**Solid State Light Sources in Capillary Electrophoresis** DAN XIAO, Sichuan University N/A
4:20	(80-6)	**Quantum Cascade Lasers: How to Revolutionize Mid-Infrared Gas and Liquid Phase Diagnostics** BORIS MIZAIKOFF, Ulm University

ORGANIZED CONTRIBUTED SESSIONS **Session 90**

Advances in Mass Spectrometry of RNA - Half Session
arranged by Norman Chiu, University of North Carolina at Greensboro

Sunday Afternoon, Room B403
Norman Chiu, University of North Carolina at Greensboro, Presiding

1:30	(90-1)	**Toward Detection of Isomeric/Isobaric MicroRNA Biomarkers** NORMAN CHIU, University of North Carolina at Greensboro
1:50	(90-2)	**New LC-MS/MS Strategies for Modified Oligonucleotides and RNAs** PATRICK A LIMBACH, University of Cincinnati, Robert L Ross
2:10	(90-3)	**Mass Spectrometry, RNA and Infectious Disease: tRNA Reprogramming and a Second Genetic Code Control Mycobacterial Dormancy** PETER DEDON, Massachusetts Institute of Technology
2:30	(90-4)	**Identification and Quantitation of RNA Post-Transcriptional Modifications by Nano-flow Liquid Chromatography - Tandem Mass Spectrometry and Database Searching.** HIROSHI NAKAYAMA, RIKEN CSRS, Masato Taoka, Nobuhiro Takahashi, Toshiaki Isobe

ORGANIZED CONTRIBUTED SESSIONS **Session 100**

Ionophore-Based Chemical Sensors I
arranged by Philippe Buhlmann, University of Minnesota and Eric Bakker, University of Geneva

Sunday Afternoon, Room B404
Philippe Buhlmann, University of Minnesota, Presiding

1:30	(100-1)	**pH Independent Fluorescent Ion Sensors** ERIC BAKKER, University of Geneva
1:50	(100-2)	**Measurement of Carbon Dioxide in Urine to Guide the Treatment of Patients in Severe Sepsis or Septic Shock** ERNO LINDNER, The University of Memphis, James G Atherton, Marcin Guzinski, Jasinski Artur, Bradford D Pendley
2:10	(100-3)	**Potentiometric Nanosensors Towards Direct Detection of Nucleic Acids** RÓBERT E GYURCSÁNYI, Budapest University of Technology and Economics, Istvan Makra, Gyula Jágerszki, Alexandra Brajnovits, Peter Fürjes, Márton Bojtár
2:30	(100-4)	**Potentiometric Detection of Biomacromolecules Based on Surface Molecular Imprinting** RONGNING LIANG, Yantai Institute of Coastal Zone Research, CAS, Wei Qin N/A
2:50		**Recess**
3:05	(100-5)	**Voltammetric Ionophore-Based Electrode for Protamine in Human Blood** SHIGERU AMEMIYA, University of Pittsburgh
3:25	(100-6)	**Conducting Polymer-Coated Electrode and Its Application to the Thin Layer Electrolysis Cell for Coulometric Determination** YUMI YOSHIDA, Kyoto Institute of Technology, Mao Fukuyama, Kohji Maeda
3:45	(100-7)	**Biofouling of Ionophore-Doped Ion-Selective Electrode Membranes Revisited** PHILIPPE BUHLMANN, University of Minnesota, Adam Dittmer
4:05	(100-8)	**Metastable Photoacids Towards Activatable and Controllable Ion Sensing Bulk Membranes for Cations Detection** KARIN Y CHUMBIMUNI-TORRES, University of Central Florida, Parth K Patel

ORGANIZED CONTRIBUTED SESSIONS **Session 110**

R&D to QC: Bridging the Gap - Half Session
arranged by Justin Shearer, Dow AgroSciences

Sunday Afternoon, Room B312
Justin Shearer, Dow AgroSciences, Presiding

1:30	(110-1)	**Back to the Future: Bringing Life to an Old Crop Protection Product** JENNIFER HOLLIE JONES, Dow AgroSciences, Tammy Hamilton, Grant Von Wald, Maciej Turowski, Binghe Gu, Chengli Zu, Scott Kelso, Edward Bellinger, Justin Shearer N/A
1:50	(110-2)	**Effective Method Transfer: Ensuring End Results** MARY ELLEN P MCNALLY, DuPont Crop Protection, Stephen Platz
2:10	(110-3)	**Transferring Analytical Methods to a Quality Control Laboratory** AMANDA DARLAND, Dow AgroSciences N/A
2:30	(110-4)	**R&D to QC: An R&D Perspective in Chromatographic Method Development** JUSTIN SHEARER, Dow AgroSciences, Suresh Annangudi Palani, Daniel Knueppel, Rose Nelson

ORAL SESSIONS **Session 120**

Bioanalytical: LC Techniques - Half Session

Sunday Afternoon, Room B315
Sean M Burrows, Oregon State University, Presiding

1:30	(120-1)	**Characterization of New 2 µm Particle Size, 25 nm Pore Size Analytical Size Exclusion Chromatography Column with Larger Exclusion Limit Useful for the Separation of Biomolecules Using UHPLC and HPLC** CRYSTAL BENNER, Tosoh Bioscience LLC, Atis Chakrabarti
1:50	(120-2)	**Studies of Drug Interactions with $Alpha_1$-Acid Glycoprotein by Using On-Line Immunoextraction and High Performance Affinity Chromatography** CONG BI, University of Nebraska-Lincoln, Ryan Matsuda, Zitha Isingizwe, Chenhua Zhang, David Hage
2:10	(120-3)	**HPAE-PAD Applications for Stepwise Biosimilar Development Processes: Monosaccharide and Sialic Acid Determinations** HUA YANG, Thermo Fisher Scientific, Linda Lopez
2:30	(120-4)	**Analysis of Hormone-Protein Binding in Solution by Ultrafast Affinity Extraction** XIWEI ZHENG, University of Nebraska-Lincoln, Cong Bi, Marrisa Brooks, David Hage

Author and presider lists as of January 20, 2016, are available at www.pittcon.org

TECHNICAL PROGRAM

ORAL SESSIONS **Session 130**

Bioanalytical: LC/MS Techniques - Half Session

Sunday Afternoon, Room B315
Sean M Burrows, Oregon State University, Presiding

3:05	(130-1)	**Discovery of Brown Recluse Spider Sex Pheromones Using Advanced Ultrafast Liquid Chromatography – Tandem Mass Spectrometry Methodologies** CASEY BURTON, Missouri University of Science and Technology, Ariel Donovan, Jennifer Parks, William Stoecker, Honglan Shi
3:25	(130-2)	**Analysis of Herceptin Oxidation Variants Using a Supermacroporous Reverse Phase Column Coupled with an Orbitrap Mass Spectrometer** SHANHUA LIN, Thermo Fisher Scientific, Ilze Birznieks, Christopher Pohl, Xiaodong Liu, Jessica Wang, Terry Zhang, Jonathan Josephs
3:45	(130-3)	**Quantitation of Endogenous Adenosine in Mouse Blood, Cell Lysate and Lysosomes by LC-MS/MS Using SurrogateMatrix Method** XIAO DING, Genentech
4:05	(130-4)	**Isomeric Separation of Glycopeptides Using Porous Graphitic Carbon (PGC) LC at High Temperature** RUI ZHU, Texas Tech University, Jingfu Zhao, Yehia Mechref

ORAL SESSIONS **Session 140**

Chemical Methods - Half Session

Sunday Afternoon, Room B316
Richard Bormett, Renishaw, Presiding

1:30	(140-1)	**Bromide-Assisted Anisotropic Growth of Surfactant-Free Gold Nanoparticles** MELISSA KERR, North Carolina Central University, Fei Yan
1:50	(140-2)	**Simultaneous Reduction of Metal Ions by Multiple Reducing Agents Initiate the Asymmetric Growth of Metallic Nanocrystals** MAHMOUD MAHMOUD, Georgia Institute of Technology
2:10	(140-3)	**Novel Optical Properties of Segmented Au-Ag Nanocylinders: Effects of Metallic Junctions on Surface Plasmon Resonance** VINEET KUMAR, North Carolina State University, Gufeng Wang
2:30	(140-4)	**Measurement of Oxidative Potential of Particulate Matter by DTT assay** SHIORI OTA, Tokai University, Kazuhiro Misawa, Yoshika Sekine

ORAL SESSIONS **Session 150**

Gas Chromatography Innovations

Sunday Afternoon, Room B310
Olujide T Akinbo, Butler University, Presiding

1:30	(150-1)	**Method Translation in Gas Chromatography to Get the Same Chromatogram** JAAP DE ZEEUW, Restek, Chris English, Chris Nelson, Jack Cochran
1:50	(150-2)	**Chemometric Treatment of GC-VUV Data - Samples in a New Light** JAMES J HARYNUK, University of Alberta, Keisean Stevenson, Seo Lin Nam, Lawrence A Adutwum
2:10	(150-3)	**All 5 Phases are Not the Same! Considerations for Method Development and Selection** RAMKUMAR DHANDAPANI, Phenomenex, Tim Anderson, Kristen Parnell
2:30	(150-4)	**Flow-through Microfluidic Photoionization Detectors for Rapid and Highly Sensitive Vapor Detection** HONGBO ZHU, University of Michigan, Robert Nidetz, Menglian Zhou, Jiwon Lee, Sanketh Buggaveeti, Katsuo Kurabayashi, Xudong Fan
2:50		Recess
3:05	(150-5)	**Safeguarding a Mass Detector from Difficult Sample Components** AMANDA B DLUGASCH, Waters Corporation, Thomas E Wheat, Patricia R McConville
3:25	(150-6)	**Wide Range Inert Dilution System for Gas Standard Generation** STANLEY D STEARNS, Valco Instruments Co. Inc., Alex Plistil, Sunny Sinlapadech, David McCharthy, Huamin Cai
3:45	(150-7)	**Multimode Plasma Emission Detector** YVES GAMACHE, Analytical Flow Products
4:05	(150-8)	**Polyionic Ionic Liquid GC Stationary Phase Evaluations** LEONARD M SIDISKY, Supelco/Sigma-Aldrich, Greg A Baney, James L Desorcie, Gustavo Serrano, Daniel Shollenberger

ORAL SESSIONS **Session 160**

GC Optimization - Half Session

Sunday Afternoon, Room B309
David Benanou, Veolia, Presiding

1:30	(160-1)	**Evaluation and Application of Sample Preparation Techniques for the Determination of Residual Volatiles in Biological Matrices Using GC-FID** ADRIENE MALSBURY, Bristol-Myers Squibb, William Fish, Frank Tomasella
1:50	(160-2)	**A New Web-Based Application for Modelling Gas Chromatographic Separations** REBECCA STEVENS, Restek Corporation, Jaap de zeeuw, Amanda Rigdon, Linx Waclaski, Dan Li
2:10	(160-3)	**Overcoming New Problems With Old Solutions: Enhanced Separations With GC Retention Gap Columns** RAMKUMAR DHANDAPANI, Phenomenex, Tim Anderson, Kristen Parnell
2:30	(160-4)	**Improved Inertness Performance for Polyethylene Glycol GC Columns** KENNETH G LYNAM, Agilent Technologies, Ngoc-A Dang, Allen Vickers, Yun Zou

ORAL SESSIONS **Session 170**

Instrument Innovations - Half Session

Sunday Afternoon, Room B316
Richard Bormett, Renishaw, Presiding

3:05	(170-1)	**Rapid Analysis of SO_2 to Determine Catalyst Efficiency** DEBBIE ALCORN, INFICON
3:25	(170-2)	**Portable Gas Analyzer for Continuous Monitoring of Sulfur Dioxide in Gas Streams** SAYED A MARZOUK, UAE University, Mohamed Alnaqbi, Muna Bufaroosha, Mohamed Al-Marzouqi
3:45	(170-3)	**Triple Mode of Action of L-tyrosine Derived Probes: Solvent Mediated Flip-Flop Halide (iodide/fluoride) Sensors and Reversible Chromogenic pH Indicators** SANJAY KUMAR MANDAL, IISER Mohali
4:05	(170-4)	**Saving Time and Improving Accuracy by Eliminating the Need for Standards and Calibration in GC/FID Analyses** JONES ANDREW, Activated Research Company, Charlie Spanjers

ORAL SESSIONS **Session 180**

Measurement Strategies - Sensors and Spectroscopy

Sunday Afternoon, Room B313
Stephanie Archer-Hartmann, Complex Carbohydrate Research Center, Presiding

1:30	(180-1)	**Real-Time Voltammetric Characterization of Non-Electroactive Metal Complexation** THUSHANI SIRIWARDHANE, University of South Carolina, Shawn McElmurry, Parastoo Hashemi
1:50	(180-2)	**Phthalocyanine Based Microfluidic Sensors for the Detection of Oxidative Stress** KEVIN J KLUNDER, Colorado State University
2:10	(180-3)	**Multiple Light Scattering to Characterize Emulsions with Polymers** JONATHAN DENIS, Formulaction, Yoann Lefeuvre, Pascal Bru, Christelle Tisserand, Gérard Meunier
2:30	(180-4)	**A Practical Approach for Maximizing ICP-MS Data: A Closer Look at Microwave Sample Prep, The Steps that Precede It, and How to Minimize External Factors that Affect Data Quality** JOHAN NORTJE, Milestone Inc.
2:50		Recess
3:05	(180-5)	**Obtaining Maximum Information from Fast Chemical Reactions Using a Photodiode Array (PDA) UV-Visible Spectrophotometer** IAN ROBERTSON, PerkinElmer Limited, Steve Upstone, Christopher Lynch
3:25	(180-6)	**Microfluidic Visual Rheometer** CHRISTELLE TISSERAND, Formulaction, Patrick Abgrall, Patrycja Adamska, Gérard Meunier
3:45	(180-7)	**Real-time Investigation of Antibiotics-induced Oxidative Stress and Superoxide Release in Bacteria Using an Electrochemical Biosensor** XIAOBO LIU, Clarkson University, Mouna Marrakchi, Michael Jahne, Shane Rogers, Silvana Andreescu
4:05	(180-8)	**Fabrication of Micro Ir/IrO_x pH Sensor for Dental Applications** CHINDANAI RATANAPORNCHAROEN, Tokyo Medical and Dental University, Miyuki Tabata, Yuji Miyahara, Tatsuro Goda, Akira Matsumoto, Junji Tagami, Yuichi Kitasako, Masaomi Ikeda

Author and presider lists as of January 20, 2016, are available at www.pittcon.org

ORAL SESSIONS — Session 190

Process Analytical Techniques - Half Session

Sunday Afternoon, Room B309
David Benanou, Veolia, Presiding

3:05 (190-1) **A Model Study of Pseudo-Absolute Quantitative Analysis Using Gas Chromatography – Vacuum Ultraviolet Spectroscopy** LING BAI, The University of Texas at Arlington, Jonathan Smuts, Phillip Walsh, Kevin A Schug

3:25 (190-2) **Rapid Process Control Using GC-ION Mobility Spectrometry** CHANDRASEKHARA HARIHARAN, ION-GAS GmbH, Wolfgang Vautz

3:45 (190-3) **Facilitating Complex Analysis Using Multiple Fast Temperature Programming Zones** DALE ASHWORTH, Valco Instruments Co. Inc., Huamin Cai, Martin Brisbin, Chris Bishop, William Coontz, Andrew Rochon, Steve Werner, Stanley D Stearns

4:05 (190-4) **Analyzing Trace Level Nitric Oxide in a Flammable Gas Matrix** KENNETH WONG, Air Liquide

SUNDAY POSTER SESSION — Session 200

Sunday posters will be on display from 3:30 PM to 7:30 PM with authors present from 5:30 PM to 7:30 PM. The location for the Sunday posters is Room A412.

Sunday Poster Session
Sunday Afternoon, Room A412

(200-1 P) **Influence of Particle Size Distribution on Protein Digestibility and Functionality** ALBERTA ARYEE, Agriculture & Agri-Food Canada, Joyce I Boye N/A

(200-2 P) **Increasing Stability of a Core-Shell Particle** MARK WOODRUFF, Fortis Technologies, Ken Butchart

(200-3 P) **Rapid Screening of Virgin and Recycled Polymer Resins Using FTIR and Raman Libraries of Pre-Computed Mixture Spectra** WILLIAM COSTA, Fiveash Data Management, Inc., Bill McCarthy, Todd Strother

(200-4 P) **The Complimentary Nature of NMR and SPME GCMS for the Analysis of Fermented Beverages** NEIL FITZGERALD, Marist College, Samantha E Soprano, Sarah R Johnson, John C Edwards

(200-5 P) **Evaluation of an Integrated Optic-Fiber-Microfluidic Analyzer for Polyphenols Measurement** MARIA CAÑIZARES-MACÍAS, Universidad Nacional Autónoma de México, Oscar Sandoval-Ventura, Luis F Olguín-Contreras N/A

(200-6 P) **Selective Separation and Detection of Catecholamines with Capillary Electrophoresis** MAOJUN GONG, Wichita State University, Qiyang Zhang

(200-7 P) **Cold EI – Approaching the Ideal GC-MS Interface and Ion Source** AVIV AMIRAV, Tel Aviv University, Bogdan Belgorodsky, Uri Keshet, Alexander Fialkov, Tal Alon

(200-8 P) **New Sampling Device for the Analysis of Lung Cancer Biomarkers in Exhaled Breath** MOHAMED ABDEL-REHIM, Stockholm University N/A

(200-9 P) **Ceramic Ion Source for Catalytic Pyrolysis and Catalytic Combustion Ionization Detection of Selected Constituents in Complex Petroleum and Biofuel Samples** PAUL L PATTERSON, Detector Engineering & Technology, Inc., Jennifer Seroy

(200-10 P) **Using Microscopy to Measure Rates of Heterogeneous Reactions** WALTER BOWYER, Hobart and William Smith Colleges, Gabriella Mylod

(200-11 P) **Crowding in Cell-Like Environments Alters Diffusion and Enzyme Kinetics** KRISTIN M SLADE, Hobart and William Smith Colleges, Michael Conroy, Sophia Melvin

(200-12 P) **Effects of Macromolecular Crowding on YADH Enzyme Mechanism** KRISTIN M SLADE, Hobart and William Smith Colleges, Allison Wilcox

(200-13 P) **Mixed Valence Mn,La,Sr-oxide based Magnetic Nanoparticles Coated with Silica Nanoparticles for Immunosensor Fabrication** AMOS MUGWERU, Rowan University

(200-14 P) **Hydrophilic Interaction Chromatography (HILIC) and Enzymatic/Spectrometric Methods for the Determination of Uric Acid and Creatinine in Human Biofluids** YUEGANG ZUO, University of Massachusetts Dartmouth, Si Zhou, Xiaofei Lu, Ningning Zhang, Faten Albalawi

(200-15 P) **Profiling/Monitoring Activated Carbons for Drinking Water** HENRY G NOWICKI, PACS Inc

(200-16 P) **Profiling/Monitoring Gas Phase Activated Carbon Systems** HENRY G NOWICKI, PACS Inc

(200-17 P) **Lignin Waste Transformed to High Quality Activated Carbons** HENRY G NOWICKI, PACS Inc

(200-18 P) **A Rapid Hydrophilic Interaction Liquid Chromatography (HILIC) Method for Determination of Trace Nitrate and Nitrite in Snow and Rain Samples** XIAOFEI LU, University of Massachusetts Dartmouth, Yuegang Zuo

(200-19 P) **Occurrence and Identification of Bisphenol A and Other Alkylphenols in Sea Shore Crabs** JOSEPH MICHAEL, University of Massachusetts Dartmouth, Yuegang Zuo

(200-20 P) **Evaluation of Porous Layer Thickness of Core Shell Particle for Separation of Proteins** NORIKAZU NAGAE, ChromaNik Technologies Inc., Tomoyasu Tsukamoto

ACS POSTER — Session 210

ACS posters will be on display from 3:30 PM to 7:30 PM with authors present from 5:30 PM to 7:30 PM. The location for the ACS posters is Room A412.

ACS Division of Analytical Chemistry Poster Session
Sunday Afternoon, Room A412

(210-1 P) **Phospholipid/Aromatic Thiol Hybrid Bilayers** CHAO LI, Auburn University, Mingming Wang, Matthew Ferguson, Wei Zhan

(210-2 P) **Dynamic Records of Spatiotemporal Pb: Lichens versus Trees and Sediments** NATHAN W BOWER, Colorado College, Ben Greydanus, Sam Kramer, Eric Wolatz, Stephen Getty, Craig Lundstrom

(210-3 P) **Dating Paintings and Prints Using ATR-FTIR: A Philatelic Case Study from Lodz Ghetto** NATHAN W BOWER, Colorado College, Conor J Blanchet, Michael S Epstein

(210-4 P) **SPME/GC-MS Analysis of Dynamic Compositional Changes in the Volatile Components in Red Wine After "Breathing"** AMBER I HILLS, East Stroudsburg University, Jon S Gold, Richard S Kelly

(210-5 P) **Development of Portable Elisa System for Infectious Disease Diagnosis** KAZUHIRO MORIOKA, Tokyo Metropolitan University, Harpal Singh, Hizuru Nakajima, Akihide Hemmi, Masayuki Shimojima, Le Van An, Sazaly AbuBakar, Hulie Zeng, Shungo Kato, Masami Sugamata, Ming Yang, Katsumi Uchiyama

(210-6 P) **Applications of a Quartz Crystal Microbalance for Monitoring Bacterial Biofilm Growth and Removal** HUNTER J SISMAET, Northeastern University, Pegah N Abadian, Edgar D Goluch

(210-7 P) **A Capillary Electrophoresis Approach to the Characterization and Application of Graphene Quantum Dots as Sensing Agents** LEONA SIRKISOON, Wake Forest University, Honest Makamba, Shingo Saito, Christa Colyer

(210-8 P) **A pH Sensor for Non-Invasive In Vivo Detection and Imaging of Implant Associated Infection** UNAIZA UZAIR, Clemson University

(210-9 P) **Scanning Electrospray Microscopy with Nanopipettes** ELIZABETH M YUILL, Indiana University, Wenqing Shi, Lane A Baker

(210-10 P) **Phosphoproteomics Studies of Human Immunodeficiency Virus-1** KEVIN MARK, LaGuardia Community College City University of NY, Pratikkumar Rathod, Emmanuel Chang, Hsin-Pin Ho, Xu Yu, Mathias Lichterfeld

(210-11 P) **Segmented Flow Sampling with Theta Pipettes** ANUMITA SAHA-SHAH, Indiana University, Curtis M Green, Lane A Baker

(210-12 P) **Mapping Local Permeability Change During Degradation with Scanning Ion Conductance Microscopy-Scanning Electrochemical Microscopy (SICM-SECM)** WENQING SHI, Indiana University, Lane A Baker

(210-13 P) **Profiling N, N'-Dibutylbenzimidazolium Salt and Its Derivatives Using CYCLIC Voltammetry** HUGGINS Z MSIMANGA, Kennesaw State University, Andrew Montalvo, Daniela Tapu

(210-14 P) **Comparative Study of Elemental Nutrients in Organic and Conventional Vegetables by Laser Induced Breakdown Spectroscopy (LIBS)** CHET R BHATT, Mississippi State University

(210-15 P) **Photoacoustic Spectroscopy with SF6, An Optically Thick Greenhouse Gas** HAN PARK, University of Tennessee at Chattanooga

(210-16 P) **Integrating Authentic Research Experiences into Undergraduate Analytical Chemistry** FEI YAN, North Carolina Central University, Melissa Kerr

(210-17 P) **Proteomic Analysis of Tetrahymena Thermophila Using MALDI-TOF/TOF** DOUGLAS BEUSSMAN, St. Olaf College, Harrison VanDolah

(210-18 P) **Possible Age and Location Differences in Human Scent Profiles** DOUGLAS BEUSSMAN, St. Olaf College, Laura Muehlbauer

(210-19 P) **LC-QTOF Detection of Methamphetamine in Packaging Residue** DOUGLAS BEUSSMAN, St. Olaf College, Matthew Bock

(210-20 P) **Environmental Toxins from Decorative Candles** DOUGLAS BEUSSMAN, St. Olaf College, Allison Sager

(210-21 P) **Analysis of Fibers Via Isotope Ratio Mass Spectrometry** DOUGLAS BEUSSMAN, St. Olaf College, Elaine Macon, Hannah Brown

(210-22 P) **Withdrawn**

(210-23 P) **HPLC Method Development for Analysis of Dissolution Samples of a Highly Soluble Drug Substance in a Hydrophobic Matrix** MATTHEW LOUCKS, Banner Life Sciences, Angela Moore, Wayne Craig, Sara Draper

(210-24 P) **Confocal Raman Microscopy Investigation of the Kinetic Barrier to PAH Partitioning into Individual C_{18}–Silica Particles** DAVID A BRYCE, University of Utah, Jay P Kitt, Joel M Harris

(210-25 P) **Development of a Flow-Based Electrogenerated Chemiluminescent Detection of Biogenic Amines on a Microfluidic Chip** ERIN GROSS, Creighton University, Leah Schaffer, Emily Lowry, Charles Henry, Rachel M Feeny, John Wydallis

TECHNICAL PROGRAM

MONDAY, MARCH 7, 2016
MORNING

AWARDS **Session 220**

Chromatography Forum of Delaware Valley Dal Nogare Award
arranged by Mary Ellen McNally, E.I. DuPont de Nemours and Company

Monday Morning, Room B312
Mary Ellen McNally, E.I. DuPont de Nemours and Company, Presiding

8:30		**Introductory Remarks - Mary Ellen McNally**
8:35		**Presentation of the 2016 Dal Nogare Award to Stephen Weber, University of Pittsburgh, by Mary Ellen McNally, E.I. DuPont de Nemours and Company**
8:40	(220-1)	**Opportunities and Challenges with Capillary Liquid Chromatography** STEPHEN WEBER, University of Pittsburgh
9:15	(220-2)	**How Much Performance is Lost in LC by Optimizing Only Velocity and Column Length Using Limited Number of Particle Size?** PETER W CARR, University of Minnesota, Adam Matula, Dwight R Stoll
9:50	(220-3)	**UHPLC-MS for Multiplexed Neurochemical Analysis** ROBERT T KENNEDY, University of Michigan
10:25		**Recess**
10:40	(220-4)	**Integrated Microfluidic Separations Devices Interfaced to Mass Spectrometry** J MICHAEL RAMSEY, University of North Carolina at Chapel Hill, Erin Redman, William Black
11:15	(220-5)	**The Second Dimension is a Strange Place - Fundamental Aspects of the Optimization of the Second Dimension in Two-Dimensional Liquid Chromatography** DWIGHT R STOLL, Gustavus Adolphus College, John Halvorson, Carston Dammann, Eli Larson, Alex Wilson, David C Harmes, Monika Dittmann, Sarah C Rutan, Abraham Lenhoff

AWARDS **Session 230**

The Pittsburgh Conference Achievement Award
arranged by Amit Ghosh, Bayer Materials Science LLC

Monday Morning, Room B314
Amit Ghosh, Bayer Materials Science LLC, Presiding

8:30		**Introductory Remarks - Amit Ghosh**
8:35		**Presentation of the 2016 Pittsburgh Conference Achievement Award to Jared L Anderson, Iowa State University, by Elias S Absey, Chair, Society for Analytical Chemists of Pittsburgh**
8:40	(230-1)	**Exploiting Ionic Liquids, Magnetic Ionic Liquids, and Polymeric Ionic Liquids in Sample Preparation and Multidimensional Gas Chromatography** JARED L ANDERSON, Iowa State University
9:15	(230-2)	**The Impact and Evolution of High Efficiency Chiral and Achiral Separations** DANIEL W ARMSTRONG, University of Texas at Arlington
9:50	(230-3)	**Use of Metal-Organic Frameworks in Sample Preparation Analytical Schemes: An Overview of Their Performance in Solid-Phase Extraction Modes** VERÓNICA PINO, University of La Laguna, Priscilla Rocío-Bautista, Juan H Ayala, Jorge Pasán, Catalina Ruiz-Pérez, Ana M Afonso
10:25		**Recess**
10:40	(230-4)	**Achiral LSER Investigation of HPLC Retention on Cinchona Alkaloid-Based Chiral Stationary Phases** APRYLL M STALCUP, Dublin City University
11:15	(230-5)	**Applications of Conductive Ionic Liquids in Chemical Analysis** JON R KIRCHHOFF, University of Toledo, Joshua A Young, Amila M Devasurendra, LM Viranga Tillekeratne, Jared L Anderson, Cheng Zhang

SYMPOSIUM **Session 240**

ACS-ANYL - BRAIN Initiative Advancements in Neurochemical and Physiological Measurements
arranged by Michael L Heien, University of Arizona

Monday Morning, Room B308
Michael L Heien, University of Arizona, Presiding

8:30		**Introductory Remarks - Michael L Heien**
8:35	(240-1)	**New Platforms for Multiplexing Electrochemical Measurements In Vivo** MICHAEL L HEIEN, University of Arizona
9:10	(240-2)	**Simultaneous Detection of Dopamine Release and Multiple Single-Unit Activity in Awake and Behaving Rats** STEPHEN COWEN, University of Arizona, Michael L Heien, Kate L Parent, Daniel F Hill, Jean-Paul Wiegand, Michael A Miller, Christopher W Atcherley
9:45	(240-3)	**New Nano Tools for Molecular Sensing and Imaging of Single Neuron-Neuron Communication** X NANCY XU, Old Dominion University, Pavan K Cherukuri, Preeyaporn Songkiatisak
10:20		**Recess**
10:35	(240-4)	**In Vivo Multiphoton Microscopy of Mouse Brain** CHRIS XU, Cornell University
11:10		**Open Discussion**

SYMPOSIUM **Session 250**

Advances in Raman Spectroscopy
arranged by Sanford A Asher, University of Pittsburgh

Monday Morning, Room B304
Sanford A Asher, University of Pittsburgh, Presiding

8:30		**Introductory Remarks - Sanford A Asher**
8:35	(250-1)	**Micro-Scale Spatially Offset Raman Spectroscopy (Micro-SORS)** PAVEL MATOUSEK, Rutherford Appleton Laboratory, Claudia Conti, Marco Realini, Chiara Colombo, Giuseppe Zerbi
9:10	(250-2)	**Recent Advances in Raman Optical Activity (ROA) Methodology and Applications** LAURENCE A NAFIE, Syracuse University, Rina K Dukor
9:45	(250-3)	**Fast Raman Imaging Using Optimized Binary Compressive Detection** DOR BEN-AMOTZ, Purdue University, Owen G Rehrauer, Bharat R Mankani, Gregery T Buzzard, Bradley J Lucier, Stanley Chan
10:20		**Recess**
10:35	(250-4)	**Stimulated Raman Scattering Microscopy of Vibrational Tags** WEI MIN, Columbia University
11:10	(250-5)	**Nanoparticle Based Analysis of Cells, Molecules and Tissue by SERS** DUNCAN GRAHAM, University of Strathclyde, Samuel Mabbott, Lee Barrett, Steven Asiala, Kirsten Gracie, Karen Faulds

SYMPOSIUM **Session 260**

Innovative Applications of Mass Spectrometry in Biopharmaceutical and Diagnostic Development
arranged by William Christopher Lamanna, Sandoz

Monday Morning, Room B305
William Christopher Lamanna, Sandoz, Presiding

8:30		**Introductory Remarks - William Christopher Lamanna**
8:35	(260-1)	**Characterization of Filgrastim Using Intact and Top-Down MS** WILLIAM CHRISTOPHER LAMANNA, Sandoz, Johann Holzmann, Hansjoerg Toll
9:10	(260-2)	**Tandem Mass Spectrometry as an Excellent Quantification Tool to Guide Process Development for Optimal Host Cell Protein Removal in Biopharmaceutical Products** DONALD WALKER, Genentech
9:45	(260-3)	**Antibody Characterization with Ultrahigh-Resolution Mass Spectrometry: Selecting Well Behaved Molecules for Optimal Product Quality Attributes** LISA MARZILLI, Pfizer, Inc., Mellissa Ly, Elaine Stephens, Heather S DeGruttola, Andrew E Saati, Jason C Rouse
10:20		**Recess**

10:35	(260-4)	**Advanced Process Control Using Fast Analytics to Monitor Multiple Critical Quality Attributes of a Monoclonal Antibody** LI ZANG, Biogen
11:10	(260-5)	**Analysis of Glycosaminoglycan Non-Reducing End Structures by Liquid Chromatography Tandem Mass Spectrometry and Its Uses in Discovering New Disease-Specific Biomarkers** ROGER LAWRENCE, BioMarin, Heather Prill, Evelyn Wang, Raymond Y Wang, Jeffery D Esko, Brett Crawford, Marzia Pasquali, Toni Pollock, William Christopher Lamanna, Nancy Pryer

SYMPOSIUM **Session 270**

Neurotransmission at Single and Nano-Resolved Bio-Structures
arranged by Mei Shen, University of Illinois at Urbana-Champaign and Christian A Amatore, CNRS & ENS

Monday Morning, Room B303
Mei Shen, University of Illinois at Urbana-Champaign, Presiding

8:30		**Introductory Remarks - Mei Shen and Christian Amatore**
8:35	(270-1)	**Electrochemical Cytometry of the Contents of Nanometer Vesicles Out and Inside Cells** ANDREW EWING, University of Gothenburg and Chalmers University
9:10	(270-2)	**Detection of Ionic Neurotransmitters at Biological Nanostructures via Versatile Nanosensor Probes** MEI SHEN, University of Illinois at Urbana-Champaign
9:45	(270-3)	**Electrochemical Approaches to Monitoring Neural Activity** R MARK WIGHTMAN, University of North Carolina at Chapel Hill
10:20		**Recess**
10:35	(270-4)	**Full Fusion During Vesicular Exocytosis: A Real Release Stage or a Hoax?** CHRISTIAN A AMATORE, CNRS & ENS
11:10	(270-5)	**Coulter Counter Voltage Trapping of Nanoparticles with Sub-Nanometer Size Resolution** HENRY WHITE, University of Utah, Sean German, Martin Edwards, Yulun Zhang

SYMPOSIUM **Session 280**

Omics for Environmental and Public Health Protection
arranged by Jeanette Van Emon, U.S. EPA and Hercules Moura, CDC

Monday Morning, Room B309
Jeanette Van Emon, U.S. EPA, Presiding

8:30		**Introductory Remarks - Jeanette Van Emon and Hercules Moura**
8:35	(280-1)	**Applications of Accelerator Mass Spectrometry to Human Health Research** TED OGNIBENE, Lawrence Livermore National Laboratory
9:10	(280-2)	**Proteomic Approaches for Molecular Epidemiological Studies** ROLF U HALDEN, Arizona State University, Nicole Hansmeier, Tzu-Chiao Chao, Julie B Herbstman, Lynn R Goldman, Frank R Witter
9:45	(280-3)	**Proteomic Applications in Microbiology** HERCULES MOURA, CDC NCEH-DLS, John R Barr
10:20		**Recess**
10:35	(280-4)	**Development of Metagenomic Tools for Identification and Characterization of Infectious Agents from Clinical Samples and for Microbiome Association Studies** JAMES POSEY, CDC
11:10	(280-5)	**Proteomics for Adverse Outcome Pathway Discovery for Environmental Contaminants** JEANETTE VAN EMON, U.S. EPA

SYMPOSIUM **Session 290**

SEAC - Nano-Electroanalysis for a Sustainable World
arranged by Joaquin Rodriguez Lopez, University of Illinois

Monday Morning, Room B310
Joaquin Rodriguez Lopez, University of Illinois, Presiding

8:30		**Introductory Remarks - Joaquin Rodriguez Lopez**
8:35	(290-1)	**Chemically Modified 3D Energy-Storage Architectures** DEBRA R ROLISON, U.S. Naval Research Laboratory, Jeffrey W Long, Christopher N Chervin, Joseph F Parker, Megan B Sassin
9:10	(290-2)	**Sunlight-Driven Hydrogen Formation by Membrane-Supported Photoelectrochemical Water Splitting** NATHAN S LEWIS, Caltech
9:45	(290-3)	**Elucidating Charge Transfer on Polymer Nano-Structures for a New Concept in Energy Storage** JOAQUIN RODRIGUEZ LOPEZ, University of Illinois
10:20		**Recess**
10:35	(290-4)	**Ion Transport in 1D and 3D Networked Porous Nanostructure Electrodes** SANG BOK LEE, University of Maryland, Eleanor Gillette
11:10	(290-5)	**Operando Methods for the Characterization Energy Materials** HECTOR D ABRUNA, Cornell University

ORGANIZED CONTRIBUTED SESSIONS **Session 300**

Infrared Spectroscopy Beyond the Diffraction Limit
arranged by Ellen Miseo, Hamamatsu Corporation and Craig Prater, Anasys Instruments

Monday Morning, Room B311
Ellen Miseo, Hamamatsu Corporation, Presiding

8:30	(300-1)	**Advancing the Field of Infrared Nanospectroscopy for Materials and Life Sciences** CRAIG PRATER, Anasys Instruments
8:50	(300-2)	**Resonance Enhanced AFM-IR Induced by Quantum Cascade Laser** ALEXANDRE DAZZI, University Paris-Sud, Jeremie Mathurin, Ariane Deniset-Besseau, Johanna Saunier, Najet Yagoubi, Kevin Kjoller
9:10	(300-3)	**Unprecedented Nanoscale Chemical Characterization of Materials Using AFM-Based Infrared Spectroscopy** CURTIS MARCOTT, Light Light Solutions, Dillon Eoghan, Kevin Kjoller
9:30	(300-4)	**Withdrawn**
9:50		**Recess**
10:05	(300-5)	**AFM-IR Spectroscopy and Imaging of Biodegradable Polymers** JOHN F RABOLT, University of Delaware, Liang Gong, Isao Noda, Bruce Chase
10:25	(300-6)	**Nanoscale Molecular Imaging of Polymer Systems Using AFM-IR** MARK RICKARD, Dow Chemical, Gregory Meyers, Carl Reinhardt, Jamie Stanley
10:45	(300-7)	**Absorption Spectroscopy and Imaging from the Visible through Mid-IR with 20 nm Resolution Using AFM Probes** ANDREA CENTRONE, NIST
11:05	(300-8)	**Open Discussion**

ORGANIZED CONTRIBUTED SESSIONS **Session 310**

Ionophore-Based Chemical Sensors II
arranged by Philippe Buhlmann, University of Minnesota and Eric Bakker, University of Geneva

Monday Morning, Room B404
Philippe Buhlmann, University of Minnesota, Presiding

8:30	(310-1)	**Magnetic Nanoparticles as Dispersible Electrodes** JUSTIN J GOODING, University of New South Wales, Roya Tavallaie, Elizabeth Morago, Saimon M Silva, Kyloon Chuah
8:50	(310-2)	**Development of an Ultra-Selective Optode Nanosensor for Potassium Imaging** ALI SAHARI, Northeastern University, Tim Ruckh, Richard Hutchings, Heather Clark
9:10	(310-3)	**Low Detection Limit of Ion-Selective Electrodes; Is the Story Really Over?** ALEKSANDAR RADU, Keele University, Lukasz K Mendecki, Sergio Granados-Focil, Christina McGraw, Peter Dillingham
9:30	(310-4)	**Multifunctional Detection and Delivery** ELIZABETH (LISA) HALL, University of Cambridge, Nadia Tsao
9:50		**Recess**

10:05 (310-5) **Printed Paper-Based Ion-Selective Optode Devices** DANIEL CITTERIO, Keio University, Hiroyuki Shibata, Terence G Henares, Nobutoshi Komuro, Koji Suzuki

10:25 (310-6) **Conducting Polymers Nanospheres for Sensors – Receptor or Transducers?** AGATA MICHALSKA, University of Warsaw, Katarzyna Klucinska, Anna Kisiel, Krzysztof Maksymiuk N/A

10:45 (310-7) **Signal Transduction Based on Constant-Potential Coulometry for Solid-Contact ISEs** JOHAN BOBACKA, Åbo Akademi University, Ulriika Vanamo, Elisa Hupa, Ville Yrjänä

11:05 (310-8) **Plasticizer-Free Paper-Based Ion-Selective Optodes** XUEWEI WANG, University of Michigan, Mark Meyerhoff, Yu Qin

ORGANIZED CONTRIBUTED SESSIONS — Session 320

New Perspectives on the History of Chromatography

arranged by Richard Ulrych, Chemical Heritage Foundation

Monday Morning, Room B407

Sarah Reisert, Chemical Heritage Foundation, Presiding

8:30 (320-1) **Building the Market for Ion Chromatography: The Early Days** ARTHUR WILLIAM FITCHETT, Thermo Fisher Scientific

8:50 (320-2) **Early Days in GC 1957-1962** HAROLD MONROE MCNAIR, Virginia Tech

9:10 (320-3) **Evolution of Capillary Chromatography and Capillary Electrophoresis: A Personal Perspective** MILOS V NOVOTNY, Indiana University

9:30 (320-4) **The Development of HPLC Method Development** LLOYD R SNYDER, LC Resources Inc.

9:50 **Recess**

10:05 (320-5) **Pioneering Days of Chromatography in Lab Automation** JACK GILL, Chemical Heritage Foundation

10:25 (320-6) **The Role of Temperature Programming in F&M's Success** AARON J MARTIN, Marlabs, Retired

10:45 **Liquid Chromatography Before It Became HPLC** JAMES WATERS, Chemical Heritage Foundation N/A

11:05 (320-7) **Chromatography In, or Next to Chemistry? Discipline Formation and Identity of Chromatography Practitioners in the 1960s and 1970s** APOSTOLOS GERONTAS, Coburg University of Applied Sciences

ORAL SESSIONS — Session 330

Biomedical: Advances in Point-of-Care Technologies

Monday Morning, Room B313

Mustafa Culha, Yeditepe University, Presiding

8:30 (330-1) **Low-Cost Molecular Diagnostics on a Smartphone** HYUNGSOON IM, Massachusetts General Hospital, Cesar M Castro, Huilin Shao, Monty Liong, Jun Song, Divay Pathania, Lioubov Fexon, Changwook Min, Rosemary H Tambouret, Misha Pivovarov, Ralph Weissleder, Hakho Lee

8:50 (330-2) **Smartphone Based Detection of Stress Biomarkers in Saliva** AADHAR JAIN, Cornell University, Elizabeth Rey, David Erickson, Dakota O'Dell, Seoho Lee

9:10 (330-3) **A Smartphone Platform for Quantitative Point-of-Care Detection of Micronutrient Deficiencies** DAKOTA O'DELL, Cornell University, Seoho Lee, Jessica Hohenstein, David Erickson

9:30 (330-4) **Bed-Side Detection of Volatile Human Metabolites for Medical Applications** CHANDRASEKHARA HARIHARAN, ION-GAS GmbH, Wolfgang Vautz, Sascha Liedtke

9:50 **Recess**

10:05 (330-5) **Improved Optical Cavity Based Biosensor with Differential Detection Method Through Simultaneous Detection** TONY BUJANA, Letourneau University, DongGee Rho, Cody Joy, Peter Cowles, Seunghyun Kim

10:25 (330-6) **A PDMS/Paper Hybrid Microfluidic Biochip for Multiplexed Instrument-Free Bacterial Meningitis Diagnosis** MAOWEI DOU, University of Texas El Paso, Sanjay Sharma Timilsina, Juan Sanchez, Delfina Dominguez, Xiujun James Li

10:45 (330-7) **Measuring Biochemical Effects of Pulmonary Rehabilitation and OMT on COPD Patients through LC-MS-MS Analysis of Plasma Metabolites** CHEN ZHANG, Michigan State University, Sherman Gorbis, John Wang, A Daniel Jones

11:05 (330-8) **Surface-Enhanced Raman Scattering for In Situ Analysis of Living Systems** MUSTAFA CULHA, Yeditepe University

ORAL SESSIONS — Session 340

Capillary Electrophoresis

Monday Morning, Room B302

Christopher R Harrison, San Diego State University, Presiding

8:30 (340-1) **Quantitation of Kinase Activity in a Social Amoeba Using Capillary Electrophoresis and a Peptide Substrate Reporter** MICHELLE L KOVARIK, Trinity College, Kunwei Yang, Allison J Tierney

8:50 (340-2) **Development of a Degradation Resistant Peptide Reporter for Monitoring E3 Ligase and Proteasome Activity** GREGERY WOSS, University of North Carolina at Chapel Hill, Nancy Allbritton, Marcey Waters, Adam Melvin, Kaiulani Houston

9:10 (340-3) **A New Analytical Technique to Profile Multiple Steroids in Individual Female Zebrafish to Study Endocrine Disruption** VINCENT T NYAKUBAYA, West Virginia University, Paige A Reed, William J Feeney, Regina Rockwell, Amber Kantes, Jennifer Ripley-Stueckle, Lisa A Holland

9:30 (340-4) **A Miniaturized Immunoaffinity Capillary Electrophoresis Based-Biomarker Analyzer for Bioanalytical Applications** NORBERTO A GUZMAN, Princeton Biochemicals, Inc., Daniel E Guzman

9:50 **Recess**

10:05 (340-5) **Electrophoretic Technologies for Catecholamine Detection** MAOJUN GONG, Wichita State University, Qiyang Zhang, Maddukuri Naveen

10:25 (340-6) **Integrating Microscale Enzyme Reactions Into Capillary Separation** SRIKANTH GATTU, West Virginia University, Cassandra Crihfield, Lisa A Holland, Anthony Moncrief

10:45 (340-7) **Investigation of Oxidative Metabolism by Electrochemistry/Capillary Electrophoresis/Mass Spectrometry Using a Novel Sheathless Interface Design** NHAN TO, University of Kansas, Ryan T Johnson, John Stobaugh, Craig E Lunte

11:05 (340-8) **Manipulation of the EOF in Phospholipid Coated Capillaries Through the Incorporation of Different Metal Cations** CHRISTOPHER R HARRISON, San Diego State University, Shane Wells, Eduardo De La Toba, Srilatha Vydha

ORAL SESSIONS — Session 350

Data Analysis and Manipulation

Monday Morning, Room B301

A Peter Snyder, Retired Government, Presiding

8:30 (350-1) **Quantitative Evaluation of Spectral Interferences in Atomic Emission Spectroscopy** MATTHIEU BAUDELET, University of Central Florida, Jessica Chappell, Brandon Seesahai, Martin Richardson, Michael Sigman

8:50 (350-2) **Comparison of Various Methods for the Determination of Uncertainty due to Long Term Stability** DANIEL BIGGERSTAFF, o2si Smart Solutions, Huichen Stavros, Mark Filla

9:10 (350-3) **Lifecycle of Analytical Methods: Development of Equivalent Dissolution Methods for Immediate-Release Oral Dosage Forms Post-Approval** IVELISSE COLON, Vertex Pharmaceuticals, Taryn Ryan, Joseph Medendorp

9:30 (350-4) **Investigating Robustness and Ruggedness of Analytical Methods Employing aQBD Principles** PETER ANDREAS LUND JACOBSEN, Fertin Pharma

9:50 **Recess**

10:05 (350-5) **Chemometrics for Big Data** ROBERT A LODDER, University of Kentucky, Anne Brooks

10:25 (350-6) **Air Quality Networks- Results from Validation and Lessons on Calibration** JOHN R SAFFELL, Alphasense Ltd., Roderic L Jones

10:45 (350-7) **Using Prior Probabilities to Increase the Confidence of Chemical Identification** TYLER A ZIMMERMAN, NIST, Tytus D Mak, W Gary Mallard, Nirina R Andriamaharavo, Dmitrii V Tchekhovskoi, Stephen E Stein

11:05 (350-8) **Applied Analytics: Using Performance-Based Analytical Test Methodology for Monitoring Laboratory Methods Required by EPA Tier III Standards** JOHN MAURER, Valero

ORAL SESSIONS — Session 360

Environmental LC

Monday Morning, Room B408
Ariel Donovan, Missouri University Science and Technology, Presiding

8:30 (360-1) **Intrigues of Analyte Peak Distortion in Ion Chromatography by Overloaded Matrix Ions –Trace Analysis of Bromate in High Ionic Strength Samples** MICHAEL K PAPPOE, University of Alberta, Mohammad H Naeeni, Charles A Lucy

8:50 (360-2) **Multi-Dimensional Ion Chromatography for Trace Ion Analysis** RONG LIN, Thermo Fisher Scientific, Kannan Srinivasan, Herb Wagner

9:10 (360-3) **Assessment of Amino Acid Stability in the Early Oceans by Liquid Chromatography-Mass Spectrometry** ERIC T PARKER, Georgia Institute of Technology, Karen L Britnon, Aaron S Burton, Daniel P Glavin, Jason P Dworkin, Jeffrey L Bada

9:30 (360-4) **Increase Sample Productivity in the Environmental Lab with High Pressure Ion Chromatography** FRANK HOEFLER, Thermo Fisher Scientific, David G Moore, Yan Liu

9:50 **Recess**

10:05 (360-5) **A Novel Cation Exchange Stationary Phase for Analysis of Common Cations and Amines Using Ion Chromatography** MANI JAYARAMAN, Thermo Fisher Scientific, Christopher Pohl, Charanjit Saini, Yan Liu

10:25 (360-6) **Effects of Titanium Dioxide Nanoparticles on Endocrine Disruption in Zebrafish** MARRIAH ELLINGTON, West Virginia University, Cassandra Crihfield, Sara Melow, Lisa A Holland

10:45 (360-7) **Trace Analysis of Guanidine Compounds in Surface Water with Resorcinarene-Based Ion Chromatography Columns** ROGER G HARRISON, Brigham Young University, Tayyabeh Panahi, John Lamb

11:05 (360-8) **Analysis of Molecular Markers of Animal Waste by LC-MS/MS** SREE HARSHITHA VELAGA, Tennessee Technological University, Sreedharan Lakshmi Narayanan, John Harwood

ORAL SESSIONS — Session 370

Fluorescence and Luminescence Advances

Monday Morning, Room B315
Chenzhong Li, Florida International University, Presiding

8:30 (370-1) **Measuring Displacement on the Micron Scale: Novel Luminescent Spectral Rulers for Orthopedic Applications** MELISSA M ROGALSKI, Clemson University, Donald Benza, Hunter Pelham, Nakul Ravikumar, Dakotah Anderson, Johnathan Heath, John D DesJardins, Jeffrey N Anker

8:50 (370-2) **Simplified Two-Photon Synchronous Scanning Fluorescence for Resolution of Co-Localized Emitters** CHRISTOPHER K ALMLIE, Oregon State University, Sean M Burrows, Kuan-Jen Chen, Karan A Patel

9:10 (370-3) **Development of Chitosan-Modified Near Infrared Fluorescent Graphene Oxide Nanocomposite** MINH H DUONG, University of North Dakota, Steve Xu Wu, Julia Xiaojun Zhao

9:30 (370-4) **Studying Chemical Reaction, Mass Transport and Their Coupling in 3D Multi-layer Catalysts at Single-Molecule Level** BIN DONG, Georgia State University, Chen Y Pei, Xiao X Chao, Wen Y Huang, Ning Fang

9:50 **Recess**

10:05 (370-5) **Slowing Down of Nanoparticle Diffusion on Sub-Micrometer Oil Droplet-Aqueous Buffer Interface Studied with Three Dimensional Tracking** YANING ZHONG, North Carolina State University, Luyang Zhao, Gufeng Wang

10:25 (370-6) **Two-Photon Metal Enhanced Fluorescence Properties of Gold Nanostars** LIXIA ZHOU, Oregon State University, Sean M Burrows

10:45 (370-7) **DNA-Aligner Controlled Nicking and Extension for Isothermal Amplification of Nucleic Acids** TAO ZHANG, Zhejiang University N/A

11:05 (370-8) **Construction of Convenient Biosensors Based on Optical Techniques** QIAOLI YUE, Liaocheng University, Chenzhong Li

ORAL SESSIONS — Session 380

FTIR and Terahertz Applications

Monday Morning, Room B316
Chin I Shyr, The Pittsburgh Conference, Presiding

8:30 (380-1) **EPA Methods 320 and 18 by Portable GC/FTIR for Total Source Emission Determination** MARTIN L SPARTZ, Prism Analytical Technologies, Inc., Anthony S Bonanno, Charles Mark Phillips, Peter P Behnke, Kelly R McPartland

8:50 (380-2) **A Novel Infrared Interferometer Suitable for 3D Infrared Hyperspectral Imaging** RYUJI TAO, Kagawa University, Akira Nishiyama, Kenji Wada, Ichiro Ishimaru

9:10 (380-3) **New Methodology for Finding Optimal Spectral Matches in Reference Databases** GREGORY BANIK, Bio-Rad Laboratories, Inc., Ty Abshear, Karl Nedwed

9:30 (380-4) **Experimental Optimization of IR pMAIRS Using A New Analytical Concept** TAKESHI HASEGAWA, Kyoto University, Nobutaka Shioya, Takafumi Shimoaka, Miyako Hada

9:50 **Recess**

10:05 (380-5) **ATR-FTIR Spectroscopic Imaging of Pharmaceutical Formulations Under Continuous Flow** SERGEI KAZARIAN, Imperial College London, Andrew Ewing

10:25 (380-6) **Advances in Pattern Recognition for the Remote Detection of Sulfur Dioxide by Passive Infrared Spectrometry** BRIAN W DESS, University of Iowa, Gary W Small

10:45 (380-7) **Demonstration of Detection of Hidden Persons and Illegal Substances with an Array of Quantum Cascade Lasers and Cantilever Enhanced Photoacoustic Spectroscopy** SAULI SINISALO, Gasera Ltd., Ismo Kauppinen

11:05 (380-8) **Measurement of Trace HF in Clean Rooms and Ambient Air** ISMO KAUPPINEN, Gasera Ltd., Tuomas Hieta, Timo Rajamäki, Sauli Sinisalo

ORAL SESSIONS — Session 390

GCMS of Environmental Analysis

Monday Morning, Room B401
Archana Kumar, Genentech, Presiding

8:30 (390-1) **Continuous Monitoring of Polycyclic Aromatic Hydrocarbons Using Automatic Thermal Desorption-Gas Chromatography** FRANCK AMIET, Chromatotec Inc., Michel Robert, Damien Bazin

8:50 (390-2) **Method Development for Evaluation of Pesticides Residue in Lake Lanier-Georgia Gwinnett County's Drinking Water Resource Using Disposable Pipette Extraction (DPX) and Gas Chromatography–Mass Spectrometry (GC-MS)** HONGXIA GUAN, Georgia Gwinnett College, Huang Wenlin, Xiaoping Li, Simon Mwongela, Rashad Simmons

9:10 (390-3) **Overcoming Cost and Supply: Let's Use Nitrogen** LEE MAROTTA, PerkinElmer, Jacob Rebholz, Roger Bardsley, Tom Hartlein

9:30 (390-4) **Analytical Method Development for 2,4,6,8-Tetrachlorodibenzothiohio-phene (TCDT) in River Sediments Utilizing GCxGC-TOFMS and APGC-TQS** MAURA K MCGONIGAL, Penn State, Frank Dorman, Kari Organtini, Robert Parette, Wendy Pearson, Doug Stevens, Adam Ladak

9:50 **Recess**

10:05 (390-5) **A Kendrick Mass Defect Approach Towards the Characterization of Hydraulic Fracturing Fluids** PAULINA PIOTROWSKI, The Pennsylvania State University, Frank Dorman, Jonathan D Byer, Joseph E Binkley

10:25 (390-6) **Collaboration of Government (EPA), Industry, Academia to Update EPA Method 625 to Solid Phase Extraction for Drinking and Wastewater Analysis Combining Stir-Bar Sorptive Extraction and EPA Method 6800** ANIL SRINIVAS CHAITANYA VISHNUVAJJHALA, Duquesne University, Weier Hao, Matt Pamuku, Andrew Boggess, Mizanur Rahman, Skip Kingston, David Singer, Ed Pfannkoch

10:45 (390-7) **Analysis of Vapor Mercury in Ambient Air and Flue Gas Emissions Using Thermal Desorption Trap Hyphenated with Gas Chromatography/Mass Spectrometry** PAOLO BENEDETTI, IIA - CNR, Carlo Crescenzi, Ettore Guerriero

11:05 (390-8) **Soil Gas: Are Targets Being Missed?** LEE MAROTTA, PerkinElmer, Roberta Provost

ORAL SESSIONS — Session 400

Innovative Approaches to Science Education - Half Session

Monday Morning, Room B409
Hubert MacDonald, The Pittsburgh Conference, Presiding

8:30	(400-1)	**Teaching Through Research (Mini) Projects of Industrial Relevance** JURICA BAUER, Inholland, Rosalba Bellini, Gertjan Heijne, Mark Jansen, Iris Kuiper, Maarten Kuiper, Lieke van Hemert, Mark Verheij, John Vessies, Niek Persoon
8:50	(400-2)	**Promote Science Education Through Collaborative Learning** YI HE, John Jay College/CUNY, Sandra Swenson
9:10	(400-3)	**Open Source Instruments for Environmental Monitoring and Science Education** JACK SUMMERS, Western Carolina University, Benjamin Hickman
9:30	(400-4)	**Application of Environmental Data for Public Health Response Actions** JAMES S HOLLER, ATSDR

ORAL SESSIONS — Session 410

Microfluidics/Lab-on-a-Chip - Bioanalytical and Others

Monday Morning, Room B406
Vishal Sahore, Brigham Young University, Presiding

8:30	(410-1)	**Droplet Microfluidic Device: Application of Nucleosome Preparation to Nucleosome Analysis** YI XU, University of Illinois at Urbana-Champaign, Richard M Graybill, Mallika V Modak, Jeong-Heon Lee, Tamas Ordog, Ryan C Bailey
8:50	(410-2)	**3D-Printed Analytical Devices Facilitate Investigation of Stored Erythrocytes Used in Transfusion Medicine** CHENGPENG CHEN, Saint Louis University, Dana Spence, R Scott Martin
9:10	(410-3)	**Investigation of Bacterial Behavior in Water Filtration Processes by Using Nanofluidic Devices** NIL TANDOGAN, Northeastern University, Kai-Tak Wan, Edgar D Goluch
9:30	(410-4)	**A Novel Chemiluminescence Signal Amplification Strategy Based on Microchip Electrophoresis Platform for Highly Sensitive MicroRNA Detection** SHULIN ZHAO, Guangxi Normal University, Jian Li, Jingjin Zhao, Yong Huang, Yi-Ming Liu
9:50		Recess
10:05	(410-5)	**Singe Molecule Nanoelectrophoresis within Thermoplastics** COLLEEN ONEIL, University of North Carolina at Chapel Hill, Swathi Pullagurla, Steven A Soper
10:25	(410-6)	**Parallel Functionality Determination of shRNA Knockdown Constructs Using Microfluidic Technologies** KRISTEN ENTWISTLE, Michigan State University, Dana Spence
10:45	(410-7)	**A Universal Droplet Microfluidic Strategy for On-Chip Operations Including Reagent Injection, Sample Washing, and Droplet Tagging** STEVEN R DOONAN, University of Illinois at Urbana-Champaign, Richard M Graybill, Dongkwan Lee, Yi Xu, Ryan C Bailey
11:05	(410-8)	**Reversible Wettability Switching in Fabric-Based Microfluidic Devices** TANYA NARAHARI, Northeastern University, Dhananjaya Dendukuri, Tripurari Choudhary, Shashi K Murthy

ORAL SESSIONS — Session 420

Neurochemistry

Monday Morning, Room B402
Rose Ann Clark, Saint Francis University, Presiding

8:30	(420-1)	**Fast-Scan Cyclic Voltammetry Measurements of Serotonin Release and Reser Pools in R6/2 Huntington's Disease Model Mice** RACHEL C GEHRINGER, The University of Kansas, Sam V Kaplan, Sarah Fantin, Michael A Johnson
8:50	(420-2)	**Combining Voltammetry, Mathematical Modeling and 2-Photon Microscopy to Correlate In Vivo Serotonin Chemistry to Physiology** AYA ABDALLA, University of South Carolina, Christopher W Atcherley, Yunju Jin, Michael L Heien, Janet Best, Michael C Reed, David Linden, Parastoo Hashemi
9:10	(420-3)	**Coregulation of Serotonin and Histamine in the Mouse Premammilary Nucleus** SRIMAL A SAMARANAYAKE, University of South Carolina, Aya Abdalla, Rhiannon Robke, H Frederick Nijhout, Michael C Reed, Janet Best, Parastoo Hashemi
9:30	(420-4)	**Application of In Vivo Fast-scan Cyclic Voltammetry to Examine the Impact of (+)-Methamphetamine of the Regulation of Norepinephrine and Dopamine** KEN WAKABAYASHI, State University of New York at Buffalo, Jinwoo Park
9:50		Recess
10:05	(420-5)	**Characterization of Spontaneous Transient Adenosine in Rat Brain Slices** SCOTT T LEE, University of Virginia, B Jill Venton
10:25	(420-6)	**Comparing Contralateral Catecholamine Release with Simultaneous Voltammetric Measurements** MEGAN E FOX, University of North Carolina at Chapel Hill, R Mark Wightman
10:45	(420-7)	**Insights into the Effects of Non-stimulant ADHD Drugs on Catecholamine Transmission in the Rat Brain Using In Vivo Fast-scan Cyclic Voltammetry** ROHAN BHIMANI, The State University of New York at Buffalo, Jinwoo Park
11:05	(420-8)	**Investigation of Enkephalin-Evoked Catecholamine Secretion in Adrenal Tissue by Fast-Scan Cyclic Voltammetry** LARS DUNAWAY, North Carolina State University, Leslie A Sombers

ORAL SESSIONS — Session 430

Pharmaceutical-GC and LC

Monday Morning, Room B403
Sam Subramaniam, Miles College, Presiding

8:30	(430-1)	**Analyte Stability and Side Reactions During Pharmaceutical Analysis** LEAH XIONG, Merck Co., Nathan D Contrella, Leih-Shan Yeung, Cory Bottone, Brian Regler, Justin Pennington, Eugenia Muschajew
8:50	(430-2)	**A New Tool for Residual Solvent Analysis of Technical Grade Hexanes** KENNETH G LYNAM, Agilent Technologies, John Oostdijk, Johan Kuipers, Ramaprasad Ganni
9:10	(430-3)	**Development and Validation of a Fast Ion-Paired Reversed Phase Stability-Indicating Method for the Assay of Thiabendazole and Estimation of its Related Compounds** PENG ZHANG, Merial, A Sanofi Company, Jingzhi Tian, Abu Rustum
9:30	(430-4)	**Determining Lithium in Pharmaceutical Products** SACHIN P PATIL, Thermo Fisher Scientific, Jeffrey Rohrer
9:50		Recess
10:05	(430-5)	**Two Dimensional Liquid Chromatography (2D-LC), A "Must-Have" Tool in Pharmaceutical Development, When 1D-LC is Inadequate** IMAD A HAIDAR AHMAD, Novartis, Adrian Clarke, James Tam, Xue Li, Thomas Tarara, Andrei Blasko
10:25	(430-6)	**Enantioresolution of Several Amino Alcohol Drugs Containing Multiple Stereogenic Centers Using Immobilized Polysaccharide-Based HPLC Chiral Stationary Phases** MOHAMED HEFNAWY, King Saud University N/A
10:45	(430-7)	**Introducing Modern LC Column Technology into a Research-Led Pharmaceutical Teaching Environment** WILLIAM J LOUGH, University of Sunderland
11:05	(430-8)	**Improved Gas Chromatographic (GC) Quantification of Acidic and Neutral Cannabinoids** AMANDA RIGDON, Restek Corporation, Jack Cochran, Joan Serdar, Corby Hilliard, Linx Waclaski, Rebecca Stevens

ORAL SESSIONS — Session 440

Synthesis and Characterization of Nano Particles

Monday Morning, Room B405
Ashish Tripathi, Leidos, Inc., Presiding

8:30	(440-1)	**Antibacterial Activities and Cytotoxicity of Green Synthesized Stable Gold Nanoparticles from Flavonoid Derivatives** FRANCIS JUMA OSONGA, Binghamton University, Idris Yazgan, David C Luther, Apryl P Jimenez, Phuong N Lee, Omowunmi Sadik
8:50	(440-2)	**Wavelength Selective Photocatalysis Using Gold-Platinum Nano-Rattles** MAHMOUD MAHMOUD, Georgia Institute of Technology, Batyr Garlyyev, Mostafa A El-Sayed
9:10	(440-3)	**Reversible Electron Delocalization of Molecule-Like CdSe Nanoclusters Using Z-Type Ligand Functionalization** KATIE N LAWRENCE, Indiana University-Purdue University Indianapolis
9:30	(440-4)	**Characterization Techniques for Nanomaterials – An Overview** CHADY STEPHAN, PerkinElmer
9:50		Recess

10:05 (440-5) **A Novel Nanoparticle Tracking Analysis System for Improved Determinations of Nanoparticle Concentration and Size Distribution** DARIUSZ STRAMSKI, Manta Instruments Inc, Kuba Tatarkiewicz, Rick A Reynolds, Monette Karr, Rick Cooper

10:25 (440-6) **Coupled Calorimetric-Manometric Technique for the Study of Sorption and Thermodynamic Properties of Macroscopic and Nanosized Materials** KRISTINA LILOVA, Setaram Inc., Link Brown

10:45 (440-7) **Exploring the Effects of Surface Ligand Structural Parameters on Exciton Delocalization of CdSe Nanocrystals** MEGHAN TEUNIS, Indiana University-Purdue University Indianapolis

11:05 (440-8) **Metallic and Hybrid Nanomaterials: Fabrication and Applications** SIMONA HUNYADI MURPH, Savannah River National Lab

POSTER SESSION — Session 450

All posters are to be mounted by 10:00 AM and remain on display until 4:00 PM. Authors must be at their posters from 10:00 AM to 12:00 PM. Location of the morning posters is on the Exposition Floor, 400 Aisle. PLEASE NOTE: You cannot get onto the Exposition Floor until after 9:00 AM.

Agriculture

Monday Morning, Exposition Floor, 400 Aisle

(450-1 P) **Growth Performance of Chicory to Plant Growth Regulators** RAMESHBHAI P DABHI, J & J College of Science, Alpesh M Patel, Maheshkumar B Chauhan N/A

(450-2 P) **Analysis of Sugarcane Growing Soils at Kakamega North District Kenya for Micro and Macro Nutrients** ONDITI O ANAM, Jomo Kenyatta University of Agriculture and Technology N/A

(450-3 P) **Fast and Precise Nitrogen and Carbon Determination Using an Elemental Analyzer** GUIDO GIAZZI, Thermo Fisher Scientific, Liliana Krotz, Francesco Leone, Dominique Chevalier

(450-4 P) **Carbon/Nitrogen Ratio in Soils and Plants Using an Elemental Analyzer** GUIDO GIAZZI, Thermo Fisher Scientific, Liliana Krotz

(450-5 P) **Development of Electrochemical Detection System Combining with Nitrocellulose Membrane for Quantitative Immunochromatography** WATARU IWASAKI, Natl. Inst. Adv. Ind. Sci. Technol. (AIST), Mizuki Ryu, Ramachandra Rao Sathuluri, Ryoji Kurita, Osamu Niwa, Masaya Miyazaki

POSTER SESSION — Session 460

All posters are to be mounted by 10:00 AM and remain on display until 4:00 PM. Authors must be at their posters from 10:00 AM to 12:00 PM. Location of the morning posters is on the Exposition Floor, 400 Aisle. PLEASE NOTE: You cannot get onto the Exposition Floor until after 9:00 AM.

Clinical/Toxicology

Monday Morning, Exposition Floor, 400 Aisle

(460-1 P) **Benzo[a]pyrene Levels in Mainstream Smoke from Spectrum Research Cigarettes** JARED HUGHES, Centers for Disease Control and Prevention, Wayne Chen, Bryan Hearn, Shirley Ding, Clifford Watson

(460-2 P) **Selected Carbonyl Levels in Mainstream Smoke from Spectrum Research Cigarettes** MICHELE CHAN, Centers for Disease Control and Prevention, Shirley Ding, Clifford Watson, Xizheng Yan

(460-3 P) **Absolute Quantification of Apolipoproteins in Serum and the Efficacy of Trypsin While Utilizing Ultra Performance Liquid Chromatography - Isotope Dilution Mass Spectrometry (UPLC-IDMS)** MICHAEL L ANDREWS, Centers for Disease Control and Prevention, Bryan A Parks, Christopher Toth, Jeffrey Jones, Kuklenyik Zsuzsanna, Michael S Gardner, Jon Rees, David Schieltz, John R Barr

(460-4 P) **Mainstream Smoke Deliveries of Tobacco Specific Nitrosamines in Spectrum Clinical Research Cigarettes** PATRICK CHEN, Centers for Disease Control and Prevention, Clifford Watson, Shirley Ding, Liqin Zhang, Josh Wong

(460-5 P) **Separation of Vitamin D2 and D3 for Clinical Application** MARK WOODRUFF, Fortis Technologies, Ken Butchart

(460-6 P) **Measurement of Ammonia Emanating from Human Skin as a Possible Biomarker for Physical/Mental Stress Responses** SHOTA FURUKAWA, Tokai University, Minami Takahashi, Keita Kimura, Shiro Ikeda, Yoshika Sekine, Asai Satomi, Umezawa Kazuo, Hayato Miyachi

(460-7 P) **Measurement of Acetic Acid Emanating from Human Skin as a Potential Biomarker for Quality of Sleep** MINAMI TAKAHASHI, Tokai University, Furukawa Shota, Keita Kimura, Yoshika Sekine, Satomi Asai, Kazuo Umezawa, Miyachi Hayato

(460-8 P) **LC-MS/MS Determination of Interactions Between Sunitinib and Green Tea Polyphenol by Equilibrium Dialysis** MATTHEW VERGNE, Lipscomb University, Lincoln Shade

(460-9 P) **Identification of Hypocretin-1 in Cerebrospinal Fluid: A Potential Diagnostic Biomarker for Narcolepsy** HEMASUDHA CHATRAGADDA, Duquesne University, Skip Kingston, Birgitte R Kornum, Matt Pamuku

(460-10 P) **ICP-MS – A Perfect Tool for the Bio-monitoring of Trace Elements in Body Fluids** EWA M PRUSZKOWSKI, PerkinElmer, Inc.

(460-11 P) **Incorporation of Amphiphilic Dendrimers in Supported Lipid Bilayers to Enhance Stability and Functionalization** CHARLES J RUIZ, University of California, Riverside, Samuel S Hinman, Quan Cheng, Ling Peng

(460-12 P) **Chiral Capillary Electrophoresis-Mass Spectrometry: Turning an Analytical Technique into High Throughput Screening of Chiral Compounds Using Novel Polymerized beta-D-Glucopyranoside Surfactants** LIU YIJIN, Georgia State University, Shahab S Shamsi

(460-13 P) **Measurement of Diacetyl and 2-Nonenal Emanating from Human Skin by Passive Flux Sampler** KEITA KIMURA, Tokai University, Shota Furukawa, Minami Takahashi, Yoshika Sekine, Kazuo Umezawa, Satomi Asai, Hayato Miyachi

(460-14 P) **Aptamer-Modified Gold Nanoparticles Coupled with Nitrocellulose Membranes for Detection of Thrombin by LDI-MS** CHIA-YIN CHANG, National Taiwan Ocean University

(460-15 P) **On-line Membrane Assisted Distillation Coupled with Ion Chromatography: A Novel Approach to Determine Trace Fluoride in Serum** LOU CHAOYAN, Zhejiang University

(460-16 P) **SERS Probe for Rapid Detection of Erythropoietin in Urine** UGUR TAMER, Gazi University, Yesim Selbes, Gokhan M Caglayan, Nejdet Saglam N/A

(460-17 P) **Electrochemical Measurement of Thyroid Hormone for Rapid Diagnostics Technology** BARBARA CATA, Northern Kentucky University, Celeste A Morris

(460-18 P) **Forensics Meets Green Chemistry: Removing Known Carcinogens from Blood Alcohol Content Protocols for Safer Applied Spectroscopy Laboratories** SARAH E GRAY, Armstrong State University, Mathew Holmes

(460-19 P) **Preparation and Evaluation of Nifedipine-Cyclodextrin Complex Microspheres with In-Vitro Studies** SUNILKUMAR H CHAUDHARI, Mantri Health Care, Shailesh Jain N/A

(460-20 P) **A Rapid and Sensitive SERS Based Measles Detection** RAMESH KATTUMENU, Argent Diagnostics Inc

(460-21 P) **The Efficiency of Multi-Sample Analysis Using Dual Gradient LCMS System** WATANABE SATORU, Shimadzu Corporation, Nakayama Daisuke, Yamaguchi Tadayuki, Inohana Yusuke

POSTER SESSION — Session 470

All posters are to be mounted by 10:00 AM and remain on display until 4:00 PM. Authors must be at their posters from 10:00 AM to 12:00 PM. Location of the morning posters is on the Exposition Floor, 400 Aisle. PLEASE NOTE: You cannot get onto the Exposition Floor until after 9:00 AM.

Drug Discovery

Monday Morning, Exposition Floor, 400 Aisle

(470-1 P) **Automated Mass-Directed Purification of Bioactive Peptides** LAINE STEWART, Gilson, Inc., Luke Roenneburg, Karen Kleman N/A

(470-2 P) **Synthesis of Nanosized Poly B-aminoester Holding Chlorambucil Drug as Slow Release Drug System for Antitumer** FAHIMA MOSAAD HELALY, National Research Centre (NRC)

(470-3 P) **Synthesis, Characterization and Antibacterial Studies of Cobalt Complexes of Isomeric Aminophenol Schiff Bases** TOLULOPE M FASINA, University of Lagos, Felicia N Ejiah, Oluwole B Familoni, Neerish Revaprasadu

(470-4 P) **Mass Spectral and Chromatographic Studies on Substituted Cathinones: Bath Salt-type Aminoketone Designer Drugs** YOUNIS ABIEDALLA, Auburn University, Randall Clark, Jack DeRuiter, Karim Abdel-Hay

(470-5 P) **An In Vitro Microfluidic Platform to Unravel Mechanisms of Action of Drug Therapies used in Multiple Sclerosis** TIFFANY BELL, Michigan State University, Dana Spence

(470-6 P) **Streamlining Compound Isolation Automatically with UPLC to Prep Chromatography Using Mass-Directed Auto Purification** JO-ANN M JABLONSKI, Waters Corporation, Andrew J Aubin, Wendy Harrop, Thomas E Wheat

(470-7 P) **Phytochemical Screening and Antimicrobial Activity of *Boerhavia Verticillata Poir.* and *Fagonia Schweinfurthii Hadidi*** YAMINI A JOSHI, M G Science Institute N/A

(470-8 P) **Aluminum-Based MOF Composite as Polymer Monoliths for Microextraction of Sulfonamides** YUNG-HAN SHIH, Chung Yuan Christian University, Kuen-Yun Wang, Hsi-Ya Huang N/A

(470-9 P) **Preparation and Study of Tramadol Imprinted Micro-and Nanoparticles by Precipitation Polymerization: Microwave Irradiation and Conventional Heating Method** MARYAM HASSANPOUR MOGHADAM, Pharmaceutical Research Center, School of Pharmacy N/A

(470-10 P) **Spectral Characterization of Cytochrome P450cam Active Site Using NMR Methods Including ^{13}C-Doubled Filtered ^{1}H-^{1}H Noesy Experiments for Mapping Distances** REMIGIO USAI, Marquette University, James R Kincaid, Daniel S Sem, Daniel Kaluka

(470-11 P) **Synthesis, Anti-TB and Crystallography of Benzothiazole Analogues** KATHARIGATTA NARAYANASWAMY VENUGOPALA, King Faisal University, Al-Dhubiab Bander N/A

(470-12 P) **Inhibition of Human Cytochrome P450 Enzymes by Rottlerin, A Naturally Occurring Constituent of Mallotus Philippensis Using an In-Vitro Cocktail Approach and Liquid Chromatography-Tandem Mass Spectrometry** ATUL S RATHORE, Bharati Vidyapeeth Deemed University, L Sathiyanarayanan, Kakasaheb R Mahadik N/A

(470-13 P) **Photostability of Pharmaceutical Drug Substance as Free Acid and Salt** JENNY WANG, Genentech Inc., Geoffrey Yeh, Lulu Dai, Christine Gu, Kelly Zhang

POSTER SESSION — Session 480

All posters are to be mounted by 10:00 AM and remain on display until 4:00 PM. Authors must be at their posters from 10:00 AM to 12:00 PM. Location of the morning posters is on the Exposition Floor, 400 Aisle. PLEASE NOTE: You cannot get onto the Exposition Floor until after 9:00 AM.

Food Safety

Monday Morning, Exposition Floor, 400 Aisle

(480-1 P) **Analysis of Local Brews in Jua, Kiambu County, Kenya, for Methanol and Ethanol by Gas Chromatography** ONDITI O ANAM, Jomo Kenyatta University of Agriculture and Technology, Momanyi M Moraa N/A

(480-2 P) **Cannabinoids and Residual Solvents by Headspace GC** TIM ANDERSON, Phenomenex, Kristen Parnell, Ramkumar Dhandapani

(480-3 P) **Melamine in Pet Food** TIM ANDERSON, Phenomenex, Kristen Parnell, Ramkumar Dhandapani

(480-4 P) **Impact of HPLC Stationary Phase Selection on Matrix Effects During LC-MS/MS Analysis of Multiple Mycotoxins in Corn** EMILY R BARREY, Supelco/Sigma-Aldrich, Lynne Perez-Blanco, Olga I Shimelis, Michael Ye, Jennifer Claus

(480-5 P) **UHPLC/MS/MS Analysis of Lipophilic Marine Toxins from Homogenized Shellfish** EMILY R BARREY, Supelco/Sigma-Aldrich, Olga I Shimelis, Michael Ye, Jennifer Claus

(480-6 P) **Utilizing Mass Spectrometry for Gluten Detection for Use in Gluten-Free Foods** SOPHIE BROMILOW, University of Manchester, Lee Gethings, Peter Shewry, Michael Buckley, Michael Bromley, Phil Johnson, Clare Mills

(480-7 P) **Easy Development of an MRM Method for Analysis of Environmental Contaminants in Milk Using SPME and GC-MS/MS** NICOLE M LOCK, Shimadzu Scientific Instruments, Di Wang, Laura Chambers, Brahm Prakash, Robert Clifford, Shilpi Chopra N/A

(480-8 P) **Analysis of Target Pesticides in Essential Oils Using a Novel GC/MS/MS System** THOMAS DILLON, PerkinElmer, Samuel Tolley, Adam J Patkin, Sharanya Reddy

(480-9 P) **A Raman Spectroscopic Method for Determination of Erucic Acid in Canola Oils** ELIF ERCIOGLU, Hacettepe University, Ismail H Boyaci, Tumay H Temiz, Serap Durakli Velioglu, Murat H Velioglu

(480-10 P) **Multiple Foodborne Pathogen Analysis Using a 96-Well Assay in Less than 5 Hours** STUART FARQUHARSON, Real-Time Analyzers, Inc, Chetan Shende, Kathryn Dana

(480-11 P) **Stand-Off Raman Detection of Adulteration in Honey** KENNETH GARCIA, Alabama A&M University, Carlton Farley, Aschalew Kassu, Anup Sharma

(480-12 P) **Evaluating the Extraction Efficiency of Food Borne Pathogens on Automated Homogenizer Platforms** SHARI GARRETT, Omni International, James Atwood, Brandon Easparro

(480-13 P) **Nitrogen/Protein Determination in Infant Food by Dumas Combustion Method in Alternative to Kjeldahl Method** GUIDO GIAZZI, Thermo Fisher Scientific, Liliana Krotz

(480-14 P) **Non-Targeted Screening of Nutritional Supplements with GC, GC×GC, TOFMS, and HR-TOFMS** ELIZABETH M HUMSTON-FULMER, Leco Corporation, David E Alonso, Jonathan D Byer, Joseph E Binkley, Lorne E Fell

(480-15 P) **Synthesis, Physico-Chemical Characterization and Potential Biological Activity of Newly Synthetised Cu(II) and Ni(II) Complexes of N-(Benzyloxycarbonyl)-1H-pyrazOle-1-Carboxamidine (HL) ligand** ZELJKO JACIMOVIC, Faculty of Metallurgy and Technology, Milica Kosovic, Goran Bogdanovic, Sladjana Novakovic, Nedeljko Latinovic, Gerald Giester N/A

(480-16 P) **Combining SERS with Liquid-Liquid Extraction Method for Simple, Rapid Detection of Rhodamine B in Raw Food** HUAIZHI KANG, Xiamen University

(480-17 P) **Concurrent High Sensitivity Conductometric Detection of Sulfide and Cyanide in a Suppressed Anion Chromatography System** HONGZHU LIAO, University of Texas at Arlington, Purnendu Dasgupta, Akinde Kadjo

(480-18 P) **A Computational Study Assisted Nano-Aptasensor for Detection of Tetracycline in Honey with a DNA Aptamer** SAI WANG, Beijing Univerisity of Chemical Technology

(480-19 P) **Simple, Rapid Extraction of Chlorinated Pesticides in Poultry Fat by Solid Phase Extraction and GC/ECD** ALLEN MISA, Phenomenex, Ramkumar Dhandapani, Tim Anderson

(480-20 P) **Approaches to Measuring both Trace and Nutritional Elements in Food in a Single Analysis by ICP-MS** KENNETH NEUBAUER, PerkinElmer, Stan Smith

(480-21 P) **Comparative Evaluation of the Antibacterial Activities of the Essential Oils of Rosmarinus Officinalis L. Obtained by Hydrodistillation and Solvent Free Microwave Extraction Methods** OMOBOLA OLURANTI OKOH, University of Fort Hare N/A

(480-22 P) **Impact of Thermal Processing on the Solubility and Detection of Peanut Allergens Using LC-MS/MS based Targeted Proteomics with Multiple-Reaction Monitoring (MRM)** REBEKAH L SAYERS, University of Manchester, Phil E Johnson, Justin T Marsh, Clare Mills, Helen Brown

(480-23 P) **Measurement of Arsenic in Wine by Hydride Generation – Flame Atomic Absorption** NICK SPIVEY, PerkinElmer Inc., Kenneth Neubauer, Stan Smith

(480-24 P) **Determination of Organic Tin Pesticides in Fruits and Vegetables by Gas Chromatography Coupled to Tandem Mass Spectrometry** XIAOBO LIU, Shimadzu

(480-25 P) **Determination of 198 Pesticide Residues in Eggplant Using Gas Chromatography Tandem Mass Spectrometry/Mass Spectrometry (GCMS/MS)** WANG YAN, Shimadzu

(480-26 P) **A Direct Ultra-Performance Liquid Chromatography-Mass Spectrometry Method for the Simultaneous Quantitation of Fatty Acids in Olive Oils of Different Origins** ZEID ABDULLAH ALOTHMAN, King Saud University N/A

(480-27 P) **Development of On-Line SFE-SFC System and Its Application for Food Contaminant Analysis** SHIN-ICHI KAWANO, Shimadzu (China) Co., Ltd., Xiaohua Liu, Lingling Shen, Yan Wang, Taohong Huang, Naoki Hamada, Yuki Hashi

(480-28 P) **High Resolution Accurate Mass (HRAM) Collision Energy Profile of Residues of Concern for Food Safety** DANIEL BIGGERSTAFF, o2si Smart Solutions, Huichen Stavros, Min Cai

(480-29 P) **Quantification of Pesticide Residues in Fruits and Vegetables by Gas Chromatography - Mass Spectrometry** KAELYB SUCHEVITS, California University of PA

(480-30 P) **Study of Sugar and Humectant Profiles in Smokeless Tobacco Products Using an LC-ESI-MS/MS Method** LIQUN WANG, Battelle, Roberto Bravo, Stephen Stanfill, Liza Valentin, Clifford Waston

(480-31 P) **High-Throughput Screening of Domoic Acid in Shellfish by Laser Ablation Electrospray Ionization (LAESI)-MS** PEARSE MCCARRON, National Research Council, Kelley Reeves, Callee M Walsh, Pamela Cantrell, Wade A Rourke, Sinead O'Brien, Daniel Beach

(480-32 P) **Development and Implementation of a Fast, Reliable and Sensitive Analytical Test for Determining Methylmercury in Fish** ANA MARIA MUÑOZ, Lasallian University , Claudio Jiménez, Daniel E León

(480-33 P) **Characterization of Adulterated Argan Oil Using a Portable Gas Chromatography Mass Spectrometer** PRESHIOUS REARDEN, 1st Detect Corporation, Parminder Kaur N/A

(480-34 P) **A New Insight Into Fish Meat Freshness: ZnO /PPy Modified Biosensor** BUKET SAHYAR (YALCIN), Indesit Company, Merve Kaplan, Ahmet Yavas, Mehmet Ozsoz, Erdal Celik, Semih Otles, Ömer Mindivanli, Metin Ozgul

POSTER SESSION — Session 490

All posters are to be mounted by 10:00 AM and remain on display until 4:00 PM. Authors must be at their posters from 10:00 AM to 12:00 PM. Location of the morning posters is on the Exposition Floor, 400 Aisle. PLEASE NOTE: You cannot get onto the Exposition Floor until after 9:00 AM.

Microfluidics/Lab-on-a-Chip

Monday Morning, Exposition Floor, 400 Aisle

(490-1 P) **A Simple and Sensitive Paper-Based Device Coupling Electrochemical Sample Pretreatment and Colorimetric Detection** WILLIAM REIS DE ARAUJO, USP, Thalita G Silva, Thiago Paixao

(490-2 P) **Mixing Reaction on Paper-Based Analytical Devices** CHING MAN CHOY, California State Polytechnic University, Pomona, Yan Liu

(490-3 P) **Combining Vibrational Spectroscopy with Microfluidics** KATHLEEN E BERG, Colorado State University, Scott D Noblitt, Monpichar Srisa-Art, Amber T Krummel, Charles Henry

(490-4 P) **Prostate Cancer Biomarker Detection Using a 16-Sensor Electrochemical Microfluidic Immunoarray** ABBY JONES, University of Connecticut, Brunah A Otieno, Colleen Krause, Mohammed Sherafeldin, Amit Joshi, James F Rusling

(490-5 P) **Real-Time Profiling of Pancreatic Hormone Secretion Dynamics Using an in Flow Fluorescence Polarization Immunoassay** NIKITA MUKHITOV, Florida State University, Adrian M Schrell, Michael G Roper

(490-6 P) **Alternating Current Driven Electroosmotic Pumping through Conical Pore Membrane** XIAOJIAN WU, University of Florida, Charles R Martin

(490-7 P) **Two-Fold Control of Pressure and Flow-Rate for Flow Control and Quality Management in Fluidic Processes** ANNE LE NEL, Fluigent, Thibaut Thupnot, Benjamin Rouffet, Nicolas Petit

(490-8 P) **A 3D-printed Device for High-Throughput Membrane-Based Cellular Analysis** RUIPENG MU, Michigan State University, Dana Spence

(490-9 P) **3D-Printed Tools to Enhance Targeted Drug Therapy** CODY WAYNE PINGER, Michigan State University, Dana Spence

(490-10 P) **Fluorescence-Based Quantification of Oxygen in Paper-Based Cultures of Mammalian Cells** MATTHEW W BOYCE, University of North Carolina at Chapel Hill, Andrew S Truong, Rachael M Kenney, Matthew R Lockett

(490-11 P) **Methods for Quantifying Hypoxia and the Hypoxic Responses of Cells in Paper-Based Invasion Assays** ANDREW S TRUONG, University of North Carolina at Chapel Hill, Rachael M Kenney, Matthew W Boyce, Christian A Lochbaum, C Chad Lloyd, Matthew R Lockett

(490-12 P) **A Microfluidic Copper Detection System Incorporating a Ratiometric Fluorescent Quantum Dot Pair** SUMATE PENGPUMKIAT, Oregon State University, Yuanyuan Wu, Anukul Boonloed, Chandima Bandara, Vincent T Remcho

(490-13 P) **Ultra-Thin Layer Chromatography with Integrated Silver Colloid-based SERS Detection** RYAN A WALLACE, University of Tennessee, Knoxville, Nickolay V Lavrik, Michael J Sepaniak

(490-14 P) **Affinity Cell Separation Based on Surface Antigen Expression Difference in a Sequential Concentration Microfluidic Chip** YE ZHANG, Texas Tech University, Dimitri Pappas

(490-15 P) **Microscale Size-Based Sorting with Capillary Electrophoresis and Phospholipid Additives** CASSANDRA CRIHFIELD, West Virginia University, Lisa A Holland

(490-16 P) **Kinetic Characterization of Neurotransmitters Release from Neurons Cells Using MicroChip Electrophoresis--ESI/Mass Spectrometry** XIANGTANG LI, Jackson State University, Shulin Zhao, Yiming Liu N/A

(490-17 P) **High-Throughput Microfluidic Isolation and Analysis of Exosomes** KRISTINA M HERRERA, University of North Carolina at Chapel Hill, Steven A Soper

(490-18 P) **Quick Production of Microfluidic Devices by Laser Engraving of Wax-Coated Glass Slides** MAURO SERGIO FERREIRA SANTOS, Clemson University, Eric T da Costa, Hong Jiao, Claudimir L do Lago, Ivano G R Gutz, Carlos D Garcia

(490-19 P) **Microfluidic Platform for Mass Spectrometry-Based Monitoring of Protein-ligand Binding Dynamics** YONGZHENG CONG, Pacific Northwest National Laboratory, Cameron Trader, Ryan T Kelly, Daniel Orton, Erin S Baker, Tao Geng

(490-20 P) **Low-Cost Microfluidic Devices for the Determination of Renal Health** CHRISTOPHER A HEIST, Oregon State University, Vincent T Remcho, Joel C Pommerenck

POSTER SESSION — Session 500

All posters are to be mounted by 10:00 AM and remain on display until 4:00 PM. Authors must be at their posters from 10:00 AM to 12:00 PM. Location of the morning posters is on the Exposition Floor, 400 Aisle. PLEASE NOTE: You cannot get onto the Exposition Floor until after 9:00 AM.

Process Analytical Techniques

Monday Morning, Exposition Floor, 400 Aisle

(500-1 P) **Characterization and Monitoring of Cavitation Through Its Acoustic Emission** NOEMIE CAILLOL, IDEEL, Franck Baco-Antoniali, Sebastien Leinardi, Sylvain Charquet, Serge Henrot, Pascal Pitiot, Davide Zonca, Christine Villard

(500-2 P) **Monitoring of Tablet Coating Using Raman Spectroscopy** HOEIL CHUNG, Hanyang University, Jaejin Kim, Young Ah Woo

(500-3 P) **Simple and Rapid Determination of Polyanion-Polycation Binding Ratio Using Pulsed Chronopotentiometry with Polyion-Selective Electrodes** EMMA GORDON, Northern Kentucky University, Kebede L Gemene

(500-4 P) **How Can Your Process Benefit From External Flow Cells and Optimized Pump Heads?** KATHRYN E MONKS, Knauer, Ingo Piotrowski

(500-5 P) **Applications of a New Wear Resistant, Chemically Inert Coating that Improves Reliability, Lifetime and Accuracy of Process, Analytical and Sampling Systems** LUKE PATTERSON, SilcoTek Corporation, Min Yuan, David Smith

(500-6 P) **Withdrawn**

(500-7 P) **Online Analysis Using LIBS for Industrial Process** RONALD BERGER-LEFÉBURE, IVEA

(500-8 P) **Non-Porous, No Glass, Leak-Free Reference Electrode** ZIAD H TAHA, Innovative Instruments, Inc.

(500-9 P) **Suggested QC Practices for On-line Analyzers** WILLIAM LIPPS, Shimadzu Scientific Instruments

(500-10 P) **A Novel Approach to Cleaning Validation for Pharmaceutical Manufacturing by Online SFE-SFC** KENICHIRO TANAKA, Shimadzu Scientific Instruments, Inc., William Hedgepeth, Daisuke Nakayama, Hidetoshi Terada, Minori Nakashima, Tadayuki Yamaguchi

(500-11 P) **Advances in Raman Analyzers for In Situ Studies of Small Volume Liquid-phase Reactors** LISA GANSTER, Kaiser Optical, Ian Lewis

(500-12 P) **Establishing Raman Spectroscopy for the Process Environment** LISA GANSTER, Kaiser Optical, Karen Esmonde-White, Harry Owen, Ian Lewis

(500-13 P) **In situ Raman Measurements of Pharmaceutical Solids During Process Unit Operations** LISA GANSTER, Kaiser Optical, Karen Esmonde-White, Carsyen Uerpmann, Sean Gilliam, Ian Lewis

(500-14 P) **Recovery of Challenging Compounds in Cleaning Validation Using Total Organic Carbon (TOC) Analysis** JENNY G WATSON, GE Analytical Instruments, David Wayne

POSTER SESSION — Session 510

All posters are to be mounted by 10:00 AM and remain on display until 4:00 PM. Authors must be at their posters from 10:00 AM to 12:00 PM. Location of the morning posters is on the Exposition Floor, 400 Aisle. PLEASE NOTE: You cannot get onto the Exposition Floor until after 9:00 AM.

Safety

Monday Morning, Exposition Floor, 400 Aisle

(510-1 P) **A Cheap and Simple Approach to Monitor Sun Exposure based on Photocatalytic Properties of Titanium Dioxide** PARISA SOWTI, University of New South Wales

(510-2 P) **A Wide Selection of New Psychoactive Substances Investigated with Proton Transfer Reaction – Mass Spectrometry within the Marie Curie Training Network Proton Ionization Molecular Mass Spectrometry (PIMMS)** MATTEO LANZA, IONICON Analytik, W Joe Acton, Philipp Sulzer, Kostiantyn Breiev, Simone Jürschik, Alfons Jordan, Eugen Hartungen, Gernot Hanel, Jens Herbig, Lukas Märk, Christian Lindinger, Chris A Mayhew, Tilmann D Maerk

POSTER SESSION — Session 520

All posters are to be mounted by 10:00 AM and remain on display until 4:00 PM. Authors must be at their posters from 10:00 AM to 12:00 PM. Location of the morning posters is on the Exposition Floor, 400 Aisle. PLEASE NOTE: You cannot get onto the Exposition Floor until after 9:00 AM.

Sampling and Sample Preparation - Pharmaceutical, Clinical/Toxicology, Food Safety, and Others

Monday Morning, Exposition Floor, 400 Aisle

(520-1 P) **Preserving the Free-radical Scavenger Activity of Key Bioactives During Extraction and Purification Processes** VALERIE DESYROY, SiliCycle, Vincent Bedard, Genevieve Gingras, Denis Boudriau, Aurelie Meng

(520-2 P) **Determination of Quinolones in Bovine Kidney Using Hollow-Fiber Supported Liquid Membrane (HF-SLM) and LC-ESI+MS/MS** SIMISO DUBE, University of South Africa N/A

(520-3 P) **Working with Challenging Samples, Sampling Systems** YVES GAMACHE, Analytical Flow Products

(520-4 P) **Analytical Method Development for Cleaning Verification of Manufacturing Equipment - Exploring the Effect of the Cleanliness of Stainless Steel Coupons on Sample Recovery** IMAD A HAIDAR AHMAD, Novartis, James Tam, Xue Li, Thomas Tarara, Andrei Blasko

(520-5 P) **Evaluation of Antibacterial and Wound Healing Properties of Hydro-ethanolic Extracts of Gossypium Barbadense Leaves** NWAMAKA H IGBOKWE, University of Lagos, Eugene E Ikobi, Cecilia I Igwilo

(520-6 P) **Evaluating the Potential for Cross Contamination when Performing 96-Well Sample Preparation Prior to LC-MS/MS Analysis** HELEN LODDER, Biotage GB Limited, Lee Williams, Victor Vandell N/A

(520-7 P) **Acrylamide from Coffee Using a Simplified Liquid** ALLEN MISA, Phenomenex, Matthew Brusius, Zeshan Aqeel

(520-8 P) **Analysis of Food Grains Using Automated Block Digestion** MICHAEL A RUTZKE, Cornell University, Suhas Narkhede, Nick McLeod

(520-9 P) **Does Weather Effect Pipetting?** GEORGE W RODRIGUES, Artel, Emily Avis, Alf Price, Verena Gerbe

MONDAY, MARCH 7, 2016
AFTERNOON

AWARDS — **Session 530**

LCGC Lifetime Achievement and Emerging Leader in Chromatography Award
arranged by Laura Bush, LCGC & Spectroscopy

Monday Afternoon, Room B314
Laura Bush, LCGC & Spectroscopy, Presiding

1:30		**Introductory Remarks - Laura Bush**
1:35		**Presentation of the LCGC 2016 Lifetime Achievement in Chromatography Award to Milton L Lee, Brigham Young University, by Laura Bush, LCGC & Spectroscopy**
1:40	(530-1)	**Columns in Small-Scale Chromatography** MILTON L LEE, Brigham Young University
2:15	(530-2)	**Recent Chromatographic and Mass Spectrometric Developments Applied to the Characterization of Recombinant Proteins, Monoclonal Antibodies and Antibody-Drug Conjugates** PAT SANDRA, RIC, Koen Sandra
2:50	(530-3)	**Analytical Glycoscience: Quo Vadis?** MILOS V NOVOTNY, Indiana University
3:25		**Recess**
3:40		**Presentation of the LCGC Emerging Leader in Chromatography Award to Debby Mangelings, Vrije Universiteit Brussel, by Laura Bush, LCGC**
3:45	(530-4)	**Generic Chiral Separation Strategies for Pharmaceutical Compounds Using Chromatographic and Electrophoretic Techniques** DEBBY MANGELINGS, Vrije Universiteit Brussel, Yvan Vander Heyden
4:20	(530-5)	**Recent Trends in High-Performance Liquid Chromatographic Separation of Enantiomers** BEZHAN CHANKVETADZE, Tbilisi State University N/A

AWARDS — **Session 540**

SEAC - Charles N Reilley and Royce W Murray Award
arranged by Richard M Crooks, University of Texas at Austin

Monday Afternoon, Room B312
Shelley Minteer, University of Utah, Presiding

1:30		**Introductory Remarks - Shelley Minteer**
1:35		**Presentation of the 2016 SEAC - Charles N Reilley Award to Reginald M Penner, University of California, Irvine by Shelley Minteer, SEAC President**
1:40	(540-1)	**Electrodeposited Nanowires for Faster and More Sensitive Hydrogen Gas Detection** REGINALD M PENNER, University of California, Irvine
2:15	(540-2)	**A Membrane-Based AC Electroosmotic Pump** CHARLES R MARTIN, University of Florida, Pradeep Ramiah Rajasekaran, Xiaojian Wu
2:50	(540-3)	**Electrocatalytic Amplification of Single Nanoparticle Collisions Using DNA-Modified Surfaces** RICHARD M CROOKS, University of Texas at Austin, Timothy M Alligrant, Radhika Dasari, Keith J Stevenson
3:25		**Recess**
3:40		**Presentation of the 2016 SEAC - Royce W Murray Award to Ryan J White, University of Maryland Baltimore County, by Shelley Minteer, SEAC President**
3:45	(540-4)	**Imaging Molecular Flux Using Protein Channel Based Scanned Probe Microscopy** RYAN J WHITE, University of Maryland Baltimore County, Florika Caling Macazo
4:20	(540-5)	**Imaging with Nanopipettes** LANE A BAKER, Indiana University

SYMPOSIUM — **Session 550**

Emerging Platforms for Lab-on-a-Chip Analyses
arranged by Ryan T Kelly, Pacific Northwest National Laboratory

Monday Afternoon, Room B301
Ryan T Kelly, Pacific Northwest National Laboratory, Presiding

1:30		**Introductory Remarks - Ryan T Kelly**
1:35	(550-1)	**Rapid Screening for Infectious Diseases Using Paper-Analytic Devices** CHARLES HENRY, Colorado State University
2:10	(550-2)	**High-Throughput Microfluidic Experimentation One Drop at a Time** ANDREW J DEMELLO, ETH Zürich
2:45	(550-3)	**Simple, Microfluidic Flow Distance-Based Determination of Biomolecule Concentrations** ADAM T WOOLLEY, Brigham Young University, Chatterjee Debolina, Sahore Vishal
3:20		**Recess**
3:35	(550-4)	**Automated Droplet Manipulation, Analysis and Screening Based on Sequential Operation Droplet Array Technique** QUN FANG, Zhejiang University, Zhu Ying
4:10	(550-5)	**Microfluidic Platforms with Integrated Microvalves for Novel Biochemical Analyses** RYAN T KELLY, Pacific Northwest National Laboratory, Yongzheng Cong, Tao Geng, Sachin Jambovane, Katipamula Shanta, Spencer Prost, Michael Russcher N/A

SYMPOSIUM — **Session 560**

Innovative Applications of Surface-Enhanced Raman Spectroscopy
arranged by Bhavya Sharma, University of Tennessee, Knoxville

Monday Afternoon, Room B303
Bhavya Sharma, University of Tennessee, Knoxville, Presiding

1:30		**Introductory Remarks - Bhavya Sharma**
1:35	(560-1)	**Ultrafast Surface-Enhanced Raman Spectroscopy** RENEE R FRONTIERA, University of Minnesota, Nathaniel C Brandt, Emily L Keller, Alyssa A Cassabaum, James L Brooks
2:10	(560-2)	**Detecting Small Molecules Using Probe-Mediated SERS Schemes** JON CAMDEN, University of Notre Dame
2:45	(560-3)	**Utilization of SERS Nanoparticles as Contrast Agents for Molecular Imaging in Cancer** CRISTINA ZAVALETA, Stanford University
3:20		**Recess**
3:35	(560-4)	**Gels, Inkjet, and Laser Ablation: A New Toolkit for Surface-Enhanced Raman Spectroscopy** MARCO LEONA, The Metropolitan Museum of Art N/A
4:10	(560-5)	**Development of Surface Enhanced Spatially Offset Raman Spectroscopy (SESORS) for Neuroscience** BHAVYA SHARMA, University of Tennessee

SYMPOSIUM — **Session 570**

Miniature Mass Spectrometers
arranged by R Graham Cooks and Zheng Ouyang, Purdue University

Monday Afternoon, Room B304
R Graham Cooks, Purdue University, Presiding

1:30		**Introductory Remarks - R Graham Cooks and Zheng Ouyang**
1:35	(570-1)	**Introduction of a Multi-Optic Coaxial Ring Ion Trap (MoCRIT) for External Ionization in Portable Mass Spectrometry** GUIDO F VERBECK, University of North Texas
2:10	(570-2)	**From 2G to 3G - Quantitative Analysis and Biomarker Profiling Using Miniature Mass Spectrometry System** ZHENG OUYANG, Purdue University, Ren Yue, Ran Zou, Yuan Su, Graham Cooks, Yu Xia
2:45	(570-3)	**Describing and Optimizing Toroidal Trapping Fields for the Development of Miniature Mass Spectrometers** STEPHEN A LAMMERT, PerkinElmer, Edgar D Lee, Daniel E Austin, Karl R Warnick
3:20		**Recess**
3:35	(570-4)	**Practical Applications of Outside-the-Lab Mass Spectrometry** MITCH WELLS, FLIR Detection, Inc.
4:10	(570-5)	**The Development and Performance Enhancement of A Mini-Mass Spectrometer with Continuous Atmospheric Pressure Interface** WEI XU, Beijing Institute of Technology, YanBing Zhai, Muyi He

SYMPOSIUM **Session 580**

Nanomedicine, From Diagnostics to Large Animal Therapy
arranged by Raoul Kopelman, University of Michigan

Monday Afternoon, Room B305
Raoul Kopelman, University of Michigan, Presiding

1:30		**Introductory Remarks - Raoul Kopelman**
1:35	(580-1)	**The Plasmonic Photo-Thermal Death of Cancer in Cells(1) and in Animals(2,3) Using Gold Nano-rod** MOSTAFA A EL-SAYED, Georgia Institute of Technology
2:10	(580-2)	**Platinum(II) and Gold(III) Compounds for Cancer Diagnosis and Therapy** CHI-MING CHE, The University of Hong Kong N/A
2:45	(580-3)	**Nanophotonics-Based Diagnostics and Therapy: From Deep In-Vivo Photo-Acoustic Chemical Imaging of Tumors to Photodynamic Treatment of Heart Disease** RAOUL KOPELMAN, University of Michigan
3:20		**Recess**
3:35	(580-4)	**Interfacial Assembly of Functional Mesoporous Nanospheres with Multi-Level Architectures for Bioapplications** DONGYUAN ZHAO, Fudan University
4:10	(580-5)	**Stickyflares: Tracking the Amount and Location of RNA in Single Cells** CHAD A MIRKIN, Northwestern University

SYMPOSIUM **Session 590**

Non-Traditional Human Biometrics for Threat Assessment: Using Chemical Forensics for National Security and Intelligence Applications
arranged by Joachim Dieter Pleil, US EPA and Kevin O'Connell, InQTel

Monday Afternoon, Room B308
Joachim Dieter Pleil, US EPA, Presiding

1:30		**Introductory Remarks - Joachim Dieter Pleil and Kevin O' Connell**
1:35	(590-1)	**Human Biomonitoring and In Vitro Toxicity Testing Applications for Covert Threat Analysis and Security Applications** JOACHIM DIETER PLEIL, US EPA, William E Funk
2:10	(590-2)	**Noninvasive Infectious Disease Monitoring for Health Applications** MICHAEL SCHIVO, University of California, Davis, Cristina E Davis
2:45	(590-3)	**Comprehensive Analysis of the Chemicals Within: The Human Exposome and Applications for Threat Assessment** GARY W MILLER, Emory University
3:20		**Recess**
3:35	(590-4)	**The US Army Edgewood Chemical Biological Center's Chemical Threat Analysis Within the Framework of the Organization for the Prohibition of Chemical Weapons** STANLEY A OSTAZESKI, US Army ECBC, Ethan A Jestel N/A
4:10	(590-5)	**Feasibility of Using Breath to Predict Exposure to Ionizing Radiation** TERENCE H RISBY, Johns Hopkins University

SYMPOSIUM **Session 600**

Novel Mass Spectrometric Approaches and Applications to Polymer Analysis
arranged by Charles L Wilkins, University of Arkansas

Monday Afternoon, Room B309
Charles L Wilkins, University of Arkansas, Presiding

1:30		**Introductory Remarks - Charles L Wilkins**
1:35	(600-1)	**Characterization of Synthetic Polymers Using as Little as a Small Molecule Matrix and the Vacuum of the Mass Spectrometer for Ionization** SARAH TRIMPIN, Wayne State University, Casey D Foley, Joshua Fischer, Zachary Devereaux, Sashiprabha M Vithanarachchi, Matthew J Allen, Barbara S Larsen
2:10	(600-2)	**Model Polymer Systems: Studies by Mass Spectrometry, Ion Mobility, and Computational Strategies** DAVID M HERCULES, Vanderbilt University, Sarah M Stow, Tiffany M Onifer, John A McLean
2:45	(600-3)	**Shape Selective Studies of Macromolecular Systems** JAMES HOWARD SCRIVENS, Teesside University
3:20		**Recess**
3:35	(600-4)	**Shape Sensitive Multidimensional Mass Spectrometry of Synthetic Polymers and Hybrid Materials** CHRYS WESDEMIOTIS, The University of Akron
4:10	(600-5)	**MALDI-TOF and MALDI-FTICR of Challenging Polymer Analysis Problems** CHARLES L WILKINS, University of Arkansas, Evegenia Tisdale

SYMPOSIUM **Session 610**

The Twenty-Seventh James L Waters Symposium on Super-Resolution Microscopy
arranged by W Richard Howe, University of Pittsburgh

Monday Afternoon, Room B405
W Richard Howe, University of Pittsburgh, Presiding

1:30		**Introductory Remarks - W Richard Howe**
1:35	(610-1)	**Imaging Life at High Spatiotemporal Resolution** ERIC BETZIG, Janelia Research Campus
2:10	(610-2)	**Accessing the Emerging Imaging Technologies at HHMI Janelia** TENG-LEONG CHEW, Janelia Research Campus
2:45	(610-3)	**Closing the Gap Between First Implementation and Product – Update on Lattice Light Sheet Commercialization Effort** ALEX SOELL, Carl Zeiss Microscopy, LLC
3:20		**Recess**
3:35	(610-4)	**Life Inside the Cell: STORM, CRISPR and Imagenomics** BO HUANG, University of California, San Francisco
4:10	(610-5)	**Structured Scanned Plane Bessel Microscopy for Super-Resolution Neuroanatomy** TIMOTHY HARRIS, HHMI Janelia Research Campus

SYMPOSIUM **Session 620**

Trials and Tribulations of Dietary Supplement Analysis: Authentication, Adulteration and Contaminant Testing
arranged by Lowri DeJager, U.S. Food and Drug Administration

Monday Afternoon, Room B311
Lowri DeJager, U.S. Food and Drug Administration, Presiding

1:30		**Introductory Remarks - Lowri DeJager**
1:35	(620-1)	**Protecting Consumers One Analysis at a Time: Identifying Harmful Adulterants in Dietary Supplements** TRAVIS M FALCONER, U.S. Food & Drug Administration
2:10	(620-2)	**Authentication of Foods and Botanical Supplements Using Chemometric Methods** JAMES HARNLY, USDA
2:45	(620-3)	**DNA Authentication and Adulterant Detection: Dispelling the Myths and Facing the Facts** DANICA HARBAUGH REYNAUD, Authen Technologies N/A
3:20		**Recess**
3:35	(620-4)	**Authenticity of Herbal Dietary Supplements: Comparison of Chemical and DNA Barcoding Methods** RAHUL PAWAR, CFSAN/FDA, Sara M Handy, Erich Grundel, Raymond Cheng, Nicole Shyong
4:10	(620-5)	**Spectroscopic Detection of Adulteration in Botanical Dietary Supplements: What is Licorice?** CHARLOTTE SIMMLER, University of Chicago, Guido Pauli, Shao-Nong Chen

ORGANIZED CONTRIBUTED SESSIONS — Session 630

Cell Phone Spectroscopy - Handheld Spectroscopy for the Citizen

arranged by Mark Druy, Druy Consulting and Richard Crocombe, PerkinElmer

Monday Afternoon, Room B402

Mark Druy, Druy Consulting, Presiding

1:30	(630-1)	**Ready When You Are: Cell Phone Spectrometry When Cellcams Yield RAW Data** ALEXANDER SCHEELINE, SpectroClick
1:50	(630-2)	**Biomedical Applications of Cellphone Spectroscopy** ANSHUMAN DAS, MIT Media Lab
2:10	(630-3)	**Optical Smartphone Biosensing Techniques** BRIAN T CUNNINGHAM, University of Illinois at Urbana-Champaign, Kenneth Long
2:30	(630-4)	**Withdrawn**
2:50		**Recess**
3:05	(630-5)	**Quantum Dots and Smartphone Fluorescence Imaging: A Perfect Marriage for Bioassays** RUSS ALGAR, University of British Columbia, Eleonora Petryayeva
3:25	(630-6)	**Point-of-Care Colorimetric Detection with a Cell Phone** IAN PAPAUTSKY, University of Cincinnati
3:45	(630-7)	**Mobile Technologies for Personalized Diagnostics and Global Health** DAVID ERICKSON, Cornell University
4:05	(630-8)	**A Handheld Optoelectronic Nose** KENNETH S SUSLICK, University of Illinois at Urbana-Champaign, Zheng Li, Maria K LaGasse, Jon R Askim

ORGANIZED CONTRIBUTED SESSIONS — Session 640

Isolation and Characterization of Impurities/Degradation Product: Understanding Your Impurity Profile Throughout the Development Process

arranged by Andy Miles and Paul Wrezel, Regis Technologies

Monday Afternoon, Room B313

Andy Miles, Regis Technologies, Presiding

1:30	(640-1)	**Risk Analysis of Impurities in Drug Substances** DAN WEISSMUELLER, Regis Technologies
1:50	(640-2)	**Low level Impurity Isolations for Impurity Profiling and Structure Elucidation** TONY (QI) YAN, Pfizer
2:10	(640-3)	**Case Studies Involving Method Development for Trace-Level Impurities** PAUL WREZEL, Regis Technologies
2:30	(640-4)	**Impurity Identification for API Process Development** ZHAO YANQUN, AbbVie, Inc., Wayne Pritts
2:50		**Recess**
3:05	(640-5)	**Emerging Techniques for the Identification of Impurities and Degradation Products in Agricultural Research** JEFFRIE A GODBEY, Dow AgroSciences, Jesse L Balcer, Jeffrey R Gilbert, Yelena A Adelfinskaya, Mary D Evenson
3:25	(640-6)	**The Development of SFC Stationary Phases for the Optimized Separation of Chemical Mixtures Containing a Wide Range of Polarities** MATTHEW PRZYBYCIEL, ES Industries
3:45	(640-7)	**Isolation and Characterization of Impurities in the Synthesis of Drug Substance to Support Drug Development** QIFENG XUE, Theravance Biopharma US, Inc., Kanaka Hettiarachchi, Ken Ngim, Zhengtian (Titan) Gu
4:05	(640-8)	**Open Discussion**

ORAL SESSIONS — Session 650

Biomedical: Advances in Detection and Therapeutics of Cancer

Monday Afternoon, Room B302

Rabih E Jabbour, US Army Edgewood Chemical Biological Center, Presiding

1:30	(650-1)	**KS-Detect: A Portable, Solar-Thermal, Polymerase Chain Reaction System for the Point-of-Care Diagnosis of Kaposi's Sarcoma** RYAN SNODGRASS, Cornell University, Andrea Gardner, Li Jiang, Ethel Cesarman, David Erickson
1:50	(650-2)	**Nano-Plasmonic Exosome Platform (nPLEX) for Label-Free Detection and Molecular Profiling of Exosomes** HYUNGSOON IM, Massachusetts General Hospital, Huilin Shao, Park Yongil, Vanessa Peterson, Cesar M Castro, Ralph Weissleder, Hakho Lee
2:10	(650-3)	**Generate DNA Aptamers Against Glypican 3 with Expanded Genetic Systems** LIQIN ZHANG, University of Florida
2:30	(650-4)	**Systematic and Quantitative Analysis of Surface N-Sialoglycoproteins in Cancer Cells with Distinct Invasiveness** RONGHU WU, Georgia Institute of Technology
2:50		**Recess**
3:05	(650-5)	**DNA "AND" Logic Platform Integrated on Nanoparticle for Programmed Recognition and Therapy of Cancer** ZHENBAO LIU, University of Florida, Weihong Tan N/A
3:25	(650-6)	**A Core/Shell Structure of Reduced Graphene Oxide/ Mesoporous Silica with Oligonucleotide Gates for Cancer Treatment** XIAO LIU, University of North Dakota, Xu Wu, Xuefei Zhang, Yuqian Xing, Julia Xiaojun Zhao
3:45	(650-7)	**Simultaneous Photothermo-/Chemotherapy Using Reduced Graphene Oxide Based Nanocomposites** YUQIAN XING, University of North Dakota, Xu Wu, Xiao Liu, Julia Xiaojun Zhao
4:05	(650-8)	**Imaging of Cancer Protein-Protein Interactions and Small Molecule Inhibitions by a Surface Plasmon Resonance Microarray** CHARUKSHA WALGAMA, Oklahoma State University, Zainab Hussain Al Mubarak, Bing Zhang, Mayowa Akinwale, Anuruddha Pathiranage, Junpeng Deng, Darrell K Berlin, Doris M Benbrook, Sadagopan Krishnan

ORAL SESSIONS — Session 660

Environmental Applications of Elemental Analysis

Monday Afternoon, Room B315

Eugene F Barry, University of Massachusetts Lowell, Presiding

1:30	(660-1)	**Analyzing Mercury from Contaminated Mining Sites Using a Direct Mercury Analyzer** SUMEDH PHATAK, Milestone Inc.
1:50	(660-2)	**ICP-MS for the Analysis of High Salinity Samples** ERICA CAHOON, PerkinElmer, Daniel H Jones
2:10	(660-3)	**Nanoparticle Removal During Alum and Ferric Coagulation Characterized by Single Particle ICP-MS** ARIEL DONOVAN, Missouri University of Science and Technology, Honglan Shi, Yinfa Ma, Craig Adams, Chady Stephan, Todd Eichholz
2:30	(660-4)	**Applications of Single Particle ICP-MS to Environmental Matrices** CHADY STEPHAN, PerkinElmer, Kenneth Neubauer
2:50		**Recess**
3:05	(660-5)	**Issues in Deep Ocean LIBS: Internal Calibration and the Effect of Suspended Particles** JOSEPH BONVALLET, University of South Carolina, S Michael Angel
3:25	(660-6)	**Selective Measurement of Metal Ions at Covalently Functionalized Carbon-Fiber Microelectrodes** YUANYUAN YANG, Wayne State University, Ahmad A Ibrahim, Jennifer L Stockdill, Parastoo Hashemi
3:45	(660-7)	**MWCNT/Bi Composite Film Modified PGE for Voltammetric Determination of Lead and Cadmium** YESIM TUGCE YAMAN, Hacettepe University, Serdar Abaci N/A
4:05	(660-8)	**Application of Mössbauer spectrometry in Environmental Studies of Fly Ashes and Road Dusts** TADEUSZ SZUMIATA, RADWAG Balances and Scales, Tadeusz Szumiata, Malgorzata Gzik-Szumiata, Katarzyna Brzozka, Bogumil Gorka, Michal Gawronski, Ryszard Swietlik, Marzena Trojanowska

ORAL SESSIONS — Session 670

Environmental GC

Monday Afternoon, Room B406
Walter B Wilson, National Institute of Standards and Technology, Presiding

1:30	(670-1)	**Multi-Dimensional Micro Gas Chromatography Device for the Rapid and Sensitive Analysis of Volatile Organic Compounds** JIWON LEE, University of Michigan, Menglian Zhou, Hongbo Zhu, Robert Nidetz, Katsuo Kurabayashi, Xudong Fan
1:50	(670-2)	**SPME On-Fiber Derivatization Using a Stable and Reusable Pentafluorophenyl Hydrazine Standard Gas Generating Vial** JUSTEN J POOLE, University of Waterloo, Jonathan J Grandy, German A Gomez-Rios, Emanuela Gionfriddo, Janusz Pawliszyn
2:10	(670-3)	**Gas Chromatography-Vacuum Ultraviolet Detection (GC-VUV) for Analysis of Polychlorinated Biphenyls (PCBs)** CHANGLING QIU, University of Texas at Arlington, Jonathan Smuts, Kevin A Schug
2:30	(670-4)	**Gas Chromatographic Retention Behavior on Select Groups of Isomeric Polycyclic Aromatic Compounds and Their Alkyl-Substituted Derivatives on Stationary Phases of Different Selectivity** WALTER B WILSON, National Institute of Standard and Technology, Federica Nalin, Lane C Sander, Leonard M Sidisky, Stephen A Wise
2:50		**Recess**
3:05	(670-5)	**Optimization of Phthalate Analysis by Gas Chromatography-Mass Spectrometry Using Computer Modeling Software** DAN LI, Restek Corporation, Rebecca Stevens, Jack Cochran, Amanda Rigdon, Chris English
3:25	(670-6)	**Analysis of Greenhouse Gases Using the Dielectric Barrier Discharge Detector** MATTHEW MONAGLE, AIC LLC N/A
3:45	(670-7)	**Fast, Very Fast and Ultra-Fast Forensic and Homeland Security GC-MS** AVIV AMIRAV, Tel Aviv University, Uri Keshet, Alexander Fialkov, Tal Alon
4:05	(670-8)	**Passive Monitoring – A Guide to Sorbent Tube Sampling for EPA Method 325** CAROLINE WIDDOWSON, Markes International

ORAL SESSIONS — Session 680

Food Product Quality and Component Characterization

Monday Afternoon, Room B316
Ed Guthrie, Agilent Technologies, Presiding

1:30	(680-1)	**Analysis of Fat Crystallization Thanks to Microrheology** MAXIME BAZIN, Formulaction, Giovanni Brambilla, Roland Ramsch, Mathias Fleury, Gérard Meunier
1:50	(680-2)	**Atomic Absorption and Potentiometric Analysis of Electrolytes in Food Substances Used for Rapid Muscle Cramp Relief by Athletes** STEPHANIE HOOPER MAROSEK, Methodist University, Taylor Tipton
2:10	(680-3)	**Microrheological Analyses for Dairy Formulations** ROLAND RAMSCH, Formulaction, Giovanni Brambilla, Gérard Meunier
2:30	(680-4)	**Analysis of Aroma Compounds in Beer by TD–GC–TOF MS with Soft Electron Ionization** LAURA MCGREGOR, Markes International Ltd, Caroline Widdowson, Massimo Santoro, Chris Hall, Ken Umbarger
2:50		**Recess**
3:05	(680-5)	**Food Analysis Using Laser Desorption GS-Ion Mobility Spectrometry – Olive Oil as an Example** WOLFGANG VAUTZ, ISAS, Sascha Liedtke, Joachim Franzke
3:25	(680-6)	**Monitoring of Protein Changes in Pasteurized Liquid Egg Using Capillary Electrophoresis** REYHAN S UYSAL, Hacettepe University, Ismail H Boyaci, Esra Acar, Nusret Ertas
3:45	(680-7)	**Exploiting Polymeric Ionic Liquids-Based SPME Sorbents Coupled to Gas-Chromatography/Mass Spectrometry for Food Quality and Safety Assessment** ERICA A SOUZA-SILVA, Universidade Federal do Rio Grande do Sul, Emanuela Gionfriddo, Nathaly Reyes-Garces, German A Gomez-Rios, Jared L Anderson, Janusz Pawliszyn
4:05	(680-8)	**Detection, Identification, and Pattern Recognition of Microbial Volatile Organic Compounds from Virulent and Hypo-Virulent Cryphonectria Parasitica Species by Headspace-SPME-GC-MS and Chemometrics** JINYAN SHE, Mississippi State University

ORAL SESSIONS — Session 690

LC Method Development - Half Session

Monday Afternoon, Room B403
William E Barber, Agilent Technologies, Presiding

1:30	(690-1)	**Method Transfer and Column Scalability with Superficially Porous Particles** WILLIAM LONG, Agilent Technologies, Anne Mack, Stephen Luke, Jason Link
1:50	(690-2)	**Isolation of Impurities in Biopharmaceutical Formulations** THOMAS E WHEAT, Waters Corporation, Amanda B Dlugasch, Patricia R McConville
2:10	(690-3)	**New Technological Solutions to Maximizing Uptime on an Ion Chromatography System** DAVID G MOORE, Thermo Fisher Scientific, Pranathi Perati, Sally Eastman
2:30	(690-4)	**Accurate Measurement of Analyte Dispersion through Connecting Tubes used in Fast Very High-Pressure Liquid Chromatography** FABRICE GRITTI, Waters Corporation, Martin Gilar, Thomas McDonald

ORAL SESSIONS — Session 700

LC Optimization - Half Session

Monday Afternoon, Room B404
Rachael Wilson, University of Pittsburgh, Presiding

1:30	(700-1)	**Dynamic Temperature Control in Capillary Liquid Chromatography: Increasing Analysis Sensitivity, Speed, and Peak Capacity** STEPHEN R GROSKREUTZ, University of Pittsburgh, Rachael Wilson, Stephen Weber
1:50	(700-2)	**Towards the Development of a Refractive Index-Based Optical Microcavity Mass Concentration Detector Compatible with Gradient Elution Liquid Chromatography** ZACHARY S WIERSMA, University of Illinois at Urbana-Champaign, Todd O Pangburn, Alexandria L Stanton, James H Wade, David Mark Meunier, Ryan C Bailey
2:10	(700-3)	**Increasing the Peak Capacity of Quantitative Liquid Chromatography Using Chemometric Curve Resolution** DANIEL W COOK, Virginia Commonwealth University, Sarah C Rutan, Dwight R Stoll
2:30	(700-4)	**Improving Sensitivity, Resolution, and Peak Capacity of Gradient Elution Capillary Liquid Chromatography with Temperature-Assisted On-Column Solute Focusing** RACHAEL WILSON, University of Pittsburgh, Stephen R Groskreutz, Stephen Weber

ORAL SESSIONS — Session 710

Neurochemistry

Monday Afternoon, Room B401
Chi Leng Leong, Imperial College London, Presiding

1:30	(710-1)	**Measurement of Phasic Dopamine Signals in the Rat Nucleus Accumbens Core and Shell in Response to Noxious Stimuli** CHRISTOPHER W ATCHERLEY, Mayo Clinic, Edita Navratilova, Yanhua Xie, Levi Lazarus, Michael L Heien, Frank Porreca N/A
1:50	(710-2)	**Oxygen Changes and Dopamine Release during Spreading Depression** CADDY N HOBBS, University of North Carolina at Chapel Hill, Justin A Johnson, R Mark Wightman
2:10	(710-3)	**Spontaneous, Transient Adenosine Release from Brain During Ischemia-Reperfusion Injury** MALLIKARJUNARAO GANESANA, University of Virginia, B Jill Venton
2:30	(710-4)	**Treatment on Carbon Nanotube Yarn Microelectrode for Sensitive and Rapid Dopamine Detection In Vivo** CHENG YANG, University of Virginia, B Jill Venton, Ilia N Ivanov, Michael D Nguyen, Christopher B Jacobs
2:50		**Recess**
3:05	(710-5)	**Neurochemical Investigation of Epilepsy Using Microdialysis Sampling to Study Multiple Seizure Rat Models** AMANDA M FURNESS, University of Kansas
3:25	(710-6)	**PKC-b Inhibitors Attenuate Amphetamine and Cocaine Stimulated Dopamine Release** ALEXANDROS G ZESTOS, University of Michigan, Robert T Kennedy, Margaret E Gnegy

3:45 (710-7) **Clinical Measurements at the Bedside: Dynamic Neurochemical Changes in the Injured Human Brain Monitored Using an Online Microdialysis System** MICHELLE L ROGERS, Imperial College London, Chi Leng Leong, Sally A Gowers, Sharon Jewell, Shumaila Khan, Anthony J Strong, Martyn G Boutelle

4:05 (710-8) **Online Clinical Microdialysis: Detecting the Neurochemical Consequences of Spreading Depolarization** CHI LENG LEONG, Imperial College London, Michelle L Rogers, Sally A Gowers, Sharon Jewell, Anthony J Strong, Shumaila Khan, Martyn G Boutelle

ORAL SESSIONS — Session 720

Unique Developments in Spectroscopy - Half Session

Monday Afternoon, Room B403
William E Barber, Agilent Technologies, Presiding

3:05 (720-1) **Simultaneous Measurement of N_2O and CH4 Emissions from Agriculture Using Photoacoustic Detection and QCL Laser** ARTO BRANDERS, Gasera Ltd., Ismo Kauppinen, Jaakko Lehtinen

3:25 (720-2) **Infrared Spectroscopic Remote Sensing of Pulsed Signals from Nearby Stars** ROBERT A LODDER, University of Kentucky, Anne Brooks

3:45 (720-3) **High Resolution Coherent Multidimensional Spectroscopy** PETER CHEN, Spelman College

4:05 (720-4) **Biologics Starting Materials Identified through Opaque Containers by Spatially Offset Raman Spectroscopy (SORS)** MATTHEW BLOOMFIELD, Cobalt Light Systems, Darren Andrews, Pavel Matousek

POSTER SESSION — Session 730

All posters are to be mounted by 10:00 AM and remain on display until 4:00 PM. Authors must be at their posters from 1:00 PM to 3:00 PM. Location of the afternoon posters is on the Exposition Floor, 400 Aisle. PLEASE NOTE: You cannot get onto the Exposition Floor until after 9:00 AM.

Art and Archaeology

Monday Afternoon, Exposition Floor, 400 Aisle

(730-1 P) **Maximizing the Information Obtained from Small Archaeological Samples by Sequencing DART-MS and Plasma-Chemical Oxidation for AMS Radiocarbon Dating** RUTH ANN ARMITAGE, Eastern Michigan University, Kathryn Jakes, Suzanne Baker

(730-2 P) **The Dose Makes the Poison: Quantitation of Pollutant VOCs From Materials Used in A Museum Environment** MICHAEL J SAMIDE, Butler University, Gregory D Smith, Jericha Mill

(730-3 P) **Using Reflectance Spectroscopy to Determine the Rate of Formation of Prussian Blue Pigment** JACOB APPLEGARTH, Butler University, Michael J Samide

(730-4 P) **Elemental Profiling of Archaeometallurgical Artefacts by ICP Spectrometry** SANDA RONCEVIC, University of Zagreb, Ivan Nemet N/A

POSTER SESSION — Session 740

All posters are to be mounted by 10:00 AM and remain on display until 4:00 PM. Authors must be at their posters from 1:00 PM to 3:00 PM. Location of the afternoon posters is on the Exposition Floor, 400 Aisle. PLEASE NOTE: You cannot get onto the Exposition Floor until after 9:00 AM.

Chemical Methods

Monday Afternoon, Exposition Floor, 400 Aisle

(740-1 P) **Single Nanoparticle to 3D Supercage: Framing for an Artificial Enzyme System** REN CAI, University of Florida, Weihong Tan

(740-2 P) **Synthesis and Characterization of Superparamagnetic Manganese Ferrite Nanoparticles** SIMONAS RAMANAVICIUS, Vilnius University, Marija Kurtinaitiene, Kestutis Mazeika, Vidas Pakstas, Arunas Jagminas

(740-3 P) **Optimization of Ibuprofen Micronization by the Rapid Expansion of a Supercritical Solution (RESS)** ROLF SCHLAKE, Applied Separations, Al Kaziunas, Brian Day

(740-4 P) **DNA-Aligner-Controlled Nicking-Based Isothermal Exponential Amplification Reaction for High Sensitive Detection of Nucleic Acids** WU WANGHUA, Zhejiang University, Zhou Jianguang, Zhang Tao, Yu Dongdong

(740-5 P) **Electrochemical, Spectroscopic and Chromatographic Techniques for Monitoring the Active Pharmaceutical Ingredients Degradation Kinetics: Which Methodology Fulfills More Principles of the Green Analytical Chemistry?** MOHAMED K ABD EL-RAHMAN, Cairo University, Amr M Mahmoud

(740-6 P) **Investigation of Heterogeneous Reaction Mechanism Between Formaldehyde and MnO_2/CeO_2 at Room Temperature by Gas Analysis Approach** HAYASHI HIROKI, Tokai University, Sekine Yoshika

(740-7 P) **Automated, Low-Level Distillation of Phenolics for Use in Environmental and Manufacturing Applications** BRANT HOEKSTRA, OI Analytical, Jonathan Howerton N/A

(740-8 P) **Identification of 6-chlorotestosterone and Other Designer Anabolic Steroids in Dietary Supplements with Semi-Quantitative Content Determination Using Surrogate Compounds** SARAH ELIZABETH VOELKER, U.S. FDA, Forensic Chemistry Center, Lisa M Lorenz, Mary B Jones, Travis M Falconer, Rick A Flurer

POSTER SESSION — Session 750

All posters are to be mounted by 10:00 AM and remain on display until 4:00 PM. Authors must be at their posters from 1:00 PM to 3:00 PM. Location of the afternoon posters is on the Exposition Floor, 400 Aisle. PLEASE NOTE: You cannot get onto the Exposition Floor until after 9:00 AM.

Commercial Products Characterization

Monday Afternoon, Exposition Floor, 400 Aisle

(750-1 P) **Considerations in Sample Preparation and Method Development Using ICP-OES** KENNETH NEUBAUER, PerkinElmer, Nick Spivey, Stan Smith, Laura Thompson

(750-2 P) **Chemical Fingerprinting of Tobacco and Related Products by TD–GC–TOF MS** LAURA MCGREGOR, Markes International Ltd, Chris Hall, Caroline Widdowson, Nicola Watson, Ken Umbarger

(750-3 P) **Levels of Mercury and Methylmercury in Fish (Prochilodus Magdalenae) Ayapel Marsh (Colombia)** EDINELDO LANS CEBALLOS, University of Cordoba, Mauricio Lora Agamez, Amira Padilla Jimenez

(750-4 P) **Mercury Residual and Methylmercury in Canned Tuna Distributed in Monteria –Colombia** EDINELDO LANS CEBALLOS, University of Cordoba, Mauro Lombana Agamez, Basilio Diaz Pogutá

(750-5 P) **Automated SPME for the Analysis of Environmental Contaminants in Milk** NICOLE M LOCK, Shimadzu Scientific Instruments, Robert Clifford, Di Wang, Laura Chambers, Brahm Prakash, Shilpi Chopra

(750-6 P) **Estimation of Dietary Intake of Essential and Non-Essential Metals through the Consumption of Dietary Supplements Available on Nigerian Retail Market** IMAOBONG UDOUSORO, University of Uyo, Ikem Abua, Olujide T Akinbo

(750-7 P) **Analysis of Aflatoxins in Pet Food by UHPLC Using PDA and Fluorescence Detection** CATHARINE LAYTON, PerkinElmer, Jason Weisenseel, Wilhad M Reuter

(750-8 P) **Antioxidant Activity of Flavored Tea and Its Content of Phenol, Flavonoid and Tannins** ABD EL-MONEIM M AFIFY, Cairo University

(750-9 P) **High Sensitive LC/MS Analysis of Stevia Sweeteners Using Polymer-Based Amino Column** RONALD BENSON, Shodex/Showa Denko America, Inc., Junji Sasuga, Satoko Sakai, Tomokazu Umezawa

(750-10 P) **Assigning Unknown Fingernail Polish to Known Manufacturers by Raman Spectroscopy and Multivariate Statistics** GARY H NAISBITT, Utah Valley University, Kelsey Hartt N/A

(750-11 P) **Characterizing the Rheological Properties of Wax Emulsions used as Carriers for Biopesticides in Agricultural Pest Management** KRISTEN JORDAN, Western Carolina University, Cynthia Atterholt

(750-12 P) **Multivariate Quantification of API Release from Combination Tablets in the Presence of Matrix Effects Using Fiber Optic Dissolution** JOSEPH MEDENDORP, Vertex Pharmaceuticals, Taryn Ryan, Mahidhar Shapally, Ivelisse Colon

(750-13 P) **Simultaneous Analysis of Methylisothiazolinone, Salicylic Acid and Parabens in Antidandruff Hair Shampoos by Monolithic Silica High-performance Liquid Chromatographic Column** ABDULRAHMAN ALMAJED, King Saud Univesriy, College of Pharmacy

(750-14 P) **Simultaneous Analysis Red Wine Absorbance, CIE Lab Color Indices and Fluorescence Excitation-Emission Matrices** ADAM MATTHEW GILMORE, Horiba, Sakiko Akaji

(750-15 P) **Extractables and Leachables Analysis of IV Bag Systems Using Thermal Desorption and Stir Bar Sorptive Extraction with GC Single Quad and GC Time-of-Flight Mass Spectrometric Detection** ANDREAS HOFFMANN, Gerstel GmbH & Co.KG, Thomas Albinus, Kurt Thaxton, Elizabeth Almasi

POSTER SESSION — Session 760

All posters are to be mounted by 10:00 AM and remain on display until 4:00 PM. Authors must be at their posters from 1:00 PM to 3:00 PM. Location of the afternoon posters is on the Exposition Floor, 400 Aisle. PLEASE NOTE: You cannot get onto the Exposition Floor until after 9:00 AM.

Data Analysis and Manipulation

Monday Afternoon, Exposition Floor, 400 Aisle

(760-1 P) **Automatic, High-Throughput Motion Classification of Surface-Attached E. Coli Cells Using Bright Field Microscopy Data** SIMON SHEN, Arizona State University, Karan Syal, Nongjian Tao, Shaopeng Wang N/A

(760-2 P) **Ranking of Variables in the Dataset Using the Degree of Separation: Pure Statistical Method** WALEED MASWADEH, US Army, Jason A Guicheteau, A Peter Snyder

(760-3 P) **Procurement and Distribution Channels of Commonly Used Drugs in Nigeria: A Case Study of the Pharmaceutical Industry in Abia State** LILIAN I OGUGUO, National Malaria Elimination Programme FMoH, Chidiebere A Odike-Aduaka, Ifeoma Agwo

(760-4 P) **Effective Data Management in the Analytical Laboratory** TOSHINOBU YANAGISAWA, Shimadzu Corporation, Kazuhito Wakabayashi, Masayuki Shibata, Keisuke Yoshizawa, Ryuji Nishimoto

(760-5 P) **Combining PLS Classification and Regression Analyses for Robust Monitoring of Nuclear Materials Reprocessing** ROBERT LASCOLA, Savannah River National Laboratory, Patrick O'Rourke

POSTER SESSION — Session 770

All posters are to be mounted by 10:00 AM and remain on display until 4:00 PM. Authors must be at their posters from 1:00 PM to 3:00 PM. Location of the afternoon posters is on the Exposition Floor, 400 Aisle. PLEASE NOTE: You cannot get onto the Exposition Floor until after 9:00 AM.

Education/Teaching

Monday Afternoon, Exposition Floor, 400 Aisle

(770-1 P) **Development of Interactive Learning Modules used in Teaching Instrumental Analysis** YI HE, John Jay College/CUNY, Sandra Swenson, Wong Tiffany, Colleen McNamara

(770-2 P) **Conservation Science Tutorial Website: Analysis of VOCs Emitted from Museum Construction Materials** JERICHA MILL, Butler University, Michael J Samide, Gregory D Smith

(770-3 P) **Analytical Chemistry Course Embedded STEM Projects with Trolox Equivalent Antioxidant Capacity Assay** XIAOPING LI, Georgia Gwinnett College, Rashad Simmons

(770-4 P) **Inexpensive Teaching Instruments for Atomic Emission and Molecular Spectroscopy** ABD AL-KARIM ALI, Miami University, Taryn L Winner, Neil D Danielson

(770-5 P) **The Antibacterial Effects of Nanoparticles** NEYDA BRIDGITTE CHACON, SUNO N/A

(770-6 P) **Strategies for Designing Calibration Curves that Extend over Three Decades of Concentration for Ultraviolet-Visible Absorption Spectroscopy** LAUREN GRABOWSKI, University of South Carolina, Scott Goode

(770-7 P) **Analytical Chemistry Experiments with a Forensic Flavor** ROBERT Q THOMPSON, Oberlin College

(770-8 P) **HPLC Refurbishment** CONNOR PUTNAM, St. John Fisher College

(770-9 P) **Analytical Chemistry 2.1: An Open-Access Digital Resource for Undergraduate Education in Analytical Chemistry** DAVID T HARVEY, DePauw University

(770-10 P) **Introducing Undergraduate Chemists to Chemometrics: Using Microsoft Excel to Illustrate the NIPALS Algorithm used in Principal Component Analysis and Partial Least Squares Regression** MARK T STAUFFER, University of Pittsburgh - Greensburg

(770-11 P) **Applying Analytical Chemistry to Solve Problems in the Developing World** RHONDA L GROSSE, Chemists Without Borders, Bego Gerber, Marya Lieberman, Steve Chambreau, Satinder Ahuja

(770-12 P) **Interdisciplinary Undergraduate Research in Chemometrics: The Faculty Perspective** HELEN M BOYLAN, Westminster College, Carolyn Cuff, Stephanie Homitz, Christopher Caroff, Keilah Ireland, Domenic DiSanti

(770-13 P) **Development of an Undergraduate Laboratory Experiment to Determine Arsenic in Sinus Wash and Tap Water by Inductively Coupled Plasma-Mass Spectrometry** ANNA M DONNELL, University of Cincinnati, Keaton Nahan, Dawone Holloway, Anne P Vonderheide

POSTER SESSION — Session 780

All posters are to be mounted by 10:00 AM and remain on display until 4:00 PM. Authors must be at their posters from 1:00 PM to 3:00 PM. Location of the afternoon posters is on the Exposition Floor, 400 Aisle. PLEASE NOTE: You cannot get onto the Exposition Floor until after 9:00 AM.

Electrochemistry

Monday Afternoon, Exposition Floor, 400 Aisle

(780-1 P) **Laser-based Heating for Nanopore DNA Duplex Analysis** CHRISTOPHER E ANGEVINE, Virginia Commonwealth University, Sarah Seashols-Williams, Joseph E Reiner

(780-2 P) **A Novel Sandwiched Electrochemiluminescence Immunosensor for Detection of Carcinoembryonic Antigen Based on Carbon Quantum Dots** WANG HUAI-SHENG, Liaocheng University N/A

(780-3 P) **Label-Free DNA Sensors for Detection of BRCA1 Mutation** BRUNO C JANEGITZ, Federal University of Sal Carlos, Lais Ribovski, Valtencir Zucolotto N/A

(780-4 P) **Conductivity as Sensor for Real-time Monitoring of the Solution Soluble Corrosion Products under Cell Culture Conditions During Corrosion of High Purity Magnesium** KOLADE O OJO, University of Cincinnati, Tracy Hopkins, Sarah Pixley, William R Heineman, Vesselin Shanov, Madhura Joshi, Pravahan Salunke, Keaton Nahan, Guangqi Zhang

(780-5 P) **Gold Nanoparticles Enzymatic Sensors for Determination of Glucose** ARUNAS RAMANAVICIUS, Vilnius University, Povilas Genys, German Natalija, Almira Ramanaviciene

(780-6 P) **RC Constant Based Label Free Biomarkers Detection** PRADEEP RAMIAH RAJASEKARAN, University of Florida, Charles R Martin, Jennifer Ottoy

(780-7 P) **Electrochemical Detection of Evoked Dopamine Release in Zebrafish** MIMI SHIN, University of Kansas, Thomas Field, Michael A Johnson, Mia Furgurson

(780-8 P) **Quantification of Ion Selectivity of Single Asymmetric Nanopore-Channels for Better Energy Harvesting from Salinity Gradients** WARREN D BROWN, Georgia State University, Yan Li, Dengchao Wang, Maksim Kvetny, Gangli Wang

(780-9 P) **Charge Transfer Mechanism of Organic Molecules Associated with Dendrimers at Polarized Liquid/Liquid Interfaces** HIROKI SAKAE, Kanazawa University, Hirohisa Nagatani, Hisanori Imura

(780-10 P) **Study of Electrochemical Hydrogen Evolution and Oxygen Evolution Reactions at Ir and Ru Oxide Alloys with Scanning Electrochemical Microscopy** YUN-BIN CHO, Ewha Womans University, Chongmok Lee, Youngmi Lee

(780-11 P) **Automated, Accurate pH and Conductivity Measurements Using a Discrete Photometric Analyzer with an ECM Module** MARI KIVILUOMA, Thermo Fisher Scientific, Annu Suoniemi-Kahara

(780-12 P) **Atmospheric Corrosion Study of Metals in an Industrial Environment of Ahmedabad** PAREKH P SUNILKUMAR PUNAMBHAI, CU Shah Science College N/A

(780-13 P) **Synthesis and Characterization of Silver Coated Electrospun Cobalt Nanotubes and Their Electrocatalytic Activity** AREUM YU, Ewha Womans University, Chongmok Lee, Myung Hwa Kim, Youngmi Lee

(780-14 P) **Single Microwire Electrodeposition by Electrochemical Liquid-Liquid-Solid Growth** TIM ZHANG, University of Michigan

(780-15 P) **Simultaneous Detection Heavy Metals by Anodic Stripping Voltammetry Using Carbon Nanotube Thread** DAOLI ZHAO, University of Cincinnati, David R Siebold, Tingting Wang, Noe R Alvarez, Vesselin Shanov, William R Heineman

(780-16 P) **Low Concentration Detection by Electrogenerated Chemiluminescence Using Bipolar Electrochemistry in a Thin Layer Manner** SONGYAN YU, Auburn University, Mark D Holtan, Sanjun Fan, Curtis Shannon

(780-17 P) **Atypical Induction of Membrane Potential in Ion Selective Membranes** DEMETRA M PANTELIS, University of Florida, Pradeep R Rajasekaran

(780-18 P) **Characterization and Spectroelectrochemical Sensing with a Boron Doped Diamond Optically Transparent Electrode Coated with a Charge Selective Polymer Film** CORY A RUSINEK, University of Cincinnati, William R Heineman, Michael Becker, Robert Rechenberg, Daoli Zhao, Necati Kaval

(780-19 P) **Simultaneous Sensing of α-Tocopherol and Retinol in Micellar Media by Using Poly(2,21-(1,4 phenylenedivinylene) bis-8-hydroxyquinaldine)/MWCNTs Modified Electrode** HAYATI FILIK, Istanbul University

(780-20 P) **Electrochemical Investigation of a Series of Uranyl Salen Complexes: Effect of Ligand Conjugation on the U(V)/U(VI) Redox Couple** EMILY E HARDY, Auburn University, Madeleine A Eddy, Anne E Gorden

(780-21 P) **Predictable Standard Potential of Solid-Contact Ion-Selective Electrodes by Using Prussian Blue Analogues as Solid Contacts** YU ISHIGE, Hitachi Ltd., Stefan Klink, Wolfgang Schuhmann

(780-22 P) **Electrochemical Behaviour of Highly- and Poorly-doped p-Si-AuNP Electrodes** MEHRAN KASHI, University of New South Wales, Yanfang Wu, Vinicius Goncales, Moinul Choudhury, Simone Ciampi, Justin J Gooding

(780-23 P) Withdrawn

(780-24 P) **The Effect of Excited Fluorophore on Vesicle Fusion at the Surface of the Electrode** NEDA NAJAFINOBAR, Chalmers University of Technology, Jelena Lovric, Johan Dunevall, Hoda M Fathali, Andrew Ewing, Ann-Sofie Cans N/A

(780-25 P) **Novel Electrode PtCr/PAA (Polyamic Acid) for Efficient Ethanol Oxidation Reaction** JING ZHANG, Binghamton University

(780-26 P) **Voltammetric Serotonin Measurements in Mouse Models of Depression** RACHEL A SAYLOR, University of South Carolina, Aya Abdalla, Parastoo Hashemi

(780-27 P) **Kinetic Size-Spectra of Gas Molecules at Ionic Liquid (IL)-Metal Interface and Its Application for Highly Selective Gas Sensing** ZHE WANG, Xavier University

(780-28 P) **New Arginine-Acetaminophen Incorporation for Selective Determination in Serum with Graphene-based Biosensor** ZHE WANG, Xavier University

(780-29 P) **Core-shell Nanoparticles Supported on Carbon Nanotubes as Promising Catalyst for Methanol Electro-oxidation** MANZAR SOHAIL, Center Of Excellence In Nanotechnology, KFUPM N/A

POSTER SESSION — Session 790

All posters are to be mounted by 10:00 AM and remain on display until 4:00 PM. Authors must be at their posters from 1:00 PM to 3:00 PM. Location of the afternoon posters is on the Exposition Floor, 400 Aisle. PLEASE NOTE: You cannot get onto the Exposition Floor until after 9:00 AM.

Materials Science

Monday Afternoon, Exposition Floor, 400 Aisle

(790-1 P) **Non-Toxic Corrosion Inhibitor from Tagetes Patula L. for Aluminum in Acid Fluids used in Industrial Operations** OLUSEGUN KEHINDE ABIOLA, Federal University of Petroleum Resources, A O Aliyu, E E Elemike

(790-2 P) **Synthesis of Fluorine Substituted $LiFeBO_3$ Composite Material as a Cathode Material for a Lithium Secondary Battery and Characterization Using the Solid State NMR** YOUNGIL LEE, University of Ulsan, JaeMin Bak, Hansol Lee

(790-3 P) **Innovative Instrument Design for Highly Precise Determination of Sulfur and Carbon Concentration** MARKUS JUNG, Elementar Analysensysteme GmbH, David Nennstiel, Christian Schmidt N/A

(790-4 P) **Analytical Methods to Estimate Biodurability of Engineered Nanomaterials** MARY-LUYZA AVRAMESCU, Health Canada, Pat E Rasmussen, Marc Chénier, H David Gardner

(790-5 P) **Intensity Ratio of Resonant Raman Modes for Chirality Enriched Carbon Nanotubes** YANMEI PIAO, National Institute of Standards and Technology, Jeffrey Simpson, Jason Streit, Stephanie Lam, Geyou Ao, Jeffrey Fagan, Angela Hight Walker

(790-6 P) **Liquid-Phase Laser Ablation as a Controlled Method to Produce Graphene Quantum Dots** ROSEMARY LYNN EASTERDAY, University of Kentucky, Wenjin Cao, Yiyang Liu, Doo Young Kim, Dong-Sheng Yang

(790-7 P) **Physicochemical Characterization and Nanotoxicity of Polishing Slurries During CMP Process** EDUARD DUMITRESCU, Clarkson University, Dinusha Karunaratne, Kenneth Wallace, Silvana Andreescu

(790-8 P) **Polycapillary X-Ray Optics to Play a Key Role in NASA Mars 2020 Mission** NING GAO, XOS, Jared Sachs, George Allen, Genevieve DeMarco, George Allen, Larry Wade, Peng Lu, Igor Ponomarev, Jay Burdett, Robert Sharrow, Jaime Luna, Douglas Dawson

(790-9 P) **Preparation and Characterization of Bleached Ground Peanut Hulls** HOLLY TRULUCK, Western Carolina University, Melisa J Glatte, Carmen L Huffman

(790-10 P) **Real Time Measurement of Layer Thickness, Erosion Rates and Crater Depth in Glow Discharge Optical Emission Spectrometry** SOFIA GAIASCHI, Horiba Jobin Yvon, Simon Richard, Patrick Chapon, Kayvon Savadkouei, Philippe Hunault

(790-11 P) **HH-XRF and HH-LIBS for Alloy Analysis** JIYAN GU, Bruker

(790-12 P) **X-Ray Photoelectron Spectroscopy, Low Energy Ion Scattering, and Time-of-Flight Secondary Ion Mass Spectrometry, including Chemometrics Analysis, of Display Glass Surfaces** MATTHEW R LINFORD, Brigham Young University, Cody V Cushman, Barry M Lunt, Philipp Brüner, Julia Zakel, Thomas Grehl, Nicholas J Smith

(790-13 P) **The Particle Size Paradox** JACK G SAAD, Micromeritics, Paul A Webb

(790-14 P) **Gas Chromatography-Vacuum Ultraviolet Absorbance Spectroscopy for Quantitation of Trace and Bulk Water in Organic Solvents: An Emerging Alternative to Karl Fischer Titration** LINDSEY NICHOLE SHEAR, VUV Analytics, Leonard M Sidisky, Dale Harrison

(790-15 P) **Complete Size Characterization of Diatomaceous Earth** JACK G SAAD, Micromeritics

POSTER SESSION — Session 800

All posters are to be mounted by 10:00 AM and remain on display until 4:00 PM. Authors must be at their posters from 1:00 PM to 3:00 PM. Location of the afternoon posters is on the Exposition Floor, 400 Aisle. PLEASE NOTE: You cannot get onto the Exposition Floor until after 9:00 AM.

Physical Measurements

Monday Afternoon, Exposition Floor, 400 Aisle

(800-1 P) **Structural Studies of Co-Spinel Ferrite Synthesized by an Auto Combustion Method** DIPAL P CHAUDHARY, Mehsana Urban Institute of Sciences

(800-2 P) **Fabrication of A Novel Fiber-Optic based Single-Cell Temperature Sensor** QINGBO YANG, Missouri University of Science and Technology, Ke Li, Hai Xiao, Honglan Shi, Yinfa Ma

(800-3 P) Withdrawn

POSTER SESSION — Session 810

All posters are to be mounted by 10:00 AM and remain on display until 4:00 PM. Authors must be at their posters from 1:00 PM to 3:00 PM. Location of the afternoon posters is on the Exposition Floor, 400 Aisle. PLEASE NOTE: You cannot get onto the Exposition Floor until after 9:00 AM.

Polymer Characterization and Analysis

Monday Afternoon, Exposition Floor, 400 Aisle

(810-1 P) **Analyzing the Vibration-Rotation Spectrum of Hydrochloric Acid Generated with a Thermogravimetric Analyzer Coupled to a Fourier Transform Infrared Spectrophotometer** ANTHONY J LANG, PerkinElmer, Jack Botting

(810-2 P) **Analysis of PLGA Molecular Weight and Structure by the Latest Advanced Multi-Detector GPC Systems** MARK RICHARD POTHECARY, Malvern Instruments, Carrie Schindler, Ulf Nobbmann

(810-3 P) **Molecular Level Understanding of Etching Chemistry at Polymer Interfaces** JOHN MYERS, University of Michigan

(810-4 P) **Dissolution Dynamic Nuclear Polarization Study of Living Anionic Polymerization Reaction** YOUNGBOK LEE, Hanyang University

(810-5 P) **Extractables Analysis of SPE Frits** KIRSTEN KINNEBERG, Porex Corporation, Gary Li

(810-6 P) **Automated Microwave Sample Preparation of Difficult Petroleum and Polymer Matrices** ROBERT LOCKERMAN, CEM Corporation, Dan Iversen, Tina Restivo, Ian Goldstein, Austin Thornton

(810-7 P) **Spectroelectrochemical and Electrochemical Impedance Spectroscopy Analysis of Coronene and Perylene Dimide Copolymers** SAMIR C PAUL, Auburn University, Vince Cammarata N/A

(810-8 P) **Microwave, r0 Structural Parameters, Conformational Stability and Vibrational Assignment of (Chloromethyl)fluorosilane** DATTATRAY K SAWANT, University of Missouri Kansas City, James R Durig

(810-9 P) **Micro/Nano-Structured Flexible Foils for Anti-Counterfeiting Purposes** NASTASIA OKULOVA, Danapak Flexibles A/S and Technical University of Denmark, Rafael Taboryski, Lars Christensen

(810-10 P) **Self-Cleaning Properties of Nanostructured Polypropylene Foils Fabricated by Roll-to-Roll Extrusion Coating** AGNIESZKA TELECKA, Danish Technical University, Rafael Taboryski, Ling Sun

(810-11 P) **Synthesis and Characterization of Poly(p-methylstyrene) Spiropyran Conjugates** MATTHEW J PRICE, California University of PA, Zachary Sullenberger

(810-12 P) **Advantages of Ion Mobility Mass Spectrometry for Extractables Testing** BAIBA CABOVSKA, Waters Corporation, Eleanor Riches

(810-13 P) **A Novel Device for DART-MS System** MICHAEL J CHURCHILL, BioChromato, Chikako Takei

(810-14 P) **Characterization of Inorganic Fillers in Complex Polymer Matrices Using Mid-IR and Far-IR Spectroscopy** WILLIAM WIHLBORG, Thermo Fisher Scientific, Ronald Rubinovitz

POSTER SESSION **Session 820**

All posters are to be mounted by 10:00 AM and remain on display until 4:00 PM. Authors must be at their posters from 1:00 PM to 3:00 PM. Location of the afternoon posters is on the Exposition Floor, 400 Aisle. PLEASE NOTE: You cannot get onto the Exposition Floor until after 9:00 AM.

Portable Instruments

Monday Afternoon, Exposition Floor, 400 Aisle

(820-1 P) **Portable Optical Detection System for Determining Physical and Chemical Changes on Bioimplant Surfaces** DONALD BENZA, Clemson University, Fenglin Wang, Jeffrey N Anker

(820-2 P) **Field-Portable LED Array Based Multi-Wavelength Photothermal Lens Spectrometer** MICAH W ELLER, Tennessee Technological University, Andrew Callender

(820-3 P) **Translating Molecular Recognition into Pressure Signal for Rapid, Sensitive, and Portable Biomedical Analysis** ZHI ZHU, Xiamen University, Yang Chaoyong N/A

(820-4 P) **Development of Optical "Clamp-meter" Using Silicone Optical Technology for In-Situ Absorption Spectroscopy** HIROAKI NOMADA, Kyushu University, Hirokazu Higuchi, Hiroaki Yoshioka, Morita Kinichi, Yuji Oki

(820-5 P) **Development of a Portable Optical Cavity to Enhance 1-GHz Mode-Locked Laser Pulses for Broadband Absorption Spectroscopy** YUTARO ITO, Kyushu University, Zaitsu Shin-ichi, Imasaka Totaro

(820-6 P) **Characterization of a Miniaturized Benchtop Fast Gas Chromatograph Mass Spectrometer** COREY STEDWELL, 1st Detect Corporation, Daniel DeBord N/A

(820-7 P) **Enzymes and Photometers** ELLEN RUTH CAMPBELL, NECi Superior Enzymes, Bill Campbell, Justin Walbeck, David A Squires

(820-8 P) **Personal Monitoring of Ozone Exposure: A Fully Portable Device for Under $150 USD Cost** TINGTING CAO, Texas Tech University, Jonathan E Thompson

(820-9 P) **Producing a Miniaturized High-Performance Liquid Chromatography System for Exploration-Based Research** KYLE B LYNCH, University of Oklahoma

POSTER SESSION **Session 830**

All posters are to be mounted by 10:00 AM and remain on display until 4:00 PM. Authors must be at their posters from 1:00 PM to 3:00 PM. Location of the afternoon posters is on the Exposition Floor, 400 Aisle. PLEASE NOTE: You cannot get onto the Exposition Floor until after 9:00 AM.

Quality/QA/QC

Monday Afternoon, Exposition Floor, 400 Aisle

(830-1 P) **A Novel Approach to Specialty Gas Certifications by Means of GC/FTIR Analysis** PETER P BEHNKE, Prism Analytical Technologies, Inc., Martin L Spartz, Charles Mark Phillips, Anthony S Bonanno, Kelly R McPartland

(830-2 P) **Evaluation of the Flavor of Strawberry Preparation Using Gas Chromatography Electronic Nose** ANDREW COWELL, Alpha MOS, Jean-Christophe Mifsud, Herve Lechat, Valérie Vabre, Fatma Ayouni, Marion Bonnefille

(830-3 P) **Developments in the Automatism for CHNS and Oxygen Determination Using an Elemental Analyzer for Chemical Characterization** GUIDO GIAZZI, Thermo Fisher Scientific, Liliana Krotz, Francesco Leone, Walter Galotta

(830-4 P) **Running Legacy HPLC and Optimized UPLC Methods on a Single UPLC Platform, Comparative Studies and Steps Towards Implementation in a QC Environment** CHRISTOPHER HENRY, Waters Corporation, Mark Wrona, Richard Ladd, Andy Boughey

(830-5 P) **Quantitative Analysis of Calcite in Desulfurization Gypsum Using Raman Spectroscopy** YOUNG TAEK MA, Hanyang University

(830-6 P) **The Use of a Triple Detection System, (UV, ELSD, MS) for Pharmaceutical Degradation Studies** AARON D PHOEBE, Waters Corporation, Paula Hong, Patricia R McConville

(830-7 P) **Fast Analysis of Short Chain Fatty Acids in Feeds By SFC/MS** JINCHUAN YANG, Waters Corporation, Carrie Snyder, BJ Bench, Jessica Lance, Jayant Shringarpure

(830-8 P) **Application of FTIR Spectroscopy to Study the Effect of Processing on the Secondary Structure of Lentil Proteins** ALBERTA ARYEE, Agriculture & Agri-Food Canada, Joyce I Boye N/A

(830-9 P) **Accelerated Oxidation Tests on Olive Oil Stored in Plastic Packaging and Submitted to Autoclaving** STEFANO CASIRAGHI, VELP Scientifica, Stefania Corti, Claudia Mancinelli

(830-10 P) **Fast Analytical Method for Essential Oil Flavor Characterization Using a Polar Stationary Phase for Optimized Selectivity** RAMKUMAR DHANDAPANI, Phenomenex, Tim Anderson, Kristen Parnell

(830-11 P) **Steep Time and Temperature Effects on Flavor and Flavonoid Extraction of Black Tea** ANNE JUREK, EST Analytical, Lindsey Pyron, Kelly Cravenor, Justin Murphy

(830-12 P) **A Novel Approach to the Analysis of Multivitamin in Foodstuff by Online Supercritical Fluid Sample Extraction/Supercritical Fluid Chromatography** QIANG LI, Shimadzu (China) Co., LTD

(830-13 P) **Simultaneous Measurement of L-Lactate and Ethanol in Tomato-Based Products** WILLIAM MILLER, YSI, Inc, June Klingensmith

(830-14 P) **Iron Translocation in Adzuki Beans Sprouts: Enrichment Effects** ALINE P OLIVEIRA, Federal University of São Paulo, Juliana Naozuka, Alexandrina C Carvalho, Cassiana S Nomura

(830-15 P) **The Qualitative and Quantitative Analysis of α-Acids in Hops and Beers by UHPLC with UV Detection** WILHAD M REUTER, PerkinElmer, Jason Weisenseel

(830-16 P) **SFC Analytical Method Development for Vitamin D3 and Related Compounds** TAKASHI SATO, YMC Co., Ltd., Roland Spaegele, Junko Kawabata, Saoko Nozawa, Toshikazu Adachi, Noritaka Kuroda

(830-17 P) **Fast and Cost-effective Sugar Analysis Using HPAE-PAD** HUA YANG, Thermo Fisher Scientific, Linda Lopez, David G Moore

(830-18 P) **Improving Analytical Performance with Enhanced Matrix Removal - Lipid** DERICK LUCAS, Agilent Technologies, Limian Zhao, Bruce Richter, Megan Juck, Joan Stevens

(830-19 P) **Food Inspection Using NIR Reflectance and Transmittance Imaging** SATORU TSUCHIKAWA, Nagoya University, Hikaru Kobori, Te Ma, Norihisa Katayama N/A

(830-20 P) **A Fast Method for Predicting the Phenolic Content of Whisky Malts** ANDREW COWELL, Alpha MOS, Jean-Christophe Mifsud, Herve Lechat, Valérie Vabre, Fatma Ayouni, Marion Bonnefille

(830-21 P) **Combining FTIR and Mass Spectrometric Detection in the Gas Chromatographic Analysis of Fragrances** TRACY PHILLPOTT, DANI Instruments, Roberta Lariccia, Matthew S Klee, Daniele Recenti

(830-22 P) **The Effect of Food Additives on Hippocampus NMDA Receptor Subunits and Oxidative Stress** ZAFER YONDEN, Mustafa Kemal University ZAFER YONDEN, Mustafa Kemal University, Oguzhan Ozcan, Yeşim Göçmen, Namık Delibaş N/A

(830-23 P) **Holographic Characterization of Multicomponent Colloidal Suspensions** JAROSLAW M BLUSEWICZ, Spheryx, Inc., David B Ruffner, Laura Philips

(830-24 P) **Mitigating Electrostatic Effects Improves Measurement Accuracy** GREG GUMKOWSKI, NRD LLC, Arnold Steinman

(830-25 P) **Safe Approach to Gas Purification** BRIAN WARRICK, ARM, Inc, Daniel Spohn

(830-26 P) **Advanced UHPLC Instrument to Instrument Method Transfer** GREGORY HUNLEN, Agilent Technologies, Michael Woodman

POSTER SESSION **Session 840**

All posters are to be mounted by 10:00 AM and remain on display until 4:00 PM. Authors must be at their posters from 1:00 PM to 3:00 PM. Location of the afternoon posters is on the Exposition Floor, 400 Aisle. PLEASE NOTE: You cannot get onto the Exposition Floor until after 9:00 AM.

Sensors

Monday Afternoon, Exposition Floor, 400 Aisle

(840-1 P) **An Ultrasensitive Impedance Biosensor based on Immunomagnetic Separation and Urease Catalysis for Rapid Detection of Foodborne Pathogens Using an Immobilization-Free Interdigitated Microelectrode** LIN JIANHAN, China Agricultural University N/A

(840-2 P) **Enabling Tools to Combat Antibiotic Resistant Bacteria** ANDREW HELLER, Michigan State University, Dana Spence

(840-3 P) **Optimizing the Scintillation Cascade in Nano-Scintillation Proximity Assay (nanoSPA) for Multiplexed Detection of Small Biomolecules** ZEINAB MOKHTARI, University of Arizona, Isen Andrew C Calderon, Colleen Janczak, Craig A Aspinwall

(840-4 P) **Dendritic Gold Structures for Glucose Biosensor Design** ALMIRA RAMANAVICIENE, Vilnius University, Laura Sakalauskiene, Anton Popov, Natalija German, Asta Kausaite-Minkstimiene, Arunas Ramanavicius

(840-5 P) **High-Throughput Thiamine Quantification in Environmental Matrices Using Periplasmic-Binding Protein Biorecognition** KATIE A EDWARDS, Cornell University, Seth Feder, Antje J Baeumner, Cliff Kraft

(840-6 P) **Multi-Purpose Individual Air Monitor – Conception to Proof of Concept Laboratory Prototype** NICHOLAS FITZGERALD, Defence Science and Technology Group, Karl Pavey

(840-7 P) **Characterization of Self-Oscillating Reaction Catalyzed by Metal Porphyrin Complexes** TAKASHI ARIMURA, AIST Interdisciplinary Research Center, Masaru Mukai, Jung Hee Do, Kenichi Tominaga

(840-8 P) **Simultaneous Detection of Dopamine Release and Neural Activity** KATE L PARENT, University of Arizona, Daniel F Hill, Jean-Paul Wiegand, Michael A Miller, Christopher W Atcherley, Stephen Cowen, Michael L Heien

(840-9 P) **Development of a Novel Position-Sensitive MCP Detector** BLAKE WIGGINS, Indiana University, Davinder Siwal, Romualdo T deSouza

(840-10 P) **Real Time Monitoring Magnesium Alloys Corrosion by Electrochemical H2 Sensor In Vivo** DAOLI ZHAO, University of Cincinnati, Tingting Wang, Zhongyun Dong, William R Heineman, Vesselin Shanov

(840-11 P) **A Cellulose Acetate Membrane-Based Colorimetric Device to Discriminate Bacteria** LIGIA BUENO, Instituto de Química - Universidade de Sao Paulo, Subrayal M Reddy, Alison Cottel, Thiago Paixao

(840-12 P) **A Smart Wearable and Autonomous Negative Pressure Device for Wound Monitoring** FABIO DI FRANCESCO, Università di Pisa, Bernardo Melai, Pietro Salvo, Andrea Pucci, Vincenzo Mollica, Anna Maria Raspolli, Valentina Dini, Marco Romanelli, Beatrice Lazzerini, Aldo Paolicchi, Valter Castelvetro, Roger Fuoco N/A

(840-13 P) **A Smart Polymer Hydrogel as a Chemical Sensor on Biomedical Implant** MOHAMMED ARIFUZZAMAN, Clemson University, Caleb Behrend, Jeffrey N Anker

(840-14 P) **Oxygen Sensitive Probes** PETER GENNARO, Clemson University

(840-15 P) **Graphene Oxide as an Efficient Antibacterial Agent in Macrophages and in Mice** XU WU, University of North Dakota, Xiao Liu, Yuqian Xing, Julia Xiaojun Zhao, Min Wu

(840-16 P) **pH Indicator to Enhance Surface Plasmon Resonance Imaging Detection of Small Organic Molecules** ZAINAB HUSSAIN AL MUBARAK, Oklahoma State University, Gayan Premaratne, Cassandra Rodenbaugh, Lucy Lehoczky, Sadagopan Krishnan

(840-17 P) **Membrane Based Electrodes for Electrochemical Applications in Biology** TANYA BAKMAND, DTU Nanotech, Dorota Kwasny, Fatima Al-Zahraa Al Atraktchi, Helle Waagepetersen, Winnie E Svendsen, Maria Dimaki

(840-18 P) **Paper-Based Colorimetric Glucose Determination Using Smartphone** HAKAN CIFTCI, Kirikkale University, Nazlı Ayyıldız, Ugur Tamer

POSTER SESSION **Session 850**

All posters are to be mounted by 10:00 AM and remain on display until 4:00 PM. Authors must be at their posters from 1:00 PM to 3:00 PM. Location of the afternoon posters is on the Exposition Floor, 400 Aisle. PLEASE NOTE: You cannot get onto the Exposition Floor until after 9:00 AM.

Thermal Analysis

Monday Afternoon, Exposition Floor, 400 Aisle

(850-1 P) **New TGA-FT-IR Library for Polymers** EKKEHARD POST, NETZSCH Geraetebau GmbH, Carolin Fischer, Bob Fidler

(850-2 P) **Beyond Classical Dynamic Mechanical Analysis - Using A New High-Force, High-Temperature DMA to Characterize Advanced Materials** BOB FIDLER, NETZSCH Instruments NA LLC, Horst Deckmann, Juergen Blumm, Ekkehard Post, Tobias Pflock

(850-3 P) **Exploring the Properties of Textiles with Thermal Analysis, Infrared and UV/Vis Spectroscopy** ANTHONY J LANG, PerkinElmer, Aaron H Adams, Jack Botting

TUESDAY, MARCH 8, 2016
MORNING

AWARDS **Session 860**

Pittsburgh Analytical Chemistry Award

arranged by Joanne Smith, The Pittsburgh Conference

Tuesday Morning, Room B312

Joanne Smith, The Pittsburgh Conference, Presiding

8:30		**Introductory Remarks - Joanne Smith**
8:35		**Presentation of the 2016 Pittsburgh Analytical Chemistry Award to Sanford A Asher, University of Pittsburgh, by Elias S Absey, Chair, Society for Analytical Chemists of Pittsburgh**
8:40	(860-1)	**Photonic Crystal Hydrogel and Organogel Sensors for Chemical and Biological Analytes** SANFORD A ASHER, University of Pittsburgh, Natasha L Smith, Zhongyu Cai, Andrew E Coukouma
9:15	(860-2)	**Structural Characterization of Methylenedianiline Regioisomers by Ion Mobility-Mass Spectrometry, Tandem Mass Spectrometry, and Computational Strategies - MALDI Spectra** DAVID M HERCULES, Vanderbilt University, Sarah M Stow, Tiffany M Onifer, John A McLean
9:50	(860-3)	**Liposomal Spherical Nucleic Acids: A New Approach to Immunomodulatory Therapy** CHAD A MIRKIN, Northwestern University
10:25		**Recess**
10:40	(860-4)	**Measuring DNA Hybridization Kinetics by Single-Molecule Fluorescence Imaging** JOEL M HARRIS, University of Utah, Eric M Peterson, Michael Manhart, Frances Morris
11:15	(860-5)	**SERS-Based Metabolic Profiling for Diagnostics and Forensics** LAWRENCE ZIEGLER, Boston University

AWARDS **Session 870**

The Coblentz Society/ABB - Bomem-Michelson Award

arranged by Brooks Pate, University of Virginia

Tuesday Morning, Room B314

Brooks Pate, University of Virginia, Presiding

8:30		**Introductory Remarks - Brooks Pate**
8:35		**Presentation of the 2016 Coblentz Society/ABB- Bomem-Michelson Award to Shaul Mukamel, University of California Irvine, by Henry L. Buijs, ABB**
8:40	(870-1)	**Coherent Ultrafast Multidimensional Spectroscopy of Molecules with Optical, X-Ray, and Quantum Light** SHAUL MUKAMEL, University of California, Irvine
9:15	(870-2)	**Photosynthetic Light Harvesting from Individual Complexes to the Grana Membrane** GRAHAM RICHARD FLEMING, University of California at Berkeley N/A
9:50	(870-3)	**A Few Lessons from Non-Adiabatic Excited State Dynamics Simulations of Large Molecules** SERGEI TRETIAK, Los Alamos National Laboratory
10:25		**Recess**
10:40	(870-4)	**Photosynthetic Light Harvesting and Ultrafast Energy Transfer** GREG D SCHOLES, Princeton University
11:15	(870-5)	**Elucidation of Chemical Reactions by Two-Dimensional Resonance Raman Spectroscopy** ANDREW MORAN, University of North Carolina

SYMPOSIUM **Session 880**

ACS-ANLY - Chemometrics: A New Dimension in Chromatography
arranged by Amber M Hupp, College of the Holy Cross

Tuesday Morning, Room B308
Amber M Hupp, College of the Holy Cross, Presiding

8:30		**Introductory Remarks - Amber M Hupp**
8:35	(880-1)	**Applying the Hotelling Trace Criterion to Optimize Chromatogram Alignment of Biodiesel Diesel Blended Fuels** AMBER M HUPP, College of the Holy Cross, Gopal Yalla, John O'Connor, Kevin Walsh, Edward Soares
9:10	(880-2)	**Forensic Signatures for the Source Attribution of Chemical Threat Agents Using Chemical Profiling, Stable Isotope Ratios and Chemometrics** CARLOS FRAGA, Pacific Northwest National Laboratory
9:45	(880-3)	**Chemometrics: An Old Dimension in Chromatography – Application to New Dimensions** SARAH C RUTAN, Virginia Commonwealth University, Daniel W Cook, Melanie M Sinanian
10:20		**Recess**
10:35	(880-4)	**Data Reduction and Processing Tools for GCxGC-TOFMS** JAMES J HARYNUK, University of Alberta, Lawrence A Adutwum, A Paulina de la Mata
11:10	(880-5)	**Chemometric Approaches to Maximize Interpretation of GCxGC - TOFMS Data for Discovery-Based Analyses** ROB SYNOVEC, University of Washington

SYMPOSIUM **Session 890**

Advances in Analytical Methodologies for the Detection of Food Allergens and Gluten
arranged by Lowri DeJager and Christine Parker, U.S. Food and Drug Administration

Tuesday Morning, Room B302
Christine Parker, U.S. Food and Drug Administration, Presiding

8:30		**Introductory Remarks - Lowri DeJager and Christine Parker**
8:35	(890-1)	**Approaches to Multianalyte Allergen Analysis in Food** CLARE MILLS, The University of Manchester, Karine Adel-Patient, Sabine Baumgartner
9:10	(890-2)	**Effects of Cross-Reactivity in Food Allergy Detection and Diagnosis** SOHEILA J MALEKI, USDA-Agricultural Research Service
9:45	(890-3)	**Novel Approaches to Identifying Amadori Products in Peanut Extract** GEOFFREY MUELLER, National Institute of Environmental Health Sciences, Jason Williams, Katina Johnson
10:20		**Recess**
10:35	(890-4)	**Overview of Analytical Methods for Food Allergen and Gluten Analysis: Challenges and Trends from a Regulatory Perspective** TERRY KOERNER, Health Canada
11:10	(890-5)	**Effects of a Proline Endopeptidase on the Detection and Quantification of Gluten During the Fermentation of Beer** RAKHI PANDA, Food and Drug Administration, Katherine L Fiedler, Chung Y Cho, Eric A Garber

SYMPOSIUM **Session 900**

Advances in Two-Dimensional Liquid Chromatography Separations of Biopharmaceuticals
arranged by Dwight R Stoll, Gustavus Adolphus College

Tuesday Morning, Room B303
Dwight R Stoll, Gustavus Adolphus College, Presiding

8:30		**Introductory Remarks - Dwight R Stoll**
8:35	(900-1)	**Column Technology for the Chromatographic Characterization of Biopharmaceuticals** DAVY GUILLARME, University of Geneva, Jean-Luc Veuthey, Szabolcs Fekete
9:10	(900-2)	**Optimization of Two-Dimensional Liquid Chromatography Separations of Therapeutic Monoclonal Antibodies Involving Ion-Exchange and Reversed-Phase Separation Modes** DWIGHT R STOLL, Gustavus Adolphus College, David C Harmes, Jacob Bush, Matthew Sorensen, Gregory Staples, Szabolcs Fekete, Davy Guillarme, Alain Beck
9:45	(900-3)	**Application of a Multi-Dimensional Approach for Quantification of Free Drug in Antibody Drug Conjugates** BROOKE M KOSHEL, Waters Corporation, Robert Birdsall, Alain Beck, Sean M McCarthy
10:20		**Recess**
10:35	(900-4)	**2-Dimensional Liquid Chromatography for Rapid Parallel Separation of Biopharmaceuticals** YAN HE, Pfizer, Ian Hartzel, Michael T Jones
11:10	(900-5)	**Antibody-Drug-Conjugates (ADC) Drug Product Profiling by Multi-dimensional HPLC** KELLY ZHANG, Genentech

SYMPOSIUM **Session 910**

Graphene Nanomaterials for Bio/Sensing Applications
arranged by Ashok Mulchandani, University of California, Riverside and Chenzhong Li, Florida International University

Tuesday Morning, Room B304
Ashok Mulchandani, University of California, Riverside, Presiding

8:30		**Introductory Remarks - Ashok Mulchandani and Chenzhong Li**
8:35	(910-1)	**New Concepts in Biosensing Using Single Walled Carbon Nanotubes and Graphene** MICHAEL S STRANO, Massachusetts Institute of Technology
9:10	(910-2)	**Graphene Electrodes for Bio/chemical Sensors** ASHOK MULCHANDANI, University of California, Riverside
9:45	(910-3)	**A Novel Wireless Biosensing Platform Enabled by Graphene Varactors** STEVEN J KOESTER, University of Minnesota
10:20		**Recess**
10:35	(910-4)	**Graphene-Enabled Nano/Bio Hybrids for Chemical Detection and Medical Diagnostics** A T CHARLIE JOHNSON, University of Pennsylvania
11:10	(910-5)	**Impedance Sensing of Nanotoxicity of Graphene at the Cellular and Tissue Level** CHENZHONG LI, Florida International University

SYMPOSIUM **Session 920**

JAIMA - The State-of-the-Art Technologies from Japan: Analytical Instruments with/for Nano-Chemistry Technology and Advanced Diagnosis (I)
arranged by Koichiro Matsuda, JAIMA

Tuesday Morning, Room B305
Koichiro Matsuda, JAIMA, Presiding
Takeshi Kawamoto, JAIMA, Presiding

8:30		**Introductory Remarks - Gon-emon Kurihara, President of JAIMA**
8:35	(920-1)	**Creation of Bio/Chemical Sensing Probes** KOJI SUZUKI, Keio University, Daniel Citterio
9:10	(920-2)	**Ab Initio Powder Structure Determination Opening Up New Research Fields** MASAKI KAWANO, Pohang University of Science and Technology
9:45	(920-3)	**Nanodroplet Formation and Chemical Analysis in Microfluidic Devices** AKIHIDE HIBARA, Tokyo Institute of Technology
10:20		**Recess**
10:35	(920-4)	**Development of Mass Microscope and Applications in Drug and Oncometabolite Visualization** SHUICHI SHIMMA, Osaka University
11:10	(920-5)	**Plasma Phospholipids and Prevalence of Mild Cognitive Impairment/ Dementia** DANNI LI, University of Minnesota

SYMPOSIUM — Session 930

Ultrahigh-Resolution Mass Spectrometry: A New Frontier
arranged by Alan G Marshall, Florida State University

Tuesday Morning, Room B301
Alan G Marshall, Florida State University, Presiding

8:30		**Introductory Remarks - Alan G Marshall**
8:35	(930-1)	**High-Resolution TOFMS via a Folded Flight Path: You Can Have Your Cake and Eat It Too** VIATCHESLAV ARTAEV, Leco Corporation, Michael Mason, Peter Willis, Anatoly Verenchikov
9:10	(930-2)	**21 Tesla FT-ICR Mass Spectrometer for Top-Down Protein Identification and Characterization** CHRISTOPHER L HENDRICKSON, National High Magnetic Field Laboratory, Lissa C Anderson, John P Quinn, Nathan K Kaiser, Donald F Smith, Greg T Blakney, Alan G Marshall
9:45	(930-3)	**Exploring the Microbiome via Advanced Mass Spectrometry** LJILJANA PASA-TOLIC, EMSL, PNNL, Jared B Shaw, Malak Tfaily, Nikola Tolic, Errol Robinson, David W Koppenaal, Nancy J Hess
10:20		**Recess**
10:35	(930-4)	**Complex Mixture Analysis by Ultra-High Resolution Mass Spectrometry** RYAN P RODGERS, Florida State University, Amy C Clingenpeel, Steven M Rowland, Vladislav V Lobodin, David C Podgorski, Yuri E Corilo, Priscila M Lalli, Jonathan C Putman, Winston K Robbins
11:10	(930-5)	**Orbitrap Mass Analyzers: The Teenage Years** MICHAEL W SENKO, Thermo Fisher Scientific, Jesse D Canterbury, Eduard Denisov, Alexander Makarov

ORGANIZED CONTRIBUTED SESSIONS — Session 940

High Performance SFC for the Analysis of Pharmaceuticals, Nutraceuticals, Natural Products and Metabolomics
arranged by David Kohler, ES Industries and Patrick G Fromal, Shimadzu Scientific Instruments Inc

Tuesday Morning, Room B309
David Kohler, ES Industries, Presiding

8:30	(940-1)	**Analysis of Mycotoxins in Various Food Matrices via SFE/SFC/MS** TODD ANDERSON, Shimadzu Scientific
8:50	(940-2)	**Evaluation of a Liquid Carbon Dioxide Based Flash Chromatography System** RAY MCCLAIN, Merck
9:10	(940-3)	**Fluoro-Methylphenylcarbamates of Cellulose as Chiral Stationary Phases for Supercritical Fluid Chromatography** WILLIAM FARRELL, Pfizer, Inc., Matthew Przybyciel
9:30	(940-4)	**Potential of Supercritical Fluid Extraction and Separation Technologies in Metabolomics** BAMBA TAKESHI, Kyushu University
9:50		**Recess**
10:05	(940-5)	**SFC – An Essential "Green" Separation Technique for the Chromatographic Analysis of a Nutraceuticals** MATTHEW PRZYBYCIEL, ES Industries
10:25	(940-6)	**SFC, A Walk on the Wild Side** BRENT MURPHY, Genentech, Mengling Wong, Joseph H Pease, Michael Hayes, Amber Guillen, Christopher Hamman
10:45	(940-7)	**Understanding the Value of Preparative Chiral Separations in Advancing Pharmaceutical R&D Projects** MICKEY REGO, Averica Discovery Services, Jeffrey Kiplinger, Paul Lefebvre, Keith Galyan, Emily Showell-Rouse
11:05	(940-8)	**Open Discussion**

ORGANIZED CONTRIBUTED SESSIONS — Session 950

Molecular Modelling and Quantum Mechanical Calculations: From Small Molecules to Large Multimeric Protein Complexes
arranged by Iain Campuzano and Michael D Bartberger, Amgen

Tuesday Morning, Room B310
Iain Campuzano, Amgen, Presiding

8:30	(950-1)	**Vibrational Optical Activity of Helical Peptides** JAMES R CHEESEMAN, Gaussian, Inc.
8:50	(950-2)	**Interplay of Theory and Experiment: Small Molecule Ion Mobility and Spectroscopic Characterization of Metabolites** MICHAEL D BARTBERGER, Amgen
9:10	(950-3)	**Efficient Photostitching of Peptide Ion Complexes in the Gas Phase with the Photo-Leucine Zipper** FRANTISEK TURECEK, University of Washington, Christopher J Shaffer, Akis P Andrikopoulos, Lubomir Rulisek
9:30	(950-4)	**Chiroptical Spectroscopy for Molecular Structure Determination** PRASAD L POLAVARAPU, Vanderbilt University
9:50		**Recess**
10:05	(950-5)	**The Use of Molecular Modelling for Ion Mobility Drift Time and Fragment Ion Prediction in Ion Mobility and Mass Spectrometry** CRIS LAPTHORN, University of Greenwich, Frank S Pullen, Babur Chowdhry, Patricia Wright, George Perkins, Yanira Heredia, Trevor Dines
10:25	(950-6)	**Dynamical Networks of Protein Residue-Residue Contacts Provide Insights into Enzyme Function** DONALD HAMELBERG, Georgia State University
10:45	(950-7)	**A Molecular Dynamics and Ion Mobility Study of Protein Structure Collapse in the Gas-phase** IAIN CAMPUZANO, Amgen, Morgan Lawrenz, Carlos Larriba-Andaluz
11:05	(950-8)	**Hell Bent on Opening: Structural Basis for Estrogen Modulation of Potassium Recycling During Epithelial Secretion** BRETT KRONCKE, Vanderbilt University

ORGANIZED CONTRIBUTED SESSIONS — Session 960

SEAC Young Investigator Session
arranged by Hitomi Mukaibo, University of Rochester

Tuesday Morning, Room B311
Hitomi Mukaibo, University of Rochester, Presiding

8:30	(960-1)	**Fluorescence-Enabled Electrochemistry** BO ZHANG, University of Washington
8:50	(960-2)	**Designing Functional Nanostructured Gels for Electrochemical Energy Storage and Biosensors** GUIHUA YU, University of Texas at Austin
9:10	(960-3)	**Inhibiting Electrochemical Processes in Li-ion Batteries at High Temperatures Using Responsive Polymers** MARK E ROBERTS, Clemson University, Jesse C Kelly, Nick L DeGrood
9:30	(960-4)	**Revealing the Mechanism of Electron Uptake in Methanogenic Biofilm Community Using a Solid-Phase Electron Donor** SOFIA BABANOVA, JCVI, Kayla Carpenter, Sujal Phadke, Tony Phan, Shunichi Ishii, Shino Suzuki, Michael Salvacion, Michael Flynn, John Hogan, Orianna Bretschger
9:50		**Recess**
10:05	(960-5)	**Reagentless and Reusable Electrochemical Metal Ion Sensors** REBECCA Y LAI, University of Nebraska-Lincoln
10:25	(960-6)	**Electrochemical Imaging of Ionic Reactivity at Operating Ion Batteries** JOAQUIN RODRIGUEZ LOPEZ, University of Illinois
10:45	(960-7)	**In Vivo Serotonin Chemistry and Local Cytoarchitecture: A Combined Voltammetric, Mathematical and Microscopy Study** PARASTOO HASHEMI, University of South Carolina, Aya Abdalla
11:05	(960-8)	**Ec-LLS at the Micro- and Macroscale: Electrodeposition of Semiconductor Films and Nanocrystals** STEPHEN MALDONADO, University of Michigan

ORGANIZED CONTRIBUTED SESSIONS **Session 970**

Supercritical CO_2 – SFE/SFC: Advances in Extraction and Purification for Pharmaceutical and Natural Products

arranged by Andy Miles, Regis Technologies and Gregory K. Webster, AbbVie

Tuesday Morning, Room B313
Andy Miles, Regis Technologies, Presiding

8:30	(970-1)	**Supercritical Fluid Chromatography in Support of Pharmaceutical – A Study of Scale-Up from Analytical to Preparative Scale with Isocratic Conditions** MIRLINDA BIBA, Merck, Jinchu Liu, Lindsey Jacobs, Judy Morris, Ingrid Mergelsberg
8:50	(970-2)	**Strategies for Increasing Throughput of Chiral Separations by Supercritical Fluid Chromatography** ERIN JORDAN, AbbVie, Philip A Searle
9:10	(970-3)	**Withdrawn**
9:30	(970-4)	**SFC-MS as the Technique of Choice for Small Molecules Purification** GERARD ROSSE, Dart Neuroscience
9:50		**Recess**
10:05	(970-5)	**Practical Approaches to SFE and SFC in Drug Discovery** JOSEPH H PEASE, Genentech, Brent Murphy, Mengling Wong, Michael Hayes, Amber Guillen
10:25	(970-6)	**One Phase, Three Techniques: Utilizing One SFC Stationary Phase Across Three Chromatographic Techniques** CHRISTINE AURIGEMMA, Pfizer, Inc., Perrine Hoerter, William Farrell
10:45	(970-7)	**Online SFE-SFC Purification Method Development Using SFC Solubility Determination** WES BARNHART, Amgen, Kyung Gahm
11:05	(970-8)	**Utilizing SFE and SFC for Extraction and Isolation of Cannabinoids** CHRISTOPHER HUDALLA, ProVerde Laboratories

ORAL SESSIONS **Session 980**

Biomedical: Advances in Glucose Monitoring and Therapeutics of Diabetes - Half Session

Tuesday Morning, Room B408
Barbara Manner, The Pittsburgh Conference, Presiding

8:30	(980-1)	**Skin Interface and Spectrometer Development for Noninvasive Glucose Measurements in People over Combination Near-Infrared Wavelengths** ARIEL BOHMAN, University of Iowa, Mark A Arnold, Gary W Small, Michael J Miller
8:50	(980-2)	**Advances in Development of Glucose Biosensors** ARUNAS RAMANAVICIUS, Vilnius University, Natalija German, Asta Kausaite-Minkstimiene, Inga Vilkonciene, Povilas Genys, Jaroslav Voronovic, Jurate Petroniene, Almira Ramanaviciene
9:10	(980-3)	**A Microfluidic Cell Culture Device for Automated Sample Preparation and Improved Biomimetic Modeling in Diabetes Metabolomics** LAURA FILLA, Saint Louis University, James L Edwards
9:30	(980-4)	**Analysis of Drug-Protein Interactions During Diabetes by High-Performance Affinity Chromatography** ZHAO LI, University of Nebraska-Lincoln, Ryan Matsuda, David Hage

ORAL SESSIONS **Session 990**

Capillary Electrophoresis

Tuesday Morning, Room B315
Elizabeth Harris, Mannkind Corporation, Presiding

8:30	(990-1)	**Monitoring Neurotransmitter Secretion from Islets of Langerhans** KIMBERLY EVANS, Florida State University, Xue Wang, Michael G Roper
8:50	(990-2)	**Observing Peptide Folding Intermediates Using Capillary Electrophoresis** ALISON E HOLLIDAY, Moravian College, John D Barr, Liuqing Shi, David E Clemmer
9:10	(990-3)	**PDMS Micro-cross Junction for Online Nanoliter Heart-cut for Two Dimensional CE Separation** VITALY AVILOV, University of Illinois at Chicago, Scott A Shippy
9:30	(990-4)	**Real Time Monitoring of Amino Acid Secretions from Islets of Langerhans Using a Microfluidic Device** XUE WANG, Florida State University, Michael G Roper, Lian Yi
9:50		Recess
10:05	(990-5)	**Single-Cell Chemical Cytometry of Akt Activity within Primary Rheumatoid Arthritis Synovial Fibroblasts Illustrates Heterogeneity in Patient Responses to TNFα** EMILIE R MAINZ, University of North Carolina at Chapel Hill, Stephen Serafin, David Lawrence, Teresa Tarrant, Christopher Sims, Nancy Allbritton
10:25	(990-6)	**The Determination of Oxidative Stress Biomarkers of Lipid Peroxidation Using a Novel Capillary Electrophoresis-Mass spectrometry Sheathless Interface Design** RYAN T JOHNSON, University of Kansas, Craig E Lunte, Nhan To, John Stobaugh
10:45	(990-7)	**Analysis of Metabolite Biomarkers in Prostrate Cancer Tissues by Capillary Electrochromatography Mass Spectrometry** YANG LU, Georgia State University, Shahab S Shamsi, Dean Troyer
11:05	(990-8)	**Rapid Determination of Cyanide in Human Urine by Capillary Electrophoresis Coupled with Laser-Induced Fluorescence Detection** QIYANG ZHANG, Wichita State University, Naveen Maddukuri, Maojun Gong

ORAL SESSIONS **Session 1000**

Electrochemistry - New Approaches and Techniques

Tuesday Morning, Room B316
Lawrence A Bottomley, Georgia Institute of Technology, Presiding

8:30	(1000-1)	**Stochastic Electrochemistry of TiO_2 Nanoparticles and Their Agglomerates** MARIO ALPUCHE-AVILES, University of Nevada, Reno, Pushpa Chhetri, Krishna K Barakoti, Ganesh Rana, Andrew Recinos
8:50	(1000-2)	**The Development of an Automated NanoElectrode Array Sensor to Detect Evaporation and Changes in Cellular Bioenergetics in a Submicroliter Chamber from an Organ-on-a-Chip System** ANNA N DAVIS, Vanderbilt University, John Wikswo, David Cliffel
9:10	(1000-3)	**Scanning Electrochemical Microscopy (SECM): A New Tool to Study Microbial Metabolism** DIPANKAR KOLEY, Oregon State University, Vrushali Joshi
9:30	(1000-4)	**Drug Metabolism Assays and Metabolite Synthesis Using Microsomes Based Bioreactor** RAJASEKHARA NERIMETLA, Oklahoma State University, Sadagopan Krishnan
9:50		**Recess**
10:05	(1000-5)	**High-Density Microelectrode Arrays as Electrochemical Imaging Platforms** RACHEL M FEENY, Colorado State University, John Wydallis, Lang Yang, Stacy Willett, Tom Chen, Stuart A Tobet, Charles Henry
10:25	(1000-6)	**Imaging of Immobilized Enzymes and Yeast Cells by Scanning Electrochemical Microscopy** ARUNAS RAMANAVICIUS, Vilnius University, Inga Vilkonciene, Rita Sareikaite, Aura Kisieliute, Povilas Genys, Jurate Petroniene, Almira Ramanaviciene
10:45	(1000-7)	**Rapid Temperature Control of Electrochemical and Biochemical Systems Using Microfabricated Heaters** NICHOLAS M CONTENTO, Biomolecular Measurement Division, Sarah M Robinson, Kurt D Benkstein, Herman O Sintim, Steve Semancik
11:05	(1000-8)	**A Simple, High Yield Method to Microfabricate Planar Microelectrode Arrays** MOHAMED M MAREI, University of Louisville, Mark Crain, Richard P Baldwin, Robert S Keynton

ORAL SESSIONS **Session 1010**

Enhancements in Pharmaceutical and Environmental Separations

Tuesday Morning, Room B402
Richard Henry, Consultant, Presiding

8:30	(1010-1)	**Sub-ppm Level Formaldehyde Measurement in Complex Sample Matrices Using a Variety of Analytical Methods – Method Comparison and Practical Considerations** PEILIN YANG, The Dow Chemical Company, Francois Huby, James N Alexander IV
8:50	(1010-2)	**Optimization of a Single-Stage, Consumable Free Thermal Modulator for GC x GC** HALEIGH BOSWELL, University of Waterloo, Tadeusz Gorecki
9:10	(1010-3)	**Thermo Tuning of Redox Potential on Nanostructured Adsorbent for a Reagent-less Recovery of Pollutant Oxoanions** MANUEL VALIENTE, Universitat Autonoma de Barcelona, He Liu, Liu Tong, Gustavo Perez
9:30	(1010-4)	**Super/subcritical Fluid Chromatographic Chiral Separations with High Efficiency Chiral Stationary Phases** ZACHARY BREITBACH, The University of Texas at Arlington, Chandan L Barhate, M Farooq Wahab, Daniel W Armstrong

9:50		Recess
10:05	(1010-5)	**Novel Solid-Phase Microextraction and Capillary Electrochromatographic Column Techniques for Pharmaceutical Analysis** ZILIN CHEN, Wuhan University
10:25	(1010-6)	**Retention in Porous Layer Pillar Array Planar Separation Platforms** DANIELLE R LINCOLN, University of Tennessee, Knoxville, Nickolay V Lavrik, Michael J Sepaniak
10:45	(1010-7)	**Carbon Nanofibers Decorated with Magnetic Nanoparticles as a New Sorbent for the Magnetic Solid Phase Extraction of Selected Polycyclic Aromatic Hydrocarbons from Water Samples** ALI SARAFRAZ YAZDI, Ferdowsi University of Mashhad N/A
11:05	(1010-8)	**Analytical Method Development: Are We Solving the Right Problem? A Systematic Approach to Select an Appropriate RPLC Column and to Optimize Separation** IMAD A HAIDAR AHMAD, Novartis, James Tam, Xue Li, Thomas Tarara, Andrei Blasko

ORAL SESSIONS — Session 1020

Environmental Air Quality and Analysis

Tuesday Morning, Room B403
John Kokosa, Research Consultant, Presiding

8:30	(1020-1)	**On-site Determination of Formaldehyde Using SPME and a Portable GC-TMS** JUSTEN J POOLE, University of Waterloo, Jonathan J Grandy, German A Gomez-Rios, Janusz Pawliszyn
8:50	(1020-2)	**Fast, Accurate, and Precise: How to Comply with EPA Method 325b (Fence Line Monitoring for Benzene)** LEE MAROTTA, PerkinElmer, Roberta Provost, Amy Jacobson, Mariah Peronto
9:10	(1020-3)	**Investigation and Modelling of the Sampling Process in a PDMS-Based Permeation Passive Sampler** FATEN SALIM, University of Waterloo, Marios Ioannidis, Tadeusz Gorecki
9:30	(1020-4)	**QCL-Based Perfluorocarbon Emission Monitoring** LUIS H ESPINOZA-NAVA, Alcoa Inc
9:50		Recess
10:05	(1020-5)	**Laser Derivitization for Soot Source Identification** RANDY VANDER WAL, Penn State University, Chethan K Gaddam
10:25	(1020-6)	**Analysis of Damaged Floor Coverings Emissions in Indoor Air Quality with Cantilever-Enhanced Photoacoustic Spectroscopy** JUSSI RAITTILA, Gasera Ltd., Ismo Kauppinen, Sauli Sinisalo, Jaakko Lehtinen
10:45	(1020-7)	**Ship Emissions Monitoring with Laser-Based Cantilever-Enhanced Photoacoustic Detection** JAAKKO LEHTINEN, Gasera Ltd., Ismo Kauppinen, Jussi Raittila
11:05	(1020-8)	**Hand Held Detector Based on an Ion Mobility Spectrometer and an Additional Detector (Electrochemical Cell or Alternatively a Photo Ionization Detector) for the Detection of Toxic Gases in Chemical Industries** ANDREAS WALTE, Airsense Analytics, Bert Ungethuem, Wolf Muenchmeyer

ORAL SESSIONS — Session 1030

Magnetic Resonance - Half Session

Tuesday Morning, Room B408
Barbara Manner, The Pittsburgh Conference, Presiding

10:05	(1030-1)	**Quantitative Determination of Short-Chain Branching and Co-Monomer Content in High Density Polyethylene by NMR Spectroscopy** RAVINDRA KUMAR, Indian Oil Corporation Ltd, Veena Bansal, Sujit Mondal, GS Kapur, V Kagdiyal, Deepak Saxena N/A
10:25	(1030-2)	**GPC-NMR Analysis of Polymeric Mixtures** NIKKI H LAFEMINA, Arkema N/A
10:45	(1030-3)	**Quantitative Analysis of Phosphorous Containing Drug Molecules Encapsulated in pH-Sensitive Nanoparticle Formulations to Establish Drug Loading and Drug Releasing Profile by Utilizing 31P Solid State and Solution State NMR Spectroscopy** SUDHAUNSHU S PUROHIT, University of Missouri Kansas City, Jianing Meng, Vivek Agrahari, Bi Botti Celestin Youan, Nathan Oyler
11:05	(1030-4)	**Nanoparticle-Assisted Removal of Proteins in Human Serum for Metabolomics Studies** BO ZHANG, The Ohio State University, Mouzhe Xie, Lei Bruschweiler-Li, Rafael Brüschweiler

ORAL SESSIONS — Session 1040

Mass Spectrometry - Bioanalytical

Tuesday Morning, Room B404
Daniel E Austin, Brigham Young University, Presiding

8:30	(1040-1)	**Electrospray Ionization-High Pressure Mass Spectrometry for Peptide and Protein Analysis** RUSSELL E BORNSCHEIN, University of North Carolina at Chapel Hill, William M Gilliland, J Michael Ramsey
8:50	(1040-2)	**Current Analytical Techniques for Glycoprotein Characterization by Mass Spectrometry** PARASTOO AZADI, Complex Carbohydrate Research Center
9:10	(1040-3)	**Nanopatterning Ligands to Enable Cell-Based Assays Using SAMDI-Mass Spectrometry** MARIA D CABEZAS, Northwestern University, Milan Mrksich, Chad A Mirkin
9:30	(1040-4)	**Shotgun Lipidomic Analysis of Human Meibum by MS/MS^{all} with Successive Switching between Positive and Negative Detection Modes** JIANZHONG CHEN, University of Alabama at Birmingham, Kelly Nichols
9:50		Recess
10:05	(1040-5)	**Development, Characterization, and Application of Coated Blade Spray Ionization** GERMAN A GOMEZ-RIOS, University of Waterloo, Nathaly Reyes-Garces, Ezel Boyaci, Janusz Pawliszyn
10:25	(1040-6)	**Discrimination of Carbohydrate Isomers as Transition Metal Adducts Using Ion Mobility Spectrometry and Tandem Mass Spectrometry** YUTING HUANG, University of Nebraska-Lincoln, Eric D Dodds, Lauren M Petrosh
10:45	(1040-7)	**Scan-by-Scan Analysis of Orbitrap Fine Isotope Structures for Unique Elemental Composition Determination** YONGDONG WANG, Cerno Bioscience, Ming Gu
11:05	(1040-8)	**Mass Spectra of Analytical Derivatives of Amino Acids and Small Peptides** NINO G TODUA, NIST, Stephen E Stein, Anzor I Mikaia

ORAL SESSIONS — Session 1050

Materials Characterization and Engineering

Tuesday Morning, Room B405
Cecil Dybowski, University of Delaware, Presiding

8:30	(1050-1)	**Chemical Analysis Applications and Optical Properties of 3D Printed <100 µm Dimension Microfluidic Channels** MICHAEL BEAUCHAMP, Brigham Young University, Hua Gong, Steven Perry, Greg Nordin, Adam T Woolley
8:50	(1050-2)	**Crack-Free Three-Dimensionally Ordered Macroporous (3DOM) Structure in Microfluidic Reactor** XIAORAN ZHANG, Michigan State University, Gary Blanchard
9:10	(1050-3)	**Interfacial Structure-Function Correlation of Perovskite Solar Cell** MINYU XIAO, University of Michigan
9:30	(1050-4)	**Optical Spectroscopy Analyses of Perovskite Nanomaterials** DANIEL J FREPPON, Iowa State University, Long Men, Ujjal Bhattacharjee, Feng Zhu, Jacob W Petrich, Emily A Smith, Javier Vela
9:50		Recess
10:05	(1050-5)	**Utilizing Thermal Analysis Methods to Develop and Optimized Porous Structural Ceramics** JOHN P SANDERS, Clemson University, Nathaniel Huygen N/A
10:25	(1050-6)	**Carbazole-Dye Conjugate - Derived Group of Uniform Materials Based on Organic Salts (GUMBOS) for Optoelectronic Applications** DEEPTHIKA DE SILVA, Louisiana State University, Noureen Siraj, Isiah M Warner
10:45	(1050-7)	**Molecular-Scale IR Thermometer Reveals Sub-Molecular Photo-Plasticization in Azomaterials** CHRISTIAN PELLERIN, University of Montreal, Audrey Laventure, Jaana Vapaavuori, Geraldine Bazuin, Olivier Lebel
11:05	(1050-8)	**Complete Characterization of the UV-Visible Properties of Optical Materials Using a Total Absolute Measurement System (TAMS)** IAN ROBERTSON, PerkinElmer Limited, Steve Upstone, Christopher Lynch

Author and presider lists as of January 20, 2016, are available at www.pittcon.org

ORAL SESSIONS **Session 1060**

Pharmaceutical-MS, UV-VIS and Others

Tuesday Morning, Room B406
Jason N Payne, Western Kentucky University, Presiding

8:30	(1060-1)	**Speciation of Elemental Impurities for Compliance with USP <232>** JON L SIMS, Perkin Elmer, Helmut Ernstberger, Kenneth Neubauer
8:50	(1060-2)	**Oxidative Degradation in Pharmaceuticals: Mechanism and Stabilization of Spray Dried Amorphous Drug - A Case Study** ARCHANA KUMAR, Genentech, Raghendhar Kotha
9:10	(1060-3)	**Evaluation of Antimicrobial and Neutraceutical Properties of Plukenetia Conophora (Walnut) Leaves** CHUKWUEMEKA PAUL AZUBUIKE, University of Lagos, Cecilia I Igwilo, Karamot O Suara
9:30	(1060-4)	**Novel Self-Patented Gold Nanoparticles for Antineoplastic Activity** JASON N PAYNE, Western Kentucky University, Rajalingam Dakshinamurthy
9:50		**Recess**
10:05	(1060-5)	**Expanding the Analytical Toolbox for Material Verification: Spectroscopic Screening of Raw Ingredients Using Portable Spectrometers** JASON D RODRIGUEZ, FDA Division of Pharmaceutical Analysis, Hirsch Srivastava, Fabiola Semidei Ortiz
10:25	(1060-6)	**Image Directed Identification of Sub-visible Particles in Protein Based Therapeutics, Classification According USP<787> of Intrinsic, Inherent and Extrinsic Particulate Matter on the Sub-visible Level** OLGA LASKINA, rap.ID, Kathryn A Lee, Markus Lankers, Oliver Valet
10:45	(1060-7)	**Coupling Chemical Analysis to High Resolution Dark Field Microscopy for Enhanced Physicochemical Characterization of Complex Drug Formulations** KATHERINE TYNER, Food and Drug Administration, Sheetal D'Mello, Sau (Larry) Lee
11:05	(1060-8)	**Therapeutic Potential of Polyamine-Tethered Low Generation PAMAM Dendrimer Derivatives for Nucleic Acid Delivery** RUBY BANSAL, CSIR-Institute of Genomics and Integrative Biology N/A

ORAL SESSIONS **Session 1070**

Sensors - Biomedical

Tuesday Morning, Room B407
Emil Ciurczak, Doramaxx Consulting, Presiding

8:30	(1070-1)	**Disposable Sensors for the Remote Monitoring of Chronic Wounds** FABIO DI FRANCESCO, Università di Pisa, Nicola Calisi, Bernardo Melai, Pietro Salvo, Clara Paoletti, Letizia Moni, Consuelo Politino, Alessio Ceccarini
8:50	(1070-2)	**Real-Time Monitoring Urinary Encrustation Using Quartz Crystal Microbalance Sensor** PEGAH N ABADIAN, Northeastern University, Jun Li, Edgar D Goluch, John Victor, Jonathan Zhang
9:10	(1070-3)	**A Fiber Optic Biosensor for Noninvasive Transdermal Glucose Sensing Based on the Glucose Binding Protein** CRISTINA E TIANGCO, University of Maryland Baltimore County, Dieudonne Fon, Yordan Kostov, Govind Rao, Leah Tolosa, Fortunato Sevilla, Dayanand Bagdure
9:30	(1070-4)	**Direct Measurement of Total Concentration of Major Physiological Anions, Chloride and Bicarbonate, Using Pulsed Chronopotentiometry with Ion-Selective Electrodes (Pulstrode)** KEBEDE L GEMENE, Northern Kentucky University, Adaeze Stella Iloegbunam, Sara Keshtvarz, Simon Segal
9:50		**Recess**
10:05	(1070-5)	**Non-Invasively Interrogating Chemical and Mechanical Sensors on Implanted Medical Devices** JEFFREY N ANKER, Clemson University, Jeremy Tzeng, Fenglin Wang, Donald Benza, Peter Gennaro, Yash Raval, Mohammed Arifuzzaman
10:25	(1070-6)	**Detection of MicroRNA Presence or Absence with Dual Functioning Signal-On/Off Fluorescent Biosensors** NICHOLAS E LARKEY, Oregon State University, Lulu Zhang, Sean M Burrows
10:45	(1070-7)	**Single Nanoparticle Plasmonic Spectroscopy and Biosensors for Imaging of Efflux Functions of Single Live Cells** X NANCY XU, Old Dominion University, Kerry J Lee, Tao Huang, Prakash d Nallathamby, Feng Ding
11:05	(1070-8)	**Detection of Chemotherapeutic-Induced Damage in Genomic DNA Using Integrated Thermoplastic Nanofluidic Sensor Devices** KUMUDITHA MADUSHANKA WEERAKOON-RATNAYAKE, University of North Carolina at Chapel Hill, Franklin I Uba, Robert Schotzinger, Steven A Soper

POSTER SESSION **Session 1080**

All posters are to be mounted by 10:00 AM and remain on display until 4:00 PM. Authors must be at their posters from 10:00 AM to 12:00 PM. Location of the morning posters is on the Exposition Floor, 400 Aisle. PLEASE NOTE: You cannot get onto the Exposition Floor until after 9:00 AM.

Environmental Air Quality and Analysis

Tuesday Morning, Exposition Floor, 400 Aisle

(1080-1 P)	**Tea Polyphenols Reduces Toxicity of PM2.5 in Human Alveolar Epithelial A549 Cells** YING ZHANG, Shijiazhuang CDC, Jie Jiang, Yan He, Yanzhong Chang
(1080-2 P)	**Testing of Gas Purifiers for VOC Removal Down to the PMOL/MOL Level** ANNARITA BALDAN, VSL, Janneke van Wijk, Jianrong Li, Rina Wortman, Jennifer Englert, Christian Plass-Duelmer, Stefan Persijn, Guido Sassi, Alessia Demichelis, Mariapaola Sassi
(1080-3 P)	**An Analytical Method for the Measurement of Reduced Sulfur Gases** JÜRGEN MICHAEL LOBERT, Entegris, Inc., Charles M Miller, Emily C Zaloga N/A
(1080-4 P)	**Developing and Field Tests of an Automatic Impinger System for Continuous Sampling of Volatile Amines in the Environment** CHIA-JUNG LU, National Taiwan Normal University, Wang Chih-Chia, Rih-Sheng Jian, Sung Lung-Yu
(1080-5 P)	**SIFT-MS: A One-Stop Analytical Tool for Detection of Fumigation Chemicals** DANIEL MILLIGAN, Syft Technologies Ltd, Vaughan Langford, Barry Prince, Thomas McKellar, David Hera, Murray McEwan
(1080-6 P)	**SIFT-MS: A Complete Solution for Analysis of Ambient Air** DANIEL MILLIGAN, Syft Technologies Ltd, Vaughan Langford, Thomas McKellar, David Hera, Barry Prince, Murray McEwan
(1080-7 P)	**Real Time Detection and Identification of Chemical Releases via GCMS** PARMINDER KAUR, 1st Detect Corporation, Corey Stedwell, Daniel DeBord
(1080-8 P)	**Continuous Fenceline Monitoring Using a Miniature Mass Spectrometer** PRESHIOUS REARDEN, 1st Detect Corporation, Madonna Marie Mamerow, Parminder Kaur
(1080-9 P)	**Changes in Tobacco-Specific Nitrosamine Cigarette Smoke Deliveries in Unburned, Recycled Portions of Roll-Your-Own Cigarettes** BENJAMIN WADE ALVERSON, Centers for Disease Control and Prevention, Jose J Perez, Morgan E Larango, Roberto Bravo, Liqin Zhang, Patrick Chen, Shirley Ding, Mary Halstead, R Steven Pappas, Clifford Watson, Liza Valentin-Blasini
(1080-10 P)	**A Method of Performing In-Trap Photoionization in a Miniature Ion Trap Mass Spectrometer** COREY STEDWELL, 1st Detect Corporation, Daniel DeBord, Michael Spencer, David Rafferty

POSTER SESSION **Session 1090**

All posters are to be mounted by 10:00 AM and remain on display until 4:00 PM. Authors must be at their posters from 10:00 AM to 12:00 PM. Location of the morning posters is on the Exposition Floor, 400 Aisle. PLEASE NOTE: You cannot get onto the Exposition Floor until after 9:00 AM.

Environmental and Geochemical Analysis: Soils, Minerals, and Agriculture

Tuesday Morning, Exposition Floor, 400 Aisle

(1090-1 P)	**Automated Analysis of Explosives in Soil Samples** WILLIAM HEDGEPETH, Shimadzu Scientific Instruments, Kenichiro Tanaka
(1090-2 P)	**Mechanisms for Controlling Soil Organic Matter Decompositions: An Application of Pyrolysis –Cryogenic –GC/MS to Molecular Characterizations of Organic Matter in Grass and Forestry Soils** XIANZHI (AMANDA) SONG, Young Harris College
(1090-3 P)	**A New Analytical Technique for the Determination of Carbonate in Soil Samples** TACETTIN OZTURK, Lita Analytical, Onur Iscan N/A
(1090-4 P)	**Ion Selective Electrodes – A Cheaper, Simpler and more Robust Analytical Method for Monitoring of Nitrate and Ammonium in Water and Soil** TOLULOPE A FAYOSE, Keele University
(1090-5 P)	**Multiple Surface-Science Techniques to Elucidate the Reactive Nature of a Metal Phosphide Mineral** DANNA QASIM, Kennesaw State University, Aaron Pital, Thomas Beckman, Heather Abbott-Lyon
(1090-6 P)	**A Systems Chemistry Investigation into Nucleoside Phosphorylation Mechanisms by Prebiotic Meteoritic Materials** MIKE A MOJICA, Georgia Institute of Technology, Maheen Gull, Matthew Pasek, Charles Liotta, Thomas Orlando, Aaron McKee, Facundo M Fernandez N/A
(1090-7 P)	**Current Advances in Instrumentation of Soil Elemental Analyzers** DOMINIK MARGRAF, Elementar Analysensysteme GmbH, Christian Schmidt, Tony Szuppa, Sabine Kraus, Lutz Lange, Hans-Peter Sieper N/A
(1090-8 P)	**Analysis of PAHs in Soil by Online SFE-SFC** KENICHIRO TANAKA, Shimadzu Scientific Instruments, Inc., William Hedgepeth N/A
(1090-9 P)	**Determination of PAHs in Durban City Road Dusts** ABDULMUMEEN ABDULKADIR, University of Kwazulu-Natal N/A

(1090-10 P) **Speciation of Metal-Binding Proteins in Marine Environment** PILAR BERMEJO-BARRERA, University of Santiago de Compostela, Spain, Maria del Carmen Barciela-Alonso, Natalia García-Otero, Olga Cristina Vázquez-Padín, Antonio Moreda-Piñeiro, Raquel Domínguez-Gonzalez, Elena Pena-Vazquez N/A

(1090-11 P) **A New Advancement in the Automated Preparation of Pressed Pellets for XRF Analysis** DAVID COLER, FLSmidth, Lukas Bruzenak, Ian Campbell

(1090-12 P) **Multifunctional Ligand Platform for Detection, Capturing and Removal of Cerium Oxide Nanoparticles** ALI OTHMAN, Clarkson University, Silvana Andreescu

POSTER SESSION — Session 1100

All posters are to be mounted by 10:00 AM and remain on display until 4:00 PM. Authors must be at their posters from 10:00 AM to 12:00 PM. Location of the morning posters is on the Exposition Floor, 400 Aisle. PLEASE NOTE: You cannot get onto the Exposition Floor until after 9:00 AM.

Environmental Applications of Elemental Analysis and Speciation

Tuesday Morning, Exposition Floor, 400 Aisle

(1100-1 P) **Rare Earth Elements – How to Accurately Determine Contamination Levels and Remove Spectral Interferences Created by Them** EWA M PRUSZKOWSKI, PerkinElmer, Inc., Cynthia Bosnak

(1100-2 P) **Determination of Heavy Metals In and Around Lake Ontario** SIMRAN SANDHU, St. John Fisher College, Kimberly Chichester

(1100-3 P) **The Analysis of Flue Gas Desulfurization Fluids by ICP-MS Using Universal Cell and FastFIAS Technology** MICHELLE M COKER, SCE&G, Daniel H Jones, George W Eargle

(1100-4 P) **Determination of Cadmium in Environmental Water Samples Collected in Superfund Sites in New York City** YI HE, John Jay College/CUNY, Kathleen Lopez, Sandra Swenson, Kate Good

(1100-5 P) **Withdrawn**

(1100-6 P) **Evaluation of Chromium Stability on Filters** TAMUTSIWA M MUTUTUVARI, High Purity Standards, Svetlana Uzunova, Kim-Phuong Tran

(1100-7 P) **Withdrawn**

(1100-8 P) **Online Analysis and Speciation of Antimony in Various Wastewater Streams Using Hydride Generation-AFS** BIN CHEN, P S Analytical, Warren T Corns, Peter B Stockwell

(1100-9 P) **An Efficient Recovery of Rare Metal Ions with Calix[4]arene Derivatives from Acidic Media Using Droplet-Based Microreactor System** MASAYA MIYAZAKI, National Institute of Advanced Industrial Science and Technology, Keisuke Ohto, Masatoshi Maeki, Ramachandra Rao Sathuluri

(1100-10 P) **Breakthrough Development for Quantitative Analysis of Total Metals in Soil by Portable High Definition X-ray Fluorescence** ZEWU CHEN, XOS, Shenghua Song, Danhong Li, Kyle Kuwitzky

(1100-11 P) **Chromogenic and Fluorogenic Chemosensors for the Selective Detection of Nd(III) Ion and Their Electrochemical Properties** ASHOK K SINGH, IIT Roorkee, Neha Gupta N/A

(1100-12 P) **Collaborative Certification of a New Low-Level Hexavalent Chromium Standard Reference Material in a Soil Matrix** JAMES E HENDERSON, Duquesne University, Patrick Benecewicz, Anil Srinivas Chaitanya Vishnuvajjhala, Weier Hao, Logan T Miller, Matt Pamuku, Jennifer Crawford, Teresa Switzer, Vasile Furdui, Pam Wee, Francine Walker, Bob O'Brien, Skip Kingston

(1100-13 P) **Impact of Coexistent Elements and Its Concentration in the Quantification of Strontium-90 Using ICP-MS with Cascade Separation System** YOSHITAKA TAKAGAI, Fukushima University, Takahiro Suzuki, Aya Yokoyama, Makoto Furukawa, Yutaka Kameo, Katz Suzuki

POSTER SESSION — Session 1110

All posters are to be mounted by 10:00 AM and remain on display until 4:00 PM. Authors must be at their posters from 10:00 AM to 12:00 PM. Location of the morning posters is on the Exposition Floor, 400 Aisle. PLEASE NOTE: You cannot get onto the Exposition Floor until after 9:00 AM.

Environmental Organic Analysis: VOCs, Pesticides, and Others

Tuesday Morning, Exposition Floor, 400 Aisle

(1110-1 P) **Determination of VOCs by US EPA Method 8260 with Extended Dynamic Range Using Fast, Sensitive Capillary GC/MS** BRAHM PRAKASH, Shimadzu Scientific Instruments, Inc, William Lipps, Di Wang, Nicole M Lock, Shilpi Chopra, Laura Chambers

(1110-2 P) **Determination of Organochlorine Pesticides and Polychlorinated Biphenyls Using GC-MS/MS Operated in the MRM Mode** BRAHM PRAKASH, Shimadzu Scientific Instruments, Inc., William Lipps, Di Wang, Nicole M Lock, Shilpi Chopra, Laura Chambers

(1110-3 P) **Analysis of Terpenes Using Gas Chromatography with Vacuum Ultraviolet Detection** CHANGLING QIU, University of Texas at Arlington, Jonathan Smuts, Phillip Walsh, Kevin A Schug

(1110-4 P) **Use of a Boron Doped Diamond Electrode Sensor for Carbamate Pesticide Classification Using a Chemometric Approach** THIAGO SELVA, University of Sao Paulo, Thiago Paixao

(1110-5 P) **PCBs and DDTs in Bluefin Tuna From the Adriatic Sea** DARIJA KLIN I, Institute for Med Research and Occupational Health, Snjažana Herceg Romanić, Zorana Kljakovi -Gašpi , Vjekoslav Tičina

(1110-6 P) **From Freon to PAHs - A New Generation of Multipurpose Thermal Desorption Tubes** PAOLO BENEDETTI, IIA - CNR, Carlo Crescenzi, Ettore Guerriero

(1110-7 P) **Search for Organic Substance Sources in the 1B Mine Pool of Sydney Coalfield Abandoned Mines** ALLEN BRITTEN, Cape Breton University, Ceilidh MacDonald, Judy MacInnis, Martin Mkandawire N/A

(1110-8 P) **Analysis of Volatile Organic Compounds in Wastewater by Purge and Trap GC/TOF-MS According to EPA Method 624** MOIRA ZANABONI, DANI Instruments, Roberta Lariccia, Matthew S Klee

(1110-9 P) **Degradation of Environmental Contaminants Using Chlorine Dioxide** SUSHMA APPALA, Middle Tennessee State University, Samanwi Munagala, Megan Z Chong, Ooi G Beng, Ngee Sing Chong

(1110-10 P) **Automating Solid Phase Extraction and Florisil Clean-Up for Organichlorine Pesticides and PCB Aroclors** PHILIP BASSIGNANI, Fluid Management Systems, Rudolf Addink

(1110-11 P) **Evaluation of a Novel Hand-Held and Easy-to-Use GC-PID Prototype for Fast and Selective On-site Analysis of Benzene and VOC** MATTHIAS SCHMITTMANN, Bentekk GmbH

(1110-12 P) **Direct Determination of Glyphosate, Glufosinate, and AMPA in Egg by Liquid Chromatography/Tandem Mass Spectrometry** NARONG CHAMKASEM, FDA, Cynthia Morris, Krystle L Hargrove

(1110-13 P) **Direct Coupling of Active and Passive Samplings of Organics with Microwave Assisted Thermal Desorption as an Innovative Solvent-Free Method** WILLIAMS ESTEVE, INRS

(1110-14 P) **Deuterated Monitoring Compounds for Better Accuracy and Precision Measurement of GC/MS Environmental Data** CHARLES G APPLEBY, U.S. Environmental Protection Agency

(1110-15 P) **Withdrawn**

(1110-16 P) **Wastewater Monitoring by Fluorescence Excitation and Emission Matrix with Parallel Factor Analysis** SAM LI, NUS, Baisheng Chen

(1110-17 P) **New Plastic In-Syringe Based Ultrasound Assisted Salt-induced Liquid-liquid Microextraction Technique for the Rapid Analysis of Triclosan in Aqueous Samples** VINOTH KUMAR PONNUSAMY, National Chung Hsing University, Jen Fon Jen

POSTER SESSION — **Session 1120**

All posters are to be mounted by 10:00 AM and remain on display until 4:00 PM. Authors must be at their posters from 10:00 AM to 12:00 PM. Location of the morning posters is on the Exposition Floor, 400 Aisle. PLEASE NOTE: You cannot get onto the Exposition Floor until after 9:00 AM.

Environmental Water Quality and Analysis

Tuesday Morning, Exposition Floor, 400 Aisle

(1120-1 P) **Electrochemical Determination of the Levels of Some Potentially Toxic Metals in Groundwater and Leachate Samples from Solous II and III Landfill, Lagos Nigeria** WESLEY OHIFEME OKIEI, University of Lagos, Modupe M Ogunlesi, Jeffery Undiandeye N/A

(1120-2 P) **Determination of N-Nitrosamines by USEPA Method 521 Using Triple Quadrupole Mass Spectrometry** BRAHM PRAKASH, Shimadzu Scientific Instruments, Inc, William Lipps, Di Wang, Nicole M Lock, Shilpi Chopra, Laura Chambers

(1120-3 P) **Optimizing Treatment of Reclaimed Water at a Drinking Water Plant by Online Monitoring of Organic Carbon Levels** MARK MULLET, GE Analytical Instruments, Dondra Biller

(1120-4 P) **Development of an Arduino Shield for Water Quality Analysis Probes** MICHAEL CHIA, Northern Kentucky University, Kelley Weigman, Grant Foreman, Richard Durtsche, Celeste A Morris

(1120-5 P) **Analysis of Peanut Hulls as an Alternative to Ion Exchange Resins** CARMEN L HUFFMAN, Western Carolina University, Kanika O Davis, Melisa J Glatte, Holly Truluck, Tyler S Cook

(1120-6 P) **Analytical Strategies for Monitoring Ionic Liquids Breakdown by Electro-Fenton Process** ELISA GONZÁLEZ-ROMERO, University of Vigo, Elvira Bocos, Jessica Meijide, Aida Díez, Marta Pazos, María Ángeles Sanromán

(1120-7 P) **A Study of Quality of Water and Effect of Various Coagulants on Water of Lodra Lake of Gandhinagar District** JAIMIN K DESAI, Gujarat University, Parekh P Paresh N/A

(1120-8 P) **Determination of the Toxins Found in Lake Ontario** MARTA LABECKI, St. John Fisher College, Kimberly Chichester

(1120-9 P) **EPA Method 557 Quantitation of Haloacetic Acids, Bromate and Dalapon in Drinking Water Using Ion Chromatography and Tandem Mass Spectrometry** JONATHAN BECK, Thermo Fisher Scientific, Terri T Christison, Charles Yang, Hans Schweingruber

(1120-10 P) **Assessment of Water Quality Parameters from the Lowber Abandoned Mine Drainage Treatment Facility, Part 2: Further Studies and Results** MARK T STAUFFER, University of Pittsburgh - Greensburg, Tell M Lovelace, Aaron K Hirshka, Luke J Metzler

(1120-11 P) **Breaking the Biofouling Code: Towards Reliable In-Pipe Water Quality Sensors** ROBERT E WILSON, Imperial College London, Ivan Stoianov, Danny O'Hare

(1120-12 P) **Analysis of the Residual Oil in Water with Excitation Emission Matrix and Parallel Factor Analysis** KAWAGUCHI YOSHIHIKO, HORIBA Advanced Techno, Co., Ltd., Kojima Reiji

(1120-13 P) **Sensitive Determination of Arsenate and Phosphate by Molybdenum Blue Method with Membrane Filter Extraction Using a Portable 8-Channel LED-Based Reflective Photometer** YASUTADA SUZUKI, University of Yamanashi, Susumu Kawakubo

(1120-14 P) **Assessment of Physico-Chemical Analysis of Drinking Water of Kheda District Gujarat (INDIA) and its Impact on Human Health** MAHESHKUMAR B CHAUHAN, J & J College of Science, Dipakkumar K Bhoi, Rameshbhai P Dabhi N/A

(1120-15 P) **Nano-Sensing Approach for the Determination of Phosphate Ions** MADELEINE JOHNSON, University of Central Florida N/A

(1120-16 P) **Study of Enhancement Effects of Functionalized Gold Nanorods in Quantitative Analysis of 1-H Benzotriazole by Surface Enhanced Raman Spectroscopy (SERS)** UTTAM SHARMA PHUYAL, Tennessee Technological University, Andrew Callender

(1120-17 P) **Seasonal Variations in Water Quality Parameters at Thol Bird Sanctuary, Gujarat India** AMRUTLAL B PARMAR, J & J College of Science, Rameshkumar V Parmar, Arunkumar H Dholakia N/A

(1120-18 P) **A Study of Water Quality Index (W.Q.I) of Pariyej Lake District: Kheda - Gujarat** BUNTY R PATEL, J & J College of Science, Alpesh M Patel, Fulaji J Thakor N/A

(1120-19 P) **Colorimetric and Electrochemical Sensor for Monohydrogen Phosphate Ion Based on 1,3-bis(2-formylphenoxy) Propane Bis(2,4-dinitrophenyl hydrazine)** DIVYA SINGHAL, IIT Roorkee, Ashok K Singh N/A

(1120-20 P) **The Effect of Slaughter House Waste on the Water Quality of Okpu River Aba** CHIDIEBERE A ODIKE-ADUAKA, Abia State Teaching Hospital Aba, Lilian I Oguguo

(1120-21 P) **Biosensor for Toxic Compounds in Wastewater Based on Microbial Electrochemistry** SAM LI, NUS

(1120-22 P) **Rapid Determination of Endocrine Disrupting Bisphenol A (BPA) in Drinking Water by Solid Phase Nano-Extraction and Room-Temperature Fluorescence Spectroscopy** MAHA AL-TAMEEMI, University of Central Florida, Bassam Alfarhani, Andres Campiglia, Jung Jong Seok

(1120-23 P) **GC/MS Screening of Water Samples for Organic Pollutants by Stir Bar Sorptive Extraction (SBSE)** OLIVER LERCH, Gerstel GmbH & Co. KG, Jasmin Zboron, Andreas Hoffmann, Chris Sandy

POSTER SESSION — **Session 1130**

All posters are to be mounted by 10:00 AM and remain on display until 4:00 PM. Authors must be at their posters from 10:00 AM to 12:00 PM. Location of the morning posters is on the Exposition Floor, 400 Aisle. PLEASE NOTE: You cannot get onto the Exposition Floor until after 9:00 AM.

Environmental, Food and Elemental Analyses - Atomic Spectroscopy

Tuesday Morning, Exposition Floor, 400 Aisle

(1130-1 P) **Analysis of Radium-226 in Shale-Gas Wastewater Using Inductively Coupled Plasma Mass Spectrometry** YUQIANG BI, University of Michigan, Wenjia Fan, Thomas P Yavaraski, Brian R Ellis, Kim F Hayes

(1130-2 P) **Speciation of Organic Mercury in Water Samples by Alkylation, Organic Solvent Extraction and GC-AFS Detection – A Comparison of Ethylation, Propylation and Phenylation** CORNELIUS BROMBACH, P S Analytical, Bin Chen, Warren T Corns, Jun Yoshinaga

(1130-3 P) **Models for Predicting Atmospheric Mercury Concentrations Using Meteorological Data and Mercury Concentrations in *Salix* (Willow) Leaves** DAVID LEHMPUHL, Colorado State University Pueblo, Lauren Bartolo

(1130-4 P) **Assessment of Titanium Dioxide Nanoparticles in Aquatic Tanks** SARA MELOW, Elmira College, Marriah Ellington, Lisa A Holland

(1130-5 P) **Feasibility Use of Ceramics as Solid Support for Cr(III) Measurement in Water by LIBS** CASSIANA S NOMURA, Institute of Chemistry - University of Sao Paulo, Alexandrina C Carvalho, Daniel M Silvestre, Flávio O Leme, Danielle P Intima, Juliana Naozuka

(1130-6 P) **Lead Determination in Soil from a Recreational Shooting Range Built on a Reclaimed Strip Mine: Effects of Oxidant Flow Rate on PB Measurements Obtained by FAAS, and Other Considerations** MARK T STAUFFER, University of Pittsburgh - Greensburg, Luke J Metzler

(1130-7 P) **Determination of Metals in Three Types of Loose-Leaf Tea: Can Metal Content Indicate the Type of Tea?** MARK T STAUFFER, University of Pittsburgh - Greensburg, Aaron K Hirshka

(1130-8 P) **Determination of Selected Metals and Nonmetals in Pre-Workout Supplements: Results from Initial and Ongoing Investigations** MARK T STAUFFER, University of Pittsburgh - Greensburg, Kelly M Boyles, Nicholas E Glotfelty N/A

(1130-9 P) **Speciation of Mercury in Rice with a New Online Pre-Concentration HPLC-CV-AFS Method** CORNELIUS BROMBACH, P S Analytical, Piumi K Dona, Bin Chen, Warren T Corns, Eva Krupp, Joerg Feldman

(1130-10 P) **Optimizing a Total Protein Combustion Instrument for Maximum Sample Throughput and Lowest Cost-Per-Analysis** JEFFERY GAST, LECO Corporation, Mason Marsh

(1130-11 P) **Laser Induced Breakdown Spectroscopy (LIBS) of Food Samples: Case Study of Tortillas** CHARLES GHANY, Mississippi State University, Herve Sanghapi, Chet R Bhatt, Bader Alfarraj, Fang Y Yueh, Jagdish P Singh

(1130-12 P) **Development of a Method For the Determination of Titanium Dioxide Nanoparticles in Food Products Using SP-ICPMS** ANTONIO MOREDA-PIÑEIRO, University of Santiago de Compostela, María del Carmen Barciela-Alonso, Olga Cristina Vázquez-Padín, Francisco Javier Vilariño-Páxaro, Manuel Aboal-Somoza, Pilar Bermejo-Barrera, Elena Pena-Vazquez

(1130-13 P) **Evaluation of Optical Depths of Ca Emission Lines in Laser Induced Breakdown Spectroscopy (LIBS)** BADER ALFARRAJ, Mississippi State University, Herve Sanghapi, Chet R Bhatt, Charles Ghany, Fang Y Yueh, Jagdish P Singh

(1130-14 P) **Standard Dilution Analysis for the Determination of Calcium by Flame Atomic Emission Spectrophotometry** CLIFTON P CALLOWAY, Winthrop University, Katja A Hall, Emily A Watson

(1130-15 P) **Cyanide Detection in Blood Using Indirect Atomic Absorption Spectroscopy** JEFFREY ROSENTRETER, Idaho State University, Matt Kirkham, Jeff Kuhlmeier

(1130-16 P) **Advanced Application of Speciation Analysis Using ICP-MS detection** DANIEL KUTSCHER, Thermo Fisher Scientific, Monika Verma, Shona McSheehy Ducos

(1130-17 P) **Multi-trophic Analysis of Lead Using a Flame Atomic Absorption Spectrometer** MICHAEL DECAROLIS, St. John Fisher College, Christopher Collins, Kimberly Chichester

(1130-18 P) **Atomic and Molecular Laser Induced Breakdown Spectroscopy for Detection of Chlorine in Concrete** WILL B JONES, University of Florida, Ebo Ewusi-Annan, Tobias Guenther, Ben Smith, Nico Omenetto

(1130-19 P) **A Modern Multi-Excitation Concept for Total Reflection X-Ray Spectrometry** ARMIN GROSS, Bruker Nano GmbH, Ulrich Waldschlaeger N/A

(1130-20 P) **Determination of Lead in Keratin Tissues: Method Development and Candidate Reference Material Production Using Caprine Horns** MINA TEHRANI, Wadsworth Center/ SUNY Albany, Patrick J Parsons N/A

(1130-21 P) **Lead Concentrations in Lagos Groundwater** AMALACHUKWU OLIVE ANADI, Hydrochrom Resources Ltd, Chimezie A Anyakora N/A

(1130-22 P) **Fast Monitoring Processed Manure Using WD-XRF Spectroscopy for Nutrients and Metals** ALEXANDER SEYFARTH, Rock River Axs LLC, Aicado Roa-Espinosa

(1130-23 P) **Onsite Monitoring of Manure Using XRF for Nutrients and Metals** ALEXANDER SEYFARTH, Bruker Nano Analytics, Aicado Roa-Espinosa

POSTER SESSION **Session 1140**

All posters are to be mounted by 10:00 AM and remain on display until 4:00 PM. Authors must be at their posters from 10:00 AM to 12:00 PM. Location of the morning posters is on the Exposition Floor, 400 Aisle. PLEASE NOTE: You cannot get onto the Exposition Floor until after 9:00 AM.

Sampling and Sample Preparation - Environmental

Tuesday Morning, Exposition Floor, 400 Aisle

(1140-1 P)	**The Measurement of Formaldehyde in Drinking Water Using Automated SPE and HPLC** ALICIA CANNON, Horizon Technology, Chris Shevlin, Michael Ebitson
(1140-2 P)	**Analysis of Extract Drying Criteria for Oil and Grease Method 1664A/B** MICHAEL EBITSON, Horizon Technology, David Gallagher, William Jones
(1140-3 P)	**Fundamentals and Comparisons for Organic Sample Extract Evaporation** ZOE GROSSER, Horizon Technology, Robert Johnson
(1140-4 P)	**Determination of 2-Methylisoborneol and Geosmin in Water Using Solid Phase Micro Extraction** ANNE JUREK, EST Analytical
(1140-5 P)	**The Advantages of Automated Sample Preparation** ANNE JUREK, EST Analytical, Lindsey Pyron, Kelly Cravenor, Justin Murphy
(1140-6 P)	**Highly Robust Polymeric Ionic Liquid Coatings for Solid Phase Microextraction: Multiclass Determinations with Application of Direct-Immersion-Headspace Mode Using Gas Chromatography-Mass Spectrometry** JOSIAS MERIB, Iowa State University, Honglian Yu, Carasek Eduardo, Jared L Anderson
(1140-7 P)	**Development of Dispersive Liquid-Liquid Microextraction for the Determination of Six Steroidal Hormones in Wastewater Using High Pressure Liquid Chromatography-Charged Aerosol Detector** MATHEW MUZI NINDI, Unisa, Cecilia O Osunmakinde, Simiso Dube
(1140-8 P)	**Extraction of Analytes of Forensic Toxicological Interest from Plasma with Enhanced Matrix Removal-Lipid Material** JOAN STEVENS, Agilent Technologies, Derick Lucas, Limian Zhao, Megan Juck, William Long
(1140-9 P)	**Modified Sample Clean-up for Combined POPs Using Automated Multi-Column Fractionation and Analytical Optimization** PHILIP BASSIGNANI, Fluid Management Systems, Rudolf Addink
(1140-10 P)	**Ice Concentration Linked with Extractive Stir Bar** NUJUD O MASLAMANI, South Dakota State University
(1140-11 P)	**The Best Sample Preparation for High Throughput Quantitative X-Ray Diffraction of Mineral Mixtures** ROGER MEIER, FLSmidth A/S, Ian Campbell, Lukas Bruzenak, Detlev Götz
(1140-12 P)	**Nanoporous Solid Phase Microextraction (SPME) Fibers by Sputtering Silicon** MASSOUD KAYKHAII, Brigham Young University, Anubhav Diwan, Bhupinder Singh, Matthew R Linford

TUESDAY, MARCH 8, 2016
AFTERNOON

AWARDS **Session 1150**

Pittsburgh Spectroscopy Award

arranged by Joseph Grabowski, University of Pittsburgh

Tuesday Afternoon, Room B312

Fu-Tyan Lin, The Pittsburgh Conference, Presiding

1:30		**Introductory Remarks - Fu-Tyan Lin**
1:35		**Presentation of the 2016 Pittsburgh Spectroscopy Award to Jürgen Popp, Friedrich-Schiller University Jena, by Joseph Grabowski, Chair, Spectroscopy Society of Pittsburgh**
1:40	(1150-1)	**Overcoming Unmet Medical Needs: Advances in Raman Spectroscopy** JÜERGEN POPP, Friedrich-Schiller University Jena
2:15	(1150-2)	**Developing Deep UV Raman Standoff Spectrometers for Trace Explosives** SANFORD A ASHER, University of Pittsburgh, Sergei V Bykov, Katie L Gares, Kyle T Hufziger
2:50	(1150-3)	**Enhanced Vibrational Circular Dichroism: Moving Beyond Established Applications of Vibrational Circular Dichorism (VCD)** LAURENCE A NAFIE, Syracuse University
3:25		**Recess**
3:40	(1150-4)	**Raman Big Data Analysis for Automatic and Objective Living Cell Discrimination/Diagnosis** HIROO HAMAGUCHI, National Chiao Tung University, Masahiro Ando
4:15	(1150-5)	**Raman Scattering from Single, Laser-Trapped Microparticles: A Review** WOLFGANG KIEFER, University of Würzburg

AWARDS **Session 1160**

RSC - JAAS Emerging Investigator Lectureship Award

arranged by Philippa Ross, Royal Society of Chemistry

Tuesday Afternoon, Room B314

Philippa Ross, Royal Society of Chemistry, Presiding

1:30		**Introductory Remarks - Philippa Ross**
1:35		**Presentation of the 2016 RSC - JAAS Emerging Investigator Lectureship Award to Gerardo Gamez, Texas Tech University, by Philippa Ross, Royal Society of Chemistry**
1:40	(1160-1)	**Revealing Surface Elemental Landscapes with Ultra-High Throughput via GDOES** GERARDO GAMEZ, Texas Tech University
2:15	(1160-2)	**Laser Ablation-Based Chemical Analysis Techniques: A Short Review** JHANIS JOSE GONZALEZ, Lawrence Berkeley National Laboratory
2:50	(1160-3)	**Direct Determination of Trace Antimony and Arsenic in Natural Waters by Photochemical Vapor Generation ICPMS** LU YANG, National Research Council Canada
3:25		**Recess**
3:40		**Presentation of the 2016 RSC - JAAS Emerging Investigator Lectureship Award to Lara Lobo Revilla, University of Oviedo, by Philippa Ross, Royal Society of Chemistry**
3:45	(1160-4)	**Pulsed Glow Discharge Time-of-Flight Mass Spectrometry (Positive and Negative Ionization Modes) for Elemental Depth Profiling of Innovative Materials and Polymer Fingerprinting** LARA LOBO REVILLA, University of Oviedo, R Muniz, B Fernandez, R Pereiro, A Sanz-Medel
4:20	(1160-5)	**What is XRF Doing in a Mass Spectrometry Award Symposium?** GEORGE HAVRILLA, Los Alamos National Laboratory

SYMPOSIUM **Session 1170**

ACS-ANYL - New Approaches to Nuclear Safeguards and Forensics Analysis
arranged by Douglas Duckworth, Pacific Northwest National Laboratory

Tuesday Afternoon, Room B308
Douglas Duckworth, Pacific Northwest National Laboratory, Presiding

1:30		**Introductory Remarks - Douglas Duckworth**
1:35	(1170-1)	**Advances in Online Spectroscopic Monitoring for Process Control and Safeguarding of Radiochemical Streams** SAM A BRYAN, Pacific Northwest National Laboratory, Amanda J Casella, Amanda M Lines, Gilbert L Nelson, Job M Bello
2:10	(1170-2)	**New XRF Applications to Nuclear Safeguards and Nuclear Forensics** GEORGE HAVRILLA, Los Alamos National Laboratory
2:45	(1170-3)	**Fieldable Mass Spectrometry: Sample Preparation and Rapid Field Analysis for Nuclear Safeguards** PETER STARK, Los Alamos National Laboratory, Elizabeth Judge, Keri Campbell, Lisa Meyers, Chris Leibman, Ning Xu, Thomas Yoshida, Matthew R Dirmyer
3:20		**Recess**
3:35	(1170-4)	**New Analytical Methods for Trace Elemental and Isotopic Analysis of Nuclear Fuel Cycle Materials** GREGORY C EIDEN, Pacific Northwest National Laboratory, Andrew M Duffin, Michael P Dion, April J Carman, Orville T Farmer, Jesse D Ward, Carmen S Menoni, Martin Liezers
4:10	(1170-5)	**Contribution and Impact of Laser Ablation Inductively Coupled Plasma Mass Spectrometry (LA-ICP-MS) and Laser Induced Breakdown Spectroscopy (LIBS) to Forensic Analysis** JHANIS JOSE GONZALEZ, Lawrence Berkeley National Laboratory

SYMPOSIUM **Session 1180**

Emerging Mass Spectrometry-Based Techniques for Biomolecular Analysis
arranged by Ronghu Wu, Georgia Institute of Technology

Tuesday Afternoon, Room B302
Ronghu Wu, Georgia Institute of Technology, Presiding

1:30		**Introductory Remarks - Ronghu Wu**
1:35	(1180-1)	**Searching for Biomarkers Using Ambient Ionization Mass Spectrometry** GRAHAM COOKS, Purdue University
2:10	(1180-2)	**An Inside-Outside Strategy to Study Cell Communication** CATHERINE FENSELAU, University of Maryland, Sitara Chauhan, Lucia Geis-Asteggiante, Avantika Dhabaria, Nathan Edwards, Suzanne Ostrand-Rosenberg
2:45	(1180-3)	**Developing Cross-linking Mass Spectrometry to Define Protein-Protein Interactions** LAN HUANG, University of California, Irvine
3:20		**Recess**
3:35	(1180-4)	**Innovative Instrumentation and Methods for the Identification of Intact Proteins in Mixtures and for Sequence Analysis of Antibodies and Posttranslationally-Modified, Intact Proteins on a Chromatographic Time-Scale** DONALD F HUNT, University of Virginia
4:10	(1180-5)	**Charge Detection Mass Spectrometry for Single Ions** EVAN WILLIAMS, University of California, Berkeley, Andrew Elliot, Zijie Xia

SYMPOSIUM **Session 1190**

Emerging Pollutants in the Environment – from Sources to Effects
arranged by Rudolf J Schneider, BAM Federal Institute for Materials Research and Testing

Tuesday Afternoon, Room B303
Rudolf J Schneider, BAM Federal Institute for Materials Research and Testing, Presiding

1:30		**Introductory Remarks - Rudolf J Schneider**
1:35	(1190-1)	**Using High Resolution Mass Spectrometry to Uncover New, Emerging Iodinated and Nitrogen-Containing Disinfection Byproducts** SUSAN D RICHARDSON, University of South Carolina, Cristina Postigo, Christina M Joseph, Hannah K Liberatore, Jessie Kadlec, Amy Cuthbertson, Friedrich Wendel, Christian Luetke-Eversloh, Thomas A Ternes, Edward Machek, Stephen Duirk, Elizabeth Wagner, Michael J Plewa
2:10	(1190-2)	**Measuring Plant Uptake and Effects of Pharmaceuticals Using Liquid Chromatography/Mass Spectrometry** DIANA S AGA, University at Buffalo, Rachel Mullen
2:45	(1190-3)	**Analysis of Hydraulic Fracturing Additives by LC/Q-TOF-MS** IMMA FERRER, University of Colorado, Michael Thurman
3:20		**Recess**
3:35	(1190-4)	**Uptake and Disposition of Pharmaceuticals by Bluegill Exposed at Constant Concentrations in a Flow-Through Aquatic Exposure System** EDWARD FURLONG, U.S. Geological Survey, Jian-Liang Zhao, Heiko L Schoenfuss, Dana W Kolpin, Kyle L Bird, David J Feifarek, Eric A Schwab, Guang-Guo Ying
4:10	(1190-5)	**Pharmaceuticals in Surface Waters - Analysis and Effects** RUDOLF J SCHNEIDER, BAM

SYMPOSIUM **Session 1200**

Emerging Technologies for Disease Biomarker Detection
arranged by Xiujun James Li, University of Texas at El Paso

Tuesday Afternoon, Room B304
Xiujun James Li, University of Texas at El Paso, Presiding

1:30		**Introductory Remarks - Xiujun James Li**
1:35	(1200-1)	**Phase Separated Droplets Enable Multiplexing of Difficult ELISA Panels** SHUICHI TAKAYAMA, University of Michigan
2:10	(1200-2)	**Looking for Rare Cells via High-throughput Single Cell Mass Spectrometry Profiling** JONATHAN V SWEEDLER, University of Illinois
2:45	(1200-3)	**Biomarker Detection Using Paper/PDMS Hybrid Microfluidic Platforms for Low-cost Disease Diagnosis** XIUJUN JAMES LI, University of Texas at El Paso, Maowei Dou, Sharma T Sanjay
3:20		**Recess**
3:35	(1200-4)	**'Cytology-on-a-Chip' Based Sensors for Monitoring of Potentially Malignant Oral Lesions** JOHN T MCDEVITT, New York University
4:10	(1200-5)	**Biomarker Discovery Using DNA Aptamers** WEIHONG TAN, University of Florida

SYMPOSIUM **Session 1210**

JAIMA - The State-of-the-Art Technologies from Japan: Analytical Instruments with/for Nano-Chemistry Technology and Advanced Diagnosis (II)
arranged by Koichiro Matsuda, JAIMA

Tuesday Afternoon, Room B305
Koichiro Matsuda, JAIMA, Presiding
Koji Suzuki, Keio University, Presiding

1:30		**Introductory Remarks - Koji Suzuki, President of JSAC**
1:35	(1210-1)	**Electrochemical Application of Boron-Doped Diamond Electrodes** YASUAKI EINAGA, Keio University
2:10	(1210-2)	**Plasmonic Nanomaterials** TETSU TATSUMA, University of Tokyo
2:45	(1210-3)	**The Unique Combination of Nanotechnology with Raman and SPRi Platforms Offers Innovative and Ultrasensitive Solutions for Diagnostics** MARINELLA SANDROS, HORIBA Scientific
3:20		**Recess**
3:35	(1210-4)	**Medicinal Cannabinomics and Mass Spectrometry Applications to Cannabis Testing Laboratories** SCOTT KUZDZAL, Shimadzu Scientific Instruments, Di Wang, Jonathan Edwardsen, William Lipps, Jeff H Dahl
4:10	(1210-5)	**Application of Laser/desorption Ionization Mass Spectrometry as a Novel Surface Analytical Tool** TAKAYA SATOH, JEOL Ltd.

TECHNICAL PROGRAM

SYMPOSIUM — Session 1220

New Advances in Analytical Mass Spectrometry
arranged by Gary Martin Hieftje, Indiana University

Tuesday Afternoon, Room B309
Gary Martin Hieftje, Indiana University, Presiding

1:30		**Introductory Remarks - Gary Martin Hieftje**
1:35	(1220-1)	**21 Tesla Fourier Transform Ion Cyclotron Resonance Mass Spectrometry: New Capabilities for Ultrahigh-Resolution Mass Analysis** ALAN G MARSHALL, Florida State University, John P Quinn, Nathan K Kaiser, Donald F Smith, Greg T Blakney, Tong Chen, Steven C Beu, Christopher L Hendrickson
2:10	(1220-2)	**The Unique Analytical Capabilities of Distance-of-Flight Mass Spectrometry** STEVEN J RAY, SUNY-Buffalo, Elise Dennis, David W Koppenaal, Christie G Enke, Gary Martin Hieftje
2:45	(1220-3)	**Plasma Mass Spectrometry: A Tool for and a Source of Chemical Reactions** JACOB T SHELLEY, Kent State University, Sunil P Badal, Yi You, Kelsey Williams, Garett MacLean
3:20		**Recess**
3:35	(1220-4)	**New Paths for Mass Spectrometry based upon Structures for Lossless Ion Manipulations (SLIM)** RICHARD D SMITH, Pacific Northwest National Laboratory
4:10	(1220-5)	**Improvements in Velocity-Based Mass Analysis by Use of Constant-Momentum Acceleration** GARY MARTIN HIEFTJE, Indiana University, Elise Dennis, Steven J Ray, Christie G Enke, Alexander W Gundlach-Graham

SYMPOSIUM — Session 1230

SEAC - New Trends in Electrochemical Neurochemistry
arranged by Adrian C Michael, University of Pittsburgh

Tuesday Afternoon, Room B310
Ann-Sofie Cans, Chalmers University, Presiding

1:30		**Introductory Remarks - Ann-Sofie Cans**
1:35	(1230-1)	**Fast-scan Cyclic Voltammetry Reveals Dopamine Spikes to Food Reward that are Tuned by Physiological State and Its Proxies** MITCHELL ROITMAN, University of Illinois at Chicago
2:10	(1230-2)	**Electroenzymatic Detection of Basolateral Amygdala Glutamate Release During Reward Seeking** KATE M WASSUM, University of California Los Angeles, Melissa Malvaez, Venuz Y Greenfield, Allison M Yorita, Lili Feng, Harold G Monbouquette
2:45	(1230-3)	**Improving Temporal Resolution of Enzyme Based Electrochemical Sensors for Detection of Non-Electroactive Analytes Important in Brain Chemistry** ANN-SOFIE CANS, Chalmers University of Technology, Yuanmo Wang, Jenny Bergman, Joakim Wigström
3:20		**Recess**
3:35	(1230-4)	**Building a 'Well-tempered' Biosensor for Real-Time Neurochemical Monitoring in the Intensive Care Unit** MARTYN G BOUTELLE, Imperial College London, Isabelle C Samper, Chu Wang, Thomas Watts, Michelle L Rogers, Chi Leng Leong, Sally A Gowers
4:10	(1230-5)	**Optogenetic-Control of Glutamate Release in the Rat Hippocampus and Frontal Cortex Measured Using Enzyme-Coated Ceramic Based Microelectrode Arrays** GREG A GERHARDT, University of Kentucky Medical Center

SYMPOSIUM — Session 1240

The Challenge of Detection for Drugged Driving
arranged by Herbert H Hill, Washington State University

Tuesday Afternoon, Room B311
Herbert H Hill, Washington State University, Presiding

1:30		**Introductory Remarks - Herbert H Hill**
1:35	(1240-1)	**Police Officer Difficulties with Drug-Impaired Driver Arrests** NICHOLAS PETER LOVRICH, Washington State University
2:10	(1240-2)	**Human Cannabinoid Metabolism and Disposition in Biological Matrices after Controlled Cannabis Administration** MARILYN ANN HUESTIS, NIDA
2:45	(1240-3)	**Analytical Methods for the Detection of Marijuana in Biological Fluids in a Forensic Toxicology Laboratory** BRIANNA PETERSON, Washington State Patrol
3:20		**Recess**
3:35	(1240-4)	**Detection of Drug Consumption in Europe** WOLFGANG VAUTZ, ISAS
4:10	(1240-5)	**Detection of Marijuana from Human Breath by Breathalyzer-IMS** HERBERT H HILL, Washington State University, Jessica A Tufariello

WORKSHOPS — Session 1250

Core-Shell versus Fully Porous HPLC Particles – The Current State of the Art in HPLC Columns
arranged by Jason Anspach and Lawrence Loo, Phenomenex

Tuesday Afternoon, Room B313
Jason Anspach, Phenomenex, Presiding

1:30		**Introductory Remarks - Jason Anspach and Lawrence Loo**
1:35	(1250-1)	**Perspectives on the Development and Future of Monodisperse Fully-Porous Silica Supports** DAVID S BELL, Supelco/Sigma-Aldrich
2:05	(1250-2)	**Fully-Porous vs. Core-shell Particles — The Past, the Present and the Future** LAWRENCE LOO, Phenomenex, Jason Anspach, Tivadar Farkas, Mike Chitty, Thuylinh Tran
2:35	(1250-3)	**Advantages and Limitations of Superficially Porous Particles** KEN BROECKHOVEN, Vrije Universiteit Brussel, Gert Desmet
3:05		**Recess**
3:20	(1250-4)	**Next Generation Superficially Porous Particle Technology: Highly Ordered Pore Structures Formed by Pseudomorphic Transformation** WILLIAM E BARBER, Agilent Technologies, Ta-Chen Wei, Wu Chen, Anne Mack, Jia Liu, Monika Dittmann, Xiaoli Wang
3:50	(1250-5)	**Clarifying the Difference Between Columns Packed with Solid Core and Fully Porous Particles** JACOB FAIRCHILD, Waters Corporation, Jonathan Turner, Bonnie Alden, Kevin Wyndham, Babajide Okandeji

ORGANIZED CONTRIBUTED SESSIONS — Session 1260

LIMS Live @ Pittcon: Best Practices and Lessons Learned From The Laboratory
arranged by Ken Ochi and Devender Gandhi, Accelerated Technology Laboratories

Tuesday Afternoon, Room B315
Ken Ochi, Accelerated Technology Laboratories, Presiding

1:30	(1260-1)	**Where Do We Start? A Roadmap to LIMS Success** CHRISTINE PASZKO, Accelerated Technology Laboratories
1:50	(1260-2)	**The LIMS Needs Assessment: Your Secret Weapon to a Successful Deployment** ALAN SERRERO, Gwinnett County Department of Public Utilities
2:10	(1260-3)	**Finding the Perfect LIMS: Keys to a Successful RFP** ASTER TEKLE, Alexandria Renew Enterprises
2:30	(1260-4)	**Preparing for a LIMS – The Importance of Proper Planning** KEITH KEESEE, Oklahoma Department of Agriculture, Cassandra Kontas
2:50		**Recess**
3:05	(1260-5)	**Implementing a LIMS is a Project – Treat It Like One** ROY D JONES, Duke Energy
3:25	(1260-6)	**Life with a LIMS: What It's Meant for the City of Clearwater** MARIA DE LA CANTERA, City of Clearwater (FL)
3:45	(1260-7)	**Leveraging LIMS for Streamlining Next Generation Sequencing Data** JENNIFER WELLER, University of North Carolina - Charlotte
4:05	(1260-8)	**What Have We Learned? Final Thoughts On the Road to LIMS Success (Getting the Most From Your LIMS)** DEVENDER GANDHI, Accelerated Technology Laboratories

Tuesday Afternoon

Author and presider lists as of January 20, 2016, are available at www.pittcon.org

ORGANIZED CONTRIBUTED SESSIONS — Session 1270

Quantifying the Tumor Microenvironment
arranged by Matthew R Lockett, University of North Carolina at Chapel Hill

Tuesday Afternoon, Room B316
Matthew R Lockett, University of North Carolina at Chapel Hill, Presiding

1:30	(1270-1)	**Phosphoproteomics in Prostate Extracellular Vesicles** W ANDY TAO, Purdue University
1:50	(1270-2)	**Exploring the Permissive Stromal Microenvironment** AMANDA B HUMMON, University of Notre Dame, Pinar Zorlutuna, Eric M Weaver
2:10	(1270-3)	**Modulating Drug Resistance in Hypoxia Tumors** DIMITRI PAPPAS, Texas Tech University
2:30	(1270-4)	**Paper-Based Assays for the Study of Cancer Cell Biology, Invasion, and Metastasis** MATTHEW R LOCKETT, University of North Carolina at Chapel Hill
2:50		**Recess**
3:05	(1270-5)	**Examining Small Molecule Cellular Signaling in Complex Environments with Microscale Systems** ASHLEIGH B THEBERGE, University of Washington
3:25	(1270-6)	**Microengineered Physiological Biomimicry: Human Organ-on-Chips** D DAN HUH, University of Pennsylvania
3:45	(1270-7)	**Direct Optical Microscopy of Biological Interfaces** CHARLES R MACE, Tufts University, Jenna A Walz, Irene Lui, Daniel J Wilson
4:05	(1270-8)	**Deconvolving Glycans in Metastasis** LARA K MAHAL, New York University

ORGANIZED CONTRIBUTED SESSIONS — Session 1280

SEAC - The Student Session in Electroanalysis
arranged by Stephen Maldonado, University of Michigan

Tuesday Afternoon, Room B401
Stephen Maldonado, University of Michigan , Presiding

1:30	(1280-1)	**Developing an Innovative Bio-Inspired Scanning Probe Microscopy (Bio-SPM) Approach to Map Specific Molecular Flux** FLORIKA CALING MACAZO, University of Maryland Baltimore County, Ryan J White
1:50	(1280-2)	**Electrodeposition of Semiconductor Thin Films Using Electrochemical Liquid-Liquid-Solid (ec-LLS) Deposition** JOSHUA DEMUTH, University of Michigan
2:10	(1280-3)	**Alternating Current Potentiometric Scanning Ion Conductance Microscopy (AC-PSICM)** LUSHAN ZHOU, Indiana University Bloomington, Yi Zhou, Wenqing Shi, Lane A Baker
2:30	(1280-4)	**Toward the Electrochemical Detection of Single Atoms and Ions** JEFFREY E DICK, The University of Texas at Austin
2:50		**Recess**
3:05	(1280-5)	**Miniaturized Potentiometric Ion-Sensing Systems: From Bulk Electrodes to Paper-Based Ion-Sensing Devices** JINBO HU, University of Minnesota, Andreas Stein, Philippe Buhlmann
3:25	(1280-6)	**Spontaneous Photoelectrochemical Growth of Nanopatterned Semiconductor Films Driven by Anisotropic Interfacial Light Collection** AZHAR I CARIM, California Institute of Technology, Nicolas A Batara, Anjali Premkumar, Harry A Atwater, Nathan S Lewis
3:45	(1280-7)	**New Enzymes for the Hybrid Enzymatic and Organic Electrocatalytic Cascade for the Complete Oxidation of Glycerol** SOFIENE ABDELLAOUI, University of Utah, Shelley D Minteer
4:05	(1280-8)	**The Unique Electrochemical Reactivity of Small Metal Nanoparticles** RAFAEL MASITAS, University of Louisville, Francis Zamborini

ORGANIZED CONTRIBUTED SESSIONS — Session 1290

Specialty Gas Analysis
arranged by Tracey Jacksier, Air Liquide

Tuesday Afternoon, Room B402
Tracey Jacksier, Air Liquide, Presiding

1:30	(1290-1)	**Real-Time, Selective Analysis of Air and Specialty Gases** DANIEL MILLIGAN, Syft Technologies Ltd, Vaughan Langford, Barry Prince, Murray McEwan
1:50	(1290-2)	**Cavity Ring-Down Spectroscopy Analyzer for Trace Moisture Detection in Ultra-Pure Ammonia** HELEN WAECHTER, Tiger Optics, Florian Adler, Marten Beels, Brian Siller, Bill West, Yu Chen
2:10	(1290-3)	**Novel FTIR/GC Detector for Analyzing Impurities in Gas Standards** MARTIN L SPARTZ, Prism Analytical Technologies, Inc., Charles Mark Phillips, Anthony S Bonanno, Peter P Behnke, Kelly R McPartland
2:30	(1290-4)	**Optimization of a Cavity Ring-Down Spectrometer for the Measurement of Trace Ammonia Contamination in Semicondutor Cleanroom Environments** GRAHAM A LEGGETT, Picarro, Mark Camenzind
2:50		**Recess**
3:05	(1290-5)	**Analytical Challenges of Measuring Impurities in Biogas** JANNEKE VAN WIJK, VSL, Annarita Baldan, Adriaan van der Veen, Stefan Persijn, Jeanrong Li
3:25	(1290-6)	**Analysis and Stability of Low Concentration HCl Standards** NATHALIE LUU , Air Liquide, Steve Hagen, Anthony Schleisman
3:45	(1290-7)	**Analytical QA/QC Measures to Validate Your Analysis Method Using FTIR** SYLVIE BOSCH-CHARPENAY, MKS Instruments
4:05	(1290-8)	**Open Discussion**

ORAL SESSIONS — Session 1300

Biomedical: Nanotechnology - Half Session

Tuesday Afternoon, Room B301
Jagdish Singh, Mississippi State University, Presiding

1:30	(1300-1)	**Green Synthesis, Characterization of Saccharide Coated Gold Nanoparticles for Catalytic Applications** HARSH MOOLANI, Western Kentucky University, Jason N Payne, Rajalingam Dakshinamurthy
1:50	(1300-2)	**NanoCluster Beacons for Detection of a Single N6-Methyladenine Epigenetic Modification** TIM YEH, University of Texas at Austin
2:10	(1300-3)	**Self-assembly Approach to Integrated Nanozymes: Rational Design and Biomedical Applications** HUI WEI, Nanjing University
2:30	(1300-4)	**High Resolution Separation of Oligonucleotides and DNA Fragments Using a New Polymer-Based Reversed Phase Column** JULIA BAEK, Thermo Fisher Scientific, Shanhua Lin, Xiaodong Liu, Jessica Wang

ORAL SESSIONS — Session 1310

Consumer Products Characterization - Half Session

Tuesday Afternoon, Room B408
William R LaCourse, University of Maryland Baltimore County, Presiding

1:30	(1310-1)	**Characterization of Metallic Nanoparticles in Tattoo Ink Using Asymmetrical Flow Field-Flow Fractionation Coupled with ICP-MS** SOHEYL TADJIKI, Postnova Analytics Inc., Evelin Moldenhauer, Tony Pfaffe, Trevor Havard, Thorsten Klein
1:50	(1310-2)	**Developing a Color Matching Database in Supercritical CO_2 for Waterless Textile Dyeing** ROLF SCHLAKE, Applied Separations, Susan Crowe, Madhu Anand, Rob Dorrycott
2:10	(1310-3)	**Detection and Quantification of Allergens in Personal Care Products by GC and GC×GC Paired with TOFMS** ELIZABETH M HUMSTON-FULMER, Leco Corporation, David E Alonso, Jonathan D Byer, Joseph E Binkley, Lorne E Fell
2:30	(1310-4)	**Carbohydrate-Rich Microalgae Characterization Using High-Performance Anion-Exchange Chromatography** PARUL ANGRISH, Thermo Fisher Scientific N/A

ORAL SESSIONS **Session 1320**

Detection of Illicit Drugs - Half Session

Tuesday Afternoon, Room B407
Nathaniel R Gomer, ChemImage Sensor Systems, Presiding

1:30	(1320-1)	**The Detection of Illicit Drugs and Cutting Agents Using Shortwave Infrared Hyperspectral Imaging** NATHANIEL R GOMER, ChemImage Sensor Systems, Jeffrey Beckstead, Oksana Olkhovyk, Matthew P Nelson
1:50	(1320-2)	**Characterization of Synthetic Phenethylamines Using High-Resolution GC-TOFMS and Mass Defect Filters** RUTH SMITH, Michigan State University, Alexandria Anstett, Fanny Chu, David E Alonso
2:10	(1320-3)	**The Development of a Novel Color Test for Improved Detection of Synthetic Cathinones** TSUNGHSUEH WU, University of Wisconsin-Platteville, Charles Cornett, Brook Tashner, Nicole Kloepfer
2:30	(1320-4)	**Potentiometric Sensor for Forensic Analysis: The Detection of the 'Undetectable Poison' Succinylcholine and Study of Its Enzymatic Degradation Kinetics** MOHAMED K ABD EL-RAHMAN, Cairo University, Amr M Mahmoud

ORAL SESSIONS **Session 1330**

Environmental and Instrumentation Application of LC/MS - Half Session

Tuesday Afternoon, Room B408
William R LaCourse, University of Maryland Baltimore County, Presiding

3:05	(1330-1)	**A New Type of Electron Ionization LC-MS and Its Applications** AVIV AMIRAV, Tel Aviv University, Boaz Seemann, Svetlana Tsizin, Alexander Fialkov, Tal Alon
3:25	(1330-2)	**Assessing the Occurrence and Fate of UV Filters in Seawater Swimming Pools by UPLC-Q-ToF-MS** TAREK MANASFI, Aix-Marseille University, Ravier Sylvain, Bruno Coulomb, Jean-Luc Boudenne
3:45	(1330-3)	**Expanded Analysis of Human Hormones in Drinking Water Using Solid Phase Extraction and Liquid Chromatography Tandem Mass Spectrometry** CARL FISHER, Thermo Fisher Scientific, Claudia Martins, Pranathi Perati
4:05	(1330-4)	**Multiclass Determination of New Psychoactive Substances in Municipal Wastewater** IVAN SENTA, Rudjer Boskovic Institute, Ivona Krizman, Marijan Ahel, Senka Terzic

ORAL SESSIONS **Session 1340**

Environmental Applications of Electrochemistry and Sensors - Half Session

Tuesday Afternoon, Room B409
Susan S Marine, Miami University Middletown, Presiding

1:30	(1340-1)	**Pretreatment and Spectroelectrochemical Sensing of Re(I)-Carbonyl Complexes** SHIRMIR D BRANCH, University of Cincinnati, Amanda D French, Amanda M Lines, Brian M Rapko, Sam A Bryan, William R Heineman
1:50	(1340-2)	**Wearable Gas Sensors: Shrinking Electrochemical Cells** JOHN R SAFFELL, Alphasense Ltd., Joseph R Stetter
2:10	(1340-3)	**Cloud Point Extraction for Electroanalysis: Anodic Stripping Voltammetry of Lead** CORY A RUSINEK, University of Cincinnati, William R Heineman, Adam Bange, Ian Papautsky, Mercedes Warren
2:30	(1340-4)	**Development of an Electrochemical Sensor for Detection of Dissolved Polycyclic Aromatic Hydrocarbons in Water** ABRA PENEZIC, Rudjer Boskovic Institute, Blazenka Gasparovic, Andrew Nelson, Drazenka Stipanicev

ORAL SESSIONS **Session 1350**

Food Product Quality and Component Characterization II

Tuesday Afternoon, Room B403
Alexander J Krynitsky, US FDA, Presiding

1:30	(1350-1)	**A Green Sample Preparation Device for Complex Biological, Environmental, Food, Pharmaceutical and Toxicological Samples** ABUZAR KABIR, Florida International University, Rodolfo Mesa, Rayma Blanko, Kenneth G Furton
1:50	(1350-2)	**Cannabinoids and Terpenes in Food** TIM ANDERSON, Phenomenex, Kristen Parnell, Ramkumar Dhandapani
2:10	(1350-3)	**Determination of Whey Adulteration in Milk Powder by Using Laser Induced Breakdown Spectroscopy** GONCA BILGE, Hacettepe University, Banu Sezer, Kemal E Eseller, Halil Berbero lu, Ali Topcu, Ismail H Boyaci
2:30	(1350-4)	**Bioactive Compounds and Antioxidant Activity of Guava Processed Byproducts and Wastes** NEELA EMANUEL, NIFTEM, Sao Khushbu, Aman Kaushik
2:50		**Recess**
3:05	(1350-5)	**Analysis of Non Volatile Congeners in Spirits by Creation of an Ion Fragmentation Database for Use with Time of Flight LC MS** RITA STEED, Agilent Technologies, Joni Stevens, Tarun Anumol, Gregory Hunlen, Luke Adam, Sue Dantonio
3:25	(1350-6)	**A Simple Field Test Kit for the Detection of Iodine in Food Grade (Table) Salt** RUFUS SHA'ATO, University of Agriculture, Nasiru L Usman
3:45	(1350-7)	**Chromatography Advancements in Nutraceuticals and Dietary Supplements Testing** ALLEN MISA, Phenomenex, Zeshan Aqeel
4:05	(1350-8)	**Raman Microspectroscopy and Chemometrics: A Combined Approach to Perform A Rapid Untargeted Screening of Bacteria Present in Food Samples** ALI ASSAF, University of Nantes, Emilie Grangé, Christophe Cordella, Douglas Rutledge, Michele Lees, Gerald Thouand N/A

ORAL SESSIONS **Session 1360**

Forensic Trace Analysis - Half Session

Tuesday Afternoon, Room B407
Nathaniel R Gomer, ChemImage Sensor Systems, Presiding

3:05	(1360-1)	**Analysis and Comparison of Fatty Acid Compositions in Latent and Smudged Fingermarks via Gas Chromatography-Mass Spectrometry (GC-MS) and Comprehensive Two-Dimensional Gas Chromatography-Time-of-Flight Mass Spectrometry (GCxGC-MS)** CAITLIN COBERN, Penn State, Frank Dorman, Seth Michalski
3:25	(1360-2)	**Mathematical Model to Predict Evaporation of Ignitable Liquids for Forensic Applications** VICTORIA L MCGUFFIN, Michigan State University, Rebecca Brehe, John W McIlroy, Ruth Smith
3:45	(1360-3)	**Analysis of Smoke Residues from Illicit Drugs as a Potential Source of Forensic Evidence** JULIE BITTER, National Institute of Standards and Technology, Matthew Staymates
4:05	(1360-4)	**LIBS Instrumentation for Fast Quantitative Analysis of Soil and Forensic Investigation of Nuclear Materials** ALAIN BLOUIN, National Research Council Canada, Paul Bouchard, Josette El Haddad, Aissa Harhira, Mohamad Sabsabi

ORAL SESSIONS **Session 1370**

Laser Induced Breakdown Spectroscopy (LIBS) and Glow Discharge in Atomic Spectroscopy - Half Session

Tuesday Afternoon, Room B301
Jagdish Singh, Mississippi State University, Presiding

3:05 (1370-1) **Standoff LIBS Using a Spatial Heterodyne Spectrometer with Sub-Microsteradian Collection Optics** PATRICK D BARNETT, University of South Carolina, Nirmal Lamsal, S Michael Angel

3:25 (1370-2) **Study of Matrix Effects for Reproducible LIBS Analysis of Powders** MATTHIEU BAUDELET, University of Central Florida, Sudeep Jung Pandey, Richard Locke, Brandon Seesahai, Romain Gaume, Martin Richardson

3:45 (1370-3) **Particulate Identification Using Image Directed Laser Induced Breakdown Spectroscopy (LIBS) with Enhanced Spectral and Spatial Resolution for Pharmaceutical and Industrial Applications** MARK SULLIVAN, rap.ID Inc, Oliver Valet

4:05 (1370-4) **Spectroscopy and Imaging Studies of a Solution-Cathode Glow Discharge** MICHAEL R WEBB, University of North Carolina Wilmington, Christian G Decker, Denise E Moon

ORAL SESSIONS **Session 1380**

LC/MS Biological Applications

Tuesday Afternoon, Room B404
Mary Ellen McNally, E.I. DuPont de Nemours and Company, Presiding

1:30 (1380-1) **Distinguishing Isomeric Acylsugar Metabolites: Strategies for Labeling Using ^{13}C-amino Acid Precursors and LC-MS/MS** XIAOXIAO LIU, Michigan State University, Banibrata Ghosh, A Daniel Jones

1:50 (1380-2) **Oxidative Techniques for Pteridine Bioanalysis and Implications for ESI-MS Applications** CASEY BURTON, Missouri University of Science and Technology, Honglan Shi, Yinfa Ma

2:10 (1380-3) **Intact Histones Separation by Using Submicron Particles with RPLC-MS** XIMO ZHANG, Purdue University, Mary J Wirth

2:30 (1380-4) **High Throughput Analysis of TCA Metabolites Using Column Switching and IC-HRAM Mass Spectroscopy** TERRI T CHRISTISON, Thermo Fisher Scientific, Junhua Wang, John E Madden, Monika Verma

2:50 **Recess**

3:05 (1380-5) **Unified Drug Testing by Online SPE-LC/MS/MS with Focus on Productivity Achieved Through Ease of Use by Lab Technicians: One Totally Automated Method Measures ALL the Drugs in Urine and/or Oral Fluids** MARK HAYWARD, ITSP Solutions, Rick Youngblood, Kim Gamble, Martin Johnson, Matthew Hardison

3:25 (1380-6) **High-Throughput Mass Spectrometric Analysis of Covalent Protein-Inhibitor Adducts for the Discovery of Irreversible Inhibitors: A Complete Workflow** IAIN CAMPUZANO, Amgen, Tisha San Miguel, Todd Rowe, Daniel Onea, Victor Cee, Tara Arvedson, John McCarter

3:45 (1380-7) **Efficient Use of pH Control in Developing LC/UV/MS Methods** THOMAS E WHEAT, Waters Corporation, Amanda B Dlugasch, Patricia R McConville

4:05 (1380-8) **Selectivity and Column Choices in HPLC Method Development** WILLIAM LONG, Agilent Technologies, Anne Mack, Stephen Luke, Jason Link

ORAL SESSIONS **Session 1390**

Polymer Characterization and Applications

Tuesday Afternoon, Room B405
Christopher Henry, Waters Corporation, Presiding

1:30 (1390-1) **Transferring a GPC Method to A More Efficient SEC Method for Zoladex Co-Polymer Using an Advanced Polymer Chromatography Based U(H)PLC System Coupled with RI Detection** CHRISTOPHER HENRY, Waters Corporation, Jeanette Bowden, Mark Wrona, Richard Ladd, Andy Boughey

1:50 (1390-2) **Characterization of Polyacrylamide at Different Ionic Strength and pH Conditions Using Asymmetrical Flow FFF and Multi-Angle Light Scattering Detector** SOHEYL TADJIKI, Postnova Analytics Inc., Trevor Havard, Japan Trivedi, Thorsten Klein

2:10 (1390-3) **Patterning of Polycaprolactone-Impregnated Glass Microfiber Membranes: A Novel Approach to Fabrication of Microfluidic Devices** GAYAN C BANDARA, Oregon State University, Vincent T Remcho

2:30 (1390-4) **Advanced Polymer Chromatography - Method Development Tools for SEC Analysis of PEG** MICHAEL OLEARY, Waters Corporation, Damian Morrison

2:50 **Recess**

3:05 (1390-5) **Determination of Minor Component Differences and Additives in Polyethylene Using Thermal Desorption, Heart-Cutting EGA, Reactive Pyrolysis and GC/MS Techniques** TERRY RAMUS, Diablo Analytical, Dave Randle, Itsuko Iwai, Robert R Freeman

3:25 (1390-6) **Applying Automatic Polymer Identification Capability to DSC Thermograms** BOB FIDLER, NETZSCH Instruments NA LLC, Ekkehard Post, Tobias Pflock, Stefan Schmoelzer, Gabriele Kaiser, Alexander Schindler

3:45 (1390-7) **Characterization and Determination of Irganox 1076 and 1010 in Polyethylene Using Thermal Desorption and Reactive Pyrolysis – GC/MS** DAVE RANDLE, Frontier Lab USA, Itsuko Iwai, Terry Ramus, Robert R Freeman, Aki Hosaka, Ichi Watanabe

4:05 (1390-8) **Simple Analysis Method of Degradation for The Molded Articles Polymer by FTIR and SEM** YOKO KATO, Advantest N/A

ORAL SESSIONS **Session 1400**

Sampling and Sample Preparation-Environmental and Food (ID, Safety and Contaminants)

Tuesday Afternoon, Room B406
Garry Lynch, Bechtel Marine Propulsion Corporation, Presiding

1:30 (1400-1) **Sampling Condensed Hookah Smoke with C-18 SPME Fibers** AMBERLIE A CLUTTERBUCK, University of Cincinnati, Joseph Caruso, William Wetzel, Julio Landero N/A

1:50 (1400-2) **Bacterial DNA Analysis Using Solid-Phase Microextraction by Different Polymeric Ionic Liquid-Based Sorbent Coatings** OMPRAKASH NACHAM, Iowa State University, Kevin D Clark, Matthew P Bommarito, Andrea E Tsatalis, Jared L Anderson

2:10 (1400-3) **ICE Concentration Linked with Extractive Stir Bar (ICECLES): A Novel Sample Preparation Technique for Ultratrace Analysis** BRIAN LOGUE, South Dakota State University

2:30 (1400-4) **Analysis of Pesticides in Foods by Direct Immersion SPME Using an Overcoated Fiber** LEONARD M SIDISKY, Supelco/Sigma-Aldrich, Katherine Stenerson, Robert Shirey, Yong Chen, Tyler Young

2:50 **Recess**

3:05 (1400-5) **Recent Advances in Sample Preparation for Extraction and SPE** SM RAHMAT ULLAH, Thermo Fisher Scientific, Kannan Srinivasan, Mike McAdams, Glenn Kuse, Aaron Kettle

3:25 (1400-6) **Development of a Carbon Mesh Supported Thin Film Microextraction Membrane as a Means to Lower the Detection Limits of Bench-top and Portable GC-MS Instrumentation** JONATHAN J GRANDY, University of Waterloo, Janusz Pawliszyn

3:45 (1400-7) **Investigation of Polymeric Ionic Liquid Sorbent Coatings in Solid-Phase Microextraction Coupled to High-Performance Liquid Chromatography for the Analysis of Polar Compounds** HONGLIAN YU, Iowa State University, Josias Merib, Jared L Anderson

4:05 (1400-8) **Justification of Kinetic Calibration in Pre-Equilibrium Solid Phase Microextraction with a Mathematical Model and Computational Simulation** MD NAZMUL ALAM, University of Waterloo, Fardin Ahmadi, Luis Ricardez-Sandoval, Janusz Pawliszyn

POSTER SESSION — Session 1410

All posters are to be mounted by 10:00 AM and remain on display until 4:00 PM. Authors must be at their posters from 1:00 PM to 3:00 PM. Location of the afternoon posters is on the Exposition Floor, 400 Aisle. PLEASE NOTE: You cannot get onto the Exposition Floor until after 9:00 AM.

Advances in Biomedical Applications

Tuesday Afternoon, Exposition Floor, 400 Aisle

(1410-1 P) **Strategies for Glycomics, Glycoproteomics, and Glycosaminoglycan Research at the Complex Carbohydrate Research Center** STEPHANIE A ARCHER-HARTMANN, Complex Carbohydrate Research Center, Roberto Sonon, Parastoo Azadi, Christian Heiss, Mayumi Ishihara

(1410-2 P) **Down Regulation of Smad-2 and VEGF Transcription and TGF-B1 Signaling in Nano Sized Titanium Dioxide-Induced Liver Injury in Mice by Potent Antioxidants** SAMY A ABDEL AZIM, Cairo University, Abd El-Moneim M Afify

(1410-3 P) **One Hydrothermal Processing of 1D Hydroxyapatite for Biomedical Application** ZORAN S STOJANOVIC, Institute of Technical Sciences of SASA, Nenad L Ignjatovic, Vuk D Uskokovic, Miroslav M Miljkovic, Vojka Zunic

(1410-4 P) **DNA Micelle Flares: Thermodynamic Stability and Cellular Internalization** YANYUE WANG, University of Florida, Weihong Tan

(1410-5 P) **Study of Various Cationic CPEs' Interaction with Mammalian Cells** SHANSHAN WANG, University of Florida, Zhiliang Li, Yun Huang, Kirk S Schanze

(1410-6 P) **Generation of T-cell Specific Aptamers Using a Novel Cell-Selex Method: Antibody Guided Cell-SELEX Technology** HASAN E ZUMRUT, City University of New York, The Graduate Center, Mallikaratchy Prabodhika, Shomi Chakrabarti, Mst Naznin Ara, George Maio

(1410-7 P) **Adeno-Associated Virus** YUAN WU, University of Florida

(1410-8 P) **Continuous Electroporation Through a Mesoporous Gold Membrane** JULIETTE EXPERTON, University of Florida, Aaron Wilson, Charles R Martin

(1410-9 P) **Analyzing the Effect of Beverages and Fluoride on Tooth Enamel with a FAAS and Dissolution** ANDREA GERCHMAN, St. John Fisher College, Kimberly Chichester

(1410-10 P) **Profiling Volatile Organic Compounds in Exhaled Breath by TD–GC–TOF MS** LAURA MCGREGOR, Markes International Ltd, Caroline Widdowson, Nicola Watson, Chris Hall, Ken Umbarger

(1410-11 P) **Purification of Pharmaceutical Proteins Including Antibody and Peptides Using Ion-Exchange Bulk Media Designed for High-Throughput Purification** TAKASHI SATO, YMC CO., Ltd., Munehiro Shoda, Chiaki Iwata, Noriko Shoji, Takatomo Takai

(1410-12 P) **An Analysis of the Protective Effects of Selenium on Porcine Jejunal Epithelial Cells Following Cadmium-induced Oxidative DNA Damage** SARAH JOANNE LYNCH, Dublin City University, Blánaid White, Dermot Walls, Karina Horgan

(1410-13 P) **Imaging Mass Spectrometry Reveals the Chemistry in Chemically Fixed Adrenal Cells Prepared for Transmission Electron Microscopy Analysis** JELENA LOVRIC, Chalmers University of Technology, Per Malmberg, Bengt R Johansson, John S Fletcher, Andrew Ewing

(1410-14 P) **Measurement of the Correlation Between Type of Mutation and Conditions that Alter Aging in Yeast** ANDREEA P MUSTEATA, Rensselaer Polytechnic Institute, Patrick Maxwell

(1410-15 P) **Self-Assembly Aptamer–Graphene Oxide Nanosheets as an Anticoagulant** PEI-XIN LAI, National Taiwan Ocean University

(1410-16 P) **Synthesis of Protein-Capped Gold Nanoparticles with Specific Protein Orientation as a DNA Transfection Vehicle** JU-YI MAO, National Taiwan Ocean University

(1410-17 P) **Recovering the Electrocatalytic Activity of Pt Nanoparticle-DNA Collisions via Nuclease Digestion for Sensing Applications** ALMA CASTANEDA, University of Texas at Austin, Donald Robinson, Richard M Crooks, Keith Stevenson

(1410-18 P) **Study of Stressed Monoclonal Antibody (mAb) Pharmaceuticals by Using Deep-UV Resonance Raman (DUVRR) Spectroscopy** CHEN QIU, US Food and Drug Administration, Sergey Arzhanstev, John Kauffman

(1410-19 P) **Paper Membrane Based SERS Platform for Rapid Bacteria Enumeration** UGUR TAMER, Gazi University, Aysen Gumustas, Bozkurt Akif, Merve Eryilmaz, Esra Acar, Demet Cetin, Zekiye Suludere, Ismail H Boyaci

(1410-20 P) **Microfluidics for the Detection of Minimal Residual Disease in Acute Myeloid Leukemia Patients Using Circulating Leukemic Cells Selected from Blood** JAMES TAYLOR, University of North Carolina at Chapel Hill, Joshua Jackson, Malgorzata Witek, Paul Armistead, Steven A Soper

POSTER SESSION — Session 1420

All posters are to be mounted by 10:00 AM and remain on display until 4:00 PM. Authors must be at their posters from 1:00 PM to 3:00 PM. Location of the afternoon posters is on the Exposition Floor, 400 Aisle. PLEASE NOTE: You cannot get onto the Exposition Floor until after 9:00 AM.

Advances in Metabolomics, Proteomics, Lipidomics

Tuesday Afternoon, Exposition Floor, 400 Aisle

(1420-1 P) **Universal Derivatization of Metabolites for Improved Sensitivity in Electrospray Ionization Mass Spectrometry** TIANJIAO HUANG, Saint Louis University, Maria Toro, James L Edwards

(1420-2 P) **Identification of a Biological Fear Cue in Blue Crab Urine Via 1H NMR-Based Metabolomics** REMINGTON X POULIN, Georgia Institute of Technology, Marc Weissburg, Julia Kubanek

(1420-3 P) **Analysis of Urine SRMs Using Solid Phase Micro Extraction, Dynamic Headspace and Liquid Injection with Comprehensive Two-Dimensional Gas Chromatography (GC×GC)-High Resolution Time-of-Flight Mass Spectrometry** DAVID E ALONSO, Leco Corporation, Joseph E Binkley, Elizabeth M Humston-Fulmer, Lorne E Fell, Jonathan D Byer

(1420-4 P) **Feasibility of Early Detection of Acute Pulmonary Exacerbations by Exhaled Breath Condensate Metabolomics** XIAOLING ZANG, Georgia Institute of Technology, Maria E Monge, Nael A McCarty, Arlene Stecenko, Facundo M Fernandez

(1420-5 P) **Prebiotic Peptidomics: An Ultra-Performance Liquid Chromatography-Ion Mobility-Tandem Mass Spectrometry (UPLC-IM-MS/MS) Workflow Applied to Origins-of-Life Chemistry** JAY G FORSYTHE, Georgia Institute of Technology, Sheng-Sheng Yu, Ramanarayanan Krishnamurthy, Martha A Grover, Nicholas V Hud, Facundo M Fernandez

(1420-6 P) **Development of a High Throughput Organelle Extraction Procedure from Rat Tissues** BRANDON EASPARRO, Omni International, Shari Garrett, James Atwood

(1420-7 P) **Serum Lipidomics Identifies Biomarkers of Acute Traumatic Brain Injury** SCOTT HOGAN, Georgia Institute of Technology, David A Gaul, Melissa A Velez, Michelle C LaPlaca, Facundo M Fernandez

(1420-8 P) **An Accelerated Protein Sample Preparation Method for LC-MS-Based Proteomics** SUJATHA CHILAKALA, Cleveland State University, Yan Xu

(1420-9 P) **High-Throughput Proteomics Analysis by LC-MS with AJS-CESI Technology** SUJATHA CHILAKALA, Cleveland State University, Yan Xu

(1420-10 P) **Response of NEIL1 to Oxidatively Damaged G-Quadruplexes via Affinity Purification-Mass Spectrometry** JONATHAN ASHBY, University of California, Davis, Brittany Anderson, Sheila David N/A

(1420-11 P) **Expression and Spectroscopic Characterization of Allene Oxide Synthase: A Cytochrome P450 for the Rearrangement of Small Molecule Hydroperoxides** JULIE C MCINTOSH, University of North Carolina at Chapel Hill, Nathan A Whitman, Matthew R Lockett

(1420-12 P) **Practical Considerations for Quantifying Protein Therapeutics in Biological Matrices by LC-MS/MS** ERIN E CHAMBERS, Waters Corporation, Mary Lame, Paula Orens, Hua Yang N/A

POSTER SESSION — Session 1430

All posters are to be mounted by 10:00 AM and remain on display until 4:00 PM. Authors must be at their posters from 1:00 PM to 3:00 PM. Location of the afternoon posters is on the Exposition Floor, 400 Aisle. PLEASE NOTE: You cannot get onto the Exposition Floor until after 9:00 AM.

Bioanalytical and Neurochemistry

Tuesday Afternoon, Exposition Floor, 400 Aisle

(1430-1 P) **Using Effective Conductivity to Study Brain Tissue Morphology** JENNA DEVIVO, University of Pittsburgh, Yangguang Ou, Erika Varner, Adrian C Michael, Stephen Weber

(1430-2 P) **Serotonin and Histamine Coregulation in the Mouse Premammilary Nucleus** RHIANNON ROBKE, University of South Carolina, Srimal A Samaranayake, Aya Abdalla, H Frederick Nijhout, Michael C Reed, Janet Best, Parastoo Hashemi

(1430-3 P) **Development of a Nanoscale Calcium-Selective Electrode** THERESA M RUWE, Northern Kentucky University, Edward A Dobrzykowski, Teri Rae Armstrong, Celeste A Morris

(1430-4 P) **Optimized Determination of L-Dopa at a Glassy Carbon Electrode Modified with Electrodeposited Films of Caffeic Acid** AHMAD ROHANI FAR, The University of Toledo, Amila M Devasurendra, Joshua A Young, Jon R Kirchhoff

(1430-5 P) **Mass Spectrometry Imaging for Targeted Metabolomics of Medulloblastoma** MARTIN R PAINE, Georgia Institute of Technology, Jingbo Liu, Danning Huang, Tobey MacDonald, Facundo M Fernandez

(1430-6 P) **Ultra-High Performance Liquid Chromatography Mass Spectrometry Metabolic Fingerprinting of a Medulloblastoma Mouse Model** DANNING HUANG, Georgia Institute of Technology, Martin R Paine, Jingbo Liu, Tobey MacDonald, Facundo M Fernandez

(1430-7 P) **Biophysical Evaluation of Surfactant Effects on Nanoparticle Toxicity a Lipid Model of the Blood-Brain Barrier** ADAM L HOFFMANN, Northern Kentucky University, Darcy Poor, Rolf Fowee, Marcus Jones, Andrew Hall, Kristi L Haik, Celeste A Morris

(1430-8 P) **Novel Graphene-Modified Graphite Pencil Electrode for the Trace Quantification of L-Tyrosine in Human Urine** ABDEL-NASSER KAWDE, King Fahd University of Petroleum and Minerals

(1430-9 P) **Online Liquid Chromatography - Surface Enhanced Raman Detection for Metabolic Profiling** ANH H NGUYEN, University of Notre Dame, Zachary D Schultz

(1430-10 P) **Combining Microchip Electrophoresis, Mass Spectrometry, and Standard Addition to Identify N-glycan Structures in Serum** XIAOMEI ZHOU, Indiana University, Christa Snyder, Margit I Campos, Milos V Novotny, Stephen C Jacobson

(1430-11 P) **Multichannel Impedance-Based Biosensing Using Virus-Polyethylenimine Films for Bladder Cancer Detection** ALANA OGATA, University of California Irvine, Reginald M Penner, Crystin Eggers N/A

(1430-12 P) **Development of a Wearable Device for Neurochemical Monitoring of Energy Availability in the Injured Brain Using On-Line Microdialysis** ISABELLE C SAMPER, Imperial College London, Sally A Gowers, Chu Wang, Martyn G Boutelle

(1430-13 P) **Label-Free Profiling of O-Linked Glycans by HPLC with Charged Aerosol Detection** IAN N ACWORTH, Thermo Fisher Scientific, David Thomas, Rainer Bauder, William Kopaciewicz

(1430-14 P) **Amperometric Quantitative Measurements of Individual Vesicular Transmitters in Single Adrenal Chromaffin Cells with Nano-Tip Conical Carbon Fiber Microelectrodes** SOODABEH MAJDI, Chalmers University of Technology N/A

(1430-15 P) **In Vitro Electrochemical Investigation of ATP: Catecholamine Interactions** ZAHRA TALEAT, Chalmers University, Johan Dunevall, Ricardo Borges, Judith Estevez, Andrew Ewing

(1430-16 P) **Quantifying the Progression of Amyotrophic Lateral Sclerosis** AIDAN P WICKHAM, Imperial College London, Christopher E Shaw, Kerry R Mills, James Bashford, Emmanuel M Drakakis, Martyn G Boutelle

POSTER SESSION — Session 1440

All posters are to be mounted by 10:00 AM and remain on display until 4:00 PM. Authors must be at their posters from 1:00 PM to 3:00 PM. Location of the afternoon posters is on the Exposition Floor, 400 Aisle. PLEASE NOTE: You cannot get onto the Exposition Floor until after 9:00 AM.

Bioanalytical: Miscellaneous Analytical Techniques

Tuesday Afternoon, Exposition Floor, 400 Aisle

(1440-1 P) **An Analytic Approach to the Laser-Assisted Microscopy of Erythrocyte Deformability** FARAMARZ RAHNAMA, University of Calcutta, Azamat Vaseghi N/A

(1440-2 P) **Fabrication of Passive Microfluidic Diodes with Tunable Breakthrough Junctions** MARK D HOLTAN, Auburn University, Christopher J Easley

(1440-3 P) **Microfluidic Thermofluorimetric Analysis (μTFA) for Protein Quantification in Nanoliter and Picoliter Volumes** JUAN HU, Auburn University, Joonyul Kim, Mark D Holtan, Christopher J Easley

(1440-4 P) **Analysis of Fluorinated Lidocaine Derivatives Under Varying Conditions and Its Application in the Body** CYANN CICCONI, Seton Hill University

(1440-5 P) **Non-Enzymatic Modification of Human Serum Albumin: A Study Focusing On Advanced Glycation End Products by D-Galactose and D-Glucose** MENASHI A COHENFORD, Marshall University

(1440-6 P) **Study Interactions Between 21 Proteins and Nanoparticles** YAOKAI DUAN, University of California Riverside, Wenwan Zhong

(1440-7 P) **Development of Conductive Polymeric Ionic Liquid-Based Electrodes** AMILA M DEVASURENDRA, The University of Toledo, Joshua A Young, Ahmad Rohani Far, LM Viranga Tillekeratne, Jon R Kirchhoff

(1440-8 P) **Plasmonic Applications in the Mid-IR: Spectroscopic Surface Plasmon Resonance (SPR) Detection of N_2O Gas and Hexadecanethiol Self-Assembled Monolayer on a Low Loss, Plasmonic Tunable Novel Material Dy:CdO** HNIANG KHAMH, North Carolina State University

(1440-9 P) **Orientation of Membrane-Bound Cytochrome P450's in Nanodiscs** IVAN LENOV, University of Illinois, Stephen Sligar

(1440-10 P) **A Comparison of Binding Constants for FITC-Labeled Single Stranded DNA with Anti-FITC Antibody** QIAN LIU, Wake Forest University, Keith Bonin, Jason Gagliano, Kathryn Riley

(1440-11 P) **A Micro Pathogenic Microorganism Detector Applied for Mobile Phone Devices** LIANG SIJIA, Zhejiang University, Yu Dongdong, Zhou Jianguang

(1440-12 P) **The Preservation of DNA Using Magnetic Ionic Liquids** MATTHEW SORENSEN, Gustavus Adolphus College, Kevin D Clark, Omprakash Nacham, Jared L Anderson

(1440-13 P) **Automatic Data Analysis for Nanopore Detection** ZHEN GU, East China University of Science and Technology, Chan Cao, Yi-Lun Ying, Yi-Tao Long N/A

(1440-14 P) **Experiments and Modeling of Sequential Enzyme Activity in Biphasic Reaction Media** BRADLEY W DAVIS, Waynesburg University, William M Aumiller, Negar Hashemian, Antonios Armaou, Christine D Keating

(1440-15 P) **Fast Multipoint Immobilized MOF Bioreactor** HSI-YA HUANG, Chung Yuan Christian University, Wan-Ling Liu, Stephen Lirio, Chia-Her Lin

(1440-16 P) **HPLC Determination of the Rosmarinic Acid Content of Common Household Herbs** BRANDT A WOOD, California University of Pennsylvania N/A

(1440-17 P) **Molecular Modeling of Binding of Coumarins to DNA Gyrase** SARA MARIYAM ABDULLA, College of Chemical Sciences N/A

(1440-18 P) **Monitoring Cluster Ions Derived from Aptamer-Modified Gold Nanofilms under Laser Desorption/Ionization for the Detection of Circulating Tumor Cells** WEI-JANE CHIU, National Taiwan Ocean University, Chih-Ching Huang N/A

(1440-19 P) **Same Nucleotide Composition and High Sequence Similarity of MicroRNAs - An Analytical Challenge to RNA Studies?** JOSEPH N MWANGI, University of North Carolina at Greensboro, Norman Chiu

(1440-20 P) **Travelling Wave Ion Mobility Mass Spectrometry of Isomeric MicroRNA Biomarkers** NORMAN CHIU, University of North Carolina at Greensboro, Joseph Mwangi N/A

(1440-21 P) **Ultrafast Spectroscopic Studies of Molecular Interactions and Vibrational Energy Relaxation Dynamics in Binary Solvents** CHEN QIU, Michigan State University, Gary Blanchard N/A

(1440-22 P) **Surface-enhanced Raman Spectroscopy Based Total Protein Assay** MERVE ERYILMAZ, Gazi University, Adem Zengin, Ugur Tamer

(1440-23 P) **Ionic Liquid Crosslinkers for Chiral Imprinted NanoGUMBOS** SUZANA HAMDAN, Louisiana State University, Leonard Moore, Jason LeJeune, Farhana Hasan, Trevor K Carlisle, Jason E Bara, Douglas Gin, Andrew L LaFrate, Richard D Noble, David Spivak, Isiah M Warner

(1440-24 P) **Fabricated Multi-functional DNA Nanogel to Delivery Drug to Target Cells** XIGAO CHEN, University of Florida N/A

(1440-25 P) **Withdrawn**

(1440-26 P) **A Novel Bioassay Platform Using Silica Core-Stabilized Liposome Shell Microparticles for Ligand Discovery** KENDALL E SANDY, University of Arizona, Jinyan Wang, Mark T Agasid, Craig A Aspinwall

(1440-27 P) **Detection of Disease Associated MicroRNA Combinations with a Smart AND Sensor** LULU ZHANG, Oregon State University, Sean M Burrows

(1440-28 P) **Withdrawn**

(1440-29 P) **E-spun Collagen-CNT/Silk-CNT Composite Fibers to Transmit Electrical Signals for Cell Stimulation** NAIWEI CHI, Illinois Institute of Technology

(1440-30 P) **Sensitive Quantification of Protease Activity in Biological Samples by Using MCE-LIF** LI PAN, Jackson State University, Yiming Liu, Shulin Zhao N/A

POSTER SESSION — Session 1450

All posters are to be mounted by 10:00 AM and remain on display until 4:00 PM. Authors must be at their posters from 1:00 PM to 3:00 PM. Location of the afternoon posters is on the Exposition Floor, 400 Aisle. PLEASE NOTE: You cannot get onto the Exposition Floor until after 9:00 AM.

Bioanalytical: Separation Techniques

Tuesday Afternoon, Exposition Floor, 400 Aisle

(1450-1 P) **Surface Modified Nylon Capillary-Channeled Polymer (C-CP) Fibers for Protein Ion-Exchange Separations** LIUWEI JIANG, Clemson University, R Kenneth Marcus

(1450-2 P) **SEC Analysis of a Monoclonal Antibodies Using a Hybrid Silica Based Stationary Phase** JEFFREY KAKALEY, YMC America Inc., Ernest Sobkow N/A

(1450-3 P) **HPLC Method Transfer for Biopharmaceutical Analysis** BROOKE M KOSHEL, Waters Corporation, Sean M McCarthy

(1450-4 P) **Ratio of Different Fatty Acids Determined by GC-MS in Exosomes Purified Through Size Exclusion Chromatography** RUI XU, Jackson State University, Yiming Liu, Joseph Fernandes, Radhika Pochampally

(1450-5 P) **Analysis of Chromium Species in Dietary Supplements Using ICP-MS and Speciated Isotope Dilution Mass Spectrometry (SIDMS)** KAITLIN MILLER, Duquesne University, Logan T Miller, Jennifer Crawford, Stuart Procter, Matt Pamuku, Skip Kingston

(1450-6 P) **Metabolomic Signatures from Early Stage Ovarian Cancer Patients** DAVID A GAUL, Georgia Institute of Technology, Christina M Jones, Long Q Tran, John F McDonald, Facundo M Fernandez

(1450-7 P) **Reducing Adhesion of Proteins on Stainless Steel Components by the Application of a Carboxysilane Coating** LUKE PATTERSON, SilcoTek Corporation, Alfredo Narvaez, David Daghfal, Vaidya Shyam, Min Yuan, David Smith

(1450-8 P) **Optimizations of Proteomic Sample Preparation Method for *Xenopus Laevis* Embryonic Proteomics** ELIZABETH H PEUCHEN, University of Notre Dame, Liangliang Sun, Norman J Dovichi

(1450-9 P) **Determination of the Constituent Compounds in the Essential Oil from the Stem Bark of Ficus Capensis, A Multipurpose Phytomedicine, by GCMS and their Relevance to the Bioactivity of the Plant** MODUPE M OGUNLESI, University of Lagos, Christianah T Aleshinloye

(1450-10 P) **GC-MS Analysis of the Essential Oil from the Stem Bark of Tetrapleura Tetrapetra, a Multipurpose Medicinal Plant, and Bioactivities of some Constituent Compounds** MODUPE M OGUNLESI, University of Lagos, Christianah T Aleshinloye

(1450-11 P) **Quantification of Trehalose and Other Sugars in Submergence Resistant Rice** ELIZABETH N MARTINEZ, California State Polytechnic University, Pomona, Rejbana Alam, Julia Bailey-Serres, Endang M Septiningsih, Gregory A Barding

(1450-12 P) **Exploring SFC for the Separation of Peptides and Small Proteins** CECILIA MAZZA, AkzoNobel PPC AB, Joakim Högblom, Peter Gidlund

(1450-13 P) **Automated Solid Phase Extraction Method for the Assessment of Human Exposure to Polycyclic Aromatic Hydrocarbons Using the Biomarker Metabolite 1-Hydroxypyrene in Urine** MICHAEL JOE TANNER, J2 Scientific, Jeff Wiseman

(1450-14 P) **High Speed SDS–PAGE of Proteins** PARUL MODI, Thermo Fisher Scientific, Stephen Roemer

(1450-15 P) **Withdrawn**

(1450-16 P) **Optimized Wide Pore Superficially Porous Particles by One-Step Coating Process for Fast and Efficient Separation of Large Biomolecules** WU CHEN, Agilent Technologies, Anne Mack, Xiaoli Wang

(1450-17 P) **Fast Quantification of Immunoglobulin G Using A New Protein A Analytical HPLC Column** KOSUKE ARAKI, Tosoh, Satoshi Fujii, Shigeru Nakatani, Atis Chakrabarti

(1450-18 P) **Charging YOYO-1 on Capillary Wall for Online DNA Intercalation and Integrating This Approach with Multiplex PCR and Bare Narrow Capillary–Hydrodynamic Chromatography for Online DNA Analysis** HUANG CHEN, University of Oklahoma, Zaifang Zhu, Joann Lu, Shaorong Liu

(1450-19 P) **From Peptide Fractions to Pure, Dry Powders: Development of a Novel Automated Chromatographic Purification Process Supported by Solid-Phase Trapping** YAMAZAKI TOMOYUKI, Shimadzu Corporation, Okoba Tsutomu, Matsuo Eiichi, Masuda Junichi, Iwata Yosuke, Robert E Buco, Nishimura Masayuki

(1450-20 P) **Coupling Ion Exchange Chromatography with Reverse Phase Liquid Chromatography for High-Throughput Analysis of Intact Proteins** ZAIFANG ZHU, University of Oklahoma, Joann Lu, Huang Chen, Shaorong Liu

POSTER SESSION — Session 1460

All posters are to be mounted by 10:00 AM and remain on display until 4:00 PM. Authors must be at their posters from 1:00 PM to 3:00 PM. Location of the afternoon posters is on the Exposition Floor, 400 Aisle. PLEASE NOTE: You cannot get onto the Exposition Floor until after 9:00 AM.

High-Throughput Chemical Analysis

Tuesday Afternoon, Exposition Floor, 400 Aisle

(1460-1 P) **Withdrawn**

(1460-2 P) **Characterization and Use of a Microspectrophotometer for Quantitative Bio-Applications** THOMAS M SPUDICH, Maryville University, Bradley Postier, Ronald Mills

(1460-3 P) **California Chlor-Alkali Production Facility Monitors Organic Carbon for Increased Reliability and Equipment Protection** MARK MULLET, GE Analytical Instruments, Gary Erickson

(1460-4 P) **New Analysis Technology of Ultra-Trace Yellow Components in a Transparent Film** HOKO SUTO, Hitachi Chemical Co., Ltd., Kosuke Iwamoto, Akihiro Unnno

(1460-5 P) **Leaning out Stage 1 Conductivity** JENNY G WATSON, GE Analytical Instruments

(1460-6 P) **Theoretical Simulation of a Helium DC Glow Discharge Used as an Ambient Desorption/Ionization Source for Mass Spectrometry** WADE C ELLIS, Brigham Young University, Paul B Farnsworth, Ross L Spencer

(1460-7 P) **Mathematical Modeling and Computational Simulation of Matrix Effect on Uptake Kinetics in Solid Phase Microextraction** MD NAZMUL AALM, University of Waterloo, Luis Ricardez-Sandoval, Janusz Pawliszyn

(1460-8 P) **Simple Imager for Multi-Well Plates** THAYUMANASAMY SOMASUNDARAM, Florida State University, Michael Zawrotny

POSTER SESSION — Session 1470

All posters are to be mounted by 10:00 AM and remain on display until 4:00 PM. Authors must be at their posters from 1:00 PM to 3:00 PM. Location of the afternoon posters is on the Exposition Floor, 400 Aisle. PLEASE NOTE: You cannot get onto the Exposition Floor until after 9:00 AM.

Surface and Microscopic Characterization of Nanostructures and Biological Materials

Tuesday Afternoon, Exposition Floor, 400 Aisle

(1470-1 P) **Characterizing Nanoparticle Size and Particle-Surface Interactions Using Nanophotonic Force Microscopy** DAKOTA O'DELL, Cornell University, Perry Schein, Summer Saraf, David Erickson

(1470-2 P) **Tuning Localized Surface Plasmon Resonance Wavelengths of Nanoparticles by Mechanical Deformation** FATHIMA S AMEER, Clemson University, Shilpa Varahagiri, Fenglin Wang, Hannah Mack, Marian Kennedy, Jeffrey N Anker

(1470-3 P) **Comparison of Color Pigment Removal between Graphitized Carbon Black and Zirconia-Based Adsorbents for QuEChERS Process** PATRICK MYERS, Supelco/Sigma-Aldrich, Katherine Stenerson, Tyler Young, Jennifer Claus, Michael Ye

(1470-4 P) **Potential-Dependent Adsorption of Water-Soluble Porphyrins at Liquid/Liquid Interfaces Studied by Polarization-Modulation Total Internal Reflection Fluorescence Spectroscopy** SHO YAMAMOTO, Kanazawa University, Hirohisa Nagatani, Hisanori Imura, Kotaro Morita

(1470-5 P) **The Concept of Lipobeads in the Context of Encapsulated Drug Delivery: Technological Challenges vs. Potential Advantages** SERGEY V KAZAKOV, Pace University

(1470-6 P) **Combination of Surface Plasmon Resonance - Surface Enhanced Raman Scattering Spectroscopy in the Kretschmann Configuration** JU-YOUNG KIM, University of Notre Dame, Zachary D Schultz

(1470-7 P) **Preparation and Characterization of Photo-Patterned Amorphous Carbon Films with Thiol-Click Reactions** CATHERINE G MCKENAS, University of North Carolina at Chapel Hill, Matthew R Lockett

(1470-8 P) **Electrophoretic Separation of Carbon Dots** KARINA M TIRADO-GONZÁLEZ, University at Buffalo, SUNY, Zuqin Xue, Luis A Colón

(1470-9 P) **Preparation and Separation of Highly Fluorescent Carbon Dots** ZUQIN XUE, University at Buffalo, SUNY, Luis A Colón, Karina M Tirado-González

(1470-10 P) **Microscopy in Analysis of Erythrocyte Shape Changes** VALIEV HAMMAT, Institute Applied Mechanics RAS, Karnet Yulia, Yumashev Oleg, Snegireva Nataliya N/A

(1470-11 P) **High Temperature In-Situ Reaction Monitoring of CdS Quantum Dots Using Spectrophotometers with Peltier Cell Holders** KYUNBAE LEE, Scinco R&D Center, In-Sung Kang, Kyung-Won Ro

(1470-12 P) **Determination of Airborne Concentration of Single-Wall Carbon Nanotubes and Metals by Wet Electrostatic Precipitation and Inductively Coupled Plasma Mass Spectroscopy** PETER ANDERSEN, Elemental Scientific, John Aumen, Matt Anderson, Grant Josh

(1470-13 P) **Impacts of Mesoporous Silica Shells on Reactivity of Metal-Semiconductor Hybrid Nanocatalysts** FEI ZHAO, Georgia State University, Bin Dong

(1470-14 P) **Investigating Chemical Reactivity of Nanoparticles Using Nano-Impact Electrochemistry** ANAHITA KARIMI, Clarkson University

(1470-15 P) **Advanced Analysis of LIB and Related Materials** KEIJI SUMIYA, Hitachi Chemical Co., Ltd., Hiroki Hirano

(1470-16 P) **Evaluation of Apples Browning Using a Camera-Imaging Visual Analyzer** ANDREW COWELL, Alpha MOS, Jean-Christophe Mifsud, Herve Lechat, Marion Bonnefille

(1470-17 P) **Ligand-Coated Zinc Oxide Nanoparticle Adsorption to Cellulose** STEPHEN R PRINTZ, Western Carolina University, Carmen L Huffman N/A

(1470-18 P) **Gemcitabine Conjugated Cysteine Modified Gold Nanoparticles: Improved Potency and Targeted Nanocarriers for Cancer Treatment** NIKUNJKUMAR NARAYANBHAI VALAND, Gujarat University, Manish B Patel, Kalpesh B Solanki, Shobhana K Menon N/A

(1470-19 P) **Investigation of Nitrogen-Doped Graphene Quantum Dots: Temperature Dependent Nitrogen Incorporations and their Effect on Optical Properties** TIMOTHY PILLAR-LITTLE, University of Kentucky, Doo Young Kim

(1470-20 P) **Industrial Characterization of Nano-scale Roughness on Polished Surfaces** NIKOLAJ A FEIDENHANS'L, Danish Fundamental Metrology, Poul-Erik Hansen, Lukas Pilny, Morten H Madsen, Giuliano Bissacco, Jan C Petersen, Rafael Taboryski

(1470-21 P) **High Throughput Fabrication of Nano and Micro Structured Polymer Foils by Roll-to-Roll-Extrusion Coating** SWATHI MURTHY, Danish Tchnical University/ Inmold, Maria Matschuk, Henrik Pranov, Henrik Pedersen, Guggi Kofod, Rafael Taboryski N/A

(1470-22 P) **Holographic Characterization of Particles in Complex Suspensions** DAVID B RUFFNER, Spheryx, Inc., Jaroslaw M Blusewicz, Priya Kasimbeg, David G Grier, Laura Philips

(1470-23 P) **Withdrawn**

(1470-24 P) Impacts of Mesoporous Silica Shells on Reactivity of Metal-Semiconductor Hybrid **Nanocatalysts** FEI ZHAO, Georgia State University, Bin Dong N/A

(1470-25 P) **Characterization of Carbon Nanomaterial Dispersions for Printed Electronics** QIHUA WU, Brewer Science, Carissa Jones, Kay Mangelson, Christopher Landorf, Stephen Gibbons

(1470-26 P) **Quantum Dots Embedded Silica Nanoparticles Functionalized for Interface Partitioning** SAFYAN AKRAM KHAN, King Fahd University of Petroleum and Minerals, Zain H Yamani, Mohammed H Aljabri N/A

WEDNESDAY, MARCH 9, 2016
MORNING

AWARDS **Session 1480**

Satinder Ahuja Award for Young Investigators in Separation Sciences
arranged by Neil D Danielson, Miami University of Ohio

Wednesday Morning, Room B312
Neil D Danielson, Miami University of Ohio, Presiding

8:30		**Introductory Remarks - Neil D Danielson**
8:35		**Presentation of the 2016 ACS Division of Analytical Chemistry Award for Young Investigators in Separation Sciences to Matthew D Miller, The Dow Chemical Company, by Neil D Danielson, Miami University of Ohio**
8:40	(1480-1)	**Fully Unlocking Polyolefin Chemical Composition Distributions: Break-through Separations Using Graphitic Carbon** MATTHEW D MILLER, The Dow Chemical Company, Rongjuan Cong, Willem deGroot, Chanda Klinker, Dean Lee, John W Lyons, David Mark Meunier, Abhishek Roy, Freddy Van Damme, Bill Winniford, Zhe Zhou
9:15	(1480-2)	**Demystifying Flow Modulated Comprehensive Two Dimensional Gas Chromatography (GC x GC) as a Practical Problem Solving Tool** BILL WINNIFORD, Dow Chemical, James Griffith, Anna Sandlin, Jim Luong, Chris Siegler, Kefu Sun
9:50	(1480-3)	**Separation of Building Blocks from Block Copolymers by High Performance Liquid Chromatography with Preloaded Adsorption Barriers** DAVID MARK MEUNIER, Dow Chemical Company, Yongfu Li, Tirtha Chatterjee, Todd O Pangburn, Eric Pearce, Mark Rickard, John W Lyons
10:25		**Recess**
10:40	(1480-4)	**Macromolecule and Nanoparticle Analyses: Beyond Molecular Weight and Size Measurements** S KIM R WILLIAMS, Colorado School of Mines
11:15	(1480-5)	**Uncertainties in Analyte Measurements in Dried Blood Spots** PURNENDU DASGUPTA, University of Texas at Arlington, Brian Stamos, Jordan Berg

SYMPOSIUM **Session 1490**

ACS-ANYL - Supported Bilayers in Bio/Chemical Analysis
arranged by Craig A Aspinwall, University of Arizona

Wednesday Morning, Room B308
Craig A Aspinwall, University of Arizona, Presiding

8:30		**Introductory Remarks - Craig A Aspinwall**
8:35	(1490-1)	**Spectroscopic Studies of the Formation, Structure, and Applications of Hybrid Supported Phospholipid Bilayers** JOEL M HARRIS, University of Utah, Jay P Kitt
9:10	(1490-2)	**Surface-Sensitive Imaging of Supported Membranes and Single Lipid Vesicles for Medical Applications** FREDRIK HÖÖK, Chalmers University of Technology
9:45	(1490-3)	**Mobile, Oriented Proteinaceous Supported Bilayers Made Directly from Cell Plasma Membranes for Bioanalytical Assays** SUSAN DANIEL, Cornell University
10:20		**Recess**
10:35	(1490-4)	**Stabilized Lipid Bilayers as a Platform for Fabrication of Ion Channel Functionalized Sensors** CRAIG A ASPINWALL, University of Arizona, S Scott Saavedra, Leonard K Bright, Mark T Agasid
11:10	(1490-5)	**Supported Lipid Bilayers and Scanning Ion Conductance Microscopy** LANE A BAKER, Indiana University

TECHNICAL PROGRAM

SYMPOSIUM **Session 1500**

Analytical Applications of Terahertz Time Domain Spectroscopy (THz-TDS)
arranged by Mark A Arnold, University of Iowa

Wednesday Morning, Room B302
Mark A Arnold, University of Iowa, Presiding

8:30		**Introductory Remarks - Mark A Arnold**
8:35	(1500-1)	**Exploration of Interlayer Chemistry in Clay Minerals by Terahertz Spectroscopy** INGRID WILKE, Rensselaer Polytechnic Institute
9:10	(1500-2)	**Analytical Applications of Terahertz Spectroscopy in Nanotechnology and Biotechnology** ANIS RAHMAN, Applied Research & Photonics
9:45	(1500-3)	**PAT Measurements to Enhance Small-Molecule Drug Development and Manufacturing** HUIQUAN WU, Food and Drug Administration
10:20		**Recess**
10:35	(1500-4)	**Understanding Bonding in Cocrystals and Identifying Polymorphs in Small-Molecule Drugs** TIMOTHY MICHAEL KORTER, Syracuse University
11:10	(1500-5)	**Analytical Measurements and Dielectric Properties of Organic Cocrystals by Time-Domain Terahertz Spectroscopy** MARK A ARNOLD, University of Iowa

SYMPOSIUM **Session 1510**

Frontiers of Plasmonics
arranged by Zachary D Schultz, University of Notre Dame and Jean-Francois Masson, Université de Montréal

Wednesday Morning, Room B303
Zachary D Schultz, University of Notre Dame, Presiding

8:30		**Introductory Remarks - Zachary D Schultz and Jean-Francois Masson**
8:35	(1510-1)	**Real-Time Tunable Emission from Plasmonic Nanolasers** TERI W ODOM, Northwestern University
9:10	(1510-2)	**Nanoplasmonic Sensors for Rapid Concentration and Sensitive Detection of Biomolecules** SANG-HYUN OH, University of Minnesota
9:45	(1510-3)	**Plasmonic Rectification** ZACHARY D SCHULTZ, University of Notre Dame
10:20		**Recess**
10:35	(1510-4)	**Trends and Challenges of Nanoplasmonic Biosensors for Clinical Use in Diagnostics** LAURA M LECHUGA, ICN2. CSIC & CIBER-BBN
11:10	(1510-5)	**Plasmonic Nanobiosensors: From Therapeutic Drug Monitoring to the Detection of Molecules Secreted by Cells** JEAN-FRANCOIS MASSON, Universite de Montreal

SYMPOSIUM **Session 1520**

IAEAC: International Association of Environmental Analytical Chemistry - Upconverting Nanocrystals: Near Infrared Excitable Probes for Background-Free Luminescent Sensing
arranged by Antje J Baeumner, University of Regensburg

Wednesday Morning, Room B304
Antje J Baeumner, University of Regensburg, Presiding

8:30		**Introductory Remarks - Antje J Baeumner**
8:35	(1520-1)	**Optical Nanotransformers for In-Situ Upconversion: From Design to Functional Imaging and Sensing** PARAS N PRASAD, SUNY Buffalo, Guanying Chen, Tymish Y Ohulchanskyy
9:10	(1520-2)	**Photodynamic Therapy, Drug Delivery, Persistent and Photo-Stimulated Emission Using Low Excitation Photons** JOHN A CAPOBIANCO, Concordia University
9:45	(1520-3)	**Printing Enhanced Upconverting Nanocrystals on Solid Supports** STANLEY MAY, University of South Dakota, Aravind Baride, Jeevan Meruga, Mary Berry, Jon Kellar, William Cross, Grant Crawford
10:20		**Recess**
10:35	(1520-4)	**Surface Modifications of Upconverting Nanoparticles and Their (Bio)analytical Applications** THOMAS HIRSCH, University of Regensburg
11:10	(1520-5)	**Upconversion Nanoparticles as Active Elements in Optical Sensing for Development of Protein and Oligonucleotide Bioassays** ULRICH J KRULL, University of Toronto Mississauga, Samer Doughan, Anna Shahmuradyan, Feng Zhou, Yi Han

SYMPOSIUM **Session 1530**

Ion Mobility/Mass Spectrometry for Metabolomics and Clinical Analysis
arranged by Richard Alan Yost, University of Florida

Wednesday Morning, Room B305
Richard Alan Yost, University of Florida, Presiding

8:30		**Introductory Remarks - Richard Alan Yost**
8:35	(1530-1)	**Structural Mass Spectrometry for Systems, Synthetic, and Chemical Biology** JOHN A MCLEAN, Vanderbilt University
9:10	(1530-2)	**Enhancing Ion Mobility-Mass Spectrometry Metabolomic Analyses with High Throughput Front End Separations** ERIN S BAKER, Pacific Northwest National Laboratory, Xing Zhang, Kristin E Burnum-Johnson, Jennifer E Kyle, Cameron P Casy, Young-Mo Kim, Erika M Zink, Dennis Mehinagic, Matthew E Monroe, Yehia M Ibrahim, Daniel Orton, Justin G Teeguarden, Thomas O Metz, Richard D Smith
9:45	(1530-3)	**Ion Mobility and Metabolomics, Two New Tools for Current Drug Discovery and Drug Development** ROB J VREEKEN, Janssen Pharmaceutica
10:20		**Recess**
10:35	(1530-4)	**Cystic Fibrosis Breathomics by Transmission-Mode Direct Analysis in Real Time-Traveling Wave Ion Mobility-Mass Spectrometry** FACUNDO M FERNANDEZ Georgia Institute of Technology, Facundo M Fernandez, Christina M Jones, Maria E Monge, Nael A McCarty, Arlene Stecenko, Jose J Perez
11:10	(1530-5)	**Ion Mobility/Mass Spectrometry for Metabolomics and Clinical Analysis: Progress and Prospects** RICHARD ALAN YOST, University of Florida, Christopher D Chouinard, Christopher R Beekman, Michael T Costanzo, Jared J Boock, Robin Kemperman, Wei S Michael, Timothy J Garrett, Christopher A Beecher

SYMPOSIUM **Session 1540**

Precision Bioanalytical Measurements
arranged by Steven A Soper, St. Louis University and Susan M Lunte, University of Kansas

Wednesday Morning, Room B309
Steven A Soper, St. Louis University, Presiding

8:30		**Introductory Remarks - Steven A Soper and Susan M Lunte**
8:35	(1540-1)	**Microdialysis Sampling and Separations: A Tribute** SUSAN M LUNTE, University of Kansas
9:10	(1540-2)	**Expanding the Application Space of In Vivo Microdialysis Sampling in the Areas of Drug Metabolism, Free Radical Chemistry, Neurochemistry, and Tissue Engineering** JULIE STENKEN, University of Arkansas
9:45	(1540-3)	**The Development of New Tools Based on Whispering Gallery Mode Sensing** ROBERT C DUNN, University of Kansas, Sarah Wildgen, Daniel Kim, Judith Flores
10:20		**Recess**
10:35	(1540-4)	**Utilizing Oxidative DNA Damage to Explore the Mode of Action of Oxidative Events and Antioxidative Responses** BLÁNAID WHITE, Dublin City University, Dermot Walls, Sarah Joanne Lynch, Karina Horgan, Sinead Loughran, Roya Hakimjavadi
11:10	(1540-5)	**Precision Medicine: Enabled by Single Cell Analysis** STEVEN A SOPER, University of North Carolina

Author and presider lists as of January 20, 2016, are available at www.pittcon.org

SYMPOSIUM — **Session 1550**

Standoff Detection Methods for Security Applications

arranged by Sanford A Asher, University of Pittsburgh
and Michael Shepard, Dept. Homeland Security, S&T Directorate

Wednesday Morning, Room B310
Michael Shepard, Dept. Homeland Security, S&T Directorate, Presiding

8:30		**Introductory Remarks - Michael Shepard**
8:35	(1550-1)	**Novel Approaches to Standoff Hyperspectral Imaging Based Detection of Explosives and Other Threats** CHARLES W GARDNER, ChemImage Sensor Systems, Matthew P Nelson, Nathaniel R Gomer, Oksana P Klueva
9:10	(1550-2)	**Standoff Detection of Explosives Residues Using Integrated Quantum Cascade Laser Arrays** MARK F WITINSKI, Pendar Technologies, Romain Blanchard, Daryoosh Vakhshoori
9:45	(1550-3)	**Standards for Optical Based Standoff Detection Fabricated Using Inkjet Printing** GREG GILLEN, NIST
10:20		**Recess**
10:35	(1550-4)	**Recent Advances in Standoff Chemical Threat Detection Using Deep-Ultraviolet Raman Spectroscopy** ADAM J HOPKINS, Alakai Defense Systems, Kenneth R Pohl, Rob Waterbury, Edwin Dottery
11:10	(1550-5)	**Deep UV Standoff Raman Detection of Explosives: Fundamentals and Methodologies** SERGEI V BYKOV, University of Pittsburgh, Katie L Gares, Kyle T Hufziger, Sanford A Asher

SYMPOSIUM — **Session 1560**

Sum Frequency Generation (SFG) Vibrational Spectroscopic Studies on Proteins and Peptides at Interfaces

arranged by Zhan Chen, University of Michigan

Wednesday Morning, Room B311
Zhan Chen, University of Michigan, Presiding

8:30		**Introductory Remarks - Zhan Chen**
8:35	(1560-1)	**Determining the Structure of Surface Bound Proteins** DAVID G CASTNER, University of Washington
9:10	(1560-2)	**The Interactions of Ions with Peptides and Lipid Bilayers** PAUL CREMER, Penn State University
9:45	(1560-3)	**Proteins at Interfaces: Structure and Platform for Novel Biomaterials** MISCHA BONN, Max Planck Institute for Polymer Research, Tobias Weidner
10:20		**Recess**
10:35	(1560-4)	**Proteins at Interfaces Probed by Chiral Sum Frequency Generation** ELSA CY YAN, Yale University
11:10	(1560-5)	**Structure-Function Relationships of Surface Immobilized Peptides and Enzymes** ZHAN CHEN, University of Michigan

WORKSHOPS — **Session 1570**

Analytical Information Markup Language (AnIML) Data Standards

arranged by Burkhard Schaefer, BSSN Software GmbH

Wednesday Morning, Room B313
Burkhard Schaefer, BSSN Software GmbH, Presiding

8:30		**Introductory Remarks - Burkhard Schaefer**
8:35	(1570-1)	**Generating AnIML Technique Definitions** MARK FAULKNER BEAN, GSK
9:05	(1570-2)	**Filling the Automation and Enterprise Gap with Data and Device Standardization** CARMEN CONDRAU, SiLA Consortium, Devon Johnston
9:35	(1570-3)	**Instant JChem and AnIML: Curation, Searching, and Visualization of MS Library and Empirical Data** MICHAEL BRADEN, ChemAxon N/A
10:05		**Recess**
10:20	(1570-4)	**Use of AnIML and Other Methods for Software Visualization and Automation** DAVID M COX, SCIEX, John Gibbons
10:50	(1570-5)	**Cloud-Based Analytical Data Management Using the AnIML Standard** BURKHARD SCHAEFER, BSSN Software GmbH

ORGANIZED CONTRIBUTED SESSIONS — **Session 1580**

Food Contaminant Methods

arranged by Wendy Young, Food and Drug Administration

Wednesday Morning, Room B315
Wendy Young, Food and Drug Administration, Presiding

8:30	(1580-1)	**Quantitative HPLC-MS/MS Analysis of Metabolites of Hypoglycin A and Methylenecyclopropylglycine (MCPG) in Human Urine** SAMANTHA L ISENBERG, Centers for Disease Control and Prevention, Melissa D Carter, Leigh A Graham, Thomas P Mathews, Darryl Johnson, Jerry D Thomas, James Pirkle, Rudolph C Johnson
8:50	(1580-2)	**LC-MS/MS Detection of Tetrodotoxin in Fresh/Frozen and Salt-Dried Fish Products** SARA C MCGRATH, FDA/CFSAN, Jonathan R Deeds
9:10	(1580-3)	**Fast Liquid Chromatography - Tandem Mass Spectrometry Analysis of >150 Drug Residues including Aminoglycosides in Food Animal Tissues** STEVEN J LEHOTAY, USDA Agricultural Research Service, Alan R Lightfield
9:30	(1580-4)	**Determination of Diglycolic Acid in Foods Containing Carboxymethyl Cellulose** WENDY YOUNG, Food and Drug Administration, Lowri DeJager
9:50		**Recess**
10:05	(1580-5)	**Sulfite Determination in Food by Liquid Chromatography-Mass Spectrometry** KATHERINE S ROBBINS, US FDA, Shaun MacMahon, Lowri de Jager
10:25	(1580-6)	**Pesticide Screen Method Development Using Micro Flow LC and High Resolving Power Mass Spectrometry** MARK CROSSWHITE, Florida Department of Agriculture, Walter Hammack, Gerard Ghislain, Fadi Aldeek, Matthew Standland, Daniel Canzani N/A
10:45	(1580-7)	**Application of Raman Imaging for the Analysis of Food Packaging Stability** RIC GONZALEZ, ConAgra Foods, Inc., Indarpal Singh
11:05	(1580-8)	**Open Discussion**

ORGANIZED CONTRIBUTED SESSIONS — **Session 1590**

On-Site Detection of THC and Related Drugs

arranged by Maggie Tam, Canada Border Services Agency and Charles S Harden, US Army ECBC

Wednesday Morning, Room B316
Maggie Tam, Canada Border Services Agency, Presiding

8:30	(1590-1)	**Challenges of Drug Detection in the Field from the Enforcement Perspective** MAGGIE TAM, Canada Border Services Agency
8:50	(1590-2)	**Forensic Screening Using High-Performance Ion Mobility Spectrometry** CHING WU, Excellims Corporation, Mark Osgood, Anthony Midey
9:10	(1590-3)	**Handheld Differential Mobility Spectrometry for Drug Interdiction** PAUL J RAUCH, Chemring Detection Systems, Eric Lynch
9:30	(1590-4)	**Detection of Tetrahydrocannabinol and Related Compounds in Human Breath Using High-Field Asymmetric Waveform Ion Mobility Spectrometry** JARED J BOOCK, University of Florida, Michael T Costanzo, Richard Alan Yost
9:50		**Recess**
10:05	(1590-5)	**Detection of Illicit Drugs of Abuse Using Existing Military Chemical Detection Equipment** CHARLES S HARDEN, US Army ECBC
10:25	(1590-6)	**Innovative and Rapid Detection of Marihuana Consumption from Direct Breath Analysis** CHANDRASEKHARA HARIHARAN, ION-GAS GmbH, Oliver Kayser
10:45	(1590-7)	**Understanding Canine Detection of Explosives and Narcotics Using a 3D Printed Artificial Dog Nose** MATTHEW STAYMATES, NIST, William MacCrehan, Jessica Staymates, Greg Gillen, Brent Craven
11:05	(1590-8)	**Open Discussion**

TECHNICAL PROGRAM

ORGANIZED CONTRIBUTED SESSIONS — Session 1600

PAI-NET - Characterization of Micro/Nano Liquid Phases
arranged by Akihide Hibara, Tokyo Institute of Technology and Kenji Kojima, PAI-NET

Wednesday Morning, Room B301
Akihide Hibara, Tokyo Institute of Technology, Presiding

8:30	(1600-1)	**Development of Microfluidic Lattices for High-Performance Cell/Particle Separations** MASUMI YAMADA, Chiba University, Minoru Seki
8:50	(1600-2)	**Microfluidic-Based Approach for Producing Diffraction Quality Protein Crystals** MAEKI MASATOSHI, Hokkaido University, Pawate Ashtamurthy, Sugishima Masakazu, Watanabe Keiichi, Tokeshi Manabu, Kenis J Paul, Miyazaki Masaya
9:10	(1600-3)	**Sample Pretreatments for Droplet-Based Microanalytical System by Using Nanodroplet Formation** MAO FUKUYAMA, Kyoto Institute of Technology, Akihide Hibara, Yumi Yoshida, Kohji Maeda
9:30	(1600-4)	**Development of DNA and/or RNA Extraction Method from Single Cell** YUKIHIRO OKAMOTO, Osaka University
9:50		**Recess**
10:05	(1600-5)	**In Vivo Detection and Quantification of Circulating Cells and Nanoparticles in Whole Blood Using Photoacoustic-Fluorescence Flow Cytometry** DMITRY A NEDOSEKIN, University of Arkansas for Medical Sciences, Vladimir P Zharov
10:25	(1600-6)	**Microfluidic Construction of Artificial Cell-Like Reactors** MASAHIRO TAKINOUE, Tokyo Technology
10:45	(1600-7)	**Parallel Lipid Bilayer Formation by Microfabrication and Its Applications** RYUJI KAWANO, Tokyo University of Agriculture and Technology
11:05	(1600-8)	**Unique Liquid Properties by Surface Silanol Groups in Extended-Nano Spaces** YUTAKA KAZOE, The University of Tokyo, Keisuke Ikeda, Kazuma Mawatari, Takehiko Kitamori

ORAL SESSIONS — Session 1610

Biomedical: New Technologies for Breath Analysis (Half Session)

Wednesday Morning, Room B409

8:30	(1610-1)	**Heart-Cutting Multidimensional Micro Gas Chromatography (μGC) System for Breath Analysis** MENGLIAN ZHOU , University of Michigan, Jiwon Lee, Hongbo Zhu, Robert Nidetz, Kevin Ward, Xudong Fan, Katsuo Kurabayashi
8:50	(1610-2)	**Real-time PTR-TOF-MS Measurements Reveal Effects of Respiratory Maneuvers onto Exhaled Breath Biomarker Profiles** PRITAM SUKUL, University Medicine of Rostock, Peter Oertel, Phillip Trefz, Khushman Taunk, Jochen K Schubert, Wolfram Miekisch, Svend Kamysek
9:10	(1610-3)	**High Altitude Respiratory Research Using Quadrupole Mass Spectrometry** CHARLES W DE CARLO, Extrel CMS, James R Brenner, Zbigniew Krieger, Frank DeThomas
9:30	(1610-4)	**VOC Detection in Animal Models for Medical Research** WOLFGANG VAUTZ, ISAS, Liedtke Sascha, Nils Kunze, Thorsten Perl, Ursula Telgheder

ORAL SESSIONS — Session 1620

Clinical/Toxicology

Wednesday Morning, Room B402
Tyler Davis, West Virginia University, Presiding

8:30	(1620-1)	**Determination of Manganese Using Cathodic Stripping Voltammetry on a Platinum Electrode** WENJING KANG, University of Cincinnati, Cory A Rusinek, Adam Bange, Erin Haynes, William R Heineman, Ian Papautsky
8:50	(1620-2)	**Impact of Amphiphilic Poly(amido)amine Dendrimers on the Biophysical and Biorecognition Properties of Bilayer Membranes** SAMUEL S HINMAN, University of California, Riverside, Charles J Ruiz, Ling Peng, Quan Cheng
9:10	(1620-3)	**Analysis of Drugs in Saliva During Treatment of Post-Traumatic Stress Disorder Patients** KATHRYN DANA, Real-Time Analyzers, Inc, Chetan Shende, Stuart Farquharson, Albert J Arias
9:30	(1620-4)	**G-Quadruplex: A Biocompatible Additive for Enhancing the Antibacterial Activity of H_2O_2** YUQIAN XING, University of North Dakota, Xiao Liu, Minh H Duong, Julia Xiaojun Zhao
9:50		**Recess**
10:05	(1620-5)	**Capillary Electrophoresis Investigation of Multiwalled Carbon Nanotube-Biomolecule Binding** TYLER DAVIS, West Virginia University, Lisa A Holland, Julia Ann Mouch
10:25	(1620-6)	**Determining Variability in Potency of Marijuana for Potential Medical Application** REBECCA PLESSEL, Pennsylvania State University, Frank Dorman, Amanda Rigdon
10:45	(1620-7)	**Design and Development of a Portable Aptasensor for Toxicity Monitoring of Field Samples** GONCA BULBUL, Clarkson University, Akhtar Hayat, Silvana Andreescu
11:05	(1620-8)	**Quantitative Assessment of Nanoparticles Exposure Potential Effects in Embryonic Zebrafish** XIAOBO LIU, Clarkson University, Eduard Dumitrescu, Kenneth Wallace, Silvana Andreescu

ORAL SESSIONS — Session 1630

Electrochemistry - Biological Applications

Wednesday Morning, Room B403
Yinfa Ma, Missouri University of Science and Technology, Presiding

8:30	(1630-1)	**Development of an Optrode for In Vivo Neurochemical Studies** THOMAS FIELD, University of Kansas, Meng Sun, Peter M Ruggles, Michael A Johnson
8:50	(1630-2)	**Adenosine Transiently Modulates Vasodilation in Caudate-Putamen** YING WANG, University of Virginia, B Jill Venton
9:10	(1630-3)	**Amperometric and Voltammetric Measurements in the Cell Cytosol Using Conical Carbon Nanoelectrodes** EDWIN MITCHELL, North Carolina State University, James Roberts, Leslie A Sombers
9:30	(1630-4)	**Tunable Electroosmotic Push-Pull Perfusion Shows Higher Aminopeptidase N Activity in CA1 than CA3 of the Rat Hippocampus** YANGGUANG OU, University of Pittsburgh, Bocheng Yin, Jenna DeVivo, German Barrionuevo, Stephen Weber
9:50		**Recess**
10:05	(1630-5)	**Optogenetic Control of Octopamine Release in Drosophila Melanogaster Larval Ventral Nerve Cord and Detection with Fast Scan Cyclic Voltammetry (FSCV)** POOJAN PYAKUREL, University of Virginia, B Jill Venton
10:25	(1630-6)	**Clarifying the Complex Chemical Mechanisms that Underlie the Voltammetric Detection of Hydrogen Peroxide** LESLIE A SOMBERS, North Carolina State University, Samantha K Smith, Leyda Z Lugo-Morales, James Roberts, Maxim A Voinov, Tatyana I Smirnova
10:45	(1630-7)	**FSCV Measurements of Neurotransmitters in Daphnia** MATT N JACKSON, Wayne State University, Srimal A Samaranayake, David Pitts, Shawn McElmurry, Annette R Tremonti, Parastoo Hashemi
11:05	(1630-8)	**Glutamate Receptor Influence on Localized Oxygen Metabolism** LINDSAY WALTON, University of North Carolina at Chapel Hill, Nick Boustead, Susan Carroll, R Mark Wightman

ORAL SESSIONS **Session 1640**

Environmental, Pharm and Nano Methods Development in Atomic Spectroscopy

Wednesday Morning, Room B404
David Clarke, Teledyne CETAC Technologies, Presiding

8:30	(1640-1)	**Novel Automation Streamlining Microwave Digestion to Detection for Elemental Analysis** DAVID CLARKE, Teledyne CETAC Technologies, Matthew Nigro, James Block
8:50	(1640-2)	**Use of Atomic Fluorescence Spectrometry (AFS) for Element Specific Hg Detection Combined with Combustion** WARREN T CORNS, P S Analytical
9:10	(1640-3)	**Development of Online Method for Simultaneous Preconcentration of Cd, Cu, Ni and Zn in Environmental Samples Using Modified Alumina** ZAHEER A KHAN, SBBU, SBA
9:30	(1640-4)	**Dual Analyte Analysis of Bimetallic and Monometallic Nanoparticle Mixtures Using Field Flow Fractionation Separation Coupled to spICP-MS** CHADY STEPHAN, PerkinElmer, Ruth Merrifield
9:50		**Recess**
10:05	(1640-5)	**Analysis of Trace Metals in Pharmaceutical Formulations - Issues with USP 232 and Q3D Regulations** PHILIP RIBY, Liverpool John Moores University, Samar Thiab, Philip Salmon, Matt Roberts, Philip Riby N/A
10:25	(1640-6)	**Development and Validation of a New Method to Measure Activity of the Na+, K+ ATPase Using ICP-MS QQQ** CORY A STINER, University of Cincinnati, Julio Landero, Judith Heiny
10:45	(1640-7)	**Arsenic Speciation in Water Samples – Development of a New ISO/CEN Method** CORNELIUS BROMBACH, P S Analytical, Bin Chen, Warren T Corns, K C Thompson
11:05	(1640-8)	**Exploiting the Limits of Single Particle ICP-MS – From Particle Size to Particle Number** CHADY STEPHAN, PerkinElmer

ORAL SESSIONS **Session 1650**

Food Contaminants

Wednesday Morning, Room B314
Joan M Stevens, Agilent Technologies, Inc., Presiding

8:30	(1650-1)	**The Analysis of Chlorinated Dioxins and Difurans in Pet Food** PHILIP BASSIGNANI, Fluid Management Systems, Rudolf Addink
8:50	(1650-2)	**Selective Lipid Removal from Complex Samples for Multi-Residue Analysis** DERICK LUCAS, Agilent Technologies, Limian Zhao, Bruce Richter, Joan Stevens, Megan Juck
9:10	(1650-3)	**Arsenic Contamination and the Emergence of Speciation in the Food Chain** HELMUT ERNSTBERGER, Perkin Elmer, Kenneth Neubauer
9:30	(1650-4)	**Elemental Profile of Tobacco used in Counterfeit Cigarettes** YI HE, John Jay College/CUNY, Carrie Green, Rufus Chaney, Fidelis Tan, Ye Hua, Victoria Mei, Marin Kurti, Klaus von Lampe
9:50		**Recess**
10:05	(1650-5)	**Determination of Antimony in Food Samples by Slurry Sampling Hydride Generation Atomic Absorption Spectrometry** JERZY MIERZWA, Tennessee State University
10:25	(1650-6)	**The Analysis of Chlorinated Dioxins, Difurans and Polychlorinated Biphenyls in Edible Oils** PHILIP BASSIGNANI, Fluid Management Systems, Rudolf Addink
10:45	(1650-7)	**Application of Mid-Infrared Portable Spectrometry in Determination of Trans-fatty Acid Content in Bakery Products** MEI SHOTTS, The Ohio State University, Luis Rodriguez-Saona
11:05	(1650-8)	**Rapid Detection of Processed Uranium in Food** ABDUR-RAFAY SHAREEF, FDA, Lin Zhichao, Emanuele Kathryn, Stephanie Healey, Patrick Regan, William Cunningham, Brian Baker

ORAL SESSIONS **Session 1660**

LIMS – No One Size Fits All

Wednesday Morning, Room B405
Hubert MacDonald, The Pittsburgh Conference, Presiding

8:30	(1660-1)	**Quantum Time Savings with LIMS Deployment** DEVENDER GANDHI, Accelerated Technology Laboratories
8:50	(1660-2)	**Informatics for Externalization** GRAHAM A MCGIBBON, ACD/Labs, Pranas Japertas, Ryan Sasaki
9:10	(1660-3)	**Going Mobile with LIMS for Field Data Collection** KEN OCHI, Accelerated Technology Laboratories
9:30	(1660-4)	**Pay Now or Later: Creating Solid System Application User Requirements** KATHERINE TEMPLE, CSOLS, Inc., Daniel Freel
9:50		**Recess**
10:05	(1660-5)	**Key Factors to Consider in Transitioning to a New LIMS** SONJA STUTSMAN, Accelerated Technology Laboratories
10:25	(1660-6)	**LIMS Implementation: "Big Bang" or "Phased" Approach** HOWARD J ROSENBERG, CSols, Inc.
10:45	(1660-7)	**Migrating from a Legacy LIMS to a Modern Platform** LAURA LEE WILLIFORD, Accelerated Technology Laboratories, Devender Gandhi
11:05	(1660-8)	**Integrated Informatics: Single Vendor vs. "Best of Breed"** HOWARD J ROSENBERG, CSols, Inc.

ORAL SESSIONS **Session 1670**

Mass Spectrometry-Environmental, ICP-MS and Others

Wednesday Morning, Room B406
Yongbo Dan, Missouri University of Science and Technology, Presiding

8:30	(1670-1)	**Ultra-Trace Analysis of Mercury Species in Drinking Water Using Ion Chromatography and Speciated Isotope Dilution Mass Spectrometry (IC-SIDMS)** PATRICK BENECEWICZ, Duquesne University, Skip Kingston, Matt Pamuku, Stuart Procter, Christopher Loran
8:50	(1670-2)	**Ion Mobility-Mass Spectrometry Screening Reveals Small Molecules Capable of Chemical and Structural Modulation of Amyloidogenic Protein** RICHARD A KERR, University of Michigan, Jeffrey Derrick, Michael Beck, Hyuck Jin Lee, Mi Hee Lim, Brandon T Ruotolo
9:10	(1670-3)	**Transformation Kinetics of Metallic Nanoparticles in Environmental and Cell Culture Exposure Media as Measured by spICP-MS** CHADY STEPHAN, PerkinElmer, Ruth Merrifield
9:30	(1670-4)	**Reading Information Digitally-Encoded in Synthetic Polymers: A Sequencing Approach by Tandem Mass Spectrometry** LAURENCE CHARLES, Aix-Marseille University, Jean-François Lutz
9:50		**Recess**
10:05	(1670-5)	**Enzymatic Digestion-Single Particle ICP-MS Method to Characterize Nanoparticle Uptake by Plants** DAN YONGBO, Missouri University of Science and Technology, Weilan Zhang, Runmiao Xue, Ma Xingmao, Chady Stephan, Honglan Shi
10:25	(1670-6)	**Study of the Transmission of Megadalton-Sized Ions from Atmospheric Pressure to Vacuum in a Q-TOF Charge Detection Mass Spectrometer** STACI N ANTHONY, Indiana University, Benjamin E Draper, Martin F Jarrold
10:45	(1670-7)	**Characterization of TiO_2-Nanoparticles Using Asymmetrical Flow- and Centrifugal Field-Flow-Fractionation Coupled with MALS, DLS and ICP-MS Detection** SOHEYL TADJIKI, Postnova Analytics Inc., Florian Meier, Evelin Moldenhauer, Trevor Havard, Thorsten Klein
11:05	(1670-8)	**Enhanced Real-Time Gas Analysis with SIFT-MS Using Negative Reagent Ions** DANIEL MILLIGAN, Syft Technologies Ltd, Vaughan Langford, Thomas McKellar, Barry Prince, Murray McEwan

ORAL SESSIONS — Session 1680

Pharmaceutical Applications of Liquid Chromatography

Wednesday Morning, Room B401
Paulina Piotrowski, The Pennsylvania State University, Presiding

8:30	(1680-1)	**Method Development Strategies and Applications of 2D LC for Pharmaceutical Analysis** PANKAJ AGGARWAL, Pfizer Inc., David T Fortin, Angel R Diaz N/A
8:50	(1680-2)	**Coupling Mass Detection with UV to Improve Method Sensitivity for Esters of Benzenesulfonic and p-Toluenesulfonic Acids in Analysis of Genotoxic Impurities** MARGARET MAZIARZ, Waters Corporation, Christopher Henry, Mark Wrona
9:10	(1680-3)	**Method Development for Non-Chromophoric Pharmaceutical Analytes Using Alternative Chromatographic and Detection Techniques** ZONGYUN HUANG, Bristol-Myers Squibb, William Fish
9:30	(1680-4)	**The Impact of LC Instrument Characteristics on HPLC and UPLC Method Migration and Method Transfer** PAULA HONG, Waters Corporation, Patricia R McConville
9:50		**Recess**
10:05	(1680-5)	**Development and Validation of A Novel Stability-Indicating Reversed-Phase High-Performance Liquid Chromatography Method for Assay of Thiabendazole and Estimation of its Related Compounds** JIANGTAO HE, Merial, A Sanofi Company, Huang Junmin, Abu Rustum
10:25	(1680-6)	**Fast Centrifugal Partitioning Chromatography** ROB DRISCOLL, Robatel Inc.
10:45	(1680-7)	**Development and Validation of a Reversed Phase Chiral HPLC Method for Verification of Afoxolaner as a Racemic Mixture** NILUSHA PADIVITAGE, Merial, A Sanofi Company, Satish Kumar, Abu Rustum
11:05	(1680-8)	**Development for a Sensitive Method for the Determination of Diphenylphosphoryl Azide and Hydrogen Azide in Active Pharmaceutical Compounds** XUEJUN XU, Bristol-Myers Squibb, Martin Nunez, Yun K Ye, Thomas V Raglione

ORAL SESSIONS — Session 1690

Portable Instruments - Half Session

Wednesday Morning, Room B409
Robert W Baudoux, Sr., RWB Convention Mgt, Presiding

10:05	(1690-1)	**Development of Portable Instrumentation Using the Arduino Microcontroller Platform for Field-Ready Electrochemical Experimentation** DREW C FARRELL, University of Arizona, Michael L Heien
10:25	(1690-2)	**A Novel Instrument for Microscale IR Thermography in High Temperature Applied to Solar Salts** JUNKO MORIKAWA, Tokyo Institute of Technology, Massimiliano Zamengo, Yukitaka Kato
10:45	(1690-3)	**A New GC/FTIR Detection Method as Applied to Inline Monitoring of Siloxanes in Biogas** CHARLES MARK PHILLIPS, Prism Analytical Technologies, Inc., Martin L Spartz, Anthony S Bonanno, Peter P Behnke
11:05	(1690-4)	**Next-Generation Handheld XRF Analyzer – Smarter, Smaller and Faster** ESA NUMMI, Thermo Fisher Scientific

ORAL SESSIONS — Session 1700

Sensors - Bioanalytical and Homeland Security/Forensics

Wednesday Morning, Room B407
Katie Edwards, Cornell University, Presiding

8:30	(1700-1)	**Photonic Crystal Protein Hydrogel Sensor for Candida Albicans** ZHONGYU CAI, University of Pittsburgh, Daniel H Kwak, David Punihaole, Zhenmin Hong, Sachin S Velankar, Xinyu Liu, Sanford A Asher
8:50	(1700-2)	**Development of a Wireless Microfluidic Biosensor System for Real-Time Monitoring of Human Transplant Organs in Transit** SALLY A GOWERS, Imperial College London, Isabelle C Samper, Chu Wang, Thomas Watts, Bynvant Sandhu, Vassilios Papalois, Martyn G Boutelle
9:10	(1700-3)	**Ion-Selective Electrodes with PEDOT(PSS) as Solid Contact: Influence of the PEDOT(PSS) Thickness on the Equilibration Time** MARCIN GUZINSKI, The University of Memphis, Jennifer M Jarvis, Erno Lindner, Bradford D Pendley
9:30	(1700-4)	**Evaluating Real Time Binding Interactions in Insulin Immunoassay for Diagnosis of Type of Diabetes by Surface Plasmon Resonance** VINI SINGH, Oklahoma State University, Sadagopan Krishnan
9:50		**Recess**
10:05	(1700-5)	**Extended Nanopore Residence Times via Metallic Clusters** JOSEPH E REINER, Virginia Commonwealth University, Amy E Chavis, Kyle T Brady, Nuwan Kothalawala
10:25	(1700-6)	**Kinetics Quantification of MicroRNAs as Disease Biomarkers on Microelectrode Point-of-Care Sensors at Attomoles within Minutes** TANYU WANG, Georgia State University, Gangli Wang
10:45	(1700-7)	**Hyphenated Inverted Mesa E-QCM Sensors for Explosives Detection** ABDUL REHMAN, King Fahd University of Petroleum and Minerals N/A
11:05	(1700-8)	**A New Miniaturized Sensor for Real-Time Suit Penetration Assessment in the Man-In-Simulant-Test (MIST) Protocol** NICHOLAS FITZGERALD, Defence Science and Technology Group, Karl Pavey

ORAL SESSIONS — Session 1710

Thermal Analysis

Wednesday Morning, Room B408
Fu-mei Lin, The Pittsburgh Conference, Presiding

8:30	(1710-1)	**Using Isothermal Titration Calorimetry to Measure Thermodynamic Parameters of Adsorption on Chromatographic Media** ANTHONY R HORNER, University of Pittsburgh, Stephen R Groskreutz, Stephen Weber
8:50	(1710-2)	**Thermal Conductivity Measurement of Solar Salt in High Temperature Using the Temperature Wave Method** JUNKO MORIKAWA, Tokyo Institute of Technology, Yukitaka Kato, Massimiliano Zamengo
9:10	(1710-3)	**TGA-GC-MS Analysis of Different Tobacco Types** EKKEHARD POST, NETZSCH Geraetebau GmbH, Bob Fidler, Jan Hanss
9:30	(1710-4)	**Pyrolysis Gases of Polycarbonate Identified by TGA-FT-IR and TGA-GC-MS** EKKEHARD POST, NETZSCH Geraetebau GmbH, Bob Fidler
9:50		**Recess**
10:05	(1710-5)	**Understanding Auto-Catalysis Using Scanning, Isothermal and Adiabatic Calorimetry** PETER RALBOVSKY, Netzsch Instruments, Bob Fidler
10:25	(1710-6)	**Developing Databases and Optimized Spectral Searching from TGA-IR Hyphenation Experiments** IAN ROBERTSON, PerkinElmer Limited, Justin Lang, Jack Botting
10:45	(1710-7)	**Thermal Excitation, Optical Response: A Novel Approach to Thermal Analysis by TMOR** SARAH SCHWARZ G HENRIQUES, Anton Paar OptoTec GmbH, Tobias Husemann, Jens Kruse, Nils Bertram
11:05	(1710-8)	**Characterization of Nanomaterials with Thermal Analysis and Hyphenated Techniques** CHADY STEPHAN, PerkinElmer, Jun Wang

POSTER SESSION **Session 1720**

All posters are to be mounted by 10:00 AM and remain on display until 4:00 PM. Authors must be at their posters from 10:00 AM to 12:00 PM. Location of the morning posters is on the Exposition Floor, 400 Aisle. PLEASE NOTE: You cannot get onto the Exposition Floor until after 9:00 AM.

Applications of LC/MS

Wednesday Morning, Exposition Floor, 400 Aisle

(1720-1 P) **Determination of Fentanyl in Canine Plasma Using HPLC-MS Detection** JOAN B BAILEY, University of Tennessee, Sherry K Cox, Molly White, Kristen Gordon, Reza Seddighi

(1720-2 P) **Targeted Metabonomic Study of Plasma from Rats with Acute Colitis Using LCMS-IT-TOF Based Metabonomics** LINGLING SHEN, Shimadzu Global COE, Shimadzu Co., Ltd, Jingting Yao, Taohong Huang, Xiaojun Zhang, Qisheng Zhong

(1720-3 P) **Optimization of On-column Trypsin Digestion Coupled with LC-MS/MS for Analysis of Apolipoproteins in Serum** CHRISTOPHER TOTH, Centers for Disease Control and Prevention, Zsuzsanna Kuklenyik, Jeffrey Jones, Bryan A Parks, Michael S Gardner, Jon Rees, Yulanda Williamson, David Schieltz, Lisa McWilliams, James Pirkle, John R Barr

(1720-4 P) **A Simple and Sensitive LC-MS/MS Method for the Determination of Free 8-Hydroxy-2'-deoxyguanosine in Human Urine** ZUWEI WANG, JES Tech, Scott M Smith

(1720-5 P) **Evaluation of Streamlined SPE Processing Using Novel Column based Components prior to LC-MS/MS** LEE WILLIAMS, Biotage GB Limited, Helen Lodder, Victor Vandell

(1720-6 P) **Multi-Class Screening of Drug Abuse in Hair by Matrix Solid Phase Dispersion – Ultrasound Extraction and HPLC-MS/MS Detection** ANTONIO MOREDA-PIÑEIRO, University of Santiago de Compostela, Mercedes Saavedra-Suarez, Juan Sanchez-Gonzalez, Pilar Bermejo-Barrera, Maria del Carmen Barciela-Alonso, Elena Pena-Vazquez, Raquel Dominguez-Gonzalez

(1720-7 P) **Optimization of QuEChERS Sample Preparation Method for the Determination of Bisphenol A in Carrots** OLUJIDE T AKINBO, Butler University, Hugh Kestufskie

(1720-8 P) **Quantification of Iodine-Containing Hormones Present in Dietary Supplements by Liquid Chromatography Tandem Mass Spectrometry (LC-MS/MS)** ENRIQUE G YANES, U.S. Food and Drug Administration (FDA), Robert A Wilson, James A Turner, Ryan Saadawi

(1720-9 P) **Ultra High Performance Liquid Chromatography Tandem Mass Spectrometry Method for Simultaneous Determination of Multiple Bioactive Constituents in Fruit Extracts of Myristica Fragrans and its Marketed Poly Herbal Formulations Using a Polarity Switch Technique** RENU PANDEY, Sophisticated Analytical Instrument Facility, CSIR, Brijesh Kumar N/A

(1720-10 P) **A Targeted Multidimensional Approach with MS Detection (SPE-RPLC/MS) for the Assessment of Trace Free Drug Species in Unadulterated Antibody-Drug Conjugate (ADC) Samples with Improved Specificity and Sensitivity** ROBERT BIRDSALL, Waters Corporation, Sean M McCarthy, Scott Berger, Weibin Chen, Alain Beck

(1720-11 P) **Method Development and Optimization for the Combined Analysis of Synthetic Cannabinoids and Designer Cathinones in Urine** HOLLY CASTELLANO, Duquesne University, Stephanie Wetzel N/A

(1720-12 P) **Separation and Identification of Lentil (Lens Culinaris) Proteins by Mass Spectrometry** ALBERTA ARYEE, Agriculture & Agri-Food Canada, Joyce I Boye N/A

(1720-13 P) **Withdrawn**

(1720-14 P) **Quantitation of Proto-Peptide Building Blocks in Complex Model Prebiotic Mixtures via Liquid Chromatography-Tandem Mass Spectrometry** ERIC T PARKER, Georgia Institute of Technology, Jeffrey L Bada, Facundo M Fernandez

POSTER SESSION **Session 1730**

All posters are to be mounted by 10:00 AM and remain on display until 4:00 PM. Authors must be at their posters from 10:00 AM to 12:00 PM. Location of the morning posters is on the Exposition Floor, 400 Aisle. PLEASE NOTE: You cannot get onto the Exposition Floor until after 9:00 AM.

Capillary Electrophoresis

Wednesday Morning, Exposition Floor, 400 Aisle

(1730-1 P) **Capillary Electrophoresis Method to Detect Circulating Steroids in Individual Zebrafish Plasma** PAIGE A REED, West Virginia University, Vincent T Nyakubaya, William J Feeney, Lisa A Holland, Jennifer Ripley-Stueckle

(1730-2 P) **Fluorogenic Derivatization of Amino Acids for Laser-Induced Fluorescence Detection in Capillary Electrophoresis** NAVEEN MADDUKURI, Wichita State University, Qiyang Zhang, Maojun Gong

(1730-3 P) **Investigating Electrospray Behavior in Capillary Electrophoresis Coupled Mass Spectrometry** JARED A LAMP, University of Notre Dame

(1730-4 P) **Validated Capillary Zone Electrophoretic Method for Simultaneous Determination of Some Antihypertensive Drugs in Their Single-pill Combination Therapy** FAWZY A EL-YAZBI, Alexandria University, Hytham M Ahmed, Tarek S Belal, Rasha A Shaalan, Sohaila M Elonsy

(1730-5 P) **Single Enzyme Molecule Studies with CE-LIF and CE-MS** EMILY A AMENSON, University of Notre Dame, Norman J Dovichi

(1730-6 P) **Xenopus Laevis Metabolomic Profiling via CZE-ESI-MS/MS and MALDI-TOFMS** JENNIFER ARCEO, University of Notre Dame, Nicole Schiavone, Danielle Boley, Elizabeth H Peuchen, Norman J Dovichi

(1730-7 P) **Insights into Chiral Recognition Mechanisms for Acryloyl Terminated Polymeric Surfactants: Application of Linear Solvation Energy Relationship in Micellar Electrokinetic Chromatography and Capillary Electrochromatography** YANG LU, Georgia State University, Shahab S Shamsi

(1730-8 P) **Study of Electrooxidation Products of Primary Alcohols by EC-CE-C4D: Assessment of the Conversion Efficiency of Alcohols into Their Carboxylates on Gold and Platinum Electrodes in Different Media** MAURO SERGIO FERREIRA SANTOS, USP, Fernando S Lopes, Ivano G R Gutz

(1730-9 P) **Microscale Quantification of Nanoparticle-Biomolecule Interactions with Capillary Electrophoresis** JULIA ANN MOUCH, Bethany College, Tyler Davis, Lisa A Holland

(1730-10 P) **Comparative Validation of Sofosbuvir Determination in Pharmaceuticals by Several Chromatographic, Electrophoretic and Spectrophotometric Methods** AMIRA F EL-YAZBI, Alexandria University

(1730-11 P) **Quantification of Amino Acids in Cordyceps by MEKC** XIN WEI, Wuhan University Zhongnan Hospital, Hankun Hu, Yiming Liu, Yue Xu

(1730-12) Isolation of Intact Cell Populations by Preparative Capilary Zone Electrophoresis Sarh N. Lum, Bonnie Huge, Matthew Champion, Norman J. Dovichi

POSTER SESSION **Session 1740**

All posters are to be mounted by 10:00 AM and remain on display until 4:00 PM. Authors must be at their posters from 10:00 AM to 12:00 PM. Location of the morning posters is on the Exposition Floor, 400 Aisle. PLEASE NOTE: You cannot get onto the Exposition Floor until after 9:00 AM.

Forensics and Homeland Security

Wednesday Morning, Exposition Floor, 400 Aisle

(1740-1 P) **Fingerprinting of Falsified Artemisinin Combination Therapies (ACTs) via DART Ionization Coupled to a Compact Single Quadrupole Mass Spectrometer** MATTHEW C BERNIER, Georgia Institute of Technology, Joseph LaPointe, Brian Musselman, Facundo M Fernandez

(1740-2 P) **Lead-Free Gunshot Residues as Forensic Evidence** CHRISTOPHER R DOCKERY, Kennesaw State University, Lashaundra Fambro, Ethan Miller, Deidre VanDenbos, Wassim Abdul Khalek

(1740-3 P) **Forensic Analysis of Textile Fibers Exposed to Cigarette Smoke Using Non-Destructive Room Temperature Excitation-Emission Fluorescence Microscopy** HUGH HAYES, University of Central Florida, Andres Campiglia, Arsenio Munoz de la Pena, Hector C Goicoechea N/A

(1740-4 P) **Forensic Analysis Of Textile Fibers Exposed To Laundry Detergents Using Fluorescence Excitation-Emission Spectroscopy** NIRVANI MUJUMDAR, University of Central Florida

(1740-5 P) **Eye-safe, Wide-Area Hyperspectral Raman Imaging Using a Spatial Heterodyne Raman Spectrometer** NATHANIEL R GOMER, ChemImage Sensor Systems, Matthew P Nelson

(1740-6 P) **Development and Optimization of Solid Phase Extraction (SPE) Method for Determination of Benzodiazepines in Wastewater and Surface Water by High-Performance Liquid Chromatography (HPLC)** HONGXIA GUAN, Georgia Gwinnett College, Qingsong Cai

(1740-7 P) **Determination and Measurement of Wildland Fire Markers in Residential Structures Using TD-GCMS** MARY C MARTIN, Prism Analytical and Central Michigan University, Alice Delia, Dale LeCaptain

(1740-8 P) **Differentiate Delta-9-tetrahydrocannabinol (Δ9-THC) and Delta-8-tetrahydrocannabinol (Δ8-THC)** KEN TSENG, Nacalai USA, Toshi Ono, Tsunehisa Hirose, Kazuhiro Kimata

(1740-9 P) **Identification and Quantification of Explosives and Their Residues in Water Using a Novel Surfactant in MEKC** CHRISTINE COPPER, U.S. Naval Academy, Marlene Perez, Alexis Clark, Karen Brensinger, Christopher Rollman, Jacqueline Rine, Ashton Genzman, Ira Lurie, Mehdi Moini

(1740-10 P) **Novel Concept of Biomarker Analysis in Forensic Analysis** JAN HALAMEK, SUNY Albany N/A

(1740-11 P) **Evaluation of GC-IRD Methods for Sensitivity and Selectivity in Forensic Drug Identification** RANDALL CLARK, Auburn University, Younis Abiedalla, Jack DeRuiter N/A

(1740-12 P) **Screening of Drugs of Abuse Using DART-MS and Real Time Reverse Library Search** FREDERICK LI, IonSense, Inc., Brian Musselman, Joe Tice, Stephen Shrader

(1740-13 P) **Sol-gel Sorbent Beds for All-in-One Sampling, Preconcentration, and Separation of Trace Explosive Vapors** MICHELLE CERRETA, U.S. Naval Research Laboratory, Braden Giordano, Kevin Johnson

(1740-14 P) **Techniques for Analyzing Volatile Organic Compounds Emitted During Aerobic Decomposition of Pig Carcasses and Swine Tissues** MASOUMEH DALILIAN, Middle Tennessee State University, Ngee Sing Chong, Samantha Keene, Lydia Rickman

(1740-15 P) **High Frequency, High Pressure Tandem Mass Spectrometry** ANDREW HAMPTON, University of North Carolina at Chapel Hill, J Michael Ramsey

(1740-16 P) **Gun Residue Analysis Using Paper Microfluidics** CHASTITY PAREDES-RODRIGUEZ, Penn State Berks, James Karlinsey

POSTER SESSION — Session 1750

All posters are to be mounted by 10:00 AM and remain on display until 4:00 PM. Authors must be at their posters from 10:00 AM to 12:00 PM. Location of the morning posters is on the Exposition Floor, 400 Aisle. PLEASE NOTE: You cannot get onto the Exposition Floor until after 9:00 AM.

Liquid Chromatography

Wednesday Morning, Exposition Floor, 400 Aisle

(1750-1 P) **Determination of Pyrethrins in Pyrethrum Oil Extracts** IAN N ACWORTH, Thermo Fisher Scientific, David Thomas, Alan Wong, Jan Glinski

(1750-2 P) **Pesticide Residue Analysis of Cereal Grains Treated with Traditional Fungicides or a Tannic Acid Biopesticide** BROOKE M BIEN, Western Carolina University, Cynthia Atterholt

(1750-3 P) **Microwave-Assisted Extraction of Triketone and Pyrazole Corn Herbicides from Agricultural Soil** SANJA STIPIČEVI, Institute for Medical Research and Occupational Health, Milena Milaković, Marija Dvoršćak, Sanja Fingler, Gordana Mendaš

(1750-4 P) **Development and Validation of a Method for the Simultaneous Extraction and Separated Measurement of Diflubenzuron and Azamethiphos from the Soft Tissue of Mussel M. Chilensis** LUIS NORAMBUENA, Instituto de Fomento Pesquero, Sergio Contreras-Lynch

(1750-5 P) **Determination of Anions and Cations in Water Matrices Using Non-suppressed Ion Chromatography** JOSEPH PAUL ROMANO, Waters Corporation, Mark E Benvenuti, Kenneth Rosnack

(1750-6 P) **Application of Factorial Design for the Determination of Ammonium in Saline Waters by Ion Chromatography Using an Ultrasound-assisted Purge-and-Trap System** MANUEL C CARNEIRO, CETEM, Fernanda N Ferreira, Fernanda Veronesi M Pontes, Julio C Afonso, Arnaldo A Neto, Maria Ines Couto Monteiro N/A

(1750-7 P) **Determination of Nitrogen Compounds in Saline Waters by Matrix Interference-Free Ion Chromatography** MARIA INES COUTO MONTEIRO, CETEM, Fernanda N Ferreira, Fernanda Veronesi M Pontes, Julio C Afonso, Manuel C Carneiro, Arnaldo A Neto N/A

(1750-8 P) **Micellar and Sub-Micellar UHPLC of Aromatic Acid Geometric Isomers** JENNIFER M FASCIANO, Miami University, Neil D Danielson

(1750-9 P) **Hydrophilic Interaction Liquid Chromatography of Aromatic Acid Isomers on a Plain Silica Stationary Phase Using a Ternary Mobile Phase** ASHLEY E RICHARDSON, Miami University, Neil D Danielson

(1750-10 P) **Understanding the Importance of Instrument Design To Take Full Advantage of 1.0mm Internal Diameter (ID) Columns when Running UPLC** JENNIFER SIMEONE, Waters Corporation, Paula Hong, Patricia R McConville

(1750-11 P) **Investigation of Reverse-Phase - HILIC Continuous Analysis Using a One Column** RONALD BENSON, Shodex/Showa Denko America, Inc., Junji Sasuga, Tomokazu Umezawa, Hideyuki Kondo

(1750-12 P) **One Diode Array Detector for Analytical, Semi-Preparative, Preparative and Biocompatible HPLC** KATHRYN E MONKS, Knauer

(1750-13 P) **A Novel Diphenyl Stationary Phase for Metabolite Profiling** MARK WOODRUFF, Fortis Technologies, Ken Butchart

(1750-14 P) **Silica Hybrid Monoliths with a Carbonaceous Surface for Liquid Chromatography** AMARIS C BORGES-MUNOZ, University at Buffalo SUNY, Luis A Colón

(1750-15 P) **Evaluation of the Performance of Ultra-high Pressure Chromatography (UPLC) Systems** IMAD A HAIDAR AHMAD, Novartis, Frank Hrovat, Adrian Clarke, James Tam, Xue Li, Thomas Tarara, Andrei Blasko

(1750-16 P) **Optimizing Reagent Free Ion Chromatography; Electrolytic Water Purification** JOHN M RIVIELLO, Trovion Company, Archava Siriraks, Daniel Khor

(1750-17 P) **Functionalized Carbon Nanotubes as Pseudo-Stationary Phases in Capillary Electrokinetic Chromatography - Evaluation of Retention Energetics and Analysis of a Wide Range of Neutral and Charged Species** SARAH ALHARTHI, Oklahoma State University, Ziad El Rassi

(1750-18 P) **Determination of Hydroperoxides Using High Performance Liquid Chromatography with Reductive Electrochemical Detection** JUN CHENG, Thermo Fisher Scientific, Yan Liu, Khalil Divan

(1750-19 P) **Label-Free Analysis by HPLC with Charged Aerosol Detection of N-linked Glycans Separated by Charge** IAN N ACWORTH, Thermo Fisher Scientific, David Thomas, William Kopaciewicz

(1750-20 P) **Multi-Modal Analyte Detection of Cyclodextrin and Ketoprofen Inclusion Complex Using UV and CAD on an Integrated UHPLC System** BRUCE BAILEY, Thermo Fisher Scientific, Marc Plante, Ian N Acworth, Evert-Jan Sneekes, Frank Steiner

(1750-21 P) **RP-HPLC Method Development and Validation for the Analysis of Pharmaceutical Drugs-Linezolid** RAJESHKUMAR H CHAUDHARI, B/H Saikrishna Hospital, Vadilal G Patel N/A

(1750-22 P) **Purification of Chlorogenic Acid from Green Coffee Using Core-shell Technology in Axia Preparative Format** MARC JEAN JACOB, Phenomenex

(1750-23 P) **HILIC HPLC Separation of Oxymorphone for Assay and Purity** BRADLEY KUMAGAI, Theravance Biopharma

(1750-24 P) **Transfer of the Method for Related Substances Analysis of Metoclopramide HCl Between Different LC Systems** MARGARET MAZIARZ, Waters Corporation, Christopher Henry, Mark Wrona

(1750-25 P) **RP-HPLC Method Development and Validation for the Analysis of Pharmaceutical Drugs-60% Sodium Lactate** VADILAL G PATEL, B/H Rajpath Hotel, Rajeshkumar H Chaudhari N/A

(1750-26 P) **Polar Stationary Phases for Capillary Liquid Chromatography based on Metallic Oxides** CARLA GRAZIELI AZEVEDO DA SILVA, Unicamp, Carol H Collins, Carla B Bottoli

(1750-27 P) **Selection of Chromatographic Columns by Supercritical Fluid Chromatography** CARLA GRAZIELI AZEVEDO DA SILVA, IQ/UNICAMP, Carol H Collins, Isabel Cristina S Jardim

(1750-28 P) **Stability Evaluation of Core Shell C18 with Encapsulated Type End-Capping** NORIKAZU NAGAE, ChromaNik Technologies Inc., Tomoyasu Tsukamoto

(1750-29 P) **Development of an Interactive Counter Current Extraction Simulation** SEAN A REED, Westminster College, Erin Wilson, James Anthony

POSTER SESSION — Session 1760

All posters are to be mounted by 10:00 AM and remain on display until 4:00 PM. Authors must be at their posters from 10:00 AM to 12:00 PM. Location of the morning posters is on the Exposition Floor, 400 Aisle. PLEASE NOTE: You cannot get onto the Exposition Floor until after 9:00 AM.

Pharmaceutical-IC, LC, and SFC

Wednesday Morning, Exposition Floor, 400 Aisle

(1760-1 P) **Determination of Aluminum in OTC Products** MANALI AGGRAWAL, Thermo Fisher Scientific, Jeffrey Rohrer

(1760-2 P) **Analysis of Antibiotics Sold in Pharmaceutical Market in Idumota, Lagos Using HPLC** SIXTUS I AMADI, Hydrochrom Resources Ltd

(1760-3 P) **A Sensitive Method for Direct Analysis of Impurities in Apramycin and Other Aminoglycoside-Antibiotics Using Hydrophilic Interaction Liquid Chromatography and Charged Aerosol Detection** ZHEN LONG, Thermo Fisher Scientific, Qi Zhang, Lina Liang, Yan Jin, Ian N Acworth, Evert-Jan Sneekes, Frank Steiner

(1760-4 P) **Chemically Stable Reversed Phase Chromatography Material Scalable from UHPLC to Semi Prep and Production** CECILIA MAZZA, AkzoNobel PPC AB, Fredrik Lime

(1760-5 P) **Characterization and Lot-to-Lot Variability of Complex Surfactants by High Performance Liquid Chromatography and Charged Aerosol Detection** MARC PLANTE, Thermo Fisher Scientific, Ian N Acworth, Bruce Bailey, Evert-Jan Sneekes, Frank Steiner

(1760-6 P) **Direct-Determination of Underivatized Carbohydrates in Biopharmaceutical Formulations Using Heart-Cut, 2D HPLC-HILIC and Charged Aerosol Detection** MARC PLANTE, Thermo Fisher Scientific, Bruce Bailey, Ian N Acworth, Evert-Jan Sneekes, Frank Steiner

(1760-7 P) **Fast Method Screening for Chromatographic Separation of Enantiomers Utilizing Polysaccharides Type Chiral Stationary Phases** TAKASHI SATO, YMC CO., Ltd., Noritaka Kuroda, Saoko Nozawa, Keiko Kihara, Noriko Shoji, Takatomo Takai

(1760-8 P) **Label-Free Measurement of Sialic Acids Released From Glycoproteins, by High Performance Liquid Chromatography and Charged Aerosol Detection** QI ZHANG, Thermo Fisher Scientific, Ian N Acworth, Bruce Bailey, Evert-Jan Sneekes, Frank Steiner

(1760-9 P) **Challenges in Developing Analytical Methods for Cleaning Validations in a GMP Regulated Environment** XIAOHUI YANG, Baxter Healthcare, Robert Garber, Walter Wasylenko, Lakshmy Nair, Jane Werling, George Monen, Beena Uchil

(1760-10 P) **SFC Analysis of Nutraceuticals and Pharmaceuticals Using SFC Optimized Stationary Phases** MATTHEW PRZYBYCIEL, ES Industries

(1760-11 P) **Analysis of Metoprolol and Select Impurities Using a Hydrophilic Interaction Chromatography Method with Combined UV and Charged Aerosol Detection** BRUCE BAILEY, Thermo Fisher Scientific, Ian N Acworth, Evert-Jan Sneekes, Frank Steiner

(1760-12 P) **Benefits of Using Mass Detection for Assessing Quality and Purity of Cetrimonium Bromide Pharmaceutical Raw Material** MARGARET MAZIARZ, Waters Corporation, Christopher Henry, Mark Wrona, Dominic Moore, Chengxia O'Shea

(1760-13 P) **Characterization of a Biologic Therapeutic: Reversed Phase/HILIC Analysis of Protein and Excipients** BRUCE BAILEY, Thermo Fisher Scientific, Marc Plante, Ian N Acworth, Evert-Jan Sneekes, Frank Steiner

(1760-14 P) **Efficient, Effective and Proven Approach to Chiral Method Development for Purification Scale Up** SEAN ORLOWICZ, Phenomenex, Michael McCoy

(1760-15 P) **Guidelines for Method Transfer and Optimization of the Newest Charged Aerosol Detector** MARC PLANTE, Thermo Fisher Scientific, Bruce Bailey, Ian N Acworth, Paul Gamache, Evert-Jan Sneekes, Frank Steiner

(1760-16 P) **Chiral Separation Using SFC and HPLC** HIDETOSHI TERADA, Shimadzu, Takato Uchikata, Funada Yasuhiro, Tanaka Kenichiro, Arao Yohei

(1760-17 P) **Determination of Ascorbic Acid in Citrus Fruits and Pharmaceutical Formulations by Hydrophilic Interaction Chromatography (HILIC)** YUEGANG ZUO, University of Massachusetts Dartmouth, Ruiting Zuo, Si Zhou, Yiwei Deng

(1760-18 P) **HPLC Method Development and Validation for the Assay and Organic Impurities of Naproxen in Naproxen API, Naproxen Delayed-Release Tablets and Naproxen Oral Suspension** JENNIFER FEDOROWSKI, United States Pharmacopeial Convention, Joshua Bhattacharya, Arindam Ganguly, Natalia Kouznetsova, Jennifer Belsky

(1760-19 P) **Sample Analysis of Compounds with Multiple Chiral Centers by Two-Dimensional HPLC** CHI TSANG, Genentech, Kelly Zhang

(1760-20 P) **Mobile Phase Effects in Reversed-Phase Chromatography of Monoclonal Antibodies at High Temperature** HILLEL K BRANDES, Sigma-Aldrich/Supelco

(1760-21 P) **Separation of Aminoglycoside Antibiotics by Using Hydrophilic Interaction Liquid Chromatography** YU LONG, Dalian Institute of Chemical Physics, Wei Jie, Shen Aijin, Guo Zhimou, Liang Xinmiao N/A

(1760-22 P) **Utilization of HPTLC and Diffuse Reflectance Spectroscopy to Quickly Evaluate Product Quality of Cotrimoxazole Tablets from Tanzania** DAVID WAYNE JENKINS, FHI 360, Eliangiringa Kaale, Samuel Hope, Thomas Layloff

(1760-23 P) **Amino Acid Analysis for Qualitative and Quantitative Composition of Pharmaceutical Products** NATALIA BELIKOVA, SGS Life Science Services

(1760-24 P) **Separation and Detection of Small Molecules by Low-Flow Liquid Chromatography Mass Spectrometry for Pharmacokinetic Studies** JAMES N MARR, Merck & Co., Rena Zhang, Gary Adamson

POSTER SESSION — Session 1770

All posters are to be mounted by 10:00 AM and remain on display until 4:00 PM. Authors must be at their posters from 10:00 AM to 12:00 PM. Location of the morning posters is on the Exposition Floor, 400 Aisle. PLEASE NOTE: You cannot get onto the Exposition Floor until after 9:00 AM.

Pharmaceutical-MS, LC/MS and Others

Wednesday Morning, Exposition Floor, 400 Aisle

(1770-1 P) **A New Approach to the Automated Identification of Metabolites in Multi-Vendor Datasets** RICHARD LEE, ACD/Labs, Vitaly Lashin, Andrey Paramonov, Alexandr Sakharov, Rytis Kubilius

(1770-2 P) **Analysis of Atropine Sulfate by HPLC Using Mass Spec HPLC Mobile Phase** JEFFREY KAKALEY, YMC America Inc., Ernest Sobkow

(1770-3 P) **Alternative Oxidation Technique for Pharmaceutical Forced Degradations** KEN NGIM, Theravance Biopharma US, Inc., Claudine Ooi, Bradley Kumagai

(1770-4 P) **Investigating Transdermal Diffusion of Vitamin D and 25-Hydroxyvitamin D** MARCEL MUSTEATA, Albany College of Pharmacy, Isaac Mall

(1770-5 P) **Calibration Free, Semi-Quantitative Analysis of Defined Drug Formulations Using FTIR Pre-computed Mixture Spectra** WILLIAM COSTA, Fiveash Data Management, Inc., Bill McCarthy, Todd Strother

(1770-6 P) **Novel Self-Patented Gold Nanoparticle Synthesis, Characterization and Antibacterial Susceptibility Testing** WILLIAM HAMILTON, Western Kentucky University, Jason N Payne, Fenil Chavda, Rajalingam Dakshinamurthy

(1770-7 P) **Unattended, Representative Sampling of a Wide Range of Chemical Reactions** VASO VLACHOOS, Mettler Toledo, Jane Riley N/A

(1770-8 P) **Square Wave Adsorptive Stripping Voltammetric Determination of Ketoconazole Drug in the Saudi Market** ABDEL-NASSER KAWDE, King Fahd University of Petroleum and Minerals, Mohamed A Morsy

(1770-9 P) **Capillary Electrophoresis: MEKC Assay Method for Simultaneous Determination of Olmesartan Medoxomil, Amlodipine Besylate and Hydrochlorothiazide in Tablets** MAHESH V ATTIMARAD, College of Clinical Pharmacy, Sree N Harsha, Bander E Al Dubaib, Anroop B Nair N/A

(1770-10 P) **A Comparison of Complete Dissolution Versus Leach of the Target Analytes by Using and Omitting HF** DAN IVERSEN, CEM, Tyler Edwards

(1770-11 P) **Standardization of Experimental Conditions of USP Melting Point Reference Standards in DSC Applications** OSOMWONKEN IGBINOSUN, United States Pharmacopeial Convention, Guillermo A Casay, Antonio Hernandez-Cardoso, Steven T Rau, Kanda K Balasubramanian

(1770-12 P) **Identification of Persistent Pd-containing Impurities Using LC-MS/MS and LC-ICP-MS** RENEE K DERMENJIAN, Merck N/A

(1770-13 P) **A Co-Crystal of Febuxostat and Isonicotinamide: Synthesis and Characterization** YANLEI KANG, Zhejiang University, Xiurong Hu, Jianming Gu, Dongdong Yu, Jianguang Zhou

(1770-14 P) **Real Time Stability of NSAIDs in Aqueous Solutions by Infrared Spectroscopy** ANUMEHA P MUTHAL, Seton Hall University, Vrushali Bhawtankar, Nicholas Snow

(1770-15 P) **Headspace Sampling of Residual Solvents Per USP 467 Using a Gas Tight Syringe** ANNE JUREK, EST Analytical, Lindsey Pyron, Kelly Cravenor, Justin Murphy

(1770-16 P) **Comparative Evaluation of Physicochemical Properties of Some Commercially Available Brands of Metformin Hydrochloride Tablets in Lagos, Nigeria** ADERONKE A ADEPOJU-BELLO, University of Lagos, Olawale S Bisiriyu N/A

(1770-17 P) **A Simple and Rapid LC-MS/MS Method for the Determination of BMCL26, a Novel Anti-Parasitic Agent, in Rat Plasma** RAMAKRISHNA REDDY VOGGU, Cleveland State University, Xiang Zhou, Bin Su, Baochuan Guo

(1770-18 P) **Derivative and Chemometric Spectrophotometric Methods for the Determination of Aripiprazole in Presence of its Related Impurities** SALLY TAREK, Cairo University N/A

(1770-19 P) **Drug-Herb Interaction: A Crossover Study of the Effect of a Polyherbal Formulation on Metroinidazole Pharmacokinetic Profile** GRACE EIGBIBHALU UKPO, University of Lagos, Teddy S Ehianeta, Steve O Ogbonnia, Idris O Balogun

(1770-20 P) **Supersaturation of Spray-Dried-Dispersion (SDDs) - Development and Evaluation of a Characterization Method** BENJAMIN H WU, Bristol-Myers Squibb

(1770-21 P) **Headspace Grade Solvents for Trace Level Analyte Detection** SUBHRA BHATTACHARYA, Thermo Fisher Scientific, Eric Oliver, Stephen Roemer

(1770-22 P) **Analytical Strategies in the Development of Generic Drug Products: Excipient Quantitation** ALEXANDER WILLIAM GARNER, Mayne Pharma

(1770-23 P) **Quantitative Laser Diffraction Method for the Assessment of Subvisible Protein Particles** ROBERT E BUCO, Shimadzu Corporation, Matthew Ferrarelli, Ariadna Martos, Andrea Hawe, Michael Wiggenhorn, Shinichiro Totoki, Haruo Shimaoka, Markus Ortlieb

(1770-24 P) **3D-Printed LTP Ionization Source for the Direct Analysis of Biomolecules** SANDRA MARTINEZ JARQUIN, Centro de Investigation Y de Estudios Avanzados Del IPN, Robert Winkler

(1770-25 P) **Demonstrating the Uptake Mechanism of Cisplatin in Cells by Single Cell ICP-MS** CHADY STEPHAN, PerkinElmer

(1770-26 P) **The Practice and Challenges of Ultrafast Chiral Separations in UHPLC and Super/Subcritical Fluid Chromatography (SFC)** CHANDAN L BARHATE, University of Texas at Arlington, M Farooq Wahab, Daniel W Armstrong

(1770-27 P) **Isolation of Pharmaceutical Impurities Utilizing MS-Directed Purification Platforms** ERIC STRECKFUSS, Merck N/A

(1770-28 P) **Determining Equivalency of Generic and Name Brand Oral Suspensions Using Zeta Potential** JACK G SAAD, Micromeritics, Myke Scoggins, Danielle Sowle

(1770-29 P) **Analysis of the Photodegradation Products of Nifedipine in Photoprotective Topical Formulations by HPLC and Mass Spectrometry** ELLEN K WASAN, University of Saskatchewan, Joshua Poteet, Ed Krol, Jacqueline Cawthray, Anas El-Aneed, Kevin Soulsbury, Munawar Mohammed N/A

TECHNICAL PROGRAM

POSTER SESSION — Session 1780

All posters are to be mounted by 10:00 AM and remain on display until 4:00 PM. Authors must be at their posters from 10:00 AM to 12:00 PM. Location of the morning posters is on the Exposition Floor, 400 Aisle. PLEASE NOTE: You cannot get onto the Exposition Floor until after 9:00 AM.

Practical Chromatography in Today's Laboratory

Wednesday Morning, Exposition Floor, 400 Aisle

(1780-1 P) **A Rapid Isocratic LC Separation of Soft Drink Additives Using an Environmentally Friendly Mobile Phase with UV Detection** MARK E BENVENUTI, Waters Corporation, Joseph Paul Romano

(1780-2 P) **High Efficiency Chiral Separations Using Sub-2 µm Monodisperse Chiral Stationary Phases** ZACHARY BREITBACH, The University of Texas at Arlington, Chandan L Barhate, M Farooq Wahab, David S Bell, Daniel W Armstrong

(1780-3 P) **Advances in Chiral HPLC Column Technology: Superficially Porous Particle Based Chiral Stationary Phases** ZACHARY BREITBACH, The University of Texas at Arlington, M Farooq Wahab, Daniel W Armstrong

(1780-4 P) **A New Standard in Analytical Workflow Design** WILLIAM HEDGEPETH, Shimadzu Scientific Instruments, Kenichiro Tanaka

(1780-5 P) **Conductivity and pH Monitoring as an Effective Tool for Multi-Step Biological Separations** LUKE ROENNEBURG, Gilson, Inc., Laine Stewart, Karen Kleman N/A

(1780-6 P) **Centrifugal Partition Chromatography: A Preparative Tool for Isolation and Purification of Xylindein from Chlorociboria Aeruginosa** ANUKUL BOONLOED, Oregon State University, Genevieve L Weber, Sumate Pengpumkiat, Vincent T Remcho

(1780-7 P) **Analysis of Petroleum Products Using Comprehensive Two-Dimensional Gas Chromatography (GC×GC) with Both Time of Flight MS and Flame Ionization Detectors** JOSEPH E BINKLEY, Leco Corporation, Christina N Kelly, Jonathan D Byer, David E Alonso, Lorne E Fell

(1780-8 P) **Evaluation of Polycyclic Aromatic Hydrocarbon Standard Reference Materials 869b and 1647f on Different Stationary Phases for Liquid Chromatography** WALTER B WILSON, National Institute of Standards and Technology, Jorge O Oña-Ruales, Lane C Sander, Stephen A Wise

(1780-9 P) **Liquid Chromatographic Retention Behavior of Polycyclic Aromatic Sulfur Heterocycles and Their Alkyl-Substituted Derivatives** WALTER B WILSON, National Institute of Standard and Technology, Lane C Sander, Stephen A Wise

(1780-10 P) **Evaluation of 25-Hydroxy Vitamin D Extraction Using Phospholipid Depletion Plate Technology and Method Comparison Using Automated Sample Preparation** KERRY CHALLENGER, Biotage GB Limited, Lee Williams, Helen Lodder, Victor Vandell

(1780-11 P) **Effects of Ultracentrifugation on HDL and LDL Size Distribution** JEFFREY JONES, Centers for Disease Control and Prevention, Zsuzsanna Kuklenyik, Christopher Toth, Bryan A Parks, Michael S Gardner, Jon Rees, Yulanda Williamson, David Schieltz, Lisa McWilliams, John R Barr, James Pirkle

(1780-12 P) **Biomolecular Separations through Tunable Nanoporous Gold Membranes** DANIEL A MCCURRY, University of Illinois at Urbana-Champaign, Ryan C Bailey

(1780-13 P) **Non-Contact Pd Separation based on Laser-Induced Particle Formation for Determination of ^{107}Pd with ICP-MS** TAKUMI YOMOGIDA, Japan Atomic Energy Agency, Shiho Asai, Morihisa Saeki, Yukiko Hanzawa, Fumitaka Esaka, Hironori Ohba, Yoshihiro Kitatsuji

(1780-14 P) **Surfactant-Pluronic Gel Phases for Electrophoresis** ASHLEY E RICHARDSON, Miami University, Elise M Leonard, Neil D Danielson

(1780-15 P) **Insights into the Effect of the PDMS-layer on the Kinetics and Thermodynamics of Analytes Sorption onto the PDMS-overcoated Coating** ERICA A SOUZA-SILVA, Universidade Federal do Rio Grande do Sul, Emanuela Gionfriddo, Janusz Pawliszyn

(1780-16 P) **A New Anion Exchange Column for Fast Ion Chromatographic Separation of Monosaccharides and Disaccharides in Biofuel, Food, and Beverage Samples** YAN LIU, Thermo Fisher Scientific, Andy Woodruff, Charanjit Saini, Yury Agroskin, Christopher Pohl

(1780-17 P) **Column Robustness Challenges during HPLC Method Development** JUN WANG, Takeda Boston, Laila Kott, Elizabeth Hewitt, Scott Zugel

(1780-18 P) **There are Problems Associated with Gradient and Method Transfer in HPLC and UHPLC – Are There Explanations and Usable Workarounds? Part 1 of 2** MICHAEL WOODMAN, Agilent Technologies, Gregory Hunlen

(1780-19 P) **Understanding the Causes and Minimizing the Impact of LC Carryover in Most LC Systems** MICHAEL WOODMAN, Agilent Technologies, Gregory Hunlen

SEAC POSTER SESSION — Session 1790

All posters are to be mounted by 10:00 AM and remain on display until 4:00 PM. Authors must be at their posters from 10:00 AM to 12:00 PM. Location of the morning posters is on the Exposition Floor, 400 Aisle. PLEASE NOTE: You cannot get onto the Exposition Floor until after 9:00 AM.

SEAC Poster Session

Wednesday Morning, Exposition Floor, 400 Aisle

(1790-1 P) **New Portable Electrochemical Instrument for *In Situ* Analysis** PABLO FANJUL-BOLADO, DropSens S.L, David Hernández-Santos, Laura Fernández-Llano, Marta Neves, Pablo Bobes-Limenes, Alejandro Pérez-Junquera, Begoña González-García, Carla Navarro-Hernández

(1790-2 P) **Aqueous and Non-Aqueous Electrochemical Activities of Mercaptosuccinic Acid Stabilized Au11-13 Clusters** JONATHAN PADELFORD, Georgia State University, Gangli Wang

(1790-3 P) **Copolymerized Triazole Based Ionic Liquid as New Sensing Material in Ion-Selective Sensors** LUKASZ K MENDECKI, Keele University

(1790-4 P) **Investigation of Cloud-Point Extraction and UV-Vis for Determining Copper and Cadmium in Vegetables** ASADUZZAMAN NUR, Tennessee Technological University, Andrew Callender

(1790-5 P) **Potentiometric Ion-Selective Electrodes based on Metastable Photoacid for Cation Detection** PARTH K PATEL, University of Central Florida, Karin Y Chumbimuni-Torres

(1790-6 P) **Signal Amplification of a Highly Selective Universal MicroRNA Electrochemical Sensor for Single Nucleotide Polymorphism Detection** DAWN MILLS, University of Central Florida, Jeffer Pinzon, Percy Calvo-Marzal, Dmitry M Kolpashchikov, Karin Y Chumbimuni-Torres

(1790-7 P) **Self-Reference Single Strip Paper Based Sensors for Ion Detection** ANDREW J MANHAN, University of Central Florida, Stephanie Armas, Karin Y Chumbimuni-Torres

(1790-8 P) **Development and Characterization of an Ion Selective Microsensor for the Detection and Monitoring of Zinc Levels in Citrus Plants** COURTNEY HULCE, University of Central Florida, Jared Church, Swadeshmukul Santra, Woo Hyoung Lee, Karin Y Chumbimuni-Torres

(1790-9 P) **Poly(3-octylthiophene)-Based Solid Contact Ion-Selective Electrodes with Improved Potential Stability** JENNIFER M JARVIS, University of Memphis, Marcin Guzinski, Bradford D Pendley, Erno Lindner

(1790-10 P) **Urine Carbon Dioxide in Septic Shock** JAMES G ATHERTON, University of Memphis, Artur Jasinski, Erno Lindner, Bradford D Pendley, Marcin Guzinski, William King

(1790-11 P) **Disposable Paper-Based Electrochemical Ion-Sensing Platform** JINBO HU, University of Minnesota, Andreas Stein, Philippe Buhlmann

(1790-12 P) **Development of Calibration-Free Electrochemical Sensors Using Novel Redox Polymers** XUE ZHEN, University of Minnesota, Philippe Buhlmann

(1790-13 P) **Ultrasensitive Detection of Dopamine with Carbon Nanopipettes** KEKE HU, Queens College CUNY, Michael Mirkin, Yun Yu, Min Zhou

(1790-14 P) **Avoiding Errors in Electrochemical Measurements: Effect of Frit Material on the Performance of Reference Electrodes with Porous Frits** MARAL PS MOUSAVI, University of Minnesota, Stacey Saba, Evan Anderson, Marc A Hillmyer, Philippe Buhlmann

WEDNESDAY, MARCH 9, 2016
AFTERNOON

AWARDS **Session 1800**

Ralph N Adams Award
arranged by Norman J Dovichi, University of Notre Dame

Wednesday Afternoon, Room B312
Norman J Dovichi, University of Notre Dame, Presiding

1:30		**Introductory Remarks - R Mark Wightman**
1:35		**Presentation of the 2016 Ralph Adams Award to David R Walt, Tufts University, by Norman J Dovichi, University of Notre Dame**
1:40	(1800-1)	**Clinical Applications of Single Molecule Arrays (Simoa)** DAVID R WALT, Tufts University
2:15	(1800-2)	**Microengineered Devices for Biomedical Research** NANCY ALLBRITTON, University of North Carolina at Chapel Hill
2:50	(1800-3)	**Micro-and Nanofabricated Molecular Measurement Devices** J MICHAEL RAMSEY, University of North Carolina at Chapel Hill N/A
3:25		**Recess**
3:40	(1800-4)	**Paper-Based Microfluidic Devices for Point-of-Need Bioanalysis** CHARLES R MACE, Tufts University, Syrena C Fernandes, Samuel Berry
4:15	(1800-5)	**Developmental Proteomics** NORMAN J DOVICHI, University of Notre Dame, Liangliang Sun, Matthew M Champion, Guijie Zhu, Paul Huber

AWARDS **Session 1810**

The Coblentz Society - Williams-Wright Award
arranged by Nancy Jestel, Sabic Innovative Plastics

Wednesday Afternoon, Room B314
Nancy Jestel, Sabic Innovative Plastics, Presiding

1:30		**Introductory Remarks - Nancy Jestel**
1:35		**Presentation of the 2016 Coblentz Society's Williams-Wright Award to D Warren Vidrine, Vidrine Consulting, by Nancy Jestel, Sabic Innovative Plastics**
1:40	(1810-1)	**Industrial Analysis Utilizing Vibrational Spectrometry** D WARREN VIDRINE, Vidrine Consulting
2:15	(1810-2)	**Biological Infrared Spectroscopy: An Overnight Success Story 64 Years in the Making** BOB MESSERSCHMIDT, Nueon Inc.
2:50	(1810-3)	**Reflections of a Chemometric Spectroscopist** DAVID M HAALAND, Spectral Resolutions, David A Melgaard, Howland D Jones
3:25		**Recess**
3:40	(1810-4)	**The Laminated Card** STUART YANIGER, ITW Fluids North America N/A
4:15	(1810-5)	**Open Discussion**

SYMPOSIUM **Session 1820**

ACS-ANYL - Advances in Instrumentation for Ion Mobility Mass Spectrometry
arranged by Matthew F Bush, University of Washington

Wednesday Afternoon, Room B308
Matthew F Bush, University of Washington, Presiding

1:30		**Introductory Remarks - Matthew F Bush**
1:35	(1820-1)	**Ultra-High Resolution Ion Mobility Separations based upon Long Path Length Structures for Lossless Ion Manipulations (SLIM)** RICHARD D SMITH, Pacific Northwest National Laboratory, Ahmed Hamid, Sandilya Garimella, Yehia M Ibrahim, Liulin Deng, Ian K Webb, Aleksey V Tolmachev, Erin S Baker
2:10	(1820-2)	**Ion Mobility Mass Spectrometers for Structural Biology and Biophysics** MATTHEW F BUSH, University of Washington
2:45	(1820-3)	**Toward Protein Ion Surface Characterization with IMS/HDX-MS/MS Techniques** STEPHEN J VALENTINE, West Virginia University, Mahdiar Khakinejad, Gregory Donohoe, Samaneh Ghassabi Kondalaji
3:20		**Recess**
3:35	(1820-4)	**Maximizing Fragmentation Duty Cycle for Ion Mobility-Ion Trap Instrumentation Using the Fourier Transform and Photodissociation** BRIAN CLOWERS, Washington State University, Kelsey Morrison
4:10	(1820-5)	**A Multi-Pass Cyclic Ion Mobility Separator: Design and Performance** KEVIN GILES, Waters Corporation, Jason Wildgoose, Steve Pringle

SYMPOSIUM **Session 1830**

Advancing Strategies for Chronic In Vivo Sensing
arranged by R Mark Wightman, University of North Carolina at Chapel Hill and Adrian C Michael, University of Pittsburgh

Wednesday Afternoon, Room B302
R Mark Wightman, University of North Carolina at Chapel Hill, Presiding

1:30		**Introductory Remarks - R Mark Wightman and Adrian C Michael**
1:35	(1830-1)	**Modulating Blood-Brain Barrier Healing Around Intracortical Electrode Implants** RAVI BELLAMKONDA, Georgia Institute of Technology, Varun Yarabarla, Alexus Clark, Brianna Gresham, Robert Kretschmar, Shoba Paul, Jessica Falcone
2:10	(1830-2)	**Is Microdialysis Specifically Monitoring the Tonic Modality of Dopamine Transmission?** GAETANO DI CHIARA, University of Cagliari
2:45	(1830-3)	**Long-Term Monitoring of Dopamine** PAUL EM PHILLIPS, University of Washington
3:20		**Recess**
3:35	(1830-4)	**Calibration of In Vivo Voltammetry** R MARK WIGHTMAN, University of North Carolina at Chapel Hill
4:10	(1830-5)	**Advancing the Possibilities for Chronic Brain Microdialysis** ADRIAN C MICHAEL, University of Pittsburgh, Erika Varner, Khanh Ngo, Andrea Jaquins-Gerstl, Stephen Weber

SYMPOSIUM **Session 1840**

Analytical Challenges Relating to the Discovery, Development, Manufacturing and Use of Cancer Immunotherapy Medicines
arranged by Maribel Beaumont and Christopher Welch, Merck Research Laboratories

Wednesday Afternoon, Room B303
Maribel Beaumont, Merck Research Laboratories, Presiding

1:30		**Introductory Remarks - Maribel Beaumont**
1:35	(1840-1)	**Concepts of Cancer Immunotherapy** ROBERT KASTELEIN, Merck & Co.
2:10	(1840-2)	**Identifying and Profiling Tumor Specific T Cells Using Mass Cytometry and Highly Multiplexed Peptide-MHC Tetramer Staining** EVAN W NEWELL, Singapore Immunology Network
2:45	(1840-3)	**Analytical Challenges in the Discovery, Development and Commercialization of Keytruda** MARIBEL BEAUMONT, Merck Research Laboratories
3:20		**Recess**
3:35	(1840-4)	**Development of Microtools for Immune Therapy Applications** CHRISTOPHER LOVE, Koch Institute for Integrative Cancer Research at MIT
4:10	(1840-5)	**Analytical Challenges During Development of Monoclonal Antibody Therapeutics** JOHN T STULTS, Genentech, Inc, David Michels

TECHNICAL PROGRAM

SYMPOSIUM — Session 1850

Analytical Chemistry of Oil and Gas Prospecting in Brazil
arranged by Maira Menezes, NurnbergMesse Brasil

Wednesday Afternoon, Room B304
Jailson de Andrade, Universidade Federal da Bahia-UFBA, Presiding

1:30		**Introductory Remarks - Jailson de Andrade**
1:35	(1850-1)	**Analytical Chemistry Challenges of Oil and Gas Prospection and Exploration** JAILSON B DE ANDRADE, UFBA N/A
2:10	(1850-2)	**Compound Specific δ13C Determination of Light Hydrocarbons (n-alkanes and Olefins from C1 to C5) at Low Concentration for Oil and Gas Prospection** ARTHUR DE LEMOS SCOFIELD, PUC-RIO, Angela R Wagener, Laura R Morales, Lilian F Almeida
2:45	(1850-3)	**Mass Spectrometry by FT-ICR and Orbitrap: Analysis of Crude Oil and Its Derivatives** EDUARDO MORGADO SCHMIDT, State University Campinas

SYMPOSIUM — Session 1860

Big Data in Analytical Sciences - Challenges and Solutions
arranged by Hang Lu, Georgia Tech and Andriana San Miguel, North Carolina State University

Wednesday Afternoon, Room B305
Hang Lu, Georgia Tech, Presiding

1:30		**Introductory Remarks - Hang Lu**
1:35	(1860-1)	**Data Analysis Challenges in Plant Biology** PHILIP BENFEY, Duke University
2:10	(1860-2)	**First Steps into Big Data Chemistry- Ultra-High-Throughput Screening** SPENCER D DREHER, Merck & Co., Inc.
2:45	(1860-3)	**Big, But Small, Data: Network Analysis in Small Sample Size Systems Biology** MARK P STYCZYNSKI, Georgia Institute of Technology
3:20		**Recess**
3:35	(1860-4)	**Clustering and Differential Alignment Algorithm: Identification of Early Stage Regulators in the Arabidopsis Thaliana Iron Deficiency Response** CRANOS WILLIAMS, North Carolina State University, Terri A Long, Alexandr Koryachko, Anna Matthiadis, James Tuck, Durreshahwar Muhammad, Joel Ducoste, Siobhan Brady, Jessica Foret
4:10	(1860-5)	**Machine Learning for Big Nonlinear Problems in Science and Engineering** LE SONG, Georgia Institute of Technology

SYMPOSIUM — Session 1870

Electrical and Electrochemical Sensing and Detection based on Nucleic Acid Recognition
arranged by Rebecca Lai, University of Nebraska-Lincoln
and Ryan J White, University of Maryland Baltimore County (UMBC)

Wednesday Afternoon, Room B309
Rebecca Lai, University of Nebraska-Lincoln, Presiding

1:30		**Introductory Remarks - Rebecca Lai and Ryan J White**
1:35	(1870-1)	**A New Approach to POC Serology** KEVIN W PLAXCO, University of California Santa Barbara
2:10	(1870-2)	**Detection of Hepatitis B Virus DNA with a Paper Electrochemical Sensor** RICHARD M CROOKS, University of Texas at Austin, Xiang Li, Karen Scida
2:45	(1870-3)	**Using Widely Available Electrochemical Device for Sensing or Diagnostics** YI LU, University of Illinois at Urbana-Champaign, JingJing Zhang, Yu Xiang
3:20		**Recess**
3:35	(1870-4)	**Electrochemical Analysis of Clinically-Relevant Biomolecular Analytes Using Nanostructured Microelectrodes** SHANA KELLEY, University of Toronto
4:10	(1870-5)	**Signal Amplification for Biosensing Based on Nucleic Acid Recognition** JU HUANGXIAN, Nanjing University

SYMPOSIUM — Session 1880

Vibrational Spectroscopy of Biodegradable Plastics: Evolution, Revolution or Back to the Future
arranged by John F Rabolt, University of Delaware

Wednesday Afternoon, Room B310
John F Rabolt, University of Delaware, Presiding

1:30		**Introductory Remarks - John F Rabolt**
1:35	(1880-1)	**Spectroscopic Study of Structural Evolution Dynamics of Bio-Based and Biodegradable Poly(hydroxyalkanoate) Copolymers** ISAO NODA, University of Delaware, Brian Sobieski, Liang Gong, Bruce Chase, John F Rabolt
2:10	(1880-2)	**Degradation Mechanisms of Poly(lactic acid)-Nanoparticle Composite and Phthalate Plasticized Poly(vinyl chloride)** ZHAN CHEN, University of Michigan
2:45	(1880-3)	**Vibrational Spectroscopic Studies on Biodegradable Polymer Microstructures** SHAW L HSU, University of Massachusetts
3:20		**Recess**
3:35	(1880-4)	**2D IR Correlation Study of Thin Films of Biodegradable PHA/PEG Blend** YOUNG MEE JUNG, Kangwon National University, Yeonju Park, Yujing Chen, Isao Noda
4:10	(1880-5)	**Characterization of Single Electrospun Biopolymer Nanofibers Using AFM-IR and Selected Area Electron Diffraction** JOHN F RABOLT, University of Delaware, Liang Gong, Isao Noda, Bruce Chase, Curtis Marcott, C J McBrin, David Martin, Jinglin Liu, Chao Ni

WORKSHOPS — Session 1890

CACA - How to be Successful in Your Career
arranged by Tao Jiang, Mallinckrodt Pharmaceuticals and Michael Ye, Supelco/Sigma-Aldrich

Wednesday Afternoon, Room B311
Tao Jiang, Mallinckrodt Pharmaceuticals, Presiding

1:30		**Introductory Remarks - Tao Jiang and Michael Ye**
1:35	(1890-1)	**Insights on Job Searching** MICHAEL W DONG, MWD Consulting
2:05	(1890-2)	**Recognizing and Managing Career Passover** ROBERT L STEVENSON, American Laboratory
2:35	(1890-3)	**Advancing Academic Career as a Faculty** YI HE, John Jay College/CUNY
3:05		**Recess**
3:20	(1890-4)	**Career Development Workshop** NAIDONG WENG, JNJ
3:50	(1890-5)	**From a Bench Chemist to an Entrepreneur** BRUCE LIU, ACB Scitech, Inc. N/A
4:20	(1890-6)	**Go Wider and Higher** CHUPING LUO, Waters Corporation

WORKSHOPS — Session 1900

Natural Health Products: Scientific Approaches to Securing Product Quality and Safety
arranged by Rob O'Brien, ISURA and Bob Chapman, National Research Council of Canada

Wednesday Afternoon, Room B313
Rob O'Brien, ISURA, Presiding

1:30		**Introductory Remarks - Rob O'Brien and Bob Chapman**
1:35	(1900-1)	**Development of a Chemical Barcoding Methodology to Identify and Support the Quality and Safety of Functional Ingredients** BOB CHAPMAN, National Research Council of Canada, Fabrice Berrue, Junzeng Zhang, Ian Burton, Joseph Hui, Sabrena MacKenzie, Camilo Martinez-Farina, Aissa Harhira, El Haddad Josette, Mohamad Sabsabi, Alain Blouin, Yuan-Chun Ma, Rob O'Brien
2:05	(1900-2)	**Utility and Limitations of DNA Barcoding and Next Generation Sequencing for Herbal Product Authentication** JONATHAN VAN HAMME, Thompson Rivers University
2:35	(1900-3)	**Developing Metabolomics Approaches for Natural Product Authentication** LIANG LI, University of Alberta
3:05		**Recess**
3:20	(1900-4)	**Natural Health Product Quality Control: Addressing the Current Issue of Adulterants** YUAN-CHUN MA, Canadian Phytopharmaceuticals Corporation
3:50	(1900-5)	**Practical Gas Chromatographic based Approaches to Confirm Identity of Powdered Herbal Products** ROB O'BRIEN, ISURA, Anderson Smith
4:20:50	(1900-6)	**TLC and HPLC Fingerprinting for Authentication of Natural Health Products** RUDOLF BAUER, University of Graz

ORGANIZED CONTRIBUTED SESSIONS — **Session 1910**

High Throughput Analysis for Food Safety and Cosmetics: Challenges and Validation
arranged by Perry G Wang, US FDA and Xiaogang Chu, China Academy of Inspection and Quarantine

Wednesday Afternoon, Room B315
Perry G Wang, US FDA, Presiding

1:30	(1910-1)	**Rapid Determination of Non-Allowed Active Pharmaceutical Ingredients for the Treatments of Hair Loss in Cosmetics Using UHPLC-HRMS** WANLONG ZHOU, US FDA, Perry G Wang, James B Wittenberg, Maria A Dionisio De Sousa, Alexander J Krynitsky
1:50	(1910-2)	**Advances in High-Throughput Analysis for Determination of Marine Biotoxins in Seafood** PEARSE MCCARRON, National Research Council, Daniel Beach
2:10	(1910-3)	**Determination of Prostaglandin Analogs in Eye Area Cosmetic Products by High Performance Liquid Chromatography with Tandem Mass Spectrometry** JAMES B WITTENBERG, Food and Drug Administration, Wanlong Zhou, Perry G Wang, Alexander J Krynitsky
2:30	(1910-4)	**Withdrawn**
2:50		**Recess**
3:05	(1910-5)	**Analysis of Color Additives (Permitted and Non-Permitted) in Different Food Matrices by a HPLC Method** SNEH D BHANDARI, Merieux NutrSciences, Tiffany Gallegos-Peretz
3:25	(1910-6)	**A Comprehensive Approach on Food Safety Analysis, Screening and Quantitation by Using Data Independent Acquisition (DIA) and DDMS2 on HR/AM Q Exactive System** JAMES S CHANG, Thermo
3:45	(1910-7)	**Separation of Aminoglycoside Antibiotics by Using Hydrophilic Interaction Liquid Chromatography** YU LONG, Dalian Institute of Chemical Physics, Wei Jie, Shen Aijin, Guo Zhimou, Liang Xinmiao
4:05	(1910-8)	**Open Discussion**

ORGANIZED CONTRIBUTED SESSIONS — **Session 1920**

Precision Bioanalytical Measurements
arranged by Robert Dunn, University of Kansas

Wednesday Afternoon, Room B316
Robert Dunn, University of Kansas, Presiding

1:30	(1920-1)	**Analytical Precision in the Age of Metabolomics – Focus on the Fundamentals** HOWARD HENDRICKSON, University of Arkansas for Medical Sciences, Lin Song
1:50	(1920-2)	**Biocompatible Self-Tuning Nanomaterials Improve the Precision of Biomolecule Processing and Separation** LISA A HOLLAND, West Virginia University, Srikanth Gattu, Brandon C Durney, Cassandra Crihfield
2:10	(1920-3)	**Analytical Approaches for Environmental Metabolomics and Ecotoxicity Modeling** CYNTHIA K LARIVE, University of California - Riverside, Corey M Griffith, Melissa M Morgan
2:30	(1920-4)	**Extending the Free Drug Hypothesis: Physical Properties Driving Asymmetric Tissue Distribution** DENNIS O SCOTT, Pfizer
2:50		**Recess**
3:05	(1920-5)	**Barbeques and Tornadoes: Analytical Strategies for Metal Ion Determinations Using Separations and Sensors** FIONA REGAN, Dublin City University
3:25	(1920-6)	**A Mass Spectrometry Based High Throughput Screening Approach with Exquisite Selectivity** HEATHER R DESAIRE, University of Kansas, Imaduwage Kasun
3:45	(1920-7)	**Microfluidic Devices with Integrated Electrodes for Monitoring Cellular Systems** R SCOTT MARTIN, Saint Louis University
4:05	(1920-8)	**Novel Applications of Microdialysis Sampling: Where No Probe has Gone Before!** SARA R THOMAS, Kansas University, Craig E Lunte, Susan M Lunte

ORGANIZED CONTRIBUTED SESSIONS — **Session 1930**

Technology Strategies for Explosives Sensing
arranged by Michael Shepard, Dept. Homeland Security, S&T Directorate and Samar K Guharay, The MITRE Corporation

Wednesday Afternoon, Room B401
Samar K Guharay, The MITRE Corporation, Presiding

1:30	(1930-1)	**Overview of Operational Challenges to Domestic Security Screening** MICHAEL SHEPARD, US Dept. of Homeland Security
1:50	(1930-2)	**Generalized Systems Analysis Framework** SAMAR K GUHARAY, The Mitre Corporation
2:10	(1930-3)	**Statistical Methods for Constructing Chemometric Signatures in Noisy Environments** CHRISTOPHER P SAUNDERS, South Dakota State University
2:30	(1930-4)	**Operational Outlook for Remote Trace Explosives Detection** RODERICK KUNZ, MIT Lincoln Laboratory
2:50		**Recess**
3:05	(1930-5)	**Pushing the Limits of Trace Detection Technology** STEFAN LUKOW, Morpho Detection
3:25	(1930-6)	**Optimized Sampling and Analysis Strategies for Trace Explosives Detection** GREG GILLEN, NIST
3:45	(1930-7)	**Chemical Attribution Signatures of Homemade Explosives** JOE CHIPUK, Signature Science N/A
4:05	(1930-8)	**Open Discussion**

ORAL SESSIONS — **Session 1940**

Advances in Fuel and Petrochemical Analyses

Wednesday Afternoon, Room B402
John Baltrus, US Department of Energy-NETL, Presiding

1:30	(1940-1)	**Chemical Fingerprinting of Crude Oils: Gaining an Extra Dimension from GCxGC–TOF MS** LAURA MCGREGOR, Markes International Ltd, Nicola Watson, Kevin Collins, Chris Hall, Ken Umbarger
1:50	(1940-2)	**Fast and Accurate Analysis of Extended Natural Gas Composition and Physical Properties Using a Temperature Programmable Gas Analyzer** DEBBIE ALCORN, INFICON
2:10	(1940-3)	**Application of Polymeric Ionic Liquids as Highly Robust and Selective Stationary Phases for Comprehensive Two-Dimensional Gas Chromatography** CHENG ZHANG, Iowa State University, Rodney A Park, Jared L Anderson
2:30	(1940-4)	**Solving Industrial Problems by Determining Compound Classes in Refinery Streams and Products** CHRIS GOSS, Alberta Innovates Technology Futures, Dan Wispinski, Lee Marotta
2:50		**Recess**
3:05	(1940-5)	**Optical Sensors for the Detection of Nitrogen Compounds in Aviation Fuels** ROBERTO A FEDERICO-PEREZ, University of Tennessee, Ziling (Ben) Xue
3:25	(1940-6)	**Fast Profiling of Petrochemical Samples by Thermal Analysis- Soft Ionization Mass Spectrometry: From Source Rock and Kerogen via Crude Oil to Petrochemical Products** RALF ZIMMERMANN, University Rostock /HMGU, Thorsten Streibel, Mohammad Saraji, Michael Fischer , Sebastian Wohlfahrt, Andreas Walte, Thomas Denner
3:45	(1940-7)	**Micro-Hollow Discharge Analysis of Carbonaceous Materials** RANDY VANDER WAL, Penn State University, Chethan K Gaddam N/A
4:05	(1940-8)	**On-Site Fuel Analysis Using a Portable Near-Infrared Spectrometer** STUART FARQUHARSON, Real-Time Analyzers, Inc, Carl Brouillette, Wayne Smith, Chetan Shende

TECHNICAL PROGRAM

ORAL SESSIONS — Session 1950

Bioanalytical: Using Microfluidics/Lab-on-a-Chip Techniques

Wednesday Afternoon, Room B404
Casey Burton, Missouri University of Science and Technology, Presiding

1:30	(1950-1)	**Pressure-Actuated Integrated Microfluidic Devices for Biomarker Analysis** VISHAL SAHORE, Brigham Young University, Suresh Kumar, Adam T Woolley
1:50	(1950-2)	**Real-Time Imaging of Cancer Cell Chemotaxis in Paper-Based Scaffolds** RACHAEL M KENNEY, University of North Carolina at Chapel Hill, Matthew W Boyce, Andrew S Truong, Matthew R Lockett
2:10	(1950-3)	**Observation of Rapid Changes in Drug Susceptibility of Tumor Cells in a Hypoxia Microfluidic Culture Device** TODD GERMAIN, Texas Tech University, Dimitri Pappas
2:30	(1950-4)	**Magnetic Microbead Based Capture and Labeling of DNA from Carbapenemase Resistance Genes** RILEY K MILLS, Brigham Young University, Radim Knob, Adam T Woolley N/A
2:50		Recess
3:05	(1950-5)	**Screening Small Molecule Modulators of Cellular Chemotaxis in Paper-Based Invasion Assays** C CHAD LLOYD, University of North Carolina at Chapel Hill, Matthew R Lockett, Rachael M Kenney, Andrew S Truong, Matthew W Boyce, Christian A Lochbaum
3:25	(1950-6)	**Microchip Affinity Monoliths for Solid Phase Extraction of DNA for Bacteria Infection Detection** RADIM KNOB, Brigham Young University, Riley K Mills, Adam T Woolley
3:45	(1950-7)	**MALDI-IMS Evaluation of 3D Cell Cultures Treated with Combination Chemotherapeutics by a 3D Printed In-Vitro PK/PD Microfluidic Platform** GABRIEL J LABONIA, University of Notre Dame, Amanda B Hummon
4:05	(1950-8)	**Fully Automated Microfluidic Input/Output Multiplexer for Endocrine Tissue Culturing and Hormone Secretion Sampling** XIANGPENG LI, Auburn University, Jessica C Brooks, Katarena Ford, Christopher J Easley

ORAL SESSIONS — Session 1960

Environmental Water Quality and Analysis

Wednesday Afternoon, Room B403
John P Auses, University of Pittsburgh, Presiding

1:30	(1960-1)	**Targeted Discovery of Disinfection By-Products in Swimming Pools and Hot Tubs** SUSAN D RICHARDSON, University of South Carolina, Jonathan D Byer, Eric J Daiber, Sridevi A Ravuri, Joseph E Binkley, Christina M Joseph
1:50	(1960-2)	**Evaluation of Iodo-, Bromo-, and Chloro-Acetic Acids Formation by Peracetic Acid Disinfection Using a Newly Developed Rapid HPIC-MS/MS Method** RUNMIAO XUE, Missouri University of Science and Technology, Honglan Shi, Yinfa Ma, John Yang, Bin Hua, Enos Inniss, Craig Adams, Todd Erichholz
2:10	(1960-3)	**Removal of Pharmaceutical Products from Wastewater Using Magnetic Bio-Char** AKILA G KARUNANAYAKE, Mississippi State University, Olivia A Todd, Todd E Mlsna
2:30	(1960-4)	**Determination of Priority Water Contaminants by Solid-Phase Extraction and UFLC-MS/MS Method** HAITING ZHANG, Missouri University of Science and Technology, Danielle West, Honglan Shi, Yinfa Ma, Craig Adams, Todd Erichholz
2:50		Recess
3:05	(1960-5)	**Characterization of Oil-based Pollutants Using Webcam-based Spectrometer** YAGIZ SUTCU, InfoScope Research, Aysegul Ergin
3:25	(1960-6)	**Coupling of Thin Film Microextraction Techniques to Portable GC-TMS Instrumentation for the On-Site, Sub-ppb, Detection of Pesticides From Environmental Waters** JONATHAN J GRANDY, University of Waterloo, Janusz Pawliszyn
3:45	(1960-7)	**SPE in US EPA Method 625, A Step Closer to Reality after Good Performance in Two Round Robins** ZOE GROSSER, Horizon Technology, William Jones, David Gallagher, Michael Ebitson, Alicia Cannon
4:05	(1960-8)	**Adsorption of Pb^{2+} from Aqueous Solution Using Low-Cost Chitosan-Modified Biochar, A Green Adsorbent** NARADA W BOMBUWALA DEWAGE, Mississippi State University, Todd E Mlsna

ORAL SESSIONS — Session 1970

GC Fuels, Energy and Petrochemical

Wednesday Afternoon, Room B405
Patricia Ranaivo, Indiana University Southeast, Presiding

1:30	(1970-1)	**Zeolite-Loaded Metal-Catalyzed Hydrotreatment of Lignin to Aromatic Monomers Using Subcritical Water** ERIC A BOAKYE, South Dakota State University, Douglas Raynie
1:50	(1970-2)	**Olefins in Refinery Streams by GC-VUV** DAN WISPINSKI, Alberta Innovates Technology Futures, Chris Goss, Phillip Walsh
2:10	(1970-3)	**GC-Ion Mobility Spectrometry for Determination of Ageing of Mineral Oil Impregnated Presspaper Isolation Systems** WOLFGANG VAUTZ, ISAS, Liedtke Sascha, Torben Muth, Frank Jenau
2:30	(1970-4)	**High Level Fixed Gas Analysis, Including Hydrogen, with one Detector and One Carrier** MATTHEW MONAGLE, AIC LLC N/A
2:50		Recess
3:05	(1970-5)	**Analysis of Dissolved Hydrocarbon Gases in Water – Pitfalls and Improvements** MARK L BRUCE, TestAmerica
3:25	(1970-6)	**Performance Evaluation of Modern Stainless Steel Capillary GC Columns** REBECCA STEVENS, Restek Corporation, Corby Hilliard, Amanda Rigdon, Linx Waclaski, Jaap de zeeuw
3:45	(1970-7)	**Determination of Polycyclic Aromatic Sulfur Heterocycles and Their Alkyl-Substituted Derivatives in Standard Reference Material 1597a** WALTER B WILSON, National Institute of Standards and Technology, Stephen A Wise
4:05	(1970-8)	**Vehicle Interior Air Quality - (S)VOC Emission from Materials: Regulation, Standard Methods and Analytical Implementation** CAROLINE WIDDOWSON, Markes International

ORAL SESSIONS — Session 1980

Mass Spectrometry - Bioanalytical and Omics

Wednesday Afternoon, Room B406
Logan T Miller, Duquesne University, Presiding

1:30	(1980-1)	**Mass Spectrometry in Discovery of Lipid Markers for Alzheimer's Disease** SATYA GIRISH CHANDRA AVULA, Cleveland State University, Christine Reece, Jagan A Pillai, Yan Xu
1:50	(1980-2)	**Novel Cationization Strategies for Improved Separation of Metabolite Isomers with Ion Mobility – Mass Spectrometry** CHRISTOPHER D CHOUINARD, University of Florida, Robin Kemperman, Harrison King, Christopher R Beekman, Richard Alan Yost
2:10	(1980-3)	**Development of Dried Matrix Card Quantification by Speciated Isotope Dilution Mass Spectrometry Using Elution and Laser Desorption Techniques** LOGAN T MILLER, Duquesne University, Kaitlin Miller, Sarah Sheffield, Skip Kingston, Matt Pamuku, Scott Faber, Mark Little, Silverio Iacono
2:30	(1980-4)	**Protease-Containing Membranes for Rapid Antibody Digestion Prior to Mass Spectrometry Analysis** YONGLE PANG, Michigan State University, Wei-Han Wang, Gavin Reid, Donald F Hunt, Merlin Bruening
2:50		Recess
3:05	(1980-5)	**Quantitative Environmental Human Health Assessment of Inorganic Elements in Dried Blood Spots Using Direct Isotope Dilution Laser Ablation and Elution Mass Spectrometry** SARAH SHEFFIELD, Duquesne University, Logan T Miller, Scott Faber, Matt Pamuku, Skip Kingston
3:25	(1980-6)	**Identification of the Cell Surface N-Glycoproteome by MS-based Proteomics** JOHANNA SMEEKENS, Georgia Institute of Technology, Weixuan Chen, Ronghu Wu
3:45	(1980-7)	**Probing the Cell-Surface N-Glycoproteome with Metabolic Chemical Reporters (MCRs)** HAOPENG XIAO, Georgia Tech
4:05	(1980-8)	**Quantitative Characterization of Protein Content from HDL and LDL Size Fractions** BRYAN A PARKS, Centers for Disease Control and Prevention, Zsuzsanna Kuklenyik, David Schieltz, Michael S Gardner, Jon Rees, Lisa McWilliams, Yulanda Williamson, John R Barr

ORAL SESSIONS **Session 1990**

Metabolomics, Proteomics, and Genomics

Wednesday Afternoon, Room B407

Qingbo (Roger) Yang, Missouri University of Science and Technology, Presiding

1:30	(1990-1)	**A Combination of Multidimensional Chromatography and High Resolution Mass Spectrometry for Chemical Exposure Analysis** DAVID E ALONSO, Leco Corporation, Joseph E Binkley, Elizabeth M Humston-Fulmer, Jonathan D Byer, Lorne E Fell
1:50	(1990-2)	**Metabolomics of Nonalcoholic Fatty Liver Disease via LC-MS** RAINEY E PATTERSON, University of Florida, Richard Alan Yost, Timothy J Garrett, Nishanth E Sunny, Srilaxmi Kalavalapalli, Kenneth Cusi
2:10	(1990-3)	**Application of PDMS-Overcoated Matrix-Compatible Solid Phase Microextraction Fiber Coupled to Comprehensive Two-Dimensional Gas Chromatography–Time-of-Flight Mass Spectrometry for Improved Exploitation of Chromatographic Space in Global Chemical Profiling of Brazilian Cachaça** ERICA A SOUZA-SILVA, Universidade Federal do Rio Grande do Sul, Fernando C Fontanive, Claudia A Zini
2:30	(1990-4)	**Application of the Isotopic Ratio Outlier Analysis Phenotypic Protocol for Metabolomic Biomarker Discovery in Type 1 Diabetes Using T Cells** CANDICE Z ULMER, University of Florida, Richard Alan Yost, Timothy J Garrett, Christopher A Beecher, Jing Chen, Clayton Matthews
2:50		**Recess**
3:05	(1990-5)	**Secretome of Murine Islets of Langerhans** ANDREW W SCHMUDLACH, University of Notre Dame, Jeremy Felton, Robert T Kennedy, Norman J Dovichi
3:25	(1990-6)	**Optimizing Sampling Protocols for the Identification and Quantitation of Neuropeptides from Brain Tissues** NING YANG, University of Illinois at Urbana-Champaign, Stanislav S Rubakhin, Jonathan V Sweedler, Krishna D Anapindi
3:45	(1990-7)	**A Study of Protein Degradation and the Effect of Proteasome Inhibition by Combining Biorthogonal Noncanonical Amino Acid Labeling with Click Chemistry** ZHENYU ZHOU, Georgia Institute of Technology N/A
4:05	(1990-8)	**Benchmarking of DNA-seq and RNA-seq Variant Detection Software** GANG FENG, Northwestern University, Zhujun Huang, Lei Huang, Riyue Bao, Hongmei Jiang, Yuchen Bai, Xiaoyong Sun N/A

ORAL SESSIONS **Session 2000**

Raman, SERS, UVRR Applications

Wednesday Afternoon, Room B408

Matt Cerutti, Thermo Scientific, Presiding

1:30	(2000-1)	**UV Resonance Raman (UVRR) Structural Studies of Polyglutamine (polyQ) Side Chains and Fibrils** DAVID PUNIHAOLE, University of Pittsburgh, Zhenmin Hong, Ryan Jakubek, Elizabeth Dahlburg, Riley Workman, Jeffry Madura, Sanford A Asher
1:50	(2000-2)	**Sheath-Flow Microfluidic Approach for Combined Surface Enhanced Raman Scattering and Electrochemical Detection** MATTHEW R BAILEY, University of Notre Dame, Amber Pentecost, Asmira Selimovic, R Scott Martin, Zachary D Schultz
2:10	(2000-3)	**Raman Spectroscopy Monitors Glutamine and Asparagine Side Chain OCCC Dihedral Angles** RYAN JAKUBEK, University of Pittsburgh, David Punihaole, Zhenmin Hong, Elizabeth Dahlburg, Steven Geib, Sanford A Asher
2:30	(2000-4)	**Assessment of the Protein–Protein Interactions in a Highly Concentrated Antibody Solution by Using Raman Spectroscopy** CHIKASHI OTA, Horiba, Ltd., Shintaro Noguchi, Satoru Nagatoishi, Kouhei Tsumoto
2:50		**Recess**
3:05	(2000-5)	**Raman Imaging of Samples with Complex Surface Topographies** TIM BATTEN, Renishaw plc, Tim Smith
3:25	(2000-6)	**Improving Raman and Surface Enhanced Raman Spectroscopy Through Effective Sampling** DONGMAO ZHANG, Mississippi State University N/A
3:45	(2000-7)	**Spatially Offset vs Conventional Raman for Through-Barrier Material Identification** ROBERT STOKES, Cobalt Light Systems, Paul Loeffen, Darren Andrews, Ken Mann, Oliver Presly, Pavel Matousek
4:05	(2000-8)	**Degradation Analysis of International Space Station Medications by Raman Spectroscopy** STUART FARQUHARSON, Real-Time Analyzers, Inc, Chetan Shende, Alexander May, Carl Brouillette, Joseph Cosgrove

ORAL SESSIONS **Session 2010**

Surface and Microscopic Characterization of Nanostructures and Biological Materials

Wednesday Afternoon, Room B409

Brian R Strohmeier, United States Steel Corporation, Presiding

1:30	(2010-1)	**Fundamental Understanding of the Synergy Between Electroactive Poly (amic) Acid Membranes and Their Interaction with Nanoparticles** VICTOR KARIUKI, Binghamton University
1:50	(2010-2)	**Optical Property and Catalytic Activity of Gold Nanorods End-capped with a Second Metal** GUFENG WANG, North Carolina State University, Nathalia Ortiz, Vineet Kumar
2:10	(2010-3)	**Autocorrelation Function Analysis of Rotational Dynamics of Gold Nanorod** KUANGCAI CHEN, Iowa State University/Georgia State University, Ning Fang
2:30	(2010-4)	**Antibody-Like Biorecognition Sites for Proteins from Surface Imprinting on Nanoparticles** SNEHASIS BHAKTA, University of Connecticut, Saiful Seraji, James F Rusling, Steven L Suib
2:50		**Recess**
3:05	(2010-5)	**Selective Raman Imaging of Integrin Receptors Through Coupled Plasmonic Nanostructures** LIFU XIAO, University of Notre Dame, Zachary D Schultz, Hao Wang
3:25	(2010-6)	**Deep and High-Resolution Three-Dimensional Tracking of Single Particles Using Nonlinear and Multiplexed Illumination** TIM YEH, University of Texas at Austin
3:45	(2010-7)	**Development of a Dual Microscope System for Integration of Intracellular Calcium Imaging with Monitoring Insulin Secretion from Islets of Langerhans** LIAN YI, Florida State University, Xue Wang, Michael G Roper
4:05	(2010-8)	**Establishing Statistical Criteria for Detecting Molecular Motion in Single-molecule Diffusional Trajectories** MOUSSA BARHOUM, University of Utah, Karl-Heinz Gericke, Joel M Harris N/A

ORAL SESSIONS **Session 2020**

Trace Explosives Detection - Half Session

Wednesday Afternoon, Room B301

Maria K Ferguson, PA Department of Environmental Protection, Presiding

1:30	(2020-1)	**Analysis of the Decomposition of Hexamethylene Triperoxide Diamine (HMTD) as Determined by SPME-GC/MS and LC/MS** LAURYN DEGREEFF, Naval Research Laboratory, Christopher Katilie, Frank L Steinkamp
1:50	(2020-2)	**Separation of Inorganic Ions and Neutral Organic Nitroaromatic Compounds by Electrokinetic Chromatography** JULIE R MCGETTRICK, University of Montana, Christopher P Palmer
2:10	(2020-3)	**Comparison of Inter-Instrument Relative Response Factors with Thermal Desorption Internal Standards** H MITCHELL RUBENSTEIN, USAF, Maomian Fan, Claude C Grigsby, Kathy Fullerton, Brian Geier, Darrin Ott
2:30	(2020-4)	**Through-Barrier Explosives and Hazardous Material Detection Using a Handheld Spatially Offset Raman Spectrometer** PAUL LOEFFEN, Cobalt Light Systems, Robert Stokes, Ken Mann, Darren Andrews, Oliver Presly, Pavel Matousek

UNDERGRADUATE POSTER SESSION — Session 2030

All posters are to be mounted by 10:00 AM and remain on display until 4:00 PM. Authors must be at their posters from 1:00 PM to 3:00 PM. Location of the afternoon posters is on the Exposition Floor, 400 Aisle. PLEASE NOTE: You cannot get onto the Exposition Floor until after 9:00 AM.

Undergraduate Poster Session

Wednesday Afternoon, Exposition Floor, 400 Aisle

(2030-1 P) **Methods for Improving Cytokine in Vivo Calibration During Microdialysis Sampling** PATRICK M PYSZ, University of Arkansas, Tina M Poseno, Julie Stenken N/A

(2030-2 P) **Analysis of Nicotine Levels in Electronic Cigarettes** MARTIN ERIC MILLER, Austin Peay State University, Jacob Williams, Dillon Burrow

(2030-3 P) **Thermal Behavior of Barium and Strontium Carbonates** CHARLES MANSFIELD EARNEST, Berry College, Ethan Miller

(2030-4 P) **Improved Quantification of Gibbsite in Bauxite Ores** CHARLES MANSFIELD EARNEST, Berry College, Karla Gann, Britney Stong

(2030-5 P) **Signal Enhancement Compensation in ICP-MS Analysis for Arsenic in the Presence of Organics** ADAM KAGEL, University of California, Davis, Carla Kagel, Richard Kagel, Nels Worden

(2030-6 P) **Isotope Labeling in Astrobiology: Ethanol as a Carbon Source** NICOLE G PERKINS, California State Polytechnic University, Pomona, Gregory A Barding, Rakesh Mogul

(2030-7 P) **Sulfidation of Silver Nanoparticles** NATHANIEL D FLETCHER, College of Charleston, Katherine M Mullaugh

(2030-8 P) **Photodegradation and Ecotoxicity Studies of Sertraline, Fluoxetine, and their Photodegradants** SYLVIA C DAVILA, College of Charleston, Jessica R Hinson, Allison Welch, Wendy C Cory

(2030-9 P) **Photodegradation of Bupropion and Gabapentin** NEHA V MUPPALA, College of Charleston, Kristina K Tran, Wendy C Cory

(2030-10 P) **Accelerated Degradation of Diphenhydramine** JAMES L SOLOMON, College of Charleston, Michael T Blanton, Wendy C Cory N/A

(2030-11 P) **Preliminary Investigation of the Geographical Distribution of Synthetic UV Filters in Bahamian Surface Waters** AEJIN KIM, Elmira College, Christopher B Hall, Emily J Feldpausch, Betsy A Smith, Jared S Baker

(2030-12 P) **Surface Plasma Polaritons from Voltage Charged Gold Nanotubes** SAM KONCHAN, University of Florida, Pradeep Ramiah Rajasekaran, Charles R Martin, Alec Talin

(2030-13 P) **Electroporation Facilitation via Gold-Plated Membranes** AARON WILSON, University of Florida, Juliette Experton, Charles R Martin

(2030-14 P) **The Analysis of Electronic Cigarette Solutions by ICP-MS for USP Regulated Contaminates** CHLOE E FERNANDES, Georgia Gwinnett College, Daniel H Jones

(2030-15 P) **Comparison of Cellular Response to Atorvastatin Among Selected Cell Lines by MS-Based Proteomics** GEORGE TANG, Georgia Institute of Technology N/A

(2030-16 P) **Discovery Metabolomics of Early-Stage Ovarian Cancer in a Dicer-Pten Double Knockout Murine Model** LAURA C WINALSKI, Georgia Institute of Technology, Christina Jones, Maria E Monge, Jaeyeon Kim, Martin M Matzuk, Facundo M Fernandez

(2030-17 P) **Colorimetric Detection of Pyrocatechol as a Model for Urushiol Analysis in Poison Ivy** COLLIN J STEEN, Kalamazoo College, Kari Anderson

(2030-18 P) **Quantification of Salicylates in Stomach Relief Aids Using GC-MS and UV-Vis** MARINA CHRISTINA KOETHER, Kennesaw State University, Kimberly C Powers

(2030-19 P) **Correlation Between Different Extraction Methods and the Ratio of Neral to Geranial in the Essential Oils of Lemongrass by GC-MS** MARINA CHRISTINA KOETHER, Kennesaw State University, Skyler Mize

(2030-20 P) **Determining Dissolution Testing Time of Potassium in Potassium Gluconate Tablets by Conductivity with Confirmation by Flame Atomic Absorption Spectrometry** MARINA CHRISTINA KOETHER, Kennesaw State University, Minwoo Lee

(2030-21 P) **Study of the Implementation of the Systematic Method in General Chemistry II** VICTORIA Y REINDERS, Maryville University Saint Louis, Thomas M Spudich, Natalie Ulrich

(2030-22 P) **Characterization and Identification of Biosurfactants for Oil Remediation** KAYDREN B ORCUTT, Mercer University, Justis E Ward, Joseph W Kloepper, Garland L Crawford, Kathryn D Kloepper

(2030-23 P) **Green Diesel and Biodiesel Fuel Additives and Their Effects on VOC Emissions** CHRISTOPHER MOORE, Middle Tennessee State University, Joseph Close N/A

(2030-24 P) **Investigating Dopamine Fluctuations Associated with Impulsive Decision Making Using Fast Scan Cyclic Voltammetry** BRENNEN GUZIK, North Carolina State University, Xiaohu Xie, Leslie A Sombers

(2030-25 P) **Tracking Cellular Invasion and Death in Three Dimensional Paper-Based Cultures with Quantitative PCR** CHRISTIAN A LOCHBAUM, University of North Carolina at Chapel Hill, Andrew S Truong, Matthew R Lockett

(2030-26 P) **Real-Time Measurements of Oxidative Stress During Chronic L-DOPA Treatment for Parkinson's Disease** CATHERINE F MASON, North Carolina State University

(2030-27 P) **Pulsed Chronopotentiometry with Asymmetric Cellulose Triacetate Membrane-Based Ion-Selective Electrodes for Kinetic Discrimination of Lipophilic Ions During Measurement of Chloride** SIMON SEGAL, Northern Kentucky University, Jeremy Meyers, Kebede L Gemene

(2030-28 P) **Water Quality Analysis of the Chattahoochee River** KIZGEL DAVIS-DESOUZA, Oglethorpe University, Kelly Jacobson, Md H Kabir

(2030-29 P) **Physicochemical Characterization of Biodiesel from Different Vegetable Oils and Seeds** JOSEPH GOODWIN, Oglethorpe University, Michelle Huang, Grace Djokoto, Markus Germann, Md H Kabir N/A

(2030-30 P) **Colloidal CdSe Quantum Dots: Synthesis, Characterization, and Applications** KELLY JACOBSON, Oglethorpe University, Md H Kabir, Michael Rulison

(2030-31 P) **Nanoparticle Toxicity on the Development of Brine Shrimp and Zebrafish** CAITLIN MAY, Oglethorpe University, Gregory Gabriel, Michael Rulison, Md H Kabir

(2030-32 P) **Evaluation of Phthalate Wiping Protocols for Estimation of Dermal Exposure from Consumer Products** ALEXANDRIA VAN GROUW, University of Portland, Richard Kagel, Carla Kagel, Stephen McWeeney

(2030-33 P) **Gaseous Molecular Analysis by High Resolution Coherent Multidimensional Spectroscopy (HRCMDS)** ANGELAR KANINI MUTHIKE, Spelman College, Peter Chen, Jessica Robinson

(2030-34 P) **Spectral Imaging of Plasma Optical Emission Via Compressed Sensing** JOHN D USALA, Texas Tech University, Gerardo Gamez

(2030-35 P) **Analysis of Melamine in Solid Pet Food Samples Using Gold Nanoparticles and Reversed Phase Chromatography** AARON HUMMERT, Washburn University, Seid Adem

(2030-36 P) **Using a Solvent-Free Synthesis-Grafting Method to Attach Phenanthroline and Dimethoxybenzene Groups to Glassy Carbon Electrodes** SOPHIA L MELNYK, Wittenberg University, Kristin K Cline N/A

(2030-37 P) **Does Grinding Glassy Carbon Electrodes in Diazonium Salts Lead to Covalently Bonded Groups?** CHELSEA L HORVATH, Wittenberg University, Kristin K Cline

(2030-38 P) **Comparison of Grafted and Untreated Activated Carbon as Solid Phase Extraction Media for Preconcentrating Copper and Lead Ions** MARGARET COLE, Wittenberg University, Kristin K Cline

(2030-39 P) **Barium Leaching in Ceramic Glazes** JASON HALMO, Hampden-Sydney College, Joshua Chamberlin, Paul Mueller

(2030-40 P) **Spectroelectrochemical Urinalysis: A Kinetic Assay for Uric Acid** PAUL FLOWERS, University of North Carolina at Pembroke, Sean Downes, Sonvia Brown

(2030-41 P) **Determination of Formaldehyde Concentration in Electronic Cigarettes** ROLAND LANDERS, Cumberland University, Sarah Pierce

(2030-42 P) **Matrix Targeting Peptide Impact on Tertiary Folding of Cargo Proteins** TYLER J SMITH, Truman State University, Brian P Adams, Bethany P Manning

(2030-43 P) **Electrospray Mass Spectrometry and Density Functional Theory Studies of the Estrone Fragmentation Mechanisms** YASSIN JEILANI, Spelman College, Daphney Sihwa, Gabrielle Webb, Nasrin Aweis

(2030-44 P) **Metals in Mushrooms of Western Pennsylvania** KAELYN MARGARET GRESKO, California University of PA, Kimberly A Woznack N/A

(2030-45 P) **Method Development for the Analysis of Pesticide Degradates by GC-ECD** JESSICA REILLY, Saint Francis University, Samantha Radford

(2030-46 P) **Investigation of Iron Dissolution from Pyrite Electrodes Using Electrochemical and Atomic Absorption Methods** KATELYN NUSBAUM, Saint Francis University, Brandyn Pryce, Rose A Clark

(2030-47 P) **The Efficacy of Duckweed in Reducing the Concentration of Manganese in Abandoned Mine Drainage (AMD) through Phytoremediation** REBECCA ANNE BRADNAM, Westminster College, Helen M Boylan

(2030-48 P) **Analysis of Water Quality in Western Pennsylvania near Hydraulic Fracturing Sites** KELSEY ANN KILBANE, Westminster College, Helen M Boylan, Jamie Linderman, Christina Mauri

(2030-49 P) **Electrophoretic Character of Borate Buffers in Capillary and Microfluidic Channels** LAUNICK SAINT-FORT, Pennsylvania State University-Berks Campus, James Karlinsey

(2030-50 P) **Interdisciplinary Undergraduate Research in Chemometrics: The Students' Perspective** STEPHANIE HOMITZ, Westminster College, Helen M Boylan, Christopher Caroff, Keilah Ireland, Carolyn Cuff

(2030-51 P) **Analysis of the Toluene Efflux Pumps in Microorganisms Through Bioinformatics** MATHILDA WILLOUGHBY, Westminster College, Samantha Tower, Sarah Kennedy

(2030-52 P) **Analysis of Urine Organic Acids via GC/MS-based Metabolomics to Determine the Effect of Diet on Urine Composition** JESSICA MINNICK, Georgia College and State University, Catrena Lisse N/A

(2030-53 P) **A Cascade SERS Signal Amplification Approach for Telomerase Activity at Single-cell Level** MULING SHI, Hunan University

THURSDAY, MARCH 10, 2016 MORNING

SYMPOSIUM **Session 2040**

ACS-ANLY - Advances in Electrokinetic Methods for Bioanalysis
arranged by Alexandra Ros, Arizona State University

Thursday Morning, Room B308
Alexandra Ros, Arizona State University, Presiding

8:30		**Introductory Remarks - Alexandra Ros**
8:35	(2040-1)	**Nanofluidic Devices for Single-Particle Analysis of Virus Assembly** STEPHEN C JACOBSON, Indiana University, Zachary D Harms, Lisa Selzer, Adam Zlotnick
9:10	(2040-2)	**Amyloid Oligomers Analysis Using Microchannel Electrophoresis** CHRISTA HESTEKIN, University of Arkansas, Sadia Paracha, Melissa Moss
9:45	(2040-3)	**Surface Isoelectric Focusing (SIEF) for Therapeutic Protein Separations** ADRIENNE R MINERICK, Michigan Technological University, Zhichao Wang
10:20		**Recess**
10:35	(2040-4)	**Microfluidic-Based Electrokinetic Methods for Protein Separation and Sensing and Manipulation of Particles and Droplets** CAROLYN L REN, University of Waterloo
11:10	(2040-5)	**Towards Organelle Separation Exploiting Deterministic Absolute Negative Mobility** ALEXANDRA ROS, Arizona State University, Jinghui Luo, Edgar Arriaga, Katherine Muratore N/A

SYMPOSIUM **Session 2050**

Advances in Vibrational Spectroscopy for Medical Diagnostics
arranged by Igor K Lednev, University at Albany, SUNY

Thursday Morning, Room B302
Igor K Lednev, University at Albany, SUNY, Presiding

8:30		**Introductory Remarks - Igor K Lednev**
8:35	(2050-1)	**Stimulated Raman Scattering Microscopy as a Tool for Brain and Skin Cancer Tissue Diagnoses** SUNNEY XIE, Harvard University
9:10	(2050-2)	**Raman Spectroscopy for Clinical Cell Analysis** JÜERGEN POPP, Friedrich-Schiller University Jena
9:45	(2050-3)	**Classification of Lung Cancers by Infrared Spectral Histopathology (SHP)** MAX DIEM, Northeastern University
10:20		**Recess**
10:35	(2050-4)	**Label-Free Spectroscopic Imaging for Molecular Diagnosis** JI-XIN CHENG, Purdue University
11:10	(2050-5)	**Raman Spectroscopy of Blood for Alzheimer's Disease Diagnostics** IGOR K LEDNEV, University at Albany, SUNY, Elena Ryzhikova, Oleksandr Kazakov, Lenka Halamkova, Eric Molho, Earl Zimmerman

SYMPOSIUM **Session 2060**

Computational Chemistry Coupled to Analytical Measurements: A Synergistic Relationship
arranged by Bruce Chase, University of Delaware

Thursday Morning, Room B303
Bruce Chase, University of Delaware, Presiding

8:30		**Introductory Remarks - Bruce Chase**
8:35	(2060-1)	**Advancing the Understanding of Rigid Rod Polymers with Statistical Mechanics and Analytical Chemistry** STEVE LUSTIG, DuPont, Steven Allen, Juan David Londono, Christopher Seay
9:10		**Binding Free Energy of DNA on Graphite and Carbon Nanotubes by Single-Molecule Peeling** ANAND JAGOTA, Lehigh University N/A
9:45	(2060-3)	**Theory and Simulations of Macromolecular Soft Materials: Linking Molecular Design to Macroscale Morphology and Function** ARTHI JAYARAMAN, University of Delaware, Tyler B Martin
10:20		**Recess**
10:35	(2060-2)	**Recent Computational Studies Related to the Use of Plasmonic Materials for Analytical Applications** GEORGE C SCHATZ, Northwestern University
11:10	(2060-4)	**Differentiation of Alpha and Beta Crystalline Polymorphs in Biodegradable Poly Hydroxybutyrate** BRUCE CHASE, University of Delaware, Brian Sobieski, Isao Noda

SYMPOSIUM **Session 2070**

Identification and Analysis for Food Safety
arranged by Perry G Wang, US FDA

Thursday Morning, Room B304
Perry G Wang, US FDA, Presiding

8:30		**Introductory Remarks - Perry G Wang**
8:35	(2070-1)	**Validation and Challenge for the Determination of Chemical Components in Cosmetic Products Using LC-MS and GC-MS** PERRY G WANG, US FDA, Wanlong Zhou, Alexander J Krynitsky
9:10	(2070-2)	**Development of Multi-Functional Ambient Mass Spectrometry for Food Safety Screening and Characterizing Polymers in Packing Materials** JENTAIE SHIEA, National Sun Yat-sen University
9:45	(2070-3)	**Emerging Disinfection Byproducts Halobenzoquinones in Treated Drinking Water** XING-FANG LI, University of Alberta
10:20		**Recess**
10:35	(2070-4)	**Effects of Different Dietary Doses of Copper and High Fructose Feeding on Rat Fecal and Liver Metabolome** XIANG ZHANG, University of Louisville, Ming Song, Xiaoli Wei, Xinmin Yin, Aminul Prodhan, Craig McClain
11:10	(2070-5)	**Identification and Confirmation of Chemical Residues in Foods for Regulatory Purposes** STEVEN J LEHOTAY, USDA Agricultural Research Service

SYMPOSIUM **Session 2080**

Integrated Microfluidics
arranged by R Scott Martin, Saint Louis University

Thursday Morning, Room B305
R Scott Martin, Saint Louis University, Presiding

8:30		**Introductory Remarks - R Scott Martin**
8:35	(2080-1)	**Integrated Microfluidics for Forensic Analysis: Creating Simple, Portable and Cost-effective Systems** JAMES LANDERS, University of Virginia
9:10	(2080-2)	**Microfluidic Technology for Protein Crystallization and Pharmaceutical Solid form Screening** PAUL KENIS, University of Illinois at Urbana Champaign
9:45	(2080-3)	**Integrated Microfluidic Systems for Measuring Secretion from Cellular Networks** MICHAEL G ROPER, Florida State University, Lian Yi, Adrian M Schrell, Xue Wang, Nikita Mukhitov, Basel Bandak, Kimberly Evans
10:20		**Recess**
10:35	(2080-4)	**New Strategies for Enhancing Human-on-Chip Systems** DANA SPENCE, Michigan State University
11:10	(2080-5)	**Integrated Microfluidic Platform for Mass Spectrometry based Metabolomics** JAMES L EDWARDS, Saint Louis University

TECHNICAL PROGRAM

SYMPOSIUM **Session 2090**

New Bioanalytical Separations for Molecular Mechanisms of Disease
arranged by B Jill Venton, University of Virginia and Lisa A Holland, West Virginia University

Thursday Morning, Room B309
B Jill Venton, University of Virginia, Presiding

8:30		**Introductory Remarks - B Jill Venton and Lisa A Holland**
8:35	(2090-1)	**Capillary Separations that Unravel Molecular Mechanisms of Endocrine Dysfunction** LISA A HOLLAND, West Virginia University, Vincent T Nyakubaya, Jennifer R Stueckle
9:10	(2090-2)	**Separation-Based Methods for Measuring Reactive Oxygen and Nitrogen Species in Biological Samples** SUSAN M LUNTE, University of Kansas
9:45	(2090-3)	**How Separations—Both High and Low Resolution—Enable Selection of Clinically Useful Aptamers** REBECCA WHELAN, Oberlin College
10:20		**Recess**
10:35	(2090-4)	**Merging Microfluidics, Electrophoresis, and Mass Spectrometry for Protein Assays** ROBERT T KENNEDY, University of Michigan
11:10	(2090-5)	**CE-FSCV for Determining Neurotransmitter Tissue Content in Drosophila Disease Models** B JILL VENTON, University of Virginia, Madelaine E Denno, Ryan Borman

SYMPOSIUM **Session 2100**

Overcoming the Obstacles to Making Measurements in the Brain
arranged by Stephen Weber, University of Pittsburgh

Thursday Morning, Room B310
Stephen Weber, University of Pittsburgh, Presiding

8:30		**Introductory Remarks - Stephen Weber**
8:35	(2100-1)	**Probing the Spatial and Temporal Dynamics of Signaling Peptides in the Nervous Systems by a Multi-Faceted MS Approach** LINGJUN LI, University of Wisconsin
9:10	(2100-2)	**Overcoming Obstacles to Understanding Voltammetric Measurements of Dopamine in the Brain** ADRIAN C MICHAEL, University of Pittsburgh, Seth H Walters, Elaine M Robbins, Brendan P Sestokas, Andrea Jaquins-Gerstl
9:45	(2100-3)	**Simultaneous Optical Imaging of Neuronal, Glia, and Hemodynamic Waves During Seizures** HONGTAO MA, Weill Cornell Medical College, Andy Daniel, Eliza Baird-Daniel, Philippe Laffont, Mingrui Zhao, Theodore H Schwartz
10:20		**Recess**
10:35	(2100-4)	**Placing New Pieces in the Puzzle of Human Traumatic Brain Injury Using Multimodal Monitoring** MARTYN G BOUTELLE, Imperial College London, Michelle L Rogers, Sally A Gowers, Chi Leng Leong, Vassilios Kontojannis, Anthony J Strong, Sharon Jewell
11:10	(2100-5)	**Electroosmotic Perfusion in Brain Tissue for Determining Ectopeptidase Activity** STEPHEN WEBER, University of Pittsburgh, Stephen R Groskreutz, Khanh Ngo, Yangguang Ou, Rachael Wilson, Jenna DeVivo, Bocheng Yin

SYMPOSIUM **Session 2110**

SAS - Handheld Spectrometers
arranged by Richard A Crocombe, PerkinElmer and Mark A Druy, Technology Assessment & Partnering

Thursday Morning, Room B311
Richard A Crocombe, PerkinElmer, Presiding

8:30		**Introductory Remarks - Richard A Crocombe and Mark A Druy**
8:35	(2110-1)	**Handheld Laser-Induced Breakdown Spectroscopic Instruments** AMY JO RAY BAUER, TSI, Incorporated
9:10	(2110-2)	**Chemometrics in Action: Moving the Lab to the Field** SUZANNE SCHREYER, Thermo Fisher Scientific, Michael Hargreaves
9:45	(2110-3)	**Mass Spectrometry in Miniature** CHRISTOPHER BROWN, 908 Devices
10:20		**Recess**
10:35	(2110-4)	**Next Generation Portable Spectrometers: Spectroscopy Solutions Wherever You Want Them** KATHERINE A BAKEEV, B&W Tek, Inc, Ken Li, Sean Wang, Jing Li, Jack Zhou
11:10	(2110-5)	**High Sensitivity Measurements in Liquids Using Mid-IR Lasers** DON KUEHL, RedShift Systems, Rick Sharp, Eugene Ma, Jinghong Kim, Chip Marshall

SYMPOSIUM **Session 2120**

Single Cell Molecular Analysis
arranged by Milan Mrksich, Northwestern University

Thursday Morning, Room B312
Milan Mrksich, Northwestern University, Presiding

8:30		**Introductory Remarks - Milan Mrksich**
8:35	(2120-1)	**Using Single Molecule Arrays (Simoa) to Measure Proteins and Nucleic Acids in Single Cells** DAVID R WALT, Tufts University, Stephanie Walter, Stephanie Schubert, Payal Ghatak
9:10	(2120-2)	**Semiconducting Polymer Dots (Pdots) for Single-Cell Sensing and Molecular Analysis** DANIEL T CHIU, University of Washington
9:45	(2120-3)	**Assaying Single Cells as a Diagnostic Tool** NANCY ALLBRITTON, University of North Carolina at Chapel Hill
10:20		**Recess**
10:35	(2120-4)	**Using SAMDI Mass Spectrometry to Measure Enzyme Activity in Single Cell Lysates** MILAN MRKSICH, Northwestern University
11:10	(2120-5)	**A Novel Tool for Detection and Enumeration of Circulating Tumor Cells** DAVID A GILJOHANN, AuraSense

ORGANIZED CONTRIBUTED SESSIONS **Session 2130**

Recent Advances in Ion Analysis
arranged by Kannan Srinivasan, Thermo Fisher Scientific

Thursday Morning, Room B313
Kannan Srinivasan, Thermo Fisher Scientific, Presiding

8:30	(2130-1)	**Anion Analysis Using Capillary Ion Chromatography of Steam Cycle Water and Anion Analysis of Reactor Water by Matrix Elimination in a Nuclear Power Plant** RICHARD WALLWORK, Pacific Gas and Electric
8:50	(2130-2)	**New Developments in Multidimensional Analysis for Drinking Water Applications** HERB WAGNER, Independent Contractor
9:10	(2130-3)	**Recent Advances in Suppressor Technology in Ion Chromatography for Achieving Low Noise Performance** KANNAN SRINIVASAN, Thermo Fisher Scientific, Brittany Omphroy, Sheetal Bhardwaj, Rong Lin, Christopher Pohl
9:30	(2130-4)	**Indirect Pulsed Electrochemical Detection following Anion-Exchange Chromatography** BILL LACOURSE, University of Maryland Baltimore County N/A
9:50		**Recess**
10:05	(2130-5)	**Revisiting the Many Facets of Ion Exclusion Chromatography** C PHILLIP SHELOR, University of Texas Arlington, Purnendu Dasgupta
10:25	(2130-6)	**Factors Underlying Recent Advances in High Speed and High Resolution Ion Chromatography** CHARLES A LUCY, University of Alberta
10:45	(2130-7)	**Peak Shapes in High Efficiency and Fast Chromatography: Contributions from the Slurry Packing Process and Detector Settings** M FAROOQ WAHAB, University of Texas at Arlington, Purnendu Dasgupta, Daniel W Armstrong
11:05	(2130-8)	**Recent Developments in Stationary Phases for Ion Chromatography** CHRISTOPHER POHL, Thermo Fisher Scientific, Charanjit Saini, Mani Jayaraman, Maria Rey, Andy Woodruff

ORAL SESSIONS **Session 2140**

Bioanalytical: Fluorescence/Luminescence Techniques

Thursday Morning, Room B405
Nathan Whitman, University of North Carolina at Chapel Hill, Presiding

8:30	(2140-1)	**Ultrasensitive Detection of Ricin Toxin in Multiple Sample Matrixes Using Single-Domain Antibodies** TRINH DINH, Tufts University, Shonda Gaylord, David R Walt, Kevin Ngan
8:50	(2140-2)	**Unusual Red Emission of Graphene Quantum Dots at Extremely High pH** YIYANG LIU, University of Kentucky, Doo Young Kim
9:10	(2140-3)	**Bright Large Stokes' Shift NIR Fluorescent Silica Nanoparticle Labels and Probes** GABOR PATONAY, Georgia State University, Gala Chapman, Maged Henary, Kyle Emer
9:30	(2140-4)	**Development of Fluorescent Magnetic Particles for "On-Off" Switching based Detection of Various Lectin-Saccharide Interactions** SUZUKI YOSHIO, AIST
9:50		**Recess**
10:05	(2140-5)	**Cell-free Expression of Cytochrome P450-containing Liposomes for Drug Metabolism Screens** NATHAN A WHITMAN, University of North Carolina at Chapel Hill, Julie C McIntosh, Jeffrey B Penley, Matthew R Lockett
10:25	(2140-6)	**Tethered Cationic Lipoplex Nanoparticle Biochip for Fast Disease Detection** JIAMING HU, The Ohio State University, L J Lee N/A
10:45	(2140-7)	**Cell Surface Engineering with Lipid-Molecular Beacon Aptamer for Real Time Probing of Proteins in Cellular Microenvironment** WEIJIA HOU, University of Florida
11:05	(2140-8)	**Ratiometric Fluorescence Assay based Quantum Dots for Recognition of RRE RNA Ligands Using A New Fluorescence Indicator** ZHIQI ZHANG, Shaanxi Normal University N/A

ORAL SESSIONS **Session 2150**

Electrochemistry - New Methods and Applications

Thursday Morning, Room B314
Leslie A Sombers, North Carolina State University, Presiding

8:30	(2150-1)	**Elucidating the Structure/Function Relationship of Conductive Polymer Microelectrodes for Use in Fast-Scan Cyclic Voltammetry of Neurotransmitters** ADAM R MEIER, University of Arizona, William Bahureksa, Michael L Heien
8:50	(2150-2)	**Electrochemically Reduced Graphene Oxide as an Electrocatalyst Support for H_2S Detection** JASON A BENNETT, Penn State Erie - The Behrend College
9:10	(2150-3)	**Amperometric Determination of Aurocyanide for Hydrometallurgical Gold Processing** WAYNE DICKINSON, Kemira
9:30	(2150-4)	**Method for Removal of Non-Faradaic Contributions to Fast-Scan Cyclic Voltammetry Recordings** JUSTIN A JOHNSON, University of North Carolina at Chapel Hill, R Mark Wightman
9:50		**Recess**
10:05	(2150-5)	**Quantitative Analysis of Microiontophoresis Drug Delivery** DOUGLAS KIRKPATRICK, University of North Carolina at Chapel Hill, R Mark Wightman
10:25	(2150-6)	**Fast Scan Cyclic Voltammetry of Metals at Carbon-Fiber Microelectrodes: Correlation Between FSCV Response and Solution Dynamics** PAVITHRA PATHIRATHNA, Wayne State University, Thushani Siriwardhane, Shawn McElmurry, Parastoo Hashemi
10:45	(2150-7)	**Construction of Training Sets for Valid Calibration Using Principal Component Analysis** NATHAN T RODEBERG, University of North Carolina at Chapel Hill, Justin A Johnson, R Mark Wightman
11:05	(2150-8)	**Electrochemical Characterization and Catalytic Activity of Ultrasmall Gold Nanoparticles** FRANCIS ZAMBORINI, University of Louisville, Rafael Masitas, Stacy Allen, Dhruba Pattadar

ORAL SESSIONS **Session 2160**

Food Safety Evaluations - Half Session

Thursday Morning, Room B407
Alice Chen, The Pittsburgh Conference, Presiding

8:30	(2160-1)	**Characterization of the Adulteration, Counterfeiting and Contamination of Spices, Spice Products and Supplements by the Detection of Toxic and Banned Organic Chemicals in Commercial Botanical Products** PATRICIA ATKINS, SPEX CertiPrep
8:50	(2160-2)	**Using SPE to Adjust Sensitivity to Analytical Requirements for Food Safety** CHRIS SHEVLIN, Horizon Technology, William Jones
9:10	(2160-3)	**Ensuring a Safe and Stable Food Supply Using ICP-OES for Elemental Monitoring** NICK SPIVEY, PerkinElmer Inc., Kenneth Neubauer, Stan Smith, Laura Thompson
9:30	(2160-4)	**Automated Solid Phase Extraction and Quantitative UHPLC Analysis of Cannabis Compounds in Food Matrices** CHRIS SHEVLIN, Horizon Technology, Robert E Buco

ORAL SESSIONS **Session 2170**

Liquid Chromatography Column Chemistry

Thursday Morning, Room B406
A Carl Sanchez, Phenomenex, Presiding

8:30	(2170-1)	**Functionalized Octatetrayne as Novel Carbon Media for Capillary Liquid Chromatography** JIAYI LIU, The Ohio State University, Susan Olesik
8:50	(2170-2)	**Stationary Phases Based on the Thiol-ene Reaction on Mercaptopropylsilane-Modified Nonporous Silica** ERIN SHIELDS, University of Pittsburgh, Kayla Thomas, Stephen Weber
9:10	(2170-3)	**Using 5 Micron Superficially Porous Particles in Capillary and Microfluidic LC Columns** JAMES P GRINIAS, University of Michigan, Robert T Kennedy
9:30	(2170-4)	**Selectivity in Reversed-Phase Liquid Chromatography: Impact of Stationary Phase Chemistry** DAVID S BELL, Supelco/Sigma-Aldrich
9:50		**Recess**
10:05	(2170-5)	**A Specialty Column for High Resolution Separation of Aminoglycoside Antibiotics by HPLC** XUEFEI SUN, Thermo Fisher Scientific, Xiao Cui, Yoginder Singh, Xiaodong Liu
10:25	(2170-6)	**Liquid Chromatography, Hydrophilic Interaction Chromatography, HILIC** DAVID S BELL, Supelco/Sigma-Aldrich
10:45	(2170-7)	**The Benefits of 1mm ID UHPLC Columns Made Real** STEPHEN LUKE, Agilent Technologies, Norwin Von Doehren, William Long, Jason Link
11:05	(2170-8)	**Comparison of the Practical Kinetic Performance Limits of Core-Shell and Fully Porous (U)HPLC Sorbents Using Commercially Available Column Formats** A CARL SANCHEZ, Phenomenex, Gareth Friedlander, Jason Anspach, Tivadar Farkas

ORAL SESSIONS **Session 2180**

Microfluidics/Lab-on-a-Chip - Bioanalytical I

Thursday Morning, Room B403
Hubert MacDonald, The Pittsburgh Conference, Presiding

8:30	(2180-1)	**A Method for Measurement of Temporally Resolved Insulin Secretion from Islets of Langerhans in Response to Fatty Acid Hydroxy Fatty Acids** BASEL BANDAK, Florida State University, Lian Yi, Nikita Mukhitov, Michael G Roper
8:50	(2180-2)	**Macro-to-Microfluidic Interfacing for Primary Endocrine Cell Culture and Sampling Using 3D Printed Device Templates and Fluidic Manifolds** JESSICA C BROOKS, Auburn University, Mark D Holtan, Katarena Ford, Dylan Holder, Christopher J Easley
9:10	(2180-3)	**A Simple Droplet Microfluidic Capillary Viscometer Based on Droplet Frequency for Rheological Measurements of Proteins** MICHAEL F DELAMARRE, University of Illinois at Chicago, Scott A Shippy
9:30	(2180-4)	**Improving Detection Sensitivity in Microchip Electrophoresis-Laser Induced Fluorescence Assays by Target-Induced Exonuclease Assisted Strand Circle Signal Amplification** SHULIN ZHAO, Guangxi Normal University, Yingfeng Qin, Liangliang Zhang, Yong Huang, Yi-Ming Liu

9:50 Recess

10:05 (2180-5) **A Complementary Method to CD4 Counting: Measurement of CD4+/CD8+ T Lymphocytes Ratio in a Serial Microfluidic System** WENJIE LI, Texas Tech University, Dimitri Pappas

10:25 (2180-6) **PDMS-Based Injection Valves for SPE-MS Analysis of Biomolecules** JAMES P GRINIAS, University of Michigan, Colleen E Dugan, Robert T Kennedy

10:45 (2180-7) **Development of an Online Microchip Electrophoresis with LED-Induced Fluorescence System for In Vivo Detection of Excitatory Amino Acid Neurotransmitters Following a Traumatic Brain Injury** MICHAEL L HOGARD, University of Kansas, Nathan Oborny, Elton E Melo Costa, Susan M Lunte, Craig E Lunte

11:05 (2180-8) **Determination of Amplification of Cellular Effects by Hormones Derived from Different Tissues** KESHAVARZ HAMIDEH, Michigan State University, Dana Spence

ORAL SESSIONS — Session 2190

Novel Applications with Gas Chromatography Mass Spectrometry

Thursday Morning, Room B315
Rudolf Addink, Toxic Report, Presiding

8:30 (2190-1) **How GC-MS with Cold EI Improves NIST Library Identification** AVIV AMIRAV, Tel Aviv University, Uri Keshet, Tal Alon

8:50 (2190-2) **Use of Automated Column Chromatography Clean Up with Reduced Solvent Volume in POPs Analysis** RUDOLF ADDINK, Toxic Report, Philip Bassignani

9:10 (2190-3) **Effective QuEChERS Cleanup and Quantitation of Planar Pesticides from Spinach and Other Food Matrices Using a Novel Graphitized Carbon Black and a Zirconia-Based Adsorbent** PATRICK MYERS, Supelco/Sigma-Aldrich, William Betz, Bill Ozanich, Jennifer Claus, Michael Ye

9:30 (2190-4) **Use of Micro Scale Solid Phase Extraction and Automated Clean Up in POPs Analysis of Human Milk and Serum** RUDOLF ADDINK, Toxic Report

9:50 Recess

10:05 (2190-5) **Multidimensional Comprehensive Gas Chromatography Multireflection High Resolution Time-of-Flight Mass Spectrometry: Combining Accurate Mass Information with Ultra-High Chromatographic Resolution** RALF ZIMMERMANN, University Rostock/HMGU, Thomas Groeger, Benedikt Weggler, Juergen Wendt

10:25 (2190-6) **Rapid, New Methods for the Analysis of 3-MCPD and 1,3 DCP in Soy Sauce** SUSAN GENUALDI, US FDA, Lowri DeJager, Patsy Nyman

10:45 (2190-7) **A Method Development Software Tool for Comprehensive Two-Dimensional Gas Chromatography Evaluated for GCxGC-TOFMS** MARK F MERRICK, LECO Corporation, Viatcheslav Artaev, Leonid M Blumberg

11:05 (2190-8) **Innovative TG-GC-MS Methods for Thermal Degradation Studies of Polymers** KRISTINA LILOVA, Setaram Inc., Link Brown

ORAL SESSIONS — Session 2200

Novel Synthesis and Applications of Nanomaterials

Thursday Morning, Room B316
David Pensenstadler, The Pittsburgh Conference, Presiding

8:30 (2200-1) **Nanofibers from Hydrothermal Treatment of Cellulose Nanocrystals** YIMEI WEN, Clemson University

8:50 (2200-2) **Graphene Nanoribbons: Engineering, Characterization of Edge Defects and Sensor Applications** PANKAJ RAMNANI, University of California, Riverside, Ashok Mulchandani

9:10 (2200-3) **Modeling of Seed-Mediated Nanoparticle Growth** HANS MUSGRAVE, University of North Dakota, Julia Xiaojun Zhao N/A

9:30 (2200-4) **Single-Molecule Tracking Studies of the Effects of Solvent Swelling on the Properties of Cylindrical Block Copolymer Microdomains** TAKASHI ITO, Kansas State University, Dol R Sapkota, Khanh-Hoa Tran-Ba, Daniel A Higgins

9:50 Recess

10:05 (2200-5) **Characterization of Single and Multi-Walled Carbon Theranostic Nanovectors** MARKUS MARTINCIC, Institut de Ciencia de Materials de Barcelona, Elzbieta Pach, Belen Ballesteros, Gerard Tobias

10:25 (2200-6) **Investigating 4-Wheeled Nanocar Diffusion Kinetics on Differently Modified Solid Surface with Single Molecule Fluorescence Microscopy (SMFM)** FANG CHEN, North Carolina State University, Víctor García-López, James Tour, Gufeng Wang

10:45 (2200-7) **Sealing and Opening of Metallic Nanotubes with a Laser Beam: A Potential Drug Delivery Vehicle** NATHALIA ORTIZ, North Carolina State University, Gufeng Wang

11:05 (2200-8) **Optically Transparent Carbon Nanotube Film Electrode for Thin Layer Spectroelectrochemistry** TINGTING WANG, University of Cincinnati

ORAL SESSIONS — Session 2210

Sampling and Sample Preparation - Bioanalytical, Neurochemistry, and Material Science

Thursday Morning, Room B404
Denise Wilkins, Bechtel Bettis Inc., Presiding

8:30 (2210-1) **Rapid Protein Purification and Digestion with Membrane-Containing Pipette Tips** WENJING NING, Michigan State University, Merlin Bruening

8:50 (2210-2) **DNA Extraction and Analysis Using Magnetic Ionic Liquid Solvents** KEVIN D CLARK, Iowa State University, Melissa Yamsek, Omprakash Nacham, Jared L Anderson

9:10 (2210-3) **Localized Laser Ablation Sample Transfer for Tissue Proteomics** FABRIZIO DONNARUMMA, Louisiana State University, Michael E Pettit, Touradj Solouki, Kermit K Murray

9:30 (2210-4) **An Ultra Sensitive Sample Preparation Approach that Eliminates the Need to Dry Down and Reconstitute** SHAHANA HUQ, Phenomenex, Matthew Brusius, Jessica Detsch, Zeshan Aqeel, Ramkumar Dhandapani

9:50 Recess

10:05 (2210-5) **Pulled Low Flow Push-Pull Perfusion Probe Tips for Sampling from Tissue Slices** MARISSA R BECKER, University of Illinois at Chicago, Scott A Shippy, David E Featherstone

10:25 (2210-6) **Design of Protein-Binding Membranes through Adsorption of Star-Shaped Polyelectrolytes in Membrane Pores** WEIJING LIU, Michigan State University, Salinda Wijeratne, Merlin Bruening

10:45 (2210-7) **Direct Coupling of Solid Phase Microextraction to Mass Spectrometry Via Nano-Electrospray Ionization: Development and Applications in Bioanalysis** GERMAN A GOMEZ-RIOS, University of Waterloo, Nathaly Reyes-Garces, Ezel Boyaci, Barbara Bojko, Janusz Pawliszyn

11:05 (2210-8) **Molecularly Imprinted Polymer-Sol-Gel for Dispersive Micro-Solid Phase Extraction** MOHAMED ABDEL-REHIM, Stockholm University, Aziza El-Beqqali N/A

ORAL SESSIONS — Session 2220

Sensors - Others

Thursday Morning, Room B401
Gufeng Wang, North Carolina State University, Presiding

8:30 (2220-1) **PANi Electrospun Fibers and Drop Cast Film Sensor Array for the Detection of Small Chained Alcohols** KELVIN TRAN, University of California Riverside, Andrew J Burris, Quan Cheng

8:50 (2220-2) **Application of Thermo-Reversible Interpenetrating Poly(Vinyl Alcohol) Networks to Stabilize Mechanically Fragile Hydrogel Sensors** ANDREW E COUKOUMA, University of Pittsburgh, Sanford A Asher N/A

9:10 (2220-3) **Improving the Sustainability of Drinking Water Systems Using Nanostructured Biosensors for Escherichia Coli** HEATHER A CRAPO, Binghamton University, Melissa McDonald, Idris Yazgan, Omowunmi Sadik

9:30 (2220-4) **An Innovative Biosensor to Assess Quickly the Biodegradable Organic Fraction** SULIVAN JOUANNEAU, University of Nantes, Marie-José Durand, Gerald Thouand, Ali Boukabache, Yves Primault N/A

9:50 Recess

10:05 (2220-5) **Surface Plasmon Resonance Immunosensor Using Au Nanoparticle Modified Antibody** DULAL C KABIRAZ, Hokkaido University, Kinichi Morita, Toshikazu Kawaguchi

10:25 (2220-6) **QCM Virtual Multisensor Array for Detection of Gasoline Adulterants** NICHOLAS SPELLER, Louisiana State University, Noureen Siraj, Isiah M Warner, Stephanie Vaughan

10:45 (2220-7) **Development of an Aptamer Functionalized Electrode Array for Real-Time In Vivo Cocaine Detection Using Square Wave Voltammetry** IAN MITCHELL TAYLOR, University of Pittsburgh, Emma Bigelow, Tracy Cui

11:05 (2220-8) **Optomechanical Switching by Plasmonic Nanoparticle Monolayers on Elastic Substrate Induced by Stretching** MAHMOUD MAHMOUD, Georgia Institute of Technology

ORAL SESSIONS — Session 2230

Vibrational Spectroscopy Instrumentation and Applications

Thursday Morning, Room B402
Jane Chan, Bechtel Bettis Inc., Presiding

8:30 (2230-1) **Experimental Studies and Electromagnetic Modeling of Localized Plasmon Surface on a New SERS Platform** MATHIEU EDELY, IMMM Institut des Molécules et Matériaux du Mans, Guy Louarn, Jean-François Bardeau, Nicolas Delorme, Ludovic Douillard N/A

8:50 (2230-2) **The Effect of Molecular Polarity and Solubility on Adsorption Rates and Equilibrium Constants for Molecules on Noble Metal Surfaces Using Surface-Enhanced Raman Spectroscopy** ERIK DAVID EMMONS, US Army, Ashish Tripathi, Neal D Kline, Jerry Cabalo, Jason A Guicheteau, Augustus W Fountain, Steven D Christesen

9:10 (2230-3) **Room Temperature Freezing and Orientation Control of Surface Immobilized Biomoelcules in Air** YAOXIN LI, University of Michigan, Zhang Xiaoxian, Somayesadat Badieyan, Zhan Chen

9:30 (2230-4) **Fiber Spectroscopy for Process Control and Medical Diagnostics** VIACHESLAV ARTYUSHENKO, Art Photonics GmbH

9:50 **Recess**

10:05 (2230-5) **Characterization of Polymer/Epoxy Buried Interfaces with Silane Adhesion Promoters Before and after Hygrothermal Aging for the Elucidation of Molecular Level Details Relevant to Adhesion** NATHAN W ULRICH, The University of Michigan, John Myers, Zhan Chen

10:25 (2230-6) **Surface Interaction of Nitrogen Containing Aromatic Molecules with Gold Investigated with Surface Enhanced Raman Spectroscopy (SERS)** ASHISH TRIPATHI, Leidos, Inc., Erik David Emmons, Augustus W Fountain, Jason A Guicheteau, Steven D Christesen, Martin Moskovits

10:45 (2230-7) **Transmission Raman Spectroscopy as a Regulatory-Approved Method for Content Uniformity Analysis – Replacing HPLC** DARREN ANDREWS, Cobalt Light Systems, Julia Griffen, Matthew Bloomfield, Andrew Owen, Mark Mabry, Pavel Matousek

11:05 (2230-8) **A New Microscope for FT-IR Microspectrometry** DAVID W SCHIERING, Czitek, Gregg Ressler

POSTER SESSION — Session 2240

All posters are to be mounted by 10:00 AM and remain on display until 2:00 PM. Authors must be at their posters from 10:00 AM to 12:00 PM. Location of the posters is on the Exposition Floor, 400 Aisle. PLEASE NOTE: You cannot get onto the Exposition Floor until after 9:00 AM.

Application of Mass Spectrometry

Thursday Morning, Exposition Floor, 400 Aisle

(2240-1 P) **Variations on a Theme: The Detection of NBOMe Designer Drugs on Blotter Paper by High Resolution Time-of-flight Mass Spectrometry (TOFMS) with and without Chromatography** DAVID BARAJAS, Boston University School of Medicine, Frank A Kero, Noelle Elliot, Bogdan Bogdanov, Craig Young, Jason Weisenseel, Sabra Botch

(2240-2 P) **Proton Transfer Reaction – Mass Spectrometry: Automated Measurement and Evaluation** JENS HERBIG, IONICON Analytik, Klaus Winkler, Johann Seehauser, Lukas Mäerk, Christian Lindinger, Alfons Jordan

(2240-3 P) **Functionalized Gold Surface for SPR-MS Determination of Enzymatic Activities and Specificity of Lectins** HYOJIK YANG, University of California, Riverside, Quan Cheng

(2240-4 P) **High-Resolution Atmospheric Pressure Drift Tube Ion Mobility Spectrometry Coupled with High-Resolution Accurate Mass Orbitrap Mass Spectrometry** JOEL D KEELOR, Georgia Institute of Technology, Facundo M Fernandez, Brian Clowers

(2240-5 P) **Real-Time Metabolome Analysis by Probe Electrospray Ionization-Tandem Mass Spectrometry (PESI-MS/MS): Preliminary Challenge to Real-Time Metabolomics** ZAITSU KEI, Nagoya University, Hayashi Yumi, Murata Tasuku, Nakajima Hiroki, Nakajima Tamie, Kusano Maiko, Tsuchihashi Hitoshi, Ishikawa Tetsuya, Ishii Akira

(2240-6 P) **Use of High Speed/High Resolution Size Based Chromatographic Separation of Surfactants and Oligomeric Materials with Single Quadrupole Mass Spectrometry Detection** MICHAEL OLEARY, Waters Corporation

(2240-7 P) **Coupling Surface Acoustic Wave Nebulization (SAWN) with Vacuum-Assisted Plasma Ionization (VaPI) Mass Spectrometry for Enhanced Ionization and Transmission Efficiency** STEPHEN C ZAMBRZYCKI, Georgia Institute of Technology, Matthew C Bernier, Joel D Keelor, Fernandez M Facundo, Sung H Yoon, David Goodlett

(2240-8 P) **Withdrawn**

(2240-9 P) **Withdrawn**

(2240-10 P) **Visualizing the Distribution of Volatile and Semi-Volatile Compounds by Low Temperature Plasma Mass Imaging (LTP-MSI)** ROBERT WINKLER, CINVESTAV Unidad Irapuato, Sandra Martinez Jarquin, Abigail Moreno Pedraza

POSTER SESSION — Session 2250

All posters are to be mounted by 10:00 AM and remain on display until 2:00 PM. Authors must be at their posters from 10:00 AM to 12:00 PM. Location of the posters is on the Exposition Floor, 400 Aisle. PLEASE NOTE: You cannot get onto the Exposition Floor until after 9:00 AM.

Fluorescence and Luminescence

Thursday Morning, Exposition Floor, 400 Aisle

(2250-1 P) **Determination of Thiamine by Flow Analysis System Based on the Chemiluminescence Inhibition Using Multicommutation** DEBORAH AZZI, Federal University of São Carlos, Marina Batistão, Bruno C Janegitz, Marina Baccarin, Geiser Oliveira, Orlando Fatibello-Filho

(2250-2 P) **Selection of Aptamers Targeting B-Cell Receptor (BCR) Using Antibody Guided Cell-Selex Method - A Novel Approach** SHOMI CHAKRABARTI, City University of New York, The Graduate Center, Hasan E Zumrut, George Maio, Prabodhika Mallikaratchy, Mst Naznin Ara

(2250-3 P) **Spectral Encoders: Detecting Position-Dependent Luminescent Spectra Through Tissue** MELISSA M ROGALSKI, Clemson University, Bobby Smith, Hunter Pelham, Nakul Ravikumar, John D DesJardins, Jeffrey N Anker

(2250-4 P) **Photoluminescence of Novel Osmium and Ruthenium Complexes in the Presence of Polyanions** MEHRUN UDDIN, St. John's University, Cody Piotrowski, Besiana Kurti, Armando Seitllari, Elise Megehee, Enju Wang

(2250-5 P) **Quantifying Free Fatty Acid Uptake Dynamics in Primary Adipocytes Using Customized Micro-Wells Made with 3D-Printed Templates** TESFAGEBRIEL M HAGOS, Auburn University, Jessica C Brooks, Christopher J Easley

(2250-6 P) **Toward Non-Invasively Detecting Radiolabeled Analytes Near Implanted Medical Devices Coated in Radioluminescent Phosphors** GRETCHEN B SCHOBER, Clemson University, Jeffrey N Anker

(2250-7 P) **Development of Silicone Filled Optical Module for Laser Fluorescence Trace Molecular Detection** HIROKAZU HIGUCHI, Kyushu University, Hiroaki Nomada, Hiroaki Yoshioka, Kinichi Morita, Yuji Oki

(2250-8 P) **Recombinant Tobacco Peroxidase: A 100-Fold More Effective Luminescent Label Than Horseradish Peroxidase** IRINA GAZARYAN, Pace University, Galina Zakharova, Andrey Poloznikov, Dmitry Hushpulian, Vladimir Tishkov

(2250-9 P) **Using Diffusional Motion to Gauge Fluidity and Interfacial Adhesion Strength of Supported Octadecylphosphonic Acid (ODPA) Monolayers** STEPHEN BAUMLER, Michigan State University

(2250-10 P) **Improving Selectivity of Fluorimetric Water Sensing in Aprotic Solvents** KATARZYNA KŁUCIŃSKA, University of Warsaw, Patryk Rzepiński, Michał Cyrański, Krzysztof Maksymiuk, Agata Michalska N/A

(2250-11 P) **Irradiation of Gold Nanodots by Ultraviolet Light: Modulation of Ligand Density and Photoluminescence** YU-TING TSENG, National Taiwan University

(2250-12 P) **A Preliminary Study of Factors Affecting Quenching or Enhancement of Fluorescence of 1,10-Phenantrololine Toward Quantitative Determination of Selected Metals in mine Drainage and Related Samples** MARK T STAUFFER, University of Pittsburgh - Greensburg, Tell M Lovelace N/A

(2250-13 P) **Traceable Mercury Gas Phase Calibrations Based Upon Gravimetry** ANNARITA BALDAN, VSL, Janneke van Wijk, Hugo Ent

(2250-14 P) **Detection of Caffeine Using a Field-Portable Fluorescence Device** HALEY CURTIS, Tennessee Technological University, Andrew Callender

(2250-15 P) **The Quenching of Riboflavin Fluorence by Nicotine in Bicontinuous Microemulsion** MAURICE O IWUNZE, Morgan State University

(2250-16 P) **Optimizing Fluorescence Tagging Strategy to Study Single Molecule Diffusion on Air-Solid Surface** TAO JIN, North Carolina State University, Fang Chen, James Tour, Victor Lopez, Gufeng Wang

TECHNICAL PROGRAM

POSTER SESSION — Session 2260

All posters are to be mounted by 10:00 AM and remain on display until 2:00 PM. Authors must be at their posters from 10:00 AM to 12:00 PM. Location of the posters is on the Exposition Floor, 400 Aisle. PLEASE NOTE: You cannot get onto the Exposition Floor until after 9:00 AM.

Fuels, Energy and Petrochemical

Thursday Morning, Exposition Floor, 400 Aisle

(2260-1 P) **Increasing Resolution of Propylene Glycol Impurities with High-Efficiency GC Columns** RAMKUMAR DHANDAPANI, Phenomenex, Tim Anderson, Kristen Parnell

(2260-2 P) **Analysis of Impurities in Propane/Propylene Streams Using a Pulsed Flame Photometric Detector (PFPD)** CYNTHIA ELMORE, OI Analytical, Michael Duffy, Brian Mistovich, J Garrett Slaton

(2260-3 P) **High C6+ Analysis** MATTHEW MONAGLE, AIC LLC N/A

(2260-4 P) **High Performance Chromatographic Diatomaceous Earth (DE)** KATARINA ODEN, Restek, Jaap de zeeuw, Barry Burger, Linx Waclaski, Rebecca Stevens

(2260-5 P) **Methanizer – A Simple Solution for CO/CO2 Gas Analysis by Gas Chromatography (GC)** KATARINA ODEN, Restek, Jaap de zeeuw, Barry Burger, Linx Waclaski, Rebecca Stevens

(2260-6 P) **Analysis of A Sulfur Mixture in Hydrocarbon Standard Cylinder in ppb Level** YANG QIN, Air Liquide Specialty Gases

(2260-7 P) **Reactive Pyrolysis-GC/MS of Polymers in a Steam Environment Used to Study Potential Bio-oil** KAREN SAM, CDS Analytical, Gary Deger, Steve Wesson

(2260-8 P) **Extended Lower Detection Range for Hydrogen Sulfide and Carbonyl Sulfide with Metal Surface Deactivated Sample Inlet for Micro Gas Chromatography** REMKO VAN LOON, Agilent Technologies, Thomas Szakas, Coen Duvekot

(2260-9 P) **The Improvement of ASTM D3612 TOGA Analysis** MAX WANG, Shimadzu Scientific Instrument Inc., Clifford M Taylor, Marty Smith

(2260-10 P) **Accurate and Reproducible Determination of Halogens in Coal Using Combustion Ion Chromatography** CARL FISHER, Thermo Fisher Scientific, Daniel Khor, Mark Manahan, Adelon Augustin, Kirk Chassaniol

(2260-11 P) **Nitrogen Determination in Lubricant by Elemental Analyzer** GUIDO GIAZZI, Thermo Fisher Scientific, Liliana Krotz, Francesco Leone

(2260-12 P) **New Methodology for the Analysis of Silicon in Petrochemical Samples by ICP-MS** ANTHONY PALERMO, PerkinElmer, Kenneth Neubauer, Stan Smith

(2260-13 P) **Artificial Photosynthesis** MINGMING WANG, Auburn University, Wei Zhan, Chao Li

(2260-14 P) **Rapid Measurement of Xylose and Glucose for Monitoring Corn Stover Fermentation in Bioethanol Production** WILLIAM MILLER, YSI, Inc, June Klingensmith

(2260-15 P) **Safety and Performance Testing of Li-ion Cells Using Thermal Analysis** PETER RALBOVSKY, Netzsch Instruments, Bob Fidler

(2260-16 P) **The Determination of Mercury in Liquefied Petroleum Gas – A Comparison of Sampling Techniques** MATTHEW A DEXTER, P S Analytical, C A Rogers, Warren T Corns

(2260-17 P) **Salen Quinoxalinol Ligands for Selective Coordination and Sensors** ANNE E GORDEN, Auburn University

(2260-18 P) **A New Hydroxide Selective Anion Exchange Phase for Ion Chromatography** CHARANJIT SAINI, Thermo Fisher Scientific, Christopher Pohl, Yan Liu

(2260-19 P) **Physicochemical Characterization of Microalgal Biodiesel** KIZGEL DAVIS-DESOUZA, Oglethorpe University, Grace Djokoto, Md H Kabir

(2260-20 P) **Graphene Quantum Dots Immobilized Nanoporous N-TiO2 Thin Films for Efficient Photocatalytic Water Splitting** NAMAL WANNINAYAKE, University of Kentucky, Syed Z Islam, Allen Reed, Stephen E Rankin, Doo Young Kim

(2260-21 P) **Techniques for Polysaccharide Research at the Complex Carbohydrate Research Center** ROBERTO SONON, Complex Carbohydrate Research Center, Stephanie A Archer-Hartmann, Zhirui Wang, Ian Black, Radnaa Naran, Mayumi Ishihara, Christian Heiss, Dandan Zhou, Artur Muszynski, Scott Forsberg, Asif Shajahn, Justyna Dobruchowska, Parastoo Azadi

(2260-22 P) **Application of Laser Induced Breakdown Spectroscopy (LIBS) in Analysis of Out Crop Samples from the Marcellus Shale** JINESH JAIN, AECOM, Alexander Bol'shakov, Herve Sanghapi, Christina Lopano, Dustin McIntyre, Richard Russo

(2260-23 P) **Direct Detection of Hydrocarbons from Microalgae Using Low Temperature Plasma – Mass Spectrometry (LTP-MS)** ABIGAIL MORENO PEDRAZA, Cinvestav, Robert Winkler

(2260-24 P) **Comprehensive Online Real-Time Analysis of Natural Gas Using VUV Absorption Spectroscopy** JAMES ANTHONY DIEKMANN III, VUV Analytics, Jonathan Smuts, Michael Roecker

(2260-25 P) **IR-Spectroscopic Analyses of Chemical Adaptation Dynamics of Live Microalgal Biomass to Shifting Nutrient Conditions** FRANK VOGT, University of Tennessee N/A

(2260-26 P) **Inorganic Salt Doped with Nanoparticulate Additives for Thermal Energy Storage with Improved Radiative Heat Transfer** PHILIP D MYERS, University of South Florida Clean Energy Research Center, D Yogi Goswami N/A

(2260-27 P) **Transformer Oil Gas Analysis Using Gas Chromatography – Vacuum Ultraviolet Absorbance Spectroscopy** JONATHAN SMUTS, VUV Analytics, Lindsey Nichole Shear, James Anthony Diekmann III, Jeff Tenney, Andy Shkolnik

(2260-28 P) **Ionic Liquids as Electrolytes for Electrochemical Double-Layer Capacitors: Structures that Optimize Specific Energy** MARAL PS MOUSAVI, University of Minnesota, Benjamin E Wilson, Sadra Kashefolgheta, Evan Anderson, Andreas Stein, Philippe Buhlmann

(2260-29 P) **Using Time Resolved FT-IR-ATR Spectroscopy to Study Biodiesel Fuel Diffusion in Flexible Elastomer Materials** JAMES M SLOAN, U.S. Army Research Laboratory N/A

(2260-30 P) **Engine State Monitoring Technology based on Engine Exhaust Test** WEIKUI WANG, Beijing Research Institute of Telemetry, Xiantao Yang, Yi Zheng N/A

POSTER SESSION — Session 2270

All posters are to be mounted by 10:00 AM and remain on display until 2:00 PM. Authors must be at their posters from 10:00 AM to 12:00 PM. Location of the posters is on the Exposition Floor, 400 Aisle. PLEASE NOTE: You cannot get onto the Exposition Floor until after 9:00 AM.

New Developments in GC

Thursday Morning, Exposition Floor, 400 Aisle

(2270-1 P) **Introduction of a New, High-Quality, Cost Efficient Headspace GC Autosampler** MAX WANG, Shimadzu Scientific Instrument Inc., Clifford M Taylor, Marty Smith

(2270-2 P) **Unique Selectivity: The Power of Ionic Liquid GC Columns** LEONARD M SIDISKY, Supelco/Sigma-Aldrich, Greg A Baney, James L Desorcie, Gustavo Serrano, Daniel Shollenberger, Michael D Buchanan

(2270-3 P) **Use of Traditional/Apex Track Integration in Commercially Available Software for GC Data Generated for EPA Methods 8082 and 8015** TOM KWOKA, PerkinElmer, Bill Hahn, Sharanya Reddy

(2270-4 P) **Evaluation of GC Conditions Using Inert Nano Stationary Phase GC Columns** ALLEN BRITTEN, Cape Breton University, Krishnat Naikwadi, Amy Clemens, Robyn Novorolsky, Allison Clarke N/A

(2270-5 P) **Evaluation of Gas Chromatographic Liner Deactivation when Exposed to Various Solvents and Extracts** LINX WACLASKI, Restek, Jack Cochran, Scott Adams, Rebecca Stevens, Dan Li

(2270-6 P) **Development of an Innovative New Thermal Modulator for Comprehensive Multidimensional Gas Chromatography** MATTHEW S KLEE, DANI Instruments, Roberta Lariccia, Vincenzo Casilli N/A

(2270-7 P) **Modification of a GC for High Level Tritium Exposure** WILLIAM SPENCER, SRNL, Jose Cortes Concepcion, Jacob Schaufler, Robert Lascola

(2270-8 P) **Micro-Scale Vapor Extractor for Micro-GC Analysis of VOCs in Biofluids** JUNQI WANG, University of Michigan, Joseph A Potkay, Edward T Zellers

POSTER SESSION — Session 2280

All posters are to be mounted by 10:00 AM and remain on display until 2:00 PM. Authors must be at their posters from 10:00 AM to 12:00 PM. Location of the posters is on the Exposition Floor, 400 Aisle. PLEASE NOTE: You cannot get onto the Exposition Floor until after 9:00 AM.

UV/VIS

Thursday Morning, Exposition Floor, 400 Aisle

(2280-1 P) **Qualitative Colorimetric and Quantitative Flow Injection Determination of Alkyl Nitrite** ABD AL-KARIM ALI, Miami University, Neil D Danielson

(2280-2 P) **Multi-order Visible Absorption and Reflectance Spectrometry: Parallels to Atomic Emission Line Interferences** ALEXANDER SCHEELINE, SpectroClick, Mark D Ginsberg

(2280-3 P) **Nano-Embedded Optochemical Sensors for In-Vivo Photo-acoustic Chemical Imaging of Potassium Ions** WULIANG ZHANG, University of Michigan, Chang Lee, Jeffery Folz, Xueding Wang, Raoul Kopelman

(2280-4 P) **Colorimetric Determination of Sulfate Using Barium Ion and the Chromate/Dichromate Equilibrium: Preliminary Results and Comparison with the Classical Gravimetric Method** MARK T STAUFFER, University of Pittsburgh - Greensburg, Jarrod W Qualk, Jeremiah C Jamrom N/A

(2280-5 P) **Derivatives of 4,5-Diazafluorene for Chelation and Colorimetric Determination of FE(II) and CU(I): Better FE(II) and CU(I) Chelators Than 1,10-Phenanthroline?** MARK T STAUFFER, University of Pittsburgh - Greensburg, Luke J Metzler, Matthew R Luderer N/A

(2280-6 P) **Simultaneous Determination of Iron and Aluminum by Spectrophotometry and Partial Least Squares Regression: Comparison of Two Potential Ligands and Application to Mine Drainage and Related Water Samples** MARK T STAUFFER, University of Pittsburgh - Greensburg

(2280-7 P) **A Highly Selective Naked Eye Anion Detector Based on Chromone Derivatives and their Electroanalytical Studies** NEHA GUPTA, IIT Roorkee, Ashok K Singh N/A

(2280-8 P) **TBAF and It's Spectral Interference** IAN ADAMS, The University of Alabama

POSTER SESSION — Session 2290

All posters are to be mounted by 10:00 AM and remain on display until 2:00 PM. Authors must be at their posters from 10:00 AM to 12:00 PM. Location of the posters is on the Exposition Floor, 400 Aisle. PLEASE NOTE: You cannot get onto the Exposition Floor until after 9:00 AM.

Various Applications of GCMS

Thursday Morning, Exposition Floor, 400 Aisle

(2290-1 P) **Evaluation of Methylisothiazolinone (MI) Extraction from Sunscreen Using Supported Liquid Extraction prior to GC/MS Analysis** RHYS JONES, Biotage GB Limited, Lee Williams, Victor Vandell

(2290-2 P) **Extending the Range of GC-MS Applications with Cold EI** AVIV AMIRAV, Tel Aviv University, Alexander Fialkov, Uri Keshet , Tal Alon

(2290-3 P) **New Sampling Device for Early Cancer Screening by Non-Invasive Detection of VOCs Biomarkers in Exhaled Breath** PAOLO BENEDETTI, IIA - CNR, Ettore Guerriero, Federico Marini, Mark Ragusa, Maria Cristina Zappa, Carlo Crescenzi

(2290-4 P) **Quantitative and Qualitative Multi-Residue Analysis of Chemical Contaminants in Food and Feed by GC-HRTOFMS** JONATHAN D BYER, Leco Corporation, Joseph E Binkley, David E Alonso N/A

(2290-5 P) **Chemical Warfare Agents (CWAVX) by Large Volume Injection/Programmable Temperature Vaporization (LVI/PTV) - Gas Chromatography/Quadrupole Mass Spectrometry** TOM FOWLER, CSS-Dynamac, Kelly Head, Julia Capri

(2290-6 P) **Quantification of Persistent Organic Pollutants in Dietary Supplements Using Stir-bar Sorptive Extraction- Thermal Desorption- GC/MS and Isotope Dilution Mass Spectrometry** WEIER HAO, Duquesne University, Andrew Boggess, Skip Kingston

(2290-7 P) **Automated Sampling of Methanol Extractions** ANNE JUREK, EST Analytical, Lindsey Pyron, Kelly Cravenor, Justin Murphy

(2290-8 P) **Complex Sample Characterization by GC×GC-TOF** ROBERTA LARICCIA, DANI Instruments, Vincenzo Casilli, Matthew S Klee N/A

(2290-9 P) **Pesticide Analysis in Drinking Water and Beverages Using Multiple Techniques** XIAOPING LI, Georgia Gwinnett College, Zoe Goldstein, Hongxia Guan, Michelle Huang, Rashad Simmons, Simon Mwongela

(2290-10 P) **Identification of Contaminants in Powdered Foods by Direct Extraction Thermal Desorption GC/MS** RONALD SHOMO, Scientific Instrument Services, Christopher Baker, John Manura

(2290-11 P) **Methods Development for Sampling and Analysis of Biogenic Volatile Organic Compounds Released from Plants** PRITHVIRAJ SRIPATHI, Middle Tennessee State University, Christopher Moore, Kathleen Kuklewicz, James G Milstead, Beng Guat Ooi, Ngee Sing Chong

(2290-12 P) **Pyrolysis-GC/MS as a Screening Tool for Phthalate Esters and Brominated Flame Retardants in the RoHS Directive** NICOLE M LOCK, Shimadzu Scientific Instruments, Shilpi Chopra, Di Wang, Laura Chambers, Mark Janeczko

(2290-13 P) **Withdrawn**

(2290-14 P) **Improved Flavor Profile of Italian Wine and Scotch Whiskey Using an Aqueous-Stable Polyethylene Glycol Stationary Phase** RAMKUMAR DHANDAPANI, Phenomenex, Tim Anderson, Kristen Parnell

(2290-15 P) **Separation Solutions for Triglycerides in Food Fat and Oil by High Temperature GC** RAMKUMAR DHANDAPANI, Phenomenex, Kristen Parnell, Tim Anderson

(2290-16 P) **Volatile Organic Compounds in Energy Drinks as Determined by GC/MS with Purge and Trap Sample Concentration** CYNTHIA ELMORE, OI Analytical, J Garrett Slaton

(2290-17 P) **What's in Your Morning Drink? Comprehensive Characterization of Coffee and Tea Extracts by GCxGC--TOF MS** LAURA MCGREGOR, Markes International Ltd, Nicola Watson, Massimo Santoro, Chris Hall, Ken Umbarger

(2290-18 P) **Research of Polychlorinated Biphenyls (PCBs) in Vegetables by GC-MS/MS** XIZHI WANG, Shimadzu, Shiheng Luo, Feifei Tian, Jun Fan, Guixiang Yang, Taohong Huang, Shin-ichi Kawano, Yuki Hashi

(2290-19 P) **Analysis of FAMEs Using Cold EI GC/MS for Improved Molecular Ion Information** ADAM J PATKIN, PerkinElmer, Sharanya Reddy

(2290-20 P) **Basmati or Not Basmati? That is the Question** KENNETH ROSNACK, Waters Corporation, Gareth Cleland, Adam Ladak, Jennifer Burgess, Steven Lai

POSTER SESSION — Session 2300

All posters are to be mounted by 10:00 AM and remain on display until 2:00 PM. Authors must be at their posters from 10:00 AM to 12:00 PM. Location of the posters is on the Exposition Floor, 400 Aisle. PLEASE NOTE: You cannot get onto the Exposition Floor until after 9:00 AM.

Vibrational Spectroscopy Advances

Thursday Morning, Exposition Floor, 400 Aisle

(2300-1 P) **Development of a Non-Invasive Probing Method for Pharmaceutical Analysis Using Spatially Offset Raman Spectroscopy** HYUNG MIN KIM, Kookmin University

(2300-2 P) **Spatial Heterodyne Raman Spectrometer with LED Sources** WILLIAM HUNTINGTON, University of South Carolina, S Michael Angel

(2300-3 P) **Silver-Coated Self-Assembled Polystyrene Nanospheres for Surface-enhanced Raman Spectroscopy** LARA MIKAC, Rudjer Boskovic Institute, Mile Ivanda, Marijan Gotic, Vesna Janicki, Hrvoje Zorc, Tibor Janci, Sanja Vidacek N/A

(2300-4 P) **Second Harmonic Generation of Gold Nanorods Coupled with Atomic Force Microscopy** DARBY NELSON, University of Notre Dame, Zachary D Schultz

(2300-5 P) **Chemical Detection and Tracking of c-RGD Peptide-Conjugated Gold Nanoparticles Interacting with αVβ3 Integrins in Living Cells** HAO WANG, University of Notre Dame, Lifu Xiao, Zachary D Schultz

(2300-6 P) **Molecular Interactions between Nanoparticles and Model Cell Membranes Determined via Combined Vibrational Spectroscopic Studies** PEIPEI HU, The University of Michigan at Ann Arbor

(2300-7 P) **A Stable, Disposable Nanostructured Substrate for Surface Enhanced Raman Scattering (SERS) Detection of Drugs with Environmental Applications** HONEY MADUPALLI, Central Michigan University, Brandon Russell, Mary M Tecklenburg N/A

(2300-8 P) **A New Diamond ATR Video Microscopy Accessory** DAVID W SCHIERING, Czitek

(2300-9 P) **From Mars to Mission Critical Process Control** DAN WOOD, Keit Ltd

(2300-10 P) **Measurement of Component Distribution in "Soft Chew" Formulations by ATR FTIR Imaging** RONALD RUBINOVITZ, Thermo Fisher Scientific, William Wihlborg

TECHNICAL PROGRAM

THURSDAY, MARCH 10, 2016 AFTERNOON

SYMPOSIUM **Session 2310**

ACS-ANLY - Ultrasensitive Bioanalysis on the Pico-to Femtoliter Scales
arranged by Yong Zeng, University of Kansas

Thursday Afternoon, Room B308
Yong Zeng, University of Kansas, Presiding

1:30		**Introductory Remarks -** **Yong Zeng**
1:35	(2310-1)	**Bioanalysis in Ultrasmall Volumes** DANIEL T CHIU, University of Washington N/A
2:10	(2310-2)	**Engineering Hydrogels for Sensitive miRNA Assays** PATRICK DOYLE, Massachusetts Institute of Technology
2:45	(2310-3)	**Microfluidic Devices with Integrated Nanochannel Arrays to Study Development and Aging of Individual Bacteria** STEPHEN C JACOBSON, Indiana University, Joshua D Baker, David T Kysela, Yves V Brun
3:20		**Recess**
3:35	(2310-4)	**Single Cell Genomic and Proteomic Analysis** ANUP K SINGH, Sandia National Laboratories
4:10	(2310-5)	**Ultrasensitive and Broad-Range Microfluidic Immunoassays** YONG ZENG, University of Kansas

SYMPOSIUM **Session 2320**

Analytical Techniques in Neuroscience
arranged by Hang Lu, Georgia Tech and Andriana San Miguel, North Carolina State University

Thursday Afternoon, Room B309
Hang Lu, Georgia Tech, Presiding

1:30		**Introductory Remarks -** **Hang Lu**
1:35	(2320-1)	**The Chemical Characterization of the Brain: From New MS-Based Measurement Tools to New Insights** JONATHAN V SWEEDLER, University of Illinois
2:10	(2320-2)	**Scalable Proteomic Imaging of Intact Biological Systems** KWANGHUN CHUNG, MIT
2:45	(2320-3)	**In Situ RNA Profiling by FISH SCALYS (Sequential Coding anALYSis)** LONG CAI, Caltech N/A
3:20		**Recess**
3:35	(2320-4)	**Expansion Sequencing (ExSEQ): Comprehensive In Situ Transcriptome Characterization Throughout Intact Brain Circuits** SHAHAR ALON, Massachusetts Institute of Technology
4:10	(2320-5)	**Genetic Approaches to Brain Circuit Mapping and Cell Type Characterization** HONGKUI ZENG, Allen Institute for Brain Science

SYMPOSIUM **Session 2330**

Bioanalytical Chemistry Using the Next Generation of Nanomaterials
arranged by Isiah M Warner, Louisiana State University and Chieu D Tran, Marquette University

Thursday Afternoon, Room B310
Isiah M Warner, Louisiana State University, Presiding

1:30		**Introductory Remarks -** **Isiah M Warner and Chieu D Tran**
1:35	(2330-1)	**GUMBOS at the Nanoscale: Size Control, Characterization, and Applications** ISIAH M WARNER, Louisiana State University, Suzana Hamden
2:10	(2330-2)	**Discovering "Genetic Codes" for Nanomaterials Morphologies and Employing the DNA-Encoded Nanomaterials for Sensing and Imaging** YI LU, University of Illinois at Urbana-Champaign, Li Huey Tan, Peiwen Wu, Nitya Sai Reddy Satyavolu
2:45	(2330-3)	**Free-Standing Gold and Silver Nanoparticles Films as Flexible Sensing Platforms** JENNIFER S SHUMAKER-PARRY, University of Utah
3:20		**Recess**
3:35	(2330-4)	**Ultrafast and Nonlinear Spectroscopy of Plasmonic Nanoparticles for Drug Delivery and Photothermal Applications** LOUIS H HABER, Louisiana State University, Raju R Kumal, Tony E Karam, Holden T Smith, Corey R Landry, Mohammad Abu-Laban, Daniel J Hayes
4:10	(2330-5)	**Biocompatible Nanoparticle Composite Materials: Green Synthesis and Applications** CHIEU D TRAN, Marquette University

SYMPOSIUM **Session 2340**

Bioinformatics: Metabolite Identification and Quantification
arranged by Xiang Zhang, University of Louisville

Thursday Afternoon, Room B311
Xiang Zhang, University of Louisville, Presiding

1:30		**Introductory Remarks -** **Xiang Zhang**
1:35	(2340-1)	**New Methods for Improved NMR Quantitation and Reliable MS Coverage in Metabolomics** DANIEL RAFTERY, University of Washington
2:10	(2340-2)	**Point Matching Global Peak Alignment Algorithms for Comprehensive Two-Dimensional Gas Chromatography Coupled with Mass Spectrometry (GC×GC-MS)** KIM SEONGHO, Wayne State University/Karmanos Cancer Institute
2:45	(2340-3)	**Computational and Database Assisted Structure Identification Tools for Untargeted Metabolomics** DAVID GRANT, University of Connecticut
3:20		**Recess**
3:35	(2340-4)	**Mass Informatics of Quantitative Metabolomics by Integrating LCxLC-MS and GCxGC-MS Data** XIANG ZHANG, University of Louisville, Xiaoli Wei, Imhoi Koo, Pawel Lorkiewicz
4:10	(2340-5)	**Translating Big Data from HR Imaging MS Data into Molecular Knowledge** THEODORE ALEXANDROV, EMBL

SYMPOSIUM **Session 2350**

Micro and Nano-Scale Optofluidic Lasers for Biological Applications
arranged by Xudong Fan, University of Michigan and Seok-Hyun Yun, Harvard Medical School

Thursday Afternoon, Room B312
Xudong Fan, University of Michigan, Presiding

1:30		**Introductory Remarks -** **Xudong Fan and Seok-Hyun Yun**
1:35	(2350-1)	**Optofluidic Laser as a New Bioanalytical Tool** XUDONG FAN, University of Michigan
2:10	(2350-2)	**Latest Progress in Spasers** MARK I STOCKMAN, Georgia State University
2:45	(2350-3)	**Recent Advances in Biolasers: Stimulated Emission from Solid-State Fluorescent Protein and Lasing Inside Live Cells** MALTE C GATHER, University of St Andrews
3:20		**Recess**
3:35	(2350-4)	**Speckle-Free Lasers** HUI CAO, Yale University
4:10	(2350-5)	**Bio-Lasers** S H ANDY YUN, Harvard University N/A

SYMPOSIUM **Session 2360**

Nanofiber Materials Overcome Enduring (Bio) Analytical Challenges
arranged by Margaret W Frey, Cornell University and Antje J Baeumner, University of Regensburg

Thursday Afternoon, Room B313
Margaret W Frey, Cornell University, Presiding

1:30		**Introductory Remarks - Margaret W Frey and Antje J Baeumner**
1:35	(2360-1)	**Nanofiber Chemistry and Synthesis and the Impact on Analytical Systems** MARGARET W FREY, Cornell University, Larissa M Shepherd, Edurne Gonzalez, Nidia Trejo
2:10	(2360-2)	**Electrospun Fibers for Electrochemical Analysis** GREGORY RUTLEDGE, Massachusetts Institute of Technology, Xianwen Mao, Yuxi Zhang, T Alan Hatton, Harry Tuller
2:45	(2360-3)	**Carbon Nanotube–Nanocrystalline Diamond Hybrid Electrodes: A Route for Development of a Highly Sensitive Neurochemical Microsensor** PRABHU ARUMUGAM, Louisiana Tech University
3:20		**Recess**
3:35	(2360-4)	**Piezoelectric Nanofiber Platform for Cell Monitoring** CAROLINE L SCHAUER, Drexel University
4:10	(2360-5)	**New Concepts for Lab-on-a-Chip Systems Using Electrospun Nanofibers** ANTJE J BAEUMNER, University of Regensburg

SYMPOSIUM **Session 2370**

SEAC - Nanoengineered Biosensors
arranged by Adrian C Michael, University of Pittsburgh

Thursday Afternoon, Room B314
Reginald M Penner, University of California Irvine, Presiding

1:30		**Introductory Remarks - Reginald M Penner**
1:35	(2370-1)	**Biosensors for Early Cancer Detection Based Upon an Electrical Interface to Virus Particles** REGINALD M PENNER, University of California Irvine
2:10	(2370-2)	**Organic Electronics Biosensors for Label-Free Femtomolar Protein Detection** LUISA TORSI, Università degli Studi di Bari "A. Moro", Maria Magliulo, Gerdo Palazzo
2:45	(2370-3)	**Single Nanoparticle SPR Imaging and Plasmonic Nanocone Arrays: Smart Materials and Smart Chemistries for Advanced Optical Biosensors and Biomimetic Devices** ROBERT M CORN, University of California Irvine, Adam M Maley, H W Millie Fung
3:20		**Recess**
3:35	(2370-4)	**Ultrasensitive Biomolecular Detection Using Nanostructured Microelectrodes** SHANA KELLEY, University of Toronto
4:10	(2370-5)	**DNA Nanostructures and Networks** WEIHONG TAN, University of Florida

ORGANIZED CONTRIBUTED SESSIONS **Session 2380**

Biosensing Devices for Neuron Mapping
arranged by Chenzhong Li, Florida International University

Thursday Afternoon, Room B315
Chenzhong Li, Florida International University, Presiding

1:30	(2380-1)	**Integration of CNS and PNS Components with Silicon Devices via Surface Microengineering for Neuronal Mapping Applications** JAMES J HICKMAN, University of Central Florida
1:50	(2380-2)	**Biomimetic Strategies for Seamless Integration of Neural Interface Technology** TRACY CUI, University of Pittsburgh
2:10	(2380-3)	**Automated Micro- and Nanoscale Systems for Single Neuronal Activity** JIT MUTHUSWAMY, Arizona State University, Swathy Sampath Kumar, Sindhu Anand
2:30	(2380-4)	**Electronic Biosensing Devices for Recording Neuron Transmitter Expression at Single Cell and 3D Tissue Level** CHENZHONG LI, Florida International University
2:50		**Recess**
3:05	(2380-5)	**Visualization of Nanoscale Neuron Surface Topography and Detection of Neurotransmitter Release Using Nano Electrochemical Microscopy** YASUFUMI TAKAHASHI, Tohoku University, Hitoshi Shiku, Tomokazu Matsue
3:25	(2380-6)	**Trans-Synaptic In Vitro Mapping Using Microfluidic Approaches for Neuroscience Discovery** ANNE MARION TAYLOR, University of North Carolina at Chapel Hill and North Carolina State
3:45	(2380-7)	**CMOS Technology Enabled Brain Machine Interface (BMI) For Chronic Neuronal Mapping** MUHAMMAD MUSTAFA HUSSAIN, King Abdullah University of Science and Technology (KAUST), Aftab M Hussain, Amir N Hanna
4:05	(2380-8)	**Wireless Stimulation and Recording for In-Vivo Electrophysiology** JAMES MORIZIO, Triangle BioSystems Inc.

ORAL SESSIONS **Session 2390**

Bioanalytical: Electrochemical Techniques

Thursday Afternoon, Room B305
Leslie Wilson, North Carolina State University, Presiding

1:30	(2390-1)	**Compatibility of Nitric Oxide Release Coatings with Implantable Enzymatic Glucose Sensors Based on Osmium(III/II) Mediated Electrochemistry** KYOUNG HA CHA, University of Michigan
1:50	(2390-2)	**Application of Nanopipette Electrodes for Real-Time Measurement of Thyroid Hormones to Evaluate Thyrotoxic Storm** CELESTE A MORRIS, Northern Kentucky University, Barbara Cata, Theresa M Ruwe, Edward A Dobrzykowski, Teri Rae Armstrong
2:10	(2390-3)	**Neurochemical and Behavioral Analysis of Post-Chemotherapy Cognitive Impairment** MICHAEL A JOHNSON, University of Kansas, Sam V Kaplan, David P Jarmolowicz, Rachel C Gehringer, Michael J Sofis, Ryan Limbocker, Mimi Shin, Meng Sun
2:30	(2390-4)	**In-Vitro Amperometric Sensing of Dynamic Changes of Endogenous NO and CO Gases for Co-cultured Endothelial and Neuronal Cells** HA YEJIN, Ewha Womans University, Heo Chaejeong, Woo Juhyun, Suh Minah, Lee Youngmi
2:50		**Recess**
3:05	(2390-5)	**Real-Time Measurements of Oxidative Stress During Chronic L-DOPA Treatment For Parkinson's Disease** LESLIE WILSON, North Carolina State University, Christie Lee, Catherine F Mason, Leslie A Sombers
3:25	(2390-6)	**Dual Function Ion Selective Microelectrodes for Scanning Electrochemical Microscopy** GANESH UMMADI, Oregon State University, Corey Downs, Dipankar Koley
3:45	(2390-7)	**Label-Free Electrochemical microRNA Detection based on Different Modifier: Conducting Polymer and Graphene on the Surface of Pencil Graphite Electrode** MEHMET OZSOZ, Gediz University, Merve Kaplan
4:05	(2390-8)	**Quantitative Measurement of Transmitters in Individual Vesicles with Microelectrodes** XIANCHAN LI, University of Gothenburg, Soodabeh Majdi, Andrew Ewing

ORAL SESSIONS **Session 2400**

Bioanalytical: Sampling and Sample Preparation - Half Session

Thursday Afternoon, Room B304
Jinesh Jain, NETL, Presiding

1:30	(2400-1)	**Aptamer Functionalized Solid Phase Microextraction for Selective Enrichment of Thrombin** MD NAZMUL ALAM, University of Waterloo, Fuyou Du, Janusz Pawliszyn
1:50	(2400-2)	**New Generation of Solid SPME Coatings for Complementary Gas- and Liquid- Phase Separation: A Step Toward Integration of Metabolomics Platforms** EMANUELA GIONFRIDDO, University of Waterloo, Ezel Boyaci, Janusz Pawliszyn
2:10	(2400-3)	**Solid Phase Microextraction as Sample Preparation Tool in Brain Tumors Analysis** NATHALY REYES-GARCES, University of Waterloo, Barbara Bojko, Janusz Pawliszyn
2:30	(2400-4)	**MEMS Based Pre-Concentrator GC-Ion Mobility Spectrometry for Trace Gas Analysis** WOLFGANG VAUTZ, ISAS, Sascha Liedtke, Stefano Zampolli, Chandrasekhara Hariharan

TECHNICAL PROGRAM

ORAL SESSIONS — Session 2410

Bioanalytical: Techniques Using Sensors

Thursday Afternoon, Room B302
Ke Li, Missouri University of Science and Technology, Presiding

1:30	(2410-1)	**Core-Shell Nanoparticle Scintillator Probes for Low-Energy Radionuclide Quantification in Aqueous Media** COLLEEN JANCZAK, University of Arizona, Isen Andrew C Calderon, Zeinab Mokhtari, Craig A Aspinwall
1:50	(2410-2)	**Use of Silicon Photonic Microring Resonators for the High-Throughput Analysis of Multi-Protein Complex Formation in the Blood Coagulation Cascade** ELLEN MUEHL, University of Illinois at Urbana-Champaign, Josh M Gajsiewicz, Ivan Lenov, Jim H Morrissey, Ryan C Bailey
2:10	(2410-3)	**Label-Free RNA Probes for Live Cell Dual-Color Imaging of EGFR** XIAOHONG TAN, Carnegie Mellon University
2:30	(2410-4)	**Integration of Whispering Gallery Mode Detectors into Fluidic Platforms for Clinical Diagnostics** DANIEL KIM, University of Kansas, Robert C Dunn
2:50		**Recess**
3:05	(2410-5)	**Ultra-High Spatial Resolution Detection of Localized pH within a Single Live Cell** QINGBO YANG, Missouri University of Science and Technology, Hai Xiao, Honglan Shi, Xiaobei Zhang, Yinfa Ma
3:25	(2410-6)	**Real Time Analysis of Hepatitis B Virus Assembly with Multi-Pore Nano-fluidic Devices for Enhanced Resolution of Particle Size and Electrophoretic Mobility** PANAGIOTIS KONDYLIS, Indiana University, Jinsheng Zhou, Zachary D Harms, Lisa Selzer, Adam Zlotnick, Stephen C Jacobson
3:45	(2410-7)	**Graphene-Based Chemiresistive Nanobiosensors for Detection of Citrus Greening Disease** THIEN-TOAN H TRAN, University of California, Riverside, Clark Kelley, Jinxia Shi, Wenbo Ma, Ashok Mulchandani
4:05	(2410-8)	**A Quantum Dot Based Fiber-Optic Micro-Sensor for Niche-Environment Temperature Monitoring** KE LI, Missouri University of Science and Technology, Qingbo Yang, Honglan Shi, Yinfa Ma, Hai Xiao, Qi Zhang

ORAL SESSIONS — Session 2420

Computers in Chemistry - Half Session

Thursday Afternoon, Room B301
Dean Tzeng, The Pittsburgh Conference, Presiding

1:30	(2420-1)	**An Inexpensive, Programmable System for Prototyping Instruments, Making Short Run Specialty Measurement Systems, and Computerizing Outdated Hardware** SCOT D ABBOTT, Phoenix Frist Response
1:50	(2420-2)	**Is SAP the Only System You Need to Operate Your QC Lab?** GEOFF R TURNBULL, CSols, Inc. N/A
2:10	(2420-3)	**Maximizing the Information from the Infrared Spectra of Mixtures Using Advanced Software Algorithms** IAN ROBERTSON, PerkinElmer Limited, Jerry Sellors, Justin Lang
2:30	(2420-4)	**Driving Governance and Organizational Change in Large and Complex Informatics Projects** ADAM S BORENSTEIN, LabAnswer, Brian Brunner, Terryl Kibodeaux

ORAL SESSIONS — Session 2430

Glycan Analysis - Half Session

Thursday Afternoon, Room B303
Christa M Snyder, Indiana University, Presiding

1:30	(2430-1)	**Identification of Serum N-Glycans as Cancer-Specific Biomarkers by Microchip Electrophoresis and Mass Spectrometry** CHRISTA SNYDER, Indiana University, Xiaomei Zhou, Margit I Campos, Milos V Novotny, Stephen C Jacobson
1:50	(2430-2)	**Improved Separation of Saccharide Standards by EFL-HILIC** RAFFEAL BENNETT, The Ohio State University, Susan Olesik
2:10	(2430-3)	**Receptor for Advanced Glycation End Products Diffusion and Ligand-Binding Events Studied by Fluorescence Recovery after Photobleaching and Surface Plasmon Resonance** QIAOCHU ZHU, Iowa State University, Aleem Syed, Chamari S Wijesooriya, Emily A Smith
2:30	(2430-4)	**Comprehensive Quantitative and Structural Analysis of Permethylated N-glycan Using PGC-LC-MS/MS** SHIYUE ZHOU, Texas Tech University, Yehia Mechref

ORAL SESSIONS — Session 2440

LC and Sample Matrix Solutions - Half Session

Thursday Afternoon, Room B304
Jinesh Jain, NETL, Presiding

3:05	(2440-1)	**Analysis of Biofluids Using Solid Phase Microextraction Devices Made on Plastic Support** NATHALY REYES-GARCES, University of Waterloo, Barbara Bojko, Janusz Pawliszyn
3:25	(2440-2)	**Studies of Matrix Effects on the Determination of Hydrogen Peroxide by Using High Performance Anion Exchange Chromatography with Pulsed Amperometric Detection** JUN CHENG, Thermo Fisher Scientific, Yan Liu, Kannan Srinivasan, Christopher Pohl
3:45	(2440-3)	**HPLC Purity Method Development Using Low pH Mobile Phase and Ion-Pairing Reagent: Application to GMP Tert-Leucine Analysis** JOHN VINCI, AbbVie, Inc., Clifford Mitchell
4:05	(2440-4)	**Combining Orthogonal Separation Modes for Analysis of Multiple Sample Components** THOMAS E WHEAT, Waters Corporation, Amanda B Dlugasch, Patricia R McConville

ORAL SESSIONS — Session 2450

Microfluidics/Lab-on-a-Chip - Bioanalytical II

Thursday Afternoon, Room B316
Michael Woodman, Agilent Technologies, Presiding

1:30	(2450-1)	**Separation of Preterm Birth Biomarkers Using Capillary and Microchip Electrophoresis** ANNA V NIELSEN, Brigham Young University, Radim Knob, Adam T Woolley
1:50	(2450-2)	**Analysis of Nitrosative Stress in Macrophage Cells Using Microchip Electrophoresis with Electrochemical Detection** JOSEPH M SIEGEL, University of Kansas, Manjula B Wijeshinghe, Kelci M Schilly, Susan M Lunte
2:10	(2450-3)	**Ultrasensitive Electrochemical Microfluidic Immunoarray for Assessment of Aggressive vs Indolent forms of Prostate Cancer Biomarkers** BRUNAH A OTIENO, University of Connecticut, Colleen E Krause, Abby Jones, Amit Joshi, Mohammed Sherafeldin, James F Rusling
2:30	(2450-4)	**Microfluidic and 3-D Printed Devices for Near-Real-Time and Simultaneous Detection of Neurotransmitters** ALEXANDRA D TOWNSEND, Saint Louis University, R Scott Martin
2:50		**Recess**
3:05	(2450-5)	**Electrokinetically Operated Integrated Microfluidic Platform for Preterm Birth Biomarker Analysis** MUKUL SONKER, Brigham Young University, Radim Knob, Vishal Sahore, Adam T Woolley
3:25	(2450-6)	**Discrete Stimulation of Lymph Node Slices on Chip** ASHLEY E ROSS, University of Virginia, Jacob F Woodroof, Rebecca R Pompano
3:45	(2450-7)	**Ultrasensitive ELISA for Detection of Infectious Diseases on Surface Modified PMMA Microfluidic Microplates** SANJAY SHARMA TIMILSINA, University of Texas at El Paso, Maowei Dou, Xiujun James Li
4:05	(2450-8)	**Optical Formulation of 3D Printer Resin for Minimum Microfluidic Flow Channel Size** GREG NORDIN, Brigham Young University, Hua Gong, Michael Beauchamp, Steven Perry, Adam T Woolley

Session # 1280 **Abstract # 1280-4** **Organized Contributed Sessions**

Session Title SEAC - The Student Session in Electroanalysis

Abstract Title **Toward the Electrochemical Detection of Single Atoms and Ions**

Primary Author
Jeffrey E. Dick
The University of Texas at Austin

Date: Tuesday, March 08, 2016 - Afternoo
Time: 02:30 PM
Room: B401

Co-Author(s)

Abstract Text

Historically, methods of analysis have entailed ensemble measurements over thousands to billions of molecules and nanoparticles in an attempt to quantify and understand their properties. This means that minute peculiarities between molecules are averaged out in ensemble measurements. Thus, studying the single molecule/single nanoparticle regime is necessary to probe properties of individual analyte species and perhaps learn something a bit different from bulk measurements.

Electrocatalytic amplification has been used as an electrochemical technique to detect single nanoparticles colliding on a relatively inert ultramicroelectrode (UME) surface. When a metal nanoparticle collides onto an UME poised at a potential where the catalytic reaction occurs on the nanoparticle but not the UME, faradaic current flows, and the current increase can be detected in the amperometric i-t response. The limits of detection for this technique will be discussed as well as future prospects.

Keywords: Electrochemistry
Application Code: Nanotechnology
Methodology Code: Electrochemistry

Session # 1280 Abstract # 1280-5 **Organized Contributed Sessions**

Session Title SEAC - The Student Session in Electroanalysis

Abstract Title **Miniaturized Potentiometric Ion-Sensing Systems: From Bulk Electrodes to Paper-Based Ion-Sensing Devices**

Date: Tuesday, March 08, 2016 - Afternoo
Time: 03:05 PM
Room: B401

Primary Author Jinbo Hu
University of Minnesota

Co-Author(s) Andreas Stein, Philippe Buhlmann

Abstract Text

Potentiometric sensors, comprising ion-selective electrodes (ISEs) and reference electrodes, are electrochemical sensors used to determine the concentration of target ions. In view of affordable and portable analytical devices with small sample volumes, all-solid-state ISEs and reference electrodes, in which a solid contact is used as an ion-to-electron transducer, are highly desirable. The first half of this talk will focus on the development of all-solid-state potentiometric cells utilizing colloid-imprinted mesoporous (CIM) carbon as a novel solid contact. CIM carbon exhibits desirable properties as a solid contact material, including a low content of redox-active impurities and a high double layer capacitance. Therefore, potentiometric sensors based on CIM carbon can be constructed with superior electrochemical performance, including excellent ionic response, reproducibility and signal stability [1,2]. Strategies of developing calibration-free ion sensors are discussed. To achieve miniaturized ion-sensing systems with low cost, the second half of this talk will focus on disposable ion-sensing devices based on paper. Previously, a three-dimensional paper-based ion sensor with conventional ISEs and reference electrodes was developed [3]. To simplify the use of these sensors, CIM carbon-based all-solid-state ISEs and reference electrodes are integrated on paper to construct a two-dimensional disposable ion-sensing platform. For a measurement, only one droplet of sample is needed. These devices exhibit linear responses towards different concentrations of electrolyte ions. They are disposable, simple to use, and do not require any supply reagents to function.

[1] Hu, J.; Zou, X. U.; Stein, A.; Bühlmann, P. Anal. Chem. 2014, 86, 7111-7118.
[2] Hu, J.; Ho, K. T.; Zou, X. U.; Smyrl, W. H.; Stein, A.; Bühlmann, P. Anal. Chem. 2015, 87, 2981-2987.
[3] Lan, W.-J.; Zou, X. U.; Hamedi, M. M.; Hu, J.; Parolo, C.; Maxwell, E. J.; Bühlmann, P.; Whitesides, G. M. Anal. Chem. 2014, 86, 9548-9553.

Keywords: Clinical Chemistry, Electrochemistry, Integrated Sensor Systems, Ion Selective Electrodes
Application Code: Clinical/Toxicology
Methodology Code: Electrochemistry

Session # 1280 Abstract # 1280-6 **Organized Contributed Sessions**

Session Title SEAC - The Student Session in Electroanalysis

Abstract Title **Spontaneous Photoelectrochemical Growth of Nanopatterned Semiconductor Films Driven by Anisotropic Interfacial Light Collection**

Date: Tuesday, March 08, 2016 - Afternoo
Time: 03:25 PM
Room: B401

Primary Author Azhar I. Carim
California Institute of Technology

Co-Author(s) Anjali Premkumar, Harry A. Atwater, Nathan S. Lewis, Nicolas A. Batara

Abstract Text
Photoelectrodeposition of SeTe films spontaneously produces ordered, nanoscale lamellar morphologies with periodicities and orientations tuneable by varying the illumination wavelength and polarization, respectively. These control mechanisms have been investigated by examining the morphologies of deposits in response to tailored illumination inputs including multimodal spectral profiles and multiple polarizations. Films grown with different sources, having similar average wavelengths but highly differing spectral bandwidths, all produced structures with a single, common periodicity. Deposition using simultaneous illumination from two narrowband sources, which differed in wavelength by several hundred nm, resulted in structures with only a single periodicity intermediate between the periods observed when either source alone was used; this periodicity could be varied by changing the relative source intensities. Simulations of light concentration in idealized lamellar arrays indicated that a self-optimization of the pattern periodicity, resulting in the maximization of the anisotropy of interfacial light absorption in the 3D structure, is consistent with the observed growth process.

Additionally, growth using light from two non-orthogonally polarized same-wavelength sources generated morphologies in which the long axes of the lamellae were oriented parallel to the intensity-weighted average polarization orientation. Deposition using orthogonally polarized sources produced structures that consisted of two intersecting sets of orthogonally oriented lamellae in which the relative heights of the two sets were proportional to the relative source intensities. Simulations of light absorption performed in analogous, idealized structures and revealed that the lamellae preferentially absorbed light polarized along their long axes. These data sets also indicate that the observed structures are a result of maximization of interfacial light concentration in the growing structures.

Keywords: Electrochemistry, Material Science, Nanotechnology, Semiconductor
Application Code: Nanotechnology
Methodology Code: Electrochemistry

Session # 1280 Abstract # 1280-7 **Organized Contributed Sessions**

Session Title SEAC - The Student Session in Electroanalysis

Abstract Title **New Enzymes for the Hybrid Enzymatic and Organic Electrocatalytic Cascade for the Complete Oxidation of Glycerol**

Date: Tuesday, March 08, 2016 - Afternoo
Time: 03:45 PM
Room: B401

Primary Author Sofiene Abdellaoui
University of Utah

Co-Author(s) Shelley D. Minteer

Abstract Text

In bioelectrocatalysis, oxidases and dehydrogenases are the most frequently used enzymes for bioelectrode design. The redox centers of these enzymes usually operate with sophisticated catalytic mechanisms. In the living cell, enzymes are not optimized for bioelectrocatalysis. It is therefore important to tailor biocatalysts to improve their stability, to expand substrate specificity and to control enzyme orientation on the electrode. Enzyme engineering offers the possibility to create and tailor new structural designs that could improve the enzymes at the enzyme–electrode interfaces. General approaches in protein engineering include rational design involving targeted modification carried out by site-directed mutagenesis and directed evolution based on the screening of mutant libraries obtained by random mutagenesis. Here, I present the application and tools for directed evolution for the improvement of a hybrid enzymatic and organic catalytic system for the complete electrochemical oxidation of the glycerol, to CO2. This system combines an organic catalyst, (2,2,6,6-Tetramethylpiperidin-1-yl)oxy (TEMPO) with oxalate decarboxylase (Oxdc), resulting in the complete electrochemical oxidation of glycerol at a carbon electrode (Figure 1). This hybrid approach consists of five initial oxidative steps (by TEMPO), resulting in the oxidation of glycerol to mesoxalic acid (1 □6, Figure 1). A combination of Oxdc and TEMPO then facilitates the oxidation of mesoxalic (6) acid to glyoxalic acid (7), oxalic acid (8), formic acid (9) and finally, CO2.

Keywords: Biofuels, Chemically Modified Electrodes, Enzyme Assays, Protein
Application Code: Fuels, Energy and Petrochemical
Methodology Code: Electrochemistry

Session # 1280 Abstract # 1280-8 **Organized Contributed Sessions**

Session Title: SEAC - The Student Session in Electroanalysis

Abstract Title: **The Unique Electrochemical Reactivity of Small Metal Nanoparticles**

Date: Tuesday, March 08, 2016 - Afternoo
Time: 04:05 PM
Room: B401

Primary Author: Rafael Masitas
University of Louisville

Co-Author(s): Francis Zamborini

Abstract Text

In this study we describe 1) Oxidation of highly unstable <4 nm diameter gold nanoparticles at 800 mV negative of the bulk oxidation potential, 2) Effect of surface charge and electrode material on the size-dependent oxidation of surface-attached metal nanoparticles, 3) An electrophoretic low pass filter for assembling ultra-small gold nanoparticles onto an indium tin oxide electrode, and 4) Galvanic exchange reaction between PtCl[sub]4[/sub][sup]2-[/sup] and Au nanoparticles as a function of size. The motivation of this work is to better understand the unique electrochemical reactivity of small metal nanoparticles. We described the oxidation of < 4 nm diameter Au nanoparticles (NPs) attached to indium tin oxide-coated glass electrodes (glass/ITO) in Br[sup]-[/sup] and Cl[sup]-[/sup] solution. The size-dependent oxidation of Au NPs electrodeposited or drop-coated directly on glass/ITO electrodes is different compared to those chemically synthesized and electrostatically or drop-cast deposited onto aminopropyltriethoxysilane (APTES)-modified glass/ITO electrodes. The Ep of 9 nm diameter citrate-capped Ag NPs attached to Au, Pt, glassy carbon (GC), and glass/ITO electrodes, following the order (vs Ag/AgCl) of Au □Pt > GC > glass/ITO. We studied the effect of voltage and time on the electrophoretic deposition of different-sized Au NPs onto glass/ITO electrodes. The applied electric field served as a low pass filter for the deposition of Au NPs onto glass/ITO. We studied the galvanic exchange reaction between PtCl[sub]4[/sub][sup]2-[/sup] and 128 nm, 50 nm, 15 nm and 4 nm diameter citrate-coated Au NPs chemically attached to glass/ITO/APTES. The level of galvanic exchange reaction between PtCl[sub]4[/sub][sup]2-[/sup] and citrate-coated Au NPs is size dependent.

Keywords: Electrochemistry, Electrophoresis, Nanotechnology, Stripping Analysis
Application Code: Nanotechnology
Methodology Code: Electrochemistry

Session # 1290 Abstract # 1290-1 **Organized Contributed Sessions**

Session Title Specialty Gas Analysis

Abstract Title **Real-Time, Selective Analysis of Air and Specialty Gases**

Date: Tuesday, March 08, 2016 - Afternoo
Time: 01:30 PM
Room: B402

Primary Author Daniel Milligan
Syft Technologies Ltd

Co-Author(s) Barry Prince, Murray McEwan, Vaughan Langford

Abstract Text

Selected Ion Flow Tube Mass Spectrometry (SIFT-MS) (Smith & Spanel, 2005) is a real-time analytical technique that offers rapid analysis of volatile organic compounds (VOCs) and many inorganic gases to ultra-trace levels in air (Prince et al., 2010). Quantitation limits in the low part-per-trillion range (by volume; pptv) can be achieved without sample preparation or preconcentration and results compare well with those obtained at an accredited laboratory using the United States' Environmental Protection Agency (US EPA) TO-15 Compendium Method (Langford et al., 2014).

Historically SIFT-MS has been able to analyze most VOCs and a small number of inorganic gases using soft chemical ionization with positive ion reagents, H[sub]3[/sub]O[sup]+[/sup], NO[sup]+[/sup], and O[sub]2[/sub][sup]+[/sup]. A very recent development involving negative reagent ions (OH[sup]-[/sup], O[sup]-[/sup], O[sub]2[/sub][sup]-[/sup], and NO[sub]2[/sub][sup]-[/sup]) expands SIFT-MS detection capability to many more inorganic volatiles, including carbon dioxide, hydrogen chloride, hydrogen fluoride, nitrous oxide, and sulfur dioxide, plus some small halogenated VOCs.

This paper describes how SIFT-MS can be applied to detection of compounds present at trace levels in a wide variety of matrices, including air and specialty gases, by appropriate choice from the seven reagent ions. Quantitation performance for specific applications will be described.

Acknowledgement: This work was funded by Syft Technologies Ltd, New Zealand.

Langford, V.S., Graves, I., & McEwan, M.J. (2014) Rapid Commun. Mass Spectrom., 28, 10–18.
Prince, B.J., Milligan, D.B., & McEwan, M.J. (2010). Rapid Commun. Mass Spectrom., 24, 1763-1769.
Smith, D., & Spanel, P. (2005). Mass Spec. Rev., 24, 661– 700.

Keywords: Chemical Ionization MS, Gas, High Throughput Chemical Analysis, Specialty Gas Analysis
Application Code: High-Throughput Chemical Analysis
Methodology Code: Mass Spectrometry

Session # 1290 Abstract # 1290-2 **Organized Contributed Sessions**

Session Title: Specialty Gas Analysis

Abstract Title: **Cavity Ring-Down Spectroscopy Analyzer for Trace Moisture Detection in Ultra-Pure Ammonia**

Primary Author: Helen Waechter
Tiger Optics

Date: Tuesday, March 08, 2016 - Afternoo
Time: 01:50 PM
Room: B402

Co-Author(s): Bill West, Brian Siller, Florian Adler, Marten Beels, Yu Chen

Abstract Text

Use of high-brightness light emitting diodes (HB-LEDs) has been increasing over the last several years as a result of their high energy efficiency, compact footprint, and long lifetime. The energy saving potential of LEDs has motivated government-enacted legislation and regulation that will eventually lead to elimination of the inefficient incandescent light bulb. The key to maximizing manufacturing yield and quality is the purity of the ammonia gas (NH[sub]3[/sub]) used for growing the devices. In particular, moisture is a critical impurity in the gas which directly affects the LED's efficiency and therefore its maximum brightness. Thanks to recent gas manufacturing improvements, gas companies now specify moisture at low single-digit parts-per-billion (ppb) levels, compared to 12 ppb a few years ago. This, of course, calls for a measurement technology that can keep up with these developments.

Continuous-Wave Cavity Ring-Down Spectroscopy (CW-CRDS) is a highly sensitive, field-proven technique for measurement of a variety of analytes in a suite of gas matrices, ranging from inert to specialty gases. CW-CRDS derives the analyte concentration from monitoring light decay inside a high-finesse optical cavity caused by direct optical absorption of the target molecule. Providing excellent sensitivity and selectivity, CW-CRDS can be applied to even the most challenging applications, such as moisture analysis in gases that are corrosive and spectrally interfering. We present recent developments of a CW-CRDS analyzer for the real-time monitoring of trace moisture impurity in ammonia, with an unprecedented detection limit of less than 2 ppb. Key improvements and novel facets of the instrument will be presented.

Keywords: Molecular Spectroscopy, Quality Control, Specialty Gas Analysis, Trace Analysis
Application Code: Quality/QA/QC
Methodology Code: Molecular Spectroscopy

Session # 1290 Abstract # 1290-3 **Organized Contributed Sessions**

Session Title: Specialty Gas Analysis

Abstract Title: **Novel FTIR/GC Detector for Analyzing Impurities in Gas Standards**

Primary Author: Martin L. Spartz
Prism Analytical Technologies, Inc.

Date: Tuesday, March 08, 2016 - Afternoo
Time: 02:10 PM
Room: B402

Co-Author(s): Anthony S. Bonanno, Charles M. Phillips, Kelly R. McPartland, Peter P. Behnke

Abstract Text

Analyzing some gas standards for trace level impurities can at times be challenging due to interferences arising from the bulk gas which is being certified. Some analysis methods use traditional gas chromatography-mass spectrometry (GC/MS) which requires daily or even hourly calibration checks. Furthermore, large splits must be done to reduce or remove the bulk gas which could overwhelm the MS, thus reducing the sensitivity.

Coupling gas chromatography (GC) with an FTIR analyzer in an innovative way combines the compound separation of the GC with the qualitative and quantitative capabilities of optical spectroscopy. Unlike mass spectrometry, the FTIR detection technology calibration is constant so a one-time characterization of the compound is the only requirement, saving analysis time. In addition, due to a unique aspect of the FTIR detection scheme and controlling software, the gases to be analyzed are sealed within the FTIR gas cell and repeatedly probed for up to minutes; this long integration time gives rise to very low detection limits. The separation power of the GC allows any bulk gas to be removed from the background, either by mathematically ratioing it out or letting it pass through the gas cell before analysis of the impurities begins, thus removing any interferences which could reduce sensitivity. Since we also have the spectroscopic power of the FTIR, all impurities can be individually analyzed and speciated.

Work performed on measurement of impurities will be presented to demonstrate the power of this GC/FTIR technique, both in sensitivity and speciation.

Keywords: FTIR, Gas, Gas Chromatography, Vibrational Spectroscopy
Application Code: Validation
Methodology Code: Integrated Sensor Systems

Session # 1290 Abstract # 1290-4 **Organized Contributed Sessions**

Session Title: Specialty Gas Analysis

Abstract Title: **Optimization of a Cavity Ring-Down Spectrometer for the Measurement of Trace Ammonia Contamination in Semicondutor Cleanroom Environments**

Date: Tuesday, March 08, 2016 - Afternoo
Time: 02:30 PM
Room: B402

Primary Author: Graham A. Leggett
Picarro

Co-Author(s): Mark Camenzind

Abstract Text

The measurement and control of trace level contaminants is critical in a variety of industrial applications, perhaps none more so than semiconductor fabrication and the manufacture of integrated circuits (ICs). The measurement of critical contaminants at parts per billion (ppb) levels and below is required to model the impact of source air, factory recirculated air, troubleshooting/detection of leaks, chemical filter exhaustion and to predict filter lifetimes. The production of ICs is a lengthy, multi-step process that exposes the evolving wafer to a variety of chemicals and their associated contaminants. IC performance is severely affected by contamination of electrically active impurities, and it is very difficult to guarantee a contaminant-free process due to the number of fabrication steps involved. Furnace operations expose the wafer to contamination present in ambient gases, while etching and cleaning cycles contribute to further contamination. Cavity rong-down spectroscopy provides both the sensitivity and speed needed for the most demanding trace contaminant monitoring applications. Delivering continuous measurements with parts per trillion (ppt) sensitivity, CRDS analyzers are ideal for semiconductor fabrication applications, including airborne molecular contamination (AMC) cleanroom monitoring. We report the optimization of a commercial trace ammonia CRDS analyzer for AMC cleanroom monitoring applications, with particular attention paid to the sampling system and particulate control measures upstream of the optical cavity. Air/gas monitoring instruments require filters be used to protect cavity optics, but particle loading of filters varies depending upon the cleanliness of the source and particles can build up and act as a sorbent that can affect the response time of the CRDS for low level ammonia. We assess filter materials and pre-treatments to optimize and maintain response /clear-down time for trace ammonia for continuous monitoring applications.

Keywords: Air, Contamination, Trace Analysis, Ultratrace Analysis
Application Code: Process Analytical Chemistry
Methodology Code: Molecular Spectroscopy

Session # 1290 Abstract # 1290-5 **Organized Contributed Sessions**

Session Title Specialty Gas Analysis

Abstract Title **Analytical Challenges of Measuring Impurities in Biogas**

Date: Tuesday, March 08, 2016 - Afternoo
Time: 03:05 PM
Room: B402

Primary Author Janneke v. Wijk
VSL

Co-Author(s) Adriaan van der Veen, Annarita Baldan, Jeanrong Li, Stefan Persijn

Abstract Text

VSL is among the leading national metrology institutes worldwide that provide metrological traceability and quality assurance for measurements of all kinds of energy gases. It offers a wide range of primary gaseous reference materials (PRMs) related to biogas: biogenic gas, biosyngas, sulphur-containing components in methane and biogas, among others. In the ongoing research and development programme, the focus is on developing such PRMs for impurities such as siloxanes, hydrogen chloride, ammonia, carbon monoxide, and halogenated hydrocarbons. Such PRMs, and related services for, e.g., calibrating gas mixtures, proficiency testing are urgently needed to support to proliferation of the use of biomethane and upgraded biogas. Given the wide diversity of feedstocks, which give rise to a very wide array of impurities potentially present in biomethane and upgraded biogas, there is a need to determine these contents to meet specifications for grid injection and use in refuelling stations. Challenges in developing such PRMs range from stability issues, the development of dynamic gas mixture preparations, addressing the infrared activity of the matrix when using infrared spectroscopy. The paper gives an overview over the most recent achievements and the work in progress.

Keywords: Biofuels, Energy, Gas, Quality
Application Code: Fuels, Energy and Petrochemical
Methodology Code: Chemical Methods

Session # **1290** **Abstract #** **1290-6** **Organized Contributed Sessions**

Session Title: Specialty Gas Analysis

Abstract Title: **Analysis and Stability of Low Concentration HCl Standards**

Date: Tuesday, March 08, 2016 - Afternoo
Time: 03:25 PM
Room: B402

Primary Author: Nathalie Luu
Air Liquide

Co-Author(s): Anthony Schleisman, Steve Hagen

Abstract Text

The National Institute of Standards and Technology (NIST) and the US Environmental Protection Agency (EPA) have requested the manufacture of low level of hydrogen chloride (HCl) mixtures because of changing environmental regulations on HCl emissions from cement production and other sources.

Hydrogen chloride mixtures, even at low ppm concentrations, can be a challenging analysis. A reliable sampling system is necessary to achieve accurate and repeatable results.

The objective of this study was to evaluate four different existing Air Liquide cylinder preparations for their effect on HCl stability over a 12-month period. The final aim is to determine a global standardized preparation technique for low concentration HCl standards. Hydrogen chloride mixtures were blended at 10 ppm and 1 ppm in balance nitrogen and were analyzed by Fourier Transform Infrared spectroscopy (FTIR).

Keywords: FTIR, Sample Preparation, Specialty Gas Analysis, Standards
Application Code: Environmental
Methodology Code: Sampling and Sample Preparation

Session # 1290 Abstract # 1290-7 **Organized Contributed Sessions**

Session Title Specialty Gas Analysis

Abstract Title **Analytical QA/QC Measures to Validate Your Analysis Method Using FTIR**

Primary Author Sylvie Bosch-Charpenay
MKS Instruments

Date: Tuesday, March 08, 2016 - Afternoo
Time: 03:45 PM
Room: B402

Co-Author(s)

Abstract Text

There are many calculations available to estimate the accuracy and detection limits of spectroscopic analysis methods, but figuring out which ones are needed and what they represent can be a daunting challenge. EPA uses a number of different methodologies depending upon whether the analysis is for gases, liquids or solids. In the case of gas analysis using FTIR spectroscopy, EPA (through ASTM D6348) typically prescribes the use of three different Minimum Detectable Concentrations (MDC) calculations named MDC1, MDC2, and MDC3. These are based on a noise-only spectrum (MDC1), statistics on interferent-only spectra (MDC2) and spectral fit on an actual sample spectrum (MDC3). The values calculated for MDC1 ("noise-limited" MDC) are not very good at determining the analyzer detection limit as it does not account for the power of spectral analysis, but it can provide a minimal level of "goodness-of-fit" parameter. MDC2 and MDC3 are labeled as "analytical algorithm error" MDCs. In the case of MDC3 it provides a goodness-of-fit parameter on the sample spectrum that encompasses accuracy information, including biases due to other gases that may be present in the sample stream. MDC2 provides the best option for the detection limit of the analyzer in the presence of known interferents. This talk will focus on these calculations in order to provide users with an estimate of both accuracy and detection limit when using FTIR spectroscopy.

Keywords: Data Analysis, FTIR, Spectroscopy, Statistical Data Analysis
Application Code: Quality/QA/QC
Methodology Code: Physical Measurements

Session # 1300 **Abstract # 1300-1** **Oral Sessions**

Session Title: Biomedical: Nanotechnology - Half Session

Abstract Title: **Green Synthesis, Characterization of Saccharide Coated Gold Nanoparticles for Catalytic Applications**

Date: Tuesday, March 08, 2016 - Afternoo
Time: 01:30 PM
Room: B301

Primary Author: Harsh Moolani
Western Kentucky University

Co-Author(s): Jason N. Payne, Rajalingam Dakshinamurthy

Abstract Text

Gold nanoparticles (AuNPs) have gained an immense interest due to their wide applications in the fields of biomedical and pharmaceutical, which is due to their unique physico-chemical properties when they are reduced to their nanoscale size range. Here, we present a novel single step bio-friendly process for synthesis of fructose (monosaccharide), sucrose (disaccharide) and raffinose (trisaccharide) capped AuNPs, wherein saccharides are directly capped onto gold without the use of any secondary capping/stabilizing agent. Our study is mainly focused on the effect of various lengths of the saccharides in the formation and catalytic reduction activity of saccharide capped AuNPs. Characterization of synthesized AuNPs was accomplished using various analytical characterization techniques such as TEM, SEM-EDS, FTIR, and UV-Vis spectroscopy. A 4-nitrophenol reduction assay was utilized to evaluate the catalytic reduction activity of various saccharide capped AuNPs at different temperatures using UV-Vis spectroscopy. Using the spectroscopic data, rate constant for three saccharide capped AuNPs were determined followed by its activation energy and exponential factor using different equations. Utilizing the exponential factor data we were also able to calculate the kinetics of the change in entropy of the catalysis. From the kinetic data, the catalytic reduction activity for three saccharides was determined as, in the descending order: fructose, sucrose and raffinose AuNPs respectively. This difference in the catalytic activity is hypothesized to be inversely proportionally to the size of ligand on gold surface which greatly influences the surface/volume ratio.

Keywords: Carbohydrates, Characterization, Nanotechnology, UV-VIS Absorbance/Luminescence
Application Code: Biomedical
Methodology Code: UV/VIS

Session # 1300 Abstract # 1300-2 **Oral Sessions**

Session Title Biomedical: Nanotechnology - Half Session

Abstract Title **NanoCluster Beacons for Detection of a Single N6-Methyladenine Epigenetic Modification**

Primary Author Tim Yeh
University of Texas at Austin

Co-Author(s)

Date: Tuesday, March 08, 2016 - Afternoo
Time: 01:50 PM
Room: B301

Abstract Text

NanoCluster Beacons (NCBs) are a new type of activatable molecular probes that are low cost, easy to prepare and have high fluorescence enhancement ratios. NCBs employ DNA-templated, few-atom silver nanoclusters (DNA/Ag NCs, with about 2~20 silver atoms per cluster) as reporters which can significantly "light up" through interactions with a nearby DNA sequence (called an enhancer). Taking advantage of this fluorescence tunability by altering the surrounding ligands, a property that is not commonly seen among existing reporters, NCB soon evolved to a multicolor probe, termed chameleon NanoCluster Beacon (cNCB), for single-nucleotide polymorphism (SNP) detection. Here we bring the NCB detection to the next level by designing a new NCB specifically for N6-methyladenine (hereafter denoted as m6A) detection. m6A is a methylation modification abundant in prokaryotic genomes, and also found in lower eukaryotes and higher plants. So far detection of m6A relies on methods such as TLC, HPLC, MS, and enzymatic reactions, which are often laborious, expensive, with low specificity and varying reactivity. Whereas high-resolution melting analysis is able to detect a single m6A modification within a target DNA via the destabilizing effect of m6A, HRM cannot pinpoint the location of m6A in the sequence. A simple and cost-effective way to identify single m6A at any specific sites is therefore highly desired. Here, we developed a robust, simple, enzyme-free and hybridization-based method for m6A detection with "pinpoint specificity", using a new type of silver cluster-based DNA probe which we term methyladenine-specific NanoCluster Beacon (maNCB). To date, there is no hybridization technique that has the potential to reach these remarkable results.

Keywords: Biosensors, Fluorescence, Nanotechnology, Nucleic Acids
Application Code: Biomedical
Methodology Code: Sensors

Session # 1300 Abstract # 1300-3 **Oral Sessions**

Session Title: Biomedical: Nanotechnology - Half Session

Abstract Title: **Self-Assembly Approach to Integrated Nanozymes: Rational Design and Biomedical Applications**

Primary Author: Hui Wei
Nanjing University

Date: Tuesday, March 08, 2016 - Afternoo
Time: 02:10 PM
Room: B301

Co-Author(s)

Abstract Text

Nanozymes are nanomaterials with enzymatic activities. They have received great attention recently due to their wide applications and distinct advantages over natural enzymes, even to the conventional artificial enzymes. Herein, we developed an self-assembly approach to integrated nanozymes. By adaptively encapsulating catalytic guests within a zeolitic imidazolate framework (ZIF) through a one-pot synthesis under mild conditions. The as-prepared integrated nanozymes enabled one-step glucose detection through the cascade enzymatic reactions. The integrated nanozymes showed numerous merits, such as considerable high catalytic activity, enhanced stability, and recyclability. The integrated nanozymes were used to determine cerebral glucose in living rats' brains. Most importantly, when further combined with microdialysis and microfluidic techniques, an integrative analytical platform was built for real time monitoring dynamic changes of striatum glucose in living brains. The current studies not only demonstrate new ways to design high-performance nanozymes, but also provide effective platform for biomedical applications.

Keywords: Biomedical, Biosensors, Nanotechnology, Bioanalytical
Application Code: Biomedical
Methodology Code: Sensors

Session # 1300 Abstract # 1300-4 **Oral Sessions**

Session Title: Biomedical: Nanotechnology - Half Session

Abstract Title: **High Resolution Separation of Oligonucleotides and DNA Fragments Using a New Polymer-Based Reversed Phase Column**

Primary Author: Julia Baek
Thermo Fisher Scientific

Date: Tuesday, March 08, 2016 - Afternoo
Time: 02:30 PM
Room: B301

Co-Author(s): Jessica Wang, Shanhua Lin, Xiaodong Liu

Abstract Text

Synthetic oligonucleotides are widely used for various applications including DNA amplification and sequencing, in situ hybridization, gene silencing, molecular diagnostics and treatment of cancer and viral infections. These applications require highly pure oligonucleotides and therefore quality control of synthetic oligonucleotides is crucial. High performance liquid chromatography (HPLC) and mass spectrometry (MS) are valuable tools for the purity assessment and characterization of oligonucleotides. Among the chromatographic modes, ion-pair reversed-phase HPLC have been shown to be most suitable for LC-MS analysis. Here we introduce a new reversed phase column for the analysis of nucleic acids including synthetic oligonucleotides and double-stranded DNA and RNA. This column is based on hydrophobic, polymer resin which is stable at high pH and high temperature. Separation of failure sequences, methylated oligonucleotide, and phosphorothioate diastereoisomers were achieved using the new column coupled to an orbitrap mass spectrometer. In addition, the resin has large pore size (1,500 Å) which provided high resolution separation of large double-stranded DNA fragments. This feature may potentially be used for characterization and preparation of next-generation sequencing (NGS) library.

Keywords: Bioanalytical, Biopharmaceutical, Liquid Chromatography/Mass Spectroscopy, Nucleic Acids
Application Code: Bioanalytical
Methodology Code: Liquid Chromatography/Mass Spectrometry

Session # 1310 Abstract # 1310-1 **Oral Sessions**

Session Title: Consumer Products Characterization - Half Session

Abstract Title: **Characterization of Metallic Nanoparticles in Tattoo Ink Using Asymmetrical Flow Field-Flow Fractionation Coupled with ICP-MS**

Primary Author: Soheyl Tadjiki
Postnova Analytics Inc.

Date: Tuesday, March 08, 2016 - Afternoo
Time: 01:30 PM
Room: B408

Co-Author(s): Evelin Moldenhauer, Thorsten Klein, Tony Pfaffe, Trevor Havard

Abstract Text

Field-Flow Fractionation-ICP-MS is a powerful analytical technique that can be used for characterization of nanomaterials. As a hyphenated system, FFF-ICP-MS generates elemental-based size distribution over a broad size range. It can also provide elemental molar ratio distribution which helps to study particles chemical composition as a function of particle size [1-6].

TiO2 nanoparticles are added to the tattoo inks to enhance color strength. Commercial Ink products containing non-metal colorants can be contaminated by traces of toxic metals. For example cutaneous allergies may occur due to the presence of Ni in inks.

This presentation demonstrates the use of Asymmetrical Flow FFF coupled with ICP-MS to study various kinds of tattoo ink. Particle size and elemental distributions of various metals are measured to study the composition of ink ingredients as a function particle size.

References:

1.Giddings, J. C., Ratanathanawongs, S. Kim, Barman, Bhajendra N., Moon, Myeong Hee, Liu, Guangyue, Tjelta, Brenda L, Hansen, Marcia E., Adv. Chem. Ser., 1994, Characterization of colloidal and particulate silica by field-flow fractionation
2. C. Giddings: Science, 260, 1456-1465 (1993)
3.Chittleborough, David J.; Tadjiki, Soheyl; Ranville, J. F.; Shanks, Finlay; Beckett, R., SuperSoil 2004: 3rd Australian New Zealand Soils Conference, 5 – 9 December 2004, University of Sydney, Australia, Soil colloid analysis by Flow Field-Flow Fractionation
4.Andreas F. Thünemann,* Jenny Kegel, Jörg Polte, and Franziska Emmerling, ASAP Anal. Chem., 10.1021/ac8004814, 2008, Superparamagnetic Maghemite Nanorods: Analysis by Coupling Field-Flow Fractionation and Small-Angle X-ray Scattering
5.Jessica Lohrke, Andreas Briel, Karsten Mäder, Nanomedicine, 3(4), 437-452, 2008, Characterization of supramagnetic iron oxide nanoparticles by asymmetrical flow-field-flow-fractionation
6.Stolpe, B., Guo, L., Schiller, A. M. Hassellöv, M. (2010) Mar. Chem., 118, 119-128

Keywords: Characterization, ICP-MS, Materials Characterization, Nanotechnology
Application Code: Consumer Products
Methodology Code: Mass Spectrometry

Session # 1310 Abstract # 1310-2 **Oral Sessions**

Session Title Consumer Products Characterization - Half Session

Abstract Title **Developing a Color Matching Database in Supercritical CO2 for Waterless Textile Dyeing**

Primary Author Rolf Schlake
Applied Separations

Date: Tuesday, March 08, 2016 - Afternoo
Time: 01:50 PM
Room: B408

Co-Author(s) Madhu Anand, Rob Dorrycott, Susan Crowe

Abstract Text

The use of supercritical carbon dioxide (scCO2) is getting increased attention from the textile industry. Several companies have set up commercial facilities and are offering scCO2dyed garments to the market. The drivers for this change:

- very large consumption of potable water in the process
- impending shortage of water
- discharge of polluted water into the environment.

Environmentally-friendly scCO2 based dyeing is waterless consequently releasing no dye or polluted water into the environment.

Despite these production concerns, fashion and consumer demand today still dictate that the industry provide an almost infinite pallet of colors of textiles. As color development and color match capability are essential to this industry, most current textile plants have developed color databases which allow them to mix dyes to produce a desired color.. However, these databases are all aqueous based. Unless an analogous database is developed for scCO2 dyeing, this environmentally-friendly, green technology will have a difficult time succeeding in the marketplace.

We have developed a commercial color matching database for scCO2 textile dyeing. Under continuing development, we have approved several commercial disperse dyes that successfully dye polyester in a myriad of colors under our dyeing conditions and with our scCO2 dyeing equipment. The dye selection involves many parameters including:

- knowledge of dye color components
- knowledge of dye solubility in scCO2,
- dyeing uniformity
- affinity of the dye for the polyester.

Our expanding database includes primary colors as well as navy, black and violet dyes. Polyester fabric was dyed at different dye concentrations and dyeing conditions. The color of the dyed fabric was characterized using a DatacolorTM colorimeter. The measured K/S correlated well with dye uptake in the fabric for both primary colors and binary dye mixtures. Similar success has been demonstrated even in color matching with multi-dye mixtures.

Keywords: Consumer Products, Polymers & Plastics, SFE, UV-VIS Absorbance/Luminescence
Application Code: Consumer Products
Methodology Code: UV/VIS

Session # **1310** **Abstract #** **1310-3** **Oral Sessions**

Session Title: Consumer Products Characterization - Half Session

Abstract Title: **Detection and Quantification of Allergens in Personal Care Products by GC and GC×GC Paired with TOFMS**

Primary Author: Elizabeth M. Humston-Fulmer
Leco Corporation

Date: Tuesday, March 08, 2016 - Afternoo
Time: 02:10 PM
Room: B408

Co-Author(s): David E. Alonso, Jonathan D. Byer, Joseph E. Binkley, Lorne E. Fell

Abstract Text

Most cosmetics and personal care products contain fragrances and other analytes, some of which are potential allergens or skin irritants to some consumers. It is important for manufacturers to understand the composition of their products in order to provide this information to consumers and also to be compliant with rules and regulations. In particular, the European Cosmetics Directive has 26 regulated contact allergens that must be reported if present in a product above certain levels. Gas Chromatography (GC) paired with mass spectrometry (MS) is well suited for screening these target analytes and we present calibration data for regulated analytes as well as representative examples of product screening for quantification of these analytes with GC-TOFMS instrumentation. TOFMS inherently provides non-targeted analyses at the same time with full-mass range data acquisition. We present additional information on other non-targeted analytes that were detected within these samples. With the complexity of cosmetic and personal care products, there are some coelutions that exceed the peak capacity of GC. In these cases, GC×GC provides important benefits that are demonstrated here. Examples of one-dimensional chromatographic coelutions that are resolved in the second dimension are shown. Sensitivity enhancements through thermal modulation are also shown for analytes detectable by GC×GC that were below the LOD with GC. Representative benefits and examples of these analytical technologies for allergen detection and personal care product characterization are demonstrated.

Keywords: Calibration, Consumer Products, Gas Chromatography/Mass Spectrometry, Time of Flight MS
Application Code: Consumer Products
Methodology Code: Gas Chromatography/Mass Spectrometry

Session # 1320 **Abstract # 1320-1** **Oral Sessions**

Session Title: Detection of Illicit Drugs - Half Session

Abstract Title: **The Detection of Illicit Drugs and Cutting Agents Using Shortwave Infrared Hyperspectral Imaging**

Date: Tuesday, March 08, 2016 - Afternoo
Time: 01:30 PM
Room: B407

Primary Author: Nathaniel R. Gomer
ChemImage Sensor Systems

Co-Author(s): Jeffrey Beckstead, Matthew P. Nelson, Oksana Olkhovyk

Abstract Text

The detection and identification of illicit drugs is a vital need for law enforcement, especially in correctional facilities where illicit materials concealed in the mail are a known problem. The introduction of these materials into correctional facilities creates security issues and increases the costs for the correctional facilities to continue to inhibit their introduction. In order to combat this problem, ChemImage Sensor Systems has developed a shortwave infrared hyperspectral imaging sensor (VeroVision[registered]), capable of detecting concealed illicit drugs hidden in the mail. Utilizing liquid crystal tunable filter technology, the sensor is capable of providing wide-area, hyperspectral imagery and detection analysis in real time, made possible by its advanced sensing algorithms and processing methods. In addition, the sensor boasts a library of drugs of interest and common cutting materials that are most commonly introduced into these facilities. This paper will present an overview of shortwave infrared hyperspectral imaging and the design of the sensor, and provide examples of various drug detection scenarios.

Keywords: Drugs, Imaging, Infrared and Raman, Instrumentation
Application Code: Homeland Security/Forensics
Methodology Code: Sensors

Session # 1320 Abstract # 1320-2 **Oral Sessions**

Session Title: Detection of Illicit Drugs - Half Session

Abstract Title: **Characterization of Synthetic Phenethylamines Using High-Resolution GC-TOFMS and Mass Defect Filters**

Date: Tuesday, March 08, 2016 - Afternoo
Time: 01:50 PM
Room: B407

Primary Author: Ruth Smith
Michigan State University

Co-Author(s): Alexandria Anstett, David E. Alonso, Fanny Chu

Abstract Text

The popularity of synthetic designer drugs has increased rapidly, with more than 240 such drugs identified since 2009. Despite legislation regulating these drugs, they remain a substantial problem, primarily due to the continual emergence of new analogs. These new analogs are often structurally similar to regulated compounds, with similar psychoactive effects, but are designed to circumvent current legislation.

Identification of these new analogs is challenging using the conventional gas chromatography-mass spectrometry (GC-MS) instruments typically available in forensic laboratories, which are equipped with electron ionization sources and single quadrupole mass analyzers. The nominal mass data are often not sufficient for definitive identification, primarily due to the high incidence of isomers and the lack of suitable reference standards for spectral comparison.

In this research, a set of synthetic phenethylamines was analyzed by GC-time-of-flight (TOF) MS. This instrument also uses electron ionization but, because a high-resolution mass analyzer is used, accurate mass data are obtained. These data were used to calculate mass defects for the molecular and fragment ions for each phenethylamine. The mass defects were used to define a series of filters that were investigated for characterization of the phenethylamines. With absolute mass defect filters, phenethylamines were distinguished from other designer drug classes while Kendrick mass defect filters were used to further characterize the phenethylamines according to structural subclass. This presentation will describe the development of each mass defect filter and the utility of mass defects for characterization of synthetic designer drugs.

Keywords: Forensics, Forensic Chemistry, GC-MS, Time of Flight MS
Application Code: Homeland Security/Forensics
Methodology Code: Gas Chromatography/Mass Spectrometry

Session # 1320 Abstract # 1320-3 **Oral Sessions**

Session Title: Detection of Illicit Drugs - Half Session

Abstract Title: **The Development of a Novel Color Test for Improved Detection of Synthetic Cathinones**

Date: Tuesday, March 08, 2016 - Afternoo
Time: 02:10 PM
Room: B407

Primary Author: Tsunghsueh Wu
University of Wisconsin-Platteville

Co-Author(s): Brook Tashner, Charles Cornett, Nicole Kloepfer

Abstract Text

Designer and emergent illicit drugs have entered the market at a rapid pace in the past five years, and synthetic cathinones (aka: "bath salts) are one example of these drugs. Synthetic cathinones are the beta-keto phenethylamine derivatives, that produce pharmacological effects similar to the Schedule I substances such as cathinone, methcathinone, and 3,4- methylenedioxymethamphetamine (MDMA). The parent cathinone structure is easily derivatized at any of four sites to generate analogues not yet regulated by state legislation or the DEA Scheduled substance list. Unfortunately, the derivatization of the cathinone structure has resulted in presumptive color tests that are often not detecting new bath salts or are providing different colors for different cathinone derivatives or require several vials of various component reagents to be effective. It stands to reason that a reliable color test for synthetic cathinones capable of operating with great sensitivity over a wide range of environmental conditions and packaged in two or fewer ampules would constitute a significant advancement in the field.

Our research and development has produced a two-stage color test. These reagents utilize organic dyes containing groups such as a sulfonic acid along with organic solvents and buffers presented in this poster. Experiments conducted demonstrate that the reagents are well suited for deployment and use in the field by law enforcement as well as the bench top as a presumptive test. Our two stage color test provides a consistent yellow color in the organic layer, and there are no impediments to its use in the field within a two-ampule system. In addition the twenty synthetic cathinones tested can be detected in amounts ranging from microgram quantities to over 30 milligram per testing package. This presentation details the 100% effectiveness in presumptive indication of a variety of synthetic cathinones.

Keywords: Forensics
Application Code: Homeland Security/Forensics
Methodology Code: Chemical Methods

Session # 1320 Abstract # 1320-4 **Oral Sessions**

Session Title Detection of Illicit Drugs - Half Session

Abstract Title **Potentiometric Sensor for Forensic Analysis: The Detection of the 'Undetectable Poison' Succinylcholine and Study of Its Enzymatic Degradation Kinetics**

Date: Tuesday, March 08, 2016 - Afternoo
Time: 02:30 PM
Room: B407

Primary Author Mohamed K. Abd El-Rahman
Cairo University

Co-Author(s) Amr M. Mahmoud

Abstract Text
In Forensic science, Succinylcholine (SUC) has a long reputation as an undetectable, "perfect" poison. It is widely used as murder weapon particularly in cases involving sudden, unexpected, and unexplained death with a medical professional as a potential suspect. Pharmacologically, SUC is a short-acting muscle relaxant favored for both its fast onset of action and short duration as it is degraded rapidly by plasma cholinesterase. From an analytical chemistry perspective, direct determination of SUC is a challenging analytical task due to the lack of a detectable chromophore and sensitive detection techniques as well as difficulty in isolation from biological specimens. In this work, an ion selective electrode (ISE) with a nanomolar detection limit for monitoring of SUC in a wide variety of forensic sample matrices is developed. In order to push the detection limit lower, we have optimized the ISE inner filling solution by buffering the activity of SUC and consequently reducing the transmembrane ion fluxes. This has been performed through utilizing the outstanding complexation properties of water-soluble p-sulfonatocalixarenes towards the quaternary ammonium SUC cation to form host-guest inclusion complex. The performance characteristics of the developed ISE revealed a linear range from 1 mM to 14 nM with LOD of 1 nM. To investigate the ability of the ISE to detect SUC in real forensic specimens, SUC has been spiked at a concentration comparable to its anticipated level and the proposed ISE shows an excellent platform for forensic analytical investigations. Moreover, the developed ISE is successfully used for measurement of the rate of in-vitro metabolism of SUC as substrate by serum cholinesterase enzyme and clearly identifies patients with increased risk of SUC sensitivity.

Keywords: Electrochemistry, Forensic Chemistry, Potentiometry, Sensors
Application Code: Homeland Security/Forensics
Methodology Code: Sensors

Session # 1330 Abstract # 1330-1 **Oral Sessions**

Session Title Environmental and Instrumentation Application of LC/MS - Half Session

Abstract Title **A New Type of Electron Ionization LC-MS and Its Applications**

Date: Tuesday, March 08, 2016 - Afternoo
Time: 03:05 PM
Room: B408

Primary Author Aviv Amirav
Tel Aviv University

Co-Author(s) Alexander Fialkov, Boaz Seemann, Svetlana Tsizin, Tal Alon

Abstract Text

Electron ionization can significantly benefit LC-MS through the provision of automated library identification and extensive fragment information. Thus, bringing back EI to LC-MS is highly valuable, if a reliable and robust EI interface can be developed. We developed a new type of EI-LC-MS system based on interfacing LC and MS with supersonic molecular beams (SMB), sample ionization with electrons as vibrationally cold compounds in the SMB and using a new Agilent 5977 MS of GC-MS as the based mass spectrometer. The output of the LC is pneumatically sprayed, followed by thermal vaporization of the sample compounds in a vaporization oven inside an inert GC liner. The vaporized sample, solvent vapor and nebulizing helium gas pass fused silica capillary flow restrictor and expand from a supersonic nozzle into the vacuum system at about 0.2 Bar pressure to suppress cluster formation while obtaining efficient vibrational cooling. The details of our new combination of EI-LC-MS with Agilent 5977 MSD will be described and its few recent applications will be demonstrated and explained including the analysis of Canabinol, Tetrahydrocannabinol (THC) and THC acid in Cannabis flower. EI-LC-MS with SMB provides several important benefits including:

A. Library based identification with names and structures at the isomer level.
B. Enhanced molecular ions and MS structural information with EI of cold molecules in SMB.
C. No ion suppression or enhancement effects are exhibited hence facilitating faster LC-MS analysis.
D. Uniform semi-quantitative ionization yields including for fully non polar compounds.

Keywords: Liquid Chromatography/Mass Spectroscopy, Mass Spectrometry
Application Code: Other
Methodology Code: Liquid Chromatography/Mass Spectrometry

Session # 1330 Abstract # 1330-2 **Oral Sessions**

Session Title: Environmental and Instrumentation Application of LC/MS - Half Session

Abstract Title: **Assessing the Occurrence and Fate of UV Filters in Seawater Swimming Pools by UPLC-Q-ToF-MS**

Date: Tuesday, March 08, 2016 - Afternoo
Time: 03:25 PM
Room: B408

Primary Author: Tarek Manasfi
Aix-Marseille University

Co-Author(s): Bruno Coulomb, Jean-Luc Boudenne, Ravier Sylvain

Abstract Text

Chemical UV filters belong to emergent organic compounds that are turning out to be a sticking point for both environmental and health concerns. Today, these compounds are used in increasing quantities and can be found in various environmental media. However, little is known about the occurrence of UV filters in seawater swimming pools and about their reactivity in presence of disinfectants such as chlorine. Determining the concentrations of UV filters in swimming pools and their transformation products is paramount to any risk assessment consideration, especially that their potential halogenated byproducts could be toxic. For this reason, seawater swimming pool samples from Southern France were analyzed to determine the concentrations of 5 widely used UV filters, namely dioxybenzone, oxybenzone, avobenzone, octocrylene, and 2-ethylhexyl-4-methoxycinnamate. Different sample preparation techniques were tested and analytical methods were developed. Analyses were performed using ultra performance liquid chromatography-quadrupole time-of-flight mass spectrometer (UPLC/Q-ToF-MS). Additionally, the reactivity of the UV filters in the presence of different molar ratios of chlorine in seawater was examined and their byproducts were identified through lab experiments. Identification of byproducts was conducted by accurate mass measurements. MS-MS experiments were also performed to elucidate structures of the found compounds. The levels of UV filters in swimming pool samples varied considerably from one compound to another. The reactivity/stability of the UV filters also varied depending on the compound's structure. Chlorination by-products of the reactive UV filters were identified for the first time in this study and transformation pathways were proposed based on the identified byproducts.

Acknowledgement: This work was supported by the Doctoral School of "Environmental Sciences" at Aix-Marseille University and the French Ministry of Higher Education and Research.

Keywords: Environmental Analysis, Liquid Chromatography/Mass Spectroscopy, Trace Analysis, Water
Application Code: Environmental
Methodology Code: Liquid Chromatography/Mass Spectrometry

Session # 1330 Abstract # 1330-3 **Oral Sessions**

Session Title: Environmental and Instrumentation Application of LC/MS - Half Session

Abstract Title: **Expanded Analysis of Human Hormones in Drinking Water Using Solid Phase Extraction and Liquid Chromatography Tandem Mass Spectrometry**

Date: Tuesday, March 08, 2016 - Afternoo
Time: 03:45 PM
Room: B408

Primary Author: Carl Fisher
Thermo Fisher Scientific

Co-Author(s): Claudia Martins, Pranathi Perati

Abstract Text

The presence of hormones in drinking water is a human health concern with several being routinely monitored as part of the U.S. Environmental Protection Agency (EPA) Unregulated Contaminant Monitoring Rule 3 (UCMR3). Various forms of estrogen are prescribed as a hormonal contraceptive device, for estrogen deficiency syndromes, and to counter the negative effects associated with the natural decline in estrogen levels, such as accelerated bone loss, in postmenopausal women. Due to the widespread use of hormone pharmaceuticals, these often end up in the sewage system as a result of excretion and disposal of unwanted quantities. There is evidence that hormones may not be effectively removed during wastewater treatment, and as a result, significant amounts of these hormones may be present in drinking water sources. To monitor the levels of the seven most common hormones in drinking water, EPA method 539 was developed. The work presented here updates this method to include five additional hormones and describes the use of an automated solid phase extraction (SPE) system containing high surface area reversed-phase (HRPHS) cartridges, followed by liquid chromatography electrospray ionization tandem mass spectrometry (LC-ESI-MS/MS) with timed selected-reaction monitoring (TSRM) mode for detection and quantification. Data acquisition and processing was automatically performed using a quantitation software package. The results of chromatographic separation, SPE recoveries, and method detection limits (MDL) will be presented.

Keywords: Environmental/Water, HPLC, Mass Spectrometry, Solid Phase Extraction
Application Code: Environmental
Methodology Code: Liquid Chromatography/Mass Spectrometry

Session # 1330 Abstract # 1330-4

Oral Sessions

Session Title Environmental and Instrumentation Application of LC/MS - Half Session

Abstract Title **Multiclass Determination of New Psychoactive Substances in Municipal Wastewater**

Date: Tuesday, March 08, 2016 - Afternoo
Time: 04:05 PM
Room: B408

Primary Author Ivan Senta
Rudjer Boskovic Institute

Co-Author(s) Ivona Krizman, Marijan Ahel, Senka Terzic

Abstract Text

Beside the "traditional" illicit drugs, such as cocaine, heroin, and marijuana, there is a growing concern about the use of modern "designer drugs" that have emerged in large numbers over the past few years. In this work, a sensitive and selective method for simultaneous determination of 25 synthetic psychoactive compounds, including amphetamine, substituted amphetamines and cathinones, in raw wastewater (RW) and secondary effluent (SE) was developed. Samples were enriched by solid phase extraction (SPE) on mixed-mode reversed-phase/strong cation-exchange sorbent and analysed by reversed-phase liquid chromatography coupled to electrospray ionisation tandem mass spectrometry (LC-MS/MS). The target compounds were separated on a Synergi Polar column and detected using multiple reaction monitoring (MRM) in positive polarity. Accurate quantification was achieved by using several deuterated analogues as surrogate standards. Careful optimisation and validation of the procedure resulted in a reliable determination of all target analytes in low ng/L range for both RW and SE, which makes the method suitable for the application in sewage epidemiology. The method was applied for assessment of selected compounds in municipal wastewater from Croatia. It was shown that the well-known synthetic illicit drugs, amphetamine and MDMA still dominate in all samples (concentrations up to 545 ng/L and 55 ng/L in RW, respectively), while the new psychoactive substances were detected very rarely and in low concentrations.

This work was funded by Croatian Ministry of Science, Education and Sports (Project No. 098-982934-2712).

Keywords: Drugs, Environmental/Waste/Sludge, Liquid Chromatography/Mass Spectroscopy, Solid Phase Extrac
Application Code: Environmental
Methodology Code: Liquid Chromatography/Mass Spectrometry

Session # 1340 Abstract # 1340-1 **Oral Sessions**

Session Title: Environmental Applications of Electrochemistry and Sensors - Half Session

Abstract Title: **Pretreatment and Spectroelectrochemical Sensing of Re(I)-Carbonyl Complexes**

Primary Author: Shirmir D. Branch
University of Cincinnati

Date: Tuesday, March 08, 2016 - Afternoo
Time: 01:30 PM
Room: B409

Co-Author(s): Amanda D. French, Amanda M. Lines, Brian M. Rapko, Sam A. Bryan, William R. Heineman

Abstract Text

A method for the spectroelectrochemical detection of rhenium(I)-tricarbonyl complexes is being investigated. The [Re(CO)3]+ complexes are first introduced to a pretreatment solution. Pretreatment involves conversion of [Re(CO)3]+ into a spectroscopically detectable species by the incorporation of a sensitizing ligand; adjusting solution conditions, such as selection and concentration of the sensitizing ligand, pH, and ionic strength; and preconcentration of the optically active species into a selective film. Spectroelectrochemical detection comprises three modes of selectivity: selective partitioning, electrochemistry, and detection by spectroscopy. The chemically selective polymer thin film allows for preconcentration. An electrochemical potential is applied to modulate the oxidative state of the preconcentrated complexes, providing selectivity by effectively turning the optical response “on” and “off.” Fluorescence spectroscopy is used to monitor this electrochemically modulated signal.

Keywords: Nuclear Analytical Applications, Optimization, Sensors, Spectroelectrochemistry
Application Code: Environmental
Methodology Code: Sensors

Session # 1340 Abstract # 1340-2 **Oral Sessions**

Session Title: Environmental Applications of Electrochemistry and Sensors - Half Session

Abstract Title: **Wearable Gas Sensors: Shrinking Electrochemical Cells**

Primary Author: John R. Saffell
Alphasense Ltd.

Date: Tuesday, March 08, 2016 - Afternoo
Time: 01:50 PM
Room: B409

Co-Author(s): Joseph R. Stetter

Abstract Text

Amperometric electrochemical cells are a mature technology, used worldwide to protect workers in mines, industrial plants, sewers and other confined spaces. Recent advances have pushed the limit of detection from ppm to low ppb concentrations for gases such as CO, H2S, NO, NO2 and O3. These gas sensors are now being used in fixed site air quality networks and recently in handheld air quality monitors.

A radical new design for amperometric gas cells has greatly reduced the size and cost of this sensor technology, approaching the requirements for smart phone and wearable applications. We will discuss the effect on performance as electrochemical cells shrink in size, especially the limit of detection; not all properties scale linearly. Opportunities for lifestyle, personal air quality monitoring and CO protection will be discussed.

Keywords: Consumer Products, Electrochemistry, Environmental/Air, Portable Instruments
Application Code: Environmental
Methodology Code: Electrochemistry

Session # 1340 Abstract # 1340-3 **Oral Sessions**

Session Title: Environmental Applications of Electrochemistry and Sensors - Half Session

Abstract Title: **Cloud Point Extraction for Electroanalysis: Anodic Stripping Voltammetry of Lead**

Primary Author: Cory A. Rusinek
University of Cincinnati

Date: Tuesday, March 08, 2016 - Afternoo
Time: 02:10 PM
Room: B409

Co-Author(s): Adam Bange, Ian Papautsky, Mercedes Warren, William R. Heineman

Abstract Text

Lead (Pb2+) is a common heavy metal ion that can cause lasting health complications. Thus, pre-concentration and quantitative detection of Pb2+ is of profound importance. Cloud point extraction (CPE) is a well-known method for pre-concentrating analytes to detectable levels with subsequent analysis typically completed via atomic absorption spectroscopy (AAS), UV – vis spectroscopy, or HPLC. However, little is known about the applicability of CPE for electroanalytical techniques such as stripping voltammetry. We developed CPE method using anodic stripping voltammetry (ASV) analysis for trace detection of Pb2+. Pb2+ was neutralized and extracted in basic pH using a common chelating agent, dithizone. Because of the critical temperature needed to form the micellar phase in CPE, Triton X-114 was chosen as the surfactant for extraction. A mercury-coated glassy carbon (Hg-GC) electrode was used for the CPE-ASV analysis due to its excellent sensitivity for Pb2+ relative to other electrodes (Bare GC, Pt, and bismuth coated GC). CPE-ASV offered excellent sensitivity for Pb2+ with a LOD of 0.27 ppb at a deposition time of just 2 min, a LOD 55x lower than the Pb2+ maximum contaminant level in drinking water set by the EPA. This is also a 6.2x increase in sensitivity and a 37x lower limit of detection compared to Pb2+ ASV without CPE. The applicability of the method for the determination of Pb2+ in environmental samples was also demonstrated. This inexpensive, environmentally friendly method of extraction can potentially be used with other heavy metal ions and organic compounds.

Keywords: Environmental/Water, Extraction, Lead, Stripping Analysis
Application Code: Environmental
Methodology Code: Electrochemistry

Session # 1340	Abstract # 1340-4	Oral Sessions

Session Title: Environmental Applications of Electrochemistry and Sensors - Half Session

Abstract Title: **Development of an Electrochemical Sensor for Detection of Dissolved Polycyclic Aromatic Hydrocarbons in Water**

Primary Author: Abra Penezic
Rudjer Boskovic Institute

Date: Tuesday, March 08, 2016 - Afternoo
Time: 02:30 PM
Room: B409

Co-Author(s): Andrew Nelson, Blazenka Gasparovic, Drazenka Stipanicev

Abstract Text

The aquatic environment is exposed to an increasing amount of pollution from various sources. In order to protect the Earths diverse ecosystems, close monitoring and early detection of potentially dangerous substances is crucial. Polycyclic aromatic hydrocarbons (PAHs) present a group of hydrophobic organic pollutants which can have cargenogenic and mutagenic effects on the aquatic organisms. More hydrophobic PAHs usually adsorb and associate with organic particles in the aquatic environment, while less hydrophobic PAHs may be present in the dissolved form, making them more available for uptake by aquatic organisms. We are developing an electrochemical sensor which could be used for early detection of the dissolved fraction of PAHs present in waters. As it has been found, PAHs interact with phospholipid monolayers adsorbed on a mercury electrode surface, causing a disruption of the monolayers' fluidity and structure. This interaction is monitored electrochemically by fast cyclic voltammetry using a semi-automated flow cell system which incorporates a chip based mercury film microelectrode. A mixed layer of a phospholipid palmitoyl-2-oleoyl-sn-glicero-3-phosphocholine and triolein, an oleic acid triglyceride, in 3:1 molar ratio, adsorbed onto a mercury film microelectrode, was used as a sensing element for PAH detection. The system proved sensitive to the presence of four different PAH molecules, anthracene, phenanthrene, pyrene and fluoranthene, in different matrices, with limits of detection down to 0.2 [micro]g/L [1]. The performance of the system was tested on a natural river sample, and its sensitivity compared to a conventional GC – MS method used for determination of PAHs.

Reference:
1. Penezic,A., Gasparovic, B., Stipanicev, D., Nelson, A. In-situ electrochemical method for detecting freely dissolved polycyclic aromatic hydrocarbons in water, Environmental Chemistry 2014, 11(2), 173-180.

Keywords: Electrochemistry, Environmental Analysis, PAH, Sensors
Application Code: Environmental
Methodology Code: Electrochemistry

Session # **1350** **Abstract #** **1350-1** **Oral Sessions**

Session Title: Food Product Quality and Component Characterization II

Abstract Title: **A Green Sample Preparation Device for Complex Biological, Environmental, Food, Pharmaceutical and Toxicological Samples**

Date: Tuesday, March 08, 2016 - Afternoo
Time: 01:30 PM
Room: B403

Primary Author: Abuzar Kabir
Florida International University

Co-Author(s): Kenneth G. Furton, Rayma Blanko, Rodolfo Mesa

Abstract Text

Despite the phenomenal advancements in analytical instrument during the last couple of decades, leading to miniaturized, highly sensitive and more rugged systems, sample preparation still remains as a major bottleneck in the overall analytical workflow.

Real-life analytical samples including food, pharmaceutical, biological, toxicological, and environmental samples often contain high volume of matrix interferents such as particulates, debris, biomasses, cells, proteins, lipids etc. and inevitably require a pre-treatment regimen prior to applying extraction and preconcentration techniques. Sample pre-treatment involves filtration and/or protein precipitation. Sample preparation also frequently includes solvent evaporation and sample reconstitution. This multi-step procedure potentially leads to significant loss of analytes and compromise the data quality.

⍰

To completely eliminate sample pre-treatment and post-treatment operations from sample preparation work-flow, herein, we introduce capillary microextraction (CME) as a green sample preparation device. The device utilizes a porous polypropylene tube with 0.55 mm internal diameter and variable length to house sol-gel derived sorbent inside the tube. A small magnetic metal rod is also integrated into the tubular device. The microporous tube serves as a built-in filter and only allows rapid permeation of the aqueous solution containing the target analyt(s). The magnetic metal rod helps diffusing the analyte(s) under the influence of a magnetic stirrer and catalyzes the extraction kinetics. After reaching the extraction equilibrium, a small volume of organic solvent can be used for solvent mediated back-extraction.

Analytical data collected from a number of biological samples will be presented to demonstrate the advantages and performance superiority of the new device.

Keywords: Clinical/Toxicology, Environmental/Biological Samples, Environmental/Water, Sample Preparation
Application Code: Bioanalytical
Methodology Code: Sampling and Sample Preparation

Session # **1350** **Abstract #** **1350-2** **Oral Sessions**

Session Title: Food Product Quality and Component Characterization II

Abstract Title: **Cannabinoids and Terpenes in Food**

Primary Author: Tim Anderson
Phenomenex

Date: Tuesday, March 08, 2016 - Afternoo
Time: 01:50 PM
Room: B403

Co-Author(s): Kristen Parnell, Ramkumar Dhandapani

Abstract Text

As states adopt laws legalizing recreational and medicinal cannabis use, there is an increase in demand for reliable and robust preparation and analytical techniques. In particular, edible cannabis products are very challenging from preparation to injection. Components of interest can range from contaminants, such as pesticides, to a screen for cannabinoids and terpenes. Labs and manufacturers have different analytes of interest, and as such a reliable general purpose approach is of value to testing laboratories. Due to the complex nature of food matrices, a less active column can overcome some of these chromatographic challenges. In this paper, we demonstrate how improved GC column technology can be used to identify a range of cannabinoids and terpenes in edible cannabis products. A capillary column with less activity reduces peak tailing, increases sensitivity, and produces better injection to injection reproducibility.

Keywords: Food Safety, Gas Chromatography, Gas Chromatography/Mass Spectrometry, Sample Preparation
Application Code: Food Identification
Methodology Code: Gas Chromatography/Mass Spectrometry

Session # 1350 Abstract # 1350-3 **Oral Sessions**

Session Title: Food Product Quality and Component Characterization II

Abstract Title: **Determination of Whey Adulteration in Milk Powder by Using Laser Induced Breakdown Spectroscopy**

Primary Author: Gonca Bilge
Hacettepe University

Date: Tuesday, March 08, 2016 - Afternoo
Time: 02:10 PM
Room: B403

Co-Author(s): Ali Topcu, Banu Sezer, Halil Berbero□u, Ismail H. Boyaci, Kemal E. Eseller

Abstract Text

Dairy products have high nutritional value and are widely consumed food groups for public nutrition. Addition of cheese whey to milk or whey solids to dairy products is one of the most frequently applied adulteration method. Whey is a low cost by-product of cheese making process. There are many detection methods for this type of adulteration such as electrophoretic analysis, reversed phase high performance liquid chromatography (HPLC), cation exchange chromatography, coupled to mass spectrometry, immunochemical assays, ELISA. However, these methods are not suitable for routine analysis because they require a long analysis time and expensive equipment and materials. Many studies have showed the differences in mineral composition levels between milk powder and whey products. The aim of this study was to evaluate the potential of LIBS for the quantitative and qualitative determination of whey in milk powder. The raw cow's milk samples from five different reliable sources were obtained from different dairies in Ankara, Turkey. Milk powder, sweet, acid whey powders were produced as standard samples and milk adulteration was performed at different ratios. In addition commercial milk products were collected from market. All of them were analyzed with LIBS. Elemental composition of skim milk powders and whey powders were also measured using inductively coupled plasma-mass spectrometry (ICP-MS) as reference method. Based on LIBS spectra of standard samples and commercial products, species was identified using principle component analysis (PCA) method, and discrimination rate of milk and whey powders was found as 80.5%. Calibration curves were obtained with partial least squares regression (PLS). Correlation coefficient (R[sup]2[/sup]) and limit of detection (LOD) values were 0.981 and 1.55% for adulteration with sweet whey powder, and 0.985 and 0.55% for adulteration with acid whey powder, respectively.

Keywords: Chemometrics, Food Identification, Plasma Emission (ICP/MIP/DCP/etc.), Spectroscopy
Application Code: Food Identification
Methodology Code: Atomic Spectroscopy/Elemental Analysis

Session # 1350 Abstract # 1350-4 **Oral Sessions**

Session Title Food Product Quality and Component Characterization II

Abstract Title **Bioactive Compounds and Antioxidant Activity of Guava Processed Byproducts and Wastes**

Primary Author Neela Emanuel
NIFTEM

Date: Tuesday, March 08, 2016 - Afternoo
Time: 02:30 PM
Room: B403

Co-Author(s) Aman Kaushik, Sao Khushbu

Abstract Text

Fruit Processing Industries generate lots of waste and byproducts which can be a valuable source of bioactive compounds. Gauva fruit's seed and pomace along with the peel were analyzed for antioxidant activity and bioactive compounds. In the present study the total phenolic content (TPC), evaluation of antioxidant activities using 2,2-diphenyl-1-picrylhydrazyl (DPPH) scavenging activity, Ferric reducing Antioxidant Power (FRAP), Total Flavonoid content (TFC) and quantification of some bioactive phenolics were carried out using HPLC. The pomace along with the peel extract contained highest phenolic content 432 mg GAE/g and possessed strong antioxidant activity in antioxidant assay. The compound extracted from these byproducts and wastes are excellent source of natural antioxidants and can be utilized as nutraceuticals in the food and pharmaceutical industries.

Keywords: Extraction, Food Identification, Food Science, HPLC Detection
Application Code: Food Identification
Methodology Code: Liquid Chromatography

Session # 1350 Abstract # 1350-5 **Oral Sessions**

Session Title: Food Product Quality and Component Characterization II

Abstract Title: **Analysis of Non Volatile Congeners in Spirits by Creation of an Ion Fragmentation Database for Use with Time of Flight LC MS**

Primary Author: Rita Steed
Agilent Technologies

Date: Tuesday, March 08, 2016 - Afternoo
Time: 03:05 PM
Room: B403

Co-Author(s): Gregory Hunlen, Joni Stevens, Luke Adam, Sue Dantonio, Tarun Anumol

Abstract Text

Novel Aspect Profiling of Spirits with Time of Flight and confirmation of the 10 non volatile congeners which indicate barrel ageing in spirits using fragment ion confirmation.

Introduction

Many types of spirits are aged in oak barrels for the absorbance of color and flavors. The barrels have a life moving from Whiskies, Bourbons, to Sherries and tequila. These 10 compounds must be present for a spirit to be considered barrel aged. Bourbon is a 1.5 billion dollar a year industry it is estimated that 10 % of the world's Bourbon is fake. In order to prosecute these counterfeiters, there must be proof that a product is fake. We have added another level on confirmation to the presence of the congeners. With the addition of fragment confirmation, we have now have ways to prove the authenticity of spirits: Retention time, isotopic fidelity and spacing and accurate mass.

Methods

We have a standard of the 10 nonvolatile congeners which we ran by by LC MS MS QTOF. We created an accurate mass data base containing spectral data. We added the 10 compounds to the data base. This data base was shipped to several sites with only Time of Flight instrumentation. This allowed users to easily search the data base to confirm by ion fragmentation pattern.

Keywords: Food Contaminants, Food Identification, Liquid Chromatography/Mass Spectroscopy, Time of Flight
Application Code: Food Identification
Methodology Code: Liquid Chromatography/Mass Spectrometry

Session # 1350 Abstract # 1350-6 **Oral Sessions**

Session Title: Food Product Quality and Component Characterization II

Abstract Title: **A Simple Field Test Kit for the Detection of Iodine in Food Grade (Table) Salt**

Primary Author: Rufus Sha'Ato
University of Agriculture

Date: Tuesday, March 08, 2016 - Afternoo
Time: 03:25 PM
Room: B403

Co-Author(s): Nasiru L. Usman

Abstract Text

A simple, sensitive and quick-to-operate test for iodine (as iodate) in food grade (table) salt has been developed. The test is based on the reduction of the IO3- in iodated salt with the liberation of free iodine at ordinary temperatures. The free iodine generated subsequently reacts with aqueous starch in situ to produce the well-known starch-iodide complex, revealed by an intense blue black colored solution, which thus, serves as a positive test for iodine in the salt. The sensitivity of the test was evaluated using laboratory grade sodium chloride samples spiked with potassium iodate (KIO3). Results showed that the test is sensitive to KIO3 levels as low as 0.03mg/g of salt. Samples of seven brands of food grade salt were obtained from the open market and tested for their iodine content using this test. All but one of the samples tested positive. The market samples of table salt were also analyzed for iodine using the titrimetric method recommended by the Standard Organization of Nigeria (SON). Results obtained confirmed the validity of the test. On these bases and for convenience of field application, a user-friendly field test kit was invented, with simple instructions to users. The simple kit for field use is a single solution contained in a 5- or 10-mL bottle and has a shelf-life of up to 12 months. We believe this kit will be a very useful tool in the Universal Salt Iodization (USI) campaign aimed at combating Iodine Deficiency Disorder (IDD), especially in low-income countries.

Keywords: Analysis, Detection, Food Identification, Quality
Application Code: Food Identification
Methodology Code: Chemical Methods

Session # 1350 Abstract # 1350-7 **Oral Sessions**

Session Title: Food Product Quality and Component Characterization II

Abstract Title: **Chromatography Advancements in Nutraceuticals and Dietary Supplements Testing**

Date: Tuesday, March 08, 2016 - Afternoo
Time: 03:45 PM
Room: B403

Primary Author: Allen Misa
Phenomenex

Co-Author(s): Zeshan Aqeel

Abstract Text

Chromatography – particularly HPLC – is an analytical technique that serves as the basis for many analytical methods. The widespread use of HPLC stems from its powerful ability to unravel complex mixtures of analytes. Current widely used analytical methods for Nutraceuticals and Dietary Supplements are based upon HPLC technology that was developed decades ago. Since then, a variety of improvements in column particle morphology and chemistries have been introduced which have significantly improved the resolution, speed and sensitivity of complex natural product composition testing.

In this session we will present a diverse selection of recently developed chromatographic methods that demonstrate the benefits to be gained from employing the best that separation science currently has to offer. In particular, we will demonstrate how Core-Shell Technology can significantly improve complex natural product composition testing and provide a superior basis for detecting fraud or ensuring authenticity.

Keywords: Chromatography, Food Science, Liquid Chromatography/Mass Spectroscopy, Natural Products
Application Code: Food Safety
Methodology Code: Liquid Chromatography

Session # 1360 Abstract # 1360-1 **Oral Sessions**

Session Title: Forensic Trace Analysis - Half Session

Abstract Title: **Analysis and Comparison of Fatty Acid Compositions in Latent and Smudged Fingermarks via Gas Chromatography-Mass Spectrometry (GC-MS) and Comprehensive Two-Dimensional Gas Chromatography-Time-of-Flight Mass Spectrometry (GCxGC-MS)**

Primary Author: Caitlin Cobern
Penn State

Date: Tuesday, March 08, 2016 - Afternoo
Time: 03:05 PM
Room: B407

Co-Author(s): Frank Dorman, Seth Michalski

Abstract Text

Secretions that make up the fingerprints mainly come from the Eccrine glands – concentrated in the palms of the hands and soles of the feet – and the sebaceous glands – concentrated on the face and around the scalp. Previous research has indicated that ratios of the fatty acids that compose fingerprints differ between genders and cultural background. Smudged fingerprints left behind at a crime scene can be limited in their value because minutiae are indistinguishable. The goal of this project was to develop a new method of fingerprint analysis for forensic labs to utilize in criminal investigations so suspects could be narrowed down. Fingerprint samples were collected from Caucasian, African/African-American, and Asian/Asian-American, male and female between the ages of 18 and 25. The samples were then extracted using a chloroform/methanol solution and prepared for analysis on a GC-MS and a GCxGC-TOFMS. Before samples were extracted the extraction technique was analyzed for effectiveness using solutions containing deuterated lauric acid. A Food Industry FAME solution, ranging from four (4) carbons to twenty-four (24) carbons, was used to compare the retention times to the chromatogram of the fingerprint samples. Samples analyzed with GC-MS were then examined for fatty acid identification and ratio determination which was done using hexadecanoic acid methyl ester as a normalizer for peak areas. Comprehensive Two-Dimensional Gas Chromatography-Time-of-Flight Mass Spectrometry (GCxGC-TOFMS) was also used to analyze samples in order to see a separation of similar length fatty acid methyl esters in the second dimension of the chromatogram. The distinct ratios of fatty acids associated with gender and ethnic background determined from this project will help law enforcement narrow down possible suspects in an ongoing criminal investigation.

Keywords: Forensics, Gas Chromatography/Mass Spectrometry, GC-MS, Time of Flight MS
Application Code: Homeland Security/Forensics
Methodology Code: Gas Chromatography/Mass Spectrometry

Session # 1360 Abstract # 1360-2 **Oral Sessions**

Session Title: Forensic Trace Analysis - Half Session

Abstract Title: **Mathematical Model to Predict Evaporation of Ignitable Liquids for Forensic Applications**

Date: Tuesday, March 08, 2016 - Afternoo
Time: 03:25 PM
Room: B407

Primary Author: Victoria L. McGuffin
Michigan State University

Co-Author(s): John W. McIlroy, Rebecca Brehe, Ruth Smith

Abstract Text

Forensic fire debris analysis involves the identification of any ignitable liquid present in samples collected from the scene of a suspected arson. Samples are analyzed by gas chromatography-mass spectrometry (GC-MS) and the resulting chromatograms and spectra are compared to a suitable reference database for identification. To account for chemical changes occurring during the fire, these databases typically include ignitable liquid reference standards evaporated to different levels.

In this work, a kinetic model has been developed to mathematically generate chromatograms of ignitable liquids corresponding to different evaporation levels. The model is used to calculate the evaporation rate constant for each compound within the liquid, which is then used to predict the fraction remaining for a specified evaporation level. The fraction remaining ranges from 0, indicating complete evaporation, to 1, indicating no evaporation. The fraction of each compound remaining is plotted versus retention index and the resulting curve is multiplied by the chromatogram of the unevaporated liquid to generate the mathematical chromatogram.

The model has been applied to various petroleum distillates and gasoline. Each liquid was evaporated to four different levels (ranging from 25% to 90% v/v) and analyzed by GC-MS. The experimentally derived chromatograms were compared to the mathematically derived chromatograms corresponding to the appropriate evaporation level. Pearson product-moment correlation (PPMC) coefficients indicated strong correlation between the experimental and mathematical chromatograms, irrespective of liquid or evaporation level. This presentation will discuss these results in more detail and describe applications of such a model for fire debris applications.

Keywords: Data Analysis, Forensics, Forensic Chemistry, GC-MS
Application Code: Homeland Security/Forensics
Methodology Code: Data Analysis and Manipulation

Session # 1360 Abstract # 1360-3 **Oral Sessions**

Session Title Forensic Trace Analysis - Half Session

Abstract Title **Analysis of Smoke Residues from Illicit Drugs as a Potential Source of Forensic Evidence**

Date: Tuesday, March 08, 2016 - Afternoo
Time: 03:45 PM
Room: B407

Primary Author Julie Bitter
National Institute of Standards and Technology

Co-Author(s) Matthew Staymates

Abstract Text

Smoke aerosols and vapors, like those from cigarettes, deposit onto surfaces like clothing, countertops, tables, and chairs, in the form of residues which contaminate indoor environments. Studies have demonstrated that cigarette smoke residues and their decomposition products can be collected from various surfaces and identified in areas of habitual smoking. In this work, 500µg samples of dried cocaine and methamphetamine solutions were volatilized at 200°C to produce an illicit drug smoke that was left exposed to ambient conditions for up to four weeks. These residues were collected after discreet time intervals and analyzed by scanning electron microscopy, electrospray-ionization time of flight mass spectrometry, and secondary ion mass spectrometry. Chemical analyses of drug residues collected on various surfaces (plastic, laminate, artificial leather, etc.) indicated low recovery that decreased rapidly in the first 24 hours, but was detectable out through four weeks. Decomposition products were also identified, which formed as a result of both volatilization and exposure to ambient laboratory conditions. Particles were collected by inertial impaction to examine the range of sizes produced by volatilization. Predominantly diameters less than 2µm were seen, indicating particles of respirable sizes. The persistence and size data suggests that smoke residues could simultaneously be a useful form of forensic trace evidence and a environmental hazard to occupants where smoking takes place. Our most recent work has focused on creating realistic smoking scenarios using glass pipes to examine temperature effects and the chemical species that are generated. Additional new interest is of how adulterants and cutting agents impact product formation and sample recovery, with the goal of identifying limits of recovery and determining the usefulness of smoke residues as a viable source of forensic evidence.

This work was funded in part by a National Research Council Fellowship.

Keywords: Time of Flight MS, Electrospray, Forensic Chemistry, Microscopy
Application Code: Homeland Security/Forensics
Methodology Code: Mass Spectrometry

Session # **1360** **Abstract #** **1360-4** **Oral Sessions**

Session Title Forensic Trace Analysis - Half Session

Abstract Title **LIBS Instrumentation for Fast Quantitative Analysis of Soil and Forensic Investigation of Nuclear Materials**

Date: Tuesday, March 08, 2016 - Afternoo
Time: 04:05 PM
Room: B407

Primary Author Alain Blouin
National Research Council Canada

Co-Author(s) Aissa Harhira, Josette El Haddad, Mohamad Sabsabi, Paul Bouchard

Abstract Text

Laser-Induced Breakdown Spectroscopy (LIBS) is a method of optical emission spectroscopy that uses laser-generated plasma as the source of vaporization, atomization and excitation. LIBS can provide on-site real-time fast analysis with no contact and no or minimal sample preparation. These characteristics are key features for the application to soil analysis and nuclear material assessment.

In the realm of nuclear safety, first responders and regulators need to adapt to the challenge of safeguarding the use of uranium, plutonium and thorium worldwide, avoiding their diversion use in weapons of mass destruction or explosive devices. In that perspective, there is a need for technologies that can provide quick and accurate information, in order to prevent clandestine activities or initiate rapid responses to them. The LIBS approach has several advantages for this purpose; in particular, this technique allows performing on-site / real-time measurements remotely and without contact, thus avoiding contamination by radioactive materials. The performance obtained with such a LIBS sensor coupled to data analytics methods and tools applied to nuclear and other radioactive materials will be discussed. More specifically, the determination of isotopic ratios using LIBS, the assessment of yellow cake origin as well as the identification of different compounds found in the uranium ore refining process will be presented.

For the analysis of soils, the nutrients bioavailability concentration (phosphorus, calcium, aluminum, etc...), the pH and the percentage of organic matter (CO (%)) are very important to evaluate soil and crops nutrient requirements. A specific LIBS system for agricultural soil analysis was built. LIBS spectral analysis by chemometric methods allowed not only the quantification of nutrient concentrations but also the quantification of the pH value and the percentage of organic matter in soils. The results of different chemometric methods will be presented.

Keywords: Atomic Emission Spectroscopy, Chemometrics, Instrumentation
Application Code: Homeland Security/Forensics
Methodology Code: Atomic Spectroscopy/Elemental Analysis

Session # 1370 Abstract # 1370-1 **Oral Sessions**

Session Title: Laser Induced Breakdown Spectroscopy (LIBS) and Glow Discharge in Atomic Spectroscopy - Half Sessi

Abstract Title: **Standoff LIBS Using a Spatial Heterodyne Spectrometer with Sub-Microsteradian Collection Optics**

Date: Tuesday, March 08, 2016 - Afternoo
Time: 03:05 PM
Room: B301

Primary Author: Patrick D. Barnett
University of South Carolina

Co-Author(s): Nirmal Lamsal, S Michael Angel

Abstract Text

In this paper we describe standoff laser-induced breakdown spectroscopy (LIBS) measurements using a spatial-heterodyne LIBS spectrometer (SHLS). The wide field of view of the SHLS relaxes requirements for laser pointing stability and alignment of the collection optics. Also, because spectral resolution is not a strong function of entrance aperture size, high spectral resolution can be achieved in a relatively small spectrometer using small diffraction gratings. The high light throughput of the SHLS allows LIBS measurements to be made for samples at distances up to 20 meters, with no collection optics, other than the 10 mm diffraction gratings, which corresponds to a collection solid angle of less than one microsteradian. In previous work, our group described a spatial heterodyne Raman spectrometer (SHRS) for standoff Raman measurements at both visible and UV wavelengths, however, a standoff spatial heterodyne LIBS spectrometer (SHLS) has not been previously described. In the described work the 10 mm diffraction gratings provide ~0.3 nm spectral resolution and ~1x10[sup]-3[/sup] cm[sup]2[/sup] sr Etendue throughput, a factor of 10-100 larger than a typical dispersive monochromator. In this paper, stand-off LIBS spectra of solid Cu, Mg, Ca, and Mn, collected at distances up to 20 m, will be shown using the SHLS alone and with a small telescope to improve light collection.

Keywords: Atomic Spectroscopy, Instrumentation, Plasma Emission (ICP/MIP/DCP/etc.), Spectrometer
Application Code: General Interest
Methodology Code: Atomic Spectroscopy/Elemental Analysis

Session # 1370 Abstract # 1370-2 **Oral Sessions**

Session Title: Laser Induced Breakdown Spectroscopy (LIBS) and Glow Discharge in Atomic Spectroscopy - Half Sessi

Abstract Title: **Study of Matrix Effects for Reproducible LIBS Analysis of Powders**

Primary Author: Matthieu Baudelet
University of Central Florida

Date: Tuesday, March 08, 2016 - Afternoo
Time: 03:25 PM
Room: B301

Co-Author(s): Brandon Seesahai, Martin Richardson, Richard Locke, Romain Gaume, Sudeep Jung Pandey

Abstract Text

The fast chemical analysis of powders has become more and more important with the reliance of modern manufacturing on powders (additive manufacturing, ceramics) as well as the need for examination of particulate materials such as soil for environmental and forensic analysis. However, LIBS being a laser-ablation based technique, its results are usually matrix-dependent. And powders are perhaps the most complex media from this viewpoint. LIBS results from powders show a dependence on moisture, temperature and texture without mentioning concentration of major elements.

This study is showing a fundamental study of the effect of the laser interaction with powders for laser-ablation based techniques, with an emphasis on LIBS. Powders of known stoichiometry (Al_2O_3) and different particle sizes are sampled under different laser conditions (spot size and wavelength) in order to determine an adaptive regime of identical laser energy deposition and plasma creation. This study is based on Monte Carlo simulation of laser interaction with powders including scattering, absorption and reflection. The simulations are based as well on the analytical Kubelka-Munk theory of diffuse reflectance.

The comparison between simulations and experimental results will be discussed to establish the conditions for existence of such a reproducible regime of laser ablation and plasma creation from powders.

Keywords: Atomic Emission Spectroscopy, Forensics, Laser, Material Science
Application Code: Material Science
Methodology Code: Atomic Spectroscopy/Elemental Analysis

Session # 1370 Abstract # 1370-3 **Oral Sessions**

Session Title: Laser Induced Breakdown Spectroscopy (LIBS) and Glow Discharge in Atomic Spectroscopy - Half Sessi

Abstract Title: **Particulate Identification Using Image Directed Laser Induced Breakdown Spectroscopy (LIBS) with Enhanced Spectral and Spatial Resolution for Pharmaceutical and Industrial Applications**

Date: Tuesday, March 08, 2016 - Afternoo
Time: 03:45 PM
Room: B301

Primary Author: Mark Sullivan
rap.ID Inc

Co-Author(s): Oliver Valet

Abstract Text

The integration of LIBS with light microscopy provides a powerful tool for particle characterization (color, size, morphology) and elemental identification of inorganic materials and metals. In this work we demonstrate the analytical advantages for LIBS due to a) increasing spectroscopic resolution from 1 nm to 0.1 nm and b) reducing the focused laser beam diameter from a nominal 10 μm down to 2 μm. The spectroscopic resolution improvement significantly lowers the detection limits for mixed elements down to the ppm range. Enhanced spatial resolution provides the capability to identify smaller particles and also to probe smaller features of bulk specimens, such as inclusions in metals, while minimizing destruction of the sampled area. Sample preparation procedures are straightforward for isolating and immobilizing particles on a suitable substrate. Automated instrument control and analysis software makes it practical to obtain size and shape distributions on thousands of particles and sequentially apply that information to selectively direct the LIBS analysis based on particle characteristics. By constructing databases of elemental profiles, LIBS becomes a viable forensic analysis tool, even in complicated matrices. Examples of diverse applications in the energy and the pharmaceutical industries will be presented.

Keywords: Atomic Spectroscopy, Forensic Chemistry, Pharmaceutical, Soil
Application Code: High-Throughput Chemical Analysis
Methodology Code: Atomic Spectroscopy/Elemental Analysis

Session # 1370 Abstract # 1370-4 **Oral Sessions**

Session Title: Laser Induced Breakdown Spectroscopy (LIBS) and Glow Discharge in Atomic Spectroscopy - Half Sessi

Abstract Title: **Spectroscopy and Imaging Studies of a Solution-Cathode Glow Discharge**

Primary Author: Michael R. Webb
University of North Carolina Wilmington

Co-Author(s): Christian G. Decker, Denise E. Moon

Date: Tuesday, March 08, 2016 - Afternoo
Time: 04:05 PM
Room: B301

Abstract Text

Solution-cathode glow discharge - optical emission spectrometry is an emerging technique for low-power, low-cost atomic spectroscopy with a small-footprint instrument. Some progress has been made towards understanding the operational mechanism of the instrument, but much remains uncertain. Organic modifiers, particularly formic acid, have been shown to significantly improve the detection limits of the technique for several elements, but the mechanism behind this improvement is also not well understood. We will present recent studies performed to improve understanding of solution-cathode glow discharge and related techniques. These studies include measurements of sample properties before and after contact with the discharge, spectroscopic measurements, images of the solution-plasma interface, and images of the plasma's emission.

Keywords: Atomic Emission Spectroscopy, Atomic Spectroscopy, Instrumentation, Plasma
Application Code: Environmental
Methodology Code: Atomic Spectroscopy/Elemental Analysis

Session # 1380 Abstract # 1380-1 **Oral Sessions**

Session Title: LC/MS Biological Applications

Abstract Title: **Distinguishing Isomeric Acylsugar Metabolites: Strategies for Labeling Using [sup]13[/sup]C-amino Acid Precursors and LC-MS/MS**

Primary Author: Xiaoxiao Liu
Michigan State University

Date: Tuesday, March 08, 2016 - Afternoo
Time: 01:30 PM
Room: B404

Co-Author(s): A Daniel Jones, Banibrata Ghosh

Abstract Text

Chemical diversity presents one of the greatest challenges in comprehensive profiling of plant specialized metabolites. Such profiling and metabolite annotation are keys to identification of gene and enzyme functions. In this study, a rapid LC-MS/MS method is introduced to distinguish isomeric plant metabolites, particularly with regard to the lengths and branching of aliphatic chains derived from metabolism of amino acids. Acylated sugar derivatives (named "acylsugar") in wild tomato [i]S. pennellii[/i] LA0716 and LA2560 plants were germinated from seeds and cultivated on agar medium in the presence of 13C-amino acids. 13C5-valine, 13C6-leucine and 13C6-isoleucine were evaluated as precursors in acylsugar related branched aliphatic chain biosynthesis. Leaflets of labeled plants were extracted using isopropanol: acetonitrile: water (3:3:2, v/v/v) and specialized metabolites were resolved by UHPLC separation. Heavy isotope labeling patterns were traced using both multiplexed CID and MS/MS mode, in which precursor ions [M+formate+4]- and [M+formate+5]- were selected as acylsugars labeled by 13C5-Val and 13C6-Leu or Ile, respectively. The presence of 13C-labeled acylsugar fragment ions in MS/MS mode provided evidence that 13C-amino acids were incorporated within plants. Additionally, 13C-anteiso-branched aliphatic chains were only observed in 13C-Ile labeled samples, whereas 13C-Val and 13C-Leu specifically incorporated in iso-branched chains. Isomers of acylsugar with different branched acyl chain types were distinguished based on this result. LC-MS/MS analysis of 13C-amino acid labeling plants provided a simplified strategy to investigate structural diversity of plant specialized metabolites without purification.

Keywords: Liquid Chromatography/Mass Spectroscopy, Metabolomics, Metabonomics, Natural Products
Application Code: Bioanalytical
Methodology Code: Liquid Chromatography/Mass Spectrometry

Session # 1380 Abstract # 1380-2 **Oral Sessions**

Session Title LC/MS Biological Applications

Abstract Title **Oxidative Techniques for Pteridine Bioanalysis and Implications for ESI-MS Applications**

Date: Tuesday, March 08, 2016 - Afternoo
Time: 01:50 PM
Room: B404

Primary Author Casey Burton
Missouri University of Science and Technology

Co-Author(s) Honglan Shi, Yinfa Ma

Abstract Text

Pteridines are endogenous metabolites involved with nucleotide synthesis, redox regulation, DNA methylation and photodegradation, and more. Their biological and clinical significance has led to the rapid development of numerous and varied analytical techniques in recent years. An equally diverse array of oxidative and antioxidative pretreatments have been employed to counter complex pteridine redox chemistry which is highlighted by the existence of three oxidation states. However, many pretreatments predate electrospray ionization - mass spectrometry (ESI-MS) methodologies, posing significant compatibility and ionization suppression concerns. Moreover, oxidative techniques can result in adverse formation of 7,8-dihydroxanthopterin from unrelated pteridine derivatives. In this study, we systematically evaluated common oxidative and antioxidative pretreatments using an improved high-performance liquid chromatography – tandem mass spectrometry (HPLC-MS/MS) urinary pteridine assay. Pretreatment techniques were evaluated under a range of conditions including protic and aprotic solvents, solvent pH, dissolved oxygen levels, reagent concentration, and reaction time. Enhanced mass spectrum scans coupled with information dependent acquisition of enhanced product ion scans (EMS-IDA-EPI) were used to identify pretreatment byproducts from chemical standards. Our results indicate that conventional permanganate and triiodide oxidative procedures lead to pyrazine ring opening and substantial ionization suppression, respectively, whereas antioxidative procedures fail to account for in vivo pteridine autooxidation. We additionally show post-column addition of 100 µM ascorbic acid limits in-source oxidation of tetrahydro- and dihydro-pteridines. The detailed HPLC-MS/MS and EMS-IDA-EPI methods and results will be presented at the conference.

This study was supported by University of Missouri FastTrack Research funding and a National Science Foundation Graduate Research Fellowship.

Keywords: Bioanalytical, Liquid Chromatography/Mass Spectroscopy, Method Development, Sample Preparatio
Application Code: Bioanalytical
Methodology Code: Liquid Chromatography/Mass Spectrometry

Session # **1380** **Abstract #** **1380-3** **Oral Sessions**

Session Title LC/MS Biological Applications

Abstract Title **Intact Histones Separation by Using Submicron Particles with RPLC-MS**

Date: Tuesday, March 08, 2016 - Afternoo
Time: 02:10 PM
Room: B404

Primary Author Ximo Zhang
Purdue University

Co-Author(s) Mary J. Wirth

Abstract Text

Here we present a method of RPLC-MS with sub-0.5 [micro]m nonporous silica particles to separate intact histones. The levels of post-translational modifications (PTMs) of histones are known to correlate with different stages of cancer. In order to explore the use of histone modification as biomarkers, efficient separation of histones is needed. However, due to the lack of resolution on intact histones separation, conventional methods for histones analysis require time-consuming digestion, which can cause the loss of PTM information. With slip-flow chromatography, high efficiency of protein separation can be achieved because of minimized eddy diffusion and reduced resistance to mass transfer by orderly packed stationary phase and slip flow enhancement. In this work, a 4 cm capillary with 470 nm particle size and C18 bonded phase was used. In a 20 min gradient, resolution was greatly improved for the isoforms of both linker histones and core histones. The levels of histone phosphorylation and other PTMs were verified by deconvoluted MS spectra. Meanwhile, the injection amount of sample was as low as 1 picomolar. This high sensitivity suggests great potential of slip-flow capillary for future clinical analysis.

This work is supported by grant NIH R01-GM105874

Keywords: Liquid Chromatography, Liquid Chromatography/Mass Spectroscopy, Pharmaceutical, Proteomics
Application Code: Genomics, Proteomics and Other 'Omics
Methodology Code: Liquid Chromatography/Mass Spectrometry

Session # 1380 Abstract # 1380-4 **Oral Sessions**

Session Title LC/MS Biological Applications

Abstract Title **High Throughput Analysis of TCA Metabolites Using Column Switching and IC-HRAM Mass Spectroscopy**

Primary Author Terri T. Christison
Thermo Fisher Scientific

Date: Tuesday, March 08, 2016 - Afternoo
Time: 02:30 PM
Room: B404

Co-Author(s) John E. Madden, Junhua Wang, Monika Verma

Abstract Text

It was previously reported that capillary ion chromatography (CapIC) separations had 10 to 100-fold higher sensitivities than other separation methods (HILIC and RP) thanks to the low chemical background and suppressor technology. Additionally, 11 isomeric monophosphate sugars and nine isomeric diphosphate sugars in cell lysates were resolved by CapIC but not by HILIC and RP. These results were reproduced using an analytical IC system running 20-min gradient. It demonstrates the potential of high throughput separation of IC. However, shorter runs were needed to fulfill the need of a fast assay for targeted analysis of large sample sets. We optimized the gradient to allow a 9-min separation with additional 5-min equilibration time. A mixture of six stable isotopic labeling (SIL) standards was well retained with acceptable reproducibility. Isomeric pairs like citrate and isocitrate; trans- and cis-aconitate were baseline resolved. Minimal resolution loss was observed for the isomeric mono- and di-phosphate sugars in the cell lysate samples. We then reconfigured the ICS-5000+ HPIC system to run column switching through an additional 10-port valve. The column switching or called multiplexing is achieved by the new interfacing software SII which couples Xcalibur with IC operating system, Chromeleon 7. Both pumps had the same flow rate (0.38 mL/min) through separate eluent generator cartridges. When one column finishes the 9 min separation gradient delivered by pump 1, it switches to the equilibration gradient delivered by pump 2, and the eluent flows to waste. The second column switches inline to separation gradient delivered by pump 1. This multiplexing is automated by programmed commands in the script; no other changes, such as additional pumps or suppressors are needed. In doing so, the total time was further shortened but the actual equilibration time for each column was extended, resulting in further improved reproducibility.

Keywords: High Throughput Chemical Analysis, Ion Chromatography, Liquid Chromatography/Mass Spectroscop
Application Code: Genomics, Proteomics and Other 'Omics
Methodology Code: Liquid Chromatography/Mass Spectrometry

Session # 1380 Abstract # 1380-5 **Oral Sessions**

Session Title LC/MS Biological Applications

Abstract Title **Unified Drug Testing by Online SPE-LC/MS/MS with Focus on Productivity Achieved Through Ease of Use by Lab Technicians: One Totally Automated Method Measures ALL the Drugs in Urine and/or Oral Fluids**

Date: Tuesday, March 08, 2016 - Afternoo
Time: 03:05 PM
Room: B404

Primary Author Mark Hayward
ITSP Solutions

Co-Author(s) Kim Gamble, Martin Johnson, Matthew Hardison, Rick Youngblood

Abstract Text

While the continued growth of the use of LC/MS/MS for the measurement of drugs of abuse in urine and OF seems certain, there still are several technical challenges that need to be met. These needs include being able to easily measure low dose drugs at or near 1 ng/g concentration (for medical purposes, Pesce, et. al. 2012 AACC conference, as well as zero tolerance testing), simplicity for performing measurements with lab technicians with relatively little training, and the ability to achieve high productivity for all work while minimizing the labor and number of workflows required.

In an effort to meet these needs, we have developed the automated on-line SPE-LC/MS/MS method. It uses SPE to clean and pre-concentrate samples so that low dose drugs at or near 1 ng/g concentration are easily measured at S/N 20. At the same time, the method's design is balanced to address (identify / measure) all of the drugs (acidic and basic drugs as well as polar and non-polar drugs), as well as either urine or OF samples, all in one method, all in one workflow. Urine and OF samples even can be measured together on the same LC/MS/MS in the same run list. The method is simple, robust, and can be readily performed with a minimum amount of labor by lab technicians with MS familiarity. It is completely automated from sample plates/vials to results (with no change in work flow while still using only the native MS software) and can process two 96-well plates of samples overnight per LC/MS/MS. The results will be waiting for you in the morning.

Total automation is achieved using the PAL system LC autosampler which performs all sample preparation, including the SPE in parallel to LC/MS/MS analysis, and injects the sample into the LC/MS/MS. The cycle time achieved for on-line SPE-LC/MS/MS was 4.5 minutes for 71 drugs (opiates, metabolites, illicits, opioids, barbs, benzos, and THCA) in urine.

Keywords: Automation, Clinical Chemistry, Liquid Chromatography/Mass Spectroscopy, Solid Phase Extraction
Application Code: Clinical/Toxicology
Methodology Code: Liquid Chromatography/Mass Spectrometry

Session # 1380 Abstract # 1380-6 **Oral Sessions**

Session Title LC/MS Biological Applications

Abstract Title **High-Throughput Mass Spectrometric Analysis of Covalent Protein-Inhibitor Adducts for the Discovery of Irreversible Inhibitors: A Complete Workflow**

Date: Tuesday, March 08, 2016 - Afternoo

Time: 03:25 PM

Room: B404

Primary Author Iain Campuzano
Amgen

Co-Author(s) Daniel Onea, John McCarter, Tara Arvedson, Tisha San Miguel, Todd Rowe, Victor Cee

Abstract Text

We have implemented a solid-phase extraction (SPE)-based RapidFire(Agilent) time-of-flight mass spectrometer (ToF-MS) system in combination with novel informatics to rapidly screen and characterize the covalent binding of different irreversible inhibitors to intact proteins. This highthroughput screening (HTS)-MS platform can be used to accurately detect and quantitate the extent of formation of different covalent proteininhibitor adducts between electrophilic inhibitors and nucleophilic residues such as cysteine, lysine, or histidine of intact proteins.

A key element of the workflow is the automated identification and quantitation of the expected masses of covalent protein-inhibitor adducts using a custom PipeLine Pilot (Accelrys) script which obviates the need to manually inspect each individual spectra.

Together this high-throughput SPE-MS and data analysis system has the capacity to test and analyze thousands of samples in a relatively short time enabling the screening of libraries of reactive compounds or use in a weekly medicinal chemistry covalent inhibitor lead optimization cycle.

Parallel screens were performed on a library of approximately 1,000 acrylamide-containing compounds of different structures and reactivities using both this SPE-MS based assay and a fluorescence-based thiol-reactive probe assay enabling comparison of false positives and false negatives between these orthogonal screening approaches.

Keywords: Automation, Bioinformatics, Drug Discovery, Liquid Chromatography/Mass Spectroscopy
Application Code: Drug Discovery
Methodology Code: Liquid Chromatography/Mass Spectrometry

Session # 1380 Abstract # 1380-7 **Oral Sessions**

Session Title LC/MS Biological Applications

Abstract Title **Efficient Use of pH Control in Developing LC/UV/MS Methods**

Primary Author Thomas E. Wheat
Waters Corporation

Date: Tuesday, March 08, 2016 - Afternoo
Time: 03:45 PM
Room: B404

Co-Author(s) Amanda B. Dlugasch, Patricia R. McConville

Abstract Text

Reliable chromatographic methods are developed to separate all sample components. Reversed-phase methods are usually based on manipulating organic solvents, column chemistries, and gradient slope as the steps in developing a separation method. In addition, adjustment of mobile phase pH can also provide important and useful changes in selectivity for ionizable analytes. These experiments are underutilized in method development because the preparation of buffered mobile phases is time consuming and labor-intensive. In common practice, only widely separated pH conditions are tested to maximize retention. For groups of related compounds, however, small changes in pH can be more effective because the analytes behave as though partially charged. We have developed techniques and software to facilitate the preparation of buffered mobile phases from concentrated stocks using the solvent proportioning capabilities of the UPLC pump. In using these techniques for method development, the selected detectors in the LC system add constraints to choice of buffers for pH control. In the common combination of UV and MS detection, the mobile phase must be both volatile and UV-transparent. The usual buffers to meet these conditions, ammonium formate and ammonium acetate, can only be used over narrow pH ranges. They also do not provide any buffering capacity in the pH range from pH 6 to 8 where many chemical functionalities can be discriminated. New volatile, transparent buffer systems have been developed to cover the pH range from 2-12. Electrospray MS detection can, therefore, be combined with UV spectral detection using a photodiode array detector. This buffers will be usedfor developing separations of pharmaceutical formulations as well as industrial chemicals. The automated preparation of mobile phases with different pH properties helps to ensure the optimum selectivity for reversed-phase liquid chromatography with UV and MS detection.

Keywords: HPLC, Liquid Chromatography, Liquid Chromatography/Mass Spectroscopy, Separation Sciences
Application Code: Pharmaceutical
Methodology Code: Liquid Chromatography/Mass Spectrometry

Session # **1380** **Abstract #** **1380-8** **Oral Sessions**

Session Title LC/MS Biological Applications

Abstract Title **Selectivity and Column Choices in HPLC Method Development**

Primary Author William Long
Agilent Technologies

Co-Author(s) Anne Mack, Jason Link, Stephen Luke

Date: Tuesday, March 08, 2016 - Afternoo
Time: 04:05 PM
Room: B404

Abstract Text

Modern superficially porous particle (SPP) based columns were introduced in 2006, and since then, many phases have been introduced from a variety of manufacturers, allowing easy transition from totally porous to superficially porous particles. These columns deliver high efficiency as well as very long lifetimes at half the pressure of totally porous sub-2 micron particles. Short highly efficient SPP columns can be used to replace longer 5 micron TPP columns with the resulting separation often having more resolution and efficiency while requiring significantly less time. Endcapped C18 SPP columns are an excellent first choice in separations because they typically result in successful separations. When performing method development, selectivity is the most powerful parameter to adjust, influencing resolution far more than efficiency or k', and can be driven by solvent strength adjustments, bonded phase changes, or modifications in the mobile phase composition or pH.In many cases an alternative column to a C18 should be chosen early in the method development cycle based on the type of compounds to be analyzed. For example separations involving closely related compounds such as isomers can be quickly separated using an SPP PFP while highly polar compounds can be separated using HILIC or an AQ phase. Examples of successful and unsuccessful separations will be made as well as a discussion of relating Tanaka and Hydrophobic Subtraction models to help explain the differences in these phases.

Keywords: Chromatography, HPLC, HPLC Columns
Application Code: Pharmaceutical
Methodology Code: Liquid Chromatography/Mass Spectrometry

Session # 1390 Abstract # 1390-1 **Oral Sessions**

Session Title Polymer Characterization and Applications

Abstract Title **Transferring a GPC Method to A More Efficient SEC Method for Zoladex Co-Polymer Using an Advanced Polymer Chromatography Based U(H)PLC System Coupled with RI Detection**

Primary Author Christopher Henry
Waters Corporation

Date: Tuesday, March 08, 2016 - Afternoo
Time: 01:30 PM
Room: B405

Co-Author(s) Andy Boughey, Jeanette Bowden, Mark Wrona, Richard Ladd

Abstract Text

We will discuss the workflow, optimization and steps towards validation of a compound which is routinely analysed by quality control departments using a registered Pharmacopeial Gel Permeation Chromatography (GPC) coupled with Refractive index RI detection method for molecular weight average.

GPC is a well established method for the characterization of polymers but has limitations associated with low resolution due to large particle size and long equilibration times due to the relatively unstable packing material in traditional GPC columns. Within this body of work we present a faster, more efficient method for determination of the molecular weight average of the co-polymer. The method was developed using a low dispersion Waters Advanced Polymer Chromatography (APC) system using sub 3µm Acquity APC columns reducing run time by over three-fold and improving resolution. Columns evaluated employed hybrid silica technology which was found to greatly improve method reliability and robustness due to the rigid silica based packing's lack of susceptibility to the shrinking/swelling inherent with traditional gel-based columns. The advantages of this technology are reduced equilibration time and more consistent chromatography.

We will also demonstrate efficiency savings in instrument time of approximately 70% with column equilibration time savings of over 80% while detailing the corresponding solvent usage savings using this method and technology. This will be achieved whilst comfortably satisfying the existing system suitability criteria laid down in the GPC method currently used by Astra Zeneca. The methodology developed provides a robust and reliable alternative to a challenging legacy GPC method.

AstraZeneca are now working towards full validation ahead of implemeting for routine use for batch release.

Keywords: Chromatography, HPLC, Polymers & Plastics, Quality Control
Application Code: Polymers and Plastics
Methodology Code: Liquid Chromatography

Session # 1390 Abstract # 1390-2 **Oral Sessions**

Session Title Polymer Characterization and Applications

Abstract Title **Characterization of Polyacrylamide at Different Ionic Strength and pH Conditions Using Asymmetrical Flow FFF and Multi-Angle Light Scattering Detector**

Date: Tuesday, March 08, 2016 - Afternoo
Time: 01:50 PM
Room: B405

Primary Author Soheyl Tadjiki
Postnova Analytics Inc.

Co-Author(s) Japan Trivedi, Thorsten Klein, Trevor Havard

Abstract Text

Polyacrylamide (PAM) is a highly water-absorbent polymer which has been used widely in different industrial applications such as mining processes, swage treatments and oil recovery industry.

Different characterization methods have been used to study PAM in dilute solutions such as gel permeation chromatography, viscometry, light scattering and ultracentrifugation [1].

In this presentation the Asymmetrical Flow FFF system (AF4) was used to fractionate high molecular weight PAM samples in different carrier pH values ranging from 12 to 3. The FFF system was equipped with a multi-angle light scattering and refractive index detectors to measure molar mass and R.M.S radius. For all samples the molecular mass increased substantially from 105 g/mole to 107 g/mole as the pH of the carrier solution decreased from 12 to 3. The samples' R.M.S radii showed the same trend increasing from 50 nm to 200 nm as the pH of the carrier solution changed from basic to acidic.

References

[1] W. M. Kulicke, R. Kniewske and J. Klein, Prog. Polym. Sci. 1982, 8, 373-468

Keywords: Analysis, Characterization, Petroleum, Polymers & Plastics
Application Code: Polymers and Plastics
Methodology Code: Separation Sciences

Session # 1390 Abstract # 1390-3 **Oral Sessions**

Session Title: Polymer Characterization and Applications

Abstract Title: **Patterning of Polycaprolactone-Impregnated Glass Microfiber Membranes: A Novel Approach to Fabrication of Microfluidic Devices**

Date: Tuesday, March 08, 2016 - Afternoo
Time: 02:10 PM
Room: B405

Primary Author: Gayan C. Bandara
Oregon State University

Co-Author(s): Vincent T. Remcho

Abstract Text

Lateral flow microfluidics has gained popularity in recent years as a miniaturized, inexpensive and rapid analytical platform. The method of fabrication and the materials of manufacture largely determine the quality and utility of these devices. The basic fabrication method for devices of this kind involves patterning of a support substrate (an organic or inorganic filtration membrane) into hydrophilic channels defined by hydrophobic barriers comprised of various polymers or waxes. Polar hydrophilic membranes, such as glass microfiber (GMF) membranes, hold great potential in microfluidic device fabrication as they are inexpensive, chemically inert and stable – yet GMF membranes are largely unexplored in microfluidic applications. Impregnation of these membranes with non-polar polymers such as polycaprolactone (PCL) will convert the hydrophilic GMF into a hydrophobic medium. Controlled alteration of the surface chemistry of PCL/GMF substrates allows for the fabrication of microfluidic patterns on the surface.

Using this approach, PCL-impregnated GMF media were selectively exposed to oxygen radicals so that the exposed surface became permanently superhydrophilic while the unexposed area remained hydrophobic. This patterning process can be used as a simple, rapid, single-step microfluidic device fabrication approach. Selective exposure was done using a mask made up of an inexpensive material, wherein the required microfluidic pattern was cut into the mask prior to assembly and exposure. Combining different channel geometries (flow-through, flow-through + lateral flow, and surface-lateral flow) allows for fabrication of complex multidimensional (2D and 3D) microfluidic devices on the same polymer-impregnated membrane with unique properties and applications.

Keywords: Environmental Analysis, Lab-on-a-Chip/Microfluidics, Plasma, Polymers & Plastics
Application Code: Polymers and Plastics
Methodology Code: Microfluidics/Lab-on-a-Chip

Session # 1390 Abstract # 1390-4 **Oral Sessions**

Session Title Polymer Characterization and Applications

Abstract Title **Advanced Polymer Chromatography - Method Development Tools for SEC Analysis of PEG**

Primary Author Michael OLeary
Waters Corporation

Co-Author(s) Damian Morrison

Date: Tuesday, March 08, 2016 - Afternoo
Time: 02:30 PM
Room: B405

Abstract Text

Traditionally, size exclusion or gel permeation chromatography (SEC/GPC) is used for the characterization of polymeric material, specifically their molecular weight distribution. In order to resolve polymeric species, long column lengths and banked column configurations are commonly used, resulting in lengthy analytical test cycle times as well as the associated consumption of costly and often hazardous solvents. Additionally, many test sets suffer from minimal replicate data points due to the typical analysis time resulting data with limited statistical weighting.

In this paper the benefits of a comprehensive systematic approach for polymer molecular weight characterization will be presented. Waters ACQUITY® Advanced Polymer Chromatography® (APC™) System, with its innovative and robust ACQUITY APC™ column technology, allows for improved resolution of polymer distributions with significantly shorter chromatographic total analysis cycle times. Additionally, we will describe some of the tools that are available to develop stable and impactful test methods that result in richer data sets based on more stable operating conditions, and replicate analyses that are easily obtained within minutes and not hours. [registered][registered][registered]

Keywords: Chromatography, HPLC, Polymers & Plastics
Application Code: Polymers and Plastics
Methodology Code: Liquid Chromatography

Session # 1390 Abstract # 1390-5 **Oral Sessions**

Session Title: Polymer Characterization and Applications

Abstract Title: **Determination of Minor Component Differences and Additives in Polyethylene Using Thermal Desorption, Heart-Cutting EGA, Reactive Pyrolysis and GC/MS Techniques**

Date: Tuesday, March 08, 2016 - Afternoo
Time: 03:05 PM
Room: B405

Primary Author: Terry Ramus
Diablo Analytical

Co-Author(s): Dave Randle, Itsuko Iwai, Robert R. Freeman

Abstract Text

The laboratory chemist is often asked to find chemical differences between two materials thought to be the same. While there are a host of analytical techniques used to characterize various materials, the combined results are often confusing or are inconsistent. This report will introduce a methodical approach that is simple, fast and, in most cases, provides sufficient data to either chemically differentiate the two samples or suggest a follow-up technique. This strategic approach, the so-called "method map" (see Fig. 1), utilizes the multi-mode capability of modern pyrolyzers directly interfaced to a GC/MS.

The sample is profiled in a very broad manner using evolved gas analysis (EGA). The EGA total ion thermogram and related selected ion chromatograms will often be sufficient to differentiate the two samples. At a minimum, the EGA results can be used to select the analytical method and conditions which will yield additional specificity and detail about the composition of the sample.

The "method map" approach will be described utilizing multiple techniques such as EGA (Evolved Gas Analysis), thermal desorption (TD), Heart-cutting EGA (HC-EGA) and pyrolysis (PY) using a multi-functional pyrolyzer and GC/MS. This report will show how these techniques were used to find and identify a minor component in a polyethylene sample that was causing a product performance issue. In addition, an introduction to reactive pyrolysis (RxPy) will demonstrate how this technique can be used for identifying additional additives in these same polyethylene samples

Keywords: Gas Chromatography/Mass Spectrometry, Polymers & Plastics, Pyrolysis, Quantitative
Application Code: Polymers and Plastics
Methodology Code: Gas Chromatography/Mass Spectrometry

Session # 1390 Abstract # 1390-6 **Oral Sessions**

Session Title: Polymer Characterization and Applications

Abstract Title: **Applying Automatic Polymer Identification Capability to DSC Thermograms**

Primary Author: Bob Fidler
NETZSCH Instruments NA LLC

Date: Tuesday, March 08, 2016 - Afternoo
Time: 03:25 PM
Room: B405

Co-Author(s): Alexander Schindler, Ekkehard Post, Gabriele Kaiser, Stefan Schmoelzer, Tobias Pflock

Abstract Text

'Identify' is a unique tool for automatic identification and interpretation of DSC curves with only a single click. Designed for material identification and quality assurance, it analyzes properties of the DSC curve for a given material being investigated and compares with the integrated database, allowing for the automatic identification of plastic types. Such a database comparison is unique in the area of DSC technology. The Identify database contains a library for typical polymers and can also be extended by adding the user's own polymers or mixtures. User-specific quality criteria can additionally be applied in order to define categories. For the first time, individual batches can be objectively compared with one another – an ability which is particularly relevant in the fields of quality assurance and failure analysis.

Keywords: DSC, Identification, Polymers & Plastics, Thermal Analysis
Application Code: Polymers and Plastics
Methodology Code: Thermal Analysis

Session # 1390 Abstract # 1390-7 **Oral Sessions**

Session Title: Polymer Characterization and Applications

Abstract Title: **Characterization and Determination of Irganox 1076 and 1010 in Polyethylene Using Thermal Desorption and Reactive Pyrolysis – GC/MS**

Date: Tuesday, March 08, 2016 - Afternoo
Time: 03:45 PM
Room: B405

Primary Author: Dave Randle
Frontier Lab USA

Co-Author(s): Aki Hosaka, Ichi Watanabe, Itsuko Iwai, Robert R. Freeman, Terry Ramus

Abstract Text

Irganox 1076 (MW=531) and 1010 (MW=1178), are sterically hindered phenolic antioxidants used to provide long-term thermal stability to a large and diverse array of polymeric materials. In addition, both antioxidants are approved for use in food contact applications. Each is highly resistant to conventional solvent based extraction techniques and each has a very low vapor pressure and so conventional GC analysis has proven inadequate for routine analysis. This report details a simple, GC/MS-based analytical method for the qualitative and quantitative determination of Irganox 1076 and 1010 in polyethylene (PE) using a temperature programmable multi-functional pyrolyzer.

Both 1076 and 1010 have an ester linkage which can be thermally hydrolyzed and methylated using tetramethylammonium hydroxide (TMAH). The amount of TMAH reagent used and the optimal temperature for reactive pyrolysis must be determined experimentally, and these results will be discussed. Each PE sample is analyzed twice: Methylation using a reactive organic alkali yields a value for the sum of [1076] + [1010]; thermal desorption is used to determine the concentration of 1076. Simply subtracting the two numbers is all that is necessary to find the amount of Irganox 1010. Standard addition quantitation is used because it reduces or eliminates non-target compound interferences in complex polymeric materials. 1076 can be determined directly using thermally desorption. For example, a four point standard addition calibration (PE sample, 100ng, 250ng and 500ng)[R[sup]2[/sup]=0.999] illustrates the precision of the method. The concentration of 1076 was 476 ppm (%RSD [n=5]=3.6). The precision and accuracy of the method will be reported.

Keywords: Gas Chromatography/Mass Spectrometry, Polymers & Plastics, Pyrolysis, Quantitative
Application Code: Polymers and Plastics
Methodology Code: Gas Chromatography/Mass Spectrometry

Session # 1400 Abstract # 1400-2 **Oral Sessions**

Session Title: Sampling and Sample Preparation-Environmental and Food (ID, Safety and Contaminants)

Abstract Title: **Bacterial DNA Analysis Using Solid-Phase Microextraction by Different Polymeric Ionic Liquid-Based Sorbent Coatings**

Date: Tuesday, March 08, 2016 - Afternoo
Time: 01:50 PM
Room: B406

Primary Author: Omprakash Nacham
Iowa State University

Co-Author(s): Andrea E. Tsatalis, Jared L. Anderson, Kevin D. Clark, Matthew P. Bommarito

Abstract Text

The extraction and preconcentration of DNA plays a central role in the analysis of microorganisms. Conventional DNA purification approaches often involve large sample volumes, the use of organic solvents, time-consuming and laborious centrifugation steps or multiple sample transfer steps that increase the risk of contamination. Solid-phase microextraction (SPME) is a non-exhaustive and solvent-free sample preparation technique. Due to its simplicity and amenability toward automation, it represents a very interesting platform for performing the preconcentration of biomolecules. In this study, a simple and relatively easy sample preparation method, polymeric ionic liquid (PIL)-based SPME was applied for the preconcentration of plasmid DNA (pDNA) from bacterial cells. Ultraviolet photoinitiated polymerization technique was utilized for the preparation of PIL-based SPME devices on a nitinol support. pDNA was extracted from buffered aqueous solution using the PIL-based sorbent coating followed by the amplification of a target gene by polymerase chain reaction (PCR). In comparison to a commercial polyacrylate sorbent coating, the PIL sorbent coating extracted greater quantities of pDNA. With an extraction time of 5 min, the PIL-based SPME technique was capable of preconcentrating a sufficient amount of template pDNA from a 20 ng/mL solution. Tuning and modification of PIL-sorbent coating composition such as incorporation of benzyl moieties in PIL structure resulted in enhanced extraction efficiency. The developed method was successfully employed for the analysis of pDNA from two different E. coli transformants in a dilute aqueous solution. Applicability of the PIL-based SPME technique was also investigated for the determination of bacterial contamination in processed food products such as orange juice samples.

The authors acknowledge funding from Chemical Measurement and Imaging Program at the National Science Foundation (Grant number CHE-1413199).

Keywords: Biological Samples, Food Contaminants, Nucleic Acids, SPME
Application Code: Food Contaminants
Methodology Code: Sampling and Sample Preparation

Session # 1400 Abstract # 1400-3 **Oral Sessions**

Session Title: Sampling and Sample Preparation-Environmental and Food (ID, Safety and Contaminants)

Abstract Title: **ICE Concentration Linked with Extractive Stir Bar (ICECLES): A Novel Sample Preparation Technique for Ultratrace Analysis**

Date: Tuesday, March 08, 2016 - Afternoo

Time: 02:10 PM

Room: B406

Primary Author: Brian Logue
South Dakota State University

Co-Author(s)

Abstract Text

The ability to determine contaminants at "ultratrace" concentrations is a critically important, but currently challenging, aspect of many types of analyses, including ensuring safe drinking water. Our recently discovered technique, coined ICE Concentration Linked with Extractive Stirrer (ICECLES), is a significant step towards routine ultratrace analysis. ICECLES combines the highly complementary techniques of stir bar sorptive extraction (SBSE) and a little-known concentration technique called freeze concentration (FC). In FC, solutes are concentrated based on the direct relationship between freezing point depression and solute molality. Specifically, if a solution is slowly frozen (typically with vigorous stirring), local regions of solvent with a low solute concentration are frozen first and the solution left behind is more concentrated. In ICECLES, a solution is stirred vigorously with a sorbent-coated stir bar as it is slowly frozen. Currently, ICECLES has produced ng/L limits of detection and over 1000-fold signal increases compared to SBSE. Here, we will introduce our work to develop ICECLES into an alternative analytical sample preparation technique especially suited for trace analysis of aqueous samples.

Keywords: Environmental Analysis, Extraction, Trace Analysis, Ultratrace Analysis

Application Code: Environmental

Methodology Code: Sampling and Sample Preparation

Session # 1400 Abstract # 1400-4 **Oral Sessions**

Session Title Sampling and Sample Preparation-Environmental and Food (ID, Safety and Contaminants)

Abstract Title **Analysis of Pesticides in Foods by Direct Immersion SPME Using an Overcoated Fiber**

Primary Author Leonard M. Sidisky
Supelco/Sigma-Aldrich

Date: Tuesday, March 08, 2016 - Afternoo
Time: 02:30 PM
Room: B406

Co-Author(s) Katherine Stenerson, Robert Shirey, Tyler Young, Yong Chen

Abstract Text

Analysis of pesticides in many food commodities is typically done using QuEChERS methodologies. While this technique offers many advantages, it produces a dilute extract that often requires a concentration step or use of large volume injection for GC analysis. In addition, despite the use of cleanup sorbents, some extracts can contain a great deal of background, which interferes with detection in GC/MS. The use of solid phase microextraction (SPME) for pesticide analysis offers several advantages over QuEChERS, including better sensitivity, and less background and hands-on sample preparation time. In addition, it is a greener technique in that no organic solvents are required. SPME is often used as for headspace extraction, however many pesticides cannot be sampled in this manner due to their low vapor pressures. Direct immersion of standard SPME fibers into complex matrices such as foods can prematurely foul the fiber, leading to loss in analyte response. In this work, we utilized a new polydimethylsiloxane (PDMS) overcoated version of a PDMS/divinylbenzene (PDMS/DVB) SPME fiber for analysis of pesticides from food matrices. The use of a PDMS overcoating offers protection for the DVB, allowing for direct immersion into complex matrices. Data will be presented in which this new fiber was used for extraction of select pesticides from baby food and spaghetti sauce. Fiber performance will be compared with a standard DVB fiber (i.e. no overcoating) for accuracy, reproducibility and durability. In addition, a comparison will be presented between results obtained using QuEChERS and the SPME methods for accuracy, reproducibility and GC/MS background.

Keywords: Pesticides, Sample Preparation, SPME

Application Code: Food Contaminants
Methodology Code: Sampling and Sample Preparation

Session # 1400 **Abstract # 1400-5** **Oral Sessions**

Session Title: Sampling and Sample Preparation-Environmental and Food (ID, Safety and Contaminants)

Abstract Title: **Recent Advances in Sample Preparation for Extraction and SPE**

Date: Tuesday, March 08, 2016 - Afternoo
Time: 03:05 PM
Room: B406

Primary Author: SM Rahmat Ullah
Thermo Fisher Scientific

Co-Author(s): Aaron Kettle, Glenn Kuse, Kannan Srinivasan, Mike McAdams

Abstract Text

Accelerated solvent extraction (ASE) is a high-temperature and high-pressure extraction technique that is widely used in environmental, chemical and food analysis. Extractions at high temperatures and pressures allow faster extraction of analytes relative to conventional solid-liquid or liquid-liquid based extraction techniques. The sample is mixed with dispersant and loaded into a metallic ASE cell followed by extraction in the ASE systems. The extraction cell consists of end caps and cell body. Some components of the end cap such as frit and seal need periodical replacement after several extractions. A user needs to disassemble the entire end cap assembly to replace either the seal or the frit in the current cell design. In the first half of this presentation we show a new design of the cell that is more users friendly. Example applications with new design will also be presented here. The AutoTrace 280 Solid-Phase Extraction (SPE) instrument performs automated SPE of large-volume liquid samples. Liquid–liquid extractions that normally take a long time to achieve complete extraction can be automated using an AutoTrace 280 SPE instrument. In the second half of this presentation we discuss new methods for in-line drying and in-line cleanup using AutoTrace 280 SPE instrument.

Keywords: Accelerated Solvent Extraction, Environmental, Sample Preparation, Solid Phase Extraction

Application Code: Environmental

Methodology Code: Sampling and Sample Preparation

Session # 1400 Abstract # 1400-6 **Oral Sessions**

Session Title: Sampling and Sample Preparation-Environmental and Food (ID, Safety and Contaminants)

Abstract Title: **Development of a Carbon Mesh Supported Thin Film Microextraction Membrane as a Means to Lower the Detection Limits of Bench-top and Portable GC-MS Instrumentation**

Primary Author: Jonathan J. Grandy
University of Waterloo

Date: Tuesday, March 08, 2016 - Afternoo
Time: 03:25 PM
Room: B406

Co-Author(s): Janusz Pawliszyn

Abstract Text

Since being introduced in 1989 Solid phase microextraction has continually grown amongst the analytical chemistry community due to a clean, portable and easy to handle design. However, the same miniaturization that has allowed for such positive characteristics has inherently limited the surface area, and therefor sensitivity, of the SPME fiber. Henceforth, this work proposes a durable, high surface area, and easy to handle thin film microextraction (TFME) device. The membrane is comprised of poly-divinylbenzene resin particles suspended in a high density polydimethylsiloxane glue spread onto a carbon mesh support. This novel design was shown to exhibit a substantially lesser amount of siloxane bleed during thermal desorption while providing a statistically similar extraction efficiency towards a broad spectrum of compounds when compared to an unsupported DVB/PDMS membrane of similar size that had been prepared with former methods. Membranes cut to 4 cm long, 4.85 mm wide and 30-40 µm thick (per side), were shown to extract 21.2, 19.8, 18.5, 18,4, 26.8, and 23.7 times the amount of 2,4 dichlorophenol 2,4,6 trichlorophenol, phorate D10, fonofos, chloropyrifos, and parathion respectively, from a 10 ppb aqueous solution than a comparable 65 µm DVB/PDMS SPME fiber. Indeed, by increasing extraction efficiencies by a factor of 20 or more, TFME may prove to be a promising future architecture of microextraction technology, further pushing down detection limits of benchtop and portable GC-MS instrumentation alike.

This work was supported by the Natural Sciences and Engineering Research Council of Canada, Supelco Co. and Torion Technologies of Pelkin Elmer Inc.

Keywords: Environmental Analysis, GC-MS, Portable Instruments, SPME
Application Code: Environmental
Methodology Code: Sampling and Sample Preparation

Session # 1400 Abstract # 1400-7 **Oral Sessions**

Session Title: Sampling and Sample Preparation-Environmental and Food (ID, Safety and Contaminants)

Abstract Title: **Investigation of Polymeric Ionic Liquid Sorbent Coatings in Solid-Phase Microextraction Coupled to High-Performance Liquid Chromatography for the Analysis of Polar Compounds**

Primary Author: Honglian Yu
Iowa State University

Date: Tuesday, March 08, 2016 - Afternoo
Time: 03:45 PM
Room: B406

Co-Author(s): Jared L. Anderson, Josias Merib

Abstract Text

Solid-phase microextraction (SPME) is a non-exhaustive extraction technique based on the distribution of analytes between the sample system and the micro-volume extraction phase. There are various commercial SPME fibers with sorbent coatings covering a wide range of polarity currently available in the market. However, the selectivity of these commercial SPME coatings towards specific target analytes needs significant improvement. The interests on ionic liquids (ILs)-based SPME coatings rapidly grew since their introduction, due to their tunable structures and multiple solvation capabilities. Polymeric ionic liquid (PILs) were proven to be substantially more robust as SPME coatings, compared to their monomeric counterparts. In this study, a number of crosslinked PIL-based coatings were synthesized and chemically bonded to derivatized nitinol supports by UV-initiated polymerization. The PIL coatings were then applied for the analysis of several classes of selected polar analytes, including phenolic compounds, nonsteroidal anti-inflammatory drugs and insecticides by direct immersion SPME coupled to HPLC. The roles of various cations and anions comprised in each PIL were intimately studied to gain insight on their selectivity towards the chosen analytes. Extraction and desorption parameters were optimized by design of experiment to achieve the best analytical results. Analytical data were also compared to commercially available SPME fibers to demonstrate the applicability of the novel materials for trace-level analysis of the selected polar compounds. Furthermore, the structural integrity and robustness of the coatings were closely monitored, in various matrix conditions, to determine the usefulness of PIL-based SPME when coupled to HPLC.

Keywords: Environmental Analysis, HPLC, Pesticides, SPME
Application Code: Environmental
Methodology Code: Sampling and Sample Preparation

Session # 1400 Abstract # 1400-8 **Oral Sessions**

Session Title: Sampling and Sample Preparation-Environmental and Food (ID, Safety and Contaminants)

Abstract Title: **Justification of Kinetic Calibration in Pre-Equilibrium Solid Phase Microextraction with a Mathematical Model and Computational Simulation**

Date: Tuesday, March 08, 2016 - Afternoo
Time: 04:05 PM
Room: B406

Primary Author: Md Nazmul Alam
University of Waterloo

Co-Author(s): Fardin Ahmadi, Janusz Pawliszyn, Luis Ricardez-Sandoval

Abstract Text

A calibration approach based on standard chemicals loaded onto extraction phases has gained popularity in various areas of sample analysis such as environmental, toxicological and tissue sampling. The kinetics of the calibrant release and analyte uptake between the sample and extraction phase were studied with a finite-element method (FEM) using COMSOL Multiphysics® software package. Effect of various sample environment parameters such as fluid flow velocity, temperature, presence of a binding matrix component on the performance of calibrant loaded extraction phase (CL-EP) was investigated in details with the model. The simulation results demonstrates the suitability of the CL-EP method for target analytes in different sample environments. Although the outcome of this study is applicable to any sampler based on calibrant-loaded liquid or solid extraction phase, experimental data using solid-phase microextraciton (SPME) sampler was used to fit our simulation results. Combined together with matrix matched calibration the CL-SPME approach provides total concentration, while with matrix free calibration it provides free concentration. The numerical results are in very good agreement with the experimental data reported previously. The mechanistic model and numerical simulation presented will aid in the optimization of sampler design and sampling parameters prior to laboratory experiment to achieve maximum sampler efficiency, which will save both time and expensive chemical reagents.

Keywords: Computers, Extraction, Sampling, Water
Application Code: Environmental
Methodology Code: Computers, Modeling and Simulation

Session # 1410 Abstract # 1410-1 **Poster Sessions**

Session Title: Advances in Biomedical Applications

Abstract Title: **Strategies for Glycomics, Glycoproteomics, and Glycosaminoglycan Research at the Complex Carbohydrate Research Center**

Primary Author: Stephanie A. Archer-Hartmann
Complex Carbohydrate Research Center

Date: Tuesday, March 08, 2016 - Afternoo
Time:
Room: Exposition Floor, 400 Aisle

Co-Author(s): Christian Heiss, Mayumi Ishihara, Parastoo Azadi, Roberto Sonon

Abstract Text

The Analytical Services Laboratory of the Complex Carbohydrates Research Center (CCRC) at The University of Georgia is a non-profit entity that offers services for structural characterization of glycoconjugates derived from animal, plant, or microbial origin. The Analytical Services Group routinely analyzes samples from a wide variety of institutions including universities, federal agencies, and industry groups from the US and other countries. The CCRC service laboratory is well complemented with instruments such as LTQ-Orbi-MS, MALDI-TOF, and high field NMRs in addition to SAX-HPLC, CE, HPAEC-PAD, and GC-MS.

Here we will show several examples of current projects in our laboratory that highlight a combination of analytical techniques for the structural elucidation of N- and O- released glycans, glycopeptides, and proteoglycans by liquid chromatography and LC-MS. Finally, this poster highlights the recent collaborative training events at the CCRC in glycomics, glycoproteomics, and glycosaminoglycan research.

Keywords: Biopharmaceutical, Carbohydrates, Liquid Chromatography/Mass Spectroscopy, Mass Spectrometry
Application Code: Biomedical
Methodology Code: Mass Spectrometry

Session # 1410 Abstract # 1410-2 **Poster Sessions**

Session Title: Advances in Biomedical Applications

Abstract Title: **Down Regulation of Smad-2 and VEGF Transcription and TGF-B1 Signaling in Nano Sized Titanium Dioxide-Induced Liver Injury in Mice by Potent Antioxidants**

Date: Tuesday, March 08, 2016 - Afternoo

Time:

Room: Exposition Floor, 400 Aisle

Primary Author: Samy A. Abdel Azim
Cairo University

Co-Author(s): Abd El-Moneim M. Afify

Abstract Text

Background/Aim: The present study investigated the in vitro and in vivo effects of individual and combined doses of idebenone, carnosine and vitamin E on ameliorating some of the biochemical indices of nano-sized titanium dioxide (n-TiO2) in mice liver. Methods: The in vitro cytotoxic effect of nano-sized anatase TiO2 (21 nm) on hepatic cell lines (HepG 2) was investigated. Additionally, n-TiO2 was orally administered (150 mg/kg/day) for 2 weeks, followed by a daily intragastric gavage of the aforementioned antioxidants for 1 month. Results: n-TiO2 induced significant cytotoxicity in hepatic cell lines and elevated the levels of serum alanine aminotransferase (ALT), aspartate aminotransferase (AST), hepatic total antioxidant capacity (TAC) and nitrite/nitrate (NOx) levels. Meanwhile, glutathione-S-transferase (GST) activity was significantly reduced. Moreover, RT-PCR and western blot analysis showed that n-TiO2 significantly altered the mRNA and protein expressions of transforming growth factor-beta (TGF-□1) and Smad-2, as well as vascular endothelium growth factor (VEGF). Histopathological examination of hepatic tissue reinforced these results. Conclusion: our findings, suggest that the different antioxidants; Idebenone, carnosine and vitamin E exert a therapeutic protective effect in n-TiO2 toxicity by decreasing oxidative stress, mRNA gene expression and liver fibrosis and angiogenesis which might be implicated in n-TiO2-induced liver toxicity.

Keywords: Bioanalytical, Biological Samples, Biomedical, Biotechnology

Application Code: Biomedical

Methodology Code: Sampling and Sample Preparation

Session # 1410 Abstract # 1410-3 **Poster Sessions**

Session Title: Advances in Biomedical Applications

Abstract Title: **One Hydrothermal Processing of 1D Hydroxyapatite for Biomedical Application**

Primary Author: Zoran S. Stojanovic
Institute of Technical Sciences of SASA

Date: Tuesday, March 08, 2016 - Afternoo
Time:
Room: Exposition Floor, 400 Aisle

Co-Author(s): Miroslav M. Miljkovic, Nenad L. Ignjatovic, Vojka Zunic, Vuk D. Uskokovic

Abstract Text

One dimensional (1D) hydroxyapatite (HA) nano and microstructures such as wires, ribbons and tubes exert a broad array of functions and have role as substrates, reinforcing phases in composites, active substance bearing agents, and thus, they are highly used in bone tissue engineering. Herein we synthesized such 1D HA structures on gram scale by hydrothermal batch processes using urea as source of hydroxide ions and calcium phosphates as precursor. The "one pot" process produced 1D HA aforementioned structures. Since conversion of starting precipitate to HA in this particular system is rather complex than straightforward process, involving many simultaneously occurring intermediate reactions, "one pot" process fail to meet uniformity criteria over particle morphology and size. Dividing procedure in two stages we succeeded to get uniform HA nanowires and, at the same time, to maintain same synthesis condition and get high yields. In this case, dicalcium phosphate dihydrate (DCPD) platelets, synthesized apart in flask, were used as precursor with or without dimethylformamide (DMF) modification. We made effort to reveal the details about mechanism, effect of DMF and drying on DCPD microstructure which finally led to formation of different HA morphology. The washed and dried products were characterized by X – ray powder diffraction (XRD), electron diffraction, scanning electron microscopy (SEM), field emission scanning electron microscopy (FE SEM), and laser diffraction (LD). For determination of size distributions and XRD diffraction peaks, components of EM algorithm were used (normalEM, mixtools R package). Our results showed that for complete transformation of DCPD to uniform HA nanowires, certain structural configuration and drying step are required. Biological response to DCPD platelets and HA nanowires, assessed using MTT assay and immunofluorescence (IF) microscopy, showed high biocompatibility of synthesized materials.

Keywords: Biomedical, Materials Characterization, Microscopy, Nanotechnology
Application Code: Biomedical
Methodology Code: Microscopy

Session # 1410 Abstract # 1410-4 **Poster Sessions**

Session Title: Advances in Biomedical Applications

Abstract Title: **DNA Micelle Flares: Thermodynamic Stability and Cellular Internalization**

Primary Author: Yanyue Wang
University of Florida

Co-Author(s): Weihong Tan

Date: Tuesday, March 08, 2016 - Afternoo
Time:
Room: Exposition Floor, 400 Aisle

Abstract Text

DNA micelle flares are sphere-shaped DNA-lipid structures formed by hydrophobic interactions between the lipid tails coupled to single DNA strands. Compared to other sphere-shaped nucleic acids, DNA micelle flares have high cellular permeabilities and low critical micelle concentrations, making them potential supporters for drug delivery and gene therapy. For this purpose, the particles must retain high binding affinity with mRNA in various conditions, including in the presence of heat and high ion concentrations. They should also be able to enter cancer cells, not just fuse into the cell membranes. In this work, the melting properties of DNA micelles towards their complementary DNAs (cDNAs) have been studied using fluorescence for systematic investigations of their thermostability. Besides, the properties and pathway for the cell internalization of DNA micelle flares have been studied. Results have shown that DNA micelles flares have sharper melting transition and higher melting temperatures compared with molecular beacons containing the same base pairs. The □S and □H of the melting procedure were also characterized using different concentrations of DNA micelle flares. The properties of the DNA micelle flares/cell membrane fusion have also demonstrated how they entered the cells.

Keywords: Bioanalytical, Fluorescence, Microscopy, Thermal Analysis
Application Code: Biomedical
Methodology Code: Fluorescence/Luminescence

Session # 1410 Abstract # 1410-5 **Poster Sessions**

Session Title: Advances in Biomedical Applications

Abstract Title: **Study of Various Cationic CPEs' Interaction with Mammalian Cells**

Date: Tuesday, March 08, 2016 - Afternoo
Time:
Room: Exposition Floor, 400 Aisle

Primary Author: Shanshan Wang
University of Florida

Co-Author(s): Kirk S. Schanze, Yun Huang, Zhiliang Li

Abstract Text

Conjugated polyelectrolytes (CPEs) are fluorescent water soluble polymers. Recently, CPEs are extensively studied as materials used in biomedical applications, for example, tumor cell diagnosis, fluorescence imaging, gene/drug delivery, etc. Understanding how CPEs interact with mammalian cells are important in ensuring their safe usage and suggesting further application opportunities. The work presented is a systematic study on how CPEs' structure differences influence their interactions with mammalian cells. Eleven CPEs with varying backbones and side groups are tested on their interactions with MCF-7 cells (women breast cancer cells). These CPEs show varying but low toxicity to MCF-7 cells; CPEs bearing more charged side groups show relatively higher toxicity than CPEs bearing less charged side groups. Laser scanning confocal microscope (LSCM) images show that all these CPEs can be taken in and located to the cells' lysosomes (an organelle in the cytoplasm), the photon bleaching of CPEs in cells are very low compared to the current commercial fluorescent probes for lysosome labeling and tracking, suggesting the CPEs' application in cellular fluorescence imaging. More interestingly, two CPEs with high bacterial killing activity are found nontoxic to MCF-7 cells, implicating their potential antibacterial medical applications where bacterial is selectively killed but mammalian cells' growth isn't interfered, such as wound dressing, photodynamic antimicrobial therapy, etc.

Keywords: Biomedical, Fluorescence, Imaging, Microspectroscopy
Application Code: Biomedical
Methodology Code: Fluorescence/Luminescence

Session # 1410 Abstract # 1410-6 **Poster Sessions**

Session Title: Advances in Biomedical Applications

Abstract Title: **Generation of T-cell Specific Aptamers Using a Novel Cell-Selex Method: Antibody Guided Cell-SELEX Technology**

Date: Tuesday, March 08, 2016 - Afternoo
Time:
Room: Exposition Floor, 400 Aisle

Primary Author: Hasan E. Zumrut
City University of New York, The Graduate Center

Co-Author(s): George Maio, Mallikaratchy Prabodhika, Mst Naznin Ara, Shomi Chakrabarti

Abstract Text

Nucleic acid aptamers (nucleic acid based antibody analogues) are synthetic DNA or RNA molecules that specifically bind to target molecules. Due to their advantages over antibodies, aptamers are being investigated to develop therapeutic molecules for cancer. In order to develop aptamers as successful therapeutic molecules, it is necessary to generate aptamers for known specific targets. Currently, target specific aptamers are selected using over-expressed protein either on a cell or as a purified protein, which may not recognize the targeting epitope when the protein is expressed at its native levels, in its native environment. Recently introduced cell-SELEX method allows the selection of aptamers towards membrane targets at their native state. Here we utilized a novel cell-SELEX approach to generate T-cell receptor (TCR) specific aptamers using antibody-antigen interactions as a guide. The T-cell receptor (TCR) consists of the highly variable and chains expressed at the cell membrane as a complex with the invariant CD3 chains. Using an antibody against TCR complex we have identified two specific aptamers against TCR complex expressed in Jurkat E6 cells. The specificity binding the selected aptamers towards TCR positive cells is validated utilizing flowcytometric assays against cell lines that express TCR complex. The epitope specificity of the aptamers is further confirmed by investigating the competition of the aptamers against the cognate antibody towards TCR complex.

Keywords: Bioanalytical, Biomedical, Biotechnology, Drug Discovery
Application Code: Biomedical
Methodology Code: Other (Specify)

Session # 1410 Abstract # 1410-7 **Poster Sessions**

Session Title: Advances in Biomedical Applications

Abstract Title: **Adeno-Associated Virus**

Primary Author: Yuan Wu
University of Florida

Date: Tuesday, March 08, 2016 - Afternoo
Time:
Room: Exposition Floor, 400 Aisle

Co-Author(s)

Abstract Text

Adeno-associated virus (AAV)-based vectors are becoming increasingly popular for gene therapy because of non-pathogenic, long-term expression, genomic integration and low immunogenicity. Recombinant AAV type 2 (rAAV2) vectors bind primarily to heparin sulfate proteoglycan (HSPG) receptor on the membrane of cell types, this in turn limits the transduction of AAV2 vectors. Thus, development of methods to achieve targeted transduction will have profound benefit in the future application of these vectors. Aptamers offer significant potential as convenient and evolvable targeting groups for cargo delivery. We design a novel conjugation-based targeting method to achieve cell-specific transduction of AAV2 containing GFP gene-based vectors. We utilized an amino group-modified sgc8 aptamers to conjugate with the amino group on the capsid of AAV2 by disuccinimidyl suberate (DSS) crosslinkers. Conjugation of specific targeting aptamers to the vectors was achieved by chemical crosslinking without abolishing the capsid structure, internalization, and subsequent GFP gene expression of AAV2 vectors. The sgc8 aptamers-modified vectors, targeted via PTK7 receptor, resulted in a significant increase in transduction efficiency of PTK7-positive CEM cells. Further optimization of this targeting method enhance the potential of AAV2 vectors in vitro and in vivo gene therapy and may form the basis for developing targeting methods for other AAV serotype capsids.

Keywords: Gene Therapy
Application Code: Biomedical
Methodology Code: Chemical Methods

Session # 1410 Abstract # 1410-8 **Poster Sessions**

Session Title: Advances in Biomedical Applications

Abstract Title: **Continuous Electroporation Through a Mesoporous Gold Membrane**

Primary Author: Juliette Experton
University of Florida

Date: Tuesday, March 08, 2016 - Afternoo
Time:
Room: Exposition Floor, 400 Aisle

Co-Author(s): Aaron Wilson, Charles R. Martin

Abstract Text

Cell permeation methods are critical. They are widely used in a context where gene therapy was demonstrated to cure many diseases. In this regard, electroporation stood out as a very efficient and versatile technique. Conventional methods consist on applying a short electrical pulse to a cuvette containing the cells in order to create temporary pores in their membrane and allow the introduction of foreign molecules. However, this approach suffers from the need of high voltage, up to 2kV, leading to cell lysis and limited control over the amount of molecular uptake.

Our study focuses on the design and characterization of a mesoporous gold plated polycarbonate membrane for flow-through electroporation. As [i]Escherichia coli[/i] travel through the membrane, they experience 30ms electric pulses at a voltage of 1 to 4V inducing a high electric field inside the pores. This allows a DNA staining dye in solution to penetrate the bacteria and be detected by fluorescence spectroscopy and microscopy. After an hour, another impermeable dye is added to the solution in order to determine the fluorescence intensity of lysed bacteria. The flow of bacteria through the membrane is detected by a UV-Visible spectrophotometer.

Subsequently, about 50% of the bacteria go through the membrane and up to 40% of them are electroporated, while almost no changes were observed if no voltage was applied. Hence, we showed evidences that we could develop an effective and robust design for the continuous electroporation of [i]E. coli[/i] under low voltage.

Keywords: Electrode Surfaces, Fluorescence, Membrane, UV-VIS Absorbance/Luminescence
Application Code: Biomedical
Methodology Code: Electrochemistry

Session # 1410 Abstract # 1410-9 **Poster Sessions**

Session Title Advances in Biomedical Applications

Abstract Title **Analyzing the Effect of Beverages and Fluoride on Tooth Enamel with a FAAS and Dissolution**

Date: Tuesday, March 08, 2016 - Afternoo

Time:

Room: Exposition Floor, 400 Aisle

Primary Author Andrea Gerchman
St. John Fisher College

Co-Author(s) Kimberly Chichester

Abstract Text

A Distek Dissolution System 2100B was used to evaluate enamel erosions on teeth after being exposed to a Monster Energy drink, Snapple ice-tea, Svedka Vodka, Coca-Cola soft drink, Gatorade sports drink, and coffee. Research has shown that the chemical structure of fluoride can bind with the tooth enamel to form an extremely thin layer of densely packed particles to protect the tooth. In this study, enamel will be exposed to different toothpastes and the tooth will be exposed to the six different liquid conditions to determine which toothpaste is best at stopping tooth erosion. The teeth will be analyzed with two different methods to compare the effect of fluoride on the tooth enamel under each condition. A caliper will be used to determine the width, length and height of the tooth to compare the erosion on the tooth before and after it was exposed to each condition and a flame atomic absorption spectroscopy will used to measure the calcium content in the tooth. Determining the concentration of calcium in the media will show how much the tooth eroded. The enamel layer before and after the addition and exposer to fluoride will be studied. Figures 1 and 2 shown below are part of my control, where the teeth were soaked in the beverages without any fluoride.

Keywords: Atomic Absorption, Biomedical, Dissolution, Surface Analysis

Application Code: Biomedical

Methodology Code: Atomic Spectroscopy/Elemental Analysis

Session # 1410 Abstract # 1410-10 **Poster Sessions**

Session Title: Advances in Biomedical Applications

Abstract Title: **Profiling Volatile Organic Compounds in Exhaled Breath by TD–GC–TOF MS**

Date: Tuesday, March 08, 2016 - Afternoo

Time:

Room: Exposition Floor, 400 Aisle

Primary Author: Laura McGregor
Markes International Ltd

Co-Author(s): Caroline Widdowson, Chris Hall, Ken Umbarger, Nicola Watson

Abstract Text

The accurate identification of biomarkers in human breath is a non-invasive approach that can potentially flag a range of physiological and pathological conditions. Until now, breath analysis in the pptv range has required time-consuming sample preparation to achieve the required sensitivity. Pre-concentration techniques have been improved by using solid adsorbents to entrain exhaled breath, followed by analysis by thermal desorption–gas chromatography–time-of-flight mass spectrometry (TD–GC–TOF MS). To identify biomarkers in breath, it is essential that sensitivity is maintained whilst recording full spectral information. The nature of the TOF MS approach means that, unlike traditional quadrupole technology, there is no loss of ion signal due to analyser mass filtering. This fundamental advantage yields dramatic sensitivity improvements, ensuring no compromise in response for target species. As in other metabolomics matrices, highly diagnostic compounds are rarely of high abundance, and by adopting a comprehensive approach, measurement of the maximum possible number of compounds is achieved in a realistic run time. In order to link exhaled substances to diseases, it is important that trace VOCs are reliably identified and accurately measured. This study uses an innovative TOF design that produces 'classical' spectra, allowing direct comparison with established libraries. Novel data processing software is also demonstrated, for background subtraction, deconvolution and library-searching on-the-fly, to allow near-real-time perception of results. This poster will show how breath sampling, performed with TD tubes at point-of-care, followed by prompt, sensitive TOF MS analysis, could enable the earliest possible diagnosis of a range of serious conditions.

Keywords: Biomedical, Gas Chromatography/Mass Spectrometry, Time of Flight MS, Volatile Organic Compound
Application Code: Biomedical
Methodology Code: Gas Chromatography/Mass Spectrometry

Session # **1410** **Abstract #** **1410-11** **Poster Sessions**

Session Title: Advances in Biomedical Applications

Abstract Title: **Purification of Pharmaceutical Proteins Including Antibody and Peptides Using Ion-Exchange Bulk Media Designed for High-Throughput Purification**

Primary Author: Takashi Sato
YMC Co., Ltd.

Date: Tuesday, March 08, 2016 - Afternoo
Time:
Room: Exposition Floor, 400 Aisle

Co-Author(s): Chiaki Iwata, Munehiro Shoda, Noriko Shoji, Takatomo Takai

Abstract Text

The recent development of bio-pharmaceutical industry has been remarkable, and shortening the development time and reducing the cost become increasingly important. The development of efficient, economical and selective separation method is required for successful commercialization of biopharmaceutical products. Especially on intermediate purification and polishing step, high resolution is required. To meet this requirement, smaller particle such as 10 to 30 [micro]m is widely applied. However, there are some difficulties on commercially available products; for example, low mechanical stability, significant decrease in binding capacity/resolution when increasing flow rate and low salt tolerance. These are hurdles for increasing purification throughput.

We recently developed a novel purification media designed for high-throughput purification media with 10 and 30 [micro]m particles. The new material has a balanced mechanical stability, binding capacity and separation ability. In this poster, major characteristics of this new media will be introduced through some examples which include antibody purification and insulin purification.

Keywords: Biological Samples, Biopharmaceutical, Ion Exchange, Prep Chromatography
Application Code: Biomedical
Methodology Code: Liquid Chromatography

Session # 1410 Abstract # 1410-12 **Poster Sessions**

Session Title: Advances in Biomedical Applications

Abstract Title: **An Analysis of the Protective Effects of Selenium on Porcine Jejunal Epithelial Cells Following Cadmium-induced Oxidative DNA Damage**

Date: Tuesday, March 08, 2016 - Afternoo
Time:
Room: Exposition Floor, 400 Aisle

Primary Author: Sarah J. Lynch
Dublin City University

Co-Author(s): Blánaid White, Dermot Walls, Karina Horgan

Abstract Text

Oxidative DNA damage has been linked with several human and animal pathologies including Alzheimer's disease, Parkinson's disease and cancer. Oxidative DNA damage occurs in response to the production of reactive oxygen species (ROS). ROS may be generated as a consequence of normal metabolic processes or exposure to adverse environmental factors. DNA damage can ensue and DNA repair mechanisms become impaired when the production of ROS exceeds cellular antioxidant levels. The essential trace element selenium (Se), which can be present in various forms including organoselenium in animal feed supplements, has been reported to possess antioxidant properties. In this study, the antioxidant properties and molecular basis of the protective effect of both organic and inorganic selenium was investigated using Porcine Jejunal Epithelial cells. The latter were cultured in medium supplemented with organic or inorganic Se, and subsequently exposed to the heavy metal cadmium (Cd). The extent to which each form of Se lead to a reduction of Cd-induced DNA damage and/or improvement in cell recovery was subsequently analysed. Levels of nuclear DNA damage were determined using single cell gel electrophoresis (Comet assay), while the TUNEL assay was utilised to analyse the extent to which cell apoptosis resulted from Cd exposure. Overall cellular health was determined by analysing membrane integrity (LDH assay) and cell metabolic activity (MTT assay). Good correlation was observed between results obtained using the Comet assay, TUNEL assay and LDH assay. Inorganic Se was shown to enhance the deleterious effect of Cd insult, while organic Se was shown to exhibit a statistically significant protective effect, reducing the extent of DNA damage after Cd exposure.

Keywords: Biopharmaceutical, Electrophoresis, Food Contaminants, UV-VIS Absorbance/Luminescence
Application Code: Bioanalytical
Methodology Code: UV/VIS

Session # 1410 Abstract # 1410-13 **Poster Sessions**

Session Title: Advances in Biomedical Applications

Abstract Title: **Imaging Mass Spectrometry Reveals the Chemistry in Chemically Fixed Adrenal Cells Prepared for Transmission Electron Microscopy Analysis**

Primary Author: Andrew Ewing
Chalmers University of Technology

Date: Tuesday, March 08, 2016 - Afternoo
Time:
Room: Exposition Floor, 400 Aisle

Co-Author(s): Bengt R. Johansson, Jelena Lovric, John S. Fletcher, Per Malmberg

Abstract Text

Here we use imaging mass spectrometry to study the neurochemistry in small environments such as large dense core vesicles (LDCVs). The major goals are to investigate the impact of pharmaceuticals like L-3,4-dihydroxyphenylalanine (L-DOPA) and reserpine onto metabolic pathways of neurotransmitter dopamine and to measure the dopamine molecule distribution between sub-vesicular compartments. Our work is focused on mapping individual LDCVs in PC12 cells with average vesicle diameters of 150 nm. Cells were treated with 13C isotopically labeled L-DOPA, a metabolic precursor for neurotransmitter dopamine in PC12 cells. 13C isotopically enriched cells were chemically fixed, resin embedded and sectioned for transmission electron micrososcopy (TEM) and following nano-scale imaging measurements were performed. Imaging inside vesicles was done with a Cameca NanoSIMS 50L ion microprobe using Cs+ primary ions with approx. 50 nm spatial resolution. High spatial resolution negative secondary ion images of isotopic ratios 14N13C/14N12C and 13C12C/12C2 were acquired from micrometer-sized cell regions of the corresponding TEM sections. The overlays of isotopic ratios images and TEM images were used to localize the transmitters in the cells as well to reveal the possible transmitter vesicular regions. The use of NanoSIMS opens the great possibilities for the study of neurochemistry in single transmitter vesicles and allows us to identify vesicle compartments as well as to relatively quantify transmitters in them. The major goals of this research are to understand new mechanisms of exocytosis and transmitter storage in vesicles as well as to investigate the impact of pharmacology on regulation of neurotransmitter release.

Keywords: Biological Samples, Mass Spectrometry, Microscopy
Application Code: Bioanalytical
Methodology Code: Mass Spectrometry

Session # 1410 Abstract # 1410-14 **Poster Sessions**

Session Title: Advances in Biomedical Applications

Abstract Title: **Measurement of the Correlation Between Type of Mutation and Conditions that Alter Aging in Yeast**

Primary Author: Andreea P. Musteata
Rensselaer Polytechnic Institute

Co-Author(s): Patrick Maxwell

Date: Tuesday, March 08, 2016 - Afternoo
Time:
Room: Exposition Floor, 400 Aisle

Abstract Text

Background and Significance: DNA damage and mutations frequently accumulate in aging cells and organisms. Microorganisms offer useful models for investigating mutation rates and types because they offer the possibility of selecting for mutant phenotypes from large populations. In Saccharomyces cerevisiae (yeast), two common mutation selections are for loss of function of the URA3 gene or loss of function of the CAN1 gene. Loss of function of URA3 or CAN1 results in resistance to 5-fluoroorotic acid or canavanine, respectively. Typically, these selections result in mainly small point mutations.

Methods and Results: Inserting the CAN1 gene at a new site in the yeast genome near the end of chromosome VIII gave rise to a much higher frequency of mutations than would normally be expected for this type of selection assay. PCR analysis identified that most mutation events were the result of recombination between chromosome VIII sequences flanking the site of CAN1 or recombination between chromosome VIII and a portion of chromosome I that is nearly identical in sequence to the relevant part of chromosome VIII.

Conclusions: Saccharomyces cerevisiae has proven a useful model for understanding cell and molecular-level factors associated with aging. A stronger correlation was observed between growth conditions that alter yeast lifespan and changes in the rate of the recombination events detected when CAN1 is present on chromosome VIII than changes in the rate of point mutations. Our results are consistent with changes in the rate of recombination being more relevant to aging than changes in the rate of point mutations. This particular mutation selection assay and analysis is useful for measuring and analyzing various types of complex recombination events instead of mainly point mutations.

Keywords: Biological Samples, Biomedical, Electrophoresis, Medical
Application Code: Biomedical
Methodology Code: Biospectroscopy

Session # 1410 Abstract # 1410-15 **Poster Sessions**

Session Title: Advances in Biomedical Applications

Abstract Title: **Self-Assembly Aptamer�Graphene Oxide Nanosheets as an Anticoagulant**

Date: Tuesday, March 08, 2016 - Afternoo
Time:
Room: Exposition Floor, 400 Aisle

Primary Author: Pei-Xin Lai
National Taiwan Ocean University

Co-Author(s)

Abstract Text

Graphene oxide (GO) possesses many unique physical properties such as high surface area, superior electric conductivity and excellent mechanical strength. Graphene oxide has been discovered that can effectively adsorb single-strand DNA mainly through �-�stacking and hydrogen-bonding interactions interactions. Herein, we demonstrated that thrombin-binding aptamer-conjugated GO (TBA–GO), prepared from a self-assembled hybrid monolayer (SAHM) of triblock aptamers on GO (~230 nm), can effectively inhibit thrombin activity toward fibrinogen. The first block poly(adenine) at the end of the triblock TBA was used for the self-assembly on GO surface. The second block, in the middle of TBA, was composed of oligonucleotides that could hybridize with each other. The third block, containing TBA15 (15-base, binding to the exosite I of thrombin) and TBA29 (29-base, binding to the exosite II of thrombin) provided bivalent interaction with thrombin. The self-assembly aptamers have optimal distances between TBA15 and TBA29, aptamer density, and orientation on the GO surfaces. These properties strengthen the interactions with thrombin, resulting in an extremely high anticoagulant potency. The dose-dependence of thrombin clotting time (TCT) delay caused by TBA–GO nanosheets is 1000 times higher than commercially available drugs (heparin, argatroban, hirudin or warfarin). In addition, the rat-tail bleeding assay time further demonstrated the self-assembly TBA–GO nanosheets were superior (>2-fold) to heparin. Our results suggested the TBA–GO nanosheets possess good potential for the treatment of various diseases related to blood-clotting disorders.

Keywords: Biomedical, Biopharmaceutical, Drugs, Medical
Application Code: Nanotechnology
Methodology Code: Biospectroscopy

Session # 1410 **Abstract # 1410-16** **Poster Sessions**

Session Title: Advances in Biomedical Applications

Abstract Title: **Synthesis of Protein-Capped Gold Nanoparticles with Specific Protein Orientation as a DNA Transfection Vehicle**

Date: Tuesday, March 08, 2016 - Afternoo

Time:

Room: Exposition Floor, 400 Aisle

Primary Author: Ju-Yi Mao
National Taiwan Ocean University

Co-Author(s)

Abstract Text

Proteins are biological materials with multiple specific functions that make them as ideal components to functionalize nanomaterials used in biological systems. However, many proteins will lose their activity while anchoring onto nanomaterials, due to their fragile nature and blocking the activity sites by attaching nanomaterials with wrong orientation. In this work, we tried different strategies to immobilize a heat-resistant DNA binding protein from Sulfolobus solfataricus (Sso7d) to gold nanoparticles (Au NPs). Such Sso7d-capped Au NPs (Sso7d􀀀Au NPs) may be used as a vehicle for therapeutic nucleotide reagents. First, we prepared a recombinant Cys-Sso7d with an extra cysteine (Cys) residue in the N-terminal. Since there is no other Cys residue in Sso7d, the only Cys at the N-terminal may provide strong thiol-gold interaction to anchor Sso7d onto Au NPs. We further developed a one step for preparation of Sso7d􀀀Au NPs through reduction of HAuCl4 with Sso7d, which acts as both a reducing and capping agent. After optimization of reaction buffer, pH, and temperature, stable Sso7d􀀀Au NPs can be prepared in the sodium phosphate buffer (5 mM, pH 12) without addition of reducing reagents at room temperature. Comparing with Sso7d without Cys-residue, Cys-Sso7d displayed better synthetic efficiency and the Cys-Sso7d were stable at 4 oC for at least two months. In this work, we successful synthesized novel protein-capped nanomaterials. By maintaining the orientation of protein molecules on the nanomaterials, we expected they would have better biological activity and biological applications such as a DNA carrier and gene transfection.

Keywords: Adsorption, Amino Acids, Biotechnology, Spectrometer

Application Code: Nanotechnology

Methodology Code: Biospectroscopy

Session # 1410 Abstract # 1410-17 **Poster Sessions**

Session Title: Advances in Biomedical Applications

Abstract Title: **Recovering the Electrocatalytic Activity of Pt Nanoparticle-DNA Collisions via Nuclease Digestion for Sensing Applications**

Primary Author: Alma Castaneda
University of Texas at Austin

Date: Tuesday, March 08, 2016 - Afternoo
Time:
Room: Exposition Floor, 400 Aisle

Co-Author(s): Donald Robinson, Keith Stevenson, Richard M. Crooks

Abstract Text
We report the electrochemical detection of electrocatalytic amplification (ECA) collisions due to hydrazine oxidation on gold (Au UMEs) and mercury/platinum (Hg/Pt UMEs) ultramicroelectrodes from DNA-modified platinum nanoparticles (Pt NPs) by enzymatic digestion of the surface-bound DNA strands. This work demonstrates significant developments towards the application of ECA collisions for the detection of biochemical events. Specifically, we have previously demonstrated that modifying the surface of metallic nanoparticles with thiolated DNA shuts off electrocatalytic activity. Our development, which utilizes the enzymatic activity of Exonuclease I to digest the anchored DNA strands, presents a simple solution to this problem. Second, ECA collision systems implementing enzymes for detection of bio-analytes have not been previously reported. Lastly, the colloidal stability of these digested Pt NPs is remarkably good. We observed an improvement in monodispersity via Nanoparticle Tracking Analysis (NTA) when compared to unmodified, citrate-capped Pt NPs. These results, which demonstrate the ability to recover electrochemistry from passivated Pt NPs, set the stage for the detection of specific DNA and RNA sequences.

Keywords: Electrochemistry, Microelectrode, Nanotechnology, Nucleic Acids
Application Code: Bioanalytical
Methodology Code: Electrochemistry

Session # 1410 Abstract # 1410-18 **Poster Sessions**

Session Title: Advances in Biomedical Applications

Abstract Title: **Study of Stressed Monoclonal Antibody (mAb) Pharmaceuticals by Using Deep-UV Resonance Raman (DUVRR) Spectroscopy**

Date: Tuesday, March 08, 2016 - Afternoo
Time:
Room: Exposition Floor, 400 Aisle

Primary Author: Chen Qiu
US Food and Drug Administration

Co-Author(s): John Kauffman, Sergey Arzhanstev

Abstract Text

Therapeutic proteins and, especially, monoclonal antibodies (mAb) has become the new trend in the pharmaceutical industry because of their target selectivity, least side effects and ability to treat rare diseases. Comparing to small-molecule drugs, these macromolecules have highly complex structure, which is still a challenge to fully characterize and easy perform quality assessments.

Deep ultraviolet resonance Raman (DUVRR) spectroscopy serves as a useful tool of assessing the higher order structure of proteins due to high sensitivity of the amide vibrational bands to the structure. In this work, DUVRR spectroscopy is applied to formulated Avastin (bevacizumab) to study the changes of protein higher order structure when stressed conditions are applied. By using UV excitation, DUVRR spectroscopy is extremely sensitive to the changes in protein secondary structure, and is capable of measurements with sample concentration as low as 1 mg/ml in formulation with minimal sample preparation. The effect of thermal stress, environmental pH, and addition of surfactant on Avastin has been studied by DUVRR. This work illustrates the promise of DUVRR as a powerful tool of probing mAb secondary structure, as well as its potential of adding to current methods in quality control of macro-biomolecules and protein drugs.

Keywords: Biopharmaceutical, Chemometrics, Quality Control, Raman
Application Code: Pharmaceutical
Methodology Code: Vibrational Spectroscopy

Session # 1410 Abstract # 1410-19 **Poster Sessions**

Session Title Advances in Biomedical Applications

Abstract Title **Paper Membrane Based SERS Platform for Rapid Bacteria Enumeration**

Date: Tuesday, March 08, 2016 - Afternoo
Time:
Room: Exposition Floor, 400 Aisle

Primary Author Ugur Tamer
Gazi University

Co-Author(s) Aysen Gumustas, Bozkurt Akif, Demet Cetin, Esra Acar, Ismail H. Boyaci, Merve Eryilmaz, Zekiye Suludere

Abstract Text

The diagnostic methods for the detection of bacteria are time consuming because they include culturing and isolation steps before microscopic enumeration. However, diagnostic, food and pharmaceutical sector need rapid detection methods to test the sampling or products. There are numerous methods for the detection of bacteria, however, early diagnosis can not be achieved with these methods. One of the alternative detection methods can be Surface enhanced Raman scattering (SERS) to detect bacteria using modified nanoparticles and paper as support material. In this work, we proposed a paper-based immunuoassays. Here, antibody modified paper surface are employed for capture probe of bacteria and antibody modified gold nanoparticles are used for Raman labeling of bacteria. We tested different kind of paper such as Whatman 1, nitrocelluloce, Whatman LF1, Nylon, and PVDF. After interacting between the paper capture probe with solutions of E. coli having different initial cell concentrations (1,5 x101-1x108 cfu/mL), SERS measurements are taken. After preparation of the paper membrane based SERS platform, optimization studies were also carried out to find the amount of antibody, the interaction time, and the amount of nanoparticle. Measured intense Raman peak is used for quantitative detection of E. coli by using plotted calibration curve. Total detection time was less than three hours which indicates clear advantage of the proposed method compared with plate counting classical methods.

Keywords: Bioanalytical, Biological Samples, Food Safety, Surface Enhanced Raman

Application Code: Bioanalytical

Methodology Code: Vibrational Spectroscopy

Session # 1410 Abstract # 1410-20 **Poster Sessions**

Session Title: Advances in Biomedical Applications

Abstract Title: **Microfluidics for the Detection of Minimal Residual Disease in Acute Myeloid Leukemia Patients Using Circulating Leukemic Cells Selected from Blood**

Primary Author: James Taylor
University of North Carolina at Chapel Hill

Date: Tuesday, March 08, 2016 - Afternoo
Time:
Room: Exposition Floor, 400 Aisle

Co-Author(s): Joshua Jackson, Malgorzata Witek, Paul Armistead, Steven A. Soper

Abstract Text

The leading cause of death for acute myeloid leukemia (AML) patients is disease relapse, which creates a need to monitor disease progression, characterized as the minimum residual disease (MRD), or the number of leukemic cells present in the blood or bone marrow. Flow cytometry can assess MRD in ~90% of AML patients but its poor sensitivity requires bone marrow biopsies, which are too invasive for frequent sampling. Polymerase chain reaction (PCR) is highly sensitive in detecting leukemia-associated mutations, but requires assay development to detect patient-specific mutations, limiting widespread implementation. We designed a microfluidic assay for detection of circulating leukemic cells (CLCs) from the peripheral blood as a minimally invasive alternative to bone marrow biopsies. Cells were enriched via positive affinity selection targeting antigens commonly expressed on AML cells (i.e., CD34, CD33 and CD117) for ~90% patient coverage. 3 parallel microfluidic chips, one for each antigen, were infused with blood to isolate the CLCs. Following isolation, selected cells were stained with fluorescently-labeled antibodies specific for normal leukocytes and the aberrant markers indicative of AML CLCs. Cells were released, visualized, and CLCs were identified using immunofluorescence. Data from a pilot study showed high concordance of the microfluidic assay with therapeutic treatment and overall outcome. Two patients were found to relapse; in one case, our assay detected MRD several months prior to PCR analysis with MRD proliferating until relapse was confirmed by a blood smear. Our microfluidic assay enables frequent patient testing and the early detection of MRD, which can influence clinical decisions.

Keywords: Bioanalytical, Isolation/Purification, Lab-on-a-Chip/Microfluidics, Medical
Application Code: Bioanalytical
Methodology Code: Microfluidics/Lab-on-a-Chip

Session # 1420 Abstract # 1420-1 **Poster Sessions**

Session Title: Advances in Metabolomics, Proteomics, Lipidomics

Abstract Title: **Universal Derivatization of Metabolites for Improved Sensitivity in Electrospray Ionization Mass Spectrometry**

Primary Author: Tianjiao Huang
Saint Louis University

Date: Tuesday, March 08, 2016 - Afternoo
Time:
Room: Exposition Floor, 400 Aisle

Co-Author(s): James L. Edwards, Maria Toro

Abstract Text

Metabolomics, the study of small molecules involved in cellular processes, offers the potential for investigating disease pathogenesis. Electrospray ionization mass spectrometry (ESI-MS) is widely used in metabolomics due to its high sensitivity and ability to generate qualitative information. Metabolites are a diverse group of compounds with a variety of functionalities including hydroxyl, amine, carboxyl, phosphoryl, and thiol groups. However, the structural diversity of metabolites results in differential signal response with ESI and consequently varying degrees of sensitivity. These limitations may prevent the detection of molecules present in low abundance in biological samples. This work undertakes a derivatization approach to improve electrospray by tagging multiple functional groups to boost metabolite sensitivity. By tagging most functional groups, adduct formation and in-source fragmentation is dramatically diminished. This project is to use two distinct tags to label hydroxyl, amine, carboxyl, phosphoryl, and thiol groups on various metabolites and improve sensitivity in ESI-MS. This work will make the simultaneous detection of a diverse group of metabolites possible, leading to a more complete picture of the metabolic system of interest.

Keywords: Derivatization, Liquid Chromatography/Mass Spectroscopy, Mass Spectrometry, Metabolomics, Met
Application Code: Genomics, Proteomics and Other 'Omics
Methodology Code: Mass Spectrometry

Session # 1420 Abstract # 1420-2 **Poster Sessions**

Session Title: Advances in Metabolomics, Proteomics, Lipidomics

Abstract Title: **Identification of a Biological Fear Cue in Blue Crab Urine Via 1H NMR-Based Metabolomics**

Date: Tuesday, March 08, 2016 - Afternoo
Time:
Room: Exposition Floor, 400 Aisle

Primary Author: Kathryn Martin
Georgia Institute of Technology

Co-Author(s): Facundo M. Fernandez, John F. McDonald, Julia Kubanek, Marc Weissburg, Remington X. Poulin

Abstract Text

For years, chemists have searched for techniques and mathematical approaches to simplify spectral data of complex mixtures and allow for the identification of individual compounds, particularly biologically important unknowns, such as novel antibiotics. The proverbial "needle-in-a-haystack" is all too common and recent advances in spectral deconvolution are still lacking, however modest progress is being made (Bingol and Bruschweiler, 2011). Identification of biologically active compounds becomes increasingly difficult when observed biological activity of a sample is dependent upon multiple components acting together. Bioassay-guided fractionation of complex mixtures using chromatographic separations coupled to bioassays after each step results in decomposition of some compounds. By linking the biological activity of a complex mixture to specific spectral features using partial least square regression (PLS-R), a multivariate statistical approach, we identified individual components of a biologically important fear-inducing cue from urine of the blue crab, Callinectes sapidus. A synthetic cue made of compounds identified by the PLS-R model elicit a statistically similar level of biological activity. We hope to test our synthetic cue in field based assays to address the feasibility of using the cue to artificially reduce predation rates on commercially important organisms in the Southeastern reef system. We believe the ability to identify single and multicomponent cues from complex mixtures using NMR-based analyses will reduce the amount of time, materials, and cost of biologically and commercially important research involving chemically mediated ecological interactions.

Keywords: Metabolomics, Metabonomics, Natural Products, NMR
Application Code: Genomics, Proteomics and Other 'Omics
Methodology Code: Magnetic Resonance

Session # 1420 Abstract # 1420-3 **Poster Sessions**

Session Title: Advances in Metabolomics, Proteomics, Lipidomics

Abstract Title: **Analysis of Urine SRMs Using Solid Phase Micro Extraction, Dynamic Headspace and Liquid Injection with Comprehensive Two-Dimensional Gas Chromatography (GC×GC)-High Resolution Time-of-Flight Mass Spectrometry**

Primary Author: David E. Alonso
Leco Corporation

Date: Tuesday, March 08, 2016 - Afternoo
Time:
Room: Exposition Floor, 400 Aisle

Co-Author(s): Elizabeth M. Humston-Fulmer, Jonathan D. Byer, Joseph E. Binkley, Lorne E. Fell

Abstract Text

In this study, various sample introduction techniques were used for the identification of volatiles and semi-volatiles in urine. The applied methodology included a combination of complementary hard and soft ionization methods coupled to high resolution time of flight mass spectrometry, as well as, effective data processing methods for characterization of samples. Analyses resulted in confident identification of a wide variety of materials (e.g., polyaromatic hydrocarbons) and their metabolites. GCxGC-TOFMS chromatograms were highly structured showing clustered classes of compounds and provided high quality spectral data that were searched against large, well-established databases. High resolution time-of-flight mass spectrometry (HRT) resulted in additional benefits such as robust formula determinations for fragment, molecular and adduct ions, as well as, increased selectivity that reduced background interferences. Comprehensive HRT data was probed multiple times via targeted and/or untargeted processing methods to identify important classes of compounds.

Keywords: Reference Material, Semi-Volatiles, Time of Flight MS, Volatile Organic Compounds
Application Code: Genomics, Proteomics and Other 'Omics
Methodology Code: Mass Spectrometry

Session # 1420 Abstract # 1420-4 **Poster Sessions**

Session Title: Advances in Metabolomics, Proteomics, Lipidomics

Abstract Title: **Feasibility of Early Detection of Acute Pulmonary Exacerbations by Exhaled Breath Condensate Metabolomics**

Primary Author: Xiaoling Zang
Georgia Institute of Technology

Date: Tuesday, March 08, 2016 - Afternoo
Time:
Room: Exposition Floor, 400 Aisle

Co-Author(s): Arlene Stecenko, Facundo M. Fernandez, Maria E. Monge, Nael A. McCarty

Abstract Text

Progressive lung function decline in cystic fibrosis (CF) is punctuated by acute pulmonary exacerbations (APEs). The frequency of APEs is a crucial factor of death in CF patients and the diagnosis remains challenging. The objective of this research is to develop reliable methods to predict oncoming APEs in order to prevent associated lung function loss, mortality and morbidity. In this study, untargeted metabolomics profiling of exhaled breath condensate (EBC) samples from 4 pre-APE (CF patients 1 to 3 months before an APE) and 19 stable CF patients (CF subjects who are clinically stable without an APE for □3 months) was performed using ultra performance liquid chromatography coupled to quadrupole-time-of-flight mass spectrometry. A supervised classification model, orthogonal partial least squares discriminant analysis (OPLS-DA), distinguished pre-APE from stable CF samples with good accuracy, sensitivity and specificity. Tentatively identified discriminant metabolites include lactic acid and hydroxyacetone. Elevated lactic acid in pre-APE compared to stable CF EBC samples was found in this study, in agreement with a previous work reporting an increased level of lactate in bronchoalveolar lavage fluid from CF patients with higher inflammation [1]. Hydroxyacetone is an intermediate in the metabolism of glycine, serine and threonine. There is evidence in literature about higher amino acid content in sputum extracts from CF patients compared to non-CF patients [2]. However, direct relationship between hydroxyacetone and CF or other respiratory diseases remains to be elucidated.

1. Wolak, J. E.; Esther, C. R., Jr.; O'Connell, T. M. Biomarkers 2009, 14 (1), 55-60.
2. Barth, A. L.; Pitt, T. L. J Med Microbiol 1996, 45 (2), 110-9.

Keywords: Biological Samples, Detection, Liquid Chromatography/Mass Spectroscopy
Application Code: Genomics, Proteomics and Other 'Omics
Methodology Code: Liquid Chromatography/Mass Spectrometry

Session # 1420 Abstract # 1420-5 **Poster Sessions**

Session Title: Advances in Metabolomics, Proteomics, Lipidomics

Abstract Title: **Prebiotic Peptidomics: An Ultra-Performance Liquid Chromatography-Ion Mobility-Tandem Mass Spectrometry (UPLC-IM-MS/MS) Workflow Applied to Origins-of-Life Chemistry**

Primary Author: Jay G. Forsythe
Georgia Institute of Technology

Date: Tuesday, March 08, 2016 - Afternoo
Time:
Room: Exposition Floor, 400 Aisle

Co-Author(s): Facundo M. Fernandez, Martha A. Grover, Nicholas V. Hud, Ramanarayanan Krishnamurthy, Sheng-Sheng Yu

Abstract Text

One of the most perplexing questions in origins-of-life research is how amino acids on the early Earth polymerized into functional proteins. Recently, we discovered a prebiotically-plausible reaction which may explain the first stage of this process: amino acid polymerization into peptide chains. This reaction utilizes alpha-hydroxy acids, compounds which differ from amino acids in that the amine group is replaced with a hydroxyl group. These compounds were likely present along with amino acids in many prebiotic settings. Hydroxy and amino acids link together to form depsipeptides, or peptides which contain a mixture of amide and ester linkages in the polymer backbone. In previous work, depsipeptide reaction products were primarily analyzed by mass spectrometry (MS). However, for more realistic (and complex) reaction mixtures, it quickly became clear that a proteomics-inspired higher resolution method was necessary to determine hydroxy/amino acid sequences and to monitor their chemical evolution toward peptides. Here, we introduce an ultra-performance liquid chromatography-ion mobility-tandem mass spectrometry (UPLC-IM-MS/MS) workflow for separating and sequencing depsipeptides. Samples were initially separated using reverse-phase UPLC (Waters BEH C18 column, 2.1 x 150 mm, 1.7 [micro]m particles) and subsequently infused via electrospray ionization (ESI) into a Waters Synapt G2 HDMS system with traveling-wave ion mobility (TWIM) gas-phase separation and quadrupole - time-of-flight (QTOF) mass analysis. Additionally, we describe data analysis strategies for incorporating non-traditional monomers and TWIM separation data into software packages designed for traditional LC-MS/MS proteomics.

Keywords: Method Development, Peptides, Proteomics, Tandem Mass Spec
Application Code: Genomics, Proteomics and Other 'Omics
Methodology Code: Mass Spectrometry

Session # 1420 Abstract # 1420-6 **Poster Sessions**

Session Title: Advances in Metabolomics, Proteomics, Lipidomics

Abstract Title: **Development of a High Throughput Organelle Extraction Procedure from Rat Tissues**

Date: Tuesday, March 08, 2016 - Afternoo
Time:
Room: Exposition Floor, 400 Aisle

Primary Author: Brandon Easparro
Omni International

Co-Author(s): James Atwood, Shari Garrett

Abstract Text

Comprehensive bottom up profiling of tissue proteomes is limited by the large dynamic range of protein expression and the ability of modern LC-MS/MS instrumentation to adequately separate and detect peptides in such a complex mixture. Tissue proteomes are diverse and a common method for segregating the proteome is subcellular fractionation; isolating organelles which contain smaller subsets of proteins that can be more readily detected. Isolating tissues organelles is commonly performed manually using either a dounce or rotor stator homogenizer. However, this process is not amendable to high throughput quantitative proteomic studies as the reproducibility is low and the processing time is significant.

In this study, we evaluate the potential for automated bead mill disruption of tissues for isolation of nuclei from rat tissues and compare nuclei purity, nuclear proteome coverage and protein extraction reproducibility to the extracts isolated by dounce and rotor stator homogenization. Following each extraction, nuclei were isolated by centrifugation. Purity was then assessed by both microscopy and western blotting for known nuclear and cytoplasmic protein markers. Proteins from each extraction method were then analyzed by LC-MS/MS and protein extraction yields were quantified by spectral counting.

Keywords: Isolation/Purification, Laboratory Automation, Liquid Chromatography/Mass Spectroscopy, Method
Application Code: Genomics, Proteomics and Other 'Omics
Methodology Code: Laboratory Informatics

Session # 1420 Abstract # 1420-7 Poster Sessions

Session Title: Advances in Metabolomics, Proteomics, Lipidomics

Abstract Title: **Serum Lipidomics Identifies Biomarkers of Acute Traumatic Brain Injury**

Primary Author: Scott Hogan
Georgia Institute of Technology

Date: Tuesday, March 08, 2016 - Afternoo
Time:
Room: Exposition Floor, 400 Aisle

Co-Author(s): David A. Gaul, Facundo M. Fernandez, Melissa A. Velez, Michelle C. LaPlaca

Abstract Text

Mild traumatic brain injury (mTBI), commonly known as concussion, is currently diagnosed by self-reported symptoms, which are subjective in nature and may not be noticeable until after the acute phase (1-24 hours following injury). The lack of a rigorous mTBI detection method underscores the need for an objective diagnostic tool. Previously, brain specific proteins were investigated as potential biomarkers of mTBI to facilitate recognition of injury. However, the diffusion of these proteins can be prevented by an intact blood-brain barrier (BBB), limiting their potential for utility in the diagnosis of mild injuries. This is in contrast to lipids, which can more readily permeate the BBB. As a result, increased efforts in the fields of biomarker research and serum lipidomics provide promising avenues for the discovery of novel markers of the secondary injury cascade.

In this study, adult male Sprague-Dawley rats were injured using a controlled cortical impact (CCI) device. Subsequently, blood samples were collected at 1 and 24 hours post-injury. Following protein precipitation, samples were analyzed in duplicate by ultra performance liquid chromatography mass spectrometry (UPLC-MS) in positive and negative ion modes. Mass spectral detection was achieved using a Waters Xevo G2-QTOF mass spectrometer. Using a discovery metabolomics workflow, serum samples from injured animals were compared to both naïve and time-matched sham controls. Preliminary results showed evidence of significant changes in lipid concentrations as early as 1 hour post-injury. This methodology is suitable for clinical use, thereby, the detected molecular species could have far-reaching implications as potential TBI biomarkers as it pertains to development of an objective concussion diagnostic tool.

Keywords: Chemometrics, Lipids, Liquid Chromatography/Mass Spectroscopy, Neurochemistry
Application Code: Genomics, Proteomics and Other 'Omics
Methodology Code: Liquid Chromatography/Mass Spectrometry

Session # 1420 Abstract # 1420-8 **Poster Sessions**

Session Title Advances in Metabolomics, Proteomics, Lipidomics

Abstract Title **An Accelerated Protein Sample Preparation Method for LC-MS-Based Proteomics**

Primary Author Sujatha Chilakala
Cleveland State University

Date: Tuesday, March 08, 2016 - Afternoo
Time:
Room: Exposition Floor, 400 Aisle

Co-Author(s) Yan Xu

Abstract Text

Sample preparation for proteomics by LC-MS usually entails two sequential steps: (a) denaturation, reduction and alkylation of proteins; and (b) digestion of proteins to peptides by protease such as trypsin, where the resultant peptides fall in the preferred mass range with basic residue at the carboxyl terminus suitable for LC-MS analysis and produce information-rich mass spectra for protein sequencing, identification and quantitation. Although high-throughput and automated LC-MS methods are available for protein characterization, sample preparation oftentimes is the rate-limiting step for many proteomic workflows. For instance, reduction and alkylation of proteins usually take ca. 2h, and digestion of proteins takes >16 h. To reduce the time consumed in protein sample preparation, new technologies (e.g., infrared-, ultrasound- and microwave-assisted enzyme digestion) have emerged. However, these new approaches focus only on the enzyme digestion step, and still need about 2 h for reduction and alkylation of proteins.

In this study, we demonstrated an accelerated method for protein sample preparation by combined use of ultrasound and microwave technologies. The entire sample preparation took less than 30 min to complete using bovine serum albumin as a model protein and trypsin as a protease. In this work, ultrasonic water bath was used for protein reduction and alkylation steps (5 min each) and followed by microwave-assisted enzyme digestion (15 min). The resultant peptides were separated and analyzed by Agilent 6540 Accurate-Mass Q-TOF LC/MS system, and protein sequencing and identification were accomplished using MASCOT database. Comparison data on protein sample throughput and recovery among the conventional, microwave-assisted and our method are presented. The data show that our method has comparable protein recovery with those of the existing methods but much higher sample throughput.

Keywords: Liquid Chromatography/Mass Spectroscopy, Method Development, Microwave, Proteomics
Application Code: Genomics, Proteomics and Other 'Omics
Methodology Code: Liquid Chromatography/Mass Spectrometry

Session # **1420** Abstract # **1420-9** **Poster Sessions**

Session Title: Advances in Metabolomics, Proteomics, Lipidomics

Abstract Title: **High-Throughput Proteomics Analysis by LC-MS with AJS-CESI Technology**

Date: Tuesday, March 08, 2016 - Afternoo
Time:
Room: Exposition Floor, 400 Aisle

Primary Author: Sujatha Chilakala
Cleveland State University

Co-Author(s): Yan Xu

Abstract Text

Although LC-MS techniques are well developed for proteomics workflow, there will always be needs for high sensitive and reproducible methods when sample amount is limited. Nano LC/MS system may be a remedy; but it requires the use of a nano-pump (e.g., 0.2-0.3 μL/min) and a nano-column (e.g., 75 μm, i.d.). Moreover, the maintenance of a nano LC/MS could be difficult and column overloading may result in skewed peaks with shifted retention times. Therefore, a compromise would be the use of smaller column (e.g., 1 mm, i.d.) with a flow rate of □50 μL/min.

Agilent jet stream technology with electrospray ionization (AJS-ESI) was developed by improving electrospray droplets. Compared to the conventional ESI, AJS-ESI improves MS and MS/MS signal-to-noise ratio of 5 to 10 folds. Capillary electrospray nebulizer (CE-ESI) was developed to handle microliter flow rates. The combined use of these two technologies (AJS-CE-ESI) may provide a high-throughput platform for proteomics analysis. In this pioneering study, we have tested the use of AJS-CESI technology without significant changes of conventional LC-MS system and method to improve the detection limit and sensitivity of small amount of protein sample.

In our experiments, tryptic digested bovine serum albumin was used as model protein, and peptides were separated on a Zobrax C18 LC column (1 x 150 mm, 3.5 μm) at a flow rate of 50 μL/min and analyzed by Agilent 6540 Accurate-Mass Q-TOF LC/MS system. Our data showed that on average a 50-fold improvement in the sensitivity with a LOD at 15 nmol/L by 3.0 μL injection (or 45 femtomoles) could be attained using AJS-CESI technology

Keywords: Capillary Electrophoresis, High Throughput Chemical Analysis, Liquid Chromatography/Mass Spectros
Application Code: High-Throughput Chemical Analysis
Methodology Code: Liquid Chromatography/Mass Spectrometry

Session # 1420 Abstract # 1420-11 **Poster Sessions**

Session Title: Advances in Metabolomics, Proteomics, Lipidomics

Abstract Title: **Expression and Spectroscopic Characterization of Allene Oxide Synthase: A Cytochrome P450 for the Rearrangement of Small Molecule Hydroperoxides**

Primary Author: Julie C. McIntosh
University of North Carolina at Chapel Hill

Co-Author(s): Matthew R. Lockett, Nathan A. Whitman

Date: Tuesday, March 08, 2016 - Afternoo
Time:
Room: Exposition Floor, 400 Aisle

Abstract Text

Cytochrome P450s (CYPs) are a superfamily of structurally conserved heme-thiolate enzymes that selectively oxidize small organic molecules involved in the synthesis of biomolecules such as steroids and lipids or the primary metabolism of xenobiotics. Allene oxide synthase (AOS) catalyzes the first "committed" step in the synthesis of jasmonates as part of the signaling pathway for plant defense responses. AOS is an atypical CYP because it: rearranges hydroperoxides without the need for external oxygen or NADPH reductase and lacks key amino acids in the binding site needed to bind carbon monoxide, a hallmark of most CYPs. In this work, we express and characterize functional AOS from [i]Arabidopsis thaliana[/i] in [i]E. coli[/i] and with a cell-free system that utilizes [i]E. coli[/i] lysate. This particular strain of AOS was selected because there is a documented crystal structure, which provides vital structural information about the binding pocket. The enzyme has not, however, been well characterized in terms of its binding affinity, activity, and redox potential to carry out rearrangements on substrate hydroperoxides. We characterized binding of small molecules to AOS expressed in both systems with absorption and electron paramagnetic resonance spectroscopy. We also developed screening assays to monitor the activity of AOS and structural mutants with key amino acids substituted into the active site using HPLC, CE, and UV/Vis. A platform capable of readily expressing and screening the activity of enzymes with engineered active sites that target organic reactions important in the synthesis of lipids and other aromatic molecules will not only provide insight into structure-function relationships of enzyme catalysis but also lead to optimized structures for the large-scale enzymatic synthesis of industrially interesting starting materials.

Keywords: Bioanalytical, Characterization, Protein, UV-VIS Absorbance/Luminescence
Application Code: Bioanalytical
Methodology Code: UV/VIS

Session # 1430 Abstract # 1430-1 **Poster Sessions**

Session Title: Bioanalytical and Neurochemistry

Abstract Title: **Using Effective Conductivity to Study Brain Tissue Morphology**

Date: Tuesday, March 08, 2016 - Afternoo

Time:

Room: Exposition Floor, 400 Aisle

Primary Author: Jenna DeVivo
University of Pittsburgh

Co-Author(s): Adrian C. Michael, Erika Varner, Stephen Weber, Yangguang Ou

Abstract Text

Neuronal injury causes swelling of the cells and elicits an immune response that results in increased density of glia near the injury site. These events alter the formation factor (fF) of the extracellular space (ECS). The formation factor is a dimensionless number that measures the degree to which a porous medium impedes transport and is a ratio of porosity to tortuosity. Currently, there is no method that can monitor these events in vivo over long periods of time and can do so with high spatial resolution without the method itself causing injury. We have developed an electrochemical technique to determine fF of porous media using micro-conductivity measurements. Excitation of a 7-□m carbon fiber microelectrode in a solution or porous medium with an AC voltage (10 – 100 kHz) gives a frequency-dependent AC current. This frequency-dependent current depends on the electrical resistance of the solution, which itself depends on the fF of the porous medium. We developed a simple circuit model and verified it via nonlinear curve fitting to experimental data from measurements of KCl solutions. We demonstrated the validity of the technique by measuring fF of simulated brain: a bed of 15-□m borosilicate glass particles. We measured an fF of 0.4 ± 0.1, a very reasonable value for particle packing of this dimension. We will apply this technique in vivo to monitor the healing of brain tissue near the probe after its implantation and to determine the effect of anti-inflammatory drugs such as dexamethasone on the rate of healing.

Keywords: Bioanalytical, Electrochemistry, Microelectrode, Neurochemistry
Application Code: Neurochemistry
Methodology Code: Electrochemistry

Session # 1430 Abstract # 1430-2 **Poster Sessions**

Session Title Bioanalytical and Neurochemistry

Abstract Title **Serotonin and Histamine Coregulation in the Mouse Premammilary Nucleus**

Primary Author Rhiannon Robke
University of South Carolina

Date: Tuesday, March 08, 2016 - Afternoo
Time:
Room: Exposition Floor, 400 Aisle

Co-Author(s) Aya Abdalla, H Frederick Nijhout, Janet Best, Michael C. Reed, Parastoo Hashemi, Srimal A. Samaranayake

Abstract Text

Serotonin and histamine's chemistries have long thought to be intertwined. Previous studies have explored this interrelationship by conducting experiments on each neurotransmitter individually, frequently failing to conclude simultaneous results of histamine and serotonin modulation in real-time. There are many neurological diseases and disorders that are speculated to involve both serotonin and histamine. These include, but are not limited to, Alzheimer's, Parkinson's, Schizophrenia, Huntington's, Down's Syndrome, anxiety, addiction, and anorexia. Many drugs have also targeted these two molecules in the potential treatment of these disorders. In this research, fast-scan cyclic voltammetry (FSCV) was utilized to simultaneously measure serotonin and histamine in real-time, for the first time. Using our novel histamine waveform, it has been shown that electrically evoked histamine rapidly and potently inhibits ambient serotonin activity in a concentration dependent manner. These findings, along with FSCV techniques, are the building blocks for deciphering the full extent of the delicate interconnectivity between histamine and serotonin and will ultimately reveal fundamentally important mechanistic information.

Keywords: Electrochemistry, Neurochemistry
Application Code: Neurochemistry
Methodology Code: Electrochemistry

Session # 1430 **Abstract # 1430-3** **Poster Sessions**

Session Title: Bioanalytical and Neurochemistry

Abstract Title: **Development of a Nanoscale Calcium-Selective Electrode**

Date: Tuesday, March 08, 2016 - Afternoo

Time:

Room: Exposition Floor, 400 Aisle

Primary Author: Theresa M. Ruwe
Northern Kentucky University

Co-Author(s): Celeste A. Morris, Edward A. Dobrzykowski, Teri Rae Armstrong

Abstract Text

Calcium (Ca[sup]2+[/sup]) plays an integral role in physiology and healthy cellular function to control neurotransmitter release, heart and muscular contractions, and bone health. The determination of calcium in blood is valuable for diagnostic clues to indicate bone, kidney, and liver diseases, as well as disorders of the parathyroid gland. The normal range of total calcium concentration in blood is 8.5-10.2 mg/dL. Ca[sup]2+[/sup] detection in the blood requires blood draw followed by a laboratory test with a delay time between analysis and results for both medical professionals and patients. The proposed nanoscale sensor could be contained within a phlebotomy needle utilized for blood draw and would record a Ca[sup]2+[/sup] analysis in vivo within several seconds of measurement time (real-time analysis). We present a nanoscale Ca[sup]2+[/sup] sensor that utilizes potentiometric measurement of calcium in phosphate buffer with a response of 25.4 mV/log[Ca[sup]2+[/sup]].

Keywords: Electrode Surfaces, Ion Selective Electrodes, Nanotechnology, Sensors

Application Code: Bioanalytical

Methodology Code: Electrochemistry

Session # **1430** **Abstract #** **1430-4** **Poster Sessions**

Session Title: Bioanalytical and Neurochemistry

Abstract Title: **Optimized Determination of L-Dopa at a Glassy Carbon Electrode Modified with Electrodeposited Films of Caffeic Acid**

Primary Author: Ahmad Rohani Far
The University of Toledo

Date: Tuesday, March 08, 2016 - Afternoo
Time:
Room: Exposition Floor, 400 Aisle

Co-Author(s): Amila M. Devasurendra, Jon R. Kirchhoff, Joshua A. Young

Abstract Text

L-Dopa is used as a drug for the treatment of Parkinson's disease. It is an amino acid which is the precursor for dopamine and is produced biologically from L-tyrosine. In this work, determination of L-dopa at a glassy carbon electrode (GCE) modified with electrodeposited films of caffeic acid was optimized in the presence of common biological interferences. An electroactive thin layer of poly caffeic acid was deposited on the surface of a GCE under potentiostatic conditions in aqueous solution. Analysis of L-dopa was determined by square wave voltammetry (SWV) after the experimental parameters of pH, step, amplitude, and frequency were studied by a full factorial design to identify the significant parameters and their interactions. The results showed that pH, step, and amplitude were significant whereas amplitude did not significantly impact response. Subsequently, a central composite design was performed for the three significant factors in order to further optimize the method. The optimal conditions were pH: 4; step potential: 6 mV; amplitude: 92 mV. Under these optimized conditions, the modified electrode demonstrated high selectivity for L-dopa in the presence of common biological interferences such as ascorbic acid by the observation of two distinctive and well resolved oxidation peaks. These results infer the possibility of electrochemical detection of L-dopa in the presence of common neurotransmitter interferences. The optimized electrode showed simplicity, short analysis time and high sensitivity for analysis of L-dopa.

Keywords: Chemically Modified Electrodes, Chemometrics, Detection, Electrochemistry
Application Code: Bioanalytical
Methodology Code: Electrochemistry

Session # 1430 Abstract # 1430-5 **Poster Sessions**

Session Title: Bioanalytical and Neurochemistry

Abstract Title: **Mass Spectrometry Imaging for Targeted Metabolomics of Medulloblastoma**

Primary Author: Martin R. Paine, Georgia Institute of Technology

Date: Tuesday, March 08, 2016 - Afternoo

Time:

Room: Exposition Floor, 400 Aisle

Co-Author(s): Danning Huang, Facundo M. Fernandez, Jingbo Liu, Tobey MacDonald

Abstract Text

Medulloblastoma (MB) is the most common and malignant brain tumor in children with a high propensity to metastasize throughout the central nervous system (CNS). The malignant nature of this cancer translates to 30-40% of children relapsing with terminal metastatic CNS disease, even after the use of aggressive radiation therapy of the brain and spinal cord. However, those that don't relapse still suffer permanent and debilitating neurocognitive impairment as a result of the treatment. By investigating matched primary and metastatic tumor tissue samples from the mouse sonic hedgehog MB models, SmoA1 and [i]Ptch[sup]+/-[/sup][/i], metabolomic biomarkers may be identified that can be used as a screening tool to predict the risk of malignancy and help direct lower toxicity treatments to patients at less risk of CNS metastasis. Mass spectrometry imaging was employed to investigate sagittal tissue sections of the whole brain and spinal cord from MB mice, allowing simultaneous detection and spatial co-localization of tumor specific metabolites. Thin tissue sections (10 µm) mounted on glass slides were analyzed using desorption electrospray ionization using a Thermo Scientific Exactive Plus Orbitrap mass spectrometer and by matrix-assisted laser desorption/ionization on a Bruker Autoflex III time-of-flight mass spectrometer after applying 1,5-diaminonaphthalene to the surface. The two techniques provide complementary information and will be used to assist in the discovery of metabolites from a concomitant LC-MS metabolomics study of homogenized MB mouse tissues. Together, these two studies will help provide novel insights into the biological pathways leading from primary tumor progression to metastasis throughout the CNS.

Keywords: Imaging, Lipids, Mass Spectrometry, Metabolomics

Application Code: Bioanalytical

Methodology Code: Mass Spectrometry

Session # 1430 Abstract # 1430-6 **Poster Sessions**

Session Title: Bioanalytical and Neurochemistry

Abstract Title: **Ultra-High Performance Liquid Chromatography Mass Spectrometry Metabolic Fingerprinting of a Medulloblastoma Mouse Model**

Date: Tuesday, March 08, 2016 - Afternoo

Time:

Room: Exposition Floor, 400 Aisle

Primary Author: Danning Huang
Georgia Institute of Technology

Co-Author(s): Facundo M. Fernandez, Jingbo Liu, Martin R. Paine, Tobey MacDonald

Abstract Text

Medulloblastoma (MB) is the most common malignant pediatric brain tumor. Currently, the standard treatment is radiation administered to the entire brain and spine (craniospinal irradiation, CSI). Although 60-70% of children with MB will be long-term survivors, the majority of patients will be left with permanent and debilitating neurocognitive impairment as a result of the aggressive CSI treatment. Therefore, better therapeutic strategies and reduction of CSI are needed, where possible. Understanding the biology that driving tumor metastasis may allow us to molecularly predict cancer aggressiveness and therefore develop personalized therapies. In this study, ultra-high performance liquid chromatography mass spectrometry (UHPLC-MS) was applied to homogenized MB tumor specimens from sonic hedgehog MB mouse models, SmoA1 and [i]Ptch[sup]+/-[/sup][/i]. Both the primary tumor and matched metastatic tumor tissues within the central nervous system (CNS) from these mice were examined, providing comprehensive and rapid global metabolic fingerprinting of MB tumor samples. Small amounts of each tissue sample were homogenized, extracted in isopropanol, and centrifuged to remove protein precipitate. The supernatant was separated on an Acquity UHPLC BEH C18 column and analyzed with a Waters Xevo Q-ToF mass spectrometer. In parallel, these experiments were augmented by tissue-based mass spectrometry imaging (MSI) to localize metabolites of interest. Together, this multi-pronged approach will provide new insights into the changes occurring in the CNS tumor metabolic microenvironment during MB progression, identifying biomarkers for tumor classification and enabling the rational development of tailored therapeutic strategies.

Keywords: HPLC Detection, Lipids, Liquid Chromatography/Mass Spectroscopy
Application Code: Bioanalytical
Methodology Code: Mass Spectrometry

Session # 1430 Abstract # 1430-7 **Poster Sessions**

Session Title: Bioanalytical and Neurochemistry

Abstract Title: **Biophysical Evaluation of Surfactant Effects on Nanoparticle Toxicity a Lipid Model of the Blood-Brain Barrier**

Primary Author: Adam L. Hoffmann
Northern Kentucky University

Date: Tuesday, March 08, 2016 - Afternoo
Time:
Room: Exposition Floor, 400 Aisle

Co-Author(s): Andrew Hall, Celeste A. Morris, Darcy Poor, Kristi L. Haik, Marcus Jones, Rolf Fowee

Abstract Text

The blood-brain barrier is a highly selective membrane capable of blocking 98% of drugs used to treat central nervous system conditions. Recently, polymeric nanoparticles coated in various surfactants have shown enhanced drug delivery and efficacy. However, toxicity of these nanoparticle-surfactant complexes on healthy brain tissue remains unclear. This study used Langmuir-Blodgett trough techniques to deposit a dipalmitoylphosphatidylcholine monolayer on a mica substrate. Atomic force microscopy was then used to evaluate lipid monolayer morphology when exposed to poly(butylcyanoacrylate) nanoparticals coated with polysorbate 80, pluronic 188, or cetrimonium bromide surfactants. The images obtained were analyzed to determine percent liquid condensed and liquid expanded phases. It was hypothesized that each of these surfactants would exhibit a dose-dependent decrease in healthy, liquid condensed monolayer morphology. Our results support this hypothesis and are further confirmed by resazurin cytotoxicity bioassays. Taken together, the data presented validate this method of biophysical toxicity evaluation.

Keywords: Biopharmaceutical, Microscopy, Nanotechnology, Neurochemistry
Application Code: Bioanalytical
Methodology Code: Microscopy

Session # 1430 Abstract # 1430-8 **Poster Sessions**

Session Title Bioanalytical and Neurochemistry

Abstract Title **Novel Graphene-Modified Graphite Pencil Electrode for the Trace Quantification of L-Tyrosine in Human Urine**

Primary Author Abdel-Nasser Kawde
King Fahd University of Petroleum and Minerals

Date: Tuesday, March 08, 2016 - Afternoo
Time:
Room: Exposition Floor, 400 Aisle

Co-Author(s)

Abstract Text

A simple and novel method of detecting L-tyrosine in urine was introduced using a graphene-modified graphite pencil electrode (GR-modified GPE). Graphene oxide (GO) was directly reduced using cyclic voltammetry on the surface of the GPE. Synthesized graphene oxide was characterized by FTIR and Raman spectroscopy. The morphology of the electrode surface was characterized by field emission scanning electron microscopy (FE-SEM) and the electrochemical properties were characterized by cyclic voltammetry (CV), electrochemical impedance spectroscopy (EIS), and square wave voltammetry (SWV). The graphene layer on the GPE dramatically enhanced the electroactive surface area and electrochemical oxidation of the L-tyrosine. A satisfactory linear response was obtained in the square wave voltammogram from 0.8 μM to 60 μM, with a regression constant (R2) of 0.9995. The modified electrode yielded a low L-tyrosine limit of detection of 0.07 μM and a high sensitivity of 0.012 AmM-1. The modified electrode surface was free from interfering species. Furthermore, the developed electrode was successfully applied for the determination of the L-tyrosine in human urine. The modified electrode was low in cost and easy to modify. It displayed excellent sensitivity, selectivity, reproducibility, a low limit of detection and a wide linear response range.

Keywords: Biomedical, Chemically Modified Electrodes, Electrochemistry, Voltammetry

Application Code: Bioanalytical

Methodology Code: Electrochemistry

Session # 1430 Abstract # 1430-9 **Poster Sessions**

Session Title: Bioanalytical and Neurochemistry

Abstract Title: **Online Liquid Chromatography - Surface Enhanced Raman Detection for Metabolic Profiling**

Date: Tuesday, March 08, 2016 - Afternoo
Time:
Room: Exposition Floor, 400 Aisle

Primary Author: Anh H. Nguyen
University of Notre Dame

Co-Author(s): Zachary D. Schultz

Abstract Text

Metabolic profiling in biological fluids is key for understanding biological systems and their interconnected biochemical pathways. In particular, measuring levels of metabolites will provide new insights into pathological cellular processes which are very useful in biomedical diagnosis and drug alterations in cells. Our laboratory had developed sheath flow detector, motivated by the idea of hydrodynamic focusing, for surface enhanced Raman (SERS) detection which is compatible with chemical separation techniques. Here we will discuss the development of SERS detection in flow following capillary liquid chromatography (LC) separations for identification of metabolites. The combination of capillary LC-SERS provides several advantages such as taller peaks, better resolution, improved limits of detection, and less sample consumption. Successful utilization of this approach will result in complementary characterization to the more common LC-MS while offering better coverage of the metabolome.

Keywords: Bioanalytical, Capillary LC, Metabolomics, Metabonomics, Surface Enhanced Raman
Application Code: Bioanalytical
Methodology Code: Vibrational Spectroscopy

Session # 1430 Abstract # 1430-10 **Poster Sessions**

Session Title: Bioanalytical and Neurochemistry

Abstract Title: **Combining Microchip Electrophoresis, Mass Spectrometry, and Standard Addition to Identify N-glycan Structures in Serum**

Primary Author: Xiaomei Zhou
Indiana University

Date: Tuesday, March 08, 2016 - Afternoo
Time:
Room: Exposition Floor, 400 Aisle

Co-Author(s): Christa Snyder, Margit I. Campos, Milos V. Novotny, Stephen C. Jacobson

Abstract Text

The relative abundances of N-glycans in serum vary among disease-free individuals and patients with cancer and can potentially be used as biomarkers to determine the type of disease and disease state. We are developing screening methods based on microchip electrophoresis to identify N-glycan structures in human serum through standard addition of known N-glycans from common glycoproteins. Electrophoretic analysis easily resolves N-glycans and their structural isomers and provides complementary information to mass spectrometric (MS) methods. N-Glycans were enzymatically cleaved from glycoproteins in serum and common glycoproteins, treated with methylamine to neutralize the charge on sialylated glycans, and labeled with 8-aminopyrene-1,3,6-trisulfonic acid (APTS) as an electrophoretic and fluorescent tag. The amidation and APTS labeling steps ensure that all glycans have a uniform -3 charge and, thus, allow direct correlation of the electrophoretic mobilities of glycans to their masses. The glycan samples were analyzed on microfluidic devices with 22-cm long separation channels operated at 900 V/cm and coupled with fluorescence detection. N-Glycans from ribonuclease B, immunoglobulin G, lactoferrin, fibrinogen, □1-acid glycoprotein, and haptoglobin were added to serum N-glycan samples in various ratios to determine what glycan structures were present in the serum N-glycan profiles. Through comparison of electrophoretic mobilities with MALDI-MS data and standard addition of N-glycans from common glycoproteins, we have identified over 30 unique N-glycan structures including 40 positional and linkage isomers present in serum.

Keywords: Bioanalytical, Biological Samples, Carbohydrates, Lab-on-a-Chip/Microfluidics
Application Code: Bioanalytical
Methodology Code: Microfluidics/Lab-on-a-Chip

Session # 1430 Abstract # 1430-12 **Poster Sessions**

Session Title: Bioanalytical and Neurochemistry

Abstract Title: **Development of a Wearable Device for Neurochemical Monitoring of Energy Availability in the Injured Brain Using On-line Microdialysis**

Primary Author: Isabelle C. Samper
Imperial College London

Date: Tuesday, March 08, 2016 - Afternoo
Time:
Room: Exposition Floor, 400 Aisle

Co-Author(s): Chu Wang, Martyn G. Boutelle, Sally A. Gowers

Abstract Text

In the recent consensus statement from the 2014 International Microdialysis Forum [1], clinicians have identified the need for continuous real-time chemical monitoring of traumatic brain injury (TBI) patients. The reason for this is that important secondary insults such as Spreading Depolarisation (SD) waves place an extreme dynamic challenge on the energy supply to the tissue. These events not only happen when patients are being monitored in ICU but also earlier, during the initial intensive periods of treatment in A&E; and up to 12 days later at a stage where patients are awake. However, current systems used to monitor patients are only operating in ICU because they require wires / tubing that prevent patient mobility and obstruct routine chemical care. Therefore, there is a need to build a wearable device for neurochemical monitoring of patients.

We recently developed a first generation 3D-printed microfluidic manifold for cyclist monitoring [2]. We are now adapting this device for clinical use and we are integrating the electronics to the system. The device consists of a 3D-printed microfluidic flow-cell that holds two amperometric biosensors for glucose and for lactate. It is designed to make effective connection to a clinical microdialyisis probe whilst facilitating a robust electrical link to the ZigBee wireless potentiostat module. The whole device will then be enclosed in a secondary 3D-printed casing.

An automated microfluidic board will be used to reproduce changes in glucose and lactate in the CSF dialysate of an injured brain. Results using the novel wearable wireless device will be presented.

References:

1. Hutchinson et al. Intensive care Med. DOI: 10.1007/s00134-015-3839-5 (2015)
2. Gowers et al., Analytical Chem. 87, 7763–7770 (2015)

Keywords: Electrochemistry, Integrated Sensor Systems, Lab-on-a-Chip/Microfluidics, Neurochemistry
Application Code: Biomedical
Methodology Code: Integrated Sensor Systems

Session # **1430** **Abstract #** **1430-13** **Poster Sessions**

Session Title: Bioanalytical and Neurochemistry

Abstract Title: **Label-Free Profiling of O-Linked Glycans by HPLC with Charged Aerosol Detection**

Primary Author: Ian N. Acworth, Thermo Fisher Scientific

Date: Tuesday, March 08, 2016 - Afternoo
Time:
Room: Exposition Floor, 400 Aisle

Co-Author(s): David Thomas, Rainer Bauder, William Kopaciewicz

Abstract Text

The goal of this work was to develop a quantitative profiling assay for O-linked glycans released from glycoproteins by reductive beta elimination. Reduction helps prevent peeling side reactions; however, the resulting O-linked glycan alditols cannot be derivatized easily with a fluorescent label. Charged aerosol detection (CAD) does not require a fluorophore or chromophore for sensitive, accurate quantification, and so HPLC-CAD provides a simple, direct approach to separate and quantify native glycans. The method uses a volatile mobile phase fully compatible with mass spectrometry, if further characterization is desired. O-linked glycan pools released from various proteins were analyzed including those from bovine fetuin, bovine submaxillary mucin, hemocyanin, and IgG. Quantitative performance including precision, detection limits and dynamic range is presented. Figures of merit include sensitivity at the low-nanogram on-column level, dynamic range over two orders of magnitude, and peak area precision averaging less than three percent RSD. By responding directly to any non-volatile compound, charged aerosol detection is able to quantify unlabeled O-linked glycans. The uniform response of charged aerosol detection also provides simple, accurate, and precise estimates of relative concentration even in the absence of pure primary standards. The use of mixed mode chromatography and gradient optimization allowed for assessment of the heterogeneity of glycan structures on the complex glycoprotein mucin.

Keywords: Bioanalytical, Biopharmaceutical, HPLC, Protein
Application Code: Bioanalytical
Methodology Code: Liquid Chromatography

Session # 1430 Abstract # 1430-15 **Poster Sessions**

Session Title Bioanalytical and Neurochemistry

Abstract Title **In Vitro Electrochemical Investigation of ATP: Catecholamine Interactions**

Primary Author Zahra Taleat
Chalmers University

Date: Tuesday, March 08, 2016 - Afternoo
Time:
Room: Exposition Floor, 400 Aisle

Co-Author(s) Andrew Ewing, Johan Dunevall, Judith Estevez, Ricardo Borges

Abstract Text

Secretory vesicles are key organelles in the exocytotic release of neurotransmitters. The vesicle matrix is mostly formed by the aggregation of catecholamines with some soluble components such as chromogranins, calcium, ascorbate, protons and ATP. Interactions of catecholamines with these componets are likely to be involved in storage and transport mechanisms. ATP, a moiety present in all kind of secretory vesicles and in all living species, can bind catecholamines and it results in a reduction of osmotic forces crucial to generate very high concentration of catecholamines in vesicles.

In this study, fast-scan cyclic voltammetry (FSCV) and chronoamperometry with disk-shaped carbon fiber microelectrodes have been used to detect the effect of ATP on catecholamine oxidation in solutions similar to that inside a vesicle (pH 5.5). We demonstrated that the redox behavior of catecholamines is influenced in the presence of different ratios of ATP. In order to understand the molecular mechanism of complexation we have examined the electrochemistry of different compounds with similar structures to ATP including adenosine, phosphoric acid, sodium phosphate salts, and catechol. The results show that ATP and adenosine have an important effect on the oxidation of catecholamines, while the changes upon adding phosphoric acid and catechol are much less significant. Thus the complexation between dopamine and ATP appears to be an electrostatic interaction between the side chains of each molecule. This work helps us to understand the storage and release of catecholamines during exocytosis, and gives hits about how to control or regulate these processes.

Keywords: Bioanalytical, Electrochemistry, Electrodes, Voltammetry
Application Code: Bioanalytical
Methodology Code: Electrochemistry

Session # **1430** **Abstract #** **1430-16** **Poster Sessions**

Session Title: Bioanalytical and Neurochemistry

Abstract Title: **Quantifying the Progression of Amyotrophic Lateral Sclerosis**

Primary Author: Aidan P. Wickham
Imperial College London

Date: Tuesday, March 08, 2016 - Afternoo
Time:
Room: Exposition Floor, 400 Aisle

Co-Author(s): Christopher E. Shaw, Emmanuel M. Drakakis, James Bashford, Kerry R. Mills, Martyn G. Boutelle

Abstract Text

Amyotrophic Lateral Sclerosis (ALS) is a degenerative disorder that causes motor neurons to progressively die, resulting in muscle atrophy, loss of motor function, and eventually death. ALS currently lacks a biomarker for measuring disease activity, molecular or electrophysiological, making it difficult to assess the rate and stage of disease progression. Current methods of assessment include functional rating scales, such as the ALS-FRS, which are coarse and inconsistent and thus hamper the development of drugs and therapeutics for treatment.

As the motor neurons become diseased, they struggle to hold their membrane potential causing them to discharge spontaneously. This spontaneous discharge is known as a Fasciculation Potential (FP) and can cause visible muscle twitches in shallow motor units. The waveform of an FP can be captured using electromyography (EMG). Evidence has been found to show that variability in the FP waveform increases in patients at later stages of the disease [1]. Quantitative assessment of the FP waveform may be of use as a surrogate biomarker to determine disease progression and therapeutic response to drug trials.

Developing a quantifiable measure of ALS progression will require large amounts of data. Due to the stochastic nature of a fasciculation, recording sessions may have to be several hours long to ensure enough FPs have been captured for quantification. An algorithm to automatically process EMG data, detect and group FPs originating from the same motor unit has been developed to quantify the variability of FPs. Preliminary data has been captured from ALS patients using a commercial wireless system, and processed with the described algorithm. A multiple channel, wireless device is in development in order to capture both temporal and spatial EMG data using high-density surface EMG grid arrays.

References:
1. De Carvalho M, Swash M. J Neurol Neurosurg Psychiatry 2013; 84:963–968

Keywords: Data Analysis, Electrodes, Sensors
Application Code: Other
Methodology Code: Data Analysis and Manipulation

Session # 1440 **Abstract # 1440-2** **Poster Sessions**

Session Title: Bioanalytical: Miscellaneous Analytical Techniques

Abstract Title: **Fabrication of Passive Microfluidic Diodes with Tunable Breakthrough Junctions**

Primary Author: Mark D. Holtan
Auburn University

Date: Tuesday, March 08, 2016 - Afternoo
Time:
Room: Exposition Floor, 400 Aisle

Co-Author(s): Christopher J. Easley

Abstract Text

Many microfluidic applications (e.g. reagent mixing, cell sampling, directional pumping) rely on anisotropic fluid flow for proper operation, for which an ideal remedy is a passive fluidic diode. Although, microfluidic diodes have been demonstrated by several groups, reliable methods for consistent diode fabrication are elusive. Key hurdles are the lack of consistent approaches for interlayer via formation and the problem of breakthrough pressure reproducibility. As others have recently demonstrated, CO[sub]2[/sub] laser drilling of vias is automatable and consistent, making it an attractive solution for diode fabrication. However, breakthrough pressure reproducibility remains an unmet need. Using multilayer soft lithography, we have fabricated functional diode structures with tunable breakthrough junctions. Borrowing concepts developed by Devaraju and Unger (Lab Chip, 2012, 12, 4809-4815.), tunable breakthrough junctions were made with post-assembly photocuring of pressurized channels in layers adjacent to flow layers. Sacrificial control channels were selectively pressurized with a photo-curable acrylate monomer (ditrimethylolpropane tetraacrylate, 10% catalyst KT-046), where photocuring effectively stored the applied pressure and permitted normally-closed channel architectures. Vias were fabricated by CO[sub]2[/sub] laser drilling using an in-house built apparatus, which is capable of holes 170 [micro]m in diameter. These diodes operate in the forward direction at <2 psi, and reverse breakdown requires at least 50 psi. Furthermore, we show that fluorescence microscopy at the junction is highly informative for dynamic characterization and pressure dependence of the components. Our unique diode architecture is amenable to fabrication of both push-up and push-down junctions, which will be the subject of future work.

Keywords: Bioanalytical, Biosensors, Lab-on-a-Chip/Microfluidics, Laser

Application Code: Bioanalytical

Methodology Code: Microfluidics/Lab-on-a-Chip

Session # 1440 Abstract # 1440-3 **Poster Sessions**

Session Title Bioanalytical: Miscellaneous Analytical Techniques

Abstract Title **Microfluidic Thermofluorimetric Analysis (µTFA) for Protein Quantification in Nanoliter and Picoliter Volumes**

Primary Author Juan Hu
Auburn University

Date: Tuesday, March 08, 2016 - Afternoo
Time:
Room: Exposition Floor, 400 Aisle

Co-Author(s) Christopher J. Easley, Joonyul Kim, Mark D. Holtan

Abstract Text

Compared to heterogeneous assays such as ELISA, homogeneous protein assays possess simple workflow and low cost, yet they are not able to differentiate signal from background directly, resulting in interferences and low signal-to-noise ratios (S/N). We recently showed that thermofluorimetric analysis (TFA) could be applied to protein quantification using two target binding mechanisms: single aptamer-protein binding (TFA) and bivalent aptamer or antibody-oligo probe assembly (TFAB). Through thermal scanning during fluorescent readout, both TFA and TFAB permit discrimination of signal from background analytically based on the distinct thermal stabilities of signal and background components (dF/dT analysis). Both methods are homogenous proteins assays with simple workflows on a standard qPCR instrument, effectively translating protein input quantities into DNA melting curves for analysis. Background autofluorescence could be negated in minimally diluted biological matrices such as human serum and plasma, greatly increasing S/N. Here, we exploit the homogeneous nature of TFA and TFAB by miniaturizing the assays to a microfluidic channel format. We used a microfluidic device with 7 parallel channels designed around the field of view of a 20x objective on a fluorescence microscope, and an in-house built Peltier control module was used for feedback temperature control using LabVIEW. This µTFA system was shown function with thrombin TFAB, where merely 1 amol of protein was easily detected in dF/dT signals within a microchannel of 100 pL volume. This result represents a miniaturization of the assay by 5 orders of magnitude, showing great promise for future applications in ultrasmall-volume, homogeneous protein quantification.

Keywords: Bioanalytical, Fluorescence, Lab-on-a-Chip/Microfluidics, Microscopy
Application Code: Bioanalytical
Methodology Code: Microfluidics/Lab-on-a-Chip

Session # 1440 Abstract # 1440-4 **Poster Sessions**

Session Title: Bioanalytical: Miscellaneous Analytical Techniques

Abstract Title: **Analysis of Fluorinated Lidocaine Derivatives Under Varying Conditions and Its Application in the Body**

Primary Author: Cyann Cicconi
Seton Hill University

Date: Tuesday, March 08, 2016 - Afternoo

Time:

Room: Exposition Floor, 400 Aisle

Co-Author(s)

Abstract Text

Recently, fluorine has been introduced in to various commercial drugs such as anti-anxiety, steroids, and anesthetics. The addition of fluorine has been useful because it increases the rate in which the drug metabolizes in the body. To better understand reactions of drugs containing fluorine bonds, the reaction kinetics and were observed using fluorinated lidocaine derivatives. The derivatives included fluorine in the ortho, para, and meta positions. All derivatives were used in various experiments testing the effects of temperature, concentration, acid concentration, and type of acid on the rate of the reaction. Each experiment was monitored using both fluorine and proton nuclear magnetic resonance spectroscopy. The data from the spectra proved that the reaction order was one. The way the fluorinated lidocaine was created and the mechanism in which the reaction proceeds will also be highlighted through the presentation.

Keywords: Analysis, Biomedical, Chemical, NMR
Application Code: Bioanalytical
Methodology Code: Magnetic Resonance

Session # 1440 Abstract # 1440-5 **Poster Sessions**

Session Title Bioanalytical: Miscellaneous Analytical Techniques

Abstract Title **Non-Enzymatic Modification of Human Serum Albumin: A Study Focusing On Advanced Glycation End Products by D-Galactose and D-Glucose**

Date: Tuesday, March 08, 2016 - Afternoo
Time:
Room: Exposition Floor, 400 Aisle

Primary Author Menashi A. Cohenford
Marshall University

Co-Author(s)

Abstract Text

The Advanced Glycation Endproducts (AGEs) of D-glucose with proteins have been extensively studied for two major reasons. First, because these molecules can readily form in vivo and, in conditions such as diabetes, contribute to many of the chronic complications of the disease. Second, because AGEs of D-glucose can form in vitro; a situation facilitating the characterization of these complex structures. Studies in our lab have revealed that akin to D-glucose, D-galactose can also interact with proteins and in Classical Galactosemia contribute to the formation of glycated proteins. Also, that the Amadori products of D-galactose prompted the formation of protein AGEs inducing changes in their structure and function. The aim of this study was to employ fluorescence spectroscopy, UV and visible spectrometry to: 1) evaluate under physiological conditions of temperature and pH, the levels of Amadori products formed between human serum albumin (HSA) and D-galactose; i.e., relative to when D-glucose was employed as the glycating agent, 2) determine the total amount of AGEs formed with D-galactose compared to D-glucose, 3) relate the vulnerability of albumin's arginine side chains to nonenzymatic attack by D-galactose, and 4) use circular dichroism to determine if HSA's glycation influenced its interaction with bilirubin; i.e., in view of the high levels of free bilirubin in infants with Classical Galactosemia. Our results revealed three important things. First, that the formation of Amadori products was more pronounced with D-galactose than D-glucose. Second, that the formation of AGEs depended on glycose concentration and the time albumin was exposed to each sugar. Finally, that the non-enzymatic binding of D-galactose with albumin primarily occurred via HSA's arginine residues. The galactation of proteins and their subsequent AGE formation warrants further evaluation, as this phenomenon may account for some of the pathophysiology of Classical Galactosemia.

Keywords: Fluorescence, UV-VIS Absorbance/Luminescence
Application Code: Bioanalytical
Methodology Code: Fluorescence/Luminescence

Session # 1440 Abstract # 1440-6 **Poster Sessions**

Session Title: Bioanalytical: Miscellaneous Analytical Techniques

Abstract Title: **Study Interactions Between 21 Proteins and Nanoparticles**

Primary Author: Yaokai Duan
University of California Riverside

Date: Tuesday, March 08, 2016 - Afternoo
Time:
Room: Exposition Floor, 400 Aisle

Co-Author(s): Wenwan Zhong

Abstract Text

The interaction between proteins and nanoparticles are very important when nanoparticles is applied in biological environment. Proteins can undergo structural change and show different binding affinity upon interacting with specific nanoparticles. Here we choose 21 proteins and two kinds of nanoparticles including silica and polystyrene nanoparticles, and can rapidly screen the interaction between those proteins and nanoparticles. The tool used here is fluorescamine, which is one fluorogenic dye and can target the surface primary amines on proteins. After protein binds to nanoparticles, its conformation will change in different degree, and the number of surface primary amine will change accordingly. Thus different fluorescence single will be observed for labeled protein before or after incubating with nanoparticles. Our results show polystyrene nanoparticles can induce fluorescence increase for most of proteins which is caused by unfolding, while silica nanoparticles can induce fluorescence decrease for some proteins because of blockage. Moreover, smaller nanoparticles will induce more fluorescence increase, but bigger ones decrease the fluorescence. Besides, the fluorescence profile could also be helpful for the study of which protein property can play a crucial role in the binding to nanoparticles. Our results indicate the size of protein will influence the binding. Specifically, bigger proteins will show more fluorescence increase for polystyrene nanoparticles than smaller ones. Overall, our study provides some preliminary information for protein-nanoparticle interactions.

Keywords: Bioanalytical, Fluorescence, Nanotechnology
Application Code: Bioanalytical
Methodology Code: Fluorescence/Luminescence

Session # 1440 Abstract # 1440-7 **Poster Sessions**

Session Title: Bioanalytical: Miscellaneous Analytical Techniques

Abstract Title: **Development of Conductive Polymeric Ionic Liquid-Based Electrodes**

Primary Author: Amila M. Devasurendra
The University of Toledo

Date: Tuesday, March 08, 2016 - Afternoo
Time:
Room: Exposition Floor, 400 Aisle

Co-Author(s): Ahmad Rohani Far, Jon R. Kirchhoff, Joshua A. Young, LM Viranga Tillekeratne

Abstract Text

An ionic liquid (IL) is a salt which possesses a melting point less than or equal to 100 [degree]C. Most commonly, it is a combination of an organic cation and an organic or inorganic anion. Therefore, physical properties of ILs such as thermal stability, vapor pressure, solubility and viscosity can be tuned by regulating its cation/anion composition. Further, the nature of different chemical functionalities of ILs can be modulated to impart unique interactions for selective detection in complex matrices. Due to the above mentioned characteristics, applications of ILs have thus become an active research field among the analytical research community.

In this research, ILs will be covalently bonded to electropolymerizable organic heterocycles such as pyrrole and thiophene to create novel electrically conductive and polymerizable ILs. The newly synthesized monomers will be deposited onto the surface of electrode substrates by electropolymerization, creating conductive polymeric ionic liquid chemically modified electrodes (CPIL-CMEs), which possess a variety of unique properties including ion exclusion. For their characterization, more specific electrochemical and other independent methodologies will be employed. The CPIL-CMEs will be then exploited for use as novel immobilization matrices for creating selective and sensitive sensors aimed at the detection and quantitation of important biomolecules.

Keywords: Chemically Modified Electrodes, Detection, Electrochemistry, Sensors
Application Code: Bioanalytical
Methodology Code: Electrochemistry

Session # 1440 Abstract # 1440-8 **Poster Sessions**

Session Title: Bioanalytical: Miscellaneous Analytical Techniques

Abstract Title: **Plasmonic Applications in the Mid-IR: Spectroscopic Surface Plasmon Resonance (SPR) Detection of N[sub]2[/sub]O Gas and Hexadecanethiol Self-Assembled Monolayer on a Low Loss, Plasmonic Tunable Novel Material Dy:CdO**

Date: Tuesday, March 08, 2016 - Afternoo

Time:

Room: Exposition Floor, 400 Aisle

Primary Author: Hniang Khamh
North Carolina State University

Co-Author(s)

Abstract Text

Since the realization that plasmonics can be detected in conducting metal oxides (CMOs), heavy investment was put forth to discover ideal CMO materials that can support plasmonic in mid-IR energies. The requirement for a good surface plasmon polaritons (SPP) material is for the mobility to be sufficiently high to produce a narrow SPP. By doping CMO, the carrier concentration increases which subsequently increases mobility and the plasma frequency changes. This shifts the resulting SPP to higher energies. A material that can support a low loss SPP in the mid-IR was recently developed. This novel material, cadmium oxide doped with dysprosium (CdO:Dy), possesses extreme peak of mobility along with high carrier density. This gateway material for mid-infrared is also tunable to the entire IR region by varying the carrier concentration of CdO substrate.

This new development in mid-IR plasmonic materials prompted further research regarding potential applications. Utilizing Dy-doped CdO as a substrate, molecular binding to the surface can be detected using surface plasmon resonance (SPR) technique. SPR is known to be very sensitive to its environment and monitors minute changes in the refractive index of chemically functionalized surfaces. The primary goal of this work is to explore the use of Dy:CdO as a possible biosensor in conjunction with SPR technique.

SPR detection of gas molecules and self-assembled monolayer (SAM), on CdO:Dy substrate with high carrier densities above 10[sup]20[/sup] cm[sup]-3[/sup], mobility values 500 cm[sup]2[/sup]V[sup]-1[/sup]s[sup]-2[/sup] at mid-IR are presented. Data procured through simulations and experimentation of plasmonic coupling between the substrate (Dy:CdO) and N[sub]2[/sub]O gas at 20, 40, and 60 psi suggest the pressure dependence of Plasmonic splitting; the Lorentzian peak splitting increases proportional to pressure. An angle shift of 0.4 degrees was observed with SPR detection of a hexadecanethiol (HDT) SAM on the substrate Dy:CdO. The spectroscopic data shows IR stretches of HDT that are evident when the signal is plotted as r[sub]p[/sub]/r[sub]s[/sub]. The surface plasmon modes were characterized using spectroscopic IR-ellipsometry and a sample holder specifically designed for gas molecule detection. Current research effort involves adding a capping layer (SiO[sub]2[/sub]) to protect water soluble CdO, the attachment of SAMs to the resulting surface, and exploring potential applications in the biosensor field.

Keywords: Material Science, Semiconductor, Spectroscopy, Surface Analysis
Application Code: Bioanalytical
Methodology Code: Surface Analysis/Imaging

Session # 1440 Abstract # 1440-9 **Poster Sessions**

Session Title Bioanalytical: Miscellaneous Analytical Techniques

Abstract Title **Orientation of Membrane-Bound Cytochrome P450's in Nanodiscs**

Date: Tuesday, March 08, 2016 - Afternoo
Time:
Room: Exposition Floor, 400 Aisle

Primary Author Ivan Lenov
University of Illinois

Co-Author(s) Stephen Sligar

Abstract Text

Cytochrome P450 (CYP) is a family of heme enzymes that are present in a range of organisms, involved in drug and hormone metabolism. The most abundant isoform in humans, cytochrome P450 3A4 (CYP3A4), is responsible for the metabolism of over 50% of the currently available drugs, while others, such as cytochrome P450 17A1 (CYP17) and aromatase cytochrome P450 (CYP19) are responsible for steroidogenesis. However, despite published crystal structures, the mechanism of active site entry is not clearly understood, as the crystal structures are of CYPs outside of the membrane. In order to better understand this mechanism, and to determine membrane orientation, CYP3A4, CYP17, and CYP19 were incorporated into Nanodiscs of various lipid composition – a membrane system that is water-soluble and preserves the structure of membrane protein. A series of linear dichroism measurements were then carried out by first adsorbing the protein-Nanodisc complex onto a glass substrate then measuring absorbance values of the heme at orthogonal polarizations of light. Angles of the heme enzymes were calculated for the different lipid compositions, illustrating that active site access channels are buried within the membrane.

Keywords: Bioanalytical, Biopharmaceutical, Nanotechnology, Protein
Application Code: Bioanalytical
Methodology Code: Physical Measurements

Session # 1440 Abstract # 1440-10 **Poster Sessions**

Session Title: Bioanalytical: Miscellaneous Analytical Techniques

Abstract Title: **A Comparison of Binding Constants for FITC-Labeled Single Stranded DNA with Anti-FITC Antibody**

Date: Tuesday, March 08, 2016 - Afternoo
Time:
Room: Exposition Floor, 400 Aisle

Primary Author: Qian Liu
Wake Forest University

Co-Author(s): Jason Gagliano, Kathryn Riley, Keith Bonin

Abstract Text

A capillary electrophoresis-based method is developed to allow for the determination of the binding constant for FITC with its corresponding anti-FITC antibody. This ligand-target couple serves as a model system by which to compare differences in binding under free-solution conditions relative to systems with immobilized ligand or target. In these studies, the fluorescent probe FITC (□ex = 495 nm, □em = 519 nm) is covalently bound to a single-stranded DNA construct (26 nucleotides plus primers). Such a construct is representative of a member of a random DNA library for aptamer selection studies, and so it is anticipated that this binding study can be translated more broadly to the evaluation of aptamer selectivity (according to a method developed in our lab that combines CE-based selection and Next Generation Sequencing, with ssDNA aptamer immobilized on Ion Sphere ParticlesTM). Specifically, we describe a capillary transient isotachophoresis (ctITP) method with laser-induced fluorescence (LIF) detection, which leads to a sharp signal for FITC-DNA alone, which is diminished after binding with the anti-FITC antibody. To overcome the quenching of the FITC emission upon antibody binding, the commercially available, noncovalent DNA tag "SYBR Gold" was added to the buffers. Based on the areas of the complex and free FITC-DNA peaks, we can calculate the binding constant of FITC-labeled DNA and anti-FITC antibody according to the theory of non-equilibrium capillary electrophoresis of equilibrium mixtures (NECEEM).

Keywords: Separation Sciences
Application Code: Bioanalytical
Methodology Code: Capillary Electrophoresis

Session # 1440 Abstract # 1440-11

Poster Sessions

Session Title: Bioanalytical: Miscellaneous Analytical Techniques

Abstract Title: **A Micro Pathogenic Microorganism Detector Applied for Mobile Phone Devices**

Primary Author: Liang Sijia
Zhejiang University

Date: Tuesday, March 08, 2016 - Afternoo
Time:
Room: Exposition Floor, 400 Aisle

Co-Author(s): Yu Dongdong, Zhou Jianguang

Abstract Text

With the improvement of people's living condition, demands for hygienic food are increasing. Foodborne pathogen is the main factor leading to food safety incidents. However the detection of foodborne pathogen is usually time-consuming, rapid detection of foodborne pathogen has attracted more and more attention. In this study, we developed a rapid pathogenic microorganism detector which is suitable for the mobile phone devices. Up conversion luminescent materials can emit visible light when excited by infrared light, and always be used in immune detection. This detector contents a piece of up-converting phosphor technology based lateral flow (UPT-LF) test paper, a 980nm micro laser source, a piece of optical filter and a wireless communication module. With its own mechanical structure , this detector is able to connect with mobile phone through the micro-USB interface which also supplies power for it. The UPT-LF test paper use up conversion luminescent materials as a marker, foodborne pathogenic microorganisms reacted with up conversion luminescent materials in infrared light then emitted visible phosphorescence. We took a picture of that UPT-LF test paper with the mobile phone camera after the reaction. Using machine vision image algorithm inside the mobile phone's APP to analysis phosphorescent information in the photo, we can get the final test results. This method has the advantages of simple, fast, portable, low-cost, and so on. It will greatly reduce the detection time and cost, and promote the healthy development of China food industry.

Supported by National Key Technology Support Program (Grants 2012BAB19B07)

Keywords: Detection, Imaging, Portable Instruments, Sensors
Application Code: Bioanalytical
Methodology Code: Portable Instruments

Session # 1440 **Abstract # 1440-12** **Poster Sessions**

Session Title: Bioanalytical: Miscellaneous Analytical Techniques

Abstract Title: **The Preservation of DNA Using Magnetic Ionic Liquids**

Primary Author: Matthew Sorensen
Gustavus Adolphus College

Date: Tuesday, March 08, 2016 - Afternoo
Time:
Room: Exposition Floor, 400 Aisle

Co-Author(s): Jared L. Anderson, Kevin D. Clark, Omprakash Nacham

Abstract Text

Deoxyribonucleic acid (DNA) is an extensively studied biomolecule that contains unique genetic information for every living organism. Due to its importance in research across the scientific community, DNA samples must be handled with care and stored under conditions that minimize degradation or mutation. Although low temperature and low moisture conditions are commonly utilized for the long-term storage of DNA, contamination of samples with endonucleases poses a threat to the stability of the nucleic acid. In this study, magnetic ionic liquids (MILs) were employed as novel solvents for the storage and preservation of DNA. Two hydrophobic MILs, namely trihexyl(tetradecyl)phosphonium tetrachloroferrate(III) ([P66614+][FeCl4-]) and benzyltrioctylammonium bromotrichloroferrate(III) ([(C8)3BnN+][FeCl3Br-]), were used as solvents for DNA storage. Both MILs successfully preserved genomic and plasmid DNA (pDNA) for up to 1 week at room temperature, as determined by agarose gel electrophoresis and polymerase chain reaction (PCR). Furthermore, the [P66614+][FeCl4-] and the [(C8)3BnN+][FeCl3Br-] MILs were investigated for their ability to protect DNA from enzymatic cleavage by deoxyribonuclease I (DNase I). No degradation of genomic DNA or pDNA was observed for up to 1 week upon spiking DNA-enriched MIL with 20 units of DNase 1, demonstrating a convenient and effective method for storing DNA in the presence of an endonuclease. The MIL preservation medium eliminates the need for refrigeration of DNA samples, providing an economical advantage over conventional storage methods In the future, MILs may also be used as DNA extraction solvents to diminish the risk of sample contamination during sample transfer.

This research was supported by the National Science Foundation (CHE-1413199).

Keywords: Biological Samples, Nucleic Acids, Sample Preparation

Application Code: Bioanalytical

Methodology Code: Sampling and Sample Preparation

Session # 1440 **Abstract # 1440-14** **Poster Sessions**

Session Title: Bioanalytical: Miscellaneous Analytical Techniques

Abstract Title: **Experiments and Modeling of Sequential Enzyme Activity in Biphasic Reaction Media**

Primary Author: Bradley W. Davis
Waynesburg University

Date: Tuesday, March 08, 2016 - Afternoo
Time:
Room: Exposition Floor, 400 Aisle

Co-Author(s): Antonios Armaou, Christine D. Keating, Negar Hashemian, William M. Aumiller

Abstract Text

The intracellular compartmentalization of biomolecular activity is a distinguishing feature of cellular function. This colocalization of biomolecules provides an avenue for controlling the kinetics of metabolic pathways. Enzymes can be compartmentalized on cellular scaffolds, in multiprotein scaffolds, or inside subcellular organelles. Phase separation induced by macromolecular crowding may also serve as a means of intracellular organization. Experiments and computational modeling were used to investigate the kinetic consequences related to the compartmentalization of sequential enzymes within a macromolecularly crowded, biphasic polymer solution. Two enzymes of the de novo purine biosynthesis pathway, adenylosuccinate lyase (ASL, step 8) and 5-aminoimidazole-4-carboxamide ribonucleotide transformylase/inosine monophosphate cyclohydrolase (ATIC, steps 9 and 10) were studied in a polyethylene glycol/dextran aqueous two-phase system. Biomolecules partition to the dextran-rich phase droplets, which serve as a model of subcellular compartmentalization. In this colocalized system we did not observe significant rate enhancements under experimental conditions. Experimental results were used to adapt a computational model to quantitatively describe the sequential reaction kinetics in the biphasic environment. The mathematical model was then used to systematically explore various, experimentally inaccessible conditions to determine when increased local concentrations of enzymes and substrates can (or cannot) be expected to increase metabolic rates. Results indicate that compartmentalization within this biphasic system can lead to enhanced sequential product formation under some conditions, but very strong partitioning into the phase compartments is required. In vivo, this could be induced by specific binding affinities between proteins that are within or compose the phase compartments.

Keywords: Bioanalytical, Biotechnology, Enzyme Assays, HPLC
Application Code: Bioanalytical
Methodology Code: Liquid Chromatography

Session # 1440   Abstract # 1440-15   **Poster Sessions**

Session Title   Bioanalytical: Miscellaneous Analytical Techniques

Abstract Title   **Fast Multipoint Immobilized MOF Bioreactor**

Date: Tuesday, March 08, 2016 - Afternoo
Time:
Room: Exposition Floor, 400 Aisle

Primary Author   Hsi-Ya Huang
Chung Yuan Christian University

Co-Author(s)   Chia-Her Lin, Stephen Lirio, Wan-Ling Liu

Abstract Text

In this study, a trypsin bioreactor with exemplary proteolytic performance, even after 5th reuse, was produced through a novel, rapid and simple multipoint immobilization technique without any chemical of solid support. Herein, a 4-chloro-7-nitrobenzofurazan (NBD), as fluorescent dye, was used to bind and stably anchor trypsin molecules onto the MOFs (trypsin-NBD@MOFs). We applied this strategy to several common MOFs with characteristic pore sizes between 0.5-2.1nm. The digestion efficiency of bioreactor was evaluated by bovine serum albumin digestion. The results showed that the stability of the resultant bioreactor can be affected based on the MOF's pore sizes. Among all the trypsin-NBD@MOFs, the trypsin-NBD@UiO-66 exhibited the highest BSA proteolytic performance and can be reused for at least 5 cycles with storage stability until 30 days.Metal-organic frameworks (MOFs), which are porous materials constructed from metal ions and organic ligands, have extended their applications to serve as bioreactors or biocatalysts for specific reactions.

?

Enzyme immobilization has been useful due its advantages such as stability and cost-effectiveness that overcome the time-consuming digestion and production of side-reaction products. In addition, the immobilized enzymes are regarded to be reusable with almost the same catalytic or digestion performance. Enzyme immobilization could be carried out through chemical bonding or physical adsorption. However, the former requires more time than the latter, which could be a contribution to green chemistry and thus, possesses economical advantages.

?

Therefore, in this study, physical adsorption of trypsin was established by utilizing a fluorescent dye, 4-chloro-7-nitrobenzofurazan (NBD), to bind with and stably anchor trypsin molecules on the MOFs. Different MOFs (MIL-101(Cr), MIL-100(Cr), UiO-66(Zr) and CYCU-4(Al)) were also used as immobilizer for trypsin.

Keywords: Enzyme Assays, Immobilization, Materials Characterization, Proteomics
Application Code: Bioanalytical
Methodology Code: Chemical Methods

Session # 1440 Abstract # 1440-19 **Poster Sessions**

Session Title: Bioanalytical: Miscellaneous Analytical Techniques

Abstract Title: **Same Nucleotide Composition and High Sequence Similarity of MicroRNAs - An Analytical Challenge to RNA Studies?**

Primary Author: Joseph N. Mwangi
University of North Carolina at Greensboro

Co-Author(s): Norman Chiu

Date: Tuesday, March 08, 2016 - Afternoo
Time:
Room: Exposition Floor, 400 Aisle

Abstract Text

Introduction

There are >2500 reported human microRNAs (miRNA) in the freely accessible miRBase repository. With the small size of mature miRNAs (19-25 nt), sequence similarity among these important biomarkers is expected. However, the extent of sequence similarity remains unclear. The RNA sequence and/or the nucleotide composition of miRNA are the characteristics that may determine the outcome on measuring specific miRNA in biological samples. For example, isomeric miRNAs with identical nucleotide composition could be co-eluted from a chromatography column, thus posing a challenge to the end point measurement including tandem mass spectrometry. The aim of this study is to investigate the extent of similarity on RNA sequence and nucleotide composition among human miRNAs.

Experimental Approach

Database

All human mature miRNA sequences used in this study were downloaded from miRBase (http://mirbase.org). In addition, human miRNAs that are linked to specific diseases are identified using the mir2disease database.

Nucleotide composition

Nucleotide composition of each miRNA was determined by using Mongo Oligo calculator in addition to our in-house developed Excel-based method.

Sequence similarity

For determining the sequence similarity among miRNAs, the MAFFT software was employed (http://www.ebi.ac.uk/Tools/msa/mafft/).

Results

Among all the human miRNAs, 55% are isomers i.e. same nucleotide composition. The differences between some miRNAs sequences are only one nucleotide. Among miRNAs that are linked to our selected diseases, the average sequence similarity is 32%.

Conclusion

For the first time, the extent of similarity in RNA sequence and nucleotide composition among human miRNAs are determined.

Keywords: Bioanalytical, Biological Samples, Liquid Chromatography/Mass Spectroscopy, Sample & Data Manag

Application Code: Genomics, Proteomics and Other 'Omics

Methodology Code: Data Analysis and Manipulation

Session # 1440 Abstract # 1440-22 **Poster Sessions**

Session Title: Bioanalytical: Miscellaneous Analytical Techniques

Abstract Title: **Surface-Enhanced Raman Spectroscopy Based Total Protein Assay**

Primary Author: Merve Eryilmaz
Gazi University

Date: Tuesday, March 08, 2016 - Afternoo
Time:
Room: Exposition Floor, 400 Aisle

Co-Author(s): Adem Zengin, Ugur Tamer

Abstract Text

The quantification of total protein is an important analytical task for the life sciences, ranging from food quality to the diagnosis of diseases. Parallel to development in spectroscopy, there is a demand for quick, interference-free total protein assays with accurate results. For this purpose, a new method for total protein quantitation was developed using Surface-enhanced Raman Spectroscopy (SERS). In this method, o-phtalaldehyde (OPA) complete reagent was used to form moieties which give SERS signals in the presence of protein and gold nanoparticles (AuNPs). A working curve was obtained by plotting the intensity of the SERS signal at a specific wavelength versus the concentration of protein standard and the correlation was found to be linear within the range of 0.054 mg/mL – 0.72 mg/mL. The limit of detection for the method was determined 0.07 mg/mL. The ability of determination of total protein amount in real samples was also investigated. For instance, milk was only diluted and there was not any other sample preparation. The measurement time was ten seconds and the sample volume was 100 L. Besides these advantages, this method offers less chemical consumption and waste with elimination of labor-consuming operations. Finally, obtained results were compared to conventional spectroscopic methods and the Kjeldahl assay.

This study has a patent number of 2014/15773.

Keywords: Protein, Quantitative, Surface Enhanced Raman
Application Code: Nanotechnology
Methodology Code: Vibrational Spectroscopy

Session # 1440 Abstract # 1440-23 **Poster Sessions**

Session Title: Bioanalytical: Miscellaneous Analytical Techniques

Abstract Title: **Ionic Liquid Crosslinkers for Chiral Imprinted NanoGUMBOS**

Date: Tuesday, March 08, 2016 - Afternoo
Time:
Room: Exposition Floor, 400 Aisle

Primary Author: Suzana Hamdan
Louisiana State University

Co-Author(s): Andrew L. LaFrate, David Spivak, Douglas Gin, Farhana Hasan, Isiah M. Warner, Jason LeJeune, Jason E. Bara, Leonard Moore, Richard D. Noble, Trevor K. Carlisle

Abstract Text

Vinylimidazole ionic liquid crosslinkers were explored as matrix supports for the synthesis of molecularly imprinted polymeric nanoGUMBOS (nanoparticles derived from a group of uniform materials based on organic salts). Each crosslinker incorporated a unique functional spacer between the vinylimidazole groups, which resulted in different imprinting performance among the synthesized polymeric nanoparticles. In this study, chiral recognition properties toward the amino acid L-tryptophan were crafted in the imprinted polymeric nanoparticles. The strong interactions between the functional groups of the ionic liquid, and the template (L-tryptophan) allowed for a successful imprinting under aqueous conditions, which is considered a challenge due to the competitive interruption of the non-covalent polymer-template interactions by water. Herein, high uptake values for L-tryptophan were found in the 13-87 umol/g range, and the binding ratios of L-tryptophan over D-tryptophan were ranging from 5 to 13:1 for different type of crosslinker. Our data imply that this novel class of ionic liquid crosslinkers is ideal for selectively imprinting aqueous templates of biological nature and for application of the synhesized nanoGUMBOS as theranostic agents.

Keywords: Characterization, Chiral, Nanotechnology, Polymers & Plastics
Application Code: Material Science
Methodology Code: Microscopy

Session # 1440 Abstract # 1440-26 **Poster Sessions**

Session Title: Bioanalytical: Miscellaneous Analytical Techniques

Abstract Title: **A Novel Bioassay Platform Using Silica Core-Stabilized Liposome Shell Microparticles for Ligand Discovery**

Primary Author: Kendall E. Sandy
University of Arizona

Date: Tuesday, March 08, 2016 - Afternoo
Time:
Room: Exposition Floor, 400 Aisle

Co-Author(s): Craig A. Aspinwall, Jinyan Wang, Mark T. Agasid

Abstract Text

Multiple disease states are caused by dysregulation of biochemical pathways through ligand-receptor interactions. Discovery of ligands that target transmembrane proteins is limited to platforms that support protein function. There is a need for rapid and highly specific assay platforms for identifying novel ligand receptor interactions while minimizing crosstalk and non-specific binding. To address this need, we have developed a novel microparticle architecture that utilizes silica core particle that is functionalized with receptors within stabilized liposomes. This particle architecture is then used to perform pulldown assays in complex solutions with subsequent analysis by electrospray ionization (ESI) or matrix-assisted laser desorption ionization (MALDI) mass spectrometry. As a proof-of-concept, recovery of serotonin via binding to 5-HT1A receptors within CHO-K1 cell membranes was evaluated. CHO-K1 cell membrane fractions were isolated through homogenization and centrifugation, and were extruded to form vesicles, which could be subsequently stabilized using polymer scaffold stabilization approaches. The vesicles were then immobilized to the particle surface to yield silica core-cell membrane vesicle shell particles. Particles were characterized using flow cytometry, to verify attachment of cell membrane vesicles with and without 5-HT1A receptors to modified particles. Serotonin was incubated with the silica core-cell membrane vesicle shell particles containing the serotonin receptor, and centrifugation was used to pull down the particles. ESI-MS confirmed the successful pull down of the serotonin ligand. This new platform shows promise for the discovery of unknown ligand/receptor pairs with minimal sample preparation.

Keywords: Bioanalytical, Drug Discovery, Mass Spectrometry
Application Code: Bioanalytical
Methodology Code: Mass Spectrometry

Session # 1440 **Abstract # 1440-27** **Poster Sessions**

Session Title: Bioanalytical: Miscellaneous Analytical Techniques

Abstract Title: **Detection of Disease Associated MicroRNA Combinations with a Smart AND Sensor**

Date: Tuesday, March 08, 2016 - Afternoo
Time:
Room: Exposition Floor, 400 Aisle

Primary Author: Lulu Zhang
Oregon State University

Co-Author(s): Sean M. Burrows

Abstract Text

Research will be presented on the fundamental biorecognition properties of a smart in situ fluorescent-based biosensor to detect microRNA (miR) combinations. MicroRNA often works in groups to interact with messenger-RNA and subsequently interrupt protein synthesis. Several studies are finding links between disease and changes in the expression of certain groups of miRs. In breast cancer, evidence suggests the Forkhead Box O1 (FOXO1) messenger-RNA is co-regulated by three microRNA: miR-27a, miR-96, and miR-182. Suppression of FOXO1 protein by these miRs disrupts apoptotic pathways, allowing the cancer cell to live and grow uncontrollably. Here we present an innovative Smart AND Sensor to detect the co-expression of miR-27a, miR-96, and miR-182. The sensor consists of three parts: (1) a donor reporter strand (R-96) partially bound to a sensing strand (S-96) for miR-96, (2) an acceptor reporter strand (R-182) partially bound to another sensing strand (S-182) for miR-182, and (3) a linker with two hairpin regions for each reporter and a hybridization site (HS-27a) for miR-27a. We will present signal to background ratio data from the smart AND sensor to determine: (1) maximum excitation wavelength for different combinations of donor and acceptor strand orientations, (2) influence of the molecular environment on reporter strands, (3) cross reactivity of the sensors components, and (4) sensing ability for a specific combination of miRs.

Keywords: Bioanalytical, Biosensors, Detection, Spectroscopy
Application Code: Bioanalytical
Methodology Code: Biospectroscopy

Session # 1440 **Abstract # 1440-29** **Poster Sessions**

Session Title: Bioanalytical: Miscellaneous Analytical Techniques

Abstract Title: **E-spun Collagen-CNT/Silk-CNT Composite Fibers to Transmit Electrical Signals for Cell Stimulation**

Primary Author: Naiwei Chi
Illinois Institute of Technology

Date: Tuesday, March 08, 2016 - Afternoo
Time:
Room: Exposition Floor, 400 Aisle

Co-Author(s)

Abstract Text

Fibrous collagen and synthetic silk are both novel biomaterials due to their outstanding mechanical properties, flexibility and bio-compatibility which enable a wide range of applications particularly in tissue engineering. Carbon nanotube (CNT) was found to interface with biological materials and markedly reinforce the scaffolds due to its high tensile strength. In this study, we incorporated CNT in collagen and silk fibers, and explored the utility of CNT's electrical conductivity rendering the nano-biocomposite materials a dimension of new functionality. To mimic the native extracellular matrix, we applied electrospinning technique to fabricate well-aligned composite fibers. The addition of proper amount of functionalized CNT was found to enhance the fiber alignment, mechanical strength and conductivity. While collagen-CNT fibers are superior in fiber alignment and stretchability, silk-CNT fibers offer higher tensile strength, elasticity and stability. Our results from cell proliferation and viability tests suggest these composite fibers are biocompatible. Using them as cell culture matrices, we grew primary fibroblast cells extracted from human connective tissues. In response to the applied low-frequency potential delivered by a home-built device, the fibroblast cells synthesized substantially higher levels of collagen and cytoskeleton protein genes, which are critical for wound healing. Interestingly, cells on silk-CNT synthesized much more collagen than cells on collagen-CNT, implying silk-CNT can more effectively transmit the electrical signal for cell stimulation. The capability of stimulating cells through the CNT incorporated biopolymers inspires the design of multi-functional biomatrices for more effective tissue regeneration.

Keywords: Bioanalytical, Biomedical, Microscopy, Nanotechnology
Application Code: Biomedical
Methodology Code: Surface Analysis/Imaging

Session # 1450 Abstract # 1450-1 **Poster Sessions**

Session Title: Bioanalytical: Separation Techniques

Abstract Title: **Surface Modified Nylon Capillary-Channeled Polymer (C-CP) Fibers for Protein Ion-Exchange Separations**

Primary Author: Liuwei Jiang
Clemson University

Date: Tuesday, March 08, 2016 - Afternoo
Time:
Room: Exposition Floor, 400 Aisle

Co-Author(s): R Kenneth Marcus

Abstract Text

Capillary-channeled polymer (C-CP) fibers have been studied as stationary phases for protein separations in recent years. C-CP fibers are made from melt-extruding of commonly used polymers such as polypropylene (PP), polyamide (Nylon) and polyester (PET). Each C-CP fiber has eight capillary channels on the fiber surface extending the whole length of the fiber. These channels provide C-CP fiber-packed bed excellent fluid transportation properties and approximately 3x more surface area than the circular cross-sectional fiber packed bed. Previously, surface modified PET fibers have been used for weak anion-exchange chromatography but suffered from peak tailing and broadening. In this study, nylon C-CP fibers were modified and used for ion-exchange protein separations. In comparison to modified PET C-CP fibers, the modified nylon C-CP fiber-packed bed provide better protein separation resolutions and higher stability in the column regeneration process in which strong acid and base are applied. Compared with the native nylon C-CP fiber, the modified fiber packed bed has higher protein dynamic binding capacity as well as better protein separation resolutions. The modified nylon fiber column has excellent hydraulic permeability and can be operated at high solvent flow rates without generating high column pressure. The simple modification method reported here, along with the modified nylon C-CP fibers, hold a great potential in analytical protein separations and protein downstream processing.

Keywords: Chromatography, HPLC Columns, Ion Exchange, Liquid Chromatography
Application Code: Bioanalytical
Methodology Code: Liquid Chromatography

Session # 1450 Abstract # 1450-3 **Poster Sessions**

Session Title: Bioanalytical: Separation Techniques

Abstract Title: **HPLC Method Transfer for Biopharmaceutical Analysis**

Date: Tuesday, March 08, 2016 - Afternoo
Time:
Room: Exposition Floor, 400 Aisle

Primary Author: Brooke M. Koshel
Waters Corporation

Co-Author(s): Sean M. McCarthy

Abstract Text

Many of the best-selling pharmaceuticals on the market as well as those making their way through the pipeline are protein-based. Unlike small molecule drugs which are synthesized through a controlled chemical process, biologics are produced from live cell cultures, which can complicate analysis. A host of chromatographic techniques are generally needed throughout the lifecycle of a drug product to ensure product quality and consistency. Because analysis can be carried out throughout various in-house laboratories and in some cases contract organizations, it is important that methods can be successfully transferred from one laboratory to the next without compromising results or requiring method revalidation.

The purpose of this work is to evaluate HPLC method transfer between two different chromatographic systems. Established methods for monoclonal antibody separations (SEC, CEX, and peptide mapping) will first be replicated on a modern UHPLC platform. Novel multi-flow path technology allows emulation of HPLC instrumentation as well as transferability to UHPLC operation on the same UHPLC platform. After assessing method transfer, we will show system reproducibility. Differences in system dispersion between the two chromatographic systems and the impact system dispersion has on resolution will also be addressed. Finally, the UHPLC platform will be used to update an SEC-HPLC method to a UHPLC method which takes advantage of smaller particle size and lower system dispersion to yield better resolution, shorter run time, and narrower peak profiles.

Keywords: Bioanalytical, Biopharmaceutical, HPLC
Application Code: Bioanalytical
Methodology Code: Liquid Chromatography

Session # 1450 Abstract # 1450-4 **Poster Sessions**

Session Title: Bioanalytical: Separation Techniques

Abstract Title: **Ratio of Different Fatty Acids Determined by GC-MS in Exosomes Purified Through Size Exclusion Chromatography**

Date: Tuesday, March 08, 2016 - Afternoo
Time:
Room: Exposition Floor, 400 Aisle

Primary Author: Rui Xu
Jackson State University

Co-Author(s): Joseph Fernandes, Radhika Pochampally, Yiming Liu

Abstract Text

Exosomes are derived from endosomal vesicles. They can be recognized and fused with cells. Fatty acids are main components of the membrane of exosomes and cells. In this presentation, gas chromatography mass spectrometry (GC-MS) technique coupled with size exclusion chromatography is demonstrated advantages to purify and quantification of fatty acids (oleic acid, linoleic acid, docosahexaenoic acid and araachidonic acid) in exosomes. Pre-column derivatization of fatty acids before GC-MS is necessary. We compare two derivatization methods to achieve the best assay sensitivity. One is derivatization by trimethylsilyldiazomethane(TMSDAM). Samples are extracted by hexan and the add 2.0 M TMSDAM. The other is derivatization by methanolysis/methylation using conc. HCl under mild condition.

The application of this method will help us reveal the relationships between the exosomes (bone marrow mesenchymal stem cells derived/PC-12 cells derived) and the original cells. Finally, the future perspective of the application can be used as a prognosis suggestions by characterization the cells pathological states from the exosomes from bloods or urea.

Acknowledgement: The research is partially supported by US National Institutes of Health (GM089557 to YML).

Keywords: Bioanalytical, Biological Samples, Characterization, GC-MS
Application Code: Bioanalytical
Methodology Code: Gas Chromatography/Mass Spectrometry

Session # 1450 Abstract # 1450-5 **Poster Sessions**

Session Title: Bioanalytical: Separation Techniques

Abstract Title: **Analysis of Chromium Species in Dietary Supplements Using ICP-MS and Speciated Isotope Dilution Mass Spectrometry (SIDMS)**

Date: Tuesday, March 08, 2016 - Afternoo

Time:

Room: Exposition Floor, 400 Aisle

Primary Author: Kaitlin Miller
Duquesne University

Co-Author(s): Jennifer Crawford, Logan T. Miller, Matt Pamuku, Skip Kingston, Stuart Procter

Abstract Text

Chromium exists in two primary oxidation forms: trivalent (Cr[sup]+3[/sup]) and hexavalent (Cr[sup]+6[/sup]). Hexavalent chromium is a potent and harmful known human carcinogen. However, trivalent chromium is a necessary trace mineral for human insulin function, and is therefore included in many dietary supplements. While many approaches are taken by pharmaceutical producers to prevent the oxidation of Cr[sup]+3[/sup] to Cr[sup]+6[/sup], the chemistry involved in the formation of pills presents oxidative potential for chromium. Final dietary supplement product analysis is therefore critical to ensuring their safety. Government requirements mandate the correct labeling of products for content and therefore necessitate testing of dietary supplements for hexavalent chromium content. Due to the extreme variation in risk factors associated with each species, analysis of substances for total chromium is inconclusive as to the health effects it may elicit. By using Inductively-Coupled Plasma Mass Spectroscopy (ICP-MS) according to EPA method 6800, update V, we are able to quantifiably measure each individual chromium species, and transformations of species, in dietary supplement samples. This study also identified the presence of several previously unknown chromium species, including possible Cr[sup]+6[/sup] anomalies. Investigation of the molecular composition of these complexes is being performed with ion chromatography fraction collection and subsequent quadrupole time-of-flight mass spectrometry to identify the chromium species present in the dietary supplement samples. Results and findings in commercial products will be discussed.

Keywords: ICP, Ion Chromatography, Mass Spectrometry, Time of Flight MS

Application Code: Bioanalytical

Methodology Code: Mass Spectrometry

Session # 1450 Abstract # 1450-6 **Poster Sessions**

Session Title: Bioanalytical: Separation Techniques

Abstract Title: **Metabolomic Signatures from Early Stage Ovarian Cancer Patients**

Primary Author: David A. Gaul
Georgia Institute of Technology

Date: Tuesday, March 08, 2016 - Afternoo
Time:
Room: Exposition Floor, 400 Aisle

Co-Author(s): Christina M. Jones, Facundo M. Fernandez, John F. McDonald, Long Q. Tran

Abstract Text

Lack of symptoms as well as the deficiency of highly specific biomarkers has resulted in only a quarter of ovarian cancer (OC) cases being diagnosed in stage I. Early detection combined with conventional therapies has resulted in 5-year survival rates up to 92%; while, 5-year overall survival is less than 30% for women with advanced stage OC. Investigation into characteristic metabolomic patterns for disease has the potential to detect changes in cells, tissues, and biofluids that can aid in diagnosis at an early stage, which is particularly advantageous for OC patients.

Serum samples were collected from early-stage papillary serous epithelial ovarian cancer (EOC) (n=46) and normal (n=49) patients and analyzed using ultra performance liquid chromatography coupled with high resolution mass spectrometry (UPLC-MS) and tandem mass spectrometry (MS/MS). Untargeted multivariate statistical analysis employing support vector machine learning methods and recursive feature elimination selected a panel of 16 metabolites that differentiated between the age-matched samples with very high accuracy, sensitivity, and specificity. The dominate classes of metabolites in the panels were lipids and fatty acids; this correlated well with the literature in which the metabolomes of EOC patients exhibit disruption of lipid metabolism and profiles. Our preliminary work demonstrated that metabolites in serum samples may be useful for detecting early-stage EOC and support conducting larger, more focused studies.

Keywords: Bioinformatics, Liquid Chromatography/Mass Spectroscopy, Metabolomics, Metabonomics
Application Code: Bioanalytical
Methodology Code: Liquid Chromatography/Mass Spectrometry

Session # 1450 Abstract # 1450-7 **Poster Sessions**

Session Title: Bioanalytical: Separation Techniques

Abstract Title: **Reducing Adhesion of Proteins on Stainless Steel Components by the Application of a Carboxysilane Coating**

Date: Tuesday, March 08, 2016 - Afternoo

Time:

Room: Exposition Floor, 400 Aisle

Primary Author: Luke Patterson

SilcoTek Corporation

Co-Author(s): Alfredo Narvaez, David Daghfal, David Smith, Min Yuan, Vaidya Shyam

Abstract Text

Protein binding and carryover in analytical systems is a challenge that can reduce accuracy and throughput when transferring and analyzing biomedical fluids. This work presents data on the effect of three surfaces on analytical accuracy, reduction of carryover and increase of throughput when they are in contact with a range of proteins. The areas of application are broad, including all sample-contacting components of medical diagnostic equipment, HPLC columns, fittings, and transfer tubing. The protein-resistant properties of a carboxysilane coating (deposited via chemical vapor deposition) was studied using quartz crystal microbalance with dissipation monitoring (QCM-D) and compared to that of bare stainless steel and a Teflon-like fluoropolymer coating (AF1600). With the assistance of a nonionic surfactant-containing wash solution, the carboxysilane coating was found to facilitate 100% removal of adsorbed proteins (BSA, mouse IgG and normal human plasma), whereas these proteins remain adsorbed on the bare stainless steel surface under the same conditions. Compared to the carboxysilane coating, AF1600 showed similar resistance to plasma protein adhesion at initial use, but the AF1600 performance degraded due to mechanical wear-induced surface delamination. The carboxysilane coating, on the other hand, maintained the same level of protein resistance through harsh chemical washes and multiple sonication cycles, demonstrating excellent chemical stability and physical durability of the CVD coating. The carboxysilane coating on stainless steel components improved reliability, reduced carryover and cycle time, and increased the durability of analytical systems.

Keywords: Adsorption, Liquid Chromatography, Material Science, Protein

Application Code: Bioanalytical

Methodology Code: Liquid Chromatography/Mass Spectrometry

Session # 1450 **Abstract # 1450-8** **Poster Sessions**

Session Title: Bioanalytical: Separation Techniques

Abstract Title: **Optimizations of Proteomic Sample Preparation Method for [i]Xenopus Laevis[/i] Embryonic Proteomics**

Date: Tuesday, March 08, 2016 - Afternoo
Time:
Room: Exposition Floor, 400 Aisle

Primary Author: Elizabeth H. Peuchen
University of Notre Dame

Co-Author(s): Liangliang Sun, Norman J. Dovichi

Abstract Text

[i]Xenopus laevis[/i] has been an important model organism in developmental biology. Landmark advances in biology, using this animal include nuclear transfer experiments, the first isolation of a gene from any organism, the first complete nucleotide sequence of a gene, and the purification of the first eukaryotic transcription factor.

There is a large literature base on this organism's transcriptome, but the study of its proteome is at an embryonic state. Only three [i]Xenopus laevis[/i] proteomics papers are published. This organism is approximately 90% yolk protein by weight at the single stage embryo, making the proteome extraction and the following mass spectrometry analysis challenging. Unfortunately, there is still no report about optimizations of proteomic sample preparation for this model system. Here, we systematically compared three common buffer systems for protein extraction from [i]Xenopus laevis[/i] embryos: SDS, 8M urea and NP40.

We employed BCA protein concentration measurement, SDS-PAGE, and nano-liquid chromatography-tandem mass spectrometry for comparing those three buffer systems. Initial findings show SDS extracts the largest amount of protein in comparison to 8M urea and NP40, with a large portion of it being yolk protein, which can be observed on a gel and with a BCA assay, lowering the identifications of peptides with SDS. NP40 extracts considerably less protein, but allows for overall higher peptide identifications since little yolk protein is extracted. SDS and NP40 samples were fractionated to uncover masked proteins in 1D analysis by yolk protein interferences. Identification of protein was improved for SDS by 165%. For complete analysis, the FASP procedure was found to have the top sample recovery and purity. This study will be valuable for [i]Xenopus laevis[/i] proteomics.

Keywords: Bioanalytical, Biological Samples, Liquid Chromatography/Mass Spectroscopy, Proteomics
Application Code: Bioanalytical
Methodology Code: Liquid Chromatography/Mass Spectrometry

Session # 1450 Abstract # 1450-9 **Poster Sessions**

Session Title Bioanalytical: Separation Techniques

Abstract Title **Determination of the Constituent Compounds in the Essential Oil from the Stem Bark of Ficus Capensis, A Multipurpose Phytomedicine, by GCMS and their Relevance to the Bioactivity of the Plant**

Date: Tuesday, March 08, 2016 - Afternoo

Time:

Room: Exposition Floor, 400 Aisle

Primary Author Modupe M. Ogunlesi
University of Lagos

Co-Author(s) Christianah T. Aleshinloye

Abstract Text

Ficus capensis, a multipurpose medicinal plant, is applied in the management of dysentery, leprosy, epilepsy, rickets, infertility, respiratory disorders, gonorrhea, threatened abortion and skin infections. The aim of the study is to relate the constituents of the essential oil of the stem bark with some of its medicinal applications. The stem bark was cut into small pieces, dried and pulverized. The essential oil was extracted into hexane by hydro distillation using a Clevenger apparatus. Collection was in two modes: continuous collection over a four-hour period and hourly collection over the same period. Analysis was carried out on a gc-ms fitted with HP5MS column using a temperature program of 80 deg. C (2min) increased at 5 deg. / min to 120 deg. (2min) increased at 3.5 deg./min to 240 deg. (5min). The constituents were saturated fatty acids; C10 (0-0.6%), C12 (0.3-1.3%), C14 (2.4-5.4%), C15 (4.8-6.5%), C16 (40.1-71.0%), C17 (0.8-1.3%), C18 (1.0-2.2%), unsaturated fatty acids; cis-13-octadecenoic acid (6.5-7.3%), and 9,12-octadecadienoic acid (9.1 and 9.5%) in the first and second hourly collections, hydrocarbons; undecane (3.1%), dodecane (1.5%) and tridecane (4.0%) present in the 4-hr. collection. Fatty acids exhibit anti-inflammatory and antimicrobial activities and are involved in cell signal and may therefore be relevant in the management of infectious diseases and epilepsy. Thus the essential oil from the stem bark contains compounds which are relevant in the management of some diseases for which it is employed.

Funding is by the University of Lagos and the authors.

Keywords: Flavor/Essential Oil, GC-MS, Natural Products
Application Code: Bioanalytical
Methodology Code: Gas Chromatography/Mass Spectrometry

Session # 1450 Abstract # 1450-10 **Poster Sessions**

Session Title Bioanalytical: Separation Techniques

Abstract Title **GC-MS Analysis of the Essential Oil from the Stem Bark of Tetrapleura Tetrapetra, a Multipurpose Medicinal Plant, and Bioactivities of some Constituent Compounds**

Date: Tuesday, March 08, 2016 - Afternoo
Time:
Room: Exposition Floor, 400 Aisle

Primary Author Modupe M. Ogunlesi
University of Lagos

Co-Author(s) Christianah T. Aleshinloye

Abstract Text

Tetrapleura tetrapetra is used for the management of fractured bones, female infertility, convulsion, sexually – transmitted diseases, asthma, rheumatism, insomnia, and as a preservative. The aim of this study is to correlate the constituent compounds in the essential oil from the stem bark with its medicinal uses. The essential oil was extracted into hexane by hydro distillation for 4 hours using a Clevenger apparatus and analyzed on GC-MS fitted with an HP5MS capillary column. The temperature program employed was 70 deg. C (4min) increased at 3.5 deg. / min to 240 deg. C (5min). The major component was n-hexadecanoic acid (40.1%). Other components include tetradecanoic- (1.5%), pentadecanoic- (1.0%), octadecanoic- (1.6%), acid. Unsaturated fatty acids present were 9,12-octadecadienoic (Z,Z) (4.2%) – and cis and trans -13-octadecenoic- (5.8%) acid. Fatty acid methyl esters present were those from hexadecanoic- (2.9%), 9,12- octadecadienoic - (7.6%), 9- octadecenoic - (Z) (10.5%) and 16-methyl heptadecanoic- acid (2.3%). Saturated fatty acids have been known to exhibit antimicrobial and anti-inflammatory activity hence may be useful in the management of infertility, asthma, rheumatism and sexually transmitted diseases. Saturated and unsaturated fatty acids can serve as preservatives. The fatty acid methyl esters may be metabolized in the human system to give fatty acids, thus increasing the fatty acid available to 77.4%. Fatty acids are involved in cell signal and may be useful in the management of convulsion. Thus the essential oil from the stem bark is relevant to the associated medicinal uses.

Funding is by University of Lagos and the authors.

Keywords: Flavor/Essential Oil, GC-MS, Natural Products
Application Code: Bioanalytical
Methodology Code: Gas Chromatography/Mass Spectrometry

Session # 1450 Abstract # 1450-11 **Poster Sessions**

Session Title Bioanalytical: Separation Techniques

Abstract Title **Quantification of Trehalose and Other Sugars in Submergence Resistant Rice**

Date: Tuesday, March 08, 2016 - Afternoo

Time:

Room: Exposition Floor, 400 Aisle

Primary Author Elizabeth N. Martinez
California State Polytechnic University, Pomona

Co-Author(s) Endang M. Septiningsih, Gregory A. Barding, Julia Bailey-Serres, Rejbana Alam

Abstract Text

Trehalose, a disaccharide found in many organisms, is an important sugar often associated with increased tolerance to a variety of stressors. In plants, trehalose is present in exceedingly low quantities making it difficult to detect and quantify, especially due to its low relative abundance comparted to sucrose, fructose and glucose. Gas chromatography – mass spectrometry (GC-MS) equipped with a quadrupole detector is a powerful technique for the sensitive detection of metabolites after derivatization with a suitable reagent, such as MSTFA. In this study, trehalose, fructose, sucrose, and glucose were evaluated in two different varieties of rice, one submergence tolerant and the other submergence intolerant. The LOD and LOQ for trehalose was found to be 4.4 nM 14.7 nM, respectively, with a 1 uL injection. Because of the excessive amount of sucrose in rice, a separate measurement was taken after a 1:100 dilution to ensure detection of the lower abundant saccharides. The analysis was first ran using a 60 – 600 m/z scan to identify each sugar using library matching as well as an in-house library generated with standards. Subsequently, selected ion monitoring (SIM) was after selecting ions specific to the sugars to further reduce the background associated with unwanted signals and increase sensitivity relative to full scan mode. In unstressed rice tissue, fructose had an abundance of approximately 0.01 g/g tissue, glucose had an abundance of 0.09 g/g tissue, Sucrose had an abundance of 19.40 mg/g tissue, and trehalose had an abundance of 0.01 g/g tissue.

Keywords: Bioanalytical, Biological Samples, Gas Chromatography/Mass Spectrometry

Application Code: Bioanalytical

Methodology Code: Gas Chromatography/Mass Spectrometry

Session # 1450 Abstract # 1450-12 **Poster Sessions**

Session Title Bioanalytical: Separation Techniques

Abstract Title **Exploring SFC for the Separation of Peptides and Small Proteins**

Date: Tuesday, March 08, 2016 - Afternoo
Time:
Room: Exposition Floor, 400 Aisle

Primary Author Cecilia Mazza
AkzoNobel PPC AB

Co-Author(s) Joakim Högblom, Peter Gidlund

Abstract Text

SFC is a chromatography technique that is in resurgence due to lower solvent consumption compared to HPLC as the SFC eluent is a mixture of mainly CO2 and small percentages of organic solvents. In addition to SFC being a low cost tool to run on day-to-day basis due to the eluent used, it is also considered a quick sample turnaround technique. Further, when moving to drug development and purification, where the fractions collected require dry down, the reduced time of solvent evaporation increases the overall productivity and efficiency in the lab. Larger bio-molecules such as peptides and small proteins are commonly analyzed and purified using reversed-phase chromatography. Recent studies show that SFC can be a tool for the separation of peptides. Here we present a comparative study of the separation of peptides and small proteins under SFC conditions.

Keywords: Analysis, Biopharmaceutical, Biotechnology, Prep Chromatography
Application Code: Bioanalytical
Methodology Code: Supercritical Fluid Chromatography

Session # 1450 Abstract # 1450-13 **Poster Sessions**

Session Title: Bioanalytical: Separation Techniques

Abstract Title: **Automated Solid Phase Extraction Method for the Assessment of Human Exposure to Polycyclic Aromatic Hydrocarbons Using the Biomarker Metabolite 1-Hydroxypyrene in Urine**

Primary Author: Michael J. Tanner
J2 Scientific

Co-Author(s): Jeff Wiseman

Date: Tuesday, March 08, 2016 - Afternoo
Time:
Room: Exposition Floor, 400 Aisle

Abstract Text

Monitoring human exposure to polycyclic aromatic hydrocarbons (PAHs) from environmental sources is an important tool in the assessment of occupational and environmental health. Following exposure, through dietary, inhalation or dermal pathways, PAHs undergo a two phase biological transformation, first to hydroxylated form (Phase I), followed by conversion to the more water soluble glucuronide or sulfate conjugates (Phase II) for elimination from the body. Routine measurements of the urinary levels of 1-hydroxypyrene have been shown to be a reliable biological indicator of the total exposure to parent PAHs in risk assessments.

Following enzymatic hydrolysis of buffered urine samples, to convert the conjugated metabolite to the hydroxylated form, the samples were subjected to automated solid phase extraction (SPE) isolation and cleanup using a J2 Scientific PrepLincTM system. The automated method allows the samples to be transferred directly from the hydrolysis containers through SPE cartridges, washing of interferences and elution of the metabolite from the SPE substrate, concentration of the eluent to final volume and transfer to autosampler vials for analysis. All aspects of the SPE method (i.e. conditioning, loading, washing, elution and concentration) were automated and the operational demands for the procedure were reduced to loading the hydrolyzed samples on to the instrument and selecting a preprogrammed method for the isolation of the anaylte of interest. High performance liquid chromatography with fluorescence detection (HPLC-FL) was used for quantification of the 1-hydroxypyrene present in urine samples. The automation of the procedure allowed for less variability between replicate samples and faster sample preparation times when compared to traditional, manual methods.

Keywords: Automation, Biological Samples, HPLC, Solid Phase Extraction
Application Code: Bioanalytical
Methodology Code: Sampling and Sample Preparation

Session # 1450 Abstract # 1450-14 **Poster Sessions**

Session Title: Bioanalytical: Separation Techniques

Abstract Title: **High Speed SDS–PAGE of Proteins**

Primary Author: Parul Modi
Thermo Fisher Scientific

Co-Author(s): Stephen Roemer

Date: Tuesday, March 08, 2016 - Afternoo
Time:
Room: Exposition Floor, 400 Aisle

Abstract Text

Sodium dodecyl sulfate–polyacrylamide gel electrophoresis (SDS–PAGE) is a widely used method for assessing in a reproducible manner the relative mass of denatured polypeptide chains and the purity of a protein preparation. For more than 40 years, the Laemmli SDS–discontinuous system based on Tris–Glycine buffer has been used with slab gels for high-resolution fractionation of protein mixtures under dissociating conditions. However, the electrophoresis run time is long for a standard 10 x 10 cm gel (~60 min) since the gel cannot be run at higher voltage without generating excess heat and compromising the resolution of separated protein bands. Through the years, various gel chemistries and buffering systems have been developed to improve protein band resolution at higher speed, but always with increased cost. In this work, we show that a simple change to the running buffer can provide high separation speed with enhanced resolution of protein bands. Tris-glycine mini-gels (precast and homemade) cast with the traditional Laemmli recipe can be run 50-60% faster (run time < 25 min) using our new proprietary Tris-SDS buffer. Gels are run at higher voltage without generating excessive heat and impacting the clarity of protein bands. Faster electrophoresis time improves the overall throughput of the protein lysate to Western blot work flow.

Keywords: Biotechnology, Electrophoresis, Protein, Proteomics
Application Code: Bioanalytical
Methodology Code: Electrochemistry

Session # 1450 Abstract # 1450-16 **Poster Sessions**

Session Title: Bioanalytical: Separation Techniques

Abstract Title: **Optimized Wide Pore Superficially Porous Particles by One-Step Coating Process for Fast and Efficient Separation of Large Biomolecules**

Primary Author: Wu Chen
Agilent Technologies

Co-Author(s): Anne Mack, Xiaoli Wang

Date: Tuesday, March 08, 2016 - Afternoo
Time:
Room: Exposition Floor, 400 Aisle

Abstract Text

Superficially porous particles (SPPs) with wide pore size of 300Å provide even better benefits over totally porous particles for separation of large biomolecules such as proteins and monoclonal antibodies (mAbs) because the analyte diffusion length inside a particle plays a more important role for fast separation in large molecule separation than in small molecule separation. Wide pore SPPs in 5 µm size have been used for protein separations for a decade. And more recently, wide pore SPPs in smaller size ranging from 2.6 µm to 3.6 µm have been developed to meet the need of continuing interest in larger therapeutic molecules by biopharmaceutical companies.

We have developed one step coating approach called "coacervation method" for synthesis of SPPs, in which the surface modified solid silica spheres, urea, formaldehyde, and silica colloidal sol under acidic conditions form a coating of coacervate of urea-formaldehyde polymer and ultra-pure silica sol particles on the solid spheres in one step. This method had been successfully applied to synthesis of commercial product, Poroshell 120 particles, for small molecule separation.

In this report, we would like to report synthesis of a series of new wide pore SPPs with different particle size, pore size, and shell thickness using this one step coating coacervation method. By evaluating chromatographic performance of those particles using van Deemtor plot, protein mixtures, larger monoclonal antibodies, the 3.6 µm, 450 Å pores SPPs were chosen as optimized for separation of large biomolecules, and were compared with other wide pore SPPs and totally porous particles.

Keywords: Bioanalytical, HPLC Columns, Liquid Chromatography, Separation Sciences
Application Code: Bioanalytical
Methodology Code: Liquid Chromatography

Session # 1450 Abstract # 1450-17 **Poster Sessions**

Session Title: Bioanalytical: Separation Techniques

Abstract Title: **Fast Quantification of Immunoglobulin G Using A New Protein A Analytical HPLC Column**

Primary Author: Atis Chakrabarti
Tosoh Bioscience LLC

Date: Tuesday, March 08, 2016 - Afternoo
Time:
Room: Exposition Floor, 400 Aisle

Co-Author(s): Kosuke Araki, Satoshi Fujii, Shigeru Nakatani

Abstract Text

Monoclonal antibodies (mAbs) and their derivatives are widely used as biopharmaceuticals and will continue to be a focus of the industry for years to come. In the development of therapeutic mAbs, a high-throughput and robust analytical method is needed for quantifying them in harvested cell culture fluids, downstream product pools, and so on. Protein A is an antibody binding protein generally used in affinity chromatography to purify and determine antibodies. High affinity and selectivity of Protein A for antibodies lead to an effective removal of host cell proteins. Here, we will report on a fast analysis of immunoglobulin G (IgG) titer in the cell culture supernatant using a newly developed Protein A affinity column. This column was designed based on a novel polymeric resin with a recombinant Protein A ligand. Due to a high dynamic loading capacity, it was able to quantify IgG in a range of 0.1 to 10 mg/mL. At the flow rate of 2.0 mL/min, the total analysis time was 2.0 min (including the time required for sample loading, wash, elution and re-equilibration). Column lifetime tests showed that the column could be use over 2,000 cycles with very little loss of performance.

Keywords: Bioanalytical, Biopharmaceutical, HPLC Columns, Separation Sciences
Application Code: Bioanalytical
Methodology Code: Liquid Chromatography

Session # 1450	Abstract # 1450-18	Poster Sessions

Session Title: Bioanalytical: Separation Techniques

Abstract Title: **Charging YOYO-1 on Capillary Wall for Online DNA Intercalation and Integrating This Approach with Multiplex PCR and Bare Narrow Capillary–Hydrodynamic Chromatography for Online DNA Analysis**

Date: Tuesday, March 08, 2016 - Afternoo

Time:

Room: Exposition Floor, 400 Aisle

Primary Author: Huang Chen
University of Oklahoma

Co-Author(s): Joann Lu, Shaorong Liu, Zaifang Zhu

Abstract Text

Multiplex polymerase chain reaction (PCR) has been widely utilized for high-throughput pathogen identification. A dye is usually used to intercalate the PCR products, and identifications of the pathogens are accomplished by DNA melting curve analysis or gel electrophoresis. Logically, coupling DNA amplification with identification can maximize the benefit of multiplex PCR. PCR and gel electrophoresis have been integrated, but replenishing the gels after each run is tedious and time-consuming. Recently, we develop an approach to address this issue. We perform multiplex PCR inside a capillary and on-line transfer the amplified fragments to a bare narrow capillary, where we measure their lengths online using bare narrow capillary-hydrodynamic chromatography (BaNC-HDC), a new technique developed in our laboratory for free-solution DNA separation. To intercalate the DNA with YOYO-1(a fluorescent dye) for BaNC-HDC, we flush the capillary column with a YOYO-1 solution; positively charged YOYO-1 is charged onto the negatively charged capillary inner wall. As DNA molecules are driven down the column for separation, they are online-intercalated with the YOYO-1 stored on the capillary wall. With a single YOYO-1 charging, the column can be used for more than 40 runs, although the fluorescence signal intensities of the DNA peaks decrease gradually. Although the dye-DNA intercalation takes place during the separation, it does not affect the retention times or separation efficiencies.

Keywords: Bioanalytical
Application Code: Bioanalytical
Methodology Code: Microfluidics/Lab-on-a-Chip

Session # 1450 Abstract # 1450-19 **Poster Sessions**

Session Title Bioanalytical: Separation Techniques

Abstract Title **From Peptide Fractions to Pure, Dry Powders: Development of a Novel Automated Chromatographic Purification Process Supported by Solid-Phase Trapping**

Primary Author Yamazaki Tomoyuki
Shimadzu Corporation

Date: Tuesday, March 08, 2016 - Afternoo
Time:
Room: Exposition Floor, 400 Aisle

Co-Author(s) Iwata Yosuke, Masuda Junichi, Matsuo Eiichi, Nishimura Masayuki, Okoba Tsutomu, Robert E. Buco

Abstract Text

There are many commercially available instruments that can be used to process preparative liquid chromatography (LC) fractions, recovering the pure compound of interest and discarding background from the mobile phase.Some of these technologies utilize solid-phase trapping media to retain the compounds of interest while allowing the water, organic solvents and salt additives to be flushed to waste. Nearly two years ago, Shimadzu introduced a new instrument that leveraged this concept of solid-phase trapping and implemented a trapping column packed with a generically applicable resin that was appropriate for the retention of most small organic molecules. In addition to removing the mobile phase matrix background from the LC fraction, this instrument provided a mechanism to treat the retained compound on-column and recover it in a specific salt form as a highly pure and very dry powder, achieving better quality than traditional fraction dry-down techniques in significantly less time. In this presentation, we focus on the development of a new resin that expands the capability of this instrument to include peptides.

Keywords: Liquid Chromatography, Pharmaceutical
Application Code: Pharmaceutical
Methodology Code: Liquid Chromatography

Session # 1450 **Abstract # 1450-20** **Poster Sessions**

Session Title: Bioanalytical: Separation Techniques

Abstract Title: **Coupling Ion Exchange Chromatography with Reverse Phase Liquid Chromatography for High-Throughput Analysis of Intact Proteins**

Primary Author: Zaifang Zhu
University of Oklahoma

Date: Tuesday, March 08, 2016 - Afternoo

Time:

Room: Exposition Floor, 400 Aisle

Co-Author(s): Huang Chen, Joann Lu, Shaorong Liu

Abstract Text

Multi-dimensional liquid chromatography is an effective technique for top-down proteomic research. In this work, we report a micro-scale (250 m i.d.) monolithic column for protein separation first with anion-exchange and then reversed-phase liquid chromatography. With NaCl in 10 mM Tris-HCl (pH~7.6) as the gradient elution for anion-exchange chromatography, the column was capable of separating protein standards with high efficiencies. The monolith pore size was controlled by adjusting the monomer percentage and the porogen ratio in the formula of making columns. We investigated the effect of monolith pore size on protein separations. We also optimized other separation conditions, including linear flow rate, gradients, and injection volume. Under the optimized conditions, protein lysates from E. Coli cells were analyzed and more than 70 peaks were identified with UV detection. We then coupled ion-exchange chromatography with reversed-phase liquid chromatography for high-throughput separations of intact separations. The resolved proteins can be directly identified by mass spectrometry.

Keywords: Bioanalytical, Ion Chromatography, Liquid Chromatography, Proteomics
Application Code: Bioanalytical
Methodology Code: Liquid Chromatography/Mass Spectrometry

Session # 1460 Abstract # 1460-2 **Poster Sessions**

Session Title: High-Throughput Chemical Analysis

Abstract Title: **Characterization and Use of a Microspectrophotometer for Quantitative Bio-Applications**

Date: Tuesday, March 08, 2016 - Afternoo

Time:

Room: Exposition Floor, 400 Aisle

Primary Author: Thomas M. Spudich
Maryville University

Co-Author(s): Bradley Postier, Ronald Mills

Abstract Text

Spectrophotometers are a ubiquitous tool for biological research as it provides a means to quantify and qualify biological materials of interest including nucleic acids, proteins, cell density, and many small molecules. However limitations to the size of a traditional spectrophotometer limit its use efficiency. We have integrated the capabilities of a spectrophotometer into the barrel of a micropipette to alleviate problems with traditional spectrophotometric operations. The authors will present the newest design for this instrument as well as initial characterization for different solutions to include red and blue dyes as well as completing a protein analysis using the Bradford assay. This design has several benefits over a traditional instrument such as: reduced manufacturing costs, small size, reduced to no sample loss, and utility in enclosed environments. This instrument utilizes disposable tips like a traditional micropipette, can be used or even dedicated to clean environments including a PCR/RNA clean areas, glove bag, BSL3 hood, fume hood, or taken into the field. Data can be presented on the device or sent wirelessly to a server, computer, tablet, phone, or wearable device. In combination, the benefits of these instruments make spectrophotometric analysis much more user friendly and efficient.

Keywords: Analysis, Fluorescence, Spectrometer, Spectroscopy

Application Code: High-Throughput Chemical Analysis

Methodology Code: Portable Instruments

Session # 1460 Abstract # 1460-3 **Poster Sessions**

Session Title High-Throughput Chemical Analysis

Abstract Title **California Chlor-Alkali Production Facility Monitors Organic Carbon for Increased Reliability and Equipment Protection**

Date: Tuesday, March 08, 2016 - Afternoo
Time:
Room: Exposition Floor, 400 Aisle

Primary Author JP Pasterczyk
GE Analytical Instruments

Co-Author(s) Gary Erickson, Mark A. Mullet

Abstract Text
Chlor-Alkali manufacturing is one of many challenging industries for online analytics. These plants utilize saturated brine (saltwater) and electricity, with advanced membrane technology, to produce chlorine for water treatment, pure bleach, caustic soda and other, chlorine-related products. Excess organic contamination of the brine can destroy membrane systems and cause plant shutdowns. Of the available technologies for chlor-alkali production, (membrane cell, diaphragm cell, and mercury cell), advanced membrane systems are becoming the leading solution.

A Northern California facility chose to move the organics monitoring, critical to protecting the membranes, from the laboratory to the process floor by implementing online analysis. Based on past experience with the high maintenance required to support operation of their laboratory analyzer, the lab manager knew online operation would present exceptional challenges.

The instrument was installed to protect the membrane system by monitoring the ultrapure brine supply from the Brine Ion Exchange unit going into the Ultrapure Brine Storage. During the evaluation, the facility maintained the laboratory measurements to verify online performance was comparable to or better than, laboratory results. The analyzer output was connected to the plant SCADA system using the 4-20mA outputs to provide real-time control room visibility with alarm implementation.

Based on historical plant operations, the operators knew safe levels of organics for their expensive advanced membranes were below 5ppm of organic carbon. Continuous monitoring and reporting allowed for detection of excursions which resulted in avoidance of damage to the membranes within the first few months of operation.

Keywords: Chemical, Membrane, Process Control, Total Organic Carbon
Application Code: High-Throughput Chemical Analysis
Methodology Code: Process Analytical Techniques

Session # 1460 Abstract # 1460-4 **Poster Sessions**

Session Title High-Throughput Chemical Analysis

Abstract Title **New Analysis Technology of Ultra-Trace Yellow Components in a Transparent Film**

Date: Tuesday, March 08, 2016 - Afternoo
Time:
Room: Exposition Floor, 400 Aisle

Primary Author Hoko Suto
Hitachi Chemical Co., Ltd.

Co-Author(s) Akihiro Unnno, Kosuke Iwamoto

Abstract Text

Recently, smartphones and tablets are spreading rapidly, and the market of touch panel displays are increasing. The visibility is one of the most important qualities of the touch panel displays. Therefore, the material used on the top of the display must be transparent. An organic transparent film is often used to protect displays and to reduce reflection of light. In many cases this film becomes yellow with the passage of time. The amount of yellowing compounds is often so small that the identification is difficult. In this report, we will show two analysis technologies of yellow compounds generated under ageing test in transparent films. In the case that the yellowing compound can be extracted by solvents, the LC-TOF/MS having high-separation LC and the high-resolution MS are suitable for the analysis. And the important feature of system is to arrange a UV detector and a MS detector in series. The time difference occurs between the UV detector and the MS detector in this way. Using this time difference which comes from selected ion chromatogram peaks and 400-500 nm chromatogram peaks, the yellow compound is identified. In another case that the yellowing compound cannot be extracted by solvents, Py-GC X GC-MS and molecular simulation are effective ways to analyze it. Comparing with the transparent film and the yellowing film, differences become obvious in ageing test. Then the origin of those differences is derived from the resin compositions. Using the molecular simulation, the yellow compound can be specified from many detected differences.

Keywords: Analysis, GC-MS, High Throughput Chemical Analysis, Liquid Chromatography/Mass Spectroscopy
Application Code: High-Throughput Chemical Analysis
Methodology Code: Mass Spectrometry

Session # 1460 Abstract # 1460-5 **Poster Sessions**

Session Title: High-Throughput Chemical Analysis

Abstract Title: **Leaning out Stage 1 Conductivity**

Date: Tuesday, March 08, 2016 - Afternoo
Time:
Room: Exposition Floor, 400 Aisle

Primary Author: Andy Young
GE Analytical Instruments

Co-Author(s): Dondra Biller, Jenny Watson

Abstract Text

Pharmaceutical and biopharmaceutical companies across the US are being tasked with leaning out their processes and looking for opportunities to save money. Testing of water samples for total organic carbon (TOC) and conductivity is a requirement in the US dictated by the United States Pharmacopeia (USP). There have been a variety of instrument improvements that increase the speed of analysis for TOC, but conductivity has been left behind. Many companies have abandoned using USP 645 Stage 1 testing due to the risk of failure. These companies have implemented USP 645 Stage 2 testing instead, which can take as long as 4 hours. This talk will discuss how a major pharmaceutical company implemented the use of a single instrument to perform simultaneous TOC and Stage 1 conductivity leading to faster results and real-time analysis of water and cleaning validation samples. Discussion will include the steps taken in method transfer, a presentation of the data analyzed, considerations taken to enable the transfer from manual measurement to automated measurement, and the projected cost savings. Leaning out the process used to take these required measurements will help companies run more efficiently.

Keywords: Biopharmaceutical, Quality Control, Sample & Data Management, Total Organic Carbon
Application Code: High-Throughput Chemical Analysis
Methodology Code: Chemical Methods

Session Title High-Throughput Chemical Analysis

Abstract Title **Theoretical Simulation of a Helium DC Glow Discharge Used as an Ambient Desorption/Ionization Source for Mass Spectrometry**

Date: Tuesday, March 08, 2016 - Afternoo

Time:

Room: Exposition Floor, 400 Aisle

Primary Author Wade C. Ellis
Brigham Young University

Co-Author(s) Paul B. Farnsworth, Ross L. Spencer

Abstract Text

The purpose of this study is to simulate the chemistry that occurs in a DC glow discharge and its afterglow region using helium as the plasma gas. The simulation is two dimensional and axisymmetric. It is divided into two parts, a flow model and a kinetic model. Star-CCM+ (CD-Adapco, New York, USA) is used to construct the glow discharge geometry and simulate the flow of helium gas into air in the afterglow region. The kinetic model is being developed using Matlab (MathWorks, Natick, MA, USA), and takes into account the reactive species present in the plasma based on findings published in many experimental papers. In particular, we take into account measurements of the absolute number densities of helium metastable atoms taken in our lab. Other reactive species considered include nitrogen molecules and ions; water molecules, ions, and clusters; and electrons. Preliminary findings show the fast depletion of helium metastable ions to form nitrogen ions (N2+ and N4+) which in turn react with water to ultimately form protonated water clusters (H(H2O)n+) that react with analyte molecules. The primary source for nitrogen and water was found to be the impurities in the helium gas stream from the cylinder or the tubing. Atmospheric nitrogen and water also played an important role. This study builds on previous similar studies by including water in the reactions and by considering the afterglow region rather than just the active plasma.

Keywords: Computers, Mass Spectrometry, Plasma
Application Code: High-Throughput Chemical Analysis
Methodology Code: Computers, Modeling and Simulation

Session # **1460** Abstract # **1460-7** **Poster Sessions**

Session Title: High-Throughput Chemical Analysis

Abstract Title: **Mathematical Modeling and Computational Simulation of Matrix Effect on Uptake Kinetics in Solid Phase Microextraction**

Date: Tuesday, March 08, 2016 - Afternoo

Time:

Room: Exposition Floor, 400 Aisle

Primary Author: Md Nazmul Alam
University of Waterloo

Co-Author(s): Janusz Pawliszyn, Luis Ricardez-Sandoval

Abstract Text

Solid phase microextraction (SPME) is a well-known sampling and sample preparation technique used for a wide variety of analytical applications. As there are various complex processes taking place at the time of extraction that influence the parameters of optimum extraction, a mathematical model and computational simulation describing the SPME process is required for experimentalists to gain more insight on this process and implement the technique without performing multiple costly and time-consuming experiments in the laboratory. In this study, a mechanistic mathematical model for the processes occurring in SPME extraction of analyte(s) from an aqueous sample medium is presented. The model integrates analyte diffusion in the coating and diffusion, convection and reversible binding in the analyte sample. The proposed mechanistic model has been validated with previously reported experimental data taken from three different sources. Several key factors that affect the extraction kinetics, such as sample agitation, fiber coating thickness, and the presence of other undesirable species (matrix), are discussed. Moreover, insights regarding enhancement or retardation of extraction kinetics in the presence of the matrix are provided with the help of an asymptotic analysis. To the authors' knowledge, this is the first study that addresses this point. The parameters that contribute to the different types of observed matrix effects on the uptake kinetics are also discussed. Numerical simulation results show that the proposed model captures the phenomena occurring in SPME, leading to a better understanding of this process. Therefore, the currently presented model can be used to identify optimum experimental parameters without the need to perform a large number of experiments in the laboratory.

Keywords: Bioanalytical, Environmental, Extraction, High Throughput Chemical Analysis

Application Code: Bioanalytical

Methodology Code: Computers, Modeling and Simulation

Session # 1460   Abstract # 1460-8   **Poster Sessions**

Session Title High-Throughput Chemical Analysis

Abstract Title **Simple Imager for Multi-Well Plates**

Primary Author Thayumanasamy Somasundaram
Florida State University

Date: Tuesday, March 08, 2016 - Afternoo
Time:
Room: Exposition Floor, 400 Aisle

Co-Author(s) Michael Zawrotny

Abstract Text

We have built a simple and low-cost imager that is suitable for capturing images of multi-well plates (96-, 192-, and 288-wells) used for crystallization screening and cell culture. The imager is assembled from commercially available variable zoom USB microscope and a home-built 2-dimensional stage. The software that controls the movement and captures the images is a modified open source version and the microprocessor control is built on Arduino platform. Stage movement is driven by small stepper motors controlled by an Arduino microprocessor running the GRBL firmware commonly found in 3D printers. Instructions are sent through the USB serial connection to the Arduino by a python program developed in-house. The same program utilizes the python bindings for the OpenCV to perform the image acquisition. The program was developed under both Mac OS X and Linux. The imager reduces the time and fatigue involved in manual inspection while providing image history of the multi-well plates during the course of an experiment. Since the design is flexible further modifications to suit other multi-well plates and experiments are possible.

Keywords: Biological Samples, Microscopy, Protein, X-ray Diffraction
Application Code: Biomedical
Methodology Code: Microscopy

Session # 1470 Abstract # 1470-1 **Poster Sessions**

Session Title: Surface and Microscopic Characterization of Nanostructures and Biological Materials

Abstract Title: **Characterizing Nanoparticle Size and Particle-Surface Interactions Using Nanophotonic Force Microscopy**

Primary Author: Dakota O'Dell
Cornell University

Date: Tuesday, March 08, 2016 - Afternoo
Time:
Room: Exposition Floor, 400 Aisle

Co-Author(s): David Erickson, Perry Schein, Summer Saraf

Abstract Text

The long-term stability of nanoparticle suspensions is governed by a complex interplay of particle-particle and particle-surface force interactions. Although understanding suspension dynamics and stability is critical for quality control in many industries, there is currently no analytic model which can handle the diversity and range of forces involved, and there is little hope of reliable prediction from first principles. To accurately probe the suspension stability, therefore, it is desirable to make a direct measurement of the net forces in solution without having to predict the magnitude of each individual component.

Here, we present an optofluidic technique for directly measuring and characterizing interactions between colloidal particles and surfaces using the light that they scatter as they interact with the evanescent field above an optical waveguide. Within the evanescent field, particles experience an attractive optical gradient force towards the center of the waveguide. By using this optical force to counterbalance surface interaction forces on the particle, we can measure the intensity of the scattered light from the particle as it undergoes Brownian motion and directly determine the magnitude of the repulsive surface forces. We can also analyze the Brownian dynamics of the nanoparticles under the influence of the applied optical forces to simultaneously obtain an estimate the particle's hydrodynamic radius. Using our near-field light scattering technique, we report high-throughput measurements of particle size and sub-pN interaction forces for particles as small as 50 nm.

Keywords: Characterization, Light Scattering, Nanotechnology, Surface Analysis
Application Code: Nanotechnology
Methodology Code: Microscopy

Session # 1470 Abstract # 1470-2 **Poster Sessions**

Session Title: Surface and Microscopic Characterization of Nanostructures and Biological Materials

Abstract Title: **Tuning Localized Surface Plasmon Resonance Wavelengths of Nanoparticles by Mechanical Deformation**

Primary Author: Fathima S. Ameer
Clemson University

Date: Tuesday, March 08, 2016 - Afternoo
Time:
Room: Exposition Floor, 400 Aisle

Co-Author(s): Fenglin Wang, Hannah Mack, Jeffrey N. Anker, Marian Kennedy, Shilpa Varahagiri

Abstract Text

When the size of noble metal nanoparticles is reduced to less than the wavelength of light, the particles intensely absorb and scatter light at wavelengths that depends on the particle size, shape, and local dielectric environment due to Localized Surface Plasmon Resonance (LSPR) modes. Recently, plasmonic particles have been used in a wide variety of sensor and optical device applications including immunoassays and surface enhanced Raman spectroscopy substrates. Controlling nanoparticle shape is essential for tuning these nanoparticles for a given application; however, controlling shape and exposing new surface area without affecting the particle volume is challenging. Reported here is a simple technique to mechanically alter the shape of gold and silver nanoparticles by manually rolling a glass rod over them. This shape change in turn induces a red-shift in the LSPR wavelength of both gold and silver nanoparticles. The flattened particles were characterized by both optical and electron microscopy, single nanoparticle scattering spectroscopy, and surface enhanced Raman spectroscopy. Our quantitative SERS measurements show ~30% increase in surface area of nanoparticles after deformation. The simple technique employed here requires no lithographic templates and has potential for rapid, inexpensive and scalable tuning of nanoparticle shape, surface area, and resonance while preserving particle volume.

Keywords: Method Development, Microscopy, Spectroscopy, Surface Enhanced Raman
Application Code: Other
Methodology Code: Microscopy

Session # 1470 Abstract # 1470-3 **Poster Sessions**

Session Title: Surface and Microscopic Characterization of Nanostructures and Biological Materials

Abstract Title: **Comparison of Color Pigment Removal between Graphitized Carbon Black and Zirconia-Based Adsorbents for QuEChERS Process**

Primary Author: Patrick Myers
Supelco/Sigma-Aldrich

Date: Tuesday, March 08, 2016 - Afternoo
Time:
Room: Exposition Floor, 400 Aisle

Co-Author(s): Jennifer Claus, Katherine Stenerson, Michael Ye, Tyler Young

Abstract Text

The clean-up of fruit and vegetable samples for pesticide analysis is complicated by the presence of plant pigments in the matrix. Pigment molecules are generally large, non-volatile compounds that tend to get trapped in the inlet liner of the GC/MS. Use of the QuEChERS clean-up method does not necessarily remove all pigments from the matrix. Method EN 15662 recommends the addition of Graphitized Carbon Black (GCB) to remove plant pigments from samples. However, in addition to removing plant pigments, GCBs also retain some pesticides, especially planar pesticides. It would be of interest to look at other sorbents that may be complementary to GCB in removal of color pigments during QuEChERS process. In this presentation, a range of plant pigments was chosen to be representative of the pigments found in various fruits and vegetables. In addition to GCB, zirconia-containing adsorbents Z-Sep and Z-Sep+ were used to remove the pigments from acetonitrile solutions designed to simulate dispersive SPE extracts. It is found that zirconia-containing adsorbent Z-Sep+ overall removed between 40% and 80% of the plant pigment from the solutions, and is comparable with GCB in removal of crocin and xanthophyll, which contribute to the red and yellow pigments, respectively, in plants. The Z-Sep+ material removed more betanin, a red color compound, than GCB. The effect of surface area of the GCB material as well as a zirconia-coated GCB material on the color removal will also be discussed.

Keywords: Pesticides
Application Code: Food Safety
Methodology Code: Surface Analysis/Imaging

Session # 1470 Abstract # 1470-4 Poster Sessions

Session Title: Surface and Microscopic Characterization of Nanostructures and Biological Materials

Abstract Title: **Potential-Dependent Adsorption of Water-Soluble Porphyrins at Liquid/Liquid Interfaces Studied by Polarization-Modulation Total Internal Reflection Fluorescence Spectroscopy**

Primary Author: Sho Yamamoto
Kanazawa University

Date: Tuesday, March 08, 2016 - Afternoo

Time:

Room: Exposition Floor, 400 Aisle

Co-Author(s): Hirohisa Nagatani, Hisanori Imura, Kotaro Morita

Abstract Text

An interface between two immiscible electrolyte solutions (ITIES) is a two-dimensional specific reaction field, where charge transfer processes are significantly affected by the Galvani potential difference across the interface. The molecular orientation of reactants adsorbed at ITIES plays a crucial role in the heterogeneous reaction mechanisms. In this study, we developed a novel in situ spectroscopic method, polarization-modulation total internal reflection fluorescence (PM-TIRF) spectroscopy, for dye species adsorbed with a certain orientation at ITIES. In PM-TIRF experiments, the fluorescence signal from the interfacial region is analyzed as a function of periodic modulation of linear-polarizations ([i]p[/i] and [i]s[/i] polarizations) of excitation beams. PM-TIRF spectroscopy can effectively extract the fluorescence signals of interfacially oriented species from the signals arising throughout the optical path in the incident medium because no PM-TIRF signal from randomly oriented bulk solution species.

The adsorption behavior of fluorescent dyes, water-soluble porphyrins at the polarized water|1,2-dichloroethane interface was investigated by PM-TIRF. PM-TIRF spectroscopy clearly demonstrated that, for example, [i]meso[/i]-substituted free base porphyrins were adsorbed with relatively lying orientation at the interface. The wavelength dependence of PM-TIRF intensity, i.e. PM-TIRF spectra, indicated that the spectral features of adsorbed species is different from that of the bulk solution species and the porphyrins are adsorbed with a modified hydration state at the interface. Furthermore, an anionic porphyrin, protoporphyrin IX (H[sub]2[/sub]PP[sup]2□[/sup]) exhibited that its adsorption states were affected by the potential-induced J-aggregation of H[sub]2[/sub]PP[sup]2□[/sup] occurred only at ITIES.

Keywords: Adsorption, Fluorescence, Spectroelectrochemistry, Surface Analysis

Application Code: Bioanalytical

Methodology Code: Surface Analysis/Imaging

Session # 1470 **Abstract # 1470-5** **Poster Sessions**

Session Title: Surface and Microscopic Characterization of Nanostructures and Biological Materials

Abstract Title: **The Concept of Lipobeads in the Context of Encapsulated Drug Delivery: Technological Challenges vs. Potential Advantages**

Primary Author: Sergey V. Kazakov
Pace University

Date: Tuesday, March 08, 2016 - Afternoo
Time:
Room: Exposition Floor, 400 Aisle

Co-Author(s)

Abstract Text

The concept of lipobeads has been proposed about 30 years ago, however, lipobead-based drug delivery systems are still largely experimental. A possible reason for a limited number of studies on lipobeads published could be in the lack of a comprehensive understanding of the advantages of these drug carriers versus the feasibility of their production. The time has come to explore analytically, microscopically, and spectroscopically a combination of lipid bilayer and cross-linked polymer network, which Nature uses to achieve workability, multifunctionality, and dynamism in living cells of different types. Lipobeads exhibit the properties attractive for the next generation of drug delivery systems: (i) retaining all the important benefits of polymeric and liposomal drug carriers, a hydrogel core brings mechanical stability and environmental responsiveness to the formulation in one construct, (ii) lipobead-delivered combination therapy shows no toxicity on intravenously administered mice, accumulation of drug-loaded lipobeads both in the area surrounding tumor and within the tumor itself outside the vasculature, high therapeutic activity at the targeted site, significantly greater reduction in both tumor growth rate and tumor mass, and drastically increased survival, (iii) bipartite structure of lipobeads can provide a number of novel and unique options (new schemes of drug release, consecutive multi-step triggering, and combined drug delivery systems). Here we present microscopic (optical, fluorescence, confocal, AFM) data on nano- and giant lipobeads demonstrating their technological achievability and supporting the expectations that additional expenses on their production will be reimbursed by the potential advantages of their use. In addition, the ideas on the conceptually new drug delivery systems, new mechanisms of lipobead internalization into the cell and mechanisms of drug release regulated by specific signaling are discussed.

Keywords: Biomedical, Biotechnology, Drugs, Nanotechnology
Application Code: Biomedical
Methodology Code: Microscopy

Session # 1470 Abstract # 1470-6 **Poster Sessions**

Session Title: Surface and Microscopic Characterization of Nanostructures and Biological Materials

Abstract Title: **Combination of Surface Plasmon Resonance - Surface Enhanced Raman Scattering Spectroscopy in the Kretschmann Configuration**

Primary Author: Ju-Young Kim
University of Notre Dame

Co-Author(s): Zachary D. Schultz

Date: Tuesday, March 08, 2016 - Afternoo
Time:
Room: Exposition Floor, 400 Aisle

Abstract Text

Surface plasmon resonance (SPR) spectroscopy is a surface sensitive bioanalytical tool that has been widely used for molecular binding studies. As a label-free technique, SPR spectroscopy can detect the protein-ligand binding event occurring on a metallic surface (i.e. gold) in real-time by detecting the changes in refractive indices.1 In this research, we combine surface enhanced Raman scattering (SERS) and SPR spectroscopy in a Kretschmann configuration.2 We have designed an instrument that has an objective lens in addition to the SPR spectroscopy in order to collect both the Raman scattered light and the monitor the change in reflectivity associated with SPR from the sample in a flow cell. We are using this to study proteins deposited on a gold surface that interact with ligand functionalized gold nanoparticles. When the protein and ligand bind together, the electric field arising from the nanoparticles and the surface plasmon-polaritons (SPPs) of the gold film can excite Raman scattering of the protein-ligand complex. Simultaneously, the reflected light can be detected by a photodiode, and then the reflectivity provides information about the binding interaction. This design enables us to complement binding studies in SPR with chemical specifics from the Raman scattering. Therefore, we believe that the simultaneous measurement will be the useful tool for surface analysis and biomolecule interaction.

(1)■. G. Patching, Surface plasmon resonance spectroscopy for characterisation of membrane protein–ligand interactions and its potential for drug discovery, Biochimica et Biophysica Acta(BBA)-Biomembranes, Elsevier, 1838 (2014), 43-55
(2)■. A. Meyer, E. C. Le Ru, P. G. Etchegoin, Combining Surface Plasmon Resonance (SPR) Spectroscopy with Surface-Enhanced Raman Scattering (SERS), Analytical Chemistry, (2011), 83, 2337-2344

Keywords: Bioanalytical, Spectroscopy, Surface Analysis, Surface Enhanced Raman
Application Code: Bioanalytical
Methodology Code: Surface Analysis/Imaging

Session # 1470 Abstract # 1470-7 **Poster Sessions**

Session Title: Surface and Microscopic Characterization of Nanostructures and Biological Materials

Abstract Title: **Preparation and Characterization of Photo-Patterned Amorphous Carbon Films with Thiol-Click Reactions**

Date: Tuesday, March 08, 2016 - Afternoo
Time:
Room: Exposition Floor, 400 Aisle

Primary Author: Catherine G. McKenas
University of North Carolina at Chapel Hill

Co-Author(s): Matthew R. Lockett

Abstract Text

Amorphous carbon films are an ideal substrate to prepare optical and electrochemical sensors or arrays of biomolecules because the carbon surface is chemically inert, but can be functionalized. These films are readily deposited at room temperature, have a high adhesion, and can thus provide a chemically stable surface while maintaining the physical properties of the underlying substrate. Here we present two chemical strategies to prepare surfaces, which can be patterned with thiol-click reactions, by adding terminal thiol or vinyl groups on the amorphous carbon films. These films were then further functionalized or patterned with photo-initiated thiol-ene/yne click reactions. In order to understand how each of these reactions alters the films, we characterized the surfaces with: atomic force microscopy, contact angle measurements, infrared and photoelectron spectroscopies, cyclic voltammetry, and capacitance measurements. The specificity of the thiol-ene/yne reaction on the amorphous carbon films was investigated, and surfaces were patterned with small molecules, protein, and DNA imaged with a number of techniques. We also investigated the potential of these amorphous carbon films as electrochemical sensors and characterized the surfaces electrochemically through potential window, capacitance, and polarization measurements. The rate constants of ferrocene in solution and chemically bound to the surface were obtained and demonstrate the thiol-ene/yne chemistry is an effective means of attaching electroactive molecules to the surface.

Keywords: Electrochemistry, Elemental Analysis, Microscopy, Surface Analysis
Application Code: Material Science
Methodology Code: Surface Analysis/Imaging

Session # **1470** **Abstract #** **1470-8** **Poster Sessions**

Session Title: Surface and Microscopic Characterization of Nanostructures and Biological Materials

Abstract Title: **Electrophoretic Separation of Carbon Dots**

Date: Tuesday, March 08, 2016 - Afternoo
Time:
Room: Exposition Floor, 400 Aisle

Primary Author: Karina M. Tirado-González
University at Buffalo, SUNY

Co-Author(s): Luis A. Colón, Zuqin Xue

Abstract Text

Carbon Dots (C-dots) have generated considerable attention during recent years, mainly because of their promising applications in optoelectronic devices, bioimaging, drug delivery, and metal sensing. In addition to their intrinsic photoluminescence, low chemical reactivity, stability, and good biocompatibility, their potential to replace the more toxic and less environmentally friendly metal-based quantum dots is being explored. After dialysis and filtration, these materials can still exist as a mixture with different surface functionalities, sizes, and surface charge. We have used capillary electrophoresis with laser induced fluorescence detection to monitor C-dots prepared under different reaction conditions in an effort to establish conditions that produce less heterogeneous C-dots samples. The C-dots have been prepared using the bottom-up approach using citric acid and diethylenetriamine as the molecular precursors. Photoluminescence and electrophoretic methods were used to characterize the mixtures. We will present how electrophoretic capillary parameters affect the separation of C-dots and how the separation profile correlates to synthetic parameters.

Keywords: Capillary Electrophoresis, Fluorescence, Nanotechnology, Separation Sciences
Application Code: Nanotechnology
Methodology Code: Capillary Electrophoresis

Session # 1470 Abstract # 1470-9 **Poster Sessions**

Session Title: Surface and Microscopic Characterization of Nanostructures and Biological Materials

Abstract Title: **Preparation and Separation of Highly Fluorescent Carbon Dots**

Date: Tuesday, March 08, 2016 - Afternoo

Time:

Room: Exposition Floor, 400 Aisle

Primary Author: Zuqin Xue
University at Buffalo, SUNY

Co-Author(s): Karina M. Tirado-González, Luis A. Colón

Abstract Text

Carbon dots (C-dots) have been the rising star among nanomaterials as they provide unique chemical and optical properties. Bottom-up synthesis of C-dots has gained popularity due to its facile synthesis, abundant raw materials, and relative high quantum yields. However, the often-neglected problem is the purification of the C-dots. Previous study in our research group indicated that a complex mixture was obtained after reaction and it is necessary to use high-resolution separation techniques for further purification and fractionation. In this study, C-dots were prepared from citric acid and amine using a hydrothermal route. The reaction conditions were varied to obtain different C-dots products by changing reaction time, reaction temperature, and molar ratio. Dialysis was performed to obtain fractions containing certain C-dots sizes. Fourier transform infrared spectroscopy and transmission electron microscopy were employed to characterize all these products. Mass spectrometry was used to characterize the C-dots fractions. High performance liquid chromatography and capillary electrophoresis was used to separate and analyze the C-dots mixture. The complexity of the products and the uniqueness of some fractions were evaluated. The focus of this presentation will be the formation mechanism of the C-dots and the separation of unique fractions from the complex C-dots mixture.

Keywords: Fluorescence, Liquid Chromatography, Nanotechnology, Separation Sciences
Application Code: Nanotechnology
Methodology Code: Separation Sciences

Session # **1470** **Abstract #** **1470-11** **Poster Sessions**

Session Title: Surface and Microscopic Characterization of Nanostructures and Biological Materials

Abstract Title: **High Temperature In-Situ Reaction Monitoring of CdS Quantum Dots Using Spectrophotometers with Peltier Cell Holders**

Date: Tuesday, March 08, 2016 - Afternoo

Time:

Room: Exposition Floor, 400 Aisle

Primary Author: Kyunbae Lee
Scinco R&D Center

Co-Author(s): In-Sung Kang, Kyung-Won Ro

Abstract Text

The particle diameters of CdS quantum dots (QDs) affect the optical properties such as absorbance and photoluminescence. In general, CdS QDs are synthesized at high temperature because of nucleation and the particle sizes of QDs can be controlled by changing heating duration [1]. However, in-situ monitoring of the optical properties of QDs during synthesis reaction was difficult due to high temperature reaction condition.

In this study, we propose a new monitoring method of optical properties of QDs at high temperature by using peltier cell holders installed in UV-Vis and fluorescence spectrophotometers. This method eliminates tedious and time consuming process, such as sampling, quenching and re-dispersion, after synthesis of QDs for the measurement of optical properties. By using the peltier cell holder in a fluorescence spectrophotometer (FS-2, SCINCO Ltd, South Korea), we could successfully monitor the change of photoluminescence spectra of CdS QDs in seconds at high reaction temperature (110□) (Figure 1). In addition, we could find the shift of absorption wavelength of CdS QDs to long wavelength at a rate of 0.1 nm/sec for 5 min in a peltier cell holder of UV-Vis Spectrophotometer (Lambda 465, Perkin Elmer, USA) (Figure 2). Absorbance can be converted to band gap energy and this band gap energy is used to estimate the size of CdS QDs by Brus equation [2,3]. We found that band gap energy increased to 2.8 eV from 3.13 eV and CdS QDs grew to 2.11 nm from 1.63 nm from absorbance spectra acquired through in-situ measurement (Figure 3).

In conclusion, this new in-situ monitoring method is very useful to analyze the optical properties for QDs more quickly and simply than the conventional analysis methods and will be applicable to the in-situ reaction monitoring of synthesis process of optic probes.

Keywords: Analysis, Fluorescence, Material Science, Spectrophotometry
Application Code: Nanotechnology
Methodology Code: Fluorescence/Luminescence

Session # 1470 Abstract # 1470-12 **Poster Sessions**

Session Title Surface and Microscopic Characterization of Nanostructures and Biological Materials

Abstract Title **Determination of Airborne Concentration of Single-Wall Carbon Nanotubes and Metals by Wet Electrostatic Precipitation and Inductively Coupled Plasma Mass Spectroscopy**

Primary Author Peter Andersen
Elemental Scientific

Date: Tuesday, March 08, 2016 - Afternoo
Time:
Room: Exposition Floor, 400 Aisle

Co-Author(s) Grant Josh, John Aumen, Matt Anderson

Abstract Text

The industrial production of single-wall carbon nanotubes (SWCNT) involves the release of some SWCNT into the air. Airborne SWCNT is known to be hazardous to the health of workers. Therefore it is vital that a technique be developed to measure airborne concentrations of SWCNT in order to certify that the air is of acceptable purity. Earlier work has been done by Health Canada (Rasmussen et. al 2013 J. Phys.: Conf. Ser. 429 012007) trying to use wet electrostatic precipitators (WESP) to determine airborne SWCNT concentrations. SWCNT contain a number of metal impurities that are left over from the manufacturing process. Two elements, yttrium and nickel, were selected for use as tracers due to their elevated concentration in the SWCNT material when compared to the background air. Using a desolvation nebulizer, aqueous samples of SWCNT and metal tracers were aerosolized and introduced to the collection stack of the WESP. Analysis of the recirculation solution by inductively coupled plasma mass spectroscopy (ICP-MS) determined the yttrium and nickel concentration. Based on these concentrations the overall SWCNT concentration and SWCNT collection efficiency were calculated. Using the same testing procedure solutions containing various metals were also collected by the WESP. Additionally, the WESP was integrated with an auto sampler. This enables the WESP to be automated, allowing a user to create and run a method with samples being loaded and dispensed by the auto sampler.

Keywords: Elemental Mass Spec, Environmental/Air, ICP-MS, Nanotechnology
Application Code: Nanotechnology
Methodology Code: Mass Spectrometry

Session # 1470 **Abstract # 1470-13** **Poster Sessions**

Session Title: Surface and Microscopic Characterization of Nanostructures and Biological Materials

Abstract Title: **Impacts of Mesoporous Silica Shells on Reactivity of Metal-Semiconductor Hybrid Nanocatalysts**

Primary Author: Fei Zhao
Georgia State University

Co-Author(s): Bin Dong

Date: Tuesday, March 08, 2016 - Afternoo

Time:

Room: Exposition Floor, 400 Aisle

Abstract Text

Mesoporous silica shells are widely used for stabilization and better dispersion of single catalyst particles. However the impact of mesoporous silica shells on the reactivity of nanocatalysts is not well understood. Photocatalytic Au-CdS hybrid nanorods with and without mesoporous silica shells are used as nanocatalysts to study the real-time oxidation reaction of nonfluorescent amplex red to highly fluorescent resorufin at the single-particle level. Prism-type total internal reflection fluorescence microscope is used to monitor the catalytic reaction. By analyzing fluorescence intensity trajectories from single catalyst particles, two durations are analyzed: □on: the dissociation time of the resorufin product molecules from the Au-CdS nanorod surface, and □off : the waiting time before the oxidation reaction happens on the nanocatalyst. Using the superlocalization imaging method, different kinds of surface reactive sites on both kinds of nanocatalysts are mapped on individual nanocatalysts. Au-CdS nanocatalysts with different specifically designed geometries are also studied.

Keywords: Imaging, Microscopy

Application Code: Nanotechnology

Methodology Code: Microscopy

Session # 1470 Abstract # 1470-14 **Poster Sessions**

Session Title: Surface and Microscopic Characterization of Nanostructures and Biological Materials

Abstract Title: **Investigating Chemical Reactivity of Nanoparticles Using Nano-Impact Electrochemistry**

Primary Author: Anahita Karimi
Clarkson University

Date: Tuesday, March 08, 2016 - Afternoo
Time:
Room: Exposition Floor, 400 Aisle

Co-Author(s)

Abstract Text

The electrochemical study of particle impacts on electrode is a rapidly developing field which provides the extensive capabilities of nanoparticle detection and characterization. It has demonstrated significant promise for quantification of nanoparticles and characterization in terms of nanoparticle coatings and catalytic activity. In this presentation, we describe the development of an electroanalytical collision technique to characterize the fundamental surface properties, functionalization and redox reactivity of metal and metal oxide nanoparticles by nano-impact electrochemistry. We will demonstrate the potential of this method: (1) as a screening tool of particle reactivity, (2) study of the adsorption/desorption of environmental contaminants with single particle resolution, and (3) extract mechanistic information that would be predictive of the chemical reactivity of nanoparticles for various applications. We will discuss the potential of this approach to complement or replace costly characterization techniques and enable routine study of nanoparticles and their reactivity.

Keywords: Electrochemistry, Environmental/Water, Microelectrode, Nanotechnology
Application Code: Environmental
Methodology Code: Electrochemistry

Session # 1470 Abstract # 1470-15 **Poster Sessions**

Session Title: Surface and Microscopic Characterization of Nanostructures and Biological Materials

Abstract Title: **Advanced Analysis of LIB and Related Materials**

Date: Tuesday, March 08, 2016 - Afternoo
Time:
Room: Exposition Floor, 400 Aisle

Primary Author: Keiji Sumiya
Hitachi Chemical Co., Ltd.

Co-Author(s): Hiroki Hirano

Abstract Text

To develop advanced Li-ion batteries, elucidating the influence of functional components on battery performance is crucial.However, Li-ion batteries are very difficult to analyze, because they comprise various organic-inorganic or liquid-solid materials. Accordingly, in this study, we developed a new analytical method to elucidate the two- and three dimensional nanostructure and crystalline distribution as well as a method to visualize the quantified dispersion state of the ingredient for Li-ion batteries.

Key Features of Analytical Techniques
New advanced analytical technique applicable to an analytical area.
Clarification of unidentified performance characteristics and the beneficial effect of lithium-ion batteries and battery related materials.

As specific successful case examples:
1) 3D image-surface observation: Visualization of coating formation, acicular structure and fibrous (mesh) structure in nanoscale.
2) Visualization of images starting from the spatial distribution of functional groups to the distribution of binders inside the electrode.
3) In-situ observation of the microstructure inside a battery used for analytical study during charge/discharge cycles using a high-resolution X-ray CT apparatus.

Keywords: Characterization, Raman, Surface Analysis
Application Code: Material Science
Methodology Code: Surface Analysis/Imaging

Session # 1470 Abstract # 1470-16 **Poster Sessions**

Session Title: Surface and Microscopic Characterization of Nanostructures and Biological Materials

Abstract Title: **Evaluation of Apples Browning Using a Camera-Imaging Visual Analyzer**

Date: Tuesday, March 08, 2016 - Afternoo
Time:
Room: Exposition Floor, 400 Aisle

Primary Author: Andrew Cowell
Alpha MOS

Co-Author(s): Herve Lechat, Jean-Christophe Mifsud, Marion Bonnefille

Abstract Text

Some apples, such as Gala apples, sometimes suffer from color change during storage. To evaluate the efficiency of various processes aimed at avoiding this degradation, a precise measurement of browning extent and color is required. The objective is to compare the visual aspect of ten apples submitted to different treatments with IRIS visual analyzer. The method consists of taking a picture of cut apples, then to analyse the colors on the pictures. For that, various picture processings are applied. First, the pictures background is removed in order to focus on apples color. Then, a filter is applied to only select the colours linked with the browning of apples and not the seeds. This allows to measure the shape and surface of brown areas for each apple, and thus the proportion of brown compared to the whole apple surface. With this method it is possible to rapidly calculate the percentage of brown color in an apple and to objectively compare the efficiency of various treatments. In addition, a Quality Control card can be set up to decide whether to accept or reject batches based on determined quality criteria (e.g.1% maximum of brown color in the apple).

Keywords: Food Science, Imaging, Quality Control
Application Code: Food Science
Methodology Code: Surface Analysis/Imaging

Session # **1470** Abstract # **1470-19** **Poster Sessions**

Session Title Surface and Microscopic Characterization of Nanostructures and Biological Materials

Abstract Title **Investigation of Nitrogen-Doped Graphene Quantum Dots: Temperature Dependent Nitrogen Incorporations and their Effect on Optical Properties**

Primary Author Timothy Pillar-Little
University of Kentucky

Date: Tuesday, March 08, 2016 - Afternoo
Time:
Room: Exposition Floor, 400 Aisle

Co-Author(s) Doo Young Kim

Abstract Text

Graphene quantum dots (GQDs) are an emerging class of carbon nanomaterials with attractive physicochemical and optical properties. Advantages including cheap synthesis and tunable absorption/emission make GQDs ideal candidates for bio-imaging, electrocatalysis, and inorganic and biological sensing. Recently, it was reported that the insertion of heteroatoms such as N, B, and S greatly influences emission position and quantum yield of GQDs. However, the relationship between synthetic conditions, chemical states of heteroatoms, and their effect on optical behaviors remain unclear. The present study aims to differentiate the role of specific N chemical states in GQDs. GQDs were synthesized by chemically oxidizing carbon nano-onions. Temperature-dependent hydrothermal treatment with aqueous ammonia produced nitrogen-doped GQDs (N-GQDs). The structure and morphology of GQDs and N-GQDs were characterized by AFM and TEM. The chemical nature of nitrogen was analyzed by X-ray photoelectron spectroscopy (XPS). XPS results demonstrated that the total nitrogen content is sensitive to hydrothermal temperature. Amine-N was the dominant chemical state at low temperature. The relative content of pyridinic-N, pyrrolic-N and quaternary-N increased with higher temperatures. Fluorescence studies show that amine-N and pyridinic-N cause weak, red-shifted emission while pyrrolic-N and quaternary-N exhibit blue-shifted emission with enhanced intensity. Furthermore, pH-dependent emission behaviors of N-GQDs were studied to probe surface functionalities. In summary, the localization of N into specific chemical states was tuned by varying hydrothermal temperature and the resultant effect on optical behaviors was investigated.

This research was supported by the NSF KY EPSCoR grant and the Kentucky Science & Engineering Foundation (KSEF) grant.

Keywords: Electron Spectroscopy, Material Science, Nanotechnology, UV-VIS Absorbance/Luminescence
Application Code: Nanotechnology
Methodology Code: Fluorescence/Luminescence

Session # 1470 Abstract # 1470-20 **Poster Sessions**

Session Title: Surface and Microscopic Characterization of Nanostructures and Biological Materials

Abstract Title: **Industrial Characterization of Nano-Scale Roughness on Polished Surfaces**

Primary Author: Nikolaj A. Feidenhans'l
Danish Fundamental Metrology

Date: Tuesday, March 08, 2016 - Afternoo
Time:
Room: Exposition Floor, 400 Aisle

Co-Author(s): Giuliano Bissacco, Jan C. Petersen, Lukas Pilny, Morten H. Madsen, Poul-Erik Hansen, Rafael Taboryski

Abstract Text

Manufacturing of nanostructured polymer devices set strict demands on the smoothness of the mold master. High quality polishing of the manufacturing master is obtainable with present industrial polishing methods, but only few instruments allow in-process characterization of the samples.

In this work, we compare three optical methods for characterizing nanoscale roughness: a laboratory scatterometer measuring the angular distribution of light scattered from a surface, a confocal 3D optical profiler, and an industrial scatterometer called an OptoSurf, which is designed for in-situ characterization. The roughness parameters evaluated are: the root-mean-square height (Rq), defined in the ISO 4287 standard, and the variance of the scattered light distribution (Aq), defined in the German VDA2009 standard.

We present a relation between the Aq value measured by the OptoSurf and the ISO standardized Rq parameter. Such relation has not been shown before, but it enables a fast characterization of surface roughness, with an industrial applicable tool. The analyzed samples are made using different methods and therefore have different roughness profiles, hence the presented relations are expected to be valid for most steel surfaces made by a directional processes, such as milling, grinding, or unidirectional polishing. An overview illustrating the conversion from Aq to Rq is seen in the attached figure.

Additionally we present a study of the theoretical and empirical changes to the angular scattering distribution, if the sample surface is not properly cleaned but contains a thin liquid film.

Keywords: Characterization, Instrumentation, Nanotechnology, Process Monitoring
Application Code: Material Science
Methodology Code: Surface Analysis/Imaging

Session # 1470 Abstract # 1470-22 **Poster Sessions**

Session Title: Surface and Microscopic Characterization of Nanostructures and Biological Materials

Abstract Title: **Holographic Characterization of Particles in Complex Suspensions**

Primary Author: David B. Ruffner, Spheryx, Inc.

Date: Tuesday, March 08, 2016 - Afternoo
Time:
Room: Exposition Floor, 400 Aisle

Co-Author(s): David G. Grier, Jaroslaw M. Blusewicz, Laura Philips, Priya Kasimbeg

Abstract Text

Holographic microscopy is a powerful technology capable of characterizing colloidal microspheres with a high level of precision and accuracy, using Lorenz-Mie theory to give the size, index of refraction, and 3D position of particles in suspensions. Although this technique approximates particles as spheres, it produces meaningful results even for non-spherical particles. Holographic microscopy is used to analyze a wide array of samples that challenge current available technologies. We will present results demonstrating the application of the technology to waster water samples that contain oil emulsion droplets. A second application we will present, is in the area of polishing slurries, where the occasional presence of larger agglomerates can result in damage in manufacturing. Holographic microscopy can detect these larger agglomerates in the presence of the background signal from the slurry itself, without dilution or special sample preparation. The
size and index of refraction data from these studies demonstrate the ability of this technique to detect subtle changes in the distribution of non-ideal particles in suspension in each of these cases. These results suggest that holographic characterization could be a valuable tool for monitoring and quality control in a variety of applications of industrial interest.

Keywords: Contamination, Environmental/Water, Particle Size and Distribution, Semiconductor
Application Code: Quality/QA/QC
Methodology Code: Microscopy

Session # 1470 Abstract # 1470-25 **Poster Sessions**

Session Title Surface and Microscopic Characterization of Nanostructures and Biological Materials

Abstract Title **Characterization of Carbon Nanomaterial Dispersions for Printed Electronics**

Primary Author Qihua Wu
Brewer Science

Date: Tuesday, March 08, 2016 - Afternoo
Time:
Room: Exposition Floor, 400 Aisle

Co-Author(s) Carissa Jones, Christopher Landorf, Kay Mangelson, Stephen Gibbons

Abstract Text

Carbon nanomaterials have been actively explored for both fundamental research and commercial applications due to their exceptional chemical, electrical, thermal, and mechanical properties. Brewer Science has been working on modification and characterization of carbon nanomaterials for transducer and high-speed sensor applications.

In this work, carbon nanomaterial dispersions including functionalized carbon nanotubes (CNTs) and graphene were developed as materials for printed electronics. Material-specific properties were characterized by Thermogravimetric analysis, UV-Vis-NIR, fluorescence and Raman spectroscopy. The particle morphology and stability of the dispersions were evaluated through measuring particle size, size distribution and surface charge as zeta-potential by dynamic light scattering technique. It was found that solvent composition plays an important role on the stability of carbon nanomaterial dispersions. Variation in pHs resulted in significant changes of particle hydrodynamic size and surface charge. The detailed results and conclusions will be presented at the conference.

Keywords: Light Scattering, Materials Characterization, Nanotechnology, Sensors
Application Code: Nanotechnology
Methodology Code: Physical Measurements

Session # 1480 Abstract # 1480-1 **Awards**

Session Title Satinder Ahuja Award for Young Investigators in Separation Sciences

Abstract Title **Fully Unlocking Polyolefin Chemical Composition Distributions: Breakthrough Separations Using Graphitic Carbon**

Date: Wednesday, March 09, 2016 - Morn
Time: 08:40 AM
Room: B312

Primary Author Matthew D. Miller
The Dow Chemical Company

Co-Author(s) Abhishek Roy, Bill Winniford, Chanda Klinker, David M. Meunier, Dean Lee, Freddy Van Damme, John W. Lyons, Rongjuan Cong, Willem deGroot, Zhe Zhou

Abstract Text

Polyolefins are critical materials that are utilized in a broad range of applications and are differentiated by properties such as comonomer content and distribution, as well as, molecular weight and molecular weight distribution. The comonomer distribution for semi-crystalline polyolefins is accessible by crystallinity-based separations such as temperature rising elution fractionation (TREF) and crystallization elution fractionation (CEF). However, TREF and CEF fail with amorphous polyolefins as there is no discrimination mechanism. The invention of graphitic-carbon based separations changed this reality. Graphitic carbon-based stationary phases interact with polyolefins on the basis of the backbone composition and separations are independent of crystallizability, enabling understanding of the full range of ethylene and alpha-olefin comonomer content.

This presentation will detail the development of graphitic-carbon based separations of polyolefins and their implementation in both solvent and thermal gradient systems. Multidimensional separations that deconvolute chemical composition and molecular weight distributions as well as quantitative IR-based detection systems will be presented. Specific applications of the technology to different polymer systems and architectures will also be examined, highlighting the impact of this technology.

Keywords: Characterization, High Temperature, HPLC, Polymers & Plastics
Application Code: Polymers and Plastics
Methodology Code: Separation Sciences

Session # 1480 Abstract # 1480-2 **Awards**

Session Title Satinder Ahuja Award for Young Investigators in Separation Sciences

Abstract Title **Demystifying Flow Modulated Comprehensive Two Dimensional Gas Chromatography (GC x GC) as a Practical Problem Solving Tool**

Primary Author Bill Winniford
Dow Chemical

Date: Wednesday, March 09, 2016 - Morn
Time: 09:15 AM
Room: B312

Co-Author(s) Anna Sandlin, Chris Siegler, James Griffith, Jim Luong, Kefu Sun

Abstract Text

Comprehensive two dimensional gas chromatography has reached its 25th year since invention and has had significant growth as a technique in the last 15 years. But it tends to still be a technique practiced by a relatively small number of experts and has not seen widespread adoption. This talk emphasizes the simplicity of implementing flow modulated GC x GC on a lab GC and the learnings from the past 10 years how to quickly achieve useful quantitative results. Unlike cryogenically modulated GC x GC, flow modulated GC x GC has no limitations on component volatility. The entire range from permanent gases to high boiling components can be modulated equally well. Though TOF/MS and FID are the most common detectors used, others such as the vacuum UV detector can be used to identify isomers when mass spectra are ambiguous. A set of examples related to polymer production: monomer purity, byproducts, additives and polymer microstructure illustrate the selection of operating parameters and data handling.

Keywords: Capillary GC, Gas Chromatography/Mass Spectrometry, GC Detectors
Application Code: Polymers and Plastics
Methodology Code: Gas Chromatography/Mass Spectrometry

Session # 1480 Abstract # 1480-3 **Awards**

Session Title: Satinder Ahuja Award for Young Investigators in Separation Sciences

Abstract Title: **Separation of Building Blocks from Block Copolymers by High Performance Liquid Chromatography with Preloaded Adsorption Barriers**

Date: Wednesday, March 09, 2016 - Morn
Time: 09:50 AM
Room: B312

Primary Author: David M. Meunier
Dow Chemical Company

Co-Author(s): Eric Pearce, John W. Lyons, Mark Rickard, Tirtha Chatterjee, Todd O. Pangburn, Yongfu Li

Abstract Text

Block copolymers, such as diblocks (e. g., polyA-b-polyB), are widely used as compatibilizers for blends and as tie layers in multilayer packaging. Because block copolymers undergo nanophase separation, they can also be used for microdevice patterning. Most block copolymer materials contain unattached building blocks, and their presence can have an influence on properties and performance. Additionally, complete characterization of the block polymer composition and molecular weight can be difficult when unattached building blocks are (unknowingly) present. The separation of both polyA and polyB from polyA-b-polyB is often not possible by ordinary gradient liquid chromatography, because the most retained block of a diblock and the corresponding unattached building block often have very similar retention characteristics.

A fundamental study of the separation of building blocks from diblock copolymers by liquid chromatography with preloaded discrete adsorption barriers was performed and will be reported in this talk. The diblock copolymers used in this study were polystyrene-b-polymethylmethacrylate and polystyrene-b-polylactide. UV absorbance and evaporative light scattering detectors were used for monitoring barrier composition and polymer elution, respectively. Control experiments were performed by blending unattached building blocks with diblock copolymer samples to demonstrate complete separation and quantitation of all three components (polyA, polyB and polyA-b-polyB). The influence of the chemical composition of diblock copolymer on the separation was studied. Results showed a diblock copolymer having 90wt% of the more retained block co-eluted with any unattached building block, to illustrate a limitation of this approach for separating block copolymer components. Application of the discrete barrier approach for separating olefin block copolymers, made from polydisperse building blocks, will also be presented and discussed.

Keywords: Adsorption, Characterization, HPLC, Polymers & Plastics
Application Code: Polymers and Plastics
Methodology Code: Liquid Chromatography

Session # **1480** **Abstract #** **1480-4** **Awards**

Session Title: Satinder Ahuja Award for Young Investigators in Separation Sciences

Abstract Title: **Macromolecule and Nanoparticle Analyses: Beyond Molecular Weight and Size Measurements**

Date: Wednesday, March 09, 2016 - Morn
Time: 10:40 AM
Room: B312

Primary Author: S Kim R. Williams
Colorado School of Mines

Co-Author(s)

Abstract Text

Macromolecules and nanoparticles are analytically challenging sample systems because of their inherent complexity. Polymers typically possess a distribution of molecular weights (MW), compositions, microstructures, degrees of branching, and architectures. Similarly, synthesized and functionalized nanoparticles can have heterogeneities in the form of distributions in size, shape, and surface chemistry and coverage. These distributions can produce remarkably different properties and end product performances. Determinations of MW and size, which are now commonly done for many macromolecules and nanomaterials, are clearly insufficient as they present only the tip-of-the-iceberg with respect to key characteristics that affect observed properties and behaviors.

Field-flow fractionation (FFF), particularly asymmetrical flow FFF (AF4), is hitting its stride as a mainstream analytical technique for analyte species that range in size from the low nanometers to tens of micrometers. Another technique, thermal FFF (ThFFF), is particularly interesting because of its ability to separate analytes on the basis of composition and microstructure. We have taken a combined theory and experimental approach to understanding thermal diffusion and to developing new capabilities for complex polymers and nanoparticles analyses. The role of solvents has been examined and a universal calibration approach has been established for determining composition distributions of copolymers. A route towards obtaining the number of polymer chain ends has also been devised and applied to different polymer architectures such as stars and bottlebrushes. Very importantly, these ThFFF chain end determinations are accomplished without the use of linear polymer analogues. ThFFF is also being developed to probe the composition and structure of nanoparticles and nanohybrids. Details of these polymer and nanoparticle studies and ThFFF capabilities will be presented.

Keywords: Chromatography, Nanotechnology, Particle Size and Distribution, Polymers & Plastics
Application Code: Polymers and Plastics
Methodology Code: Separation Sciences

Session # **1480** **Abstract #** **1480-5** **Awards**

Session Title: Satinder Ahuja Award for Young Investigators in Separation Sciences

Abstract Title: **Uncertainties in Analyte Measurements in Dried Blood Spots**

Primary Author: Purnendu Dasgupta
University of Texas at Arlington

Date: Wednesday, March 09, 2016 - Morn
Time: 11:15 AM
Room: B312

Co-Author(s): Brian Stamos, Jordan Berg

Abstract Text

In 1963 Robert Guthrie first introduced screening for metabolic diseases from blood spots collected on paper filters from heel or fingerpricks. By the end of the decade screening for phenylketonuria in neonates was used nationwide in Scotland. This is now practiced in over two dozen countries to screen for similar screening and more recently for congenital hypothyroidism, sickle cell disorders and HIV infection. Many analytes are stabilized in the dried blood spot (DBS) form and unlike neat body fluid samples, DBS samples in appropriate packaging are not considered hazardous greatly facilitating transport, storage and archival. Traditionally the analytical horizon of DBS was limited as it represents only a limited amount of analyte. But improvements in the sensitivities attainable by present day mass spectrometry as well as the possibilities to amplify an appropriate analyte by the polymerase chain reaction, etc. are greatly widening the horizon of DBS applications. Typically in DBS analysis, subsamples are punched out of a larger spot for extraction and subsequent determination. If quantitative accuracy is important, it becomes necessary to know how much blood the given spot represents. Because hematocrit content of blood varies widely and this in turn controls how it spreads on the filter. Because of the near constancy of the NaCl content of blood, sodium is determined in an aliquot of the abstract. This is typically a consumptive measurement by atomic spectrometry and wasteful considering how little of the abstract is available. In this presentation I will discuss the use of a nondestructive 2-dimensional conductance measurement microprobe and the underlying theory that relates to minimum liquid depth. We have generally come to think of chromatography as a friend. Sometimes, however, it may not be: Are analytes distributed uniformly in a blood spot?

Keywords: Biological Samples, Small Samples, Trace Analysis
Application Code: Clinical/Toxicology
Methodology Code: Sampling and Sample Preparation

Session # 1490 Abstract # 1490-1 **Symposia**

Session Title ACS-ANYL - Supported Bilayers in Bio/Chemical Analysis

Abstract Title **Spectroscopic Studies of the Formation, Structure, and Applications of Hybrid Supported Phospholipid Bilayers**

Date: Wednesday, March 09, 2016 - Morn
Time: 08:35 AM
Room: B308

Primary Author Joel M. Harris
University of Utah

Co-Author(s) Jay P. Kitt

Abstract Text

Measuring lipid-membrane partitioning of small molecules is critical to developing drug delivery strategies and understanding mechanisms of cell signaling. Historically, these measurements have been approximated by octanol/water partitioning, which probes the relative solubility in bulk octanol and aqueous phases and represents a compromise for molecules having interfacial activity. A stable model membrane interface has been developed through the non-covalent interaction between a phospholipid and n-alkane modified surface or hybrid-bilayer membrane. These structures have been generated by adsorption of phospholipids onto n-alkane-thiol self-assembled monolayers on gold and used for electrochemical, infrared reflection, and surface-plasmon resonance measurements. The concept has also been adapted to producing hybrid-bilayers on C18-modified porous silica, where chromatographic retention of membrane-active peptides is comparable to their relative affinities for liposomes. Spectroscopic probing of lipid bilayers using Raman spectroscopy can provide insight into their formation, structure, and functioning; this technique is challenged, however, by the sensitivity needed to observe a small number of lipid molecules in the bilayer of a vesicle or planar supported bilayer. In this work, the sensitivity challenge is met by depositing hybrid-lipid bilayers on n-alkyl modified porous silica chromatographic particles. The highly porous silica provides a large surface area, where 1,2-dimyristoyl-sn-glycero-3-phosphocholine (DMPC) forms a stable monolayer on covalently-bound C18 chains on the interior surface. The coverage of the hybrid bilayer was quantified by Raman microscopy on individual particles. By monitoring acyl chain conformation versus temperature, it was possible to observe the melting transition, which is shifted to slightly higher temperatures compared to a DMPC vesicle. To understand the nature of melting transitions of hybrid-bilayers, Raman scattering from lipid acyl chains was resolved from that of the surface C18-chains by using deuterated DMPC. Applications of these structures in studies of small-molecule affiliation with hybrid lipid membranes will be presented. The affiliation of solutes with these model membranes will be compared to octanol/water partitioning and sorption to C18 stationary phases.

Keywords: Lipids, Microscopy, Microspectroscopy, Raman
Application Code: Bioanalytical
Methodology Code: Surface Analysis/Imaging

Session # 1490 Abstract # 1490-2 Symposia

Session Title ACS-ANYL - Supported Bilayers in Bio/Chemical Analysis

Abstract Title **Surface-Sensitive Imaging of Supported Membranes and Single Lipid Vesicles for Medical Applications**

Date: Wednesday, March 09, 2016 - Morn
Time: 09:10 AM
Room: B308

Primary Author Fredrik Höök
Chalmers University of Technology

Co-Author(s)

Abstract Text

Measurements of ligand-binding events to membrane-protein receptors in a near-natural environment display an opportunity in mechanistic studies of membrane receptors. Furthermore, the residence time of drug-target interactions is being increasingly recognized as a key parameter in evaluating drug efficacy, but is hampered by the technical challenge to perform such studies for membrane proteins. However, with membrane proteins embedded in nanoscale lipid vesicles and detection methods with single molecule sensitivity, such information can be gained in a broad dynamic range, as requested in both drug-screening and diagnostic applications. A diverse set of tools with single-nanoparticle sensitivity is now available, to which we recently contributed a concept that enables simultaneous fluorescent and scattering-based label-free imaging of thousands of surface-bound nanoscale entities [Agnarsson B et al., in revision]. The principle is based on the use of lipid vesicles as enhancer elements in optical waveguide based fluorescence and label-free evanescent-wave scattering microscopy, making the concept compatible with analysis of both water-soluble and cell-membrane bound receptors. The concept is currently evaluated as a diagnostic assay for biomarker detection and in drug-screening applications, previously explored by us using conventional total internal reflection fluorescence (TIRF) microscopy [Gunnarsson et al., Anal Chem (2015)]. The use of scattering microscopy in the context of single-enzyme detection in complex biological fluids will be presented, with focus on single-molecule biomarker detection in cerebrospinal fluid from individuals suffering from Alzheimer's disease [Angew Chemie, 2015]. A new means to utilize the two-dimensional fluidity of supported cell-membrane derived lipid bilayers in microfluidic designs for nanoparticle size determination and sorting applications will also be discussed [Simonnson et al., JACS, 2011 and Pace et al., Anal Chem, 2015].

Keywords: Imaging, Medical, Membrane, Surface Analysis
Application Code: Bioanalytical
Methodology Code: Sensors

Session # 1490 Abstract # 1490-3 **Symposia**

Session Title ACS-ANYL - Supported Bilayers in Bio/Chemical Analysis

Abstract Title **Mobile, Oriented Proteinaceous Supported Bilayers Made Directly from Cell Plasma Membranes for Bioanalytical Assays**

Date: Wednesday, March 09, 2016 - Morn
Time: 09:45 AM
Room: B308

Primary Author Susan Daniel
Cornell University

Co-Author(s)

Abstract Text

Understanding the functions of membrane proteins is important for combating disease and designing therapeutics, but they are a complicated class of biomolecules to incorporate into bioanalytical assay platforms. The supported lipid bilayer (SLB) platform is planar and compatible with many surface characterization tools. However, its full potential has yet to be reached because of the challenges associated with integration of membrane proteins, namely maintaining protein fluidity, orientation, and function. A significant bottleneck is the method in which membrane proteins are introduced into the SLB. A prevalent method uses detergent membrane disruption and proteoliposome reconstitution. This method involves tedious optimization and alters the protein orientation and structure. We developed a method for the delivery of proteins to SLBs via cell blebs. Cell blebs are sections of the cell membrane that bud off into a proteoliposome during local detachment of the membrane from the cystoskeleton. Native membrane travels with the proteins to the SLB, so crucial lipid interactions can be preserved, however interactions between the proteins and the underlying support can cause immobility. By mixing cell blebs with polyethylene glycol (PEG) containing vesicles, we generate a cushioned bleb bilayer. We examined a GPI-linked protein and a multi-pass transmembrane protein, the P2X receptor, an ATP-gated ion channel with 6 transmembrane helices. Individual proteins were tracked and analyzed to distinguish diffusion modes. Orientation assays determined that the rupture process for both types of proteins results in predominantly outward facing membrane proteins, implying a "parachute" mechanism of bleb rupture. This platform containing mobile, oriented proteins preserved with lipids from the plasma membrane is a critical intermediate platform that bridges whole cell to in vitro systems and will play a key role in the development of novel membrane bioanalytical assays in the future

Keywords: Bioanalytical, Biological Samples, Biosensors, Biotechnology
Application Code: Bioanalytical
Methodology Code: Fluorescence/Luminescence

Session # 1490 Abstract # 1490-4 **Symposia**

Session Title ACS-ANYL - Supported Bilayers in Bio/Chemical Analysis

Abstract Title **Stabilized Lipid Bilayers as a Platform for Fabrication of Ion Channel Functionalized Sensors**

Date: Wednesday, March 09, 2016 - Morn
Time: 10:35 AM
Room: B308

Primary Author Craig A. Aspinwall
University of Arizona

Co-Author(s) Leonard K. Bright, Mark T. Agasid, S Scott Saavedra

Abstract Text

Suspended lipid membranes, also known as black lipid membranes, (BLMs) provide a synthetic environment that facilitates measurement of ion channel activity in diverse analytical platforms. The limited electrical, mechanical and temporal stabilities of BLMs pose a significant challenge to development of highly stable measurement platforms. To overcome these limitations, we have devised a series of approaches that involve: surface chemical modification of the underlying support; formation of BLMs using synthetic, reactive lipids; and/or c) polymer scaffolding of natural lipid membranes. Using these approaches, we have developed BLMs that are stable for days to weeks, compared to hours for traditional BLMs. Furthermore, these stabilized BLMs support the function of ion channels that can be used in the preparation of biochemical sensors and other ion channel-functionalized measurement platforms. These key enabling results provide the foundation upon which our future efforts in the development of ion channel-functionalized sensors, sequencers, diagnostic and discovery platforms and beyond will be based. In this presentation we will discuss current strategies to further improve BLM stability and our efforts to integrate ion channels to develop ligand responsive ion channel sensors.

Keywords: Biosensors, Electrochemistry, Lipids, Sensors
Application Code: Bioanalytical
Methodology Code: Sensors

Session # 1490 Abstract # 1490-5 Symposia

Session Title ACS-ANYL - Supported Bilayers in Bio/Chemical Analysis

Abstract Title **Supported Lipid Bilayers and Scanning Ion Conductance Microscopy**

Date: Wednesday, March 09, 2016 - Morn
Time: 11:10 AM
Room: B308

Primary Author Lane A. Baker
Indiana University

Co-Author(s)

Abstract Text
The sensitivity and selectivity of ion channels provides an appealing opportunity for sensor development. Here, we describe ion channel probes (ICPs) which consist of multiple ion channels reconstituted into lipid bilayers suspended across the opening of perflourinated glass micropipets. When incorporated with a scanning ion conductance microscope (SICM), ICPs displayed a distance dependent current response that was depended on the number of ion channels in the membrane. With distance dependent current as feedback, probes were translated laterally, to demonstrate the possibility of imaging with ICPs. The ICP platform yields several potential advantages for SICM that will enable exciting opportunities for incorporation of chemical information into imaging and for high resolution imaging.

Keywords: Biosensors, Electrochemistry, Imaging
Application Code: Bioanalytical
Methodology Code: Electrochemistry

Session # 1500 Abstract # 1500-1 **Symposia**

Session Title: Analytical Applications of Terahertz Time Domain Spectroscopy (THz-TDS)

Abstract Title: **Exploration of Interlayer Chemistry in Clay Minerals by Terahertz Spectroscopy**

Date: Wednesday, March 09, 2016 - Morn
Time: 08:35 AM
Room: B302

Primary Author: Ingrid Wilke
Rensselaer Polytechnic Institute

Co-Author(s)

Abstract Text

Among clay minerals, montmorillonites are known to catalyze numerous chemical reactions relevant to the chemical and pharmaceutical industries. Since montmorillonites are a natural resource abundantly present on Earth with no reported toxicity to humans, animals and plants, they are regarded as a promising candidate for the development of chemical processes that reduce or eliminate the generation of hazardous waste. Synthesis of organic molecules using montmorillonite as catalysts is empirically known to work but not clearly understood at the molecular level. Catalysis by montmorillonite is the result of the layered structure of the material. In montmorillonite, two tetrahedral sheets of silicon oxides sandwich a central octahedral sheet of aluminum oxides. For catalysis, the molecules of interest are "inserted" into the nanometer-sized space between the layers. This is the location where the chemical reaction occurs and the reaction products are formed. Spectroscopic probes for the investigation of interlayer adsorbate molecular structure are well established in surface chemistry, e.g. x-ray absorption, infrared -, electron spin resonance - and nuclear magnetic resonance spectroscopy. THz spectroscopy is emerging as an approach to study the dynamics of polar molecules intercalated in clay minerals because it provides unique complementary information about the adsorbate molecular dynamics in comparison with state-of-the art spectroscopic probes.

In my talk I present experimental results illustrating that THz spectroscopy meets the key physical criteria for accurate spectroscopic identification of adsorbate molecular dynamics. The approach is selective, i.e. the octa- and tetrahedral sheets of the montmorillonite absorb THz radiation very weakly whereas polar molecules, e.g. water, absorb strongly in the THz frequency band. Also THz spectroscopy is sufficiently sensitive to detect adsorbed molecules at relevant concentrations, e.g. monomolecular layers.

Keywords: Adsorption, Molecular Spectroscopy, Surface Analysis, Vibrational Spectroscopy
Application Code: General Interest
Methodology Code: Vibrational Spectroscopy

Session # 1500 Abstract # 1500-2 **Symposia**

Session Title: Analytical Applications of Terahertz Time Domain Spectroscopy (THz-TDS)

Abstract Title: **Analytical Applications of Terahertz Spectroscopy in Nanotechnology and Biotechnology**

Primary Author: Anis Rahman
Applied Research & Photonics

Date: Wednesday, March 09, 2016 - Morn
Time: 09:10 AM
Room: B302

Co-Author(s)

Abstract Text

Accessing the terahertz region of the electromagnetic spectrum (~0.1 THz to ~30 THz) has enabled a number of analytical applications utilizing the energy within this band. T-ray is non-ionizing and can penetrate almost all non-metallic objects without causing radiation damage. Recent discovery of dendrimer dipole excitation (DDE) allows one to generate continuous wave T-ray without requiring a femto-second pulsed laser. This allows an opportunity to investigate intrinsic properties of biomaterials in their native environment without damage via ionization. A wide broadband terahertz spectrometer was utilized to illustrate unique applications of terahertz spectroscopy as an analytical tool. Measurements are conducted in time-domain to capture the time evolution of an interferogram that is generated by scanning the probe arm of the spectrometer by the pump arm that interrogates the sample. Fourier transform techniques are used to analyze the time-domain signal that generates an absorbance spectrum or magnitude spectrum. Other analysis options are also available. Illustrative examples include analysis of chemical compounds, nanoparticles with ligand, analysis of emulsifying additives, contamination in gasoline, and others. It was found that the technique is suitable for interrogating molecular motions that are not easily visible by traditional means. The time-domain terahertz spectrometer (TeraSpectra) not only reproduces the known absorbance peaks of standard materials, it generates additional peaks that were not visible before. The difficulty, however, is that interpretation of these newly observed peaks may not be readily available. The technique thus provides a unique opportunity for discovering critical molecular signatures of materials of practical importance.

Keywords: Biotechnology, Nanotechnology, Spectroscopy, Ultra Fast Spectroscopy
Application Code: Nanotechnology
Methodology Code: Vibrational Spectroscopy

Session # 1500 Abstract # 1500-3 **Symposia**

Session Title Analytical Applications of Terahertz Time Domain Spectroscopy (THz-TDS)

Abstract Title **PAT Measurements to Enhance Small-Molecule Drug Development and Manufacturing**

Date: Wednesday, March 09, 2016 - Morn
Time: 09:45 AM
Room: B302

Primary Author Huiquan Wu
Food and Drug Administration

Co-Author(s)

Abstract Text
FDA's PAT initiative created unprecedented opportunities for innovative pharmaceutical development and manufacturing. Terahertz (THz) Spectroscopy has found wide application in the pharmaceutical area, from drug discovery to drug development and has potential applications to drug manufacturing. As an emerging technology, THz spectroscopy may be adopted as a PAT tool applied to pharmaceutical polymorphic form identification, structural differentiation, and real time process monitoring and control of extended release dosage form manufacturing. Based on recent FDA research as well as other published literature, the following THz PAT applications will be discussed: 1) THz spectroscopy used for pharmaceutical product characterization and pharmaceutical development, such as polymorph screening and solid-state assessment, assessing weak interaction and intra-molecular interaction, and mapping chemical compositions and tablet density; 2) THz spectroscopy and THz Pulse Imaging (TPI) used for process monitoring of pharmaceutical tablet film coating, including coating layer thickness, coating uniformity, and dissolution performance; 3) chemometric aspects, such as impact of THz data quality and spectral data preprocessing algorithms on predictive power and robustness of the THz PAT model; and 4) First principles modeling approaches to address model validation challenges such as in situ THz monitoring for examining solvent diffusion in polymers and ex-situ THz monitoring for examining moisture content in test disk of pharmaceuticals. Challenges and opportunities in adopting THz spectroscopy as a PAT tool will be discussed from a technical perspective.

Keywords: On-line, Pharmaceutical, Process Monitoring, Quality Control
Application Code: Pharmaceutical
Methodology Code: Process Analytical Techniques

Session # 1500	Abstract # 1500-4	Symposia

Session Title: Analytical Applications of Terahertz Time Domain Spectroscopy (THz-TDS)

Abstract Title: **Understanding Bonding in Cocrystals and Identifying Polymorphs in Small-Molecule Drugs**

Primary Author: Timothy M. Korter
Syracuse University

Date: Wednesday, March 09, 2016 - Morn
Time: 10:35 AM
Room: B302

Co-Author(s)

Abstract Text

Terahertz spectroscopy is rapidly becoming an established analytical tool for the characterization of the three-dimensional arrangements of organic molecules in the solid-state. This sensitivity to molecular shape and crystal packing arises from the nature of the lattice vibrations (torsions, translations, and rotations) that occur in the terahertz spectral region (e.g. sub-100 cm[sup]-1[/sup]). It is precisely this global vibrational character that makes terahertz spectra difficult to assign to particular molecular motions, since group frequency tables are not applicable here. To meet this challenge, solid-state density functional theory has been utilized with great success, enabling accurate simulations of molecular crystal structures and dynamics to be achieved.

In this seminar, experimental terahertz spectra of several pure and binary crystalline samples will be shown, with primary focus being the interpretation of the spectral data by quantum mechanical techniques. For example, detailed analyses of the enantiotropic and monotropic relationships of d-mannitol polymorphs will be presented, including the calculation of Gibbs free energy curves. The importance of third-generation London force corrections (DFT-D3) and continuous basis set superposition error evaluation will be highlighted. Finally, as part of the ongoing effort to enhance understanding of intermolecular bonding in pharmaceutical co-crystals, the computational techniques of crystal orbital overlap population (COOP) and crystal orbital Hamiltonian population (COHP) calculations will be described for use in such systems, including aspects concerning their recent inclusion into atom-centered solid-state density functional theory software for the first time.

Keywords: Infrared and Raman, Pharmaceutical, Spectroscopy, Vibrational Spectroscopy
Application Code: Pharmaceutical
Methodology Code: Vibrational Spectroscopy

Session # 1500 Abstract # 1500-5 **Symposia**

Session Title Analytical Applications of Terahertz Time Domain Spectroscopy (THz-TDS)

Abstract Title **Analytical Measurements and Dielectric Properties of Organic Cocrystals by Time-Domain Terahertz Spectroscopy**

Primary Author Mark A. Arnold
University of Iowa

Date: Wednesday, March 09, 2016 - Morn
Time: 11:10 AM
Room: B302

Co-Author(s)

Abstract Text

Terahertz time domain spectroscopy (THz-TDS) offers the opportunity to obtain high signal-to-noise spectra over optical frequencies associated with the far infrared region of the electromagnetic spectrum (0.2-4.0 THz or 5-130 cm-1). For THz-TDS, a pulse of coherent THz radiation is generated by acceleration of electrons in response to a femtosecond pulse of an excitation laser (800 nm) onto a polarized region of a solid-state semiconductor. After propagating through the sample, gated detection is used to record the amplitude of the electric-field vector of the transmitted radiation. A time-domain signal results and represents the superposition of the resulting THz frequency waves, thereby encoding both phase and power information. This time-domain signal can be Fourier transformed to provide a frequency-domain spectrum that encodes absorption and scattering information associated with the sample. The dielectric properties of the sample can be determined directly from analysis of the "time of flight" of the coherent THz radiation, while concentration information can be obtained from analysis of the measured absorption strength at specific frequencies.

This presentation will explore the analytical utility of THz-TDS for characterization of organic cocrystals. Selectivity and quantitative properties of cocrystals will be presented for THz absorption spectra collected for resourcinol-templated cocrystals of 4,4'-bispyridylethylene. Quantitation properties are further established by demonstrating our ability to monitor the single-crystal-to-single-crystal [2+2] intermolecular photodimerization reaction of 5-cyanoresourcinol 4,4'-bispyridylethylene cocrystals to form 5-cyanoresourcinol tertakis(4-pyridylcyclobutane) cocrystals. In addition, the ability to measure both dielectric and polarizability properties directly from THz time-domain signals will be detailed for a family of cocrystals.

Keywords: Material Science, Monitoring, Quantitative
Application Code: Material Science
Methodology Code: Chemical Methods

Session # 1510 Abstract # 1510-1 **Symposia**

Session Title: Frontiers of Plasmonics

Abstract Title: **Real-Time Tunable Emission from Plasmonic Nanolasers**

Date: Wednesday, March 09, 2016 - Morn
Time: 08:35 AM
Room: B303

Primary Author: Teri W. Odom
Northwestern University

Co-Author(s)

Abstract Text

Plasmon lasers can support ultrasmall mode confinement and ultrafast dynamics with device sizes below the diffraction limit. However, most spasers or plasmon-based nanolasers rely on solid gain materials that preclude the possibility of dynamic tuning. This talk will discuss how to achieve real-time, tunable lattice plasmon lasing based on 2D arrays of gold nanoparticles and liquid gain. Optically pumped gold nanoparticle arrays surrounded by dye solutions exhibited lasing emission that could be tuned as a function of dielectric environment. Wavelength-dependent, time-resolved experiments showed distinct lifetime characteristics below and above lasing threshold. Dynamic tuning of the plasmon lasing wavelength can be achieved by integrating the nanoparticle arrays in a microfluidic device. Tunable lattice plasmon lasers offer new prospects for applications such as on-chip light sources and bio-sensing.

Keywords: Nanotechnology
Application Code: Material Science
Methodology Code: Mass Spectrometry

Session # 1510 **Abstract # 1510-2** **Symposia**

Session Title: Frontiers of Plasmonics

Abstract Title: **Nanoplasmonic Sensors for Rapid Concentration and Sensitive Detection of Biomolecules**

Date: Wednesday, March 09, 2016 - Morn
Time: 09:10 AM
Room: B303

Primary Author: Sang-Hyun Oh
University of Minnesota

Co-Author(s)

Abstract Text

For surface-based biosensors, diffusive transport of analyte molecules into the sensing "hot spot" is typically a slow process that limits the sensor performance. We discuss new fabrication techniques and sensing strategies to overcome the diffusion limit and enable rapid and sensitive detection of biomolecules. We use template stripping and atomic layer lithography to engineer noble metal nanostructures with ultra-high patterning resolution. After integration with microfluidic chips, these patterned nanostructures can concurrently act as trapping sites for biological analytes.

Keywords: Bioanalytical, Biosensors, Spectroscopy, Vibrational Spectroscopy
Application Code: Bioanalytical
Methodology Code: Sensors

Session # **1510** **Abstract #** **1510-3** **Symposia**

Session Title: Frontiers of Plasmonics

Abstract Title: **Plasmonic Rectification**

Date: Wednesday, March 09, 2016 - Morn
Time: 09:45 AM
Room: B303

Primary Author: Zachary D. Schultz
University of Notre Dame

Co-Author(s)

Abstract Text

The excitation of plasmon resonances results in electric fields that can be used for a variety of applications from sensing to catalysis and photovoltaics. The intense electric fields are associated with surface enhanced Raman scattering (SERS). High sensitivity SERS sensing often requires plasmon coupling between nanostructures, to form hotspots or junctions with the greatest electric fields. However, the observed Raman scattering can change with different geometric arrangements of nanoparticles, excitation wavelengths, and chemical environments; suggesting differences in the local electric field. Experimental results we have obtained suggest that the intense electric fields on nanostructures can give rise to second order nonlinear phenomena, such as optical rectification. The rectified plasmonic field is evident as a shift in the vibrational frequency of reporter molecules on the nanostructures. This rectified field has not been widely explored, but provides new insight into the control of electric fields on nanostructures. Here we present a strategy that utilizes the controlled formation of coupled plasmonic structures to experimentally measure both the magnitude of the electric fields and the observed Raman scattering. Preliminary results suggest that quantum tunneling between nanostructures may have significant consequences for sensing applications.

Keywords: Biosensors, Nanotechnology, Surface Enhanced Raman
Application Code: Nanotechnology
Methodology Code: Surface Analysis/Imaging

Session # 1510 Abstract # 1510-4 **Symposia**

Session Title: Frontiers of Plasmonics

Abstract Title: **Trends and Challenges of Nanoplasmonic Biosensors for Clinical Use in Diagnostics**

Primary Author: Laura M. Lechuga
ICN2. CSIC & CIBER-BBN

Date: Wednesday, March 09, 2016 - Morn
Time: 10:35 AM
Room: B303

Co-Author(s)

Abstract Text

Motivated by the benefits such as user-friendly, multiplexing capabilities and high sensitivities, Plasmonic biosensors have profiled themselves as an excellent alternative to traditional analytical techniques. Both SPR and LSPR sensors have attracted a remarkable interest for label-free biosensing. But the main challenge for the Plasmonic sensors is the demonstration of its utility in the clinical practise by handling and directly analyse biomarkers in minimum amounts of body fluids without previous treatment.

During last years, we have developed fully-automated and portable plasmonic sensor platforms. Surpassing the challenges, we have demonstrated the suitability of our plasmonic sensors for the clinical diagnostics, as for example:(i)Drug allergy diagnosis (antibiotic) using dendrimer receptors which enable the detection of IgE antibodies directly in patients´serum. (ii)Detection of gluten-derivative peptides in urine as a rapid and non-invasive technique for dietary control of celiac patients.(iii)Detection of specific tumor-related autoantibodies in serum, which indicate the onset of colorectal cancer in its preliminary stage. In all cases, our plasmonic sensors has shown excellent robustness with high reproducibility and sensitivity, rendering in a valuable tool for the diagnostics of bodily fluids.

In addition, we have recently demonstrate the use of plasmonic sensors as an unconventional strategy for studying gene expression pathways in cells such as: (i)alternative mRNA splicing variants, (ii) epigenetics modifications as DNA methylation; (iii)interaction with non-coding RNA regulators such as microRNAs. This approach can be relevant for the prediction of patient response to cancer treatments by identifying defective pathways leading to apoptosis-resistance of tumors.

Our next challenge is to achieve a stand-alone plasmonic biosensor for the clinical practice at the point of care with competitive levels of multiplexing, robustness and integration.

Keywords: Bioanalytical, Biomedical, Biosensors, Lab-on-a-Chip/Microfluidics
Application Code: Bioanalytical
Methodology Code: Sensors

Session # 1510 Abstract # 1510-5 **Symposia**

Session Title: Frontiers of Plasmonics

Abstract Title: **Plasmonic Nanobiosensors: From Therapeutic Drug Monitoring to the Detection of Molecules Secreted by Cells**

Date: Wednesday, March 09, 2016 - Morn
Time: 11:10 AM
Room: B303

Primary Author: Jean-Francois Masson
Universite de Montreal

Co-Author(s)

Abstract Text

In this presentation, the properties of different plasmonic nanostructures will be discussed in the context of clinical and biological sensing. The optical properties of nanohole arrays were optimized for sensing proteins and antibodies with a sensitivity approaching classical SPR sensors. The nanohole arrays of 1.2 micron periodicity and 800 nm hole duiameter were fabricated with photolithography on a 4" wafer. A 96-well plate reader competent for full speactra acquisition of transmission measurements at an angled incidence was constructed and integrated with acquisition software. This platform was used to screen antibodies for PSA sensing. Using a Kretschmann SPR sensing platform based on a small and portable instrument, competition assays were validated for therapeutic drug quantitation. Au nanoparticle competitors were synthesized based on an analogue molecule to the therapeutic drug and validated in the clinical range for methotrexate and antibiotics. A sensor was also designed for monitoring therapeutic responses of patients undergoing leukemia treatments. In this assay, antibodies expressed as an allergic response to the treatment was measured in the sera of young patients during ongoing treatments. Lastly, we are currently exploring the concept of plasmonic nanopipettes for monitoring cell secretion events. Monitoring cell secretion events remains a challenge to overcome in chemical analysis. Plasmonic nanopipettes were developed based on the decoration of patch clamp nanocapillaries with Au nanoparticles. The plasmonic nanopipette is thus competent for dynamic SERS measurements in the liquid environment near cells. This nanobiosensor was tested with the detection of small metabolites near live cells.

Keywords: Biosensors, Raman, Spectroscopy
Application Code: Biomedical
Methodology Code: Sensors

Session # **1520** Abstract # **1520-1** **Symposia**

Session Title IAEAC: International Association of Environmental Analytical Chemistry - Upconverting Nanocrystals:

Abstract Title **Optical Nanotransformers for In-Situ Upconversion: From Design to Functional Imaging and Sensing**

Primary Author Paras N. Prasad
SUNY Buffalo

Date: Wednesday, March 09, 2016 - Morn
Time: 08:35 AM
Room: B304

Co-Author(s) Guanying Chen, Tymish Y. Ohulchanskyy

Abstract Text

Optical nanotransformers, converting a photon of one specific wavelength to another, open up possibilities for a broad range of biomedical applications1-3. A core-multiple shell nanoarchitecture provides nanophotonic control of excitation dynamics that allows to produce very efficient and selective upconversion (up to five-photon processes), whereby IR photons can be transformed to blue or even UV photons under NIR irradiance of low power (~ 10-1 W/cm2) accessible from low-cost coherent or incoherent light sources. The in-situ upconversion is capable to overcome the major hurdle of limited penetration of light of shorter wavelengths into a biological medium. In addition, an appropriate core–multiple shell design can produce simultaneous upconversion at a number of wavelengths, allowing for multiplexed optical imaging, and combine different modalities of imaging (e.g., photoacoustic and MRI). Furthermore, "see, treat and see" theranostic approaches combining diagnostic with therapy can be readily introduced. In this talk, we present various types of core-multiple shell designs of our optical nanotransformers. We demonstrate successful applications of these newly developed nanoparticles for high-contrast in vivo optical imaging, through-bone imaging, photoacoustic bioimaging, NIR light-activated in situ uncaging and optogenetics. Functional imaging and sensing for pH, local temperature, cell membrane potential as well as cell morphological changes during a disease progression or a therapeutically induced bioprocess will also be discussed. This talk will conclude with a discussion of future outlook and new opportunities.

1.P. N. Prasad "Nanophotonics" John Wiley & Sons, (2004).
2.P. N. Prasad "Introduction to Biophotonics" John Wiley & Sons, (2003).
3.P. N. Prasad "Introduction to Nanomedicine and Nanobioengineering" John Wiley & Sons,(2012).

Keywords: Imaging, Laser, Luminescence, Microscopy
Application Code: Biomedical
Methodology Code: Near Infrared

Session # 1520 **Abstract # 1520-2** **Symposia**

Session Title: IAEAC: International Association of Environmental Analytical Chemistry - Upconverting Nanocrystals:

Abstract Title: **Photodynamic Therapy, Drug Delivery, Persistent and Photo-Stimulated Emission Using Low Excitation Photons**

Primary Author: John A. Capobianco
Concordia University

Date: Wednesday, March 09, 2016 - Morn
Time: 09:10 AM
Room: B304

Co-Author(s)

Abstract Text

Lanthanide doped nanoparticles have the ability to undergo upconversion. Upconversion is a non-linear anti-Stokes process that efficiently converts two or more low-energy excitation photons, which are generally near infrared (NIR) light, into a higher energy photon (e.g., NIR, visible, ultraviolet) through the use of long lifetime and real ladder-like energy levels of trivalent lanthanide ions embedded in an appropriate inorganic host lattice. Thus, these materials are quickly emerging as candidates in novel biological applications such for PDT, photoswitching and as imaging probes for cancer cells. This stems from their unique optical and chemical properties, such as non-blinking, non-photobleaching, absence of autofluorescence, low-toxicity, low photodamage to cells, and their remarkable ability of NIR light to efficiently penetrate tissues. We will report on studies using upconverting nanoparticles in photodynamic therapy and drug delivery and on the synthesis, the characterization and the optical properties of CaS:Eu2+/Dy3+ nanophosphors, which demonstrate strong red light emission following NIR excitation.

Keywords: Biomedical, Nanotechnology, Near Infrared, UV-VIS Absorbance/Luminescence

Application Code: Nanotechnology

Methodology Code: Fluorescence/Luminescence

Session # 1520 **Abstract #** 1520-3 **Symposia**

Session Title IAEAC: International Association of Environmental Analytical Chemistry - Upconverting Nanocrystals:

Abstract Title **Printing Enhanced Upconverting Nanocrystals on Solid Supports**

Date: Wednesday, March 09, 2016 - Morn
Time: 09:45 AM
Room: B304

Primary Author Stanley May
University of South Dakota

Co-Author(s) Aravind Baride, Grant Crawford, Jeevan Meruga, Jon Kellar, Mary Berry, William Cross

Abstract Text

Recent advances in producing highly resolved, pre-defined patterns of upconversion (UC) nanophosphors via printing and other techniques present new opportunities for the use of these materials in sensing, theranostic, and security applications. UC phosphors convert long-wavelength excitation light to shorter-wavelength luminescence. The UC materials discussed here are invisible under ambient light and UV excitation, but become visible (or otherwise detectable) under NIR excitation. We have demonstrated that covert upconversion QR codes printed using aerosol-jet technology (and subsequently adapted to inkjet technology), are readable using a near-IR laser, and can be successfully scanned using a smart phone. This research demonstrates that UC inks can be used to provide selected access to encoded information. We have also demonstrated an RGB additive-color printing system that produces highly resolved pre-defined patterns that are invisible under ambient lighting, but which are viewable as luminescent multi-color images under NIR excitation. Most recently, we have developed a print-and-read system based on NIR-to-NIR (980 nm-to-800 nm) upconversion luminescence. Remaining in the NIR spectral region for both excitation and emission has distinct advantages, because both excitation and emission wavelengths are able to penetrate highly scattering and / or visibly opaque media. Moreover, inexpensive CCD-based detectors and imaging devices exhibit peak sensitivity at the 800 nm emission wavelength of these nanocrystals. NIR-to-NIR images are easily captured, for example, using an inexpensive, modified point-and-shoot CCD camera, even at modest excitation power densities (1 W/cm[sup]2[/sup]). Finally, we will discuss the potential for using metal-pattern-array substrates for significant enhancement of upconversion brightness through both plasmonic and optical mechanisms. (See Fig. 1)

Keywords: Bioanalytical, Imaging, Luminescence, Spectroscopy
Application Code: Nanotechnology
Methodology Code: Fluorescence/Luminescence

Session # **1520** **Abstract #** **1520-4** **Symposia**

Session Title IAEAC: International Association of Environmental Analytical Chemistry - Upconverting Nanocrystals:

Abstract Title **Surface Modifications of Upconverting Nanoparticles and Their (Bio)analytical Applications**

Date: Wednesday, March 09, 2016 - Morn

Time: 10:35 AM

Room: B304

Primary Author Thomas Hirsch
University of Regensburg

Co-Author(s)

Abstract Text

Upconverting nanoparticles (UCNPs) have emerged as a promising new class of nanomaterials due to their ability to convert near-infrared (NIR) light into visible luminescence, narrow emission bands, high chemical stability, and photostability. The highest upconversion efficiency has been observed in hexagonal-phase NaYF[sub]4[/sub]:Yb,Er nanocrystals. The effect of surface ligands on the luminescence properties and colloidal stability of water-dispersible UCNPs is presented. The ratio of the 545 nm and 658 nm emission bands of these UCNPs determined at a constant excitation power density clearly depends on the surface chemistry. Modifications relying on the deposition of additional (amphiphilic) layer coatings show reduced non-radiative quenching by water as compared to UCNPs that are rendered water-dispersible via ligand exchange. The brightness of the upconversion luminescence is strongly affected by the type of surface modification, yet hardly by the chemical nature of the ligand. Modifications using an additional shell provide a higher dynamic range of the green/red ratio favorable for sensing schemes based on inner filter effects, while ligand exchange seems to be better suited for the design of sensors utilizing fluorescence resonance energy transfer (FRET). Since FRET is distance dependent, the diameter of the UCNPs should affect the FRET efficiency. To study this influence UCNPs of different diameters ranging from 10 to 42 nm have been investigated in order to determine the optimum particle size for FRET based sensing applications. The surface of the monodisperse, oleate-capped UCNPs was modified with fluorescent dyes via a two-step ligand exchange assisted by NOBF[sub]4[/sub], resulting in the shortest possible distance between donor and acceptor for highest FRET efficiencies. Optimized FRET processes can be used for the development of hybridization assays, optical sensors for e.g. pH, or for efficient photodynamic therapy using NIR excitation.

Keywords: Bioanalytical, Nanotechnology, Sensors

Application Code: Bioanalytical

Methodology Code: Sensors

Session # **1520** **Abstract #** **1520-5** **Symposia**

Session Title IAEAC: International Association of Environmental Analytical Chemistry - Upconverting Nanocrystals:

Abstract Title **Upconversion Nanoparticles as Active Elements in Optical Sensing for Development of Protein and Oligonucleotide Bioassays**

Primary Author Ulrich J. Krull
University of Toronto Mississauga

Date: Wednesday, March 09, 2016 - Morn
Time: 11:10 AM
Room: B304

Co-Author(s) Anna Shahmuradyan, Feng Zhou, Samer Doughan, Yi Han

Abstract Text

Luminescence from upconversion nanoparticles (UCNPs) can be used to interrogate selective interactions at the surface of the nanoparticles by means of resonance energy transfer and by direct excitation. Selectivity can be established by use of immobilized oligonucleotide probes for hybridization, or for aptamer-ligand interaction. Transduction of selective binding can be determined from changes of emission of a fluorescent label that is excited by the UCNPs. Of interest is whether these systems can be made to have physical and chemical stability, whether a high signal-to-noise in the optical channels can be achieved by using near infrared excitation of the UCNPs to reduce background, and whether effective multiplexing is possible. Covalent immobilization of UCNPs would offer opportunity for preparation of stable bioassay and biosensor platforms. UCNPs were modified by ligand exchange of oleic acid with o-phosphorylethanolamine (PEA) to place amine groups on the surface for covalent conjugation. PEA-UCNPs were covalently immobilized on aldehyde functionalized coverslips and on paper substrates. The immobilized films had a density of coverage that was similar to that observed using other forms of nanoparticles. Analytical functionality of the immobilized UCNPs was demonstrated by an aptamer-based sandwich assay for the detection of thrombin that made use of quantum dots as acceptors. Using UCNPs adsorbed to paper substrates, dye-labeled oligonucleotide targets have been used to develop a duplex bioassay using two concurrent emission bands of the UCNPs. In further work, use of intercalating dyes to transduce hybridization of target oligonucleotides by single-stranded oligonucleotide probes has been achieved. Single-stranded probes that integrate the transducing dye into the probe structure have been developed to eliminate the need for labeled target.

Keywords: Bioanalytical, Fluorescence, Luminescence, Nanotechnology
Application Code: Bioanalytical
Methodology Code: Fluorescence/Luminescence

Session # 1530 Abstract # 1530-1 **Symposia**

Session Title Ion Mobility/Mass Spectrometry for Metabolomics and Clinical Analysis

Abstract Title **Structural Mass Spectrometry for Systems, Synthetic, and Chemical Biology**

Primary Author John A. McLean
Vanderbilt University

Date: Wednesday, March 09, 2016 - Morn
Time: 08:35 AM
Room: B305

Co-Author(s)

Abstract Text

One of the predominant challenges in systems-wide analyses is the broad-scale characterization of the molecular inventory in cells, tissues, and biological fluids. Advances in computational systems biology rely heavily on the experimental capacity to make omics measurements, i.e. integrated metabolomics, proteomics, lipidomics, glycomics, etc., accompanied with fast minimal sample preparation, fast measurements, high concentration dynamic range, low limits of detection, and high selectivity. This confluence of figures-of-merit place demanding challenges on analytical platforms for such analyses. Ion mobility-mass spectrometry (IM-MS) provides rapid (ms) gas-phase electrophoretic separations on the basis of molecular structure and is well suited for integration with rapid (us) mass spectrometry detection techniques. Furthermore, the timescales of this multi-dimensional separation are well suited for combination with fast condensed-phase separations such as GC, SFC, and UPLC (min) for enhanced separation selectivity as the sample complexity becomes ever more challenging. This report will describe recent advances in IM-MS omics measurement strategies in the analyses of complex biological samples of interest in systems, synthetic, and chemical biology. Specifically, avenues for rapidly characterizing organotypic cultures in "human-on-a-chip" endeavors, microbiome and commensal communities, and drug discovery from hypogean organisms will be discussed. New advances in bioinformatics and biostatistics will also be described to approach biological queries from an unbiased and untargeted perspective and to quickly mine the data gathered to provide targeted and actionable information.

Keywords: Bioanalytical, Bioinformatics, Mass Spectrometry, Metabolomics
Application Code: Bioanalytical
Methodology Code: Mass Spectrometry

Session # 1530 Abstract # 1530-2 **Symposia**

Session Title: Ion Mobility/Mass Spectrometry for Metabolomics and Clinical Analysis

Abstract Title: **Enhancing Ion Mobility-Mass Spectrometry Metabolomic Analyses with High Throughput Front End Separations**

Date: Wednesday, March 09, 2016 - Morn
Time: 09:10 AM
Room: B305

Primary Author: Erin S. Baker
Pacific Northwest National Laboratory

Co-Author(s): Cameron P. Casy, Daniel Orton, Dennis Mehinagic, Erika M. Zink, Jennifer E. Kyle, Justin G. Teeguarden, Kristin E. Burnum-Johnson, Matthew E. Monroe, Richard D. Smith, Thomas O. Metz, Xing

Abstract Text

Metabolomic analyses of complex plasma and urine samples present numerous analytical challenges such as isomeric indistinguishability and inadequate throughput of measurements. Ion mobility separations (IMS) minimize these limitations by providing high throughput structurally informative analyses, and when combined with mass spectrometry (MS) measurements, the multidimensional IMS-MS analyses provide in depth characterization of each metabolite. However, ionization suppression is readily observed in IMS-MS direct injection studies of complex samples due to the numerous components in plasma and the high salt concentrations in urine. Thus, a rapid separation is often necessary prior to IMS-MS analyses for high molecular coverage and sample cleanup. This presentation will report on the analysis of plasma and urine metabolomic samples utilizing RapidFire solid phase extraction (SPE) and liquid chromatography (LC) prior to IMS-MS measurements in order to reduce ionization suppression, quickly remove salts, and detect endogenous and exogenous metabolites from picomolar to millimolar concentration levels. The multiple dimensions in the LC-IMS-MS study provided added separation power for distinguishing lipid isomers but were not as high throughput as the RapidFire-IMS-MS platform, which was able to analyze six biological samples per minute. Results from both approaches will be illustrated in the presentation.

Keywords: Mass Spectrometry
Application Code: High-Throughput Chemical Analysis
Methodology Code: Mass Spectrometry

Session # 1530 Abstract # 1530-3 **Symposia**

Session Title: Ion Mobility/Mass Spectrometry for Metabolomics and Clinical Analysis

Abstract Title: **Ion Mobility and Metabolomics, Two New Tools for Current Drug Discovery and Drug Development**

Date: Wednesday, March 09, 2016 - Morn
Time: 09:45 AM
Room: B305

Primary Author: Rob J. Vreeken
Janssen Pharmaceutica

Co-Author(s)

Abstract Text

In drug discovery and drug development we are continuously pushed to the boundaries of the analytical techniques in view of challenges like, e.g. required sensitivity, advanced pharmacodynamics, analysis of reactive (endogenous and exogenous) metabolites, small sample volumes and increased throughput in target engagement analysis. Currently used UPLC-MS/MS or UPLC-HRMS techniques often run into their limits in view of sensitivity and/or selectivity and additional techniques are required to overcome these issues. Amongst various techniques, ion mobility offers an orthogonal technique to current UPLC-HRMS strategies. Separation based on size, shape and mass to charge allows for signal intensity improvements in complex sample analysis. Examples will be given where Ion Mobility significantly enhances sensitivity and selectivity of the applied UPLC-HR-MS analysis, resulting in better understanding the 'mode of action' of specific interventions for single target and/or multi-target approaches. Especially in combination with targeted and non-targeted metabolomics methodologies, ion mobility can be used to 'pull-out' relevant endogenous and/or exogenous metabolites responsible for adverse effects. This results in better understanding of actual metabolic or off-target reactions. In conjunction with Quan-Qual methods metabolomics and ion mobility offer the opportunity to zoom in into group/pathway specific pattern changes related to specific pheno-/metabo-types in various intervention strategies.

Keywords: Bioanalytical, Mass Spectrometry, Metabolomics, Metabonomics
Application Code: Drug Discovery
Methodology Code: Liquid Chromatography/Mass Spectrometry

Session # 1530 Abstract # 1530-4 **Symposia**

Session Title: Ion Mobility/Mass Spectrometry for Metabolomics and Clinical Analysis

Abstract Title: **Cystic Fibrosis Breathomics by Transmission-Mode Direct Analysis in Real Time-Traveling Wave Ion Mobility-Mass Spectrometry**

Primary Author: Facundo M. Fernandez
Georgia Institute of Technology

Date: Wednesday, March 09, 2016 - Morn
Time: 10:35 AM
Room: B305

Co-Author(s): Arlene Stecenko, Christina M. Jones, Jose Perez, Maria E. Monge, Nael A. McCarty

Abstract Text

Cystic fibrosis (CF) is an autosomal recessive disorder caused by mutations in the cystic fibrosis transmembrane conductance regulator (CFTR) gene affecting many organs, and particularly damaging effects to the lungs. To better understand CF pathology and pathogenesis at the systems biology level, breath discovery metabolomics (breathomics) investigations are gaining remarkable importance. In particular, investigation of simple-to-collect biofluids such as exhaled breath condensate (EBC) provide a non-invasive way of studying metabolic alterations associated with lung inflammatory processes.

In this study, we evaluated the applicability of positive and negative-ion transmission-mode direct analysis in real time-traveling wave ion mobility spectrometry-time-of-flight mass spectrometry (TM-DART-TWIMS-TOF MS) for rapid, high-throughput discovery CF breathomics investigations. The EBC metabolome coverage yielded by DART was compared with equivalent direct-infusion electrospray ionization (ESI) and atmospheric pressure chemical ionization (APCI) techniques based on sensitivity, [i]m/z[/i]and drift time range, type of ionic species detected, and metabolome coverage. Using TM-DART-TWIMS-TOF MS in combination with multivariate analysis, EBC samples from CF patients and healthy subjects were screened for metabolic differences. Nine salient spectral features selected by a genetic algorithm were used to build a classification model via orthogonal projections to latent structures-discriminant analysis (oPLS-DA), which successfully discriminated between sample classes with 100% cross-validated accuracy, sensitivity, and specificity.

Keywords: Bioanalytical, Biological Samples, Mass Spectrometry, Metabolomics
Application Code: Genomics, Proteomics and Other 'Omics
Methodology Code: Mass Spectrometry

Session # 1530 Abstract # 1530-5 **Symposia**

Session Title: Ion Mobility/Mass Spectrometry for Metabolomics and Clinical Analysis

Abstract Title: **Ion Mobility/Mass Spectrometry for Metabolomics and Clinical Analysis: Progress and Prospects**

Date: Wednesday, March 09, 2016 - Morn
Time: 11:10 AM
Room: B305

Primary Author: Richard A. Yost
University of Florida

Co-Author(s): Christopher A. Beecher, Christopher D. Chouinard, Christopher R. Beekman, Jared J. Boock, Michael T. Costanzo, Robin Kemperman, Timothy J. Garrett, Wei S. Michael

Abstract Text

Ion mobility/mass spectrometry has tremendous potential for metabolomics and clinical analysis. Ion mobility can resolve compounds unresolved by LC/MS/MS, provide additional structural information not available from mass spectrometry, and reduce or eliminate the need for chromatographic separation. These features offer significant improvements for quantitative targeted metabolomics and clinical analysis, as well as for untargeted (global) metabolomics studies.

This presentation will explore innovations in ion mobility/mass spectrometry for metabolomics and clinical analysis. Topics to be covered include both classic drift tube ion mobility (IMS) and high-field asymmetric-waveform ion mobility (FAIMS), in conjunction with MS, MS/MS, and LC/MS. Applications will include a range of metabolomics and targeted clinical analyses.

Keywords: Bioanalytical, Clinical Chemistry, Mass Spectrometry, Metabolomics, Metabonomics
Application Code: Genomics, Proteomics and Other 'Omics
Methodology Code: Liquid Chromatography/Mass Spectrometry

Session # 1540 Abstract # 1540-1 **Symposia**

Session Title: Precision Bioanalytical Measurements

Abstract Title: **Microdialysis Sampling and Separations: A Tribute**

Date: Wednesday, March 09, 2016 - Morn
Time: 08:35 AM
Room: B309

Primary Author: Susan M. Lunte
University of Kansas

Co-Author(s)

Abstract Text

Craig Lunte pioneered the use of microdialysis sampling for drug metabolism studies in the late 1980s. His group developed several novel probes that are used for monitoring drug metabolism and disposition in a variety of tissues. To effectively perform pharmacokinetic and pharmacodynamic studies (PK/PD) studies, it is critical to incorporate a separation method and a selective detector as part of the analytical protocol. This makes it possible to monitor the drug, its metabolites, and relevant biomarkers simultaneously in near real-time. This presentation will review some of the important criteria that need to be taken into consideration when combining microdialysis sampling with a separation method. It will review some of the early work from the C. Lunte lab regarding the combination of microdialysis sampling with fast LC to monitor drug metabolism, as well as current projects using capillary and microchip electrophoresis as the separation methods.

Keywords: Capillary Electrophoresis, Liquid Chromatography, Pharmaceutical, Sampling
Application Code: Pharmaceutical
Methodology Code: Sampling and Sample Preparation

Session # 1540 Abstract # 1540-2 **Symposia**

Session Title: Precision Bioanalytical Measurements

Abstract Title: **Expanding the Application Space of In Vivo Microdialysis Sampling in the Areas of Drug Metabolism, Free Radical Chemistry, Neurochemistry, and Tissue Engineering**

Primary Author: Julie Stenken
University of Arkansas

Date: Wednesday, March 09, 2016 - Morn
Time: 09:10 AM
Room: B309

Co-Author(s)

Abstract Text

Research in Professor Craig Lunte's group always focused on creative and novel uses of in vivo microdialysis sampling in a variety of different disease contexts as well as the actual tissue being probed. Most of his research was well outside the traditional realm of microdialysis sampling which has focused on neurochemistry. Reflecting on both my time as a Lunte group member and my academic career, I trust that I and my research group have carried on this tradition. This talk will highlight work that started in the Lunte lab 25 years ago and will describe how it leads up to present research interests and needs in my research group. In the 1990s, we were concerned about quantitative applications of microdialysis sampling and this led to attempts to better understand the role of the membrane in the transport processes. These needs are still important as we embark on sampling proteins which have non-ideal interactions in the tissue space as well as with the dialysis membranes employed for microdialysis sampling. Early work with drug metabolism led to interesting studies with in situ and localized metabolism and experiments designed to trap oxygen-derived free radicals. This work has expanded into local biochemical studies of matrix metalloproteinases (MMPs). All of this past work has evolved into important studies of cytokine signaling in both the brain and in applications of localized modulation of macrophage cells. My group has now come "full circle" with very recent research combining microdialysis sampling with microelectrode array measurements for important neurochemical measurements. This and other research over the years will be described in this session honoring the life, research, and memory of Professor Craig Lunte.

Keywords: Bioanalytical, Biomedical, Protein, Sampling
Application Code: Bioanalytical
Methodology Code: Sampling and Sample Preparation

Session # 1540 Abstract # 1540-3 **Symposia**

Session Title: Precision Bioanalytical Measurements

Abstract Title: **The Development of New Tools Based on Whispering Gallery Mode Sensing**

Date: Wednesday, March 09, 2016 - Morn
Time: 09:45 AM
Room: B309

Primary Author: Robert C. Dunn
University of Kansas

Co-Author(s): Daniel Kim, Judith Flores, Sarah Wildgen

Abstract Text

Whispering gallery mode (WGM) resonators are small, axially symmetric dielectric structures that can efficiently trap and store light. Light evanescently coupled into a WGM resonator, such as a glass microsphere, undergoes continuous total internal refection at the sphere interface. When the light circumnavigates the resonator and returns in phase, sharp resonances are observed that are characterized by large quality factors (Q factors). These resonances shift in response to changes in the effective refractive index, which has led to many applications in biosensing using functionalized resonators. The small footprint (10-50 micron diameters) and large Q factors (10[sup]4[/sup]-10[sup]9[/sup]) of WGM resonators offer intriguing capabilities for label-free sensing and the development of new analytical tools. For the latter, we have recently introduced a new scanning probe method based on WGM sensing. Scanning resonator microscopy (SRM) uses a small WGM resonator attached to the end of an atomic force microscopy (AFM) tip to simultaneously map surface refractive index and topography with high spatial resolution. We have also begun exploring the integration of WGM resonators with separation platforms to develop new detection capabilities. Initial studies have coupled WGM sensing with capillary electrophoresis (CE) to detect the separation of small ions. As part of this development, we introduce a modulation method for tracking resonant shifts using phase sensitive detection. This approach enables real-time monitoring of bands eluting from the CE column with detection limits of 10[sup]-7[/sup] refractive index units. These capabilities are currently being extended to detect the separation of serum proteins for applications in the diagnosis and treatment of patients with multiple myeloma. These and other applications using high-Q WGM resonators for novel sensing will be discussed.

Keywords: Biosensors, Microscopy, Sensors, Surface Analysis
Application Code: Bioanalytical
Methodology Code: Sensors

Session # 1540 Abstract # 1540-4 **Symposia**

Session Title: Precision Bioanalytical Measurements

Abstract Title: **Utilizing Oxidative DNA Damage to Explore the Mode of Action of Oxidative Events and Antioxidative Responses**

Primary Author: Blánaid White
Dublin City University

Date: Wednesday, March 09, 2016 - Morn
Time: 10:35 AM
Room: B309

Co-Author(s): Dermot Walls, Karina Horgan, Roya Hakimjavadi, Sarah J. Lynch, Sinead Loughran

Abstract Text

By monitoring oxidative DNA damage, the mode of action of proposed oxidants, and antioxidant therapies, can be investigated. Analytical platforms such as CE-EC, HPLC-UV-EC and HPLC-MS, in tandem with bioanalytical techniques such as flow cytometry and the Comet assay have been used to monitor oxidative DNA damage at both a molecular and cellular level. In this presentation, the mode of action of metal based oxidants is explored. A number of antioxidant therapies, from UV protecting sunscreens to nutrient enriched food stuffs will also be discussed in terms of their ability to protect against oxidative DNA damage. The analytical and bioanalytical platforms utilised for their determinations evaluated in terms of the accuracy and precision of the resulting measurements. For example, a series of novel therapeutics evaluated using evaluated using the oestrogen-positive MCF-7 breast cancer cell line. Guanine oxidation studies using HPLC-UV-EC confirmed that one derivative was capable of generating oxidative damage via a reactive oxygen species-mediated mechanism. During sunscreen analysis, it was determined that particular ingredients resulted in an increase in skin protection, as evidenced by a decrease in oxidative damage to the nuclear DNA within the cells. Using FACS and comet assay, a placebo cream was found not to reproducibly affect the level of cellular oxidative DNA damage where it was applied. However, an antioxidant cream resulted in a decrease in the concentration of oxidative DNA damage relative to control sites where no cream was applied. Selenium enriched protein digests were also explored to determine their antioxidant properties. Using a number of bioanalytical platforms, they were compared to organic and inorganic selenium sources and their antioxidant properties evaluated, for a number of different types of oxidative stress insult.

Keywords: Bioanalytical, Environmental/Biological Samples, Nucleic Acids, Separation Sciences
Application Code: Bioanalytical
Methodology Code: Separation Sciences

Session # 1540 Abstract # 1540-5 **Symposia**

Session Title: Precision Bioanalytical Measurements

Abstract Title: **Precision Medicine: Enabled by Single Cell Analysis**

Date: Wednesday, March 09, 2016 - Morn
Time: 11:10 AM
Room: B309

Primary Author: Maggie Witek
University of North Carolina

Co-Author(s): Steven A. Soper

Abstract Text

Improved therapies that yield more cures and better overall survival for cancer patients are needed. For example, women with breast cancer have a 5-year survival rate of 22% (Stage IV) and 72% (Stage III). Doxorubicin, cisplatin, paclitaxel, and tamoxifen are examples of drugs used for treating breast cancer with selection of therapy typically based on the classification and staging of the patient's cancer. While treatment regimens assigned to some patients may be optimal using the current classification model, others within certain breast cancer sub-types fail therapy. New assays must be developed to determine how a patient's physiology and genetic makeup affects drug efficacy. In this presentation, a novel SMARTChip™ design will be discussed for the isolation and processing of circulating tumor cells (CTCs). The SMARTChip™ quantifies response to therapy using three pieces of information secured from the CTCs; (1) CTC number; (2) CTC viability; and (3) the frequency of DNA damage (abasic (AP) sites) in genomic DNA (gDNA) harvested from the CTCs. The SMARTChip consists of task-specific modules integrated to a fluidic motherboard. Micro-scale modules are used for CTC selection, CTC enumeration and viability determinations, lysing CTCs, and purifying gDNA. The module to read AP sites is a nanosensor made via nano-imprinting in plastics and contains a nanochannel with dimensions less than the persistence length of double-stranded DNA (~50 nm). Labeling AP sites with fluorescent dyes and stretching the gDNA in the nanochannel to near its full contour length allows for the direct readout of the AP sites, even from a single CTC. This information can be used to determine how a patient is responding to therapy.

Keywords: Clinical Chemistry, Electrophoresis, Fluorescence, Medical
Application Code: Biomedical
Methodology Code: Microfluidics/Lab-on-a-Chip

Session # 1550 **Abstract # 1550-1** **Symposia**

Session Title: Standoff Detection Methods for Security Applications

Abstract Title: **Novel Approaches to Standoff Hyperspectral Imaging Based Detection of Explosives and Other Threats**

Date: Wednesday, March 09, 2016 - Morn
Time: 08:35 AM
Room: B310

Primary Author: Charles W. Gardner
ChemImage Sensor Systems

Co-Author(s): Matthew P. Nelson, Nathaniel R. Gomer, Oksana P. Klueva

Abstract Text

Hyperspectral Imaging is the marriage of spectroscopy with the power of digital imaging to reveal chemical-based details of the world. This talk will provide examples of how CISS has developed standoff hyperspectral imaging systems to detect explosives at entry control points, chemicals in the environment and details of how one standoff hyperspectral mode can be used to target another in order to dramatically improve the area search rate for the detection of chemicals on environmental surfaces. Details of the instrumentation and software that enable these applications of hyperspectral imaging will be presented.

Keywords: Forensics, Molecular Spectroscopy
Application Code: Homeland Security/Forensics
Methodology Code: Molecular Spectroscopy

Session # 1550 **Abstract # 1550-2** **Symposia**

Session Title: Standoff Detection Methods for Security Applications

Abstract Title: **Standoff Detection of Explosives Residues Using Integrated Quantum Cascade Laser Arrays**

Primary Author: Mark F. Witinski, Pendar Technologies

Co-Author(s): Daryoosh Vakhshoori, Romain Blanchard

Date: Wednesday, March 09, 2016 - Morn
Time: 09:10 AM
Room: B310

Abstract Text

This presentation introduces the spectroscopic concepts and results enabled by arrays of Distributed Feedback (DFB) QCLs, with each element at a slightly different wavelength than its neighbor. In portable optical systems, such as standoff threat detectors and in situ gas analyzers, this increases analyte sensitivity and selectivity by broadening spectral source coverage while also allowing for extremely fast all-electronic wavelength tuning with no moving parts.

This talk will first present the QCL array and its packaging, then move into the description of an integrated prototype standoff detection system, and finally show standoff detection results from a handheld system from over 2 meters.

The data show how monolithic and all-electronic tuning enables next-generation spectroscopes that are not only more robust and miniature than those that utilize external cavity-tuned lasers, but that are inherently more stable in terms the shot-to-shot amplitude and wavelength parameters. This enhanced stability increases signal to noise for a given configuration (pathlength, averaging time, concentration, etc...). Some discussion of how to maximize the benefits of high speed, highly reproducible tuning is presented, including detector, preamplifier, and digitization considerations for both backscattered and closed path configurations.

Keywords: Chemometrics, Infrared and Raman, Molecular Spectroscopy, Sensors
Application Code: Homeland Security/Forensics
Methodology Code: Molecular Spectroscopy

Session # 1550 Abstract # 1550-3 **Symposia**

Session Title: Standoff Detection Methods for Security Applications

Abstract Title: **Standards for Optical Based Standoff Detection Fabricated Using Inkjet Printing**

Primary Author: Greg Gillen
NIST

Date: Wednesday, March 09, 2016 - Morn
Time: 09:45 AM
Room: B310

Co-Author(s)

Abstract Text

Optimization and evaluation of deployed and next generation trace detection systems is a key component of the NIST program in Trace Explosives Detection conducted in collaboration with the DHS Science and Technology Standards and Explosives Divisions. To support development of emerging optical based trace detection methods, a procedure has been developed to fabricate patterned arrays of micrometer sized monodisperse solid particles of homemade and military explosives on hydrophobic/oleophobic surfaces using inkjet printing. The method relies on dispensing a precise quantity of a concentrated analyte solution onto a modified silicon surface. Depending on the contact angle of the solution and the solute concentration, a spherical shaped cap droplet is formed. During the evaporation process, the droplet maintains a constant geometry until converting to a dried solid particle. Environmental scanning electron microscopy is used to determine the diameter and height of the particles. Particle size and shape can be modified by changing the number of drop/spot, the contact angle, or the solution concentration. The fill factor - the ratio of sample coverage to surface area of the sample - can be modified to conform to the characteristics of actual trace residues. The primary application of this fabrication approach is to prepare test materials for evaluation of spatially-resolved optical and mass spectrometry -based chemical analysis techniques. Results will be shown demonstrating the use of these materials for characterizing the performance of laboratory-based Raman, IR and imaging MS systems as function of particle size. The approach may also be extended to a variety of explosives such as AN, PC, HMX, TNT, RDX and PETN. Strategies for using these test materials will be presented. A review will also be provided of other approaches being used for fabrication of optical standoff detection standards.

Keywords: Forensics, Forensic Chemistry, Infrared and Raman, Instrumentation
Application Code: Homeland Security/Forensics
Methodology Code: Molecular Spectroscopy

Session # 1550 Abstract # 1550-4 **Symposia**

Session Title: Standoff Detection Methods for Security Applications

Abstract Title: **Recent Advances in Standoff Chemical Threat Detection Using Deep-Ultraviolet Raman Spectroscopy**

Date: Wednesday, March 09, 2016 - Morn
Time: 10:35 AM
Room: B310

Primary Author: Adam J. Hopkins
Alakai Defense Systems

Co-Author(s): Edwin Dottery, Kenneth R. Pohl, Rob Waterbury

Abstract Text

Alakai Defense Systems has created two systems, CPEDS (Check Point Explosives Detection System) and PRIED (Portable Raman Improvised Explosives Detector), to detect bulk to near-trace quantities of explosives and HMEs at standoff ranges from one to hundreds of meters. We present a short description of the instruments and CONOPS for these two designs, as well as data for a variety of explosives, precursors, and hazardous materials. We also provide an overview of the improvements in our explosives detection technologies, as well as a discussion of some challenges to detection of trace chemicals at standoff ranges.

Keywords: Detection, Portable Instruments, Raman, Sensors
Application Code: Homeland Security/Forensics
Methodology Code: Sensors

Session # 1550 Abstract # 1550-5 **Symposia**

Session Title: Standoff Detection Methods for Security Applications

Abstract Title: **Deep UV Standoff Raman Detection of Explosives: Fundamentals and Methodologies**

Date: Wednesday, March 09, 2016 - Morn
Time: 11:10 AM
Room: B310

Primary Author: Sergei V. Bykov
University of Pittsburgh

Co-Author(s): Katie L. Gares, Kyle T. Hufziger, Sanford A. Asher

Abstract Text

We report on the development of Deep UV Standoff Raman instrumentation and methodologies for standoff explosive detection. Deep UV Raman, excited < 250 nm, enables trace explosives detection at a distance due to the resonance enhancement of Raman band intensities, much stronger light scattering at shorter wavelengths and negligible florescence interference.

We are developing deep UV Raman detection methodology by investigating resonance enhancement of explosives excited in the deep UV, determined the optimal excitation wavelengths, investigating explosives UV-photochemistry, characterizing the photochemically induced spectral changes and determining UV-photodecomposition quantum yields.

Successful application of UV Raman for standoff trace explosive detection requires development of state-of-the-art UV Raman instrumentation. In collaboration with UVisIR Inc. We recently developed two novel, compact, acousto-optically Q-switched diode pumped solid state (DPSS) deep UV lasers capable of producing up to 100 mW of 213 nm and up to 20 mW of 228 nm quasi-CW light. We also developed and built novel high throughput, high efficiency deep UV Raman spectrographs both dispersive and imaging. Dispersive spectrograph utilizes Echelle grating in quasi-Littrow optical configuration while imaging spectrograph utilizes novel deep UV photonic crystals for wavelength selection.

We used these novel deep UV Raman instrumentation and methodology for explosive detection at a distance.

Keywords: Detection, Laser, Raman, Spectroscopy
Application Code: Homeland Security/Forensics
Methodology Code: Vibrational Spectroscopy

Session # **1560** **Abstract #** **1560-1** **Symposia**

Session Title: Sum Frequency Generation (SFG) Vibrational Spectroscopic Studies on Proteins and Peptides at Interfa

Abstract Title: **Determining the Structure of Surface Bound Proteins**

Primary Author: David G. Castner
University of Washington

Date: Wednesday, March 09, 2016 - Morn
Time: 08:35 AM
Room: B311

Co-Author(s)

Abstract Text

Molecular level Information about protein structure and function at interfaces is crucial in drug design, biosensor applications and biomaterial engineering. Proteins on surfaces are an integral part of many biomedical applications. This importance has stimulated research towards developing techniques to assess the structure, activity, and surface interactions of immobilized proteins. Recent advances in SFG spectroscopy now provide detailed information about surface bound proteins and peptides. Combining SFG measurements with NEXAFS spectroscopy and ToF-SIMS provide a particularly powerful approach for investigating protein structures on surfaces. Short, well-defined leucine-lysine (LK) peptides were used as model systems to develop the methods for detailed analysis of protein-surface interactions at the molecular level. By deuerating the iso-propyl group of a given leucine side and collecting SFG spectra at different polarization combinations the orientation of that side chain with respect to the surface can be determined. Similarly SFG spectra of the amide I band can be used to determine the overall orientation of the peptide with respect to the surface. On hydrophobic surfaces the backbone of the helical LK peptides were found to be oriented parallel to the surface. For the beta-strand LK peptides NEXAFS showed the backbones were parallel to the surface. We have also successfully expanded this approach to the B1 domain of the Protein G to determine secondary structure and backbone orientation of this protein immobilized onto appropriately functionalized surfaces. SFG is a powerful tool to probe proteins on surfaces, but its full potential can only be realized when combined with complementary techniques. However, the combination of surface analytical tools alone can still not provide atomic structures of entire proteins. This requires the integration of computer modeling and stimulations with the experimental methods.

Keywords: Mass Spectrometry, Protein, Spectroscopy, Surface Analysis
Application Code: Bioanalytical
Methodology Code: Surface Analysis/Imaging

Session # 1560 Abstract # 1560-2 **Symposia**

Session Title: Sum Frequency Generation (SFG) Vibrational Spectroscopic Studies on Proteins and Peptides at Interfa

Abstract Title: **The Interactions of Ions with Peptides and Lipid Bilayers**

Primary Author: Paul Cremer
Penn State University

Date: Wednesday, March 09, 2016 - Morn
Time: 09:10 AM
Room: B311

Co-Author(s)

Abstract Text

Hofmeister cations and anions can both interact with biomolecules, but typically do so for different reasons. Specifically, weakly solvated anions usually interact with the hydrophobic portions of biomolecules including methyl and methylene moieties. For this type of interaction to occur, both the anion and the biointerface must shed hydration waters. By contrast, cations usually interact with charge groups like carboxylates, phosphates, or uncharged hydrophilic moieties like the amide oxygen of the backbone of peptides. Of course, cations must also shed their hydration shells for these interactions to occur. However, only the most strongly hydrated cations typically do this. It is energetically more unfavorable for strongly hydrated cations to shed waters of hydration, but the new interactions with the biomolecules are typically much stronger than for more weakly hydrated cations. Information about the role that water plays in these binding events can be obtained by using vibrational sum frequency spectroscopy. Moreover, reorientation of the biomolecules by the ions can also be observed.

Keywords: Bioanalytical, Lipids, Surface Analysis, Vibrational Spectroscopy
Application Code: Bioanalytical
Methodology Code: Biospectroscopy

Session # **1560** Abstract # **1560-3** **Symposia**

Session Title: Sum Frequency Generation (SFG) Vibrational Spectroscopic Studies on Proteins and Peptides at Interfa

Abstract Title: **Proteins at Interfaces: Structure and Platform for Novel Biomaterials**

Primary Author: Mischa Bonn
Max Planck Institute for Polymer Research

Date: Wednesday, March 09, 2016 - Morn
Time: 09:45 AM
Room: B311

Co-Author(s): Tobias Weidner

Abstract Text

Using surface-specific, label-free vibrational spectroscopy in conjunction with other surface spectroscopies, we investigate the structure and conformation of peptides and proteins at interfaces, for two purposes. Firstly, we aim to understand the structure-function relation for biological relevant membrane proteins, specifically the protein IM30, which is involved in the generation of the photosynthetic membrane systems. We show the orientation of this protein at the membrane, and correlate that with its putative function. In a second line of research, we use synthetic peptides to control the growth of biomimetic minerals at various interfaces. Specifically, we demonstrate that vaterite, the least stable polymorph of calcium carbonate, can be stabilized directly at surfaces – by engineered oligo(glutamic acid) peptides. Our data show that the peptide-induced vaterite mineralization process occurs via a 'self templating' process where calcium ions restructure the peptide backbone, which in turn allows for effective vaterite precipitation. Similarly, we show that interfacial biomimetic silica sheets can be synthesized by surface-active synthetic leucine–lysine peptides with differently folded structures. The nanometer thin and micrometer-size silica sheets are stable and self-supported. Moreover, the detailed morphology of the sheets can be controlled by the secondary structure of the peptides when interacting with the silica.

Keywords: Bioanalytical, Spectroscopy, Vibrational Spectroscopy
Application Code: Bioanalytical
Methodology Code: Vibrational Spectroscopy

Session # 1560 Abstract # 1560-4 Symposia

Session Title: Sum Frequency Generation (SFG) Vibrational Spectroscopic Studies on Proteins and Peptides at Interfa

Abstract Title: **Proteins at Interfaces Probed by Chiral Sum Frequency Generation**

Date: Wednesday, March 09, 2016 - Morn
Time: 10:35 AM
Room: B311

Primary Author: Elsa C. Yan
Yale University

Co-Author(s)

Abstract Text

Characterization of protein secondary structures at interfaces in situ and in real time is important to understand biological processes associated with cell membranes and solve problems in bioengineering and biomedical sciences. However, such characterization remains challenging because it requires techniques that are both selective to interfaces and protein secondary structures. Here, we demonstrate that chiral sum frequency generation spectroscopy (SFG) can provide peptide amide I and N-H stretch vibrational signals that are free of water background and characteristic to parallel beta-sheet, anti-parallel beta-sheet, alpha-helix, 3-10 helix and random coil. This enables chiral SFG to distinguish protein secondary structures at interfaces, similar to circular dichroism spectroscopy for protein characterization in solution. Using chiral SFG, we followed the kinetics of misfolding of an amyloid protein and perform structure-function studies of a biofilm protein. The results demonstrate and the potential of chiral SFG in addressing a wide range of problems related to proteins at interfaces.

Keywords: Chiral, Molecular Spectroscopy, Protein, Vibrational Spectroscopy
Application Code: Biomedical
Methodology Code: Vibrational Spectroscopy

Session # 1560 Abstract # 1560-5 **Symposia**

Session Title: Sum Frequency Generation (SFG) Vibrational Spectroscopic Studies on Proteins and Peptides at Interfa

Abstract Title: **Structure-Function Relationships of Surface Immobilized Peptides and Enzymes**

Primary Author: Zhan Chen
University of Michigan

Date: Wednesday, March 09, 2016 - Morn
Time: 11:10 AM
Room: B311

Co-Author(s)

Abstract Text

Surface immobilized peptides have been widely used for antimicrobial coatings, and many biosensors and biochips use surface immobilized enzymes. In this study, we applied sum frequency generation (SFG) vibrational spectroscopy to study conformation and orientation of a variety of surface immobilized peptides such as cecropin P1, MSI-78, and cecropin-melittin hybrid peptide, and surface immobilized enzymes such as nitro-reductase, □-galactosidase, and haloalkane dehalogenase. ATR-FTIR and CD spectroscopies were used as supplemental tools in the study, along with molecular dynamics simulations. We also measured the activities of these surface immobilized biological molecules. We successfully elucidated the structure-function relationships of surface immobilized peptides and proteins.

Keywords: Bioanalytical, Peptides, Protein, Spectroscopy
Application Code: Bioanalytical
Methodology Code: Vibrational Spectroscopy

Session # **1570** **Abstract #** **1570-1** **Workshops**

Session Title: Analytical Information Markup Language (AnIML) Data Standards

Abstract Title: **Generating AnIML Technique Definitions**

Primary Author: Mark F. Bean
GSK

Date: Wednesday, March 09, 2016 - Morn
Time: 08:35 AM
Room: B313

Co-Author(s)

Abstract Text

Different approaches to generation of AnIML Technique Definitions will be summarized with particular attention to the sample separation technique of Chromatography and also to the detector technique of Mass Spectrometry. Hints and shortcuts to speed the process will be covered along with intentional convergence with existing vocabularies.

Keywords: Automation, Data Analysis, Informatics, Liquid Chromatography/Mass Spectroscopy
Application Code: Other
Methodology Code: Data Analysis and Manipulation

Session # 1570 **Abstract # 1570-2** **Workshops**

Session Title: Analytical Information Markup Language (AnIML) Data Standards

Abstract Title: **Filling the Automation and Enterprise Gap with Data and Device Standardization**

Date: Wednesday, March 09, 2016 - Morn
Time: 09:05 AM
Room: B313

Primary Author: Del Ray Jackson
Hamilton Company

Co-Author(s): Carmen Condrau, Devon Johnston

Abstract Text

With the increased efficiency brought to labs by automated systems, the amount of data being collected, extracted, analyzed and manipulated is growing at an unprecedented rate. Likewise, the technology of automated devices, software platforms and information management systems is becoming more sophisticated to handle this surge in Big Data. However, further scientific advances and innovations critically depend on more complex, increasingly integrated, and highly adaptive systems.

In an effort to match these requirements, the lab automation's information and communications technology industry has recognized the need for a flexible, modular approach to both the hardware and software of system integrations. However, modularity poses its own challenges. Laboratories still find themselves with a heterogeneous mix of different vendor hardware and software technologies, none of which easily lend themselves to seamless integration. In order for laboratory automation systems to adopt a modular, plug-and-play architecture, vendors must agree upon standards, both in the hardware as well as in the software, such that the system and the data the system produces can withstand reconfiguration and the test of time.

While over the years, several hardware standards have become adopted by laboratory automation device manufacturers (e.g. USB, Ethernet, SBS micro plate footprint), industry-wide software standards have been slower to emerge. The SiLA Consortium is uniquely focused on providing standardization solutions for both the horizontal communication between devices (e.g. status data, commands) and the vertical integration of result data and workflows. Join us to learn more about how SiLA satisfies the entire value proposition chain from device integration to data collection, transmission and analysis.

Keywords: Laboratory Automation, Laboratory Informatics, Sample & Data Management, Standards
Application Code: Bioanalytical
Methodology Code: Laboratory Informatics

Session # 1570 Abstract # 1570-4 **Workshops**

Session Title: Analytical Information Markup Language (AnIML) Data Standards

Abstract Title: **Use of AnIML and Other Methods for Software Visualization and Automation**

Date: Wednesday, March 09, 2016 - Morn
Time: 10:20 AM
Room: B313

Primary Author: David M. Cox
SCIEX

Co-Author(s): John Gibbons

Abstract Text

Modern mass spectrometers produce a wealth of analytical information capable of identifying compounds in complex mixtures. As the amount of data increases, the importance of software for interpreting the results becomes even more important. Software is particularly critical for understanding large amounts of data, by improving visualization of data, and automation of processing. Visualization is important for understanding data. It is much easier to grasp the meaning of large amounts of data when it is displayed graphically and interactively, than when it is simply a table of numbers. Automation of processing is another area where software plays a critical role. Existing solutions include the ability to automatically identify unknown compounds through library searching and formula finding. Where we, as a community, go next is very exciting. Today's results files are increasingly open formats (such as AnIML) or convertible to other formats. Simple, command line driven, tools exist for processing data and generating reports. Combining open formats with simple tools can empower a variety of custom workflows. Underlying visualization and automation, is the data itself. The amount of data, the number of compounds (known or unknown at analysis time) and the speed at which answers can be mined from it, all become important. Data independent acquisition (SWATH) ensures that MS/MS for all possible compounds is collected, but it does increase the amount of the data being collected. As the amount of data being collected increases, the speed of extracting chromatograms and MS/MS becomes a critical step. Visualization, automation, and speed of processing are critical factors for analysing and understanding the wealth of data produced by modern mass spectrometry analysis.

Keywords: Data Analysis, Laboratory Automation, Mass Spectrometry, Sample & Data Management
Application Code: General Interest
Methodology Code: Mass Spectrometry

Session # 1570 Abstract # 1570-5 **Workshops**

Session Title Analytical Information Markup Language (AnIML) Data Standards

Abstract Title **Cloud-Based Analytical Data Management Using the AnIML Standard**

Date: Wednesday, March 09, 2016 - Morn
Time: 10:50 AM
Room: B313

Primary Author Burkhard Schaefer
BSSN Software GmbH

Co-Author(s)

Abstract Text

This presentation highlights the current state of the ASTM AnIML Data Standard. It demonstrates how the standard can be used to deliver analytical data through web and cloud technologies.

Keywords: Chemometrics, Laboratory Informatics, LIMS, Software
Application Code: General Interest
Methodology Code: Laboratory Informatics

Session # 1580 Abstract # 1580-1 **Organized Contributed Sessions**

Session Title Food Contaminant Methods

Abstract Title **Quantitative HPLC-MS/MS Analysis of Metabolites of Hypoglycin A and Methylenecyclopropylglycine (MCPG) in Human Urine**

Primary Author Samantha L. Isenberg
Centers for Disease Control and Prevention

Date: Wednesday, March 09, 2016 - Morn
Time: 08:30 AM
Room: B315

Co-Author(s) Darryl Johnson, James Pirkle, Jerry D. Thomas, Leigh A. Graham, Melissa D. Carter, Rudolph C. Johnson, Thomas P. Mathews

Abstract Text

Hypoglycin A and methylenecyclopropylglycine (MCPG) are naturally occurring amino acids found in some members of the soapberry family. Both hypoglycin A and MCPG have been shown to induce encephalopathy and hypoglycemia in animal studies, and several illness outbreaks have been linked to the ingestion of these toxins. Jamaican Vomiting Sickness (JVS) was linked to hypoglycin A in ackee fruit, and the ingestion of maple seeds containing hypoglycin A has been determined to cause seasonal pasture myopathy (SPM) in horses. MCPG has been hypothesized as a possible cause of an Acute Enchephalitis Syndrome (AES) in Asia. A specific method for the identification and quantification of urine metabolites for exposure to hypoglycin A and MCPG is presented in this work. The urine metabolites of hypoglycin A and MCPG are glycine adducts methylenecyclopropylacetyl-glycine (MCPA-Gly) and methylenecyclopropylformyl-glycine (MCPF-Gly), respectively. Samples consisting of 10 µL of urine were processed by isotope-dilution, separated by reverse-phase liquid chromatography, and monitored by electrospray-ionization tandem mass spectrometry. The reportable range of metabolite concentration was from the lowest reportable limit 0.100 µg/mL to 20.0 µg/mL, with a correlation coefficient of r □0.99. This method was further applied for laboratory analysis for cases of suspected hypoglycin A and/or MCPG toxicity and provided the first reported direct measurement of MCPA-Gly and MCPF-Gly in human urine.

Keywords: Clinical/Toxicology, Food Contaminants, Liquid Chromatography/Mass Spectroscopy, Toxicology
Application Code: Food Contaminants
Methodology Code: Liquid Chromatography/Mass Spectrometry

Session # 1580 Abstract # 1580-2 **Organized Contributed Sessions**

Session Title: Food Contaminant Methods

Abstract Title: **LC-MS/MS Detection of Tetrodotoxin in Fresh/Frozen and Salt-Dried Fish Products**

Date: Wednesday, March 09, 2016 - Morn
Time: 08:50 AM
Room: B315

Primary Author: Sara C. McGrath, FDA/CFSAN

Co-Author(s): Jonathan R. Deeds

Abstract Text

Tetrodotoxin (TTX) is a small, non-protein, alkaloid toxin most commonly associated with puffer fish poisoning. Despite tight regulation of both imported and domestic puffer fish, there have been multiple instances of puffer fish poisoning in the US in the last decade from illegally imported fish in various product forms. Our goal is to establish a TTX detection method that will provide a fast and sensitive estimation of toxicity to facilitate a rapid response to TTX-related illnesses and reinforce the safety of the food supply. It is critical that analytical methods for detection and quantification of TTX be applicable to multiple types of food matrices and include robust sample preparation protocols.

This presentation will discuss the development of an optimized method for detection and quantification of TTX analogues in fish using liquid chromatography-tandem mass spectrometry (LC-MS/MS). Optimization of the LC-MS/MS method conditions include column stationary phase chemistry, mobile phase buffer system, and selection of product ion transitions for TTX-related analytes using standard reference materials. The extraction of TTX from naturally-contaminated samples has been optimized for selected organs from fresh/frozen fish as well as fish that have been preserved by salting or dehydration. Sample cleanup protocols include a combination of solvent extraction, liquid-liquid partitioning and solid phase extraction. While the US has not imposed a regulatory limit for TTX in food products, this optimized method will allow consistent detection of TTX in fish at a level 10x below the anticipated safety level.

Keywords: Food Contaminants, Food Safety, Liquid Chromatography/Mass Spectroscopy
Application Code: Food Contaminants
Methodology Code: Liquid Chromatography/Mass Spectrometry

Session # 1580 Abstract # 1580-3 **Organized Contributed Sessions**

Session Title: Food Contaminant Methods

Abstract Title: **Fast Liquid Chromatography - Tandem Mass Spectrometry Analysis of >150 Drug Residues including Aminoglycosides in Food Animal Tissues**

Date: Wednesday, March 09, 2016 - Morn
Time: 09:10 AM
Room: B315

Primary Author: Steven J. Lehotay
USDA Agricultural Research Service

Co-Author(s): Alan R. Lightfield

Abstract Text

The USDA Food Safety and Inspection Service (FSIS) currently employs several methods to monitor veterinary drug residues in slaughtered food animals. Many of their methods use UHPLC-MS/MS analysis to screen, quantify, identify, and confirm violative residues for regulatory enforcement purposes. A goal of our work is to increase the number of drug analytes into one simple and easy workflow protocol to achieve acceptable quality of results and high sample throughput at low cost. This can be done in a single method for >150 veterinary drugs, but due to their different chemical nature, highly polar aminoglycoside antibiotics require different extraction conditions. Conventionally, they also need different chromatographic conditions, too, but we have devised one method to separate and analyze all analytes including aminoglycosides in 10 min using ultrahigh performance liquid chromatography (UHPLC) - triple quadrupole tandem mass spectrometry (MS/MS). An ion-pairing reagent is added to the combined extracts from two streamlined sample preparation methods, which also improved peak shapes for other drugs, such as tetracyclines and macrolides. The new method was validated at regulatory levels in muscle tissues from cattle, chicken, and pork according to FSIS standards.

Keywords: Agricultural, High Throughput Chemical Analysis, Liquid Chromatography/Mass Spectroscopy, Metho
Application Code: Food Safety
Methodology Code: Liquid Chromatography/Mass Spectrometry

Session # 1580 Abstract # 1580-4 **Organized Contributed Sessions**

Session Title Food Contaminant Methods

Abstract Title **Determination of Diglycolic Acid in Foods Containing Carboxymethyl Cellulose**

Primary Author Wendy Young
Food and Drug Administration

Date: Wednesday, March 09, 2016 - Morn
Time: 09:30 AM
Room: B315

Co-Author(s) Lowri DeJager

Abstract Text

Carboxymethyl cellulose (i.e. CMC or cellulose gum) is a fluid absorbent used in food packaging and food contact materials. CMC can also be used as a direct additive for foods and drugs in order to change texture and act as a binder. CMC and other carboxymethyl starches are synthesized by condensing glycolic acid with monochloroacetic acid. Diglycolic acid (DGA) is a byproduct produced by this condensation which cannot be completely removed. Currently there are no analytical methods to accurately detect and quantify DGA in foods and food packaging materials. Because DGA has been shown to be a potential renal toxicant, the determination of DGA available from foods is needed. A simple methanol/water extraction was developed and a novel LC-MS/MS method used to determine the DGA concentration in food contact materials, direct additive CMC, as well as over the counter drugs and supplements. This talk will discuss the development and validation of preparation and cleanup of various food matrices and LC-MS/MS analysis for the presence of DGA. Matrices investigated include but are not limited to cereals, snacks, bake mixes, dressings, milk drinks and ice cream.

Keywords: Food Science, Liquid Chromatography/Mass Spectroscopy, Method Development, Solid Phase Extrac
Application Code: Food Safety
Methodology Code: Chemical Methods

Session # 1580   Abstract # 1580-5   **Organized Contributed Sessions**

Session Title: Food Contaminant Methods

Abstract Title: **Sulfite Determination in Food by Liquid Chromatography-Mass Spectrometry**

Date: Wednesday, March 09, 2016 - Morn
Time: 10:05 AM
Room: B315

Primary Author: Katherine S. Robbins
US FDA

Co-Author(s): Lowri de Jager, Shaun MacMahon

Abstract Text

Sulfites are food additives used to limit browning and microbial growth. Sensitive individuals have reported severe allergic-type reactions following consumption of sulfite treated foods. In 1986, the US FDA mandated that sulfites be declared on the label of any product containing in excess of 10 ppm SO[sub]2[/sub]. Currently, the optimized Monier-Williams method is the most common approach for determining sulfite concentrations in foods. However, this method is tedious and time consuming. This research describes the development of a LC-MS/MS method for determining sulfite in food matrices. The method was validated according to the FDA Foods Program Guidelines for Chemical Methods. In these validation studies, spiked recoveries ranging from 84-115% were observed for dried fruit, jam, vinegar, and juice products. Commercially sulfited products were analyzed using both the MW and LC-MS/MS methods to provide a comparison. The current regulatory method produces false positive results with vegetables from the [i]Allium[/i] (garlic) and [i]Brassica[/i] (cabbage) genera due to the extraction conditions causing endogenous sulfur compounds to release SO[sub]2[/sub]. Because of this bias, special consideration is needed for regulatory analyses. Vegetables were selected from these genera and were analyzed using multiple sulfite methods to determine the false positive rate. Sulfite concentrations greater than 10 ppm SO[sub]2[/sub] were observed with the MW analyses. The LC-MS/MS method had concentrations below 10 ppm for the [i]Brassica[/i] samples but was not successful with the [i]Allium[/i] matrices. The ability to eliminate false positives will enable accurate determination of added sulfite to ensure compliance with sulfite labeling requirements.

Keywords: Food Contaminants, Food Safety, Liquid Chromatography/Mass Spectroscopy
Application Code: Food Safety
Methodology Code: Liquid Chromatography/Mass Spectrometry

Session # 1580 Abstract # 1580-7 **Organized Contributed Sessions**

Session Title: Food Contaminant Methods

Abstract Title: **Application of Raman Imaging for the Analysis of Food Packaging Stability**

Primary Author: Ric Gonzalez
ConAgra Foods, Inc.

Date: Wednesday, March 09, 2016 - Morn
Time: 10:45 AM
Room: B315

Co-Author(s): Indarpal Singh

Abstract Text

The use of appropriate packaging is crucial to ensure compatibility with food and the avoidance of food contamination. The determination of the stability of packaging under various conditions is necessary to ensure its suitability for food application. Raman spectroscopy has reached a level of maturity to recently become a key analytical tool in the analysis of packaging and food, including package-food interactions. Most recently, the integration of Raman spectroscopy with optical microscopy has allowed for a very powerful means to identify and map molecular structure at the micron scale, including impurities, with imaging. This presentation will demonstrate the application of confocal Raman microscopy to the analysis of multi-layer packaging structures and their suitability for various food matrices.

Keywords: Food Contaminants, Microscopy, Polymers & Plastics, Raman
Application Code: Food Contaminants
Methodology Code: Molecular Spectroscopy

Session # 1590 Abstract # 1590-1 **Organized Contributed Sessions**

Session Title: On-Site Detection of THC and Related Drugs

Abstract Title: **Challenges of Drug Detection in the Field from the Enforcement Perspective**

Primary Author: Maggie Tam
Canada Border Services Agency

Date: Wednesday, March 09, 2016 - Morn
Time: 08:30 AM
Room: B316

Co-Author(s)

Abstract Text

When there is sufficient sample quantity, a clean operating environment and ready access to power and supplies, an analyst has ample time and resources to extract, analyze and confirm the presence of illicit substances. Field detection often lacks the luxury of some, if not all of the above parameters.

This presentation shall explore the unique challenges faced by enforcement agencies in regards to field detection of drugs, the constantly evolving target materials, the issues of non-seizure alarms, and more importantly, to initiate a much needed international discussion on trace drug detection.

Keywords: Detection, Drugs, Forensics, Portable Instruments
Application Code: Homeland Security/Forensics
Methodology Code: Portable Instruments

Session # **1590** **Abstract #** **1590-2** **Organized Contributed Sessions**

Session Title: On-Site Detection of THC and Related Drugs

Abstract Title: **Forensic Screening Using High-Performance Ion Mobility Spectrometry**

Primary Author: Ching Wu
Excellims Corporation

Co-Author(s): Anthony Midey, Mark Osgood

Date: Wednesday, March 09, 2016 - Morn
Time: 08:50 AM
Room: B316

Abstract Text

The National Institute for Drug Abuse (NIDA) reports marijuana is the most commonly used illegal drug in the United States [HHS Publication No. (SMA) 14-4887; NSDUD Series H-49, 2014]. Usage statistics in the coming years will reflect the increasing number of states allowing medical marijuana prescriptions or decriminalization/legalization for recreational use. The well-known psychoactive ingredient Δ9-tetrahydrocannabinol (THC) found in the plant is fat soluble after consumption, ultimately metabolized and oxidized to the inactive 11-nor- Δ9-THC acidic form that is pharmacologically inactive. This is the main metabolite detected in biological matrices for user drug screening [Moffat et al., Clarke's Analysis of Drugs and Poisons, 3rd Ed., London: Pharma. Press; 2004. p.740-743.]. High-performance Ion Mobility Spectrometry (HPIMS) using electrospray ionization (ESI) has proven adept at detecting a wide range of illegal drug compounds, even in a urine-type matrix for potential forensic or clinical use [Midey et al., Talanta, 2013, 116, 77; Joshi et al., For. Sci. Int. 2014, 255, 196]. HPIMS uses the rapid ion mobility based separation of IMS and gives resolving power comparable to chromatography. Therefore, these initial studies have been extended to focus on detection of THC and related compounds in matrices using ESI-HPIMS with direct ionization from a sample syringe aimed at an alternative rapid quantitative screening for cannabis use.

Keywords: Detection, Drugs, Electrospray, Trace Analysis
Application Code: High-Throughput Chemical Analysis
Methodology Code: Portable Instruments

Session # 1590 Abstract # 1590-3 **Organized Contributed Sessions**

Session Title: On-Site Detection of THC and Related Drugs

Abstract Title: **Handheld Differential Mobility Spectrometry for Drug Interdiction**

Date: Wednesday, March 09, 2016 - Morn
Time: 09:10 AM
Room: B316

Primary Author: Paul J. Rauch
Chemring Detection Systems

Co-Author(s): Eric Lynch

Abstract Text

Differential ion mobility spectrometry (DMS) is capable of detecting and identifying a wide range of chemical vapors by leveraging the nonlinear dependence of an ion's mobility within high electric field conditions. This allows DMS to separate ions that have similar mobilities under the low field conditions typically used in ion mobility spectrometers (IMS) or similar molecular weights in mass spectrometers. Ions within a DMS flow through an analyzer region where they are subjected to two electrical fields a high frequency, high potential asymmetric waveform and a low potential DC field. These two fields create a tunable filter that permits ions to be sorted, allowing high selectivity and sensitivity. Chemring Detection Systems has developed a handheld DMS system that detects and identifies narcotic compounds directly and in a breath matrix, results from this system will be discussed.

Keywords: Drugs, Instrumentation, Spectrophotometry
Application Code: Homeland Security/Forensics
Methodology Code: Sensors

Session # 1590 Abstract # 1590-4 **Organized Contributed Sessions**

Session Title: On-Site Detection of THC and Related Drugs

Abstract Title: **Detection of Tetrahydrocannabinol and Related Compounds in Human Breath Using High-Field Asymmetric Waveform Ion Mobility Spectrometry**

Primary Author: Jared J. Boock
University of Florida

Date: Wednesday, March 09, 2016 - Morn
Time: 09:30 AM
Room: B316

Co-Author(s): Michael T. Costanzo, Richard A. Yost

Abstract Text

Data show that THC can be found in detectable limits in human breath for a period of two hours after cannabis consumption. This short amount of time necessitates an instrument be available onsite. High-field asymmetric waveform ion mobility spectrometry (FAIMS) is a highly selective atmospheric-pressure separation technique that has the potential to provide a portable, field capability for the detection of THC in breath.

Keywords: Drugs, Forensics, Instrumentation, Mass Spectrometry
Application Code: Homeland Security/Forensics
Methodology Code: Mass Spectrometry

Session # 1590 **Abstract #** 1590-5 **Organized Contributed Sessions**

Session Title: On-Site Detection of THC and Related Drugs

Abstract Title: **Detection of Illicit Drugs of Abuse Using Existing Military Chemical Detection Equipment**

Date: Wednesday, March 09, 2016 - Morn
Time: 10:05 AM
Room: B316

Primary Author: Charles S. Harden
US Army ECBC

Co-Author(s)

Abstract Text

U.S. Army research and development at Edgewood Chemical Biological Center is developing and demonstrating detection of illicit drugs of abuse, e.g., marijuana, cocaine, heroin, and designer drugs, using existing chemical detection systems and integrating illicit drug detection into the family of Chemical Biological Radiological Nuclear and Explosives (CBRNE) sensors. This work will integrate hazard detection equipment and reduce the overall number of types of detection equipment fielded; saving purchase and maintenance cost, as well as training time.

This report is an extension of recent explosives detection work at ECBC that resulted in development of a "Chemical Explosives Detector" (CED)1 based on the fielded M4A1-JCAD (Joint Chemical Agent Detector) and development and fielding of a "Colorimetric Reconnaissance Explosive Squad Screening" (CRESS) kit2 – these systems have been demonstrated to be effective for detection and identification of trace and bulk levels of explosives and their precursors.

Test results and the gas phase ion molecule chemistry of the CED ion mobility spectrometry (IMS) and the CRESS kit specific colorimetric chemistry will be presented and discussed.

1http://www.ecbc.army.mil/mobile/news/2014/Teaching-JCAD-new-tricks-ECBC-scientists-turn-handheld-JCAD-dual-use-chemical-explosives-detector.html
2http://www.army.mil/article/125449/Explosives_detection_kit_ready_to_enter_new_acquisition_phase/

Keywords: Detector, Drug Discovery, Drugs, Trace Analysis
Application Code: Drug Discovery
Methodology Code: Portable Instruments

Session # 1590 Abstract # 1590-6 **Organized Contributed Sessions**

Session Title: On-Site Detection of THC and Related Drugs

Abstract Title: **Innovative and Rapid Detection of Marihuana Consumption from Direct Breath Analysis**

Date: Wednesday, March 09, 2016 - Morn
Time: 10:25 AM
Room: B316

Primary Author: Chandrasekhara Hariharan
ION-GAS GmbH

Co-Author(s): Oliver Kayser

Abstract Text

In several clinical studies, the potential of ion mobility spectrometry coupled to rapid gas-chromatographic pre-separation (GC-IMS) for a comprehensive analysis of human breath was demonstrated in the past decade. Exploring the exhaled metabolic profile enables medical diagnosis and therapy control e.g. in nephrology or diabetes. Furthermore, the quantification of various remedies with sufficient correlation to plasma concentrations was shown e.g. for anaesthetics such as Propofol or Fluranes. Such methods could be applied for on-line anaesthesia control.

Encouraged from those findings, we developed a method for the on-site quantification of the consumption of Cannabis sativa, one of the most common illicit drugs world-wide with the objective to provide a tool for the regulatory authorities e.g. to control if the drivers vigilance is affected. A characteristic pattern of metabolites exclusively caused by Cannabis consumption was developed and quantified for the correlation with the THC plasma concentration which is at least the measure for the effects of the drug. The method was validated successfully with regard to false-positives e.g. caused by other hemp products and plant-based commodities. Presently, the method enables the on-site non-invasive detection of Marihuana consumption even after 3-4 hours which is in the range required by the authorities.

Keywords: Drugs, Forensics, Metabolomics, Metabonomics
Application Code: Safety
Methodology Code: Chemical Methods

Session # 1590 Abstract # 1590-7 **Organized Contributed Sessions**

Session Title: On-Site Detection of THC and Related Drugs

Abstract Title: **Understanding Canine Detection of Explosives and Narcotics Using a 3D Printed Artificial Dog Nose**

Primary Author: Matthew Staymates
NIST

Date: Wednesday, March 09, 2016 - Morn
Time: 10:45 AM
Room: B316

Co-Author(s): Brent Craven, Greg Gillen, Jessica Staymates, William MacCrehan

Abstract Text

An investigation of the external aerodynamics of canine olfaction is presented. Extending upon the previous work done by Settles (2002) and Craven (2010), we have developed an anatomically-correct artificial dog nose. The nose is modeled from detailed MRI imaging of a female Labrador retriever and fabricated using a 3D printer. Realistic sniffing flow rate and frequency is facilitated via custom piston/cylinder system. Flow visualization experiments using schlieren imaging enable real-time examination of the dogs remarkable ability to attract and sample vapors from extended distances. During exhale, a turbulent air jet emanates from each nostril and entrains fluid from ahead of the nose, sometimes at a distance of many tens of centimeters. This vapor is now readily available for inhalation, during which the nose now acts as a potential-flow inlet. During active sniffing, this exhale/inhale cycle is repeated at a frequency of around 5Hz. We have learned that the dog is an active aerodynamic sampling system, utilizing fluid dynamics to increase its aerodynamic reach to sample vapors at increasingly large distances.

As a form of biomimicry, we are now utilizing bio-inspired design principles from the dog and applying them to current- and next-generation vapor sampling technology. This presentation will show many flow visualization examples of canine olfaction, and results from biomimicry experiments that demonstrate improvements in vapor sampling of commercially-available explosives and narcotics vapor samplers by making them “sniff” rather than continuously inhale. Applications of this effort include potential optimization of real-time detection of THC and related drugs from a vapor sampling perspective.

GS Settles, et.al. A chapter in Sensors and Sensing in Biology and Engineering, ed. F.G. Barth, J.A.C. Humphrey, and T.W. Secomb, Springer, Vienna & NY, 2002.

BA Craven, et.al. J.R. Soc. Interface, 7, 2010

Keywords: Biosensors, Detection, Portable Instruments, Sampling
Application Code: Homeland Security/Forensics
Methodology Code: Sensors

Session # 1600 Abstract # 1600-1 **Organized Contributed Sessions**

Session Title PAI-NET - Characterization of Micro/Nano Liquid Phases

Abstract Title **Development of Microfluidic Lattices for High-Performance Cell/Particle Separations**

Primary Author Masumi Yamada
Chiba University

Co-Author(s) Minoru Seki

Date: Wednesday, March 09, 2016 - Morn
Time: 08:30 AM
Room: B301

Abstract Text

In the last decade, microfluidics-based particle/cell sorting systems have been developed by many researchers. Although most of previously developed microfluidic devices enable highly precise cell sorting with relatively simple experimental procedures, there are several problems in terms of the low throughput of sample processing and the microchannel clogging. Here we present a new size-based particle/cell sorting system using lattice-patterned microfluidic channels. The lattice region was composed of two types of microchannels, which are perpendicularly crossing and slanted against the overall flow direction. Because there was a significant difference in the densities of these two types of microchannels, only a small amount of the flow is split at every intersection. Large particles/cells flow along the streamline whereas small ones are separated from the stream, resulting in the continuous size-dependent cell sorting. This system is robust against the problem of microchannel clogging, and enables relatively high-throughput sample processing. We successfully demonstrate the sorting of model particles with size of several micrometers, and examined factors affecting the separation behaviors of particles. As a biological application, we performed direct sorting of blood cells from a diluted blood samples. In addition, dual-height microfluidic systems was proposed, which would further improve the sorting throughput. The presented scheme of particle/cell sorting would be advantageous because of the simplicity in operation, high precision of the sorting performance, and the relatively high-speed processing.

Keywords: Biomedical, Isolation/Purification, Lab-on-a-Chip/Microfluidics, Separation Sciences
Application Code: Biomedical
Methodology Code: Microfluidics/Lab-on-a-Chip

Session # 1600 Abstract # 1600-2 **Organized Contributed Sessions**

Session Title PAI-NET - Characterization of Micro/Nano Liquid Phases

Abstract Title **Microfluidic-Based Approach for Producing Diffraction Quality Protein Crystals**

Date: Wednesday, March 09, 2016 - Morn
Time: 08:50 AM
Room: B301

Primary Author Maeki Masatoshi
Hokkaido University

Co-Author(s) Kenis J. Paul, Miyazaki Masaya, Pawate Ashtamurthy, Sugishima Masakazu, Tokeshi Manabu, Watanabe Keiichi

Abstract Text

Protein crystallization and its crystal structure analysis provide essential information in the field of drug discovery. High diffraction protein crystals must be prepared to obtain the three dimensional structure data at high resolution. However, preparation of high quality protein crystals is bottleneck of the protein three-dimensional structure analysis. We demonstrated a single crystallization method by using microdroplet to obtain high quality protein crystal. In this study, we investigated the protein crystal growth behavior in microspace with the aim of obtaining the high diffraction quality of protein crystal. Lysozyme, glucokinase from Pseudoalteromonas sp. AS-131 (PsGK), and heme oxygenase (HO) complex were used as model proteins. To confirm the effect of the microspace on protein crystal growth, two types of microfluidic device were fabricated. Lysozyme crystallization experiments were performed by using 10 and 50 [micro]m depth of crystallization chamber. We found that the (1 1 0) face was preferentially grown in the 10 [micro]m depth crystallization chamber. On the other hand, the (1 1 0) and (1 0 1) faces of lysozyme crystal were randomly grown in the 50 [micro]m depth crystallization chamber. Typically, the difference in the growth rate of the (1 1 0) and (1 0 1) faces of lysozyme crystal was dominated the supersaturation of crystallization solution. In the case of PsGK and HO-complex, the depth of crystallization chamber also affected the crystal growth. Fig. 1 (a) shows the PsGK crystal obtained by conventional crystallization method. Conversely, the microfluidic-based crystallization can produce single PsGK crystal in each crystallization chamber. However, the growth behavior of protein crystal was dramatically changed by depth of the crystallization chamber, as shown in Fig. 1 (c) and (d). Consequently, our microfluidic-based approach provides a simple preparation method of the high quality protein crystals.

Keywords: Bioanalytical, Lab-on-a-Chip/Microfluidics, X-ray Diffraction
Application Code: Bioanalytical
Methodology Code: Microfluidics/Lab-on-a-Chip

Session # 1600 Abstract # 1600-3 **Organized Contributed Sessions**

Session Title PAI-NET - Characterization of Micro/Nano Liquid Phases

Abstract Title **Sample Pretreatments for Droplet-Based Microanalytical System by Using Nanodroplet Formation**

Date: Wednesday, March 09, 2016 - Morn
Time: 09:10 AM
Room: B301

Primary Author Mao Fukuyama
Kyoto Institute of Technology

Co-Author(s) Akihide Hibara, Kohji Maeda, Yumi Yoshida

Abstract Text

We report the selective concentration of the microdroplets' contents using spontaneous emulsification and its application to the biochemical analytical processes. Recently, droplet microfluidics involving aqueous microdroplets formed in microfluidic devices have been studied in the fields of biochemical and chemical analyses for the applications to high-throughput analyses such as single cell assays. However, the lack of the separation method for water-soluble solutes in microdroplets has limited the applicability of the droplet microfluidics to the biochemical analyses.

In order to overcome this difficulty, we have been developed the selective condensation method for microdroplet contents using spontaneous emulsification at the interface of the microdroplets. When an aqueous microdroplet was exposed to the organic phase containing Span 80, a nonionic surfactant, nanodroplets were formed at the interface of the microdroplet as the result of the spontaneous emulsification. Consequently, hydrophilic and large solutes were concentrated in the microdroplet and the other solutes partitioned to the nanodroplets.

In the presentation, the control of the selectivity and the application of this method to biochemical analyses will be presented.

Keywords: Lab-on-a-Chip/Microfluidics, Sample Preparation, Surfactants
Application Code: High-Throughput Chemical Analysis
Methodology Code: Microfluidics/Lab-on-a-Chip

Session # 1600 Abstract # 1600-4 **Organized Contributed Sessions**

Session Title PAI-NET - Characterization of Micro/Nano Liquid Phases

Abstract Title **Development of DNA and/or RNA Extraction Method from Single Cell**

Primary Author Yukihiro Okamoto
Osaka University

Date: Wednesday, March 09, 2016 - Morn
Time: 09:30 AM
Room: B301

Co-Author(s)

Abstract Text

The significance of single cell analysis in diagnosis and medical researches has been increasing because present cell analysis cannot find and analyze rare cells in the cell group. However, despite the progress of high performance apparatus, present single cell analysis has still been hampered by troublesome pretreatment such as single cell lysis and extraction of biomolecules in cell. Therefore, for true attainment and practical application of single cell, we must overcome the problems related to pretreatments.

Microfluidic device has attracted many researchers interested in single cell analysis and pretreatment because micro space is suitable for single cell treatment because sample loss, dilution and contamination could be avoided. However, proposed single cell pretreatments using microfluidic device adopt conventional pretreatments and never overcome the problems of single cell analysis. Furthermore, the integration of several pretreatment procedures into single microfluidic device is still difficult. Therefore, we must propose a new principle for single cells pretreatments, which utilize advantages of microspace.

In this talk, in view of present state of single cell pretreatment, I propose some techniques and ideas combined with the advantages of microfluidic device. At first, I will talk about single cell sorting with microfluidic device and optical trapping. Subsequently, the extraction method of DNA or RNA from single cell with microfluidic device will be briefly explained. Finally, I will introduce in situ formation of single cell pretreatments and the performance of our method for single cell pretreatments.

Keywords: Bioanalytical, Biomedical, Lab-on-a-Chip/Microfluidics, Nucleic Acids
Application Code: Bioanalytical
Methodology Code: Microfluidics/Lab-on-a-Chip

Session # 1600 Abstract # 1600-5 **Organized Contributed Sessions**

Session Title: PAI-NET - Characterization of Micro/Nano Liquid Phases

Abstract Title: **In Vivo Detection and Quantification of Circulating Cells and Nanoparticles in Whole Blood Using Photoacoustic-Fluorescence Flow Cytometry**

Date: Wednesday, March 09, 2016 - Morn
Time: 10:05 AM
Room: B301

Primary Author: Dmitry A. Nedosekin
University of Arkansas for Medical Sciences

Co-Author(s): Vladimir P. Zharov

Abstract Text

<i>In vivo</i> molecular analysis of live tissues and bio-fluids may provide crucial information for optimization and development of novel drugs and monitoring of progression for various diseases. However, in the most cases repeated biopsy for <i>ex vivo</i> analysis is impossible, as it compromises integrity of the analyzed system. In this case noninvasive analysis based on a variety of biomedical optical techniques may provide required data in real time without the need to extract tissue sample. Still, light scattering and absorption by live tissues hinder applications of most light based techniques and requires carefull approach toward developing <i>in vivo</i> analysis method. We have developed integrated photoacoustic (PA) fluorescence flow cytometry (PAFFC) system for <i>in vivo</i> liquid biopsy of bio-fluids including monitoring of micro blood and lymph vessels. The molecular analysis based on absorbance and fluorescence properties of circulating cells and nanoparticles (NPs) provides label-free identification for a numbers of compounds including cancer cells. Moreover, molecular specific targeting of circulating cells using dyes and NPs further extend the range of possible applications for PAFFC system. <i>In vivo</i> PAFFC platform was used to study pharmacokinetics of dyes and nanoparticles in live animals. PAFFC provides a flexible approach for cancer research by allowing identification and quantification of circulating tumor cells (CTCs) in real time in blood circulatory. In PA mode label-free detection of pigmented cancer cells is possible even in human patients. Fluorescence module of the PAFFC provides detection of cancer cells genetically engineered to produce fluorescent proteins. We also proposed and demonstrated labeling and later identification of individual cancer cells in flow using photoswitchable fluorescent proteins. Finally, integrated <i>in vivo</i> PAFFC detection may reveal interactions of NPs with CTCs in blood flow.

Keywords: Biomedical, Lab-on-a-Chip/Microfluidics, Nanotechnology, Photoacoustic
Application Code: Biomedical
Methodology Code: Biospectroscopy

Session # 1600 **Abstract # 1600-6** **Organized Contributed Sessions**

Session Title: PAI-NET - Characterization of Micro/Nano Liquid Phases

Abstract Title: **Microfluidic Construction of Artificial Cell-Like Reactors**

Date: Wednesday, March 09, 2016 - Morn
Time: 10:25 AM
Room: B301

Primary Author: Masahiro Takinoue
Tokyo Technology

Co-Author(s)

Abstract Text

Understanding the essence of dynamical behaviors of living systems is one of the most important issues in science. Although our understanding of the molecular basis of living systems has dramatically increased, the whole picture of living systems as autonomous integrated molecular systems has not been revealed yet. Recently, artificial cell-like systems as simplified models of living cells have been proposed. The artificial cell-like systems have helped us to characterize living systems as autonomous integrated molecular systems. However, most of proposed artificial cell-like systems have limitations caused by the difficulty in controlling nonequilibrium conditions in micrometer-sized systems, and it is required to develop experimental methods to control nonequilibrium conditions in micrometer-sized systems by realizing sustained matter and energy flows into/out of cell-sized systems. In this presentation, we first introduce microfluidic method to control micrometer-sized nonequilibrium open conditions of chemical microreactors toward the construction of artificial cell-like systems. The method is based on the fusion and fission of water-in-oil (W/O) microdroplets. Second, we briefly show the study of molecular robots based on artificial cell-like reactors. We believe that these technologies will promote the studies of applied physics of soft matter and biomedical engineering in the future.

Keywords: Biomedical, Biotechnology, Lab-on-a-Chip/Microfluidics
Application Code: Biomedical
Methodology Code: Microfluidics/Lab-on-a-Chip

Session # 1600 Abstract # 1600-7 **Organized Contributed Sessions**

Session Title PAI-NET - Characterization of Micro/Nano Liquid Phases

Abstract Title **Parallel Lipid Bilayer Formation by Microfabrication and Its Applications**

Primary Author Ryuji Kawano
Tokyo University of Agriculture and Technology

Date: Wednesday, March 09, 2016 - Morn
Time: 10:45 AM
Room: B301

Co-Author(s)

Abstract Text

Artificial cell membranes have emerged as a biomimetic tool in such areas as membrane protein study, synthetic biology, and drug discovery. Planar lipid bilayers are used for functional studies of ion channel proteins using electrophysiological techniques. However, the stability of lipid bilayers and the reproducibility of bilayer formation remain challenging. To improve these issues, we are trying to use microfabrication technology that has a major advantage: easy to handle lipid molecules or solution at micron scale using microfluidics. Applying this advantage, we propose a stable and reproducible preparation procedure for the planar lipid bilayers using "droplet contact method", and they are applying to ion channel measurements and portable biosensings.

Keywords: Electrochemistry, Lab-on-a-Chip/Microfluidics, Lipids, Membrane
Application Code: Bioanalytical
Methodology Code: Microfluidics/Lab-on-a-Chip

Session # 1600 Abstract # 1600-8 **Organized Contributed Sessions**

Session Title PAI-NET - Characterization of Micro/Nano Liquid Phases

Abstract Title **Unique Liquid Properties by Surface Silanol Groups in Extended-Nano Spaces**

Primary Author Yutaka Kazoe
The University of Tokyo

Date: Wednesday, March 09, 2016 - Morn
Time: 11:05 AM
Room: B301

Co-Author(s) Kazuma Mawatari, Keisuke Ikeda, Takehiko Kitamori

Abstract Text

Microfluidics has developed to realize various chemical processes at small volumes, by integrating chemical unit operations in micro fluidic channels. Recently, the research field is extended to extended-nano spaces (10-1000 nm). Exploiting aL-fL volumes and dominant surface effects of extended-nanochannels, novel chemical devices such as aL-chromatography and single cell/single molecule analysis are expected. Previously, we have revealed various unique liquid properties in fused-silica extended-nanochannels such as higher viscosity, lower dielectric constant and higher proton mobility. Based on these findings, we hypothesized proton transfer phase of loosely coupled water within 50 nm by interaction between surface silanol groups and water molecules. Therefore, for engineering using extended-nano spaces, verification of the liquid model is required to optimally design fluidic devices and develop novel applications. In this study, we verified the hypothesis indicating that the interaction between surface silanol groups and water molecules induce the unique properties in extended-nano spaces.

We controlled the density of surface silanol groups in fused-silica extended-nanochannel by dehydration reaction. Using the surface-controlled extended-nanochannels, we investigated molecular motion and proton transfer in pure water by nuclear magnetic resonance. The water molecular motion was slower and the activation energy of proton transfer through water molecules was lower, with increasing the surface silanol density. The results suggest that the water molecules are loosely structured and make hydrogen bond networks at high surface silanol density. Therefore, we could verify our hypothesis of water confined in extended-nano spaces. The knowledge obtained from this study suggests importance of surface chemical groups interacted with water molecules for design of nanofluidic analytical devices.

Keywords: Bioanalytical, Lab-on-a-Chip/Microfluidics, Sample Preparation, Water
Application Code: Bioanalytical
Methodology Code: Microfluidics/Lab-on-a-Chip

Session # 1610 Abstract # 1610-1 **Oral Sessions**

Session Title: Biomedical: New Technologies for Breath Analysis (Half Session)

Abstract Title: **Heart-Cutting Multidimensional Micro Gas Chromatography ([micro]GC) System for Breath Analysis**

Primary Author: Menglian Zhou
University of Michigan

Date: Wednesday, March 09, 2016 - Morn
Time: 08:30 AM
Room: B409

Co-Author(s): Hongbo Zhu, Jiwon Lee, Katsuo Kurabayashi, Kevin Ward, Robert Nidetz, Xudong Fan

Abstract Text

Human's exhaled breath is known to contain more than 300 volatile organic compounds (VOC). Some of these compounds, e.g., ethane, pentane, isoprene, and acetone are produced by metabolic/pathologic processes. Therefore, monitoring those VOC markers can help us predict the onset and severity of certain diseases, and guide therapy accordingly. Gas chromatography coupled with mass spectrometry (GC-MS) is the most established method for breath analysis. Standard GC provides results with high sensitivity and resolution, but its application is limited by high cost and low mobility; on the other hand, a [micro]GC system is compact, portable, and rapid but its separation capability is inadequate for complex samples like human breath. To address the need for a portable GC for rapid and quantitative detection of VOC markers in human breath, we constructed and optimized a heart-cutting multi-dimensional [micro]GC system that consisted of two separation columns coated with different stationary phases. Flow-through microfluidic photoionization detectors were installed after first and second dimension columns and a lab-built Deans switch was utilized for transferring the effluent from the first dimensional column to second dimensional columns for further separation and detection. The system was fully automated and can either be attached to a mechanical ventilator or use a tedlar bag to gather breath samples.

Keywords: Biomedical, Gas Chromatography
Application Code: Biomedical
Methodology Code: Gas Chromatography

Session # 1610 Abstract # 1610-2 **Oral Sessions**

Session Title Biomedical: New Technologies for Breath Analysis (Half Session)

Abstract Title **Real-Time PTR-TOF-MS Measurements Reveal Effects of Respiratory Maneuvers onto Exhaled Breath Biomarker Profiles**

Date: Wednesday, March 09, 2016 - Morn
Time: 08:50 AM
Room: B409

Primary Author Pritam Sukul
University Medicine of Rostock

Co-Author(s) Jochen K. Schubert, Khushman Taunk, Peter Oertel, Phillip Trefz, Svend Kamysek, Wolfram Miekisch

Abstract Text

Analysis of breath VOC biomarkers (VOCs) is promising in the field of noninvasive diagnosis. Hemodynamic changes due to different breathing patterns or postures cause immediate substance specific effects. In our study we investigated the effect of forced expiratory maneuvers onto exhaled VOC profiles.

A PTR-ToF-MS-8000 (PDrift=2.3mbar, TDrift=75°C, VDrift=610Volt, E/N=139Td, Time-resolution=200ms) was used in continuous side stream mode (Sampling flow=20ml/min) for breath resolved measurements of VOC profiles in 15 healthy volunteers with parallel monitoring of lung-function and hemodynamics. After the first minute of paced breathing (12/min) a maneuver for spirometric determination of FEV1 (forced expiratory volume/1s) was performed. 30 selected VOCs were quantified in alveolar and inspiratory air by using a custom made data processing algorithm.

Profound changes of hemodynamic and respiratory parameters and, consecutively, of exhaled VOC concentrations happened within seconds. Normalized mean isoprene concentration increased by 13% (from 114ppb to 129ppb) during full exhalation (before the forced expiration) and decreased by 46% (to 62ppb) after forced exhalation. Exhaled isoprene concentrations then increased again (by 18% up to 135ppb) and even exceeded baseline levels, mirroring time profiles of cardiac output and pET-CO2. Normalised mean tidal-volume and minute-ventilation were increased to 257% and 143% respectively during forced exhalation. Acetone, H2S (from oral bacteria) and benzene (exogenous) concentrations remained almost constant.

Blood borne exhaled VOC concentrations changed during forced expiration. Changes depended on both respiratory and hemodynamic parameters, on origin and physico-chemical properties of the substances. In a perspective breath VOCs may be used to gain additional information on lung function.

Keywords: Bioanalytical, Biomedical, Mass Spectrometry, Sampling

Application Code: Biomedical

Methodology Code: Mass Spectrometry

Session # **1610** **Abstract #** **1610-3** **Oral Sessions**

Session Title Biomedical: New Technologies for Breath Analysis (Half Session)

Abstract Title **High Altitude Respiratory Research Using Quadrupole Mass Spectrometry**

Date: Wednesday, March 09, 2016 - Morn
Time: 09:10 AM
Room: B409

Primary Author Charles W. De Carlo
Extrel CMS

Co-Author(s) Frank DeThomas, James R. Brenner, Zbigniew Krieger

Abstract Text

For several decades quadrupole mass spectrometry has been the preferred technique for respiratory gas analysis. There are three primary advantages to using the mass spectrometer in a breath analysis application. The analyzer has the ability to provide high sensitivity measurements of all of the necessary components (nitrogen, oxygen, carbon dioxide, argon, water, etc.), and it is capable of performing the analysis using a very small amount of sample flow relative to the gas in a typical exhalation. In addition, the speed of analysis ensures that the analyzer is able to accurately quantify the entire profile of the exhalation by performing several measurements per second.

Biomedical and biophysical researchers have employed the approach for years in a number of applications to determine how various conditions impact the respiration of the test subject. One of the more challenging variables to test is the effect of high elevation breathing, due to the associated decrease in atmospheric pressure. Normally this results in decreased sample pressure and reduced quality of response at the mass spectrometer. To counter this effect a variable pressure sample inlet was developed to maintain constant sample pressures, and high speed air sampling was conducted under conditions simulating several elevations. Using this configuration the mass spectrometer performed the air analysis at a rate of <10 milliseconds per component, with no loss of precision under high altitude conditions.

Keywords: Bioanalytical, Biomedical, Mass Spectrometry, Quadrupole MS
Application Code: Biomedical
Methodology Code: Mass Spectrometry

Session # 1610 Abstract # 1610-4 **Oral Sessions**

Session Title: Biomedical: New Technologies for Breath Analysis (Half Session)

Abstract Title: **VOC Detection in Animal Models for Medical Research**

Date: Wednesday, March 09, 2016 - Morn
Time: 09:30 AM
Room: B409

Primary Author: Wolfgang Vautz
ISAS

Co-Author(s): Liedtke Sascha, Nils Kunze, Thorsten Perl, Ursula Telgheder

Abstract Text

The feasibility of the application of ion mobility spectrometry coupled to rapid gas-chromatographic pre-separation (GC-IMS) in animal models e.g. breath analysis for providing information on the metabolome was already demonstrated for an asthma mouse model. In general, this non-invasive method could lead to more relevant information from every single animal experiment, thus consequently enabling a reduction of the number of experiments. Recently, the method was applied for the investigation of two substantial medical problems.

Infections are still a major contributor to death in intensive care medicine. Survival in sepsis patients depends on the immediate administration of an adequate antibiotic therapy. A faster identification of pathogens in sepsis patients could be an important step towards a more effective use of antibiotic substances. Specific VOC patterns of pathogens have been successfully used for their differentiation in in-vitro experiments. The investigation of exhaled air from sepsis patients could allow the on-site identification of pathogen at the intensive care unit. Therefore, we developed a suitable method by MCC-IMS breath analysis in a rabbit model.

Acute respiratory distress syndrome (ARDS) is a severe reaction of lung tissue on different trigger including pneumonia, sepsis and trauma. There is an urgent need for studies unravelling pathogenetic mechanisms to optimise the prevention and treatment of ARDS. We conducted a pilot study with breath gas analysis in domestic pigs under sedation before and after induction of an ARDS. In this limited pilot study several compounds with significant differences between native and ARDS samples can be described.

Keywords: Biomedical, Clinical Chemistry, Medical, Metabolomics
Application Code: Biomedical
Methodology Code: Chemical Methods

Session # 1620 Abstract # 1620-1 **Oral Sessions**

Session Title: Clinical/Toxicology

Abstract Title: **Determination of Manganese Using Cathodic Stripping Voltammetry on a Platinum Electrode**

Primary Author: Wenjing Kang
University of Cincinnati

Date: Wednesday, March 09, 2016 - Morn
Time: 08:30 AM
Room: B402

Co-Author(s): Adam Bange, Cory A. Rusinek, Erin Haynes, Ian Papautsky, William R. Heineman

Abstract Text

Manganese (Mn) is an essential trace metal and plays an important role in metabolic enzyme activation. Yet it is harmful to neurological systems at high concentrations, and has been associated with Parkinson's disease or neurological malfunction in children. Thus it is critical to monitor Mn levels in human's blood, and in the environment due to various sources of exposure. Since the well-established spectroscopic methods (ICP-MS or AAS) are inconvenient for point-of-care (POC) applications, we are developing a portable system with disposable sensors based on stripping voltammetry that offer efficient and accurate measurements of Mn in different samples. The three-electrode sensor contains platinum working and auxiliary electrodes, and an electroplated Ag/AgCl reference electrode. Another advantage of this Pt-based miniaturized sensor is the small size (~10 mm on side) and need for only 1-10 µL sample. We explored the potential window and located the Mn reduction peaks using cyclic voltammetry in acetate and borate buffer. Then we optimized the parameters for cathodic stripping voltammetry followed by calibration of the sensor in acetate buffer (0.2 M, pH 5.5), with concentrations of Mn from 5-50 ppb. With optimal parameters, we achieved a calculated detection limit as 0.89 ppb (16.3 nM) and 1.394 nA/nM sensitivity. Currently we are able to measure 1960 ppb spiked Mn from digested bovine blood, and a more-efficient extraction procedure is under development. Eventually we believe this sensor system could act as an alternative for POC applications in local clinics or resource-limited areas.

Keywords: Biomedical, Clinical/Toxicology, Electrochemistry, Sensors
Application Code: Clinical/Toxicology
Methodology Code: Electrochemistry

Session # 1620 Abstract # 1620-2 **Oral Sessions**

Session Title Clinical/Toxicology

Abstract Title **Impact of Amphiphilic Poly(amido)amine Dendrimers on the Biophysical and Biorecognition Properties of Bilayer Membranes**

Date: Wednesday, March 09, 2016 - Morn
Time: 08:50 AM
Room: B402

Primary Author Samuel S. Hinman
University of California, Riverside

Co-Author(s) Charles J. Ruiz, Ling Peng, Quan Cheng

Abstract Text

Amphiphilic dendrimers have proven to be promising candidates for drug delivery and cancer therapy; these synthetic lipids are capable of assembling into supramolecular micelles, which may encapsulate drugs and deliver them to specific sites and locations. In cultured cell systems, the high treatment effectiveness has been attributed to promoting cellular uptake and decreasing efflux of the chosen drug. However, no effort has yet been made to determine what effects these synthetic lipids may have on the plasma membrane. In this study, we utilize supported bilayer membrane systems that incorporate various poly(amido)amine (PAMAM) dendrimers of different headgroup sizes to investigate how these molecules interact with and alter the bilayer properties. We first show how amphiphilic PAMAM dendrimer micelles may disrupt the POPC bilayer, possibly via a surfactant effect. However, under controlled conditions, the dendrimers may be integrated into the bilayer for tailored supported bilayer properties. Using surface plasmon resonance (SPR), we show how increasing concentration and headgroup size of the PAMAM dendrimers leads to an increase in bilayer stability. This is supported by confocal fluorescence microscopy investigations, which show a decrease in POPC bilayer fluidity under similar conditions. The amine terminals of the dendrimer headgroups allow for facile, in situ derivitization of complexes that incorporate these molecules, and we exemplify this through biotin tagging and streptavidin binding to dendrimer containing bilayers. Taken together, we expect these results to influence future drug carrier designs, as the supramolecular assembly process plays an important role in the efficacy and potential toxicity of carrier structures.

Keywords: Bioanalytical, Biospectroscopy, Lipids, Membrane

Application Code: Clinical/Toxicology

Methodology Code: Biospectroscopy

Session # 1620 **Abstract # 1620-3** **Oral Sessions**

Session Title: Clinical/Toxicology

Abstract Title: **Analysis of Drugs in Saliva During Treatment of Post-Traumatic Stress Disorder Patients**

Date: Wednesday, March 09, 2016 - Morn
Time: 09:10 AM
Room: B402

Primary Author: Kathryn Dana
Real-Time Analyzers, Inc

Co-Author(s): Albert J. Arias, Chetan Shende, Stuart Farquharson

Abstract Text

Medication is frequently prescribed as a means to relieve the stresses and traumas of combat. Some 476,000 military personnel with post-traumatic stress disorder (PTSD) have developed drug or alcohol dependence. Treatments are available for drug and alcohol dependence including medications to reduce their use. Veterans Affairs hospital physicians must frequently test patients to identify the discontinuation of medications or any recurrence of drug use and then adjust treatment appropriately. These tests most often involve collecting and sending a urine sample to a clinical lab for analysis by gas chromatography coupled mass spectrometers. Results are usually returned in 1-2 weeks, a delay that makes timely adjustment of treatment difficult. In an effort to provide an analyzer that can determine drug use at the time of VA hospital or clinic visits so that treatment can be adjusted, we have been developing a surface-enhanced Raman spectroscopy (SERS) based assay that can detect and identify numerous drugs in saliva at ng/mL concentrations within 10 minutes. This technology represents a major advancement in drug analysis, allowing for real-time at-site analysis, which will likely lead to improvements in clinical care as well as cost benefits. Measurements of lab and patient samples will be presented.

Keywords: Biomedical, Clinical Chemistry, Drugs, Surface Enhanced Raman
Application Code: Clinical/Toxicology
Methodology Code: Vibrational Spectroscopy

Session # **1620** **Abstract #** **1620-4** **Oral Sessions**

Session Title Clinical/Toxicology

Abstract Title **G-Quadruplex: A Biocompatible Additive for Enhancing the Antibacterial Activity of H2O2**

Date: Wednesday, March 09, 2016 - Morn
Time: 09:30 AM
Room: B402

Primary Author Yuqian Xing
University of North Dakota

Co-Author(s) Julia Xiaojun Zhao, Minh H. Duong, Xiao Liu

Abstract Text

G-quadruplexes have been broadly applied to bioanalytical chemistry with peroxidase-mimicking activity. This property makes G-quadruplexes possible to catalyze the decomposition of H2O2. The hydroxyl radical (·OH) generated during the procedure has a higher antibacterial activity than the original H2O2. Herein, an efficient and biocompatible antibacterial system, which provides the same antibacterial efficiency at lower H2O2 concentration to avoid the H2O2 toxicity, has been demonstrated based on the conversion of H2O2 to ·OH. First, the mechanism of ·OH generation was verified by monitoring the reaction between ·OH and terephthalic acid. Under the optimized condition, the disk diffusion assay and growth inhibition assay showed the obvious enhancement of the antibacterial activity against E. coli of the single H2O2. Meanwhile, the excellent biocompatibility is observed based on the human-derived molecules as additive. This system can be suitable for in actual wound disinfection in vivo.

Keywords: Biotechnology, Nucleic Acids
Application Code: Clinical/Toxicology
Methodology Code: Chemical Methods

Session # 1620 Abstract # 1620-5 **Oral Sessions**

Session Title: Clinical/Toxicology

Abstract Title: **Capillary Electrophoresis Investigation of Multiwalled Carbon Nanotube-Biomolecule Binding**

Primary Author: Tyler Davis
West Virginia University

Date: Wednesday, March 09, 2016 - Morn
Time: 10:05 AM
Room: B402

Co-Author(s): Julia A. Mouch, Lisa A. Holland

Abstract Text

With increased use of multiwalled carbon nanotube in medical and commercial products it is imperative to determine toxicity. To better understand toxicity is it important to understand how carbon nanotubes interaction with biomolecules, however, the toxicity of carbon nanotubes has varied in published data. Current methods for studying these interactions require large amounts of time, animal models, and cell lines. To rapidly determine carbon nanotubes toxicity, an affinity measuring techniques in capillary is used. In capillary affinity analysis determines binding for single stranded DNA to functionalized multiwalled carbon nanotubes in a single run. Binding studies were performed across several factors to include DNA sequence, DNA length and carbon nanotube functionalization, to determine effects on affinity. For method validation these studies were compared to traditional affinity binding measurements. Several reports have shown that poly(GT) DNA had a high affinity for single-walled carbon nanotubes (SWCNTs) for dispersion and DNA transportation. Yet, little reporting has been done on multiwalled carbon nanotubes (MWCNT) with DNA. To date this is the first study done to determine binding affinity for multiwalled carbon nanotubes to biomolecules.

Keywords: Analysis, Capillary Electrophoresis, Clinical/Toxicology
Application Code: Clinical/Toxicology
Methodology Code: Capillary Electrophoresis

Session # 1620 **Abstract # 1620-6** **Oral Sessions**

Session Title: Clinical/Toxicology

Abstract Title: **Determining Variability in Potency of Marijuana for Potential Medical Application**

Date: Wednesday, March 09, 2016 - Morn
Time: 10:25 AM
Room: B402

Primary Author: Rebecca Plessel
Pennsylvania State University

Co-Author(s): Amanda Rigdon, Frank Dorman

Abstract Text

The legalization of marijuana is controversial, and popular opinion seems to drive it forward rather than scientific verification of its safety. Marijuana's suspected medical properties have also pushed the legislation for its legalization, but, due to its prior illegality, research in this area was difficult to conduct. Finding that marijuana has some medicinal effects, research expanded to further determine the compounds in the plant and their various effects on the human body. The hallucinogen, [delta]-9-tetrahydrocannabinol (THC), is the most pharmacologically active substance in the plant, yet the other cannabinoids have shown positive results as potential treatments for immunocompromised patients, multiple sclerosis, and Crohn's disease, to name a few. For those who have been administered medical marijuana, they now have the concern about potency. The preferred method of administration is ingestion by a cannabinoid-injected food, where the user breaks pieces off and eats the food over time. Unfortunately, this leads to a potency gradient in the food, from high on the injected end to almost nothing on the opposite. One solution is a pill. To determine the viability, we will test the cannabinoid concentrations across various plants of the same genotype, using Gas Chromatography Flame Ionization Detector. In addition, we will determine the variance in the matrix and the instrument itself, counteracting any potential variables. If the variation in concentration between plants remains minimal, the viability of a pill would be high. If the variance is high, the pill would be no better, or potentially worse, than the current ingestion method.

Keywords: Drugs, Forensic Chemistry, Gas Chromatography
Application Code: Clinical/Toxicology
Methodology Code: Gas Chromatography

Session # 1620 Abstract # 1620-7 **Oral Sessions**

Session Title: Clinical/Toxicology

Abstract Title: **Design and Development of a Portable Aptasensor for Toxicity Monitoring of Field Samples**

Date: Wednesday, March 09, 2016 - Morn
Time: 10:45 AM
Room: B402

Primary Author: Gonca Bulbul, Clarkson University

Co-Author(s): Akhtar Hayat, Silvana Andreescu

Abstract Text

The adoption of nanoparticle based analytical technologies has gained increasing acceptance in various fields of chemical analysis in the environmental, clinical, food and biomedical sectors. Different types of methodologies which employ electrochemical, fluorimetric, piezo-electric output signals are currently developed for the detection of various molecules. In this respect, colorimetric assays are very promising since they enable rapid visual detection on the spot without any need for elaborate laboratory equipment. Additionally, they provide specific advantages such as portability, ease-to-operate, and low cost which make them very attractive from an application view of point. Nanoceria particles have gained significant interest due to their catalytic and free radical scavenging properties. Here, we report a newly discovered phenomenon for quantifying molecular recognition based on the reversible assembly of single-stranded DNA aptamers on redox active nanoceria particles. The method involves target tunable electrostatic and steric repulsion phenomena of the ssDNA to the surface of nanoceria which changes its spectral and functional catalytic properties upon binding of the target analyte. As a proof of concept, the proposed strategy was employed to construct an aptaswitch for the colorimetric sensing of OchratoxinA (OTA), providing a detection limit of 0.15 nM OTA. This approach is generally applicable for sensitive and specific detection of a wide spectrum of analytes, since any aptamer–target binding event can in principle be translated to conformational transition and be detected based on this strategy.

Keywords: Biosensors, Food Contaminants, Sensors
Application Code: Environmental
Methodology Code: Sensors

Session # 1620 Abstract # 1620-8 **Oral Sessions**

Session Title: Clinical/Toxicology

Abstract Title: **Quantitative Assessment of Nanoparticles Exposure Potential Effects in Embryonic Zebrafish**

Date: Wednesday, March 09, 2016 - Morn
Time: 11:05 AM
Room: B402

Primary Author: Xiaobo Liu
Clarkson University

Co-Author(s): Eduard Dumitrescu, Kenneth Wallace, Silvana Andreescu

Abstract Text

The exponential increase of nanoparticles (NPs) containing consumer products require a comprehensive assessment of the risks associated with the use of these materials. In this presentation, we described an integrated methodology to assess the potential toxic effects of silver nanoparticles (Ag NPs) in embryonic zebrafish. [i]In situ[/i] electrochemical experiments were used to establish the relationship between NP exposure and the release of nitric oxide (NO), as an indicator of oxidative stress. RNA [i]in situ[/i] hybridization probes for inducible nitric oxide synthase (NOS2a and NOS2b) was synthesized and used to determine the expression pattern of NOS2a and NOS2b. These results were correlated with the amount of released silver ions and embryonic viability after NP exposure, and related to the NP size and environmental transformation. This methodology can be used to establish mechanisms of nanotoxicity at organ levels and accelerate testing and screening of nanomaterials for their potential effects on the environment and biological systems.

Keywords: Bioanalytical, Biosensors, Biotechnology, Electrochemistry
Application Code: Clinical/Toxicology
Methodology Code: Electrochemistry

Session # 1630 Abstract # 1630-1 **Oral Sessions**

Session Title Electrochemistry - Biological Applications

Abstract Title **Development of an Optrode for In Vivo Neurochemical Studies**

Date: Wednesday, March 09, 2016 - Morn
Time: 08:30 AM
Room: B403

Primary Author Thomas Field
University of Kansas

Co-Author(s) Meng Sun, Michael A. Johnson, Peter M. Ruggles

Abstract Text

Photochemical methods such as optogenetics and caged compound photoactivation are potentially useful for controlling specific neurotransmitter systems with a high level of spatial and temporal resolution. When these photochemical methods are combined with fast analytical methods, such as fast scan cyclic voltammetry (FSCV), they can be powerful tools to understand how the brain functions in vivo. One of the difficulties with combining these methods is the need to position the analytical probe and the light guide accurately in the location of interest because of space constraints. In order to minimize this difficulty, our group developed a probe that combine a working electrode, an optical fiber and a silica capillary in one unit to simplify placement of the analytical tools in the desired location. The probe consists of a 7 □M diameter working electrode for FSCV measurements, a 60 □M fiber optic to act as a light guide, and a 70 □M silica capillary for introduction of compounds. These three components are enclosed in a 200 □M silica capillary and sealed with epoxy. The 200 □M capillary is then incased in a 1.2 mm glass capillary and sealed with epoxy. We demonstrated each application of the probe in vitro by measuring dopamine release with FSCV in a flow cell, imaging dye delivery from the capillary, and by performing uncaging reactions in a microliter vessel.

Keywords: Bioanalytical, Electrochemistry, Microelectrode, Neurochemistry
Application Code: Neurochemistry
Methodology Code: Electrochemistry

Session # 1630 Abstract # 1630-2 **Oral Sessions**

Session Title: Electrochemistry - Biological Applications

Abstract Title: **Adenosine Transiently Modulates Vasodilation in Caudate-Putamen**

Primary Author: Ying Wang, University of Virginia

Co-Author(s): B Jill Venton

Date: Wednesday, March 09, 2016 - Morn
Time: 08:50 AM
Room: B403

Abstract Text

Adenosine is an endogenous nucleoside that modulates important physiological processes, such as vasodilation, in the central nervous system. A rapid mode of adenosine signaling on the second time scale has been recently discovered, but the ability of this type of adenosine signaling to rapidly modulate blood flow has not been characterized. In this study, the effect of transient adenosine release to cause vasodilation was evaluated by simultaneously measuring adenosine and oxygen using fast-scan cyclic voltammetry. Oxygen changes occur when there is an increase in local cerebral blood flow and thus are a measure of vasodilation. About 33% of adenosine transients in the rat caudate-putamen caused a subsequent transient change in oxygen. The amount of oxygen was correlated with concentration of adenosine release and larger adenosine transients (over 0.4 uM) always caused oxygen changes. The duration of adenosine and oxygen transients were 3.24 ± 0.03 seconds and 3.52 ± 0.06 seconds, respectively. On average, spontaneous adenosine peaks 0.2 seconds prior to the peak in oxygen. The A[sub]1[/sub] antagonist, DPCPX, significantly decreased the concentration of spontaneous adenosine release and consequently oxygen release also decreased. The A[sub]2a[/sub] antagonist, SCH442416, decreased the frequency of oxygen release while the frequency of adenosine release was unchanged. These results suggested that adenosine-induced vasodilation was modulated via A[sub]2a[/sub] receptors. Adenosine can modulate blood flow on a rapid, sub-second time scale, a novel finding that rapid adenosine changes can induce rapid, but transient, vasodilation.

Keywords: Bioanalytical, Electrochemistry, Microelectrode, Voltammetry

Application Code: Bioanalytical

Methodology Code: Electrochemistry

Session # 1630 Abstract # 1630-3 **Oral Sessions**

Session Title: Electrochemistry - Biological Applications

Abstract Title: **Amperometric and Voltammetric Measurements in the Cell Cytosol Using Conical Carbon Nanoelectrodes**

Primary Author: Edwin Mitchell
North Carolina State University

Date: Wednesday, March 09, 2016 - Morn
Time: 09:10 AM
Room: B403

Co-Author(s): James Roberts, Leslie A. Sombers

Abstract Text

Carbon microelectrodes of a planar geometry have proven to be indispensible tools for examining exocytotic events at single cells. While these traditional electrodes are ideal for measurements of neurotransmitter released into the extracellular space, their size and disk geometry preclude study of intracellular chemical dynamics. To address this limitation we have developed a simple method to construct conical carbon-fiber nanoelectrodes. A masking procedure was devised that uses polymer insulation to reproducibly control the length of exposed carbon surface. Classical voltammetric techniques and microscopy were used to characterize the shape and size of these electrodes. The diameter at the tip of an electrochemically etched electrode ranged from approximately 200-400 nm. This work presents the first implementation of fast scan cyclic voltammetry (FSCV) measurements for intracellular analysis, and the data were compared with amperometric measurements in the cytosol of bovine adrenal chromaffin cells. Amperometry offers the highest temporal resolution, allowing the recording of single vesicular events, but this approach lacks chemical selectivity. FSCV provides qualitative identification of intracellular analytes, but operates on a slower time scale. Fortunately, once the electrode is positioned within the cell, both electrochemical methods can be exploited. The data demonstrate that FSCV and amperometry are complementary techniques for single cell experiments.

Keywords: Characterization, Electrochemistry, Electrodes
Application Code: Bioanalytical
Methodology Code: Electrochemistry

Session # 1630 **Abstract # 1630-4** **Oral Sessions**

Session Title: Electrochemistry - Biological Applications

Abstract Title: **Tunable Electroosmotic Push-Pull Perfusion Shows Higher Aminopeptidase N Activity in CA1 than CA3 of the Rat Hippocampus**

Date: Wednesday, March 09, 2016 - Morn
Time: 09:30 AM
Room: B403

Primary Author: Yangguang Ou
University of Pittsburgh

Co-Author(s): Bocheng Yin, German Barrionuevo, Jenna DeVivo, Stephen Weber

Abstract Text

Ectopeptidases are membrane-bound enzymes whose catalytic domains face the extracellular space (ECS). These enzymes represent a largely unexplored mechanism for the control of neurochemical activity and are critical therapeutic targets for pharmacological treatment of certain disease states, including inflammatory diseases and ischemia. Our lab has developed a novel tool that can study ectopeptidase activity in intact tissue cultures. This technique, called electroosmotic push-pull perfusion (EOPPP, Fig. 1), uses electroosmosis to introduce peptide substrates and collect hydrolysis products from the rat organotypic hippocampal slice cultures (OHSCs). The determination of the concentration of hydrolysis products and/or the unhydrolyzed peptide by capillary LC gives the enzyme activity. Using this tunable approach with leucine enkephalin as substrate, we measured a six-fold higher activity of the bestatin-sensitive aminopeptidase N (APN, EC 3.4.11.2) in the CA1 than in the CA3. Using an oxygen glucose deprivation (OGD) model, we show that inhibition of this enzyme significantly reduces damage in CA1 under ischemic conditions compared to that in the CA3 (as measured by propidium iodide fluorescence). We hypothesize that this is due to the role of APN in cleaving endogenous neuroprotective peptides, such as enkephalins, and seek to further understand the pathway by which this neuroprotection occurs.

Keywords: Bioanalytical, Electrochemistry, Method Development, Neurochemistry
Application Code: Bioanalytical
Methodology Code: Electrochemistry

Session # 1630 Abstract # 1630-5 **Oral Sessions**

Session Title: Electrochemistry - Biological Applications

Abstract Title: **Optogenetic Control of Octopamine Release in Drosophila Melanogaster Larval Ventral Nerve Cord and Detection with Fast Scan Cyclic Voltammetry (FSCV)**

Date: Wednesday, March 09, 2016 - Morn
Time: 10:05 AM
Room: B403

Primary Author: Poojan Pyakurel
University of Virginia

Co-Author(s): B Jill Venton

Abstract Text

Octopamine is an endogenous biogenic amine that plays important roles as a neurotransmitter, neurohormone, and neuromodulator in invertebrates, and has functional analogy with norepinephrine in vertebrates. Fast scan cyclic voltammetry (FSCV) is an excellent technique to detect the rapidly changing levels of octopamine in the brain, however, an FSCV waveform has not been optimized for octopamine detection [i]in situ[/i]. The FSCV waveform was optimized so that the potential for octopamine oxidation would not be near the switching potential and that the secondary peak would be observed, which is important to consistently detect and quantify octopamine. Endogenous octopamine release was stimulated with two new stimulation methods, the ATP sensitive channel P2X2, and a red-light sensitive channelrhodopsin CsChrimson. For CsChrimson mediated release, the evoked current decreases in the presence of octopamine synthesis inhibitor, disulfiram, which confirms that the current is due to octopamine and not its precursor, tyramine. On average, a 2 s stimulation with CsChrimson evokes 0.15 ± 0.07 μM of octopamine release in the larval VNC. Current due to evoked octopamine is stable upon repeated stimulation with 2 and 5 minutes interstimulation times, and the release is dependent on the frequency of applied light pulse. The ability to study this important neurotransmitter in Drosophila will allow studying the effects of drugs and mutations on octopamine release.

Keywords: Electrochemistry, Electrodes, Voltammetry
Application Code: Neurochemistry
Methodology Code: Electrochemistry

Session # 1630 Abstract # 1630-6 **Oral Sessions**

Session Title: Electrochemistry - Biological Applications

Abstract Title: **Clarifying the Complex Chemical Mechanisms that Underlie the Voltammetric Detection of Hydrogen Peroxide**

Date: Wednesday, March 09, 2016 - Morn
Time: 10:25 AM
Room: B403

Primary Author: Leslie A. Sombers
North Carolina State University

Co-Author(s): James Roberts, Leyda Z. Lugo-Morales, Maxim A. Voinov, Samantha K. Smith, Tatyana I. Smirnova

Abstract Text

Hydrogen peroxide (H[sub]2[/sub]O[sub]2[/sub]) is a molecule that is of broad interest in multiple disciplines. In the field of biosensing, oxidase-based electrochemical sensors are developed to combine the exquisite selectivity of H[sub]2[/sub]O[sub]2[/sub]-producing enzymes with the rapid response time and low detection limits inherent to electrochemistry. This union has resulted in the development of powerful analytical tools that are used in many fields, including clinical diagnostics, drug discovery, biodefense, environmental monitoring, military, and security applications. We have found that cyclic voltammetry is particularly useful for monitoring rapid fluctuations of H[sub]2[/sub]O[sub]2[/sub], as it enables real-time measurements that boast electrochemical selectivity. This talk will describe experiments that combine voltammetry and electron paramagnetic resonance spectroscopy to identify and define the electrochemical role of a hydroxide radical liberated during the oxidation of H[sub]2[/sub]O[sub]2[/sub]. These data demonstrate that the hydroxide radical is a principal contributor to the voltammetry of H[sub]2[/sub]O[sub]2[/sub] when using carbon or platinum microelectrodes and a variety of scan rates. They incorporate a missing, fundamental element to our knowledge base that will inform the intelligent design of optimized biosensors for a range of applications.

This work is supported by the NSF (CAREER CHE7151264).

Keywords: Biosensors, Electrodes, Magnetic Resonance, Voltammetry

Application Code: Bioanalytical
Methodology Code: Electrochemistry

Session # 1630 Abstract # 1630-7 **Oral Sessions**

Session Title: Electrochemistry - Biological Applications

Abstract Title: **FSCV Measurements of Neurotransmitters in Daphnia**

Date: Wednesday, March 09, 2016 - Morn
Time: 10:45 AM
Room: B403

Primary Author: Matt N. Jackson
Wayne State University

Co-Author(s): Annette R. Tremonti, David Pitts, Parastoo Hashemi, Shawn McElmurry, Srimal A. Samaranayake

Abstract Text

Cladocera (Daphnia) respond to light stimulus by moving away from it. This is thought to be a defense mechanism developed by the Daphnia to avoid predators. It has been speculated that the chemistry underlying this mechanism is mediated by serotonin and histamine. As the global usage of selective serotonin reuptake inhibitors has increased over the last decade, low but significant levels of several SSRIs are now found in many natural water systems. Given that SSRIs change daphnia behavior in experimental settings, this SSRI contamination of natural waters may be ecologically hazardous. We seek to understand if Daphnia's behaviors are serotonin and histamine mediated and to assess the effects of SSRIs on these neurochemicals. We therefore utilize in vivo FSCV to assess the neurochemical response of daphnia to light. Daphnia were immobilized on the head of a pin and a carbon fiber microelectrode was implanted into its brain. Serotonin and histamine were measured simultaneously as light stimuli were delivered to the Daphnia. We found stimulation locked, frequency dependent serotonin signals in response to light stimulation. The same stimulations were administered in the presence of an SSRI and significant differences were noted. Our preparation can help define the fundamental chemistry of the daphnia's defense mechanism and assess the effects of low level SSRI exposure. Understanding this fundamental chemistry is important for understanding the ecological impact of anthropogenic behaviors.

Keywords: Electrochemistry, Electrodes, Environmental Analysis, Environmental/Water
Application Code: Environmental
Methodology Code: Electrochemistry

Session # 1630 Abstract # 1630-8 **Oral Sessions**

Session Title: Electrochemistry - Biological Applications

Abstract Title: **Glutamate Receptor Influence on Localized Oxygen Metabolism**

Primary Author: Lindsay Walton
University of North Carolina at Chapel Hill

Date: Wednesday, March 09, 2016 - Morn
Time: 11:05 AM
Room: B403

Co-Author(s): Nick Boustead, R Mark Wightman, Susan Carroll

Abstract Text

Functional hyperemia is the biological mechanism of maintaining homeostasis in the brain through mediating local cerebral blood flow (CBF) supply based on neuronal activity. When neurons fire, CBF increases in a biphasic manner to provide blood rich with oxygen and glucose and renew locally depleted energy sources. Dysregulation of this system is noted in many disease states, including Alzheimer's and stroke. Irregular neurotransmitter supply is also prevalent in neuropathies, making the relationship between CBF and neurotransmission important to understand. The largest percentage of neurons in the brain release glutamate, an excitatory neurotransmitter that elicits cell firing. Its behavior is best characterized in the cortex or in slices, but technical issues have limited its study subcortically. We aim to better understand glutamatergic neurovascular influence deeper within an intact brain.

This research explores the role of glutamate on highly localized metabolic activity through simultaneous detection of cell firing and oxygen in response to glutamate. Iontophoresis, a local drug delivery technique, ejects glutamate and other drugs in close proximity to a neuron as opposed to less precise, systemic injections. Cell firing is detected using single unit recording electrophysiology and oxygen responses are recorded with fast-scan cyclic voltammetry, both at a single carbon fiber microelectrode. Our method of glutamate delivery is gentle enough to not provoke CBF changes, and instead probes the intensity and duration of increased oxygen metabolism demands that follow action potentials. We present glutamate-elicited neuronal activation and subsequent metabolic changes as cells are exposed to a number of glutamatergic receptor antagonists in both the nucleus accumbens and somatosensory cortex.

Keywords: Electrochemistry, Neurochemistry, Voltammetry
Application Code: Neurochemistry
Methodology Code: Electrochemistry

Session # 1640 Abstract # 1640-1 **Oral Sessions**

Session Title: Environmental, Pharm and Nano Methods Development in Atomic Spectroscopy

Abstract Title: **Novel Automation Streamlining Microwave Digestion to Detection for Elemental Analysis**

Date: Wednesday, March 09, 2016 - Morn
Time: 08:30 AM
Room: B404

Primary Author: David Clarke
Teledyne CETAC Technologies

Co-Author(s): James Block, Matthew Nigro

Abstract Text

Progressively lower detection limits for ICP-MS analysis have allowed for new frontiers in element research. The earth's crust, however, contains both high concentration major elements and low concentration trace elements and most natural samples are representative of this fact. Many testing laboratories split samples between ICP-OES and ICP-MS, with a separate suite of elements tested on each instrument. Microwave digestion is well proven for complete dissolution of some of the most difficult matrices. The only time consuming part is weighing the samples before digestion and diluting the dissolution after digestion and prior to analysis. In this work, we show a method which streamlines the process from microwave to detection. The medium which links digestion and detection is a novel automation platform. The SDX High Performance Liquid Dilution (HPLD) system combines both prescriptive and intelligent dilution of up to 40,000X. With this technology accurate quantitation ranging from weight percent concentration to part per billion concentration is readily achievable as well as a clear path to weight percent to part per trillion. Method performance metrics are discussed as well as recoveries of several pertinent standard reference materials.

Keywords: Atomic Spectroscopy, ICP-MS, Metals, Microwave
Application Code: Environmental
Methodology Code: Atomic Spectroscopy/Elemental Analysis

Session # 1640 Abstract # 1640-2 **Oral Sessions**

Session Title: Environmental, Pharm and Nano Methods Development in Atomic Spectroscopy

Abstract Title: **Use of Atomic Fluorescence Spectrometry (AFS) for Element Specific Hg Detection Combined with Combustion**

Primary Author: Warren T. Corns
P S Analytical

Co-Author(s)

Date: Wednesday, March 09, 2016 - Morn
Time: 08:50 AM
Room: B404

Abstract Text

The determination of Hg using cold vapour atomic fluorescence spectrometry (CV-AFS) is well established technique that offers low part per detection limits. The CV-AFS approach requires that samples are digested prior to analysis irrespective if the samples are liquids or solids. This is time consuming operation requiring some knowledge of the best digestion approach for particular sample types without introducing losses of Hg or elevated blanks from contaminated reagents. More recently direct combustion in pure oxygen coupled to atomic absorption spectrometry (AAS) with and without gold amalgamation has become commercially available and widely adopted by labs worldwide. A catalyst is typically used to scrub acid gas combustion products to avoid downstream contamination of poisoning of gold trapping systems. In this presentation we will review the analytical performance characteristics of amalgamation AFS when coupled to thermal combustion. Results will be presented for a wide range of environmental, clinical food and industrial samples. Two methods of calibration will be compared including in the injection of saturated Hg vapour from a bell jar at known temperature and aqueous phase calibrations. The system accuracy was validated using certified reference materials. The presentation will highlight the performance enhancements of AFS compared to AAS.

Keywords: Atomic Spectroscopy, Mercury, Pyrolysis, Trace Analysis
Application Code: Environmental
Methodology Code: Atomic Spectroscopy/Elemental Analysis

Session # 1640 Abstract # 1640-3 **Oral Sessions**

Session Title: Environmental, Pharm and Nano Methods Development in Atomic Spectroscopy

Abstract Title: **Development of Online Method for Simultaneous Preconcentration of Cd, Cu, Ni and Zn in Environmental Samples Using Modified Alumina**

Date: Wednesday, March 09, 2016 - Morn

Time: 09:10 AM

Room: B404

Primary Author: Zaheer A. Khan
SBBU, SBA

Co-Author(s)

Abstract Text

A sensitive and simple method for online simultaneous preconcentration of trace heavy metal ions in some environmental samples by flame atomic absorption spectrometry has been reported. In present study ammonium lauryl sulphate is used as a new surfactant for modification of alumina with dithizone as a adsorbent for simultaneous preconcentration of Cd, Cu, Ni and Zn. The adsorbed metals on modified alumina were eluted using 1 mL of 1 mol L-1 hydrochloric acid. The influences of the analytical parameters including pH, flow rate of sample, type and volume of eluent and sample volume were investigated. The calibration graph was linear in the range of 0.05-0.5 ng mL-1 for Cd, Cu, Ni and Zn. The limits of detection were 1, 0.5, 1.1 and 0.61 ng mL-1 for Cd, Cu, Ni and Zn respectively. The preconcentration factor of 100 for Cu and 150 for Cd, Ni and Zn was achieved. The developed method was successfully applied for the determination of trace metal ions in soil, water and plant samples.

Keywords: Adsorption
Application Code: Environmental
Methodology Code: Atomic Spectroscopy/Elemental Analysis

Session # 1640 Abstract # 1640-4 **Oral Sessions**

Session Title Environmental, Pharm and Nano Methods Development in Atomic Spectroscopy

Abstract Title **Dual Analyte Analysis of Bimetallic and Monometallic Nanoparticle Mixtures Using Field Flow Fractionation Separation Coupled to spICP-MS**

Date: Wednesday, March 09, 2016 - Morn
Time: 09:30 AM
Room: B404

Primary Author Chady Stephan
PerkinElmer

Co-Author(s) Ruth Merrifield

Abstract Text

The production and of nano-enabled products for consumer and industrial applications/processes is increasing. The extensive use of these manmade NPs results in an increase in potential release and subsequent risk to the environment. A major difficulty in assessing the fate, behavior and effects of these nanoparticles is the limitations of characterization and metrology methods in relation to the complex media and low nanoparticle concentrations found in the environment. The recent advances in spICP-MS allow for short dwell times, increasing the signal to noise ratio and allowing the measurement of dual analytes in the same nanoparticle. Here we have combined this technique with FFF, which allows us to further reduce the ionic background noise as well as separate complex NP suspensions into size fractions before analyzing their size and composition directly with spICP-MS. Different suspensions containing low concentrations of mixtures of Au, Ag and Au@Ag NPs all with diameters of 60 nm (NanoComposix) were analyzed using the FFF-spICP-MS combination and show the ability of the FFF-spICP-MS to quickly and efficiently distinguish between single and dual analyte NPs in complex mixes.

Keywords: Detection, ICP-MS, Metals, Nanotechnology
Application Code: Nanotechnology
Methodology Code: Atomic Spectroscopy/Elemental Analysis

Session # 1640 **Abstract #** 1640-6 **Oral Sessions**

Session Title: Environmental, Pharm and Nano Methods Development in Atomic Spectroscopy

Abstract Title: **Development and Validation of a New Method to Measure Activity of the Na+, K+ ATPase Using ICP-MS QQQ**

Date: Wednesday, March 09, 2016 - Morn
Time: 10:25 AM
Room: B404

Primary Author: Cory A. Stiner
University of Cincinnati

Co-Author(s): Judith Heiny, Julio Landero

Abstract Text

The sodium-potassium pump (Na+, K+ ATPase) is a vital enzyme in all eukaryotic cells. The Na+, K+ ATPase is also the receptor for cardiotonic steroids, a class of compounds that are used for the treatment of congestive heart failure. Endogenous cardiotonic steroids are also present at low levels in various tissues, organs, and blood. These compounds play an important role in blood pressure regulation. The presence of un-characterized cardiotonic steroids in tissues present a relevant area for research within this field of interest.

One of the tools to assess the activity of the Na+, K+ ATPase is the measurement of rubidium uptake, as a substitute for its natural potassium target. The current method used to quantify Rb/measure the Na+, K+ ATPase activity in tissues and cells require the use of radioactive materials. The cells are bathed in a physiological saline that contains a trace amount of 86Rb, and the Na+, K+ ATPase is activated by various physiological stimuli. The amount of 86Rb transported into the cells by the Na+, K+ ATPase per unit time is quantified to determine the turnover rate of the enzyme.

The instrument used for the current method to detect and measure the ionizing radiation from 86Rb is called a scintillation counter. The new method developed using ICP-MS QQQ will not require the use of radioactive 86Rb. Instead natural abundance 85Rb and 87Rb, can be used to quantify Rb within the cells to determine the activity of the Na+, K+ ATPase. The new method provides greater precision and statistical power with conventional tracer flux measurements. This method is also widely applicable to studies of other metal ion transporters and metal-dependent processes in a range of cell types and conditions.

Keywords: Bioanalytical, Biological Samples, Elemental Analysis, ICP-MS
Application Code: Bioanalytical
Methodology Code: Atomic Spectroscopy/Elemental Analysis

Session # 1640 Abstract # 1640-7 **Oral Sessions**

Session Title Environmental, Pharm and Nano Methods Development in Atomic Spectroscopy

Abstract Title **Arsenic Speciation in Water Samples – Development of a New ISO/CEN Method**

Date: Wednesday, March 09, 2016 - Morn
Time: 10:45 AM
Room: B404

Primary Author Cornelius Brombach
P S Analytical

Co-Author(s) Bin Chen, K C. Thompson, Warren T. Corns

Abstract Text

A method was developed for the speciation of arsenic in water samples by using HPLC-HG-AFS which is intended for the use as an ISO/CEN technical standard in the future. The four arsenic species arsenite (As(III)), arsenate (As(V)), monomethylarsonic acid (MMA) and dimethylarsinic acid (DMA) can be separated by various mobile phase compositions and the HG-AFS detection makes the method sensitive, cost-effective and element-specific. This presentation will present an overview of the proposed method and will summarize the analytical performance found. One of the keys areas of discussion within the method development working group was in relation to the stability of arsenic species and best methods of sample storage. This is not only an important consideration for the final method but also for inter laboratory collaborative trial preparation. A stability test of different arsenic species in bottled drinking water and seawater was conducted to test the best storage options for arsenic speciation. Three different types of bottles (polyethylene, fluorinated polyethylene and amber polyethylene), two different storage types (fridge or room temperature) and also the effect of acid stabilisation was evaluated.

Keywords: Atomic Spectroscopy, Environmental/Water, Ion Chromatography, Liquid Chromatography
Application Code: Environmental
Methodology Code: Atomic Spectroscopy/Elemental Analysis

Session # 1640 Abstract # 1640-8 **Oral Sessions**

Session Title: Environmental, Pharm and Nano Methods Development in Atomic Spectroscopy

Abstract Title: **Exploiting the Limits of Single Particle ICP-MS – From Particle Size to Particle Number**

Date: Wednesday, March 09, 2016 - Morn
Time: 11:05 AM
Room: B404

Primary Author: Chady Stephan
PerkinElmer

Co-Author(s)

Abstract Text

Single Particle ICP-MS is a new advancement in ICP-MS devoted to the analysis of individual particles ranging from single digit nm up to a few µm. It is element specific, that allows the differentiation between ionic (M+) and particulate signals (particles) in a wide variety of matrices without any prior separation. In one sample analysis, SP-ICP-MS provides ionic and particle concentration, particle composition, size and size distribution.

In the following, I share my 5 years of experience working with SP-ICP-MS. I detail the various parameters to consider when analyzing for particles in various matrices from sample flow rate and transport efficiency to the more advanced parameters such dwell time and Dynamic Reaction Cell (DRC) for the analysis of Iron and silicon dioxide particles. Factors affecting particle size detection limit, how small of a particle can we detect? How large of a particle can we measure? What is the lowest particle number that we could realistically work with? What is the particle size resolution of the technique? What is the lowest mass base concentration that we can achieve, could it be sub-ppq?

Keywords: Environmental, ICP-MS, Nanotechnology, Particle Size and Distribution
Application Code: Nanotechnology
Methodology Code: Atomic Spectroscopy/Elemental Analysis

Session # **1650** **Abstract #** **1650-1** **Oral Sessions**

Session Title: Food Contaminants

Abstract Title: **The Analysis of Chlorinated Dioxins and Difurans in Pet Food**

Date: Wednesday, March 09, 2016 - Morn
Time: 08:30 AM
Room: B314

Primary Author: Philip Bassignani
Fluid Management Systems

Co-Author(s): Rudolf Addink

Abstract Text

Screening for chlorinated dioxins and furans in human foods is well-established. Both EU and US have protocols for testing of food for human consumption and additives to feed for commercial livestock. Given recent contamination issues in pet food there is growing concern for safety. To screen pet foods and additives, development of analytical techniques is a high priority. With high lipid content in canned and dried pet foods, Persistent Organic Pollutants (POPs) are likely to be found.

Three brands canned dog food were analyzed. 10g of sample was spiked with 13C surrogates and mixed with diatomaceous earth for drying. Samples were put in cells for Pressurized Liquid Extraction (PLE) with 50/50 dichloromethane/hexane, followed by volume reduction and solvent exchange to hexane. Cleanup was done using automated column chromatography (acid-base-neutral silica, alumina, carbon columns). Samples were loaded onto ABN silica columns, eluted onto alumina, and then onto carbon columns using dichloromethane. The eluate was collected (fraction # 1). The carbon column was eluted with toluene collected (fraction # 2). Fractions were reduced in volume and analyzed with high res GC/MS.

Canned dog food was extracted and cleaned up with the techniques described and analyzed for PCDD/Fs. Recoveries were (averages): tetra-CDD/F 88%; penta-CDD/F 74%; hexa-CDD/F 93%; hepta-CDD/F 81%; octa-CDD/F 73%. Detection limits for all native analytes were < 0.1 pg/g. Review of the 3 pet food matrices showed labeled recoveries well within EPA1613 limits. The method blank showed no background above the CS 0.1 calibration standard level except for OCDD. Combining the clean background with good recoveries demonstrates the ability of PLE and automated column chromatography to handle wet pet food of various types. With total processing time < 5h, same day analysis is possible. When factoring in the need to qualify batches for release and product delays, the value of rapid testing become critical.

Keywords: Food Science, Gas Chromatography/Mass Spectrometry, Sample Handling/Automation, Trace Analysi
Application Code: Food Contaminants
Methodology Code: Gas Chromatography/Mass Spectrometry

Session # **1650** Abstract # **1650-2** **Oral Sessions**

Session Title: Food Contaminants

Abstract Title: **Selective Lipid Removal from Complex Samples for Multi-Residue Analysis**

Date: Wednesday, March 09, 2016 - Morn
Time: 08:50 AM
Room: B314

Primary Author: Derick Lucas
Agilent Technologies

Co-Author(s): Bruce Richter, Joan Stevens, Limian Zhao, Megan Juck

Abstract Text

Sample preparation methodologies for multi-residue analysis often implement a general extraction followed by analysis with instrumentation including LC/MS/MS and GC/MS/MS. Ideally the techniques are simple and ideal for the extraction of diverse analyte groups, however this also extracts large amounts of matrix and traditional cleanup materials struggle to effectively and selectively remove unwanted interferences. Samples high in fat are particularly problematic as lipid co-extractives cause poor reproducibility, ion suppression/enhancement, changes in analyte response over time, and instrument maintenance. Agilent Bond Elut Enhanced Matrix Removal – Lipid (EMR-Lipid) represents the next generation of sample preparation technology; providing selective lipid removal for complex samples without analyte retention. EMR-Lipid is available in a convenient dispersive solid phase extraction (dSPE) format and is amenable to widely accepted workflows such as QuEChERS and protein precipitation. Data will demonstrate the performance benefits achieved by cleaner sample extracts using this new material in applications involving multi-class, multi-residue analysis for pesticides, veterinary drugs, and mycotoxins in complex, high fat samples. Dramatic improvements are obtained for matrix removal, analyte recovery, and reproducibility compared to currently available cleanup sorbents. The high performance and selectivity of EMR-Lipid make it an attractive option for laboratories seeking to simplify sample preparation for fatty samples, while enhancing analytical and instrumental integrity.

Keywords: Gas Chromatography/Mass Spectrometry, Lipids, Liquid Chromatography/Mass Spectroscopy, Sampl
Application Code: Food Contaminants
Methodology Code: Sampling and Sample Preparation

Session # 1650 **Abstract # 1650-3** **Oral Sessions**

Session Title: Food Contaminants

Abstract Title: **Arsenic Contamination and the Emergence of Speciation in the Food Chain**

Date: Wednesday, March 09, 2016 - Morn
Time: 09:10 AM
Room: B314

Primary Author: Helmut Ernstberger
Perkin Elmer

Co-Author(s): Kenneth Neubauer

Abstract Text

The assessment of health hazards due to the arsenic content of food and drink products has received widespread attention due to its potential toxicity. However, not all forms of arsenic are toxic: organic arsenic compounds are generally considered non-toxic, while inorganic arsenic is toxic. Therefore, just measuring the total arsenic content is not an accurate assessment of the toxic arsenic content: the need exists to differentiate organic from inorganic arsenic.

An initial focus of arsenic speciation in foods has been fruit juice, specifically apple juice. The reason is that children tend to drink large quantities of fruit juice and are more susceptible to the effects of inorganic arsenic than adults. For apple juice, this led to the proposal of an action level for inorganic arsenic concentration by the FDA. Speciation analysis is required if the total level exceeds the set limit in order to differentiate inorganic from organic species. Currently, much attention focuses on arsenic speciation in rice due to its elevated arsenic content.

Here we present novel analysis methods for both apple juice and rice using LC-ICP-MS with ion interaction chromatography to achieve separation. The methods are validated on a wide variety of samples, including certified reference materials. The results demonstrate that both reliable and fast speciation analysis can be achieved, meeting the prerequisites for routine application in speciation monitoring.

Keywords: Food Contaminants, HPLC, ICP-MS
Application Code: Food Contaminants
Methodology Code: Atomic Spectroscopy/Elemental Analysis

Session # 1650 Abstract # 1650-4 **Oral Sessions**

Session Title: Food Contaminants

Abstract Title: **Elemental Profile of Tobacco used in Counterfeit Cigarettes**

Primary Author: Yi He
John Jay College/CUNY

Date: Wednesday, March 09, 2016 - Morn
Time: 09:30 AM
Room: B314

Co-Author(s): Carrie Green, Fidelis Tan, Klaus von Lampe, Marin Kurti, Rufus Chaney, Victoria Mei, Ye Hua

Abstract Text

Elemental profile of a plant sample provides important information of on the growing environment and the general health of the plant studied. This project investigated multiple elements including Ca, Cd, Co, Cu, Fe, K, Mg, Mn, Ni, P, Pb and Zn in tobacco leaves used in counterfeited Marlboro Red, Marlboro Gold and Newport, the three most counterfeited cigarette brands in the US. All counterfeit samples, provided by law enforcement agencies, were prepared by either dry ashing followed by hotplate acid digestion or microwave digestion (HNO3-H2O2), and analyzed using inductively coupled plasma- atomic emission spectroscopy (ICP-AES) using yttrium as an internal standard. The two sample preparation methods were compared and evaluated by analyzing In-house tobacco standards and NIST 1573a: Tomato leaves. Preliminary results revealed that elevated Pb and Cd were commonly found in the counterfeit cigarettes. Because cadmium in tobacco products already contributes about 50% of lifetime Cd accumulation in the kidney of smokers, contaminated tobacco products are a significant additional Cd risk to smokers.

Keywords: Atomic Spectroscopy, Elemental Analysis, Food Contaminants, Forensic Chemistry
Application Code: Food Contaminants
Methodology Code: Atomic Spectroscopy/Elemental Analysis

Session # 1650 Abstract # 1650-5 **Oral Sessions**

Session Title: Food Contaminants

Abstract Title: **Determination of Antimony in Food Samples by Slurry Sampling Hydride Generation Atomic Absorption Spectrometry**

Primary Author: Jerzy Mierzwa
Tennessee State University

Date: Wednesday, March 09, 2016 - Morn
Time: 10:05 AM
Room: B314

Co-Author(s)

Abstract Text

A method for the determination of antimony in some food samples based on simplified sample preparation procedure was developed. Antimony hydride was generated directly from slurried samples and subsequently directed to atomic absorption spectrometer. Slurried samples were prepared in a mixture of 4% (w/w) hydrochloric acid containing 0.02% of the surfactant Triton X-100.

The repeatability of this analytical procedure expressed in term of relative standard deviation (RSD) was typically better than 7.5 % at the slurry concentration of approx. 80 mg/ml. The characteristic mass for peak absorption of antimony (at the wavelength of 217.6 nm) was 24 pg and the limit of detection (LOD) was about 0.12 ug/kg. The results of the above analytical procedure were very similar to the results obtained after full digestion/dissolution of the tested samples. A Certified Reference Material (BCR-679, White Cabbage) was also employed and the results of the antimony determination by this technique are in very close agreement with the certified value.

Keywords: Atomic Absorption, Food Contaminants, Hydride, Trace Analysis
Application Code: Food Contaminants
Methodology Code: Atomic Spectroscopy/Elemental Analysis

Session # 1650 Abstract # 1650-6 **Oral Sessions**

Session Title Food Contaminants

Abstract Title **The Analysis of Chlorinated Dioxins, Difurans and Polychlorinated Biphenyls in Edible Oils**

Date: Wednesday, March 09, 2016 - Morn
Time: 10:25 AM
Room: B314

Primary Author Philip Bassignani
Fluid Management Systems

Co-Author(s) Rudolf Addink

Abstract Text

Polychlorinated dibenzo-p-dioxins and furans, and polychlorinated biphenyls, are of great concern to human health. They accumulate in adipose tissue and end up in food supplies. For this reason, the FDA and EU have established strict regulations for monitoring of food products for human consumption, in particular edible oils. Manual extraction of oils is a time consuming procedure with long turnaround times. By automation, food oil samples can be reliably processed within 24 hours. The following procedure utilizes an automated sample clean-up system.

5 g of various oil matrices (lard, and oils (olive, corn, cod, red palm, unrefined pumpkin, and unrefined vegetable oil) were spiked with 13C labeled standards, diluted into hexane and drawn up in a gas-tight syringe. Columns (acid-base-neutral silica, alumina and carbon) of the automated chromatography clean up system were conditioned. Samples were loaded across ABN silica columns and eluted onto alumina columns. The samples were eluted onto the carbon column using dichloromethane. The eluate was collected as fraction # 1. The carbon column was eluted with toluene collected as fraction # 2. The fractions were reduced in volume and analyzed with high res GC/MS.

Edible oils were cleaned up with the technique described and analyzed for PCBs and PCDD/Fs. Recoveries for PCBs (averages) were: tetra-CBs 69%; penta-CBs 76%; hexa-CBs 64%; hepta-CBs 79%. Recoveries for PCDD/F (averages): tetra-CDD/F 74%; penta-CDD/F 82%; hexa-CDD/F 71%; hepta-CDD/F 78%; octa-CDD/F 70%. Detection limits for all native species measured were < 0.5 pg/g oil. Analysis of the 6 matrices processed yielded good recoveries for all analytes. Analysis of n-hexane as blank gave no detectable target analytes measured in the calibration range of each respective compound. With a total processing time of less than 2.5 hours, the automated column chromatography system and concentrator deliver efficient, completely automated sample prep for edible oils.

Keywords: Food Science, Gas Chromatography/Mass Spectrometry, Sample Preparation, Trace Analysis
Application Code: Food Contaminants
Methodology Code: Gas Chromatography/Mass Spectrometry

Session # 1650 Abstract # 1650-7 **Oral Sessions**

Session Title: Food Contaminants

Abstract Title: **Application of Mid-Infrared Portable Spectrometry in Determination of Trans-Fatty Acid Content in Bakery Products**

Date: Wednesday, March 09, 2016 - Morn
Time: 10:45 AM
Room: B314

Primary Author: Mei Shotts
The Ohio State University

Co-Author(s): Luis Rodriguez-Saona

Abstract Text

In 2013, the US Food and Drug Administration announced plans to remove trans-fats and/or partially hydrogenated oils (PHOs) from the list of food ingredients the are generally regarded as safe (GRAS). Our objective was to develop a predictive model to quantify trans-fat concentrations in bakery food products using portable mid-infrared (MIR) spectrometers coupled with attenuated total reflectance (ATR) to aid in the monitoring of the usage of trans-fats. Isolation of trans-fats were obtained with AOAC Method #2000.10 and spectra collection was done using a zinc selenide (ZnSe) ATR crystal coupled with a temperature control (65ºC). The approach was tested using 35 standards and 89 different bakery products by blending products into powders with liquid nitrogen and used gravimetrically prepared, extracted fat via AOAC method #960.39. Linear regression models were developed using the unique trans-fat fingerprint region at 966cm-1, developing a SEP (Standard Error of Prediction) of 0.80% for a validation set of samples. PLSR (Partial Least Square Regression) of trans-fat gave R2>0.998 and developing a SEP of 0.9% for a validation set of samples, respectively. This shows the determination of trans-fat using portable ATR-MIR spectrometers shows rapid throughput and high accuracy making it ideal for regulatory applications and well suited to quality control applications.

Keywords: Analysis, Portable Instruments, Quantitative, Vibrational Spectroscopy
Application Code: Food Science
Methodology Code: Vibrational Spectroscopy

Session # 1650   Abstract # 1650-8   **Oral Sessions**

Session Title Food Contaminants

Abstract Title **Rapid Detection of Processed Uranium in Food**

Date: Wednesday, March 09, 2016 - Morn
Time: 11:05 AM
Room: B314

Primary Author Abdur-Rafay Shareef
FDA

Co-Author(s) Brian Baker, Emanuele Kathryn, Lin Zhichao, Patrick Regan, Stephanie Healey, William Cunningham

Abstract Text

The US Food and Drug Administration (FDA) maintains an active, robust program for routine monitoring of food for radionuclide contamination and stands ready to respond to emergencies. Due to on-going interest in increasing our analytical capabilities, a focus of recent work has been applying isotope ratio mass spectrometry (IR-MS) to the determination of radionuclides in food. In this current study, a rapid, high-throughput method for the detection of processed uranium in food was developed. This technique is particularly useful for uranium isotopes because the isotope ratios for processed uranium, as used for nuclear energy and weaponry, differ significantly from natural uranium. In response to an incident, the Agency may need means to differentiate between natural and processed uranium. Since radioactive tracers and spikes are not used, non-radiochemical laboratories can perform this analysis. Several foods were examined with excellent results: sub-picogram quantities of U-235 can be quantified and errors were below 4 % for all food types examined to date.

Keywords: Chemical, Food Science, ICP, Mass Spectrometry
Application Code: Food Safety
Methodology Code: Mass Spectrometry

Session # 1660 **Abstract # 1660-1** **Oral Sessions**

Session Title LIMS-No One Size Fits All

Abstract Title **Quantum Time Savings with LIMS Deployment**

Primary Author Devender Gandhi
Accelerated Technology Laboratories

Date: Wednesday, March 09, 2016 - Morn
Time: 08:30 AM
Room: B405

Co-Author(s)

Abstract Text

As laboratories adopt automation (auto samplers, automatic data uploads and calculations, etc.), the price per test remains consistent. The competition among contract analytical laboratories is fierce, so laboratories that leverage automation to save time, will have a competitive and financial advantage. Contract testing is a mature market and once managers have optimized their operations, wise managers and leaders turn their attention to automating the human resource aspect of operations. ATL's subject matter experts have over two decades of expertise in laboratory automation and have performed hundreds of gap analysis or needs assessments. We have learned that areas that draw the most resources and frequently cause bottlenecks are sample login, importing instrument data, and quoting/invoicing/reporting. In addition, customer service, which includes calls about sample status, re-sending invoices, questions on requested testing, understanding results, and IT concerns. Part of the needs assessment that we perform includes detailed workflow analysis with recommendations for maximizing resources and productivity. This talk describes tools that we deploy which maximize our client's technology investment and enhance productivity. These tools include an integrated LIMS/ERP solution, secure web portal for 24/7 data access, a LIMS Kiosk for clients to submit samples, Instrument and Enterprise integration, and leveraging Software as a Service (SaaS). Outsourcing IT makes financial sense, allowing clients to focus on their core competency – enhancing data quality, growing laboratory business and maximizing productivity and profitability. To learn how your laboratory can become more efficient and experience quantum time savings, request our free LIMS Guide and a private consultation.

Keywords: Data Analysis, Laboratory Automation, Laboratory Informatics, LIMS
Application Code: Laboratory Management
Methodology Code: Laboratory Informatics

Session # 1660 Abstract # 1660-2 **Oral Sessions**

Session Title LIMS-No One Size Fits All

Abstract Title **Informatics for Externalization**

Primary Author Graham A. McGibbon
ACD/Labs

Date: Wednesday, March 09, 2016 - Morn
Time: 08:50 AM
Room: B405

Co-Author(s) Pranas Japertas, Ryan Sasaki

Abstract Text

Challenges in data sharing and collaboration are already apparent throughout most chemically involved R&D industries and being exacerbated by a trend towards outsourcing core and non-core scientific activities. It has been estimated that pharmaceutical and biotech R&D spending outside of company boundaries is now >40% [1] and Contract Research Organizations (CROs) in China, e.g. Wuxi and ChemPartner, are among the largest employers of synthetic chemists. As externalization and research virtualization evolve and business critical tasks of library synthesis, process chemistry, metabolism, toxicology, and intermediates manufacturing are included, the demand for information accessibility intensifies. Typical data sharing, by email on demand or via document management systems like SharepointTM between each Sponsor Company and their CROs isn't yet optimal. Formats are diverse and many, even spreadsheets and PDF files, may be less apt for future leveraging in a scientifically meaningful way.

This presentation will highlight a new laboratory informatics externalization model based on analytical data sharing as a use case. A standardized way to collaborate systematically and efficiently via accessing share data is suggested. A software platform ensures data is managed in a way that enables data mining for the purpose of identifying raw materials, impurities, metabolites and other chemical ingredients. Collaborative workspaces enable creating analytical knowledge packages that can have 'live' analytical data, metadata, and chemistry information independent of instrument source for seamless sharing. The value of fingertip access to this 'live' information that can be searched, shared, re-processed, re-purposed, and re-analyzed will be discussed. Platform knowledge content can be easily accessed via web or mobile client interfaces and is amenable to integration and cloud-based deployments.

[1] M.E. Elliot, The De-Evolution of Informatics, Scientific Computing, October 2012

Keywords: Informatics, Laboratory Informatics, Scientific Data Management, Software
Application Code: General Interest
Methodology Code: Laboratory Informatics

Session # 1660	Abstract # 1660-3	Oral Sessions

Session Title LIMS-No One Size Fits All

Abstract Title **Going Mobile with LIMS for Field Data Collection**

Date: Wednesday, March 09, 2016 - Morn
Time: 09:10 AM
Room: B405

Primary Author Ken Ochi
Accelerated Technology Laboratories

Co-Author(s)

Abstract Text

Although there has been a proliferation of the use of smart phones and tablets over a variety of industries, field data collectors are still carrying around a lot of paper forms to capture data. Many laboratories are limping along with old laboratory data management technology that limits the laboratory's ability to work efficiently, quickly generate the best quality data and maximize their resources. Over time, business needs changes and the technology must change to ensure that those needs are met and the laboratory remains efficient and productive. Although tablets are in use, they are often only used for checking email, browsing the web or sharing photos. Using tablets to let field collectors know what the schedule of collected samples is for the day is not only smart, but more efficient and provides better data quality and faster data management. This also ensures that no sampling events are missed, especially those for regulatory compliance purposes.

Leveraging tablets and smart phones to replace paper-based data capture in the field means that organizations can:
- Increase their efficiency by conducting more collections/inspections/visits in shorter time frames.
- Save on mailing, data-entry costs and errors made in transcription.
- Send real-time information (data and images) to the LIMS from the field for reporting and analysis.
- Provider quicker services and responses to customers, leading to higher customers satisfaction and also increased revenue.
- No need to remember schedule collections or inspections as they are sent from the LIMS.
- Quickly investigate customer complaints and upload data to the laboratory in real-time.

This presentation will incorporate customer testimonials to validate the benefits that mobile technology brings to the laboratory.

Keywords: Data Analysis, Laboratory Automation, Laboratory Informatics, LIMS
Application Code: Laboratory Management
Methodology Code: Laboratory Informatics

Session # 1660 Abstract # 1660-4 **Oral Sessions**

Session Title LIMS-No One Size Fits All

Abstract Title **Pay Now or Later: Creating Solid System Application User Requirements**

Date: Wednesday, March 09, 2016 - Morn
Time: 09:30 AM
Room: B405

Primary Author Katherine Temple
CSOLS, Inc.

Co-Author(s) Daniel Freel

Abstract Text

A laboratory informatics solution (e.g. LIMS, ELN, CDS) is only as strong as the weakest part of its foundation. The foundation for the implementation of a laboratory informatics solution is the user requirements document that is generated to identify exactly what the system is intended to do. The requirements document is used extensively throughout the selection, development, and implementation of the solution and in the end, is the yardstick used to measure its success. Unfortunately, many times this document is not created with the thoroughness and attention to detail that its importance deserves. Too often it is looked at as a checkbox exercise where many in the business take the approach that they will have plenty of time to correct any inconsistencies as the project evolves. Unfortunately, it is often the case that many decisions have already been made based on the document contents before the realization occurs that the requirements documented were not detailed enough or entirely accurate to actual business needs. The result of such neglect can be time and cost overrun, a system that is inefficient, users who are unhappy, and the possibility of total project failure.

This presentation will take a look at proven, thorough, yet sensible approaches to ensuring that your business is in the best position to create solid user requirements for your informatics system. Overcoming different obstacles such as lack of resources, inconsistent work processes, lack of informatics "experts", and lack of business motivation within the company will also be addressed. In addition to generating a solid, well thought out user requirements document, the presentation will provide ways to also leverage the requirements generation exercise to improve communication, morale, and work processes, all while having a little bit of fun.

Keywords: Laboratory Informatics, LIMS, Quality, Sample & Data Management
Application Code: Laboratory Management
Methodology Code: Laboratory Informatics

Session # 1660 Abstract # 1660-5 Oral Sessions

Session Title LIMS-No One Size Fits All

Abstract Title **Key Factors to Consider in Transitioning to a New LIMS**

Date: Wednesday, March 09, 2016 - Morn
Time: 10:05 AM
Room: B405

Primary Author Sonja Stutsman
Accelerated Technology Laboratories

Co-Author(s)

Abstract Text

It is no secret that laboratories that automate their operations not only increase efficiency and productivity, but also increase market share and profitability. Many laboratories are limping along with old laboratory data management technology that lacks support or hasn't kept up with advances in technology. An older system limits the laboratory's ability to work efficiently, quickly generate the best quality data or maximize resources. Older data management systems can also affect a laboratory's ability to achieve the certifications or to electronically submit data to its customers.

During this presentation we will identify key factors to consider when transitioning to a new LIMS. One important factor is considering key functionalities provided by the new LIMS to ensure the lab is planning for future growth and integrating field data, like those set by ISO 17025, NELAC and state regulatory agencies. Another factor to consider is data migration from the old system: how much data will need to be migrated and what's the validation process? Lastly, there needs to be a plan to manage the transition from the older system to the new LIMS with training, an appropriate level of vendor support, and documentation.

Keywords: Data Analysis, Laboratory Automation, Laboratory Informatics, LIMS
Application Code: Laboratory Management
Methodology Code: Laboratory Informatics

Session # 1660 Abstract # 1660-6 **Oral Sessions**

Session Title LIMS-No One Size Fits All

Abstract Title **LIMS Implementation: "Big Bang" or "Phased" Approach**

Primary Author Howard J. Rosenberg
CSols, Inc.

Co-Author(s)

Date: Wednesday, March 09, 2016 - Morn
Time: 10:25 AM
Room: B405

Abstract Text

Congratulations! Your laboratory organization has gone through the process of documenting your lab processes, workflows, and data flows as well as your lab and business needs. You have leveraged all this information to drive your laboratory informatics system selection process and you have chosen a commercial off the shelf Laboratory Information Management System (LIMS). The hard part is over. No more big decisions to make, right? Unfortunately this is not the case. There will be many more decisions to make and plenty of work to be done before your LIMS is up and running and everyone in the lab organization is using and benefiting from the system.

One of the first decisions you are going to need to make is how you are going to go about implementing your new LIMS. There are many good methodologies that you can utilize to successfully implement your LIMS but regardless of the method you choose, you will also need to decide on your implementation approach. There are two main approaches for implementing a LIMS: the "Big Bang" approach and the "Phased" approach.

This talk will compare and contrast the "Big Bang" and the "Phased" LIMS implementation approaches. We will discuss the advantages/benefits and disadvantages/challenges with each approach. A framework for assisting you in determining which approach would be best for your organization will also be shared.

Keywords: Lab Management, Laboratory Informatics, LIMS, Sample & Data Management
Application Code: Laboratory Management
Methodology Code: Laboratory Informatics

Session # 1660 Abstract # 1660-7 **Oral Sessions**

Session Title LIMS-No One Size Fits All

Abstract Title **Migrating from a Legacy LIMS to a Modern Platform**

Primary Author
Laura Lee Williford
Accelerated Technology Laboratories

Date: Wednesday, March 09, 2016 - Morn
Time: 10:45 AM
Room: B405

Co-Author(s) Devender Gandhi

Abstract Text

Migration from one LIMS to another is a challenge Laboratory Leaders face. Successful migrations can be ensured with a strategy. Factors leading to a migration include unsupported LIMS, dated hardware, changing requirements, and lack of support resources. Migrations in the past were complicated and painful, however modern commercial tools help make the transition much easier. ATL has migrated many LIMS platforms to our modern workflow-centric LIMS. Steps for migrations include:

- Migrating static and dynamic data
- Moving customizations and configurations
- Creating new reports
- Validation effort and time
- Impact on current practices along with user acceptance and training

These challenges required a lot of time, costs, planning, configurations, training, and validation. New processes make migrations faster and not as cost prohibitive. A regimented methodology used by ATL helps reduce the time required for migrations with significant cost savings and faster ROI. Migration consists of a multi-step process. These include:

- Detailed analysis of current LIMS or as-is workflows
- Create To-Be workflows
- Establish a Project Plan or a Gantt chart
- Map from old schema to new LIMS schema
- Identify and document migration workflow steps
- Conduct a "Pilot" migration of a subset of data and validate
- Conduct total migration and verify
- Configure workflows, and conduct validation and training and Go LIVE

While migrations are based on case-by-case review of current LIMS, transitioning to a new LIMS has become faster and easier resulting in cost savings and many benefits overall. New LIMS deployments will pay off sooner than later and reduce annual support costs incurred with legacy LIMS.

Keywords: Data Analysis, Laboratory Automation, Laboratory Informatics, LIMS
Application Code: Laboratory Management
Methodology Code: Laboratory Informatics

Session # **1660** Abstract # **1660-8** **Oral Sessions**

Session Title LIMS-No One Size Fits All

Abstract Title **Integrated Informatics: Single Vendor vs. "Best of Breed"**

Primary Author Howard J. Rosenberg
CSols, Inc.

Date: Wednesday, March 09, 2016 - Morn
Time: 11:05 AM
Room: B405

Co-Author(s)

Abstract Text

Your organization has actively embraced laboratory informatics tools to help increase your lab's efficiency and effectiveness, information availability across the enterprise, innovation and discovery, and collaboration amongst your scientists and partners. While the laboratory informatics solutions you currently have implemented have enabled you to make great strides, in order to get to the next level, you need to implement a fully Integrated Informatics Solution (IIS). The goal of an IIS is to connect all your laboratory information systems (LIMS, ELN, SDMS, etc.), instrument systems (CDS, MS, balances, etc.), business and manufacturing systems, and reporting and analytics tools such that the data and information flows seamlessly between them.

Fully integrated informatics solutions can be implemented by following two distinct approaches: Single Vendor or "Best of Breed". There are advantages/benefits and disadvantages/challenges with either approach.

This talk will compare and contrast Single Vendor and Best of Breed Integrated Informatics approaches. We will also discuss the advantages/benefits and disadvantages/challenges with each approach. A framework for assisting you in determining which approach would be best for your organization will also be shared.

Keywords: Laboratory Automation, Laboratory Informatics, Sample & Data Management, Scientific Data Manag
Application Code: Laboratory Management
Methodology Code: Laboratory Informatics

Session # 1670 **Abstract # 1670-1** **Oral Sessions**

Session Title: Mass Spectrometry-Environmental, ICP-MS and Others

Abstract Title: **Ultra-Trace Analysis of Mercury Species in Drinking Water Using Ion Chromatography and Speciated Isotope Dilution Mass Spectrometry (IC-SIDMS)**

Primary Author: Patrick Benecewicz
Duquesne University

Date: Wednesday, March 09, 2016 - Morn
Time: 08:30 AM
Room: B406

Co-Author(s): Christopher Loran, Matt Pamuku, Skip Kingston, Stuart Procter

Abstract Text

Mercury is a known biological toxin that causes many devastating side effects when humans are exposed to the several common ionic and molecular species. Mercury exists in several common forms in the environment and each species has different toxicological and physiological risk factors. The most common two forms are organic (usually methyl mercury) and inorganic mercury (usually doubly charged). Organic mercury species are especially hazardous to human health due to its ability to cross the blood-brain and placental barrier. Recently, agencies such as the California Department of Toxic Substances Control (DTSC) and the Environmental Protection Agency (EPA) have been monitoring the levels of total mercury in drinking water. Because toxicity and risk factors differ greatly depending on species, measurements of the metals' species are now being requested and demanded. California would like to establish lower allowable concentration of methyl mercury in drinking water to protect sensitive populations, including pregnant mothers and their unborn children. These populations are highly susceptible to exposure of methyl mercury due to the risk of Minamata disease in the fetus. The state of California would like to reduce the lower limit of quantification of inorganic and organic mercury to 0.2ppt and 0.02ppt respectively. This study creates an IC instrument for the pre-concentration of drinking water mercury species, and couples it to an ICP-MS using EPA method 6800, update V, for quantification. Method 6800 allows for the correction of species conversion that can occur during the analysis enabling legally defensible and actionable monitoring.

Keywords: ICP-MS, Ion Chromatography, Mercury, Water
Application Code: Environmental
Methodology Code: Mass Spectrometry

Session # 1670	Abstract # 1670-2	Oral Sessions

Session Title Mass Spectrometry-Environmental, ICP-MS and Others

Abstract Title **Ion Mobility-Mass Spectrometry Screening Reveals Small Molecules Capable of Chemical and Structural Modulation of Amyloidogenic Protein**

Date: Wednesday, March 09, 2016 - Morn
Time: 08:50 AM
Room: B406

Primary Author Richard A. Kerr
University of Michigan

Co-Author(s) Brandon T. Ruotolo, Hyuck Jin Lee, Jeffrey Derrick, Mi Hee Lim, Michael Beck

Abstract Text

Between 2002 and 2012, 244 small molecule drugs targeting the causes or effects of Alzheimer's disease (AD) were brought to trial in the US, with only Memantine progressing for FDA approval. Given an aging population and projected future increases in AD diagnoses, developing new methodologies capable of screening drug scaffolds that target the causative factors of AD are critically important. Based on these concerns we have developed a multi-faceted, ion mobility-mass spectrometry (IM-MS) centric, screening methodology capable of identifying small molecules that bind and induce structural and/or covalent modifications within the target, Amyloid □(A[beta]), that can be correlated with downstream [i]in vivo[/i] efficacy.

Our methodology combines native IM-MS, a gas phase technique that separates mass resolved protein-small molecule complex ions according to their size to charge ratio, with tandem mass spectrometry (MS/MS) analysis. If target-ligand binding is observed, analysis of IM-MS data can be used to correlate changes in the ligand bound arrival times with the formation of downstream off-pathway aggregates. If ligand dependent covalent modifications are observed however, MS/MS can be used to identify their source. Using this method we have identified several new, rationally targeted, inhibitors of metal free and metal-associated A[beta] amyloidosis. Continued development of this method is enabling us to screen larger ligand libraries, with important mechanistic insight into their modes of action.

In this presentation we will discuss our multi-faceted IM-MS and MS/MS screening methodology in greater detail. Recent data sets will be discussed in order to highlight the ability of our approach to detect potential inhibitors of A[beta] amyloidosis, as well as their modes of action. Ideally placed, our methodology looks to improve the throughput of the AD drug discovery process by removing candidates that lack key traits early in the development cycle.

Keywords: Drug Discovery, Mass Spectrometry, Method Development, Protein
Application Code: Drug Discovery
Methodology Code: Mass Spectrometry

Session # 1670 Abstract # 1670-3 **Oral Sessions**

Session Title: Mass Spectrometry-Environmental, ICP-MS and Others

Abstract Title: **Transformation Kinetics of Metallic Nanoparticles in Environmental and Cell Culture Exposure Media as Measured by spICP-MS**

Date: Wednesday, March 09, 2016 - Morn
Time: 09:10 AM
Room: B406

Primary Author: Chady Stephan
PerkinElmer

Co-Author(s): Ruth Merrifield

Abstract Text

The environmental fate and behavior and biological effects of manmade nanoparticles is currently the subject of great debate. Detection and characterization of these nanoparticles under relevant conditions and concentrations remains a significant challenge due to limitations in characterization techniques. Recent advances in spICP-MS has provided very short dwell times, decreasing the ionic background and increasing the particle detection limit potentially allowing measurement of pristine and transformed nanoparticles at low concentrations. Here we have exposed two NPs (60 nm Ag and Au) to a synthetic surface water and a cell culture media at a mass concentration of 0.3 ppb (100,000 NPs / ml). The effect of temperature, UV exposure and turbulence of the suspensions were altered and controlled to mimic different OECD toxicity exposure conditions for typical algae, daphnia, and snail and cell culture conditions. Changes in number and mass concentration, size distribution and other properties were measured with spICP-MS over 72 hours and the results are reported. Increases in temperature and UV exposure both effect the size and particle number of the suspensions with increased levels of natural organic matter (NOM) slowing these affects.

Keywords: Detection, Environmental, ICP-MS, Nanotechnology
Application Code: Environmental
Methodology Code: Mass Spectrometry

Session # 1670 Abstract # 1670-4 **Oral Sessions**

Session Title Mass Spectrometry-Environmental, ICP-MS and Others

Abstract Title **Reading Information Digitally-Encoded in Synthetic Polymers: A Sequencing Approach by Tandem Mass Spectrometry**

Date: Wednesday, March 09, 2016 - Morn
Time: 09:30 AM
Room: B406

Primary Author Laurence Charles
Aix-Marseille University

Co-Author(s) Jean-François Lutz

Abstract Text

Strings of information can be implemented in synthetic polymers using two comonomers intentionally defined as 0-bit and 1-bit, as long as polymeric chains are strictly monodisperse and exhibit controlled sequences of comonomers. Storage of binary-coded messages in poly(alkoxyamine amide)s [1] and polyphosphates [2] was recently demonstrated in our group. Decoding of molecular messages stored in such synthetic copolymers is typically a sequencing task which can be readily achieved using tandem mass spectrometry.

Poly(alkoxyamine amide)s contained coded amide moieties, defined as 0 (71.0 Da) and 1 (85.1 Da), linked by TEMPO (T) alkoxyamine spacers, and were obtained by successive orthogonal coupling of bromo-anhydrides (a-0 and a-1) and amino-TEMPO building-blocks on a solid support. Interchangeable use of a-0 and a-1 in the iterative synthesis allows encoding of binary messages. So-obtained tercopolymer chains were protonated in positive mode electrospray ionization and readily dissociated upon collisional activation via competitive homolytic cleavages of alkoxyamine linkages between any encoding comonomers and a T moiety. Simple dissociation rules based on two series of complementary fragments were then established for poly(alkoxyamine amide)s, allowing their sequencing in a straightforward manner.

Non-natural sequence-encoded polyphosphates were prepared on a DNA synthesizer using two phosphoramidite monomers (with a mass difference of 28 Da) to form binary-coded sequences. These monodisperse polymers were ionized as multiply charged deprotonated molecules using electrospray in the negative mode. Dissociation reactions in MS/MS were mainly observed to occur at bonds in the phosphate groups, leading to eight fragment series which complementarities allowed full sequence coverage by MS/MS.

[1] R.K. Roy, A. Meszynska, C. Laure, L. Charles, C. Verchin, J.F. Lutz Nat. Commun. 2015, 6, 7237.
[2] A. Al Ouahabi, L. Charles, J.F. Lutz J. Am. Chem. Soc. 2015, 137, 5629.

Keywords: Characterization, Mass Spectrometry, Polymers & Plastics, Tandem Mass Spec
Application Code: Polymers and Plastics
Methodology Code: Mass Spectrometry

Session # 1670 **Abstract # 1670-5** **Oral Sessions**

Session Title: Mass Spectrometry-Environmental, ICP-MS and Others

Abstract Title: **Enzymatic Digestion-Single Particle ICP-MS Method to Characterize Nanoparticle Uptake by Plants**

Date: Wednesday, March 09, 2016 - Morn
Time: 10:05 AM
Room: B406

Primary Author: Dan Yongbo
Missouri University of Science and Technology

Co-Author(s): Chady Stephan, Honglan Shi, Ma Xingmao, Runmiao Xue, Weilan Zhang

Abstract Text

Plant uptake and accumulation of nanoparticles (NPs) represent an important pathway for potential human expose to NPs. Consequently, it is imperative to understand the uptake and accumulation of NPs in plant tissues and their unique physical and chemical properties within plant tissue. Current technologies are limited in revealing the unique characteristics of NPs after they enter plant tissues. Enzymatic digestion followed by single particle inductively coupled plasma-mass spectrometry (SP-ICP-MS) analysis methods were developed for simultaneous determination of NP size, size distribution, particle concentration, and dissolved analyte concentration in plant tissues. A PerkinElmer NexION 350D ICP-MS with Syngistix™ Nano Application module the market only dedicated software for SP-ICP-MS was used for the high throughput analysis. The experimental results showed that Macerozyme R-10 enzyme was capable of extracting NPs from tomato plants without causing dissolution or aggregation of NPs. The detection limit for quantification of AuNP size was 20 nm and the AuNPs particle concentration detection limit was 1000 NPs/mL. The particle concentration recoveries of spiked AuNPs were high (79%-96%) in quality control samples. The developed SP-ICP-MS method was able to accurately measure NP size, size distribution, and particle concentration in plant matrix. The dosing study indicated that tomato can uptake AuNPs as intact particles without alternating the AuNPs properties. The enzyme dosage and digestion time was optimized. The applicability of the developed method to different plant species will also be presented.

Keywords: Agricultural, Food Safety, ICP-MS, Nanotechnology
Application Code: Agriculture
Methodology Code: Mass Spectrometry

Session # **1670** **Abstract #** **1670-6** **Oral Sessions**

Session Title: Mass Spectrometry-Environmental, ICP-MS and Others

Abstract Title: **Study of the Transmission of Megadalton-Sized Ions from Atmospheric Pressure to Vacuum in a Q-TOF Charge Detection Mass Spectrometer**

Primary Author: Staci N. Anthony
Indiana University

Date: Wednesday, March 09, 2016 - Morn
Time: 10:25 AM
Room: B406

Co-Author(s): Benjamin E. Draper, Martin F. Jarrold

Abstract Text

The vast majority of ion trajectory simulations disregard gas flow throughout the various differentially pumped regions of a mass spectrometer and assume only a static gas. Those few simulations that do account for the presence of gas flow only study the transmission of small (m/z < 3,000) ions. As the size of the ion increases, the effect of gas flow relative to that of the ion optics becomes more pronounced. To study these effects, a Fortran program was written to simulate ion motion based on gas flow and both DC and RF electric fields through various stages of a home-built mass spectrometer.

A multi-physics solver, STAR-CCM+, was used to simulate gas flow from atmospheric pressure, through a capillary interface, to the first differentially pumped stage of the mass spectrometer. At pressures below 1 Torr, a Parallel Interactive Direct Simulation Monte Carlo (PI-DSMC) program was used to model gas flow through the rest of the instrument. SIMION 8.1 was used to generate the electric fields from both DC and RF potentials applied throughout the instrument. Information from these three programs were then incorporated into a Fortran program to model ion motion due to these electric fields and drag. Simulations suggest that the incorporation of gas flow dramatically reduces transmission throughout the instrument as compared to simulations with a static gas, especially for larger ions. In particular, these simulations have identified key areas of ion loss due to gas flow, and have allowed the design of a much more sensitive mass spectrometer.

Keywords: Electrospray, Instrumentation, Mass Spectrometry, Tandem Mass Spec
Application Code: Bioanalytical
Methodology Code: Mass Spectrometry

Session Title Mass Spectrometry-Environmental, ICP-MS and Others

Abstract Title **Characterization of TiO2-Nanoparticles Using Asymmetrical Flow- and Centrifugal Field-Flow-Fractionation Coupled with MALS, DLS and ICP-MS Detection**

Date: Wednesday, March 09, 2016 - Morn
Time: 10:45 AM
Room: B406

Primary Author Soheyl Tadjiki
Postnova Analytics Inc.

Co-Author(s) Evelin Moldenhauer, Florian Meier, Thorsten Klein, Trevor Havard

Abstract Text

TiO2 nanoparticles have become an essential part of our everyday life. It has applications in a vast range of consumer products such as white wall paint, cleansers or sunscreen formulations. Comprehensive characterization of TiO2 nanomaterials in consumer products is important for good product quality and safety and risk assessments.

In this presentation, we demonstrate the application of Asymmetrical Flow- and Centrifugal Field-Flow-Fractionation coupled with MALS, DLS and ICP-MS detection to characterize TiO2 nanoparticles in various commercial products such as Tego® Sun T805, Aeroxide® P25 and AeroDisp® w740x. In addition, CO2-dried sunscreen formulations containing TiO2 nanoparticles are also evaluated.

The measurements of the radii of gyration, hydrodynamic radii and the elemental composition facilitated the direct comparison of the applied TiO2 nanoparticles with particular respect toward agglomeration. Hence, the obtained results not only clearly indicate the excellent applicability of FFF as a reliable technique suitable for the routine analysis of TiO2 nanoparticles during and after the manufacturing process, but also in the sensitive determination of TiO2 nanoparticles in CO2-dried sunscreen formulations. The capability of determining TiO2 nanoparticles with hydrodynamic radii below 100 nm in real samples renders the FFF technology ideally suitably for the monitoring of existing and upcoming EU regulations, where nanoparticle containing consumer products have to be specifically marked.

Keywords: Characterization, ICP-MS, Laser, Nanotechnology
Application Code: Consumer Products
Methodology Code: Mass Spectrometry

Session # 1670 Abstract # 1670-8 **Oral Sessions**

Session Title Mass Spectrometry-Environmental, ICP-MS and Others

Abstract Title **Enhanced Real-Time Gas Analysis with SIFT-MS Using Negative Reagent Ions**

Date: Wednesday, March 09, 2016 - Morn
Time: 11:05 AM
Room: B406

Primary Author Daniel Milligan
Syft Technologies Ltd

Co-Author(s) Barry Prince, Murray McEwan, Thomas McKellar, Vaughan Langford

Abstract Text

Selected Ion Flow Tube Mass Spectrometry (SIFT-MS) (Smith & Spanel, 2005) is a real-time analytical technique that rapidly analyzes volatile organic compounds (VOCs) to ultra-trace levels in air (Prince et al., 2010). Traditional SIFT-MS utilizes three positively charged reagent ions, H[sub]3[/sub]O[sup]+[/sup], NO[sup]+[/sup], and O[sub]2[/sub][sup]+[/sup] that are created from a microwave discharge through moist air and subsequently mass-selected using a quadrupole mass filter. This means that reagent ions can be switched in 10 milliseconds, rather than minutes when different source gases are used. The addition of negatively charged reagents enables analysis of chemicals previously inaccessible and provides even greater selectivity.

This paper describes a new development of the SIFT-MS ion source that enables four negatively charged reagent ions to be generated in addition to the existing positive ions. OH[sup]-[/sup] is created and mass-selected from the same stream of moist air that is used to generate the positive ions. The remaining three ions, O[sup]-[/sup], O[sub]2[/sub][sup]-[/sup], and NO[sub]2[/sub][sup]-[/sup] are generated from dry air.

The negatively charged reagent ions greatly extends the applicability of SIFT-MS to compounds not previously accessible to the positive reagent ions, because their ionization properties were unsuitable (ionization energies were too high or proton affinities too low). Examples of compounds now detectable include ozone, nitrous oxide, and hydrogen fluoride. Detection limits are often similar in both positive- and negative-ion modes (part-per-trillion by volume).

Acknowledgement: This work was funded by Syft Technologies Ltd, New Zealand.

Prince, B.J., Milligan, D.B., & McEwan, M.J. (2010). Rapid Commun. Mass Spectrom., 24, 1763-1769.
Smith, D., & Spanel, P. (2005). Mass Spec. Rev., 24, 661– 700.

Keywords: Chemical Ionization MS, Gas, High Throughput Chemical Analysis, Volatile Organic Compounds
Application Code: High-Throughput Chemical Analysis
Methodology Code: Mass Spectrometry

Session # 1680 Abstract # 1680-2 **Oral Sessions**

Session Title: Pharmaceutical Applications of Liquid Chromatography

Abstract Title: **Coupling Mass Detection with UV to Improve Method Sensitivity for Esters of Benzenesulfonic and p-Toluenesulfonic Acids in Analysis of Genotoxic Impurities**

Date: Wednesday, March 09, 2016 - Morn
Time: 08:50 AM
Room: B401

Primary Author: Margaret Maziarz
Waters Corporation

Co-Author(s): Christopher Henry, Mark Wrona

Abstract Text

Genotoxic impurities (GTI's) are intermediates or reactants that can develop during the synthesis of a drug substance. In addition to process impurities, certain drugs may generate GTI's via degradation during formulation or storage of the pharmaceutical drug products. The genotoxic compounds have the potential to react with DNA, consequently produce a carcinogenic response and tumor. It is therefore essential to identify presence of these impurities early in the drug development process and to have reliable and highly sensitive methods for accurate determination in both drug substance and drug product.

Alkyl esters of methanesulfonic acid (mesylate), benzenesulfonic acid (besylate) and p-toluenesulfonic acid (tosylate) are commonly used as alkylation agents in chemical synthesis or manufacturing processes. The sulfonic acids are commonly used as counter ions to form Active Pharmaceutical Ingredients (API) salts and can interact with residual alcohols to generate alkyl esters, which are considered potential genotoxic impurities.

In this work, we present a robust and quick dual detection UPLC method for analysis of methyl and ethyl esters of benzenesulfonic and p-toluenesulfonic acids. The UPLC method utilizes both the UV and mass detection (with an ACQUITY QDa detector) for fast and accurate monitoring of genotoxic impurities. We will demonstrate the linearity, sensitivity, and specificity of achievable with both UV and mass detection. We will show that by employing mass detection we improve sensitivity of the method, which is essential for analysis of low level genotoxic impurities in pharmaceutical samples.

Keywords: Chromatography, Liquid Chromatography, Liquid Chromatography/Mass Spectroscopy, Pharmaceuti
Application Code: Pharmaceutical
Methodology Code: Liquid Chromatography

Session # 1680 Abstract # 1680-3 **Oral Sessions**

Session Title: Pharmaceutical Applications of Liquid Chromatography

Abstract Title: **Method Development for Non-Chromophoric Pharmaceutical Analytes Using Alternative Chromatographic and Detection Techniques**

Primary Author: Zongyun Huang
Bristol-Myers Squibb

Date: Wednesday, March 09, 2016 - Morn
Time: 09:10 AM
Room: B401

Co-Author(s): William Fish

Abstract Text

Reversed phase HPLC (RPLC) coupled with UV detection is the most common chromatographic technique in pharmaceutical analysis. However, the determination and quantitation of analytes which lack a suitable chromopore offer unique challenges. Usually, these analytes are hydrophilic are poorly retained using RPLC. This presentation, utilizing real case studies, will discuss various practical method development approaches to quantitate these types of analytes with a focus on using alternative detection (Corona CAD, ESI-MS and low wavelength UV) and alternative chromatographic (HILIC, mix-mode) techniques. For each case study, the advantages, disadvantages and method performance will be discussed with an emphasis on guiding scientists to apply these techniques in a systematic fashion.

Keywords: HPLC, HPLC Columns, HPLC Detection, Liquid Chromatography
Application Code: Pharmaceutical
Methodology Code: Liquid Chromatography

Session # 1680 Abstract # 1680-4 **Oral Sessions**

Session Title: Pharmaceutical Applications of Liquid Chromatography

Abstract Title: **The Impact of LC Instrument Characteristics on HPLC and UPLC Method Migration and Method Transfer**

Date: Wednesday, March 09, 2016 - Morn
Time: 09:30 AM
Room: B401

Primary Author: Paula Hong
Waters Corporation

Co-Author(s): Patricia R. McConville

Abstract Text

Transfer of established reversed-phase methods across HPLC, UHPLC and UPLC chromatographic instrumentation requires careful consideration of the operating parameters and design of each instrument. For example, gradient formation can be influenced by binary or quaternary mixing, dwell volume, viscosity changes in the mobile phase, the gradient shape, residual volumes, and many other factors that can vary across different pumps. The mechanism for sample injection, as well as the detector, can influence linearity and quantitation. Extra column dispersion can impact the resolution and efficiency of a separation. To understand the effect these factors may have on methods transfer, both instrument characteristics and specific method conditions must be factored and evaluated when transferring HPLC and UHPLC methods.

In this presentation, studies will evaluate methods transfer, for a variety of samples and methods, including USP assays. The first set of studies will evaluate the same column and method across multiple instruments from different manufacturers. The impact of instrument attributes- including solvent delivery and temperature control - on the separation fidelity will be evaluated. The transferability of the method will be assessed through system suitability criteria (relative retention, % area, etc.). An additional set of experiments will evaluate the impact of system attributes on method scaling, in which the L/dp ratio remains constant, and the stationary phase ranges from 5 [micro]m to –sub-2- [micro]m particles. In both sets of examples, consideration will be made to conduct method transfer in accordance with regulatory guidelines for allowable adjustments to compendial methods.

Keywords: HPLC, HPLC Detection, Liquid Chromatography, Pharmaceutical
Application Code: Pharmaceutical
Methodology Code: Liquid Chromatography

Session # 1680 Abstract # 1680-5 **Oral Sessions**

Session Title: Pharmaceutical Applications of Liquid Chromatography

Abstract Title: **Development and Validation of A Novel Stability-Indicating Reversed-Phase High-Performance Liquid Chromatography Method for Assay of Thiabendazole and Estimation of its Related Compounds**

Date: Wednesday, March 09, 2016 - Morn
Time: 10:05 AM
Room: B401

Primary Author: Jiangtao He
Merial, A Sanofi Company

Co-Author(s): Abu Rustum, Huang Junmin

Abstract Text

Thiabendazole is a broad-spectrum antihelmintic agent used predominantly in treatment of intestinal pinworm and strongyloides infection. It is similar in structure and mechanism of action to albendazole and mebendazole, which act by selective binding to beta-tubulin of parasitic worms, causing their immobilization and death. A stability-indicating reverse-phase high performance liquid chromatographic method was developed and validated for assay of thiabendazole and estimation of both of its degradation compounds and process impurities. The stability-indicating capability of the method was demonstrated through adequate separation of all potential milbemycin oxime related compounds and also from each other that are present in aged and stressed degradation thiabendazole samples. Chromatographic separation of thiabendazole and its related compounds was achieved by using a gradient elution at a flow rate of 1.0 mL/min on a Waters XBridge C18 column (50 mm × 4.6 mm, 3.5 µm particle size) at 35°C. Mobile phase A of the gradient was 20 mM ammonium hydroxide in aqueous solution, and mobile phase B was acetonitrile. UV detection at 300 nm was employed to monitor the analytes. The new method has been demonstrated to be accurate, linear, precise, reproducible, specific and robust for its intended purpose according to International Conference on Harmonization guidelines and proved to be suitable for routine QC use.

Keywords: HPLC, Method Development, Pharmaceutical, Validation
Application Code: Pharmaceutical
Methodology Code: Liquid Chromatography

Session # 1680 Abstract # 1680-6 **Oral Sessions**

Session Title Pharmaceutical Applications of Liquid Chromatography

Abstract Title **Fast Centrifugal Partitioning Chromatography**

Date: Wednesday, March 09, 2016 - Morn
Time: 10:25 AM
Room: B401

Primary Author Rob Driscoll
Robatel Inc.

Co-Author(s)

Abstract Text

Centrifugal Chromatography instrumentation is used in the natural products, pharmaceutical, biological, and in academia as a means of fractionating complex mixtures of organic components. Fast centrifugal partitioning chromatography [FCPC] offers several advantages versus traditional methods such as HPLC in that no solid packing material is required, which in itself can cross contaminate fractions. The FCPC is a discrete stage-wise device that uses the specific partitioning coefficients of the individual components for isolation of the product fractions. Due to the large quantity of extraction stages in the rotating column, components with similar molecular structures can be easily isolated. Recent improvements in the stage cell design, such as diamond shaped profiles versus traditional Z-cell profiles improve diffusion and reduce mobile phase holdup within the cells. Twin cell designs have been developed that improve resolution and degree of fractionation. FCPC technology is commonly used in isolation of botanicals such as Cannabis, Tobacco, Opiate Derivatives, and Nutraceuticals. Capacities range from the milligram scale for analytical scale quantities, up to preparative scale quantities in the gram scale, and production scale capacities are available for kilogram scale fraction collection. Developments have been made in the sample collection system software. Manual systems are available or they can be used in conjunction with a PLC. Commonly used solvent systems include the Arizona range, HeMWat, aqueous biphasics, and both polar and non-polar solvent pairs.

Keywords: Liquid Chromatography
Application Code: Pharmaceutical
Methodology Code: Liquid Chromatography

Session # 1680 Abstract # 1680-7 **Oral Sessions**

Session Title: Pharmaceutical Applications of Liquid Chromatography

Abstract Title: **Development and Validation of a Reversed Phase Chiral HPLC Method for Verification of Afoxolaner as a Racemic Mixture**

Date: Wednesday, March 09, 2016 - Morn
Time: 10:45 AM
Room: B401

Primary Author: Nilusha Padivitage
Merial, A Sanofi Company

Co-Author(s): Abu Rustum, Satish Kumar

Abstract Text

Afoxolaner is a new antiparasitic molecule from the isoxazoline family that acts on insect and acarine gamma-aminobutyric acid (GABA) receptors. Isoxazoline family of compounds have been employed as active pharmaceutical ingredient in drug products prescribed for control of fleas and ticks in dogs. Afoxolaner, with a chiral center at isoxazoline ring, exists as a racemic mixture of two enantiomers. A chiral reversed phase high performance liquid chromatography (HPLC) method has been developed to verify that Afoxolaner is a racemic mixture as demonstrated by specific rotation. A Chiralpak® AD-RH column (150 mm × 4.6 mm I.D.) maintained at 45 °C was used in the method. Analytes were analyzed with an isocratic elution using 40%Water/50%IPA/10%ACN (v/v/v) as the mobile phase with a detection wavelength of 312 nm. Desired separation of the two enantiomers was achieved within 10 minutes. The resolution and selectivity factors of the two enantiomers were 2.3 and 1.24 respectively. The Limit of Quantitation (LOQ) of the method is 1.6 [micro]g/mL of Afoxolaner. This method was appropriately validated according to the ICH guidelines for its intended use. Since this is a reversed phase HPLC method, it is more easily adaptable to analyze afoxolaner chiral purity in the drug products compared to a normal phase method. ® All marks are the property of their respective owners.

Significance: Effective separation/quantitation of a new chiral API's enantiomers was achieved within 10 minutes.

Keywords: Chiral Separations, Drugs, HPLC, Validation
Application Code: Pharmaceutical
Methodology Code: Liquid Chromatography

Session # 1680 Abstract # 1680-8 **Oral Sessions**

Session Title: Pharmaceutical Applications of Liquid Chromatography

Abstract Title: **Development for a Sensitive Method for the Determination of Diphenylphosphoryl Azide and Hydrogen Azide in Active Pharmaceutical Compounds**

Date: Wednesday, March 09, 2016 - Morn
Time: 11:05 AM
Room: B401

Primary Author: Xuejun Xu
Bristol-Myers Squibb

Co-Author(s): Martin Nunez, Thomas V. Raglione, Yun K. Ye

Abstract Text

Diphenylphosphoryl azide (DPPA) is a common reagent in many pharmaceutical processes. It is often utilized in the Curtius rearrangement reaction to synthesize active pharmaceutical compounds of interest. Under certain conditions DPPA may hydrolyze to generate hydrogen azide (HN3), a genotoxic impurity (GTI). Therefore, a sensitive method was needed to determine this GTI and its precursor, DPPA, at the 1 ppm level in the final drug substance. The main challenges encountered during method development were 1) the instability of DPPA in aqueous solutions due to hydrolysis and 2) the detection of HN3, which has minimal response in GC-MS and no response in LC-MS (electrospray ionization mode). The derivatization of HN3, as reported in the literature, requires complicated reaction conditions and a long reaction time. The method development that will be presented includes the investigation of DPPA hydrolysis conditions, liquid-liquid extraction of HN3 to enrich the sample concentration, and HPLC analysis. A simple HPLC-UV method was successfully developed to determine these GTIs in the final drug substance at the 1 ppm level. The method was demonstrated to be sensitive, robust, and may be utilized to analyze DPPA and/or HN3 in other matrices.

Keywords: HPLC, Method Development, Sample Preparation, Trace Analysis
Application Code: Pharmaceutical
Methodology Code: Liquid Chromatography

Session # 1690 Abstract # 1690-1 **Oral Sessions**

Session Title: Portable Instruments - Half Session

Abstract Title: **Development of Portable Instrumentation Using the Arduino Microcontroller Platform for Field-Ready Electrochemical Experimentation**

Primary Author: Drew C. Farrell
University of Arizona

Date: Wednesday, March 09, 2016 - Morn
Time: 10:05 AM
Room: B409

Co-Author(s): Michael L. Heien

Abstract Text

Developments in consumer electronics, such as the operational amplifier, charge coupled device (CCD), and microcontroller have revolutionized the field of measurement science. The Arduino[registered] platform is an open source prototyping board aimed at building the power of modern microcontrollers into a consumer friendly environment, thus making it an excellent platform for the simplified design of new instrumentation. We have built two handheld, portable instruments using the Arduino[registered] platform. The first is designed to monitor the respiratory rate of sedated lab animals during in-vivo experimentation, and the second is a portable, battery powered potentiostat for “in the field” electrochemical experimentation. This potentiostat is capable of performing several different electrochemical experiments including cyclic voltammetry and anodic stripping voltammetry. The device runs on four 9V batteries, which afford several hours of active experimentation time. The potentiostat is capable of saving data on an onboard data storage card, displaying it on an on-board display, as well as exporting it to a computer for further analysis. This device is not only portable, but also very inexpensive to produce, with a total production cost of just $150. These devices are powerful, but also showcase the power of the Arduino[registered] platform for the development of new analytical instrumentation.

Keywords: Electrochemistry, Forensic Chemistry, Portable Instruments, Trace Analysis
Application Code: Other
Methodology Code: Portable Instruments

Session # 1690 Abstract # 1690-2 **Oral Sessions**

Session Title: Portable Instruments - Half Session

Abstract Title: **A Novel Instrument for Microscale IR Thermography in High Temperature Applied to Solar Salts**

Date: Wednesday, March 09, 2016 - Morn
Time: 10:25 AM
Room: B409

Primary Author: Junko Morikawa
Tokyo Institute of Technology

Co-Author(s): Massimiliano Zamengo, Yukitaka Kato

Abstract Text

Thermal imaging with a spatial resolution of micro-scale is required in increasing numbers of applications of thermal functioning materials, monitoring of small temperature changes and its gradients in localized micro-scale area. We proposed the microscopic IR thermography instruments with original optics' designs and original signal processing systems based on the superimpose technique, which was applied to some novel measuring methods such as (1) pseudo acceleration of temporal resolution of periodic thermal events, (2) pixel by pixel emissivity-temperature conversion, (3) micro-scale flying spot method.

In this study, we newly propose an instrumentation of micro-scale infrared thermal imaging system which is applicable to high temperature measurements up to 700 degC or higher, in order to observe the micro-scale heat transfer and the heat exchange of solar salts and related heat exchange materials.

Considering the handy size portable instrumentation for a practical use in industry with a smaller amount of specimens, the un-cooled micro-bolometer (thermal detector) is chosen as an IR-FPA sensor, with a frame rate of 60Hz, pixel numbers 640x512, pixel pitch 17 micron. The newly designed optics gives the 10 micrometer spatial resolution in the spectrum band 7-14 micron. In order to avoid the thermal radiation effect on the optics, the shading and cooling setup is designed.

Combined with the superimpose technique, the micro-scale thermal imaging in high temperature was achieved and the molten flow of the solar salts was successfully observed. The microscale heat transfer properties of molten salts were also determined.

This work was supported by Council of Science, Technology and Innovation(CSTI), Cross-ministerial Strategic Innovation Program (SIP), "Energy Carrier"(Foundation agency SIP).

Keywords: Energy, Materials Characterization, Portable Instruments, Temperature
Application Code: Material Science
Methodology Code: Portable Instruments

Session # 1690 Abstract # 1690-3 **Oral Sessions**

Session Title: Portable Instruments - Half Session

Abstract Title: **A New GC/FTIR Detection Method as Applied to Inline Monitoring of Siloxanes in Biogas**

Date: Wednesday, March 09, 2016 - Morn
Time: 10:45 AM
Room: B409

Primary Author: Charles M. Phillips
Prism Analytical Technologies, Inc.

Co-Author(s): Anthony S. Bonanno, Martin L. Spartz, Peter P. Behnke

Abstract Text

The mesophillic and thermophillic action of anaerobic microbes in wastewater and landfill waste produce methane which can be burned as a fuel for power generating engines or turbines. Alternatively, the filtered biogas can be introduced into the natural gas pipeline for use by households and businesses or processed as compressed natural gas (CNG) for transportation. In all these situations, the biogas must be rid of siloxanes, which are normally found in the raw sources, as they will form silicates when burnt, causing abrasion to any moving parts. Most facilities use periodic monitoring of the filtered biogas stream to insure that the siloxane levels are acceptable. However, both the methodology of grab sampling using impingers or bags as well as the time frame for analysis do not guarantee that the siloxane levels measured are representative of the actual sample. We have developed an inline instrument in which we are able to statically trap and monitor (via FTIR detection) the gas chromatographic eluants and use a unique algorithm to analyze for each eluant. Since the eluants are trapped within the gas cell, their spectra can be integrated for a period of time up to minutes; this results in analytical detection limits from the single digit ppb to pptr levels (through the use of thermal desorption tubes). The data from actual wastewater and landfill inline samples will be discussed, demonstrating how the two dimensions of this technique (chromatographic separation and spectral deconvolution via classical least squares) produces these detection characteristics without the need for recalibration.

Keywords: FTIR, Gas Chromatography, Portable Instruments, Vibrational Spectroscopy
Application Code: Fuels, Energy and Petrochemical
Methodology Code: Portable Instruments

Session # 1690 Abstract # 1690-4 **Oral Sessions**

Session Title: Portable Instruments - Half Session

Abstract Title: **Next-Generation Handheld XRF Analyzer – Smarter, Smaller and Faster**

Primary Author: Esa Nummi
Thermo Fisher Scientific

Date: Wednesday, March 09, 2016 - Morn
Time: 11:05 AM
Room: B409

Co-Author(s)

Abstract Text

Field portable X-ray fluorescence (XRF) spectrometry is a mature technology for elemental chemical analysis, dating from the 1960s. The last two decades were signified by the rapid pace of development in handheld XRF analyzers, large isotope based two-component systems were replaced by the lightweight; X-ray tube based one-piece handheld devices.

The majority of handheld XRF instruments are used for analysis and identification of metal alloys, but handheld XRF technology has become an important tool also in many other diverse applications ranging from ore exploration to screening toys for hazardous elements. Speed of a single analysis, which translates into a large number of tests, its non-destructive character and the economics of use, are key features that have made handheld XRF analyzer a tool of choice for so many applications.

Handheld XRF is today clearly a mature analytical method, but this doesn't mean that pace of development would have slowed down. In this presentation, we briefly review the evolution of handheld XRF and look at the science and technology of the latest generation of handheld XRF analyzers, their new applications and new analytical capabilities. In addition to ever improving analytical performance, latest technology provides smart connectivity and other new features in smaller and lighter package, making the handheld XRF technology more powerful and versatile analytical tool than ever before.

Keywords: Material Science, Metals, Spectroscopy, X-ray Fluorescence
Application Code: Material Science
Methodology Code: Portable Instruments

Session Title: Sensors - Bioanalytical and Homeland Security/Forensics

Abstract Title: **Photonic Crystal Protein Hydrogel Sensor for Candida Albicans**

Date: Wednesday, March 09, 2016 - Morn
Time: 08:30 AM
Room: B407

Primary Author: Zhongyu Cai
University of Pittsburgh

Co-Author(s): Daniel H. Kwak, David Punihaole, Sachin S. Velankar, Sanford A. Asher, Xinyu Liu, Zhenmin Hong

Abstract Text

We report two-dimensional (2-D) photonic crystal (PC) sensing materials that selectively detect Candida albicans (C. albicans). These sensors utilize Concanavalin A (Con A) protein hydrogels that multivalently and selectively bind to mannan on the C. albicans cell surface to form cross-links. The cross-links formed shrink the Con A protein hydrogel, decrease the 2-D PC particle spacing and blue shift the light diffracted from the 2-D PC embedded on the Con A protein hydrogel surface. Increasing C. albicans concentrations increasingly blue shift the diffraction. The diffraction shifts can be visually monitored, or measured with a spectrometer, or determined from the Debye diffraction ring diameter. Our unoptimized Con A protein hydrogel sensor has a detection limit of ~32 CFU/mL for C. albicans. This sensor distinguishes between C. albicans and gram negative bacteria such as E. coli. This sensor demonstrates proof-of-concept for utilizing recognition between lectins and microbial cell surface carbohydrates to detect microorganisms in aqueous environments.

This work was financially supported by HDTRA under grant no. 1-10-1-0044.

Keywords: Biosensors, Carbohydrates, Material Science, Protein

Application Code: Bioanalytical

Methodology Code: Sensors

Session # 1700 Abstract # 1700-2 **Oral Sessions**

Session Title: Sensors - Bioanalytical and Homeland Security/Forensics

Abstract Title: **Development of a Wireless Microfluidic Biosensor System for Real-Time Monitoring of Human Transplant Organs in Transit**

Date: Wednesday, March 09, 2016 - Morn
Time: 08:50 AM
Room: B407

Primary Author: Sally A. Gowers
Imperial College London

Co-Author(s): Bynvant Sandhu, Chu Wang, Isabelle C. Samper, Martyn G. Boutelle, Thomas Watts, Vassilios Papalois

Abstract Text

Kidney transplantation is the preferred treatment for patients with end-stage renal failure. As a result of the severe shortage of donor organs, kidneys from extended criteria donors are increasingly being used to meet the demand. These organs are under-utilised as they may have incurred more ischaemic damage prior to recovery, which could impair their function. Therefore it is vital to be able to assess the viability of these organs after donation in order increase the number of successful transplants.

We have previously shown that tissue metabolite levels can provide valuable information about renal health in a laboratory setting (1). However, it is also necessary to be able to monitor the organ immediately after donation and in transit between donor and recipient sites in order to gain time-critical viability information. Therefore, we are developing a portable analysis system that uses biosensors to quantify key metabolic markers of tissue health in real time. Microdialysis is used to sample the tissue during storage and the resulting dialysate is monitored for changes in glucose and lactate levels using integrated needle electrodes (2) coated with a layer of entrapped enzyme (3). The biosensors are also coated with an additional diffusion-limiting polyurethane film to extend the dynamic range of the sensors to include the potentially high initial levels. These sensors are housed within a 3D-printed microfluidic device (4) and are coupled to potentiostats that wirelessly link with a laptop. The system also incorporates an autocalibration board, which allows calibration of the system in transit.

Preliminary results will be presented showing system validation as well as results of initial proof-of-concept experiments.

References:
1. Hamaoui et al. J. Surg. Res. In press (2015)
2. Rogers et al. ACS Chem. Neurosci. 4, 799-807 (2013)
3. Vasylieva et al. Biosens. Bioelectron. 26, 3993-4000 (2011)
4. Gowers et al. Anal. Chem. 87, 7763-7770 (2015)

Keywords: Bioanalytical, Biosensors, Electrochemistry, Portable Instruments
Application Code: Bioanalytical
Methodology Code: Sensors

Session Title Sensors - Bioanalytical and Homeland Security/Forensics

Abstract Title **Ion-Selective Electrodes with PEDOT(PSS) as Solid Contact: Influence of the PEDOT(PSS) Thickness on the Equilibration Time**

Date: Wednesday, March 09, 2016 - Morn
Time: 09:10 AM
Room: B407

Primary Author Marcin Guzinski
The University of Memphis

Co-Author(s) Bradford D. Pendley, Erno Lindner, Jennifer M. Jarvis

Abstract Text

Solid contact ion-selective electrodes (ISEs) are multilayer structures, in which an intermediate layer is utilized between the electron conducting substrate and the ion-selective membrane. This intermediate layer serves as ion-to-electron conductor to improve reproducibility and stability of ISEs compared to the coated wire electrodes. In our recent paper we showed that the equilibration times of solid contact ISEs are significantly longer when a PEDOT(PSS) (poly(3,4-ethylenedioxythiophene) polystyrene sulfonate) intermediate layer is used over platinum surface instead gold or glassy carbon.[sup]1[/sup]

In this work we report the influence of thickness and deposition technique of the intermediate layer on the equilibration times and long term stability of ISEs with PEDOT(PSS) as solid contact. The equilibration time is defined as the time interval from the first solution contact of the electrode and the time when the potential drift drops below a threshold value (0.3 mV/min). The equilibration time is uniquely important for single use sensors and sensors implemented in point-of-care devices.

We have studied the equilibration time for solid contact potassium ion-selective electrodes with 20 nm, 0.1µm, 0.5 µm, 1 µm, 2 µm and 4 µm thickness of PEDOT(PSS) layer. In addition we have compared the equilibration time of K+ ISEs in which the PEDOT(PSS) film has been deposited by cyclic voltammetry, chronopotentiometry and chronoamperometry.

(1) Guzinski, M.; Jarvis, J. M.; Pendley, B. D.; Lindner, E. Equilibration Time of Solid Contact Ion-Selective Electrodes. Analytical Chemistry 2015, 87, 6654-6659.

Keywords: Biosensors, Electrochemistry, Ion Selective Electrodes, Potentiometry
Application Code: Bioanalytical
Methodology Code: Sensors

Session # 1700 Abstract # 1700-4 **Oral Sessions**

Session Title: Sensors - Bioanalytical and Homeland Security/Forensics

Abstract Title: **Evaluating Real Time Binding Interactions in Insulin Immunoassay for Diagnosis of Type of Diabetes by Surface Plasmon Resonance**

Date: Wednesday, March 09, 2016 - Morn
Time: 09:30 AM
Room: B407

Primary Author: Vini Singh
Oklahoma State University

Co-Author(s): Sadagopan Krishnan

Abstract Text

Diabetes, an autoimmune disorder, is caused by the inability of pancreas to produce enough insulin. Diabetes has reached alarming rates as reported by the latest 2014 statistics from the American Diabetes Association. About 86 million people are projected to have prediabetic conditions. Such a high population suggests the need for better management and treatment outcomes. Our objective is to develop novel ultrasensitive optical and electrochemical detection tools, which can diagnose and identify the type of diabetes (Type 1-insulin dependent and Type 2-insulin independent), based on insulin levels. Complex clinical matrices such as human serum and whole blood pose huge non-specific interactions on the sensor surface and hence detecting clinically relevant picomolar insulin levels in these matrices still remains a challenge.

Our prior work involved developing electrochemical mass and voltammetric immunosensors using magnetic nanoparticles conjugated to serum insulin and detection of insulin levels by capture antibodies immobilized on the sensor surface with a limit of detection (LOD) 5 pM serum insulin. We present herein the development of a sandwich immunosensor that can measure insulin levels in whole blood samples, captured by magnetic microbeads carrying detection antibodies, and detected on an insulin-antibody microarray surface. Additionally, we monitored the real time binding interactions of insulin to surface anti-insulin antibody by change in reflectivity with time using a Surface Plasmon Resonance imager (SPRi) offering better throughput and sensitivity. The microarray imager developed was successful in establishing the kinetic parameters, ka (association constant) and kd (dissociation constant) and thus can be a promising tool in quickly diagnosing diabetes at clinical sites.

Acknowledgements. Financial support by the National Institute of Diabetes and Digestive and Kidney Diseases of the National Institutes of Health is gratefully acknowledged.

Keywords: Bioanalytical, Biosensors, Charge Transfer Devices (CID CCD), Detection
Application Code: Bioanalytical
Methodology Code: Sensors

Session # 1700 Abstract # 1700-5 **Oral Sessions**

Session Title Sensors - Bioanalytical and Homeland Security/Forensics

Abstract Title **Extended Nanopore Residence Times via Metallic Clusters**

Date: Wednesday, March 09, 2016 - Morn
Time: 10:05 AM
Room: B407

Primary Author Joseph E. Reiner
Virginia Commonwealth University

Co-Author(s) Amy E. Chavis, Kyle T. Brady, Nuwan Kothalawala

Abstract Text

Resistive-pulse nanopore sensing utilizes the Coulter counting principle where molecules enter a nanoscale pore separating two electrolyte solutions and reduce the flow of ionic current, yielding short-lived current blockades. An important goal of this field is to develop techniques to extract as much information as possible from the current blockades. One way to do this is to keep molecules in the pore for extended periods of time. Increasing the blockade duration improves the nanopore sensor in two ways; it enables more accurate discrimination of different sized molecules and it extends the time over which one can analyze the fluctuations within each current blockade. Previously, researchers have increased the time a molecule spends in the pore by changing the physical or chemical properties in or around the pore. However, these methods are limited because the residence time enhancement has been too small or the pore's physical properties were irreversibly altered.

This presentation will describe the use of charged metallic clusters to increase the residence time of analyte in an alpha hemolysin pore. Anionic clusters (Au25(SG)18) enter the pore from one side and increase the residence time of cationic molecules entering from the opposite side. Initial studies demonstrated that the cluster greatly improves the pore's ability to discriminate between different-sized PEG molecules. More recent studies have focused on detecting the biologically relevant peptides angiotensin and neurotensin with the cluster enhanced residence time mechanism. We will discuss these results and the implications of cluster enhancement for peptide detection with nanopore sensors.

Keywords: Bioanalytical, Electrochemistry, Nanotechnology, Peptides
Application Code: Bioanalytical
Methodology Code: Sensors

Session # 1700 **Abstract # 1700-6** **Oral Sessions**

Session Title: Sensors - Bioanalytical and Homeland Security/Forensics

Abstract Title: **Kinetics Quantification of MicroRNAs as Disease Biomarkers on Microelectrode Point-of-Care Sensors at Attomoles within Minutes**

Date: Wednesday, March 09, 2016 - Morn
Time: 10:25 AM
Room: B407

Primary Author: Tanyu Wang
Georgia State University

Co-Author(s): Gangli Wang

Abstract Text

DNA-/aptamer-based electrochemical sensors have become promising analytical tools for rapid and quantitative folding-based detection recently. The incorporation of micro- and nanoelectrodes in sensor fabrication is meritorious for miniaturized quantity/volume detection at high throughput. An intrinsic limit for such approaches is the low signal for detection in general. We have established a current amplification mechanism and developed a one-step label-free electrochemical sensor for the detection of microRNAs that is challenged by the low analyte abundance in physiological samples. The folding-based DNA probe functions as signal-on sensing mechanism upon binding with target. A chemical reductant tris-(2-carboxyethyl) phosphine hydrochloride is introduced in the bulk solution to cyclically reduce the electro-oxidized MB thus enables enzyme-less signal amplification for ultrasensitive detection. The detected current is governed by sequential kinetic processes including mass transport of miRNA from solution to the surface, hybridization with surface DNA sequence, ET among the TCEP, MB, and the electrode, and the diffusion of TCEP. A detection limit of miR-122 of 0.1 fM via direct readout within minutes, with a wide detection range from sub fM to nM is being further optimized. The sensor demonstrates excellent discrimination against 2-mismatch sequences and superb stability when stored in dry form. Mechanistic understanding of the governing kinetics at different time/concentration domains enables improved linearity of calibration curve and broadened detection range for robust practical applications.

Keywords: Bioanalytical, Biosensors, Electrochemistry, Microelectrode
Application Code: Bioanalytical
Methodology Code: Sensors

Session # 1700 Abstract # 1700-8 **Oral Sessions**

Session Title: Sensors - Bioanalytical and Homeland Security/Forensics

Abstract Title: **A New Miniaturized Sensor for Real-Time Suit Penetration Assessment in the Man-In-Simulant-Test (MIST) Protocol**

Date: Wednesday, March 09, 2016 - Morn
Time: 11:05 AM
Room: B407

Primary Author: Nicholas Fitzgerald
Defence Science and Technology Group

Co-Author(s): Karl Pavey

Abstract Text

Military personnel and civilian first responders regularly put themselves in harm's way to serve the public interest. One of the most insidious threats that they face is that of toxic chemical vapours either released deliberately or by accident. The ability to effectively and efficiently evaluate the protective capability of new protective suit ensembles that are used in such situations is of vital importance.

The gold standard for evaluating protective ensembles against chemical vapour is the Man-In-Simulant-Test (MIST) which involves human participants performing prescribed activities in a chemical warfare agent simulant environment whilst wearing protective clothing. Measurement of Methyl Salicylate (MeS – simulant for Sulfur Mustard) under the ensemble during a MIST is achieved by using passive adsorbent PADs attached to the skin of the participants and subsequent analysis by liquid extraction/HPLC. This method provides no time domain data and limited information about the spread of penetrated vapours beneath the ensemble.

The authors have developed a real-time, ultra-miniaturised, wireless sensor which is able to be attached to the body during the MIST. Using miniaturised single-wavelength spectroscopy tuned to a specific colorimetric reaction, the new sensor can detect MeS at concentrations below 50 ppb. Smart chemical treatments and electronics design allow the sensor to work effectively across a wide range of temperature and humidity conditions (20-35ºC, 30-95% RH) while retaining ultra-low sensitivity in a compact lightweight package. This sensor has the potential to radically increase MIST throughput and provide important time resolved data for better protective ensemble design.

Keywords: Detection, Portable Instruments, Sensors, Spectrophotometry
Application Code: Homeland Security/Forensics
Methodology Code: Sensors

Session # 1710 **Abstract # 1710-1** **Oral Sessions**

Session Title: Thermal Analysis

Abstract Title: **Using Isothermal Titration Calorimetry to Measure Thermodynamic Parameters of Adsorption on Chromatographic Media**

Date: Wednesday, March 09, 2016 - Morn
Time: 08:30 AM
Room: B408

Primary Author: Anthony R. Horner
University of Pittsburgh

Co-Author(s): Stephen Weber, Stephen R. Groskreutz

Abstract Text

Thermodynamic parameters of adsorption/retention between analyte and stationary phase were studied using isothermal titration calorimetry (ITC). These parameters are typically studied using liquid chromatography, and there is limited literature measuring adsorption thermodynamics using ITC. Typically, ITC is used to study the thermodynamic parameters of proteins and small molecules by measuring the heat generated or absorbed during chemical interactions and reactions. In this work, unfunctionalized 1.8 [micro]m fully porous silica particles are titrated with sodium hydroxide with and without sodium chloride to evaluate substrate effects and to study electrostatic effects of bare silica. C18 modified silica was evaluated using 1-phenyloctan-1-one to determine the thermodynamics of adsorption. Accurate measurements of enthalpy have been determined using ITC and compared to those obtained from a chromatogram. Enthalpy is calculated in chromatography using retention factors at varied experimental temperatures. Chromatographically, the enthalpy is -12.69±0.01 kJ/mol and calorimetrically, the enthalpy is -12.09±0.59 kJ/mol. In chromatography, free energy is calculated from retention factor. In theory it can be measured by increasing the analyte adsorbed on the stationary phase, however we have shown that this measurement cannot be made in this experimental set up. Experiments are modeled allowing for experimental sensitivity to unknown thermodynamic parameters to be predicted and for accurate fitting of experimental data.

Keywords: HPLC, HPLC Columns, Liquid Chromatography, Thermal Analysis
Application Code: General Interest
Methodology Code: Thermal Analysis

Session # 1710 Abstract # 1710-2 **Oral Sessions**

Session Title: Thermal Analysis

Abstract Title: **Thermal Conductivity Measurement of Solar Salt in High Temperature Using the Temperature Wave Method**

Primary Author: Junko Morikawa
Tokyo Institute of Technology

Date: Wednesday, March 09, 2016 - Morn
Time: 08:50 AM
Room: B408

Co-Author(s): Massimiliano Zamengo, Yukitaka Kato

Abstract Text

The precise control of thermal conductivity of molten salt in high temperature is one of the key technology in power-saving technology. The principle of the double lock-in method using the originally manufactured high-sensitivity micro-thermopiles and the FPGA controlled signal processing systems are presented in this study. The thermopile sensors are located in the different distances of a few micro-meters from the modulated heat source controlled by a Peltier thermo-module. A normalized gain (an attenuation ratio of the temperature wave) gives the thermal impedance of the sample in relation with the environmental conditions. Thermal conductivity of the sample is determined by the relationship between the gain and the thermal impedance of the sample, in which the experimental heat loss factor can be calibrated with standard materials. The numerical simulations of the influence of the thermal impedance on the gain is precisely compared with the experimental results to confirm the accuracy and the validity of this method. By using this technique, the temperature dependence of solar salt is precisely determined. The handy size instrumentation for a practical use in industry with a smaller amount of specimens in a quick time measurement is proposed.

This work is supported by Counsil of Science, Technology, and Inovation(CSTI), Cross-ministerial Strategic Innovation Promotion Program(SIP), "ENERGY CAREER"(Funding agency:JST).

Keywords: Energy, Materials Characterization, Sensors, Thermal Analysis
Application Code: Material Science
Methodology Code: Thermal Analysis

Session # **1710** **Abstract #** **1710-3** **Oral Sessions**

Session Title Thermal Analysis

Abstract Title **TGA-GC-MS Analysis of Different Tobacco Types**

Date: Wednesday, March 09, 2016 - Morn
Time: 09:10 AM
Room: B408

Primary Author Ekkehard Post
NETZSCH Geraetebau GmbH

Co-Author(s) Bob Fidler, Jan Hanss

Abstract Text

Smoking tobacco involves a low level oxygen combustion of the tobacco leaves. It is of course indisputable that during that burning process – apart from the nicotine - several other organic products are in the smoke present. With TGA coupled to a GC-MS system these gaseous products can be identified. Applying different measurement modes like cryo or quasi-continuous mode allow either a very good identification of the gaseous products due to an optimum separation by the GC or a better temperature resolution with the drawback of poorer GC separation. In this contribution two different tobacco types were analyzed by GC-MS and compared.

Keywords: GC-MS, Pyrolysis, Thermal Analysis, Toxicology
Application Code: Material Science
Methodology Code: Thermal Analysis

Session # 1710 **Abstract # 1710-4** **Oral Sessions**

Session Title: Thermal Analysis

Abstract Title: **Pyrolysis Gases of Polycarbonate Identified by TGA-FT-IR and TGA-GC-MS**

Date: Wednesday, March 09, 2016 - Morn
Time: 09:30 AM
Room: B408

Primary Author: Ekkehard Post
NETZSCH Geraetebau GmbH

Co-Author(s): Bob Fidler

Abstract Text

Polycarbonate (PC) is a widely used thermoplastic polymer. PC is amorphous and shows a high mechanical stability over a broad temperature range. PC is used for high-strength technical parts, medical devices, dishes, etc. Due to the chemical structure (aromatic components), numerous toxic compounds are evolved during pyrolysis or even combustion with an oxygen deficit. A thermogravimetric analyzer (TGA) coupled to evolved gas analysis instruments can help simulating such pyrolysis/combustion processes by identifying the evolved gases versus mass loss and temperature. Both EGA methods complement each other and/or confirm the results. The TGA-GC-MS coupling clearly shows advantages in the identification of the gases due to the previous separation of the gases by means of the GC prior to the actual identification with the mass spectrometer. The FTIR, on the other hand, allows a better time/temperature resolution along with the mass loss. In this paper, the different coupling systems will be presented and the resulting pyrolysis gases will be discussed.

Keywords: GC-MS, Polymers & Plastics, Pyrolysis, Thermal Analysis
Application Code: Material Science
Methodology Code: Thermal Analysis

Session # 1710 Abstract # 1710-5 **Oral Sessions**

Session Title Thermal Analysis

Abstract Title **Understanding Auto-Catalysis Using Scanning, Isothermal and Adiabatic Calorimetry**

Date: Wednesday, March 09, 2016 - Morn
Time: 10:05 AM
Room: B408

Primary Author Peter Ralbovsky
Netzsch Instruments

Co-Author(s) Bob Fidler

Abstract Text

Autocatalytic reactions can be the most dangerous types of reactions, as the autocatalytic nature of the reaction can go undetected even after typical hazards screening methods have been employed. Also there are many instances were multiple reactions can complicate what would otherwise be a straightforward analysis approach. There are many methodologies which have been developed to systematically screen for reactive hazards. Those chemicals, alone or in combination with other chemicals, which meet a threshold of concern are often then subjected to DSC or similar tests. From these tests the thermodynamics and the kinetics can be determined and can be used directly for modeling or applied to any number of ranking methods. Ranking methods are designed to be fairly easy to implement and definitely easy to understand and communicate. With simplicity sometimes robustness is sacrificed. Adiabatic calorimetry testing will provide test data which includes pressure data and also provides indirectly the effect of heat capacity with the determination of the adiabatic temperature rise (reaction heat x heat capacity) corrected for thermal inertia. Furthermore, in a single test the ARC can also provide kinetic information as well as an indication as to whether or not the reaction is autocatalytic. However as a screening method, adiabatic testing can be time-consuming. DSC testing on the other hand is much faster but lacks the pressure data. Multiple tests, run at different rates can be used to determine the existence of an autocatalytic reaction and kinetic parameters can be determined. The effect of autocatalysis on test data, and the application of this test data to simple analysis techniques, hazards ranking will be reviewed and summarized.

Keywords: Chemical, Data Analysis, DSC, Thermal Analysis
Application Code: Material Science
Methodology Code: Thermal Analysis

Session # 1710 Abstract # 1710-6 **Oral Sessions**

Session Title: Thermal Analysis

Abstract Title: **Developing Databases and Optimized Spectral Searching from TGA-IR Hyphenation Experiments**

Primary Author: Ian Robertson
PerkinElmer Limited

Co-Author(s): Jack Botting, Justin Lang

Date: Wednesday, March 09, 2016 - Morn
Time: 10:25 AM
Room: B408

Abstract Text

Hyphenated techniques, such as combining Thermogravimetric Analysis (TGA) with an FT-IR (IR), can generate significant amounts of data. The information from the individual techniques is extremely valuable, but the data from the combined techniques yields additional and complementary information. However, the hyphenated techniques can generate large datasets that can take considerable time to examine and extract the valuable information. Spectral library searching on TGA-IR data is complicated by the fact that the breakdown products often consist of multiple species being evolved simultaneously. The use of a mixture search algorithm capable of identifying multiple components from a single mixture spectrum greatly enhances the identification of the species present. Use of Informatics with data visualization tools, including PCA analysis of the entire hyphenated experiment, offers a rapid method for extracting the maximum information in the easiest and fastest way. TGA-IR applications will be presented demonstrating the wealth of information available from these experiments and the ability to extract the most important information in an easy manner.

Keywords: FTIR, Informatics, Other Hyphenated Techniques, Thermal Analysis
Application Code: Material Science
Methodology Code: Thermal Analysis

Session # 1710 Abstract # 1710-7 **Oral Sessions**

Session Title: Thermal Analysis

Abstract Title: **Thermal Excitation, Optical Response: A Novel Approach to Thermal Analysis by TMOR**

Primary Author: Sarah Schwarz G Henriques
Anton Paar OptoTec GmbH

Co-Author(s): Jens Kruse, Nils Bertram, Tobias Husemann

Date: Wednesday, March 09, 2016 - Morn
Time: 10:45 AM
Room: B408

Abstract Text

Phase transitions play a key role in many industrial applications: While for example glass transitions govern the production of hard candy and also determine the mechanical properties of polymers and adhesives, melting transitions of fats are essential for the perceived quality of chocolate. Several phase transitions are easily measurable with commercially available instruments. Other phase transitions – typically second order phase transitions or glass transitions – are more difficult to detect. In particular calorimetrists are faced with experimental challenges such as base-line instability, choice of matching crucibles for sample and reference, thermal contact of crucibles to the sensor plate and costly equipment.

We present a novel technique, TMOR (temperature-modulated optical refraction), for monitoring both simple and more complex phase transitions, which is not subject to the above obstacles. The TMOR principle is based on a modulation of the sample temperature and the detection of the response in refractive index [1]. This grants access to one of the fundamental susceptibilities, namely the thermal expansion coefficient. Further analysis sheds light onto temperature- and time-dependent processes, e.g. melting and glass transitions as well as polymerisations. Recent data on edible fats indicate that even structural information on polymorphisms can be obtained. By discussing both food and polymer applications of the innovative measuring principle at the example of cocoa butter (melting process), caramel and PVAC (glass transitions) we demonstrate the potential of the technique for research and development, as well as quality management.

[1] Müller et al. Thermochimica Acta, 2013

Keywords: Food Science, Polymers & Plastics, Thermal Analysis
Application Code: Food Safety
Methodology Code: Thermal Analysis

Session # 1710 **Abstract # 1710-8** **Oral Sessions**

Session Title: Thermal Analysis

Abstract Title: **Characterization of Nanomaterials with Thermal Analysis and Hyphenated Techniques**

Date: Wednesday, March 09, 2016 - Morn
Time: 11:05 AM
Room: B408

Primary Author: Chady Stephan
PerkinElmer

Co-Author(s): Jun Wang

Abstract Text

With the rapid growth of the nanotechnology market, the need for modern characterization techniques for nanomaterial analysis is increasing. Thermal Analysis has been heavily explored as a standard-alone technique, and connected to other instrumentation such as GC/MS and FTIR as well for evolved gas analysis. This presentation illustrates some typical applications of thermal analysis and such hyphenations with various examples.

Thermal analysis is a group of analytical techniques that measure the physical properties of a material (i.e., heat flow, mass, dimension) as a function of temperature. The application of these techniques on both polymer nanocomposites, where nanoparticles are used as a filler material to modify the mechanical and thermal properties, and pure nanomaterials are presented. The effects of particle size and concentration of nanofiller in polymer composite material on storage modulus, glass transition, crystallinity can be quantified by Dynamic Mechnical Analysis (DMA) and Differential Scanning Calorimetry (DSC). The purity, thermal stability and composition are measured with Thermogravimetric Analysis (TGA). The application of hyphenated techniques is demonstrated by the characterization of some nanomaterials using TG-GC/MS.

Keywords: DSC, Nanotechnology, Thermal Desorption
Application Code: Nanotechnology
Methodology Code: Thermal Analysis

Session # 1720 **Abstract # 1720-1** **Poster Sessions**

Session Title: Applications of LC/MS

Abstract Title: **Determination of Fentanyl in Canine Plasma Using HPLC-MS Detection**

Date: Wednesday, March 09, 2016 - Morn

Time:

Room: Exposition Floor, 400 Aisle

Primary Author: Joan B. Bailey
University of Tennessee

Co-Author(s): Kristen Gordon, Molly White, Reza Seddighi, Sherry K. Cox

Abstract Text

A simple, easy, and accurate high-performance liquid chromatographic method for the determination of fentanyl concentrations in plasma samples has been developed and validated. Following a liquid extraction of acetonitrile:water (90:10), samples were separated by reversed-phase high-performance liquid chromatography on an XBridge C18 3.5µm column (2.1 x 50 mm) and detected by mass spectroscopy. The mobile phase was a mixture of water with 0.1% formic acid, and acetonitrile with 0.1% formic acid (90:10), with a flow rate of 0.6 mL/min. The procedure produced a linear curve over the concentration range of 0.1-25 ng/mL for fentanyl in canine plasma with an LOD of 0.05 ng/mL and LOQ of 0.1 ng/mL. Intra- and inter-assay variability ranged from 2.6%-8.2% and the average recovery for fentanyl was 100%.

Keywords: Biological Samples, HPLC, Liquid Chromatography/Mass Spectroscopy, Sample Preparation
Application Code: Bioanalytical
Methodology Code: Liquid Chromatography/Mass Spectrometry

Session # 1720 Abstract # 1720-2 **Poster Sessions**

Session Title: Applications of LC/MS

Abstract Title: **Targeted Metabonomic Study of Plasma from Rats with Acute Colitis Using LCMS-IT-TOF Based Metabonomics**

Date: Wednesday, March 09, 2016 - Morn
Time:
Room: Exposition Floor, 400 Aisle

Primary Author: Lingling Shen
Shimadzu Global COE, Shimadzu Co., Ltd

Co-Author(s): Jingting Yao, Qisheng Zhong, Taohong Huang, Xiaojun Zhang

Abstract Text

According to published estimates, there is tens of thousands of all kinds of endogenous metabolites in the human body. Differences in the concentration range and physical & chemical properties of various metabolites in body fluids, there is no one analysis technique of all metabolites in body fluids. NMR, mass spectrometry and various separation methods became to the important analysis tools in metabonomics, such as: the liquid chromatography-mass spectrometry (LC-MS), gas chromatography-mass spectrometry (GC-MS) etc. By using multivariable model identify methods,it is found potential markers from massive metabonomics data analysis. The diagnosis of IBD primarily depends on invasive methods, such as endoscope, however, which usually limits early diagnosis and treatments. Based on metabonomics, this paper will provide useful information for developing non-invasive and sensitive diagnosis methods for IBD. As an example of serum samples of rats with acute colitis, this paper established a method of screening potential biomarkers from the serum metabolic fingerprint by using LCMS-IT-TOF. This method described systematically the process from data collection, data preprocessing, pattern recognition and finally screening potential biomarker. The PLS model showed that there are significant differences in the metabolites of TNBS rats and healthy rats , and 8 compounds screened were identified as potential biomarkers, such as henylacetylglycine󠀠p-cresol glucuronide󠀠Butyryl-L-carnitine󠀠8Z,11Z,14Z,18Z-eicosatetraenoic acid and LysoPCs (PC16:0, PC18:0, PC18:2, PC18:1).

Keywords: Bioanalytical, HPLC Columns, Liquid Chromatography/Mass Spectroscopy, Time of Flight MS
Application Code: Bioanalytical
Methodology Code: Liquid Chromatography/Mass Spectrometry

Session # 1720 Abstract # 1720-3 **Poster Sessions**

Session Title: Applications of LC/MS

Abstract Title: **Optimization of On-Column Trypsin Digestion Coupled with LC-MS/MS for Analysis of Apolipoproteins in Serum**

Date: Wednesday, March 09, 2016 - Morn

Time:

Room: Exposition Floor, 400 Aisle

Primary Author: Christopher Toth
Centers for Disease Control and Prevention

Co-Author(s): Bryan A. Parks, David Schieltz, James Pirkle, Jeffrey Jones, John R. Barr, Jon Rees, Lisa McWilliams, Michael S. Gardner, Yulanda Williamson, Zsuzsanna Kuklenyik

Abstract Text

Accurate isotope dilution LC-MS/MS quantification of proteins through trypsin digestion requires preferably the analysis of multiple target peptides per each protein. It is assumed that each of the target peptides is cleaved completely with the same rate from the protein(s) during a given digestion time while both native cleavage products and corresponding isotope labeled internal standards remain stable. When targeted quantitative measurement of multiple protein analytes is performed, finding an optimal digestion time that produces reproducible results for all analytes becomes increasingly difficult. On-column trypsin digestion can alleviate these problems because digestion time is controlled with high precision through flow rate and column volume. Furthermore, because of the short on-column digestion time (typically 4-6 minutes) and direct on-line coupling to LC-MS analysis, potential degradation of the target peptides can be minimized. Using a design of experiment (DoE), the effect of digestion time, temperature, detergent type and detergent concentration on the digestion efficacy were examined. Using these optimized conditions we show the application of on-column trypsin digestion for the analysis of apolipoproteins in serum. We evaluated all possible tryptic cleavage products by linear regression of LC-MS/MS signal count vs. concentration curves, providing information on the most structurally accessible target peptides for quantitative analysis. Using serum calibrators with known protein concentration and isotopically labeled peptides as internal standards, we applied the method to the high throughput analysis of apolipoproteins in high and low density lipoproteins separated from serum and plasma.

Keywords: Liquid Chromatography/Mass Spectroscopy, On-line, Protein, Quantitative

Application Code: Bioanalytical

Methodology Code: Liquid Chromatography/Mass Spectrometry

Session # 1720 Abstract # 1720-4 **Poster Sessions**

Session Title Applications of LC/MS

Abstract Title **A Simple and Sensitive LC-MS/MS Method for the Determination of Free 8-Hydroxy-2'-deoxyguanosine in Human Urine**

Primary Author Zuwei Wang
JES Tech

Date: Wednesday, March 09, 2016 - Morn
Time:
Room: Exposition Floor, 400 Aisle

Co-Author(s) Scott M. Smith

Abstract Text

Urinary free 8-hydroxy-2'-deoxyguanosine (8OHdG), an oxidized product of DNA, is frequently chosen as a biomarker of oxidative stress in humans, including studies of oxidative DNA damage during spaceflight. It is challenging to accurately and efficiently quantify urinary free 8OHdG in large scale human studies. LC-MS/MS is emerging as a preferable analytical technique owing its high sensitivity, selectivity and efficiency, compared to some traditional methods such as ELISA and HPLC.

A simple and sensitive LC-MS/MS method has been developed for the determination of free 8OHdG in human urine. Sample preparation was done by solid phase extraction with a Waters' Oasis HLB 96 well plate. A Waters' Alliance 2795 HT Separation Module combined with a Quattro Micro tandem mass spectrometer was used as the LC-MS/MS system. The runtime of 1 injection can be less than 5 minutes using a reversed phase C18 column and an isocratic flow of methanol/water. ESI positive ions were quantified in the multiple reaction modes (MRM) using m/z 284 [greater than] 168 for 8OHdG and m/z 289 [greater than] 173 for stable isotope labeled internal standard [15N5] 8OHdG.

With this method for 8OHdG analyses, a lower limit of quantitation of 2 nM (0.56 ng/ml) has been achieved using 100 [micro]l urine sample. The analytical range is between 1.0 and 100 nM with a correlation coefficient [greater than] 0.99. Good reproducibility can be obtained with intra-assay and inter-assay CVs less than 10% for 8OHdG spiked urine QC samples. This method can be used in high-throughput routine analysis of free 8OHdG in human urine.

Keywords: Biological Samples, Liquid Chromatography/Mass Spectroscopy, Solid Phase Extraction
Application Code: Bioanalytical
Methodology Code: Liquid Chromatography/Mass Spectrometry

Session # 1720 Abstract # 1720-5 **Poster Sessions**

Session Title: Applications of LC/MS

Abstract Title: **Evaluation of Streamlined SPE Processing Using Novel Column based Components prior to LC-MS/MS**

Date: Wednesday, March 09, 2016 - Morn

Time:

Room: Exposition Floor, 400 Aisle

Primary Author: Lee Williams
Biotage GB Limited

Co-Author(s): Helen Lodder, Victor Vandell

Abstract Text

Introduction

Solid phase extraction is regarded as the gold standard of sample preparation approaches. However, SPE processing can be time consuming. In most cases you are required to condition the phase with an organic solvent, followed by equilibration with an aqueous solvent prior to sample loading. Many modern polymer-based SPE sorbents are water-wettable. EVOLUTE EXPRESS combines sorbent wettability with optimized SPE components allowing better flow consistency and in many cases eliminating the need for conditioning.

Methodology

SPE was performed on spiked human urine using various chemistries and respective generic extraction protocols using 50-500 mg sorbent bed weights: polymer-based hydrophobic SPE; strong and weak mixed-mode cation and anion exchange SPE (ABN, CX, WCX, AX and WAX). Multiple suites incorporating a number of acidic, basic and neutral analytes were investigated. The resultant extracts were evaporated to dryness and reconstituted in mobile phases for subsequent LC-MS/MS analysis. Positive and negative ions were acquired using electrospray in the MRM mode.

Results

The use of novel column components led to substantial improvements in flow rates when using viscous matrices over conventional SPE technologies. Overall flow rates differed between column chemistries in line with known water wettability of the sorbents: mixed-mode anion exchange SPE columns exhibit slower flows than the equivalent cation exchange sorbents. It was possible to eliminate column conditioning and equilibration for all column chemistries whist observing good flow characteristics. Overall all chemistries demonstrated equivalent recoveries and reproducibility when using both processing protocols. The results demonstrated that it is possible to perform SPE without conditioning and equilibration steps when using optimized SPE column components and water wettable sorbent chemistries. As a result substantial time and solvent savings were observed.

Keywords: Clinical/Toxicology, Forensic Chemistry, Liquid Chromatography/Mass Spectroscopy, Solid Phase Extr
Application Code: Clinical/Toxicology
Methodology Code: Liquid Chromatography/Mass Spectrometry

Session # **1720** Abstract # **1720-6** **Poster Sessions**

Session Title: Applications of LC/MS

Abstract Title: **Multi-Class Screening of Drug Abuse in Hair by Matrix Solid Phase Dispersion – Ultrasound Extraction and HPLC-MS/MS Detection**

Date: Wednesday, March 09, 2016 - Morn

Time:

Room: Exposition Floor, 400 Aisle

Primary Author: Antonio Moreda-Piñeiro
University of Santiago de Compostela

Co-Author(s): Elena Pena-Vazquez, Juan Sanchez-Gonzalez, Maria del Carmen Barciela-Alonso, Mercedes Saavedra-Suarez, Pilar Bermejo-Barrera, Raquel Dominguez-Gonzalez

Abstract Text

Recent data from the United Nations Office on Drugs and Crime (UNODC) state cocaine, as well as other recreational drugs, as one of the most widely used illicit substances worldwide (United Nations Office on Drugs and Crime, 2013). Rapid and low-cost methodologies for assessing cocaine abuse are therefore needed as screening and confirmative methods. The objective of this work is the development of a multi-class method for detecting/determining drugs of abuse in hair (cocaine, amphetamines, opiates, designer drugs, and metabolites). Hair as an alternative clinical sample offers a number of advantages over the analysis of urine and blood. Fist, hair sampling is a less invasive procedure and it does not violate the privacy of the person. In addition, sample contamination and/or adulteration is more difficult. Furthermore, hair analysis allows the detection of drugs up to 1 or 2 years.

The developed sample pretreatment is based on the abrasion of the sample (50 mg hair) with 180 mg of alumina (dispersant) in the presence of 20 mL of dithiothreitol (DTT) for 5 minutes. The dispersed sample is then transferred into an Eppendorf tube, and extraction is performed with 1.4 mL of 2mM NH4Ac in MeOH mixture under ultrasound irradiation (37 kHz) for 40 min. The extracted compounds (cocaine, benzoylecgonine, ecgonine methyl ester, cocaethylene, morphine, 6-monoacetylmorphine, codeine, amphetamine, methamphetamine, buprenorphine, norephedrine, etc.) are determined by high resolution liquid chromatography - tandem mass spectrometry (HPLC-MS/MS). The described methodology has been validated according to the guidelines proposed by the Food and Drug Administration (FDA), and has been successfully applied to the multi-class screening of drugs of abuse in hair.

Keywords: Drugs, HPLC, Mass Spectrometry, Sample Preparation
Application Code: Clinical/Toxicology
Methodology Code: Liquid Chromatography/Mass Spectrometry

Session # **1720** **Abstract #** **1720-7** **Poster Sessions**

Session Title: Applications of LC/MS

Abstract Title: **Optimization of QuEChERS Sample Preparation Method for the Determination of Bisphenol A in Carrots**

Primary Author: Olujide T. Akinbo
Butler University

Co-Author(s): Hugh Kestufskie

Date: Wednesday, March 09, 2016 - Morn
Time:
Room: Exposition Floor, 400 Aisle

Abstract Text

Bisphenols are used in the polymer industry as additives or intermediates to impart properties such as rigidity, softness, and resistance to tear, heat, and flame. Some of these polymers are used for coating the inner surface of cans used for food storage and can leach into the food. Given its potential adverse health impacts, the goal of this study is to optimize, validate, and apply QuEChERS sample preparation method with LC-ESI-MS/MS detection to determine human exposure to BPA through the consumption of canned foods. The MS was validated in FIA and chromatographic modes. In the FIA mode the system was successfully calibrated between 2-10 ng BPA/ mL range with a correlation coefficient of 0.9960 for m/z 227□132.5 and 0.9966 for m/z 227□212. LOD and LOQ are lower for the 212 transition compared to the 132.5. Correspondingly sensitivity is higher for the 212 transition. In the chromatographic mode sensitivity was higher for the 132.5 transition but the LOD and LOQ were still lower for the 212 transition. This is explained by the better signal to noise ratio observed with the 212 transition. The m/z-212 was subsequently used for quantitation. Significant signal suppression was observed with the QuEChERS extract both with and without food due to the presence of metal ions (Na+ and Mg2+). Utility of cation exchange (PK228) and anion exchange (HPA25) resins for desalting the extract prior to the LC-MS analysis were investigated for signal improvement. Preliminary results indicate that the cation exchange resin facilitated some signal improvement.

Keywords: Analysis, Chromatography, Environmental, Method Development

Application Code: Food Contaminants

Methodology Code: Liquid Chromatography/Mass Spectrometry

Session # 1720 Abstract # 1720-8 **Poster Sessions**

Session Title: Applications of LC/MS

Abstract Title: **Quantification of Iodine-Containing Hormones Present in Dietary Supplements by Liquid Chromatography Tandem Mass Spectrometry (LC-MS/MS)**

Primary Author: Enrique G. Yanes
U.S. Food and Drug Administration (FDA)

Date: Wednesday, March 09, 2016 - Morn
Time:
Room: Exposition Floor, 400 Aisle

Co-Author(s): James A. Turner, Robert A. Wilson, Ryan Saadawi

Abstract Text

Thyroid hormones are produced by the thyroid gland and are essential in a number of important metabolic functions. They play a key role in regulating cellular activity, growth and brain development. Diseases related to thyroid hormones include hyperthyroidism and hypothyroidism which have been linked to excessive iodine intake. In the US, a recommended dietary allowance (RDA) of 150 µg/day and a tolerable upper intake level (UL) of 1,100 µg/day for iodine has been established for adults.

In this study, commercially available dietary supplements suspected to contain thyroglobulin have been analyzed for the presence of thyroid hormones using liquid chromatography tandem mass spectrometry following a treatment with the enzyme pronase to digest/liberate the bound iodine-containing compounds. The analysis was performed on supplement products in the form of tablets and capsules. For the enzymatic hydrolysis, 50 mg of the ground sample was treated with a 0.5 mL of a proteolytic enzyme solution consisting of 8 mg/mL pronase in 50 mM ammonium bicarbonate buffer, and incubated for 26 hours at ~37 °C. Results/figures of merit will be presented for the quantification of the individual iodine-containing hormones, specifically 3,5,3'-triiodothyronine (T3, commonly known as liothyronine) and 3,5,3',5'-tetraiodothyronine (T4; commonly known as thyroxine). In addition, the total iodine concentration determined by ICP-MS following an alkaline extraction will be discussed.

Keywords: Food Identification, Food Safety, Liquid Chromatography/Mass Spectroscopy
Application Code: Food Safety
Methodology Code: Liquid Chromatography/Mass Spectrometry

Session # 1720 **Abstract # 1720-10** **Poster Sessions**

Session Title: Applications of LC/MS

Abstract Title: **A Targeted Multidimensional Approach with MS Detection (SPE-RPLC/MS) for the Assessment of Trace Free Drug Species in Unadulterated Antibody-Drug Conjugate (ADC) Samples with Improved Specificity and Sensitivity**

Date: Wednesday, March 09, 2016 - Morn

Time:

Room: Exposition Floor, 400 Aisle

Primary Author: Robert Birdsall
Waters Corporation

Co-Author(s): Alain Beck, Scott Berger, Sean M. McCarthy, Weibin Chen

Abstract Text

Incomplete conjugation processes associated with the production of ADCs can result in free drug species increasing the risk to patients and reducing the efficacy of the ADC. Despite stringent purification steps, trace levels of free drug species may be present in formulated ADCs, reducing the therapeutic window. The reduction of sample preparation steps through the incorporation of multidimensional techniques has afforded analysts more efficient methods in the assessment of trace drug species. However SEC-RPLC/UV methods have limited sensitivity and do not offer analysts a high degree of method control when using SEC, a potential dilemma considering the diversity of biological substrates and drug candidates currently under investigation. The current study addresses these challenges through the development of an SPE-RPLC/MS approach that is specific, sensitive, and enables method control in both dimensions. The proposed method was evaluated using a clinically relevant valine-citrulline surrogate molecule based on brentuximab vedotin. The method was determined to be fit-for-purpose with accuracy and precision both determined to be within 5% of regulatory recommendations. Assays sensitivity was two orders more sensitive using MS detection in comparison to UV based detection with an LOQ of 0.30 ng/mL. Free-drug species were determined to be present in an unadulterated ADC surrogate sample at concentrations below 7.0 ng/mL, levels not detectable by UV alone. The proposed 2DLC method provides an assay with improved specificity and sensitivity in the assessment of trace free drug species with improved control over each dimension enabling straightforward integration into existing or novel workflows.

Keywords: Automation, Biopharmaceutical, Characterization, Liquid Chromatography/Mass Spectroscopy
Application Code: Pharmaceutical
Methodology Code: Liquid Chromatography/Mass Spectrometry

Session # 1720 Abstract # 1720-14 **Poster Sessions**

Session Title: Applications of LC/MS

Abstract Title: **Quantitation of Proto-Peptide Building Blocks in Complex Model Prebiotic Mixtures via Liquid Chromatography-Tandem Mass Spectrometry**

Date: Wednesday, March 09, 2016 - Morn

Time:

Room: Exposition Floor, 400 Aisle

Primary Author: Eric T. Parker
Georgia Institute of Technology

Co-Author(s): Facundo M. Fernandez, Jeffrey L. Bada

Abstract Text

Prebiotic simulation experiments produce complex mixtures that have been analyzed for amino [Miller Science 1953]- and hydroxy [Miller JACS 1955]-acid proto-biopolymer building blocks. Detection of polymerization products (i.e. peptides [Parker ACIE 2014]) continues to be of great interest to the origins of life field. Peptide synthesis is limited by formation of diketopiperazines (DKPs) [Long TFS 1971; Steinberg Science 1981; Qian GCA 1993], stable cyclic dipeptides.

Polyesters have been synthesized by environmental cycling of [alpha]-hydroxy acids [Mamajanov Macromolec. 2014] (AHAs) and [alpha]-amino acids. This forms mixed amide/ester linkages, yielding core-enriched depsipeptides with up to n=10 units [Forsythe ACIE 2015]. To further understand this chemistry, it is necessary to establish identity and quantity of AHAs in model prebiotic mixtures. Previous analytical approaches were time-intensive [Cronin Science 1997], used expensive derivatization agents [Fransson JCA 1998], or exhibited limited detection capabilities [Ehling JAFC 2014].

A method for direct analysis of AHAs was developed by ultrahigh performance liquid chromatography with triple quadrupole mass spectrometry (UHPLC/QqQ-MS). A C18 column was used for chromatography and multiple reaction monitoring mode (MRM) for MS detection.

Targeted AHAs were glycolic, lactic, malic, 2-hydroxybutyric, 2-hydroxyisobutyric, 2-hydroxy-2-methylbutyric, 2-hydroxyisovaleric, 2-hydroxyglutaric, 2-hydroxycaproic, and 2-hydroxyisocaproic acids. Chromatographic separation was achieved using hexylamine as an ion-pairing agent. MRM transitions were identified for all 10 analytes, 7 of which had a confirmatory transition in addition to one for quantitation.

Combining ion-pairing chromatography and MRM-MS yielded rapid, direct analysis of many AHAs within a single run. Experiments demonstrated lactic, 2-hydroxybutyric, and malic acids were detected in complex mixtures from Miller-type spark discharge experiments.

Keywords: Liquid Chromatography, Mass Spectrometry, Peptides, Tandem Mass Spec
Application Code: Environmental
Methodology Code: Liquid Chromatography/Mass Spectrometry

Session # 1730 Abstract # 1730-1 Poster Sessions

Session Title Capillary Electrophoresis

Abstract Title **Capillary Electrophoresis Method to Detect Circulating Steroids in Individual Zebrafish Plasma**

Date: Wednesday, March 09, 2016 - Morn

Time:

Room: Exposition Floor, 400 Aisle

Primary Author Paige A. Reed
West Virginia University

Co-Author(s) Jennifer Ripley-Stueckle, Lisa A. Holland, Vincent T. Nyakubaya, William J. Feeney

Abstract Text

By combining capillary electrophoresis,UV-visible absorbance detection ,and sample stacking, detection limits ranging from 0.2 to 2 ng/mL (0.8 to 6 nM) can be achieved for circulating steroid hormones. Stacking is accomplished using a negatively charged cyclodextrin, as a steroid-carrier, at a pH interface between the basic reconstituted plasma sample and the acidic separation phosphate buffer. Steroids are then separated in under 5 minutes using a capillary electrophoresis method incorporating a secondary equilibria with sodium dodecyl sulfate and cyclodextrin. Using this recently reported method, seven steroidal compounds are quantified from five microliters or less of zebrafish blood. Looking at the changes in circulating steroid levels, provides more information about potential mechanisms of endocrine

Keywords: Biological Samples, Capillary Electrophoresis, Environmental, Toxicology
Application Code: Bioanalytical
Methodology Code: Capillary Electrophoresis

Session # 1730 Abstract # 1730-2 **Poster Sessions**

Session Title: Capillary Electrophoresis

Abstract Title: **Fluorogenic Derivatization of Amino Acids for Laser-Induced Fluorescence Detection in Capillary Electrophoresis**

Primary Author: Naveen Maddukuri
Wichita State University

Co-Author(s): Maojun Gong, Qiyang Zhang

Date: Wednesday, March 09, 2016 - Morn
Time:
Room: Exposition Floor, 400 Aisle

Abstract Text

Capillary Electrophoresis (CE) is a powerful separation technique in terms of separation efficiency, analysis speed, and ease of use. However, the detection sensitivity is incomparable to that of HPLC when optical detection methods are used because of the small light path in the dimension of the inner diameter of the separation capillary. Laser-induced fluorescence (LIF) detection is one of the most sensitive methods commonly used in CE detection. However, most of the targeted analytes have no inherent fluorophores thus fluorogenic derivatization is often required before being detected by LIF. Orthophthalaldehyde (OPA) and 2-mercaptoethanol (2-ME) could be fluorogenic reagents for the detection of primary amines. However, the optimal excitation of the derivatives requires the ultra-violet range (340-360 nM), and the quantum yield is low. Here we report a method for derivatizing amino acids based on the reaction of OPA, fluorescein-thiol, and primary amines. The reaction could be completed in less than 30 s, and the fluorescein-thiol was highly fluorescent with the excitation at 491 nm and emission at 520 nm. Our experimental results show that 10 amino acids including Glutamate, Aspartate, Glycine, Glutamine and Serine etc. could be separated and detected in less than 60 s, and the detection limit was typically at 200 pM. This method was applied to the detection of amino acid neurotransmitters in complex biological samples such as cerebrospinal fluid (CSF), and it will be used for in vivo neurotransmitter monitoring.

Keywords: Amino Acids, Capillary Electrophoresis, Derivatization, Fluorescence
Application Code: Bioanalytical
Methodology Code: Capillary Electrophoresis

Session # 1730 **Abstract # 1730-3** **Poster Sessions**

Session Title: Capillary Electrophoresis

Abstract Title: **Investigating Electrospray Behavior in Capillary Electrophoresis Coupled Mass Spectrometry**

Primary Author: Jared A. Lamp
University of Notre Dame

Date: Wednesday, March 09, 2016 - Morn
Time:
Room: Exposition Floor, 400 Aisle

Co-Author(s)

Abstract Text

Recent reports demonstrate a large sensitivity increase in CZE-ESI-MS systems that employ etched capillaries, potentially due to ion distribution alterations within the electrospray. Here, ion distribution in electrospray generated from an electrokinetically driven sheath flow CZE-ESI-MS interface was mapped. The sheathing and separation buffers contained either a 25 [micro]g/mL solution of dimethylated bradykinin or a 25 [micro]g/mL solution of isotopically labeled dimethylated bradykinin (+6 Da deuterium), respectively, or vice versa. An electrospray deposition setup was developed by suspending the sheath flow interface above a grounded ITO/DHB-coated microscope slide. Bradykinins were then electrosprayed from a distance such that no solvent accumulated on the slide. After spraying, sections of the analyte spots could be interrogated with a MALDI laser in order to determine spatial distribution of peptides within the spot, effectively highlighting the density of ions entering through the capillary as well as any spatial inhomogeneity in the spray. Mapping was completed using either a standard capillary (150 [micro]m o.d.) or a capillary where the outer diameter had been etched with HF (60 [micro]m o.d.). Our results may suggest a more focused beam of peptides present while using an etched capillary, leading to a greater number of ions entering the mass spectrometer, and hence a more sensitive measurement.

Keywords: Bioanalytical, Capillary Electrophoresis, Electrospray, Mass Spectrometry
Application Code: Bioanalytical
Methodology Code: Capillary Electrophoresis

Session # 1730 Abstract # 1730-4 **Poster Sessions**

Session Title: Capillary Electrophoresis

Abstract Title: **Validated Capillary Zone Electrophoretic Method for Simultaneous Determination of Some Antihypertensive Drugs in Their Single-pill Combination Therapy**

Primary Author: Fawzy A. El-Yazbi
Alexandria University

Date: Wednesday, March 09, 2016 - Morn
Time:
Room: Exposition Floor, 400 Aisle

Co-Author(s): Hytham M. Ahmed, Rasha A. Shaalan, Sohaila M. Elonsy, Tarek S. Belal

Abstract Text

Recently several single-pill combinations (SPC) have been used as a promising choice for the treatment of hypertension. However, such combinations increased the challenge to analysts in Pharmaceutical Quality Control Laboratories. In the present work, a simple, sensitive, selective and fast capillary zone electrophoresis (CZE) method coupled with diode array detection (DAD) was developed for the determination of different combinations of antihypertensive drugs in one pharmaceutical preparation. For this purpose, our method was used for the simultaneous determination of two different antihypertensive combinations containing amlodipine besylate (AML), benazepril (BEN), hydrochlorothiazide (HCT) and valsartan (VAL) in their combined formulations. Electrophoretic conditions were optimized to improve the sensitivity, and rapidity of separation. The separation was performed using a fused silica capillary (effective length 70 cm × 75 µm id), with 10 mM phosphate buffer as background electrolyte adjusted at pH 7.5 and 12.0 s injection time. The applied voltage was 30 kV. AML, BEN, and VAL were detected at 210 nm, while HCT was detected at 225 nm. The four compounds were resolved in less than 10 minutes. Migration times were 4.7, 5.6, 6.9 and 8.9 min for AML, HCT, BEN, and VAL, respectively. The described method was linear for the four drugs with correlation coefficients > 0.9994 and limits of detection less than 2 µg/mL and limits of quantitation less than 7 µg/mL for the four drugs. Peak identity and purity were confirmed by DAD. The validated CZE method was successfully applied for the analysis of two combinations of AML, HCT, BEN, and VAL in their pharmaceutical tablets (Exforge HCT® for AML, HCT, and VAL) and (Loadless® for AML, and BEN). Therefore, the proposed method can be used for the estimation of large number of quality control SPC samples of these drugs in a short time.

Keywords: Capillary Electrophoresis, Quality Control, Quantitative
Application Code: Quality/QA/QC
Methodology Code: Capillary Electrophoresis

Session # 1730 Abstract # 1730-5 **Poster Sessions**

Session Title: Capillary Electrophoresis

Abstract Title: **Single Enzyme Molecule Studies with CE-LIF and CE-MS**

Date: Wednesday, March 09, 2016 - Morn

Time:

Room: Exposition Floor, 400 Aisle

Primary Author: Emily A. Amenson
University of Notre Dame

Co-Author(s): Norman J. Dovichi

Abstract Text

A capillary electrophoresis with laser induced fluorescence (CE-LIF) instrument was constructed with a low limit of detection. A calibration curve for this instrument was created using fluorescein. The individual activity of single enzyme molecules will be measured through CE-LIF. The calf intestinal alkaline phosphatase (ciAP) enzyme will be studied first in these experiments; a single enzyme molecule will be injected into a capillary filled with substrate, and several incubations will be performed at different points along the capillary. Each of these incubations with the AttoPhos substrate produces a pool of highly fluorogenic product known as AttoFluor, which will migrate through a highly sensitive laser-induced fluorescence detection system. The areas of the peaks produced will be used to create a kinetic plot to determine the activity of the enzyme with high precision. These experiments will be adapted for use with mass spectrometry and different enzymes in the future.

Keywords: Capillary Electrophoresis, Electrophoresis, Enzyme Assays, Mass Spectrometry
Application Code: Bioanalytical
Methodology Code: Capillary Electrophoresis

Session # 1730 Abstract # 1730-6 **Poster Sessions**

Session Title: Capillary Electrophoresis

Abstract Title: **Xenopus Laevis Metabolomic Profiling via CZE-ESI-MS/MS and MALDI-TOFMS**

Date: Wednesday, March 09, 2016 - Morn
Time:
Room: Exposition Floor, 400 Aisle

Primary Author: Jennifer Arceo
University of Notre Dame

Co-Author(s): Danielle Boley, Elizabeth H. Peuchen, Nicole Schiavone, Norman J. Dovichi

Abstract Text

Xenopus Laevis is a pivotal animal model used to study early development. The majority of research efforts have been focused on the genome and transcriptome. There is a growing interest in expression changes of small molecules and lipids during early development. In our study we extract lipids and small molecules during four stages of Xenopus embryonic development and analyze them using Capillary electrophoresis coupled to tandem mass spectrometry to produce metabolic profiles. We observed increased expression of metabolite features from the 1 cell stage to stage 22 when early organ differentiation begins. Our data is consistent with key biological events and provides complementary data to genomic and proteomic studies.

Keywords: Capillary Electrophoresis, Imaging, Lipids, Metabolomics
Application Code: Bioanalytical
Methodology Code: Capillary Electrophoresis

Session # 1730 Abstract # 1730-7 **Poster Sessions**

Session Title: Capillary Electrophoresis

Abstract Title: **Insights into Chiral Recognition Mechanisms for Acryloyl Terminated Polymeric Surfactants: Application of Linear Solvation Energy Relationship in Micellar Electrokinetic Chromatography and Capillary Electrochromatography**

Primary Author: Yang Lu
Georgia State University

Co-Author(s): Shahab S. Shamsi

Date: Wednesday, March 09, 2016 - Morn
Time:
Room: Exposition Floor, 400 Aisle

Abstract Text

Linear solvation energy relation (LSER) is used to understand the types and relative strengths of chemical interactions that control retention and selectivity in various separation techniques [1]. In this study, LSER was applied to characterize polymerizable acryloyl terminated amino acid based surfactants in capillary electrochromatograpy (CEC) using monolithic column, and micellar electrokinetic chromatography (MEKC) using molecular micelles. Using monolithic column as a true stationary phase and molecular micelles as pseudostationary phases we varied the linkers (amide and carbamate), head groups (alanine, leucine and valine) and chain lengths (C8, C10 and C12) of the acryloyl terminated amino acid based surfactants, to understand the factors governing retention and enantioselectivity in solution phase versus solid phase Preliminary data reveals differences in system parameters among surfactant head group, linker and chain length affecting the separation selectivity of both achiral and chiral compounds. In general, CEC phase are more retentive than MEKC phases even when comparing the surfactant of same head groups, linker and chain length. This study showed the predictive capability of LSER to understand the aforementioned intermolecular processes controlling retention and enantioselecitivity and by doing so will be able to quantitatively predict the experimental conditions to achieve acceptable chiral or achiral separations.

Reference

[1] M. Vitha and P. W. Carr, “The chemical interpretation and practice of linear solvation energy relationships in chromatography,” Journal of Chromatography A, vol. 1126, no. 1-2, pp. 143–194, 2006.

Keywords: Capillary Electrophoresis, Chiral Separations, Surfactants
Application Code: Clinical/Toxicology
Methodology Code: Capillary Electrophoresis

Session # 1730 Abstract # 1730-8 **Poster Sessions**

Session Title Capillary Electrophoresis

Abstract Title **Study of Electrooxidation Products of Primary Alcohols by EC-CE-C4D: Assessment of the Conversion Efficiency of Alcohols into Their Carboxylates on Gold and Platinum Electrodes in Different Media**

Primary Author Mauro S. Santos
Clemson University

Co-Author(s) Fernando S. Lopes, Ivano G R. Gutz

Date: Wednesday, March 09, 2016 - Morn
Time:
Room: Exposition Floor, 400 Aisle

Abstract Text

Kinetic studies of electrooxidation of alcohols are widely undertaken on different electrode materials and the complete oxidation to carbon dioxide is aimed whenever the purpose is to develop fuel cells, although variable proportions of intermediates like aldehydes, ketones and carboxylic acids may be also formed. Ionic or ionizable electrooxidation products can be measured online with a recently developed EC-CE-C4D flow system[1] and the formation of carboxylates from some alcohols on gold and platinum electrodes in different media was assessed in this work .

Experiments were performed using a three-electrodes electrochemical cell coupled to a lab-made CE equipment with C4D detector. Double gold (and platinum) disk electrodes embedded in Kel-F were used as working and auxiliary electrodes and a silver wire covered with AgCl served as quasi-reference electrode. A silica capillary of 50 µm i.d. and 45 cm length (20 +/- 2 cm to the detector) was used. The injection was performed during 3 s under 25 kPa and the separation voltage was 30 kV. The separation BGE used was 30 mM Tris + 10 mM HCl (pH 8.6). The electrooxidation of alcohols with 2 to 5 C and at a concentration of 1 mM was evaluated in the following electrolytes: 10 mM HNO3 + 1 mM HCl, 10 mM KNO3 + 1 mM KCl and 10 mM NaOH + 1 mM NaCl.

The best electrooxidation time period and potential found were 50 s and 1.5 V (vs. Ag/AgCl) for Au and Pt electrodes in acid and neutral medium. On Au electrode in alkaline solution the C4D peak signal intensity was lower than in acidic and neutral ones, however, they appeared at 0.85 V instead of 1.35 V. On both electrodes the signals obtained in acidic solution were higher than in neutral medium. In acidic solution a noteworthy selectivity effect in terms of carboxylate yield was observed on similarly polished polycrystalline Au and Pt electrodes, as shown in the figure. While the pentanoate formation was identical on Pt and Au, by shortening the alcohols' carbon chain, the conductivity peak areas increase on Au and decrease gradually on Pt up to a ratio of four for acetate.

Acknowledgments

To CNPq (Brazil) for fellowships and a grant.

References

[1] Santos, M. S. F., at. al. Anal. Chem. 2012, 84, 7599–7602.

Keywords: Capillary Electrophoresis, Derivatization, Electrochemistry

Application Code: Fuels, Energy and Petrochemical

Methodology Code: Capillary Electrophoresis

Session # 1730 **Abstract # 1730-9** **Poster Sessions**

Session Title: Capillary Electrophoresis

Abstract Title: **Microscale Quantification of Nanoparticle-Biomolecule Interactions with Capillary Electrophoresis**

Primary Author: Julia A. Mouch
Bethany College

Co-Author(s): Lisa A. Holland, Tyler Davis

Date: Wednesday, March 09, 2016 - Morn
Time:
Room: Exposition Floor, 400 Aisle

Abstract Text

The uses of nanotechnology are becoming more prevalent in everyday commercial products; however any potential health risks or harmful side effects of nanoparticles must be understood. Nanoparticle toxicity has been demonstrated in animals and cells, but fundamental studies are required to elucidate specific interactions between nanoparticles and physiologically relevant biomolecules. Conventional affinity methods require excessive quantities of homogeneous nanoparticles. These barriers are overcome with rapid capillary electrophoresis affinity analyses of nanoparticle-biomolecule interactions. Methods based on capillary electrophoresis separate nanoliter sample volumes of nanoparticles in minutes based on velocity in an electric field. Under pre-equilibrium conditions affinity binding is determined by directly assessing bound and free concentrations of biomolecules. Under nonequilibrium conditions the nanoparticle-equilibrium mixtures can be analyzed in a single capillary electrophoresis separation. This research outlines the suitability of capillary electrophoresis for different nanoparticles.

Keywords: Bioanalytical, Capillary Electrophoresis, Nanotechnology
Application Code: Nanotechnology
Methodology Code: Capillary Electrophoresis

Session # 1730 Abstract # 1730-10 **Poster Sessions**

Session Title: Capillary Electrophoresis

Abstract Title: **Comparative Validation of Sofosbuvir Determination in Pharmaceuticals by Several Chromatographic, Electrophoretic and Spectrophotometric Methods**

Primary Author: Amira F. El-Yazbi
Alexandria University

Date: Wednesday, March 09, 2016 - Morn
Time:
Room: Exposition Floor, 400 Aisle

Co-Author(s)

Abstract Text

The analysis of pharmaceuticals is an integral part of an overall drug development process. Due to the increasing importance of sofosbuvir after it has been approved by the FDA, in 2013, for the treatment of hepatitis C viral infection, rapid and simple methods for routine analysis and quality control of commercial formulations are very desirable. It should also be noted that all the current editions of various Pharmacopoeia still don't have any analytical methods for sofosbuvir quantification. In this work five accurate methods for the determination of sofosbuvir in tablets: reversed phase high pressure liquid chromatography (RP-HPLC), capillary zone electrophoresis (CZE), high performance thin layer chromatography (HPTLC) with densitometric detection, UV spectrophotometric and derivative spectrometry methods, were developed and validated. The HPLC was carried out using C18 Thermo stationary phase and mobile phase consisted of 0.1% formic acid-acetonitrile (60 : 40 v/v) with flow rate 1 mL min□1 and UV detection at 260 nm. CZE was performed using 75 □m × 82 cm fused silica capillary. Detection was carried out at 230 nm with 10 mM phosphate buffer pH 7.50, 30 kV voltage and 25 □C temperature. NP-HPTLC was carried out using HPTLC silica F254 plates, developed with methanol–chloroform (70:30, v/v) through 19 cm distance. Analysis were scanned with densitometer at 260 nm. UV spectrophotometry was carried out using 260 nm for direct assay and 215 and 245 nm for the first derivative assay. The proposed methods proved to be rapid, simple, sensitive, selective and accurate analytical procedures, suitable for reliable determination of sofosbuvir in tablets for routine quality control. It should be noted that all methods gave acceptable results with respect to precision and accuracy. The ANOVA test confirmed that there is no significant differences between the proposed assays, Thus any of these methods can be used for routine analysis of sofosbuvir in commercial tablets.

Keywords: Capillary Electrophoresis, HPLC, Quality Control, Thin Layer Chromatography
Application Code: Pharmaceutical
Methodology Code: Capillary Electrophoresis

Session # 1730 Abstract # 1730-11 **Poster Sessions**

Session Title: Capillary Electrophoresis

Abstract Title: **Quantification of Amino Acids in Cordyceps by MEKC**

Date: Wednesday, March 09, 2016 - Morn
Time:
Room: Exposition Floor, 400 Aisle

Primary Author: Xin Wei
Wuhan University Zhongnan Hospital

Co-Author(s): Hankun Hu, Yiming Liu, Yue Xu

Abstract Text

Quality assessment of Cordyceps sinensis, a precious and pricey natural product that offers a variety of health benefits is highly significant. In this work, the profiles of amino acids and their glucosides in C. sinensis and in cultured Cordyceps militaris are comparatively investigated by micellar electrokinetic capillary chromatography (MEKC). Samples are cryogenic milled into powders and extracted with methanol /water (2+1) containing 0.1M HCl. Amino acids extracted out are derivatized with NBD-F.

The derivatized sample is then separated by MEKC with a 25 mM borate buffer containing 20 mM SDS and 10% methanol. It is demonstrated that the proposed MEKC separation is very effective for NBD-amino acids present tin these samples. The separation is completed within 15 min. Amino acids, including lysine, glutamic acid, proline, and phenylalanine are detected both in C. sinensis and in cultured C. militaris. Interestingly, however, the levels of the glucosides of these amino acids in these samples are significantly different, indicating a variation in enzymatic activity.

Acknowledgement: The research is partially supported by US National Institutes of Health (GM089557 to YML).

Keywords: Amino Acids, Method Development, Pharmaceutical, Sample Preparation

Application Code: Quality/QA/QC

Methodology Code: Capillary Electrophoresis

Session # 1730 Abstract # 1730-12 **Poster Sessions**

Session Title: Capillary Electrophoresis

Abstract Title: **Isolation of Intact Cell Populations by Preparative Capillary Zone Electrophoresis**

Primary Author: Sarah N. Lum
University of Notre Dame

Date: Wednesday, March 09, 2016 - Morn
Time:
Room: Exposition Floor, 400 Aisle

Co-Author(s): Bonnie Huge, Matthew M. Champion, Norman J. Dovichi

Abstract Text

An estimated one percent of all microbial life in the environment can be cultivated using standard laboratory conditions. The remaining ninety-nine percent require complex atmospheres that are not as easily mimicked in the laboratory but are essential to promote bacterial survival and reproduction. Furthermore, effective separation and characterization of individual microbial components of these complex atmospheres would be of great value in both environmental and medical fields; for example, isolated bacterial species can be sequenced and screened for potential use as alternative energy sources or pharmaceuticals. In this work, we address the challenge of isolating a single strain of bacteria from a complex environment while maintaining cellular vitality. We use capillary zone electrophoresis (CZE) for the separation of bacteria within a mixture. We isolate each separated component with an automated fraction collector developed in house. Viability, and to what degree, will be accomplished in culture. Finally, species identification will be confirmed by deep sequencing analysis.

Keywords: Bioanalytical, Capillary Electrophoresis
Application Code: Bioanalytical
Methodology Code: Capillary Electrophoresis

Session # 1740 Abstract # 1740-1 **Poster Sessions**

Session Title: Forensics and Homeland Security

Abstract Title: **Fingerprinting of Falsified Artemisinin Combination Therapies (ACTs) via DART Ionization Coupled to a Compact Single Quadrupole Mass Spectrometer**

Date: Wednesday, March 09, 2016 - Morn
Time:
Room: Exposition Floor, 400 Aisle

Primary Author: Matthew C. Bernier
Georgia Institute of Technology

Co-Author(s): Brian Musselman, Facundo M. Fernandez, Joseph LaPointe

Abstract Text

Artemisin-based combination therapies (ACTs), such as those containing artemether-lumefantrine, have proved to be a successful treatment of plasmodium falciparum malaria. The effective treatment of vector-borne infections, however, has been hindered by an influx of poor quality (falsified, substandard, degraded) drugs into many parts of Africa. Initial detection of artemether-lumefantrine (AL) counterfeits involve the use of Raman spectroscopy, which can confirm whether or not AL is present but provides little information on all components. For better understanding of the complete make-up of the counterfeits, mass spectrometry has emerged as an important tool. The information MS provides on counterfeit composition can allow for tracing and immediate understanding of their health effects, and hence, is vitally important. Currently, much of the work in fingerprinting via MS analysis must be done outside of the regions directly impacted by counterfeits as these regions have few facilities with such technology. However, with the increased use of more cost efficient low-resolution benchtop instruments, the option of providing full information on a fake tablet upon immediate discovery is becoming possible. The work here involves the direct comparison of just such an MS, a Waters QDa, compared to a higher resolution Bruker QTOF instrument for determining whether deployment of these low-resource bench-tops could compete with laboratory standard instruments and be advantageous on the front-lines of counterfeit enforcement in affected areas. Preliminary results have shown that direct analysis in real time (DART) ionization with the QDa MS offers suitable fingerprinting of counterfeits comparable to higher resolution MS that could provide a significant advantage in quickly characterizing seized counterfeit ACTs.

Keywords: Drugs, Mass Spectrometry, Plasma Emission (ICP/MIP/DCP/etc.), Portable Instruments
Application Code: Homeland Security/Forensics
Methodology Code: Mass Spectrometry

Session # 1740 Abstract # 1740-2 **Poster Sessions**

Session Title Forensics and Homeland Security

Abstract Title **Lead-Free Gunshot Residues as Forensic Evidence**

Primary Author Christopher R. Dockery
Kennesaw State University

Date: Wednesday, March 09, 2016 - Morn
Time:
Room: Exposition Floor, 400 Aisle

Co-Author(s) Deidre VanDenbos, Ethan Miller, Lashaundra Fambro, Wassim Abdul Khalek

Abstract Text

Gunshot residue (GSR) analyses have been forensically relevant for many years and are well characterized in the literature. However, the manufacture and distribution of lead-free alternatives to small caliber firearms ammunition is increasing rapidly as they are environmentally friendly and safer for use in indoor ranges. With this emerging market, the forensic community must develop and validate methods aimed at the detection of lead-free GSR on the hands of suspected shooters. This study investigates the use of laser-induced breakdown spectroscopy (LIBS) and scanning electron microscopy energy-dispersive x-ray spectroscopy (SEM-EDX) as means of characterizing gunshot residues originating from the discharge of lead-free blank training rounds. The scope of this project is to provide investigators with a chemical fingerprint for lead-free GSR by LIBS, to fully characterize the rates of error for shooters and non-shooters, and to determine the amount of time that forensically relevant quantities of lead-free GSR can be recovered from the hands of shooter. Additionally, this study will serve as a comparison for the analysis of lead-free GSR via LIBS as a rapid and relatively non-destructive screening method followed by confirmation with SEM-EDX on the preserved evidence.

Keywords: Atomic Emission Spectroscopy, Forensic Chemistry, Microscopy, Plasma Emission (ICP/MIP/DCP/etc.
Application Code: Homeland Security/Forensics
Methodology Code: Atomic Spectroscopy/Elemental Analysis

Session # 1740 Abstract # 1740-4 **Poster Sessions**

Session Title: Forensics and Homeland Security

Abstract Title: **Forensic Analysis Of Textile Fibers Exposed To Laundry Detergents Using Fluorescence Excitation-Emission Spectroscopy**

Date: Wednesday, March 09, 2016 - Morn

Time:

Room: Exposition Floor, 400 Aisle

Primary Author: Nirvani Mujumdar
University of Central Florida

Co-Author(s)

Abstract Text

Textile fibers are a major form of trace evidence, and the ability to reliably classify them is useful for forensic scientists. Before becoming an important source of evidence, fibers discovered at a crime scene are likely to be exposed to multiple launderings, which can possibly add characteristic fluorescence components on to them. This can be a useful tool for the identification and analysis of forensic fiber evidence, since detergents usually contain fluorescence whitening agents, which can modify the original spectra of these fibers if accumulated. The main goal was to distinguish between fibers exposed to multiple launderings versus those never exposed. A non-destructive technique called fluorescence excitation-emission spectroscopy was used to examine alteration in fluorescence spectral fingerprint of single fibers due to exposure to detergents containing fluorescent whitening agents. 7 top-selling detergents were used to wash fibers composed of diverse classes of dyes, colors and compositions. Undyed acrylic, cotton and nylon fabrics dyed respectively with basic green 4, direct blue 1 and acid yellow 17 dyes were washed up to 6 times, and the spectral contribution of each detergent on single fibers was quantified and shown to reach a maximum after 5 serial washes. However in certain cases, some detergents even showed statistically meaningful differences to the fiber spectra after only a single wash. Fluorescence emission spectral profiles were collected and principle component cluster analysis was employed for additional statistical comparisons and differentiations, which facilitated to determine that the spectra of washed fibers were distinct from those of dyed, unwashed fibers. We attempt to explain the effects of repetitive laundering on total fluorescence of fibers, and to determine the number of washes needed to make distinctions between washed and unwashed fibers. Results indicate that washed and unwashed fibers were distinguishable with 95% accuracy.

Keywords: Fluorescence, Forensic Chemistry, Trace Analysis, UV-VIS Absorbance/Luminescence
Application Code: Homeland Security/Forensics
Methodology Code: UV/VIS

Session # 1740 Abstract # 1740-5 **Poster Sessions**

Session Title Forensics and Homeland Security

Abstract Title **Eye-safe, Wide-Area Hyperspectral Raman Imaging Using a Spatial Heterodyne Raman Spectrometer**

Primary Author Nathaniel R. Gomer
ChemImage Sensor Systems

Date: Wednesday, March 09, 2016 - Morn
Time:
Room: Exposition Floor, 400 Aisle

Co-Author(s) Matthew P. Nelson

Abstract Text

Raman spectroscopy is a valuable tool for the investigation and analysis of threat materials because it provides a unique molecular fingerprint that allows for unambiguous target identification. Raman can highly beneficial for this application, since it is capable of providing in situ surface analysis, can work at standoff distances, and is capable of being conducted remotely. Current generation Raman systems typically offer very low throughput, are physically large and heavy, and can only probe an area the size of a tightly focused laser beam, eliminating their ability to be eye-safe and severely hindering the ability of the system to investigate larger areas efficiently.

The majority of these limitations are directly related to a system's spectrometer, typically dispersive grating based. Typical slit-based UV Raman systems require very narrow slit widths and very long focal length optics to accurately disperse and resolve all light that enters the spectrometer, which tends to make them large and heavy. These limitations with UV Raman make implementation for applications with SWaP constraints difficult. In order for UV Raman to become a viable tool for threat detection, an innovative Raman system is needed that addresses these shortcomings.

CISS has begun development on an eye-safe, wide-area hyperspectral UV Raman sensor that overcomes the limitations of current generation UV Raman systems. The proposed innovation couples a spatial heterodyne spectrometer (SHS), a novel slit-less spectrometer that operates similar to Michelson interferometer, with a fiber array spectral translator (FAST) fiber array, a two-dimensional imaging fiber for hyperspectral imagery, to create a novel wide-area hyperspectral Raman sensor capable of yielding very high spectral resolution in a small form factor while using defocused laser excitation. This paper will provide an overview of spatial heterodyne spectroscopy and FAST hyperspectral imaging, and discuss preliminary results.

Keywords: Instrumentation, Raman, Spectrometer, Vibrational Spectroscopy
Application Code: Homeland Security/Forensics
Methodology Code: Vibrational Spectroscopy

Session # 1740 Abstract # 1740-6 **Poster Sessions**

Session Title Forensics and Homeland Security

Abstract Title **Development and Optimization of Solid Phase Extraction (SPE) Method for Determination of Benzodiazepines in Wastewater and Surface Water by High-Performance Liquid Chromatography (HPLC)**

Primary Author Hongxia Guan
Georgia Gwinnett College

Co-Author(s) Qingsong Cai

Date: Wednesday, March 09, 2016 - Morn
Time:
Room: Exposition Floor, 400 Aisle

Abstract Text

Benzodiazepines are among the most common emerging drugs of abuse, and the presence of benzodiazepines in surface water and waste water has been reported. A reliable analytical method needs to be developed for the analysis of benzodiazepines. In the current research, different types of SPE sorbents and solvent system were investigated for selective extraction and preconcentration of seven common benzodiazepines from water samples surface water and drinking water. The benzodiazepines studied were alpha-hydroxy alprazolam, alprazolam, clonazepam, diazepam, lorazepam, Nordiazepam, and Temazepam. Combination of three different SPE sorbents (C18, SDVB, and WAX) and three solvents (methanol, acetonitrile, and ethyl acetate) were compared to obtain the best sorbent and solvent for all benzodiazepines studied. Extraction efficiency was determined by HPLC/UV. The SDVB and WAX adsorbent were found to provide good accuracy (recoveries >85%) and precision (%RSDs < 10%) for analysis of these drugs at ppb level.

Keywords: Forensic Chemistry, Liquid Chromatography, Solid Phase Extraction

Application Code: Homeland Security/Forensics

Methodology Code: Liquid Chromatography

Session # 1740 Abstract # 1740-7 **Poster Sessions**

Session Title: Forensics and Homeland Security

Abstract Title: **Determination and Measurement of Wildland Fire Markers in Residential Structures Using TD-GCMS**

Date: Wednesday, March 09, 2016 - Morn
Time:
Room: Exposition Floor, 400 Aisle

Primary Author: Mary C. Martin
Prism Analytical and Central Michigan University

Co-Author(s): Alice Delia, Dale LeCaptain

Abstract Text

Each year thousands of wildland fires blaze across the United States causing secondary ("smoke") damage to numerous businesses and personal property. Currently there are no specific industry standards or guidelines for determining wildfire combustion residues. Remediation decisions often rely on anecdotal evidence from occupants. A variety of particulate methods are used to assess surface contamination but there are few methods for evaluating organic chemical residues that encompass the wide range of chemical classes produced during wildland fires. The most common laboratory analysis method for chemical wildland fire damage is solvent extraction for semi-volatile organic compounds such as polycyclic aromatic hydrocarbons. However it is labor intensive, uses harsh chemicals, and does not account for other classes of chemicals or particulate matter from wildland fires. A new method is being developed employing a thermal desorption gas chromatography mass spectrometry (TD-GCMS) method. TD-GCMS using novel sorbent beds decreases the sample preparation substantially and enables sampling of bulk materials by off-gassing. Furthermore, the method developed is specific to wildland fire events.

Keywords: Environmental/Air, Gas Chromatography/Mass Spectrometry, Pyrolysis, Thermal Desorption
Application Code: Environmental
Methodology Code: Gas Chromatography/Mass Spectrometry

Session # 1740 Abstract # 1740-8 **Poster Sessions**

Session Title: Forensics and Homeland Security

Abstract Title: **Differentiate Delta-9-tetrahydrocannabinol (□9-THC) and Delta-8-tetrahydrocannabinol (□8-THC)**

Primary Author: Ken Tseng
Nacalai USA

Date: Wednesday, March 09, 2016 - Morn
Time:
Room: Exposition Floor, 400 Aisle

Co-Author(s): Kazuhiro Kimata, Toshi Ono, Tsunehisa Hirose

Abstract Text

Of the roughly 60 cannabinoids, delta-9-tetrahydrocannabinol (□9-THC) is the primary psychoactive molecule found in cannabis plants. □9-THC and its metabolites have been widely studied.

Delta-8-tetrahydrocannabinol (□8-THC) is an isobaric isomer of □9-THC that differs by the position of a double bond. It has lower psychoactive potency, more chemically stable, and potentially better medicinal properties than □9-THC.

Cannabinol (CBN) is used to monitor the freshness of the sample since □9-THC easily oxidizes to CBN. Cannabidiol (CBD) has no psychoactive activity but it has many potent medicinal properties.

These four cannabinoids, CBD, CBN, □9-THC, and □8-THC were analyzed by core-shell HPLC columns. The C18 core-shell column produced co-eluting peak of □9-THC and □8-THC. The peak shapes in this column were fronting under the MS-friendly experimental condition.

Cosmocore Cholester is a core-shell HPLC column that has similar hydrophobicity as C18. The rigid cholesterol functional group provides it with higher steric selectivity to resolve □9-THC and □8-THC peaks. The peak shapes were symmetrical using MS-compatible solvents as the mobile phase.

Other cannabinoids and metabolites are also successfully separated and identified to the baseline level.

Keywords: Agricultural, HPLC, HPLC Columns, Separation Sciences
Application Code: Agriculture
Methodology Code: Liquid Chromatography

Session # 1740 Abstract # 1740-9 **Poster Sessions**

Session Title: Forensics and Homeland Security

Abstract Title: **Identification and Quantification of Explosives and Their Residues in Water Using a Novel Surfactant in MEKC**

Date: Wednesday, March 09, 2016 - Morn
Time:
Room: Exposition Floor, 400 Aisle

Primary Author: Christine Copper
U.S. Naval Academy

Co-Author(s): Alexis Clark, Ashton Genzman, Christopher Rollman, Ira Lurie, Jacqueline Rine, Karen Brensinger, Marlene Perez, Mehdi Moini

Abstract Text

Micellar Electrokinetic Capillary Chromatography (MEKC) has been used to separate numerous classes of compounds, including explosives and their residues. The most popular micellar system employed in these studies is sodium dodecyl sulfate (SDS) and detection is almost always performed using an absorbance system. While MEKC using SDS micelles results in adequate resolution of explosives and their residues, these micelles are not compatible with mass spectrometric detection due to their lack of volatility. The goal of this work was to prove the usefulness of MEKC with a volatile micellar system, perfluorooctanoic acid (PFOA), for the separation of the explosives and residues in EPA Mixes 8330A and B along with other related compounds. Further, the method reported herein is shown to have utility in identifying and quantifying several of these compounds in water, sand and soil samples.

Keywords: Capillary Electrophoresis, Environmental/Water, Water
Application Code: Environmental
Methodology Code: Capillary Electrophoresis

Session # 1740 Abstract # 1740-12 **Poster Sessions**

Session Title: Forensics and Homeland Security

Abstract Title: **Screening of Drugs of Abuse Using DART-MS and Real Time Reverse Library Search**

Date: Wednesday, March 09, 2016 - Morn

Time:

Room: Exposition Floor, 400 Aisle

Primary Author: Frederick Li
IonSense, Inc.

Co-Author(s): Brian Musselman, Joe Tice, Stephen Shrader

Abstract Text

Routine screenings for drugs of abuse are often performed with either presumptive techniques such as color tests or confirmatory techniques such as GC-MS and LC-MS. Although these techniques still represent the gold standard for routine analysis and identification of drugs of abuse, presumptive tests are designed to identify the presence of a set of possible drugs and chromatography-based methods are time-consuming. Direct analysis in real time – mass spectrometry (DART-MS) has been shown as an attractive method for rapid, real time detection and identification of drugs of abuse. In an effort to improve the level of confidence for drug identification, the utility of in-source fragmentation and reverse search libraries was investigated. For our detection system, a DART source was interfaced to a single quadrupole mass spectrometer as a means of providing a low cost instrument with the capability of in-source fragmentation. The voltage applied to the mass spectrometer sample cone was varied to induce various degrees of fragmentation. A reverse search library of the most characteristic precursor and fragment ions for a collection of drugs of abuse was created and evaluated to determine performance. Mass spectral data, including in-source fragmentation data, were obtained in seconds per sample. The reverse library search resulted in 100% identification with lower limits of detection of 0.5 ppm with our particular instrument. DART-MS in combination with reverse library search provides an attractive method for rapid, confirmatory screening of drugs of abuse.

Keywords: Detection, Drugs, Forensics, Mass Spectrometry
Application Code: Homeland Security/Forensics
Methodology Code: Mass Spectrometry

Session # 1740 Abstract # 1740-13 **Poster Sessions**

Session Title: Forensics and Homeland Security

Abstract Title: **Sol-gel Sorbent Beds for All-in-One Sampling, Preconcentration, and Separation of Trace Explosive Vapors**

Primary Author: Michelle Cerreta
U.S. Naval Research Laboratory

Co-Author(s): Braden Giordano, Kevin Johnson

Date: Wednesday, March 09, 2016 - Morn
Time:
Room: Exposition Floor, 400 Aisle

Abstract Text

Homemade and improvised explosive devices (IEDs), have become a popular weapon for terrorist action at home and abroad, necessitating research to develop new analytical methods for trace explosives vapor. A typical trace vapor analysis involves sorbent trapping, followed by desorption and gas-chromatographic analysis. Importantly, both sorbent trapping and gas chromatographic separation involve similar fundamental analyte-stationary phase interactions, but with different optimization criteria. In this work, we examine the potential of an approach combining sorbent sampling and preconcentration with partial separation achieved through temperature-programmed thermal desorption to eliminate the need for a separate gas chromatographic separation and thus reduce analytical complexity and analysis time of trace explosive vapors. A novel sorbent trap was formulated in which glass tubes were coated with a sol-gel based polymer stationary phase and installed in a programmable-temperature gas chromatograph inlet that was directly connected to a mass-selective detector. Methods for polymer synthesis and coating were developed and optimized in order to create a thermally stable polymer apparatus viable for analyte desorption and separation. The capability of the system to separate trace vapor mixtures was evaluated, as were the run-to-run reproducibility of sorption, sampling, and separation characteristics.

Keywords: Forensics, High Throughput Chemical Analysis, Polymers & Plastics, Separation Sciences
Application Code: High-Throughput Chemical Analysis
Methodology Code: Separation Sciences

Session # 1740 Abstract # 1740-14 **Poster Sessions**

Session Title: Forensics and Homeland Security

Abstract Title: **Techniques for Analyzing Volatile Organic Compounds Emitted During Aerobic Decomposition of Pig Carcasses and Swine Tissues**

Primary Author: Masoumeh Dalilian
Middle Tennessee State University

Date: Wednesday, March 09, 2016 - Morn
Time:
Room: Exposition Floor, 400 Aisle

Co-Author(s): Lydia Rickman, Ngee Sing Chong, Samantha Keene

Abstract Text

Studying the emission of volatile compounds from the decomposition of animal carcasses will be beneficial to the forensic community by providing important information about chemical compounds that are emitted from various tissues of animal carcasses and their relationship to the different stages of the decomposition process. In this study, the decomposition of pig carcasses and the relative decomposition rates of various swine tissues were studied. The influence of environmental media in the decomposition process was minimized by conducting the study of decomposing tissues in glass vessels that are configured with a leak-proof design for efficient collection of compounds emitted from the degradation of tissues. Samples were collected approximately 3 time a week over a period of ten weeks using pre-evacuated bottles or canisters that were analyzed by gas chromatography-mass spectrometry (GC-MS) with analyte enrichment on a 3-trap preconcentrator. The cryofocusing GC-MS technique was able to detect compounds with carbon numbers ranging from 3 to 14 down to 0.1 parts per billion by volume or ppbv levels. The emissions from the decomposition of porcine tissues include alkanes, alkenes, ketones, aldehydes, alcohols, and organic sulfides. The presence of sulfur-containing compounds, such as carbon disulfide, methanethiol, dimethylsulfide, dimethyldisulfide, and dimethyltrisulfide, could be linked to the biochemical degradation of sulfur-containing amino acids like methionine, cystine, and cysteine. Lipids or fatty tissues in the swine parts are also related to the production of ketones and alcohols that are potentially useful for predicting post-mortem intervals.

Keywords: Analysis, Forensics, Gas Chromatography/Mass Spectrometry, Organic Mass Spectrometry

Application Code: Homeland Security/Forensics
Methodology Code: Mass Spectrometry

Session # 1740 Abstract # 1740-15 **Poster Sessions**

Session Title: Forensics and Homeland Security

Abstract Title: **High Frequency, High Pressure Tandem Mass Spectrometry**

Date: Wednesday, March 09, 2016 - Morn
Time:
Room: Exposition Floor, 400 Aisle

Primary Author: Andrew Hampton
University of North Carolina at Chapel Hill

Co-Author(s): J Michael Ramsey

Abstract Text

Mass spectrometry is a gold-standard for chemical analysis despite a traditional constraint to laboratory settings. Price, size, and the needs for significant power and trained operators have limited conventional instrumentation. Portable mass spectrometry is possible by operating at higher pressures of background gas (1 Torr) and by using air instead of helium as the buffer gas. Pumping requirements are significantly diminished, leading to fully hand-portable devices. Field-portable instruments are crucial for time-sensitive applications such as rapid threat detection. Fast information enables appropriate response procedures, mitigating harm. Many response protocols require significant resources and are only effective for a specific threat. Analytical selectivity is therefore extremely important for threat detection; false positives must be eliminated while maintaining low limits of detection.

Mass spectrometry is inherently selective by measuring a compound's mass-to-charge ratio. This selectivity can be improved with tandem mass spectrometry. Collision induced dissociation (CID) can be achieved in an ion trap with no extra electrodes or power supplies, making it attractive for portable instruments. Identification of analytes is aided with analysis of the fragmentation patterns of trapped ions. CID has been demonstrated in microscale cylindrical ion traps (r0 = 170-500 µm) at pressures up to and above 1 Torr of air. A wide variety of compounds have been fragmented, and studies have explored the most effective conditions for efficient fragmentation. Drive frequency and critical trap dimensions are factors in determination of the pseudopotential well depth, which is important for stably trapping a fragmented ion. Increasing frequency, up to 25 MHz, is shown to result in more efficient charge conversion.

Collision-induced dissociation is studied in microscale ion traps with high drive rf frequency for improving the selectivity of high pressure mass spectrometers.

Keywords: Ion Trap, Mass Spectrometry, Portable Instruments, Tandem Mass Spec
Application Code: Homeland Security/Forensics
Methodology Code: Mass Spectrometry

Session # 1740 Abstract # 1740-16 **Poster Sessions**

Session Title Forensics and Homeland Security

Abstract Title **Gun Residue Analysis Using Paper Microfluidics**

Date: Wednesday, March 09, 2016 - Morn
Time:
Room: Exposition Floor, 400 Aisle

Primary Author Chastity Paredes-Rodriguez
Penn State Berks

Co-Author(s) James Karlinsey

Abstract Text

At borders, security check-points, and crime scenes, officials need an efficient test for explosive and gun powder residues. This work describes the development and evaluation of a microfluidic paper-based analytical device ([micro]PAD) that is intended to be used for the rapid detection of aromatic compounds present in gunpowder residue. The [micro]PAD offers benefits of rapid prototyping, reduced cost, disposability, and ease of transport, which is important for an application that is designed for use in the field. Similar work has been reported in the literature for a [micro]PAD-based colorimetric assay that detects aromatics from explosives (Pesenti et al., Analytical Chemistry, 2014, 86, 4707-14). The [micro]PAD presented here was designed using AutoCAD software and patterned onto chromatography paper of various grades with a dedicated printer using wax-based ink to define fluidic flow paths. A panel of common residues was evaluated on paper in order to better understand experimental limitations, prior to the acquisition of real samples from local authorities (e.g., campus police) that are able to provide substrate containing residue. Two different types of analysis were explored: a colorimetric response on paper and the extraction of sample from the paper for traditional benchtop analysis using in-house instrumentation (e.g., infrared or absorbance spectroscopy). The latter analysis could prove beneficial for follow-up analysis after an initial colorimetric screening “in the field”.

Keywords: Detector, Forensics, Lab-on-a-Chip/Microfluidics
Application Code: Homeland Security/Forensics
Methodology Code: Microfluidics/Lab-on-a-Chip

Session # 1750 Abstract # 1750-1 **Poster Sessions**

Session Title: Liquid Chromatography

Abstract Title: **Determination of Pyrethrins in Pyrethrum Oil Extracts**

Date: Wednesday, March 09, 2016 - Morn
Time:
Room: Exposition Floor, 400 Aisle

Primary Author: Ian N. Acworth
Thermo Fisher Scientific

Co-Author(s): Alan Wong, David Thomas, Jan Glinski

Abstract Text

Pyrethrins are terpenoid esters derived from the flowers of Chrysanthemum cinerariifolium. The pyrethrin family includes six similar compounds containing a cyclopropane core, named pyrethrin 1 and 2, cinerin 1 and 2 and jasmolin 1 and 2. Pyrethrins both repel and kill insects by delaying the closure of voltage-gated sodium ion channels in the nerve cells. The insecticidal and insect repellent properties of these compounds have been known for millennia and Chrysanthemum species have long been cultivated for this purpose. Interest in using pyrethrin insecticides is growing because of their low toxicity to humans (allowing home use) and favorable, fast biodegradability. On the negative side, pyrethrum oil can trigger allergic reactions in susceptible people. It is also toxic to bees with fatal doses as low as 0.02 micrograms, thus requiring very cautious application. With their increasing use in agricultural and consumer products, there is a need for improved analytical techniques both to assure product quality and to monitor the fate of pyrethrins in the environment. Current reversed phase HPLC methods using UV detection fail to fully resolve the closely eluting components without excessively long run times. In this work, we show that a new UHPLC system comprising a binary parallel pump capable of operating at pressures up to 1500 bar, a high sensitivity diode-array detector with new light-pipe technology and an integrated charged aerosol detector, enables the resolution and detection of more compounds in pyrethrum oil in less time than previously possible.

Keywords: Agricultural, HPLC Detection, Natural Products
Application Code: Agriculture
Methodology Code: Liquid Chromatography

Session # **1750** **Abstract #** **1750-2** **Poster Sessions**

Session Title: Liquid Chromatography

Abstract Title: **Pesticide Residue Analysis of Cereal Grains Treated with Traditional Fungicides or a Tannic Acid Biopesticide**

Date: Wednesday, March 09, 2016 - Morn

Time:

Room: Exposition Floor, 400 Aisle

Primary Author: Brooke M. Bien
Western Carolina University

Co-Author(s): Cynthia Atterholt

Abstract Text

Tannic acid wax emulsions were developed for the spray application of tannic acid to agricultural plants. Research has shown that tannic acid can be used as a biopesticide to protect crops such as wheat and barley from Fusarium head blight (FHB) disease caused by the Fusarium fungus. Commercial fungicides used to protect crops from FHB include metconazole, tebuconazole, and prothiconazole. However, recent research indicated that tannic acid, a natural product, could be used to effectively protect crops from FHB. One objective of this research was to develop analytical techniques to measure residual fungicides or tannic acid that had been sprayed on crops for protection against FHB. Measuring the residual tannic acid on treated plants would help determine whether this could be an effective alternate treatment for protection against FHB for agricultural crops. HPLC was used to quantify residual tannic acid, and gas chromatography was used to quantify the residual commercial fungicides on treated plants.

Keywords: Agricultural, Gas Chromatography/Mass Spectrometry, HPLC, Pesticides
Application Code: Agriculture
Methodology Code: Liquid Chromatography

Session # 1750 Abstract # 1750-3 **Poster Sessions**

Session Title: Liquid Chromatography

Abstract Title: **Microwave-Assisted Extraction of Triketone and Pyrazole Corn Herbicides from Agricultural Soil**

Primary Author: Sanja Stipi□evi□
Institute for Medical Research and Occupational Health

Date: Wednesday, March 09, 2016 - Morn
Time:
Room: Exposition Floor, 400 Aisle

Co-Author(s): Gordana Mendaš, Marija Dvorš□ak, Milena Milakovi□, Sanja Fingler

Abstract Text

The benzoylcyclohexanedione (triketone) herbicides mesotrione and tembotrione and benzoylpyrazole herbicide topramezone have recently been released for commercial use in corn production. The objective of this study was to develop and validate a new method for the simultaneous extraction of mesotrione with its two degradation products 4-methylsulfonyl-2-nitrobenzoic acid (MNBA) and 2-amino-4-methylsulfonyl-benzoic acid (AMBA), tembotrione and topramezone from agricultural soil. Our analytical method is based on microwave-assisted extraction (MAE) followed by reversed-phase liquid chromatography, using gradient elution with acetonitrile and acidified water (pH 2.2) and diode array detection at 205 and 220 nm. MAE operational parameters (solvent type and volume, extraction temperature and time) were optimized with respect to the extraction efficiency of the target compounds from dry soils fortified at 0.2 and 0.1 mg kg[sup]-1[/sup]. The highest recoveries (70-94 % for herbicides and MNBA and <30 % for AMBA), with satisfactory method precision (RSD □20 %), were achieved by extraction of soil with solvent mixture [Psi] (methanol, 0.1 mol L[sup]-1[/sup] HCl) = 9:1 at 60 [degree]C for 5 min. The method has proven suitable for residue analysis of target herbicides and MNBA in neutral, low humic silt loam soils, with an analyte detection limit of 0.002 mg kg[sup]-1[/sup] of soil dry mass. This study has been supported in full by the Croatian Science Foundation under project No. 8366.

Keywords: Environmental/Soils, Liquid Chromatography, Pesticides, Trace Analysis
Application Code: Environmental
Methodology Code: Liquid Chromatography

Session # 1750 Abstract # 1750-4 **Poster Sessions**

Session Title Liquid Chromatography

Abstract Title **Development and Validation of a Method for the Simultaneous Extraction and Separated Measurement of Diflubenzuron and Azamethiphos from the Soft Tissue of Mussel M. Chilensis**

Date: Wednesday, March 09, 2016 - Morn

Time:

Room: Exposition Floor, 400 Aisle

Primary Author Luis Norambuena
Instituto de Fomento Pesquero

Co-Author(s) Sergio Contreras-Lynch

Abstract Text

A simple and rapid method for the extraction and detection of the antiparasitics Diflubenzuron and Azamethiphos in the soft tissue of the mussel species M. chilensis was developed and validated. The compounds were extracted from the mussel using the QuEChERS method (quick, easy, cheap, effective, rugged, and safe), to later be detected and quantified by high performance liquid chromatography (HPLC) using an diode array detector (Diflubenzuron) and fluorescence (Azamethiphos). The limit of quantification (LOQ) was 100 ng/g for both analytes. The coefficient of variation for repeatability and intermediate precision was less than 10%. The calibration curve was
linear between 20 and 400 ng/ml (Azamethiphos) and between 25 and 400 ng/ml (Diflubenzuron).The recovery of the analytes was over 80%.

Keywords: Environmental, HPLC, Sample Preparation

Application Code: Environmental

Methodology Code: Liquid Chromatography

Session # 1750 Abstract # 1750-5 **Poster Sessions**

Session Title: Liquid Chromatography

Abstract Title: **Determination of Anions and Cations in Water Matrices Using Non-suppressed Ion Chromatography**

Date: Wednesday, March 09, 2016 - Morn

Time:

Room: Exposition Floor, 400 Aisle

Primary Author: Joseph P. Romano
Waters Corporation

Co-Author(s): Kenneth Rosnack, Mark E. Benvenuti

Abstract Text

Ion Chromatography (IC) is the preferred method for the determinations of anions and cations in aqueous samples using conductivity detection. Improvements in chromatographic hardware, column and detection technology have increased the range of solutes which can be analyzed and also lowered the levels at which they can be detected. Despite its widespread use, by far the most significant application of IC is the routine determination of the common inorganic anions, fluoride, chloride, nitrite, bromide, nitrate, phosphate and sulfate, in drinking water and wastewater samples.

In this work, a borate/gluconate eluent was used with a high resolution anion-exchange column in the non-suppressed mode. This eluent has shown to give the best overall separation selectivity for the common inorganic anions. For determination of monovalent and divalent cations, lithium, sodium, ammonium, potassium, magnesium, calcium, strontium and barium, an EDTA/nitric acid eluent was used with a high resolution cation-exchange column in the non-suppressed mode. Samples of drinking water and waste water were analyzed to determine the levels of both anions and cations in these water matrices.

Keywords: Chromatography, Environmental Analysis, Environmental/Water, Ion Chromatography
Application Code: Environmental
Methodology Code: Liquid Chromatography

Session # 1750 **Abstract # 1750-8** **Poster Sessions**

Session Title: Liquid Chromatography

Abstract Title: **Micellar and Sub-Micellar UHPLC of Aromatic Acid Geometric Isomers**

Primary Author: Jennifer M. Fasciano
Miami University

Date: Wednesday, March 09, 2016 - Morn
Time:
Room: Exposition Floor, 400 Aisle

Co-Author(s): Neil D. Danielson

Abstract Text

Micellar liquid chromatography (MLC) has been used primarily for the separation of neutral analytes of varying polarities, most commonly phenols and polyaromatic hydrocarbons, but has not been used to study aromatic carboxylic acids. We have studied the separation of hydroxybenzoic acid mixtures, including monohydroxybenxoic and dihydroxybenzoic acid geometric isomers by MLC. SDS was investigated as the modifying surfactant on a C18 UHPLC column (100 x 2.1 mm, 1.8 m). The addition of only SDS (no organic solvent) to the mobile phase reduced the influence of hydrophobic interactions while improving the retention times, resolution, and peak shapes, even at concentrations below the CMC of SDS (0.23%). The separation of seven hydroxybenzoic acids, including six dihydroxybenzoic acid geometric isomers and one trihydroxybenzoic acid, was achieved with high efficiency using 0.1% SDS in 1.84 mM sulfuric acid (pH = 2.43) mobile phase, in less than six minutes with a flow rate of 0.3 mL/min, and in less than four minutes with a flow rate of 0.7 mL/min. Six monohydroxybenzoic acid isomers were also effectively separated by MLC, using a 0.35% SDS mobile phase modifier, in less than twenty minutes with a flow rate of 0.3 mL/min, and in less than fourteen minutes with a flow rate of 0.7 mL/min. Solute-micelle equilibrium constants and partition coefficients were calculated for six monohydroxybenzoic acids based on a plot of MLC retention factor vs. mobile phase micelle concentration. All aromatic carboxylic acid isomers studied can be classified as binding solutes in the MLC retention mechanism. Less effective separations were observed with shorter chain anionic surfactants (C6, C9, C10), leading to higher retention times and poor peak shapes. It was concluded that SDS is the preferred modifying surfactant for the examined separations.

Keywords: Liquid Chromatography, Surfactants
Application Code: Food Science
Methodology Code: Liquid Chromatography

Session # 1750 Abstract # 1750-9 **Poster Sessions**

Session Title: Liquid Chromatography

Abstract Title: **Hydrophilic Interaction Liquid Chromatography of Aromatic Acid Isomers on a Plain Silica Stationary Phase Using a Ternary Mobile Phase**

Primary Author: Ashley E. Richardson
Miami University

Co-Author(s): Neil D. Danielson

Date: Wednesday, March 09, 2016 - Morn
Time:
Room: Exposition Floor, 400 Aisle

Abstract Text

Hydrophilic interaction liquid chromatography (HILIC) has become increasingly popular as an alternative to reversed phase LC due to its ease of separating complex polar compound mixtures and the compatibility of the mobile phase with mass spectrometry. The variation of a third solvent such as a series of alcohols from methanol to isopropanol in the mobile phase has been shown to be effective at increasing the retention factors for polar analytes. A HILIC simulation study has predicted a non-water miscible third solvent will markedly increase polar solute retention over that with a water miscible third solvent. This is because the non-water miscible solvent will remain in the acetonitrile (MeCN) layer providing a strong polarity contrast to the water rich stationary phase. We have found the addition of such a third solvent to the mobile phase of MeCN and ammonium acetate buffer has proven effective at separating aromatic acid geometric isomers (particularly m- and p-). With a plain silica column (150mm x 4.6mm) and a mobile phase comprised of MeCN/ammonium acetate (pH 7)/pentane (90/5/5), the separation of three hydroxybenzoic acid isomers, three hydroxycinnamic acid isomers, syringic acid, and vanillic acid can be achieved in about 60 min with close to baseline resolution. The presence of both the buffer and organic solvent are important and their concentration and type need to be studied. Variation of the acid, base buffer components is underway. We plan to characterize pairs of aprotic ternary solvents such as 2-butanone/ethyl acetate and pentane/heptane which have similar polarities but quite different solubilities in water. The retention trend may vary as a function of both solvent polarity and water solubility since aromatic acids are structurally both polar and nonpolar.

Keywords: Liquid Chromatography
Application Code: Food Identification
Methodology Code: Liquid Chromatography

Session # 1750 **Abstract # 1750-10** **Poster Sessions**

Session Title: Liquid Chromatography

Abstract Title: **Understanding the Importance of Instrument Design To Take Full Advantage of 1.0mm Internal Diameter (ID) Columns when Running UPLC**

Date: Wednesday, March 09, 2016 - Morn
Time:
Room: Exposition Floor, 400 Aisle

Primary Author: Jennifer Simeone
Waters Corporation

Co-Author(s): Patricia R. McConville, Paula Hong

Abstract Text

The goal of this work is to show the important link between instrument characteristics and resulting chromatography. Specifically, the internal diameter of a column should be matched to instrument characteristics such as system volume and dispersion in order to produce optimal chromatography. If the instrument is not designed to run smaller ID columns, the resulting chromatography may show peak broadening, loss of resolution, and decreased peak capacity. Any peak broadening or loss of resolution will be especially problematic for complex separations that contain many peaks, or for separations which require a minimum resolution between critical pairs.

In order to achieve high quality separations using 1.0 mm ID columns, the user must be cognizant of the LC instrument characteristics and the potential negative effects it may have on the chromatography. In order to maximize resolution between critical pairs, it is important to have the lowest possible dispersion in the instrument system. If this volume cannot be decreased to achieve chromatographic resolution, column length can be increased. However, this comes at a price of increased system pressure, increased run time, increased solvent usage, and lower overall throughput.

Keywords: Chromatography, Liquid Chromatography, Separation Sciences
Application Code: General Interest
Methodology Code: Liquid Chromatography

Session # 1750 Abstract # 1750-11 **Poster Sessions**

Session Title Liquid Chromatography

Abstract Title **Investigation of Reverse-Phase - HILIC Continuous Analysis Using a One Column**

Date: Wednesday, March 09, 2016 - Morn
Time:
Room: Exposition Floor, 400 Aisle

Primary Author Ronald Benson
Shodex/Showa Denko America, Inc.

Co-Author(s) Hideyuki Kondo, Junji Sasuga, Tomokazu Umezawa

Abstract Text

A sample matrix that includes both hydrophobic and hydrophilic substances are difficult to analyze by reversed phase chromatography (RPC) or hydrophilic interaction chromatography (HILIC) alone. The analysis of such matrices requires the separation using both an RPC column and a HILIC column. Herein, we describe the method for analyzing both hydrophobic and hydrophilic compounds simultaneously with one column by applying RPC/HILIC continuous analysis method.

First, the transition in the retention behavior of hydrophilic compounds (Cytosine, Uridine and Uracil) and hydrophobic compounds (Toluene and Naphthalene) corresponding to changing CH3CN ratio was examined by using CH3CN / H2O as eluent. Chromatographic separation was conducted using RPC column (4.6mmI.D. ×150mmL) packed with polyhydroxymethacrylate. As a result, the retention order was reversed in two conditions, 30% CH3CN and 90% CH3CN. It is suggested that the RPC mode acts in low CH3CN concentration and the HILIC mode act in high CH3CN concentration. By taking account of other conditions, it is confirmed that the RCP mode appears in under 50% CH3CN condition and HILIC mode appears in above 70% CH3CN condition.

Second, the analysis for a commercially available energy drink by LC/MS with gradient condition from RPC mode to HILIC mode continuously was attempted. As a result, the method that both hydrophobic and hydrophilic compound can be separated in a single measurement was established with twice sample injection. This method realizes quick equilibration as switching the mode and rapid analysis.

The polymer-based RPC column containing hydroxyl group which was considered this time has a unique performance that can be applied to HILIC mode as well. This column realizes efficient analysis for hydrophobic compound and hydrophilic compound in a single measurement.

Keywords: HPLC Columns
Application Code: General Interest
Methodology Code: Liquid Chromatography

Session # 1750 **Abstract # 1750-12** **Poster Sessions**

Session Title: Liquid Chromatography

Abstract Title: **One Diode Array Detector for Analytical, Semi-Preparative, Preparative and Biocompatible HPLC**

Date: Wednesday, March 09, 2016 - Morn

Time:

Room: Exposition Floor, 400 Aisle

Primary Author: Kathryn E. Monks
Knauer

Co-Author(s)

Abstract Text

KNAUER has been manufacturing world-class HPLC instruments for the past five decades. One of our key competencies is the development of robust and highly sensitive diode array detectors. The latest devices to come to market are characterized by an optimized optical bench, active temperature control and wide range of flow cells covering numerous application fields. Furthermore, with the aid of a cartridge adapter cell, remote flow cells can be installed on these detectors via fiber optic cables. The fiber optics flow cells have the advantage that they can be decoupled from the detector and mounted for instance directly onto the column outlet. This eliminates the delay volume between column outlet and detector. Other remote flow cells applications include detection in high temperature and hazardous environments. The DAD detectors can be controlled via various Chromatographic Data Systems.

Keywords: Detection, HPLC, HPLC Detection, Liquid Chromatography

Application Code: General Interest

Methodology Code: Liquid Chromatography

Session # **1750** **Abstract #** **1750-13** **Poster Sessions**

Session Title Liquid Chromatography

Abstract Title **A Novel Diphenyl Stationary Phase for Metabolite Profiling**

Date: Wednesday, March 09, 2016 - Morn
Time:
Room: Exposition Floor, 400 Aisle

Primary Author Mark Woodruff
Fortis Technologies

Co-Author(s) Ken Butchart

Abstract Text

In chromatography selectivity is often hard to achieve in a simple manner, complex samples often require a complex mobile phase system to be produced that is not seen as productive, reproducible and easily transferable between sites. We discuss the use of a novel di-phenyl bonded phase chemistry which allows the use of simple mobile phase systems even for complex sample types, such as positional isomers, loss of functional groups and metabolites which are closely related.

We discuss the use of this unique stationary phase in terms of its physical and chemical characteristics and stability, particularly in MS where the unique bonding function allows for a stable baseline to be achieved. No "MS-bleed" is achieved and this in turn allows sensitivity and resolution to be optimised.
We compare to both traditional C18 chemistry and also to other phenol functionalities in terms of the extra selectivity and extra stability that is available.

Applications highlighting the unique selectivity that can be achieved with simple mobile phases and unique di-phenyl functionality are shown, examples of positional isomers, metabolites and other closely related species are referenced. Which mobile phases are best and how simplicity is retained due to the unique nature of the stationary phase.

Keywords: Chromatography, HPLC, HPLC Columns, Liquid Chromatography/Mass Spectroscopy
Application Code: General Interest
Methodology Code: Liquid Chromatography

Session # **1750** **Abstract #** **1750-14** **Poster Sessions**

Session Title: Liquid Chromatography

Abstract Title: **Silica Hybrid Monoliths with a Carbonaceous Surface for Liquid Chromatography**

Date: Wednesday, March 09, 2016 - Morn
Time:
Room: Exposition Floor, 400 Aisle

Primary Author: Amaris C. Borges-Munoz
University at Buffalo SUNY

Co-Author(s): Luis A. Colón

Abstract Text

Organo-silica hybrid monoliths have proven to be hydrolytically stable materials for liquid chromatography. The preparation of such materials follows the typical sol-gel process, but using at least one silane precursor that will impart the organic character to the otherwise inorganic silica network. To achieve chromatographic selectivity, the surface of the monolithic column can contain a moiety of interest incorporated through bulk modification during synthesis or by further attachment to the surface. Herein, we report the synthesis of an allyl-silica hybrid monolith, which has been modified to produce a carbonaceous surface. The monolith has been characterized by infrared spectroscopy (IR), thermogravimetric analysis (TGA), and surface gas adsorption. The monolithic column was tested under liquid chromatographic conditions to assess the retention behavior of probe compounds. Details of the synthesis, modification of the surface, and the physicochemical as well as the preliminary results for the chromatographic characterization will be presented.

Keywords: Liquid Chromatography, Materials Characterization, Modified Silica

Application Code: Material Science

Methodology Code: Liquid Chromatography

Session # 1750 Abstract # 1750-15 **Poster Sessions**

Session Title: Liquid Chromatography

Abstract Title: **Evaluation of the Performance of Ultra-high Pressure Chromatography (UPLC) Systems**

Primary Author: Imad A. Haidar Ahmad
Novartis

Date: Wednesday, March 09, 2016 - Morn
Time:
Room: Exposition Floor, 400 Aisle

Co-Author(s): Adrian Clarke, Andrei Blasko, Frank Hrovat, James Tam, Thomas Tarara, Xue Li

Abstract Text

High performance liquid chromatography is one of the most used analytical techniques. Therefore, testing the performance of a chromatography system is fundamental to verifying its suitability to produce reliable results. Here, we tested three ultra-high pressure liquid chromatography (UPLC) instruments by examining the key functions of each system, each consisting of a pump, auto-injector, column thermostat, and absorbance detector. The three UPLC systems used in this study are the Agilent 1290, the Vanquish from Thermo Fisher, and the Nexera X2 from Shimadzu. The following performance factors were evaluated: injector linearity and carryover, accuracy and precision of temperature control, pressure ripple, pump mixing accuracy, retention time fluctuation, extra-column volume and dwell volume determination, gradient accuracy, flow rate accuracy, and detector linearity. The largest variability was found for temperature control accuracy, pressure ripple, and pump mixing accuracy. Such tests were designed to identify any problems that adversely affect system performance when using these UPLC systems, including retention time irreproducibility, low sensitivity, or potential method transfer failure or issues.

Keywords: Chromatography, Liquid Chromatography
Application Code: Other
Methodology Code: Liquid Chromatography

Session # 1750 Abstract # 1750-16 **Poster Sessions**

Session Title: Liquid Chromatography

Abstract Title: **Optimizing Reagent Free Ion Chromatography; Electrolytic Water Purification**

Primary Author: John M. Riviello
Trovion Company

Date: Wednesday, March 09, 2016 - Morn
Time:
Room: Exposition Floor, 400 Aisle

Co-Author(s): Archava Siriraks, Daniel Khor

Abstract Text

Over the past decade, ion chromatography using electrolytic technologies for eluent production and suppression have emerged as the standard for ion analysis. Referred to as Reagent-Free Ion Chromatography[circumflex O](RFIC), deionized water is used as the pumped phase to electrochemically generate an acid or base eluent. For anion analysis, deionized water is pumped to the electrolytic eluent generator where potassium hydroxide is produced. The purity of the potassium hydroxide is largely dependent on the purity of the deionized water. Deionized water with a resistance of at least 18.0 M[omega]-cm is recommended for RFIC[circumflex O], but in practice, the resistance of the deionized water in the eluent bottle is typically 1 M[omega]-cm as a result of the absorption of ambient carbon dioxide. The potassium hydroxide eluent now contains potassium carbonate which causes an increase in the suppressed background conductivity, a higher gradient blank and will effect quantitation of analytes. Other common anions in deionized water such as chloride and sulfate can also cause an increase in the gradient blank.

In this poster, we will describe an electrolytic water purifier which integrates directly into the ion chromatograph. This point-of-use approach ensures that the water used for electrolytic eluent generation is essentially ion-free. Since the pressure drop through the electrolytic water purifier is very low, no additional pump or pressurized reservoir is required. Data will be presented which shows a reduction in the suppressed background conductivity and the gradient blank for carbonate and other interfering anions. The ability to maintain consistent system performance even as the deionized water is contaminated with carbonate and trace ions will also be demonstrated.

Keywords: Ion Chromatography, Ion Exchange, Ultratrace Analysis, Water
Application Code: General Interest
Methodology Code: Liquid Chromatography

Session # 1750 Abstract # 1750-17 **Poster Sessions**

Session Title Liquid Chromatography

Abstract Title **Functionalized Carbon Nanotubes as Pseudo-Stationary Phases in Capillary Electrokinetic Chromatography - Evaluation of Retention Energetics and Analysis of a Wide Range of Neutral and Charged Species**

Primary Author Sarah Alharthi
Oklahoma State University

Date: Wednesday, March 09, 2016 - Morn
Time:
Room: Exposition Floor, 400 Aisle

Co-Author(s) Ziad El Rassi

Abstract Text

Functionalized multiwalled carbon nanotubes (MWCNTs) exhibit unique chemical and physical properties that significantly enhance separation in capillary electrokinetic Chromatography (EKC). In this investigation, MWCNTs have been functionalized with hydroxyl, carboxylic and sulfonic groups and evaluated over a wide range of electrolyte composition with various neutral and charged species, e.g., alkylbenzenes, phenylalkyl alcohols, dansyl amino acids, barbiturates, urea herbicides and some aromatics. Functionalized CNTs have been characterized by spectroscopic methods. In all cases, the results on the functionalized MWCNTs were compared to those obtained on unmodified MWCNTs in the presence of SDS in the running electrolyte. The ratio of MWCNTs to SDS affected the electrokinetic systems under investigation significantly in terms of migration time window and in turn system resolution. Plots of log k' of the same solutes on the various MWCNTs were used to evaluate the retention energetics as well as the hydrophobic phase ratios. While the various MWCNTs showed homoenergetic retention behaviors, they differed in terms of their hydrophobicity with the sulfonated ones being the least hydrophobic toward all solutes examined. The migration time window of the functionalized MWCNTs was quite wide allowing the separation of the various neutral and charged species investigated with high selectivity and resolution, and yielded a plate count reaching as high as 184,000 plates/m.

Keywords: Amino Acids, Capillary Electrophoresis, Environmental Analysis, Pharmaceutical
Application Code: Bioanalytical
Methodology Code: Capillary Electrophoresis

Session # 1750 Abstract # 1750-18 **Poster Sessions**

Session Title: Liquid Chromatography

Abstract Title: **Determination of Hydroperoxides Using High Performance Liquid Chromatography with Reductive Electrochemical Detection**

Date: Wednesday, March 09, 2016 - Morn

Time:

Room: Exposition Floor, 400 Aisle

Primary Author: Jun Cheng
Thermo Fisher Scientific

Co-Author(s): Khalil Divan, Yan Liu

Abstract Text

The majority of applications using electrochemical detection in liquid chromatography have been performed with oxidative electrochemical detection. There are relatively few reports on the use of reductive electrochemical detection in liquid chromatography. The liquid chromatography-electrochemical detection (LC-ED) applications would become much more popular if a sensitive and reliable reductive electrochemical detection method can be developed. Advances in this field have been relatively slow due in part to the difficulties of making experimental measurements with the reductive detection with dissolved oxygen in the mobile phase. It often requires a special procedure to eliminate the oxygen to achieve sensitive and reliable results.

In this paper, a reductive electrochemical detection method was developed for the detection of organic hydroperoxides after liquid chromatographic separation without a special step for removing dissolved oxygen. The mobile phase and detection potential were found to be critical to achieve good analytical performance. With the optimized mobile phase and detection potential, the detection limits of organic hydroperoxides were determined to be 19.3 μM (t-butyl hydroperoxide) and 22.5 μM (cumene hydroperoxides) using a 20 μl injection volume. Calibration plots for both hydroperoxides are linear from 30 μM to 10,000 μM.

Keywords: Biopharmaceutical, Electrochemistry, HPLC Detection, Method Development
Application Code: Pharmaceutical
Methodology Code: Liquid Chromatography

Session # 1750 Abstract # 1750-19 **Poster Sessions**

Session Title Liquid Chromatography

Abstract Title **Label-Free Analysis by HPLC with Charged Aerosol Detection of N-Linked Glycans Separated by Charge**

Date: Wednesday, March 09, 2016 - Morn

Time:

Room: Exposition Floor, 400 Aisle

Primary Author Ian N. Acworth
Thermo Fisher Scientific

Co-Author(s) David Thomas, William Kopaciewicz

Abstract Text

The purpose of this work was to develop a quantitative glycan profiling assay that is simple, fast, and eliminates the hassle of labeling glycans with a fluorophore. The method is intended for characterization and quality control of glycoproteins biotherapeutics. High performance liquid chromatography with charged aerosol detection (HPLC-CAD) uses a volatile mobile phase fully compatible with mass spectrometry in case further characterization is desired. The oligosaccharide component of glycoproteins is a key determinant of their function. Changes in the number, type, composition or linkage pattern of these glycans may serve as a biomarker of disease or influence the efficacy of a biotherapeutic product. For this reason, the ability to correctly identify and measure these glycans is of scientific interest, and to do so reliably, quickly and inexpensively is of practical benefit. This work explores direct detection of native glycans as an alternative to the common techniques for glycan analysis that rely on labeling reactions to render glycans detectable. The lack of a detectable chromophore in native glycans is overcome by using HPLC with charged aerosol detection, which can quantitatively measure any non-volatile compound. Glycans released from various proteins were analyzed including those from bovine fetuin and alpha acid glycoprotein. Quantitative performance including precision, detection limits and dynamic range is presented. Figures of merit include sensitivity at the low nanogram on-column level, dynamic range over two orders of magnitude, and peak area precision averaging less than three percent RSD. By responding directly to any non-volatile compound, charged aerosol detection is able to quantify unlabeled N-linked glycans. The uniform response of charged aerosol detection also provides simple, accurate and precise estimates of relative concentration even in the absence of pure primary standards.

Keywords: Biopharmaceutical, Biotechnology, HPLC Detection, Protein
Application Code: Bioanalytical
Methodology Code: Liquid Chromatography

Session # 1750 Abstract # 1750-20 **Poster Sessions**

Session Title: Liquid Chromatography

Abstract Title: **Multi-Modal Analyte Detection of Cyclodextrin and Ketoprofen Inclusion Complex Using UV and CAD on an Integrated UHPLC System**

Primary Author: Bruce Bailey
Thermo Fisher Scientific

Date: Wednesday, March 09, 2016 - Morn
Time:
Room: Exposition Floor, 400 Aisle

Co-Author(s): Evert-Jan Sneekes, Frank Steiner, Ian N. Acworth, Marc Plante

Abstract Text

Several drug candidates have limited bioavailability or stability due to their inherent chemical properties. The formation of a non-covalent inclusion complex with hydrophobic “guest” molecules can be used to alter the bioavailability of these drugs. An integrated UHPLC system with orthogonal and complementary diode array detector and charged aerosol detector (CAD) is used to separate and detect an inclusion complex containing cyclodextrin and ketoprofen. Beta cyclodextrin has a ring structure containing seven sugars but since it is comprised of carbohydrate molecules it does not possess a suitable chromophore for UV detection and is only detected by a universal detector. The drug, ketoprofen can be measured by both UV and a universal detector. A reverse phase HPLC method was developed, using a long porous C18 UHPLC column and gradient elution chromatography. The Corona Veo CAD provides the sensitivity and linearity of correlation that is required for quantification of compounds with non-chrompohore characteristics. This method has several advantages since the integrated system offers high precision performance for drug formulations. The %RSD on retention time was quite low and calibration curves for ketoprofen also exhibit excellent correlations (R2 of 0.999 for both the UV and charged aerosol detectors).

Keywords: Cyclodextrin, HPLC Detection, Liquid Chromatography
Application Code: Pharmaceutical
Methodology Code: Liquid Chromatography

Session # 1750 Abstract # 1750-22 **Poster Sessions**

Session Title Liquid Chromatography

Abstract Title **Purification of Chlorogenic Acid from Green Coffee Using Core-shell Technology in Axia Preparative Format**

Date: Wednesday, March 09, 2016 - Morn
Time:
Room: Exposition Floor, 400 Aisle

Primary Author Marc J. Jacob
Phenomenex

Co-Author(s)

Abstract Text

High performance HPLC/UHPLC core-shell material is the latest technological advancement in chromatographic media. When used under analytical conditions, core-shell particles show improved efficiency and performance over fully porous particles of equivalent particle size.1,2
With the recent commercialization of a lower pressure 5 µm core-shell media, it is now possible to offer core-shell media in a preparative format (>20 mm ID) that's compatible with standard prep LC systems.

In this poster, we will demonstrate that this new 5 µm core-shell particle size, available in a variety of bonded phases, can be packed efficiently in preparative format with internal diameter greater than 2 centimeters. We will highlight the advantage of such product for the isolation of antioxidant chlorogenic acid from Green Coffee extract.

Reference;
1. F. Gritti and G. Guiochon LC-GC, 2012, 30, 586-595.
2. F. Gritti and G. Guiochon J. Chromatogr. A, 2013, 1230, 35–50

Keywords: HPLC, Natural Products, Pharmaceutical, Prep Chromatography
Application Code: Pharmaceutical
Methodology Code: Liquid Chromatography

Session # 1750 **Abstract # 1750-23** **Poster Sessions**

Session Title: Liquid Chromatography

Abstract Title: **HILIC HPLC Separation of Oxymorphone for Assay and Purity**

Date: Wednesday, March 09, 2016 - Morn
Time:
Room: Exposition Floor, 400 Aisle

Primary Author: Bradley Kumagai
Theravance Biopharma

Co-Author(s)

Abstract Text

Developing an HPLC assay and impurity method for oxymorphone was of interest to support drug product development. A survey of the available published and compendial methods was limited for oxymorphone, particularly for impurity methods, prompting development efforts. Reversed phase HPLC on multiple columns was attempted, but exhibited issues with peak shape and retention. This led to the evaluation of a hydrophobic interaction liquid chromatography (HILIC) retention mechanism, which improved these performance aspects and led to the development of an improved assay and impurity method for oxymorphone.

Keywords: HPLC, Liquid Chromatography, Method Development, Pharmaceutical
Application Code: Pharmaceutical
Methodology Code: Liquid Chromatography

Session # 1750 **Abstract # 1750-24** **Poster Sessions**

Session Title: Liquid Chromatography

Abstract Title: **Transfer of the Method for Related Substances Analysis of Metoclopramide HCl Between Different LC Systems**

Primary Author: Margaret Maziarz, Waters Corporation

Co-Author(s): Christopher Henry, Mark Wrona

Date: Wednesday, March 09, 2016 - Morn
Time:
Room: Exposition Floor, 400 Aisle

Abstract Text

Analytical methods used for testing of pharmaceutical raw materials and finished products are often transferred across organizations or to contract partners that utilize instrument from different vendors. It is therefore essential for the analytical laboratories to successfully transfer these methods to ensure product consistency and compliance with the regulations. Effective method transfer generates equivalent results for the same analysis independent of the instrument, laboratory, or the resources. This is important as it eliminates the need to revalidate the method, which is time consuming and costly.

In this work, we present transfer of the HPLC method for related substances analysis of metoclopramide HCl between different LC systems. Additionally, we will demonstrate a method improvement by scaling this method to a column with a smaller particle size. The HPLC method will be scaled to smaller particle size using the Waters ACQUITY Columns Calculator. The success of the method transfer to the new instrumentation will be measured by examining chromatographic separation for comparable results and verifying that the system suitability results of replicate injections meet the specifications defined in the USP General Chapter <621> Chromatography.

As shown in this study, the transfer of the method between LC systems, as well as method improvements by migration to a column with smaller particle size was successful. The chromatographic separation between all the peaks were comparable on both systems. The system suitability results for each component were in excellent agreement with the USP specifications and comparable on both systems.

Keywords: HPLC, Pharmaceutical
Application Code: Pharmaceutical
Methodology Code: Liquid Chromatography

Session # 1750 Abstract # 1750-26 **Poster Sessions**

Session Title: Liquid Chromatography

Abstract Title: **Polar Stationary Phases for Capillary Liquid Chromatography based on Metallic Oxides**

Date: Wednesday, March 09, 2016 - Morn

Time:

Room: Exposition Floor, 400 Aisle

Primary Author: Carla Grazieli A. da Silva
Unicamp

Co-Author(s): Carla B. Bottoli, Carol H. Collins

Abstract Text

Monolithic stationary phases (SP) that are used in capillary columns are prepared with either organic or inorganic materials. The porous body is synthesized inside the capillary by an organic free radical reaction or by the sol-gel process with inorganic precursors.

The goal of this work was the preparation and characterization of new monolithic stationary phases for cLC using a mixture of polyethylene glycol (PEG) (M = 10 000 g/mol), acetic acid, water and then modified with zirconium or titanium alkoxides. The SP were compared to silica monoliths prepared by the sol-gel process using tetramethoxysilane, PEG, urea and acetic acid.

The new SP were submitted to physical-chemical characterization (thermogravimetric analysis, Fourier transform infrared spectroscopy (FTIR), elemental analysis, X-ray fluorescence and imaging techniques (scanning electron microscopy (SEM) and 3D structural characterization using synchrotron radiation). The capillary columns were submitted to chromatographic characterization by the separation of two mixtures of aromatic hydrocarbons in the HILIC (hydrophilic-interaction chromatography) mode. These results were compared with commercial silica-based C18 SP (5 µm of particle size; 150 mm) and with monolithic silica-based SP prepared by the sol-gel process. The results showed that these new stationary phases based in metallic oxides have the potential to be used in cLC for the separation of small molecules in the HILIC mode.

Keywords: Chromatography, Imaging, Liquid Chromatography, Material Science
Application Code: Material Science
Methodology Code: Liquid Chromatography

Session # 1750 Abstract # 1750-27 **Poster Sessions**

Session Title Liquid Chromatography

Abstract Title **Selection of Chromatographic Columns by Supercritical Fluid Chromatography**

Primary Author Carla Grazieli A. da Silva
Unicamp

Co-Author(s) Carol H. Collins, Isabel Cristina S. Jardim

Date: Wednesday, March 09, 2016 - Morn
Time:
Room: Exposition Floor, 400 Aisle

Abstract Text

Supercritical fluid chromatography (SFC) is a powerful and interesting tool used to characterize chromatographic columns. The method used for stationary phases (SP) classification and comparison in SFC uses the key-solutes quantitative structure-retention relationship (QSRR), based on the linear solvation energy relationship (LSER) model and in the Abraham descriptors [1,2] obtained by multilinear regression of retention factors of specific groups of aromatic solutes.

In this work, SP for high performance liquid chromatography (HPLC) and SFC (type C8, C18, phenyl, etc.) were evaluated using LSER model using subcritical conditions (CO2:organic modifier (90:10, v/v); flow: 3 mL min-1; temperature: 25 °C; outlet pressure: 15 MPa/60oC and UV detection at 210 nm).

The results reveleaded different profiles of separation based on the retention factors and different selectivities depending of the organic modifier used (Figure 1; k-k plot for C18 SP using methanol (MeOH) and acetonitrile (ACN) as organic modifier).

References
[1] C.F. Poole, S. K. Poole, J. Chromatogr. A 965 (2002) 263.
[2] M. Vitha, P. W. Carr, J. Chromatogr. A 1126 (2006) 143.

Keywords: Characterization, Data Analysis, SFC, Supercritical Fluid Chromatography
Application Code: General Interest
Methodology Code: Supercritical Fluid Chromatography

Session # 1750 Abstract # 1750-28 **Poster Sessions**

Session Title Liquid Chromatography

Abstract Title **Stability Evaluation of Core Shell C18 with Encapsulated Type End-Capping**

Primary Author Norikazu Nagae
ChromaNik Technologies Inc.

Co-Author(s) Tomoyasu Tsukamoto

Date: Wednesday, March 09, 2016 - Morn
Time:
Room: Exposition Floor, 400 Aisle

Abstract Text

A column packed with core shell particles has been widely used on HPLC and UHPLC, because it showed not only excellent column efficiency but also lower back pressure than sub-2 um column. More than 20 kinds of core shell column are available in the market. Recently high stability under a basic pH condition has been requested for a core shell reversed-phase as well as a fully porous C18. In this study, dense end-capping using a difunctional silyl-reagent was evaluated as an encapsulated type end-capping.

1,5-dichlorohexamethyltrisiloxane or/and 1.2-bis(dimethoxymethyl)ethane and trimethoxyoctadecylsilane were bonded. These reagents were bonded on 2.6 um core shell silica at three kinds of mixture ratio. The column life was measured to elute the mobile phase adjusted at pH10 or pH 11. The peak shape of acidic, basic and metal chelating compounds was compared with the other brand C18s.

The Stability of core shell C18s with seven kinds of coverage of the end-capping was tested using methanol/50 mM phosphate buffer pH 11=10/90 as a mobile phase. The most stable core shell C18 was synthesized with a mixture of two reagents with 1:1 ratio. In comparison with three other brand C18s including hybridized core shell C18 not only regarding a column life under pH 10.5 condition at 60 degree Celsius but also regarding peak shape of acidic, basic and metal chelating compounds, proposed most stable C18 showed the highest stability and the best peaks.

The proposed encapsulated dense end-capping could make the stability under basic pH conditions high without deterioration of a peak shape.

Keywords: HPLC Columns
Application Code: General Interest
Methodology Code: Liquid Chromatography

Session # 1750 Abstract # 1750-29 **Poster Sessions**

Session Title Liquid Chromatography

Abstract Title **Development of an Interactive Counter Current Extraction Simulation**

Date: Wednesday, March 09, 2016 - Morn
Time:
Room: Exposition Floor, 400 Aisle

Primary Author Sean A. Reed
Westminster College

Co-Author(s) Erin Wilson, James Anthony

Abstract Text

Counter current extraction is a method of step-wise extractions involving the movement of analytes by partitioning through a series of stationary phase fractions through step-wise mobile phase transfers. This allows analytes in the sample to be separated based on their affinity to the stationary and mobile phase (i.e. their partition coefficient). Counter current extraction can serve as a conceptual bridge for students between liquid-liquid chromatography and column chromatography. The interactive program written simulates counter current extraction of either two or three analytes with user control of the volume of the mobile phase and stationary phase, partition coefficient of each analyte, and how many Stationary phase fractions are used and the number of extraction step. The program was written using Processing software (available at Processing.org) and the program is run through Java. This program can be used as a way to visualize the concept of chromatography and counter current extraction, and how each variable alters the separation of the analytes.

Keywords: Chromatography, Education, Teaching/Education

Application Code: Other

Methodology Code: Computers, Modeling and Simulation

Session # 1760 Abstract # 1760-1 **Poster Sessions**

Session Title Pharmaceutical-IC, LC, and SFC

Abstract Title **Determination of Aluminum in OTC Products**

Primary Author Manali Aggrawal
Thermo Fisher Scientific

Co-Author(s) Jeffrey Rohrer

Date: Wednesday, March 09, 2016 - Morn
Time:
Room: Exposition Floor, 400 Aisle

Abstract Text

Aluminum containing compounds are used extensively in cosmetics, pharmaceuticals and over-the-counter (OTC) products. A number of them are used as active ingredients in antiperspirant products. Most OTC antiperspirants contain aluminum chloride or aluminum chlorohydrate as being the main active ingredient. In USA, the FDA allows OTC sale of antiperspirants containing 15%-25% aluminum (with the amount varying based on the specific compound used). Aluminum-based compounds are also used in antacid formulations. Aluminum hydroxide and magnesium hydroxide act as antacids by neutralizing stomach acid that result in an increasing pH in the stomach. United States Pharmacopeia (USP) has adopted several different assays for aluminum in various OTC products. These analytical techniques include complexometric titration, chelatometric titration, ion-exclusion, and reversed-phase liquid chromatography. Aluminum can be determined using Ion Chromatography (IC) with an easy setup and fast run time.

Here, we report the validation of an IC method for the determination of aluminum in OTC formulations with a cation-exchange column and post column derivatization. We demonstrated the results for two OTC products; antacid and antiperspirant. Figure 1 shows the chromatogram of antacid and antiperspirant. For antacid, a suspension formulation was used and for antiperspirant, solid sticks were used. This method uses an IonPac CS10 (4 × 250 mm) column and post column addition of Tiron, a colorimetric reagent, followed by UV absorbance. The method was evaluated in terms of linearity, precision, accuracy, ruggedness, and limit of quantitation for aluminum. The calibration results show that the detection is linear over the concentration range (1.5- 45 mg/L), with a coefficient of determination >0.9999. Spike recoveries for Aluminum in antacid sample are within 99-101%, for the antiperspirants is 75-110%. The estimated LOD and LOQ for Aluminum found to be 0.032 and 0.107 mg/L respectively.

Keywords: Cosmetic, Ion Chromatography, Pharmaceutical, UV-VIS Absorbance/Luminescence
Application Code: Pharmaceutical
Methodology Code: Liquid Chromatography

Session # 1760 Abstract # 1760-2 **Poster Sessions**

Session Title: Pharmaceutical-IC, LC, and SFC

Abstract Title: **Analysis of Antibiotics Sold in Pharmaceutical Market in Idumota, Lagos Using HPLC**

Primary Author: Sixtus I. Amadi
Hydrochrom Resources Ltd

Date: Wednesday, March 09, 2016 - Morn
Time:
Room: Exposition Floor, 400 Aisle

Co-Author(s)

Abstract Text

The study was undertaken with the objectives of determining the percentage purity of the active components of thirty two brands of five categories of antibiotics sold in Idumota pharmaceutical market, Lagos, Nigeria. Metronidazole, Tetracycline, Ciprofloxacin, Erythromycin Stearate and Cotrimoxazole (TMP and SMZ) tablets were studied.
The assay was carried out using HPLC with variable wavelength detector and Xbridge C18, 150x4.6mm at wavelengths 276nm, 355nm, 254nm, 200nm and 254nm as in the order of drugs listed above. The mobile phase compositions comprise; Metronidazole (0.1% H3PO4/ACN–60:40), Tetracycline (H20/ACN–50:50), Ciprofloxacin (0.1% TFA/ACN-87:13), Erythromycin Stearate (0.02M KH2PO4/ACN-60:40) and Cotrimoxazole (0.025M NaH2PO4- gradient elution).
The result of the nine brands of Metronidazole gave purity, 98.33, 97.79, 94.46, 69.86, 95.47, 94.99, 103.71, 127.85 111.59%, three of the samples were not within USP range. One of five Tetracycline HCl brands gave purity 143.25% while the others were within range. Six brands of Ciprofloxacin HCl assayed only one, 92.01 was within USP range while others had 81.40, 85.05, 82.64, 88.33, 87.34%. Five brands of Erythromycine Stearate were assayed only one of the brands had purity 93.01% while the others were outside the USP range, 90.0-120.0%. All the seven brands assayed for Cotrimoxazole, none was within the specified USP range.
In conclusion, the thirty two brands of antibiotics assayed, 62.5% did not meet the standard required dosage.

Keywords: Detector, Drugs, HPLC, HPLC Columns
Application Code: Pharmaceutical
Methodology Code: Liquid Chromatography

Session # 1760 **Abstract # 1760-3** **Poster Sessions**

Session Title: Pharmaceutical-IC, LC, and SFC

Abstract Title: **A Sensitive Method for Direct Analysis of Impurities in Apramycin and Other Aminoglycoside-Antibiotics Using Hydrophilic Interaction Liquid Chromatography and Charged Aerosol Detection**

Date: Wednesday, March 09, 2016 - Morn

Time:

Room: Exposition Floor, 400 Aisle

Primary Author: Zhen Long
Thermo Fisher Scientific

Co-Author(s): Evert-Jan Sneekes, Frank Steiner, Ian N. Acworth, Lina Liang, Qi Zhang, Yan Jin

Abstract Text

Aminoglycosides are a group of structurally similar antibiotics used to treat infections caused by aerobic gram-negative bacteria. Analytical methods are required for rapid assessment of drug purity and minor degradants. As they lack a chromophore, these compounds are not amenable to UV detection. Apramycin is an antibiotic used in veterinary medicine. Reported methods for apramycin and impurities analysis usually involves pre- or post column derivatization with UV detection. Such approaches are tedious and time consuming, and may not be able to detect all impurities. Charged aerosol detection directly measures any nonvolatile impurities of aminoglycoside antibiotics, allowing for accurate degradation studies and improved assessments of product purity. A sensitive gradient method for analysis of apramycin and impurities using hydrophilic interaction liquid chromatography with charged aerosol detection (HILIC-CAD) was developed. More than 16 impurities were detected with this method and four were identified with MS. The SCX-UV method, recommened by British Pharmacopoeia (veterinary) 2008, only detected seven impurities, as not all of them could be derivatized by the SCX-UV approach. The HILIC-CAD method is much more sensitive than ELSD. ELSD is less sensitive due to the sigmoidal nature of its response curve, and cannot detect low-level impurities. After an injection of 49.6 mg/mL apramycin, more than 16 impurities (S/N >3) were detected by CAD, while only 12 impurities were detected by ELSD. This method, with or without slight modification, could be used for impurity measurement of an additional eleven aminoglycoside antibiotics.

Keywords: Biopharmaceutical, HPLC, HPLC Detection, Pharmaceutical
Application Code: Pharmaceutical
Methodology Code: Liquid Chromatography

Session # 1760 Abstract # 1760-4 **Poster Sessions**

Session Title Pharmaceutical-IC, LC, and SFC

Abstract Title **Chemically Stable Reversed Phase Chromatography Material Scalable from UHPLC to Semi Prep and Production**

Date: Wednesday, March 09, 2016 - Morn
Time:
Room: Exposition Floor, 400 Aisle

Primary Author Cecilia Mazza
AkzoNobel PPC AB

Co-Author(s) Fredrik Lime

Abstract Text

Classic reversed phase chromatography silica materials are often limited to pH between 2 and 8 due to breakage of the siloxane bonds at low pH or dissolution of silica at high pH. Merged organic/inorganic materials offer the opportunity to work in a wider pH range due to their organosilane reinforcement, therefore analytical, medicinal and drug discovery laboratories adopt these phases when analyzing and purifying mixtures that require harsh pH conditions. The merged silica materials are useful in laboratories where there is a need for repeated injections without shift in retention times operating in an extended pH range, they also serve for fast sample screening and in multi-purpose purifications under same conditions. This new organosilane reinforced materials with C8, C18 and phenyl hexyl derivatizations have a superior mechanical stability, pore size and pore structure. To use the same chemistry across all particle sizes (1.8, 2.5, 5 and 10 μm) with for example reinforced C18 material allows organizations to scale their chromatography from time savings UHPLC all the way to semi prep and full scale production.

Keywords: Chromatography, Drug Discovery, HPLC, Prep Chromatography
Application Code: Pharmaceutical
Methodology Code: Liquid Chromatography

Session # 1760 Abstract # 1760-5 **Poster Sessions**

Session Title: Pharmaceutical-IC, LC, and SFC

Abstract Title: **Characterization and Lot-to-Lot Variability of Complex Surfactants by High Performance Liquid Chromatography and Charged Aerosol Detection**

Primary Author: Marc Plante
Thermo Fisher Scientific

Date: Wednesday, March 09, 2016 - Morn
Time:
Room: Exposition Floor, 400 Aisle

Co-Author(s): Bruce Bailey, Evert-Jan Sneekes, Frank Steiner, Ian N. Acworth

Abstract Text

Surfactants are present in biopharmaceutical products, pharmaceuticals, over-the-counter products, and many processed foods. Polymeric surfactants, such a polysorbate 20, polysorbate 80, Triton X-100, Brij 35, and poloxamer 128 are used to promote the solubility of APIs and the solubility and stability of proteins and possibly other sparingly-soluble components in their formulations. Raw material testing and qualification of these surfactants are an important criterion of quality control measures, and the ICH Q6B Guidelines recommends "liquid chromatographic patterns ... for identity, homogeneity, and purity."

A method (and variants of) was created using a C-18 column and gradient elution chromatography with detection by charged aerosol detector to provide means of analyzing a sample. Data was then processed using a mathematical means of comparing a polymeric surfactant sample to a known standard or reference compound, which serves as a basis of "similarity."

Complete characterization of polysorbate 80 was complete in 34 minutes, providing a chromatogram containing the different subsets of this complex analyte. Peak areas integrated by retention time windows when subtracted from areas obtained from the reference analyte help determine the level of differences that may exist. The sum of these absolute differences can then be used to determine important, actionable decisions concerning product quality so that unacceptable products are not used in final production. Two different lots were compared against a reference lot, and the results are discussed.

Keywords: HPLC, HPLC Detection, Pharmaceutical, Surfactants

Application Code: Pharmaceutical
Methodology Code: Liquid Chromatography

Session # 1760 Abstract # 1760-6 **Poster Sessions**

Session Title: Pharmaceutical-IC, LC, and SFC

Abstract Title: **Direct-Determination of Underivatized Carbohydrates in Biopharmaceutical Formulations Using Heart-Cut, 2D HPLC-HILIC and Charged Aerosol Detection**

Primary Author: Marc Plante
Thermo Fisher Scientific

Date: Wednesday, March 09, 2016 - Morn
Time:
Room: Exposition Floor, 400 Aisle

Co-Author(s): Bruce Bailey, Evert-Jan Sneekes, Frank Steiner, Ian N. Acworth

Abstract Text

The stability of biopharmaceutical protein formulations is often enhanced by the addition of sugar based excipients. Carbohydrates used in these biopharmaceutical products generally consist of simple sugars or sugar alcohols. The determination of carbohydrates in these formulations is complex, with the inclusion of the protein API which may affect the HILIC chromatography. To prevent this and to provide a method for the determination of these sugars, a "heart-cut" reversed-phase HPLC-HILIC method was developed and used to analyse the six carbohydrates: sorbitol, mannitol, sucrose, lactose, maltose, and trehalose in combination with bovine serum albumin as the protein surrogate.

To address detection of the carbohydrates, the Corona charged aerosol detector was used, which is a universal mass-sensitive detector requiring no derivatization of the analytes. The six carbohydrates were separated and calibrated on a HILIC HPLC column, from 15.6 – 2000 ng on column with a quadratic regression. The correlation coefficients were > 0.999 for all six sugars, and limits of quantitation of 15 ng on column. The method injected the sample on to a C4 column, with the hydrophilic analytes transferred to the HILIC column through a heart-cut and mixing with additional acetonitrile. The protein was analyzed by ultraviolet and CAD from the C4 column, while the sugars were analyzed from the HILIC column and detected by CAD during an 18-minute analysis time.

Keywords: Biopharmaceutical, Carbohydrates, HPLC, HPLC Detection
Application Code: Pharmaceutical
Methodology Code: Liquid Chromatography

Session # **1760** **Abstract #** **1760-7** **Poster Sessions**

Session Title: Pharmaceutical-IC, LC, and SFC

Abstract Title: **Fast Method Screening for Chromatographic Separation of Enantiomers Utilizing Polysaccharides Type Chiral Stationary Phases**

Date: Wednesday, March 09, 2016 - Morn
Time:
Room: Exposition Floor, 400 Aisle

Primary Author: Takashi Sato
YMC Co., Ltd.

Co-Author(s): Keiko Kihara, Noriko Shoji, Noritaka Kuroda, Saoko Nozawa, Takatomo Takai

Abstract Text

Mechanism of chiral separation on liquid chromatography is very complicated, and the separation is made by complex combination of various interactions, such as hydrophobic, hydrogen bonding, dipole-dipole, and [pi]-[pi]. This makes method development of chiral separation difficult. Therefore, the column screening is commonly recognized as the first stage of separation method development. The fast column screening is the key driver for the rapid establishment of separation method.

Recently, we developed chiral stationary phases consisting of polysaccharides derivatives coated/immobilized on 3 [micro]m silica gel particle. These new materials are ideal for the fast method screening due to the high column efficiency across a wide range of flow rate. Moreover, they show separation selectivity identical to conventional materials in 5, 10, and 20 [micro]m particle sizes. This feature enables predictable method transfer from ultrafast separation method to conventional method using 5 m material, and even to preparative method.
In this poster, we will present an fast method screening of chiral separation utilizing the 3 [micro]m chiral separation columns through some applications.

Keywords: Chiral Separations, High Throughput Chemical Analysis, HPLC, Pharmaceutical
Application Code: Pharmaceutical
Methodology Code: Liquid Chromatography

Session # 1760 Abstract # 1760-8 **Poster Sessions**

Session Title: Pharmaceutical-IC, LC, and SFC

Abstract Title: **Label-Free Measurement of Sialic Acids Released From Glycoproteins, by High Performance Liquid Chromatography and Charged Aerosol Detection**

Primary Author: Qi Zhang
Thermo Fisher Scientific

Date: Wednesday, March 09, 2016 - Morn
Time:
Room: Exposition Floor, 400 Aisle

Co-Author(s): Bruce Bailey, Evert-Jan Sneekes, Frank Steiner, Ian N. Acworth

Abstract Text

Sialic acids comprise a large family of N- and O-substituted neuraminic acids. Sialic acids can influence the biological and physiochemical properties of glycoproteins, and play an important role in circulatory half-life, biological activity, and solubility of therapeutic glycoproteins. Two important sialic acids, N-acetyl-neuraminic acid (Neu5Ac) and N-glycolyl-neuraminic acid (Neu5Gc), are the most routinely determined glycoprotein products. Released sialic acids can be measured by a number of analytical methods. Spectrometric methods, although easy to perform, tend to overestimate sialic acid content due to the presence of numerous interferences. High performance liquid chromatography (HPLC) with fluorescent labeling has been widely used for accurate and sensitive determination of sialic acids, but derivatization procedures can be time consuming. Presented here is a simple approach for the direct determination of sialic acids using HPLC with charged aerosol detection. The lack of a detectable chromophore in sialic acids is overcome by the universal nature of the charged aerosol detector (CAD) that can quantitatively measure any non-volatile compound. An isocratic HPLC -CAD method was developed for direct determination of sialic acids from glycoproteins following their release by neuraminidase digestion. N-acetyl-neuraminic acid and N-glycolyl-neuraminic acid were measured in less than 12 min with detection limits of approximately 3pmol on column. Response was linear in the range of 15 to 320 pmol on column. The method was evaluated using human and bovine transferrin.

Keywords: Drugs, HPLC Detection, Liquid Chromatography/Mass Spectroscopy, Pharmaceutical
Application Code: Pharmaceutical
Methodology Code: Liquid Chromatography

Session # 1760 Abstract # 1760-9 **Poster Sessions**

Session Title: Pharmaceutical-IC, LC, and SFC

Abstract Title: **Challenges in Developing Analytical Methods for Cleaning Validations in a GMP Regulated Environment**

Primary Author: Xiaohui Yang
Baxter Healthcare

Date: Wednesday, March 09, 2016 - Morn
Time:
Room: Exposition Floor, 400 Aisle

Co-Author(s): Beena Uchil, George Monen, Jane Werling, Lakshmy Nair, Robert Garber, Walter Wasylenko

Abstract Text

Cleanliness of pharmaceutical manufacturing equipment is a GMP requirement for drug substance and drug product manufacturers. Cleaning validation is one of the critical control processes to ensure the effectiveness of the cleaning procedures. Selection of an analytical method for measuring drug residues depends on the chemical nature of the residues and the analytical method sensitivity. For certain highly potent drug products, the limit for drug residue in the manufacturing equipment can be very low (low ppb), which pose challenges in developing analytical methods.

Two types of cleaning methods are required by regulatory agencies. One is to measure the drug residue in rinse water samples and the second one is to measure residue in swab samples. Typical analytical techniques used for cleaning methods are Ultraviolet Spectroscopy (UV), Total Organic Carbon (TOC) analyzer, and High Performance Liquid Chromatography/UV (HPLC/UV) depending on the sensitivity & specificity required for the method.

In this presentation, two cleaning methods will be discussed using HPLC for the measurement of active in rinse and swab samples. Both of these two methods required low ppb levels of sensitivity, were recently developed and validated in our laboratory. HPLC methods used sub-2 micron columns, large injection volume and photodiode array (PDA) detection with extended path length UV flow cell (60mm) to meet such difficult method requirements. The methodology, experimental design and the results from validation studies will be presented.

Keywords: Liquid Chromatography, Pharmaceutical, Validation
Application Code: Pharmaceutical
Methodology Code: Liquid Chromatography

Session # **1760** **Abstract #** **1760-10** **Poster Sessions**

Session Title: Pharmaceutical-IC, LC, and SFC

Abstract Title: **SFC Analysis of Nutraceuticals and Pharmaceuticals Using SFC Optimized Stationary Phases**

Primary Author: Matthew Przybyciel
ES Industries

Date: Wednesday, March 09, 2016 - Morn
Time:
Room: Exposition Floor, 400 Aisle

Co-Author(s)

Abstract Text

Both normal and reversed HPLC is widely used for separation and analytical analysis of many nutraceuticals and pharmaceutical mixtures. Unfortunately, there are mixtures that are not well separated by HPLC leading to incomplete analytical results. An alternative separation technique maybe required such as SFC (supercritical fluid chromatography) to effect a complete separation of many mixtures. SFC provides many unique features including producing high pressure/high speed separations. These features suit SFC well to utilizing columns packed with small particles and SFC optimized stationary phases. It is the purpose of the work to develop high performance columns that have been engineered to specifically for SFC. We will provide examples and applications on how the chromatographer can benefit from these types of stationary phases using the high performance columns. We will demonstrate how these SFC columns can provide for the high resolution separations over a wide variety flow rate conditions and mobile compositions.

Keywords: Natural Products, Pharmaceutical, SFC, Supercritical Fluid Chromatography
Application Code: Pharmaceutical
Methodology Code: Supercritical Fluid Chromatography

Session # 1760 Abstract # 1760-11 **Poster Sessions**

Session Title Pharmaceutical-IC, LC, and SFC

Abstract Title **Analysis of Metoprolol and Select Impurities Using a Hydrophilic Interaction Chromatography Method with Combined UV and Charged Aerosol Detection**

Date: Wednesday, March 09, 2016 - Morn
Time:
Room: Exposition Floor, 400 Aisle

Primary Author Bruce Bailey
Thermo Fisher Scientific

Co-Author(s) Evert-Jan Sneekes, Frank Steiner, Ian N. Acworth

Abstract Text

The drug, metoprolol succinate USP, is a selective beta1-adrenoreceptor blocking agent which reduces chest pain and lowers high blood pressure. Some impurities of metoprolol, such as M and N are non-aromatic □-hydroxyamines, and thus do not possess a detectable UV chromophore. The European Pharmacopeia (EP) indicates that impurities M and N are analyzed by TLC which does not provide good quantitative data. The USP monograph modernization program has indicated that a chromatographic method is more desirable. Charged Aerosol Detection (CAD) is a mass sensitive technique for determining levels of any non-volatile and many semi-volatile analytes after separation and has been used to measure metoprolol and its non-chromophore impurities at levels of 0.1% compared to the active drug.
A HILIC method with UV and charged aerosol detection is described for measuring metoprolol and impurities A, M and N. The method was developed using a tri-modal HPLC column that provides HILIC, anion-exchange and cation-exchange mixed-mode retention mechanisms. The UHPLC system integrates both diode array and charged aerosol detectors so that chromophore and non-chromophore compounds can be simultaneously detected. This provides a suitable technique for quantitative data of both metoprolol and its non-chromophore impurities at levels of 0.1% compared to the active drug.

Keywords: HPLC Detection, Pharmaceutical, Quality Control
Application Code: Pharmaceutical
Methodology Code: Liquid Chromatography

Session # 1760 Abstract # 1760-12

Poster Sessions

Session Title Pharmaceutical-IC, LC, and SFC

Abstract Title **Benefits of Using Mass Detection for Assessing Quality and Purity of Cetrimonium Bromide Pharmaceutical Raw Material**

Date: Wednesday, March 09, 2016 - Morn

Time:

Room: Exposition Floor, 400 Aisle

Primary Author Margaret Maziarz
Waters Corporation

Co-Author(s) Chengxia O'Shea, Christopher Henry, Dominic Moore, Mark Wrona

Abstract Text

Pharmaceutical raw materials are substrates or elements used for manufacturing different drug products. Raw materials include active pharmaceutical ingredients (API), excipients, and other inactive ingredients. The quality and purity of raw materials are critical to the safety of the final drug product and must be controlled. Materials with low purity or containing contaminants may often compromise the safety and efficacy of the end pharmaceutical product, therefore it is essential to have robust and reliable methods for accurate assessment of quality and purity.

Rapid and accurate analysis of components that lack UV chromophores or have low UV-extinction coefficient can be challenging. As these compounds cannot be directly detected by UV, alternative methods must be employed to accurately identify and measure them. In case of cetrimonium bromide, a non-chromophoric material, the current assay methodology listed in the United States Pharmacopeia (USP) Monograph is based on a complex and time consuming titrimetric analysis. Mass detection enables quick, direct and accurate analysis of non-chromophoric compounds, eliminating the need for complex sample-preparation procedure.

In this work, we present a robust and quick UPLC method for analysis of cetrimonium bromide raw material. The UPLC method utilizes a mass detector for fast and accurate assessment of quality and purity. We will demonstrate the system suitability and linearity of the method achievable with mass detection. Then, we will apply this method to confirm identity, assess purity and assay of cetrimonium bromide raw material purchased from three different suppliers.

Keywords: Liquid Chromatography, Liquid Chromatography/Mass Spectroscopy, Pharmaceutical
Application Code: Pharmaceutical
Methodology Code: Liquid Chromatography

Session # **1760** **Abstract #** **1760-13** **Poster Sessions**

Session Title: Pharmaceutical-IC, LC, and SFC

Abstract Title: **Characterization of a Biologic Therapeutic: Reversed Phase/HILIC Analysis of Protein and Excipients**

Primary Author: Bruce Bailey
Thermo Fisher Scientific

Date: Wednesday, March 09, 2016 - Morn
Time:
Room: Exposition Floor, 400 Aisle

Co-Author(s): Evert-Jan Sneekes, Frank Steiner, Ian N. Acworth, Marc Plante

Abstract Text

Therapeutic proteins (e.g., antibodies and vaccines) vary considerably due to the nature and dose of the protein molecule. Aggregation is a major degradation pathway of protein therapeutics during storage. Stabilization of these protein formulations can be enhanced through the addition of specific excipients such as surfactants, amino acids and sugars. An integrated UHPLC system with a UV and universal charged aerosol detection offers multi-mode detection for the simultaneous analysis of both non-chromophore and chromophore containing compounds.

The chromatographic separation of therapeutic protein and amino acid excipients was performed using a parallel RP/HILIC separation. Proteins were separated using a C4 column followed by detection via diode array detector. A "heart cut" containing the polar amino acids from the sample injection was transferred to a second column via a switching valve. The separation of underivatized amino acid excipients and several ions was achieved using HILIC with charged aerosol detection. The mixed mode column provides cation, anion and reversed phase separation characteristics so gradient conditions could be adjusted by selecting appropriate ionic buffer strength, pH and level of organic solvents. A biologic formulation containing a mixture of surfactant, amino acids, ions and protein was successfully analyzed demonstrating the capability of the described RP/HILIC method.

Keywords: Amino Acids, Biopharmaceutical, HPLC Detection, Protein
Application Code: Pharmaceutical
Methodology Code: Liquid Chromatography

Session # 1760 Abstract # 1760-14 **Poster Sessions**

Session Title: Pharmaceutical-IC, LC, and SFC

Abstract Title: **Efficient, Effective and Proven Approach to Chiral Method Development for Purification Scale Up**

Date: Wednesday, March 09, 2016 - Morn
Time:
Room: Exposition Floor, 400 Aisle

Primary Author: Sean Orlowicz
Phenomenex

Co-Author(s): Michael McCoy

Abstract Text

With the majority of today's pharmaceuticals being chiral compounds, and a renewed focus on the differences in biological activity of isomers, chiral separations and isolations have continued to grow in importance. Our lab screened hundreds of compounds on various Chiral Stationary Phases (CSPs), and over the years has developed an efficient experimental strategy for identifying the appropriate CSP for a unique chiral separation. Due to the success this chiral screening service, many customers have returned asking for the isolation of pure enantiomer. As a result, we have explored the additional challenges of chiral separations as we scale up to preparative HPLC. In this poster, we demonstrate through case study, a unique and effective approach to both CSP screening and subsequent scale up for chiral small molecules. When screening compounds for the appropriate CSP, the analyte's physical properties often define the mode of chromatography most likely to be successful for any given CSP. Screening Normal Phase, Reversed Phase and Polar Organic modes of chromatography on a selection of 6 orthogonal polysaccharide-type phases has proven to be successful for more than 80% of the unique submissions. Mobile phase composition in each mode is optimized to achieve resolution. When isolating enantiomers, we must further consider the physical and chemical properties of the analytes. Preparative isolations on various CSPs can be drastically affected by scale up phenomena such as; loading capacity of the CSP itself, stability of the sorbent and nonlinear chromatography. In summary, though real world examples we have developed an effective strategy for identifying chromatographic conditions for chiral separation of small molecules with moderate chiral complexity. We have successfully demonstrated the ability to transition those analytical separations to preparative isolations.

Keywords: Chiral Separations, Liquid Chromatography, Method Development, Pharmaceutical
Application Code: Pharmaceutical
Methodology Code: Liquid Chromatography

Session # 1760 Abstract # 1760-15 **Poster Sessions**

Session Title Pharmaceutical-IC, LC, and SFC

Abstract Title **Guidelines for Method Transfer and Optimization of the Newest Charged Aerosol Detector**

Date: Wednesday, March 09, 2016 - Morn
Time:
Room: Exposition Floor, 400 Aisle

Primary Author Marc Plante
Thermo Fisher Scientific

Co-Author(s) Bruce Bailey, Evert-Jan Sneekes, Frank Steiner, Ian N. Acworth, Paul Gamache

Abstract Text

No single liquid chromatography (LC) detector delivers ideal results. Often one analyte responds more strongly than another, or may not respond at all. What is most desired is the ability to accurately measure a wide range of analytes with consistent response, simultaneously. Charged aerosol detection (CAD) is a mass sensitive technique for determining levels of any non-volatile and many semi-volatile analytes after separation by liquid chromatography. This technique provides consistent analyte response independent of chemical characteristics, and gives greater sensitivity as well as having a wider dynamic range than other nebulizer-based detectors. The response to an analyte does not depend on optical properties, as with UV-vis absorbance, or the ability to ionize, as with mass spectrometry (MS). There is no need for the presence of chromophores, radiolabels, ionizable moieties, or chemical derivatization for detection. HPLC and UHPLC methods using CAD have limits of detection of between mid-picograms to low nanograms (on column) and have a wide dynamic range from nanogram to microgram levels, with high reproducibility.

Since the introduction of this technology in 2004, the charged aerosol detector is now a mature, fourth-generation product. The version of this detector shows a number of improvements over its predecessors: a new concentric nebulizer design forms a stable aerosol, which increases both assay reproducibility and sensitivity, the use of evaporation temperature enables an expanded range of mobile phase constituents to be used, including basic mobile phases, and a linearity function can also be used for linear calibration curves. Comparative data is presented and certain guidelines for method transfer from the previous generation product are highlighted. Superior performance of the newest charged aerosol detector versus evaporative light scattering detection is also discussed.

Keywords: HPLC, HPLC Detection, Instrumentation, Optimization
Application Code: Pharmaceutical
Methodology Code: Liquid Chromatography

Session # 1760 Abstract # 1760-16 **Poster Sessions**

Session Title: Pharmaceutical-IC, LC, and SFC

Abstract Title: **Chiral Separation Using SFC and HPLC**

Primary Author: Hidetoshi Terada
Shimadzu

Date: Wednesday, March 09, 2016 - Morn
Time:
Room: Exposition Floor, 400 Aisle

Co-Author(s): Arao Yohei, Funada Yasuhiro, Takato Uchikata, Tanaka Kenichiro

Abstract Text

It is important that the efficacy and safety of compounds are evaluated as enantiomers especially in pharmaceutical formulations and its related industries. Chiral separation using SFC and HPLC is one of the typical methods for purifying enantiomers from racemic mixtures. In this method, the suitable column and mobile phase for targeted chiral separation have to be evaluated before starting the analysis. Evaluating analysis conditions for chiral compounds requires trial and error in order to find suitable combinations of columns and mobile phases to separate the component of interest. This process requires a lot of work and time.

We have developed a method screening system and workflow using SFC and HPLC to evaluate chiral separation more efficiently. This system have two solvent delivery pumps and one carbon dioxide delivery pump, and can be used for SFC and HPLC with the single instrument by only switching the pumps. The system is configured by installing a column switching valve inside the oven and a solvent switching valve within solvent delivery pumps, thereby permitting comprehensive data collection while continuously switching through multiple combinations of columns and mobile phases automatically by using dedicated control software. Here we report the process of high efficiency method development of chiral compounds by using SFC and HPLC.

Keywords: Chiral, HPLC, Method Development, Supercritical Fluid Chromatography
Application Code: Pharmaceutical
Methodology Code: Supercritical Fluid Chromatography

Session # **1760** **Abstract #** **1760-17** **Poster Sessions**

Session Title: Pharmaceutical-IC, LC, and SFC

Abstract Title: **Determination of Ascorbic Acid in Citrus Fruits and Pharmaceutical Formulations by Hydrophilic Interaction Chromatography (HILIC)**

Primary Author: Yuegang Zuo
University of Massachusetts Dartmouth

Date: Wednesday, March 09, 2016 - Morn

Time:

Room: Exposition Floor, 400 Aisle

Co-Author(s): Ruiting Zuo, Si Zhou, Yiwei Deng

Abstract Text

Ascorbic acid (AA), also known as vitamin C, is a natural antioxidant and plays an important role in various physiological processes such as biosynthesis of collagen, norepinephrine, peptide hormones, and tyrosine, intestinal absorption of iron, defense against cellular oxidation, quenching of free radicals, and prevention of cardiovascular disease and cancer. Since humans cannot synthesize AA, they must take this essential nutrient from foods and beverages. Citrus fruits are a major source of vitamin C, many analytical methods such as spectrophotometry, voltammetry, GC, CE, and HPLC have been developed for the determination of ascorbic acid in fruit juices. Due to the complex matrices with numerous nonspecific interferences for the AA determination in fruits by other techniques, HPLC has gained popularity [Zuo, Y. Ed., High-Performance Liquid Chromatography (HPLC): Principles, Procedures and Practices. Nova Science Publishers, Inc., New York (2014)]. As AA is very polar small compound, it is difficult to be retained in conventional RP-HPLC and separated from void volume. A rapid and sensitive hydrophilic interaction chromatography (HILIC) method has been developed here for the determination of AA in citrus fruits and pharmaceutical formulations. Fresh fruits were hand-squeezed, pharmaceutical pills were dissolved in the HILIC mobile phase, and then centrifuged, filtered, diluted and separated on a Waters Spherisoro S5NH2 column fitted with a C18 guard column using an isocratic elution program consisting of a 50% acetonitrile and 50% aqueous phosphate buffer solution. The UV detection was made at the wavelength of 205 nm. AA had a good retention on the amino column and was well separated from other components in citrus juices and pharmaceutical matrices. A typical chromatogram of an orange sample is presented in Fig. 1. The developed HILIC technique has been also successfully applied to examine the absorption and metabolism of AA in human fluids.

Keywords: Agricultural, Bioanalytical, Clinical Chemistry, Drugs
Application Code: Food Science
Methodology Code: Liquid Chromatography

Session # 1760 Abstract # 1760-18 **Poster Sessions**

Session Title Pharmaceutical-IC, LC, and SFC

Abstract Title **HPLC Method Development and Validation for the Assay and Organic Impurities of Naproxen in Naproxen API, Naproxen Delayed-Release Tablets and Naproxen Oral Suspension**

Primary Author Jennifer Fedorowski
United States Pharmacopeial Convention

Date: Wednesday, March 09, 2016 - Morn
Time:
Room: Exposition Floor, 400 Aisle

Co-Author(s) Arindam Ganguly, Jennifer Belsky, Joshua Bhattacharya, Natalia Kouznetsova

Abstract Text

A stability-indicating HPLC-UV method was developed and validated for the Assay and Organic Impurities of naproxen in drug substance and two drug products. The method utilizes a 4.6 × 150 mm, 3.5 µm C18 (L1) column maintained at 35º, gradient elution with a mobile phase consisting of 0.1% acetic acid in water and 0.01% acetic acid in methanol and a flow rate of 1.0 mL/min. UV detection was executed at 272 nm for Assay and 240 nm for Organic Impurities. The method separates naproxen from the 14 naproxen related compounds (RC) listed by the European Pharmacopeia (EP) with a resolution of more than 2.0. All inactive ingredients of the drug products were well separated from the naproxen peak and four specified impurities. Forced degradation studies resulted in the partial degradation (NMT 15%) of naproxen under oxidative and photolytic stress conditions. The naproxen peak was observed to be spectrally pure in all stressed samples. In addition, the method was shown to be robust to small changes in flow rate (±10%), mobile phase composition (±5%), and column temperature (±10º).

The Assay of naproxen was validated in the range of 80% –120% (drug substance) and 70% -130% (drug product) of the nominal concentration of 0.1 mg/mL. The specified impurities were validated at the ICH limit of 0.1% (range of 0.05% - 1.0%) for drug substance and at the ICH limit of 0.2% (range of 0.1% - 1.0%) for drug product. Both procedures were found to be specific, linear, accurate and precise.

Keywords: Chromatography, Method Development, Pharmaceutical, Validation

Application Code: Pharmaceutical

Methodology Code: Liquid Chromatography

Session # 1760	Abstract # 1760-19	Poster Sessions

Session Title: Pharmaceutical-IC, LC, and SFC

Abstract Title: **Sample Analysis of Compounds with Multiple Chiral Centers by Two-Dimensional HPLC**

Date: Wednesday, March 09, 2016 - Morn

Time:

Room: Exposition Floor, 400 Aisle

Primary Author: Chi Tsang
Genentech

Co-Author(s): Kelly Zhang

Abstract Text

More than half of the drugs currently in use are chiral compounds. Although they have the same chemical formula, most isomers of chiral drugs exhibit marked differences in biological activities such as pharmacology, toxicology, pharmacokinetics, and metabolism etc. Therefore, an analytical method to promote chiral separation and assess the purity of the correct drug form is highly critical. By utilizing a heart-cutting two-dimensional liquid chromatography technique, a reversed phase achiral HPLC method in the first dimension with time-triggered fraction collection and subsequent chiral reversed phase HPLC analysis in the second dimension, complete resolution of eight enantiomers in the mixture was possible.

A 2D UHPLC method was developed to separate eight enantiomers with multi-chiral centers. The final chromatographic separation was achieved on Ascentis Express C18 achiral column in the first dimension. Our in-house customized UHPLC-2DLC gives us the flexibility to analyze the collected fractions with multiple columns Chiralcel OD-3 and Chiralpak AS-3 chiral stationary phase in the second dimension. In addition, UHPLC method was developed in both the 1D and 2D HPLC analysis which gives a complete analysis in less than 20 minutes.

Keywords: Chiral Separations, Liquid Chromatography, Method Development, Pharmaceutical
Application Code: Pharmaceutical
Methodology Code: Liquid Chromatography

Session # 1760   Abstract # 1760-20   **Poster Sessions**

Session Title   Pharmaceutical-IC, LC, and SFC

Abstract Title   **Mobile Phase Effects in Reversed-Phase Chromatography of Monoclonal Antibodies at High Temperature**

Primary Author   Hillel K. Brandes
Sigma-Aldrich/Supelco

Date: Wednesday, March 09, 2016 - Morn
Time:
Room: Exposition Floor, 400 Aisle

Co-Author(s)

Abstract Text

Non-specific protein adsorption onto an HPLC column has been shown to affect protein recovery, peak shape and resolution. Thus in order to get complete recovery and improve chromatographic performance, it is necessary to select a mobile phase that reduces interaction between the protein and the accessible surface area in the column. We previously identified that optimal reversed-phase (RP) chromatography of monoclonal antibodies is generally best performed with conventional water-acetonitrile gradients at high temperature (60 - 80° C); this observation is supported throughout the literature. This presentation extends this concept to examine the impact of secondary mobile phase modifiers on protein recovery and chromatographic performance. Briefly, the impact of the addition of several aliphatic primary alcohols on monoclonal antibodies (mAbs) recovery, peak shape and resolution was examined as a function of temperature. Interestingly, it was found that adding certain, but not all, aliphatic primary alcohols dramatically reduced the temperature required to achieve optimum recovery and chromatography of mAbs. Thus, the mechanisms by which the mobile phase modifiers operate are not entirely straight forward. This data will benefit analysts assessing optimum conditions for the separation of proteins or mAbs during development or biomarker research.

Keywords: Biopharmaceutical, HPLC
Application Code: Other
Methodology Code: Liquid Chromatography

Session # 1760 Abstract # 1760-22 **Poster Sessions**

Session Title: Pharmaceutical-IC, LC, and SFC

Abstract Title: **Utilization of HPTLC and Diffuse Reflectance Spectroscopy to Quickly Evaluate Product Quality of Cotrimoxazole Tablets from Tanzania**

Primary Author: David W. Jenkins
FHI 360

Date: Wednesday, March 09, 2016 - Morn
Time:
Room: Exposition Floor, 400 Aisle

Co-Author(s): Eliangiringa Kaale, Samuel Hope, Thomas Layloff

Abstract Text

In order to improve access to life saving medications under the U.S. President's Emergency Plan for AIDS Relief (PEPFAR), the "Partnership for Supply Chain Management" (PFSCM) has established a program to procure locally manufactured product within the same country as the destination health facilities / clinics. Tanzania has been the prime country for implementation, where PFSCM in support of the "Supply Chain Management System" (SCMS) project purchases on a consignment basis, large amounts of cotrimoxazole tablets produced by a Tanzanian manufacturer. The manufacturer is not prequalified by the World Health Organization (WHO) or approved by a Stringent Regulatory Authority, but is registered in Tanzania and has been qualified through the PFSCM's internal quality assurance program. In order to assure that the products are consistently produced under good process controls, PFSCM has instituted a diffuse reflectance spectroscopy procedure with periodic assessment by assay and dissolution testing using validated HPTLC techniques (with weight variation and disintegration testing). The implementation of this program has demonstrated improvements of the ongoing product quality from this local manufacturer. Based on data from primary test methods, the first group of product yielded less than 80% compliance, where subsequent groups improved to more than 99% compliance. Being faster than traditional pharmaceutical testing techniques, this approach provides a mechanism for assessing product quality and could be applied to other products.

Keywords: Analysis, Chemical, Chromatography, Near Infrared
Application Code: Pharmaceutical
Methodology Code: Portable Instruments

Session # **1760** **Abstract #** **1760-23** **Poster Sessions**

Session Title: Pharmaceutical-IC, LC, and SFC

Abstract Title: **Amino Acid Analysis for Qualitative and Quantitative Composition of Pharmaceutical Products**

Primary Author: Natalia Belikova
SGS Life Science Services

Date: Wednesday, March 09, 2016 - Morn
Time:
Room: Exposition Floor, 400 Aisle

Co-Author(s)

Abstract Text

The United States Pharmacopoeial Convention (USP), European Pharmacopoeia (Ph. Eur.) and Japanese Pharmacopoeia (JP) define requirements for the qualitative and quantitative composition of medicines, as well as the tests to be carried out on medicines and on substances and materials used in their production. Amino Acid Analysis can be used for identification testing for biopharmaceutical active ingredients, impurities and related substances determination and single or total amino acid quantitation in drug products. The objective of the project was to develop a liquid chromatography method for simultaneous determination ~20 amino acids in simple and complex mixtures. Method utilizes mobile phase gradient program as well as column temperature gradient program for separating amino acids of interest. After separation, eluent underwent post-column ninhydrin derivatization with subsequent detection at 570 nm and 440 nm for amino and imino acids, respectively. The developed methods were optimized to comply with system suitability requirements per Ph. Eur and JP monographs. Methods were successfully verified and used for analysis of ninhydrin-positive substances in individual amino acid samples per Ph. Eur. as well as in de-formulation projects to monitor excipient levels in biopharmaceutical products.

Keywords: Amino Acids, Analysis, Liquid Chromatography, Pharmaceutical
Application Code: Pharmaceutical
Methodology Code: Liquid Chromatography

Session # 1760 Abstract # 1760-24 **Poster Sessions**

Session Title: Pharmaceutical-IC, LC, and SFC

Abstract Title: **Separation and Detection of Small Molecules by Low-Flow Liquid Chromatography Mass Spectrometry for Pharmacokinetic Studies**

Primary Author: James N. Marr
Merck & Co.

Date: Wednesday, March 09, 2016 - Morn

Time:

Room: Exposition Floor, 400 Aisle

Co-Author(s): Gary Adamson, Rena Zhang

Abstract Text

Standard LC flow rates of 500 µL/min or higher coupled with SRM detection is the primary analytical platform used in small molecule bioanalytical labs, allowing for high sample throughput. Nano or low micro-flow chromatography is generally reserved for proteomic analysis where throughput is not a priority. We evaluated low-flow chromatography to understand the relationship between flow rates, peak resolution, and ionization efficiency to determine the feasibility of low-flow LC for small molecule bioanalysis.

Using an API 4500 triple quadrupole MS utilizing electrospray ionization for our detector, a nanoACQUITY run at 100 µL/min with a 50 x 1 mm ID column was compared to a Thermo LX-2 run at 750 µL/min with a 50 x 2.5 mm ID column. The nanoACQUITY showed an increase in signal to noise for 7 of 10 structurally diverse compounds in plasma extract. To achieve the same peak resolution for all 10 compounds, it took the nanoACQUITY 10 minutes compared to 3 minutes on the LX2. Research has shown that equivalent response is achieved when scaling linear velocities and injection volumes with the square of the column radius, limiting column injection volume. While the main benefits of low-flow chromatography are a gain in sensitivity and the ability to inject from low sample volumes, the high cost of sample throughput to achieve the same peak resolution makes utilizing low flow systems less than ideal. The key is to find a low-flow system capable of performing analyte separations at the same efficiency as the standard flow systems.

Keywords: HPLC Columns, Liquid Chromatography/Mass Spectroscopy, Pharmaceutical, Plasma
Application Code: Bioanalytical
Methodology Code: Liquid Chromatography/Mass Spectrometry

Session # 1770 Abstract # 1770-1 **Poster Sessions**

Session Title: Pharmaceutical-MS, LC/MS and Others

Abstract Title: **A New Approach to the Automated Identification of Metabolites in Multi-Vendor Datasets**

Primary Author: Richard Lee
ACD/Labs

Date: Wednesday, March 09, 2016 - Morn
Time:
Room: Exposition Floor, 400 Aisle

Co-Author(s): Alexandr Sakharov, Andrey Paramonov, Rytis Kubilius, Vitaly Lashin

Abstract Text

For scientists involved in the study of drug metabolism the challenge of accurately and rapidly identifying metabolites is met using a variety of techniques and software packages. Whilst most instruments vendors offer software to help identify metabolites they are of little use in a multi-vendor laboratory, and they lack the flexibility and customisation required. Most users have to then resort to purchasing additional software and using Microsoft Excel to get a complete workflow.

Here we present a new approach for the automated identification of metabolites, which allows data from nearly all the mass spectrometry vendors to be processed, reviewed and databased. The automated file capture and processing capabilities makes it suitable for high-throughput environment whilst the biotransformation scientist can check the results and make any changes such as modifying an assignment. The batch processing allows multiple time point samples to be processed and the calculation of pharmacokinetic parameters such as area under the cure (AUC). To help reduce the number of false positives a structure based prediction approach is used. Confirmation of the site of biotransformation is checked using the available MSMS data. The metabolite fragment mass shifts, relative to the parent MSMS spectrum, help localise the site of biotransformation. In the cases where there is not sufficient evidence to support a single site of biotransformation, then the metabolite structure can be represented using the Markush notation. Unexpected metabolites are also be identified using a combination of mass defect filtering, control sample comparison and component profiling. All the metabolites and metadata can then be stored in a database for future use. This allows for a greater degree of collaboration between the discovery and development departments which can save a huge amount of time and effort. Answering the question 'Have I seen this metabolite before?' then becomes very easy.

Keywords: Informatics, Mass Spectrometry, Metabolomics, Metabonomics
Application Code: Pharmaceutical
Methodology Code: Mass Spectrometry

Session # 1770 **Abstract # 1770-2** **Poster Sessions**

Session Title: Pharmaceutical-MS, LC/MS and Others

Abstract Title: **Analysis of Atropine Sulfate by HPLC Using Mass Spec HPLC Mobile Phase**

Primary Author: Jeffrey Kakaley
YMC America Inc.

Co-Author(s): Ernest Sobkow

Date: Wednesday, March 09, 2016 - Morn
Time:
Room: Exposition Floor, 400 Aisle

Abstract Text

When developing modern analytical HPLC methods, today's scientist typically tries to avoid using ion-pairing agents in their mobile phase(s). This becomes especially important should the method need to be used with mass spec detection. This poster investigates a MS-friendly HPLC method for Atropine Sulfate as an alternative to the traditional USP method that incorporates ion-pairing agent.

Keywords: Chromatography, HPLC, Liquid Chromatography/Mass Spectroscopy, Pharmaceutical
Application Code: Pharmaceutical
Methodology Code: Liquid Chromatography/Mass Spectrometry

Session # 1770 Abstract # 1770-3 **Poster Sessions**

Session Title Pharmaceutical-MS, LC/MS and Others

Abstract Title **Alternative Oxidation Technique for Pharmaceutical Forced Degradations**

Primary Author Bradley Kumagai
Theravance Biopharma

Date: Wednesday, March 09, 2016 - Morn
Time:
Room: Exposition Floor, 400 Aisle

Co-Author(s) Claudine Ooi, Ken Ngim

Abstract Text

The forced degradation of pharmaceutical actives is performed routinely as an aid for developing and validating stability-indicating chromatography methods. Pharmaceutical oxidations are typically performed using hydrogen peroxide, which reacts as a nucleophile or electrophile, and free radical initiators. In this study, solutions of a developmental pharmaceutical active (~ 0.1 mg/mL in water) with dilute hydrogen peroxide (5:1 to 45:1 molar ratios of hydrogen peroxide:active) are irradiated by UV light to facilitate the reaction of generated hydroxyl radical with the active. The desired decomposition for pharmaceutical forced degradations (~10%) is achieved within 3 hours, and the reactivity is approximately proportional to the reagent content used and is quenched by turning off the light source. The resultant oxidation products of the pharmaceutical active, which were characterized by LC/MS, qualitatively approximated those observed in actual stability samples. The use of UV-irradiated hydrogen peroxide for forced degradation proved to be a rapid, controlled and relevant means of oxidizing a developmental active.

Keywords: HPLC, Mass Spectrometry, Method Development, Validation
Application Code: Pharmaceutical
Methodology Code: Liquid Chromatography/Mass Spectrometry

Session # 1770 **Abstract # 1770-4** **Poster Sessions**

Session Title: Pharmaceutical-MS, LC/MS and Others

Abstract Title: **Investigating Transdermal Diffusion of Vitamin D and 25-Hydroxyvitamin D**

Date: Wednesday, March 09, 2016 - Morn
Time:
Room: Exposition Floor, 400 Aisle

Primary Author: Marcel Musteata
Albany College of Pharmacy

Co-Author(s): Isaac Mall

Abstract Text

Objective. Transdermal extraction is proposed for vitamin D3 and its metabolites to overcome problems associated with blood analysis. The main challenge for this analytical method is to overcome the barrier properties of skin, especially for very lipophilic compounds such as vitamin D and its metabolites.

Significance. In the last decade, the scientific and medical community was confronted with a renewed interest in vitamin D and its metabolites, interest prompted by the new discoveries regarding the association between the members of the vitamin D family and a great number of physiological functions and pathological states. However, the concentration of vitamin D and its metabolites in biological samples is very difficult to measure. Since currently used procedures are lengthy and result in analyte degradation, we are investigating a simpler vitamin D analysis procedure based on transdermal extraction patches.

Methods. Extraction patches with a thickness of 127µm were placed for various time intervals on the stratum corneum side of skin patches placed in Franz diffusion cells. After extraction, the patches were analyzed by LC-MS/MS to determine the amount of vitamin D and 25(OH)D extracted, using a gradient of water:methanol with modifiers as mobile phase. The experiments were repeated at various concentrations of analytes in the donor phase section of the diffusion cells. Furthermore, the influence of several penetration enhancers on transdermal diffusion was investigated.

Results. The flux of vitamin through the skin was found to be 2.3 pg/cm2/hr for untreated skin and 12 pg/cm2/hr for skin treated with penetration enhancers, suggesting that amounts of vitamin D high enough for detection by LC-MS/MS can be extracted in a matter of hours.
Conclusions. The research shows promise that transdermal extraction could be an effective analysis route for vitamin D3, especially when ethanol and dodecylamine are used as penetration enhancers.

Keywords: Bioanalytical, Biological Samples, Extraction, Pharmaceutical
Application Code: Pharmaceutical
Methodology Code: Liquid Chromatography/Mass Spectrometry

Session # 1770 Abstract # 1770-5 **Poster Sessions**

Session Title Pharmaceutical-MS, LC/MS and Others

Abstract Title **Calibration Free, Semi-Quantitative Analysis of Defined Drug Formulations Using FTIR Pre-computed Mixture Spectra**

Primary Author William Costa
Fiveash Data Management, Inc.

Co-Author(s) Bill McCarthy, Todd Strother

Date: Wednesday, March 09, 2016 - Morn
Time:
Room: Exposition Floor, 400 Aisle

Abstract Text

The significant advantages of Attenuated Total Reflectance (ATR) Fourier Transform Infrared (FTIR) spectroscopy (e.g., non-destructive and fast analysis, reproducible results with absorbance linear to concentration, availability of spectral libraries, and the ability to provide quantitative results) have made this technique the method of choice for many industrial and research applications. ATR is a particularly important technique for pharmaceutical and forensic analysis. Previous ATR libraries for pharmaceuticals has been limited to primarily neat (>90% pure) ingredients but most drug formulations are mixtures of varying concentrations of active and inactive ingredients that contribute to the overall spectral signature. Creating samples of all likely concentrations of these drugs is often not feasible because of the high cost and intensive labor involved.

This presentation will describe the utilization of a new ATR spectral library of pre-computed mixture spectra to provide a semi-quantitative result of relative concentrations. The use of such libraries does not require the creation of calibration or validation standards. Semi-quantitative results for defined drug mixtures will be evaluated. The advantages and disadvantages of this rapid screening method for the forensic science community and the pharmaceutical industry will be discussed.

Keywords: Drugs, Forensics, FTIR, Pharmaceutical
Application Code: Pharmaceutical
Methodology Code: Molecular Spectroscopy

Session # 1770 Abstract # 1770-6 **Poster Sessions**

Session Title: Pharmaceutical-MS, LC/MS and Others

Abstract Title: **Novel Self-Patented Gold Nanoparticle Synthesis, Characterization and Antibacterial Susceptibility Testing**

Primary Author: William Hamilton
Western Kentucky University

Date: Wednesday, March 09, 2016 - Morn
Time:
Room: Exposition Floor, 400 Aisle

Co-Author(s): Fenil Chavda, Jason N. Payne, Rajalingam Dakshinamurthy

Abstract Text

With soaring increase in the cases of multi-drug resistant (MDR) bacteria all over the world, we are on the verge of entering post-antibiotic era if no immediate action is taken against this global crisis. As an alternative route to modify current commercial antibiotics, we made an attempt to design an array of effective antibacterial agents involving gold nanoparticles (AuNPs) conjugated to an antibiotic, like those of the aminoglycoside, cephalosporin, and carbapenem drug classes. Due to recent emergence of infections due to both Gram-positive and Gram-negative bacterial strains with advanced patterns of antimicrobial resistance bactericidal agents such as these are being view as a prime candidates for further development and augmentation. Unlike conventional methods, a unique self-patented green process was used for AuNPs synthesis wherein the antibiotic assists in both reducing and stabilizing the AuNPs resulting in antibiotic conjugated gold nanoparticles (Ab-AuNPs) which were morphologically characterized using transmission electron microscope (TEM), UV-Vis spectroscopy, scanning electron microscopy/energy-dispersive X-ray spectroscopy (SEM-EDS), and dynamic light scattering (DLS). The presence of ligand (antibiotic) onto AuNPs was confirmed using TGA analysis. Antibacterial efficiency was evaluated on Gram-positive and Gram-negative bacterial strains using turbidmetric and spread plate assay. AuNPs activity was further confirmed with propidium iodide assay. Super-thin cross-sections of bacteria treated with Ab-AuNPs observed under TEM showed bactericidal activity by causing perforations and disturbing the cellular environment leading to cell lysis and apoptosis. The minimum inhibitory concentrations (MIC) of Ab-AuNPs was significantly less when compared to pure antibiotic drugs which proves the synergistic activity of Ab-AuNPs.

Keywords: Drug Discovery, Drugs, Nanotechnology, Pharmaceutical
Application Code: Pharmaceutical
Methodology Code: UV/VIS

Session # 1770 Abstract # 1770-8 **Poster Sessions**

Session Title Pharmaceutical-MS, LC/MS and Others

Abstract Title **Square Wave Adsorptive Stripping Voltammetric Determination of Ketoconazole Drug in the Saudi Market**

Primary Author Abdel-Nasser Kawde
King Fahd University of Petroleum and Minerals

Date: Wednesday, March 09, 2016 - Morn
Time:
Room: Exposition Floor, 400 Aisle

Co-Author(s) Mohamed A. Morsy

Abstract Text

The widespread use of the Ketoconazole compound in its three formulations, cream, gel and shampoo, and the need for clinical and pharmacological study require fast and simple techniques for determination of its presence in biological fluids. Electrochemical investigations of these drugs in particular and developing new electrochemical methods will help in monitoring the quality of the commercially available drugs in the Saudi market. A simple, rapid and reliable electrochemical method, square wave adsorptive stripping voltammetry, for the determination of Ketoconazole (KTZ) in pure and drug formulation forms is developed. Development of such methods helps in monitoring the quality of the available commercial medicinal products in the Saudi market, and thus will have a great impact on human health. Square-wave adsorptive stripping voltammetry of KTZ measured at glassy carbon electrode (GCE) surfaces at different pHs, 0.1 M phosphate buffer solution, pH 7.0, showed the best anodic peak For KTZ oxidation. Under optimum conditions, a dynamic calibration curve (R2 = 0.97) at KTZ concentrations of 10–25000 ppb is obtained with a limits of detection (3) and quantification of 3 ppb and 10 ppb KTZ, respectively.

Keywords: Electrochemistry, Pharmaceutical, Quantitative, Voltammetry
Application Code: Pharmaceutical
Methodology Code: Electrochemistry

Session # 1770 Abstract # 1770-10 **Poster Sessions**

Session Title: Pharmaceutical-MS, LC/MS and Others

Abstract Title: **A Comparison of Complete Dissolution Versus Leach of the Target Analytes by Using and Omitting HF**

Primary Author: Dan Iversen
CEM

Date: Wednesday, March 09, 2016 - Morn
Time:
Room: Exposition Floor, 400 Aisle

Co-Author(s): Tyler Edwards

Abstract Text

The completed dissolution of pharmaceuticals required by the upcoming USP method 232/233 utilizes Hydrofluoric Acid(HF). Many labs shy away from the use of HF due to increased safety hazards and additional handling procedures. This study will compare a complete dissolution and a leach of the target analytes by using and omitting HF. The types of products to be tested will range from gel caps, safety caps, and pressed pills. The results will be compared and presented.

Keywords: Atomic Spectroscopy, High Temperature, Pharmaceutical, Sample Preparation
Application Code: Pharmaceutical
Methodology Code: Atomic Spectroscopy/Elemental Analysis

Session # 1770 Abstract # 1770-11 **Poster Sessions**

Session Title: Pharmaceutical-MS, LC/MS and Others

Abstract Title: **Standardization of Experimental Conditions of USP Melting Point Reference Standards in DSC Applications**

Primary Author: Osomwonken Igbinosun
United States Pharmacopeial Convention

Date: Wednesday, March 09, 2016 - Morn
Time:
Room: Exposition Floor, 400 Aisle

Co-Author(s): Antonio Hernandez-Cardoso, Guillermo A. Casay, Kanda K. Balasubramanian, Steven T. Rau

Abstract Text

The pharmaceutical industry has seen a considerable increase in the use of differential scanning calorimetry (DSC) for melting point determinations. This led to the publication of studies exploring parameters from DSC measurements and providing melting points for various materials. A literature search was initiated to determine the spread of values reported for melting point standards. The result of the search found that the melt onset and the peak temperatures are used interchangeably for the same material. For consistency, USP is proposing to provide system suitability conditions to ensure reproducible reporting. USP laboratories evaluated the effects of heating rate, pulverization, and amount of sample on the melting point measurement using established USP Melting Point Standards (MPS) using Thermal Analysis <891>. The melt onset temperature of dried material, gently pulverized, at 0.1 °/min, 0.5 °/min, 1.0 °/min, 3.0 °/min, 5.0 °/min, 10.0 °/min and 20.0 °/min heating rates, remained constant. DSC values gathered for 4 consecutive days on two different lots of the same material showed small standard deviations (< 0.0500). USP will propose experimental conditions for the use of USP MPS in system suitability testing of the DSC apparatus. USP proposes to add "melt onset temperature by DSC" to the label text of all the USP MPS.

Keywords: Pharmaceutical, Thermal Analysis
Application Code: Pharmaceutical
Methodology Code: Thermal Analysis

Session # 1770 Abstract # 1770-13 **Poster Sessions**

Session Title: Pharmaceutical-MS, LC/MS and Others

Abstract Title: **A Co-Crystal of Febuxostat and Isonicotinamide: Synthesis and Characterization**

Date: Wednesday, March 09, 2016 - Morn

Time:

Room: Exposition Floor, 400 Aisle

Primary Author: Yanlei Kang
Zhejiang University

Co-Author(s): Dongdong Yu, Jianguang Zhou, Jianming Gu, Xiurong Hu

Abstract Text

Febuxostat was investigated for its good effect for the treatment of gout, however, its low solubility which seriously affected the drug efficacy became the bottleneck of its development. In recent years, co-crystal has increasingly being applied to enhance the drug solubility by connecting the active pharmaceutical ingredient (API) and the coformers through non-covalent bonds. So the co-crystal of Febuxostat (FX) and Isonicotinamide (ISON) was synthesized to improve the solubility of the drug through the method of cooling crystallization. In addition, the prepared co-crystal was studied by microscope, Powder X-Ray Diffraction, Single Crystal X-ray Diffraction, Differential Scanning Calorimetry, infrared spectrometers. At the same time, the co-crystal solubility and dissolution rate were explored. Among those experimental results, the DSC study and PXRD study presented the successful preparation of a new solid form, which is subsequently verified by Single Crystal X-ray Diffraction. Furthermore, the co-crystal exhibited a higher solubility compared to the pure drug. Besides, it can be obvious to draw conclusion from the infrared spectrometer results that the carbonyl groups both exist in the FX and ISON molecules involved in the formation of hydrogen bond. In summary, the current experiment can conclude that co-crystallization can be a better way to enhance the solubility of the poorly water soluble drug.

Supported by National Key Technology Support Program (No. 2012BAB19B07)

Keywords: DSC, Pharmaceutical, Solution, X-ray Diffraction
Application Code: Pharmaceutical
Methodology Code: X-ray Techniques

Session # 1770 **Abstract # 1770-14** **Poster Sessions**

Session Title: Pharmaceutical-MS, LC/MS and Others

Abstract Title: **Real Time Stability of NSAIDs in Aqueous Solutions by Infrared Spectroscopy**

Primary Author: Anumeha P. Muthal
Seton Hall University

Date: Wednesday, March 09, 2016 - Morn
Time:
Room: Exposition Floor, 400 Aisle

Co-Author(s): Nicholas Snow, Vrushali Bhawtankar

Abstract Text

Attenuated total reflectance fourier transform infrared spectrometry (ATR-FTIR) has been used to study the dissolution profile of drugs using an in-situ ATR-FTIR system to monitor the behavior of aspirin tablets under simulated physiological conditions, where aspirin and the hydrolysis product (salicylic acid) were detected as two separate signals.[1] This poster discusses monitoring the hydrolysis of several NSAIDs, including aspirin, ibuprofen and narpoxen in real time using in-situ ATR-FTIR to monitor the reaction. In-situ ATR-FTIR provides rapid real-time and selective monitoring of both the parent compound and hydrolysis product(s). This real time analysis can be used to determine the kinetics of hydrolysis, which contributes to understanding of degradation processes. NSAIDs provide excellent model compounds for general use in pharmaceutical analysis. In situ ATR-FTIR is envisioned as a complement to more traditional analysis by HPLC and GC.

References

1. Abe Kassis, Vrushali M. Bhawtankar, John R. Sowa Jr. , Attenuated total reflection infrared spectroscopy (ATR-IR) as an in situ technique for dissolution studies, Journal of Pharmaceutical and Biomedical Analysis, 53 (2010) 269–273.

Keywords: Dissolution, Drugs, FTIR, Near Infrared
Application Code: Pharmaceutical
Methodology Code: Near Infrared

Session # **1770** Abstract # **1770-15** **Poster Sessions**

Session Title Pharmaceutical-MS, LC/MS and Others

Abstract Title **Headspace Sampling of Residual Solvents Per USP 467 Using a Gas Tight Syringe**

Date: Wednesday, March 09, 2016 - Morn
Time:
Room: Exposition Floor, 400 Aisle

Primary Author Anne Jurek
EST Analytical

Co-Author(s) Justin Murphy, Kelly Cravenor, Lindsey Pyron

Abstract Text

During the synthesis of some pharmaceuticals it is sometimes necessary to use solvents in order to increase yield or purity of the product. After the pharmaceutical is produced, the solvent(s) are removed to the greatest extent possible. The products are then tested for any residual solvents in order to limit patient exposure. Residual solvents are separated into classes according to the risk to patient health. Class 1 solvents are to be avoided at all costs as they are known to be human carcinogens or have adverse effects on the environment. Class 2 solvents have less severe toxicities but should be avoided because of the potential of adverse effects. Finally, Class 3 solvents have low toxicity and can be used when needed. Pharmaceutical manufacturers are required to test for residual solvents in order to ensure that any residual solvents in the product are below established exposure limits. United States Pharmacopeia (USP) general chapter <467> describes a static headspace gas chromatography procedure for the determination of residual solvents. This application will demonstrate the USP <467> procedure using an autosampler configured with a Gas Tight Syringe for static headspace sampling.

Keywords: GC, Headspace, Pharmaceutical
Application Code: Pharmaceutical
Methodology Code: Gas Chromatography

Session # 1770 Abstract # 1770-17 **Poster Sessions**

Session Title Pharmaceutical-MS, LC/MS and Others

Abstract Title **A Simple and Rapid LC-MS/MS Method for the Determination of BMCL26, A Novel Anti-Parasitic Agent, in Rat Plasma**

Date: Wednesday, March 09, 2016 - Morn
Time:
Room: Exposition Floor, 400 Aisle

Primary Author Ramakrishna Reddy Voggu
Cleveland State University

Co-Author(s) Baochuan Guo, Bin Su, Xiang Zhou

Abstract Text

BMCL26 is a potential drug derived from nimesulide, which has exhibited the substantial anti-parasitic activity in various cell lines. To conduct various pharmacological and toxicological properties of this drug. We developed and validated a rapid LC-MS/MS method for its quantification in accordance with the FDA guidelines. Protein precipitation with 0.1 % formic acid in acetonitrile was used to extract the analyte along with the internal standard (JCC76) from rat plasma. It was found that the calibration curve of the method had an excellent linearity (r2 □0.9993) for the analyte concentration ranging from 0.5 to 1000 ng/mL with acceptable inter- and intra-assay, precision, accuracy and stability. the matrix effect and extraction recovery were in the range of 101.3 - 110.1 % and 90.2 - 105.0 %, respectively. This LC-MS/MS method is simple and rapid and can be used in the further pharmaceutical studies of BMCL26.

Keywords: Liquid Chromatography/Mass Spectroscopy, Method Development, Sample Preparation, Validation
Application Code: Pharmaceutical
Methodology Code: Liquid Chromatography/Mass Spectrometry

Session # 1770 Abstract # 1770-19 **Poster Sessions**

Session Title: Pharmaceutical-MS, LC/MS and Others

Abstract Title: **Drug-herb Interaction: A Crossover Study of the Effect of a Polyherbal Formulation on Metroinidazole Pharmacokinetic Profile**

Primary Author: Grace E. Ukpo
University of Lagos

Date: Wednesday, March 09, 2016 - Morn
Time:
Room: Exposition Floor, 400 Aisle

Co-Author(s): Idris O. Balogun, Steve O. Ogbonnia, Teddy S. Ehianeta

Abstract Text
The effect of the polyherbal formulation, Katoka mixture (KTK) on the pharmacokinetics of metronidazole was studied using rats in a two phase cross over study method. The study arose due to the wide use of the herbal preparation claimed to treat over 100 ailments and metronidazole commonly used for all gastrointestinal complains in our communities without prescription.

Though not statistically significant, the male rats had higher clearance and elimination half lives in the presence of the herbal formulation suggesting that gender is likely to affect the elimination profile of metronidazole, especially when co-administered with KTK. There was a decrease in the absorption and plasma half-life showing that both elimination and absorption were reduced in females. The AUC0□18h was observed to increase in both male and female rats which was not statistically significant ($p<0.05$) on concomitant administration of MTZ with KTK extract. The derived pharmacokinetic results show a statistical significance in AUC0□□ thus suggesting the susceptibility of males to MTZ toxicity than females when MTZ is co-administered with KTK. The significant rise suggests that males are more prone to the adverse effects of metronidazole in the presence of Katoka herbal formulation, than females.

This is further supported by the significant increase in bioavailability (F) of MTZ on co-administration.

Keywords: Biopharmaceutical, Chromatography, Liquid Chromatography, Natural Products
Application Code: Pharmaceutical
Methodology Code: Liquid Chromatography

Session # 1770 Abstract # 1770-20 **Poster Sessions**

Session Title: Pharmaceutical-MS, LC/MS and Others

Abstract Title: **Supersaturation of Spray-Dried-Dispersion (SDDs) - Development and Evaluation of a Characterization Method**

Primary Author: Benjamin H. Wu
Bristol-Myers Squibb

Date: Wednesday, March 09, 2016 - Morn
Time:
Room: Exposition Floor, 400 Aisle

Co-Author(s)

Abstract Text

Use of spray-dried-dispersions (SDDs) to improve the bioavailability of poorly water-soluble compounds has become a common practice in supporting early phase clinical studies. However, evaluation of their performance, whether in solid dosage forms or alone, still presents significant challenges. A microcentrifuge dissolution method has been reported to quickly assess the dissolution performance of SDDs [1]. In this work, two poorly water-soluble compounds (indomethacin and ketoconazole) and two commonly used polymers (PVP and HPMC-AS) were chosen to prepare SDDs. A typical micro-centrifuge dissolution procedure as devised in [1] was followed. In addition, after separation of the supernatant from precipitation, some of samples were filtered through filters of various sizes to investigate the particulate nature of the supernatant. Furthermore, the centrifuge speed was varied to study sedimentation of API, SDD or polymer particles. Results indicated that the SDDs of four drug-polymer pairs behaved differently in micro-centrifuge dissolution, depending on the polymer and the drug used. The SDDs of indomethacin with either PVP or HPMC-AS showed a reproducible dissolution with minimum variability even after filtration and with different centrifuge speeds, suggesting that the supernatant behaved like a solution. However, ketoconazole-PVP and ketoconazole-HPMCAS SDDs displayed a significant variation in concentration as the centrifuge speed and the pore sizes of filters were altered, indicating that their supernatant was heterogeneous with the presence of particulates.

[1] Curatolo W, Nightingale JA, Herbig SM 2009, Utility of Hydroxypropylmethylcellulose acetate succinate (HPMCAS) for initiation and maintenance of drug supersaturation in the GI milieu Pharm. Res., 26, 1419-1431

Keywords: Dissolution, HPLC, Pharmaceutical
Application Code: Pharmaceutical
Methodology Code: Liquid Chromatography

Session # 1770 Abstract # 1770-21 Poster Sessions

Session Title: Pharmaceutical-MS, LC/MS and Others

Abstract Title: **Headspace Grade Solvents for Trace Level Analyte Detection**

Date: Wednesday, March 09, 2016 - Morn

Time:

Room: Exposition Floor, 400 Aisle

Primary Author: Subhra Bhattacharya
Thermo Fisher Scientific

Co-Author(s): Eric Oliver, Stephen Roemer

Abstract Text

Headspace gas chromatography is the preferred technique for the analysis of highly volatile organics which are partitioned efficiently into the gas phase from solid or liquid matrices. Detection of residual solvents in pharmaceuticals is performed by headspace GC analysis. The residual solvents in pharmaceuticals are defined as organic volatile chemicals that are used or produced in the manufacturing of drug substances or excipients, or in the preparation of drug products per United States Pharmacopoeia (USP) and International Conference on Harmonization (ICH)). Residual solvents are difficult to remove completely from drug products and therefore the amount of residual solvent should be evaluated by headspace GC analysis. USP and ICH guidelines recommend the acceptable amounts of residual solvents in pharmaceuticals for the safety of the patient. Residual solvents are classified as Class 1 (most toxic), Class 2 and Class 3 according to the toxicity effects. In headspace GC analysis, the pharmaceutical compound of interest is often dissolved in a high boiling solvent to evaluate the amount of volatile components. Although water is the most common solvent for this type of analysis, other solvents are frequently used when the pharmaceutical compounds are not soluble in water.

We have evaluated five different solvents for headspace GC analysis. The solvents are water, dimethyl sulfoxide (DMSO), N,N-dimethylformamide (DMF), N,N-dimethylacetamide (DMAC) and 1-methyl-2-pyrrolidinone (NMP). Allergy medications such as Claritin, Zyrtech, Advil (allergy), Wal-Phed and Wal-Dry were purchased from Walgreens. Residual solvent analysis from these allergy medications was performed using five different headspace solvents. Our results showed that all five solvents are suitable for detection of trace level volatile organic components from the pharmaceutical compounds.

Keywords: Gas Chromatography/Mass Spectrometry, GC-MS, Headspace
Application Code: Pharmaceutical
Methodology Code: Gas Chromatography/Mass Spectrometry

Session # 1770 Abstract # 1770-22 **Poster Sessions**

Session Title Pharmaceutical-MS, LC/MS and Others

Abstract Title **Analytical Strategies in the Development of Generic Drug Products: Excipient Quantitation**

Date: Wednesday, March 09, 2016 - Morn
Time:
Room: Exposition Floor, 400 Aisle

Primary Author Alexander W. Garner
Mayne Pharma

Co-Author(s)

Abstract Text

As an early part of the development process involved in producing a generic drug, it is advantageous to determine the amounts of key excipients in the innovator drug (reference listed drug, RLD). The advantage in obtaining this information is to provide a foundation for producing a viable formulation for the submission of an Abbreviated New Drug Application (ANDA), that matches the performance of the innovator drug both in-vitro and in-vivo. Reverse engineering of the innovator drug formulation is commonly referred to as "deformulation." Excipients are used in many, if not all, drug products found on the pharmaceutical market. Excipients range from small molecules to large polymers that are used as surfactants, binders, flavors, colorants, extended release mechanisms, lubricants, glidants, preservatives, coatings, disintegrants, etc., some of which contain chromophores while most do not. The process of deformulation is not well described in the current analytical chemistry literature, and therefore requires use of some innovative techniques. Novel methods are also required because excipients are not typically evaluated by the pharmaceutical industry outside of USP identification and purity type testing. The greatest challenge however is the quantitation of an individual excipient in the presence of other excipients and the pharmaceutical active, especially when it lacks a visible chromophore. Presented is an approach to deformulation, which discusses various aspects of reverse engineering innovator drug products; like background research on the individual excipients, sample preparation, methods of testing and detection, and critical quality attributes used to determine the suitability of the method to be used.

Keywords: HPLC Detection, Method Development, Pharmaceutical, Separation Sciences
Application Code: Pharmaceutical
Methodology Code: Process Analytical Techniques

Session # 1770 Abstract # 1770-23 **Poster Sessions**

Session Title: Pharmaceutical-MS, LC/MS and Others

Abstract Title: **Quantitative Laser Diffraction Method for the Assessment of Subvisible Protein Particles**

Date: Wednesday, March 09, 2016 - Morn

Time:

Room: Exposition Floor, 400 Aisle

Primary Author: Robert E. Buco
Shimadzu Corporation

Co-Author(s): Andrea Hawe, Ariadna Martos, Haruo Shimaoka, Markus Ortlieb, Matthew Ferrarelli, Michael Wiggenhorn, Shinichiro Totoki

Abstract Text

Biopharmaceuticals (e.g. antibody drugs) have been widely used in the treatment of autoimmune diseases and cancers. During the production, purification, and storage of most antibody drugs, protein aggregation and particle formation is observed to a certain extent, which has been suggested to impact the risk of in-vivo immunogenicity. Because of this, proper monitoring and assessment of protein aggregates in biopharmaceuticals is highly important and is required by some regulatory authorities. Of particular interest is the development of robust analytical methods for the quantitative analysis of particles between 0.2-2[micro]m in diameter.

Aggregates are divided into four categories according to their diameter: <0.2[micro]m, 0.2-2[micro]m, 2-10[micro]m, and 10-25[micro]m. Quantification of those with diameters <0.2[micro]m can be achieved by employing orthogonal methods including size-exclusion chromatography, analytical ultracentrifugation, and field flow fractionation, and those with diameters between 1-25[micro]m can be assessed by light obscuration and dynamic imaging analysis. However, accurate quantification of protein particles with diameters in the subvisible particle size range, especially between 0.2–1[micro]m, remains a challenge.

Laser Diffraction (LD) is widely recognized as a method for estimating particle size distribution. This presentation describes the application of a newly developed quantitative LD (qLD) system, which combines a LD method with extensive deconvolution analysis, toward the quantitative analysis of subvisible protein particles between 0.2-10[micro]m in diameter. The robustness of the method, interference by excipients and silicone oil, and its performance in relevant formulations of therapeutic proteins (e.g. at high protein concentration) are discussed, and the qLD analysis is compared with other available particle methods.

Keywords: Biopharmaceutical, Particle Size and Distribution, Protein, Quantitative
Application Code: Other
Methodology Code: Physical Measurements

Session # 1770 Abstract # 1770-24 **Poster Sessions**

Session Title Pharmaceutical-MS, LC/MS and Others

Abstract Title **3D-Printed LTP Ionization Source for the Direct Analysis of Biomolecules**

Primary Author Sandra Martinez Jarquin
Centro de Investigation Y de Estudios Avanzados Del IPN

Co-Author(s) Robert Winkler

Date: Wednesday, March 09, 2016 - Morn
Time:
Room: Exposition Floor, 400 Aisle

Abstract Text

Analysis of compounds directly from samples is a promising technique for in vivo analysis of organisms. Ambient ionization sources for mass spectrometry make this possible. Analysis by low-temperature plasma (LTP) ionization enables to directly monitor metabolic processes on organisms, because no sample preparation is necessary and minimal harm is caused to the studied object. However, currently no LTP sources are commercial available. Therefore, devices have to be built in-house to have access to the technology (Martínez-Jarquín & Winkler, 2013)

About five conceptually different configurations of LTP sources have been reported in the literature. But the lack of a standardized device affects the reproducibility of results. 3D printing represents an emerging technology for building prototypes for a great number of applications. Repositories with open laboratory devices are available on internet webpages such as Thingiverse (http://www.thingiverse.com).

Here, we report the construction of a 3D printed LTP ionization source and present experimental data on the performance of the probe. Interested researchers may download our public template files, print, modify and use our LTP probe design.

Martínez-Jarquín, S., & Winkler, R. (2013). Design of a low-temperature plasma (LTP) probe with adjustable output temperature and variable beam diameter for the direct detection of organic molecules. Rapid Communications in Mass Spectrometry RCM, 27(5), 629–34. doi:10.1002/rcm.6494

Keywords: Biological Samples, Instrumentation, Mass Spectrometry, Plasma
Application Code: Process Analytical Chemistry
Methodology Code: Mass Spectrometry

Session # 1770 **Abstract # 1770-25** **Poster Sessions**

Session Title Pharmaceutical-MS, LC/MS and Others

Abstract Title **Demonstrating the Uptake Mechanism of Cisplatin in Cells by Single Cell ICP-MS**

Date: Wednesday, March 09, 2016 - Morn
Time:
Room: Exposition Floor, 400 Aisle

Primary Author Chady Stephan
PerkinElmer

Co-Author(s)

Abstract Text

Platinum chemotherapeutic agents have a broad range of activity in malignant disease and are used to treat many types of cancer (platinum compounds include: cisplatin, carboplatin, and oxaliplatin). Initially patients respond well to treatment but later relapse and display resistance to platinum compounds. This work focuses on the use of Single Cell ICP-MS technique exploring the uptake mechanism for cisplatin in cells based on individual cell information.

Keywords: Bioanalytical, Biomedical, ICP-MS, Nanotechnology
Application Code: Bioanalytical
Methodology Code: Atomic Spectroscopy/Elemental Analysis

Session # 1770 Abstract # 1770-26 **Poster Sessions**

Session Title: Pharmaceutical-MS, LC/MS and Others

Abstract Title: **The Practice and Challenges of Ultrafast Chiral Separations in UHPLC and Super/Subcritical Fluid Chromatography (SFC)**

Date: Wednesday, March 09, 2016 - Morn

Time:

Room: Exposition Floor, 400 Aisle

Primary Author: Chandan L. Barhate
University of Texas at Arlington

Co-Author(s): Daniel W. Armstrong, M Farooq Wahab

Abstract Text

With the introduction of sub-2 µm narrow particle size distribution particles it has become possible to do chiral separations in highly efficient 2-5 cm short columns. Recently, we have directed our efforts towards achieving sub-minute chiral separations. However, with high speed separations, not only the column packing but the instrument itself can be a limiting factor in getting the fastest separations. In this talk we will show the nuances associated with high speed separations and discuss how to overcome the challenges associated with packing short columns and detector settings. SFC is inherently more prone to noise. More importantly, the concept of "response time" and "data sampling intervals" becomes very significant in the SFC mode. We will also focus on the pros and cons of different types of digital filters and how they affect retention time, peak shapes and efficiencies.

Keywords: Chiral, Chromatography, Liquid Chromatography, Supercritical Fluid Chromatography

Application Code: Pharmaceutical

Methodology Code: Separation Sciences

Session # 1770 Abstract # 1770-28 **Poster Sessions**

Session Title: Pharmaceutical-MS, LC/MS and Others

Abstract Title: **Determining Equivalency of Generic and Name Brand Oral Suspensions Using Zeta Potential**

Date: Wednesday, March 09, 2016 - Morn
Time:
Room: Exposition Floor, 400 Aisle

Primary Author: Jack G. Saad
Micromeritics

Co-Author(s): Danielle Sowle, Myke Scoggins

Abstract Text

Suspension stability plays a role in the taste and texture of an oral suspension containing one or more active pharmaceutical ingredients (API). The effects of aggregating or flocculating nanoparticles or emulsions can impact this key component where generic oral suspension must demonstrate equivalency to name brand oral suspensions. Zeta potential can be used to quantify and control suspension stability and the formation of aggregates and flocculants, thereby optimizing the taste and texture of a generic oral suspension to match the name brand product. A generic and a name brand oral suspension are visually compared as a qualitative test. The electrophoretic mobility of each suspension is then determined using electrophoretic light scattering (ELS) technology. From the electrophoretic mobility, the electrical potential of a particle or emulsion at the shear or slipping plane or zeta potential, is determined. The larger the absolute millivolt (mV) value of zeta potential, the more stable a suspension will be and the more likely the particles or emulsions will repel each other rather than flocculate. The generic formulation shows zeta potential values that are much smaller than the name brand oral suspension. Visually, the generic oral suspension shows signs of settling and requires shaking before sampling. The generic formulation may require the addition of additives that can increase the zeta potential to the levels of the name brand and aid in stabilizing the suspension.

Keywords: Electrochemistry, Light Scattering, Process Analytical Chemistry, Separation Sciences
Application Code: Pharmaceutical
Methodology Code: Process Analytical Techniques

Session # 1780 Abstract # 1780-1 **Poster Sessions**

Session Title: Practical Chromatography in Today's Laboratory

Abstract Title: **A Rapid Isocratic LC Separation of Soft Drink Additives Using an Environmentally Friendly Mobile Phase with UV Detection**

Date: Wednesday, March 09, 2016 - Morn
Time:
Room: Exposition Floor, 400 Aisle

Primary Author: Mark E. Benvenuti
Waters Corporation

Co-Author(s): Joseph P. Romano

Abstract Text

The soft drink market is an important revenue source for many major food and beverage producers. Such beverages include traditional carbonated soft drinks, energy drinks, and wellness beverages such as vitamin waters and teas. These products often contain caffeine as an energy booster, benzoate and sorbate for preservation, and non-nutritive sweeteners such as aspartame, saccharin, and acesulfame K for diet formulations. For quality control purposes, the conformance of target concentrations of analytes to specified ranges is critical. This presentation will demonstrate a rapid isocratic LC separation of these analytes using an environmentally compatible mobile phase with UV detection. Applicability to several varieties of beverages will be shown . We will also show how analyte interferences can be eliminated by using selective UV wavelengths.

Keywords: Food Identification, Food Science, HPLC Detection, UV-VIS Absorbance/Luminescence
Application Code: Food Identification
Methodology Code: Separation Sciences

Session # 1780 Abstract # 1780-2 **Poster Sessions**

Session Title Practical Chromatography in Today's Laboratory

Abstract Title **High Efficiency Chiral Separations Using Sub-2 □m Monodisperse Chiral Stationary Phases**

Primary Author Zachary Breitbach
The University of Texas at Arlington

Co-Author(s) Chandan L. Barhate, Daniel W. Armstrong, David S. Bell, M Farooq Wahab

Date: Wednesday, March 09, 2016 - Morn
Time:
Room: Exposition Floor, 400 Aisle

Abstract Text

Achiral HPLC column technologies and hardware have seen tremendous advances over the past decade through the use of more efficient packing materials. The use of sub-2□m silica particles in combination with high pressure, low system volume, UPLC instrumentation has allowed for routine achiral separations to yield >250,000 plates per meter. Conversely, chiral column technology has remained relatively unchanged in the last ten years, with 5□m silica based columns resulting in enantiomeric separations which exhibit □30,000 plates per meter being considered acceptable. In this work, highly efficient chiral phases were produced using novel, sub-2□m, Titan silica particles. Titan particles have very narrow particle size distributions, unique porous structures, and have proven to provide some of the highest efficiency achiral packing materials to date. Herein, three chiral selectors (teicoplanin, teicoplanin aglycone, and vancomycin) were chemically boned to 1.9□m Titan silica and packed into 4.6 mm i.d columns. Their UPLC chromatographic performance was compared to state of the art 5□m commercial columns, as well as, 1.7□m chiral phases made in house. The latter comparison gives a clear picture of the true advantages of the Titan material over other sub-2□m silica. The resulting Titan chiral phases were applied to UPLC and SFC chiral separations, resulting in unprecedented enantiomeric separation efficiencies. Such high efficiencies allow for the use of short columns, which in turn, provide the ability to produced ultra-fast, sub-1minute, enantiomeric separations.

Keywords: Chiral, Chiral Separations, HPLC, HPLC Columns
Application Code: Pharmaceutical
Methodology Code: Separation Sciences

Session # 1780 Abstract # 1780-3 **Poster Sessions**

Session Title: Practical Chromatography in Today's Laboratory

Abstract Title: **Advances in Chiral HPLC Column Technology: Superficially Porous Particle Based Chiral Stationary Phases**

Primary Author: Zachary Breitbach
The University of Texas at Arlington

Date: Wednesday, March 09, 2016 - Morn
Time:
Room: Exposition Floor, 400 Aisle

Co-Author(s): Daniel W. Armstrong, M Farooq Wahab

Abstract Text

In this report, chemically bonded brush type chiral selectors on fully and superficially porous particles (SPPs) were developed. The chromatographic performance of superficially porous CSP based columns is compared with columns packed with 5 µm, 3 m, and 1.7 m fully porous particles (FPPs). Theoretical treatments have indicated that SPP based CSPs may show increases in resolution. Herein, true gains in efficiency and resolution are obtained using a variety of brush-type chiral stationary phases (CSPs) developed using SPPs. For example, when using a cyclofructan based CSP, at a flow rate of 3.0 ml/min, the number of plates on column afforded by the SPP column was ~ 7x greater than the number of plates on column (same length) obtained when using the 5 µm FPP based column. The enantiomeric selectivities were comparable and sometimes better for the SPP based columns compared to the FPP based columns even though the SPP columns contained lower absolute amounts of chiral selector. Under constant retention conditions, the SPP based CSPs greatly improved resolution compared to FPP based columns. Given their high efficiencies and relatively low back pressures, columns containing these particles were particularly advantageous for ultrafast "chiral" separations in the 4 to 40 seconds range. It is demonstrated these separations can be performed in any mobile phase conditions or mode, i.e., reversed phase, normal phase, polar organic, HILIC. Finally, the practice of ultrafast chiral LC often produces interesting and unusual consequences that must be recognized, dealt with, and/or properly understood for optimal performance.

Keywords: Chiral, Chiral Separations, Liquid Chromatography, Pharmaceutical
Application Code: Pharmaceutical
Methodology Code: Separation Sciences

Session # 1780 Abstract # 1780-4 **Poster Sessions**

Session Title: Practical Chromatography in Today's Laboratory

Abstract Title: **A New Standard in Analytical Workflow Design**

Date: Wednesday, March 09, 2016 - Morn
Time:
Room: Exposition Floor, 400 Aisle

Primary Author: William Hedgepeth
Shimadzu Scientific Instruments

Co-Author(s): Kenichiro Tanaka

Abstract Text
Current analytical methodology typically consists of a separate off-line sample preparation technique that is followed by a chromatographic analysis. Sample preparation and manual transfer to the analytical instrument often consumes a majority of the analyst's time and effort. Recently, an innovative new concept was introduced that greatly reduces sample preparation times and the variability associated with manual procedures. This new technique automates the sample preparation and analysis of samples by supercritical fluid extraction of compounds from the sample matrix, which are then transported to the analytical column for analysis without any human intervention. A number of applications that include food, environmental, and pharmaceutical areas will be shown that show the flexibility of this technique.

Keywords: Laboratory Automation, Sample Preparation, Separation Sciences, SFC
Application Code: General Interest
Methodology Code: Separation Sciences

Session # 1780 Abstract # 1780-6 **Poster Sessions**

Session Title Practical Chromatography in Today's Laboratory

Abstract Title **Centrifugal Partition Chromatography: A Preparative Tool for Isolation and Purification of Xylindein from Chlorociboria Aeruginosa**

Primary Author Anukul Boonloed
Oregon State University

Date: Wednesday, March 09, 2016 - Morn
Time:
Room: Exposition Floor, 400 Aisle

Co-Author(s) Genevieve L. Weber, Sumate Pengpumkiat, Vincent T. Remcho

Abstract Text

A centrifugal partition chromatography (CPC) method was developed for the preparative-scale isolation and purification of xylindein from the wood-staining fungi, Chlorociboria aeruginosa. Xylindein, a blue-green pigment naturally secreted from the hyphae and fruiting bodies of the fungus, has great value in the decorative wood industry and textile coloration. Xylindein has great potential for use as a fluorescent labeling agent as well as in organic semiconductor applications. However, a primary limitation of xylindein is its poor solubility in most common organic solvents and aqueous solutions. It is thus arduous to purify using preparative liquid chromatography or solid-phase extraction (SPE). Support-free liquid-liquid chromatographic methods, including CPC, where solutes are separated based on their different partition coefficients between two immiscible solvent systems, are promising alternatives for the purification the compound on a preparative scale. In this work, a new biphasic solvent system suitable for CPC separation of xylindein was developed. Various groups of solvents were assessed for their suitability as xylindein extractants. Appropriate solvent systems for CPC were sought and identified based on their partition coefficients. A CPC system equipped with a fraction collector was then used for the isolation of xylindein from crude extracts. Qualitative characterization and purity determination for various xylindein fractions were then carried out by HPLC.

Keywords: HPLC, Isolation/Purification, Prep Chromatography, Separation Sciences

Application Code: Other

Methodology Code: Separation Sciences

Session # 1780 Abstract # 1780-7 **Poster Sessions**

Session Title: Practical Chromatography in Today's Laboratory

Abstract Title: **Analysis of Petroleum Products Using Comprehensive Two-Dimensional Gas Chromatography (GC×GC) with Both Time of Flight MS and Flame Ionization Detectors**

Date: Wednesday, March 09, 2016 - Morn
Time:
Room: Exposition Floor, 400 Aisle

Primary Author: Joseph E. Binkley
Leco Corporation

Co-Author(s): Christina N. Kelly, David E. Alonso, Jonathan D. Byer, Lorne E. Fell

Abstract Text

Comprehensive Two-Dimensional Gas Chromatography (GC×GC) has proven to be an extremely valuable analytical technique for the petroleum industry due to its ability to substantially increase the chromatographic peak capacity beyond that of traditional single dimension gas chromatography. Pairing GC×GC with Time of Flight Mass Spectrometry (TOFMS) provides unsurpassed characterization capabilities due to the separation power of GC×GC and the ability of TOFMS to provide rich data to deconvolution algorithms which help unravel the complexity of difficult petroleum matrices. Petrochemical labs also often utilize flame ionization detectors for GC×GC to provide quantitative results via area percent calculations. The combination of data from these two analytical platforms would provide a wealth of information for both the characterization and quantitation of petroleum samples. The goal of this poster is to present results from several possible configurations to develop a workflow which would allow either subsequent or simultaneous acquisitions of GC×GC TOFMS and FID data on the same GC×GC instrument. The following configurations will be explored and results reported: 1. Installing independent GC×GC column sets in separate GC inlets (Front and Back) and terminating at respective detectors (TOFMS and FID). This option will only allow data collection via subsequent injections. 2. Connecting independent GC×GC column sets in a single inlet by way of a dual-hole ferrule and terminating at respective detectors (TOFMS and FID). 3. Installing a single primary column splitting to two second dimension columns terminating at respective detectors. Options 2 and 3 will allow simultaneous acquisition of TOFMS and FID data.

Keywords: Fuels\Energy\Petrochemical, Gas Chromatography, Gas Chromatography/Mass Spectrometry, Time
Application Code: Fuels, Energy and Petrochemical
Methodology Code: Separation Sciences

Session # 1780 Abstract # 1780-8 **Poster Sessions**

Session Title: Practical Chromatography in Today's Laboratory

Abstract Title: **Evaluation of Polycyclic Aromatic Hydrocarbon Standard Reference Materials 869b and 1647f on Different Stationary Phases for Liquid Chromatography**

Date: Wednesday, March 09, 2016 - Morn
Time:
Room: Exposition Floor, 400 Aisle

Primary Author: Walter B. Wilson
National Institute of Standard and Technology

Co-Author(s): Jorge O. Oña-Ruales, Lane C. Sander, Stephen A. Wise

Abstract Text

Polycyclic aromatic hydrocarbons (PAHs) are important environmental pollutants originating from a wide variety of natural and anthropogenic sources. PAHs are generally formed during incomplete combustion of organic matter containing carbon and hydrogen. Due to the carcinogenic nature of some PAHs, their chemical analysis is of great environmental and toxicological importance. Among the hundreds of PAHs present in the environment, the U.S. Environmental Protection Agency (EPA) have included sixteen in their priority pollutants list. Liquid chromatography (LC) is a standard analysis technique used for determining the 16 EPA-PAHs. Currently at the National Institute of Standard Technology (NIST), standard reference materials (SRM) 869b and 1647f are available for evaluating current and new LC columns. SRM 869b is a mixture of three PAHs for characterizing a LC column selectivity for separation of PAHs. Depending on the elution order of the three PAHs, column selectivity can be predicted for complex PAH mixtures. SRM 1647f is a calibration solution for use in LC methods for the determination of the 16 EPA-PAHs. In this study, multiple LC columns consisting of C18, C30, phenyl-hexyl and penta-fluoro-phenyl stationary phases are evaluated with SRM 869b and SRM 1647f.

Keywords: Liquid Chromatography, PAH, Separation Sciences, Standards
Application Code: Environmental
Methodology Code: Separation Sciences

Session # 1780 Abstract # 1780-9 **Poster Sessions**

Session Title: Practical Chromatography in Today's Laboratory

Abstract Title: **Liquid Chromatographic Retention Behavior of Polycyclic Aromatic Sulfur Heterocycles and Their Alkyl-Substituted Derivatives**

Primary Author: Walter B. Wilson
National Institute of Standard and Technology

Date: Wednesday, March 09, 2016 - Morn
Time:
Room: Exposition Floor, 400 Aisle

Co-Author(s): Lane C. Sander, Stephen A. Wise

Abstract Text

Polycyclic aromatic compounds comprise a complex class of condensed multi-ring benzenoid compounds originating from a wide variety of natural and anthropogenic sources. The parent homocyclic species, which contain only carbon and hydrogen, are the familiar polycyclic aromatic hydrocarbons (PAHs). Along with PAHs, heterocyclic compounds containing at least one heteroatom such as polycyclic aromatic sulfur heterocycles (PASH) are largely present in petroleum products. The total number of possible isomeric structures for PASH is greatly increased compared with the corresponding PAH because both ring arrangement and position of the heteroatom substitution within the rings give rise to unique isomers. Similar to PAHs, alkylated-PASH isomers increase the sample complexity due to the increased number of structural isomers. Gas chromatography coupled to mass spectrometry (GC/MS) is the primary separation technique currently used for the determination of PASHs in complex samples. With the exception of a couple of publications, very limited data have been reported concerning liquid chromatographic (LC) retention characteristic of PASHs even though LC is a commonly used analytical technique for analyzing PAHs. In this study, the retention behavior of both PASHs and alkylated-PASHs on monomeric and polymeric C18 phases are reported. Molecular descriptors (length, breadth, thickness (T) and length-to-breadth (L/B) ratio were calculated for all the compounds studied. Correlations for retention on both stationary phases and PASH geometry (L/B and T) ratios were investigated.

Keywords: Environmental Analysis, Liquid Chromatography, PAH, Separation Sciences
Application Code: Environmental
Methodology Code: Separation Sciences

Session # 1780 Abstract # 1780-10 **Poster Sessions**

Session Title Practical Chromatography in Today's Laboratory

Abstract Title **Evaluation of 25-Hydroxy Vitamin D Extraction Using Phospholipid Depletion Plate Technology and Method Comparison Using Automated Sample Preparation**

Date: Wednesday, March 09, 2016 - Morn
Time:
Room: Exposition Floor, 400 Aisle

Primary Author Kerry Challenger
Biotage GB Limited

Co-Author(s) Helen Lodder, Lee Williams, Victor Vandell

Abstract Text

Introduction

Vitamin D deficiency can result in health issues such as osteoporosis, liver, kidney and even increased risk of cancers and MS. Many sample preparation approaches have been employed prior to LC-MS/MS analysis. This poster demonstrates the use of a novel protein and phospholipid depletion plate. The extraction protocol was ultimately transferred to an SPE automation platform and method performance versus manual processing was compared.

Methodology

25-hydroxy vitamin D2/D3 was spiked into charcoal stripped human serum at various concentrations. A solvent first crash methodology was employed using ISOLUTE PLD+ phospholipid depletion plates in conjunction with repeat aspirate/dispense steps for optimal mixing. The optimized method was transferred to the Extrahera automated sample preparation platform. Extracts were evaporated and reconstituted for LC/MS analysis.

Results

Initial method development focussed on optimization of protein crash ratio and optimal organic solvent composition. Various proportions of MeOH and ACN were investigated. Serum precipitation with ACN in a 1:4 ratio demonstrated recoveries greater than 70%, while MeOH and MeOH/ACN solvent combinations demonstrated lower recoveries and higher RSDs. Good protein removal was afforded by the depth filter effect provided by the frit combinations. Optimal matrix crash ratios of 1:4 serum:ACN demonstrat excellent phospholipid removal. Linearity was observed over the conccentration range, correlation coefficients greater than 0.99 for both 25-hydroxy vitamin D2 and D3. The optimized extraction protocol was then transferred onto the Extrahera. The positive pressure processing unit provided very precise flow control. Final method performance of analyte recovery, precision, accuracy, linearity and coefficients of determination of calibration curves were compared between manual and automated processing methods. The Extrahera provided better precision and accuracy compared to manual processing.

Keywords: Bioanalytical, Food Science, Liquid Chromatography/Mass Spectroscopy, Solid Phase Extraction
Application Code: Bioanalytical
Methodology Code: Separation Sciences

Session # 1780 Abstract # 1780-11 **Poster Sessions**

Session Title: Practical Chromatography in Today's Laboratory

Abstract Title: **Effects of Ultracentrifugation on HDL and LDL Size Distribution**

Primary Author: Jeffrey Jones
Centers for Disease Control and Prevention

Date: Wednesday, March 09, 2016 - Morn
Time:
Room: Exposition Floor, 400 Aisle

Co-Author(s): Bryan A. Parks, Christopher Toth, David Schieltz, James Pirkle, John R. Barr, Jon Rees, Lisa McWilliams, Michael S. Gardner, Yulanda Williamson, Zsuzsanna Kuklenyik

Abstract Text

Lipoproteins are natural liposome carriers of lipids and proteins that are central to numerous physiological functions throughout the body. Of particular interest are two common classes of extracellular lipoproteins, high density lipoproteins (HDL) and low density lipoproteins (LDL), that are established risk factors in the development of coronary heart disease. Recent epidemiologic research has established links between the heterogeneity of HDL and LDL particles and cardiovascular risk, which led to the development of various methods for their fractionation into sub-classes based on physicochemical properties. The most prevalent method of sub-fractionation is still traditional ultracentrifugation, where either serum or plasma is separated on the basis of density into the various subclasses. However, during the separation the particles undergo extreme forces as well as being subjected to a high salt concentration, resulting in several studies showing alteration in lipid/protein composition. This work utilizes Asymmetric Flow Field-Flow Fractionation (AF4), a comparatively gentle technique, to physically separate the different lipoprotein sub-classes and collect fractions for targeted liquid chromatography coupled mass spectrometry analysis. AF4 uses laminar flow dynamics as the separation mechanism at physiological pH and salt conditions without sheer forces, minimizing the risk of sample degradation, which makes it an ideal technique to show changes in sub-class distribution before and after ultracentrifugation. By using AF4, the results show that significant changes occur during ultracentrifugation, which may confound quantification of the sub-class distribution of HDL and LDL particles and subsequent understanding of their atherogenic properties.

Keywords: Bioanalytical, Biological Samples, Lipids, Protein
Application Code: Bioanalytical
Methodology Code: Separation Sciences

Session # 1780 Abstract # 1780-12 Poster Sessions

Session Title Practical Chromatography in Today's Laboratory

Abstract Title **Biomolecular Separations through Tunable Nanoporous Gold Membranes**

Primary Author Daniel A. McCurry
University of Illinois at Urbana-Champaign

Date: Wednesday, March 09, 2016 - Morn
Time:
Room: Exposition Floor, 400 Aisle

Co-Author(s) Ryan C. Bailey

Abstract Text

The analysis of biomolecules, including proteins and DNA, is rapidly moving towards lab-on-a-chip devices in an effort to minimize sample volume and sample loss. Synthetic nanopores have been developed in order to discriminate between different base pairs of DNA for sequencing or characterize proteins as they pass through the pore on a very small scale. Unfortunately, these devices often demonstrate low throughput and must be fabricated on a per-application basis as the diameter and surface charge of the pore dictates its selectivity. In the present study, a nanoporous gold (NPG) membrane is used as an alternative separations device. The three-dimensional porosity allows for higher throughput than traditional single pores and its inherent conductivity along with facile surface modification provide a tunable surface charge. NPG was produced through free-corrosion de-alloying in concentrated nitric acid. Pore sizes were measured to be 50 ± 20 nm via scanning electron microscopy. The wide pore size distribution was important for maintaining size-selectivity while still allowing multiple sizes of analyte through. The NPG was then functionalized following standard self-assembled monolayer techniques with alkane thiols. Lysozyme (pI=11.4), bovine hemoglobin (pI=6.8), and bovine serum albumin (pI=4.7) were used to model transport of differently charged proteins through the NPG as a function of the alkane thiol functionalization. The transport rate was monitored in real time using a fiber-optically coupled UV-Visible spectrometer. An electric potential applied directly to the gold was shown to gate transport against specifically charged proteins. Support for this work was provided through the University of Illinois at Urbana-Champaign.

Keywords: Electrochemistry, Nanotechnology, Protein, Separation Sciences
Application Code: Bioanalytical
Methodology Code: Separation Sciences

Session # 1780 **Abstract # 1780-13** **Poster Sessions**

Session Title: Practical Chromatography in Today's Laboratory

Abstract Title: **Non-Contact Pd Separation based on Laser-Induced Particle Formation for Determination of [sup]107[/sup]Pd with ICP-MS**

Date: Wednesday, March 09, 2016 - Morn

Time:

Room: Exposition Floor, 400 Aisle

Primary Author: Takumi Yomogida
Japan Atomic Energy Agency

Co-Author(s): Fumitaka Esaka, Hironori Ohba, Morihisa Saeki, Shiho Asai, Yoshihiro Kitatsuji, Yukiko Hanzawa

Abstract Text

Palladium-107, one of the Pd isotopes, can be found in high-level radioactive wastes (HLW). Owing to its long half-life (6.5 million years), determination of the [sup]107[/sup]Pd contents in HLW is essential to evaluate the long-term safety of HLW repositories. Measurement reports on the [sup]107[/sup]Pd, however, have not been established because of difficulties in separation. Pd ions are, in general, strongly retained onto adsorbents, such as anion-exchange resins. This makes the quantitative elution of the Pd ions hard to achieve. In the handling samples for [sup]107[/sup]Pd determination may cause radioactive contamination and radiation exposure to workers. Therefore, a simple and contamination-free procedure is desired.

In this study, we developed a novel separation technique based on laser-induced particle formation. This technique enables us to perform non-contact separation of the Pd species. A simulated HLW solution, comprised of 14 major elements (Rb, Sr, Zr, Mo, Ru, Rh, Pd, Cs, Ba, La, Ce, Pr, Nd, Sm) in a 3 M HNO[sub]3[/sub] solution, was used to evaluate the separation performance. A portion of the solution (0.1 mL) was added to 1.0 M HNO[sub]3[/sub], ultrapure water and ethanol mixed solution to adjust the volume and concentrations to 2 mL of 0.5 M HNO[sub]3[/sub]-40% ethanol. The resultant solution including 4.8 □g of Pd was irradiated with 355 nm pulsed laser for 20 min. The Pd particles were formed by the photoreduction of Pd ions induced by the irradiation. They were isolated from the solution by centrifugation and subsequently dissolved with 0.05 mL of aqua regia. The Pd solution was diluted with 1.0 M HCl solution. The concentration of Pd in the resultant solution was measured with ICP-MS. The results showed that more than 60% of Pd were recovered, while about 99.9% of the other 13 elements were removed. This indicates that non-contact and high-selective separation of Pd is achievable with the proposed separation technique.

Keywords: ICP-MS, Laser, Nuclear Analytical Applications, Separation Sciences

Application Code: Nuclear

Methodology Code: Separation Sciences

Session # 1780 **Abstract # 1780-14** **Poster Sessions**

Session Title: Practical Chromatography in Today's Laboratory

Abstract Title: **Surfactant-Pluronic Gel Phases for Electrophoresis**

Date: Wednesday, March 09, 2016 - Morn
Time:
Room: Exposition Floor, 400 Aisle

Primary Author: Ashley E. Richardson
Miami University

Co-Author(s): Elise M. Leonard, Neil D. Danielson

Abstract Text

Previously polyacrylamide gel electrophoresis (PAGE) has not been amenable for the separation of low molecular mass proteins and peptides. Pluronic polymers are triblock co-polymers of ethylene oxide (EO) and propylene oxide (PO) having the general formula H[OCH2CH2][sub]a[/sub][OCH(CH3)CH2][sub]b[/sub][OCH2CH2][sub]a[/sub]OH where a=106 and b=70. Below room temperature, the hydration layer keeps the polymer in solution. As the temperature is raised, the hydrophilic chains become desolvated due to breaking of hydrogen bonds favoring hydrophobic interaction of the exposed PO groups resulting in a network of micellar strands and gel formation. The goals of our study are to first ascertain the effect of gel formation when a surfactant is added to the polymer solution at low temperature and then characterize these polymers for the separation of different charged dye mixtures. We have found that Pluronic gel formation at 30% can be maintained at room temperature below, at, or above the critical micelle concentration (CMC) of sodium dodecylsulfate (SDS). Using 1mm ID x 7.3 cm glass tubes filled with 30% Pluronic F127 gel and SDS either above or below the CMC, electrophoretic separations of a dye test mixture containing an anionic dye (Alizarin Red), zwitterionic dye (Rhodamine B), and a cationic dye (methylene blue) have been carried out with the application of +300 V. In the presence of SDS above the CMC, the cationic and zwitterionic dyes did not migrate but the anionic dye SDS did migrate about 35 mm. This is expected since the upper buffer solution is cathodic and the lower buffer solution is anodic. We plan to vary the SDS concentration to tailor the separation of a mixture of anionic dyes with charges from -1 to -4. In addition, the cationic surfactant cetyltrimethylammonium bromide will be used to facilitate the separation of cationic and zwitterionic dyes. This work will hopefully provide insight into predicting the separation of a series of short chain peptides.

Keywords: Electrophoresis
Application Code: Bioanalytical
Methodology Code: Separation Sciences

Session # 1780 Abstract # 1780-15 **Poster Sessions**

Session Title: Practical Chromatography in Today's Laboratory

Abstract Title: **Insights into the Effect of the PDMS-Layer on the Kinetics and Thermodynamics of Analytes Sorption onto the PDMS-Overcoated Coating**

Date: Wednesday, March 09, 2016 - Morn

Time:

Room: Exposition Floor, 400 Aisle

Primary Author: Erica A. Souza-Silva
Universidade Federal do Rio Grande do Sul

Co-Author(s): Emanuela Gionfriddo, Janusz Pawliszyn

Abstract Text

In any DI-SPME method dealing with highly complex matrices, it is important to assure that matrix components do not impair the performance of the method due to non-specific attachment of matrix components onto the coating surface. The implementation of a thin outerlayer of PDMS onto the commercial SPME coating has led to the achievement of a matrix-compatible coating surface. The developed configuration can be seen as a built-in membrane, utilizing a non-porous membrane, i.e. PDMS, placed between the sample and the DVB coating.

In the present work, one of the main premises behind the choice of PDMS as an antifouling material to produce a matrix-compatible fibre for food analysis is attributed to its hydrophobicity, which lessens the attachment of sugars and charged macromolecules to its surface. This process, in turn, significantly decreases the formation of side products and artifacts, due to reactions occurring between carbohydrates and other matrix components, especially for thermal desorption.

In addition, being PDMS materials widely regarded as hydrophobic,the ability of given compound to permeate through PDMS must be carefully investigated. Understanding the role of the PDMS layer in the extraction process employing the PDMS-overcoated solid coatings is particularly important when considering the following questions: (1) how does the PDMS layer affect the uptake of analytes for kinetic extractions (under non-equilibrium conditions)? (2) Would the PDMS layer impose a bias on the representativeness of sampling (polar vs non-polar analytes)? (3) Does the PDMS layer affect the coating capacity towards target analytes as compared to the original coating?
To address these questions, eleven analytes were selected to model and discuss the mass transfer of analytes within the PDMS-modified coating during the mass uptake process. In addition, the thermodynamic parameters, here associated with the fibre constant, were also investigated.

Keywords: Environmental, Food Contaminants, Sample Preparation, SPME

Application Code: Food Contaminants

Methodology Code: Sampling and Sample Preparation

Session # 1780 Abstract # 1780-16 **Poster Sessions**

Session Title: Practical Chromatography in Today's Laboratory

Abstract Title:

A New Anion Exchange Column for Fast Ion Chromatographic Separation of Monosaccharides and Disaccharides in Biofuel, Food, and Beverage Samples

Primary Author: Yan Liu
Thermo Fisher Scientific

Date: Wednesday, March 09, 2016 - Morn
Time:
Room: Exposition Floor, 400 Aisle

Co-Author(s): Andy Woodruff, Charanjit Saini, Christopher Pohl, Yury Agroskin

Abstract Text

The biofuel, food, and beverage industries process very high volumes of carbohydrate-containing samples. These industries require high throughput determination of monosaccharides and disaccharides in various sample matrices. We have recently developed a new anion exchange column designed specifically to provide exceptionally fast, high-resolution separation of monosaccharides and disaccharides. In this paper, we will describe the characteristics of the anion exchange stationary phase used in this new colum and discuss the performance of this new anion exchange column in the determination of monosaccharides and disaccharides using high pressure ion chromatography systems with electrolytic eluent generation and electrochemical detection capabilities. We will demonstrate the rapid and high-resolution separations of carbohydrates such as fucose, sucrose, arabinose, galactose, glucose, xylose, mannose and fructose in biofuel samples as well as carbohydrates such as sucrose, glucose, fructose, lactose, cellibiose, and maltose in food and beverage samples.

Keywords: Beverage, Biofuels, Ion Chromatography, Liquid Chromatography
Application Code: Food Identification
Methodology Code: Liquid Chromatography

Session # 1780 Abstract # 1780-17 **Poster Sessions**

Session Title: Practical Chromatography in Today's Laboratory

Abstract Title: **Column Robustness Challenges during HPLC Method Development**

Primary Author: Jun Wang
Takeda Boston

Date: Wednesday, March 09, 2016 - Morn
Time:
Room: Exposition Floor, 400 Aisle

Co-Author(s): Elizabeth Hewitt, Laila Kott, Scott Zugel

Abstract Text

Column robustness issues rarely occur in reversed-phase liquid chromatography, but they can represent a great challenge for HPLC method development. In this study, the compound itself is complex, containing 4 chiral centers and a sulfamate group. It also has 8 known related impurities with similar structure as the API but very different pKas. All the impurities are extremely sensitive to pH variation and column surface changes. Column robustness issues were observed for a validated method involving 5 impurities on columns from different manufacturing lots. To avoid sourcing a specific column lot for the method, extensive development work was conducted to re-develop a robust method.

With special attention to inert columns, over 20 columns from different manufacturers and different stationary phases such as C8, C18, Phenyl were investigated and screened at a wide range of pH values. Fourteen conditions (from 12 different columns at pH 3.2, 3.5, 4.0, 8.5 and 10) effectively separated all impurities. Unfortunately, column reproducibility issues were observed consistently at every condition. Using a Poroshell PFP column at the ideal pH range of 3.5, a short and sensitive analytical method was developed. The method is simpler and more robust than the current method; however, it also suffers from a column reproducibility issue involving a critical impurity pair - one of them is a low level process impurity and the other is a major drug product degradant. With strategic considerations for impurity reporting, this method can be used as assay & impurity method for both drug substance and drug product.

Keywords: HPLC, HPLC Columns, Method Development, Pharmaceutical
Application Code: Pharmaceutical
Methodology Code: Liquid Chromatography

Session # 1780 Abstract # 1780-18 **Poster Sessions**

Session Title: Practical Chromatography in Today's Laboratory

Abstract Title: **There are Problems Associated with Gradient and Method Transfer in HPLC and UHPLC – Are There Explanations and Usable Workarounds? Part 1 of 2**

Date: Wednesday, March 09, 2016 - Morn
Time:
Room: Exposition Floor, 400 Aisle

Primary Author: Michael Woodman
Agilent Technologies

Co-Author(s): Gregory Hunlen

Abstract Text

UHPLC has allowed many users to explore practical examples of what was in the past only possible under theoretical consideration. We observe, however, that considerable difficulty may be found when a demanding HPLC method is transferred to UHPLC columns and/or systems, or the reverse is attempted. Is it due to dramatic increases in operating pressure and viscous heating, dwell volume disparities or is it that many methods are simply not translated with sufficient calculation precision? For every remarkable example of perfect translation, generally prepared by an instrument or column vendor, as many as five examples of catastrophic results may be found. The answer is no doubt complex and is more and more the subject of discussion now that it is quite clear that UHPLC concepts and practice are here to stay and growing in use.

We have interrogated systems in the UHPLC classification, with a selection of columns generally meeting the definition of UHPLC in design and performance, and a set of HPLC systems and conditions for comparison. Gradient delay volumes were carefully measured by various approaches and used to evaluate the effectiveness of simply offsetting the time of the gradient to achieve resolution and retention time parity. Extra-column dispersive effects, while of considerable interest, are ignored as the focus of this work is on retention time replication. The nature of systematic distinctions between gradient delay, gradient dwell volume and the intriguing transition volume effect are explored as we assess how well this time-offset approach can perform with typical reversed phase gradients.

Keywords: HPLC, Instrumentation, Liquid Chromatography, Method Development
Application Code: General Interest
Methodology Code: Liquid Chromatography

Session # 1780 Abstract # 1780-19 **Poster Sessions**

Session Title: Practical Chromatography in Today's Laboratory

Abstract Title: **Understanding the Causes and Minimizing the Impact of LC Carryover in Most LC Systems**

Date: Wednesday, March 09, 2016 - Morn
Time:
Room: Exposition Floor, 400 Aisle

Primary Author: Michael Woodman
Agilent Technologies

Co-Author(s): Gregory Hunlen

Abstract Text

As liquid chromatography detector sensitivities have improved and the demand for lower impurity and analyte detection has been increased, today's users have become keenly aware of the problems associated with sample carryover. While carryover is not a new phenomenon, having been covered in some detail in doubtless hundreds of publications, application notes and user operating manuals, it has been our experience that the solutions are inconsistently applied. The most practical solutions, given instruments with varying capability to address the causes without undue difficulty or vendor design bias, are relatively consistent for most systems. Carryover is the observance of measurable quantities of material from prior injections appearing in subsequent injections. While commonly attributed to the injection process, there are some additional possibilities worth including in this work. The degree to which carryover impairs the intended use of the system, as one might expect, is directly related to normal analytical processes that would be strongly affected by the otherwise insignificant amount of carryover typical of normal system operation. One such example might be the intentional injection of a lower limit of detection (LLOD) standard immediately following the injection of a high level standard. If the expected accuracy of the LLOD standard is strictly controlled, the contribution of even minor carryover from a preceding high level standard would have disastrous effects. Diagrams of the primary autosampler designs are presented to help the reader identify common contamination points, and wash features associated with most autosampler designs can be used to effectively counteract the problem.

Keywords: HPLC, Instrumentation, Sample Handling/Automation, Sample Introduction

Application Code: General Interest

Methodology Code: Liquid Chromatography

Session # 1790 Abstract # 1790-1 **SEAC Poster Session**

Session Title SEAC Poster Session

Abstract Title **New Portable Electrochemical Instrument for [i]In Situ[/i] Analysis**

Primary Author Pablo Fanjul-Bolado
DropSens S.L

Date: Wednesday, March 09, 2016 - Morn
Time:
Room: Exposition Floor, 400 Aisle

Co-Author(s) Alejandro Pérez-Junquera, Begoña González-García, Carla Navarro-Hernández, David Hernández-Santos, Laura Fernández-Llano, Marta Neves, Pablo Bobes-Limenes

Abstract Text

Analytical chemistry is moving towards the development of sensitive, miniaturized and cost-effective strategies, within the concept of in situ analysis. The use of sensors for routine decentralized measurements has been encouraging the development of new technologies and more suitable methodologies.

In this work a hand-held battery powered potentiostat has been developed. This small electrochemical device works with disposable screen-printed electrodes. Accordingly to the developed application, it is customized with a voltammetric, amperometric or open circuit potential (OCP) technique and its specific optimized parameters, as well as the calibration curve. The results, besides being displayed in the instrument, can be downloaded via USB to a PC.

The relevance of the described device is presented with two different analytical applications: pH control in sea water and determination of ethanol in alcoholic beverages. Measurements of pH through the difference of potential between reference and working electrode were done using OCP technique and antimony sensors. The calibration of the pH remained linear over the range 2 – 10.9 (Britton- Robinson 0.1M solution, r2 > 0.99) with repeatability between same pH values □4%. The analytical method described was successfully tested to measure the pH of North Atlantic Ocean sea water. A non-enzymatic ethanol sensor, based on a platinum nanostructured electrode, responded linearly between 0 – 4.7 g L-1 of ethanol (in a KOH 1.0 M solution, r2 > 0.99), employing a linear sweep voltammetry technique. Moreover, the sensor was used to determine the alcoholic strength in different beer and wine brands, showing an adequate analytical performance.

This work has being supported by the seventh framework programme under the project acronym "Common Sense".

Keywords: Electrochemistry, Instrumentation, Portable Instruments, Sensors
Application Code: Other
Methodology Code: Portable Instruments

Session # 1790 Abstract # 1790-2 **SEAC Poster Session**

Session Title: SEAC Poster Session

Abstract Title: **Aqueous and Non-Aqueous Electrochemical Activities of Mercaptosuccinic Acid Stabilized Au11-13 Clusters**

Date: Wednesday, March 09, 2016 - Morn

Time:

Room: Exposition Floor, 400 Aisle

Primary Author: Jonathan Padelford
Georgia State University

Co-Author(s): Gangli Wang

Abstract Text

Tremendous progresses have been achieved in determining the molecular composition and atomic structures of thiolate stabilized gold clusters recently. This category of nanomaterials display exciting optical and electrochemical properties and have shown promising in applications such as biological imaging, drug delivery and catalysis. While aqueous soluble Au clusters are extensively characterized for optical properties, studies of the corresponding electrochemical properties are rare. Several limiting factors include the cluster monodispersity, small potential window limited by water reactivity, and the low charging energy due to the high dielectric constant of water. Through further optimization in synthesis and isolation, a composition of Au11MSA8 is proposed from ongoing MALDI mass spectrometry analysis. Several electron transfer peaks in the negative potential range were resolved in aqueous solution via cyclic and square wave voltammetry. The impacts of ligand charges on the electron transfer activity were evaluated by variations of the solution pH. To combat water limitations, the clusters were phase transferred into a non-aqueous medium using tetraoctylammonium bromide. Excitingly, multiple electron transfer peaks separated by a HOMO-LUMO gap of 2.6 V were observed for the same clusters in non-aqueous solvent. Similar voltammetric features and a larger gap in comparison to the highly investigated Au25 are consistent with smaller core size. The size/composition dependent electrochemistry features are further correlated with optical activities including absorption and near infrared photoluminescence.

Keywords: Electrochemistry
Application Code: Nanotechnology
Methodology Code: Electrochemistry

Session # 1790 Abstract # 1790-3 **SEAC Poster Session**

Session Title SEAC Poster Session

Abstract Title **Copolymerized Triazole Based Ionic Liquid as New Sensing Material in Ion-Selective Sensors**

Primary Author Lukasz K. Mendecki
Keele University

Date: Wednesday, March 09, 2016 - Morn
Time:
Room: Exposition Floor, 400 Aisle

Co-Author(s)

Abstract Text

Novel solid contact iodide selective electrodes based on covalently attached 1,2,3 triazole ionic liquid (IL) were prepared and investigated in this study. Triazole based IL moieties were synthesized using click chemistry and were further copolymerized with lauryl methacrylate via a simple one step free radical polymerization to produce a self-plasticized copolymer. The mechanical properties of obtained polymer were found to be suitable for the fabrication of plasticizer-free ion-selective membrane electrodes and bulk optode. The influence of the membrane compositions and pH, the effect of ionophore and lipophilic ionic sites on the response properties of the ISEs were investigated in this study. The most optimal responses were observed for both ionophore and ionic sites free membranes demonstrating that covalently attached IL moieties provide adequate functionality to the ion selective membrane. This also indicates that triazole based IL can be directly involved in binding and stabilization of iodide ions in the membrane bulk and therefore have direct influence on membrane's selectivity. Further potentiometric experiments revealed that each electrode displays high selectivity towards iodide anions over a number of inorganic anions. The electrodes exhibited a near Nernstian behavior with a theoretical slope of -58.2 mV per decade across large concentration range with lower detection limits found at approximately 10[sup]-7.5[/sup] M. Moreover, the potentiometric responses of these electrodes were independent of pH changes over the range of 3.0 – 10.0 with satisfactory reproducibility. These all-solid state sensors were utilized for the selective potentiometric determination of iodide ions in natural water and human urine samples in the nanomolar concentration range.

Keywords: Electrochemistry, Potentiometry, Sensors
Application Code: Clinical/Toxicology
Methodology Code: Sensors

Session # 1790 **Abstract # 1790-4** **SEAC Poster Session**

Session Title SEAC Poster Session

Abstract Title **Investigation of Cloud-Point Extraction and UV-Vis for Determining Copper and Cadmium in Vegetables**

Primary Author Asaduzzaman Nur
Tennessee Technological University

Date: Wednesday, March 09, 2016 - Morn
Time:
Room: Exposition Floor, 400 Aisle

Co-Author(s) Andrew Callender

Abstract Text

I used cloud point extraction method for the analysis of cadmium and copper in vegetables (lettuce and spinach), soft drinks and water. I presented a comparison among three different ligands pyruvaldehyde thiosemicarbazone (PTSM) and the commonly used reagent APDC. UV-vis absorbanmce was used as the detection method. Extraction conditions were optimized for pH, salt concentration, and ligand concentration, amount of extraction solvent, extraction time and temperature.

Keywords: Extraction, Metals, Trace Analysis, UV-VIS Absorbance/Luminescence
Application Code: Environmental
Methodology Code: UV/VIS

Session # 1790 Abstract # 1790-5 **SEAC Poster Session**

Session Title SEAC Poster Session

Abstract Title **Potentiometric Ion-Selective Electrodes based on Metastable Photoacid for Cation Detection**

Date: Wednesday, March 09, 2016 - Morn
Time:
Room: Exposition Floor, 400 Aisle

Primary Author Parth K. Patel
University of Central Florida

Co-Author(s) Karin Y. Chumbimuni-Torres

Abstract Text

Liquid contact ion-selective electrodes (LC-ISE) have been used for many practical applications to detect various ions.[sup]1[/sup] These sensors based of an ionophore (selective to the ion of interest) and an ion-exchanger (to keep electroneutrality of the membrane) within a plasticized polymer matrix, requires pre-condition step before measurements can be performed. During the pre-condition step, the ionophore is doped with the ion of interest based on the principle of mass transfer equilibria which takes hours.[sup]2[/sup] Herein, we present a 'one of a kind' LC-ISE based on metastable photoacid ([i]m[/i]PAH) that is used as a photoactive ion-exchanger which could reduce the time required for the pre-condition step of the LC-ISE. The [i]m[/i]PAH photodissociate its protons upon irradiation and undergoes thermal reassociation in order of minutes, after activation with visible light.[sup]3[/sup] When this [i]m[/i]PAH is incorporated in LC-ISE membrane, the photodissociated proton is exchanged with the cation of interest after activation, subsequently doping the ionophore within minutes. Furthermore, the diffusion coefficient of cations can be determined from kinetic studies prior to potentiometric analysis by absorption spectroscopy using the same membrane. With this work, we expect to present for the first time the use of a photoactive ion-exchanger that could behave the same as traditional non-photoactive ion-exchangers.

1. P. Buhlmann, E. Pretsch, E. Bakker, Chem. Rev., 1998, 98, 1593.
2. E. Bakker, P. Buhlmann, E. Pretsch, Chem. Rev., 1997, 97, 3083.
3. V. K. Johns, P. Peng, J. DeJesus, Z. Z. Wang, Y. Liao, Chem.-Eur. J., 2014, 20, 689.

Keywords: Detection, Ion Selective Electrodes, Potentiometry, UV-VIS Absorbance/Luminescence
Application Code: General Interest
Methodology Code: Electrochemistry

Session # 1790 Abstract # 1790-6 SEAC Poster Session

Session Title SEAC Poster Session

Abstract Title **Signal Amplification of a Highly Selective Universal MicroRNA Electrochemical Sensor for Single Nucleotide Polymorphism Detection**

Primary Author Dawn Mills
University of Central Florida

Date: Wednesday, March 09, 2016 - Morn
Time:
Room: Exposition Floor, 400 Aisle

Co-Author(s) Dmitry M. Kolpashchikov, Jeffer Pinzon, Karin Y. Chumbimuni-Torres, Percy Calvo-Marzal

Abstract Text

Detection of microRNAs (miRNAs) has gained tremendous interest due to their excellent capability to function as biomarkers for human malignancies such as cancer and neurological disorders.[sup]1-2[/sup] Herein, we report the design and optimization of a four-way junction (4J) universal electrochemical sensor that consists of an electrode-immobilized DNA stem-loop(SL) probe and two adaptor strands (m and f). Adaptor strand m was labeled with a methylene blue (MeB) redox marker. Two adaptor strands can hybridize to SL probe and target miRNA to form a quadripartite associate, which is stabilized by 4J structure. In this complex MeB is brought close to the electrode surface thus enabling electrochemical output signal. The sensor shows a linear range from 5 to 50 nM with a detection limit of 3.2 nM miRNA122 using square wave voltammetry. The sensor was shown to regenerate under mild conditions, allowing at least 6 measurements without loss of efficiency. The sensor was further characterized using ellipsometric and electrochemical techniques. In order to detect lower amounts of miRNA by signal amplification, MeB redox marker was replaced with hexaamine ruthenium (III) (RuHex), which interacts electrostatically with anionic phosphates on the DNA backbone. Potassium ferrocyanide (II) was introduced to enable turnover of RuHex[sup]3[/sup] and was shown to amplify the signal >150x.

1. Bartel, D., 2004, Cell. 116, 281–297.
2. Li, F., Peng, J., Zheng, Q., Guo, X.,Tang, H., Yao, S., 2015, Anal. Chem. 87, 4806□4813.
3. Boon, E., Jackson, N., Wightman, M., Kelley, S., Hill, M., Barton, 2003, J. Phys. Chem. 107, 11805-11812.

Keywords: Biosensors, Electrochemistry, Nucleic Acids, Voltammetry
Application Code: Bioanalytical
Methodology Code: Electrochemistry

Session # **1790** **Abstract #** **1790-7** **SEAC Poster Session**

Session Title: SEAC Poster Session

Abstract Title: **Self-Reference Single Strip Paper Based Sensors for Ion Detection**

Primary Author: Andrew J. Manhan
University of Central Florida

Date: Wednesday, March 09, 2016 - Morn
Time:
Room: Exposition Floor, 400 Aisle

Co-Author(s): Karin Y. Chumbimuni-Torres, Stephanie Armas

Abstract Text

There is a growing interest in the development of low-cost, portable, versatile, and reliable sensors for ion detection. Therefore, we propose to create a sensor that incorporates the ion-selective membrane (ISM) and a reference membrane (RM) in a single strip paper-based device. This device will eliminate the need of an external reference electrode, allowing portability, and reliability by achieving a Nernstian Response through the coupled ISM and RM. The RM will be based on the copolymer methyl methacrylate-co-decyl methacrylate (MMA-DMA) (support matrix), combined with ionic liquids (ILs) to create and maintain a stable potential that is unaffected by an increase of ionic activity. The chosen matrix for the RM is preferable over polyvinyl chloride, as it does not require a plasticizer, avoiding the risk of plasticizer leaching, leading to better electrode lifetime (2). The functionality of the RM is dictated by limited partitioning of the IL into a sample solution, which then dictates the potential of the reference electrode (1). Preliminary experiments show promising results with near Nernstian response to potassium and sodium ions.

[1] Cicmil, D.; Anastasova, S.; Kavanagh, A.; Diamond, D.; Mattinen, U.; Bobacka, J.; Lewenstam, A.; Radu, A. Electroanalysis 2011, 23 (8), 1881–1890.
[2] Y. Qin, S. Peper, E. Bakker, Plasticizer-free polymer membrane ion-selective electrodes containing a methacrylic copolymer matrix, Electroanal, 14(2002) 1375-81.

Keywords: Ion Exchange, Polymers & Plastics, Potentiometry
Application Code: Environmental
Methodology Code: Sensors

Session # 1790 Abstract # 1790-8 **SEAC Poster Session**

Session Title SEAC Poster Session

Abstract Title **Development and Characterization of an Ion Selective Microsensor for the Detection and Monitoring of Zinc Levels in Citrus Plants**

Date: Wednesday, March 09, 2016 - Morn

Time:

Room: Exposition Floor, 400 Aisle

Primary Author Courtney Hulce
University of Central Florida

Co-Author(s) Jared Church, Karin Y. Chumbimuni-Torres, Swadeshmukul Santra, Woo Hyoung Lee

Abstract Text

Ion selective sensors (ISSs) have become an invaluable tool for biological applications because of their low energy requirements and ability to be miniaturized while maintaining low limits of detection (LOD). Commercialized microsensors have been developed for alkali, alkaline earth metals, and some anions, however, more efforts are required to develop ISSs for heavy metal ions, zinc (Zn[sup]2+[/sup]) in particular, which is critical to many cellular functions[sup]1[/sup], but is toxic beyond a certain concentration level[sup]2[/sup]. This study seeks to develop a zinc ion selective microsensor to monitor zinc levels in citrus plants. Preliminary results in developing the zinc ion-selective membrane showed an ISS with Nernstian response of 30 ± 2 mV decade[sup]-1[/sup] and LOD of 1.30×10-7 mol L[sup]-1[/sup]. Using the zinc ion-selective membrane, a micro-ISS was fabricated. The miniaturized ISS designed had a 400 µm long ion-selective membrane with a tip diameter of 15 µm. The calibration curve showed good correlation between the potential measurements at different concentrations of zinc ions. The successful development of the microsensor will allow for the measurement of ion concentration profiles without perturbing the biological environment in citrus plants[sup]3[/sup].

1.Tomoya Hirano,Kazuya Kikuchi,Yasuteru Urano,Tsunehiko Higuchi, Tetsuo Nagano J. Am. Chem. Soc., 2000, 122 (49), pp 12399–12400.
2.Clemens, S. Biochimie, 2006. 88(11): p. 1707-1719.
3.Schramm A, Larsen LH, Revsbech NP, Ramsing NB, Amann R, Schleifer KH. Appl. Microbiol. Biotechnol. 1996, 62 (12) 4641-4647.

Keywords: Characterization, Detection, Ion Selective Electrodes, Microelectrode
Application Code: Agriculture
Methodology Code: Electrochemistry

Session # 1790 Abstract # 1790-9 SEAC Poster Session

Session Title SEAC Poster Session

Abstract Title **Poly(3-octylthiophene)-Based Solid Contact Ion-Selective Electrodes with Improved Potential Stability**

Primary Author Jennifer M. Jarvis
University of Memphis

Date: Wednesday, March 09, 2016 - Morn
Time:
Room: Exposition Floor, 400 Aisle

Co-Author(s) Bradford D. Pendley, Erno Lindner, Marcin Guzinski

Abstract Text

Solid contact ion-selective electrodes (SC ISEs) are attractive for analyzing small volumes of clinical samples (e.g. blood samples of newborn babies) because of the possibility of sensor miniaturization. In SC ISEs, a conductive polymer (CP) serves as an ion-to-electron transducer between an electron-conducting metal substrate and an ion-conducting ion-selective membrane. However, SC ISEs often have poor potential stability and irreproducible standard potentials. Due to these drawbacks, SC ISEs require frequent calibrations, which is a serious limitation for sensors aimed for continuous monitoring or for single-use. The irreproducible standard potentials and the poor stability of SC ISEs are thought to be due to gradual oxidation of the CP, i.e., due to changes of the CP redox potential over time. To control the redox potential of the CP, we have implemented a redox couple (7,7,8,8-tetracyanoquinodimethane, (TCNQ)) into the CP layer (poly(3-octylthiophene), (POT)) at a high concentration. TCNQ has a standard redox potential below that of POT so its reduced form is more easily oxidized than POT. When the TCNQ concentration in the POT film is sufficiently high and the concentration ratio of its oxidized and reduced forms is electrochemically set close to 1:1 it is expected to stabilize the potential of the POT film and thus improve the potential stability of the SC ISE. SC ISEs with TCNQ in the POT film showed improved potential stability and improved standard potential reproducibility versus those without TCNQ: -1.1 mV/hr versus -2.5 mV/hr and 397±10 mV versus 431±33 mV, respectively.

Keywords: Biosensors, Electrochemistry, Ion Selective Electrodes, Potentiometry
Application Code: Biomedical
Methodology Code: Sensors

Session # 1790 Abstract # 1790-10 **SEAC Poster Session**

Session Title: SEAC Poster Session

Abstract Title: **Urine Carbon Dioxide in Septic Shock**

Date: Wednesday, March 09, 2016 - Morn

Time:

Room: Exposition Floor, 400 Aisle

Primary Author: James G. Atherton, University of Memphis

Co-Author(s): Artur Jasinski, Bradford D. Pendley, Erno Lindner, Marcin Guzinski, William King

Abstract Text

Since 2001 intensive care doctors have used “early, goal directed therapy” for the management of shock. The protocol's essence is careful monitoring of metabolic parameters (hematocrit, venous carbon dioxide (CO2) and adapting treatment to their changes. Emanuel Rivers demonstrated mortality decrease from 47 to 31% by following this protocol. We hypothesize that urine carbon dioxide may also indicate septic shock prognosis and therefore provide benefit to monitoring. Scientists have shown large differences in urine CO2 between healthy controls and hemodynamically unstable patients. Increased CO2 production and decreased CO2 clearance may contribute to the difference.
As current detected, researchers have difficulty knowing urine carbon dioxide. We needed to determine if urine carbon dioxide as a prognostic indicator in septic shock has enough promise to justify the efforts for us to streamline the detection of urine carbon dioxide. To investigate the utility of urine CO2 as a prognostic tool for septic shock we validated a sampling protocol and we collected samples in an IRB approved pilot study to know if developing a more refined sensor system is justified. During the initial phase of our project we built a wall-jet flow through manifold for measuring urine carbon dioxide and we have established that our system provides accurate values. We validated it for varying urine environments. During the second phase of our study we used a syringe with acceptable CO2 impermeability characteristics to collect samples from patients' Foley catheter. In an IRB approved study, we successfully monitored urine for carbon dioxide levels in 12 intensive care unit patients both during and after shock. During analysis of our data we recognized the need for better urine CO2 monitoring. In this phase we establish that urine CO2 may indeed correlate with patient hemodynamic status and have the justification for the third phase of our project, fabricating a better urine CO2 sensor.

Keywords: Biomedical, Electrochemistry, Ion Selective Electrodes

Application Code: Biomedical

Methodology Code: Sensors

Session # 1790 Abstract # 1790-11 **SEAC Poster Session**

Session Title SEAC Poster Session

Abstract Title **Disposable Paper-Based Electrochemical Ion-Sensing Platform**

Primary Author Jinbo Hu
University of Minnesota

Date: Wednesday, March 09, 2016 - Morn
Time:
Room: Exposition Floor, 400 Aisle

Co-Author(s) Andreas Stein, Philippe Buhlmann

Abstract Text

Better healthcare is a global challenge, and countless lives could be saved if affordable diagnostic devices were available, especially in the developing world. Previously, a three-dimensional paper-based ion sensor with conventional ion-selective and reference electrodes was developed to quantitatively determine the concentrations of clinically relevant ions (K+, Na+, Ca2+, and Cl-) [1]. In this work, we simplify the use of paper-based ion sensor by developing a low-cost disposable two-dimensional ion-sensing platform. This ion-sensing platform contains miniaturized all-solid-state ion-selective and reference electrodes integrated on paper with microfluidic channels. The all-solid-state electrodes are based on nanostructured colloid-imprinted mesoporous carbon, which serves as an ion-to-electron transducer. For a measurement, only one droplet of sample is needed. These devices exhibit linear responses towards different concentrations of electrolyte ions. They are disposable, simple to use, and do not require any supply reagents to function. They provide a promising affordable solution to the quantitative analysis of electrolytes in blood.

[1] Lan, W.-J.; Zou, X. U.; Hamedi, M. M.; Hu, J.; Parolo, C.; Maxwell, E. J.; Bühlmann, P.; Whitesides, G. M. Anal. Chem. 2014, 86, 9548-9553.

Keywords: Clinical Chemistry, Electrochemistry, Ion Selective Electrodes, Lab-on-a-Chip/Microfluidics
Application Code: Clinical/Toxicology
Methodology Code: Electrochemistry

Session Title SEAC Poster Session

Abstract Title **Development of Calibration-Free Electrochemical Sensors Using Novel Redox Polymers**

Date: Wednesday, March 09, 2016 - Morn
Time:
Room: Exposition Floor, 400 Aisle

Primary Author Xue Zhen
University of Minnesota

Co-Author(s) Philippe Buhlmann

Abstract Text

Good electrode-to-electrode reproducibility is a necessity for calibration-free electrochemical sensors. Previously, our group introduced cobalt(II)/cobalt(III) redox buffers as a component of the transducer layer of solid-contact ion-selective electrodes in order to produce a well-defined phase boundary potential at the interface between the sensing membrane and the underlying electron conductor. The electrode-to-electrode deviation of the calibration curve was minimized to as low as 0.70 mV, making calibration-free measurements possible [1]. However, due to their low lipophilicity, leaching of the redox buffer from the ion-selective membrane into the aqueous sample could not be prevented, resulting in potential drifts. In this work, we develop a new class of polymeric redox buffers by attaching redox couples to polymers through covalent bonds, and using them for the fabrication of solid-contact ion-selective electrodes that exhibit exceptional long-term stability by avoiding leaching of the redox buffer from the ion-selective membrane into the aqueous sample. Another application of the new redox buffer polymers is their use for other modes of electro¬chemical sensing, such as ionophore-assisted ion transfer voltammetry and coulometry.

[1] Zou, X. U.; Zhen, X. V.; Cheong, J. H.; Bühlmann, P., Calibration-Free Ionophore-Based Ion-Selective Electrodes With a Co(II)/Co(III) Redox Couple-Based Solid Contact. Anal. Chem. 2014, 86, 8687-8692.

Keywords: Electrochemistry, Electrodes, Polymers & Plastics, Potentiometry
Application Code: Environmental
Methodology Code: Sensors

Session # 1790 Abstract # 1790-13 **SEAC Poster Session**

Session Title SEAC Poster Session

Abstract Title **Ultrasensitive Detection of Dopamine with Carbon Nanopipettes**

Date: Wednesday, March 09, 2016 - Morn

Time:

Room: Exposition Floor, 400 Aisle

Primary Author Keke Hu
Queens College CUNY

Co-Author(s) Michael Mirkin, Min Zhou, Yun Yu

Abstract Text

Micrometer- and nanometer-sized carbon fiber electrodes have been extensively used for measuring dopamine and other neurotransmitters in biological systems. While the radius of some reported probes was <<1 μm, the length of the exposed carbon was typically on the micrometer scale, thus, limiting the spatial resolution of electroanalytical measurements. More recent attempts to determine neurotransmitters in single cells and vesicles provided additional impetus for decreasing the probe dimensions. The disk-type nanoelectrodes are not suitable for such experiments because a larger surface area is required for sensitive detection of dopamine. Here we report three types of dopamine sensors based on carbon nanopipettes (CNPs) prepared by chemical vapor deposition of carbon into the pre-pulled quartz capillary. These include 5-100 nm radius CNPs with either a shallow or a deep cavity near the orifice and CNPs with an open path in the middle, in which the volume of sampled solution is controlled by the applied pressure. Because of the relatively large surface area of carbon exposed to solution inside the pipette, all three types of CNPs exhibited greatly improved voltammetric responses to dopamine with the attainable detection limit <100 pM. The sensor properties (e.g., orifice radius, pipette angle, and carbon roughness) were characterized by high resolution TEM and SEM imaging to demonstrate the agreement between the experimental and simulated voltammograms.

Keywords: Analysis, Biosensors, Electrochemistry, Sampling
Application Code: Bioanalytical
Methodology Code: Electrochemistry

Session # 1790 Abstract # 1790-14 **SEAC Poster Session**

Session Title: SEAC Poster Session

Abstract Title: **Avoiding Errors in Electrochemical Measurements: Effect of Frit Material on the Performance of Reference Electrodes with Porous Frits**

Date: Wednesday, March 09, 2016 - Morn

Time:

Room: Exposition Floor, 400 Aisle

Primary Author: Maral PS Mousavi
University of Minnesota

Co-Author(s): Evan Anderson, Marc A. Hillmyer, Philippe Buhlmann, Stacey Saba

Abstract Text

Reference electrodes are designed to provide a constant and sample-independent reference potential and are used in almost every electrochemical measurement. In many reference electrodes, a nanoporous Vycor glass frit is used to contain the electrolyte solution that forms a salt bridge between the sample and the reference solutions. It was recently discovered that in samples with low ionic strength, the half-cell potentials of reference electrodes with nanoporous Vycor frits are affected by sample composition and can shift by more than 50 mV (which can cause up to 900% error in the measurement). It was confirmed that such large potential variations result from electrostatic screening of ion transfer into the glass nanopores that have negative surface charges, and not by the liquid junction potential at the interface of sample and reference solutions.[sup]1[/sup] Vycor glass frits in reference electrodes have been recently replaced by new materials, mainly two porous glasses with brand names of CoralPor and Electro-porous KT and two porous polymers of Teflon and polyethylene. We show that glass frits with larger pore sizes and polymer frits nearly eliminate the potential variations caused by screening of ion transport through the pores of the frit, however, much larger flow rates of reference solution through the pores and into the test solution was observed for larger pores.

1. M. P. S. Mousavi and P. Buhlmann Anal. Chem., 2013, 85, 8895-8901

Keywords: Electrochemistry, Electrodes, Potentiometry, Voltammetry

Application Code: General Interest

Methodology Code: Electrochemistry

Session # 1800 Abstract # 1800-1 **Awards**

Session Title: Ralph N Adams Award

Abstract Title: **Clinical Applications of Single Molecule Arrays (Simoa)**

Primary Author: David R. Walt
Tufts University

Co-Author(s)

Date: Wednesday, March 09, 2016 - After
Time: 01:40 PM
Room: B312

Abstract Text

My laboratory has developed a digital immunoassay method based on single molecule ELISAs that we call single molecule arrays (Simoa). In this method, single immunocomplexes are formed on paramagnetic beads and sealed in femtoliter microwell arrays. An enzyme label on the immunocomplex catalyzes the conversion of thousands of substrate molecules to a fluorescent product. The resulting concentrated fluorescent product enables visualization of the signal such that fluorescent microwells can easily be counted. In this way digital counting of individual molecules can be achieved, enabling ultra-sensitive measurements of both proteins and nucleic acids. The Simoa technology has been applied to a variety of clinical applications including early cancer diagnostics and infectious disease detection.

Keywords: Bioanalytical, Fluorescence, Immunoassay, Proteomics
Application Code: Bioanalytical
Methodology Code: Fluorescence/Luminescence

Session # 1800 Abstract # 1800-2 Awards

Session Title: Ralph N Adams Award

Abstract Title: **Microengineered Devices for Biomedical Research**

Date: Wednesday, March 09, 2016 - After
Time: 02:15 PM
Room: B312

Primary Author: Nancy Allbritton
University of North Carolina at Chapel Hill

Co-Author(s)

Abstract Text

The ability to monitor and manipulate the microenvironment of cells and tissues is one of the most promising applications for microengineered systems. The laboratory is developing a suite of technologies based on microengineered platforms and microfluidics to manipulate and analyze living cells and organoids. We have developed simple, inexpensive fabrication methods utilizing photoresists, plastics, and hydrogels to array cells and organoids. The fabricated devices include detachable, deformable, or biodegradable array elements designed for cell/organoid analysis and sorting. Cells cultured in these engineered microenvironments yield high-density arrays for high through put analyses and assays followed by isolation of individual cells and organoids. The arrays bring the power of cell sorting capabilities to high-content-screening microscopy. Applications for this technology are envisioned for the sorting of single cells for mRNA expression analysis, identification and capture of rare immune cells for immunotherapy, and screening of primary samples using complex phenotypes as separation criteria.

Keywords: Bioanalytical, Biological Samples, Biomedical, Biotechnology
Application Code: Biomedical
Methodology Code: Microfluidics/Lab-on-a-Chip

Session # **1800** **Abstract #** **1800-4** **Awards**

Session Title: Ralph N Adams Award

Abstract Title: **Paper-Based Microfluidic Devices for Point-of-Need Bioanalysis**

Primary Author: Charles R. Mace
Tufts University

Date: Wednesday, March 09, 2016 - After
Time: 03:40 PM
Room: B312

Co-Author(s): Samuel Berry, Syrena C. Fernandes

Abstract Text

We have developed a suite of three-dimensional paper-based microfluidic devices that enable diagnostic assays to be performed directly at the point-of-need. Paper offers a number of attractive characteristics that supports its use as a foundational platform for bioanalytical sensing in resource-limited settings: it is inexpensive, disposable, and compatible with an extensive range of chemical and biochemical reactions. The use of paper-based devices significantly reduces the burden of effort on an end-user, as complex biological samples (e.g., urine and blood) can be applied to the device directly without additional preparation and all biochemical reactions needed for an analysis (e.g., incubation with secondary antibodies) are performed autonomously within the device. Paper-based microfluidic devices thus permit many critical functions used by more sophisticated lab-on-a-chip technologies, but do not rely on power or additional instrumentation to provide valuable information on the health status of a patient. We have demonstrated our method by designing a general device architecture that is capable of performing immunoassays broadly, and we have expanded this approach to include multiplexed immunoassays. Further, we have demonstrated that the careful tuning of the physical and chemical properties of paper can enable entirely new classes of bioanalytical assays. Diagnostic assays created using simple and inexpensive materials have the potential to transform healthcare management systems in resource-limited settings.

Keywords: Bioanalytical, Biological Samples, Chromatography, Lab-on-a-Chip/Microfluidics
Application Code: Bioanalytical
Methodology Code: Microfluidics/Lab-on-a-Chip

Session # 1800 Abstract # 1800-5 **Awards**

Session Title Ralph N Adams Award

Abstract Title **Developmental Proteomics**

Date: Wednesday, March 09, 2016 - After
Time: 04:15 PM
Room: B312

Primary Author Norman J. Dovichi
University of Notre Dame

Co-Author(s) Guijie Zhu, Liangliang Sun, Matthew M. Champion, Paul Huber

Abstract Text
Modern mass spectrometry-based methods provide an exciting opportunity to characterize protein expression in the developing embryo. We employed an isotopic labeling technology to quantify the expression dynamics of nearly 6,000 proteins across six stages of development in Xenopus laevis from single stage zygote through the mid-blastula transition and the onset of organogenesis (1). The expression of ~40% of the proteins showed significant expression changes across the development stages; the expression changes for these proteins naturally falls into six clusters that correspond to major events that mark early Xenopus development. A subset of experiments in this study quantified protein expression differences between single embryos at the same stage of development; within experimental error, embryos at the same stage of development have identical protein expression levels.

(1) Sun L, Bertke MM, Champion MM, Zhu G, Huber PW, Dovichi NJ. Sci Rep. 2014; 4: 4365

Keywords: Liquid Chromatography/Mass Spectroscopy, Proteomics
Application Code: Biomedical
Methodology Code: Mass Spectrometry

Session # 1810 Abstract # 1810-1 **Awards**

Session Title The Coblentz Society - Williams-Wright Award

Abstract Title **Industrial Analysis Utilizing Vibrational Spectrometry**

Primary Author D Warren Vidrine
Vidrine Consulting

Date: Wednesday, March 09, 2016 - After
Time: 01:40 PM
Room: B314

Co-Author(s)

Abstract Text

Vibrational industrial spectroscopy began in fits and starts, with IR photometers used to control industrial processes in World War II, and dispersive spectrometers gaining a role in analytical lab & offline quality control through the 1970's. Online filter photometers continued to be used in industry, but for the most part these photometers were unique ad hoc solutions to exigent problems.

Karl Norris kicked off the modern era of industrial vibrational analysis in 1977, based on newly available filter NIR instrumentation & computational methods. That same year I left academia and joined the fledgling Nicolet Analytical FTIR group. Since then, I have been involved with the development of vibrational industrial instrumentation & analysis, particularly real-time FTIR-based methods, for my whole career.

This development of industrial process analysis & control has progressed incrementally, and three major types of advances have set the pace: robust spectrometers, sampling interfaces, and chemometric methods. Overlying all of this has been the ancient & recurrent problem of defining the analysis: what IS the critical measurement desired?

This presentation will explore the interaction of these factors in influencing the development of real-time vibrational industrial process analysis, and attempt to answer the question of where we are, how we got here, and where might we be going.

Keywords: Infrared and Raman, Instrumentation, Process Analytical Chemistry, Spectrophotometry
Application Code: Process Analytical Chemistry
Methodology Code: Vibrational Spectroscopy

Session # **1810** **Abstract #** **1810-2** **Awards**

Session Title The Coblentz Society - Williams-Wright Award

Abstract Title **Biological Infrared Spectroscopy: An Overnight Success Story 64 Years in the Making**

Primary Author Bob Messerschmidt
Nueon Inc.

Date: Wednesday, March 09, 2016 - After
Time: 02:15 PM
Room: B314

Co-Author(s)

Abstract Text

Infrared spectroscopy is a natural choice to gain molecular understanding of biological materials. Indeed some of the earliest applications of infrared spectrometers were in the qualitative analysis of biological samples(1). So what has changed? Nothing and everything. The progression of instrumentation in infrared spectroscopy has been entirely evolutionary and not revolutionary, despite the marketing claims. The outputted data is still an infrared spectrum which is a fundamental physical property. Developments have however completely changed the time commitment needed to obtain a high quality spectrum, and this in itself has allowed the development of new sampling modalities and computational methods. In this paper I hope to give a bit of a history of the key developments that made infrared spectroscopy a modern day powerhouse for biological insight.

(1) Bird, G. R. and Blout, E. R. Laboratory Investigation 1,266 (1952).

Keywords: Biomedical, FTIR, Instrumentation, Medical
Application Code: General Interest
Methodology Code: Vibrational Spectroscopy

Session # 1810 Abstract # 1810-3 Awards

Session Title: The Coblentz Society - Williams-Wright Award

Abstract Title: **Reflections of a Chemometric Spectroscopist**

Primary Author: David M. Haaland
Spectral Resolutions

Co-Author(s): David A. Melgaard, Howland D. Jones

Date: Wednesday, March 09, 2016 - After
Time: 02:50 PM
Room: B314

Abstract Text

After more than three decades as a chemometrician and spectroscopist, and after serving as the North American editor of the Journal of Chemometrics, I have a number of insights that I would like to share for those interested in performing chemometric research and applying chemometric methods to the analysis of spectral data. A portion of the talk will be devoted to outlining good chemometric practices and recommendations for authors and reviewers of chemometric research articles. In addition, the significant advantages of the little known Augmented Classical Least Squares (ACLS) calibration and prediction methods will be highlighted. These ACLS methods are able to achieve significant advances and remove all the barriers to the use of CLS approaches to quantitative spectral analyses. With these advances, ACLS methods demonstrate quantitative prediction abilities that have always been as good as or better than achieved with PLS or PCR. However, the most significant advantage of ACLS calibrations over PLS and PCR is that they can be readily updated, without the need for recalibration, for spectrometer drift, changes in spectrometers and the introduction of new chemical species not present in the original calibration samples. In addition, CLS methods form the basis of multivariate curve resolution (MCR) which can be applied to the analysis of spectral images to discover and quantify all the independently varying spectral species in the imaged sample without any a priori information being required. Examples of the new capabilities of the ACLS methods will be presented using multicomponent mixtures and spectral images.

Keywords: Chemometrics, Fluorescence, Imaging, Near Infrared
Application Code: Process Analytical Chemistry
Methodology Code: Chemometrics

Session # 1820 Abstract # 1820-1 **Symposia**

Session Title ACS-ANYL - Advances in Instrumentation for Ion Mobility Mass Spectrometry

Abstract Title **Ultra-High Resolution Ion Mobility Separations based upon Long Path Length Structures for Lossless Ion Manipulations (SLIM)**

Primary Author Richard D. Smith
Pacific Northwest National Laboratory

Date: Wednesday, March 09, 2016 - After
Time: 01:35 PM
Room: B308

Co-Author(s) Ahmed Hamid, Aleksey V. Tolmachev, Erin S. Baker, Ian K. Webb, Liulin Deng, Sandilya Garimella, Yehia M. Ibrahim

Abstract Text

Ion mobility separations are of increasing importance in conjunction with analytical applications of mass spectrometry, not only providing additional structure-related information, but potentially more complete analysis of complex samples, detection of lower level constituents, and greater measurement throughput than feasible with on-line liquid phase separations. The benefits of mobility-based separations generally increase as separation power increases, but high resolution mobility separations to this point have only been achieved in conjunction with significant ion losses, substantially limiting their practicality and range of applications. This presentation will describe new approaches using Structures for Lossless Ion Manipulations (SLIM) for enabling very long path length high resolution ion mobility separations, and initial SLIM design implementations and experimental results will be presented. The presentation will conclude with consideration of pending developments enabled by SLIM and their potential impacts for mass spectrometry-based measurements.

Keywords: Instrumentation, Mass Spectrometry, Separation Sciences
Application Code: Bioanalytical
Methodology Code: Mass Spectrometry

Session # **1820** **Abstract #** **1820-2** **Symposia**

Session Title ACS-ANYL - Advances in Instrumentation for Ion Mobility Mass Spectrometry

Abstract Title **Ion Mobility Mass Spectrometers for Structural Biology and Biophysics**

Date: Wednesday, March 09, 2016 - After
Time: 02:10 PM
Room: B308

Primary Author
Matthew F. Bush
University of Washington

Co-Author(s)

Abstract Text

Native ion mobility mass spectrometry is an emerging approach for characterizing the stoichiometry, assembly, and shapes of noncovalent complexes in solution. These technologies are especially useful for investigating proteins and protein complexes that are challenging to characterize using condensed-phase experiments, including those that are heterogeneous, have large mass, and are membrane bound. I will discuss how my lab uses radio frequency (RF) confining drift cells and structures for lossless ion manipulation (SLIM) to analyze intact proteins and protein complexes. RF confining drift cells use a series of ring electrodes to establish a constant electric field along the axis of transmission, similar to conventional drift tubes, but also superimpose RF potentials applied to all electrodes (alternating phases applied to neighboring electrodes) that confine ions along the transverse axes. SLIM devices also use a combination of direct-current and RF potentials, but apply those potentials to electrodes deposited directly onto parallel pairs of printed circuit boards. I will discuss the implementation and application of these ion mobility architectures in the context of structural biology and biophysics.

Keywords: Bioanalytical, Instrumentation, Mass Spectrometry
Application Code: Bioanalytical
Methodology Code: Mass Spectrometry

Session # **1820** **Abstract #** **1820-3** **Symposia**

Session Title ACS-ANYL - Advances in Instrumentation for Ion Mobility Mass Spectrometry

Abstract Title **Toward Protein Ion Surface Characterization with IMS/HDX-MS/MS Techniques**

Date: Wednesday, March 09, 2016 - After
Time: 02:45 PM
Room: B308

Primary Author Stephen J. Valentine
West Virginia University

Co-Author(s) Gregory Donohoe, Mahdiar Khakinejad, Samaneh Ghassabi Kondalaji

Abstract Text

Although gas-phase hydrogen deuterium exchange (HDX) techniques were presented for protein ion characterization more than 20 years ago, the approach received little attention in the ensuing decades. In part, this was a result of the fact that the sites of deuterium incorporation could not be located due to the translocation of the deuterium label upon collisional activation of protein ions. Recently several research groups have investigated the combination of ion mobility spectrometry (IMS) and HDX by employing non-ergodic fragmentation techniques. This presentation will describe the first IMS/HDX-MS/MS experiments in which electron transfer dissociation (ETD) was used to designate deuterium incorporation at specific sites for select gas-phase conformers. The instrumentation used to perform these experiments and the approach for mapping the HDX accessibility of protein and protein complex ions will be presented.

Keywords: Instrumentation, Mass Spectrometry, Protein, Proteomics
Application Code: Bioanalytical
Methodology Code: Mass Spectrometry

Session # 1820 Abstract # 1820-4 Symposia

Session Title ACS-ANYL - Advances in Instrumentation for Ion Mobility Mass Spectrometry

Abstract Title **Accurate Measurements of Ion Mobilities**

Date: Wednesday, March 09, 2016 - After
Time: 03:35 PM
Room: B308

Primary Author Herbert H. Hill
Washington State University

Co-Author(s) Brian Clowers, Kelsey Morrison

Abstract Text
To be discussed

Keywords: Instrumentation, Mass Spectrometry
Application Code: Homeland Security/Forensics
Methodology Code: Mass Spectrometry

Session # 1820 Abstract # 1820-5 **Symposia**

Session Title ACS-ANYL - Advances in Instrumentation for Ion Mobility Mass Spectrometry

Abstract Title **A Multi-Pass Cyclic Ion Mobility Separator: Design and Performance**

Date: Wednesday, March 09, 2016 - After
Time: 04:10 PM
Room: B308

Primary Author Kevin Giles
Waters Corporation

Co-Author(s) Jason Wildgoose, Steve Pringle

Abstract Text

The combination of ion mobility (IM) separation with mass spectrometry (MS) is becoming a more widely used approach, in part due to relatively recent advances in the hyphenated technology. With increasing acceptance comes the driver to improve both performance and functionality; one of the more recent concepts in IM separation involves use of non-linear devices. Here, the design and performance of a novel cyclic IM (cIM) separator will be presented. The purpose of the cIM device is multi-fold: the circular path minimises instrument footprint whilst providing a longer, higher resolution separation path; a multi-pass capability facilitates mobility 'zoom' type operation, providing significantly higher resolution; the device can be enabled for mobility separation or by-passed if not required and, the multifunctional ion entry/exit array can selectively eject species within a range of mobilities, providing additional functionality. The cIM device comprises a stacked plate ion guide with a path length of 100cm around which t-waves circulate to provide mobility separation. The cIM device replaces the standard t-wave mobility separator in a Synapt G2-S quadrupole-IM-ToF system. A mobility resolution of around 60 is indicated for a single pass around the cIM separator with values significantly in excess of this for multiple passes.

Keywords: Chromatography, Instrumentation, Mass Spectrometry, Time of Flight MS
Application Code: Other
Methodology Code: Mass Spectrometry

Session # 1830 **Abstract # 1830-1** **Symposia**

Session Title: Advancing Strategies for Chronic In Vivo Sensing

Abstract Title: **Modulating Blood-Brain Barrier Healing Around Intracortical Electrode Implants**

Primary Author: Ravi Bellamkonda
Georgia Institute of Technology

Date: Wednesday, March 09, 2016 - After
Time: 01:35 PM
Room: B302

Co-Author(s): Alexus Clark, Brianna Gresham, Jessica Falcone, Robert Kretschmar, Shoba Paul, Varun Yarabarla

Abstract Text

Electrically interfacing with the brain is the next frontier, which requires successful biological integration of recording electrodes for eventual closed-looped prosthetics. However, inflammation and neurodegeneration at the electrode-tissue interface greatly inhibit the reliability of intracortical recording electrodes. When the electrodes are inserted, blood vessels in the blood brain barrier (BBB) are ruptured to make room for the implant. After recovery from the initial injury, the presence of a chronic implant within the brain can lead to endothelial cell inflammation. Monocytes and macrophages then attach to the activated endothelial cells and infiltrate the brain. We hypothesize that the presence of these inflammatory leukocytes activates a cytokine cascade that eventually leads to local neurodegeneration. Here we propose to reduce chronic neurodegeneration by modulating the BBB. A main target to reduce BBB leakiness is endothelial cells. Imatinib, a tyrosine kinase inhibitor, has been used in previous studies to reduce permeability of endothelial cells and improve neural survival in neurodegenerative disease models. For this study, to evaluate the state of the BBB, electrodes were implanted into the barrel cortex of the rat. By administering imatinib daily for 2 weeks, IgG leakage was reduced, which correlates with reduced BBB leakage. Further studies will be conducted to evaluate the chronic effects of imatinib treatment in the electrode implant model.

Keywords: Biomedical, Biosensors, Microelectrode, Pharmaceutical
Application Code: Biomedical
Methodology Code: Sensors

Session # 1830 **Abstract # 1830-2** **Symposia**

Session Title: Advancing Strategies for Chronic In Vivo Sensing

Abstract Title: **Is Microdialysis Specifically Monitoring the Tonic Modality of Dopamine Transmission?**

Primary Author: Gaetano Di Chiara
University of Cagliari

Date: Wednesday, March 09, 2016 - After
Time: 02:10 PM
Room: B302

Co-Author(s)

Abstract Text

Dopamine neurons are currently thought to transmit neural information to units they project to by two distinct modalities, tonic and phasic. These two terms however are too generic and might refer to different time scales. Therefore, an operational definition of tonic and phasic modalities of dopamine neuron activity can be conveniently referred to the two modalities of dopaminergic neuron activity, single spike and burst firing, that can be demonstrated in vivo by extracellular recording. According to A. Grace, while microdialysis estimates the overflow of dopamine released by tonic, single spike dopamine neuron activity, voltammetry estimates phasic, burst firing activity. This assumption however contrasts with a number of considerations and observations. First, after discounting for the difference in time-scale, a high level of concordance is observed between the effects of various stimuli on the in vivo release of dopamine as estimated by microdialysis and voltammetry. Second, due to the low efficacy in promoting calcium influx, the ability of single spike firing to release dopamine exocytotically is likely to be minor as compared to burst firing. Third, doses of gamma hydroxy butyrate that selectively depress burst firing decrease dialysate dopamine (Nyssbrandt et al, 1994). Burst firing of dopamine neurons is known to depend on the activity of small-conductance calcium-activated K+-(SK) channels. In order to investigate the contribution of burst firing to in vivo dopamine release as estimated by microdialysis, we have studied the effect of apamin and of CyPPA, respectively an antagonist and an allosteric activator of SK channels, on dialysate dopamine in the n.accumbens shell. The results are consistent with the hypothesis that dopamine release as estimated by microdialysis is largely accounted by burst firing of dopamine neurons.

Keywords: Drugs, Electrochemistry, HPLC Columns, Liquid Chromatography
Application Code: Neurochemistry
Methodology Code: Liquid Chromatography

Session # **1830** **Abstract #** **1830-3** **Symposia**

Session Title Advancing Strategies for Chronic In Vivo Sensing

Abstract Title **Long-Term Monitoring of Dopamine**

Date: Wednesday, March 09, 2016 - After
Time: 02:45 PM
Room: B302

Primary Author Paul E. Phillips
University of Washington

Co-Author(s)

Abstract Text

We have used fused-silica-insulated carbon-fiber microelectrodes for chronic implantation into the striatum of rodents for long-term tracking of extracellular dopamine concentrations with sub-second temporal resolution. Implantation of these electrodes produces minimal neuroinflammatory responses without glial encapsulation; and the recording sites appear unperturbed at the optical microscopy level, with normal levels of tyrosine-hydroxylase-positive staining around the implantation site. As assessed by both in-vitro calibration and the presentation of positive-control stimuli in vivo, electrode sensitivity is stable over months of implantation. Recordings from these electrodes faithfully reproduce the concentrations and kinetics observed with acutely implanted electrodes, but have the distinct advantage of being able to track within-in subject changes in neurotransmission over days, or even weeks, that result from psychological and/or pathological processes. The electrodes are sensitive to impulse-dependent dopamine neurotransmission as indicated by the detection of electrically stimulated changes in extracellular dopamine, by the attenuation of behaviorally evoked responses following the inactivation of midbrain dopamine neurons, and they have replicated many of the patterns of activity observed by electrophysiological recordings of dopamine neurons. Compared to glass-pulled electrodes, the epoxy-sealed fused-silica electrode design of the chronic electrode has lower and, importantly, less variable capacitance, making them more reproducible.

Keywords: Neurochemistry, Voltammetry
Application Code: Neurochemistry
Methodology Code: Electrochemistry

Session # 1830 **Abstract # 1830-4** **Symposia**

Session Title: Advancing Strategies for Chronic In Vivo Sensing

Abstract Title: **Calibration of In Vivo Voltammetry**

Date: Wednesday, March 09, 2016 - After
Time: 03:35 PM
Room: B302

Primary Author: R Mark Wightman
University of North Carolina at Chapel Hill

Co-Author(s)

Abstract Text

Principal component regression, a multivariate calibration technique, is an invaluable tool for the analysis of voltammetric data collected in vivo with acutely implanted microelectrodes. This method utilizes training sets to separate cyclic voltammograms into contributions from multiple electroactive species. The introduction of chronically implanted microelectrodes permits longitudinal measurements at the same electrode and brain location over multiple recordings. The reliability of these measurements depends on a consistent calibration methodology. One published approach has been the use of training sets built with data from separate electrodes and animals to evaluate neurochemical signals in multiple subjects. Alternatively, responses to unpredicted rewards have been used to generate calibration data. This study addresses these approaches using voltammetric data from three different experiments in freely-moving rats obtained with acutely implanted microelectrodes. The findings demonstrate critical issues arising from the misuse of principal component regression that result in significant underestimates of concentrations and improper statistical model validation that, in turn, can lead to inaccurate data interpretation.

Keywords: Calibration, Chemometrics, Electrochemistry, Neurochemistry
Application Code: Neurochemistry
Methodology Code: Electrochemistry

Session # 1830 Abstract # 1830-5 **Symposia**

Session Title: Advancing Strategies for Chronic In Vivo Sensing

Abstract Title: **Advancing the Possibilities for Chronic Brain Microdialysis**

Primary Author: Adrian C. Michael
University of Pittsburgh

Date: Wednesday, March 09, 2016 - After
Time: 04:10 PM
Room: B302

Co-Author(s): Andrea Jaquins-Gerstl, Erika Varner, Khanh Ngo, Stephen Weber

Abstract Text

Microdialysis probes are widely used to sample the chemical content of the brain extracellular space. The probes both collect the sample and prepare it for analysis by eliminating high molecular weight substances and other cellular debris that might be incompatible with high performance analytical tools, such as HPLC, CE, MS, etc. Thus, especially when the analysis is performed on-line, microdialysis offers the capability of near real-time, highly sensitive, and highly selective chemical monitoring of the brain processes. However, some evidence in the published literature suggests that the performance of brain microdialysis is affected by the damage to the brain tissue that occurs when the probe is implanted. This so-called penetration injury, left to its own devices, triggers a wound healing process (also sometimes called a foreign body response) that will lead to the encapsulation of the microdialysis probe in scar tissue within a few days. The formation of scar tissue might alter the neurochemical properties and activity of the tissue surrounding the probe. We hypothesize that the retrodialysis delivery of anti-inflammatory and neuroprotective agents directly to the sampling site might be an affective strategy for mitigating the disruptive effects of the penetration injury and ensuing scar formation. Here we report the effects of retrodialysis of dexamethasone, a powerful anti-inflammatory glucocorticoid, on measures of dopamine, a highly significant neurotransmitter, in the striatum.

Keywords: Bioanalytical, Biological Samples, Electrochemistry, Neurochemistry
Application Code: Neurochemistry
Methodology Code: Sampling and Sample Preparation

Session # 1840 Abstract # 1840-1 **Symposia**

Session Title Analytical Challenges Relating to the Discovery, Development, Manufacturing and Use of Cancer Immu

Abstract Title **Concepts of Cancer Immunotherapy**

Date: Wednesday, March 09, 2016 - After

Time: 01:35 PM

Room: B303

Primary Author Robert Kastelein

Merck & Co.

Co-Author(s)

Abstract Text

The past few decades have seen a groundswell of research on the immune system yielding a deeper understanding of how cancer progresses and offering new ways to stop it. These new efforts in cancer immunotherapy are built on the concept of "re-activating" the immune system to target tumor cells through manipulation of co-inhibitory or co-stimulatory molecules involved in regulation of the immune response. Proof of concept for this approach has been demonstrated with the recent success of anti-CTLA-4 antibody ipilimumab (YervoyTM) and, more recently, with the approval of two anti-PD-1 antibodies, pembrolizumab (KeytrudaTM) and nivolumab (OptivoTM), for use in several cancer indications. The unprecedented clinical activity observed by targeting the PD-1 pathway in many different tumor types has initiated a broad effort across industry and academics alike to take advantage of newly gained insights. Blockade of the PD-1 pathway appears to be central to immune cell re-activation within tumors and provides a general paradigm for targeting immune-modulatory molecules (IMRs). The focus of cancer immunotherapy research over the foreseeable future will be on 'raising the survival tail' further, beyond what is achievable with the current checkpoint inhibitors and other immune therapies. There is general agreement that this will require combination therapies that target both cancer cell pathways as well as additional immune cell pathways. This presentation will introduce the concepts of cancer immunotherapy.

Keywords: Biopharmaceutical, Biotechnology, Drug Discovery

Application Code: Pharmaceutical

Methodology Code: Education/Teaching

Session # 1840 Abstract # 1840-2 **Symposia**

Session Title: Analytical Challenges Relating to the Discovery, Development, Manufacturing and Use of Cancer Immu

Abstract Title: **Identifying and Profiling Tumor Specific T Cells Using Mass Cytometry and Highly Multiplexed Peptide-MHC Tetramer Staining**

Primary Author: Evan W. Newell
Singapore Immunology Network

Date: Wednesday, March 09, 2016 - After
Time: 02:10 PM
Room: B303

Co-Author(s)

Abstract Text

Antigen specific T cells are critical initiators and orchestrators of the adaptive immune response. To better understand how T cells participate (appropriately or inappropriately) in the immune response to cancer, more meaningful categorizations of these cells are needed. For this purpose, a very large number of cellular parameters can be considered. Focusing on human CD8+ T cells, the utility of high dimensional mass cytometry (i.e., Cytometry by Time-Of-Flight, CyTOF) analysis will be discussed. By this approach, T cells can be probed with unprecedented detail by simultaneously evaluating surface marker expression, functional capacity and antigen-specificity. For antigen-specificity, this analysis can be performed in conjunction with a highly multiplexed method based on peptide-MHC tetramers, which allows simultaneous assessment of >500 different antigen specificities in a single sample. To visualize and understand the diversity of these cells, several computational data analysis approaches are being used. Application of these approaches to identify and phenotypically profile various cancer specific T cells in humans as well as in mouse models will be described. In addition to the possibility of finding novel therapeutic targets, a long-term goal of this work is to discover more accurate biomarkers of clinical outcomes (e.g., responsiveness to immunotherapy) based on the high-dimensional characteristics of antigen-specific T cells.

Keywords: Bioinformatics, Immunoassay, Mass Spectrometry, Method Development
Application Code: Biomedical
Methodology Code: Mass Spectrometry

Session # 1840 **Abstract # 1840-3** **Symposia**

Session Title Analytical Challenges Relating to the Discovery, Development, Manufacturing and Use of Cancer Immu

Abstract Title **Analytical Challenges in the Discovery, Development and Commercialization of Keytruda**

Date: Wednesday, March 09, 2016 - After
Time: 02:45 PM
Room: B303

Primary Author Maribel Beaumont
Merck Research Laboratories

Co-Author(s)

Abstract Text

With their unique specificity, potential for agonistic or antagonistic responses, moldable effector function and optimal pharmacokinetic properties, monoclonal antibodies and antibody-based biologics have delivered an impressive success rate in the clinic. Specifically, antibodies targeting immune checkpoints such as programmed cell death protein 1 (PD1), programmed cell death 1 ligand 1 (PDL1) and cytotoxic T lymphocyte antigen 4 (CTLA4) have resulted in the commercialization of new medicines that are having a profound effect on survival rates and patient outcomes.

Keytruda (MK-3475) is a humanized IgG4 antibody against human PD-1. IgG4 molecules have the uncanny ability to exchange Fab arms by swapping a heavy chain and attached light chain (half molecule) with a heavy-light chain pair from another molecule, resulting in bispecific antibodies. This intriguing property is not desirable for a therapeutic mAb and mutating S228 at the hinge region to a proline (so it resembles IgG1) has been shown to confer stability to IgG4 antibodies. However, regulatory agencies still require the Biopharmaceutical industry to demonstrate that S228P modified IgG4 are unable to engage in arm exchange. A series of orthogonal analytical methods were applied to study and understand the half molecule exchange in buffer and Human serum (ex-vivo). Lastly, half molecule swapping was also examined in vivo using both immunoassay and affinity-purification coupled to LC-MS (AP-LC-MS). Results from these studies will be highlighted.

Keywords: Biopharmaceutical, Characterization, Mass Spectrometry
Application Code: Bioanalytical
Methodology Code: Liquid Chromatography/Mass Spectrometry

Session # 1840 Abstract # 1840-4 **Symposia**

Session Title: Analytical Challenges Relating to the Discovery, Development, Manufacturing and Use of Cancer Immu

Abstract Title: **Development of Microtools for Immune Therapy Applications**

Primary Author: Christopher Love
Koch Institute for Integrative Cancer Research at MIT

Date: Wednesday, March 09, 2016 - After
Time: 03:35 PM
Room: B303

Co-Author(s)

Abstract Text

Over the past decade, monoclonal antibodies have emerged as an important class of therapeutics. More recently, antibodies targeting immune checkpoint have resulted in important clinical advances. Developing and using microtools for single cell multiparametric analysis to accelerate and streamline biologics development or to use in monitoring the immune system in cancer and with drug treatment is critical to transform our understanding of human disease and in turn improve human health.

Keywords: Biopharmaceutical, Biotechnology, Drug Discovery, Nanotechnology
Application Code: Pharmaceutical
Methodology Code: Microfluidics/Lab-on-a-Chip

Session # 1840 Abstract # 1840-5 **Symposia**

Session Title: Analytical Challenges Relating to the Discovery, Development, Manufacturing and Use of Cancer Immu

Abstract Title: **Analytical Challenges During Development of Monoclonal Antibody Therapeutics**

Date: Wednesday, March 09, 2016 - After
Time: 04:10 PM
Room: B303

Primary Author: John T. Stults
Genentech, Inc

Co-Author(s): David Michels

Abstract Text

As a clinical antibody-based project progresses through the development stages on the path to commercialization, a number of analytical challenges are encountered during establishment of both the manufacturing process and the control strategy, including method development, molecule characterization, and Critical Quality Attribute assessment. This presentation will provide examples of these challenges and the strategies used to address them.

Keywords: Biopharmaceutical, Characterization, Mass Spectrometry, Protein
Application Code: Bioanalytical
Methodology Code: Mass Spectrometry

This page intentionally left blank.

Session # 1850 Abstract # 1850-2 **Symposia**

Session Title: Analytical Chemistry of Oil and Gas Prospecting in Brazil

Abstract Title: **Compound Specific 13C Determination of Light Hydrocarbons (n-alkanes and Olefins from C1 to C5) at Low Concentration for Oil and Gas Prospection**

Date: Wednesday, March 09, 2016 - After
Time: 02:10 PM
Room: B304

Primary Author: Arthur D. Scofield
PUC-RIO

Co-Author(s): Angela R. Wagener, Laura R. Morales, Lilian F. Almeida

Abstract Text

The concentration and isotopic composition of hydrocarbons C1-C5 has revealed as a powerful tool for interpretation of interesting processes in the oil exploration industry, studies of global climate change and carbon cycle elucidation. In this study a gas pre-concentration device, dedicated to the determination of methane and nitrous oxide of atmospheric samples, was modified allowing the measuring of carbon isotope ratio of alkanes mixtures from C1 to C5 at low concentration by GC-IRMS. The system allows the trapping of condensable gases (C2+, CO2, Ar, N2, O2, N2O, H2O) at low temperature using just a few milliliters of sample. In addition to that, the methane is oxidized, cryo-focused as CO2 and then injected in the GC-IRMS system. After the methane analysis, the condensed gases are sent to the GC-IRMS system. The pre-concentration and the release of analytes were tested by evaluation of standards in GC-IRMS system. After the release the determination of C2+ was made by the GC-IRMS system. To ensure the selective oxidation of methane, a few modifications were tested in the cryogenic traps and in the analytical method. The introduction of PLOT column in the traps, and the use of others cryogenic temperatures demonstrated the quantitative collection of gas, the separation of non-methane hydrocarbons and the decrease of interferences collected in liquid nitrogen (N2 and O2). In addition, interferences were minimized with the use of a cleaning chemical trap. The analytical system was later modified to allow the analysis of olefins with the introduction of a PLOT column of alumina, and simplification of the cleaning trap. The new method enables the analysis of concentrated samples (<100 µmol.mol-1) by direct injection without the interference of air (N2 and O2) on methane. Also the same facilities of the pre-concentration system can be applied for diluted samples. The methods were fully tested with real geological samples.

Keywords: Isotope Ratio MS, Petroleum, Sample Preparation, Specialty Gas Analysis
Application Code: Fuels, Energy and Petrochemical
Methodology Code: Mass Spectrometry

Session # 1850 Abstract # 1850-3 **Symposia**

Session Title: Analytical Chemistry of Oil and Gas Prospecting in Brazil

Abstract Title: **Mass Spectrometry by FT-ICR and Orbitrap: Analysis of Crude Oil and Its Derivatives**

Primary Author: Eduardo M. Schmidt
State University Campinas

Date: Wednesday, March 09, 2016 - After
Time: 02:45 PM
Room: B304

Co-Author(s)

Abstract Text

With world-wide petroleum reserves ending, effective processing of petroleum becomes increasingly important. The mass spectrometry came to be a fundamental tool in crude oil analysis since GC-MS until FT-ICR-MS [1] and now with Orbitrap [2-4]. Crude oil is a complex mixture of hydrocarbons, containing multiple aromatic rings including heterocycles of N, S and O and crude acids in its polar fraction. Mass spectrometry is an analytical technique that currently allows from the analysis of a single molecular ion to a profile of the distribution of ions in complex samples. This has been possible through direct analysis with different sources of ionization. For complex samples such as oil, the most appropriate source of ionization is the electrospray (ESI) due to its stability and consequently reproducibility. The mass analyzer FT-ICR is already being used for direct analysis of crude oil and allows monitoring and investigating crude oil and its derivatives. These new data about polar compounds may be useful to add information starting from the crude oil exploration until refining. Additionally, it may also be useful in forensic analysis such as in crude oil spills and in the investigation of the compliance of different motor oils (mineral, semi-synthetic and synthetic) with their labels. Thus, this results offers an enormous potential for the analysis and characterization of oil and its derivatives, through a simple and fast procedure, without any pretreatment of samples.

[1] Qian et al., Energy&Fuels, 15, 2001

[2] Smith et al., Energy&Fuels, 24, 2010

[3] Denisov et al., Int. J. of Mass Spectrom. 325– 327, 2012

[4] Lababidi S, PhD thesis, 2013

Keywords: Electrospray, Forensics, Mass Spectrometry, Petroleum
Application Code: Fuels, Energy and Petrochemical
Methodology Code: Mass Spectrometry

Session # 1860 **Abstract # 1860-1** **Symposia**

Session Title: Big Data in Analytical Sciences - Challenges and Solutions

Abstract Title: **Data Analysis Challenges in Plant Biology**

Date: Wednesday, March 09, 2016 - After
Time: 01:35 PM
Room: B305

Primary Author: Philip Benfey
Duke University

Co-Author(s)

Abstract Text

To understand the progression from stem cells to differentiated tissues we are exploiting the simplifying aspects of root development. We have profiled mRNA, small RNAs, alternative splicing and DNA methylation at cell-type specific resolution within the Arabidopsis root. We are developing new experimental, analytical and imaging methods to identify networks functioning within different cell types and developmental stages. Because, root systems are high value targets for crop improvement due to their potential to boost yields, improve drought tolerance, and reduce the need for fertilizers, we are using a variety of approaches to enhance root system architecture. We have developed a semi-automated 3D imaging and phenotyping system to identify the genetic basis of root architecture. The integrated system combines hardware, imaging, software and analysis. We automatically reconstructed and phenotyped a well-studied rice mapping population, identifying QTLs for RSA traits that control the extent, shape, distribution, and surface size of root networks. We have extended this work to maize where we mapped large effect QTLs for RSA. We are also using X-ray imaging to phenotype roots grown in soil. Thus, our approach can directly aid breeding efforts as well as identify important genes underlying environmentally robust QTLs.

Keywords: Agricultural, Bioinformatics, Genomics, Imaging
Application Code: Genomics, Proteomics and Other 'Omics
Methodology Code: Data Analysis and Manipulation

Session # 1860 **Abstract #** 1860-2 **Symposia**

Session Title: Big Data in Analytical Sciences - Challenges and Solutions

Abstract Title: **First Steps into Big Data Chemistry- Ultra-High-Throughput Screening**

Primary Author: Spencer D. Dreher
Merck & Co., Inc.

Date: Wednesday, March 09, 2016 - After
Time: 02:10 PM
Room: B305

Co-Author(s)

Abstract Text

The advent of miniaturized ultra-high-throughput chemistry has allowed chemists to enter into the unfamiliar world of big-data science. In principle, informatics approaches that have proven immensely valuable to other disciplines should also enable dramatic improvements to the rate and quality of synthetic chemistry. The real barrier to turning on big data chemistry is rapid, effective reaction analysis. New analytical approaches will be required that can keep up with a modern chemist who can run upwards of 5,000 chemistry experiments per day in search of new drug candidates or vastly improved syntheses of known valuable molecules.

Keywords: Nanotechnology, Pharmaceutical, Sample & Data Management, Scientific Data Management
Application Code: Pharmaceutical
Methodology Code: Chemical Methods

Session # 1860	Abstract # 1860-3	Symposia

Session Title Big Data in Analytical Sciences - Challenges and Solutions

Abstract Title **Big, But Small, Data: Network Analysis in Small Sample Size Systems Biology**

Primary Author
Mark P. Styczynski
Georgia Institute of Technology

Date: Wednesday, March 09, 2016 - After
Time: 02:45 PM
Room: B305

Co-Author(s)

Abstract Text

While genome-scale systems biology experimental techniques provide valuable data to understand and allow engineering of biological systems, the large number of measured variables cause a curse of dimensionality: finding statistically significant relationships between variables requires more and more samples as the number of observed variables grows. Nonetheless, the application of systems biology approaches in complex animal model systems is an exciting area of research, often with a goal of learning about underlying regulatory network structures (e.g., gene regulatory networks). These complex animal model systems typically have significant limitations on cohort sizes, number of samples, and the ability to perform follow-up and validation experiments, which is exactly the situation where the curse of dimensionality has its greatest impact. These constraints are particularly problematic for many current network learning approaches, which require large numbers of samples and may predict many more regulatory relationships than actually exist. Here, we will discuss our recent work to learn regulatory networks from systems-scale, but small sample size, data. We will present approaches using existing tools, as well as new algorithms to supplement those approaches, that help to enable such analyses. Common themes across the approaches we discuss include resampling and permutation-based analysis of statistical significance in order to identify the strongest and most robust relationships between variables. We will discuss this work as applied to data generated by recent studies of host-pathogen interactions in non-human primate models of malaria: a system where sample size is quite limited, but where biological insight would be particularly valuable. Our approaches have helped to identify biological relationships that were not expected and that were not otherwise found using standard univariate or multivariate methods.

Keywords: Bioinformatics, Data Analysis, Data Mining
Application Code: Genomics, Proteomics and Other 'Omics
Methodology Code: Data Analysis and Manipulation

Session # 1860 **Abstract # 1860-4** **Symposia**

Session Title: Big Data in Analytical Sciences - Challenges and Solutions

Abstract Title: **Clustering and Differential Alignment Algorithm: Identification of Early Stage Regulators in the Arabidopsis Thaliana Iron Deficiency Response**

Date: Wednesday, March 09, 2016 - After
Time: 03:35 PM
Room: B305

Primary Author: Cranos Williams
North Carolina State University

Co-Author(s): Alexandr Koryachko, Anna Matthiadis, Durreshahwar Muhammad, James Tuck, Jessica Foret, Joel Ducoste, Siobhan Brady, Terri A. Long

Abstract Text

Time course transcriptome datasets are commonly used to predict key gene regulators associated with stress responses and to explore gene functionality. Techniques developed to extract causal relationships between genes from high throughput time course expression data are limited by low signal levels coupled with noise and sparseness in time points. We deal with these limitations by proposing the Cluster and Differential Alignment Algorithm (CDAA). This algorithm was designed to process transcriptome data by first grouping genes based on stages of activity and then using similarities in gene expression to predict influential connections between individual genes. Regulatory relationships are assigned based on pairwise alignment scores generated using the expression patterns of two genes and some inferred delay between the regulator and the observed activity of the target. We applied the CDAA to an iron deficiency time course microarray dataset to identify regulators that influence 7 target transcription factors known to participate in the Arabidopsis thaliana iron deficiency response. A set of 931 possible regulatory relationships between 133 differentially expressed transcription factors and the 7 chosen targets was reduced by the CDAA to a very testable subset of 7 regulator transcription factors 32 connections. The 7 regulators identified by the CDAA were previously unlinked to iron homeostasis. We validated over half of predicted influential relationships using qRT-PCR expression analysis in mutant backgrounds. One predicted regulator-target relationship was shown to be a direct binding interaction according to yeast one-hybrid (Y1H) analysis. These results serve as a proof of concept emphasizing the utility of the CDAA for identifying unknown or missing nodes in regulatory cascades, providing the fundamental knowledge needed for constructing predictive gene regulatory networks.

Keywords: Bioinformatics, Data Mining, Genomics, Pattern Recognition
Application Code: Other
Methodology Code: Data Analysis and Manipulation

Session # 1860 Abstract # 1860-5 Symposia

Session Title: Big Data in Analytical Sciences - Challenges and Solutions

Abstract Title: **Machine Learning for Big Nonlinear Problems in Science and Engineering**

Date: Wednesday, March 09, 2016 - After
Time: 04:10 PM
Room: B305

Primary Author: Le Song
Georgia Institute of Technology

Co-Author(s)

Abstract Text

Modern science and engineering problems are generating datasets with increasing volume, velocity and variety. The complexity and scale of big data impose tremendous challenges for their analysis. Yet, big data also offer us great opportunities. Some nonlinear phenomena or relations, which are not clear or cannot be inferred reliably from small and medium data, now become clear and can be learned robustly from big data. Typically, the form of the nonlinearity is unknown to us, and needs to be learned from data as well. Being able to harness the nonlinear structures from big data could allow us to tackle problems which are impossible before or obtain results which are far better than previous state-of-the-arts. In this talk, I will discuss how to use large scale deep learning and kernel methods, both are nonlinear machine learning methods, to address challenges arising from large scale images classification, time-series analysis, and materials discovery problems.

Keywords: Laboratory Informatics
Application Code: Laboratory Management
Methodology Code: Laboratory Informatics

Session # 1870 Abstract # 1870-1 Symposia

Session Title Electrical and Electrochemical Sensing and Detection based on Nucleic Acid Recognition

Abstract Title **A New Approach to POC Serology**

Date: Wednesday, March 09, 2016 - After
Time: 01:35 PM
Room: B309

Primary Author Kevin W. Plaxco
University of California Santa Barbara

Co-Author(s)

Abstract Text
To date fluorescence polarization (FP) is the only quantitative, general molecular diagnostic to achieve any significant penetration at the point of care. It is not, however, without significant limitations associated with the difficulty of performing high-precision fluorimetry in blood serum, much less whole blood. In response, we are developing an electrochemical equivalent to FP that circumvents many of the difficulties associated with the latter and which thus may prove of particular value in the point-of-care measurement of specific analytes.

Keywords: Bioanalytical, Biomedical, Biosensors, Electrochemistry
Application Code: Bioanalytical
Methodology Code: Electrochemistry

Session # 1870 Abstract # 1870-2 **Symposia**

Session Title Electrical and Electrochemical Sensing and Detection based on Nucleic Acid Recognition

Abstract Title **Detection of Hepatitis B Virus DNA with a Paper Electrochemical Sensor**

Date: Wednesday, March 09, 2016 - After
Time: 02:10 PM
Room: B309

Primary Author Richard M. Crooks
University of Texas at Austin

Co-Author(s) Karen Scida, Xiang Li

Abstract Text

In this presentation we show that a simple paper-based electrochemical sensor, fabricated by paper folding, is able to detect a 30-base nucleotide sequence characteristic of DNA from the hepatitis B virus (HBV) with a detection limit of 85 pM. This device is based on design principles we have reported previously for detecting proteins via a metalloimmunoassay. It has four desirable attributes. First, its design combines simple origami (paper folding) assembly, the open structure of a hollow-channel paper analytical device to accommodate micron-scale particles, and a convenient slip layer for timing incubation steps. Second, two stages of amplification are achieved: silver nanoparticle labels provide a maximum amplification factor of 250,000 and magnetic microbeads, which are mobile solid-phase supports for the capture probes, are concentrated at a detection electrode and provide an additional ~25-fold amplification. Third, there are no enzymes or antibodies used in the assay, thereby increasing its speed, stability, and robustness. Fourth, only a single sample incubation step is required before detection is initiated.

Keywords: Bioanalytical, Electrochemistry
Application Code: Bioanalytical
Methodology Code: Electrochemistry

Session # 1870 Abstract # 1870-3 Symposia

Session Title: Electrical and Electrochemical Sensing and Detection based on Nucleic Acid Recognition

Abstract Title: **Using Widely Available Electrochemical Device for Sensing or Diagnostics**

Primary Author: Yi Lu
University of Illinois at Urbana-Champaign

Date: Wednesday, March 09, 2016 - After
Time: 02:45 PM
Room: B309

Co-Author(s): JingJing Zhang, Yu Xiang

Abstract Text

On site and real time detection is very important for environmental monitoring, food safety and medical diagnostics. Despite much effort, only a few sensors are widely available to the public. Among them, electrochemical devices based on personal glucose meters (PGMs) and pH meters are shining examples. To translate the success of PGM/pH meters into sensors for other targets, both scientific and technological barriers need to be overcome.

In scientific research, designing sensors based on a single class of molecules for a broad range of targets remains a challenge, where successes in designing sensors for one target can be difficult to translate for others. To meet the challenge, we need to develop general strategies to obtain sensing molecules for any targets, improve selectivity, transform the binding into detectable signals, and tune the dynamic range to match the levels of the targets. Toward these goals, we have used in vitro selection to obtain DNAzymes and aptamers for a wide range of targets, parcitually small molecular targets (e.g., metal ions, organic toxins, and biomarkers for diseases) that are difficult to detect using other methods, and used negative selection strategy to improve the selectivity.

In technological development, it is difficult to adopt new devices developed in laboratories into markets. We are exploring ways to overcome this barrier by taking advantages of the wide availability and low cost of pocket-sized PGM/pH meters to detect many non-glucose targets, ranging from recreational drugs to biological cofactors, to disease markers and metal ions. We achieved the success by using target-induced release of conjugates between the DNAzyme/aptamers and an enzyme that can convert sucrose into glucose or cause pH changes through enzymatic turnovers. Since in vitro selection can be used to obtain DNAzyme/aptamers to bind a wide range of targets, this approach can be readily used by the public to detect many targets at home and in the field.

Keywords: Bioanalytical, Biomedical, Electrochemistry, Environmental

Application Code: Bioanalytical

Methodology Code: Sensors

Session # 1870 Abstract # 1870-4 **Symposia**

Session Title: Electrical and Electrochemical Sensing and Detection based on Nucleic Acid Recognition

Abstract Title: **Electrochemical Analysis of Clinically-Relevant Biomolecular Analytes Using Nanostructured Microelectrodes**

Date: Wednesday, March 09, 2016 - After
Time: 03:35 PM
Room: B309

Primary Author: Shana Kelley
University of Toronto

Co-Author(s)

Abstract Text

The analysis of panels of molecular biomarkers offers valuable diagnostic and prognostic information for clinical decision making. Robust, practical platforms that detect low levels of biomolecules (< 1000 copies) are urgently needed to advance medical care by diagnosing and predicting the progression of cancer and other disease states. Electrochemical methods providing low cost and direct biomarker read-out have attracted a great deal of attention for this application, but have, to date, failed to provide clinically-relevant sensitivity. We exploit controlled nanostructuring of electrode surfaces to promote surface accessibility and enhance capture rate and efficiency to solve this long-standing problem, and showed that the nanoscale morphologies of electrode surfaces control their sensitivities. This presentation will highlight our efforts to use these components to detect markers in clinical samples to develop tests for infectious disease diagnosis, oncological management and transplant medicine.

Keywords: Electrochemistry, Genomics, Integrated Sensor Systems, Lab-on-a-Chip/Microfluidics
Application Code: Biomedical
Methodology Code: Electrochemistry

Session # 1870 **Abstract # 1870-5** **Symposia**

Session Title: Electrical and Electrochemical Sensing and Detection based on Nucleic Acid Recognition

Abstract Title: **Signal Amplification for Biosensing Based on Nucleic Acid Recognition**

Primary Author: Ju Huangxian
Nanjing University

Date: Wednesday, March 09, 2016 - After
Time: 04:10 PM
Room: B309

Co-Author(s)

Abstract Text

As the development of life science and biomedical science, the detection of biomolecules with low abundance and the acquisition of ultra weak biological signals have been become a bottleneck of these fields. High sensitive analytical methods coupling with the specificity of biological recognition are in urgent need. In recent years, our group focusses on the design of signal amplification strategies for highly sensitive biosensing. The first type of strategies is nano signal amplification, which includes 5 ways. Here I introduce the second type of signal amplification strategies, which is based on nuclei acid recognition, such as PCR, RCA, target-induced repeated primer extension, hybridization chain reaction, loop-mediated amplification, target DNA recycling, proximity hybridization and DNA assembly. They have been used for electrochemical, electrochemiluminescent, and photoelectrochemical detections; optical such as chemiluminescent, fluorescent and Raman analysis; mass spectrometric analysis and the development of imaging technologies such as chemiluminescence imaging, fluorescence imaging and Raman spectral imaging. The established methods can conveniently be used in the detections of small biomolecules, DNA, proteins, cells, carbohydrate sites on cell surfaces, intracellular microRNA, telomerase, ATP and sialyltransferase activity. Some methods can even realize quasi-single-molecule detection. The designed nanoprobes have been used for highly selective and sensitive biosensing, cell–subtype specific siRNA delivery, precise near-infrared cancer therapy, therapeutic monitoring, and monitoring of the evolution of intracellular caspase family.

Keywords: Bioanalytical, Biosensors, Imaging, Nucleic Acids
Application Code: Bioanalytical
Methodology Code: Sensors

Session # 1880 Abstract # 1880-1 **Symposia**

Session Title Vibrational Spectroscopy of Biodegradable Plastics: Evolution, Revolution or Back to the Future

Abstract Title **Spectroscopic Study of Structural Evolution Dynamics of Bio-Based and Biodegradable Poly(hydroxyalkanoate) Copolymers**

Date: Wednesday, March 09, 2016 - After
Time: 01:35 PM
Room: B310

Primary Author Isao Noda
University of Delaware

Co-Author(s) Brian Sobieski, Bruce Chase, John F. Rabolt, Liang Gong

Abstract Text

Poly(hydroxylalkanoate)s or PHAs are a class of aliphatic polyesters accumulated within cellular bodies of a number of microorganisms as a carbon and energy storage medium in a manner similar to lipids in higher organisms. Among many types of bacterial PHAs, poly[(R)-3-hydroxybutyrate-co-(R)-3-hydroxyhexanoate] or PHBHx, now manufactured under the tradename of Nodax™, has shown a promising set of physical properties potentially capable of replacing many conventional plastics derived from petroleum. PHBHx exhibits rich multi-phase morphological structures consisting of both amorphous and crystalline domains. Furthermore, it has recently discovered that PHBHx has at least two distinct crystalline polymorphs: most commonly observed □form with 21 helical structure and □form with planar zigzag structure. Because of the characteristic IR absorption band assignable to each constituent structure, one can monitor the dynamics of the supramolecular level transformation of PHBHx from the melt or solution to solid or gel in a real time measurement as function of time or temperature. A series of spectral data thus obtained during the transformation process are then examined with two-dimensional correlation analysis to obtain the significant insight into the multi-phase evolution dynamics of the product made of these bioplastics.

Keywords: FTIR, Infrared and Raman, Polymers & Plastics, Spectroscopy
Application Code: Polymers and Plastics
Methodology Code: Molecular Spectroscopy

Session # 1880   Abstract # 1880-2   **Symposia**

Session Title: Vibrational Spectroscopy of Biodegradable Plastics: Evolution, Revolution or Back to the Future

Abstract Title: **Degradation Mechanisms of Poly(lactic acid)-Nanoparticle Composite and Phthalate Plasticized Poly(vinyl chloride)**

Date: Wednesday, March 09, 2016 - After
Time: 02:10 PM
Room: B310

Primary Author: Zhan Chen
University of Michigan

Co-Author(s)

Abstract Text

Poly(lactic acid) (PLA) – nanoparticle composites have great potential in biomedical and materials applications such as drug delivery. We investigated the degrade mechanisms of such composites using a variety of analytical tools such as sum frequency generation (SFG) vibrational spectroscopy, scanning electron microscopy, atomic force microscopy, and contact angle goniometry. The results obtained from the in vitro drug release studies can be well explained by the deduced degrade mechanisms. It was found that the drug release behavior is determined by the PLA hydrolysis rate, which is medicated by the surface hydrophobicity and surface crystallization. Phthalate molecules are extensively used for plasticizers, but they may have negative impact on environment. We applied a combination of vibrational spectroscopic techniques including SFG, ATR-FTIR, and CARS spectroscopies to study degradation mechanisms of phthalate plasticized poly(vinyl chloride) (PVC) after UV irradiation. We also studied the effects of plasma treatment on such degradation mechanisms.

Keywords: Polymers & Plastics, Surface Analysis, Vibrational Spectroscopy
Application Code: Polymers and Plastics
Methodology Code: Vibrational Spectroscopy

Session # 1880 Abstract # 1880-3 Symposia

Session Title Vibrational Spectroscopy of Biodegradable Plastics: Evolution, Revolution or Back to the Future

Abstract Title **Vibrational Spectroscopic Studies on Biodegradable Polymer Microstructures**

Date: Wednesday, March 09, 2016 - After
Time: 02:45 PM
Room: B310

Primary Author Shaw L. Hsu
University of Massachusetts

Co-Author(s)

Abstract Text

We have relied on vibrational spectroscopy (infrared and Raman) to characterize the structures and their changes of various biodegradable polymers, with emphasis on poly(lactic acid) (PLA). The merits of these spectroscopic techniques are simple sample preparation, modest cost instrumentation, but most importantly, extremely useful composition and structural information can be obtained that complement other characterization techniques well. This is particularly true with the robust computation capability commonly available nowadays. Unlike techniques such as X-ray diffraction, where long range structural coherence is required, optical activity governing vibrational transitions is characteristic of localized morphological features. Therefore, detailed information such as functional group composition can be analyzed extremely accurately. Polymers obviously are formed when strong covalent bonds connect the monomers. Therefore, the infrared and Raman spectra often exhibit additional features that are different from monomers directly reflecting the connectivity along the backbone. As previous studies have shown, the perturbing effects of this connectivity are also seen in many polymer systems with various secondary forces "connecting" the monomers of different chains. In this presentation, I shall highlight the merits of using vibrational spectroscopy in determining the crucial structural element that gives PLA rise its stability. Vibrational spectroscopy is also able to elucidate the relaxation mechanism for PLA to seek the most stable crystalline phase. Since our research primarily deal with biomedical applications, the unusual water-PLA interactions have also been characterized. There is no question that vibrational spectroscopy has advanced our understanding of PLA microstructure making it an attractive candidate for replacing petroleum based polymers.

Keywords: Characterization, Infrared and Raman, Molecular Spectroscopy, Polymers & Plastics
Application Code: Polymers and Plastics
Methodology Code: Vibrational Spectroscopy

Session # 1880 Abstract # 1880-4 **Symposia**

Session Title Vibrational Spectroscopy of Biodegradable Plastics: Evolution, Revolution or Back to the Future

Abstract Title **2D IR Correlation Study of Thin Films of Biodegradable PHA/PEG Blend**

Date: Wednesday, March 09, 2016 - After
Time: 03:35 PM
Room: B310

Primary Author Young Mee Jung
Kangwon National University

Co-Author(s) Isao Noda, Yeonju Park, Yujing Chen

Abstract Text

Biodegradable poly(3-hydroxybutyrate-co-3-hydroxyhexanoate) (P(HB-co-HHx)) has recently received considerable attention owing to its potential applications in the environmental protection fields. P(HB-co-HHx) copolymers are fully miscible with biodegradable polyethylene glycol (PEG). The study of P(HB-co-HHx)/PEG blend is of great practical interest. We investigated thermal behavior of thin films of P(HB-co-HHx)/PEG blend upon heating process by infrared reflection absorption spectroscopy. Further understanding was obtained using 2D correlation spectroscopy. In this presentation, details of thermal behavior of thin films of biodegradable P(HB-co-HHx)/PEG blend investigated by 2D IR correlation analysis will be discussed.

Keywords: Chemometrics, Polymers & Plastics, Spectroscopy, Vibrational Spectroscopy
Application Code: Polymers and Plastics
Methodology Code: Vibrational Spectroscopy

Session # **1880** **Abstract #** **1880-5** **Symposia**

Session Title: Vibrational Spectroscopy of Biodegradable Plastics: Evolution, Revolution or Back to the Future

Abstract Title: **Characterization of Single Electrospun Biopolymer Nanofibers Using AFM-IR and Selected Area Electron Diffraction**

Date: Wednesday, March 09, 2016 - After
Time: 04:10 PM
Room: B310

Primary Author: John F. Rabolt
University of Delaware

Co-Author(s): Bruce Chase, C J. McBrin, Chao Ni, Curtis Marcott, David Martin, Isao Noda, Jinglin Liu, Liang Gong

Abstract Text

A metastable □-crystalline form of biodegradable and biocompatible poly[(R)-3-hydroxybutyrate-co-(R)-3-hydroxyhexanoate] (PHBHx) was fabricated using a combination of solution electrospinning and high-speed collection on a rotary disk with a tapered edge. Using an AFM-IR instrument and low dose selected area electron diffraction (SAED), we have explored the correlation between structure, processing and chain orientation/crystallinity in these single nanofibers and tested the hypothesis that different processing protocols can alter the concentration of the stable □-crystalline form and the metastable □-crystalline form. The ability to obtain IR spectra at high spatial resolutions has allowed us to map crystalline populations as a function of nanofiber diameter and as a function of location within a single nanofiber and we have observed, for the first time, the existence of a core-shell structure in a single electrospun nanofiber.

Keywords: Infrared and Raman, Instrumentation, Polymers & Plastics, Vibrational Spectroscopy
Application Code: Polymers and Plastics
Methodology Code: Vibrational Spectroscopy

Session # 1890 Abstract # 1890-1 **Workshops**

Session Title CACA - How to be Successful in Your Career

Abstract Title **Insights on Job Searching**

Primary Author Michael W. Dong
MWD Consulting

Date: Wednesday, March 09, 2016 - After
Time: 01:35 PM
Room: B311

Co-Author(s)

Abstract Text

Job searching is a fundamental skill particularly important in this changing job market for analytical chemists. An overview on job searching will be presented including how to develop a good understanding of your own skill set and the requirement of various industries, as well as the changing trend in today's job market for different career stages. Topics discussed are: networking, dealing with headhunters and outplacement organizations, the use of internet sites (monster.com and LinkedIn), developing interviewing skills and dealing with relocation issues.

Keywords: Biopharmaceutical, Chromatography, Education, HPLC
Application Code: Pharmaceutical
Methodology Code: Separation Sciences

Session # 1890	Abstract # 1890-2	Workshops

Session Title CACA - How to be Successful in Your Career

Abstract Title **Recognizing and Managing Career Passover**

Date: Wednesday, March 09, 2016 - After
Time: 02:05 PM
Room: B311

Primary Author Robert L. Stevenson
American Laboratory

Co-Author(s)

Abstract Text

Passover is a career problem usually affecting scientists in mid career. It refers to the practice of managers, particularly in industry, to favor young recent graduates for assignments to explore or bring in new technologies. Passover is a fork in one's career path that can lead to very adverse outcomes. Unfortunately, employers are not motivated to help experienced employees avoid the career problem.

Let me explain: The time cycle of many concepts and technologies in the sciences is about 20 to 30 years. Look at HPCE, SFC, solvolitic displacement reactions, ligand field theory, hard vs. soft acids, etc. Each were trendy topics during my early career. Each had an interest cycle that lasted only few decades.

For scientists that specialized in these and similar fields. The career path is very favorable during the ascendancy but deadly during the declining years.

Scientists, especially those employed in industry, need to critically monitor the status of their specialty. If interest and growth is positive, and you are an respected technical leader, then things may be good. But if this is not the case, it is prudent to look for something that one could move into. This can be a technology extension such as HPLC moving into ESI or TOF Mass Spectrometry. Or, if the future of one's particular specialty is looking bleak, then jump on another train that is fast tracked such as regulatory affairs or formulation.

Remember, you must manage your own career, since no one else will.

Keywords: Education
Application Code: Other
Methodology Code: Computers, Modeling and Simulation

Session # 1890 Abstract # 1890-3 Workshops

Session Title: CACA - How to be Successful in Your Career

Abstract Title: **Advancing Academic Career as a Faculty**

Primary Author: Yi He
John Jay College/CUNY

Date: Wednesday, March 09, 2016 - After
Time: 02:35 PM
Room: B311

Co-Author(s)

Abstract Text

Using personal examples, this presentation shares the experiences of how to prepare to become a faculty, the tenure process, and how to be effective when pursuing a faculty career. For audience who is a Ph.D. student or a postdoc, this presentation will introduce how to prepare an application package for a tenure-track faculty position and requirements and expectations of different types of universities/colleges. For audience who has already been a faculty, the discussion will focus on research, teaching and service – the three important parts in tenure process and promotion. Finally, this presentation will introduce "seven habits of highly effective people", which has been found useful in developing a successful career.

Keywords: Education
Application Code: Other
Methodology Code: Education/Teaching

Session # 1890 **Abstract # 1890-4** **Workshops**

Session Title CACA - How to be Successful in Your Career

Abstract Title **Career Development Workshop**

Primary Author Naidong Weng
JNJ

Date: Wednesday, March 09, 2016 - After
Time: 03:20 PM
Room: B311

Co-Author(s)

Abstract Text

You can have career development at any places, in any companies, and doing any work – big or small. Career development is not always a promotion or even financial reward. It is what your passion leads you. Find a good boss who can help your career development – those give you candid feedbacks. Recruit and retain talents and develop their careers – lead from behind and compassionate about their success. Give them your candidate feedbacks too. Involve in community services and be a good citizen at and outside workplaces.

Keywords: Education
Application Code: Pharmaceutical
Methodology Code: Education/Teaching

Session # 1890 Abstract # 1890-6 Workshops

Session Title: CACA - How to be Successful in Your Career

Abstract Title: **Go Wider and Higher**

Primary Author: Chuping Luo
Waters Corporation

Date: Wednesday, March 09, 2016 - After
Time: 04:20 PM
Room: B311

Co-Author(s)

Abstract Text

Many Chinese scientists have the talent to be creative in their career. However, a successful career is far beyond that. In this presentation, the author will present how to be successful in their career in terms of planning, learning, thinking, collaboration, developing good habits, and networking etc.

Keywords: Chromatography, Gas Chromatography, Liquid Chromatography, Supercritical Fluid Chromatography
Application Code: Other
Methodology Code: Education/Teaching

Session # 1900 Abstract # 1900-1 **Workshops**

Session Title Natural Health Products: Scientific Approaches to Securing Product Quality and Safety

Abstract Title **Development of a Chemical Barcoding Methodology to Identify and Support the Quality and Safety of Functional Ingredients**

Date: Wednesday, March 09, 2016 - After
Time: 01:35 PM
Room: B313

Primary Author Bob Chapman
National Research Council of Canada

Co-Author(s) Aissa Harhira, Alain Blouin, Camilo Martinez-Farina, El Haddad Josette, Fabrice Berrue, Ian Burton, Joseph Hui, Junzeng Zhang, Mohamad Sabsabi, Rob O'Brien, Sabrena MacKenzie, Yuan-Chun Ma

Abstract Text

One of the greatest challenges finish product manufactures of dietary supplements face is sourcing the needed high quality functional ingredient inputs for their products. For manufactures there is an inherent risk in the functional ingredients hitting their loading dock as ingredients derived from natural plant and herbal raw materials can vary considerably from different suppliers. The lack of ingredient standards, modernized analytical methodologies and industry oversight creates the potential for low quality and in some cases deliberately adulteration of ingredients. The growing consumer demand for safe and efficacious products is driving leading companies to find solutions to the ingredient quality issue. DNA barcoding has emerged as one tool but its suitability to NHPs has not been established. The NRC in partnership with our clients have been developing analytical chemistry methods based on LC-HRMS and NMR characterization to fingerprint raw materials to generate a “chemical barcode” to evaluate ingredients in great detail. Unlike DNA barcoding that provides only one single type of information, chemical barcoding allows quantification of the bioactive chemical entities that can determine the authenticity of functional ingredients. A key advantage of a non-directed chemical approach to evaluate ingredients quality is the ability to detect in the same analysis the presence of both the expected bioactives as well as any potential adulterants such as herbicides and/or pesticides that are presumed to be absent. The NRC presentation will introduce the concept of chemical barcoding that utilizes multiple analytical approaches including NMR spectroscopy, LC-HRMS analysis and laser induced breakdown spectroscopy (LIBS) followed by statistical analysis of the datasets. The development of the methodology has been demonstrated on multiple ingredients including two important Canadian grown medicinal crops, American ginseng and goldenseal.

Keywords: Identification, Mass Spectrometry, Natural Products, NMR
Application Code: Quality/QA/QC
Methodology Code: Mass Spectrometry

Session # 1900 Abstract # 1900-2 Workshops

Session Title: Natural Health Products: Scientific Approaches to Securing Product Quality and Safety

Abstract Title: **Utility and Limitations of DNA Barcoding and Next Generation Sequencing for Herbal Product Authentication**

Date: Wednesday, March 09, 2016 - After
Time: 02:05 PM
Room: B313

Primary Author: Jonathan Van Hamme
Thompson Rivers University

Co-Author(s)

Abstract Text

Recent high profile cases involving mass recalls of herbal products has resulted in increased political, legal and public pressure for companies to provide DNA barcoding data as a means to guarantee accurate product labeling. DNA barcoding in this context refers to the specific amplification and sequencing of gene fragments from genomes, followed by bioinformatic analysis to assign identities to organisms within a sample. It has become clear that there is no universal barcode sequence that can be used to differentiate between all organisms on the planet and, as such, specific barcodes must be identified for different groups. For plants, the most popular barcode regions are currently <i>rbcL, matK, trnH-psbA</i> (chloroplast genome) and a nuclear ITS region. Unfortunately, while DNA barcoding is a powerful tool for plant identification, improper experimental design and oversimplified data interpretation can easily lead to both false positive and false negative results with significant economic and legal consequences. For each plant species of interest, it is critical to optimize the DNA barcoding workflow from sample collection, DNA extraction, choice of barcode, DNA amplification conditions, choice of sequencing technology and bioinformatics tools. As more plant genomes are sequenced, the reliability of DNA barcoding will increase and, as sequencing costs continue to decline there will be a shift to sequencing larger barcode targets to improve accuracy. Even with reliable DNA barcode information, it is critical to realize that the presence of plant DNA in a sample will not correlate to the presence of bioactive compounds in a product. As such, it is not advisable to rely solely on DNA barcoding for authentication of herbal products.

Keywords: Food Science, Genomics, Identification, Natural Products
Application Code: Food Identification
Methodology Code: Data Analysis and Manipulation

Session # 1900 Abstract # 1900-3 **Workshops**

Session Title: Natural Health Products: Scientific Approaches to Securing Product Quality and Safety

Abstract Title: **Developing Metabolomics Approaches for Natural Product Authentication**

Primary Author: Liang Li
University of Alberta

Date: Wednesday, March 09, 2016 - After
Time: 02:35 PM
Room: B313

Co-Author(s)

Abstract Text

Metabolomics involves the characterization of all the small molecules in a biological system. New analytical tools, primarily based on liquid chromatography mass spectrometry (LC-MS), are being developed to generate comprehensive metabolomic profile of a sample. For example, chemical isotope labeling (CIL) of metabolites, in combination with LC-MS, can be used to determine metabolomic differences of comparative samples with a coverage of several thousands of metabolites. In this presentation, we will describe the development of CIL LC-MS for large scale profiling of natural products such as fruits, plants and herbal extracts. We will demonstrate the use of metabolomic signatures of plant species for natural product authentication.

Keywords: Food Identification, Liquid Chromatography, Mass Spectrometry, Metabolomics, Metabonomics
Application Code: Genomics, Proteomics and Other 'Omics
Methodology Code: Liquid Chromatography/Mass Spectrometry

Session # 1900 Abstract # 1900-4 **Workshops**

Session Title: Natural Health Products: Scientific Approaches to Securing Product Quality and Safety

Abstract Title: **Natural Health Product Quality Control: Addressing the Current Issue of Adulterants**

Date: Wednesday, March 09, 2016 - After
Time: 03:20 PM
Room: B313

Primary Author: Yuan-Chun Ma
Canadian Phytopharmaceuticals Corporation

Co-Author(s)

Abstract Text

The natural health product industry has been plagued for years with issues ranging from problematic fillers to the complete adulteration. Manufacturers of products sold in Canada are required to comply with strict standards set out by Health Canada and the United States, Chinese, and European Pharmacopeias. These international Pharmacopeias have monographs which specify methods of analysis and standards for many herbal products. Unfortunately, the current monograph methods have short comings which allow for the adulteration of a number of popular herbal products. These deficiencies recently gathered attention in the United States when the New York State Attorney General's Office ordered four national retailers, GNC, Walgreens, Target, and Wal-Mart to stop selling a number of their dietary supplements, due to adulteration. Also, this May, the Chinese SFDA announced a new temporary testing method to try to effectively identify Ginkgo biloba adulterants in the market. This method is capable of identifying many of the spiked products; however, some of the adulterants remain elusive. Based on these results, early this June, the Chinese Government ordered 247 manufacturers to remove all Ginkgo biloba products from their shelves.

To help eliminate these flaws, we have done extensive research into new methods for the authentication of a number of herbal products, and the analytical methods developed for on Rhodiola rosea, Panax notoginseng, and Ganoderma lucidum have been published as monographs and the standard extracts are being used as standard reference materials in the United States Pharmacopoeia. Most recently we have established an effective and rapid HPLC method for the analysis of Ginkgo biloba plants, extracts, and finished products. The new analytical method is currently under review by the United States and Chinese Pharmacopeias.

Keywords: HPLC, Identification, Natural Products, Quality Control
Application Code: Quality/QA/QC
Methodology Code: Liquid Chromatography

Session # 1900 Abstract # 1900-5 **Workshops**

Session Title Natural Health Products: Scientific Approaches to Securing Product Quality and Safety

Abstract Title **Practical Gas Chromatographic based Approaches to Confirm Identity of Powdered Herbal Products**

Date: Wednesday, March 09, 2016 - After
Time: 03:50 PM
Room: B313

Primary Author Rob O'Brien
ISURA

Co-Author(s) Anderson Smith

Abstract Text
The proper identification of herbal medicines and Natural Health Products (NHP) is a significant issue as incorrect identification of species can lead to serious allergenic reactions and result in products with limited or no effectiveness. Many suppliers in North America receive dried powdered bulk material and it is critical for them to have cost effective solutions to determine authentic products so that they do not inadvertently introduce fraudulent products into the market place. With these objectives in mind we have evaluated the effectiveness of a series of Gas Chromatographic based approaches to see which is the most effective approach to "Chemically Barcode" products. We have evaluated a wide variety of techniques including simple extraction with GC/FID detection, GCMS (full scan) approaches from both solvent extracts and Head Space – Solid Phase Micro Extraction (SPME). The data sets generated in this matter were compared using a wide variety of statistical models. We also collected a wide variety of orthogonal data from high resolution LCMS, NMR and DNA Sequencing approaches and these GC results were compared to those other techniques.

GC approaches offer a low per sample cost approach that can also achieve relative high sample throughput with limited technical infrastructure requirements. All of the benefits and limitations of this approach will be discussed.

Keywords: Chromatography, Food Identification, Gas Chromatography/Mass Spectrometry, GC
Application Code: Food Identification
Methodology Code: Gas Chromatography/Mass Spectrometry

Session # 1900 Abstract # 1900-6 **Workshops**

Session Title Natural Health Products: Scientific Approaches to Securing Product Quality and Safety

Abstract Title **TLC and HPLC Fingerprinting for Authentication of Natural Health Products**

Primary Author Rudolf Bauer
University of Graz

Date: Wednesday, March 09, 2016 - After
Time: 04:20 PM
Room: B313

Co-Author(s)

Abstract Text

In order to guarantee efficacy and safety of natural health products, quality control is a basic requirement. Testing for identity is obligatory in order to avoid adulterations and mixing up plant material [1]. Adulterations can not only negatively influence efficacy of a product, but can also lead to toxicological implications. The methods most widely used for authentication of natural health products are TLC and HPLC fingerprinting [2]. While TLC fingerprinting is used regularly in pharmacopoeias [3], HPLC fingerprints and chemometric analysis are more found in research and may be the concept for the future [4,5].

Several examples, like Echinacea, Ginkgo, and Angelica containing herbs will be presented to demonstrate the power of chromatographic fingerprint analysis. Compared the DNA barcoding, TLC and HPLC fingerprinting have the big advantage that they provide also information on the content of constituents. Therefore they can also be used for authentication ofr extracts and processed material, which do not contain intact DNA.

References

[1] Zhao ZZ, Hu Y, Liang Z, Yuen JP, Jiang Z, Leung KS. Planta Med. 72(10):865-74 (2006).
[2] Wagner, H., Bauer, R., Melchart, D., Xiao, P.-G., StaudingerA. (Eds). Chromatographic Fingerprint Analysis of Herbal Medicines - Thin-layer and High Performance Liquid Chromatography of Chinese Drugs, Springer-Verlag, Wien (2011).
[3] Bauer, R., Franz, G. Planta Med. 76(17):2004-11 (2010).
[4] Tistaert C, Dejaegher B, Vander Heyden Y. Analytica chimica acta. 690:148–61 (2011).
[5] Chen Z, Liao L, Yang Y, Zhang Z, Wang Z. J Sep Sci. 38(2):231-8 (2015).

Keywords: HPLC, Identification, Quality Control, Thin Layer Chromatography
Application Code: Quality/QA/QC
Methodology Code: Liquid Chromatography

Session # 1910 Abstract # 1910-1 **Organized Contributed Sessions**

Session Title: High Throughput Analysis for Food Safety and Cosmetics: Challenges and Validation

Abstract Title: **Rapid Determination of Non-Allowed Active Pharmaceutical Ingredients for the Treatments of Hair Loss in Cosmetics Using UHPLC-HRMS**

Date: Wednesday, March 09, 2016 - After
Time: 01:30 PM
Room: B315

Primary Author: Wanlong Zhou
US FDA

Co-Author(s): Alexander J. Krynitsky, James B. Wittenberg, Maria A. Dionisio De Sousa, Perry G. Wang

Abstract Text

Cosmetic products containing non-allowed active pharmaceutical ingredients (APIs) for the treatments of hair loss have appeared on the internet. Some therapeutic functions have been claimed for the products in order to attract consumers. These APIs may have potential side effects and possibly cause adverse health effects to consumers, especially when used without medical control. A rapid and sensitive UHPLC-HRMS method using Q Orbitrap mass spectrometry has been developed to determine these APIs which include finasteride, dutasteride, minoxidil, ketoconazole, spironolactone, and flutamide. A high-resolution product ion scan spectrum library has been established and could reliably be used to identify these APIs in complicated cosmetics matrices. The preliminary data show that the most of the analytes have a linear range from 1 to 1,000 ng/mL with a correlation coefficients (r) greater than 0.995. Ongoing work includes validation of the UHPLC-HRMS method on cosmetic products and to perform a limited survey on commercial cosmetic products to determine the levels of these APIs in cosmetic products.

Keywords: Cosmetic, Drugs, Liquid Chromatography/Mass Spectroscopy, Method Development
Application Code: Consumer Products
Methodology Code: Liquid Chromatography/Mass Spectrometry

Session # **1910** **Abstract #** **1910-2** **Organized Contributed Sessions**

Session Title: High Throughput Analysis for Food Safety and Cosmetics: Challenges and Validation

Abstract Title: **Advances in High-Throughput Analysis for Determination of Marine Biotoxins in Seafood**

Date: Wednesday, March 09, 2016 - After
Time: 01:50 PM
Room: B315

Primary Author: Pearse McCarron
National Research Council

Co-Author(s): Daniel Beach

Abstract Text

Marine biotoxins include a variety of compounds that accumulate in shellfish to levels that pose serious risks to human health and the seafood industry. Early techniques used for shellfish safety testing were based on animal bioassays. More recently, faster analytical chemistry based techniques have been developed and are now used almost exclusively. To help deal with the increasing demands for analysis, work is being conducted on the development and validation of higher-throughput methods for toxin detection and quantitation.

LC-MS methods have been validated for quantitating the full suite of regulated lipophilic toxins including okadaic acid, dinophysistoxins, azaspiracids, pectenotoxins and yessotoxins, in a single run. Non-regulated toxins of interest such as gymnodimines, spirolides and pinnatoxins can also be included. This has enabled the monitoring of lipophilic toxins by LC MS in most testing labs. Our recent work to develop a multitoxin shellfish matrix certified reference material helps to further enable multitoxin analysis in a quality-controlled environment.

We have also investigated emerging analytical techniques to enhance throughput by eliminating, or significantly reducing, sample preparation or chromatography steps in toxin analysis. Laser Ablation Electrospray Ionization (LAESI) with MS detection was evaluated for determination of the neurotoxin domoic acid directly from shellfish tissue without sample preparation or chromatography. After a preliminary study, a validation exercise was conducted to compare the results from LAESI screening to those from traditional LC-UV for a variety of shellfish samples obtained from regulatory partners. Differential mobility spectrometry (DMS) is a rapid gas phase separation technique that operates after electrospray ionization and before MS detection. This technique was investigated as an alternative to chromatography in high-throughput screening of hydrophilic and lipophilic marine toxins.

Keywords: Food Safety, Liquid Chromatography/Mass Spectroscopy, Quantitative, Reference Material
Application Code: Food Safety
Methodology Code: Liquid Chromatography/Mass Spectrometry

Session # 1910 **Abstract # 1910-3** **Organized Contributed Sessions**

Session Title: High Throughput Analysis for Food Safety and Cosmetics: Challenges and Validation

Abstract Title: **Determination of Prostaglandin Analogs in Eye Area Cosmetic Products by High Performance Liquid Chromatography with Tandem Mass Spectrometry**

Date: Wednesday, March 09, 2016 - After
Time: 02:10 PM
Room: B315

Primary Author: James B. Wittenberg
Food and Drug Administration

Co-Author(s): Alexander J. Krynitsky, Perry G. Wang, Wanlong Zhou

Abstract Text

A method was developed and validated for the determination of 16 prostaglandin analogs in eye area cosmetic products. The objective of this work was spurred by the FDA's regulatory concerns over the eyelash-enhancing properties of bimatoprost and its analogs in eye area products being marketed as cosmetics. The QuEChERS (Quick, Easy, Cheap, Efficient, Rugged, Safe) liquid-liquid extraction method, typically used for pesticide residue analysis, was utilized as the sample preparation technique. The prostaglandin analogs were chromatographically separated and quantified using high performance liquid chromatography with tandem mass spectrometry (HPLC-MS/MS). Thirty-one cosmetic products were surveyed, and 13 products were determined to contain a prostaglandin analog with amounts ranging from 27.4 to 297 [micro]g/g. The calculated concentrations for the cosmetic products were in a similar range when compared to the concentrations of three different prostaglandin analog-containing prescription products.

Keywords: Cosmetic, Liquid Chromatography/Mass Spectroscopy, Method Development
Application Code: Other
Methodology Code: Liquid Chromatography/Mass Spectrometry

Session # 1910 **Abstract # 1910-5** **Organized Contributed Sessions**

Session Title: High Throughput Analysis for Food Safety and Cosmetics: Challenges and Validation

Abstract Title: **Analysis of Color Additives (Permitted and Non-Permitted) in Different Food Matrices by a HPLC Method**

Date: Wednesday, March 09, 2016 - After
Time: 03:05 PM
Room: B315

Primary Author: Sneh D. Bhandari
Merieux NutrSciences

Co-Author(s): Tiffany Gallegos-Peretz

Abstract Text

There is a need for accurate and precise method applicable to a wide variety of food matrices for analysis of artificial color additives in food. A method to quantitate fourteen permitted and non-permitted color additives in food products was developed. The method involves extraction of color additives by a multi-step solvent extraction method involving pH adjustments at different steps. The color additives are analyzed by a HPLC method. The chromatographic method provides a good resolution of all of the fourteen color additives at selected wave lengths (420-620 nm) of detection. The method has been applied to a variety of food matrices (beverages, beverage powders, gelatins, sauces, bakery, and dairy). The investigated food matrices did not interfere in resolution of the color additives by the method . The precision of the analysis for all of the analyzed color substances was found to be satisfactory (%RSD range = 0.6-21). The spike recovery of the analysis of all of the color additives except FD&C Blue 2 was found to be in range of 81-121% in all of the tested matrices. FD&C Blue-2 in some of the matrices provided a lower (39-57%) spike recovery because of its instability in the extracts. Estimated LOD of the method was found to be around 1 ppm. The method has been applied in high through put mode particularly for analysis of FD&C colors.

Keywords: Food Science, High Throughput Chemical Analysis, HPLC, Liquid Chromatography
Application Code: Regulatory
Methodology Code: Liquid Chromatography

Session # 1910 Abstract # 1910-6 **Organized Contributed Sessions**

Session Title High Throughput Analysis for Food Safety and Cosmetics: Challenges and Validation

Abstract Title **A Comprehensive Approach on Food Safety Analysis, Screening and Quantitation by Using Data Independent Acquisition (DIA) and DDMS2 on HR/AM Q Exactive System**

Primary Author James S. Chang

Thermo

Date: Wednesday, March 09, 2016 - After

Time: 03:25 PM

Room: B315

Co-Author(s)

Abstract Text

Multiple reaction monitoring (MRM), two stages of mass filtering are employed on a triple quadruple mass spectrometer. In the first stage the precursor ion is preselected in Q1 then into a collision chamber and collides with a neutral gas in a pressurized collision cell (Q2) which will results of induced to fragment by collisional excitation. In the second stage, instead of obtaining full scan ms/ms where all the possible fragment ions derived from the precursor are mass analyzed in Q3, only selected ions are mass analyzed in Q3. This targeted MS analysis using MRM enhances the S/N ration which results the sensitivity increase. DDMS2 method consists of a generic chromatographic method and a full-scan data-dependent MS/MS (FS-ddMS2) mass spectrometric method which including a list of accurate mass on target compounds, the system will automatically triggered for a ms/ms scan according to the list if it find under the full scan data. Data Independent Acquisition (DIA) has been widely employed in shotgun proteomic workflow, not only has it been used in identification and characterization of proteins in a complex biological matrix, the method is also used in quantitative proteomics, particularly with the use of isobaric labeling or tagging. The system was set to sequentially isolate and fragment precursor windows of preset isolation windows by collision-activated dissociation (CAD) until a desired range was covered. Furthermore, the method provided time-consistent ion sampling and was able to confirm the results in MS1 and with the simplicity which doesn't have to know the nature of compounds need to be analyzed.

Keywords: Calibration, Food Science, Mass Spectrometry, Tandem Mass Spec

Application Code: High-Throughput Chemical Analysis

Methodology Code: Mass Spectrometry

Session # 1910 Abstract # 1910-7 **Organized Contributed Sessions**

Session Title: High Throughput Analysis for Food Safety and Cosmetics: Challenges and Validation

Abstract Title: **Separation of Aminoglycoside Antibiotics by Using Hydrophilic Interaction Liquid Chromatography**

Date: Wednesday, March 09, 2016 - After
Time: 03:45 PM
Room: B315

Primary Author: Yu Long
Dalian Institute of Chemical Physics

Co-Author(s): Guo Zhimou, Liang Xinmiao, Shen Aijin, Wei Jie

Abstract Text

Aminoglycoside is an important category of antibiotics that has been extensively applied in veterinary infection treatment or as feed additive. Residue of aminoglycoside antibiotics in animal-originated foods may pose potential threat to human health. Therefore, it is of great significance to develop analytical techniques for the detection and residue monitoring of this type of antibiotics. Herein, we introduced a customized hydrophilic interaction liquid chromatography (HILIC) column named Click TE-Cys for the effective separation of aminoglycoside compounds. Conferred by the unique zwitterionic feature of the bonded functionalities, Click TE-Cys had exhibited superior hydrophilic selectivity for the ten model aminoglycosides, which could be well resolved on this column under the chromatographic condition of low pH and high salt concentration. Compared with five other HILIC columns commonly used in aminoglycoside separation, Click TE-Cys also displayed much better separation effectiveness, even for the structurally similar aminoglycosides. The results indicated the great potential of Click TE-Cys column in the detection and residue level monitoring of the aminoglycoside antibiotics.

Keywords: Chromatography, Food Safety, HPLC, HPLC Columns
Application Code: Food Safety
Methodology Code: Liquid Chromatography

Session # 1920 Abstract # 1920-1 **Organized Contributed Sessions**

Session Title: Precision Bioanalytical Measurements

Abstract Title: **Analytical Precision in the Age of Metabolomics – Focus on the Fundamentals**

Date: Wednesday, March 09, 2016 - After
Time: 01:30 PM
Room: B316

Primary Author: Howard Hendrickson
University of Arkansas for Medical Sciences

Co-Author(s): Lin Song

Abstract Text

Efforts in my research group are directed toward the discovery and validation of disease markers, through targeted metabolomics. Precise bioanalytical measurement of endogenous metabolites and subsequent correlation of this measurement with a unique physiological state, presents several challenges not normally encountered with therapeutic drug monitoring or preclinical development of a potential therapeutic agent. A rather obvious challenge of metabolite determinations is the lack of a suitable blank sample while a less obvious challenge is how to establish a typical concentration for a given metabolite under normal physiological conditions. Overlooking one or both of these factors can adversely affect the precision of the measurement and ultimate conclusions drawn from the data collected. Collaboration with multiple research groups or regulatory labs is a key to our success in this area. Preclinical research into the effects of ionizing radiation exposure on amino acid metabolism is a key area of current interest in our group. Data from these studies and efforts toward the development a research grade botanical product will be discussed within the context of optimizing analytical precision.

Keywords: Bioanalytical, Biological Samples, Liquid Chromatography/Mass Spectroscopy, Metabolomics, Metab
Application Code: Validation
Methodology Code: Liquid Chromatography/Mass Spectrometry

Session # 1920 Abstract # 1920-2 **Organized Contributed Sessions**

Session Title: Precision Bioanalytical Measurements

Abstract Title: **Biocompatible Self-Tuning Nanomaterials Improve the Precision of Biomolecule Processing and Separation**

Primary Author: Lisa A. Holland
West Virginia University

Date: Wednesday, March 09, 2016 - After
Time: 01:50 PM
Room: B316

Co-Author(s): Brandon C. Durney, Cassandra Crihfield, Srikanth Gattu

Abstract Text

Self-assembled lipid nanophases are significant to chemical separations because they extend the size range of the separation, expand the precision of size discrimination, and are used to precisely pattern enzymes, antibodies, and lectins in microscale channels for nanoscale biomolecule processing. The phospholipid nanophases are used to reconstitute and non-covalently pattern proteins in fluid channels. Patterned lipid nanophases enhance enzyme activity and increase protein stability. Each molecular affinity step in a patterned nanophase utilizes a small (~5 nL) plug of protein in the separation channel. Once patterned, nanophases also serve as a tunable separation material for separations based on hydrodynamic volume or molecular sieving. As a separation medium, these nanomaterials form temporary gels that break and reform and are self-tuning. Alternatively, nanophases may be directly tuned to provide an appropriate viscosity to accentuate differences in electrophoretic mobility without increasing longitudinal diffusion. These multifunctional materials serve as an enabling tool to address complex biological analyses. The research directly impacts separations chemistry, self-tuning smart phases, and multifunctional microscale analyses.

Keywords: Bioanalytical, Biotechnology, Capillary Electrophoresis
Application Code: Bioanalytical
Methodology Code: Capillary Electrophoresis

Session # 1920 **Abstract # 1920-3** **Organized Contributed Sessions**

Session Title: Precision Bioanalytical Measurements

Abstract Title: **Analytical Approaches for Environmental Metabolomics and Ecotoxicity Modeling**

Date: Wednesday, March 09, 2016 - After
Time: 02:10 PM
Room: B316

Primary Author: Cynthia K. Larive
University of California - Riverside

Co-Author(s): Corey M. Griffith, Melissa M. Morgan

Abstract Text

Understanding the impacts of contaminants on environmental health depends on advances in analytical technologies and methods for ecotoxicity evaluation. Metabolomics offers possibility as a new biomonitoring tool to assess environmental health by monitoring the response of organisms to toxicants via the toxicometabolome. In this work, earthworms (Eisenia fetida) and brine shrimp (Artemia salina) are used as ecotoxicity models to evaluate the metabolomic impacts of contaminants, including dose-response and bioaccumulation. Proton and two dimensional nuclear magnetic resonance is used in tandem with mass spectrometry to characterize the metabolomic impact of contaminants such as agrochemicals on these non-target organisms.

Keywords: Gas Chromatography/Mass Spectrometry, Liquid Chromatography/Mass Spectroscopy, Magnetic Res
Application Code: Environmental
Methodology Code: Magnetic Resonance

Session # 1920 Abstract # 1920-4 **Organized Contributed Sessions**

Session Title: Precision Bioanalytical Measurements

Abstract Title: **Extending the Free Drug Hypothesis: Physical Properties Driving Asymmetric Tissue Distribution**

Primary Author: Dennis O. Scott
Pfizer

Date: Wednesday, March 09, 2016 - After
Time: 02:30 PM
Room: B316

Co-Author(s)

Abstract Text

A mathematical framework of underlying physicochemical properties of tissues and drug molecules will be described along with examples of it's application to the understanding of drug cellular and sub-cellular disposition which may lead to deviations from the traditional understanding of the free drug hypothesis. Passive distribution is modeled via a Fick-Nernst-Planck approach, using in vitro experimental data to estimate the permeability of both the neutral and ionized species. This framework allows for the derivation of a single set of parameters which govern the distribution of drug molecules across multiple local sub-cellular conditions both in vitro and in vivo. A case study using this approach will be discussed in which the hepatoselective GKa activator achieves hepatoselectivity via a combination of active uptake into the liver and exclusion from other tissues.

Keywords: Pharmaceutical
Application Code: Pharmaceutical
Methodology Code: Liquid Chromatography/Mass Spectrometry

Session # 1920 Abstract # 1920-5 **Organized Contributed Sessions**

Session Title: Precision Bioanalytical Measurements

Abstract Title: **Barbeques and Tornadoes: Analytical Strategies for Metal Ion Determinations Using Separations and Sensors**

Date: Wednesday, March 09, 2016 - After
Time: 03:05 PM
Room: B316

Primary Author: Fiona Regan
Dublin City University

Co-Author(s)

Abstract Text

The accurate determination of metal ions at trace levels in environmental matrices is a complex problem and as a result, methods for determination and preconcentration of trace metals are continuously under investigation. Selection of suitable ligands for sensitive and selective detection of metal ions is seeing much attention. This work presents a historical perspective of methods used for metal speciation from separation techniques to several novel sensing strategies for developing optical chemical sensors. Determination of metal ions by capillary electrophoresis (CE) using oncolumn complexation with 4,2-pyridylazo resorcinol (PAR) followed by peak stacking is described. This method of trace enrichment achieved a limit of detection of 1.10‘8 M for Co(II), Zn(II) and Fe(II). For Pb(II) determination, a 100 fold improvement in sensitivity was achieved using on-column complexation as opposed to precolumn derivitization. The optical sensors are comprised of metal chelating reagents, together with an ion carrier immobilised within polymeric thin films, i.e. hybrid sol-gel thin films, PVC membranes, ORMSOLs, and functionalised cellulose membranes. The developed test strips based on 2-(5-bromo-2-2pyridydylazo)-5-diethylaminophenol (Br-PADAP) immobilised on hybrid nafion/sol-gel membranes are capable of selectively monitoring Ni2+ in water samples. A novel series of double armed spirocyclic calix[4]arene compounds has been investigated for their binding abilities with heavy metal ions. This compound demonstrated potential for the use as selective-ionophore for the development of a Pb2+ selective sensing system. In recent times autonomous sensors are a desired approach and we will present current work on the development of a centrifugal lab-on-a-disc platform for metal ion determination. This platform includes a fully integrated optical detection with long path length, on-disc mixing, in-built motor and 3-D printed casing.

Keywords: Bioanalytical, Capillary Electrophoresis, Environmental Analysis, Lab-on-a-Chip/Microfluidics
Application Code: Environmental
Methodology Code: Capillary Electrophoresis

Session # 1920 Abstract # 1920-6 **Organized Contributed Sessions**

Session Title: Precision Bioanalytical Measurements

Abstract Title: **A Mass Spectrometry Based High Throughput Screening Approach with Exquisite Selectivity**

Primary Author: Heather R. Desaire
University of Kansas

Co-Author(s): Imaduwage Kasun

Date: Wednesday, March 09, 2016 - After
Time: 03:25 PM
Room: B316

Abstract Text

High throughput screening of potential drug candidates is now a necessary prerequisite in almost all drug discovery workflows. When a screen based on a fluorescent or chemiluminescent read-out is feasible, tens of thousands of compounds can be assayed per day. Many druggable targets, however, cannot be probed in these standard assays because the drug action of interest is a protein binding event that does not induce a change in light emission or absorbance. In these cases, much lower-throughput assays based on NMR or SPR are routinely employed. We developed a screening method using LC-MS to identify potential protein binding partners. The method competes very favorably against existing technology like solution based indirect Affinity Selection Mass Spectrometry (ASMS). Our method can be used to screen thousands of compounds per day, and its key advantage over existing approaches is that virtually zero false positive hits are detected. The method was demonstrated to be effective with two different protein examples; in both cases, a single known inhibitor could be picked out of a background of hundreds of non-binding compounds.

Keywords: Bioanalytical, High Throughput Chemical Analysis, Mass Spectrometry, Pharmaceutical
Application Code: Pharmaceutical
Methodology Code: Mass Spectrometry

Session # **1920** **Abstract #** **1920-7** **Organized Contributed Sessions**

Session Title Precision Bioanalytical Measurements

Abstract Title **Microfluidic Devices with Integrated Electrodes for Monitoring Cellular Systems**

Date: Wednesday, March 09, 2016 - After
Time: 03:45 PM
Room: B316

Primary Author R Scott Martin
Saint Louis University

Co-Author(s)

Abstract Text

We have recently developed methods to fabricate polystyrene-based microfluidic devices where encapsulation of materials (tubing and electrodes) can be used to integrate fluidic interconnects and electrochemical detection with other processes such as cell culture and microchip electrophoresis. This general encapsulation approach allows the electrode material to be polished before each use and also enables the use of different electrode materials. This includes a palladium electrode for coupling microchip electrophoresis with electrochemical detection as well as any working electrode material of choice (such as carbon or platinum). This talk will show some of this recent work and demonstrate how the devices are robust for both the immobilization of cells (both PC 12 and endothelial cells) and the analysis of molecules (such as catecholamines and nitric oxide) released from these cells upon stimulation. We will also show that a unique feature of this approach is the ability to create planar membranes from 3-dimensional pillar electrodes. Work towards using this type of device to monitor cell-to-cell interactions will be discussed. In addition, this talk will also feature related work using 3D printers to create highly functional devices that can be used to easily transfer technology to biomedical laboratories.

Keywords: Bioanalytical, Electrochemistry, Electrophoresis, Lab-on-a-Chip/Microfluidics
Application Code: Bioanalytical
Methodology Code: Microfluidics/Lab-on-a-Chip

Session # 1920 Abstract # 1920-8 **Organized Contributed Sessions**

Session Title: Precision Bioanalytical Measurements

Abstract Title: **Novel Applications of Microdialysis Sampling: Where No Probe has Gone Before!**

Primary Author: Sara R. Thomas
Kansas University

Date: Wednesday, March 09, 2016 - After
Time: 04:05 PM
Room: B316

Co-Author(s): Craig E. Lunte, Susan M. Lunte

Abstract Text

Monitoring pharmacokinetics/pharmacodynamics (PK/PD) in various tissues using microdialysis sampling was a central component of the research performed in the Craig Lunte Research Group. The group was mainly divided into two parts: a subgroup focused on the development of instrumentation and methods for analyzing microdialysis samples and a second subgroup concentrated on implementing novel microdialysis sampling strategies [i]in vivo [/i]. The instrumental subgroup was focused primarily on minimizing sample volumes required for analysis, improving the limits of detection, and/or determining the appropriate separation parameters to ensure that the optimal analytical method was employed for the microdialysis experiments. Concurrently, the microdialysis subgroup expanded the applicability of microdialysis sampling by developing a new microdialysis probe for intravenous sampling, and by perfecting the implantation of microdialysis probes in a variety of tissues, including heart, lung, and GI tract. This was done in the quest to gain a better understanding of the PK/PD of drugs/biomarkers in these organs/tissues.

Utilizing microdialysis sampling for PK/PD studies, it is possible to continuously monitor [i]in vivo [/i] changes in drug disposition and pharmacological biomarkers in a site-specific fashion. Examples of novel applications of microdialysis developed in Craig Lunte's lab will be presented, including the implantation of a microdialysis probe in the submucosa layer of the gastrointestinal tract and in the apex of the heart. The information obtained from these experiments would not have been possible without Craig Lunte's guidance, perseverance, and enthusiasm to expand the capabilities of microdialysis sampling.

Keywords: Biological Samples, Biopharmaceutical, HPLC Detection, Sample Handling/Automation
Application Code: Bioanalytical
Methodology Code: Separation Sciences

Session # 1930 Abstract # 1930-1 **Organized Contributed Sessions**

Session Title Technology Strategies for Explosives Sensing

Abstract Title **Overview of Operational Challenges to Domestic Security Screening**

Primary Author Michael Shepard
US Dept. of Homeland Security

Date: Wednesday, March 09, 2016 - After
Time: 01:30 PM
Room: B401

Co-Author(s)

Abstract Text

The focus of many Government-sponsored R&D initiatives is typically on a specific core technology. While this is appropriate, it often leads to sub-optimal solution sets, and imposes costly spiral development and re-purposing activities. In the context of domestic explosives screening, this presentation will highlight operational and logistical challenges that should be considered early in the development cycle. Areas of concern may include adaptability to emerging threats, user-imposed "excursion" applications, common interface designs, and operational safety validation.

Keywords: Calibration, Forensic Chemistry, Identification, Sampling
Application Code: Homeland Security/Forensics
Methodology Code: Sensors

Session # 1930 Abstract # 1930-2 Organized Contributed Sessions

Session Title Technology Strategies for Explosives Sensing

Abstract Title **Generalized Systems Analysis Framework**

Primary Author Samar K. Guharay
The Mitre Corporation

Date: Wednesday, March 09, 2016 - After
Time: 01:50 PM
Room: B401

Co-Author(s)

Abstract Text

A generalized framework based on information structures is discussed. This enables combining information from multiple sources and deriving decision rules for optimal utility of resources. Combining orthogonal information is a natural corollary of this generalized approach. This generalized framework is agnostic of any specific application and system components. It has merits to be applied to diverse mission objectives, including optimal combination of information from multiple sensing modalities under risk constraints, communications, business practice, and many others.

Keywords: Analysis, Informatics, Integrated Sensor Systems, Method Development
Application Code: Homeland Security/Forensics
Methodology Code: Integrated Sensor Systems

Session # 1930 Abstract # 1930-3 **Organized Contributed Sessions**

Session Title Technology Strategies for Explosives Sensing

Abstract Title **Statistical Methods for Constructing Chemometric Signatures in Noisy Environments**

Date: Wednesday, March 09, 2016 - After
Time: 02:10 PM
Room: B401

Primary Author Austin P. O'Brien
South Dakota State University

Co-Author(s) Christopher Saunders

Abstract Text

In this presentation we will discuss various statistical methods for constructing signatures and profiles for various compounds in the presence of outliers. Our focus is on atypicality measures and methods for extending these measures to high dimensional data arising from raman spectra.

Keywords: Chemometrics, Forensics, Gas Chromatography/Mass Spectrometry, Raman
Application Code: Homeland Security/Forensics
Methodology Code: Chemometrics

Session # 1930 Abstract # 1930-4 **Organized Contributed Sessions**

Session Title: Technology Strategies for Explosives Sensing

Abstract Title: **Operational Outlook for Remote Trace Explosives Detection**

Primary Author: Roderick Kunz
MIT Lincoln Laboratory

Date: Wednesday, March 09, 2016 - After
Time: 02:30 PM
Room: B401

Co-Author(s)

Abstract Text

The ability to inspect a suspect item for the presence of concealed explosives is an important capability for both military and civilian safety. One means of doing so relies of the detection of explosive traces exterior to the device, which are good indicators for the likelihood of a concealed explosive. The current methods of detection are based on either collection of surface contamination and introduction of the collected sample into an instrument, or by use of trained animals such as canines. Both methods require the detector to be in close proximity to the inspected object.

In recent years, significant attention has been made to the development of trace explosive detectors that can work from a greater distance than is currently possible. Such a capability would allow larger areas to be searched more quickly and more safely. Put another way, is it possible to do from 100 meters what a canine can do from one meter? The answer to this question lies not only in the technical capabilities of the proposed solutions (which is the usual focus), but in the very nature of the explosive signature itself. In fact, the signature characteristics play a lead role in both defining the technical requirements and determining the ultimate operational utility of even the best detection system.

This talk will provide an overview of the role that trace signature science plays in informing the design and potential capabilities of remote trace detection systems, and will draw on previous work [1,2] to provide a framework for how the operational potential of any remote trace detection system can be informed by a detailed knowledge of the signature science.

[1] Kunz, R. R., et al., Anal. Bioanal. Chem. 395(2), 357-369 (2009).
[2] Kunz R. R., et al., J. Phys. Chem A 116, 3611-3624 (2012).

Keywords: Infrared and Raman, Spectroscopy, Trace Analysis
Application Code: Homeland Security/Forensics
Methodology Code: Molecular Spectroscopy

Session # 1930 Abstract # 1930-5 **Organized Contributed Sessions**

Session Title: Technology Strategies for Explosives Sensing

Abstract Title: **Pushing the Limits of Trace Detection Technology**

Primary Author: Stefan Lukow
Morpho Detection

Date: Wednesday, March 09, 2016 - After
Time: 03:05 PM
Room: B401

Co-Author(s)

Abstract Text

An industry perspective on the development of Explosive Trace Detection systems will be presented highlighting the challenges faced with increasingly strict requirements for a variety of key metrics. Pushing operational limits to improve both sensitivity and breadth of detection while maintaining, or even lowering, a given false alarm level is of prime concern. However, increases in library size, lowering power, weight, and footprint requirements and reducing environmental health and safety concerns are also major focus efforts toward the development of next generation systems. With an ever-expanding list of explosive threat materials, a more dynamic set of requirements is imposed on industry, driving the development of more flexible detection platforms intended to be upgraded as threat priorities shift. The current state of trace detection will be discussed in light of foreseen challenges toward these desired improvements.

Keywords: Instrumentation, Sampling, Trace Analysis
Application Code: Homeland Security/Forensics
Methodology Code: Portable Instruments

Session # 1930 Abstract # 1930-6 **Organized Contributed Sessions**

Session Title: Technology Strategies for Explosives Sensing

Abstract Title: **Optimized Sampling and Analysis Strategies for Trace Explosives Detection**

Primary Author: Greg Gillen
NIST

Date: Wednesday, March 09, 2016 - After
Time: 03:25 PM
Room: B401

Co-Author(s)

Abstract Text

Effective sampling of explosive particle residues is the critical front end process for successful trace chemical detection and forensic analysis of trace contraband materials. The chemical and physical characteristics of these particles must be understood in order to develop optimized sampling and analysis strategies. This presentation will focus on recent studies aimed at understanding the characteristics of contraband particles in residues deposited on model surfaces. A variety of micro analytical tools including micro CT, cathodoluminescence, optical and electron microcopies, atmospheric pressure mass spectrometry and secondary ion mass spectrometry have been employed to study the morphology and chemical composition of explosive particles. Based on this information, optimized sampling protocols and sampling materials have been developed for swipe- based particle collection. These studies have demonstrated the critical nature of applied force, swiping direction and swipe collection material. Each of these variables will be discussed. Of particular interest for swipe sampling is the incorporation of force sensitive resistor imaging technology into sampling wands to evaluate effectiveness of swipe sampling. Assuming successful collection of targeted trace particle residues, subsequent chemical identification is typically achieved by rapidly heating the particles to produce vapors which are characterized by ion mobility or mass spectrometry. The use of high speed video microscopes, temperature programmed desorption, particle counting and laser based scattering measurements has allowed us to develop a better understanding of the desorption process leading to analytical conditions with enhanced analytical sensitivity.

Keywords: Forensic Chemistry, Materials Characterization
Application Code: Homeland Security/Forensics
Methodology Code: Chemical Methods

Session # 1940 Abstract # 1940-1 **Oral Sessions**

Session Title: Advances in Fuel and Petrochemical Analyses

Abstract Title: **Chemical Fingerprinting of Crude Oils: Gaining an Extra Dimension from GCxGC–TOF MS**

Date: Wednesday, March 09, 2016 - After
Time: 01:30 PM
Room: B402

Primary Author: Laura McGregor
Markes International Ltd

Co-Author(s): Chris Hall, Ken Umbarger, Kevin Collins, Nicola Watson

Abstract Text

The enhanced separation offered by comprehensive two-dimensional gas chromatography with time-of-flight mass spectrometry (GCxGC–TOF MS) has made the technique a popular choice for petrochemical analyses. Despite this enhanced separation, the identification of individual compounds in complex samples may become complicated when similar mass spectral characteristics are evident across entire chemical classes. Branched alkanes are a prime example, with weak molecular ions that further complicate the process. Spectral similarity can be addressed by the use of soft ionization to reduce the degree of ion fragmentation, but this approach has been cumbersome to implement until now. Select-eV ion-source technology addresses this problem through the ability to switch effortlessly between hard and soft electron ionization without loss in sensitivity. The novel ion source provides enhanced molecular ions whilst retaining structurally-significant fragment ions, thus simplifying the identification of isomeric compounds. The enhanced sensitivity and selectivity stemming from the dramatic reduction in fragmentation at low energies also greatly increases the number of compounds confidently identified, permitting the robust statistical comparisons that are essential for successful chemical fingerprinting.

Keywords: Gas Chromatography/Mass Spectrometry, Hydrocarbons, Petrochemical, Time of Flight MS
Application Code: Fuels, Energy and Petrochemical
Methodology Code: Gas Chromatography/Mass Spectrometry

Session # 1940 Abstract # 1940-2 **Oral Sessions**

Session Title: Advances in Fuel and Petrochemical Analyses

Abstract Title: **Fast and Accurate Analysis of Extended Natural Gas Composition and Physical Properties Using a Temperature Programmable Gas Analyzer**

Primary Author: Debbie Alcorn
INFICON

Date: Wednesday, March 09, 2016 - After
Time: 01:50 PM
Room: B402

Co-Author(s)

Abstract Text

Precise and fast measurement of natural gas composition is critical for producers, gatherers, and natural gas appliance and engine manufacturers. Physical property calculations, such as heating value (measured in BTU), provide valuable information to the end user. Due to variations in natural gas composition, gas samples must be analyzed frequently, as small changes in physical properties can have a significant financial impact.

INFICON Micro GC Fusion builds upon proven MEMS based technology and introduces temperature programming, which allows for faster, more accurate analyses. Compared to isothermal operation, temperature programming provides sharper peaks and extends application capability. With a single column, C1-C8+ components can be analyzed within 4 minutes. For extended analysis up to C12, a second module may be utilized. Micro GC Fusion's small size and portability makes it an ideal instrument either in the field or in the lab. Combined with Diablo EZReporter for physical property calculation, Micro GC Fusion is a powerful tool for natural gas composition analysis.

Keywords: Energy, Gas Chromatography, GC, Portable Instruments
Application Code: Fuels, Energy and Petrochemical
Methodology Code: Gas Chromatography

Session # 1940 Abstract # 1940-3 **Oral Sessions**

Session Title Advances in Fuel and Petrochemical Analyses

Abstract Title **Application of Polymeric Ionic Liquids as Highly Robust and Selective Stationary Phases for Comprehensive Two-Dimensional Gas Chromatography**

Date: Wednesday, March 09, 2016 - After
Time: 02:10 PM
Room: B402

Primary Author Cheng Zhang
Iowa State University

Co-Author(s) Jared L. Anderson, Rodney A. Park

Abstract Text

Structurally tuned mono and dicationic imidazolium-based ionic liquids (ILs) have been applied in comprehensive two-dimensional gas chromatography (GC×GC) as second dimension columns and have been shown to exhibit high selectivity for the separation of aliphatic hydrocarbons from kerosene. However, the low maximum allowable operating temperatures (MAOTs) of the ILs have largely limited their applications at high operating temperature. To address this issue, a series of polymeric ionic liquid (PIL)-based stationary phases were prepared via in-column free radical polymerization using imidazolium-based IL monomers and crosslinkers. The IL monomers were functionalized with long alkyl chain substituents to provide the selectivity required for the separation of aliphatic hydrocarbons. Columns with different film thicknesses were prepared to determine the best column dimension for the GC×GC separation of kerosene. The bis[(tri□uoromethyl)sulfonyl]imide ([NTf2]-)-based PIL with high film thickness (0.28 μm) exhibited high selectivity and MAOT of 300 □C. Mixtures of IL monomer and IL-based crosslinkers were examined in this study. Better column efficiency and selectivity was obtained when a long chain crosslinker was applied. It was also observed that the addition of crosslinker did not show a significant effect on the thermal stability of the stationary phases. However, better resolution of aliphatic hydrocarbons was obtained when 50% (w/w) of crosslinker was incorporated into the stationary phase. The best performing PIL-based stationary phases exhibited better selectivity and higher thermal stability compared to commercial poly ethylene glycol (PEG) columns.

The authors acknowledge funding from Chemical Measurement and Imaging Program at the National Science Foundation (Grant number CHE-1413199).

Keywords: Fuels\Energy\Petrochemical, Gas Chromatography, Petroleum, Separation Sciences
Application Code: Fuels, Energy and Petrochemical
Methodology Code: Gas Chromatography

Session # 1940 Abstract # 1940-4 **Oral Sessions**

Session Title Advances in Fuel and Petrochemical Analyses

Abstract Title **Solving Industrial Problems by Determining Compound Classes in Refinery Streams and Products**

Primary Author Chris Goss
Alberta Innovates Technology Futures

Date: Wednesday, March 09, 2016 - After
Time: 02:30 PM
Room: B402

Co-Author(s) Dan Wispinski, Lee Marotta

Abstract Text

This presentation discusses a technique that measures compound classes problematic in the petroleum refining process, including cyclics, saturates, aromatics, olefins, heteroatomic (oxygen, nitrogen, sulfur, phosphorous, halogens), and volatile organometallic compounds across the boiling point distribution. A few target classes are looked at here that are known to be either corrosive or destabilizing to the process.

Olefins in condensate streams used as diluents can have a negative impact on the catalyst. Olefins in products can cause destabilization. Naphthenic acids may cause operational problems such as foaming, and are corrosive especially at high temperatures. Odor problems in products may be caused by their prescence. They may cause deactivation of catalyst. The basic nitrogen in combination with phenols may form colorbodies and promote gum formation in the naptha and kerosene fractions. Sulfur, may cause corrosiveness and has a negative impact on our environemnt. All of these classes have the potential to cause the product to fail specification.
In this research, olefins were tested in jet fuels and condensates. Naphthenic acids, sulfur and nitrogen compounds were investigated in light and heavy crudes.

Chromatographic co-elution and similarity of spectra in a traditional electron ionization mass (EI) spectrometry present an analytical challenge to identifying these classes and compounds. This challenge was overcome using a Cold EI source which enhances the molecular ion relative to the fragmentation ions. Cold EI while enhancing the molecular ion also allows for NIST searchable spectra, if the peaks can be deconvoluted. The extracted enhanced molecular ion paired with boiling point distribution is used to attain improved compound identification in complex mixtures Enhanced molecular ion also allows for enhanced MS/MS to confirm structural and compositional identification on the same instrument.

Keywords: Fuels\Energy\Petrochemical, Gas Chromatography/Mass Spectrometry, Petrochemical, Petroleum
Application Code: Fuels, Energy and Petrochemical
Methodology Code: Gas Chromatography/Mass Spectrometry

Session # 1940 Abstract # 1940-5 **Oral Sessions**

Session Title Advances in Fuel and Petrochemical Analyses

Abstract Title **Optical Sensors for the Detection of Nitrogen Compounds in Aviation Fuels**

Date: Wednesday, March 09, 2016 - After
Time: 03:05 PM
Room: B402

Primary Author Roberto A. Federico-Perez
University of Tennessee

Co-Author(s) Ziling (Ben) Xue

Abstract Text

Aromatic nitrogen compounds have impacts on thermal and storage stability of aviation fuels. Presence of nitrogenated compounds is largely represented by indoles, anilines, pyrroles, quinolines, and carbazoles. These compounds are traditionally determined by MS, X-ray, IR, HPLC, and GC, using specialized equipment, sample transportation and handling. Optical sensors represent a different approach that can potentially overcome these drawbacks.

The Ehrlich reaction has been applied for the detection of indole compounds. A red product is formed by condensation of p-dimethylaminobenzaldehyde (DMAB) with indole species in acidic media. We have fabricated a sensor that encloses an Ehrlich-type system into a polymer matrix. Exposure to indole solutions led to the development of red color. This change corresponds to an increasing absorbance at 540 nm that shows a fair correlation with the concentration of indole (R[superscript 2] = 0.988). When aniline is present in the solution, yellow color developed with a peak at 432 nm (R[superscript 2] = 0.990). The detection limits are ppb for indole or aniline. Our work poses an easy and inexpensive alternative for the detection of indole and aniline in aviation fuels. By utilizing a portable spectrophotometer, on-site detection can be achieved in a shorter time in comparison to traditional methods.

Keywords: Fuels\Energy\Petrochemical, Sensors, UV-VIS Absorbance/Luminescence
Application Code: Fuels, Energy and Petrochemical
Methodology Code: UV/VIS

Session # 1940 Abstract # 1940-6 **Oral Sessions**

Session Title: Advances in Fuel and Petrochemical Analyses

Abstract Title: **Fast Profiling of Petrochemical Samples by Thermal Analysis- Soft Ionization Mass Spectrometry: From Source Rock and Kerogen via Crude Oil to Petrochemical Products**

Date: Wednesday, March 09, 2016 - After
Time: 03:25 PM
Room: B402

Primary Author: Ralf Zimmermann
University Rostock /HMGU

Co-Author(s): Andreas Walte, Michael Fischer, Mohammad Saraji, Sebastian Wohlfahrt, Thomas Denner, Thorsten Streibel

Abstract Text

Mass Spectrometry (MS) with soft ionisation methods (i.e. fragmentation-free ionisation methods) such as photoionisation (PIMS) enables the direct detection of traces of intact molecules from complex organic vapours. PIMS is a perfect tool for Evolved Gas Analysis (EGA) in Thermal Analysis (TA). In particular it allows the determination of the molecular organic signatures (i.e. intact organic molecules) of desorption-, pyrolysis- and combustion-processes. A newly developed TA-PIMS system was applied for detection of the evolved signatures of intact organic molecules from thermogravimetic analyses of petrochemical samples (source rock, keroegn, crude oils etc.). In the lecture initially a briefly explanation of the principles of the used soft ionisation mass spectrometry approaches is given. The main part of the lecture comprehends applications results of the Thermal Analysis - Photoionisation Mass Spectrometry (TG-PIMS). This approach allows the detailed analysis of the thermochemical behaviour of petrochemically relevant samples. In the case of crude oil, e.g., the yield and chemical composition of the different distillation fractions can be determined (50-400°C). At higher temperatures the cracking products of the non-volatile residue (asphalthenes etc.) are analysed. Further examples are the source rock samples. Here the composition and formation temperature of shale-oils and -gases can be measured. The comparison of TG-PIMS laboratory results and industrial pilot-plant process studies finally suggest the applicability of the laboratory instruments for simulation and optimization of industrial processes. In summary the potential of TA hyphenated to soft photoionisation mass spectrometry for industrial- as well as for fundamental-research applications is demonstrated and discussed.

Keywords: Fuels\Energy\Petrochemical, Gasoline, Thermal Analysis, Time of Flight MS
Application Code: Fuels, Energy and Petrochemical
Methodology Code: Thermal Analysis

Session # 1940 **Abstract # 1940-8** **Oral Sessions**

Session Title: Advances in Fuel and Petrochemical Analyses

Abstract Title: **On-Site Fuel Analysis Using a Portable Near-Infrared Spectrometer**

Date: Wednesday, March 09, 2016 - After
Time: 04:05 PM
Room: B402

Primary Author: Stuart Farquharson, Real-Time Analyzers, Inc

Co-Author(s): Carl Brouillette, Chetan Shende, Wayne Smith

Abstract Text

Fuel quality is becoming increasingly important in many countries where fuel theft occurs in the form of diluting shipments using less expensive petroleum products. Diluted fuels with inferior chemical and physical properties lead to decreased engine performance and ultimately engine failure. These properties include aromatic, fuel system icing inhibitor, olefinic, naphthene, oxygenate, and saturate content; cetane and octane numbers; cloud, distillation, flash, freezing, and pour points; density (API gravity), Reid vapor pressure, and viscosity. Complete analysis of a suspect fuel requires multiple analyzers, typically one per property, and as much as 8 hours. Such analysis is not always performed. Consequently, there exists a need to rapidly verify fuel quality upon receipt. Often, such verification must be performed far from a lab containing standard fuel analysis instruments and apparatus. To satisfy this need, we have developed an inexpensive, compact (7x13x16", 14.5 lb) fuel analyzer based on near-infrared spectroscopy that employs chemometrics to determine the 16 properties listed above in 10 seconds, using only 2 mL of sample. This presentation describes the analyzer and the chemometric analysis.

Keywords: Chemometrics, Fuels\Energy\Petrochemical, Gasoline, Portable Instruments

Application Code: Fuels, Energy and Petrochemical

Methodology Code: Near Infrared

Session # 1950 Abstract # 1950-1 **Oral Sessions**

Session Title: Bioanalytical: Using Microfluidics/Lab-on-a-Chip Techniques

Abstract Title: **Pressure-Actuated Integrated Microfluidic Devices for Biomarker Analysis**

Primary Author: Vishal Sahore
Brigham Young University

Date: Wednesday, March 09, 2016 - After
Time: 01:30 PM
Room: B404

Co-Author(s): Adam T. Woolley, Suresh Kumar

Abstract Text

We are developing pressure-actuated integrated microfluidic devices to diagnose the risk of a pre-term birth (PTB). The novelty of our work lies in the integration of pressure-driven immunoaffinity extraction, enrichment, labeling, and injection followed by electrophoretic separation, on a single platform. A microchip electrophoresis (□CE) module has been developed in a three-layer poly(dimethylsiloxane) (PDMS) system. Each device has pneumatic valves that surround the injection intersection to capture a sample plug containing PTB biomarkers, which are subsequently separated using □CE. An on-chip labeling and enrichment module has been developed with a reversed-phase porous polymer monolith in a cyclic olefin copolymer (COC) thermoplastic layer. The labeling module is integrated with a □CE module to allow the retention, labeling, injection, and separation of a PTB biomarker. We have successfully demonstrated ~80-fold enrichment of a PTB protein biomarker with this integrated on-chip labeling-□CE device. Furthermore, for the selective capture, injection, and separation of PTB biomarkers, an integrated immunoaffinity extraction-□CE device has been manufactured with a porous polymer monolith in the COC layer. Currently, studies are being performed to optimize conditions for loading, extraction, and elution of PTB biomarkers on the monolith; subsequently, the affinity-extracted sample will be injected in the □CE module for separation and quantitation. Finally, we plan to combine all three modules and construct a fully automated, integrated microfluidic device for pre-term birth diagnosis.

Keywords: Bioanalytical, Capillary Electrophoresis, Lab-on-a-Chip/Microfluidics, Solid Phase Extraction
Application Code: Bioanalytical
Methodology Code: Microfluidics/Lab-on-a-Chip

Session # 1950 Abstract # 1950-2 **Oral Sessions**

Session Title: Bioanalytical: Using Microfluidics/Lab-on-a-Chip Techniques

Abstract Title: **Real-Time Imaging of Cancer Cell Chemotaxis in Paper-Based Scaffolds**

Date: Wednesday, March 09, 2016 - After
Time: 01:50 PM
Room: B404

Primary Author: Rachael M. Kenney
University of North Carolina at Chapel Hill

Co-Author(s): Andrew S. Truong, Matthew R. Lockett, Matthew W. Boyce

Abstract Text

Chemokines, factors secreted by cells for intercellular signaling, are believed to play an important role in directing differentiation, proliferation, and the migration of cells in solid tumors. Two dimensional migration assays are amenable to high-throughput screening of cellular migration in the presence of chemokines; these assays have ill-defined concentration gradients, oversimplify [i]in vivo[/i] conditions, and cannot incorporate the three-dimensional nature of a tissue. Microfluidic devices support 2D and 3D cultures, offer exquisite control over the spatial and temporal gradients of soluble molecules and are compatible with real-time optical imaging. Unfortunately, microfluidic devices often require specialized equipment and engineering experience, as they are difficult to setup and maintain. Here we describe a paper-based invasion assay that provides an easy to assemble 3D tissue-like environment. The assay utilizes a single sheet of paper with wax-patterned channels to seed cells; a gradient is generated down the length of the channel and cellular movement is observed in real-time with fluorescence microscopy. Specifically, we have demonstrated that MDA-MB-231 cells selectively migrate toward regions containing higher concentrations of different chemoattractants (e.g., oxygen, CXCL12). We will discuss the response of breast cancer lines in the invasion channels, and how altering these gradients affect the extent of their invasion. This three-dimensional invasion assay could provide a platform that is more representative of [i]in vivo[/i] cellular responses than current chemotaxis assays, and screen potential chemoattractants or small molecules inhibiting invasion with a system that is easily prepared and analyzed in any tissue culture laboratory.

Keywords: Bioanalytical, Biotechnology, Imaging, Lab-on-a-Chip/Microfluidics
Application Code: Bioanalytical
Methodology Code: Microfluidics/Lab-on-a-Chip

Session # 1950 Abstract # 1950-3 **Oral Sessions**

Session Title: Bioanalytical: Using Microfluidics/Lab-on-a-Chip Techniques

Abstract Title: **Observation of Rapid Changes in Drug Susceptibility of Tumor Cells in a Hypoxia Microfluidic Culture Device**

Date: Wednesday, March 09, 2016 - After
Time: 02:10 PM
Room: B404

Primary Author: Todd Germain
Texas Tech University

Co-Author(s): Dimitri Pappas

Abstract Text

In this work we studied prostate cancer drug susceptibility in a hypoxic environment observing rapid adaptation to hypoxia. In a previous study we have looked at the response of prostate cancer cells to hypoxia at preconditioning times of an hour or greater. In this study we observe the response of PC-3 cells to staurosporine while in a hypoxic environment at preconditioning times from 0 min to 30 min in a low-shear microfluidic cell culture device. The cells were assayed using Annexin-V-Alexa Fluor 647 and Sytox Green after being cultured for up to 9 hours. Cells cultured in normoxia in the microfluidic device that were not exposed to staurosporine were found to be 84.5% viable, a viability that was similar to cells exposed to 2 □M staurosporine with 30 min of hypoxia preconditioning (78.0% viability). Cells exposed to the same concentration of staurosporine but with a preconditioning time of 0 min were found to be 48.6% viable. Cell viability was on a similar level for all preconditioning times below 30 min and the staurosporine control experiment that did not involve hypoxia, as well as the 30 min normoxia preconditioning experiment (51.6% viability). The shear force experienced by the cells in the culture chamber was studied by observing the diffusion of a fluorescent dye flowing into a channel that had been photobleached. The shear stress in the culture chambers was found to be 0.021 dynes/cm[sup]2[/sup], which was 0.62% of the shear force in the medium-supply channel which was found to be 3.42 dynes/cm[sup]2[/sup]. This microfluidic device is capable of culturing cells in a low shear stress environment, and this method is capable of looking at different cell types with a minimal preconditioning time of 30 min under hypoxic conditions. The observation that cell adaptation to hypoxia is rapid has broad implications in the study of solid, hypoxic tumors.

Keywords: Bioanalytical, Lab-on-a-Chip/Microfluidics
Application Code: Bioanalytical
Methodology Code: Microfluidics/Lab-on-a-Chip

Session # 1950 Abstract # 1950-5 **Oral Sessions**

Session Title: Bioanalytical: Using Microfluidics/Lab-on-a-Chip Techniques

Abstract Title: **Screening Small Molecule Modulators of Cellular Chemotaxis in Paper-Based Invasion Assays**

Primary Author: C Chad Lloyd
University of North Carolina at Chapel Hill

Date: Wednesday, March 09, 2016 - After
Time: 03:05 PM
Room: B404

Co-Author(s): Andrew S. Truong, Christian A. Lochbaum, Matthew R. Lockett, Matthew W. Boyce, Rachael M. Kenney

Abstract Text

There are various methods to test the effects of small molecules on cancer cells. Most of these methods investigate the effects of small drug molecules on the invasiveness of cancer cells in a two-dimensional assay. Using paper based scaffolds, one can analyze the effects of various small molecules in a three-dimensional environment. We prepare paper-based invasion assays and subject them to different gradients of small molecules that could modulate invasiveness. With this application, we can better determine whether a small molecule acts as an agonist, antagonist, or inverse agonist toward cellular movement. We compare the invasiveness of breast and prostate cancer cell lines in the presence of small molecules that target cannabinoid receptors. These receptors are overexpressed in tumors when compared to normal tissue. Paper-based invasion assays are useful in testing a variety of concentrations of a specific small molecule on a single culture or exposing a single culture to several different small molecules at once.

Keywords: Bioanalytical, Biomedical, Drugs
Application Code: Bioanalytical
Methodology Code: Microfluidics/Lab-on-a-Chip

Session # 1950 **Abstract # 1950-6** **Oral Sessions**

Session Title: Bioanalytical: Using Microfluidics/Lab-on-a-Chip Techniques

Abstract Title: **Microchip Affinity Monoliths for Solid Phase Extraction of DNA for Bacteria Infection Detection**

Date: Wednesday, March 09, 2016 - After
Time: 03:25 PM
Room: B404

Primary Author: Radim Knob
Brigham Young University

Co-Author(s): Adam T. Woolley, Riley K. Mills

Abstract Text

Antibiotic resistant bacteria present a significant threat that brings difficulties in diagnosis, which causes complications in treatment. Analysis of DNA presents an essential tool in diagnosis of many diseases, and detection of specific DNA sequences could be utilized in identification of specific bacteria or resistance genes. Current methods for identification of antibiotic resistance rely on a slow culture step to grow sufficient quantities of bacteria for detection.

In our approach, a device for capturing nucleic acid sequences representing specific bacterial strains or genes related to antibiotic resistance is under development. One component of the device is aimed at selective extraction of DNA from bacterial lysate.

For this purpose, a porous polymer monolith stationary phase was developed and functionalized with a DNA sequence for selective capture of a complementary target DNA. Glycidyl methacrylate based monoliths were prepared in a specific part of polymethylmethacrylate microfluidic devices using photo-induced polymerization. Properties of the porous polymer were optimized regarding porosity and flow-through characteristics, allowing use of the stationary phase with a low-pressure pump. Conditions for solid-phase extraction of a model fluorescein-labeled DNA sequence were studied for capture and elution. Immobilization conditions and composition of rinsing and eluting solutions were tested. Extraction was visualized by monitoring fluorescence on the monolith by a CCD camera, and elution was recorded by laser-induced fluorescence using a photomultiplier tube for detection.

These monolithic stationary phases will extract target DNA from bacterial lysate and deliver it concentrated in a small volume to a single-molecule detection module in the final device. This should significantly reduce the time needed for identification of bacteria type and resistance genes, compared to current methods.

Keywords: Bioanalytical, Chromatography, Immobilization, Lab-on-a-Chip/Microfluidics
Application Code: Bioanalytical
Methodology Code: Microfluidics/Lab-on-a-Chip

Session # 1950 Abstract # 1950-7 **Oral Sessions**

Session Title: Bioanalytical: Using Microfluidics/Lab-on-a-Chip Techniques

Abstract Title: **MALDI-IMS Evaluation of 3D Cell Cultures Treated with Combination Chemotherapeutics by a 3D Printed In-Vitro PK/PD Microfluidic Platform**

Date: Wednesday, March 09, 2016 - After
Time: 03:45 PM
Room: B404

Primary Author: Gabriel J. LaBonia
University of Notre Dame

Co-Author(s): Amanda B. Hummon

Abstract Text

Combination chemotherapies are commonly employed by oncologists for the treatment of metastatic colorectal cancer. Aside from ethical concerns, typical in-vivo research on these toxic chemotherapeutic drugs is low throughput, time consuming and costly. With this project, a novel in-vitro platform is employed to better understand the effects of chemotherapeutic treatments on cancer cells. This in-vitro alternative to typical animal models is 3-Dimensional cell culture tumor spheroids. These multi-cellular aggregates, or spheroids, of colorectal cancer cells have been shown to exhibit many of the physiological and chemical gradients of a typical clinical tumor, while still maintaining the flexibility of a cell culture system. This project utilizes a 3D printed microfluidic device that facilitates high-throughput treatment of these tumor spheroids across a semipermeable membrane. The device also allows for a dynamic drug-dosing gradient similar to a typical in-vivo pharmacokinetic profile. The treated tumor spheroids are evaluated with MALDI-Imaging Mass Spectrometry to elucidate the spatial distribution of the drug molecules and corresponding metabolites. This innovative system enables both pharmacokinetic and pharmacodynamic evaluations for a wide range of compounds, and can have a transformative impact on the pre-clinical evaluation of drug candidates. The described analytical platform, in tandem with the 3D cell culture environment is cost effective, enables higher throughput, and is readily available for the exploration of new and creative cancer treatments.

Keywords: Bioanalytical, Drugs, Imaging, Mass Spectrometry
Application Code: Bioanalytical
Methodology Code: Mass Spectrometry

Session # **1950** Abstract # **1950-8** **Oral Sessions**

Session Title: Bioanalytical: Using Microfluidics/Lab-on-a-Chip Techniques

Abstract Title: **Fully Automated Microfluidic Input/Output Multiplexer for Endocrine Tissue Culturing and Hormone Secretion Sampling**

Date: Wednesday, March 09, 2016 - After
Time: 04:05 PM
Room: B404

Primary Author: Xiangpeng Li
Auburn University

Co-Author(s): Christopher J. Easley, Jessica C. Brooks, Katarena Ford

Abstract Text

Our group has previously developed microfluidic systems for culture and sampling of primary murine tissue, specifically pancreatic islets and adipocytes. Tissue culture reservoirs were directly built, or "landscaped," into the polydimethylsiloxane (PDMS) above the microchannels using manually fabricated templates. We have recently begun using 3D printed templates designed to match device channel patterns. However, the devices have been based entirely on passive flow control, which limits capabilities for temporally-resolved hormone sampling and controlled stimulation. Here, we report a fully automated, pneumatically valved microfluidic system for endocrine tissue culture as well as temporally resolved stimulation and hormone sampling. The device consists of a central reservoir for tissue culture interfaced to a 16-channel input/output microfluidic multiplexer (□MUX-16) for automated perfusion and sampling of cells. The reservoir is fabricated with 3D printed templates, and the valved microchannels are made with multilayer soft lithography. As in analogous electronic components, the □MUX-16 can address large numbers (N=16) of fluid channels with a smaller number (2 log2N=8) of pneumatic controls, allowing up to 4-bit resolution on input or output timing—or a combination of both. The device is automated by an in-house written LabVIEW application that controls solenoids for valving, detects reservoir filling with conductivity sensors, and precisely determines reservoir emptying with transmitted light imaging through real-time image analysis. The system is proven functional with dynamic glucose inputs and insulin secretion outputs from primary islets. Overall, our novel □MUX-16 is shown to be a highly flexible, high resolution fluidic interface to primary tissue culture, which serves as a mimic of the circulatory system and should be directly applicable to other endocrine tissues or even organ-on-a-chip platforms.

Keywords: Automation, Bioanalytical, Lab-on-a-Chip/Microfluidics, Sampling
Application Code: Bioanalytical
Methodology Code: Microfluidics/Lab-on-a-Chip

Session # **1960** **Abstract #** **1960-1** **Oral Sessions**

Session Title: Environmental Water Quality and Analysis

Abstract Title: **Targeted Discovery of Disinfection By-Products in Swimming Pools and Hot Tubs**

Primary Author: Susan D. Richardson
University of South Carolina

Date: Wednesday, March 09, 2016 - After
Time: 01:30 PM
Room: B403

Co-Author(s): Christina M. Joseph, Eric J. Daiber, Jonathan D. Byer, Joseph E. Binkley, Sridevi A. Ravuri

Abstract Text

Swimming pools are treated with disinfectants to protect swimmers from pathogens and prevent illness. The water used to fill a swimming pool is also often treated with disinfectants such as chlorine. Disinfectants will react with naturally occurring organic matter in water and, in the case of swimming pools, they can also react with chemicals introduced to the water by the swimmers themselves to produce by-products that can be potentially harmful. It is important to treat water while minimizing the risk of disinfection by-products (DBPs). One of the first steps is to chemically characterize the DBPs in swimming pools and hot tubs, very complex matrices, using discovery techniques because a lot of the contaminants are unknowns. Comprehensive two-dimensional gas chromatography high resolution time-of-flight mass spectrometry (GC×GC-HR-TOF-MS) was used for the tentative identification of "known unknowns" and "unknown unknowns" in swimming pool and hot tub water. The "known unknowns" were identified by library database searching deconvoluted spectra. The "unknown unknowns" were tentatively identified using a combination of electron ionization (EI) and chemical ionization (CI) accurate mass data for chemical formulae determination and structural elucidation by leveraging the accurate mass fragments.

Keywords: Environmental Analysis, Environmental/Water, Gas Chromatography/Mass Spectrometry
Application Code: Environmental
Methodology Code: Mass Spectrometry

Session # 1960 Abstract # 1960-2 **Oral Sessions**

Session Title: Environmental Water Quality and Analysis

Abstract Title: **Evaluation of Iodo-, Bromo-, and Chloro-Acetic Acids Formation by Peracetic Acid Disinfection Using a Newly Developed Rapid HPIC-MS/MS Method**

Primary Author: Runmiao Xue
Missouri University of Science and Technology

Date: Wednesday, March 09, 2016 - After
Time: 01:50 PM
Room: B403

Co-Author(s): Bin Hua, Craig Adams, Enos Inniss, Honglan Shi, John Yang, Todd Erichholz, Yinfa Ma

Abstract Text

Haloacetic acids (HAAs), including chloroacetic acids (CAAs) and bromoacetic acids (BAAs), are toxic water disinfection byproducts (DBPs) regulated by the U.S. Environmental Protection Agency (USEPA). Iodoacetic acids (IAAs) are the emerging DBPs that are more toxic than the corresponding CAAs and BAAs. In this study, a novel rapid and sensitive high performance ion chromatography-tandem mass spectrometry (HPIC – MS/MS) method has been developed for simultaneous analysis of all these HAAs, bromate, bromide, iodide, and iodate, seventeen compounds in total, without any tedious sample preparation. The analytes were separated with a ion chromatography column coupled with a guard column and detected by negative electrospray ionization-tandem mass spectrometry. The detection limits were mostly in the range of 0.02 to 0.05 μg/L except a couple of them are higher. The method has been used for analyzing water samples of various matrixes, including river water, groundwater, swimming pool water, and finished drinking water. Appropriate dilutions are necessary in order to achieve good accuracy for some samples with complex matrices. Peracetic acid (PAA) has been demonstrated to be a possible green disinfectant. The formation potential of HAAs from PAA disinfection has been investigated and compared with the HAA formation from other disinfectants. The detailed experimental conditions and the results will be presented at the conference.

This study is supported by US EPA STAR program (grant # 83517301) and Missouri Department of Natural Resource.

Keywords: Contamination, Environmental Analysis, Liquid Chromatography/Mass Spectroscopy
Application Code: Environmental
Methodology Code: Chemical Methods

Session # 1960 **Abstract # 1960-3** **Oral Sessions**

Session Title: Environmental Water Quality and Analysis

Abstract Title: **Removal of Pharmaceutical Products from Wastewater Using Magnetic Bio-Char**

Date: Wednesday, March 09, 2016 - After
Time: 02:10 PM
Room: B403

Primary Author: Akila G. Karunanayake
Mississippi State University

Co-Author(s): Olivia A. Todd, Todd E. Mlsna

Abstract Text

Salicylic acid, 4-nitroaniline, benzoic acid and phthalic acid are pharmaceutical products that find their way into the environment in industrial and municipal wastewater. In high concentrations, these pharmaceuticals can have detrimental environmental effects. Therefore, it is imperative that these contaminants be removed before treated wastewater is returned to the environment. A novel method of these pharmaceutical removal from wastewater is through the use of magnetic bio-char, which is an environmentally-friendly and cheap. Magnetic bio-char was prepared by iron oxide precipitation onto the surface of commercially available Rinsed Ultra bio-char using an aqueous $Fe3+/Fe2+$ solution followed by NaOH treatment. The surface chemistry and composition of magnetic bio-char were examined by SEM, SEM-EDX, TGA, PZC, elemental analysis, and surface area measurements. For both magnetic and non-magnetic bio-char, batch sorption studies were carried out between pH values from 2 to 10, adsorbate concentrations from 25 to 100 ppm and temperatures of 25 and 35°C. Sorption performance at different temperatures was evaluated using the Langmuir, Freundlich, Redlich–Peterson, Toth, Sips, and Radke–Prausnitz adsorption isotherm models. Results indicate that magnetic bio-char removes the pharmaceuticals and remediated the solutions more effectively than the non-magnetic bio-char.

Keywords: Analysis, Environmental/Water, Pharmaceutical, UV-VIS Absorbance/Luminescence
Application Code: Environmental
Methodology Code: UV/VIS

Session # 1960 Abstract # 1960-4 **Oral Sessions**

Session Title: Environmental Water Quality and Analysis

Abstract Title: **Determination of Priority Water Contaminants by Solid-Phase Extraction and UFLC-MS/MS Method**

Primary Author: Haiting Zhang
Missouri University of Science and Technology

Date: Wednesday, March 09, 2016 - After
Time: 02:30 PM
Room: B403

Co-Author(s): Craig Adams, Danielle West, Honglan Shi, Todd Erichholz, Yinfa Ma

Abstract Text

Recent studies indicate the potential widespread occurrence of pharmaceuticals and personal care products, hormones, and other organic contaminants in the aquatic environment. In this study, seven high-priority drinking water contaminants were accessed in Missouri drinking water systems. The compounds include two pharmaceuticals (fluoxetine and gemfibrozil), three pesticides (3-hydroxycarbofuran, azinphos-methyl, and chlorpyrifos), and two hormones (progesterone and testosterone). A solid phase extraction (SPE) method has been developed for this study. Each 500 mL water samples was filtered through 0.45 μm nylon membrane filter, adjusted to pH 7.0 ± 0.5, and extracted through the conditioned HLB SPE cartridges at a flow rate of 10-15 mL/min. The analytes were eluted with 7 mL of methanol and then evaporated down to 1 mL at 32°C for analysis. An ultra-fast liquid chromatography-tandem mass spectrometry (UFLC-MS/MS) method was developed to detect the concentrations of these contaminants. The method was used for assessment of these contaminants in eighteen Missouri drinking water facilities. The water samples include source and finished water samples from different types of source water of surface, ground, groundwater influenced by surface water. The concentrations of the studied contaminants were bellow detection limits in most of the selected facilities, only a few source water samples got the contamination level approximately 0.02 μg/L, close the detection limits.

Keywords: Environmental Analysis, Liquid Chromatography/Mass Spectroscopy, Solid Phase Extraction
Application Code: Environmental
Methodology Code: Chemical Methods

Session # 1960 Abstract # 1960-5 **Oral Sessions**

Session Title: Environmental Water Quality and Analysis

Abstract Title: **Characterization of Oil-based Pollutants Using Webcam-based Spectrometer**

Primary Author: Yagiz Sutcu
InfoScope Research

Co-Author(s): Aysegul Ergin

Date: Wednesday, March 09, 2016 - After
Time: 03:05 PM
Room: B403

Abstract Text

Spectroscopic techniques such as absorbance and fluorescence/luminescence spectroscopy are common analytical tools used in many areas including environmental monitoring, pharmaceuticals, bio-imaging and food analysis etc. Fortunately, spectroscopic instruments are now available in handheld/portable form and do not require specialized laboratory facilities and personnel to operate. These portable devices are easy to use with minimal user training, and once the spectral signatures are extracted and validated, characterization of unknown samples can be performed almost in real time. In addition to these advantages, accessibility of free online platforms enables widespread use of portable spectroscopic instruments for many applications. More recently, due to the availability of very low-cost sensors and 3D printers, do-it-yourself (DIY) spectrometers have become much more accurate, reliable and accessible.

In this study, we investigated the capabilities and limitations of an extremely affordable, portable, webcam-based spectrometer in environmental pollution monitoring applications. More specifically, we focused on oil-based pollutant characterization using fluorescence spectroscopy. Our preliminary results showed that despite its simple design and low-cost, this webcam-based DIY spectrometer is a very promising tool which can be used by citizen scientists to investigate environmental pollution. Furthermore, our results indicate that similar approach can be used to detect not only other types of environmental pollutants in water and soil, but also can be used to analyze edible oils to detect adulteration. Moving forward, it is crucial to collect and maintain a database of high quality spectral data and an efficient and scalable algorithm in order to be able to achieve acceptable accuracy to use this approach for real-world application.

Keywords: Environmental, Fluorescence, Identification, Instrumentation
Application Code: Environmental
Methodology Code: Fluorescence/Luminescence

Session # 1960 Abstract # 1960-6 **Oral Sessions**

Session Title Environmental Water Quality and Analysis

Abstract Title **Coupling of Thin Film Microextraction Techniques to Portable GC-TMS Instrumentation for the On-Site, Sub-ppb, Detection of Pesticides From Environmental Waters**

Date: Wednesday, March 09, 2016 - After
Time: 03:25 PM
Room: B403

Primary Author Jonathan J. Grandy
University of Waterloo

Co-Author(s) Janusz Pawliszyn

Abstract Text

Hand-portable GC-MS technology has been slowly growing for application in completely on-site environmental analysis. However, these instruments are known to exhibit much less sensitivity than their benchtop counterparts. With this in mind, more efficient sample preparation techniques may prove key in decreasing the limits of detection for on-site analytical methods. Hench, in this work it is shown that a DVB/PDMS/Carbon mesh, thin film microextraction membrane may be coupled directly to a hand portable GC-TMS instrument by use of a prototype, high volume desorption module thus, increasing the sensitivity of on-site methods by a factor of 20 or more. This prototype high volume desorption module was shown to effectively transfer all analytes loaded onto the TFME membrane to the GC-MS instrument with no detectable sample loss completely on-site. These membranes, which possess a surface area of 3.88 cm 2 allow for the detection of organochlorine and organophosphorus based pesticides from an aqueous solution, below the ppb level in less than 15 minutes. In-fact, it was shown that 2,4 dichlorophenol 2,4,6 trichlorophenol, phorate D10, fonofos, and parathion could be detected from water on portable GC-TMS instrumentation at 100 ppt using these membranes. The coupling of these TFME devices to hand portable GC-MS instrumentation can push detection limits for these pesticides down to levels expected for standard benchtop instruments.

This work was supported by the Natural Sciences and Engineering Research Council of Canada, Supelco Co. and Torion Technologies of Pelkin Elmer Inc.

Keywords: Environmental Analysis, Pesticides, Portable Instruments, SPME

Application Code: Environmental

Methodology Code: Portable Instruments

Session # 1960 **Abstract # 1960-7** **Oral Sessions**

Session Title: Environmental Water Quality and Analysis

Abstract Title: **SPE in US EPA Method 625, A Step Closer to Reality after Good Performance in Two Round Robins**

Date: Wednesday, March 09, 2016 - After
Time: 03:45 PM
Room: B403

Primary Author: Zoe Grosser
Horizon Technology

Co-Author(s): Alicia Cannon, David Gallagher, Michael Ebitson, William Jones

Abstract Text

US EPA method 625 is a general semivolatile method for wastewater analysis applied to a large suite of target analytes. Although method 625 was developed a number of years ago, through the EPA Office of Water, Office of Science and Technology, the method has been updated several times. As new technology is developed either for the determinative measurement or, earlier in the analysis process, for the sample preparation, data must be collected to demonstrate that the new technology is compliant and reproducible.

Two small round-robin studies using multiple vendor products and solid phase extraction (SPE) materials in a variety of laboratories have been run to demonstrate the compliance of SPE with method requirements. The first study relied on the quality control in the existing method to catch when the equipment or sorbent did not work properly. In the second study, the choice of surrogates was enlarged to ensure that errors not caught with the criteria in the older version of 625 would be identified in this version.

This paper will discuss the performance of SPE in general for method 625 and the specific performance of a disk used with a single pass of acidified water through it rather than a pass with basified water and a second pass with the same water, now acidified, which is typical for liquid-liquid extraction. Recoveries of a large suite of compounds from a variety of matrices and laboratories will be examined and the effect of surrogates will be considered. The results from the first round robin demonstrated recoveries from 70-130% of most all the acid/base/neutral/pesticides chosen for the study using a multi-mode disk adsorbent. The second study used a TCLP matrix as the challenge and these results will be compared to results from the first study. The implications of larger and smaller sample volumes will be discussed.

Keywords: Environmental/Water, Sample Handling/Automation, Sample Preparation, Solid Phase Extraction
Application Code: Environmental
Methodology Code: Sampling and Sample Preparation

Session # 1960 Abstract # 1960-8 **Oral Sessions**

Session Title: Environmental Water Quality and Analysis

Abstract Title: **Adsorption of Pb[sup]2+[/sup] from Aqueous Solution Using Low-Cost Chitosan-Modified Biochar, A Green Adsorbent**

Primary Author: Narada W. Bombuwala Dewage
Mississippi State University

Date: Wednesday, March 09, 2016 - After
Time: 04:05 PM
Room: B403

Co-Author(s): Todd E. Mlsna

Abstract Text

Pine wood biochar, the waste product produced from fast pyrolysis during bio-oil production, was investigated as potential green adsorbents for lead remediation from aqueous solution. Biochar was obtained from bio-oil production at 698 K with a residence time of 20-30 s in an auger-fed reactor and then modified by mixing with chitosan acetate solution, followed by treatment with NaOH. The characterization of both chitosan-modified and non-modified biochars were studied using Fourier transform infrared spectroscopy, scanning electron microscopy, scanning electron microscopy/energy dispersive X-ray spectroscopy, surface area measurement, elemental analysis, thermogravimetric analysis, and □potential measurements. Characterization results showed that the application of chitosan on biochar surface could improve its performance as an adsorbent. Batch sorption studies were performed at pH values from 2 to 5, and temperatures from 298 to 318 K. The total amount of Pb[sup]2+[/sup] adsorbed was determined quantitatively using atomic absorption spectrophotometry. Maximum lead removal occurred at pH 5. The chitosan-modified biochar showed enhanced removal of Pb[sup]2+[/sup] from solution, as opposed to non-modified biochar, suggesting that the modification of biochar with chitosan generates different adsorption sites on their solid surface for metal ion adsorption. Pseudo-second order kinetics provided the best fit with regression coefficients of 0.998 or greater. Sorption was evaluated from 298 to 318 K using the Freundlich and Langmuir isotherm models. The mechanisms of Pb[sup]2+[/sup] adsorption on chitosan-modified biochar were studied by using the Fourier transform infrared spectroscopy and X- ray photoelectron spectroscopy. The N atoms in chitosan are believed to play a major role in lead adsorption with the studied pH range.

Keywords: Adsorption, Environmental, Lead, Metals
Application Code: Environmental
Methodology Code: Atomic Spectroscopy/Elemental Analysis

Session # 1970 **Abstract # 1970-1** **Oral Sessions**

Session Title: GC Fuels, Energy and Petrochemical

Abstract Title: **Zeolite-Loaded Metal-Catalyzed Hydrotreatment of Lignin to Aromatic Monomers Using Subcritical Water**

Date: Wednesday, March 09, 2016 - After
Time: 01:30 PM
Room: B405

Primary Author: Eric A. Boakye
South Dakota State University

Co-Author(s): Douglas Raynie

Abstract Text

Lignin forms about 30% of lignocellulosic material, which is the second most abundant non-fossil organic carbon source in the biosphere. However, it is often treated as waste or, in some instances, burned to supply energy. Developing an efficient and environmentally benign method to convert lignin to high value-added aromatic monomers (guaiacyl, vanillin, and syringyl moieties) as a feedstock for chemical polymers has become necessary. Mineral bases, such as NaOH and CsOH, or supported-metal catalysts (Pt, Ru, Pd, Ni on C) have been used to form aromatic monomers; but associated drawbacks are corrosion, catalyst recovery, sintering of metals, and loss of activity.

?

Therefore, we are currently using zeolite-supported metal catalysts (CoO, LaO, and MoO) with subcritical water at 200°C and 240°C to develop a method to convert alkali lignin to high value-added aromatic monomers. Separations of organic and aqueous phases were done by liquid-liquid extraction using ethyl acetate as solvent. Our results indicate the formation of guaiacol, homovanillic acid, vanillic acid, 3-methoxyacetophenone, acetovanillone, and vanillin moieties. Gas chromatography-mass spectrometry analysis of the organic extracts shows 2-4.5 wt% and 3-8 wt% formation of phenolic compounds for temperatures at 200°C and 240°C respectively using 25MPa pressure. MoO catalyst gave the highest %wt yield phenolic monomers for both temperatures. The presence of aromatic nature in the products has been confirmed by FTIR analysis.

Keywords: Chemical, Energy, Extraction, Gas Chromatography/Mass Spectrometry
Application Code: Fuels, Energy and Petrochemical
Methodology Code: Gas Chromatography

Session # **1970** **Abstract #** **1970-2** **Oral Sessions**

Session Title GC Fuels, Energy and Petrochemical

Abstract Title **Olefins in Refinery Streams by GC-VUV**

Date: Wednesday, March 09, 2016 - After
Time: 01:50 PM
Room: B405

Primary Author Dan Wispinski
Alberta Innovates Technology Futures

Co-Author(s) Chris Goss, Phillip Walsh

Abstract Text

The VGA 100 vacuum UV (VUV) detector has many potential applications for the petroleum/refining industry. This presentation will discuss several solutions that this new technology can bring to this industry.

This research will describe a new gas chromatography (GC) -VUV technique which will speciate and quantitate individual olefins, including reactive conjugated diolefins of particular interest in refinery streams. Olefins may pose a risk to refineries by fouling heat exchangers and catalysts. The presence of olefins in finished products negatively impacts quality. Incoming feedstock may contain olefins from partial upgrading of produced bitumen (thermal cracking processes) or from the use of cracked stock as a diluent for bitumen.

The Canadian CRW specification for condensate limits olefins to a maximum of 1.0 mass%. This specification cites a proton NMR method which detects total olefin content as 1-decene equivalent, without differentiation of olefin type. Comparisons to proton NMR, ASTM D1319, detailed hydrocarbon analysis by Canadian CGSB 3.0 No. 14.3 method matrix will also be discussed.

Comparison to UOP 326 Diene Value by Maleic Anhydride Addition Reaction will also be investigated. This industry accepted test method for the detection of conjugated diolefins in petroleum products is a very complicated time consuming titration method.

With the use of a VUV detector, a GC method for olefin and diolefins has advantages over current techniques.

Keywords: Analysis, Fuels\Energy\Petrochemical, Gas Chromatography
Application Code: Fuels, Energy and Petrochemical
Methodology Code: Gas Chromatography

Session # 1970 Abstract # 1970-3 **Oral Sessions**

Session Title GC Fuels, Energy and Petrochemical

Abstract Title **GC-Ion Mobility Spectrometry for Determination of Ageing of Mineral Oil Impregnated Presspaper Isolation Systems**

Date: Wednesday, March 09, 2016 - After
Time: 02:10 PM
Room: B405

Primary Author Wolfgang Vautz
ISAS

Co-Author(s) Frank Jenau, Liedtke Sascha, Torben Muth

Abstract Text

The ageing of transformers is mainly determined by the status of the mineral oil impregnated presspaper isolation system which is strongly depending on the level of strain. This makes a valid determination of measures characteristic for the ageing difficult and elaborate sampling is an additional handicap. Therefore, an appropriate determination of the correlation of ageing with temperature and electric field is indispensable for a valid estimation of the residual term.

An experiment was designed for accelerated aging of isolation systems under different temperature and electric field conditions. The samples from this experiment were investigated by different established methods and by ion mobility spectrometry coupled to rapid gas-chromatography (GC-IMS) as innovative method in addition. Using GC-IMS, the headspace of few mL of the samples was analysed for their volatile compounds – in general a non-invasive method.

Several signals detected with GC-IMS showed a correlation with the ageing process, in particular Pentanal und 2-Pentanone. Pentanal furthermore shows an additional influence of the electric field. However, it has to be kept in mind, that both analytes are fragments of the oil decomposition but no related to the ageing of the presspaper itseld. Concluding, the innovative method using GC-IMS for headspace analysis of oil samples shows high potential for characterisation of the ageing process of isolation systems. However, further work has to be done, in particular the identification of all compounds correlated to ageing and the description of their origin and relevance.

Keywords: Energy, Gas Chromatography, Quality Control, Spectrometer
Application Code: Fuels, Energy and Petrochemical
Methodology Code: Chemical Methods

Session # 1970 Abstract # 1970-5 **Oral Sessions**

Session Title GC Fuels, Energy and Petrochemical

Abstract Title **Analysis of Dissolved Hydrocarbon Gases in Water – Pitfalls and Improvements**

Date: Wednesday, March 09, 2016 - After
Time: 03:05 PM
Room: B405

Primary Author Mark L. Bruce
TestAmerica

Co-Author(s)

Abstract Text
Analysis of methane and other light hydrocarbons is frequently done as part of predrill background studies to support hydraulic fracturing activities in the various shale plays in the United States. Gaseous hydrocarbon analysis in water also is used to monitor degradation of environmental contaminants. There is no fully validated USEPA method for this type of analysis. Commercial environmental laboratories currently reference an SOP (RSK-175) from an EPA lab. There is significant variability in how this SOP is interpreted and implemented at various laboratories. There are no commercially available proficiency test samples. Hence the method performance is not documented as well as other common environmental parameters. Some internal studies indicate there are opportunities to improve accuracy and reproducibility and provide data more suitable for the oil & gas industry.

Loss of these very volatile analytes during sample preparation by too much exposure to air prior to headspace-gas chromatographic analysis is a primary concern. In addition analytical process differences between handling of water samples and gas phase based standards can introduce bias. Several different options for producing water based standards have been investigated. Development and use of water based calibration standards and proficiency testing samples will facilitate other method improvements and documentation of method performance. Also, several different sample preparation options have been evaluated for both accuracy and precision.

Keywords: Environmental/Water, Gas Chromatography, Headspace, Hydrocarbons
Application Code: Environmental
Methodology Code: Gas Chromatography

Session # **1970** **Abstract #** **1970-6** **Oral Sessions**

Session Title GC Fuels, Energy and Petrochemical

Abstract Title **Performance Evaluation of Modern Stainless Steel Capillary GC Columns**

Date: Wednesday, March 09, 2016 - After
Time: 03:25 PM
Room: B405

Primary Author Rebecca Stevens
Restek Corporation

Co-Author(s) Amanda Rigdon, Corby Hilliard, Jaap de zeeuw, Linx Waclaski

Abstract Text

Steel capillary columns for gas chromatography (GC) have several specialized characteristics compared with polyimide coated fused silica capillaries. These include high mechanical strength, temperature stability, and smaller minimum coiling diameter. Because of these qualities steel capillaries are often preferred to fused silica in high temperature and process GC applications, where ruggedness is essential.

One drawback is that historically steel capillary columns have suffered from lower efficiency and more activity towards polar analytes than their fused silica counterparts. The steel surface is inherently more difficult to coat with liquid stationary phase and requires special deactivation. With modern deactivation and coating techniques, however, steel capillary columns can be produced that rival the chromatographic performance of fused silica.

Our lab critically evaluated the performance of a large cross section of commercially available steel capillary columns including 4 different stationary phases in both 0.25mm ID and 0.53mm ID. Data was collected in triplicate under isothermal conditions for an expanded Grob type test mix. Most steel columns showed similar inertness and efficiency to fused silica columns of the same phase ratio. We propose that some long held perceptions of steel capillary columns for GC may not be entirely correct. Our studies have indicated that the tradeoff of ruggedness for chromatographic performance presented by steel capillaries is not as drastic as once thought.

Keywords: Capillary GC, GC Columns, Petrochemical, Process Monitoring
Application Code: Process Analytical Chemistry
Methodology Code: Gas Chromatography

Session # 1970 Abstract # 1970-7 **Oral Sessions**

Session Title GC Fuels, Energy and Petrochemical

Abstract Title **Determination of Polycyclic Aromatic Sulfur Heterocycles and Their Alkyl-Substituted Derivatives in Standard Reference Material 1597a**

Date: Wednesday, March 09, 2016 - After
Time: 03:45 PM
Room: B405

Primary Author Walter B. Wilson
National Institute of Standard and Technology

Co-Author(s) Stephen A. Wise

Abstract Text

Standard reference material (SRM) 1597a is a combustion-related mixture of polycyclic aromatic hydrocarbons (PAHs) that has been previously isolated from a coal tar sample and dissolved in toluene. This SRM is widely used for the evaluation and validation of analytical methods for the determination of PAHs. In addition to PAHs, incomplete combustion of organic matter leads to numerous heterocyclic compounds containing at least one heteroatom such as polycyclic aromatic sulfur heterocycles (PASHs). The total number of possible isomeric structures for PASH is greatly increased compared with the corresponding PAH because both ring arrangement and position of the heteroatom substitution within the rings give rise to unique isomers. In this study, an analytical method was developed for the separation and identification of PASHs in SRM 1597a. The compounds measured include the following groups of PASHs: (1) non-substituted PASHs with molecular mass (MM) 184, 234, 258, 284 and 334 Da; (2) methyl-substituted PASHs with MM 198 Da; (3) ethyl- and dimethyl-substituted PASHs with MM 212 Da; (4) trimethyl-substituted PASHs with MM 226 Da. Because of the occurrence of PAHs and PASHs together with their large number of possible alkyl-substituted isomers, the analytical method described requires multiple cleanup steps. The PASHs are isolated from the PAHs and then the PASHs are isolated based on the number of aromatic carbons. These aromatic ring fractions are analyzed via liquid chromatography/ultraviolet-visible (LC-UV) and gas chromatography/mass spectrometry (GC/MS).

Keywords: Chromatography, Method Development, PAH, Petroleum
Application Code: Environmental
Methodology Code: Separation Sciences

Session # 1970 **Abstract # 1970-8** **Oral Sessions**

Session Title: GC Fuels, Energy and Petrochemical

Abstract Title: **Vehicle Interior Air Quality - (S)VOC Emission from Materials: Regulation, Standard Methods and Analytical Implementation**

Primary Author: Caroline Widdowson
Markes International

Date: Wednesday, March 09, 2016 - After
Time: 04:05 PM
Room: B405

Co-Author(s)

Abstract Text

Emissions from vehicle trim components can adversely affect vehicle interior air quality (VIAQ). There has a development of harmonized methods to quantitate the release of chemicals released from the materials used, and assess the overall quality of in-vehicle air. Thermal desorption and related sampling methods are stipulated in these methods, to aid understanding of both the volatile content and emission profiles of car trim, as well as for analysis of in-vehicle air. This paper will give an overview of the national regulations, standard methods, correlation and differences between differing national strategies, followed by a focused study on Micro-scale Chamber - (S)VOCs Extraction Methodology.

Keywords: Sample Introduction, Semi-Volatiles, Thermal Desorption, Volatile Organic Compounds
Application Code: Regulatory
Methodology Code: Gas Chromatography/Mass Spectrometry

Session # 1980 Abstract # 1980-1 — Oral Sessions

Session Title: Mass Spectrometry - Bioanalytical and Omics

Abstract Title: **Mass Spectrometry in Discovery of Lipid Markers for Alzheimer's Disease**

Primary Author: Satya Girish Chandra Avula
Cleveland State University

Co-Author(s): Christine Reece, Jagan A. Pillai, Yan Xu

Date: Wednesday, March 09, 2016 - After
Time: 01:30 PM
Room: B406

Abstract Text

Alzheimer's disease (AD) is the most common form of progressive dementia that currently affects 35 million individuals worldwide and is expected to triple in the next three decades. By the current diagnostic tools, AD can't be diagnosed until the disease progresses into the final stage of dementia. Thus, there is an urgent need in developing diagnostic tools for early detection of AD. Lipids are known to be critical in brain function and have been implicated in number of neurological disorders. Growing evidences suggest that classes of lipids such as the phospholipids play a major role in disease progression. Hence, we developed an effective shotgun lipidomics method to profile the various classes of lipids that are up- and down-regulated in the plasma of normal & AD patient samples, as a mean to discover biomarkers for prognosis of the disease.

In this work, plasma samples were first extracted by modified Bligh-Dyer method, then fractionated into various lipid classes by amino-propyl solid-phase-extraction cartridges. Each class of lipids was subjected to an array of multiple precursor-ion and neutral-loss scans through direct sample infusion. Lipid identifications accomplished using LipidView™ software which showed that more than 300 glycerolipids and phospholipids were present in the plasma samples. Semi-quantitative analyses of these identified glycerolipids and phospholipids were achieved using internal standards for each sub-classes. Statistical analysis of plasma lipids (e.g., principal component analysis and t-test) between the two comparison groups were done using MarkerView™ software, which revealed that more than 60 phospholipid species had significant differences between the normal controls and AD patients (p□k□0.005) with concentration changes □fivefolds. Thus, mass spectrometry may be an effective technical platform for discovery of biomarkers for early detection of AD.

Keywords: Bioanalytical, Lipids, Mass Spectrometry, Plasma

Application Code: Genomics, Proteomics and Other 'Omics
Methodology Code: Mass Spectrometry

Session # 1980 Abstract # 1980-2 **Oral Sessions**

Session Title: Mass Spectrometry - Bioanalytical and Omics

Abstract Title: **Novel Cationization Strategies for Improved Separation of Metabolite Isomers with Ion Mobility – Mass Spectrometry**

Date: Wednesday, March 09, 2016 - After
Time: 01:50 PM
Room: B406

Primary Author: Christopher D. Chouinard
University of Florida

Co-Author(s): Christopher R. Beekman, Harrison King, Richard A. Yost, Robin Kemperman

Abstract Text

Ion mobility spectrometry (IMS) uses a uniform electric field drift tube to separate gas-phase ions based on differences in size, shape, and charge. This offers the potential for separation of isomers that cannot be resolved with mass spectrometry alone. To improve this separation potential, complexation with metal ions can produce significant differences in gas-phase structure between isomers. This study has investigated the effects of alkali, alkaline earth, and first row transition metal ions for improvements in separation of biologically relevant isomers such as steroids, bile acids, and vitamin D metabolites.

Preliminary experiments have focused on a steroid epimer pair, androsterone (Andro) and trans-androsterone (t-Andro), as a model for evaluating the effects with alkali and alkaline earth metals. Results were obtained with an Agilent 6560 IM-QTOF instrument (Santa Clara, CA), which allows measurement of collision cross section (CCS) for each species identified. Monomer ions for these steroids displayed minor differences in their CCS, but more significant differences were seen in the dimers. For example, alkali metals lithium and sodium improved separation, yielding the following CCS for Andro and t-Andro, respectively: [2M+Li]+, m/z 587.465 – 243.0 Å2 vs. 257.4 Å2; [2M+Na]+, m/z 603.439 – 242.6 Å2 vs. 256.3 Å2. Similarly, dimers formed with alkaline earth metals magnesium and calcium (with acetate ion) yielded the following CCS for Andro and t-Andro, respectively: [2M+Mg+Ac]+, m/z 663.448 – 257.2 vs. 265.1 Å2; [2M+Ca+Ac]+, m/z 679.425 – 260.9 vs. 266.4 Å2. Studies are underway with first row transition metals, plus palladium, silver, cadmium, and gold to investigate the species formed (especially for those with +3 charge state) and their effects on CCS. Additionally, density functional theory calculations have been performed to model the lowest energy gas-phase structures of the dimer metal complexes for comparison with experimentally obtained CCS.

Keywords: Chiral Separations, Mass Spectrometry, Metabolomics, Metabonomics, Time of Flight MS
Application Code: Genomics, Proteomics and Other 'Omics
Methodology Code: Mass Spectrometry

Session # 1980 Abstract # 1980-3 **Oral Sessions**

Session Title: Mass Spectrometry - Bioanalytical and Omics

Abstract Title: **Development of Dried Matrix Card Quantification by Speciated Isotope Dilution Mass Spectrometry Using Elution and Laser Desorption Techniques**

Primary Author: Logan T. Miller
Duquesne University

Date: Wednesday, March 09, 2016 - After
Time: 02:10 PM
Room: B406

Co-Author(s): Kaitlin Miller, Mark Little, Matt Pamuku, Sarah Sheffield, Scott Faber, Silverio Iacono, Skip Kingston

Abstract Text

Quantitative and effective techniques for blood sampling and analysis are required to meet the growing needs of the medical and bioanalytical communities. Traditional blood draws are inherently used because of the amount of blood typically desired to be withdrawn from the patient. Dried Matrix Spots (DMS) greatly reduce the volume of collected blood, thus enabling more frequent and easier procurement with a finger stick instead of a venous blood draw. Additionally, DMS cards can be legally transferred across international borders. Development of the ability to quantitate DMS analytes using mass spectrometry is of importance to bioanalytical research and medicine. Traditional DMS workflows require tedious sample preparation and is therefore more time consuming than the methodology used in this study. This study presents a unique approach to characterizing DMS by using a combination of on card "clamp-and-elute" technology and direct analysis using laser-enhanced ionization. This technique is applied to analytes with the benefit that having a multi-pronged analytical approach allows new diagnostic and patient monitoring for many classes of medical intervention. Initially, this study focuses on the ratio and quantification of reduced and oxidized glutathione using Speciated Isotope Dilution Mass Spectrometry (SIDMS). This research was performed on an experimentally designed automated DMS elution instrument system combining a tunable laser system mounted to a quadrupole time-of-flight mass spectrometer and ICP mass spectrometer. By coupling the DMS cards to both liquid-elution and solid-state desorption instruments with SIDMS quantification, high quality measurements can be performed.

Keywords: Laser Desorption, Liquid Chromatography/Mass Spectroscopy, Mass Spectrometry, Method Develop
Application Code: Bioanalytical
Methodology Code: Mass Spectrometry

Session # 1980 **Abstract # 1980-4** **Oral Sessions**

Session Title: Mass Spectrometry - Bioanalytical and Omics

Abstract Title: **Protease-Containing Membranes for Rapid Antibody Digestion Prior to Mass Spectrometry Analysis**

Date: Wednesday, March 09, 2016 - After
Time: 02:30 PM
Room: B406

Primary Author: Yongle Pang
Michigan State University

Co-Author(s): Donald F. Hunt, Gavin Reid, Merlin Bruening, Wei-Han Wang

Abstract Text

Monoclonal antibodies (mAbs) have emerged as important biotherapeutic drugs with high specificities. Facile characterization of mAb post-translational modifications is essential for quality control, and mass spectrometry (MS) is the most powerful tool for antibody characterization. This research uses pepsin-modified membranes as proteolysis reactors that rapidly (<1 min) digest mAbs prior to MS analysis. Peptic digestion is particularly convenient because the acidic conditions enable antibody proteolysis without urea denaturation and alkylation. After a 15-min antibody reduction with TCEP under acidic conditions, passage of the antibody-containing solution through a pepsin-containing membrane yields proteolytic peptides whose length depends on the flow rate through the membrane. Without further purification, the membrane effluent is suitable for direct infusion into an Orbitrap Velos mass spectrometer.

Variation of the residence time (3-sec to 3-msec) of reduced-antibody solutions in the membrane yields bottom-up (1-2 kDa) to middle-down sized peptides (5-15 kDa) for both the light and heavy chains, and these peptides cover essentially the entire antibody sequence. A peptide with light-chain amino acids 1-140 covers the three complementarity-determining regions to facilitate their characterization. MS and MS/MS analysis of the proteolytic peptides reveals oxidation, deamidation, pyroglutamate formation, and glycosylation on the light and heavy chains. MS/MS with various dissociation methods gives 99% sequence coverage of the light chain compared to 98% coverage for in-solution digestion and 55% coverage for "top-down" analysis. With minimal preparation time, membrane digestion leads to high peptide and sequence coverages for identification of PTMs by MS.

Keywords: Analysis, Mass Spectrometry, Membrane, Protein
Application Code: Bioanalytical
Methodology Code: Mass Spectrometry

Session # 1980   Abstract # 1980-5   **Oral Sessions**

Session Title   Mass Spectrometry - Bioanalytical and Omics

Abstract Title   **Quantitative Environmental Human Health Assessment of Inorganic Elements in Dried Blood Spots Using Direct Isotope Dilution Laser Ablation and Elution Mass Spectrometry**

Primary Author   Sarah Sheffield
Duquesne University

Date: Wednesday, March 09, 2016 - After
Time: 03:05 PM
Room: B406

Co-Author(s)   Logan T. Miller, Matt Pamuku, Scott Faber, Skip Kingston

Abstract Text

There is a need for more accessible blood analysis in the United States and internationally as testing demands are increasing. By utilizing finger sticks and dried blood spot (DBS) cards, a smaller volume of blood is required and can be taken and shipped by the patient for analysis almost anywhere on the globe. This allows testing to be performed on a smaller volume while retaining or improving the analytical measurement quality. The application of DBS analysis can be completed by the patient at home by either a finger or heel stick. Previously, metals analysis underwent extensive sample preparation including microwave digestion for inorganic analytes in blood using inductively coupled plasma–mass spectrometer (ICP-MS). However, by using a laser, the blood can be ablated and directly analyzed, saving sample preparation time. Additionally, on card elution using "clamp-and-elute" technology can be used to extract the analyte of interest directly from the card. Quantification in laser ablation (LA)–ICP-MS and liquid elution can be achieved by utilizing isotope dilution mass spectrometry (IDMS), which eliminates the use of calibration curves and provides a direct mathematical calculation for the analyte's concentration. This method has been standardized and is covered in EPA method 6800 update V, 2015. The quantification of analytes enables enhanced patient monitoring for doctors to better follow their patients and improves treatment decisions. Examples demonstrating method 6800's quantitative ability will be discussed.

Keywords: Elemental Mass Spec, Environmental/Biological Samples, Laser Desorption, Method Development
Application Code: Bioanalytical
Methodology Code: Mass Spectrometry

Session # 1980 Abstract # 1980-6 **Oral Sessions**

Session Title: Mass Spectrometry - Bioanalytical and Omics

Abstract Title: **Identification of the Cell Surface N-Glycoproteome by MS-based Proteomics**

Date: Wednesday, March 09, 2016 - After
Time: 03:25 PM
Room: B406

Primary Author: Johanna Smeekens
Georgia Institute of Technology

Co-Author(s): Ronghu Wu, Weixuan Chen

Abstract Text

Cell surface N-glycoproteins are essential in cell-cell communication, cell-matrix interactions, and cellular response to environmental cues. Many N-glycoproteins are present at low abundances within the cell, resulting in challenges with separation and comprehensive analysis. Here, we have developed a novel method integrating metabolic labeling, copper-free click chemistry, and mass spectrometry (MS)-based proteomics to analyze cell surface N-glycoproteins comprehensively and site-specifically. First, glycoproteins were labeled with an azidosugar analog, N-azidoacetylgalactosamine (GalNAz). Glycoproteins with the azido group on the cell surface were then bound to dibenzocyclooctyne (DBCO)-sulfo-biotin via copper-free click chemistry under physiological conditions. Proteins were extracted and digested, and glycopeptides containing the biotin tag were enriched by NeutrAvidin conjugated beads. Glycans were removed from enriched peptides with peptide-N-glycosidase F (PNGase F) in heavy-oxygen water; in the process, asparagine was converted to aspartic acid and tagged with [sup]18[/sup]O for MS analysis. This method was applied to HEK 293T cells and 144 unique N-glycopeptides containing 152 N-glycosylation sites were identified in 110 proteins. Membrane proteins comprised 95% of identified glycoproteins, and many sites were located on key receptors, transporters, and cluster of differentiation proteins. The current method proved to be very effective for the comprehensive and site-specific identification of the cell surface N-glycoproteome and can be extensively applied to other cell surface protein studies.

Keywords: Carbohydrates, Liquid Chromatography/Mass Spectroscopy, Proteomics
Application Code: Genomics, Proteomics and Other 'Omics
Methodology Code: Mass Spectrometry

Session # 1980	Abstract # 1980-7	Oral Sessions

Session Title Mass Spectrometry - Bioanalytical and Omics

Abstract Title **Probing the Cell-Surface N-Glycoproteome with Metabolic Chemical Reporters (MCRs)**

Primary Author Haopeng Xiao
Georgia Tech

Date: Wednesday, March 09, 2016 - After
Time: 03:45 PM
Room: B406

Co-Author(s)

Abstract Text

The frequent modification of cell-surface proteins by N-linked glycans is known to be correlated with many biological processes. Aberrant glycosylation on surface proteins is associated with different cellular statuses and disease progression. However, it is extraordinarily challenging to globally and site-specifically identify glycoproteins located only on the cell surface. Currently mass spectrometry (MS)-based proteomics provides us the possibility for the global analysis of the N-glycoproteome, though the effective separation and enrichment of surface glycoproteins have yet to be developed. Since metabolic labeling-based methods are commonly used in glycoprotein studies, we have devised strategies to comprehensively evaluate the performance of different metabolic chemical reporters (MCRs) in the identification and quantification of cell-surface N-glycoproteins by combining copper-free click chemistry and MS-based proteomics. Three MCRs, namely, GalNAz, GlcNAz and ManNAz, were utilized and compared. Labeling using GalNAz allowed us to obtain the best results, and duplicate experiments resulted in the identification of 591 unique sites from 274 surface N-glycoproteins. In the quantification experiment for statin-treated HepG2 liver cells, 311 unique singly N-glycosylated sites were profiled among 945 peptides. Many glycopeptides were down-regulated in treated cells compared to untreated cells because statin prevents the synthesis of dolichol, which is essential for the formation of the dolichol-linked precursor oligosaccharide. The quantification results demonstrated that many glycosylation sites on surface proteins were down-regulated in statin-treated HepG2 cells. Several glycosylation sites in proteins belonging to the pathway of Alzheimer's disease were down-regulated, and 35 sites on CD molecules were also down-regulated. This method can be extensively applied for the global analyses of the cell-surface N-glycoproteome in multiple biological events.

Keywords: Biopharmaceutical, High Throughput Chemical Analysis, Mass Spectrometry, Proteomics
Application Code: Genomics, Proteomics and Other 'Omics
Methodology Code: Mass Spectrometry

Session # 1980 Abstract # 1980-8 **Oral Sessions**

Session Title Mass Spectrometry - Bioanalytical and Omics

Abstract Title **Quantitative Characterization of Protein Content from HDL and LDL Size Fractions**

Primary Author Bryan A. Parks
Centers for Disease Control and Prevention

Date: Wednesday, March 09, 2016 - After
Time: 04:05 PM
Room: B406

Co-Author(s) David Schieltz, John R. Barr, Jon Rees, Lisa McWilliams, Michael S. Gardner, Yulanda Williamson, Zsuzsanna Kuklenyik

Abstract Text

Cardiovascular disease (CVD) is one of the leading cause of mortality in the world. In recent years, the efficacy of the traditional approaches to CVD risk assessment based on the use of high density and low density lipoprotein cholesterol measurements (HDL-C and LDL-C) is highly questioned based on epidemiological studies which suggest the use of stronger risk indicators, such as apolipoproteins A-1 and B-100 (ApoA-I and ApoB-100). Isotope dilution mass spectrometry (IDMS) is potentially the most direct and selective alternative to measure ApoA-1 and ApoB-100 concentrations in total sera.

In the first part of this presentation, we discuss the importance of choosing appropriate peptides (i.e. limit peptides) for quantification to avoid protein reduction and alkylation. The effect of key parameters (pH, digestion temperature, detergent type and concentration) on digestion efficacy and method ruggedness will be discussed along with quantitation accuracy. The IDMS method has been validated on ApoA-1 and ApoB-100 serum primary reference materials (accuracy>95% and CVs □10%).

In the second part of this presentation, we show application of the IDMS method to the analysis of HDL and LDL sub-fractions separated by hydrodynamic size using asymmetric flow field-flow fractionation (AF4). The IDMS analysis of the AF4 size fractions showed that ApoA-II, ApoA-IV and ApoM bind to ApoA-I containing HDL particles. ApoC-II, ApoC-III and ApoE bind to both ApoA-I and ApoB-100 containing lipoproteins. The apolipoprotein profiles varied significantly between samples with low and high total cholesterol/triglyceride serum levels (low vs. high CVD risk samples).

Keywords: Lipids, Liquid Chromatography/Mass Spectroscopy, Peptides, Proteomics
Application Code: Genomics, Proteomics and Other 'Omics
Methodology Code: Liquid Chromatography/Mass Spectrometry

Session # 1990 Abstract # 1990-1 **Oral Sessions**

Session Title: Metabolomics, Proteomics, and Genomics

Abstract Title: **A Combination of Multidimensional Chromatography and High Resolution Mass Spectrometry for Chemical Exposure Analysis**

Primary Author: David E. Alonso
Leco Corporation

Date: Wednesday, March 09, 2016 - After
Time: 01:30 PM
Room: B407

Co-Author(s): Elizabeth M. Humston-Fulmer, Jonathan D. Byer, Joseph E. Binkley, Lorne E. Fell

Abstract Text

Environmental exposure plays an important role in the development of chronic diseases such as atherosclerosis, cancer and diabetes. Unfortunately, the relationship between exposure and disease is not fully understood. The arrival of untargeted profiling methods are critical for effective investigation of endogenous and exogenous environmental exposure. Metabolomics is particularly useful due to its proximity to system phenotype and its relatively quick insight into system perturbations from drugs, food, smoke and persistent organic pollutants. Challenges associated with metabolomics include the enormous number of chemically diverse metabolites present in a wide range of concentrations. The main bottleneck in metabolomics is identification and structural characterization of compounds. The main objective in this study was to develop and apply a workflow for rapid and confident identification of compounds in urine. While no single instrument is capable of fully profiling all compounds in these samples, we have developed a workflow that utilizes high resolution time-of-flight (HRT) instrumentation for rapid and confident identification of metabolites. Excellent mass accuracy values (MA < 1.0 ppm) allowed for confident elemental composition determinations for molecular, fragment and adduct ions. Furthermore, enhanced chromatographic resolution improves spectral data resulting in excellent matches to well-established large databases. Our methodology included untargeted data acquisition using different ionization methods to obtain comprehensive profiles of derivatized biological samples. In addition, software tools were used to quickly reinterrogate rich data sets in a targeted manner after compounds of interest (e.g., acids, Table 1) were identified using comprehensive Peak Find processing.

Keywords: Mass Spectrometry, Metabolomics, Metabonomics, Time of Flight MS
Application Code: Genomics, Proteomics and Other 'Omics
Methodology Code: Gas Chromatography/Mass Spectrometry

Session # 1990 Abstract # 1990-2 **Oral Sessions**

Session Title: Metabolomics, Proteomics, and Genomics

Abstract Title: **Metabolomics of Nonalcoholic Fatty Liver Disease via LC-MS**

Date: Wednesday, March 09, 2016 - After
Time: 01:50 PM
Room: B407

Primary Author: Rainey E. Patterson
University of Florida

Co-Author(s): Kenneth Cusi, Nishanth E. Sunny, Richard A. Yost, Srilaxmi Kalavalapalli, Timothy J. Garrett

Abstract Text

Approximately one-third of the US population is affected by a nonalcoholic fatty liver (NAFL), with half advancing to nonalcoholic steatohepatitis (NASH). Investigation of mouse models and humans via LC-MS reveals metabolic differences indicative of disease progression. Metabolomics analysis may allow biomarker identification in human blood, enabling faster detection and more frequent monitoring. Mice (C57/BL6) were fed either a control diet or a high fructose, trans-fat diet (TFD) for 8 weeks or 24 weeks to induce NAFL or NASH, respectively. Both human and mouse samples had disease status verified with liver histology. Investigation of diacylglycerols (DAG) and ceramides (Cer), specifically, shows contribution to disease progression, especially among DAGs with 18:1, 18:0, and 20:4 fatty acid chains. Not only are the abundance of these compounds significantly different from age-matched controls ($p<0.05$), but 14 DAGs targeted significantly changed from NAFL to NASH. Statistical separation of DAG (TFD vs control) was represented by 91% of the first principal component. Ceramide levels in general increased from NAFL to NASH, and principal components analysis indicated that the first component explained 84% of the variation comparing TFD mice and controls. Hexosylceramides, however, decrease in NAFL mice and must be investigated further. Triacylglycerols (TAG), although analytically more difficult to identify, are increased by at least an order of magnitude in the TFD mice. Noticeably, TAGs with a higher number of unsaturations (>5) are more concentrated in the NAFL mice compared to the NASH counterparts. Acylcarnitines with long carbon chains (C□6) have been shown to increase in NASH but not in NAFL, indicating impaired fat oxidation in NASH. Comparisons with the aqueous metabolites and lipids of human blood plasma indicate the role of TCA cycle induction, believed to impair fat oxidation, which leads to the large increase in lipids, creating lipotoxicity.

Keywords: Bioanalytical, Liquid Chromatography/Mass Spectroscopy, Mass Spectrometry, Metabolomics
Application Code: Genomics, Proteomics and Other 'Omics
Methodology Code: Liquid Chromatography/Mass Spectrometry

Session # 1990 Abstract # 1990-3 **Oral Sessions**

Session Title Metabolomics, Proteomics, and Genomics

Abstract Title **Application of PDMS-Overcoated Matrix-Compatible Solid Phase Microextraction Fiber Coupled to Comprehensive Two-Dimensional Gas Chromatography–Time-of-Flight Mass Spectrometry for Improved Exploitation of Chromatographic Space in Global Chemical Profiling of Brazilian Cachaça**

Date: Wednesday, March 09, 2016 - After
Time: 02:10 PM
Room: B407

Primary Author Erica A. Souza-Silva
Universidade Federal do Rio Grande do Sul

Co-Author(s) Claudia A. Zini, Fernando C. Fontanive

Abstract Text

Although advanced technologies such as GCxGC can expand the space for separation and/or selectivity, it is also true that such approaches are often underutilized. Indeed, the thorough utilization of the chromatographic space requires not only careful evaluation of the chromatographic conditions, but also sample preparation/extraction. In response to the ever-increasing interest in the development of comprehensive methods competent with obtaining a global chemical profiling of food and beverages, the purpose of the current investigation is to test the feasibility of solid phase microextraction employing the recently introduced matrix compatible fiber for profiling of volatile and semivolatile compounds in Brazilian cachaça. Firstly, the GCxGC-ToFMS method was evaluated as per the utilization of the chromatographic space utilizing a model Brazilian cachaça sampled by solid phase microextraction (SPME) utilizing both a commercial DVB/Car/PDMS fiber, and a PDMS-overcoated DVB/Car/PDMS fiber (also known as matrix compatible fiber). Moreover, eighteen samples of artisanal cachaças aged in different types of wood barrels were sampled and the feasibility of each SPME mode to identify and classify such samples was assessed. In the case of the commercial fiber, HS-SPME was utilized giving the limitation of solid coatings to be exposed directly to matrix of certain complexity. Conversely, in the case of the matrix-compatible fiber, DI-SPME was employed leading to the much broader coverage of metabolites. In fact, the employment of matrix-compatible fiber in DI mode led to a superior exploitation of the chromatographic space, therefore providing a more broad screening of samples' constituents.

Keywords: GC-MS, Metabolomics, Metabonomics, SPME
Application Code: Genomics, Proteomics and Other 'Omics
Methodology Code: Sampling and Sample Preparation

Session # 1990 Abstract # 1990-4 **Oral Sessions**

Session Title: Metabolomics, Proteomics, and Genomics

Abstract Title: **Application of the Isotopic Ratio Outlier Analysis Phenotypic Protocol for Metabolomic Biomarker Discovery in Type 1 Diabetes Using T Cells**

Date: Wednesday, March 09, 2016 - After
Time: 02:30 PM
Room: B407

Primary Author: Candice Z. Ulmer
University of Florida

Co-Author(s): Christopher A. Beecher, Clayton Matthews, Jing Chen, Richard A. Yost, Timothy J. Garrett

Abstract Text

Type 1 Diabetes (T1D) is an incurable, auto-immune disease resulting from the destruction of pancreatic beta cells by pathogenic T lymphocytes. Of the few experimental designs targeting metabolic and lipidomic dysregulation in T1D, many incorporate animal models that fail to account for pathophysiological differences in humans. There is a need to better understand the metabolic and lipidomic signatures of this disease using human samples. This work employs the phenotypic IROA protocol, an isotopic labeling LC-HRMS method, to identify the metabolic and lipidomic trends of immune dysregulation using primary T cells obtained from T1D patients.

Jurkat T-human leukemia cells were grown in 95% 13C-IROA media for 2-3 passages. Primary T cells were collected from Type 1 diabetics, their 1st-degree relatives, and healthy control patients. Equal aliquots of 95% 13C-labeled Jurkat cells (IROA standards) were spiked into the patient samples using the total protein content for normalization. The metabolites were isolated using ice cold 80% MeOH and the lipids were extracted from the same cell pellet using the Folch method. Samples were analyzed using LC-HRMS.

The ratio of the 12C/13C peak intensity was calculated for each sample within all three groups using the ClusterFinder software. Multivariate and univariate analyses were performed on the primary T cell patient data to distinguish m/z peaks with the greatest and most significant variation between the T1D and healthy control/1st-degree relative groups. Many metabolites, including as valine and arginine, show a higher peak intensity ratio in the T1D samples compared to the control.

Keywords: Liquid Chromatography/Mass Spectroscopy, Method Development, Metabolomics, Metabonomics
Application Code: Genomics, Proteomics and Other 'Omics
Methodology Code: Liquid Chromatography/Mass Spectrometry

Session # 1990 Abstract # 1990-5 **Oral Sessions**

Session Title: Metabolomics, Proteomics, and Genomics

Abstract Title: **Secretome of Murine Islets of Langerhans**

Primary Author: Andrew W. Schmudlach
University of Notre Dame

Date: Wednesday, March 09, 2016 - After
Time: 03:05 PM
Room: B407

Co-Author(s): Jeremy Felton, Norman J. Dovichi, Robert T. Kennedy

Abstract Text

Diabetes mellitus, a disease characterized by chronic high blood sugar, is one of the primary health concerns for global human health. In the United States alone an estimated 29.1 million adults have diabetes, many of which are undiagnosed. The islets of Langerhans are miniorgans, which secrete different hormones to properly balance blood sugar via carbohydrate and lipid metabolism. This goal of this project is to characterize the secretome, or the total secreted protein content, of the islets of Langerhans using bottom up proteomics.

Keywords: Bioanalytical, Liquid Chromatography/Mass Spectroscopy, Proteomics, Tandem Mass Spec
Application Code: Genomics, Proteomics and Other 'Omics
Methodology Code: Liquid Chromatography/Mass Spectrometry

Session # 1990 Abstract # 1990-6 **Oral Sessions**

Session Title: Metabolomics, Proteomics, and Genomics

Abstract Title: **Optimizing Sampling Protocols for the Identification and Quantitation of Neuropeptides from Brain Tissues**

Primary Author: Ning Yang
University of Illinois at Urbana-Champaign

Date: Wednesday, March 09, 2016 - After
Time: 03:25 PM
Room: B407

Co-Author(s): Jonathan V. Sweedler, Krishna D. Anapindi, Stanislav S. Rubakhin

Abstract Text

Neuropeptides are signaling molecules synthesized and released by neurons acting as neurotransmitters, neuromodulators, and hormones. They participate in many physiological processes such as pain, memory and homeostasis. In the past decade, study of neuropeptides has been greatly aided by mass spectrometry due to its good sensitivity, multiplexity, and ability of peptide sequencing. However, issues with sampling still limit a number of neuropeptide studies. First, neuropeptides are quickly degraded by enzymes after being released from a cell. Secondly, many neuropeptides are expressed at low levels so that their signals can be overwhelmed by protein degradation products formed ex vivo. Heat treatment of samples is a good approach to deactivate the peptidases and proteases and stabilize chemical contents of the tissues. Here, we investigated the effects of different sample stabilization method involving heat treatment in the identification and quantitation of neuropeptides. Two groups of fresh rat hypothalami and two groups of snap frozen rat hypothalami were utilized in the experiments. One of each type of groups was stabilized with boiling water or an energy input controllable laser. Peptide extracts from each group were subjected to nanoLC-MS/MS and database searches. 78 and 77 neuropeptides were identified from fresh and snap frozen tissues stabilized with laser. In contrast, 16 and 41 neuropeptides were detected in the samples heated with boiling water. The amounts of three known neuropeptides in the rat habenula stabilized with boiling water and laser were determined by LC-QqQ mass spectrometery. Signal intensities of the three neuropeptides in the laser stabilized samples were higher than those in the boiling water stabilized samples. Lesser chemical complexity of the laser treated sample extracts; therefore, lesser signal suppression and interference in analyte detection can be one of possible causes of the observed effects.
Acknowledgement: NIH2P30DA018310B

Keywords: Mass Spectrometry, Peptides
Application Code: Genomics, Proteomics and Other 'Omics
Methodology Code: Liquid Chromatography/Mass Spectrometry

Session # 2000 Abstract # 2000-1 **Oral Sessions**

Session Title: Raman, SERS, UVRR Applications

Abstract Title: **UV Resonance Raman (UVRR) Structural Studies of Polyglutamine (polyQ) Side Chains and Fibrils**

Primary Author: David Punihaole
University of Pittsburgh

Date: Wednesday, March 09, 2016 - After
Time: 01:30 PM
Room: B408

Co-Author(s): Elizabeth Dahlburg, Jeffry Madura, Riley Workman, Ryan Jakubek, Sanford A. Asher, Zhenmin Hong

Abstract Text

There is currently little that is known about the structure of polyglutamine (polyQ) fibrils, which are involved in at least ten neurodegenerative diseases. New spectroscopic markers and biophysical methods need to be developed in order to obtain high-resolution structures of polyQ and other amyloid-like fibrils. Here, we discuss our recent, deep understanding of the UVRR spectra of polyQ peptides and fibrils that enables us to gain molecular-level structural information. For example, our recent studies have determined how the hydrogen bonding and local dielectric environment of glutamine (Gln) and asparagine (Asn) side chains impact the Raman cross sections and frequencies of the AmI[sup]P[/sup] and AmII[sup]P[/sup] vibrations. In addition, we also discovered a novel spectroscopic marker, the AmIII[sup]P[/sup] vibration, which sensitively reports on the OCCC dihedral angle of Gln and Asn side chains. We show how these UVRR bands can be used to report on the local hydrogen bonding, dielectric environment, and side chains of polyQ peptides in solution-state and in fibril aggregates. We also show how UVRR can be used to examine the peptide backbone structure and hydrogen bonding in solution-state and in fibrils. Finally, we detail how the structural information that we obtain from UVRR can be synergistically utilized to guide Molecular Dynamics (MD) simulations in order to obtain a structural model for polyQ peptides in solution-state and in fibrils.

Keywords: Bioanalytical, Biospectroscopy, Molecular Spectroscopy, Raman
Application Code: Bioanalytical
Methodology Code: Vibrational Spectroscopy

Session # 2000 Abstract # 2000-2 **Oral Sessions**

Session Title Raman, SERS, UVRR Applications

Abstract Title **Sheath-Flow Microfluidic Approach for Combined Surface Enhanced Raman Scattering and Electrochemical Detection**

Primary Author Matthew R. Bailey
University of Notre Dame

Date: Wednesday, March 09, 2016 - After
Time: 01:50 PM
Room: B408

Co-Author(s) Amber Pentecost, Asmira Selimovic, R Scott Martin, Zachary D. Schultz

Abstract Text

The combination of hydrodynamic focusing with embedded capillaries in a microfluidic device is shown to enable both surface enhanced Raman scattering (SERS) and electrochemical characterization of analytes at nanomolar concentrations in flow. Linking multiple detection methods allows for the unique advantage of obtaining complimentary information about analytes. The approach utilizes a versatile polystyrene chip that contains an encapsulated microelectrode and fluidic tubing with a polydimethylsiloxane (PDMS) microchannel positioned over both to generate a sheath-flow that confines and increases the interaction between the analyte and the surface to improve detection. The microfluidic device was characterized using finite element simulations, amperometry, and Raman experiments. An examination of riboflavin (vitamin B12) and catechol demonstrated a SERS and amperometric detection limit near 1 and 25 nM, respectively. Combining SERS and amperometry onto a single platform provides an improved method to identify and quantify electroactive analytes over either technique independently and suggests a straightforward route to improving trace detection both spectroscopically and electrochemically. The device is currently being used to investigate a number of different species, including a series of neurotransmitters.

Keywords: Electrochemistry, Lab-on-a-Chip/Microfluidics, Spectroelectrochemistry, Surface Enhanced Raman
Application Code: Bioanalytical
Methodology Code: Vibrational Spectroscopy

Session # 2000	Abstract # 2000-3	Oral Sessions

Session Title Raman, SERS, UVRR Applications

Abstract Title **Raman Spectroscopy Monitors Glutamine and Asparagine Side Chain OCCC Dihedral Angles**

Primary Author: Ryan Jakubek, University of Pittsburgh

Date: Wednesday, March 09, 2016 - After
Time: 02:10 PM
Room: B408

Co-Author(s) David Punihaole, Elizabeth Dahlburg, Sanford A. Asher, Steven Geib, Zhenmin Hong

Abstract Text

We discovered vibrational spectroscopic markers that are able to monitor the OCCC dihedral angles of glutamine (Gln) and asparagine (Asn) side chains. Density functional theory (DFT) calculations predict that the frequency of the Amide III[sup]P[/sup] (AmIII[sup]P[/sup]) Raman bands of Gln and Asn are correlated to their □[sub]3[/sub] and □[sub]2[/sub] dihedral angles, respectively. We use Raman and UV Resonance Raman (UVRR) spectroscopy to experimentally verify this correlation. DFT calculations show that the C[sub]□[/sub]-C[sub]□[/sub] (C[sub]□[/sub]-C[sub]□[/sub]) bond length of Gln (Asn) is inversely correlated to the AmIII[sup]P[/sup] frequency. Thus, a change in the force constant of the C[sub]□[/sub]-C[sub]□[/sub] (C[sub]□[/sub]-C[sub]□[/sub]) bond with Gln (Asn) □[sub]3[/sub](□[sub]2[/sub]) dihedral angle induces a frequency shift in the AmIII[sup]P[/sup] vibration. We show that the change in C[sub]□[/sub]-C[sub]□[/sub] (C[sub]□[/sub]-C[sub]□[/sub]) bond length is caused by hyperconjugation between the C[sub]□[/sub]-C[sub]□[/sub] (C[sub]□[/sub]-C[sub]□[/sub]) □and C[sub]□[/sub]=O[sub]□[/sub] (C[sub]□[/sub]=O[sub]□[/sub]) □* orbitals of Gln (Asn). We derived a series of equations that can be used to calculate the □[sub]3[/sub] and □[sub]2[/sub] dihedral angle distributions of Gln and Asn side chains from the inhomogeneous bandwidth of the AmIII[sup]P[/sup] band. We demonstrate the use of these equations by determining the □[sub]3[/sub] angle distribution for two Gln peptides with PPII-like secondary structures. Using a peptide backbone dependent rotamer library we show that the preferred □[sub]3[/sub] and □[sub]2[/sub] dihedral angles of Gln and Asn depend on the residue's backbone Ramachandran angles. The AmIII[sup]P[/sup] spectral marker enables us to obtain important new structural information about solution-state peptides as well as amyloid-like and prion aggregates.

Keywords: Bioanalytical, Biospectroscopy, Raman, Vibrational Spectroscopy

Application Code: Bioanalytical

Methodology Code: Vibrational Spectroscopy

Session # 2000 Abstract # 2000-4 **Oral Sessions**

Session Title: Raman, SERS, UVRR Applications

Abstract Title: **Assessment of the Protein–Protein Interactions in a Highly Concentrated Antibody Solution by Using Raman Spectroscopy**

Date: Wednesday, March 09, 2016 - After
Time: 02:30 PM
Room: B408

Primary Author: Chikashi Ota
Horiba, Ltd.

Co-Author(s): Kouhei Tsumoto, Satoru Nagatoishi, Shintaro Noguchi

Abstract Text

Recently, many biopharmaceuticals such as antibody drugs have been developed due to their high efficacy. In the biopharmaceutical industry, highly concentrated liquid formulations (>100 mg/mL) are required. To investigate the protein–protein interactions of a highly concentrated antibody solution, we measured concentration dependence over a wide range of concentrations (10–200 mg/mL). Our analysis of the amide I band, the band width of Trp at 1555 cm-1, I856/I830 of Tyr showed that across this wide range of concentrations, the secondary structure of the IgG molecules did not change; however, short-range attractive interactions around Trp and Tyr residues occurred as the distance between the IgG molecules decreased with increasing concentration. Our data show that Raman spectroscopy can provide valuable information based on conformational approaches to support conventional colloidal approaches, especially for analyses of highly concentrated solutions.

Keywords: Biopharmaceutical, Pharmaceutical, Protein, Raman
Application Code: Pharmaceutical
Methodology Code: Vibrational Spectroscopy

Session # 2000 Abstract # 2000-5 **Oral Sessions**

Session Title: Raman, SERS, UVRR Applications

Abstract Title: **Raman Imaging of Samples with Complex Surface Topographies**

Date: Wednesday, March 09, 2016 - After
Time: 03:05 PM
Room: B408

Primary Author: Tim Batten
Renishaw plc

Co-Author(s): Tim Smith

Abstract Text

In this work, we discuss recent advances in hardware and software that enable micro-Raman focus to be maintained over large areas during data collection. These developments allow the analysis of samples that in the past were impractical or even impossible because of variations in surface topography.

When collecting micro-Raman data it is vital to maintain the focus of the microscope objective as it determines the location of the collection volume and therefore signal strength. This is particularly important when conducting Raman imaging as the sample moving in and out of focus may result in artefacts in the images or even erroneous data. In an ideal world all Raman samples would be perfectly flat negating the need for any form of surface tracking. However, in reality we need to analyse a wide range of samples that may demand complex focus adjustment throughout the measurement. Here we discuss and present data on a range of extremely difficult samples including:

•Graphene on a Cu foil, a sample that is inherently rough on a micrometre length scale

•Unprepared pharmaceutical tablets where the sample surface is both curved and has a complex surface geometry which contains indented logos and lettering

•A snapped pharmaceutical tablet section with an extremely rough fracture surface

Keywords: Imaging, Materials Characterization, Pharmaceutical, Raman
Application Code: General Interest
Methodology Code: Vibrational Spectroscopy

Session # 2000 Abstract # 2000-7 **Oral Sessions**

Session Title Raman, SERS, UVRR Applications

Abstract Title **Spatially Offset vs Conventional Raman for Through-Barrier Material Identification**

Date: Wednesday, March 09, 2016 - After
Time: 03:45 PM
Room: B408

Primary Author Darren Andrews
Cobalt Light Systems

Co-Author(s) Ken Mann, Oliver Presly, Paul Loeffen, Pavel Matousek, Robert Stokes

Abstract Text

Raman spectroscopy has the advantages of very high chemical specificity and being a non-invasive, non-destructive technique. It is a widely used technique for material identification applications with many commercial variants available. Spatially Offset Raman Spectroscopy (SORS) is a novel variant that allows rapid and efficient through-barrier material identification without any prior knowledge of the container. Conventional Raman systems rely on a single spectrum, typically obtained from the surface of a sample. A SORS system, in contrast, relies on two measurements. The first is collected from a "zero" position (similar to conventional Raman) and a second at an offset position. By mathematically processing these two measurements, including a resultant scaled subtraction of one from the other, the SORS spectrum of the contents can be obtained. SORS thus has the capability to reject the signal component from the container and isolate the Raman signature from the contents. This paper will describe in detail the capability advantage gained by this technique in through-barrier detection over conventional Raman. Using a novel and recently developed handheld SORS system, specific capability will be addressed for a range of materials such as coloured and opaque plastics, paper, card, sacks, fabric and glass. Performance factors include speed of measurement, suppression of container fluorescence, signal/noise and quality of measurement enabling direct matching to an on-board spectroscopic library. Mixtures analysis using Raman is a challenging area. SORS strongly benefits mixture analysis as residual signal from the container is removed and does not influence the matching algorithm.

Keywords: Detection, Identification, Portable Instruments, Raman
Application Code: Homeland Security/Forensics
Methodology Code: Vibrational Spectroscopy

Session # 2000 Abstract # 2000-8

Oral Sessions

Session Title: Raman, SERS, UVRR Applications

Abstract Title: **Degradation Analysis of International Space Station Medications by Raman Spectroscopy**

Primary Author: Stuart Farquharson
Real-Time Analyzers, Inc

Date: Wednesday, March 09, 2016 - After
Time: 04:05 PM
Room: B408

Co-Author(s): Alexander May, Carl Brouillette, Chetan Shende, Joseph Cosgrove

Abstract Text

Astronauts have available pharmaceutical drugs to overcome the deleterious effects of weightlessness, sickness and injuries. Unfortunately, recent studies have shown that some of the drugs currently used may degrade more rapidly in space, losing their potency well before their expiration dates. To complicate matters, the degradation products of some drugs can be toxic, such as p-aminophenol formed from acetaminophen (Tylenol®), which can cause liver damage. Consequently there is a need for a space-worthy analyzer that can determine if a drug is safe at the time of use, as well as to monitor and understand space-induced degradation, so that drug types, formulations, and packaging can be improved. To address this need we have been developing a Raman spectroscopy system to monitor and quantify drug degradation and determine if a drug is suitable for use based on the presence of 90% or more of the original API concentration. Raman spectroscopy measures the vibrational modes of molecules, which allows identification of virtually any substance, including drugs. Furthermore, it is a "point-and-shoot" technology, in that the sample is simply placed at the laser focal point to make a measurement, typically in 1 minute or less. Here we present measurements of medications recently returned from ISS by Space X using a compact, low mass, Raman spectrometer.

Keywords: Consumer Products, Pharmaceutical, Raman, Toxicology
Application Code: Pharmaceutical
Methodology Code: Vibrational Spectroscopy

Session # **2010** Abstract # **2010-1** **Oral Sessions**

Session Title Surface and Microscopic Characterization of Nanostructures and Biological Materials

Abstract Title **Fundamental Understanding of the Synergy Between Electroactive Poly (amic) Acid Membranes and Their Interaction with Nanoparticles**

Primary Author Victor Kariuki
Binghamton University

Date: Wednesday, March 09, 2016 - After
Time: 01:30 PM
Room: B409

Co-Author(s)

Abstract Text

Certain metal nanoparticles exhibit excellent synergistic properties with conjugated polymers (CPs) and distinct electrochromic properties have been exploited in sensing, optics and electrochemical switches. Our laboratory at SUNY-Binghamton has discovered a new class of nanostructured, -conjugated, poly (amic) acid - PAA. The uniqueness of PAA lies in its excellent chromatic, electronic1, biodegradable and mechanical properties. PAA membranes showed remarkable potential as sensors for engineered nanoparticles. However its conductivity, conjugation and electroactivity have been hampered by the structure of one of the monomers. This talk specifically focuses on understanding the properties of different forms of poly (amic) acid, their interaction with nanoparticles (Gold and silver) and the applications of both PAA polymer and PAA-nanoparticle composites.

Keywords: Electrochemistry, Membrane, Nanotechnology, Polymers & Plastics
Application Code: Nanotechnology
Methodology Code: Microscopy

Session # 2010 Abstract # 2010-2 **Oral Sessions**

Session Title: Surface and Microscopic Characterization of Nanostructures and Biological Materials

Abstract Title: **Optical Property and Catalytic Activity of Gold Nanorods End-capped with a Second Metal**

Primary Author: Gufeng Wang
North Carolina State University

Date: Wednesday, March 09, 2016 - After
Time: 01:50 PM
Room: B409

Co-Author(s): Nathalia Ortiz, Vineet Kumar

Abstract Text

Recent studies show that plasmonic bimetal nanoparticles have potentials to catalyze chemical reactions efficiently due to their surface plasmon resonance (SPR) properties. This study explores the synthesis of new bimetallic nanoparticles and investigates their optical properties and catalytic potentials. New capping agents are used to synthesize gold nanorods, which facilitates the modification of a second metal on Au nanorod ends. Pd-end capped and Ag-end capped Au nanorods are synthesized. Their optical property is studied with FDTD simulation and individual particle spectroscopy. The synthesized Ag- and Pd-end capped Au nanorods retain the gold nanorod SPR peaks, which shows that they potentially can generate hot electrons by light excitation. Their catalytic activity is tested using the resazurin reduction reaction.

Keywords: Material Science, Microscopy, Nanotechnology, Spectroscopy
Application Code: Material Science
Methodology Code: Surface Analysis/Imaging

Session # 2010 **Abstract # 2010-3** **Oral Sessions**

Session Title: Surface and Microscopic Characterization of Nanostructures and Biological Materials

Abstract Title: **Autocorrelation Function Analysis of Rotational Dynamics of Gold Nanorod**

Primary Author: Kuangcai Chen
Iowa State University/Georgia State University

Date: Wednesday, March 09, 2016 - After
Time: 02:10 PM
Room: B409

Co-Author(s): Ning Fang

Abstract Text

Single particle tracking (SPT) is proven to be powerful in investigating the rotational dynamics of complex systems. With the development of light microscope and sensitive camera, vast time dependent imaging data can be recorded in single particle orientation and rotational tracking (SPORT) experiments. It is of importance to use the appropriate data analysis method to extract accurate and precise information from these recorded movies. Autocorrelation function (ACF) analysis was reported to be an effective tool to study the correlation of the fluctuation of intensities in a time series. In this study, the rotational dynamics of PEG-modified gold nanorods (AuNRs) on synthetic lipid bilayers were recorded with high temporal resolution in differential interference contrast (DIC) microscopy without suffering photobleaching. Binning was used to study the influence of the exposure time on AuNR rotation. ACF analysis coupled with computer simulations was used to investigate rotational dynamics based on the correlation of the fluctuations between the DIC bright and dark intensity. The comprehensive study on the data treatment and the effect of experimental parameters on rotational dynamics will be beneficial to data interpretation and experiment design in DIC-based SPORT experiment.

Keywords: Data Analysis, Imaging, Lipids, Nanotechnology
Application Code: Nanotechnology
Methodology Code: Surface Analysis/Imaging

Session # 2010 Abstract # 2010-4 **Oral Sessions**

Session Title: Surface and Microscopic Characterization of Nanostructures and Biological Materials

Abstract Title: **Antibody-Like Biorecognition Sites for Proteins from Surface Imprinting on Nanoparticles**

Date: Wednesday, March 09, 2016 - After
Time: 02:30 PM
Room: B409

Primary Author: Snehasis Bhakta
University of Connecticut

Co-Author(s): James F. Rusling, Saiful Seraji, Steven L. Suib

Abstract Text

Natural antibodies are used widely in important areas such as biomedical analysis, cancer therapy and directed drug delivery, but are expensive and can have limited stability. This paper describes synthesis of antibody-like binding sites by molecular imprinting on silica nanoparticles (SiNP) using a combination of four organosilane monomers with amino acid-like side chains providing hydrophobic, hydrophilic and H-bonding interactions with target proteins. This approach provided artificial antibody (AA) nanoparticles with good selectivity and specificity to binding domains on target protein a relatively low cost synthesis. We made AAs by polymer grafting onto SiNPs for human serum albumin (HSA) and glucose oxidase (GOx), and tested binding affinity, selectivity and specificity vs. several other proteins using adsorption isotherms and surface plasmon resonance (SPR). The Langmuir-Freundlich adsorption model was used to obtain apparent binding constants (KLF) from binding isotherms of HSA (6.7x104) and GOx (4.7x104) to their respective AAs. These values were 4-300 fold larger compared to a series of non-template proteins. SPR binding studies of AAs with proteins attached to a gold surface confirmed good specificity and revealed faster binding for the target proteins compared to non-target proteins. Target proteins retained their secondary structures upon binding.

Keywords: Bioanalytical, Material Science, Surfactants, Surface Analysis
Application Code: Bioanalytical
Methodology Code: Surface Analysis/Imaging

Session # 2010 Abstract # 2010-5 **Oral Sessions**

Session Title: Surface and Microscopic Characterization of Nanostructures and Biological Materials

Abstract Title: **Selective Raman Imaging of Integrin Receptors Through Coupled Plasmonic Nanostructures**

Date: Wednesday, March 09, 2016 - After
Time: 03:05 PM
Room: B409

Primary Author: Lifu Xiao
University of Notre Dame

Co-Author(s): Hao Wang, Zachary D. Schultz

Abstract Text

Integrin receptors are a key class of trans-membrane proteins that regulate intracellular signaling and a host of other down-stream events. The expression and function of these receptors are closely related to various types of tumor angiogenesis and metastasis. Currently the chemical structures and binding activities of these receptors are accessible mainly through molecular biology techniques such as crystallography and NMR. Developing molecular imaging tools that not only allow direct monitoring in cells but also provide chemical characteristics of the receptor is highly desirable. Surface-enhanced Raman scattering (SERS) imaging of live cells has proven to be robust and versatile, enabling chemical imaging studies of many cellular components. Meanwhile, Tip-enhanced Raman scattering (TERS), a near-field Raman approach, has matured into a routine super-resolution imaging technique and has already been employed to image specific receptors on fixed cells. Despite these advances, interpretation of SERS/TERS signals obtained in cells is challenging and sometimes impossible, due to a lack of control in probe specificity and the complexity of the cell itself. Using bio-orthogonal Raman reporters can conveniently avoid the complexity of interpreting Raman signals. In this study, we use peptide functionalized nanoparticle probes to improve the targeting specificity to the receptors and combine SERS monitoring, SEM, and TERS to resolve two integrin receptors that have similar structures but different functions and associated diseases. The binding and endocytosis of the nanoparticle probes will be captured via dark-field microscopy and SEM/TEM. We expect these results to provide insights on the binding events of different integrin receptors on cancer cells and potentially their uses in identifying drug-targets.

Keywords: Biospectroscopy, Protein, Vibrational Spectroscopy
Application Code: Nanotechnology
Methodology Code: Surface Analysis/Imaging

Session # 2010 Abstract # 2010-6 **Oral Sessions**

Session Title Surface and Microscopic Characterization of Nanostructures and Biological Materials

Abstract Title **Deep and High-Resolution Three-Dimensional Tracking of Single Particles Using Nonlinear and Multiplexed Illumination**

Date: Wednesday, March 09, 2016 - After
Time: 03:25 PM
Room: B409

Primary Author Tim Yeh
University of Texas at Austin

Co-Author(s)

Abstract Text

Molecular trafficking within cells, tissues, and engineered three-dimensional multicellular models is critical to the understanding of the development and treatment of various diseases including cancer. However, current tracking methods are either confined to two dimensions or limited to an interrogation depth of ~15 μm. Here we present a new 3D tracking method capable of quantifying rapid molecular transport dynamics in highly scattering environments at depths up to 200 μm. The system has a response time of 1 ms with a temporal resolution down to 50 s in high signal-to-noise conditions, and a spatial localization precision as good as 35 nm. Built upon spatiotemporally multiplexed two-photon excitation, this approach requires only one detector for 3D particle tracking and allows for two-photon, multi-color imaging. 3D tracking of epidermal growth factor receptor (EGFR) complexes at a depth of ~100 μm in tumor spheroids is demonstrated. Our 3D tracking microscope is built upon spatiotemporally multiplexed two-photon excitation and uses time-gated analysis via a photon counting histogram to discern the molecular 3D position. Feedback control then steers the excitation to lock-on to the single molecule as it travels at a high speed. The molecular trajectories are reconstructed from the recorded actuator positions from the feedback control loop operating at 1-5 ms. Dynamics down to 50 μs can be inferred from analysis of the photon counting histogram. In our method, the first PMT channel is used for particle tracking while the second and the third PMT channels can be used for two-photon scanning microscopy, colocalization analysis, and energy transfer studies. We have coined this technique TSUNAMI (Tracking Single particles Using Nonlinear And Multiplexed Illumination).

Keywords: Fluorescence, Imaging, Laser, Microscopy
Application Code: Biomedical
Methodology Code: Microscopy

Session # **2010** Abstract # **2010-7** **Oral Sessions**

Session Title Surface and Microscopic Characterization of Nanostructures and Biological Materials

Abstract Title **Development of a Dual Microscope System for Integration of Intracellular Calcium Imaging with Monitoring Insulin Secretion from Islets of Langerhans**

Date: Wednesday, March 09, 2016 - After
Time: 03:45 PM
Room: B409

Primary Author Lian Yi
Florida State University

Co-Author(s) Michael G. Roper, Xue Wang

Abstract Text

Glucose stimulated insulin secretion is of great interest since this process is disrupted in type 2 diabetes, but the mechanism is still unknown. Multiple signaling molecules are involved in this process, including reactive oxygen species, NADPH, mitochondrial membrane potential, and intracellular Ca[sup]2+[/sup]. To better understand the mechanism of impaired insulin secretion in diabetes, it would be ideal to have a system that can correlate intracellular events with insulin secretion from islets in real-time. The goal of this work is to integrate fluorescence imaging of intracellular Ca[sup]2+[/sup] with the monitoring of insulin secretion.

To accomplish this goal, a dual microscope system was developed that allowed simultaneous fluorescence imaging of Ca[sup]2+[/sup] and laser-induced fluorescence detection of insulin immunoassays. A gravity-driven perfusion system was used to deliver different glucose concentrations to an islet chamber on a microfluidic device. Islets were loaded with Fura-2 dye, which was excited at both 340 and 380 nm using a filter wheel. The ratio of fluorescence intensity excited at the two wavelengths was monitored every 20 s and converted to [Ca[sup]2+[/sup]][sub]i[/sub]. The secreted insulin from islets was monitored simultaneously with a competitive immunoassay on the same microfluidic device every 10 s. The peak height ratios of bound and free Cy5 labeled insulin (B/F) were inversely proportional to the secreted insulin concentrations. The insulin immunoassay had a detection limit of 6 nM. The B/F at different insulin concentrations was reproducible with RSDs <3%.

This system will be used to monitor the intracellular Ca[sup]2+[/sup] and insulin secretion from islets of Langerhans simultaneously. The developed system can be extended to other signaling molecules involved in the process of glucose stimulated insulin secretion in islets by use of other fluorescent probes.

Keywords: Bioanalytical, Fluorescence, Method Development, Microscopy
Application Code: Bioanalytical
Methodology Code: Microscopy

Session # 2020 Abstract # 2020-1 **Oral Sessions**

Session Title: Trace Explosives Detection - Half Session

Abstract Title: **Analysis of the Decomposition of Hexamethylene Triperoxide Diamine (HMTD) as Determined by SPME-GC/MS and LC/MS**

Date: Wednesday, March 09, 2016 - After
Time: 01:30 PM
Room: B301

Primary Author: Lauryn DeGreeff
Naval Research Laboratory

Co-Author(s): Christopher Katilie, Frank L. Steinkamp

Abstract Text

Hexamethylene triperoxide diamine (HMTD) is an explosive, nitrogen-containing, organic peroxide compound consisting of to nitrogen atoms each attached to three methyl groups in a planar arrangement. Like triacetone triperoxide (TATP), another organic peroxide explosive, HMTD is highly sensitive to shock, friction, heat, and static, making it very difficult to handle or transport safely. It thus has no military or commercial applications. Unlike TATP and other explosive compounds, detection of HMTD has proven to be challenging, most notably due to its thermal instability and tendency to decompose at relatively low temperatures. The goal of this research was to determine these decomposition products and to suggest possible mechanistic pathways. This knowledge will ultimately be applied to improve HMTD detection in the field. The decomposition of HMTD yields several small volatile compounds detectable in the headspace and in solution. This decomposition was monitored over time and under several conditions, including temperature, presence of water/humidity, and formulation. Headspace analysis was carried out above the bulk material using solid phase microextraction (SPME) with GC/MS, while the solution analysis was carried out by LC/MS. Degradation products detected included acetic acid, formic acid, formamide, and trimethylamine. The results suggested that the decomposition of HMTD is complicated, dynamic process involving the decomposition of HMTD and subsequent decomposition products. Solution analysis suggests that the decomposition mechanism may involve the interaction between individual HMTD molecules, and may be driven by the presence of water.

Keywords: Chromatography, Forensic Chemistry, Gas Chromatography, Volatile Organic Compounds
Application Code: Homeland Security/Forensics
Methodology Code: Separation Sciences

Session # 2020	Abstract # 2020-2	Oral Sessions

Session Title: Trace Explosives Detection - Half Session

Abstract Title: **Separation of Inorganic Ions and Neutral Organic Nitroaromatic Compounds by Electrokinetic Chromatography**

Date: Wednesday, March 09, 2016 - After
Time: 01:50 PM
Room: B301

Primary Author: Julie R. McGettrick
University of Montana

Co-Author(s): Christopher P. Palmer

Abstract Text

Explosives residues are complex mixtures that can contain both neutral organic molecules and inorganic ions. Most forensic protocols use two techniques for explosives analysis: ion chromatography for ions and HPLC for organic compounds. Electrokinetic chromatography (EKC) is a powerful analytical technique that combines the instrumentation of capillary electrophoresis and the principles of chromatography and makes it possible to analyze both anions and neutral organic compounds in the same analytical run. In this study AB diblock copolymers that form latex nanoparticles with a cationic shell and hydrophobic core were synthesized and used as pseudostationary phases. Capillaries were coated with a cationic polymer to create anodic electroosmotic flow. A contactless conductivity detector and UV detector were used to detect ions and nitroaromatic compounds, respectively. A mixture of anions and nitroaromatic compounds can be separated in less than ten minutes.The effect of different cationic shells and hydrophobic cores on the separation was investigated.

Keywords: Capillary Electrophoresis, Forensic Chemistry, Method Development, Separation Sciences

Application Code: Homeland Security/Forensics

Methodology Code: Capillary Electrophoresis

Session # 2020 Abstract # 2020-3 **Oral Sessions**

Session Title: Trace Explosives Detection - Half Session

Abstract Title: **Comparison of Inter-Instrument Relative Response Factors with Thermal Desorption Internal Standards**

Primary Author: H MItchell Rubenstein
USAF

Date: Wednesday, March 09, 2016 - After
Time: 02:10 PM
Room: B301

Co-Author(s): Brian Geier, Claude C. Grigsby, Darrin Ott, Kathy Fullerton, Maomian Fan

Abstract Text

Experiments to establish inter-instrument correlations were conducted with three units of the HAPSITE ER (ER), a portable GCMS. Specifically, detection and quantification was accomplished using thermal desorption (TD). Furthermore, we examined the use of thermal desorption internal standards (TDIS). Data will be presented to demonstrate the stability of both retention times (RTs) and relative response factors (RRFs) over time. While portable gas chromatographic mass spectrometry has achieved huge success in the qualitative identification of unknowns, quantitative analysis remains elusive. Primarily, field technicians do not always have the materials to prepare standard curves and instead rely on calibrations generated by the manufacturer and programmed into the instruments. These pre-established curves have considerable variability among instruments. For this reason we have examined approaches to tighten the data. We accomplished this by spiking the TD tubes with known concentrations of TDIS compounds that would behave similarly to the target analytes. Tubes prepared in this manner were examined on three ER instruments over a six week period and will be presented. The data demonstrates that tubes prepared in this manner are stable at room temperature over this period for select compounds. An experimental design determined that compounds are well retained when an additional 3.6L volume of nitrogen is passed through the tubes at 50[degree]C. This establishes the use of these compounds for incorporation onto TD tubes prior to field sampling. Based on our data, we propose that pre-established calibration curves can be corrected by RRF correction factors obtained from internal standards spiked onto the thermal desorption tubes prior to sampling. The data objective of <30% Relative Standard Deviation can be met significantly improving the quality of data attained in field analyses.

Keywords: Gas Chromatography/Mass Spectrometry, Semi-Volatiles, Thermal Desorption
Application Code: Homeland Security/Forensics
Methodology Code: Gas Chromatography/Mass Spectrometry

Session # **2020** **Abstract #** **2020-4** **Oral Sessions**

Session Title: Trace Explosives Detection - Half Session

Abstract Title: **Through-barrier Explosives and Hazardous Material Detection Using a Handheld Spatially Offset Raman Spectrometer**

Date: Wednesday, March 09, 2016 - After
Time: 02:30 PM
Room: B301

Primary Author: Steve Wood
Cobalt Light Systems

Co-Author(s): Darren Andrews, Ken Mann, Oliver Presly, Paul Loeffen, Pavel Matousek, Robert Stokes

Abstract Text

Raman spectroscopy allows the acquisition of molecularly specific signatures of pure compounds and mixtures making it a popular method for material identification applications. Spatially Offset Raman Spectroscopy (SORS) is a novel variant of Raman spectroscopy whereby multiple measurements at differing positions are used to separate the spectrum arising from the sub layers of a sample from the spectrum at the surface. Crucially this method requires no prior knowledge of the surface layer. This paper will demonstrate the application of this technique to rapid identification of hazardous materials that are concealed by a wide variety of sealed containers such as coloured and opaque plastics, paper, card, sacks, fabric and glass. The range of potential target materials includes toxic industrial chemicals, explosives, narcotics, chemical warfare agents and biological materials. Recent development of a handheld SORS system has allowed the technique to be successfully deployed in applications such as: improvised explosive materials within oil jerry cans, liquid explosives and precursor materials within bottles and narcotics within sealed envelopes / postage packs. Typical scan times in normal ambient light conditions are around 30s. A major advantage of SORS in this context is that the operator is able to obtain a positive identification without opening or disturbing the container to gain access to the material or to take a sample. This is important as it has the potential to improve the safety and efficiency of operating procedures in explosive ordinance disposal, first response, hazmat incident management, illicit lab search and rapid screening at security checkpoints.

Keywords: Detection, Portable Instruments, Raman, Vibrational Spectroscopy
Application Code: Homeland Security/Forensics
Methodology Code: Portable Instruments

Session # 2030 Abstract # 2030-2 **Undergraduate Poster Session**

Session Title: Undergraduate Poster Session

Abstract Title: **Analysis of Nicotine Levels in Electronic Cigarettes**

Primary Author: Martin E. Miller
Austin Peay State University

Co-Author(s): Dillon Burrow, Jacob Williams

Date: Wednesday, March 09, 2016 - After

Time:

Room: Exposition Floor, 400 Aisle

Abstract Text

Electronic cigarettes are gaining popularity across the globe. Marketed as a smoking cessation aid and as a healthier alternative to tobacco products, electronic cigarettes deliver nicotine to users as a water-based vapor without the typical toxic chemicals found from the combustion of tobacco products. Currently, electronic cigarettes are not under the authority of the Food and Drug Administration, since they do not contain tobacco. Due to a lack of regulation, the physiological effects and chemical content of these products has not undergone close analytical scrutiny. Presented here are the efforts towards the quantification of nicotine content in electronic cigarette liquid. Using HPLC, high performance liquid chromatography, nicotine content in several electronic cigarette liquid samples is being assessed to compare to the values listed on the products. Calibration curves are being utilized to determine the amount of nicotine in each sample, variation in batches of the same brand, and degradation of the nicotine in the solutions over time.

Keywords: Chromatography, HPLC, HPLC Columns, HPLC Detection
Application Code: Consumer Products
Methodology Code: Liquid Chromatography

Session # **2030** Abstract # **2030-3** **Undergraduate Poster Session**

Session Title: Undergraduate Poster Session

Abstract Title: **Thermal Behavior of Barium and Strontium Carbonates**

Primary Author: Charles M. Earnest
Berry College

Co-Author(s): Ethan Miller

Date: Wednesday, March 09, 2016 - After
Time:
Room: Exposition Floor, 400 Aisle

Abstract Text

Both Barium and Strontium Carbonates are commercially important in Material Science. Both of these carbonates have also been sold as temperature standards by NIST for use in calibrating Differential Thermal Analysis (DTA) instruments. Both of these carbonates undergo polymorphic transitions from the orthorhombic to hexagonal forms when heated to temperatures above 800 deg C. On further heating, the Barium Carbonate undergoes an additional polymorphic transition where the hexagonal form transforms to a face-centered cubic form near 976 deg C. On continued heating, this cubic form of Barium Carbonate decomposes to Barium Oxide liberating carbon dioxide gas. Strontium Carbonate, on the other hand, tends to be thermally less stable than Barium Carbonate. It is reported to lose mass over a broad range of temperatures beginning at temperatures below the observed polymorphic transition. Our studies, presented here, employ both differential thermal analysis (DTA) and Thermogravimetric Analysis (TGA) in an attempt to answer questions of uncertainty of the reported temperatures of transition, as well as, heats of transition for these industrially important compounds. The effects of the composition of the dynamic puge atmosphere employed in the thermal anaysis were also observed.

Keywords: Analysis, Materials Characterization, Standards, Thermal Analysis
Application Code: Material Science
Methodology Code: Thermal Analysis

Session # **2030** Abstract # **2030-4** **Undergraduate Poster Session**

Session Title: Undergraduate Poster Session

Abstract Title: **Improved Quantification of Gibbsite in Bauxite Ores**

Primary Author: Charles M. Earnest
Berry College

Co-Author(s): Britney Stong, Karla Gann

Date: Wednesday, March 09, 2016 - After
Time:
Room: Exposition Floor, 400 Aisle

Abstract Text

Bauxite ores are composed of a mixture of minerals. The components of the mixture vary with location of the Bauxite. Generally speaking, the two most important components are Gibbsite, Al(OH)3, and Boehmite, AlO(OH). It is these components that determine the commercial value of these ores since it is aluminum metal or aluminum oxide that is subsequently derived from them. The Gibbsite and Boehmite content are usually assigned by the use of X-ray diffraction spectroscopy(XRD). The use of the methods of Thermogravimetric Analysis(TGA)and Derivative Thermogravimetry (DTG) have been used to a lesser extent due to less than stoichiometric dehydroxylation of the Gibbsite component which liberates water vapor and a small amount of Boehmite. In this work, we describe the use of an "empirical gravimertic factor" rather than the usual theoretical gravimetric factor employed for thermogravimetric quantification assignments. As will be seen from the data presented here, by employing our "empirical gravimetric factor" for the quantification of the Gibbsite content of a number of Bauxites (that were previously analyzed by XRD), we were able to obtain results for the Gibbsite content within plus or minus one percent of that obtained by the method of XRD.The improvement in TGA methodology exhibed here is the result of the experimental development an "empirical gravimetric factor" for use in quantitative assignment rather than using a gravimetric factor calculated from the balanced chemicak equation. This eliminates the error associated with less than complete dehydroxylation of the Gibbsite component on heating in the TGA instrumentation.

Keywords: Optimization, Quantitative, Thermal Analysis
Application Code: Material Science
Methodology Code: Thermal Analysis

Session # 2030 Abstract # 2030-5 **Undergraduate Poster Session**

Session Title: Undergraduate Poster Session

Abstract Title: **Signal Enhancement Compensation in ICP-MS Analysis for Arsenic in the Presence of Organics**

Date: Wednesday, March 09, 2016 - After

Time:

Room: Exposition Floor, 400 Aisle

Primary Author: Adam Kagel
University of California, Davis

Co-Author(s): Carla Kagel, Nels Worden, Richard Kagel

Abstract Text

Significant signal enhancement was observed for arsenic measurements using ICP-MS due to the presence of various carbon sources in samples and sample digests. Enhancements ranged to greater than 300 percent depending on the level of carbon present. This relationship between carbon concentration and signal enhancement was studied along with methods for compensating for this effect during quantitative analysis. An approach to compensate for the observed large positive biases as well as extremely high calculated recoveries on matrix spike results will be presented.

Keywords: Atomic Spectroscopy, Elemental Analysis, Environmental Analysis, ICP-MS

Application Code: Environmental

Methodology Code: Atomic Spectroscopy/Elemental Analysis

Session # 2030 Abstract # 2030-6 **Undergraduate Poster Session**

Session Title: Undergraduate Poster Session

Abstract Title: **Isotope Labeling in Astrobiology: Ethanol as a Carbon Source**

Primary Author: Nicole G. Perkins
California State Polytechnic University, Pomona

Co-Author(s): Gregory A. Barding, Rakesh Mogul

Date: Wednesday, March 09, 2016 - After
Time:
Room: Exposition Floor, 400 Aisle

Abstract Text

Originally isolated from the Mars Odyssey prior to launch, Acinetobacter radioresistens 50v1, a gram negative bacteria, was found to exhibit extremophile capabilities as it can resist multiple decontamination techniques within spacecraft facilities. In particular, this strain of Acinetobacter is believed to not only survive ethanol cleaning but also utilize it as its sole carbon source. To verify and understand the unique characteristics of Acinetobacter radioresistens 50v1 to resist and exploit the ethanol-based decontamination process, several different growth conditions were compared and analyzed by GC-MS. First, the capability of the bacterium to utilize ethanol as a sole carbon source in a nutrient-poor environment was evaluated. After collecting the cells from an inoculated sample, GC-MS analysis was done in order to identify if ethanol was present and incorporated into the cell. To confirm and explore the switch to ethanol as an energy source, the experiment was repeated using U-[sup]13[/sup]C and deuterium labeled ethanol and compared to growth under nutrient rich conditions using glucose as the carbon source. Isotopic labeling was extensively observed and included metabolites such as proline, lysine, isoleucine, valine, and trehalose. Because of the lack of other compounds as carbon sources in the growth media, the labeling of metabolites and disaccharides strongly suggest that the organism is capable of using ethanol as its sole carbon source.

Keywords: Bioanalytical, Gas Chromatography/Mass Spectrometry, Metabolomics, Metabonomics
Application Code: Genomics, Proteomics and Other 'Omics
Methodology Code: Gas Chromatography/Mass Spectrometry

Session # 2030 **Abstract # 2030-7** **Undergraduate Poster Session**

Session Title: Undergraduate Poster Session

Abstract Title: **Sulfidation of Silver Nanoparticles**

Primary Author: Nathaniel D. Fletcher
College of Charleston

Co-Author(s): Katherine M. Mullaugh

Date: Wednesday, March 09, 2016 - After
Time:
Room: Exposition Floor, 400 Aisle

Abstract Text

Silver nanoparticles (Ag NPs) are one of the most widely used forms of nanoparticles in consumer goods due to their anti-microbial properties. Their ubiquity has raised concerns about possible environmental harm due to their potential to release toxic silver ions (Ag[sup]+[/sup]) to natural waters. To explore the potential hazard of Ag NPs in natural waters, we studied chemical transformations that Ag NPs may undergo as they pass through sulfide-rich conditions common in waste water treatment plants, which may limit the release of Ag[sup]+[/sup] from Ag NPs due to the formation of low-solubility silver sulfide (Ag[sub]2[/sub]S). However, it is uncertain whether sulfidation of Ag NPs is complete and if sulfidized Ag NPs can continue to release Ag[sup]+[/sup]. To simulate these conditions, we monitored the reaction of Ag NPs with various levels of sulfide with an ion selective electrode and UV/visible spectrophotometry. We additionally characterized the products of the sulfidation reactions with a purge-and-trap acid volatile sulfide (AVS) analysis, which served as a measure of the extent of Ag NP sulfidation. Having optimized the AVS method for the analysis of Ag NP sulfidation products, we are studying how sulfidized Ag NPs change over time to assess their potential as a source of Ag[sup]+[/sup]. Changes in Ag NP absorbance several days after reaction with sulfide suggest a reversal of Ag NP sulfidation is possible, especially when silver is initially in excess of sulfide.

Keywords: Environmental Analysis, Nanotechnology, Purge and Trap, Sulfur

Application Code: Environmental

Methodology Code: UV/VIS

Session # 2030 Abstract # 2030-8 **Undergraduate Poster Session**

Session Title: Undergraduate Poster Session

Abstract Title: **Photodegradation and Ecotoxicity Studies of Sertraline, Fluoxetine, and their Photodegradants**

Primary Author: Sylvia C. Davila
College of Charleston

Date: Wednesday, March 09, 2016 - After
Time:
Room: Exposition Floor, 400 Aisle

Co-Author(s): Allison Welch, Jessica R. Hinson, Wendy C. Cory

Abstract Text

Selective serotonin reuptake inhibitors (SSRIs) are a commonly prescribed class of antidepressants Their ubiquity suggests that they, as unchanged parent compounds or metabolites, may be present in waste water via human excretion products. After sewage treatment these compounds may, upon exposure to sunlight, be further degraded or transformed into structurally similar compounds that may pose greater toxicity to humans and other non-target organisms. Photodegradation kinetics were studied for sertraline (SER) and fluoxetine (FLX) in Holtfreter's solution. The effects of added humic acid and fulvic acid on the photodegradation rates were studied. Aliquots were obtained at multiple time points during photoexposure and were analyzed via HPLC and LC-MS. Peak photodegradant concentrations were observed to occur at the half-lives of the parent compounds. Southern Toad tadpoles were exposed to cocktails of these compounds at environmentally relevant concentrations to test for effects on acute toxicity as well as swimming behavior and development.

Keywords: Environmental Analysis, Liquid Chromatography/Mass Spectroscopy, Pharmaceutical, Toxicology
Application Code: Environmental
Methodology Code: Liquid Chromatography/Mass Spectrometry

Session # 2030 Abstract # 2030-9 **Undergraduate Poster Session**

Session Title: Undergraduate Poster Session

Abstract Title: **Photodegradation of Bupropion and Gabapentin**

Date: Wednesday, March 09, 2016 - After
Time:
Room: Exposition Floor, 400 Aisle

Primary Author: Neha V. Muppala
College of Charleston

Co-Author(s): Kristina K. Tran, Wendy C. Cory

Abstract Text

Prescribed medications that act as maintenance drugs can affect the environment through landfills and the water system. These drugs eventually degrade and can form products that can potentially pose a greater threat to human health and the environment than when in their original composition. To study the degradation of commonly prescribed maintenance drugs, we observed and analyzed the photodegradation rate of bupropion (BUP) and gabapentin (GAB) in aqueous solutions that mimic the water system of the natural environment. Samples of BUP and GAB with and without humic acid were exposed to light in a solar simulator. The resultant solutions were analyzed via HPLC and LC-MS. Rates of degradation and degradant identities were identified.

Keywords: Environmental/Water, HPLC, Liquid Chromatography/Mass Spectroscopy, Pharmaceutical
Application Code: Pharmaceutical
Methodology Code: Liquid Chromatography/Mass Spectrometry

Session # 2030 Abstract # 2030-11 **Undergraduate Poster Session**

Session Title: Undergraduate Poster Session

Abstract Title: **Preliminary Investigation of the Geographical Distribution of Synthetic UV Filters in Bahamian Surface Waters**

Primary Author: Aejin Kim
Elmira College

Date: Wednesday, March 09, 2016 - After
Time:
Room: Exposition Floor, 400 Aisle

Co-Author(s): Betsy A. Smith, Christopher B. Hall, Emily J. Feldpausch, Jared S. Baker

Abstract Text

Certain organic UV filters, commonly found in personal care products such as sunscreen, have been associated with a potential for endocrine disruption in biological systems and are often found in low levels in a variety of surface waters. Consequently, there is considerable interest in understanding the environmental transport and fate of these chemical species. San Salvador Island, The Bahamas, represents a unique opportunity to study UV filters in natural surface waters in that it is a sparsely-populated outer Bahamian island whose interior lakes experience little human contact. Conversely, there exists a point source for the delivery of UV filters into the ocean waters around the periphery of the island as guests at a popular tourist resort frequent the beach and ocean. An initial sampling of numerous surface and sub-surface water sources was conducted using fabric-phase sorptive extraction (FPSE) devices. The FPSE device was constructed by polymerizing a sol-gel derived hybrid organic-inorganic sorbent onto a cellulose fabric. This sorbent provides for an effective means to extract a wide range of UV filters from aqueous solutions, however, it has yet to be used in this capacity. This presentation will discuss the extraction procedure and the preliminary analysis of a number of real samples for the presence of commonly encountered UV filters using FPSE coupled with gas chromatography-mass spectrometry.

Keywords: Environmental/Water, Extraction, GC-MS, Sampling
Application Code: Environmental
Methodology Code: Sampling and Sample Preparation

Session # 2030 Abstract # 2030-12 Undergraduate Poster Session

Session Title: Undergraduate Poster Session

Abstract Title: **Surface Plasma Polaritons from Voltage Charged Gold Nanotubes**

Primary Author: Sam Konchan
University of Florida

Date: Wednesday, March 09, 2016 - After
Time:
Room: Exposition Floor, 400 Aisle

Co-Author(s): Alec Talin, Charles R. Martin, Pradeep Ramiah Rajasekaran

Abstract Text

The field of surface plasmons has been gaining a lot of interest in the recent years due to it's wide range of applicability. Studying and controlling the surface plasmons will have potential applications in the area of sensors, photonics and opto electronics. Surface plasmons can also give information about the atomic and molecular properties of materials adjacent to the surface layer. We have recently developed a technique to investigate the behavior of voltage-charged template-synthesized gold nanotube membranes electrochemically. The electrochemical characterization reveled the charge distribution and localization behavior in these gold nanotube membranes upon charging. Here we extend this method and employ photophysical methods to study voltage charged gold nanotube membranes. Extinction spectra of theses gold nanotube membranes are recorded as a function of the applied voltage. Depending upon the polarity of the voltage applied, the gold nanotubes will become electron rich or depleted. As the electron population has a direct impact on the surface plasmons, the extinction spectra will be reflection of the charge on the gold nano tubes. This interesting phenomenon will provide monumental information about the materials adjacent to the metallic layer. This could be used as an analytical tool to investigate the electrical double layer and any other materials adjacent to the gold surface. Apart from this, it is also a basic science phenomenon of fundamental interest

Keywords: Biosensors, Electrochemistry, Electrode Surfaces, Microspectroscopy
Application Code: Material Science
Methodology Code: UV/VIS

Session Title Undergraduate Poster Session

Abstract Title **Electroporation Facilitation via Gold-Plated Membranes**

Primary Author Aaron Wilson
University of Florida

Date: Wednesday, March 09, 2016 - After
Time:
Room: Exposition Floor, 400 Aisle

Co-Author(s) Charles R. Martin, Juliette Experton

Abstract Text

Electroporation is a convenient tool used for gene transfection in cells. It allows for the control of the permeability of cellular membranes to facilitate the uptake of molecules, such as plasmids, from the extracellular medium. Current electroporation methods require large voltages, upwards of 2,000V, that can be dramatically reduced using the facilitation of a gold-plated mesoporous membrane. Our model uses [i]Escherichia coli[/i] bacteria as the target of electroporation. While the bacteria flowed through the membrane, electrical pulses from 1 to 4V were applied to it inducing a high electric field inside the mesopores. To measure electroporation, a combination of two impermeable dyes, a yellow-fluoresced dye and red-fluoresced propidium iodide, were selectively added to the medium. They are known to independently and significantly increase their fluorescence intensity when staining the bacteria DNA. The use of those two dyes allows the determination of the electroporation efficiency while maintaining the exclusion of lysed bacteria. Using a UV-Vis Spectrophotometer a linear increase in absorbance was measured over time showing bacteria passing through the membrane. The fluorescence intensity of the bacterial solution was also measured versus time and showed an increasing amount of fluorescent bacteria as electroporation occurred. Microscopy techniques were used to identify bacteria that exhibited fluorescence in order to demonstrate successful electroporation. Up to 40% electroporation efficiency was calculated using the combination of these data.

Keywords: Electrode Surfaces, Fluorescence, Membrane, UV-VIS Absorbance/Luminescence
Application Code: Biomedical
Methodology Code: Electrochemistry

Session # 2030 Abstract # 2030-14 **Undergraduate Poster Session**

Session Title: Undergraduate Poster Session

Abstract Title: **The Analysis of Electronic Cigarette Solutions by ICP-MS for USP Regulated Contaminates**

Primary Author: Chloe E. Fernandes
Georgia Gwinnett College

Date: Wednesday, March 09, 2016 - After
Time:
Room: Exposition Floor, 400 Aisle

Co-Author(s): Daniel H. Jones

Abstract Text

Electronic cigarettes are portable, battery-operated devices that transform a liquid solution containing flavorings and often nicotine into a vapor that gets inhaled via an electronic atomizer. These electronic devices are often promoted as a safer alternative to traditional cigarettes; however regulations on the safety of the e-cigarette solutions have not yet been imposed by any federal FDA guidelines. Consequently, many of these solutions have not been vigorously tested for contaminates or other harmful components, including heavy metals. The United States Pharmacopeia does not regulate e-cigarettes; however, it does set guidelines for daily inhalation exposure in pharmaceutical products. These limits are referred to as Permissible Daily Exposure or PDE. In this study five solutions were evaluated by ICP-MS against these limits of daily exposure that are normally reserved for therapeutic drugs. These limits are based upon not only the toxicity of the element but also the concentration and the amount inhaled into the body. It is estimated that smoking 100ul of an e-cigarette solution is equivalent to smoking one traditional cigarette. Therefore, two milliliters of an e-cigarette solution correlates to 20 traditional cigarettes or one pack. A calculation of the daily exposure of heavy metals from the e-cigarette solution can be determined based upon a typical usage of one pack a day of traditional cigarettes. Results reported in this study are in micrograms per 2mls of e-cigarette solution or total micrograms per daily dose.

Keywords: Atomic Spectroscopy, Consumer Products, ICP-MS, Method Development
Application Code: Consumer Products
Methodology Code: Atomic Spectroscopy/Elemental Analysis

Session # **2030** Abstract # **2030-16** **Undergraduate Poster Session**

Session Title: Undergraduate Poster Session

Abstract Title: **Discovery Metabolomics of Early-Stage Ovarian Cancer in a Dicer-Pten Double Knockout Murine Model**

Primary Author: Laura C. Winalski
Georgia Institute of Technology

Date: Wednesday, March 09, 2016 - After
Time:
Room: Exposition Floor, 400 Aisle

Co-Author(s): Christina M. Jones, Facundo M. Fernandez, Jaeyeon Kim, Maria E. Monge, Martin M. Matzuk

Abstract Text

A Dicer-Pten double-knockout (DKO) mouse model of high-grade serous ovarian cancer (HGSC) has proved to be a valuable means for understanding the molecular basis of the deadliest gynecological cancer among women, with a global 5-year survival rate of only 44.2%. Characterized by rapid growing tumors, most cases remain undiagnosed until the cancer is late-stage. Using ultra-performance liquid chromatography (UPLC) coupled to mass spectrometry (MS), our team is studying early metabolome changes identifiable in DKO mouse serum which could aid in early detection of this highly aggressive disease.

Blood serum samples were collected from control and early-stage DKO mice. Metabolites were extracted after protein precipitation using methanol in a 3:1 (v/v) dilution ratio to serum. Metabolite extracts were then lyophilized and reconstituted in the initial composition of the chromatographic mobile phase. High resolution mass spectra were acquired in positive electrospray ionization mode for the m/z 50-1200 range. Analysis was conducted using a Waters ACCQUITY H Class system coupled to a Xevo G2 QTOF mass spectrometer. Metabolic features were extracted using MZmine 2.10 software. With 17 metabolites selected from these profiles using genetic algorithms for variable selection, orthogonal projection to latent structures-discriminant analysis (oPLS-DA) separated early-stage DKO mice from control mice with 98% accuracy and sensitivity, and 100% specificity. Continuing work involves identifying these 17 discriminating features utilizing UPLC tandem MS, retention time, and ion mobility collisional cross section matching to chemical standards for further validation, thereby providing insight into the metabolic alterations associated with the developments and proliferation of HGSC.

Keywords: Bioanalytical, Liquid Chromatography/Mass Spectroscopy
Application Code: Genomics, Proteomics and Other 'Omics
Methodology Code: Liquid Chromatography/Mass Spectrometry

Session # 2030 Abstract # 2030-17 **Undergraduate Poster Session**

Session Title: Undergraduate Poster Session

Abstract Title: **Colorimetric Detection of Pyrocatechol as a Model for Urushiol Analysis in Poison Ivy**

Primary Author: Collin J. Steen
Kalamazoo College

Date: Wednesday, March 09, 2016 - After
Time:
Room: Exposition Floor, 400 Aisle

Co-Author(s): Kari Anderson

Abstract Text

Since poison ivy can vary in physical appearance, it is often difficult to recognize in nature. Consequently, a simple colorimetric detection method could successfully aid in not only identifying the plant, but also quantifying urushiol. Consisting of an aromatic ring with two hydroxyl groups and a long hydrophobic chain, urushiol is the active compound responsible for the allergenic response to poison ivy. Pyrocatechol, a similar compound, only lacks the hydrophobic chain and thus provides a potential model system for analysis of urushiol. Tyrosinase, an enzyme normally found in both plants and animals, oxidizes the hydroxyl groups resulting in the formation of benzoquinone, a yellow-brown colored product. Using UV-visible absorbance spectroscopy ([lambda]=386 nm), the concentration of pyrocatechol in a solution of known 233 [micro]M was quantified as 234 ± 1 [micro]M (SEM, n=3) with an average detection limit of 3 [micro]M. Another colorimetric technique involves sodium molybdate which, in the presence of trichloroacetic acid, complexes two catechol molecules forming a red-colored product. Using UV-visible absorbance spectroscopy ([lambda]=505 nm), the concentration of pyrocatechol in a solution of known 49.7 [micro]M was quantified as 48.9 ± 0.2 [micro]M (SEM, n=3) with an average detection limit of 0.4 [micro]M. Although extraction of urushiol from the plant itself has proven difficult, these two colorimetric techniques represent potential ways to identify and quantify urushiol from poison ivy.

Keywords: Environmental, Extraction, Method Development, UV-VIS Absorbance/Luminescence
Application Code: General Interest
Methodology Code: UV/VIS

Session # 2030 **Abstract # 2030-18** **Undergraduate Poster Session**

Session Title: Undergraduate Poster Session

Abstract Title: **Quantification of Salicylates in Stomach Relief Aids Using GC-MS and UV-Vis**

Primary Author: Marina C. Koether
Kennesaw State University

Date: Wednesday, March 09, 2016 - After

Time:

Room: Exposition Floor, 400 Aisle

Co-Author(s): Kimberly C. Powers

Abstract Text

Some stomach relief aids have bismuth subsalicylate and, as a flavoring, methyl salicylate as ingredients. A variety of methods will be described to quantify the salicylates using GC-MS, UV-Vis and the internal standard method.

Keywords: GC-MS, Method Development, Pharmaceutical, UV-VIS Absorbance/Luminescence
Application Code: Pharmaceutical
Methodology Code: Gas Chromatography/Mass Spectrometry

Session # 2030 Abstract # 2030-19 Undergraduate Poster Session

Session Title: Undergraduate Poster Session

Abstract Title: **Correlation Between Different Extraction Methods and the Ratio of Neral to Geranial in the Essential Oils of Lemongrass by GC-MS**

Date: Wednesday, March 09, 2016 - After

Time:

Room: Exposition Floor, 400 Aisle

Primary Author: Marina C. Koether
Kennesaw State University

Co-Author(s): Skyler Mize

Abstract Text

Citral is a compound that is found in the essential oils of the lemongrass plant. Citral contains two different isomers, neral(Z) and geranial(E). The purpose of these experiments is to investigate the effects that different extraction methods, and dilutions have on the citral in a sample of lemongrass and extraction efficiencies. The three different extraction types analyzed were steam distillation, solvent soak with exposure to the sun, and solvent soak in a cool, dark, well ventilated place. The results from the GC-MS supports the fact that as the energy in the extraction increases, more geranial is converted to neral. However, diluting a sample results in a larger ratio of E/Z.

Keywords: Calibration, Data Analysis, Extraction, GC-MS
Application Code: Food Science
Methodology Code: Separation Sciences

Session # 2030 Abstract # 2030-20 Undergraduate Poster Session

Session Title Undergraduate Poster Session

Abstract Title **Determining Dissolution Testing Time of Potassium in Potassium Gluconate Tablets by Conductivity with Confirmation by Flame Atomic Absorption Spectrometry**

Primary Author Marina C. Koether
Kennesaw State University

Date: Wednesday, March 09, 2016 - After
Time:
Room: Exposition Floor, 400 Aisle

Co-Author(s) Minwoo Lee

Abstract Text

The potassium in potassium gluconate tablets can be monitored during dissolution testing by conductivity. A correlation can be made with flame atomic absorption spectroscopy to ensure that the conductivity measured is due to potassium. Results indicate that different brands have different dissolution rates. Some are 100% dissolved in less than 30 minutes while others take over 40 minutes to completely dissolve.

Keywords: Atomic Absorption, Atomic Spectroscopy, Data Analysis, Pharmaceutical
Application Code: Pharmaceutical
Methodology Code: Atomic Spectroscopy/Elemental Analysis

Session # 2030 Abstract # 2030-21 **Undergraduate Poster Session**

Session Title: Undergraduate Poster Session

Abstract Title: **Study of the Implementation of the Systematic Method in General Chemistry II**

Date: Wednesday, March 09, 2016 - After

Time:

Room: Exposition Floor, 400 Aisle

Primary Author: Victoria Y. Reinders
Maryville University Saint Louis

Co-Author(s): Natalie Ulrich, Thomas M. Spudich

Abstract Text

The objective of this study is to examine how the systematic method affords helpful opportunities for learning equilibrium concepts in a General Chemistry course. The systematic method approaches chemical equilibrium by modeling systems using conservation of mass and charge principles rather than the traditional "ICE" table method. Students in two general chemistry courses – one taught with the systematic method and one taught using the ICE table method – were observed during the Equilibrium unit of their courses. Identical tests were administered and test scores were compared to determine which equilibrium problem solving method allowed for a better understanding of the concepts of equilibrium. The first data set was collected in the Spring of 2015, and a second set of data is currently being collected under the direction of the same course instructors. This study's ultimate goal is to test whether the use of the systematic method in general chemistry education improves student performance and understanding, and subsequently implement the concepts and tools currently taught in the Quantitative Analysis course into the General Chemistry curriculum. By contrasting abilities to understand equilibrium that stem from using the systematic and traditional methods, this research is working to uncover the potential of the systematic method to help students better understand and describe chemical equilibrium. We are illustrating that a basic understanding of the systematic method in general chemistry can lead to more in-depth discussions built from the basic knowledge of different analyses currently found in a traditional analytical chemistry course.

Keywords: Education
Application Code: Other
Methodology Code: Education/Teaching

Session # 2030 Abstract # 2030-22 **Undergraduate Poster Session**

Session Title: Undergraduate Poster Session

Abstract Title: **Characterization and Identification of Biosurfactants for Oil Remediation**

Primary Author: Kaydren B. Orcutt, Mercer University

Date: Wednesday, March 09, 2016 - After Time:

Room: Exposition Floor, 400 Aisle

Co-Author(s): Garland L. Crawford, Joseph W. Kloepper, Justis E. Ward, Kathryn D. Kloepper

Abstract Text

This study examines the potential of an environmentally-safe, biosurfactant-producing Bacillus strain for the bioremediation of oil spills. To better understand biosurfactant production by this strain, cells were grown in controlled conditions, and biosurfactant production was evaluated by determining the emulsifying ability of the cell-free broth across specific time points during the growth cycle. We have developed bioanalytical protocols that mimic environmental conditions to evaluate biosurfactant activity and stability as a function of pH, salinity, and oil type. Results indicate that the produced biosurfactants are comparable to biological and chemical surfactants and thus may have the potential to provide a method for the environmentally-friendly remediation of oil spills. Continued work aims to separate and identify the individual biosurfactants using a combination of RP-HPLC, FT-IR, and NMR.

Keywords: Bioanalytical, Environmental, Petrochemical, Petroleum

Application Code: Bioanalytical

Methodology Code: Separation Sciences

Session # 2030 Abstract # 2030-24 **Undergraduate Poster Session**

Session Title: Undergraduate Poster Session

Abstract Title: **Investigating Dopamine Fluctuations Associated with Impulsive Decision Making Using Fast Scan Cyclic Voltammetry**

Primary Author: Brennen Guzik
North Carolina State University

Date: Wednesday, March 09, 2016 - After
Time:
Room: Exposition Floor, 400 Aisle

Co-Author(s): Leslie A. Sombers, Xiaohu Xie

Abstract Text

Impulsivity is prevalent in numerous neurological disorders, including attention deficit hyperactive disorder and substance abuse. For the purposes of this study, impulsivity was quantified in a rat animal model using delayed discounting paradigms, which can measure impulsive decision making. In the delayed discounting task, animals were trained in an operant conditioning chamber equipped with visual stimuli, retractable levers and a sucrose pellet dispenser. Animals were allowed to choose either an immediate, small reward or delayed, large reward. Lever availability was indicated by a cue light, and each animal developed a preference for one reward type over the other. Dopamine, an electroactive neurotransmitter in the brain, is highly implicated in motivated behavior and is elicited in the ventral striatum by cues that predict reward availability. Using this animal model, we investigated the role of dopamine in individual difference in delay discounting behavior. Fast scan cyclic voltammetry was used to measure sub-second dopamine release in the shell of the nucleus accumbens in rats performing the delay discounting task. This approach quantified real time changes in concentrations of dopamine elicited in the vicinity of the carbon-fiber microelectrode as the animal made decisions based on learned associations and preference. Based on current data, we observed a positive correlation between the magnitude of cue-induced dopamine release and the preferred choice of reward. In the future, we intend to repeat these experiments under the effects acute nicotine exposure, which has been implicated in enhancing impulsive decision making.

Keywords: Bioanalytical, Electrochemistry, Neurochemistry
Application Code: Neurochemistry
Methodology Code: Electrochemistry

Session Title Undergraduate Poster Session

Abstract Title **Tracking Cellular Invasion and Death in Three Dimensional Paper-Based Cultures with Quantitative PCR**

Date: Wednesday, March 09, 2016 - After

Time:

Room: Exposition Floor, 400 Aisle

Primary Author Christian A. Lochbaum
University of North Carolina at Chapel Hill

Co-Author(s) Andrew S. Truong, Matthew R. Lockett

Abstract Text

Paper–based scaffolds support the prolonged culture of mammalian cells in three-dimensional environments, are easily prepared, and have been used to monitor cellular: growth, response to chemotherapeutics, and invasion. The invasion of cancer cells in the presence of gradients of oxygen, soluble chemokines, or both has been quantified in paper-based invasion assays. This quantification has largely relied on low-resolution fluorescence images, which have a limit of detection of approximately 1000 cells and requires the cells to express a fluorescent protein or be labeled with a small fluorescent molecule. The environment of a solid tumor contains multiple cell types as well as live and dead cells, to accurately mimic this environment in vitro invasion assays must also be able to accommodate these complex environments. The fluorescence-based readouts currently utilized can only differentiate a limited number of fluorophore-labeled cell lines and cannot readily distinguish live and dead cells. We are developing qPCR-based methods to quantify small numbers (< 100) of invasive cancer cells, which have been pre-labeled with unique DNA barcodes, in paper-based invasion assays in the presence of multiple cell types. We also incorporate propidium monoazide, which has been used to eliminate the amplification of genomic DNA from dead bacteria in qPCR assays, into our assays to quantify the number of dead cells in the paper scaffolds. This qPCR-based method of monitoring cell invasion within paper-based scaffolds eliminates many of the downfalls of conventional fluorescence-based methods, while maintaining the high throughput and cost-effective nature of the paper scaffolds.

Keywords: High Throughput Chemical Analysis, Nucleic Acids, Quantitative, Small Samples
Application Code: Bioanalytical
Methodology Code: Process Analytical Techniques

Session # 2030 Abstract # 2030-26 **Undergraduate Poster Session**

Session Title Undergraduate Poster Session

Abstract Title **Real-Time Measurements of Oxidative Stress During Chronic L-DOPA Treatment for Parkinson's Disease**

Primary Author Catherine F. Mason
North Carolina State University

Date: Wednesday, March 09, 2016 - After
Time:
Room: Exposition Floor, 400 Aisle

Co-Author(s)

Abstract Text

Parkinson's disease (PD) is a neurodegenerative disease characterized by the slow degeneration of dopaminergic neurons found in a region of the midbrain called the substantia nigra (SN). Dopamine (DA) plays a key role in regulating motor function. Thus, the destruction of these neurons and the consequential decrease in DA concentrations in the striatum leads to the deterioration of motor control. The drug Levodopa (L-DOPA) has been used to treat PD by helping to increase the concentration of DA in the brain. This drug has been proven to alleviate the motor symptoms of PD; however, after a short period of time, dyskinetic symptoms can develop. It is thought that oxidative stress is a principal contributor to the destruction of dopaminergic neurons, and possibly to the development of dyskinesias, in PD and its treatment. This experiment uses fast-scan cyclic voltammetry (FSCV) coupled with carbon-fiber microelectrodes to monitor the generation of hydrogen peroxide (H2O2) in the SN as an indicator of the presence of oxidative stress. FSCV allows for rapid, real-time, simultaneous measurement of DA and H2O2 in the dorsal striatum. Neurochemical fluctuations can be time-locked to dyskinetic episodes. Overall, these studies will aid in our understanding of how oxidative stress modulates nigrostriatal DA signaling, as well as the behavioral consequences of this interaction. The results will inform improved therapeutic strategies for the treatment of PD.

Keywords: Electrochemistry, Neurochemistry
Application Code: Neurochemistry
Methodology Code: Electrochemistry

Session Title: Undergraduate Poster Session

Abstract Title: **Pulsed Chronopotentiometry with Asymmetric Cellulose Triacetate Membrane-Based Ion-Selective Electrodes for Kinetic Discrimination of Lipophilic Ions During Measurement of Chloride**

Primary Author: Simon Segal
Northern Kentucky University

Date: Wednesday, March 09, 2016 - After
Time:
Room: Exposition Floor, 400 Aisle

Co-Author(s): Jeremy Meyers, Kebede L. Gemene

Abstract Text

Chloride is the most prevalent anion in human blood, at an extracellular concentration of 100-110 mM. Measurement of chloride concentration in blood is of high demand since it is related to prolonged vomiting and respiratory distress [1]. However, lipophilic anions that can be found in blood such as salicylate and thiocyanate, from aspirin metabolism and tobacco smoking, respectively, interfere with chloride measurements with ion-selective electrodes. Therefore, the objective of this project was to alleviate the interference of lipophilic anions on the measurement of chloride by kinetic discrimination using asymmetric cellulose triacetate (CTA) membranes. This is done by casting a thin layer of CTA membrane without sensing components followed by hydrolysis of this barrier layer with NaOH to produce a hydrophilic porous layer. A second layer of CTA with sensing components is casted atop the hydrolyzed layer to create an asymmetric membrane. The hydrolyzed porous barrier layer will allow for the faster passage of small ions like chloride into the sensing layer of the membrane, while retarding the movement of larger ions like salicylate. The small concentration of the lipophilic anions that reaches at the sensing surface will be depleted by extraction into the membrane under pulsed chronopotentiometric measuring mode and will not interfere with the measurement of the abundant hydrophilic chloride ions[2]. In addition to the hydrophilic porous layer, the magnitude and duration of the applied current pulse under pulsed chronopotentiometry can play an important role for kinetic selectivity of hydrophilic anions. Here we have demonstrated very low detection limit of chloride and considerable kinetic discrimination of the interference of lipophilic anions using our sensing protocol.

References

1.Lab Tests Online. https://labtestsonline.org/understanding/analytes/chloride/tab/test/ (accessed on 8/14/2015).
2.Gemene, K. L; Bakker, E. Anal. Chim. Acta. 2007, 583, 190-196

Keywords: Biosensors, Electrochemistry, Ion Selective Electrodes, Potentiometry
Application Code: Biomedical
Methodology Code: Electrochemistry

Session # 2030 Abstract # 2030-28 **Undergraduate Poster Session**

Session Title: Undergraduate Poster Session

Abstract Title: **Water Quality Analysis of the Chattahoochee River**

Primary Author: Kizgel Davis-deSouza
Oglethorpe University

Date: Wednesday, March 09, 2016 - After
Time:
Room: Exposition Floor, 400 Aisle

Co-Author(s): Kelly Jacobson, Md H. Kabir

Abstract Text

The Chattahoochee River which is one of the major sources of water to the Metro Atlanta Area and North Georgia, starts in North Georgia and creates part of the Georgia-Alabama border and some of the Alabama-Florida border. This River is an important part of the ecosystem for humans, land organisms, and aquatic life. Water quality parameters such as temperature, pH, alkalinity, hardness, chemical oxygen demand (COD), biological oxygen demand (BOD), phosphate concentration, lead, arsenic, cadmium, and mercury are determined using chemical and spectrophotometric methods. The data collected over a range of years. The analysis of the current and historical water quality conditions of the Chattahoochee will be reported in this presentation.

Keywords: Environmental/Water
Application Code: Environmental
Methodology Code: UV/VIS

Session # 2030 Abstract # 2030-30 **Undergraduate Poster Session**

Session Title: Undergraduate Poster Session

Abstract Title: **Colloidal CdSe Quantum Dots: Synthesis, Characterization, and Applications**

Primary Author: Kelly Jacobson
Oglethorpe University

Date: Wednesday, March 09, 2016 - After

Time:

Room: Exposition Floor, 400 Aisle

Co-Author(s): Md H. Kabir, Michael Rulison

Abstract Text

Colloidal nanocrystal quantum dots are gaining increasing interest in the recent decades because of their great potential in non-invasive fluorescence imaging, light emitting devices, and photovoltaic technologies. Quantum dot are semiconductors that are small enough in size to display unique optical properties due to the combination of their material band gap energy and quantum well phenomenon. Colloidal cadmium selenide (CdSe) nanocrystals were synthesized and characterized in order to understand their structure and to use them to harvest solar energy. Quantum dots were characterized using UV-Visible spectroscopy and transmission electron microscopy. Quantum dots of different sizes ranging from 1.5-2.7 nanometers were synthesized. Increasing size and a corresponding red shift in the absorbance spectrum was observed for a series of CdSe samples synthesized at progressively higher growth temperatures. The characterization of CdSe quantum dots and the efficiency of the synthesized CdSe-TiO2 photovoltaic cell in harvesting solar energy will be presented in this presentation.

Keywords: Fuels\Energy\Petrochemical, Spectroscopy
Application Code: Fuels, Energy and Petrochemical
Methodology Code: UV/VIS

Session # 2030 Abstract # 2030-31 Undergraduate Poster Session

Session Title: Undergraduate Poster Session

Abstract Title: **Nanoparticle Toxicity on the Development of Brine Shrimp and Zebrafish**

Primary Author: Caitlin May
Oglethorpe University

Date: Wednesday, March 09, 2016 - After
Time:
Room: Exposition Floor, 400 Aisle

Co-Author(s): Gregory Gabriel, Md H. Kabir, Michael Rulison

Abstract Text

Nanoparticles are known to have unique and surprising physicochemical properties that differ from equivalent bulk substances. Despite their increasing presence in commercially produced goods and their applications in medicine, optical, and electronics fields, nanoparticles and their toxicity to humans and the environment are not yet well understood. In this study, the effects of silver nanoparticles on the hatching of brine shrimp eggs were evaluated at different concentrations. First, the nanoparticles were synthesized, then diluted to various concentrations and finally added to brine shrimp eggs, which were monitored closely to track hatching over four days. It was found that brine shrimp vitality decreased at higher concentrations of nanoparticles application. Toxicity effects of silver nitrate and coated silver nanoparticles on brine shrimp eggs and zebrafish will also be presented. The role of environmental conditions (water, plants, sediment) on the toxicity effect of nanoparticles will be discussed in this presentation.

Keywords: Clinical/Toxicology, Environmental/Biological Samples, Toxicology
Application Code: Clinical/Toxicology
Methodology Code: Chemical Methods

Session # 2030 Abstract # 2030-32 Undergraduate Poster Session

Session Title Undergraduate Poster Session

Abstract Title **Evaluation of Phthalate Wiping Protocols for Estimation of Dermal Exposure from Consumer Products**

Primary Author Alexandria Van Grouw
University of Portland

Date: Wednesday, March 09, 2016 - After
Time:
Room: Exposition Floor, 400 Aisle

Co-Author(s) Carla Kagel, Richard Kagel, Stephen McWeeney

Abstract Text
Standard protocols for estimating human exposure to phthalate plasticizers due to dermal contact are not well established and for this reason various approaches have been used without supporting data. GC-MS analysis was utilized in this study to evaluate the efficiency of various wiping procedures as applied to vinyl plastic and how these protocols relate to actual dermal exposure to the common plasticizer, diethyl hexyl phthalate (DEHP).

The relationship between total content, wiping efficiency and correlation to dermal exposure were evaluated for the three most common wiping protocols. Results were found to vary significantly depending on which protocol was used and in some cases vastly overestimated actual dermal exposure.

Keywords: Consumer Products, GC-MS, Sampling, Toxicology
Application Code: Consumer Products
Methodology Code: Gas Chromatography/Mass Spectrometry

Session # 2030 Abstract # 2030-33 **Undergraduate Poster Session**

Session Title: Undergraduate Poster Session

Abstract Title: **Gaseous Molecular Analysis by High Resolution Coherent Multidimensional Spectroscopy (HRCMDS)**

Primary Author: Angelar K. Muthike
Spelman College

Date: Wednesday, March 09, 2016 - After
Time:
Room: Exposition Floor, 400 Aisle

Co-Author(s): Jessica Robinson, Peter Chen

Abstract Text

Polyatomic molecules tend to produce very dense peaks whose interpretation is rather challenging. Different techniques (like Laser Induced Florescence (LIF) and Supersonic Molecular Jet (SMJ)) have been used to extract useful information from the spectra produced by different molecules. However, some gas molecules produce heavily congested spectra due to perturbations. LIF and SMJ are used in order to reduce the congestion, but SMJ reduces rotational information and LIF is not universal. In HRCMDS, the second and the third dimensions of the technique distribute peaks over a large space and create recognizable patterns that can be used for the analysis of gaseous polyatomic molecules. Four Wave Mixing (FWM) signals are produced by selecting certain input laser beam frequencies and an output frequency is detected depending on whether it is a 2D or 3D HRCMDS process.

Keywords: Analysis, Gas, Laser, Spectroscopy
Application Code: Other
Methodology Code: Molecular Spectroscopy

Session # 2030 **Abstract # 2030-34** **Undergraduate Poster Session**

Session Title: Undergraduate Poster Session

Abstract Title: **Spectral Imaging of Plasma Optical Emission Via Compressed Sensing**

Date: Wednesday, March 09, 2016 - After

Time:

Room: Exposition Floor, 400 Aisle

Primary Author: John D. Usala
Texas Tech University

Co-Author(s): Adrian Maag, Gerardo Gamez

Abstract Text

Spectral imaging currently requires array detectors that are very expensive. Further, the acquired image files are typically subjected to compression to keep data manageable which is achieved without loss of critical information, showing the sparsity of the data. Imaging through compressed sensing, however, aims at performing compression during acquisition which results in a more efficient use of experimental resources, such as time and cost.

In this study, a single pixel photon detector utilizing compressive sensing will be constructed and used to obtain optical emission spectral images from atmospheric pressure plasmas. The setup consists of a digital micro-mirror array and a monochromator which is an order of magnitude less expensive than traditional setups using CCD cameras. Spectral images of optical emission from plasma species of interest (He, O, N) will be obtained. The spatial resolution will be characterized as a function of experimental and image processing parameters such as sensing matrix selection and recovery algorithm choice. Further, the spectral images will be compared to a more expensive push-broom hyperspectral imaging system featuring an ICCD camera.

Keywords: Atomic Emission Spectroscopy, Charge Transfer Devices (CID CCD), Sampling
Application Code: Other
Methodology Code: Atomic Spectroscopy/Elemental Analysis

Session # 2030 Abstract # 2030-35 **Undergraduate Poster Session**

Session Title: Undergraduate Poster Session

Abstract Title: **Analysis of Melamine in Solid Pet Food Samples Using Gold Nanoparticles and Reversed Phase Chromatography**

Primary Author: Aaron Hummert
Washburn University

Co-Author(s): Seid Adem

Date: Wednesday, March 09, 2016 - After
Time:
Room: Exposition Floor, 400 Aisle

Abstract Text

Melamine is a small organic compound rich in nitrogen and was implicated in the recalls of pet and infant foods in 2007, as it was deliberately added to boost the apparent protein content of such foods. In this study, a sensitive, repeatable and a rather simple method was developed based on gold nanoparticles (AuNPs) for detection of melamine in pet foods. The method of detection in this study is based on the measurement of AuNP's aggregation in the presence of melamine. The color of the well dispersed AuNPs is wine red and [lambda][sub]max[/sub] for its surface plasmon resonance (SPR) is centered at about 521 nm. The AuNPs undergo aggregation in the presence of melamine and there is a clear shift of the absorption peak to longer wavelengths and broadening of the SPR band. The color of gold nanoparticles changed from wine red to blue in the presence of melamine enabling visual detection and no color change was observed before melamine was introduced into the food sample. The method involves less sample preparation and is thus suitable as a portable method for detection of melamine in food products. Furthermore, the method is so sensitive that a limit of detection about 0.56 ppm was obtained, which is well below the threshold set by the United States Food and Drug Administration. The results of the new method were also verified using reversed phase chromatography with a diode-array detection at 236 nm with a commonly used mobile phase (38% methanol/ 61% water/1% glacial acetic acid) and resorcinol as internal standard. A working calibration curve of melamine was constructed to determine the limit of detection in aqueous solution and food matrix samples and a good correlation between instrument response and melamine concentration was obtained with R[sup]2[/sup] values of 0.999 and 0.997, respectively. The percent recovery of melamine in pet food samples that have been spiked with a known amount of melamine ranges from 92.7 ± 1.9% - 107.6 ± 2.8% (n = 6).

Keywords: Analysis, Chromatography, Food Contaminants, HPLC
Application Code: Food Contaminants
Methodology Code: Liquid Chromatography

Session # 2030 Abstract # 2030-37 **Undergraduate Poster Session**

Session Title: Undergraduate Poster Session

Abstract Title: **Does Grinding Glassy Carbon Electrodes in Diazonium Salts Lead to Covalently Bonded Groups?**

Primary Author: Chelsea L. Horvath
Wittenberg University

Date: Wednesday, March 09, 2016 - After
Time:
Room: Exposition Floor, 400 Aisle

Co-Author(s): Kristin K. Cline

Abstract Text

Diazonium ions are known to readily react with carbon electrodes to yield grafted surfaces. Typically this grafting is effected with electrochemical reduction or occurs spontaneously in solution. Previous work in our lab showed that grinding glassy carbon electrodes in dry diazonium salts yields a modified surface. This solid-state grafting strategy minimizes solvent waste and eliminates solution reactions that produce multilayers on the surface. The objective of the present study is to optimize conditions for this process and investigate the nature of the modifying groups.

Glassy carbon electrodes were conventionally polished and sonicated before being ground in dry diazonium salt—either anthraquinone-1-diazonium hemi(zinc chloride) salt or 4-nitrobenzenediazonium tosylate. Control electrodes were exposed to diazonium salt without grinding. Electrodes were rinsed and sonicated to remove physisorbed groups.

Cyclic voltammetry of modified electrodes showed peaks characteristic of attached nitrophenyl or anthraquinone groups, which were integrated to calculate surface coverage. For electrodes that underwent sonication in acetone and toluene, the surface coverage of anthraquinone groups is 0.36 ±0.05 nmol/cm[sup]2[/sup] and of nitrophenyl groups is 0.35 ± 0.02 nmol/cm[sup]2[/sup]; with further sonication in toluene or voltammetric scanning in alkaline solution, values reach a lower limit of 0.3 and 0.2 nmol/cm[sup]2[/sup], respectively. The larger decrease in apparent coverage of nitrophenyl groups may be due to irreversible conversion to aminophenyl groups. We conclude that the grinding process is complicated by the physical adsorption of byproducts; however, with sufficient sonication, strongly bonded groups remain, at submonolayer coverage.

This work was made possible by the Virginia Ellis Franta Fund for Chemistry.

Keywords: Chemically Modified Electrodes, Electrochemistry, Electrode Surfaces, Electrodes
Application Code: Material Science
Methodology Code: Electrochemistry

Session # 2030 Abstract # 2030-38 **Undergraduate Poster Session**

Session Title: Undergraduate Poster Session

Abstract Title: **Comparison of Grafted and Untreated Activated Carbon as Solid Phase Extraction Media for Preconcentrating Copper and Lead Ions**

Date: Wednesday, March 09, 2016 - After

Time:

Room: Exposition Floor, 400 Aisle

Primary Author: Margaret Cole
Wittenberg University

Co-Author(s): Kristin K. Cline

Abstract Text

Activated carbon has been used as solid phase extraction (SPE) media, typically for organic compound extraction. Organic ligands have been adsorbed to carbon particles to yield an extraction phase for metal ions. Since diazonium ions are known to react readily with carbon materials to yield modified surfaces, this chemistry may provide a convenient and versatile route to producing new SPE phases for a variety of applications. The objective of this study is to produce diazonium ion-grafted carbon particles for use as SPE media for the extraction of metal ions. To our knowledge, this strategy has not yet been reported.

In this study, we synthesized carboxybenzenediazonium and nitrobenzenediazonium ions in the presence of carbon particles. Cyclic voltammetry of the product solution showed characteristic diazonium reduction peaks. Carbon paste electrodes made from modified particles produced voltammetric signatures for attached nitrophenyl groups.

Untreated and carboxyphenyl-grafted carbon particles were packed into tubes for extraction of aqueous copper (II) and lead (II) solutions. Eluted metal ions were quantified with flame atomic absorption spectroscopy. Initial results showed no significant difference for copper and lead extraction on unmodified vs. carboxybenzene-grafted carbon particles. Both effectively trapped metal ions from dilute solutions, with tested concentration factors of 5-20 yielding 90% or better recovery.

This work was made possible by the Virginia Ellis Franta Fund for Chemistry.

Keywords: Atomic Absorption, Electrochemistry, Solid Phase Extraction
Application Code: Environmental
Methodology Code: Separation Sciences

Session Title Undergraduate Poster Session

Abstract Title **Barium Leaching in Ceramic Glazes**

Primary Author Jason Halmo
Hampden-Sydney College

Date: Wednesday, March 09, 2016 - After
Time:
Room: Exposition Floor, 400 Aisle

Co-Author(s) Joshua Chamberlin, Paul Mueller

Abstract Text

Systematic investigations into Barium-based glazes and their leaching properties that have been reported in the chemical literature are scarce. Through the use of Microwave Plasma Atomic Emission Spectroscopy (MP-AES), this study, using statistical sampling, provides numerical concentration values of leached Barium ions for several glazes prepared with varying levels of Barium and Copper Carbonates and fired at several temperatures. Barium ions were leached by utilizing 4% citric and 4% acetic Acid. General trends showed that values for leached Barium ions are less at higher temperatures then when compared to lower temperatures. These numbers will be compared to known allowable values for Barium to determine if these glazes present a health hazard. Additionally, a colorimetric study was conducted to determine the degree of color changes due to leaching in glazes exposed to citric and acetic acids for extended time periods.

Keywords: Atomic Spectroscopy, Paint/Coatings
Application Code: Art/Archaeology
Methodology Code: Atomic Spectroscopy/Elemental Analysis

Session # 2030 Abstract # 2030-40 **Undergraduate Poster Session**

Session Title Undergraduate Poster Session

Abstract Title **Spectroelectrochemical Urinalysis: A Kinetic Assay for Uric Acid**

Date: Wednesday, March 09, 2016 - After
Time:
Room: Exposition Floor, 400 Aisle

Primary Author Paul Flowers
University of North Carolina at Pembroke

Co-Author(s) Sean Downes, Sonvia Brown

Abstract Text

Spectroelectrochemistry (SEC) refers to measurements involving the simultaneous application of spectral and electrochemical techniques. First reported in the early 1960s, SEC methods have become well established tools for fundamental studies of redox chemistry. In the more recent past, increased attention has been given to the use of SEC techniques for purely analytical applications, i.e., as the basis for quantitative assays. Research in our laboratory has been focused on the development of SEC-based assays for various small molecules of biomedical significance. Potential benefits of the SEC approach relative to existing assays include decreased cost due to the elimination of expensive reagents; increased sample throughput due to the relatively short analysis time; and the possibility of simultaneous multi-analyte determinations due to the selective control of electrolysis potential and measurement wavelength. In this poster, we describe a novel kinetic SEC assay for uric acid in human urine. The relatively slow post-electrolysis decomposition of the uric acid electro-oxidation product permits discrimination of its signal from that of common interfering species, for example, ascorbic acid. The kinetic SEC assay exhibits a sensitivity and precision comparable to standard clinical assays, and it effectively eliminates interference from ascorbic acid at clinically relevant concentrations. Results obtained for split urine samples subjected to both the SEC analysis and a standard colorimetric assay show a significant positive correlation, suggesting the SEC approach could be successfully implemented in clinical settings.

Keywords: Clinical Chemistry, Spectroelectrochemistry
Application Code: Clinical/Toxicology
Methodology Code: Electrochemistry

Session # 2030 **Abstract # 2030-41** **Undergraduate Poster Session**

Session Title: Undergraduate Poster Session

Abstract Title: **Determination of Formaldehyde Concentration in Electronic Cigarettes**

Primary Author: Roland Landers
Cumberland University

Co-Author(s): Sarah Pierce

Date: Wednesday, March 09, 2016 - After
Time:
Room: Exposition Floor, 400 Aisle

Abstract Text

Electronic cigarettes are often marketed as a safer alternative to smoking and have now become a part of popular culture appealing even to non-smokers. Recent studies in various academic journals have reported high amounts of formaldehyde in electronic cigarette vapor. Formaldehyde, a cancerous compound used in embalming fluid, is generated in electronic cigarettes through the partial combustion of the oil-based reagents that constitute the greater part of "e-juice." Visible spectroscopy was used to detect the concentration of formaldehyde present in electronic cigarette vapor at various degrees of voltage by condensing the vapor. A colorimetric pentane-2,4-dione assay was used to determine concentration. Additionally, the effects of nicotine concentration in relation to formaldehyde concentration were also tested. The detection of formaldehyde at toxic levels in electronic cigarette vapor is an important step to debunking the belief in electronic cigarettes as a safer alternative to smoking.

Keywords: Consumer Products, Detection, Drugs, UV-VIS Absorbance/Luminescence
Application Code: Consumer Products
Methodology Code: UV/VIS

Session # 2030 Abstract # 2030-42 **Undergraduate Poster Session**

Session Title: Undergraduate Poster Session

Abstract Title: **Matrix Targeting Peptide Impact on Tertiary Folding of Cargo Proteins**

Primary Author: Tyler J. Smith
Truman State University

Co-Author(s): Bethany P. Manning, Brian P. Adams

Date: Wednesday, March 09, 2016 - After
Time:
Room: Exposition Floor, 400 Aisle

Abstract Text

Proteins that are destined for the mitochondria are typically synthesized in the cytosol of the cell and must be transported into the organelle via carrier proteins, chaperones, translocases, or a combination thereof. N-terminal mitochondrial targeting sequences (MTSs) are responsible for the trafficking of these proteins into the mitochondria. MTSs range from 20 to 40 amino acids in size and unlike other targeting peptides, MTSs have a very low homology. Our previous work has suggested that the MTS may affect the folding of its cargo protein. This effect could keep the protein in a partially folded, molten globule state which would assist in the translocation of the protein across the outer and inner mitochondrial membrane. To investigate this effect, we fused MTS sequences to green fluorescent protein (GFP), a protein that fluoresces based on its overall tertiary fold. This allows for measurement of change in GFP folding by monitoring of fluorescence. Preliminary results show that the MTS of aldosterone synthase, a protein targeted to the mitochondrial matrix, impacts the folding of AS in comparison to scrambled control sequences.

Keywords: Amino Acids, Fluorescence, Protein
Application Code: Biomedical
Methodology Code: Fluorescence/Luminescence

Session # 2030 Abstract # 2030-43 **Undergraduate Poster Session**

Session Title: Undergraduate Poster Session

Abstract Title: **Electrospray Mass Spectrometry and Density Functional Theory Studies of the Estrone Fragmentation Mechanisms**

Primary Author: Yassin Jeilani
Spelman College

Date: Wednesday, March 09, 2016 - After
Time:
Room: Exposition Floor, 400 Aisle

Co-Author(s): Daphney Sihwa, Gabrielle Webb, Nasrin Aweis

Abstract Text

Estrone has been shown to influence growth, body development, behavior and regulation of reproductive cycles in animals. It has been often targeted in biological samples by liquid chromatography coupled with tandem mass spectrometry (MS/MS). Understanding the MS/MS fragmentation pathways is important for proper identification of estrone in complex mixtures as well as in complex biological matrices. In this project, we study the fragmentation pathways of estrone by collision induced dissociation and density functional theory (DFT) at B3LYP/6-311G(d,p) level of theory. The product ion spectrum of the [M+H]+ peak of estrone is complex and shows major peaks at m/z: 253, 225, 213, 197, 183, 173, 157, 145, 133, and 121. The pathways involve rearrangements and neutral eliminations. The rearrangements are based on both charge-induced and charge-remote steps. DFT results were used to determine energy barriers and structures of the product ions. These pathways provide proper identification of product ions which are important for proper identification of estrone.

Keywords: Bioanalytical, Mass Spectrometry, Organic Mass Spectrometry, Tandem Mass Spec
Application Code: Bioanalytical
Methodology Code: Mass Spectrometry

Session Title: Undergraduate Poster Session

Abstract Title: **Method Development for the Analysis of Pesticide Degradates by GC-ECD**

Primary Author: Jessica Reilly
Saint Francis University

Co-Author(s): Samantha Radford

Date: Wednesday, March 09, 2016 - After
Time:
Room: Exposition Floor, 400 Aisle

Abstract Text

Pesticides are commonly used worldwide to keep insects from destroying agricultural crops. Even though these pesticides help to keep food costs down, they may be harmful to the human body. Children are at the greatest risk of being negatively affected by pesticides, because the brain is still developing and especially susceptible to neurotoxicants. Further, due to the small size of a child and their greater food intake per kilogram body weight, children receive more concentrated doses of these compounds. Because of the developmental concerns surrounding pesticides, we need robust measurements of human exposure to them. A common way of measuring exposure to pesticides is by tracking the pesticide metabolites in urine. However, if pesticide degradates are preformed in the environment, tracking exposure through urinary metabolites may cause overestimation of exposure. Therefore, we need to know if these pesticides break down in food before it is ever eaten. The pesticides investigated in this research are chlorpyrifos and diazinon. Chlorpyrifos breaks down to form the specific metabolite 3,5,6-trichloro-2-pyridinol (TCPy) and diazinon breaks down to form 2-isopropyl-6-methyl-4-pyrimidinol (IMP). A previously developed method for the extraction of pesticide degradates in baby food and juice was modified with a derivatization step to allow for analysis by gas chromatography with electron capture detection. In the future, this method will be used to study pesticide breakdown in fruit juices, which are popular foods for children. This research will lead to a better understanding of pesticide breakdown in the environment and to a more complete assessment of human exposure to these toxic compounds.

Keywords: Food Safety, Gas Chromatography, Method Development, Pesticides
Application Code: Food Contaminants
Methodology Code: Gas Chromatography

Session Title Undergraduate Poster Session

Abstract Title **Investigation of Iron Dissolution from Pyrite Electrodes Using Electrochemical and Atomic Absorption Methods**

Date: Wednesday, March 09, 2016 - After

Time:

Room: Exposition Floor, 400 Aisle

Primary Author Katelyn Nusbaum
Saint Francis University

Co-Author(s) Brandyn Pryce, Rose A. Clark

Abstract Text

Pyrite is a common metal found in coal mines and contaminates local acid mine drainage (AMD) sites. The environmental conditions where the pyrite is found will affect the dissolution rate of the mineral. To better understand how pyrite dissolution is affected by the environment, pyrite electrodes were created and examined electrochemically under varying solution conditions. The electrodes were tested in 20 mM H2SO4 and 33 mM Na2H2SO4 solution at pH 1.7 and the electrochemical response was found to be stable and reproducible. The electrodes were then studied in 20 mM H2SO4 and 33 mM Na2H2SO4 solutions at varying pHs to evaluate how pyrite dissolution is affected. The pyrite dissolution rate increases with increasing acidity of the aqueous solution. Solutions were also investigated from a local AMD remediation site from the top and bottom of the weir. Cyclic voltammetry was used to examine electrode performance and reproducibility during the pyrite dissolution. Tafel plots were implemented to find the dissolution rates (corrosion rates) for each electrode during the varied solution condition experiments.

?

Bulk electrolysis of the pyrite electrodes at varying potentials was also conducted. Iron concentrations in the bulk solution were determined using Atomic Absorption with a Graphite Furnace. The relationship between iron concentration, potential, and solution conditions will be discussed.

Keywords: Atomic Absorption, Electrochemistry, Electrode Surfaces
Application Code: Environmental
Methodology Code: Electrochemistry

Session # 2030 Abstract # 2030-47 **Undergraduate Poster Session**

Session Title: Undergraduate Poster Session

Abstract Title: **The Efficacy of Duckweed in Reducing the Concentration of Manganese in Abandoned Mine Drainage (AMD) through Phytoremediation**

Primary Author: Rebecca A. Bradnam, Westminster College

Co-Author(s): Helen M. Boylan

Date: Wednesday, March 09, 2016 - After
Time:
Room: Exposition Floor, 400 Aisle

Abstract Text

Coal mines from the late 1800's and early 1900's have left stream water contaminated because of mine drainage, which is often characterized by low pH and high metal content. Western Pennsylvania has been particularly hard hit by this pollution source, with "yellow boy" coating numerous stream beds. While passive treatment of AMD has been successful in improving most water quality parameters, the removal of manganese from AMD through passive treatment remains challenging. Duckweed, a common, small aquatic plant that floats on the surface of slow-moving water, has been previously demonstrated to remove metal contamination, including manganese, from water through the process of phytoremediation. In this research, a particular species of duckweed, Lemna minor, is being investigated. A preliminary study was performed to determine viable manganese concentrations and timeline for manganese exposure. In this study, duckweed was exposed to 0, 5, 15, and 25 µg/mL Mn over 9 days. The water was collected on day 0, 3, 6, and 9, and on day 9, the plants were collected to test for manganese uptake. Manganese analyses were performed by inductively coupled plasma-optical emission spectroscopy (ICP-OES) following sample preparation. The results from this preliminary study have been used to establish the conditions for a more complete experiment with replication, in which duckweed is exposed to real-world AMD. The goals of the study are to investigate the efficacy of duckweed in decreasing manganese concentration in contaminated water and to determine at what concentration this phytoremediation process would be viable.

Keywords: Atomic Spectroscopy, Education, Environmental/Biological Samples, Environmental/Water

Application Code: Environmental

Methodology Code: Atomic Spectroscopy/Elemental Analysis

Session # 2030 Abstract # 2030-48 **Undergraduate Poster Session**

Session Title: Undergraduate Poster Session

Abstract Title: **Analysis of Water Quality in Western Pennsylvania near Hydraulic Fracturing Sites**

Primary Author: Kelsey A. Kilbane
Westminster College

Date: Wednesday, March 09, 2016 - After
Time:
Room: Exposition Floor, 400 Aisle

Co-Author(s): Christina Mauri, Helen M. Boylan, Jamie Linderman

Abstract Text

Lawrence and Mercer counties, rural areas in western Pennsylvania along the Ohio border, have recently been the target of expanded development of unconventional oil and gas wells. Although the risk of surface water contamination from fracking is low, our community is committed to being vigilant of our stream quality. Six stream locations were chosen to be monitored based on their proximity to an existing or planned unconventional well and accessibility. These streams are being monitored routinely during the academic year, and for a 6-week duration, samples were collected on a weekly basis at upstream and downstream locations of each site. The tests conducted in the field on these streams were pH, total dissolved solids (TDS), conductivity, chloride, alkalinity, iron, hardness, total, calcium and turbidity, and a visual assessment was performed each time the sites were monitored. The water samples were also analyzed for metal content, including barium, strontium, calcium, and iron, by inductively coupled plasma-optical emission spectroscopy (ICP-OES). One of the seven sites that was monitored is being used to establish baseline water quality data near a planned well location. This planned fracking site is located in Wilmington Township (Lawrence County), a township that has recently considered controversial zoning ordinances for the oil and gas industry. Its location, in an agricultural zone close to Westminster College (and its pristine Field Station), is a great cause for concern to the community. The collected data is being shared with the public via the Lawrence and Mercer ALLARM webpage, www.lawmerallarm.org.

Keywords: Environmental Analysis, Environmental/Water, Fuels\Energy\Petrochemical, ICP
Application Code: Environmental
Methodology Code: Atomic Spectroscopy/Elemental Analysis

Session # 2030 Abstract # 2030-49 **Undergraduate Poster Session**

Session Title: Undergraduate Poster Session

Abstract Title: **Electrophoretic Character of Borate Buffers in Capillary and Microfluidic Channels**

Primary Author: Launick Saint-Fort
Pennsylvania State University-Berks Campus

Date: Wednesday, March 09, 2016 - After
Time:
Room: Exposition Floor, 400 Aisle

Co-Author(s): James Karlinsey

Abstract Text

This work describes the investigation of electroosmotic flow (EOF) in glass capillaries and microfluidic channels. EOF is a phenomenon that exists when voltage is applied to electrolytic buffers in microscale flow channels, resulting in bulk flow of buffer through the channel in response to the electric field. EOF is evaluated and reported here in both glass and poly(dimethyl siloxane) (PDMS) substrates, including glass capillary coated with PDMS and hybrid glass-PDMS microfluidic devices. Devices with various parameters (i.e., cross-section, length) were designed using AutoCAD software and fabricated in-house using a CNC mill to create molds onto which PDMS was cast. EOF was evaluated because it is the dominant flow employed in capillary electrophoresis (CE) separations and an important consideration in microchip electrophoresis (ME). Many reports of integrated microfluidic analysis, with applications including protein and DNA analysis, have featured ME, but there is significantly less attention paid to EOF in favor of the more complex analysis steps. In this work, borate buffer was employed due to its widespread use in both CE and ME, monitoring EOF in response to varied electrophoretic parameters (i.e., composition, ionic strength). EOF was evaluated using current measurements and collecting fluorescent signal from charged and uncharged analytes, providing valuable information regarding the different factors affecting separations in capillaries and microfluidic devices.

Keywords: Capillary Electrophoresis, Lab-on-a-Chip/Microfluidics, Separation Sciences
Application Code: General Interest
Methodology Code: Capillary Electrophoresis

Session # 2030 Abstract # 2030-50 **Undergraduate Poster Session**

Session Title Undergraduate Poster Session

Abstract Title **Interdisciplinary Undergraduate Research in Chemometrics: The Students' Perspective**

Primary Author Stephanie Homitz
Westminster College

Date: Wednesday, March 09, 2016 - After
Time:
Room: Exposition Floor, 400 Aisle

Co-Author(s) Carolyn Cuff, Christopher Caroff, Helen M. Boylan, Keilah Ireland

Abstract Text

At Westminster College we have established an ongoing undergraduate research project in the field of chemometrics. This research is a collaboration involving an analytical chemist, a statistician, undergraduate chemistry students, and undergraduate mathematics students. The language barrier between the fields of chemistry and math presents a significant challenge for cross-communication between the two disciplines, but regular interdisciplinary meetings allow us to overcome this difficulty. The statistical analysis of data sets that both disciplines can make sense of (body fat, solar energy production) have been key to our ability to progress to complex chemical data sets. This presentation will focus on the elemental analysis of a wide range of pet food samples by ICP-OES and laser induced breakdown spectroscopy and the subsequent chemometric analysis of the data.

Keywords: Chemometrics, Consumer Products, Education, Plasma Emission (ICP/MIP/DCP/etc.)
Application Code: Other
Methodology Code: Education/Teaching

Session # 2030 **Abstract # 2030-51** **Undergraduate Poster Session**

Session Title: Undergraduate Poster Session

Abstract Title: **Analysis of the Toluene Efflux Pumps in Microorganisms Through Bioinformatics**

Primary Author: Mathilda Willoughby
Westminster College

Co-Author(s): Samantha Tower, Sarah Kennedy

Date: Wednesday, March 09, 2016 - After
Time:
Room: Exposition Floor, 400 Aisle

Abstract Text

Some bacteria exhibit tolerance to organic solvents such as toluene because of the presence of a toluene efflux pump that removes toluene from the cell. Efflux pumps are classified by five different categories and act as transporters located in the cytoplasmic membrane that require chemical energy for their performance. Identifying and analyzing the genes responsible for these pumps can contribute to future work in environmental and biotechnology industries. In this work, the genes for toluene efflux pumps from multiple microorganisms were explored using an online bioinformatics program called GENI-Act through the Microbial Genome Annotation Network (MGAN).

Keywords: Bioinformatics, Biotechnology, Genomics, Protein
Application Code: Genomics, Proteomics and Other 'Omics
Methodology Code: Computers, Modeling and Simulation

Session # 2030 Abstract # 2030-53 **Undergraduate Poster Session**

Session Title: Undergraduate Poster Session

Abstract Title: **A Cascade SERS Signal Amplification Approach for Telomerase Activity at Single-Cell Level**

Primary Author: Muling Shi
Hunan University

Date: Wednesday, March 09, 2016 - After
Time:
Room: Exposition Floor, 400 Aisle

Co-Author(s)

Abstract Text

As an important biomarker and therapeutic target, telomerase has attracted extensive attention concerning its detection and monitoring. Recently, enzyme-assisted amplification approaches have provided useful platforms for the telomerase activity detection, however, further improvement in sensitivity is still hindered by the single-step signal amplification. Herein, we develop a quadratic signal amplification strategy for ultrasensitive surface-enhanced Raman scattering (SERS) detection of telomerase activity. The central idea of our design is using telomerase-induced silver nanoparticles (AgNPs) assembly and silver ions (Ag+)-mediated cascade amplification. In our approach, each telomerase-aided DNA sequence extension could trigger the formation of a long double-stranded DNA (dsDNA), making numerous AgNPs assembling along with this long strand through specific Ag–S bond, to form a primary ampli□cation element. For secondary ampli□cation, each conjugated AgNP was dissolved into Ag+, which can effectively induce the 4-aminobenzenethiol (4-ABT) modified gold nanoparticles (AuNPs@4-ABT) to undergo aggregation to form numerous "hot-spots". Through quadratic ampli□cations, a limit of detection down to single HeLa cell was achieved. More importantly, this method demonstrated good performance when applied to tissues from colon cancer patients, which exhibits great potential in the practical application of telomerase-based cancer diagnosis in early stages. To demonstrate the potential in screening the telomerase inhibitors and telomerase-targeted drugs, the proposed design is successfully employed to measure the inhibition of telomerase activity by 3'-azido-3'-deoxythymidine.

Keywords: Analysis, Biological Samples, Biosensors, Raman
Application Code: Bioanalytical
Methodology Code: Chemical Methods

Session # 2040 **Abstract # 2040-1** **Symposia**

Session Title ACS-ANLY - Advances in Electrokinetic Methods for Bioanalysis

Abstract Title **Nanofluidic Devices for Single-Particle Analysis of Virus Assembly**

Date: Thursday, March 10, 2016 - Mornin
Time: 08:35 AM
Room: B308

Primary Author Stephen C. Jacobson
Indiana University

Co-Author(s) Adam Zlotnick, Lisa Selzer, Zachary D. Harms

Abstract Text

We are using resistive-pulse sensing as a label-free, nondestructive technique to characterize the assembly of Hepatitis B Virus Cp149 dimers into T = 3 and T = 4 symmetry capsids. This single-particle counting technique permits real-time detection of both capsid formation and intermediate depletion and has a sufficient sensitivity to monitor assembly at dimer concentrations as low as 50 nM, well below the pseudo-critical dimer concentration. In 1 M NaCl, assembly reactions below, near, and above the pseudo-critical dimer concentration reveal three distinct phases of assembly. Below the pseudo-critical dimer concentration, the ratio of T = 3 to T = 4 capsids increases with decreasing dimer concentration. Above the pseudo-critical dimer concentration where kinetic traps form, pre-T = 4 particles assemble rapidly and slowly anneal into T = 4 capsids. At all dimer concentrations tested, the T = 3 capsids formed more rapidly than T = 4 capsids, suggesting distinct pathways for the two forms.

Keywords: Bioanalytical, Lab-on-a-Chip/Microfluidics, Nanotechnology
Application Code: Bioanalytical
Methodology Code: Microfluidics/Lab-on-a-Chip

Session # **2040** Abstract # **2040-2** **Symposia**

Session Title ACS-ANLY - Advances in Electrokinetic Methods for Bioanalysis

Abstract Title **Amyloid Oligomers Analysis Using Microchannel Electrophoresis**

Date: Thursday, March 10, 2016 - Mornin
Time: 09:10 AM
Room: B308

Primary Author Christa Hestekin
University of Arkansas

Co-Author(s) Melissa Moss, Sadia Paracha

Abstract Text

The aggregation of proteins into amyloids is involved in a number of diseases including Alzheimer's Disease and type 2 diabetes. A number of technical challenges exist in performing quantitative analysis of the formation of different sizes of amyloid oligomers which are present during the early stages of aggregation, therefore requiring new techniques to be developed. Microchannel electrophoretic techniques have emerged as powerful tools for the quantitative analysis of proteins. In this study, we investigated the ability of capillary and microchip electrophoresis to monitor the early stages of amyloid aggregation using amyloid beta (A) and amylin as a model amyloid-forming proteins. Microchannel electrophoresis demonstrated differences in the aggregation based on sample preparation as well as the use of a fluorescent label. In addition, a coating polymer was found to significantly decrease the amount of protein adsorption. Field amplified sample stacking was used to increase amount of sample detected. Due to the highly unstable nature of the small oligomers, photo-crosslinking was also used to stabilize the oligomers and compared with uncrosslinked samples.

Keywords: Analysis, Protein
Application Code: Biomedical
Methodology Code: Capillary Electrophoresis

Session # 2040 Abstract # 2040-3 **Symposia**

Session Title: ACS-ANLY - Advances in Electrokinetic Methods for Bioanalysis

Abstract Title: **Surface Isoelectric Focusing (SIEF) for Therapeutic Protein Separations**

Date: Thursday, March 10, 2016 - Mornin
Time: 09:45 AM
Room: B308

Primary Author: Adrienne R. Minerick
Michigan Technological University

Co-Author(s): Zhichao Wang

Abstract Text

Our previously reported surface isoelectric focusing (sIEF) technique illustrated the potential to resolve protein molecules from a complex mixture at micrometer scale, which allows sIEF to be integrated with protein array libraries. This property allows sIEF to characterize proteins on surfaces in situ via surface accessible spot arrays. Glycoengineering of functional proteins has been shown to improve molecular stability, regulate physicochemical and pharmacological properties, and improve pharmacokinetics with better absorption and longer circulation times. Degree of glycosylation can be accessed with sIEF in situ, in parallel, with <10 min run times enabling automation, high-resolution, low cost, and rapid analysis for pharmaceutical and clinical glycoprotein analysis.

The sIEF devices utilize PharmalyteTM solutions in either pH 3 - 10 or 6.7 - 7.7 ranges are co-printed with an acrylamide monomer solution and then polymerized in-situ spanning two microfabricated electrodes. Gel-resolving power is maximized with coprinted additive chemicals and dielectric surface coatings over electrodes. Glycosylated therapeutic monoclonal antibodies (mAb) are printed on the gel, the separation electrodes are energized, and focusing is optically observed over 6 to 9 minutes. Our results illustrate optimized sIEF with a continuous 3-10 pH gradient to focus and detect mAb; protein transfer from adjacent protein arrays is demonstrated using a surface-patterning tool. The advantages of sIEF, such as lower applied voltages, smaller sample requirements and device reusability will be reviewed, which make sIEF an attractive tool for glycoprotein analysis within pharmaceutical and clinical settings.

Keywords: Biopharmaceutical, Electrophoresis, Proteomics
Application Code: Pharmaceutical
Methodology Code: Capillary Electrophoresis

Session # 2040 Abstract # 2040-4 **Symposia**

Session Title: ACS-ANLY - Advances in Electrokinetic Methods for Bioanalysis

Abstract Title: **Microfluidic-Based Electrokinetic Methods for Protein Separation and Sensing and Manipulation of Particles and Droplets**

Primary Author: Carolyn L. Ren
University of Waterloo

Date: Thursday, March 10, 2016 - Mornin
Time: 10:35 AM
Room: B308

Co-Author(s)

Abstract Text

Protein separation is instrumental for many applications such as disease diagnosis and drug screening. There have been many benchtop systems developed for protein separation using different mechanisms. Microfluidic-based methods offer unique advantages over traditional systems owing to its small scale that results in reduced reagent consumption and thus low cost, and its capability for multiplexing and integration which is promising for multidimensional protein separation. This talk consists of two sections. The first half of the talk focuses on introducing microfluidic-based electrokinetic methods for protein separation using isoelectric focusing, temperature gradient focusing and two-dimensional separation techniques. The second half of the talk will mainly discuss sensing and manipulation of (bio)particles and droplets using microfluidic-based electrokinetic methods.

Keywords: Biosensors, High Throughput Chemical Analysis, Lab-on-a-Chip/Microfluidics, Microwave
Application Code: High-Throughput Chemical Analysis
Methodology Code: Microfluidics/Lab-on-a-Chip

Session # 2050 Abstract # 2050-1 **Symposia**

Session Title: Advances in Vibrational Spectroscopy for Medical Diagnostics

Abstract Title: **Stimulated Raman Scattering Microscopy as a Tool for Brain and Skin Cancer Tissue Diagnoses**

Date: Thursday, March 10, 2016 - Mornin
Time: 08:35 AM
Room: B302

Primary Author: Wenlong Yang
Harvard University

Co-Author(s): Sunney Xie

Abstract Text

Stimulated Raman scattering (SRS) microscopy is a label-free and noninvasive imaging technique using vibration spectroscopy as the contrast mechanism. SRS has opened a wide range of biomedical applications since it provides instant tissue examination without the need of previous histological staining, does not affect cell function, and is best suited for imaging small metabolite molecules. The ability to distinguish normal versus cancer cells is key during surgery to avoid removal of healthy tissue and secure complete elimination of cancer cells, especially important during brain tumor surgery. Because tumor cells are rich in proteins, we have shown that SRS microscopy allows for clear distinction between tumor and non-tumor cells and visualization of the localization of tumor infiltration in areas that appeared normal by eye, making SRS a valuable tool to be used during surgery.

We have also applied SRS to study the distribution of DNA and the changes on DNA condensation in the nucleus of a single cell. SRS is a fast, reliable in vivo and in real time label-free histology method that provides comparable results to other conventional tissue staining methods such as H&E. We expect that in vivo counting of the mitotic rate using SRS may not only speed up Mohs surgery by on-site label-free imaging of tumor tissue with margins, but it could also have the potential for in vivo noninvasive detection and progress evaluation of skin lesions in real time. Thus, SRS microscopy allows a label-free, noninvasive and rapid tissue evaluation during cancer surgery.

Keywords: Biomedical, Molecular Spectroscopy, Raman, Vibrational Spectroscopy

Application Code: Biomedical

Methodology Code: Vibrational Spectroscopy

Session # **2050** Abstract # **2050-2** **Symposia**

Session Title Advances in Vibrational Spectroscopy for Medical Diagnostics

Abstract Title **Raman Spectroscopy for Clinical Cell Analysis**

Primary Author Juergen Popp
Friedrich-Schiller University Jena

Date: Thursday, March 10, 2016 - Mornin
Time: 09:10 AM
Room: B302

Co-Author(s)

Abstract Text

Raman spectroscopy has shown to be a powerful tool to study biological cells. The focus of this presentation is concerned with Raman studies on eukaryotic cells for biomedical applications like e.g. better surveillance of cancer onset and its treatment by detecting tumor cells circulating in body liquids. We will highlight the potential of Raman spectroscopy for a label-free discrimination between normal and tumor cells based on their biochemical composition or towards establishing a Raman spectroscopic hemogram i.e. characterizing leukozytes. Thereby cellular Raman spectra were recorded after drying, in laser tweezers or trapped in a microfluidic environment. In particular we will report about recent progress we made towards Raman activated cell sorting (RACS) by coupling Raman spectroscopy with microfluidics and micromanipulation approaches. In a first step, a microfluidic chip made of quartz was introduced which integrates injection of cells, trapping by fiber lasers and sorting of cells. Second, an all-fiber Raman-on-chip setup was introduced which accommodates laser excitation fibers and multi-core single-mode collection fibers. Without microscope, this Raman-on-chip setup offers low detection limits for solutions and enables to collect Raman spectra of trapped cells. Fiber Bragg gratings were inscribed into the collection fibers to suppress elastic scattered light. Furthermore, we present lab-on-a-chip (LOC) Surface enhanced Raman spectroscopy (SERS). Here, SERS increases the Raman scattering efficiency while the combination of SERS with a droplet based microfluidic platform guarantees high sample throughput and reproducible SERS conditions. Such a LOC-SERS approach has been used to determine the enzyme activity in lysed red blood cells.

Keywords: Biomedical, Infrared and Raman, Lab-on-a-Chip/Microfluidics, Surface Enhanced Raman
Application Code: Biomedical
Methodology Code: Vibrational Spectroscopy

Session # 2050	Abstract # 2050-3	Symposia

Session Title Advances in Vibrational Spectroscopy for Medical Diagnostics

Abstract Title **Classification of Lung Cancers by Infrared Spectral Histopathology (SHP)**

Date: Thursday, March 10, 2016 - Mornin
Time: 09:45 AM
Room: B302

Primary Author Max Diem
Northeastern University

Co-Author(s)

Abstract Text

Spectral histopathology (SHP) is an approach for the classification and diagnosis of tissue sections by spectral methods, in particular, Fourier transform infrared (FTIR) microspectral imaging coupled with multivariate statistical analytical methods. We report an extension and continuation of the approach originally pioneered by Lasch et al, namely to use unsupervised methods to delineate areas of high spectral homogeneity for the training of supervised classifiers to detect tissue abnormalities in SHP. Datasets comprising hundreds of patients, in tissue micro-array as well as large tissue section format have been collected at several laboratories worldwide for different cancer types and organs. We report here recent progress on a lung cancer study that now contains well over 600 patients, and that has revealed astounding sensitivity for the classification and sub-classification of lung cancers.

Keywords: Chemometrics, Infrared and Raman, Microspectroscopy
Application Code: Biomedical
Methodology Code: Vibrational Spectroscopy

Session # **2050** **Abstract #** **2050-4** **Symposia**

Session Title Advances in Vibrational Spectroscopy for Medical Diagnostics

Abstract Title **Label-Free Spectroscopic Imaging for Molecular Diagnosis**

Date: Thursday, March 10, 2016 - Mornin
Time: 10:35 AM
Room: B302

Primary Author
Ji-Xin Cheng
Purdue University

Co-Author(s)

Abstract Text

Current medical imaging modalities including CT, MRI, Ultrasound and OCT are based on tissue properties. On the other hand, diseases are driven by altered molecular pathways, which lead to a change of chemical composition of the tissue. This gap highlights a need for molecule based early diagnosis technology. This presentation will discuss our recent efforts in pushing molecular spectroscopy towards molecular diagnosis. I will present a spectroscopic imaging pen for in situ detection of cancer and pathogens under ambient light, and a deep-tissue bond-selective imaging platform for in vivo sensing of vulnerable plaque in coronary artery.

Keywords: Bioanalytical, Biospectroscopy, Laser, Vibrational Spectroscopy
Application Code: Biomedical
Methodology Code: Biospectroscopy

Session # 2050 Abstract # 2050-5 **Symposia**

Session Title: Advances in Vibrational Spectroscopy for Medical Diagnostics

Abstract Title: **Raman Spectroscopy of Blood for Alzheimer's Disease Diagnostics**

Date: Thursday, March 10, 2016 - Mornin
Time: 11:10 AM
Room: B302

Primary Author: Igor K. Lednev
University at Albany, SUNY

Co-Author(s): Earl Zimmerman, Elena Ryzhikova, Eric Molho, Lenka Halamkova, Oleksandr Kazakov

Abstract Text

The diagnosis of Alzheimer's disease is cumbersome and often delayed by lack of a simple diagnostic tool such as a blood test. In this study, we applied near infrared (NIR) Raman microspectroscopy coupled with advanced multivariate statistics for the differential diagnosis of AD based on blood serum. We analyzed NIR Raman spectral data from patients diagnosed with AD, patients diagnosed with other types of dementia (OD) and Healthy Control (HC) subjects. Artificial neural networks (ANN) and a support vector machine (SVM) were utilized for spectral data analysis. A Raman spectrum of blood serum represents the total biochemical composition of the fluid, subtle and specific changes of which could reflect a specific disease. We found that advanced statistical analysis of the serum Raman spectra allows for differentiating AD, OD and HC subjects with more than 95% sensitivity and specificity. Further study of a much larger cohort is required for the validation of the method and for establishing its effectiveness for early disease diagnostics. When fully developed, this fast, inexpensive noninvasive method could be used for screening of at risk patient populations for AD development and progression.

Keywords: Biomedical, Chemometrics, Raman, Vibrational Spectroscopy
Application Code: Biomedical
Methodology Code: Biospectroscopy

Session # 2060 Abstract # 2060-1 **Symposia**

Session Title: Computational Chemistry Coupled to Analytical Measurements: A Synergistic Relationship

Abstract Title: **Advancing the Understanding of Rigid Rod Polymers with Statistical Mechanics and Analytical Chemistry**

Primary Author: Steve Lustig
DuPont

Date: Thursday, March 10, 2016 - Mornin
Time: 08:35 AM
Room: B303

Co-Author(s): Christopher Seay, Juan David Londono, Steven Allen

Abstract Text

Many rigid rod polymers form such highly crystalline solid and liquid phases that common analytical methods of characterization are at best difficult, but typically intractable. Two examples are described in which the combining of computational statistical mechanics and analytical measurements advanced our understanding of concentrated rigid rod systems.

The first example uncovers the mechanism in which water weakens the tensile modulus of poly(hydroquinone diimidazopyridine), M5. Two-dimensional, high resolution X-ray scattering and molecular simulation elucidate the detailed structure of M5 fiber, subsequent structural changes with different levels of water content and the locations in which water resides within the crystalline morphology. Quantum chemical calculations predict quantitatively the experimentally measured modulus decrease due intrachain hydrogen bonding with water, without having to invoke an additional chemical hydrolysis mechanism. Quantum molecular dynamics simulations uncover that the polymer actually exists in two keto-enol tautomeric forms, that interconvert dynamically. Neither form is particularly sensitive to chemical hydrolysis.

The second example uncovers a mechanism for concentrated liquid crystalline solutions of poly(para-phenylenediamine terephthaloyl), PPDT, to resist forming a single, homogeneous nematic phase and retain rheological elasticity. Rheological characterization in shear over a wide range of low molecular weights show the onset of elasticity with increasing molecular weight. New theory and simulations support the contention that PPDT chains entangle to form twist-tie knots at sufficient lengths and concentrations. The topological nature of these entanglements originates from the distribution of rotational isomeric states that create intramolecular helicities.

Keywords: Characterization, Computers, Water, X-ray Diffraction
Application Code: Polymers and Plastics
Methodology Code: Computers, Modeling and Simulation

Session # 2060 **Abstract # 2060-2** **Symposia**

Session Title: Computational Chemistry Coupled to Analytical Measurements: A Synergistic Relationship

Abstract Title: **Recent Computational Studies Related to the Use of Plasmonic Materials for Analytical Applications**

Date: Thursday, March 10, 2016 - Mornin
Time: 09:10 AM
Room: B303

Primary Author: George C. Schatz
Northwestern University

Co-Author(s)

Abstract Text

This talk will describe studies in which theoretical methods (computational electromagnetics, electronic structure calculations) are used to study plasmonic materials (silver, gold and aluminum nanostructures) that are of interest in analytical applications. Most of this discussion will focus on surface enhanced Raman scattering (SERS) and related techniques such as TERS, where we will show that high electromagnetic enhancement factors can be obtained both for nanoparticle dimers and other small aggregates, and for larger structures (film over nanospheres) which are composed of many small gaps. Factors that determine plasmon width, and which dictate the wavelength range for useful measurements are considered, and we also examine chemical contributions to the enhancement factor for certain classes of molecular adsorbates. We also consider structures where plasmons are coupled to photonic modes, as these offer opportunities for index of fraction sensing and of luminescence and stimulated emission in which the far-field properties are dictated by the photonic lattice while the near-field properties involve plasmonic enhancements.

Keywords: Computers, Method Development, Molecular Spectroscopy
Application Code: Nanotechnology
Methodology Code: Vibrational Spectroscopy

Session # **2060** **Abstract #** **2060-3** **Symposia**

Session Title: Computational Chemistry Coupled to Analytical Measurements: A Synergistic Relationship

Abstract Title: **Theory and Simulations of Macromolecular Soft Materials: Linking Molecular Design to Macroscale Morphology and Function**

Date: Thursday, March 10, 2016 - Mornin
Time: 09:45 AM
Room: B303

Primary Author: Arthi Jayaraman
University of Delaware

Co-Author(s): Tyler B. Martin

Abstract Text

We use theory and simulation techniques to connect molecular features of macromolecular materials, specifically polymers, to their morphology and macroscopic properties, thereby guiding synthesis of materials for various applications in the energy and biomedical fields.

In this talk I will present our recent theory and simulation studies of polymer functionalized nanoparticles in polymer nanocomposites. The goal of this work is to control spatial arrangement of nanoparticles in a polymer nanocomposite so as to engineer materials with target mechanical or optical properties. One can tailor the inter-particle interactions and precisely control the assembly of the particles in the polymer matrix by functionalizing nanoparticle surfaces with polymers, and systematically tuning the composition, chemistry, molecular weight and grafting density of these grafted polymers. We have developed an integrated self-consistent approach involving Polymer Reference Interaction Site Model (PRISM) theory and molecular simulations to study polymer grafted nanoparticles in polymer matrix, and understand the effect of heterogeneity, such as monomer chemistry, monomer sequence, and polydispersity, in the polymer functionalization on the effective interactions, and dispersion/assembly of functionalized nanoparticles in a polymer matrix.

Keywords: Computers, Material Science
Application Code: Material Science
Methodology Code: Computers, Modeling and Simulation

Session # 2060 Abstract # 2060-4 **Symposia**

Session Title: Computational Chemistry Coupled to Analytical Measurements: A Synergistic Relationship

Abstract Title: **Differentiation of Alpha and Beta Crystalline Polymorphs in Biodegradable Poly Hydroxybutyrate**

Date: Thursday, March 10, 2016 - Mornin
Time: 10:35 AM
Room: B303

Primary Author: Bruce Chase
University of Delaware

Co-Author(s): Brian Sobieski, Isao Noda

Abstract Text

Crystalline polymorphs can present significant problems in understanding vibrational spectra of polymers. Small changes in conformational states can have a major effect on how the polymer chains pack. While vibrational spectroscopy is sensitive to conformations, it is often difficult to assign specific vibrational bands to specific conformations. The use of computational chemistry to calculate the vibrational spectra is now a valuable tool for analysis and assignment of spectral bands. We will demonstrate the power of this approach to the analysis of alpha and beta crystalline forms of a biodegradable polymer, poly hydroxybutyrate.

Keywords: Computers, FTIR
Application Code: General Interest
Methodology Code: Computers, Modeling and Simulation

Session # 2070 Abstract # 2070-1 **Symposia**

Session Title: Identification and Analysis for Food Safety

Abstract Title: **Validation and Challenge for the Determination of Chemical Components in Cosmetic Products Using LC-MS and GC-MS**

Primary Author: Perry G. Wang
US FDA

Date: Thursday, March 10, 2016 - Mornin
Time: 08:35 AM
Room: B304

Co-Author(s): Alexander J. Krynitsky, Wanlong Zhou

Abstract Text

Ingredients in cosmetics may cause adverse reactions in consumers. The FDA needs reliable analytical methods to identify and quantify the ingredients. This presentation will discuss the FDA's activities in developing both gas chromatography- mass spectrometry (GC-MS) and liquid chromatography-mass spectrometry (LC-MS) methods for the determination of chemical components in the complex matrices of cosmetic products. The projects that will be discussed include: the determination of parabens by GC-MS/MS and CAPB by LC-MS/MS in cosmetic and personal care products. For both studies, respective stable isotopically labeled analogues were used as internal standards to correct for matrix effects and recovery. Sample preparation, matrix effects and validation will be discussed. These methods have been fully validated and successfully used in limited surveys of cosmetics and personal care products to determine the prevalence of specific chemical components.

Keywords: Gas Chromatography/Mass Spectrometry, High Throughput Chemical Analysis, Liquid Chromatograp
Application Code: Food Safety
Methodology Code: Gas Chromatography/Mass Spectrometry

Session # 2070 Abstract # 2070-2 **Symposia**

Session Title Identification and Analysis for Food Safety

Abstract Title **Development of Multi-Functional Ambient Mass Spectrometry for Food Safety Screening and Characterizing Polymers in Packing Materials**

Date: Thursday, March 10, 2016 - Mornin

Time: 09:10 AM

Room: B304

Primary Author Jentaie Shiea
National Sun Yat-sen University

Co-Author(s)

Abstract Text

An ESI+APCI dual source was developed to simultaneously ionize both polar and nonpolar compounds. The ESI+APCI ion source can be operated in ESI-only, APCI-only, and ESI+APCI modes. The ESI+APCI source was constructed by inserting a fused silica capillary into a stainless steel tube which was enclosed in a glass tube. An AC high voltage was applied to the electrode on the glass tube to produce plasma from the nitrogen flowing through the tube and a DC high voltage was applied to the electrode in the source to generate an ESI plume from the methanol flowing out of the fused silica capillary. For ESI-only mode, polar analytes are ionized by interacting with the primary species generated in the ESI plume. For APCI-only mode, less polar or nonpolar analytes are ionized by interacting with the primary species generated in the APCI plasma. For ESI+APCI mode, the analytes are ionized by interacting with the primary species in the ESI+APCI plume.

The combination of ESI+APCI/MS with different sampling methods including thermal desorption and laser desorption makes the technique suitable for rapid analysis of the compounds over wide polarity and molecular weight range. For TD-ESI+APCI/MS analysis, a metallic sample probe was inserted into a preheated oven to thermally desorb the analytes on the probe. For CW LD-ESI+APCI/MS analysis, a diode laser was used to desorb the analytes on a sampling probe. The desorbed analytes moved upward to join in the ESI, APCI, or ESI+APCI plume for ionization, the analyte ions were subsequently detected by an ion trap mass analyzer attached to the source. TD-ESI+APCI/MS was used to rapidly characterize residual pesticides on fruits and vegetables' surface, preservatives in sauces, fatty acids in cooking oils, bisphenol A in water bottle, and phthalates on common commodities. CW LD-ESI+APCI/MS was used to rapidly characterize the polymeric components (such as PE, PP, PS, PLA, PVC and PVDC) and additives in various food packing

Keywords: Food Safety, Mass Spectrometry, Material Science, Polymers & Plastics

Application Code: Food Safety

Methodology Code: Mass Spectrometry

Session # **2070** **Abstract #** **2070-3** **Symposia**

Session Title: Identification and Analysis for Food Safety

Abstract Title: **Emerging Disinfection Byproducts Halobenzoquinones in Treated Drinking Water**

Primary Author: Xing-Fang Li
University of Alberta

Date: Thursday, March 10, 2016 - Mornin
Time: 09:45 AM
Room: B304

Co-Author(s)

Abstract Text

Water disinfection is necessary for killing pathogens, but it creates an unintended chemical risk from the formation of disinfection byproducts (DBPs). Epidemiological studies have shown an association of water disinfection byproducts with increased risk of bladder cancer. The DBPs responsible for the observed adverse health effects remain unknown. The quantitative structure and toxicity relationship analysis predicts that halobenzoquinones (HBQs) are potential bladder cancer carcinogens. The objectives of this study are to characterize the occurrence, formation, transformation, removal, and toxicity of HBQs as DBPs. Analysis of water samples from over ten drinking water treatment plants (DWTPs) show that HBQs occur widely in treated water, supported by the detection of dichlorobenzoquinone (DCBQ) in all samples. Analysis of HBQ formation potential of source waters collected from nine DWTPs support the widespread presence of precursors of HBQs. We further investigated the current treatment processes, including coagulation, filtration, granular activated carbon, ozonation, and UV irradiation, for removal of HBQ precursors. The results demonstrate that the combined treatments cannot substantially reduce HBQ precursor levels in water. In vitro experiments have shown that HBQs induce greater cytotoxicity and/or greater neurotoxicity than most of the regulated DBPs. HBQs as an emerging class of DBPs warrant monitoring because of their widespread presence and high toxicity.

Keywords: Characterization, Liquid Chromatography, Mass Spectrometry, Water
Application Code: Other
Methodology Code: Liquid Chromatography/Mass Spectrometry

Session # 2070 **Abstract # 2070-4** **Symposia**

Session Title: Identification and Analysis for Food Safety

Abstract Title: **Effects of Different Dietary Doses of Copper and High Fructose Feeding on Rat Fecal and Liver Metabolome**

Primary Author: Xiang Zhang
University of Louisville

Date: Thursday, March 10, 2016 - Mornin
Time: 10:35 AM
Room: B304

Co-Author(s): Aminul Prodhan, Craig McClain, Ming Song, Xiaoli Wei, Xinmin Yin

Abstract Text

Increased fructose consumption and inadequate copper intake are two critical risk factors in the development of nonalcoholic fatty liver disease (NAFLD). To investigate whether copper deficiency plays a role in fructose-induced fatty liver, male weanling Sprague-Dawley rats (35–45 g) were housed in stainless steel cages in a temperature and humidity controlled room with a 12:12 h light:dark cycle. All rats were allocated randomly into four groups. Each group of rats was fed either an adequate copper or marginally copper deficient diet. At the same time, distilled water or distilled water containing 30% fructose (w/v) was given ad lib for 4 weeks. Fructose enriched drinking water was changed twice a week. At the end of the experiment, all the animals were killed under anesthesia with pentobarbital (50mg/kg I.P. injection). Liver and feces were collected and snap-frozen with liquid nitrogen for metabolomics study.

Metabolites were extracted using a solvent mixture of methanol and water. The extracted metabolites were derivatized using N-(tert-butyldimethylsilyl)-N-methyltrifluoroacetamide (MTBSTFA). The derivatized samples were analyzed on LECO Pegasus 4D GC×GC□TOF MS instrument. Experimental data were processed by ChromaTOF software for peak selection and compound identification, followed by MetPP software for retention index matching and metabolite quantitification.

Preliminary analysis of rat fecal samples shows a significant metabolite profile difference between sample groups. Among the metabolites detected with significant abundance alteration between groups, short chain fatty acids were markedly decreased in groups with excessive fructose intake irrespective of copper levels. C15:0 and C17:0 long chain fatty acids, produced only by bacteria, were increased by either high copper level or high fructose intake. In addition, increased fecal urea and malic acid paralleled the increased hepatic fat accumulation.

Keywords: Food Safety, Food Science, Gas Chromatography/Mass Spectrometry, Statistical Data Analysis

Application Code: Food Safety

Methodology Code: Gas Chromatography/Mass Spectrometry

Session # 2070 **Abstract # 2070-5** **Symposia**

Session Title: Identification and Analysis for Food Safety

Abstract Title: **Identification and Confirmation of Chemical Residues in Foods for Regulatory Purposes**

Primary Author: Steven J. Lehotay
USDA Agricultural Research Service

Date: Thursday, March 10, 2016 - Mornin
Time: 11:10 AM
Room: B304

Co-Author(s)

Abstract Text

Despite much focus during method validation and quality control to address the accuracy in quantitative determinations of chemical residues in food for regulatory purposes, the issues surrounding qualitative identification and confirmation are often taken for granted. The rates of false positives and negatives in real-world monitoring are usually untested and unknown, even when using the most powerful analytical tools. Many notable examples of erroneous chemical identifications in foods have occurred in the past. Nearly every misidentification originates from spurious error (human mistakes), which can never be eliminated because humans are ultimately liable and responsible for any result, even if automated systems and fixed identification criteria are implemented. Just as analytical methods must be validated for quantitative purposes, they should also be validated qualitatively by assessing rates of false positives and negatives. For regulatory purposes in which economic losses and personal reputations are at stake, all violative findings should be confirmed via re-analysis of a duplicate sample portion using a different method of sample preparation and/or analysis involving a different chemical mechanism.

Keywords: Chromatography, Mass Spectrometry, Method Development, Quality Control

Application Code: Food Safety

Methodology Code: Mass Spectrometry

Session # 2080 **Abstract # 2080-1** **Symposia**

Session Title: Integrated Microfluidics

Abstract Title: **Integrated Microfluidics for Forensic Analysis: Creating Simple, Portable and Cost-effective Systems**

Primary Author: James Landers
University of Virginia

Date: Thursday, March 10, 2016 - Mornin
Time: 08:35 AM
Room: B305

Co-Author(s)

Abstract Text

The last decade has seen an exponential growth in the development of integrated microfluidic systems for specific applications. Inherent in the development of technologies that will either be applied to new applications or supplant existing ones, is that they be simple to operate (unskilled personnel) and cost-effective (affordable). While COC-like substrates have dominated polymeric materials for microchip fabrication, we will discuss the use of standard overhead transparencies (polyester film, Pe) for creating microfluidic devices with complex multilevel architecture. Standard printing processes are utilized as a fundamental part of the fabrication with toner (T) functioning for multilayer adhesion as well as providing passive valves. In addition, with a view to simple fluidic flow control, the microdevices are designed to fit a CD-sized format so that centrifugal force can drive flow, metering, mixing, reaction and detection. A variety of fluidic architectures in these rotation-driven microdevices (RDMs) provides analytical functionality in a broad application space, including protein and DNA quantification, white blood cell counting, infectious pathogen detection, DNA purification, CD4 cell counting, cancer mutation detection, DNA amplification through thermal and isothermal processes, explosives detection, drug detetion and, finally, separations with fluorescence or optical detection. Discussion of select applications will demonstrate that PeT RDMs provide one approach for obtaining cost-effective and versatile analytical functionality when coupled simple external hardware that has the potential to provide truly portable (handheld even) systems for field use.

Keywords: Biological Samples, Biomedical, Forensics, Genomics
Application Code: Biomedical
Methodology Code: Microfluidics/Lab-on-a-Chip

Session # 2080 Abstract # 2080-2 **Symposia**

Session Title: Integrated Microfluidics

Abstract Title: **Microfluidic Technology for Protein Crystallization and Pharmaceutical Solid form Screening**

Primary Author: Paul Kenis
University of Illinois at Urbana Champaign

Date: Thursday, March 10, 2016 - Mornin
Time: 09:10 AM
Room: B305

Co-Author(s)

Abstract Text

This presentation will focus on the development of microfluidic technology for (a) crystallization of membrane proteins for structure determination, and (b) the screening for solid forms (co-crystals, polymorphs, ...) of active pharmaceutical ingredients. For both applications one wants to screen many conditions using only a minimal amount of material, hence the suitability of microfluidic platforms (high throughput screening). Furthermore, our work focuses on the use of Raman Spectroscopy and/or X-ray diffraction to analyze the crystals or pharmaceutical solid forms on-chip, eliminating the need to harvest and analyze many small crystals one-by-one. This presentation will highlight the design, fabrication and successful application of different suitable microfluidic platforms for the two aforementioned applications.

Keywords: Pharmaceutical, Protein, Raman, X-ray Diffraction
Application Code: Drug Discovery
Methodology Code: Microfluidics/Lab-on-a-Chip

Session # 2080 Abstract # 2080-3 **Symposia**

Session Title Integrated Microfluidics

Abstract Title **Integrated Microfluidic Systems for Measuring Secretion from Cellular Networks**

Date: Thursday, March 10, 2016 - Mornin
Time: 09:45 AM
Room: B305

Primary Author Michael G. Roper
Florida State University

Co-Author(s) Adrian M. Schrell, Basel Bandak, Kimberly Evans, Lian Yi, Nikita Mukhitov, Xue Wang

Abstract Text

Secretion of small molecules, peptides, and proteins is key to communication between different cells, tissues, and organs. Disrupted secretion has been linked to numerous disease states. Over the last 10 years, our group has focused on secretion from islets of Langerhans, the endocrine portion of the pancreas responsible for maintaining glucose homeostasis. These cellular networks consist of multiple cell types that secrete different small molecules and peptides, which can act in both autocrine and paracrine manners.

This talk will center on the development and characterization of analytical systems that can deliver various stimuli to these cell networks while simultaneously measuring the secretory output. Implementation of electrophoretic separations and optical measurements allow the secretion to be measured in a time-resolved manner. Part of the presentation will focus on how these cells respond to dynamically changing stimuli, which is an important facet for understanding how they interact in vivo.

Keywords: Bioanalytical, Capillary Electrophoresis, Electrophoresis, Lab-on-a-Chip/Microfluidics
Application Code: Bioanalytical
Methodology Code: Microfluidics/Lab-on-a-Chip

Session # 2080 Abstract # 2080-4 Symposia

Session Title Integrated Microfluidics

Abstract Title **New Strategies for Enhancing Human-on-Chip Systems**

Primary Author Dana Spence
Michigan State University

Date: Thursday, March 10, 2016 - Mornin
Time: 10:35 AM
Room: B305

Co-Author(s)

Abstract Text

Microfluidic devices have been successfully used in biological assays, especially for assays involving bacteria and cells. While important studies involving bacteria and cells are still in progress in many research laboratories, many devices are now being developed such that tissues, and even organs, can be cultivated on these devices for more realistic biological studies. For nearly a decade, the PI has been culturing endothelial cells in microfluidic devices and investigating the interactions of the endothelium (as a vessel wall tissue) with a flowing stream of blood components such as red blood cells (RBCs) and platelets. In this application, the investigative team wishes to expand upon these tissues to include components of pancreatic tissue (beta-cells, the cells responsible for secreting the well-known biological peptide, insulin) and even smooth muscle cells. One of the major objectives of this application is to measure and understand cell-to-cell and tissue-to-tissue communication between the beta-cells, the bloodstream, and the vessel wall (endothelium). Unfortunately, it has been shown that affecting cells beta-cells by knockout or knockdown approaches often adversely affects the cell as a whole, thus making it difficult to distinguish key factors in the biological pathways. In this presentation, the we report on a novel platform based on 3D-printing of fluidic devices hosting membrane inserts with modified pores that selectively enables, or inhibits, molecular messengers traveling through the bloodstream from one tissue to another without affecting the secreting cell. We anticipate that this platform will represent a novel, powerful and nearly universal technology for use in biological mechanisms without the use of knockouts or silencing techniques.

Keywords: Biological Samples, Biotechnology, Lab-on-a-Chip/Microfluidics, Membrane
Application Code: Bioanalytical
Methodology Code: Microfluidics/Lab-on-a-Chip

Session # 2080 Abstract # 2080-5 **Symposia**

Session Title Integrated Microfluidics

Abstract Title **Integrated Microfluidic Platform for Mass Spectrometry based Metabolomics**

Date: Thursday, March 10, 2016 - Mornin

Primary Author James L. Edwards
Saint Louis University

Time: 11:10 AM

Room: B305

Co-Author(s)

Abstract Text

Investigations of metabolite profiles from endothelial cells cultured for multiple days on a polydimethylsioxane (PDMS)-glass chip will be discussed. To mimic the cardiovascular environment, aortic endothelial cells are cultured in a PDMS based microfluidic platform with continuous flow imparting shear stress >15 dynes/cm2. Cells are viable and maintain physiological shape for greater than 7 days under these conditions. Novel cell lysing methods are employed to optimize reproducibility, reduce artifacts from sample preparation, and increase automation. Removal of enzymes and lipids from the metabolite lysate will be explored. Metabolites relevant to glucose metabolism and oxidative stress will be evaluated as a function of external glucose stimuli. Metabolite investigations will employ both MALDI and ESI based mass spectrometry.

Keywords: Bioanalytical, Capillary LC, Lab-on-a-Chip/Microfluidics, Mass Spectrometry
Application Code: Bioanalytical
Methodology Code: Liquid Chromatography/Mass Spectrometry

Session # 2090 Abstract # 2090-1 **Symposia**

Session Title New Bioanalytical Separations for Molecular Mechanisms of Disease

Abstract Title **Capillary Separations that Unravel Molecular Mechanisms of Endocrine Dysfunction**

Primary Author Lisa A. Holland
West Virginia University

Date: Thursday, March 10, 2016 - Mornin
Time: 08:35 AM
Room: B309

Co-Author(s) Jennifer R. Stueckle, Vincent T. Nyakubaya

Abstract Text

Zebrafish are integral to assessing the effects of endocrine-disrupting chemicals on human health because genes, proteins, hormones, and the hypothalamic-pituitary-gonadal axis are similar to humans. Estrogenic endocrine disrupting chemicals bind to estrogen receptors and activate, inhibit, or have no effect on gene expression. Circulating steroid hormones are a more direct measure of endocrine disruption than monitoring transcription, but a set of steroid hormones must be quantified in an individual to elucidate the physiological response to chemical exposure because estrogens, androgens, and progestogens are interrelated through a complex pathway. Steroids are difficult to detect in individual zebrafish using conventional methods due to the limited blood volume of these animals. An innovative capillary electrophoresis is used to separate five endogenous steroids and one synthetic steroid in 5 microliters of plasma in under 5 minutes. The separation method utilizes stacking to generate steroid detection limits of 0.2 to 2 ng/mL (0.8 to 6 nM) using only UV-visible absorbance detection. The steroids: 17[alpha],20beta-dihydroxy-pregn-4-en-3-one, testosterone, 11-ketotestosterone, estrone, 17beta-estradiol and 17[alpha]-ethinyl estradiol are monitored in individual zebrafish subject to exposure studies mandated by the Organisation for Economic Co-operation and Development. Test No. 229: Fish Short Term Reproduction Assay. Monitoring multiple steroids in individual fish demonstrates that pooling plasma confounds analyses of circulating steroids. The results of this work, reveal that estrone is produced in zebrafish in response to estrogenic activity. This capillary electrophoresis technology demonstrates that levels of different circulating steroids distinguish endocrine disruptors with different modes of action (17beta-estradiol, trenbolone, prochloraz) as well as those with similar physiological outcomes (17beta-estradiol vs 17[alpha]-ethinyl estradiol).

Keywords: Capillary Electrophoresis, Environmental/Biological Samples, Toxicology
Application Code: Bioanalytical
Methodology Code: Capillary Electrophoresis

Session # 2090 Abstract # 2090-2 **Symposia**

Session Title: New Bioanalytical Separations for Molecular Mechanisms of Disease

Abstract Title: **Separation-Based Methods for Measuring Reactive Oxygen and Nitrogen Species in Biological Samples**

Date: Thursday, March 10, 2016 - Mornin
Time: 09:10 AM
Room: B309

Primary Author: Susan M. Lunte
University of Kansas

Co-Author(s)

Abstract Text

Reactive oxygen and nitrogen species (RNOS) such as nitric oxide, peroxynitrite, and superoxide have been implicated in a number of diseases, including cardiovascular and neurodegenerative disorders. Approaches for the simultaneous detection and quantitation of multiple RNOS in a single sample are lacking, especially for single-cell analysis applications. In this presentation, methods for the detection of RNOS using microchip electrophoresis with electrochemical and laser-induced fluorescence detection will be described. Many RNOS, including nitrite, peroxynitrite, hydrogen peroxide, and nitric oxide, are electroactive and can be separated and detected using microchip electrophoresis with electrochemical detection (MEEC). In addition, intracellular oxidants such as glutathione and ascorbate can be measured using the same technique. Alternatively, laser-induced fluorescence (LIF) can be used for the indirect detection of RNOS such as nitric oxide, superoxide, and peroxynitrite. ME-LIF requires derivatization of the analytes with specific fluorophores prior to analysis. In this case, the electrophoretic separation is used to isolate the desired product from reaction side products, making more precise quantitation possible. Examples of the detection of RNOS in macrophage cells using both ME-EC and ME-LIF will be presented.

Keywords: Biological Samples, Capillary Electrophoresis, Lab-on-a-Chip/Microfluidics, Microelectrode
Application Code: Bioanalytical
Methodology Code: Microfluidics/Lab-on-a-Chip

Session # 2090 Abstract # 2090-3 **Symposia**

Session Title New Bioanalytical Separations for Molecular Mechanisms of Disease

Abstract Title **How Separations—Both High and Low Resolution—Enable Selection of Clinically Useful Aptamers**

Date: Thursday, March 10, 2016 - Mornin
Time: 09:45 AM
Room: B309

Primary Author Rebecca Whelan
Oberlin College

Co-Author(s)

Abstract Text

Affinity probes—either protein or nucleic acid based—form the basis of diagnostic tests and therapeutic approaches. Our lab focuses on the development of nucleic acid aptamers with affinity for ovarian cancer biomarkers. This talk will describe two versions of separations used in the service of aptamer selection. In one case, the protein target is CA125, the clinical gold standard for ovarian cancer monitoring and detection of recurrence. In our novel "one-pot" approach, CA125 is allowed to adsorb onto the inner surface of a PCR tube. Unselected DNA library is added, and those sequences with high affinity remain on the PCR wall. Direct addition of amplification reagents enables the enrichment of the population with the CA125 binding sequences of interest, with no material transfer. This low-resolution separation has yielded a set of aptamers whose affinity and specificity will be described. In the second case, the protein target of the aptamer selection process is the secreted protein HE4, whose abundance in early stage ovarian cancer patients complements that of CA125. In this case, the high-resolution separation of capillary electrophoresis is used to separate DNA-protein complexes from unbound DNA. This technique has also yielded a set of aptamers. The endpoint of both selection processes is the sequencing of selected DNA on an Illumina platform, followed by bioinformatics analysis of the abundant resulting data. Our bioinformatics pipeline, which is composed of freely available online tools, will be described.

We acknowledge support from the National Cancer Institute.

Keywords: Bioanalytical, Capillary Electrophoresis, Immunoassay, Nucleic Acids
Application Code: Bioanalytical
Methodology Code: Capillary Electrophoresis

Session # 2090 Abstract # 2090-4 Symposia

Session Title New Bioanalytical Separations for Molecular Mechanisms of Disease

Abstract Title **Merging Microfluidics, Electrophoresis, and Mass Spectrometry for Protein Assays**

Date: Thursday, March 10, 2016 - Mornin
Time: 10:35 AM
Room: B309

Primary Author Robert T. Kennedy
University of Michigan

Co-Author(s)

Abstract Text

Microchip electrophoresis enables high speed separations on small samples which opens new possibilities for chemical analysis. We have explored new ways of using microchip electrophoresis for protein analysis related to high-throughput screening (HTS) and western blotting. The high speed of electrophoresis allows intact non-covalent complexes to be separated and detected. This capability has been extensively used for immunoassay and aptamer assay. We have recently demonstrated that we can separate protein-protein complexes involved in intracellular signaling to generate quantitative information on binding and selectivity. We have also demonstrated fast (< 1 s) separations of substrates and products from enzymes allowing rapid enzyme assays. The high speed of the assays suggests they could be used in HTS for drug discovery. To achieve this, we have developed a droplet microfluidic system that allows discrete samples to be rapidly loaded in sequence onto a chip for injection and separation. Using this method over 1000 injections have been achieved in 17 min. Preliminary results suggest the method is stable enough for HTS opening the door to robust HTS of protein-protein interactions and enzyme activity. The droplet system has also been interfaced to mass spectrometry for assays at rates up to 5/s. A third approach that we have investigated is a microchip western blot. Western blotting is one of the most widely used protein assays and the lack of a miniaturized counterpart has likely held back the use of microfluidics in routine biochemistry. We have interfaced chip separations to blotting membranes for rapid westerns. Potential for multi-protein analysis is greatly enhanced by this method.

Keywords: Bioanalytical, Electrophoresis, Mass Spectrometry
Application Code: Bioanalytical
Methodology Code: Microfluidics/Lab-on-a-Chip

Session # 2090 Abstract # 2090-5 **Symposia**

Session Title: New Bioanalytical Separations for Molecular Mechanisms of Disease

Abstract Title: **CE-FSCV for Determining Neurotransmitter Tissue Content in Drosophila Disease Models**

Primary Author: B Jill Venton
University of Virginia

Date: Thursday, March 10, 2016 - Mornin
Time: 11:10 AM
Room: B309

Co-Author(s): Madelaine E. Denno, Ryan Borman

Abstract Text

Drosophila are a model organism for studying many mechanisms of basic neurotransmission. Our lab has pioneered methods to use electrochemistry to measure stimulated dopamine or serotonin release in larvae. However, in order to understand neurotransmitter regulation, it is also necessary to have a measure of tissue content. Therefore we have developed capillary electrophoresis with fast-scan cyclic voltammetry detection (CE-FSCV) for the analysis of neurotransmitter tissue content in single fly brains. Studying fly brains is challenging because each sample is less than 8 nL in volume and contains only femtomoles of sample. We have applied this method to measure dopamine, serotonin, octopamine, and tyramine throughout fly development. We have also compared histamine, carcinine, dopamine, and beta-alanyl-dopamine levels in brains, eyes and cuticles of adult flies. Finally, in order to better understand disease models, we have measured dopamine content in fly models of Parkinson disease. These results have been correlated with results from stimulated electrical release. CE-FSCV is a convenient method for understanding the tissue content of small organisms.

Keywords: Bioanalytical, Capillary Electrophoresis, Voltammetry
Application Code: Neurochemistry
Methodology Code: Capillary Electrophoresis

Session # 2100 Abstract # 2100-1 **Symposia**

Session Title: Overcoming the Obstacles to Making Measurements in the Brain

Abstract Title: **Probing the Spatial and Temporal Dynamics of Signaling Peptides in the Nervous Systems by a Multi-Faceted MS Approach**

Date: Thursday, March 10, 2016 - Mornin
Time: 08:35 AM
Room: B310

Primary Author: Lingjun Li
University of Wisconsin

Co-Author(s)

Abstract Text

Signaling peptides regulate neuronal circuits and a wide range of physiological processes. Over the past several years, mass spectrometry (MS)-based strategies have revolutionized the discovery of neuropeptides in numerous model organisms, especially in decapod crustaceans. Despite substantial acceleration of the rate of neuropeptide discovery, it remains challenging to map the spatial distribution of neuropeptides and monitor their temporal dynamics in body fluids of live animals. In this presentation, I will discuss our recent advances in the use of MS-based techniques to explore neuropeptidome in the spatial and temporal domains. These MS-enabled investigations provide valuable information about the distribution, secretion and potential function of neuropeptides with high molecular specificity and sensitivity. Specifically, we report on a multiplex-MSI method, which combines high resolution accurate mass (HRAM) mass spectrometry imaging (MSI) technology with data dependent acquisition (DDA) tandem MS analysis in a single experiment. To enhance the dynamic range and efficiency of in situ DDA, we introduced a novel gas-phase fractionation strategy prior to MS/MS scans, to decrease molecular complexity of tissue samples for enhanced peptidome coverage. Furthermore, we combined the HRAM multiplexed MSI method with in vivo microdialysis sampling technique to allow monitoring of trace-level neuropeptide secretion, small signaling molecule dynamic changes in the crustacean hemolymph. Finally, the dynamic degradation profiles of neuropeptides were investigated via in vivo microdialysis coupled to HRAM multiplexed MSI on a MALDI LTQ Orbitrap platform.

Keywords: Bioanalytical, Mass Spectrometry, Neurochemistry, Peptides
Application Code: Neurochemistry
Methodology Code: Mass Spectrometry

Session # 2100 Abstract # 2100-2 **Symposia**

Session Title: Overcoming the Obstacles to Making Measurements in the Brain

Abstract Title: **Overcoming Obstacles to Understanding Voltammetric Measurements of Dopamine in the Brain**

Date: Thursday, March 10, 2016 - Mornin
Time: 09:10 AM
Room: B310

Primary Author: Adrian C. Michael
University of Pittsburgh

Co-Author(s): Andrea Jaquins-Gerstl, Brendan P. Sestokas, Elaine M. Robbins, Seth H. Walters

Abstract Text

Fast-scan cyclic voltammetry (FSCV) in conjunction with carbon fiber microelectrodes permits the in vivo, real time chemical monitoring of dopamine, an immensely important neurotransmitter deeply involved in numerous aspects of brain function and heavily implicated in several neurological and psychiatric disorders. Many studies, including studies from our laboratory, have made use of FSCV to monitor so-called evoked dopamine, i.e. dopamine that is initially released into the extracellular space from terminals upon the electrical stimulation of dopamine axons and then returned to dopamine terminals by the dopamine reuptake mechanism involving the transmembrane dopamine transporter (DAT). An obstacle to thoroughly understanding the evoked responses is their inherent heterogeneity. However, recent results confirm that the responses are not randomly heterogeneous but rather are organized into sub-types that reflect local kinetic diversity within the dopamine terminal field. The full range of features of the kinetically diverse responses only become clearly understandable upon a correction to account for the technical aspects of FSCV, which includes the consequences of dopamine's tendency to adsorb to carbon fiber electrodes: a correction to account for adsorption effects is proposed. Application of the correction and a kinetic analysis to characterize the effects of bupropion, a popular anti-depressant medication, is described. Histochemical analysis of the electrode implantation site shed light on the underlying anatomical basis of the kinetic diversity.

Keywords: Electrochemistry, Microelectrode, Neurochemistry, Voltammetry
Application Code: Neurochemistry
Methodology Code: Electrochemistry

Session # 2100 Abstract # 2100-3 **Symposia**

Session Title Overcoming the Obstacles to Making Measurements in the Brain

Abstract Title **Simultaneous Optical Imaging of Neuronal, Glia, and Hemodynamic Waves During Seizures**

Date: Thursday, March 10, 2016 - Mornin
Time: 09:45 AM
Room: B310

Primary Author Hongtao Ma
Weill Cornell Medical College

Co-Author(s) Andy Daniel, Eliza Baird-Daniel, Mingrui Zhao, Philippe Laffont, Theodore H. Schwartz

Abstract Text

"Neurovascular coupling" is a term used to describe the close relationship between neuronal activation, increased metabolism, and the resulting increase in cerebral blood flow. Neurovascular coupling based brain imaging techniques, such as functional magnetic resonance imaging and positron emission tomography, have been widely used in the clinic for brain mapping. However, the precise details of neurovascular coupling are not well understood, especially during non-repeating brain activities, such as seizures. A method that can simultaneously record hemodynamic change as well as underlying brain activity is necessary in elucidating these processes. We employed simultaneous multiple-wavelength wide field optical imaging to record the neuronal, glia, and hemodynamic changes during acute seizures in rats. Brain activity was recorded with calcium dyes, either OGB-1, whose data can be filtered to reflect glia (< 1 Hz) or neuronal activity (> 1Hz), or the calcium dye Rhod-2, which stains only astrocytes, in addition to intrinsic optical imaging, which record the hemodynamic changes. This method can offer a spatial resolution of 50 μm, a temporal resolution of 55 Hz, and a spatial coverage of 14 x 14 mm.

We found clear glial wave which began focally and spread across the cortex occurred simultaneously with each ictal event. However, glial waves propagated further than cerebral blood volume (CBV) and neuronal activity. Despite widely varying seizure durations, the duration of astrocytic activation remained relatively constant. The hemodynamic change, on the other hand, lasted longer than both the astrocytic and neuronal activity. Moreover, glial waves were significantly delayed in onset compared to CBV changes. Our results suggest that during ictal events, each compartment in the neurovascular unit displays a unique spatiotemporal onset and evolution and call into question the putative essential role of astrocytes in ictal neurovascular coupling.

Keywords: Imaging, Medical, Neural Network
Application Code: Biomedical
Methodology Code: Biospectroscopy

Session # 2100 Abstract # 2100-4 **Symposia**

Session Title: Overcoming the Obstacles to Making Measurements in the Brain

Abstract Title: **Placing New Pieces in the Puzzle of Human Traumatic Brain Injury Using Multimodal Monitoring**

Date: Thursday, March 10, 2016 - Mornin
Time: 10:35 AM
Room: B310

Primary Author: Martyn G. Boutelle
Imperial College London

Co-Author(s): Anthony J. Strong, Chi Leng Leong, Michelle L. Rogers, Sally A. Gowers, Sharon Jewell, Vassilios Kontojannis

Abstract Text

TBI contributes to 31% of all injury related deaths in the US. Each year in the US 1.7 million people suffer a traumatic brain injury (TBI) of whom 52,000 die, 275,000 are hospitalised and 1.3865 million are treated in the emergency room before release [1]. New clinical strategies for treating TBI are desperately needed. New analytical instrumentation has a vital role in devising these strategies.

At the heart of the injury caused by TBI is a disruption of normally tight coupling between neuronal activity and energy metabolism. Hence injury causes transient excessive cellular depolarisation at the same time as low, unreactive blood flow. This leads to local failure of the energy delivery to neurons necessary for repolarisation. We are developing a multimodal monitoring approach in TBI patients to monitor the pieces of this complex puzzle as the injury evolves dynamically. Electrical activity is determined using electrical contacts placed onto the brain via a craniotomy or bolt. Blood flow changes can be detected non-invasively by use of near infrared (at 700 and 900nm) to detect changes in oxy and deoxyhaemaglobin. To detect chemical changes in the brain tissue requires implanted probes. Changes in tissue oxygen levels are detected using a clinically approved Clark electrode, while changes in extracellular neurochemicals are detected by use of a microdialysis probe that feeds into a novel microfluidic analysis chip coupled to amperometric microelectrode biosensors (glucose, lactate, pyruvate) and micro ion selective electrodes (potassium).

Interpretation of any one of these measured signals in isolation is challenging. However, by carefully aligning the real-time signals with respect to each other clear patterns of changes can be identified that both allow the rapid detection of 'adverse' events, and which suggest underlying mechanisms of secondary brain injury.

[1] US Centre for Disease Control http://www.cdc.gov/traumaticbraininjury/

Keywords: Biosensors, Electrochemistry, Near Infrared, Neurochemistry
Application Code: Bioanalytical
Methodology Code: Electrochemistry

Session # 2100 Abstract # 2100-5 **Symposia**

Session Title: Overcoming the Obstacles to Making Measurements in the Brain

Abstract Title: **Electroosmotic Perfusion in Brain Tissue for Determining Ectopeptidase Activity**

Date: Thursday, March 10, 2016 - Mornin
Time: 11:10 AM
Room: B310

Primary Author: Stephen Weber
University of Pittsburgh

Co-Author(s): Bocheng Yin, Jenna DeVivo, Khanh Ngo, Rachael Wilson, Stephen R. Groskreutz, Yangguang Ou

Abstract Text

While small molecule neurotransmitters have a specific function to carry a signal, neuropeptides have a variety of functions in the brain. While the production of neuropeptides and their interactions with receptors on neurons has been the focus of much attention, the fate of peptides in the extracellular space has been less studied. We have developed a method to determine the activity of enzymes, ectopeptidases, that hydrolyze neuropeptides in the extracellular space of cultured brain slices. The method reveals for the first time significant differences in the activity of certain ectopeptidases that inactivate enkephalins in different brain areas. A major effort is underway to improve the throughput of this measurement by investigating multiple ectopeptidases at the same time.

Keywords: Bioanalytical, Capillary LC, Liquid Chromatography/Mass Spectroscopy, Neurochemistry
Application Code: Neurochemistry
Methodology Code: Sampling and Sample Preparation

Session # 2110 Abstract # 2110-1 Symposia

Session Title SAS - Handheld Spectrometers

Abstract Title **Handheld Laser-Induced Breakdown Spectroscopic Instruments**

Primary Author Amy J. Bauer
TSI, Incorporated

Date: Thursday, March 10, 2016 - Mornin
Time: 08:35 AM
Room: B311

Co-Author(s)

Abstract Text
This talk will begin with a brief history of laser induced breakdown spectroscopy (LIBS). This discussion will set the stage for the technological challenges that had to be overcome before miniaturization of the technique was possible. These impressive achievements include a reduction of laser dimensions from the size of a refrigerator to the size of a pack of playing cards. We will then address the major measurement approaches taken by companies now striving to commercialize handheld LIBS instruments, segregated by laser repetition rate and power deposition specifications. Lastly, the most important of the applications addressed by handheld LIBS, from geological analysis to scrap metal sorting, will be addressed.

Keywords: Atomic Spectroscopy, Instrumentation, Laser, Portable Instruments
Application Code: General Interest
Methodology Code: Atomic Spectroscopy/Elemental Analysis

Session # 2110 Abstract # 2110-2 **Symposia**

Session Title SAS - Handheld Spectrometers

Abstract Title **Chemometrics in Action: Moving the Lab to the Field**

Primary Author Michael Hargreaves
Thermo Fisher Scientific

Co-Author(s) Suzanne Schreyer

Date: Thursday, March 10, 2016 - Mornin
Time: 09:10 AM
Room: B311

Abstract Text

In recent years, there is an increasing trend of moving analytical chemistry testing from laboratory environments to field based material identification. The use of analyzers in the field has characteristics which clearly distinguishes itself from the use in a traditional laboratory setting. Portable instruments in the field need to be able to give rapid, consistent results over a range of conditions: from testing in a warehouse to high stress situations involving emergency personnel and unknown conditions. This puts additional requirements on the software and underlying data treatment. For field based applications, merely presenting a spectrum is not enough. Rather the instrument is required to present a clear and unambiguous result that can be quickly interpreted by the user. This puts greater emphasis on analytical method development and is a big hurdle for wide adoption of new analyzers. Intelligent chemometric algorithms must be used in these cases to give users clear answers for their intended use in the field. In this presentation, an overview will be given on recent developments of chemometrics algorithms for handheld analyzers based on Raman, Infrared and Near Infrared spectroscopy. Specific examples will be given of field based applications for safety and security applications, recent developments in agricultural methods and also pharmaceutical applications.

Keywords: Chemometrics, Portable Instruments, Spectroscopy

Application Code: Other

Methodology Code: Portable Instruments

Session # 2110 Abstract # 2110-3 Symposia

Session Title: SAS - Handheld Spectrometers

Abstract Title: **Mass Spectrometry in Miniature**

Primary Author: Christopher Brown
908 Devices

Date: Thursday, March 10, 2016 - Mornin
Time: 09:45 AM
Room: B311

Co-Author(s)

Abstract Text
Novel micro-scale ion-trap mass spectrometers operating at high pressures have been the foundation for the first truly handheld mass spectrometers. The applications of these purpose-built systems will be discussed in the context of safety & security needs, as well as the logical extension of the core miniature MS platform for more traditional hyphenated techniques with gas and liquid-phase separations coupled to the MS backend. Applied examples will illustrate the opportunities and challenges with these new frontiers in ultra-miniature instrumentation.

Keywords: Gas Chromatography/Mass Spectrometry, Liquid Chromatography/Mass Spectroscopy, Mass Spectro
Application Code: Safety
Methodology Code: Portable Instruments

Session # 2110 Abstract # 2110-4 **Symposia**

Session Title SAS - Handheld Spectrometers

Abstract Title **Next Generation Portable Spectrometers: Spectroscopy Solutions Wherever You Want Them**

Primary Author Katherine A. Bakeev
B&W Tek, Inc

Date: Thursday, March 10, 2016 - Mornin
Time: 10:35 AM
Room: B311

Co-Author(s) Jack Zhou, Jing Li, Ken Li, Sean Wang

Abstract Text
Portable instrumentation has been an important trend over the last decade. With continued developments in the technology, purpose-built instruments targeted for particular applications have grown in popularity. We will present information on handheld spectrometers, including Raman, near-infrared and LIBS. We'll also provide information on the technology, application needs and emerging solutions.

Keywords: Atomic Spectroscopy, Near Infrared, Portable Instruments, Raman
Application Code: General Interest
Methodology Code: Portable Instruments

Session # 2110 Abstract # 2110-5 **Symposia**

Session Title SAS - Handheld Spectrometers

Abstract Title **High Sensitivity Measurements in Liquids Using Mid-IR Lasers**

Primary Author Don Kuehl
RedShift Systems

Date: Thursday, March 10, 2016 - Mornin
Time: 11:10 AM
Room: B311

Co-Author(s) Chip Marshall, Eugene Ma, Jinghong Kim, Rick Sharp

Abstract Text

Mid-Infrared Quantum Cascade and Interband Cascade Lasers (QCL, ICL) are an extremely bright, high resolution, tunable source of mid-IR radiation suitable for a variety of spectroscopy applications. Their high resolution (<0.001 cm-1) provides significant sensitivity advantages over FTIR for gas sensing applications while their high power makes them attractive for stand-off detection of analytes on surfaces. However, the high brightness of mid-IR lasers also holds promise for higher sensitivity measurements of optically dark samples such as analytes in water. Some early results had been somewhat disappointing, which may be attributed in part to the immaturity of the first generation of commercial tunable QC lasers and to low frequency instabilities in the lasers and optical systems. More recent results have shown significant progress in overcoming some of these challenges by providing sensitivity equivalent to FTIR while also improving sampling methods (1).

In this presentation, we will describe a new approach for the measurement of analytes in aqueous solutions using low noise, continuous wave (CW) mid-IR lasers coupled with novel continuous flow microfluidic sampling techniques. The system not only provides the advantage of longer pathlength transmission cells, it exhibits a one to two order of magnitude improvement in sensitivity compared to an FTIR . By utilizing microfluidic techniques, the system is highly immune to short and long-term system noise, which also improves the accuracy for both quantitative and qualitative analysis. Some example measurements have been made on both chemical and biological systems and the results will be compared and contrasted to traditional FTIR measurements.

[1] M. Alcaraz B. Lendl, et al., "External-Cavity Quantum Cascade Laser Spectroscopy for Mid-IR Transmission Measurements of Proteins in Aqueous Solution", Anal. Chem. 2015, 87, 6980 6987

Keywords: Biospectroscopy, Environmental/Water, Infrared and Raman, Protein
Application Code: General Interest
Methodology Code: Vibrational Spectroscopy

Session # 2120 Abstract # 2120-1 **Symposia**

Session Title: Single Cell Molecular Analysis

Abstract Title: **Using Single Molecule Arrays (Simoa) to Measure Proteins and Nucleic Acids in Single Cells**

Primary Author: David R. Walt, Tufts University

Date: Thursday, March 10, 2016 - Mornin
Time: 08:35 AM
Room: B312

Co-Author(s): Payal Ghatak, Stephanie Schubert, Stephanie Walter

Abstract Text

Measuring proteins in cells typically entails lysing thousands of cells to obtain a detectable concentration. Such ensemble measurements provide for average concentrations of the protein of interest but the individual cell concentrations are lost. We have developed single molecule arrays (Simoa) that can measure proteins at 100-100x lower concentrations than ELISAs. We have used Simoa to measure the protein concentration in lysed individual cells. By measuring hundreds of cells, we can determine the protein distribution of cell populations. We have also applied the Simoa technology to the detection of nucleic acids in single cells. Such methods may one day enable rare cells to be identified in tissue samples such as biopsies, to search for particularly invasive tumor cells in a large background of relatively benign cells. In such cases, rare cells would be lost by making bulk measurements of homogenized tissue.

Keywords: Bioanalytical, Fluorescence, Genomics, Proteomics
Application Code: Bioanalytical
Methodology Code: Fluorescence/Luminescence

Session # 2120 Abstract # 2120-2 Symposia

Session Title: Single Cell Molecular Analysis

Abstract Title: **Semiconducting Polymer Dots (Pdots) for Single-Cell Sensing and Molecular Analysis**

Date: Thursday, March 10, 2016 - Mornin
Time: 09:10 AM
Room: B312

Primary Author: Daniel T. Chiu
University of Washington

Co-Author(s)

Abstract Text

Semiconducting polymer dots (Pdots) have attracted considerable attention in recent years because of their outstanding characteristics as fluorescent probes. In this presentation, I will describe our work in Pdot-based biosensors, encoded Pdots for single-cell tracking, and photoswitchable Pdots for single-cell analysis.

Keywords: Bioanalytical, Biosensors, Fluorescence, Nanotechnology
Application Code: Bioanalytical
Methodology Code: Fluorescence/Luminescence

Session # 2120 Abstract # 2120-3 **Symposia**

Session Title: Single Cell Molecular Analysis

Abstract Title: **Assaying Single Cells as a Diagnostic Tool**

Date: Thursday, March 10, 2016 - Mornin
Time: 09:45 AM
Room: B312

Primary Author: Nancy Allbritton
University of North Carolina at Chapel Hill

Co-Author(s)

Abstract Text

Cancer cells are characterized by substantial heterogeneity in basal signaling and drug response. Consequently technologies to analyze single cells are greatly needed to fully elucidate cellular behavior. To address this demand, the laboratory is developing a suite of technologies based on microengineered platforms to analyze single living cells. A key aspect to this research is the design of single-cell biochemical probes that report the enzymatic activity of kinases, lipases, and proteases with the end goal of creating clinical diagnostic and prognostic assays in patients. An automated microelectrophoresis system facilitates rapid separation and quantification of enzyme probes and their metabolic products from a cell. Single-cell analysis rates of 4 cells per min enable sufficient data collection to achieve statistically significant results. Measurements in primary tumor cells from patients demonstrate significant intercellular heterogeneity in enzymatic activities. This variability in enzyme signaling has important implications for cancer therapeutics and demonstrates the value of single-cell analysis in characterizing the nature of oncogenic signaling in cancer.

Keywords: Biomedical, Capillary Electrophoresis, Enzyme Assays, Medical
Application Code: Biomedical
Methodology Code: Capillary Electrophoresis

Session # 2120 Abstract # 2120-4 **Symposia**

Session Title Single Cell Molecular Analysis

Abstract Title **Using SAMDI Mass Spectrometry to Measure Enzyme Activity in Single Cell Lysates**

Primary Author Milan Mrksich
Northwestern University

Date: Thursday, March 10, 2016 - Mornin
Time: 10:35 AM
Room: B312

Co-Author(s)

Abstract Text

The measurement of enzyme activities in individual cells typically relies on GFP-based fluorescent reagents. The approach has the limitations that the development of the reagents requires an extensive effort and many enzyme activities are not easily addressable with this approach. This talk will illustrate how SAMDI mass spectrometry can be used to measure a broad range of enzyme activities in cell lysates. SAMDI uses self-assembled monolayers that present peptide, carbohydrate or small molecule substrates for an enzyme of interest. Treatment of the monolayer with a sample containing the enzyme results in a structural change of the substrate. This change can be directly observed with laser desorption ionization mass spectrometry, since irradiation of the monolayer results in release of the alkanethiolates, but with little fragmentation. This talk will describe a method that can assay lysates isolated from individual cells. The method uses monolayers that are patterned with a cell adhesive ligand (here, the common RGD tripeptide) and also with a peptide that is a substrate for a phosphatase enzyme. Cells are allowed to attach and spread on these features. The media is then removed and a lysis buffer is applied to individual cells, such that the lysate that is generated is in direct contact with the substrate for the phosphatase enzymes. After the lysate is removed, the monolayer is analyzed by SAMDI to determine the extent of enzyme activity. Hence, this work provides a general strategy to combine cell-based assays with a molecular assay of enzyme activity.

Keywords: Biological Samples, Drug Discovery, Enzyme Assays, Method Development

Application Code: Bioanalytical

Methodology Code: Microfluidics/Lab-on-a-Chip

Session # 2120 Abstract # 2120-5 **Symposia**

Session Title Single Cell Molecular Analysis

Abstract Title **A Novel Tool for Detection and Enumeration of Circulating Tumor Cells**

Date: Thursday, March 10, 2016 - Mornin
Time: 11:10 AM
Room: B312

Primary Author David A. Giljohann
AuraSense

Co-Author(s)

Abstract Text
"NanoFlares" are the basis for a new class of live cell intracellular RNA detection assays. NanoFlare constructs exhibit high stability, high binding specificity, and unparalleled transfection efficiency into more than 60 cell and tissue types tested. The unique and proven abilities of NanoFlares to detect genetic information in living cells can be leveraged to create innovative new detection capabilities for circulating tumor cells from patient samples. For example, prior to metastatic tumor formation, cells from a primary tumor must first invade and travel through the blood stream. At any given time, only a small portion of tumor cells, known as circulating tumor cells (CTCs) are present in the blood stream. CTCs have the potential to extravasate, seed the growth of metastases, and ultimately spread. Recent studies indicate that the presence of CTCs before treatment is an independent predictor of poor survival in patients with metastatic cancer, while the presence of residual CTCs after treatment is associated with treatment failure. CTC molecular expression profiles, however, can be largely heterogeneous within the context of a single cancer type or even a single patient. As a result, evaluation of disease progression and treatment efficacy is often impaired or inaccurate. Current methods for CTC detection rely on only a single expressed protein signal, and these processes miss as much as 40% of CTC subpopulations that do not express the appropriate surface proteins necessary for either isolation or detection, and do not offer any additional information about isolated cells at the single-cell level. The ability of NanoFlares to enter cells and fluorescently detect RNA for any gene target of interest will open up the entire transcriptome for analysis at the single cell level.

NanoFlares were developed at Northwestern University and are licensed to and marketed by life sciences giant EMD Millipore under the trade name SmartFlare for laboratory research use.

Keywords: Biotechnology, Imaging, Nanotechnology, Nucleic Acids
Application Code: Biomedical
Methodology Code: Fluorescence/Luminescence

Session # **2130** Abstract # **2130-1** **Organized Contributed Sessions**

Session Title: Recent Advances in Ion Analysis

Abstract Title: **Anion Analysis Using Capillary Ion Chromatography of Steam Cycle Water and Anion Analysis of Reactor Water by Matrix Elimination in a Nuclear Power Plant**

Date: Thursday, March 10, 2016 - Mornin
Time: 08:30 AM
Room: B313

Primary Author: Richard Wallwork
Pacific Gas and Electric

Co-Author(s)

Abstract Text

Ion chromatography (IC) is a primary technique for monitoring water quality with respect to corrosive ions in the nuclear power industry. Diablo Canyon Power Plant is a Nuclear power plant located on the central coast of California. In this presentation, we will discuss two methods of anion analysis using Ion Chromatography. In the first method anion analysis in the Steam Cycle Water using a Capillary IC setup with an in-line sample capability will be discussed. The focus will be on method development considerations particularly protocols geared to address the challenges in quantifying trace (ppt) levels of anions present in a matrix containing hydrazine and ethanolamine. We will also comment on our experience with operating a capillary IC system with in-line sample capability. In the second method we will discuss analysis of anions in the Reactor Water after pursuing a matrix elimination method. The focus again will be on method development aspects particularly protocols geared to address the challenges of quantitating ppb level anion concentration in a matrix containing ppm level of boron and lithium.

Keywords: Capillary Ion Analysis, Ion Chromatography, Nuclear Analytical Applications, Sample Handling/Autom
Application Code: General Interest
Methodology Code: Liquid Chromatography

Session # 2130 Abstract # 2130-2 **Organized Contributed Sessions**

Session Title: Recent Advances in Ion Analysis

Abstract Title: **New Developments in Multidimensional Analysis for Drinking Water Applications**

Date: Thursday, March 10, 2016 - Mornin
Time: 08:50 AM
Room: B313

Primary Author: Herb Wagner
Independent Contractor

Co-Author(s)

Abstract Text

The quality of drinking water is essential to public health. To ensure that the drinking water is safe for consumption, it has to meet stringent regulatory requirements. Ion chromatography with suppressed conductivity detection has been well adopted as a reliable method for compliance monitoring. Over the years, as we become more aware of the various potential contaminants in the drinking water, the demand for trace analysis increased substantially. This presentation will discuss the new developments in ion chromatography technology with a focus on improving method reporting limits for trace contaminants in drinking water. Specifically the discussion will focus on multi-dimensional ion chromatography. The combination of different column chemistry significantly expands the resolution power and quantitation capability, thereby providing an ideal platform for trace analysis. The application of this technology for bromate. perchlorate and haloacetic acids analysis in drinking waters will be discussed.

Keywords: Environmental Analysis, Environmental/Water, Ion Chromatography
Application Code: Environmental
Methodology Code: Liquid Chromatography

Session # 2130 Abstract # 2130-3 **Organized Contributed Sessions**

Session Title: Recent Advances in Ion Analysis

Abstract Title: **Recent Advances in Suppressor Technology in Ion Chromatography for Achieving Low Noise Performance**

Date: Thursday, March 10, 2016 - Mornin
Time: 09:10 AM
Room: B313

Primary Author: Kannan Srinivasan
Thermo Fisher Scientific

Co-Author(s): Brittany Omphroy, Christopher Pohl, Rong Lin, Sheetal Bhardwaj

Abstract Text

The technique of suppressed ion chromatography is synonymous with ion chromatography (IC) and the suppressor has become an essential component of the ion chromatograph. Suppression as implemented with an electrolytic membrane suppressor offers continuous operation with the highest dynamic suppression capacity thus enabling the use of high capacity columns and gradient mode of operation. When pursuing anion analysis with hydroxide eluents, the product from the suppressor is water and results in a low background and low noise using conductivity detection. When pursuing anion analysis using carbonate and/or bicarbonate eluents, the suppressor product is carbonic acid which results in a greater than 10 fold higher background than hydroxide eluents and relatively high noise. In the chemical mode of operation, the noise is not impacted by the suppressed conductivity background; however leakage of the chemical reagent can compromise the operational dynamic capacity of the suppressor and the detection sensitivity.

We investigated various design configurations of the electrolytic membrane suppressor to understand the underlying cause of the higher noise phenomena with carbonate and/or bicarbonate eluents. In this presentation we provide highlights of these investigations. Additionally we discuss a new design of the suppressor that resulted in low noise performance with carbonate and/or bicarbonate eluents. We show results from comparing the performance of the conventional membrane suppressor design with the performance of the new design. Example applications with drinking water will also be shown here that demonstrates the improved signal to noise performance with the new suppressor design.

Keywords: Chromatography, Environmental, Ion Chromatography, Ion Exchange
Application Code: Environmental
Methodology Code: Liquid Chromatography

Session # **2130** **Abstract #** **2130-5** **Organized Contributed Sessions**

Session Title: Recent Advances in Ion Analysis

Abstract Title: **Revisiting the Many Facets of Ion Exclusion Chromatography**

Date: Thursday, March 10, 2016 - Mornin
Time: 10:05 AM
Room: B313

Primary Author: C Phillip Shelor
University of Texas Arlington

Co-Author(s): Purnendu Dasgupta

Abstract Text

There are few substitutes for Ion Exclusion Chromatography (ICE) if the task is to determine a number of weak organic acids. Gradient elution is possible with a nonabsorbing strong acid eluent where a decreasing acid eluent concentration over time provides the gradient. But this approach is applicable only if the analyte acids are absorbing. With a variety of analytes of interest, even at 210 nm the absorption is weak and leads to relatively poor sensitivity. Suppressed conductometric ICE dominates current practice. A large low mobility anion strong acid (e.g., heptafluorobutyric acid) is used as the eluent and the proton is exchanged in the suppressor for a large low mobility cation (e.g. tetrabutylammonium). While this leads to a lower conductivity background than if the eluent is not suppressed, the usual benefits of suppression are not attained – analyte signal also decreases by salt formation and it is not really possible to run gradients. Optical detection that does not rely on the absorption properties of the analyte but is translated via pH changes of a pH-sensitive indicator has also been used but this also does not permit eluent gradients. It is also common that weak acids do not represent the entire horizon of interest, in recent years interesting combinations of IC and ICE have been reported. In this presentation, we will cover our recent efforts to perform conductometric ICE of weak acids in an unsuppressed mode, yet with a low conductance background, also permitting gradient elution.

Keywords: Ion Chromatography
Application Code: General Interest
Methodology Code: Liquid Chromatography

Session # 2130 Abstract # 2130-6 **Organized Contributed Sessions**

Session Title Recent Advances in Ion Analysis

Abstract Title **Factors Underlying Recent Advances in High Speed and High Resolution Ion Chromatography**

Primary Author Charles A. Lucy
University of Alberta

Date: Thursday, March 10, 2016 - Mornin
Time: 10:25 AM
Room: B313

Co-Author(s)

Abstract Text

Separation speeds and efficiencies in ion chromatography (IC) have lagged behind those of HPLC and UHPLC. Traditionally, IC columns were based on 7-13 μm particles. In recent years, dramatic strides have been made in reducing separation times and increasing efficiency in ion chromatography by using particles as small as 2 μm.

This presentation will review technological developments such as higher pressure PEEK flow components, new eluent generators, and membrane suppressors with increased pressure and lower volumes that enable faster separations.

But the heart of a faster IC separation lies in the use of smaller ion exchange particles. Such particles allow shorter columns to be used with correspondingly faster separations, or if packed in longer columns, yield higher efficiency columns. However, the use of small IC columns come with challenges such as increased back pressure, greater demands on extra column components, complex overload behavior, and overcoming the colloidal nature of the particles when packing columns. Means of overcoming these challenges will be discussed. Finally, recent investigations of new materials for IC separations will also be presented.

Keywords: Chromatography, Instrumentation, Ion Chromatography, Liquid Chromatography
Application Code: General Interest
Methodology Code: Liquid Chromatography

Session # 2130 Abstract # 2130-7 **Organized Contributed Sessions**

Session Title: Recent Advances in Ion Analysis

Abstract Title: **Peak Shapes in High Efficiency and Fast Chromatography: Contributions from the Slurry Packing Process and Detector Settings**

Date: Thursday, March 10, 2016 - Mornin
Time: 10:45 AM
Room: B313

Primary Author: M Farooq Wahab
University of Texas at Arlington

Co-Author(s): Daniel W. Armstrong, Purnendu Dasgupta

Abstract Text

Peak shapes in chromatography provide a fundamental insight into the processes, which take place inside the column and within the HPLC instrument. Ideally, it is desired to obtain a Gaussian peak with very small variance to achieve highest detection sensitivity and maximum peak capacity. With ever-improving column technology, it is becoming imperative to understand the processes, which affect the peak shape. This talk will focus on the fundamental aspects, which affect the peak shape, firstly from the slurry packing process and secondly from the electronic effects on modern HPLCs.

Although slurry-packing process is considered an art, scientifically, the whole packing process can be considered as a fundamental problem in suspension rheology in confined cylindrical tubes. The nature of slurry, wall effects and correlation between optical microscopy with rheology, all of which control the final peak shape will be discussed. In the second part, the effect of detector-sampling rate and the detector response time will be considered. In this talk, we revisit the concept of data sampling from the fundamental sampling theorem. It is shown by Fast Fourier Transform (FFT) of simulated chromatograms that even for extremely narrow peaks; the minimum sampling frequency required by the sampling theorem is concentrated at very low frequencies (< 10 Hz). However, in real life instrumentation, the sampling frequency may be coupled with data bunching or integration and even with response times depending on the manufacturer. Using an ultra-efficient column (40,000 plates) one can evaluate how different instruments convolute a Gaussian peak by their proprietary digital filtering. We give examples of digital filtering on two state of the art HPLCs and show its effect on peak shape, peak width, noise amplitude, and retention time. Square wave experiments with a 1 Hz light emitting diode reveal the nature of the digital filter.

Keywords: HPLC Detection, Instrumentation, Ion Chromatography, Liquid Chromatography
Application Code: General Interest
Methodology Code: Liquid Chromatography

Session # 2130 Abstract # 2130-8 **Organized Contributed Sessions**

Session Title: Recent Advances in Ion Analysis

Abstract Title: **Recent Developments in Stationary Phases for Ion Chromatography**

Primary Author: Christopher Pohl
Thermo Fisher Scientific

Co-Author(s): Andy Woodruff, Charanjit Saini, Mani Jayaraman, Maria Rey

Date: Thursday, March 10, 2016 - Mornin
Time: 11:05 AM
Room: B313

Abstract Text

2015 marked the 40th anniversary of the introduction to the world of the technique that came to be known as: Ion Chromatography. In spite of the relative maturity of this technique, research into the development of new stationary phases for ion chromatography continues at an active pace. The focus of most research in this area in recent years has been in the development of stationary phases that maximize the utility of the preferred eluent species in ion chromatography: hydronium and hydroxide. In conventional ion exchange media hydronium and hydroxide suffer from the disadvantage of being relatively weak eluents that are incapable of eluting strongly retained analytes. Furthermore these reagents tend to be corrosive to all but a few stationary phase architectures. In this work I will review the latest developments in new ion exchange phases developed specifically for ion chromatography, with a focus on the use of such eluent systems. We will cover a variety of different stationary phase architectures and include examples illustrating their application to water quality and environmental analytical challenges.

Keywords: Chromatography, Environmental, Ion Chromatography, Ion Exchange
Application Code: Environmental
Methodology Code: Liquid Chromatography

Session # 2140 **Abstract # 2140-1** **Oral Sessions**

Session Title: Bioanalytical: Fluorescence/Luminescence Techniques

Abstract Title: **Ultrasensitive Detection of Ricin Toxin in Multiple Sample Matrixes Using Single-Domain Antibodies**

Date: Thursday, March 10, 2016 - Mornin
Time: 08:30 AM
Room: B405

Primary Author: Trinh Dinh
Tufts University

Co-Author(s): David R. Walt, Kevin Ngan, Shonda Gaylord

Abstract Text

Ricin is an extreme biological toxin in which a single ricin molecule can inactivate 1,500 ribosomes per minute and lead to cell necrosis. Although ricin intoxication is not contagious, there is no currently available vaccine or therapeutic treatment for it but only the supportive care. Therefore, it is urgently needed of a highly sensitive detection method of ricin toxins in human biological fluids to increase the survival chance of intoxicated patients. Here we present the detection of ricin toxins in human urine and serum using the single molecule array (Simoa) in a fully automatic, high throughput system that generates the first sample result in as little as 64 min, compared to 9-12 hours assay time needed for Immuno-PCR. The Simoa assay formed the sandwich immunocomplex on the paramagnetic beads by using the smallest engineered binding fragments, single domain antibodies (sdAbs). The ricin toxin is detectable at levels of 1 pg/ mL in buffer, urine and serum using Simoa assay, while demonstrating no cross-reactivity with other non-related toxins or toxoids.

Keywords: Bioanalytical, Biological Samples, Clinical/Toxicology, Enzyme Assays
Application Code: Bioanalytical
Methodology Code: Fluorescence/Luminescence

Session # 2140 Abstract # 2140-2 **Oral Sessions**

Session Title Bioanalytical: Fluorescence/Luminescence Techniques

Abstract Title **Unusual Red Emission of Graphene Quantum Dots at Extremely High pH**

Primary Author Yiyang Liu
University of Kentucky

Date: Thursday, March 10, 2016 - Mornin
Time: 08:50 AM
Room: B405

Co-Author(s) Doo Young Kim

Abstract Text

Graphene quantum dots (GQDs) are small pieces of graphene nano-sheets usually with the lateral dimension of 3-20 nm and are rich in surface functional groups. Unlike bulk carbon materials, GQDs are highly fluorescent due to their non-zero band gap. This, combined with their excellent biocompatibility, makes GQDs a potent material for bio-imaging. However, most of the reported GQDs fluoresce in the shorter wavelength end of visible spectrum, exhibiting either blue or green color. This greatly limits their utility in their application for bioimaging because shorter wavelength fluorescence is susceptible to absorbance by tissues. Fluorescence at longer wavelength, such as red or near-infrared (NIR) emissions, is therefore highly desirable in GQDs. In this work we report an unusual observation of red (630 nm) emission of GQDs induced by high concentration of alkaline media (pH □13). The GQDs in this work are synthesized through a typical top-down method using carbon nano-onions as precursors. The as-prepared GQDs exhibit green fluorescence at neutral pH condition and switch to red fluorescence upon adding KOH. The process was found to be instantaneous and reversible by neutralization. To determine the origin of such phenomenon, we performed carefully controlled solution-phase FT-IR and cyclic voltammetry to track the change of surface functional groups associated with the fluorescence shifting process. From the data collected, we concluded that the red emission was due to the structural transformation of GQDs from quinone structure to hydroxyl groups. This study will be inspiring in the design of red or NIR fluorescent quantum dots for bioimaging.

This research was supported by the NSF KY EPSCoR grant and the Kentucky Science & Engineering Foundation (KSEF) grant.

Keywords: Biosensors, FTIR, Imaging, Luminescence
Application Code: Material Science
Methodology Code: Fluorescence/Luminescence

Session # 2140 Abstract # 2140-3 **Oral Sessions**

Session Title Bioanalytical: Fluorescence/Luminescence Techniques

Abstract Title **Bright Large Stokes' Shift NIR Fluorescent Silica Nanoparticle Labels and Probes**

Date: Thursday, March 10, 2016 - Mornin
Time: 09:10 AM
Room: B405

Primary Author Gabor Patonay
Georgia State University

Co-Author(s) Gala Chapman, Kyle Emer, Maged Henary

Abstract Text

There have been continuous research efforts to develop highly fluorescent reporting labels and probes for bioanalytical applications. One approach to achieve high fluorescence intensity is to incorporate several fluorophores into a single reporting or sensor entity. Silica nanoparticles can be made of using almost any kind of fluorophores that can be modified to have suitable functional moiety for covalent binding to silicates. Although fluorescent silica nanoparticles can be made by simply saturating porous particles with NIR dyes; solid silica nanoparticles containing covalently copolymerized dyes have much superior properties. Using NIR dyes in these structures have significant spectral advantages especially in bioanalytical applications as the NIR spectral region (650-900 nm) offers lower background interference and large penetration depths. NIR dyes often have relatively lower fluorescent quantum yield and they are more prone to self quenching. This can significantly be reduced by using NIR dyes that have large Stokes' shift achieved by substituting meso position halogens in the NIR fluorescent carbocyanine dye with a linker containing amino moiety. This substitution lowers the excitation wavelength and has very little influence on the fluorescence wavelength and simultaneously can serve as a linker to covalently attach the dye molecule to the nanoparticle backbone. This presentation describes a systematic study of several NIR carbocyanines covalently synthesized into silica nanoparticles backbones. The surface of these silica nanoparticles then can be functionalized for use as covalent labels. Several applications will be discussed of these bright, large Stokes' shift labels, such as immunochemistry, flow cytometry, capillary electrophoresis, etc. The covalently immobilized NIR dye retains its environmental sensitivity which can be utilized for high sensitivity probe applications, such as metal ion sensors or solvent polarity measurements.

Keywords: Bioanalytical, Molecular Spectroscopy, Near Infrared, Spectrophotometry

Application Code: Bioanalytical

Methodology Code: Fluorescence/Luminescence

Session # **2140** Abstract # **2140-4** **Oral Sessions**

Session Title: Bioanalytical: Fluorescence/Luminescence Techniques

Abstract Title: **Development of Fluorescent Magnetic Particles for "On-Off" Switching based Detection of Various Lectin-Saccharide Interactions**

Date: Thursday, March 10, 2016 - Mornin
Time: 09:30 AM
Room: B405

Primary Author: Suzuki Yoshio
AIST

Co-Author(s)

Abstract Text

The detection of various glycoconjugates, as biomarkers of various biological events, is important not only for the advancement of basic scientific research but also for diagnostic applications, and model systems such as self-assembled monolayers and gold glyconanoparticles have been usually used to study carbohydrate-protein binding events. However, limit of detection in previous systems were poor, and more sensitive model systems for the detection of carbohydrate-protein interaction should be needed.

We have developed a fluorescence-based monitoring system for the detection of lectin-saccharide interactions using a novel fluorescence emitter and quenching pair. The emitter was composed of a dansyl fluorophore and Con A (I-D), and quencher was based on a cyanopyranyl group and maltose (II-P). The fluorescence intensities of I-D decreased specifically in the presence of II-P. These changes were caused by a fluorescence quenching of dansyl fluorophore by cyanopyranyl moiety due to the formation of a stable complex between Con A and maltose. Further, to demonstrate the application of this detection method, the emitter and Con A were immobilized on the surface of magnetic particles, and the detections of various lectin-saccharide interactions was successfully performed in highly sensitive (limit of detection was 0.1 nM) way. Our results clearly indicate that the novel emitter-quencher pair is a good indicator for a highly-sensitive detection of lectin-saccharide interactions.

Keywords: Biosensors, Fluorescence, Material Science, Protein
Application Code: Bioanalytical
Methodology Code: Fluorescence/Luminescence

Session # 2140 **Abstract # 2140-5** **Oral Sessions**

Session Title: Bioanalytical: Fluorescence/Luminescence Techniques

Abstract Title: **Cell-Free Expression of Cytochrome P450-Containing Liposomes for Drug Metabolism Screens**

Date: Thursday, March 10, 2016 - Mornin
Time: 10:05 AM
Room: B405

Primary Author: Nathan A. Whitman
University of North Carolina at Chapel Hill

Co-Author(s): Jeffrey B. Penley, Julie C. McIntosh, Matthew R. Lockett

Abstract Text

Cytochrome P450 enzymes (CYPs) are a superfamily of heme-containing enzymes responsible for the oxidation of steroids, bile acids, fatty acids, prostaglandins, and leukotrienes as well as the primary metabolism of biogenic amines or retinoids, various carcinogens, and environmental pollutants. In humans, CYPs metabolize ~ 75% of marketed drugs and are widely used in drug metabolism screens in the form of human liver microsomes. The over-expression of full-length CYPs in bacterial or insect cells is difficult due to low expression yields and the challenge of solubilizing and purifying the membrane-anchored CYPs. Here we present a high-throughput method of expressing full-length human CYPs in a cell-free system of [i]E. coli[/i] extract containing an energy regeneration system, cofactors to assist with enzyme expression and folding, and synthetic liposomes. The expression and capture of functional CYPs in the synthetic liposomes was confirmed with UV/vis spectroscopy and the Michaelis-Menten kinetic parameters obtained from fluorescence-based activity assays. These CYP-containing liposomes are produced in large enough quantities for well plate-based screens of drug activity. We have also co-expressed multiple CYPs to study drug-induced CYP inhibition. This cell-free expression system provides a high-throughput method to prepare purified liposomes containing specific CYPs or CYP-isoforms, and will allow drug metabolism screens to be tailored to the individual.

Keywords: Bioanalytical, Characterization, Protein
Application Code: Bioanalytical
Methodology Code: Fluorescence/Luminescence

Session # 2140 Abstract # 2140-7 **Oral Sessions**

Session Title: Bioanalytical: Fluorescence/Luminescence Techniques

Abstract Title: **Cell Surface Engineering with Lipid-Molecular Beacon Aptamer for Real Time Probing of Proteins in Cellular Microenvironment**

Primary Author: Weijia Hou
University of Florida

Date: Thursday, March 10, 2016 - Mornin
Time: 10:45 AM
Room: B405

Co-Author(s)

Abstract Text

we propose a one-step self-construction of a cell-surface protein sensor by anchoring diacyllipid-aptamer conjugates onto the cell membrane. The diacyllipid synthesized by our group contains two C18H37 hydrocarbon tails and can be coupled to the 5 end of the DNA sequence using solid phase synthesis by the automated DNA synthesizer.With the hydrophilic DNA head segment and the hydrophobic diacyllipid tail, this amphiphilic molecule can be functionalized on the cell surface by inserting the lipid tail into the lipid bilayer cell membrane with high efficiency. Compared to traditional methods for cell membrane modification such as genetic engineering and covalent conjugation, this self-inserting strategy is extremely rapid, highly efficient, and amount controllable. The most important consideration is that the diacyllipid-aptamer functionalizes the cell membrane with additional nucleic acid molecules without changing biological and chemical property of the cell.

Keywords: Bioanalytical, Biosensors, Fluorescence
Application Code: Bioanalytical
Methodology Code: Fluorescence/Luminescence

Session # 2150 Abstract # 2150-1 **Oral Sessions**

Session Title Electrochemistry - New Methods and Applications

Abstract Title **Elucidating the Structure/Function Relationship of Conductive Polymer Microelectrodes for Use in Fast-Scan Cyclic Voltammetry of Neurotransmitters**

Date: Thursday, March 10, 2016 - Mornin
Time: 08:30 AM
Room: B314

Primary Author Adam R. Meier
University of Arizona

Co-Author(s) Michael L. Heien, William Bahureksa

Abstract Text

Our understanding of rapid chemical neurotransmission has been edified by the use of fast-scan cyclic voltammetry (FSCV). The majority of this research has used the same sensor platform, the carbon-fiber microelectrode. While effective, this material suffers from limitations in geometry. The creation of new sensor materials that overcome this limitation will allow researchers to measure chemical events that are currently inaccessible due to sensor size and number. Poly(3,4-ethylenedioxythiophene) (PEDOT) is a conductive polymer that has been used in a variety of applications such as implantable biosensors and solar cells. PEDOT is an attractive alternative to traditional electrode materials for biosensor applications because it is inexpensive, mechanically robust, biocompatible, and can be easily processed. Unfortunately, commercial PEDOT is highly capacitive (>2000 μF/cm2) making it unsuitable for rapid electrochemical applications. A vapor deposition synthesis was used to produce PEDOT films with low capacitance (<100 μF/cm2). The CVD method results in a PEDOT film with higher electrical conductivity and more uniform film thickness than commercially available PEDOT. It has faster electron transfer kinetics compared to other PEDOT systems. Scanning electron microscopy and x-ray photoelectron spectroscopy were used to study surface morphology and film composition. These measurements, combined with electrochemical data, show our CVD-synthesized PEDOT is a much "denser" film, containing a significantly higher concentration of PEDOT than dopant. This yields a better conductor and electrode material compared to commercial PEDOT. Optimizing the electrochemical properties of PEDOT has resulted in the first viable PEDOT-based platform for FSCV measurements of neurotransmitters.

Keywords: Electrochemistry, Electrodes, Neurochemistry, Polymers & Plastics
Application Code: Neurochemistry
Methodology Code: Electrochemistry

Session # 2150 Abstract # 2150-2 **Oral Sessions**

Session Title: Electrochemistry - New Methods and Applications

Abstract Title: **Electrochemically Reduced Graphene Oxide as an Electrocatalyst Support for H[sub]2[/sub]S Detection**

Primary Author: Jason A. Bennett
Penn State Erie - The Behrend College

Date: Thursday, March 10, 2016 - Mornin
Time: 08:50 AM
Room: B314

Co-Author(s)

Abstract Text

Electrochemically Reduced Graphene Oxide (ERGO) has attracted much attention for various applications, including as a sensing and electrocatalyst support material, due to its relative simplicity, speed, potential for controllability, and benign environmental impact. Most studies focus on ERGO deposited on glassy carbon electrodes from an aqueous suspension, but of those that utilize an underlying metallic electrode, few ascertain whether the underlying electrode surface is completely covered by the ERGO layer. The exposure of the underlying metal electrode can severely impact the observed electrochemistry, and specifically hinder the selectivity in sensing applications. Moreover, the complete coverage of the metal electrode is quite difficult to achieve.

This work will present a new protocol for depositing complete ERGO coatings onto a Pt electrode and evaluate its use as an electrocatalyst support material for hydrogen sulfide detection. Hydrogen sulfide is a gaseous signaling molecule known to be physiologically significant in the central nervous and cardiovascular systems. Independently detecting H[sub]2[/sub]S in the presence of its two main interferences, NO and CO, will allow for the full understanding of both its individual physiological roles as well as its interactions with NO and CO. The objective of this research is to utilize ERGO to support cyanide-coordinated electropolymerized ferriprotoporphyrin to achieve a selective and stable H[sub]2[/sub]S sensing material that can be incorporated into future amperometric sensors.

This research is sponsored by an award from the National Science Foundation.

Keywords: Biosensors, Electrochemistry, Sensors
Application Code: Bioanalytical
Methodology Code: Electrochemistry

Session # 2150 **Abstract # 2150-3** **Oral Sessions**

Session Title: Electrochemistry - New Methods and Applications

Abstract Title: **Amperometric Determination of Aurocyanide for Hydrometallurgical Gold Processing**

Date: Thursday, March 10, 2016 - Mornin
Time: 09:10 AM
Room: B314

Primary Author: Wayne Dickinson
Kemira

Co-Author(s)

Abstract Text

Determination of dissolved gold concentration is a fundamental need in hydrometallurgical gold extraction. An analytical method suitable for on-line measurement was developed that can enable more frequent gold analysis compared to standard batch AA methods; enabling more rapid response to changing conditions within the gold mill. DC amperometry was used to measure aurocyanide, Au(CN)[sub]2[/sub][sup]-[/sup] according to a modified Levich equation. A three-electrode glass cell with gold working, carbon counter and calomel reference electrodes was filled with 100 mL of a 0.05M borate-buffered, 50 µM Au(CN)[sub]2[/sub][sup]-[/sup] pH 12 solution adjusted to 0.02M CN[sup]-[/sup] and 14 µM Pb[sup]2+[/sup] with NaCN and PbCl[sub]2[/sub] respectively. Pb[sup]2+[/sup] was used to cathodically polarize hydrogen reduction, minimizing interference with the Au(CN)[sub]2[/sub][sup]-[/sup] reduction. The solution was aggressively sparged with nitrogen for 10 minutes prior to analysis to eliminate oxygen and sparging was continued throughout the test to maintain diffusion-limited conditions. The potential was stepped to -1.2 V[sub]sce[/sub] using a potentiostat and current was recorded over an 8 minute period. The analytical signal both as terminal current and total coulombs was linear in response to gold concentration with a detection limit of 0.5 mg/L. The rate of gold uptake onto activated carbon used in gold processing was determined for native carbon and for carbon treated with a polymeric material to reduce carbon abrasion. A 110 mL volume of 10 mg/L gold solution was tumbled in glass jars containing borate buffer and 4.5 g/L native or polymer-treated carbon. At intervals over 90 minutes, the solution was sampled, amended with CN[sup]-[/sup] and Pb[sup]2+[/sup] then analyzed. The treatment had no effect on gold uptake indicating compatibility with the gold adsorption process. Concentrations were in excellent agreement with separate analysis by atomic adsorption.

Keywords: Electrochemistry, Monitoring, Process Analytical Chemistry, Voltammetry
Application Code: Process Analytical Chemistry
Methodology Code: Electrochemistry

Session # 2150 Abstract # 2150-4 **Oral Sessions**

Session Title: Electrochemistry - New Methods and Applications

Abstract Title: **Method for Removal of Non-Faradaic Contributions to Fast-Scan Cyclic Voltammetry Recordings**

Date: Thursday, March 10, 2016 - Mornin
Time: 09:30 AM
Room: B314

Primary Author: Justin A. Johnson
University of North Carolina at Chapel Hill

Co-Author(s): R Mark Wightman

Abstract Text

Fast-scan cyclic voltammetry (FSCV) at carbon-fiber microelectrodes has proven to be an essential tool for [i]in vivo[/i] monitoring of subsecond fluctuations in catecholamines (e.g. dopamine and norepinephrine) and other electroactive neurotransmitters. The large background current that arises at the high voltammetric scan rates employed for FSCV necessitates the use of digital background subtraction in order to resolve the analytical signal of interest. Such an approach depends critically on the stability of the background both before and during the neurobiological phenomenon under study. However, ionic fluctuations, which occur both naturally (e.g. potassium waves in spreading depression) and artificially (e.g. iontophoretic ejections), alter the electrode capacitance and, consequently, the background signal, complicating analytical quantification. While multivariate data analysis (e.g. principal component analysis) may be used for separation of the ionic and analytically relevant signals, this approach necessitates the recording of multiple CVs of 'pure' ionic changes at the site of recording for training of the chemometric model, increasing experimental complexity.

Here, we describe and characterize an alternative measurement protocol for the removal of these non-Faradaic contributions to the FSCV signal. This approach relies on the use of a modified waveform to measure the capacitive response of the electrode, which is then used for the prediction and removal of the non-Faradaic component of the voltammetric response. In addition to discussing the waveform characteristics that enable accurate prediction of these ionic contributions, we will demonstrate the successful application of this methodology for the removal of such signals from FSCV recordings both [i]in vitro[/i] (e.g. during flow-cell analysis of mixtures of catecholamines and nonelectroactive species) and [i]in vitro[/i] (e.g. during iontophoretic ejections in anesthetized animals).

Keywords: Bioanalytical, Electrochemistry, Neurochemistry, Voltammetry
Application Code: Neurochemistry
Methodology Code: Electrochemistry

Session # 2150 Abstract # 2150-5 **Oral Sessions**

Session Title: Electrochemistry - New Methods and Applications

Abstract Title: **Quantitative Analysis of Microiontophoresis Drug Delivery**

Date: Thursday, March 10, 2016 - Mornin
Time: 10:05 AM
Room: B314

Primary Author: Douglas Kirkpatrick
University of North Carolina at Chapel Hill

Co-Author(s): R Mark Wightman

Abstract Text

Microiontophoresis describes drug administration from a micropipette delivered by a small current. It has been utilized as a tool in neurochemical investigations to administer drugs to targeted brain regions. This technique offers advantages over alternative drug delivery methods due to its rapid application time and spatial selectivity, making it ideal to study individual cell and receptor properties. However microiontophoresis is limited by the inability to precisely determine ejected concentrations and dosages. To address these shortcomings, this approach investigates factors underlying drug delivery rates in microiontophoresis. Chromatographic, electrochemical, and fluorescence methods are employed to reveal ejection rates under different conditions. Analyte concentration, molecular charge, and ionic strength are all considered. Through this process, practical guidelines are established for how microiontophoresis parameters should be chosen to anticipate ejected concentration ranges. In addition, relative scaling of the dosage by the ejection current is demonstrated. Lastly, applications for the established principles are displayed in traditional neurochemical protocols.

Keywords: Chromatography, Fluorescence, Liquid Chromatography, Neurochemistry
Application Code: Neurochemistry
Methodology Code: Electrochemistry

Session # 2150   Abstract # 2150-6   **Oral Sessions**

Session Title: Electrochemistry - New Methods and Applications

Abstract Title: **Fast Scan Cyclic Voltammetry of Metals at Carbon-fiber Microelectrodes: Correlation Between FSCV Response and Solution Dynamics**

Primary Author: Pavithra Pathirathna
Wayne State University

Date: Thursday, March 10, 2016 - Mornin
Time: 10:25 AM
Room: B314

Co-Author(s): Parastoo Hashemi, Shawn McElmurry, Thushani Siriwardhane

Abstract Text

The behavior of trace metals is the environment is controlled by speciation. For example, metal complexation with organic/inorganic ligands reduces the impact of trace metals. On the flip side, trace metals are mobilized during dynamic environmental events such as storms, which increases their toxicity. Rapid, real-time characterization of metal complexation would allow a better understanding of metals in the environment. We recently described an ultrafast, Hg-free method to detect copper and lead at carbon fiber microlectrodes (CFMs) using fast scan cyclic voltammetry (FSCV). Moreover, we explored the surface adsorption as the underlying mechanism of our fast FSCV signal. In this work, we study copper binding with a model set of ligands illustrating a wide spectrum of thermodynamic equilibrium constants expected to be found in natural waters. We identify mathematical relationships between thermodynamic equilibrium constants (K) for copper complexation and the FSCV signal. We utilize fast scan controlled adsorption voltammetry (FSCAV) to quantify ambient Cu2+ levels in real environmental samples and develop a model that relates the FSCV signal to free copper in solution to the solution K. We, hence, showcase the power of FSCV as a speciation sensor.

Keywords: Electrochemistry, Environmental, Metals, Microelectrode
Application Code: Environmental
Methodology Code: Electrochemistry

Session # 2150 **Abstract # 2150-7** **Oral Sessions**

Session Title: Electrochemistry - New Methods and Applications

Abstract Title: **Construction of Training Sets for Valid Calibration Using Principal Component Analysis**

Date: Thursday, March 10, 2016 - Mornin
Time: 10:45 AM
Room: B314

Primary Author: Nathan T. Rodeberg
University of North Carolina at Chapel Hill

Co-Author(s): Justin A. Johnson, R Mark Wightman

Abstract Text

Principal component regression (PCR) is a useful multivariate calibration tool for fast-scan cyclic voltammetry that permits the separation and quantitation of multiple overlapping neurophysiological signals. For measurements made at acutely implanted carbon-fiber microelectrodes, training sets are typically built for both dopamine and pH changes using representative cyclic voltammograms (CVs) from electrically evoked dopamine release at the same brain location and electrode. However, in some experiments it is impractical to implant stimulating electrodes. In these cases, training sets built from CVs acquired at separate electrodes have been used to analyze data. However, this method has not been investigated and it has been previously suggested that this approach would fail.

Here we investigate this approach using data from behaving animals acquired with acutely implanted electrodes, and compare it to a previously established PCR protocol. These data demonstrate differences in CV characteristics across electrodes lead to the failure of this approach, including misestimated and inconsistent concentration predictions and, more importantly, an inability to validate PCR models. Alternative strategies to this approach are also investigated.

Keywords: Calibration, Chemometrics, Neurochemistry, Voltammetry
Application Code: Neurochemistry
Methodology Code: Electrochemistry

Session # 2150 **Abstract # 2150-8** **Oral Sessions**

Session Title: Electrochemistry - New Methods and Applications

Abstract Title: **Electrochemical Characterization and Catalytic Activity of Ultrasmall Gold Nanoparticles**

Primary Author: Francis Zamborini
University of Louisville

Date: Thursday, March 10, 2016 - Mornin
Time: 11:05 AM
Room: B314

Co-Author(s): Dhruba Pattadar, Rafael Masitas, Stacy Allen

Abstract Text

In this talk we describe a method to synthesize ultrasmall citrate-coated gold nanoparticles estimated to be about 1 nm in diameter and demonstrate their unique electrochemical reactivity. UV-vis spectroscopy provides spectroscopic characterization of the bulk solution of gold nanoparticles, which were synthesized by the borohydride reduction of tetrachloroaurate in the presence of citrate. After synthesis, the gold nanoparticles were attached electrostatically to amine-functionalized indium-tin-oxide electrodes. Linear sweep voltammetry provided the oxidation potential and coverage of the gold nanoparticles on the electrode surface. Two main peaks appeared in the voltammetry as a result of two major size populations. One size population was estimated to be about 1 nm in diameter based on its large negative shift from the bulk oxidation potential of gold. Samples with this size population showed strong electrocatalytic activity towards carbon dioxide reduction, which is important as a possible route to prepare chemical fuels. Carbon dioxide reduction did not occur after removal of the ultrasmall gold nanoparticles or with 2-4 nm diameter nanoparticles attached to the surface. This research importantly shows a very large negative shift in the oxidation potential and strong catalytic activity for carbon dioxide reduction for very small ~1 nm diameter gold nanoparticles. The altered synthesis allows a large coverage of these small nanoparticles to be attached to the electrode surface. The nanoparticles are highly active due to their small size and presence of the weak citrate stabilizer, which does not block their catalytic surface.

Keywords: Electrochemistry, Nanotechnology, Particle Size and Distribution, Voltammetry
Application Code: Nanotechnology
Methodology Code: Electrochemistry

Session # 2160 Abstract # 2160-1 **Oral Sessions**

Session Title Food Safety Evaluations - Half Session

Abstract Title **Characterization of the Adulteration, Counterfeiting and Contamination of Spices, Spice Products and Supplements by the Detection of Toxic and Banned Organic Chemicals in Commercial Botanical Products**

Primary Author Patricia Atkins
SPEX CertiPrep

Date: Thursday, March 10, 2016 - Mornin
Time: 08:30 AM
Room: B407

Co-Author(s)

Abstract Text
The consumption of botanical products has increased over the past two decades as consumers trend to what are perceived to be natural and high quality botanical products. The primary regions of spice and tea production around the have often been cited as having less stringent safety and quality standards in regards to consumer products. Products from these regions have been noted to contain a variety of adulterants and contaminants.

Common spices and botanicals in the US (Black Pepper, Red Pepper, Cinnamon, Mustard Seed, Cumin, and Turmeric) in various forms (i.e. spices, teas, condiments and supplements) were purchased at dollar stores, farmer's markets, chain stores, and online vitamin outlets. Products selected covered the range of preparations including organic products. Cryogenic grinding and microwave extraction were employed in sample processing. Physical and chemical screening methods were used to detect gross adulteration and counterfeiting. GC-MS and LC-MS were used to determine the possible adulteration or contamination of these products with compounds such as restricted azodyes, processing solvents and banned pesticides. High levels of many restricted or banned compounds were found in spices and spice brands which were previously believed to be counterfeited or adulterated through ICP and ICP-MS analysis of these same spices for elemental content. The presence of banned and toxic organic compounds added further weight to the toxic elemental content some spices creating chemical and elemental profiles of counterfeit and adulterated spice products being sold within the United States.

Keywords: Food Contaminants, Food Safety, Liquid Chromatography/Mass Spectroscopy, Natural Products
Application Code: Food Safety
Methodology Code: Gas Chromatography/Mass Spectrometry

Session # 2160 Abstract # 2160-2 **Oral Sessions**

Session Title: Food Safety Evaluations - Half Session

Abstract Title: **Using SPE to Adjust Sensitivity to Analytical Requirements for Food Safety**

Date: Thursday, March 10, 2016 - Mornin
Time: 08:50 AM
Room: B407

Primary Author: Chris Shevlin
Horizon Technology

Co-Author(s): William Jones

Abstract Text

More and more compounds that may be carcinogens are found in food and beverages. The compound 4-Methylimidizole (4-MEI) is formed as a byproduct in some foods and beverages. Caramel coloring type III and Type IV in beverages is one of the ingredients which may contain 4-MEI.

Products that potentially contain 4-MEI include certain colas, beers, soy sauces, breads, coffee, ammoniated livestock feed and other products. Sensitive measurement of 4-MEI is required to ensure safe levels are not exceeded.

Laboratories are equipped with a variety of equipment, but smaller laboratories or those in less developed parts of the world may not be able to afford the most expensive analytical equipment, such as mass spectrometry or triple quad mass spectrometry. In this case, sample preparation using solid phase extraction (SPE) can aid in concentrating the analytes of interest so a less expensive analytical technique, such as HPLC with UV or Fluorescence detection can be used. If a mass spectrometry instrument is available, solid phase extraction may not be needed for concentrating the analytes, but may prove useful for cleaning up the extract to provide a cleaner sample for injection.

The capabilities of SPE and application to water and soft drinks will be discussed. The tradeoffs for different analytical techniques will be illustrated with data on a variety of samples and spiked samples.

Keywords: Food Safety, Liquid Chromatography/Mass Spectroscopy, Pesticides, Solid Phase Extraction
Application Code: Food Safety
Methodology Code: Liquid Chromatography/Mass Spectrometry

Session # **2160** Abstract # **2160-3** **Oral Sessions**

Session Title: Food Safety Evaluations - Half Session

Abstract Title: **Ensuring a Safe and Stable Food Supply Using ICP-OES for Elemental Monitoring**

Date: Thursday, March 10, 2016 - Mornin
Time: 09:10 AM
Room: B407

Primary Author: Nick Spivey
PerkinElmer Inc.

Co-Author(s): Kenneth Neubauer, Laura Thompson, Stan Smith

Abstract Text

Consumers have become ever more aware of the global nature of the food supply chain and the fragmented nature of global regulatory oversight and the weaknesses of regional safety and quality control. Food manufacturers have become acutely sensitized to the potential fallout if a food product, fresh or processed, were to cause health issues or illness with the public even if the problem can be traced back to a supplier. Both food producers and consumers alike are concerned with the ability of suppliers to maintain the global food supply and minimize the negative impact on the soil and environment.

Elemental analysis of fresh and processed raw materials and food products can determine the nutritional content of the food and screen for harmful elemental contaminants. Understanding the elemental nutrient content of the soil allows for targeted fertilizer application without waste and environmentally damaging fertilizer runoff. Fertilizers can be analyzed to eliminate mislabeled or falsely labelled material.

With a number of techniques available for elemental analysis, analytical measurement by ICP-OES provides a strong balance between cost, speed and simplicity when compared with flame AA and ICP-MS.

Effective sample preparation of such a wide array of samples can be a challenge. The use of microwave digestion minimizes this while providing safe and simple preparation methods.

This work focuses on the preparation and analysis of food, fertilizer and soil samples using microwave digestion followed by ICP-OES analysis.

Keywords: Agricultural, Elemental Analysis, Food Safety, ICP
Application Code: Food Safety
Methodology Code: Atomic Spectroscopy/Elemental Analysis

Session # 2160 Abstract # 2160-4 Oral Sessions

Session Title Food Safety Evaluations - Half Session

Abstract Title **Automated Solid Phase Extraction and Quantitative UHPLC Analysis of Cannabis Compounds in Food Matrices**

Date: Thursday, March 10, 2016 - Mornin
Time: 09:30 AM
Room: B407

Primary Author Chris Shevlin
Horizon Technology

Co-Author(s) Robert E. Buco

Abstract Text

With the recent legislation that legalized the recreational use of marijuana in some states, and the medicinal use of marijuana in many more, there is a fast-growing need to develop standardized analytical techniques to assess the potency of cannabis plants and edibles. Sample preparation is one of the most significant challenges, especially as it pertains to the analysis of edibles such as brownies, cookies, chocolate bars, and gummy bears that have been prepared with cannabis plant extracts. This presentation describes the development and optimization of a sample preparation and solid phase extraction method to quantitatively analyze cannabis compounds in edibles, focusing on automated procedures to increase precision, and exploring the sensitivity and accuracy of the complete workflow when combined with UHPLC analysis using a PDA detector.

Keywords: Food Safety, HPLC, Sample Preparation, Solid Phase Extraction
Application Code: Food Safety
Methodology Code: Sampling and Sample Preparation

Session # 2170 **Abstract # 2170-1** **Oral Sessions**

Session Title: Liquid Chromatography Column Chemistry

Abstract Title: **Functionalized Octatetrayne as Novel Carbon Media for Capillary Liquid Chromatography**

Date: Thursday, March 10, 2016 - Mornin

Time: 08:30 AM

Room: B406

Primary Author: Jiayi Liu
The Ohio State University

Co-Author(s): Susan Olesik

Abstract Text

Open tubular liquid chromatography was developed for over 40 years due to the inherent advantages of high plate numbers, low solvent consumption and low sample injection. The small sample volumes and high detection limit of capillary HPLC have limited its use. Numerous stationary phases have been proposed but with limited success. The ideal stationary phase for capillary HPLC must be highly efficient with high surface area to improve the sample capacity and decrease the detection limit. A new carbon stationary phase will be illustrated that has excellent performance. The uniform carbon film was covalently coated onto silica capillary columns with high stability and unique absorption properties for both polar and non polar analytes. This new carbon capillary column provides improved separation performance with reversed phase properties.

Keywords: Capillary LC, Hydrocarbons, Liquid Chromatography, UV-VIS Absorbance/Luminescence

Application Code: Other

Methodology Code: Liquid Chromatography

Session # 2170 **Abstract # 2170-2** **Oral Sessions**

Session Title: Liquid Chromatography Column Chemistry

Abstract Title: **Stationary Phases Based on the Thiol-ene Reaction on Mercaptopropylsilane-Modified Nonporous Silica**

Date: Thursday, March 10, 2016 - Mornin
Time: 08:50 AM
Room: B406

Primary Author: Erin Shields
University of Pittsburgh

Co-Author(s): Kayla Thomas, Stephen Weber

Abstract Text
The thiol-ene reaction is a robust widely used "click chemistry" reaction. The lack of metal catalysts and other species that can be detrimental to chromatography make the reaction an excellent step to create countless new stationary phases. To study the effect of the embedded sulfur and new stationary phases, a method was developed to produce monodisperse nonporous silica particles with a diameter of 1.2 micrometers to use as the stationary phase support. The coupling of the thiol-ene reaction with nonporous silica helps remove the ambiguity from the diffusion in porous silica, and allows easier studies of new stationary phases. With a greater understanding of the thiol-ene reaction's effect on separations, new stationary phases were custom built and studied using the nonporous silica support. A mixed-mode weak cation exchange reverse phase (WCX-RP) stationary phase was created to separate monoamine neurotransmitters in brain dialysate. The cation exchange moiety separates the positively charged monoamines from neutral and negatively charged compounds without the need for ion-pairing reagents. When used with small samples having low solute concentrations to remain below the loading capacity, we anticipate using the WCX-RP stationary phase to allow faster separations of monoamine neurotransmitters.

Keywords: Capillary LC, HPLC, HPLC Columns, Liquid Chromatography
Application Code: Bioanalytical
Methodology Code: Liquid Chromatography

Session # 2170 Abstract # 2170-3 **Oral Sessions**

Session Title Liquid Chromatography Column Chemistry

Abstract Title **Using 5 Micron Superficially Porous Particles in Capillary and Microfluidic LC Columns**

Date: Thursday, March 10, 2016 - Mornin
Time: 09:10 AM
Room: B406

Primary Author James P. Grinias
University of Michigan

Co-Author(s) Robert T. Kennedy

Abstract Text

Large-size (4–5 micron) superficially porous particles yield lower plate heights (with hmin = 1.5) than fully porous particles of a similar size when packed into large-bore columns. This property (thought to be due to reduced stagnant mobile phase effects and eddy dispersion) allows for better chromatographic performance without the higher pressures required for smaller particles. This study explored the use of such particles in microfluidic LC columns where materials and fitting pressure limits can constrain the size of particle used. The theoretically predicted performance improvements compared to fully porous particles were not demonstrated in capillary columns (with hmin = 2 for both particle types), in agreement with previous studies that examined smaller superficially porous particles. Microfluidic columns were then compared to capillary columns. Capillary columns significantly outperformed microfluidic columns due to imperfections imposed by microfluidic channel asymmetry and the world-to-chip connection at the optimal flow rate; however, superficially porous particles packed in microfluidic LC columns had flatter plate height versus flow rate curves indicating potential for better performance at high reduced velocities. The use of these particles in capillary LC columns for the separation of in vivo neurotransmitter samples was also investigated.

Keywords: Capillary LC, HPLC Columns, Lab-on-a-Chip/Microfluidics, Liquid Chromatography
Application Code: General Interest
Methodology Code: Liquid Chromatography

Session # 2170 Abstract # 2170-4 **Oral Sessions**

Session Title: Liquid Chromatography Column Chemistry

Abstract Title: **Selectivity in Reversed-Phase Liquid Chromatography: Impact of Stationary Phase Chemistry**

Date: Thursday, March 10, 2016 - Mornin
Time: 09:30 AM
Room: B406

Primary Author: David S. Bell
Supelco/Sigma-Aldrich

Co-Author(s)

Abstract Text

Many of the problems encountered executing HPLC methods are a result of the initial choice of stationary phase chemistry. In practice, many analysts will reach for their favorite C18 upon commencement of method development; however alkyl (Cn) phases are often not the most appropriate tool for a given set of separation requirements (molecules with varying pKa values, solubilities, molecular weights, etc.). When retention or resolution are not readily achieved, analysts will often resort to the addition of ion-pair reagents or other exotic and complicated mobile phase preparations to force them to work. This common practice often results in the development of highly complex methods that suffer from lack of robustness and ruggedness and are not as easily transferred to other laboratories or other analysts.
There are many choices for alternative stationary phase chemistries that render the phase decision difficult. In this work alternative stationary phase classes which are highly complementary to alkyl phases from a fundamental molecular interaction point of view are discussed. An understanding of the contrasting interactions that these different classes of stationary phase chemistries provide then leads to more accurate decisions regarding the choice of phase that may be most appropriate for a given set of analytes. This critical information promises to facilitate method development and generate simpler, more reproducible separations.

Keywords: HPLC, Liquid Chromatography, Method Development
Application Code: General Interest
Methodology Code: Liquid Chromatography

Session # 2170 Abstract # 2170-5 **Oral Sessions**

Session Title Liquid Chromatography Column Chemistry

Abstract Title **A Specialty Column for High Resolution Separation of Aminoglycoside Antibiotics by HPLC**

Date: Thursday, March 10, 2016 - Mornin
Time: 10:05 AM
Room: B406

Primary Author Xuefei Sun
Thermo Fisher Scientific

Co-Author(s) Xiao Cui, Xiaodong Liu, Yoginder Singh

Abstract Text

Aminoglycoside is a family of antibiotics with amino modified sugar structures. They are widely used as clinical and veterinary medicines to treat the infections mainly caused by Gram-negative bacteria. However, these antibiotics can cause varying degree of ototoxicity and nephrotoxicity. Therefore, it is important to develop sensitive and reliable analytical methods to characterize and quantify the drug purity and impurities, determine and monitor aminoglycosides residue in different matrices. Ion-pairing reversed phase liquid chromatography (IP-RPLC) is widely utilized to analyze aminoglycosides because of their hydrophilic and positive charged nature. In addition, they have limited solubility in many organic solvents, which makes HILIC separation challenging. The conventional C18 column is usually implemented for IP-RPLC separation of aminoglycosides using high concentration of ion-pairing reagents, such as trifluoroacetic acid (TFA).However, most C18 columns are not stable under such low pH condition (e.g. pH ~1 in 0.1 M TFA) due to hydrolysis. To address this challenge, here we present a specifically designed C18 column for analysis of various aminoglycoside antibiotics by IP-RPLC. The stationary phase is based on polymer encapsulated silica technology that ensures unusual stability when exposure to the extreme conditions. In addition, this specialty column provides excellent selectivity and high resolution for the aminoglycoside analysis and it is easy to use. In this work, a couple of aminoglycoside samples have been analyzed using this column and the factors affecting the separation have also been systematically studied.

Keywords: Drugs, HPLC, HPLC Columns
Application Code: Pharmaceutical
Methodology Code: Liquid Chromatography

Session # 2170 Abstract # 2170-6 **Oral Sessions**

Session Title Liquid Chromatography Column Chemistry

Abstract Title **Liquid Chromatography, Hydrophilic Interaction Chromatography, HILIC**

Date: Thursday, March 10, 2016 - Mornin
Time: 10:25 AM
Room: B406

Primary Author David S. Bell
Supelco/Sigma-Aldrich

Co-Author(s)

Abstract Text

HILIC chromatography is a complex system involving partition, polar and ion-exchange interactions. Method development can be greatly facilitated by understanding the interactions that the different stationary phases provide and applying that knowledge to the separation task at hand. A straight forward approach toward determining the relative ion-exchange contribution for various HILIC stationary phases is described and used to classify HILIC columns by their predominant mechanisms of interaction. The investigation demonstrates that hydroxyl, zwitterionic and amide phases provide relatively little ion-exchange interactions and relatively high partitioning and polar interactions. Conversely, pentafluorophenyl and cyano phases exhibit primarily ion-exchange and relatively little partitioning character. Bare silica and several variations of HILIC columns are shown to provide different blends of ion-exchange and partitioning/polar interactions. Several examples are provided that demonstrate facilitated and predictable method development based on knowledge of stationary phase retention mechanisms. Through an understanding of the main physiochemical differences between a set of analytes intended to be resolved, a column or set of columns can be judiciously chosen to screen for effective retention and selectivity.

Keywords: HPLC Columns, Liquid Chromatography
Application Code: General Interest
Methodology Code: Liquid Chromatography

Session # 2170 Abstract # 2170-7 **Oral Sessions**

Session Title Liquid Chromatography Column Chemistry

Abstract Title **The Benefits of 1mm ID UHPLC Columns Made Real**

Date: Thursday, March 10, 2016 - Mornin
Time: 10:45 AM
Room: B406

Primary Author Stephen Luke
Agilent Technologies

Co-Author(s) Jason Link, Norwin Von Doehren, William Long

Abstract Text

In theory, UHPLC columns with a 1 mm internal diameter allow LC and LC/MS users to improve the sensitivity of their analysis compared to that achieved with wider bore columns by reducing the dilution of compounds in column and running with more MS compatible flow rates. Additionally, because of the lower flow rates used, 1mm ID columns allow chromatographers to save costs in the purchase and disposal of mobile phase solvents. However, these potential advantages provide little benefit if the chromatographic performance and, importantly, lifetime of UHPLC columns is compromised when the internal diameter is reduced. This work will show that the limitations of narrow bore columns mentioned above are overcome with an optimized 1mm UHPLC column hardware design packed with superficially porous Agilent Poroshell 120 particles. We will detail lifetimes under UHPLC conditions substantially longer than those achievable with other 1mm ID columns plus efficiencies and peak capacities that are at least equivalent to those obtained with the columns of the current industry standard of 2.1 mm ID. The work will also show the importance of the correct UHPLC instrument configuration when working with 1mm ID. The benefits of 1mm UHPLC columns can be made real.

Keywords: HPLC, HPLC Columns, Liquid Chromatography, Liquid Chromatography/Mass Spectroscopy
Application Code: General Interest
Methodology Code: Liquid Chromatography

Session # 2170 Abstract # 2170-8 **Oral Sessions**

Session Title: Liquid Chromatography Column Chemistry

Abstract Title: **Comparison of the Practical Kinetic Performance Limits of Core-Shell and Fully Porous (U)HPLC Sorbents Using Commercially Available Column Formats**

Primary Author: A Carl Sanchez
Phenomenex

Date: Thursday, March 10, 2016 - Mornin
Time: 11:05 AM
Room: B406

Co-Author(s): Gareth Friedlander, Jason Anspach, Tivadar Farkas

Abstract Text

Various approaches for comparing the separation power of different HPLC and UHPLC sorbents have been used over the years such as Van Deemter, Poppe and Kinetic plots. Core-shell and fully porous sorbents of various particle sizes have been compared previously [1-2] over very wide operational ranges. However, to date, a comprehensive comparison of the kinetic performance of core-shell and fully porous sorbents spanning the full particle size range currently available (1.3/1.7 – 5 μm) using strictly practical operational conditions has not been shown.

In this work, the kinetic plot approach was used to determine the optimum particle morphology and size for various practical separation goals. Practical limits were imposed on analysis time, column length, flow rate and column maximum operating pressure to focus results on the most relevant operational ranges. Additionally, results were restricted to combinations of column formats (length and diameter) and operational ranges (flow rate and pressure) compatible with currently available HPLC and UHPLC systems and columns. The kinetic performance of columns packed with core shell particles having diameters of 1.3, 1.7, 2.6, 3.6 and 5 μm were compared to each other and to that obtained with fully porous particles with diameters of 1.7, 3 and 5 μm. Optimum, practical combinations of column length, column I.D., particle size, particle morphology, etc. for various common chromatographic scenarios such as minimum required peak capacity (or chromatographic efficiency), maximum analysis time and combinations thereof are proposed for both HPLC and UHPLC systems.

References

[1] Fekete S.; Guillarme, D. J. Chromatogr. A 2013, 1308, 104-113
[2] Vanderheyden, Y.; Cabooter, D.; Desmet, G.; Broeckhoven, K. J. Chromatogr. A 2013, 1312, 80-86

Keywords: High Throughput Chemical Analysis, HPLC Columns, Method Development, Optimization
Application Code: High-Throughput Chemical Analysis
Methodology Code: Liquid Chromatography

Session # 2180 Abstract # 2180-1 **Oral Sessions**

Session Title: Microfluidics/Lab-on-a-Chip - Bioanalytical I

Abstract Title: **A Method for Measurement of Temporally Resolved Insulin Secretion from Islets of Langerhans in Response to Fatty Acid Hydroxy Fatty Acids**

Date: Thursday, March 10, 2016 - Mornin
Time: 08:30 AM
Room: B403

Primary Author: Basel Bandak
Florida State University

Co-Author(s): Lian Yi, Michael G. Roper, Nikita Mukhitov

Abstract Text

The release of peptides, such as insulin and glucagon, from the pancreatic islets of Langerhans regulates glucose homeostasis. Abnormal secretions of these peptides can cause a failure in glucose homeostasis. A recently discovered class of mammalian lipids, known as fatty acid hydroxy fatty acids (FAHFAs), has been found to lower glycemia and improve glucose tolerance in mice. In this report, we aimed to develop an analytical method to temporally-resolve the secretory response of individual pancreatic islets in response to these FAHFAs using a microfluidic immunoassay system coupled with a gravity-driven perfusion system.

The microfluidic device consisted of a two-layer glass chip that contained a perfusion system for delivery of FAHFAs, an islet incubation chamber, and a separation channel for measurement of secreted insulin. Insulin was electrophoretically sampled and mixed online with Cy5-labeled insulin and anti-insulin antibodies. Injections into the separation channel were performed every 10 s and allowed baseline resolution of bound (B) and free (F) Cy5-labeled insulin. Calibration curves were obtained for the range of 0 -200 nM insulin. This assay demonstrated good reproducibility of the bound-to-free ratio (B/F). Relative standard deviations of B/F were < 4% and the detection limit was 5 nM.

This device provides a method for stimulating islets, and for real time monitoring of insulin secretion with high temporal resolution. The application of this method to study the stimulated secretion of insulin by FAHFAs will elucidate the ameliorative effects that these recently discovered compounds have on elevated glycemia.

Keywords: Fluorescence, Immunoassay, Lab-on-a-Chip/Microfluidics, Separation Sciences
Application Code: Bioanalytical
Methodology Code: Microfluidics/Lab-on-a-Chip

Session # 2180 Abstract # 2180-2 **Oral Sessions**

Session Title Microfluidics/Lab-on-a-Chip - Bioanalytical I

Abstract Title **Macro-to-Microfluidic Interfacing for Primary Endocrine Cell Culture and Sampling Using 3D Printed Device Templates and Fluidic Manifolds**

Primary Author Jessica C. Brooks
Auburn University

Date: Thursday, March 10, 2016 - Mornin
Time: 08:50 AM
Room: B403

Co-Author(s) Christopher J. Easley, Dylan Holder, Katarena Ford, Mark D. Holtan

Abstract Text

Interfacing laboratory scale fluidic systems—pipettes, pressure lines, cell culture chambers—to microfluidic channels is a constant challenge in microchip design and fabrication. In typical polydimethylsiloxane (PDMS) devices, although above-channel PDMS material accounts for 95.5% of the device height, it is rarely utilized for integrating additional design features other than cylindrically punched interfaces that span the entire device height. In this work, we introduce novel methods to topologically landscape this above-channel PDMS using 3D printed templates. With an inexpensive printer (MakerBot Replicator 2), a 3D printed positive mold (polylactic acid, PLA) can be placed in uncured PDMS to form negative impressions as small as 0.5 mm in diameter in the cured device. Using these techniques, microfluidic devices were fabricated to accommodate on-chip culture of primary, murine islets and adipocytes. Both cell types were cultured above our 8-channel design, where rapid design and prototyping of interface templates allowed simple adaptation to varied culture constraints with the different tissues. These devices accommodated both temporally resolved or unresolved sampling into the channels, depending on the application. Higher throughput sampling was also achieved using a customized 3D printed fluidic manifold. Vacuum was applied to the manifold and samples were collected in 8-strip PCR tubes, where the strip lid was replaced with a PLA lid containing input and output ports. Overall, we show that rapid prototyping of 3D printed interface templates and fluidic manifolds provides a novel and highly flexible approach for interfacing to microfluidic channels, which we applied to endocrine cell sampling.

Keywords: Bioanalytical, Biological Samples, Lab-on-a-Chip/Microfluidics, Protein
Application Code: Bioanalytical
Methodology Code: Microfluidics/Lab-on-a-Chip

Session # 2180 Abstract # 2180-3 **Oral Sessions**

Session Title: Microfluidics/Lab-on-a-Chip - Bioanalytical I

Abstract Title: **A Simple Droplet Microfluidic Capillary Viscometer Based on Droplet Frequency for Rheological Measurements of Proteins**

Date: Thursday, March 10, 2016 - Mornin
Time: 09:10 AM
Room: B403

Primary Author: Michael F. DeLaMarre
University of Illinois at Chicago

Co-Author(s): Scott A. Shippy

Abstract Text
Droplet microfluidics is a powerful technique capable of performing many laboratory sample handling and analysis techniques on the nanoliter scale in an automated fashion. Despite the wide variety of droplet techniques available, relatively few capable of characterizing rheological properties have been reported. Considering the importance of rheology to a wide variety of disciplines, including proteomics, development of reliable droplet techniques for characterizing rheological properties like viscosity is important. We report the development of a droplet viscometer consisting of only a microfluidic T-junction, which utilizes droplet frequency to determine the viscosity of aqueous samples.
The viscometer is capable of processing samples as small as 19 nL, and was found to have both an accuracy and precision better than ±1%. Measurements can be performed in less than10 seconds on samples up to 1000 cP and shear rates between 10 and 1.4×104 s[sup]-1[/sup]. A fluorinated surfactant was added to the oil phase to prevent protein adsorption to polydimethylsiloxane surfaces, which allowed bovine serum albumin solutions up to 26 mg/mL to be analyzed without adverse effects on device surface properties. Finally, we demonstrate the ability to measure sample viscosity using off-chip fluorescence detection after using a T-junction for addition of reagent to droplets. This could allow for simultaneous spectroscopic assay and rheological characterization of samples, which would be particularly useful in proteomics and other biomedical applications.

Keywords: Lab-on-a-Chip/Microfluidics, Protein, Rheology, Small Samples
Application Code: Bioanalytical
Methodology Code: Microfluidics/Lab-on-a-Chip

Session # 2180 Abstract # 2180-4 **Oral Sessions**

Session Title: Microfluidics/Lab-on-a-Chip - Bioanalytical I

Abstract Title: **Improving Detection Sensitivity in Microchip Electrophoresis-Laser Induced Fluorescence Assays by Target-Induced Exonuclease Assisted Strand Circle Signal Amplification**

Primary Author: Shulin Zhao
Guangxi Normal University

Date: Thursday, March 10, 2016 - Mornin
Time: 09:30 AM
Room: B403

Co-Author(s): Liangliang Zhang, Yi-Ming Liu, Yingfeng Qin, Yong Huang

Abstract Text

A new method was developed for simple, sensitive, highly specific detection of IFN-□based on T7 exonuclease (T7 Exo) assisted signal amplification using microchip electrophoresis-laser induced fluorescence detection (MCE-LIF). In this method, the T7 Exo was used to assist fluorescence signal amplification; MCE-LIF was used for fast and efficient separation and sensitve detection of IFN-□ The peak height of FAM-5-nucleoside monophophates was shown to be good linear with the concentration of IFN-□ranging from 1.5×10-11 to 2.5×10-9 M. The detection limit (S/N = 3) was estimated to be 4.5×10-12 M. The present method was successfully applied for the detection of IFN-□in human plasma with satisfactory results. Furthermore, this detection system appears to be a universal approach for the detection of target molecules by simply changing the aptamer sequence of the hairpin probe. And bying design different length of the fluorescence probe can achieve multiple detection of target under single excitation. Thus it could be greatly expanded the scope of application of the method.

Keywords: Capillary Electrophoresis, Lab-on-a-Chip/Microfluidics, Ultratrace Analysis

Application Code: Bioanalytical

Methodology Code: Microfluidics/Lab-on-a-Chip

Session # 2180 Abstract # 2180-5 **Oral Sessions**

Session Title: Microfluidics/Lab-on-a-Chip - Bioanalytical I

Abstract Title: **A Complementary Method to CD4 Counting: Measurement of CD4+/CD8+ T Lymphocytes Ratio in a Serial Microfluidic System**

Date: Thursday, March 10, 2016 - Mornin
Time: 10:05 AM
Room: B403

Primary Author: Wenjie Li
Texas Tech University

Co-Author(s): Dimitri Pappas

Abstract Text
The CD4+ T lymphocytes are related to human immunodeficiency virus (HIV) infection. Besides the absolute number of CD4+ T lymphocytes in per microliter of adult blood, the measurement of the CD4+/CD8+T lymphocyte ratio also plays an important role in HIV diagnosis and antiretroviral treatment (ART) assessment. We describe a microfluidic system that isolates and enumerates CD4+ and CD8+ T lymphocytes on two serial linked affinity regions, then the CD4+/CD8+ T lymphocyte ratio is calculated. A drop of human blood is injected to a serial microfluidic channel, and blood cells are captured on monoclonal antibody coated affinity surfaces using stop flow; fluorescence-labeling antibodies are then injected with optimized shear stress to retain the CD4+ /CD8+ T lymphocytes on the corresponding affinity regions while elute background cells. Different fluorescent signals are detected to identify and enumerate CD4+/CD8+ T lymphocytes respectively. A linear relationship is reported between the CD4+/CD8+ T lymphocyte ratio achieved using conventional flow cytometry and our serial microfluidic system with high coefficient of determination (R[superscript 2]=0.97). This inexpensive and effective method will potentially provide an alternative to direct CD4 counting in HIV implication in point-of-care. This project was funded by Texas Research Incentive Program.

Keywords: Bioanalytical, Lab-on-a-Chip/Microfluidics
Application Code: Bioanalytical
Methodology Code: Microfluidics/Lab-on-a-Chip

Session # 2180 Abstract # 2180-6 **Oral Sessions**

Session Title: Microfluidics/Lab-on-a-Chip - Bioanalytical I

Abstract Title: **PDMS-Based Injection Valves for SPE-MS Analysis of Biomolecules**

Primary Author: James P. Grinias
University of Michigan

Date: Thursday, March 10, 2016 - Mornin
Time: 10:25 AM
Room: B403

Co-Author(s): Colleen E. Dugan, Robert T. Kennedy

Abstract Text

Using the inherently elastomeric PDMS as a substrate for microfluidic devices allows for the integration of pressure driven valves to selectively close fluidic channels. These valves can be arranged in such a way to integrate on-chip the function of four- or six-port valves, which are commonly used for injection in LC. Here, biologically relevant samples containing high salt concentrations that were generated or collected on the chip were manipulated into a solid phase extraction bed using a PDMS six-port valve. With the extraction bed, samples were de-salted and eluted to a mass spectrometer using electrospray ionization for sample identification and quantification. In one example, non-esterified fatty acids (NEFAs) secreted from adipocytes loaded on-chip were measured under basal and stimulated conditions. The SPE-MS assay improves upon a previous fluorescent assay used for adipocyte analysis by enabling the identification of individual NEFAs compared to total FA concentration. In another application, in vivo microdialysis samples (to study neurotransmitter concentrations in rat brains) were collected, derivatized using on-chip reagent addition, trapped on an SPE bed, washed and eluted from the bed using a PDMS six-port valve, and quantified using MS-MS analysis. Both examples demonstrate the potential of PDMS valves connected to SPE beds as a low-cost analytical technique that can greatly improve MS analysis compared to direct injection.

Keywords: Bioanalytical, Lab-on-a-Chip/Microfluidics, Mass Spectrometry, Solid Phase Extraction
Application Code: Bioanalytical
Methodology Code: Microfluidics/Lab-on-a-Chip

Session # 2180 Abstract # 2180-7 **Oral Sessions**

Session Title Microfluidics/Lab-on-a-Chip - Bioanalytical I

Abstract Title **Development of an Online Microchip Electrophoresis with LED-Induced Fluorescence System for In Vivo Detection of Excitatory Amino Acid Neurotransmitters Following a Traumatic Brain Injury**

Date: Thursday, March 10, 2016 - Mornin

Time: 10:45 AM

Room: B403

Primary Author Michael L. Hogard
University of Kansas

Co-Author(s) Craig E. Lunte, Elton E. Melo Costa, Nathan Oborny, Susan M. Lunte

Abstract Text

Traumatic brain injury (TBI) is one of the leading causes of death and disability worldwide. Damage from the primary trauma event can induce deleterious secondary issues such as ischemia, inflammation, and cell death. The development of analytical techniques that can directly detect TBI biomarkers in vivo will assist in medical treatments and interventions for TBI patients. A method was developed for the detection of excitatory amino acid (EAA) neurotransmitters using microchip electrophoresis (ME) with laser-induced fluorescence (LIF) detection. Detection was achieved via a derivatization reaction with the fluorescent tag naphthalene-2,3-dicarbozaldedye (NDA). The EAAs glutamate and aspartate, as well as the amino acids citrulline, arginine, taurine, and histamine, were separated using this method and subsequently detected in intercranial microdialysis (MD) samples. However, traditional analytical equipment (e.g. LIF) is too cumbersome for use in a medical setting, such as a hospital room. With this in mind, a mobile and fully integrated light-emitting diode-induced fluorescence (LED-IF) detection system was designed and built in-house to achieve the same results. The method was optimized for online MD to ME-LED-IF analysis with on-chip NDA derivatization. An internal standard was added to the separation to account for the loss of signal response over the duration of the MD experiment. This system will be used to monitor in vivo EAA concentration changes in rats following TBI induction. The ultimate goal is to use the MD-ME-LED-IF system for bedside monitoring of biomarkers in TBI patients to help medical professionals identify secondary issues before they can cause further injuries.

Keywords: Capillary Electrophoresis, Fluorescence, Lab-on-a-Chip/Microfluidics, Neurochemistry

Application Code: Bioanalytical

Methodology Code: Microfluidics/Lab-on-a-Chip

Session # **2180** **Abstract #** **2180-8** **Oral Sessions**

Session Title Microfluidics/Lab-on-a-Chip - Bioanalytical I

Abstract Title **Determination of Amplification of Cellular Effects by Hormones Derived from Different Tissues**

Date: Thursday, March 10, 2016 - Mornin
Time: 11:05 AM
Room: B403

Primary Author Keshavarz Hamideh
Michigan State University

Co-Author(s) Dana Spence

Abstract Text

Leptin, a hormone produced mainly by adipose tissue, is believed to balance in vivo energy level by regulation of food intake and body weight. Higher concentrations of leptin are observed in obese people, who are thought to be "leptin resistant". Leptin resistance is common in patients with diabetes and studies show leptin can have glucoregulatory effect on patients with type 1 and type 2 diabetes. In spite of this, obese diabetic patients do not exhibit strong response to exogenous leptin. Thus, a detailed understanding of the glucose regulatory function of leptin is of crucial importance in order to overcome the shortcomings of leptin therapy and its use as a potential therapeutic for humans. Previously, our group has reported that C-peptide, a 31 amino acid peptide secreted from pancreatic beta cells, binds to red blood cells (RBCs) and has cellular energetic effects. Here, we show that leptin amplifies these C-peptide effects. Using a 3D printed microfluidic device, measurements were performed to determine ATP release from RBCs that had been incubated with C-peptide, zinc, and leptin and combinations thereof. In the presence of C-peptide and zinc, a significant increase in RBC-derived ATP is measured; this signal is further significantly enhanced in the presence of leptin. Importantly, leptin alone has no effect on the RBC. To improve the measurement scheme, we have also explored the design of a fluidic device that will contain adipose tissue, insulin and C-peptide secreting INS-1 cells and a flowing stream of blood cells to confirm our preliminary findings in an enhanced organ-on-chip format.

Keywords: Biological Samples, Biomedical, Biotechnology, Lab-on-a-Chip/Microfluidics
Application Code: Bioanalytical
Methodology Code: Microfluidics/Lab-on-a-Chip

Session # 2190 Abstract # 2190-1 **Oral Sessions**

Session Title Novel Applications with Gas Chromatography Mass Spectrometry

Abstract Title **How GC-MS with Cold EI Improves NIST Library Identification**

Date: Thursday, March 10, 2016 - Mornin
Time: 08:30 AM
Room: B315

Primary Author Aviv Amirav
Tel Aviv University

Co-Author(s) Tal Alon, Uri Keshet

Abstract Text

Cold EI is electron ionization of vibrationally cold molecules in supersonic molecular beams. GC-MS with Cold EI provides mass spectra with all the standard EI fragment ions combined with enhanced molecular ions and high mass fragments. We found that despite the above MS changes NIST library identification actually improves with Cold EI, as library identification probabilities for the correct library mass spectra are increased despite the lower matching factors. Computer simulations support measurements and show that while enhanced molecular ion and high mass fragments lower the matching factor of the correct library compound, the matching factors of the incorrect library candidates are lowered even more, resulting in a rise of the identification probability for the correct compound. This behavior emerges from the fact that molecular ions characterize sample compounds more than low mass ions and therefore carry more weight in library search identification algorithms. Furthermore, enhanced molecular ions in Cold EI serve by themselves to confirm or reject the library identification and enable the use of TAMI software for the provision of elemental formula based on improved quadrupole MS mass accuracy and isotope abundance analysis. In addition, GC-MS with Cold EI can serve for the analysis of significantly increased range of large, polar and labile compounds. Thus, GC-MS with Cold EI and TAMI software is superior in sample identification to GC-HR-TOF since greater range of compounds can be eluted and exhibit trustworthy molecular ion with it, while the TAMI software confirms NIST identification and/or provides elemental formula.

Keywords: Gas Chromatography/Mass Spectrometry, GC-MS, Identification
Application Code: Other
Methodology Code: Gas Chromatography/Mass Spectrometry

Session # 2190 Abstract # 2190-2 **Oral Sessions**

Session Title: Novel Applications with Gas Chromatography Mass Spectrometry

Abstract Title: **Use of Automated Column Chromatography Clean Up with Reduced Solvent Volume in POPs Analysis**

Primary Author: Rudolf Addink
Toxic Report

Date: Thursday, March 10, 2016 - Mornin
Time: 08:50 AM
Room: B315

Co-Author(s): Philip Bassignani

Abstract Text

Persistent Organic Pollutants have been studied since the 1970s. They include polychlorinated dibenzo-p-dioxins and furans (PCDD/Fs), polychlorinated biphenyls (PCBs), and brominated flame retardants (PBDEs). Sample prep has traditionally involved multi-day Soxhlet extraction of the sample (up to 48h) and manual preparative multistep column chromatography.

Automating the sample prep process has been the focus of work ongoing since the 1980s. Automation will, among others, result in faster turnaround time of samples, lower and control the cost of analysis, and improve the quality of the data generated, e.g., as part of method validation.

Automated Pressurized Liquid Extraction (PLE) which can be done in a total processing time of ~ 1h has been developed as an alternative to Soxhlet extraction. Similarly automated column chromatography (acid-base-neutral silica – alumina – carbon) can be done in up to 40 min for even complex matrices such as lipids and oils. The original program for automated clean up – based on the existing manual methods - used up to 1 L of solvent, such as hexane, dichloromethane and toluene (or various mixes of those). Recently an attempt has been made to reduce solvent consumption for the automated sample clean up. The reduced volume programs have total solvent consumption of 250-400 mLs depending on the type of silica column used (roughly between 5-40 g of silica per column). More demanding matrices will require use of higher amounts of silica (both acidified and basic). With these reduced volume programs excellent recoveries of the analytes have been achieved for matrices such as fish oil, peanut butter and soil.

Keywords: Gas Chromatography/Mass Spectrometry, Mass Spectrometry, Prep Chromatography, Trace Analysis
Application Code: Other
Methodology Code: Gas Chromatography/Mass Spectrometry

Session # 2190 Abstract # 2190-3 **Oral Sessions**

Session Title: Novel Applications with Gas Chromatography Mass Spectrometry

Abstract Title: **Effective QuEChERS Cleanup and Quantitation of Planar Pesticides from Spinach and Other Food Matrices Using a Novel Graphitized Carbon Black and a Zirconia-Based Adsorbent**

Date: Thursday, March 10, 2016 - Mornin
Time: 09:10 AM
Room: B315

Primary Author: Patrick Myers
Supelco/Sigma-Aldrich

Co-Author(s): Bill Ozanich, Jennifer Claus, Michael Ye, William Betz

Abstract Text

Pesticides with a planar structure, such as hexachlorobenzene and chlorothalonil, are among those commonly used during the cultivation of spinach and other green, leafy crops. Extraction and analysis of these residues is complicated by the presence of chlorophyll in the food matrices. Large pigment molecules like chlorophyll are deleterious to both GC-MS and LC-MS/MS analyses; accumulating in the inlet and degrading column performance in GC and contaminating the source in LC. Method EN15662 recommends the addition of Graphitized Carbon Black (GCB) to cleanup tubes to remove chlorophyll and other pigment interferences. In addition to removing pigment molecules, however, GCB also retains the planar pesticides because of □-□interactions between the graphitic carbon and the planar structures of the pesticides. Previous work shows a near linear inverse relationship between pigment removal and recovery of planar pesticides. Suggestions to overcome this problem include adding toluene to the extraction solvent to minimize binding to the GCB and using smaller amounts of GCB to balance recovery of planar pesticides with color removal. Data show that neither of these solutions is completely satisfactory. This presentation, will show a QuEChERS cleanup mix, Supel QuE Verde, that comprises a proprietary GCB and a zirconia-containing adsorbent, Z-Sep+, to remove chlorophyll while maintaining high recovery of planar as well as other pesticide residues in spinach, green pepper, kiwi, strawberry and oregano. Color removal was determined using a UV-Vis spectrophotometer. Planar pesticide recovery was determined using a GC-MS/MS Triple Quad. The effects of quantity and surface areas of GCB on the color removal and pesticide recovery will also be discussed.

Keywords: GC-MS, Pesticides
Application Code: Food Safety
Methodology Code: Gas Chromatography/Mass Spectrometry

Session # 2190 Abstract # 2190-4 **Oral Sessions**

Session Title: Novel Applications with Gas Chromatography Mass Spectrometry

Abstract Title: **Use of Micro Scale Solid Phase Extraction and Automated Clean Up in POPs Analysis of Human Milk and Serum**

Date: Thursday, March 10, 2016 - Mornin
Time: 09:30 AM
Room: B315

Primary Author: Rudolf Addink
Toxic Report

Co-Author(s)

Abstract Text

Laboratory analysis of human milk and serum for Persistent Organic Compounds (POPs) has become increasingly important. Biomonitoring in which levels of toxic chemicals are assessed in humans (e.g., in breast milk and serum) has received much attention in the last decade. Automation of the sample prep process can result in faster turn around time of samples, lower costs, and improved quality of the data generated. As the amount of sample used in bio-monitoring with Solid Phase Extraction is typically lower than in, e.g., water analysis, a micro system was used.

PCBs/PBDEs analysis serum: 2 g serum was added to 2g water and treated with 4g formic acid twice; 100 uL methanol, 100 uL HCl (pH~2) and 13C labeled standards were added. The sample was loaded on an HLB-500 cartridge (conditioned with dichloromethane, methanol, water) with positive pressure and dried (N2). Elution with 12mLs DCM, volume reduction to 3mLs, solvent-exchange with hexane. Cleanup over 2g acidified silica (in same system). Elution with 20 mLs hexane, followed by volume reduction to 10 uL in 24 position vial evaporator.

PCDD/Fs analysis serum: 20g serum was mixed with 20g water and 2 x 40g formic acid. No methanol or HCl used; C18 cartridge used; 30 mLs DCM for elution; reduce to 5mLs and exchange to 5 mLs hexane. Subsequent clean up in automated column chromatography system.

PCBs/PBDEs analysis milk: 1g milk (as received, not freeze dried) spiked with 13C labeled standards; absorbed into 1g Hydromatrix™ cartridge; dried with N2 (positive pressure, no conditioning); elution with 12mLs DCM; subsequent steps same as for serum.

PCDD/Fs analysis milk: 5g milk plus 13C labeled standards absorbed into 5 g of Hydromatrix™; dried with N2; elution 20 mLs DCM; volume reduction to 5 mLs; exchange with hexane. Subsequent cleanup in automated chromatography system.

All analyses were done with high res GC/MS. Milk and serum gave excellent recoveries (all POPs) with detection levels at the low pg level.

Keywords: Gas Chromatography/Mass Spectrometry, Solid Phase Extraction
Application Code: Other
Methodology Code: Gas Chromatography/Mass Spectrometry

Session # 2190 Abstract # 2190-5 **Oral Sessions**

Session Title: Novel Applications with Gas Chromatography Mass Spectrometry

Abstract Title: **Multidimensional Comprehensive Gas Chromatography Multireflection High Resolution Time-of-Flight Mass Spectrometry: Combining Accurate Mass Information with Ultra-High Chromatographic Resolution**

Date: Thursday, March 10, 2016 - Mornin
Time: 10:05 AM
Room: B315

Primary Author: Ralf Zimmermann
University Rostock /HMGU

Co-Author(s): Benedikt Weggler, Juergen Wendt, Thomas Groeger

Abstract Text

Kendrick mass defect plots etc. are common graphical tools to visualize data from high resolution mass spectrometers (HRMS). Even plots from very complex mixtures (e.g. petrochemical samples) show structured features, indicating homologous series for various substance classes. However, a drawback is the inability of HRMS to resolve the isomeric composition behind the elemental sum-formulas. This motivates the hyphenation of HRMS to high resolution chromatography. Especially comprehensive two-dimensional gas chromatography (GCxGC) is able to resolve the isomeric composition of complex mixtures. Compound classes and isomeric compositions depict a structured order in the separation plane. Complementary information from both techniques could be combined to build up novel chromatographic resolved mass defect plots. A GCxGC-HRMS system was used to analyze petrochemical fractions. Mass spectra can be acquired by the high resolving multi-reflection time-of-flight MS with up to 200 Hz at mass resolutions/accuracies better than 25.000/1 ppm, respectively. For demonstration of the novel information space, different petrochemical matrices (e.g. Diesel fuel with FAME, Heavy Fuel Oil or Fischer-Tropsch condensates (FTC)) were analyzed. While e.g. the Diesel Fuel represent a pure hydrocarbon matrix (beside of the oxygenated FAME content), HFO also comprises N- and S-containing compound classes. The mean mass defect plots were coherent with direct MS data obtained from ultra-high resolution instrumentation (FT-ICR). Variations in the isomeric composition, however, can be observed by adding one or two chromatographic dimensions to the mass defect plots. The technique was then applied to a series of FTC samples. Different FTC generated by a FT model-reactor operated at different reaction temperatures where analyzed. A shift in the isomeric composition (degree of branching) as well as oxidation level due to the temperature differences could be readily detected by GCxGC-HRMS.

Keywords: Capillary GC, Gas Chromatography, Petrochemical, Time of Flight MS
Application Code: Fuels, Energy and Petrochemical
Methodology Code: Gas Chromatography/Mass Spectrometry

Session # 2190 Abstract # 2190-6 **Oral Sessions**

Session Title Novel Applications with Gas Chromatography Mass Spectrometry

Abstract Title **Rapid, New Methods for the Analysis of 3-MCPD and 1,3 DCP in Soy Sauce**

Date: Thursday, March 10, 2016 - Mornin
Time: 10:25 AM
Room: B315

Primary Author Susan Genualdi
US FDA

Co-Author(s) Lowri DeJager, Patsy Nyman

Abstract Text

Acid hydrolyzed vegetable protein (aHVP) is used for flavoring a wide variety of foods and also in the production of non-fermented soy sauce. During the production of aHVP, chloropropanols including 3-monochloropropane-1,2-diol (3-MCPD) and 1,3 dichloropropane-2-ol (1,3 DCP) can be formed through the reaction of the hydrochloric acid catalyst and residual fat and the reaction of 3-MCPD with acetic acid, respectively. 3-MCPD is a carcinogen and a suspected genotoxin in humans. The European Union (EU) has set a maximum level of 0.02 ppm of 3-MCPS in aHVP, and the Food and Drug Administration (FDA) set a guidance limit of 1 ppm of 3-MCPD in aHVP. Prior to the guidance level being set, a survey of 55 samples performed by the FDA found 33% of samples to have concentrations greater than 1 ppm. An AOAC method was used for this analysis, which is time consuming, labor intensive, and requires excessive solvents. A new survey of 60 sauces was performed in 2015 to determine if concentrations have changed since 2008 using newer, more rapid methods. Alternative methods were investigated, including derivatization of 3-MCPD using phenylboronic acid and headspace analysis for 1,3 DCP. Additionally, a new technique using microvial thermal desorption coupled with GCMS was developed involving minimal sample preparation and the rapid assessment of 3-MCPD contamination in soy sauce samples.

Keywords: Food Science, Headspace, Sample Preparation, Thermal Desorption
Application Code: Food Safety
Methodology Code: Gas Chromatography/Mass Spectrometry

Session # 2190 **Abstract # 2190-7** **Oral Sessions**

Session Title: Novel Applications with Gas Chromatography Mass Spectrometry

Abstract Title: **A Method Development Software Tool for Comprehensive Two-Dimensional Gas Chromatography Evaluated for GCxGC-TOFMS**

Primary Author: Mark F. Merrick
LECO Corporation

Date: Thursday, March 10, 2016 - Mornin
Time: 10:45 AM
Room: B315

Co-Author(s): Leonid M. Blumberg, Viatcheslav Artaev

Abstract Text

Comprehensive two-dimensional gas chromatography/time-of-flight mass spectrometry (GCxGC-TOFMS)has been commercially available since 2002; however, the development of methods, the selection of column dimensions and GC conditions, continues to be a challenge. Generally, analysts have relied on examples in the published literature which typically use a 30 m x 0.25 mm primary column, a 1 to 2 m x 0.1 mm secondary column and a long modulation period (5 - 10 sec). A methodology for optimized GCxGC has been described in the literature, but is has not yet been widely accepted. In this lecture we present results of an evaluation of a GCxGC method development software tool utilizing the theory of optimized GCxGC. In a fully automatic mode, the tool calculates the optimal lengths of the primary and secondary column, the optimal flow rate, and the optimal heating rate for the user selected column diameters, transfer line dimensions, carrier gas, modulator-generated peak width, and second dimension peak capacity. The tool can also operate in a partially automatic mode or a totally manual mode in which the user can select all parameters.

The tool was evaluated by comparing experimental results (two-dimensional peak capacity) to the predicted results of the tool for a range of conditions including first and second column dimensions, column flow, heating rate, and modulation period. The experimental data were collected on a GCxGC-TOFMS system with a quad jet, LN2-cooled modulator. Preliminary data indicate that the experimental results are sufficiently accurate for providing good initial conditions for GCxGC method development.

Keywords: Gas Chromatography/Mass Spectrometry, Method Development, Optimization, Time of Flight MS
Application Code: Other
Methodology Code: Gas Chromatography/Mass Spectrometry

Session # 2190 Abstract # 2190-8 **Oral Sessions**

Session Title: Novel Applications with Gas Chromatography Mass Spectrometry

Abstract Title: **Innovative TG-GC-MS Methods for Thermal Degradation Studies of Polymers**

Date: Thursday, March 10, 2016 - Mornin
Time: 11:05 AM
Room: B315

Primary Author: Kristina Lilova
Setaram Inc.

Co-Author(s): Link Brown

Abstract Text

The application of coupled techniques in the field of thermogravimetric analysis is well established, particularly for the investigation of the chemistry of thermal decomposition and identification of the evolved species. However, these techniques have limitations, especially when a large number of molecules are evolved simultaneously. For that reason a more comprehensive TG-GC-MS technique, which involves a separation of the evolving species by GC and identification by MS, became a necessity. SETARAM and Automation worked at designing a TG-GC-MS line allowing to coordinate TGA measurements with injections on the GC column. The system includes an appropriate automated sampling loop that allows either injection of the gas stream in the GC column or releasing it outside the system. All gases evolving from the TGA furnace also flow through the sample loop so their concentration in the loop reflects the exact gas ratio in the furnace. The whole line, including the sampling loop is designed to withstand high temperatures and can be heated up to 350 °C. Switching between injections and venting can be done during the sample decomposition process. Three main modes of operation are available: Examples related to the study of the decomposition of polymers (including polystyrene) illustrate how this technique that can give more in-depth understanding of the degradation of complex substances. The presence of small quantities of heavy molecules with high boiling points (e.g. styrene dimer) can be detected in a gas flow rich in other substances.

Keywords: FTIR, Gas Chromatography/Mass Spectrometry, Materials Characterization, Thermal Analysis
Application Code: Polymers and Plastics
Methodology Code: Thermal Analysis

Session # 2200 Abstract # 2200-1 **Oral Sessions**

Session Title: Novel Synthesis and Applications of Nanomaterials

Abstract Title: **Nanofibers from Hydrothermal Treatment of Cellulose Nanocrystals**

Primary Author: Yimei Wen
Clemson University

Date: Thursday, March 10, 2016 - Mornin
Time: 08:30 AM
Room: B316

Co-Author(s)

Abstract Text

There is an increasing demand for materials made from sustainable and renewable resources such as cellulose. Natural cellulose contains crystalline domains and amorphous domains. In a controlled sulfuric acid hydrolysis, amorphous domains decompose and the remaining crystalline domains are commonly referred as cellulose nanocrystals (CNC). CNC are among the most exciting cutting edge materials currently used for liquid crystals and reinforcing polymers. They also have potential for many new applications such as antibacterial films, biomedical implants, etc.Here, a hydrothermal synthesis of 1-2 nm carbonaceous fibers from CNC is reported. CNC used in this work had 20±6 nm width with 107±55 nm length and were treated in a Teflon® lined autoclave at 240°C. The nanofibers were characterized by AFM, TEM, Raman spectroscopy and XRD. AFM measurements of the fibers on different substrates revealed their hydrophilic nature. TEM images indicated strong van der Waals forces between individual nanofibers leading to the formation of bundles of various shapes and sizes. The comparison of Raman spectra before and after the hydrothermal treatment of CNC, revealed the formation of graphitic hydrochar. The demonstration of the nanofiber synthesis via the hydrothermal treatment of CNC can lead to the development of environmentally friendly technology for producing carbon nanofibers.

Keywords: Material Science, Raman, X-ray Diffraction
Application Code: Nanotechnology
Methodology Code: Microscopy

Session # 2200 Abstract # 2200-2 **Oral Sessions**

Session Title: Novel Synthesis and Applications of Nanomaterials

Abstract Title: **Graphene Nanoribbons: Engineering, Characterization of Edge Defects and Sensor Applications**

Primary Author: Pankaj Ramnani
University of California, Riverside

Date: Thursday, March 10, 2016 - Mornin
Time: 08:50 AM
Room: B316

Co-Author(s): Ashok Mulchandani

Abstract Text

Large area graphene is a zero band gap semi-metal with the valence and conduction bands intersecting at the K points in the Brillouin zone. Band gap engineering of graphene by lateral confinement of charge carriers into pseudo one-dimensional nanostructures like graphene nanoribbons (GNRs) and nanomesh has been demonstrated both experimentally and theoretically. The most commonly used methods for patterning graphene, including electron-beam lithography (EBL), nano-imprint lithography (NIL) and block co-polymer lithography (BCP) use an etch mask to selectively protect a certain region of graphene and the exposed graphene region is removed using O2 plasma. However, the GNRs fabricated using these methods typically have a high degree of edge roughness and edge functionalization of the C atoms which introduces doping and significantly alters the charge carrier mobility and band gap of graphene. In this work, large-grain single-layer graphene films were grown using chemical vapor deposition (CVD). GNRs with widths varying from 30 – 200 nm were patterned using electron-beam lithography (EBL) and O2 plasma treatment. We use Raman spectroscopy and x-ray photoelectron spectroscopy (XPS) for detailed characterization of the GNR edges. Thermal annealing and electrochemical reduction was used to partially repair the edge defects created by the O2 plasma treatment. The electrical properties of GNRs, such as charge carrier mobility, extent of doping and band gap were characterized by fabricating field-effect transistors (FETs). Finally, we demonstrate the enhanced sensitivity of the GNR-FET devices by investigating their performance as a chemical and biological sensor.

Keywords: Materials Characterization, Raman, Semiconductor, Sensors
Application Code: Nanotechnology
Methodology Code: Sensors

Session # 2200 Abstract # 2200-4 **Oral Sessions**

Session Title: Novel Synthesis and Applications of Nanomaterials

Abstract Title: **Single-Molecule Tracking Studies of the Effects of Solvent Swelling on the Properties of Cylindrical Block Copolymer Microdomains**

Date: Thursday, March 10, 2016 - Mornin
Time: 09:30 AM
Room: B316

Primary Author: Takashi Ito
Kansas State University

Co-Author(s): Daniel A. Higgins, Dol R. Sapkota, Khanh-Hoa Tran-Ba

Abstract Text

Thorough understanding of solvent swelling of block copolymer microdomains will help control microdomain morphology/alignment via solvent vapor annealing and the chemical separation efficiency of block copolymer membranes. This presentation will discuss the effects of solvent swelling on the molecular permeability, dimensions and alignment of individual cylindrical microdomains in polystyrene-block-poly(ethylene oxide) (PS-b-PEO) thin films (ca. 50 µm thick). These films were prepared from 30 wt% PS-b-PEO solutions in benzene or THF between two glass coverslips, as reported previously (JPCB 2014, 118, 11406). The cylindrical PEO microdomains were doped with sulforhodamine B (SRB) dye molecules, and aligned in the direction of solution flow. The diffusional motions of individual SRB molecules were measured using single-molecule tracking (SMT) as a function of film drying time. The radii and orientation of the individual microdomains were determined using orthogonal regression methods, and the microdomain permeability was assessed from diffusion coefficients of individual SRB molecules. We have found that the microdomain orientation was similar regardless of drying time, as expected for shear-induced alignment of PS-b-PEO assembles in the fairly concentrated solution. The radius and permeability gradually decreased with increased drying time. Both solvents exhibited similar drying-induced changes in these microdomain properties, suggesting that these solvents similarly swell the PS and PEO microdomains as expected from their solubility parameters. Importantly, no clear correlations were observed between microdomain radius and diffusion coefficient for individual microdomains, likely due to the wide distribution of microdomain properties in the PS-b-PEO films.

This work was funded by the US-DOE (DE-FG02-12ER16095).

Keywords: Imaging, Materials Characterization, Nanotechnology, Polymers & Plastics
Application Code: Nanotechnology
Methodology Code: Microscopy

Session # **2200** **Abstract #** **2200-5** **Oral Sessions**

Session Title: Novel Synthesis and Applications of Nanomaterials

Abstract Title: **Characterization of Single and Multi-Walled Carbon Theranostic Nanovectors**

Primary Author: Markus Martincic
Institut de Ciencia de Materials de Barcelona

Date: Thursday, March 10, 2016 - Mornin
Time: 10:05 AM
Room: B316

Co-Author(s): Belen Ballesteros, Elzbieta Pach, Gerard Tobias

Abstract Text

Closed-ended carbon nanotubes (CNTs) filled with different inorganic materials have potential theranostic applications [1]. The spontaneous closure of CNT tips [2] has also been demonstrated for multi-walled CNTs under different conditions. Therefore, filling experiments were performed on both single and multi-walled CNTs with different inorganic materials including samarium(III) chloride, gadolinium(III) chloride, iodine, sodium iodide and barium iodide. Multi-walled CNTs have more attractive properties due to their larger cavity, ease of functionalization and dispersion when compared to single-walled CNTs, which may provide increased biocompatibility and theranostic efficacy of the material at hand [3]. After the removal of external non-encapsulated material, different qualitative and quantitative analytical techniques were employed to describe filled material within CNTs including thermogravimetric analysis, inductively coupled plasma and scanning and transmission electron microscopy. The presence of the chosen payloads has been confirmed within carbon nanotubes. The novel material can find its use in in-vitro and in-vivo imaging and therapy [3].

[1] S.Y. Hong, G. Tobias, K.T. Al-Jamal, B. Ballesteros, H. Ali-Boucetta, S.L. Perez, P.D. Nellist, R.B. Sim, C. Finucane, S.J. Mather, M.L.H. Green, K. Kostarelos, B.G. Davis, Nat Mater, 9, 485 (2010).
[2] L. Shao, G. Tobias, Y. Huh and M.L.H. Green, Carbon, 44, 2849 (2006).
[3] M. Martincic, G. Tobias, Expert Opin Drug Deliv, 12(4), 563 (2015).

Keywords: Nanotechnology, Quantitative, Thermal Analysis, UV-VIS Absorbance/Luminescence
Application Code: Nanotechnology
Methodology Code: UV/VIS

Session # 2200 Abstract # 2200-6 **Oral Sessions**

Session Title: Novel Synthesis and Applications of Nanomaterials

Abstract Title: **Investigating 4-Wheeled Nanocar Diffusion Kinetics on Differently Modified Solid Surface with Single Molecule Fluorescence Microscopy (SMFM)**

Primary Author: Fang Chen
North Carolina State University

Date: Thursday, March 10, 2016 - Mornin
Time: 10:25 AM
Room: B316

Co-Author(s): Gufeng Wang, James Tour, Víctor García-López

Abstract Text

Driven by the motivation of manipulating microscopic objects to even molecules, a series of structures composed of aromatic chassis, axles, and wheels, named "nanocars" have been synthesized to translate on a solid surface. The ultimate goal is to incorporate "motor", a structure that could convert external energy such as electricity, heat, or light to mechanical motion, to achieve controlled movement of the nanocar. Monitoring and understating molecular movement at the single molecule level is the first step. Scanning tunneling microscopy (STM), which provides atomic scale resolution, is the most used techniques in probing molecular surface diffusion. However, the requirement of conductive substrates limits the applicable surfaces. Furthermore, due to the electrical beam, the imaged structure and the movement of the molecule are perturbed in the image collection process. To overcome this problem, we use non-perturbing single molecule fluorescence microscopy (SMFM) to study diffusion of 4-adamantane-wheeled nanocars on different surfaces. We found that both the molecule mobility and their diffusion trajectory show a time-dependence. That provides us the diffusion kinetics of 4-wheeled nanocar on different surfaces. With understanding different kinetics behaviors associating with surface properties, it extends our knowledge to design better system to manipulate molecules on solid surface.

Keywords: Analysis, Detection, Imaging, Surface Analysis
Application Code: Nanotechnology
Methodology Code: Microscopy

Session # 2200	Abstract # 2200-7	Oral Sessions

Session Title: Novel Synthesis and Applications of Nanomaterials

Abstract Title: **Sealing and Opening of Metallic Nanotubes with a Laser Beam: A Potential Drug Delivery Vehicle**

Date: Thursday, March 10, 2016 - Mornin
Time: 10:45 AM
Room: B316

Primary Author: Nathalia Ortiz
North Carolina State University

Co-Author(s): Gufeng Wang

Abstract Text

Metallic nanostructures with hallow interior have been studied for their potential applications in nano-electronics and nano-medicine. In this study, we propose to use metallic nanotubes as carriers for controlled delivery of medicines with light. High aspect ratio tungsten nanotubes were synthesized by combing nanopore templates and vapor-phased atomic layer deposition (ALD). The tungsten nanotubes can have a dimension from 70 nm to 350 nm outer diameter (OD) with a thickness of 20 nm and a length of 5~40 micrometers. We focus on their response to laser induced heating. Specifically, tungsten nanotubes exposed to a laser beam respond to light differently depending on the surroundings medium and the laser power. Finally, their potential as a drug delivery carrier is studied using bodipy dye as the probe. We demonstrate that we are able to encapsulate and release the dye molecules from the nanotubes in a controlled manner.

Keywords: Biomedical, Material Science, Nanotechnology, Spectroscopy
Application Code: Nanotechnology
Methodology Code: Microscopy

Session # **2200** **Abstract #** **2200-8** **Oral Sessions**

Session Title: Novel Synthesis and Applications of Nanomaterials

Abstract Title: **Optically Transparent Carbon Nanotube Film Electrode for Thin Layer Spectroelectrochemistry**

Primary Author: Tingting Wang
University of Cincinnati

Date: Thursday, March 10, 2016 - Mornin
Time: 11:05 AM
Room: B316

Co-Author(s): Daoli Zhao, Noe R. Alvarez, Vesselin N. Shanov, William R. Heineman

Abstract Text

Carbon nanotube (CNT) film was evaluated as an optically transparent electrode (OTE) for thin layer spectroelectrochemistry. Chemically inert CNT arrays were synthesized by chemical vapor deposition (CVD) using thin films of Fe and Co as catalysts. Vertically aligned CNT arrays were drawn onto a quartz slide to form CNT films that constituted the OTE. Good Adequate conductivity and transparency make this material a good OTE for spectroelectrochemistry. These properties could be varied by the number of layers of CNTs used to form the OTE. Detection in the UV/near UV region down to 200 nm can be achieved using these transparent CNT films on quartz. The OTE was characterized by transmission electron microscopy, scanning electron microscopy, Raman spectroscopy, UV-visible spectroscopy, cyclic voltammetry, electrochemical impedance spectroscopy, and thin layer spectroelectrochemistry. Ferricyanide, tris-(2,2'-bipyridine) ruthenium(II) chloride, and cytochrome c were used as representative redox probes for thin layer spectroelectrochemistry using the CNT film OTE and the results correlated well with their known properties. Direct electron transfer of cytochrome c was achieved on the CNT film electrode.

Keywords: Electrochemistry, Materials Characterization, Sensors, Voltammetry
Application Code: Nanotechnology
Methodology Code: Electrochemistry

Session # 2210 Abstract # 2210-1 Oral Sessions

Session Title: Sampling and Sample Preparation - Bioanalytical, Neurochemistry, and Material Science

Abstract Title: **Rapid Protein Purification and Digestion with Membrane-Containing Pipette Tips**

Primary Author: Wenjing Ning
Michigan State University

Date: Thursday, March 10, 2016 - Mornin
Time: 08:30 AM
Room: B404

Co-Author(s): Merlin Bruening

Abstract Text

Protein isolation and digestion are vital steps in mass spectrometry (MS) studies of protein structures, interactions and post-translational modifications. However, conventional bead-based affinity isolation suffers from long incubation times because of slow diffusion within bead pores, and in-solution digestion usually takes hours because of the low protease concentration required to avoid self-digestion. Porous membranes are an attractive alternative platform for these operations because convective flow rapidly transports proteins to functional sites in membrane pores, and protease immobilization avoids self-digestion. Integrating membrane technology into pipette tips should make isolation and digestion especially convenient. For example, pushing a protein-containing solution through a protease-containing membrane at the end of a pipette tip enables proteolysis in 30 s or less. Passage of an apomyoglobin solution through a trypsin-modified membrane in 30 s achieves more complete proteolysis than in-solution digestion for 30 min. Similar digestion of the monoclonal antibody Herceptin in a pepsin-containing membrane and subsequent direct infusion electrospray ionization-MS analysis gives 100% peptide coverage for the light chain and 90% for the heavy chain. In protein isolation, pipetting a His-tagged protein-containing solution through a membrane modified with Ni[sup]2+[/sup] complexes and subsequent rinsing and elution yield purified His-tagged protein in 2 minutes. Stacking membranes or increasing membrane area can increase protein-binding capacity if necessary. These results demonstrate that membrane-based devices connected directly to pipette tips enable convenient proteolysis or protein purification, potentially for high-throughput studies when coupled to robotic systems.

Keywords: Mass Spectrometry, Membrane, Protein, Sample Preparation
Application Code: Bioanalytical
Methodology Code: Sampling and Sample Preparation

Session # 2210 Abstract # 2210-2 **Oral Sessions**

Session Title Sampling and Sample Preparation - Bioanalytical, Neurochemistry, and Material Science

Abstract Title **DNA Extraction and Analysis Using Magnetic Ionic Liquid Solvents**

Date: Thursday, March 10, 2016 - Mornin
Time: 08:50 AM
Room: B404

Primary Author Kevin D. Clark
Iowa State University

Co-Author(s) Jared L. Anderson, Melissa Yamsek, Omprakash Nacham

Abstract Text

The extraction and purification of DNA represents a significant bottleneck in nucleic acid analysis. Very recently, hydrophobic magnetic ionic liquids (MILs) were employed as solvents for the rapid and efficient extraction of DNA from aqueous solution and bacterial cell lysate. By application of a magnetic field, the DNA-enriched MIL microdroplets were readily isolated from bulk aqueous sample medium. Although the MIL-based procedure allowed for a rapid extraction step, isolation of sufficiently pure nucleic acid from the MIL extraction phase proved to be a time-consuming and labor-intensive challenge. In this study, a novel strategy for the MIL-based extraction of plasmid DNA (pDNA) followed by immediate polymerase chain reaction (PCR) amplification of a target gene is described. Two hydrophobic MILs including trihexyl(tetradecyl)phosphonium tetrachloroferrate(III) ([P6,6,6,14+][FeCl4□]) and benzyltrioctylammonium bromotrichloroferrate(III) ([(C8)3BnN+][FeCl3Br□]) were investigated as solvents for rapid nucleic acid analysis. Both the paramagnetic anion and the hydrophobic cation components of the two MILs studied were found to inhibit PCR. However, careful design of the components within a PCR buffer enabled the direct addition of pDNA-enriched MIL to a PCR tube for successful amplification of a target gene without further sample purification. Sequence analysis of the PCR amplicon obtained from MIL-facilitated DNA analysis revealed an identical sequence when compared to a standard. The optimized PCR conditions provided a suitable medium for the amplification of pDNA extracted from bacterial cell lysate, demonstrating the feasibility of integrating MIL solvents with biochemical assays for rapid DNA analysis.

Keywords: Bioanalytical, Extraction, Nucleic Acids, Sample Preparation
Application Code: Bioanalytical
Methodology Code: Sampling and Sample Preparation

Session # 2210 Abstract # 2210-3 **Oral Sessions**

Session Title: Sampling and Sample Preparation - Bioanalytical, Neurochemistry, and Material Science

Abstract Title: **Localized Laser Ablation Sample Transfer for Tissue Proteomics**

Primary Author: Fabrizio Donnarumma
Louisiana State University

Date: Thursday, March 10, 2016 - Mornin
Time: 09:10 AM
Room: B404

Co-Author(s): Kermit K. Murray, Michael E. Pettit, Touradj Solouki

Abstract Text

The goal of this work is to develop a comprehensive approach for ambient laser ablation sampling for mass spectrometry and other bioanalytical techniques which enables localized extraction of tissue material for high-throughput tandem mass spectrometry proteomics analysis. An optical parametric oscillator laser system operating in the 3 [micro]m wavelength region of the mid-infrared is used to ablate material from tissue samples that is drawn into a capillary by means of suction or captured by interception with a solvent. The vacuum capture approach displays higher material recovery at the expense of an increased downstream sample process time and it is indicated for sampling experiments aimed to obtain the largest amount of information. The liquid capture approach has higher multiplex sampling capabilities and lower sample-to-sample processing time which result in a method suitable for fast profiling techniques. In vacuum capture mode, the material is drawn to a filter cartridge. The collected material is brought into solution and further processed, for example by filter aided sample preparation (FASP) bottom-up proteomic workflow, before being analyzed with nanoLC coupled to tandem MS/MS. Alternatively, the solubilized material can be deposited on a target for MALDI analysis. Liquid capture can be performed using 96 or 384 format well plates and the ablated material can be either further processed as previously described or stored. We have found that it is possible to transfer large intact proteins (MW>100 kDa) from thin tissue samples and that the captured samples can be processed without addition of surfactants. Multi-compartmental proteomic analyses have been achieved from sampling area as small as 500x500 [micro]m. Evaluation of high throughput proteomics performances, both in data dependent as well as independent modes, are currently being conducted using high resolution quadrupole time of flight mass spectrometry with ion mobility capabilities.

Keywords: Laser, Liquid Chromatography/Mass Spectroscopy, Proteomics, Sampling

Application Code: Bioanalytical

Methodology Code: Sampling and Sample Preparation

Session # 2210 Abstract # 2210-4 **Oral Sessions**

Session Title Sampling and Sample Preparation - Bioanalytical, Neurochemistry, and Material Science

Abstract Title **An Ultra Sensitive Sample Preparation Approach that Eliminates the Need to Dry Down and Reconstitute**

Date: Thursday, March 10, 2016 - Mornin
Time: 09:30 AM
Room: B404

Primary Author Shahana Huq
Phenomenex

Co-Author(s) Jessica Detsch, Matthew Brusius, Ramkumar Dhandapani, Zeshan Aqeel

Abstract Text

Solid Phase Extraction (SPE) is an excellent cleanup solution for bioanalytical samples because it is selective, reproducible, and results in ultra-clean and concentrated samples. While the technique is effective, traditional formats such as 10 mg 96-well plates or tubes pose challenges for small or limited sample volumes and can result in dilute eluents if the sample is not dried down and reconstituted after the extraction procedure. This dry down step can require 30 or more minutes, adding a significant amount of time to the procedure. To overcome these challenges, the μElution 96-well SPE plate format was developed. The μElution SPE plates contain significantly less sorbent as compared to a traditional 10 mg 96-well SPE plate, allowing analysts to elute in volumes as low as 25 μL. These low elution volumes result in ultra-concentrated samples that do not need to be blown down. In this talk, we will demonstrate the time savings and increase in sensitivity that can be achieved by moving a 10 mg SPE method to the μElution format.

Keywords: Biological Samples, Liquid Chromatography/Mass Spectroscopy, Sample Preparation, Solid Phase Ext
Application Code: Bioanalytical
Methodology Code: Sampling and Sample Preparation

Session # 2210 Abstract # 2210-5 **Oral Sessions**

Session Title: Sampling and Sample Preparation - Bioanalytical, Neurochemistry, and Material Science

Abstract Title: **Pulled Low Flow Push-Pull Perfusion Probe Tips for Sampling from Tissue Slices**

Date: Thursday, March 10, 2016 - Mornin
Time: 10:05 AM
Room: B404

Primary Author: Marissa R. Becker
University of Illinois at Chicago

Co-Author(s): David E. Featherstone, Scott A. Shippy

Abstract Text

[i]In vivo[/i] sampling methodologies are commonly used to describe chemical compositions for a wide range of analytes; however, most sampling methods are relatively large, and measurements are susceptible to bias from tissue damage. This is particularly problematic for brain tissue slice models, commonly used in neuroscience experiments. Low-flow push-pull perfusion sampling (LFPS) utilizes a gentle perfusion of tissue at the tip of a capillary and is well suited for brain slices. Nonetheless, the relatively large dimensions of LFPS probe may still bias sample composition due to tissue damage. The objective of this work is to build, characterize, and demonstrate the use of a pulled LFPS probe that will collect extracellular fluid from smaller tissues volumes and reduce the probability of damage-induced tissue bias.

The probes are fabricated using fused-silica capillaries in a concentric design similar to literature reports by us. The ends of the probe are secured into a vertical puller where a hanging weight is used to provide the axial force necessary to pull the probe. A small section (1-4mm) of the probe is heated with a butane flame. Softened fused-silica is pulled by gravity, and a fine tip is produced. This method provides an average 50 µm outer diameter probe tip while delivering consistent shape and length confirmed by optical microscopy. [i]In vitro[/i] calibration was performed to characterize stable infusion and withdrawal flow rates at 10-15 nL/min. Relative recovery of amino acids, glycine and arginine, with the natively fluorescent molecule, riboflavin, from quiescent solutions showed 70-80% recovery. To demonstrate the applicability of these probes, they were used to sample extracellular fluid from mouse hippocampal tissue slices. Collected samples were analyzed to determine amino acid content using capillary electrophoresis with laser induced fluorescence. Although the probe size is similar to those in electrophysiology, the probes differentiate by allowing direct chemical information which is likely to be significant for modern neurochemistry.

Keywords: Amino Acids, Bioanalytical, Capillary Electrophoresis, Small Samples
Application Code: Neurochemistry
Methodology Code: Sampling and Sample Preparation

Session # 2210 Abstract # 2210-6 **Oral Sessions**

Session Title Sampling and Sample Preparation - Bioanalytical, Neurochemistry, and Material Science

Abstract Title **Design of Protein-Binding Membranes through Adsorption of Star-Shaped Polyelectrolytes in Membrane Pores**

Primary Author Weijing Liu
Michigan State University

Date: Thursday, March 10, 2016 - Mornin
Time: 10:25 AM
Room: B404

Co-Author(s) Merlin Bruening, Salinda Wijeratne

Abstract Text

Porous membranes are attractive for rapid protein isolation because convection efficiently transports proteins to affinity sites. However, creation of membranes with high binding capacities requires modification of pore walls with films that rapidly adsorb multilayers of protein. This work employs layer-by-layer adsorption of star-polyelectrolytes to create coatings with large pores that may enhance the rate of protein capture in these films. Specifically, we synthesized star-poly(acrylic acid) (star-pAA) and star-poly(2-dimethylaminoethyl methacrylate) (star-pDMAEMA) with 3, 4 and 6 arms and examined the morphology and protein-binding capacities of films prepared through adsorption of these polymers. AFM images of multilayer (star-pDMAEMA/star-pAA)n films on gold-coated wafers reveal pore diameters ranging from 200-550 nm, depending on the number of arms in the polymer and the number of bilayers, n, in the films. Layer-by-layer adsorption of star-pAA and star-pDMAEMA in porous hydroxylated nylon gives membranes that bind proteins via ion-exchange interactions, and the binding capacity increases with the number of adsorbed bilayers and varies with the deposition pH and number of arms on the polymers. Such membranes capture as much as 120 mg of lysozyme per cm3 of membrane, which is more than twice the capacity of commercial ion-exchange membranes. Derivatization of the star-pAA side chains with aminobutyl nitrilotriacetate□Ni2+ complexes allows formation of membranes that selectively bind polyhistidine-tagged proteins. Such membranes capture 50 mg of polyhistidine-tagged ubiquitin per cm3 of membrane, which is similar to the capacity of commercial beads.

Keywords: Adsorption, Membrane, Polymers & Plastics, Sample Preparation

Application Code: Material Science

Methodology Code: Sampling and Sample Preparation

Session Title: Sampling and Sample Preparation - Bioanalytical, Neurochemistry, and Material Science

Abstract Title: **Direct Coupling of Solid Phase Microextraction to Mass Spectrometry Via Nano-Electrospray Ionization: Development and Applications in Bioanalysis**

Primary Author: German A. Gomez-Rios
University of Waterloo

Date: Thursday, March 10, 2016 - Mornin
Time: 10:45 AM
Room: B404

Co-Author(s): Barbara Bojko, Ezel Boyaci, Janusz Pawliszyn, Nathaly Reyes-Garces

Abstract Text

Different geometrical configurations of SPME have been directly coupled to mass spectrometry aiming to diminish matrix effects, improving detection limits, and enhancing the analysis throughput. Although SPME fibres have been used for decades, its maximum quantitative potential has remained fully unexplored. Herein, we present the direct coupling of biocompatible-SPME (Bio-SPME) fibres to mass spectrometry via nano-ESI emitters as a powerful tool for fast quantitative analysis of target analytes in biofluids. The total sample preparation time does not exceed 2 minutes and, by selecting the appropriate fibre length and sample vessel, sample volumes ranging between 10 and 1500 □L can be scrutinized. Limits of detection in the sub-nanogram per millilitre, great accuracy (□90 %), and outstanding linearity (r □ 0.99) were achieved for all the studied probes in PBS, urine, plasma, and whole blood. Given that Bio-SPME-nano-ESI efficiently integrates sample clean-up, analyte extraction/enrichment, and ionization, our results demonstrated that it is an advantageous configuration for bioanalytical applications such as therapeutic drug monitoring, doping in sports and pharmacological studies. In addition, this study thoroughly evaluates the effect of the coating characteristics (e.g. thickness, coating chemistry, and coating length) on the performance of Bio-SPME-nano-ESI.

Keywords: Bioanalytical, Clinical/Toxicology, Mass Spectrometry, SPME
Application Code: Bioanalytical
Methodology Code: Sampling and Sample Preparation

Session # 2220 Abstract # 2220-1 **Oral Sessions**

Session Title Sensors - Others

Abstract Title **PANi Electrospun Fibers and Drop Cast Film Sensor Array for the Detection of Small Chained Alcohols**

Date: Thursday, March 10, 2016 - Mornin
Time: 08:30 AM
Room: B401

Primary Author Kelvin Tran
University of California Riverside

Co-Author(s) Andrew J. Burris, Quan Cheng

Abstract Text
Electrospun nanofibers can serve as a great sensing platform as it is facile process and has great versatility as these fibers can be functionalized to detect a wide assortment of analytes. Substrates made from electrospun microfiber and drop-cast films of polyaniline (PANi)/ (+)camphor-10-sulfonic acid (HCSA)/polyethylene oxide (PEO) composite doped with variants of graphene oxide were fabricated and evaluated as chemiresistor for sensing gas analytes. This new approach of improving PANi gas sensor performance is shown using varied GO dopants that have been reduced by various methods to different degree of oxidation. The GO were reduced thermally at 500 C (trGO) and also chemically reduced by citrate reduction for 6 hours (crGO-6) and 24 hours (crGO-24) which gave tunable enhancement for the sensor. Upon exposure of the sensor substrate to small chain alcohol vapors, the crGO-6 dopant exhibited higher sensitivity than the other dopants which suggest that this dopant improves sensor performance via increase electrical conductivity that surpasses the non-reduced GO, and enhanced hydrogen bonding capability over that of the crGO-24. The electrospun fibers outperformed the drop-cast films in terms of response time, linear range, and sensitivity due to the increased specific surface area of the sensing membrane. Sensor array consisting of all three reduced GO dopant in the PANi/HSCA/PEO substrate successfully identified methanol, ethanol, and propanol vapors using principal component analysis (PCA).

Keywords: Gas, Material Science, Sensors
Application Code: Material Science
Methodology Code: Sensors

Session # 2220 Abstract # 2220-3 **Oral Sessions**

Session Title: Sensors - Others

Abstract Title: **Improving the Sustainability of Drinking Water Systems Using Nanostructured Biosensors for Escherichia Coli**

Primary Author: Heather A. Crapo
Binghamton University

Date: Thursday, March 10, 2016 - Mornin
Time: 09:10 AM
Room: B401

Co-Author(s): Idris Yazgan, Melissa McDonald, Omowunmi Sadik

Abstract Text

The US EPA's Safe Drinking Water Act (SDWA) requires that all surface water be filtered and disinfected before consumption. This requires the development of low-cost, innovative nanotechnologies that can efficiently detect and remove microbial contaminants from drinking water. Microorganisms are in sizes of nanometesr to hundreds of nanometers, which limits the choice of method in decontamination. We have developed nano-structured poly(amic) acid membranes integrated with chitosan by glutaraldehyde in order to decontaminate tap-water [1, 2]. The synthesized semi-interpenetrating membrane showed improved mechanical properties and excellent solvent resistance against buffers and many organic solvents, which makes them eligible for tap water decontamination. Filtration capacity of the membranes were tested against three water contaminants, namely Escherichia coli, Citrobacter freundii and Staphylococcus epidermidis with absolute removal achieved using through dead end and tangential flow filtrations. In this work, we are exploiting the strong affinity of mannose-containing oligosaccharides with fimbrial lectin for E.coli detection [3]. E. coli, being one of the most widely used indicators for fecal contamination, is a rational target choice for determining whether or not drinking water is safe for consumption. We seek to improve the sustainability of drinking water and wastewater treatment facilities through the development of nano-structured sensor/membranes with the capabilities for rapid, ultrasensitive, and highly selective detection by utilizing the high affinity of mannose for the FimH lectin of E. coli type 1 pili. This work combines the selective ligands with electrochemical quartz crystal microbalance for ultrasensitive detection of E.coli at sub-infectious doses.

Keywords: Biosensors, Contamination, Sensors, Water
Application Code: Food Contaminants
Methodology Code: Sensors

Session # 2220 Abstract # 2220-5 **Oral Sessions**

Session Title: Sensors - Others

Abstract Title: **Surface Plasmon Resonance Immunosensor Using Au Nanoparticle Modified Antibody**

Date: Thursday, March 10, 2016 - Mornin
Time: 10:05 AM
Room: B401

Primary Author: Dulal C. Kabiraz
Hokkaido University

Co-Author(s): Kinichi Morita, Toshikazu Kawaguchi

Abstract Text

Surface Plasmon Resonance (SPR) biosensor is a highly sensitive signal transducer from a biochemical reaction. Hence, many applications such as affinity evaluation of protein, monitoring of environmental pollutants, medical diagnosis, detection of explosives, have been studied. However, know to date, the miniaturized SPR sensor is not commercialized yet. Since SPR can detect the mass change in pg cm-2 order at the interface, it is used for monitoring of the initial process of biochemical reaction. As a result, the obtained data includes huge error, because the observed signal change is quite small. Thus, the signal amplification is essentially necessary for a practical use. In this study, we have studied the indirect competitive inhibition method for immunosensing that used for detection of antibody (Mw > 10 kDa) instead of small analyte (illegal compound in food: clenbuterol: Mw = 277). However, it could not be achieved the commercialization, because enough SPR signal change was not obtained. In order to further enhance SPR signal change, the antibody was devised with Au nanoparticle in this study. Here, primary antibody was modified with Au nanoparticle, and it was applied to immunosensing. The size dependence of Au nanoparticle and the comparison of secondary antibody modification were studied. We also report here the signal amplification mechanism based on kinetics and the characterization of immunosensing by using STM, XPS, FT-IR and electrochemical methods. Briefly, it was found that the signal change and the sensitivity increased by 10 times (8 mdeg. to 80 mdeg.) and 40 times (2 ppt (pg/mL) to 50 ppq (fg/mL)), respectively.

Keywords: Biosensors, Food Safety, Particle Size and Distribution, Sensors
Application Code: Food Safety
Methodology Code: Sensors

Session # 2220 Abstract # 2220-6 **Oral Sessions**

Session Title: Sensors - Others

Abstract Title: **QCM Virtual Multisensor Array for Detection of Gasoline Adulterants**

Primary Author: Nicholas Speller
Louisiana State University

Date: Thursday, March 10, 2016 - Mornin
Time: 10:25 AM
Room: B401

Co-Author(s): Isiah M. Warner, Noureen Siraj, Stephanie Vaughan

Abstract Text

Gasoline adulteration is a persistent problem that can have significant impact on the economy, environment and consumer vehicle reliability. Unfortunately, traditional detection methods are either expensive and require expert knowledge (e.g. GC-MS, Dye markers, etc.) or inexpensive yet not very robust (e.g. evaporation /distillation techniques, density based techniques etc). As a result, there is a need for accurate detection methods that are simplistic yet cost effective. One approach that could satisfy these criteria are the use of Quartz crystal microbalance (QCM) based cross-reactive sensor arrays. Herein, we present a simple method for detection of common gasoline adulterants using a novel QCM sensing approach based on chemical affinity, viscoelasticity, and harmonics. In this regard, we present an ionic liquid based QCM virtual multisensor array for highly accurate identification and quantification of gasoline adulterants.

This material is based upon work supported by the National Science Foundation Graduate Research Fellowship Program under grant number DGE-1247192; National Science Foundation under grant numbers, CHE-1243916 and CHE-1307611; and funds form the Phillip W. West Endowment to IMW.

Keywords: Analysis, Gasoline, Sensors, Volatile Organic Compounds

Application Code: Fuels, Energy and Petrochemical

Methodology Code: Sensors

Session # 2220 Abstract # 2220-7 **Oral Sessions**

Session Title: Sensors - Others

Abstract Title: **Development of an Aptamer Functionalized Electrode Array for Real-Time In Vivo Cocaine Detection Using Square Wave Voltammetry**

Date: Thursday, March 10, 2016 - Mornin
Time: 10:45 AM
Room: B401

Primary Author: Ian M. Taylor
University of Pittsburgh

Co-Author(s): Emma Bigelow, Tracy Cui

Abstract Text

Cocaine is a well-known and highly addictive competitive inhibitor of the dopamine transporter (DAT) shown to up-regulate dopamine (DA) signaling in the dorsal striatum and nucleus accumbens regions of the brain. While the concentrations of IV and IP injection dosages are predetermined, identifying actual drug concentration in specific regions of the brain presents a challenge. Microdialysis has commonly been used to sample the extracellular space of the brain, but fails to accurately determine local analyte concentration due to spatial averaging, unknown extracellular volume, and probe-triggered immune response. We have developed an aptamer-based selective cocaine sensor designed to directly quantify cocaine concentration in discrete regions of the extracellular space in vivo. The probe consists of an array of 16 individual gold electrodes (38 µm diameter) positioned along a single 60 µm x 15 mm (w x l) silicon shank. After electrodeposition of a fresh gold layer, a monolayer of 3' methylene blue (MB) functionalized cocaine-binding aptamer was covalently bound to the electrode surface and remaining regions of unbound gold were treated with 6-mercapto-1-hexanol (MCH). The cocaine-binding aptamer robustly exhibits a concentration dependent conformation shift in the presence of cocaine, bringing the electroactive MB closer to the electrode surface. Using 100Hz square wave voltammetry detection, the sensor exhibits sub second temporal resolution and selectivity for cocaine over dopamine, GABA, pH and several other neurochemically relevant interferents. The aptasensor currently exhibits a 10 µM lower limit of detection and sensitivity over at least three orders of magnitude. The cocaine sensor exhibits stable baseline signal in the rat dorsal striatum over an 8 hour sampling period and immediately detects local injection of a 1 µL bolus of 250mM cocaine solution in vivo.

Keywords: Electrochemistry, Neurochemistry, Nucleic Acids, Sensors
Application Code: Neurochemistry
Methodology Code: Sensors

Session # 2220 Abstract # 2220-8 **Oral Sessions**

Session Title Sensors - Others

Abstract Title **Optomechanical Switching by Plasmonic Nanoparticle Monolayers on Elastic Substrate Induced by Stretching**

Date: Thursday, March 10, 2016 - Mornin
Time: 11:05 AM
Room: B401

Primary Author Mahmoud Mahmoud
Georgia Institute of Technology

Co-Author(s)

Abstract Text

A monolayer assembly of silver nanodisks (AgNDs) was fabricated on the surface of a polydimethylsiloxane (PDMS) polymer substrate using the Langmuir-Blodgett technique. Upon stretching the PDMS substrate, the localized surface plasmon resonance (LSPR) spectrum of the AgND monolayer is blue shifted when the incident light excitation is polarized parallel to the stretching direction. Conversely, a red shift in the LSPR spectrum of the AgND monolayer is observed in the case of light polarization orthogonal to the stretching direction. The magnitude of the shift in the LSPR spectrum is proportional to the degree of stretching of the PDMS substrate. Stretching PDMS in one direction causes its shrinking in the orthogonal direction. Consequently, the interparticle distance between individual AgNDs on the PDMS surface increases in the same direction as the mechanical stretching and simultaneously decreases in the orthogonal direction. The different optical responses of the AgND assembly on the surface of stretched PDMS when excited with different polarization directions is due to the changing strength of the plasmon field coupling, which is inversely proportional to the separation gap between the AgNDs. The experimentally measured LSPR spectra upon stretching the PDMS substrate to different lengths and varying the incident light polarization were confirmed using the discrete dipole approximation calculation technique. The same optical response was obtained for an AgND monolayer sandwiched between two PDMS substrates. Covering the surface of the AgND monolayer on the PDMS substrate with another PDMS layer on top eliminates their deformation after multiple stretching-shrinking cycles and increases its chemical stability.

Keywords: Metals, Nanotechnology, Sensors
Application Code: Nanotechnology
Methodology Code: Sensors

Session # 2230 Abstract # 2230-2 **Oral Sessions**

Session Title: Vibrational Spectroscopy Instrumentation and Applications

Abstract Title: **The Effect of Molecular Polarity and Solubility on Adsorption Rates and Equilibrium Constants for Molecules on Noble Metal Surfaces Using Surface-Enhanced Raman Spectroscopy**

Date: Thursday, March 10, 2016 - Mornin
Time: 08:50 AM
Room: B402

Primary Author: Erik D. Emmons
US Army

Co-Author(s): Ashish Tripathi, Augustus W. Fountain, Jason A. Guicheteau, Jerry Cabalo, Neal D. Kline, Steven D. Christesen

Abstract Text

Adsorption of molecules on surfaces plays an important role in many different technical applications, and can be influenced by many factors. Molecular parameters such as polarity and solubility are expected to play an important role in adsorption. We are studying the effect of polarity and solubility on the adsorption of aromatic molecules on noble metal surfaces using the technique of surface-enhanced Raman spectroscopy (SERS). Detection of low concentrations of analytes of interest such as explosives and toxic chemicals is possible using SERS, but only if the molecule adsorbs to the surface. To reduce the relative lack of information on what causes molecules to adsorb, we have studied different series of aromatic thiol and nitrogen-containing molecules with different substituents that lead to varying levels of overall molecular polarity and solubility. These factors have been showed to have an effect on adsorption rates, equilibrium constants, and binding energies. These measurements show trends which can provide predictive power for determining the SERS response.

Keywords: Infrared and Raman, Ultratrace Analysis, Vibrational Spectroscopy
Application Code: Homeland Security/Forensics
Methodology Code: Vibrational Spectroscopy

Session # 2230 Abstract # 2230-3 **Oral Sessions**

Session Title: Vibrational Spectroscopy Instrumentation and Applications

Abstract Title: **Room Temperature Freezing and Orientation Control of Surface Immobilized Biomoelcules in Air**

Primary Author: Yaoxin Li
University of Michigan

Date: Thursday, March 10, 2016 - Mornin
Time: 09:10 AM
Room: B402

Co-Author(s): Somayesadat Badieyan, Zhan Chen, Zhang Xiaoxian

Abstract Text

A biological molecule such as a peptide or protein usually requires an aqueous environment to fold into the native structure to be active. For a biomolecule immobilized on a surface, its activity may be optimized via maintaining its native structure as well as controlling its orientation. Such native structure and orientation usually collapse in air, therefore the biological function is greatly reduced or completely lost. In this study, a simple and reversible method to stabilize and control the "native" structure and orientation of surface immobilized biomolecules in air using sugar coating was developed. The immobilized biomolecule structure and orientation were examined using sum frequency generation vibrational spectroscopy and circular dichroic spectroscopy. We believe that this method is general and can be applied to other more complicated molecules, providing new potential on the improvement of heterogeneous bio-catalysis in harsh or "water-free" conditions.

Keywords: Analysis, Peptides, Protein, Spectroscopy
Application Code: Bioanalytical
Methodology Code: Vibrational Spectroscopy

Session # 2230 Abstract # 2230-4 **Oral Sessions**

Session Title Vibrational Spectroscopy Instrumentation and Applications

Abstract Title **Fiber Spectroscopy for Process Control and Medical Diagnostics**

Primary Author Viacheslav Artyushenko
Art Photonics GmbH

Date: Thursday, March 10, 2016 - Mornin
Time: 09:30 AM
Room: B402

Co-Author(s)

Abstract Text

Various fiber spectroscopy methods can be used for process control and medical diagnostics: transmission/absorption, diffuse reflection, fluorescence and Raman scattering. They can be used alone or combined - to enhance sensitivity and accuracy in media composition analysis. Review of research fiber spectrometers will be done to compare them with promising fiber sensors designed for customized applications. Portable and robust fiber sensors may accelerate development of various process-control solutions because of their low price, small size, high speed and easy reading. Their industrial application will assist in a better yield of products with enhanced quality. Fiber spectroscopy used for reaction monitoring in labs and for process control in industry provides well known advantages:

1)eliminates the need to take samples for lab analysis from a running process as fiber probes enables direct media analysis in-line/in-citu;

2)remote process-control can be done at long distance - up to 300m in UV-Vis-NIR range, where silica fibers possess by high transmission;

3)robust design of industrial spectral probes can withstand to toxic, aggressive or radioactive media and to the harsh reaction conditions: high pressure, high or low temperature, vibrations, etc.

As silica fibers could transmit light till 2.2µm only, the new Mid IR-fiber probes are used for 2-17µm finger-print range - the most reach by specific bands used for molecular vibration analysis. Process control by spectral changes of media can be done with complex spectral systems and optimal chemometric model, but total cost of such a solution can be too high for volume applications. Customized fiber sensors may eliminated this trouble to expand PAT methods into industrial applications and medical diagnostics. This possibility will be presented in review of promising fiber sensors based on LED and tunable Fabry-Perot MEMS filters - with examples of promising applications in industry and for oncology diagnostics.

Keywords: Fiber Optics, Fluorescence, Infrared and Raman, Spectroscopy
Application Code: Biomedical
Methodology Code: Vibrational Spectroscopy

Session # 2230 Abstract # 2230-5 **Oral Sessions**

Session Title: Vibrational Spectroscopy Instrumentation and Applications

Abstract Title: **Characterization of Polymer/Epoxy Buried Interfaces with Silane Adhesion Promoters Before and after Hygrothermal Aging for the Elucidation of Molecular Level Details Relevant to Adhesion**

Date: Thursday, March 10, 2016 - Mornin
Time: 10:05 AM
Room: B402

Primary Author: Nathan W. Ulrich
The University of Michigan

Co-Author(s): John Myers, Zhan Chen

Abstract Text

Buried interfacial structures containing epoxy underfills are important in the microelectronics industry and their structures determine the interfacial adhesion properties of electronic devices and ultimately their lifetime. Weak adhesion and delamination at such interfaces lead to premature failure of microelectronic devices. In this work, sum frequency generation (SFG) vibrational spectroscopy, an intrinsically surface sensitive technique, was utilized to investigate the molecular structure of buried epoxy interfaces before and after accelerated stress testing. This technique was used in order to relate the molecular-level structural changes of the epoxy systems to the macroscopic adhesion strength and determine if silane adhesion promoters can affect a polymer/epoxy system. Two controls systems were used; a hydrophilic and hydrophobic surface, interfacial water was detected after hygrothermal aging on the hydrophilic surface but was not detected on the hydrophobic surface. Lap shear analysis was performed and it was found that the hydrophilic surface had greatly reduced in adhesion strength after aging, much more reduced than the hydrophobic surface. To determine if the adhesion strength could be increased after aging, silane adhesion promoters, (3-aminopropyl)trimethoxysilane (ATMS), (3-glycidoxypropyl)trimethoxysilane (GPS), and octadecyltrimethoxysilane (OTMS), were introduced into the system. For the hydrophilic model surface, each silane prevented interfacial water from forming and increased adhesion strength after aging. For the hydrophobic model surface, ATMS and GPS increased the adhesion strength, while OTMS did not change the adhesion strength after aging. This research demonstrates that molecular structural studies of buried epoxy interfaces during hygrothermal aging using SFG vibrational spectroscopy can greatly contribute to the overall understanding of moisture-induced failure mechanisms of organic adhesives found in microelectronic packaging.

Keywords: Laser, Light Scattering, Polymers & Plastics, Raman
Application Code: Polymers and Plastics
Methodology Code: Vibrational Spectroscopy

Session # **2230** Abstract # **2230-6** **Oral Sessions**

Session Title: Vibrational Spectroscopy Instrumentation and Applications

Abstract Title: **Surface Interaction of Nitrogen Containing Aromatic Molecules with Gold Investigated with Surface Enhanced Raman Spectroscopy (SERS)**

Primary Author: Ashish Tripathi
Leidos, Inc.

Date: Thursday, March 10, 2016 - Mornin
Time: 10:25 AM
Room: B402

Co-Author(s): Augustus W. Fountain, Erik D. Emmons, Jason A. Guicheteau, Martin Moskovits, Steven D. Christesen

Abstract Text

Surface-enhanced Raman scattering (SERS) is potentially a capable tool for detecting trace concentrations of analytes. In order to predict the SERS activity of relevant toxic compounds, it is important to understand the nature of the binding of these compounds to the noble metal surface. In this effort we have studied the binding of several nitrogen containing aromatic chemicals to nanostructured gold surfaces from aqueous media. These studies have revealed that a complex equilibrium state exists between the molecules and the gold surface that cannot be explained by a simple Langmuir isotherm. The peak area of selected Raman features verses concentration data of each of the chemical analyzed suggests two equilibrium states. The two equilibrium states could be attributed to concentration depended arrangements/orientation and/or two types of adsorption sites . The Raman spectral data also shows observable differences between the two observed equilibrium states. Our studies were expanded to investigate the causation of the observed complex equilibrium state between these molecules and nanostructured gold surfaces.

Keywords: Adsorption, Nanotechnology, Raman, Spectroscopy
Application Code: Homeland Security/Forensics
Methodology Code: Molecular Spectroscopy

Session # 2230 **Abstract # 2230-7** **Oral Sessions**

Session Title: Vibrational Spectroscopy Instrumentation and Applications

Abstract Title: **Transmission Raman Spectroscopy as a Regulatory-Approved Method for Content Uniformity Analysis – Replacing HPLC**

Date: Thursday, March 10, 2016 - Mornin
Time: 10:45 AM
Room: B402

Primary Author: Darren Andrews
Cobalt Light Systems

Co-Author(s): Andrew Owen, Julia Griffen, Mark Mabry, Matthew Bloomfield, Pavel Matousek

Abstract Text

Recent developments in transmission Raman spectroscopy (TRS) have culminated in regulatory approval for routine content uniformity (CU) testing in pharmaceutical manufacturing. TRS is now used around the world as an alternative to costly HPLC methods. This talk discusses best-practice method development, example CU method creation and the key regulatory issues for a successful TRS CU method.

TRS as a technique for quantitative pharmaceutical analysis has been discussed since 2005 and commercially available since 2010. It has several benefits over other spectroscopic methods, such as mid-IR NIR and even ssNMR and XRPD. TRS, like transmission NIR, samples the bulk of the tablet; however, TRS spectra are sharp and feature-rich and directly assignable to the chemical ingredients in the tablet, not the physical properties related to scattering. This makes the analysis and method development simpler, the results easier to interpret, with a corresponding reduction in the time and resource needed to build a model. As a practical non-invasive QC-laboratory technique TRS requires no sample preparation and works with capsules and coated tablets – through as much as 10mm of powdered/tableted material.

As well as quantifying APIs in intact tablets TRS can also quantify the excipients, without adding greatly to the development work, which is useful for formulation development and process development/validation. Measuring polymorph content and residual crystallinity at levels of 0.1-1% w/w polymorph content in intact tablets makes it an ideal alternative to X-ray and NMR methods for many applications.

Keywords: Molecular Spectroscopy, Pharmaceutical, Quality Control, Raman
Application Code: Quality/QA/QC
Methodology Code: Vibrational Spectroscopy

Session # 2230 Abstract # 2230-8 **Oral Sessions**

Session Title Vibrational Spectroscopy Instrumentation and Applications

Abstract Title **A New Microscope for FT-IR Microspectrometry**

Primary Author David W. Schiering
Czitek

Date: Thursday, March 10, 2016 - Mornin
Time: 11:05 AM
Room: B402

Co-Author(s) Gregg Ressler

Abstract Text

For the last thirty years, reflecting microscopes interfaced to Fourier transform infrared (FT-IR) spectrometers have been the method of choice to characterize microsamples. These so called FT-IR microscopes have grown in sophistication and performance. The advantages of FT-IR microscopes lie in the ability to view, manipulate, and isolate the microsample. Areas of disparate chemical properties can be measured separately by isolating the area with an aperture mask while observing under magnification. This presentation shall concern the design and performance of a novel, low cost microscope for FT-IR microspectrometry. The system operates in surface reflection, attenuated total reflection (ATR), and transmission spectroscopy modes and interfaces to the sample compartment of commercial FT-IR spectrometers. The microscope employs off-axis ellipsoidal mirrors with nominal 5X magnification. The observation of samples is accomplished with a 2X magnification, parfocal optical microscope objective that images the specimen onto a 5 megapixel CMOS camera. A series of fixed aperture masks are used to isolate microsamples. These masks are controlled automatically with a software program that is also used to display video images from the camera. The microscope accommodates various sample supports and substrates including low-E glass slides, IR windows, and compression cells. The microscope supports diamond, ZnSe, and Ge single reflection internal reflection elements (IRE) in the ATR mode. Specimens can be observed through the IRE, if it is optically transparent. The performance of the system will be demonstrated with spectra collected from microsamples including small diameter single fibers, paint chips, and surface contaminants.

Keywords: Forensics, FTIR, Infrared and Raman, Microspectroscopy
Application Code: Homeland Security/Forensics
Methodology Code: Vibrational Spectroscopy

Session # 2240 Abstract # 2240-1 **Poster Sessions**

Session Title: Application of Mass Spectrometry

Abstract Title: **Variations on a Theme: The Detection of NBOMe Designer Drugs on Blotter Paper by High Resolution Time-of-Flight Mass Spectrometry (TOFMS) with and without Chromatography**

Primary Author: David Barajas
Boston University School of Medicine

Date: Thursday, March 10, 2016 - Mornin
Time:
Room: Exposition Floor, 400 Aisle

Co-Author(s): Bogdan Bogdanov, Craig Young, Frank A. Kero, Jason Weisenseel, Noelle Elliot, Sabra Botch

Abstract Text

Novel Psychoactive Substances (NPS) have been associated with the cause of death in a number of cases in the United States and have led investigators to rethink traditional drug monitoring protocols. Of particular interest to this investigation are the variable phenethylamine chemical structures known as NBOMes', which pose an emerging threat to public health with incidence steadily growing over the past decade. In the culture of abuse, NBOMes are commonly applied to blotter paper and administered sublingually to induce episodes of hallucinations (similar but more potent effects when compared to LSD). This study considers two approaches for screening confiscated blotter paper to determine the presence of NBOMes using high resolution mass spectrometry in forensic case studies. The first approach is an extraction prior to UPLC-ESI-TOFMS. The second is DSA-TOFMS, a direct measurement using ambient ionization mass spectrometry without chromatographic separation. The key advantage of the second approach would reduce the analysis time per sample from minutes to seconds. Additional value added considerations in the reduction of consumable cost and solvent waste should also be noted. Samples were prepared at Boston University-School of Medicine Department of Biomedical Forensic Sciences (Boston, MA). These samples were analyzed at PerkinElmer's Tech Center (Oak Brook, IL). Feasibility of both approaches will be presented.

Keywords: Chromatography, Drugs, Forensic Chemistry, Mass Spectrometry
Application Code: Homeland Security/Forensics
Methodology Code: Mass Spectrometry

Session # 2240 Abstract # 2240-2 **Poster Sessions**

Session Title: Application of Mass Spectrometry

Abstract Title: **Proton Transfer Reaction – Mass Spectrometry: Automated Measurement and Evaluation**

Primary Author: Jens Herbig, IONICON Analytik

Date: Thursday, March 10, 2016 - Mornin
Time:
Room: Exposition Floor, 400 Aisle

Co-Author(s): Alfons Jordan, Christian Lindinger, Johann Seehauser, Klaus Winkler, Lukas Mäerk

Abstract Text

PTR-MS is a well-established direct injection trace gas analysis method with various advantages: high sensitivity, low detection limits, no sample preparation and online quantification. The latter is possible because of well-known conditions in the PTR drift tube and particularly simple because of low fragmentation caused by the chemical ionization via H[sub]3[/sub]O[sup]+[/sup]. However, especially when quadrupole mass spectrometer based PTR-MS instruments are used (instead of high mass resolution TOF analyzers), overlaps of product ions (protonated parent and fragment ions) may occur and make data evaluation somewhat complicated.

Recently, we have published a series of studies which demonstrate that by changing the ionization conditions, such as the reduced electric field strength in the PTR drift tube (E/N) or by switching the reagent ions (SRI) from H[sub]3[/sub]O[sup]+[/sup] to NO[sup]+[/sup] or O[sub]2[/sub][sup]+[/sup], respectively, the selectivity of a PTR-MS instrument can be considerably increased. Stimulated by these results we have developed a new software tool, "Automated Measurement and Evaluation" (AME), which performes measurements at various ionization conditions and subsequently processes the data with real-time output. We demonstrate this process using several examples. For instance, isopropyl alcohol and acetic acid, two isobaric compounds, are both detected at m/z 61 (protonated parent) and m/z 43 (fragment) in PTR-MS, which hinders identification and quantification. Using a substance library containing the fragmentation ratios of both compounds at different E/N levels in combination with an ordinary least squares (OLS) model, the AME software separates and quantifies the two isobars independently. Additionally, AME also averages the data, to reduce noise, exports and displays the processed data. All these features have been implemented with industrial monitoring applications in mind, where an automated process is required.

Keywords: Automation, Chemical Ionization MS, Mass Spectrometry, Time of Flight MS
Application Code: General Interest
Methodology Code: Mass Spectrometry

Session # 2240 Abstract # 2240-3 **Poster Sessions**

Session Title: Application of Mass Spectrometry

Abstract Title: **Functionalized Gold Surface for SPR-MS Determination of Enzymatic Activities and Specificity of Lectins**

Primary Author: Hyojik Yang
University of California, Riverside

Date: Thursday, March 10, 2016 - Mornin
Time:
Room: Exposition Floor, 400 Aisle

Co-Author(s): Quan Cheng

Abstract Text

In this work, we report a method that combines MALDI mass spectrometry (MS) and Surface Plasmon Resonance (SPR) spectroscopy for determination of enzymatic activities and specificity of carbohydrate-lectin interactions.Functionalized surfaces are established by immobilization of lipoic acid derivatives through gold-sulfur bonds. Specifically, lipoic amido (LA)-octaethylene glycol (PEG8)-2,3,5,6-tetrafluorophenol (TFP) was used as a selected ligand in this study. Construction and quality control of self-assembled monolayers of the ligands was monitored through SPR spectroscopy. Direct measurement of molecular mass of the ligands on the surface was performed by MALDI-MS. The TFP ester group allowed for facile immobilization of various bio-molecules, such as neurotensin and 4-aminophenyl glucopyranoside. The use of the labeled biomolecules as substrates for measurement of enzyme activities has been demonstrated. Furthermore, patterned resonance plasmonic microarrays were functionalized with LA-PEG8-TFP for fabrication of carbohydrates chips. These chips were used for SPR imaging/MALDI-MS determination of the selectivity of lectins, the carbohydrate binding proteins. From the results we have showed that the lipoic acid derivatives are ideal for functionalization of various biomolecules on the surface. SPR in combination of MALDI-MS technique is effective for characterization of the ligands and subsequently interactions on the gold surface, show high-quality performance. The biomolecule functionalized surfaces reported here could find broad applications in enzyme activity study as well as investigation of lectin-carbohydrate interactions.

Keywords: Bioanalytical, Enzyme Assays, Laser Desorption, Surface Analysis
Application Code: Bioanalytical
Methodology Code: Surface Analysis/Imaging

Session # 2240 Abstract # 2240-4 **Poster Sessions**

Session Title: Application of Mass Spectrometry

Abstract Title: **High-Resolution Atmospheric Pressure Drift Tube Ion Mobility Spectrometry Coupled with High-Resolution Accurate Mass Orbitrap Mass Spectrometry**

Primary Author: Joel D. Keelor
Georgia Institute of Technology

Date: Thursday, March 10, 2016 - Mornin
Time:
Room: Exposition Floor, 400 Aisle

Co-Author(s): Brian Clowers, Facundo M. Fernandez

Abstract Text

Ion mobility spectrometry (IMS) has become routinely integrated with mass spectrometry (MS), affording enhanced selectivity and separation power for multidimensional analysis of complex mixtures while also providing molecular structure information derived from ion collision cross-section measurements. The majority of modern ion mobility-mass spectrometry platforms incorporate drift tube or traveling wave mobility stages between the ion source (or first quadrupole) and a time-of-flight (ToF) mass analyzer. In these instruments, the resolving power of mobility separations is severely limited by the need for operating at the reduced pressures required for ToF compatibility. Moreover, sensitivity and resolving power are typically constrained by the ToF duty cycle and its effective flight path length. In an effort to circumvent such drawbacks and improve the overall analytical performance of IM-MS instrumentation, we present work coupling atmospheric pressure drift tube ion mobility spectrometry (AP-DTIMS) with a Thermo Scientific Orbitrap Q ExactiveTM mass spectrometer. This new instrument enables achieving high-resolving power in the mobility separation dimension together with ultra-accurate mass analysis. Experiments aimed at improving the duty cycle of the mobility stage using digitally multiplexing will also be presented.

Keywords: Electrospray, Mass Spectrometry, Separation Sciences
Application Code: High-Throughput Chemical Analysis
Methodology Code: Mass Spectrometry

Session # 2240 Abstract # 2240-5 **Poster Sessions**

Session Title: Application of Mass Spectrometry

Abstract Title: **Real-Time Metabolome Analysis by Probe Electrospray Ionization-Tandem Mass Spectrometry (PESI-MS/MS): Preliminary Challenge to Real-Time Metabolomics**

Primary Author: Zaitsu Kei
Nagoya University

Date: Thursday, March 10, 2016 - Mornin
Time:
Room: Exposition Floor, 400 Aisle

Co-Author(s): Hayashi Yumi, Ishii Akira, Ishikawa Tetsuya, Kusano Maiko, Murata Tasuku, Nakajima Hiroki, Nakajima Tamie, Tsuchihashi Hitoshi

Abstract Text

Introduction: Real-time metabolome analysis of live animals must be the ultimate goal for metabolomics, because it can achieve to understand exact metabolome alteration in animals. Probe electrospray ionization (PESI) is a novel ionization technique that enables direct endogenous compounds analysis in biological tissues without invasion and sample preparation. We have proposed intact metabolome analysis of dissected mice liver samples by PESI/tandem mass spectrometry (MS/MS). In this presentation, we applied PESI/MS/MS to real-time hepatic metabolome analysis of mice. Analytical methods: An LCMS-8040 tandem mass spectrometer with probe electrospray ion source (Shimadzu) was used. Small plastic pot were directly attached to the surface of liver of 129/sv mice under isoflurane anesthesia. After 1-day recovery period, treated mice were set in front of the mass spectrometer and 50% EtOH aqueous solution were poured into the pot. Under isoflurane anesthesia, real-time analysis for metabolites like ATP, sugars and intermediates of TCA cycle was executed. Preliminary results: Loop time of the mass spectrometer and cycle time of probe movement of the PESI ion source needed to be synchronized and thus were also optimized: loop time and the cycle time were optimized at 10 sec and 0.1 Hz, respectively. Under above-mentioned analytical condition, we could monitor real-time variation of metabolites in live mice up to 15 min at 10-sec intervals at this time. In conclusion, real-time metabolome analysis by PESI/MS/MS was achieved demonstrating its high potential of becoming the novel approach to real-time monitoring of metabolome alteration in live animals.

Keywords: Bioanalytical, Biological Samples, Mass Spectrometry, Metabolomics
Application Code: Genomics, Proteomics and Other 'Omics
Methodology Code: Mass Spectrometry

Session # 2240 Abstract # 2240-6 **Poster Sessions**

Session Title: Application of Mass Spectrometry

Abstract Title: **Use of High Speed/High Resolution Size Based Chromatographic Separation of Surfactants and Oligomeric Materials with Single Quadrupole Mass Spectrometry Detection**

Primary Author: Michael OLeary
Waters Corporation

Date: Thursday, March 10, 2016 - Mornin
Time:
Room: Exposition Floor, 400 Aisle

Co-Author(s)

Abstract Text

Recent developments in polymerization processes have utilized a wide array of strategies. The development has evolved from simple polymer chains to complex polymers capable of performing multiple functions within a single molecular chain. As these new materials evolve their control and understanding has come under intense scrutiny utilizing a wide range of analytical technology ranging from chromatographic separation to advanced mass spectrometry.

Addressing the challenges of material characterization has often been focused on hyphenated detection techniques coupled with separation. This approach utilizes a concentration detector such as a refractive index (RI) detector as well as a viscosity detector and a multi angle light scattering detector and more recently the use of Mass Spectrometry (MS) detectors.

With the introduction of high speed-high resolution size based separation techniques, a novel approach to the design of the separation equipment including the separation column as well as the entire flow path are used to yield a high speed / resolution separation maintained from injection to detection with traditional detector options such as RI and UV detection. However, the use of this high speed high resolution separation technique has seen limited pairing with MS detection due to the need to control material ionization and solvent matrix effects.

In this study the expansion of the APC approach to the size based separation is presented. A single system control platform is evaluated to pair the chromatographic system to the MS detector system allowing for a high through put/ high resolution evaluation of size based separation of polymeric material while allowing for controlled MS ionization without interfering with the chromatographic separation.

Keywords: HPLC, Mass Spectrometry, Polymers & Plastics, Surfactants
Application Code: Polymers and Plastics
Methodology Code: Mass Spectrometry

Session # 2240 Abstract # 2240-7 **Poster Sessions**

Session Title: Application of Mass Spectrometry

Abstract Title: **Coupling Surface Acoustic Wave Nebulization (SAWN) with Vacuum-assisted Plasma Ionization (VaPI) Mass Spectrometry for Enhanced Ionization and Transmission Efficiency**

Primary Author: Stephen C. Zambrzycki
Georgia Institute of Technology

Date: Thursday, March 10, 2016 - Mornin
Time:
Room: Exposition Floor, 400 Aisle

Co-Author(s): David Goodlett, Fernandez M. Facundo, Joel D. Keelor, Matthew C. Bernier, Sung H. Yoon

Abstract Text

Surface acoustic wave nebulization (SAWN) is an atmospheric pressure ionization technique for mass spectrometry (MS) that has been shown to successfully ionize proteins, phospholipids, and drugs with minimal fragmentation compared to electrospray ionization (ESI) [1-3]. Sampling efficiency with SAWN, however, is potentially restricted by dispersive plume effects during nebulization that can effectively diminish material collection at the spectrometer ambient interface. We here propose a simple remedy to this limitation based on the coupling of SAWN with a new vacuum-assisted plasma ionization (VaPI) approach. VaPI is an new ion generation motif that uses the vacuum draw of an extended inlet capillary to entrain volatilized samples into a high-power helium glow discharge enclosed before, and directly connected to, the mass spectrometer inlet [4]. Combining SAWN and VaPI is hypothesized to increase ion source performance and sensitivity in several ways compared to SAWN alone. Better control of the fluid dynamics at the interface using the VaPI gas inflow facilitates sample collection, thereby improving mass-transport and ion transmission. The temperature-regulated collection region also serves to accelerate solvent droplet evaporation and ion declustering while reactive species in the excited plasma afterglow bolster ionization efficiency and combat possible ion suppression during direct analysis. The first set of experiments presented is aimed at basic ion source parameter space characterization, using small molecule standards to evaluate the hybrid SAWN-VaPI operation. Following these experiments, we use benzylpyridium salts as “thermometer ions” to determine the energies imparted onto analytes using the SAWN-VaPI versus individual SAWN and VaPI operation [5].

Keywords: Mass Spectrometry, Optimization, Plasma, Sampling

Application Code: Other

Methodology Code: Mass Spectrometry

Session # 2240 Abstract # 2240-10 **Poster Sessions**

Session Title: Application of Mass Spectrometry

Abstract Title: **Visualizing the Distribution of Volatile and Semi-Volatile Compounds by Low Temperature Plasma Mass Imaging (LTP-MSI)**

Date: Thursday, March 10, 2016 - Mornin
Time:
Room: Exposition Floor, 400 Aisle

Primary Author: Robert Winkler
CINVESTAV Unidad Irapuato

Co-Author(s): Abigail Moreno Pedraza, Sandra Martinez Jarquín

Abstract Text

Mapping organic compounds with low vapor pressure by mass spectrometry (MS) requires an ionization method, which operates at ambient pressure and temperature. Therefore, we developed a low temperature plasma (LTP) probe, which allows the detection of molecules from organic surfaces [1]. Moving the sample in x- and y- direction enables the creation of molecular images [2]. From 2D coordinates and their respective mass spectra, .imzML data files are generated, which can be analyzed for the distribution of known and unknown molecules [3].

To improve the spatial resolution of our initial prototypes, we constructed a sampling/ imaging robot in RepRap (3D printer) technology (Fig.1 A). The robot can be controlled by open source software (Pronterface) and simple G-code programs (Fig.1 B). Further, we use 3D printed components to optimize the LTP probe (Fig.1 C). The modular and open design of our system allows various configurations and can be adjusted to different mass analyzers (Fig. 1) and (bio)analytical questions.

[1] S. Martínez-Jarquín, R. Winkler, Rapid Commun. Mass Spectrom. 27 (2013) 629–634. [2] M. Maldonado-Torres, J.F. López-Hernández, P. Jiménez-Sandoval, R. Winkler, J. Proteomics. 102 (2014) 60–65. [3] R. Gamboa-Becerra, E. Ramírez-Chávez, J. Molina-Torres, R. Winkler, Anal. Bioanal. Chem. 407 (2015) 5673–5684.

We acknowledge the support by the CINVESTAV Agency 3C, MakerMex and the funding by the CONACYT scholarships, grants I0017/CB 2010 01/151596 and FINNOVA I010/260/2014.

Keywords: Biological Samples, Imaging, Organic Mass Spectrometry, Surface Analysis

Application Code: Bioanalytical

Methodology Code: Surface Analysis/Imaging

Session # 2250	Abstract # 2250-1	Poster Sessions

Session Title: Fluorescence and Luminescence

Abstract Title: **Determination of Thiamine by Flow Analysis System Based on the Chemiluminescence Inhibition Using Multicommutation**

Primary Author: Deborah Azzi
Federal University of São Carlos

Date: Thursday, March 10, 2016 - Mornin
Time:
Room: Exposition Floor, 400 Aisle

Co-Author(s): Bruno C. Janegitz, Geiser Oliveira, Marina Baccarin, Marina Batistão, Orlando Fatibello-Filho

Abstract Text

The development of analytical procedures using chemiluminescence inhibition for the determination of thiamine by flow analysis system is described. The first experiment is based on the inhibition of the chemiluminescent signal from the reaction between luminol and sodium hypochlorite by adding thiamine to react with luminol. Under optimal experimental conditions, the analytical curve was linear in the thiamine concentration range from 5.0×10^{-5} to 5.0×10^{-4} mol L^{-1}, with a detection limit of 2.3×10^{-5} mol L^{-1} and an analytical frequency of 144 h^{-1}. A second flow-injection method for thiamine based on the inhibition of chemiluminescent reaction between the luminol and hydrogen peroxide solution, catalyzed by potassium hexacyanoferrate (III) (preliminarily reacted to the thiamine) is proposed. The analytical curve presented a linear range from 1.0×10^{-5} to 1.0×10^{-4} mol L^{-1}, with a detection limit of 7.4×10^{-6} mol L^{-1} and an analytical frequency of 100 h^{-1}. The proposed methods were success applied for the thiamine determination in pharmaceutical samples, achieving recovery values between 97.6 and 106 % that indicated no matrices effect. The developed procedures presented high selectivity, low reagent consumption, high analytical frequency; besides, it used deionized water as carrier solution, generating low toxic waste and economy.

Keywords: Chemical, Optimization, Quality, Quantitative
Application Code: Pharmaceutical
Methodology Code: Fluorescence/Luminescence

Session # 2250 Abstract # 2250-2 **Poster Sessions**

Session Title Fluorescence and Luminescence

Abstract Title **Selection of Aptamers Targeting B-Cell Receptor (BCR) Using Antibody Guided Cell-Selex Method - A Novel Approach**

Primary Author Shomi Chakrabarti
City University of New York, The Graduate Center

Date: Thursday, March 10, 2016 - Mornin
Time:
Room: Exposition Floor, 400 Aisle

Co-Author(s) George Maio, Hasan E. Zumrut, Mst Naznin Ara, Prabodhika Mallikaratchy

Abstract Text

Nucleic acid aptamers are single stranded DNA or RNA molecules that have stable three-dimensional structures. Aptamers show specific binding to wide-range of target molecules. Due to high specificity and affinity of aptamer molecules to their targets, aptamer molecules are considered as antibody analogues. Aptamers are generated by a process named "systematic evolution of ligands by exponential enrichment" (SELEX). We have developed a new method to select aptamers towards specific target receptors. To demonstrate this novel method, we utilized monoclonal antibody binding to membrane IgM (mIgM) as a guide to select aptamers specific for mIgM. The target, mIgM is highly expressed on a Burkitt's lymphoma cell line and play a significant role in B-cell development. Aptamers selected by mAb-guided-SELEX show high specificity towards mIgM expressed in cultured B-cell lymphomas and in primary B-cells obtained from healthy individuals. While these aptamers show superior specificity, the affinity of the aptamers are in the comparable range of affinities with the aptamers evolved from traditional SELEX methods. Also, a cocktail containing four of the identified aptamers specific for mIgM, can block the cognate monoclonal antibody binding to its target, which demonstrates that aptamers can be selected using this strategy against a pre-determined entity on cell surface receptors.

Keywords: Bioanalytical, Biomedical, Fluorescence, Method Development
Application Code: Biomedical
Methodology Code: Fluorescence/Luminescence

Session # 2250 **Abstract # 2250-3** **Poster Sessions**

Session Title: Fluorescence and Luminescence

Abstract Title: **Spectral Encoders: Detecting Position-Dependent Luminescent Spectra Through Tissue**

Primary Author: Melissa M. Rogalski
Clemson University

Date: Thursday, March 10, 2016 - Mornin
Time:
Room: Exposition Floor, 400 Aisle

Co-Author(s): Bobby Smith, Hunter Pelham, Jeffrey N. Anker, John D. DesJardins, Nakul Ravikumar

Abstract Text

We have developed a non-invasive method for monitoring local chemical and physical changes including pH and strain through tissue on the surface of implanted orthopedic devices. Our sensors monitor change by measuring position dependent luminescence through tissue. The sensors contain a substrate patterned with luminescent materials that is overlaid with an analyzer mask containing opaque regions and transparent windows or holes. Position of the analyzer mask above the luminescent layer controls the resultant wavelength of light visible through the openings. The position-modulated signal is collected by spectrometer and a spectral intensity ratio from closely spaced peaks is calculated. We have demonstrated this concept using x-ray excited optical luminescence and fluorescence spectral encoders through 6 mm of chicken breast. We have also extended this methodology to upconversion luminescence which provides an essentially background-free signal through tissue. In addition, we demonstrate proof of concept for the use of magnetic actuation to control position of the mask and modulate optical signals. We demonstrate how this approach can be used to reject background for pH measurements utilizing a fluorescent spectral encoder.

Keywords: Biomedical, Fluorescence, Luminescence, Spectroscopy
Application Code: Biomedical
Methodology Code: Fluorescence/Luminescence

Session # 2250 Abstract # 2250-4 **Poster Sessions**

Session Title: Fluorescence and Luminescence

Abstract Title: **Photoluminescence of Novel Osmium and Ruthenium Complexes in the Presence of Polyanions**

Primary Author: Mehrun Uddin
St. John's University

Date: Thursday, March 10, 2016 - Mornin
Time:
Room: Exposition Floor, 400 Aisle

Co-Author(s): Armando Seitllari, Besiana Kurti, Cody Piotrowski, Elise Megehee, Enju Wang

Abstract Text

DNA detection is important to many biological processes. DNA biosensors are increasingly used in hybridization reactions, mutation detection, genomic sequencing, and identification of pathogens. The macromolecular polysaccharide-based polyanions, including heparin salts, dextran sulfate, carrageenan, and chondroitin sulfate have unique properties and functions in physiology, microbiology and food technology. The quantity of polyanion reflecting the exact number of charges in samples administered in biological procedures has to be strictly controlled. Thus the detection of these polyanions in clinical or commercial samples is key in the diagnostic and quality control processes of related fields.

A series of Os(II) carbonyl complexes with two phenanthroline, and a variety of the sixth ligand exhibit moderate to high luminescent emission intensity in the visible region. The emission intensity of these complexes is significantly altered with the addition of heparin or other polysaccharide based polyanions. The complexes were successfully applied for the detection of injectable heparin preparations and chondroitin in joint support supplement tablets. Our recent results show that the luminescence intensity of these osmium complexes can be reduced or enhanced by different DNA strands. DNA sensing using these complexes are highly promising.

In the interests of reducing the cost of the luminescence markers for Heparin and DNA sensing, we synthesized the Ru(II)CO complex analogs and evaluated their luminescence characteristics. This presentation compares the luminescence behaviors of the Os(II) and Ru(II) complexes in the presence of various polyanions, in the effort of developing heparin and DNA detection markers.

Keywords: Bioanalytical, Spectroscopy, UV-VIS Absorbance/Luminescence
Application Code: Bioanalytical
Methodology Code: Fluorescence/Luminescence

Session # 2250 Abstract # 2250-5 **Poster Sessions**

Session Title: Fluorescence and Luminescence

Abstract Title: **Quantifying Free Fatty Acid Uptake Dynamics in Primary Adipocytes Using Customized Micro-Wells Made with 3D-Printed Templates**

Primary Author: Tesfagebriel M. Hagos
Auburn University

Co-Author(s): Christopher J. Easley, Jessica C. Brooks

Date: Thursday, March 10, 2016 - Mornin
Time:
Room: Exposition Floor, 400 Aisle

Abstract Text

Free fatty acid (FFA) uptake by adipocytes in response to nutritional or hormonal stimuli is a key process in the complex metabolic cascade leading to diabetes and obesity. Improved methods for measuring FFA uptake dynamics will enhance our understanding of these disease states. We employed a fluorescently labeled FFA analog, BODIPY-hexadecanoic acid (i.e. BODIPY-palmitic acid), to directly follow FFA uptake by adipocytes under various treatments using a multi-well plate reader. First, we validated the FFA uptake assay using manual pipetting of solutions incubated with primary murine adipocytes, and we observed a rapid uptake in <30 min. These results confirmed that cellular mechanisms for FFA uptake were intact and functional, and the dynamics agreed with recent work by others. However, to achieve a real-time FFA uptake assay, manual pipetting steps must be excluded. Toward this goal, a modification of micro-well architecture was required to block optical excitation of adipocytes while preserving interrogation of solution fluorescence. 3D-printed templates were used as molds for polydimethylsiloxane (PDMS) micro-wells, allowing rapid prototyping of various architectures. Optimized PDMS microplates included 24 wells (7.5 mm diameter) with tissue culture moats on well floors, 3D-printed caps (polylactic acid, or PLA) with 2-mm diameter pinholes, and a 3D-printed microplate base replica. These hybrid PDMS/PLA microplates allowed interrogation of the solution but prevented excitation of moat regions, which should allow real-time quantification of FFA uptake dynamics, along with minimizing photo toxicity of the cells. The system is also well-poised for future integration with microfluidic channels.

Keywords: Fluorescence, Lipids, Microspectroscopy, UV-VIS Absorbance/Luminescence
Application Code: Bioanalytical
Methodology Code: Fluorescence/Luminescence

Session # 2250 Abstract # 2250-6 Poster Sessions

Session Title: Fluorescence and Luminescence

Abstract Title: **Toward Non-Invasively Detecting Radiolabeled Analytes Near Implanted Medical Devices Coated in Radioluminescent Phosphors**

Primary Author: Gretchen B. Schober
Clemson University

Date: Thursday, March 10, 2016 - Mornin
Time:
Room: Exposition Floor, 400 Aisle

Co-Author(s): Jeffrey N. Anker

Abstract Text

The objective of this project is to provide proof of principle for a modified scintillation proximity method to visualize and quantify antibiotic penetration of bacterial biofilms. The modified scintillation proximity method consists of an immobilized Gd[sub]2[/sub]O[sub]2[/sub]S:Eu[sup]3+[/sup] scintillator film that emits visible light when excited by subatomic particles of decaying radioactive atoms in close proximity. Emitted light intensity correlates with the concentration of radioisotope within a calculable distance of the scintillator film, which depends on the subatomic particle decay range of the radioisotope used. The signal intensity and location are simultaneously recorded using a Spectrum Preclinical In Vivo Imaging System (IVIS). In application, radiolabelled antibiotics will be used to quantify location and depth of antibiotic penetration through a bacterial biofilm to an underlying scintillator film. As an extension of this study, the method will be altered to monitor antibiotic delivery and release by coupling Gd[sub]2[/sub]O[sub]2[/sub]S:Eu[sup]3+[/sup] microparticles to radiolabelled antibiotics. Antibiotic release from microparticles will be quantified by detecting a decrease in emission signal intensity. While antibiotic penetration of bacterial biofilms has been investigated, quantitative data regarding concentration, penetration depth, and pharmacokinetics has yet to surface. As a result, reduced antibiotic susceptibility in bacterial biofilms remains unexplained and biofilm formation on biomedical implants continues to be difficult to eradicate non-invasively. Ultimately this modified scintillation proximity method provides a way to acquire quantitative data both [i]in vitro[/i] and [i]in vivo[/i]. This will reveal whether limited antibiotic penetration is a factor reducing bacterial biofilm antibiotic susceptibility.

Keywords: Bioanalytical, Luminescence, Quantitative
Application Code: Bioanalytical
Methodology Code: Fluorescence/Luminescence

Session # 2250 Abstract # 2250-7 **Poster Sessions**

Session Title: Fluorescence and Luminescence

Abstract Title: **Development of Silicone Filled Optical Module for Laser Fluorescence Trace Molecular Detection**

Primary Author: Hirokazu Higuchi
Kyushu University

Date: Thursday, March 10, 2016 - Mornin
Time:
Room: Exposition Floor, 400 Aisle

Co-Author(s): Hiroaki Nomada, Hiroaki Yoshioka, Kinichi Morita, Yuji Oki

Abstract Text

Portable spectroscopy systems have been studied by many research groups[1,2]. The LIF (Laser Induced Fluorescence) method is useful for immunoassay[3], quench bodies [4] and so on. However, high sensitivity requires a very dark background, so reduction of laser stray lights and background radiation is the most important developing point. Though the blank light pass was normally adopted, the non-blank optical system such as microfluidic chips or fiber optical circuits, causes severe internal scattering noise.

Recently, we proposed novel concept SOT (silicone optical technology) based on PDMS matrix. SOT provides good noise suppression due to low and homogeneous scattering at PDMS / absorbing PDMS composite[5]. An absorption measurement palmtop systems was also demonstrated[6]. In this report, LIF module based on SOT was proposed and demonstrated.

The figure shows the schematic of the SOT-LIF module. The module size was only 54x60x15 mm3. A battery driven Nd:YVO4 laser and a battery driven photomultiplier tube were integrated into hybrid PDMS chamber. The chamber was composed of carbon particle dispersed PDMS and conclude optical path folded to reduce the internal reflection noise. A notch filter and a green cut filter were inserted to remove the laser stray light.

As an evaluation, the fluorescence of Rhodamine6G water solution was assumed as detection sample. The detection procedure is as follows: pipette the sample solution in PCR tube and put the tube in the mounting spot on the module. From the calibration curve, the sample of the concentration of 1nM showed a clear difference from the black sample.
1: R. H. Byrne et al.:Talanta 50 (2000) 1307
2: F. J. Fortes, J.J. Laserna.: Spectrochimica Acta Part B 65 (2010) 975
3: A. J. Ozinskas et al.: Anal Biochem. 213(1993) 264
4: R. Abe et al.: J. Am. Chem. Soc. 133, (2011)17386
5: K. Morita et al.: Flow Analysis XIII (2015) P25
6: H. Nomada et al.: Pittcon (2014) 1450-1

Keywords: Bioanalytical, Biospectroscopy, Fluorescence, Raman
Application Code: Bioanalytical
Methodology Code: Fluorescence/Luminescence

Session # 2250 Abstract # 2250-8 **Poster Sessions**

Session Title: Fluorescence and Luminescence

Abstract Title: **Recombinant Tobacco Peroxidase: A 100-Fold More Effective Luminescent Label Than Horseradish Peroxidase**

Date: Thursday, March 10, 2016 - Mornin
Time:
Room: Exposition Floor, 400 Aisle

Primary Author: Irina Gazaryan
Pace University

Co-Author(s): Andrey Poloznikov, Dmitry Hushpulian, Galina Zakharova, Vladimir Tishkov

Abstract Text

Anionic tobacco peroxidase (TOP) is superior over horseradish peroxidase (HRP) in the chemiluminescent assay: HRP needs enhancer substrates whereas TOP is more active than HRP just with luminol as the only substrate. A newly optimized refolding protocol permits production of 60 mg active TOP enzyme per 1L of E.coli culture, with the refolding yield as high as 85%. Such quantity is comparable with that from transgenic plants overproducing tobacco peroxidase. The cost of production of recombinant TOP is comparable to that for native HRP. Availability of large quantities of TOP permits its direct comparison with HRP in various assay formats. The performance of TOP-IgG conjugates was compared to that of HRP-IgG using a commercial kit for IgG assay (FastELISA, RD-Biotech, France). Enzyme immunoassay kits based on HRP contain native HRP conjugated with antibodies via modification of oligosaccharide chains in HRP. To conjugate IgG with recombinant TOP that has no an "oligosaccharide fur", the method based on TOP amino group derivatization with sulfo-SMCC followed by binding to SATA-activated antibodies has been used. The conjugation method preserves TOP activity unchanged. Assay of IgG content with TOP-IgG and HRP-IgG shows that TOP-based assay results in a 3 fold increase in colorimetric signal, and a 110-fold increase in a luminescent signal. An additional advantage is that the linear range of the calibration plot is so wide that there is no need to dilute the sample before the measurements. The use of recombinant TOP warrants both high sensitivity and lower detection limits than the assay based on HRP. Thus, the recombinant TOP now available in large quantities is a much better chemiluminescent label than HRP, providing the better assay characteristics and avoiding the need in a specific formulation of the assay mixture, since it does not need an enhancer substrate.

Keywords: Bioanalytical, Enzyme Assays, Luminescence, UV-VIS Absorbance/Luminescence
Application Code: Bioanalytical
Methodology Code: Fluorescence/Luminescence

Session # 2250 Abstract # 2250-9 **Poster Sessions**

Session Title: Fluorescence and Luminescence

Abstract Title: **Using Diffusional Motion to Gauge Fluidity and Interfacial Adhesion Strength of Supported Octadecylphosphonic Acid (ODPA) Monolayers**

Date: Thursday, March 10, 2016 - Mornin

Time:

Room: Exposition Floor, 400 Aisle

Primary Author: Stephen Baumler
Michigan State University

Co-Author(s)

Abstract Text

We report on the use of diffusion measurements to gauge the fluidity and surface binding properties of a molecular monolayer. The monolayer film consists of octadecyl-1-phosphonic acid (ODPA) and controlled amounts of a lyso-phosphatidic acid tagged with the fluorescent probe BODIPY (BLPA). The monolayer films were formed using a Langmuir-Blodgett (LB) trough and deposited onto a glass slide. Monolayer morphology was characterized during film formation using Brewster angle microscopy (BAM). Fluorescence Recovery After Photobleaching (FRAP) microscopy was used to measure translational diffusion of BLPA and Time-Resolved Fluorescence Anisotropy Imaging (TR-FAIM) was used to measure rotational diffusion of the BLPA chromophore. These results provide information on the motional freedom of the probe and, importantly, on the strength of interaction between the probe and the support. Compositional variations in the monolayer give rise to changes in constituent dynamics that reflect intermolecular interactions.

Keywords: Fluorescence, Imaging, Spectroscopy, Surfactants
Application Code: Material Science
Methodology Code: Fluorescence/Luminescence

Session # 2250 Abstract # 2250-11 **Poster Sessions**

Session Title: Fluorescence and Luminescence

Abstract Title: **Irradiation of Gold Nanodots by Ultraviolet Light: Modulation of Ligand Density and Photoluminescence**

Primary Author: Yu-Ting Tseng
National Taiwan University

Date: Thursday, March 10, 2016 - Mornin
Time:
Room: Exposition Floor, 400 Aisle

Co-Author(s)

Abstract Text

In this study, we developed a novel, simple and convenient method to tune the photoluminescence (PL) properties of gold nanodots (Au NDs). The (11-mercaptoundecyl)-N, N, N-trimethylammonium bromide (11-MUTAB)–Au NDs (11-MUTAB–Au NDs) obtained from the reaction of gold nanoparticles (ca. 3 nm) and 11-MUTAB; they exhibited weak emission [quantum yield (QY): 0.01%] at 700 nm upon excitation at 365 nm. We found 11-mercaptoundecanoic acid (11-MUA) could rapidly self-assemble (<30 min) on 11-MUTAB–Au NDs under the irradiation by ultraviolet (UV) light. The as-prepared 11-MUTAB–Au NDs@11-MUA exhibited superior PL intensity (QY: 2.83%) at 520 nm. The dramatic changes of PL properties of Au NDs were due to its smaller particle sizes, higher surface ligand densities as well as surface Au oxidation states. Our findings revealed that UV-mediated the self-assembly of thiol lignands on Au NDs can rapid modulate the PL properties of Au NDs.

Keywords: Fluorescence, Material Science, Nanotechnology, Surface Analysis
Application Code: Material Science
Methodology Code: Fluorescence/Luminescence

Session # 2250 Abstract # 2250-13 **Poster Sessions**

Session Title: Fluorescence and Luminescence

Abstract Title: **Traceable Mercury Gas Phase Calibrations Based Upon Gravimetry**

Primary Author: Annarita Baldan
VSL

Date: Thursday, March 10, 2016 - Mornin
Time:
Room: Exposition Floor, 400 Aisle

Co-Author(s): Hugo Ent, Janneke van Wijk

Abstract Text

Within the EMRP (European Metrology Research Programme) project MeTra a mercury vapor generator is being developed to establish traceability of mercury vapor measurement results, based upon gravimetry, for ambient air levels as well as higher concentrations. Current measurement capabilities are maintained at levels of 0,25 – 350 µg Hg/m3, whereas the aim of the new gravimetric primary standard is to realize traceability for the range 5 ng Hg/m3 - 60 µg Hg/m3, covering key requirements for ambient air monitoring (1 - 2 ng Hg/m3), health-based exposure standards (50 ng Hg/m3), concentrations relevant to stationary source emissions (upwards of 1 µg Hg/m3), the minimum alveolar concentration value (20 µg Hg/m3), and also affording a comparison with the currently used mercury vapor equation based calibration concentrations. In order to realize traceability starting at this low mercury vapor content level, we developed a strongly modified type of diffusion tube and a new measurement method to weigh the loss in (mercury) mass of these diffusion tubes during use (ca. 6-8 µg mass difference between successive weighings). The specifics of the newly introduced capabilities, to calibrate mercury monitors at levels of 10 ng Hg/m3 and higher, will be highlighted. These new capabilities are especially important for measurement services testing indoor and work place related mercury content levels according to health standards, but will also contribute to higher safety standards and cost reductions in e.g. the LNG field, where aluminium main cryogenic heat exchangers are used which are particular prone to corrosion caused by mercury.

Keywords: Calibration, Mercury, Monitoring, Validation
Application Code: Environmental
Methodology Code: Fluorescence/Luminescence

Session # 2250 Abstract # 2250-14 **Poster Sessions**

Session Title: Fluorescence and Luminescence

Abstract Title: **Detection of Caffeine Using a Field-Portable Fluorescence Device**

Primary Author: Haley Curtis
Tennessee Technological University

Date: Thursday, March 10, 2016 - Mornin
Time:
Room: Exposition Floor, 400 Aisle

Co-Author(s): Andrew Callender

Abstract Text

Caffeine is an interesting molecule due to its inexpensive and prevalent nature. We will present our work thus far on a technique to detect caffeine in coffee and tea. Our approach is based on using a simple extraction, along with a fluorescent dye to be used in a simple field-portable device. The fluorescent dye, acridine orange forms aggregates in water that show reduced fluorescence intensity. Caffeine disrupts the aggregates, increasing the fluorescence. This allows us to detect caffeine with millimolar sensitivity, suitable for determination of caffeine in beverages. The device consists of an LED source, a simple plastic casing, and a detector; the detector can be either a biased silicon photodiode or a CdS photocell resistor. We will compare the performance of the two detectors to confirm that the inexpensive photocell gives acceptable performance. The application of this device will be in an educational setting to help promote the interest of high-school students in the wide array of possibilities that arise from the study of chemistry, and specifically analytical chemistry.

Keywords: Beverage, Education, Food Science, Portable Instruments
Application Code: Consumer Products
Methodology Code: Fluorescence/Luminescence

Session # 2250 Abstract # 2250-15 **Poster Sessions**

Session Title Fluorescence and Luminescence

Abstract Title **The Quenching of Riboflavin Fluorence by Nicotine in Bicontinuous Microemulsion**

Date: Thursday, March 10, 2016 - Mornin
Time:
Room: Exposition Floor, 400 Aisle

Primary Author Maurice O. Iwunze
Morgan State University

Co-Author(s)

Abstract Text
Steady-state fluorescence spectroscopy was used to study the quenching of riboflavin fluorescence by nicotine in a bicontinuous microemulsion system made up of 42:13.7:21.34:22.85 % w/w of water:oil:surfactant:co-surfactant. The surfactant used is cetyltrimethylammonium bromide (CTAB), the oil is tetradecane and the co-surfactant is 1-pentanol. In this medium the observed rate constant, kq, of the riboflavin fluorescence quenching by nicotine was observed as 5.14 x 109 M-1s-1. The electron transfer rate constant, kET, was calculated as 2.79 x 109 s-1 with an activation rate constant, ka, of 4.51 x 109 s-1. The calculated solvent reorganization energy, s, of the reaction was 0.76 eV. The mechanism of the reaction is proposed to be activated electron transfer in a solvent separated (outer-sphere) scheme within a diffusion limited regime.

Keywords: Spectroscopy
Application Code: Bioanalytical
Methodology Code: Fluorescence/Luminescence

Session # 2250 **Abstract # 2250-16** **Poster Sessions**

Session Title: Fluorescence and Luminescence

Abstract Title: **Optimizing Fluorescence Tagging Strategy to Study Single Molecule Diffusion on Air-Solid Surface**

Primary Author: Tao Jin
North Carolina State University

Date: Thursday, March 10, 2016 - Mornin
Time:
Room: Exposition Floor, 400 Aisle

Co-Author(s): Fang Chen, Gufeng Wang, James Tour, Victor Lopez

Abstract Text

Single molecule machines such as nanocars and nanomotors perform their namesake functions at the molecular level. Our eventual goal of developing nanocars is to utilize these molecular machines to transport cargos from one place to another. To achieve this goal, the first step is to monitor their movement at single molecule level. Single molecule fluorescence microscopy (SMFM) is a non-intrusive method that provides sufficient spatial and time resolution to detect and track single molecule machines on air-solid interface. However, single molecule fluorescence microscopy suffers from photobleaching. In this study, we investigate the photostability of different molecules tagged with two bodipy dyes on air-glass surface. Interestingly, we found that the molecules show different blinking and bleaching behaviors when the bodipy fluorophores are conjugated or not. Singlet-triplet annihilation and Forster energy transfer model between the homo-dye pairs were proposed and compared to explain the self-quenching mechanism. Distance relied oxygen radical diffusion was proposed to explain the bleaching kinetics. With the better understanding of the photostability for the dye-tagged molecules, we are able to better design fluorophores in single molecule tracking.

Keywords: Fluorescence, Laser, Method Development, Spectroscopy
Application Code: Process Analytical Chemistry
Methodology Code: Fluorescence/Luminescence

Session # 2260 Abstract # 2260-1 **Poster Sessions**

Session Title: Fuels, Energy and Petrochemical

Abstract Title: **Increasing Resolution of Propylene Glycol Impurities with High-Efficiency GC Columns**

Date: Thursday, March 10, 2016 - Mornin

Time:

Room: Exposition Floor, 400 Aisle

Primary Author: Ramkumar Dhandapani
Phenomenex

Co-Author(s): Kristen Parnell, Tim Anderson

Abstract Text

Changes in GC column efficiency can significantly impact separations, even when selectivity is held constant. This work explores the impact of changes in column dimensions and therefore efficiency on column resolution, using propylene glycol impurity testing as an example. The experiments were performed on a Zebron™ ZB-WAXplus™ GC column with large internal diameter and thick film, and compared with a smaller internal diameter and thin film. Enhanced resolution was observed for 2,4-pentanediol-1, 1,2-butanediol, and 2,4-pentanediol-2 using the thin film and smaller ID column. Stationary phase interactions, mass transfer, and the impact of variations in these parameters on resolution are discussed. The study illustrated that careful selection of high efficiency column dimension would help in enhancing the resolution of close eluting analytes; aqueous stability and its benefits Additionally, maintaining high resolution without altering column efficiency over time for aqueous propylene glycol samples.

Keywords: Fuels\Energy\Petrochemical, Gas Chromatography, GC, GC Columns

Application Code: Fuels, Energy and Petrochemical

Methodology Code: Gas Chromatography

Session # 2260 Abstract # 2260-2 **Poster Sessions**

Session Title Fuels, Energy and Petrochemical

Abstract Title **Analysis of Impurities in Propane/Propylene Streams Using a Pulsed Flame Photometric Detector (PFPD)**

Date: Thursday, March 10, 2016 - Mornin
Time:
Room: Exposition Floor, 400 Aisle

Primary Author Cynthia Elmore
OI Analytical

Co-Author(s) Brian Mistovich, J Garrett Slaton, Michael Duffy

Abstract Text

Some of the key processes in the petrochemical industry are conversions of high-grade ethylene (C2) and propane/propylene (C3) feedstocks into end products (polyethylene , polypropylene) and intermediates such as 1-butene . These , of course , are the building blocks for plastics and a wide range of products , and are a large industry , with 55 million metric tons of polypropylene produced in 2013 . Unfortunately , even trace levels of sulfur species H2S and COS , which are often entrained in C2 and C3 feedstocks , corrode pipes and equipment , inhibit or damage catalyst beds , and lower product yield and purity . The need for a fast , reliable method for H2S and COS in C2 and C3 feedstocks is obvious , but sulfur in C3 is a difficult application , owing to the poor separation of the impurities from the matrix when coupled with the quenching of the PFPD detector signal by propane/propylene . We present here a fast , reliable and robust method for the analysis of sulfur contaminants in C3 feedstocks that makes use of an automated gas loop injection system , separation by gas chromatography , and pulsed-flame photometric detection that can detect sulfur at better than 0.1ppmv .

Keywords: Fuels\Energy\Petrochemical, Gas Chromatography, GC Detectors, Sulfur
Application Code: Fuels, Energy and Petrochemical
Methodology Code: Gas Chromatography

Session # 2260 Abstract # 2260-4 **Poster Sessions**

Session Title Fuels, Energy and Petrochemical

Abstract Title **High Performance Chromatographic Diatomaceous Earth (DE)**

Primary Author Katarina Oden
Restek

Date: Thursday, March 10, 2016 - Mornin
Time:
Room: Exposition Floor, 400 Aisle

Co-Author(s) Barry Burger, Jaap de zeeuw, Linx Waclaski, Rebecca Stevens

Abstract Text

The use of Diatomaceous Earth (DE) solid support media for packed columns was predicted to be obsolete in the 1980's with the introduction of glass capillary columns, followed by fused silica capillary columns. However, the packed column is still thriving today in the petroleum and pharmaceutical markets. Within the last decade, the availability of inert, chromatographic grade DE has been limited. We evaluated several commercially available DE packed columns. Comparison of surface area, particle size distribution, density and pore size of these highly efficient packed columns was performed. This poster will highlight the chromatographic advantages of this material as a packed column support.

Keywords: Fuels\Energy\Petrochemical, Gas Chromatography, Materials Characterization, Solution
Application Code: Fuels, Energy and Petrochemical
Methodology Code: Gas Chromatography

Session # **2260** **Abstract #** **2260-5** **Poster Sessions**

Session Title: Fuels, Energy and Petrochemical

Abstract Title: **Methanizer – A Simple Solution for CO/CO2 Gas Analysis by Gas Chromatography (GC)**

Primary Author: Katarina Oden
Restek

Date: Thursday, March 10, 2016 - Mornin
Time:
Room: Exposition Floor, 400 Aisle

Co-Author(s): Barry Burger, Jaap de zeeuw, Linx Waclaski, Rebecca Stevens

Abstract Text

CO, CO_2 and other permanent gases are common components in refinery gases, natural gases, fuel cell gases, greenhouse gases and are present in many other industrial processes. Accurate and precise determination of the final components can save industries millions of dollars per year.

A novel approach has been taken towards the analysis of CO and/or CO2. A modern, user friendly and accurately controlled Methanizer was developed that can be easily integrated within any existing GC/FID system. The Methanizer consists of a Ni-catalyst in a heated chamber that converts CO and/or CO2 in presence of hydrogen into a FID detectable gas, methane. The GC independent controller runs a PID algorithm to proportionately control the heating of the Ni-catalyst to ±1°C, assuring controlled and reliable conversion of CO and/or CO2. CO and/or CO2 analytical solutions, robustness and system performance will be presented.

Keywords: Environmental Analysis, Fuels\Energy\Petrochemical, Gas, Gas Chromatography
Application Code: Fuels, Energy and Petrochemical
Methodology Code: Gas Chromatography

Session # 2260 Abstract # 2260-6 **Poster Sessions**

Session Title: Fuels, Energy and Petrochemical

Abstract Title: **Analysis of A Sulfur Mixture in Hydrocarbon Standard Cylinder in ppb Level**

Primary Author: Yang Qin
Air Liquide Specialty Gases

Date: Thursday, March 10, 2016 - Mornin
Time:
Room: Exposition Floor, 400 Aisle

Co-Author(s)

Abstract Text

The demands for sulfur mixture in hydrocarbon in low ppm and ppb level has increased rapidly from refinery and petroleum company. Most are multi sulfurs balance with propylene or C2 and C4, and are in gas phase. Because of the natural absorption and reaction of sulfurs, we have a difficulty to give our customer a confident certification number on sulfur mixture, especially in ppb level and in hydrocarbon matrix.

Plus, our in-house standards only has <100 pounds pressure for a gas phase. Very often, before we can give a confident certified number to the product, the standard runs out. Additional, sulfur molecules in standard absorbed on cylinder wall and reacted with moisture on the wall make our in-house standards unstable or short of shelf life.

I develop a sulfur liquid standard to certify the customer blends. By using a vaporizer and response factor, a stable and a long shelf life standards can give a sulfur mixture a confident and consistent certified number, especially in ppb level (US patent# 8,993,336B1)

Keywords: Calibration, Gas Chromatography, Standards, Trace Analysis
Application Code: Fuels, Energy and Petrochemical
Methodology Code: Gas Chromatography

Session # 2260 Abstract # 2260-7 **Poster Sessions**

Session Title Fuels, Energy and Petrochemical

Abstract Title **Reactive Pyrolysis-GC/MS of Polymers in a Steam Environment Used to Study Potential Bio-oil**

Date: Thursday, March 10, 2016 - Mornin

Time:

Room: Exposition Floor, 400 Aisle

Primary Author Karen Sam
CDS Analytical

Co-Author(s) Gary Deger, Steve Wesson

Abstract Text

The decrease of crude oil reserves, and the environmental burden of using fossil fuels has driven the need to discover new ways of producing liquid feedstock that can be used for fuels. Pyrolysis of other organic materials, like biomass, produces bio-oil, which can be used as a feedstock. However, bio-oil is often different from conventional oil. For example, it is high in oxygenated compounds, causing difficulties in equipment and refineries. Pyrolysis under reactive atmospheres and catalysts have been studied as methods of converting biomass to a more useable fuel, and pyrolysis under the presence of steam, has also been studied. In addition, reactant gases, including water influence fixed gas composition (gasification.)

Before scaling up to a bed reactor, it is already possible to investigate the conversion of biomass to bio-oil on a microscale, using reactant gases, like CO2 and H2, catalysts, and pressures up to 500psi with an analytical pyrolyzer interfaced to a GC/MS. A recently developed pyrolyzer has introduced the capability of using steam as a reactant gas. Additionally, a fixed gas analyzer can be added to study fixed gas composition.

In this study, analytical pyrolysis GC/MS and FGA (Fixed Gas Analysis) using steam as a reactant gas will be studied on a variety synthetic and natural materials (potential feedstock) to see if its presence changes pyrolysis products, creating more favorable fuel or gases.

Keywords: Biofuels, Fuels\Energy\Petrochemical, Gas Chromatography/Mass Spectrometry, Pyrolysis
Application Code: Fuels, Energy and Petrochemical
Methodology Code: Gas Chromatography/Mass Spectrometry

Session # 2260 Abstract # 2260-8 **Poster Sessions**

Session Title: Fuels, Energy and Petrochemical

Abstract Title: **Extended Lower Detection Range for Hydrogen Sulfide and Carbonyl Sulfide with Metal Surface Deactivated Sample Inlet for Micro Gas Chromatography**

Date: Thursday, March 10, 2016 - Mornin

Time:

Room: Exposition Floor, 400 Aisle

Primary Author: Thomas Szakas
Agilent Technologies

Co-Author(s): Coen Duvekot, Remko van Loon

Abstract Text

Hydrogen sulfide (H2S) and carbonyl sulfides (COS) are common sulfur gases in light hydrocarbon streams. The determination of these compounds in gaseous samples, such as natural gas or biogas, is required for multiple reasons. Next to their undesired smell, these sulfur gases have corrosive and toxic properties. Potentially causing damage to industrial pipelines. In additional, their presence can affect the performance of industrial gases, such as unwanted chemical reaction and loss of catalyst activity, ultimately resulting in lower yield.

Gas Chromatography (GC) is a commonly used technique to quantify sulfur gases like H2S and COS. On a PLOT U type capillary columns, these compounds are easily separated from C2 and C3 hydrocarbons when present at nominal concentration levels. When analyzing these sulfur gases, it's important to use deactivated sample lines because low concentration levels tend to adsorb to active spots on untreated stainless steel surfaces. This paper will highlight fast analysis of hydrogen sulfide and carbonyl sulfide using Micro Gas Chromatography (Micro GC) utilized with metal-surface deactivated sample lines. Chromatograms and repeatability figures show reliable, low ppm level analysis in under 60 seconds.

The Micro GC's compact, modular design and low carrier gas consumption enables easy implementation in process applications and mobile laboratories. Direct, on-site analysis secures the integrity of the sample. Moreover, it leads to fast availability of results. Out-of-spec values can directly be communicated and corrective actions can be taken accordingly.

Keywords: Fuels\Energy\Petrochemical, Gas Chromatography, Portable Instruments, Process Monitoring
Application Code: Fuels, Energy and Petrochemical
Methodology Code: Gas Chromatography

Session # 2260 Abstract # 2260-9 **Poster Sessions**

Session Title: Fuels, Energy and Petrochemical

Abstract Title: **The Improvement of ASTM D3612 TOGA Analysis**

Date: Thursday, March 10, 2016 - Mornin
Time:
Room: Exposition Floor, 400 Aisle

Primary Author: Max Wang
Shimadzu Scientific Instrument Inc.

Co-Author(s): Clifford M. Taylor, Marty Smith

Abstract Text

ASTM D-3612 is a GC method to extract and determine hydrogen, oxygen, nitrogen, methane, carbon monoxide, C2-C3 in hydrocarbon-based transformer oil. Three procedures have been described based on a gas chromatogram with TCD, and Methanizer FID. In this work, two improvements have been made by employing a universal Barrier Ionization Discharge (BID) detector in place of the TCD: (a) hydrogen detection levels have been lowered from 10ppm to 100ppb; and (b) propane and butane have been incorporated in the chromatograms.

Keywords: Energy, Gas Chromatography, GC, Instrumentation
Application Code: Fuels, Energy and Petrochemical
Methodology Code: Gas Chromatography

Session # 2260 Abstract # 2260-10 **Poster Sessions**

Session Title: Fuels, Energy and Petrochemical

Abstract Title: **Accurate and Reproducible Determination of Halogens in Coal Using Combustion Ion Chromatography**

Primary Author: Carl Fisher
Thermo Fisher Scientific

Date: Thursday, March 10, 2016 - Mornin
Time:
Room: Exposition Floor, 400 Aisle

Co-Author(s): Adelon Augustin, Daniel Khor, Kirk Chassaniol, Mark Manahan

Abstract Text

When coal is burned, the exhaust contains halogens that can be released into the atmosphere as acidic gases (such as hydrochloric acid (HCl) and hydrofluoric acid (HF)), which can result in the production of acid rain. Mercury (Hg), another toxic pollutant, is also present in coal exhaust and the ability of pollution control devices to capture this element from flue gases is enhanced in the presence of halogens. The adverse environmental impacts from coal burning by power plants resulted in the U.S. Environmental Protection Agency of establishing the Mercury and Air Toxics Standards (MATS) rule to reduce emissions of Hg and other toxic pollutants. Combustion Ion Chromatography (CIC) methods have been developed to measure trace halogens and sulfur (as sulfate) in non-water soluble solids and semi-solids such as organic solvents, plastics, polymers, rubber, wood chips, and petroleum liquids. This presentation describes the determination of total fluorine, chlorine, and bromine in coal using an automated approach to combustion sample preparation in combination with ion chromatography to deliver results with excellent accuracy and reproducibility, while eliminating the tedium of manual combustion methods.

Keywords: Energy, Fuels\Energy\Petrochemical, Ion Chromatography, Sample Preparation
Application Code: Fuels, Energy and Petrochemical
Methodology Code: Liquid Chromatography

Session # 2260 Abstract # 2260-11 **Poster Sessions**

Session Title: Fuels, Energy and Petrochemical

Abstract Title: **Nitrogen Determination in Lubricant by Elemental Analyzer**

Date: Thursday, March 10, 2016 - Mornin
Time:
Room: Exposition Floor, 400 Aisle

Primary Author: Guido Giazzi
Thermo Fisher Scientific

Co-Author(s): Francesco Leone, Liliana Krotz

Abstract Text

In a typical production process of mineral oils, the nitrogen content is periodically monitored and tested for quality control. The reproducibility of data, which is measured as deviation of results from the mean value, is one of the main targets in all analytical tests for the alternative techniques accepted. The method for nitrogen analysis in lubricants, based on combustion, is described in ASTM D5291. For this reason the use of an accurate instrumental analytical techniques is required. As the demand for improved sample throughput, reduction of operational costs and minimization of human errors is becoming every day more notable, it is very important apply a simple and automatic technique which allows the fast analysis with excellent reproducibility. The FLASH 2000 Analyzer, using typically Helium gas carrier and based on the dynamic flash combustion of the sample, copes effortlessly with the wide array of laboratory requirements such as accuracy, day to day reproducibility and high sample throughput. However as in the last years there is a possible worldwide shortage and large cost increasing for Helium, it is necessary to test as alternative gas, Argon which is readily available. This paper presents data on Nitrogen determination in lubricants with different Nitrogen concentration to show the performance of the system using Argon as carrier gas and the reproducibility of the results obtained.

Keywords: Characterization, Elemental Analysis, Petrochemical
Application Code: Fuels, Energy and Petrochemical
Methodology Code: Atomic Spectroscopy/Elemental Analysis

Session # **2260** Abstract # **2260-12** **Poster Sessions**

Session Title Fuels, Energy and Petrochemical

Abstract Title **New Methodology for the Analysis of Silicon in Petrochemical Samples by ICP-MS**

Primary Author Anthony Palermo
PerkinElmer

Date: Thursday, March 10, 2016 - Mornin
Time:
Room: Exposition Floor, 400 Aisle

Co-Author(s) Kenneth Neubauer, Stan Smith

Abstract Text

The analysis of trace levels of silicon by ICP-MS is challenging for several reasons: contamination, interferences, and sample-to-sample matrix variations. The main isotope of Si (m/z 28 at 92.2% abundance) suffers primarily from two common polyatomic interferences: N2+ and CO+. In aqueous matrices, the N2+ interference is most abundant, while in organic matrices, CO+ dominates.

In the petrochemical industry, the need exists to reduce the CO+ interference to allow accurate silicon measurements in organic matrices. Further complicating the measurement is the variety of middle distillates in which silicon must be measured. Because of the volatility of the organic samples, CO+ levels are elevated and can vary between samples, making it difficult to obtain accurate analyses.

In addition, silicon is present everywhere, meaning that the background can fluctuate greatly as the result of silicon background/contamination. Certain organic solvents leach organo-Si compounds from plastic, while Si from the ICP-MS torch and sample introduction system must also be considered.

To overcome these challenges, new methodology was developed which both limits the sample-to-sample variations in CO+ levels and significantly reduces both the N2+ and CO+ interferences. This work will discuss method development and show initial results for the analysis of silicon in a variety of samples and solvents commonly used in the petrochemical industry.

Keywords: Elemental Analysis, ICP-MS, Petrochemical
Application Code: Fuels, Energy and Petrochemical
Methodology Code: Atomic Spectroscopy/Elemental Analysis

Session # 2260 Abstract # 2260-13 **Poster Sessions**

Session Title: Fuels, Energy and Petrochemical

Abstract Title: **Artificial Photosynthesis**

Date: Thursday, March 10, 2016 - Mornin

Time:

Room: Exposition Floor, 400 Aisle

Primary Author: Mingming Wang
Auburn University

Co-Author(s): Chao Li, Wei Zhan

Abstract Text

We report here a new protein-free photosynthetic mimicking strategy that relies on supercomplexed lipid nanoassemblies to organize small organic species for coordinated light harvesting, energy/electron transfer and photo-to-electrochemical energy conversion. Specifically, we demonstrate that efficient photoinduced electron transfer can be achieved between rhodamine and fullerene assembled in these nanoassemblies. Effect of lipid phases including liquid disordered (ld) phase, liquid ordered (lo) phase and gel phase of these nanoassemblies on photoconversion efficiency was investigated. Atomic force microscopy, UV-Vis spectroscopy, fluorescence spectroscopy and ultrafast transient absorption spectroscopy were used to characterize the architecture and photodynamics behavior of these lipid complexes.

Keywords: Membrane, Spectroscopy, Ultra Fast Spectroscopy, UV-VIS Absorbance/Luminescence
Application Code: Fuels, Energy and Petrochemical
Methodology Code: Fluorescence/Luminescence

Session # 2260 Abstract # 2260-14 **Poster Sessions**

Session Title Fuels, Energy and Petrochemical

Abstract Title **Rapid Measurement of Xylose and Glucose for Monitoring Corn Stover Fermentation in Bioethanol Production**

Date: Thursday, March 10, 2016 - Mornin
Time:
Room: Exposition Floor, 400 Aisle

Primary Author William Miller
YSI, Inc

Co-Author(s) June Klingensmith

Abstract Text

Acid-treated corn stover produces a variety of fermentable sugars that include glucose, xylose, mannose, arabinose, cellobiose, and galactose. HPLC is commonly used to measure glucose and xylose, which are important markers for evaluating the progress and efficiency of ethanol fermentation processes. The HPLC method is accurate and precise; however, it requires significant sample pretreatment and analytical cycle times □30 minutes. A YSI Biochemistry Analyzer, configured with glucose oxidase and pyranose oxidase immobilized enzyme membranes, was evaluated for its ability to simultaneously measure glucose and xylose during fermentation of pretreated corn stover in a bioethanol production process. In this study filtered corn stover samples were periodically measured for glucose and xylose over a 48-hour period during a lab-scale bioethanol fermentation. Samples were analyzed on both a YSI Biochemistry Analyzer and an HPLC. Comparability of the two analytical methods were evaluated with regards to precision and analysis time. A strong, positive correlation of the two methods was demonstrated. The YSI Biochemistry analyzer performed simultaneous analysis of glucose and xylose within one minute, providing a rapid, precise analytical alternative to the HPLC analytical method.

Keywords: Biofuels, Biosensors, Fuels\Energy\Petrochemical, Process Monitoring
Application Code: Fuels, Energy and Petrochemical
Methodology Code: Process Analytical Techniques

Session # 2260 Abstract # 2260-15 **Poster Sessions**

Session Title Fuels, Energy and Petrochemical

Abstract Title **Safety and Performance Testing of Li-ion Cells Using Thermal Analysis**

Date: Thursday, March 10, 2016 - Mornin

Time:

Room: Exposition Floor, 400 Aisle

Primary Author Peter Ralbovsky
Netzsch Instruments

Co-Author(s) Bob Fidler

Abstract Text

Li-Ion batteries have gained wide spread use in numerous portable power applications including HEV and PHEV systems. Different applications have different requirements on key operating parameters such as cycle life and power needs. Of continuing concern with Li-Ion batteries is safety. Special materials are being developed to provide the unique operating conditions to meet the use scenario. Thermal analysis is one of the key methods to use in order to characterize the performance, safety and reliability of battery components and cells. Calorimetry of a complete cell can provide information on thermal degradation and thermal runaway. It can also be used to look at efficiencies at different temperatures and operating conditions. Isothermal testing can be used to look at entropic changes needed for improved lifetime and operation. ?

Keywords: Chemical, Electrochemistry, Energy, Thermal Analysis
Application Code: Fuels, Energy and Petrochemical
Methodology Code: Thermal Analysis

Session # **2260** Abstract # **2260-16** **Poster Sessions**

Session Title: Fuels, Energy and Petrochemical

Abstract Title: **The Determination of Mercury in Liquefied Petroleum Gas – A Comparison of Sampling Techniques**

Primary Author: Matthew A. Dexter
P S Analytical

Co-Author(s): C A. Rogers, Warren T. Corns

Date: Thursday, March 10, 2016 - Mornin
Time:
Room: Exposition Floor, 400 Aisle

Abstract Text

Liquefied petroleum gas (LPG) is a mixture of C3 and C4 compounds, primarily propane and butane. LPG is pressurised to maintain it in the liquid phase to facilitate storage and transport and vaporised when used. The mercury concentration in LPG is typically measured by atomic spectroscopy-based methods, following vaporisation of a representative sample. Various sampling techniques are available, many of which use amalgamation-atomic-spectroscopy based analysis techniques:

1. Vaporising the LPG sample in the field using a heated regulator, followed by collection of mercury from a known volume of sample by amalgamation onto gold for subsequent analysis.
2. Collecting a liquid LPG sample in a sample bomb for laboratory analysis. A low flow of gas is then extracted from the vessel into a Tedlar bag for analysis as a gas. Mercury from a known volume of gas in the bag is then collected onto gold for subsequent analysis.
3. Collecting a liquid LPG sample in a sample bomb for laboratory analysis. A sub-sample is then vaporised using a heated pressure regulator and sampled by amalgamation onto gold.

Approaches 1 and 3 are based on the ISO6978 and ASTM 6350 methods for determination of mercury in natural gas. Approach 2 is described in JLPGA-S-07. Other sampling methods are also available, including direct injection of a liquid LPG into a GC-AFS system.

There are benefits and challenges associated with each sampling approach. This poster presents a laboratory comparison of sampling approaches 1-3 with sample determination of mercury by amalgamation-atomic fluorescence spectrometry.

Keywords: Atomic Spectroscopy, Mercury, Sampling, Trace Analysis
Application Code: Fuels, Energy and Petrochemical
Methodology Code: Sampling and Sample Preparation

Session # 2260 Abstract # 2260-17 **Poster Sessions**

Session Title: Fuels, Energy and Petrochemical

Abstract Title: **Salen Quinoxalinol Ligands for Selective Coordination and Sensors**

Primary Author: Anne E. Gorden
Auburn University

Date: Thursday, March 10, 2016 - Mornin
Time:
Room: Exposition Floor, 400 Aisle

Co-Author(s)

Abstract Text

Increasing the use of nuclear energy to provide civilian electrical power is one proposal to limit our ever increasing global dependence on fossil fuels; however, the environmental and health concerns that surround the use of nuclear fuels limit the acceptance of nuclear power by the public. Because of this, new materials are required that can coordinate, sense, and isolate actinides both for use in sensors and isolation or decontamination applications. In an effort to develop inexpensive ligands for colorimetric sensing of actinides, we have incorporated quinoxalines or imine aza-donors into an ethylene backbone framework for rapid identification or detection of uranyl. Using TDDFT calculations, we found that the hypsochromic shift upon UO22+ ion complexation of ligand is due to ligand to metal charge transfer, while the bathochromic shift observed with the addition of Cu2+ ion is metal to ligand charge transfer. In addition, it was found that the addition of Cu2+ to either ligand resulted in rapid, complete quenching of the ligand fluorescence. Here, we describe the further expansion of these systems to include phenazine based systems and naphthalene based systems to further decrease detection limits and reduce signal to noise. By incorporating UV/Vis microscopy into our system, we can follow some of these reactions in real time to observe reactions with very small sample sizes.

Keywords: Fluorescence, Lab-on-a-Chip/Microfluidics, Microspectroscopy, UV-VIS Absorbance/Luminescence
Application Code: Nuclear
Methodology Code: Fluorescence/Luminescence

Session # 2260 Abstract # 2260-18 **Poster Sessions**

Session Title Fuels, Energy and Petrochemical

Abstract Title **A New Hydroxide Selective Anion Exchange Phase for Ion Chromatography**

Primary Author Charanjit Saini
Thermo Fisher Scientific

Co-Author(s) Christopher Pohl, Yan Liu

Date: Thursday, March 10, 2016 - Mornin
Time:
Room: Exposition Floor, 400 Aisle

Abstract Text

In the area of ion chromatography of anions, hydroxide based eluent systems provide the highest detection sensitivity as well as the lowest noise, resulting in substantially lower detection limits. Over the years, this evolution of stationary phases designed for use with hydroxide eluents has preceded in a number of different directions. Hydroxide eluent is ideally suited to the application of gradients in anion exchange ion chromatography in terms of detection properties. As a result, we have developed a series of new hydroxide selective phases over the past 15 years aimed at producing materials well suited to the chemical and chromatographic properties of hydroxide eluents. This column family includes the IonPac AS11, AS11-HC, AS15, AS16, AS17, AS18, AS19, AS20, AS24, AS25, AS26, and AS27 columns. Through the evolutionary process of developing these columns, there has been a steady improvement in hydroxide elution power, selectivity, solvent compatibility and overall column ruggedness.

In this presentation, we will demonstrate recent advances in hydroxide selective stationary phases. First, we will demonstrate a high performance hydroxide selective stationary phase designed for analysis of high purity water for the semiconductor industry. Second, we will demonstrate recent developments in resin technology have allowed the use of 4□m resin particles in ion exchange columns. The benefits of columns packed with smaller particles include more efficient peaks, better resolution, faster analysis time, easier integration and more reliable analytical results.

Keywords: Environmental Analysis, Fuels\Energy\Petrochemical, Ion Chromatography, Liquid Chromatography
Application Code: Environmental
Methodology Code: Liquid Chromatography

Session # **2260** **Abstract #** **2260-19** **Poster Sessions**

Session Title: Fuels, Energy and Petrochemical

Abstract Title: **Physicochemical Characterization of Microalgal Biodiesel**

Date: Thursday, March 10, 2016 - Mornin
Time:
Room: Exposition Floor, 400 Aisle

Primary Author: Kizgel Davis-deSouza
Oglethorpe University

Co-Author(s): Grace Djokoto, Md H. Kabir

Abstract Text

Conventional sources of energy, such as oil, natural gas, and coal is now widely recognized as unsustainable because of increasing energy demand and depleting sources, and their contribution to the accumulation of greenhouse gases in the environment that are causing disastrous climate change. Renewable, carbon neutral, transportation fuels are necessary for environmental and economic sustainability. In recent decades, microalgae have emerged as an attractive feedstock for the large production of renewable bio-fuel, biodiesel because of high biomass and lipid production. To evaluate the potential of microalgal biodiesel as a renewable transportation fuels in this work, we focused to study the effect of nutrients on algal growth, efficiency of lipid extraction, efficient transesterification, amount of biodiesel production, and physicochemical properties of biodiesel extracted from laboratory grown microalgae.

Keywords: Biofuels, Fuels\Energy\Petrochemical
Application Code: Fuels, Energy and Petrochemical
Methodology Code: Chemical Methods

Session # 2260 Abstract # 2260-20 **Poster Sessions**

Session Title: Fuels, Energy and Petrochemical

Abstract Title: **Graphene Quantum Dots Immobilized Nanoporous N-TiO2 Thin Films for Efficient Photocatalytic Water Splitting**

Primary Author: Namal Wanninayake
University of Kentucky

Date: Thursday, March 10, 2016 - Mornin

Time:

Room: Exposition Floor, 400 Aisle

Co-Author(s): Allen Reed, Doo Young Kim, Stephen E. Rankin, Syed Z. Islam

Abstract Text

Our group recently reported that N2/Ar plasma-induced treatment is an efficient method to promote visible-light absorption and photocatalytic activities of TiO2. The nitrogen-doped TiO2 (N-TiO2) showed the significant reduction of band-gap (3.5 eV to 2.9 eV) and the unprecedented enhancement (200 times) of photocatalytic hydrogen production. This talk will address our continuous effort to enhance visible-light-driven photoactivity of N-TiO2 by the modification with graphene quantum dots (GQDs). The cubic ordered mesoporous TiO2 films were prepared via a surfactant templated sol-gel method and followed by brief calcination. Then, TiO2 films were treated with N2/Ar plasma for the incorporation of substitutional N atoms. GQDs were prepared by chemically oxidizing carbon nano-onions. The immobilization of GQDs was accomplished by reacting carboxyl groups of GQDs with amine groups of N-TiO2 developed by the prior immobilization of (3-aminopropyl)triethoxysilane (APTES). Successful immobilization of GQDs onto N-TiO2 was probed by UV-Vis, FT-IR, and scanning electron microscopy. Further, zeta potential and contact angle measurements showed enhanced surface charge and hydrophilicity, confirming the successful immobilization of GQDs. Linear-sweep-voltammetry and chronoamperometry were employed to determine their photocatalytic performance. GQD-immobilized N-TiO2 showed about 3X enhancement compared to unmodified N-TiO2, showing the important role of GQDs. We conclude that this enhancement comes from the synergistic effect of N-TiO2 and GQDs, including enhanced visible light absorption, efficient charge-separation at their interface, and improved water wettability and catalytic effect by GQDs.

This research was supported by the NSF KY EPSCoR grant and the Kentucky Science & Engineering Foundation (KSEF) grant.

Keywords: Electrochemistry, FTIR

Application Code: Fuels, Energy and Petrochemical

Methodology Code: Electrochemistry

Session # 2260 Abstract # 2260-21 **Poster Sessions**

Session Title: Fuels, Energy and Petrochemical

Abstract Title: **Techniques for Polysaccharide Research at the Complex Carbohydrate Research Center**

Primary Author: Roberto Sonon
Complex Carbohydrate Research Center

Date: Thursday, March 10, 2016 - Mornin
Time:
Room: Exposition Floor, 400 Aisle

Co-Author(s): Artur Muszynski, Asif Shajahn, Christian Heiss, Dandan Zhou, Ian Black, Justyna Dobruchowska, Mayumi Ishihara, Parastoo Azadi, Radnaa Naran, Scott Forsberg, Stephanie A. Archer-Hartmann,

Abstract Text

The Analytical Services Laboratory is a non-profit unit of the Complex Carbohydrate Research Center (CCRC) at The University of Georgia. We offer routine and specialty services for structural characterization of polysaccharides of plant or microbial origin that are coming from university, federal agencies, and industry researchers and from the US and other countries. The laboratory is equipped with state-of-the art instruments, which include Orbitrap Fusion MS, Velos Orbitrap-Elite MS, LTQ-Orbitrap MS, MALDI TOF/TOF MS, Capillary Electrophoresis MS, High Field NMRs, GC-MS, HPLC, and HPAEC.

On this poster, we will present several examples of recent research results on polysaccharide characterization that highlight a combination of analytical techniques using various instruments. Also, we will show recent collaborative trainings conducted at CCRC on analytical techniques designed for scientists who are engaged in the area of polysaccharide research.

Keywords: Biofuels, Carbohydrates, Food Science, Mass Spectrometry
Application Code: Fuels, Energy and Petrochemical
Methodology Code: Gas Chromatography/Mass Spectrometry

Session # 2260 Abstract # 2260-22 **Poster Sessions**

Session Title: Fuels, Energy and Petrochemical

Abstract Title: **Application of Laser Induced Breakdown Spectroscopy (LIBS) in Analysis of Out Crop Samples from the Marcellus Shale**

Primary Author: Jinesh Jain
AECOM

Co-Author(s): Alexander Bol'shakov, Christina Lopano, Dustin McIntyre, Herve Sanghapi, Richard Russo

Date: Thursday, March 10, 2016 - Mornin
Time:
Room: Exposition Floor, 400 Aisle

Abstract Text

Organic rich shale formations are unique in the production of natural gas because those not only serve as a source for the gas but also form the reservoir. In addition, the formations that are depleted of the gas are potential candidates for geologic storage of CO2 accompanied by an enhanced gas recovery (EGR). Both CO2 and natural gas adsorption/desorption seem to have a correlation with mineral composition of the shale rocks. Also the rocks that contain higher amount of organic material have greater ability to generate natural gas and potentially a greater capacity of CO2 storage. In order to ascertain the utility of laser induced breakdown spectroscopy (LIBS) for elemental characterization of shale rocks we analyzed the outcrop samples from the Marcellus Shale. It is advantageous to use LIBS because it enables a rapid in situ sample analysis with little or no sample preparation and can perform multi-element analysis including total carbon. In this study, a powdered sample was pressed to form a pellet and used for LIBS analysis. Laser pulse energy 25 mJ, a pulse duration 4 ns, gate width 1.05 ms, and gate delay 0.2 µs were used and the data were collected using 10 laser shots per spot of 150 µm diameter. Between partial least squares regression (PLS-R) and simple linear regression (SLR) calibration methods the PLS-R yielded better accuracy error. The samples were also analyzed using ICP-OES and a comparison between LIBS and ICP-OES results showed that the both results were comparable within 10%. Development of a LIBS method would provide rapid analysis of shale samples and would potentially benefit the depleted gas shale carbon sequestration research.

Keywords: Atomic Spectroscopy, Environmental Analysis, Laser, Plasma Emission (ICP/MIP/DCP/etc.)

Application Code: Environmental

Methodology Code: Atomic Spectroscopy/Elemental Analysis

Session # 2260 Abstract # 2260-23 **Poster Sessions**

Session Title: Fuels, Energy and Petrochemical

Abstract Title: **Direct Detection of Hydrocarbons from Microalgae Using Low Temperature Plasma –Mass Spectrometry (LTP-MS)**

Primary Author: Abigail Moreno Pedraza
Cinvestav

Co-Author(s): Robert Winkler

Date: Thursday, March 10, 2016 - Mornin
Time:
Room: Exposition Floor, 400 Aisle

Abstract Text

Botryococcus braunii produces up to 86% of its dry weight as oils, which are easily converted in biofuels (1). A major problem in process development and monitoring is the quantification of hydrocarbons, which usually requires a complicated and time-consuming extraction process prior analysis. Low-temperature plasma (LTP) is an ambient ionization technique for the direct measurement of compounds without an extraction procedure (2). Coupled to mass spectrometry (MS), LTP allows to monitor the biosynthesis of relevant compounds, or to map the distribution of molecules (3).

Here, we present a rapid and direct methodology for detection and/or imaging of hydrocarbon from the surface of microbial cultures using LTP-MS. Main advantages of LTP-MS compared to conventional ionization methods include the operation under ambient conditions, the absence of organic solvents and the low energy consumption. Thus, LTP-MS demonstrates high potential for the on-line monitoring of hydrocarbon bioprocesses.

(1) Cornejo-Corona, et al., (2015) "The biofuel potential of the green colonial microalga Botryococcus braunii" Chapter 3 in Microalgae and other phototrophic bacteria.

(2) Martínez-Jarquín and Winkler, (2013) "Design of a Low-Temperature Plasma (LTP) Probe with Adjustable Output Temperature and Variable Beam Diameter for the Direct Detection of Organic Molecules."

(3) Maldonado-Torres et al., (2014) "'Plug and Play' Assembly of a Low-Temperature Plasma Ionization Mass Spectrometry Imaging (LTP-MSI) System"

Keywords: Bioanalytical, Biofuels, Instrumentation, Plasma
Application Code: Fuels, Energy and Petrochemical
Methodology Code: Process Analytical Techniques

Session # 2260 Abstract # 2260-24 **Poster Sessions**

Session Title: Fuels, Energy and Petrochemical

Abstract Title: **Comprehensive Online Real-Time Analysis of Natural Gas Using VUV Absorption Spectroscopy**

Date: Thursday, March 10, 2016 - Mornin
Time:
Room: Exposition Floor, 400 Aisle

Primary Author: James A. Diekmann III
VUV Analytics

Co-Author(s): Jonathan Smuts, Michael Roecker

Abstract Text

Vacuum ultraviolet (VUV) absorption spectroscopy has recently received considerable attention as a new detection technology for use with gas chromatography. The growing interest in this novel technique has been driven to a large extent by the unique capabilities afforded by probing molecular electronic transitions. Virtually all molecules absorb in this energy range and provide unique absorbance spectra that can be used to readily distinguish even closely related species like isomers providing unique qualitative information. Quantitative analysis follows well established Beer-Lambert Law principles and renders accurate quantitation via spectral deconvolution simple and predictable. The combination of these characteristics results in a detection technology which for certain applications, can be used with little or no separation. Comprehensive analysis of natural gas using VUV absorption has the potential to provide a more accurate determination of the BTU desired for custody transfer. The real-time online measurement capability will provide better control and monitoring than is currently available, enabling process variations to be detected and corrected to almost perfect efficiency. We present here a robust, rapid, and repeatable method to perform a detailed analysis of all of the components in natural gas in less than one minute. This method has the potential to transform the way online real-time analysis of natural gas is conducted.

Keywords: Fuels\Energy\Petrochemical, Molecular Spectroscopy, On-line, Process Control
Application Code: Fuels, Energy and Petrochemical
Methodology Code: Gas Chromatography

Session # 2260 Abstract # 2260-27 **Poster Sessions**

Session Title: Fuels, Energy and Petrochemical

Abstract Title: **Transformer Oil Gas Analysis Using Gas Chromatography – Vacuum Ultraviolet Absorbance Spectroscopy**

Primary Author: Jonathan Smuts
VUV Analytics

Date: Thursday, March 10, 2016 - Mornin
Time:
Room: Exposition Floor, 400 Aisle

Co-Author(s): Andy Shkolnik, James A. Diekmann III, Jeff Tenney, Lindsey N. Shear

Abstract Text

Vacuum ultraviolet absorbance (VUV) spectroscopy is a powerful tool for studying the electronic transitions of virtually all molecules. Recent advances in instrumentation have allowed for the first ever application of this technology to chromatographic systems. The GC-VUV detector produces highly characteristic absorbance spectra for nearly all chemical species in the wavelength region of 125-240 nm. This system allows not only for identification but also for robust quantitation of a variety of compounds, even in the presence of coelution. A routine dissolved gas analysis (DGA) of the decomposition products and primary constituents comprising transformer oil gases is critical to predicting and preventing transformer failure. Traditionally such measurements are determined using a combination of TCD/FID/Methanizer. When used to analyze bio-based transformer oil gases this technique requires regular methanizer replacement. GC-VUV offers a powerful, yet simple alternative for the characterization of bio-based transformer oil gases without the use of methanizers. This new approach allows for unique spectral identification of all chemical components in the sample via the measurement of characteristic VUV absorbance spectra. We present here a robust, rapid and repeatable method that has the potential to transform the way transformer oil gas analysis (TOGA) is conducted.

Keywords: Gas Chromatography, GC Detectors, Molecular Spectroscopy, Process Control
Application Code: Fuels, Energy and Petrochemical
Methodology Code: Gas Chromatography

Session # 2260 Abstract # 2260-28 **Poster Sessions**

Session Title Fuels, Energy and Petrochemical

Abstract Title **Ionic Liquids as Electrolytes for Electrochemical Double-Layer Capacitors: Structures that Optimize Specific Energy**

Primary Author Maral PS Mousavi
University of Minnesota

Date: Thursday, March 10, 2016 - Mornin
Time:
Room: Exposition Floor, 400 Aisle

Co-Author(s) Andreas Stein, Benjamin E. Wilson, Evan Anderson, Philippe Buhlmann, Sadra Kashefolgheta

Abstract Text

The double-layer capacitance and the operating potential of the cell are the parameters that determine the specific energy of electrochemical double-layer capacitors (EDLCs). The operating potential of the cell is generally limited by the electrochemical window of the electrolyte solution, that is, the potential range within which electro-decomposition of electrolyte solution does not occur. Because ionic liquids offer wide potential windows, they can provide larger specific energies in EDLCs. In this work, we quantified the electrochemical stability and capacitance of several ionic liquids with structurally diverse anions and cations and showed that the cation size has a significant effect on the electrochemical capacitance. Imidazolium- and pyridinium-based ionic liquids were shown to provide the highest cell capacitance, and ammonium-based ionic liquids offered the largest potential windows. Increasing the chain length of the alkyl substituents in 1-alkyl-3 methylimidazolium trifluoromethanesulfonimide did not widen the potential window of the ionic liquid. We identified the ionic liquids that maximize the specific energies of EDLCs by combining the effects of their potential windows and the double-layer capacitance. The highest specific energies were obtained with ionic liquid electrolytes that had moderate electrochemical stability, but small ionic volumes, low viscosity, and high conductivity, and that was ionic liquid of 1-ethyl-3-methylimidazolium bis(trifluoromethylsulfonyl)imide.

Keywords: Electrochemistry, Electrodes, Energy, Voltammetry
Application Code: Fuels, Energy and Petrochemical
Methodology Code: Electrochemistry

Session # 2270 **Abstract # 2270-1** **Poster Sessions**

Session Title: New Developments in GC

Abstract Title: **Introduction of a New, High-quality, Cost Efficient Headspace GC Autosampler**

Date: Thursday, March 10, 2016 - Mornin
Time:
Room: Exposition Floor, 400 Aisle

Primary Author: Max Wang
Shimadzu Scientific Instrument Inc.

Co-Author(s): Clifford M. Taylor, Marty Smith

Abstract Text

HS-10, a new low cost version of the HS-20 headspace autosampler, has been developed and recently released for sale. The HS-10 is a transfer line, sample loop type, static headspace sampler. It has many of the same features of the high end model but is offered at a lower price. The features of this new headspace unit will be described along with applications.

Keywords: Chemical, Gas Chromatography, GC, Sample Handling/Automation
Application Code: General Interest
Methodology Code: Gas Chromatography

Session # 2270 Abstract # 2270-2 **Poster Sessions**

Session Title New Developments in GC

Abstract Title **Unique Selectivity: The Power of Ionic Liquid GC Columns**

Primary Author Leonard M. Sidisky
Supelco/Sigma-Aldrich

Date: Thursday, March 10, 2016 - Mornin
Time:
Room: Exposition Floor, 400 Aisle

Co-Author(s) Daniel Shollenberger, Greg A. Baney, Gustavo Serrano, James L. Desorcie, Michael D. Buchanan

Abstract Text

Choosing a stationary phase is the most critical step in column selection, more important than the column's I.D., film thickness, or length. This is because the stationary phase determines the selectivity of the column, and that selectivity influences resolution. Changing stationary phase may be an effective way to increase resolution.1

Beginning in 2006, extensive evaluations of columns manufactured with ionic liquid stationary phases have occurred. Their main strength was discovered to be unique selectivity. These columns have the ability to perform many of the same applications as columns made with polysiloxane polymer or polyethylene glycol stationary phases of similar polarity, but with slight elution order changes. Many times this results in increased resolution and/or shorter run times.

Recent solvation parameter model (SPM) evaluations indicate that only ionic liquid columns are capable of simultaneously providing intense H-acceptor and intense H-donor interactions, along with dipolar and □-□interactions.2 The effect of these interactions is illustrated through multiple chromatograms showing an ionic liquid column providing unique selectivity resulting in better resolution and/or faster analysis when compared to a non-ionic liquid column with similar polarity.

References

1.□Barry EF. Columns: packed and capillary; column selection in gas chromatography. In: Grob RL, Barry EF, editors. Modern Practice of Gas Chromatography, Fourth Edition. New Jersey: John Wiley & Sons, Inc.; 2004. p 65–191.
2.□Rodríguez-Sánchez S, Galindo-Iranzo P, Soria AC, Sanz ML, Quintanilla-López JE, Lebrón-Aguilar R. Principle component analysis (PCA) evaluation of seven commercial ionic liquid capillary GC columns. Supelco Reporter/2015; 33.1: 3–4.

Keywords: Capillary GC, Gas Chromatography
Application Code: Other
Methodology Code: Gas Chromatography

Session # 2270 Abstract # 2270-3 **Poster Sessions**

Session Title: New Developments in GC

Abstract Title: **Use of Traditional/Apex Track Integration in Commercially Available Software for GC Data Generated for EPA Methods 8082 and 8015**

Primary Author: Amanda Prior
PerkinElmer

Co-Author(s): Bill Hahn, Sharanya Reddy, Tom Kwoka

Date: Thursday, March 10, 2016 - Mornin
Time:
Room: Exposition Floor, 400 Aisle

Abstract Text

Peak detection and integration are fundamental to good data quality for accuracy and precision. There are several challenges to accurate peak determination such as poor resolution and baseline drift that can be compounded by challenging/dirty sample matrix. The co-elution of isomers, presence of shoulder peaks along with hundreds of components present in a single sample can make data processing very difficult. The data generated with EPA methods 8082 and 8015 which, involves multicomponent analysis of complex mixtures that are not fully resolved offers the desired level of complexity for testing peak detection and integration with Empower software. Two distinct approaches were taken for peak identification using data from the two methods. In case of EPA 8082, Apex track integration which uses a second derivative function of the chromatogram was used to identify changes in the chromatogram signifying peak start, stop and additional features such as shouldering. Using Apex track the overall integration of a very complex sample improved dramatically, with rapid peak detection and reduced the need for manual adjustment thereby making the process more robust. The data from the EPA 8015 method was integrated using the traditional peak integration algorithm where peak resolution is unobtainable and baseline shifts are expected. With Empower software, the user can analyze their data using two different integration algorithms for both simple and complex samples.

Keywords: Environmental Analysis, Fuels\Energy\Petrochemical, GC
Application Code: Environmental
Methodology Code: Gas Chromatography

Session # 2270 Abstract # 2270-5 **Poster Sessions**

Session Title New Developments in GC

Abstract Title **Evaluation of Gas Chromatographic Liner Deactivation when Exposed to Various Solvents and Extracts**

Date: Thursday, March 10, 2016 - Mornin
Time:
Room: Exposition Floor, 400 Aisle

Primary Author Linx Waclaski
Restek

Co-Author(s) Dan Li, Jack Cochran, Rebecca Stevens, Scott Adams

Abstract Text

Inlet liners are a critical part of the gas chromatographic system, in which the liquid sample must vaporize reproducibly and without discrimination or sample loss to ensure precise and accurate analyses. In addition, the sample is exposed to high temperatures, which require a highly inert surface. Interactions between the analytes and the liner can affect the integrity of results. It is critical to use inert, deactivated liners to minimize these adverse interactions, especially when using splitless injections. Despite choosing a highly inert liner to begin with, there are several conditions that can affect the integrity of even the best liner deactivations, such as solvent exposure and "dirtiness" of extracts. It is important for the analyst to maintain awareness of the condition of an inlet liner and perform maintenance when necessary.

The following experiments will examine different liner deactivations, as well as types of wool, for ruggedness to a variety of solvents and extracts. Performance is monitored by looking at several sensitive compounds, including organochlorine, organophosphorus, organonitrogen, and carbamate pesticides. Solvents such as acetified acetonitrile, commonly used in AOAC QuEChERS methods, can affect the integrity of liner deactivations, decreasing responses for certain compounds. Even non-polar solvents like hexane, after repeatedly injected in a hot inlet, can cause degradation of liner performance. The matrices of sample extracts, whether environmental, food, etc., can also leave behind non-volatile components in the liner, causing issues. Liner inertness following exposure to solvents and a variety of matrices will be presented.

Keywords: Environmental Analysis, Food Safety, Gas Chromatography, Sample Introduction
Application Code: General Interest
Methodology Code: Gas Chromatography

Session Title New Developments in GC

Abstract Title **Modification of a GC for High Level Tritium Exposure**

Primary Author William Spencer
SRNL

Date: Thursday, March 10, 2016 - Mornin
Time:
Room: Exposition Floor, 400 Aisle

Co-Author(s) Jacob Schaufler, Jose Cortes Concepcion, Robert Lascola

Abstract Text

An Inficon 3000 micro gas chromatograph with molecular sieve and PlotU columns was selected for an application involving analysis of tritium gas streams. The system will primarily monitor common gas impurities like nitrogen, argon, water, and trace C1-C4 organics. The Inficon GC has the advantages of being compact, requiring low amperage line voltage, consuming minimal carrier gas, and being designed for external control with a simple Ethernet jump cable while performing a standard atmospheric gas sweep in less than two minutes. These properties made the system suitable for installation inside an inert glovebox for process monitoring applications. However, on close examination the chromatograph had several features that made it incompatible with high levels of tritium gas. The chromatographic modules had to be significantly modified to remove both the plastic venting tubing and the diaphragm based sampling pump system. The system was found to be challenged by less than positive pressure samples and was not well designed for sample recapture. The modifications to the system with steel tubing, tip seal-less mini scroll pump, and additive manufacturing brackets will be described.

Re:SRNL-L4110-2015-00005

Keywords: Gas Chromatography, Nuclear Analytical Applications, Nuclear Energy
Application Code: Nuclear
Methodology Code: Gas Chromatography

Session # 2270 Abstract # 2270-8 **Poster Sessions**

Session Title New Developments in GC

Abstract Title **Micro-Scale Vapor Extractor for Micro-GC Analysis of VOCs in Biofluids**

Primary Author Junqi Wang
University of Michigan

Date: Thursday, March 10, 2016 - Mornin
Time:
Room: Exposition Floor, 400 Aisle

Co-Author(s) Edward T. Zellers, Joseph A. Potkay

Abstract Text

We describe a new passive, membrane-mediated microfabricated vapor extractor ([micro]VE) for extraction of aqueous volatile organic compounds (VOCs), which we interfaced with a new micro-scale gas chromatograph prototype ([micro]GC). VOC extractions are normally achieved by purge-and-trap or solid-phase microextraction, which, can suffer from low extraction efficiency, long cycle times, and large sample volumes, and thus are not well-suited for use with on-site portable analyzers. Our 2.8 × 2.3 cm [micro]VE chip has a lower Si substrate containing etched conduction and extraction [micro]channels, an upper glass substrate with an etched gas [micro]channel, and a cast PDMS membrane bonded at the interface. An aqueous sample is pumped through the lower µchannels and VOCs diffuse through the PDMS and are swept by N[sub]2[/sub] to the downstream analyzer. Initial characterization of a [micro]VE with a 100-[micro]m thick PDMS membrane challenged with several VOCs gave time-to-steady-state ranging from ~4-12 mins, and steady-state permeation rates within +/-50% of model predictions. The [micro]VE was fluidically interfaced to our [micro]GC, which contains a [micro]preconcentrator, [micro]column, and chemiresistor array detector. A mixture of 2-butanone, trichloroethylene, toluene, and tetrachloroethylene spiked into synthetic urine at ppm concentrations was extracted from a 700 [micro]L sample in 3.5 min and passed directly into the [micro]preconcentrator and injected. Separation and analysis took 1.3 min with the lowest LOD = 660 ppb. These preliminary results suggest that the hybrid [micro]VE-[micro]GC microsystem may provide rapid field/clinical analysis of VOC water contaminants and urinary biomarkers in ultra-small sample volumes.

Funding: University of Michigan and NIOSH-CDC.

Keywords: Clinical Chemistry, Environmental/Biological Samples, Extraction, GC
Application Code: Bioanalytical
Methodology Code: Microfluidics/Lab-on-a-Chip

Session # 2280 Abstract # 2280-1 **Poster Sessions**

Session Title UV/VIS

Abstract Title **Qualitative Colorimetric and Quantitative Flow Injection Determination of Alkyl Nitrite**

Date: Thursday, March 10, 2016 - Mornin

Time:

Room: Exposition Floor, 400 Aisle

Primary Author Abd al-karim Ali
Miami University

Co-Author(s) Neil D. Danielson

Abstract Text

Alkyl nitrites such as isopentyl- or butyl-nitrite are volatile organic compounds with the general formula of R-O-N=O. Isopentyl nitrite is a legitimate drug prescribed for angina or as an antidote for cyanide poisoning. Butyl- or other short chain alkyl nitrites formulated as concentrated solutions called "poppers" have alleged aphrodisiac and euphoric properties when inhaled and are illegal in the USA. Most analytical methods for alkyl nitrites are based on gas chromatography (GC) or the colorimetric Griess reaction, the latter method used for the determination of inorganic nitrite. However a cheap, portable, facile, and fast qualitative test for alkyl nitrites is still needed and would be important for forensic purposes. We have investigated analytically the rapid reaction of isopentyl nitrite in 0.1 M sulfuric acid with 3-mercaptopropionic acid (MCPA) to produce the pink nitroso product compound. For the qualitative test for alkyl nitrites, a cotton wool plug soaked with acidic MCPA is placed at the bottom of a plastic syringe. The syringe is placed over the sample and the headspace vapor is withdrawn through the cotton wool plug. The presence of a pink color on the plug surface is indicative of an alkyl nitrite. We have been able to detect easily 0.1% solution of isopentyl nitrite which is well below the expected concentration of about 50% in poppers. Using a dual valve flow injection analysis method, we have quantitatively determined trace levels of isopentyl nitrite. Injection of the MCPA reagent and the alkyl nitrite solution using separate valves in the sulfuric acid carrier permit the reaction to take place in a mixing tee and the product is monitored continuously at 335 nm using a HPLC UV-VIS detector. A linear range of 0.6–10 mM for eight points gave a correlation coefficient of 0.9992. These data are comparable to our prior work using the same method to determine inorganic nitrite.

Keywords: Drugs, Forensic Chemistry, UV-VIS Absorbance/Luminescence

Application Code: Clinical/Toxicology

Methodology Code: UV/VIS

Session Title UV/VIS

Abstract Title **Multi-order Visible Absorption and Reflectance Spectrometry: Parallels to Atomic Emission Line Interferences**

Date: Thursday, March 10, 2016 - Mornin
Time:
Room: Exposition Floor, 400 Aisle

Primary Author Alexander Scheeline
SpectroClick

Co-Author(s) Mark D. Ginsberg

Abstract Text

Stacked, mutually-rotated transmission diffraction gratings allow generation of low resolution spectra that can be exploited by 8-bit cameras for performing visible absorption or reflectance spectrometry. Exploiting this capability requires several steps, some of which have been previously reported: accurate identification of diffraction pattern center, identification of each order, dispersion calibration, extraction of individual orders, normalization of throughput among orders, and combination/averaging of orders to obtain a precise spectrum. Order throughput varies over a wide range, allowing a single exposure to measure light intensity over 3.5 orders of magnitude despite the 2.5 order dynamic range of our spectrometer's camera. While not obvious upon qualitative inspection of raw data, quantitation demands compensation for order overlap in ways resembling problems with line overlaps in atomic emission spectrometry.

Our poster shows the chemometrics for extracting a spectrum from raw SpectroBurst™ images. Examples of spectra with inadequate compensation for order overlap and with optimal compensation are presented.

Keywords: Instrumentation, Spectrometer, Spectroscopy, UV-VIS Absorbance/Luminescence
Application Code: General Interest
Methodology Code: UV/VIS

Session # **2280** Abstract # **2280-3** **Poster Sessions**

Session Title UV/VIS

Abstract Title **Nano-Embedded Optochemical Sensors for In-Vivo Photo-Acoustic Chemical Imaging of Potassium Ions**

Primary Author Wuliang Zhang
University of Michigan

Date: Thursday, March 10, 2016 - Mornin
Time:
Room: Exposition Floor, 400 Aisle

Co-Author(s) Chang Lee, Jeffery Folz, Raoul Kopelman, Xueding Wang

Abstract Text

Potassium is very important in cellular biochemical reactions and metabolism; it participates in the intracellular protein synthesis and carbohydrate metabolism.[sup]1[/sup] Given the importance of potassium, probes that are able to visualize the intracellular K[sup]+[/sup] behavior are necessary. To achieve this goal, potassium ion-selective optical nanosensors were prepared for photoacoustic imaging in-vivo, for future non-invasive, spatially and time-resolved biochemical analysis. They are based on the incorporation of Potassium Ionophore III and an indicative dye, Chromoionohore I.[sup]2, 3[/sup] They were embedded in Pluronic micelles and another layer of polybutyl methacrylate nanoparticles within the micelles. The polybutyl methacrylate nanoparticles were formed via free-radical polymerization. The nanoparticles obtained were typically on the order of 60 nm. The potassium sensors were characterized by both UV-VIS and photoacoustic spectroscopy. Notably, photoacoustic spectroscopy uses the same principles as that of UV-VIS spectroscopy,[sup]4[/sup] but it has much deeper tissue penetration than optical spectroscopies.[sup]5 [/sup]Work is in progress on other ions, such as Na[sup]+[/sup] and Ca[sup]2+[/sup].

References

(1) Sigel, A.; Sigel, H.; Sigel, R.; Pohl, H.; Wheeler, J.; Murray, H.[i] Met ions Life Sci[/i][b] 2013[/b].
(2) Brasuel, M.; Kopelman, R.; Miller, T.; Tialkens, R.; Philbert, M.[i] Anal. Chem.[/i][b] 2001[/b],[i] 73[/i], 22212228.
(3) Xie, X.; Zhai, J.; Bakker, E.[i] Analytical Chemistry[/i][b] 2014[/b], [i]86[/i], 28532856.
(4) Cash, K.; Li, C.; Xia, J.; Wang, L.; Clark, H. [i]ACS Nano [/i][b]2015[/b], [i]9[/i], 16921698.
(5) Ray, A.; Yoon, H.; Lee, Y.; Kopelman, R.; Wang, X.[i] Analyst[/i] [b]2013[/b], [i]138[/i], 3126-3130.

Acknowledgement
The authors acknowledge the National Institute of Health for financial support. Grant number: NIH/NIBIB 1R01CA186769.

Keywords: Imaging, Nanotechnology, Photoacoustic, Sensors
Application Code: Nanotechnology
Methodology Code: UV/VIS

Session # 2280 Abstract # 2280-6 **Poster Sessions**

Session Title: UV/VIS

Abstract Title: **Simultaneous Determination of Iron and Aluminum by Spectrophotometry and Partial Least Squares Regression: Comparison of Two Potential Ligands and Application to Mine Drainage and Related Water Samples**

Primary Author: Mark T. Stauffer
University of Pittsburgh - Greensburg

Date: Thursday, March 10, 2016 - Mornin
Time:
Room: Exposition Floor, 400 Aisle

Co-Author(s)

Abstract Text

This presentation will examine the use of visible (VIS) spectrophotometry and partial least squares regression (PLSR) for simultaneous determination of iron and aluminum, toward applications to analysis of abandoned mine drainage (AMD) and related water samples for these two metals. Abandoned mine drainage is the result of collection of water in abandoned coal mines, causing oxidation of pyrite (iron disulfide) in and near coal seams and subsequent formation of sulfate and ferrous iron. Along with Fe, Al may also be a primary metallic constituent of AMD. Fe and Al in AMD are determined by atomic spectrometry, and less expensively by VIS spectrophotometry. Many colorimetric methods exist for determination of iron and aluminum in water samples, using chelators such as xylenol orange (XO) and Eriochrome cyanine R (ECR). For a number of these colorimetric methods, though, either iron or aluminum will interfere with the other analyte due to severe spectral overlap produced by chelates of these two metals with the same ligand. This is especially true with XO and ECR, which yield strongly absorbing chelates with both Fe and Al and also exhibit severe spectral overlap. In such cases, use of PLSR makes possible simultaneous determination of iron and aluminum using the same chelator and elimination of extra procedural steps for isolation and separate determination of each analyte. Thus, the focus of this presentation is on simultaneous colorimetric determination of Fe and Al and PLSR, using XO and ECR under appropriate conditions that promote chelation of both Fe and Al by each ligand. Comparison of results for Fe and Al using each ligand will be made. Experimental details, significance of the results, potential application of these methods toward characterization of water hardness of AMD and related samples, and future directions for this work, will be presented and discussed.

Keywords: Calibration, Chemometrics, Metals, Spectrophotometry
Application Code: Environmental
Methodology Code: UV/VIS

Session # 2280 Abstract # 2280-8 **Poster Sessions**

Session Title UV/VIS

Abstract Title **TBAF and It's Spectral Interference**

Primary Author Ian Adams
The University of Alabama

Date: Thursday, March 10, 2016 - Mornin
Time:
Room: Exposition Floor, 400 Aisle

Co-Author(s)

Abstract Text

Tetrabutylammonium fluoride (TBAF) is a frequently used source for fluoride in laboratory settings which measure fluoride sensing capabilities, especially of novel compounds. However, when using some commercial sources of TBAF in high concentrations, a pale yellow visible impurity is present at high concentrations. While this yellow impurity has been noted to occur, even by commercial suppliers, it can cause spectral interference in both UV/vis and fluorescence spectroscopy. This interference has the potential to alter some fluoride sensory findings which have been published. Our research which we will present on is focused on identification, removal, and spectroscopic influence of this impurity.

Keywords: Biosensors, Integrated Sensor Systems, Sensors, UV-VIS Absorbance/Luminescence
Application Code: General Interest
Methodology Code: UV/VIS

Session # 2290 Abstract # 2290-1 **Poster Sessions**

Session Title: Various Applications of GCMS

Abstract Title: **Evaluation of Methylisothiazolinone (MI) Extraction from Sunscreen Using Supported Liquid Extraction prior to GC/MS Analysis**

Date: Thursday, March 10, 2016 - Mornin

Time:

Room: Exposition Floor, 400 Aisle

Primary Author: Rhys Jones
Biotage GB Limited

Co-Author(s): Lee Williams, Victor Vandell

Abstract Text

Introduction

Methylisothiazolinone (MI) is an antimicrobial preservative that is used in a variety of personal care products, such as sunscreens, lotions cosmetics etc. MI is a cytotoxin and as a result there is major concern because of sensitization and allergic reactions as well as cell and nerve damage. From July 2015, the European Commission will adopt a ban of MI as an ingredient in leave-on cosmetic products. This poster demonstrates the development of a simple sample preparation protocol using supported liquid extraction columns prior to GC/MS analysis.

Methodology

Sunscreen was weighed and extracted with 50/50 MeOH/1M NaCl aq. Homogenization was performed using vortex mixing or an OMNI Bead Ruptor. Samples were centrifuged and the supernatant applied to the ISOLUTE SLE+ columns. Analyte extraction was performed and extracts were evaporated and reconstituted in EtOA. Samples were analyzed by GC/MS.

Results

Optimization of the evaporation condition involved the use of 0.2M HCl in IPA added to the collection tubes prior to blowdown. It was necessary to avoid over-drying to minimize variability. The complex nature of the matrix in terms of various additives provided an interesting challenge. Optimization of the supported liquid extraction protocol involved investigation of loading volume and elution solvent combination. The optimal extraction solvent of 50/50 hexane/EtOAc demonstrated high analyte recoveries, greater than 90%, RSDs below 10%, good visual extract cleanliness and no interference on the SIM trace. Method performance was compared using vortex mixing to the use of an OMNI bead ruptor instrument. The latter provided a far more homogenous extract compared to vortex mixing. Calibration curves were constructed spiking MI into sunscreen from 50-750 ng/mL. Although no internal standard was used, good coefficients of determination (r2) greater than 0.99 were demonstrated.

Keywords: Consumer Products, Cosmetic, GC-MS, Separation Sciences
Application Code: Consumer Products
Methodology Code: Gas Chromatography/Mass Spectrometry

Session # **2290** Abstract # **2290-2** **Poster Sessions**

Session Title: Various Applications of GCMS

Abstract Title: **Extending the Range of GC-MS Applications with Cold EI**

Date: Thursday, March 10, 2016 - Mornin
Time:
Room: Exposition Floor, 400 Aisle

Primary Author: Aviv Amirav
Tel Aviv University

Co-Author(s): Alexander Fialkov, Tal Alon, Uri Keshet

Abstract Text

GC-MS suffer from a few limitations that restrict its use and growth. A major limitation is its inability to analyze relatively non-volatile and thermally labile compounds. Another downside is the weakness or absence of molecular ions. Furthermore, its typical >30 min analysis time impedes its use in areas that require real time analysis. GC-MS with Cold EI is based on electron ionization of vibrationally cold molecules in supersonic molecular beams (SMB) and on interfacing the GC and MS with SMB while replacing the EI ion source with Cold EI fly-through ion source. In GC-MS with Cold EI the GC elution temperatures can be significantly lowered upon the reduction of the column length and increase of carrier gas flow rate. Furthermore, via using high column flow rate, lower injector temperatures can be used and sample degradation at the Cold EI fly-through ion source is inherently eliminated. Thus, extended range of thermally labile, polar and low volatility compounds are amenable for analysis. Furthermore, the ionization of cold molecules results in enhancement of the molecular ions and GC-MS with Cold EI facilitates much shorter analysis time up to real time analysis with low thermal mass fast GC inlet. Extended range of GC-MS with Cold EI applications will be demonstrated in the analysis of large hydrocarbons, full range of all the organic explosives, large polar drugs, thermally labile pesticides, lipidomics analysis including free fatty acids, cholesterol cholesterilesters and triglycerides, service GC-MS for synthetic organic compounds and real time forensic analysis.

Keywords: Gas Chromatography/Mass Spectrometry, GC-MS
Application Code: Other
Methodology Code: Gas Chromatography/Mass Spectrometry

Session # 2290 Abstract # 2290-3 **Poster Sessions**

Session Title Various Applications of GCMS

Abstract Title **New Sampling Device for Early Cancer Screening by Non-Invasive Detection of VOCs Biomarkers in Exhaled Breath**

Primary Author Paolo Benedetti
IIA - CNR

Date: Thursday, March 10, 2016 - Mornin
Time:
Room: Exposition Floor, 400 Aisle

Co-Author(s) Carlo Crescenzi, Ettore Guerriero, Federico Marini, Maria Cristina Zappa, Mark Ragusa

Abstract Text

In the last decade, exhaled breath analysis demonstrated a huge potential as screening technique for cancer biomarkers. On the other hand lacks in standardization of sampling methods and data normalization still affect most of the studies. A new simplified sampling methodology was explored in order to develop a robust and reliable procedure. The method is based on new sampling device derived from a modification of the geometry of commonly used thermal desorption tubes. The device allows a direct sampling and delivers accurate information by GC-MS analysis. Major problem such as high air flow impedance and water condensation are overstepped due the combination of a recently develop sorbent and the new geometry. Reduced sample manipulation and trapping on condensate allow collection of more representative samples. Preliminary results of a study aimed to the enlargement of potential target analytes are presented.

Keywords: Biomedical, Chemometrics, Gas Chromatography/Mass Spectrometry, Volatile Organic Compounds
Application Code: Biomedical
Methodology Code: Gas Chromatography/Mass Spectrometry

Session # 2290 Abstract # 2290-5 **Poster Sessions**

Session Title Various Applications of GCMS

Abstract Title **Chemical Warfare Agents (CWAVX) by Large Volume Injection/Programmable Temperature Vaporization (LVI/PTV) - Gas Chromatography/Quadrupole Mass Spectrometry**

Primary Author Tom Fowler
CSS-Dynamac

Co-Author(s) Julia Capri, Kelly Head

Date: Thursday, March 10, 2016 - Mornin
Time:
Room: Exposition Floor, 400 Aisle

Abstract Text

The purpose of this study was to develop a method for the analysis of chemical warfare agent compounds using GC-Quadrupole Mass Spectrometry (GCQMS) in full scan mode to achieve the low picogram detection levels typically seen when analyzing these compounds by GC/Time-of-flight mass spectrometry (GCTOFMS).

This procedure allows for the detection of chemical warfare agents, in the low picogram range as opposed to the low nanogram range, which is typically achieved using GCQMS instrumentation in full scan mode. The benefits of the analysis in full scan mode include the generation of more definitive spectra and decreased incidence of false positives that are common when analyses are performed by SIM.

A GCQMS instrument equipped with a Large Volume Injection/Programmable Temperature Vaporization (LVI/PTV) injector was utilized to develop a full scan method in which 10uL of sample extract was injected by way of a solvent vent injection program.

By optimizing the LVI/PTV parameters, the low pictogram detection levels common to GC-time-of-flight instruments were achieved by GC quadrupole MS in full scan mode.

Data from calibration curve and MDL studies analyzed by the developed method validate the method as a viable alternative to analysis of chemical warfare agents by GCTOFMS.

Keywords: Environmental Analysis, Gas Chromatography/Mass Spectrometry, Mass Spectrometry, Quadrupole
Application Code: Environmental
Methodology Code: Gas Chromatography/Mass Spectrometry

Session # 2290 Abstract # 2290-6 **Poster Sessions**

Session Title: Various Applications of GCMS

Abstract Title: **Quantification of Persistent Organic Pollutants in Dietary Supplements Using Stir-bar Sorptive Extraction- Thermal Desorption- GC/MS and Isotope Dilution Mass Spectrometry**

Primary Author: Weier Hao
Duquesne University

Date: Thursday, March 10, 2016 - Mornin
Time:
Room: Exposition Floor, 400 Aisle

Co-Author(s): Andrew Boggess, Skip Kingston

Abstract Text

A method using Stir-bar Sorptive Extraction (SBSE) coupled with Thermal Desorption- Gas Chromatography- Mass Spectrometry (TD-GC-MS) and Isotope Dilution Mass Spectrometry (IDMS, EPA Method 6800 Update V 2015) has been developed and applied to this study. This method has previously been applied to quantification of persistent organic pollutants (POPs) in human blood, wastewater, and drinking water and has been proven accurate, precise, reproducible, and relatively green. Besides drinking water, dietary supplements and foods are significant pathways that people are exposed to POPs. Dietary supplements have been found to have incorrect labels with missing ingredients or incorrect ingredient concentrations and they have become a source of xenobiotics that can have adverse impact on human health. In this study, SBSE-TD-GC-MS-IDMS is adapted and applied to quantification of POPs in dietary supplements. Selected commercially available dietary supplements from different manufacturers sold ubiquitously across the United States and internationally are analyzed. Compared with classical extraction methods such as liquid-liquid extraction and traditional solid phase extraction, SBSE is more environmental friendly requiring much less solvent from beginning to end. Thermal desorption is fully automated, which makes the process efficient and enables a high level of quality assurance and quality control. Instead of traditional calibration curve, IDMS is used for quantification. It provides higher precision and accuracy in less time and eliminates the error from recovery. This green and effective method, along with the instruments and products, will also be demonstrated in medical and human health assessment.

Keywords: Gas Chromatography/Mass Spectrometry, Isotope Ratio MS, Solid Phase Extraction, Thermal Desorpt
Application Code: Food Contaminants
Methodology Code: Gas Chromatography/Mass Spectrometry

Session # 2290 Abstract # 2290-7 **Poster Sessions**

Session Title: Various Applications of GCMS

Abstract Title: **Automated Sampling of Methanol Extractions**

Primary Author: Anne Jurek
EST Analytical

Co-Author(s): Justin Murphy, Kelly Cravenor, Lindsey Pyron

Date: Thursday, March 10, 2016 - Mornin
Time:
Room: Exposition Floor, 400 Aisle

Abstract Text

The United States Environmental Protection Agency (USEPA) Method 8260 is used in order to ascertain volatile organic compounds in waters, soils and solid waste samples. Often times, soil and solid waste samples are so highly contaminated the sample needs to be dispersed in methanol. Sample collection for contaminated soils can be obtained in two ways. One, dispersing a bulk soil sample into a 40ml vial and adding methanol in the lab or two, sending pre-weighed vials with a septum sealed cap that already contains the pre-requisite methanol out in the field for soil sampling. No matter how the soil sample is dispersed in methanol, an aliquot of the methanol extract needs to be added to water and purged using USEPA Method 5030. This application will investigate automated sampling of methanol soil extractions.

Keywords: Environmental/Soils, Extraction, Volatile Organic Compounds
Application Code: Environmental
Methodology Code: Gas Chromatography/Mass Spectrometry

Session # 2290   Abstract # 2290-9   **Poster Sessions**

Session Title: Various Applications of GCMS

Abstract Title: **Pesticide Analysis in Drinking Water and Beverages Using Multiple Techniques**

Primary Author: Xiaoping Li
Georgia Gwinnett College

Date: Thursday, March 10, 2016 - Mornin
Time:
Room: Exposition Floor, 400 Aisle

Co-Author(s): Hongxia Guan, Michelle Huang, Rashad Simmons, Simon Mwongela, Zoe Goldstein

Abstract Text

Pesticides have become ubiquitous environmental and human health hazards. There is clear evidence that long term exposure to pesticides can cause serious diseases. This project involves a qualitative and quantitative analysis of one of the most widely applied pesticide (atrazine) and its degradation products in various real-world samples (water and soft drink) with multiple separation/detection techniques, including Gas Chromatography-Mass Spectrometer (GC-MS), High-Performance Liquid Chromatography (HPLC), and Enzyme-linked Immunosorbent Assay (ELISA). The detection limit, quantitation limit, linearity and spike recovery from different techniques will be compared and discussed.

Keywords: Environmental Analysis, HPLC, Water
Application Code: Environmental
Methodology Code: Gas Chromatography/Mass Spectrometry

Session # 2290 Abstract # 2290-10 **Poster Sessions**

Session Title: Various Applications of GCMS

Abstract Title: **Identification of Contaminants in Powdered Foods by Direct Extraction Thermal Desorption GC/MS**

Date: Thursday, March 10, 2016 - Mornin
Time:
Room: Exposition Floor, 400 Aisle

Primary Author: Ronald Shomo
Scientific Instrument Services

Co-Author(s): Christopher Baker, John Manura

Abstract Text

The ability to identify volatile and semi-volatile contaminants present in food products without the use of solvent extractions has several advantages including improving sample throughput, reducing the chance of a volatile component being "lost" in the extraction process and eliminating the need for solvent disposal. This study utilizes the advantages of direct thermal extraction GC/MS to identify contaminants in powdered food products. Direct Thermal Extraction GC/MS provides for fast analysis with no carryover problems that can be associated with other GC/MS techniques.

Keywords: Contamination, Food Contaminants, Gas Chromatography/Mass Spectrometry, GC-MS
Application Code: Food Contaminants
Methodology Code: Gas Chromatography/Mass Spectrometry

Session # 2290 Abstract # 2290-11 **Poster Sessions**

Session Title Various Applications of GCMS

Abstract Title **Methods Development for Sampling and Analysis of Biogenic Volatile Organic Compounds Released from Plants**

Primary Author Prithviraj Sripathi
Middle Tennessee State University

Date: Thursday, March 10, 2016 - Mornin
Time:
Room: Exposition Floor, 400 Aisle

Co-Author(s) Beng Guat Ooi, Christopher Moore, James G. Milstead, Kathleen Kuklewicz, Ngee Sing Chong

Abstract Text

Plants emit biogenic volatile organic compounds (BVOCs) including terpenes and terpenoid compounds that play a vital role in the atmospheric chemistry. These BVOCs will react with the hydroxyl radicals that will result in the formation of tropospheric ozone and secondary organic aerosols. This study investigates the emission of terpenes in a forest setting under the diurnal variation of sunlight or darkness, temperature, and humidity. Biogenic terpenes released by the Eastern Red Bud, Eastern Red Cedar, Shagbark Hickory, and Winged Elm trees located in Stones River Battle National Field are sampled using different sampling methodologies including sorbent tubes, solid phase microextraction, 6-liter canisters, and passive sampling devices. Ground-based sampling schemes requiring the placement of intact branches inside a conical cage coupled to a canister with a 1-hour passive sampler, balloon-based air sampling where a canister or sorbent based passive sampler is tied to the tethered balloon, and drones carrying air sampling accessories are explored and their performance characteristics compared. The GC-MS methods based on scan mode as well as the dual mode of both scan and selected ion monitoring will be compared in terms of their ability to monitor low levels of BVOCs at the 10-100 parts per trillion levels. The retention time indices of terpenes along with the selection of their common mass spectrometric ion fragments are used to enhance the confidence in correct identification of structurally similar terpenes.

Keywords: GC-MS, SPME, Thermal Desorption
Application Code: Environmental
Methodology Code: Gas Chromatography/Mass Spectrometry

Session # 2290 **Abstract # 2290-12** **Poster Sessions**

Session Title: Various Applications of GCMS

Abstract Title: **Pyrolysis-GC/MS as a Screening Tool for Phthalate Esters and Brominated Flame Retardants in the RoHS Directive**

Date: Thursday, March 10, 2016 - Mornin
Time:
Room: Exposition Floor, 400 Aisle

Primary Author: Nicole M. Lock
Shimadzu Scientific Instruments

Co-Author(s): Di Wang, Laura Chambers, Mark Janeczko, Shilpi Chopra

Abstract Text

The Restriction of Hazardous Substances (RoHS) Directive controls six hazardous substances commonly used in electronic and electrical equipment. Two of the restricted substances are compound classes commonly used in flame retardants: polybrominated biphenyls (PBB) and polybrominated diphenyl ethers (PBDE), both known to cause serious health concerns due to their high halogen content. Furthermore, four types of phthalate esters are expected to be regulated in 2019 in Europe, which adds phthalate esters to the substances restricted under RoHS. To quantitate these substance in a polymer matrix, the traditional approach involves extraction of PBBs, PBDEs and phthalate esters from the sample matrix, followed by detection and quantitation by GC/MS. This method is time consuming and poses the risk to exposure to multiple toxic solvents.

Pyrolysis followed by gas chromatography/mass spectrometry (PY-GC/MS) has been well established for detection of volatile and semi-volatile compounds in both natural and synthetic polymers. Using the pyrolysis technique, a temperature programed micro-furnace provides thermal desorption and pyrolysis of the polymer matrix, releasing the PBBs, PBDEs and phthalate esters for GC/MS analysis.

In this poster, PY-GC/MS method has been used to screen for 7 phthalate esters and 11 brominated flame retardants. A commercially available method package was used, which contains phthalate and PBDE standards, pre-registered methods with acquisition and data processing parameters, and calibration curves for semi-quantitative calculation of compound concentration. Quantitation results were generated with minimal sample preparation requiring no organic solvents. A software program for efficient multi-analyte data confirmation and QAQC review is discussed in the detail.

Keywords: Environmental, Gas Chromatography/Mass Spectrometry, Polymers & Plastics, Pyrolysis
Application Code: Environmental
Methodology Code: Gas Chromatography/Mass Spectrometry

Session # 2290 Abstract # 2290-14 **Poster Sessions**

Session Title: Various Applications of GCMS

Abstract Title: **Improved Flavor Profile of Italian Wine and Scotch Whiskey Using an Aqueous-Stable Polyethylene Glycol Stationary Phase**

Date: Thursday, March 10, 2016 - Mornin

Time:

Room: Exposition Floor, 400 Aisle

Primary Author: Ramkumar Dhandapani
Phenomenex

Co-Author(s): Kristen Parnell, Tim Anderson

Abstract Text

Monitoring the flavor profile of both high boiling and low boiling components in fermented beverages is very important for health safety and quality control. During the fermentation process in distilled spirit production, compounds called congeners are formed. These congeners contribute to the distinct flavor and aroma profile of the spirit and can include varying amounts of fusel alcohols, esters and acids; when present in excess however, congeners can cause harmful health effects and are additionally suspected to contribute to the well-known "hangover". Some spirits such as vodka therefore undergo extra processing steps to eliminate these compounds. Aside from health concerns, overabundance of a specific congener can signify production or improper storage problems. Analysis is therefore routinely performed using GC; however, fermented beverage samples including distilled spirit congeners are historically challenging to analyze by neat injection with the polar stationary phases required for selectivity due to their instability with high-aqueous matrices.

Presented is a study of the congener flavor profile in Italian wine and Scotch whiskey, which contains 60% water. The methods employed in the study were performed using an aqueous stable Zebron™ ZB-WAXplus™ GC column, and demonstrates reproducibility of the method over time to repeated injections of highly water-based alcoholic beverage samples. In particular, no changes in retention times or peak intensities even after consecutive injections are observed. In addition to stability, the methods offer symmetric peak shape for polar analytes and good peaks shapes for late eluting fatty acids in the same run. Overall, the study achieves accurate, reproducible methods for high-aqueous distilled spirit samples, without compromising resolution.

Keywords: Beverage, Food Identification, Gas Chromatography, GC Columns
Application Code: Food Science
Methodology Code: Gas Chromatography

Session # 2290 Abstract # 2290-15 **Poster Sessions**

Session Title: Various Applications of GCMS

Abstract Title: **Separation Solutions for Triglycerides in Food Fat and Oil by High Temperature GC**

Primary Author: Ramkumar Dhandapani, Phenomenex

Co-Author(s): Kristen Parnell, Tim Anderson

Date: Thursday, March 10, 2016 - Mornin
Time:
Room: Exposition Floor, 400 Aisle

Abstract Text

Triglycerides are esters of glycerol with three fatty acids, and are naturally occurring in food. These compounds have relatively high molecular weights that increase with the degree of unsaturation, and are typically analyzed using low polarity GC stationary phases. Because the boiling point of these compounds are very low, high oven temperatures are essential to elute the analyte out of the stationary phase. Traditional columns used for this testing are limited to maximum oven ramp temperatures of approximately 370 °C. This could cause problems such as carryover of un-eluted high boiling compound in the following injection, excess degradation of the external column coating, and increased phase bleed, which leads to both reduced sensitivity and increased cost due to consumable replacement.

The present work focuses on the high temperature analysis of triglycerides in butter, olive oil, peanut oil and canola oil using GC oven temperatures as high as 400 °C. Methods for high boiling compounds utilized optimized ramp procedures and a thin film, low-polarity Zebron™ ZB-5HT Inferno™ GC column, which provided stability to 430 °C for the high oven ramp programs used. Considering the high boiling point of the triglycerides from different food sources, the method was optimized individually using thin film high temperature Zebron ZB-5HT Inferno columns. The study revealed preferred materials and methods that avoided external film degradation, prevented carryover, and improved sensitivity.

Keywords: Food Identification, Food Science, GC, GC Columns

Application Code: Food Science

Methodology Code: Gas Chromatography

Session # **2290** **Abstract #** **2290-16** **Poster Sessions**

Session Title: Various Applications of GCMS

Abstract Title: **Volatile Organic Compounds in Energy Drinks as Determined by GC/MS with Purge and Trap Sample Concentration**

Date: Thursday, March 10, 2016 - Mornin
Time:
Room: Exposition Floor, 400 Aisle

Primary Author: Cynthia Elmore
OI Analytical

Co-Author(s): J Garrett Slaton

Abstract Text

Energy drinks and shots are a roughly $30 billion dollar industry with major beverage companies , such as PepsiCo , and focused energy drink manufacturers , like Red Bull GmbH , participating . They are consumed by a large number of people , with a 2014 WHO report estimating the consumer base to include 30% of adults , 68% of adolescents , and 18% of children under the age of 10 . Energy drinks may instill a feeling of increased energy or focus . They contain a broad mix of active ingredients as well as various flavorings , preservatives , and coloring agents . The overuse of energy drinks can be problematic , as the number of emergency hospital visits attributed to overconsumption of energy drinks doubled over the period from 2007 to 2010 . In an effort to learn more about the composition of energy drinks , we have analyzed a selection of them using gas chromatography-mass spectrometry with a new purge and trap concentrator and will present these results, identifying volatile organic compounds found in the National Primary Drinking Water Regulations present at detectable levels .

Keywords: Beverage, Food Safety, Food Science, GC-MS
Application Code: Food Science
Methodology Code: Gas Chromatography/Mass Spectrometry

Session # 2290 Abstract # 2290-17 **Poster Sessions**

Session Title: Various Applications of GCMS

Abstract Title: **What's in Your Morning Drink? Comprehensive Characterization of Coffee and Tea Extracts by GCxGC–TOF MS**

Primary Author: Laura McGregor
Markes International Ltd

Date: Thursday, March 10, 2016 - Mornin
Time:
Room: Exposition Floor, 400 Aisle

Co-Author(s): Chris Hall, Ken Umbarger, Massimo Santoro, Nicola Watson

Abstract Text

Almost one thousand different compounds have been identified in roast coffee extracts, with chemical composition varying due to a range of factors, such as coffee bean origin and degree of roasting. The overall flavour and aroma results from a complex combination of chemical classes, including terpenes, oxygenates (aldehydes, esters and ketones) and thiophenes, as well as a range of nitrogen-containing compounds (pyrazines, pyridines and thiazoles). Comprehensive two-dimensional gas chromatography coupled with time-of-flight mass spectrometry (GCxGC–TOF MS) is ideal for the analysis of complex samples such as coffee and tea. The enhanced separation capacity of this technique allows characterisation of the entire sample within a single run. In this study, key flavour compounds, which would have been subject to extensive co-elution in a conventional GC–MS system, were quickly and confidently identified using automated search tools. For example, simple scripting functions were applied to allocate the pyrazines, pyridines and thiazoles based on their characteristic fragmentation patterns, allowing fast characterisation of the sample.

Keywords: Flavor/Essential Oil, Gas Chromatography/Mass Spectrometry, Time of Flight MS, Volatile Organic Co
Application Code: Food Science
Methodology Code: Gas Chromatography/Mass Spectrometry

Session # 2290 **Abstract # 2290-18** **Poster Sessions**

Session Title: Various Applications of GCMS

Abstract Title: **Research of Polychlorinated Biphenyls (PCBs) in Vegetables by GC-MS/MS**

Primary Author: Xizhi Wang, Shimadzu

Date: Thursday, March 10, 2016 - Mornin
Time:
Room: Exposition Floor, 400 Aisle

Co-Author(s): Feifei Tian, Guixiang Yang, Jun Fan, Shiheng Luo, Shin-ichi Kawano, Taohong Huang, Yuki Hashi

Abstract Text

Polychlorinated biphenyls (PCBs), one of the most famous "dirtydozen" persistent organic pollutants (POPs) with carcinogenicity, teratogenicity and mutagenicity, are used to be produced and commercially used as mixtures. Even PCBs were banned by most countries as early as 1970s, they can still be detected in air, soil, water, sediment and biota at a global scale, even in remote areas such as the polar regions, deep seas and high mountains. In this research, Carrot, Potato and Ginger were selected as representative samples analyzed by triple quadrupole tandem mass spectrometry(GC-MS/MS) with MRM mode. The results of this research indicated that the relative coefficients of the 7 kinds of PCBs ranged from 1 to 500μg/L are above 0.999. Precision of this method was measured by analyzing the same sample (1μg/L) six times. The overall RSD% of analysis were below 5%. The limit of detection (LOD; S/N=3) of most compounds were below 0.09μg/L. Commercially available Carrot, Potato and Ginger were used for recovery test, spiked concentration was 10μg/kg and the recovery rate of carrot was 72~108% and that of Potato and Ginger were 76~110% and 93~120%. The developed method in this study was proved to be reliable and accurate, and permits rapid determination of PCBs can be easily applied for quality control of vegetables.

Keywords: Food Contaminants, Gas Chromatography/Mass Spectrometry, Tandem Mass Spec
Application Code: Food Contaminants
Methodology Code: Gas Chromatography/Mass Spectrometry

Session # 2290 Abstract # 2290-19 **Poster Sessions**

Session Title: Various Applications of GCMS

Abstract Title: **Analysis of FAMEs Using Cold EI GC/MS for Improved Molecular Ion Information**

Primary Author: Charles Schmidt
PerkinElmer

Date: Thursday, March 10, 2016 - Mornin
Time:
Room: Exposition Floor, 400 Aisle

Co-Author(s): Adam J. Patkin, Sharanya Reddy

Abstract Text
[b]Introduction[/b]
Characterization of fatty acid methyl esters (FAMEs) is used in several important fields, ranging from fat content in foods and blood to biofuels. They are generally derivatized from free fatty acids and mono-, di-, and triglycerides. FAMEs may be saturated, mono- or polyunsaturated, linear or branched, and of variable chain lengths.

Electron Ionization Gas Chromatography / Mass Spectrometry (EI GC/MS) is often used to characterize FAMEs, but may fail to produce a useful molecular ion for short, unsaturated, or branched chains, making identification more difficult.

Contrastingly, Cold Electron Ionization GC/MS (Cold EI GC/MS) can substantially increase the molecular ion intensity of these compounds, retain EI fragmentation patterns for spectral library searching, and not require modification to established GC methodologies.

[b]Experimental[/b]
A mixture of FAMEs was characterized using EI and Cold EI GC/MS.

GC column effluent with added makeup gas was expanded through a nozzle, creating a supersonic molecular beam. Adiabatic expansion cools the analytes to ~15 K, reducing the molecular vibrational energy and thus fragmentation caused by the ionizing electron. This results in higher molecular ion intensities than for conventional EI, allowing use of this ion for more selective compound analysis of FAME isomers which may have unstable molecular ions, resulting in low or no intensity in EI.

[b]Results[/b]
Very large molecular ion relative intensity enhancements are noted for mid-length unsaturated linear chains. Other compounds have smaller enhancements. The average relative intensity increased a factor of 33.

This enhanced molecular ion in Cold EI provides improved selectivity and less ambiguous molecular weight determination in complex mixtures. Cold EI molecular ion mass chromatograms can be successfully used for quantification with better selectivity than a lower mass fragment which might otherwise have to be used for EI.

Keywords: Biomedical, Food Identification, Fuels\Energy\Petrochemical, Gas Chromatography/Mass Spectrome
Application Code: Food Identification
Methodology Code: Gas Chromatography/Mass Spectrometry

Session # 2290 Abstract # 2290-20 **Poster Sessions**

Session Title: Various Applications of GCMS

Abstract Title: **Basmati or Not Basmati? That is the Question**

Primary Author: Kenneth Rosnack
Waters Corporation

Co-Author(s): Adam Ladak, Gareth Cleland, Jennifer Burgess, Steven Lai

Date: Thursday, March 10, 2016 - Mornin
Time:
Room: Exposition Floor, 400 Aisle

Abstract Text

There is increased concern on the authenticity of basmati rice sold throughout the world. There is reason to believe that economic adulteration, of the commodity being imported from India, is occurring. According to many reports, basmati rice is being adulterated with the cheaper, easier to grow and less aromatic rice, CSR30. More and more sophisticated techniques are required to detect food fraud across a range of commodities that demand a higher price than their cheaper alternatives. Owing to the volatile nature of compounds responsible for the aromatic smell and flavor of basmati rice, dry rice was heated, agitated and compounds of interest extracted from the headspace by SPME. Separations were performed using a 30 m DB5 column. Atmospheric pressure ionization gas chromatography (APGC) coupled to an IMS QTof was used for the analysis of several samples. HDMSE data was collected for replicate samples analyzed as part of a randomized sample set. All data was processed using a novel statistical analysis package which first aligns all experimental runs before peak picking, deconvoluting and performing statistical analysis using various models. The integration of several publically available databases into the processing method made preliminary identification of significant ions identified by the statistical analysis facile and automated calculation of collision cross section (CCS) values for all detected compounds was performed using the drift times obtained from the ion mobility separation. These CCS values can be added to database entries for use as an additional level of selectivity in future experiments. A proof-of-principle method for the investigation of basmati rice authenticity and potential food fraud was devised.

Keywords: Agricultural, Food Identification, Gas Chromatography/Mass Spectrometry, Time of Flight MS
Application Code: Food Identification
Methodology Code: Gas Chromatography/Mass Spectrometry

Session # 2300 Abstract # 2300-1 **Poster Sessions**

Session Title: Vibrational Spectroscopy Advances

Abstract Title: **Development of a Non-Invasive Probing Method for Pharmaceutical Analysis Using Spatially Offset Raman Spectroscopy**

Primary Author: Hyung Min Kim
Kookmin University

Date: Thursday, March 10, 2016 - Mornin
Time:
Room: Exposition Floor, 400 Aisle

Co-Author(s)

Abstract Text

Vibrational spectroscopies, such as Infrared spectroscopy and Raman spectroscopy, should be a better solution for obtaining all information from large batch, since they only examine small volume of the sample. Furthermore, these spectroscopic methods are available for gathering molecular information including functional groups, chirality and tautomerism. They provide much valuable information, comparing with other sample selection – and detecting methods, such as high performance liquid chromatography.

Recently, an emerging new technique, spatially offset Raman spectroscopy (SORS) is a promising tool for the pharmaceutical analysis. SORS can provide more information than conventional Raman spectroscopy because it covers much larger sample size and wide depth. Also, it is non-destructive analysis tool through containers, such as bottles, sacks, and polymers. In this work, we developed SORS setup for probing mixing process of pharmaceutical ingredients. We improved the sampling and analysis speed and presented spectra with better signal to noise ratio, since this technique increases Raman signals. Also, for pharmaceutical industry, uniformity of all tablet components is essential to keep the patients' health. So, we demonstrated this technique as an analysis tool for proving the uniformity of the tablets.

1. Griffen. Julia A., Andrew W. Owen, and Pavel Matousek. Analyst, 140, 107-112 (2014)
2. P. Matousek, I. P. Clark, E. R. C. Draper, M. D. Morris, A. E. Goodship, N. Everall, M.Towrie, W. F. Finney, A. W. Parker, Appl. Spectrosc. 59, 393-340 (2005).
3. H. M. Kim, H. Park, Y. Cho, S. M. Jin, K. T. Lee, Y. M. Jung, Y. D. Suh, J. Mol. Struct., 1069, 223–228 (2014)
4. H. S. Lee, Y. Cho, B. Kim, S. Song, Y. M. Jung, C. R. Park, and H. M. Kim, Bull. Kor. Chem. Soc, 36, 1032–1036 (2015)

Keywords: Chemometrics, Pharmaceutical, Raman
Application Code: Pharmaceutical
Methodology Code: Vibrational Spectroscopy

Session # 2300 Abstract # 2300-2 **Poster Sessions**

Session Title: Vibrational Spectroscopy Advances

Abstract Title: **Spatial Heterodyne Raman Spectrometer with LED Sources**

Primary Author: William Huntington
University of South Carolina

Co-Author(s): S Michael Angel

Date: Thursday, March 10, 2016 - Mornin
Time:
Room: Exposition Floor, 400 Aisle

Abstract Text

The use of LED light sources offers the potential to decrease the size, weight, and power consumption of Raman instruments. These considerations are particularly important for instruments that may be used in planetary exploration. For many Raman applications the output power of LED sources is sufficiently high, but the emission is several nm wide, making it too broad for use. The spatial radiation pattern of powerful multi-element LED sources extends over a very large angle, making it difficult to concentrate all the emitted light onto a sample. Optical bandpass filters can be used to decrease the line width of an LED source, but at the expense of transmitted power. A more efficient approach is to disperse the LED emission using a diffraction grating and focus the dispersed light onto the sample in a line, so that the excitation wavelength changes across it. An imaging detector can then be used to collect the Raman scattered light and the Raman spectrum reconstructed with no loss of resolution or intensity. However, focusing the diffuse source image onto a narrow spectrometer slit still leads to large signal losses. The spatial heterodyne Raman spectrometer (SHRS) has a wide acceptance angle and there is no slit, making it much easier to collect light efficiently from the large illumination spots provided by an LED. This paper will demonstrate multiple wavelength selection schemes for using the LED as a Raman excitation source with the SHRS, and LED Raman spectra of solid and liquid samples.

Keywords: Instrumentation, Molecular Spectroscopy, Raman, Spectrometer
Application Code: Other
Methodology Code: Vibrational Spectroscopy

Session # 2300 Abstract # 2300-4 **Poster Sessions**

Session Title: Vibrational Spectroscopy Advances

Abstract Title: **Second Harmonic Generation of Gold Nanorods Coupled with Atomic Force Microscopy**

Primary Author: Darby Nelson
University of Notre Dame

Date: Thursday, March 10, 2016 - Mornin
Time:
Room: Exposition Floor, 400 Aisle

Co-Author(s): Zachary D. Schultz

Abstract Text
Combining SHG and AFM provides new capabilities that are beneficial to the study of plasmonic interactions. Results suggest that plasmon resonances have significant impact on nonlinear processes, and these nonlinear processes can result in altered electric fields on the nanostructure surface. Second Harmonic Generation (SHG) is a powerful technique for examining surface and interfacial properties but the spatial resolution is limited by the excitation volume (on the order of hundreds of nanometers). Atomic Force Microscopy (AFM) exhibits improved lateral resolution, (on the order of tens of nanometers), along with providing topographic information. Combining these techniques enables improved correlation of the optical response to the surface morphology. This project aims to couple SHG and AFM to study the changing electric fields at the surface of gold nanorods. Our instrument consists of a femtosecond-pulsed super continuum for broadband SHG excitation from 694 nm to 945 nm after filtration coupled to shear force feedback AFM. The super continuum probes the plasmon resonances of the nanorod to reveal how the number of indices of the nanorod changes the resonance of the local surface plasmon with a corresponding AFM image to illustrate the orientation of the nanorods. By understanding the correlation between morphology and optical response, we aim to provide new understanding of the electric fields surrounding plasmonic structures.

Keywords: Imaging, Instrumentation, Vibrational Spectroscopy
Application Code: Other
Methodology Code: Vibrational Spectroscopy

Session # 2300	Abstract # 2300-5	Poster Sessions

Session Title: Vibrational Spectroscopy Advances

Abstract Title: **Chemical Detection and Tracking of c-RGD Peptide-Conjugated Gold Nanoparticles Interacting with αVβ3 Integrins in Living Cells**

Date: Thursday, March 10, 2016 - Mornin
Time:
Room: Exposition Floor, 400 Aisle

Primary Author: Hao Wang
University of Notre Dame

Co-Author(s): Lifu Xiao, Zachary D. Schultz

Abstract Text

Integrins are membrane receptor proteins that have functions associated with a range of cellular activities like cell adhesion and proliferation. Thus, these heterodimer receptors are common drug targets. However, the process of drug screening is inefficient and costly. A key challenge is the lack of ability to monitor the molecular interactions between drug molecules and receptors on the plasma membrane. Here we demonstrate an approach that correlates specific ligand-receptor binding chemistry with binding dynamics by using high-speed dark-field imaging for dynamic tracking of nanoparticle-conjugated c-RGD peptide ligands, and using tip-enhanced Raman scattering (TERS) for specific chemical characterization of the ligands binding with □V□ integrins in living cells. 50 nm spherical gold nanoparticles (AuNPs) have been conjugated with cyclic-arginine (R)-glycine (G)-aspartic acid (D)-phenylalanine (f)-cysteine (C) (c-RGDfC, or c-RGD) peptides, which specifically bind to □V□ receptors, to enable dark-field imaging and generate TERS enhancement. The interactions of these ligand functionalized nanoparticles in two colon cancer cell lines SW480 and SW620, derived from a cancer patient before and after tumor metastasis, are being studied. Real-time tracking of nanoparticles interacting with cells was achieved using a high-speed CCD camera. The controlled plasmonic interaction between the TERS tip and the peptide-conjugated nanoparticles generates significantly enhanced Raman signal, which provides chemical-specific information reflecting binding between c-RGD and □V□ integrins. Our work develops a platform to study the molecular interaction of ligand-receptor recognition in cells.

Keywords: Bioanalytical, Imaging, Vibrational Spectroscopy
Application Code: Biomedical
Methodology Code: Vibrational Spectroscopy

Session # 2300 Abstract # 2300-6 **Poster Sessions**

Session Title: Vibrational Spectroscopy Advances

Abstract Title: **Molecular Interactions between Nanoparticles and Model Cell Membranes Determined via Combined Vibrational Spectroscopic Studies**

Primary Author: Peipei Hu
The University of Michigan at Ann Arbor

Date: Thursday, March 10, 2016 - Mornin
Time:
Room: Exposition Floor, 400 Aisle

Co-Author(s)

Abstract Text

In recent years, nanoparticles (NPs) have attracted the attention of researchers. NPs are widely applied in biomedical areas such as medical diagnosis, targeted drug delivery, and treatment of many diseases. For these applications, a majority of NPs have potential to cross cell membranes and reach the cellular nucleus. Many efforts have been made to determine the cytotoxicity of NPs and to elucidate NP cellular uptake mechanisms, it remains undetermined about how different parameters (size, surface properties, and nature of the NPs, cell types etc.) affect the entry of NPs through cell membrane. In this study, sum frequency generation (SFG) vibrational spectroscopy, a surface/interface-sensitive technique, was applied to investigate the interaction of lipid bilayers (to mimic model mammalian cell membranes) with surface-modified gold (Au) and silver (Ag) NPs. It was determined that flip-flop of the cell membrane was induced by Au NPs with amine and carboxyl coatings. However, the mechanism of Ag NP-cell membrane interaction is the combination of flip-flop and NP accumulation onto the cell membrane. The process of flip-flop dominated during the initial state of the Ag NP-cell membrane interaction, while the NP accumulation process increased as time increased. This study provides dynamic observation of how different NPs (Au and Ag) interact with model mammalian cell membranes on a molecular level.

Keywords: Biomedical, FTIR, Membrane, Spectroscopy
Application Code: Nanotechnology
Methodology Code: Molecular Spectroscopy

Session # 2300 Abstract # 2300-8 **Poster Sessions**

Session Title Vibrational Spectroscopy Advances

Abstract Title **A New Diamond ATR Video Microscopy Accessory**

Primary Author David W. Schiering
Czitek

Date: Thursday, March 10, 2016 - Mornin
Time:
Room: Exposition Floor, 400 Aisle

Co-Author(s)

Abstract Text

Type IIA diamond is a near perfect material for infrared (IR) spectroscopy applications. The combination diamond's IR transmission, hardness, and chemical resistivity are unsurpassed by alternative IR materials. Diamond offers an additional advantage – the optical transmission in the visible range – that allows the combination of IR spectroscopy with visible imaging of the specimen. This presentation will concern the application of MicromATR Vision, a new diamond attenuated total reflection (ATR) – video microscopy accessory for Fourier transform infrared (FT-IR) instruments. The accessory employs a single reflection diamond internal reflection element (IRE) in a "roof top" prism configuration. Microscopical observation is accomplished by imaging the specimen plane through the diamond "roof top" onto a megapixel camera. Live images are observed on a computer monitor via USB interface and video microscopy software. Images can be captured, analyzed, and documented using this software. An external reflection interface can also be used in place of the diamond ATR on MicromATR Vision. The system has been used to characterize a variety of surfaces in ATR and external reflection modes. Comparative external reflection and ATR data has been analyzed from spectra recorded from aluminum surfaces coated with polymers. Surface oxidation on metals has also been measured. Polymer coatings on electrical wires were found to have various compositions including polyethylene, polyvinylidene fluoride, and polyvinyl chloride. Finally contaminants and inks on documents and currencies have been measured and analyzed.

Keywords: Forensics, FTIR, Infrared and Raman, Vibrational Spectroscopy
Application Code: Homeland Security/Forensics
Methodology Code: Vibrational Spectroscopy

Session # 2300 Abstract # 2300-9 **Poster Sessions**

Session Title: Vibrational Spectroscopy Advances

Abstract Title: **From Mars to Mission Critical Process Control**

Date: Thursday, March 10, 2016 - Mornin
Time:
Room: Exposition Floor, 400 Aisle

Primary Author: Dan Wood
Keit Ltd

Co-Author(s)

Abstract Text

Keit's revolutionary new FTIR spectrometer will be presented. Tracings its origins in space instrumentation to development and launch as a rugged in-line process measurement spectrometer. By using an enhanced Sagnac interferometer Keit has been able to deliver pharmaceutical-grade sensitivity in a device that is small and rugged enough to be directly attached to a reaction vessel or continuous reactor. Results from field trials will be presented.

Keywords: FTIR, Infrared and Raman, Pharmaceutical, Process Analytical Chemistry
Application Code: Process Analytical Chemistry
Methodology Code: Vibrational Spectroscopy

Session # 2300 Abstract # 2300-10 **Poster Sessions**

Session Title: Vibrational Spectroscopy Advances

Abstract Title: **Measurement of Component Distribution in "Soft Chew" Formulations by ATR FTIR Imaging**

Primary Author: Ronald Rubinovitz
Thermo Fisher Scientific

Co-Author(s): William Wihlborg

Date: Thursday, March 10, 2016 - Mornin
Time:
Room: Exposition Floor, 400 Aisle

Abstract Text

In this study, the benefits of attenuated total reflectance (ATR) imaging microscopy are applied to "soft chew" formulations in order to obtain distribution information related to their key components. This increasingly popular formulation type presents sampling challenges for micro FTIR measurements by alternate methods. Imaging in transmission mode is challenging since samples are difficult to cut thin enough and simply pressing them flat destroys the distribution information within the sample. Measuring these samples by reflectance mode tends to result in spectra of poor quality. The ATR measurement of a sample cross section offers a viable alternative since sample thickness is not an issue. However, as these sample formulations tend to be sticky and contain oil, the occurrence of "sample carry over" make single point ATR mapping problematic. Also, maps acquired by ATR typically require longer measurement times due to the time required to raise and lower the sample stage at each measurement point. However, the use of an imaging ATR accessory which makes contact with the sample just once while still permitting measurements across the sample surface removes these issues. Further, the utilization of an array detector enables the collection of thousands of spectra in relatively short time periods. The distributions of active and other key components within these products are revealed in chemical maps. Results will show that strong spectral features are readily imaged by univariate methods, such as peak height, while weaker spectral features are detected and mapped by multivariate methods such as multivariate curve resolution.

Keywords: Imaging, Infrared and Raman, Microspectroscopy, Spectroscopy
Application Code: Pharmaceutical
Methodology Code: Vibrational Spectroscopy

Session # **2310** **Abstract #** **2310-2** **Symposia**

Session Title ACS-ANLY - Ultrasensitive Bioanalysis on the Pico-to Femtoliter Scales

Abstract Title **Engineering Hydrogels for Sensitive miRNA Assays**

Date: Thursday, March 10, 2016 - Afterno
Time: 02:10 PM
Room: B308

Primary Author Patrick Doyle
Massachusetts Institute of Technology

Co-Author(s)

Abstract Text
MicroRNAs (miRNAs) are short non-coding RNAs that mediate protein translation and are known to be dysregulated in diseases including diabetes, Alzheimer's, and cancer. With greater stability and predictive value than mRNA, this relatively small class of biomolecules has become increasingly important in determining disease diagnosis and prognosis. Although miRNAs have become increasingly important in disease diagnosis, clinicians still lack proper tools for high-confidence quantification in complex media due to low abundance and sequence homology. In this talk, I will discuss technologies being developed by our group which exploit structured hydrogels for the detection of miRNA. By carefully engineering the physico-chemical properties of the gel and the development of lithographic processes to physically pattern gel modules, we are able to directly detect DNA in both tissue and serum samples. Extensions of these concepts towards single cell assays will also be discussed.

Keywords: Biomedical, Biosensors, Nucleic Acids, Polymers & Plastics
Application Code: Bioanalytical
Methodology Code: Microfluidics/Lab-on-a-Chip

Session # 2310 Abstract # 2310-3 **Symposia**

Session Title ACS-ANLY - Ultrasensitive Bioanalysis on the Pico-to Femtoliter Scales

Abstract Title **Microfluidic Devices with Integrated Nanochannel Arrays to Study Development and Aging of Individual Bacteria**

Date: Thursday, March 10, 2016 - Afterno
Time: 02:45 PM
Room: B308

Primary Author Stephen C. Jacobson
Indiana University

Co-Author(s) David T. Kysela, Joshua D. Baker, Yves V. Brun

Abstract Text

We are developing microfluidic devices with integrated nanochannel arrays to trap individual bacteria and monitor growth and reproduction of lineages over multiple generations. Our poly(dimethylsiloxane) based device comprises a pneumatically actuated nanochannel array with >1200 channels with widths from 400 to 1000 nm to actively trap diverse bacteria. Integrated pumps and valves perform on-chip fluid and cell manipulations that provide dynamic control of cell loading and nutrient flow, permitting chemostatic growth for extended periods of time. Nanochannels confine bacterial growth to a single dimension, facilitating high-resolution time-lapse imaging and tracking of individual cells. We use the device to monitor the growth of single bacterial cells that undergo symmetric (Bacillus subtilis) and asymmetric (Caulobacter crescentus) division and reconstruct their lineages to correlate growth measurements through time and among related cells. Furthermore, we monitor the motility state of single B. subtilis cells through multiple generations by the expression of a fluorescently labeled motility gene and observe the state of the epigenetic switch that is correlated over four generations. Our device allows imaging of cellular lineages with high spatiotemporal resolution to facilitate the analysis of biological processes spanning multiple generations.

Keywords: Bioanalytical, Lab-on-a-Chip/Microfluidics, Nanotechnology
Application Code: Bioanalytical
Methodology Code: Microfluidics/Lab-on-a-Chip

Session # 2310 **Abstract # 2310-4** **Symposia**

Session Title: ACS-ANLY - Ultrasensitive Bioanalysis on the Pico-to Femtoliter Scales

Abstract Title: **Single Cell Genomic and Proteomic Analysis**

Date: Thursday, March 10, 2016 - Afterno
Time: 03:35 PM
Room: B308

Primary Author: Anup K. Singh
Sandia National Laboratories

Co-Author(s)

Abstract Text

We have developed microfluidic platforms for genomic, transcriptomic and proteomic analysis of single cells. It integrates cell handling and analysis with high-resolution imaging and flow cytometry to provide spatio-temporal measurement of signaling pathways with single-cell resolution. The platform is capable of imaging single cells to obtain dynamic translocation data as well as acquisition of quantitative protein expression, phosphorylation and secretion information of selected cell populations. The platform consists of multiple modules such as single-cell array, cell sorter, and phosphoflow chips to provide confocal imaging, cell sorting, and flow cytometry-based phosphorylation assays. The platform allows high-resolution imaging as well as quantitative proteomic measurements with high sensitivity (<pM) and time-resolution (~15 s) in the same population of cells, a feat not achievable by current techniques.

I will also discuss another microfluidic platform that permits digital droplet multiple displacement amplification (MDA) for genome sequencing of samples with limited DNA amounts. While it is routine to sequence an organism when large amount of purified DNA is available (e.g., from cultured cells), genome sequencing from limited number of cells or single cell is still challenging. Multiple displacement amplification (MDA) is the most commonly used technique for whole genome amplification from samples containing single or few cells but suffers from high amplification bias and low specificity. We present a droplet digital MDA (ddMDA) system in which we use droplets to perform MDA with each droplet containing very few DNA templates. The ddMDA approach enables more uniform coverage of amplification over the entire length of the genome, with significantly greater specificity and coverage than conventional tube MDA or microfluidic nanoliter chamber systems.

Keywords: Bioanalytical, Biological Samples, Genomics, Lab-on-a-Chip/Microfluidics
Application Code: Genomics, Proteomics and Other 'Omics
Methodology Code: Microfluidics/Lab-on-a-Chip

Session # 2310 Abstract # 2310-5 **Symposia**

Session Title ACS-ANLY - Ultrasensitive Bioanalysis on the Pico-to Femtoliter Scales

Abstract Title **Ultrasensitive and Broad-Range Microfluidic Immunoassays**

Primary Author Yong Zeng
University of Kansas

Date: Thursday, March 10, 2016 - Afterno
Time: 04:10 PM
Room: B308

Co-Author(s)

Abstract Text

Ultrasensitive analytical measurements becomes increasingly important to biological and clinical applications, ranging from single cell analysis of phenotypic/functional heterogeneity of cells to early detection of diseases. Many of these cases requires quantitative detection of biomolecules that are present in the concentration range beyond the sensitivity of current standard methods (e.g., 10□18 to 10□15 M). In comparison with nucleic acids that can be enzymatically amplified, quantitative detection of low-level proteins remains very challenging. Here we report the development of microfluidics-based platforms for ultrasensitive protein immunoassay with broad dynamic range. The core component of these microfluidic system is an integrated microwell array that are pneumatically controllable. The microstructure allows for solid-state sandwich immunoassay and enzymatic fluorogenic detection in the different manners: measuring ensemble intensity fluorescence signal of the analyte (analog) and single-copy counting of the absolute copy number of protein molecules (digital) confined in femtoliter-scale reactors. Microfluidic engineering of standard immunoassay substantially enhances the analytical performance, such as sensitivity and dynamic range, while reducing the sample demand and analysis time. We will demonstrate the applications of these microfluidic platforms to ultrasensitive and quantitative analysis of protein biomarkers relevant to cancer diagnostics and therapeutics, as well as their potential for single-cell analysis. Overall, these microfluidic systems would provide useful bioanalytical technologies that promote quantitative measurement of complex biological systems and clinical disease diagnosis.

Keywords: Bioanalytical, Immunoassay, Lab-on-a-Chip/Microfluidics, Protein
Application Code: Bioanalytical
Methodology Code: Microfluidics/Lab-on-a-Chip

Session # 2320 Abstract # 2320-1 **Symposia**

Session Title: Analytical Techniques in Neuroscience

Abstract Title: **The Chemical Characterization of the Brain: From New MS-Based Measurement Tools to New Insights**

Date: Thursday, March 10, 2016 - Afterno
Time: 01:35 PM
Room: B309

Primary Author: Jonathan V. Sweedler
University of Illinois

Co-Author(s)

Abstract Text
In the postgenomic era, one expects the suite of chemical players in a brain region to be known and their functions uncovered. However, many cell-to-cell signaling molecules remain poorly characterized and for those that are known, their localization and dynamics are oftentimes unknown. A suite of mass spectrometry-based approaches are described that allow us to assay individual neurons and small brain regions; these approaches include capillary scale separations coupled to mass spectrometry, direct mass spectrometric-based profiling, and mass spectrometry imaging. A key to successful measurement involves optimized tissue and cell sampling protocols. Depending on the sample being assayed and metabolites being measured, we use mechanical isolation, optical tweezers, patch pipettes, dialysis probes and microfluidics, all of which have advantages for specific sample types. Several applications of single cell mass spectrometry are highlighted including the discovery of unusual metabolites to characterizing the peptides in single cells. Specifically, new serotonin-related compounds, the cellular redox state, and literally hundreds of new neuropeptides have been characterized in well-defined neuronal networks, and in several cases, the functional roles of these molecules described. Imaging mass spectrometry and dynamic sampling of the extracellular environment are used for elucidating novel cell to cell signaling molecules in a range of neuronal model systems. Current analytical technology efforts involve extending the depth of metabolome coverage and adapting these analytical approaches to higher throughput single cell assays. Our overarching biological goal is to uncover the complex chemical mosaic of the brain and pinpoint key cellular players in physiological and pathological processes.

Keywords: Capillary Electrophoresis, Mass Spectrometry, Neurochemistry, Time of Flight MS
Application Code: Neurochemistry
Methodology Code: Mass Spectrometry

Session # 2320 Abstract # 2320-2 Symposia

Session Title Analytical Techniques in Neuroscience

Abstract Title **Scalable Proteomic Imaging of Intact Biological Systems**

Primary Author Kwanghun Chung
MIT

Date: Thursday, March 10, 2016 - Afterno
Time: 02:10 PM
Room: B309

Co-Author(s)

Abstract Text

Combined measurement of diverse molecular and anatomical traits that span multiple levels, from molecules to cells to an entire system, remains a major challenge in biology. In this talk, I will introduce a simple method that enables proteomic imaging for scalable, integrated, high-dimensional phenotyping of both animal tissues and human clinical samples. This method, termed SWITCH, uniformly secures tissue architecture, native biomolecules, and antigenicity across an entire system by synchronizing the tissue preservation reaction. The heat- and chemical-resistant nature of the resulting framework permits virtually unlimited rounds of relabeling of a single tissue with precise co-registration of multiple datasets. Furthermore, SWITCH synchronizes labeling reactions to improve probe penetration depth and uniformity of staining. With SWITCH, we demonstrated combinatorial protein expression profiling and high-dimensional quantitative analysis of the human cortex. Such integrated high-dimensional information may accelerate our understanding of biological systems at multiple levels.

Keywords: Biotechnology
Application Code: Biomedical
Methodology Code: Sampling and Sample Preparation

Session # 2320 Abstract # 2320-4 **Symposia**

Session Title: Analytical Techniques in Neuroscience

Abstract Title: **Expansion Sequencing (ExSEQ): Comprehensive In Situ Transcriptome Characterization Throughout Intact Brain Circuits**

Primary Author: Shahar Alon
Massachusetts Institute of Technology

Date: Thursday, March 10, 2016 - Afterno
Time: 03:35 PM
Room: B309

Co-Author(s)

Abstract Text

Enabling the mapping of the cell types of the brain, as well as the systematic analysis of cell types in complex behavioral and disease states, would benefit greatly from a technology for the comprehensive analysis of gene expression patterns in neurons throughout a neural circuit. Ideally we could perform this analysis in intact brain circuits, so that these transcriptional profiles could be combined with morphological and circuit topology information, resulting in unified pictures of the cell types and cell states of the brain. Current tools do not permit this: optical methods maintain the spatial location of molecules, but the number of molecules that can be studied simultaneously is limited. On the other hand transcriptomic approaches allow the multiplexed measurement of potentially all the RNA molecules, but spatial information is lost in the process. Here we devise a new method for in situ sequencing of nucleic acids throughout all the neurons of an intact brain circuit, by creating new forms of expansion microscopy (ExM), a technology that physically magnifies brain tissues while preserving nanoscale isotropy (Science 347:543-548), as well as fluorescent in situ sequencing (FISSEQ; Science 343:1360-1363). Using this new technology, which we call expansion sequencing (ExSEQ), users can expand brain circuits, then sequence the RNAs within the expanded tissue, resolving transcripts throughout entire neurons and neural circuits, enabling systematic cell type and cell state classification in health and disease.

Keywords: Biotechnology, Microscopy
Application Code: Genomics, Proteomics and Other 'Omics
Methodology Code: Microscopy

Session # 2320 Abstract # 2320-5 **Symposia**

Session Title: Analytical Techniques in Neuroscience

Abstract Title: **Genetic Approaches to Brain Circuit Mapping and Cell Type Characterization**

Primary Author: Hongkui Zeng
Allen Institute for Brain Science

Date: Thursday, March 10, 2016 - Afterno
Time: 04:10 PM
Room: B309

Co-Author(s)

Abstract Text

The brain circuit is an intricately interconnected network of numerous cell types. At the Allen Institute, we are developing a comprehensive program to combine molecular, genetic, anatomical and physiological approaches to unravel the diversity and connectivity of the neuronal cell types that compose of neural circuits. To build ground-laying technologies, we have generated transgenic mouse lines that target sensors and effectors to specific types of cells to enable identification, labeling, monitoring and manipulation of these cells. In our cell types program, we use the mouse visual cortex as a model to characterize the transcriptomic, morphological, and electrophysiological properties of different kinds of neurons in a standardized way, towards a systematic taxonomy of cell types for this circuit. Finally, we wish to gain a comprehensive and detailed understanding of how different types of neurons are connected to each other. The Allen Mouse Brain Connectivity Atlas represents the first of such large-scale efforts, in which axonal projections from different regions and different cell types within these regions are systematically mapped throughout the brain to generate a 3D whole-brain projectome.

Keywords: Database, Genetic Engineering, Neural Network
Application Code: Genomics, Proteomics and Other 'Omics
Methodology Code: Microscopy

Session # 2330 Abstract # 2330-1 **Symposia**

Session Title Bioanalytical Chemistry Using the Next Generation of Nanomaterials

Abstract Title **GUMBOS at the Nanoscale: Size Control, Characterization, and Applications**

Date: Thursday, March 10, 2016 - Afterno
Time: 01:35 PM
Room: B310

Primary Author Isiah M. Warner
Louisiana State University

Co-Author(s) Suzana Hamden

Abstract Text

We have been exploring unique applications of room-temperature ionic liquids (RTILs) for several years. More recently, my group has extended the range of these materials to include applications of similar solid phase materials, i.e. organic salts with melting points of frozen ionic liquids (25 °C to 100 °C) up to melting points of 250 °C. To contrast and better define these new materials from RTILs, we have created an acronym, GUMBOS (Group of Uniform Materials Based on Organic Salts). Our GUMBOS have the tunable properties frequently associated with RTILs, including tunable solubility, melting point, viscosity, thermal stability, and functionality. Thus, when taken in aggregate, these properties allow production of solid phase materials which have a wide range of properties, and thus also applications. In this talk, I will highlight select applications of GUMBOS at the nanoscale which we have recently explored. We have designated these GUMBOS-based nanomaterials as nanoGUMBOS. In regard to nanoGUMBOS, we believe that our methodology represents an extremely useful approach to production of nanomaterials since our materials are designed and assembled for specific uses, rather than adapted for use as is true for many nanomaterials. In this talk, I will highlight selected applications, including sensor applications and cancer therapy. Particular emphasis will be placed on the unique advantages acquired by use of nanoGUMBOS and on materials relevant to the general areas of analytical chemistry and the biomedical sciences.

Keywords: Bioanalytical, Chemical, Fluorescence, Material Science
Application Code: Bioanalytical
Methodology Code: Fluorescence/Luminescence

Session # 2330 Abstract # 2330-2 Symposia

Session Title: Bioanalytical Chemistry Using the Next Generation of Nanomaterials

Abstract Title: **Discovering "Genetic Codes" for Nanomaterials Morphologies and Employing the DNA-Encoded Nanomaterials for Sensing and Imaging**

Date: Thursday, March 10, 2016 - Afterno
Time: 02:10 PM
Room: B310

Primary Author: Yi Lu
University of Illinois at Urbana-Champaign

Co-Author(s): Li Huey Tan, Nitya Sai Reddy Satyavolu, Peiwen Wu

Abstract Text

It has been recognized that different shapes and surface properties of nanomaterials can exhibit different optical, electrical, magnetic properties that can be ideal choices for sensing and imaging [1]. However, it has been difficult to control these morphologies and most of these materials lack selectivity toward analytic targets. Inspired by the discovery of genetic codes in biology, we have discovered DNA codes for fine control of the morphologies of nanomaterials, when they are synthesized from the seeds of gold nanospheres, gold nanoprism, gold nanorods and silver nanocubes [2]. Rules of shape control by difference DNAs and their combinations are summarized. These new DNA codes and rules can play an important role in rational design and synthesis of novel nanomaterials with predictive shape control. As a result of these investigations, we have obtained novel nanomaterials display a wide variety of morphologies with high yields and possessing many interesting and tunable properties, such as near-IR surface plasmon resonance. Because the DNA were found to be embedded into the nanomaterials, these DNA-nanomaterials possesses much higher stability than DNA nanomaterials using other immobilization methods, including thio-gold chemistry. The combination of these novel nanomaterials with DNAzymes and aptamers have resulted in new classes of colorimetric and fluorescent sensors, and smart MRI contrast agents with high sensitivity and selectivity for many analytical targets, including metal ions, organic and biomolecules [3].

1. a) Hang Xing, et al., Curr. Opin. Chem. Biol. 16, 429 (2012); b) Li Huey Tan, et al., Acc. Chem. Res., 47, 1881 (2014).
2. a) Zidong Wang, et al., Angew. Chem., Int. Ed. 51, 9078 (2012); b) Jiangjiexing Wu, et al., J. Am. Chem. Soc., 136, 15195 (2014); (c) Tingjie Song, et al., Angew. Chem., Int. Ed. 54, 8114 (2015).
3. a) Peiwen Wu, et al., J. Am. Chem. Soc. 135, 5254 (2013); b) Lele Li & Yi Lu, J. Am. Chem. Soc., 137, 5272 (2015).

Keywords: Bioanalytical, Biomedical, Biosensors, Nanotechnology
Application Code: Bioanalytical
Methodology Code: Sensors

Session # 2330 Abstract # 2330-3 **Symposia**

Session Title: Bioanalytical Chemistry Using the Next Generation of Nanomaterials

Abstract Title: **Free-Standing Gold and Silver Nanoparticles Films as Flexible Sensing Platforms**

Primary Author: Jennifer S. Shumaker-Parry
University of Utah

Date: Thursday, March 10, 2016 - Afterno
Time: 02:45 PM
Room: B310

Co-Author(s)

Abstract Text

Plasmonic architectures offer an opportunity to control the electromagnetic near field response to enhance signals for spectroscopic detection of analytes. A number of approaches based on top down fabrication and bottom up assembly have been used to create plasmonic sensing platforms. We present a simple approach to controlled assembly of gold and silver nanoparticles to produce a plasmonic sensing platform based on asymmetric functionalization, or creation of Janus-like particles. Both gold nanoparticle dimers and silver nanocube dimers prepared using this process will be discussed. An unexpected outcome of the functionalization procedure has been the formation of free-standing gold nanopshere films and silver nanocube films. By modifying the asymmetric functionalization process, single layer nanoparticle films were prepared through a simple adsorption and delamination process. The free-standing films have been transferred to different substrates including polystyrene beads, paper, and plastic. Electron microscopy analysis shows the conformal coverage of the films on these different surfaces and the close spacing of the nanoparticles within the film. The gold and silver nanoparticle films demonstrate good stability based on surface enhanced Raman spectroscopy analysis. Surface characterization using x-ray photoelectron spectroscopy indicates presence of silanes in the nanoparticle films. Interestingly, aged nanoparticles exhibit limited adsorption and do not form films. The mechanism for nanoparticle film formation and applications for detection of bio-analytes will be discussed.

Keywords: Bioanalytical, Nanotechnology, Raman, Spectroscopy
Application Code: Bioanalytical
Methodology Code: Vibrational Spectroscopy

Session # 2330 Abstract # 2330-4 **Symposia**

Session Title: Bioanalytical Chemistry Using the Next Generation of Nanomaterials

Abstract Title: **Ultrafast and Nonlinear Spectroscopy of Plasmonic Nanoparticles for Drug Delivery and Photothermal Applications**

Date: Thursday, March 10, 2016 - Afterno
Time: 03:35 PM
Room: B310

Primary Author: Louis H. Haber
Louisiana State University

Co-Author(s): Corey R. Landry, Daniel J. Hayes, Holden T. Smith, Mohammad Abu-Laban, Raju R. Kumal, Tony E. Karam

Abstract Text

Properties of plasmonic colloidal nanoparticles and their interactions with ions, biological molecules, and light are investigated using ultrafast and nonlinear spectroscopy. The surface charge density, electrostatic potentials, and ion adsorptions on 50 nm colloidal gold nanoparticles coated with mercaptosuccinic acid in water are determined using second harmonic generation (SHG) measurements. Numerical solutions to the Poisson-Boltzmann equation are fit using the SHG signal of gold nanoparticles in water as a function of added NaCl and MgCl2 concentrations to include for effects from nanoparticle surface curvature and ion adsorption to the Stern layer, showing excellent agreement with corresponding electrophoretic measurements. In addition, the photocleaving dynamics of microRNA-functionalized gold nanoparticles in colloidal suspension in water are monitored in real time using SHG. MicroRNA is functionalized to 70 nm spherical colloidal gold nanoparticles using a nitrobenyzl linker that cleaves upon ultraviolet irradiation. Photocleaving rates measured at different irradiation wavelengths show a maximum photocleaving efficiency on resonance at 365 nm. The photocleaving kinetics are measured as a function of the irradiation power to demonstrate that the photocleaving is a one-photon process. Lastly, ultrafast energy transfer and heating dynamics are studied in colloidal gold nanoparticles as well as gold-silver-gold core-shell-shell nanoparticles using transient absorption spectroscopy. The plasmon resonance of these core-shell-shell nanoparticles can be controlled by varying the core and shell dimensions for enhanced photothermal effects in the near infrared for potential cancer therapies and drug-delivery applications. These nonlinear and ultrafast spectroscopic investigations provide important information on the surface chemistry, photocleaving rates, and enhanced heating dynamics of plasmonic nanoparticles for potential biological applications.

Keywords: Nanotechnology, Spectroscopy, Ultra Fast Spectroscopy
Application Code: Nanotechnology
Methodology Code: Biospectroscopy

Session # **2330** **Abstract #** **2330-5** **Symposia**

Session Title: Bioanalytical Chemistry Using the Next Generation of Nanomaterials

Abstract Title: **Biocompatible Nanoparticle Composite Materials: Green Synthesis and Applications**

Primary Author: Chieu D. Tran
Marquette University

Date: Thursday, March 10, 2016 - Afterno
Time: 04:10 PM
Room: B310

Co-Author(s)

Abstract Text

A novel, green and recyclable methods has been developed for the synthesis of biocompatible and biodegradable composite materials containing cellulose (CEL), chitosan (CS) and keratin (KER), from wool, hair and chicken feather. Gold, silver and copper oxide nanoparticles will be encapsulated into the composites to enhance their properties. Various spectroscopy and imaging techniques including FT-IR, NIR, CP-MAS-NMR, XRD, SEM, TGA, DSC were used to monitor the synthetic process, to characterize the materials and to determine their chemical and properties. Novel applications of the nanoparticle composites including antimicrobial activities, drug delivery, hemostasis and would healing, removal of pollutants (organic and heavy metal ions) and toxins will be described.s

Keywords: Bioanalytical, Biomedical, Environmental/Water, Imaging
Application Code: Bioanalytical
Methodology Code: Molecular Spectroscopy

Session # **2340** **Abstract #** **2340-1** **Symposia**

Session Title: Bioinformatics: Metabolite Identification and Quantification

Abstract Title: **New Methods for Improved NMR Quantitation and Reliable MS Coverage in Metabolomics**

Primary Author: Daniel Raftery
University of Washington

Date: Thursday, March 10, 2016 - Afterno
Time: 01:35 PM
Room: B311

Co-Author(s)

Abstract Text

The high complexity of biological samples provides a challenging analysis problem for the field of metabolomics. Ideally, broad metabolome coverage provides the opportunity for deep insights into biological problems, while excellent quantitation allows high reproducibility and an ability to compare across studies. However, these goals are difficult to achieve on a routine basis because the highly complex sample matrix often precludes reliable measurements of many metabolites and complicates quantitation efforts. To improve quantitation in NMR-based metabolomics, we have evaluated a simple protein precipitation procedure that allows the absolute quantitation of over 70 metabolites using a single standard compound. These metabolites, including some at even sub-micromolar concentrations, span a broad range of classes and pathways, including organic and amino acids, as well as energy metabolites and co-enzymes (NAD+, NADH, NADP+, NADPH) and many others. This approach is even useful for calibrating quantitative MS measurements, and has lead to some unusual findings about the stability of well-known metabolites such as glutamine and others. To improve metabolome coverage, we have developed a novel approach that optimizes the number of metabolites measureable using a triple quad MS instrument, which we are calling GOT (globally optimized targeted)-MS. In this approach we can profile over 1000 metabolites using short chromatographic runs (4x9 min each) with excellent reproducibility (average CV <8% for over 1900 MRM transitions). Over half of these transitions match those in known metabolite databases, while the other half are unknown, indicating that GOT-MS can detect unknown metabolites, unlike targeted assays. GOT-MS has high dynamic range (~5 orders of magnitude), with excellent sensitivity and reproducibility, and very few missing values. Application of the new methods for biomarker discovery and cancer biology studies will be discussed.

Keywords: Chemometrics, Mass Spectrometry, Metabolomics, Metabonomics, NMR
Application Code: Genomics, Proteomics and Other 'Omics
Methodology Code: Mass Spectrometry

Session # 2340 Abstract # 2340-2 **Symposia**

Session Title: Bioinformatics: Metabolite Identification and Quantification

Abstract Title: **Point Matching Global Peak Alignment Algorithms for Comprehensive Two-Dimensional Gas Chromatography Coupled with Mass Spectrometry (GC×GC-MS)**

Date: Thursday, March 10, 2016 - Afterno
Time: 02:10 PM
Room: B311

Primary Author: Kim Seongho
Wayne State University/Karmanos Cancer Institute

Co-Author(s)

Abstract Text

Comprehensive two-dimensional gas chromatography coupled with mass spectrometry (GC×GC-MS) has been used to analyze multiple samples in a metabolomics study. However, due to some uncontrollable experimental conditions, such as the differences in temperature or pressure, matrix effects on samples, and stationary phase degradation, there is always a shift of retention times in the two GC columns between samples. In order to correct the retention time shifts in GC×GC-MS, the peak alignment is a crucial data analysis step to recognize the peaks generated by the same metabolite in different samples. Two approaches have been developed for GC×GC-MS data alignment: profile alignment and peak matching alignment. However, these existing alignment methods are all based on a local alignment, resulting that a peak is not correctly aligned in a dense chromatographic region where many peaks are present in a small region. False alignment will result in false discovery in the downstream statistical analysis. We, therefore, develop a global comparison based peak alignment method using point matching algorithm (PMA-PA). The developed algorithm PMA-PA first extracts feature points in the chromatography and then searches globally the matching peaks in the consecutive chromatography by adopting the projection of rigid and non-rigid transformation. Simulation studies show that PMA-PA outperforms the existing alignment algorithms in terms of F1 score.

Keywords: Bioinformatics, GC-MS, Liquid Chromatography/Mass Spectroscopy, Metabolomics
Application Code: Genomics, Proteomics and Other 'Omics
Methodology Code: Chemometrics

Session # 2340 Abstract # 2340-3 **Symposia**

Session Title Bioinformatics: Metabolite Identification and Quantification

Abstract Title **Computational and Database Assisted Structure Identification Tools for Untargeted Metabolomics**

Date: Thursday, March 10, 2016 - Afterno
Time: 02:45 PM
Room: B311

Primary Author David Grant
University of Connecticut

Co-Author(s)

Abstract Text

Despite having free access to multiple metabolomics databases (HMDB, Metlin, PubChem), most non-targeted discovery metabolomics studies report the identification of <20% of all detectable metabolites. HMDB and Metlin are state-of-the-art metabolomics databases that include nearly all known human biochemical compounds. It is therefore alarming that our understanding of the small molecule composition of typical human biofluids (e.g. urine or serum) is so profoundly incomplete. This leads to a current paradox within the field of metabolomics: we cannot identify the structures of unknown metabolites if they are not present in databases and we cannot add new metabolites to these databases without first identifying them. Although metabolomics researches have access to the largest freely available chemical database in the world (PubChem, with >30x106 chemical compounds), there is currently no way to effectively search this resource. In order for metabolomics research to make a sustained and lasting contribution to human health, we need: i) better methods for efficiently searching (and filtering) large biochemical/chemical databases, ii) biochemical databases that contain [nearly] all possible biochemical compounds (known and unknown), and, iii) innovative approaches that allow us to identify the structures of metabolites that are not yet known to exist in humans. In this talk I will provide a brief overview of: 1) current methods of using databases for identifying the structures of unknown metabolites; 2) current methods of augmenting metabolomics databases with unknown biochemical compounds; and, 3) approaches for identifying compounds even if they are not in databases.

Keywords: Bioinformatics, High Throughput Chemical Analysis, Mass Spectrometry, Software
Application Code: High-Throughput Chemical Analysis
Methodology Code: Liquid Chromatography/Mass Spectrometry

Session # 2340 Abstract # 2340-4 **Symposia**

Session Title Bioinformatics: Metabolite Identification and Quantification

Abstract Title **Mass Informatics of Quantitative Metabolomics by Integrating LCxLC-MS and GCxGC-MS Data**

Date: Thursday, March 10, 2016 - Afterno
Time: 03:35 PM
Room: B311

Primary Author Xiang Zhang
University of Louisville

Co-Author(s) Imhoi Koo, Pawel Lorkiewicz, Xiaoli Wei

Abstract Text

Currently, multiple analytical platforms such as liquid chromatography–mass spectrometry (LC-MS), gas chromatography–mass spectrometry (GC-MS) and nuclear magnetic resonance spectroscopy (NMR) are widely used in metabolomics for global metabolic profiling and targeted metabolomics to reveal metabolite abundance alteration in metabolome. To further increase metabolite coverage and to detect low abundance metabolites, multi-dimensional separation methods have been developed, including two dimensional liquid chromatography mass spectrometry (LCxLC-MS) and comprehensive two-dimensional gas chromatography mass spectrometry (GCxGC-MS). However, these analytical platforms generate massive information-rich mass spectrometry data. In order to accurately decipher the biological knowledge from the experimental data, we developed a bioinformatics package that is able to analyze the metabolomics data acquired on LCxLC-MS and GC-xGC-MS platforms for quantitative metabolomics.

In the developed software package, experimental data acquired on each analytical platform are respectively analyzed for spectrum deconvolution, metabolite identification, peak list alignment, normalization and statistical significance tests. The analysis results are then integrated to investigate for metabolite association network and pathway analysis. A suite of data analysis algorithms are developed to assess the quality of experiment data acquired on different analytical platforms. The developed bioinformatics platform has been validated by analysis of experimental data of mixtures of metabolite standards as well as biological samples.

Keywords: Bioinformatics, Chemometrics, Metabolomics, Metabonomics
Application Code: Other
Methodology Code: Data Analysis and Manipulation

Session # 2340 Abstract # 2340-5 **Symposia**

Session Title Bioinformatics: Metabolite Identification and Quantification

Abstract Title **Translating Big Data from HR Imaging MS Data into Molecular Knowledge**

Primary Author Andrew Palmer
European Molecular Biology Laboratory (EMBL)

Date: Thursday, March 10, 2016 - Afterno
Time: 04:10 PM
Room: B311

Co-Author(s) Theodore Alexandrov

Abstract Text

Spatial metabolomics is emerging as a powerful approach to localize hundreds of metabolites directly from sections of biological samples with the grand challenge to be in the molecular annotation of big data generated. Existing bioinformatics tools cannot be applied directly because of the sheer data size and high complexity of spectra. We developed algorithms for molecular annotation for High Resolution Imaging Mass Spectrometry that integrate both spectral and spatial filters and map the results onto metabolic pathways. We will present our efficient implementation using modern big data technologies and apply it to 3D cell spheroids, microbial agar plates, and biological tissues. We will discuss the mass spectrometry as well as algorithmic challenges. One major challenge that we addressed is the formulation of a False Discovery Rate that would be applicable in the spatial metabolomics setting. We also present the evaluation of the proposed FDR. Finally, we present how these algorithms and online engine will be integrated into METASPACE, a novel European project on Bioinformatics for Spatial Metabolomics.

Keywords: Chemometrics, Imaging, Informatics, Metabolomics
Application Code: General Interest
Methodology Code: Data Analysis and Manipulation

Session # 2350 Abstract # 2350-1 **Symposia**

Session Title: Micro and Nano-Scale Optofluidic Lasers for Biological Applications

Abstract Title: **Optofluidic Laser as a New Bioanalytical Tool**

Date: Thursday, March 10, 2016 - Afterno
Time: 01:35 PM
Room: B312

Primary Author: Xudong Fan
University of Michigan

Co-Author(s)

Abstract Text

The optofluidic laser is an emerging technology that integrates microfluidics, miniaturized laser cavity, and laser gain medium in liquid. Due to its unique biocompatibility, bio-interaction and bio-process can take places within the optical cavity mode volume, and thus modulate the gain medium of the laser. Rather than using fluorescence, the optofluidic laser based detection uses laser emission, i.e., stimulated emission, as the sensing signal, which takes advantage of optical amplification provided by the laser cavity to achieve much higher sensitivity. Besides intensity and polarization, the optofluidic laser has many other characteristics that can be used monitored to reveal the details of the bio-interaction and bio-process inside a cavity, such as lasing threshold, lasing efficiency, and lasing mode spatial profile, etc. Therefore, the optofluidic laser provides a powerful complementary technology to fluorescence in bioanalysis.

In this presentation, I will briefly review the history of the optofluidic laser and introduce the concept of optofluidic laser based bioanalysis. Then I will discuss the three important components in the optofluidic laser, i.e., cavity, gain medium, and microfluidics, as well as excitation mechanisms. Finally, I will describe current applications of the optofluidic laser in bioanalysis at the molecular, cellular, and tissue level. If time permits, I will discuss briefly the future research and application directions.

Keywords: Analysis, Bioanalytical, Fluorescence, Lab-on-a-Chip/Microfluidics
Application Code: Bioanalytical
Methodology Code: Microfluidics/Lab-on-a-Chip

Session # 2350 Abstract # 2350-2 Symposia

Session Title Micro and Nano-Scale Optofluidic Lasers for Biological Applications

Abstract Title **Latest Progress in Spasers**

Date: Thursday, March 10, 2016 - Afterno
Time: 02:10 PM
Room: B312

Primary Author Mark I. Stockman
Georgia State University

Co-Author(s)

Abstract Text

Nanoplasmonics deals with collective electron excitations at the surfaces of metal nanostructures, called surface plasmons. The surface plasmons localize and nano-concentrate optical energy creating highly enhanced local fields. Nanoplasmonics has numerous applications in science, technology, biomedicine, environmental monitoring, and defense.

There is an all-important need in active devices capable of generating and amplifying coherent optical fields on the nanoscale analogous to lasers and amplifiers of the conventional optics or transistors of microelectronics. Such an active device is the spaser (surface plasmon amplification by stimulated emission of radiation), also called plasmonic nanolaser. We will focus on the newest ideas and review the latest experimental progress in spasers, which presently cover a wide optical spectrum from near-IR to near-UV.

We will present two new theoretical ideas in the field of spasers: spaser with electric pumping via quantum wire and quantum-cascade graphene spaser. We will consider an example of the latest progress in spasers and some applications of spasers. Among them is a recent breakthrough in ultrasensitive detection of explosives using the spaser. Another recent breakthrough to be presented is an application of the spaser as an ultrabright nanolabel and an efficient theranostic agent in biomedicine (cancer diagnostics and treatment).

Keywords: Bioanalytical, Biomedical, Nanotechnology, Plasma
Application Code: Bioanalytical
Methodology Code: Vibrational Spectroscopy

Session # 2350 Abstract # 2350-3 **Symposia**

Session Title: Micro and Nano-Scale Optofluidic Lasers for Biological Applications

Abstract Title: **Recent Advances in Biolasers: Stimulated Emission from Solid-State Fluorescent Protein and Lasing Inside Live Cells**

Primary Author: Malte C. Gather
University of St Andrews

Date: Thursday, March 10, 2016 - Afterno
Time: 02:45 PM
Room: B312

Co-Author(s)

Abstract Text

An exciting area of recent progress is the integration of small lasers with biological structures[1]. Here, recent work in two areas of biolasers will be summarized; the use of the green fluorescent protein (GFP) as a solid-state gain material and the demonstration of lasing within live cells.

We recently found that the biologically produced GFP represents an exciting solid-state emitter which offers unique optical properties[2]. In contrast to many organic dyes, the special molecular structure of GFP ensures that in solid-state the protein fluorophores have a fixed interspacing. This suppresses concentration quenching and enables strong optical amplification in thin protein films. Using this effect, we fabricate efficient solid-state vertical cavity surface emitting protein lasers. We also find indication that strong exciton-photon coupling may be observed in fluorescent proteins if suitably designed resonators are used [3].

The second part of the talk reports on a laser that is fully embedded within a live cell[4]. Our earlier single cell lasers required extracellular optical feedback structures with limited potential for bio-integration[5]. By harnessing natural endocytosis, we now introduce a whispering gallery mode (WGM) microresonator into the cell. On pumping with nanojoule light pulses, cells with WGM resonators generate green laser emission. The characteristics of the lasing spectrum provide each cell with a barcode-type label which enables identifying and tracking individual migrating cells. Our approach can be applied to different cell types, and cells with microresonators remain viable for weeks under standard conditions.

1. M.T. Hill, M.C. Gather, Nat. Photon. 8, 908-918 (2014)
2. M.C. Gather, S.H. Yun, Nat. Comm. 5, 5722 (2014)
3. C.P. Dietrich, S. Höfling, M.C. Gather, Appl. Phys. Lett. 105, 233702 (2014)
4. M. Schubert et al, Nano Lett. 15, 5647 (2015)
5. M.C. Gather, S.H. Yun, Nat. Photon. 5, 406-410 (2011)

Keywords: Biological Samples, Biospectroscopy, Fluorescence, Nanotechnology
Application Code: Biomedical
Methodology Code: Microscopy

Session # **2350** **Abstract #** **2350-4** **Symposia**

Session Title: Micro and Nano-Scale Optofluidic Lasers for Biological Applications

Abstract Title: **Speckle-Free Lasers**

Date: Thursday, March 10, 2016 - Afterno

Time: 03:35 PM

Room: B312

Primary Author: Hui Cao
Yale University

Co-Author(s)

Abstract Text

High spatial coherence is a defining characteristic of traditional lasers, but in the setting of imaging it has one notorious consequence: speckle. Coherent artifacts such as speckle have precluded the use of lasers in full-field imaging applications. Instead, low spatial coherence sources (e.g. thermal sources and LEDs) are still used in most full-field imaging applications, despite having lower power per mode, inferior efficiency, and less spectral control than laser sources. These limitations are particularly pronounced in applications requiring high-speed imaging, or imaging in lossy or scattering media, prompting the use of raster-scanning based laser imaging systems. This represents a trade-off due to the lack of an appropriate source (combining the low spatial coherence of a thermal source with the high power per mode of a laser).

To address this issue, we have designed special types of lasers, e.g., random lasers and chaotic microcavity lasers, to provide low spatial coherence while maintaining high power per mode. Using such lasers, we demonstrate speckle-free full-field imaging in the setting of intense optical scattering. The images exhibit superior quality than those generated with spatially coherent illumination. These special lasers are well suited for a host of full-field imaging applications from full-field microscopy to digital light projector systems.

Keywords: Imaging, Laser, Light Scattering

Application Code: Biomedical

Methodology Code: Portable Instruments

Session # 2360 **Abstract # 2360-1** **Symposia**

Session Title: Nanofiber Materials Overcome Enduring (Bio) Analytical Challenges

Abstract Title: **Nanofiber Chemistry and Synthesis and the Impact on Analytical Systems**

Primary Author: Margaret W. Frey
Cornell University

Co-Author(s): Edurne Gonzalez, Larissa M. Shepherd, Nidia Trejo

Date: Thursday, March 10, 2016 - Afterno
Time: 01:35 PM
Room: B313

Abstract Text

Nanofibers can provide large functional surfaces for lab-on-chip and lateral flow assay devices without occluding significant volume. These fibers with sub-micron diameters are produced via an easily scaled up process that does not require a clean room. Based on the needs of specific analysis systems, we have developed a broad range of nanofiber functionalities including hydrophilic, hydrophobic, positive and negative surface charges, nitrate, biotinylated, proteins, conductive and piezoelectric and demonstrated these in model devices.

In our research, micro and nanofibers are produced primarily by electrospinning with a focus on using the process variables to drive final morphology of individual fibers and non-woven membranes. The strong elongational flow field, electrical gradient and thermodynamics of solvent evaporation and polymer phase separation all contribute to production of fibers with fine diameters, high concentration of active components at the fiber surface and membrane structures ranging from random to well aligned. Utilizing phase separation, functionalities based on small molecules, expensive and non-fiber forming polymers and proteins have been achieved by co-dissolution or suspension of the active material with a good fiber former to produce a fiber combining sufficient mechanical properties with the desired surface chemistry. Nanofiber membranes have been incorporated into devices by directly spinning into microfluidic channels or cutting shapes from larger membranes for use in microfluidic channels and lateral flow assay devices. Additionally, it is straightforward to interface and combine nanofiber components with other types of components on a device.

Keywords: Material Science, Membrane, Nanotechnology, Polymers & Plastics
Application Code: Polymers and Plastics
Methodology Code: Sampling and Sample Preparation

Session # 2360 Abstract # 2360-2 **Symposia**

Session Title Nanofiber Materials Overcome Enduring (Bio) Analytical Challenges

Abstract Title **Electrospun Fibers for Electrochemical Analysis**

Primary Author Xiamwen Mao
Massachusetts Institute of Technology

Date: Thursday, March 10, 2016 - Afterno
Time: 02:10 PM
Room: B313

Co-Author(s) Gregory Rutledge, Harry Tuller, T Alan Hatton, Yuxi Zhang

Abstract Text

Due to their high specific surface area, electrospun fibers offer the potential for high sensitivity and wide dynamic range in sensor and device applications. With their small diameters, these fibers can provide relatively fast response times. By formulating conductive materials whose electrical properties are sensitive to specific reactions with compounds in the surrounding environment, effective electrochemical sensors based on electrospun polymeric and carbon fibers can be obtained. Applications of these sensors include detection of vapors and of biologically relevant compounds for medical applications. In this talk we demonstrate the fabrication of electrospun fibers of the intrinsically conducting polymer (ICP) polyaniline (PAni), optionally doped to varying degrees with (+)-camphor-10-sulfonic acid (HCSA), both as blends with polyethylene oxide and as the neat ICP fiber by two-fluid electrospinning. Electrical conductivities as high as 150 S/cm were observed for these organic fibers. We show that these fibers can serve as effective low temperature chemoresistive sensors to detect small concentrations of electron-donating or electron-withdrawing gases, like ammonia or nitrogen dioxide. We also discuss the fabrication of carbon fibers from electrospun polyacrylonitrile (PAN), and focus on the electrochemical activities that can be realized through the modulation of electronic densities of states (DOS) and conductivities with carbonization at intermediate temperature (1000-1200°C) and levels of oxidation. These carbon fibers can then serve as the active components of electrochemical devices such as (bio)sensors and supercapacitors.

Funding by the US Army, through the Institute for Soldier Nanotechnologies, and by the US Department of Energy is gratefully acknowledged.

Keywords: Electrochemistry, Energy, Material Science, Sensors

Application Code: Bioanalytical

Methodology Code: Electrochemistry

Session # 2360 **Abstract # 2360-3** **Symposia**

Session Title: Nanofiber Materials Overcome Enduring (Bio) Analytical Challenges

Abstract Title: **Carbon Nanotube–Nanocrystalline Diamond Hybrid Electrodes: A Route for Development of a Highly Sensitive Neurochemical Microsensor**

Primary Author: Prabhu Arumugam
Louisiana Tech University

Date: Thursday, March 10, 2016 - Afterno
Time: 02:45 PM
Room: B313

Co-Author(s)

Abstract Text

Chronic neurochemical monitoring is critical for identifying the neural basis of human behavior and treating brain disorders. Studies have already shown that any abnormal neurochemical signaling cause brain disorders such as epilepsy, Parkinson's disease, traumatic brain injury and drug addiction. To treat such disorders, it is important to understand neurochemical dynamics over long-term, preferably in all areas of the brain. Currently, the preferred detection method is fast-scan cyclic voltammetry (FSCV) and the preferred electrode material is carbon fiber microelectrode (CFM). Unfortunately, CFM's increased sensitivity (sub-micromolar levels) is at the expense of increased surface fouling and chemical etching, which limits electrode lifetime to few days. Emerging carbon nanomaterials have spurred renewed interest in investigating new electrode material technology. Among them, boron-doped ultrananocrystalline diamond (UNCD) with its high chemical inertness, electrochemical and dimensional stability has the potential to be an ideal electrode for sensing. But they lack sensitivity and kinetics at physiologically-relevant chemical concentrations. In this talk, we present a novel carbon nanotube (CNT)-UNCD hybrid electrode for neurochemical detection. CNT is selectively patterned using electrophoretic deposition to achieve highest sensitivity, excellent electrode kinetics and comparable S/N ratio to UNCD. Long-term electrode surface stability and changes to electrochemical properties when exposed to buffer solutions and biological fluids will be reported. Post-surface treatment strategies to improve CNT-UNCD's chemical selectivity and sensitivity towards dopamine, a neurochemical linked to several neurodegenerative diseases will be presented. Finally, development of electrochemical models to explain the progression of surface fouling using impedance techniques will be presented.

Keywords: Electrochemistry, Electrodes, Neurochemistry, Sensors
Application Code: Bioanalytical
Methodology Code: Sensors

Session # 2360 Abstract # 2360-4 **Symposia**

Session Title: Nanofiber Materials Overcome Enduring (Bio) Analytical Challenges

Abstract Title: **Piezoelectric Nanofiber Platform for Cell Monitoring**

Primary Author: Caroline L. Schauer
Drexel University

Date: Thursday, March 10, 2016 - Afterno
Time: 03:35 PM
Room: B313

Co-Author(s)

Abstract Text

Cell-based biosensors can provide new diagnostic or therapeutic tools for numerous diseases. Piezoelectrics are a unique class of materials that are typically utilized in electronic, energy, sensor technology, and electromechanical applications. Combining these two technologies into a nanofibers platform can create a contractile cell-based biosensor. To demonstrate this, a piezoelectric polymer, PVDF-TrFe, was electrospun into aligned nanofibers and interfaced with a flexible plastic substrate, termed a nanogenerator. We have previously demonstrated the strong voltage signal obtained from electrospun aligned PVDF-TrFe nanogenerators in response to cellular scale deformations. Using a coaxial setup, a core-shell polymer fiber can be created using collagen as the shell. These nanogenerators have been successfully characterized for their piezoelectric response, which was an average of $\pm$ 0.1 V. The nanogenerators can be used as a contractile analysis lab-on-a-chip based device as demonstrated using HeLa cell contraction, which was induced with potassium chloride. The nanogenerator system was able to detect both the resting state of HeLa cells, a contraction state, and a hyperpolarized state, proving their potential use as contractile analysis microdevices.

Keywords: Biomedical, Biosensors, Material Science
Application Code: Biomedical
Methodology Code: Sensors

Session # 2360 Abstract # 2360-5 **Symposia**

Session Title Nanofiber Materials Overcome Enduring (Bio) Analytical Challenges

Abstract Title **New Concepts for Lab-on-a-Chip Systems Using Electrospun Nanofibers**

Date: Thursday, March 10, 2016 - Afterno
Time: 04:10 PM
Room: B313

Primary Author Antje J. Baeumner
University of Regensburg

Co-Author(s)

Abstract Text

Microfluidic biosensors, labs-on-a-chip and lateral flow assays for the detection of viable organisms, toxins, and clinically relevant markers have been successfully developed in our research group including analytes such as B. anthracis, C. parvum, dengue virus, E. coli, S. pyogenes, cholera toxin, CD4+ T-lymphocytes, thrombin and myoglobin.

Recently, we initiated the study of electrospun nanofibers and their potential to enhance bioassays in paper-based lateral-flow assays (LFA) and in polymer-based microfluidic devices by adding functionalities to the formats otherwise not available. In the case of the LFA format we successfully demonstrated the de novo fabrication of nanofiber-mats as membrane material enabling immobilization of biorecognition elements, adding novel surface chemistries and preventing non-specific binding without the use of blocking reagents.

Nanofiber-enhanced microfluidic devices provide additional degrees of freedom for bioassay designs, as nanofibers with various surface chemistries are electrospun into distinct locations in the microfluidic channels. As the resulting fiber mats can be of varying density and size and hence generate a 3D-structure (see figure of a 3D model of a typical nanofiber mat) within the channels intimate contact with the sample is guaranteed throughout the channel volume. Current investigations study these systems for sample preparation, as mixers, and as concentrators where, for example, a concentration of E. coli cells by a factor of 20,000 has already been demonstrated. We also develop a nanofiber modeling software that enables fluid dynamic studies to investigate mixing capabilities within our microfluidic devices.

Keywords: Biosensors, Environmental, Food Safety, Lab-on-a-Chip/Microfluidics
Application Code: Bioanalytical
Methodology Code: Microfluidics/Lab-on-a-Chip

Session # 2370 **Abstract # 2370-1** **Symposia**

Session Title: SEAC - Nanoengineered Biosensors

Abstract Title: **Biosensors for Early Cancer Detection Based Upon an Electrical Interface to Virus Particles**

Date: Thursday, March 10, 2016 - Afterno
Time: 01:35 PM
Room: B314

Primary Author: Reginald M. Penner
University of California, Irvine

Co-Author(s)

Abstract Text

In this talk I'll describe a new approach to biosensors that has as its objective the development of ultra-cheap, disposable biosensors that are able to detect cancer markers directly in urine. The realization of this biosensor is made possible by two new developments in our laboratory and that of our collaborator – Professor Gregory Weiss of UCI: The first is a nanowire fabrication technique called Lithographically Patterned Nanowire Electrodeposition (LPNE) that permits very long (> 1 cm), very uniform noble metal nanowires as small as 6 nm x 20 nm to be patterned on glass surfaces. Previously, such nanowires could only be obtained using electron beam lithography – a tedious and expensive fabrication method. The second is the demonstration that filamentous bactiophage particles, that have been engineered using phage display to selectively recognize and bind a particular analyte molecule, can immobilized onto electrode surfaces. The resulting "virus surfaces" retain the ability to recognize and bind molecules from a buffer solution. In fact, these surfaces show kinetic and thermodynamic binding properties for selected analyte molecules that are comparable to immobilized monoclonal antibodies, the gold standard receptors for biosensing. How can LPNE and virus particles be used in conjunction to prepare a biosensor? Three experimental approachs will be described. In one of these, nanowires of the conductive polymer PEDOT (polypoly(3,4-ethylenedioxythiophene)) are fabricated in which virus particles are entrained. These composite nanowires show a change in their electrical impedance upon exposure to peptides – upregulated by cancers - that selectively bind to the entrained virus particles.

Literature References.

1. Synthesis of a Virus Electrode for Measurement of Prostate Specific Membrane Antigen
J.E. Diaz, L.C. Yang, J.A. Lamboy, R.M. Penner*, G.A. Weiss* Methods in Molecular Biology: Biosensors and Biodeection, Vol 504, Humana Press, Chapter 16 (2008) p. 255. 10.1007/978-1-60327-569-9_16
2. Direct Electrical Transduction of Antibody Binding to a Covalent Virus Layer Using Electrochemical Impedance, L.C. Yang, J.J. Diaz, T.M. McIntire, G.A. Weiss*, R.M. Penner* Analytical Chemistry 80 (2008) 5695. 10.1021/ac8008109
3. A Covalent Virus Surface for Mass-Based Biodetection, L.C. Yang, J.J. Diaz, T.M. McIntire, G.A. Weiss*, R.M. Penner* Analytical Chemistry 80 (2008) 933. 10.1021/ac071470f
4. Virus-PEDOT Nanowires for Biosensing, Jessica A. Arter , David K. Taggart , Theresa M. McIntire, Reginald M. Penner,* and Gregory A. Weiss* Nano Letters 10 (2010) 4858. 10.1021/nl1025826
5. Virus-PEDOT Composite Films for Impedance-Based Biosensing, Keith C. Donavan, Jessica A. Arter, Rosa Pilolli, Nicola Cioffi, Gregory A. Weiss*, and Reginald M. Penner*Analytical Chemistry 83 (2011) 2420. 10.1021/ac2000835
6. Virus-PEDOT Biocomposite Films, Keith C. Donavan, Jessica A. Arter, Gregory A. Weiss*, Reginald M. Penner* Langmuir 28 (2012) 12581. 10.1021/la302473j.
7. Virus-Polymer Hybrid Nanowires Tailored to Detect Prostate-Specific Membrane Antigen.
Jessica A. Arter, Juan E. Diaz, Keith C. Donavan, Reginald M. Penner* and Gregory A. Weiss*
Analytical Chemistry 84 (2012) 2776. 10.1021/ac203143y
8. Chemically Synthesized and Genetically Encoded Ligands for Synergistic Binding and Detection of Prostate Specific Membrane Antigen, Kritika Mohan, Keith C. Donavan, Jessica A. Arter, Reginald M. Penner*, and Gregory A. Weiss* J. Am. Chem. Soc. 135 (2013) 7761 10.1021/ja4028082

Keywords: Biomedical, Biosensors, Electrochemistry
Application Code: Bioanalytical
Methodology Code: Sensors

Session # 2370 Abstract # 2370-2 **Symposia**

Session Title SEAC - Nanoengineered Biosensors

Abstract Title **Organic Electronics Biosensors for Label-free Femtomolar Protein Detection**

Date: Thursday, March 10, 2016 - Afterno
Time: 02:10 PM
Room: B314

Primary Author Luisa Torsi
Università degli Studi di Bari "A. Moro"

Co-Author(s) Gerdo Palazzo, Maria Magliulo

Abstract Text

Point-of-care (POC) biosensors are integrated diagnostic systems employed for the detection of clinically relevant analytes in biological fluids such as blood, urine and saliva. These devices offer the advantage to provide rapid results directly where the information is needed (e.g. patient's home, doctor's office or emergency room), thus facilitating an earlier diagnosis and a prompt patient's treatment. Various technologies have been proposed for the realization of POC biosensors including label-free techniques based on optical, mechanical and electrochemical transducers. However, reliable, quantitative and ultrasensitive devices have been not yet commercialized. Electronic biosensors based on organic thin-film transistors (OTFTs) are a promising choice for the development of the next generation of POC devices. These biosensors can be combined with integrated electrical circuits, microfluidic systems and wireless technologies. Furthermore, they offer high sensitivity, biocompatibility and possibility to produce all-printed low-cost biosensors in flexible and disposable formats. Among them, electrolyte-gated (EG)-OTFTs have been identified as ideal candidates for biosensors development as they operate at low voltages directly in aqueous buffer solutions. Using these configurations ultrasensitive label-free immunosensors for the detection of C-reactive protein (CRP), a specific biomarker of inflammatory and infection diseases, at the femtomolar concentration level have been developed. The devices are also able to perform chiral differential detection of odorant molecules. The specific features of the proposed EGOTFT biosensors as well as their analytical performances will be discussed.

- L. Torsi, M. Magliulo, K. Manoli, G. Palazzo, Chemical Society Review 42 (2013) 8612-8628.
- M.Y. Mulla, E. Tuccori, M. Magliulo, G. Lattanzi, G. Palazzo, K. Persaud and L. Torsi. Nature Communications 6 (2015) 6010.
- K. Manoli et al. Angewandte Chemie International Edition, 2015, 54, 12562–12576.

Keywords: Biosensors, Chiral Separations, Material Science, Sensors
Application Code: Bioanalytical
Methodology Code: Sensors

Session # 2370 Abstract # 2370-3 **Symposia**

Session Title SEAC - Nanoengineered Biosensors

Abstract Title **Single Nanoparticle SPR Imaging and Plasmonic Nanocone Arrays: Smart Materials and Smart Chemistries for Advanced Optical Biosensors and Biomimetic Devices**

Date: Thursday, March 10, 2016 - Afterno
Time: 02:45 PM
Room: B314

Primary Author Robert M. Corn
University of California Irvine

Co-Author(s) Adam M. Maley, H W Millie Fung

Abstract Text

In this talk, I will describe how we use Single Nanoparticle SPRI for: (i) the detection and characterization of single hydrogel and glycopolymer nanoparticles, (ii) detection and characterization of single surface RNA enzyme reactions and (iii) the detection of single microRNA molecules. I will also discuss how we fabricate and use large area plasmonic nanocone array surfaces that exhibit: (i) wideband anti-reflection properties, (ii) superhydrophobicity, (iii) enhanced photocatalysis and (iv) rapid biosensing capabilities.

Keywords: Bioanalytical, Spectroscopy, Surface Analysis
Application Code: Bioanalytical
Methodology Code: Sensors

Session # 2370 Abstract # 2370-4 **Symposia**

Session Title SEAC - Nanoengineered Biosensors

Abstract Title **Ultrasensitive Biomolecular Detection Using Nanostructured Microelectrodes**

Date: Thursday, March 10, 2016 - Afterno
Time: 03:35 PM
Room: B314

Primary Author
Shana Kelley
University of Toronto

Co-Author(s)

Abstract Text

The analysis of panels of molecular biomarkers offers valuable diagnostic and prognostic information for clinical decision making. Robust, practical platforms that detect low levels of biomolecules (< 1000 copies) are urgently needed to advance medical care by diagnosing and predicting the progression of cancer and other disease states. Electrochemical methods providing low cost and direct biomarker read-out have attracted a great deal of attention for this application, but have, to date, failed to provide clinically-relevant sensitivity. We exploit controlled nanostructuring of electrode surfaces to promote surface accessibility and enhance capture rate and efficiency to solve this long-standing problem, and showed that the nanoscale morphologies of electrode surfaces control their sensitivities. In addition, we have worked towards integrating nanomaterials-based electrodes into a chip-based platform to facilitate multiplexed analysis in a robust, practical format. Recently, we have developed assays that are able to detect nucleic acids, proteins and small molecules, with universally high sensitivity levels. This presentation will highlight how the nanostructured electrodes were engineered to exhibit high levels of sensitivity and specificity, and how they have been integrated with devices for sample preparation and data analysis.

Keywords: Biosensors, Electrochemistry, Lab-on-a-Chip/Microfluidics, Nucleic Acids
Application Code: Biomedical
Methodology Code: Electrochemistry

Session # 2370	Abstract # 2370-5	Symposia

Session Title SEAC - Nanoengineered Biosensors

Abstract Title **DNA Nanostructures and Networks**

Date: Thursday, March 10, 2016 - Afterno
Time: 04:10 PM
Room: B314

Primary Author Weihong Tan
University of Florida

Co-Author(s)

Abstract Text

Biological systems use complex 'information processing cores' composed of molecular networks to coordinate their external environment and internal states. An example of this is the acquired, or adaptive, immune system (AIS), which is composed of both humoral and cell-mediated components. Here we report the step-by-step construction of a prototype mimic of the AIS which we call Adaptive Immune Response Simulator (AIRS). DNA and enzymes are used as simple artificial analogues of the components of the AIS to create a system which responds to specific molecular stimuli in vitro. We show that this network of reactions can function in a manner which is superficially similar to the most basic responses of the vertebrate acquired immune system, including reaction sequences that mimic both humoral and cellular responses. As such, AIRS provides guidelines for the design and engineering of artificial reaction networks and molecular devices. We will also discuss other DNA nanostructures for molecular medicine.

Keywords: Bioanalytical, Biomedical, Biotechnology
Application Code: Nanotechnology
Methodology Code: Chemical Methods

Session # 2380 Abstract # 2380-1 **Organized Contributed Sessions**

Session Title: Biosensing Devices for Neuron Mapping

Abstract Title: **Integration of CNS and PNS Components with Silicon Devices via Surface Microengineering for Neuronal Mapping Applications**

Date: Thursday, March 10, 2016 - Afterno
Time: 01:30 PM
Room: B315

Primary Author: James J. Hickman
University of Central Florida

Co-Author(s)

Abstract Text
One of the major obstacles hindering the development of an effective therapy for Amyotrophic Lateral Sclerosis, Alzheimer's and other neurodegenerative diseases is the lack of valid, functional in vitro models. Our advancements in culturing human and animal neurons in a defined serum-free medium, suggest outstanding potential for answering questions related to maturation, aging, neurodegeneration and injury, as well as to screen different novel and putative drug candidates. The long-term research goal of our group is to learn how to handle and prepare cells to serve as components for microdevices and engineered tissues, and then to demonstrate the practicality of this approach by manipulating them to build hybrid systems and engineer functional tissues. The idea is to integrate microsystems fabrication technology and surface modifications with protein and cellular components, with the aim of initiating and maintaining self-assembly and growth into biologically, mechanically and electronically interactive functional multi-component systems. The HSL is using this ability to manipulate the biological components and integrate it with silicon-based systems to create cell-based sensors for high throughput drug discovery and functional genomic assays as well as for hybrid neuronal/silicon systems to study biological computation and neuronal mapping. We are also using what we learn for a more fundamental understanding of cellular development and neuronal regeneration. Finally, we believe our research is at the forefront of the next generation of Systems Biology tools necessary to establish the field as the predominant method utilized in drug discovery.

Keywords: Biosensors, Drugs, Lab-on-a-Chip/Microfluidics, Neural Network
Application Code: Bioanalytical
Methodology Code: Microfluidics/Lab-on-a-Chip

Session # 2380 Abstract # 2380-2 **Organized Contributed Sessions**

Session Title: Biosensing Devices for Neuron Mapping

Abstract Title: **Biomimetic Strategies for Seamless Integration of Neural Interface Technology**

Date: Thursday, March 10, 2016 - Afterno
Time: 01:50 PM
Room: B315

Primary Author: Tracy Cui
University of Pittsburgh

Co-Author(s)

Abstract Text

Micro-fabricated neural electrode arrays, placed in the nervous system to directly interface with neurons, have tremendous research and clinical significance. However, current intracortical neural electrodes arrays experience recording failure including signal drift and degradation due to biochemical, mechanical and electrical mismatch between the artificial device and brain tissue. To understand the cellular and tissue response at the neural electrode-tissue interface, multimodal analysis was done using a combination of postmortem immunohistochemistry, live-animal multi-photon imaging and neurophysiological recording. It was observed that implantation of microelectrodes cause immediate vascular damage and microglial activation followed by inflammatory gliosis and chronic neural degeneration in the brain tissue. Furthermore, several bioengineering strategies have been developed to modulate the brain tissue response towards a seamless and stable neural electrode-tissue interface. The first strategy is to immobilize biomolecules onto the implant surface to promote the growth and attachment of neurons while suppressing the glial cell response. Neural electrodes coated with neural adhesion molecule L1 showed enhanced neuronal ingrowth and minimized microglia reaction on and around the implant. Secondly, an on-demand release coating that can actively deliver anti-inflammatory or neuroprotective drugs is being developed. This coating takes advantage of the electrically conductive and electroactive conducting polymer and carbon nanotubes, which allow the drug molecules to be loaded and electrically released while enhancing the recording and stimulation capabilities of the electrodes. Thirdly, electrode arrays that are ultra-small and/or ultra-compliant have been developed to minimize the brain tissue response.

Keywords: Biosensors, Electrode Surfaces, Microelectrode, Sensors
Application Code: Neurochemistry
Methodology Code: Sensors

Session # 2380 Abstract # 2380-3 **Organized Contributed Sessions**

Session Title: Biosensing Devices for Neuron Mapping

Abstract Title: **Automated Micro- and Nanoscale Systems for Single Neuronal Activity**

Primary Author: Jit Muthuswamy
Arizona State University

Date: Thursday, March 10, 2016 - Afterno
Time: 02:10 PM
Room: B315

Co-Author(s): Sindhu Anand, Swathy Sampath Kumar

Abstract Text

Current technologies for sensing single neuronal activity (both extracellular and intracellular) in vivo suffer from limitations that hinder a complete characterization of neural circuits particularly during behavior in long-term experiments. For example, current fixed microelectrode arrays do not allow for optimal tuning of the placement of individual electrodes, which is critical to target neurons that are specific to a given behavior. Technologies for intracellular recording in vivo suffer from even more serious limitations such as (a) bulky form factors (b) requiring highly skilled personnel (c) not scalable for brain mapping applications and (d) not readily usable in behaving animals. We present here a versatile, scalable micro- and nano-scale system that allows for automated extracellular and intracellular recordings from single neurons in vivo. The automated system consists of microscale actuators, closed-loop controls (regulator) and nanoscale interfaces. In n=9 rodent experiments, we determined that an optimal regulator for the neural sensing devices in in vivo experiments resulted in significant reduction in the number of microelectrode movements (0.23 movements/min) and longer periods of stable signal-to-noise ratio (53% of the time) compared to a non-optimal regulator (1.48 movements/min and 23% of the time respectively). In this study, we also demonstrate the ability of this technology to autonomously isolate, penetrate and record from single neurons in abdominal ganglion of aplysia californica. The performance of the system was assessed in n= 50 attempts of single neuron penetrations. Membrane potentials ranging from 40-55 mV were consistently recorded in every attempt.

Keywords: Biosensors, Microelectrode, Neurochemistry, Robotics
Application Code: Biomedical
Methodology Code: Sensors

Session # 2380 Abstract # 2380-4 **Organized Contributed Sessions**

Session Title Biosensing Devices for Neuron Mapping

Abstract Title **Electronic Biosensing Devices for Recording Neuron Transmitter Expression at Single Cell and 3D Tissue Level**

Primary Author Chenzhong Li
Florida International University

Date: Thursday, March 10, 2016 - Afterno
Time: 02:30 PM
Room: B315

Co-Author(s)

Abstract Text

Neurons secrete neurotransmitters to communicate with other cells and coordinate functions. A Neuron's inability to generate and release neurotransmitters is associated with many neurodegenerative diseases, such as ALS, Parkinson's, and Alzheimer's' disease. In fact, the change in neurotransmitters release from neuronal cells is often recorded to identify materials' toxicity, substance of abuse effects, and drug efficacy. Hence, neurotransmitter measurement is a vital task to understand neuronal health. Our group has focused on development of novel biosensing platforms which offer high sensitivity, reliability of use, economical and provide higher throughput as compared to the traditional approaches. We have developed biosensing devices to record neurotransmitter expression at single cell level when exposed to nanomaterials to identify nanotoxicity in the earliest possible stage. We have used a novel large scale integrated array system along with electrochemical mapping to study the effects of dopaminergic drugs on 3D spheroids of neuronal cells. These novel platforms offer promising solutions to the need of reliable, sensitive and high throughput analytical devices to accelerate the neuronal research.

Keywords: Biosensors, Electrochemistry, Lab-on-a-Chip/Microfluidics, Neurochemistry
Application Code: Neurochemistry
Methodology Code: Electrochemistry

Session # 2380 Abstract # 2380-5 **Organized Contributed Sessions**

Session Title Biosensing Devices for Neuron Mapping

Abstract Title **Visualization of Nanoscale Neuron Surface Topography and Detection of Neurotransmitter Release Using Nano Electrochemical Microscopy**

Date: Thursday, March 10, 2016 - Afterno
Time: 03:05 PM
Room: B315

Primary Author Yasufumi Takahashi
Tohoku University

Co-Author(s) Hitoshi Shiku, Tomokazu Matsue

Abstract Text

The dynamics of local chemical concentration and topography changes at cell surface underpins a wide range of neuroscience phenomena. For instance, cell-cell communication is intermediated by neurotransmitter in synapse. Therefore, it is vital to perform measurements of chemical flux with nanoscale spatial resolution. One technique with the potential to measure chemically specific fluxes on the nanoscale is scanning electrochemical microscopy (SECM), but a lack of reliable distance control and difficulties in fabricating small-scale electrodes have largely restricted the technique to the microscale.

We have developed scanning ion conductance microscopy (SICM) and SECM hybrid system to overcome these problems. SICM uses a nanopipette as a probe and imaging cell surface nanoscale topography without contact (Figure 1(a)). We developed SECM-SICM nanoprobes fabrication method, which is an extremely quick (<2 min) and simple process with a high success rate. The double-barrel pipette was pulled from a "theta" quartz capillary (Figure 1(b)). The nanoprobes were fabricated with one barrel filled with pyrolytic carbon for use as the SECM nanoelectrode, and the other barrel filled with electrolyte for SICM.

One of the great advantages of SECM-SICM is that the SICM barrel can be used to apply voltage-driven local chemical ejector. We performed voltage-driven application of K+ ions by using the nanoprobe to achieve both the local depolarization of the cell membrane and simultaneous detection of the neurotransmitter (Figure 1(c)). With local stimulation we always detected either a low frequency of current spikes compared with whole cell stimulation. This finding suggests that the SECM-SICM can be used to induce and detect localized release of the neurotransmitter over the cell surface, thus opening up possibilities to perform the mapping of neurotransmitter release sites.

Keywords: Electrochemistry, Microelectrode
Application Code: Neurochemistry
Methodology Code: Electrochemistry

Session # 2380 Abstract # 2380-6 **Organized Contributed Sessions**

Session Title: Biosensing Devices for Neuron Mapping

Abstract Title: **Trans-Synaptic In Vitro Mapping Using Microfluidic Approaches for Neuroscience Discovery**

Primary Author: Anne M. Taylor
University of North Carolina at Chapel Hill and North Caroli

Date: Thursday, March 10, 2016 - Afterno
Time: 03:25 PM
Room: B315

Co-Author(s)

Abstract Text

Distal injury of long pyramidal tracts remodels cortical circuitry by enhancing excitability, thus influencing recovery following injury; the mechanisms underlying this plasticity are unknown. We developed a novel microfluidics-based in vitro model system to examine synaptic remodeling following distal axon injury in long projection pyramidal neurons. We found that distal axotomy of pyramidal neurons caused dendritic spine loss at synapses onto injured neurons followed by a delayed and persistent retrograde trans-synaptic enhancement in presynaptic excitability. This hyper-excitability involved the elimination of presynaptically "silent" and inhibitory presynaptic terminals without a change in the number of glutamatergic terminals. Further, we found that these changes required differential gene expression and axotomy decreased mRNA expression of the secreted factor netrin-1 coinciding with spine loss. Exogenous netrin-1 applied two days after injury normalized presynaptic hyper-excitability by restoring the excitatory/ inhibitory balance of inputs onto injured neurons.

Keywords: Biological Samples, Drug Discovery, Lab-on-a-Chip/Microfluidics, Method Development
Application Code: Biomedical
Methodology Code: Microfluidics/Lab-on-a-Chip

Session # 2380 Abstract # 2380-7 **Organized Contributed Sessions**

Session Title: Biosensing Devices for Neuron Mapping

Abstract Title: **CMOS Technology Enabled Brain Machine Interface (BMI) For Chronic Neuronal Mapping**

Date: Thursday, March 10, 2016 - Afterno
Time: 03:45 PM
Room: B315

Primary Author: Muhammad M. Hussain
King Abdullah University of Science and Technology (KAUST

Co-Author(s): Aftab M. Hussain, Amir N. Hanna

Abstract Text

According to the World Health Organization (WHO), nearly 20% of the world's population suffers from various neurological illnesses including but not limited to Alzheimer's Disease, multiple sclerosis, autism, dementia, paralysis, epilepsy, traumatic brain injury and such. Neuropsychiatric disorders are among the leading causes of worldwide disability in young people. It is thus imperative to study and understand the neurological condition of our brain. According to US National Academy of Engineering, one of the grandest engineering challenges of the 21st century is development of high-resolution high-performance implantable electronics and their placement in the intracranial space and on the brain's soft tissue. By gathering neurological signals generated in the brain, such systems can offer invaluable information. If we enhance the functionality of such systems for stimulation then they can provide critical clinical therapeutic to Parkinson's disease and other neurological disorders. Still, a convenient mobile system for chronic use is yet to be developed that can monitor the neurological activities. Due to the fact that complementary metal oxide semiconductor (CMOS) can provide ultra-large-scale-integration of ultra-scaled devices, therefore, I will discuss a Brain Machine Interface (BMI) system which is (i) implantable but non-penetrating, eliminating possibility of hemorrhage and inflammatory tissue responses; (ii) physically flexible to be compliant on asymmetric soft surfaces of the brain; (iii) ultra-dense high-performance, ultra-low-power CMOS electronics integrated for large area mapping and high-resolution data acquisition; (iv) fully implantable (wireless data transmission and power supply) and (v) for chronic use with ease (portable and no wire out after implant).

Keywords: Biomedical, Semiconductor, Sensors
Application Code: Biomedical
Methodology Code: Sensors

Session # 2380 Abstract # 2380-8 **Organized Contributed Sessions**

Session Title: Biosensing Devices for Neuron Mapping

Abstract Title: **Wireless Stimulation and Recording for In-Vivo Electrophysiology**

Primary Author: James Morizio
Triangle BioSystems Inc.

Date: Thursday, March 10, 2016 - Afterno
Time: 04:05 PM
Room: B315

Co-Author(s)

Abstract Text

I will present wireless technology advancements for neural recording and stimulation used for in-vivo electrophysiology that can be used with freely moving small rodents species(mice) to large non-human primates. System level concepts will be described that explain the design challenges and tradeoffs of these technologies to acquire EEG, EMG, ECG and single units or spikes signals from brain, central nerve or peripheral nerves. Sub-system components and accessories will also be described that include electrodes or neural interfaces, low noise integrated headstage electronics, RF circuitry, DAQ hardware and analysis software used with electrical and optogenetic Stimulation, neural Recording and combo Stimulation/Recording headstage technologies. In this presentation I will show how these technologies scale with number of input channels, battery life and weight. In addition a variety experiment paradigms will be introduced and described to illustrate correlating animal behavior with electrophysiology application areas or disease models.

Keywords: Biotechnology, Instrumentation, Sample & Data Management, Sensors
Application Code: Drug Discovery
Methodology Code: Integrated Sensor Systems

Session # 2390 Abstract # 2390-1 **Oral Sessions**

Session Title: Bioanalytical: Electrochemical Techniques

Abstract Title: **Compatibility of Nitric Oxide Release Coatings with Implantable Enzymatic Glucose Sensors Based on Osmium(III/II) Mediated Electrochemistry**

Date: Thursday, March 10, 2016 - Afterno
Time: 01:30 PM
Room: B305

Primary Author: Kyoung Ha Cha
University of Michigan

Co-Author(s)

Abstract Text

The development of continuous glucose monitoring systems (CGMS) is an important area of bioanalytical research.1 However, recent commercial CGMS systems involve sensors placed subcutaneously, not within the bloodstream. This leads to a significant lag time in observed sensor signal with respect to actual changes in blood glucose levels. CGMSs that could accurately monitor glucose within the bloodstream would be advantageous especially for ICU patients, where tight glycemic control can greatly improve patient outcome.1 Intravascular (IV) CGMSs typically exhibit thrombus formation on the surface of the implanted sensor that can lead to inaccurate results.2

Nitric oxide (NO) is well known to prevent platelet activation and subsequent thrombosis.3 Needle-type IV amperometric glucose sensor, based on hydrogen peroxide detection at + 0.65 V vs. Ag/AgCl, with an outer NO releasing polymer coating have been reported.4 However, larger background currents were detected due to oxidation of NO. Herein, we examine the compatibility of NO release coatings with glucose sensors prepared with “wired” glucose oxidase (GOX). In this system, GOX is immobilized within a hydrogel of poly(1-vinylimiidazole) and Os(II/III) complex.5 The potential required for oxidation Os (II/III) complex is very low, +400 mV vs. SCE, so that typical interferences present in blood cannot be readily oxidized. It will be shown that this low potential avoids the larger background current from the oxidation of NO. It will be further shown that the NO does not negatively perturb the redox chemistry of Os(II/III) species, resulting in glucose sensors that yield rapid and fully reversible response in the range of 0-20 mM glucose.

1.Van Den Berghe, G. et al. N. Engl. J. Med. 2001, 345(19), 1359.
2.Pronovost, P. et al. Engl. J. Med. 2006, 355(26), 2725.
3.Nguyen, B. L. et al. Am. J. Physiol. 1991, 261, H1043.
4.Yan, Q. et al. Anal. Chem. 2011, 83, 8341.
5.Ohara, T. J. et al. Anal. Chem. 1993, 65, 3512.

Keywords: Bioanalytical, Biomedical, Biosensors, Electrochemistry
Application Code: Bioanalytical
Methodology Code: Electrochemistry

Session # 2390 **Abstract # 2390-2** **Oral Sessions**

Session Title: Bioanalytical: Electrochemical Techniques

Abstract Title: **Application of Nanopipette Electrodes for Real-Time Measurement of Thyroid Hormones to Evaluate Thyrotoxic Storm**

Date: Thursday, March 10, 2016 - Afterno
Time: 01:50 PM
Room: B305

Primary Author: Celeste A. Morris
Northern Kentucky University

Co-Author(s): Barbara Cata, Edward A. Dobrzykowski, Teri Rae Armstrong, Theresa M. Ruwe

Abstract Text

Development of rapid-analysis technology for thyroid hormones in blood is essential for immediate diagnosis of thyrotoxic storm. We assessed the hypothesis that electrochemical reduction of thyroxine would be a viable analytical technique for real-time measurement of thyroxine in blood. Nanopipette electrodes utilized for electrochemical measurement of thyroxine are especially suited for point-of-care testing due to rapid analysis times, simple calibration, and ability to provide hormone concentration levels in blood for patients experiencing thyrotoxicosis. The selective detection and measurement of thyroxine in blood serum has been demonstrated via cyclic voltammetry with □M sensitivity. Improvements in sensitivity and dynamic range of thyroxine measurement was achieved through chemical modification with (11-mercaptoundecyl)-N,N,N-trimethylammonium bromide on gold micro and nanoelectrodes. In conclusion, we developed nanoscale electrodes for real-time thyrotoxic storm evaluation.

Keywords: Biosensors, Biotechnology, Electrochemistry, Nanotechnology
Application Code: Bioanalytical
Methodology Code: Electrochemistry

Session # 2390 Abstract # 2390-3 **Oral Sessions**

Session Title: Bioanalytical: Electrochemical Techniques

Abstract Title: **Neurochemical and Behavioral Analysis of Post-Chemotherapy Cognitive Impairment**

Primary Author: Michael A. Johnson
University of Kansas

Date: Thursday, March 10, 2016 - Afterno
Time: 02:10 PM
Room: B305

Co-Author(s): David P. Jarmolowicz, Meng Sun, Michael J. Sofis, Mimi Shin, Rachel C. Gehringer, Ryan Limbocker, Sam V. Kaplan

Abstract Text

Post-chemotherapy cognitive impairment, also known as chemobrain, is a medical complication of cancer treatment that is characterized by a general decline in cognition affecting visual and verbal memory, attention, complex problem solving skills, and motor function. Previously, we have shown that dopamine release is impaired in rats treated with carboplatin, a widely-used chemotherapeutic agent that has been associated with cognitive impairment. Here, we show that enhanced hydrogen peroxide production persists after carboplatin treatment has concluded. Additionally, we present a novel behavioral paradigm that not only reveals cognitive impairment in these rats but also is amenable to in vivo neurochemical measurements. Finally, the use of zebrafish as a neurochemical model for chemobrain will be discussed.

Keywords: Bioanalytical, Electrochemistry, Microelectrode, Voltammetry
Application Code: Bioanalytical
Methodology Code: Electrochemistry

Session # 2390 Abstract # 2390-4 **Oral Sessions**

Session Title Bioanalytical: Electrochemical Techniques

Abstract Title **In-Vitro Amperometric Sensing of Dynamic Changes of Endogenous NO and CO Gases for Co-cultured Endothelial and Neuronal Cells**

Primary Author Ha Yejin
Ewha Womans University

Date: Thursday, March 10, 2016 - Afterno
Time: 02:30 PM
Room: B305

Co-Author(s) Heo Chaejeong, Lee Youngmi, Suh Minah, Woo Juhyun

Abstract Text

Nitric oxide (NO) and carbon monoxide (CO) are known to play significant roles in broad range of biological phenomena– such as neurotransmission, vasodilation, inflammation, platelet aggregation, etc. Mediating neurotransmission and vasodilation with these gases in the brain is one of the important responsibilities for maintaining brain functions. Nevertheless of their importance, behaviors of NO and CO have not been clearly revealed due to the methodological limitation to study these gases which have chemically analogous properties, exist in minute concentrations, and are easily oxidized in biological environments. To investigate these gases in biological condition, improved amperometric NO/CO dual microsensor, recently developed in our research group, is used. Two platinum (Pt) disks (76 and 50 m in diameter each) modified with selected second metal nanoparticles, and gas permeable membrane enable the sensor to measure NO and CO gases separately. In fact, with an Ag/AgCl counter/reference electrode, NO or CO are selectively oxidized at the working electrode disks. The appropriate modifications of the Pt microdisks allow high sensitivity, selectivity and short response time for NO and CO measurements. Simultaneous detection of the gases with micro-sized electrodes is also an advantage for detecting spontaneously generated gases. In this study, we simply developed the neurovascular coupling unit at cellular level in vitro by co-culturing neuronal cells and brain endothelial cells. This system is optimized to study well-defined NO/CO responses clearly. The results suggest that NO/CO regulation for neurovascular coupling dynamics immediately following the neural stimulation by chemicals.

This research was supported by Basic Science Research Program through the National Research Foundation of Korea (NRF) funded by the Ministry of Science, ICT & Future Planning (2014R1A2A2A05003769), and IBS-R015-D1.

Keywords: Biosensors, Electrochemistry, Microelectrode, Neurochemistry
Application Code: Bioanalytical
Methodology Code: Electrochemistry

Session # 2390 Abstract # 2390-5 **Oral Sessions**

Session Title Bioanalytical: Electrochemical Techniques

Abstract Title **Real-Time Measurements of Oxidative Stress During Chronic L-DOPA Treatment For Parkinson's Disease**

Primary Author Leslie Wilson
North Carolina State University

Date: Thursday, March 10, 2016 - Afterno
Time: 03:05 PM
Room: B305

Co-Author(s) Catherine F. Mason, Christie Lee, Leslie A. Sombers

Abstract Text

Parkinson's disease (PD) is a chronic neurodegenerative disorder characterized by the preferential loss of dopaminergic neurons stemming from the midbrain's substantia nigra pars compacta and innervating the dorsal striatum. The substantial decreases in striatal dopamine (DA) result in devastating hypokinetic movements and motor disturbances. One potential contributor to Parkinsonian symptoms is increased generation of reactive oxygen species, such as hydrogen peroxide (H2O2). However, the precise role of H2O2 in the the initiation, progression, and maintenance of the disease remains unclear, as reactive oxygen species are difficult to monitor in brain tissue. Further, several lines of evidence suggest that the standard treatment strategy of dopaminergic replacement therapy via administration of Levodopa (L-DOPA; L-3,4 dihydroxyphenylalanine) may serve to increase oxidative stress and potentiate cell death. We aim to investigate how striatal H2O2 and DA dynamics underlie behavioral changes that result from chronic L-DOPA administration in a rodent model of PD (unilateral 6-OHDA lesion) using fast-scan cyclic voltammetry, an electrochemical technique that affords precise spatial and temporal resolution, as well as selective detection of these neurochemicals. Specifically, carbon-fiber microelectrodes are used to simultaneously quantify rapid H2O2 and DA fluctuations at single recording sites in the dorsal striatum over several weeks of L-DOPA administration. The chemical fluctuations are correlated with behavioral abnormalities that develop over the course of treatment. These studies will aid in our understanding of how oxidative stress modulates nigrostriatal DA signaling, and will demonstrate how these signals correspond with the development of dyskinetic movements in the treatment of PD.

Keywords: Bioanalytical, Electrochemistry, Voltammetry
Application Code: Bioanalytical
Methodology Code: Electrochemistry

Session # 2390 Abstract # 2390-6 **Oral Sessions**

Session Title Bioanalytical: Electrochemical Techniques

Abstract Title **Dual Function Ion Selective Microelectrodes for Scanning Electrochemical Microscopy**

Date: Thursday, March 10, 2016 - Afterno
Time: 03:25 PM
Room: B305

Primary Author Ganesh Ummadi
Oregon State University

Co-Author(s) Corey Downs, Dipankar Koley

Abstract Text

Solid-state ion-selective microelectrodes have an inherent advantage of having very fast response time and are used as probes for scanning electrochemical microscope (SECM). New dual function calcium ion selective microelectrodes (Ca[sup]2+[/sup]-ISME) were developed and used as a SECM probes in this study to quantitatively map the chemical microenvironment produced by a model substrate, bioactive glass (BAG). This carbon powder based Ca[sup]2+[/sup]-ISME showed broad linear response range with Nernstian slope. The selectivity coefficients of this Ca[sup]2+[/sup]-ISME were found to be log K[sub]Ca[sup]2+[/sup],A[/sub] = -5.5, -6.7, and -6.1 for Mg[sup]2+[/sup], Na[sup]+[/sup], and K[sup]+[/sup], respectively. The SECM experiments with the Ca[sup]2+[/sup]-ISME showed that the BAG releases calcium ions in acidic conditions and neutralizes the pH in the process. This was an important finding because there was no previous characterization of the 3D chemical environment at the surface of the BAG. The Ca[sup]2+[/sup]-ISME had a tip diameter of 25 [micro]m and the inclusion of carbon powder made it redox sensitive. The amperometric function of the sensor was used to fix the tip-substrate distance by using approach curve technique with the SECM. New developments in SECM based 3D chemical imaging would be presented in the meeting.

Keywords: Electrochemistry, Electrodes, Imaging, Sensors
Application Code: Bioanalytical
Methodology Code: Electrochemistry

Session # 2390 Abstract # 2390-7 **Oral Sessions**

Session Title Bioanalytical: Electrochemical Techniques

Abstract Title **Label-Free Electrochemical microRNA Detection based on Different Modifier: Conducting Polymer and Graphene on the Surface of Pencil Graphite Electrode**

Date: Thursday, March 10, 2016 - Afterno
Time: 03:45 PM
Room: B305

Primary Author Mehmet Ozsoz
Gediz University

Co-Author(s) Merve Kaplan

Abstract Text

MicroRNAs (miRNAs) are small non-protein-coding single-stranded RNAs that are 19-23 nucleotides and endogenously expressed . Since miRNA functions are involved in regulating genes , it is considered that aberrant levels of miRNAs are related to many cancer. miRNAs bind to 3' untranslated regions of target mRNA and causes gene inactivation. The voltammetric and impidimetric detection of miRNA as pure form or miRNA from cell lysates has been investigated by using graphene modified disposable pencil graphite electrodes .[1]

Doped nucleic acid probes could also be obtained as a film of electropolymerized polypyrrole (PPy) on the surface of pencil graphite electrode (PGE). The polypyrrole electropolymerization has been done by six voltammetric scanning (between 0.00 and +0.85V; 25 mV/s) using a 0.10 M pyrrole containing 5mg/L oligonucleotide probe solution. Experiments done with and without nucleic acid probes as dopant have been compared. Doping nucleic acid electrodes gave better response than without doping. miRNAs detection has also been done by immobilization of complementary antimir to the surface of the graphene (GRP) modified pencil graphite electrodes then followed the solid phase hybridization with either synthetic miRNA or miRNA included total RNA isolated cell line

1.Kilic, T., et al., Electrochemical Detection of a Cancer Biomarker mir-21 in Cell Lysates Using Graphene Modified Sensors. Electroanalysis, 2015. 27(2): p. 317-326.

Keywords: Bioanalytical, Biosensors, Electrochemistry, Nucleic Acids
Application Code: Bioanalytical
Methodology Code: Electrochemistry

Session # 2390 Abstract # 2390-8 **Oral Sessions**

Session Title Bioanalytical: Electrochemical Techniques

Abstract Title **Quantitative Measurement of Transmitters in Individual Vesicles with Microelectrodes**

Primary Author Xianchan Li
University of Gothenburg

Co-Author(s) Andrew Ewing, Soodabeh Majdi

Date: Thursday, March 10, 2016 - Afterno
Time: 04:05 PM
Room: B305

Abstract Text

During neuronal transmission, vesicles are the major organelles involved in the storage and release of chemical messengers. Quantification of vesicular transmitter content is important in order to study mechanisms of neurotransmission and malfunction in disease and yet this it is incredibly difficult to measure these small quantities in the attoliter volume of a single vesicle.

Recently, we developed a new approach to characterize the contents of mammalian vesicles isolated from adrenal gland. In this approach, nanoscale mammalian vesicles are allowed to adsorb to electrodes and subsequently rupture thereby expelling their contents eliciting an oxidation current that can be used to quantify the catecholamine contents of the vesicles.

With this method that is effective to quantify the contents of individual isolated vesicles it is an exciting prospect to measure the contents of vesicles directly in the intracellular environment. This eliminates the vesicle isolation procedure and allows direct comparison to exocytotic release. Thus, we have developed another method, intracellular vesicle electrochemical cytometry. A nano-tip conical carbon fiber microelectrode is used to electrochemically measure the total contents of electroactive neurotransmitters from individual nanoscale vesicles in single cells as these vesicles lyse on the electrode inside the living cell. The results demonstrate that only a fraction of the quantal contents of neurotransmitter is released during exocytosis. These data support the intriguing hypothesis that the vesicle does not open all the way during the normal exocytosis process. In addition, these levels can be changed with pharmacological manipulation.

Acknowledgement: Funding for this project was provided by the European Research Council (Advanced Grant), Knut and Alice Wallenberg Foundation, the Swedish Research Council (VR), and the National Institutes of Health.

Keywords: Electrochemistry, Microelectrode, Nanotechnology, Neurochemistry
Application Code: Neurochemistry
Methodology Code: Electrochemistry

Session # 2400 Abstract # 2400-1 **Oral Sessions**

Session Title: Bioanalytical: Sampling and Sample Preparation - Half Session

Abstract Title: **Aptamer Functionalized Solid Phase Microextraction for Selective Enrichment of Thrombin**

Primary Author: Md Nazmul Alam
University of Waterloo

Co-Author(s): Fuyou Du, Janusz Pawliszyn

Date: Thursday, March 10, 2016 - Afterno
Time: 01:30 PM
Room: B304

Abstract Text

A novel solid phase microextraction (SPME) coating functionalized with a DNA aptamer for selective enrichment of a low abundance protein from diluted human plasma is described. This approach is based on the covalent immobilization of an aptamer ligand on electrospun microfibers made with the hydrophilic polymer poly(acrylonitrile-co-maleic acid) (PANCMA) on stainless steel rods. A plasma protein, human -thrombin, was employed as a model protein for selective extraction by the developed Apt-SPME probe, and the detection was carried out with liquid chromatography/tandem mass spectrometry (LC–MS/MS). The SPME probe exhibited highly selective capture, good binding capacity, high stability and good repeatability for the extraction of thrombin. The protein selective probe was employed for direct extraction of thrombin from 20-fold diluted human plasma samples without any other purification. The Apt-SPME method coupled with LC–MS/MS provided a good linear dynamic range of 0.5–50 nM in diluted human plasma with a good correlation coefficient (R2 = 0.9923), and the detection limit of the proposed method was found to be 0.30 nM. Finally, the Apt-SPME coupled with LC–MS/MS method was successfully utilized for the determination of thrombin in clinical human plasma samples. One shortcoming of the method is its reduced efficiency in undiluted human plasma compared to the standard solution. Nevertheless, this new aptamer affinity-based SPME probe opens up the possibility of selective enrichment of a given targeted protein from complex sample either in vivo or ex vivo.

Keywords: Bioanalytical, Biological Samples, Sampling, Solid Phase Extraction
Application Code: Bioanalytical
Methodology Code: Sampling and Sample Preparation

Session # 2400 Abstract # 2400-2 **Oral Sessions**

Session Title: Bioanalytical: Sampling and Sample Preparation - Half Session

Abstract Title: **New Generation of Solid SPME Coatings for Complementary Gas- and Liquid- Phase Separation: A Step Toward Integration of Metabolomics Platforms**

Date: Thursday, March 10, 2016 - Afterno
Time: 01:50 PM
Room: B304

Primary Author: Emanuela Gionfriddo
University of Waterloo

Co-Author(s): Ezel Boyaci, Janusz Pawliszyn

Abstract Text

The research focused on the development of new sorbents for SolidPhase Mircoextraction (SPME) has been active in the past decade with the purpose of providing extraction phases suitable to a vast variety of applications. In particular, considering the recent advances of the technique in metabolomics and untargeted analysis, the need for coatings able to extract a broad range of analytes is of outmost importance. In this work we present a new generation of SPME coatings constituted of hydrophilic-lipophilic balanced polymeric particles (HLB) immobilized on a fiber by a fluorocarbon-based polymer. The main novelty of this coating is its suitability for both thermal and solvent desorption, characteristic that allows its use in both liquid- and gas-chromatographic applications. Considered the versatility of the new coatings and the capability of HLB sorbent to extract a broad range of analytes, it is possible to consider their use for integrated metabolomics approaches involving multiple analytical platforms. Additionally, the biocompatibility gained by the selected glue allows extraction from complex matrices where biofouling of macromolecules from the sample matrix may constitute an issue. The extraction performances of the new HLB/F-polymer coatings were tested and compared to conventional coatings previously used for metabolomics investigation by gas- and liquid- chromatography. The probe analytes chosen to carry out the evaluation of coating performances were metabolites occurring in fruits and drugs bearing broad ranges of molecular weights, polarities and functionalities. An attentive study on adsorption kinetic was also carried out in order to fully characterize the coating.

Keywords: Bioanalytical, Extraction, Sample Preparation
Application Code: Bioanalytical
Methodology Code: Sampling and Sample Preparation

Session # 2400 Abstract # 2400-3 **Oral Sessions**

Session Title: Bioanalytical: Sampling and Sample Preparation - Half Session

Abstract Title: **Solid Phase Microextraction as Sample Preparation Tool in Brain Tumors Analysis**

Date: Thursday, March 10, 2016 - Afterno
Time: 02:10 PM
Room: B304

Primary Author: Nathaly Reyes-Garces
University of Waterloo

Co-Author(s): Barbara Bojko, Janusz Pawliszyn

Abstract Text

Solid phase microextraction (SPME) has demonstrated to be a well-suited tool for tissue analysis. The integration of sampling and sample preparation in a single step, and the feasibility of doing in vivo analysis, make of SPME an ideal technique for such type of applications. As a matter of fact, recent studies involving pig and murine models have reported the outstanding capabilities of this technology to monitor metabolic changes in lung, liver and brain. Considering this scientific evidence, in this work we present a preliminary investigation of the potential of SPME for the chemical characterization of human brain tumors. For this purpose, 7 mm mixed mode SPME fibers were used to sample from six different brain tumors as soon as they were resected (in vitro). Depending on each tumor size, three or more fibers were inserted into the tissue to perform the extraction. SPME fibers were exposed to the tissue for 30 min at room temperature. Afterwards, all fibers were quickly rinsed using nanopure water and then stored at -30 °C for further analysis. Extracts obtained after desorbing the SPME fibers in appropriate solvent were run using liquid chromatography coupled to high resolution mass spectrometry (Orbitrap). Statistical analysis showed satisfactory separation among the fibers according to the tumor from which extraction was performed. Interestingly, tumors classified as gliomas (2) showed clear clustering after performing partial least square (PLS) analysis. Among the main metabolites influencing data separation, creatine, inosine, glutamic acid and arachidonic acid were tentatively identified.

Keywords: Clinical Chemistry, Metabolomics, Metabonomics, Sample Preparation
Application Code: Bioanalytical
Methodology Code: Sampling and Sample Preparation

Session # **2400** Abstract # **2400-4** **Oral Sessions**

Session Title: Bioanalytical: Sampling and Sample Preparation - Half Session

Abstract Title: **MEMS Based Pre-Concentrator GC-Ion Mobility Spectrometry for Trace Gas Analysis**

Primary Author: Wolfgang Vautz
ISAS

Date: Thursday, March 10, 2016 - Afterno
Time: 02:30 PM
Room: B304

Co-Author(s): Chandrasekhara Hariharan, Sascha Liedtke, Stefano Zampolli

Abstract Text

Modern analytical techniques are already quite powerful. Nevertheless, industrial and technical developments require continuously a further improvement of the analytical power and sensitivity. In particular multi-dimensional techniques – e.g. mass spectrometry or ion mobility spectrometry coupled to gas or liquid chromatography – with already considerable analysis time would suffer from additional pre-concentration steps, in the worst case in an off-line setup.

We validated the potential of a MEMS based pre-concentrator under these conditions by coupling to an ion mobility spectrometer with gas-chromatographic pre-separation. This technique already requires an off-line sample introduction. Samples cannot be introduced continuously. Therefore, in general a sample volume is flushed with the sample and by switching a valve, the sample volume is introduced into the GC. This procedure also requires considerable time in the range of seconds, depending on the sample volume and the experimental setup.

By applying the MEMS based pre-concentrator, the sampling time could even be reduced while maintaining sensitivity. Furthermore, by variation of the sample volume, the sensitivity can be adapted in a very flexible way to the application still in few seconds. The design of the pre-concentrator moreover enables the easy exchange of the adsorption cartridges, thus enabling the flexible application of different adsorption materials. Summarising, the MEMS based pre-concentrator has high potential for application in many analytical instruments.

Keywords: Adsorption, Portable Instruments, Sampling, Thermal Desorption
Application Code: Bioanalytical
Methodology Code: Sampling and Sample Preparation

Session # 2410 Abstract # 2410-1 **Oral Sessions**

Session Title: Bioanalytical: Techniques Using Sensors

Abstract Title: **Core-Shell Nanoparticle Scintillator Probes for Low-Energy Radionuclide Quantification in Aqueous Media**

Primary Author: Colleen Janczak
University of Arizona

Date: Thursday, March 10, 2016 - Afterno
Time: 01:30 PM
Room: B302

Co-Author(s): Craig A. Aspinwall, Isen Andrew C. Calderon, Zeinab Mokhtari

Abstract Text

[beta]-particle emitting radionuclides such as [sup]3[/sup]H, [sup]33[/sup]P, and [sup]35[/sup]S are useful molecular labels due to their small size and ubiquity, but are challenging to detect and quantify with temporal resolution in biological samples due to their low energies (E[sub]max[/sub] □300 keV) and short penetration depths (□0.6 mm) in aqueous media. Activity measurements for these [beta]-emitters are usually made in milliliter volumes of liquid scintillation cocktail (LSC), a mixture of energy-absorbing organic solvents, surfactants, and scintillant fluorophores, which is incompatible with living cells and therefore dynamic biological measurements. Solid polymer or inorganic crystal scintillator fabrics, papers, sheets, and particles are an alternative to LSC, but the relatively large diameters (□1 μm), the high density of inorganic scintillators, and the orientation of fabrics, papers, and sheets can result in inaccurate counting. We have developed polystyrene-core silica-shell nanoparticle scintillators (nanoScint) which avoid much of the toxicity of LSC as well as many of the limitations of polymer and inorganic crystal scintillators. The polystyrene acts as an absorber for energy from emitted [beta]-particles, and can be loaded with a range of scintillant fluorophores to which the energy is transferred, leading to photon emission at visible wavelengths. The silica shell serves as a hydrophilic shield for the polystyrene core, and facilitates functionalization and attachment of specific binding molecules for target specific imaging or scintillation proximity assays. Furthermore, nanoScint particles have been recovered and re-used for [sup]3[/sup]H activity measurements in bulk aqueous samples, demonstrating the potential for significantly reduced waste in radionuclide quantification.

Keywords: Biosensors, Radiochemical Methods, Sensors
Application Code: Bioanalytical
Methodology Code: Sensors

Session # 2410 Abstract # 2410-2 **Oral Sessions**

Session Title: Bioanalytical: Techniques Using Sensors

Abstract Title: **Use of Silicon Photonic Microring Resonators for the High-Throughput Analysis of Multi-Protein Complex Formation in the Blood Coagulation Cascade**

Date: Thursday, March 10, 2016 - Afterno
Time: 01:50 PM
Room: B302

Primary Author: Ellen Muehl
University of Illinois at Urbana-Champaign

Co-Author(s): Ivan Lenov, Jim H. Morrissey, Josh M. Gajsiewicz, Ryan C. Bailey

Abstract Text

The blood coagulation cascade is governed by interactions of soluble proteins with the cell membrane and membrane-bound proteins. The critical initiating step involves the interaction of soluble protein factor X (FX) with the complex of the membrane bound tissue factor (TF) and soluble activated factor VII (FVIIa). This interaction is influenced not only by the binding domains on the proteins but also by the lipid composition of the membrane. Much is known about the binding of FVIIa to TF but the binding of FX to this complex has up till now been studied only indirectly. To further elucidate these complex interactions, we developed a multiplex assay incorporating TF and TF mutants into nanodiscs of varied lipid composition which were spatially arrayed on silicon photonic microring resonator sensor chips. After formation of the TF-FVIIa complexes, FX binding to up to eight different lipid compositions or TF mutants can be monitored simultaneously. Nanodiscs are lipid bilayer mimics. They are easily fabricated and offer a high degree of control over lipid composition and incorporation of integral membrane proteins. Silicon photonic microring resonators constitute a highly multiplexible, label free technology that can monitor binding interactions in real time. By creating nanodisc arrays on microring resonator sensor ships, kinetic rates of FX binding were determined and analyzed to provide unique insight into the formation of this key multi-protein interaction of the blood clotting cascade.

Keywords: Bioanalytical, Biosensors, Lipids, Protein
Application Code: Bioanalytical
Methodology Code: Sensors

Session # 2410 Abstract # 2410-3 **Oral Sessions**

Session Title Bioanalytical: Techniques Using Sensors

Abstract Title **Label-Free RNA Probes for Live Cell Dual-Color Imaging of EGFR**

Date: Thursday, March 10, 2016 - Afterno
Time: 02:10 PM
Room: B302

Primary Author Xiaohong Tan
Carnegie Mellon University

Co-Author(s)

Abstract Text

The discovery that single-chain antibody fragments exhibit considerable promiscuity in their recognition of fluorogenic cyanines, leading to a catalogue of protein fluoromodules with colors spanning the visible spectrum in which the protein component is constant but the dye varies, motivated our efforts to explore the promiscuity of DIR-binding aptamers. We reported the selection of a RNA aptamer that activates fluorescence from dimethyl indole red (DIR), establishing the unsymmetrical cyanine dyes as attractive components for development of RNA fluoromodules for bioimaging and sensing. Unlike our or others' previous reports in which the pools were symmetric with the stem-loop in the center, the RNA pool used for this selection was asymmetric. After SELEX and sequence minimization and optimization, we obtained a short anti-DIR RNA aptamer (DIR2s-Apt) which can bind and activate DIR fluorescence. Interestingly, this aptamer showed high fluorescence activation ability towards another class of fluorogenic cyanines based on the oxazole thiazole blue core. By constructing EGFR-DIR2s fusing aptamer, we successfully imaged EGFR on the surface of living cells by a dual-color labeling approach. In addition, we demonstrated that cell-surface and internalized EGFR can be discriminated using blue or red fluorogenic dyes.

Keywords: Bioanalytical, Biosensors, Imaging
Application Code: Bioanalytical
Methodology Code: Sensors

Session # 2410 Abstract # 2410-4 **Oral Sessions**

Session Title: Bioanalytical: Techniques Using Sensors

Abstract Title: **Integration of Whispering Gallery Mode Detectors into Fluidic Platforms for Clinical Diagnostics**

Primary Author: Daniel Kim
University of Kansas

Date: Thursday, March 10, 2016 - Afterno
Time: 02:30 PM
Room: B302

Co-Author(s): Robert C. Dunn

Abstract Text

Whispering gallery mode (WGM) resonators enable the label-free detection of analytes based on refractive index sensing. We recently demonstrated a large scale multiplexed imaging platform where hundreds of resonators are simultaneously characterized by coupling a fluorescent dye to the resonator surface. This scheme was used to quantify several biomarkers of ovarian cancer with detection limits comparable to ELISA. Recently, we extended this technique by using an evanescent scattering approach for characterizing WGM resonances. Given the large scattering signals and simplified optics, this approach offers promise for developing fast, inexpensive, and sensitive assays of disease biomarkers. Here, the integration of WGM detectors with separation platforms such as capillary electrophoresis (CE) and microfluidics will be presented. In particular, progress towards coupling WGM immunosensing with serum protein electrophoresis will be discussed to improve diagnostics for multiple myeloma. Integration of WGM immunosensing with CE enables specific detection of biomarkers eluting from the CE column while minimizing interference from non-specific binding.

Keywords: Biological Samples, Biosensors, Capillary Electrophoresis, Immunoassay
Application Code: Bioanalytical
Methodology Code: Sensors

Session # 2410 Abstract # 2410-5 **Oral Sessions**

Session Title Bioanalytical: Techniques Using Sensors

Abstract Title **Ultra-High Spatial Resolution Detection of Localized pH within a Single Live Cell**

Date: Thursday, March 10, 2016 - Afterno
Time: 03:05 PM
Room: B302

Primary Author Qingbo Yang
Missouri University of Science and Technology

Co-Author(s) Hai Xiao, Honglan Shi, Xiaobei Zhang, Yinfa Ma

Abstract Text

The fundamental understanding of a single cell behavior is an essential prerequisite for mechanism study of many biological and pathological processes, such as early-stage carcinogenesis as well as pluripotent cell differentiation. Intracellular pH homeostasis play a key role at the very early phases of cell response to many external and internal stimuli, and mediate profound downstream signaling processes. Thus it can strongly regulate the cell fate, and work as a representative parameter of status quo of a cell. However, there is a lack of sensitive and high pH resolution techniques to measure the pH in a single live cell without permanent damages. In present study, a novel single cell pH sensor with ultra-high spatial resolution was developed with unique tip-cut, hexagonal 1-in-6 fiber configured sensing probe. An ultra-thin, porous aerogel layer based on organic-modified silicate (OrMoSils) technique with pH sensing ability was deposited onto the tip surface. Only 1 - 3 m tip-region pH condition can actively be detected. An excellent linear correlation between fluorescent peak area and pH within biologically relevant range was obtained with 0.03 pH resolution. The fabricated probe can provide real-time sensing ability, with minimum invasiveness and without chronic cell growth disruption. Hence, the new probe can be used in detecting intracellular pH variation with high spatial resolution. The detailed experimental designs and results will be presented at the conference.

This project was supported by National Institute of Health (1R21GM104696-01).

Keywords: Bioanalytical, Biosensors, Detector, Sensors
Application Code: Bioanalytical
Methodology Code: Sensors

Session # 2410 Abstract # 2410-6 **Oral Sessions**

Session Title: Bioanalytical: Techniques Using Sensors

Abstract Title: **Real Time Analysis of Hepatitis B Virus Assembly with Multi-Pore Nanofluidic Devices for Enhanced Resolution of Particle Size and Electrophoretic Mobility**

Date: Thursday, March 10, 2016 - Afterno
Time: 03:25 PM
Room: B302

Primary Author: Panagiotis Kondylis
Indiana University

Co-Author(s): Adam Zlotnick, Jinsheng Zhou, Lisa Selzer, Stephen C. Jacobson, Zachary D. Harms

Abstract Text

Understanding virus assembly pathways and developing a general model for self-assembly will accelerate the development of potential antiviral drugs. To study virus assembly, we are using resistive-pulse sensing for real time analysis of reactions in solution. In our experiments, better size discrimination is needed to fully resolve complete capsids from reaction intermediates. Measurement of particles with multi-pore devices improves not only the signal-to-noise ratio but also the size resolution by a factor equal to the square root of the number of pores. In addition, the pore-to-pore time (or electrophoretic mobility) can be used to further characterize particles. To demonstrate improved measurement precision, we fabricated nanofluidic devices with 2, 4, and 8 nanopores connected in series and compared the resolution of particle size and pore-to-pore time for the assembly of empty T = 3 (90 dimers) and T = 4 (120 dimers) Hepatitis B Virus capsids. With the 8-pore devices, the amplitude (size) resolution improved by a factor of 2.2, and the pore-to-pore time resolution improved by 3.2 when compared to measurements made on 2-pore devices. Results from these devices provide a more complete picture of capsid formation and the annealing of late-stage intermediates into fully formed capsids.

Keywords: Analysis, Bioanalytical, Electrophoresis, Nanotechnology
Application Code: Bioanalytical
Methodology Code: Sensors

Session # 2410   Abstract # 2410-7   **Oral Sessions**

Session Title: Bioanalytical: Techniques Using Sensors

Abstract Title: **Graphene-Based Chemiresistive Nanobiosensors for Detection of Citrus Greening Disease**

Primary Author: Thien-Toan H. Tran
University of California, Riverside

Date: Thursday, March 10, 2016 - Afterno
Time: 03:45 PM
Room: B302

Co-Author(s): Ashok Mulchandani, Clark Kelley, Jinxia Shi, Wenbo Ma

Abstract Text

Citrus greening disease, also known as Huanglongbing (HLB), is posing a worldwide threat to the multi-billion dollars citrus industry. Containment of the disease is heavily dependent on early detection of infected hosts for quarantine. One of the major pathogens responsible is the bacteria [i]Candidatus[/i] Liberibacter asiaticus ([i]C[/i]Las). Current methods for detection of HLB are based on qualitative assessment of disease symptoms and nucleic acid assays which are susceptible to error and inaccuracies and suffer from lack of portability and ease-of-use making them unsuitable for onsite applications due to variable latent time and sporadic distribution of the pathogens in infected plants.

Here, we report various chemiresistive immunosensors that use graphene allotropes for quantitative detection of HLB biomarkers in tree phloem extract by targeting proteins secreted by [i]C[/i]Las. These secreted proteins can systematically distribute in the infected trees, which affords reliable and selective diagnosis of infected plants. Using these biomarkers, we have generated custom polyclonal antibodies which were then used to functionalize our biosensors for specific detection of HLB biomarkers. The biosensor utilizes semiconducting graphene-based nanomaterials, such as reduced graphene oxide (rGO) platelets and single-walled carbon nanotubes (SWNTs), as electrical transducers. Antigen-antibody binding at the surface of rGO platelets and SWNTs lead to changes in the local electrostatic environment and consequently leads to proportionate modulation of electrical resistance of the nanomaterials and the sensing device. In summary, this biosensor provides a viable analytical tool for the citrus industry for management of HLB.

Keywords: Biosensors, Biotechnology, Immunoassay, Nanotechnology

Application Code: Bioanalytical

Methodology Code: Sensors

Session # 2410 **Abstract # 2410-8** **Oral Sessions**

Session Title: Bioanalytical: Techniques Using Sensors

Abstract Title: **A Quantum Dot Based Fiber-Optic Micro-Sensor for Niche-Environment Temperature Monitoring**

Primary Author: Ke Li
Missouri University of Science and Technology

Co-Author(s): Hai Xiao, Honglan Shi, Qi Zhang, Qingbo Yang, Yinfa Ma

Date: Thursday, March 10, 2016 - Afterno
Time: 04:05 PM
Room: B302

Abstract Text

Temperature may regulate or reflect immediate bio-reactions during cell-environment interactions and play a key role in fields like drug delivery, cancer diagnosis, nanotoxicity, and so on. It's still a great challenge, however, to detect the thermal oscillations of a biological/biochemical system within a confined and miniaturized environment. In this study, a novel 80-100 □m sized fiber-optic sensor was designed and fabricated based on the temperature sensitive core-shell CdSe/ZnS quantum dots (QDs). Twisted-fused dual-fiber tips were flat-cut and sealed together with nano-liter QDs within the end cap of a tapered hollow glass tube. The fluorescence intensity of the QDs was found proportionally correlated with the temperature change and a good linearity was obtained between fluorescent peak area and temperature within a narrow, biologically relevant range of ~30□C to 45□C (r2=0.9901). Moreover, the equilibrating time was fast (less than five seconds) with good, reproducible detecting accuracy of about ± 0.1□C. Finally, this newly developed temperature sensor probe was successfully used in measuring artificially adjusted temperature gradients as well as real-time monitoring of thermal behaviors of cell colony under environmental stimuli. In summary, this novel temperature probe can potentially be used for rapid monitoring of local temperature change with ultra-small scale variations, and thus may greatly contribute to future biological researches and applications

More detailed experimental designing and results will be presented at the conference, and this project was supported by the National Institute of Health (1R21GM104696-01).

Keywords: Fiber Optics, Sensors, Temperature
Application Code: Bioanalytical
Methodology Code: Sensors

Session # 2420 Abstract # 2420-1 **Oral Sessions**

Session Title: Computers in Chemistry - Half Session

Abstract Title: **An Inexpensive, Programmable System for Prototyping Instruments, Making Short Run Specialty Measurement Systems, and Computerizing Outdated Hardware**

Primary Author: Scot D. Abbott
Phoenix First Response

Date: Thursday, March 10, 2016 - Afterno
Time: 01:30 PM
Room: B301

Co-Author(s)

Abstract Text

A major challenge for developing instrumentation and measurement systems is the cost (time and money) of high quality computerization for projects. Usual routes have been either (1) expensive software linked to expensive i/o devices, (2) low quality i/o devices, and/or (3) require significant programming. This stopped many development efforts because the of timescale, programming costs, and inflexibility.

In our laboratories, we needed to computerize several different kinds of devices and develop flexible user interfaces at modest cost. We also have old instruments that needed to be computerized. We needed to have the ability to take in several forms of data, provide real time graphics and use PC's running various versions of Windows® (32 and 64 bit) and Office®. We have developed a high performance instrumentation-oriented system which is very easy to program, inexpensive, takes many forms of input (analog, digital and frequency), and has several forms of output. We have found this approach very workable in our labs. It has now been used developing and prototyping new instrument personalities, and reviving old instrumentation. Examples and data will be provided.

Keywords: Computers, Flow Injection Analysis, Instrumentation, Process Analytical Chemistry
Application Code: General Interest
Methodology Code: Computers, Modeling and Simulation

Session # 2420   Abstract # 2420-3   **Oral Sessions**

Session Title   Computers in Chemistry - Half Session

Abstract Title   **Maximizing the Information from the Infrared Spectra of Mixtures Using Advanced Software Algorithms**

Date: Thursday, March 10, 2016 - Afterno
Time: 02:10 PM
Room: B301

Primary Author   Ian Robertson
PerkinElmer Limited

Co-Author(s)   Jerry Sellors, Justin Lang

Abstract Text

Infrared (IR) spectroscopy is considered to be one of the primary analytical techniques for the identification of materials. The IR spectrum gives a unique fingerprint for different chemical species. This can be utilized for the identification of materials against substantial spectral libraries or for the qualification of materials against databases of known "good" materials. These approaches work well for single or pure component materials. However, multicomponent or adulterated/contaminated materials can offer a more significant challenge. In the case of multicomponent mixtures it is necessary to be able to identify all of the components present. In the case of adulterated materials it is necessary to first recognize that the material is adulterated, and secondly identify the adulterant material. Selection of the appropriate algorithm allows for the extraction of the maximum amount of information for the specific requirements of the analysis. Several applications will be shown to demonstrate the applicability of IR spectroscopy and advanced algorithms to multicomponent mixtures and contaminated/adulterated materials.

Keywords: FTIR, Molecular Spectroscopy, Software
Application Code: General Interest
Methodology Code: Molecular Spectroscopy

Session # 2420 **Abstract # 2420-4** **Oral Sessions**

Session Title: Computers in Chemistry - Half Session

Abstract Title: **Driving Governance and Organizational Change in Large and Complex Informatics Projects**

Primary Author: Adam S. Borenstein
LabAnswer

Date: Thursday, March 10, 2016 - Afterno
Time: 02:30 PM
Room: B301

Co-Author(s): Brian Brunner, Terryl Kibodeaux

Abstract Text

For decades, large organizations have tried and failed to deliver complex enterprise-scale laboratory informatics projects. Some that claim success have left user communities with extreme implementation fatigue and mountains of data that cannot be leveraged as actionable knowledge. Damaged relationships between traditional IT and Laboratory Operations have opened the door to "shadow IT" within the business. Today, emerging technologies and the promise of critical analytics on holistic sets of scientific data are drawing large organizations back to considering enterprise-scale informatics projects. This presentation focuses on the non-technical keys to success in highly technical programs. Namely, how program governance and organizational change management will help both business and laboratory operations executives achieve their operational and strategic goals, while avoiding "bottlenecks" that hamper solution acceptance and adoption.

Keywords: Laboratory Informatics, LIMS, Quality, Sample & Data Management
Application Code: Laboratory Management
Methodology Code: Laboratory Informatics

Session # 2430	Abstract # 2430-1	Oral Sessions

Session Title: Glycan Analysis - Half Session

Abstract Title: **Identification of Serum N-Glycans as Cancer-Specific Biomarkers by Microchip Electrophoresis and Mass Spectrometry**

Primary Author: Christa Snyder
Indiana University

Date: Thursday, March 10, 2016 - Afterno
Time: 01:30 PM
Room: B303

Co-Author(s): Margit I. Campos, Milos V. Novotny, Stephen C. Jacobson, Xiaomei Zhou

Abstract Text

Glycosylation patterns of proteins can be highly sensitive to the local biological environment and often become aberrant in response to diseases, e.g., cancer. Microchip electrophoresis of serum N-glycans generates N-glycan profiles specific to disease states and permits quantitative differentiation between samples from disease-free (control) individuals and samples from patients with various cancers. N-glycans were extracted from 5-[micro]L aliquots of serum, methylamidated to neutralize native charges associated with the presence of sialic acids, fluorescently labeled with 8-aminopyrene-1,3,6-trisulfonic acid (APTS), separated on microfluidic devices, and detected with laser-induced fluorescence detection.

Analysis of serum N-glycan profiles differentiated samples from disease-free individuals and colorectal and ovarian cancer patients after their first and third treatment cycles. Separations of the neutralized serum samples had aligned migration time reproducibilities better than 0.5% and separation efficiencies up to 700,000 plates. The 40-most intense peaks from the electropherograms of cancer samples and healthy individuals were used to create principal component analysis (PCA) scores plots that show differentiation between healthy and cancer sample groups for both types of cancer studied. The profiles for ovarian and colorectal cancers were also compared to one another, which resulted in differentiation between the two cancer types, in addition to the differentiation already seen between healthy and cancer samples. Structural identification was previously done through standard addition of known N-glycans derived from glycoproteins. Methylamidated serum N-glycans in this study were also analyzed by electrospray ionization (ESI) and matrix-assisted laser desorption ionization (MALDI) mass spectrometry to obtain unambiguous identification of structures associated with disease progression.

Keywords: Bioanalytical, Carbohydrates, Electrophoresis, Lab-on-a-Chip/Microfluidics
Application Code: Bioanalytical
Methodology Code: Microfluidics/Lab-on-a-Chip

Session # 2430 Abstract # 2430-2 **Oral Sessions**

Session Title Glycan Analysis - Half Session

Abstract Title **Improved Separation of Saccharide Standards by EFL-HILIC**

Date: Thursday, March 10, 2016 - Afterno
Time: 01:50 PM
Room: B303

Primary Author Raffeal Bennett
The Ohio State University

Co-Author(s) Susan Olesik

Abstract Text

Oligosaccharides and other sugars are the building blocks for glycans which affect important biological processes like signaling, gene expression, protein trafficking, and stem cell differentiation. Many of the methods in development for the comprehensive profiling of these oligosaccharides either require a non-“green,” expensive (i.e. acetonitrile) solvent or derivatization of the analytes at the expense of time, sample loss, and loss of quantitative information. Enhanced-fluidity liquid chromatography (EFLC) is an alternative separation method that involves the use of an alcohol/water mixture (a green solvent) as the primary mobile phase component and liquid CO[sub]2[/sub] as the modifier in subcritical conditions. The liquid CO[sub]2[/sub] affords a tunable property to the mobile phase in terms of polarity, diffusivity, and viscosity. As a proof-of-concept, EFLC separations were performed using a HILIC (hydrophilic interaction chromatography) amide column and a LC ELSD to show that higher performance can be obtained as opposed to HPLC separations. This work demonstrates that EFLC-HILIC markedly lowers the analysis time for a test mixture of oligosaccharides compared to an optimized HPLC analysis while maintaining equivalent or better efficiencies. This work was sponsored by the Department of Chemistry and Biochemistry at the Ohio State University.

Keywords: Carbohydrates, Chromatography, HPLC, SFC
Application Code: Other
Methodology Code: Liquid Chromatography

Session # **2430** **Abstract #** **2430-3** **Oral Sessions**

Session Title: Glycan Analysis - Half Session

Abstract Title: **Receptor for Advanced Glycation End Products Diffusion and Ligand-Binding Events Studied by Fluorescence Recovery after Photobleaching and Surface Plasmon Resonance**

Date: Thursday, March 10, 2016 - Afterno
Time: 02:10 PM
Room: B303

Primary Author: Qiaochu Zhu
Iowa State University

Co-Author(s): Aleem Syed, Chamari S. Wijesooriya, Emily A. Smith

Abstract Text

The Receptor for Advanced Glycation Endproducts (RAGE) is a pattern-recognition receptor that can interact with a broad range of ligands including advanced glycation endproducts and S100 proteins. Upon ligand binding, RAGE signaling pathways are initiated that affect gene expression. The signaling downstream events are associated with several diseases such as cancer, Alzheimer's disease, diabetes, and viral infections. Diffusion is a key mechanism for regulating signal transduction events that are initiated by membrane components. We use fluorescence recovery after photobleaching (FRAP) to elucidate the diffusion properties of RAGE in cultured mammalian cells and quantify the differences in RAGE diffusion upon binding to a variety of ligands. Additional binding studies between RAGE and its ligands were performed by surface plasmon resonance (SPR) and isothermal titration calorimetry (ITC). Many previous RAGE binding studies used chemically-heterogeneous ligands. In contrast, we have used self-assembled monolayers of chemically-synthesized, and well characterized, ligands on gold substrates to study specific RAGE-ligand binding interactions. Our work aims to provide key information on the biophysical mechanism of RAGE through the study of its diffusion and ligand-binding events.

This work was supported by National Science Foundation Award Number: 1412084.

Keywords: Bioanalytical, Biospectroscopy, Fluorescence, Protein
Application Code: Bioanalytical
Methodology Code: Biospectroscopy

Session # 2430 Abstract # 2430-4 **Oral Sessions**

Session Title Glycan Analysis - Half Session

Abstract Title **Comprehensive Quantitative and Structural Analysis of Permethylated N-glycan Using PGC-LC-MS/MS**

Primary Author Shiyue Zhou
Texas Tech University

Date: Thursday, March 10, 2016 - Afterno
Time: 02:30 PM
Room: B303

Co-Author(s) Yehia Mechref

Abstract Text

Glycosylation as one of the most common post-translational modification (PTM) plays critical roles in various biological processes. Development of quantitative and structural glycomic profiling method is in great importance for understanding biofunction of glycans and discovering potential glycan biomarkers for diseases. Although high resolution mass spectrometry facilitates accurate sequential identification of glycans, identification of glycan isomers is still relying on efficient LC separation. Permethylation is a popular derivation method that enhances glycan MS signal intensity. Moreover, permethylation can stabilize sialic acid and eliminate fucose migration during ionization, contributing to the acquisition of more reliable glycan structural information. However, the isomeric separation of permethylated glycans is always not satisfactory due to the increased intramolecular interaction after permethylation. In this study, we have achieved efficient isomeric separation for permethylated glycans by using high temperature porous graphite carbon (PGC) LC. Meanwhile, more informative fragments of glycans can be obtained in MS2 for the structural elucidation of glycans benefiting from permethylation. Glycans released from model glycoproteins (Ribonuclease B, Fetuin and IgG) and various biological sources including human blood serum, cell line and milk were analyzed utilizing this strategy. Isomers resulting from different unit sites and linkages got base-line separated in PGC LC and diagnostic ions for core fucosylation and different branch galactose can be found in collision-induced dissociation (CID) MS2. Hence, comprehensive quantitative and structural glycomic profiling can be achieved by using PGC-LC-MS/MS.

Keywords: Carbohydrates, Liquid Chromatography/Mass Spectroscopy
Application Code: Bioanalytical
Methodology Code: Liquid Chromatography/Mass Spectrometry

Session # 2440 Abstract # 2440-1 **Oral Sessions**

Session Title LC and Sample Matrix Solutions - Half Session

Abstract Title **Analysis of Biofluids Using Solid Phase Microextraction Devices Made on Plastic Support**

Date: Thursday, March 10, 2016 - Afterno
Time: 03:05 PM
Room: B304

Primary Author Nathaly Reyes-Garces
University of Waterloo

Co-Author(s) Barbara Bojko, Janusz Pawliszyn

Abstract Text

The availability of solid phase microextraction (SPME) devices with biocompatible extraction phases has allowed for new opportunities to apply this technology in fields such as the bioanalytical. Indeed, these biocompatible coatings enable the direct immersion of SPME devices in complex matrices while permitting a selective extraction of small molecules. Some of the most important advantages of biocompatible SPME samplers include negligible protein attachment, reduction of possible matrix effects, and the possibility of using such samplers for in vivo analysis. Although in SPME the compatibility and chemistry of the coating are some of the most critical factors to take into consideration, having a biocompatible support able to provide good coating stability also plays an important role. In this work, thin film SPME samplers made of hydrophilic-lipophilic balanced particles and polyacrylonitrile (PAN) were used for the analysis of a diverse set of drugs in urine, plasma and whole blood. Particularly, the devices herein used were manufactured utilizing polybutylene terephthalate (PBT) as a support material. Rewarding results in terms of coating stability, absence of absolute matrix effects and interferences, linearity, accuracy, and precision were attained when the proposed samplers were evaluated in the three listed biofluids. Furthermore, by comparing water contact angles measured on HLB-PAN and C18-PAN coatings, it was verified that the presence of HLB in the extraction phase favored the coating wettability characteristics. Overall, the introduction of alternative materials, such as PBT, may represent an important opportunity for new advances in the commercialization and acceptance of SPME and in the development of innovative samplers.

Keywords: Bioanalytical, Biological Samples, Sample Preparation, SPME
Application Code: Bioanalytical
Methodology Code: Sampling and Sample Preparation

Session # 2440 Abstract # 2440-2 **Oral Sessions**

Session Title LC and Sample Matrix Solutions - Half Session

Abstract Title **Studies of Matrix Effects on the Determination of Hydrogen Peroxide by Using High Performance Anion Exchange Chromatography with Pulsed Amperometric Detection**

Date: Thursday, March 10, 2016 - Afterno
Time: 03:25 PM
Room: B304

Primary Author Jun Cheng
Thermo Fisher Scientific

Co-Author(s) Christopher Pohl, Kannan Srinivasan, Yan Liu

Abstract Text

The concentration of hydrogen peroxide in various samples can be determined by using high performance anion exchange chromatography with pulsed amperometric detection. However, there have been reports that the sample matrices may dramatically affect the analytical results in the determination of hydrogen peroxide more recently. We recently performed studies to investigate the effects of various matrices such as water, sodium hydroxide and sodium chloride on the determination of hydrogen peroxide by using high performance anion exchange chromatography with pulsed amperometric detection. The response, the retention time and the linear calibration were identified to vary from matrix to matrix. In order to eliminate the matrix effect, we optimized the experimental conditions, including injection volume and degassing efficiency. In addition, Sample pretrement cartridges such as OnGuard II H and OnGuard II Ag/H were used to remove hydroxide and chloride, respectively. Excellent limits of detection and linear calibration were obtained for hydrogen peroxide in all the three different matrices. The recovery was determined to be in the range of 101% to 117%.

Keywords: Bioanalytical, Electrochemistry, Ion Chromatography, Sample Preparation
Application Code: Bioanalytical
Methodology Code: Liquid Chromatography

Session # 2440 **Abstract # 2440-3** **Oral Sessions**

Session Title: LC and Sample Matrix Solutions - Half Session

Abstract Title: **HPLC Purity Method Development Using Low pH Mobile Phase and Ion-Pairing Reagent: Application to GMP Tert-Leucine Analysis**

Date: Thursday, March 10, 2016 - Afterno
Time: 03:45 PM
Room: B304

Primary Author: John Vinci
AbbVie, Inc.

Co-Author(s): Clifford Mitchell

Abstract Text

Research and advances in liquid chromatography continue to be centered on particle technology, including particle construction and novel stationary phases. Developing these new separation media is essential to meet the evolving needs of many industries, pharmaceuticals in particular. As regulatory expectations increase with advancing technology and stricter control requirements, it is essential to evaluate and implement new technology and strategies. Further, due to recent trends of more regulatory pushback of starting materials, especially by the EMA, it is becoming increasingly common to have smaller structures as starting materials that are frequently more polar than their respective later-stage intermediates and API, and less amenable to traditional reversed-phase separations. However, developing scientifically sound, robust HPLC methods for Good Manufacturing Practices (GMP) release and understanding of starting materials and their impurities remains of paramount importance for developing new processes and bringing innovative therapies to commercialization. As a result, novel analytical methods must be developed to address these challenges. Amino acids typically have carboxylic acid pKa around 2-2.5 and protonated amine group pKa around 9-10. As a result, amino acids are zwitterionic at a majority of the pH range and singly charged at each pH extreme. This ionic behavior of amino acids equates to low or no retention by traditional reversed-phase methods, even with 100% aqueous mobile phase. Further, the lack of strong chromophore makes detectability an issue. While amino acids have been separated by polar-polar separation mechanisms in the past, the presence of non-polar impurities complicates matters and makes developing an HPLC method for the release of regulatory starting material a significant challenge. This paper highlights an innovative approach of low pH mobile phase and ion-pairing reagent that provides adequate quantitation limit and scientific soundnesss for the GMP analysis of the purity and assay of tert-leucine.

Keywords: HPLC, HPLC Columns, Ion Exchange, Pharmaceutical
Application Code: Pharmaceutical
Methodology Code: Liquid Chromatography

Session # **2440** Abstract # **2440-4** **Oral Sessions**

Session Title LC and Sample Matrix Solutions - Half Session

Abstract Title **Combining Orthogonal Separation Modes for Analysis of Multiple Sample Components**

Primary Author Thomas E. Wheat
Waters Corporation

Date: Thursday, March 10, 2016 - Afterno
Time: 04:05 PM
Room: B304

Co-Author(s) Amanda B. Dlugasch, Patricia R. McConville

Abstract Text

Analytical methods that incorporate separations technology are required to resolve all sample components to provide unequivocal identification with the possibility of reliable quantitation. Many powerful tools and procedures have been developed for meeting this objective. All chromatographic techniques, however, reach a fundamental limit in the number of components that can be resolved in a single analysis. This limitation can be addressed using multi-dimensional chromatography, where a particular band or peak is transferred to a second separation column. There are specific challenges in applying this principle. To derive the full benefit of the additional column, the modes of separation should be orthogonal to one another. While appropriate method combinations can be selected, the mobile phases for the two modes are often incompatible. The most desirable multi-dimensional system will include provision for adjusting the mobile composition. It may also be necessary to subject multiple peaks to the second mode of chromatography without moving to full comprehensive 2-dimensional chromatography. We have assembled an automated system to address both of these needs. A set of valves provides multiple positions for isolating multiple peaks or segments of a chromatogram. The isolated analyte can be kept in solution or it can be adsorbed to a solid support. When the first chromatographic separation is complete, the several components can be analyzed with the second mode of chromatography. The system incorporates an At-column Dilution stage for adjusting the isolated peak composition before the second mode of chromatography. While this most often uses water to reduce the organic solvent content, we have also used the technique for pH adjustment. We will evaluate the performance of this system using both biopharmaceutical protein preparations and small molecule pharmaceuticals with impurity.

Keywords: Chromatography, HPLC, Instrumentation, Liquid Chromatography
Application Code: Pharmaceutical
Methodology Code: Liquid Chromatography

Session Title: Microfluidics/Lab-on-a-Chip - Bioanalytical II

Abstract Title: **Separation of Preterm Birth Biomarkers Using Capillary and Microchip Electrophoresis**

Date: Thursday, March 10, 2016 - Afterno
Time: 01:30 PM
Room: B316

Primary Author: Anna V. Nielsen
Brigham Young University

Co-Author(s): Adam T. Woolley, Radim Knob

Abstract Text

Preterm birth (PTB) is currently the leading cause of pregnancy-related complications and infant mortality. Esplin, et al.1 recently discovered a panel of biomarkers consisting of three peptides and six proteins which can be used to determine PTB risk with 86.5% selectivity and 80.6% specificity up to four weeks in advance of the onset of early labor. We are developing a point-of-care analysis microchip device which will allow for early diagnosis of PTB via the combined detection of these nine biomarkers. A key part of this analysis system involves electrokinetic separation and subsequent laser-induced fluorescence detection of these biomarkers. As many of the qualities of microchip electrophoresis are analogous to conventional capillary electrophoresis (CE), we have been studying the separation of PTB biomarkers using conventional CE. In order to optimize the separation pH, the effective mobilities of each of the biomarkers in their native and fluorescently-labelled states are being measured over a broad pH range. From these mobilities, an ideal separation pH can then be determined, and the separation can be optimized in conventional CE before it is miniaturized in a microchip device. Once a separation of all nine biomarkers has been performed in the microchip device, it will be possible to integrate this separation with the other sample preparation functions in a miniaturized system.

Reference

1. Esplin, M.S.; Merrell, K.; Goldenberg, R.; et al. Proteomic identification of serum peptides predicting subsequent spontaneous preterm birth. Am. J. Obstet. Gynecol. 2011. 204, 391.e1-8.

Acknowledgement

National Institutes of Health (R01 EB006124)

Keywords: Bioanalytical, Capillary Electrophoresis, Lab-on-a-Chip/Microfluidics, Separation Sciences
Application Code: Bioanalytical
Methodology Code: Microfluidics/Lab-on-a-Chip

Session # 2450 **Abstract # 2450-2** **Oral Sessions**

Session Title: Microfluidics/Lab-on-a-Chip - Bioanalytical II

Abstract Title: **Analysis of Nitrosative Stress in Macrophage Cells Using Microchip Electrophoresis with Electrochemical Detection**

Date: Thursday, March 10, 2016 - Afterno
Time: 01:50 PM
Room: B316

Primary Author: Joseph M. Siegel
University of Kansas

Co-Author(s): Kelci M. Schilly, Manjula B. Wijeshinghe, Susan M. Lunte

Abstract Text

Nitric oxide is naturally produced in macrophage cells during the immune response. However, chronic inflammation and over-stimulation can lead to excess NO production and the formation of dangerous reactive nitrogen species (RNS) such as peroxynitrite. These RNS can react with lipids, proteins, and DNA to inhibit their function and have been linked to neurodegenerative and cardiovascular disorders. The research presented here describes the progress regarding the development of a microchip electrophoresis with electrochemical detection (ME-EC) method to monitor nitrosative stress in macrophage cells. Previously, we reported a ME-EC method to separate and detect nitrite, the primary degradation product of NO, from cellular interferences in bulk macrophage cell lysates following stimulation with lipopolysaccharide (LPS). However, this method suffers from high LODs, and NO is detected indirectly as nitrite. To improve the LODs, a platinum working electrode was modified with platinum black, which enhances the electrochemical signal for RNS. Platinum black was deposited using a constant current density. To investigate the signal enhancement for specific RNS, NO and peroxynitrite were generated using the PROLI/NONOate and SIN-1 systems, respectively. A greater than three-fold signal enhancement was observed for both species. In addition to improving the LOD, conditions were modified to achieve a baseline separation of nitrite, NO, and peroxynitrite. This method was then applied to the analysis of bulk macrophage cells stimulated with LPS and interferon-□to further induce NO production. In the future, this method will be utilized to monitor the nitrosative stress in macrophages due to other stimulants, such as beta amyloid.

Keywords: Bioanalytical, Biological Samples, Electrochemistry, Lab-on-a-Chip/Microfluidics
Application Code: Bioanalytical
Methodology Code: Microfluidics/Lab-on-a-Chip

Session # 2450 Abstract # 2450-3 **Oral Sessions**

Session Title: Microfluidics/Lab-on-a-Chip - Bioanalytical II

Abstract Title: **Ultrasensitive Electrochemical Microfluidic Immunoarray for Assessment of Aggressive vs Indolent forms of Prostate Cancer Biomarkers**

Date: Thursday, March 10, 2016 - Afterno
Time: 02:10 PM
Room: B316

Primary Author: Brunah A. Otieno
University of Connecticut

Co-Author(s): Abby Jones, Amit Joshi, Colleen E. Krause, James F. Rusling, Mohammed Sherafeldin

Abstract Text

Prostate cancer is the most common cause of cancer-related death in men in the US and throughout the world. Current practices for detection and staging of prostate cancer often fall short in terms of sensitivity, specificity, limited predictive power and inability to distinguish between aggressive and indolent forms of prostate cancer. These limitations lead to unnecessary treatments that adversely affect the patients' quality of life with minimal or no gain. Measurement of small panels of signature molecular biomarkers in serum holds tremendous potential for prostate cancer diagnostics and personalized therapy. Here we describe a simple, low-cost, multiple-biomarker based microfluidic system for on-line capture and detection of prostate cancer protein biomarkers. The protein panel includes PSA, CD-14, ERG, GOLM-1, PEDF-1, IGF-1, VEGF-D and IGFBP-3, many of which are thought to be specific for aggressive prostate cancer. The system features a small chamber for on-line protein capture from serum by magnetic beads labeled with many copies of analyte-specific antibodies and signal-transducing enzyme labels, positioned upstream of a detection chamber housing a nanostructured 8-electrode sensor array. Gold immunoarrays fabricated by ink-jet printing ($0.2) or commercial screen printed carbon arrays ($5) are fitted into the microfluidic detection chamber to achieve high sensitivity. Detection limit in the low fM range was achieved for multiplexed detection of the cancer biomarker proteins from as little as 5 uL sample. Measurements of this panel of selected biomarkers will be tested with prostate cancer patient samples in future to assess its diagnostic capability.

Keywords: Biosensors, Electrochemistry, Immunoassay, Lab-on-a-Chip/Microfluidics
Application Code: Bioanalytical
Methodology Code: Microfluidics/Lab-on-a-Chip

Session # 2450 **Abstract # 2450-4** **Oral Sessions**

Session Title Microfluidics/Lab-on-a-Chip - Bioanalytical II

Abstract Title **Microfluidic and 3-D Printed Devices for Near-Real-Time and Simultaneous Detection of Neurotransmitters**

Date: Thursday, March 10, 2016 - Afterno
Time: 02:30 PM
Room: B316

Primary Author Alexandra D. Townsend
Saint Louis University

Co-Author(s) R Scott Martin

Abstract Text

There is a need in the biomedical community to develop easy-to-use devices to measure changes in neurotransmitter levels with high temporal resolution. This talk will outline two different approaches with this overarching theme. The first approach involves microchip-based amperometric detection of nitric oxide released from endothelial cells. Nitric oxide is a difficult molecule to detect due to its reactivity and short half-life. This talk will describe the creation of a device that utilizes a PDMS chip designed to contain immobilized endothelial cells in one channel and detect the released NO in a separate channel via a planar membrane. The microfluidic device contains a Nafion and platinum-black modified glassy carbon detection electrode embedded in a polystyrene base which makes it suitable for cell culture. The second approach involves the use of microchip-based and 3-D printed devices to study co-transmission of neurotransmitters within the sympathetic nervous system. This talk will describe the simultaneous detection of basal and stimulated levels of norepinephrine and ATP perfused from the mesenteric bed of a rat. A 3-D printed droplet splitter allows the sample to enter a luminometer for ATP detection with a luciferin-luciferase reaction while simultaneously flowing through a PDMS-PS microfluidic device for amperometric detection of norepinephrine. The microfluidic device consists of a straight channel and a nafion-coated gold pillar array embedded within the polystyrene. Both of these approaches enable sensitive and selective detection of neurotransmitters in close-to real time.

Keywords: Bioanalytical, Electrochemistry, High Throughput Chemical Analysis, Lab-on-a-Chip/Microfluidics
Application Code: Bioanalytical
Methodology Code: Microfluidics/Lab-on-a-Chip

Session # 2450 Abstract # 2450-5 Oral Sessions

Session Title Microfluidics/Lab-on-a-Chip - Bioanalytical II

Abstract Title **Electrokinetically Operated Integrated Microfluidic Platform for Preterm Birth Biomarker Analysis**

Primary Author Mukul Sonker
Brigham Young University

Date: Thursday, March 10, 2016 - Afterno
Time: 03:05 PM
Room: B316

Co-Author(s) Adam T. Woolley, Radim Knob, Vishal Sahore

Abstract Text

Preterm Birth (PTB), the most common complication in pregnancy, affects more than 500,000 children every year in USA alone and is the leading cause of newborn deaths and illnesses. Due to the lack of clinical methods that can predict PTB at an early stage, a simple, inexpensive and sensitive detection system for assessing risk of PTBs is highly desired. Our goal is to develop an electrokinetically operated microfluidic platform that can analyze a PTB biomarker panel characterized by Esplin et al.,[sup]1[/sup] that includes 3 peptides and 6 proteins. A microchip electrophoresis (μCE) module has been developed in a thermoplastic material and was used to successfully separate four PTB biomarkers (3 peptides and 1 protein). A solid-phase extraction module with a reversed-phase monolith was also developed to retain, enrich and elute PTB biomarkers, with a sample enrichment efficiency of >25 fold. Furthermore, we have developed a reactive porous polymer monolith immunoaffinity extraction module to selectively capture, enrich, and elute ferritin, a PTB biomarker. Currently we are integrating the solid-phase extraction and μCE modules to enrich and separate PTB biomarkers, with initial results showing ~8 fold enrichment. We are also working on integration of the immunoaffinity and μCE modules to further automate analysis. We believe that such a platform offers great opportunities as a diagnostic tool for various potential analytes.

Reference:

1.Esplin, S., et al., Am. J. Obstet. Gynecol., 204, 391.e1-8 (2011).

Acknowledgements:
We acknowledge National Institutes of Health (R01 EB006124) for the financial support.

Keywords: Electrophoresis, Fluorescence, Lab-on-a-Chip/Microfluidics, Solid Phase Extraction
Application Code: Bioanalytical
Methodology Code: Microfluidics/Lab-on-a-Chip

Session # 2450 Abstract # 2450-6 **Oral Sessions**

Session Title Microfluidics/Lab-on-a-Chip - Bioanalytical II

Abstract Title **Discrete Stimulation of Lymph Node Slices on Chip**

Date: Thursday, March 10, 2016 - Afterno
Time: 03:25 PM
Room: B316

Primary Author Ashley E. Ross
University of Virginia

Co-Author(s) Jacob F. Woodroof, Rebecca R. Pompano

Abstract Text

Protective immune responses depend on exquisitely organized chemical communications inside tiny organs such as the lymph nodes. The ability to chemically target discrete regions of a lymph node would allow local analysis of inflammatory processes, but traditional methods for studying the immune system cannot target specific regions of the node. Controlled stimulation would improve tissue analysis by providing quantifiable monitoring of immune responses. Microfluidics, a technology to control fluidic delivery at the 10 – 1000 µm scale, provides a unique solution to the spatial analysis problem, and has already been implemented in neuroscience to deliver stimuli to brain on chip. Here, we show for the first time, a microfluidic device capable of discretely stimulating intact mouse lymph node slices with 200-□m resolution, the size of critical lymph node sub-structures. Fluorescein and FITC-dextrans of varying molecular weight were delivered to a slice for optimization of stimulus delivery. The delivery time, flow rate, and concentration were varied. Preliminary results indicate that for short pulse times, spread and intensity of stimulus was linearly dependent on pulse time, making delivery predictable. The target resolution was achieved using protein-sized molecules; e.g., a 5-s stimulation with 0.1 mg/mL 40-kDa FITC-dextran resulted in 210 ± 42 µm initial spread (mean ± SEM). As a proof of concept for analysis on-chip, a 42-kDa antigen, ovalbumin, was locally applied on-chip and T cell migration was monitored. In the future, monitoring of local immune responses will provide a wealth of information about the spatially organized signaling of the lymph node.

Keywords: Bioanalytical, Lab-on-a-Chip/Microfluidics, Monitoring
Application Code: Bioanalytical
Methodology Code: Microfluidics/Lab-on-a-Chip

Session # 2450 Abstract # 2450-7 **Oral Sessions**

Session Title: Microfluidics/Lab-on-a-Chip - Bioanalytical II

Abstract Title: **Ultrasensitive ELISA for Detection of Infectious Diseases on Surface Modified PMMA Microfluidic Microplates**

Date: Thursday, March 10, 2016 - Afterno
Time: 03:45 PM
Room: B316

Primary Author: Sanjay Sharma Timilsina
University of Texas at El Paso

Co-Author(s): Maowei Dou, Xiujun James Li

Abstract Text

Unspecific absorption of protein often leads to high background and low sensitivity in enzyme linked immunosorbent assay (ELISA). Covalent binding of proteins can enhance the binding efficiency and improve the immunoassay sensitivity. Herein, we have developed a simple, miniaturized poly(methyl methacrylate) (PMMA) ELISA microfluidic microplate, where the protein is covalently bound to carboxylated PMMA surface. Unlike ELISA in traditional microplates, which is often limited by long incubation and blocking time, rapid and ultrasensitive detection of disease biomarkers can be completed within 90 min in this microplate with much less reagent consumption. Immunoassays do not require expensive and sophisticated equipment and results can even be observed by the naked eye. Quantitative analysis can be achieved by calculating the brightness of images scanned by a desktop scanner. Although no specialized ELISA equipment was used, the limits of detection of 51 pg/mL for Immunoglobulin G (IgG) and 135 pg/mL for hepatitis B surface antigen (HBsAg) have been achieved using this PMMA microplate, which is around 30 fold more sensitive as compared to commercial ELISA kits.

Keywords: Bioanalytical, Enzyme Assays, Immobilization, Lab-on-a-Chip/Microfluidics
Application Code: Bioanalytical
Methodology Code: Microfluidics/Lab-on-a-Chip

Session # 2450 **Abstract # 2450-8** **Oral Sessions**

Session Title Microfluidics/Lab-on-a-Chip - Bioanalytical II

Abstract Title **Optical Formulation of 3D Printer Resin for Minimum Microfluidic Flow Channel Size**

Primary Author Greg Nordin
Brigham Young University

Date: Thursday, March 10, 2016 - Afterno
Time: 04:05 PM
Room: B316

Co-Author(s) Adam T. Woolley, Hua Gong, Michael Beauchamp, Steven Perry

Abstract Text
3D printing is a potentially attractive fabrication and prototyping method for microfluidic devices. In particular, there has been great interest in using commercially available 3D printers or 3D printing services to fabricate microfluidic devices. To date, however, most efforts using stereolithgraphic 3D printers have been based on the use of commercially available resins, which are not necessarily optimized to fabricate small flow channels. In this presentation we describe an optical approach to formulate 3D printer resin to reliably fabricate flow channels as small as 60 microns tall and 4 pixels (~100 microns) wide. Moreover, we elucidate rules-of-thumb that are applicable to any resin, commercial or custom, for the minimum achievable flow channel size based on readily measureable optical properties of the resin. We also develop a mathematical model to calculate the local optical dose delivered to any position within a 3D printed microfluidic device during fabrication. We use the model to develop guidelines for choice of build layer thickness and layer exposure dose for any resin given its optical properties, and experimentally demonstrate the validity of these guidelines using a variety of resins.

Keywords: Bioanalytical, Lab-on-a-Chip/Microfluidics
Application Code: Bioanalytical
Methodology Code: Microfluidics/Lab-on-a-Chip